G222.
CONSTANT EVOLUTION.

The G222 is a versatile, tough and spacious military tactical transport aircraft that does its job, whenever and wherever it's needed. No matter what the mission, the G222's sturdy design and state-of the-art systems allow it to operate in remote areas and on short and roughly prepared runways, with no ground support required. It's the perfect aircraft for troop and cargo transport, paradropping, aeromedical transport and airdropping of materials into inaccessible areas. Leading air forces on four continents, including those of the United States and other NATO members, have discovered the G222's secret: it's an aircraft that's in constant evolution. And that's why it's the right choice for today and for the future.

Alenia

A F I N M E C C A N I C A C O M P A N Y

The power of information, the key to success

DEFENCE   **AEROSPACE**   **TRANSPORT**

 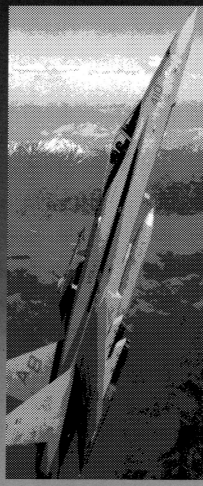 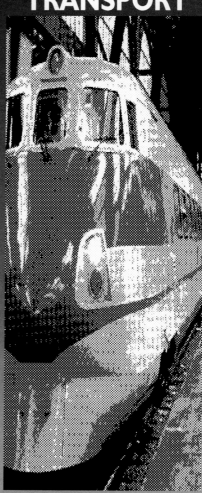

YEARBOOKS • BINDERS • DIRECTORIES

Jane's
INFORMATION GROUP

Call or fax us NOW for your free Jane's catalogue:

UK and Rest of World:
Dept. DSM, Sentinel House, 163 Brighton Road, Coulsdon, Surrey, CR5 2NH, UK
Tel: +44 (0)81 763 1030 Fax: +44 (0)81 763 1006

USA, Canada and South America:
Marketing Dept., 1340 Braddock Place, Suite 300, Alexandria, Virginia 22314 – 1651, USA
Tel: (703) 683 3700 Fax: (703) 836 0029

CONTENTS

This Edition has been compiled by:

Mark Lambert	AIRCRAFT SECTION: GERMANY, INTERNATIONAL (PART), ITALY, UNITED STATES OF AMERICA (DOUGLAS AIRCRAFT AND BOEING COMMERCIAL AIRPLANE GROUP)
Kenneth Munson	AIRCRAFT SECTION: ARGENTINA TO BELGIUM, BRAZIL TO FINLAND, GREECE TO ISRAEL, INTERNATIONAL (PART), JAPAN TO TURKEY; LIGHTER THAN AIR; FIRST FLIGHTS
John Taylor	AIRCRAFT SECTION: RUSSIA, UKRAINE, UZBEKISTAN
Paul Jackson	AIRCRAFT SECTION: BOSNIA-HERCEGOVINA, FRANCE, INTERNATIONAL (PART), UNITED KINGDOM, UNITED STATES OF AMERICA (MILITARY AIRCRAFT), YUGOSLAVIA (FORMER)
Michael Taylor	AIRCRAFT SECTION: FRANCE, UNITED KINGDOM, UNITED STATES OF AMERICA (CIVIL AIRCRAFT)
Bill Gunston	AERO ENGINES; GLOSSARY

Foreword: Into the pit and out again .... [13]
Some first flights made between
1 April 1992 and 30 June 1993 .... [19]
Glossary .... [22]
Airline fleet list .... [27]

Aircraft
Argentina 1
Australia 4
Austria 11
Belgium 11
Bosnia-Hercegovina 13
Brazil 13
Canada 20
Chile 41
China, People's Republic 44
Colombia 60
Czech Republic 63
Egypt 72
Ethiopia 72
Finland 73
France 74
Germany 95
Greece 105
India 106
Indonesia 112
International Programmes 114
Iran 181
Iraq 181
Israel 181
Italy 184
Japan 203
Korea, South 212
Malaysia 214
Netherlands 215
New Zealand 221
Nigeria 224
Norway 224
Pakistan 225
Peru 226
Philippines 226
Poland 228
Portugal 247
Romania 247
Russia 255
Singapore 338
South Africa 338
Spain 343
Sweden 348
Switzerland 357
Taiwan 363
Turkey 365
Ukraine 366
United Kingdom 376
United States of America 410
Uzbekistan 586
Yugoslavia (former) 586

Lighter than Air
Canada 591
China, People's Republic 591
France 591
Germany 591
Russia 592
Ukraine 594
United Kingdom 595
United States of America 598

Aero Engines
Australia 603
Austria 603
Belgium 603
Brazil 603
Canada 604
China, People's Republic 608
Czech Republic 612
Egypt 614
France 614
Germany 619
India 621
International Programmes 621
Israel 627
Italy 627
Japan 628
Poland 629
Romania 632
Russia 632
Singapore 649
South Africa 649
Spain 649
Sweden 649
Turkey 649
Ukraine 650
United Kingdom 654
United States of America 660
Yugoslavia (former) 684

Addenda 685

Indexes 699

ADMINISTRATION

Publishing Director: Robert Hutchinson

Managing Editor: Keith Faulkner

Publishing Supervisor: Ruth Simmance

Publishing Assistant: Lynn Morse

Product Group (Marketing) Manager: Sandra Dawes

EDITORIAL OFFICES

Jane's Information Group Limited, Sentinel House,
163 Brighton Road, Coulsdon, Surrey CR5 2NH, United Kingdom

Tel: 081 763 1030 International +44 81 763 1030
Telex: 916907 Janes G
Fax: 081 763 1006 International +44 81 763 2582

SALES OFFICES

Send enquiries to:
Fabiana Angelini, Sales Manager,
Jane's Information Group Limited, UK address as above

Send USA enquiries to:
Joe McHale, Senior Vice-President Product Sales,
Jane's Information Group Inc, 1340 Braddock Place, Suite 300,
Alexandria, VA 22314-1651

Tel: +1 703 683 3700
Telex: 6819193
Fax: +1 703 836 0029

ADVERTISEMENT SALES OFFICES

Advertisement Sales Manager: Barbara Urry

Australia: Brendan Gullifer, Havre & Gullifer (PTY) Ltd, 253 Richardson Street, Middle Park, Victoria 3206, Australia

Tel: +61 (3) 6960288
Fax: +61 (3) 6966951

Benelux: Peter McSherry, Jane's Information Group (see United Kingdom)

Brazil: L Bilyk, Brazmedia International S/C Ltda, Alameda Gabriel Monteiro da Silva, 366 CEP, 01442, São Paulo

Tel: +55 11 853 4133
Telex: 32836 BMED BR
Fax: +55 11 852 6485

France: Patrice Février, Jane's Information Group – France, BP 418, F75824 Paris, France

Tel: +33 1 45 72 3311
Fax: +33 1 45 72 1795

Germany and Austria: Rainer Vogel, Media Services International, Schwabenbergstrasse 12, D-8089 Emmering

Tel: +49 8141 42534
Fax: +49 8141 6706

Greece: Anwar Aswad, A&M Advertising & Marketing Consultants, Zaimi 7-9, Apt 1, Palaio-Faliron

Tel: +30 1 982 2577
Telex: 218947 GNM
Fax: +30 1 723 2990

Hong Kong: Jeremy Miller, Major Media Ltd, Room 142, 14F Capitol Centre, 5-19 Jardine's Bazaar, Causeway Bay

Tel: +852 5 890 3110
Fax: +852 5 576 3397

Israel: Oreet Ben-Yaacov, Oreet International Media, 15 Kineret Street, 51201 Bene-Berak

Tel: +972 3 570 6527
Fax: +972 3 570 6526

Italy and Switzerland: Ediconsult Internazionale Srl, Piazza Fontane Marose 3, I-16123 Genoa

Tel: +39 10 583684
Telex: 281197 EDINT I
Fax: +39 10 566578

Japan: Hiro Sakagami, Intermart/E.A.C., Inc, 1-7 Akasaka, 9-Chome, Minato-ku, Tokyo 107

Tel: +81(3)5474-7835
Fax: +81(3)5474-7837

Korea: Young Seoh Chinn, JES Media International, KPO Box 576, Seoul

Tel: +82 2 545 8001/2
Fax: +82 2 549 8861

Scandinavia: Gillian Thompson, First Call International, 11 Chardmore Road, London N16 6JA, UK

Tel: +44 81 806 2301, 3538
Fax: +44 81 806 8137

Singapore, Indonesia, Malaysia, Philippines, Taiwan and Thailand: Hoo Siew Sai, Major Media (Singapore) Pte Ltd, 6th Floor, 52 Chin Swee Road, Singapore 0316

Tel: +65 738 0122
Telex: RS 43370 AMPLS
Fax: +65 738 2108

Spain: Jesus Moran Iglesias, Varex SA, Modesto Lafuente 4, E-28010 Madrid

Tel: +34 1 448 7622
Fax: +34 1 446 0198

United States and Canada: Kimberly S. Hanson, Director of Advertising Sales and Marketing, Jane's Information Group Inc, 1340 Braddock Place, Suite 300, Alexandria, VA 22314-1651

Tel: +1 703 683 3700
Telex: 6819193
Fax: +1 703 836 0029

USA South Eastern Region: Kristin Schulze, Regional Advertising Manager
(see United States and Canada)

USA North Eastern Region and Canada: Melissa C Gunning, Regional Advertising Manager

USA Western Region and Canada: Anne Marie St. John-Brooks, Regional Advertising Manager, Jane's Information Group, 1523 Rollins Road, Burlingame, CA 94010

Tel: +1 415 259 9982
Fax: +1 415 259 9751

United Kingdom/Rest of World: Peter McSherry, Barbara Urry, Jane's Information Group, Sentinel House, 163 Brighton Road, Coulsdon, Surrey CR5 2NH

Tel: 081 763 1030 International +44 81 763 1030
Telex: 916907 Janes G
Fax: 081 763 1006 International +44 81 763 1006

Administration: Tara Betts, Jane's Information Group
(see United Kingdom)

AIR SUPERIORITY IS NOW WITHIN YOUR REACH

ELBIT'S AIRCRAFT UPGRADES TURN OUTDATED PLATFORMS INTO FIRST-CLASS FIGHTERS

Today's flight missions demand performance capabilities that aging aircraft can't deliver. Aircraft upgrades are the logical answer, and Elbit is the address. With across-the-board expertise in aircraft retrofits and advanced avionics, Elbit is in a unique position to provide cost-effective solutions to assure that current platforms match future requirements. Elbit upgrades "rejuvenate" aircraft. With structural improvements for extended operational life. Advanced cockpits and next-generation capabilities that equal, and often exceed those of today's fighters, by incorporating the following high performance systems:

■ Weapon delivery and navigation ■ Mission computers ■ Sight helmet ■ Day and night vision ■ Tactical systems ■ Multifunction displays (color and monochromatic) ■ Stores management ■ Radar ■ EW.

Elbit's upgrade experience has been proven in major programs all over the globe. Team with Elbit. Together we can reach new heights.

 **A STRONG SENSE OF DIRECTION**

Elbit Ltd.
P.O.Box 539, Haifa 31053, Israel
Tel: (972) 4-315315
Fax: (972)4-550002/551623

Elbit Systems Of America, Inc.
5519 Glenwood Hills Parkway, S.E.
Grand Rapids, Michigan 49512, U.S.A.
Tel: (1)616-957-3337 Fax: (1)616-957-1183

Alphabetical list of advertisers

A

Aermacchi SpA
Via P. Foresio, I-21040 Venegono Superiore,
Italy .... *Facing Page* [1]

Alenia SpA
81 Via Vitorchiano, I-00189 Rome,
Italy .... *Facing Inside Front Cover*

AMX International
Gruppo Velivoli da Combattimento,
Corso Marche 41, I-10146 Turin,
Italy .... [9]

Atlas Aviation
A Division of Denel (Pty) Ltd
PO Box 11, Kempton Park 1620, Transvaal,
South Africa .... [14]

E

Elbit Ltd
PO Box 539, Haifa 31053,
Israel .... [5]

Elettronica SpA
Via Tiburtina Km 13,700, I-00131 Rome,
Italy .... [7]

L

Lockheed Fort Worth
PO Box 748, Fort Worth, Texas 76101,
USA .... [12]

Lockheed Aeronautical Systems Group
86 South Cobb Drive, Marietta, Georgia 30063,
USA .... *Inside Front Cover*

Jane's
INFORMATION GROUP

Leading suppliers of impartial, factual, professional information to the defence, aerospace and transport industries.

The Group's unique capabilities for research and analysis enables it to provide the most comprehensive information from a single source.

Products and Services
CD-ROM · Subscription Services
Yearbooks and Directories · On-Line Services
Magazines · Confidential Market Research
Electronic Databases

Jane's Information Group
the unique answer to your intelligence requirements

Jane's Information Group
Sentinel House, 163 Brighton Road
Coulsdon, Surrey CR5 2NH, United Kingdom
Tel: +44 (0)81 763 1030

Jane's Information Group
1340 Braddock Place, Suite 300
Alexandria VA 22314-1651, United States
Tel: (703) 683 3700

AIRBORNE ELECTRONIC WARFARE SYSTEMS are from ELETTRONICA S.p.A.

ELETTRONICA PRODUCES A FULL RANGE OF EW EQUIPMENT AND SYSTEMS TO PROVIDE AIRCRAFT WITH COMPLETE PROTECTION AGAINST MODERN WEAPON SYSTEMS.

ELETTRONICA S.p.A.
Via Tiburtina km 13.700
ROME - ITALY
Phone +39.6. 41541 - Fax +39.6. 4192869
Telex 626527 ELT

Classified list of advertisers

The companies listed advertising in this publication have informed us that they are involved in the fields of manufacture indicated below.

Aero engine controls and accessories
Atlas Aviation
Pratt & Whitney

Aero engine test plant
Atlas Aviation
Pratt & Whitney

Aero engines
Atlas Aviation
Pratt & Whitney

Aircraft, combat
Aermacchi
AMX International
Atlas Aviation
Boeing
Lockheed

Aircraft, commercial trainers
Atlas Aviation

Aircraft construction
Aermacchi
Alenia
Atlas Aviation
Boeing
Lockheed

Aircraft, military
Aermacchi
Alenia
AMX International
Atlas Aviation
Boeing
Lockheed

Aircraft, military trainers
Aermacchi

Aircraft, military transport
Alenia
Atlas Aviation
Boeing
Lockheed

Aircraft modifications
Aermacchi
Alenia
Atlas Aviation
Boeing
Elbit
Elettronica
Lockheed

Aircraft product support
Atlas Aviation

Aircraft upgrades
Alenia
Atlas Aviation
Boeing
Elbit
Lockheed

Airframes
Aermacchi
Alenia
Atlas Aviation
Boeing
Lockheed

Antennas
Elbit
Elettronica

Antennas, aircraft
Elbit
Elettronica

Carbon fibre components
Alenia
Atlas Aviation
Boeing
Lockheed

Civil and military aircraft
Aermacchi
Alenia
AMX International
Atlas Aviation
Boeing
Lockheed

Combat command and control
Boeing
Elbit
Elettronica
Lockheed

Composite structures
Alenia
Atlas Aviation
Boeing
Lockheed

Defence contractors
Aermacchi
Alenia
AMX International
Atlas Aviation
Boeing
Elbit
Elettronica
Lockheed

Design services
Aermacchi
Alenia
AMX International
Atlas Aviation
Boeing
Elbit
Elettronica
Lockheed

Electrical equipment and components
Atlas Aviation
Elbit
Elettronica

Electronic countermeasures (ECM)
Elbit
Elettronica

Engine design and manufacture
Atlas Aviation
Pratt & Whitney

Engine parts fabrication
Atlas Aviation
Pratt & Whitney

Engine starting equipment
Aermacchi
Atlas Aviation

Engine testing equipment
Aermacchi
Atlas Aviation
Pratt & Whitney

Gas turbines
Atlas Aviation

Helicopter parts and components
Atlas Aviation

Helicopter support
Atlas Aviation
Elbit

Helicopters, Military — Naval
Atlas Aviation

Infra-red equipment
Elbit
Elettronica

Infra-red materials
Elbit
Elettronica

Infra-red systems
Elbit
Elettronica

Instruments, aircraft
Atlas Aviation

Jet engine parts
Atlas Aviation
Pratt & Whitney

Maintenance and overhaul — airframe
Alenia
Atlas Aviation

Manufacturing services
Aermacchi
Alenia
Atlas Aviation
Boeing
Lockheed

Night vision equipment
Elbit
Elettronica

Optical infra-red detectors
Elbit
Elettronica

Reconnaissance, airborne
Aermacchi
Alenia
Boeing
Elbit
Elettronica
Lockheed

Repair and maintenance of aircraft
Alenia
Atlas Aviation
Boeing
Lockheed

Repair and overhaul of aero engines
Atlas Aviation
Pratt & Whitney

Repair and maintenance of aircraft wheels and brakes
Atlas Aviation

Repair of aircraft instruments
Atlas Aviation

RPV electronics
Alenia

Sheet metal work
Aermacchi
Alenia
Atlas Aviation
Boeing
Lockheed

Space systems
Alenia
Boeing

Turboprops and turboshafts
Atlas Aviation

ADVANCED COMBAT TRAINER

MISSION: LO-LO-LO-LO

Low level. The two-seater AMX is both an advanced trainer for OCU AND a dedicated surface attack aircraft, designed from the outset for effective battlefield and maritime operations, by day or night.

Low effort. Superb low altitude performance, manoeuvrability and survivability, plus advanced avionics and ECM keep crew fatigue to a minimum.

Low cost. With very high parts commonality with the single-seater, identical engine and low-maintenance airframe, the AMX costs less to operate.

Further savings are made from the ease of pilot conversion, and from opportunities to reduce expensive multi-role inventory via a mixed fleet.

Low lead times. The AMX is already in squadron service, and unlike combat aircraft converted from trainers, the two-seater AMX is available for early delivery.

AMX. For potency, survivability, affordability and availability.

JAS 39 Gripen takes pride of place in combining small size, high performance and a wide range of air-to-air and air-to-surface weapons

Between 16 and 18 June, the Airbus World Ranger A340-200 flew 10,307 nm (19,100 km; 11,853 miles) from Paris to Auckland in 21 h 32 min arriving with 14,500 kg (31,967 lb) of fuel remaining; it flew the 10,392 nm (19,258 km; 11,950 miles) back to Paris in 21 h 46 min with 6,500 kg (14,329 lb) of fuel remaining. These were the longest flights by any airliner and the fastest in their class

JANE'S ALL THE WORLD'S AIRCRAFT 1993-94

Jane's Information Group Limited, Sentinel House, 163 Brighton Road, Coulsdon, Surrey CR5 2NH, UK
Jane's Information Group Inc, 1340 Braddock Place, Suite 300, Alexandria, VA 22314-1651, USA

Times Change. So Do F-16s.

The world has seen some dramatic changes since the first F-16 was introduced. The Berlin Wall has come down. The Soviet Union and Warsaw Pact have been dissolved. And new potential trouble spots have emerged.

Dramatic changes in weapon technology have also taken place. Fighter aircraft have improved radar capabilities, faster computers and more advanced weapons.

Through the years the F-16 has proven it can truly stay ahead of the threat.

Its ability to continually adapt new avionics and weaponry has led to an incredible service record, including 65 aerial dogfight victories, with no losses.

F-16 (Night Attack) Cockpit

The F-16 was the workhorse of Desert Storm. It flew 13,500 missions and had the highest readiness rate of any fighter in theater. With LANTIRN and GPS, F-16s were the premiere scud hunters.

The F-16 we're building today incorporates literally hundreds of new state-of-the-art technologies. The entire cockpit has been modernized. Engine thrust has been increased 25%, and there is a choice of the world's two best fighter engines manufactured by Pratt & Whitney and General Electric.

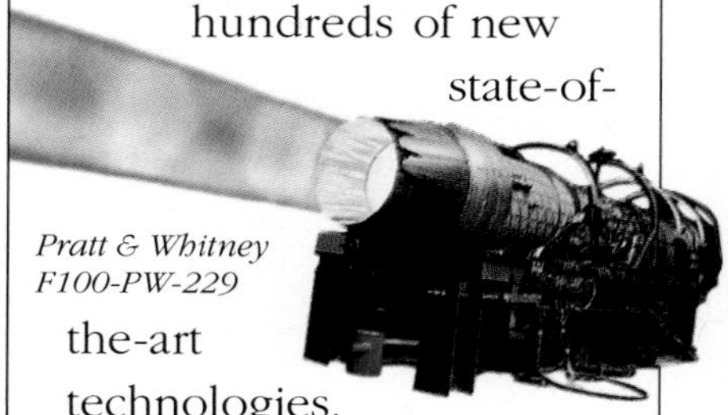

Pratt & Whitney F100-PW-229

General Electric F110

We've added beyond-visual-range firepower with Sparrow and AMRAAM radar missiles, night/all weather attack and autonomous precision attack with LANTIRN, IIR Mavericks, and laser guided bombs; anti-radar attack with HARM; and anti-ship with Penguin.

While the F-16's combat capability has been significantly enhanced, it was not done at the expense of operation and support costs. In terms of reliability, maintainability, readiness and lifecycle cost, the F-16 remains the best frontline fighter in the world.

And that's something we never intend to change.

Lockheed
Fort Worth Company

All three of the new Airbus airliners which made their first flights between the 1991 and 1993 Paris Air Shows showed off their fly-by-wire manoeuvre protection features by flying past with the side-stick controller held hard back

FOREWORD: Into the pit and out again

"Yesterday is a memory, tomorrow is a vision, but today is just miserable."
I don't know who originally said this, but it was quoted by Boeing Company President Philip Condit during the 1993 Paris Air Show.

Aerospace manufacturers today face four main problems: how to achieve the most promising alliances, how to focus on the most appropriate product lines, how to cope with growing competition from newcomers and, finally, whether to stay in the aerospace business. I have not known so much turmoil since the late 1950s and early 1960s in Britain.

The downward slide

The downturn in military production has really begun to bite in every country. Even the useful crop of new military export orders, F-15Ss and Tornados for Saudi Arabia, Mirage 2000-5s for Taiwan, Hawks for the Middle East and Indonesia, and F/A-18s for Switzerland, Finland, Kuwait and Malaysia, have not prevented several production lines from coming to a standstill. There will probably be an 11-month gap between the last US Air Force F-15E and the first Saudi F-15S. The Tornado line was being restarted in late June 1993 and South Korea was seen as a possible new customer for it. F/A-18 production has been trimmed to four per month. Production deliveries of the F/A-18E/F are not expected until 1998.

At the same time, the airlines have run into an unprecedented slump. It has been made worse in the

The new look *Jane's*

This year, we have introduced a major change of emphasis in *Jane's All the World's Aircraft*. We have greatly expanded our coverage of the commercially significant aircraft and ceased covering the range of purely recreational aircraft and balloons. The one balances the other and we believe that the readers for whom the book is now mainly produced will find *Jane's* even more informative and much more useful.

Each of the major aircraft is more thoroughly covered. Instead of providing a single photograph, a three-view drawing and a structured text on each aircraft, we give several photographs to show also the disposition of landing gear, high-lift devices and other special features. These are supported by pictures of flight decks, illustrations of internal structural layout and diagrams showing cabin cross-sections and seating layouts. For the Airbus A320 and A330/340 we show for the first time block diagrams of their fly-by-wire systems. Orders for the main aircraft are listed in the form of tables showing, where practicable, customers and versions.

At the front of the book we have added a listing of airline fleets compiled from Jane's Information Group's airline and airliner database. This will allow readers to search for information by locating the operator first and then consulting the appropriate aircraft entry in the main body of the book.

For some years our coverage of the little aircraft, sailplanes, microlights, hang gliders and the homebuilt sector has been steadily reduced and compressed. Last year we gave no more than basic outline information. At the same time, as the price of the book has been increased, our readership has become more professional and official. The time had to come when we accepted the inevitable and ceased covering the very small and largely recreational end of the spectrum. We have set aside balloons as similarly strictly recreational, but we have retained airships in our Lighter Than Air section and will include with them professionally used balloons or those of special technical interest.

We have conserved from last year's Private Aircraft section those light aircraft that are already substantive commercial ventures or are likely to become so. Those we have dropped we judge to be unlikely to achieve some form of certification or extensive public recognition. The small end of the market has been very fluid and new technology and certification requirements have left many grey areas, but we will continue to ensure that significant small aircraft remain in *Jane's*.

Sport/private aircraft, sailplanes, microlights, hang gliders and balloons continue to feature in the index, because this covers the preceding decade. Readers will still be able to find reference to entries in those preceding editions.

IN OUR BUSINESS SOMETIMES THINGS GET UGLY

The CSH-2 Rooivalk.

Designed to be mean, Rooivalk is a sophisticated, high performance combat support helicopter with the high power fighting capabilities you need at your command.

Born from active combat experience, Rooivalk flies as Atlas Aviation's milestone aviation achievement - comparable to the world's best combat helicopters.

A symbol of the strength and reliability Atlas Aviation provides in everything - from aircraft design and development to manufacture and maintenance.

It's this dedication to aerospace perfection that made the Rooivalk possible.

And it's the kind of dedication that ensures you'll always have the advantage. Even when things get ugly.

ATLAS AVIATION
MAINTAINING YOUR EDGE

Atlas Aviation, a Division of Denel (Pty) Ltd. P O Box 11 Kempton Park 1620, South Africa. Contact: Marketing Manager Tel: 27-11-927 4117 Fax: 27-11-973 5353.

FOREWORD

USA by reckless fare discounting and remorseless increases in capacity which have blocked any return to profitability. Many of the US airlines are operating in the never-never land of Chapter 11 protection, which powerfully distorts the market.

IATA sees little hope of the airlines getting out of the red during 1993. Even a 22 per cent growth in international traffic over five years has been offset by 33 per cent growth in capacity; costs fell one per cent in 1992 and yields rose 0.7 per cent, but overcapacity wiped out the gains.

Tales of woe could go on for ever. As far as the manufacturers are concerned, $15 billion of orders for new aircraft have been cancelled. New orders have been coming in at the rate of 300 or so a year, but production has been running at 800 a year. So suddenly did the downturn follow the order rush of the late 1980s.

Boeing is unquestionably the strongest aerospace company in the world, ending 1992 with a backlog of $87.9 billion, but it took prompt and massive action, including 28,000 redundancies, to overcome the effects of simultaneous airline and defence declines. Boeing's Defense and Space Group now has major shares in the F-22, Comanche, V-22 Osprey, 707 and 767 AWACS, Chinook, maritime avionics, Avenger missile system, Inertial Upper Stage and NASA Space Station, and returned to profitability in 1992.

The airlines have ordered more than 450 Boeing airliners since the last Paris Air Show, but all the manufacturers together will be selling 40 per cent fewer passenger seats by the Farnborough Air Show in 1994. The big three, Boeing, Airbus and Douglas, are all severely cutting back production for the next few years, but their forecasts agree that output will pick up in the late 1990s and again during the first decade of the next century. Even then, it will only return to pre-recession level rather than hit a new high. World air travel will strongly increase, though mainly in the Asia-Pacific region.

There are various ways of coping with redundancies. Boeing can fire thousands who, it seems, go back to family farms in the region. Aerospatiale workers and suppliers will have from two to five weeks of enforced unpaid holiday during 1993, mainly as a result of order cancellations by Northwest and GPA. Instead of reaching an output of 350 Airbus and ATR airliners during 1993, as originally forecast, Aerospatiale now foresees 205 in 1993 and 240 or fewer in 1994. Airbus had orders for 92 airliners cancelled during 1992 and retained orders for 41. Boeing lost only 13, but had a large number of deliveries postponed.

Japan's aircraft industry, an increasingly significant player in the market, lost out in both exports and exchange rates in 1992-93. Volume was down by only a few per cent, but the weak dollar forced the value of exports down by 28 per cent.

The engine companies cannot escape. Pratt & Whitney spares sales declined by about one-third in the first quarter of 1993 and are now $700 million a year less than two years ago. The workforce is being reduced by 28 per cent to 30,000 by the end of 1994 and the company is aiming for 26,000 in 1995. Factory floor area has been reduced by 35 per cent. Output has declined from about 1,700 large military and civil engines in 1981 to 575 in 1993.

General Electric, reacting quickly, has reduced its workforce by 13 per cent to 29,100. Rolls-Royce will have reduced its workforce by 17 per cent to 24,500 by the end of 1994. By contrast Rolls-Royce has doubled its civil engine deliveries during the past five years from 200 to more than 400 and expects to move up from a 23 per cent share of the civil market to about 30 per cent by the end of the decade.

The ruthless logic of cost has now forced the three engine giants to recognise that none of them might be able to finance the development of a new big engine alone. They are already discussing co-operation.

The recovery

Now the train has hit the buffers. The immediate remedy, as always, has been to reduce capacity, but when redundancies reach the scale of the present wave (117,000 in the USA alone during 1992 and probably not far off 100,000 in 1993) manufacturers start worrying about whether they can stay in the business and whether the industry can retain a full development ca-

There will probably be an 11-month hiatus on the McDonnell Douglas production line between the delivery of the last US Air Force F-15E and the first F-15S for Saudi Arabia

One of the two new worldwide super groups will be formed around the advanced supersonic transport. This is the Aerospatiale/BAe Alliance proposal

pability. General Dynamics decided early and sold its whole Fort Worth fighter manufacturing business to Lockheed and the associated electronics business to GM Hughes.

There are some positive features. Factories are being reorganised and lead times reduced. For example, Douglas Aircraft took the first step of reducing its workforce from between 40,000 and 43,000 in 1990, before the military C-17 was hived off to McDonnell Douglas Aerospace, to around 19,000 by the end of 1992 and 15,000 by the end of 1993. At the same time, it streamlined its assembly process so that the man-hours required to produce an MD-11 are reducing from 300,000 in 1991 to 150,000 by the end of 1993. Man-hours required for a twin-jet are similarly being reduced from 100,000 to 50,000. The greatly streamlined MD-90 manufacturing system, which replaces fuselage assembly from long panels by assembly in modular barrel sections, is being applied to the MD-80

line during 1993. Douglas can therefore remain viable while producing fewer than 30 twin-jets a year (compared with 136 needed in 1989) and around 30 tri-jets a year.

Among European manufacturers, ATR says it can now survive on fewer than 60 ATR 42/72 deliveries a year and Saab has reduced its assembly time for the 340 in two stages from nearly 200 days to well under 100. Fast response to orders is now a prime objective and will become extremely important in riding out the peaks and troughs in orders.

BAe saw substantial profits from its defence business absorbed by losses in the civil aircraft enterprises. It set up those activities as substantive and therefore saleable enterprises and put them on the market. After successive false starts, the regional jet business has become a joint enterprise with Taiwan Aerospace and revived the old Avro name with the title Avro International Aerospace.

[15]

FOREWORD

Jetstream Aircraft was formed to press on with the two Jetstream turboprops, which have suffered the same sales downturn as all the other commuter airliners. Jetstream also picked up the old Avro/Hawker Siddeley/BAe ATP and is appropriately expanding it into a family of new Jetstreams.

BAe's Corporate Jets found a good home with Raytheon, starting an instant family to go with Beechcraft's ex-Mitsubishi Beechjet. The ex-BAe 800 and 1000 might revive another old British name, because their predecessors were known in the USA as Hawkers when Beechcraft was a sales agent during the 1960s.

Reorganisation

Last year I expressed surprise that the US airframe industry showed so few signs of rationalising to match the seemingly permanent reduction in sales. This year, the companies are well into the realignment process and there is certainly more to come, either voluntarily or forced by market pressures. Planning must be greatly complicated by yet another reshuffle of US military programmes introduced by the incoming Clinton administration, but at least the US aerospace industry is targeted as a focal point for economic recovery.

The main realignment has come in the avionics industry, with Hughes being bought by General Motors, GE Aerospace by Martin Marietta and LTV avionics interests going to Loral. European avionics companies have followed a similar rationalisation, but by means of alliances rather than acquisitions.

While Boeing and Lockheed seem to be safe and sound, I wonder very much how the airframe-building sections of McDonnell Douglas, Rockwell, Northrop, Vought and Grumman will look this time next year. McDonnell Douglas has taken its helicopter subsidiary off the market and will keep it.

Chaos still reigns in the regional airliner sector: too many suppliers, too many types of aircraft, too many outdated designs, too many unrealistic ambitions and too many new entrants in the market. At least the manufacturers are bringing some order into the market. DASA has finally taken over Fokker, but will let Fokker hold onto its name and status until a decision on the 130-passenger airliner has to be taken. It is just possible that ATR will join Fokker/DASA. Bombardier has taken over Canadair, de Havilland and Shorts. It remains to rationalise the range of types. It is difficult to see the logic of the CASA 3000 and the IPTN 250-100 although they are at least aimed at the viable 70-passenger end of the market. By the same token, the Fokker 70 looks well aimed. By contrast, the Dornier 328 is uncomfortably placed in the market. It is too early to judge the prospects of Avro International Aerospace and Jetstream.

Does the regional future lie with the jet or the turboprop? Can active noise control and excess power allow the turboprop to hold onto its cost advantage over the jet?

Here come the super groups

It is clear that we are about to see the emergence of two super groups to undertake the new programmes that no single manufacturer can finance. They will take shape around the advanced supersonic transport and the very large commercial transport (VLCT) otherwise known as the ultra-high capacity airliner (UHCA). The international effort between Boeing and the Airbus partners to examine the potential for a VLCT seems to be progressing more positively and quickly than expected.

Passenger traffic between Asia and Europe is likely to grow by 7.5 per cent a year until 2010: between Asia and North America it will grow slightly faster and over the North Atlantic it will grow at 4.5 per cent a year.

For those long over-ocean routes and in the face of increasing airport congestion, the manufacturers believe they can now see a market for between 400 and 500 VLCTs, which is commercially worthwhile, but too small to justify more than one type of aircraft. Despite the economic situation, there seems to be a good chance that an industrial team for the VLCT could be formed. Firm plans must be made fairly soon if the VLCT is to be ready at the turn of the century.

On the technical side, the companies seem to agree that the right initial capacity will be 600 passengers with a range of about 7,000 nm (13,000 km; 8,000 miles). The aircraft must fit existing airports and respect the environment. For optimum work-sharing, subassemblies should be as large as possible, but they must still be transportable across the Atlantic in one direction or the other.

The advanced supersonic transport is taking shape as the other super group-forming programme. Many more companies and countries are involved in the initial studies (see SCTICSG in the International section). R&D is well under way and money is already being spent. In the USA, NASA is increasing its civil SST expenditure from $117 million in 1993 to $187 million in 1994 and a fair proportion of this is going to Boeing and Douglas in the form of study contracts. Many technical breakthroughs remain to be made, including laminar flow and variable cycle engines, but all the manufacturers seem to believe that these will be achieved.

There must have been half a dozen SST models at the 1993 Paris Air Show, including those from Japan and Russia. Whereas Concorde was a major technical success and a social disaster, it is now taken for granted that the SST must be genuinely commercially viable, environmentally acceptable and able to operate economically over land as well as the oceans. The 14 Concordes which have been operating for 20 years are still the most prestigious airliners in the world, but the lessons have been well and truly learned.

How can Russia/Ukraine join the club?

How can the former Soviet aircraft industry, for so many years a secret giant in the Soviet state, rejoin the mainstream of international aeronautical activity? Is it possible at all?

The industry has virtually been privatised and told to rely on its own resources. Its reaction seems to have been to diversify by forming numerous new groupings and letting the bigger manufacturing centres set up their own sales activities. An amazing variety of new designs has been proposed in the apparent expectation of a sudden outburst of flying in all its forms. But it is hard to believe that this will occur or that the export market will absorb more than a tiny fraction of the products proposed. There is no lack of manufacturing skill, but there seems to be little real commercial judgement.

The solid airliner programmes will succeed, because CIS airlines need them and will not be able to

One of the two expected super groups will form around the so-called very large commercial transport. The Deutsche Aerospace Airbus A 2000 proposal is shown here (*Brian M. Service*)

The Russian Il-96M powered by P&W turbofans flew during the Paris Air Show, as promised two years ago, but will probably first fly in the West as a freighter (*Brian M. Service*)

buy more than a few foreign airliners. A hard core of military programmes, like Su-27 developments, the new MiG (with SNECMA) or Yak (with Aermacchi) advanced trainers, the A-50 AWACS and a new military tactical freighter will take up some capacity.

Foreign suppliers, including East European companies, are demanding hard currency for necessary aircraft like the Czech Let 410/610 and Polish-made An-28, and their spares, and they have become unaffordable. Pratt & Whitney has invested heavily in its Russian programme, but requires returns in hard currency.

Russia and Ukraine have had a strong traditional export market, not least in helicopters, but that has been invaded by Boeing and the others during the years of chaos. To tap into the general world market, the Russians have to win the vital membership ticket of Western certification.

Theoretically, that should be easy. Russian manufacturers are perfectly capable of meeting Western engineering standards, but there will inevitably be Western resistance to a big new competitor in the world market. It has always been easy to put up those subtle barriers of the sort that have already held up some certification validations within the West. Eurocopter and Mil plans to develop the Mi-38 together seem eminently sound.

Another problem is that Western certification demands an immense volume of supporting documentation, which must be presented in English. With the best will in the world, Russia does not have the translation capacity to get through this before 1995 or even later. Another snag is that fire-retardant furnishing fabrics have never been used there. They will have to be acquired from the West and integrated into the system.

Low labour rates will keep Russian airliners cheap in the West, even if the manufacturers work out a realistic commercial price in Roubles. A Rolls-Royce-powered Tu-204 is offered at about $38 million, some 30 per cent less than the equivalent Boeing 757. The Il-96M is going for a mere $68 million. But price is not everything and Western customers are also concerned about after-sales support. One story going the rounds at the Paris Show was that a company proposing to buy Il-96Ms asked about product support and was offered three extra aircraft free of charge as a spare parts pool.

Another factor is that Eastern operators buying Russian airliners powered by Western engines will have to produce hard currency. To earn that money, the aircraft must be operated on high-density hard currency-earning routes. Netherlands leasing company Partnair started the ball rolling by signing initial documents in June 1993 to buy five Il-96Ms, plus options on five, but still has to obtain a certifiable passenger cabin interior. Virtually all the Il-96s and other types sold outside the CIS will initially be supplied as freighters and will prove themselves in that trade before being accepted for passenger flying.

In the end, the airlines will dictate developments. Will they trust the product and do they want to see another supplier in the marketplace?

In the military field, the Malaysian fighter decision, announced as we go to press, may be a pointer: a small batch of eight F/A-18Ds delivered in 1996 and 18 MiG-29s delivered almost immediately. The cheap purchase combined with the high-cost insurance policy.

One of the surprises of the Paris Air Show was the appearance of a newly painted example of the ancient Beriev Be-32 commuter airliner. The original Be-30 flew in 1967 and the Be-32 in 1970 and the aircraft at Paris had literally been taken out of a museum and refurbished. Production is being restarted at Irkutsk. The reason is that the Be-32 is the best answer to the pressing need for a replacement of the L-410 and An-28, now starved of spares.

China

China has become a greater uncertainty than Russia/Ukraine. The industry is now a major supplier of military aircraft and missiles and has broken into the space launcher market. But civil production and exports have hardly exploded and the Trunkliner programme is cautious in its scope. Douglas is keen to publicise the programme, but Boeing has sold well over 100 of its airliners in the country. As tourism grows, the internal airlines have multiplied and new aircraft are being introduced quite rapidly. But where will the industry be in five years' time?

China may in fact be seen as less of a threat to the stability of the Western aircraft market than South Korea, Taiwan, Indonesia and similar countries.

Gripen is go

The Saab Gripen must be the most significant military aircraft on the market today and we have given it pride of place in our frontispiece. If one scans the available military aircraft from the lightest upwards, remembering that weight has a direct bearing on cost, the Gripen is the first one to offer the full range of capabilities an air force might require. It can make interceptions beyond visual range, it can deliver large air-to-surface missiles, it has built-in reconnaissance capability and can operate from short stretches of road. It is designed to be serviced by simply trained reservist groundcrews.

Now, the Gripen's time has come. Performance has exceeded expectations and the first aircraft has been delivered to the Swedish Air Force, which has ordered 130. It has launched a wide selection of its intended range of weapons. Finally, the Swedish government is supporting export sales much more seriously.

The Gripen is the fifth successive combat aircraft that Sweden has successfully developed. Not one has been cancelled. That achievement is deeply impressive.

Eurofighter 2000

A thoroughly regrettable feature of the Paris Air Show was the absence of the Eurofighter 2000. All three examples of France's Rafale were flying in public, demonstrating once again the success of single-minded determination. Sweden's Gripen was not there in the flesh, but not out of mind.

Where was Eurofighter 2000? Apparently glued to the concrete more than a year after its initially programmed first flight date and suffering grievously from more or less justified political interference and absolutely appalling publicity. German Defence Minister Volker Rühe finally released next year's government funding, but not this year's. He won a reduction in price and a stretch-out in the programme. The manufacturers declined to amend their contracts. The price went down 13 per cent and Mr Rühe learned something about the reality of contracts and programmes, but no-one was happy. Now, we have to see

JAS 39 Gripen was absent from the 1993 Paris Air Show, but evidently progressing well. All three prototypes of France's Dassault Rafale flew in public. Eurofighter 2000 was conspicuously still grounded 18 months after its original first flight date, bogged down in political, financial and technical problems. Shown above is the two-seat French Air Force Rafale B01 (*F. Robineau*)

how Eurofighter will fare in the marketplace against the powerful, but keenly priced Sukhoi Su-27/30 Flanker.

It is strange that an aircraft that flies is so much more credible (and uncancellable) than one stuck on the ground. It has to be admitted, nevertheless, that in these days of fly-by-wire, the publicity resulting from over-hasty first flights, as we have seen, can be worse than a prudent delay.

Technology: still progressing

It is tempting to think that little progress remains to be made in aircraft design and that avionics are taking over as the driving force. That is true to some extent if one considers fly-by-wire, flight management systems, so-called autonomous approaches using the Global Positioning System, integrated helmet displays for night low-level flight, countermeasures and a host of virtually indispensable systems. But aerodynamic innovation and new propulsion technologies continue to improve the performance of aircraft.

"A single 767 can carry as many passengers as two 727s plus half as much cargo as a 707 freighter while consuming two per cent less fuel than a single 727", remarked Boeing President Phil Condit during the Paris Air Show. Aircraft technology is still advancing and it is becoming more friendly, because such reductions in fuel consumption directly benefit the environment by reducing both consumption and emissions.

Airbus is not letting people forget that the promised cross-qualification all the way from A319 to A340, with difference courses lasting only a few days, has significant implications for operators of Airbus fleets. As 15-hour non-stop flights with double crews become routine, pilots can go for many weeks without making a landing or take-off and suffer the effects of boredom. With cross-qualification, long-range pilots can be periodically transferred to narrow-body aircraft and keep in touch with intensive short-range operations.

Boeing plans to obtain clearance for extended-range twin-engined over-water operations (ETOPS) for the 777 at entry into service. ETOPS is already a dominant airline operating technique and more passengers now cross the Atlantic in twin-engined than in three- or four-engined airliners, but there are professionals who regard a ditching as inevitable. It remains to be proved in the long run whether ETOPS will have been a clearance too far.

The light aircraft scene is still in some confusion and

FOREWORD

depression. The US top two, Cessna and Beechcraft, moved out of the small single-engined business years ago and the trade is still menaced by product liability law as well as minimal sales. Strangely, the largest number of piston singles was produced last year by slowly reviving Piper under the shelter of Chapter 11 protection from bankruptcy.

European Joint Airworthiness Regulations for Very Light Aircraft (JAR/VLA) and the FAA equivalent, are giving a new lease of life to light aircraft by reducing the cost of development and operation. At the same time, better aerodynamics and more innovative structures are improving performance. Precisely built composites aerofoils and light structures are making it possible to carry two people in comfort on as little as 52 kW (70 hp).

It is too early to say where this might eventually lead worldwide club and private aviation. For us, it makes it all the harder to decide whether a new small aircraft deserves a place in our Aircraft section or not.

Sevenoaks, July 1993　　　　　　　　　Mark Lambert

Acknowledgements

Our editorial team, too, has had a hard year and been reduced by one. After 24 years' stalwart service, Assistant Editor Michael Taylor is leaving to head a new venture. The decision to eliminate most private/sport aircraft, sailplanes, microlights and hang gliders left him virtually no compiling to do. He took over the civil sections of France, UK and USA in place of John Cook, who left us last Summer, but decided to sever his connection altogether after the completion of this production cycle.

For the last time, Mike Taylor and Deputy Editor Ken Munson have shared the reading of all the proofs and maintained the style and standard of the book. In future, Ken will carry the full burden of editing and continue to compile virtually all the countries from Argentina to Turkey and the Lighter Than Air section. Editor Emeritus John Taylor still contributes his unique expertise on Russia/Ukraine. Paul Jackson is our infallible source for the military side of USA, UK and France and for Yugoslavia. Bill Gunston is our authority on aero engines. I now compile most of the International section, Germany, Italy and the Boeing and Douglas entries, which require a different approach. Ian Strachan, fresh from an unusually multifarious career in the British Royal Air Force, has joined us as a compiler.

Our hundreds of three-view drawings, this year much expanded, are the work, as usual, of Mike Keep and Dennis Punnett. Sadly, Dennis Punnett died suddenly on 12 July. This is the end of an era in the art and technique of the three-view drawing. Our two full-page cutaways are by Mike Badrocke and Robert Roux. Maurice Allward compiles our 10-year rolling index. Incidentally, it continues to list private and sport aircraft and the sailplanes and small aircraft that have featured in past editions.

Of our many outside contributors and photographers, those who are willing to be mentioned include Andrea Artoni of *Volare*, the *Avio Data* team, Peter Bowers, Piotr Butowski, John Cook, Peter J. Cooper, Paul R. Duffy, *Flight International*, John Fricker, Lutz Freundt, Mike Gradidge, Geoffrey P. Jones, Vaclav Jukl of *Letectvi + Kosmonautika*, Howard Levy, Neil A. Macdougall, Ryszard Jaxa-Malachowski, Jacques Marmain, Brian and Margaret Service, Ivo Sturzenegger and Jim Thorn of *Australian Aviation*.

At Coulsdon, Ruth Simmance continues as Publishing Supervisor, leading VDU typesetters Sarah Erskine and Kathryn Jones; Lynn Morse in the same department co-ordinates all the illustrations and finally reads all the proofs.

A new publishing system allowing us to typeset copy electronically, scan illustrations into a data bank and make up pages has been installed in the Coulsdon offices this year. Neil Beirne made up all the pages to create the book as we see it. Chrissie Richards, Jane's Book Production Controller, watches over that activity.

The managing directorship of Jane's Information Group passed from Michael Goldsmith to Alfred Rolington at the beginning of 1993. The whip is still being cracked over Jane's Data Division by Publishing Director Bob Hutchinson, assisted now by Gina Dawkins, and by Managing Editor Keith Faulkner.

First Flights

Some of the first flights made during the period 1 April 1992 to 30 June 1993

APRIL 1992
- 1 Airbus A340-200 (F-WWBA) (International)
- 9 USAF/General Dynamics variable stability (VISTA) NF-16D (86-0048) (USA)
- 11 Guimbal G2 Cabri (F-PILA) (France)
- 11 WAI Sentinel 1000 airship, first flight with production-standard fly-by-light control system (USA)
- 15 McDonnell Douglas F/A-18 Hornet, first flight with Hughes AN/APG-73 upgraded radar (USA)
- 15 McDonnell Douglas AH-64D Longbow Apache (89-0192) (USA)
- 17 Northrop B-2, fourth prototype (82-1069) (USA)
- 18 Dee Howard (Boeing) 727-100, first refitted with Tay engines (USA)
- 24 Zlin 143 L (OK-074) (Czech Republic)
- 26 Sequoia Model 300 Sequoia homebuilt (N48BL) (USA)

MAY 1992
- 5 WSK-PZL Mielec I-22 Iryda, first pre-production (103) (Poland)
- 18 McDonnell Douglas C-17A Globemaster III, first production (P-1/88-0265) (USA)

JUNE 1992
- 4 Dornier 328, second prototype (D-CATI) (Germany)
- 4 IAI (Sikorsky) CH-53D, first Yasur 2000 upgrade (044) (Israel/USA)
- 11 Grob G 115T (D-EMGT) (Germany)
- 12 Eurocopter/Vought AS 565 Panther 800 (N6040T) (France/USA)
- 15 Grumman EA-6B Prowler, first VEP upgrade aircraft (USA)
- 15 Global Helicopter Huey 800 (N800NT, formerly 69-15092) (USA)
- 15 Learjet 60, first production (N601LJ) (USA)
- 15 LTV F-8E(FN) Crusader, first upgraded aircraft for Aéronavale (35) (USA/France)
- 21 McDonnell Douglas C-17A Globemaster III, second production (P-2/88-0266) (USA)
- 26 SAMF (McDonnell Douglas) MD-83, first Chinese-built (China/USA)

JULY 1992
- 3 Saab 2000, second prototype (SE-002) (Sweden)
- 6 Air Tractor AT-802A (N1546H) (USA)
- 7 Rockwell/General Dynamics F-111F, first Pacer Strike upgrade (USA)
- 8 BAe Jetstream 41, fourth aircraft (first demonstrator) (G-JMAC) (UK)
- 8 Bede BD-10J homebuilt (N2BJ) (USA)
- 10 Dassault Falcon 20 testbed, first flight with RBE2 radar for Rafale (104) (France)
- 20 Panavia Tornado ECR, first Italian conversion (MM7079) (International)
- 25 Teledyne Ryan Model 410, first unmanned flight (N53578) (USA)

AUGUST 1992
- 7 Ilyushin Il-114, first production (Russia)
- 14 Tupolev Tu-204-220, first flight with Rolls-Royce RB211-535E4 engines (RA-64004) (Russia/UK)
- 18 Panavia Tornado F. Mk 2, first flight as TIARA avionics research testbed (ZD902) (International)
- 20 Hindustan Aeronautics ALH (Z 5182) (India)
- 21 Cessna Citation VII, first flight as testbed for Allison GMA 3007 engine (N650) (USA)
- 21 UNC (Bell) UH-1H Ultra Huey (N700UH) (USA)
- 28 Saab 2000, third development aircraft (SE-003) (Sweden)

SEPTEMBER 1992
- 7 McDonnell Douglas C-17A Globemaster III, third production (P-3/89-1189) (USA)
- 10 IG JAS 39A Gripen, first production (39101) (Sweden)
- 13 AASI Jetcruzer 450/500P, first flight in production configuration (N102JC) (USA)
- 15 Tridair 206L-3ST Gemini ST, second prototype (N100TH) (USA)
- 17 SIAI-Marchetti S.211A (I-PATS) (Italy)
- 22 McDonnell Douglas/BAe AV-8B Harrier II Plus (164129) (International)
- 24 Airtech CN-235 M, first assembled by Turkish Aerospace Industries (International/Turkey)

OCTOBER 1992
- 2 Eurocopter BK 117 P5 (Kawasaki fly-by-wire flight control system testbed) (JQ0003) (International)
- 5 Northrop B-2, fifth aircraft (82-1070) (USA)
- 20 Dornier 328, first pre-production (D-COOL) (Germany)
- 20 Yakovlev Yak-112 (RA-00001) (Russia)

707 with a difference: reconfigured by Israel Aircraft Industries with Elta's Phalcon AEW system, it flew for the first time on 12 May 1993 and was to be delivered later in the year to the Chilean Air Force

FIRST FLIGHTS

NOVEMBER 1992
- 2 Airbus A330 (F-WWKA) (International)
- 6 EAI Eagle XTS, production prototype (VH-XEP) (Australia)
- 10 Airbus A320, first flight with IAE V2500-A5 engines (F-WWAI) (International)
- 13 McDonnell Douglas AH-64D Longbow Apache, second prototype (USA)
- 18 PAC Cresco 08-600-34AG (ZK-TMN) (New Zealand)
- 22 BHEL LT-1 Swati, first production (VT-XSR) (India)
- 27 Avro International Aerospace RJ85 (G-CROS, first production RJ) (UK)

DECEMBER 1992
- 8 Eurocopter AS 532UL Cougar, first flight with full Horizon radar (F-ZVLJ) (France)
- 9 McDonnell Douglas C-17A Globemaster III, fourth production (P-4/89-1190) (USA)
- 10 Fokker 50 Maritime Enforcer Mk 2 (PH-LXX) (Netherlands)
- 17 Molniya-1 (Russia)
- 18 Let L-610G (OK-134/OK-XZA) (Czech Republic)
- 18 McDonnell Douglas Explorer (N900MD) (USA)
- 22 WSK-PZL Mielec I-22 Iryda, fourth prototype, first flight with K-15 engines (SP-PWD) (Poland)
- 23 Beechcraft (Pilatus) PC-9 Mk II, first production prototype (N8284M) (USA/Switzerland)
- 28 Sabca/Dassault Mirage 5BA, first MIRSIP upgrade (BA60) (Belgium/France)

JANUARY 1993
- 15 Rockwell/DASA Ranger 2000 (D-FANA) (International)
- 17 Let L-33 Solo sailplane, first production (c/n 930101) (Czech Republic)
- 23 Dornier 328, first production (c/n 4, D-CITI) (Germany)
- 26 Grob G 115C (Germany)
- 31 McDonnell Douglas C-17A Globemaster III, fifth production (P-5/89-1191) (USA)
- 31 Sikorsky UH-60Q (86-24560) (USA)

FEBRUARY 1993
- 2 Northrop B-2, sixth aircraft (82-1071) (USA)
- 8 Rocky Mountain Helicopters (Eurocopter) SA 315B Lama, first flight with PT6B-35H engine (N220RM) (International/USA)
- 12 Eurocopter AS 355 Ecureuil/TwinStar, first flight with Arrius 2C engine (F-WGTM) (International)
- 22 McDonnell Douglas MD-90 (N901DC) (USA)

MARCH 1993
- 4 IG JAS 39A Gripen, second production, first for Swedish Air Force (39102) (Sweden)
- 4 Dassault Falcon 2000 (F-WNAV) (France)
- 11 Airbus A321 (F-WWIA) (International)
- 17 Saab 2000, first production (SE-004) (Sweden)
- 22 Freewing Aircraft Freebird Mk 5 (N13FW) (USA)

APRIL 1993
- 2 Fokker 70 (PH-MKC) (Netherlands)
- 6 Ilyushin Il-96MO (RA-96000) (Russia)
- 7 Embraer EMB-312F Tucano, second pre-series aircraft for French Air Force (but first to fly) (PP-ZVD/439) (Brazil)
- 8 Embraer EMB-312F Tucano, first pre-series aircraft for French Air Force (PP-ZVC/438) (Brazil)
- 9 Fournier RF-47 (F-WNDF) (France)
- 18 Hindustan Aeronautics ALH, second prototype (Z 5183) (India)
- 21 Grob G 520T (D-FDST) (Germany)
- 22 Eurocopter HAP Gerfaut (second Tiger prototype) (PT2/F-ZWWY) (International)
- 30 Dassault Rafale B (first two-seat Rafale) (B01) (France)

Hindustan Aeronautics' Advanced Light Helicopter made its first flight on 20 August 1992, followed by the second prototype in April 1993

First production example of the Fokker 50 Maritime Enforcer Mk 2, which flew for the first time on 10 December 1992

The second, third and fourth Dornier 328s join the prototype (farthest from camera) for a formation flypast

FIRST FLIGHTS

MAY 1993
- 7 Boeing 747-400F (N6005C) (USA)
- 8 McDonnell Douglas C-17A Globemaster III, sixth production (P-6/89-1192) (USA)
- 10 Aero L-139 Albatros (5501) (Czech Republic)
- 12 Boeing 707, first with Elta Phalcon AEW radar (4X-JYT) (Israel)
- 14 Saab 2000, second production (SE-005) (Sweden)
- 15 Embraer EMB-312H Super Tucano (PT-ZTV) (Brazil)
- 25 Vought/FMA Pampa 2000, first 'missionised' prototype (E-812) (Argentina/USA)
- 28 Pilatus PC-12, second prototype (HB-FOB) (Switzerland)
- 29 Panavia Tornado GR. Mk 4 (converted Mk 1), first fully equipped trials aircraft for mid-life update (ZD708) (International)

JUNE 1993
- 2 WSK-PZL Warszawa-Okecie PZL-130TC Turbo-Orlik (SP-PCE) (Poland)
- 9 Saab 2000, third production (SE-006) (Sweden)
- 18 Rockwell/DASA Ranger 2000, second prototype (D-FANB) (International)*
- 24 Saab 2000, fourth production (SE-007) (Sweden)
- 30 McDonnell Douglas AH-64D Longbow Apache, third prototype (USA)

*Aircraft destroyed in an accident 27 July

First example of the Garrett-engined Aero L-139, appearing at the Paris Air Show soon after its first flight on 10 May 1993 *(Paul Jackson)*

First (right) and second prototypes of Rockwell/DASA's JPATS candidate, the Ranger 2000, first flown on 15 January and 18 June 1993

Glossary of aerospace terms in this book

AAM Air-to-air missile.
AATH Automatic approach to hover.
AC Alternating current.
ACE Actuator control electronics.
ACLS (1) Automatic carrier landing system; (2) Air cushion landing system.
ACMI Air combat manoeuvring instrumentation.
ACN Aircraft classification number (ICAO system for aircraft pavements).
ADAC Avion de décollage et atterrissage court (STOL).
ADAV Avion de décollage et atterrissage vertical (VTOL).
ADC (1) US Air Force Aerospace Defense Command (no longer active); (2) air data computer.
ADF Medium frequency automatic direction finding (equipment).
ADG Accessory-drive generator.
ADI Attitude/director indicator.
aeroplane (N America, airplane) Heavier-than-air aircraft with propulsion and a wing that does not rotate in order to generate lift.
AEW Airborne early warning.
AFB Air Force Base (USA).
AFCS Automatic flight control system.
AFRP Aramid fibre reinforced plastics.
afterburning Temporarily augmenting the thrust of a turbofan or turbojet by burning additional fuel in the jetpipe.
AGM Air-to-ground missile.
AGREE Advisory Group on Reliability in Electronic Equipment.
Ah Ampère-hours.
AHRS Attitude/heading reference system.
AIDS Airborne integrated data system.
aircraft All man-made vehicles for off-surface navigation within the atmosphere, including helicopters and balloons.
airstair Retractable stairway built into aircraft.
AIS Advanced instrumentation subsystem.
ALCM Air-launched cruise missile.
AM Amplitude modulation.
AMAD Airframe mounted accessory drive.
anhedral Downward slope of wing seen from front, in direction from root to tip.
ANVIS Aviator's night vision system.
AP Ammonium perchlorate.
APFD Autopilot flight director.
approach noise Measured 1 nm from downwind end of runway with aircraft passing overhead at 113 m (370 ft).
APS Aircraft prepared for service; a fully equipped and crewed aircraft without usable fuel and payload.
APU Auxiliary power unit (part of aircraft).
ARINC Aeronautical Radio Inc, US company whose electronic box sizes (racking sizes) are the international standard.
ARV Air recreational vehicle or air reconnaissance vehicle, according to context.
ASE (1) Automatic stabilisation equipment; (2) Aircraft survivability equipment.
ASI Airspeed indicator.
ASIR Airspeed indicator reading.
ASM Air-to-surface missile.
aspect ratio Measure of wing (or other aerofoil) slenderness seen in plan view, usually defined as the square of the span divided by gross area.
ASPJ Advanced self-protection jammer.
AST Air Staff Target (UK).
ASV (1) Air-to-surface vessel; (2) Anti-surface vessel.
ASW Anti-submarine warfare.
ATA Air Transport Association of America.
ATC Air traffic control.
ATDS Airborne tactical data system.
ATHS Airborne target handover (US, handoff) system.
ATR Airline Transport Radio ARINC 404 black box racking standards.
attack, angle of (alpha) Angle at which airstream meets aerofoil (angle between mean chord and free-stream direction). Not to be confused with angle of incidence (which see).
augmented Boosted by afterburning.
autogyro Rotary-wing aircraft propelled by a propeller (or other thrusting device) and lifted by a freely running autorotating rotor.
AUW All-up weight (term meaning total weight of aircraft under defined conditions, or at a specific time during flight). Not to be confused with MTOGW (which see).
avionics Aviation electronics, such as communications radio, radars, navigation systems and computers.
AVLF Airborne very low frequency.
AWACS Airborne warning and control system (aircraft).

bar Non-SI unit of pressure adopted by this yearbook pending wider acceptance of Pa. 1 bar = 10^5 Pa. ISA pressure at S/L is 1013.2 mb or just over 1 bar. ICAO has standardised hectopascal for atmospheric pressure, in which ISA S/L pressure is 101.32 hPa.
bare weight Undefined term meaning unequipped empty weight.
basic operating weight MTOGW minus payload (thus, including crew, fuel and oil, bar stocks, cutlery etc).
BCAR British Civil Airworthiness Requirements (see JAR).
bearingless rotor Rotor in which flapping, lead/lag and pitch change movements are provided by the flexibility of the structural material and not by bearings. No rotor is rigid.
Beta mode Propeller or rotor operating regime in which pilot has direct control of pitch.
birotative Having two components rotating in opposite directions.
BITE Built-in test equipment.
bladder tank Fuel (or other fluid) tank of flexible material.
BLC Boundary-layer control.
bleed air Hot high-pressure air extracted from gas turbine engine compressor or combustor and taken through valves and pipes to perform useful work such as pressurisation, driving machinery or anti-icing by heating surfaces.
blisk Blade plus disc (of turbine engine) fabricated in one piece.
blown flap Flap across which bleed air is discharged at high (often supersonic) speed to prevent flow breakaway.
BOW Basic operating weight (which see).
BPR Bypass ratio.
BRW Brake release weight, maximum permitted weight at start of T-O run.
BTU Non-SI unit of energy (British Thermal Unit) = 0.9478 J.
bulk cargo All cargo not packed in containers or on pallets.
bus Busbar, main terminal in electrical system to which battery or generator power is supplied.
BVR Beyond visual range.
bypass ratio Airflow through fan duct (not passing through core) divided by airflow through core.

C^3 Command, control and communications.
C^3CM Command, control, communications and countermeasures.
CAA Civil Aviation Authority (UK).
cabin altitude Height above S/L at which ambient pressure is same as inside cabin.
CAD/CAM Computer-assisted design/computer-assisted manufacture.
CAM Cockpit-angle measure (crew field of view).
canards Foreplanes, fixed or controllable aerodynamic surfaces ahead of CG.
CAN 5 Committee on Aircraft Noise (ICAO) rules for new designs of aircraft.
CAR Civil Airworthiness Regulations.
carbonfibre Fine filament of carbon/graphite used as strength element in composites.
CAS (1) Calibrated airspeed, ASI calibrated to allow for air compressibility according to ISA S/L; (2) close air support; (3) Chief of the Air Staff (also several other aerospace meanings).
CBR California bearing ratio, measure of ability of airfield surface (paved or not) to support aircraft.
CBU Cluster bomb unit.
CCV Control configured vehicle.
CEAM Centre d'Expériences Aériennes Militaires.
CEAT Centre d'Essais Aéronautiques de Toulouse.
CEP Circular error probability (50/50 chance of hit being inside or outside) in bombing, missile attack or gunnery.
CEV Centre d'Essais en Vol.
CFRP Carbonfibre-reinforced plastics.
CG Centre of gravity.
chaff Thin slivers of radar-reflective material cut to length appropriate to wavelengths of hostile radars and scattered in clouds to protect friendly aircraft.
chord Distance from leading-edge to trailing-edge measured parallel to longitudinal axis.
CKD Component knocked down, for assembly elsewhere.
clean (1) In flight configuration with landing gear, flaps, slats etc retracted; (2) Without any optional external stores.
c/n Construction (or constructor's) number.
COINS Computer operated instrument system.
combi Civil aircraft carrying both freight and passengers on main deck.
comint Communications intelligence.
composite material Made of two constituents, such as filaments or short whiskers plus adhesive binding matrix.

CONUS Continental USA (ie, excluding Hawaii, etc).
convertible Transport aircraft able to be equipped to carry passengers or cargo.
core Gas generator portion of turbofan comprising compressor(s), combustion chamber and turbine(s).
C/R Counter-rotating (propellers).
CRT Cathode-ray tube.
CSAS Command and stability augmentation system (part of AFCS).
CSD Constant-speed drive (output shaft speed held steady, no matter how input may vary).
CSRL Common strategic rotary launcher (for air-launched missiles of various types).

DADC Digital air data computer.
DADS Digital air data system.
daN Decanewtons (Newtons force × 10); thus, torque measured in daN-metres.
DARPA Defense Advanced Research Projects Agency.
databus Electronic highway for passing digital data between aircraft sensors and system processors, usually MIL-STD-1553B or ARINC 419 (one-way) and 619 (two-way) systems.
dB Decibel.
DC Direct current.
DECU Digital engine control unit.
derated Engine restricted to power less than potential maximum (usually such engine is flat rated, which see).
design weight Different authorities have different definitions; weight chosen as typical of mission but usually much less than MTOGW.
DF Direction finder, or direction finding.
DGAC Direction Générale à l'Aviation Civile.
dibber bomb Designed to cause maximum damage to concrete runways.
dihedral Upward slope of wing seen from front, in direction from root to tip.
DINS Digital inertial navigation system.
disposable load Sum of masses that can be loaded or unloaded, including payload, crew, usable fuel etc; MTOGW minus OWE.
DLC Direct lift control.
DME UHF distance-measuring equipment; gives slant distance to a beacon; DME element of Tacan.
dog-tooth A step in the leading-edge of a plane resulting from an increase in chord (see also saw-tooth).
Doppler Short for Doppler radar – radar using fact that received frequency is a function of relative velocity between transmitter or reflecting surface and receiver; used for measuring speed over ground or for detecting aircraft or moving vehicles against static ground or sea.
double-slotted flap One having an auxiliary aerofoil ahead of main surface to increase maximum lift.
dP Maximum design differential pressure between pressurised cabin and ambient (outside atmosphere).
DRA Defence Research Agency, Farnborough and Bedford.
drone Pilotless aircraft, usually winged, following preset programme of manoeuvres.
DS Directionally solidified.

EAA Experimental Aircraft Association (divided into local branches called Chapters).
EAS Equivalent airspeed, RAS minus correction for compressibility.
EB welding Electron beam welding.
ECCM Electronic counter-countermeasures.
ECM Electronic countermeasures.
EFIS Electronic flight instrument(ation) system, in which large multifunction CRT displays replace traditional instruments.
EGT Exhaust gas temperature.
ehp Equivalent horsepower, measure of propulsive power of turboprop made up of shp plus addition due to residual thrust from jet.
EICAS Engine indication (and) crew alerting system.
EIS Entry into service.
ekW Equivalent kilowatts, SI measure of propulsive power of turboprop (see ehp).
elevon Wing trailing-edge control surface combining functions of aileron and elevator.
ELF Extreme low frequency.
elint electronics intelligence.
ELT Emergency locator transmitter, to help rescuers home on to a disabled or crashed aircraft.
EMP Electromagnetic pulse of nuclear or electronic origin.
EO Electro-optical.
EPA Environmental Protection Agency.
EPNdB Effective perceived noise decibel, SI unit of EPNL.

Jane's Readership Survey

FREE GIFT

In the run-up to our centenary, we are endeavouring to discover more about you - our Yearbook user. In return for your help, we would like to send you the FREE gift which is described at the bottom of this page.

Please photocopy this page now . . . .

. . . . and answer the six simple questions listed below. Then complete your name & address details and return this photocopied page to the address below.

Thank you for your assistance.

1. Were YOU the person responsible for PURCHASING this Jane's Yearbook? Y / N
2. Have you referred to THIS Jane's Yearbook, previously? Y / N
3. If "YES", for HOW MANY YEARS have you used it? . . . .
4. Do YOU use OTHER Jane's Yearbooks or Products? Y / N
5. Within which of the following categories do you have the most interest?

DEFENCE	NAVAL	MILITARY AEROSPACE
LAND WARFARE	CIVIL AEROSPACE	MARITIME TRANSPORTATION
LAND TRANSPORTATION	AIRPORTS	OTHER (please specify):

. . . .

6. Do YOU have access to a CD-ROM reader? Y / N

Your Name: Job Title:

Company / Organisation:

Address:

                                                                    Tel No:
Country: Postcode: Fax No:

As a "thank you" for your assistance in our survey, we will send you a **FREE** copy of
JANE'S DEFENCE GLOSSARY . . . .
Over 10,000 defence - related acronyms & abbreviations are arranged alphabetically with each entry giving a brief description and, where appropriate, country of origin.

OFFICER RANKS FOR NATO COUNTRIES
NATO REPORTING NAMES FOR FORMER SOVIET MISSILES & AIRCRAFT
JETDS CODING SYSTEM - UNITS OF MEASUREMENT
CONVERSION FACTORS - FREQUENCY BANDS & DESIGNATORS
THE INTERNATIONAL PHONETIC ALPHABET - MEMBERSHIP OF INTERNATIONAL ORGANISATIONS

AWA   *Normally priced £30*

Please return this completed page to Jane's Readership Survey, Department DSM, Jane's Information Group, 163 Brighton Road, Coulsdon, Surrey CR5 2NH, United Kingdom. Or, fax this page to +44 (0) 81 763 1006.

EPNL Effective perceived noise level, measure of noise effect on humans which takes account of sound intensity, frequency, character and duration, and response of human ear.
EPU Emergency power unit (part of aircraft, not used for propulsion).
EROPS Alternative abbreviation (favoured by airlines) for ETOPS.
ERP Effective radiated power.
ESA European Space Agency.
ESM (1) Electronic surveillance (or support) measures; (2) Electronic signal monitoring.
ETOPS Extended-range twin (engine) operations, routing not more than a given flight time (120 min or 180 min) from a usable alternative airfield.
EVA Extra-vehicular activity, ie outside spacecraft.
EW Electronic warfare.
EWSM Early-warning support measures.

FAA Federal Aviation Administration.
factored Multiplied by an agreed number to take account of extreme adverse conditions, errors, design deficiencies or other inaccuracies.
FADEC Full authority digital engine (or electronic) control.
FAI Fédération Aéronautique Internationale.
fail-operational System which continues to function after any single fault has occurred.
fail-safe Structure or system which survives failure (in case of system, may no longer function normally).
FAR Federal Aviation Regulations.
FAR Pt 23 Defines the airworthiness of private and air taxi aeroplanes of 5670 kg (12,500 lb) MTOGW and below.
FAR Pt 25 Defines the airworthiness of public transport aeroplanes exceeding 5670 kg (12,500 lb) MTOGW.
FBW Fly by wire (which see).
FDS Flight director system.
feathering Setting propeller or similar blades at pitch aligned with slipstream to give resultant torque (not tending to turn shaft) and thus minimum drag.
FEL Fibre elastomeric rotor head.
fence A chordwise projection on the surface of a wing, used to modify the distribution of pressure.
fenestron Helicopter tail rotor with many slender blades rotating in short duct.
ferry range Extreme safe range with zero payload.
FFAR Folding-fin (or free-flight) aircraft rocket.
FFVV Fédération Française de Vol à Voile (French gliding authority).
field length Measure of distance needed to land and/or takeoff; many different measures for particular purposes, each precisely defined.
flaperon Wing trailing-edge surface combining functions of flap and aileron.
flat-four Engine having four horizontally opposed cylinders; thus, flat-twin, flat-six etc.
flat rated Propulsion engine capable of giving full thrustor power for take-off to high airfield height and/or high ambient temperature (thus, probably derated at S/L).
FLIR Forward-looking infra-red.
FLOT Forward line of own troops.
fly by light Flight control system in which signals pass between computers and actuators along fibre optic leads.
fly by wire Flight control system with electrical signalling (ie, without mechanical interconnection between cockpit flying controls and control surfaces).
FM Frequency modulation.
FMCS Flight management computer system.
FMS (1) Foreign military sales (US DoD); (2) Flight management system.
FOL Forward operating location.
footprint (1) A precisely delineated boundary on the surface, inside which the perceived noise of an aircraft exceeds a specified level during take-off and/or landing; (2) Dispersion of weapon or submunition impact points.
Fowler flap Moves initially aft to increase wing area and then also deflects down to increase drag.
free turbine Turbine mechanically independent of engine upstream, other than being connected by rotating bearings and the gas stream, and thus able to run at its own speed.
frequency agile (frequency hopping) Making a transmission harder to detect by switching automatically to a succession of frequencies.
FSD Full scale development.
FSW Forward-swept wing.
FY Fiscal year; 1 October to 30 September in US government affairs (1 October 1992 starts FY 1993).

g Acceleration due to mean Earth gravity, ie of a body in free fall; or acceleration due to rapid change of direction of flight path.
gallons Non-SI measure; 1 Imp gallon (UK) = 4.546 litres, 1 US gallon = 3.785 litres.
GCI Ground-controlled interception.
GfK Glassfibre-reinforced plastics (German).
GFRP Glassfibre-reinforced plastics.
glide ratio Of a sailplane, distance travelled along track divided by height lost in still air.
glideslope Element giving descent path guidance in ILS.
glove (1) Fixed portion of wing inboard of variable sweep wing; (2) additional aerofoil profile added around normal wing for test purposes.
GPS Global Positioning System, US military/civil satellite-based precision navaid.
GPU Ground power unit (not part of aircraft).
GPWS Ground-proximity warning system.
green aircraft Aircraft flyable but unpainted, unfurnished and basically equipped.
gross wing area See wing area.
GS Glideslope, of ILS.
GSE Ground support equipment (such as special test gear, steps and servicing platforms).
GTS Gas turbine starter (ie starter is miniature gas turbine).
gunship Helicopter designed for battlefield attack, normally with slim body carrying pilot and weapon operator only.

h Hour(s).
hardened Protected as far as possible against nuclear explosion.
hardpoint Reinforced part of aircraft to which external load can be attached, eg weapon or tank pylon.
HASC US House Armed Services Committee.
HBPR High bypass ratio (engine).
hectopascal (hPa) Unit of pressure, Pa × 100.
helicopter Rotary-wing aircraft both lifted and propelled by one or more power-driven rotors turning about substantially vertical axes.
HF High frequency.
HMD Helmet-mounted display; hence HMS = sight.
homebuilt Experimental category aircraft built/assembled from plans or kits.
hot and high Adverse combination of airfield height and high ambient temperature, which lengthens required TOD.
hovering ceiling Ceiling of helicopter (corresponding to air density at which maximum rate of climb is zero), either IGE or OGE.
HP High pressure (HPC, compressor; HPT, turbine).
hp Horsepower.
HSI Horizontal situation indicator.
HUD Head-up display (bright numbers and symbols projected on pilot's aiming sight glass and focused on infinity so that pilot can simultaneously read display and look ahead).
HVAR High-velocity aircraft rocket.
Hz Hertz, cycles per second.

IAS Indicated airspeed, ASIR corrected for instrument error.
IATA International Air Transport Association.
ICAO International Civil Aviation Organisation.
ICNIA Integrated communications, navigation and identification avionics.
IFF Identification friend or foe.
IFR Instrument flight rules (ie, not VFR).
IGE In ground effect: helicopter performance with theoretical flat horizontal surface just below it (eg mountain).
ILS Instrument landing system.
IMC Instrument meteorological conditions. Meteorological conditions too poor for pilot to fly without reference to blind-flying instruments.
IMK Increased manoeuvrability kit.
Imperial gallon 1.20095 US gallons; 4.546 litres.
IMS Integrated multiplex system.
INAS Integrated nav/attack system.
incidence Strictly, the angle at which the wing is set in relation to the fore/aft axis. Wrongly used to mean angle of attack (which see).
inertial navigation Measuring all accelerations imparted to a vehicle and, by integrating these with respect to time, calculating speed at every instant (in all three planes) and by integrating a second time calculating total change of position in relation to starting point.
INEWS Integrated electronic warfare system.
INS Inertial navigation system.
integral construction Machined from solid instead of assembled from separate parts.
integral tank Fuel (or other liquid) tank formed by sealing part of structure.
intercom Wired telephone system for communication within aircraft.
inverter Electric or electronic device for inverting (reversing polarity of) alternate waves in AC power to produce DC.
IOC Initial operational capability.
IP (1) Intermediate pressure; (2) initial point in attack manoeuvre.
IPC, IPT Intermediate-pressure compressor, turbine.
IR Infra-red.
IRAN Inspect and repair as necessary.
IRLS Infra-red linescan (builds TV-type picture showing cool regions as dark and hot regions as light).
IRS Inertial reference system.
IRST Infra-red search and track.
ISA International Standard Atmosphere.
ISIS (1 Boeing Vertol) Integral spar inspection system; (2 Ferranti) integrated strike and interception sight.
ITE Involute throat and exit (rocket nozzle).
IVSI Instantaneous VSI.

J Joules, SI unit of energy.
JAR Joint Airworthiness Requirements, agreed by all major EC countries (JAR.25 equivalent to FAR Pt 25).
JASDF Japan Air Self-Defence Force.
JATO Jet-assisted take-off (actually means rocket-assisted).
JCAB Japan Civil Airworthiness Board.
JDA Japan Defence Agency.
JGSDF Japan Ground Self-Defence Force.
JMSDF Japan Maritime Self-Defence Force.
joined wing Tandem wing layout in which forward and aft wings are swept so that the outer sections meet.
J-STARS US Air Force/Navy Joint Surveillance Target Attack Radar System in Boeing E-8A.
JTIDS Joint Tactical Information Distribution System.

K One thousand bits of memory.
Kevlar Aramid fibre used as basis of high-strength composite material.
km/h Kilometres per hour.
kN Kilonewtons (the Newton is the SI unit of force; 1 lbf = 4.448 N).
knot 1 nm per hour.
Krueger flap Hinges down and then forward from below the leading-edge.
kVA Kilovolt-ampères.
kW Kilowatt, SI measure of all forms of power (not just electrical).

LABS Low-altitude bombing system designed to throw the bomb up and forward (toss bombing).
LANTIRN Low-altitude navigation and targeting infra-red, night.
LARC Low-altitude ride control.
LBA Luftfahrtbundesamt (German civil aviation authority).
lbf Pounds of thrust.
LCD Liquid crystal display, used for showing instrument information.
LCN Load classification number, measure of 'flotation' of aircraft landing gear linking aircraft weight, weight distribution, tyre numbers, pressures and disposition.
LED Light-emitting diode.
Lidar Light detection and ranging (laser counterpart of radar).
LITVC Liquid-injection thrust vector control.
LLTV Low-light TV (thus, LLLTV, low-light-level).
load factor (1) percentage of max payload; (2) design factor for airframe.
LOC Localiser (which see).
localiser Element giving steering guidance in ILS.
loiter Flight for maximum endurance, such as supersonic fighter on patrol.
longerons Principal fore-and-aft structural members (eg in fuselage).
Loran Long range navigation; family of hyperbolic navaids based on ground radio emissions, now mainly Loran C.
LOROP Long-range oblique photography.
low observables Materials and structures designed to reduce aircraft signatures of all kinds.
lox Liquid oxygen.
LP Low pressure (LPC, compressor; LPT, turbine).
LRMTS Laser ranger and marked-target seeker.
LRU Line-replaceable unit.

m Metre(s), SI unit of length.
M or Mach number The ratio of the speed of a body to the speed of sound (1116 ft; 340 m/s in air at 15°C) under the same ambient conditions.
MAC (1) US Air Force Military Airlift Command; (2) mean aerodynamic chord.
MAD Magnetic anomaly detector.
Madar Maintenance analysis, detection and recording.
Madge Microwave aircraft digital guidance equipment.
marker, marker beacon Ground beacon giving position guidance in ILS.
mb Millibars, bar × 10^{-3}.
MBR Marker beacon receiver.
MEPU Monofuel emergency power unit.
METO Maximum except take-off.
MF Medium frequency.
MFD Multi-function (electronic) display.

mg Milligrammes, grammes × 10^{-3}.
MLS Microwave landing system.
MLW Maximum landing weight.
mm Millimetres, metres × 10^{-3}.
MMH Monomethyl hydrazine.
MMO Max operating Mach number.
MMS Mast-mounted sight.
MNPS Minimum navigation performance specification.
MO Maximum permitted operating Mach number.
monocoque Structure with strength in outer shell, devoid of internal bracing (semi-monocoque, with some internal supporting structure).
MoU Memorandum of understanding.
MPA Man-powered aircraft.
mph Miles per hour.
MRW Maximum ramp weight.
MSIP US armed forces multi-staged improvement programme.
MTBF Mean time between failures.
MTBR Mean time between removals.
MTI Moving-target indication (radar).
MTOGW Maximum take-off gross weight (MRW minus taxi/run-up fuel).
MTTR Mean time to repair.
MZFW Maximum zero-fuel weight.

NACA US National Advisory Committee for Aeronautics (now NASA).
Nadge NATO air defence ground environment.
NAS US Naval Air Station.
NASA National Aeronautics and Space Administration.
NASC US Naval Air Systems Command (also several other aerospace meanings).
NATC US Naval Air Training Command or Test Center (also several other aerospace meanings).
NBAA US National Business Aircraft Association.
NDB Non-directional beacon.
NDT Non-destructive testing.
NGV Nozzle guide vane.
NH_4ClO_4 Ammonium perchlorate.
nm nautical mile, 1.8532 km, 1.15152 miles.
NOAA US National Oceanic and Atmospheric Administration.
NOE Nap-of-the-Earth (low flying in military aircraft, using natural cover of hills, trees etc).
NOS Night observation surveillance.
NO_x Generalised term for oxides of nitrogen.
Ns Newton-second (1 N thrust applied for 1 second).
NVG Night vision goggles.

OAT Outside air temperature.
OBOGS Onboard oxygen generating system.
OCU Operational Conversion Unit.
OEI One engine inoperative.
offset Workshare granted to a customer nation to offset the cost of an imported system.
OGE Out of ground effect; helicopter hovering, far above nearest surface.
Omega Long-range hyperbolic radio navaid.
OMI Omni-bearing magnetic indicator.
omni Generalised word meaning equal in all directions (as in omni-range, omni-flash beacon).
on condition maintenance According to condition rather than at fixed intervals.
optronics Combination of optics and electronics in viewing and sighting systems.
OSTIV Organisation Scientifique et Technique Internationale du Vol à voile (international gliding authority).
OTH Over the horizon.
OTPI On-top position indicator (indicates overhead of submarine in ASW).
OWE Operating weight empty. MTOGW minus payload, usable fuel and oil and other consumables.

PA system Public or passenger address.
pallet (1) for freight, rigid platform for handling by forklift or conveyor; (2) for missile, mounting and electronics box outside aircraft.
payload Disposable load generating revenue (passengers, cargo, mail and other paid items), in military aircraft loosely used to mean total load carried of weapons, cargo or other mission equipment.
PD radar Pulse Doppler.
penaids Penetration aids, such as jammers, chaff or decoys to help aircraft fly safely through hostile airspace.
PFA Popular Flying Association (UK).
PFCS Primary flight computer system.
phased array Radar in which the beam is scanned electronically in one or both axes without moving the antenna.
PHI Position and heading (or homing) indicator.
plane A lifting surface (eg wing, tailplane).
plug door Door larger than its frame in pressurised fuselage, either opening inwards or arranged to retract parts prior to opening outwards.
plume The region of hot air and gas emitted by a helicopter jetpipe.
pneumatic de-icing Covered with flexible surfaces alternately pumped up and deflated to throw off ice.
port Left side, looking forward.
power loading Aircraft weight (usually MTOGW) divided by total propulsive power or thrust at T-O.
prepreg Glassfibre cloth or rovings pre-impregnated with resin to simplify layup.
pressure fuelling Fuelling via a leakproof connection through which fuel passes at high rate under pressure.
pressure ratio In gas turbine engine, compressor delivery pressure divided by ambient pressure (in supersonic aircraft, divided by ram pressure downstream of inlet).
primary flight controls Those used to control trajectory of aircraft (thus, not trimmers, tabs, flaps, slats, airbrakes or lift dumpers etc).
primary flight display Single screen bearing all data for aircraft flight path control.
propfan A family of new technology propellers characterised by multiple scimitar-shaped blades with thin sharp-edged profile. Single and contra-rotating examples promise to extend propeller efficiency up to about Mach 0.8. See also UDF.
pulse Doppler Radar sending out pulses and measuring frequency-shift to detect returns only from moving target(s) in background clutter.
pylon Structure linking aircraft to external load (engine nacelle, drop tank, bomb etc). Also used in conventional sense in pylon racing.

radius In terms of performance, the distance an aircraft can fly from base and return without intermediate landing.
RAE See DRA.
RAI Registro Aeronautico Italiano.
RAM Radar absorbent material.
ram pressure Increased pressure in forward-facing aircraft inlet, generated by converting (relative) kinetic energy to pressure.
ramp weight Maximum weight at start of flight (MTOGW plus taxi/run-up fuel).
range Too many definitions to list, but essentially the distance an aircraft can fly (or is permitted to fly) with specified load and usually whilst making allowance for specified additional manoeuvres (diversions, stand-off, go-around etc).
RANSAC Range surveillance aircraft.
RAS Rectified airspeed, IAS corrected for position error.
raster Generation of large-area display, eg TV screen, by close-spaced horizontal lines scanned either alternately or in sequence.
RAT Ram air turbine.
redundant Provided with spare capacity or data channels and thus made to survive failures.
refanned Gas turbine engine fitted with new fan of higher BPR.
RFP Request(s) for proposals.
rigid rotor (see bearingless rotor).
RLD Rijksluchtvaartdienst. Netherlands civil aviation department.
RMI Radio magnetic indicator; combines compass and navaid bearings.
R/Nav Calculates position, distance and time from groups of airways beacons.
RON Research octane number.
rotor-kite Rotary-wing aircraft with no internal power, lifted by a freely running autorotating rotor and towed by an external vehicle.
roving Multiple strands of fibre, as in a rope (but usually not twisted).
RPV Remotely piloted vehicle (pilot in other aircraft or on ground).
RSA Réseau du Sport de l'Air.
ruddervators Flying control surfaces, usually a V tail, that control both yaw and pitch attitude.
RVR Runway visual range.

s Second(s)
SAC US Air Force Strategic Air Command.
safe-life A term denoting that a component has proved by testing that it can be expected to continue to function safely for a precisely defined period before replacement.
salmon (French saumon) Streamlined fairings, usually at wingtip of sailplane, serving same function as end plate and acting also as tip-skid.
SAR (1) Search and rescue; (2) synthetic aperture radar.
SAS Stability augmentation system.
SATS (1) Small airfield for tactical support; (2) Small Arms Target System.
saw-tooth Same as dog-tooth.
SCAS Stability and control augmentation system.
second-source Production of identical item by second factory or company.
semi-active Homing on to radiation reflected from target illuminated by radar or laser energy beamed from elsewhere.
service ceiling Usually height equivalent to air density at which maximum attainable rate of climb is 100 ft/min.
servo A device which acts as a relay, usually augmenting the pilot's efforts to move a control surface or the like.
SFAR Special Federal Aviation Regulation(s).
sfc Specific fuel consumption.
SGAC Secrétariat Général à l'Aviation Civile (now DGAC).
shaft Connection between gas turbine and compressor or other driven unit. Two-shaft engine has second shaft, rotating at different speed, surrounding the first (thus, HP surrounds inner LP or fanshaft).
shipment One item or consignment delivered (by any means of transport) to customer.
Shoran Short range navigation (radio).
shp Shaft horsepower, measure of power transmitted via rotating shaft.
sideline noise EPNdB measure of aircraft landing and taking off, at point 0.25 nm (2- or 3-engined) or 0.35 nm (4-engined) from runway centre line.
sidestick Control column in the form of a short hand-grip beside the pilot.
SIF Selective identification facility.
sigint Signals intelligence.
signature Characteristic 'fingerprint' of all electromagnetic radiation (radar, IR etc).
single-shaft Gas turbine in which all compressors and turbines are on common shaft rotating together.
S/L Sea level.
SLAR Side-looking airborne radar.
snap-down Air-to-air interception of low-flying aircraft by AAM fired from fighter at a higher altitude.
soft target Not armoured or hardened.
specific fuel consumption Rate at which fuel is consumed divided by power or thrust developed, and thus a measure of engine efficiency. For jet engines (air-breathing, ie not rockets) unit is mg/Ns, milligrams per Newton-second; for shaft engines unit is μg/J, micrograms (millionths of a gram) per Joule (SI unit of work or energy).
specific impulse Measure of rocket engine efficiency; thrust divided by rate of fuel/oxidant consumption per second, the units for mass and force being the same so that the answer is expressed in seconds.
SPILS Stall protection and incidence-limiting system.
spool One complete axial compressor rotor; thus a two-shaft engine may have a fan plus an LP spool.
SSB Single-sideband (radio).
SSR Secondary surveillance radar.
SST Supersonic transport.
st Static thrust.
stabiliser Fin (thus, horizontal stabiliser = tailplane).
stall strips Sharp-edged strips on wing leading-edge to induce stall at that point.
stalling speed Airspeed at which aircraft stalls at 1*g*, ie wing lift suddenly collapses.
standard day ISA temperature and pressure.
starboard Right side, looking forward.
static inverter Solid-state inverter of alternating wave-form (ie, not rotary machine) to produce DC from AC.
stick-pusher Stall-protection device that forces pilot's control column forward as stalling angle of attack is neared.
stick-shaker Stall-warning device that noisily shakes pilot's control column as stalling angle of attack is neared.
STOL Short take-off and landing. (Several definitions, stipulating allowable horizontal distance to clear screen height of 35 or 50 ft or various SI measures.)
store Object carried as part of payload on external attachment (eg bomb, drop tank).
strobe light High-intensity flashing beacon.
substrate The underlying layer on which something (such as a solar cell or integrated circuit) is built up.
supercritical wing Wing of relatively deep, flat-topped profile generating lift right across upper surface instead of concentrated close behind leading-edge.
sweepback Backwards inclination of wing or other aerofoil, seen from above, measured relative to fuselage or other reference axis, usually measured at quarter-chord (25%) or at leading-edge.
synchronous satellite Geostationary.

t Tonne, 1 Megagram, 1000 kg.
tabbed flap Fitted with narrow-chord tab along entire trailing-edge which deflects to greater angle than main surface.
tabs Small auxiliary surfaces hinged to trailing-edge of control surfaces for purposes of trimming, reducing hinge moment (force needed to operate main surface) or in other way assisting pilot.
TAC US Air Force Tactical Air Command.
Tacan Tactical air navigation UHF military navaid giving bearing and distance to ground beacons; distance element (see DME) can be paired with civil VOR.

GLOSSARY

taileron Left and right tailplanes used as primary control surfaces in both pitch and roll.
tailplane Main horizontal tail surface, originally fixed and carrying hinged elevator(s) but today often a single 'slab' serving as control surface.
TANS Tactical air navigation system; Decca Navigator or Doppler-based computer, control and display unit.
TAS True airspeed, EAS corrected for density (often very large factor) appropriate to aircraft height.
TBO Time between overhauls.
t/c ratio Ratio of the thickness (aerodynamic depth) of awing or other surface to its chord, both measured at the same place parallel to the fore-and-aft axis.
TET Turbine entry temperature (of the gas); also turbine inlet temperature (TIT), inter-turbine temperature (ITT) and turbine gas temperature (TGT).
TFR Terrain-following radar (for low-level attack).
thickness Depth of wing or other aerofoil; maximum perpendicular distance between upper and lower surfaces.
tilt-rotor Aircraft with fixed wing and rotors that tilt up for hovering and forward for fast flight.
T-O Take-off.
T-O noise EPNdB measure of aircraft taking off, at point directly under flight path 3.5 nm from brakes-release (regardless of elevation).
TOD Take-off distance.
TOGW Take-off gross weight (not necessarily MTOGW)
ton Imperial (long) ton = 1.016 t (Mg) or 2240 lb, US (short) ton = 0.9072 t or 2000 lb.
track Distance between centres of contact areas of main landing wheels measured left/right across aircraft (with bogies, distance between centres of contact areas of each bogie).
transceiver Radio transmitter/receiver.
transponder Radio transmitter triggered automatically by a particular received signal as in civil secondary surveillance radar (SSR).
TRU Transformer/rectifier unit.
TSFC Thrust specific fuel consumption of jet engine (turbojet, turbofan, ducted propfan or ramjet).
TSO Technical Standard Order (FAA).
turbofan Gas-turbine jet engine generating most thrust by a large-diameter cowled fan, with small part added by jet from core.
turbojet Simplest form of gas turbine comprising compressor, combustion chamber, turbine and propulsive nozzle.
turboprop Gas turbine in which as much energy as possible is taken from gas jet and used to drive reduction gearbox and propeller.
turboshaft Gas turbine in which as much energy as possible is taken from gas jet and used to drive high-speed shaft (which in turn drives external load such as helicopter transmission).
TVC Thrust vector control (rocket).
TWT Travelling-wave tube.
tyre sizes In simplest form, first figure is rim diameter (in or mm) and second is rim width (in or mm). In more correct three-unit form, first figure is outside diameter, second is max width and third is wheel diameter.

UAV Unmanned air vehicle.
UBE, Ubee Ultra bypass engine, alternative terminology (Boeing) for UDF.
UDF Unducted fan, one form of advanced propulsion system in which gas turbine blading directly drives large fan (propfan) blades mounted around the outside of the engine pod. (GE registered abbreviation.)
UHF Ultra-high frequency.
unfactored Performance level expected of average pilot, in average aircraft, without additional safety factors.
upper surface blowing Turbofan exhaust vented over upper surface of wing to increase lift.
usable fuel Total mass of fuel consumable in flight, usually 95-98 per cent of system capacity.
useful load Usable fuel plus payload.
US gallon 0.83267 Imperial gallon; 3.785 litres.

variable geometry Capable of grossly changing shape in flight, especially by varying sweep of wings.
V_D Maximum permitted diving speed.
VDU Video (or visual) display unit.
vernier Small thruster, usually a rocket, for final precise adjustment of a vehicle's trajectory and velocity.
VFR Visual flight rules.
VHF Very high frequency.
VLF Very low frequency (area-coverage navaid).
V_{MO} Maximum permitted operating flight speed (IAS, EAS or CAS must be specified).
VMS Vehicle management system.
V_{NE} Never-exceed speed (aerodynamic or structural limit).
VOR VHF omni-directional range (VHF radio beacons providing bearing to or from beacon).
vortex generators Small blades attached to wing and tail surfaces to energise local airflow and improve control.
vortillon Short-chord fence ahead of and below leading-edge.
VSI Vertical speed (climb/descent) indicator.
VTOL Vertical take-off and landing.

washout Inbuilt twist of wing or rotor blade reducing angle of incidence towards the tip.
WDNS Weapon delivery and navigation system.
wheelbase Minimum distance from nosewheel or tailwheel (centre of contact area) to line joining mainwheels (centres of contact areas).
wing area Total projected area of clean wing (no flaps, slats etc) including all control surfaces and area of fuselage bounded by leading- and trailing-edges projected to centreline (inapplicable to slender-delta aircraft with extremely large leading-edge sweep angle). Described in *Jane's* as gross wing area; net area excludes projected areas of fuselage, nacelles, etc.
wing loading Aircraft weight (usually MTOGW) divided by wing area.
winglet Small auxiliary aerofoil, usually sharply upturned and often sweptback, at tip of wing.

zero-fuel weight MTOGW minus usable fuel and other consumables, in most aircraft imposing severest stress on wing and defining limit on payload.
zero/zero seat Ejection seat designed for use even at zero speed on ground.
ZFW Zero-fuel weight.
μg Microgrammes, grammes × 10^{-6}.

Airline fleet list

This listing of the fleets of world airlines was extracted from the *Jane's World Airlines* database, which is compiled by Paul Portnoi. Updates are published quarterly. The list includes information up to 30 June 1993. The purpose of including it here is to allow the reader a different approach to locating information in the main part of the book. Readers often need to know which types an airline operates before seeking aircraft details.

Figures after the engine type denote the number of aircraft in operation and the subsequent figures show how many are on firm order. These are actual airline strengths and do not indicate whether they were acquired direct from the manufacturer, through a leasing organisation or second hand. Our customer records in the main Aircraft section entries, on the other hand, list sales to leasing organisations, but not the airlines which subsequently leased them. Neither do they indicate when the aircraft passed from the first to a second or subsequent owner.

Airline and type	Engine	In operation	On order
ACES (AEROLINEAS CENTRALES DE COLOMBIA SA) (Colombia)			
727-100	P&W JT8D-7B	4	
727-200	P&W JT8D-17R	2	
ATR 42-320	P&WC PW120	3	1
DHC-6 Twin Otter 300	P&WC PT6A-27	10	
ACS OF CANADA - AIR CHARTER SYSTEMS (Canada)			
DC-8-55F	P&W JT3D-3B (Q)	3	
ADRIA AIRWAYS (Slovenia)			
A320-200	IAE V2500-A1	3	
DC-9-30	P&W JT8D-9A	3	
DHC-7 Dash 7-100	P&WC PT6A-50	2	
MD-80	P&W JT8D-217	5	
AER LINGUS PLC (Ireland)			
737-200 Advanced	P&W JT8D-9A	3	
737-300	CFMI CFM56-3B1	1	
737-400	CFMI CFM56-3B2	6	
737-500	CFMI CFM56-3B1	8	
747-100	P&W JT9D-3A	2	
747-100	P&W JT9D-7	1	
Fokker 50	P&WC PW125B	6	
Saab 340B	GE CT7-9B	4	
AERO CARIBBEAN (Cuba)			
An-26	ZMKB AI-24T	3	
DC-3C	P&W R-1830	3	
Il-14	Aviadvigatel ASh-82T	1	
Il-18D	ZMKB AI-20M	4	
Il-18V (F)	ZMKB AI-20K	1	
AEROCARIBE (Mexico)			
CV 340	P&W R-2800	1	
CV 440 Metropolitan	P&W R-2800	1	
Fairchild F-27F	RR Dart 529-7E	2	
Fairchild FH-227B	RR Dart 532-7L	2	
Fairchild FH-227D	RR Dart 532-7L	2	
AERO COZUMEL SA (Mexico)			
BN-2A Trislander	Lyc O-540-E4C5	2	
Fairchild F-27	RR Dart 514-7	1	
Fairchild F-27A	RR Dart 528-7	1	
AEROFLOT - SOVIET AIRLINES (Russia)			
A310-300	GE CF6-80C2	5	
Il-62M	Aviadvigatel D-30KU	28	
Il-76	Aviadvigatel D-30KP	19	
Il-86	KKBM NK-86	19	
Tu-154B	KKBM NK-8-2U	4	
Tu-154C	KKBM NK-8-2U	3	
Tu-154M	Aviadvigatel D-30KU-154-II	23	

Airline and type	Engine	In operation	On order
AEROLINEAS ARGENTINAS (Argentina)			
727-200 Advanced	P&W JT8D-17R	1	
737-200	P&W JT8D-9A	4	
737-200 Advanced	P&W JT8D-9A	6	
737-200C	P&W JT8D-9A	2	
747-200B	P&W JT9D-7Q	6	
767-300ER			4
A310-300			1
Fokker F28-1000	RR Spey 555-15	3	
MD-83	P&W JT8D-217	1	
MD-88	P&W JT8D-219	4	3
AERO LLOYD (Germany)			
DC-9-32	P&W JT8D-9A	3	
MD-11	P&W PW4460		
MD-83	P&W JT8D-219	13	
MD-87	P&W JT8D-219	4	
AEROMEXICO (Mexico)			
757-200	P&W PW2040	1	5
767-200ER	P&W PW4060	2	
767-300ER	P&W PW4060	2	
DC-9-31	P&W JT8D-7B	3	
DC-9-32	P&W JT8D-17	15	
DC-10-15	GE CF6-50C2F	2	
DC-10-30	GE CF6-50C2F	4	
MD-82	P&W JT8D-217A	12	
MD-83	P&W JT8D-219	2	
MD-87	P&W JT8D-219	2	
MD-88	P&W JT8D-219	10	
AEROPERU (Peru)			
727-100	P&W JT8D-7B	2	
727-100C	P&W JT8D-7A	1	
DC-8-62	P&W JT3D-3B	1	
DC-8-62	P&W JT3D-3B (Q)	3	
DC-8-62	P&W JT3D-7	1	
DC-8-63	P&W JT3D-7 (Q)	1	
Fokker F28-1000	RR Spey 555-15	2	
AEROPOSTAL - LINEAS AEROPOSTAL VENEZOLANA (Venezuela)			
DC-9-15	P&W JT8D-7	1	
DC-9-32	P&W JT8D-17	3	
DC-9-34CF	P&W JT8D-17	1	
DC-9-51	P&W JT8D-17	7	
MD-83	P&W JT8D-219	6	
AEROVIAS NACIONALES DE COLOMBIA SA - AVIANCA (Colombia)			
707-300B	P&W JT3D-3B (Q)	3	
727-100	P&W JT8D-7	5	
727-100	P&W JT8D-7A	1	
727-100	P&W JT8D-7B	5	

One of the first western airliners operated by Aeroflot is this Airbus A310-300

AIRLINE FLEET LIST

Five DC-3s from Air Atlantique took part in the anti-pollution operations after the tanker *Braer* ran aground near Sumburgh, Shetlands, in January 1993 *(Peter J.*

Airline and type	Engine	In operation	On order
AEROVIAS NACIONALES DE COLOMBIA SA - AVIANCA (Colombia) continued...			
727-200	P&W JT8D-17R	1	
727-200 Advanced	P&W JT8D-15	1	
727-200 Advanced	P&W JT8D-17R	7	
747-100 (SCD)	P&W JT9D-7AH	1	
747-200 (SCD)	P&W JT9D-7Q	1	
757-200	RR RB211-535E4	2	
767-200ER	P&W PW4056	1	2
Fokker 50	P&WC PW126	1	5
MD-83	P&W JT8D-219	1	10
AER TURAS - IRISH CARGO AIRLINES (Ireland)			
DC-8-63	P&W JT3D-7 (Q)	3	
AFFRETAIR (PRIVATE) LTD (Zimbabwe)			
DC-8-55	P&W JT3D-3B	2	
DC-8-71F	CFMI CFM56-2	1	
AFRICAN AIRLINES INTERNATIONAL LTD (Kenya)			
An-32	ZMKB AI-20M	2	
707-300B	P&W JT3D-3B	1	
707-300B	P&W JT3D-7 (Q)	1	
707-300C	P&W JT3D-3B (Q)	1	
AFRICAN INTERNATIONAL AIRWAYS (PTY) LIMITED (Swaziland)			
737-200QC	P&W JT8D-9A	2	
DC-8-54F	P&W JT3D-3B (Q)	2	
AFRICAN SAFARI AIRWAYS LTD (Kenya)			
DC-8-63	P&W JT3D-7 (Q)	1	
DC-10-30	GE CF6-52	1	
AHK AIR HONG KONG (Hong Kong)			
707-300C (F)	P&W JT3D-3B (Q)	1	
747-100F	P&W JT9-7A	1	
747-100F (SCD)	P&W JT9D-7A	2	
747-200F	TBD		1
AIR 2000 LTD (UK)			
757-200	RR RB211-535E4	15	
A320-200	IAE V2500-A1	4	
AIR AFRIQUE (Cote d'Ivoire)			
A300B4-203	GE CF6-50C2	3	
A310-300	GE CF6-80C2A2	4	1
DC-10-30	GE CF6-50C	3	
AIR ALGERIE (Algeria)			
727-200	P&W JT8D-9	2	
727-200 Advanced	P&W JT8D-15	9	
737-200 Advanced	P&W JT8D-15	10	
737-200 Advanced	P&W JT8D-17A	3	
737-200C Advanced	P&W JT8D-9	2	
737-200C Advanced	P&W JT8D-15	1	
767-300	GE CF6-80C2	3	
A310-203	GE CF6-80A3	2	
Fokker F27-400M	RR Dart 536-7R	8	
L382G Hercules	Allison 501-D22A	2	
AIR ARUBA (Aruba)			
727-200	P&W JT8D	1	
737-300	CFMI CFM56-3B2	1	
757-200	RR RB211-535E4	1	
EMB-120RT Brasilia	P&WC PW118	1	
MD-83	P&W JT8D-219	1	
MD-88	P&W JT8D-219	1	

Airline and type	Engine	In operation	On order
AIR ATLANTIC LTD (CANADIAN PARTNER) (Canada)			
BAe 146-200	Lyc ALF502R-5	3	
DHC-8 Dash 8-100	P&WC PW120A	11	
AIR ATLANTIQUE (UK)			
BN-2A Islander	Lyc O-540-E4C5	2	
Cessna 310L	Cont IO-470-V	1	
Cessna 310R II	Cont IO-520-M	2	
Cessna 402B	Cont TSIO-520-E	2	
Cessna 404 Titan	Cont GTSIO-520-M	1	
Cessna 406	Cont TSIO-520	1	
DC-3C	P&W R-1830-92	10	
DC-6A	P&W R-2800-CB17	2	
AIR ATLANTIS (Portugal)			
737-200 Advanced	P&W JT8D-15	3	
737-300	CFMI CFM56-3B2	5	1
AIR AUSTRAL (Réunion)			
737-500	CFMI CFM56-3C1	1	
BAe 748 Srs 2A	RR Dart 534-2	1	
Fokker F28-1000C	RR Spey 555-15	1	
AIR BC LTD (Canada)			
BAe 146-200A	Lyc ALF502R-5	5	
BAe Jetstream 31	Ga TPE331-10UG-513H	6	
DHC-6 Twin Otter 200	P&WC PT6A-20	4	
DHC-8 Dash 8-100	P&WC PW120A	12	
DHC-8 Dash 8-300	P&WC PW123	6	
AIR BELGIUM (Belgium)			
737-400	CFMI CFM56-3C1	2	
757-200	RR RB211-535E4	1	
AIRBORNE EXPRESS (ABX AIR, INC) (USA)			
DC-8-61	P&W JT3B-3D (Q)	10	
DC-8-62	P&W JT3B-3D (Q)	6	
DC-8-63	P&W JT3B-7 (Q)	9	2
DC-9-15	P&W JT8D-7A	2	
DC-9-31	P&W JT8D-7B	14	
DC-9-31	P&W JT8D-9/9A	4	
DC-9-32	P&W JT8D-7B	16	
DC-9-32F	P&W JT8D-7B	3	
DC-9-33F	P&W JT8D-9	6	
DC-9-41	P&W JT8D-11	7	8
DC-9-41	P&W JT8D-15	3	2
NAMC YS-11A-200	RR Dart 542-10K	11	
AIR BOTSWANA (Botswana)			
ATR 42	P&WC PW121	2	
BAe 146-100	Lyc ALF502R-5	2	
Dornier 228-100	Ga TPE331-5-252D	1	
AIR BURKINA (Burkino Faso)			
EMB-110P2 Bandeirante	P&WC PT6A-34	1	
Fokker F28-4000	RR Spey 555-15P	1	
AIR CALEDONIE (New Caledonia)			
ATR 42	P&WC PW120	3	
Dornier 228-212	Ga TPE331-5-252D	2	
AIR CALEDONIE INTERNATIONAL (New Caledonia)			
737-300	CFMI CFM56-3B2	1	
DHC-6 Twin Otter 300	P&WC PT6A-27	1	

Notes: SCD Side cargo door   TBD To be decided   EUD Extended upper deck   Q Hush kitted   SR Short range   LR Long range   F Freighter

AIRLINE FLEET LIST

Airline and type	Engine	In operation	On order
AIR CANADA (Canada)			
727-200	P&W JT8D-15	3	
747-100	P&W JT9D-7	3	
747-200C	P&W JT9D-7	3	
747-400	P&W PW4056	3	
767-200	P&W JT9D-7R4D	21	
767-300ER	P&W PW4060		6
A320-200	CFMI CFM56-5A1	33	7
DC-8-73	CFMI CFM56-2C5	5	
DC-9-30	P&W JT8D-7A	35	
AIR CAPE (South Africa)			
ATR 72	P&WC PW125	2	
CV 580	Allison 501-D13	3	
Partenavia AP68TP Victor	Lyc IO-360-A1B6	1	
AIR CHARTER (France)			
727-200	P&W JT8D-7B	2	
727-200 Advanced	P&W JT8D-15	4	
737-200	P&W JT8D-15A	2	
737-400	CFMI CFM56-3C1	1	
A300B4-203	GE CF6-50C2	2	
AIR CHINA INTERNATIONAL (People's Republic of China)			
707-300	P&W JT3D-7	5	
737-200 Advanced	P&W JT8D-17A	3	
737-300	CFMI CFM56-3	6	
747-200B	P&W JT9D-7R4G2	1	
747-200C	P&W JT9D-7R4G2	2	
747-200F	P&W JT9D-7R4G2	1	
747-400	P&W PW4056	1	4
747-400C	P&W PW4256	3	2
747SP	P&W JT9D-7J	4	
767-300ER	P&W PW4052	6	
BAe 146-100	Lyc ALF502R-5	4	
XAC Y-7	WJ 5A-1	8	
AIR COLUMBUS - TRANSPORTE AEREO NAO REGULAR SA (Portugal)			
727-200RE	P&W JT8D-17 / 217C outers	3	
737-300	CFMI CFM56-3B2	2	
AIR CREEBEC INC (Canada)			
BAe 748-200 Srs 2	RR Dart 534-2	8	
Beech Super King Air 200	P&WC PT6A-41	1	
Cessna 402C Utililiner II	Cont TSIO-520-VB	2	
DHC-8 Dash 8-100	P&WC PW120A	2	
AIR DJIBOUTI (Djibouti)			
727-200	P&W JT8D-15	1	
DC-9-32	P&W JT8D-9A	1	
DHC-6 Twin Otter 200	P&WC PT6A-20	2	
AIR EUROPA (AIR ESPANA, SA) (Spain)			
737-300	CFMI CFM56-3B1 / 3B2	10	
757-200	RR RB211-535E4	3	
AIRFAST INDONESIA (Indonesia)			
737-200	P&W JT8D-7A	1	
737-200	P&W JT8D-9	1	
737-200	P&W JT8D-9A	1	
Agusta-Bell 204B	Lyc T53-11A	1	
BAe 748 Srs 2A	RR Dart 534-2	3	
Beech Queen Air 65-B80	Lyc IGSO-540-A1D	1	
Bell 204B	Lyc T53-11A	3	
Bell 206B JetRanger II	Allison 250-C20	3	
Bell 212	P&WC PT6T-3 TwinPac	1	
Bell 412	P&WC PT6T-3 TwinPac	2	
CASA 212-200 Aviocar	Ga TPE331-10-501C	1	
DC-3	P&W R-1830-92	1	
PA-23-250 Aztec D	Lyc IO-540-C4B5	1	
Sikorsky S-58ET	P&WC PT6T-3 TwinPac	2	
Sikorsky S-58ET	P&WC PT6T-6 TwinPac	2	
AIR FOYLE LIMITED (UK)			
737-300	CFMI CFM56-3B2	1	
An-124 Ruslan	ZMKB D-18T	3	
BAe 146-200QT	Lyc ALF502R-5	3	
BAe 146-300QT	Lyc ALF502R-5	4	
BAe Jetstream 31	Ga TPE331-10UF-513H	1	
Beech Super King Air 200	P&WC PT6A-41	1	
AIR FRANCE (France)			
727-200 Advanced	P&W JT8D-15	7	
737-200	P&W JT8D-15A	19	
737-300	CFMI CFM56-3B1	6	
737-500	CFMI CFM56-3C1	16	
747-100	P&W JT9D-7	13	
747-200	GE CF6-50E	2	
747-200B (SCD)	GE CF6-50E2	10	
747-200F (SCD)	GE CF6-50E2	10	
747-300 (SCD)	GE CF6-50E2	2	
747-300 (SCD/EUD)	GE CF6-50E2	2	
747-400	GE CF6-80C2B1F	7	
747-400 (SCD)	GE CF6-80C2B1F	3	4
747-400C	GE CF6-80C2B1F	1	
767-300ER	P&W PW4060	3	
A300B2-101	GE CF6-50C	4	
A300B4-203	GE CF6-50C2	11	
A310-200	GE CF6-80A3	7	
A310-300	GE CF6-80C2A2	4	
A320-100	CFMI CFM56-5A1	6	
A320-200	CFMI CFM56-5A1	19	
A321	CFMI CFM56-5B		8
A340-300	CFMI CFM56-5C2	1	6
Concorde	RR Olympus 593-610	7	
DC-10-30	GE CF6-50C2R	5	
Fokker F27-500	RR Dart 532-7	15	
AIR GABON (Gabon)			
737-200	P&W JT8D-17	1	
747-200 (SCD)	GE CF6-50E2	1	
ATR 72	P&WC PW120		
Fokker 100	RR Tay 620-15	1	
Fokker F28-1000	RR Spey 555-15	1	
Fokker F28-2000	RR Spey 555-15	2	
L382G Hercules	Allison 501-D22A	1	
AIR GAMBIA (The Gambia)			
707-300	P&W JT3D-3B (Q)	1	
AIR GUADELOUPE (Guadeloupe)			
737-300	CFMI CFM56-3	1	
ATR 42	P&WC PW120	2	
DHC-6 Twin Otter 300	P&WC PT6A-27	4	
Dornier 228-202K	Ga TEP331-5-252D	2	
Fairchild F-27J	RR Dart 532-7	2	
AIR HOLLAND CHARTER BV (Netherlands)			
737-300	CFMI CFM56-3	1	
757-200	RR RB211-535E4	2	
AIR INDIA (India)			
747-200B	P&W JT9D-7J	4	
747-200B	P&W JT9D-7Q	5	
747-300C	GE CF6-80C2B1	2	
747-400	P&W PW4056		4
A300B4-203	GE CF6-50C2	3	
A310-300	GE CF6-80C2A2	8	
Il-62	Aviadvigatel D-30KV	1	
Il-76	Aviadvigatel D-30KP	1	
AIR INTER (France)			
A300B2-1	GE CF6-50C2R	13	
A300B2-K3C	GE CF6-50C2R	1	
A300B4-2C	GE CF6-50C2R	6	

In February 1993, Air France became the second airline to receive Airbus A340s

[29]

AIRLINE FLEET LIST

Airline and type	Engine	In operation	On order
AIR INTER (France) continued...			
A320-100	CFMI CFM56-5A1	6	
A320-200	CFMI CFM56-5A1	25	1
A321-100	CFMI CFM56-5B		7
A330	GE CF6-80E1	2	13
Dassault Mercure 100	P&W JT8D-15	8	
AIR INTER GABON SA (Gabon)			
ATR 42-300	P&WC PW120	1	
BN-2A Islander	Lyc O-540-E4C5	2	
Cessna 404 Titan II	Cont GTSIO-520-M	2	
DHC-6 Twin Otter 300	P&WC PT6A-27	2	
AIR INUIT (1985) LTD (Canada)			
BAe 748 Srs 2A	RR Dart 534-2	4	
Beech Super King Air 200C	P&WC PT6A-41	1	
CV 580	Allison 501-D13	4	
DHC-6 Twin Otter 200	P&WC PT6A-27	1	
DHC-6 Twin Otter 300	P&WC PT6A-27	4	
AIR IVOIRE (Cote d'Ivoire)			
Beech Super King Air 200	P&WC PT6A-41	2	
Fokker 100	RR Tay 620-15	2	
Fokker F27-400M	RR Dart 536-7B	1	
Fokker F27-600	RR Dart 536-7B	1	
Fokker F28-1000C	RR Spey 555-15	1	
Fokker F28-4000	RR Spey 555-15H	2	
AIR JAMAICA LIMITED (West Indies)			
A300B4-203	GE CF6-50C2	4	
727-200 Advanced	P&W JT8D-15	4	
AIR JET (France)			
BAe 146-200QC	Lyc ALF502R-5	1	
Fokker F27-600	RR Dart 532-7	2	
AIR KENYA AVIATION LIMITED (Kenya)			
Beech Baron 58	Cont IO-520-C	1	
Beech Baron 95-E55	Cont IO-520-C	2	
DC-3C	P&W R-1830	2	
DHC-6 Twin Otter 200	P&WC PT6A-20	1	
DHC-6 Twin Otter 300	P&WC PT6A-27	6	
Fokker F27-200	RR Dart 528-7E	1	
PA T-1040	P&WC PT6A-27	1	
PA-31-310 Navajo	Lyc TSIO-540-A2C	1	
AIR LANKA (Sri Lanka)			
A320-200	IAE V2500-A1	2	
A340-300	CFMI CFM56-5C2		5
L1011 TriStar 50	RR RB211-22B-02	1	
L1011 TriStar 100	RR RB211-22B-02	1	
L1011 TriStar 200	RR RB211-524B-02	2	
L1011 TriStar 500	RR RB211-524B4	2	
AIR LIBERTE (France)			
A300-600R	P&W PW4158	2	
A310-300	P&W PW4158	1	
MD-83	P&W JT8D-219	6	
AIRLINK AIRLINE (PTY) LTD (South Africa)			
ATR 42-320	P&WC PW120	2	
Beech King Air B100	P&WC PT6A-28	2	
Cessna 421	Cont GTSIO-520-D	2	
Fairchild Metro II	Ga TPE331-3UW-303G	3	
AIR LITTORAL (France)			
ATR 72	P&WC PW123	1	1
EMB-110P2 Bandeirante	P&WC PT6A-34	1	
EMB-120RT Brasilia	P&WC PW118	9	
Fokker 100	RR Tay 620-15	6	
Fairchild Metro II	Ga TPE331-3UW-303G	1	
AIR MADAGASCAR (Madagascar)			
737-200	P&W JT8D-9	1	
737-200 Advanced	P&W JT8D-15	1	
747-200B (SCD)	P&W JT9D-70A	1	
BAe 748 Srs 2B	RR Dart 536-2	2	
DHC-6 Twin Otter 300	P&WC PT6A-27	4	
AIR MALAWI (Malawi)			
737-300	CFMI CFM56-3C1	1	
ATR 42	P&WC PW121	1	
BAe 748 Srs 2A	RR Dart 534-2	1	
BAe One-Eleven/400	RR Spey 512-14DW	1	
BAe One-Eleven/500	RR Spey 512-14DW	1	
Dornier 228	Ga TPE331		1
AIR MALTA CO LTD (Malta)			
737-200A	P&W JT8D-15A	6	
737-300	CFMI CFM56-3C1		3
A320-200	CFMI CFM56-5A1	2	
AIR MANITOBA LTD (Canada)			
BAe 748-2A Cargo	RR Dart 534-2	1	
BAe 748-2A Combi	RR Dart 534-2	4	
Curtiss C46 Commando	P&W R-2800	4	
DC-3C	P&W R-1830	3	
AIR MARSHALL ISLANDS, INC (Pacific Islands Trust Territory)			
BAe 748 Srs 2B	RR Dart 536-2	1	
DC-8-62	P&W JT3D-7	1	
Dornier 228-100	Ga TPE331-5-252D	1	
Dornier 228-201	Ga TPE331-5-252D	1	
Saab 2000	Allison GMA 2100		2
AIR MARTINIQUE - CAAA (Martinique)			
ATR 42	P&WC PW120	2	
ATR 72	P&WC PW123		1
Dornier 228-202K	Ga TPE331-5-252D	3	
AIR MAURITANIE (Mauritania)			
BN-2T Islander	Allison 250-B17C	1	
Fokker F28-4000	RR Spey 555-15P	2	
PA-31T Cheyenne II	P&WC PT6A-28	1	
AIR MAURITIUS LIMITED (Mauritius)			
747SP	P&W JT9D-7FW	4	
767-200ER	GE CF6-80C2B4	2	
A340-300	CFMI CFM56-5C3		5
ATR 42	P&WC PW120	2	
Bell 206B JetRanger	Allison 250-C20	2	
AIR MIDWEST, INC (USAIR EXPRESS) (USA)			
Beech 1900 Airliner	P&WC PT6A-65B	14	
Fairchild Metro II	Ga TPE331-10UA-511G	1	
AIR MOOREA (Polynesia)			
BN-2A-8 Islander	Lyc O-540-E4C5	4	
Dornier 228-212	Ga TPE331-5A-252D	1	
PA-23-250 Aztec C	Lyc IO-540-C4B5	2	
AIR NAMIBIA (Namibia)			
737-200 Advanced	P&W JT8D-15	1	
747SP	P&W JT9D-7FW	1	
Beech 1900C Airliner	P&WC PT6A-65B	3	
AIR NAURU (Nauru)			
727-100QC	P&W JT8D-7A	1	
737-200 Advanced	P&W JT8D-17	2	
737-200C	P&W JT8D-17	1	
737-400	CFMI CFM56-3C1		2
AIR NEW ZEALAND LTD (New Zealand)			
737-200 Advanced	P&W JT8D-15A	12	
737-300	CFMI CFM56-3		6
747-200	RR RB211-524D4	4	
747-400	RR RB211-524G	3	1
767-200ER	GE CF6-80A2	6	
767-200ER	P&W JT9D-7R4D	2	
767-300ER	GE CF6-80C2	2	4
AIR NIPPON CO LTD (Japan)			
737-200 Advanced	P&W JT8D-9A	1	
737-200 Advanced	P&W JT8D-17	9	
737-500	CFMI CFM56-3C1		7
A320-200	CFMI CFM56-5A1	1	
DHC-6 Twin Otter 300	P&WC PT6A-27	2	
NAMC YS-11-100	RR Dart 543-10J/K	5	
NAMC YS-11-200	RR Dart 543-10J/K	14	
NAMC YS-11A-500	RR Dart 543-10J/K	7	
AIR NIUGINI (Papua New Guinea)			
A310-300	P&W PW4152	2	
DHC-7 Dash 7-100	P&WC PT6A-50	2	
Fokker F28-1000	RR Spey 555-15	7	
AIR NOVA (Canada)			
BAe 146-200A	Lyc ALF502R-5	5	
Canadair Regional Jet	GE CF34-3A		
DHC-8 Dash 8-100	P&WC PW120A	10	
AIR ONTARIO INC (Canada)			
DHC-8 Dash 8-100	P&WC PW120A	15	
DHC-8 Dash 8-300	P&WC PW123	6	
AIR PACIFIC LIMITED (Fiji)			
737-500	CFMI CFM56-3B1	1	
747-200	P&W JT9D-7FW	1	
767-200ER	P&W JT9D-7R4D	1	
767-300ER	GE CF6-80C2B4		2
AIR RESORTS AIRLINES (USA)			
CV 580	Allison 501-D13H	4	
AIR RWANDA (Rwanda)			
707-300C	P&W JT3D-3B (Q)	1	
DHC-6 Twin Otter 300	P&WC PT6A-27	1	
PA-23-250 Aztec E	Lyc IO-540-C485	1	
AIR SAINT-PIERRE (St Pierre and Miquelon Isles)			
BAe 748 Srs 2A	RR Dart 534-2	2	
PA-23-250 Aztec C	Lyc IO-540-C4B5	1	
PA-31-350 Chieftain	Lyc TIO-540-J2BD	1	

Notes: SCD Side cargo door   TBD To be decided   EUD Extended upper deck   Q Hush kitted   SR Short range   LR Long range   F Freighter

AIRLINE FLEET LIST

Airline and type	Engine	In operation	On order
AIR SENEGAL, SOCIETE NATIONALE DE TRANSPORT AERIEN (Senegal)			
BAe 748 Srs 2A	RR Dart 543-2	2	
DHC-6 Twin Otter 300	P&WC PT6A-27	2	
AIR SERVICE NANTES (France)			
737	TBD		1
Beech King Air 200	P&WC PT6A-41	2	
Caravelle 10B1R	P&W JT8D-7B	1	
Dassault Falcon 20C	GE CF700-2C	1	
Dassault Falcon 20C	GE CF700-2D2	1	
AIR SEYCHELLES (Seychelles)			
757-200ER	RR RB211-535E4	1	
767-200ER	GE CF6-80C2B4	1	
BN-2A Islander	Lyc O-540-E4C5	1	
DHC-6 Twin Otter 300	P&WC PT6A-27	4	
AIR SINAI (Egypt)			
737-200 Advanced	P&W JT8D-17	5	
Fokker F27-500	RR Dart 532-7	2	
AIRSUL - TRANSPORTE AEREO SA (Portugal)			
737-200 Advanced	P&W JT8D-17A	2	
AIR SWAZI CARGO (PTY) LTD (Swaziland)			
707-300C	P&W JT3D-3B (Q)	1	
AIR TAHITI (French Polynesia)			
ATR 42-300	P&WC PW120	4	
ATR 72	P&WC PW124	1	1
DHC-6 Twin Otter 300	P&WC PT6A-27	1	
Dornier 228-200	Ga TPE331-5A-252D	1	
AIR TANZANIA (Tanzania)			
737-200C Advanced	P&W JT8D-17	2	
DHC-6 Twin Otter 300	P&WC PT6A-27	2	
Fokker F27-600CF	RR Dart 536-7R	2	
AIRTOURS INTERNATIONAL AVIATION (GUERNSEY) LTD (UK)			
MD-83	P&W JT8D-219	8	
AIR TRANSAT (Canada)			
727-200 Advanced	P&W JT8D-15	3	
757-200	RR RB211-535E4	2	
L1011 TriStar 100	RR RB211-22B	4	
AIR TRANSPORT INTERNATIONAL (USA)			
DC-8-61F	P&W JT3D-3B (Q)	2	
DC-8-62F	P&W JT3D-3B (Q)	3	
DC-8-63F	P&W JT3D-7 (Q)	4	
AIR TUNGARU CORP (Kiribati)			
BN-2A Trislander	Lyc O-540-E4C5	2	
CASA 212-200 Aviocar	Ga TPE331-10R-511C	1	
AIR UK (LEISURE) LIMITED (UK)			
737-400	CFMI CFM56-3C1	7	
767-300ER	GE CF6-80C2	2	
AIR UK LTD (UK)			
BAe 146-100	Lyc ALF502R-5	2	
BAe 146-200	Lyc ALF502R-5	2	
BAe 146-300	Lyc ALF502R-5	6	
Fokker 100	RR Tay 620-15	5	
Fokker F27-100	RR Dart 514-7	1	
Fokker F27-200	RR Dart 528-7E	10	
Fokker F27-500	RR Dart 532-7	2	
Fokker F27-600	RR Dart 528-7E	1	
Shorts 360	P&WC PT6A-65R	1	
AIR VANUATU (OPERATIONS) LIMITED (South Pacific)			
737-400	CFMI CFM56-3C1	1	
EMB-110P6 Bandeirante	P&WC PT6A-34	1	

Airline and type	Engine	In operation	On order
AIR WISCONSIN (UNITED EXPRESS) (USA)			
BAe 146-100A	Lyc ALF502R-5	1	
BAe 146-200A	Lyc ALF502R-5	5	
BAe 146-300A	Lyc ALF502R-5	5	
BAe ATP	P&WC PW124A	9	5
DHC-8 Dash 8-100	P&WC PW120A	5	
DHC-8 Dash 8-300	P&WC PW121	8	
Fokker F27-500	RR Dart 552-7R	3	
AIR ZAIRE, SOCIETE (Zaïre)			
737-200 Advanced	P&W JT8D-15	2	
DC-8-63AF	P&W JT3D-7	2	
DC-10-30	GE CF6-50C	1	
AIR ZIMBABWE (Zimbabwe)			
707-300B	P&W JT3D-7	1	
707-300B	P&W JT3D-7 (wet)	2	
737-200 Advanced	P&W JT8D-15A	3	
767-200ER	P&W PW4056	2	
BAe 146-200	Lyc ALF502R-5	1	
ALASKA AIRLINES, INC (USA)			
727-100	P&W JT8D-7B	2	
727-200 Advanced	P&W JT8D-17	8	
727-200 Advanced	P&W JT8D-17R	7	
737-200C Advanced	P&W JT8D-17	7	
737-400	CFMI CFM56-3C1	10	12
MD-82	P&W JT8D-217	4	
MD-82	P&W JT8D-217A	4	
MD-82	P&W JT8D-219	2	
MD-83	P&W JT8D-219	28	
MD90-30	IAE V2525-D5		10
ALITALIA - LINEE AEREE ITALIANE, SPA (Italy)			
747-200B	GE CF6-50E2	7	
747-200B (SCD)	GE CF6-50E2	5	
747-200F (SCD)	GE CF6-50E2	2	
A300B2-200	GE CF6-50C2	1	
A300B4-100	GE CF6-50C2	4	
A300B4-200	GE CF6-50C2	8	
A321-100	CFMI CFM56-5B		40
DC-9-32	P&W JT8D-9A	27	
MD-11C	GE CF6-80C2D1F	6	2
MD-82	P&W JT8D-217A	9	
MD-82	P&W JT8D-217C	23	3
ALL NIPPON AIRWAYS CO, LTD (Japan)			
737-200 Advanced	P&W JT8D-17	9	
747-200B	GE CF6-50E2	6	
747-400	GE CF6-80C2	13	11
747SR	GE CF6-45A2	17	
767-200	GE CF6-80C2	25	
767-300	GE CF6-80C2B2	22	11
767-300ER	GE CF6-80C2B6	5	1
777	P&W PW4073		15
A320-200	CFMI CFM56-5A1	14	2
A340-300	CFMI CFM56-5C2		5
Beech 58 Baron	Cont	6	1
Beech A36 Bonanza	Cont IO-550-B	6	
L1011 TriStar 1	RR RB211-22B	9	
NAMC YS-11A-500	RR Dart 543-10J/K	7	
ALM ANTILLEAN AIRLINES (Netherlands Antilles)			
DHC-8 Dash 8-300	P&WC PW123	2	2
MD-82	P&W JT8D-217	2	
MD-82	P&W JT8D-217A	1	
ALOHA AIRLINES (USA)			
737-200	P&W JT8D-9A	2	
737-200 Advanced	P&W JT8D-9A	9	
737-200C	P&W JT8D-9A	2	

All Nippon Airways Boeing 747-400 takes off from Boeing's Paine Field at Everett, north of Seattle

AIRLINE FLEET LIST

Airline and type	Engine	In operation	On order
ALL NIPPON AIRWAYS CO, LTD (Japan) continued...			
737-200 Advanced	P&W JT8D-15	1	
737-200C	P&W JT8D-15	1	
737-200C Advanced	P&W JT8D-17	1	
737-300	CFMI CFM56-3B1	3	
737-400	CFMI CFM56-3B2		8
ALOHA ISLANDAIR, INC (USA)			
DHC-6 Twin Otter 300	P&WC PT6A-27	8	
AMERICANA DE AVIACION (Peru)			
727-100	P&W JT8D-7B	4	
AMERICAN AIRLINES, INC (USA)			
727-100	P&W JT8D-7	16	
727-200 Advanced	P&W JT8D-9A	102	
727-200 Advanced	P&W JT8D-15A	22	
757-200	RR RB211-535E4B	69	22
767-200	GE CF6-80A	8	
767-200ER	GE CF6-80A2	22	
767-300ER	GE CF6-80C2B6	28	13
A300-605R	GE CF6-80C2A5	34	1
DC-10-10	GE CF6-6K	46	
DC-10-10ER	GE CF6-6K2	3	
DC-10-30	GE CF6-50C2	10	
Fokker 100	RR Tay 620-15	41	34
MD-11	GE CF6-80C2D1F	12	7
MD-82	P&W JT8D-217A	234	
MD-83	P&W JT8D-219	26	
AMERICAN TRANS AIR, INC (USA)			
727-100	P&W JT8D-7B	7	
757-200	P&W PW2037	6	
L1011 TriStar 1	RR RB211-22B	13	
AMERICA WEST AIRLINES (USA)			
737-100	P&W JT8D-9A	1	
737-200 Advanced	P&W JT8D-9A	6	
737-200 Advanced	P&W JT8D-15	15	
737-200 Advanced	P&W JT8D-15A	1	
737-300	CFMI CFM56-3B1	21	
737-300	CFMI CFM56-3B2	12	
747-400	TBD		4
757-200	RR RB211-535E4	11	10
A320-200	IAE V2500-A1	18	
AMR EAGLE (USA)			
ATR 42	P&WC PW120	46	
ATR 72-200	P&WC PW124	12	
BAe Jetstream 41	Ga TPE331-14GR/HR		
BAe Jetstream Super 31	Ga TPE331-10UR-513H	89	
Saab 340A	GE CT7-5A2	16	
Saab 340B	GE CT7-9B	99	
Saab 2000	Allison GMA 2100		
Shorts 360-200	P&WC PT6A-65AR	31	
Shorts 360-300	P&WC PT6A-67R	6	
ANSETT AUSTRALIA (Australia)			
727-200 Advanced	P&W JT8D-15	5	
727-200F Advanced	P&W JT8D-15	1	
737-300	CFMI CFM56-3B1	16	
767-200	GE CF6-80A	5	
A320-200	CFMI CFM56-5A1	12	
A321	CFMI CFM56-5B		10
BAe 146-200QT	Lyc ALF502R-5	2	
Fokker 50	P&WC PW125B	5	
ANSETT EXPRESS AIRLINES (Australia)			
Fokker 50	P&WC PW125B	6	
Fokker F28-3000	RR Spey 555-15	2	
Fokker F28-4000	RR Spey 555-15	4	
ANSETT NEW ZEALAND (New Zealand)			
BAe 146-200	Lyc ALF502R-5	2	
BAe 146-200QC	Lyc ALF502R-5	1	
BAe 146-300	Lyc ALF502R-5	6	
DHC-8 Dash 8-100	P&WC PW120A	2	
ANSETT WA (Australia)			
BAe 146-200A	Lyc ALF502R-5	6	
Fokker F28-1000	RR Spey 555-15	4	
Fokker F28-4000	RR Spey 555-15P	2	
AOM FRENCH AIRLINES (France)			
DC-10-30	GE CF6-50C	3	
DC-10-30	GE CF6-50C2	2	
DC-10-30	GE CF6-50C2B	2	
DC-10-30	GE CF6-50C2R	1	
MD-11	P&W PW4460		3
MD-83	P&W JT8D-219	5	
APA INTERNACIONAL AIRWAYS (Dominican Republic)			
707-300C	P&W JT3D-3B	1	
DC-6F	P&W R-2800-CB16	2	
APISA - AEROTRANSPORTES PERUANOS INTERNACIONALES SA (Peru)			
707-300C	P&W JT3D-3B	1	
DC-8-33F	P&W JT4A-9	1	
ARIANA AFGHAN AIRLINES (Afghanistan)			
727-100C	P&W JT8D-7	1	
727-100C	P&W JT8D-9	1	
727-200	P&W JT8D-7	3	
An-24RV	ZMKB AI-24	2	
An-26	ZMKB AI-24T	2	
Tu-154M	Aviadvigatel D-30KU-154-II	2	
Yak-40	ZMKB AI-25	2	
ARKIA ISRAELI AIRLINES LTD (Israel)			
707-300C	P&W JT3D-3B (Q)	1	
737-200 Advanced	P&W JT8D-17A	2	
747-100 (SCD)	P&W JT9D-7J	1	
757-200	RR RB211-535E4	1	
BN-2A Islander	Lyc O-540-E4B5	1	
Cessna 337B Super Skymaster	Cont IO-360-C	4	
DHC-7 Dash 7-100	P&WC PT6A-50	6	
PA-31-350 Chieftain	Lyc TIO-540-J2BD	5	
Rockwell Grand Commander	Lyc IGSO-540-B1A	1	
ARROW AIR INC (USA)			
DC-8-62F	P&W JT3D-3B (Q)	1	
DC-8-62F	P&W JT3D-7 (Q)	8	
DC-8-63F	P&W JT3D-7 (Q)	4	
ASIANA AIRLINES (Korea)			
737-400	CFMI CFM56-3C1	13	
737-500	CFMI CFM56-3C1	4	
747-400C (SCD)	GE CF6-80C2B1F	5	6
767-300	GE CF6-80C2BF	2	8
767-300ER	GE CF6-80C2B6F	3	2
ATI - LINEE AEREE NAZIONALI (Italy)			
ATR 42	P&WC PW120	7	
DC-9-32	P&W JT8D-9A	14	
MD-82	P&W JT8D-217A	13	1
MD-82	P&W JT8D-217C	15	3
ATLANTIC AIRWAYS (FAROE ISLANDS) (Faroe Islands)			
BAe 146-200A	Lyc ALF502R-5	1	
ATLANTIC SOUTHEAST AIRLINES, INC (USA)			
ATR 72-210	P&WC PW124		8
DHC-7 Dash 7-100	P&WC PT6A-50	2	
EMB-110P1 Bandeirante	P&WC PT6A-34	11	
EMB-120RT Brasilia	P&WC PW118	53	5
AURIGNY AIR SERVICES LTD (UK)			
BN-2A Trislander	Lyc O-540-E4C5	9	
Shorts 360	P&WC PT6A-67R	1	
AUSTRAL LINEAS AEREAS (Argentina)			
BAe One-Eleven/500	RR Spey 512-14DW	8	
MD-81	P&W JT8D-209	2	
AUSTRIAN AIR SERVICES (Austria)			
Fokker 50	P&WC PW125B	6	1
AUSTRIAN AIRLINES, OSTERREICHISCHE LUFTVERKEHRS AG (Austria)			
A310-300	P&W PW4152	4	
A320-200	CFMI CFM56-5B		7
A321	CFMI CFM56-5B		6
A340-200	CFMI CFM56-5C2		2
MD-81	P&W JT8D-209	7	
MD-82	P&W JT8D-217C	6	
MD-83	P&W JT8D-219	1	
MD-87ER	P&W JT8D-219	2	
MD-87SR	P&W JT8D-217C	3	
AVENSA, AEROVIAS VENEZOLANAS, SA (Venezuela)			
727-100	P&W JT8D-7B	3	
727-200	P&W JT8D-9A	1	
727-200 Advanced	P&W JT8D-9A	2	
727-200 Advanced	P&W JT8D-17	4	
757-200	RR RB211-535C	2	
Beech King Air 90	P&WC PT6A-28	1	
CV 580	Allison 501-D13D	1	
DC-9-31	P&W JT8D-7B	1	
DC-9-51	P&W JT8D-17	4	
AVIACO - AVIACION Y COMERCIO, SA (Spain)			
DC-9-32	P&W JT8D-9A	17	
DC-9-34	P&W JT8D-17	3	
DC-9-34F	P&W JT8D-17	2	
MD-88	P&W JT8D-217C	13	
AVIANCA — See AEROVIAS NACIONALES DE COLOMBIA			
AVIANOVA SPA (Italy)			
ATR 42-300	P&WC PW120	11	
AVIATECA - AEROLINEAS DE GUATEMALA (Guatemala)			
737-200	P&W JT8D-15	1	
737-300	CFMI CFM56-3B2	3	
AVIOGENEX (Serbia)			
727-200	P&W JT8D-15	3	
737-200	P&W JT8D-15	4	

Notes: SCD Side cargo door   TBD To be decided   EUD Extended upper deck   Q Hush kitted   SR Short range   LR Long range   F Freighter

AIRLINE FLEET LIST

Airline and type	Engine	In operation	On order
BAHAMASAIR HOLDINGS LTD (Bahamas)			
737-200 Advanced	P&W JT8D-9A	1	
737-200 Advanced	P&W JT8D-17	2	
BAe 748 Srs 2A	RR Dart 535-2	1	
Cessna 402C Utililiner II	Cont TSIO-520-VB	3	
DHC-8 Dash 8-300	P&WC PW123	4	4
BALAIRCTA (formerly Balair and CTA) (Switzerland)			
A310-300ER	P&W JT9D-7R4E1	1	
A320-300	P&W PW4156	3	
MD-82	P&W JT8D-217C	3	
MD-83	P&W JT8D-219	2	
MD-87	P&W JT8D-217C	4	
BALKAN-BULGARIAN AIRLINES (Bulgaria)			
737-500	CFMI CFM56-3C1	3	
767-200ER	P&W PW4060	2	
A320-200	IAE V2500-A1	4	
An-12 (F)	ZMKB AI-20K	3	
An-24V-2	ZMKB AI-24	15	
Il-18D	ZMKB AI-20	1	
Il-18V	ZMKB AI-20	2	
Tu-134A-3	Aviadvigatel D-30	5	
Tu-154B-2	KKBM NK-8-2U	17	
Tu-154M	Aviadvigatel D-30KU-154-II	6	
BALTIC INTERNATIONAL AIRLINES (Latvia)			
Tu-134A	Aviadvigatel D-30	2	
Tu-154M	Aviadvigatel D-30KU-154-II	2	
BANGKOK AIRWAYS CO LTD (Thailand)			
DHC-8 Dash 8-100	P&WC PW121	1	
DHC-8 Dash 8-300	P&WC PW123	3	
EMB-110P2 Bandeirante	P&WC PT6A-34	1	
Shorts 330-200	P&WC PT6A-45R	4	
Shorts 360 Advanced	P&WC PT6A-67R	2	
BARNAULSKOYE AVIAPREDPRIYATIYE (Russia)			
Il-76	Aviadvigatel D-30KP		
Mil Mi-8	Klimov TV2-117A		
PZL An-2	Aviadvigatel ASh-621R		
Tu-154	KKBM NK-8-2U		
Yak-40	ZMKB AI-25		
BERING AIR, INC (USA)			
Beech 18	P&W R-985	4	
Cessna 207 Stationair 8 II	Cont IO-520-F	4	
PA-31-350 Chieftain	Lyc TIO-540-J2BD	4	
PA-31-350 T1020	Lyc TIO-540-J2B	3	
BIG SKY AIRLINES (USA)			
Cessna 402C Utililiner II	Cont TSIO-520-VB	1	
Fairchild Metro II	Ga TPE331-10UA-511G	5	
BIMAN BANGLADESH AIRLINES (Bangladesh)			
BAe ATP	P&WC PW126A	2	
DC-10-30	GE CF6-50C2	4	
Fokker F28-4000	RR Spey 555-15H	2	
BINTER (Spain)			
ATR 72	P&WC PW124	6	
CASA/IPTN CN-235-10	GE CT7-7A	4	
CASA/IPTN CN-235-100	GE CT7-9C	5	
DC-9-30	P&W JT8D-17	2	
BOP AIR (Bophuthatswana)			
BAe 748 Srs 2B	RR Dart 535-2	1	
Cessna 550 Citation II	P&WC JT15D-4	1	
EMB-110P2 Bandeirante	P&WC PT6A-34	2	
EMB-120RT Brasilia	P&WC PW118A	2	
PA-31-350 Chieftain	Lyc TIO-540-J2BD	1	
BOURAQ INDONESIA AIRLINES PT (Indonesia)			
BAe 748 Srs 2A	RR Dart 534-2	9	
BAe 748 Srs 2B	RR Dart 536-2	6	
BAe Viscount	RR Dart 525F	4	
IPTN N-250-100	Allison GMA 2100		62
BRAATHENS S.A.F.E. A/S (Norway)			
737-200 Advanced	P&W JT8D-17A	2	
737-200 Heavy Advanced	P&W JT8D-17	3	
737-400	CFMI CFM56-3C1	6	1
737-500	CFMI CFM56-3C1	15	3
BRADLEY AIR SERVICES LIMITED (FIRST AIR) (Canada)			
727-100	P&W JT8D-7B	1	
727-100C	P&W JT8D-7A	2	
727-100C	P&W JT8D-7B	1	
BAe 748 Srs 2A	RR Dart 534-2	6	
BAe 748 Srs 2A	RR Dart 535-2	1	
BAe 748 Srs 2B	RR Dart 535-2	1	
DHC-2 Beaver	P&W R-985	2	
DHC-6 Twin Otter 300	P&WC PT6A-27	5	
DHC-7 Dash 7-100	P&WC PT6A-50	1	
BRIT AIR (France)			
ATR 42-300	P&WC PW120	13	
ATR 72-101	P&WC PW125	2	2
EMB-110P2 Bandeirante	P&WC PT6A-34	2	
PA-31T Cheyenne II	P&WC PT6A-28	1	
Saab 340A	GE CT7-5A2	6	
BRITANNIA AIRWAYS LTD (UK)			
737-200	P&W JT8D-9	2	
737-200 Advanced	P&W JT8D-15	13	
757-200	RR RB211-535E5	1	1
757-200ER	RR RB211-535E4	11	
767-200	GE CF6-80A	3	
767-200ER	GE CF6-80A	7	4
BRITISH AIRWAYS (UK)			
737-200	P&W JT8D-15A	43	
737-300	CFMI CFM56-3B2	3	
737-400	CFMI CFM56-3C1	32	6
747-100	P&W JT9D-7A	15	
747-200	RR RB211-524D4U	13	
747-200B (SCD)	RR RB211-524DX	3	
747-400	RR RB211-524G	25	36
757-200	RR RB211-535C	34	
757-200	RR RB211-535E4	8	3
767-300ER	RR RB211-524H	20	8
777-200 / 200ER	GE GE90-B4		15
A320-100	CFMI CFM56-5A1	5	
A320-200	CFMI CFM56-5A1	5	
BAe ATP	P&WC PW126	14	
BAe One-Eleven/500	RR Spey 512-14DW	6	
Concorde	RR Olympus 593-610	7	
DC-10-30	GE CF6-50C	7	
L1011 TriStar 1/100	RR RB211-22B	5	

Concorde, operated only by British Airways and Air France, still turns heads wherever it is seen

AIRLINE FLEET LIST

Canadian Airlines International Airbus A320, one of 12 in service and 10 on order

Airline and type	Engine	In operation	On order
BRITISH MIDLAND AIRWAYS LTD (UK)			
737-300	CFMI CFM56-3B1	7	
737-300	CFMI CFM56-3B2	1	
737-400	CFMI CFM56-3C1	6	
737-500	CFMI CFM56-3C1	3	
BAe ATP	P&WC PW126	3	
DC-9-14	P&W JT8D-7A/B	2	
DC-9-15	P&W JT8D-7A/B	3	
DC-9-32	P&W JT8D-11	2	
DC-9-32	P&W JT8D-9/9A	6	
DHC-7 Dash 7-100	P&WC PT6A-50	2	
BRITISH WORLD AIRLINES LTD (UK)			
BAe 146-300	Lyc ALF502R-5	2	
BAe Freight Master	RR Dart 510	5	
BAe One-Eleven/500	RR Spey 512-14DW	5	
BAe Viscount	RR Dart 510	5	
Handley Page Herald 214	RR Dart 532-9	1	
BRYMON EUROPEAN AIRWAYS (UK)			
BAe Jetstream 31	Ga TPE331-10UF-511H	3	
BAe One-Eleven/400	RR Spey 512-14	4	
BAe One-Eleven/500	RR Spey 512-14DW	1	
DHC-7 Dash 7-100	P&WC PT6A-50	4	
DHC-8 Dash 8-100	P&WC PW120A	2	
DHC-8 Dash 8-300	P&WC PW123	2	
BUFFALO AIRWAYS (USA)			
707-300C	P&W JT3D-3B (Q)	3	
BUSINESS EXPRESS (USA)			
Avro IAe RJ70	Lyc LF508		10
BAe 146-200	Lyc ALF502R-5	5	
Beech 1900C Airliner	P&WC PT6A-65B	22	
Saab 340A	GE CT7-5A2	17	
Saab 340B	GE CT7-9B	18	30
Saab 2000	Allison GMA 2100		
Shorts 360	P&WC PT6A-65R	8	
CAIRO AVIATION (Egypt)			
Il-76TD	Aviadvigatel D-30KP	2	
Tu-154M	Aviadvigatel D-30KU-154-II	2	
CALEDONIAN AIRWAYS (UK)			
757-200ER	RR RB211-535E4	6	
DC-10-30	GE CF6-50C2	1	
L1011 TriStar 1	RR RB211-22B	1	
L1011 TriStar 100	RR RB211-22B	4	
CALM AIR INTERNATIONAL LTD (CANADIAN PARTNER) (Canada)			
BAe 748 Srs 2A	RR Dart 534-2	4	
Beech Super King Air 200	P&WC PT6A-41	1	
DHC-6 Twin Otter 100	P&WC PT6A-20	1	
DHC-6 Twin Otter 200	P&WC PT6A-20	1	
PA-31-350 Chieftain	Lyc TIO-540-J2BD	1	
CAMEROON AIRLINES (Cameroon)			
737-200 Advanced	P&W JT8D-15	2	
737-200 Advanced	P&W JT8D-15A	1	
747-200B (SCD)	P&W JT9D-7Q	1	
BAe 748 Super 2B	RR Dart 536-2	1	
CANADA 3000 AIRLINES LIMITED (Canada)			
757-200ER	RR RB211-535E4	7	
CANADIAN AIRLINES INTERNATIONAL LTD (Canada)			
737-200 Advanced	P&W JT8D-9A	19	
737-200 Advanced	P&W JT8D-15	1	
737-200 Advanced	P&W JT8D-17	9	
737-200 Advanced	P&W JT8D-17A	5	
737-200 Advanced (ER)	P&W JT8D-17	6	

Airline and type	Engine	In operation	On order
737-200C	P&W JT8D-9A	2	
737-200C Advanced	P&W JT8D-9A	3	
737-200C Advanced	P&W JT8D-17A	2	
747-400	GE CF6-80C2B1F	3	
767-300ER	GE CF6-80C2B6	12	
A320-200	CFMI CFM56-5	12	10
DC-10-30	GE CF6-50C2	3	
DC-10-30ER	GE CF6-50C2	5	
CANADIAN REGIONAL AIRLINES (Canada)			
ATR 42-300	P&WC PW120	7	
DHC-8 Dash 8-100	P&WC PW120	6	5
DHC-8 Dash 8-300	P&WC PW123	14	
EMB-120 Brasilia	P&WC PW118		3
Fokker F28-1000	RR Spey 555-15	7	
CANAIR CARGO (Canada)			
CV 580	Allison 501-D13	12	
CARGOLUX AIRLINES INTERNATIONAL SA (Luxembourg)			
747-200F (SCD)	GE CF6-50E2	4	
747-200F (SCD)	P&W JT9D-7	1	
747-400F	GE CF6-80C2B1F		3
CARGOSUR - COMPANIA DE EXPLOTACION DE AVIONES CARGUEROS SA (Spain)			
DC-8-62F	P&W JT3D-3B (Q)	4	
CARIBBEAN INTERNATIONAL AIRWAYS (Venezuela)			
707-300C	P&W JT3D-3B	4	
DC-3	P&W R-1830	2	
CARNIVAL AIR LINES INC (USA)			
727-200	P&W JT8D-1A	2	
727-200	P&W JT8D-9A	2	
737-200	P&W JT8D-9A	4	
737-400	CFMI CFM56-3B2	3	
CATA (Argentina)			
Fairchild F-27J	RR Dart 532-7	4	
IAI Arava 102	P&WC PT6A-34	1	
Rockwell Turbo Cmdr 690B	Ga TPE331-5-251K	2	
CATHAY PACIFIC AIRWAYS LTD (Hong Kong)			
747-200	RR RB211-524C2	2	
747-200	RR RB211-524D4	5	
747-200F	RR RB211-524D4	4	
747-300	RR RB211-524C2	6	
747-400	RR RB211-524H	15	3
747-400F	RR RB211-524H		2
777	RR Trent 800		11
A330	RR Trent 700		10
L1011 TriStar 1	RR RB211-22B	17	
L1011 TriStar 100	RR RB211-22B	2	
CAYMAN AIRWAYS LTD (West Indies)			
737-200 Advanced	P&W JT8D-15A	2	
CCAIR, INC (USAIR EXPRESS) (USA)			
BAe Jetstream 31	Ga TPE331-5-252D	12	
DHC-8 Dash 8-100	P&WC PW120A	4	
Shorts 360-300	P&WC PT6A-67R	9	
CHALLENGE AIR CARGO (USA)			
707-300C	P&W JT3D-3B (Q)	1	
757-200PF	RR RB211-535E4	3	
CHANNEL EXPRESS (AIR SERVICES) LTD (UK)			
Handley Page Herald	RR Dart 532-9	9	
L188C Electra	Allison 501-D13A	4	

Notes: SCD Side cargo door   TBD To be decided   EUD Extended upper deck   Q Hush kitted   SR Short range   LR Long range   F Freighter

AIRLINE FLEET LIST

Airline and type	Engine	In operation	On order
CHINA AIRLINES LTD (Taiwan)			
737-200 Advanced	P&W JT8D-9A	3	
747-200B	P&W JT9D-7A	1	
747-200B	P&W JT9D-7Q	2	
747-200B (SCD)	P&W JT9D-7A	1	
747-200F (SCD)	P&W JT9D-7R4G2	2	
747-400	P&W PW4056	4	1
747SP	P&W JT9D-7A	4	
A300-600R	P&W PW4158	6	
A300B4-200	P&W JT9D-59A	6	
MD-11	P&W PW4460	2	2
CHINA EASTERN AIRLINES (People's Republic of China)			
A300-600R	GE CF6-80C2A5	3	4
A310-300	GE CF6-80C2A2	2	
A340-300	CFMI CFM56-5C2		5
BAe 146-100	Lyc ALF502R-5	3	
Fokker 100	RR Tay 650-15	1	9
MD-11	GE CF6-80C2	3	1
MD-11F	GE CF6-80C2	1	
MD-82	P&W JT8D-217	6	
MD-82 / MD/SAIC MD-82	P&W JT8D-217A	8	
SAC Y-8	WJ 6	2	
Shorts 360	P&WC PT6A-65R	4	
XAC Y7-100	WJ 5A-1	10	
CHINA NORTHERN AIRLINES (People's Republic of China)			
A300	P&W PW4000		6
MD-82 / MD/SAIC MD-82	P&W JT8D-217A	2/15	2
XAC Y7	WJ 5A-1	5	
XAC Y7-100	WJ 5A-1	3	
CHINA NORTHWEST AIRLINES (People's Republic of China)			
A300-600R	GE CF6-80C2	2	6
An-24	ZMKB AI-24A	3	
BAe 146-100	Lyc ALF502R-5	3	
BAe 146-300	Lyc ALF502R-5		8
SAP Y-5	HS 5	12	
Tu-154M	Aviadvigatel D-30KU-154-II	10	
XAC Y7-100	WJ 5A-1	6	
CHINA SOUTHERN AIRLINES (People's Republic of China)			
737-200 Advanced	P&W JT8D-17A	9	
737-300	CFMI CFM56-3	4	3
737-500	CFMI CFM56-3B1	5	4
757-200	RR RB211-535E4	12	4
777	GE GE90		6
A340	CFM1 CFM56-5C2		6
An-24	ZMKB AI-24A	5	
BN-2B-20 Islander	Lyc IO-540-K1B5	5	
Saab 340B	GE CT7-9B	1	3
Shorts 360	P&WC PT6A-65R	3	
XAC Y7-100	WJ 5A-1	5	
CHINA SOUTHWEST AIRLINES (People's Republic of China)			
707-300	P&W JT3D-7	5	
737-300	CFMI CFM56-3B1	5	
757-200	RR RB211-535E4	1	4
HAMC Y-12	P&WC PT6A-27	4	
Saab 340B	GE CT7-9B		4
Tu-154M	Aviadvigatel D-30KU-154-II	5	
XAC Y7-100	WJ 5A-1	5	
CIMBER AIR DENMARK A/S (Denmark)			
AS 262	TU Bastan VIC	2	
ATR 42	P&WC PW120	8	
Cessna Citation I/SP	P&WC JT15D-1A	1	
CITYFLYER EXPRESS LTD (UK)			
ATR 42	P&WC PW120	2	
Shorts 360	P&WC PT6A-65AR	4	
COMAIR (COMMERCIAL AIRWAYS (PTY) LTD) (South Africa)			
DC-3	P&W R-1830-92	1	
Fokker F27-200	RR Dart 532-7R	4	
COMAIR (USA)			
Canadair Regional Jet	GE CF34-3A		20
EMB-120RT Brasilia	P&WC PW118A	40	
Fairchild Metro III	Ga TPE33One-ElevenU-611G	4	
Fairchild Metro III	Ga TPE33One-ElevenU-612G	6	
Saab 340A	GE CT7-5A2	19	
Saab 2000	Allison GMA 2100		
COMPANIA PANAMENA DE AVIACION, SA (COPA) (Panama)			
737-100	P&W JT8D-9A	1	
737-200 Advanced	P&W JT8D-15	4	
737-200QC	P&W JT8D-9A	1	
CONAIR A/S (Denmark)			
A300B4-100	P&W JT9D-59A	3	
A320-200	IAE V2500-A1	6	
CONDOR FLUGDIENST GMBH (Germany)			
737-300	CFMI CFM56-3B2	8	
757-200	P&W PW2040	12	6
757-200	RR RB211-535E4	1	
767-300ER	P&W PW4060	5	8
DC-10-30	GE CF6-50C2	3	
CONNIE KALITTA SERVICES (USA)			
727-100 (F)	P&W JT8D-7B	2	
Cessna 310L (F)	Cont IO-470-V	1	
DC-8-51F	P&W JT3D-3B (Q)	2	
DC-8-54	P&W JT3D-3B (Q)	2	
DC-8-55	P&W JT3D-3B (Q)	3	
DC-8-73	CFMI CFM56-2	1	
Hamilton Westwind IISTD	Ga TPE331-1-101B	1	
Learjet 23 (F)	GE CJ610-4	2	
Learjet 24 (F)	GE CJ610-4	1	
Learjet 25 (F)	GE CJ610-6	1	
Learjet 25B (F)	GE CJ610-6	2	
Learjet 25D (F)	GE CJ610-8A	1	
Volpar Turboliner (F)	Ga TPE331-1-101B	10	
CONTINENTAL AIR LINES, INC (USA)			
727-100	P&W JT8D-7	3	
727-200 Advanced	P&W JT8D-9A	62	
727-200 Advanced	P&W JT8D-15	23	
737-100	P&W JT8D-7A	14	
737-200	P&W JT8D-9A	24	
737-200 Advanced	P&W JT8D-9A	1	
737-300	CFMI CFM56-3B1	55	50
747-100	P&W JT9D-7	2	
747-200B	P&W JT9D-7	6	
757-200	RR RB211-535E4		25
767-300ER	GE CF6-80C2		12
777A	GE GE90-B4		5
A300B4-103	GE CF6-50C	6	
A300B4-203	GE CF6-50C2	17	
DC-9-30	P&W JT8D-9A	34	
DC-10-10	GE CF6-6D	7	
DC-10-30	GE CF6-50C2	10	
MD-81	P&W JT8D-217	5	
MD-82	P&W JT8D-217	18	
MD-82	P&W JT8D-217A	32	
MD-82	P&W JT8D-219	4	
CONTINENTAL EXPRESS (USA)			
ATR 42	P&WC PW120	37	
Beech 1900C Airliner	P&WC PT6A-65B	5	
Beech 1900C-1 Airliner	P&WC PT6A-65B	10	
DHC-7 Dash 7-100	P&WC PT6A-50	5	
EMB-120 Brasilia	P&WC PW118	33	
EMB-120RT Brasilia	P&WC PW118	20	10
Fairchild F-27	RR Dart 514-7	2	
CONTINENTAL MICRONESIA INC (Pacific Islands Trust Territory)			
727-100C	P&W JT8D-9	1	
727-200 Advanced	P&W JT8D-15	9	
747-200B	P&W JT9D-7	1	
DC-10-10	GE CF6-6D	5	
DC-10-30	GE CF6-50C2	1	
CORSAIR INTERNATIONAL (France)			
737-300	CFMI CFM56-3B1	1	
737-400	CFMI CFM56-3C1	2	
747-100	P&W JT9D-7A	2	
747-300M	GE CF-50C	1	1
CORSE MEDITERRANEE (France)			
ATR 72-300	P&WC PW124B	4	1
Fokker 100	RR Tay 620-15	1	1
CROATIA AIRLINES (Croatia)			
737-200 Advanced	P&W JT8D-15	5	
ATR 42-300	P&WC PW120	2	
Cessna 310R2	Cont IO-520-MB	1	
Cessna 402C II	Cont TSIO-520-VB	3	
Cessna 550 Citation II	P&WC JT15D-4	1	
CROSSAIR (Switzerland)			
BAe 146-300	Lyc ALF502R-5	1	
AIAe RJ85	Lyc LF507-1F	4	
Fokker 50	P&WC PW125B	5	
Saab 340A	GE CT7-5A2	10	
Saab 340B	GE CT7-9B	15	
Saab 2000	Allison GMA 2100		20
CROWN AIRWAYS (USAIR EXPRESS) (USA)			
Beech 1900C Airliner	P&WC PT6A-65B	3	
DHC-6 Twin Otter 300	P&WC PT6A-27	2	
Shorts 330-200	P&WC PT6A-45R	3	
Shorts 360-300	P&WC PT6A-67R	2	
CRUZEIRO (Brazil)			
737-200 Advanced	P&W JT8D-17	6	
CSA - CESKOSLOVENSKE AEROLINIE (Czech Republic and Slovakia)			
737-500	CFMI CFM56-3B1	2	3
A310-300	GE CF6-80C2A2	2	
ATR 72	P&WC PW125	4	
Il-62	KKBM NK-8-4	2	
Il-62M	Aviadvigatel D-30KU	6	
Tu-134A	Aviadvigatel D-30-II	5	
Tu-154M	Aviadvigatel D-30KU-154-II	7	
CTA - COMPAGNIE DE TRANSPORT AERIEN (Switzerland) see BALAIRCTA			

AIRLINE FLEET LIST

Czech and Slovak airline CSA has been operating seven Tu-154M airliners *(Peter J. Cooper)*

Airline and type	Engine	In operation	On order
CUBANA DE AVIACION (Cuba)			
A310-300	GE CF6-80C2A2		
An-24RV	ZMKB AI-24	7	
An-24V	ZMKB AI-24	5	
An-26	ZMKB AI-24T	24	
Il-62M	Aviadvigatel D-30KU	10	
Il-76MD	Aviadvigatel D-30KP	2	
Il-96-300	Aviadvigatel PS-90A		5
Tu-154B-2	KKBM NK-8-2U	5	
Tu-154M	Aviadvigatel D-30KU-154-II	4	
Yak-40	ZMKB AI-25	8	
CYPRUS AIRWAYS LTD (Cyprus)			
A310-203	GE CF6-80A3	3	
A310-204	GE CF6-80C2A2	1	
A320-200	IAE V2500-A1	8	
BAe One-Eleven/500	RR Spey 512-14DW	3	
CYPRUS TURKISH AIRLINES / KIBRIS TURK HAVA YOLLARI (Turkish Northern Cyprus)			
727-200 Advanced	P&W JT8D-15	2	
DAS AIR CARGO (DAIRO AIR SERVICES) (Uganda)			
707-300C	P&W JT3D-3B (Q)	2	
DELTA AIR LINES, INC (USA)			
727-200 Advanced	P&W JT8D-9	4	
727-200 Advanced	P&W JT8D-15	137	
737-200 Advanced	P&W JT8D-15	10	
737-200 Advanced	P&W JT8D-15A	46	
737-200 Advanced	P&W JT8D-17	2	
737-300	CFMI CFM56-3B1	13	57
757-200	P&W PW2037	84	6
767-200	GE CF6-80A	13	
767-200	GE CF6-80A2	2	
767-300	GE CF6-80A2	24	6
767-300ER	P&W PW4060	12	5
A310-200	P&W JT9D-7R4E1	3	
L1011 TriStar 1	RR RB211-22B	32	
L1011 TriStar 200	RR RB211-524B	1	
L1011 TriStar 250	RR RB211-524B	6	
L1011 TriStar 500	RR RB211-524B	17	
MD-11	P&W PW4360	9	8
MD-88	P&W JT8D-219	110	15
MD90-30	IAE V2525-D5		50
DELTA AIR TRANSPORT NV (Belgium)			
BAe 146-200	Lyc ALF502R-5	7	2
EMB-120RT Brasilia	P&WC PW118	9	
EMB-145	Allison GMA 3007		
Fokker F28-3000	RR Spey 555-15H	1	
Fokker F28-4000	RR Spey 555-15P	1	
DEUTSCHE BA LUFTFAHRTGESELLSCHAFT MBH (Germany)			
737-300	CFMI CFM56-3	3	
Dornier 228-200	Ga TPE331-5-252D	1	
Saab 340A	GE CT7-5A2	9	
Saab 2000	Ga GMA 2100		5
DOMINICANA DE AVIACION (Dominican Republic)			
707-300C	P&W JT3D-3B	1	
727-100	P&W JT8D-9A	1	
727-100C	P&W JT8D-7B	1	
727-200 Advanced	P&W JT8D-9A	1	
DC-6A (F)	P&W R-2800-CB16	1	
DRAGONAIR (HONG KONG DRAGON AIRLINES LIMITED) (Hong Kong)			
737-200 Advanced	P&W JT8D-17	3	
A320-200	IAE V2500-A1	3	3
L1011 TriStar 1	RR RB211-22B	2	
DRUK AIR CORP (Bhutan)			
BAe 146-100	Lyc ALF502R-5	1	
Dornier 228-201	Ga TPE331-5-252D	2	
EAGLE AIRWAYS LTD (New Zealand)			
EMB-110P1 Bandeirante	P&WC PT6A-34	6	
Fairchild Metro III	Ga TPE33One-ElevenU-611G	2	
EASTERN AUSTRALIA AIRLINES (Australia)			
BAe Jetstream 31	Ga TPE331-10UG-513H	3	
BAe Jetstream Super 31	Ga TPE331-12UAR-701H	1	
DHC-8 Dash 8-100	P&WC PW120A	5	
EASTWEST AIRLINES LTD (Australia)			
BAe 146-300	Lyc ALF502R-5	6	
ECUATORIANA (Ecuador)			
707-300B	P&W JT3D-3B	1	
707-300B	P&W JT3D-3B (Q)	2	
707-300C (F)	P&W JT3D-3B (Q)	1	
A310-300	P&W PW4152		2
DC-10-30	GE CF6-50C	1	
EGYPTAIR (Egypt)			
707-300	P&W JT3D-7	6	
737-500	CFMI CFM56-3C1	3	2
747-300 (SCD)	P&W JT9D-7R4G2	2	
767-200ER	P&W JT9D-7R4E	3	
767-300ER	P&W JT9D-7R4G2	2	
A300-600R	P&W PW4158	9	
A300B4-203	GE CF6-50C2	4	
A320-200	IAE V2500-A1	5	
EL AL - ISRAEL AIRLINES LTD (Israel)			
707-300C	P&W JT3D-3B (Q)	1	
737-200 Advanced	P&W JT8D-17A	2	
747-200	P&W JT9D-7J	5	
747-200C	P&W JT9D-7J	2	
747-200F	P&W JT9D-7J	1	
747-400	P&W PW4056		2
757-200	RR RB211-535E4	7	
767-200	P&W JT9D-7R4D	2	
767-200ER	P&W JT9D-7R4D	2	
EMERY WORLDWIDE AIRLINES, INC (USA)			
727-100C	P&W JT8D-7B	22	
727-100F	P&W JT8D-7B (Q)	5	
727-200F	P&W JT8D-7B	3	
DC-8-54F	P&W JT3D-3B (Q)	2	
DC-8-62F	P&W JT3D-3B (Q)	7	
DC-8-63F	P&W JT3D-7 (Q)	12	
DC-8-73F	CFMI CFM56-2C	7	
DC-9-15F	P&W JT8D-7B	7	

Notes: SCD Side cargo door   TBD To be decided   EUD Extended upper deck   Q Hush kitted   SR Short range   LR Long range   F Freighter

AIRLINE FLEET LIST

Airline and type	Engine	In operation	On order
EMIRATES (United Arab Emirates)			
727-200 Advanced	P&W JT8D-17	3	
777A/B	RR Trent 880		7
A300-600R	GE CF6-80C2A5	4	1
A310-300	GE CF6-80C2A2	8	
EQUATORIAL - INTERNATIONAL AIRLINES OF SAO TOME E PRINCIPE			
(Sao Tome and Principe)			
AS 262	TU Bastan VIC1	1	
ETHIOPIAN AIRLINES CORPORATION (Ethiopia)			
727-200 Advanced	P&W JT8D-17R	4	
737-200	P&W JT8D-17A	1	
757-200	P&W PW2040	3	1
757-200F	P&W PW2040	1	
767-200ER	P&W JT9D-7R4E	3	
ATR 42	P&WC PW121	2	
DHC-5A Buffalo	GE CT64-820-4	1	
DHC-6 Twin Otter 300	P&WC PT6A-27	5	
L382G Hercules	Allison 501-D22A	2	
EURALAIR INTERNATIONAL (France)			
737-200	P&W JT8D-7B/-9A	5	
737-500	CFMI CFM56-3B1	3	3
777	GE GE90-B4		2
A321-100	CFMI CFM56-5B		2
A330-320	P&W PW4164		2
BAe 146-200QT	Lyc ALF502R-5	2	
Cessna 525 CitationJet	Williams/RR Int FJ44		3
Cessna 550 Citation I	P&WC JT15D-1A	2	
Cessna 560 Citation V	P&W JT15D-5A	2	
Dassault Falcon 10	Ga TFE731-2-1C	3	
Dassault Falcon 2000	GE/Garrett CFE738 / PW305		1
Dassault Falcon 20F5	Ga TFE731-5AR-2C	2	
Swearingen SJ30	Williams/RR Int FJ44		2
EUROBERLIN FRANCE (Germany)			
737-300	CFMI CFM56-3B1	6	
EUROCYPRIA AIRLINES LTD (Cyprus)			
A320-200	IAE V2500-A1	3	
EUROPE AERO SERVICE - EAS (France)			
727-200 Advanced	P&W JT8D-9A	2	
737-200 Advanced	P&W JT8D-17	3	
737-300	CFMI CFM56-3B1	1	
737-500	CFMI CFM56-3C1	1	
A310-300	GE CF6-80C2A2		2
Caravelle 10B1R	P&W JT8D-7A	1	
Caravelle 10B3	P&W JT8D-7A	7	
EUROPEAN AIRLIFT (Belgium)			
707-300	P&W JT3D-7	1	
DC-10-10	GE CF6-D1A	2	
Fokker F27-500	RR Dart 532-7	1	
EUROPEAN AIR TRANSPORT NV/SA (Belgium)			
727-100F	P&W JT8D-7	5	
CV 580 (F)	Allison 501-D13	11	
Fairchild Metro II	Ga TPE331-3UW-303G	3	
EUROWINGS (Germany)			
ATR 42	P&WC PW120	19	2
ATR 72	P&WC PW124	6	
ATR 72-210	P&WC PW127	2	2
BAe 146-200QT	Lyc ALF502R-5	1	

Airline and type	Engine	In operation	On order
EVA AIRWAYS CORPORATION (Taiwan)			
747-400	GE CF6-80C2B1F	2	6
767-200ER	GE CF6-80C2		4
767-300ER	GE CF6-80C2	1	
767-300ER	GE CF6-80C2B6F	4	
MD-11	GE CF6-80C2D1F		6
EVERGREEN INTERNATIONAL AIRLINES INC (USA)			
727-100C	P&W JT8D-7	1	
727-100C	P&W JT8D-7A	1	
727-100C	P&W JT8D-7B	1	
727-100C (F)	P&W JT8D-7	2	
727-100C (F)	P&W JT8D-7A	4	
727-100C (F)	P&W JT8D-7B	2	
747-100	P&W JT8D-7A	1	
747-100 (F)	P&W JT9D-7	1	
747-100 (F)	P&W JT9D-7A	1	
747-100F	P&W JT8D-7AH	2	
747-100SR (F)	P&W JT8D-7A	1	
747-100SR (F)	P&W JT8D-7AW	1	
747-200C (F)	P&W JT9D-7F	2	
DC-8-62F	P&W JT3D-3B (Q)	1	
DC-8-73F	CFMI CFM56-2	2	
DC-9-15	P&W JT8D-7B	1	
DC-9-15F	P&W JT8D-7A	1	
DC-9-32F	P&W JT8D-7	2	
DC-9-33F	P&W JT8D-9	3	
DC-9-33F	P&W JT8D-11	1	
EXCALIBUR AIRWAYS LTD (UK)			
A320-200	CFMI CFM56-5A3	4	
EXPRESS AIRLINES I, INC (NORTHWEST AIRLINK) (USA)			
BAe Jetstream 31	Ga TPE331-10UG-513H	25	
EMB/FMA CBA-123 Vector	Ga TPF351-20		
Saab 340A	GE CT7-5A2	18	
Saab 340B	GE CT7-9B		15
Saab 2000	Allison GMA 2100		10
EXPRESSO AEREO (Peru)			
Fokker F27-200	RR Dart 528-7E	1	
Fokker F27-500	RR Dart 532-7	2	
EXPRESS ONE (USA)			
727-100	P&W JT8D-7	5	
727-100	P&W JT8D-7B	8	
727-100 (F)	P&W JT8D-7	6	
727-100C	P&W JT8D-7	1	
727-200C	P&W JT8D-9A	1	
FALCON AVIATION AB (Sweden)			
737-300QC	CFMI CFM56-3C1	3	
FAR EASTERN AIR TRANSPORT (Taiwan)			
737-100	P&W JT8D-7A	2	
737-200	P&W JT8D-7A	1	
737-200	P&W JT8D-9A	3	
737-200 Advanced	P&W JT8D-9A	2	
MD-82	P&W JT8D-219	5	
MD-83	P&W JT8D-219	1	
FARNER AIR TRANSPORT AG (Switzerland)			
DHC-6 Twin Otter 300	P&WC PT6A-27	1	
EMB-110P1 Bandeirante	P&WC PT6A-34	1	
Fokker F27-400	RR Dart 532-7R	1	
Fokker F27-600	RR Dart 532-7	1	
PC-6 Turbo Porter	P&WC PT6A-27	2	

Boeing 767-332 belonging to major US operator Delta Air Lines on final approach *(Peter J. Cooper)*

AIRLINE FLEET LIST

Beech 1900C-1 of Flandre Air in TAT markings at Tours airport *(Peter J. Cooper)*

Airline and type	Engine	In operation	On order
FAUCETT PERU (Peru)			
727-200	P&W JT8D-15	1	
737-100	P&W JT8D-9A	1	
737-200	P&W JT8D-9	1	
737-200	P&W JT8D-9A	1	
DC-8-51	P&W JT3D-3B (Q)	1	
DC-8-52	P&W JT3D-3B	1	
DC-8-52	P&W JT3D-3B (Q)	2	
DC-8-61	P&W JT3D-3B (Q)	1	
L1011 TriStar 1	RR RB211-22B	1	
FEDERAL EXPRESS CORPORATION (USA)			
727-100C (F)	P&W JT8D-7	39	
727-100F	P&W JT8D-7B	46	
727-200F Advanced	P&W JT8D-9A	14	
727-200F Advanced	P&W JT8D-15	29	11
727-200F Advanced	P&W JT8D-17	6	
727-200F Advanced	P&W JT8D-17A	5	
727-200F Advanced (RE)	P&W JT8D-17 / 217C outers	12	
747-100	P&W JT9D-7A	4	
747-200F	P&W JT9D-7Q	4	
747-200F	P&W JT9D-70A	5	
A300-600F	GE CF6-80C2A5		25
Cessna 208A Caravan I	P&WC PT6A-114	10	
Cessna 208B Caravan I Super	P&WC PT6A-114/114A	206	
DC-10-10CF	GE CF6-6D	9	
DC-10-10F	GE CF6-6D	2	
DC-10-30CF	GE CF6-50C2	17	
Fokker F27-500	RR Dart 532-7	32	
MD-11F	GE CF6-80C2D1F	4	9
FIJI AIR LTD (Fiji)			
Beech Baron 95-C55	Cont IO-520-C	1	
BN-2A-27 Islander	Lyc O-540-E4C5	2	
BN-2B-20 Islander	Lyc IO-540-K1P5	1	
DH 114 Riley Heron	Lyc IO-540-K1C5	2	
DHC-6 Twin Otter 200	P&WC PT6A-20	1	
DHC-6 Twin Otter 300	P&WC PT6A-27	1	
FINNAIR OY (Finland)			
A300B4-203	GE CF6-50C2	2	
DC-9-41	P&W JT8D-17A	5	
DC-9-51	P&W JT8D-17A	12	
DC-10-30	GE CF6-50C2	4	
DC-10-30ER	GE CF6-50C2B	1	
MD-11	GE CF6-80C2D1BF	3	1
MD-82	P&W JT8D-219	9	
MD-83	P&W JT8D-219	5	
MD-87	P&W JT8D-219	3	
FINNAVIATION O/Y (Finland)			
Saab 340A	GE CT7-5A2	5	
Saab 340B	GE CT7-9B	1	
FLANDRE AIR (France)			
Beech 1900C-1	P&WC PT6A-65B	5	
Beech King Air A100	P&WC PT6A-28	1	
Beech Super King Air 200	P&WC PT6A-41	4	
FLORIDA WEST AIRLINES INC (USA)			
707-300C (F)	P&W JT3D-3B (Q)	6	
FORMOSA AIRLINES (Taiwan)			
BN-2A Islander	Lyc O-540-E4C5	3	
Dornier 228-201	Ga TPE331-5-252D	7	
Hiller UH-12E	VO-540	3	
Saab 340A	GE CT7-5A2	3	
Saab 340B	GE CT7-9B	2	

Airline and type	Engine	In operation	On order
FOUR STAR AIR CARGO (USA)			
BN-2A-26 Islander	Lyc O-540-E4C5	1	
DC-3C	P&W R-1830	5	
DHC-6 Twin Otter	P&WC PT6A-20	2	
FRED OLSEN AIR TRANSPORT LTD (Norway)			
BAe 748 Srs 2A	RR Dart 532-2L	1	
L188 Electra	Allison 501-D13	4	
GARUDA INDONESIA PT (Indonesia)			
737-300	CFMI CFM56-3B1	5	16
747-200	P&W JT9D-7Q	6	
747-400	GE CF6-80C2		9
A300-600R	GE CF6-80C2A5	1	
A300-600R	P&W PW4158	5	11
A300B4-220	P&W JT9D-59A	9	
A330-300	RR Trent 700		9
DC-9-32	P&W JT8D-9	15	
DC-10-30	GE CF6-50C	6	
Fokker F28-3000/4000	RR Spey 555-15H	14	
MD-11	GE CF6-80C2	3	6
GB AIRWAYS (UK)			
737-200	P&W JT8D-15	4	
GERMAN CARGO SERVICES (Germany)			
737-200F	P&W JT8D-15 (Q)	2	
747-200F (SCD)	GE CF6-50E2	3	
DC-8-73F	CFMI CFM56-2C5	5	
GERMANIA FLUGGESELLSCHAFT MBH (Germany)			
737-300	CFMI CFM56-3B2	10	
GHANA AIRWAYS CORPORATION (Ghana)			
DC-9-50	P&W JT8D-17	1	
DC-10-30	GE CF6-50C2	1	
Fokker F28-4000	RR Spey 555-15P	1	
GREAT CHINA AIRLINES (Taiwan)			
DHC-8 Dash 8-100	P&WC PW120A	4	
DHC-8 Dash 8-300	P&WC PW123	2	4
MD-90-30	IAE V2500-D5		3
GROENLANDSFLY A/S (Denmark)			
AS 350B1 Ecureuil	TU Arriel 1D1	2	
AS 350B2 Ecureuil	TU Arriel 1D1	3	
Bell 206B JetRanger II	Allison 250-C20	1	
Bell 206B JetRanger III	Allison 250-C20	2	
Bell 212	P&WC PT6T-3B TwinPac	6	
DHC-6 Twin Otter 300	P&WC PT6A-27	2	
DHC-7 Dash 7-100	P&WC PT6A-50	3	
Sikorsky S-61N	GE CT58-140-1	4	
GUIZHOU AIRLINES (Taiwan)			
HAMC Y-12	P&WC PT6A-27	1	
XAC Y7-100	WJ 5A-1	2	
GULF AIR (Bahrain)			
737-200 Advanced	P&W JT8D-15	10	
767-300	GE CF6-80C2	9	3
767-300ER	GE CF6-80C2B4	16	4
A320-200	CFMI CFM56-5A1	1	11
A321	CFMI CFM56-5A1		12
A340	CFMI CFM56-5C2		6
L1011 TriStar 200	RR RB211-524B4	8	

Notes: SCD Side cargo door   TBD To be decided   EUD Extended upper deck   Q Hush kitted   SR Short range   LR Long range   F Freighter

AIRLINE FLEET LIST

Airline and type	Engine	In operation	On order
GUYANA AIRWAYS CORP (Guyana)			
707-300B	P&W JT3D-3B	1	
BAe 748 Srs 2A	RR Dart 535-2L	2	
DHC-6 Twin Otter 300	P&WC PT6A-27	1	
HAMBURG AIRLINES GMBH & CO KG (Germany)			
DHC-8 Dash 8-100	P&WC PW120A	3	
DHC-8 Dash 8-300	P&WC PW123	4	
HAPAG-LLOYD FLUG GMBH (Germany)			
737-400	CFMI CFM56-3C1	9	
737-500	CFMI CFM56-3C1	5	
A310-200	GE CF6-80C2A2	4	
A310-300	GE CF6-80C2A2	3	
HARBOR AIRLINES, INC (USA)			
PA-31-350 Chieftain	Lyc TIO-540-J2BD	4	
HAWAIIAN AIRLINES (USA)			
DC-8-62	P&W JT3D-7 (Q)	2	
DC-8-62F (CF)	P&W JT3D-3B (Q)	2	
DC-8-63	P&W JT3D-7 (Q)	2	
DC-9-51	P&W JT8D-17	9	
DC-9-51	P&W JT8D-17A	5	
DHC-7 Dash 7-100	P&WC PT6A-50	4	
L1011 TriStar 50	RR RB211-22B	5	
HAZELTON AIRLINES (Australia)			
Cessna 310R II	Cont IO-520-M	2	
PA-31-350 Chieftain	Lyc TIO-540-J2BD	9	
Saab 340B	GE CT7-5A2	3	
Shorts 360-300	P&WC PT6A-67R	2	
HEAVYLIFT CARGO AIRLINES (UK)			
707-300C	P&W JT3D-3B (Q)	1	
An-124-100 Ruslan	ZMKB D-18T	3	
CL44-0 Guppy	RR Tyne 515/10	1	
Il-76TD	Aviadvigatel D-30KP	1	
L382G Hercules	Allison 501-D22A	4	
Shorts Belfast	RR Tyne 515-101W	5	
HENSON AVIATION INC (USAIR EXPRESS) (USA)			
DHC-7 Dash 7-100	P&WC PT6A-50	5	
DHC-8 Dash 8-100	P&WC PW120A	33	
HOLMSTROM AIR (Sweden)			
Beech King Air A100	P&WC PT6A-28	1	
Beech Super King Air 200	P&WC PT6A-42	1	
Cessna Turbo Centurion II	Cont TSIO-520-R	1	
Dornier 228-100	Ga TPE331-5-252D	2	
Dornier 228-201	Ga TPE331-5-252D	1	
HORIZON AIR (USA)			
DHC-8 Dash 8-100	P&WC PW120A	22	
Dornier 328	P&WC PW119		35
Fairchild Metro III	Ga TPE33One-ElevenU-611G	31	
Fokker F28-1000	RR Spey 555-15	3	
HUNTING CARGO AIRLINES LTD (UK)			
BAe Merchantman	RR Tyne 506-10	4	
L188 Electra	Allison 501-D13A	3	
L188CF Electra	Allison 501-D13A	4	
IBERIA - LINEAS AEREAS DE ESPANA, SA (Spain)			
727-200 Advanced	P&W JT8D-9	35	
737-300	CFMI CFM56-3B1	5	
747-200B	P&W JT9D-7A	1	
747-200B	P&W JT9D-7Q	3	
747-200B	P&W JT9D-7Q3	3	
757-200	RR RB211-535E4		16
A300B4-100	P&W JT9D-59A	6	
A300B4-200	GE CF6-50C	2	
A320-200	CFMI CFM56-5A1	15	12
A321	CFMI CFM56-5B		8
A340	CFMI CFM56-5C2		8
DC-9-32	P&W JT8D-7	18	
DC-9-32	P&W JT8D-9	3	
DC-10-30	GE CF6-50C	8	
MD-83	P&W JT8D-219	2	
MD-87	P&W JT8D-217C	24	
ICELANDAIR (Iceland)			
737-400	CFMI CFM56-3C1	4	
757-200ER	RR RB211-535E4	3	
Fokker 50	P&WC PW125B	4	
INDIAN AIRLINES CORPORATION (India)			
737-200	P&W JT8D-9A	4	
737-200 Advanced	P&W JT8D-9A	4	
737-200 Advanced	P&W JT8D-17A	14	
A300B2-101	GE CF6-50C	9	
A300B4-203	GE CF6-50C2	2	
A320-200	IAE V2500-A1	20	10
Fokker F27-100	RR Dart 514	2	
Fokker F27-400	RR Dart 532-7	1	
INTER-CANADIEN (Canada)			
ATR 42	P&WC PW120	8	
INTER EUROPEAN AIRWAYS LTD (UK)			
737-300	CFMI CFM56-3B2	3	
737-400	CFMI CFM56-3C1	1	
757-200	RR RB211-535E4	2	
A320-200	CFMI CFM56-5A1	2	
IRAN AIR: THE AIRLINE OF THE ISLAMIC REPUBLIC OF IRAN (Iran)			
707-300	P&W JT3D-3B	4	
727-100	P&W JT8D-7B	3	
727-200 Advanced	P&W JT8D-15	5	
737-200	P&W JT8D-15	3	
747-100	P&W JT9D-7F	1	
747-200	P&W JT9D-7F	2	
747-200F	P&W JT9D-7F	2	
747SP	P&W JT9D-7F	4	
A300-600R	GE CF6-80C2A5		2
A300B2-203	GE CF6-50C2	5	
Fokker 100	RR Tay 650-15	6	
Tu-134A	Aviadvigatel D-30	4	
IRAN ASSEMAN AIRLINES (Iran)			
ATR 42-320	P&WC PW121	1	4
ATR 72-210	P&WC PW124		4
BN-2A-3 Islander	Lyc IO-540-K1B5	2	
Dassault Falcon 20E	GE CF700-2D2	3	
Dassault Falcon 20F	GE CF700-2D2	1	
Fairchild FH-227B	RR Dart 536-7P	2	
Fokker F27-600	RR Dart 536-7R	1	
Fokker F28-1000	RR Spey 555-15	2	
Fokker F28-4000	RR Spey 555-15H	2	
PA-31-350 Chieftain	Lyc TIO-540-J2BD	3	
Rockwell Shrike Commander	Lyc IO-540-E1B5	11	

Groenlandsfly (Greenlandair) de Havilland Dash 7 at Kangerlussuaq airport *(Peter J. Cooper)*

AIRLINE FLEET LIST

Iran Air was operating six Fokker 100s in mid-1993

Airline and type	Engine	In operation	On order
IRAQI AIRWAYS (Iraq)			
707-300C	P&W JT3D-3B	1	
707-300C	P&W JT3D-3B (Q)	1	
727-200 Advanced	P&W JT8D-17	6	
737-200C Advanced	P&W JT8D-15	2	
747-200C (SCD)	P&W JT9D-7FW	3	
A310-300	P&W PW4152		5
An-12 (F)	ZMKB AI-20	5	
An-24	ZMKB AI-24	1	
An-24T	ZMKB AI-24	2	
An-24TV	ZMKB AI-24	1	
An-24V	ZMKB AI-24	2	
Dassault Falcon 20F	GE CF700-2D2	2	
Dassault Falcon 50	Ga TFE731-3-1C	3	
Il-76M (F)	Aviadvigatel D-30KP	12	
Il-76MD (F)	Aviadvigatel D-30KP	20	
L1329 JetStar II	Ga TFE731-3	6	
Piaggio P166-DL2	Lyc IGSO-540-A1H	4	
ISTANBUL AIRLINES - ISTAIR (Turkey)			
727-200	P&W JT8D-7B	2	
727-200 Advanced	P&W JT8D-15	5	
737-400	CFMI CFM56-3C1	1	
Caravelle 10B1R	P&W JT8D-7B	3	
ITAPEMIRIM TRANSPORTES AEREOS SA (Brazil)			
727-100F	P&W JT8D-7B	2	2
JAGSON AIRLINES (India)			
Dornier 228-201	Ga TPE331-5-252D	2	
JANES AVIATION 748 LTD (UK)			
BAe 748 Srs 1	RR Dart 534-2	3	
BAe 748 Srs 2A	RR Dart 534-2	4	
JAPAN AIR COMMUTER CO LTD (NIHON AIR COMMUTER) (Japan)			
Dornier 228-200	Ga TPE331-5-252D	3	
NAMC YS-11A-500	RR Dart 542-10J/K	6	
Saab 340B	GE CT7-9B	2	6
JAPAN AIRLINES COMPANY LTD - JAL (Japan)			
747-100 (LR)	P&W JT9D-7A	2	
747-100B (SR)	P&W JT9D-7A	3	
747-100B (SR/EUD)	P&W JT9D-7A	2	
747-200B	P&W JT9D-7A	1	
747-200B	P&W JT9D-7Q	8	
747-200B	P&W JT9D-7R4G2	3	
747-200B (F)	P&W JT9D-7Q	1	
747-200B (F)	P&W JT9D-7AW	10	
747-200F (SCD)	P&W JT9D-7AW	1	
747-200F (SCD)	P&W JT9D-7Q	6	
747-200F (SCD)	P&W JT9D-7R4G2	2	
747-300 (EUD)	RR RB211-524D4	3	
747-300 (LR)	P&W JT9D-7R4G2	9	
747-300 (SR)	P&W JT9D-7R4G2	4	

Airline and type	Engine	In operation	On order
747-400	GE CF6-80C2B1F	21	13
747-400D	GE CF6-80C2B1F	5	
767-200	P&W JT9D-7R4D	2	
767-300	P&W JT9D-7R4D	13	5
777-200	TBD		10
DC-10-40	P&W JT9D-59A	15	
MD-11	P&W PW4460		10
JAPAN AIR SYSTEM CO LTD (NIHON AIR SYSTEM) (Japan)			
767	TBD		7
A300B2-K3C	GE CF6-50C2R	9	
A300B4-2C	GE CF6-50C2R	8	
A300-622R	P&W PW4158	4	11
Beech Super King Air 200	P&WC PT6A-42	2	
DC-9-41	P&W JT8D-15	10	
DC-10-30	GE CF6-50C2	2	
MD-81	P&W JT8D-209	8	
MD-81	P&W JT8D-217A	12	6
MD-87	P&W JT8D-217C	8	
MD-90-30	IAE V2525-D5		10
NAMC YS-11A-500	RR Dart 542-10J/K	15	
Saab 340B	GE CT7-9B	1	7
JAPAN ASIA AIRWAYS (NIHON ASIA KOKU) (Japan)			
747-100	P&W JT9D-7A	1	
747-200B	P&W JT9D-7AW	1	
747-200B	P&W JT9D-7Q	1	
747-300	P&W JT9D-7R4G2	1	
DC-10-40	P&W JT9D-59A	4	
JERSEY EUROPEAN AIRWAYS LTD (UK)			
BAe 146-200	Lyc ALF502R-5	1	
BAe 146-300	Lyc ALF502R-5	2	
Fokker F27-500	RR Dart 532-7	7	
Shorts 360 Advanced	P&WC PT6A-65AR	4	
JETALL CORPORATION (Canada)			
Cessna 401A	Cont TSIO-520-E	1	
Cessna 402B II	Cont TSIO-520-E	1	
CV 580F (SCD)	Allison 501-D13	4	
Fairchild Merlin IV	Ga TPE331-10UA-511G	1	
Fairchild Merlin IVA	Ga TPE331-3UW-304G	1	
Fairchild Merlin IVC	Ga TPE33One-ElevenU-611G	4	
Fairchild Metro II	Ga TPE331-3UW-303G	1	
Fairchild Metro II	Ga TPE331-10UA-511G	3	
JET ALSACE (France)			
MD-83	P&W JT8D-219	2	
JUGOSLOVENSKI AEROTRANSPORT - JAT (former Yugoslavia)			
727-200 Advanced	P&W JT8D-9A	8	
737-300	CFMI CFM56-3B1	9	
ATR 72	P&WC PW124	3	
DC-10-30	GE CF6-50C1	3	
MD-11	P&W PW4460		

Notes: SCD Side cargo door   TBD To be decided   EUD Extended upper deck   Q Hush kitted   SR Short range   LR Long range   F Freighter

AIRLINE FLEET LIST

Airline and type	Engine	In operation	On order
KARAIR OY (Finland)			
ATR 72	P&WC PW124	5	1
DHC-6 Twin Otter 300	P&WC PT6A-27	1	
KELOWNA FLIGHTCRAFT AIR CHARTER LTD (Canada)			
727-100C	P&W JT8D-7B	2	
Beech Duke A60	Lyc TIO-541-E1B4	1	
Cessna 340	Cont TSIO-520-N	1	
Cessna 402B	Cont TSIO-520-E	2	
CV 580	Allison 501-D13	14	
CV 5800	Allison 501-D22G	2	
DC-3	P&W R-1830	1	
Gulfstream I	RR Dart 529-8D	1	
KEMEROVSKOE AVIAPREDPRIYATIYE (Russia)			
An-24	ZMKB AI-24T	6	
An-26	ZMKB AI-24VT	6	
PZL An-2	Aviadvigatel ASh-621R		
Tu-154	KKBM NK-8-2U	3	
KENDELL AIRLINES (AUSTRALIA) PTY LIMITED (Australia)			
Fairchild Metro II	Ga TPE331-3UW-304G	8	
Fairchild Metro 23	Ga TPE331-3UW-304G	2	
Saab 340A	GE CT7-5A2	6	
Saab 2000	Allison GMA 2100		
KENYA AIRWAYS LTD (Kenya)			
737-200 Advanced	P&W JT8D-17A	2	
757-200	RR RB211-535E4	1	
A310-300	GE CF6-80C2A2	3	
Fokker 50	P&WC PW125B	3	
KITTY HAWK GROUP, INC (USA)			
CV 240	P&W R-2800	3	
CV 440 Metropolitan	P&W R-2800	5	
Dassault Falcon 20	GE CF700-2D2	1	
DC-3	P&W R-1830	3	
KLM CITYHOPPER (Netherlands)			
Fokker 50	P&WC PW125B	10	
Fokker F28-4000	RR Spey 555-15P	4	
Saab 340B	GE CT7-9B	11	1
KLM ROYAL DUTCH AIRLINES (Netherlands)			
737-300	CFMI CFM56-3B1	15	
737-400	CFMI CFM56-3B2	10	2
747-300	GE CF6-50E2	3	
747-300M (SCD)	GE CF6-50E2	10	
747-400	GE CF6-80C2	5	
747-400F	GE CF6-80C2		2
747-400M (SCD)	GE CF6-80C2	9	2
A310-200	GE CF6-80A3	10	
DC-10-30	GE CF6-50C2	4	
Fokker 100	RR Tay 620-15	6	
MD-11	GE CF6-80C2		10
KOREAN AIR LINES CO LTD (South Korea)			
727-200	P&W JT8D-9A	3	
727-200 Advanced	P&W JT8D-9A	4	
747-200B	P&W JT9D-7A	3	
747-200C (F)	P&W JT9D-7A	1	
747-200F	P&W JT9D-7A	1	
747-200F (SCD)	P&W JT9D-7Q	4	
747-200F (SCD)	P&W JT9D-7R4G2	2	
747-300	P&W JT9D-7R4G2	2	
747-300M	P&W JT9D-7R4G2	1	
747-400	P&W PW4056	7	12
747-400C (SCD)	P&W PW4056	1	1
747SP	P&W JT9D-7	2	
A300-620	P&W JT9D-7R4H1	3	
A300-622	P&W PW4158	2	
A300-622R	P&W PW4158	15	9
A300B4-103	GE CF6-50C2	8	
A300F4-203	GE CF6-50C2	2	
A330-300	TBD		7
CASA 212-200 Aviocar	Ga TPE331-5-251C	1	
Dassault Falcon 20F	GE CF700-2D2	1	
Dassault Falcon 50	Ga TFE731-3-1C	1	
Fokker 100	RR Tay 650-15	3	9
Fokker F28-4000	RR Spey 555-15P	3	
MD-11	P&W PW4360	5	3
MD-82	P&W JT8D-217A	8	2
KUWAIT AIRWAYS CORPORATION (Kuwait)			
707-300C	P&W JT3D-3B (Q)	2	
747-200B (SCD)	P&W JT9D-7J	4	
747-400M	GE CF6-80C2B1F		3
767-200ER	P&W JT9D-7R4E4	1	
A300-600R	GE CF6-80C2A5		5
A310-222	P&W JT9D-7R4H	5	
A310-300	GE CF6-80C2A2	5	
A310-300	GE CF6-80C2A8		3
A320-200	CFMI CFM56-5A1	1	2
A321	CFMI CFM56-5B		
A340-200	CFMI CFM56-5C2		4
BAe 125-700B	Ga TFE731-3-1H	1	
Gulfstream IV	RR Tay 611-8	3	
L1011 TriStar 200	RR RB211-524B-02	2	
KYRGYZSTAN AIRLINE (Kyrgyzstan)			
PZL An-2	Aviadvigatel ASh-621R		
Tu-134	Aviadvigatel D-30		
Tu-154	KKBM NK-8-2U		
Yak-40	ZMKB AI-25		
LAM - LINHAS AEREAS DE MOCAMBIQUE (Mozambique)			
737-200	P&W JT8D-9	1	
737-200C Advanced	P&W JT8D-9	1	
737-300	CFMI CFM56-3C1	2	1
767-200ER	P&W PW4000	2	1
Beech Super King Air 200C	P&WC PT6A-41	2	
CASA 212-200 Aviocar	Ga TPE331-10-501C	4	
Cessna 402C Utililiner II	Cont TSIO-520-VB	6	
Partenavia P68C	Lyc IO-360-A1B6	4	
LANCHILE - LINEA AEREA NACIONAL CHILE (Chile)			
707-300C (F)	P&W JT3D-3B (Q)	3	
737-200 Advanced	P&W JT8D-15	1	
737-200 Advanced	P&W JT8D-17	1	
737-200 Advanced	P&W JT8D-17A	1	
767-200ER	GE CF6-80A2	3	
767-200ER	P&W PW4056	1	
767-300ER	P&W PW4060	1	
BAe 146-200	Lyc ALF502R-5	2	

Japan Air System adapted Airbus's own colour scheme for its A300-600Rs

AIRLINE FLEET LIST

BAe ATP (now to become the Jetstream 61) operated by Loganair

Airline and type	Engine	In operation	On order
LAP - LINEAS AEREAS PARAGUAYAS (Paraguay)			
707-300B	P&W JT3D-3B	3	
DC-8-61	P&W JT3D-3B (Q)	1	
DC-8-63	P&W JT3D-7	1	
DC-8-71	CFMI CFM56-2C1	1	
L188C Electra	Allison 501-D13A	1	
LAPA - LINEAS AEREAS PRIVADAS ARGENTINIAS, SA (Argentina)			
Saab 340A	GE CT7-5A2	2	
LAR TRANSREGIONAL (Portugal)			
BAe ATP	P&WC PW126	3	
Dornier 228-202	Ga TPE331-5-252D	3	
LAUDA AIR LUFTFAHRT AKTIENGESELLSCHAFT (Austria)			
737-300	CFMI CFM56-3B1	2	
737-400	CFMI CFM56-3C1	2	
767-300ER	P&W PW4056	1	
767-300ER	P&W PW4060	3	
777	GE GE90-B3		4
Learjet 36A	Ga TFE731-2-2B	1	
LESOTHO AIRWAYS CORPORATION (Lesotho)			
DHC-6 Twin Otter 300	P&WC PT6A-27	3	
Fokker F27-600	RR Dart 536-7R	1	
LIAT (1974) LTD (West Indies)			
BAe 748 Super 2B	RR Dart 536-2	4	
BN-2A Islander	Lyc O-540-E4C5	3	
DHC-6 Twin Otter 300	P&WC PT6A-27	6	
DHC-8 Dash 8-100	P&WC PW120A	5	
LIBYAN ARAB AIRLINES (Libya)			
707-300B	P&W JT3D-3B	1	
707-300C	P&W JT3D-3B	4	
727-200	P&W JT8D-9	1	
727-200 Advanced	P&W JT8D-15	8	
A310-203	GE CF6-80A3	2	
Dassault Falcon 20C	GE CF700-2C	1	
Dassault Falcon 50	Ga TFE731-3-1C	1	
Fokker F27-400	RR Dart 536-7	1	
Fokker F27-600	RR Dart 536-7	13	
Fokker F28-4000	RR Spey 555-15P	3	
Gulfstream II	RR Spey 511-8	2	
Il-76M (F)	Aviadvigatel D-30KP	1	
Il-76T (F)	Aviadvigatel D-30KP	5	
Il-76TD (F)	Aviadvigatel D-30KP	11	
L1329 JetStar 8	P&W JT12A-8	1	
LINEA AEREA DEL COBRE - LADECO (Chile)			
707-300CH	P&W JT3D-3B (Q)	1	
737-200 Advanced	P&W JT8D-9A	1	
737-200 Advanced	P&W JT8D-15	1	
737-200 Advanced	P&W JT8D-15A	1	
737-300	CFMI CFM56-3B2	2	1
757-200ER	P&W PW2040	2	
BAe One-Eleven/300	RR Spey 511-14	2	
LINEAS AEREAS COSTARRICENSES, SA - LACSA (Costa Rica)			
727-100	P&W JT8D-7A	1	
737-200 Advanced	P&W JT8D-17	3	
A310-300	GE CF6-80C2A4		2
A320-200	CFMI CFM56-5A3	5	2
DC-8-55	P&W JT3D-3B (Q)	1	
LITHUANIAN AIRLINES (Lithuania)			
737-200	P&W JT8D-15	1	
An-24RV	ZMKB AI-24	2	
An-24V	ZMKB AI-24	2	
An-26	ZMKB AI-24VT	3	
Tu-134A	Aviadvigatel D-30	8	
Tu-134A-3	Aviadvigatel D-30	1	
Yak-42	ZMKB D-36	12	
LLOYD AEREO BOLIVIANO SAM (Bolivia)			
707-300	P&W JT3D-3B	1	
727-100	P&W JT8D-9A	2	
727-100C	P&W JT8D-9A	1	
727-200 Advanced	P&W JT8D-17R	3	
A310-304	GE CF6-80C2A2	1	
Fokker F27-200	RR Dart 536-7P	1	
Fokker F27-600	RR Dart 532-7P	1	
LOGANAIR LTD (UK)			
BAe ATP	P&WC PW126	5	
BAe Jetstream 41	Ga TPE331-14GR/HR	3	
BAe Jetstream Super 31	Ga TPE331-10UGR-513H	2	
BN-2A-26 Islander	Lyc O-540-E4C5	1	
BN-2B-26 Islander	Lyc O-540-E4C5	4	
DHC-6 Twin Otter 300	P&WC PT6A-27	2	
Shorts 360-100	P&WC PT6A-65R	3	
Shorts 360-100 Advanced	P&WC PT6A-65AR	2	
LONE STAR AIRLINES (USA)			
Fairchild Metro III	Ga TPE33One-ElevenU-611G	8	
LOT - POLSKIE LINIE LOTNICZE (Poland)			
737-500	CFMI CFM56-3C1	7	3
767-200ER	GE CF6-80C2B2	2	
767-300ER	GE CF6-80C2	1	1
ATR 72	P&WC PW124	4	4
Il-18D	ZMKB AI-20	1	
Il-18E	ZMKB AI-20	2	
Tu-134A	Aviadvigatel D-30-II	7	
Tu-154M	Aviadvigatel D-30KU-154-II	13	
Yak-40	ZMKB AI-25	5	
LTE INTERNATIONAL AIRWAYS SA (Spain)			
757-200	RR RB211-535C	3	
757-200	RR RB211-535E4	2	
LTU - LUFTTRANSPORT UNTERNEHMEN GMBH (Germany)			
757-200	RR RB211-535E4	12	
A330-300	P&W PW4164		5
L1011 TriStar 100	RR RB211-22B	5	
L1011 TriStar 200	RR RB211-524B4	1	
L1011 TriStar 500	RR RB211-524B4	3	
MD-11	P&W PW4460	4	
LTU SUD INTERNATIONAL AIRWAYS (Germany)			
757-200	RR RB211-535E4	2	
757-200ER	RR RB211-535E4	5	4
767-300ER	P&W PW4060	4	
LUFTHANSA CITYLINE GMBH (Germany)			
Canadair Regional Jet	GE CF34-3A	5	20
Fokker 50	P&WC PW125B	25	

Notes: SCD Side cargo door   TBD To be decided   EUD Extended upper deck   Q Hush kitted   SR Short range   LR Long range   F Freighter

AIRLINE FLEET LIST

Airline and type	Engine	In operation	On order
LUFTHANSA GERMAN AIRLINES - DEUTSCHE LUFTHANSA AG (Germany)			
737-200 Advanced	P&W JT8D-15	35	
737-300	CFMI CFM56-3B1	31	
737-300QC	CFMI CFM56-3B1	2	5
737-400	CFMI CFM56-3C1	7	
737-500	CFMI CFM56-3B1/3C1	32	
747-200B	GE CF6-50E2	4	
747-200B (SCD)	GE CF6-50E2	10	
747-200F (SCD)	GE CF6-50E2	5	
747-400	GE CF6-80C2B1F	10	
747-400 (SCD)	GE CF6-80C2B1F	7	
A300-600	GE CF6-80C2A3	11	
A310-200	GE CF6-80A3	13	
A310-300	GE CF6-80C2A2	12	
A320-200	CFMI CFM56-5A1	34	
A321	IAE V2530-A5		20
A340-200	CFMI CFM56-5C2	3	4
A340-300	CFMI CFM56-5C2		8
Beech A36 Bonanza	Cont IO-550-B	9	
Beech Baron 58	Cont IO-540-C	19	
Beech F33A Bonanza	Cont IO-520-BA	59	
Beech T-34 Mentor	P&WC PT6A-25	2	
DC-10-30	GE CF6-50C2	11	
PA-42-720 Cheyenne IIIA	P&WC PT6A-61	7	
LUXAIR SA (Luxembourg)			
737-400	CFMI CFM56-3B2	2	
737-500	CFMI CFM56-3B1	2	
747SP	P&W JT9D-7F	1	
Cessna 441 Conquest II	Ga TPE331-8-403S	1	
EMB-120 Brasilia	P&WC PW115	3	
Fokker 50	P&WC PW124	4	
MAERSK AIR (Denmark)			
737-300	CFMI CFM56-3B2	10	
737-400	CFMI CFM56-3C1	1	
737-500	CFMI CFM56-3B1	5	
AS 332L Super Puma	TU Makila 1A	3	
AS 365N2 Dauphin 2	TU Arriel 1C2	2	
BAe 125-800	Ga TFE731-3R-1H	1	
Fokker 50	P&WC PW125B	8	
Shorts 360	P&WC PT6A-65R	1	
MALAYSIAN AIRLINE SYSTEM (Malaysia)			
737-200 Advanced	P&W JT8D-15	3	
737-200 Advanced	P&W JT8D-15A	2	
737-200C Advanced	P&W JT8D-15	1	
737-300F	CFMI CFM56-3C1	1	
737-400	CFMI CFM56-3C1	26	43
737-500	CFMI CFM56-3C1	6	
747-200B	RR RB211-524D4	2	
747-300 (SCD)	P&W JT9D-7R4G2	1	
747-400	GE CF6-80C2B1F	2	7
747-400	P&W PW4056	2	
747-400 (SCD)	GE CF6-80C2B1F	2	
A300B4-203	GE CF6-50C2	4	
A330-320	P&W PW4168		10
DC-10-30	GE CF6-50C2	6	
DHC-6 Twin Otter 300	P&WC PT6A-27	6	
Fokker 50	P&WC PW125B	11	
MALEV - HUNGARIAN AIRLINES (Hungary)			
737-200	P&W JT8D-17	3	
737-300	CFMI CFM56-3C1	3	
767-200ER	GE CF6-80C2	2	
BAe 146-200QT	Lyc ALF502R-5	1	
Tu-134A-3	Aviadvigatel D-30-III	6	
Tu-154B-2	KKBM NK-8-2U	12	
Yak-40	ZMKB AI-25	3	
MALI TIMBOUCTOU AIR SERVICE (Mali)			
An-24RV	ZMKB AI-24-2	1	
CASA 212-200 Aviocar	Ga TPE331-10-501C		1
Let 410UVP-E1 Turbolet	Mot M-601E	2	
MALMO AVIATION SCHEDULE AB (Sweden)			
AIAe RJ85	Lyc LF507		5
BAe 146-200	Lyc ALF502R-5	4	
BAe 146-200QT	Lyc ALF502R-5	1	
BAe 146-300QT	Lyc ALF502R-5	2	
MANDALA AIRLINES PT (Indonesia)			
BAe Viscount	RR Dart 525	1	
L188C Electra	Allison 501-D13A	3	
MANX AIRLINES LTD (UK)			
BAe 146-100	Lyc ALF502R-5	1	
BAe ATP	P&WC PW126	3	
BAe Jetstream 31	Ga TPE331-10UGR-514H	3	
BAe Jetstream 41	Ga TPE331-14GR/HR		5
Shorts 360-100 (F)	P&WC PT6A-65R	3	
MARKAIR, INC (USA)			
737-200C Advanced	P&W JT8D-17A	10	
DHC-7 Dash 7-100	P&WC PT6A-50	2	
DHC-8 Dash 8-300 Combi	P&WC PW123	2	
L382G Hercules	Allison 501-D22A	1	
MARTINAIR HOLLAND NV (Netherlands)			
747-200C (SCD)	GE CF6-50E2	2	
747-200F	GE CF6-50E2	1	
767-300ER	P&W PW4060	5	
A310-203C	GE CF6-80A3	1	
Cessna 404 Titan II	Cont GTSIO-520-M	1	
Cessna 550 Citation II	P&WC JT15D-4	1	
Cessna 650 Citation IV	Ga TFE731-3B-100S	1	
DC-10-30CF	GE CF6-50C	2	
Dornier 228-212	Ga TPE331-5-252D	1	
MD-11CF	P&W PW4462		4
MAYA AIRWAYS (Belize)			
BN-2A-26 Islander	Lyc O-540-E4C5	5	
MERIDIANA SPA (Italy)			
BAe 146-200	Lyc ALF502R-5	3	4
DC-9-51	P&W JT8D-17	6	
MD-82	P&W JT8D-217A	8	
MERPATI NUSANTARA AIRLINES PT (Indonesia)			
BAe ATP	P&WC PW125	5	
DC-9-32	P&W JT8D-9	9	
DHC-6 Twin Otter 300	P&WC PT6A-27	11	
Fokker 100	RR Tay 650-15		12
Fokker F27-500	RR Dart 532-7	8	
Fokker F27-500	RR Dart 536-7	6	
Fokker F28-4000	RR Spey 555-15H	28	
IPTN/CASA 212-200 Aviocar	Ga TPE331-10-511C	15	
IPTN/CASA CN-235-10	GE CT7-7A	14	
IPTN N-250-100	Allison GMA 2100		65
L382G Hercules	Allison 501-D22A	1	
MESA AIRLINES INC (USA)			
BAe 146-200A	Lyc ALF502R-5	3	
BAe Jetstream 31	Ga TPE331-10UG-513H	34	
Beech 1900 Airliner	P&WC PT6A-65B	20	
Beech 1900D Airliner	P&WC PT6A-67D	52	41
EMB-120RT Brasilia	P&WC PW118	21	

Lufthansa had taken delivery of three Airbus A340-200s by mid-1993 and had four more on order

AIRLINE FLEET LIST

Artist's impression of the Fokker 100 of which Mexicana has 10 on order

Airline and type	Engine	In operation	On order
MESABA AIRLINES (NORTHWEST AIRLINK) (USA)			
DHC-8 Dash 8-100	P&WC PW120	25	
Fokker F27-500	RR Dart 532-7R	5	
Fokker F27-600	RR Dart 532-7R	2	
Fairchild Metro III	Ga TPE33One-ElevenU-611G	21	
MEXICANA (Mexico)			
727-200 Advanced	P&W JT8D-17R	44	
A320-200	IAE V2500-A1	11	17
DC-10-15	GE CF6-50C2F	6	
Fokker 100	RR Tay 620-15		10
MGM GRAND AIR, INC (USA)			
727-100	P&W JT8D-7	3	
DC-8-62	P&W JT3D-3B (Q)	3	
MIDDLE EAST AIRLINES - AIRLIBAN (Lebanon)			
707-300C	P&W JT3D-3B (Q)	8	
720B	P&W JT3B-1	3	
747-200B	P&W JT9D-7FW	3	
A310-200	GE CF6-80A3	2	
MIDWEST EXPRESS (USA)			
DC-9-14	P&W JT8D-7B	6	
DC-9-15	P&W JT8D-7B	2	
DC-9-32	P&W JT8D-7B	6	
MD-88	P&W JT8D-217C	2	
MONARCH AIRLINES LTD (UK)			
737-300	CFMI CFM56-3B1	8	
757-200	RR RB211-535E4	8	
A300-600R	GE CF6-80C2A5	4	
A320-200	CFMI CFM56-5A3	3	2
MONGOLIAN AIRLINES - MIAT (Mongolia)			
An-24RV	ZMKB AI-24	10	
An-24V	ZMKB AI-24	2	
An-26	ZMKB AI-24T	4	
An-30	ZMKB AI-24T	1	
Mil Mi-8	Klimov TV2-117A	3	
PZL An-2	Aviadvigatel ASh-621R	30	
Tu-154M	Aviadvigatel D-30KU-154-II	1	
MOUNT COOK AIRLINE (New Zealand)			
BAe 748 Srs 2A	RR Dart 534-2	6	
BN-2A-26 Islander	Lyc O-540-E4C5	3	
Cessna 185A Skywagon	Cont IO-520-D	2	
Cessna A185F Skywagon II	Cont IO-520-D	6	
DHC-6 Twin Otter 300	P&WC PT6A-27	1	
PA-31-350 Chieftain	Lyc TIO-540-J2BD	1	
PC-6 Turbo Porter	P&WC PT6A-27	3	
MUK AIR (Denmark)			
Cessna 402B	Cont TSIO-520-VB	1	
EMB-110P2 Bandeirante	P&WC PT6A-34	4	
Fairchild Metro II	Ga TPE331-3UW-303G	1	
PA-31-350 Chieftain	Lyc TIO-540-J2BD	1	
Shorts 330-200	P&WC PT6A-45R	2	
MYANMAR AIRWAYS (BURMA AIRWAYS CORPORATION) (Myanmar)			
AS SA330J Puma	TU Turmo IVC	2	
Fokker F27-600	RR Dart 532-7	5	
Fokker F28-1000	RR Spey 555-15	1	
Fokker F28-4000	RR Spey 555-15H / -15P	2	

Airline and type	Engine	In operation	On order
NATIONAL OVERSEAS AIRLINE (Egypt)			
707-300	P&W JT3D-3B	2	
707-300B	P&W JT3D-3B	1	
An-2	ASh-62IR	5	
Bell 206B JetRanger	Allison 250-C20B	2	
Cessna 337 Super Skymaster	Cont IO-360-C	2	
Cessna 421 Golden Eagle	Cont GTSIO-520-H	1	
NIGERIA AIRWAYS LTD (Nigeria)			
707-300C	P&W JT3D-3B	3	
737-200 Advanced	P&W JT8D-15	8	
A310-222	P&W JT9D-7R4E1	4	
DC-10-30	GE CF6-50C	2	
NIPPON CARGO AIRLINES COMPANY LTD (Japan)			
747-200F (SCD)	GE CF6-50E2	6	
NORTH AMERICAN AIRLINES (USA)			
757-200	RR RB211-535E4	1	
MD-83	P&W JT8D-217	1	
NORTHERN AIR CARGO INC (USA)			
DC-6A	P&W R-2800	11	
DC-6B	P&W R-2800	1	
DC-6BF Swingtail	P&W R-2800	2	
NORTHWEST AIRLINES (USA)			
727-100	P&W JT8D-7B	8	
727-200	P&W JT8D-7B	20	
727-200 Advanced	P&W JT8D-15	27	
727-200 Advanced	P&W JT8D-17	15	
747-100	P&W JT9D-7A	12	
747-200B	P&W JT9D-7F	5	
747-200B	P&W JT9D-7Q	12	
747-200B	P&W JT9D-7R4G2	3	
747-200F (SCD)	P&W JT9D-7F	5	
747-200F (SCD)	P&W JT9D-7Q	1	
747-200F (SCD)	P&W JT9D-7R4G2	2	
747-400	P&W PW4056	10	6
757-200	P&W PW2037	33	40
A320-200	CFMI CFM56-5A1	33	17
A321	TBD		
A330	P&W PW4164		16
A340	CFMI CFM56-5C2		24
DC-9-14	P&W JT8D-7A	1	
DC-9-14	P&W JT8D-7B	20	
DC-9-15	P&W JT8D-7B	8	
DC-9-15F (RC)	P&W JT8D-7B	5	
DC-9-31	P&W JT8D-7	1	
DC-9-31	P&W JT8D-7B	23	
DC-9-31	P&W JT8D-9	26	
DC-9-31	P&W JT8D-15	7	
DC-9-32	P&W JT8D-9A	10	
DC-9-32	P&W JT8D-15	10	
DC-9-41	P&W JT8D-11	10	2
DC-9-51	P&W JT8D-17	28	
DC-10-30	GE CF6-50C	5	1
DC-10-30	GE CF6-50C2B	1	
DC-10-30ER	GE CF6-50C2B	1	
DC-10-40	P&W JT9D-20	21	
MD-82	P&W JT8D-217	8	

Notes: SCD Side cargo door   TBD To be decided   EUD Extended upper deck   Q Hush kitted   SR Short range   LR Long range   F Freighter

AIRLINE FLEET LIST

Airline and type	Engine	In operation	On order
NOVOKUZNETSKOE AVIAPREDPRIYATIYE (Russia)			
An-24	ZMKB AI-24T		
An-26	ZMKB AI-24VT	5	
Mil Mi-8	Klimov TV2-117A		
PZL An-2	Aviadvigatel ASh-621R		
Tu-154	KKBM NK-8-2U	5	
Tu-154M	Aviadvigatel D-30KU-154-II	2	
NWT AIR (NORTHWEST TERRITORIAL AIRWAYS LTD) (Canada)			
737-200	P&W JT8D-9A	3	
L382G Hercules	Allison 501-D22A	1	
OASIS INTERNATIONAL AIRLINES (Spain)			
MD-82	P&W JT8D-217C	1	
MD-83	P&W JT8D-219	5	
OLYMPIC AIRWAYS SA (Greece)			
727-200	P&W JT8D-9A	6	
727-200 Advanced	P&W JT8D-15	3	
737-200	P&W JT8D-15	11	
737-400	CFMI CFM56-3C1	6	
747-200B	P&W JT9D-7J	1	
747-200B	P&W JT9D-7Q	3	
A300-600R	GE CF6-80C2A5	1	3
A300B4-103	GE CF6-50C2	8	
OLYMPIC AVIATION SA (Greece)			
Agusta A109 II	Allison 250-C20B	1	
AS 350B Ecureuil	TU Arriel 1B	1	
ATR 42	P&WC PW121	4	
ATR 72	P&WC PW124	4	
Cessna F152 II	Lyc O-235-LC2	3	
Dornier 228-200	Ga TPE331-5-252D	7	
PA-28-140 Cherokee	Lyc IO-320-E3D	2	
Shorts 330-200	P&WC PT6A-45B /45R	5	
OMAN AVIATION SERVICES COMPANY (SAO) (Oman)			
Cessna 550 Citation II	P&WC JT15D-4	1	
DHC-6 Twin Otter 300	P&WC PT6A-27	1	
Fokker F27-500	RR Dart 536-7	4	
OMSKOYE AVIAPREDPRIYATIYE (Russia)			
An-24	ZMKB AI-24T		
An-26	ZMKB AI-24VT		
PZL An-2	Aviadvigatel ASh-621R		
Tu-154	KKBM NK-8-2U		
ONUR AIR (Turkey)			
A320-200	CFMI CFM56-5A1	3	1
PAKISTAN INTERNATIONAL AIRLINES CORPORATION (PIA) (Pakistan)			
707-300C	P&W JT3D-3B	2	
737-300	CFMI CFM56-3B2	6	
747-200B	P&W JT9D-7A	6	
747-200B (SCD)	GE CF6-50E2	2	
A300B4-200	GE CF6-50C2	7	
A310-307	GE CF6-80C2A8	4	2
DHC-6 Twin Otter 300	P&WC PT6A-27	2	
Fokker F27-200	RR Dart 532-7E	13	
Fokker F27-400	RR Dart 532-7E	1	
PARADISE ISLAND AIRLINES (USA)			
DHC-7 Dash 7-100	P&WC PT6A-50	6	
PELITA AIR SERVICE (Indonesia)			
AIAe RJ85	Lyc LF507-1F		1
BAe 125-600B	RR Viper 601-22A	1	
BAe 146-200	Lyc ALF502R-5	1	
BAe One-Eleven/400	RR Spey 511-14	1	
DHC-7 Dash 7-100	P&WC PT6A-50	5	
Fokker 70	RR Tay 620		5
Fokker 100	RR Tay 650-15	1	
Fokker F28-1000	RR Spey 555-15	2	
Fokker F28-4000	RR Spey 555-15P	5	
Gulfstream II	RR Spey 511-8	1	
Gulfstream III	RR Spey 511-8	1	
IPTN/CASA 212-100 Aviocar	Ga TPE-331-5-251C	4	
IPTN/CASA 212-200 Aviocar	Ga TPE331-10-511C	8	
IPTN/Eurocopter BO 105C	Allison 250-C20	5	
IPTN/Eurocopter NBO 105CB	Allison 250-C20B	28	
IPTN/SA 330G Puma	TU Turmo IVA/IVC	4	
IPTN/SA 330J Puma	TU Turmo IVC	11	
IPTN/SA 332C Super Puma	TU Makila 1A	2	
L382G Hercules	Allison 501-D22A	4	
S-76A II	Allison 250-C30S20	4	
Transall C-160NG	RR Tyne 522	3	
Transall C-160P	RR Tyne 522	3	
PENNSYLVANIA COMMUTER AIRLINES, INC (USA)			
Beech 1900C Airliner	P&WC PT6A-65B	13	
DHC-8 Dash 8-100	P&WC PW120A	16	
Fokker F27-500	RR Dart 535-7R	3	
Shorts 360	P&WC PT6A-65R	10	
PHILIPPINE AIR LINES, INC (PAL) (Philippines)			
737-300	CFMI CFM56-3B1	11	
747-200	GE CF6-50E2	6	
747-200B	P&W JT9D-7Q	3	
747-200B	P&W JT9D-7A	1	
747-400	GE CF6-80C2		4
A300B4-103	GE CF6-50C2	2	
A300B4-203	GE CF6-50C2	6	
A340-200	CFMI CFM56-5C2		6
DC-10-30	GE CF6-50C2	2	
Fokker 50	P&WC PW125B	10	
POLYNESIAN AIRLINES (HOLDINGS) LTD (Western Samoa)			
727-200	P&W JT8D-15	1	
737-300QC	CFMI CFM56-3B2	1	
767-200ER	GE CF6-80A2	1	
BN-2A-8 Islander	Lyc O-540-E4C5	1	
DHC-6 Twin Otter 300	P&WC PT6A-27	1	
PORTUGALIA - COMPANIA PORTUGESA DE TRANSPORTES AEREOS SA (Portugal)			
Fokker 100	RR Tay 650-15	3	
PRIMERAS LINEAS URUGUAYAS DE NAVEGACION AEREA - PLUNA (Uruguay)			
707-300C	P&W JT3D-3B	1	
737-200 Advanced	P&W JT8D-9A	3	
PROVINCIAL AIRLINES LTD (Canada)			
Beech King Air B200	P&WC PT6A-41	4	
Cessna 172 Skyhawk	Lyc O-320	2	
PA-23-250 Aztec B	Lyc O-540-A1D5	2	
PA-28-140 Cherokee	Lyc IO-360	10	
PA-28-151/161 Cherokee	Lyc IO-360	2	
PA-31-310 Navajo	Lyc TIO-540-A1A	7	
PA-31-350 Chieftain	Lyc TIO-540-J2BD	3	
PA-34-200 Seneca	Lyc IO-360-C1E6	1	
Fairchild Merlin IV	Ga TPE331-3UW-303G	1	
Fairchild Metro II	Ga TPE331-10UA-511G	3	
QANTAS AIRWAYS LTD (Australia)			
737-300	CFMI CFM56-3B2	16	
737-400	CFMI CFM56-3C1	14	3
747-200B	P&W JT9D-7FW	2	
747-200B	RR RB211-524D4	3	
747-200B (SCD)	RR RB211-524D4	2	
747-300 (EUD)	RR RB211-524D4	6	
747-400	RR RB211-542G	18	
747SP	RR RB211-524D4	2	
767-200ER	P&W JT9D-7R4E	7	
767-300ER	GE CF6-80C2B6	13	2
A300B4-203	GE CF6-50C2	4	
BAe 146-100	Lyc ALF502R-5	4	

Qantas 747-438 on the ground at London Heathrow (*Peter J. Cooper*)

AIRLINE FLEET LIST

Airline and type	Engine	In operation	On order
RATIOFLUG GMBH (Germany)			
Beech Super King Air 200	P&WC PT6A-41	2	
Cessna 404 Titan II	Cont GTSIO-520-M	1	
Dornier 228-202	Ga TPE331-5-252D	2	
Fokker F27-600	RR Dart 532-7	3	
Learjet 35	Ga TPE731-2-2B	1	
Learjet 55	Ga TPE731-3A-2B	2	
REEVE ALEUTIAN AIRWAYS, INC (USA)			
727-100C	P&W JT8D-7	2	
L188 Electra	Allison 501-D13A	3	
NAMC YS-11A Combi	RR Dart 542-10K	3	
REGIONAL AIRLINES (France)			
BAe Jetstream Super 31	Ga TPE331-12U	6	
Fairchild Metro II	Ga TPE331-10UA-511G	1	
Fairchild Metro III	Ga TPE33One-ElevenU-611G	3	
Saab 340B	GE CT7-9B	5	
REGIONAL AIR (PTY) LTD (South Africa)			
CV 580	Allison 501-D13	2	
DC-3	P&W R-18303	1	
RICH INTERNATIONAL AIRWAYS INC (USA)			
DC-8-62	P&W JT3D-3B (Q)	2	
DC-8-63	P&W JT3D-3B (Q)	1	
L1011 TriStar 1	RR RB211-22B	3	
RIO-SUL SERVICOS AEREOS REGIONAIS SA (Brazil)			
EMB-110P Bandeirante	P&WC PT6A-27	5	
EMB-120RT Brasilia	P&WC PW118A	5	
Fokker 50	P&WC PW125B	3	
Fokker F27-500	RR Dart 532-7	2	
ROYAL AIR MAROC (Morocco)			
707-300C	P&W JT3D-3B	2	
727-200	P&W JT8D-7B	2	
727-200 Advanced	P&W JT8D-15	6	
737-200 Advanced	P&W JT8D-15	5	
737-200C Advanced	P&W JT8D-15A	2	
737-400	CFMI CFM56-3C1	2	4
737-500	CFMI CFM56-3C1	1	3
747-200B (SCD)	P&W JT9D-7F	1	
747SP	P&W JT9D-7FW	1	
757-200	P&W PW2037	2	
ATR 42	P&WC PW120	3	
Beech Baron 95-B55	Cont IO-470-L	2	
Beech Super King Air 200	P&WC PT6A-41	2	
Fokker F27	RR Dart 500	2	
ROYAL BRUNEI AIRLINES (Brunei Darussalam)			
727-100	P&W JT8D-7	1	
727-200 Advanced	P&W JT8D-15	1	
757-200	RR RB211-535E4	3	
767-200ER	P&W PW4056	1	
767-300ER	P&W PW4000	1	1
ROYAL JORDANIAN (Jordan)			
707-300C	P&W JT3D-3B	3	
727-200A	P&W JT8D-17	2	
A310-300	GE CF6-80C2A2	4	
A320-200	CFMI CFM56-5A1	2	
A340	CFMI CFM56-5C2		5
L1011 TriStar 500	RR RB211-524B4	5	
ROYAL NEPAL AIRLINES CORPORATION (Nepal)			
727-100	P&W JT8D-9A	2	
757-200	RR RB211-535E4	2	
BAe 748 Srs 2A	RR Dart 543-2	3	
DHC-6 Twin Otter 300	P&WC PT6A-27	9	
PC-6B Turbo Porter	P&WC PT6A-20	1	
ROYAL SWAZI NATIONAL AIRWAYS CORPORATION (Swaziland)			
Fokker F28-3000	RR Spey 555-15H	1	
ROYAL TONGAN AIRLINES (Tonga)			
737-400	CFMI CFM56-3C1	1	
DHC-6 Twin Otter 300	P&WC PT6A-27	2	
RYANAIR LTD (Ireland)			
BAe 1-11/500	RR Spey 512-14DW	4	
Rombac 1-11/560	RR Spey 512-14DW	4	
SABENA BELGIAN WORLD AIRLINES (Belgium)			
737-200 Advanced	P&W JT8D-15A	8	
737-200C Advanced	P&W JT8D-15A	4	
737-300	CFMI CFM56-3B2	6	
737-400	CFMI CFM56-3C1	3	
737-500	CFMI CFM56-3B1	6	7
747-100 (SCD)	P&W JT9D-7A	1	
747-300C	GE CF6-50E2	2	
A310-200	P&W JT9D-7R4E1	2	
A310-300	P&W JT9D-7R4E1	1	
A340-200	CFMI CFM56-5C2		5
DC-10-30CF	GE CF6-50C2	2	
SAFAIR FREIGHTERS (PTY) LTD (South Africa)			
707-300C	P&W JT3D-7	1	
BAe 146-100QT	Lyc ALF502R-3		1
L382G Hercules	Allison 501-D22A	6	
SAM - SOCIEDAD AERONAUTICA DE MEDELLIN CONSOLIDADA SA (Colombia)			
727-100	P&W JT8D-7A	5	
727-100C	P&W JT8D-7	2	
727-200	P&W JT8D-17R	2	
SAS - SCANDINAVIAN AIRLINES SYSTEM (Sweden)			
737-500	CFMI CFM56-3B1	8	3
767-200ER	P&W PW4056	2	
767-300ER	P&W PW4050	12	2
DC-9-21	P&W JT8D-11	9	
DC-9-41	P&W JT8D-11	26	
Fokker 50	P&WC PW125B	22	
Fokker F28-1000	RR Spey 555-15N	3	
Fokker F28-4000	RR Spey 555-15P	16	
MD-81	P&W JT8D-217C	29	
MD-82	P&W JT8D-219	12	
MD-87	P&W JT8D-217C	16	
MD-90	IAE V2525-D5		6
Saab 340A	GE CT7-5A2	13	
SATA AIR ACORES - SERVICO ACOREANO DE TRANSPORTES AEREOS, E. P. (Portugal)			
BAe ATP	P&WC PW126	3	
Dornier 228-202	Ga TPE331-5-525D	1	
SAUDIA - SAUDI ARABIAN AIRLINES (Saudi Arabia)			
707-300C	P&W JT3D-3B (Q)	2	
737-200	P&W JT8D-15	20	
747-100	RR RB211-524C2-19	8	
747-200B (SCD)	P&W JT9D-7F	1	
747-200F	RR RB211-524D4	1	
747-300	RR RB211-524D4-19	11	
747SP	RR RB211-524C2-19	3	
A300-600	P&W JT9D-7R4H1	11	
Beech A36 Bonanza	Cont IO-550-B	6	
Beech King Air A100	P&WC PT6A-28	2	
Cessna 550 Citation II	P&WC JT15D-4	2	
Dassault Falcon 900	Ga TFE731-5AR-1C2	1	
DC-8-63F	P&W JT3D-7 (Q)	1	
DC-8-72	CFMI CFM56-2C5	1	
DHC-6 Twin Otter 300	P&WC PT6A-27	1	
Gulfstream II	RR Spey 511-8	4	
Gulfstream III	RR Spey 511-8	3	
Gulfstream IV	RR Tay 611-8	5	
L1011 TriStar 200	RR RB211-524B-02	17	
L1011 TriStar 500	RR RB211-524B4	1	
PA-28-181 Archer II	Lyc O-360-A4M	8	
SCANAIR LTD (Sweden)			
DC-10-10	GE CF6-6D1A	6	
MD-82	P&W JT8D-219	1	
MD-83	P&W JT8D-219	2	
SCENIC AIRLINES INC (USA)			
DHC-6 Twin Otter 300	P&WC PT6A-27	24	
SCHREINER AIRWAYS BV (Netherlands)			
AS 355 Ecureuil 2	Allison 250-C20F	1	
AS 365 Dauphin 2	TU Arriel 1A2	1	
AS 365N Dauphin 2	TU Arriel 1C	2	
Bell 206B JetRanger III	Allison 250-C20	1	
DHC-8 Dash 8-100	P&WC PW120A	1	
DHC-8 Dash 8-300	P&WC PW123		2
Fokker F27-300	RR Dart 514-7	1	
Fokker F27-400	RR Dart 532-7	3	
Fokker F27-500	RR Dart 536-7	1	
Fokker F27-600	RR Dart 532-7	1	
L382G Hercules	Allison 501-D22A	1	
ASSA 316B Alouette III	TU Artouste IIIB1	1	
ASSA 330J Puma	TU Turmo IVC	1	
SEAGREEN AIR TRANSPORT (West Indies)			
707-300C	P&W JT3D-3B	3	
707-300C (F)	P&W JT3D-3B (Q)	1	
DHC-6 Twin Otter 200	P&WC PT6A-27	1	
L1329 JetStar 8	P&W JT12A-8	1	
SERVICIOS AVENSA SA - SERVIVENSA (Venezuela)			
727-200	P&W JT8D-17	1	
DC-3	P&W R-1830	4	
DC-9-32	P&W JT8D-7B	1	
DC-9-32	P&W JT8D-17	2	
SHANGHAI AIRLINES (People's Republic of China)			
757-200	P&W PW2037	3	5
767-300	P&W PW4000		5
SHUTTLE INC (USAIR SHUTTLE) (USA)			
727-100	P&W JT8D-7B	4	
727-200	P&W JT8D-7B	11	

Notes: SCD Side cargo door   TBD To be decided   EUD Extended upper deck   Q Hush kitted   SR Short range   LR Long range   F Freighter

AIRLINE FLEET LIST

Airline and type	Engine	In operation	On order
SIBERIA AIRLINE (Russia)			
An-24	ZMKB AI-24T	6	
An-26	ZMKB AI-24VT	5	
Il-86	KKBM NK-86	6	
Tu-154	KKBM NK-8-2U	12	
Tu-154M	Aviadvigatel D-30KU-154-II	7	
SICHUAN AIRLINES (People's Republic of China)			
Tu-154M	Aviadvigatel D-30KU-154-II	2	
XAC Y7-100	WJ 5A-1	5	
SILKAIR (SINGAPORE) PRIVATE LIMITED (Singapore)			
737-300	CFMI CFM56-3B2	4	
Dornier 228-202K	Ga TPE331-5-252D	1	
SINGAPORE AIRLINES LIMITED (Singapore)			
737-300F	CFMI CFM56-3B2	1	
747-200	P&W JT9D-7Q	2	
747-200F (SCD)	P&W JT9D-7R4G2	3	
747-300 (SCD)	P&W JT9D-7R4G2	3	
747-300M	P&W JT9D-7R4G2	11	
747-400	P&W PW4056	18	18
747-400F	P&W PW4056		5
A310-200	P&W JT9D-7R4E1	6	
A310-300	P&W PW4152	12	4
A340-300	CFMI CFM56-5C4		7
Learjet 31A	Ga TFE731-2-3B	2	
SKYWAYS, AB (Sweden)			
Saab 2000	Allison GMA 2100A		2
Shorts 330-200	P&WC PT6A-45R	1	
Shorts 360	P&WC PT6A-65R	4	
SKYWEST AIRLINES INC (DELTA CONNECTION) (USA)			
Canadair Regional Jet	GE CF34-3A		20
EMB-120ER Brasilia	P&WC PW118A		8
EMB-120RT Brasilia	P&WC PW118A	20	
Fairchild Metro III	Ga TPE33One-ElevenU-611G	31	
Saab 2000	Allison GMA 2100		
SOBELAIR - SOCIETE BELGE DE TRANSPORTS AERIENS (Belgium)			
737-200 Advanced	P&W JT8D-15	2	
737-300	CFMI CFM56-3B2	1	
737-400	CFMI CFM56-3C1	2	
SOLOMON AIRLINES LTD (Solomon Islands)			
737-400	CFMI CFM56-3C1	1	
BN-2A Islander	Lyc O-540-E4C5	2	
DHC-6 Twin Otter 300	P&WC PT6A-27	1	
PA-23-250 Aztec D	Lyc IO-540-C4B5	1	
SOMALI AIRLINES (Somalia)			
707-300B	P&W JT3D-3B	1	
A310-304	GE CF6-80C2A2	1	
Dornier 228-202	Ga TPE331-5-252D	2	
Fokker F27-600	RR Dart 536-3B	1	
SOUTH AFRICAN AIRWAYS (SAA) (South Africa)			
737-200	P&W JT8D-9	4	
737-200 Advanced	P&W JT8D-15	13	
747-200B	P&W JT9D-7R4G2	5	
747-200B (SCD)	P&W JT9D-7Q	1	
747-300	P&W JT9D-7R4G2	2	
747-400	RR RB211-524G	3	1
747SP	P&W JT9D-7FW	5	
A300B2-K3C	GE CF6-50C2R	4	
A300B4-203	GE CF6-50C2	4	
A300C4-203	GE CF6-50C2	1	
A320-200	IAE V2500-A1	6	1
SOUTHERN AIR TRANSPORT, INC (USA)			
DC-8-71F	CFMI CFM56-2C1	3	
DC-8-73F	CFMI CFM56-2C1	1	
L382E/F Hercules	Allison 501-D22A	1	
L382G Hercules	Allison 501-D22A	14	
SOUTHWEST AIR LINES CO LTD (Japan)			
737-200 Advanced	P&W JT8D-17	8	
737-400	CFMI CFM56-3C1		4
DHC-6 Twin Otter 300	P&WC PT6A-27	4	
NAMC YS-11-200	RR Dart 542-10J/K	5	
SOUTHWEST AIRLINES COMPANY (USA)			
737-200 Advanced	P&W JT8D-9A	49	
737-300	CFMI CFM56-3B1	67	60
737-500	CFMI CFM56-3B1	25	
SPANAIR (Spain)			
767-300ER	P&W PW4060	2	
MD-83	P&W JT8D-219	9	
STAR AIR I/S (Denmark)			
Fokker F27-600	RR Dart 532-7	1	
Fokker F27-600	RR Dart 532-7R	1	
Fokker F27-600	RR Dart 536-7R	1	
Fokker F27-600F	RR Dart 532-7	1	
STELLAIR TA (France)			
EMB-110 Bandeirante	P&WC PT6A-34	2	
Fairchild F-27	RR Dart 532-7	2	
STERLING AIRWAYS LTD (Denmark)			
727-200 Advanced	P&W JT8D-17	5	
727-200 Advanced	P&W JT8D-17A	3	
727-200RE Advanced	P&W JT8D-17 / 217C outers	2	
757-200ER	RR RB211-535E4	2	1
SUDAN AIRWAYS (Sudan)			
707-300B	P&W JT3D-3B	1	
707-300C	P&W JT3D-3B	2	
707-300C	P&W JT3D-7	2	
737-200C Advanced	P&W JT8D-7	2	
A310-300	GE CF6-80C2A2	1	
Fokker 50	P&WC PW125B	2	
Fokker F27-500	RR Dart 532-7	1	
SUN-AIR OF SCANDINAVIA A/S (Denmark)			
BAe Jetstream 31	Ga TPE331-10UR-513H	6	
BAe Jetstream 41	Ga TPE331-14GR/HR		2
Cessna 402C Utililiner II	Cont TSIO-520-VB	1	
Cessna 500 Citation	P&WC JT15D-1A	1	
Cessna 500 Citation IA	P&WC JT15D-1A	1	
Cessna 501 Citation I/SP	P&WC JT15D-1A	1	
EMB-110P1 Bandeirante	P&WC PT6A-34	1	
Partenavia P68 Observer	Lyc IO-360-A1B6	1	
Shorts Skyvan 3	Ga TPE331-2-201A	2	
SUN COUNTRY AIRLINES (USA)			
727-200 Advanced	P&W JT8D-15	2	
727-200 Advanced	P&W JT8D-17	1	
727-200 Advanced	P&W JT8D-17R	3	
727-200RE	P&W JT8D-17A/ 217C outers	2	
DC-10-10	GE CF6-6D1A	2	
SUNEXPRESS (Turkey)			
737-300	CFMI CFM56-3B2	1	
737-300	CFMI CFM56-3C1	2	

Sudan Airways acquired two Fokker 50s

AIRLINE FLEET LIST

Airbus A310 operated by Romanian airline Tarom

Airline and type	Engine	In operation	On order
SUNFLOWER AIRLINES LTD (Fiji)			
Beech 65 Excalibur Queenaire	Lyc IO-720-A1B	1	
BN-2A Islander	Lyc O-540-E4C5	5	
DH 114 Riley Heron	Lyc IO-540-K1C5	1	
DHC-6 Twin Otter 210	P&WC PT6A-27	2	
SUNSTATE AIRLINES (Australia)			
DHC-8 Dash 8-100	P&WC PW120A	3	
EMB-110P1 Bandeirante	P&WC PT6A-34	2	
Shorts 330-200	P&WC PT6A-45R	2	
Shorts 360	P&WC PT6A-65R	3	
SURINAM AIRWAYS (Suriname)			
DC-8-63	P&W JT3D-7	1	
DHC-6 Twin Otter 300	P&WC PT6A-27	2	
SWISSAIR (Switzerland)			
747-300	P&W JT9D-7R4G2	2	
747-300 (SCD)	P&W JT9D-7R4G2	3	
A310-221	P&W JT9D-7R4D1	5	
A310-322	P&W JT9D-7R4E1	4	
A320-200	CFMI CFM56-5B4		7
A321-100	CFMI CFM56-5B		19
Fokker 100	RR Tay 620-15	10	
MD-11	P&W PW4360	12	1
MD-81	P&W JT9D-209	24	
SYRIAN ARAB AIRLINES (Syria)			
727-200	P&W JT8D-17	3	
747SP	P&W JT9D-7A	2	
An-24	ZMKB AI-24	1	
An-26	ZMKB AI-24T	4	
Caravelle 10B3	P&W JT8D-7	2	
Dassault Falcon 20F	GE CF700-2D2	2	
Il-76M	Aviadvigatel D-30KP	4	
Tu-134B-3	Aviadvigatel D-30-3	4	
Tu-154M	Aviadvigatel D-30KU-154-II	3	
Yak-40	ZMKB AI-25	6	
TAAG - LINHAS AEREAS DE ANGOLA (ANGOLA AIRLINES) (Angola)			
707-300C	P&W JT3D-3B	1	
707-300C	P&W JT3D-3B (Q)	2	
737-200 Advanced	P&W JT8D-17	2	
737-200 Advanced	P&W JT8D-17A	2	
737-200C Advanced	P&W JT8D-17	1	
An-12	ZMKB AI-20K	1	
Fokker F27-400M (F)	RR Dart 532-7R	2	
Fokker F27-500	RR Dart 532-7R	1	
Fokker F27-600	RR Dart 532-7	2	
Il-62M	Aviadvigatel D-30KU	2	
L1011 TriStar 500	RR RB211-524B4-02	1	
L382G Hercules	Allison 501-D22A	2	
TABA - TRANSPORTES AEREOS DA BACIA AMAZONICA SA (Brazil)			
DHC-8 Dash 8-300	P&WC PW123	2	
EMB-110 Bandeirante	P&WC PT6A-27/34	14	
Fairchild FH-227B	RR Dart 532-7	6	
TACA INTERNATIONAL AIRLINES SA (El Salvador)			
737-200	P&W JT8D-7	1	
737-200	P&W JT8D-9A	1	
737-200 Advanced	P&W JT8D-17	1	
737-200C Advanced	P&W JT8D-17	1	
737-300	CFMI CFM56-3B2	3	
767-200	GE CF6-80A	1	
TACV - TRANSPORTES AEREOS DE CABO VERDE (Cape Verde)			
A310-300	GE CF6-80C2A2	1	
BAe 748 Srs 2A	RR Dart 534-2	2	
CASA 212-300 Aviocar	Ga TPE331-10R-513C	2	
DHC-6 Twin Otter 300	P&WC PT6A-27	2	
EMB-120RT Brasilia	P&WC PW118	1	
TAESA - TRANSPORTES AEREOS EJECUTIVOS SA (Mexico)			
727-100	P&W JT8D-7B	10	
737-300	CFMI CFM56-3	2	
737-400	CFMI CFM56-3B1	2	
737-500	CFMI CFM56-3B1	4	
757-200	RR RB211-535E4	3	
767-200	P&W PW4060	1	
Eurocopter BO 105 LS A-3	Allison 250-C28C	2	
Gulfstream I	RR Dart 529-8X	1	
Gulfstream II	RR Spey 511-8	1	
Gulfstream IIB	RR Spey 511-8	1	
L1329 Jetstar 8	P&W JT12A-8	5	
L1329 Jetstar 731	Ga TFE731-3-1E	3	
Learjet 24D/25/25B/25D	GE CJ610	9	
Learjet 31/31A/35A	Ga TFE731	6	
Rockwell Sabreliner 40	P&W JT12A-8	1	
TAIWAN AIRLINES (Taiwan)			
BN-2A Trislander	Lyc IO-540-E4C5	2	
BN-2A-26 Islander	Lyc IO-540-E4C5	3	
BN-2B-26 Islander	Lyc IO-540-E4C5	1	
Fokker 100	RR Tay 650-15	2	
TALAIR PTY LIMITED (Papua New Guinea)			
DHC-6 Twin Otter 200	P&WC PT6A-20	7	
DHC-6 Twin Otter 300	P&WC PT6A-27	9	
EMB-110P1 Bandeirante	P&WC PT6A-34	5	
EMB-110P2 Bandeirante	P&WC PT6A-34	3	
TAM - TRANSPORTES AEREOS REGIONAIS SA (Brazil)			
EMB-110P Bandeirante	P&WC PT6A-27	6	
EMB-110P1A Bandeirante	P&WC PT6A-34	1	
EMB-110P2 Bandeirante	P&WC PT6A-34	2	
Fokker 100	RR Tay 650-15	10	
Fokker F-27	RR Dart 532-7	7	
TAME - TRANSPORTES AEREOS MILITARES ECUATORIANOS (Ecuador)			
727-100	P&W JT8D-7	1	
727-100	P&W JT8D-9A	2	
727-200 Advanced	P&W JT8D-17	1	
BAe 748 Srs 2A	RR Dart 534-2	1	
BAe 748 Srs 2A (SCD)	RR Dart 534-2	1	
Fokker F28-4000	RR Spey 555-15P	1	
TAMPA SA (TRANSPORTES AEREOS MERCANTILES PANAMERICANOS SA) (Colombia)			
707-300C	P&W JT3D-3B (Q)	5	
TAP - AIR PORTUGAL (Portugal)			
737-200 Advanced	P&W JT8D-17A	10	
737-300	CFMI CFM56-3B2	10	
A310-300	GE CF6-80C2A2	5	
A320-200	CFMI CFM56-5A1	6	
A340-300	CFMI CFM56-5C2		4
L1011 TriStar 500	RR RB211-524B4-02	7	
TAROM ROMANIAN AIR TRANSPORT (Romania)			
707-300C	P&W JT3D-3BQ	3	
737-300	CFMI CFM56-3	5	6
A310-300	P&W PW4156A	2	1
An-24RT	ZMKB AI-24	17	
BAe One-Eleven/400F	RR Spey 511-14	1	
BAe One-Eleven/500	RR Spey 511-14DW	7	
Il-18D	ZMKB AI-20	4	
Il-62	KKBM NK-8-4	2	
Il-62M	Aviadvigatel D-30KU	2	
Rombac 1-11/560	RR Spey 511-14DW	6	
Tu-154B-1	KKBM NK-8-2U	5	
Tu-154B-2	KKBM NK-8-2U	3	

Notes: SCD Side cargo door   TBD To be decided   EUD Extended upper deck   Q Hush kitted   SR Short range   LR Long range   F Freighter

AIRLINE FLEET LIST

Airline and type	Engine	In operation	On order
TAS AIRWAYS SPA (Italy)			
BAe 146-300	Lyc ALF502R-5	2	
Gulfstream I	RR Dart 529-8X	7	
TAT - TRANSPORT AERIEN TRANSREGIONAL (France)			
737-200C	P&W JT8D-9A	5	
ATR 42-300	P&WC PW120	5	6
ATR 72	P&WC P&W124	2	20
Beech King Air 200	P&WC PT6A-41	2	
Fairchild FH-227B	RR Dart 532-7	4	
Fokker 100	RR Tay 620-15	4	12/4
Fokker F28-1000	RR Spey 555-15	13	
Fokker F28-2000	RR Spey 555-15	4	
Fokker F28-4000	RR Spey 555-15H	2	
TEA BASEL (Switzerland)			
737-300	CFMI CFM56-3B2	5	
THAI AIRWAYS INTERNATIONAL (Thailand)			
737-200	P&W JT8D-15	3	
737-400	CFMI CFM56-3B2	6	1
747-200	GE CF6-50E2	6	
747-300	GE CF6-50C2B1	2	
747-400	GE CF6-80C2B1F	6	6
777-200	RR Trent 871		8
A300-601	GE CF6-80C2A1	6	
A300-605R	GE CF6-80C2A5	2	
A300-622R	P&W PW4158	7	1
A300B4-103	GE CF6-50C2	8	
A300B4-203	GE CF6-50C2	4	
A310-200	GE CF6-80C2A2	2	
A310-300	GE CF6-80C2A2	1	
A330-301	GE CF6-80E1A3F		4
A330-320	P&W PW4168		4
ATR 42	P&WC PW121	2	
ATR 72	P&WC PW124	2	
CL-601-3A/ER Challenger	GE CF34-3A	1	
MD-11	GE CF6-80C2D1F	4	3
TMA - TRANS MEDITERRANEAN AIRWAYS (Lebanon)			
707-300C (F)	P&W JT3D-3B (Q)	7	
TNT INTERNATIONAL AVIATION SERVICES (UK)			
BAe 146-200QT	Lyc ALF502R-5	13	
BAe 146-300QT	Lyc ALF502R-5	8	2
TOMSKOYE AVIAPREDPRIYATIYE (Russia)			
An-24	ZMKB AI-24T	6	
An-26	ZMKB AI-24VT	5	
PZL An-2	Aviadvigatel ASh-621R		
Tu-154	KKBM NK-8-2U	3	
TOWER AIR, INC (USA)			
747-100	P&W JT9D-7A	6	
747-100	P&W JT9D-7J	5	
TRADEWINDS INTERNATIONAL AIRLINES (WRANGLER AVIATION) (USA)			
CL44D4-2	RR Tyne 515/10	4	
CL44D4-6	RR Tyne 515/10	1	
L1011 TriStar 1 (F)	RR RB211-22B	1	
TRANSAFRIK (Sao Tome and Principe)			
Bell 206B JetRanger II	Allison 250-C20	1	
DC-8-33F	P&W JT4A-9	2	
L188C Electra	Allison 501-D13	1	
L382G Hercules	Allison 501-D22A	6	
TRANS-AIR-LINK CORPORATION - TAL (USA)			
DC-6A	P&W R-2800-CB16	3	
DC-7CF	CW R-3350	1	

Airline and type	Engine	In operation	On order
TRANSASIA AIRWAYS (Taiwan)			
A300-600R	P&W PW4158		4
A320-200	IAE V2500-A1	2	1
ATR 42-300	P&WC PW120	4	
ATR 72	P&WC PW124B	8	5
TRANSAVIA AIRLINES CV (Netherlands)			
737-200	P&W JT8D-7	1	
737-200	P&W JT8D-15A	3	
737-200 Advanced	P&W JT8D-15	2	
737-300	CFMI CFM56-3B2	8	
757-200	RR RB211-535E5	2	4
TRANSBRASIL SA - LINHAS AEREAS (Brazil)			
707-300C	P&W JT3D-3B	2	
707-300C (F)	P&W JT3D-3B	1	
737-300	CFMI CFM56-3B1	6	
737-300	CFMI CFM56-3B2	3	
737-400	CFMI CFM56-3C1	4	
767-200	GE CF6-80A	3	
767-300ER	P&W PW4060	2	5
777-200B	TBD		3
TRANS CONTINENTAL AIRLINES, INC (USA)			
CV 440F	P&W R-2800	3	
DC-6A	P&W R-2800	2	
DC-6BF	P&W R-2800	2	
DC-8-54	P&W JT3D-3B (Q)	3	2
TRANS ISLAND AIRWAYS LTD - TIA (Bahamas)			
BN-2A Islander	Lyc IO-540-E4C5	3	
PA-23-250 Aztec D	Lyc IO-540-C4B5	3	
TRANS-JAMAICAN AIRLINES LTD (West Indies)			
BN-2 Islander	Lyc O-540-E4C5	1	
BN-2A Trislander	Lyc O-540-E4C5	2	
Cessna 402C Utililiner II	Cont TSIO-520-VB	2	
Cessna U206G Stationair 6 II	Cont IO-520-F	2	
TRANSLIFT AIRWAYS (Ireland)			
DC-8-71	CFMI CFM56-2	2	
DC-8-71F	CFMI CFM56-2	1	
TRANSMED AIRLINES (Egypt)			
737-200	CFMI CFM56-3B2	3	
737-400	CFMI CFM56-3C1	2	1
TRANSWEDE AIRWAYS (Sweden)			
MD-83	P&W JT8D-219	5	2
MD-87	P&W JT8D-217	2	
TRANS WORLD AIRLINES, INC (USA)			
727-100	P&W JT8D-7	8	
727-200/200 Advanced	P&W JT8D-9A/15	51	
747-100	P&W JT9D-7	10	
747-200	P&W JT9D-7A	1	
767-200ER	P&W JT9D-7R4D	10	
A330	RR Trent 700		20
DC-9-15	P&W JT8D-7	19	
DC-9-31	P&W JT8D-9	18	
DC-9-32	P&W JT8D-9/15	14	
DC-9-33	P&W JT8D-9	1	
DC-9-34	P&W JT8D-15	3	
DC-9-41	P&W JT8D-15	3	
L1011 TriStar 1	RR RB211-22B	9	
L1011 TriStar 100	RR RB211-22B	5	
L1011 TriStar 500	RR RB211-22B	3	
MD-82	P&W JT8D-217A/C	29	
MD-83 / SAIC/MD-83	P&W JT8D-217C	10	

One of two Airbus A320s operated by TransAsia, formerly Foshing

AIRLINE FLEET LIST

One of the US airline industry's big three, United Airlines, has over 100 Boeing 737-300s in service and many more on order *(Peter J. Cooper)*

Airline and type	Engine	In operation	On order
TRANS WORLD EXPRESS, INC (USA)			
ATR 42-300	P&WC PW120	11	
BAe Jetstream 31	Ga TPE331-12UAR-701H	8	
DHC-7 Dash 7-102	P&WC PT6A-50	8	
TREK AIRWAYS (PTY) LTD (FLITESTAR) (South Africa)			
A320-200	CFMI CFM56-5A	4	
ATR 72	P&WC PW125	2	
TRINIDAD & TOBAGO (BWIA INTERNATIONAL) AIRWAYS CORPORATION (West Indies)			
L1011 TriStar 500	RR RB211-524B4	4	
MD-83	P&W JT8D-219	9	
TUNINTER (Tunisia)			
ATR 42-300	P&WC PW120	1	
ATR 72	P&WC PW124B	2	
TUNISAIR - SOCIETE TUNISIENNE DE L'AIR (Tunisia)			
727-200 Advanced	P&W JT8D-9A	8	
737-200 Advanced	P&W JT8D-9A	2	
737-200 Advanced	P&W JT8D-17	1	
737-200C Advanced	P&W JT8D-17	1	
737-500	CFMI CFM56-3B1	1	2
A300B4-203	GE CF6-50C2	1	
A320-200	CFMI CFM56-5A1	5	
TUR EUROPEAN AIRWAYS (Turkey)			
727-100	P&W JT8D-9A	1	
727-200	P&W JT8D-15	2	1
727-200	P&W JT8D-17	2	
DC-9-51	P&W JT8D-17	1	1
MD-83	P&W JT8D-219	1	
TURKISH AIRLINES INC (TURK HAVA YOLLARI AO) (Turkey)			
727-200 Advanced	P&W JT8D-15	7	
737-400	CFMI CFM56-3C1	16	
737-500	CFMI CFM56-3C1	2	
A310-200	GE CF6-80A3	7	
A310-300	GE CF6-80C2A2	7	
A340-300	CFMI CFM56-5C2		5
AIAe RJ100	Lyc LF507-1F		2
BAe ATP	P&WC PW126	4	
DC-9-32	P&W JT8D-9A	9	
TURKMENISTAN AIRLINE (TURKMENAVIA) (Turkmenistan)			
737-300	CFMI CFM56-3	1	
An-24	ZMKB AI-24T		
An-26	ZMKB AI-24VT		
Mil Mi-8	Klimov TV2-117A		
Tu-154B-2	KKBM NK-8-2U	2	
TYROLEAN AIRWAYS (TIROLER LUFTFAHRT AG) (Austria)			
DHC-7 Dash 7-100	P&WC PT6A-50	2	
DHC-8 Dash 8-100	P&WC PW120A	2	
DHC-8 Dash 8-100	P&WC PW121	6	5
DHC-8 Dash 8-300A	P&WC PW123B	3	
UGANDA AIRLINES CORPORATION (Uganda)			
707-300C	P&W JT3D-3B (Q)	1	
Fokker F27-600	RR Dart 532-7	1	
ULAN-UDENSKOYE AVIAPREDPRIYATIYE (Russia)			
An-24	AI IV-24T	7	
Mil Mi-8	Klimov TV2-117A		
An-2	Aviadvigatel ASh-621R		
Tu-154	KKBM NK-8-2U	1	

Airline and type	Engine	In operation	On order
UNITED AIRLINES, INC (USA)			
727-100	P&W JT8D-7B	1	
727-200	P&W JT8D-7B	20	
737-200	P&W JT8D-7B	45	
727-200 Advanced	P&W JT8D-15	75	
737-200 Advanced	P&W JT8D-17	18	
737-200 Advanced	P&W JT8D-9A	6	
737-300	CFMI CFM56-3B2	101	
737-300/400/500	CFMI CFM56-3C		50
737-500	CFMI CFM56-3C1	52	10
747-100	P&W JT9D-3A	3	
747-100	P&W JT9D-7A	15	
747-200B	P&W JT9D-7J	7	
747-200B	P&W JT9D-7R4G2	2	
747-400	P&W PW4056	23	1
747SP	P&W JT9D-7A	10	
757-200	P&W PW2037	78	10
767-200	P&W JT9D-7R4D	19	
767-300ER	P&W PW4000	15	3
777A	P&W PW4073		30
777B	P&W PW4073		4
A320-200	IAE V2500-A1		50
DC-10-10	GE CF6-6D1	46	
DC-10-30	GE CF6-6D1	4	
DC-10-30	GE CF6-50C2	4	
UNITED PARCEL SERVICE COMPANY UPS (USA)			
727-100 (F)	P&W JT8D-7B	44	
727-100QF	RR Tay 651-54	6	34
727-200 Advanced (F)	P&W JT8D-15	6	
727-200 Advanced (F)	P&W JT8D-17	2	
747-100 (F) (SCD)	P&W JT9D-7A	11	
757-200PF	P&W PW2040	28	7
757-200PF	RR RB211-535E4		20
767-300ER (F)	GE CF6-80C2		30
DC-8-71F	CFMI CFM56-2	23	
DC-8-73F	CFMI CFM56-2	26	
Fairchild Expediter	Ga TPE33One-ElevenU-601G	11	
USAIR, INC (USA)			
727-200 Advanced	P&W JT8D-17A	8	
737-200	P&W JT8D-9A	15	
737-200 Advanced	P&W JT8D-9A	17	
737-200 Advanced	P&W JT8D-15	24	
737-200 Advanced	P&W JT8D-15A	23	
737-300	CFMI CFM56-3B1	49	3
737-300	CFMI CFM56-3B2	52	
737-400	CFMI CFM56-3B2	54	22
757-200	RR RB211-535E4	11	14
767-200ER	GE CF6-80C2B2	11	2
DC-9-31	P&W JT8D-7	52	
DC-9-31	P&W JT8D-9	14	
DC-9-32	P&W JT8D-7B	7	
Fokker 100	RR Tay 650-15	40	
Fokker F28-1000	RR Spey 555-15P	11	
Fokker F28-4000	RR Spey 555-15P	24	
MD-81	P&W JT8D-209	19	
MD-82	P&W JT8D-217	12	8
VARIG - VIACAO AEREA RIO-GRANDENSE SA (Brazil)			
727-100	P&W JT8D-7B	3	
727-100	P&W JT8D-9A	1	
727-100C	P&W JT8D-7B	2	
727-100C	P&W JT8D-9A	3	
737-200 Advanced	P&W JT8D-17A	11	
737-300	CFMI CFM56-3B1	6	
737-300	CFMI CFM56-3B2	17	12

SCD Side cargo door   TBD To be decided   EUD Extended upper deck   Q Hush kitted   SR Short range   LR Long range   F Freighter

AIRLINE FLEET LIST

Airline and type	Engine	In operation	On order
VARIG - VIACAO AEREA RIO-GRANDENSE SA (Brazil) continued...			
747-200B	GE CF6-50E2	1	
747-200B (SCD)	GE CF6-50E2	2	
747-300	GE CF6-80C2B1	3	
747-300 (SCD)	GE CF6-50E2	2	
747-400	GE CF6-80C2B1F	2	2
767-200ER	GE CF6-80C2B2	6	
767-300ER	GE CF6-80C2	4	
DC-10-30	GE CF6-50C2	10	
DC-10-30CF	GE CF6-50C2	2	
MD-11	GE CF6-80C2D1F	4	2
VASP - VIACAO AEREA SAO PAULO SA (Brazil)			
727-200 Advanced	P&W JT8D-17R	4	
737-200	P&W JT8D-7	7	
737-200 Advanced	P&W JT8D-7	6	
737-200 Advanced	P&W JT8D-15	1	
737-200 Advanced	P&W JT8D-17	4	
737-200C (F)	P&W JT8D-7	1	
737-200C Advanced (F)	P&W JT8D-17	1	
737-300	CFMI CFM56-3B1	10	
A300B2-203	GE CF6-50C2	3	
MD-11	GE CF6-80C2D1F	2	
VAYUDOOT (India)			
Dornier 228-201	Ga TPE331-5-252D	4	
Fokker F27-100	RR Dart 514-7	4	
HAL/BAe 748 Srs 2	RR Dart 533-2	5	
HAL/BAe 748 Srs 2A	RR Dart 533-2	5	
HAL/Dornier 228-201	Ga TPE331-5-252D	5	
VIASA, VENEZOLANA INTERNACIONAL DE AVIACION, SA (Venezuela)			
A300B4-203	GE CF6-50C2	2	
DC-10-30	GE CF6-50C	6	
MD-11	GE CF6-80C2		2
VIRGIN ATLANTIC AIRWAYS (UK)			
747-100	P&W JT9D-7A	1	
747-200	P&W JT9D-7Q	2	
747-200B	P&W JT9D-7J/A	5	
747-400		4	
777			1
A340-300	CFMI CFM56-5C3		4
VIVA AIR (Spain)			
737-300	CFMI CFM56-3B2	9	
VOLGA-DNEPR CARGO AIRLINE (Russia)			
An-12	ZMKB AI-20M	7	
An-124 Ruslan	ZMKB D-18T	6	2
Il-76T	Aviadvigatel D-30KP	1	
Il-76TD	Aviadvigatel D-30KP	1	
Yak-40	ZMKB AI-25	2	
WDL AVIATION (KOLN) (Germany)			
Fokker F27-100	RR Dart 514-7E	3	
Fokker F27-200	RR Dart 532-7	1	
Fokker F27-600	RR Dart 536-7R	3	
Luftschiff WDL 1B airship	Cont IO-360-CD	2	
WESTAIR COMMUTER AIRLINES (USA)			
BAe 146-200A	Lyc ALF502R-5	3	
BAe Jetstream 31	Ga TPE331-14GR/HR	34	
EMB-120RT Brasilia	P&WC PW118	15	13
WIDEROE'S FLYVESELSKAP A/S (Norway)			
DHC-6 Twin Otter 300	P&WC PT6A-27	12	
DHC-7 Dash 7-100	P&WC PT6A-50	8	
DHC-8 Dash 8-103	P&WC PW121	1	14
WINDWARD ISLANDS AIRWAYS INTERNATIONAL NV (Netherlands Antilles)			
DHC-6 Twin Otter 300	P&WC PT6A-27	3	
WUHAN AIRLINES (People's Republic of China)			
XAC Y7-100	WJ 5A-1	2	
XIAMEN AIRLINES (People's Republic of China)			
737-200 Advanced	P&W JT8D-17	1	
737-200 Advanced	P&W JT8D-17A	3	
737-500	CFMI CFM56-3B1	3	
757-200	RR RB211-535E4	1	2
YEMEN AIRWAYS (Yemen)			
727-200 Advanced	P&W JT8D-17R	5	
737-200 Advanced	P&W JT8D-15	1	
DHC-7 Dash 7-100	P&WC PT6A-50	2	
ZAIREAN AIRLINES (Zaïre)			
707-300F	P&W JT4A-12	1	
707-400	RR Conway 508	1	
727-100	P&W JT8D-7A	1	
BAe Viscount	RR Dart 506	1	
PA-28-180	Lyc IO-360	1	
ZAMBIA AIRWAYS CORPORATION LIMITED (Zambia)			
737-200 Advanced	P&W JT8D-17	2	
757-200F	RR RB211-535E4	1	
ATR 42	P&WC PW120	2	
DC-8-71	CFMI CFM56-2C	1	
DC-10-30	GE CF6-50C2	1	
ZANTOP INTERNATIONAL AIRLINES INC (USA)			
CV 640 (F)	RR Dart 542-4	11	
DC-8-54	P&W JT8D-3B (Q)	2	
L188 Electra (F)	Allison 501-D13A	21	
ZAS AIRLINE OF EGYPT (Egypt)			
707-300C	P&W JT3D-3B	2	
A300B4-100	P&W JT9D-59A	2	
Caravelle 10R	P&W JT8D-7A	2	2
Caravelle 10R	P&W JT8D-7B	1	
L1329 JetStar 6	P&W JT12A-8	3	
MD-11	P&W PW4360		2
MD-83	P&W JT8D-219	2	
MD-87	P&W JT8D-219	1	1
ZHONGYUAN AIRLINES (People's Republic of China)			
XAC Y7-100	WJ 5A-1	2	

AIRCRAFT
ARGENTINA

AERO BOERO
AERO BOERO SA
Brasil y Alem, 2421 Morteros, Córdoba
Telephone: 54 (562) 2690 and 2121
Fax: 54 (562) 2690 and 2121
PRESIDENT: Cesar E. Boero
MANAGING DIRECTOR: Hector A. Zinoni
OTHER WORKS: Av 9 de Julio 1101, 2400 San Francisco, Córdoba
Telephone: 54 (562) 22972 and 24118

Aero Boero had produced 464 aircraft of various models by 1 January 1993, and is continuing production of AB 115 civil trainer and AB 180 RVR at combined rate of 10 aircraft per month. Indaer-Peru production of these types, under 1989 licence agreement, not yet started.

AERO BOERO 115 TRAINER

TYPE: Two/three-seat light aircraft.
PROGRAMME: Original AB 115 developed from AB 95 (1969-70 *Jane's*); subsequent variants included AB 115 BS ambulance and 112 kW (150 hp) AB 115/150 (1983-84 *Jane's*); current AB 115 Trainer ordered by Brazil in 1988 and 1989.
CURRENT VERSIONS: **AB 115 Trainer**: Standard version, *as described.*
CUSTOMERS: Total of 450 ordered by Brazilian government (100 in 1988, 350 in 1989), mainly for civilian flying club use; 237 delivered by 1 January 1993; completion due mid-1993.
DESIGN FEATURES: High-wing cabin monoplane with V bracing struts, sweptback vertical tail and fixed incidence, non-swept tailplane. Wing section modified NACA 23012; 1° 45′ dihedral; incidence 3° at root, 1° at tip.
FLYING CONTROLS: Conventional three-axis, mechanically actuated; trim tab in port elevator.
STRUCTURE: Welded steel tube (SAE 4130) fuselage and tail unit with Dacron covering; wings, including skins and flaps/ailerons, of aluminium alloy. Wings strut braced, tailplane wire braced.
LANDING GEAR: Non-retractable tailwheel type, with shock absorption by helicoidal springs inside fuselage; mainwheels carried on faired-in V struts and half-axles. Mainwheels and tyres size 6.00-6, tailwheel tyre size 2.80-2.50. Hydraulic disc brakes on main units; tailwheel steerable and fully castoring.
POWER PLANT: One 86 kW (115 hp) Textron Lycoming O-235-C2A flat-four engine, driving a Sensenich 72-CK-0-50 two-blade fixed-pitch propeller. Two aluminium fuel tanks in wings, combined capacity 128 litres (33.9 US gallons; 28.2 Imp gallons). Gravity refuelling point in top of each tank.
ACCOMMODATION: Pilot and one or two passengers in fully enclosed, heated and ventilated cabin.
SYSTEMS: 40A alternator and 12V battery.
AVIONICS: Com/nav equipment, blind-flying instrumentation and landing lights optional.

DIMENSIONS, EXTERNAL:
Wing span	10.78 m (35 ft 4½ in)
Wing chord, constant	1.615 m (5 ft 3½ in)
Wing aspect ratio	6.67
Length overall	7.08 m (23 ft 2¾ in)
Height overall	2.05 m (6 ft 8¾ in)
Tailplane span	3.20 m (10 ft 6 in)
Wheel track	1.76 m (5 ft 9¼ in)
Wheelbase	4.94 m (16 ft 2½ in)
Propeller diameter	1.93 m (6 ft 4 in)
Propeller ground clearance	0.58 m (1 ft 10¾ in)

AREAS:
Wings, gross	17.41 m² (187.4 sq ft)
Ailerons (total)	1.84 m² (19.81 sq ft)
Trailing-edge flaps (total)	1.94 m² (20.88 sq ft)
Fin	0.93 m² (10.01 sq ft)
Rudder	0.41 m² (4.41 sq ft)
Tailplane	1.40 m² (15.07 sq ft)
Elevators (total, incl tab)	0.97 m² (10.44 sq ft)

WEIGHTS AND LOADINGS:
Weight empty	556 kg (1,226 lb)
Max T-O weight	802 kg (1,768 lb)
Max wing loading	46.06 kg/m² (9.43 lb/sq ft)
Max power loading	9.33 kg/kW (15.37 lb/hp)

PERFORMANCE (at max T-O weight):
Never-exceed speed (VNE)	118 knots (220 km/h; 136 mph)
Max cruising speed	91 knots (169 km/h; 105 mph)
Stalling speed, power off:	
flaps up	41 knots (75 km/h; 47 mph)
flaps down	35 knots (64 km/h; 40 mph)
Max rate of climb at S/L	204 m (669 ft)/min
T-O run	100 m (330 ft)
T-O to, and landing from, 15 m (50 ft)	250 m (820 ft)
Landing run	80 m (265 ft)
Range with max fuel	664 nm (1,230 km; 765 miles)

Aero Boero 115 Trainer two/three-seat lightplane (*J. M. G. Gradidge*)

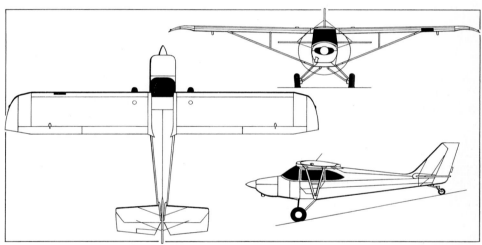

Aero Boero 115 Trainer (Textron Lycoming O-235-C2A engine) (*Jane's/Mike Keep*)

AERO BOERO 180 RVR

TYPE: Single/three-seat glider tug.
PROGRAMME: AB 180 RV, AB 180 Ag and AB 180 SP described in earlier editions; current AB 180 RVR ordered by Brazil 1987 and 1989. See also Addenda.
CURRENT VERSIONS: **AB 180 RVR**: Standard version, *as described.*
CUSTOMERS: Total of 77 for Brazil (seven ordered 1987 plus 70 in 1989), of which 58 delivered by 1 January 1993; completion due mid-1993; civil customers in Argentina.
DESIGN FEATURES: As AB 115 Trainer; stepped fuselage with added cabin rear window.
FLYING CONTROLS: As AB 115 Trainer, plus ground adjustable tab on rudder.

Aero Boero 180 Ag with ventral chemical pod and spraybars (*Photo Link*)

STRUCTURE: As AB 115 Trainer except for Ceconite covering instead of Dacron.
POWER PLANT: One 134 kW (180 hp) Textron Lycoming O-360-A1A flat-four engine, driving a Sensenich 76-EM8 fixed-pitch or Hartzell HC-92ZK-8D constant-speed two-blade propeller. Fuel capacity (two aluminium wing tanks) 200 litres (53 US gallons; 44 Imp gallons); oil capacity 8 litres (2.1 US gallons; 1.75 Imp gallons).
ACCOMMODATION: Pilot and two passengers in fully enclosed, heated and ventilated cabin.
EQUIPMENT: Glider towing hook.
DIMENSIONS, EXTERNAL: As for AB 115 Trainer
AREAS: As for AB 115 Trainer
WEIGHTS AND LOADINGS:
Weight empty	602 kg (1,327 lb)
Max T-O weight	890 kg (1,962 lb)
Max wing loading	51.12 kg/m² (10.47 lb/sq ft)
Max power loading	6.64 kg/kW (10.90 lb/hp)

PERFORMANCE (at max T-O weight except where indicated):
Never-exceed speed (VNE)	132 knots (245 km/h; 152 mph)
Max level speed at S/L	122 knots (225 km/h; 140 mph)
Max cruising speed at S/L	108 knots (201 km/h; 125 mph)
Stalling speed, flaps down	40 knots (73 km/h; 45 mph)
Max rate of climb at S/L	312 m (1,025 ft)/min
Rate of climb: with single-seat sailplane	more than 180 m (590 ft)/min
with two-seat sailplane	120 m (394 ft)/min
Time to 600 m (1,970 ft), 75% power, with Blanik two-seat sailplane	3 min 10 s
Service ceiling	more than 7,000 m (22,965 ft)
T-O run	100 m (330 ft)
T-O to 15 m (50 ft), two persons	188 m (615 ft)
Landing from 15 m (50 ft)	160 m (525 ft)
Landing run	60 m (195 ft)
Range with max fuel	636 nm (1,180 km; 733 miles)

Aero Boero 180 RVR glider tug

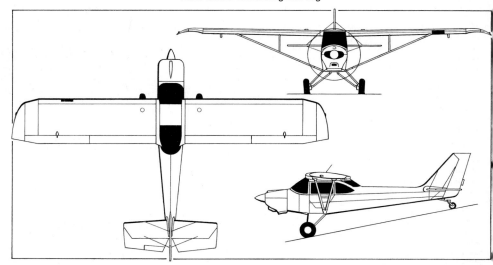

Aero Boero 180 RVR (Textron Lycoming O-360-A1A engine) (*Jane's/Mike Keep*)

CHINCUL

CHINCUL S. A. C. A. I. F. I.
25 de Mayo 489, 6° Piso, 1339 Buenos Aires
Telephone: 54 (1) 312 5671/5
Fax: 54 (1) 311 5742
Telex: 22706 MACUB AR
WORKS: Calle Mendoza entre 6 y 7, 5400 San Juan, Pcia de San Juan; and at Avenida Diaz Velez 1034, 1702 Ciudadela, Buenos Aires
PRESIDENT: José Maria Beraza
SALES DIRECTOR: Juan Pablo Beraza
EXPORT MANAGER: Oscar A. Prïeto

Wholly owned subsidiary of La Macarena (Piper's Argentine distributor), formed 1971 to assemble various Piper types; local manufacture now at Phase 3 level (more than 60 per cent of each aircraft) including wing and control surface assembly and riveting, manufacture of interiors, electrical harnesses, and other systems installation. Developed own tandem two-seat conversion of Pawnee D agricultural aircraft, and two-seat aerobatic military trainer based on Cherokee Arrow, as described in 1983-84 and earlier *Jane's*. San Juan factory has covered area of 16,500 m² (177,600 sq ft), workforce of about 300, and had delivered approx 900 aircraft by January 1990 (latest figure received); Ciudadela plant produces jigs, dies, and aircraft parts for production line and spares; Argentine built items include batteries, fabrics and upholstery, tyres, engine instruments, glassfibre components and fire extinguishers; assembles Bendix/King avionics for Chincul Pipers.

Production and marketing of Piper types by Embraer (see Brazilian section) and Chincul rationalised following May 1989 agreement. Chincul range then comprised PA-18-115 and PA-18-150 Super Cub; PA-25-235 and PA-25-260 Pawnee; PA-28-161 Warrior II; PA-28-181 Archer II; PA-28-236 Dakota; PA-28RT-201/201T Arrow/Turbo Arrow IV; PA-31-350 Chieftain; PA-31T Cheyenne I/II; PA-36-375 Pawnee Brave; and PA-42 Cheyenne III/400. Effect on Chincul of Piper suspension of single-engined types (which see) is not known.

Under July 1990 MoU with Bell Helicopter Textron, Chincul will co-produce Bell 212 and 412 models (see Canadian section), initially by assembling imported CKD kits and later (Phase 2) by manufacturing 20 per cent (by value) of each helicopter. Initial production rate to be four helicopters per year, increasing later to 12; first example (a Bell 212) was due for completion in first quarter of 1992.

Chincul built example of the Piper PA-25-235/260 Pawnee

FMA

FÁBRICA MILITAR DE AVIONES SA
Avenida Fuerza Aérea Argentina Km 5½, 5103 Córdoba
Telephone: 54 (51) 690594
Fax: 54 (51) 690698
Telex: 51965 AMCOR AR
MANAGING DIRECTOR: Brigadier Alberto H. Lindow
COMMERCIAL DIRECTOR: Comodoro Roberto Gomez
COMMERCIAL MANAGER: Ing H. Francisco Luciano

Original FMA (Military Aircraft Factory) came into operation 10 October 1927 as central organisation for aeronautical research and production; underwent several name changes (see 1987-88 and earlier *Jane's*) before reverting 1968 to original title as component of Area de Material Córdoba (AMC) of Argentine Air Force. Principal activities are aircraft design, manufacture, maintenance and repair, major current product being IA 63 Pampa jet trainer; laboratories, factories and other aeronautical division buildings occupy total covered area of 253,000 m² (2,723,265 sq ft); AMC had workforce of just over 2,900 in mid-December 1992, of whom about 2,000 engaged in design and manufacturing; Córdoba facility also accommodates Centro de Ensayos en Vuelo (Flight Test Centre), a separate division also controlled by Argentine Air Force, at which all aircraft produced in Argentina undergo certification testing.

Argentine government has legislated to privatise most national defence companies, including FMA; this required Argentine Air Force to convert FMA into a joint stock company from April 1992, Air Force buying 30 per cent of the shares to establish itself as FMA's holding and management authority; government would sell remaining 70 per cent to foreign private investors at stock exchange valuation. Privatisation expected to be completed during 1993.

EMBRAER/FMA CBA-123 VECTOR

Twin-turboprop regional transport, described in International section; FMA responsible for 33 per cent of development and manufacture.

FMA IA 58A PUCARÁ

Four sold to Sri Lanka Air Force; first delivery January 1993.

FMA IA 63 PAMPA

TYPE: Basic and advanced jet trainer and ground attack aircraft.
PROGRAMME: Initiated by Fuerza Aérea Argentina (FAA) 1979, eventual configuration being selected over six other designs early 1980, with Dornier of Germany providing technical assistance (incl manufacture of prototypes' wings and tailplanes); two static/fatigue test airframes and three flying prototypes built (first flight, by EX-01, 6 October 1984); first flight of production Pampa October 1987; first three production aircraft delivered (to IV Brigada Aérea at Mendoza) April 1988.
CURRENT VERSIONS: **Pampa:** Standard Argentine Air Force version; *detailed description applies to this version.* First six aircraft retrofitted with Argentine Air Force developed HUD (to become standard for all existing and future production Pampas); aircraft in service refitted with Elbit autonomous WDNS (weapon delivery and navigation system).
Naval version: Under development: no details yet known.
Pampa 2000: Version proposed, with Vought (ex-LTV) and Loral as FMA's partners, for USAF/USN JPATS trainer programme; second prototype EX-02 and series aircraft E-812 supplied to USA for modification to Pampa 2000 configuration; third aircraft to be delivered during 1992; EX-02 lost in crash in UK 31 August 1992.
CUSTOMERS: Firm orders for 18 for Argentine Air Force (12 delivered by 1 January 1991, remainder due by May 1992); follow-on order for 46 expected when privatisation completed. Argentine Navy interested in an initial 14. Other reported interest from Canada, Israel and Mexico may not crystallise until outcome of JPATS competition known.
COSTS: Total Argentine Air Force programme $190 million; flyaway unit cost $3.5 million.
DESIGN FEATURES: Non-swept shoulder mounted wings and anhedral tailplane, sweptback fin and rudder; single engine with twin lateral air intakes. Wing section Dornier DoA-7/-8 advanced transonic; thickness/chord ratio 14.5 per cent at root, 12.5 per cent at tip; anhedral 3°.
FLYING CONTROLS: Fully powered hydraulic actuation of all moving surfaces (ailerons, rudder, all-moving tailplane, single-slotted Fowler flaps, and door type airbrake on each side of upper rear fuselage); primary surfaces have Liebherr tandem servo-actuators and electromechanical trim; stick forces simulated by artificial feel.
STRUCTURE: Conventional all-metal semi-monocoque/stressed skin; two-spar wing box forms integral fuel tank.
LANDING GEAR: SHL (Israel) retractable tricycle type, with hydraulic extension/retraction and emergency free-fall extension. Oleo-pneumatic shock absorbers. Single Messier-Bugatti wheel and Goodrich (main) or Continental (nose) low-pressure tyre; nosewheel offset 10 cm (3.9 in) to starboard. Tyre sizes 6.50-10 (10 ply rating) on mainwheels, 380 × 150 mm (4-6 ply rating) on nosewheel, with respective pressures of 6.55 bars (95 lb/sq in) and 4.00 bars (58 lb/sq in). Nosewheel retracts rearward, mainwheels inward into underside of engine air intake trunks. Messier-Bugatti mainwheel hydraulic disc brakes incorporate anti-skid device; nosewheel steering (±47°). Gear designed for operation from unprepared surfaces.
POWER PLANT: One 15.57 kN (3,500 lb st) Garrett TFE731-2-2N turbofan installed in rear fuselage. Standard internal fuel capacity of 968 litres (255 US gallons; 213 Imp gallons) in integral wing tank of 550 litres (145 US gallons; 121 Imp gallons) and 418 litre (110 US gallon; 92 Imp gallon) flexible fuselage tank with a negative *g* chamber permitting up to 10 s of inverted flight. Additional 415 litres (109 US gallons; 91 Imp gallons) can be carried in auxiliary tanks inside outer wing panels, to give a max internal capacity of 1,383 litres (364 US gallons; 304 Imp gallons). Single-point pressure refuelling, plus gravity point in upper surface of each wing.
ACCOMMODATION: Instructor and pupil in tandem (instructor at rear, on elevated seat), on UPC (Stencel) S-III-S3IA63 zero/zero ejection seats. Ejection procedure can be pre-

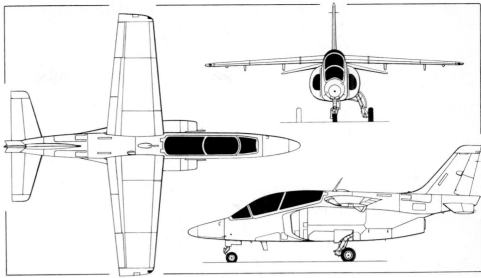

FMA IA 63 Pampa two-seat basic and advanced jet trainer *(Dennis Punnett)*

IA 63 Pampa tandem-seat jet trainer in Argentine Air Force camouflage

selected for separate single ejections, or for both seats to be fired from front or rear cockpit. Dual controls standard. One-piece wraparound windscreen. One-piece canopy, with internal screen, is hinged at rear and opens upward. Entire accommodation pressurised and air-conditioned.
SYSTEMS: AiResearch environmental control system, max differential 0.3 bar (4.35 lb/sq in), supplied by high or low pressure engine bleed air, provides a 1,980 m (6,500 ft) cockpit environment up to flight level 5,730 m (18,800 ft) and also provides ram air for negative *g* system and canopy seal. Oxygen system supplied by 10 litre (2.64 US gallon; 2.20 Imp gallon) lox converter. Engine air intakes anti-iced by engine bleed air. Two independent hydraulic systems, each at pressure of 207 bars (3,000 lb/sq in), each supplied by engine driven pump. Each system incorporates a bootstrap reservoir pressurised at 4 bars (58 lb/sq in). No. 1 system, with flow rate of 16 litres (4.2 US gallons; 3.5 Imp gallons)/min, actuates primary flight controls, airbrakes, landing gear and wheel brakes; No. 2 system, with flow rate of 8 litres (2.1 US gallons; 1.75 Imp gallons)/min, actuates primary flight controls, wing flaps, emergency and parking brakes, and nosewheel steering. AiResearch ram air turbine provides emergency hydraulic power for No. 2 system if engine shuts down in flight and pressure in this system drops below minimum. Electrical system (28V DC) supplied by Lear Siegler 400A 11.5kW engine driven starter/generator; secondary supply (115/26V AC power at 400Hz) from two Flite-Tronics 450VA static inverters and two SAFT 27Ah nickel-cadmium batteries. Thirty minutes emergency electrical power available in case of in-flight engine shutdown.
AVIONICS: Standard package comprises two redundant Collins VHF com transceivers, Becker intercom system, Collins VOR/ILS with marker beacon receiver, Collins DME, and Collins ADF. Navigation system allows complete navigation/landing training under IFR conditions. Attitude and heading information provided by Astronautics three-gyro platform, with magnetic flux valve compass for additional heading reference.
ARMAMENT: No built-in weapons. Five attachments for external stores, with max pylon load of 400 kg (882 lb) on each inboard underwing station, 250 kg (551 lb) each on fuselage centreline and outboard underwing pair. With a 30 mm gun pod containing 145 rds on the fuselage station, typical underwing loads can include six Mk 81 bombs, two each Mk 81 and Mk 82 bombs, or one 7.62 mm twin-gun pod and one practice bomb/rocket training container. Gyro-

stabilised sighting system in front cockpit (optional in rear cockpit), with recorder in front sight. Weapon management system adequate for several different tactical configurations.

DIMENSIONS, EXTERNAL:
Wing span	9.686 m (31 ft 9¼ in)
Wing aspect ratio	6.0
Length overall (excl pitot probe)	10.93 m (35 ft 10¼ in)
Height overall	4.29 m (14 ft 1 in)
Tailplane span	4.576 m (15 ft 0½ in)
Wheel track	2.663 m (8 ft 8¾ in)
Wheelbase	4.418 m (14 ft 6 in)

AREAS:
Wings, gross	15.633 m² (168.27 sq ft)
Ailerons (total)	0.894 m² (9.62 sq ft)
Trailing-edge flaps (total)	2.928 m² (31.52 sq ft)
Fin	1.863 m² (20.05 sq ft)
Rudder	0.655 m² (7.05 sq ft)
Tailplane	4.354 m² (46.87 sq ft)

WEIGHTS AND LOADINGS:
Weight empty	2,821 kg (6,219 lb)
Fuel load:	
wings (incl auxiliary tanks)	780 kg (1,719 lb)
fuselage	338 kg (745 lb)
Max underwing load with normal internal fuel	1,160 kg (2,557 lb)
T-O weight, clean configuration:	
968 litres internal fuel	3,700 kg (8,157 lb)
1,383 litres internal fuel	3,800 kg (8,377 lb)
Max T-O weight with external stores	5,000 kg (11,023 lb)
Typical landing weight	3,500 kg (7,716 lb)
Wing loading:	
at clean T-O weight:	
968 litres internal fuel	236.72 kg/m² (48.51 lb/sq ft)
1,383 litres internal fuel	243.12 kg/m² (49.82 lb/sq ft)
at max T-O weight with external stores	319.90 kg/m² (65.55 lb/sq ft)
Power loading:	
at clean T-O weight:	
968 litres internal fuel	237.8 kg/kN (2.33 lb/lb st)
1,383 litres internal fuel	244.2 kg/kN (2.39 lb/lb st)
at max T-O weight with external stores	321.4 kg/kN (3.15 lb/lb st)

PERFORMANCE (ISA, at 3,800 kg; 8,377 lb clean T-O weight with max internal fuel, except where indicated):
Max level speed at S/L   405 knots (750 km/h; 466 mph)
Approach speed at S/L, landing weight of 3,630 kg (8,000 lb)   120 knots (222 km/h; 138 mph)
Stalling speed   92 knots (171 km/h; 106 mph)
Max rate of climb: at S/L   1,560 m (5,120 ft)/min
at 4,575 m (15,000 ft)   1,215 m (3,990 ft)/min
Max rate of roll   150°/s
Service ceiling   12,900 m (42,325 ft)
T-O run at S/L, AUW of 3,700 kg (8,157 lb)   424 m (1,390 ft)
Landing run at landing weight of 3,500 kg (7,716 lb)   461 m (1,512 ft)
Typical mission radius:
air-to-air gunnery (hi-hi), T-O weight of 3,950 kg (8,708 lb) with 250 kg (551 lb) external load, 5 min allowance for dogfight, 30 min reserves   237 nm (440 km; 273 miles)
air-to-ground (hi-lo-hi), T-O weight of 4,860 kg (10,714 lb) with 1,000 kg (2,205 lb) external load, 5 min allowance for weapon delivery, 30 min reserves   194 nm (360 km; 223 miles)
Range at 300 knots (556 km/h; 345 mph) at 4,000 m (13,125 ft):
968 litres internal fuel   540 nm (1,000 km; 621 miles)
1,383 litres internal fuel   809 nm (1,500 km; 932 miles)
Ferry range at 280 knots (519 km/h; 322 mph) at 10,060 m (33,000 ft), max internal/external fuel   1,000 nm (1,853 km; 1,151 miles)
Max endurance at 300 knots (556 km/h; 345 mph) at 4,000 m (13,125 ft), 1,383 litres internal fuel   3 h 48 min
g limits   +6/−3 (+4.5 max sustained)

AUSTRALIA

AIA
AVIATION INDUSTRIES OF AUSTRALIA PTY LTD
25 Welsford Street (PO Box 814), Shepparton, Victoria 3630
Telephone: 61 (58) 31 1488
Fax: 61 (58) 31 1386
EXECUTIVE DIRECTORS:
Brian Marrison (Commercial) (Chief Executive)
Jess Smith (Technical)
CHIEF OF DESIGN: Merv Reed
AIA because responsible for Mamba aircraft programme, initiated by Melbourne Aircraft Corporation (which see in 1991-92 *Jane's*).

AIA MA-2 MAMBA
TYPE: Two-seat light aircraft.
PROGRAMME: Two years in conception and design; prototype (VH-JSA) made first flight 25 January 1989; original 86.5 kW (116 hp) Textron Lycoming O-235 later replaced by O-320-D1A; first flight of pre-production Mamba December 1989; developmental C of A received 1991; final certification under CAO 101.22 (which complies with FAR Pt 23) scheduled for 1992; commercial production began late 1991; first deliveries were expected late 1992.
CURRENT VERSIONS: **MA-2A**: Basic model, with O-320 engine. *Detailed description applies to this version except where indicated.*
MA-2C: Civil model.
MA-2M: Military model; more powerful IO-360 engine and additional instrumentation.
CUSTOMERS: Reported early orders included one for a local flying school for aerobatic training and three for forest fire patrol with Western Australia Dept of Conservation and Land Management.
DESIGN FEATURES: Braced high-wing monoplane with extensively glazed cabin, slender fuselage and slightly swept-back vertical tail surfaces. Wing section NACA 4412 (constant); dihedral 1° 30′ from roots; stores hardpoint under each wing outboard of strut.
FLYING CONTROLS: Frise type differential ailerons, rudder and elevators, all mechanically actuated; electrically actuated split flaps on wing trailing-edges; servo tab in each aileron, trim tab in one elevator.
STRUCTURE: All-metal except for marine ply belly, with 4130 chromoly steel tube fuselage frame, two-spar wings and aluminium alloy skins; wings braced by single I strut each side.
LANDING GEAR: Non-retractable tricycle type. MAC cantilever self-sprung mainwheel legs, each with Cleveland wheel and Dunlop 6.00-6 × 4 tyre. Cleveland tailwheel has 5.00-5 × 4 Dunlop tyre and is steerable ±30°. MAC mainwheel brakes. Dual brakes (toe front/hand rear) standard in MA-2M. Wheel fairings optional.
POWER PLANT: One 119 kW (160 hp) Textron Lycoming O-320-D1A flat-four engine in MA-2A (149 kW; 200 hp IO-360-A1B6 in MA-2M), driving a Hartzell two-blade propeller. Internal fuel capacity 327 litres (86 US gallons; 72 Imp gallons); provision for auxiliary fuel tank under each wing inboard of strut. Inverted flight fuel system, turbocharging and (on MA-2A) fuel injection engine optional. Oil capacity 9.1 litres (2.4 US gallons; 2 Imp gallons).
ACCOMMODATION: Two seats in tandem in heated and ventilated cabin. Dual flight and engine controls standard. Door on starboard side. Space for 20 kg (44 lb) of baggage aft of seats. Rear baggage compartment standard on MA-2M.
SYSTEMS: 14V or 28V DC electrical system (50A alternator).
AVIONICS: Com radio to customer's requirements. Basic VFR instrumentation in MA-2A, to which are added artificial horizon, directional gyro, clock, accelerometer and instrument panel lights in MA-2M. Autopilot and rear cockpit instruments optional.
EQUIPMENT: Flashing beacon and navigation/landing/taxi lights standard on MA-2M.

DIMENSIONS, EXTERNAL:
Wing span   8.36 m (27 ft 5 in)
Wing aspect ratio   7.52
Length overall   7.47 m (24 ft 6 in)
Height overall   2.74 m (9 ft 0 in)
AREAS:
Wings, gross   9.29 m² (100.0 sq ft)
WEIGHTS AND LOADINGS:
Weight empty (standard fuel):
MA-2A, MA-2M   635 kg (1,400 lb)
Max T-O weight with external stores:
MA-2A   907 kg (2,000 lb)
MA-2M   1,043 kg (2,300 lb)
Max wing loading: MA-2A   97.65 kg/m² (20.0 lb/sq ft)
MA-2M   112.30 kg/m² (23.0 lb/sq ft)
Max power loading: MA-2A   7.62 kg/kW (12.50 lb/hp)
MA-2M   7.00 kg/kW (11.50 lb/hp)
PERFORMANCE (estimated, at above max T-O weights, ISA):
Max level speed at S/L:
MA-2A   125 knots (232 km/h; 144 mph)
MA-2M   135 knots (250 km/h; 155 mph)
Max cruising speed (75% power) at S/L:
MA-2A   115 knots (213 km/h; 132 mph)
MA-2M   123 knots (228 km/h; 142 mph)
Max cruising speed (75% power) at 2,440 m (8,000 ft):
MA-2A   120 knots (222 km/h; 138 mph)
MA-2M   125 knots (232 km/h; 144 mph)
Stalling speed, power off:
flaps up:
MA-2A, MA-2M 60 knots (111 km/h; 69 mph) EAS
flaps down:
MA-2A, MA-2M 45 knots (84 km/h; 52 mph) EAS
Max rate of climb at S/L: MA-2A   335 m (1,100 ft)/min
MA-2M   610 m (2,000 ft)/min
Service ceiling: MA-2A   4,265 m (14,000 ft)
MA-2M   4,875 m (16,000 ft)
T-O run: MA-2A, 30° flap   153 m (500 ft)
MA-2M, 20° flap   92 m (300 ft)
T-O to 15 m (50 ft): MA-2A   305 m (1,000 ft)
MA-2M   214 m (700 ft)
Range at 2,440 m (8,000 ft) at max cruising speed (75% power), standard fuel, no reserves:
MA-2A   684 nm (1,267 km; 787 miles)
MA-2M   625 nm (1,158 km; 719 miles)

Prototype AIA MA-2 Mamba two-seat light aircraft

AIRCORP
AIRCORP PTY LTD
Caboolture Airfield, Brisbane, Queensland
MANAGING DIRECTOR: H. Anning
PROJECT DIRECTOR, BUSHMASTER: Brian P. Creer

AIRCORP BUSHMASTER
TYPE: Two- to four-seat aerobatic light aircraft.
PROGRAMME: First flight of two-seat B2-N prototype (VH-BOI, with 67.1 kW; 90 hp Norton Aerotor 90 two-chamber rotary engine) 28 October 1989, following four years of design and development; original plan to offer choice of Norton or Textron Lycoming engines in production versions abandoned, and airframe changes made to facilitate production (details in 1991-92 *Jane's*); not yet confirmed whether production has started.
CURRENT VERSIONS: **B2-N**: Two-seat, Norton powered prototype; described in 1991-92 *Jane's*.
B2-16: Two-seater, with 86.5 kW (116 hp) O-235 engine.

Aircorp B2-N Bushmaster two-seat prototype with original Aerotor rotary engine

B3-16: As B2-16, but with three seats. *Detailed description applies to B2-16 and B3-16 except where indicated.*
B4-16: As B2-16, but with '2 + 2' seating.
B4-60: Full four-seater, with lengthened fuselage and 119 kW (160 hp) Textron Lycoming engine.
B4-80: As B4-60, but with 134 kW (180 hp) Textron Lycoming engine.
DESIGN FEATURES: Intended for club training, private flying and rural support; conforms to FAR Pt 23 standards; conventional high-wing cabin monoplane configuration.
FLYING CONTROLS: Conventional primary control surfaces, actuated mechanically; three-stage trailing-edge flaps.
STRUCTURE: All-metal, conventional; one-piece wing (extruded spars, forming box for fuel tanks) braced by single I strut each side; welded steel tube fuselage; wire braced tailplane; light alloy skins.
LANDING GEAR: Non-retractable tailwheel type. Main units have faired-in side Vs, half-axles, and speed fairings over wheels.
POWER PLANT (B2 and B3): One 86.5 kW (116 hp) Textron Lycoming O-235-N2C flat-four engine, driving a three-blade propeller. Fuel capacity 136.4 litres (36 US gallons; 30 Imp gallons) standard.
ACCOMMODATION: Two, '2 + 2', three or four seats (see under Current Versions), in fully enclosed cabin with forward opening door on each side; B3 has pilot seat on centreline with two-person bench seat to rear (removable for carriage of cargo). Space for 55 kg (121 lb) of baggage aft of seats.
DIMENSIONS, EXTERNAL:
Wing span 10.00 m (32 ft 9¾ in)
Wing aspect ratio 7.14
Length overall 6.40 m (21 ft 0 in)
Wheel track 2.14 m (7 ft 0¼ in)
Wheelbase 4.00 m (13 ft 1½ in)
DIMENSIONS, INTERNAL:
Cabin: Length 2.70 m (8 ft 10¼ in)
Max width 1.10 m (3 ft 7¼ in)
Max height 1.10 m (3 ft 7¼ in)
AREAS:
Wings, gross 14.00 m² (150.7 sq ft)
WEIGHTS AND LOADINGS (U: Utility, N: Normal category):
Weight empty 410 kg (904 lb)
Max T-O weight: U 700 kg (1,543 lb)
N 811 kg (1,788 lb)
Max wing loading: U 50.00 kg/m² (10.24 lb/sq ft)
N 57.93 kg/m² (11.86 lb/sq ft)
Max power loading: U 8.16 kg/kW (13.42 lb/hp)
N 9.46 kg/kW (15.55 lb/hp)
PERFORMANCE:
Max cruising speed more than 100 knots (185 km/h; 115 mph)
Stalling speed 35 knots (65 km/h; 41 mph)
g limits: U +4.4/−2.2

ASTA
AEROSPACE TECHNOLOGIES OF AUSTRALIA PTY LTD

CORPORATE OFFICE: Private Box 226, Port Melbourne, Victoria 3207
Telephone: 61 (3) 647 3111
Fax: 61 (3) 646 4381 or 2253
Telex: AA 34851
CHAIRMAN: Sir Brian Inglis
MANAGING DIRECTOR: George W. Stuart
MEDIA COMMUNICATIONS: Andre R. van der Zwan
BUSINESS UNITS:
ASTA Airport, Private Bag 9, Avalon Airport, Lara, Victoria 3212
Telephone: 61 (52) 279 555
Fax: 61 (52) 823 335
AIRPORT DEVELOPMENT MANAGER: Dennis Chant
ASTA Defence, Private Bag 4, Avalon Airport, Lara, Victoria 3212
Telephone: 61 (52) 279 444
Fax: 61 (52) 823 345
GENERAL MANAGER: Rick Campbell
ASTA Engineering, PO Box 376, Port Melbourne, Victoria 3207
Telephone: 61 (3) 647 3111
Fax: 61 (3) 645 2582
GENERAL MANAGER: Noel Jenkinson
ASTA Components, Private Bag 4, Port Melbourne, Victoria 3207
Telephone: 61 (3) 647 3111
Fax: 61 (3) 645 3424
GENERAL MANAGER: Paul Mathieson
ASTA Aircraft Services Pty Ltd (ASTAAS), Private Bag 2, Avalon Airport, Lara, Victoria 3212
Telephone: 61 (52) 279 111
Fax: 61 (52) 823 892
MANAGING DIRECTOR: David Main
Pacific Aerospace Corporation Ltd (PAC): see New Zealand section

Established 1986 as private enterprise organisation to succeed former Government Aircraft Factories (GAF: see earlier editions). Group specialises in design, development, manufacture, assembly, maintenance and modification of aircraft, target drones and guided weapons; component subassembly, final assembly, modification, repair and test flying of jet and other aircraft undertaken at Avalon airfield.

ASTA Airport created October 1990 following July 1990 purchase of Avalon airfield; ASTA Defence created 1 July 1992 by combining former Military Aircraft Services, Systems and General Aviation business units. ASTA group has 75.1 per cent stake in Pacific Aerospace Corporation (see New Zealand section); HAECO of Hong Kong has 9.9 per cent interest in ASTAAS.

Recent and current aircraft activity includes component manufacture (details in 1991-92 *Jane's*), forward fuselage fitting out, final assembly and flight test of RAAF AF/ATF-18A Hornets (due for completion January 1993); 40 per cent of Australian production work on RAAF Pilatus PC-9/A programme (completed October 1991); and assembly of eight of 16 Seahawks for Royal Australian Navy. Subcontract work includes components for Boeing 747 (Krueger flaps), 757 (inspar wing ribs and rudders) and 777 (rudders: first delivery due mid-1993); Airbus A320 (flap shrouds); Sikorsky Black Hawk/Seahawk (rotor blade components); McDonnell Douglas MD-80 series; and McDonnell Douglas MD-11 (CFM56 nacelle components). ASTA also a risk-sharing contractor to Aerospatiale to design/develop/manufacture carbonfibre main and central landing gear doors and floor support panels for Airbus A330/A340.

AUSTRALIAN AUTOGYRO
THE AUSTRALIAN AUTOGYRO CO
29 Benning Avenue, Turramurra, Sydney, NSW 2074
Telephone: 61 (2) 449 9816
PROPRIETOR: E. R. Minty

AUSTRALIAN AUTOGYRO SKYHOOK
TYPE: Single-seat lightweight autogyro.
PROGRAMME: See 1982-83 and previous *Jane's* for early development history; Rotax 503 dual-ignition engine became preferred power plant from 1990, and original 1,835 cc VW engine now no longer available; other recent changes include reversion to aluminium (instead of composites) rotor blades.
CURRENT VERSIONS: **Mk I:** Basic open frame model; 38.8 kW (52 hp) Rotax 503 engine standard.
Mk II: Unpainted, with enclosed body.
Mk III: Fully enclosed and customised version. *Description applies to this version except where indicated.*
CUSTOMERS: 24 Mk Is and two Mk IIIs (all VW powered), plus numerous kits and components, sold by 1 February 1990; three more aircraft (one with Rotax) sold by end of that year. No sales in 1991 or 1992 due to Australian recession, but remains available; potential of Asian market being explored.
COSTS: A$14,000 flyaway for current Mk I; fully customised Mk III A$23,000.
DESIGN FEATURES: Two-blade rotor; fuselage, seat, engine mount, rotor mast and tail unit carried on single-beam keel; entire tail section (twin rudders, united by dihedral tailplane and braced at upper ends) operates as single unit.
FLYING CONTROLS: Joystick control actuates two push/pull cables to operate rotor head and blades; tail assembly swivels from side to side for directional stability and power-off manoeuvrability.
STRUCTURE: All-aluminium rotor blades attached to fully adjustable hub bar; push/pull cables are of nylon encased 6.35 mm (¼ in) stainless steel; keel and rotor mast are of 5.1 cm (2 in) square section 6061-T6 aluminium alloy with radiused corners; rudders of marine quality aluminium, tailplane of 2024 aluminium sheet; upper ends of rudders braced by double V triangular aluminium straps. Most structural attachments clamped to minimise potential fracture locations.
LANDING GEAR: Non-retractable type with small sprung tailwheel at rear end of keel. Fully sprung steerable nosewheel, linked to rudders. Mainwheels are ultra-lightweight 5 in nylon go-kart rims, each with a 4.00-5 tyre and tube. Disc brakes optional.
POWER PLANT: One 38.8 kW (52 hp) Rotax 503 (optionally 47 kW; 63 hp Rotax 582) two-cylinder two-stroke engine, mounted on chromoly brackets attached to mast and keel, driving a three-blade adjustable-pitch carbonfibre/Kevlar pusher propeller. Fuel in pilot's hollow seat, capacity 45.5 litres (12 US gallons; 10 Imp gallons).
ACCOMMODATION: Pilot only, in fully enclosed cockpit (Mk III only), on rotationally moulded super-strength cross-linked polyethylene seat/fuel tank located just forward of mast/keel junction, close to CG. Aircraft can be flown without fuselage shell and Plexiglas windscreen enclosure. Adjustable vents in fuselage nose provide ventilation when flown with enclosed cockpit.
DIMENSIONS, EXTERNAL:
Rotor diameter 7.01 m (23 ft 0 in)
Rotor blade chord 0.19 m (7½ in)
Length overall 3.35 m (11 ft 0 in)
Height: to top of rotor head 2.06 m (6 ft 9 in)
to top of cockpit canopy 1.42 m (4 ft 8 in)
Tailplane span (incl rudders) 0.86 m (2 ft 10 in)
Wheel track 1.68 m (5 ft 6 in)
Propeller diameter 1.37 m (4 ft 6 in)
AREAS:
Rotor disc 38.6 m² (415.5 sq ft)
WEIGHTS AND LOADINGS:
Weight empty, excl rotor blades:
Mk I (Rotax 503) 97 kg (213 lb)
Mk III 147 kg (325 lb)
Max T-O weight 271 kg (597 lb)
Max disc loading 7.03 kg/m² (1.44 lb/sq ft)
Max power loading 6.98 kg/kW (11.48 lb/hp)
PERFORMANCE:
Max level speed 69 knots (129 km/h; 80 mph)
Cruising speed 43 knots (80 km/h; 50 mph)
Max rate of climb at S/L 457 m (1,500 ft)/min
T-O run (depending on headwind) approx 122 m (400 ft)
Landing run (with disc brakes) approx 5 m (15 ft)
Range with max fuel 191 nm (354 km; 220 miles)

Skyhook Mk I single-seat autogyro with Rotax engine and lightweight open frame

BAC
BUCHANAN AIRCRAFT CORPORATION LTD
9 Production Court, Toowoomba, Queensland 4350
Telephone: 61 (76) 33 1856
Fax: 61 (76) 33 1115
CHAIRMAN: Joe Moharich
MANAGING DIRECTOR: John Buchanan
EXECUTIVE DIRECTOR, OPERATIONS: Kevin O'Brien

Company incorporated 1989; Buchanan family has major shareholding; other prime shareholder is Queensland government Industrial Development Corporation; workforce 60. Specialist in composites design and manufacture; Australian CAA approved centre for civil aircraft design, engineering, stressing and modification. Quality systems approval by Lloyds Register to ISO 9001/AS 3901.

BUCHANAN BAC-204
TYPE: Two-seat very light aircraft.
PROGRAMME: First prototype (VH-OZE) flying by early 1992; second in flight test programme by early 1993; to be certificated to JAR VLA.
DESIGN FEATURES: Designed for JAR VLA, but many aspects conform also to FAR Pt 23; laminar profile wing, low-drag, waisted fuselage; high manoeuvrability and docile stall; future adaptation possible for IMC flight, surveillance, mustering or military forward air control.
   Modified Wortmann FX-63-166 wing section; leading-edge sweptforward 4°; dihedral 4° from roots.
FLYING CONTROLS: Conventional mechanical; mass balanced elevators and rudder; wide span flaps (max travel 35°).
STRUCTURE: All-composites; first prototype built to prove wet layup process; materials and methods for second aircraft being validated early 1992.
LANDING GEAR: Non-retractable tricycle; cantilever self-sprung leg and wheel speed fairing on each unit.
POWER PLANT: One 59.7 kW (80 hp) Rotax 912-A1 flat-four engine, driving a two-blade fixed-pitch propeller. Usable fuel capacity 100 litres (26.4 US gallons; 22 Imp gallons).
ACCOMMODATION: Two seats side by side under one-piece fully transparent canopy. Space for 30 kg (66 lb) of baggage behind seats.

DIMENSIONS, EXTERNAL:
Wing span	10.50 m (34 ft 5½ in)
Wing aspect ratio	10.8
Length overall	6.40 m (21 ft 0 in)
Height overall	2.30 m (7 ft 6½ in)
Propeller diameter	1.70 m (5 ft 7 in)

AREAS:
Wings, gross	10.20 m² (109.79 sq ft)

WEIGHTS AND LOADINGS:
Weight empty	350 kg (771 lb)
Max fuel weight	71 kg (156.5 lb)
Max T-O weight	600 kg (1,323 lb)
Max wing loading	58.82 kg/m² (12.05 lb/sq ft)
Max power loading	10.06 kg/kW (16.53 lb/hp)

PERFORMANCE (estimated, at max T-O weight):
Never-exceed speed (VNE)	150 knots (278 km/h; 172 mph)
Max level speed at S/L	102 knots (189 km/h; 117 mph)
Cruising speed at 70% power at 1,980 m (6,500 ft)	100 knots (185 km/h; 115 mph)
Stalling speed: flaps up	48 knots (89 km/h; 56 mph)
flaps down	44 knots (82 km/h; 51 mph)
Max rate of climb at S/L	244 m (800 ft)/min
Service ceiling	5,490 m (18,000 ft)
Range at 70% power at 1,980 m (6,500 ft), 45 min reserves	525 nm (973 km; 604 miles)
Endurance at 70% power at 1,980 m (6,500 ft), no reserves	6 h

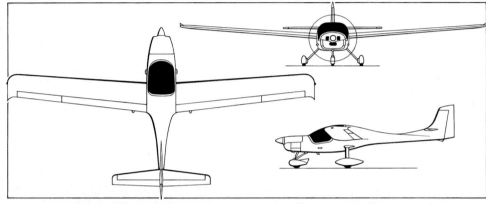

Three-view drawing *(Jane's/Mike Keep)* and photograph of the Buchanan BAC-204 two-seat lightplane

---

EAI
EAGLE AIRCRAFT INTERNATIONAL
PO Box 586, Fremantle, WA 6160
Telephone: 61 (9) 410 1077
Fax: 61 (9) 410 2430
MARKETING MANAGER: Deryck F. Graham Jr

Eagle Aircraft International is trading name of Composite Technology Pty Ltd; joint venture between Eagle Aircraft Australia Ltd (see 1991-92 *Jane's*) and Assets Accretion Snd Bhd, a corporate vehicle for the Malaysian government, came into effect 1 March 1991; entire venture purchased by Malaysian government May 1993. Plans are to build 77 aircraft in first 12 months of manufacture, and eventually to co-produce range of Eagle aircraft in Australia and Malaysia.

EAI EAGLE XTS
TYPE: Two-seat lightplane for general aviation use.
PROGRAMME: Launched 1981 with objective of producing first all-composites light aircraft in Australia; single-seat POC (proof-of-concept) aircraft (now displayed at Power House Museum, Sydney); construction of two-seat pre-production prototype started fourth quarter 1987; first flight (VH-XEG) Spring 1988, with 58 kW (78 hp) Aeropower engine, replaced later by 74.5 kW (100 hp) Continental O-200; 200 hour test programme, meeting all original design criteria, completed by October 1988; plans for 1989 production start aborted (see 1991-92 *Jane's*); component manufacture began December 1991; static testing under way; production prototype (VH-XEP) for certification to JAR VLA (first flight 6 November 1992); three production aircraft under construction January 1993, with first delivery expected in mid-1993.
CURRENT VERSIONS: **Pastoral:** Baseline model, with robust interior, basic instrumentation and larger tyres for rough-field landing. *Description applies to this version.*
   Trainer: Equipped for flying academies and clubs.
   Sports/Executive: For recreational flyer and personal business commuter; instruments and interior to customer's requirements.
CUSTOMERS: Options for approx 20 held in May 1993.
COSTS: Programme cost A$15 million.
DESIGN FEATURES: 'Three flying surface' configuration, with high mounted main wing, low mounted foreplane, and tailplane; conceived by US designer John Roncz (chief aerodynamicist on round-the-world Voyager); claimed to be very responsive at both low and high speeds, very stable in level flight, and to have forgiving low-speed stall characteristics. Production version undergoing certification to JAR VLA (very light aircraft: less than 750 kg; 1,653 lb MTOGW) standards.
FLYING CONTROLS: Flaps and small ailerons on mainplane;

Eagle XTS production prototype (93.2 kW; 125 hp IO-240 engine)

elevators on foreplane and tailplane; internally balanced rudder. All surfaces mechanically (rod) actuated. Electric elevator and rudder trim. Stall fence on each foreplane leading-edge.

STRUCTURE: Except for metal engine mounts and flight control rods, production XTS built entirely of advanced composites. Wings, fuselage and all control surfaces are Nomex honeycomb or high density foams, sandwiched between multiple layers of carbonfibre; Kevlar reinforcement around wing leading-edges and shoulder of mainplane; cockpit is impact-resistant capsule of multi-layered Kevlar and carbonfibre; carbonfibre spars. Entire structure uses EAI-designed vinylester resins for strength/longevity/impact resistance and to minimise 'wet environment' problems inherent in standard epoxies.

LANDING GEAR: Non-retractable tricycle type with glassfibre epoxy self-sprung mainwheel legs. Independent disc brakes on mainwheels; non-steerable nose unit.

POWER PLANT: One Teledyne Continental IO-240-ETS (Eagle Two-Seat) flat-four engine (93.2 kW; 125 hp at 2,850 rpm), driving a two-blade fixed-pitch wooden propeller. Fuel capacity 136 litres (36 US gallons; 30 Imp gallons).

ACCOMMODATION: Two seats side by side under forward opening one-piece canopy.

AVIONICS: Details not yet finalised.

DIMENSIONS, EXTERNAL:
Wing span	7.16 m (23 ft 6 in)
Wing chord, constant	0.75 m (2 ft 5½ in)
Wing aspect ratio	10.21
Foreplane span	4.88 m (16 ft 0 in)
Foreplane chord, constant	0.75 m (2 ft 5½ in)
Foreplane aspect ratio	6.84
Length overall	6.53 m (21 ft 5 in)
Height overall	2.26 m (7 ft 5 in)
Tailplane span	3.25 m (10 ft 8 in)
Wheel track	1.93 m (6 ft 4 in)
Wheelbase	1.52 m (5 ft 0 in)
Propeller diameter (max)	1.78 m (5 ft 10 in)
Propeller ground clearance	0.28 m (11 in)

DIMENSIONS, INTERNAL:
Cockpit: Length	1.37 m (4 ft 6 in)
Max width	1.07 m (3 ft 6 in)
Max height	0.86 m (2 ft 10 in)

AREAS:
Wings, gross	5.02 m² (54.0 sq ft)
Foreplane, gross	3.48 m² (37.5 sq ft)
Horizontal tail surfaces (total)	1.49 m² (16.0 sq ft)

WEIGHTS AND LOADINGS:
Weight empty	327 kg (720 lb)
Max T-O weight	567 kg (1,250 lb)
Max wing/foreplane loading	66.69 kg/m² (13.66 lb/sq ft)
Max power loading	6.08 kg/kW (10.00 lb/hp)

PERFORMANCE (estimated, at max T-O weight):
Never-exceed speed (V_NE)	165 knots (305 km/h; 190 mph) IAS
Max level speed	144 knots (266 km/h; 166 mph) IAS
Max cruising speed, 75% power	130 knots (241 km/h; 150 mph) IAS
Stalling speed: flaps up	61 knots (113 km/h; 71 mph) IAS
flaps down	45 knots (84 km/h; 52 mph) IAS
Max rate of climb at S/L	more than 305 m (1,000 ft)/min
Service ceiling	5,485 m (18,000 ft)
T-O run	180 m (590 ft)
Landing run	120 m (393 ft)
Range with max fuel	703 nm (1,302 km; 810 miles)
Max endurance, 60% power	6 h
g limits	+3.8/−1.9 (JAR VLA)
	+8.55/−4.27 (ultimate)

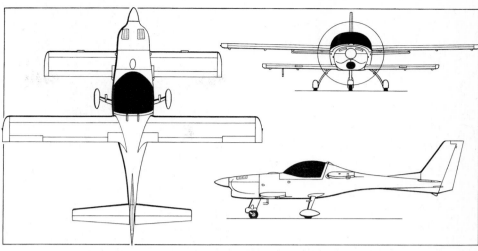

Eagle XTS two-seat light aircraft (*Jane's/Mike Keep*)

GA

GIPPSLAND AERONAUTICS PTY LTD

PO Box 881, Latrobe Valley Airfield, Morwell, Victoria 3840
Telephone: 61 (51) 74 3086
Fax: 61 (51) 74 0956

DIRECTORS:
George Morgan
Peter Furlong

DESIGN AND CERTIFICATION ENGINEER: Colin Nicholson

Involved since 1971 in modification programmes for wide range of aircraft, from wooden homebuilts to pressurised turboprops. Much of this work involved modifications to light agricultural types, prompting GA to regard these as 'prototypes' leading eventually to totally new design.

GIPPSLAND GA-200

TYPE: Two-seat agricultural aircraft.

PROGRAMME: Based on more than 10 years' experience of design innovations involving other agricultural types; full Australian CAA certification in both Normal and Agricultural categories (to CAO 101.16 and 101.22 and FAR Pt 23 at Amendment 23.36) awarded 1 March 1991.

CURRENT VERSIONS: **Standard version:** *As described in detail.* Optional 0.76 m (2 ft 6 in) wingtip extension available for short airfield or high-density operation.

Ag-trainer: Training version, with dual controls, dual rudder pedals and smaller hopper.

DESIGN FEATURES: Braced low-wing design, with large integral hopper forward of cockpit; crash-resistant, corrosion-proofed structure; gap sealed ailerons; detachable wingtips. Wing dihedral 7° from roots. Flaps and ailerons non-handed.

FLYING CONTROLS: Mechanically actuated primary surfaces; single-slotted trailing-edge wing flaps. Interconnect system applies bias to elevator trim spring when flaps extended, to avoid pitch trim changes.

STRUCTURE: Fuselage of welded SAE 4130 chromoly steel tube with removable metal side panels; wings (braced by overwing inverted V strut each side) and tail surfaces conventional all-metal; wingtips detachable.

LANDING GEAR: Non-retractable, with 6 in diameter Cleveland P/N 40-84A mainwheels mounted on tubular steel side Vs; rubber cord shock absorption with hydraulic dampers and Cleveland hydraulic disc brakes. Mainwheel tyres size 8.50 × 6 (6-ply). Scott 3200 steerable/castoring tailwheel, mounted on multi-leaf flat springs.

POWER PLANT: One 194 kW (260 hp) Textron Lycoming O-540-H2A5 flat-six engine, driving a McCauley 1A200/FA 84 52 two-blade fixed-pitch metal propeller, is standard; 186 kW (250 hp) O-540-A1D5 available optionally. Fuel in integral tank in each wing, combined usable capacity 200 litres (53 US gallons; 44 Imp gallons), plus small header tank in upper front fuselage. Oil capacity 11.4 litres (3 US gallons; 2.5 Imp gallons).

Gippsland GA-200 agricultural aircraft

ACCOMMODATION: Two seats side by side (right hand seat for loader/driver); dual controls and second set of rudder pedals for right hand seat in Ag-trainer. Four-point restraint harness(es). Cockpit doors top-hinged, opening upward, with bubble window to improve shoulder room and outward view.
SYSTEMS: 14V 55A automotive alternator and automotive or R-35 aviation battery for electrical power; 50A circuit breaker switch serves as master switch.
AVIONICS: Basic VFR instruments only.
EQUIPMENT: 800 litre (211 US gallon; 176 Imp gallon) capacity hopper in forward fuselage (approx 50 litres; 13.2 US gallons; 11 Imp gallons less in Ag-trainer). Multi-role door/hopper outlet, eliminating need to change outlet when changing between solids and liquids, can also be used for fire bombing or laying of fire retardants. Spreader vanes can be added to increase swath width. 100W landing light in each wingtip; 28V night working lights system (two retractable underwing 600W lights, powered by separate 55A alternator) also available.

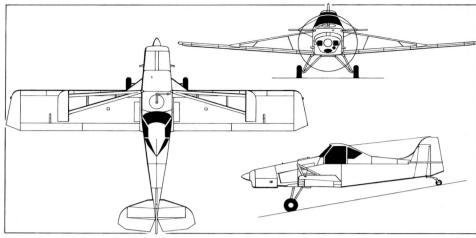

Gippsland GA-200 in standard two-seat form (*Jane's/Mike Keep*)

DIMENSIONS, EXTERNAL:
Wing span, standard	11.17 m (36 ft 7¾ in)
Wing chord, constant	1.60 m (5 ft 3 in)
Wing aspect ratio	6.82
Length overall (flying attitude)	7.48 m (24 ft 6½ in)
Height (static) over cockpit canopy	2.33 m (7 ft 7¾ in)
Tailplane span	2.90 m (9 ft 6¼ in)
Wheel track	2.337 m (7 ft 8 in)
Propeller diameter	2.13 m (7 ft 0 in)

AREAS:
Wings, gross	18.3 m² (197.0 sq ft)

WEIGHTS AND LOADINGS:
Operating weight empty	770 kg (1,698 lb)
T-O weight: certificated	1,315 kg (2,899 lb)
typical agricultural	1,700 kg (3,748 lb)
Max wing loading:	
certificated	71.86 kg/m² (14.72 lb/sq ft)
typical agricultural	92.90 kg/m² (19.03 lb/sq ft)
Max power loading:	
certificated	6.78 kg/kW (11.15 lb/hp)
typical agricultural	8.76 kg/kW (14.41 lb/hp)

PERFORMANCE (at 1,315 kg; 2,899 lb weight except where indicated):
Long-range cruising speed (clean) at 305 m (1,000 ft), ISA	100 knots (185 km/h; 115 mph)
Stalling speed:	
flaps up	54 knots (100 km/h; 62 mph) IAS
flaps down	49 knots (91 km/h; 57 mph) IAS
Stalling speed at typical landing weight:	
flaps up	45 knots (84 km/h; 52 mph) IAS
flaps down	41 knots (76 km/h; 48 mph) IAS
Max rate of climb (clean) at S/L, ISA	295 m (970 ft)/min
T-O run at 1,600 kg (3,527 lb) AUW, 15° flap	340 m (1,115 ft)

Cockpit access and flap detail of the GA-200

HDH
HAWKER DE HAVILLAND LTD (Member company of the BTR Group)
PO Box 30 (361 Milperra Road), Bankstown, NSW 2200
Telephone: 61 (2) 772 8397
Fax: 61 (2) 792 3604
Telex: AA 20719
MANAGING DIRECTOR: J. B. Hattersley
TECHNICAL DIRECTOR: R. H. Jeal
COMMERCIAL DIRECTOR: F. J. Makin

HDH primarily an aerospace and defence company engaged in design, production and support activities for civil and military customers in Australia, USA, UK and more than 20 nations in Asia, the Pacific and Middle East; early 1993 workforce, at 15 locations in Australia, Dubai, Singapore and the USA, totalled about 2,000.

Largest single element of production activity is international airframe subcontracting, including sole source for major subassemblies on Airbus A300/310/320/330/340, Boeing 737/747/757/777 and McDonnell Douglas MD-11/MD-80 airliners. HDH is a major subcontractor on RAAF P-3C ESM and F-111 avionics update programmes. It designs and manufactures the Hawker Pilot Trainer, an FAA Level 2 selectable multi-model device for flight training up to twin-engined business jet level.

Recently completed programmes (see 1992-93 and earlier *Jane's*) have included licence production of Pilatus PC-9/A turboprop trainers (last of 67 for RAAF delivered May 1992), assembly of Sikorsky Black Hawk helicopters, avionics upgrade and fitment of in-flight refuelling equipment on RAAF Boeing 707s, F404 engine production/assembly for RAAF Hornets, and assembly/test of T700 turboshaft engines for Australian Black Hawk programme. HDH has Pilatus approval to market PC-9/A to potential regional customers.

Under risk-sharing agreement of 21 February 1989, HDH is responsible for design and sole source fuselage manufacture of McDonnell Douglas Explorer light helicopter, shipping airframe sections to MDHC for final assembly and engine, transmission and other systems installation.

Australian airframe and engine overhaul facilities include Bankstown, NSW; Fishermen's Bend and Laverton, Victoria; and Perth, WA; customers include Australian defence forces and regional civil and military operators of small/medium-sized fixed-wing aircraft and helicopters. Through its Space Office, HDH is involved in a variety of space-related activities.

HUGHES
HOWARD HUGHES ENGINEERING PTY LTD
PO Box 89, 11 Smith Drive, Ballina, NSW 2478
Telephone: 61 (66) 86 3148
Fax: 61 (66) 86 8343

HUGHES LIGHTWING GR-582
TYPE: Two-seat microlight, ARV and homebuilt.
PROGRAMME: Prototype LightWing made first flight June 1986; many sold conforming to ANO 95-25 requirements; approval sought to sell kits under homebuilt or CAO 101.28 regulations, in non-retractable tailwheel and twin-float configurations.
DESIGN FEATURES: Modified Clark Y wing section. Non-retractable tailwheel landing gear; optional amphibious floats with retractable mainwheels and non-retractable tailwheels (which double as water rudders), or skis.
FLYING CONTROLS: Three-axis control.
STRUCTURE: Strut braced wings of 6061-T6 aluminium alloy and welded steel tube fuselage and tail, all Ceconite fabric covered. Glassfibre engine cowling.
POWER PLANT: One 47 kW (63 hp) Rotax 582 flat-twin engine,

Hughes LightWing GR-582 in twin-float and landplane configurations

with 2.58:1 reduction gear. Fuel capacity 60 litres (15.9 US gallons; 13.2 Imp gallons).

DIMENSIONS, EXTERNAL:
Wing span	9.70 m (31 ft 10 in)
Wing aspect ratio	6.47
Length overall	5.80 m (19 ft 0½ in)
Height overall	1.90 m (6 ft 2¾ in)
Propeller diameter	1.68 m (5 ft 6 in)

AREAS:
Wings, gross	14.55 m² (156.61 sq ft)

WEIGHTS AND LOADINGS:
Weight empty	240 kg (529 lb)
Max pilot weight	100 kg (220 lb)
Baggage capacity	11.3 kg (25 lb)
Max T-O weight	450 kg (992 lb)
Max wing loading	30.93 kg/m² (6.33 lb/sq ft)
Max power loading	9.57 kg/kW (15.75 lb/hp)

PERFORMANCE:
Max level speed at 915 m (3,000 ft)	80 knots (148 km/h; 92 mph)
Econ cruising speed	70 knots (130 km/h; 81 mph)
Stalling speed	32 knots (60 km/h; 37 mph)
Max rate of climb at S/L	366 m (1,200 ft)/min
Service ceiling (CAO 101.55)	915 m (3,000 ft)
T-O run	31 m (100 ft)
Landing run	92 m (300 ft)
Range	150 nm (278 km; 172 miles)
g limits	+6/−3

HUGHES LIGHTWING GR-912
DESIGN FEATURES: Similar to GR-582 but 59 kW (79 hp) watercooled Rotax 912 engine with dual ignition. Empty weight 275 kg (606 lb); max level speed 80 knots (148 km/h; 92 mph).

HUGHES LIGHTWING GA-55
TYPE: Two-seat homebuilt; produced under CAO 101.28 and 101.55 regulations.
PROGRAMME: Available in ready assembled and kit forms; under Australian regulations, can be owner built and registered as ultralight or registered as general aviation aircraft. First flight of prototype March 1987; first flight of kit-built aircraft 1988. Airframe details similar to those for Rotax-engined GR-582.
POWER PLANT: One 58.2 kW (78 hp) Aeropower or Limbach. Fuel capacity 60 litres (15.9 US gallons; 13.2 Imp gallons).

DIMENSIONS, EXTERNAL:
Wing span	9.11 m (29 ft 10½ in)
Length overall	5.70 m (18 ft 8½ in)
Height overall	1.90 m (6 ft 2¾ in)
Propeller diameter	1.50 m (4 ft 11 in)

WEIGHTS AND LOADINGS:
Weight empty	280 kg (617 lb)
Max pilot weight	90 kg (198 lb)
Max T-O weight	500 kg (1,102 lb)
Max power loading	8.59 kg/kW (14.13 lb/hp)

PERFORMANCE:
Max level speed	80 knots (148 km/h; 92 mph)
Max and econ cruising speed	75 knots (139 km/h; 86 mph)
Stalling speed: flaps up	38 knots (71 km/h; 44 mph)
flaps down	34 knots (63 km/h; 39 mph)
negative flaps	40 knots (74 km/h; 46 mph)
Service ceiling	3,050 m (10,000 ft)
T-O run	153 m (500 ft)
Landing run	121 m (400 ft)
Range	217 nm (402 km; 250 miles)
Endurance	3 h 30 min
g limits	+6/−3

JABIRU

JABIRU AIRCRAFT PTY LTD
PO Box 5186, Bundaberg West, Queensland 4670
Telephone: 61 (71) 55 1778
Fax: 61 (71) 52 6131
JOINT MANAGING DIRECTOR: Phil Ainsworth

JABIRU LSA (LIGHT SPORT AIRCRAFT)
TYPE: Two-seat very light aircraft.
PROGRAMME: Deliveries began late 1991.
CUSTOMERS: Twelve built and flown by February 1992; two more then under construction.
COSTS: A$45,000 (1992 flyaway).
DESIGN FEATURES: Unswept wing (braced) and tailplane, swept fin; dorsal and ventral fins; all flying surfaces square-tipped; wings detachable for storage and transportation; designed to Australian CAO 101.55 and British BCAR Section S, with reference to FAR Pt 23 and JAR VLA.
FLYING CONTROLS: Conventional mechanical by push/pull cables, with in-flight adjustable trim; wide span slotted flaps.
STRUCTURE: All-GFRP except metal wing struts and engine mount.
LANDING GEAR: Non-retractable tricycle type, with steerable nosewheel coupled to rudder. Cantilever self-sprung mainwheel legs; speed fairing on each wheel. Hydraulic disc brakes on mainwheels.
POWER PLANT: One 44.7 kW (60 hp) IAME KFM 112 flat-four engine, with electric starting; two-blade fixed-pitch wooden propeller. Fuel capacity 50 litres (13.2 US gallons; 11 Imp gallons).
ACCOMMODATION: Two seats side by side. Door on each side. Dual controls and central control stick standard. Cabin heated and ventilated.
AVIONICS: Standard VFR instrumentation and intercom; VHF antenna (inside fin) standard.

DIMENSIONS, EXTERNAL:
Wing span	8.03 m (26 ft 4 in)
Wing aspect ratio	8.16
Length overall	5.00 m (16 ft 5 in)
Height overall	2.01 m (6 ft 7 in)
Propeller diameter	1.37 m (4 ft 6 in)

AREAS:
Wings, gross	7.90 m² (85.0 sq ft)

WEIGHTS AND LOADINGS:
Weight empty	225 kg (497 lb)
Max T-O weight	429 kg (946 lb)
Max wing loading	54.34 kg/m² (11.13 lb/sq ft)
Max power loading	9.60 kg/kW (15.77 lb/hp)

PERFORMANCE (at max T-O weight):
Never-exceed speed (V_NE)	116 knots (215 km/h; 133 mph) IAS
Max level speed	98 knots (181 km/h; 113 mph) IAS
Max cruising speed (75% power)	90 knots (167 km/h; 103 mph) IAS
Stalling speed: flaps up	45 knots (84 km/h; 52 mph) IAS
flaps down	40 knots (75 km/h; 47 mph) IAS
Max rate of climb at S/L	244 m (800 ft)/min
Service ceiling	3,050 m (10,000 ft)
T-O run	275 m (903 ft)
Landing run	160 m (525 ft)
Max range	400 nm (741 km; 460 miles)
Max endurance	4 h
g limits (ultimate)	+6.6/−3.3

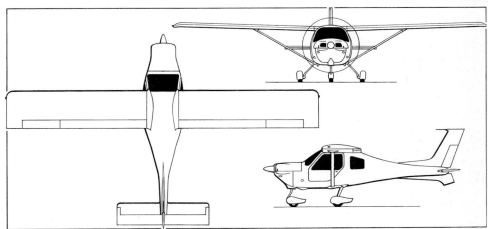

Three-view drawing (*Jane's/Mike Keep*) and photograph of the Jabiru LSA side by side two-seat light aircraft

SADLEIR

VTOL INDUSTRIES AUSTRALIA LTD
8th Floor, 160 St George's Terrace, Perth, WA 6000
Telephone: 61 (619) 481 0524
Fax: 61 (619) 481 0287
Telex: ATEPS AA 94635
MANAGING DIRECTOR: Kim V. Sadleir

SADLEIR STOVL PROTOTYPE
TYPE: Two-seat air bearing fan proof-of-concept (POC) demonstrator.
PROGRAMME: Stems from Sadleir air bearing fan (ABF) concept (last described in 1989-90 *Jane's*); modified design re-launched late 1992 for expected four-year development period. Ground test rig model has shown encouraging results, with ground effect from lift-fan and torque effect on pusher fan described as minimal.
DESIGN FEATURES: POC demonstrator to be of basically delta-winged configuration, with large-diameter fixed-pitch ABF mounted centrally and vertically, inside overall wing cross-section, upstream of single turboprop engine which also drives shrouded pusher propeller; ABF inlet doors close after transition to forward flight to form part of wing upper surface.
FLYING CONTROLS: Conventional stick and rudder controls for three-axis forward flight (elevons and rudder operating in propeller slipstream) and hover; thrust vectoring of fan airflow via four 'rotary valves' for transition between vertical and forward flight modes, lift dumping and primary control in hover; ABF de-clutched on reaching minimum forward flight speed, to increase engine power drive to propeller.
POWER PLANT: One Pratt & Whitney PT6A turboprop, installed just aft of ABF and driving both ABF and shrouded pusher propeller.

WEIGHTS AND LOADINGS:
Max T-O weight	1,300-1,500 kg (2,866-3,307 lb)

SEABIRD
SEABIRD AVIATION AUSTRALIA PTY LTD
Hervey Bay Airport, PO Box 618, Pialba, Queensland 4655
Telephone: 61 (71) 25 3144
Fax: 61 (71) 25 3123
MANAGING DIRECTOR: Donald C. Adams
CHIEF DESIGN ENGINEER: C. W. Whitney

SEABIRD SB7L SEEKER
TYPE: Two-seat lightplane for observation and reconnaissance.
PROGRAMME: Design began January 1985, construction January 1988; first flight of first (SB5N) prototype (VH-SBI, c/n 001, with Norton rotary engine) 1 October 1989, and of SB5E second prototype (VH-SBU, c/n 003, with Emdair engine) 11 January 1991 (c/n 002 was structural test airframe); fourth aircraft (c/n 004, VH-ZIG) is SB7L Seeker prototype, making first flight 6 June 1991; certificated to FAR Pt 23; version with Textron Lycoming O-360-L engine for higher performance flown early 1993 and expected to become preferred model.
DESIGN FEATURES: Braced high-wing monoplane with pod and boom fuselage, extensively glazed cabin and slightly sweptback vertical tail. Primary applications are observation/reconnaissance, offering helicopter-like view from cockpit and good low-speed handling and loiter capabilities; training and agricultural use also foreseen. Wing section NACA 63_2-215 (constant); wedge tips; twist 3°; incidence 6°; dihedral 2° 30′ from roots. Ventral and auxiliary fins added early 1993.
FLYING CONTROLS: Mechanically (rod) actuated slotted ailerons, one-piece horn balanced elevator (with trim tab) and horn balanced rudder; slotted flaps on wing trailing-edge; fixed incidence tailplane. Inboard stall strips; vortex generators on wing upper surface near leading-edge and lower surface forward of ailerons; elevator bias trim.
STRUCTURE: Front fuselage mainly of 4130 chromoly steel tube with Kevlar non-load-bearing skin; aluminium alloy tubular tailboom; wings and tail unit conventional aluminium alloy stressed skin structures, former with single bracing strut and jury strut each side.
LANDING GEAR: Non-retractable, with Cleveland 8 in mainwheels and fairings on cantilever spring steel legs; fully castoring Maule pneumatic tailwheel with oil/nitrogen oleo strut and 210 × 65 mm McCreary tyre. Mainwheels have McCreary 500 × 150 × 150 mm tyres (pressure 2.90 bars; 42 lb/sq in) and Cleveland disc brakes. Min ground turning radius 4.60 m (15 ft 1 in). Float gear to be developed.
POWER PLANT: One 86.5 kW (116 hp) Textron Lycoming O-235 flat-four engine, driving an MT three-blade variable-pitch pusher propeller. Fuel in two 96 litre (25.4 US gallon; 21.1 Imp gallon) integral wing tanks, each with overwing gravity filling point; total fuel capacity 192 litres (50.7 US gallons; 42.2 Imp gallons). Provision for auxiliary fuel tanks on underwing hardpoints. Oil capacity 7 litres (1.85 US gallons; 1.54 Imp gallons).
ACCOMMODATION: Side by side seats, adjustable fore and aft, for pilot and co-pilot or observer/passenger in extensively glazed cabin. Split (upward/downward hinged) door each side, both removable. Ram air cabin ventilation. Space for 22.7 kg (50 lb) of baggage aft of seats.
SYSTEMS: 12V electrical system, with 70A alternator and 18Ah Gill G-25 battery, for engine start, instruments and lighting; 24V system optional.
AVIONICS: Conventional instrumentation; dual VHF com, VOR, ADF, GPS and transponder mode C.
EQUIPMENT: Hardpoint for 60 kg (132 lb) of external stores beneath each wing. Quick-change photo/survey modules, stretcher or 100 litre (26.4 US gallon; 22 Imp gallon) spraytank optional in place of right hand seat.

DIMENSIONS, EXTERNAL:
Wing span	10.60 m (34 ft 9¼ in)
Wing chord, constant	1.22 m (4 ft 0 in)
Wing aspect ratio	8.8
Length overall	6.75 m (22 ft 1¾ in)
Fuselage: Length	6.45 m (21 ft 2 in)
Max width	1.22 m (4 ft 0 in)
Height overall	2.40 m (7 ft 10½ in)
Tailplane span	2.93 m (9 ft 7¼ in)
Wheel track	1.83 m (6 ft 0 in)
Wheelbase	4.50 m (14 ft 9 in)
Propeller diameter	1.66 m (5 ft 5½ in)
Propeller ground clearance	0.80 m (2 ft 7½ in)
Cabin doors (two, each): Height	0.83 m (2 ft 8¾ in)
Width	0.97 m (3 ft 2½ in)
Height to sill	0.97 m (3 ft 2½ in)

DIMENSIONS, INTERNAL:
Cabin, incl baggage space:	
Length	2.12 m (6 ft 11½ in)
Max width	1.15 m (3 ft 9¼ in)
Max height	1.09 m (3 ft 7 in)
Floor area	1.60 m² (17.22 sq ft)
Volume	1.70 m³ (60.0 cu ft)
Baggage space aft of seats	0.45 m³ (15.89 cu ft)

AREAS:
Wings, gross	12.77 m² (137.46 sq ft)
Ailerons (total)	1.22 m² (13.13 sq ft)
Trailing-edge flaps (total)	1.23 m² (13.24 sq ft)
Fin	0.91 m² (9.80 sq ft)
Rudder	0.61 m² (6.57 sq ft)
Tailplane	1.21 m² (13.02 sq ft)
Elevator, incl tab	0.90 m² (9.69 sq ft)

WEIGHTS AND LOADINGS:
Operating weight empty	550 kg (1,213 lb)
Max fuel weight	136 kg (300 lb)
Max T-O and landing weight	800 kg (1,763 lb)
Max wing loading	62.65 kg/m² (12.83 lb/sq ft)
Max power loading	9.25 kg/kW (15.20 lb/hp)

PERFORMANCE (at max T-O weight):
Never-exceed speed at S/L (V$_{NE}$)	144 knots (266 km/h; 165 mph)
Max level speed at 1,220 m (4,000 ft)	103 knots (191 km/h; 119 mph)
Max cruising speed at 1,525 m (5,000 ft)	95 knots (176 km/h; 109 mph)
Econ cruising speed at 1,525 m (5,000 ft)	85 knots (158 km/h; 98 mph)
Stalling speed at 1,220 m (4,000 ft), flaps down, power off	49 knots (91 km/h; 57 mph)
Max rate of climb at S/L	230 m (750 ft)/min
Service ceiling	6,100 m (20,000 ft)
T-O and landing run	95 m (312 ft)
T-O to 15 m (50 ft)	120 m (394 ft)
Range, 45 min reserves:	
max internal fuel	446 nm (826 km; 513 miles)
max internal and external fuel	780 nm (1,445 km; 898 miles)

Seabird SB7L Seeker prototype observation and reconnaissance aircraft before modification

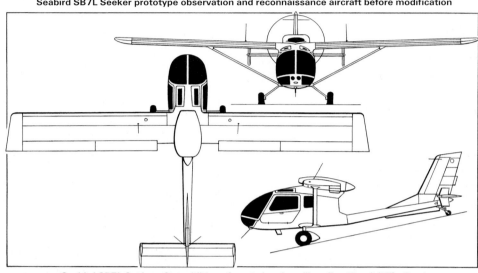

Seabird SB7L Seeker after addition of ventral and auxiliary fins (*Jane's/Mike Keep*)

SKYFOX
SKYFOX AVIATION
PO Box 910, Caloundra, Queensland 4551
Telephone: 61 (74) 91 5355
Fax: 61 (74) 91 8237
SALES MANAGER: Malcolm Fenton

SKYFOX AVIATION SKYFOX
TYPE: Two-seat cabin ARV; certificated under CAA CAO 101.55 regulations.
PROGRAMME: Skyfox first flight 6 September 1989 as modified version of Denney (now SkyStar) Kitfox, with reported 38 changes. First flight of production Skyfox 22 June 1990; 41 built by 30 January 1992.
COSTS: A$44,900 (1992), plus options that include UHF radio (720-channel VHF standard), intercom, cabin heater and deluxe interior.
DESIGN FEATURES: Similar to SkyStar Kitfox but with V

Skyfox Aviation Skyfox two-seat ARV

bracing struts attached further along wings, increased wing dihedral, and possibly larger tail surfaces. Wing folding without controls disconnection.
STRUCTURE: Similar to Kitfox except for larger size fuselage tubing and foam core/aluminium alloy spar/aluminium alloy skin ailerons.
POWER PLANT: One 60.4 kW (81 hp) Rotax 912 flat-four engine, with 2.273:1 reduction gear. Fuel capacity 52 litres (13.7 US gallons; 11.4 Imp gallons). No auxiliary tanks.
DIMENSIONS, EXTERNAL:
Wing span 9.52 m (31 ft 3 in)
Wing aspect ratio 7.82
Length overall 5.60 m (18 ft 4½ in)
Height overall 1.87 m (6 ft 1½ in)
Propeller diameter 1.73 m (5 ft 8 in)
AREAS:
Wings, gross 11.60 m^2 (124.86 sq ft)
WEIGHTS AND LOADINGS:
Weight empty 274 kg (604 lb)
Pilot weight range 55-139 kg (122-306 lb)
Baggage 10 kg (22 lb)
Max T-O weight 450 kg (992 lb)
Max wing loading 38.79 kg/m^2 (7.94 lb/sq ft)
Max power loading 7.45 kg/kW (12.25 lb/hp)
PERFORMANCE:
Max level speed at S/L 85 knots (157 km/h; 98 mph)
Max cruising speed at S/L 75 knots (139 km/h; 86 mph)
Econ cruising speed at S/L 70 knots (130 km/h; 81 mph)
Stalling speed 40 knots (75 km/h; 46 mph)
Max rate of climb at S/L 274 m (900 ft)/min
Service ceiling 3,050 m (10,000 ft)
T-O and landing run 122 m (400 ft)
Range 280 nm (518 km; 322 miles)
Endurance 4 h
g limits +6/−4

TRANSAVIA
TRANSFIELD CONSTRUCTION PTY LTD, TRANSAVIA DIVISION
73 Station Road, Seven Hills, NSW 2147
Telephone: 61 (2) 624 4400
Fax: 61 (2) 624 2548
Telex: AA 170300 TRANSAC
AVIATION SALES MANAGER: Neil McDonald
Formed 1964; division of one of Australia's largest construction companies.

TRANSAVIA SKYFARMER T-300A
TYPE: Single/two-seat agricultural aircraft.
PROGRAMME: Original model was PL-12 Airtruk (first flight 22 April 1965, certificated 10 February 1966); details of this and general purpose PL-12-U (first flight December 1970) in many earlier editions of *Jane's*. Standard model from 1978 (first flight July) was renamed Skyfarmer T-300, improved to T-300A in 1981. Production for many years lately has been at very low level, and no updated information received since 1990. See 1992-93 *Jane's* for latest detailed description and illustration.

VEETOL
VEETOL HELICOPTERS PTY LTD
PPO Box 5195C, Newcastle West, NSW 2302
Telephone: 61 (49) 43 5348
CHAIRMAN: Duan A. Phillips
Formed 1991 to carry former VTOL Aircraft Pty Ltd Phillicopter programme into production stage.

VEETOL PHILLICOPTER Mk 1
TYPE: Two-seat light helicopter.
PROGRAMME: Details of prototype (first flight 1971) last given in 1989-90 *Jane's*. Development and flight testing of modified pre-production model continuing in 1993, using existing prototype as testbed.

AUSTRIA

HOAC
HOAC AUSTRIA FLUGZEUGWERK WIENER NEUSTADT GmbH
N. A. Otto-Strasse 5, A-2700 Wiener Neustadt
Telephone: 43 (2622) 26700
Fax: 43 (2622) 26780
DIRECTOR: Christian Dries
PUBLIC RELATIONS: Michael Feinig
HOAC producing Super Dimona motor glider (see Sailplanes section of 1992-93 *Jane's*), and developing DV 20 Katana private and primary training aircraft.

HOAC DV 20 KATANA
TYPE: Two-seat private and primary training aircraft.
PROGRAMME: Revealed following first flight on 16 March 1991 of proof-of-concept LF 2000 (OE-VPX), illustrated in Addenda to 1991-92 *Jane's*; LF 2 Katana prototype (OE-CPU) made first flight December 1991; DV 20 prototype (OE-AKL, first flight November 1992) representative of production aircraft; Austrian certification April 1993; JAR VLA (very light aircraft) certification expected shortly afterwards; five aircraft completed by early May 1993; approx 20 ordered.
COSTS: Approx Sch1 million.
DESIGN FEATURES: All-composites low-wing design with T tail and side by side seats. Wing section modified Wortmann FX-63-137; sweepback 1° on leading-edge; dihedral 4°.
FLYING CONTROLS: Mechanically actuated primary surfaces, with trim tab in elevator; electrically actuated flaps.
STRUCTURE: Entirely of fibre composites.
LANDING GEAR: Non-retractable tricycle type, with cantilever self-sprung steel leg on each unit. Hydraulic disc brakes on mainwheels; streamline speed fairings on all three wheels.
POWER PLANT: One 59.7 kW (80 hp) Rotax 912A flat-four engine; 2.27 reduction gear to two-blade constant-speed propeller. Fuel capacity 80 litres (21.1 US gallons; 17.6 Imp gallons).
ACCOMMODATION: Two seats side by side.
DIMENSIONS, EXTERNAL:
Wing span 11.00 m (36 ft 1 in)
Wing chord: at root 1.252 m (4 ft 1¼ in)
at tip 0.90 m (2 ft 11½ in)
Wing aspect ratio 10.42
Length overall 7.10 m (23 ft 3½ in)
Height overall 2.10 m (6 ft 10¾ in)
Tailplane span 2.40 m (7 ft 10½ in)
Wheel track 1.90 m (6 ft 2¾ in)
Wheelbase 1.72 m (5 ft 7¾ in)

HOAC LF 2 Katana prototype two-seat private aircraft and primary trainer

AREAS:
Wings, gross 11.616 m^2 (125.03 sq ft)
Vertical tail surfaces (total) 1.066 m^2 (11.47 sq ft)
Horizontal tail surfaces (total) 1.536 m^2 (16.53 sq ft)
WEIGHTS AND LOADINGS:
Weight empty 505 kg (1,113 lb)
Fuel weight 60 kg (132 lb)
Max T-O weight 730 kg (1,609 lb)
Max wing loading 62.84 kg/m^2 (12.87 lb/sq ft)
Max power loading 12.23 kg/kW (20.12 lb/hp)
PERFORMANCE (at max T-O weight):
Max level speed 127 knots (235 km/h; 146 mph)
Cruising speed, 75% power 115 knots (214 km/h; 133 mph)
Stalling speed, flaps down 45 knots (83 km/h; 52 mph)
Max rate of climb at S/L 222 m (728 ft)/min
Service ceiling above 4,000 m (13,125 ft)
T-O run 212 m (696 ft)
T-O to 15 m (50 ft) 350 m (1,150 ft)
Range at 75% power, 30 min reserves 412 nm (765 km; 475 miles)
g limits +4.4/−2.2

BELGIUM

EPERVIER
EPERVIER AVIATION SA
Chaussée de Fleurus 179, B-6041 Gosselies
Telephone: 32 (71) 37 23 76
Fax: 32 (71) 37 28 57
MANAGING DIRECTOR: Yves Kinard

EPERVIER AVIATION EPERVIER
TYPE: Side by side two-seat cabin ARV; conforms to JAR VLA.
PROGRAMME: Construction of prototype Epervier began September 1989; first flight September 1990; first marketed in microlight form; production of ARV had been expected to begin in early 1993. Epervier also proposed for observation, with equipment on request.

Following details refer to ARV version:
COSTS: BFr1.5 million assembled.
DESIGN FEATURES: Wing section NACA 23012. Baggage area behind seats.
FLYING CONTROLS: Three-axis control, with ailerons, flaps, elevators and rudder.
STRUCTURE: Strut braced wings with CFRP/epoxy spars, PVC foam/CFRP/epoxy ribs, and GFRP/epoxy skin. Fuselage

of GFRP/CFRP/epoxy construction.
LANDING GEAR: Non-retractable tricycle type, with brakes; optional floats or skis.
POWER PLANT: One 56 kW (75 hp) Limbach L 2000 flat-four engine, driving a Hoffmann two-blade fixed-pitch aluminium propeller. Total fuel capacity in two tanks 60 litres (15.9 US gallons; 13.2 Imp gallons).

DIMENSIONS, EXTERNAL:
Wing span	11.15 m (36 ft 7 in)
Wing aspect ratio	7.10
Length overall	6.82 m (22 ft 4½ in)
Height overall	2.55 m (8 ft 4½ in)

AREAS:
Wings, gross	17.52 m² (188.58 sq ft)

WEIGHTS AND LOADINGS:
Weight empty	450 kg (992 lb)
Baggage capacity	70 kg (154 lb)
Max T-O weight	750 kg (1,653 lb)
Max wing loading	42.81 kg/m² (8.77 lb/sq ft)
Max power loading	13.39 kg/kW (22.04 lb/hp)

PERFORMANCE (estimated):
Cruising speed	98 knots (181 km/h; 112 mph)
Stalling speed	35 knots (65 km/h; 41 mph)
Max rate of climb at 49 knots (90 km/h; 56 mph)	150 m (492 ft)/min
T-O run	70-120 m (230-394 ft)
Landing run	50-100 m (164-329 ft)
g limits	+3.8/−1.9

Epervier two-seat cabin monoplane *(Geoffrey P. Jones)*

PROMAVIA
PROMAVIA SA
Chaussée de Fleurus 181, B-6041 Gosselies-Aéroport
Telephone: 32 (71) 35 08 29
Fax: 32 (71) 35 79 54
PRESIDENT: André L. Delhamende
VICE-PRESIDENT: Philippe Delhamende
MARKETING MANAGER: Joseph Bernas

Promavia formed at Charleroi-Gosselies Airport in mid-1980s by industrialists, investment companies and a bank, to initiate Jet Squalus programme following 1983 completion of a market survey. Financial backing from Belgian government in 1985 contributed towards prototype research and development. Manufacturing responsibility transferred to Canada 1991 with establishment of Promavia International Corporation (PIC) at Saskatoon, Saskatchewan, where production and developments planned to be made; Promavia SA in Belgium responsible for international marketing of PIC products. Promavia required to raise C$26 million to establish Saskatoon plant. ATTA 3000 now under joint development by Promavia and MiG.

The Promavia Jet Squalus first prototype photographed during a visit to Canada *(Neil A. Macdougall)*

PROMAVIA JET SQUALUS F1300
TYPE: Two-seat primary and basic jet trainer.
PROGRAMME: Dott Ing Stelio Frati of Italy commissioned to undertake design and prototype construction; first prototype made first flight 30 April 1987 and had completed approx 550 hours' flying by mid-1992; demonstrated to Canadian Forces April/May 1991; due to be fitted 1993 with uprated TFE109-3 of 7.12 kN (1,600 lb st); second prototype (OO-JET, not yet flown at time of going to press) being modified to airline pilot training configuration; third prototype to have pressurised cockpit. Garrett MoU February 1992 to supply initial 100 TFE109s to PIC.
CURRENT VERSIONS: **Prototypes:** Two so far built (see Programme). *Description applies to first prototype except where indicated.*
  AWS versions: Four 'air ward system' proposed production variants announced 1990 (details in 1991-92 *Jane's*); no subsequent information received.
  ATTA 3000: Tandem-seat derivative; may be proposed to USAF and USN for JPATS requirement; described under Promavia/MiG heading in International section.
DESIGN FEATURES: Jet Squalus (Latin for 'shark') designed to cover all stages of flying training, from elementary to part of advanced syllabus; side by side seating; underwing hardpoints for weapon training/light tactical missions; non-swept wings/tailplane, sweptback fin/rudder.
  Wings have supercritical section (thickness/chord ratio 13 per cent constant); incidence 1° at root, −1° 45′ at tip; dihedral 6° from roots.
FLYING CONTROLS: Primary control surfaces (Frise differential ailerons, elevators and rudder) actuated mechanically; hydraulic actuation for wing trailing-edge flaps and two-piece underfuselage airbrake; ailerons each have servo tab, port elevator electrically operated trim tab; fixed incidence tailplane.
STRUCTURE: Composites for fairings and some non-structural components, otherwise basically metal throughout (semi-monocoque/flush riveted stressed skin); large quick-disconnect panel in lower rear fuselage for rapid engine access/removal.
LANDING GEAR: Retractable tricycle type, with single wheel and oleo-pneumatic shock absorber on each unit. Mainwheels retract inward, nosewheel rearward. Hydraulic actuation, with built-in emergency system. Main gear of trailing-arm type. Nosewheel steerable ±18°. Mainwheels and tyres size 6.00-6, nosewheel 5.00-5.
POWER PLANT: One Garrett TFE109-1 turbofan mounted in rear fuselage, rated initially at 5.92 kN (1,330 lb st). Alternative turbofan under consideration is 8.45 kN (1,900 lb st) Williams-Rolls FJ44. Semi-integral metal fuel tank in centre-fuselage, max usable capacity 720 litres (190 US gallons; 158 Imp gallons). Single gravity refuelling point on top of fuselage, aft of canopy. Electric fuel pump for engine starting and emergency.
ACCOMMODATION: Two persons side by side in air-conditioned cockpit, on Martin-Baker Mk 11 lightweight ejection seats operable at altitudes up to 12,200 m (40,000 ft) and at any speed between 60 and 400 knots (111-741 km/h; 69-461 mph), including ejection through canopy. One-piece framed canopy is hinged at rear and opens upward.
SYSTEMS: Environmental control system for cockpit air-conditioning. Hydraulic system (operating pressure 117 bars; 1,700 lb/sq in) for actuation of airbrake, landing gear and flaps. System incorporates electrically driven oil pump, with two air/oil accumulators (one for normal and one for emergency operation); separate standby system for emergency lowering of landing gear. Electrical system is 28V DC, using an engine driven starter/generator and nickel-cadmium or lead-acid battery. Negretti Aviation oxygen system.
AVIONICS: Include dual Collins Pro Line II EFIS and radio equipment.
ARMAMENT: Four underwing attachment points, each of 150 kg (331 lb) capacity, capable of carrying a variety of weapons or auxiliary fuel tanks.

DIMENSIONS, EXTERNAL:
Wing span	9.04 m (29 ft 8 in)
Wing chord: at root	1.90 m (6 ft 2¾ in)
at tip	1.00 m (3 ft 3¼ in)
mean aerodynamic	1.575 m (5 ft 2 in)
Wing aspect ratio	6.0
Length of fuselage	9.36 m (30 ft 8½ in)
Height overall	3.60 m (11 ft 9¾ in)
Tailplane span	3.80 m (12 ft 5½ in)
Wheel track	3.59 m (11 ft 9¼ in)
Wheelbase	3.58 m (11 ft 9 in)

AREAS:
Wings, gross	13.58 m² (146.17 sq ft)
Ailerons (total)	1.122 m² (12.08 sq ft)
Trailing-edge flaps (total)	1.784 m² (19.20 sq ft)
Fin	1.256 m² (13.52 sq ft)
Rudder	0.782 m² (8.42 sq ft)
Tailplane	2.04 m² (22.00 sq ft)
Elevators (total, incl tab)	1.61 m² (17.33 sq ft)

WEIGHTS AND LOADINGS:
Weight empty: TFE109-1	1,300 kg (2,866 lb)
TFE109-3	1,400 kg (3,086 lb)
Max external stores load	600 kg (1,323 lb)
Max T-O weight:	
TFE109-1 (Aerobatic)	2,000 kg (4,409 lb)
TFE109-1 (Normal), TFE109-3	2,400 kg (5,291 lb)
Max wing loading: TFE109-1 (Aerobatic)	147.27 kg/m² (30.18 lb/sq ft)
TFE109-1 (Normal), TFE109-3	176.73 kg/m² (36.21 lb/sq ft)
Max power loading: TFE109-1 (Aerobatic)	337.75 kg/kN (3.31 lb/lb st)
TFE109-1 (Normal), TFE109-3	337.21 kg/kN (3.30 lb/lb st)

PERFORMANCE (at max T-O weight, TFE109-1 engine):
Never-exceed speed (V_{NE})	Mach 0.70 (345 knots; 638 km/h; 397 mph)
Max level speed at 4,265 m (14,000 ft)	280 knots (519 km/h; 322 mph)
Normal operating speed	260 knots (482 km/h; 299 mph)
Max speed for landing gear extension	150 knots (278 km/h; 173 mph)
Max speed for flap extension (landing position)	130 knots (241 km/h; 150 mph)
Stalling speed, flaps down	67 knots (124 km/h; 77 mph)
Max rate of climb at S/L	762 m (2,500 ft)/min
Service ceiling	11,275 m (37,000 ft)
Max operating ceiling	7,620 m (25,000 ft)
T-O run	335 m (1,100 ft)
T-O to 15 m (50 ft)	656 m (2,150 ft)
Landing from 15 m (50 ft)	671 m (2,200 ft)
Landing run	366 m (1,200 ft)
Ferry range at 6,100 m (20,000 ft), max internal fuel	1,000 nm (1,850 km; 1,150 miles)
g limits	+2.8 sustained, at 3,050 m (10,000 ft)
	+7/−3.5 Aerobatic

PROMAVIA ATTA 3000
This tandem-seat derivative of Jet Squalus now under joint development by Promavia and MiG: see International section.

PROMAVIA ARA 3600
TYPE: Projected single-seat light attack/reconnaissance aircraft.
PROGRAMME: Announced 1989; programme continues, but slowed down by ATTA 3000 work, from which it will use fuselage and most systems.
DESIGN FEATURES: Broadly similar configuration to ATTA 3000, but redesigned wing with different leading/trailing-edge sweep angles.
POWER PLANT: Two 7.12 kN (1,600 lb st) Garrett TFE109-3 or 8.45 kN (1,900 lb st) Williams-Rolls FJ44 turbofans; fuel capacity 757 litres (200 US gallons; 166.5 Imp gallons).
ARMAMENT: Up to 1,000 kg (2,205 lb) of external stores, including two 20 mm or four 7.62 or 12.7 mm gun pods, four 7-tube 70 mm rocket launchers, or up to four Mk 82 or smaller bombs.
DIMENSIONS, EXTERNAL: As for ATTA 3000
WEIGHTS AND LOADINGS: As for ATTA 3000
PERFORMANCE (estimated): As for ATTA 3000 except:
Max rate of climb at S/L	2,440 m (8,000 ft)/min
Service ceiling	13,715 m (45,000 ft)

SABCA
SOCIETE ANONYME BELGE DE CONSTRUCTIONS AERONAUTIQUES
Chaussée de Haecht 1470, B-1130 Brussels
Telephone: 32 (2) 729 55 11
Fax: 32 (2) 216 15 70
Telex: 21 237 SABCA B
CHAIRMAN: J. Groothaert
DIRECTOR/GENERAL MANAGER: J. Detemmerman
DEPUTY DIRECTOR: M. Humblet
MARKETING MANAGERS:
  J. E. Versmessen (Aerospace)
  R. Bhatti (Space Systems)
  P. Pellegrin (Electronic Defence)
OTHER WORKS: Aéroport de Gosselies-Charleroi, B-6041 Gosselies
Telephone: 32 (71) 25 42 11
Fax: 32 (71) 34 42 14
Telex: 51 251 SABGO B

Major aerospace company in Belgium, founded 1920; has taken part in various European aircraft programmes since 1945. Works occupy total area of approx 82,000 m² (882,640 sq ft); average workforce 1,800 in early 1993. Dassault Belgique Aviation (DBA) acquired late 1990; new factory for Sabca-Limburg NV subsidiary, operational from mid-1991, produces components in high grade composites for aircraft and aerospace applications. Sabca belongs to various European industrial consortia; Dassault Aviation (53 per cent) and Fokker (over 40 per cent) have parity holdings.

Haren factory builds major structures (eg wings, nose sections) and other structural components/equipment for Lockheed Fort Worth F-16, Alpha Jet, Mirage III/5/F1, Atlantic 1, Atlantique 2, Northrop F-5, Airbus A330/A340, Fokker 50 and 100, AS 330 Puma, Spacelab, and Ariane launchers; servo controls for F-16 and Ariane launchers also made at Haren.

Gosselies plant assembles and tests F-16s for Belgium, modifies Belgian Air Force and USAFE F-16s, and does final assembly of Agusta A 109BA helicopters for Belgium (including installing/integrating HeliTOW 2 and all other avionics).

Electronic Division produces fire-control systems, thermal imaging (infra-red) equipment, a number of electronics units, and maintains Doppler equipment. Maintenance and overhaul continues of Belgian and other armed forces' military aircraft, their electronic components and accessories, and commercial fixed-wing aircraft and helicopters; current work includes integrating ECM devices in Belgian aircraft.

MIRAGE SAFETY IMPROVEMENT PROGRAMME (MIRSIP)

Sabca is prime contractor for this Belgian Air Force upgrade programme, which involves life-extending modifications and maintenance, fitment of canard control surfaces and pressure refuelling (both Dassault systems), and new weapons delivery/navigation/reconnaissance system (SAGEM Uliss 92 INS, SAGEM Tercor terrain contour matching system with 3D digital map database, Thomson-CSF laser rangefinder, Vinten 3150/2768 colour video recording system and GEC Ferranti HUD), operating via a MIL-STD-1553B databus. (See also *Jane's Civil and Military Aircraft Upgrades*.)

Programme survived early 1993 Belgian government defence cuts; upgraded aircraft due to be redelivered to No. 42 Squadron late 1993.

SONACA
SOCIETE NATIONALE DE CONSTRUCTION AEROSPATIALE SA
Parc Industriel, Route Nationale Cinq, B-6041 Gosselies
Telephone: 32 (71) 25 51 11
Fax: 32 (71) 34 40 35
Telex: 51241
GENERAL MANAGER: M. Harmant
MANUFACTURING DIRECTOR: P. Wacquez
SALES AND MARKETING DIRECTOR: Marcel Devresse
PUBLIC RELATIONS: Claude Loriaux

Formerly Fairey SA (established 1931), Sonaca was incorporated 1 May 1978; 90 per cent of capital is held by public institutions. Built on 25 ha (61.78 acre) site near Charleroi Airport, factory covers 87,661 m² (943,574 sq ft); 1992 workforce was 1,420.

Civil and military aviation manufacturing programmes include co-production of Lockheed Fort Worth F-16 (rear fuselage, vertical fin, dorsal fairing and final mating) and wing leading-edge moving surfaces for Airbus A310/A320/A330/A340; other aircraft for which parts supplied include Eurocopter France and Agusta A 109 helicopters, Dassault Atlantique 2, Lockheed C-130, Saab 340 and Dassault/Dornier Alpha Jet.

Association in large international military and civil programmes has resulted in significant increase in capability to develop and manufacture metallic and composite aerospace structures, including structural elements for Hermès spacecraft; R&D resources include IBM CADAM (2D, 3D) and CATIA (3D) computer aided design systems.

BOSNIA-HERCEGOVINA

SOKO
VAZDUHOPLOVNA INDUSTRIJA SOKO DD
Radoč bb, 88000 Mostar
Telephone: 38 (88) 53749, 21692
Fax: 38 (88) 423049, 423205
Telex: 46322

Soko (Falcon) Aircraft Industry founded 14 October 1950, initially as subcontractor to existing aircraft industry, including Ikarus, to which Soko was informal successor; produced parts for indigenous S-49C, Aero 2, 212 and 213, Soviet-designed Yak-2 and Yak-9, and rebuilt Republic F-84G Thunderjets; prime contractor on Type 522 and G-2A Galeb trainers, plus Jastreb attack aircraft; manufactured Kraguj CO-IN type; Galeb/Jastreb built since 1963 and exported to Zambia and Libya; Galeb 3 developed late 1969, but not built. Soko privatised March 1991, 3,000 of its 3,500 workers having held 20 per cent of $110 million capital.

Claimed independence of Bosnia-Hercegovina from former Yugoslavia recognised by European Community 7 April 1992. Successor state of Yugoslavia transferred many Soko activities to Utva (which see for G-4 Super Galeb and Gazelle); Mostar factory ceased production 1 May 1992; some partly complete Super Galebs abandoned on production line; infrastructure damaged by civil war; possession of Mostar area in dispute.

International collaboration included subcontract work on de Havilland Dash 8 (emergency exits); Embraer EMB-120; Airbus A300/310/330/340 (rudder shell), A320 (cargo compartment) and A330/340 (fuselage panels) – in co-operation with Aerospatiale, SOCEA-SOGERMA, Dornier/DASA and Korean Air; ATR 42 (frames); McDonnell Douglas MD-80 and MD-11 in co-operation with Alenia and IAI; Eurocopter Super Puma (sponsons and rotor blades); Dassault Aviation; and Tupolev Tu-204 (structural elements and assemblies) in association with UAPK. On 29 July 1990, Soko delivered to Aviaexport of Moscow à half-scale Ilyushin Il-114, produced in collaboration with Utva, including electrical, pressure and hydraulic systems, for wind tunnel tests. Efforts continued until 1992 to join international collaborative ventures, including design and manufacture of structural assemblies and components in metals and composites. Mostar plant included 140,000 m² (1,507,000 sq ft) of covered floor area.

Other activities included fabrication of wings and forward fuselage, and assembly, of G-4 Super Galeb; licence manufacture of Eurocopter Gazelle, including some subassemblies for French assembly line; final assembly of Orao attack aircraft (see Soko/Avioane in International section) and wing production for both assembly lines; manufacture of front fuselage and wings of Utva Lasta; and production of aircraft ground and test equipment, rocket pods and bomb racks.

BRAZIL

EMBRAER
EMPRESA BRASILEIRA DE AERONÁUTICA SA
Caixa Postal 343, 12227-901 São José dos Campos, SP
Telephone: 55 (123) 21 8842 or 25 1529
Fax: 55 (123) 21 5339 or 8466
Telex: 1233589 EBAE BR
CHAIRMAN: Adyr da Silva
PRESIDENT AND CEO: Ozires Silva
COMMERCIAL DIRECTOR: Michel Cury
TECHNICAL DIRECTOR: Horácio Aragonés Forjaz
PRODUCTION DIRECTOR: Juarez de Siqueira Britto Wanderley
PRESS MANAGER, MARKETING DEPARTMENT:
  Mário B. de M. Vinagre

Created 19 August 1969, Embraer began operating on 2 January 1970. It has a 300,000 m² (3,229,170 sq ft) factory area, and had built more than 4,600 aircraft by 1 January 1993, when workforce was 5,800. The 4,500th aircraft to be delivered (Brasilia c/n 120204) was handed over on 27 October 1992. Neiva (which see) became a subsidiary March 1980.

Principal current own-design manufacturing programmes are EMB-120 Brasilia commuter transport, AMX tactical fighter (jointly with Alenia and Aermacchi of Italy) and EMB-312 Tucano turboprop military trainer; under development are CBA-123 Vector commuter transport (jointly with FMA of Argentina) and EMB-145 regional transport. Subsidiary Neiva manufactures EMB-201A Ipanema agricultural aircraft and, under licence from Piper Aircraft Corporation, PA-32-301 Saratoga and PA-34-220T Seneca III; production

Embraer delivered its 4,500th aircraft, an EMB-120 Brasilia, on 27 October 1992, two months after its 23rd anniversary

BRAZIL: Aircraft—EMBRAER

and marketing of Saratoga/Seneca co-ordinated with licence manufacture and sale of single-engined Piper types by Chincul of Argentina (which see).

In subcontract field, Embraer delivered 100th shipset of MD-11 outboard flaps to McDonnell Douglas, and received confirmation for third block (40 shipsets) against order placed 1987 for 200 sets, plus further 100 on option. Development continuing of tooling and production of wingtips and vertical fin fairings for Boeing 777 under 1991 contract, with first deliveries scheduled for mid-1993.

Aircraft deliveries during 1992 totalled 78 (two Bandeirantes, 22 Brasilias, 16 Tucanos, nine AMXs, 17 Ipanemas and 12 assorted light aircraft).

Brazilian government plans to privatise Embraer, retaining 20 per cent holding but offering 10 per cent of voting stock to employees, 30 per cent to local investors and 40 per cent to foreign investors. Privatisation expected to be completed during 1993.

AMX

Details of this military aircraft programme with Alenia and Aermacchi appear in International section. Over 100 now delivered. Embraer delivered nine (including first two two-seaters) to Brazilian Air Force, and 15 shipsets to Italy, during 1992.

EMBRAER/FMA CBA-123 VECTOR

Details of this twin-turboprop airliner programme with FMA of Argentina are in International section. Activities continued at slow pace during 1992; go-ahead decision still depends on Brazilian and Argentine government confirmation of respective air force orders for 40 and 20.

EMBRAER EMB-110 BANDEIRANTE

TYPE: Twin-turboprop general purpose transport.
PROGRAMME: See 1989-90 *Jane's* for history and last full description. Embraer reported March 1993 to be considering restarting production.
CUSTOMERS: Two previously unsold EMB-110P1As delivered to Colombian Air Force late 1992; three sold to Peruvian government early 1993.

EMBRAER EMB-111

Brazilian Air Force designation: P-95
TYPE: Twin-turboprop maritime surveillance aircraft.
PROGRAMME: First flight 15 August 1977; initial deliveries to Brazilian Air Force and Chilean Navy 1978; one to Gabon August 1981; follow-on Brazilian order (for P-95B) December 1987, deliveries from October 1989 to September 1991; earlier Brazilian P-95s to be brought up to P-95B standard.
CURRENT VERSIONS: **P-95**: Initial version; full description in 1985-86 and earlier *Jane's*.
P-95B: As P-95, but with structural improvements and upgraded avionics. Details in 1992-93 and earlier *Jane's*.
CUSTOMERS: Brazilian Air Force (12 P-95 and 10 P-95B); Chilean Navy (six); Gabonese Air Force (one). Brazilian operators are 1st and 2nd Squadrons of 7th Aviation Group (GAv 1°/7° and 2°/7°).

EMBRAER EMB-120 BRASILIA

Brazilian Air Force designation: VC-97
TYPE: Twin-turboprop passenger and cargo transport.
PROGRAMME: Design began September 1979; three flying prototypes, two static/fatigue test aircraft and one pre-series demonstrator built (first flight, by PT-ZBA, 27 July 1983); Brazilian CTA certification with original PW115 engines on 10 May 1985 followed by FAA (FAR Pt 25) type approval 9 July 1985, British/French/German approval 1986 and Australian in April 1990; deliveries began June 1985 (to Atlantic Southeast Airlines, USA, entering service that October); first order for corporate version (United Technologies Corporation, USA) received August 1986, delivered following month; from October 1986 (c/n 120028), all Brasilias delivered incorporate composite materials equivalent to 10 per cent of aircraft basic empty weight; also since late 1986 has been available in hot and high version (certificated 26 August 1986) with PW118A engines which maintain max output up to ISA + 15°C at S/L (first customer Skywest of USA); on 4 January 1989, first prototype began flight trials with Garrett TPE331-12B turboprop on port side of rear fuselage, as testbed for engine installation of Embraer/FMA CBA-123 (see International section); 200th Brasilia delivered 20 August 1990; one million flying hours completed January 1991; extended-range EMB-120ER announced June 1991, certificated by CTA February 1992; production rate three to four per month in 1992; two million flying hours and over 41 million passengers by January 1993.
CURRENT VERSIONS: **EMB-120RT**: Standard civil version, certificated to FAR 25 (up to Amendment 54), JAR 25 (up to Amendment 8), FAR 36 (up to Amendment 12); operational characteristics in accordance with FAR 91 and 135. *Detailed description applies to this version, except where indicated.*
EMB-120ER: Extended range version, obtained by increasing max T-O weight without major structural change; this also allows for increased standard passenger-plus-baggage weight (97.5 kg; 215 lb per person instead of 91 kg; 200 lb) and obligatory fitting of TCAS and flight data recorder. Existing EMB-120s can be retrofitted to ER standard (first two customers for retrofit: Delta Air Transport and Luxair).
Improved Brasilia: Announced 1992 for availability in 1994: based on EMB-120ER. Changes under consideration include interchangeable leading-edges on all flying surfaces; deletion of fin de-icing boot; improved door seals and redesigned interior panel joints to reduce cabin noise; new and more comfortable reclining passenger seats; new pilot seats; redesigned crew door similar to that of CBA-123; new-design overhead bins with wider doors; new passenger cabin lighting; redesigned cockpit ventilation; redesigned windscreen fairing to reduce condensation; new flight deck sun visors; new autopilot with automatic rudder trim; and main cargo capacity increased by 700 kg (1,543 lb). See also Addenda.
VC-97: VIP transport model for Brazilian Air Force.
CUSTOMERS: Total of 323 ordered, of which 266 delivered (incl four VC-97s) by 10 June 1993; further 150 then on option. Operators include airlines in Aruba (Air Aruba), Australia (Flight West), Belgium (Delta Air Transport), Bophuthatswana (Bop Air), Brazil (Air Force, Nordeste, Pantanal, Total and Rio-Sul), Canada (Ontario Express), Cape Verde (TA de Cabo Verde), France (Air Exel and Air Littoral), Luxembourg (Luxair), Norway (Norsk Air), Panama (Talais), the UK (CSE Aviation and Esquel) and the USA (ASA, Air Midwest, Comair, Midway, SkyWest, Texas Air/Britt Airways and WestAir).
DESIGN FEATURES: Low mounted unswept wings, circular-section pressurised fuselage, all-sweptback T tail unit. Wing section NACA 23018 (modified) at root, NACA

One of the two Bandeirante military transports delivered to Colombia in 1992

EMB-120 Brasilia operated as a VC-97 VIP transport by the Grupo de Transporte Especial (GTE) of the Brazilian Air Force

Embraer EMB-120 Brasilia in the insignia of Air Midwest's Trans World Express network

23012 at tip; incidence 2°; at 66 per cent chord, wings have 6° 30′ dihedral. Fixed incidence tailplane.

FLYING CONTROLS: Internally balanced ailerons, and horn balanced elevators, actuated mechanically (ailerons by dual irreversible actuators); serially hinged two-segment rudder actuated hydraulically by Bertea CSD unit; trim tabs in ailerons (one port, two starboard) and each elevator. Hydraulically actuated, electrically controlled, double-slotted Fowler trailing-edge flap inboard and outboard of each engine nacelle, with small plain flap beneath each nacelle; no slats, slots, spoilers or airbrakes. Small fence on each outer wing between outer flap and aileron; twin ventral strakes under rear fuselage.

STRUCTURE: Kevlar reinforced glassfibre for wing and tailplane leading-edges and tips, wingroot fairings, dorsal fin, fuselage nosecone and (when no APU fitted) tailcone; Fowler flaps of carbonfibre. Remainder conventional semi-monocoque/stressed skin structure of 2024/7050/7475 aluminium alloys with chemically milled skins. Fuselage pressurised between flat bulkhead forward of flight deck and hemispherical bulkhead aft of baggage compartment, and meets damage tolerance requirements of FAR Pt 25 (Transport category) up to Amendment 25-54. Wing is single continuous three-spar fail-safe structure, attached to underfuselage frames; tail surfaces also three-spar.

LANDING GEAR: Retractable tricycle type, with Goodrich twin wheels and oleo-pneumatic shock absorber on each unit (main units 12 in, nose unit 8 in). Hydraulic actuation; all units retract forward (main units into engine nacelles). Hydraulically powered nosewheel steering. Goodyear tyres, size 24 × 7.25 in (main), 18 × 5.5 in (nose); pressure 6.90-7.58 bars (100-110 lb/sq in) on main units, 4.14-4.83 bars (60-70 lb/sq in) on nose unit. Goodrich carbon brakes standard (steel optional). Hydro Aire anti-skid system standard; autobrake optional. Min ground turning radius 15.76 m (51 ft 8½ in).

POWER PLANT: Two Pratt & Whitney Canada PW118 or PW118A turboprops, each rated at 1,342 kW (1,800 shp) for T-O and max continuous power, and driving a Hamilton Standard 14RF-9 four-blade constant-speed reversible-pitch autofeathering propeller with glassfibre blades containing aluminium spars. Fuel in two-cell 1,670 litre (441 US gallon; 367.2 Imp gallon) integral tank in each wing; total capacity 3,340 litres (882 US gallons; 734.4 Imp gallons), of which 3,308 litres (874 US gallons; 728 Imp gallons) are usable. Single-point pressure refuelling (beneath outer starboard wing), plus gravity point in upper surface of each wing. Oil capacity 9 litres (2.4 US gallons; 2 Imp gallons).

ACCOMMODATION: Two-pilot flight deck. Main cabin accommodates cabin attendant and 30 passengers in three-abreast seating at 79 cm (31 in) pitch, with overhead lockable baggage racks, in pressurised and air-conditioned environment. Passenger seats of carbonfibre and Kevlar, floor and partitions of carbonfibre and Nomex sandwich, side panels and ceiling of glassfibre/Kevlar/Nomex/carbonfibre sandwich. Provisions for wardrobe, galley and toilet. Quick-change cabin attendant available optionally (first customer, Total Linhas Aéreas in 1993), for 30 passengers or 3,500 kg (7,716 lb) of cargo. Downward opening main passenger door, with airstairs, forward of wing on port side. Type II emergency exit on starboard side at rear. Overwing Type III emergency exit on each side. Pressurised baggage compartment aft of passenger cabin, with large door on port side. Also available with all-cargo interior; executive or military transport interior; or in mixed-traffic version with 24 or 26 passengers (toilet omitted in latter case), and 900 kg (1,984 lb) of cargo in enlarged rear baggage compartment.

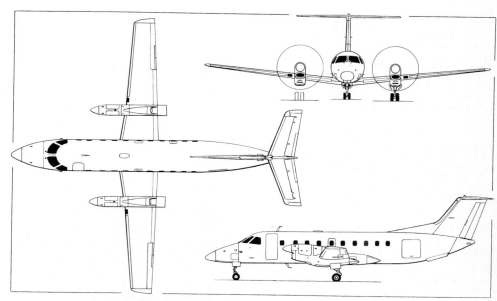

Embraer EMB-120 Brasilia twin-turboprop transport (Dennis Punnett)

SYSTEMS: AiResearch air-conditioning/pressurisation system (differential 0.48 bar; 7 lb/sq in), with dual packs of recirculation equipment. Duplicated hydraulic systems (pressure 207 bars; 3,000 lb/sq in), each powered by an engine driven pump, for landing gear, flap, rudder and brake actuation, and nosewheel steering. Emergency standby electric pumps on each system, plus single standby handpump, for landing gear extension. Main electrical power supplied by two 28V 400A DC starter/generators; two 28V 100A DC auxiliary brushless generators for secondary and/or emergency power; one 24V 40Ah nickel-cadmium battery for assisted starting and emergency power. Main and standby 450VA static inverters for 26/115V AC power at 400Hz. Single high-pressure (127.5 bars; 1,850 lb/sq in) oxygen cylinder for crew; individual chemical oxygen generators for passengers. Pneumatic de-icing for wing and tail leading-edges, and engine air intakes; electrically heated windscreens, propellers and pitot tubes; bleed air de-icing of engine air intakes. Optional Garrett GTCP36-150(A) APU in tailcone, for electrical and pneumatic power supply.

AVIONICS: Collins Pro Line II digital avionics package includes as standard dual VHF-22 com transceivers, dual VIR-32 VHF nav receivers, one ADF-60A, one TDR-90 transponder, CLT-22/32/62/92 control heads, one DME-41, one WXR-270 weather radar, dual AHRS-85 digital strapdown AHRS, dual ADI-84, dual EHSI-74, dual RMI-36, one Dorne & Margolin DMELT-81 emergency locator transmitter, dual Avtech audio/interphones, Avtech PA and cabin interphone, Fairchild voice recorder, and J. E. T. standby attitude indicator. Optional avionics include third VHF com, second transponder and DME, WXR-300 weather radar, two EFIS-86 electronic flight instrument systems, one MFD-85 multi-function display, one or two J. E. T. RNS-8000 3D or Racal Avionics RN 5000 nav, one APS-65 digital autopilot, one or two FCS-65 digital flight directors, flight entertainment music, one or two Canadian Marconi CMA-771 Alpha VLF/Omega, one or two ALT-55 radio altimeters, altitude alerter/preselect, MLS, GPWS, flight recorder, and Motorola Selcal. Alternative (Bendix/King) package available optionally. Other avionics, for special versions, as required for missions concerned.

DIMENSIONS, EXTERNAL:
Wing span	19.78 m (64 ft 10¾ in)
Wing chord: at root	2.81 m (9 ft 2¾ in)
at tip	1.40 m (4 ft 7 in)
Wing aspect ratio	9.9
Length overall: 120	20.00 m (65 ft 7½ in)
120ER	20.07 m (65 ft 10¼ in)
Fuselage: Length	18.73 m (61 ft 5½ in)
Max diameter	2.28 m (7 ft 5¾ in)
Height overall	6.35 m (20 ft 10 in)
Elevator span	6.94 m (22 ft 9¼ in)
Wheel track (c/l of shock struts)	6.58 m (21 ft 7 in)
Wheelbase	6.97 m (22 ft 10½ in)
Propeller diameter	3.20 m (10 ft 6 in)
Propeller ground clearance (min)	0.52 m (1 ft 8½ in)
Passenger door (fwd, port): Height	1.70 m (5 ft 7 in)
Width	0.774 m (2 ft 6½ in)
Height to sill	1.47 m (4 ft 10 in)
Cargo door (rear, port): Height	1.36 m (4 ft 5½ in)
Width	1.30 m (4 ft 3¼ in)
Height to sill	1.67 m (5 ft 5¾ in)
Emergency exit (rear, stbd): Height	1.37 m (4 ft 6 in)
Width	0.51 m (1 ft 8 in)
Height to sill	1.56 m (5 ft 1½ in)
Emergency exits (overwing, each):	
Height	0.91 m (3 ft 0 in)
Width	0.51 m (1 ft 8 in)
Emergency exits (flight deck side windows, each):	
Min height	0.48 m (1 ft 7 in)
Min width	0.51 m (1 ft 8 in)

DIMENSIONS, INTERNAL:
Cabin, excl flight deck and baggage compartment:
Length 9.38 m (30 ft 9¼ in)

Max width 2.10 m (6 ft 10¾ in)
Max height 1.76 m (5 ft 9¼ in)
Floor area 14.97 m^2 (161.14 sq ft)
Volume 27.40 m^3 (967.6 cu ft)
Rear baggage compartment volume:
30-passenger version 6.40 m^3 (226 cu ft)
all-cargo version 2.70 m^3 (95 cu ft)
passenger/cargo version 11.00 m^3 (388 cu ft)
Cabin, incl flight deck and baggage compartment:
total volume approx 41.8 m^3 (1,476 cu ft)
Max available cabin volume (all-cargo version)
31.10 m^3 (1,098 cu ft)

AREAS:
Wings, gross 39.43 m^2 (424.42 sq ft)
Ailerons (total) 2.88 m^2 (31.00 sq ft)
Trailing-edge flaps (total) 3.23 m^2 (34.77 sq ft)
Fin, incl dorsal fin 5.74 m^2 (61.78 sq ft)
Rudder 2.59 m^2 (27.88 sq ft)
Tailplane 6.10 m^2 (65.66 sq ft)
Elevator, incl tabs 3.90 m^2 (41.98 sq ft)

WEIGHTS AND LOADINGS:
* Weight empty, equipped: 120 7,101 kg (15,655 lb)
120ER 7,140 kg (15,741 lb)
Operating weight empty: 120 7,465 kg (16,457 lb)
120ER 7,500 kg (16,534 lb)
Max usable fuel: 120 2,600 kg (5,732 lb)
120ER 2,660 kg (5,864 lb)
* Max payload: 120 (passenger) 3,031 kg (6,682 lb)
120 (cargo) 3,500 kg (7,716 lb)
120ER 3,223 kg (7,105 lb)
Max T-O weight: 120 11,500 kg (25,353 lb)
120ER 11,990 kg (26,433 lb)
Max ramp weight: 120 11,580 kg (25,529 lb)
120ER 12,080 kg (26,632 lb)
Max landing weight: 120 11,250 kg (24,802 lb)
120ER 11,700 kg (25,794 lb)
Max zero-fuel weight: 120 10,500 kg (23,148 lb)
120ER 10,700 kg (23,589 lb)
Max wing loading: 120 292 kg/m^2 (59.81 lb/sq ft)
120ER 304.1 kg/m^2 (62.28 lb/sq ft)
Max power loading: 120 4.29 kg/kW (7.04 lb/shp)
120ER 4.47 kg/kW (7.34 lb/shp)

* *4 kg (8.8 lb) payload decrease/empty weight increase with PW118A engines*

PERFORMANCE (at max T-O weight, ISA, except where indicated):
Max operating speed:
120, 120ER 272 knots (504 km/h; 313 mph) EAS
Max level speed at 6,100 m (20,000 ft):
120 328 knots (608 km/h; 378 mph)
120ER 327 knots (606 km/h; 377 mph)
Max cruising speed at 7,620 m (25,000 ft):
120, 120ER (PW118) 300 knots (555 km/h; 345 mph)
120 (PW118A) 310 knots (574 km/h; 357 mph)
120ER (PW118A) 313 knots (580 km/h; 360 mph)
Long-range cruising speed at 7,620 m (25,000 ft):
120 260 knots (482 km/h; 299 mph)
120ER 270 knots (500 km/h; 311 mph)
Stalling speed, power off:
flaps up: 120 117 knots (217 km/h; 135 mph) CAS
120ER 120 knots (223 km/h; 138 mph) CAS
flaps down: 120 87 knots (162 km/h; 100 mph) CAS
120ER 89 knots (165 km/h; 103 mph) CAS
Max rate of climb at S/L: 120 646 m (2,120 ft)/min
120ER 762 m (2,500 ft)/min
Rate of climb at S/L, OEI: 120 206 m (675 ft)/min
120ER 168 m (550 ft)/min
Service ceiling: 120 (PW118) 9,150 m (30,000 ft)
120ER (PW118) 8,840 m (29,000 ft)
120, 120ER (PW118A) 9,750 m (32,000 ft)
Service ceiling, OEI, AUW of 11,200 kg (24,690 lb):
120 (PW118) 5,240 m (17,200 ft)
120ER (PW118 or PW118A) 4,600 m (15,100 ft)
120 (PW118A) 5,790 m (19,000 ft)
FAR Pt 25 T-O field length: 120 1,420 m (4,660 ft)
120ER 1,550 m (5,085 ft)
FAR Pt 135 landing field length, max landing weight at S/L: 120 1,370 m (4,495 ft)
120ER 1,390 m (4,560 ft)
Range at 7,620 m (25,000 ft), reserves for 100 nm (185 km; 115 mile) diversion and 45 min hold:
with max (30 passengers) payload (2,721 kg; 6,000 lb):
120 (PW118) 550 nm (1,019 km; 633 miles)
120 (PW118A) 500 nm (926 km; 575 miles)
120ER (PW118) 840 nm (1,556 km; 967 miles)
120ER (PW118A) 810 nm (1,501 km; 932 miles)
with max fuel:
120, 21 passengers (1,920 kg; 4,233 lb payload)
1,610 nm (2,983 km; 1,854 miles)
120ER, 20 passengers (1,850 kg; 4,078 lb payload)
1,510 nm (2,798 km; 1,739 miles)

OPERATIONAL NOISE LEVELS (120 and 120ER, FAR Pt 36, BCAR-N and ICAO Annex 16):
T-O 81.2 EPNdB
Approach 92.3 EPNdB
Sideline 83.5 EPNdB

EMBRAER EMB-145 JETLINER

TYPE: Twin-turbofan regional airliner.

PROGRAMME: Original development plans revealed 12 June 1989, aimed at first flight late 1991 and first deliveries mid-1993, following FAR/JAR Pt 25 and FAR Pt 36 (ICAO Annex 16) international certification; programme delayed by company cutbacks initiated in Autumn 1990. Complete redesign of wing, engine installation and landing gear completed March 1991, new wind tunnel models being tested at CTA and Boeing in following month; first flight then planned for late 1992 and certification for mid/late 1993. Further redesign announced October 1991 in which engines moved from underwing to rear of fuselage; subsonic/transonic wind tunnel tests of final configuration 1992 confirmed or bettered predicted performance; Parker Hannifin (USA) and GAMESA (Spain) announced as programme partners February 1993; Parker Hannifin will provide flight controls, fuel and steering systems, and some hydraulic components; GAMESA will develop and build wings and engine nacelles; first flight now scheduled for early 1995, certification (to FAR/JAR 25, FAR 36, ICAO Annex 16 and FAR 121) a year later.

COSTS: Estimated development costs $300 million; target sale price $13 million.

DESIGN FEATURES: Stretched EMB-120 Brasilia fuselage (with tailcone adapted for rear mounted engine installation), allied to new-design wing with Embraer supercritical section; CBA-123 nose section; T tailplane. Wing sweepback 22° 43′ 48″ at quarter-chord.

FLYING CONTROLS: Ailerons and rudder hydraulically actuated; elevator with automatic and spring tab. In-flight and ground spoilers; two pairs of double-slotted flaps.

STRUCTURE: Fuselage as for Brasilia; two-spar wing with integral fuel tanks, plus auxiliary third spar supporting landing gear; T tail unit with aluminium main boxes; wing and tailplane leading-edges aluminium, fin leading-edge composites sandwich.

LANDING GEAR: Twin-wheel main legs retract inward into wing/fuselage fairings; twin-wheel nose unit retracts forward.

POWER PLANT: Two 31.32 kN (7,040 lb st), FADEC-equipped Allison GMA 3007A turbofans, pylon mounted on rear cone of fuselage. For fuel, see Weights and Loadings.

ACCOMMODATION: Two pilots, flight observer and cabin attendant. Standard accommodation for 50 passengers, three abreast at seat pitch of 79 cm (31 in). Carry-on baggage cabinet and galley at front of cabin; toilet and main baggage compartment at rear of cabin. Cabinet plus overhead bins carry-on baggage capacity 358 kg (789 lb); underseat capacity 450 kg (992 lb); main baggage compartment capacity 1,000 kg (2,204 lb). Deletion of forward baggage cabinet enables two additional passenger seats to be installed (one each at front and rear of three-abreast rows). Outward opening plug-type door, incorporating airstair, identical to that of EMB-120; upward sliding baggage door at rear on port side; sideways opening service door at front on starboard side; inward opening emergency exit above wing on each side. Entire accommodation, including baggage compartments, pressurised and air-conditioned.

SYSTEMS: Cabin pressurisation 0.54 bar (7.8 lb/sq in). Wing and tailplane leading-edges anti-iced by engine bleed air.

DIMENSIONS, EXTERNAL:
Wing span 20.04 m (65 ft 9 in)
Wing aspect ratio 7.86
Length overall 29.87 m (98 ft 0 in)
Fuselage: Length 27.93 m (91 ft 7½ in)
Max diameter 2.28 m (7 ft 5¾ in)
Height overall 6.71 m (22 ft 0¼ in)
Elevator span 7.55 m (24 ft 9¼ in)
Wheel track (c/l of shock struts) 4.10 m (13 ft 5½ in)
Wheelbase 14.45 m (47 ft 5 in)
Passenger door (fwd, port): Height 1.70 m (5 ft 7 in)
Width 0.71 m (2 ft 4 in)
Height to sill (max) 1.69 m (5 ft 6½ in)
Baggage door (rear, port): Height 1.10 m (3 ft 7¼ in)
Width 1.00 m (3 ft 3¼ in)
Height to sill (max) 1.89 m (6 ft 2½ in)
Service door (rear, stbd): Height 1.42 m (4 ft 8 in)
Width 0.62 m (2 ft 0½ in)
Height to sill (max) 1.64 m (5 ft 4½ in)
Emergency exits (overwing, each):
Height 0.92 m (3 ft 0¼ in)
Width 0.51 m (1 ft 8 in)

DIMENSIONS, INTERNAL:
Cabin (excl flight deck, incl toilet):
Length 16.49 m (54 ft 1¼ in)
Max width 2.10 m (6 ft 10¾ in)
Baggage volume:
rear compartment 9.21 m^3 (325.25 cu ft)
front of cabin 0.70 m^3 (24.72 cu ft)
overhead bins 1.50 m^3 (52.97 cu ft)
underseat 2.20 m^3 (77.69 cu ft)

AREAS:
Wings, gross 51.08 m^2 (549.8 sq ft)
Fin 4.98 m^2 (53.60 sq ft)
Rudder 2.22 m^2 (23.90 sq ft)
Tailplane 11.20 m^2 (120.56 sq ft)
Elevators (total, incl tabs) 3.34 m^2 (35.95 sq ft)

WEIGHTS AND LOADINGS:
Basic operating weight empty 11,585 kg (25,540 lb)
Max usable fuel 4,200 kg (9,259 lb)

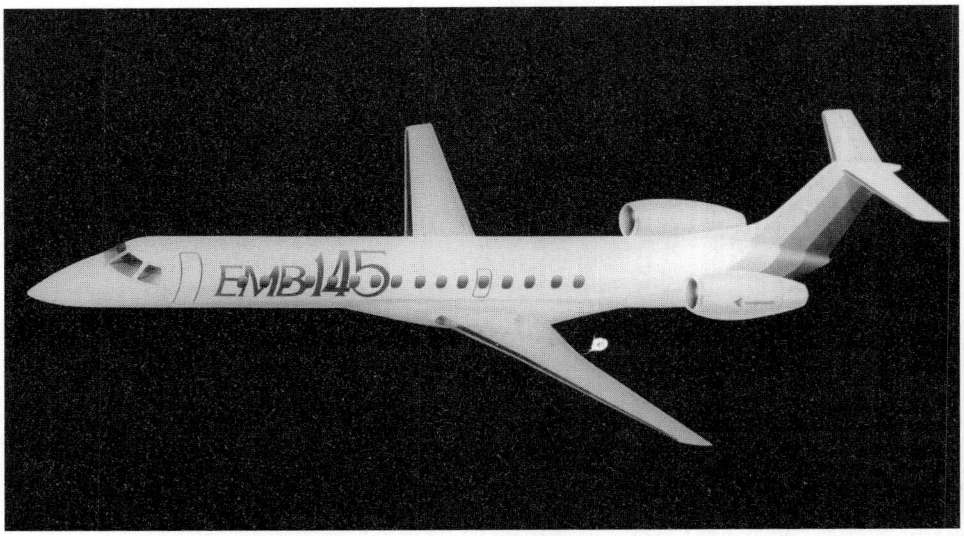

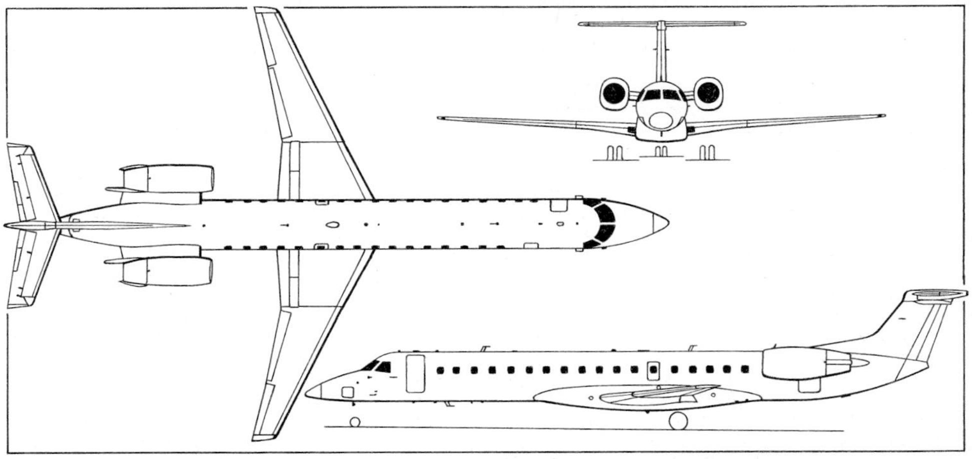

Drawing *(Jane's/Mike Keep)* and model of the EMB-145 rear-engined twin-turbofan regional airliner

Max payload	5,515 kg (12,158 lb)
Max T-O weight	19,200 kg (42,329 lb)
Max ramp weight	19,300 kg (42,549 lb)
Max zero-fuel weight	17,100 kg (37,699 lb)
Max landing weight	18,700 kg (41,226 lb)
Max wing loading	375.9 kg/m^2 (76.99 lb/sq ft)
Max power loading	306.76 kg/kN (3.01 lb/lb st)

PERFORMANCE (estimated, at max T-O weight, ISA, except where indicated):
Max operating speed
  Mach 0.78 (320 knots; 593 km/h; 368 mph CAS)
Max level speed at 9,750 m (32,000 ft)
  443 knots (821 km/h; 510 mph)
Max cruising speed at 11,275 m (37,000 ft)
  430 knots (796 km/h; 495 mph)
Long-range cruising speed at 9,750 m (32,000 ft)
  360 knots (667 km/h; 415 mph)
Stalling speed, power off:
  flaps up   117 knots (217 km/h; 135 mph)
  flaps down   97 knots (180 km/h; 112 mph)
Max rate of climb at S/L   731 m (2,400 ft)/min
Rate of climb at S/L, OEI   207 m (680 ft)/min
Service ceiling   11,275 m (37,000 ft)
Service ceiling, OEI   5,180 m (17,000 ft)
FAR Pt 25 T-O field length   1,530 m (5,020 ft)
FAR Pt 135 landing field length, typical landing weight at S/L   1,500 m (4,922 ft)
Range at 11,275 m (37,000 ft), reserves for 100 nm (185 km; 115 mile) diversion, 10% block fuel and 30 min hold:
  with max (50 passengers) payload (4,536 kg; 10,000 lb)   800 nm (1,482 km; 921 miles)
  with max fuel and 1,814 kg (4,000 lb) payload (20 passengers)   1,390 nm (2,576 km; 1,600 miles)

EMB-312F Tucano tandem two-seat trainer of the French Air Force

EMBRAER EMB-312 TUCANO (TOUCAN)
Brazilian Air Force designation: T-27

TYPE: Turboprop powered basic trainer.
PROGRAMME: Design started January 1978, Ministry of Aeronautics contract being received 6 December that year for two flying prototypes and two static/fatigue test airframes; first prototype (Brazilian Air Force serial number 1300) made first flight 16 August 1980, second (1301) on 10 December 1980; third to fly (PP-ZDK, on 16 August 1982) was to production standard; deliveries to FAB began September 1983. In same month Egypt placed order for itself (40) and Iraq (80), with options on 60 more (incl 20 for Iraq); first 10 of these built by Embraer and delivered from October 1984, remaining 110 (plus further 14 ordered early 1989) delivered to AOI in Egypt (which see) as CKD kits for local assembly; selected for RAF by UK government March 1985 (see under Shorts heading in UK section); last of 118 on original Brazilian Air Force order delivered September 1986; FAB ordered further 10 in January 1990, all of which delivered during that year; French government order for 80 placed in July 1990 came into effect October 1991, with deliveries to take place 1993-97 (first two for delivery July 1993; nine due for delivery by year-end); five more ordered by Brazilian Air Force 1992 (first two delivered December 1992, remainder early 1993); 14 for Colombia also delivered December 1992.
CURRENT VERSIONS: **EMB-312**: Standard Embraer built version. *Detailed description applies to this model.*
  EMB-312F: Version for French Armée de l'Air; increased fatigue life airframe, ventral airbrake, modifications to some systems, and French avionics.
  EMB-312H and Super Tucano: Stretched versions; described separately.
  S312: Modified version, with British equipment and more powerful 820 kW (1,100 shp) Garrett TPE331 engine, built by Shorts for Royal Air Force; described under that company's heading in UK section.
CUSTOMERS: Total firm orders 654 (Argentina 30, Brazil 133, Colombia 14, Egypt 54, France 80, Honduras 12, Iran 25, Iraq 80, Kenya 12 S312, Kuwait 16 S312, Paraguay six, Peru 30, Venezuela 31, UK 131) by January 1993, of which 510 delivered; options then held for further 96 (Brazil 35, Egypt 26, Iraq 20, UK 15). Delivery total includes kits for 25 of those ordered by UK.
DESIGN FEATURES: Meets requirements of FAR Pt 23 Appendix A, and MIL and CAA Section K specifications. Low mounted wings, stepped cockpits in tandem, fully aerobatic. NACA wing sections (63$_2$A-415 at root, 63A-212 at tip); incidence 1° 25′; geometric twist 2° 13′; dihedral 5° 30′ at 30 per cent chord; sweepback 0° 43′ 26″ at quarter-chord. Small fillet forward of tailplane root each side.
FLYING CONTROLS: Primary surfaces internally balanced and actuated mechanically; electrically actuated trim tab in, and small geared tab on, each Frise aileron; electromechanically actuated spring tab in rudder and port elevator. Electrically actuated single-slotted Fowler flaps on wing trailing-edges. Fixed incidence tailplane. Ventral airbrake on aircraft for France.
STRUCTURE: Conventional all-metal construction from 2024 series aluminium alloys; continuous three-spar wing box forms integral fuel tankage. Steel flap tracks. French aircraft strengthened to increase airframe life to 10,000 h.
LANDING GEAR: Hydraulically retractable tricycle type, with single wheel and Piper oleo-pneumatic shock absorber on each unit. Accumulator for emergency extension in the event of hydraulic system failure. Shimmy damper on nose unit. Rearward retracting steerable nose unit; main units retract inward into wings. Parker Hannifin 40-130 mainwheels, Oldi-DI-1.555-02-OL nosewheel. Tyre sizes 6.50-10 (Type III, 8-ply rating) on mainwheels, 5.00-5 (Type III, 6-ply rating) on nosewheel. Tyre pressures (±0.21 bar; 3 lb/sq in in each case) are 5.17 bars (75 lb/sq in) on mainwheels, 4.48 bars (65 lb/sq in) on nosewheel. Parker Hannifin 30-95A hydraulic mainwheel brakes.
POWER PLANT: One 559 kW (750 shp) Pratt & Whitney Canada PT6A-25C turboprop, driving a Hartzell HC-B3TN-3C/T10178-8R three-blade constant-speed fully feathering reversible-pitch propeller. Single-lever combined control for engine throttling and propeller pitch adjustment. Two integral fuel tanks in each wing, total capacity 694 litres (183.3 US gallons; 152.7 Imp gallons). Fuel tanks lined with anti-detonation plastics foam. Gravity refuelling point in each wing upper surface. Fuel system allows nominally for up to 30 s of inverted flight. (Aircraft has been test flown inverted for up to 10 min.) Provision for two underwing ferry fuel tanks, total capacity 660 litres (174.4 US gallons; 145 Imp gallons).
ACCOMMODATION: Instructor and pupil in tandem, on Martin-Baker BR8LC lightweight ejection seats, in air-conditioned cockpit. One-piece fully transparent vacuum formed canopy, opening sideways to starboard, with internal and external jettison provisions. Rear seat elevated 25 cm (9.9 in). Dual controls standard. Baggage compartment in rear fuselage, with access via door on port side. Cockpit heating and canopy demisting by engine bleed air.
SYSTEMS: Freon cycle air-conditioning system, with engine driven compressor. Single hydraulic system, consisting basically of (a) control unit, including reservoir with usable capacity of 1.9 litres (0.5 US gallon; 0.42 Imp gallon); (b) an engine driven pump with nominal pressure of 131 bars (1,900 lb/sq in) and nominal flow rate of 4.6 litres (1.22 US gallons; 1.01 Imp gallons)/min at 3,800 rpm; (c) landing gear and gear door actuators; (d) filter; (e) shutoff valve; and (f) hydraulic fluid to MIL-H-5606. Under normal operation, hydraulic system actuates landing gear extension/retraction and control of gear doors. Landing gear extension can be performed under emergency operation; emergency retraction also possible during landing and T-O with engine running. Reservoir and system are suitable for aerobatics. No pneumatic system. 28V DC electrical power provided by a 6kW starter/generator, 26Ah battery and, for 115V and 26V AC power at 400Hz, a 250VA inverter. Diluter-demand oxygen system conforms to MIL-C-5887 and is supplied individually to each occupant by six MS 21227 D2 type cylinders (total capacity approx 1,200 litres; 317 US gallons; 264 Imp gallons) at a pressure of 31 bars (450 lb/sq in).
AVIONICS: Standard avionics include two Collins VHF-20A transceivers; two Collins 387C-4 audio systems; one Embraer radio transferring system; Telephonics audio control panel; one Collins VIR-31A VOR/ILS/marker beacon receiver; one Collins TDR-90 ATC transponder; one Collins DME-40; one Collins PN-101 gyromagnetic compass; and one Collins ADF-60A. TEAM (Télécommunications Electroniques, Aéronautiques et Maritimes) responsible for Sextant/Thomson-CSF/Rockwell-Collins France/SFIM package for French Air Force Tucanos.
EQUIPMENT: Landing light in each wing leading-edge; taxying lights on nosewheel unit.
ARMAMENT: Two hardpoints under each wing, each stressed for a max load of 250 kg (551 lb). Typical loads, on GB100-20-36B pylons, include two 0.30 in C2 machine-gun pods, each with 500 rds; four 25 lb Mk 76 practice bombs; four 250 lb Mk 81 general purpose bombs; or four

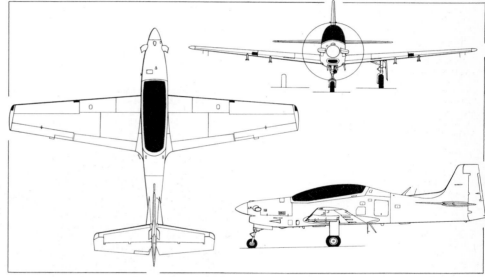

Embraer EMB-312 Tucano basic trainer *(Dennis Punnett)*

BRAZIL: AIRCRAFT—EMBRAER

First seven of 14 Tucano trainers delivered to the Colombian Air Force

LM-37/7A or LM-70/7 launchers, each with seven rockets (Avibras SBAT-37 and SBAT-70 respectively). Fixed reflex-type gunsight.

DIMENSIONS, EXTERNAL:
Wing span	11.14 m (36 ft 6½ in)
Wing chord: at root	2.30 m (7 ft 6½ in)
at tip	1.07 m (3 ft 6⅛ in)
Wing aspect ratio	6.4
Length overall	9.86 m (32 ft 4¼ in)
Fuselage: Length (excl rudder)	8.53 m (27 ft 11⅞ in)
Max width	1.00 m (3 ft 3¼ in)
Max depth	1.55 m (5 ft 1 in)
Height overall (static)	3.40 m (11 ft 1¾ in)
Tailplane span	4.66 m (15 ft 3½ in)
Wheel track	3.76 m (12 ft 4 in)
Wheelbase	3.16 m (10 ft 4½ in)
Propeller diameter	2.36 m (7 ft 9 in)
Propeller ground clearance (static)	0.33 m (1 ft 1 in)
Baggage compartment door:	
Height	0.60 m (1 ft 11½ in)
Width	0.54 m (1 ft 9¼ in)
Height to sill	1.25 m (4 ft 1¼ in)

DIMENSIONS, INTERNAL:
Cockpits: Combined length	2.90 m (9 ft 6⅛ in)
Max height	1.55 m (5 ft 1 in)
Max width	0.85 m (2 ft 9½ in)
Baggage compartment volume	0.17 m³ (6.0 cu ft)

AREAS:
Wings, gross	19.40 m² (208.82 sq ft)
Ailerons (total)	1.97 m² (21.20 sq ft)
Trailing-edge flaps (total)	2.58 m² (27.77 sq ft)
Fin, incl dorsal fin	2.29 m² (24.65 sq ft)
Rudder, incl tab	1.38 m² (14.85 sq ft)
Tailplane, incl fillets	4.77 m² (51.34 sq ft)
Elevators, incl tab	2.00 m² (21.53 sq ft)

WEIGHTS AND LOADINGS:
Basic weight empty	1,810 kg (3,991 lb)
Max internal fuel load (usable)	529 kg (1,166 lb)
Max external stores load	1,000 kg (2,205 lb)
Max T-O weight: clean	2,550 kg (5,622 lb)
with external stores	3,175 kg (7,000 lb)
Max ramp weight	3,195 kg (7,044 lb)
Max landing weight: clean	2,800 kg (6,173 lb)
Max zero-fuel weight	2,050 kg (4,519 lb)
Max wing loading: clean	131.4 kg/m² (26.92 lb/sq ft)
with external stores	163.7 kg/m² (33.52 lb/sq ft)
Max power loading: clean	4.56 kg/kW (7.50 lb/shp)
with external stores	5.68 kg/kW (9.33 lb/shp)

PERFORMANCE (at max clean T-O weight except where indicated):
Never-exceed speed (V_{NE})	280 knots (519 km/h; 322 mph) EAS
Max level speed at 3,050 m (10,000 ft)	242 knots (448 km/h; 278 mph)
Max cruising speed at 3,050 m (10,000 ft)	222 knots (411 km/h; 255 mph)
Econ cruising speed at 3,050 m (10,000 ft)	172 knots (319 km/h; 198 mph)
Stalling speed, power off:	
flaps and landing gear up	72 knots (133 km/h; 83 mph) EAS
flaps and landing gear down	67 knots (124 km/h; 77 mph) EAS
Max rate of climb at S/L	680 m (2,231 ft)/min
Service ceiling	9,150 m (30,000 ft)
T-O run	380 m (1,250 ft)
T-O to 15 m (50 ft)	710 m (2,330 ft)
Landing from 15 m (50 ft)	605 m (1,985 ft)
Landing run	370 m (1,214 ft)

Range at 6,100 m (20,000 ft) with max fuel, 30 min reserves 995 nm (1,844 km; 1,145 miles)
Ferry range at 6,100 m (20,000 ft) with underwing tanks 1,797 nm (3,330 km; 2,069 miles)
Endurance on internal fuel at econ cruising speed at 6,100 m (20,000 ft), 30 min reserves approx 5 h
g limits: fully Aerobatic category, at max clean T-O weight +6/−3
at max T-O weight with external stores +4.4/−2.2

EMBRAER EMB-312H SUPER TUCANO

TYPE: Stretched version of EMB-312.
PROGRAMME: EMB-312H announced at Paris Air Show June 1991; Embraer development aircraft PT-ZTW (c/n 312161, previously used as prototype for Garrett powered Tucano adopted by Royal Air Force) modified as Tucano H proof-of-concept (POC) prototype, making first flight in this form 9 September 1991. Embraer teamed with Northrop May 1992 to bid Super Tucano in US JPATS competition.

CURRENT VERSIONS: **EMB-312H**: Proof-of-concept prototype toured US Air Force/Navy bases August 1992 as preliminary to Super Tucano entry in JPATS competition. Two prototypes (PT-ZTV, first flight 15 May 1993, and 'ZTF, due to fly August 1993) tailored to US JPATS requirements. Will also form basis for developing specific versions for Brazilian Air Force and prospective export customers.

DESIGN FEATURES: Differs from standard EMB-312 mainly in having more powerful engine and plugs totalling 1.37 m (4 ft 6 in) inserted fore and aft of cockpit to accommodate longer engine and retain CG and stability; ventral airbrake; reinforced airframe for higher g loads and longer fatigue life; pressure refuelling; canopy and propeller de-icing; new avionics including GPS and TCAS; and onboard oxygen generating system (OBOGS). Able to cover whole primary and half of advanced pilot training syllabus, and fly precision weapons delivery and target towing missions.

POWER PLANT: One 1,193 kW (1,600 shp) Pratt & Whitney Canada PT6A-68/1 turboprop, driving a Hartzell five-blade constant-speed fully feathering reversible-pitch propeller (PT6A-67R of same rating in POC prototype).

DIMENSIONS, EXTERNAL: As EMB-312 except:
Length overall	11.42 m (37 ft 5¾ in)
Fuselage: Length (excl rudder)	10.53 m (34 ft 6½ in)
Max depth	1.86 m (6 ft 1¼ in)
Height overall (static)	3.90 m (12 ft 9½ in)
Wheelbase	3.36 m (11 ft 0¼ in)
Propeller diameter	2.39 m (7 ft 10 in)
Propeller ground clearance (static)	0.345 m (1 ft 1½ in)

WEIGHTS AND LOADINGS:
Basic weight empty	2,390 kg (5,269 lb)
Max internal fuel load (usable)	520 kg (1,146 lb)
Max external stores load	1,000 kg (2,205 lb)
Max T-O weight: clean	3,150 kg (6,944 lb)
with external stores	3,790 kg (8,356 lb)
Max ramp weight	3,180 kg (7,010 lb)
Max landing weight: clean	3,160 kg (6,967 lb)
Max zero-fuel weight	2,640 kg (5,820 lb)
Max wing loading: clean	162.9 kg/m² (33.36 lb/sq ft)
with external stores	195.4 kg/m² (40.02 lb/sq ft)
Max power loading: clean	2.65 kg/kW (4.35 lb/shp)

EMB-312H Tucano proof-of-concept prototype

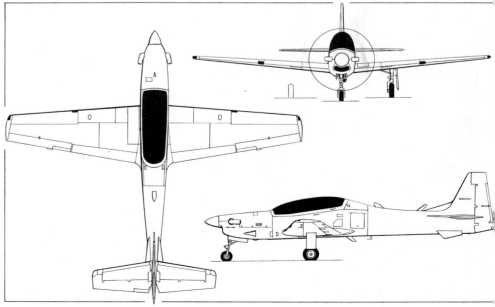

Provisional drawing of the Embraer Super Tucano stretched and higher performance version of the standard trainer (*Dennis Punnett*)

with external stores 3.18 kg/kW (5.22 lb/shp)
PERFORMANCE (at max clean T-O weight except where indicated):
Max level speed at 6,100 m (20,000 ft)
304 knots (563 km/h; 350 mph)
Max cruising speed at 6,100 m (20,000 ft)
289 knots (536 km/h; 333 mph)
Econ cruising speed at 6,100 m (20,000 ft)
234 knots (434 km/h; 269 mph)
Stalling speed, power off:
flaps and landing gear up
82 knots (152 km/h; 95 mph) EAS
flaps and landing gear down
75 knots (139 km/h; 87 mph) EAS
Max rate of climb at S/L 1,340 m (4,400 ft)/min
Service ceiling 10,670 m (35,000 ft)
T-O run 271 m (890 ft)
T-O to 15 m (50 ft) 547 m (1,795 ft)
Landing from 15 m (50 ft) 696 m (2,284 ft)
Landing run 431 m (1,414 ft)
Range at 9,150 m (30,000 ft) with max fuel, 30 min reserves 820 nm (1,519 km; 944 miles)
Ferry range at 9,150 m (30,000 ft) with underwing tanks and 30 min reserves
1,680 nm (3,113 km; 1,934 miles)
Endurance on internal fuel at econ cruising speed at 9,150 m (30,000 ft), 30 min reserves 4 h 40 min
g limits: fully Aerobatic category at 2,770 kg (6,107 lb) +7/−3.5
at 2,770 kg (6,107 lb) with external stores +4/−2.2

EMBRAER EMB-201/202 IPANEMA

TYPE: Single-seat agricultural aircraft.
PROGRAMME: First flight (EMB-200 prototype PP-ZIP) 30 July 1970; certificated 14 December 1971; see 1977-78 and earlier editions for EMB-200/200A (73 built), EMB-201 (200 built) and EMB-201R (three built); current EMB-201A first flight 10 March 1977; manufacture transferred to Neiva second half of 1981; many detail improvements (listed in 1991-92 and earlier *Jane's*) introduced late 1988; production continuing.
CURRENT VERSIONS: **EMB-201A**: Major production version until 1992, now discontinued; see 1992-93 and earlier *Jane's* for details.
EMB-202: Current model from 1992; generally similar to EMB-201A except for increased wing span, larger hopper and choice of Lycoming or Continental engine. First delivery 2 October 1992. *Description applies to this version.*
CUSTOMERS: 276 of earlier models built (see Programme); more than 425 examples of EMB-201A/202 sold by 1 January 1993, bringing overall total to more than 700.
DESIGN FEATURES: Designed to Brazilian Ministry of Agriculture specifications. Low mounted unbraced wings, with cambered leading-edges (detachable) and tips; rectangular-section fuselage; slightly sweptback vertical tail. Wing section NACA 23015 (modified); incidence 3°; dihedral 7° from roots. Fixed incidence tailplane.
FLYING CONTROLS: Conventional three-axis (ailerons Frise type), with manual/mechanical actuation; trim tab in starboard elevator, ground adjustable tab on rudder. Slotted flaps on wing trailing-edges.
STRUCTURE: All-metal safe-life fuselage frame of welded 4130 steel tube, with removable skin panels of 2024 aluminium alloy and glassfibre and specially treated against chemical corrosion. All-metal wings (single-spar) and tail surfaces (two-spar).
LANDING GEAR: Non-retractable main- and tailwheels, with rubber shock absorbers in main units. Tailwheel has tapered spring shock absorber. Mainwheels and tyres size 8.50-10. Tailwheel diameter 250 mm (10 in). Tyre pressures: main, 2.07-2.41 bars (30-35 lb/sq in); tailwheel, 3.79 bars (55 lb/sq in). Hydraulic disc brakes on mainwheels.
POWER PLANT: One 224 kW (300 hp) Textron Lycoming IO-540-K1J5D flat-six engine, driving a Hartzell two-blade (optionally three-blade) constant-speed metal pro-

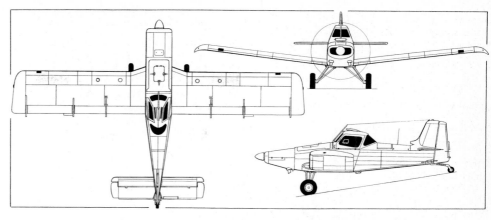

Embraer EMB-201A Ipanema single-seat agricultural aircraft *(Dennis Punnett)*

peller. Optional engine is Teledyne Continental 224 kW (300 hp) IO-550-E with McCauley two-blade constant-speed propeller. Integral fuel tanks in each wing leading-edge, with total capacity of 292 litres (77.1 US gallons; 64.2 Imp gallons), of which 264 litres (69.75 US gallons; 58 Imp gallons) are usable. Refuelling point on top of each tank. Oil capacity 12 litres (3.2 US gallons; 2.6 Imp gallons).
ACCOMMODATION: Single horizontally/vertically adjustable seat in fully enclosed cabin with bottom-hinged, triple-lock, jettisonable window/door on each side and two overhead windows. Ventilation airscoop at top of canopy front edge; second airscoop on fin leading-edge pressurises interior of rear fuselage. Inertial shoulder harness standard.
SYSTEMS: 28V DC electrical system supplied by two Cral 18EP 43Ah batteries and a CEN 240074 28V 35A alternator. Power receptacle for external battery (AN-2552-3A type) on port side of forward fuselage.
AVIONICS: Standard VFR avionics include 760-channel Bendix/King KY 96A transceiver and KR 86 ADF.
EQUIPMENT: Hopper for agricultural chemicals, capacity 950 litres (251 US gallons; 209 Imp gallons) liquid or 750 kg (1,653 lb) dry. Dusting system below centre of fuselage. Spraybooms or Micronair atomisers aft of or above wing trailing-edges respectively. Options include ram air pressure generator for use with liquid spray system; improved lightweight spraybooms; smaller/lighter Micronair AU5000 rotary atomisers; and trapezoidal spreader with adjustable inlet to improve application of dry chemicals.
DIMENSIONS, EXTERNAL:
Wing span 11.69 m (38 ft 4¼ in)
Wing chord, constant 1.71 m (5 ft 7½ in)
Wing aspect ratio 6.3
Length overall (tail up) 7.43 m (24 ft 4½ in)
Height overall (tail down) 2.22 m (7 ft 3½ in)
Fuselage: Max width 0.93 m (3 ft 0½ in)
Tailplane span 3.66 m (12 ft 0 in)
Wheel track 2.20 m (7 ft 2½ in)
Wheelbase 5.20 m (17 ft 7¼ in)
Propeller diameter 2.20 m (7 ft 2½ in)
DIMENSIONS, INTERNAL:
Cockpit: Max length 1.20 m (3 ft 11¼ in)
Max width 0.85 m (2 ft 9½ in)
Max height 1.34 m (4 ft 4¾ in)
AREAS:
Wings, gross 19.94 m² (214.63 sq ft)
Ailerons (total) 1.60 m² (17.22 sq ft)
Trailing-edge flaps (total) 2.30 m² (24.76 sq ft)
Fin 0.58 m² (6.24 sq ft)
Rudder 0.63 m² (6.78 sq ft)
Tailplane 3.17 m² (34.12 sq ft)
Elevators (total, incl tab) 1.50 m² (16.15 sq ft)
WEIGHTS AND LOADINGS (N: Normal, R: Restricted category):
Weight empty: N, R 1,040 kg (2,293 lb)

Max payload: N 530 kg (1,168 lb)
R 780 kg (1,720 lb)
Max T-O and landing weight: N 1,550 kg (3,417 lb)
R 1,800 kg (3,968 lb)
Max wing loading: N 77.75 kg/m² (15.92 lb/sq ft)
R 90.29 kg/m² (18.49 lb/sq ft)
Max power loading: N 6.92 kg/kW (11.39 lb/hp)
R 8.03 kg/kW (13.23 lb/hp)
PERFORMANCE (at max T-O weight, clean configuration, ISA):
Never-exceed speed (VNE):
N 147 knots (272 km/h; 169 mph)
R 113 knots (209 km/h; 130 mph)
Max level speed at S/L:
N 124 knots (230 km/h; 143 mph)
R 121 knots (225 km/h; 140 mph)
Max cruising speed (75% power) at 1,830 m (6,000 ft):
N 113 knots (209 km/h; 130 mph)
R 111 knots (206 km/h; 128 mph)
Stalling speed, power off (N):
flaps up 56 knots (103 km/h; 64 mph)
8° flap 54 knots (100 km/h; 62 mph)
30° flap 50 knots (92 km/h; 57 mph)
Stalling speed, power off (R):
flaps up 60 knots (110 km/h; 68 mph)
8° flap 58 knots (107 km/h; 66 mph)
30° flap 53 knots (99 km/h; 61 mph)
Max rate of climb at S/L, 8° flap:
N 283 m (930 ft)/min
R 201 m (660 ft)/min
Service ceiling, 8° flap: R 3,470 m (11,385 ft)
T-O run at S/L, 8° flap, asphalt runway:
N 200 m (656 ft)
R 354 m (1,160 ft)
T-O to 15 m (50 ft), conditions as above:
N 332 m (1,090 ft)
R 564 m (1,850 ft)
Landing from 15 m (50 ft) at S/L, 30° flap, asphalt runway:
N 412 m (1,352 ft)
R 476 m (1,562 ft)
Landing run, conditions as above: N 153 m (502 ft)
R 170 m (558 ft)
Range at 1,830 m (6,000 ft), no reserves:
N 506 nm (938 km; 583 miles)
R 474 nm (878 km; 545 miles)

EMBRAER-PIPER LIGHT AIRCRAFT PROGRAMME

Detailed descriptions of Piper aircraft built under licence by Embraer can be found in US section of this and earlier editions of *Jane's*. Manufacture undertaken by Neiva. Following types in production in 1993:
EMB-720D Minuano: Piper PA-32-301 Saratoga.
EMB-810D Cuesta: Piper PA-34-220T Seneca III. Deliveries include 35 to Brazilian Air Force for liaison duties.

HELIBRAS

HELICÓPTEROS DO BRASIL S/A (Subsidiary of Eurocopter France)

Caixa Postal 184, 37500-000 Itajubá, MG
Telephone: 55 (35) 622 3366 and 622 2455
Telex: 31 2602 HLBR BR
PRESIDENT: Bruno Boulnois
Formed 1978; now owned jointly by Bueninvest of Brazil (30 per cent), Eurocopter France (45 per cent) and state government of Minas Gerais (25 per cent). Assembly and up to 30 per cent local manufacture of SA 315B Lama, AS 350/550 single-engined and AS 355/555 twin-engined Ecureuil/Fennec helicopters and AS 565AA Panther. First assembly hall inaugurated 28 March 1980; complete facility intended to occupy 206,650 m² (2,224,360 sq ft). Workforce 275 in 1992, by which time more than 200 helicopters delivered.

HELIBRAS (AEROSPATIALE) HB 315B GAVIÃO

TYPE: Brazilian built SA 315B Lama.
PROGRAMME: First Gavião completed second half of 1979.
CUSTOMERS: Seven built by May 1989 (six for SAR and utility duties with Grupo Aereo 51 of Bolivian Air Force, one for civil customer in Chile). No later details received.

HELIBRAS (EUROCOPTER) HB 350B, HB 350B1, HB 350L1 and HB 355F2 ESQUILO

TYPE: Single-engined (HB 350) and twin-engined (HB 355) Brazilian built versions of Eurocopter France Ecureuil and Fennec (see International section).
PROGRAMME: Deliveries began 1979; aeromedical version launched February 1989.

CURRENT VERSIONS: **CH-50**: Brazilian Air Force designation of HB 350B/B1.
CH-55: Brazilian Air Force designation of HB 355F2.
HA-1: Brazilian Army designation of HB 350L1; initial 16 delivered by 1991; further 10 on order. In service with 1st Aviation Battalion at Taubaté, São Paulo; equipped for tactical support and reconnaissance.
HB 350B and B1: Single-engined civil versions.
HB 350L1: Single-engined civil version.
HB 355F2: Twin-engined civil version.
TH-50: Trainer version of CH-50.
UH-12: Brazilian Navy designation of HB 350B. In service (incl UH-12B) with 1° Esquadrão de Helicópteros de Emprego Geral (squadron of general purpose helicopters).
UH-12B: Brazilian Navy designation of HB 355F2.
VH-55: VIP transport version of CH-55.

BRAZIL/CANADA: AIRCRAFT—HELIBRAS/BELL

CUSTOMERS: More than 100 delivered by early 1993 to Brazilian Army (16 HA-1), Navy (nine UH-12, 11 UH-12B) and Air Force (30 CH/TH-50, 11 CH-55, two VH-55); Brazilian police and civilian customers; Paraguay (Air Force four HB 350B, Navy two); and civil customers in Argentina (three), Bolivia (four) and Venezuela (three). Further 10 HA-1s on order for 1992-94 delivery.
ACCOMMODATION (Aeromedical version): Doctor, nurse and two stretcher patients in addition to pilot.
EQUIPMENT (Aeromedical version): Electrocardiograph, respirator, pacemaker, stretchers, battery operated incubator, oxygen/compressed air cylinders, first aid kit, and four-way socket for 115V AC (60Hz) and 12V DC power.
ARMAMENT (HA-1): Can include a 20 mm gun and 2.75 in unguided rockets, or anti-tank missiles and HeliTOW sighting system.
ARMAMENT (UH-12/12B): Two Avibrás LM-70/7 pods each containing seven SBAT 70 mm rockets, or two FN Herstal twin 7.62 mm MAG machine-gun pods and a door mounted MAG pedestal.

HELIBRAS (EUROCOPTER) AS 565 PANTHER
Brazilian Army designation: HM-1
TYPE: Brazilian built Eurocopter Panther (see International section).
PROGRAMME: Helibras assembled final 10 of Brazilian Army order from French built CKD kits; deliveries completed January 1992.
CUSTOMERS: Brazilian Army 36 (26 delivered by Aerospatiale 1989-90, 10 assembled locally).

IPE
INDÚSTRIA PARANAENSE DE ESTRUTURAS LTDA
Caixa Postal 7931, 80011-970 Curitiba, Paraná
Telephone: 55 (41) 242 2324
MANAGER: Eng J. C. Boscardin

IPE 06 CURUCACA
TYPE: Very lightweight training/general purpose aircraft.
PROGRAMME: Developed for Brazilian club market; first flight January 1990. No further news since then; see 1992-93 *Jane's* for illustration and all known details.

NEIVA
INDÚSTRIA AERONÁUTICA NEIVA S/A (Subsidiary of Embraer)
Caixa Postal 10, 18608-900 Botucatu, SP
Telephone: 55 (149) 22 1010
Telex: 142 423 SOAN BR
PRESIDENT: Eng Antonio Garcia da Silveira
ENGINEERING MANAGER: Luiz Carlos Benetti

Formed 1954; became wholly owned Embraer subsidiary 10 March 1980; factory area 20,580 m² (221,521 sq ft), workforce 400. Substantial participation in Embraer general aviation programmes, including complete production of EMB-720D Minuano and EMB-810D Cuesta (licence built Piper twins); also responsible since 1981 for total production of EMB-202 and earlier -201A Ipanema. Other work includes subassemblies and components for Embraer Bandeirante and Tucano.

NEIVA NE-821 CARAJÁ
TYPE: High-powered version of Schafer turboprop conversion of Piper Navajo Chieftain.
PROGRAMME: Completed in 1992; see 1992-93 *Jane's* for details.
CUSTOMERS: Total of 39 delivered by 1 January 1991.

SUPER ROTOR
M. M. SUPER ROTOR INDÚSTRIA AERONÁUTICA LTDA
Rua Itapeti 541, Tatuapé, 03324-000 São Paulo, SP
Telephone: 55 (11) 295 8187
Fax: 55 (11) 941 1683
Telex: 11 941 1892
DIRECTOR: José Montalvá Perez

SUPER ROTOR AC-4 ANDORINHA
TYPE: Single-seat autogyro.
PROGRAMME: Private venture, designed 1970 by Eng Altair Coelho; first flight December 1972; modified by Francisco Mattos Jr before gaining domestic certification January 1985; produced in ready to fly or kit form.
CUSTOMERS: Domestic and foreign sales (incl kits) totalled 307 by early 1990. No subsequent information received; see 1992-93 and earlier *Jane's* for description.

SUPER ROTOR M-1 MONTALVÁ
TYPE: Tandem two-seat autogyro.
PROGRAMME: Design started June 1984; first flight March 1985; deliveries began Summer 1985.
CUSTOMERS: Total of 75 built by January 1990. No subsequent information received; see 1992-93 and earlier *Jane's* for description.

CANADA

AEROSPACE CONSORTIUM
THE AEROSPACE CONSORTIUM INC
730 Gana Court, Mississauga, Ontario L5S 1P1
Telephone: 1 (416) 670 1070
Fax: 1 (416) 670 1695

Provides single-point sourcing and project management services by drawing on combined manufacturing, engineering and management capabilities of several companies grouped under single umbrella organisation. Products and systems can include military and commercial aircraft, communications and navigation systems, radar systems, air/ground/shipboard antenna systems, control systems, electronic processing systems and surveillance systems. Consortium members have supplied services to Boeing, Bombardier, de Havilland, Garrett, Grumman, Hughes, Litton, Lockheed, McDonnell Douglas, Oerlikon and Pratt & Whitney Canada.

Currently teamed with Promavia International Corporation (which see) to set up planned jet trainer manufacturing facility in Canada; in addition to providing technical and management assistance to Promavia, will act as procurement manager for most bought-out parts and systems.

AIRTECH
AIRTECH CANADA
Peterborough Municipal Airport, PO Box 415, Peterborough, Ontario K9J 6Z3
Telephone: 1 (705) 743 9483
Fax: 1 (705) 749 0841
PRESIDENT AND CHIEF ENGINEER: James C. Mewett
GENERAL MANAGER: Bernard J. LaFrance

Specialises in fitting more powerful Polish engines to Otters and Beavers to improve performance and economy; also developed and installed auxiliary fuel tanks and medevac interiors in Cessna 401, 411, 414, 421 and 500, Piper PA-31 and PA-42, Mitsubishi MU-2, Beech King Air 100, Fairchild Metro IIB and Sikorsky S-76. More details of Beaver and Otter programmes in *Jane's Civil and Military Aircraft Upgrades 1993-94*.

AIRTECH CANADA DHC-3/1000 OTTER
Eight de Havilland Canada DHC-3 Otters re-engined with 447 kW (600 hp) Polish PZL-3S; details in 1983-84 *Jane's*. Followed by DHC-3/1000 powered by Polish 746 kW (1,000 hp) ASz-62IR radial; first flight 25 August 1983; nine flying in South and North America by May 1992; details in 1990-91 *Jane's*.

Airtech Canada's DHC-3/1000 Otter conversion, powered by a PZL Kalisz ASz-62IR radial engine

AIRTECH CANADA DHC-2/PZL-3S BEAVER
Four Beavers powered by 447 kW (600 hp) PZL-3S radial engines flying by early 1991; details in 1990-91 *Jane's*.

BELL
BELL HELICOPTER TEXTRON CANADA LTD (a Division of Textron Canada Ltd)
12800 rue de l'Avenir, Mirabel, Quebec J7J 1R4
Telephone: 1 (514) 437 2729
Fax: 1 (514) 437 6010
Telex: 05-25827
PRESIDENT: Tommy Tomason

Memorandum of understanding to start helicopter industry in Canada signed 7 October 1983; 34,560 m² (372,000 sq ft) factory opened late 1985 and employed 1,300 people at end of 1991; US civil production of 206B JetRanger and 206L LongRanger transferred to Canada by early 1987; then 212 in August 1988 and 412 in February 1989; over 800 (all models) delivered by end of 1992; about half of each helicopter made in Canada (dynamic systems supplied by Bell Fort Worth). Production rate more than 20 per month. Except for sales to Canadian government or commercial customers, all helicopters produced by Bell Canada are sold to parent company at Fort Worth (see US section) for resale to US and foreign customers.

JetRanger and 212/412 also made under licence by Agusta in Italy (which see); some cabin components for 212 made by BHK Inc in South Korea; 212 and 412 to be assembled and later part-built by Chincul in Argentina.

BELL 206B-3 JETRANGER III
US Army designation: TH-67 Creek
Canadian Forces designation: CH-139
TYPE: Five-seat turbine powered light helicopter.
PROGRAMME: Delivery of current 206B-3 JetRanger III began

Summer 1977; transferred to Mirabel, Canada, 1986; production rate in mid-1990 was six or seven a month.

CURRENT VERSIONS: **206B-3 JetRanger III:** Current civil production version. *Subject of description below.*

TH-67 Creek (Bell designation TH-206): Selected March 1993 as US Army NTH choice to replace UH-1 at pilot training school at Fort Rucker, Alabama. Instructor and one pupil in front seats, second pupil observing from rear seat sees flight instruments by closed circuit TV screen mounted on back of right hand front seat. Powered by Allison 250-C20JN engine. First batch includes nine cockpit procedures trainers outfitted by Frasca International; three configurations: VFR, IFR, and VFR with IFR provision.

CUSTOMERS: TH-206 (TH-67 Creek) declared winner of US Army NTH (New Training Helicopter) competition March 1993, with deliveries to begin October 1993; 29 aircraft (and six procedures trainers) to be delivered in time for first training course to open April 1994; total order covers 102 TH-67s and nine cockpit procedures trainers; also includes option for FY 1994 procurement of three more procedures trainers and 55 more TH-67s.

Well over 7,000 JetRangers produced by Bell and licensees by January 1993, including 4,200 Model 206Bs and about 2,000 military OH-58 series; 120 completed at Mirabel by January 1990; combined civil sales of JetRanger/LongRanger 178 in 1991 and 139 (58 Bs and 81 Ls) in 1992.

COSTS: US Army initial NTH contract $84.9 million.

DESIGN FEATURES: Two-blade teetering main rotor with preconed and underslung bearings; blades retained by grip, pitch change bearing and torsion-tension strap assembly; two-blade tail rotor; main rotor rpm 374 to 394.

FLYING CONTROLS: Hydraulic fully powered cyclic and collective controls and foot powered tail rotor control; tailplane with highly cambered inverted aerofoil section and stall strip produces appropriate nose-up and nose-down attitude during climb and descent; optional autostabiliser, autopilot and IFR systems.

STRUCTURE: Conventional light alloy structure with two floor beams and bonded honeycomb sandwich floor; transmission mounted on two beams and deck joined to floor by three fuselage frames; main rotor blades have extruded aluminium D-section leading-edge with honeycomb core behind, covered by bonded skin; tail rotor blades have bonded skin without honeycomb core.

LANDING GEAR: Aluminium alloy tubular skids bolted to extruded cross-tubes. Tubular steel skid on ventral fin to protect tail rotor in tail-down landing. Special high skid gear (0.25 m; 10 in greater ground clearance) available for use in areas with high brush. Pontoons or stowed floats, capable of in-flight inflation, available as optional kits.

POWER PLANT: One 313 kW (420 shp) Allison 250-C20J turboshaft, flat rated at 236 kW (317 shp). Transmission rating 236 kW (317 shp). Rupture resistant fuel tank below and behind rear passenger seat, capacity 344 litres (91 US gallons; 75.75 Imp gallons). Refuelling point on starboard side of fuselage, aft of cabin. Oil capacity 5.2 litres (1.38 US gallons; 1.13 Imp gallons).

ACCOMMODATION: Two seats side by side in front and three-seat rear bench. Dual controls optional. Two forward-hinged doors on each side, made of formed aluminium alloy with transparent panels (bulged on rear pair). Baggage compartment aft of rear seats, capacity 113 kg (250 lb), with external door on port side.

SYSTEMS: Hydraulic system, pressure 41.5 bars (600 lb/sq in), for cyclic, collective and directional controls. Max flow rate 7.57 litres (2 US gallons; 1.65 Imp gallons)/min. Open reservoir. Electrical supply from 150A starter/generator. One 28V 13Ah (optionally 17Ah) nickel-cadmium battery.

AVIONICS: Full range of optional kits, including VHF communications and omni navigation, ADF, DME, R/Nav, transponder and intercom and speaker system.

EQUIPMENT: Standard equipment includes cabin fire extinguisher, first aid kit, door locks, night lighting, and dynamic flapping restraints. Optional items include clock, engine hour meter, turn and slip indicator, custom seating, internal stretcher kit, rescue hoist, cabin heater, ECS, camera access door, high intensity night lights, engine fire detection system, and external cargo hook of 680 kg (1,500 lb) capacity.

DIMENSIONS, EXTERNAL:
Main rotor diameter	10.16 m (33 ft 4 in)
Tail rotor diameter	1.65 m (5 ft 5 in)
Main rotor blade chord	0.33 m (1 ft 1 in)
Tail rotor blade chord	0.122 m (4.8 in)
Distance between rotor centres	5.96 m (19 ft 6½ in)
Length: overall, rotors turning	11.82 m (38 ft 9½ in)
fuselage, incl tailskid	9.50 m (31 ft 2 in)
Height: over tail fin	2.54 m (8 ft 4 in)
to top of rotor head	2.91 m (9 ft 6½ in)
Stabiliser span	1.97 m (6 ft 5¾ in)
Width over skids	1.92 m (6 ft 3½ in)
Forward cabin doors (each): Height	1.02 m (3 ft 4 in)
Width	0.61 m (2 ft 0 in)
Rear cabin doors (each): Height	1.02 m (3 ft 4 in)
Width	0.91 m (3 ft 0 in)

DIMENSIONS, INTERNAL:
Cabin: Length of seating area	2.13 m (7 ft 0 in)
Max width	1.27 m (4 ft 2 in)
Max height	1.28 m (4 ft 3 in)
Volume	1.13 m³ (40 cu ft)
Baggage compartment volume	approx 0.45 m³ (16 cu ft)

AREAS:
Main rotor blades (each)	1.68 m² (18.05 sq ft)
Tail rotor blades (each)	0.11 m² (1.18 sq ft)
Main rotor disc	81.07 m² (872.7 sq ft)
Tail rotor disc	2.14 m² (23.04 sq ft)
Stabiliser	0.90 m² (9.65 sq ft)

WEIGHTS AND LOADINGS:
Weight empty, standard	737 kg (1,625 lb)
Max payload: internal	635 kg (1,400 lb)
external	680 kg (1,500 lb)
Max T-O weight: internal load	1,451 kg (3,200 lb)
external load	1,519 kg (3,350 lb)
Max disc loading:	
internal load	17.92 kg/m² (3.67 lb/sq ft)
external load	18.74 kg/m² (3.84 lb/sq ft)
Max power loading:	
internal load	6.15 kg/kW (10.09 lb/shp)
external load	6.43 kg/kW (10.57 lb/shp)

PERFORMANCE (at max internal load T-O weight, ISA):
Never-exceed speed (VNE) at S/L	122 knots (225 km/h; 140 mph)
Max cruising speed at S/L	115 knots (214 km/h; 133 mph)
Max rate of climb at S/L	390 m (1,280 ft)/min
Vertical rate of climb at S/L	91 m (300 ft)/min
Service ceiling	4,115 m (13,500 ft)
Hovering ceiling: IGE	3,900 m (12,800 ft)
OGE	2,680 m (8,800 ft)
Range with max fuel:	
at S/L, no reserves	365 nm (676 km; 420 miles)
at 1,525 m (5,000 ft), no reserves	395 nm (732 km; 455 miles)

Bell 206B-3 JetRanger III (Allison 250-C20J turboshaft)

BELL 206L-4 LONGRANGER IV

TYPE: Seven-seat turbine powered general purpose light helicopter (stretched JetRanger).

PROGRAMME: Announced 25 September 1973; production in Canada began January 1987. MoU early 1993 for Aymet Aerospace Industries (Turkey) to assemble 14 of 24 ordered for Turkish armed forces.

CURRENT VERSIONS: **LongRanger III:** Major version from 1973-92; Canadian production began January 1987; details in 1992-93 and earlier *Jane's*. Now superseded by LongRanger IV.

LongRanger IV: Announced March 1992 as new current standard model; transmission uprated to absorb 365 kW (490 shp) instead of 324 kW (435 shp) from same engine; gross weight raised from 1,882 kg (4,150 lb) to 2,018 kg (4,450 lb); certificated late 1992, delivered from December.

Gemini ST: Twin-engined conversion of LongRanger III/IV developed by Tridair and Soloy in USA and described under Tridair heading in US section. Also in new-build form by Bell Canada as TwinRanger.

TwinRanger: New-build equivalent of Gemini ST; described separately.

CUSTOMERS: More than 1,100 LongRangers produced by January 1990; total Canadian deliveries 120 by January 1990. No total received for 1990; see JetRanger entry for 1991-92 civil sales. LongRanger IV deliveries began (one to Mofaz Air, Malaysia) December 1992; orders include 24 for Turkish armed forces.

DESIGN FEATURES: Cabin length increased to make room for club seating and extra window; Bell Noda-Matic transmission to reduce vibration; improvements introduced on LongRanger II include new freewheel unit, modified shafting and increased-thrust tail rotor; main rotor rpm 394; rotor brake optional.

FLYING CONTROLS: As JetRanger, but with endplate fins on tailplane; single-pilot IFR with Collins AP-107H autopilot; optional SFENA autopilot with stabilisation and holds for heading, height and approach.

STRUCTURE: As JetRanger.

LANDING GEAR: As JetRanger.

POWER PLANT: One 485 kW (650 shp) Allison 250-C30P turboshaft (max continuous rating 415 kW; 557 shp). Transmission rated at 365 kW (490 shp) for take-off, with a continuous rating of 276 kW (370 shp); 340 kW (456 shp) transmission optional. Rupture resistant fuel system, capacity 419 litres (110.7 US gallons; 92.2 Imp gallons).

ACCOMMODATION: Redesigned rear cabin, more spacious than JetRanger. With a crew of two, standard cabin layout accommodates five passengers in two canted rearward facing seats and three forward facing seats. An optional executive cabin layout has four individual passenger seats. Port forward passenger seat has folding back to allow loading of a 2.44 × 0.91 × 0.30 m (8 × 3 × 1 ft) container,

British-registered Bell 206L-3 LongRanger III in Air Hanson livery

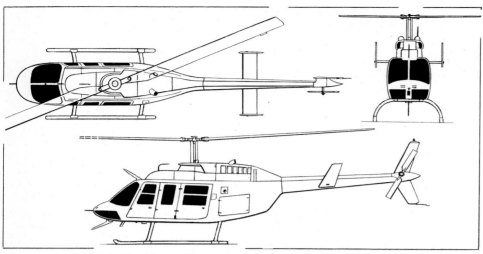

Bell 206L-3 LongRanger III general purpose light helicopter *(Dennis Punnett)*

making possible carriage of such items as survey equipment, skis and other long components. Double doors on port side of cabin provide opening 1.54 m (5 ft 1 in) wide, for straight-in loading of stretcher patients or utility cargo; in ambulance or rescue role two stretcher patients and two ambulatory patients/attendants may be carried. Dual controls optional.

SYSTEMS: Hydraulic system; 28V DC electrical power from 180A starter/generator and 17Ah battery.

AVIONICS: Standard Collins MicroLine suite includes dual nav/com, ADF, DME, transponder and marker beacon receiver. Bendix/King R/Nav, radio altimeter and encoding altimeter optional.

EQUIPMENT: Optional kits include emergency flotation gear, 907 kg (2,000 lb) cargo hook, rescue hoist, Nightsun searchlight (with high skid gear) and engine bleed air ECS.

DIMENSIONS, EXTERNAL:
Main rotor diameter	11.28 m (37 ft 0 in)
Tail rotor diameter	1.65 m (5 ft 5 in)
Main rotor blade chord	0.33 m (1 ft 1 in)
Tail rotor blade chord	0.135 m (5¼ in)
Length: overall, rotors turning	13.02 m (42 ft 8½ in)
fuselage, incl tailskid	9.81 m (32 ft 2½ in)
Height: over tail fin	3.12 m (10 ft 2¾ in)
to top of rotor head	3.14 m (10 ft 3¾ in)
Fuselage: Max width	1.32 m (4 ft 4 in)
Stabiliser span	1.98 m (6 ft 6 in)
Width over skids	2.34 m (7 ft 8 in)
Forward cabin doors (each): Height	1.02 m (3 ft 4 in)
Width	0.61 m (2 ft 0 in)
Centre cabin door (port): Height	1.02 m (3 ft 4 in)
Width	0.63 m (2 ft 1 in)
Rear cabin doors (each): Height	1.02 m (3 ft 4 in)
Width	0.91 m (3 ft 0 in)

DIMENSIONS, INTERNAL:
Cabin: Length	2.74 m (9 ft 0 in)
Max width and height	as JetRanger
Volume	2.35 m³ (83 cu ft)

AREAS:
Main rotor disc	99.89 m² (1,075.2 sq ft)
Tail rotor disc	2.13 m² (22.97 sq ft)

WEIGHTS AND LOADINGS:
Weight empty, standard	1,031 kg (2,274 lb)
Max external load	907 kg (2,000 lb)
Max T-O weight: normal	2,018 kg (4,450 lb)
external load	2,064 kg (4,550 lb)
Max disc loading: normal	20.66 kg/m² (4.23 lb/sq ft)
external load	19.29 kg/m² (3.95 lb/sq ft)
Max power loading: transmission for T-O, normal	
	5.52 kg/kW (9.08 lb/shp)
transmission for T-O, external load	
	5.65 kg/kW (9.29 lb/shp)

PERFORMANCE (at max normal T-O weight, ISA):
Never-exceed speed (V_{NE}):	
at S/L	130 knots (241 km/h; 150 mph)
at 1,525 m (5,000 ft)	133 knots (246 km/h; 153 mph)
Max cruising speed:	
at S/L	110 knots (203 km/h; 126 mph)
at 1,525 m (5,000 ft)	111 knots (206 km/h; 128 mph)
Max rate of climb at S/L	408 m (1,340 ft)/min
Service ceiling at max cruise power	3,050 m (10,000 ft)
Hovering ceiling: IGE	3,050 m (10,000 ft)
OGE	1,980 m (6,500 ft)
Range with max fuel, no reserves:	
at S/L	321 nm (595 km; 369 miles)
at 1,525 m (5,000 ft)	357 nm (661 km; 411 miles)

BELL 206LT TWINRANGER

TYPE: Seven-seat twin-turboshaft light helicopter (twin-engined LongRanger derivative).

PROGRAMME: New-build production counterpart of Tridair Gemini ST (see US section); first example in final assembly March 1993.

DESIGN FEATURES: Same airframe as LongRanger except for modified cowling contours for twin engine installation.

POWER PLANT: Two Allison 250-C20R turboshafts, each rated at 335.5 kW (450 shp) for 5 min for T-O and 283 kW (380 shp) max continuous; OEI max continuous rating 335.5 kW (450 shp). Transmission rating as for LongRanger. Fuel capacity (usable) 427 litres (112.7 US gallons; 93.8 Imp gallons).

ACCOMMODATION: As for LongRanger.

SYSTEMS: Generally as for LongRanger except electrical system uses two 150A starter/generators and 28Ah nickel-cadmium battery.

AVIONICS: Similar to those for LongRanger.
DIMENSIONS, EXTERNAL: As LongRanger IV
DIMENSIONS, INTERNAL: As LongRanger IV
AREAS: As LongRanger IV
WEIGHTS AND LOADINGS: As LongRanger IV except:
Weight empty, standard   1,246 kg (2,748 lb)
PERFORMANCE (estimated, at max normal T-O weight, ISA, except where indicated):
Max cruising speed at S/L	106 knots (196 km/h; 122 mph)
Econ cruising speed at S/L, average gross weight	108 knots (200 km/h; 124 mph)
Service ceiling	3,050 m (10,000 ft)
Service ceiling, OEI, 30 min power:	
ISA	3,050 m (10,000 ft)
ISA + 20°C	2,350 m (7,700 ft)
Hovering ceiling: IGE	3,050 m (10,000 ft)
OGE	2,100 m (6,900 ft)
Range at S/L with max fuel, long-range cruising speed, no reserves	250 nm (463 km; 288 miles)

BELL 212 TWIN TWO-TWELVE
US military designation: UH-1N
Canadian Forces designation: CH-135

TYPE: Twin-turbine utility helicopter.

PROGRAMME: Canadian government approval to develop twin-engined UH-1 with P&WC PT6T-3 Twin-Pac announced 1 May 1968; more powerful PT6T-3B introduced June 1980; manufacture transferred to Canada August 1988; production increased from one to two a month in mid-1990.

CURRENT VERSIONS: **CH-135:** Canadian version, originally CUH-1N.

UH-1N: US Air Force, Navy and Marines version.

Twin Two-Twelve: Civil version; FAA certification October 1970; FAA Category A transport certification 30 June 1971; IFR certification granted by FAA, UK's CAA, Norwegian DCA and Canadian DoT; first-ever single-pilot IFR certification with fixed floats granted June 1977. *Description applies to this version.*

CUSTOMERS: Canadian Forces received 50 CH-135s (originally CUH-1Ns) in 1971-72; delivery of 79 UH-1Ns to US Air Force began 1970 for Special Operations Force counter-insurgency, psychological warfare and unconventional warfare; delivery of 40 UH-1Ns to US Navy and 22 to Marines began 1971; further 159 delivered to US Navy and Marines 1973 to 1978.

Other military deliveries include eight to Argentine Air Force, nine to Bangladesh Air Force, 25 to Mexican Air Force, two to Panamanian Air Force and nine to Royal Thai Army; more recently 18 more ordered for Mexico and 23 more for Thailand.

Civil examples delivered to customers in Australia, China, Japan and Saudi Arabia. Recent civil sales: 18 in 1991 and eight in 1992.

DESIGN FEATURES: All-metal two-blade semi-rigid teetering main rotor with interchangeable blades; underslung feathering axis head; rotor brake optional.

FLYING CONTROLS: Fully powered hydraulic controls; gyroscopic stabiliser bar above main rotor; automatic variable incidence tailplane; IFR versions have large fin above cabin to improve roll-yaw responses during manual instrument flying.

STRUCTURE: Metal main rotor blades have extruded aluminium nose sections and laminates; glassfibre safety straps provide redundant load path; fuselage conventional light metal.

LANDING GEAR: Tubular skid type. Lock-on ground handling wheels, high skid gear, fixed floats and emergency pop-out nylon float bags optional.

POWER PLANT: Pratt & Whitney Canada PT6T-3B Turbo Twin-Pac, comprising two PT6 turboshafts coupled to combining gearbox with single output shaft. Engine rating 1,342 kW (1,800 shp) for T-O, 1,193 kW (1,600 shp) max continuous; OEI rating 764 kW (1,025 shp) for 2½ min, 723 kW (970 shp) for 30 min. Transmission rating 962 kW (1,290 shp) for T-O, 846 kW (1,134 shp) max continuous and 764 kW (1,025 shp) OEI. Five interconnected rubber fuel cells, total usable capacity 818 litres (216 US gallons; 180 Imp gallons). Two 76 or 341 litre (20 or 90 US gallon; 16.7 or 75 Imp gallon) auxiliary fuel tanks optional, to provide max possible capacity of 1,499 litres (396 US gallons; 330 Imp gallons). Single-point refuelling on starboard side of cabin. Oil capacity 11.5 litres (3 US gallons; 2.5 Imp gallons) for engines, 8.5 litres (2.25 US gallons; 1.87 Imp gallons) for transmission.

ACCOMMODATION: Pilot and up to 14 passengers. Dual controls optional. In cargo configuration, has total internal volume of 7.02 m³ (248 cu ft), including baggage space in tailboom, capacity 181 kg (400 lb). Forward opening crew door each side of fuselage. Two doors each side of cabin; forward (jettisonable) door hinged to open forward, rear door sliding aft. Accommodation heated and ventilated.

SYSTEMS: Dual hydraulic systems, pressure 69 bars (1,000 lb/sq in) each, max flow rate 22.7 litres (6 US gallons; 5 Imp gallons)/min. Open reservoir. 28V DC electrical system

Bell's 206LT TwinRanger, a twin-engined version of the LongRanger, is the new-build counterpart of this Tridair Gemini ST prototype

Bell 212 of the Sri Lanka Air Force *(Peter Steinemann)*

supplied by two completely independent 30V 200A (derated to 150A) starter/generators. Secondary AC power supplied by two independent 250VA single-phase solid state inverters. A third inverter can acquire automatically the load of a failed inverter. 40Ah nickel-cadmium battery. AiResearch air-cycle environmental control unit optional.

AVIONICS: Optional IFR avionics include dual Bendix/King KTR 908 720-channel VHF com transceivers, dual KNR 660A VOR/LOC/RMI receivers, KDF 800 ADF, KMD 700A DME, KXP 750A transponder and KGM 690 marker beacon/glideslope receiver; dual Honeywell Tarsyn-444 three-axis gyro units; stability control augmentation system; and an automatic flight control system. Flight director and weather radar also optional.

EQUIPMENT: Optional equipment includes a stretcher kit, cargo hook, cargo sling and rescue hoist.

DIMENSIONS, EXTERNAL:
Main rotor diameter	14.69 m (48 ft 2¼ in)
Tail rotor diameter	2.59 m (8 ft 6 in)
Main rotor blade chord	0.59 m (1 ft 11.4 in)
Tail rotor blade chord	0.292 m (11½ in)
Length:	
overall (main rotor fore and aft)	17.46 m (57 ft 3¼ in)
fuselage	12.92 m (42 ft 4¾ in)
Height: to top of rotor head	3.92 m (12 ft 10.2 in)
overall	4.53 m (14 ft 10¼ in)
Width: over skids	2.64 m (8 ft 8 in)
overall (main rotor fore and aft)	2.81 m (9 ft 2.6 in)
Elevator span	2.86 m (9 ft 4½ in)
Rear sliding doors (each): Height	1.24 m (4 ft 1 in)
Width	1.88 m (6 ft 2 in)
Height to sill	0.76 m (2 ft 6 in)
Baggage compartment door: Height	0.53 m (1 ft 9 in)
Width	1.71 m (2 ft 4 in)
Emergency exits (centre cabin windows, each):	
Height	0.76 m (2 ft 6 in)
Width	0.97 m (3 ft 2 in)

DIMENSIONS, INTERNAL:
Cabin, excl flight deck: Length	2.34 m (7 ft 8 in)
Max width	2.44 m (8 ft 0 in)
Max height	1.24 m (4 ft 1 in)
Floor area	4.74 m² (51.0 sq ft)
Volume	6.23 m³ (220 cu ft)
Baggage compartment volume	0.78 m³ (28 cu ft)

AREAS:
Main rotor disc	168.11 m² (1,809.6 sq ft)
Tail rotor disc	5.27 m² (56.74 sq ft)

WEIGHTS AND LOADINGS:
Weight empty, standard configuration:	
VFR	2,765 kg (6,097 lb)
IFR	2,847 kg (6,277 lb)
Max external load	2,268 kg (5,000 lb)
Max T-O weight, internal or external load	
	5,080 kg (11,200 lb)
Max disc loading	30.22 kg/m² (6.19 lb/sq ft)
Max power loading	5.28 kg/kW (8.68 lb/shp)

PERFORMANCE (at max T-O weight, ISA):
Never-exceed speed (V_{NE}) and max cruising speed at S/L	
	111 knots (206 km/h; 128 mph)
Max cruising speed at 1,525 m (5,000 ft) at max cruise power	102 knots (189 km/h; 117 mph)
Long-range cruising speed at 1,525 m (5,000 ft)	
	104 knots (193 km/h; 120 mph)
Max rate of climb at S/L	402 m (1,320 ft)/min
Service ceiling	3,960 m (13,000 ft)
Max altitude for T-O and landing	1,430 m (4,700 ft)
Hovering ceiling: IGE	1,450 m (4,750 ft)
Max range with standard fuel at 1,525 m (5,000 ft), long-range cruising speed, no reserves	
	243 nm (450 km; 280 miles)

BELL 412HP
Canadian Forces designation: CH-146 Griffon

TYPE: Four-blade, twin-engined utility helicopter.

PROGRAMME: Model 412 announced 8 September 1978 (see earlier editions of *Jane's*); FAR Pt 29 VFR approval received 9 January 1981, IFR 13 February 1981; production (SP version) transferred to Canada February 1989; first delivery (civil) 18 January 1981.

CURRENT VERSIONS: **412SP**: Special Performance version with increased max T-O weight, new seating options and 55 per cent greater standard fuel capacity. Superseded by 412HP early 1991. Details in 1991-92 *Jane's*.

Military 412: Announced by Bell June 1986; fitted with Lucas Aerospace chin turret and Honeywell Head Tracker helmet sight similar to that in AH-1S; turret carries 875 rounds, weighs 188 kg (414 lb) and can be removed in under 30 minutes; firing arcs ±110° in azimuth, +15° and −45° in elevation; other armament includes twin dual FN Herstal 7.62 mm gun pods, single FN Herstal 0.50 in pod, pods of seven or 19 2.75 in rockets, M240E1 pintle-mounted door guns, FN Herstal four-round 70 mm rocket launcher and a 0.50 in gun or two Giat M621 20 mm cannon pods.

412HP: Improved transmission giving better OGE hover; FAR Pt 29 certification 5 February 1991, first delivery (c/n 36020) later that month. Now standard current model, *to which detailed description applies*.

Griffon: Agusta developed military version (see Italian section); capable of ASW, medevac, armed support, transport, SAR and patrol.

NBell-412: Indonesia's IPTN (which see) has licence to produce up to 100 Model 412SPs, but production currently suspended.

CUSTOMERS: Total 108 Model 412 and 97 412SP delivered by end September 1990; no figures received for final quarter of 1990, but civil sales 17 in 1991 and 16 in 1992.

Military deliveries include Venezuelan Air Force (two), Botswana Defence Force (three), Public Security Flying Wing of Bahrain Defence Force (two), Sri Lankan armed forces (four), Nigerian Police Air Wing (two), Mexican government (two VIP transports), South Korean Coast Guard (one), Honduras (10), Royal Norwegian Air Force (18, of which 17 assembled by Helikopter Service, Stavanger, to replace UH-1Bs of 339 Squadron at Bardufoss and 720 Squadron at Rygge).

Canadian Forces contract for 100 Bell 412HPs (CH-146 Griffon) placed in 1992, to replace current Bell 205s (CH-118), 206s (CH-136) and 212s (CH-135); deliveries from early 1994, completion 1997.

DESIGN FEATURES: Four-blade main rotor with blades retained within central metal star fitting by single elastomeric bearings; shorter rotor mast than 212; blades can be folded; rotor brake standard; two-blade tail rotor; main rotor rpm 314.

FLYING CONTROLS: As 212 with automatic tailplane incidence control; optional IFR avionics include Bendix/King Gold Crown III and dual Honeywell AFCS.

STRUCTURE: Main rotor blade spar unidirectional glassfibre with 45° wound torque casing of glassfibre cloth; Nomex rear section core with trailing-edge of unidirectional glass-fibre; leading-edge protected by titanium abrasion strip and replaceable stainless steel cap at tip; lightning protection mesh embedded; provision for electric de-icing heater elements; main rotor hub of steel and light alloy; all-metal tail rotor.

LANDING GEAR: High skid, emergency pop-out float or non-retractable tricycle gear optional.

POWER PLANT: Pratt & Whitney Canada PT6T-3BE Turbo Twin-Pac, rated at 1,342 kW (1,800 shp) for 5 min for T-O and 1,193 kW (1,600 shp) max continuous. In event of engine failure, remaining engine can deliver up to 764 kW (1,025 shp) for 2½ min, or 723 kW (970 shp) for 30 min. Transmission rating 1,022 kW (1,370 shp) for T-O, 828 kW (1,110 shp) max continuous; OEI rating 764 kW (1,025 shp). Optional 30 kW (40 shp) for accessory drives from main gearbox. Seven interconnected rupture resistant fuel cells, with automatic shut-off valves (breakaway fittings), have a combined capacity of 1,249 litres (330 US gallons; 275 Imp gallons). Two 76 or 310 litre (20 or 82 US gallon; 16.7 or 68.3 Imp gallon) auxiliary fuel tanks, in any combination, can increase maximum total capacity to 1,870 litres (494 US gallons; 411 Imp gallons). Single-point refuelling on starboard side of cabin.

ACCOMMODATION: Generally as for Bell 212.

SYSTEMS: As Bell 212 except inverters are 450VA.

AVIONICS: Generally as for Bell 212.

EQUIPMENT: Optional equipment includes cargo sling and rescue hoist.

DIMENSIONS, EXTERNAL:
Main rotor diameter	14.02 m (46 ft 0 in)
Tail rotor diameter	2.62 m (8 ft 7 in)
Main rotor blade chord: at root	0.40 m (1 ft 3.9 in)
at tip	0.22 m (8½ in)
Tail rotor blade chord	0.29 m (11½ in)

Bell 412SP in EMS (Emergency Medical Service) configuration

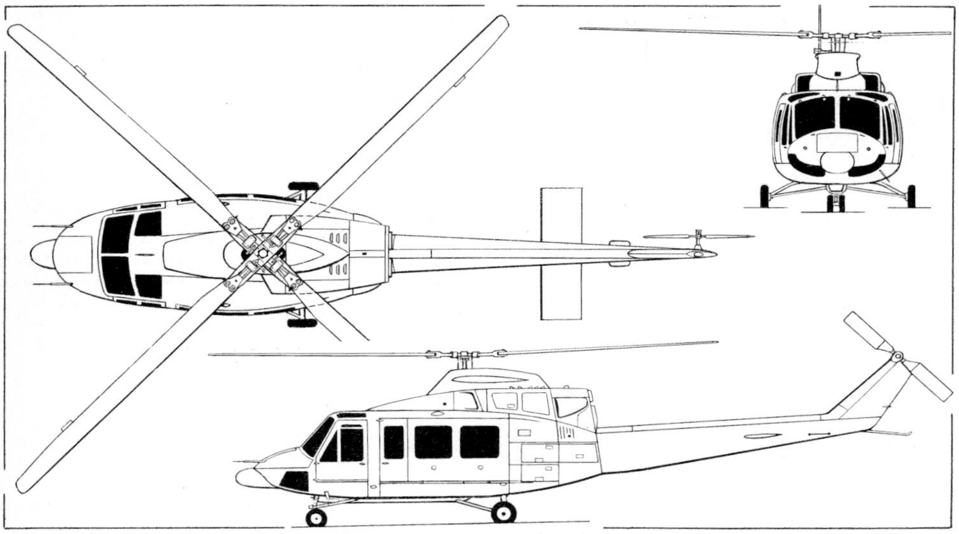

Bell 412HP twin-turboshaft utility helicopter *(Dennis Punnett)*

Length: overall, rotors turning 17.12 m (56 ft 2 in)
fuselage, excl rotors 12.92 m (42 ft 4¾ in)
Height: to top of rotor head 3.48 m (11 ft 5 in)
overall, tail rotor turning 4.57 m (15 ft 0 in)
Stabiliser span 2.87 m (9 ft 5 in)
Width over skids 2.84 m (9 ft 4 in)
Door sizes as Bell 212

AREAS:
Main rotor disc 154.40 m^2 (1,661.9 sq ft)
Tail rotor disc 5.38 m^2 (57.86 sq ft)

WEIGHTS AND LOADINGS:
Weight empty, standard equipped:
VFR 3,018 kg (6,654 lb)
IFR 3,066 kg (6,759 lb)
Max external hook load 2,041 kg (4,500 lb)
Max T-O and landing weight, internal or external load 5,397 kg (11,900 lb)
Max disc loading 34.96 kg/m^2 (7.16 lb/sq ft)
Max power loading 4.02 kg/kW (6.61 lb/shp)

PERFORMANCE (at max T-O weight, ISA):
Max cruising speed:
at S/L 122 knots (226 km/h; 140 mph)
at 1,525 m (5,000 ft) 124 knots (230 km/h; 143 mph)
Long-range cruising speed at 1,525 m (5,000 ft) 130 knots (241 km/h; 150 mph)
Service ceiling, OEI, 30 min power rating 2,070 m (6,800 ft)
Hovering ceiling: IGE 3,110 m (10,200 ft)
OGE 1,585 m (5,200 ft)
Range at 1,525 m (5,000 ft), long-range cruising speed, standard fuel, no reserves 402 nm (745 km; 463 miles)

BELL 230

TYPE: Twin-turboshaft commercial light helicopter.

PROGRAMME: Announced at 1989 NBAA Convention; two Bell 222s converted at Mirabel 1990-91; first flights 12 August (C-GEXP) and 3 October (C-GBLL) 1991; Transport Canada type approval 12 March 1992; production aircraft first flight (C-GAHJ) 23 May 1992; deliveries began (c/n 230002 for Mitsui) 16 November 1992.

CURRENT VERSIONS: **Utility** and **Executive** models, similar to corresponding versions of Model 222. **EMS** (Emergency Medical Service) versions also available. *Details are for initial production aircraft with Allison engines.*

CUSTOMERS: Launch customer Mitsui & Co (distributor) with order for 20; two others sold in 1992. One being leased by Chilean Navy 1993, equipped for SAR with wheel gear, Indal ASIST deck landing system, auxiliary fuel tanks, rescue hoist, Honeywell Primus 700 radar, Spectrolab SX-16 Nightsun searchlight, thermal imager, Honeywell EDZ-705 EFIS with SPZ-7000 AFCS, Loran/GPS receiver and HUD.

COSTS: Canadian government providing up to C$15.2 million as repayable development loan.

DESIGN FEATURES: Replaces Bell 222 (see US section of 1990-91 *Jane's*); Textron Lycoming LTS 101 turboshafts of Bell 222 replaced by Allison 250-C30G2s in first 50 aircraft, LTS 101 optional thereafter; main and tail rotors substantially same as Bell 222, former having Wortmann 090 blade section with 8 per cent thickness/chord ratio and swept tips. Independent (hydraulic) rotor brake. Short span sponson each side of fuselage houses mainwheel units and fuel tanks, and serves as work platform.

FLYING CONTROLS: Fully powered hydraulic, with elastomeric pitch change and flapping bearings; fixed tailplane with leading-edge slats and endplate fins; strakes under sponsons; single-pilot IFR system without autostabilisation approved for Bell 222.

STRUCTURE: Substantially as for Bell 222, ie two-blade main rotor with stainless steel spars and leading-edges, Nomex honeycomb trailing-edge with glassfibre skin, and glassfibre safety straps; tail rotor blades stainless steel; aluminium alloy fuselage with integral tailboom and some honeycomb panels.

LANDING GEAR: Tubular skid type on Utility version. Executive version to have hydraulically retractable tricycle gear, single mainwheels retracting forward into sponsons; forward retracting nosewheel fully castoring and self-centring; hydraulic disc brakes on main units.

POWER PLANT: First 50 aircraft each to be powered by two Allison 250-C30G2 turboshafts, each rated at 522 kW (700 shp) for 5 min for T-O, 464 kW (622 shp) max continuous, 581 kW (779 shp) OEI for 2½ min and 553 kW (742 shp) OEI for 30 min; Textron Lycoming LTS 101 optional thereafter. Main transmission rated at 690 kW (925 shp) for T-O, 652.5 kW (875 shp) max continuous and 548 kW (735 shp) for single-engined operation. Usable fuel capacity 935 litres (247 US gallons; 206 Imp gallons) in skid gear version, 710 litres (187.5 US gallons; 156 Imp gallons) in wheeled version. Optional 182 litres (48 US gallons; 40 Imp gallons) of auxiliary fuel for both versions.

ACCOMMODATION: Standard layout has forward facing seats for nine persons (2-2-2-3) including pilot(s). Options include eight-seat executive (rear six in club layout), six-seat executive (rear four in club layout with console between each pair), or 10-seat utility (2-2-3-3, all forward facing). Customised EMS (emergency medical services) versions also available, configured for pilot-only operation plus one or two pivotable stretchers and four or three medical attendants/sitting casualties respectively. Two forward opening doors each side. Entire interior ram air ventilated and soundproofed. Dual controls optional.

SYSTEMS: Dual hydraulic system (dual for main rotor collective and cyclic, single for tail rotor). Dual 28V DC electrical system, powered by two 30V 200A engine mounted starter/generators (derated to 180A) and a 24V 28Ah nickel-cadmium battery.

AVIONICS: Bendix/King Gold Crown III KTR 908 VHF com radio and KMA 24H-71 ICS standard; Sperry GH-206 attitude indicator. IFR avionics include KCS 305 compass and KPI 552B HSI with glideslope.

EQUIPMENT: Standard equipment includes rotor and cargo tie-downs, ground handling wheels for skid version, retractable 450W search/landing light. Options include dual controls, ECS, auxiliary fuel tankage, force/feel trim system, more comprehensive nav/com avionics, 136 kg (300 lb) capacity rescue hoist, 1,270 kg (2,800 lb) capacity cargo hook, emergency flotation gear, heated windscreen, particle separator and snow baffles.

DIMENSIONS, EXTERNAL:
Main rotor diameter 12.80 m (42 ft 0 in)
Tail rotor diameter 2.10 m (6 ft 10½ in)
Main rotor blade chord 0.66 m (2 ft 2 in)
Tail rotor blade chord 0.25 m (10 in)
Length, incl tailskid: overall, rotors turning (skid gear) 15.23 m (49 ft 11½ in)
overall, rotors turning (wheel gear) 15.29 m (50 ft 2 in)
fuselage (skid gear) 12.81 m (42 ft 0¼ in)
fuselage (wheel gear) 12.87 m (42 ft 2¾ in)
Height over tail fin: skid gear 3.33 m (10 ft 11 in)
wheel gear 3.39 m (11 ft 1½ in)
Height overall: skid gear 3.70 m (12 ft 1½ in)
Width over endplate fins 3.12 m (10 ft 3 in)
Width over sponsons: skid gear 3.62 m (11 ft 10½ in)
wheel gear 3.45 m (11 ft 4 in)
Width over skids 2.39 m (7 ft 10 in)
Wheel track 2.78 m (9 ft 1½ in)
Passenger doors (each): Height 1.24 m (4 ft 1 in)
Width 0.91 m (3 ft 0 in)

DIMENSIONS, INTERNAL:
Cabin (excl flight deck and rear baggage compartment):

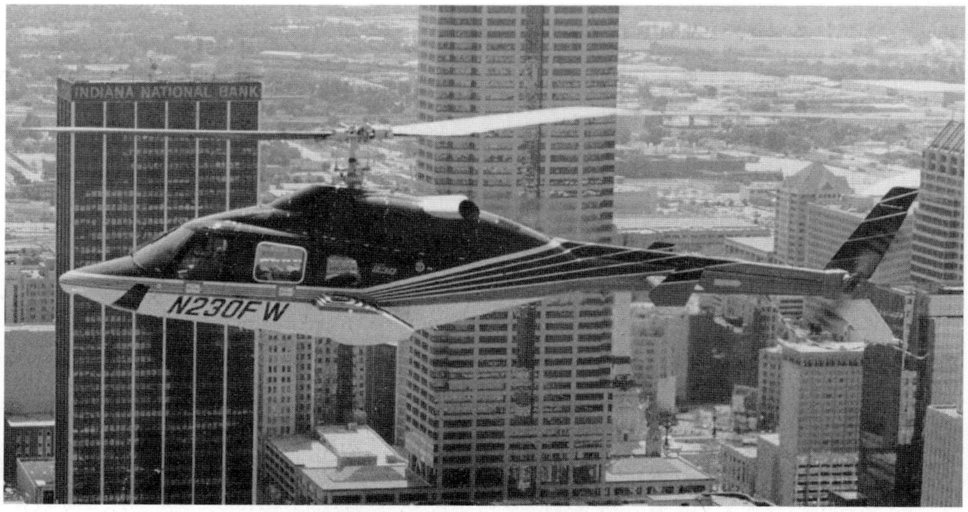

Bell 230, designed as the successor to the 222

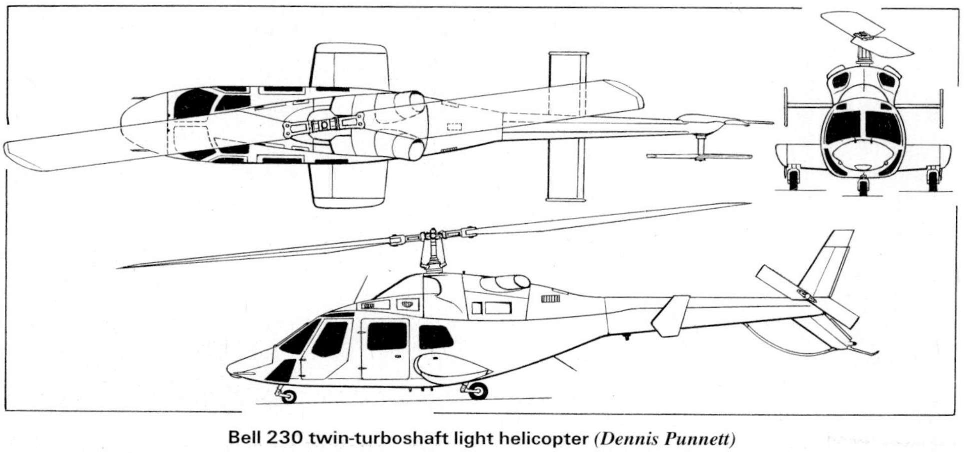

Bell 230 twin-turboshaft light helicopter *(Dennis Punnett)*

Length	2.16 m (7 ft 1 in)
Max width	1.47 m (4 ft 10 in)
Max height	1.45 m (4 ft 9 in)
Volume	3.67 m³ (129.5 cu ft)
Flight deck volume	1.87 m³ (65.9 cu ft)
Baggage compartment volume:	
aft of cabin	1.05 m³ (37.2 cu ft)
nose	0.40 m³ (14.2 cu ft)

AREAS:
Main rotor disc	128.7 m² (1,385.4 sq ft)
Tail rotor disc	3.45 m² (37.12 sq ft)

WEIGHTS AND LOADINGS:
Weight empty, standard:
skid gear	2,268 kg (5,000 lb)
wheel gear	2,312 kg (5,097 lb)
Max external sling load	1,270 kg (2,800 lb)
Max T-O weight, internal or external load	3,810 kg (8,400 lb)
Max disc loading	29.6 kg/m² (6.06 lb/sq ft)
Max power loading	5.53 kg/kW (9.08 lb/shp)

PERFORMANCE (at max T-O weight, ISA, except where indicated):
Max cruising speed at S/L:
skid gear	137 knots (254 km/h; 158 mph)
wheel gear, standard fuel	141 knots (261 km/h; 162 mph)
wheel gear, auxiliary fuel	137 knots (254 km/h; 158 mph)

Econ cruising speed at S/L at average gross weight:
skid gear, standard fuel	134 knots (248 km/h; 154 mph)
wheel gear	138 knots (256 km/h; 159 mph)

Service ceiling at max cruise power:
both	4,725 m (15,500 ft)

Service ceiling, OEI, 30 min power rating:
both	2,350 m (7,700 ft)
Hovering ceiling IGE (both): ISA	3,780 m (12,400 ft)
ISA + 20°C	2,895 m (9,500 ft)
Hovering ceiling OGE (both): ISA	2,225 m (7,300 ft)
ISA + 20°C	1,220 m (4,000 ft)

Range at S/L at econ cruising speed, standard fuel, no reserves:
skid gear	385 nm (713 km; 443 miles)
wheel gear	301 nm (558 km; 346 miles)

Range at S/L at econ cruising speed, auxiliary fuel, no reserves:
wheel gear	379 nm (702 km; 436 miles)

BELL 430
TYPE: Four-blade rotor, higher powered and stretched variant of Bell 230.
PROGRAMME: Proposed 1992; prototype completion expected by end of 1993, first flight early 1994, deliveries to begin late 1995.
DESIGN FEATURES: Bell 680 all-composites four-blade bearingless main rotor; Bell 230 fuselage lengthened by 0.46 m (18 in) plug; approx 10 per cent power increase from use of 582 kW (780 shp) Allison 250-C40 turboshafts (prototype will probably have 250-C30G2s); max T-O weight increased to 5,443 kg (12,000 lb).

BELL 442
TYPE: Twin-turboshaft medium-weight helicopter.
PROGRAMME: In design definition stage Spring 1993; same weight class as Bell 430; four-blade main rotor; intended as Bell 412 successor; international partner(s) being sought.
POWER PLANT: Two Allison/Garrett CTS800 or MTU/Turbomeca/Rolls-Royce MTR 390 turboshafts.

BELL LIGHT HELICOPTER
TYPE: Turbine-powered light helicopter.
PROGRAMME: Design definition under way early 1993 as 1,814 kg (4,000 lb) gross weight class replacement for JetRanger with four-blade main rotor; single or twin engines; international partner(s) being sought.

BOMBARDIER
BOMBARDIER AEROSPACE GROUP, NORTH AMERICA
PRESIDENT: Robert E. Brown
MEMBER COMPANIES:
Canadair: see this section
de Havilland: see this section
Learjet: see US section
BUSINESS UNITS:
Amphibious Aircraft Division
Manufacturing Division
Defence Systems Division
MARKETING, SALES & SUPPORT UNITS:
Regional Aircraft Division (Toronto)
PRESIDENT: Robert A. Wohl
VICE-PRESIDENTS:
Tom Appleton (Marketing and Sales)
Larry Dugan (Customer Support)
Paul Francoeur (Contracts)
PUBLIC RELATIONS: Colin S. Fisher
Business Aircraft Division (Montreal)
PRESIDENT: Bryan T. Moss

Created 1992 with combined workforce of nearly 15,000 (Canadair 8,200, de Havilland 3,300, Learjet 3,350). Current products and programmes comprise Canadair Challenger 601-3A, Regional Jet, Global Express and CL-215T/415; de Havilland Dash 8; and Learjet 31A, 35A, 45 and 60. Amphibious Aircraft Division responsible for CL-215T/415; Manufacturing Division for Challenger, Regional Jet, Dash 8 and Global Express, plus component manufacture for Airbus, Boeing, BAe and McDonnell Douglas; Defence Systems Division for CF-18 support, Canadian Forces flying training et al; and CL-89/289/227 UAVs. Regional Aircraft Division provides marketing, sales and support for Regional Jet and Dash 8, Business Aircraft Division for Challenger, Regional Jet, Global Express and Learjets.

The Bombardier family of regional aircraft: the 50-passenger Canadair Regional Jet (nearest camera) and the new de Havilland Dash 8 Series 200

BRISTOL AEROSPACE
BRISTOL AEROSPACE LTD
PO Box 874, 660 Berry Street, Winnipeg, Manitoba R3C 2S4
Telephone: 1 (204) 775 8331
Fax: 1 (204) 885 3195
Telex: 0757774
EXECUTIVE VICE-PRESIDENT AND GENERAL MANAGER:
Keith F. Burrows
VICE-PRESIDENT, AIRCRAFT DIVISION: James S. Butyniec

NORTHROP F-5 UPGRADE PROGRAMMES
Canadian Forces designation: CF-116
TYPE: Lead-in trainer for CF-188 Hornets.
PROGRAMME: Two 'prototypes' ordered late 1988; CF fleet fitment ordered November 1990; prototype first flown 14 June 1991; life extension changes completed on 23 CF-5s by January 1992; first service re-entry Spring 1992 with No. 419 Squadron at CFB Cold Lake, Alberta; completion due 1995. See also *Jane's Civil and Military Aircraft Upgrades*.
CUSTOMERS: Canadian Forces (13 single-seat and 33 two-seat); also contracted by CASA 1991 to undertake structural improvements on 23 Spanish Air Force F-5Bs.
COSTS: Canadian fleet fitment contract worth C$69.73 million (1990) including installation, a new database and a training simulator.
DESIGN FEATURES: Life extension modifications include repair, overhaul and re-skinning of wings and fin; replacement of dorsal longeron, rear fuselage formers and landing gear; complete rewiring and repainting.
AVIONICS: New suite includes GEC-Marconi Avionics HUD and mission computer/display processor, Litton INS, Magnavox AN/ARC-164 VHF com, J. E. T. standby attitude indicator, Conrac angle of attack sensor, Honeywell radar altimeter and Ferranti video camera.

Northrop CF-5D fighter/trainers undergoing airframe and avionics upgrade by Bristol Aerospace

CANADAIR
CANADAIR GROUP, BOMBARDIER INC

Cartierville Airport, 1800 Laurentien Boulevard, St Laurent, Quebec H4R 1K2
POSTAL ADDRESS: PO Box 6087, Station A, Montreal, Quebec H3C 3G9
Telephone: 1 (514) 744 1511
Fax: 1 (514) 744 6586
Telex: 05-826747
PRESIDENT: Robert E. Brown
PRESIDENT, BOMBARDIER BUSINESS AIRCRAFT DIVISION:
  Bryan T. Moss
PRESIDENT, BOMBARDIER REGIONAL AIRCRAFT DIVISION:
  Robert A. Wohl
PRESIDENT, AMPHIBIOUS AIRCRAFT DIVISION:
  Pierre-André Roy
VICE-PRESIDENTS/GENERAL MANAGERS:
  Roland Gagnon (Manufacturing Division)
  Walter Niemy (Defence Systems Division)
PUBLIC RELATIONS: Catherine Chase

By 1993 Canadair had manufactured some 4,200 aircraft since 1944; acquired by Bombardier Inc 23 December 1986; merged with parent company as Canadair Group of Bombardier Inc 5 August 1988; has five divisions: Business Aircraft, Regional Aircraft, Amphibious Aircraft, Manufacturing, and Defence Systems.

Three plants in St Laurent complex at Cartierville Airport; facility at Montreal (Dorval) expanded for Challenger and Regional Jet assembly; new facilities at Montreal (Mirabel) for CF-18 work and other military work; Regional Aircraft Division based at Cartierville. Total covered floor area (early 1992) 332,392 m² (3,577,834 sq ft); workforce 8,200 in September 1992.

Main subcontracts include 512 nose barrel units for McDonnell Douglas F/A-18 (471 delivered by 31 August 1992); six major fuselage subassemblies for 600 Airbus A330/340 for Aerospatiale, Airbus inboard wing leading-edge assemblies for BAe and rear fuselage sections for Boeing 767. Prime contractor for engineering support for CF-18 (with CAE Electronics and NWI); also repairs, overhauls and produces spares for other aircraft.

CANADAIR CHALLENGER
Canadian Forces designations: CC-144 and CE-144A
TYPE: Twin-turbofan business, cargo and regional transport.
PROGRAMME: First flight of first of three prototypes (C-GCGR-X) 8 November 1978; first flight production Challenger 600 with Lycoming ALF 502L-2 turbofans 21 September 1979; first customer delivery 30 December 1980; prototype Challenger 601 with GE CF34s 10 April 1982; first 601-1A delivered 6 May 1983; first 601-3A 6 May 1987 and first 601-3A/ER 19 May 1989; 601-S (1991-92 *Jane's*) launched June 1989 but not committed to production.
CURRENT VERSIONS: **Challenger 600:** Total 83 built after certification in 1980 (76 since retrofitted with winglets); 12 delivered to Canadian Department of National Defence as CC-144 (four) and CE-144A (three) (see 1989-90 and earlier *Jane's*), plus four for coastal patrol. Production completed.
  Challenger 601-1A: First production version (66 built) to have CF34 engines (see 1990-91 *Jane's*); first flight 17 September 1982. Production completed.
  Challenger 601-3A: Current standard version with 'glass cockpit' and upgraded CF34s; first flight 28 September 1986; Canadian and US certification 21 and 30 April 1987; also certificated for Cat. II and in 22 other countries; improvements include CF34-3A engines flat rated to 21°C, and fully integrated digital flight guidance and flight management systems. *Detailed description applies to this version except where indicated.*
  Challenger 601-3A/ER: Extended range option available on new 601-3As since 1989 and as retrofit to 601-1As and 601-3As; range increased to 3,585 nm (6,643 km; 4,128 miles) with NBAA IFR reserves; first flight 8 November 1988; Canadian certification 16 March 1989; tail fairing replaced with conformal tailcone fuel tank which extends fuselage length by 46 cm (18 in) and adds 118 kg (260 lb) to operating weight empty; max ramp weight increased by 680 kg (1,500 lb). Optional gross weight increase of 227 kg (500 lb).
CUSTOMERS: Total 84 Challenger 600 and 66 601-1A delivered; total 121 Challenger 601-3A delivered between 6 May 1987 and 5 January 1993; total 271 Challengers of all versions delivered by 5 January 1993.
DESIGN FEATURES: Advanced wing section; quarter-chord sweep 25°; thickness/chord ratio 14 per cent at root, 12 per cent at leading-edge sweep break and 10 per cent at tip; dihedral 2° 33'; incidence at root 3° 30'; fuselage circular cross-section, pressurised.
FLYING CONTROLS: Fully powered hydraulic controls; electrically actuated variable incidence tailplane; two-segment spoilers (outboard airbrake panels, inboard lift dumpers); two-segment double-slotted flaps.
STRUCTURE: Two-spar wing torsion box; chemically milled fuselage skin panels with riveted frames and stringers form damage-tolerant structure; multi-spar fin and tailplane.
LANDING GEAR: Hydraulically retractable tricycle type, with twin wheels and Dowty oleo-pneumatic shock absorber on each unit. Mainwheels retract inward into wing centre-section, nose unit forward. Nose unit steerable and self-centring. Mainwheels have Goodyear 25 × 6.75 tyres, pressure 14.20 bars (206 lb/sq in); nosewheels have B. F. Goodrich 18 × 4.4 tyres, pressure 10.41 bars (151 lb/sq in). ABS (Aircraft Braking Systems) hydraulically operated multiple-disc carbon brakes with fully modulated anti-skid system. Min ground turning radius 12.19 m (40 ft 0 in).
POWER PLANT: Two General Electric CF34-3A turbofans, each rated at 40.66 kN (9,140 lb st) with automatic power reserve, or 38.48 kN (8,650 lb st) without APR, pylon mounted on rear fuselage and fitted with cascade type fan-air thrust reversers. Nacelles and thrust reversers by Vought Aircraft. Integral fuel tank in centre-section (capacity 2,839 litres; 750 US gallons; 624 Imp gallons), one in each wing (each 2,725 litres; 720 US gallons; 600 Imp gallons) and auxiliary tanks (combined capacity 984 litres; 260 US gallons; 216.5 Imp gallons) beneath cabin floor; total capacity 9,278 litres (2,451 US gallons; 2,041 Imp gallons). Optional tank in tailcone (601-3A/ER), capacity 696.5 litres (184 US gallons; 153 Imp gallons). Pressure and gravity fuelling and defuelling. Oil capacity 13.6 litres (3.6 US gallons; 3 Imp gallons).
ACCOMMODATION: Two-pilot flight deck with dual controls. Blind-flying instrumentation standard. Cabin interiors to customer's specifications; maximum of 19 passenger seats approved. Typical installations include toilet, buffet, bar and wardrobe. Medevac version can carry up to seven stretcher patients, infant incubator, full complement of medical staff and comprehensive intensive care equipment. Baggage compartment, with own loading door, accessible in flight. Downward opening, power assisted door on port side, forward of wing. Overwing emergency exit on starboard side. Entire accommodation heated, pressurised and air-conditioned.
SYSTEMS: AiResearch pressurisation and air-conditioning systems, max pressure differential 0.61 bar (8.8 lb/sq in). Three independent hydraulic systems, each of 207 bars (3,000 lb/sq in). No. 1 system powers flight controls (via servo-actuators positioned by cables and pushrods); No. 2 system for flight controls and brakes; No. 3 system for flight controls, landing gear extension/retraction, brakes and nosewheel steering. Nos. 1 and 2 systems each

Canadair Challenger 601-3A twin-turbofan business aircraft

Extended rear fuselage of Challenger 601-3A/ER, housing additional fuel tank

powered by an engine driven pump, supplemented by an AC electric pump; No. 3 system by two AC pumps. Two 30kVA engine driven generators supply primary 115/200V three-phase AC electric power at 400Hz. Four transformer-rectifiers to convert AC power to 28V DC; one 43Ah nickel-cadmium battery. Alternative primary power provided by APU and/or an air driven generator, latter deployed automatically in flight if engine driven generators and APU are inoperative. Stall warning system, with stick shakers and stick pusher. Garrett GTCP-100E gas turbine APU for engine start, ground air-conditioning and other services. Electric anti-icing of windscreen, flight deck side windows and pitot heads; Sundstrand bleed air anti-icing of wing and tailplane leading-edges, engine intake cowls and guide vanes. Gaseous oxygen system, pressure 127.5 bars (1,850 lb/sq in). Continuous-element fire detectors in each engine nacelle, APU and main landing gear bays; two-shot extinguishing system for engines, single-shot system for APU.

AVIONICS: Honeywell digital avionics include SPZ-8000 five-tube electronic flight instrument system (EFIS) including a single multi-function display (MFD); dual Honeywell laser inertial reference systems (LIRS); dual flight management systems; Lasertrak navigation display unit; digital automatic flight control system, with dual channel fail-operational autopilot and flight director; Mach trim and auto trim; dual digital air data system; Honeywell Primus 650 or 870 four-colour digital weather radar (latter with turbulence detection); radio altimeter; Collins Pro Line II nav/coms, including dual VHF com; dual VHF nav; dual DME; dual ATC transponders; dual ADF; dual HF com; cockpit voice recorder; standby instruments (artificial horizon, airspeed indicator and altimeter). Systems certificated for Cat. II operations. Space provisions for flight data recorder, ELT, VLF/Omega, GPWS, and full provisions for third LIRS.

EQUIPMENT (Medevac version): Includes cardio-pulmonary resuscitation unit; physio control lifepack comprising heart defibrilator, ECG and cardioscope; ophthalmoscope; respirators and resuscitators; infant monitor; X-ray viewer; cardiostimulator; foetal heart monitor; and anti-shock suit.

DIMENSIONS, EXTERNAL:
Wing span over winglets	19.61 m (64 ft 4 in)
Wing chord: at fuselage c/l	4.89 m (16 ft 0½ in)
at tip	1.27 m (4 ft 1.9 in)
Wing aspect ratio (excl winglets)	8.5
Length overall	20.85 m (68 ft 5 in)
Fuselage: Max diameter	2.69 m (8 ft 10 in)
Height overall	6.30 m (20 ft 8 in)
Tailplane span	6.20 m (20 ft 4 in)
Wheel track (c/l of shock struts)	3.18 m (10 ft 5 in)
Wheelbase	7.99 m (26 ft 2½ in)
Passenger door (port, fwd): Height	1.78 m (5 ft 10 in)
Width	0.91 m (3 ft 0 in)
Height to sill	1.61 m (5 ft 3½ in)
Baggage door (port, rear): Height	0.84 m (2 ft 9 in)
Width	0.71 m (2 ft 4 in)
Height to sill	1.61 m (5 ft 3½ in)
Overwing emergency exit (stbd):	
Height	0.91 m (3 ft 0 in)
Width	0.51 m (1 ft 8 in)

DIMENSIONS, INTERNAL:
Cabin: Length, incl galley, toilet and baggage area, excl flight deck	8.61 m (28 ft 3 in)
Max width	2.49 m (8 ft 2 in)
Width at floor level	2.18 m (7 ft 2 in)
Max height	1.85 m (6 ft 1 in)
Floor area	18.77 m² (202 sq ft)
Volume	32.6 m³ (1,150 cu ft)

AREAS:
Wings, gross (excl winglets)	48.31 m² (520.0 sq ft)
Ailerons (total)	1.39 m² (15.0 sq ft)
Trailing-edge flaps (total)	7.80 m² (84.0 sq ft)
Fin	9.18 m² (98.8 sq ft)
Rudder	2.03 m² (21.9 sq ft)
Tailplane	6.45 m² (69.4 sq ft)
Elevators (total)	2.15 m² (23.1 sq ft)

WEIGHTS AND LOADINGS:
Manufacturer's weight empty: 3A	9,292 kg (20,485 lb)
3A/ER	9,405 kg (20,735 lb)
Operating weight empty: 3A	11,566 kg (25,500 lb)
3A/ER	11,684 kg (25,760 lb)

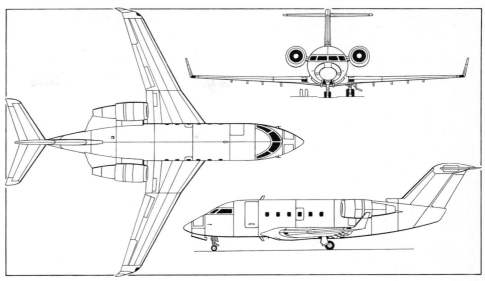

Canadair Challenger 601-3A (two General Electric CF34-3A turbofans) *(Jane's/Mike Keep)*

Early production Challenger 601-1A in service with the German Luftwaffe *(Press Office Sturzenegger)*

Max fuel: 3A	7,559 kg (16,665 lb)	Max wing loading:	
3A/ER	8,119 kg (17,900 lb)	3A, standard	404.7 kg/m² (82.88 lb/sq ft)
Max payload: 3A	1,814 kg (4,000 lb)	3A/ER, standard	418.8 kg/m² (85.77 lb/sq ft)
3A/ER	1,696 kg (3,740 lb)	3A, 3A/ER, optional	423.5 kg/m² (86.73 lb/sq ft)
Payload with max fuel: 3A	492 kg (1,085 lb)	Max power loading (without APR)	
3A/ER	494 kg (1,090 lb)	3A, standard	254.03 kg/kN (2.49 lb/lb st)
Max T-O weight: * 3A	19,550 kg (43,100 lb)	3A/ER, standard	262.86 kg/kN (2.58 lb/lb st)
* 3A/ER	20,230 kg (44,600 lb)	* 20,457 kg (45,100 lb) option available	
Max ramp weight: **3A	19,618 kg (43,250 lb)	** 20,525 kg (45,250 lb) option available	
**3A/ER	20,300 kg (44,750 lb)	*** 14,060 kg (31,000 lb) option available	
Max landing weight: 3A, 3A/ER	16,329 kg (36,000 lb)	PERFORMANCE (at max T-O weight except where indicated):	
Max zero-fuel weight:		Max cruising speed:	
***3A, 3A/ER	13,381 kg (29,500 lb)	3A, 3A/ER	476 knots (882 km/h; 548 mph)

Interior of a Challenger 601-3A operated by Thai Airways International

28 CANADA: AIRCRAFT—CANADAIR

Normal cruising speed:
 3A, 3A/ER 459 knots (851 km/h; 529 mph)
Long-range cruising speed:
 3A, 3A/ER 424 knots (786 km/h; 488 mph)
Time to initial cruise altitude: 3A 24 min 42 s
 3A/ER 22 min
Max operating altitude: 3A, 3A/ER 12,500 m (41,000 ft)
Service ceiling, OEI: 3A 6,000 m (19,700 ft)
 3A/ER 5,600 m (18,400 ft)
Balanced T-O field length (ISA at S/L):
 3A 1,645 m (5,400 ft)
 3A/ER 1,791 m (5,875 ft)
Landing distance at S/L at max landing weight:
 3A, 3A/ER 1,006 m (3,300 ft)
Range with max fuel and five passengers, NBAA IFR reserves (200 nm; 370 km; 230 mile alternate) at long-range cruising speed:
 3A 3,365 nm (6,236 km; 3,874 miles)
 3A/ER 3,585 nm (6,643 km; 4,128 miles)
Design g limit: 3A, 3A/ER +2.6
OPERATIONAL NOISE LEVELS (FAR Pt 36):
 T-O: 3A 79.4 EPNdB
 3A/ER 79.4 EPNdB
 Sideline: 3A 85.9 EPNdB
 3A/ER 85.7 EPNdB
 Approach: 3A, 3A/ER 90.8 EPNdB

CANADAIR REGIONAL JET

TYPE: Twin-turbofan regional transport.
PROGRAMME: Design studies began Autumn 1987; basic configuration frozen June 1988; formal programme go-ahead given 31 March 1989; extended range 100ER announced September 1990. Three development aircraft built (c/n 7001-7003), plus static test airframe (c/n 7991) and forward fuselage test article (7992); first flight of 7001 (C-FCRJ) 10 May 1991, following official rollout on 6 May; 7002 (C-FNRJ) first flew 2 August 1991 and 7003 on 17 November 1991; all three in 1,400 hour flight test programme in Wichita, USA. CF34-3A1 engine obtained its US type certificate 24 July 1991. Transport Canada type approval (100 and 100ER) 31 July 1992; first flight of first delivery aircraft (c/n 7004) 4 July 1992; first delivery (to Lufthansa CityLine of Germany) 29 October 1992, entering service 1 November; European JAA and US FAA certification 15 and 21 January 1993 respectively.
CURRENT VERSIONS: **Series 100**: Standard aircraft; designed to carry 50 passengers over 980 nm (1,816 km; 1,128 mile) range; max T-O weight 21,523 kg (47,450 lb).
 Series 100ER: Extended range capability with optional increase in max T-O weight to 23,133 kg (51,000 lb) and optional additional fuel capacity, for range of 1,620 nm (3,002 km; 1,865 miles).
CUSTOMERS: 26 orders and 36 options (incl one 30-seat corporate for Xerox Corporation) under contract at June 1993; further orders/options then under letters of intent. Launch customers are Lufthansa CityLine (13), MTM Aviation GmbH (two) and Comair Inc (The Delta Connection) (20). Ten delivered by 28 May 1993.

COSTS: Programme development costs C$275 million.
DESIGN FEATURES: Evolved from Challenger (which see), designed expressly for regional airline operating environment. Advanced transonic wing design, with winglets for high speed operations; fuel-efficient GE turbofans; options include higher design weights, additional fuel capacity, more comprehensive avionics, and max certificated altitude raised to 12,500 m (41,000 ft).
 Wings, designed with computational fluid dynamics (CFD), have 13.2 per cent (root) and 10 per cent (tip) thickness/chord ratios, 2° 20′ dihedral, 3° 25′ root incidence and 24° 45′ quarter-chord sweepback.
FLYING CONTROLS: Conventional three-axis primary controls with cables and push/pull rods for multiple redundancy; hydraulically actuated ailerons, elevators and rudder with at least two hydraulic power control unit actuators per surface (three on elevators); ailerons and elevators fitted with flutter dampers, rudder with dual-channel control yaw damping; artificial feel and electronic trim on all axes; variable incidence T tailplane. Double-slotted flaps with dual Datron electric motors; GEC-Marconi Avionics fly-by-wire spoiler and spoileron system, four spoilers each side, with inner two functioning as ground spoilers, outer two comprising one flight spoiler and one spoileron, both also providing lift dumping on touchdown. Avionics suite includes engine indication and crew alerting system (EICAS).
STRUCTURE: Semi-monocoque fuselage is damage tolerant FAR/JAR Pt 25 certificated airframe with chemically milled skins; flat pressure bulkheads forward of flight deck and aft of baggage compartment; extensive use of advanced composites in secondary structures (passenger compartment floor, wing/fuselage fairings, nacelle doors, wing access door covers, winglets, tailcone, avionics access doors and landing gear doors); comprehensive anti-corrosion treatment and drainage. Wing is one-piece unit mounted to underside of fuselage; two-spar box joined by ribs, covered top and bottom with integrally stiffened skin panels (three upper and three lower each side) for smooth flow; machined or built-up spars and shearweb type ribs. Short Brothers (see UK section) manufactures fuselage central section, fore and aft fuselage plugs, wing flaps, ailerons, spoilerons and inboard spoilers.
LANDING GEAR: Hydraulically retractable tricycle type, manufactured by Dowty. Inward retracting main units each have floating piston type shock absorption and 15 in Aircraft Braking System wheels with 29 × 9-15 Goodyear H type tubeless tyres, pressure 11.17 bars (162 lb/sq in) unladen. Nose unit has Dowty Canada steer-by-wire steering and unladen tyre pressure of 8.62 bars (125 lb/sq in). Aircraft Braking System steel multi-disc brakes and fully modulated Hydro Aire Mk III anti-skid system. Min ground turning radius 22.86 m (75 ft 0 in).
POWER PLANT: Two General Electric CF34-3A1 turbofans, each rated at 41.01 kN (9,220 lb st) with APR and 38.83 kN (8,729 lb st) without. Nacelles produced by Vought Aircraft. Pneumatically actuated cascade-type fan-air thrust reversers. Fuel in two integral wing tanks, combined capacity 5,300 litres (1,400 US gallons; 1,166 Imp gallons); increasable to 8,080 litres (2,135 US gallons; 1,778 Imp gallons) with optional centre wing tank. Pressure refuelling point in starboard leading-edge wingroot; two gravity points on starboard wing (one for centre tank) and one on port wing.
ACCOMMODATION: Two-pilot flight deck; one or two cabin attendants. Main cabin seats up to 50 passengers in standard configuration, four-abreast at 79 cm (31 in) pitch, with centre aisle; max capacity 52 seats. Various configurations, from 15 to 50 seats, available for corporate version. Downward opening front passenger door with integral airstairs on port side, plug type forward emergency exit/service door opposite on starboard side. Inward opening baggage door on port side at rear. Overwing Type III emergency exit each side. Entire accommodation pressurised, including rear baggage compartment.
SYSTEMS: Cabin pressurisation and air-conditioning system (max differential 0.57 bar; 8.3 lb/sq in). Primary flight control systems powered by hydraulic servo-actuators with distinct, alternate paths cable and pushrod systems. Electric trim and dual yaw dampers. Three fully independent 207 bar (3,000 lb/sq in) hydraulic systems. Three-phase 115V AC electrical primary power at 400Hz supplied by two 30kVA engine driven generators; alternative power provided by APU and air driven generator. Conversion to 28V DC by five transformer-rectifier units. Main (nickel-cadmium) battery 17Ah, APU battery 43Ah. Garrett GTCP 36-150 (RJ) APU and two-pack air-conditioning system in rear of fuselage. Wing leading-edges and engine intake cowls anti-iced by engine bleed air. Electric anti-icing of windscreen and cockpit side windows, pitot heads, air data vanes, static sources and sensors. Ice detection system standard.
AVIONICS: Collins Pro Line IV integrated all-digital avionics suite, including dual primary flight displays, dual multi-function displays, dual EICAS, dual AFCS, dual AHRS, dual nav/com radios, dual air data system and Cat. II capability. Digital weather radar system, GPWS, windshear detection system and TCAS. Loral Fairchild flight data recorder. Options include dual flight management system, dual inertial reference system in lieu of AHRS, Cat. IIIa landing capability using head-up guidance system, split-scan radar, weather radar with turbulence mode, HF radio, single Selcal, and MLS provisions.
DIMENSIONS, EXTERNAL: As for Challenger 601-3A except:
 Wing span over winglets 21.21 m (69 ft 7 in)
 Wing chord: at fuselage c/l 5.13 m (16 ft 10 in)
  at tip 1.27 m (4 ft 2 in)
 Wing aspect ratio (excl winglets) 8.85
 Length: overall 26.77 m (87 ft 10 in)
  fuselage 24.38 m (80 ft 0 in)
 Height overall 6.22 m (20 ft 5 in)
 Wheelbase 11.39 m (37 ft 4½ in)
 Service door (stbd, fwd): Height 1.22 m (4 ft 0 in)
  Width 0.61 m (2 ft 0 in)
  Height to sill 1.61 m (5 ft 3½ in)
 Baggage door (port, rear): Width 1.09 m (3 ft 7 in)
DIMENSIONS, INTERNAL: As for Challenger 601-3A except:
 Cabin (incl baggage compartment, excl flight deck):
  Length 14.76 m (48 ft 5 in)
  Max height 1.87 m (6 ft 1½ in)

Canadair Regional Jet 50/52-passenger transport in the insignia of Lufthansa CityLine

Cutaway drawing of the Canadair Regional Jet

Floor area	32.14 m² (346.0 sq ft)
Volume	57.06 m³ (2,015 cu ft)
Stowage volume:	
main (rear) baggage compartment	8.89 m³ (314.0 cu ft)
wardrobes/bins/underseat (total)	5.15 m³ (182.0 cu ft)
AREAS:	
Wings: gross (excl winglets)	54.54 m² (587.1 sq ft)
net	48.35 m² (520.4 sq ft)
Ailerons (total)	1.93 m² (20.8 sq ft)
Trailing-edge flaps (total)	10.60 m² (114.1 sq ft)
Spoilers (total)	2.26 m² (24.3 sq ft)
Winglets (total)	1.38 m² (14.9 sq ft)
Fin	9.18 m² (98.8 sq ft)
Rudder	2.03 m² (21.9 sq ft)
Tailplane	9.44 m² (101.6 sq ft)
Elevators (total)	2.84 m² (30.52 sq ft)
WEIGHTS AND LOADINGS:	
Manufacturer's weight empty:	
100, 100ER	13,236 kg (29,180 lb)
Operating weight empty: 100	13,653 kg (30,100 lb)
100ER	13,663 kg (30,122 lb)
Max payload (structural): 100	5,488 kg (12,100 lb)
100ER	6,295 kg (13,878 lb)
Max fuel: 100	4,254 kg (9,380 lb)
100ER	6,489 kg (14,305 lb)
Payload with max fuel: 100	3,728 kg (8,220 lb)
100ER	3,095 kg (6,823 lb)
Max T-O weight: 100	21,523 kg (47,450 lb)
100ER	23,133 kg (51,000 lb)
Max ramp weight: 100	21,636 kg (47,700 lb)
100ER	23,246 kg (51,250 lb)
Max zero-fuel weight: 100	19,141 kg (42,200 lb)
100ER	19,958 kg (44,000 lb)
Max landing weight: 100	20,275 kg (44,700 lb)
100ER	21,319 kg (47,000 lb)
Max wing loading: 100	394.6 kg/m² (80.82 lb/sq ft)
100ER	424.14 kg/m² (86.87 lb/sq ft)
Max power loading (APR rating):	
100	262.48 kg/kN (2.57 lb/lb st)
100ER	282.1 kg/kN (2.77 lb/lb st)
PERFORMANCE (at max T-O weight except where indicated):	
Max operating speed:	
above 9,570 m (31,400 ft)	Mach 0.85
below 7,740 m (25,400 ft)	335 knots (621 km/h; 386 mph)
High-speed cruising speed at 11,275 m (37,000 ft)	Mach 0.80 or 459 knots (851 km/h; 529 mph)
Normal cruising speed at 11,275 m (37,000 ft)	Mach 0.74 or 424 knots (786 km/h; 488 mph)
Approach speed, 45° flap, AUW of 19,504 kg (43,000 lb)	137 knots (254 km/h; 158 mph)
Max rate of climb at 457 m (1,500 ft), 250 knots CAS/ Mach 0.74 climb schedule: 100	1,128 m (3,700 ft)/min
100ER	1,036 m (3,400 ft)/min

Retouched photograph showing the Regional Jet in the colours of first corporate customer Xerox

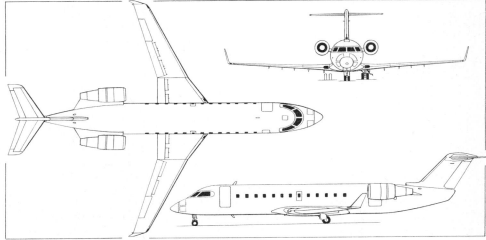

Canadair Regional Jet transport (two General Electric CF34-3A1 turbofans) *(Dennis Punnett)*

30 CANADA: AIRCRAFT—CANADAIR

Max operating altitude	12,500 m (41,000 ft)
FAR T-O field length at S/L, ISA	1,605 m (5,265 ft)
FAR landing field length at S/L, ISA, at max landing weight	1,440 m (4,725 ft)

Range with max payload at long-range cruising speed, FAR Pt 121 reserves:
100	980 nm (1,816 km; 1,128 miles)
100ER	1,620 nm (3,002 km; 1,865 miles)
Corporate (30 seats), NBAA IFR reserves	2,141 nm (3,967 km; 2,465 miles)

OPERATIONAL NOISE LEVELS:
T-O	78.6 EPNdB
Approach	92.1 EPNdB
Sideline	82.2 EPNdB

CANADAIR GLOBAL EXPRESS

TYPE: Long-range high-speed corporate transport.
PROGRAMME: Announced 28 October 1991 at NBAA Convention; preliminary design phase started early 1993 and advance design phase mid-1993; could be launched with minimum of 30 firm orders, leading to first flight 1996; launch decision expected before end of 1993.
CUSTOMERS: Deposits placed for 37 aircraft by March 1993, from operators on all five continents.
COSTS: Tentatively $25 million (1991) for 'green' aircraft.
DESIGN FEATURES: Wide-body fuselage with approximately same cabin length as Regional Jet; all-new 35° supercritical swept wings with winglets; rear mounted engines; sweptback T tail. Three-compartment passenger configuration, plus crew rest area, large galley and baggage compartment; modern flight deck may include six flat panel displays, head-up guidance system, sidestick controllers and electronic library.
POWER PLANT: Two rear mounted 65.30 kN (14,680 lb st) BMW Rolls-Royce BR710-48-C2 turbofans, flat rated to ISA + 20°C.
ACCOMMODATION: Crew of four (incl cabin attendant) and eight to 30 passengers.

Following data are provisional:

DIMENSIONS, EXTERNAL:
Wing span over winglets	27.64 m (90 ft 8 in)
Length overall	29.51 m (96 ft 10 in)
Length of fuselage	24.89 m (81 ft 8 in)
Height overall	7.42 m (24 ft 4 in)

DIMENSIONS, INTERNAL:
Cabin (excl flight deck): Length	14.63 m (48 ft 0 in)
Width at floor	2.11 m (6 ft 11 in)
Max width	2.49 m (8 ft 2 in)
Max height	1.90 m (6 ft 3 in)
Floor area	31.12 m² (335.0 sq ft)
Volume	58.81 m³ (2,077 cu ft)
Baggage compartment volume	4.96 m³ (175 cu ft)
Total baggage volume	9.17 m³ (324 cu ft)

WEIGHTS AND LOADINGS:
Max T-O weight	41,277 kg (91,000 lb)
Max landing weight	35,650 kg (78,600 lb)

PERFORMANCE (design target):
Max cruising speed	Mach 0.88
Normal cruising speed	Mach 0.85
Long-range cruising speed	Mach 0.80
Initial cruising altitude	12,500 m (41,000 ft)
Max operating altitude	15,545 m (51,000 ft)
T-O and landing field length at max T-O weight	1,830 m (6,000 ft)

Range with four crew and eight passengers:
at Mach 0.80	6,500 nm (12,045 km; 7,485 miles)
at Mach 0.85	6,000 nm (11,119 km; 6,909 miles)
at Mach 0.88	5,000 nm (9,266 km; 5,757 miles)

CANADAIR CL-215T

TYPE: Twin-turboprop amphibian.
PROGRAMME: Turboprop conversion of piston engined CL-215 (1991-92 and earlier *Jane's*), announced August 1986 and given official go-ahead January 1987; now divided into CL-215T (retrofit: see *Jane's Civil and Military Aircraft Upgrades*) and CL-415 (new production, described separately). Two Quebec CL-215s (C-FASE, c/n 1114, and C-FAWQ, c/n 1115) modified as CL-215T prototypes, latter with hydraulically powered ailerons (customer option on 215T, standard on 415); hydraulically powered rudder and elevators standard; first flights 8 June and 20 September 1989; flight test programme approx 500 hours each aircraft; certification in Restricted category (firefighting) received 28 March 1991, Utility category 24 December 1991. Retrofit kits now manufactured by Canadair in Montreal and shipped to de Havilland in Toronto for installation (from Nos. 8-13 of Spanish order; Nos. 1-7 done by Canadair, Nos. 14-15 to be done in Spain).
CUSTOMERS: Spanish government 15 conversion kits (incl powered ailerons), ordered August 1989 for aircraft of No. 43 Squadron, Spanish Air Force: five are for in-service aircraft (c/n 1056, 1057, 1061, 1080, 1109), 10 for new production CL-215s (c/n 1113, 1116-1122, 1124, 1125); first two delivered 5 June 1991; two more delivered by end of 1991, four more in 1992; remainder to be delivered by end of 1993.

Artist's impression of Canadair's projected Global Express corporate transport

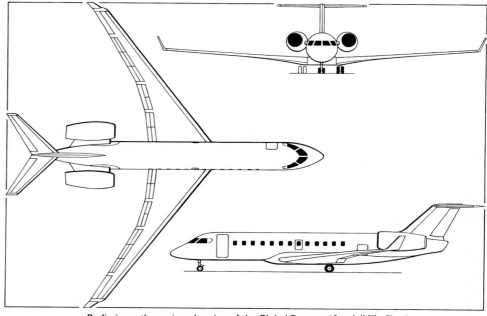

Preliminary three-view drawing of the Global Express (*Jane's/Mike Keep*)

Global Express, looking aft through the 2.29 m (7 ft 6 in) lounge/conference area to the rear stateroom

Canadair CL-215T of the Spanish government's Forestry and Nature Conservation Institute (ICONA)

Canadair CL-215T photograph, retouched to depict a CL-415 in Quebec government colours

CANADAIR CL-415

TYPE: Twin-turboprop amphibian.
PROGRAMME: Introduced as product follow-on to piston engined CL-215; given new designation CL-415 in 1991 to distinguish new production turboprop model from CL-215T retrofit; launched officially 16 October 1991 with firm orders from France and (August 1992) Quebec; first aircraft in final assembly February 1993 with first flight scheduled for October 1993; short flight test programme to be followed by first deliveries (to France) Spring 1994.
CURRENT VERSIONS: Standard aircraft and first production units will be in **firefighting** configuration; others will be developed for **maritime** and **utility/passenger** applications.
CUSTOMERS: Governments of France (12) and Quebec (eight).
COSTS: C$155 million (Quebec order).
DESIGN FEATURES: Retains well proven basic airframe of piston engined CL-215 (thick wing with zero dihedral; row of vortex generators on each wing outboard of fence; long stall strip inboard of starboard fence; leading-edge strakes and fences beside engine nacelles; water scoops behind planing step; anti-spray channels in planing bottom chine) but incorporates upgrading modifications and improvements including higher operating weights for increased firefighting productivity; pressure refuelling; winglets for lateral stability at higher engine power ratings; finlets and tailplane/fin bullet to recover longitudinal/directional stability caused by relocated thrust line, increased power and new propellers; powered rudder and elevators; new electrical system; new 'glass' cockpit with air-conditioning; enlarged four-tank firefighting drop system.
FLYING CONTROLS: Hydraulically actuated ailerons, and mass balanced elevators and rudder, standard; manual reversion in event of hydraulic failure; geared tab in each aileron, spring tab in rudder and each elevator, plus trim tab in port aileron and port elevator. Hydraulically operated single-slotted flaps, each supported by four external hinges.
STRUCTURE: No-dihedral, no-twist high wing, of constant chord, set at 2° incidence; one-piece structure with two conventional spars, extruded spanwise stringers and interspar ribs and aluminium alloy skins. All-metal, fail-safe, single-step boat-hull fuselage with numerous watertight compartments. Tail surfaces of aluminium alloy sheet and extrusions, with honeycomb panels on control surfaces.
LANDING GEAR: Hydraulically retractable tricycle type. Self-centring twin-wheel nose unit retracts rearward into hull and is fully enclosed by conformal doors. Nosewheel steering standard. Main gear support structures retract into wells in sides of hull. Plate mounted on each main gear assembly encloses bottom of wheel well. Mainwheel tyre pressure 5.31 bars (77 lb/sq in); nosewheel tyre pressure 6.55 bars (95 lb/sq in). Hydraulic disc brakes. Non-retractable stabilising floats, each carried near wingtip on pylon cantilevered from wing box structure, with break-away provision.
POWER PLANT: Two 1,775 kW (2,380 shp) Pratt & Whitney Canada PW123AF turboprops, on damage-tolerant mounts, each driving a Hamilton Standard 14SF-19 four-blade constant-speed fully feathering reversible-pitch propeller. Two fuel tanks, each of eight identical flexible cells, in wing spar box, with total usable capacity of 5,796 litres (1,531 US gallons; 1,275 Imp gallons). Single-point pressure refuelling (rear fuselage, starboard side), plus gravity points in wing upper surface. Pneumatic/electric intake de-icing system.
ACCOMMODATION: Normal crew of two side by side on flight deck, with dual controls. Additional stations in maritime patrol/SAR versions for third cockpit member, mission specialist and two observers. For water bomber cabin installation, see Equipment paragraph. With above-floor water tanks removed, transport configurations can include layout for 30 passengers plus toilet, galley and baggage area, with seat pitch of 79 cm (31 in). Combi layout offers cargo at front, full firefighting capability, plus 11 seats at rear. Other quick-change interiors available for utility/paratroop (up to 14 troop-type foldup canvas seats in cabin) or other special missions according to customer's requirements. Flush doors to main cabin on port side of fuselage forward and aft of wings. Emergency exit on starboard side aft of wing trailing-edge. Crew emergency hatch in flight deck roof on starboard side. Mooring hatch in upper surface of nose. Provision for additional cabin windows.
SYSTEMS: Vapour cycle air-conditioning system and combustion heater. Hydraulic system, pressure 207 bars (3,000 lb/sq in), utilises two engine driven pumps (max flow rate 45.5 litres; 12 US gallons; 10 Imp gallons/min) to actuate nosewheel steering, landing gear, flaps, water drop doors, pickup probes, flight controls, main gear unlocking and wheel brakes. Hydraulic fluid (MIL-H-83282) in air/oil reservoir slightly pressurised by engine bleed air. Electrically driven third pump provides hydraulic power for emergency actuation of landing gear and brakes and closure of water doors. Electrical system includes two 800VA 115V 400Hz static inverters, two 28V 400A DC engine driven starter/generators and two 40Ah nickel-cadmium batteries. Ice protection system optional.
AVIONICS: Standard installation includes four-tube Honeywell EDZ-605 EFIS for EADI and EHSI plus dual Honeywell Primus 2 digital nav/com system. Includes dual ADF, VOR/ILS, marker beacon receivers and ATC transponders, plus single DME, ELT and radio altimeter. Com

32 CANADA: AIRCRAFT—CANADAIR

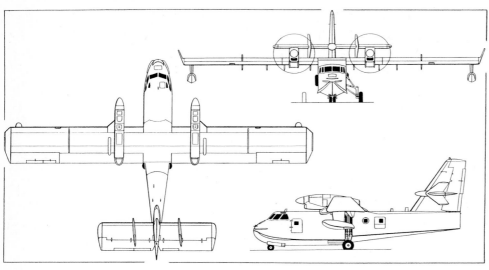

Canadair CL-415 twin-turboprop general purpose amphibian *(Dennis Punnett)*

Mockup of CL-415 flight deck with Honeywell four-tube EFIS and Primus 2 nav/com system

equipment includes dual VHF and single VHF/FM and HF. Dual Honeywell AHRS. Optional avionics include autopilot, VLF/Omega/GPS nav system and search/weather radar.

EQUIPMENT (water bomber): Four integral water tanks in main fuselage compartment, near CG (combined capacity 6,130 litres; 1,620 US gallons; 1,350 Imp gallons), plus seven inward facing seats. Tanks filled by two hydraulically actuated scoops aft of hull step, fillable also on ground by hose adaptor on each side of fuselage. Four independently openable water doors in hull bottom. Onboard foam concentrate reservoirs (capacity 680 kg; 1,500 lb) and mixing system. Improved drop pattern and drop door sequencing compared with CL-215. Optional spray kit can be coupled with firefighting tanks for large scale spraying of oil dispersants and insecticides. In a typical firefighting mission, with a fire 85 nm (157 km; 98 miles) from airbase, and a water source 10 nm (18.5 km; 11.5 miles) from the fire, aircraft could make 25 water scoops and drop circuits before having to return to base to refuel. Water tanks can be scoop-filled completely (ISA at S/L, zero wind) in an on-water distance of only 1,300 m (4,265 ft); partial water loads can be scooped on smaller bodies of water. Minimum safe water depth for scooping is only 1.40 m (4 ft 7 in).

EQUIPMENT (other versions): Stretcher kits, passenger or troop seats, cargo tiedowns, searchlight and other equipment according to mission and customer requirements. Provision for underwing pylon attachment points for auxiliary fuel tanks or other stores.

DIMENSIONS, EXTERNAL:
Wing span	28.63 m (93 ft 11 in)
Wing chord, constant	3.54 m (11 ft 7½ in)
Wing aspect ratio	8.2
Length overall	19.82 m (65 ft 0½ in)
Beam (max)	2.59 m (8 ft 6 in)
Length/beam ratio	7.5
Height overall: on land	8.98 m (29 ft 5½ in)
on water	6.88 m (22 ft 7 in)
Draught: wheels up	1.12 m (3 ft 8 in)
wheels down	2.03 m (6 ft 8 in)
Tailplane span	10.97 m (36 ft 0 in)
Wheel track	5.28 m (17 ft 4 in)
Wheelbase	7.23 m (23 ft 9 in)
Propeller diameter	3.97 m (13 ft 0¼ in)
Propeller fuselage clearance	0.59 m (1 ft 11¼ in)
Propeller water clearance	1.30 m (4 ft 3¼ in)
Propeller ground clearance	2.77 m (9 ft 1 in)
Forward door: Height*	1.37 m (4 ft 6 in)
Width	1.03 m (3 ft 4 in)
Height to sill	1.68 m (5 ft 6 in)
Rear door: Height	1.12 m (3 ft 8 in)
Width	1.03 m (3 ft 4 in)
Height to sill	1.83 m (6 ft 0 in)
Water drop door: Length	1.60 m (5 ft 3 in)
Width	0.81 m (2 ft 8 in)
Emergency exit: Height	0.91 m (3 ft 0 in)
Width	0.51 m (1 ft 8 in)

* incl 25 cm (10 in) removable sill

DIMENSIONS, INTERNAL:
Cabin, excl flight deck: Length	9.38 m (30 ft 9½ in)
Max width	2.39 m (7 ft 10 in)
Max height	1.90 m (6 ft 3 in)
Floor area	19.69 m² (212.0 sq ft)
Volume	35.59 m³ (1,257.0 cu ft)

AREAS:
Wings, gross	100.33 m² (1,080.0 sq ft)
Ailerons (total)	8.05 m² (86.6 sq ft)
Flaps (total)	22.39 m² (241.0 sq ft)
Fin	11.22 m² (120.75 sq ft)
Rudder, incl tabs	6.02 m² (64.75 sq ft)
Tailplane	20.55 m² (221.2 sq ft)
Elevators (total, incl tabs)	7.88 m² (84.8 sq ft)

WEIGHTS AND LOADINGS (A: water bomber, B: utility, land or water based):
Typical operating weight empty:	
A	12,333 kg (27,190 lb)
B	12,079 kg (26,630 lb)
Max internal fuel weight: A, B	4,649 kg (10,250 lb)
Max payload: A (disposable)	6,123 kg (13,500 lb)
B	4,431 kg (9,770 lb)
Max ramp weight: A (land)	19,890 kg (43,850 lb)
A (water), B	17,236 kg (38,000 lb)
Max T-O weight: A (land)	19,890 kg (43,850 lb)
A (water), B (land and water)	17,168 kg (37,850 lb)
Max touchdown weight for water scooping:	
A	16,420 kg (36,200 lb)
Max flying weight after water scooping:	
A	20,865 kg (46,000 lb)
Max landing weight:	
A, B (land and water)	16,783 kg (37,000 lb)
Max zero-fuel weight: A	19,051 kg (42,000 lb)
B	16,511 kg (36,400 lb)
Max wing loading:	
A (after scoop)	207.9 kg/m² (42.59 lb/sq ft)
A (land)	198.2 kg/m² (40.60 lb/sq ft)
B (land and water)	171.1 kg/m² (35.05 lb/sq ft)
Max power loading:	
A (after scoop)	5.88 kg/kW (9.66 lb/shp)
A (land)	5.60 kg/kW (9.21 lb/shp)
B (land and water)	4.84 kg/kW (7.95 lb/shp)

Fuselage/wing mating of the first CL-415 at Dorval in February 1993

PERFORMANCE (estimated at weights shown):
 Max cruising speed at 3,050 m (10,000 ft), AUW of
  14,741 kg (32,500 lb)   203 knots (376 km/h; 234 mph)
 Long-range cruising speed at 3,050 m (10,000 ft), AUW of
  14,741 kg (32,500 lb)
                          145 knots (269 km/h; 167 mph)
 Patrol speed at S/L, AUW of 14,741 kg (32,500 lb)
                          130 knots (241 km/h; 150 mph)
 Stalling speed:
  15° flap, AUW of 20,865 kg (46,000 lb)
                           79 knots (147 km/h; 91 mph)
  25° flap, AUW of 16,783 kg (37,000 lb)
                           68 knots (126 km/h; 79 mph)
 Max rate of climb at S/L, AUW of 20,865 kg (46,000 lb)
                          419 m (1,375 ft)/min
 T-O distance at S/L, ISA:
  land, AUW of 19,890 kg (43,850 lb) 844 m (2,770 ft)
  water, AUW of 17,168 kg (37,850 lb) 814 m (2,670 ft)
 Landing distance at S/L, ISA:
  land, AUW of 16,783 kg (37,000 lb) 674 m (2,210 ft)
  water, AUW of 16,783 kg (37,000 lb) 665 m (2,180 ft)
 Scooping distance at S/L, ISA (incl safe clearance
  heights)   1,293 m (4,240 ft)
 Ferry range with 499 kg (1,100 lb) payload
                          1,310 nm (2,428 km; 1,508 miles)
 Design g limits (15° flap)   +3.25/−1

CONAIR
CONAIR AVIATION LTD
PO Box 220, Abbotsford, British Columbia V2S 4N9
Telephone: 1 (604) 855 1171
Fax: 1 (604) 855 1017
Telex: 04-363529
PRESIDENT AND CEO: K. B. (Barry) Marsden
VICE-PRESIDENT: Jim Dunkel
MARKETING MANAGER: Robert M. Stitt
MANAGER, CORPORATE COMMUNICATIONS: Lorna Thomassen

Operates largest private fleet of fire-control aircraft in world with 90 aircraft (52 fixed-wing and 38 helicopters); Conair also operates oil spill control, insect control, forest fertilisation and fisheries surveillance; designs and manufactures associated systems, including fire-control belly tanks for various helicopters and Douglas DC-6B; Grumman and Canadian built Trackers (see Firecat and Turbo Firecat); spray systems produced for Lockheed C-130, Douglas DC-6B, Fokker F27 and Alenia G222. C-130 conversion includes 7,560 litre (1,997 US gallon; 1,663 Imp gallon) modular spray system; six delivered by 1991, to 356th Tactical Airlift Squadron of USAF. See also *Jane's Civil and Military Aircraft Upgrades*.

CONAIR FIRECAT
TYPE: Fire-control aircraft based on S-2A/S2F/CS2F Tracker.
PROGRAMME: In production.
CUSTOMERS: Total deliveries 32 by mid-1992.
DESIGN FEATURES: Based on Grumman S-2A (S2F-1) or de Havilland Canada CS2F-1/2 Trackers; firefighting conversion includes four-compartment retardant tank allowing choice of drop patterns, cabin floor raised 20.3 cm (8 in) to accommodate tank, reworked/updated flight instrument panels, fitting larger landing gear wheels with low pressure tyres, removing 1,361 kg (3,000 lb) of military equipment, inspecting and repairing wing spars and complete rewiring. Options include hydraulic or pneumatic discharge system with microcomputer control of drop sequence and 173 litre (45.6 US gallon; 38 Imp gallon) foam injection system.
FLYING CONTROLS: As Tracker.
STRUCTURE: As Tracker.
POWER PLANT: Two 1,100 kW (1,475 hp) Wright 982C9HE2 (R-1820-82) Cyclone nine-cylinder air-cooled radial engines, each driving a Hamilton Standard three-blade propeller. Total internal fuel capacity 1,968 litres (520 US gallons; 433 Imp gallons).
ACCOMMODATION: Minimum crew: one pilot.
WEIGHTS AND LOADINGS:
 Operating weight empty   6,895 kg (15,200 lb)
 Max payload   3,300 kg (7,275 lb)
 Max fuel   1,418 kg (3,126 lb)
 Max T-O weight   11,793 kg (26,000 lb)
 Max landing weight   11,113 kg (24,500 lb)
PERFORMANCE (at max T-O weight):
 Never-exceed speed (V_NE)
                              240 knots (444 km/h; 276 mph)
 Max level speed at 1,220 m (4,000 ft)
                              220 knots (408 km/h; 253 mph)
 Normal cruising speed   180 knots (333 km/h; 207 mph)
 Normal drop speed   120 knots (222 km/h; 138 mph)
 Stalling speed, flaps down, power off
                               82 knots (152 km/h; 95 mph)
 Max rate of climb at S/L   366 m (1,200 ft)/min
 Rate of climb at S/L, OEI   107 m (350 ft)/min
 Service ceiling   6,100 m (20,000 ft)
 T-O to 15 m (50 ft)   915 m (3,000 ft)
 Landing from 15 m (50 ft)   762 m (2,500 ft)
 Min field length   915 m (3,000 ft)
 Endurance with max payload 4 h 30 min

CONAIR TURBO FIRECAT
TYPE: Turboprop fire-control aircraft based on S-2A/S2F/CS2F Tracker.
PROGRAMME: First flight 7 August 1988; Canadian certification 22 December 1989.
CUSTOMERS: First delivery to French Sécurité Civile August 1988; total of eight for delivery by mid-1993. Conair built its own demonstrator in 1992.
DESIGN FEATURES: Turboprops to extend life of aircraft and its fire-control role; other advantages include improved performance, higher speed for greater productivity, better manoeuvring in mountainous terrain, turbine reliability, reduced operating and maintenance costs and fuel availability.
FLYING CONTROLS: As Tracker.
STRUCTURE: As Tracker.
POWER PLANT: Two 1,062 kW (1,424 shp) Pratt & Whitney Canada PT6A-67AF turboprops, each driving a Hartzell HC-B5MA-3BXI/M11296SX five-blade propeller. Total fuel capacity of 2,936 litres (775.6 US gallons; 645.8 Imp gallons), consisting of 1,972 litres (521 US gallons; 433.8 Imp gallons) internally and 964 litres (254.6 US gallons; 212 Imp gallons) in two underwing pylon tanks. Single-point refuelling station in starboard engine nacelle.
ACCOMMODATION: Minimum crew: one pilot.
AVIONICS: Include angle of attack and stall warning system.
EQUIPMENT: Four-compartment retardant tank, max capacity 3,455 litres (913 US gallons; 760 Imp gallons); max discharge rate 3,955 litres (1,045 US gallons; 870 Imp gallons)/s.
DIMENSIONS, INTERNAL:
 Retardant tank: Length   3.25 m (10 ft 8 in)
          Width   1.27 m (4 ft 2 in)
          Depth   0.86 m (2 ft 10 in)
WEIGHTS AND LOADINGS: As for Firecat except:
 Operating weight empty   6,884 kg (15,177 lb)
 Max fuel   2,339 kg (5,158 lb)
 Max T-O weight   12,473 kg (27,500 lb)
PERFORMANCE (at max T-O weight except where indicated):
 As for Firecat except:
 Normal cruising speed   220 knots (408 km/h; 253 mph)
 Rate of climb at S/L, OEI:
  at max T-O weight   61 m (200 ft)/min
  at 9,072 kg (20,000 lb) gross weight
                           227 m (745 ft)/min
 T-O to 15 m (50 ft)   1,220 m (4,000 ft)
* Landing from 15 m (50 ft) 762 m (2,500 ft)
 Min field length   1,220 m (4,000 ft)
 Endurance with max payload   5 h
* *Without available propeller reversal*

CONAIR F27 FIREFIGHTER
TYPE: Fire-control version of Fokker F27 Mk 600.
PROGRAMME: Canadian DoT approval 5 June 1986; three aircraft modified (details in 1991-92 *Jane's*); system redesigned 1992.
CURRENT VERSIONS: Alternatives include transport for cargo and fire crews, infra-red fire detection and mapping, aerial survey, aerial spraying and pararescue.
CUSTOMERS: Two operated by French Sécurité Civile.
DESIGN FEATURES: First firefighting conversion of turboprop aircraft; Conair 6,364 litre (1,681 US gallon; 1,400 Imp gallon) long-term retardant delivery system with optional 455 litre (120 US gallon; 100 Imp gallon) foam injection system; filling rate 1,514 litres (400 US gallons; 333 Imp gallons)/min; computer controlled drop door sequencing; large forward freight door retained; seating for support crew.
FLYING CONTROLS: As F27.

Conair Turbo Firecat (two P&WC PT6A-67AF turboprops) of the French Sécurité Civile

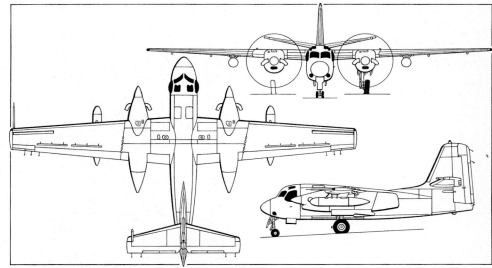

Conair Turbo Firecat twin-turboprop fire-control conversion of the Grumman Tracker (*Jane's/Mike Keep*)

34 CANADA: AIRCRAFT—CONAIR/DE HAVILLAND

STRUCTURE: As F27.
WEIGHTS AND LOADINGS:
Operating weight empty	10,646 kg (23,471 lb)
Max payload	6,731 kg (14,840 lb)
Max fuel	4,152 kg (9,153 lb)
Max T-O weight	20,411 kg (45,000 lb)
Max landing weight	18,143 kg (40,000 lb)

PERFORMANCE (at max T-O weight except where indicated):
Never-exceed speed (V_{NE})	259 knots (480 km/h; 298 mph)
Max cruising speed	230 knots (426 km/h; 265 mph)
Normal drop speed	125 knots (232 km/h; 144 mph)
Min control speed	80 knots (149 km/h; 92 mph)
Stalling speed, flaps down, power off	77 knots (143 km/h; 89 mph)
Max rate of climb at S/L	366 m (1,200 ft)/min
Rate of climb at S/L, OEI, AUW of 14,060 kg (31,000 lb)	177 m (580 ft)/min
Service ceiling	7,620 m (25,000 ft)
Service ceiling, OEI, AUW of 14,060 kg (31,000 lb)	5,640 m (18,500 ft)
T-O to 10.7 m (35 ft)	1,600 m (5,250 ft)
Landing from 15 m (50 ft) at max landing weight	987 m (3,240 ft)
Min field length	1,525 m (5,000 ft)
Max endurance	3 h 24 min

CONAIR HELITANKERS

TYPE: Generic name of helicopter fire-control conversions.
CURRENT VERSIONS: **Bell 205/212:** (1,360 litres; 359 US gallons; 299 Imp gallons); fitted with rappelling system for delivering firefighters.
  Eurocopter AS 350B, Ecureuil: (800 litres; 211 US gallons; 176 Imp gallons).
  Eurocopter SA 315B Lama: (900 litres; 238 US gallons; 198 Imp gallons).
  Eurocopter AS 330 Puma and AS 332 Super Puma: (2,355 litres; 622 US gallons; 517.5 Imp gallons). Puma has 800 litre (211 US gallon; 176 Imp gallon) two-door belly tank refilled by 261 litre (69 US gallon; 57 Imp gallon) chute from a 1,296 litre (342 US gallon; 285 Imp gallon) cabin tank; foam tank capacity 173 litres (45.6 US gallons; 38 Imp gallons).
CUSTOMERS: Sales include 26 Bell 205/212s; 19 Ecureuils; 13 Lamas; two Puma/Super Pumas.

The Conair 1,360 litre (359 US gallon; 299 Imp gallon) helitanker system on a Bell 212 self-loads in 40 seconds

DESIGN FEATURES: Helitanker system has full-length doors in rigid ventral tanks so that retardant line length and drop concentrations can be regulated; helicopter can refill while hovering over water or deliver water to a portable reservoir on ground; reversible pump allows single-point loading and offloading; foam injection system available.
FLYING CONTROLS: Unchanged.
STRUCTURE: Substantially unchanged.

DE HAVILLAND
DE HAVILLAND INC

Garrett Boulevard, Downsview, Ontario M3K 1Y5
Telephone: 1 (416) 633 7310
Fax: 1 (416) 375 4546
Telex: 0622128
PRESIDENT: Kenneth G. Laver
VICE-PRESIDENT, OPERATIONS: Carl Maggi
VICE-PRESIDENT, ENGINEERING: Carl Gerard
MARKETING AND SALES:
  See Bombardier Regional Aircraft Division
MANAGER, PUBLIC RELATIONS: Colin S. Fisher

Established 1928 as The de Havilland Aircraft of Canada Ltd, subsidiary of The de Havilland Aircraft Co Ltd, both absorbed 1961 by Hawker Siddeley Group; ownership transferred to Canadian government 26 June 1974; purchased by Boeing Company 31 January 1986 and made a division of Boeing of Canada Ltd; Boeing intention to sell announced July 1990; attempt to purchase by Aerospatiale (France) and Alenia (Italy) in April 1991 initially not accepted by Canadian government and then disapproved by EC. Sale to Bombardier Inc (51 per cent) and Government of Ontario (49 per cent), signed 22 January 1992, supported by help from Ontario and federal governments, with Canadian Export Development Corporation to provide sales financing for Dash 8; all government support conditionally repayable. Workforce (5,000 in 1991) reduced to 3,350 by September 1992, and Dash 8 production rate slowed to two per month.

DHC-8 DASH 8 SERIES 100 and 200

TYPE: Twin-turboprop short-range transport.
PROGRAMME: Launched 1980; first flight first prototype (C-GDNK) 20 June 1983, second prototype (C-GGMP) 26 October 1983, third prototype November 1983; fourth aircraft, first with production P&WC P120 engines, flying by early 1984; certificated to Canadian DoT, FAR Pts 25 and 36 and SFAR No. 27 on 28 September 1984, followed by FAA type approval; also certificated Australia, Austria, Brazil, China, Germany, Ireland, Italy, Netherlands, New Zealand, Norway, Papua New Guinea, Taiwan and UK; first delivery (NorOntair) 23 October 1984, entering service December 1984. New B model standard introduced 1992 offers reduced vibration and noise damping, improved crosswind capability, higher operating weights; Cat. IIIa weather operation to become available as option by mid-1994.
CURRENT VERSIONS: **Series 100:** Initial version, with choice of PW120A or PW121 engines; described in 1992-93 and earlier editions.
  Series 100A: Introduced 1990; PW120A engines and restyled interior with 6.35 cm (2.5 in) more headroom in aisle; first delivery to Pennsylvania Airlines July 1990. *Detailed description applies to this version except where indicated.*
  Series 100B: Improved version from 1992; PW121 engines enhance airfield and climb performance.

Austria's Tyrolean Airways, already a Dash 8 Series 100 operator, is also launch customer for the new Series 100B

DE HAVILLAND—AIRCRAFT: CANADA

Series 200A: Increased speed/payload version of Series 100A, launched 16 March 1992; increased OEI capability and greater commonality with Series 300. Same airframe as 100A/B, but PW123C engines give 30 knot (56 km/h; 35 mph) increase in cruising speed, allowing airlines to increase frequencies or operational radius. For delivery from early 1994.

Series 200B: As 200A, but with PW123D engines for full power at higher ambient temperatures.

Dash 8M and **Series 300 and 400**: described separately.

CUSTOMERS: Total 285 firm orders for civil and military (Dash 8M) Series 100/100A/100B by March 1993, of which 257 delivered.

DESIGN FEATURES: Wing has constant chord inboard section and tapered outer panels; thickness/chord ratio 18 per cent at root, 13 per cent at tip; dihedral 2° 30′ on outer panels; inboard leading-edges drooped; T tail, swept fin with large dorsal fin.

FLYING CONTROLS: Fixed tailplane; horn balanced elevator with four tabs; mechanically actuated horn balanced ailerons with inset tabs; hydraulically actuated roll spoilers/lift dumpers forward of each outer flap; two-segment serially hinged rudder, hydraulically actuated; yaw damper; lift dumper panels inboard and outboard of engine nacelles; stall strips on leading-edges outboard of engines; two-section slotted Fowler flaps. Digital AFCS.

STRUCTURE: Fuselage near-circular section, flush riveted and pressurised; adhesive bonded stringers and cutout reinforcements; wing leading-edge, radome, nose bay, wing/fuselage and wingtip fairings, dorsal fin, fin leading-edge, fin/tailplane fairings, tailplane leading-edges, elevator tips, flap shrouds, flap trailing-edges and other components of Kevlar and Nomex; wing has tip-to-tip torsion box. Wheel doors of Kevlar and other composites.

LANDING GEAR: Retractable tricycle type, by Dowty Aerospace, with twin wheels on each unit. Steer-by-wire nose unit retracts forward, main units rearward into engine nacelles. Goodrich mainwheels and brakes; Hydro-Aire Mk 3 anti-skid system. Standard tyre pressures: main 9.03 bars (131 lb/sq in), nose 5.52 bars (80 lb/sq in). Low pressure tyres optional, pressure 5.31 bars (77 lb/sq in) on main units, 3.31 bars (48 lb/sq in) on nose unit.

POWER PLANT: Two 1,491 kW (2,000 shp) Pratt & Whitney Canada PW120A turboprops in Series 100A, each driving a Hamilton Standard 14SF-7 four-blade constant-speed fully feathering aluminium/glassfibre propeller with reversible pitch. Series 100B has 1,603 kW (2,150 shp) PW121A engines; 1,603 kW (2,150 shp) PW123Cs in Series 200A are flat rated for full power at up to 26°C; PW123Ds in Series 200B maintain same power at up to 45°C. Propeller blades have Beta control. Standard usable fuel capacity (in-wing tanks) 3,160 litres (835 US gallons; 695 Imp gallons); optional auxiliary tank system increases this to 5,700 litres (1,506 US gallons; 1,254 Imp gallons). Pressure refuelling point in rear of starboard engine nacelle; overwing gravity point in each outer wing panel. Oil capacity 21 litres (5.5 US gallons; 4.6 Imp gallons) per engine.

ACCOMMODATION: Crew of two on flight deck, plus cabin attendant. Dual controls standard. Standard commuter layout provides four-abreast seating, with central aisle, for 37 passengers at 79 cm (31 in) pitch, plus buffet, toilet and large rear baggage compartment. Wardrobe at front of passenger cabin, in addition to overhead lockers and underseat stowage, provides additional carry-on baggage capacity. Alternative 39-passenger, passenger/cargo or corporate layouts available at customer's option. Movable bulkhead to facilitate conversion to mixed-traffic or all-cargo configuration. Port side airstair door at front for crew and passengers; large inward opening port side door aft of wing for cargo loading. Emergency exit each side, in line with wing leading-edge, and opposite passenger door on starboard side. Entire accommodation pressurised and air-conditioned.

SYSTEMS: Air cycle air-conditioning and pressurisation system (max differential 0.38 bar; 5.5 lb/sq in). Normal hydraulic installation comprises two independent systems, each having an engine driven variable displacement pump and an electrically driven standby pump; accumulator and handpump for emergency use. Electrical system DC power provided by two starter/generators, two transformer-rectifier units, and two nickel-cadmium batteries. Variable frequency AC power provided by two engine driven AC generators and three static inverters. Ground power receptacles in port side of nose (DC) and rear of starboard nacelle (AC). Rubber-boot de-icing of wing, tailplane and fin leading-edges by pneumatic system plus electric heating; electric de-icing of propeller blades. APU standard in corporate version. Simmonds fuel monitoring system.

AVIONICS: Standard factory installed avionics include Bendix/King Gold Crown III com/nav (KTR 908 VHF com, KNR 634 VHF nav, KDF 806 ADF, KDM 706A DME and KXP 756 transponder), Honeywell SPZ-800 dual-channel digital AFCS with integrated fail-operational flight director/autopilot system, dual digital air data system, electromechanical flight instruments, and Primus 800 colour weather radar. Honeywell electronic flight instrumentation system (EFIS) optional on commuter, standard on corporate version. Avtech audio integrating system. Telephonics PA system. Hughes/Flight Dynamics head-up guidance system (HGS) for Cat. IIIa operation to be available optionally by mid-1994.

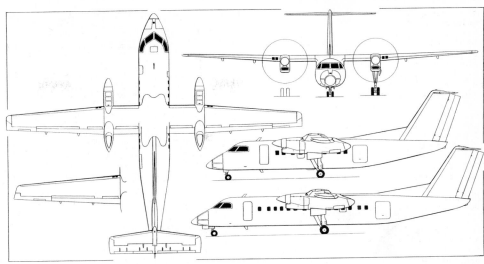

DHC-8 Dash 8 Series 100, with additional side view (bottom) and wingtip of Series 300 *(Dennis Punnett)*

Development aircraft for the Dash 8 Series 200, which is due to enter service in 1994

DIMENSIONS, EXTERNAL:
Wing span	25.91 m (85 ft 0 in)
Wing aspect ratio	12.35
Length overall	22.25 m (73 ft 0 in)
Fuselage: Max diameter	2.69 m (8 ft 10 in)
Height overall	7.49 m (24 ft 7 in)
Elevator span	7.92 m (26 ft 0 in)
Wheel track (c/l of shock struts)	7.87 m (25 ft 10 in)
Wheelbase	7.95 m (26 ft 1 in)
Propeller diameter	3.96 m (13 ft 0 in)
Propeller ground clearance	0.94 m (3 ft 1 in)
Propeller fuselage clearance	0.76 m (2 ft 6 in)
Passenger/crew door (fwd, port):	
Height	1.65 m (5 ft 5 in)
Width	0.76 m (2 ft 6 in)
Height to sill	1.09 m (3 ft 7 in)
Baggage door (rear, port):	
Height	1.52 m (5 ft 0 in)
Width	1.27 m (4 ft 2 in)
Height to sill	1.09 m (3 ft 7 in)

DIMENSIONS, INTERNAL:
Cabin (excl flight deck): Length	9.17 m (30 ft 1 in)
Max width	2.49 m (8 ft 2 in)
Width at floor	2.03 m (6 ft 8 in)
Max height	1.94 m (6 ft 4½ in)
Volume	39.36 m³ (1,390 cu ft)
Baggage compartment volume	8.5 m³ (300 cu ft)

AREAS:
Wings, gross	54.35 m² (585.0 sq ft)
Vertical tail surfaces (total)	14.12 m² (152.0 sq ft)
Horizontal tail surfaces (total)	13.94 m² (150.0 sq ft)

WEIGHTS AND LOADINGS:
Operating weight empty: 100A	10,251 kg (22,600 lb)
100B	10,273 kg (22,648 lb)
200A, 200B	10,380 kg (22,886 lb)
Max usable fuel (all): standard	2,576 kg (5,678 lb)
optional	4,646 kg (10,244 lb)
Max payload: passengers: 100A	3,810 kg (8,400 lb)
100B	4,242 kg (9,352 lb)
200A, 200B	4,134 kg (9,114 lb)
Max T-O weight: 100A	15,650 kg (34,500 lb)
100B, 200A, 200B	16,465 kg (36,300 lb)
Max landing weight: all	15,375 kg (33,900 lb)
Max zero-fuel weight: 100A	14,060 kg (31,000 lb)
100B, 200A, 200B	14,515 kg (32,000 lb)
Max wing loading: 100A	287.95 kg/m² (58.97 lb/sq ft)
100B, 200A, 200B	302.96 kg/m² (62.05 lb/sq ft)
Max power loading: 100A	5.25 kg/kW (8.62 lb/shp)
100B, 200A, 200B	5.14 kg/kW (8.44 lb/shp)

PERFORMANCE (at 95% standard MTOGW except where indicated):
Max cruising speed: 100A	265 knots (491 km/h; 305 mph)
100B	270 knots (500 km/h; 311 mph)
200A, 200B	300 knots (556 km/h; 345 mph)
Stalling speed, flaps down	72 knots (134 km/h; 83 mph)
Max rate of climb at S/L	475 m (1,560 ft)/min
Rate of climb at S/L, OEI	137 m (450 ft)/min
Certificated ceiling	7,620 m (25,000 ft)
Service ceiling, OEI	4,575 m (15,000 ft)

FAR Pt 25 T-O field length, max T-O weight, 15° flap:
100A (S/L, ISA)	942 m (3,090 ft)
100B, 200A (S/L, ISA)	960 m (3,150 ft)
200B (at 1,525 m; 5,000 ft, ISA + 30°C)	2,134 m (7,000 ft)

FAR Pt 25 landing field length at max landing weight:
all	908 m (2,980 ft)

Range with standard fuel, IFR reserves:
full passenger load	820 nm (1,519 km; 944 miles)
2,721 kg (6,000 lb) payload	1,100 nm (2,038 km; 1,266 miles)

OPERATIONAL NOISE LEVELS (FAR Pt 36 Stage 3 and ICAO Annex 16):
T-O	81 EPNdB
Sideline	86 EPNdB
Approach	95 EPNdB

DHC-8 DASH 8M
Canadian Forces designations: CC-142 and CT-142
US Air Force designation: E-9A
TYPE: Multi-mission medium utility aircraft.
CURRENT VERSIONS: **CC-142 and CT-142:** Passenger/cargo transport and navigation trainer versions of Dash 8M-100 respectively; both have long-range fuel tanks, rough-field landing gear, high-strength floors and mission avionics; Canadian Dept of National Defence operates two CC-142s and four CT-142s, latter with mapping radar in extended nose.

E-9A: US Air Force Dash 8M-100 as missile range control aircraft; relays telemetry, voice and drone and fighter control data and simultaneously observes range with radar; avionics include large electronically steered phased array radar in fuselage side and AN/APS-128D surveillance radar in ventral dome.

CUSTOMERS: Canadian Dept of Transport (two Dash 8M-100), Canadian Forces (two CC-142, four CT-142), USAF (two E-9A); all 10 included in Series 100/100A delivery total.

Dash 8M-100 in E-9A configuration as a USAF missile range control aircraft

DHC-8 DASH 8 SERIES 300
TYPE: Stretched twin-turboprop regional transport.
PROGRAMME: Announced mid-1985; launched March 1986; first flight (modified Series 100 prototype C-GDNK) 15 May 1987; Canadian DoT certification 14 February 1989; first delivery (Time Air) 27 February 1989; FAA type approval 8 June 1989; now also certificated in Brazil, China, Germany, Ireland, Italy, Netherlands and Taiwan.
CURRENT VERSIONS: **Series 300:** Initial version; differs from Series 100A in having extended wingtips; 3.43 m (11 ft 3 in) two-plug fuselage extension giving standard seating for 50 at 81 cm (32 in) pitch or 56 at 74 cm (29 in) pitch, plus second cabin attendant; also larger galley, galley service door, additional wardrobe, larger lavatory, dual air-conditioning packs and optional Turbomach T-40 APU; powered by 1,775 kW (2,380 shp) P&WC PW123s driving Hamilton Standard 14SF-15 four-blade propellers; fuel capacity as Series 100A; tyre pressures increased (mainwheels 6.69 bars; 97 lb/sq in, nosewheels 4.14 bars; 60 lb/sq in).

Series 300A: Introduced 1990; 1,775 kW (2,380 shp) PW123A engines standard (PW123B optional) and improved payload/range; same interior improvements as Series 100A; higher gross weight optional; first delivery to German Contact Air 24 August 1990. First 300A/PW123B delivered to Tyrolean Airways 1991; this version now Series 300B.

Series 300B: Introduced 1992; generally as for 300A, but payload/range and airfield performance improved by 1,864 kW (2,500 shp) PW123B engines; optional high gross weight of 300A is standard on 300B.

CUSTOMERS: Firm orders for 113 by March 1993, of which 84 delivered.

de Havilland CC-142 (Dash 8M-100) in Canadian Forces camouflage (*Press Office Sturzenegger*)

DIMENSIONS, EXTERNAL: As for Series 100A except:
Wing span	27.43 m (90 ft 0 in)
Wing aspect ratio	13.39
Length overall	25.68 m (84 ft 3 in)
Wheelbase	10.01 m (32 ft 10 in)

DIMENSIONS, INTERNAL: As for Series 100A except:
Cabin (excl flight deck): Length	12.65 m (41 ft 6 in)
Volume	50.12 m³ (1,770 cu ft)
Baggage compartment volume:	
with 50 passengers	9.06 m³ (320 cu ft)
with 56 passengers	7.93 m³ (280 cu ft)

AREAS:
Wings, gross	56.21 m² (605.0 sq ft)
Tail surfaces	as for Series 100A/200A

WEIGHTS AND LOADINGS:
Operating weight empty:	
300A, 300B	11,657 kg (25,700 lb)
Max usable fuel: 300A, 300B	as for Series 100A/200A
Max payload: 300A	5,216 kg (11,500 lb)
300B	6,259 kg (13,800 lb)
Max T-O weight: 300A	18,642 kg (41,100 lb)
300A (optional), 300B	19,504 kg (43,000 lb)
Max landing weight: 300A	18,144 kg (40,000 lb)
300B	19,050 kg (42,000 lb)
Max zero-fuel weight: 300A	16,873 kg (37,200 lb)
300B	17,917 kg (39,500 lb)
Max wing loading: 300A	331.65 kg/m² (67.93 lb/sq ft)
300B	346.99 kg/m² (71.07 lb/sq ft)
Max power loading: 300A	5.25 kg/kW (8.63 lb/shp)
300B	5.49 kg/kW (9.03 lb/shp)

PERFORMANCE (at max T-O weight except where indicated):
Max cruising speed at 95% of MTOGW:	
300A	287 knots (532 km/h; 330 mph)
300B	285 knots (528 km/h; 328 mph)
Stalling speed, flaps down	77 knots (141 km/h; 88 mph)
Max rate of climb at S/L	549 m (1,800 ft)/min
Rate of climb at S/L, OEI	137 m (450 ft)/min
Certificated ceiling	7,620 m (25,000 ft)
Service ceiling, OEI	4,115 m (13,500 ft)
FAR Pt 25 T-O field length at S/L, ISA, 15° flap:	
300A	1,085 m (3,560 ft)
300B	1,160 m (3,800 ft)
FAR Pt 25 landing field length at max landing weight:	
300A	1,021 m (3,350 ft)
300B	1,052 m (3,450 ft)
Range with standard fuel, IFR reserves:	
full passenger load	830 nm (1,538 km; 955 miles)
2,721 kg (6,000 lb) payload	870 nm (1,612 km; 1,001 miles)

Series 300 stretched version of the Dash 8 in the insignia of DLT of Germany

DHC-8 DASH 8 SERIES 400

TYPE: Further stretch of Dash 8.
PROGRAMME: Launch decision and engine choice due 1993, leading to first flight mid-1996, certification approx one year later.
Following data are provisional:
DESIGN FEATURES: Fuselage stretched to seat approx 70; new engines.
POWER PLANT: Candidates are GE/Textron Lycoming GLC38 and Allison GMA 2100, both with six-blade propellers.
DIMENSIONS, EXTERNAL: As for Series 300 except:
Wing span	28.12 m (92 ft 3 in)
Wing aspect ratio	12.74
Length overall	31.60 m (103 ft 8 in)
Height overall	8.10 m (26 ft 7 in)
Wheelbase	8.48 m (27 ft 10 in)
Baggage compartment door:	
Height to sill	1.55 m (5 ft 1 in)

DIMENSIONS, INTERNAL:
Cabin: Length	17.75 m (58 ft 3 in)
Baggage compartment volume:	
forward	2.29 m³ (81.0 cu ft)
rear	11.50 m³ (406.0 cu ft)

AREAS: As for Series 300 except:
Wings, gross	62.05 m² (668.0 sq ft)

WEIGHTS AND LOADINGS:
Operating weight empty	15,558 kg (34,300 lb)
Max usable fuel	5,352 kg (11,800 lb)
Max payload (standard passenger aircraft)	7,484 kg (16,500 lb)
Max T-O weight	25,855 kg (57,000 lb)
Max landing weight	25,628 kg (56,500 lb)
Max zero-fuel weight	23,042 kg (50,800 lb)
Max wing loading	416.6 kg/m² (85.33 lb/sq ft)

PERFORMANCE (estimated):
Max cruising speed at 95% of max T-O weight, ISA
   350 knots (648 km/h; 403 mph)
Service ceiling, OEI, at 95% of max T-O weight, ISA
   6,220 m (20,400 ft)
FAR T-O field length at S/L, ISA   1,100 m (3,609 ft)
FAR landing field length at S/L, max landing weight
   1,205 m (3,954 ft)
Range with 70 passengers and baggage, IFR reserves:
   at max cruising speed 940 nm (1,742 km; 1,082 miles)
   at long-range cruising speed
   1,110 nm (2,057 km; 1,278 miles)

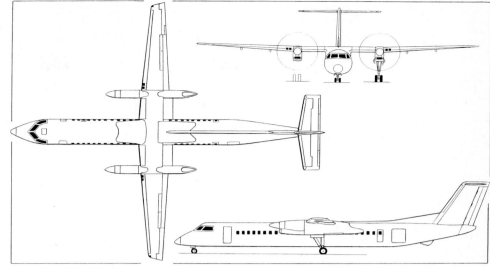
General arrangement of the Dash 8 Series 400 (Dennis Punnett)

ELMWOOD
ELMWOOD AVIATION
RR 4 Elmwood Drive, Belleville, Ontario K8N 4Z4
Telephone: 1 (613) 967 1853
DESIGNER/PROPRIETOR: Ronald B. Mason

ELMWOOD CA-05 CHRISTAVIA Mk I
TYPE: Tandem two-seat homebuilt.
PROGRAMME: First flight of Christavia Mk I (C-GENC) 3 October 1981; plans offered; aircraft considered suitable for professional flying such as missionary work; at least 475 sets of plans sold and 130 Mk Is flying; several being built for mission field.
COSTS: Plans: US$150 (1992). Average construction cost US$8,000 (1992).
DESIGN FEATURES: Computer designed wing section.
FLYING CONTROLS: Include Frise ailerons.
STRUCTURE: Braced two-spar wing of Sitka spruce, with truss ribs and Dacron covering. Aluminium can be used alternatively to fabricate spars and ribs. Wooden ailerons have Dacron covering. Fuselage of welded 4130 and 1025 steel tube, Dacron covered. Wire braced tail of steel tube, with mild steel plate ribs, Dacron covered.
LANDING GEAR: Non-retractable tailwheel type; optional floats or skis.
POWER PLANT: Prototype has a 48.5 kW (65 hp) Teledyne Continental A65. Engines of 37.25 to 112 kW (50 to 150 hp) can be fitted. Main fuel tank, capacity 57 litres (15 US gallons; 12.5 Imp gallons), in fuselage aft of firewall; auxiliary tank, capacity 23 litres (6 US gallons; 5 Imp gallons), in each wing.

DIMENSIONS, EXTERNAL:
Wing span	9.91 m (32 ft 6 in)
Wing aspect ratio	7.22
Length overall	6.30 m (20 ft 8 in)
Height overall	2.13 m (7 ft 0 in)
Propeller diameter	1.83 m (6 ft 0 in)

AREAS:
Wings, gross	13.59 m² (146.25 sq ft)

WEIGHTS AND LOADINGS (48.5 kW; 65 hp engine, 15 US gallons fuel):
Weight empty	338 kg (745 lb)
Max T-O weight	590 kg (1,300 lb)
Max wing loading	43.41 kg/m² (8.89 lb/sq ft)
Max power loading	12.16 kg/kW (20.00 lb/hp)

PERFORMANCE (at max T-O weight, engine and fuel as above):
Max level speed at S/L	104 knots (193 km/h; 120 mph)
Cruising speed at 610 m (2,000 ft), 45% power	74 knots (137 km/h; 85 mph)
Stalling speed: power off	37 knots (68 km/h; 42 mph)
power on	31 knots (57 km/h; 35 mph)
Max rate of climb at S/L	259 m (850 ft)/min
Service ceiling	3,960 m (13,000 ft)
T-O run	107 m (350 ft)
Landing from 15 m (50 ft)	198 m (650 ft)
Range with standard fuel, 45% power	382 nm (708 km; 440 miles)
g limits	+4.5/−2.8

ELMWOOD CH-8 CHRISTAVIA Mk 4
TYPE: Four-seat homebuilt cabin monoplane.
PROGRAMME: Four-seat development of Christavia Mk I: first flight 3 January 1986. Plans and information kit available; at least 200 sets of plans sold.
COSTS: Plans: $200. Average construction cost $10,000.
DESIGN FEATURES: Similar to Christavia Mk I.
POWER PLANT: One 112 kW (150 hp) Textron Lycoming O-320 or converted motorcar engine. Fuel capacity 136.4 litres (36 US gallons; 30 Imp gallons).

DIMENSIONS, EXTERNAL:
Wing span	10.67 m (35 ft 0 in)
Wing aspect ratio	7.0
Length overall	7.01 m (23 ft 0 in)
Height overall	1.98 m (6 ft 6 in)
Propeller diameter	1.78 m (5 ft 10 in)

AREAS:
Wings, gross	16.26 m² (175.0 sq ft)

WEIGHTS AND LOADINGS (112 kW; 150 hp engine):
Weight empty	522 kg (1,150 lb)
Max T-O weight	1,043 kg (2,300 lb)
Max wing loading	64.15 kg/m² (13.14 lb/sq ft)
Max power loading	9.31 kg/kW (15.33 lb/hp)

PERFORMANCE:
Max level speed at 610 m (2,000 ft)	111 knots (206 km/h; 128 mph)
Econ cruising speed, 65% power	102 knots (188 km/h; 117 mph)
Stalling speed: flaps up	35 knots (65 km/h; 40 mph)
12° flap	31 knots (57 km/h; 35 mph)
Max rate of climb at S/L	244 m (800 ft)/min
Service ceiling	5,485 m (18,000 ft)
T-O run	91 m (300 ft)
Landing run	152 m (500 ft)
Range with max fuel	548 nm (1,017 km; 632 miles)

Elmwood Aviation CA-05 Christavia Mk I two-seat homebuilt

EUROCOPTER CANADA
EUROCOPTER CANADA LTD
(Subsidiary of Eurocopter SA)
HEAD OFFICE: PO Box 250, 1100 Gilmore Road, Fort Erie, Ontario L2A 5M9
Telephone: 1 (416) 871 7772
Fax: 1 (416) 871 3320
Telex: 061-5250
VICE-PRESIDENT AND GENERAL MANAGER:
   Richard W. Harwood
GOVERNMENT MARKETING OFFICE: Suite 1202, 60 Queen Street, Ottawa, Ontario K1P 5Y7
Telephone: 1 (613) 232 1557
Fax: 1 (613) 232 5454

Started as MBB Helicopter Canada April 1984 following MBB agreement with Canadian federal and provincial governments; Fort Erie factory covering 7,897 m² (85,000 sq ft) opened mid-1986. Company holds world product mandate and design authority for high-powered BO 105 LS; sells and completes standard BO 105 CBS and Eurocopter/Kawasaki BK 117s in Canada; and is partner in new EC 135. First product BO 105 LS A-3. Workforce cut to 30 (from over 100) by mid-1993; marketing now undertaken by former Aerospatiale team based in Canada.

EUROCOPTER BO 105 LS

TYPE: Hot and high version of BO 105 CBS utility twin-engined helicopter.
PROGRAMME: First flight 23 October 1981; German LBA certification July 1984; extended for landing and take-off at 6,100 m (20,000 ft) April 1985; extended again 7 July 1986 to cover BO 105 LS A-3; FAA and Canadian DoT certification followed.
CURRENT VERSIONS: **BO 105 LS A-3**: Current production version; first delivery February 1987. *Details as for BO 105 CBS (see under Eurocopter in International section) except where indicated.*
BO 105 LS B-1: One-off testbed C-FMCL, powered by two 307 kW (412 shp) P&WC PW205B turboshafts, made first flight 13 October 1988; further details in 1989-90 *Jane's*.
CUSTOMERS: Approx 50 LS models delivered by January 1993, including 11 to Peruvian Air Force and police. Production continuing at slow rate in 1992-93.
DESIGN FEATURES: Uprated transmission and Allison 250-C28C engines produce exceptional hot and high performance.
POWER PLANT: Two Allison 250-C28C turboshafts, each rated at 410 kW (550 shp) for 2.5 min, and with 5 min T-O and max continuous power ratings of 373 kW (500 shp) and 368 kW (493 shp) respectively. Main transmission, type ZF-FS 112, rated for independent restricted input of 310 kW (416 shp) per engine at T-O power or 294 kW (394 shp) per engine for max continuous operation; or single-engine restricted input of 368 kW (493 shp) at max continuous power, or 410 kW (550 shp) for 2.5 min at T-O power. Fuel capacity as for CB/CBS. Oil capacity 5 litres (1.3 US gallons; 1.1 Imp gallons) per engine.
ACCOMMODATION: Pilot, and co-pilot or passenger, on two front seats; three or four passengers in main cabin. Cargo space behind rear seats, plus additional 20 kg (44 lb) in baggage compartment. Crew door and passengers' sliding door each side; clamshell rear cargo doors, removable for carriage of extra-long cargo. Cabin heating and air-conditioning available optionally.

Eurocopter Canada BO 105 LS four/five-passenger helicopter (two Allison 250-C28C turboshafts)

SYSTEMS: As for BO 105 CBS, except stability augmentation system is standard, bleed air anti-icing optional.
WEIGHTS AND LOADINGS:
Weight empty, basic   1,430 kg (3,152 lb)
Fuel weight   456 kg (1,005 lb)
Max T-O weight   2,600 kg (5,732 lb)
Max disc loading   34.19 kg/m² (7.00 lb/sq ft)
Max power loading (transmission restricted)
                             4.19 kg/kW (6.89 lb/shp)
PERFORMANCE (at T-O weight of 2,400 kg; 5,291 lb, ISA):
Never-exceed speed (VNE) at S/L
                             145 knots (270 km/h; 167 mph)
Max cruising speed at S/L
                             131 knots (243 km/h; 151 mph)
Max forward rate of climb at S/L   634 m (2,080 ft)/min
Max vertical rate of climb at S/L 427 m (1,400 ft)/min
Max operating altitude   6,100 m (20,000 ft)
Service ceiling, OEI, 30.5 m (100 ft)/min climb reserve
                                   2,590 m (8,500 ft)
Hovering ceiling: IGE   4,265 m (14,000 ft)
  OGE   3,385 m (11,100 ft)
Range at S/L, standard fuel, max internal payload, no reserves   281 nm (522 km; 324 miles)

IMP

IMP AEROSPACE LTD
Suite 400, 2651 Dutch Village Road, Halifax, Nova Scotia B3L 4T1
Telephone: 1 (902) 873 2250
Telex: 019 22504
MARKETING MANAGER: M. J. Garvey

Created 1970 from former Fairey Canada company when latter acquired by IMP Group. Major programmes concern weapon systems engineering, outfitting, upgrading and maintenance support of maritime aircraft used by Canadian Forces.

IMP (GRUMMAN) S-2E TRACKER UPGRADE

TYPE: Twin-turboprop anti-submarine aircraft.
PROGRAMME: Order to re-engine and upgrade 12 S-2E Tracker ASW aircraft for Brazilian Air Force (FAB designation **P-16E**) placed 1989; conversion includes 1,230 kW (1,650 shp) P&WC PT6A-67CF turboprops, Hartzell five-blade propellers and modification of fuel, hydraulic, pneumatic and electrical systems; first flight of 'prototype' 14 June 1990; redelivered to Brazil 17 December 1990 for trials in aircraft carrier *Minas Gerais*; remainder being converted in 1992. See also *Jane's Civil and Military Aircraft Upgrades*.

IMP (LOCKHEED) CP-140A ARCTURUS

TYPE: Arctic and maritime surveillance aircraft; variant of Lockheed P-3 Orion (see US section).
PROGRAMME: Outfitting and completion of three aircraft for non-ASW surface surveillance and search and rescue duties with Canadian Forces. Completed and delivered 1992-93. Details of avionics in Lockheed P-3 entry.
CUSTOMERS: Canadian Forces (CFB Greenwood, Nova Scotia): three aircraft.
COSTS: C$254 million (1992).
DESIGN FEATURES: Last three P-3A airframes from Lockheed California production line; bought 'green' and outfitted by IMP with cockpit avionics, Texas Instruments AN/APS-134 Plus surveillance radar and interior equipment and furnishing.
EQUIPMENT: SAR equipment includes liferafts and SKAD (survival kit, air-droppable).

IMP (SIKORSKY) SEA KING CONVERSION (1)

TYPE: Anti-submarine helicopter.
PROGRAMME: Six Canadian Forces CH-124A Sea Kings

Brazilian Air Force P-16E Tracker converted by IMP

converted for surface surveillance over Gulf from HMCS *Athabaskan* and *Protecteur*; one converted by IMP and remainder at Canadian Forces Base Shearwater, Nova Scotia, with IMP kits. Details in 1991-92 *Jane's*.

IMP (SIKORSKY) SEA KING CONVERSION (2)

TYPE: Anti-submarine helicopter.
PROGRAMME: Mission system avionics modification of six Canadian Forces CH-124A Sea Kings to CH-124B, for compatibility with frigates and destroyers equipped with Canadian towed array sonar system (CANTASS), under programme known as HELTAS (helicopter towed array support); due for redelivery between August 1992 and May 1993. Seventh Sea King to be equipped with MAD and Tac/Nav of CH-124B plus a 99-channel sonobuoy receiver, HAPS processor, and three competing types of acoustic recorder.

DESIGN FEATURES: Intended to evaluate equipment of, and provide training platforms for, Canadian versions of EH 101 (see International section).
ACCOMMODATION: Crew of four (two pilots and two navigators, latter acting as TACCO and SENSO).
AVIONICS: Existing dipping sonar, analog Tac/Nav, sonobuoy receiver and relay systems deleted; AN/ASQ-504 internal MAD retained. New equipment includes AN/ASN-123 Tac/Nav, AN/UYS-503 sonobuoy processor and 31-channel AN/ARR-75 sonobuoy receiver, plus 20 sonobuoys.

IMP (EHI) EH 101

TYPE: Anti-submarine helicopter.
PROGRAMME: IMP a major subcontractor in CH-148/149 Canadian versions of EH 101 helicopter (see International section); work will also include manufacture of tailplanes, gearbox casings and inter-seat consoles for all EH 101s worldwide, and rear loading ramps for all utility EH 101s.

KFC
KELOWNA FLIGHTCRAFT GROUP
No. 1, 5655 Kelowna Airport, Kelowna, British Columbia V1V 1S1
Telephone: 1 (604) 765 1481
Fax: 1 (604) 765 1489
PRESIDENT: Barry Lapointe
VICE-PRESIDENT: Jim Rogers
OPERATIONS MANAGER: Greg Carter
CV5800 PROJECT MANAGER: Bill De Meester

KFC Group comprises four companies, each specialising in one area of aviation but each complementing the others:

Kelowna Flightcraft Air Charter Ltd responds to all kinds of charter, from fire patrols to time-sensitive cargo or personnel flights. Fleet comprises two Boeing 727s, 12 Convair 580s, one Gulfstream I, two Cessna 402Bs and a DC-3.

Kelowna Flightcraft Ltd totally maintains KFC Air Charter fleet and undertakes work for outside customers; overhaul, manufacturing and engineering capabilities have Canadian MoT approval and FAA (FAR Pt 43.17) recognition.

Kelowna Flight-Comm Avionics Inc intensively involved in CV5800 rework programme, and others involving civilian, military and government aircraft.

Kelowna Flightcraft R & D Ltd involved in research and development of major aircraft modifications including STCs and STAs such as CV5800, long-range fuel system for Convair 580, cargo door modifications and AiResearch GTC 85-90 APU modification.

First prototype CV5800, a stretched conversion of the Convair 580/C-131F

KFC CV5800
TYPE: Stretched conversion of Convair 580 (applicable to any Convair 340/440/580 airframe).
PROGRAMME: Launched 1984; structural work started January 1990; first flight (N5800) 11 February 1992; second prototype 60 per cent complete December 1992 and expected to fly May 1993; first 'production' conversion also to fly 1993; certification by FAA (to FAR 25 Pt 4B) and Canadian STA expected in August 1993.
CURRENT VERSIONS: **Cargo:** As first prototype. *Detailed description applies to this version.*
Passenger: Under development; to seat 76 people at 81 cm (32 in) pitch.
COSTS: Development programme $16.5 million (1989); standard aircraft $6.5 million.
DESIGN FEATURES: Fuselage stretched by 4.34 m (14 ft 3 in) and strengthened for 2,268 kg (5,000 lb) increase in gross weight; wing also strengthened internally; standard Convair 340/580 tail unit; Honeywell four-tube EFIS and other modern avionics; Allison 501 turboprops and Hamilton Standard propellers; configured to cargo class E standard, with cargo conveyance system available optionally. Will meet Stage 3 noise and Cat. II landing standards; modification claimed to extend aircraft life by 100,000 hours.
FLYING CONTROLS: Conventional mechanical (cables), all with trim and servo tabs; vortex generators added to tailplane; control column bob-weight and increased spring tension system; Honeywell autopilot.
STRUCTURE: Conventional aluminium alloy stressed skin.
LANDING GEAR: Twin-wheel nose and main units retract forward hydraulically. Menasco oleo-pneumatic shock absorbers and Goodyear wheels and tyres on each unit; main-wheel tyres size 39 × 13-16, pressure 5.86 bars (85 lb/sq in); nosewheel tyres size 29 × 7.5-14, pressure 4.34 bars (63 lb/sq in). Goodyear disc brakes and Hytrol anti-skid units; nose unit steerable ±62°. Minimum ground turning radius about 19.81 m (65 ft 0 in).
POWER PLANT: Two Allison 501-D22G turboprops (each 3,430 kW; 4,600 shp at 13,820 rpm); Hamilton Standard 54H60-77/117 four-blade constant-speed propellers (reversible pitch optional). Standard fuel capacity 6,549 litres (1,730 US gallons; 1,440 Imp gallons) in 3,274 litre (865 US gallon; 720 Imp gallon) tank in each wing; increase to 7,874 litres (2,080 US gallons; 1,732 Imp gallons) total available optionally. Single pressure fuelling point in each lower wing skin. Oil capacity 32 litres (8.4 US gallons; 7 Imp gallons).
ACCOMMODATION: Flight crew of two. Crew/passenger door at front and cargo door at rear of cabin, both on port side. Overwing emergency exit each side. Entire accommodation pressurised and air-conditioned.
SYSTEMS: Convair/AiResearch pressurisation/air-conditioning system (max differential 0.29 bar; 4.16 lb/sq in). Dual primary hydraulic systems (pressure 207 bars; 3,000 lb/sq in), plus emergency standby. Electrical power at 400Hz supplied by two 24V DC generators, two 240V AC alternators and two 12V batteries. Gaseous oxygen system. Hot-air de-icing and anti-icing of wings, tail unit, propellers and engine air intakes. AiResearch GTC 85-90 APU for electric (DC) engine starting.
AVIONICS: Honeywell SPZ-4500 digital autopilot, dual FZ-450 flight management computers, Primus II nav/com/ident radio system, EDZ-803 four-tube EFIS, Primus 650 weather radar, dual VG-14A/C-14A AHRS, and A-A300 radio altimeter.

DIMENSIONS, EXTERNAL:
Wing span	32.10 m (105 ft 4 in)
Length overall	29.18 m (95 ft 9 in)
Height overall	8.58 m (28 ft 1⅔ in)
Tailplane span	11.15 m (36 ft 7 in)
Wheel track	7.62 m (25 ft 0 in)
Wheelbase	7.97 m (26 ft 1¾ in)
Propeller diameter	4.11 m (13 ft 6 in)
Propeller ground clearance	0.33 m (13 in)
Cabin door (fwd, port): Height	2.11 m (6 ft 11 in)
Width	0.91 m (3 ft 0 in)
Cargo door (rear, port): Height	1.83 m (6 ft 0 in)
Width	3.05 m (10 ft 0 in)
Emergency exits (each): Height	0.66 m (2 ft 2 in)
Width	0.48 m (1 ft 7 in)

DIMENSIONS, INTERNAL:
Cabin excl flight deck: Length	20.80 m (68 ft 3 in)
Max width	2.97 m (9 ft 9 in)
Max height	1.96 m (6 ft 5 in)
Floor area	48.12 m² (518 sq ft)
Volume (cargo)	87.67 m³ (3,096 cu ft)

AREAS:
Wings, gross	85.47 m² (920.0 sq ft)

WEIGHTS AND LOADINGS:
Operating weight empty (cargo version)	15,043 kg (33,166 lb)
Max payload (cargo version)	9,903 kg (21,834 lb)
Max standard fuel	5,257 kg (11,591 lb)
Max T-O weight	28,576 kg (63,000 lb)
Max landing weight	26,308 kg (58,000 lb)
Max zero-fuel weight	24,947 kg (55,000 lb)
Max wing loading	334.3 kg/m² (68.48 lb/sq ft)
Max power loading	4.17 kg/kW (6.85 lb/shp)

PERFORMANCE: Still being determined (January 1993)

NWI
NORTHWEST INDUSTRIES LIMITED
(a Division of CAE Industries Ltd)
PO Box 9864, Edmonton International Airport, Edmonton, Alberta T5J 2T2
Telephone: 1 (403) 890 6300
Fax: 1 (403) 890 7773
PRESIDENT: L. H. Prokop
VICE-PRESIDENT, MARKETING: B. McKenzie
MARKETING MANAGER: T. MacEwan

Major Canadian civil and military maintenance, repair, overhaul and modification centre; works on Lockheed C-130 and T-33, and Canadair CL-41 (CT-114); maintains mobile repair parties at CFBs Cold Lake and Bagotville to support Canadian Forces CF-5s and CF-18s: manufactures structural, mechanical and electronic components for own programmes and under subcontract for principal manufacturers.

Major structural upgrade of 22 Canadian Forces C-130Es completed 1987; two new C-130H-84s and two used C-130H-73s upgraded and modified to Canadian Forces fleet standard; progressive structural inspection (PSI) of C-130 fleet begun Autumn 1987; five new C-130s modified as flight refuelling tankers 1992; T-33A Silver Star inspections continuing; prototype CL-41 Tutor rewiring programme installed 1992; depot level inspection and repair, windscreen replacement, avionics update and operational load monitoring programme for CL-41 to start 1993.

PIC
PROMAVIA INTERNATIONAL CORPORATION
Saskatoon, Saskatchewan

Formed 1991 by Promavia SA (see Belgian section) to be responsible for production of Jet Squalus jet trainer and any derivatives; awaiting capital for factory construction and start-up.

VIVIAN
L. R. VIVIAN ASSOCIATES LIMITED
1203 Roselawn Avenue, Toronto, Ontario M6B 1C6
Telephone: 1 (416) 789 3407
Fax: 1 (416) 785 5396
MANAGING DIRECTOR: L. R. Vivian
MARKETING MANAGER: Alan Sarsons

VIVIAN HYBRID AIRCRAFT
TYPE: V/STOL heavy lift aircraft.
PROGRAMME: Design project, revealed 1991, though owes something to HTH Helitruck, last described in German section of 1985-86 *Jane's*. No information received for 1992 or 1993; all known details, and illustration, in 1992-93 *Jane's*.

ZENAIR
ZENAIR LTD
Huronia Airport, Midland, Ontario L4R 4K8
Telephone: 1 (705) 526 2871
Fax: 1 (705) 526 8022
PRESIDENT AND DESIGNER: Christophe Heintz
US WORKS: Zénith Aircraft Company, Mexico Memorial Airport, Mexico, Missouri 65265-0650
Telephone: 1 (314) 581 9000
Fax: 1 (314) 581 0011
PRESIDENT: Sebastien C. Heintz
MANAGER, ADMINISTRATION: Susan M. McCullough
MANAGER, PRODUCTION: Nicholas M. Heintz

Founded in 1974, Zenair designs, develops and manufactures light aircraft. In 1992 Zénith Aircraft Company was formed to manufacture and market Zenair aircraft in USA, which include Zodiac CH 601 and STOL CH 701. Zenair Ltd and Zénith also jointly developing Zénith CH 2000 Trainer, for FAR VLA certification. CH 601 and CH 701 are to undergo certification under FAA's new Primary Category, and production rate has been increased.

Manufacture of Mono Z-CH 100, Acro-Zénith CH 150 and Super Acro-Zénith CH-180 has ended (see 1992-93 *Jane's*); prototype development and testing of Super STOL CH 801 also discontinued. In addition to light aircraft, company constructs metal floats and amphibious floats.

Zenair has licensed representatives in Africa, South America, France, Italy, Japan, Spain and the UK.

ZENAIR ZENITH-CH 200

TYPE: Side by side two-seat, dual-control homebuilt aircraft.
PROGRAMME: Prototype first flight 22 March 1970. Many hundreds of kits and plans sold to customers in 44 countries.
COSTS: Plans: $230. Kit: $13,850. Information pack: $20. Materials and component parts also available.
DESIGN FEATURES: NACA 64A515 (modified) wing section. Plans show rudder only, with no fin. Conventional fin and rudder can be fitted if desired.
STRUCTURE: All-metal semi-monocoque, except for glassfibre engine cowling and fairings.
LANDING GEAR: Non-retractable tricycle or, optionally, tailwheel type. Floats and skis optional.
POWER PLANT: Design suitable for engines from 63.5 kW (85 hp) to 119 kW (160 hp). Fuel capacity 90 litres (24 US gallons; 20 Imp gallons). Optional fuel tanks in wing leading-edges, total capacity 72.5 litres (19.2 US gallons; 16 Imp gallons).

Zenair Zénith-CH 200 with float landing gear

DIMENSIONS, EXTERNAL:
Wing span	7.00 m (22 ft 11¾ in)
Wing aspect ratio	5.0
Length overall	6.25 m (20 ft 6 in)
Height overall	2.11 m (6 ft 11 in)
Propeller diameter	1.83 m (6 ft 0 in)

AREAS:
Wings, gross	9.80 m² (105.9 sq ft)

WEIGHTS AND LOADINGS (93 kW; 125 hp engine):
Weight empty	422 kg (930 lb)
Baggage capacity	35 kg (77 lb)
Max T-O weight	680 kg (1,500 lb)
Max wing loading	69.16 kg/m² (14.16 lb/sq ft)
Max power loading	7.31 kg/kW (12.00 lb/hp)

PERFORMANCE (at max T-O weight. 93 kW; 125 hp engine):
Max level speed at S/L	131 knots (243 km/h; 151 mph)
Cruising speed (75% power) at S/L	122 knots (227 km/h; 141 mph)
Stalling speed, flaps down	47 knots (87 km/h; 54 mph)
Max rate of climb at S/L	335 m (1,100 ft)/min
T-O and landing run	244 m (800 ft)
Range with max fuel	391 nm (724 km; 450 miles)
g limit	+9 ultimate

ZENAIR ZENITH-CH 250

TYPE: De luxe version of Zénith-CH 200.
COSTS: Plans: $260. Kit: $14,980. Information pack: $20. Materials and component parts available.
DESIGN FEATURES: Two 65 litre (16.8 US gallon; 14 Imp gallon) fuel tanks in wings and 0.71 m³ (25 cu ft) baggage area. Forward sliding canopy and rear windows similar to those of Tri-Z CH 300.
LANDING GEAR: Tailwheel or float gear can be fitted as alternatives to standard tricycle type.
POWER PLANT: Recommended engines in 86-119 kW (115-160 hp) range.

WEIGHTS AND LOADINGS (112 kW; 150 hp engine):
Weight empty	449 kg (990 lb)
Max T-O weight	730 kg (1,610 lb)
Max wing loading	74.49 kg/m² (15.26 lb/sq ft)
Max power loading	6.53 kg/kW (10.73 lb/hp)

PERFORMANCE (engine as above):
Max level speed	144 knots (267 km/h; 166 mph)
Cruising speed (75% power)	130 knots (241 km/h; 150 mph)
Stalling speed	45 knots (84 km/h; 52 mph)
Max rate of climb at S/L	518 m (1,700 ft)/min
T-O and landing run	198 m (650 ft)
Range with standard fuel	469 nm (869 km; 540 miles)

ZENAIR TRI-Z CH 300

TYPE: Three-seat stretched version of two-seat CH 200.
PROGRAMME: First flight 9 July 1977.
COSTS: Plans: $300. Kit: $16,880. Information pack: $20. Materials and component parts available.
DESIGN FEATURES: Enlarged cabin to provide room for rear bench seat able to carry third adult, two children or 95 kg (210 lb) of baggage. Construction from kits takes 1,000 working hours on average.
FLYING CONTROLS: Electrically actuated slotted flaps.
POWER PLANT: Recommended power in 93-134 kW (125-180 hp) range. Fuel carried in two 64.4 litre (17 US gallon; 14.2 Imp gallon) wing tanks. Double fuel can be carried in further wing tanks.

DIMENSIONS, EXTERNAL:
Wing span	8.08 m (26 ft 6 in)
Wing aspect ratio	5.40
Length overall	6.86 m (22 ft 6 in)
Height overall	2.08 m (6 ft 10 in)

AREAS:
Wings, gross	12.08 m² (130.0 sq ft)

WEIGHTS AND LOADINGS (93 kW; 125 hp engine):
Weight empty	476 kg (1,050 lb)
Max T-O weight	816 kg (1,800 lb)
Max wing loading	67.6 kg/m² (13.85 lb/sq ft)
Max power loading	8.77 kg/kW (14.40 lb/hp)

PERFORMANCE (engine as above):
Max level speed	130 knots (241 km/h; 150 mph)
Econ cruising speed	109 knots (201 km/h; 125 mph)
Stalling speed, flaps down	45 knots (82 km/h; 51 mph)
Max rate of climb at S/L	244 m (800 ft)/min
Range (75% power)	521 nm (965 km; 600 miles)
g limits	±5.7 ultimate

ZENAIR ZODIAC CH 601

TYPE: Side by side two-seat advanced ultralight (AUL), ARV and homebuilt; conforms to Canadian TP 10141 airworthiness standards.
PROGRAMME: First flight of CH 600 prototype June 1984. Plans and kits of parts (45 per cent premanufactured) for improved model followed. In 1991 CH 600 relaunched as CH 601HD (which see). New two-seater already developed from CH 600, as CH 601, specifically to meet TP 10141 advanced ultralight (AUL) category in Canada. Construction of prototype CH 601 began September 1990 (first flight October that year). Kit production started January 1991. Also available as 85 per cent preassembled kit, requiring only 120 working hours to complete.
COSTS: Plans: C$315. Kit: C$21,795 with engine. Information pack: C$20. Materials and component parts are available.
DESIGN FEATURES: New features included reduced empty weight, wider cabin (1.12 m; 3 ft 8 in), and standard wing lockers for baggage or extra fuel. A higher performance model became 601HDS (which see).
POWER PLANT: One 59.7 kW (80 hp) Rotax 912 flat-four engine. Fuel capacity 60 litres (16 US gallons; 13.3 Imp gallons). Two 28.4 litre (7.5 US gallon; 6.25 Imp gallon) auxiliary tanks optional.

DIMENSIONS, EXTERNAL:
Wing span	8.23 m (27 ft 0 in)
Wing aspect ratio	5.61
Length overall	5.79 m (19 ft 0 in)
Height overall	1.52 m (5 ft 0 in)
Propeller diameter	1.73 m (5 ft 8 in)

AREAS:
Wings, gross	12.08 m² (130.0 sq ft)

WEIGHTS AND LOADINGS:
Weight empty	249 kg (550 lb)
Baggage capacity	36 kg (80 lb)
Max T-O weight	476 kg (1,050 lb)
Max wing loading	39.45 kg/m² (8.08 lb/sq ft)
Max power loading	7.97 kg/kW (13.13 lb/hp)

PERFORMANCE (Rotax 912):
Max level speed	117 knots (217 km/h; 135 mph)
Econ cruising speed	104 knots (193 km/h; 120 mph)
Stalling speed	39 knots (71 km/h; 44 mph)
Max rate of climb at S/L	366 m (1,200 ft)/min
Service ceiling	more than 4,875 m (16,000 ft)
T-O run	122 m (400 ft)
Landing run	199 m (650 ft)
Range	380 nm (704 km; 437 miles)
Endurance	4 h
g limits	±6

ZENAIR ZODIAC CH 601HD

TYPE: 'Heavy duty' CH 600, available as kit and as 85 per cent preassembled kit requiring only 120 working hours to complete.
COSTS: Plans: C$340. Kit: C$21,995 with engine. Information pack: C$20. Materials and component parts are available.
DESIGN FEATURES: Wider cabin than CH 601. No fixed fin. Wing section NACA 65018.
STRUCTURE: All-metal.
LANDING GEAR: Non-retractable tricycle or tailwheel type. Skis and floats optional.
POWER PLANT: One engine of 48.5-85.75 kW (65-115 hp), including 59.7 kW (80 hp) Rotax 912, with reduction gear, 47.7 kW (64 hp) Rotax 532, 48.5-74.5 kW (65-100 hp) Teledyne Continental, 37.3 kW (50 hp) Volkswagen without electric starting or 52 kW (70 hp) Volkswagen. Fuel capacity 60 litres (16 US gallons; 13.3 Imp gallons).

DIMENSIONS, EXTERNAL: As CH 601 except:
Height overall	1.90 m (6 ft 2¾ in)
Propeller diameter	1.52 m (5 ft 0 in)

WEIGHTS AND LOADINGS (Rotax 912 for loadings):
Weight empty, equipped	268 kg (590 lb)
Max T-O weight	544 kg (1,200 lb)
Max wing loading	45.07 kg/m² (9.23 lb/sq ft)
Max power loading	9.11 kg/kW (15.00 lb/hp)

PERFORMANCE (Rotax 912, at max T-O weight):
Max level speed at S/L	117 knots (217 km/h; 135 mph)
Econ cruising speed	104 knots (193 km/h; 120 mph)
Stalling speed	41 knots (76 km/h; 47 mph)
Max rate of climb at S/L	351 m (1,150 ft)/min
Service ceiling	more than 4,875 m (16,000 ft)
T-O run	131 m (430 ft)
Landing run	138 m (450 ft)
Range with max fuel	380 nm (704 km; 437 miles)
Endurance	4 h
g limits	±6.5

ZENAIR SUPER ZODIAC CH 601HDS

TYPE: Side by side two-seat monoplane; pending Recreational category certification in Canada.
PROGRAMME: Latest member of CH 601 series, conceived to offer higher performance and true cross-country capability. Prototype first flight August 1991. Apart from recreational flying, is suitable for surveillance, patrol and training. Military version designated CH 601HDS-M.
COSTS: Plans: C$390. Kit: C$23,175 with Rotax engine.
DESIGN FEATURES: Shorter span and tapered 'speed' wing.
STRUCTURE: All-metal.
LANDING GEAR: Non-retractable tricycle or tailwheel type, with disc brakes. Optional skis and floats.

Zenair Super Zodiac CH 601HDS two-seat monoplane

POWER PLANT: One 59.7 kW (80 hp) Rotax 912, with 2.53:1 reduction gear. Optional engines include converted motorcar types, Textron Lycoming O-235 and Teledyne Continental O-200. Fuel capacity 60 litres (16 US gallons; 13.3 Imp gallons). Two 28.4 litre (7.5 US gallon; 6.25 Imp gallon) wing tanks optional.

DIMENSIONS, EXTERNAL: As CH 601 except:
Wing span	7.01 m (23 ft 0 in)
Height overall	1.57 m (5 ft 2 in)
Propeller diameter	1.73 m (5 ft 8 in)

AREAS:
Wings, gross	9.10 m² (98.0 sq ft)

WEIGHTS AND LOADINGS:
Weight empty	258 kg (570 lb)
Max T-O weight	544 kg (1,200 lb)
Max wing loading	59.78 kg/m² (12.24 lb/sq ft)
Max power loading	9.11 kg/kW (15.00 lb/hp)

PERFORMANCE:
Max level speed at S/L	130 knots (241 km/h; 150 mph)
Econ cruising speed at S/L, 60% power	113 knots (209 km/h; 130 mph)
Stalling speed, engine idling	46 knots (86 km/h; 53 mph)
Max rate of climb at S/L	314 m (1,030 ft)/min
Service ceiling	more than 4,875 m (16,000 ft)
T-O run	168 m (550 ft)
Landing run	183 m (600 ft)
Range: with standard fuel	440 nm (815 km; 506 miles)
with wing tanks	825 nm (1,528 km; 950 miles)
Endurance	7 h
g limits	±6

Artist's impression of Zenair Zénith CH 2000 Trainer

ZENAIR STOL CH 701

TYPE: Two-seat advanced ultralight or Experimental homebuilt.

PROGRAMME: Experimental category prototype made first flight Summer 1986. Plans, 49 or 85 per cent kits, or fully assembled aircraft available. Meets TP 10141 standards for advanced ultralight trainers.

COSTS: Plans: C$390. Kit: C$19,890 with engine. Information pack: C$20. Materials and component parts available.

DESIGN FEATURES: Strut braced high-wing cabin monoplane with foldable wings. In most countries, CH 701s can be registered either as advanced ultralights (AULs) or as Experimental homebuilts.

FLYING CONTROLS: Full span leading-edge flaps for STOL performance.

LANDING GEAR: Can be completed with tricycle or tailwheel gear, with floats, amphibious and ski gears as options.

POWER PLANT: One 37.3 kW (50 hp) Rotax 503 engine in prototype, but 59.7 kW (80 hp) Rotax 912, 48.5 kW (65 hp) Rotax 582, 44.7-52 kW (60-70 hp) VW or 48.5-67 kW (65-90 hp) Teledyne Continental engine optional; standard fuel capacity 60 litres (16 US gallons; 13.3 Imp gallons); optional 114 litres (30 US gallons; 25 Imp gallons).

DIMENSIONS, EXTERNAL:
Wing span	8.23 m (27 ft 0 in)
Wing aspect ratio	5.98
Length overall	6.10 m (20 ft 0 in)
Propeller diameter	1.73 m (5 ft 8 in)

AREAS:
Wings, gross	11.33 m² (122.0 sq ft)

WEIGHTS AND LOADINGS (48.5 kW; 65 hp engine for loadings):
Weight empty	208 kg (460 lb)
Max T-O weight	435 kg (960 lb)
Max wing loading	38.42 kg/m² (7.87 lb/sq ft)
Max power loading	8.97 kg/kW (14.77 lb/hp)

PERFORMANCE (48.5 kW; 65 hp engine, at max T-O weight):
Max level speed at S/L	74 knots (137 km/h; 85 mph)
Max cruising speed	65 knots (121 km/h; 75 mph)
Stalling speed	25 knots (45 km/h; 28 mph)
Max rate of climb at S/L	366 m (1,200 ft)/min
Service ceiling	4,875 m (16,000 ft)
T-O run	28 m (90 ft)
Landing run	39 m (125 ft)
Range: standard fuel	250 nm (463 km; 287 miles)
with wing tanks	450 nm (834 km; 518 miles)
Endurance	1 h 40 min
g limits	+6/−3

ZENAIR STOL CH 701-AG

TYPE: Agricultural version of CH 701.

PROGRAMME: Operated extensively in South America. Micro Ag spraying/dusting system. Aircraft are assembled by Zenair's Colombian subsidiary, Agricopteros Ltda (which see).

ZENAIR ZENITH CH 2000 TRAINER

TYPE: Two-seat multi-purpose trainer.

PROGRAMME: Under development by Zenair and Zénith Aircraft. Certification expected August 1993, with production starting early 1994.

COSTS: $59,900 (with basic equipment).

DESIGN FEATURES: Side by side two-seat and smaller span derivative of Tri-Z CH 300. Designed to conform to FAR Pt 23 (JAR VLA equivalent level of safety) and aimed at primary training market. Wing section LS (1) 0417 (mod).

FLYING CONTROLS: Mechanical, except for electrically actuated flaps. Dual controls/dual rudder pedals.

STRUCTURE: Aluminium alloy, with stressed skins.

LANDING GEAR: Non-retractable tricycle type; optional floats and skis.

POWER PLANT: One 86.5 kW (116 hp) Textron Lycoming O-235, driving metal Sensenich CKS 6-0-54 propeller. Fuel capacity 90 litres (24 US gallons; 20 Imp gallons) standard, 180 litres (48 US gallons; 40 Imp gallons) optional.

AVIONICS: Standard VFR instrumentation. IFR will be available.

DIMENSIONS, EXTERNAL:
Wing span	7.60 m (24 ft 11 in)
Wing aspect ratio	5.36
Length overall	6.81 m (22 ft 4 in)
Height overall	2.34 m (7 ft 8 in)
Propeller diameter	1.83 m (6 ft 0 in)

AREAS:
Wings, gross	10.78 m² (116.0 sq ft)

WEIGHTS AND LOADINGS:
Max payload: with max fuel	180 kg (396 lb)
with standard fuel and baggage	220 kg (485 lb)
Max T-O weight	726 kg (1,600 lb)
Max wing loading	67.34 kg/m² (13.79 lb/sq ft)
Max power loading	8.39 kg/kW (13.79 lb/hp)

PERFORMANCE (estimated):
Max level speed	128 knots (236 km/h; 147 mph)
Econ cruising speed	104 knots (193 km/h; 120 mph)
Stalling speed	45 knots (82 km/h; 51 mph)
Max rate of climb at S/L	280 m (920 ft)/min
Service ceiling	more than 3,660 m (12,000 ft)
T-O run	213 m (700 ft)
Landing run	183 m (600 ft)
Range: with max standard and optional fuel	851 nm (1,577 km; 980 miles)
with max payload and standard fuel	425 nm (788 km; 490 miles)
g limits	+4.4/−2.2

Zenair STOL CH 701 with wheel fairings

CHILE

ENAER
EMPRESA NACIONAL DE AERONÁUTICA DE CHILE

Avenida José Miguel Carrera 11087, P. 36 ½, Santiago
Telephone: 56 (2) 5282599, 5282735 and 5282823
Fax: 56 (2) 5282815
Telex: 645157 ENAER CT
CHAIRMAN: General Ramón V. Hidalgo (Ret'd)
COMMERCIAL DIRECTOR: Alejandro Vargas

State owned company, formed 1984 from IndAer industrial organisation set up 1980 by Chilean Air Force (FACh); workforce approx 1,700; aircraft manufacture started 1979 with assembly of 27 Piper PA-28 Dakota lightplanes for Chilean Air Force and flying clubs; current or recent activities, design and production of aircraft and EW equipment, upgrade programmes for Chilean Air Force (eg conversion of Beechcraft 99s to maritime surveillance role, as listed in 1987-88 *Jane's*, and retrofit of FACh Hunters with Caiquen II RWR). Present main programmes are T-35DT Turbo Pillán military trainer, Ñamcu lightplane and Pantera upgrade of FACh Mirage 50s.

ENAER T-35 PILLÁN (DEVIL)
Spanish Air Force designation: E.26 Tamiz

TYPE: Two-seat fully aerobatic basic (T-35A/C) and instrument (T-35B/D) military trainer.

PROGRAMME: First two prototypes (first flight 6 March 1981) developed by Piper; followed by three Piper kits for ENAER assembly (first flight, by FACh s/n 101, 30 January 1982); then slight redesign, replacing all-moving tailplane by electrically trimmable tailplane with conventional elevator, increasing rudder mass balance and deepening canopy; ENAER series production started September

1984; first flight of production T-35A, 28 December 1984; deliveries to FACh began 31 July 1985. Supply of ENAER T-35C kits to CASA started 27 December 1985 (first flight 12 May 1986), completed September 1987; T-35D deliveries to Panama began 1988; first flight of single-seat T-35S made 5 March 1988; T-35A/B deliveries to FACh completed by Spring 1990; export deliveries completed 1991. Full description in 1992-93 and earlier *Jane's*.

CURRENT VERSIONS: **T-35A**: Primary trainer version for Chilean Air Force. In service with Escuela de Aviación Capitan Avalos at El Bosque AB, Santiago.

T-35B: Chilean Air Force instrument trainer, with more comprehensive instrumentation.

T-35C: Primary trainer for Spanish Air Force (designation **E.26 Tamiz**), assembled by CASA from ENAER kits.

T-35D: Instrument trainer for Panama and Paraguay.

T-35S: Single-seat prototype (CC-PZB); evaluation continuing in 1990, but present status uncertain; 313 kW (420 shp) Allison 250-B17 turboprop planned for any production version. Illustrated in 1991-92 *Jane's*.

T-35DT Turbo Pillán: Turboprop version; described separately.

CUSTOMERS: Total 146 for air forces of Chile (60 T-35A, 20 T-35B), Panama (10 T-35D), Paraguay (15 T-35D) and Spain (41 T-35C/E.26), all delivered. Two converted as Aucán and Turbo Pillán prototypes (see next entry).

ENAER Chile's company demonstrator T-35 Pillán piston-engined trainer *(Brian M. Service)*

ENAER T-35DT TURBO PILLÁN

TYPE: Two-seat turboprop military trainer (converted T-35D).

PROGRAMME: Original turboprop prototype was **T-35TX Aucán** (CC-PZC: see 1988-89 *Jane's*), first flown February 1986, of which flight trials suspended 1987 after some 500 flight hours; subsequently modified to have new one-piece canopy opening sideways to starboard, oxygen system, and partial avionics upgrade. Soloy Corporation (see US section) awarded 1990 contract to develop production-ready modification kit for existing T-35s, based on Allison 250-B17D engine; marketing effort began early 1991; first flight of this **T-35DT** 'prototype' (CC-PZG) March 1991.

DESIGN FEATURES: Based on Piper Cherokee series (utilising many components of PA-28 Dakota and PA-32 Saratoga); cleared to FAR Pt 23 (Aerobatic category) and military standards for basic, intermediate and instrument flying training. Low-wing, tandem-seat design; sweptback vertical tail, non-swept horizontal surfaces.

Wing section NACA 65_2-415 on constant chord inboard panels, NACA 65_2-415 (modified) at tips; incidence 2° at root, −0° 30′ at tip; dihedral 7° from roots.

FLYING CONTROLS: Manual/mechanical actuation of primary control surfaces (ailerons slotted, elevator/rudder mass balanced); single-slotted wing flaps, aileron trim tab (port) and rudder/tailplane/elevator trim all electrically actuated; variable incidence tailplane; one-piece elevator.

STRUCTURE: Main structure of aluminium alloy and steel, with riveted skins, except for glassfibre engine cowling, wingtips and tailplane tips. Single-spar fail-safe wings, with components mainly from PA-28-236 Dakota (leading-edges) and PA-32R-301 Saratoga (trailing-edges), modified for shorter span; vertical tail virtually identical with Dakota; tailplane uses some standard components from Dakota and PA-31 (Navajo/Cheyenne); tailcone from Cherokee components, modified for narrower fuselage.

LANDING GEAR: Hydraulically retractable tricycle type, with single wheel on each unit. Main units retract inward, steerable nosewheel (±25°) rearward. Min ground turning radius 6.20 m (20 ft 4 in). Piper oleo-pneumatic shock absorber in each unit. Emergency free-fall extension. Cleveland mainwheels and McCreary tyres size 6.00-6 (8 ply), nosewheel and tyre size 5.00-5 (6 ply). Tyre pressures: 2.62 bars (38 lb/sq in) on mainwheels, 2.41 bars (35 lb/sq in) on nosewheel. Single-disc air-cooled hydraulic brake on each mainwheel. Parking brake.

POWER PLANT: One 313 kW (420 shp) Allison 250-B17D turboprop in Soloy inverted installation with Soloy reduction gearbox; Hartzell HC-B3TF-7A/T9212K-2B three-blade constant-speed reversible-pitch propeller. Total usable fuel 277.8 litres (73.4 US gallons; 61.1 Imp gallons), in two integral wing tanks; inverted-flight fuel feed system.

ACCOMMODATION: Two vertically adjustable seats, with seat belts and shoulder harnesses, in tandem beneath one-piece transparent jettisonable canopy which opens sideways to starboard. One-piece acrylic windscreen, and one-piece window in glassfibre fairing aft of canopy. Rear (instructor's) seat 22 cm (8.7 in) higher. Dual controls standard. Baggage compartment aft of rear cockpit, with external access on port side. Cockpits ventilated; cockpit heating and canopy demisting by engine bleed air.

SYSTEMS: Electrically operated hydraulic system, at 124 bars (1,800 lb/sq in) pressure for landing gear retraction and 44.8 bars (650 lb/sq in) for gear extension; separate system at 20.7 bars (300 lb/sq in) for wheel brakes. Electrical system is 28V DC, powered by 28V 70A engine driven Prestolite alternator and 24V 10Ah battery, with inverter for AC power at 400Hz to operate RMIs and attitude indicators. External power socket. Oxygen system.

AVIONICS: Nav/com system comprises dual VHF/VOR transceivers, ATC transponder, DME with dual indicators, dual HSI, ADF, dual RMI, marker beacon receiver, dual audio selector panels, dual magnetic compasses, dual directional gyros and slave meter. Blind-flying instrumentation standard.

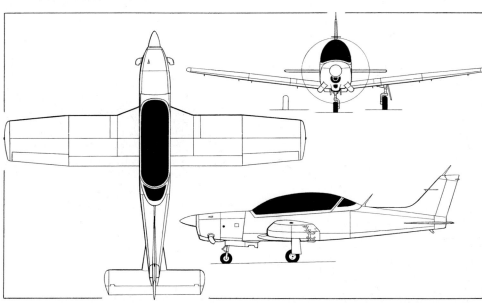

ENAER T-35DT Turbo Pillán tandem two-seat turboprop trainer *(Dennis Punnett)*

Prototype T-35DT Turbo Pillán, converted by Soloy *(Charles Bickers)*

DIMENSIONS, EXTERNAL:
Wing span	8.84 m (29 ft 0 in)
Wing chord: at root	1.88 m (6 ft 2 in)
constant portion	1.60 m (5 ft 3 in)
at tip	1.26 m (4 ft 1½ in)
Wing aspect ratio	5.71
Length overall	8.60 m (28 ft 2½ in)
Height overall	2.64 m (8 ft 8 in)
Tailplane span	3.05 m (10 ft 0 in)
Wheel track	3.01 m (9 ft 10½ in)
Wheelbase	2.00 m (6 ft 6¾ in)
Propeller diameter	1.93 m (6 ft 4 in)

DIMENSIONS, INTERNAL:

Cockpit: Length	3.24 m (10 ft 7½ in)
Max width	1.04 m (3 ft 5 in)
Max height	1.48 m (4 ft 10¼ in)

AREAS:
Wings, gross	13.69 m² (147.36 sq ft)
Ailerons (total)	1.14 m² (12.27 sq ft)
Trailing-edge flaps (total)	1.36 m² (14.64 sq ft)
Fin	0.69 m² (7.43 sq ft)
Rudder	0.38 m² (4.09 sq ft)
Tailplane	1.57 m² (16.90 sq ft)
Elevator	0.77 m² (8.29 sq ft)

WEIGHTS AND LOADINGS:
Weight empty	943 kg (2,080 lb)
Fuel weight: usable	224 kg (494 lb)
max	238 kg (524 lb)
T-O and landing weight: Aerobatic	1,315 kg (2,900 lb)
max	1,338 kg (2,950 lb)
Max wing loading	97.73 kg/m² (20.03 lb/sq ft)
Max power loading	4.27 kg/kW (7.02 lb/shp)

PERFORMANCE (at max T-O weight):
Never-exceed speed (V_{NE})	241 knots (446 km/h; 277 mph) IAS
Max manoeuvring speed (V_A)	183 knots (339 km/h; 211 mph) IAS
Max level speed at S/L	230 knots (426 km/h; 264 mph)

Cruising speed:
75% power at 2,320 m (7,610 ft)	182 knots (337 km/h; 209 mph)
65% power at 3,450 m (11,320 ft)	176 knots (326 km/h; 203 mph)
55% power at 4,630 m (15,190 ft)	169 knots (313 km/h; 194 mph)

Stalling speed, power off:
flaps and landing gear up	66 knots (123 km/h; 76 mph)
flaps and landing gear down	62 knots (115 km/h; 71 mph)
Max rate of climb at S/L	869 m (2,850 ft)/min
Service ceiling	7,620 m (25,000 ft)
Absolute ceiling	7,826 m (25,675 ft)
Time to 2,000 m (6,560 ft)	3 min 42 s
Time to 3,000 m (9,850 ft)	5 min 36 s
T-O run	195 m (640 ft)
T-O to 15 m (50 ft)	357 m (1,175 ft)
Landing from 15 m (50 ft)	555 m (1,820 ft)
Landing run	128 m (420 ft)

Range, 45 min reserves:
75% power at 2,500 m (8,200 ft)	350 nm (648 km; 403 miles)
55% power at 4,000 m (13,125 ft)	410 nm (759 km; 472 miles)
g limits	+6/−3

ENAER T-36/A-36 HALCÓN (HAWK)

TYPE: Licence built CASA C-101 (see under Spain).
PROGRAMME: Chilean Air Force (FACh) contract 1980 included licence for progressive increase from CKD kit assembly to local manufacture of part of airframe; T-36 entered service late 1983, A-36 made first flight November 1983; ENAER began phase 3B (manufacture of hydraulic and electrical systems, and small subassemblies) third quarter 1989 with aircraft s/n 417; final phase 4 involved manufacture of forward fuselages, tail units and flying control surfaces. A-36 upgrade programme reportedly under consideration in 1992.

CURRENT VERSIONS: **T-36**: Trainer version (CASA C-101BB-02), with 16.46 kN (3,700 lb st) Garrett TFE731-3 turbofan; in service with Escuela de Aviación Capitan Avalos at El Bosque AB, Santiago. Four built in Spain, 10 assembled by ENAER; production completed.

A-36: Attack version (CASA C-101CC-02), with 19.13 kN (4,300 lb st) TFE731-5 turbofan (increasable to 20.91 kN; 4,700 lb st with military power reserve); in service with 1° Esquadrão of Brigada Aérea I at Los Cóndores, northern Chile, and 12° Esquadrão of Brigada Aérea IV at Carlos Ibánez AB in extreme south. Four built in Spain, 19 by ENAER; programme completed by mid-1993 according to CASA.

CUSTOMERS: Chilean Air Force (14 T-36, 23 A-36).

ENAER PANTERA 50C (PANTHER)

TYPE: Upgraded Dassault Mirage 50.
PROGRAMME: Pantera 50C is programme to upgrade airframe and avionics of eight Mirage 50FCs, six 50CHs and two 50DCHs, with technical assistance from Israel Aircraft Industries, launched in mid-1980s. Flight testing (with foreplanes only) began 1986; first flight of fully upgraded aircraft at end of 1988; second aircraft completed by March 1992; three more then under way, but programme slowed by lack of funding.

DESIGN FEATURES: Forward fuselage untapered plug of approx 1.00 m (3 ft 3¼ in) inserted immediately ahead of windscreen; non-movable canard surfaces shoulder-mounted on engine intake trunks; in-flight refuelling probe added. Aircraft serve with 4° Esquadrão of Brigada Aérea IV at Carlos Ibánez AB, southern Chile.

AVIONICS: New suite includes INS, Elbit radar, computerised HUD, modified electrical/hydraulic/weapon control systems, ENAER Caiquen III RWR and Eclipse chaff/flare dispensing system.

ENAER A-36 Halcón in Fuerza Aérea de Chile insignia *(Denis Hughes)*

First ENAER Pantera, a modified Mirage 50 with foreplanes and upgraded nav/attack system *(Kenneth Munson)*

Prototype ENAER Ñamcu two-seat light aircraft *(Charles Bickers)*

ENAER ÑAMCU (EAGLET)

TYPE: Two-seat light aircraft.
PROGRAMME: Launched June 1986 under project name Avión Liviano (light aircraft); prototype construction started February 1987 (three to be built); first flight (CC-PZI) April 1989; first flight of second prototype (CC-PZJ) early March 1990; one prototype since lost; present programme status uncertain.

DESIGN FEATURES: First ENAER aircraft of all-Chilean design, intended as small, inexpensive club aircraft, initially for domestic market and later for export; fully aerobatic training capability; conforms to FAR Pt 23 (Utility category).
Wing section NACA 63_2-415; incidence 3° at root, 0° 30′ at tip; dihedral 5° from roots.

FLYING CONTROLS: Conventional primary surfaces (plain ailerons, balanced elevators and rudder), mechanically

actuated; trim tab in starboard elevator; plain flaps on wing trailing-edges.
STRUCTURE: All-composites (glassfibre/foam sandwich), including movable surfaces; glassfibre/carbonfibre type C main spar with carbonfibre spar caps.
LANDING GEAR: Non-retractable tricycle type. Cantilever spring steel main units; steerable and self-centring nose unit, with oleo-pneumatic shock absorber and shimmy damper. Cleveland wheel and Goodyear 6-ply tyre on each unit; all three tyres size 5.00-5. Cleveland hydraulic mainwheel disc brakes.
POWER PLANT: One Textron Lycoming O-235-N2C flat-four engine (86.5 kW; 116 hp at 2,800 rpm), driving an MT 178 R115-2C two-blade fixed-pitch wooden propeller. Integral fuel tank in each wing leading-edge, combined capacity 100 litres (26.4 US gallons; 22 Imp gallons); overwing gravity fuelling point for each tank.
ACCOMMODATION: Two seats side by side in fully enclosed cockpit, with adjustable headrests and four-piece safety harnesses. Dual controls. Two independent gull wing window/doors, hinged on centreline to open upward. Space for 10 kg (22 lb) of baggage aft of seats. Cockpit heated and ventilated. Electric defrosting of windscreen.
SYSTEMS: Hydraulic system for brakes only. Electrical power supplied by Prestolite 12V 70A alternator and 12V 35Ah lead-acid battery.
AVIONICS: VFR flight and engine instrumentation, Walter Dittel FSG 71M 760-channel VHF com transceiver, and intercom, are standard; IFR instrumentation optional.

DIMENSIONS, EXTERNAL:
Wing span	8.31 m (27 ft 3¼ in)
Wing chord: at root	1.53 m (5 ft 0¼ in)
at tip	0.84 m (2 ft 9 in)
Wing aspect ratio	7.6
Length overall	7.05 m (23 ft 1½ in)
Height overall	2.42 m (7 ft 11¼ in)
Tailplane span	3.00 m (9 ft 10 in)
Wheel track	2.54 m (8 ft 4 in)
Wheelbase	1.50 m (4 ft 11 in)
Propeller diameter	1.78 m (5 ft 10 in)

DIMENSIONS, INTERNAL:
Cockpit: Max width	1.22 m (4 ft 0 in)

AREAS:
Wings, gross	10.01 m² (107.75 sq ft)
Ailerons (total)	0.44 m² (4.74 sq ft)
Trailing-edge flaps (total)	0.92 m² (9.90 sq ft)
Fin	0.88 m² (9.47 sq ft)
Rudder	0.34 m² (3.66 sq ft)
Tailplane	2.08 m² (22.39 sq ft)
Elevators (total)	0.76 m² (8.18 sq ft)

WEIGHTS AND LOADINGS:
Basic weight empty	546 kg (1,204 lb)
Max fuel	72 kg (159 lb)
Max T-O and landing weight	800 kg (1,763 lb)
Max wing loading	79.92 kg/m² (16.38 lb/sq ft)
Max power loading	9.25 kg/kW (15.20 lb/hp)

PERFORMANCE (at max T-O weight except where indicated):
Never-exceed speed (V_{NE})	177 knots (328 km/h; 204 mph)
Max level speed at S/L	127 knots (235 km/h; 146 mph)
Max cruising speed, 75% power at 2,440 m (8,000 ft)	103 knots (191 km/h; 119 mph)
Stalling speed, power off:	
flaps up	56 knots (104 km/h; 65 mph)
flaps down	50 knots (93 km/h; 58 mph)
Max rate of climb at S/L	295 m (968 ft/min)
Service ceiling	4,270 m (14,000 ft)
T-O run	304 m (998 ft)
T-O to 15 m (50 ft)	412 m (1,352 ft)
Landing from 15 m (50 ft)	364 m (1,195 ft)
Landing run	177 m (581 ft)
Range at 75% power at 2,440 m (8,000 ft) with max fuel, 10% reserves	500 nm (926 km; 575 miles)
Endurance, conditions as above	3 h 40 min
g limits	+4.4/−2.2

CHINA, PEOPLE'S REPUBLIC

AIC
AVIATION INDUSTRIES OF CHINA
PRESIDENT: Zhu Yuli
INTERNATIONAL MARKETING:
CATIC (China National Aero-Technology Import and Export Corporation)
5 Liangguochang Road, East City District (PO Box 647), Beijing 100010
Telephone: 86 (1) 4017722
Fax: 86 (1) 4015381
Telex: 22318 AEROT CN
PRESIDENT: Liu Guomin
EXECUTIVE VICE-PRESIDENT: Tang Xiaoping

Former Ministry of Aero-Space Industry reorganised mid-1993 into AIC as economic entity to develop market economy and expand international collaboration in aviation programmes. There are design and development centres at Shenyang, Beijing, Harbin and elsewhere. Over 13,000 aircraft and 50,000 aero-engines produced since 1949, of which more than 700 exported; approx 10 per cent of exported aircraft are civil types. Xian, Chengdu, Shanghai, Shenyang, Harbin and other factories also carry out subcontract work on Airbus A300/A310/A320; ATR 42; BAe 146 and Jetstream ATP; Boeing 737/747/757; de Havilland Dash 8; Canadair CL-215; McDonnell Douglas MD-80 series; and Shorts 360. Total workforce of aerospace industry estimated at 500,000 in 1990, although most factories also have wide range of non-aerospace products. Transition from military to civil products being accelerated.

International partner being sought for 30/40-passenger commuter airliner programme (existing type or new design).

Current helicopter programmes include Changhe Z-8 (Chinese derivative of Aerospatiale Super Frelon), licence production at Harbin of Eurocopter Dauphin as Z-9, and development (see International section) with Eurocopter and Singapore Aerospace of EC 120, a small (2,500 kg; 5,500 lb class) type for agricultural and forestry work.

CAC
CHENGDU AIRCRAFT INDUSTRIAL CORPORATION
PO Box 800, Chengdu, Sichuan 610092
Telephone: 86 (28) 669629
Fax: 86 (28) 669816
Telex: 60132 CCDAC CN
PRESIDENT: Yang Baoshu
GENERAL MANAGER: Hou Jianwu
DEPUTY GENERAL MANAGER: Li Shaoming
DIRECTOR, INTERNATIONAL CO-OPERATION DIVISION:
Wang Yinggong

Major centre for fighter development and production, founded 1958; has since built over 2,000 fighters of 10 models or variants; current facility occupies site area of 510 ha (1,260 acres) and had 1992 workforce of 22,000. Production includes J-7/F-7 fighter series (several models) and limited batches of JJ-5/FT-5 fighter trainer; subcontract work includes July 1988 contract for 100 nosecones for McDonnell Douglas MD-80 series, both for Shanghai MD-82/83 programme (see SAMF entry in this section; first one delivered 13 December 1991) and for US production line. Output also includes various non-aerospace products.

CAC JJ-5
Chinese name: Jianjiji Jiaolianji-5 (Fighter training aircraft 5) or Jianjiao-5
Westernised designation: FT-5
TYPE: Tandem two-seat fighter-trainer.
PROGRAMME: Developed at Chengdu 1965; first flight 8 May 1966; certificated for mass production end of 1966.
CUSTOMERS: Standard advanced trainer of Chinese air forces, to which pupil pilots graduate after basic training on NAMC CJ-6A (which see). Production totalled 1,061 at end of 1986, with limited batch production undertaken subsequently; more than 100 exported as FT-5s to Albania (eight), Bangladesh, Pakistan, Sri Lanka (two), Sudan and Tanzania.
DESIGN FEATURES: Tandem two-seat modification of J-5A (Chinese built MiG-17PF), retaining latter's lipped intake (fairing indicates provision for radar ranging gunsight in front cockpit) but with WP5D engine and with armament reduced to single Type 23-1 (23 mm) gun in removable belly pack with barrel to starboard side of nosewheel doors.
POWER PLANT: One 26.48 kN (5,952 lb st) Xian (XAE) WP5D non-afterburning turbojet.

DIMENSIONS, EXTERNAL:
Wing span	9.628 m (31 ft 7 in)
Length overall	11.50 m (37 ft 9 in)
Height overall	3.80 m (12 ft 5¾ in)
Wheel track	3.85 m (12 ft 7½ in)

WEIGHTS AND LOADINGS:
Weight empty, equipped	4,080 kg (8,995 lb)
Normal T-O weight	5,401 kg (11,907 lb)
Max T-O weight	6,215 kg (13,700 lb)

PERFORMANCE:
Normal operating speed	418 knots (775 km/h; 482 mph)
Max rate of climb at S/L	1,620 m (5,315 ft)/min
Service ceiling	14,300 m (46,900 ft)
T-O run	760 m (2,493 ft)
Landing run	780-830 m (2,559-2,723 ft)
Range with max fuel at 12,000 m (39,370 ft)	664 nm (1,230 km; 764 miles)
Max endurance at 13,700 m (45,000 ft) with two 400 litre (105.5 US gallon; 88 Imp gallon) drop tanks	2 h 38 min

CAC J-9
Chengdu developing a new single-seat, single-engined fighter for entry into service by year 2000. Design features include sweptback wings, close-coupled canards, fly-by-wire flight controls and gross weight in 10-15 tonne (22,045-33,070 lb) class. Foreign co-operation may be sought for avionics. First flight 5 September 1991.

CAC (MIKOYAN) J-7
Chinese name: Jianjiji-7 (Fighter aircraft 7) or Jian-7
Westernised designation: F-7
TYPE: Single-seat fighter and close support aircraft.
PROGRAMME: Soviet licence to manufacture MiG-21F-13 and its R-11F-300 engine granted to Chinese government 1961, when some pattern aircraft and CKD (component knocked down) kits delivered, but necessary technical documentation not completed; assembly of first J-7 using Chinese made components began early 1964; original plan in 1964-65 was for Chengdu and Guizhou factories to become main airframe/engine production centres for J-7, backed up by Shenyang until these were fully productive, but plans affected by onset of cultural revolution. Static testing completed November 1965; first flight of Shenyang

J-7 III Chinese equivalent of the MiG-21MF 'Fishbed-J'

built J-7, 17 January 1966; Chengdu production of J-7 I began June 1967; development of J-7 II began 1975, followed by first flight 30 December 1978 and production approval September 1979; development of F-7M and J-7 III started 1981; J-7 III first flight 26 April 1984; F-7M revealed publicly October 1984, production go-ahead December 1984, named Airguard early 1986; first F-7P deliveries to Pakistan 1988; first F-7MPs to Pakistan mid-1989.

CURRENT VERSIONS: **J-7**: Initial licence version using Chinese made components, built at Shenyang; few only.

J-7 I: Initial Chengdu version for PLA Air Force (1967), with variable intake shock cone and second 30 mm gun; not accepted in large numbers, due mainly to unsatisfactory escape system (front-hinged canopy, to which ejection seat was attached).

F-7A: Export counterpart of J-7 I, supplied to Albania and Tanzania.

J-7 II: Modified and improved development of J-7 I, with WP7B turbojet of increased thrust (43.15 kN; 9,700 lb st dry, 59.82 kN; 13,448 lb st with afterburning); 720 litre (190.2 US gallon; 158.4 Imp gallon) centreline drop tank for increased range; brake-chute relocated at base of rudder to improve landing performance and shorten run; rear-hinged canopy, jettisoned before ejection seat deploys; new Chengdu Type II seat offering ejection at zero height and speeds down to 135 knots (250 km/h; 155 mph); and new Lanzhou compass system. Small batch production (eg 14 in 1989) continuing.

F-7B: Export version of J-7 II, with R550 Magic missile capability; supplied to Egypt and Iraq in 1982-83.

F-7BS: Hybrid version supplied to Sri Lanka 1991: has F-7B fuselage/tail and Chinese avionics (no HUD etc), combined with four-pylon wings of F-7M. Equips No. 5 Squadron.

F-7M Airguard: Upgraded export version, developed from J-7 II; new avionics imported from May 1979 included GEC-Marconi Avionics HUDWAC (head-up display and weapon aiming computer); new ranging radar, air data computer, radar altimeter and IFF; more secure com radio; improved electrical power generation system for the new avionics; two additional underwing stores points; improved WP7B(BM) engine; birdproof windscreen; strengthened landing gear; ability to carry PL-7 air-to-air missiles; nose probe relocated from beneath intake to top lip of intake, offset to starboard. Exported to Bangladesh, Iran and Zimbabwe. In production. *Description applies to this version except where indicated.*

F-7P Airguard: Variant of F-7M (briefly called Skybolt), embodying 24 modifications to meet specific requirements of Pakistan Air Force, including ability to carry four air-to-air missiles (Sidewinders) instead of two and fitment of Martin-Baker Mk 10L ejection seat. Delivered 1988-91.

F-7MP: Further-modified variant of F-7P; improved cockpit layout and navigation system incorporating Collins AN/ARN-147 VOR/ILS receiver, AN/ARN-149 ADF and Pro Line II digital DME-42. Avionics (contract for up to 100 sets) delivered to China from early 1989.

J-7 III: Chinese equivalent of MiG-21MF, much redesigned from J-7 II with blown flaps and all-weather, day/night capability. Main improvements are more powerful WP13 engine; additional fuel in deeper dorsal spine; JL-7 (J-band) interception radar, with correspondingly larger nose intake and centrebody radome; sideways opening (to starboard) canopy, with centrally located rearview mirror; improved HTY-4 low-speed/zero height ejection seat; more advanced fire control system; twin-barrel 23 mm gun under fuselage (with HK-03D optical gunsight); broader-chord vertical tail surfaces, incorporating antennae for LJ-2 omnidirectional RWR in hemispherical fairing each side at base of rudder; increased weapon/stores capability (four underwing stations), similar to that of F-7M; and new or additional avionics (which see). Joint development by Chengdu and Guizhou (GAIC); in production and service.

F-7 III: Improved version of J-7 III; due to enter PLA Air Force service 1992.

J-7E: Upgraded version of J-7 II with modified wing (see drawing), retaining existing leading-edge sweep angle of 57° inboard but reduced sweep of only 42° outboard; span increased by 1.17 m (3 ft 10 in) and area by 1.88 m² (20.2 sq ft), giving 8.17 per cent more wing area; four underwing stations instead of two, outer pair each plumbed for 480 litre (127 US gallon; 105.6 Imp gallon) drop tank; new WP7F version of WP7 engine, rated at 44.13 kN (9,921 lb st) dry and 63.74 kN (14,330 lb st) with afterburning; armament generally as listed for F-7M, but capability extended to include PL-8 air-to-air missiles; g limits: 8 (up to Mach 0.8) and 6.5 (above Mach 0.8); avionics include head-up display and air data computer. Reportedly due to have made first flight April 1990.

Super-7: Proposed export development of F-7M; described separately.

JJ-7/FT-7: Tandem two-seat operational trainer, based on J-7 II and MiG-21US; developed at Guizhou and described under GAIC entry.

CUSTOMERS: Several hundred built for Chinese air forces; over 400 exported to Albania (12 F-7A), Bangladesh (16 F-7M), Egypt (approx 80 F-7B?), Iran (F-7M), Iraq (approx 80 F-7B?), Pakistan (20 F-7P and 60 F-7MP delivered, all designated F-7P by PAF; more F-7MP

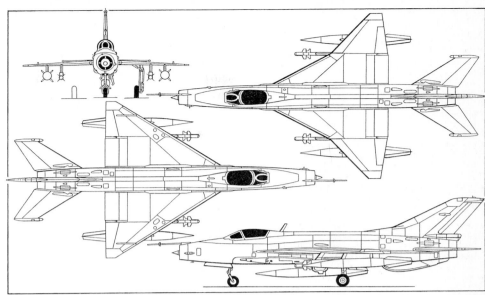

CAC F-7M Airguard single-seat fighter and close support aircraft; upper plan view shows modified outer wings of J-7E (*Jane's/Mike Keep*)

CAC F-7P fighters of the Pakistan Air Force (*Peter Steinemann*)

Cockpit layout of the F-7M Airguard

reportedly on order), Sri Lanka (four F-7BS), Tanzania (16 F-7A) and Zimbabwe (22 F-7M). Pakistan Air Force squadrons are No. 2 at Masroor, Nos. 18 and 20 at Rafiqui and No. 25 at Mianwali; F-7BSs serve with Sri Lanka's No. 5 Squadron.

DESIGN FEATURES: Diminutive tailed delta with clipped tips to mid-mounted wings; circular-section fuselage with dorsal spine; nose intake with conical centrebody; swept tail, with large vertical surfaces and ventral fin.

Wing anhedral 2° from roots; incidence 0°; thickness/chord ratio approx 5 per cent at root, 4.2 per cent at tip; quarter-chord sweepback 49° 6′ 36″; no wing leading-edge camber.

FLYING CONTROLS: Manual operation, with autostabilisation in pitch and roll; hydraulically boosted inset ailerons; plain trailing-edge flaps, actuated hydraulically; forward hinged door type airbrake each side of underfuselage below wing leading-edge; third forward hinged airbrake under fuselage forward of ventral fin; airbrakes actuated hydraulically; hydraulically boosted rudder and all-moving, trimmable tailplane.

STRUCTURE: All-metal; wings have two primary spars and auxiliary spar; semi-monocoque fuselage, with spine housing control pushrods, avionics, single-point refuelling cap and fuel tank; blister fairings on fuselage above and below each wing to accommodate retracted mainwheels.

LANDING GEAR: Inward retracting mainwheels, with 600 × 200 mm tyres (pressure 11.5 bars; 166.8 lb/sq in) and LS-16 disc brakes; forward retracting nosewheel, with 500 × 180 mm tyre (pressure 7.0 bars; 101.5 lb/sq in) and LS-15 double-acting brake. Nosewheel steerable ±47°. Min ground turning radius 7.04 m (23 ft 1¼ in). Tail braking parachute at base of vertical tail.

POWER PLANT: One Chengdu WP7B(BM) turbojet (43.15 kN; 9,700 lb st dry, 59.82 kN; 13,448 lb st with afterburning) in F-7M; WP13 turbojet (40.21 kN; 9,039 lb st dry, 64.72 kN; 14,550 lb st with afterburning) in J-7 III/F-7 III. Total internal fuel capacity 2,385 litres (630 US gallons; 524.5 Imp gallons), contained in six flexible tanks in fuselage and two integral tanks in each wing. Provision for carrying a 500 or 800 litre (132 or 211.3 US gallon; 110 or 176 Imp gallon) centreline drop tank, and/or a 500 litre drop tank on each outboard underwing pylon. Max internal/external fuel capacity 4,185 litres (1,105.3 US gallons; 920.5 Imp gallons).

ACCOMMODATION: Pilot only, on Chengdu Aircraft Industrial Corporation zero-height/low-speed ejection seat operable between 70 and 459 knots (130-850 km/h; 81-528 mph) IAS. Martin-Baker Mk 10L seat in Pakistani F-7P/MP. One-piece canopy, hinged at rear to open upward. J-7 III/F-7 III canopy opens sideways to starboard.

SYSTEMS: Improved electrical system in F-7M, using three static inverters, to cater for additional avionics. Jianghuai YX-3 oxygen system.

AVIONICS (F-7M): GEC-Marconi Avionics suite includes Type 956 HUDWAC, AD 3400 two-band UHF/VHF multi-function com system, Type 226 Skyranger ranging radar with ECCM, and an air data computer. Other avionics include Chinese Type 602 IFF transponder, Type 0101 HR A/2 radar altimeter, WL-7 radio compass, and XS-6A marker beacon receiver. HUDWAC (head-up display and weapon aiming computer) provides pilot with displays for instrument flying, with air-to-air and air-to-ground weapon aiming symbols integrated with flight-instrument symbology. It can store 32 weapon parameter functions, allowing for both current and future weapon variants. In air-to-air combat its four modes (missiles, conventional gunnery, snapshoot gunnery, dogfight and standby aiming reticle) allow for all eventualities. Navigation function includes approach mode.

AVIONICS (J-7 III): New avionics include GT-4 ECM jammer, FJ-1 flight data recorder, Type 605A ('Odd Rods' type) IFF; angle of attack vane and air data probe similar to those of F-7M; Beijing Aeronautical Instruments Factory KJ-11 twin-channel autopilot.

ARMAMENT (F-7M): Two 30 mm Type 30-1 belt-fed cannon, with 60 rds/gun, in fairings under front fuselage just forward of wingroot leading-edges. Two hardpoints under each wing, of which outer ones are wet for carriage of drop tanks. Centreline pylon used for drop tank only. Each inboard pylon capable of carrying a PL-2, -2A, -5B or -7 missile or, at customer's option, a Matra R.550 Magic; one 18-tube pod of Type 57-2 (57 mm) air-to-air and air-to-ground rockets; one Type 90-1 (90 mm) 7-tube pod of air-to-ground rockets; or a 50, 150, 250 or 500 kg bomb. Each outboard pylon can carry one of above rocket pods, a 50 or 150 kg bomb, or a 500 litre drop tank.

ARMAMENT (J-7 III): One 23 mm Type 23-3 twin-barrel gun in ventral pack. Five external stores stations can carry two to four PL-5B air-launched missiles; or four Qingan HF-16B 12-round launchers for Type 57-2 or seven-round pods of Type 90-1 rockets; or two 500 kg, four 250 kg or ten 100 kg bombs, in various combinations with 500 litre (one centreline and/or one under each wing) or 800 litre (underfuselage station only) drop tanks.

DIMENSIONS, EXTERNAL:
Wing span: except J-7E 7.154 m (23 ft 5⅝ in)
J-7E 8.32 m (27 ft 3½ in)
Wing chord: at root 5.508 m (18 ft 0¾ in)
at tip 0.462 m (1 ft 6¼ in)
Wing aspect ratio: except J-7E 2.22
J-7E 2.78
Length overall: excl nose probe 13.945 m (45 ft 9 in)
incl nose probe 14.885 m (48 ft 10 in)
Fuselage: Length 12.177 m (39 ft 11½ in)
Max diameter 1.341 m (4 ft 4¾ in)
Height overall 4.103 m (13 ft 5½ in)
Tailplane span 3.14 m (10 ft 3½ in)
Wheel track 2.692 m (8 ft 10 in)
Wheelbase 4.806 m (15 ft 9¼ in)

AREAS:
Wings, gross: except J-7E 23.00 m² (247.6 sq ft)
J-7E 24.88 m² (267.8 sq ft)
Ailerons (total): except J-7E 1.18 m² (12.70 sq ft)
Trailing-edge flaps (total) 1.87 m² (20.13 sq ft)
Fin 3.48 m² (37.46 sq ft)
Rudder 0.97 m² (10.44 sq ft)
Tailplane 3.94 m² (42.41 sq ft)

WEIGHTS AND LOADINGS:
Weight empty 5,275 kg (11,629 lb)
Normal max T-O weight with two PL-2 or PL-7 air-to-air missiles: F-7M 7,531 kg (16,603 lb)
J-7 III 8,150 kg (17,967 lb)
Wing loading at normal max T-O weight:
F-7M 327.43 kg/m² (67.10 lb/sq ft)
J-7 III 354.35 kg/m² (72.58 lb/sq ft)
Power loading at normal max T-O weight:
F-7M 125.5 kg/kN (1.230 lb/lb st)
J-7 III 125.9 kg/kN (1.234 lb/lb st)

PERFORMANCE (F-7M at normal max T-O weight with two PL-2 or PL-7 air-to-air missiles, except where indicated):
Never-exceed speed (VNE) above 12,500 m (41,010 ft)
Mach 2.35 (1,346 knots; 2,495 km/h; 1,550 mph)
Max level speed between 12,500 and 18,500 m (41,010-60,700 ft)
Mach 2.05 (1,175 knots; 2,175 km/h; 1,350 mph)
Unstick speed 167-178 knots (310-330 km/h; 193-205 mph)
Touchdown speed 162-173 knots (300-320 km/h; 186-199 mph)
Max rate of climb at S/L 10,800 m (35,435 ft)/min
Acceleration from Mach 0.9 to 1.2 at 5,000 m (16,400 ft) 35 s
Max sustained turn rate: Mach 0.7 at S/L 14.7°/s
Mach 0.8 at 5,000 m (16,400 ft) 9.5°/s
Service ceiling 18,200 m (59,710 ft)
Absolute ceiling 18,700 m (61,350 ft)
T-O run 700-950 m (2,297-3,117 ft)
Landing run with brake-chute 600-900 m (1,969-2,953 ft)
Typical mission profiles:
combat air patrol at 11,000 m (36,000 ft) with two air-to-air missiles and three 500 litre drop tanks, incl 5 min combat 45 min
long-range interception at 11,000 m (36,000 ft) at 351 nm (650 km; 404 miles) from base, incl Mach 1.5 dash and 5 min combat, stores as above
hi-lo-hi interdiction radius, out and back at 11,000 m (36,000 ft), with three 500 litre drop tanks and two 150 kg bombs 324 nm (600 km; 373 miles)
lo-lo-lo close air support radius with four rocket pods, no external tanks 200 nm (370 km; 230 miles)
Range: two PL-7 missiles and three 500 litre drop tanks 939 nm (1,740 km; 1,081 miles)
self-ferry with one 800 litre and two 500 litre drop tanks, no missiles 1,203 nm (2,230 km; 1,385 miles)
g limit +8

PERFORMANCE (J-7 III at normal max T-O weight):
Max operating Mach number 2.1
Unstick speed (with afterburning) 173 knots (320 km/h; 199 mph)
Touchdown speed (with flap blowing) 135-146 knots (250-270 km/h; 155-168 mph)
Min level flight speed 140 knots (260 km/h; 162 mph)
Max rate of climb at S/L 9,000 m (29,525 ft)/min
Service ceiling 18,000 m (59,050 ft)
Acceleration from Mach 1.2 to 1.9 at 13,000 m (42,650 ft) 3 min 27 s
Air turning radius at 5,000 m (16,400 ft) at Mach 1.2 5,093 m (16,710 ft)
T-O run (with afterburning) 800 m (2,625 ft)
Landing run (with flap blowing, drag-chute and brakes deployed) 550 m (1,805 ft)
Range: internal fuel only 518 nm (960 km; 596 miles)
with 800 litre belly tank 701 nm (1,300 km; 807 miles)
with 800 litre belly tank and two 500 litre underwing tanks 898 nm (1,664 km; 1,034 miles)
g limits: up to Mach 0.8 +8.5
above Mach 0.8 +7

CAC SUPER-7

TYPE: Proposed export development of F-7M Airguard.

PROGRAMME: Agreement between CATIC and Grumman (USA) for joint preliminary design signed 21 October 1988; Grumman participation suspended by US government mid-1989. Preliminary design and wind tunnel testing completed; CATIC seeking alternative partner to continue programme.

DESIGN FEATURES: Lateral air intakes for more powerful engine; 'solid' ogival nosecone for modern fire control radar; wings of enlarged span and area, with leading-edge slats and additional pair of hardpoints inboard for air-to-air missiles; enlarged dorsal spine housing additional fuel; single-point pressure refuelling; easier-access engine compartment; arrester hook; modified ventral fin; strengthened main landing gear with larger tyres; new straight-leg, steerable nosewheel unit; belly mounted twin-barrel 23 mm gun instead of two internal 30 mm; new cockpit, incorporating HUD and new ejection seat; revised ECS for avionics cooling.

CUSTOMERS: Pakistan regarded as main market.

DIMENSIONS, EXTERNAL:
Wing span 8.98 m (29 ft 5½ in)
Length overall 15.30 m (50 ft 2½ in)
Height overall 4.13 m (13 ft 6½ in)
Wheel track 2.79 m (9 ft 1¾ in)
Wheelbase 5.59 m (18 ft 4 in)

AREAS:
Wings, gross approx 24.62 m² (265.0 sq ft)

WEIGHTS AND LOADINGS (estimated):
Internal fuel 2,327 kg (5,130 lb)
Design gross weight 9,100 kg (20,062 lb)
Max T-O weight 11,295 kg (24,900 lb)

PERFORMANCE (estimated):
Max level speed above Mach 1.8
Service ceiling 16,765 m (55,000 ft)
T-O distance 555 m (1,821 ft)
Landing distance 860 m (2,822 ft)
Mission radius:
air-to-air (hi-hi-hi) 475 nm (880 km; 547 miles)
air-to-ground (hi-lo-hi) 329 nm (610 km; 379 miles)
g limit (max) +8.5

Model of the proposed Super-7 advanced export development of the F-7M

CAF
CHANGHE AIRCRAFT FACTORY
PO Box 109, Jingdezhen, Jiangxi 333002
Telephone: 442019
Telex: 95027 CHAF CN
DIRECTOR: Li Wanxin

Changhe Aircraft Factory, built on a 234 ha (578 acre) site at Jingdezhen, began producing coaches and commercial road vehicles in 1974. These and other automotive products still account for most of output, but since being placed under MAS jurisdiction has batch-produced helicopters and is one of three members (with Harbin Aircraft Manufacturing Co and Helicopter Design and Research Institute) of China Helicopter Industry Corporation. CAF's 1991 workforce of 6,000 included more than 800 engineers and technicians.

CAF Z-8
Chinese name: Zhishengji-8 (Vertical take-off aircraft 8) or Zhi-8

TYPE: Multi-role military and civil helicopter.
PROGRAMME: Design work begun 1976, but suspended from 1979 to mid-1984; initial flights of first prototype 11 December 1985, second prototype October 1987; domestic type approval awarded 8 April 1989; first Z-8 handed over to PLA Naval Air Force for service trials 5 August 1989; initial production has been approved. Eventual applications expected to include troop transport, ASW/ASV, search and rescue, minelaying/sweeping, aerial survey and firefighting.
CURRENT VERSIONS: Standard version, *as described*.
DESIGN FEATURES: Chinese equivalent of Aerospatiale Super Frelon (see 1982-83 *Jane's*). Six-blade main rotor and five-blade tail rotor; boat-hull fuselage with watertight compartments inside planing bottom; stabilising float at rear each side, attached to small stub-wing; small, strut braced fixed horizontal stabiliser on starboard side of tail rotor pylon.
FLYING CONTROLS: Pitch control fitting at root of each main rotor blade; drag and flapping hinges for each blade mounted on rotor head starplates; each main blade also has a hydraulic drag damper. Fully redundant flight control system, with Dong Fang KJ-8 autopilot.
STRUCTURE: Stressed skin metal fuselage, with riveted watertight compartments; gearboxes manufactured by Zhongnan Transmission Machinery Factory.
LANDING GEAR: Non-retractable tricycle type, with twin wheels and low-pressure oleo-pneumatic shock absorber on each unit. Small tripod tailskid under rear of tailboom. Boat hull and side floats permit emergency water landings and take-offs.
POWER PLANT: Three Changzhou (CLXMW) WZ6 turboshafts, each with max emergency rating of 1,156 kW (1,550 shp) and 20 per cent power reserve at S/L, ISA. Two engines side by side in front of main rotor shaft and one aft of shaft. Transmission rated at 3,072 kW (4,120 shp). Standard internal fuel capacity 3,900 litres (1,030 US gallons; 858 Imp gallons), in flexible tanks under floor of centre fuselage. Auxiliary fuel tanks can be carried inside cabin for extended range or self-ferry missions, increasing total capacity to 5,800 litres (1,532 US gallons; 1,276 Imp gallons).
ACCOMMODATION: Crew of two or three on flight deck. Accommodation in main cabin for up to 27 fully armed troops, or 39 without equipment; up to 15 stretchers and a medical attendant in ambulance configuration; a BJ-212 jeep and its crew; or other configurations according to mission. Entire accommodation heated, ventilated, soundproofed and vibration-proofed. Forward opening crew door on each side of flight deck. Rearward sliding door at front of cabin on starboard side. Hydraulically actuated rear loading ramp/door.
EQUIPMENT: Equipment for SAR role can include 275 kg (606 lb) capacity hydraulic rescue hoist and two five-person liferafts. Can also be equipped with sonar, sonobuoys, search radar, or equipment for oceanography, geological survey and forest firefighting.
ARMAMENT: Can be equipped with torpedoes, anti-shipping missiles, or gear for minelaying (eight 250 kg mines) or minesweeping.

Two views of the second prototype CAF Z-8 three-turboshaft helicopter

DIMENSIONS, EXTERNAL:
Main rotor diameter	18.90 m (62 ft 0 in)
Tail rotor diameter	4.00 m (13 ft 1½ in)
Length overall, rotors turning	23.035 m (75 ft 7 in)
Height overall, rotors turning	6.66 m (21 ft 10¼ in)
Width over main gear sponsons	5.20 m (17 ft 0¾ in)

AREAS:
Main rotor blades (each)	5.10 m² (54.90 sq ft)
Tail rotor blades (each)	0.56 m² (6.03 sq ft)
Main rotor disc	280.48 m² (3,019.1 sq ft)
Tail rotor disc	12.57 m² (135.3 sq ft)

WEIGHTS AND LOADINGS:
Weight empty, equipped	7,550 kg (16,645 lb)
Max cargo payload	5,000 kg (11,023 lb)
Max hovering weight OGE at S/L	12,480 kg (27,513 lb)
Max T-O weight: standard fuel	10,592 kg (23,351 lb)
with auxiliary fuel	12,074 kg (26,618 lb)
Max disc loading:	
standard fuel	37.76 kg/m² (7.73 lb/sq ft)
auxiliary fuel	43.05 kg/m² (8.82 lb/sq ft)
Max power loading:	
standard fuel	3.45 kg/kW (5.67 lb/shp)
auxiliary fuel	3.93 kg/kW (6.46 lb/shp)

PERFORMANCE (A at T-O weight of 9,000 kg; 19,841 lb, B at 11,000 kg; 24,251 lb, C at 13,000 kg; 28,660 lb):
Never-exceed speed (VNE):	
A	170 knots (315 km/h; 195 mph)
B	159 knots (296 km/h; 183 mph)
C	148 knots (275 km/h; 170 mph)
Max cruising speed: A	143 knots (266 km/h; 165 mph)
B	140 knots (260 km/h; 161 mph)
C	134 knots (248 km/h; 154 mph)
Econ cruising speed: A	137 knots (255 km/h; 158 mph)
B	132 knots (246 km/h; 153 mph)
C	125 knots (232 km/h; 144 mph)
Rate of climb at S/L (15° 30′ collective pitch, OEI):	
A	690 m (2,263 ft)/min
B	552 m (1,811 ft)/min
C	396 m (1,299 ft)/min
Service ceiling: A	6,000 m (19,685 ft)
B	4,900 m (16,075 ft)
C	3,050 m (10,000 ft)
Hovering ceiling IGE: A	5,500 m (18,045 ft)
B	3,600 m (11,810 ft)
C	1,900 m (6,235 ft)
Hovering ceiling OGE: A	4,400 m (14,435 ft)
B	2,300 m (7,545 ft)
Range with max standard fuel, OEI, no reserves:	
A	232 nm (430 km; 267 miles)
B	442 nm (820 km; 509 miles)
C	431 nm (800 km; 497 miles)
Ferry range with auxiliary fuel tanks, OEI, no reserves:	
C	755 nm (1,400 km; 870 miles)
Endurance with max standard fuel, OEI, no reserves:	
A	2 h 31 min
B	4 h 43 min
C	4 h 10 min

GAIC
GUIZHOU AVIATION INDUSTRY CORPORATION
PO Box 38, Anshun, Guizhou 561000
Telephone: 86 (851) 551027; or 86 (412) 22228
Fax: 86 (853) 25528
Telex: 66018 AIMGA CN
GENERAL MANAGER: Sun Ruisheng

GAIC incorporates many enterprises, factories and institutes engaged in various aerospace and non-aerospace activities; aerospace workforce is about 6,000. Aviation programmes include JJ-7/FT-7 fighter/trainer, two series of turbojets, air-to-air missiles and rocket launchers, plus participation in Chengdu (CAC, which see) production of single-seat J-7/F-7.

GAIC JJ-7
Chinese name: Jianjiji Jiaolianji-7 (Fighter training aircraft 7) or Jianjiao-7
Westernised designation: FT-7

TYPE: Tandem two-seat fighter/trainer.
PROGRAMME: Launched October 1982; first metal cut April 1985; first flight 5 July 1985; series production began February 1986; production FT-7 first flight December 1987; FT-7P first flight 9 November 1990 and received MAS production approval 13 May 1992.
CURRENT VERSIONS: **JJ-7**: Basic version, based on single-seat J-7 II and MiG-21US.
FT-7: Export version of JJ-7.
FT-7P: Version of FT-7 for Pakistan: imported fire control system; HUD and air data computer; improved instrument layout; two underwing pylons each side; increased fuel load; 25 per cent increase in operational range.
CUSTOMERS: Pakistan Air Force (15 FT-7P); Sri Lanka Air Force (one FT-7); Zimbabwe Air Force (two FT-7).
DESIGN FEATURES: Generally as J-7/F-7 and MiG-21US except for twin canopies opening sideways to starboard (rear one with retractable periscope), twin ventral strakes of modified shape, and removable saddleback fuel tank aft of second cockpit. Can provide full training syllabus for all J-7/F-7 versions, plus most of that necessary for Shenyang J-8.
FLYING CONTROLS: As J-7/F-7.
STRUCTURE: As J-7/F-7.
POWER PLANT: As F-7M. Internal fuel tank arrangement as for F-7M, but fuselage and wing total capacities are 1,880 and 560 litres respectively (496.6 and 148 US gallons; 413.5

48 CHINA: AIRCRAFT—GAIC/HAMC

Two GAIC FT-7P two-seat trainers of the Pakistan Air Force *(Peter Steinemann)*

and 123.2 Imp gallons), giving total capacity of 2,440 litres (644.6 US gallons; 536.7 Imp gallons). External drop tanks as for F-7M.

ACCOMMODATION: Second cockpit in tandem, with dual controls.

SYSTEMS: Cockpits pressurised and air-conditioned (max differential 0.3 bar; 4.35 lb/sq in). Hydraulic system pressure 207 bars (3,000 lb/sq in), flow rates 36 litres (9.51 US gallons; 7.92 Imp gallons)/min in main system, 4 litres (1.06 US gallons; 0.88 Imp gallon)/min in standby system. Pneumatic system pressure 49 bars (710.6 lb/sq in). Electrical power provided by QF-12C 12 kW engine driven starter/generator, with 400A static inverters for 28.5V DC power. YX-1 oxygen system for crew. De-icing/anti-icing standard.

AVIONICS: Include CT-3M VHF com transceiver, WL-7 radio compass, XS-6A marker beacon receiver and FJ-1 flight data recorder. Other (unidentified) Chinese nav/com avionics designated Type 222, Type 262, JT-2A and J7L.

ARMAMENT: Single underwing pylon each side (two on FT-7P) for such stores as a PL-2 or –2B air-to-air missile, an HF-5A 18-round launcher for 57 mm rockets, or a bomb of up to 250 kg size; can also be fitted with Type 23-3 twin-barrel 23 mm gun in underbelly pack and HK-03E optical gunsight.

DIMENSIONS, EXTERNAL: As F-7M except:
  Length overall, incl probe   14.874 m (48 ft 9½ in)

WEIGHTS AND LOADINGS:
  Weight empty, equipped: FT-7   5,519 kg (12,167 lb)
  FT-7P   5,330 kg (11,750 lb)

Max fuel weight: internal   1,891 kg (4,169 lb)
  external   558 kg (1,230 lb)
Max external stores load: FT-7   1,187 kg (2,617 lb)
Normal T-O weight with two PL-2 air-to-air missiles:
  FT-7P   7,590 kg (16,733 lb)
Max T-O weight: FT-7   8,555 kg (18,860 lb)
  FT-7P   8,600 kg (18,960 lb)
Max landing weight: FT-7   6,096 kg (13,439 lb)
Max zero-fuel weight: FT-7   7,300 kg (16,094 lb)
Max wing loading: FT-7   371.96 kg/m^2 (76.18 lb/sq ft)
  FT-7P   373.91 kg/m^2 (76.58 lb/sq ft)
Max power loading: FT-7   315.28 kg/kN (3.09 lb/lb st)
  FT-7P   316.94 kg/kN (3.11 lb/lb st)

PERFORMANCE (FT-7 at max T-O weight):
Never-exceed speed (V_{NE}) above 12,500 m (41,010 ft)
  Mach 2.35 (1,346 knots; 2,495 km/h; 1,550 mph)
Max level speed above 12,500 m (41,010 ft)
  Mach 2.05 (1,175 knots; 2,175 km/h; 1,350 mph)
Max cruising speed at 11,000 m (36,100 ft)
  545 knots (1,010 km/h; 627 mph)
Econ cruising speed at 11,000 m (36,100 ft)
  516 knots (956 km/h; 594 mph)
Unstick speed
  178-189 knots (330-350 km/h; 205-218 mph)
Touchdown speed
  162-181 knots (300-335 km/h; 187-209 mph)
Stalling speed, flaps down
  135 knots (250 km/h; 156 mph)
Max rate of climb at S/L   9,300 m (30,510 ft)/min
Service ceiling   17,500 m (57,415 ft)
Absolute ceiling   17,700 m (58,070 ft)
T-O run   827 m (2,714 ft)
T-O to 15 m (50 ft)   931 m (3,055 ft)
Landing from 15 m (50 ft)   1,368 m (4,489 ft)
Landing run   1,060 m (3,478 ft)
Range at 11,000 m (36,000 ft):
  max internal fuel   635 nm (1,176 km; 731 miles)
  max internal and external fuel
    798 nm (1,479 km; 919 miles)
  max external stores   708 nm (1,313 km; 816 miles)
g limit with two PL-2B missiles   +7

HAMC
HARBIN AIRCRAFT MANUFACTURING CORPORATION

PO Box 201, Harbin, Heilongjiang 150066
Telephone: 86 (451) 801122 or 802552
Fax: 86 (451) 802061
Telex: 87082 HAF CN
PRESIDENT: Yang Shouwen

HAMC is a major Chinese airframe manufacturer; established 1952, subsequently producing H-5 light bomber (Soviet designed Il-28) and Z-5 helicopter (Soviet designed Mi-4) in large numbers, as well as smaller numbers of Chinese designed SH-5 flying-boat and Y-11 agricultural light twin (see earlier editions of *Jane's* for details).

Currently producing own-design Y-12 II utility light twin and licence manufacturing Eurocopter France Dauphin 2 as Z-9A-100. Subcontract work includes doors for BAe 146, and Dauphin doors for Eurocopter France. Workforce in 1992 numbered approx 16,000.

HAMC SH-5
Chinese name: Shuishang Hongzhaji 5 (Maritime bomber 5) or Shuihong-5
Westernised designation: PS-5

TYPE: Maritime patrol and anti-submarine bomber, surveillance and SAR flying-boat.

PROGRAMME: Joint design by HAMC and Seaplane Design Institute, but development a victim of cultural revolution; detail design started February 1970, but static testing of first complete airframe not achieved until August 1974, nearly three years after its completion; first flying prototype completed December 1973, undergoing some 30 hours' water taxi trials between May 1975 and March 1976; first flight made 3 April 1976; six more SH-5s completed and flown 1984-85; four of these handed over 3 September 1986 to PLA Naval Air Force for service at Tuandao naval air station, Qingdao. ASW and avionics upgrade reportedly being sought.

CURRENT VERSIONS: Intended for wide range of duties including anti-submarine and anti-surface vessel warfare, mine-laying, search and rescue, and carriage of bulk cargo; fire-fighting water-bomber version has been evaluated.

CUSTOMERS: PLA Naval Air Force (four).

DESIGN FEATURES: Unpressurised hull, with high length/beam ratio, single-step planing bottom, curved spray suppression strakes along sides of nose, spray suppression slots in lower sides aft of inboard propeller plane; wings have constant chord centre-section and tapered, anhedral outer panels, with non-retractable stabilising float (on N struts with twin I struts inboard) beneath each wing near tip; high mounted dihedral tailplane, with oval endplate fins and rudders, on fairing above rear fuselage; thimble radome on nose; MAD in extended tail sting; small water rudder at rear of hull.

FLYING CONTROLS: Actuation details not known. Trim tab in each aileron, elevator and rudder; spoiler forward of each outer flap segment.

STRUCTURE: All-metal.

LANDING GEAR: Retractable tricycle beaching gear, with single mainwheels and twin-wheel nose unit. Oleo-pneumatic shock absorbers. Main units retract upward and rearward into wells in hull sides; nose unit retracts rearward.

POWER PLANT: Four 2,349 kW (3,150 ehp) Dongan (DEMC) WJ5A turboprops, each driving a Baoding four-blade propeller. Max fuel capacity approx 21,000 litres (5,548 US gallons; 4,620 Imp gallons).

HAMC SH-5 (PS-5) patrol and anti-submarine flying-boat, in service with the Chinese PLA Naval Air Force. Upper view shows water bombing for fighting fires

ACCOMMODATION: Standard eight-person crew includes a flight crew of five (pilot, co-pilot, navigator, flight engineer and radio operator), plus systems/equipment operators according to mission. Three freight compartments in front portion of hull. Mission crew cabin amidships, aft of which are two further compartments, one for communications and other electronic equipment and rear one for specialised mission equipment. All compartments connected by corridor, with watertight doors aft of flight deck and between each compartment.

AVIONICS: Include INS, air data computer, radio altimeter and radio compass. Doppler search radar in thimble radome

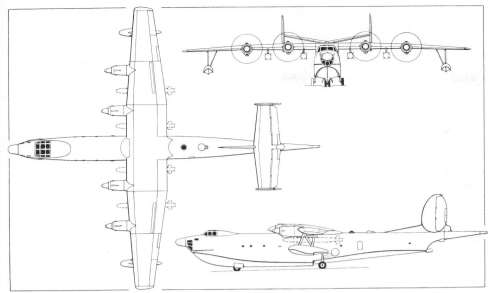

HAMC SH-5 flying-boat (four Dongan WJ5A turboprops) *(Dennis Punnett)*

HAMC SH-5 taking off from water *(Xinhua News Agency)*

forward of nose transparencies. Magnetic anomaly detector (MAD) in extended tail sting.
ARMAMENT: Two-gun dorsal turret. Four underwing hardpoints for C-101 sea skimming supersonic anti-shipping or other missiles (one on each inboard pylon), lightweight torpedoes (up to three on each outer pylon), or other stores. Depth charges, mines, bombs, sonobuoys, SAR gear or other mission equipment and stores in rear of hull, as required.

DIMENSIONS, EXTERNAL:
Wing span	36.00 m (118 ft 1¼ in)
Wing aspect ratio	9.0
Length overall	38.90 m (127 ft 7½ in)
Height overall	9.802 m (32 ft 2 in)
Span over tail-fins	10.50 m (34 ft 5½ in)
Wheel track	3.754 m (12 ft 3¾ in)
Wheelbase	10.50 m (34 ft 5½ in)
Propeller diameter	3.90 m (12 ft 9½ in)

AREAS:
Wings, gross	144.0 m² (1,550.0 sq ft)

WEIGHTS AND LOADINGS:
Weight empty, equipped:	
SAR and transport	less than 25,000 kg (55,115 lb)
ASW	26,500 kg (58,422 lb)
Fuel load (max)	13,417 kg (29,579 lb)
Max internal weapons load	6,000 kg (13,228 lb)
Max payload (bulk cargo)	10,000 kg (22,045 lb)
Normal T-O weight	36,000 kg (79,366 lb)
Max T-O weight	45,000 kg (99,208 lb)
Wing loading:	
at normal T-O weight	250.0 kg/m² (51.2 lb/sq ft)
at max T-O weight	312.5 kg/m² (64.0 lb/sq ft)
Power loading:	
at normal T-O weight	3.31 kg/kW (5.44 lb/ehp)
at max T-O weight	4.14 kg/kW (6.80 lb/ehp)

PERFORMANCE:
Max level speed	300 knots (556 km/h; 345 mph)
Max cruising speed	243 knots (450 km/h; 280 mph)
Min patrol speed	124 knots (230 km/h; 143 mph)
T-O speed (water)	87 knots (160 km/h; 100 mph)
Landing speed (water)	92 knots (170 km/h; 106 mph)
Service ceiling	10,250 m (33,630 ft)
T-O run (water)	482 m (1,582 ft)
Landing run (water)	653 m (2,143 ft)
Range with max fuel	2,563 nm (4,750 km; 2,951 miles)
Endurance (2 engines)	12 to 15 h

HAMC Y-11B (I)

Chinese name: Yunshuji-11B (I) (Transport aircraft 11B) or Yun-11B

TYPE: Agricultural and general purpose aircraft.
PROGRAMME: Initiated November 1988 to find new power plant for Y-11 able to confer required single-engine performance at 1,500 m (4,920 ft) at normal max T-O weight; two flying prototypes (first flight 25 December 1990), plus two for static test; domestic certification obtained late 1991; deliveries started 1992.
CURRENT VERSIONS: **Y-11**: Original production version, with 213 kW (285 hp) SMPMC (Zhuzhou) HS6A radial engines; more than 40 built; described in 1986-87 and earlier *Jane's*.
Y-11B (I): Upgraded current version, with turbocharged US flat-six engines, improved avionics and minor airframe changes. *Details apply to this version*. To be available in **passenger** (pilot plus seven or eight), **passenger/cargo** (optional four collapsible double seats) and **agricultural** (sowing/spraying gear and higher operating weights) configurations.

CUSTOMERS: Y-11 in service with China General Aviation Airlines (16), Flying Dragon Aviation Corporation (14), Shuangyang General Aviation Co and Xinjiang Agricultural Air Service (10), among others.
DESIGN FEATURES: Constant chord, no-dihedral high wings, braced at mid-span to small stub-wings at cabin floor level which also support mainwheel units; rectangular section fuselage, upswept at rear; non-swept tail surfaces, with dorsal fin. Wing section NACA 4412. Conforms to FAR Pt 23 and BCAR.
FLYING CONTROLS: Mechanical (rods/cables) for drooping ailerons, horn balanced rudder and elevators, all except port aileron having inset trim tab; electrically actuated two-segment double-slotted flaps on trailing-edges; automatic leading-edge slats outboard of nacelles, smaller fixed slat inboard.
STRUCTURE: Conventional aluminium alloy stressed skin; two-spar wing.
LANDING GEAR: Non-retractable tricycle type, with oleo-pneumatic shock absorber in each unit. Twin-wheel main units, attached to underside of stub-wings. Single self-centring, non-steerable nosewheel. Tyre sizes 500 × 150 mm on mainwheels, 400 × 150 mm on nosewheel; pressure 3.0 bars (43.51 lb/sq in) on all units. Hydraulic mainwheel disc brakes. Small bumper under tailcone.
POWER PLANT: Two 261 kW (350 hp) Teledyne Continental TSIO-550-B flat-six engines, each driving a Hartzell PHC-C3YF-2KUF/FC three-blade variable-pitch propeller. Fuel in two integral wing tanks, combined capacity 620 litres (163.8 US gallons; 136.4 Imp gallons); gravity refuelling point in top of each wing. Oil capacity 11.4 litres (3 US gallons; 2.5 Imp gallons).
ACCOMMODATION: Flight crew of one (VFR) or two (IFR). Seats aft of flight deck for seven passengers (three left, four right), with baggage compartment to rear. Forward opening crew door on port side, forward of wing, with main cabin door on same side aft of wing; emergency exit below wing on starboard side.
SYSTEMS: Hydraulic system (operating pressure 85 bars; 1,233 lb/sq in) for mainwheel brakes. Two 3 kW generators and a 28Ah battery for electrical power.
AVIONICS: Bendix/King com/nav (KMA 24H-70, KHF 950, dual KY 126A VHF, dual KR 87, KR 21).

DIMENSIONS, EXTERNAL:
Wing span	17.08 m (56 ft 0½ in)
Wing chord, constant	2.00 m (6 ft 6¾ in)
Wing aspect ratio	8.54
Length overall	12.15 m (39 ft 10¼ in)
Height overall	5.186 m (17 ft 0¼ in)
Wheel track (c/l of shock absorbers)	3.10 m (10 ft 2 in)
Wheelbase	3.697 m (12 ft 1½ in)
Propeller diameter	2.03 m (6 ft 8 in)
Passenger door: Height	1.23 m (4 ft 0½ in)
Width	0.988 m (3 ft 3 in)
Baggage compartment door: Height	1.30 m (4 ft 3¼ in)
Width	0.45 m (1 ft 5¾ in)
Height to sill	0.215 m (8½ in)
Emergency exit: Height	0.67 m (2 ft 2½ in)
Width	0.49 m (1 ft 7¼ in)
Height to sill	0.40 m (1 ft 3¾ in)

DIMENSIONS, INTERNAL:
Cabin: Length	3.58 m (11 ft 9 in)
Max width	1.27 m (4 ft 2 in)
Max height	1.48 m (4 ft 10¼ in)
Volume	6.73 m³ (237.7 cu ft)
Baggage compartment volume	0.48 m³ (16.95 cu ft)

AREAS:
Wings, gross	34.00 m² (365.97 sq ft)
Ailerons (total, incl tab)	2.674 m² (28.78 sq ft)
Trailing-edge flaps (total)	6.288 m² (67.68 sq ft)
Fin	5.242 m² (56.42 sq ft)
Rudder, incl tab	2.865 m² (30.84 sq ft)
Tailplane	6.973 m² (75.06 sq ft)
Elevators (total, incl tabs)	4.06 m² (43.70 sq ft)

WEIGHTS AND LOADINGS:
Weight empty, equipped	2,504 kg (5,520 lb)

Prototype HAMC Y-11B (I) (two TCM TSIO-550-B flat-six engines)

50 CHINA: AIRCRAFT—HAMC

Max fuel weight	450 kg (992 lb)
Max payload: Normal	900 kg (1,984 lb)
Restricted	1,200 kg (2,645 lb)
Max T-O and landing weight:	
Normal	3,500 kg (7,716 lb)
Restricted	3,900 kg (8,598 lb)
Max wing loading: Normal	102.94 kg/m² (21.08 lb/sq ft)
Restricted	114.71 kg/m² (23.49 lb/sq ft)
Max power loading: Normal	6.70 kg/kW (11.02 lb/hp)
Restricted	7.47 kg/kW (12.28 lb/hp)

PERFORMANCE (at Normal category max T-O weight):
Max level speed at 3,000 m (9,840 ft)	143 knots (265 km/h; 165 mph)
Max operating speed	135 knots (250 km/h; 155 mph)
Max cruising speed at 3,000 m (9,840 ft)	127 knots (235 km/h; 146 mph)
Econ cruising speed at 3,000 m (9,840 ft)	108 knots (200 km/h; 124 mph)
Stalling speed, 30° flap	57 knots (104 km/h; 65 mph)
Max rate of climb at S/L	336 m (1,100 ft)/min
Rate of climb at S/L, OEI	33 m (108 ft)/min
Service ceiling	6,000 m (19,685 ft)
Service ceiling, OEI	2,100 m (6,890 ft)
T-O run	200 m (657 ft)
T-O to 15 m (50 ft)	435 m (1,428 ft)
Landing from 15 m (50 ft)	530 m (1,739 ft)
Landing run	275 m (903 ft)
Range at 3,000 m (9,840 ft) at optimum cruising speed:	
with max payload	162 nm (300 km; 86 miles)
with max fuel	582 nm (1,080 km; 671 miles)

HAMC Y-12 (II), one of six for the Peruvian Air Force

HAMC Y-12 (II)
Chinese name: Yunshuji-12 (II) (Transport aircraft 12) or Yun-12 (II)

TYPE: Twin-turboprop STOL general purpose transport.
PROGRAMME: Initiated as uprated version of Y-12 (I) (1987-88 and earlier *Jane's*); first flight 16 August 1984; domestic certification December 1985; UK (BCAR Section K) certification received 20 June 1990; FAA type approval being sought. Production rate (1992) approx one per month.
CURRENT VERSIONS: **Y-12 (I)**: Initial version (first flight 14 July 1982), with PT6A-11 engines; three prototypes and approx 30 production examples built; described in 1987-88 *Jane's*.
Y-12 (II): Current production version, *to which detailed description applies*; higher rated engines, no leading-edge slats and smaller ventral fin.
Y-12 (IV): Improved version under development. Modifications to wingtips, control surface actuation, main gear and brakes; redesigned seating for 18-19 passengers; starboard side rear baggage door; empty and max T-O weights increased to 2,900 kg (6,393 lb) and 5,670 kg (12,500 lb).
Future versions: Plans under consideration include stretched version and one with pressurised cabin.
CUSTOMERS: Total 55 delivered by March 1992; Chinese operators include Flying Dragon Aviation (eight), China General Aviation Corporation (five), China Southwest Airlines (four) and Hainan Airlines (two). Export customers include Ethiopia, Fiji Air (one), Gabon (one), Iran (two), Lao Aviation (four), Malaysia (Berjaya Air Charter one), Mongolian Airlines (five), Nepal Airlines (three), Peruvian Air Force (six), Sri Lanka Air Force (nine), Sudan, Zambia (two) and Zimbabwe Airlines (one).
DESIGN FEATURES: Designed to standards of FAR Pts 23 and 135 Appendix A and developed to improve upon modest payload/range of original piston engined Y-11. Constant chord high braced wings, with small stub-wings at cabin floor level supporting mainwheel units; basically rectangular-section fuselage, upswept at rear; non-swept tail surfaces; large dorsal fin; ventral fin under tailcone.
Wing section LS(1)-0417; thickness/chord ratio 17 per cent; dihedral 1° 41'; incidence 4°.
FLYING CONTROLS: Drooping ailerons, horn balanced elevators and rudder, all mechanically actuated; trim tab in starboard aileron, rudder and each elevator; electrically actuated two-segment double-slotted flaps on each wing trailing-edge.
STRUCTURE: Conventional all-metal structure with two-spar fail-safe wings and stressed skin fuselage; Ziqiang-2 resin bonding on 70 per cent of wing structure and 40 per cent of fuselage; integral fuel tankage in wing spar box; bracing strut from each stub-wing out to approx one-third span.
LANDING GEAR: Non-retractable tricycle type, with oleo-pneumatic shock absorber in each unit. Single-wheel main units, attached to underside of stub-wings. Single steerable nosewheel. Mainwheel tyres size 640 × 230 mm, pressure 5.5 bars (80 lb/sq in); nosewheel tyre size 480 × 200 mm, pressure 3.5 bars (51 lb/sq in). Pneumatic brakes. Min ground turning radius 16.75 m (54 ft 11½ in).
POWER PLANT: Two Pratt & Whitney Canada PT6A-27 turboprops, each flat rated at 507 kW (680 shp) and driving a Hartzell HC-B3TN-3B/T10173B-3 three-blade constant-speed reversible-pitch propeller. All fuel in tanks in wing spar box, total capacity 1,616 litres (427 US gallons; 355.5 Imp gallons), with overwing gravity filling point each side. Domestic WJ9 turboprop (442-520 kW; 593-697 shp) under development for Y-12 and other aircraft.
ACCOMMODATION: Crew of two on flight deck, with access via forward opening door on port side. Four-way adjustable crew seats. Dual controls. Main cabin can accommodate up to 17 passengers in commuter configuration, in three-abreast layout (with aisle), at seat pitch of 79 cm (31 in). Alternative layouts for up to 16 parachutists, or all-cargo configuration with 11 tiedown rings. Passenger/cargo double door on port side at rear, rear half of which opens outward and forward half inward; foldout steps in passenger entrance. Emergency exits on each side at front of cabin and opposite passenger door on starboard side at rear. Baggage compartments in nose and at rear of passenger cabin, for 100 kg (220 lb) and 260 kg (573 lb) respectively.
SYSTEMS: Hamilton Standard R70-3WG environmental control system. Goodrich Type 29S-7D 5178 anti-icing system for wing, tailplane and fin leading-edges.
AVIONICS: VHF-251 and HF-230 com radio, AUD-251H, ADF-650A radio compass, VIR-351 and Bendix/King 1400C weather radar. Standard instrumentation includes BK-450 airspeed indicator, BDP-1 artificial horizon, BG10-1A altimeter, ZGW-3G altitude indicator, ZHZ-4A radio magnetic heading indicator, BC10 rate of climb indicator, ZWH-1 outside air temperature indicator, and ZEY-1 flap position indicator; dual engine torquemeters, interturbine temperature indicators, gas generator tachometers, oil temperature and pressure indicators, and fuel pressure and quantity indicators; HSZ-2 clock; and XDH-10B warning light box. Doppler navigation with satellite responder optional (eg in mineral detection role).
EQUIPMENT: Hopper for 1,200 litres (317 US gallons; 264 Imp gallons) of dry or liquid chemical in agricultural version. Appropriate specialised equipment for firefighting, geophysical survey (eg long, kinked sensor tailboom) and other missions.

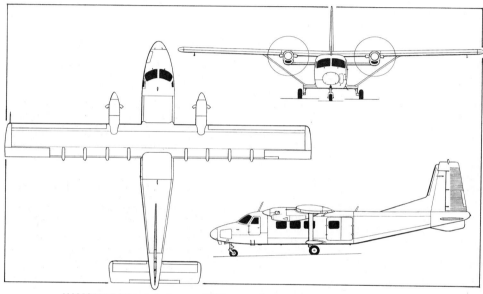

HAMC Y-12 (II) twin-turboprop STOL general purpose transport (*Dennis Punnett*)

Civil Y-12 (II) of Malaysia's Berjaya Air Charter

DIMENSIONS, EXTERNAL:
Wing span	17.235 m (56 ft 6½ in)
Wing chord, constant	2.00 m (6 ft 6¾ in)
Wing aspect ratio	8.7
Length overall	14.86 m (48 ft 9 in)

Height overall	5.575 m (18 ft 3½ in)
Elevator span	5.365 m (17 ft 7¼ in)
Wheel track	3.60 m (11 ft 9¾ in)
Wheelbase	4.698 m (15 ft 5 in)
Propeller diameter	2.49 m (8 ft 2 in)
Distance between propeller centres	4.937 m (16 ft 2⅜ in)
Fuselage ground clearance	0.65 m (2 ft 1½ in)
Crew door: Height	1.35 m (4 ft 5¼ in)
Width	0.65 m (2 ft 1½ in)
Passenger/cargo door: Height	1.38 m (4 ft 6¼ in)
Width (passenger door only)	0.65 m (2 ft 1½ in)
Width (double door)	1.45 m (4 ft 9 in)
Emergency exits (three, each):	
Height	0.68 m (2 ft 2¾ in)
Width	0.68 m (2 ft 2¾ in)
Baggage door (nose, port):	
Max height	0.56 m (1 ft 10 in)
Width	0.75 m (2 ft 5½ in)

DIMENSIONS, INTERNAL:
Cabin, excl flight deck and rear baggage compartment:

Length	4.82 m (15 ft 9¾ in)
Max width	1.46 m (4 ft 9½ in)
Max height	1.70 m (5 ft 7 in)
Volume	12.9 m³ (455.5 cu ft)
Baggage compartment volume:	
nose	0.77 m³ (27.20 cu ft)
rear	1.89 m³ (66.75 cu ft)

AREAS:

Wings, gross	34.27 m² (368.88 sq ft)
Vertical tail surfaces (total)	5.064 m² (54.51 sq ft)
Horizontal tail surfaces (total)	7.024 m² (75.61 sq ft)

WEIGHTS AND LOADINGS:

Weight empty, equipped	2,840 kg (6,261 lb)
Operating weight empty	3,000 kg (6,614 lb)
Max fuel load	1,233 kg (2,718 lb)
Max payload	1,700 kg (3,748 lb)
T-O weight for agricultural operation	4,500 kg (9,921 lb)
Max T-O and landing weight	5,300 kg (11,684 lb)
Max ramp weight	5,330 kg (11,750 lb)
Max zero-fuel weight	4,700 kg (10,362 lb)
Max cabin floor loading (cargo)	750 kg/m² (153.7 lb/sq ft)
Max wing loading	145.9 kg/m² (29.90 lb/sq ft)
Max power loading	5.23 kg/kW (8.59 lb/shp)

PERFORMANCE (at max T-O weight, ISA):

Never-exceed speed (VNE) at 3,000 m (9,840 ft)	177 knots (328 km/h; 204 mph)
Max cruising speed at 3,000 m (9,840 ft)	157 knots (292 km/h; 181 mph)
Econ cruising speed at 3,000 m (9,840 ft)	135 knots (250 km/h; 155 mph)
Max rate of climb at S/L	504 m (1,655 ft)/min
Rate of climb at S/L, OEI	101 m (331 ft)/min
Service ceiling	7,000 m (22,960 ft)
Service ceiling, OEI, 15 m (50 ft)/min rate of climb, max continuous power	3,000 m (9,840 ft)
T-O run, 15° flap	340 m (1,115 ft)
T-O to 15 m (50 ft), 15° flap	425 m (1,395 ft)
Landing from 15 m (50 ft), with braking and propeller reversal	480 m (1,575 ft)
Landing run with braking and propeller reversal	200 m (656 ft)
Range at 135 knots (250 km/h; 155 mph) at 3,000 m (9,840 ft) with max fuel, 45 min reserves	723 nm (1,340 km; 832 miles)
Endurance, conditions as above	5 h 12 min

HAMC Z-9A Haitun twin-turbine light helicopter in PLA Air Force insignia

HAMC (EUROCOPTER) Z-9 HAITUN (DOLPHIN)

Chinese name: Zhishengji-9 (Vertical take-off aircraft 9) or Zhi-9

TYPE: Licence built Eurocopter (Aerospatiale) AS 365N Dauphin 2 (which see in International section).
PROGRAMME: Licence agreement (Aerospatiale/CATIC) signed 2 July 1980; first (French built) example made initial acceptance flight in China 6 February 1982; Chinese parts manufacture began 1986; initial agreed batch of 50, last of which delivered January 1992. Production continuing as Z-9A-100 under May 1988 domestic contract, with much increased local manufacture (72.2 per cent of airframe and 91 per cent of engine).
CURRENT VERSIONS: **Z-9**: Initial Chinese licence version, equivalent to French AS 365N.
  Z-9A: Later aircraft in initial 50, to AS 365N₁ standard and with increased proportion of locally manufactured components.
  Z-9A-100: Current production version, with much increased local manufacture (see Programme). First flight 16 January 1992; flight test programme completed 20 November 1992 after almost 200 flight hours (408 flights); Chinese type approval received 30 December 1992. Weight and performance details assumed as for Z-9A.
CUSTOMERS: Serve with CAAC and all three Chinese armed services. Entered service with two PLA group armies January and February 1988 (Beijing and Shenyang Military Regions respectively); first anti-tank version (Norinco 'Red Arrow 8' missiles) first flew late 1988 or early 1989. PLA Naval Air Force reportedly seeking suitable MAD. Civil models used for various duties including offshore oil rig support and air ambulance (four stretchers/two seats or two stretchers/five seats).
STRUCTURE: Transmission manufactured by Dongan Engine Manufacturing Co at Harbin, hubs and tail rotor blades by Baoding Propeller Factory.
POWER PLANT: Arriel 1C and 1C1 turboshafts produced by SMPMC at Zhuzhou as WZ8 and WZ8A; fuel capacity 1,140 litres (301 US gallons; 251 Imp gallons).

WEIGHTS AND LOADINGS:

Weight empty, equipped: Z-9	1,975 kg (4,354 lb)
Z-9A	2,050 kg (4,519 lb)
Max payload: Z-9	1,863 kg (4,107 lb)
Z-9A	2,038 kg (4,493 lb)
Max load on cargo sling: Z-9, Z-9A	1,600 kg (3,527 lb)
Max T-O weight, internal or external load:	
Z-9	3,850 kg (8,488 lb)
Z-9A	4,100 kg (9,039 lb)

PERFORMANCE (at max T-O weight):

Max cruising speed at S/L:	
Z-9	158 knots (293 km/h; 182 mph)
Z-9A	154 knots (285 km/h; 177 mph)
Max vertical rate of climb at S/L:	
Z-9	252 m (827 ft)/min
Z-9A	246 m (805 ft)/min
Max forward rate of climb at S/L:	
Z-9	462 m (1,515 ft)/min
Z-9A	456 m (1,495 ft)/min
Service ceiling: Z-9	4,500 m (14,765 ft)
Z-9A	6,000 m (19,685 ft)
Hovering ceiling IGE: Z-9	1,950 m (6,400 ft)
Z-9A	2,600 m (8,530 ft)
Hovering ceiling OGE: Z-9	1,020 m (3,350 ft)
Z-9A	1,600 m (5,250 ft)
Max range at 140 knots (260 km/h; 161 mph) normal cruising speed, no reserves:	
standard tanks: Z-9	491 nm (910 km; 565 miles)
Z-9A	464 nm (860 km; 534 miles)
with 180 litre (47.5 US gallon; 39.6 Imp gallon) auxiliary tank:	
Z-9	572 nm (1,060 km; 658 miles)
Z-9A	539 nm (1,000 km; 621 miles)

HAMC POLE STAR

Reported early 1993 that Harbin studying future light helicopter in Bell JetRanger size category, possibly with single Allison 250-C20B or –C20R turboshaft.

NAMC
NANCHANG AIRCRAFT MANUFACTURING COMPANY

PO Box 5001-506, Nanchang, Jiangxi 330024
Telephone: 86 (791) 251833
Fax: 86 (791) 251491
Telex: 95068 NAMC CN
GENERAL MANAGER: Wu Mingwang
INFORMATION: Feng Jinghua

NAMC created 1951; built 379 CJ-5s (licence Soviet Yak-18s) between 1954 and 1958, and in 1960s shared in large production programme for J-6 fighter (Chinese development of MiG-19); also built (1957-68) 727 Y-5 (Chinese An-2) biplanes (see under SAP in this section and under NAMC in 1991-92 *Jane's*). Current programmes are CJ-6A development of CJ-5, Q-5/A-5 attack derivative of J-6 and its upgraded A-5M version, K-8 jet trainer in collaboration with PAC of Pakistan, and new N-5A dedicated agricultural aircraft. NAMC occupies a 500 ha (1,235 acre) site, with 10,000 m² (107,639 sq ft) of covered space, and had 1991 workforce of over 20,000; about 80 per cent of its activities are non-aerospace.

NAMC/PAC K-8 KARAKORUM 8

Details of this tandem-seat jet trainer can be found under above heading in International section.

NAMC Q-5

Chinese name: Qiangjiji-5 (Attack aircraft 5) or Qiang-5
Westernised designation: A-5
NATO reporting name: Fantan

TYPE: Single-seat close air support and ground attack aircraft, with air-to-air combat capability.
PROGRAMME: Derivative of J-6 fighter (see 1989-90 and earlier *Jane's* for details of changes), originating August 1958 as Shenyang design proposal; responsibility assigned to Nanchang; prototype programme cancelled 1961, but kept alive by small team and resumed officially 1963; first flight 4 June 1965; preliminary design certificate awarded and pre-production batch authorised late 1965, but further modifications (to fuel, armament, hydraulic and other systems) found necessary, leading to flight test of two much modified prototypes from October 1969; series production approved at end of 1969, deliveries beginning 1970. Improved Q-5 I proposed 1976, flight tested late 1980 and certificated for production 20 October 1981, by which time (April 1981) Pakistan had placed order for A-5C modified export version; first A-5C deliveries January 1983, completed January 1984; domestic Q-5 IA, incorporating many of A-5C improvements, certificated January 1985; design study by Flight Refuelling of UK completed in mid-1980s for receiver version (with Xian H-6 as tanker), but no go-ahead for conversion given; upgrade programmes involving Western avionics started in 1986 with France (Q-5K Kong Yun) and Italy (A-5M, which see), but Kong Yun programme terminated 1990 (details in 1990-91 *Jane's*). Batch production of all versions continuing in 1993.
CURRENT VERSIONS: **Q-5**: Initial production version, with internal fuselage bay approx 4.00 m (13 ft 1½ in) long for two 250 kg or 500 kg bombs, two underfuselage attachments adjacent to bay for two similar bombs, and two stores pylons beneath each wing; Series 6 WP6 turbojets; brake-chute in tailcone, between upper and lower pen-nib fairings. Some adapted for nuclear weapon delivery tests in early 1970s.
  Q-5 I: Extended payload/range version, with internal bomb bay blanked off and space used to enlarge main fuselage fuel tank and add a flexible tank; underfuselage stores points increased to four; improved series WP6 engines; modified landing gear; brake-chute relocated under base of rudder; improved Type I rocket ejection seat; HF/SSB transceiver added. Some aircraft, adapted for PLA Naval Air Force to carry two underfuselage torpedoes, reportedly have Doppler type nose radar and 20 m (66 ft) sea-skimming capability with C-801 anti-shipping missiles.
  Q-5 IA: Improved Q-5 I, with additional underwing hardpoint each side (increasing stores load by 500 kg; 1,102 lb), new gun/bomb sighting systems, pressure refuelling, and added warning/countermeasures systems.
  Q-5 II: As Q-5 IA, but fitted (or retrofitted) with radar warning receiver.

CHINA: AIRCRAFT—NAMC

NAMC A-5Cs of the Bangladesh Air Force (*Peter Steinemann*)

A-5C: Export version for Pakistan Air Force, involving 32 modifications from Q-5 I, notably upgraded avionics, Martin-Baker PKD10 zero/zero seats, and adaptation of hardpoints for 356 mm (14 in) lugs compatible with Sidewinder missiles and other PAF weapons; three prototypes preceded production programme; in service with Nos. 7, 16 and 26 Squadrons of PAF. Ordered also by Bangladesh. *Description applies to Q-5 IA and A-5C except where indicated.*

A-5M: Upgraded version of Q-5 II; described separately.

CUSTOMERS: Nearly 1,000 (all versions) built to date, including over 100 for export to Bangladesh (20 A-5C ordered), North Korea (40 Q-5 IA) and Pakistan (52 A-5C).

DESIGN FEATURES: Mid-mounted sweptback wings with deep, full-chord fence on each upper surface at mid-span; air intake on each side of fuselage abreast of cockpit; twin jetpipes side by side at rear with upper and lower pen-nib fairings aft of nozzles; centre-fuselage has area-ruled 'waist'; rear fuselage detachable aft of wing trailing-edge for engine access; dorsal spine fairing; shallow ventral strake under each jetpipe; all-swept tail surfaces.

Wings have 52° 30′ sweep at quarter-chord and 4° anhedral from roots; tailplane has 6° 30′ anhedral.

FLYING CONTROLS: Internally balanced ailerons and fully powered slab tailplane; mechanically actuated mass balanced rudder; hydraulically actuated Gouge flaps on inboard trailing-edges; electrically operated trim tab in port aileron and rudder; forward hinged, hydraulically actuated door type airbrake under centre of fuselage, forward of bomb attachment points; anti-flutter weight on each tailplane tip.

STRUCTURE: Conventional all-metal stressed skin structure. Multi-spar wings have three-point attachment to fuselage; fuselage built in forward and rear portions.

LANDING GEAR: Hydraulically retractable wide-track tricycle type, with single wheel and oleo-pneumatic shock absorber on each unit. Main units retract inward into wings, non-steerable nosewheel forward into fuselage, rotating through 87° to lie flat in gear bay. Mainwheels have size 830 × 205 mm tubeless tyres and disc brakes. Tail braking parachute, deployed when aircraft is 1 m (3.3 ft) above the ground, in bullet fairing beneath rudder.

POWER PLANT: Two Shenyang WP6 turbojets, each rated at 25.50 kN (5,732 lb st) dry and 31.87 kN (7,165 lb st) with afterburning, mounted side by side in rear of fuselage. Improved WP6A engines (see A-5M entry for details) available optionally. Lateral air intake, with small splitter plate, for each engine. Hydraulically actuated nozzles. Internal fuel in three forward and two rear fuselage tanks with combined capacity of 3,648 litres (964 US gallons; 802.5 Imp gallons). Provision for carrying a 760 litre (201 US gallon; 167 Imp gallon) drop tank on each centre underwing pylon, to give max internal/external fuel capacity of 5,168 litres (1,366 US gallons; 1,136.5 Imp gallons). When centre wing stations are occupied by bombs, a 400 litre (105.7 US gallon; 88 Imp gallon) drop tank can be carried instead on each outboard underwing pylon.

ACCOMMODATION: Pilot only, under one-piece jettisonable canopy which is hinged at rear and opens upward. Downward view over nose, in level flight, is 13° 30′. Low-speed seat allows for safe ejection within speed range of 135-458 knots (250-850 km/h; 155-528 mph) at zero height or above. Aircraft in Pakistan service have Martin-Baker PKD10 zero/zero seats. Armour plating in some areas of cockpit to protect pilot from anti-aircraft gunfire. Cockpit pressurised and air-conditioned.

SYSTEMS: Dual air-conditioning systems, one for cockpit environment and one for avionics cooling. Two independent hydraulic systems, each operating at pressure of 207 bars (3,000 lb/sq in). Primary system actuates landing gear extension and retraction, flaps, airbrake and afterburner nozzles; auxiliary system supplies power for aileron and all-moving tailplane boosters. Emergency system, operating pressure 108 bars (1,570 lb/sq in), for actuation of main landing gear. Electrical system (28V DC) powered by two 6 kW engine driven starter/generators, with two inverters for 115V single-phase and 36V three-phase AC power at 400Hz.

AVIONICS: Include CT-3 VHF com transceiver, WL-7 radio compass, WG-4 low altitude radio altimeter, LTC-2 horizon gyro, YD-3 IFF, Type 930 radar warning receiver (antenna in fin-tip) and XS-6 marker beacon receiver. Combat camera in small 'teardrop' fairing on starboard side of nose (not on export models). 'Odd Rods' type IFF aerials under nose on Q-5 variants, replaced on A-5C by a single blade antenna. Space provision in nose and centre-fuselage for additional or updated avionics, including an attack radar.

EQUIPMENT: Landing light under fuselage, forward of nose-wheel bay and offset to port; taxying light on nosewheel leg.

ARMAMENT: Internal armament consists of one 23 mm cannon (Norinco Type 23-2K), with 100 rds, in each wingroot. Ten attachment points normally for external stores: two pairs in tandem under centre of fuselage, and three under each wing (one inboard and two outboard of mainwheel leg). Fuselage stations can each carry a 250 kg bomb (Chinese 250-2 or 250-3, US Mk 82 or Snakeye, French Durandal, or similar). Inboard wing stations can carry 6 kg or 25 lb practice bombs, or a pod containing eight Chinese 57-2 (57 mm), seven 68 mm, or seven Norinco 90-1 (90 mm) or four 130-1 (130 mm) rockets. Centre wing stations can carry a 500 kg or 750 lb bomb, a BL755 600 lb cluster bomb, a Chinese 250-2 or -3 bomb, US Mk 82 or Snakeye, French Durandal, or similar, or a C-801 anti-shipping missile. Normal bomb carrying capacity is 1,000 kg (2,205 lb), max capacity 2,000 kg (4,410 lb). Instead of bombs, centre wing stations can each carry a 760 litre drop tank (see Power Plant paragraph) or ECM pod. Outboard wing stations can each be occupied by a 400 litre drop tank (when the larger tank is not carried on the centre wing station) or by air-to-air missiles such as Chinese PL-2, PL-2B, PL-7, AIM-9 Sidewinder and Matra R.550 Magic. Within overall max T-O weight, all stores mentioned can be carried provided that CG shift remains within allowable operating range of 31 to 39 per cent of mean aerodynamic chord; more than 22 external stores configurations possible. Aircraft carries SH-1J or ABS1A optical sight for level and dive bombing, or for air-to-ground rocket launching. Some aircraft in Chinese service can carry a single 5-20 kT nuclear bomb.

DIMENSIONS, EXTERNAL:
Wing span: Q-5 IA	9.68 m (31 ft 9 in)
A-5C	9.70 m (31 ft 10 in)
Wing chord (mean aerodynamic)	3.097 m (10 ft 2 in)
Wing aspect ratio	3.37
Length overall:	
incl nose probe: Q-5 IA	15.65 m (51 ft 4¼ in)
A-5C	16.255 m (53 ft 4 in)
excl nose probe: A-5C	15.415 m (50 ft 7 in)
Height overall: Q-5 IA	4.333 m (14 ft 2¾ in)
A-5C	4.516 m (14 ft 9¾ in)
Wheel track	4.40 m (14 ft 5¼ in)
Wheelbase	4.01 m (13 ft 2 in)

AREAS (Q-5 IA and A-5C):
Wings, gross	27.95 m² (300.85 sq ft)
Vertical tail surfaces (total)	4.64 m² (49.94 sq ft)
Horizontal tail surfaces:	
movable	5.00 m² (53.82 sq ft)

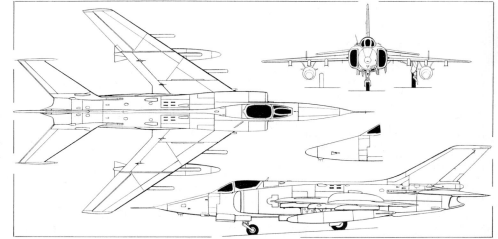

NAMC A-5C 'Fantan' single-seat twin-jet combat aircraft, with scrap view showing nose configuration of A-5M (*Dennis Punnett*)

```
total, incl projected fuselage area
                             8.62 m² (92.78 sq ft)
WEIGHTS AND LOADINGS:
  Weight empty: Q-5 IA      6,375 kg (14,054 lb)
                A-5C         6,494 kg (14,317 lb)
  Fuel: max internal         2,827 kg (6,232 lb)
    two 400 litre drop tanks   620 kg (1,367 lb)
    two 760 litre drop tanks 1,178 kg (2,597 lb)
    max internal/external    4,005 kg (8,829 lb)
  Max external stores load   2,000 kg (4,410 lb)
  Max T-O weight:
    clean: Q-5 IA            9,486 kg (20,913 lb)
           A-5C              9,530 kg (21,010 lb)
    with max external stores:
      Q-5 IA                11,830 kg (26,080 lb)
      A-5C                  12,000 kg (26,455 lb)
  Max wing loading:
    clean: Q-5 IA           339.4 kg/m² (69.5 lb/sq ft)
           A-5C             341.0 kg/m² (69.9 lb/sq ft)
    with max external stores:
      Q-5 IA                423.3 kg/m² (86.7 lb/sq ft)
      A-5C                  429.3 kg/m² (87.9 lb/sq ft)
  Max power loading:
    clean: Q-5 IA           148.8 kg/kN (1.46 lb/lb st)
           A-5C             149.5 kg/kN (1.47 lb/lb st)
    with max external stores:
      Q-5 IA                185.6 kg/kN (1.82 lb/lb st)
      A-5C                  188.3 kg/kN (1.85 lb/lb st)
PERFORMANCE (A-5C at max clean T-O weight, with after-
  burning, except where indicated):
  Max limiting Mach number                           1.5
  Max level speed:
    at 11,000 m (36,000 ft)
                    Mach 1.12 (643 knots; 1,190 km/h; 740 mph)
    at S/L          653 knots (1,210 km/h; 752 mph)
  T-O speed:
    clean, 15° flap   162 knots (300 km/h; 186 mph)
    with max external stores, 25° flap
                      178 knots (330 km/h; 205 mph)
* Landing speed:
    25° flap, brake-chute deployed
                   150-165 knots (278-307 km/h; 172-191 mph)
* Max rate of climb at 5,000 m (16,400 ft)
                      4,980-6,180 m (16,340-20,275 ft)/min
  Service ceiling            15,850 m (52,000 ft)
  T-O run:
    *clean, 15° flap          700-750 m (2,300-2,460 ft)
    with max external stores, 25° flap   1,250 m (4,100 ft)
  Landing run:
    25° flap, brake-chute deployed       1,060 m (3,480 ft)
  Combat radius with max external stores, afterburners off:
    lo-lo-lo (500 m; 1,640 ft) 216 nm (400 km; 248 miles)
    hi-lo-hi (8,000/500/8,000 m; 26,250/1,640/26,250 ft)
                             324 nm (600 km; 373 miles)
  Range at 11,000 m (36,000 ft) with max internal and
    external fuel, afterburners off
                     nearly 1,080 nm (2,000 km; 1,243 miles)
  g limits:
    with full load of bombs and/or drop tanks          5
    with drop tanks empty                            6.5
    clean                                            7.5
* depending upon airfield elevation and temperature
```

NAMC A-5M

TYPE: Upgraded export version of Q-5 II.
PROGRAMME: Begun 1 August 1986 by Alenia (Italy) and CATIC to upgrade nav/attack capability; two Q-5 IIs converted as prototypes; first one (first flight 30 August 1988) lost in crash 17 October same year; first flight of second prototype 8 March 1989; replacement for first aircraft completed subsequently; successful completion of development and flight testing announced 19 February 1991; reported mid-1992 to be intended to have further improvements including Martin-Baker zero/zero seat, new Alenia ECM pod, and WP6A III engines (same ratings, but TBO increased from 150 h to 300 h). According to chief A-5M designer Yong Zhengqiu in early 1993, ongoing upgrade programme will also include wider use of radar-absorbent composites to reduce radar signature; addition of in-flight refuelling probe; installation of IR night vision equipment (helmet mounted NVGs and electronic displays) and laser rangefinder; more powerful active jammer pod; and ability to carry laser guided bombs and anti-radiation missiles.
CUSTOMERS: None announced up to early 1993.
DESIGN FEATURES: New all-weather nav/attack system similar to that used in AMX; uprated engines; two additional underwing stores stations.
POWER PLANT: Improved WP6A turbojets with dry and afterburning ratings of 29.42 kN (6,614 lb st) and 36.78 kN (8,267 lb st) respectively; underwing drop tanks can be of 1,140 litre (301 US gallon; 251 Imp gallon) capacity.
AVIONICS: All-weather nav/attack system designed around two Singer central digital computers and dual-redundant MIL-STD-1553B databus; other new items include Pointer 2500 ranging radar, Litton LN-39A INS, Alenia HUD-35, air data computer, three-axis gyro package, RW-30 RWR, HSI, AG-5 attitude indicator, static inverters, mode controls, and interface to link these with existing AR-3201 VHF com radio, radio altimeter, radio compass, marker beacon receiver, IFF and armament system.

Prototype A-5M improved-capability attack aircraft, with new nav/attack system and digital avionics

```
EQUIPMENT: Chaff/flare dispenser added.
ARMAMENT: External stores stations increased to 12 by adding
  fourth pylon beneath each outer wing, with some redistri-
  bution of weapons on each wing station; PL-5B added to
  range of air-to-air missiles; larger underwing drop tanks
  optional (see Power Plant). External stores configurations
  otherwise generally as for Q-5 IA/A-5C.
DIMENSIONS, EXTERNAL: As for Q-5 IA/A-5C except:
  Length overall                      15.366 m (50 ft 5 in)
  Height overall                       4.53 m (14 ft 10¼ in)
WEIGHTS AND LOADINGS:
  Weight empty                         6,728 kg (14,833 lb)
* Max external stores load             2,000 kg (4,410 lb)
  Max T-O weight: clean                9,769 kg (21,537 lb)
    with max external stores          12,200 kg (26,869 lb)
  Max wing loading: clean      349.5 kg/m² (71.58 lb/sq ft)
    with max external stores   436.5 kg/m² (89.40 lb/sq ft)
  Max power loading: clean     132.8 kg/kN (1.30 lb/lb st)
    with max external stores   165.8 kg/kN (1.63 lb/lb st)
* Reportedly to be increased to 3,000 kg (6,614 lb)
PERFORMANCE (estimated):
  Max level speed at S/L at clean T-O weight
                             658 knots (1,220 km/h; 758 mph)
  Max level flight Mach number at 11,000 m (36,000 ft) at
    clean T-O weight                                1.205
  Unstick speed, with afterburning:
    no external stores, 15° flap
                             162 knots (300 km/h; 187 mph)
    full external stores, 25° flap
                             174 knots (322 km/h; 200 mph)
  Landing speed, 25° flap, brakes on and brake-chute
    deployed (depending upon AUW)
                   150-166 knots (278-307 km/h; 173-191 mph)
  Max vertical rate of climb at 5,000 m (16,400 ft) at clean
    T-O weight, with afterburning
                                     6,900 m (22,638 ft)/min
  Service ceiling at clean T-O weight  16,000 m (52,500 ft)
  T-O run, with afterburning:
    no external stores, 15° flap          911 m (2,989 ft)
    full external stores, 25° flap      1,250 m (4,101 ft)
  Landing run, 25° flap, brakes on and brake-chute
    deployed                            1,060 m (3,478 ft)
  Combat radius with full external stores:
    out at 8,000 m (26,250 ft), combat at 500 m (1,640 ft)
      and back at 11,000 m (36,000 ft)
                                280 nm (518 km; 322 miles)
    out, combat and back all at 500 m (1,640 ft)
                                174 nm (322 km; 200 miles)
  Range at 11,000 m (36,000 ft) with two 760 litre (200 US
    gallon; 167 Imp gallon) drop tanks
                             1,080 nm (2,000 km; 1,243 miles)
```

NAMC N-5A

Chinese name: Nongye Feiji 5 (Agricultural aircraft 5) or Nong-5

TYPE: Single/two-seat agricultural aircraft.
PROGRAMME: Design began November 1987, first details being revealed at Farnborough International air show September 1988; first of three prototypes (B-501L) made first flight 26 December 1989; CAAC type certificate awarded 12 August 1992; batch production then initiated in anticipation of MAS manufacturing approval.

CURRENT VERSIONS: **N-5A**: Standard version, *as described*.
CUSTOMERS: Estimated domestic market for more than 300 as Y-5 replacement; may eventually be offered for export.
DESIGN FEATURES: Designed to meet Chinese (CCAR) and US (FAR Pt 23) Normal category requirements, for specialised farming and forestry applications; crash-resistant forward fuselage with quickly removable side panels; cable cutter on windscreen, with deflector cable from this to tip of fin. Wings mainly constant chord, of LS(1)-0417 Mod section (17 per cent thickness/chord ratio), but sweptforward 18° at root; dihedral 4° 30' from roots, incidence 2°. Tailplane has inverted aerofoil section.
FLYING CONTROLS: Primary surfaces actuated mechanically, with ailerons operable differentially; ground adjustable tab on rudder and each aileron, electrically actuated trim tab in starboard elevator. Single-slotted trailing-edge flaps, also actuated electrically.
STRUCTURE: All-metal, with two-spar thin-wall wings, two-spar fin and tailplane, welded alloy steel tube forward fuselage and riveted duralumin rear fuselage; lower fuselage skin panels are of stainless steel and non-removable. Entire structure anodised before assembly and finished in polyurethane enamel paint.
LANDING GEAR: Non-retractable tricycle type, with single wheel and oleo-pneumatic shock absorber on each unit. Nose unit has a telescopic strut, a size 400 × 150 mm tyre, pressure 2.5 bars (36.3 lb/sq in), and is fitted with a shimmy damper and wire cutter. Main gear legs are of trailing-link type, with wheel tyres size 500 × 200 mm, pressure 3.0 bars (43.5 lb/sq in). Hydraulic mainwheel disc brakes and parking brake.
POWER PLANT: One 298 kW (400 hp) Textron Lycoming IO-720-D1B flat-eight engine, driving a Hartzell HC-C3YR-1RF/F8475R three-blade constant-speed metal propeller. All fuel in wing tanks with combined capacity of 315 litres (83.2 US gallons; 69.3 Imp gallons). Gravity fuelling point in upper surface of each wing at root.
ACCOMMODATION: Tandem seats and inertia reel safety harnesses for pilot and, when required, a loader/mechanic, under hard-top framed canopy with all-round field of view. Downward opening window/door on each side. Cockpit semi-sealed with ram air ventilation, slightly pressurised to prevent chemical ingress. Cockpit heating optional. Windscreen washer, wiper and demister standard. Deflector cable from top of windscreen cable cutter to tip of fin.
SYSTEMS: Hydraulic system for brakes only. No pneumatic system. Electrical system powered by Prestolite 28V 100A AC generator and 30Ah battery.
AVIONICS: Bendix/King KY 96A VHF com transceiver and stall warning system standard; KHF 950 HF/SSB com transceiver optional. Other options include Type 263 radio altimeter; LC-2 magnetic compass; WL-7A radio compass; and XS-6B marker beacon receiver.
EQUIPMENT: Glassfibre honeycomb hopper, for liquid or dry chemicals, forward of cockpit, with quick-dump system permitting release of all contents within 5 s. Solid or spray dispersal system, as appropriate. Dispersal of liquids, powered by fan driven pump, is via Y-type filter and 60-nozzle spraybars and is suitable for high, medium or low volume application. Wire cutters on main landing gear and in front of cockpit canopy.

First prototype NAMC N-5A (Textron Lycoming IO-720 flat-eight engine)

54 CHINA: AIRCRAFT—NAMC/SAC

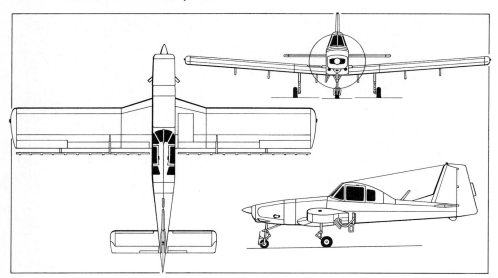

NAMC N-5A agricultural monoplane *(Jane's/Mike Keep)*

DIMENSIONS, EXTERNAL:
Wing span	13.418 m (44 ft 0¼ in)
Wing chord: at root	2.319 m (7 ft 7½ in)
constant portion	1.877 m (6 ft 2 in)
Wing aspect ratio	6.80
Length overall	10.487 m (34 ft 4⅞ in)
Height overall	3.733 m (12 ft 3 in)
Fuselage: Max width	1.01 m (3 ft 3¾ in)
Max depth	1.735 m (5 ft 8¼ in)
Tailplane span	4.59 m (15 ft 0¾ in)
Wheel track	3.528 m (11 ft 7 in)
Wheelbase	2.713 m (8 ft 10¾ in)
Propeller diameter	2.184 m (7 ft 2 in)

DIMENSIONS, INTERNAL:
Cockpit: Length	2.29 m (7 ft 6¼ in)
Max width	1.00 m (3 ft 3¼ in)
Max height	1.26 m (4 ft 1½ in)
Hopper volume	1.20 m³ (42.38 cu ft)

AREAS:
Wings, gross	26.0 m² (279.86 sq ft)
Ailerons (total)	2.08 m² (22.39 sq ft)
Trailing-edge flaps (total)	4.06 m² (43.70 sq ft)
Fin	2.28 m² (24.54 sq ft)
Rudder, incl tab	1.57 m² (16.90 sq ft)
Tailplane	4.68 m² (50.38 sq ft)
Elevators (total, incl tabs)	2.20 m² (23.68 sq ft)

WEIGHTS AND LOADINGS:
Weight empty	1,328 kg (2,928 lb)
Fuel weight: normal	85 kg (187 lb)
max	233 kg (513 lb)
Payload: normal	760 kg (1,675 lb)
max	960 kg (2,116 lb)
Max T-O weight: normal	2,250 kg (4,960 lb)
overload	2,450 kg (5,401 lb)
Max wing loading: normal	86.54 kg/m² (17.72 lb/sq ft)
overload	94.23 kg/m² (19.30 lb/sq ft)
Max power loading: normal	7.55 kg/kW (12.40 lb/hp)
overload	8.22 kg/kW (13.50 lb/hp)

PERFORMANCE (A: with, B: without dispersal equipment):
Max level speed: A	111 knots (205 km/h; 127 mph)
B	118 knots (220 km/h; 136 mph)
Normal operating speed:	
A, B	92 knots (170 km/h; 105 mph)
Stalling speed:	
A, B, flaps up	57 knots (105 km/h; 66 mph)
A, B, flaps down	47 knots (86 km/h; 54 mph)
Max rate of climb at S/L: A	257 m (845 ft)/min
B	281 m (922 ft)/min
Service ceiling: A	3,750 m (12,300 ft)
B	4,280 m (14,040 ft)
T-O run: A	303 m (995 ft)
B	296 m (971 ft)
T-O to 15 m (50 ft): A	569 m (1,867 ft)
B	553 m (1,814 ft)
Landing from 15 m (50 ft): A	373 m (1,225 ft)
B	379 m (1,243 ft)
Landing run: A	246 m (807 ft)
B	252 m (827 ft)
Min banking turn radius: A	145 m (476 ft)
B	140 m (459 ft)
Normal range with max payload, 45 min reserves:	
A	135 nm (250 km; 155 miles)
B	152 nm (282 km; 175 miles)
Endurance with max payload, 45 min reserves:	
A	1 h 48 min
B	1 h 56 min
Ferry range with max fuel:	
A	528 nm (979 km; 608 miles)
Endurance (self-ferry) with max fuel	5 h 45 min

NAMC CJ-6A

Chinese name: Chuji Jiaolianji-6A (Basic training aircraft 6A) or Chujiao-6A
Westernised designation: PT-6A

TYPE: Primary trainer.
PROGRAMME: Design initiated at Shenyang Autumn 1956 as Chinese engineered successor to CJ-5 (licence Yak-18: see 1991-92 and earlier *Jane's*); first flight of first prototype (108 kW; 145 hp Mikulin M-11ER engine) 27 August 1958, but trials disappointing; modified version with 194 kW (260 hp) Ivchenko AI-14R made first flight 18 July 1960. Responsibility subsequently transferred to Nanchang, further redesign preceding first flight of production-standard prototype 15 October 1961; production go-ahead for aircraft January 1962, for HS6 engine (Chinese AI-14R) June 1962.
CURRENT VERSIONS: **CJ-6A:** Standard version since December 1965, with uprated HS6A engine.
CJ-6B: Armed version: 10 built 1964-66.
Haiyan A: Prototype of civil agricultural version, described in 1991-92 and earlier *Jane's*. Programme superseded by N-5A (which see).
CUSTOMERS: Total of 1,796 (all versions) built by end of 1986; more than 2,000 now built for PLA Air Force and approx 200 for foreign customers. Exported to Albania, Bangladesh, Cambodia, Korea, Tanzania and Zambia.
DESIGN FEATURES: Low mounted wings with dihedral on outer panels; tandem seats in framed 'glasshouse' cockpit.
STRUCTURE: All-metal, with two-spar wings, outer panels of which are detachable.
LANDING GEAR: Retractable tricycle type with low pressure mainwheel tyres; suitable for operation from grass strips.
POWER PLANT: One 213 kW (285 hp) Zhuzhou (SMPMC) HS6A nine-cylinder air-cooled radial engine, driving a Baoding J9-G1 two-blade constant-speed propeller. Fuel capacity (two tanks) 100 litres (26.4 US gallons; 22 Imp gallons).

DIMENSIONS, EXTERNAL:
Wing span	10.22 m (33 ft 6½ in)
Length overall	8.46 m (27 ft 9 in)
Height overall	3.25 m (10 ft 8 in)

WEIGHTS AND LOADINGS:
Weight empty	1,095 kg (2,414 lb)
Max fuel	110 kg (243 lb)
Max T-O weight	1,400 kg (3,086 lb)

PERFORMANCE:
Max permissible diving speed	199 knots (370 km/h; 230 mph)
Max level speed	160 knots (297 km/h; 185 mph)
Landing speed	62 knots (115 km/h; 72 mph)
Max rate of climb at S/L	380 m (1,248 ft)/min
Service ceiling	6,250 m (20,500 ft)
T-O run	280 m (920 ft)
Landing run	350 m (1,150 ft)
Range with max fuel	372 nm (690 km; 428 miles)
Endurance	3 h 36 min

NAMC PT-6A primary trainers of the Bangladesh Air Force *(Peter Steinemann)*

SAC
SHAANXI AIRCRAFT COMPANY

PO Box 34, Chenggu, Shaanxi 723213
Telephone: 86 (916) 216301
Fax: 86 (916) 216302
Telex: 70141 SAC CN
PRESIDENT: Wang Wenfeng
MARKETING MANAGER: Li Yousen

Founded early 1970s; currently occupies a 204 ha (504 acre) site and had 1993 workforce of about 10,000; covered workspace includes largest final assembly building in China. Main aircraft programme is Y-8 transport; non-aerospace products include 36-seat coaches and small trucks.

SAC Y-8

Chinese name: Yunshuji-8 (Transport aircraft 8) or Yun-8

TYPE: Four-turboprop medium-range transport.
PROGRAMME: Redesign, as Chinese development of Antonov An-12B, started at Xian March 1969; first flight of first prototype 25 December 1974, followed by second (first built by SAC) 29 December 1975; production go-ahead given January 1980; type approval of Y-8X awarded September 1984. First Y-8B delivered 1986, first Y-8A and Y-8D 1987, first Y-8E 1989, first Y-8F early 1990. Pressurised Y-8C made first flight 17 December 1990; Y-8C certification and service entry anticipated in 1992.
CURRENT VERSIONS: **Y-8:** Prototype.
Y-8A: Helicopter carrier. Main cabin height increased by 120 mm (4.72 in) by deleting internal gantry; downward opening rear ramp/door. *Detailed description applies to this version unless otherwise indicated.*
Y-8B: Civil transport. Military equipment deleted; weight reduced by 1,720 kg (3,792 lb); some avionics differ.
Y-8C: Fully pressurised version for civil and military applications, developed with Lockheed collaboration. Pressurised volume increased from 31 m³ (1,095 cu ft) to 212 m³ (7,487 cu ft); effective length of cargo hold extended by 2.00 m (6 ft 6¾ in); other changes include redesigned cargo loading door and main landing gear; improved air-conditioning and oxygen systems; additional emergency exits. Two prototypes now flying.
Y-8D: Export version, with main avionics by Collins and Litton.
Y-8E: Drone carrier version. Forward pressure cabin accommodates drone controller's console; carrier/launch trapeze for one drone under each outer wing panel.
Y-8F: Livestock carrier, with cages to hold up to 350 sheep or goats.
Y-8H: Aerial survey version.
Y-8X: Maritime patrol version, referred to in earlier *Jane's* as Y-8MPA. Western com/nav, radar, surveillance and search equipment; larger chin radome. May be used for both ASW patrol and civilian offshore duties such as fishery patrol, pollution monitoring and oil exploration support, but not yet in production.
AEW version: Under development, with assistance from GEC-Marconi.
CUSTOMERS: More than 40 delivered by June 1993. In service in China with Changan Airlines (one) in 1992, and PLA Air Force; exports (Y-18D) include air forces of Myanmar (two), Sri Lanka (two) and Sudan (two). Sri Lankan aircraft reportedly modified locally for use as bombers.
DESIGN FEATURES: More pointed nose transparencies than An-12; high mounted wing with progressive anhedral on outer panels; circular section fuselage (forward section and tail turret pressurised), upswept at rear; angular tail surfaces with large dorsal fin.
FLYING CONTROLS: Mechanical actuation of aerodynamically balanced differential ailerons, elevators and rudder, each of which has inset trim tab; two-segment, hydraulically actuated double-slotted Fowler flaps on each wing trailing-edge; comb-shaped spoilers forward of flaps.

SAC Y-8B four-turboprop multi-purpose medium-range transport

Wing sections C-5-18 at root, C-3-16 at rib 15 and C-3-14 at tip, final two digits indicating thickness/chord ratio; incidence 4°; anhedral 1° on intermediate panels, 4° on outboard panels; 6° 50′ sweepback at quarter-chord; fixed incidence tailplane.

STRUCTURE: All-metal (aluminium alloy) conventional semi-monocoque/stressed skin; wings, tailplane and fin are two-spar box structures; landing gear and all hydraulic components manufactured by Shaanxi Aero-Hydraulic Component Factory (SAHCF).

LANDING GEAR: Hydraulically retractable tricycle type, with Shaanxi (SAHCF) nitrogen/oil shock struts on all units. Four-wheel main bogie on each side retracts inward and upward into blister on side of fuselage. Twin-wheel nose unit, hydraulically steerable to ±35°, retracts rearward. Mainwheel tyres size 1,050 × 300 mm, pressure 28.4 bars (412 lb/sq in); nosewheel tyres size 900 × 300 mm, pressure 16.7 bars (242 lb/sq in). Hydraulic disc brakes and Xingping inertial anti-skid sensor.

POWER PLANT: Four 3,169 kW (4,250 ehp) Zhuzhou (SMPMC) WJ6 turboprops, each driving a Baoding four-blade J17-G13 constant-speed propeller. Alternative power plants under consideration include GE CT7. All fuel in two integral tanks and 29 bag-type tanks in wings (20,102 litres; 5,310.5 US gallons; 4,422 Imp gallons) and fuselage (10,075 litres; 2,661.5 US gallons; 2,216 Imp gallons), giving total capacity of 30,177 litres (7,972 US gallons; 6,638 Imp gallons). Refuelling points in starboard side of fuselage (between frames 14 and 15), mainwheel fairing, and in wing upper surface.

ACCOMMODATION: Flight crew of five (pilot, co-pilot, navigator, engineer and radio operator). Forward portion of fuselage (up to frame 13) is pressurised, and can accommodate up to 14 passengers in addition to crew. Cargo compartment (between frames 13 and 43) is unpressurised. Max accommodation for up to 96 troops; or 58 paratroops; or 60 severe casualties plus 20 slightly wounded, with three medical attendants; or two 'Liberation' army trucks. Crew door and two emergency exits in forward fuselage. Three additional emergency exits in cargo compartment, access to which is via a large rear-loading ramp/door in underside of rear fuselage. Entire accommodation heated and ventilated.

SYSTEMS: Forward fuselage pressurised to maintain a differential of 0.20 bar (2.84 lb/sq in) at altitudes above 4,300 m (14,100 ft). Two independent hydraulic systems, with operating pressures of 152 bars (2,200 lb/sq in) (port) and 147 bars (2,130 lb/sq in) (starboard), plus hand and electrical standby pumps, for actuation of landing gear extension/retraction, nosewheel steering, flaps, brakes and rear ramp/door. Electrical DC power (28.5V) supplied by eight 12 kW generators, an 18 kW (24 hp) Xian Aero Engine Co APU (mainly for engine starting) and four 28Ah batteries. Four 12kVA alternators provide 115V AC power at 400Hz. Gaseous oxygen system for crew. Electric de-icing of windscreen, propellers and fin/tailplane leading-edges; hot air de-icing for wing leading-edges.

AVIONICS (Y-8X): Collins VHF, dual HF (DF-2 and DS-3) and HF/SSB radios; Litton Canada AN/APS-504(V)3 search radar; Litton Canada LTN-72 INS and LTN-211 Omega navigation system; Collins ADF, DME-42, TDR-90 ATC transponder, VOR-32, HSI-85, ADI-85A, 520-3337 RMI, IFF and autopilot. Optical and infra-red cameras, IR submarine detection system and sonobuoys.

DIMENSIONS, EXTERNAL:
Wing span	38.00 m (124 ft 8 in)
Wing chord: at root	4.73 m (15 ft 6¼ in)
at tip	1.69 m (5 ft 6½ in)
Wing aspect ratio	11.85
Length overall	34.022 m (111 ft 7½ in)
Fuselage: Max diameter of circular section	4.10 m (13 ft 5½ in)
Height overall	11.16 m (36 ft 7½ in)
Tailplane span	12.196 m (40 ft 0¼ in)
Wheel track (c/l of shock struts)	4.92 m (16 ft 1¾ in)
Wheelbase (c/l of main bogie)	9.576 m (31 ft 5 in)
Propeller diameter	4.50 m (14 ft 9¼ in)
Propeller ground clearance	1.89 m (6 ft 2½ in)
Crew door: Height	1.455 m (4 ft 9¼ in)
Width	0.80 m (2 ft 7½ in)
Rear loading hatch: Length	7.67 m (25 ft 2 in)
Width: min	2.16 m (7 ft 1 in)
max	3.10 m (10 ft 2 in)
Emergency exits (each): Height	0.55 m (1 ft 9¾ in)
Width	0.60 m (1 ft 11½ in)

DIMENSIONS, INTERNAL:
Cabin (incl flight deck, galley and toilet):	
Length	13.50 m (44 ft 3½ in)
Width: min	3.00 m (9 ft 10 in)
max	3.50 m (11 ft 5¾ in)
Height: min	2.40 m (7 ft 10½ in)
max	2.60 m (8 ft 6½ in)
Floor area	55.0 m² (592.0 sq ft)
Volume	123.3 m³ (4,354.3 cu ft)

AREAS:
Wings, gross	121.86 m² (1,311.7 sq ft)
Ailerons (total)	7.84 m² (84.39 sq ft)
Trailing-edge flaps (total)	26.91 m² (289.66 sq ft)
Rudder	6.537 m² (70.36 sq ft)
Tailplane	27.05 m² (291.16 sq ft)
Elevators (total)	7.101 m² (76.43 sq ft)

WEIGHTS AND LOADINGS:
Weight empty, equipped	35,500 kg (78,264 lb)
Max fuel load	22,909 kg (50,505 lb)
Max payload: containerised	16,000 kg (35,275 lb)
bulk cargo	20,000 kg (44,090 lb)
Max T-O weight	61,000 kg (134,480 lb)
Max taxi weight	61,500 kg (135,585 lb)
Max landing weight	58,000 kg (127,870 lb)
Max zero-fuel weight	36,266 kg (79,955 lb)
Max wing loading	500.6 kg/m² (102.5 lb/sq ft)
Max power loading	4.81 kg/kW (7.91 lb/ehp)

PERFORMANCE (at max T-O weight except where indicated):
Max level speed at 7,000 m (22,965 ft)	370 knots (685 km/h; 425 mph)
Max cruising speed at 8,000 m (26,250 ft)	297 knots (550 km/h; 342 mph)
Econ cruising speed at 8,000 m (26,250 ft)	286 knots (530 km/h; 329 mph)
Max rate of climb at S/L	473 m (1,552 ft)/min
Rate of climb at S/L, OEI	231 m (758 ft)/min
Service ceiling, AUW of 51,000 kg (112,435 lb)	10,400 m (34,120 ft)
Service ceiling, OEI, AUW of 51,000 kg (112,435 lb)	8,100 m (26,575 ft)
T-O run	1,230 m (4,035 ft)
T-O to 15 m (50 ft)	3,007 m (9,866 ft)
Landing from 15 m (50 ft)	2,174 m (7,133 ft)
Landing run	1,100 m (3,609 ft)
Range: with max payload	687 nm (1,273 km; 791 miles)
with max fuel	3,086 nm (5,720 km; 3,554 miles)
Max endurance	11 h 7 min

Black Hawk helicopter being unloaded from a Y-8A

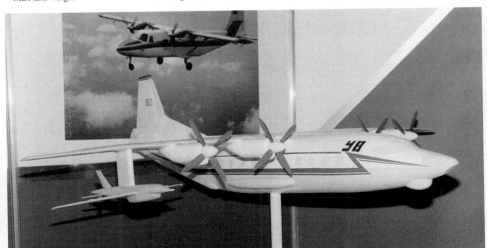

Enlarged radome of maritime Y-8X

Model of the SAC Y-8E with two WZ-5 high-altitude reconnaissance UAVs under wings

SAC

SHENYANG AIRCRAFT CORPORATION

PO Box 328, Shenyang, Liaoning 110034
Telephone: 86 (24) 462680
Fax: 86 (24) 662689
Telex: 80018 SAMC CN
GENERAL MANAGER: Tang Qiansan
DEPUTY MANAGER: Xu Guosheng

Pioneer Chinese fighter design centre; built 767 examples of J-5 (licence MiG-17F) from 1956-59 and from 1963 was major producer of J-6 series (reverse engineered MiG-19), including 634 JJ-6 tandem-seat fighter/trainers; initiated development and production of J-7 (see now under CAC). Now occupies site area of more than 800 ha (1,976 acres) and has workforce of over 20,000; only some 30 per cent of current activities are in aerospace. Principal current programme is J-8 II fighter; subcontract manufacture includes cargo doors for Boeing 757 and de Havilland Dash 8 (100th Dash 8 door delivered 12 June 1992), rudders for BAe Jetstream ATP, wing ribs and emergency hatches for Airbus A320, tailcone/landing gear door/pylon components for Lockheed C-130, and other machined parts for BAe, Boeing, Deutsche Aerospace Airbus and Saab-Scania.

SAC J-8

Chinese name: Jianjiji-8 (Fighter aircraft 8) or Jian-8
Westernised designation: F-8
NATO reporting name: Finback

TYPE: Single-seat twin-engined air superiority fighter, with secondary capability for ground attack.

PROGRAMME: Development started 1964; first flight of first of two prototypes 5 July 1969; flight trials (but no other J-8 activity) permitted to continue during 1966-76 cultural revolution, totalling 663 hours in 1,025 flights by prototypes; initial production authorised July 1979; three prototypes of J-8 I then built (one lost before flight testing); first flight 24 April 1981 by second aircraft; production go-ahead for this version given July 1985. This preceded 12 June 1984 by first flight of first of four prototypes of much redesigned J-8 II; J-8/J-8 I production ended 1987; Peace Pearl programme (see 1990-91 *Jane's*), to upgrade J-8 II with Western avionics, embargoed by US government mid-1989 and cancelled by China 1990; two prototypes supplied to Grumman for Peace Pearl returned to China 1993; recent purchase of Russian Su-27s may have obviated alternative plans for J-8 II upgrade.

CURRENT VERSIONS: **J-8 ('Finback-A'):** Initial clear-weather day fighter, powered by two Liyang (LMC) WP7B turbojets (each 43.15 kN; 9,700 lb st dry and 59.82 kN; 13,448 lb st with afterburning) and armed with two single-barrel 30 mm cannon and four wing-mounted PL-2B air-to-air missiles; single intake in nose, with conical centrebody housing ranging radar only. Built in small numbers only; production ended 1987.

J-8 I ('Finback-A'): Improved all-weather version of J-8, with same power plant but fitted from outset with Sichuan SR-4 fire control radar in intake centrebody; single twin-barrel 23 mm cannon. More than 100 now in service (including upgraded J-8s), but gradually being supplanted by J-8 II. Described in 1985-86 *Jane's.*

J-8 II ('Finback-B'): All-weather dual-role version (high altitude interceptor and ground attack), embodying some 70 per cent redesign compared with J-8 I. Main configuration change is to 'solid' nose and twin lateral air intakes, providing more nose space for fire control radar and other avionics, plus increased airflow for more powerful WP13A II turbojets. In production and service, with several dozen built by 1990, but then being manufactured in small economic batches rather than continuous production. *Detailed description applies to this version.*

F-8 II: Proposed export version: WP13B engines (uprated by 4 per cent to 68.65 kN; 15,432 lb st with afterburning), pulse Doppler lookdown radar, digital avionics (incl HUD and two HDDs) with 1553B databus, leading-edge flaps, in-flight refuelling, seven stations for 4,500 kg (9,921 lb) stores load, max speed Mach 2.2.

CUSTOMERS: China (PLA Air Force).

DESIGN FEATURES: Thin-section, mid-mounted delta wings and all-sweptback tail surfaces; fuselage has area rule 'waisting', detachable rear portion for engine access, and dorsal spine fairing; large ventral fin under rear fuselage, main portion of which folds sideways to starboard during take-off and landing, provides additional directional stability; small fence on each wing upper surface near tip; small airscoops at foot of fin leading-edge and at top of fuselage each side, above tailplane. Sweepback 60° on wing and tailplane leading-edges; wings have slight anhedral.

FLYING CONTROLS: Hydraulically boosted ailerons, rudder and low-set all-moving tailplane; two-segment single-slotted flaps on each wing trailing-edge, inboard of aileron; four door-type underfuselage airbrakes, one under each engine air intake trunk and one immediately aft of each mainwheel well.

STRUCTURE: Conventional aluminium alloy semi-monocoque/stressed skin construction, with high tensile steel for high load bearing areas of wings and fuselage and titanium in high temperature fuselage areas; ailerons, rudder and rear portion of tailplane are of aluminium honeycomb with sheet aluminium skin; dielectric skins on nosecone, tip of main fin, and on non-folding portion of ventral fin leading-edge.

LANDING GEAR: Hydraulically retractable tricycle type, with single wheel and oleo-pneumatic shock absorber on each unit. Steerable nose unit retracts forward, main units inward into centre-fuselage; mainwheels turn to stow vertically inside fuselage, resulting in slight overwing bulge. Brake-chute in bullet fairing at base of rudder.

POWER PLANT: Two Liyang (Guizhou Engine Co) WP13A II turbojets, each rated at 42.7 kN (9,590 lb st) dry and 65.9 kN (14,815 lb st) with afterburning, mounted side by side in rear fuselage with pen nib fairing above and between exhaust nozzles. Lateral, non-swept air intakes, with automatically regulated ramp angle and large splitter plates similar in shape to those of MiG-23. Internal fuel capacity (four integral wing tanks plus fuselage tanks) approx 5,400 litres (1,426 US gallons; 1,188 Imp gallons). Single-point pressure refuelling. Provision for auxiliary fuel tanks on fuselage centreline and each outboard underwing pylon.

ACCOMMODATION: Pilot only, on ejection seat under one-piece canopy hinged at rear and opening upward. Cockpit pressurised, heated and air-conditioned. Heated windscreen.

SYSTEMS: Two simple air-cycle environmental control systems, one for cockpit heating and air-conditioning and one for radar cooling; cooling air bled from engine compressor. Two 207 bar (3,000 lb/sq in) independent hydraulic systems (main utility system plus one for flight control surfaces boost), powered by engine driven pumps. 28.5V DC primary electrical power from two 12 kW engine driven starter/generators, with two 6kVA alternators for 115/200V three-phase AC at 400Hz. Pneumatic bottles for emergency landing gear extension. Pop-out ram air emergency turbine under fuselage.

AVIONICS: VHF/UHF and HF/SSB com radio, Tacan, radio compass, radar altimeter, ILS, marker beacon receiver, 'Odd Rods' type IFF, radar warning receiver and ECM; RWR antenna in fin-tip. Autopilot for attitude and heading hold, altitude hold and stability augmentation. Existing fire control system comprises a monopulse radar, optical gyro gunsight and gun camera. Enlarged avionics bays in nose and fuselage provide room for modernised fire control system and other upgraded avionics. Chaff/flare dispensers in tailcone.

ARMAMENT: One 23 mm Type 23-3 twin-barrel cannon, with 200 rds, in underfuselage pack immediately aft of nosewheel doors. Seven external stations (one under fuselage and three under each wing) for a variety of stores which can include PL-2B infra-red air-to-air missiles, PL-7 medium-range semi-active radar homing air-to-air missiles, Qingan HF-16B twelve-round pods of 57 mm Type 57-2 unguided air-to-air rockets, launchers for 90 mm air-to-surface rockets, bombs, or (centreline and outboard underwing stations only) auxiliary fuel tanks.

DIMENSIONS, EXTERNAL:
Wing span 9.344 m (30 ft $7\frac{7}{8}$ in)
Wing aspect ratio 2.1
Length overall, incl nose probe 21.59 m (70 ft 10 in)
Height overall 5.41 m (17 ft 9 in)
Wheel track 3.741 m (12 ft $3\frac{1}{4}$ in)
Wheelbase 7.337 m (24 ft $0\frac{3}{4}$ in)

AREAS:
Wings, gross 42.2 m^2 (454.2 sq ft)

WEIGHTS AND LOADINGS:
Weight empty 9,820 kg (21,649 lb)
Normal T-O weight 14,300 kg (31,526 lb)
Max T-O weight 17,800 kg (39,242 lb)
Wing loading:
at normal T-O weight 338.9 kg/m^2 (69.4 lb/sq ft)
at max T-O weight 421.8 kg/m^2 (86.4 lb/sq ft)
Power loading:
at normal T-O weight 110.5 kg/kN (1.08 lb/lb st)
at max T-O weight 137.5 kg/kN (1.35 lb/lb st)

PERFORMANCE:
Design max operating Mach number 2.2
Design max level speed 701 knots (1,300 km/h; 808 mph) IAS
Unstick speed 175 knots (325 km/h; 202 mph)
Landing speed 156 knots (290 km/h; 180 mph)
Max rate of climb at S/L 12,000 m (39,370 ft)/min
Acceleration from Mach 0.6 to 1.25 at 5,000 m (16,400 ft) 54 s
Service ceiling 20,200 m (66,275 ft)
T-O run, with afterburning 670 m (2,198 ft)
Landing run, brake-chute deployed 1,000 m (3,280 ft)
Combat radius 432 nm (800 km; 497 miles)
Max range 1,187 nm (2,200 km; 1,367 miles)
g limit in sustained turn at Mach 0.9 at 5,000 m (16,400 ft) +4.83

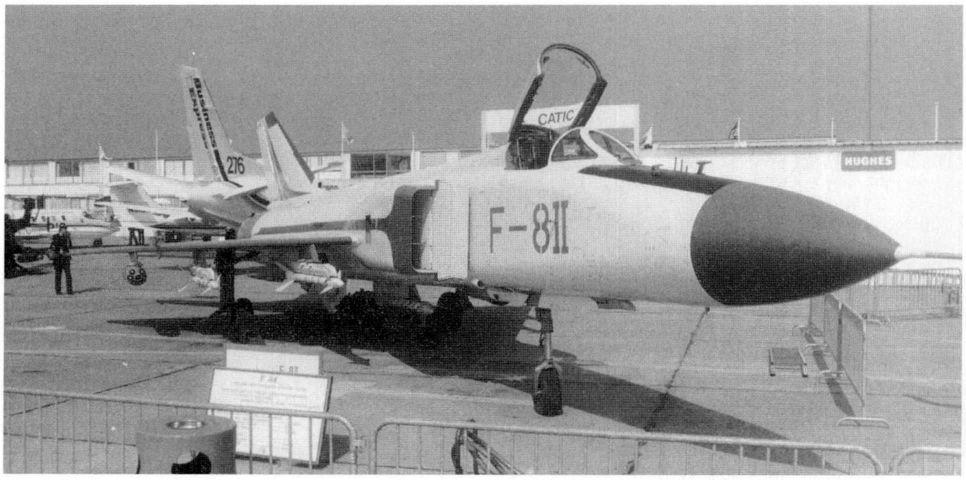

Fourth prototype of the SAC J-8 II single-seat multi-role fighter *(Paul Jackson)*

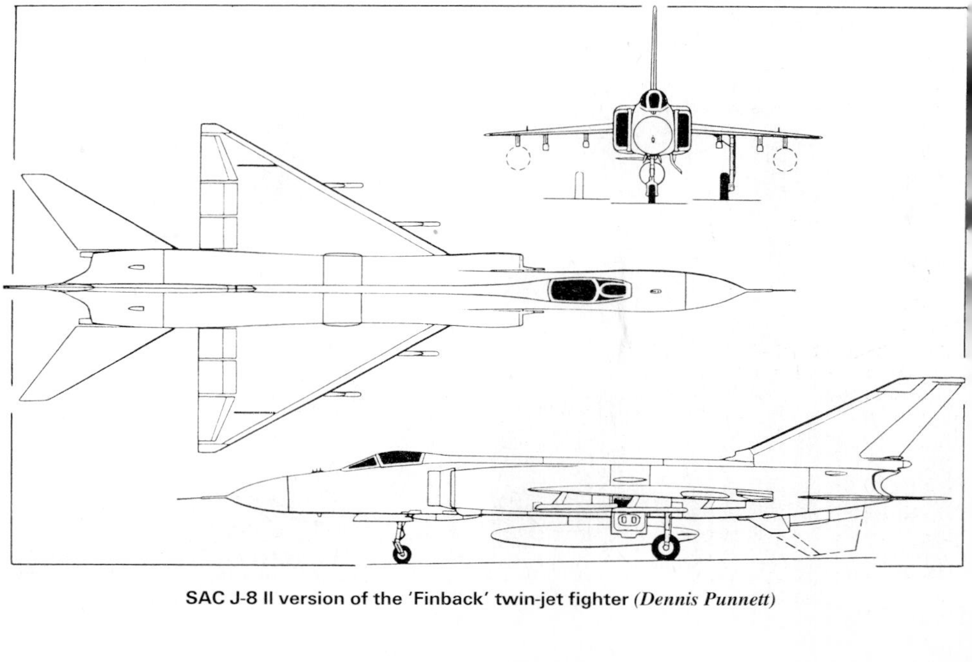

SAC J-8 II version of the 'Finback' twin-jet fighter *(Dennis Punnett)*

SAMF
SHANGHAI AIRCRAFT MANUFACTURING FACTORY (Subsidiary of Shanghai Aviation Industrial Corporation)
PO Box 232-007, Shanghai 200232
Telephone: 86 (21) 4383311
Fax: 86 (21) 6658103
Telex: 33136 SHAIR CN
PRESIDENT: Wu Zuoquan

Created 1951; occupies site area of 135 ha (333.5 acres); total workforce approx 7,000, of whom about 3,000 engaged in present MD-82/83 programme. Has produced main and landing gear doors for McDonnell Douglas MD-80 series since 1979 (1,799 sets delivered by end of 1990), and now also produces cargo and service doors, avionics bay doors and tailplanes; will produce wings for MD-95.

SAMF (MCDONNELL DOUGLAS) MD-82 and MD-83
TYPE: Commercial airliner (see US section for description).
PROGRAMME: Letter of intent (SAMF/McDonnell Douglas) for MD-82 co-production announced 11 January 1984; programme details confirmed April 1985 with announcement of sale of 26 MD-82s to China, of which one Douglas built (delivered September 1985) and 25 (all for Chinese airlines) to be assembled in China. Shanghai programme began April 1986 using major Douglas subassemblies for first three MD-82s, next 22 having gradually increasing share of Chinese manufactured components (see 1991-92 and earlier *Jane's*). First Chinese assembled MD-82 rolled out 8 June and first flight 2 July 1987, delivered (to Shenyang branch of CAAC) 31 July and entered service (with China Northern Airlines) 4 August 1987; 25th SAMF aircraft delivered 12 October 1991. Programme extended April 1990 by agreement for 10 more aircraft (five each MD-82 and MD-83, of which latter for resale to McDonnell Douglas); MD-83 work started 14 August 1991; first follow-on MD-82 delivered 30 December 1991; first MD-83 made first flight 26 June 1992, delivered to McDonnell Douglas 14 July; all five delivered by end of 1992.

Contract announced 29 June 1992 launching long-awaited Trunkliner programme with order for 20 each MD-82/82T (3 + 17) and MD-90-30T; first flight of MD-82T scheduled for 1997.
CURRENT VERSIONS: **MD-82:** As standard US version (which see); 33 to be built by SAMF (25 + 5 + 3).
MD-82T: Interim Chinese Trunkliner version, with four-wheel-bogie main landing gear for operation from runways with low bearing strength; 17 to be built by SAMF.

Shanghai-assembled McDonnell Douglas MD-82 of China Eastern Airlines

CUSTOMERS: China Eastern Airlines (11) and China Northern Airlines (19) share first 30 MD-82s.

SAMF (MCDONNELL DOUGLAS) MD-90-30T
TYPE: Commercial airliner (see US section for description).
PROGRAMME: Selected June 1992, after many years' deliberation, to fulfil Chinese Trunkliner requirement, with manufacturing contributions by Chengdu (nose section), Shenyang (vertical tail) and Xian (wings and fuselage) and final assembly at Shanghai; similar offset deal for co-production of V2500 engines; production to increase towards target of 93 per cent of airframe (by weight) being produced in China; follow-on orders expected later.

SAP
SHIJIAZHUANG AIRCRAFT PLANT
PO Box 164, Shijiazhuang, Hebei 050062
Telephone: 86 (311) 744251
Telex: 26236 HBJXC CN
DIRECTOR: Zhou Enqing

SAP has produced more than 225 An-2 general purpose biplanes since 1970, concentrating now on Y-5B customised agricultural version; other products include W-5 and W-6 microlight series (see 1992-1993 and earlier *Jane's*). Current (1992) workforce of more than 4,000 includes some 650 engineers and technicians. Became part of Xian Aircraft Industrial Group July 1992.

SAP Y-5B
Chinese name: Yunshuji-5 (Transport aircraft 5) or Yun-5
TYPE: Agricultural biplane.
PROGRAMME: Antonov An-2 (see under PZL Mielec in Polish section for full description) has been built under licence in China since 1957, chiefly at Nanchang and latterly by SAP; dedicated agricultural and forestry Y-5B version made first flight 2 June 1989, has been certificated to Chinese equivalent of FAR Pt 23 and is now in batch production; nine produced in 1990, next 12 due in 1993, total of 60 planned by end of current (8th) five-year plan period.
CURRENT VERSIONS: **Y-5N:** Standard Chinese civil transport and general purpose version.
Y-5B: Dedicated agricultural and forestry variant; *following description applies to this version.*
CUSTOMERS: Current operators of Y-5B include Xinjiang General Aviation (eight) and Jiangnan General Aviation; 12 ordered by CASC (China Aviation Supplies Corporation) in December 1992 for delivery by mid-1993.
STRUCTURE: As standard An-2/Y-5N, but specially treated to resist corrosion; cabin doors sealed against chemical ingress.
POWER PLANT: One 735.5 kW (986 hp) PZL Kalisz ASz-62IR-16 or SMPMC (Zhuzhou) HS5 nine-cylinder radial engine, driving an AW-2 or J12-G15 four-blade propeller.

Y-5B agricultural biplane produced by Shijiazhuang Aircraft Plant

ACCOMMODATION: Flight crew of one or two; cabin heating and ventilation improved by new ECS.
AVIONICS: Include Bendix/King KHF 950 HF and KY 196 VHF com radios, KR 87 ADF and KMA 24 audio control panel; some electrical and other instrument installations also improved.
EQUIPMENT: Large hopper/tank with emergency jettison of contents; high flow rate, wind driven pump; sprayers with various nozzle sizes, depending upon spray volume required.
DIMENSIONS, EXTERNAL:
Propeller diameter: AW-2 3.60 m (11 ft 9¾ in)
J12-G15 3.40 m (11 ft 2 in)
WEIGHTS AND LOADINGS (A: with dry chemical spreader, B: with liquid spray system):
Max payload: A, B 1,367 kg (3,013 lb)
Max T-O weight: A, B 5,250 kg (11,574 lb)
PERFORMANCE (A and B as for Weights):
Max level speed at S/L:
A 110 knots (205 km/h; 127 mph)
B 108 knots (200 km/h; 124 mph)
Max level speed at 1,700 m (5,575 ft):
A 119 knots (220 km/h; 137 mph)
B 116 knots (215 km/h; 133 mph)
Operating speed: A, B 86 knots (160 km/h; 99 mph)
Stalling speed: A, B 52 knots (95 km/h; 59 mph)
Max rate of climb at S/L: A 120 m (394 ft)/min
B 114 m (374 ft)/min
Rate of climb at 1,600 m (5,250 ft):
A 133 m (436 ft)/min
B 123 m (404 ft)/min
Service ceiling: A 3,460 m (11,350 ft)
B 3,250 m (10,660 ft)
Air turning radius: A, B 350 m (1,150 ft)
T-O run: A 170 m (558 ft)
B 180 m (591 ft)
Landing run: A 160 m (525 ft)
B 157 m (515 ft)
Range at S/L with fuel load of 670 litres (177 US gallons; 147 Imp gallons): A, B 456 nm (845 km; 525 miles)
Endurance, conditions as above: A, B 5 h 39 min
Swath width 20-50 m (66-165 ft)

SGAC
SHENZHEN GENERAL AVIATION COMPANY LTD
Announced 1992 as Shenzhen government development project to set up new, non-governmental production centre to manufacture light aircraft in 4- to 10-seat range, of Chinese and/or foreign design; intended to have eventual output capability of up to 180 aircraft per year.

XAC
XIAN AIRCRAFT COMPANY
PO Box 140, Xian, Shaanxi 710000
Telephone: 86 (29) 714959, 714960 and 716929
Fax: 86 (29) 715102 and 717859
Telex: 70101 XAC CN
PRESIDENT: Wang Qinping
SALES MANAGER: Wang Zhigang

Aircraft factory established at Xian 1958; current XAC has covered area of some 300 ha (741.3 acres) and 1992 workforce of 19,730, of whom about 90 per cent are engaged in aircraft production. Earlier programmes have included licence production of Tu-16 twin-jet bomber (as H-6: see 1991-92 and earlier *Jane's*); major current programmes concern JH-7 fighter-bomber and Y-7 transport series of An-24/26 derivatives. Subcontract work includes glassfibre header tanks, water float pylons, ailerons and various doors for Canadair CL-215/415 amphibians; and various components for Airbus A300, Boeing 737/747 and ATR 42.

XAC JH-7

Chinese name: Jianjiji Hongzhaji-7 (Fighter-bomber aircraft 7) or Jian Hong-7
Westernised designation: B-7

TYPE: All-weather interdictor and attack aircraft; secondary air-to-air capability.
PROGRAMME: First revealed publicly September 1988 as model at Farnborough International air show, first of two prototypes having been rolled out during previous month; first flight late 1988 or early 1989; service entry originally scheduled for 1992-93, but may have been delayed.
CURRENT VERSIONS: **JH-7:** Initial version, *as described*. Previously recorded in *Jane's* as H-7.
CUSTOMERS: Under development for PLA Air Force and, in maritime attack form, for PLA Naval Air Force.
DESIGN FEATURES: In same role and configuration class as Russian Sukhoi Su-24 'Fencer'. High-mounted wings with compound sweepback, dog-tooth leading-edges and marked anhedral; twin turbofans, with lateral air intakes; all-swept tail surfaces, comprising large main fin, single small ventral fin and low-set all-moving tailplane; small overwing fence at approx two-thirds span.
STRUCTURE: Assumed conventional all-metal except for dielectric panels.
LANDING GEAR: Retractable tricycle type.
POWER PLANT: Prototypes said in 1988 to be powered by two Xian WS9 turbofans (Rolls-Royce Spey Mk 202: each 91.2 kN; 20,515 lb st with afterburning). One Chinese source has suggested that US engines are or were envisaged but later embargoed; eventual intention believed to be to power production JH-7 with LEMC (Liming) turbofans of 71.3 kN (16,027 lb st) dry rating (138.3 kN; 31,085 lb st with afterburning).
ACCOMMODATION: Crew of two in tandem (rear seat elevated). HTY-4 ejection seats, operable at speeds from zero to 540 knots (1,000 km/h; 621 mph) and altitudes from S/L to 20,000 m (65,600 ft).
AVIONICS: Terrain-following radar and other avionics on prototypes said to be of Chinese origin.
ARMAMENT: Twin-barrel 23 mm gun in nose; two stores pylons under each wing, plus rail for close-range air-to-air missile at each wingtip. Typical underwing load for maritime attack, two C-801 sea-skimming anti-ship missiles (inboard) and two drop tanks (outboard).

DIMENSIONS, EXTERNAL:
Wing span	12.80 m (42 ft 0 in)
Length overall, incl probe	21.00 m (68 ft 10¾ in)
Height overall	6.22 m (20 ft 4⅞ in)

AREAS:
Wings, gross	52.3 m² (562.95 sq ft)

WEIGHTS AND LOADINGS (estimated):
Max T-O weight	27,500 kg (60,627 lb)
Max wing loading	525.8 kg/m² (107.7 lb/sq ft)
Max power loading (prototypes)	150.77 kg/kN (1.48 lb/lb st)

PERFORMANCE:
Max level speed at altitude	Mach 1.7
Service ceiling	16,000 m (52,500 ft)

XAC Y-7

Chinese name: Yunshuji-7 (Transport aircraft 7) or Yun-7

TYPE: Twin-turboprop short/medium-range transport.
PROGRAMME: Reverse engineering of Soviet 48/52-passenger Antonov An-24 began in mid-1970s, three prototypes (first flight 25 December 1970) and two static test airframes being completed; Chinese C of A awarded 1980; pre-production Y-7 made public debut 17 April 1982, production starting later same year; initial Y-7 entered service early 1984 (first scheduled services with CAAC, 29 April 1986); Y7-100 prototype conversion by HAECO 1985 (first flight of production Y7-100 late 1985), followed by domestic certification 23 January 1986; first flight of Y7-200B (B-528L) 23 November 1990; deliveries of Y7-200B, and first flight of Y7-200A, scheduled for 1993.
CURRENT VERSIONS: **Y-7:** Initial production version; full description in 1988-89 *Jane's*. Being retrofitted with winglet modification of Y7-100. In production.
Y-7 Freighter: Civil cargo version of Y-7, first flown late 1989; five delivered to domestic operators in 1992 (first one on 24 June).
Y7-100: Improved version, meeting BCAR standards. Changes include winglets, new three-crew (instead of five) flight deck, all-new cabin interior with 52 reclining seats, windscreen de-icing, new HF/VHF com, new nav equipment, addition of oxygen/air data/environmental control systems. One Y-7 (B-3499) converted as prototype 1985 by Hong Kong Aircraft Engineering Company (HAECO); in production. Sub-variants **Y7-100C1/C2/C3** available with alternative seating and avionics to customer's requirements. *Detailed description applies to standard Y7-100, except where indicated.*
Y7-200A: Improved Y7-100, with Pratt & Whitney Canada PW127 turboprops, Hamilton Standard 14SF four-blade propellers and Collins EFIS 85/86 avionics; fuselage 1.00 m (3 ft 3¼ in) longer than Y7-100 for 52 (standard) or 56 passengers; first flight due third quarter 1993.
Y7-200B: Improved Y7-100 for domestic market, with Dongan WJ5A IG improved turboprops, new four-blade propellers, Collins EFIS 85 avionics, higher max lift coefficient, improved stall characteristics, lower fuel consumption; winglets deleted; overall length increased by 0.74 m (2 ft 5¼ in), empty weight reduced by 500 kg (1,102 lb); Cat. II automatic landing capability; 49 hours flown by January 1993; certification due late 1993.
Y7H-500: Military and civil cargo version, derived from Antonov An-26; described separately.
CUSTOMERS: Delivered to 14 Chinese domestic airlines and armed services, including two or more Y-7s for PLA Naval Air Force. Chinese Y-7 operators at October 1992 included Changan Airlines (one), Guizhou Airlines (two), Wuhan Airlines (two) and Zhong Yuan Airlines (two); operators of Y7-100 included Air China (six), Air Great Wall (three), China Eastern (10), China General Aviation (three), China Northern (11), China Northwest (seven), China Southern (five), China Southwest (five) and Sichuan Airlines (four). Export customers include Laos Airlines (two Y7-100C). Total of 62 Y-7s and 54 or more Y7-100s sold by May 1993, of which 83 delivered.
COSTS: Y-7 $5 million, Y7-100 $6 million (September 1990).
DESIGN FEATURES: Non-swept high-mounted wings, with 2° 12' 12" anhedral on tapered outer panels; wingtip winglets standard on Y7-100, being retrofitted on Y-7; basically circular-section fuselage; sweptback fin and rudder; 9° dihedral on tailplane; ventral fin under tailcone. Wing sweepback 6° 50' at quarter-chord of outer panels; incidence 3°.

XAC Y-7 in PLA Air Force insignia

FLYING CONTROLS: Primary surfaces mechanical/manual (ailerons mass balanced); servo tab and trim tab in each aileron, balance tab in each elevator, trim tab and spring tab in rudder. Hydraulically operated Fowler flaps on wing trailing-edges, single-slotted inboard of nacelles, double-slotted outboard.
STRUCTURE: Conventional all-metal, with two-spar wing box and bonded/welded fuselage; glassfibre aileron trim tabs.
LANDING GEAR: Retractable tricycle type with twin wheels on all units. Hydraulic actuation, with emergency gravity extension. All units retract forward. Mainwheels are size 900 × 300 mm, tyre pressure 5.39-5.88 bars (78.2-85.3 lb/sq in); nosewheels size 700 × 250 mm, tyre pressure 3.92 bars (56.8 lb/sq in). (Mainwheel tyre pressures variable to cater for different types of runway.) Disc brakes on mainwheels; hydraulically steerable (±45°) and castoring nosewheel unit.
POWER PLANT: Two Dongan (DEMC) WJ5A I turboprops, each rated at 2,080 kW (2,790 shp) for T-O and 1,976 kW (2,650 shp) at ISA + 23°C; Baoding J16-G10A four-blade constant-speed fully feathering propellers. Fuel in integral wing tanks immediately outboard of nacelles, and four bag-type tanks in centre-section, total capacity 5,550 litres (1,466 US gallons; 1,220 Imp gallons). Provision for four additional tanks in centre-section. Pressure refuelling point in starboard engine nacelle; gravity fuelling point above each tank. One 8.83 kN (1,984 lb st) RU 19A-300 auxiliary turbojet in starboard engine nacelle for engine starting, to improve take-off and in-flight performance, and to reduce stability and handling problems if one turboprop fails in flight.
ACCOMMODATION: Crew of three on flight deck, plus one or two cabin attendants. Standard layout has four-abreast seating, with centre aisle, for 52 passengers in air-conditioned, soundproofed (by Tracor) and pressurised cabin. Galley (by Lermer) and toilet at rear on starboard side. Baggage compartments forward and aft of passenger cabin, plus overhead stowage bins in cabin. Alternative layouts available for 48 or 50 passengers, or 20-passenger executive interior. Passenger airstair door on port side at rear of cabin. Doors to forward and rear

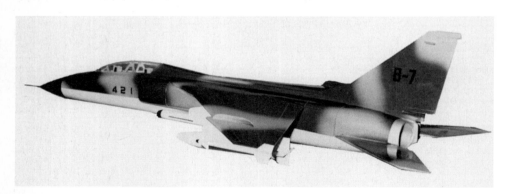

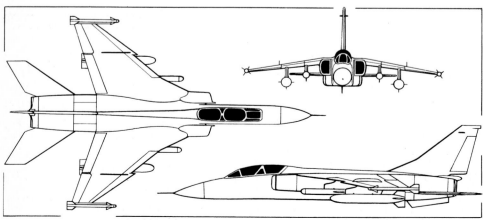

Provisional three-view *(Dennis Punnett)* and model of XAC JH-7 in maritime attack form

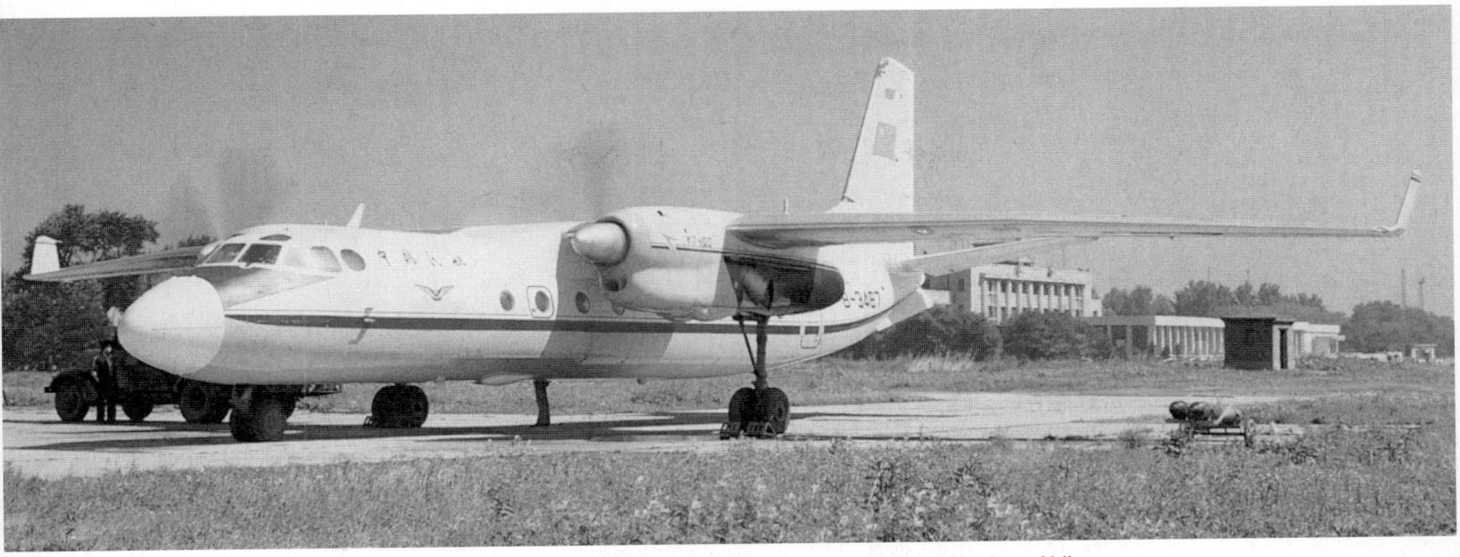

Xian Aircraft Company Y7-100 twin-turboprop transport of China Northern Airlines

baggage compartments on starboard side. All doors open inward. Electric windscreen de-icing.

SYSTEMS: Hamilton Standard environmental control system (cabin pressure differential in An-24RV is 0.29 bar; 4.27 lb/sq in). Puritan-Bennett passenger oxygen system optional. Main and emergency hydraulic systems, pressure 152 bars (2,200 lb/sq in), for landing gear actuation, nosewheel steering and braking, flaps, brakes, windscreen wipers and propeller feathering. Electrical system (28.5V DC) powered by two 18 kW QF-18 starter/generators and two 24V 28Ah 12HK-28 batteries; AC power (115V at 400Hz) supplied by two JF-30A 16kVA generators, with two 3kVA static inverters for secondary AC power (115/26V at 400Hz).

AVIONICS: Standard communications equipment comprises Collins 618M-3 dual VHF, Collins 628T-3 single HF, Becker audio selection and intercom, and Sundstrand AV-557C cockpit voice recorder. Standard navigation equipment comprises dual ADI-84A, dual EHSI-74 electronic HSI, dual RMI-36, FGS-65 flight guidance system, dual 51RV-4B VOR/ILS, dual DME-42, dual DF-206 ADF, 860F-4 radio altimeter, 621A-6A ATC transponder, 51Z-4 marker beacon receiver and CWS-80 instrument warning system, all by Collins; Litton LTN-211 VLF/Omega navigation system; Honeywell MHRS dual compass system, VG-311 dual attitude reference and Primus 90 colour weather radar; IDC air data system; Sundstrand UFDR flight data recorder; XLG-2A stall warning system with stick shaker; and KJ-6A autopilot. Gables control units. Other instrumentation by Gould, IDC, SFENA and Smiths.

DIMENSIONS, EXTERNAL (Y7-100):

Wing span (over winglets)	29.666 m (97 ft 4 in)
Wing chord: at root	3.50 m (11 ft 5¾ in)
at tip	1.095 m (3 ft 7 in)
Wing aspect ratio	11.7
Length overall	23.708 m (77 ft 9½ in)
Height overall	8.553 m (28 ft 0¾ in)
Fuselage: Max width	2.90 m (9 ft 6¼ in)
Max depth	2.50 m (8 ft 2½ in)
Tailplane span	9.08 m (29 ft 9½ in)
Wheel track (c/l of shock struts)	7.90 m (25 ft 11 in)
Wheelbase	7.90 m (25 ft 11 in)
Passenger door (port, rear):	
Height	1.40 m (4 ft 7 in)
Width	0.75 m (2 ft 5½ in)
Height to sill	1.40 m (4 ft 7 in)
Baggage compartment door (starboard, fwd):	
Height	1.10 m (3 ft 7¼ in)
Width	1.20 m (3 ft 11¼ in)
Height to sill	1.30 m (4 ft 3 in)
Baggage compartment door (starboard, rear):	
Height	1.41 m (4 ft 7½ in)
Width	0.75 m (2 ft 5½ in)

DIMENSIONS, EXTERNAL (Y7-200B):

Wing span	29.20 m (95 ft 9½ in)
Length overall	24.448 m (80 ft 2½ in)
Height overall	8.548 m (28 ft 0½ in)
Tailplane span	9.996 m (32 ft 9¾ in)
Wheelbase	8.598 m (28 ft 2½ in)
Propeller diameter	3.968 m (13 ft 0¼ in)
Propeller fuselage clearance	0.692 m (2 ft 3¼ in)

DIMENSIONS, INTERNAL (Y7-100):

Cabin:	
Length, incl flight deck	10.50 m (34 ft 5½ in)
Max width	2.80 m (9 ft 2¼ in)
Max height	1.90 m (6 ft 2¾ in)
Volume	56.0 m^3 (1,978 cu ft)
Baggage compartment volume:	
fwd	4.50 m^3 (159 cu ft)
rear	6.70 m^3 (237 cu ft)

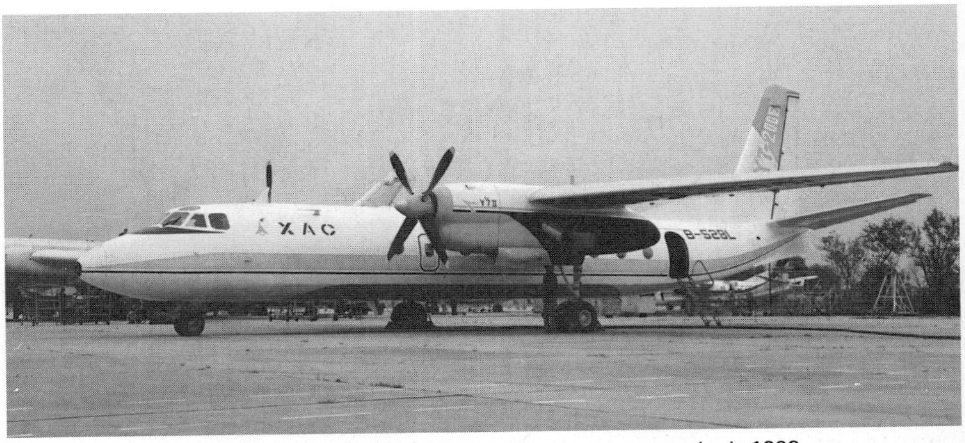

Prototype of the improved XAC Y7-200B, due to enter service in 1993

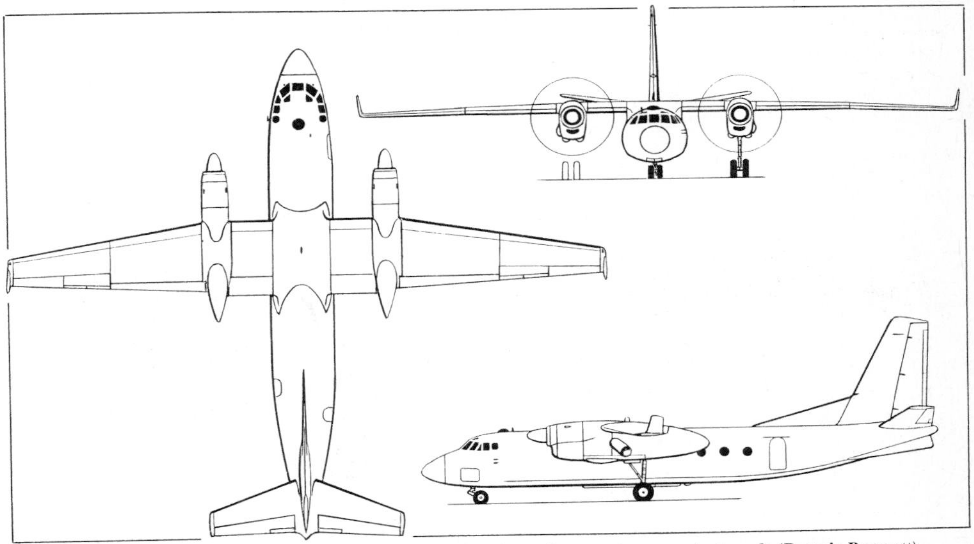

XAC Y7-100 current production version of Y-7 twin-turboprop transport aircraft *(Dennis Punnett)*

AREAS (Y7-100):

Wings, gross	75.26 m^2 (810.1 sq ft)
Vertical tail surfaces (total)	13.38 m^2 (144.0 sq ft)
Horizontal tail surfaces (total)	17.23 m^2 (185.5 sq ft)

WEIGHTS AND LOADINGS (Y7-100):

Operating weight empty	14,988 kg (33,042 lb)
Max fuel	4,790 kg (10,560 lb)
Max payload	5,500 kg (12,125 lb)
Max T-O and landing weight	21,800 kg (48,060 lb)
Max zero-fuel weight	19,655 kg (43,332 lb)
Max wing loading	289.7 kg/m^2 (59.34 lb/sq ft)
Max power loading	5.24 kg/kW (8.61 lb/shp)

PERFORMANCE (Y7-100, at max T-O weight except where indicated):

Max level speed	271 knots (503 km/h; 313 mph)
Max cruising speed at 4,000 m (13,125 ft)	262 knots (486 km/h; 302 mph)
Econ cruising speed at 6,000 m (19,685 ft)	233 knots (432 km/h; 268 mph)
Max rate of climb at S/L, AUW of 21,000 kg (46,297 lb)	458 m (1,504 ft)/min
Service ceiling, AUW of 21,000 kg (46,297 lb)	8,750 m (28,700 ft)
Service ceiling, OEI, AUW of 19,000 kg (41,887 lb)	3,850 m (12,630 ft)
T-O run at S/L, FAR Pt 25, AUW of 21,000 kg (46,297 lb):	
ISA	546 m (1,792 ft)
ISA + 20°C	1,398 m (4,590 ft)
Landing run, AUW of 21,000 kg (46,297 lb)	620 m (2,035 ft)
Range: max (52-passenger) payload	491 nm (910 km; 565 miles)
max standard fuel	1,070 nm (1,983 km; 1,232 miles)
standard and auxiliary fuel	1,296 nm (2,403 km; 1,493 miles)

XAC Y7H

TYPE: Military (Y7H) and civil cargo (Y7H-500) transport.
PROGRAMME: First flight of prototype late 1988 (originally known as Y-14-100); domestic civil certification due first half of 1993; in production.
CURRENT VERSIONS: **Y7H**: Military transport.
Y7H-500: Civil cargo version; *detailed description applies to this model except where indicated.*
DESIGN FEATURES: Chinese derivative of Antonov An-26 (see Ukraine section), with winglet modification of Y7-100, rear-loading underfuselage ramp/door, WJ5A I engines and rough-field landing gear.
FLYING CONTROLS: Generally as for An-26/Y7-100.
STRUCTURE: Generally as for An-26/Y7-100; commonality with latter includes wings and forward fuselage; flight deck windows enlarged compared with An-26.
LANDING GEAR: Mainwheel tyres size 1,050 × 400 mm; nose-wheel tyres as for Y7-100.
POWER PLANT: Two 2,081 kW (2,790 shp) Dongan (DEMC) WJ5A I(M) turboprops, each driving a Baoding J16-G10A four-blade constant-speed metal propeller. RU 19A-300 turbojet APU (max thrust 8.83 kN; 1,984 lb) for take-off assistance in hot and high conditions.
ACCOMMODATION: Flight crew of three (pilot, co-pilot and flight engineer). Up to 38 fully equipped troops or 39 paratroops in Y7H, or 24 stretcher cases and one medical attendant in medevac role.
AVIONICS: Modern navaids include AHRS and autopilot; nose mounted weather radar.
EQUIPMENT: Up to 2,000 kg (4,409 lb) of external stores, such as supply containers or weapons, on fuselage attachments.

DIMENSIONS, EXTERNAL:
Wing span	29.20 m (95 ft 9½ in)
Length overall	23.98 m (78 ft 8 in)
Height overall	8.89 m (29 ft 2 in)
Tailplane span	10.01 m (32 ft 10 in)
Wheel track (c/l of shock struts)	7.90 m (25 ft 11 in)
Wheelbase	8.036 m (26 ft 4½ in)
Propeller diameter	3.90 m (12 ft 9½ in)
Propeller fuselage clearance	0.72 m (2 ft 4¼ in)
Crew door: Height	1.40 m (4 ft 7 in)
Width	0.60 m (1 ft 11½ in)
Rear ramp/door: Max width	2.40 m (7 ft 10½ in)
Max length	3.05 m (10 ft 0 in)
Height to sill	1.74 m (5 ft 8½ in)

DIMENSIONS, INTERNAL:
Cargo hold: Length	11.43 m (34 ft 2½ in)
Width	2.78 m (9 ft 1½ in)
Height	1.91 m (6 ft 3¾ in)

AREAS:
Wings, gross	74.98 m² (807.08 sq ft)

WEIGHTS AND LOADINGS:
Operating weight empty	15,400 kg (33,950 lb)
Max fuel weight	5,500 kg (12,125 lb)
Max payload	5,500 kg (12,125 lb)
Max ramp weight	24,230 kg (53,420 lb)
Max T-O and landing weight	24,000 kg (52,910 lb)
Max wing loading	320.08 kg/m² (65.56 lb/sq ft)
Max power loading	5.55 kg/kW (9.12 lb/shp)

PERFORMANCE (at max T-O weight except where indicated):
Max level speed at 6,000 m (19,685 ft) at AUW of 22,500 kg (49,600 lb)	240 knots (445 km/h; 276 mph)
Max cruising speed at 6,000 m (19,685 ft)	236 knots (438 km/h; 272 mph)
Max rate of climb at S/L	480 m (1,575 ft)/min
Rate of climb at S/L, OEI	144 m (472 ft)/min
Service ceiling	8,200 m (26,900 ft)
Service ceiling, OEI	3,800 m (12,465 ft)
T-O run	857 m (2,812 ft)
Landing run	634 m (2,080 ft)
Range at 236 knots (438 km/h; 272 mph) at 6,000 m (19,685 ft):	
with max payload	560 nm (1,038 km; 645 miles)
with 3,300 kg (7,275 lb) payload	1,187 nm (2,200 km; 1,367 miles)
Max endurance	5 h 23 min

XAC Y7H-500 medium-range civil cargo transport

Rear-loading ramp/door of the Y7H-500

XAC Y-7 DERIVATIVE

TYPE: Design study for 60/64-seat regional transport.
PROGRAMME: In concept design phase late 1991; alternative configurations under study include high-wing twin resembling Y-7 and low-wing/dihedral type similar to Saab 340.
DESIGN FEATURES: General parameters include twin turboprops, empty weight of approx 21,200 kg (46,740 lb), range 700 nm (1,300 km; 808 miles) or 863 nm (1,600 km; 994 miles) with 60 and 56 passengers respectively; would conform to FAR Pt 25 and have Western power plant and avionics.
POWER PLANT: Pratt & Whitney Canada PW120 series and General Electric CT7-11 under consideration.

ZLAC

ZHONGMENG LIGHT AIRCRAFT CORPORATION

Wenquan, Hainan
Construction of airfield and light aircraft manufacturing facility begun Spring 1992 and planned for completion by 1995. Initial product stated to be AD-200 (development of Nanjing AD-100 described in 1987-88 *Jane's*), of advanced all-composites construction, with planned development into series including an AD-300 and AD-600. Also to produce latest (Mifeng-11) in BUAA designed microlight range. Intention is to export 90 per cent of output.

COLOMBIA

AGRO-COPTEROS

AGRO-COPTEROS LTDA

Calle 20 N 8A-18, Apartado Aéreo 1789, Cali
Telephone: 57 (3) 825110, 833519
Telex: 51138 DIEGO CO

PRESIDENT: Maximo Tedesco Kappler
Assembles North American light aircraft and rotorcraft kits, particularly for agricultural work; produced sprayplane version of the Aerosport Scamp B (see 1984-85 *Jane's*). Latest known product is agricultural version of Zenair STOL CH 701 side by side two-seat Experimental category aircraft (see Canadian section); power plant is 47.7 kW (64 hp) Rotax 532; Micro Ag dispersal system.

AICSA

AERO INDUSTRIAL COLOMBIANA SA

Aeropuerto Guaymaral, Bogotá
GENERAL MANAGER: Pedro Alberto Gil
CIAC: Entrada No. 1, Aeropuerto Eldorado, Bogotá

MANAGER: Maj Gen (retd) Alberto Guzmán M.
Has assembled Piper aircraft from kits for sale in Colombia and some Andes Pact countries since 1968 (278 single-engined and 240 twin-engined by 31 December 1991), plus two PZL Mielec (Antonov) An-2s under agreement with Pezetel. No information on subsequent production received.
AICSA is 51 per cent owned by Colombian Air Force through its subsidiary CIAC (Corporación de la Industria Aeronáutica Colombiana), which planned to merge AICSA into its own organisation.

AVIONES DE COLOMBIA

AVIONES DE COLOMBIA SA

Entrada No. 1, Aeropuerto Eldorado, Bogotá
Telephone: 57 (1) 413 8300
Fax: 57 (1) 413 8075
Telex: 45 220
WORKS: Aeropuerto Guaymaral, Apartado Aéreo 6876, Bogotá
Telephone: 57 (1) 676 0478 and 0101
Fax: 57 (1) 676 0458
GENERAL MANAGER: Rafael Urdaneta
SALES MANAGER: Mauricio Pinzon
Established as Urdaneta y Galvez Ltda in 1950s; a South American distributor for Cessna since 1961; began assembling and partly building selected Cessna aircraft in 1969; now manufactures complete airframes.
Facilities include fixed-base operation at Eldorado International Airport and 13,935 m² (150,000 sq ft) assembly plant and FAA-approved service station at Bogotá Guaymaral; workforce on 31 January 1990 was 170; total 1,047 Cessnas assembled by that date.

AVIONES DE COLOMBIA/CESSNA AGTRAINER

TYPE: Side by side two-seat agricultural pilot trainer, modified by Aviones de Colombia from Cessna Model 188 Ag Truck (see US section of 1984-85 *Jane's*).
PROGRAMME: Two prototypes; first flight 16 September 1976.
CUSTOMERS: First aircraft operated by Aeroandes cropspraying flying school; of eight Agtrainers produced by 1990, four operating in Colombia, three in Central America and one in Ecuador.
DESIGN FEATURES: As Cessna Model 188 Ag Truck, but enlarged cockpit adds 91 kg (200 lb) to empty weight. Low wing, strut braced from above, one jury strut each side; wing section NACA 2412 modified; dihedral 9°; incidence 1° 30′ at root, −1° 30′ at tip; outer strut fairings shaped to form wing fence; tailplane abrasion boots.
FLYING CONTROLS: Mechanical, with fixed tailplane; trim tab in starboard elevator; Frise ailerons with sealed gaps; single-slotted flaps.
STRUCTURE: All-metal; front fuselage steel tube with removable sheet skin panels; stressed skin aft of cockpit.
LANDING GEAR: Non-retractable tailwheel type. Land-O-Matic cantilever main legs of heavy duty spring steel. Tapered tubular tailwheel spring shock absorber. Mainwheel tyres size 22 × 8.00-8, 6-ply rating, pressure 2.41 bars (35 lb/sq in). Oversize tyres optional, size 8.50-10, 6-ply rating, pressure 1.72 bars (25 lb/sq in). Tailwheel tyre size 3.50-10, 4-ply rating, pressure 3.45-4.14 bars (50-60 lb/sq in). Steerable tailwheel. Hydraulic disc brakes and parking brake.
POWER PLANT: One 224 kW (300 hp) Teledyne Continental IO-520-D flat-six engine, driving a McCauley three-blade constant-speed propeller. Fuel capacity 204 litres (54 US gallons; 45 Imp gallons). Oil capacity 11.4 litres (3 US gallons; 2.5 Imp gallons).
ACCOMMODATION: Side by side seats for two persons, in enclosed cabin with steel overturn structure. Combined window and door on each side, hinged at bottom. Ventilation standard. Air-conditioning, heating and windscreen defrosting optional.
SYSTEMS: Electrical system powered by a 28V 60A alternator and 24V 12.75Ah battery as standard. 28V 95A alternator and 24V 15.5Ah heavy duty battery optional.
EQUIPMENT: Standard equipment includes a 1,060 litre (280 US gallon; 233 Imp gallon) hopper with shatter-resistant window, engine driven hydraulic spray system and manually controlled spray valve and gearbox without agitator, hopper side loading system on port side, pilot's four-way adjustable seat, control stick lock, wire cutters, cable deflector, navigation lights, tailcone lift handles, quick drain oil valve, remote fuel strainer drain control, and auxiliary fuel pump.

DIMENSIONS, EXTERNAL:
Wing span	12.70 m (41 ft 8 in)
Length overall	8.00 m (26 ft 3 in)
Height overall	2.44 m (8 ft 0 in)
Propeller diameter	2.03 m (6 ft 8 in)

DIMENSIONS, INTERNAL:
Cabin: Max width	1.09 m (3 ft 7 in)
Hopper volume	0.85 m³ (30.0 cu ft)

AREAS:
Wings, gross	19.05 m² (205.0 sq ft)

WEIGHTS AND LOADINGS:
Weight empty:	
without dispersal equipment	1,017 kg (2,242 lb)
with dispersal equipment	1,099 kg (2,424 lb)
Max T-O weight:	
Normal category	1,497 kg (3,300 lb)
Restricted category	1,905 kg (4,200 lb)
Max landing weight	1,497 kg (3,300 lb)
Max wing loading: Normal	78.55 kg/m² (16.10 lb/sq ft)
Restricted	99.98 kg/m² (20.49 lb/sq ft)
Max power loading: Normal	6.69 kg/kW (11.0 lb/hp)
Restricted	8.52 kg/kW (14.0 lb/hp)

PERFORMANCE (at max T-O weight):
Max level speed at S/L	105 knots (195 km/h; 121 mph)
Max cruising speed (75% power) at 1,980 m (6,500 ft)	98 knots (182 km/h; 113 mph)
Stalling speed, power off:	
flaps up	53 knots (98 km/h; 61 mph) IAS
flaps down	50 knots (92 km/h; 57 mph) IAS
Max rate of climb at S/L	210 m (690 ft)/min
Service ceiling	3,385 m (11,100 ft)
T-O run	207 m (680 ft)
T-O to 15 m (50 ft)	332 m (1,090 ft)
Landing from 15 m (50 ft)	386 m (1,265 ft)
Landing run	128 m (420 ft)
Range, reserves for start, taxi, T-O and 45 min at 45% power	256 nm (474 km; 295 miles)
Endurance, conditions as above	2 h 36 min

AVIONES DE COLOMBIA AC-05 PIJAO

TYPE: Single-seat agricultural aircraft.
PROGRAMME: Design began January 1988 and construction November 1989; prototype (HK-3631-X) first flight 10 April 1991; certificated by Colombian DAAC 5 July 1991; first production examples under construction early 1992.
COSTS: $185,000 (1992).
DESIGN FEATURES: Typical braced, unswept low-wing agricultural configuration; turned-down wingtips; STOL capability; designed for easy on-site cleaning and maintenance, even in remote areas. Wing section NACA 2412; dihedral 9°; incidence 1° 30′ at root, −1° 30′ at tip.
FLYING CONTROLS: Conventional three-axis primary surfaces, with trimmable rudder and elevators; wing flaps.
STRUCTURE: All-metal, with 4130 steel tube cabin structure, 2024-ST aluminium alloy stressed skin fuselage and two-spar wings; riveted stressed skin.
LANDING GEAR: Non-retractable cantilever self-sprung mainwheel legs, each with Cleveland 40-101 wheel and Goodyear 8.50-10-8 tyre (pressure 1.72 bars; 25 lb/sq in); Cleveland D-303380 tailwheel with Goodyear 3.5-10-4 tyre (pressure 4.14 bars; 60 lb/sq in). Cleveland 30-67C hydraulic mainwheel brakes.
POWER PLANT: One Teledyne Continental IO-520-D flat-six engine, rated at 224 kW (300 hp) at 2,850 rpm for T-O, 212.5 kW (285 hp) at 2,700 rpm max continuous; McCauley D2A34C98-N two-blade constant-speed propeller. Fuel in two wing tanks, total capacity 204 litres (54 US gallons; 45 Imp gallons). Gravity refuelling point for each tank. Oil capacity 11.4 litres (3 US gallons; 2.5 Imp gallons).
ACCOMMODATION: Pilot only. Door on each side of cockpit, with emergency quick-release. Cockpit ventilated.
SYSTEMS: 12V DC battery for electrical power. Hydraulic power for mainwheel brakes and dispersal system.
EQUIPMENT: Hopper for chemical, capacity 1,060 litres (280 US gallons; 233 Imp gallons).

DIMENSIONS, EXTERNAL:
Wing span	12.63 m (41 ft 5¼ in)
Wing chord: at root	1.63 m (5 ft 4 in)
at tip	1.12 m (3 ft 8 in)
Wing aspect ratio	8.05
Length overall	8.09 m (26 ft 6½ in)
Fuselage: Length	6.63 m (21 ft 9¼ in)
Max width	1.44 m (4 ft 8¾ in)
Height over tail fin	2.49 m (8 ft 2 in)
Tailplane span	3.35 m (11 ft 0 in)
Wheel track	3.70 m (12 ft 1½ in)
Propeller diameter	2.18 m (7 ft 2 in)
Propeller ground clearance	0.41 m (1 ft 4¼ in)
Doors (two, each): Height	0.68 m (2 ft 2¾ in)
Width	0.88 m (2 ft 10½ in)
Height to sill	0.68 m (2 ft 2¾ in)

Aviones de Colombia AgTrainer two-seat agricultural training aircraft

Two views of the prototype Aviones de Colombia AC-05 Pijao

DIMENSIONS, INTERNAL:	
Cockpit: Length	1.16 m (3 ft 9½ in)
Max width	0.77 m (2 ft 6¼ in)
Max height	1.24 m (4 ft 0¾ in)
Floor area	0.36 m² (3.86 sq ft)
Volume	1.26 m³ (44.6 cu ft)
Hopper volume	1.06 m³ (37.4 cu ft)
AREAS:	
Wings, gross	19.48 m² (209.66 sq ft)
Ailerons (total)	2.58 m² (27.74 sq ft)
Trailing-edge flaps (total)	1.97 m² (21.20 sq ft)
Rudder	0.84 m² (9.02 sq ft)
Tailplane	1.95 m² (20.94 sq ft)
Elevators (total)	1.41 m² (15.13 sq ft)

WEIGHTS AND LOADINGS (Restricted category):	
Operating weight empty	989 kg (2,180 lb)
Max fuel weight	147 kg (324 lb)
Max payload	635 kg (1,400 lb)
Max T-O weight	1,905 kg (4,200 lb)
Max landing weight	1,497 kg (3,300 lb)
Max wing loading	97.80 kg/m² (20.03 lb/sq ft)
Max power loading	8.50 kg/kW (14.0 lb/hp)

PERFORMANCE (at Restricted max T-O weight, ISA):
Never-exceed speed (V_{NE})	105 knots (194 km/h; 121 mph)
Cruising speed at 1,980 m (6,500 ft), 75% power	102 knots (188 km/h; 117 mph)
Stalling speed, power off:	
flaps up	58 knots (107 km/h; 66 mph) CAS
flaps down	48 knots (89 km/h; 55 mph) CAS
Max rate of climb at S/L	149 m (490 ft)/min
Service ceiling	2,470 m (8,100 ft)
T-O run	295 m (966 ft)
T-O to 15 m (50 ft)	596 m (1,955 ft)
Landing from 15 m (50 ft)	327 m (1,070 ft)
Landing run	128 m (420 ft)
Range with max usable fuel, max cruising speed (75% power) at 1,980 m (6,500 ft), recommended lean mixture and allowances for engine start, taxi, take-off, climb and 30 min reserves	252 nm (466 km; 290 miles)
Endurance, conditions as above	2 h 36 min

GAVILÁN
EL GAVILÁN SA
Carrera 3a No. 56-19, Apartado Aéreo 6781, Santa Fé de Bogotá
Telephone: 57 (1) 211 8100
Fax: 57 (1) 212 8952
Telex: 44 581 LAVE CO
GENERAL MANAGER: Eric C. Leaver

Formerly Aero Mercantil (see 1991-92 and earlier *Jane's*), associated with Piper Aircraft Corporation as dealer and then distributor since 1952; sold its shares in AICSA (which see) and in 1991 formed new Gavilán (Sparrowhawk) company to pursue development and manufacture of own-design EL-1 utility aircraft.

GAVILÁN EL-1 MODEL 358
TYPE: Seven-passenger utility light transport.
PROGRAMME: Launched March 1986; construction began May 1987; first flight (HK-3500-Z) 27 April 1990; after 50 hours' flying, fuselage stretched 0.305 m (1 ft 0 in) immediately forward of main spar and gross weight increased by 136 kg (300 lb); new first flight 7 November 1990; wing incidence increased 1991 to improve cruising speed, and passenger windows reshaped; damage from emergency landing in USA in February 1992 has delayed progress slightly, but company has now decided to obtain FAA (FAR Pt 23) certification from outset; this to be gained by second prototype being built by General Aviation Technical Services of Lock Haven, Pennsylvania, USA; expected by end of 1994 for first deliveries in mid-1995.
COSTS: Standard civil aircraft, VFR equipped $270,000 (1993).
DESIGN FEATURES: Constant chord, unswept braced wing of NACA 4412 section; dihedral 2°; incidence 5° 30'.
FLYING CONTROLS: All-mechanical (cables), with fixed tailplane; mass balanced ailerons on piano hinges; elevators and rudder horn balanced; trim tab with dual actuators in starboard elevator; single-slotted flaps on offset hinges.
STRUCTURE: All-metal, mainly of 2024-T3 aluminium alloy sheet; fuselage frame of 4130N steel tube skinned with 2024-T3; two-spar wing with single strut each side; two-spar fin.
LANDING GEAR: Non-retractable tricycle type, with elastomeric shock absorption and single wheel on each unit. Tyre sizes 700 × 6-6 (main) and 600 × 6-6 (nose). Cleveland hydraulic mainwheel brakes. Float/ski options to be offered later.
POWER PLANT: One Textron Lycoming TIO-540-W2A flat-six engine (261 kW; 350 hp at 2,700 rpm), driving a three-blade constant-speed Hartzell propeller. Fuel tank in each wing, combined capacity 454.2 litres (120 US gallons; 99.9 Imp gallons). Refuelling point in top of each tank. Oil capacity 11.4 litres (3 US gallons; 2.5 Imp gallons).
ACCOMMODATION: Pilot and co-pilot or one passenger at front. Two rows of three seats to rear of these, facing each other. Door at front on each side, plus larger double door at rear on port side.
SYSTEMS: Engine mounted vacuum pump for gyro instruments, driven at 0.37 bar (5.5 lb/sq in). 12V 70Ah battery. Hydraulic system for brakes only. Gaseous oxygen system optional.
AVIONICS: Standard VFR package includes Bendix/King KX 155 VHF nav/com radio and KR 87 ADF.

DIMENSIONS, EXTERNAL:	
Wing span	12.80 m (42 ft 0 in)
Wing chord, constant	1.55 m (5 ft 1 in)
Wing aspect ratio	7.84
Length overall	9.53 m (31 ft 3 in)
Fuselage: Max width	1.42 m (4 ft 8 in)
Height overall	3.35 m (11 ft 0 in)
Tailplane span	3.10 m (10 ft 2 in)
Wheel track	3.35 m (11 ft 0 in)
Wheelbase	3.35 m (11 ft 0 in)
Propeller diameter	2.13 m (7 ft 0 in)
Propeller ground clearance	0.36 m (1 ft 2 in)
Pilot's door: Height	1.17 m (3 ft 10 in)
Max width	0.86 m (2 ft 10 in)
Height to sill	1.02 m (3 ft 4 in)
Co-pilot's door: Height	1.17 m (3 ft 10 in)
Max width	0.66 m (2 ft 2 in)
Height to sill	1.02 m (3 ft 4 in)
Passenger door: Height	1.17 m (3 ft 10 in)
Width	1.24 m (4 ft 1 in)
Height to sill	1.02 m (3 ft 4 in)

DIMENSIONS, INTERNAL:	
Cabin: Length	3.40 m (11 ft 2 in)
Max width	1.37 m (4 ft 6 in)
Max height	1.37 m (4 ft 6 in)
Floor area	4.09 m² (44.0 sq ft)
Volume	5.38 m³ (190.0 cu ft)

AREAS:	
Wings, gross	18.95 m² (204.0 sq ft)
Ailerons (total)	0.97 m² (10.4 sq ft)
Trailing-edge flaps (total)	1.12 m² (12.0 sq ft)
Fin	2.16 m² (23.3 sq ft)
Tailplane	1.51 m² (16.3 sq ft)
Elevators (total, incl tab)	0.90 m² (9.7 sq ft)

WEIGHTS AND LOADINGS:	
Weight empty, equipped	1,270 kg (2,800 lb)
Max fuel	326 kg (720 lb)
Max T-O and landing weight	2,041 kg (4,500 lb)
Max wing loading	107.65 kg/m² (22.06 lb/sq ft)
Max power loading	7.82 kg/kW (12.86 lb/hp)

PERFORMANCE (at max T-O weight):
Never-exceed speed (V_{NE})	204 knots (378 km/h; 235 mph)
Max level speed at 4,575 m (15,000 ft)	140 knots (259 km/h; 161 mph)
Max cruising speed at 4,575 m (15,000 ft)	135 knots (250 km/h; 155 mph)
Econ cruising speed at 4,575 m (15,000 ft)	128 knots (237 km/h; 147 mph)
Stalling speed, 40° flap, engine idling	58 knots (108 km/h; 67 mph)
Max rate of climb at S/L	274 m (900 ft)/min
Service ceiling	7,010 m (23,000 ft)
T-O run	275 m (900 ft)
T-O to 15 m (50 ft)	457 m (1,500 ft)
Landing from 15 m (50 ft)	366 m (1,200 ft)
Landing run	183 m (600 ft)
Range, no reserves:	
with max payload	550 nm (1,019 km; 633 miles)
with max fuel	800 nm (1,482 km; 921 miles)

Gavilán EL-1 Model 358 prototype in its current configuration

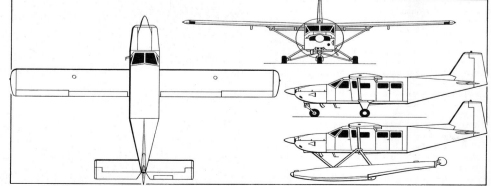

Gavilán EL-1 seven-passenger utility transport, with additional side view of future twin-float version *(Jane's/Mike Keep)*

CZECH REPUBLIC

AERO
AERO CZECH AND SLOVAK AEROSPACE INDUSTRY LTD
Beranových 130, CR-199 04 Prague 9
Telephone: 42 (2) 6838220
Fax: 42 (2) 886581
PRESIDENT AND CEO: Ing Zdeněk Pernica
COMMERCIAL DIRECTOR: Ing Vladimír Plšek

TECHNICAL DIRECTOR: Dipl Ing Jan Bartoň
MARKETING COMMUNICATIONS: Ing Zdeněk Burian

This new joint stockholding management group replaced state owned Aero Concern of Czechoslovak Aerospace Industry on 1 December 1990; airframe, engine and equipment factories, and research centres, became limited companies on 1 January 1991. Joint stock company was partly privatised (39.4 per cent) in 1992, basically changing current role of Aero Ltd from an operational holding to a financial one (100 per cent ownership of 11 subsidiaries). Activities accordingly now comprise organisation, co-ordination and financing of research, development, production and sale of aircraft and other aviation products. Share in subsidiaries scheduled to decrease during 1993 by sales to various foreign and domestic partners.

OMNIPOL
OMNIPOL COMPANY LIMITED
Nekázanka 11, CR-112 21 Prague 1
Telephone: 42 (2) 21401111
Fax: 42 (2) 21402241
Telex: 121297 or 121299 OMPO C

GENERAL DIRECTOR: Ing Stanislav Kožený
COMMERCIAL DIRECTOR: Ing Josef Stibor
PUBLICITY MANAGER: Dipl Ing Pavel Štícha

Omnipol handles export sales of some, though not all, products of the Czech and Slovak aerospace industry, and deals with customer enquiries.

AERO
AERO VODOCHODY AKCIOVÁ SPOLEČNOST (Aero Vodochody Aeronautical Works Ltd)
CR-250 70 Odolena Voda
Telephone: 42 (2) 843641
Fax: 42 (2) 823172 and 6872247
Telex: 121169 AERO C
MARKETING MANAGER: Ing Miloš Vališ

Factory established 1 July 1953; produced about 3,600 L-29 Delfin jet trainers between 1963 and 1974.

AERO L-39 ALBATROS

TYPE: Two-seat jet trainer with armed and combat variants.
PROGRAMME: First flight 4 November 1968; 10 pre-production aircraft from 1971; selected to succeed L-29 as standard trainer for USSR, Czechoslovakia and former East Germany 1972; production started 1972; service trials in USSR and Czechoslovakia 1973; in service with CzAF 1974.
CURRENT VERSIONS: **L-39 C**: Initial pilot trainer, with two underwing stations; in production. *Detailed description applies to this version except where indicated.*
L-39 V: Target towing version for Czech use; eight only.
L-39 Z0 (Z Zbrojní: armed): Reinforced wings with four underwing stations; first flight (X-09) 25 August 1975. Production completed; weight, performance and other data in 1991-92 and earlier *Jane's*.
L-39 ZA: Ground attack and reconnaissance version of Z0; four underwing stations and centreline gun pod; reinforced wings and landing gear; prototypes (X-10 and X-11) flown 1975-76; in production. Export demonstrator completed with Western avionics (HUD, mission computer, Bendix/King avionics and navigation equipment) as **L-39 ZA/MP** (for multi-purpose). Version for Thailand designated **ZE** (for Elbit avionics).
L-59: Formerly L-39 MS; new advanced training version with more powerful engine and new avionics; described separately.
L-139: Trainer powered by 18.15 kN (4,080 lb st) Garrett TFE731-4 under preliminary Aero/Garrett agreement signed June 1991; Bendix/King avionics; first flight 10 May 1993, production 1994.
CUSTOMERS: More than 2,800 produced by 1 January 1993 for Czechoslovakia (36 L-39 C, eight V and 32 ZA) and former Soviet Union (2,094 L-39 C), plus air forces of Afghanistan (12 Cs), Algeria (32 ZAs), Bulgaria (55 ZAs), Cuba (30 Cs), Ethiopia (20 Cs), former East Germany (52 Z0s), Iraq (81 Z0s), Libya (181 Z0s, of which 10 transferred to Egypt), Nigeria (51 ZAs), Romania (32 ZAs), Syria (55 Z0s and 44 ZAs) and Vietnam (24 Cs). Royal Thai Air Force has ordered 36 ZEs with Elbit avionics (deliveries to begin second half of 1993); Philippine Air Force planning to buy 18 (version not known, but possibly also ZE); Lithuanian Air Force reported to have acquired four Cs from former Soviet republic of Kyrgyzstan.
DESIGN FEATURES: Tandem two-seater with ejection seats and pressurisation; fixed tip tanks also contain navigation/landing lights; rear fuselage and tail, retained by five bolts, allow easy removal for access to engine; tapered wing has NACA 64A012 Mod.5 section; quarter-chord sweepback 1° 45′; dihedral 2° 30′ from roots; incidence 2°.
FLYING CONTROLS: Mechanical by pushrods; fixed tailplane; electrically actuated trim tab in each elevator; servo tab in rudder; mass balanced ailerons with electrically actuated servo tabs; port tab also electrically actuated for trim; two airbrake panels under fuselage just forward of wing leading-edge, operated by single hydraulic jack, extend automatically as speed approaches Mach 0.8; double-slotted flaps extended by rods from single hydraulic jack, retracting automatically as airspeed reaches 167 knots (310 km/h; 193 mph).
STRUCTURE: One-piece all-metal stressed skin wing with main and auxiliary spars; four-point wing attachment; fuselage ahead of engine break has three sections: nose containing avionics/battery/antennae/air and oxygen bottles/nose-wheel unit, next section forming pressure cabin, third section containing air intakes/fuel tanks/engine bay.
LANDING GEAR: Retractable tricycle type, with single wheel and oleo-pneumatic shock absorber on each unit; designed for touchdown sink rate of 3.4 m (11.15 ft)/s at AUW of 4,600 kg (10,141 lb). Retraction/extension operated hydraulically, with electrical control. All wheel well doors close automatically after wheels are lowered, to prevent ingress of dirt and debris. Mainwheels retract inward into wings (with automatic braking during retraction), nosewheel forward into fuselage. L-39 C has K24 mainwheels, with Barum tubeless tyres size 610 × 185 mm, and K25 castoring and self-centring nosewheel, with Barum tubeless tyre size 430 × 150 mm. L-39 ZA has K28 mainwheels with 610 × 215 mm tyres and K27 nosewheel with 450 × 165 mm tyre. Hydraulic disc brakes and anti-skid units on mainwheels; shimmy damper on nosewheel leg. Min ground turning radius (from nosewheel) 2.50 m (8 ft 2 in). Capable of operation from grass strips with bearing strength of 7 kg/cm² (99 lb/sq in) at up to 4,600 kg (10,141 lb) T-O weight, or from unprepared runways. L-39 ZA reinforced to cater for higher operating weights; at same T-O weight, this version can operate from grass strips with bearing strength of 6 kg/cm² (85 lb/sq in).

POWER PLANT: One 16.87 kN (3,792 lb st) Progress AI-25 TL turbofan in rear fuselage, with semi-circular lateral air intake, with splitter plate, on each side of fuselage above wing centre-section. Fuel in five rubber main bag tanks aft of cockpits, with combined capacity of 1,055 litres (279 US gallons; 232 Imp gallons), and two 100 litre (26.5 US gallon; 22 Imp gallon) non-jettisonable wingtip tanks. Total internal fuel capacity 1,255 litres (332 US gallons; 276 Imp gallons). Gravity refuelling points on top of fuselage and on each tip tank. Provision for two 350 litre (92.5 US gallon; 77 Imp gallon) drop tanks on inboard underwing pylons, increasing total overall fuel capacity to a maximum of 1,955 litres (517 US gallons; 430 Imp gallons). Fuel system permits up to 20 s of inverted flight.
ACCOMMODATION: Crew of two in tandem, on Czech VS-1-BRI rocket assisted ejection seats, operable at zero height and at speeds down to 81 knots (150 km/h; 94 mph); individual canopies hinge sideways to starboard and are jettisonable. Rear seat elevated. One-piece windscreen hinges forward for access to front instrument panel. Internal transparency between cockpits. Dual controls standard.
SYSTEMS: Cabin pressurised (standard differential 0.227 bar; 3.29 lb/sq in, max overpressure 0.29 bar; 4.20 lb/sq in) and air-conditioned, using engine bleed air and cooling unit.

Aero L-39 ZA of the Bulgarian Air Force *(Press Office Sturzenegger)*

Aero Vodochody's L-39 ZA/MP (multi-purpose) demonstrator, with HUD and Bendix/King avionics

64 CZECH REPUBLIC: AIRCRAFT—AERO

Air-conditioning system provides automatic temperature control from 10° to 25°C at ambient air temperatures from −55°C to +45°C. Main and standby interconnected hydraulic systems, main system having variable flow pump with operating pressure of 147 bars (2,133 lb/sq in) for actuation of landing gear, flaps, airbrakes, ram air turbine and (at 34.3 bars; 500 lb/sq in pressure) wheel brakes. Emergency system, for all of above except airbrakes, incorporates three accumulators. Pneumatic canopy seals supplied by a 2 litre compressed air bottle in nose (pressure 147 bars; 2,133 lb/sq in). Electrical system (27V DC) powered by 7.5kVA engine driven generator; if primary generator fails, V 910 ram air turbine extends automatically into airstream and generates up to 3kVA emergency power for essential services. 12V 28Ah SAM 28 lead-acid battery for standby power and APU starting. Two 800VA static inverters (first for radio equipment, ice warning lights, engine vibration measurement and air-conditioning, second for navigation and landing systems, IFF and air-to-air missiles) provide 115V single-phase AC power at 400Hz. Second circuit incorporates 500VA rotary inverter and 40VA static inverter for 36V three-phase AC power, also at 400Hz. Saphir 5 APU and SV-25 turbine for engine starting. Air intakes and windscreen anti-iced by engine bleed air; anti-icing normally sensor-activated automatically, but manual standby system also provided. Six-bottle oxygen system for crew, pressure 147 bars (2,133 lb/sq in).

AVIONICS: Standard avionics include R-832 M two-band com radio (VHF 118-140MHz, UHF 220-389MHz); SPU-9 crew intercom; RKL-41 ADF (150-1,800kHz); RV-5 radar altimeter; MRP-56 P/S marker beacon receiver; SRO-2M IFF; and RSBN-5S navigation and landing system. VOR/ILS system available at customer's option.

ARMAMENT (L-39 ZA): Underfuselage pod below front cockpit, housing a single 23 mm GSh-23 two-barrel cannon; ammunition (max 150 rds) housed in fuselage above gun pod. Gun/rocket firing and weapon release controls, including electrically controlled ASP-3 NMU-39 Z gyroscopic gunsight and FKP-2-2 gun camera, in front cockpit only. Four underwing hardpoints, inboard pair each stressed for up to 500 kg (1,102 lb) and outer pair for up to 250 kg (551 lb) each; max underwing stores load 1,100 kg (2,425 lb). Non-jettisonable pylons, each comprising a D3-57D stores rack. Typical underwing stores can include various combinations of bombs (two 500 kg, four 250 kg or six 100 kg); four UB-16-57 M pods each containing sixteen S-5 57 mm air-to-surface rockets; infra-red air-to-air missiles (outer pylons only); a five-camera day reconnaissance pod (port inboard pylon only); or (inboard stations only) two 350 litre (92.5 US gallon; 77 Imp gallon) drop tanks.

DIMENSIONS, EXTERNAL:
Wing span, incl tip tanks	9.46 m (31 ft 0½ in)
Wing chord (mean)	2.15 m (7 ft 0½ in)
Wing aspect ratio: geometric	4.4
incl tip tanks	5.2
Length overall	12.13 m (39 ft 9½ in)
Height overall	4.77 m (15 ft 7¾ in)
Tailplane span	4.40 m (14 ft 5 in)
Wheel track	2.44 m (8 ft 0 in)
Wheelbase	4.39 m (14 ft 4¼ in)

AREAS:
Wings, gross	18.80 m² (202.36 sq ft)
Ailerons (total)	1.23 m² (13.26 sq ft)
Trailing-edge flaps (total)	2.68 m² (28.89 sq ft)
Airbrakes (total)	0.50 m² (5.38 sq ft)
Fin	2.60 m² (27.99 sq ft)
Rudder, incl tab	0.91 m² (9.80 sq ft)
Tailplane	3.93 m² (42.30 sq ft)
Elevators, incl tabs	1.14 m² (12.27 sq ft)

WEIGHTS AND LOADINGS:
Weight empty, equipped: C	3,455 kg (7,617 lb)
ZA	3,565 kg (7,859 lb)
Fuel load: fuselage tanks	824 kg (1,816 lb)
wingtip tanks	156 kg (344 lb)
Max external stores load: C	284 kg (626 lb)
ZA	1,290 kg (2,844 lb)
T-O weight clean: C	4,525 kg (9,976 lb)
ZA	4,635 kg (10,218 lb)
Max T-O weight: C	4,700 kg (10,362 lb)
ZA	5,600 kg (12,346 lb)
Max wing loading: C	250.0 kg/m² (51.23 lb/sq ft)
ZA	297.9 kg/m² (61.01 lb/sq ft)
Max power loading: C	278.6 kg/kN (2.73 lb/lb st)
ZA	332.0 kg/kN (3.25 lb/lb st)

PERFORMANCE (C at clean T-O weight of 4,500 kg; 9,921 lb, ZA at max T-O weight, except where indicated):
Max limiting Mach number	0.80
Max level speed at S/L:	
C	378 knots (700 km/h; 435 mph)
ZA	329 knots (610 km/h; 379 mph)
Max level speed at 5,000 m (16,400 ft):	
C	405 knots (750 km/h; 466 mph)
ZA	340 knots (630 km/h; 391 mph)
Stalling speed: C	90 knots (165 km/h; 103 mph)
ZA	103 knots (190 km/h; 118 mph)
Max rate of climb at S/L: C	1,260 m (4,130 ft)/min
ZA	810 m (2,657 ft)/min
Time to 5,000 m (16,400 ft): C	5 min
ZA	10 min
Service ceiling: C	11,000 m (36,100 ft)
ZA	7,500 m (24,600 ft)
T-O run (concrete): C	530 m (1,740 ft)
ZA	970 m (3,182 ft)
Landing run (concrete): C	650 m (2,135 ft)
ZA	800 m (2,625 ft)

Range at 7,000 m (22,975 ft):
C, 980 kg (2,160 lb) max internal fuel
593 nm (1,100 km; 683 miles)
C, 1,524 kg (3,360 lb) max internal and external fuel
944 nm (1,750 km; 1,087 miles)

Endurance at 7,000 m (22,975 ft):
C, max internal fuel as above	2 h 30 min
C, max internal/external fuel as above	3 h 50 min

g limits:
operational: at 4,200 kg (9,259 lb) AUW	+8/−4
at 5,500 kg (12,125 lb) AUW	+5.2/−2.6
ultimate, at 4,200 kg (9,259 lb) AUW	+12

AERO L-59

TYPE: Developed version of L-39 jet trainer.
PROGRAMME: Originally known as L-39 MS; first flight of X-22 prototype (OK-184) 30 September 1986; two more prototypes (X-24, X-25) flown 26 June and 6 October 1987; first flight of production L-39 MS, 1 October 1989; first flight of prototype L-59 E April 1992; deliveries of L-59 E began (two aircraft) 29 January 1993.
CURRENT VERSIONS: **L-39 MS**: Initial production version for Czech and Slovak Air Force.
L-59 E: Production version for Egyptian Air Force, generally as L-39 MS but with Western avionics. *Detailed description applies to this version.*
CUSTOMERS: Czech and Slovak Air Force (six L-39 MS); Egyptian Air Force (48 L-59 E).
COSTS: Egyptian order reportedly worth $204 million.
DESIGN FEATURES: Main changes are a reinforced fuselage, new and more powerful engine, upgraded avionics, more pointed nose.
FLYING CONTROLS: Generally as L-39 C except that ailerons and elevators have Czech-designed irreversible power controls and no tabs.
STRUCTURE: Generally as for L-39 C except for light alloy/honeycomb sandwich ailerons and elevators, and reinforced fuselage.
LANDING GEAR: Czech design gas/oil shock absorption; K 36 mainwheels (610 × 215 mm) and K 37 nosewheel (460 × 180 mm). Mainwheel tyre pressures 6.0 bars (87 lb/sq in) on clean aircraft, 8.0 bars (116 lb/sq in) on combat-equipped version; corresponding nosewheel tyre pressures are 3.5 bars (51 lb/sq in) and 4.5 bars (65 lb/sq in). Six-piston, air-cooled hydraulic disc brakes on mainwheels, with electronic anti-skid units.
POWER PLANT: One 21.57 kN (4,850 lb st) Progress (Lotarev/ZVL) DV-2 turbofan. Internal fuel in fuselage tanks (total 1,077 litres; 284.5 US gallons; 237 Imp gallons) and two 230 litre (60.8 US gallon; 50.6 Imp gallon) non-jettisonable wingtip tanks. Provision for two underwing (inboard) 150 or 350 litre (39.6 or 92.5 US gallon; 33 or 77 Imp gallon) drop tanks.
ACCOMMODATION: Crew of two in tandem on Czech VS-2 zero/zero ejection seats. One-piece canopy, hinged at rear and opening upward.
SYSTEMS: Cockpits pressurised (max overpressure 0.30 bar; 4.35 lb/sq in) and air-conditioned, using engine bleed air (25 litres/min; 0.883 cu ft/min) and cooling unit. Automatic temperature control from 15°C to 30°C. Hydraulic system comprises first and second subsystems each with engine driven variable flow pump with operating pressure of 150 bars (2,175 lb/sq in), max flow rate 25 litres (6.6 US gallons; 5.5 Imp gallons)/min. Emergency hydraulic pump for second subsystem driven by APU. Main (9 kW) and standby (6 kW) generators for electrical power, plus 25Ah nickel-cadmium battery. Gaseous oxygen system for crew. Saphir 5M APU for engine starting and drive of standby hydraulic pump and generator.
AVIONICS (L-39 MS): Include LPR 80 VHF/UHF com radio with intercom, LUN 3524 standby radio, plus Bendix/King KNS 660 flight management system, KNR 634 VOR, KTU 709 Tacan, KDF 806 ADF, KRA 405 radar altimeter, KLN 670 GPS, KXP 756 transponder, KAH 460 AHRS, KAD 480 air data system and EFS 40 EFIS. Flight Visions FV-2000 HUD and mission computer, with video camera in front cockpit and monitor in rear cockpit.
ARMAMENT: Single twin-barrel 23 mm GSh gun in underfuselage pod below front cockpit; ammunition (150 rds) housed in fuselage. Four underwing hardpoints, inner ones each with 500 kg (1,102 lb) capacity, outer ones each 250 kg (551 lb) capacity. Underwing stores of former Soviet types, including bombs of up to 500 kg size and UB-16-57M (57 mm) rocket launchers.

DIMENSIONS, EXTERNAL: As for L-39 C except:
Wing span, incl tip tanks	9.54 m (31 ft 3½ in)
Wing chord: at root	2.80 m (9 ft 2¼ in)
at tip	1.40 m (4 ft 7 in)
Length overall	12.20 m (40 ft 0¼ in)

AREAS: As for L-39 C except:
Ailerons (total)	1.686 m² (18.15 sq ft)
Tailplane	4.15 m² (44.64 sq ft)

WEIGHTS AND LOADINGS:
Weight empty:	
trainer, incl GSh-23 gun	4,030 kg (8,885 lb)

Aero L-39 MS in Czech red, white and blue colour scheme

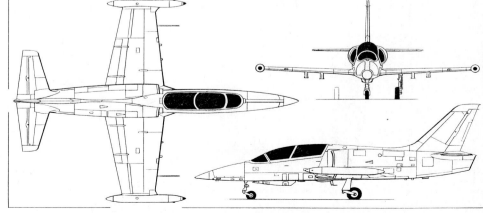

Aero L-59 two-seat basic and advanced jet trainer *(Dennis Punnett)*

Front cockpit of the L-59 E

Prototype of the L-59 E export model for Egypt in a climb

Max fuel weight:
  internal, incl wingtip tanks   1,200 kg (2,645 lb)
  external (two 350 litre drop tanks)   544 kg (1,199 lb)
Max T-O weight: trainer, clean   5,390 kg (11,883 lb)
  with external stores   7,000 kg (15,432 lb)
Max landing weight (on concrete)   6,000 kg (13,228 lb)
Max wing loading:
  clean   286.70 kg/m² (58.72 lb/sq ft)
  at 7,000 kg (15,432 lb) max T-O weight
       372.34 kg/m² (76.26 lb/sq ft)
Max power loading:
  clean   249.82 kg/kN (2.45 lb/lb st)
  at 7,000 kg (15,432 lb) max T-O weight
       324.45 kg/kN (3.18 lb/lb st)
PERFORMANCE (at max trainer clean T-O weight):
Max limiting Mach number   0.82
Max level speed at 5,000 m (16,400 ft)
       467 knots (865 km/h; 537 mph)
Stalling speed: flaps up   116 knots (215 km/h; 134 mph)
  flaps down   100 knots (185 km/h; 115 mph)
Max rate of climb at S/L   1,680 m (5,510 ft)/min
Service ceiling   11,800 m (38,725 ft)
T-O run   590 m (1,936 ft)
Landing run   770 m (2,527 ft)
Range at 7,000 m (22,975 ft) with max internal and external fuel (1,744 kg; 3,845 lb)
       1,079 nm (2,000 km; 1,243 miles)

AERO Ae 270

TYPE: Single-turboprop, nine/ten-passenger utility transport.
PROGRAMME: Announced early 1990, originally as L-270; configuration finalised 1991; first flight and certification planned for first half 1994 and third quarter 1995.
CURRENT VERSIONS: **Ae 270 U**: Basic non-pressurised model, with fixed landing gear, Motorlet M 601 E engine, Czech avionics, optional de-icing.
  Ae 270 MP: Pressurised version, with P&WC PT6A-42 engine, Bendix/King avionics, retractable gear and de-icing as standard.
DESIGN FEATURES: High aspect ratio low wing, large circular cabin windows, sweptback vertical tail.
FLYING CONTROLS: Ailerons, elevators, rudder and upper-wing spoilers actuated mechanically; wide-span Fowler flaps (70 per cent of trailing-edge) actuated hydraulically.
STRUCTURE: All-metal stressed skin, with fail-safe structural elements in fuselage and two-spar wings.
LANDING GEAR: Tricycle type (retractable on Ae 270 MP, non-retractable on Ae 270 U), with steerable nosewheel. Oleo-pneumatic shock absorbers in all units; hydraulic disc brakes on mainwheels. Inertial anti-skid system optional for both models. All wheels have tubeless tyres, pressure 3.5 or 5 bars (50.8 or 72.5 lb/sq in) on Ae 270 U, 5 bars only on Ae 270 MP.
POWER PLANT: One 589 kW (790 shp) Motorlet M 601 E turboprop in Ae 270 U, or 634 kW (850 shp) (flat rated) Pratt & Whitney Canada PT6A-42 in Ae 270 MP; constant-speed reversible-pitch propeller. Water injection system optional for both models. Integral fuel tank in each wing, combined capacity 1,200 litres (317 US gallons; 264 Imp gallons). Refuelling point in top of each wing.
ACCOMMODATION: Flight crew of two standard, but to be certificated for single-pilot operation. Cabin suitable for up to nine passengers or 1,200 kg (2,645 lb) of cargo, or combinations of both. Cabin door and baggage door on port side, emergency exit on starboard side. Heating and ventilation standard, air-conditioning and windscreen heating optional, on Ae 270 U; cockpit and cabin air-conditioning, pressurisation and windscreen heating standard on MP.
SYSTEMS: Electrical power in both models provided by 28V 7.5 kW DC engine driven starter/generator and two 25Ah nickel-cadmium batteries, with 3kVA alternator and third 25Ah battery optional. Hydraulic system, pressure 150 bars (2,175 lb/sq in), for actuation of flaps, mainwheel brakes and (on Ae 270 MP) landing gear extension/retraction. Air-conditioning and pressurisation system in Ae 270 MP maintains differential of 0.27 bar (3.92 lb/sq in) up to flight level 7,500 m (24,600 ft). Pneumatic (engine bleed air) de-icing of wing and tail leading-edges standard on Ae 270 MP, optional on Ae 270 U; electric de-icing of propeller blades, engine air intakes and pitot tube standard on both models.
AVIONICS AND EQUIPMENT: Standard instrumentation to comply with FAR Pt 23. Variety of avionics to customer's requirements for VFR and IFR operation, including auto-

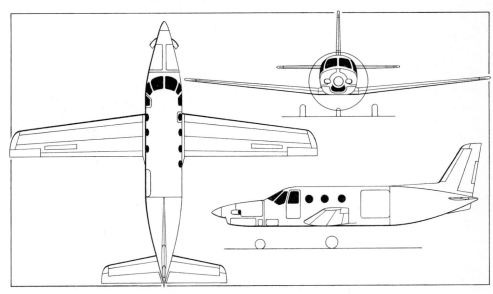

Aero Ae 270 MP single-turboprop utility transport (*Jane's/Mike Keep*)

pilot for single-pilot IFR.

DIMENSIONS, EXTERNAL:
Wing span	13.80 m (45 ft 3¼ in)
Wing aspect ratio	9.07
Length overall	12.19 m (40 ft 0 in)
Height overall	4.79 m (15 ft 8½ in)
Wheel track	2.83 m (9 ft 3½ in)
Wheelbase	3.41 m (11 ft 2¼ in)
Passenger/cargo door: Height	1.25 m (4 ft 1¼ in)
Width	1.25 m (4 ft 1¼ in)

DIMENSIONS, INTERNAL:
Cabin: Length	4.10 m (13 ft 5½ in)
Max width	1.48 m (4 ft 10¼ in)
Max height	1.37 m (4 ft 6 in)
Volume	7.5 m³ (264.9 cu ft)

AREAS:
Wings, gross	21.00 m² (226.04 sq ft)

WEIGHTS AND LOADINGS:
Weight empty	1,655 kg (3,648 lb)
Max fuel weight	900 kg (1,984 lb)
Max cargo payload	1,200 kg (2,645 lb)
Max T-O weight	3,300 kg (7,275 lb)
Max landing weight	3,135 kg (6,911 lb)
Max zero-fuel weight	2,960 kg (6,525 lb)
Max wing loading	157.14 kg/m² (32.18 lb/sq ft)
Max power loading: M 601 E	5.60 kg/kW (9.21 lb/shp)
PT6A-42	5.20 kg/kW (8.56 lb/shp)

PERFORMANCE (estimated, at max T-O weight):
Max level speed at S/L:	
U	199 knots (370 km/h; 230 mph)
MP	205 knots (380 km/h; 236 mph)
Max level speed at 6,000 m (19,685 ft):	
U	216 knots (400 km/h; 248 mph)
MP	229 knots (425 km/h; 264 mph)
Max cruising speed at 4,000 m (13,125 ft):	
U	194 knots (360 km/h; 223 mph)
MP	229 knots (425 km/h; 264 mph)
Max cruising speed at 6,000 m (19,685 ft):	
U	205 knots (380 km/h; 236 mph)
MP	226 knots (420 km/h; 261 mph)
Stalling speed, flaps up:	
U, MP	79 knots (145 km/h; 91 mph)
Stalling speed, flaps down:	
U, MP	60 knots (110 km/h; 69 mph)
Max rate of climb at S/L: U	432 m (1,417 ft)/min
MP	540 m (1,770 ft)/min
Service ceiling: U	8,200 m (26,900 ft)
MP	9,300 m (30,500 ft)
T-O run: U	250 m (821 ft)
MP	230 m (755 ft)
T-O to 15 m (50 ft): U	490 m (1,608 ft)
MP	465 m (1,526 ft)
Landing from 15 m (50 ft): U	580 m (1,903 ft)
Landing run: U	255 m (837 ft)
Range at 1,500 m (4,920 ft), 45 min reserves:	
U at 3,000 m (9,850 ft)	944 nm (1,750 km; 1,087 miles)
MP at 6,000 m (19,700 ft)	1,079 nm (2,000 km; 1,243 miles)

LET

LET AKCIOVÁ SPOLEČNOST (Let Aeronautical Works Ltd)

Uherské Hradiště, CR-686 04 Kunovice
Telephone: 42 (632) 412240
Fax: 42 (632) 61353
Telex: 60388
MANAGING DIRECTOR: Premysl Parak
TECHNICAL DIRECTOR: Josef Skrasek
CHIEF DESIGNER: Miroslav Pesak

Established at Kunovice in 1950, producing Yak-11 trainer as C-11; subsequently involved in programmes for L-200, Z-37 and L-29 Delfin. Current main aircraft programmes are L-410UVP-E and L-610; also produces L-23 Super Blanik and L-33 Solo sailplanes, plus equipment for radar and computer technology.

LET L-410UVP-E

TYPE: Twin-turboprop general-purpose light transport.
PROGRAMME: Details of prototypes and early versions in 1980-81 and earlier *Jane's*; L-410UVP first flight 1 November 1977; in 1980 became first non-Soviet aircraft to gain Soviet NLGS-2 certification; UVP production (495 built) ended late 1985. Present UVP-E version first flew (OK-120) 30 December 1984, received NLGS-2 certification March 1986. Nearly 50 aircraft unsold at February 1992 due to loss of major customer Aeroflot, resulting in suspension of assembly line for time being; future aircraft to be completed to Western equipment and certification standards.
CUSTOMERS: L-410 production (all versions) totalled 1,034 by 1 December 1992, of which 872 in service with Aeroflot by early 1992; UVP-E production had reached 364 by 1 December 1992, comprising 318 to former Soviet Union, 15 to Czechoslovakia and 31 to Bolivia/Brazil/Bulgaria/Denmark/Djibouti/Hungary/Libya/Poland; customers identified by suffix numbers (eg, –E10, –E20).
DESIGN FEATURES: Changes from UVP include baggage compartment and toilet moved aft to accommodate four more passengers in same fuselage length; wings reinforced to carry wingtip tanks increasing max range by 40 per cent; max flap deflection increased; spoiler setting 72° on ground; new vacuum sintered oil cooler; oil/fuel heat exchanger on each engine firewall to avoid use of additives; engine fire bottles moved to port rear wing/fuselage fairing; separate engine and propeller indicators for each engine; portable oxygen in cabin and improved PA system; fire extinguishing system in nose baggage compartment; operating ambient temperature range –50°C to +50°C; design life 20,000 hours or 20,000 cycles.

Wing section NACA 63A418 at root, 63A412 at tip; dihedral 1° 45′; incidence 2° at root, –0° 30′ at tip; no sweepback at front spar; tailplane dihedral 7°.
FLYING CONTROLS: Mechanical for ailerons, elevators and rudder; trim tab in port aileron, geared tab in rudder and each elevator; pop-up bank control surfaces ahead of ailerons rise automatically after engine failure to reduce lift on side of running engine. Hydraulic actuation of two-segment double-slotted flaps and lateral spoilers/lift dumpers ahead of flaps. Fixed tailplane; dorsal and ventral fins.
STRUCTURE: All-metal, two-spar torsion box wing with chemically milled skins; four wing-attachment points; one-piece tailplane; fabric covered elevators and rudder.
LANDING GEAR: Retractable tricycle type, with single wheel on each unit. Hydraulic retraction, nosewheel forward, mainwheels inward to lie flat in fairing on each side of fuselage. Technometra Radotin oleo-pneumatic shock absorbers. Non-braking nosewheel, with servo-assisted steering, fitted with 548 × 221 mm (9.00-6) tubeless tyre, pressure 4.5 bars (65 lb/sq in). Nosewheel is also steerable by rudder pedals. Mainwheels fitted with 718 × 306 mm (12.50-10) tubeless tyres, pressure 4.5 bars (65 lb/sq in). All wheels manufactured by Moravan Otrokovice, tyres by Rudy Rijen, Gottwaldov. Moravan Otrokovice K38-3200.00 hydraulic disc brakes, parking brake and anti-skid units on mainwheels. Min ground turning radius 13.40 m (43 ft 11½ in). Metal ski landing gear, with plastics undersurface, optional.
POWER PLANT: Two 559 kW (750 shp) Motorlet M 601 E turboprops, each driving an Avia V 510 five-blade constant-speed reversible-pitch metal propeller with manual and automatic feathering and Beta control. At higher ambient temperatures, engine power can be increased to 603 kW (809 ehp) for short periods by water injection into compressor. Eight bag fuel tanks in wings, total capacity 1,290 litres (341 US gallons; 284 Imp gallons), plus additional optional 200 litres (52.8 US gallons; 44 Imp gallons) of fuel in each wingtip tank. Fuel system operable after failure of electrical system. Total oil capacity (incl oil in cooler) 22 litres (5.8 US gallons; 4.8 Imp gallons). Water tank capacity (for injection into compressor) 11 litres (2.9 US gallons; 2.4 Imp gallons).
ACCOMMODATION: Crew of one or two on flight deck, with dual controls. Electric de-icing for windscreen. Standard accommodation in main cabin for 19 passengers, with pairs of adjustable seats on starboard side of aisle and single seats opposite, all at 76 cm (30 in) pitch. Baggage compartment (at rear, accessible from cabin), toilet and wardrobe standard. Cabin heated by engine bleed air. Alternative layouts include all-cargo; ambulance, accommodating six stretchers, five sitting patients and a medical attendant; accommodation for 18 parachutists and a dispatcher/instructor; firefighting configuration, carrying 16 firefighters and a pilot/observer. All-cargo version has protective floor covering, crash nets on each side of cabin, and tiedown provisions; floor is at truckbed height. Aircraft can also be equipped for aerial photography or for calibration of ground navigation aids. Double upward opening doors aft on port side, with stowable steps; right hand door serves as passenger entrance and exit. Both doors open for cargo loading, and can be removed for paratroop training missions. Rearward opening door, forward on starboard side, serves as emergency exit.
SYSTEMS: No APU, air-conditioning or pressurisation systems. Duplicated hydraulic systems, No. 1 system actuating landing gear, flaps, spoilers, automatic pitch trim surfaces, mainwheel brakes, nosewheel steering and windscreen wipers. No. 2 system for emergency landing gear extension, flap actuation and parking brake. 28V DC electrical system supplied by two 5.6 kW starter/generators, connected for autonomous starting, plus two 25V 25Ah batteries for emergency power. Two input systems for AC power (three-phase 200V/115Hz, variable frequency), incorporating two 3.7 kW alternators with

Two views of the current production L-410UVP-E twin-turboprop transport, in civil and Hungarian Air Force markings

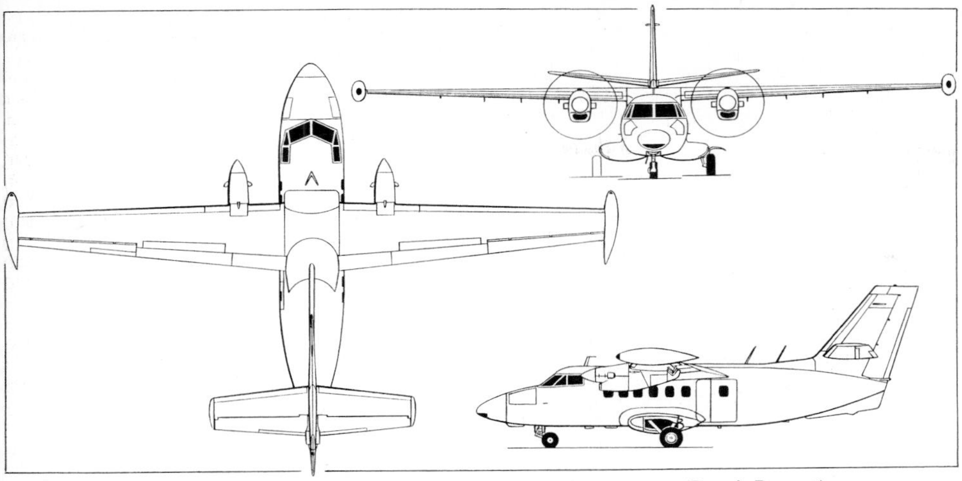

Let L-410UVP-E twin-turboprop 19-passenger light transport *(Dennis Punnett)*

alternator control unit. Port alternator provides for windscreen heating, starboard one for propeller blade de-icing. Two static inverters provide three-phase 36V/400Hz AC. Two 115V/400Hz inverters. One three-phase 36V/400Hz static inverter for standby horizon. De-icing for propeller blades (electrical) and lower intakes (bleed air); anti-icing flaps inside each nacelle. Kléber-Colombes pneumatic de-icing of wing and tail leading-edges. Two portable oxygen breathing sets on flight deck and two in passenger cabin. Fire extinguishing system for engines and nose baggage compartment.

AVIONICS: Standard instrumentation provides for flight in IMC conditions, with all basic instruments duplicated and three artificial horizons. Communications include two VHF transceivers with a range of 65 nm (120 km; 75 miles) at 1,000 m (3,280 ft) altitude, passenger address system and crew intercom. Standard instruments include LUN 1205 horizon gyros, rate of climb indicators, LUN 1215 turn and bank indicator, RMIs, gyro compasses, ILS/SP-50A instrument landing system with marker beacon receiver, dual ARK-22 ADF, A-037 radio altimeter, SO-69 SSR transponder with encoding altimeter, ASI with stall warning, magnetic compass, GMK-1GE VOR, and BUR-1-2G flight data recorder. Weather radar and VZLU autopilot optional.

EQUIPMENT: Cockpit, instrument and passenger cabin lights, navigation lights, three landing lights in nose (each with two levels of light intensity), crew and cabin fire extinguishers, windscreen wipers, and alcohol spray for windscreen and wiper de-icing, are standard.

DIMENSIONS, EXTERNAL:

Wing span: over tip tanks	19.98 m (65 ft 6½ in)
excl tip tanks	19.48 m (63 ft 11 in)
Wing chord at root	2.534 m (8 ft 3¾ in)
Length overall	14.424 m (47 ft 4 in)
Fuselage: Max width	2.08 m (6 ft 10 in)
Max depth	2.10 m (6 ft 10¾ in)
Height overall	5.83 m (19 ft 1½ in)
Tailplane span	6.74 m (22 ft 1¼ in)
Wheel track	3.65 m (11 ft 11½ in)
Wheelbase	3.67 m (12 ft 0¼ in)
Propeller diameter	2.30 m (7 ft 6½ in)
Propeller ground clearance	1.26 m (4 ft 1½ in)
Distance between propeller centres	4.82 m (15 ft 9½ in)
Passenger/cargo door (port, rear):	
Height	1.46 m (4 ft 9½ in)
Width overall	1.25 m (4 ft 1¼ in)
Width (passenger door only)	0.80 m (2 ft 7½ in)
Height to sill	0.70 m (2 ft 3½ in)
Emergency exit door (stbd, fwd):	
Height	0.97 m (3 ft 2¼ in)
Width	0.66 m (2 ft 2 in)
Height to sill	0.80 m (2 ft 7½ in)

DIMENSIONS, INTERNAL:

Cabin, excl flight deck:	
Length	6.345 m (20 ft 9¾ in)
Max width	1.95 m (6 ft 4¾ in)
Max height	1.66 m (5 ft 5¼ in)
Aisle width at 0.4 m (1 ft 3¾ in) above cabin floor	0.34 m (1 ft 1½ in)
Floor area	10.0 m² (107.6 sq ft)
Volume	17.9 m³ (632.1 cu ft)
Baggage compartment volume:	
nose	0.60 m³ (21.19 cu ft)
rear	0.77 m³ (27.19 cu ft)

AREAS:

Wings, gross	34.86 m² (375.2 sq ft)
Ailerons (total)	2.89 m² (31.11 sq ft)
Automatic bank control flaps (total)	0.49 m² (5.27 sq ft)
Trailing-edge flaps (total)	5.92 m² (63.72 sq ft)
Spoilers (total)	0.87 m² (9.36 sq ft)
Fin	4.49 m² (48.33 sq ft)
Rudder, incl tab	2.81 m² (30.25 sq ft)
Tailplane	6.41 m² (69.00 sq ft)
Elevators, incl tabs	3.15 m² (33.91 sq ft)

WEIGHTS AND LOADINGS:

Weight empty	3,985 kg (8,785 lb)
Operating weight empty, equipped	4,160 kg (9,171 lb)
Max fuel	1,300 kg (2,866 lb)
Max payload	1,615 kg (3,560 lb)
Max ramp weight	6,420 kg (14,154 lb)
Max T-O weight	6,400 kg (14,110 lb)
Max zero-fuel weight	5,775 kg (12,732 lb)
Max landing weight	6,200 kg (13,668 lb)
Max wing loading	183.6 kg/m² (37.60 lb/sq ft)
Max power loading	5.72 kg/kW (9.41 lb/shp)

PERFORMANCE (at max T-O weight):

Never-exceed speed (VNE)	192 knots (357 km/h; 222 mph) EAS
Max level speed at 4,200 m (13,780 ft)	168 knots (311 km/h; 193 mph) EAS
Max cruising speed at 4,200 m (13,780 ft)	205 knots (380 km/h; 236 mph)
Econ cruising speed at 4,200 m (13,780 ft)	197 knots (365 km/h; 227 mph)
Stalling speed:	
flaps up	84 knots (155 km/h; 97 mph) EAS
flaps down	66 knots (121 km/h; 76 mph) EAS
Max rate of climb at S/L	444 m (1,455 ft)/min
Rate of climb at S/L, OEI	108 m (354 ft)/min
Service ceiling: practical	6,320 m (20,725 ft)
theoretical	7,050 m (23,125 ft)
Service ceiling, OEI:	
practical	2,700 m (8,860 ft)
theoretical	3,980 m (13,050 ft)
T-O run	445 m (1,460 ft)
T-O to 10.7 m (35 ft)	685 m (2,250 ft)
Landing from 9 m (30 ft)	480 m (1,575 ft)
Landing run	240 m (787 ft)
Range at 4,200 m (13,780 ft), max cruising speed, 30 min reserves:	
with max payload	294 nm (546 km; 339 miles)
with max fuel and 885 kg (1,951 lb) payload	744 nm (1,380 km; 857 miles)

LET L-610

TYPE: Twin-turboprop regional transport.

PROGRAMME: First flight (OK-130) 28 December 1988; five development aircraft include two for static test; deliveries (initially to Aeroflot) began 1991; contract 18 January 1991 for General Electric to provide CT7-9D turboprops for L-610G (first two engines delivered shortly afterward); first flight of this version (OK-134) 18 December 1992; certification to FAR Pt 25 expected in 1994.

CURRENT VERSIONS: **L-610:** Basic version, with 1,358 kW (1,822 shp) Motorlet M 602 engines; weight and performance data not yet finalised.

L-610G: Version with General Electric CT7-9D turboprops and Collins digital avionics including EFIS, weather radar and autopilot. *Detailed description applies to this version except where indicated.*

CUSTOMERS: Deliveries to meet Aeroflot requirement for 600 L-610s (to replace Yak-40s and An-24s) started 1991, but held up later that year by requirement for payment in hard currency.

DESIGN FEATURES: Intended to meet Russian ENLG-S civil airworthiness requirements; wing sections MS(1)-0318D at root, MS(1)-0312 at tip; thickness/chord ratios 18.29 (root) and 12 per cent (tip); dihedral 2°; incidence 3° 8′ 38″ at root, 0° at tip; quarter-chord sweepback 1°.

FLYING CONTROLS: Ailerons, elevators and rudder actuated mechanically; trim tabs in port aileron (actuated electrically), rudder and each elevator, plus geared tab in each elevator and spring tab in rudder; ailerons horn balanced. Single-slotted Fowler flaps, and lateral control spoilers ahead of outer segments, actuated hydraulically. Fixed incidence tailplane.

STRUCTURE: All-metal, fail-safe stressed skin structure; parallel circular-section fuselage between flight deck and tail; wing contains high grade aluminium and high strength steel; honeycomb spoiler panels.

LANDING GEAR: Retractable tricycle type, with single wheel on each unit. Hydraulic actuation, mainwheels retracting inward to lie flat in fairing each side of fuselage, nosewheel retracting forward. Oleo-pneumatic shock absorber in each unit. Mainwheels are type XK 34-3000.00, with 1,050 × 390 × 480 mm tyres; type XR 25-1000.00 nosewheel has a 720 × 310 × 254 mm tyre. Hydraulic disc brakes and electronically controlled anti-skid units. Min ground turning radius 18.33 m (60 ft 1¾ in).

POWER PLANT: Two 1,305 kW (1,750 shp) General Electric CT7-9D turboprops in L-610G, each driving a Hamilton Standard HS-14RF-23 four-blade fully feathering metal propeller with reversible pitch. Fuel in two integral wing tanks, combined usable capacity 3,500 litres (925 US gallons; 770 Imp gallons). Pressure refuelling point in fuselage, gravity points in wings. Oil capacity 30 litres (7.9 US gallons; 6.6 Imp gallons).

ACCOMMODATION: Crew of two on flight deck, plus one cabin attendant. Standard accommodation for 40 passengers, four-abreast at seat pitch of 76 cm (30 in). Aisle width 51 cm (20 in). Galley, two wardrobes, toilet, freight and baggage compartment, all located at rear of cabin. Alternative mixed (passenger/cargo) and all-cargo layouts available. Passenger door at rear of fuselage, freight door at front, both opening outward on port side. Outward opening service door on starboard side, opposite passenger door, serving also as emergency exit; outward opening emergency exit beneath wing on each side. Entire accommodation pressurised and air-conditioned.

SYSTEMS: Bootstrap type air-conditioning system. Max operating cabin pressure differential 0.3 bar (4.35 lb/sq in). Duplicated hydraulic systems (one main and one standby), operating at pressure of 210 bars (3,045 lb/sq in). APU in tailcone, for engine starting and auxiliary on-ground and in-flight power. Electrical system powered by two 115/200V 25kVA variable frequency AC generators, plus a third 8kVA 115/200V three-phase AC generator driven by APU. System also includes two 115V 400Hz inverters (each 1.5kVA), two 27V DC transformer-rectifiers (each 4.5 kW), and a 25Ah nickel-cadmium battery for APU starting and auxiliary power supply. Portable oxygen

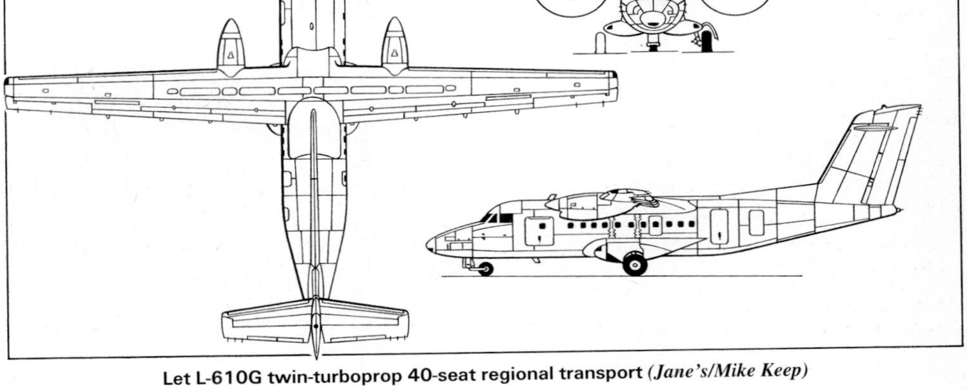

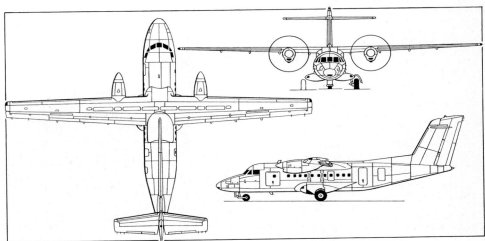

Let L-610G twin-turboprop 40-seat regional transport *(Jane's/Mike Keep)*

equipment for crew and 10 per cent of passengers. Pneumatic de-icing of wing and tail unit leading-edges, engine inlets and oil cooler; electric de-icing of propeller blade roots, windscreen, pitot static system and horn balances.

AVIONICS: Equipped with dual 760-channel VHF com, single HF com (optional), intercom, cabin address system, weather radar, blind-flying instrumentation, dual ILS with two LOC/glideslope receivers and two marker beacon receivers, single or dual ADF, Doppler radar, navigation computer, dual compasses, single or dual radio altimeters, transponder, autopilot, voice recorder, flight recorder, and Cat. II approach aids.

DIMENSIONS, EXTERNAL:
Wing span	25.60 m (84 ft 0 in)
Wing chord: at root	2.917 m (9 ft 6⅞ in)
at tip	1.458 m (4 ft 9½ in)
Wing aspect ratio	11.7
Length overall	21.72 m (71 ft 3 in)
Fuselage: Length	20.533 m (67 ft 4⅜ in)
Max diameter	2.70 m (8 ft 10¼ in)
Distance between propeller centres	7.00 m (22 ft 11½ in)
Height overall	8.19 m (26 ft 10½ in)
Tailplane span	7.908 m (25 ft 11⅓ in)
Wheel track	4.59 m (15 ft 0¾ in)
Wheelbase	6.596 m (21 ft 7¾ in)
Propeller diameter	3.50 m (11 ft 5¾ in)
Propeller fuselage clearance	0.59 m (1 ft 11¼ in)
Propeller ground clearance	1.64 m (5 ft 4½ in)
Passenger door: Height	1.625 m (5 ft 4 in)
Width	0.76 m (2 ft 6 in)
Height to sill	1.448 m (4 ft 9 in)
Freight door: Height	1.30 m (4 ft 3¼ in)
Width	1.25 m (4 ft 1¼ in)
Height to sill	1.448 m (4 ft 9 in)
Service door: Height	1.286 m (4 ft 2⅔ in)
Width	0.61 m (2 ft 0 in)
Emergency exits (underwing, each):	
Height	0.915 m (3 ft 0 in)
Width	0.515 m (1 ft 8¼ in)

DIMENSIONS, INTERNAL:
Cabin (excl flight deck): Length	11.10 m (36 ft 5 in)
Max width	2.54 m (8 ft 4 in)
Width at floor	2.02 m (6 ft 7½ in)
Max height	1.835 m (6 ft 0¼ in)
Floor area	22.4 m² (241.1 sq ft)
Volume	44.7 m³ (1,578.6 cu ft)
Wardrobe volume (total)	1.0 m³ (35.3 cu ft)
Baggage/freight hold volume (total)	7.5 m³ (264.9 cu ft)

AREAS:
Wings, gross	56.0 m² (602.8 sq ft)
Ailerons (total)	3.27 m² (35.20 sq ft)
Trailing-edge flaps (total)	11.29 m² (121.52 sq ft)
Spoilers (total)	3.54 m² (38.10 sq ft)
Fin	8.30 m² (89.34 sq ft)
Rudder, incl tabs	5.54 m² (59.63 sq ft)
Tailplane	7.68 m² (82.67 sq ft)
Elevators (total, incl tabs)	5.82 m² (62.65 sq ft)

WEIGHTS AND LOADINGS (L-610G):
Weight empty, equipped	8,730 kg (19,246 lb)
Operating weight empty	9,000 kg (19,841 lb)
Max fuel	2,700 kg (5,952 lb)
Max payload	4,200 kg (9,259 lb)
Max ramp weight	14,540 kg (32,055 lb)
Max T-O weight	14,500 kg (31,967 lb)
Max zero-fuel weight	13,200 kg (29,101 lb)
Max landing weight	14,200 kg (31,305 lb)
Max wing loading	258.9 kg/m² (53.03 lb/sq ft)
Max power loading	5.555 kg/kW (9.13 lb/shp)

PERFORMANCE (L-610G, estimated, at max T-O weight):
Never-exceed speed (V_{NE})	216 knots (400 km/h; 248 mph) EAS
Max level and max cruising speed at 7,200 m (23,620 ft)	251 knots (466 km/h; 289 mph)
Long-range cruising speed at 7,200 m (23,620 ft)	152 knots (282 km/h; 175 mph)
Approach speed	92 knots (170 km/h; 106 mph)
Stalling speed:	
flaps up	95 knots (176 km/h; 110 mph) EAS
flaps down	74 knots (136 km/h; 85 mph) EAS
Max rate of climb at S/L	504 m (1,653 ft)/min
Rate of climb at S/L, OEI	153 m (502 ft)/min
Service ceiling: theoretical	8,400 m (27,560 ft)
practical	8,050 m (26,410 ft)
Service ceiling, OEI (30.5 m; 100 ft/min rate of climb):	
theoretical	5,050 m (16,570 ft)
practical	4,400 m (14,435 ft)
T-O run	528 m (1,733 ft)
T-O to 10.7 m (35 ft)	662 m (2,172 ft)
Balanced T-O distance	1,035 m (3,396 ft)
Balanced T-O field length:	
hard runway	875 m (2,870 ft)
unpaved surface	1,030 m (3,380 ft)
Landing from 15 m (50 ft)	587 m (1,926 ft)
Landing run without propeller reversal	380 m (1,247 ft)
Range, reserves for 45 min hold and 100 nm (185 km; 115 mile) diversion:	
with max payload	469 nm (870 km; 540 miles)
with max fuel	1,333 nm (2,470 km; 1,535 miles)

Let L-610G (two General Electric CT7-9D turboprops), photographed prior to its first flight on 18 December 1992

Early production Czech (Motorlet M 602) engined L-610 prior to delivery to Aeroflot

ZLIN

MORAVAN AKCIOVÁ SPOLEČNOST (Zlin Aircraft Moravan Aeronautical Works Ltd)

CR-765 81 Otrokovice
Telephone: 42 (67) 922041 or 924351
Fax: 42 (67) 922518 or 922479
Telex: 67240
MANAGING DIRECTOR: Milan Charvat
COMMERCIAL DIRECTOR: Ing Karel Kubena
TECHNICAL DIRECTOR: Ing Antonin Partika
MARKETING MANAGER: Jiri Aron

Formed 18 September 1934 as Zlinská Letecká Akciová Společnost (Zlin Aviation Company Ltd) in Zlin; manufacture of Zlin aircraft started 1933 by Masarykova Letecká Liga (Masaryk League of Aviation); company renamed Moravan after Second World War; also now manufactures aircraft equipment.

ZLIN 142

TYPE: Two-seat fully aerobatic light training and touring aircraft.
PROGRAMME: Design begun Winter 1977/78, prototype construction April 1978; first flight 29 December 1978; FAR Pt 23 certification (Aerobatic, Utility and Normal) 1980; production started 1981; Canadian DoT certification 26 November 1991.
CURRENT VERSIONS: **Z 142 A**: Aerobatic trainer.
 Z 142 U: General trainer.
 Z 142 N: Touring aircraft.
 Z 242 L: Lycoming powered version of Z 142 for US market; described separately.
CUSTOMERS: Total 335 (all versions) built by 1 January 1992 (latest figure received). Exported to Algeria, Bulgaria, Canada, Cuba, Germany, Hungary, Poland, Romania, South Africa and Yugoslavia.
DESIGN FEATURES: Development of Zlin 42 M (1980-81 *Jane's*); basic, advanced and aerobatic training, glider towing and (with equipment) night flying and IFR training. Wing section NACA $63_2416.5$; dihedral 6° from roots; sweepforward 4° 20' at quarter-chord.
FLYING CONTROLS: Slotted, mass balanced surfaces used for ailerons and flaps; horn balanced elevator with trim tab; ground adjustable tabs in ailerons and rudder; ailerons and elevator operated by rods, rudder by cables; fixed incidence tailplane.
STRUCTURE: Mainly metal; wing has main and auxiliary spars; duralumin skins, fluted on control surfaces; metal engine cowlings; centre-fuselage is steel tube cage with composite skin panels.
LANDING GEAR: Non-retractable tricycle type, with nosewheel offset to port. Oleo-pneumatic nosewheel shock absorber. Mainwheels carried on flat spring steel legs. Nosewheel steered (±38°) by rudder pedals. Mainwheels and Barum tyres size 350 × 150 mm, pressure 1.90 bars (27.6 lb/sq in); nosewheel and Barum tyre size 350 × 135 mm, pressure 2.50 bars (36.3 lb/sq in). Hydraulic disc brakes on mainwheels can be operated from either seat. Parking brake standard.
POWER PLANT: One Avia M 337 AK inverted six-cylinder air-cooled inline engine (156.5 kW; 210 hp at 2,750 rpm), with supercharger and low-pressure injection pump, driving a two-blade Avia V 500 A constant-speed metal propeller. Fuel tanks in each wing leading-edge, with combined capacity of 125 litres (33 US gallons; 27.5 Imp gallons). Normal category version has auxiliary 50 litre (13.2 US gallon; 11 Imp gallon) tank at each wingtip, increasing total fuel capacity to 225 litres (59.4 US gallons; 49.5 Imp gallons). Fuel and oil systems permit inverted flying for up to 3 min. Oil capacity 12 litres (3.2 US gallons; 2.6 Imp gallons).
ACCOMMODATION: Side by side seats for two persons, instructor's seat to port. Both seats adjustable and permit use of back type parachutes. Baggage space aft of seats. Cabin and windscreen heating and ventilation standard. Forward sliding cockpit canopy. Dual controls standard.
SYSTEMS: Electrical system includes 600W 28V engine driven generator and 24V 25Ah Teledyne battery. External power source can be used for starting engine.
AVIONICS: Bendix/King Silver Crown or other avionics, to customer's requirements.
EQUIPMENT: Standard equipment includes cockpit, instrument and cabin lights; navigation lights; landing and taxying lights; and anti-collision light. Towing gear, for gliders of up to 500 kg (1,102 lb) weight, optional.

DIMENSIONS, EXTERNAL:
Wing span	9.16 m (30 ft 0½ in)
Wing aspect ratio	6.4
Wing chord, constant portion	1.42 m (4 ft 8 in)
Length overall	7.33 m (24 ft 0½ in)
Height overall	2.75 m (9 ft 0¼ in)
Elevator span	2.904 m (9 ft 6⅓ in)
Wheel track	2.33 m (7 ft 7¾ in)
Wheelbase	1.66 m (5 ft 5¼ in)
Propeller diameter	2.00 m (6 ft 6¾ in)
Propeller ground clearance	0.40 m (1 ft 3¾ in)

DIMENSIONS, INTERNAL:
Cabin: Length	1.80 m (5 ft 10¾ in)
Max width	1.12 m (3 ft 8 in)
Max height	1.20 m (3 ft 11¼ in)
Baggage space	0.2 m³ (7.1 cu ft)

AREAS:
Wings, gross	13.15 m² (141.5 sq ft)
Ailerons (total)	1.408 m² (15.16 sq ft)
Trailing-edge flaps (total)	1.408 m² (15.16 sq ft)
Fin	0.54 m² (5.81 sq ft)
Rudder, incl tab	0.81 m² (8.72 sq ft)
Tailplane	1.23 m² (13.24 sq ft)
Elevator, incl tabs	1.36 m² (14.64 sq ft)

WEIGHTS AND LOADINGS (A: Aerobatic, U: Utility, N: Normal category):
Basic weight empty (all versions)	730 kg (1,609 lb)
Max T-O weight: A	970 kg (2,138 lb)
U	1,020 kg (2,248 lb)
N	1,090 kg (2,403 lb)
Max landing weight: A	970 kg (2,138 lb)
U	1,020 kg (2,248 lb)
N	1,050 kg (2,315 lb)
Max wing loading: A	73.76 kg/m² (15.11 lb/sq ft)
U	77.57 kg/m² (15.89 lb/sq ft)
N	82.89 kg/m² (16.98 lb/sq ft)
Max power loading: A	6.19 kg/kW (10.17 lb/hp)
U	6.51 kg/kW (10.69 lb/hp)
N	6.96 kg/kW (11.43 lb/hp)

PERFORMANCE (at max T-O weight):
Never-exceed speed (V_{NE}) (all versions)	179 knots (333 km/h; 206 mph) IAS
Max level speed at 500 m (1,640 ft):	
A, U	125 knots (231 km/h; 143 mph)
N	122 knots (227 km/h; 141 mph)
Max cruising speed at 500 m (1,640 ft):	
A, U	106 knots (197 km/h; 122 mph)
N	102 knots (190 km/h; 118 mph)
Econ cruising speed at 500 m (1,640 ft):	
A	97 knots (180 km/h; 112 mph)
Stalling speed, flaps up:	
A	56 knots (103 km/h; 64 mph) IAS
U	58 knots (107 km/h; 67 mph) IAS
N	60 knots (110 km/h; 69 mph) IAS
Stalling speed, T-O flap setting:	
A	54 knots (99 km/h; 62 mph) IAS
U	56 knots (102 km/h; 64 mph) IAS
N	57 knots (105 km/h; 66 mph) IAS
Stalling speed, flaps down:	
A	48 knots (88 km/h; 55 mph) IAS
U	50 knots (91 km/h; 57 mph) IAS
N	52 knots (95 km/h; 60 mph) IAS
Max rate of climb at S/L, ISA:	
A	312 m (1,024 ft)/min
U	288 m (945 ft)/min
N	252 m (827 ft)/min
Service ceiling: A	4,750 m (15,580 ft)
U	4,500 m (14,765 ft)
N	4,300 m (14,100 ft)
T-O run: A	231 m (758 ft)
U	236 m (775 ft)
N	252 m (827 ft)
T-O to 15 m (50 ft): A	510 m (1,674 ft)
U	560 m (1,838 ft)
N	620 m (2,035 ft)
Landing from 15 m (50 ft): A	425 m (1,395 ft)
U	440 m (1,444 ft)
N	460 m (1,510 ft)
Landing run: A	190 m (624 ft)
U	200 m (657 ft)
N	220 m (722 ft)
Range at max cruising speed:	
A, U	276 nm (512 km; 318 miles)
N	499 nm (925 km; 575 miles)
Max range: N	552 nm (1,024 km; 636 miles)
g limits: A	+6/−3.5
U	+5/−3
N	+3.8/−1.5

ZLIN 242 L

TYPE: Lycoming powered version of Z 142.
PROGRAMME: Intended for US market; design started December 1988; first flight (OK-076) 14 February 1990; FAR Pt 23 certification (A, U and N categories) April 1992.
CUSTOMERS: No production figure received.
DESIGN FEATURES: Changes from Z 142, apart from new engine, include redesigned (and shorter) engine cowling and front fuselage, wing incidence, 0° sweep, bulged wingroot/fuselage fairing, redesigned wing and tailplane tips, redesigned fuel system and updated instruments. Spin recovery strake each side of cowling. Flying controls, structure, landing gear and accommodation details as for Z 142.
POWER PLANT: One Textron Lycoming AEIO-360-A1B6 flat-four engine (149 kW; 200 hp at 2,700 rpm), driving a Mühlbauer MTV-9-B-C/C-188-18a three-blade constant-speed wood/composites propeller. Fuel capacity 120 litres (31.7 US gallons; 26.4 Imp gallons); Normal category version wingtip tanks 55 litres (14.5 US gallons; 12.1 Imp gallons) each, bringing total capacity to 230 litres (60.7 US gallons; 50.6 Imp gallons). Inverted flight limited to 1 min. Oil capacity 8 litres (2.1 US gallons; 1.8 Imp gallons).
SYSTEMS: Electrical system includes 1.6kW 28V engine driven generator and 24V 19Ah Gill battery. External power source can be used for engine starting.
AVIONICS: To customer's specification, usually from Bendix/King Silver Crown range.
EQUIPMENT: Standard equipment includes EGT gauge, fuel flow indicator, g meter and anti-collision beacon. Optional items include cockpit/instrument/cabin lights, landing/taxying lights, anti-collision lights, and towing gear for gliders of up to 500 kg (1,102 lb) weight.

DIMENSIONS, EXTERNAL:
Wing span	9.34 m (30 ft 7¾ in)
Wing chord, constant portion	1.504 m (4 ft 11¼ in)
Wing aspect ratio	6.34
Length overall	6.94 m (22 ft 9¼ in)
Height overall	2.95 m (9 ft 8¼ in)
Elevator span	3.20 m (10 ft 6 in)
Wheel track	2.33 m (7 ft 7¾ in)
Wheelbase	1.755 m (5 ft 9 in)
Propeller diameter	1.88 m (6 ft 2 in)
Propeller ground clearance	0.41 m (1 ft 4¼ in)

DIMENSIONS, INTERNAL: As Z 142
AREAS: As Z 142 except:
Wings, gross	13.758 m² (148.09 sq ft)

WEIGHTS AND LOADINGS (A: Aerobatic, U: Utility, N: Normal category):
Basic weight empty, max T-O and landing weights as Z 142

Zlin 142 two-seat trainer/tourer, with single-seat aerobatic Zlin 50 LS beyond

Prototype Zlin 242 L (Textron Lycoming AEIO-360-A1B6 engine)

Wing aspect ratio	6.95
Length overall	7.58 m (24 ft 10½ in)
Height overall	2.91 m (9 ft 6½ in)
Elevator span	3.006 m (9 ft 10¼ in)
Wheel track	2.44 m (8 ft 0 in)
Wheelbase	1.75 m (5 ft 9 in)
Propeller diameter	1.95 m (6 ft 4¾ in)
Propeller ground clearance	0.28 m (11 in)

DIMENSIONS, INTERNAL:
Baggage compartment volume   0.25 m³ (8.83 cu ft)
AREAS: As for Z 142 except:
Wings, gross   14.776 m² (159.05 sq ft)
WEIGHTS AND LOADINGS (U: Utility, N: Normal category):
Weight empty, equipped: U, N   830 kg (1,830 lb)
Max T-O weight: U   1,080 kg (2,381 lb)
N   1,350 kg (2,976 lb)
Max landing weight: U   1,080 kg (2,381 lb)
N   1,285 kg (2,833 lb)
Max wing loading: U   73.09 kg/m² (14.97 lb/sq ft)
N   91.36 kg/m² (18.71 lb/sq ft)
Max power loading: U   6.16 kg/kW (10.13 lb/hp)
N   7.70 kg/kW (12.66 lb/hp)
PERFORMANCE (estimated, at max Normal category T-O weight, ISA at S/L):
Max level speed   143 knots (265 km/h; 165 mph)
Max rate of climb   264 m (866 ft)/min
Range at cruising speed   458 nm (850 km; 528 miles)
g limits   +3.8/−1.52 (Normal)
    +4.4/−1.76 (Utility)

ZLIN 50

TYPE: Single-seat aerobatic aircraft.
PROGRAMME: First flight (Z 50 LS) 29 June 1981; FAR Pt 23 certification 1982; won European Aerobatic Championships 1983, World Aerobatic Championships 1984 and 1986.
CURRENT VERSIONS: **Z 50 L**: Initial version, with 194 kW (260 hp) Textron Lycoming AEIO-540-D4B5; superseded (see 1982-83 *Jane's*).

Max wing loading: A   70.50 kg/m² (14.44 lb/sq ft)
U   74.14 kg/m² (15.18 lb/sq ft)
N   79.23 kg/m² (16.23 lb/sq ft)
Max power loading: A   6.50 kg/kW (10.68 lb/hp)
U   6.84 kg/kW (11.24 lb/hp)
N   7.31 kg/kW (12.01 lb/hp)
PERFORMANCE (at max T-O weight):
Never-exceed speed (V_NE):
A, U, N   172 knots (319 km/h; 198 mph) IAS
Max level speed at 500 m (1,640 ft):
A   127 knots (236 km/h; 146 mph)
U   126 knots (234 km/h; 145 mph)
N   124 knots (230 km/h; 143 mph)
Max cruising speed at 500 m (1,640 ft):
A   115.5 knots (214 km/h; 133 mph)
U   115 knots (213 km/h; 132 mph)
N   114 knots (212 km/h; 132 mph)
Stalling speed, flaps up:
A   57 knots (105 km/h; 66 mph)
U   59 knots (108 km/h; 67 mph)
N   60 knots (111 km/h; 69 mph)
Stalling speed, T-O flap setting:
A   53 knots (98 km/h; 61 mph)
U   54 knots (100 km/h; 62 mph)
N   56 knots (103 km/h; 64 mph)
Stalling speed, flaps down:
A   47 knots (87 km/h; 54 mph)
U   48 knots (89 km/h; 56 mph)
N   50 knots (93 km/h; 58 mph)
Max rate of climb at S/L: A   330 m (1,085 ft)/min
U   300 m (985 ft)/min
N   258 m (845 ft)/min
Service ceiling: A   5,150 m (16,900 ft)
U   5,000 m (16,400 ft)
N   4,800 m (15,750 ft)
T-O run: A   242 m (794 ft)
U   268 m (880 ft)
N   306 m (1,005 ft)
T-O to 15 m (50 ft): A   560 m (1,838 ft)
U   619 m (2,031 ft)
N   706 m (2,317 ft)
Landing from 15 m (50 ft): A   500 m (1,641 ft)
U   525 m (1,723 ft)
N   540 m (1,772 ft)
Landing run: A   395 m (1,296 ft)
Range at econ cruising speed at 500 m (1,640 ft):
N   558 nm (1,035 km; 643 miles)
g limits: A   +6/−3.5
U   +5/−3
N   +3.8/−1.5

ACCOMMODATION: Four seats in '2 + 2' configuration, with pilot in front left hand seat. All seats have four-point safety belts. Baggage compartment aft of rear seats, with external door on port side. One-piece forward sliding canopy. Dual controls standard. Cabin and windscreen heating and ventilation standard.
SYSTEMS: As for Z 242 L.
AVIONICS: As for Z 242 L.
DIMENSIONS, EXTERNAL:
Wing span   10.136 m (33 ft 3 in)
Wing chord, mean aerodynamic   1.4895 m (4 ft 10½ in)

The Zlin 143 L prototype four-seat trainer and touring aircraft

ZLIN 143

TYPE: Four-seat basic/advanced trainer and touring aircraft.
PROGRAMME: Launched May 1991, with prototype construction beginning that Summer and first flight (Z 143 L prototype OK-074) 24 April 1992; certification expected during 1993; production under way.
CURRENT VERSIONS: **Z 143 L**: Prototype, with Textron Lycoming engine. *Detailed description applies to this version.*
Z 143: Future production aircraft powered by 156.5 kW (210 hp) Avia M 337 AK.
DESIGN FEATURES: For basic aerobatics (except inverted flight), glider towing and (with equipment) night and IFR training. Wing section and dihedral as for Z 142.
FLYING CONTROLS: As for Z 142.
STRUCTURE: As for Z 142.
LANDING GEAR: As for Z 142.
POWER PLANT: One Textron Lycoming O-540-J3A5 flat-six engine (175 kW; 235 hp at 2,400 rpm); Mühlbauer MTV-9-B/195-45 three-blade variable-pitch propeller. Fuel capacity as for Z 242 L.

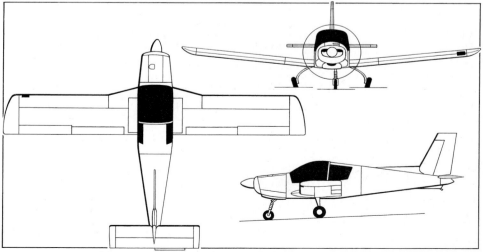

General arrangement of the Zlin 143 L (Textron Lycoming O-540-J3A5 engine) *(Jane's/Mike Keep)*

Z 50 LA: Modified L with propeller pitch control and propeller speed governor.

Z 50 LE: Short-span version with T-O weight reduced by 50 kg (110 lb); first flight 1990; only two built.

Z 50 LS: Principal version still in production; more powerful engine than earlier models; total 65 built by mid-1991. *Detailed description applies to this version.*

Z 50 M: Longer and slimmer nose than L series, with 134 kW (180 hp) Avia M 137 AZ inline engine and constant-speed propeller; first flight (OK-080) 25 April 1988; Aerobatic and Normal certification February 1989; prototype and two production aircraft built by January 1989; two more completed for 1992 delivery to South Africa. Overall length 6.96 m (22 ft 10 in), tailplane area 2.24 m² (24.1 sq ft), otherwise dimensionally as Z 50 LS; see 1991-92 *Jane's* for weights, loadings and performance data.

CUSTOMERS: Total of 75 (72 L/LA/LE/LS and three M) built by 1 January 1992 (latest figure received), with two more M then under construction. Exported to Bulgaria, Germany, Hungary, Italy, Poland, Romania, South Africa, Spain and UK.

DESIGN FEATURES: Provision for fitting range-extending tip tanks. Wing sections NACA 0018 (root), 0012 (tip); dihedral 1° 7′ 24″.

FLYING CONTROLS: Long-span, two-segment mass balanced ailerons on each wing (no flaps); inner ailerons have geared tabs, port outer has ground adjustable tab; trim tab and geared tab on elevators, geared tab on rudder; ailerons and elevators operated by rods, rudder by cables; fixed incidence tailplane.

STRUCTURE: All-metal except for fabric covered elevators and rudder; single tip-to-tip main spar with auxiliary spar; duralumin skins; stressed skin fuselage; braced tailplane.

LANDING GEAR: Non-retractable tailwheel type. Mainwheels carried on flat-spring titanium cantilever legs. Mechanical mainwheel brakes actuated by rudder pedals. Fully castoring tailwheel, with flat-spring shock absorption, has automatic locking device to maintain aircraft on a straight track during taxying, take-off and landing. Mainwheel tyres size 350 × 135 mm, pressure 2.5 bars (36 lb/sq in); tailwheel tyre size 200 × 80 mm, pressure 1.0 bar (14.5 lb/sq in). Mainwheel fairings optional.

POWER PLANT: One 224 kW (300 hp) Textron Lycoming AEIO-540-L1B5D flat-six engine, driving a Hoffmann HO-V123K-V/200AH three-blade constant-speed wooden propeller. Single main fuel tank in fuselage, aft of firewall, capacity 60 litres (15.9 US gallons; 13.2 Imp gallons). Auxiliary 50 litre (13.2 US gallon; 11 Imp gallon) tank can be attached to each wingtip for cross-country flights only. Fuel and oil systems designed for full aerobatic manoeuvres, including inverted flight. Oil capacity 12 litres (3.2 US gallons; 2.6 Imp gallons).

ACCOMMODATION: Single seat under fully transparent sideways opening (to starboard) bubble canopy, which can be jettisoned in emergency. Seat and backrest adjustable, permit use of back type parachute. Cockpit ventilated by sliding panel in canopy.

SYSTEMS: Electrical system includes 24V LUN 2111 alternator as main power source and two 12Ah batteries. External power socket in fuselage side for engine starting.

AVIONICS: VHF radio optional.

Zlin 50 M single-seat aerobatic aircraft (Avia M 137 AZ inline engine) (*Václav Jukl/Letectvi + Kosmonautika*)

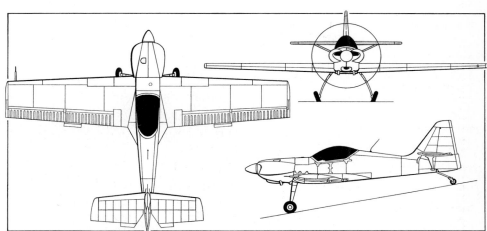

The 224 kW (300 hp) Lycoming-powered Zlin 50 LS aerobatic aircraft (*Jane's/Mike Keep*)

DIMENSIONS, EXTERNAL:
Wing span	8.58 m (28 ft 1¾ in)
Wing span over tip tanks	9.03 m (29 ft 7½ in)
Wing chord: at root	1.73 m (5 ft 8¼ in)
at tip	1.21 m (3 ft 11¾ in)
Wing aspect ratio	5.9
Length overall (tail up)	6.62 m (21 ft 8¾ in)
Height over tail (static)	2.075 m (6 ft 9¾ in)
Elevator span	3.44 m (11 ft 3½ in)
Wheel track	1.90 m (6 ft 2¾ in)
Wheelbase	5.05 m (16 ft 7 in)
Propeller diameter	2.00 m (6 ft 6¾ in)
Propeller ground clearance (tail up)	0.31 m (1 ft 0¼ in)

AREAS:
Wings, gross	12.50 m² (134.55 sq ft)
Ailerons (total)	2.80 m² (30.14 sq ft)
Fin	0.59 m² (6.35 sq ft)
Rudder, incl tab	0.81 m² (8.72 sq ft)
Tailplane	1.66 m² (17.87 sq ft)
Elevators (total, incl tabs)	1.20 m² (12.92 sq ft)

WEIGHTS AND LOADINGS (A: Aerobatic, N: Normal category):
Weight empty: A	600 kg (1,322 lb)
N	610 kg (1,345 lb)
Max T-O weight: A	760 kg (1,675 lb)
N	840 kg (1,852 lb)
Max wing loading: A	60.8 kg/m² (12.45 lb/sq ft)
N	67.2 kg/m² (13.76 lb/sq ft)
Max power loading: A	3.40 kg/kW (5.58 lb/hp)
N	3.75 kg/kW (6.17 lb/hp)

PERFORMANCE (at max Aerobatic T-O weight):
Never-exceed speed (VNE)	181 knots (337 km/h; 209 mph) CAS
Max level speed at 500 m (1,640 ft), ISA	166 knots (308 km/h; 191 mph)
Max cruising speed at 500 m (1,640 ft), ISA	148 knots (275 km/h; 171 mph)
Stalling speed, engine idling	56 knots (103 km/h; 81 mph) CAS
Max rate of climb at S/L, ISA	840 m (2,755 ft)/min
Service ceiling, ISA	8,175 m (26,820 ft)
T-O run	150 m (492 ft)
T-O to 15 m (50 ft)	300 m (985 ft)
Landing from 15 m (50 ft)	530 m (1,740 ft)
Landing run	300 m (985 ft)
g limits: A	+8/−6
N	+3.8/−1.5

ZLIN 90

TYPE: Four-seat light aircraft.
PROGRAMME: In design stage; first flight expected 1994-95.
DESIGN FEATURES: Aerobatic as single/two-seater.
POWER PLANT: One 194 kW (260 hp) Textron Lycoming O-540 flat-six engine.

ZLIN 137T AGRO TURBO

TYPE: Single-seat turboprop powered agricultural aircraft.
PROGRAMME: First flight of XZ-37T prototype (OK-146), powered by Motorlet M 601 B, 6 September 1981 (see 1981-82 *Jane's*); first flights of lower powered prototypes (OK-072 and OK-074) 12 July and 29 December 1983; certificated to BCAR Section K 1984; production as Z 137T started 1985; South African type approval 3 December 1991.
CURRENT VERSIONS: **Z 137T:** Standard production aircraft.
Z 37T-2: Two-seat training version.
CUSTOMERS: Total of 44 Z 137Ts and two Z 37T-2s built by 1 January 1992 (latest figure received); one Z 137T exported to South Africa.
DESIGN FEATURES: Unbraced low wing, with NACA 33015 (root) and 44012 (tip) sections; incidence 3° at root, 0° at tip; dihedral 7° outboard of mainwheels.
FLYING CONTROLS: Primary surfaces mechanical; ground adjustable tab on each aileron, trim tab in rudder and centre of elevator. Fixed leading-edge slats forward of ailerons; pneumatically actuated double-slotted flaps; wingtip winglets; fin offset to port to counteract engine torque.
STRUCTURE: Wing all-metal single-spar with auxiliary rear spar and metal skin; centre and outer panels built integrally with fuselage; ailerons, one-piece elevator and rudder fabric covered; fuselage steel tube with fabric and metal covering.
LANDING GEAR: Non-retractable tailwheel type, with Technometra oleo-pneumatic mainwheel shock absorbers, Moravan light alloy wheels and Barum tyres. Steerable tailwheel. Mainwheel tyres size 556 × 163 × 254 mm, tailwheel tyre size 290 × 110 mm; pressure 3.45 bars (50 lb/sq in) on all units. Moravan hydraulic drum brakes on mainwheels.
POWER PLANT: One 365 kW (490 shp) Motorlet M 601 Z turboprop, driving an Avia VJ7-508Z three-blade constant-speed propeller. Two metal fuel tanks in wing centre-section, combined capacity 350 litres (92.5 US gallons; 77 Imp gallons). Fuel can be transported to distant airstrips in four auxiliary tanks with a combined capacity of 500 litres (132 US gallons; 110 Imp gallons). Gravity refuelling point in top of each wing. Oil capacity 7 litres (1.8 US gallons; 1.5 Imp gallons). Air intake filter.
ACCOMMODATION: Pilot in enclosed cockpit, on contoured seat with headrest. Forward opening window/door, on starboard side, can be jettisoned in emergency. Auxiliary seat to rear for mechanic or loader. Cockpit heated, and provided with filtered fresh air intake. Two-seat training version available.
SYSTEMS: Pneumatic system of 50 bars (725 lb/sq in) pressure, reduced to 30 bars (435 lb/sq in) for agricultural equipment and flaps. Electrical power supplied by 28V 5.6 kW DC starter/generator.
AVIONICS: LUN 3524 VHF radio standard.
EQUIPMENT: Hopper/tank capacity (max) 1,000 litres (264 US gallons; 220 Imp gallons) of liquid or 900 kg (1,984 lb) of dry chemical. Distribution system for both liquid and dry chemicals is operated pneumatically. Chemicals can be jettisoned in 5 s in emergency. Steel cable cutter on windscreen and each mainwheel leg; steel deflector cable runs from tip of windscreen cable cutter to tip of fin. Windscreen washer and wiper standard. Other equipment includes gyro compass, clock, rearview mirror, second (mechanic's) seat, cockpit air-conditioning, ventilation and heating, and anti-collision light. Can be modified for firefighting role.

DIMENSIONS, EXTERNAL:
Wing span	13.63 m (44 ft 8½ in)
Wing chord: at root	2.39 m (7 ft 10 in)
at tip	1.224 m (4 ft 0¼ in)
Wing aspect ratio	7.0
Length overall (flying attitude)	10.46 m (34 ft 4 in)
Fuselage: Max width	1.70 m (5 ft 7 in)
Height overall	3.505 m (11 ft 6 in)
Elevator span	5.294 m (17 ft 4½ in)

Wheel track	3.30 m (10 ft 10 in)
Wheelbase	6.375 m (20 ft 11 in)
Propeller diameter	2.50 m (8 ft 2½ in)
Propeller ground clearance (min)	0.45 m (1 ft 5¾ in)

AREAS:
Wings, gross	26.69 m² (287.3 sq ft)
Ailerons (total)	2.428 m² (26.13 sq ft)
Trailing-edge flaps (total)	4.37 m² (47.04 sq ft)
Fin	1.185 m² (12.76 sq ft)
Rudder, incl tab	1.054 m² (11.35 sq ft)
Tailplane	2.776 m² (29.88 sq ft)
Elevator, incl tab	3.008 m² (32.38 sq ft)

WEIGHTS AND LOADINGS:
Weight empty with basic agricultural equipment	1,250 kg (2,756 lb)
Max payload	900 kg (1,984 lb)
Max fuel	280 kg (617 lb)
Max T-O weight: ferry flights	2,260 kg (4,982 lb)
agricultural, forestry and waterways work	2,525 kg (5,566 lb)
Max wing loading	89.9 kg/m² (18.41 lb/sq ft)
Max power loading	6.67 kg/kW (10.95 lb/shp)

PERFORMANCE (at 2,525 kg; 5,566 lb max T-O weight):
Never-exceed speed (V_{NE})	153 knots (285 km/h; 177 mph)
Max level speed at 500 m (1,640 ft)	118 knots (218 km/h; 135 mph)
Max cruising speed at 500 m (1,640 ft)	103 knots (190 km/h; 118 mph)
Working speed	78-89 knots (145-165 km/h; 90-103 mph)
Stalling speed:	
flaps up	48 knots (88 km/h; 55 mph)
flaps down	42 knots (77 km/h; 48 mph)
Max rate of climb at S/L	252 m (827 ft)/min
T-O run	265 m (870 ft)
T-O to 15 m (50 ft)	580 m (1,905 ft)
Landing from 15 m (50 ft)	720 m (2,365 ft)
Landing run	300 m (985 ft)
Range with max internal fuel	188 nm (350 km; 217 miles)
Swath width: granules	30 m (98 ft)
liquid	40 m (131 ft)
g limits	+3.2/−1.28

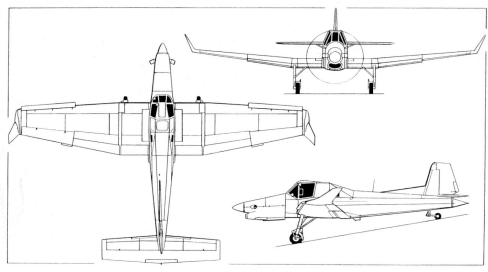

Z 137T Agro Turbo agricultural aircraft (Motorlet M 601 Z turboprop) *(Dennis Punnett)*

Zlin 137T Agro Turbo *(Martin Fricke)*

EGYPT

AOI
ARAB ORGANISATION FOR INDUSTRIALISATION
2D Abbassiya Square, PO Box 770, Cairo
Telephone: 20 (2) 932822 or 823377
Fax: 20 (2) 826010
Telex: 92090 or 92014 AOI UN
CHAIRMAN: Lt General Ibrahim Al Orabi
MANAGING DIRECTOR: Eng Mohamed El Khateeb
OPERATIONS AND MARKETING DIRECTOR:
  Eng Fathy Ibrahim El Shal

AOI founded November 1975 by Egypt, Saudi Arabia, Qatar and United Arab Emirates as basis of Arab military industry; owns five factories and shares in four joint ventures; total workforce about 20,000 including about 3,000 in the joint ventures.

Aircraft Factory, PO Box 11722, Helwan
Telephone: 20 (2) 784303 or 782407
Fax: 20 (2) 782408
Telex: 92191 NASR UN
CHAIRMAN: Eng Ahmed El Sayed

Assembly and production of fixed-wing aircraft; manufactures parts for Mirage 2000 and drop tanks for Mirage 5 (1,300 litres; 343.4 US gallons; 286 Imp gallons), Mirage 2000 (1,700 litres; 449 US gallons; 374 Imp gallons), MiG-17 (400 litres; 105.7 US gallons; 88 Imp gallons), MiG-21 (480 and 800 litres; 127 and 211.3 US gallons; 105.5 and 176 Imp gallons) and F-16 (1,173 litres; 310 US gallons; 258 Imp gallons). Latest product is multi-purpose ultralight aircraft, first flown October 1991.

Engine Factory, PO Box 12, Helwan
CHAIRMAN: Dr Mohamed El Semary

Assembly, production and overhaul of aero engines, and engine parts manufacture; overhauls Atar 9Cs, Larzac 04s and PT6As of Egyptian Mirage 5s, Alpha Jets and Tucanos.
Kader Factory, PO Box 287, Heliopolis
SAKR Factory, PO Box 33, Heliopolis
Electronics Factory, PO Box 84, Heliopolis
JOINT VENTURES:
Arab British Helicopter Co (ABH)
Arab British Engine Co (ABECo)
Arab British Dynamics Co (ABD)
Arab American Vehicle Co (AAV)

ABH founded 27 February 1978; currently undertakes partial overhaul of Gazelle helicopter, for which many of necessary tools also manufactured. ABECo founded 26 February 1978 to assemble/repair/overhaul helicopter engines, currently including TV2-117A for Mi-8, Gnome H.1400-1 for Commando and Sea King, VO-540-C20 for Hillers and Astazou XIVH for Gazelle. British Aerospace withdrew from ABD in 1992.

ETHIOPIA

EAL
ETHIOPIAN AIRLINES S. C.
PO Box 1755, Addis Ababa
Telephone: 251 (1) 612222
Fax: 251 (1) 611474
Telex: 21012 ETHAIR ADDIS
GENERAL MANAGER: Capt Zelleke Demissie
DIRECTOR, AGRO AIRCRAFT MANUFACTURING:
  Col Taddele Mekuria

Schweizer (see USA) Ag-Cat Super B Turbine being assembled and part-manufactured under name **Eshet** for Ethiopian use and export; first aircraft (ET-AIY) rolled out 20 December 1986; same as Schweizer version, except higher tyre pressure 3.83 bars (55.58 lb/sq in) and empty weight 1,500 kg (3,307 lb). Power plant is one 559 kW (750 shp) PT6A-34AG turboprop driving 2.69 m (8 ft 10 in) Hartzell HC-B3TN-3D three-blade propeller; optional engine is 507 kW (680 shp) PT6A-15AG; optional max fuel load 435 litres (115 US gallons; 95.75 Imp gallons) and oil capacity 10.6 litres (2.8 US gallons; 2.3 Imp gallons).

Ethiopian Airlines Eshet licence built version of the Schweizer Ag-Cat Super B Turbine

EAL has sole rights to build, market and service Ag-Cat throughout Africa except Algeria, Tunisia and South Africa; proportion of local manufacture being gradually increased; 12 Eshets delivered by end 1992 with more under construction.

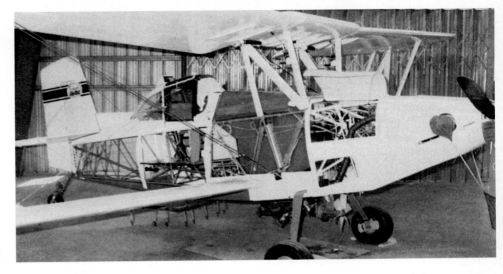

Eshet with fuselage side panels removed for cleaning

FINLAND

VALMET
VALMET AVIATION INDUSTRIES INC
Kuninkaankatu 30 (PO Box 11), SF-33201 Tampere
Telephone: 358 (31) 239444
Fax: 358 (31) 239543
PRESIDENT: Markku Valtonen
Training Aircraft Division
PO Box 53, SF-35601 Halli
Telephone: 358 (42) 8291
Fax: 358 (42) 829600
Telex: 28269
ADMINISTRATION MANAGER: Maija Aro-Suoranta

Valmet and predecessors have built 30 types of aircraft, including 19 of Finnish design, since 1922; current activities include manufacturing L-90 TP Redigo turboprop trainer, production of aircraft and aero engine parts, maintenance/overhaul/repair/modification of aircraft and accessories, aero engines and accessories, marine diesel engines, accessories and gears and foundry products. Contract signed 1989 for Valmet to produce fin, rudder, tailplane and elevators for Saab 2000; will assemble 57 F/A-18Cs for Finnish Air Force from CKD kits during period 1995-2000, and will also co-produce fuselage dorsal and side skin panels.

Facilities in Kuorevesi, Linnavuori and Tampere; workforce about 950 at 1 January 1992.

VALMET L-90 TP REDIGO
TYPE: Two/four-seat multi-stage turboprop military trainer.
PROGRAMME: Design started 1983; first flight Allison-powered prototype (OH-VTP) 1 July 1986; first flight second prototype (OH-VTM), powered by Turbomeca TP 319 derated to 313 kW (420 shp), December 1987; second prototype lost in accident 28 August 1988; Finnish certification September 1991.
CUSTOMERS: Finnish Air Force ordered 10 on 6 January 1989; deliveries completed 1992. Announced September 1992 that deliveries of 18 aircraft to three unnamed customers (one since revealed as Mexico for 10) were then in progress.
DESIGN FEATURES: Designed to FAR Pt 23 and BCAR Section K; also MIL-A-8866B fatigue spectrum and 30,000 landings in heavy military use; intended to cover training syllabus from primary to BAe Hawk level; additional roles can include search and rescue, photographic reconnaissance, weapons training and target towing.

Tapered wing with forward-swept inboard leading-edges and turned-up tips; wing section NACA 63-218 (mod B3) at root, 63-412 (mod B3) at tip; dihedral 6° from roots; incidence 3° at root; twist −3°.
FLYING CONTROLS: Elevators, ailerons and rudder cable-operated; geared tabs in elevators and rudder each also operate as trim tabs; modified Frise type mass balanced ailerons; geared tab and spring tab in each aileron; starboard geared tab also trims; slotted flaps electrically actuated.
STRUCTURE: Fail-safe all-metal structure; wing has main and auxiliary spars; CFRP and glassfibre used for engine cowlings, wingroot fairings, tailcone and dorsal fin; glassfibre wingtips; in-wing fuel tanks of Valmet load-bearing sandwich construction; fluted aluminium skin on elevators and flaps.
LANDING GEAR: AP Precision Hydraulics electrohydraulically retractable tricycle type, with single wheel and oleo-pneumatic shock absorber on each unit. Nosewheel, centred by spring, is steerable ±25° and retracts rearward; main units retract inward into wings. Spring assisted lowering of all units in event of emergency. Mainwheel tyres size 17.5 × 6.3-6.0 in, pressure 3.79 bars (55 lb/sq in); nosewheel tyre size 14.2 × 4.95-5.0 in, pressure 3.45 bars (50 lb/sq in). Differential brakes on mainwheels. Parking brake. Min ground turning radius 5.00 m (16 ft 4¾ in).
POWER PLANT: One Allison 250-B17F turboprop (max power 373 kW; 500 shp), flat rated at 313 kW (420 shp), driving a Hartzell HC-B3TF-7A/T10173-15 three-blade constant-speed reversible-pitch propeller. Fuel in four wing tanks and a fuselage collector tank, total usable capacity 378 litres (100 US gallons; 83.2 Imp gallons). Collector tank of 9 litres (2.4 US gallons; 2 Imp gallons) can be used for up to 30 s of inverted flight. Gravity refuelling point in top of each wing tank. Oil capacity 9 litres (2.4 US gallons; 2 Imp gallons).

First prototype of the Valmet L-90 TP Redigo two/four-seat military trainer, modified to production standard

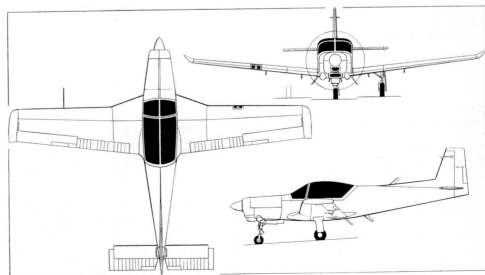

Valmet L-90 TP Redigo turboprop powered multi-stage trainer *(Dennis Punnett)*

ACCOMMODATION: Instructor and pupil, side by side, beneath one-piece rearward sliding jettisonable canopy with steel tube turnover windscreen frame. Canopy can be locked in partially open position if required. Zero/zero rocket assisted escape system optional. Dual controls standard, but instructor's or pupil's control column can be removed if desired. Both front seats adjustable longitudinally and for rake, and fitted with five-point seat belts and inertia reel shoulder harnesses. Provision for two more seats at rear, with four-point harnesses, which can be removed to make room for up to 200 kg (440 lb) of baggage. As ambulance, can accommodate one stretcher patient, and a medical attendant or sitting patient, in addition to pilot. Accommodation heated by bleed air heater. Windscreen heated and ventilated by heat exchanger, fresh air intake and mixer unit. Auxiliary fresh air intake in fin leading-edge. Air-conditioning system optional.

SYSTEMS: No hydraulic or pneumatic systems. Electrical system is 28V DC, powered by a 200A engine driven starter/generator, with a 23Ah nickel-cadmium battery for emergency supply and engine starting. Ground power receptacle. Emergency battery for main artificial horizon. Anti-icing for engine air intake, spinner and propeller blades. Oxygen system available to customer's requirements.

AVIONICS: Dual controls and instrumentation for day and night VFR and IFR operation, including VHF com radios (two), ADF, DME, transponder, RMI, HSI, and marker beacon receiver.

EQUIPMENT: Six underwing attachments, each inner point stressed for 250 kg (551 lb) and the other four for 150 kg (331 lb) each; max external stores load 800 kg (1,764 lb). When flown solo, can carry up to four photographic, TV, radar or reconnaissance pods plus two flares. As two-seater, typical loads can include five liferafts or emergency packs and one searchlight pod; and photo and TV pods. Provision for target towing with winch and hit counters. Twin landing lights in starboard wing leading-edge.

DIMENSIONS, EXTERNAL:

Wing span	10.60 m (34 ft 9¼ in)
Wing chord: at root	1.827 m (6 ft 0 in)
mean aerodynamic	1.497 m (4 ft 11 in)
at tip	1.109 m (3 ft 7⅔ in)
Wing aspect ratio	7.62
Length overall	8.53 m (27 ft 11¾ in)
Fuselage: Max width	1.22 m (4 ft 0 in)
Height overall	3.20 m (10 ft 6 in)
Elevator span	3.684 m (12 ft 1 in)
Wheel track	3.367 m (11 ft 0½ in)
Wheelbase	2.112 m (6 ft 11¼ in)
Propeller diameter	2.194 m (7 ft 2½ in)
Propeller ground clearance	0.283 m (11¼ in)

DIMENSIONS, INTERNAL:

Cockpit: Length	2.00 m (6 ft 6¾ in)
Max width	1.14 m (3 ft 9 in)
Height (seat cushion to canopy)	1.02 m (3 ft 4¼ in)

AREAS:

Wings, gross	14.748 m^2 (158.75 sq ft)
Ailerons (total, incl tabs)	1.996 m^2 (21.48 sq ft)
Trailing-edge flaps (total)	1.766 m^2 (19.01 sq ft)
Fin	0.718 m^2 (7.73 sq ft)
Rudder, incl tab	0.70 m^2 (7.53 sq ft)
Tailplane	1.609 m^2 (17.32 sq ft)
Elevators (total, incl tabs)	1.421 m^2 (15.30 sq ft)

WEIGHTS AND LOADINGS (A: Aerobatic category, U: Utility, N: Normal category):

Weight empty, equipped	950 kg (2,094 lb)
Max fuel	303 kg (668 lb)
External stores: max	800 kg (1,764 lb)
with max fuel	600 kg (1,323 lb)
Max T-O weight: A	1,350 kg (2,976 lb)
U	1,470 kg (3,241 lb)
U (with external stores)	1,900 kg (4,189 lb)
N	1,600 kg (3,527 lb)
Max landing weight	1,530 kg (3,373 lb)
Max wing loading: A	91.54 kg/m^2 (18.75 lb/sq ft)
U	99.67 kg/m^2 (20.41 lb/sq ft)
U (with external stores)	128.83 kg/m^2 (26.39 lb/sq ft)
N	108.49 kg/m^2 (22.22 lb/sq ft)
Max power loading: A	4.31 kg/kW (7.08 lb/shp)
U	4.69 kg/kW (7.70 lb/shp)
U (with external stores)	6.07 kg/kW (9.97 lb/shp)
N	5.11 kg/kW (8.40 lb/shp)

PERFORMANCE (at max T-O weight, ISA):

Never-exceed speed (VNE) 251 knots (465 km/h; 289 mph) CAS

Max level speed 224 knots (415 km/h; 258 mph) CAS

Max cruising speed at 2,400 m (7,875 ft) 190 knots (352 km/h; 219 mph)

Cruising speed (75% max continuous power) at 2,400 m (7,875 ft) 168 knots (312 km/h; 194 mph)

Stalling speed, engine idling:

flaps up	65 knots (120 km/h; 75 mph)
15° flap	60 knots (111 km/h; 70 mph)
35° flap	55 knots (101 km/h; 63 mph)
Max rate of climb at S/L	540 m (1,771 ft)/min
Time to height: 3,000 m (9,845 ft)	5 min
5,000 m (16,400 ft)	11 min
Service ceiling (engine limited)	7,620 m (25,000 ft)
T-O run	240 m (788 ft)
T-O to 15 m (50 ft)	340 m (1,116 ft)
Landing from 15 m (50 ft)	410 m (1,345 ft)
Landing run (without propeller reversal)	240 m (788 ft)

Range at 6,000 m (19,685 ft) with max internal fuel, 30 min reserves 755 nm (1,400 km; 870 miles)

Endurance, conditions as above 6 h 20 min

g limits +7/−3.5 aerobatic
+2.8 max sustained

FRANCE

AEROSPATIALE

AEROSPATIALE SNI

37 Boulevard Montmorency, F-75781 Paris Cedex 16
Telephone: 33 (1) 42 24 24 24
Fax: 33 (1) 42 24 21 32
Telex: AISPA 640025 F
PRESIDENT AND CEO: Louis Gallois
DIRECTOR OF INFORMATION AND COMMUNICATIONS:
Patrice Kreis

AIRCRAFT GROUP

PRESIDENT AND COO: Jacques Plenier
AIRBUS PROGRAMME DIRECTOR: Charles Benaben
REGIONAL TRANSPORT PROGRAMME DIRECTOR:
Jean-Paul Perrais
DIRECTOR OF COMMUNICATIONS: Patrice Prevot
WORKS AND FACILITIES:
Toulouse. PLANT MANAGER: Jean-Marie Mir
Nantes Bouguenais. PLANT MANAGER: Christian Beugnet
Saint-Nazaire. PLANT MANAGER: Jacques Crusson
Méaulte. PLANT MANAGER: Claude Berlan

AVIATION SUBSIDIARIES OF AEROSPATIALE GROUP

Maroc Aviation (part holding)
Société de Construction d'Avions de Tourisme et d'Affaires (SOCATA) (listed separately in this section)
SOGERMA-SOCEA
Société d'Exploitation et de Constructions Aéronautiques (SECA)
Sextant Avionique (50:50 with Thomson-CSF)

Aerospatiale formed 1 January 1970 by French government-directed merger of Sud-Aviation, Nord-Aviation and SEREB. Aerospatiale group sales increased 1.7 per cent in 1992 with helicopters declining, civil aircraft advancing slightly and space and defence increasing considerably. Extraordinary items of over FFr1 billion brought overall group loss to FFr2.38 billion. Order intake was constant at FFr39 billion and a backlog of FFr150 billion covers some three years' operations; workforce considerably reduced from 43,000, including 14,000 in Aircraft Group.

Besides programmes listed below, Aerospatiale has 37 per cent share in Airbus Industrie (see International section); main Airbus assembly plant is at Aerospatiale base at Toulouse. Group helicopter activities now vested in Eurocopter SA, which see in International section. Part holding in Aerospatiale (20 per cent) held by SOGEPA (which see).

AEROSPATIALE HELICOPTERS

These now listed under Eurocopter in International section, as is Eurocopter Tiger. International programmes in which Aerospatiale was a partner (EC120, formerly P120L, NH 90 and Eurofar) are listed as separate international programmes with Eurocopter as the partner.

AEROSPATIALE LIGHT AIRCRAFT (SOCATA)

Products of Aerospatiale's light aircraft manufacturing subsidiary, Socata, are listed separately in this section, as in previous years.

AEROSPATIALE/ALENIA ATR 42/72

Equally owned Groupement d'Intéret Economique formed with Aeritalia (now Alenia) 5 February 1982 to manufacture 42-passenger and 72-passenger turboprop regional transports. For details of these and developments of them, see ATR in International section.

AEROSPATIALE ATSF

Aerospatiale's own studies of advanced supersonic airliner (Avion de Transport Supersonique Futur) continue. Company has joined the other international SST studies with Britain, Germany, Japan and USA summarised under SCTICSG in International section.

VLCT/UHCA

Aerospatiale participating in US, British, German and Japanese study outside Airbus Industrie of feasibility of a very large commercial transport/ultra-high capacity airliner seating between 500 and 800 passengers. See in International section on page 126.

EUROFLAG

Aerospatiale is one of companies co-operating in development of this future military tactical transport (see International section).

EUROPATROL

Aerospatiale is one of group of European companies investigating development of a new international maritime patrol aircraft (see International section).

AVIASUD

AVIASUD ENGINEERING SA

Zone Industrielle la Palud, F-83618 Fréjus Cedex
Telephone: 33 94 53 94 00
Fax: 33 94 52 12 23
Telex: 409 926
GENERAL MANAGER: Bernard d'Otreppe

Details of the Sirocco single-seat microlight can be found in the 1992-93 *Jane's*. The company's latest aircraft is the AE 209 Albatros two-seat microlight, from which it is developing a version to be certificated under JAR-VLA as the AE 210 Alizé, which should become available in 1994.

AVIASUD AE 206 MISTRAL

TYPE: Side by side two-seat microlight and ARV aircraft; designed to FAR Pt 23.

PROGRAMME: Designed for recreational flying but also suited to professional activities normally performed by high-cost conventional aeroplanes and helicopters, such as pilot training, aerial photography, TV and surveillance, banner towing and cropspraying. Design began 1983; prototype

Aviasud AE 206 Mistral (47.7 kW; 64 hp Rotax 582)

first flown May 1985; first production aircraft flown February 1986. An agricultural example sold to a Spanish operator has 125 litre (33 US gallon; 27.5 Imp gallon) underfuselage chemical tank.

CURRENT VERSIONS: **AE 206 Mistral**: *As detailed below.*
    AE 206 US: Ultra silent version with 3.48:1 reduction drive and larger propeller.
CUSTOMERS: 190 ordered (185 delivered) by February 1993.
COSTS: FFr220,000 (excl VAT).
DESIGN FEATURES: Braced biplane. Wings have NACA 23012 section; leading-edge sweepback 6° 30'; dihedral 4°; incidence 7°; twist 0°.
FLYING CONTROLS: Manually activated; all-moving lower wings, all-moving tailplane with anti-servo tab (used also for pitch trim), and rudder.
STRUCTURE: Zicral circular wing spars and laminated wooden ribs, covered with PVF film; composites monocoque fuselage and fin, of glassfibre/polyester resin construction.
LANDING GEAR: Non-retractable tricycle landing gear, with optional floats and skis. Disc brakes.
POWER PLANT: One 47.7 kW (64 hp) Rotax 582 with reduction gear, driving three-blade fixed-pitch propeller. Fuel capacity 60 litres (16 US gallons; 13.2 Imp gallons).

DIMENSIONS, EXTERNAL:
Wing span	9.40 m (30 ft 10 in)
Wing chord, constant	0.86 m (2 ft 10 in)
Length overall	5.66 m (18 ft 6¼ in)
Height overall	2.25 m (7 ft 4½ in)
Tailplane span	2.70 m (8 ft 10¼ in)
Wheel track	1.84 m (6 ft 0½ in)
Wheelbase	1.42 m (4 ft 8 in)
Propeller diameter: normal	1.65 m (5 ft 5 in)
US version	1.75 m (5 ft 9 in)

AREAS:
Wings, gross	16.39 m² (176.42 sq ft)

WEIGHTS AND LOADINGS:
Weight empty, equipped	205 kg (452 lb)
Baggage capacity	20 kg (44 lb)
Max T-O and landing weight	390 kg (860 lb)
Max wing loading	23.8 kg/m² (4.87 lb/sq ft)
Max power loading	8.18 kg/kW (13.44 lb/hp)

PERFORMANCE (at max T-O weight):
Never-exceed speed (V_{NE})	89 knots (165 km/h; 103 mph)
Max level speed	84 knots (155 km/h; 96 mph)
Max cruising speed	72 knots (133 km/h; 83 mph)
Econ cruising speed	49 knots (90 km/h; 56 mph)
Stalling speed	33 knots (60 km/h; 38 mph)
Max rate of climb at S/L	232 m (760 ft)/min
Service ceiling	more than 4,575 m (15,000 ft)
T-O run	80 m (262 ft)
T-O to 15 m (50 ft)	200 m (657 ft)
Landing from 15 m (50 ft)	225 m (739 ft)
Range, no reserves	286 nm (530 km; 329 miles)

AVIASUD AE 207 MISTRAL TWIN

TYPE: Twin-engined derivative of Mistral.
CUSTOMERS: Twenty-five sold by February 1993.
COSTS: FFr276,000 (excl VAT).
DESIGN FEATURES: Developed specially for aerial advertising and surveillance work, this is a standard Mistral fitted with a Rotax 582 engine forward and carrying a Rotax 503 in pusher configuration above the upper wing.

WEIGHTS AND LOADINGS:
Weight empty	199-230 kg (439-507 lb)
Max T-O weight	450 kg (992 lb)

PERFORMANCE:
Max rate of climb at S/L	216 m (710 ft)/min

Aviasud AE 209 Albatros

AVIASUD AE 209 ALBATROS

TYPE: Two-seat microlight.
PROGRAMME: Design started 1988; prototype first flight March 1991.
CUSTOMERS: Sixty ordered (25 built) by February 1993.
COSTS: FFr140,000 with Rotax 503; FFr185,000 with Rotax 582.
DESIGN FEATURES: Braced high-wing monoplane; T tail; panoramic visibility; forgiving flying characteristics; folding wings (under one min). Wings NACA 63018 section; sweepback 2°; dihedral 0° 30'; twist 0°.
FLYING CONTROLS: Manual actuation; ailerons, all-moving tailplane with anti-servo tab (also used for pitch trim), rudder.
STRUCTURE: Wing spars and control surfaces of carbonfibre; remainder of airframe glassfibre/polyester or epoxy resin.
LANDING GEAR: Non-retractable tailwheel type; hydraulic disc brakes. Optional retractable skis.
POWER PLANT: One 37.3 kW (50 hp) Rotax 503 flat-twin engine, with 2.58:1 reduction drive to two-blade pusher propeller (optional 4:1 reduction); alternative 47.7 kW (64 hp) Rotax 582 (2.58:1 or 3.47:1 reduction drive) or 59.7 kW (80 hp) Rotax 912 with 2.27:1 reduction drive and three-blade propeller. Fuel capacity 60 litres (16 US gallons; 13.2 Imp gallons).

DIMENSIONS, EXTERNAL:
Wing span	9.70 m (31 ft 10 in)
Wing chord, constant	1.45 m (4 ft 9 in)
Length overall	7.36 m (24 ft 1¼ in)
Height overall	2.13 m (7 ft 0 in)
Tailplane span	2.42 m (7 ft 11¼ in)
Wheel track	1.52 m (5 ft 0 in)
Wheelbase	5.25 m (17 ft 2¾ in)
Propeller diameter: two-blade	1.62 m (5 ft 3¾ in)
three-blade	1.70 m (5 ft 7 in)

AREAS:
Wings, gross	13.90 m² (149.62 sq ft)

WEIGHTS AND LOADINGS (Rotax 582 engine):
Weight empty	230 kg (507 lb)
Fuel weight	42 kg (92.6 lb)
Max T-O and landing weight	450 kg (992 lb)
Max wing loading	32.37 kg/m² (6.63 lb/sq ft)
Max power loading	9.43 kg/kW (15.50 lb/hp)

PERFORMANCE (Rotax 582 engine):
Never-exceed speed (V_{NE})	100 knots (185 km/h; 115 mph)
Max level speed	86 knots (160 km/h; 99 mph)
Max cruising speed	76 knots (140 km/h; 87 mph)
Econ cruising speed	49 knots (90 km/h; 56 mph)
Stalling speed	35 knots (64 km/h; 40 mph)
Max rate of climb at S/L	251 m (825 ft)/min
Service ceiling	4,575 m (15,000 ft)
T-O run	75 m (246 ft)
T-O to 15 m (50 ft)	250 m (821 ft)
Landing from 15 m (50 ft)	280 m (919 ft)
Landing run	65 m (214 ft)
Range, no reserves	280 nm (520 km; 323 miles)

Aviasud AE 207 Mistral Twin

DASSAULT
DASSAULT AVIATION
9 Rond-Point Champs Elysées Marcel Dassault, F-75008 Paris
Telephone: 33 (1) 43 59 14 70
Fax: 33 (1) 42 56 22 49
PRESS INFORMATION OFFICE: 27 rue du Professeur Pauchet, F-92420 Vaucresson
   Telephone: 33 (1) 47 95 86 87
   Fax: 33 (1) 47 95 86 80/47 95 87 40
WORKS: F-92214 St Cloud, F-95101 Argenteuil,
  F-78141 Vélizy-Villacoublay, F-33127 Martignas,
  F-33701 Bordeaux-Mérignac, F-91220 Brétigny,
  F-33630 Cazaux, F-64205 Biarritz-Parme,
  F-13801 Istres, F-74371 Argonay,
  F-59472 Lille-Seclin, F-86000 Poitiers
CHAIRMAN AND CEO: Serge Dassault
DIRECTOR GENERAL, INTERNATIONAL: Pierre Chouzenoux
VICE-PRESIDENT, TECHNICAL, RESEARCH, CO-OPERATION:
  Bruno Revellin-Falcoz
VICE-PRESIDENT, ECONOMICS AND FINANCE:
  Charles Edelstenne
VICE-PRESIDENT, INDUSTRIAL AND SOCIAL AFFAIRS:
  Michel Herchin
DIRECTOR OF COMMUNICATIONS: Gen Pierre Pacalon
   Former Avions Marcel Dassault-Breguet Aviation formed from merger of Dassault and Breguet aircraft companies in December 1971; French government acquired 20 per cent of stock in January 1979, raised to 46 per cent in November 1981; present company name adopted April 1990. Some shares hold double voting rights so that French government has majority control, with 55 per cent. On 17 September 1992, Dassault joined with Aerospatiale in state-owned joint holding company, SOGEPA (which see), which has 35 per cent holding in Dassault Aviation; Dassault and Aerospatiale now pool resources and co-ordinate R&D strategy, although each remains separate entity.

Employees in 1992 totalled 11,000. Business divided 55 per cent military, 35 per cent civil and 10 per cent space. Dassault assembles and tests its civil and military aircraft in its own factories, but operates wide network of subcontractors. Other products include flight control system components, maintenance and support equipment and CAD/CAM software (CATIA). Dassault Aviation is component of Dassault Industries, which also includes Dassault Falcon Service, Dassault Belgique Aviation and Dassault Electronique.

Dassault Aviation has shared in Atlantique programme with Belgium, Germany and Italy; SEPECAT (Jaguar) with Britain; and Alpha Jet with Dornier. Offset manufacture of Dassault aircraft components arranged with Spain, Belgium, Greece, Egypt. Dassault makes fuselages for Fokker at Biarritz-Parme.

More than 6,500 aircraft produced since 1945, including 92 prototypes.

DASSAULT/DORNIER ALPHA JET
Details of latest developments in Alpha Jet programme appear in International section of 1992-93 *Jane's*.

DASSAULT/BAe JAGUAR
Details of latest developments in Anglo-French Jaguar programme appear under SEPECAT in International section. Jaguar continues in production in India (see HAL entry).

HERMES AEROSPACECRAFT
Studies for a new spaceplane have been initiated. Details of the former Hermès can be found in the 1992-93 *Jane's*.

DASSAULT MIRAGE 2000
Indian Air Force name: Vajra (Divine Thunder)
TYPE: Multi-role fighter.
PROGRAMME: Selected as main French Air Force combat aircraft 18 December 1975; first developed as interceptor with SNECMA M53 and Thomson-CSF RDM multi-mode pulse Doppler radar; M53-5 in early production aircraft

FRANCE: AIRCRAFT—DASSAULT

Dassault Mirage 2000C of Escadron de Chasse 1/5 with chaff/flare dispensers in place of braking parachute *(SIRPA 'AIR')*

Mirage 2000D No. 02 equipped with BGL 1000 laser guided bomb, Atlis laser designator and two RPL 541/542 fuel tanks *(K. Tokunaga DACT)*

succeeded by M53-P2; fitted with RDI radar from 38th French Air Force 2000C onwards; first flight of production 2000C 20 November 1982; first flight of production two-seat 2000B, 7 October 1983; first unit, EC 1/2 'Cigognes', formed at Dijon 2 July 1984; intended interceptor fleet is three wings (escadres) of three squadrons each.

CURRENT VERSIONS: **2000B**: Two-seat trainer counterpart of 2000C; Nos. 501-514 (Series) S3 with RDM radar and M53-5 power plant; Nos. 515-520, S4 with RDI J1-1 radar and M53-5, but Nos. 516 and 517 retrofitted with RDM; No. 521 S4-2 with RDI J2-4 and M53-5; No. 522 also S4-2, but M53-P2; Nos. 523-533 S5 with RDI J3-13 and M53-P2. See also Mirage 2000DA below.

2000BOB: (Banc Optronique Biplace – Two-seat Optronics Testbed). Mirage 2000B No. 504 flown 28 June 1989 after modification by CEV at Brétigny-sur-Orge; trials of Rubis FLIR, VEH-3020 holographic HUD, night vision goggles and other electro-optical systems. Fitted with OBOGS.

2000C: Standard interceptor; Nos. 1-37 built as series S1, S2 and S3 with RDM radar and M53-5 power plants; since upgraded to S3 radar standard. Loosely called Mirage **2000RDM**; Mirage 2000B and C collectively known as Mirage **2000DA** (Défense Aérienne). Surviving 30 offered to Pakistan early 1992, but not purchased; instead, FFr4,600 million ($830 million) conversion programme announced November 1992 for upgrading to **2000C-5** standard (see 2000-5 entry, below) for continued French service; conversion work by Dassault on initial 15 begins 1994, with 15 to follow in 1995. Later aircraft (loosely **2000RDI**) have RDI radar and M53-P2 power plants; Nos. 38-48 Series S4, delivered from July 1987 and later upgraded to S4-1; Nos. 49-63 S4-1; Nos. 64-74 S4-2; No. 75 upwards S5, beginning late 1990. Equipment standards of Mirage 2000B/Cs are **S3**, incapable of launching Matra 530D (530F only); **S4** RDI J1-1 radar; **S4-1** retrofit of all S4s with improved J1-2; **S4-2** further radar upgrade to J2-4; **S4-2A** retrofit of all S4-1 and S4-2 aircraft (Nos. 38-74, 515 and 518-522) with HOTAS-type throttle and improved J2-5 radar; **S5** definitive standard with J3-13 radar and, from No. 93 onwards, Spirale chaff/flare dispenser. *Detailed description applies to 2000C except where indicated.*

2000D: Two-seat conventional attack version of 2000N, lacking ASMP missile interface and nose pitot but with improved ECM and Navstar GPS; first flight (D01, ex-N01) 19 February 1991; second prototype, D02 (ex-N02), flown 24 February 1992; first 2000D delivered CEAM Mont de Marsan for trials 9 April 1993; deliveries (Nos. 601-675) due 1993 to EC 3 at Nancy for formation of first squadron on 1 January 1994.

2000E: Multi-role fighter for export; M53-P2 power plant throughout.

2000ED: Two-seat trainer counterpart of 2000E.

2000N: Two-seat low altitude penetration version to deliver ASMP nuclear stand-off missile; two prototypes; first flight 3 February 1983; first 25 production aircraft (Nos. 301-325) were 2000N-K1 with ASMP capability only; from July 1988 remaining aircraft, designated 2000N-K2, have full conventional and ASMP capability; production ended 1992; some K1s (of squadrons 3/4 and part of 2/4) retrofitted with partial air-to-ground capability for conventional attack. Equipment includes Antilope 5 terrain-following radar, two SAGEM inertial platforms, two improved TRT AHV-12 radio altimeters, Thomson-CSF colour CRT, Omera vertical camera, two Magic self-defence missiles, and ICMS (integrated countermeasures system) comprising Sabre jamming system, Serval RWR and Spirale automatic chaff/flare dispenser system.

2000N' (N Prime): Initial designation of 2000D.

2000R: Single-seat day/night reconnaissance export version of 2000E but with normal radar nose; various sensor pods possible (see Avionics paragraph).

2000-3: Private venture upgrade, begun 1987, with Rafale-type multi-function cockpit displays known as APSI (advanced pilot system interface); prototype BX1 (ex-No. 501) flew 10 March 1988.

2000-4: Private venture integration of Matra Mica AAM. First guided flight against target drone, 9 January 1992.

2000-5: Multi-role upgrade incorporating -3 and -4 improvements, plus Thomson-CSF RDY radar and new central processing unit, Thomson-CSF VEH 3020 HUD and ICMS Mk 2 countermeasures; Matra Super 530D and BAe Sky Flash as alternatives to Mica; laser-guided bombs and ASMs, or Apache stand-off dispenser in air-to-ground role. First Mirage flight of RDY radar in BY1 (ex-No. B01, later BY2) 10 March 1988; first flight of full 2000-5 (two-seat) 24 October 1990 (same aircraft; no serial number); first single-seater (conversion of trials aircraft CY1) flown 27 April 1991. Options include SNECMA M88-P20 uprated by 4 per cent to 98.06 kN (22,046 lb st); available 1995.

2000S: Export attack version of 2000D; announced April 1989; available 1994.

CUSTOMERS: France required seven prototypes and 372 production aircraft; reduced in late 1991 to 318 by abandonment of final 24 2000Cs and 30 2000Ds and transfer of 11 single-seat aircraft to trainer contract; no Mirages funded in 1992. Firm orders at 31 December 1991 totalled 33 2000Bs, 135 2000Cs, 75 2000Ns and 75 2000Ds. Some

230 delivered by early 1993; final deliveries due in 1997. Mirage 2000C equipped three squadrons of EC 2 at Dijon (1984-86), three of EC 5 at Orange (1988-90) and began conversion of EC 12 at Cambrai (squadron EC 1/12 'Cambresis' declared operational 28 April 1992). Mirage 2000N deliveries began to EC 4 at Luxeuil, 30 March 1988; EC 1/4 'Dauphine' formed 1 July 1988, now with full dual-role K2 series aircraft; second and third squadrons (K1 aircraft) operational with EC 4 on 1 July 1989 and 1 July 1990 – (latter detached to Istres); pending 2000D, EC 2/3 'Champagne' formed at Nancy 1 September 1991 with 2000N in conventional role. EC 1/3 'Navarre' operational with 2000D from 1 January 1994; EC 3/3 'Ardennes' equips 1994; EC 2/3 converts to 2000D (ex-2000N) 1995.

Egypt ordered 16 2000EMs and four BMs in December 1981; deliveries 30 June 1986 to January 1988; based at Berigat.

India first ordered 36 2000Hs and four THs in October 1982; 26 Hs and four THs temporarily powered by M53-5; final 10 by M53-P2 from outset; first flight of 2000H (KF-101) 21 September 1984; first flight of 2000TH (KT-201) early 1985; No. 7 IAF Squadron 'Battle Axe' formed at Gwalior AB 29 June 1985, coincident with first arrivals in India. Named Vajra (Divine Thunder); second Indian order for six Hs and three THs signed March 1986 and delivered April 1987 to October 1988 to complete No. 1 'Tigers' Squadron.

Peru ordered 24 2000Ps and two 2000DPs in December 1982, but reduced this to 10 2000Ps and two 2000DPs; first 2000DP handed over 7 June 1985; deliveries to Peru from December 1986; Escuadron 412 of Grupo Aéreo de Caza 4 at La Joya inaugurated 14 August 1987.

Abu Dhabi ordered 18 aircraft on 16 May 1983 and took up 18 options in 1985 for a total of 22 2000EADs, eight 2000RADs and six 2000DADs; deliveries delayed from 1986 to 1989 by provision for US weapons such as Sidewinder; deliveries to Maqatra/Al Dhafra completed November 1990 for Nos. 1 and 2 Shaheen (Warrior) Squadrons. Abu Dhabi 2000RADs carry COR2 multi-camera pod, but alternatives include Raphaël-type SLAR 2000 or Harold pods; second 18 have Elettronica ELT/158 threat warning receivers and ELT/558 self-protection jammers; all with Spirale chaff/flare system. Deliveries 7 November 1989 to November 1990.

Greece ordered 36 2000EGMs and four 2000BGMs on 20 July 1985; handed over from 21 March 1988 and delivered from 27 April 1988 for 331 'Aegeas' and 332 'Geraki' Mire Pandos Kairou within 114 Pterix Mahis at Tanagra; deliveries suspended October 1989 at 28th aircraft; resumed 1992 and completed 18 November 1992. Spirale chaff/flare system installed, plus upgraded ECM.

Jordan ordered 10 2000EJs and two 2000DJs on 22 April 1988; all cancelled in August 1991.

Taiwan ordered 60 Mirage 2000-5s with M53-P20 power plants on 17 November 1992. First export sale for -5; deliveries begin 1995.

Total 535 firm orders (excluding seven prototypes and six company-owned trials and demonstrator aircraft) by 1 January 1993 (318 French; 217 exports), of which some 375 delivered, including 157 exports. Provisional agreement signed with Pakistan in January 1992 for 44 Mirage 2000Es.

COSTS: Programme unit costs: Greece $34.5 million.
DESIGN FEATURES: Low-set thin delta wing with cambered section, 58° leading-edge sweep and moderately blended root; area ruled fuselage; cleared for 9 g and 270°/s roll at sub- and supersonic speed carrying four air-to-air missiles.
FLYING CONTROLS: Full fly-by-wire control with SFENA autopilot; two-section elevons on wing move up 16° and down 25°; inner leading-edge slat sections droop up to 17° 30' and outer sections up to 30°; fixed strakes on intake ducts create vortices at high angles of attack that help to correct yaw excursions; small airbrakes above and below wings.
STRUCTURE: Multi-spar metal wing; elevons have carbonfibre skins with AG5 light alloy honeycomb cores; carbonfibre/light alloy honeycomb panel covers avionics bay; most of fin and all rudder skinned with boron/epoxy/carbon; rudder has light alloy skin.
LANDING GEAR: Retractable tricycle type by Messier-Bugatti, with twin nosewheels; single wheel on each main unit. Hydraulic retraction, nosewheels rearward, main units inward. Oleo-pneumatic shock absorbers. Electro-hydraulic nosewheel steering (±45°). Manual disconnect permits nosewheel unit to castor through 360° for ground towing. Light alloy wheels and tubeless tyres, size 360 × 135-6, pressure 8.0 bars (116 lb/sq in) on nosewheels, size 750 × 230-15, pressure 15.0 bars (217 lb/sq in) on mainwheels. Messier-Bugatti hydraulically actuated polycrystalline graphite disc brakes on mainwheels, with anti-skid units. Compartment in lower rear fuselage for brake parachute, or arrester hook or chaff/flare dispenser.
POWER PLANT: One SNECMA M53-P2 turbofan, rated at 64.3 kN (14,462 lb st) dry and 95.1 kN (21,385 lb st) with afterburning or (available 1995) M53-P20 rated at 98.06 kN (22,046 lb st). Movable half-cone centrebody in each air intake. Internal wing fuel tank capacity 1,480 litres (391 US gallons; 326 Imp gallons); fuselage tank capacity 2,498 litres (660 US gallons; 549 Imp gallons) in single-seat aircraft, 2,424 litres (640 US gallons; 533 Imp gallons) in two-seat aircraft. Total internal fuel capacity 3,978 litres (1,050 US gallons; 875 Imp gallons) in 2000C and E, 3,904 litres (1,030 US gallons; 859 Imp gallons) in 2000B, N, D and S. Provision for one jettisonable 1,300 litre (343 US gallon; 286 Imp gallon) RPL-522 96 kg (212 lb) fuel tank under centre of fuselage, and a 1,700 litre (449 US gallon; 374 Imp gallon) RPL-501/502 210 kg (463 lb) drop tank under each wing. Total internal/external fuel capacity 8,678 litres (2,291 US gallons; 1,909 Imp gallons) in 2000C and E, 8,604 litres (2,271 US gallons; 1,892 Imp gallons) in 2000B. Detachable flight refuelling probe forward of cockpit on starboard side. Dassault type 541/542 tanks of 2,000 litres (528 gallons; 440 Imp gallons) are available for the 2000N, D and S wing attachments (and optional on 2000B/C), empty weight 240 kg (529 lb) each, increasing internal/external fuel to 9,204 litres (2,430 US gallons; 2,025 Imp gallons).
ACCOMMODATION: Pilot only in 2000C, on Hispano-Suiza licence-built Martin-Baker F10Q zero/zero ejection seat, in air-conditioned and pressurised cockpit. Pilot-initiated automatic ejection in two-seat aircraft; 500 ms delay between departures. Canopy hinged at rear to open upward and, on Mirage 2000D, covered in gold film to reduce radar signature.
SYSTEMS: ABG-Semca air-conditioning and pressurisation system. Two independent hydraulic systems, pressure

Dassault Mirage 2000DAD two-seat trainer of the Abu Dhabi Air Force

Dassault Mirage 2000-5 single-seat prototype with Matra Mica AAMs under fuselage and Matra Magic 2s outboard underwing (*Aviaplans/F. Robineau*)

Mirage 2000D conventional attack version of 2000N carrying two laser guided bombs and Atlis designator (*Dassault/Aviaplans*)

Dassault Mirage 2000N of Escadron de Chasse 3/4 'Limousin' *(Press Office Sturzenegger)*

Dassault Mirage 2000B-5 two-seat prototype *(Aviaplans/F. Robineau)*

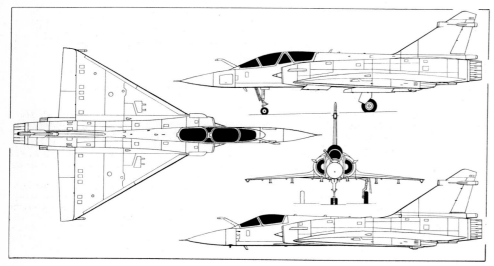
Dassault Mirage 2000N, with added side view (bottom) of Mirage 2000C *(Dennis Punnett)*

280 bars (4,000 lb/sq in) each, to actuate flying control servo units, landing gear and brakes. Hydraulic flow rate 110 litres (29 US gallons; 24 Imp gallons)/min. Electrical system includes two Auxilec 20110 air-cooled 20kVA 400Hz constant frequency alternators (25kVA in Mirage 2000-5), two Bronzavia DC transformers, a SAFT 40Ah battery and ATEI static inverter. Fly-by-wire flight control system. Eros oxygen system.

AVIONICS: Thomson-CSF RDM multi-mode radar or Dassault Electronique/Thomson-CSF RDI pulse Doppler radar, each with operating range of 54 nm (100 km; 62 miles) (Mirage 2000N has Dassault Electronique/Thomson-CSF Antilope 5 terrain-following and ground mapping radar; Antilope 50 in Mirage 2000D.) SAGEM Uliss 52 inertial platform (52E in 2000C and B; 52D for export; and two 52P in 2000N/D, plus Navstar GPS in 2000D), Dassault Electronique Type 2084 central digital computer and Digibus digital databus, Sextant TMV-980 data display system (VE-130 head-up and VMC-180 head-down) (two headdown in 2000N), SFENA 605 autopilot (606 in 2000N, 607 in 2000D, 608 in 2000-5), LMT Deltac Tacan, LMT NRAI-7A IFF transponder, SOCRAT 8900 solid-state VOR/ILS and IO-300-A marker beacon receiver, TRT radio altimeter (AHV-6 in 2000B and C, AHV-9 in export aircraft, two AHV-12 in 2000N and AHV-17 in 2000-5), TRT ERA 7000 V/UHF com transceiver, TRT ERA 7200 UHF or EAS secure voice com, Sextant Avionique Type 90 air data computer, and Thomson-CSF Atlis laser designator and marked target seeker (in pod on forward starboard underfuselage station). Defensive systems in 2000C and 2000N include Thomson-CSF Serval radar warning receiver (antennae at each wingtip and on trailing-edge of fin, near tip, plus VCM-65 cockpit display); Dassault Electronique Caméléon jammer (detector on fin leading-edge); and Matra Spirale in Karman fairings at wing trailing-edge/fuselage intersection comprising flares (left) and chaff (right) integrated with missile plume detectors in rear of Magic launch rails. Spirale fitted to 2000N-K2; retrofitted to 2000N-K1 and installed on 2000Cs from No. 93; earlier 2000Cs have Dassault Electronique Eclair system (Alkan LL5062 chaff and flare launcher) in place of braking parachute, lacking automatic operation. Spirale on some export 2000Es. Upgrade planned of 2000N with ICMS Mk 1 (integrated countermeasures system) comprising Serval, Dassault Electronique/Thomson-CSF Sabre jammers (leading-edge of fin and bullet fairing at base of rudder) and Matra Spirale; automated ICMS Mk 2 of Mirage 2000-5 adds receiver/processor in nose to detect missile command links; extra pair of antennae near top of fin and additional DF antennae scabbed to existing wingtip pods (fin and secondary wingtip antennae also on Greek Mirage 2000s); ABD2000 export version of Sabre in some Mirage 2000Es. Omera vertical camera in 2000N. Other sensors can include a 570 kg (1,257 lb) Thomson-CSF Raphaël SLAR pod, 400 kg (882 lb) Dassault COR2 multi-camera pod or 680 kg (1,499 lb) Dassault AA-3-38 Harold long-range oblique photographic (Lorop) pod; a 110 kg (243 lb) Dassault/TRT/Intertechnique Rubis FLIR pod; a 160 kg (353 lb) Thomson-CSF Atlis laser designator/marked target seeker pod on Mirage 2000D; two 182 kg (401 lb) Thomson-CSF DB 3141/3163 Remora self-defence ECM pods and one 550 kg (1,213 lb) Thomson-CSF TMV 004 (CT51J) Caiman offensive or intelligence ECM pod.

EQUIPMENT: Optional 250 kg (551 lb) Intertechnique 231-300 buddy type in-flight refuelling pod.

ARMAMENT: Two 30 mm DEFA 554 guns in 2000C and 2000E (not fitted in B, D, N or S), with 125 rds/gun. Nine attachments for external stores, five under fuselage and two under each wing. Fuselage centreline and inboard wing stations each stressed for 1,800 kg (3,968 lb) loads; other four fuselage points for 400 kg (882 lb) each, and outboard wing points for 300 kg (661 lb) each. Typical interception weapons comprise two 275 kg (606 lb) Matra Super 530D or (if RDM radar not modified with target illuminator) 250 kg (551 lb) 530F missiles (inboard) and two 90 kg (198 lb) Matra 550 Magic or Magic 2 missiles (outboard) under wings. Alternatively, each of four underwing hardpoints can carry a Magic. Matra Mica AAM (110 kg; 243 lb) optional on Mirage 2000-5. Primary weapon for 2000N is 850 kg (1,874 lb) ASMP tactical nuclear missile mounted on LM-770 centreline pylon. In air-to-surface role, the Mirage 2000 can carry up to 6,300 kg (13,890 lb) of external stores, including eighteen Matra 250 kg retarded bombs or 32.5 kg (72 lb) Thomson-Brandt BAP 100 anti-runway bombs; sixteen 219 kg (483 lb) Matra Durandal penetration bombs; one or two 990 kg (2,183 lb) Matra BGL 1000 laser guided bombs; five or six 305 kg (672 lb) Matra Belouga cluster bombs or 400 kg (882 lb) Thomson-Brandt BM 400 modular bombs; one Rafaut F2 practice bomb launcher; two 520 kg (1,146 lb) Aerospatiale AS.30L, Matra Armat anti-radar, or 655 kg (1,444 lb) Aerospatiale AM 39 Exocet anti-ship, air-to-surface missiles; four 185 kg (408 lb) Matra LR F4 rocket launchers, each with eighteen 68 mm rockets; two packs of 100 mm rockets; or a 765 kg (1,687 lb) Dassault CC 630 gun pod, containing two 30 mm guns and total 600 rounds of ammunition. For air defence weapon training, a Cubic Corporation AIS (airborne instrumentation subsystem) pod, externally resembling a Magic missile, can replace Magic on launch rail, enabling pilot to simulate a firing without carrying actual missile.

DIMENSIONS, EXTERNAL:
Wing span	9.13 m (29 ft 11½ in)
Wing aspect ratio	2.03
Length overall: 2000C, E	14.36 m (47 ft 1¼ in)
2000B, N*	14.55 m (47 ft 9 in)
Height overall: 2000C, E	5.20 m (17 ft 0¾ in)
2000B, N, D, S	5.15 m (16 ft 10¾ in)
Wheel track	3.40 m (11 ft 1¾ in)
Wheelbase	5.00 m (16 ft 4¾ in)

* 2000D, S and -5 versions lack nose pitot

AREAS:
Wings, gross	41.0 m² (441.3 sq ft)

WEIGHTS AND LOADINGS:
Weight empty: 2000C, E	7,500 kg (16,534 lb)
2000B, N, D, S	7,600 kg (16,755 lb)
Max internal fuel: 2000C	3,160 kg (6,967 lb)
2000B, N, D, S	3,100 kg (6,834 lb)
Max external fuel: 2000B, C, E	3,720 kg (8,201 lb)
2000N, D, S	3,751 kg (8,270 lb)
Max external stores load	6,300 kg (13,890 lb)
T-O weight clean: 2000C, E	10,860 kg (23,940 lb)

2000B, N, D, S 10,960 kg (24,165 lb)
Max T-O weight 17,000 kg (37,480 lb)
Max wing loading 414.63 kg/m² (84.97 lb/sq ft)
PERFORMANCE (M53-P2 power plant):
 Max level speed: at height Mach 2.2
  at S/L Mach 1.2
 Max continuous speed: 2000C, E Mach 2.2
  2000N, D, S Mach 1.4
 Min speed in stable flight
  100 knots (185 km/h; 115 mph)
 Approach speed 140 knots (259 km/h; 161 mph)
 Landing speed 125 knots (232 km/h; 144 mph)
 Max rate of climb at S/L 17,060 m (56,000 ft)/min
 Time to 15,000 m (49,200 ft) and Mach 2 4 min
 Time from brake release to intercept target flying at Mach 3 at 24,400 m (80,000 ft) less than 5 min
 Service ceiling 16,460 m (54,000 ft)
 Range: hi-hi-hi 1,000 nm (1,850 km; 1,150 miles)
  interdiction, hi-lo-hi 800 nm (1,480 km; 920 miles)
  attack, hi-lo-hi 650 nm (1,205 km; 748 miles)
  attack, lo-lo-lo 500 nm (925 km; 575 miles)
 with one 1,300 litre and two 1,700 litre drop tanks
  1,800 nm (3,335 km; 2,073 miles)
 g limits: +9.0/−4.5 normal
  +13.5/−9.0 ultimate

DASSAULT RAFALE (SQUALL)

TYPE: Two-seat Avion de Combat Tactique (French Air Force) or single-seat Avion de Combat Marine (French Navy) interceptor, multi-role fighter and reconnaissance aircraft.

PROGRAMME: Ordered to replace French Air Force Jaguars and Navy Crusaders and Super Etendards; for early development history, see 1990-91 and earlier *Jane's*; first flight of Rafale A prototype (F-ZJRE) 4 July 1986; first flight with SNECMA M88 replacing one GE F404, 27 February 1990 (was 461st flight overall); 622 sorties flown by mid-February 1992; ACE International (Avion de Combat Européen) GIE set up in 1987 by Dassault Aviation, SNECMA, Thomson-CSF and Dassault Electronique, partly to attract international partners; none found.

Single-seat Rafale C prototype, C01 (F-ZWVR), ordered 21 April 1988; flown 19 May 1991; 50th sortie 24 October 1991; officially flight tested at CEV in October 1991, two months ahead of schedule; 100th sortie 12 May 1992. First navalised prototype, M01, ordered 6 December 1988; flown 12 December 1991. Catapult trials ashore at US Naval Air Warfare Center, Patuxent River and Lakehurst, July/August 1992; second series of US trials January/February 1993 followed by deck trials on *Foch*; first deck landing, 19 April 1993. Total 1,000 Rafale sorties flown by October 1992. Two-seat, dual-control trainer Rafale B prototype, B01, ordered 19 July 1989 as first with RBE2 radar and Spectra defensive systems; planned February 1993 first flight delayed to 30 April 1993. Second naval prototype, M02, ordered 4 July 1990; to fly Autumn 1993. Second Rafale C order (C02) not placed; abandoned 1991. Test airframe, in Rafale M configuration, delivered to CEAT at Toulouse for ground trials 10 December 1991. French Air Force preference switched to operational two-seat (pilot and WSO) derivative of Rafale C in 1991; announced 1992 that 60 per cent of procurement to be two-seat, although extra cost results in cancellation of 16 aircraft.

Associated programmes include Thomson-CSF/Dassault Electronique electronic scanning RBE2 (Radar à Bayalage Electronique deux plans) multi-mode radar, ordered November 1989; test flights begun in Falcon 20, 10 July 1992; five Falcon 20s and four Mirage 2000s for RBE2 trials, plus final three Rafale prototypes. Dassault Electronique/Thomson-CSF/Matra defensive aids package named Spectra (Système pour la Protection Electronique Contre Tous les Rayonnements Adverses); wholly internal IR detection, laser warning, electromagnetic detection, missile approach warning, jamming and chaff/flare launching; nine prototypes ordered; trials in Mirage 2000 from 1992; total weight 250 kg (551 lb). Development contract awarded 1991 for Thomson-TRT/SAT OSF (Optronique Secteur Frontal) with IRST, FLIR and laser rangefinder in two modules ahead of windscreen; surveillance, tracking and lock-on by port module; target identification, analysis and optical identification by starboard module; combined output in pilot's head-level display; test flights from 1996; installation from 1999 (in 41st and subsequent Rafales).

To accelerate programme, first 20 Rafales for each service will not have ASMP, automatic terrain-following, Spectra automatic defensive subsystems, helmet-mounted sight, OSF and voice command controls; definitive S02 version will be delivered thereafter.

CURRENT VERSIONS: **Rafale B**: Originally planned dual control, two-seat version for French Air Force; weight 350 kg (772 lb) more than Rafale C; 3-5 per cent higher cost than Rafale C. Being developed into fully operational variant.

Rafale C: Single-seat combat version for French Air Force. *Detailed description applies to Rafale C, except where indicated.*

Rafale D: Original configuration from which production versions derived; now 'Rafale Discret' (stealthy) generic name for French Air Force versions.

Dassault Rafale C during an early test flight *(Aviaplans/F. Robineau)*

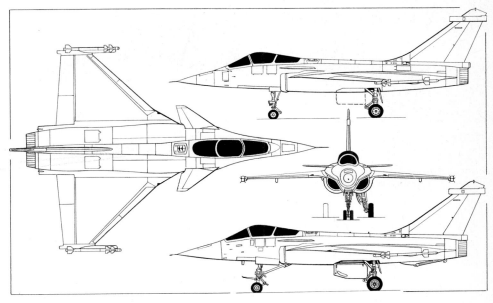

Dassault Rafale C, with lower side view of carrier-borne Rafale M *(Dennis Punnett)*

Rafale M: Single-seat carrier-borne fighter; take-off weight from existing French carrier *Foch* limited to 16,500 kg (36,376 lb); has 80 per cent structural and equipment commonality with Rafale C, 95 per cent systems commonality. Navy's financial share of French programme cut in 1991 from 25 to 20 per cent.

CUSTOMERS: Anticipated worldwide market for 500 aircraft in addition to originally planned 250 for French Air Force (225 Cs and 25 Bs) and 86 for French Navy; former service announced revised requirement for 234, comprising 95 Rafale Cs and 139 two-seat (pilot and WSO) combat versions, in 1992; naval requirement believed cut to 78 in 1991. First deliveries late 1996; 20 S1-standard in naval service by 2000 for interceptor duties with 14 Flottille aboard *Foch* and *Charles de Gaulle*, replacing Crusaders; balance for delivery as S02s 2005-2009, replacing Super Etendards. Air Force deliveries originally planned for 1996-2009, including first 20 as S01s; two year postponement announced 1992; deliveries from 1998; first squadron to form 2000 or later. Official authorisation to launch production given 23 December 1992. Initial production contract in 1993 defence budget: one aircraft each for Air Force and Navy.

COSTS: Programme estimated at FFr155 billion (1991), including FFr40 billion for R&D.

DESIGN FEATURES: Minimum weight and volume structure to hold minimum cost; thin, mid-mounted delta wing with moving canard; individual fixed, kidney-shaped intakes without shock cones.

FLYING CONTROLS: Fully fly-by-wire controls with fully modulated two-section leading-edge slats and two elevons per wing; canard incidence automatically increased to 20° when landing gear lowered; airbrake panels in top of fuselage beside leading-edge of fin; HOTAS (hands on throttle and stick) controls, with sidestick controller on starboard console and small-travel throttle lever.

STRUCTURE: Most of wing components made of carbonfibre including elevons; wing spar/fuselage attachment fittings in aluminium-lithium; slats in titanium; wingroot and tip fairings Kevlar; canard made mainly by superplastic forming and diffusion bonding; fuselage 50 per cent carbonfibre; fuselage side skins of aluminium-lithium alloy; wheel and engine doors carbonfibre; fin made primarily of carbonfibre with aluminium honeycomb core in rudder.

LANDING GEAR: Hydraulically retractable tricycle type supplied by Messier-Bugatti, with single mainwheels and twin, hydraulically steerable, nosewheels. All wheels retract forward. Designed for impact at vertical speed of 3 m (10 ft)/s, or 6 m (20 ft)/s in naval version, without flare-out. Michelin radial tyres. Mainwheel tyres size 810 × 275-15, pressure 16.0 bars (232 lb/sq in). Nosewheel tyre size 550 × 200-10. Carbon brakes on all three units, controlled by fly-by-wire system. Brake-chute for emergency use in cylindrical container at base of rudder. Rafale M has 'jump strut' nosewheel leg which releases energy stored in shock absorber at end of deck take-off run, changing aircraft's attitude for climbout without need for ski-jump ramp. 'Jump strut' advantage equivalent to 9 knots (16 km/h; 10 mph) or 900 kg (1,984 lb) extra weapon load; not to be used aboard carrier *Foch*, which to have 1° 30′ ramp giving 20 knots (37 km/h; 23 mph) or 2,000 kg (4,409 lb) advantage. Dowty Aerospace Yakima holdback fitting. Naval nosewheel steerable ±70°; or almost 360° under tow. Hydraulic (Rafale M) or tension-stored (Rafale B/C) arrester hook.

POWER PLANT: Two SNECMA M88-2 augmented turbofans, each rated at 48.7 kN (10,950 lb st) dry and 72.9 kN (16,400 lb st) with afterburning. M88-3 of 87 kN (19,558 lb st) max rating in production aircraft. Internal tanks for more than 5,325 litres (1,406 US gallons; 1,171 Imp gallons) of fuel. Fuel system by Lucas Air Equipement and Zenith Aviation; equipment by Intertechnique. One 1,700 litre (449 US gallon; 374 Imp gallon) drop tank on centreline; 2,000 litre (528 US gallon; 440 Imp gallon) drop tank on each inboard underwing pylon; and/or 1,300 litre (343 US gallon; 286 Imp gallon) tank on each centre underwing pylon. Max external fuel 6,600 litres (1,742 US gallons; 1,452 Imp gallons). Pressure refuelling in seven minutes, or four minutes for internal tanks only. Fixed (detachable) in-flight refuelling probe on all versions.

ACCOMMODATION: Pilot only, on Hispano-Suiza licence built Martin-Baker Mk 15 zero/zero ejection seat, reclined at angle of 29°. One-piece Sully Products Spéciaux blister windscreen/canopy, hinged to open sideways to starboard. Canopy gold-coated to reduce radar reflection.

SYSTEMS: Technofan bootstrap cockpit air-conditioning system. Dual hydraulic circuits, pressure 280 bars (4,000 lb/sq in), each with two Messier-Bugatti pumps and Bronzavia ancillaries. Variable frequency electrical system, with two 30/40kVA Auxilec alternators. Triplex digital plus one dual analog fly-by-wire flight control system, integrated with engine controls and linked with weapons system. EROS oxygen system. Microturbo APU.

AVIONICS: Provision for more than 780 kg (1,720 lb) of avionics equipment and racks, including GIE Radar (Thomson-CSF/Dassault Electronique) RBE2 lookdown/shootdown radar, able to track up to eight targets simultaneously, with automatic threat assessment and allocation of priority. SAGEM Uliss 52X INS (interface with carrier's navigation on Rafale M). Digital display of fuel, engine, hydraulic, electrical, oxygen and other systems information on collimated eye-level display and two lateral multi-function touch-sensitive colour LCD displays by Thomson-CSF/SFENA. Fourth cockpit screen is head-down navigation/radar display. Sextant Avionique CTH3022 wide-angle, holographic HUD. TRT com. SOCRAT VOR/ILS. Sextant Avionique voice activated radio controls and voice alarm warning system. Sextant/Intertechnique helmet-mounted sight. Thomson-CSF/CNI IFF and JTIDS. Thomson-TRT/SAT OSF electro-optical sensors. Spectra radar warning and ECM suite by Thomson-CSF, Dassault Electronique and Matra; Sextant Avionique Navstar GPS; EAS V/UHF and TRT Saturn UHF radios; Thomson-CSF helmet sight; Thomson-CSF/CNI radio altimeter and SFIM/Dassault Electronique flight recorder. Communications via SINTAC/JTIDS. Various reconnaissance, ECM, FLIR and laser designation pods.

EQUIPMENT: Integral fold-down ladder in Rafale M.

ARMAMENT: One 30 mm Giat DEFA 791B cannon in side of port engine duct. Fourteen external stores attachments: two on fuselage centreline, two beneath engine intakes, two astride rear fuselage, six under wings and two at wingtips. Forward centreline position deleted on Rafale M. Normal external load 6,000 kg (13,228 lb); max permissible, 8,000 kg (17,637 lb). In strike role, one Aerospatiale ASMP stand-off nuclear weapon. In interception role, up to eight Matra Mica AAMs (with IR or active homing) and two underwing fuel tanks; or six Micas and 5,700 litres (1,505 US gallons; 1,254 Imp gallons) of external fuel. In air-to-ground role, typically sixteen 227 kg (500 lb) bombs, two Micas and two 1,300 litre (343 US gallon; 286 Imp gallon) tanks; or two Apache stand-off weapon dispensers, two Micas and 5,700 litres of external fuel; or FLIR pod, Atlis laser designator pod, two 1,000 kg (2,205 lb) laser guided bombs, two AS.30L laser ASMs, four Micas and single 1,700 litre (449 US gallon; 374 Imp gallon) tank. In anti-ship role, two Exocet or projected ANS sea-skimming missiles, four Micas and 4,300 litres (1,135 US gallons; 946 Imp gallons) of external fuel.

DIMENSIONS, EXTERNAL:
Wing span, incl wingtip missiles	10.90 m (35 ft 9⅛ in)
Length overall	15.30 m (50 ft 2⅜ in)
Height overall (Rafale D)	5.34 m (17 ft 6¼ in)

AREAS:
Wings, gross	46.00 m² (495.1 sq ft)

WEIGHTS AND LOADINGS (estimated):
Basic weight empty, equipped:	
Rafale D	9,060 kg (19,973 lb)
Rafale M	9,800 kg (21,605 lb)
External load: normal	6,000 kg (13,228 lb)
max	8,000 kg (17,637 lb)
Max ramp weight: initial version	19,500 kg (42,990 lb)
developed version	21,500 kg (47,399 lb)

PERFORMANCE (estimated):
Max level speed: at altitude	Mach 2
at low level	750 knots (1,390 km/h; 864 mph)
Approach speed	115 knots (213 km/h; 132 mph)
T-O distance: air defence	400 m (1,312 ft)
attack	600 m (1,969 ft)

Dassault Rafale C No. 01 preparing for take-off

Dassault Rafale M (nearest) and Rafale C in formation (*Aviaplans/F. Robineau*)

Radius of action: low-level penetration with twelve 250 kg
bombs, four Mica AAMs and 4,300 litres (1,135 US
gallons; 946 Imp gallons) of external fuel in three
tanks 590 nm (1,093 km; 679 miles)
air-to-air, long-range with eight Mica AAMs and 6,600
litres (1,742 US gallons; 1,452 Imp gallons) of
external fuel in four tanks, 12,200 m (40,000 ft)
transit 1,000 nm (1,853 km; 1,152 miles)
g limits +9.0/−3.6

DASSAULT ATLANTIQUE 2

TYPE: Twin-turboprop maritime patrol aircraft. Former designation Atlantique Nouvelle Génération (ANG).
PROGRAMME: Design definition began July 1977; development began September 1978; first flight Atlantic 1 converted to ATL2 prototype 8 May 1981; first flight second converted prototype 26 March 1982; production authorised 24 May 1984; first flight production aircraft 19 October 1988; 23 Flottille declared operational with three Atlantiques 1 February 1991.
CURRENT VERSIONS: **Atlantique 2**: Standard aircraft equipped to attack surface and submarine targets, lay mines, transport personnel and freight, protect offshore interests and fly search and rescue. *Detailed description applies to this version.*
  Jet Atlantique: Proposed for RAF in 1991; two podded turbofans underwing (Garrett TFE731 considered) in addition to turboprops; main power plant change from Tyne to Allison T406 or General Electric T407 offered.
  Atlantique 3: Developed version, proposed in 1988.
  Europatrol: Dassault proposal for European maritime patrol replacement for cancelled Lockheed P-7, 1990; 'not necessarily' based on Atlantique 2.
CUSTOMERS: French Navy required 42 production Atlantique 2s, but total cut to 30 in 1992 budget; first two funded 1985; Nos. 28-30 included in 1992 budget; funding for three per year since 1990; but only 25 contracted from Dassault by January 1992; deliveries one, two, six, six and four in 1989-93, then three per year. First squadron, 23 Flottille at Lann-Bihoué, completed in 1991 with eight aircraft; co-located 24 F following from 1992; 21 and 22 F to re-equip from Atlantic 1s at Nimes-Garons, 1994-98.
COSTS: 1977 estimates of FFr3,046 million R&D and FFr21,462 million production; FFr800 million over budget by 1991.
DESIGN FEATURES: Aerodynamically conventional mid-wing twin-turboprop aircraft with double-bubble fuselage; retractable radome; weapons bay and ejectable stores stowage in lower fuselage section; avionics pods at tips of wings and fin; underwing stores stations for anti-ship or self-defence missiles.
  Wing section NASA 64 series; incidence 3°; dihedral 6° outboard of engines; sweepback 9° on leading-edge.
FLYING CONTROLS: All fully powered by SAMM tandem hydraulic power units; no trim tabs; ailerons assisted by three spoilers on each wing, some of which also act as airbrakes; fixed tailplane; three segments of slotted flaps on each wing.
STRUCTURE: Employs refined honeycomb panel bonding technique to improve corrosion resistance and maintainability of original structure; wing torsion box, pressure cabin walls, flaps and large doors made of bonded honeycomb panels; weapon bay and nose landing gear doors slide over external skin to open.
  Most members of original SECBAT Atlantic 1 consortium are sharing manufacture, including Dassault Aviation, Sabca, Sonaca, DASA (MBB and Dornier), Alenia and Aerospatiale; Tyne engines are made by Rolls-Royce, FN and DASA (MTU); propellers by Ratier and British Aerospace.
LANDING GEAR: Retractable tricycle type, by Messier-Bugatti, with twin wheels on each unit. Hydraulic retraction, nosewheels rearward, main units forward into engine nacelles. Kléber-Colombes tyres; size 39 × 13-20 on mainwheels, pressure 12 bars (170 lb/sq in), 26 × 7.75-13 on nosewheels, pressure 6.5 bars (94 lb/sq in). Messier-Bugatti disc brakes and Modistop anti-skid units.
POWER PLANT: Two 4,549 kW (6,100 ehp) Rolls-Royce Tyne RTy.20 Mk 21 turboprops, each driving a four-blade Ratier/British Aerospace constant-speed metal propeller type PD 249/476/3 on prototypes. Four pressure-refuelled integral fuel tanks in wings, with total capacity of 23,120 litres (6,108 US gallons; 5,085 Imp gallons). Updated gauging system. Oil capacity 100 litres (26.4 US gallons; 22 Imp gallons).
ACCOMMODATION: Normal flight crew of 10-12, comprising observer in glazed nose; pilot, co-pilot and flight engineer on flight deck; radio-navigator, ESM/ECM/MAD operator, radar-IFF operator, tactical co-ordinator and two acoustic sensor operators at stations on starboard side of tactical compartment; and two optional observers in beam positions at rear. Rest/relief crew compartment in centre of fuselage. Primary access via extending airstair door in bottom of rear fuselage. Emergency exits above and below flight deck and on each side, above wing trailing-edge.
SYSTEMS: Air-conditioning system supplied by two compressors driven by gearboxes. Heat exchangers and bootstrap system for cabin temperature control. Duplicated hydraulic system, pressure 186 bars (2,700 lb/sq in), to operate flying controls, landing gear, flaps, weapons bay doors and retractable radome. Hydraulic flow rate 17.85 litres (4.7 US gallons; 3.9 Imp gallons)/min. Three basic electrical systems: variable frequency three-phase 115/200V AC system, with two 60/80kVA Auxilec alternators and modernised control and protection equipment; fixed frequency three-phase 115/200V 400Hz AC system, with four 15kVA Auxilec Auxivar generators, two on each engine; 28V DC system, with four 6kW transformer-rectifiers supplied from variable frequency AC system, and one 40Ah battery. One 60kVA emergency AC generator, driven at constant speed by APU. Individual oxygen bottles for emergency use. Air Equipement/Kléber-Colombes pneumatic de-icing system on wing and tail leading-edges. Electric anti-icing for engine air intake lips, propeller blades and spinners. Turbomeca/ABG-Semca Astadyne gas turbine APU for engine starting, emergency electrical supply, and air-conditioning on ground.
AVIONICS: SAT/TRT Tango FLIR sensor in turret under nose. Thomson-CSF Iguane retractable radar immediately forward of weapons bay, with integrated LMT IFF interrogator and SECRE decoder. Sextant Avionique MAD in lengthened tail sting. Thomson-CSF Arar 13A radar detector for ESM. Thomson-CSF Sadang system for processing active and passive acoustic detection data. A distributed data processing system around a databus, with a CIMSA Mitra 125X tactical computer (512K memory), two Dassault Electronique bus computers, two SAGEM magnetic bubble mass memories and Thomson-CSF display subsystem. Other equipment includes IFF transponder and HF com by LMT, UHF AM/FM by SINTRA, Tacan and DME by Thomson-CSF, VHF AM/FM com by SOCRAT, VOR/ILS by EAS, TRT radio altimeter, Collins MF radio compass, ADF, HSI and autopilot/flight director by SFENA, dual SAGEM Uliss 53 inertial navigation systems coupled to a Navstar GPS, SAGEM high-speed printer and terminal display, Sextant Avionique navigation table and air data computer.
EQUIPMENT: More than 100 sonobuoys, with Alkan pneumatic launcher, in compartment aft of weapons bay, where whole of upper and lower fuselage provides storage for sonobuoys and 160 smoke markers and flares. Thomson-TRT

Navalised Rafale M showing revised landing gear for deck operations *(Aviaplans/F. Robineau)*

Dassault Rafale M No. 01 with Mirage 2000 chase-plane *(Aviaplans/F. Robineau)*

First production Dassault Atlantique 2 (ATL2) maritime patrol aircraft *(Patrick Biegel)*

35 cameras in starboard side of nose and in bottom of rear fuselage.

ARMAMENT: Main weapons bay in unpressurised lower fuselage can accommodate all NATO standard bombs, eight depth charges, up to eight Mk 46 homing torpedoes, seven French Murène advanced torpedoes or two air-to-surface missiles (typical load comprises three torpedoes and one AM 39 Exocet or AS 37 Martel missile). Four underwing attachments for external stores, including two or four ARMAT or Magic missiles, or future air-to-surface and air-to-air missiles or pods.

DIMENSIONS, EXTERNAL:
Wing span: incl wingtip pods	37.42 m (122 ft 9¼ in)
excl wingtip pods	37.30 m (122 ft 4½ in)
Wing aspect ratio	10.9
Length overall	33.63 m (110 ft 4 in)
Height overall	10.89 m (35 ft 8¾ in)
Fuselage: Max depth	4.00 m (13 ft 1½ in)
Tailplane span	12.31 m (40 ft 4½ in)
Wheel track (c/l of shock struts)	9.00 m (29 ft 6¼ in)
Wheelbase	9.40 m (30 ft 10 in)
Propeller diameter	4.88 m (16 ft 0 in)
Distance between propeller centres	9.00 m (29 ft 6¼ in)

DIMENSIONS, INTERNAL:
Cabin, incl rest compartment, galley, toilet, aft observers'
stations: Length	18.50 m (60 ft 8½ in)
Max width	3.60 m (11 ft 9½ in)
Max height	2.00 m (6 ft 6¾ in)
Floor area	155.0 m² (1,668 sq ft)
Volume	92.0 m³ (3,250 cu ft)
Main weapons bay: Length	9.00 m (29 ft 6¼ in)
Width	2.10 m (6 ft 10¾ in)
Depth	1.00 m (3 ft 3¼ in)

AREAS:
Wings, gross	120.34 m² (1,295.3 sq ft)
Ailerons (total)	5.26 m² (56.62 sq ft)
Flaps (total)	26.42 m² (284.38 sq ft)
Spoilers (total)	1.66 m² (17.87 sq ft)
Fin, incl dorsal fin	10.68 m² (114.96 sq ft)
Rudder	5.96 m² (64.15 sq ft)
Tailplane	24.20 m² (260.49 sq ft)
Elevators (total)	8.30 m² (89.34 sq ft)

WEIGHTS AND LOADINGS:
Weight empty, equipped, standard mission	25,700 kg (56,659 lb)
Max weapon load: internal	2,500 kg (5,511 lb)
external	3,500 kg (7,716 lb)
Max fuel	18,500 kg (40,785 lb)
Standard mission T-O weight:	
ASW or ASSW mission	44,200 kg (97,440 lb)
combined ASW/ASSW mission	45,000 kg (99,200 lb)
Max T-O weight	46,200 kg (101,850 lb)
Normal design landing weight	36,000 kg (79,365 lb)
Max landing weight	46,000 kg (101,400 lb)
Max zero-fuel weight	32,500 kg (71,650 lb)
Max wing loading	385 kg/m² (78.96 lb/sq ft)
Max power loading	5.07 kg/kW (8.34 lb/ehp)

PERFORMANCE (at T-O weight of 45,000 kg; 99,200 lb except where indicated):
Max Mach number	0.73
Max level speed:	
at optimum height	350 knots (648 km/h; 402 mph)
at S/L	320 knots (592 km/h; 368 mph)
Max cruising speed at 7,620 m (25,000 ft)	300 knots (555 km/h; 345 mph)
Normal patrol speed, S/L to 1,525 m (5,000 ft)	170 knots (315 km/h; 195 mph)
Stalling speed, flaps down	90 knots (167 km/h; 104 mph)
Max rate of climb at S/L:	
AUW of 30,000 kg (66,140 lb)	884 m (2,900 ft)/min
AUW of 40,000 kg (88,185 lb)	610 m (2,000 ft)/min
Rate of climb at S/L, OEI:	
AUW of 30,000 kg (66,140 lb)	365 m (1,200 ft)/min
AUW of 40,000 kg (88,185 lb)	213 m (700 ft)/min
Service ceiling	9,145 m (30,000 ft)
Runway LCN at max T-O weight	60
T-O to 10.7 m (35 ft)	1,840 m (6,037 ft)
Landing from 15 m (50 ft)	1,500 m (4,922 ft)
170 knot turning radius at AUW of 40,000 kg (88,185 lb)	
at: 30° bank	1,380 m (4,530 ft)
45° bank	800 m (2,625 ft)
60° bank	460 m (1,510 ft)

Typical mission profiles, with reserves of 5% total fuel, 5% of fuel consumed and 20 min hold:
anti-ship mission: T-O with max fuel and one AM 39 missile; fly 1,800 nm (3,333 km; 2,071 miles) to target area; descend for two-hour search and attack at 90 m (300 ft); return to base
anti-submarine mission: T-O at 44,300 kg (97,665 lb) AUW with 15,225 kg (33,565 lb) of fuel, four Mk 46 torpedoes, 78 sonobuoys, and a full load of markers and flares; cruise to search area at 290 knots (537 km/h; 333 mph) at 7,620 m (25,000 ft); descend for 8 h low altitude patrol at 600 nm (1,110 km; 690 miles) from base, or 5 h patrol at 1,000 nm (1,850 km; 1,150 miles) from base; return to base at 9,145 m (30,000 ft). Total mission time 12 h 31 min
Ferry range with max fuel
 4,900 nm (9,075 km; 5,635 miles)
Max endurance 18 h

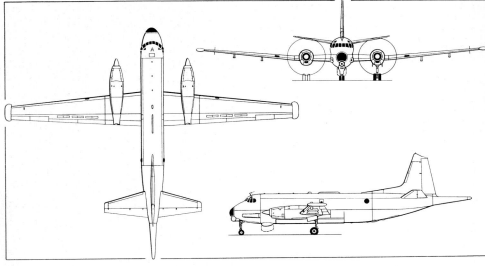

Dassault Atlantique 2 (ATL2) twin-turboprop maritime patrol aircraft *(Dennis Punnett)*

Dassault Atlantique 2 at patrol altitude *(Patrick Biegel)*

DASSAULT FALCON 20 and 200, HU-25 GUARDIAN and GARDIAN 2

Production of these Falcon versions ended in 1988. Several operators have converted aircraft for specialist roles or taken advantage of re-engine programmes. Details in *Jane's Civil and Military Aircraft Upgrades*. One of latest conversions has involved Dassault modifying Falcon 20s to perform as testbeds for Rafale's new Thomson-CSF RBE2 radar; first flown 10 July 1992.

DASSAULT FALCON 50

Spanish Air Force designation: T.16

TYPE: Three-turbofan long-range business transport.

PROGRAMME: First flight of prototype (F-WAMD) 7 November 1976, second prototype 18 February 1978, first (and only) pre-production 13 June 1978; French certification 27 February 1979; FAA certification 7 March 1979; deliveries began July 1979.

CUSTOMERS: Total 227 sold to 30 countries between July 1979 and 31 July 1992; Falcon 50 adopted by governments of Djibouti, Iraq, Jordan, Libya, Morocco, Portugal, South Africa, Spain and former Yugoslavia; three for Italian Air Force convertible to air ambulance.

DESIGN FEATURES: New design with new technology wing and sharply waisted rear fuselage and engine pod designed by computational fluid dynamics; wing has compound leading-edge sweep (24° 50'–29° at quarter-chord) and optimised section. Three-engine layout permits overflight of oceans and desert areas within public transport regulations.

FLYING CONTROLS: Fully powered controls with artificial feel; variable incidence anhedral tailplane with dual electrical actuation; drooped leading-edge inboard and slats outboard; double-slotted flaps; three two-position airbrake/spoiler panels on each wing.

STRUCTURE: All-metal, circular-section fuselage with rear baggage compartment inside pressure cabin; wing boxes are integral fuel tanks bolted to carry-through box.

Dassault announced June 1991 arrangement for MiG to produce tail surfaces for Falcons in Russia; otherwise, fuselages and tail surfaces made by Aerospatiale at Saint Nazaire and Mèaulte and cowlings by Hurel-Dubois at Vélizy-Villacoublay.

LANDING GEAR: Retractable tricycle type by Messier-Bugatti, with twin wheels on each unit. Hydraulic retraction, main units inward, nosewheels forward. Nosewheel steerable ±60° for taxying, ±180° for towing. Mainwheel tyres size 26 × 6.6-14 in, pressure 14.34 bars (208 lb/sq in). Nosewheel tyres size 14.5 × 5.5-6 in, pressure 8.96 bars (130 lb/sq in). Four-disc brakes designed for 400 landings with normal energy braking. Min ground turning radius (about nosewheels) 13.54 m (44 ft 5 in).

POWER PLANT: Three Garrett TFE731-3 turbofans, each rated at 16.5 kN (3,700 lb st) for take-off. Two engines pod mounted on sides of rear fuselage, third attached by two top mounts. Thrust reverser on centre engine. Fuel in integral tanks, with capacity of 5,787 litres (1,529 US gallons; 1,273 Imp gallons) in wings and 2,976 litres (786 US gallons; 655 Imp gallons) in fuselage tanks. Total fuel capacity 8,763 litres (2,315 US gallons; 1,928 Imp gallons). Single-point pressure fuelling.

ACCOMMODATION: Crew of two side by side on flight deck, with full dual controls and airline type instrumentation. Third seat to rear of co-pilot. Various cabin configurations available, based on two alternative toilet locations. Aft cabin toilet allows eight/nine-passenger arrangement, with four chairs in forward cabin, facing each other in pairs, and three-place sofa and two facing chairs in rear cabin. Wardrobe, galley and crew toilet located forward, in entrance area. Alternatively, forward toilet, facing door, makes possible a lounge in rear cabin, furnished with four/five-place angle sofa and chair. This rear lounge is separated from forward cabin by either a wardrobe and refreshment/recreation console, or by two additional seats, raising cabin accommodation to 12 persons. After removing forward cabin equipment (wardrobe and galley) and seats, cabin will accommodate up to three stretchers, two doctors and medical equipment, or freight. Rear baggage compartment is pressurised and air-conditioned, and has capacity of 1,000 kg (2,200 lb). Access is by separate door on port side.

SYSTEMS: Air-conditioning system utilises bleed air from all three engines. Max pressure differential 0.61 bar (8.8 lb/sq in). Pressurisation maintains a max cabin altitude of 2,440 m (8,000 ft) to a flight altitude of 13,700 m (45,000 ft). Two independent hydraulic systems, pressure 207 bars (3,000 lb/sq in), with three engine driven pumps and one emergency electric pump, actuate primary flying controls, flaps, slats, landing gear, wheel brakes, airbrakes and nosewheel steering. Plain reservoir, pressurised by bleed air at 1.47 bars (21 lb/sq in). 28V DC electrical system, with a 9 kW 28V DC starter/generator on each engine and two 23Ah batteries. Automatic emergency oxygen system. Optional 9 kW Garrett APU.

DASSAULT—AIRCRAFT: FRANCE 83

AVIONICS: Standard fit provides Collins FCS 80F autopilot in conjunction with dual five-CRT EFIS-85 system, ADC 80 air data computer, dual VHF, VOR, DME, ADF, transponders, radar altimeter and Honeywell Primus 400 weather radar. Main options provide for advanced symbology EFIS-86C, digital radio controllers used in conjunction with autotune FMS (Global GNS 1000 or UNS 1), Honeywell laser inertial systems, which may also replace gyro reference systems, and Omega. Installation of Collins digital avionics including APS-85 autopilot, EFIS-86, ADS-82, AHS-85 AHRS and Pro Line II com/nav/pulse makes possible certificated Cat. II operation.

DIMENSIONS, EXTERNAL:
Wing span	18.86 m (61 ft 10½ in)
Wing chord (mean)	2.84 m (9 ft 3¾ in)
Wing aspect ratio	7.6
Length: overall	18.52 m (60 ft 9¼ in)
fuselage	17.66 m (57 ft 11 in)
Height overall	6.97 m (22 ft 10½ in)
Tailplane span	7.74 m (25 ft 4¾ in)
Wheel track	3.98 m (13 ft 0¾ in)
Wheelbase	7.24 m (23 ft 9 in)
Passenger door: Height	1.52 m (4 ft 11¾ in)
Width	0.80 m (2 ft 7½ in)
Height to sill	1.30 m (4 ft 3¼ in)
Emergency exits (each side, over wing):	
Height	0.92 m (3 ft 0¼ in)
Width	0.51 m (1 ft 8 in)

DIMENSIONS, INTERNAL:
Cabin, incl forward baggage space and rear toilet:
Length	7.16 m (23 ft 6 in)
Max width	1.86 m (6 ft 1¼ in)
Max height	1.79 m (5 ft 10½ in)
Volume	20.15 m³ (711.6 cu ft)
Baggage space	0.75 m³ (26.5 cu ft)
Baggage compartment (rear)	2.55 m³ (90 cu ft)

AREAS:
Wings, gross	46.83 m² (504.1 sq ft)
Horizontal tail surfaces (total)	13.35 m² (143.7 sq ft)
Vertical tail surfaces (total)	9.82 m² (105.7 sq ft)

WEIGHTS AND LOADINGS:
Weight empty, equipped	9,150 kg (20,170 lb)
Max payload: normal	1,570 kg (3,461 lb)
optional	2,170 kg (4,784 lb)
with max fuel	1,130 kg (2,491 lb)
Max fuel	7,040 kg (15,520 lb)
Max T-O and ramp weight:	
standard	17,600 kg (38,800 lb)
optional	18,500 kg (40,780 lb)
Max zero-fuel weight: standard	11,000 kg (24,250 lb)
optional	11,600 kg (25,570 lb)
Max landing weight	16,200 kg (35,715 lb)
Max wing loading: standard	375.8 kg/m² (76.97 lb/sq ft)
optional	395.0 kg/m² (80.90 lb/sq ft)
Max power loading: standard	356.7 kg/kN (3.49 lb/lb st)
optional	374.9 kg/kN (3.67 lb/lb st)

PERFORMANCE:
Max operating Mach No. (M_{MO})	0.86
Max operating speed (V_{MO}):	
at S/L	350 knots (648 km/h; 402 mph) IAS
at 7,225 m (23,700 ft)	370 knots (685 km/h; 425 mph) IAS
Max cruising speed	Mach 0.82 or 475 knots (880 km/h; 546 mph)
Long-range cruising speed at 10,670 m (35,000 ft)	Mach 0.75 (430 knots; 797 km/h; 495 mph)
Approach speed, 8 passengers and 45 min long-range reserves	99.5 knots (184 km/h; 115 mph)
Max operating altitude	13,715 m (45,000 ft)
FAR 25 balanced T-O field length with 8 passengers and fuel for 3,500 nm (6,480 km; 4,025 miles)	1,365 m (4,480 ft)
FAR 121 landing distance with 8 passengers and 45 min long-range reserves	1,025 m (3,350 ft)
Range at Mach 0.75 with 8 passengers and 45 min long-range reserves	3,500 nm (6,480 km; 4,025 miles)

DASSAULT FALCON 900B
Spanish Air Force designation: T.18

TYPE: Three-turbofan intercontinental business transport.
PROGRAMME: Falcon 900 announced 27 May 1983; first flight of prototype (F-GIDE *Spirit of Lafayette*) 21 September 1984, second aircraft (F-GFJC) 30 August 1985; flew non-stop 4,305 nm (7,973 km; 4,954 miles) Paris to Little Rock, Arkansas, September 1985; returned Teterboro, New Jersey, to Istres, France, at Mach 0.84; French and US certification March 1986, including status close to FAR Pts 25 and 55 for damage tolerance of entire airframe.
CURRENT VERSIONS: **Falcon 900:** Standard version up to 1992; specifications in 1992-93 *Jane's* and earlier applied to this version; 20 kN (4,500 lb st) TFE731-5AR-1C turbofans.
Falcon 900B: French and British certification received end 1991; complies with FAR Pt 36 Stage III and ICAO 16 noise requirements; approved for Cat. II approaches, for operations from unpaved fields; re-engined with Garrett TFE731-5BR turbofans, to give 5.5 per cent power increase; initial cruising altitude 11,855 m (39,000 ft) and NBAA IFR range increased by 100 nm (185 km; 115

Dassault Falcon 50 of the Italian Air Force

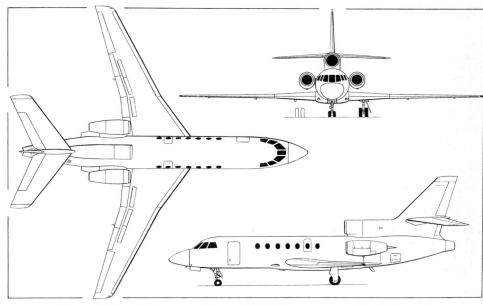
Dassault Falcon 50 long-range three-turbofan business transport (*Dennis Punnett*)

Dassault Falcon 900B operated by the Saudi government (*Press Office Sturzenegger*)

miles); retrofit to be offered. *Detailed description applies to this version, except where indicated.*

Japan ASDF: Two Falcon 900s for long-range maritime surveillance entered service September 1989; US search radar, special communications radio, operations control station, UH-25A-style search windows and drop hatch for sonobuoys, markers and flares.

CUSTOMERS: Total 117 of all versions sold to 26 countries by 31 July 1992; government/VIP versions operated by Australia, France, Malaysia, Nigeria, Saudi Arabia and Spain; F-WWFJ, completed 17 June 1988, was 1,000th of Falcon series.

DESIGN FEATURES: Larger cross-section and cabin length than Falcon 50; wing adapted from Falcon 50 but increased span and area, and optimised for Mach 0.84 cruise; quarter-chord sweepback on outer wings 24° 30′; dihedral 0° 30′. Added economy and further power increase of engines achieved by mixer compound nozzle tailpipe, mixing cold and hot flows.

FLYING CONTROLS: Fully powered flying controls with artificial feel; full span slats, combined with slotted flaps, provide very low speed capability; tailplane incidence adjustable by two independent electric screwjacks; three-position airbrakes.

STRUCTURE: Design and manufacture computer assisted; damage-tolerant structure; extensive use of carbonfibre and aramid (Kevlar); Kevlar radome, wingroot fairings and tailcone; standard 12 windows can be increased to 18; secondary rear cabin pressure bulkhead allows access to baggage in flight and additional protection against pressure loss. Nosewheel doors of Kevlar; mainwheel doors of carbonfibre. Kevlar air intake trunk for centre engine, and rear cowling for side engines. Carbonfibre central cowling around all three engines.

LANDING GEAR: Retractable tricycle type by Messier-Bugatti, with twin wheels on each unit. Hydraulic retraction, main units inward, nosewheels forward. Oleo-pneumatic shock absorbers. Mainwheels fitted with Michelin radial tyres size 29 × 7.7-15, pressure 13.0 bars (189 lb/sq in). Nosewheel tyres size 17.5 × 5.75-8, pressure 10.0 bars (145 lb/sq in). Hydraulic nosewheel steering (±60° for taxying, ±180° for towing). Messier-Bugatti triple-disc carbon brakes and anti-skid system. Min ground turning radius (about nosewheels) 13.54 m (44 ft 5 in).

POWER PLANT: Three Garrett TFE731-5BR turbofans, each rated at 21.13 kN (4,750 lb st) at ISA + 8°C. Thrust reverser on centre engine. Fuel in two integral tanks in wings, centre-section tank, and two tanks under floor of forward and rear fuselage. Total fuel capacity 10,825 litres (2,860 US gallons; 2,381 Imp gallons).

ACCOMMODATION: Type III emergency exit on starboard side of cabin permits wide range of layouts for up to 19 passengers. Flight deck for two pilots, with central jumpseat. Flight deck separated from cabin by door, with crew wardrobe and baggage locker on either side. Galley at front of main cabin, on starboard side opposite main cabin door. Passenger area is divided into three lounges. Forward zone has four 'sleeping' swivel chairs in facing pairs with tables. Centre zone is dining area, with two double seats facing a transverse table. On starboard side, storage cabinet contains foldaway bench, allowing five to six persons to be seated around table, while leaving emergency exit clear. In rear zone, inward facing three-seat settee on starboard side converts into a bed. On port side, two armchairs are separated by a table. At rear of cabin, a door leads to toilet compartment, on starboard side, and a second structural plug door to large rear baggage area. Baggage door is electrically actuated. Other interior configurations available. Alternative eight-passenger configuration has bedroom at rear and three personnel seats in forward zone. A 15-passenger layout divides a VIP area at rear from six (three-abreast) chairs forward; full fuel can still be carried with 15 passengers. The 18-passenger scheme has four rows of three-abreast airline type seats forward, and VIP lounge with two chairs and settee aft. Many optional items, including additional windows, front toilet unit, video system with one or more monitors, 'Airshow 200' navigation display system, compact disc deck, aft cabin partition, one or two couches in aft cabin convertible to bed(s), storage cabinet in baggage hold, aft longitudinal table, individual listening devices for passengers, life jackets and rafts.

SYSTEMS: Air-conditioning system uses engine bleed air or air from Garrett GTCP36-150 APU installed in rear fuselage. Softair pressurisation system, with max differential of 0.64 bar (9.3 lb/sq in), maintains sea level cabin environment to height of 7,620 m (25,000 ft), and cabin equivalent of 2,440 m (8,000 ft) at 15,550 m (51,000 ft). Cold air supply by single oversize air cycle unit. Two independent hydraulic systems, pressure 207 bars (3,000 lb/sq in), with three engine driven pumps and one emergency electric pump, actuate primary flying controls, flaps, slats, landing gear retraction, wheel brakes, airbrakes, nosewheel steering and thrust reverser. Bootstrap hydraulic reservoirs. DC electrical system supplied by three 9 kW 28V Auxilec starter/generators and two 23Ah batteries. Heated bleed air anti-icing of wing leading-edges, intakes and centre engine duct; electrically heated windscreens. Eros (SFIM/Intertechnique) oxygen system.

AVIONICS: Dual bi-directional Honeywell ASCB digital databus operating in conjunction with dual SPZ 8000 flight director/autopilot and EFIS (Honeywell EDZ 820 system).

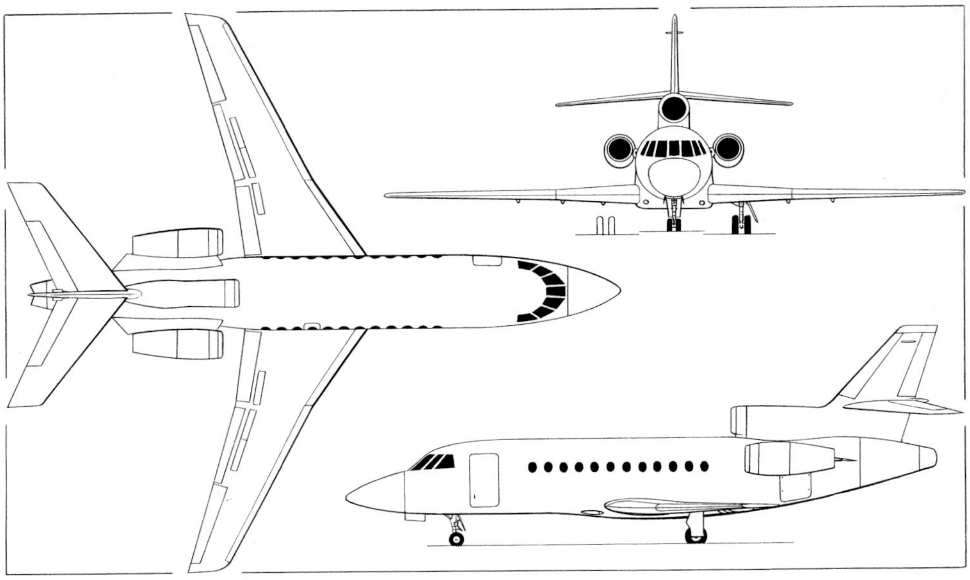

Dassault Falcon 900B (three Garrett TFE731 turbofans) ***(Dennis Punnett)***

Dual Honeywell FMZ 800 flight management system, associated with two AZ 810 air data computers and two Honeywell Laseref II laser ring inertial reference systems; magnetic heading is synthetised by the system. Standby gyro horizon with separate power supply and direct-reading magnetic compass provides standby reference. Dual Collins VIR 32 VOR/ILS/marker beacon receiver; dual Collins ADF 60B; dual Collins TDR 94D transponders; dual Collins DME 42; Sperry RT 300 radio altimeter; dual Collins VHF 22A; dual Bendix/King KHF 950 HF com; large avionics options listing. Honeywell Primus 870 colour radar.

DIMENSIONS, EXTERNAL:

Wing span	19.33 m (63 ft 5 in)
Wing chord: at root	4.08 m (13 ft 4¾ in)
at tip	1.12 m (3 ft 8 in)
Wing aspect ratio	7.62
Length overall	20.21 m (66 ft 3¾ in)
Fuselage: Max diameter	2.50 m (8 ft 2½ in)
Height overall	7.55 m (24 ft 9¼ in)
Tailplane span	7.74 m (25 ft 4¾ in)
Wheel track	4.45 m (14 ft 7¼ in)
Wheelbase	7.93 m (26 ft 0¼ in)
Passenger door: Height	1.72 m (5 ft 7¾ in)
Width	0.80 m (2 ft 7½ in)
Height to sill	1.79 m (5 ft 10½ in)
Emergency exit (overwing, stbd):	
Height	0.91 m (2 ft 11¾ in)
Width	0.53 m (1 ft 8¾ in)

DIMENSIONS, INTERNAL:

Cabin, excl flight deck, incl toilet and baggage compartments: Length	11.90 m (39 ft 0½ in)
Max width	2.34 m (7 ft 8¼ in)
Width at floor	1.86 m (6 ft 1¼ in)
Max height	1.87 m (6 ft 1½ in)
Volume	35.90 m³ (1,268 cu ft)
Rear baggage compartment volume	3.60 m³ (127 cu ft)
Flight deck volume	3.75 m³ (132 cu ft)

AREAS:

Wings, gross	49.00 m² (527.43 sq ft)
Vertical tail surfaces (total)	9.82 m² (105.7 sq ft)
Horizontal tail surfaces (total)	13.35 m² (143.7 sq ft)

WEIGHTS AND LOADINGS:

Weight empty, equipped (typical)	10,240 kg (22,575 lb)
Operating weight empty	10,545 kg (23,248 lb)
Max payload	2,185 kg (4,817 lb)
Payload with max fuel	1,385 kg (3,053 lb)
Max fuel	8,690 kg (19,158 lb)
Max T-O weight	20,640 kg (45,500 lb)
Max landing weight	19,050 kg (42,000 lb)
Normal landing weight	12,250 kg (27,000 lb)
Max zero-fuel weight	12,800 kg (28,220 lb)
Max wing loading	421.2 kg/m² (86.27 lb/sq ft)
Max power loading	325.60 kg/kN (3.19 lb/lb st)

PERFORMANCE (at AUW of 12,250 kg; 27,000 lb, except where indicated):

Max operating speed: 900, 900B:
at S/L
Mach 0.87 (350 knots; 648 km/h; 403 mph IAS)
between 3,050-7,620 m (10,000-25,000 ft)
Mach 0.84 (370 knots; 685 km/h; 425 mph IAS)

Max cruising speed at 8,230 m (27,000 ft):
900, 900B 500 knots (927 km/h; 575 mph)

Econ cruising speed: 900 Mach 0.75

Approach speed, 8 passengers and fuel reserves:
900 103 knots (191 km/h; 119 mph)

Stalling speed: 900, clean 101 knots (188 km/h; 117 mph)
900, landing configuration 79 knots (147 km/h; 91 mph)

Max cruising altitude: 900, 900B 15,550 m (51,000 ft)

Balanced T-O field length with full tanks, 8 passengers and baggage: 900 1,515 m (4,970 ft)

FAR 91 landing field length at AUW of 12,250 kg (27,000 lb): 900 700 m (2,300 ft)

Range with max payload, NBAA IFR reserves:
900 3,460 nm (6,412 km; 3,984 miles)

Range at Mach 0.75 with max fuel and NBAA IFR reserves:
900B, 15 passengers 3,840 nm (7,116 km; 4,421 miles)
900, 8 passengers 3,900 nm (7,227 km; 4,491 miles)

DASSAULT FALCON 2000

TYPE: Follow-on to Falcon 20/200 as transcontinental wide-body business transport.

PROGRAMME: Announced Paris Air Show 1989 as Falcon X; launched as Falcon 2000 on 4 October 1990, following first orders; Alenia joined as 25 per cent risk-sharing partner February 1991, with responsibility for rear fuselage sections and engine nacelles; selection of GE/Garrett CFE738 announced 2 April 1990; first flight (F-WNAV) 4 March 1993; one prototype only; second aircraft, flying about April 1994, will be demonstrator; third and fourth to be first US delivery (Summer 1994) and first European delivery (Aeroleasing, Geneva); French and US certification December 1994 to JAR 25/FAR Pt 25 Transport Category Aircraft standards, and FAR Pt 36 Stage III noise levels.

CUSTOMERS: Total 50 commitments, including 15 firm, received by early 1993 and 40 to 50 firm wanted by early 1995; total market estimated at 300 to 400 in 10 years.

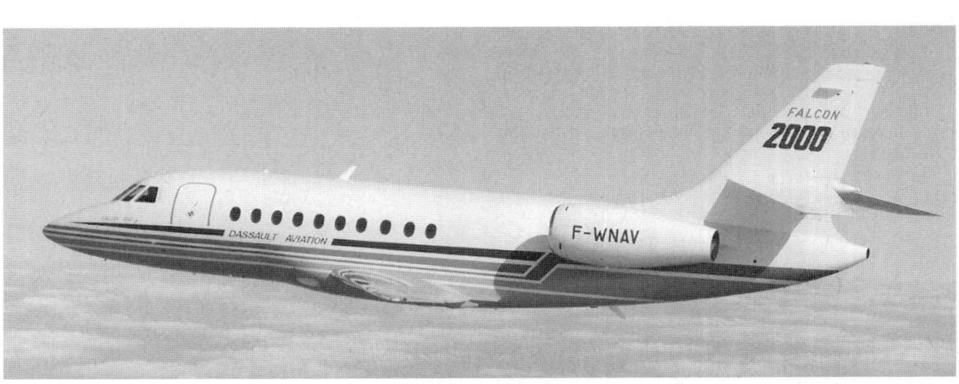

Dassault Falcon 2000 during its first flight on 4 March 1993 ***(Aviaplans/F. Robineau)***

COSTS: Initial price $13.7 million for delivery in 1994; options from November 1992 at $13.95 million.
DESIGN FEATURES: Same fuselage cross-section as Falcon 900, but 1.98 m (6 ft 6 in) shorter. Falcon 900 wing with modified leading-edge and no inboard slat; sweepback at quarter-chord 24° 50′–29°.
STRUCTURE: Largely as for Falcon 900. Rear fuselage, including engine pods and pylons, by Alenia and Piaggio; thrust reversers by Dee Howard.
POWER PLANT: Two GE/Garrett CFE738 turbofans, each rated at 26.7 kN (6,000 lb st). Fuel capacity 6,865 litres (1,814 US gallons; 1,510 Imp gallons).
ACCOMMODATION: Up to 12 passengers and two flight crew; standard passenger accommodation is four seats in forward lounge and two seats and a three-person sofa in aft lounge.
AVIONICS: Includes Collins Pro Line IV four-tube EFIS; Cat. II autopilot and digital flight director also by Collins; Sextant Avionique EICAS (engine indication and crew alerting system); Honeywell FMS.

DIMENSIONS, EXTERNAL:
Wing span 19.33 m (63 ft 5 in)
Length overall 20.23 m (66 ft 4½ in)
Height overall 6.98 m (22 ft 10¾ in)
DIMENSIONS, INTERNAL:
Cabin: Length 7.98 m (26 ft 2¼ in)
  Max width 2.34 m (7 ft 8¼ in)
  Max height 1.87 m (6 ft 1½ in)
  Volume 28.00 m³ (989.6 cu ft)
Baggage volume 4.00 m³ (141.3 cu ft)
AREAS:
Wings, gross 49.02 m² (527.65 sq ft)
WEIGHTS AND LOADINGS (estimated):
Weight empty, equipped 8,936 kg (19,700 lb)
Max payload 1,390 kg (3,064 lb)
Max fuel weight 5,513 kg (12,154 lb)
Max T-O weight 15,875 kg (35,000 lb)
Max landing weight 14,970 kg (33,000 lb)
Max zero-fuel weight 13,000 kg (28,660 lb)

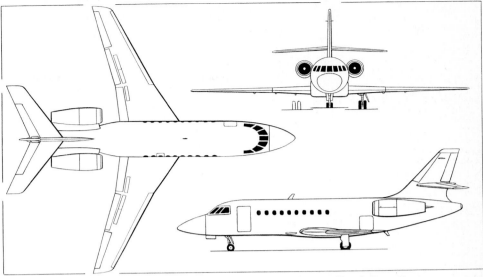

Dassault Falcon 2000 twin-engined business transport *(Dennis Punnett)*

Max wing loading 323.8 kg/m² (66.33 lb/sq ft)
Max power loading 297.6 kg/kN (2.92 lb/lb st)
PERFORMANCE (estimated, at max T-O weight, ISA, except where indicated):
Max operating Mach No. (M_{MO}) 0.85–0.87
Max operating speed (V_{MO})
 350 knots (648 km/h; 403 mph) IAS
Max cruising speed at 11,890 m (39,000 ft)
 Mach 0.83–0.85
Certificated ceiling 14,330 m (47,000 ft)
Balanced T-O field length, 8 passengers, ISA + 15°C
 1,635 m (5,365 ft)
FAR 91 landing field length, 8 passengers, ISA, at S/L
 780 m (2,560 ft)
Range with max fuel, 8 passengers and NBAA IFR reserves 2,997 nm (5,555 km; 3,451 miles)
g limits +2.64/−1

GUIMBAL
BRUNO GUIMBAL

Bruno Guimbal, an engineer with Eurocopter France at Marignane, has designed and constructed a light two-seat helicopter; Eurocopter has sponsored some of final development and flight testing; Eurocopter does not associate itself officially with project.

GUIMBAL G2 CABRI

TYPE: Two-seat light helicopter.
PROGRAMME: Privately designed and developed during the late 1980s; final pre-flight development supported by Eurocopter France; first flight (F-PILA) 11 April 1992; about 10 hours flight and full-scale ground running completed by March 1993; all initial problems solved; 100 knots (185 km/h; 115 mph) cruising speed achieved; production aircraft might be three-seater.
DESIGN FEATURES: Three-blade Spheriflex rotor designed to produce high inertia for safe autorotation, moderate control power (apparent hinge offset 4.8 per cent) and fail-safe design; design avoids mast bumping, minimises dynamic roll-over or over-control and allows wide CG travel; rotor inertia lies between that of Bell JetRanger and Bell 47 to give best practical autorotation performance; Spheriflex main rotor hub, with forked blade ends picking up single elastomeric centrifugal bearings and mast-mounted droop stops, gives low maintenance and high safety; drag dampers are simple elastomeric; patented driveshaft runs in an oblique groove in top of tail pylon; bearingless seven-blade Fenestron tail rotor has Kevlar starplate hub giving integral pitch change freedom and offers good control with protection from contact with long grass and harder obstacles; tailboom has lifting aerofoil designed to unload rotor in cruising flight and save power. Main rotor rpm 597; Fenestron rpm 5,734.
FLYING CONTROLS: Unboosted mechanical dual controls with friction damper at base of cyclic stick acting as attitude trim.
STRUCTURE: All-composite airframe shell weighs 70 kg (154.3 lb), accounting for 22 per cent of empty weight; based on central box of GFRP sandwich carrying instrument console, two integral fuel tanks, transmission, landing skids, engine and tailboom; lateral shells contain seats, windows and pylon fairing; GFRP tailboom, Fenestron and fin integral, but detachable from fuselage; high energy-absorption landing gear bows of filament-wound R-glassfibre.
LANDING GEAR: Two fixed aluminium skids and two GFRP bows giving 4 m (13.12 ft)/s impact absorption.
POWER PLANT: One 112 kW (150 shp) Textron Lycoming O-320 flat-four engine, mounted horizontally, nose forward, with 4 into 1 straight exhaust to reduce engine noise; downdraught cooling fan, fed by ram intake in pylon fairing and mounted on alternator on top of engine, driven through 90° by belt from crankshaft; engine drives transmission by four-sheave V-belt with reduction ratio 0.91:1; belt tensioned by rotating pulley bearing by eccentric lever driven by small electric actuator; main driveshaft drives

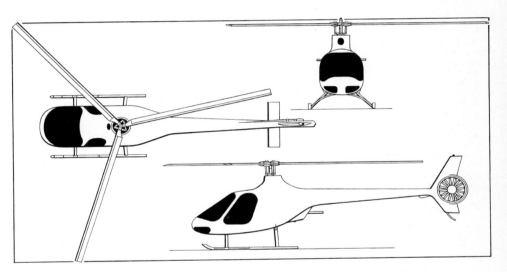

Guimbal G2 Cabri two-seat light helicopter in latest form *(Jane's/Mike Keep)*

Guimbal G2 Cabri two-seat light helicopter

Main rotor head of Guimbal Cabri follows the Spheriflex layout, with single elastomeric bearings, prominent droop stops and drag dampers

Fenestron and main transmission; reductions by simple spiral bevel gears (4:1 for main rotor and 1:2.5 for Fenestron); rotor pylon is stationary with splined driveshaft inside. Fuel capacity 154 litres (40.7 US gallons; 33.9 Imp gallons) in two interconnected integral tanks with single filler cap.

DIMENSIONS, EXTERNAL:	
Main rotor diameter	6.50 m (21 ft 4 in)
Main rotor blade chord	0.16 m (6.3 in)
Fenestron diameter	0.54 m (1 ft 9¼ in)
Fenestron blade chord	0.038 m (1½ in)
Length overall	5.75 m (18 ft 10½ in)
Fuselage: Max width	1.18 m (3 ft 10½ in)
Height to top of rotor head	2.10 m (6 ft 10¾ in)
DIMENSIONS, INTERNAL:	
Baggage volume	0.18 m³ (6.35 cu ft)
AREAS:	
Main rotor disc	33.18 m² (357.2 sq ft)
Fenestron disc	0.23 m² (2.47 sq ft)
WEIGHTS AND LOADINGS:	
Weight empty	320 kg (705.5 lb)
Max T-O weight	550 kg (1,212 lb)
Max disc loading	16.58 kg/m² (3.39 lb/sq ft)
Max power loading	4.92 kg/kW (8.08 lb/hp)
PERFORMANCE (calculated):	
Max level speed, 75% power	108 knots (200 km/h; 124 mph)
Cruising speed	97 knots (180 km/h; 112 mph)
Service ceiling	2,200 m (7,215 ft)
Range	539 nm (1,000 km; 621 miles)
Endurance	5 h

JURCA

MARCEL JURCA

3 Allées des Bordes, F-94430 Chennevieres S/M
Telephone: 33 (1) 45 94 01 38

M Marcel Jurca has for many years designed a series of high-performance light aircraft, of which two are described below. Other Jurca aircraft listed in 1992-93 *Jane's* (some available to amateur builders) are M. J. 2/M. J. 22 Tempete; M. J. 3 Dart; M. J. 4 Shadow; M. J. 5 Sirocco (Sport Wing); M. J. 14 Fourtouna; M. J. 51 Sperocco; and M. J. 53 Autan.

Jurca M. J. 5 Sirocco built in France *(P. Gaillard)*

JURCA M. J. 5 SIROCCO

TYPE: Tandem two-seat development of M. J. 2 Tempete.

PROGRAMME: Prototype first flew 3 August 1962, powered by 78.5 kW (105 hp) Potez 4 E-20 engine; factory built model subsequently awarded certificate of airworthiness in Utility category. Version for amateur construction is generally similar to factory built version, with non-retractable or retractable landing gear, including optional retractable tailwheel.

DESIGN FEATURES: Type of engine fitted indicated by suffix letter in designation. Suffix letters are A for 67 kW (90 hp) Teledyne Continental C90-8 or -14F, B for 74.5 kW (100 hp) Teledyne Continental O-200-A, C for 78.5 kW (105 hp) Potez 4 E-20, D for 86 kW (115 hp) Potez 4 E-30, E for 78.5 kW (105 hp) Hirth, F for 93 kW (125 hp) Textron Lycoming, G for 100.5 kW (135 hp) Regnier, H for 119.5 kW (160 hp) Textron Lycoming, K for 134 kW (180 hp) Textron Lycoming and L for 149 kW (200 hp) Textron Lycoming. Addition of numeral 1 indicates non-retractable landing gear; the numeral 2 indicates retractable main landing gear; 2A indicates tailwheel also retracts. One aircraft has 119 kW (160 hp) PRV modified motorcar engine and P-51 Mustang type underfuselage airscoop.

The details which follow refer to a Sirocco with an 86 kW (115 hp) Textron Lycoming O-235-C2B engine and 1.85 m (6 ft 0¾ in) diameter propeller, and a modified rudder of reduced height and greater chord:

DIMENSIONS, EXTERNAL:	
Wing span	7.00 m (23 ft 0 in)
Wing aspect ratio	4.90
Length overall	6.15 m (20 ft 2 in)
Height overall, tail up:	
with modified rudder	2.60 m (8 ft 6¼ in)
standard rudder	2.80 m (9 ft 2¼ in)
AREAS:	
Wings, gross	10.00 m² (107.64 sq ft)
WEIGHTS AND LOADINGS:	
Weight empty	430 kg (947 lb)
Max T-O weight	680 kg (1,499 lb)
Max wing loading	68.0 kg/m² (13.92 lb/sq ft)
Max power loading	7.93 kg/kW (13.04 lb/hp)
PERFORMANCE (at max T-O weight):	
Max level speed	127 knots (235 km/h; 146 mph)
Cruising speed	116 knots (215 km/h; 134 mph)
Stalling speed	44 knots (80 km/h; 50 mph)
Time to 1,000 m (3,280 ft)	4 min
Service ceiling	5,000 m (16,400 ft)
T-O run	250 m (820 ft)
Landing run	200 m (655 ft)
Endurance	4 h 20 min

JURCA M. J. 52 ZEPHYR

TYPE: Very light two-seat monoplane.

PROGRAMME: Plans available since Spring 1985. Designed to conform to FAR Pt 23 Utility category requirements, yet simple to construct in garage or similar building and inexpensive in terms of materials and working hours.

DESIGN FEATURES: Based on M. J. 5 Sirocco but using converted Volkswagen motorcar engine or Teledyne Continental engine in 30-48.5 kW (40-65 hp) range. Can have non-retractable or retractable landing gear.

DIMENSIONS, EXTERNAL:	
Wing span	9.06 m (29 ft 8¾ in)
Wing aspect ratio	6.08
Length overall	6.28 m (20 ft 7¼ in)
AREAS:	
Wings, gross	13.50 m² (145.3 sq ft)
WEIGHTS AND LOADINGS:	
Weight empty	333 kg (734 lb)
Max T-O weight	517 kg (1,140 lb)
Max wing loading	38.3 kg/m² (7.84 lb/sq ft)
PERFORMANCE (37.3 kW; 50 hp engine):	
Max level speed	59 knots (110 km/h; 68 mph)
Stalling speed	22 knots (40 km/h; 25 mph)
Endurance	5 h
g limits	+8/−4 ultimate

MUDRY

AVIONS MUDRY et CIE

Aérodrome de Bernay, BP 214, F-27300 Bernay
Telephone: 33 32 43 47 34
Fax: 33 32 43 47 90
Telex: MUDRY 180 587 F

DIRECTORS GENERAL:
Auguste Mudry (President)
Danielle Baron (Assistant Director General)

CHIEF DESIGNER: Jean Marie Klinka

Mudry Aviation Ltd

Flager County Airport, Box 18T 7, Florida 32110, USA
Telephone: 1 (904) 437 9700
Fax: 1 (904) 437 1170

PRESIDENT: Daniel Heligoin

Company established at Bernay 1958 and operated alongside CAARP of Beynes; two companies consolidated at Bernay; design office at Courcelles; total workforce 50. Plan to certify Sukhoi Su-26MX in France lapsed for administrative reasons; CAP X is no longer to be produced in Russia; Mudry has manufactured Aéronautic Baroudeur microlight in small numbers for four years.

MUDRY CAP 10 B

TYPE: Two-seat aerobatic and club aircraft.

PROGRAMME: First flight August 1968; certificated 4 September 1970; FAA certification for day and night VFR 1974.

CURRENT VERSIONS: **CAP 10 B**: Current production version with enlarged rudder and ventral fin.

CAP 10 R: Glider tug (remorqueur), but production still not begun by April 1993.

CUSTOMERS: Total 260 CAP 10/10 Bs produced by January 1992. French Air Force operates 56 at Ecole de l'Air at Salon de Provence and Ecole de Formation Initiale du Personnel Navigant (EFIPN 307) at Avord; French Navy operates eight at 51 Escadrille de Servitude at Rochefort/Soubise; Mexican Air Force operates 20 for basic training.

DESIGN FEATURES: Derived from Piel Emeraude. Wing section NACA 23012; dihedral 5°; incidence 0°.

FLYING CONTROLS: Slotted ailerons; trim tabs on both elevators; automatic rudder trim; tailplane incidence adjustable on ground; plain flaps.

STRUCTURE: All-spruce single-spar wing torsion box; rear auxiliary spar; okoumé ply skin on wing and fuselage with polyester fabric covering; double skin on forward fuselage; fin spar integral with fuselage and tailplane.

LANDING GEAR: Non-retractable tailwheel type. Mainwheel legs of light alloy, with ERAM type 9 270 C oleo-pneumatic shock absorbers. Single wheel on each main unit, tyre size 380 × 150 mm. Solid tailwheel tyre, size 6 × 200. Tailwheel is steerable by rudder linkage but can be disengaged for ground manoeuvring. Hydraulically actuated mainwheel disc brakes (controllable from port seat) and parking brake. Streamline fairings on mainwheels and legs.

POWER PLANT: One 134 kW (180 hp) Textron Lycoming AEIO-360-B2F flat-four engine, driving a Hoffmann two-blade fixed-pitch wooden propeller. Standard fuel tank aft of engine fireproof bulkhead, capacity 72 litres (19 US gallons; 16 Imp gallons). Optional auxiliary tank, capacity 75 litres (20 US gallons; 16.5 Imp gallons), beneath baggage compartment. Inverted fuel and oil (Aviat/Christen) systems permit continuous inverted flight.

ACCOMMODATION: Side by side adjustable seats for two persons, with provision for back parachutes, under rearward sliding and jettisonable moulded transparent canopy. Special aerobatic shoulder harness standard. Space for 20 kg (44 lb) of baggage aft of seats in training and touring models.

SYSTEMS: Electrical system includes Delco-Rémy 40A engine driven alternator and STECO ET24 nickel-cadmium battery.
AVIONICS: Bendix/King avionics standard.

DIMENSIONS, EXTERNAL:
Wing span	8.06 m (26 ft 5¼ in)
Wing aspect ratio	6.0
Length overall	7.16 m (23 ft 6 in)
Height overall	2.55 m (8 ft 4½ in)
Tailplane span	2.90 m (9 ft 6 in)
Wheel track	2.06 m (6 ft 9 in)

DIMENSIONS, INTERNAL:
Cockpit: Max width	1.054 m (3 ft 5½ in)

AREAS:
Wings, gross	10.85 m² (116.79 sq ft)
Ailerons (total)	0.79 m² (8.50 sq ft)
Vertical tail surfaces (total)	1.32 m² (14.25 sq ft)
Horizontal tail surfaces (total)	1.86 m² (20.0 sq ft)

WEIGHTS AND LOADINGS (A: Aerobatic, U: Utility):
Weight empty, equipped: A, U	550 kg (1,213 lb)
Fuel weight: A	54 kg (119 lb)
U	108 kg (238 lb)
Max T-O weight: A	760 kg (1,675 lb)
U	830 kg (1,829 lb)
Max wing loading: A	70.05 kg/m² (14.35 lb/sq ft)
U	76.50 kg/m² (15.67 lb/sq ft)
Max power loading: A	5.66 kg/kW (9.31 lb/hp)
U	6.19 kg/kW (10.16 lb/hp)

PERFORMANCE (at max T-O weight):
Never-exceed speed (V_{NE})	183 knots (340 km/h; 211 mph)
Max level speed at S/L	146 knots (270 km/h; 168 mph)
Max cruising speed (75% power)	135 knots (250 km/h; 155 mph)
Stalling speed: flaps up	52 knots (95 km/h; 59 mph) IAS
flaps down	43 knots (80 km/h; 50 mph) IAS
Max rate of climb at S/L	480 m (1,575 ft)/min
Service ceiling	5,000 m (16,400 ft)
T-O run	350 m (1,149 ft)
T-O to 15 m (50 ft)	450 m (1,477 ft)
Landing from 15 m (50 ft)	600 m (1,968 ft)
Landing run	360 m (1,182 ft)
Range with max fuel	539 nm (1,000 km; 621 miles)
g limits	+6/−4.5

Mudry CAP 10 B two-seat aerobatic light aircraft *(John Cook)*

MUDRY CAP 21

TYPE: Single-seat competition aerobatic aircraft.
PROGRAMME: First flight 23 June 1980; first delivery May 1982; 39 built when production ceased in January 1990; aircraft now offered as alternative to CAP 231.
DESIGN FEATURES: Symmetrical V16F wing section; thickness/chord ratio 16 per cent; dihedral 1° 30'; no twist.
FLYING CONTROLS: Mechanical controls; trim tab in each elevator and servo tab in each aileron; roll rate 180°/s at 135 knots (250 km/h; 155 mph)
STRUCTURE: All-wood airframe with plywood skinning on wings and fuselage; single-spar wing; cantilever tail surfaces.
LANDING GEAR: Non-retractable tailwheel type. Cantilever glassfibre main legs, with streamline fairings over wheels. Disc brakes.
POWER PLANT: One 149 kW (200 hp) Textron Lycoming AEIO-360-A1B flat-four engine, driving a two-blade Hartzell variable-pitch propeller. Fixed-pitch propeller optional. Normal fuel tank capacity 40 litres (10.5 US gallons; 8.8 Imp gallons). Max fuel capacity 75 litres (20 US gallons; 16.5 Imp gallons), with 15 litre (4 US gallon; 3.3 Imp gallon) gravity tank for inverted flying. Christen inverted oil system.
ACCOMMODATION: Single glassfibre seat under rearward sliding transparent canopy. Special aerobatic shoulder harness.

DIMENSIONS, EXTERNAL:
Wing span	8.08 m (26 ft 6 in)
Wing aspect ratio	7.0
Length overall	6.46 m (21 ft 2½ in)
Height overall	1.52 m (5 ft 0 in)

AREAS:
Wings, gross	9.40 m² (101.18 sq ft)
Ailerons (total)	1.42 m² (15.28 sq ft)
Fin	0.780 m² (8.40 sq ft)
Rudder	0.725 m² (7.80 sq ft)
Elevators	1.18 m² (12.70 sq ft)

WEIGHTS AND LOADINGS:
Weight empty	500 kg (1,103 lb)
Max T-O weight: Aerobatic	620 kg (1,367 lb)
Normal	750 kg (1,653 lb)
Max wing loading (Aerobatic)	65.95 kg/m² (13.5 lb/sq ft)
Max power loading (Aerobatic)	4.16 kg/kW (6.84 lb/hp)

PERFORMANCE:
Never-exceed speed (V_{NE})	200 knots (372 km/h; 231 mph)
Max level speed at S/L	151 knots (280 km/h; 174 mph)
Max cruising speed (75% power)	135 knots (250 km/h; 155 mph)
Stalling speed	49 knots (90 km/h; 56 mph)
Max rate of climb at S/L	780 m (2,560 ft)/min
Rate of roll	200°/s
Range (with supplementary tank)	431 nm (800 km; 497 miles)
g limits	+8/−6

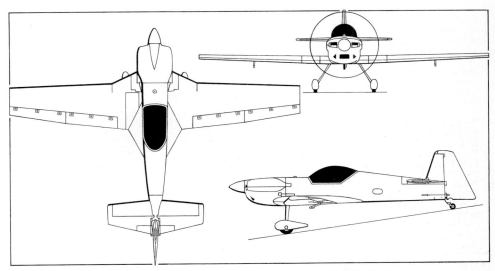

Mudry CAP 231 single-seat aerobatic aircraft *(Jane's/Mike Keep)*

Mudry CAP 231 single-seat aerobatic aircraft *(John Cook)*

MUDRY CAP 231

TYPE: Single-seat competition aerobatic aircraft.
PROGRAMME: First flight (F-WZCI) April 1990; French certification 25 July 1990. Flown by 10 pilots in 1990 world aerobatic championships.
CURRENT VERSIONS: **CAP 231**: Standard aerobatic version, *as described*.
CAP 231 EX: Improved version; described separately.
CUSTOMERS: First aircraft delivered to Tony Bianchi in Britain 1990; three to Morocco Marche Verte; 20 produced by April 1993.
DESIGN FEATURES: Some of 10 CAP 230s converted to CAP 231; main differences include elevator servo tab, forward-swept wingroots and optional Mühlbauer three-blade propeller. Wing section V16F; thickness/chord ratio 16 per cent; dihedral 1° 30'; no twist.
FLYING CONTROLS: Two elevator trim tabs, one also acting as servo tab; four-section ailerons cover three-quarters of span; servo tabs on outer ailerons.
STRUCTURE: All-wood structure; single-spar wing; wooden skins; non-flammable laminated plastics engine cowling.
LANDING GEAR: Non-retractable tailwheel type. Cantilever glassfibre main legs, with wheel fairings. Cleveland disc brakes.

POWER PLANT: One 224 kW (300 hp) Textron Lycoming AEIO-540-L1B5D flat-six engine, driving a two-blade Hartzell HC-C2YR-4CF constant-speed propeller or a three-blade Mühlbauer MTV-9BC 200-15. Fuel capacity 65 litres (17.2 US gallons; 14.3 Imp gallons). Aviat/Christen inverted oil system.
ACCOMMODATION: Single glassfibre seat under sideways opening canopy, hinged to starboard. Space for 35 kg (72 lb) baggage behind pilot. Special aerobatic shoulder harness.
DIMENSIONS, EXTERNAL:
Wing span 8.08 m (26 ft 6 in)
Wing aspect ratio 6.62
Length overall 6.75 m (22 ft 1¼ in)
Height overall 1.90 m (6 ft 2¾ in)
Tailplane span 2.82 m (9 ft 3 in)
Wheel track 2.40 m (7 ft 10½ in)
AREAS:
Wings, gross 9.86 m² (106.13 sq ft)
Ailerons (total) 0.92 m² (9.90 sq ft)
Horizontal tail surfaces (total) 3.89 m² (41.87 sq ft)
WEIGHTS AND LOADINGS (A: Aerobatic, N: Normal):
Weight empty 630 kg (1,389 lb)
Max T-O weight: A 730 kg (1,609 lb)
N 820 kg (1,807 lb)
Max wing loading: A 74.04 kg/m² (15.16 lb/sq ft)
N 83.16 kg/m² (17.03 lb/sq ft)
Max power loading: A 3.26 kg/kW (5.36 lb/hp)
N 3.67 kg/kW (6.03 lb/hp)
PERFORMANCE:
Never-exceed speed (V_{NE})
216 knots (400 km/h; 248 mph)
Max level speed at S/L 178 knots (330 km/h; 205 mph)
Max cruising speed (75% power)
162 knots (300 km/h; 186 mph)
Stalling speed 49 knots (90 km/h; 56 mph)
Max rate of climb at S/L 960 m (3,150 ft)/min
Rate of roll at 161 knots (300 km/h; 186 mph) 270°/s
T-O run 150 m (490 ft)
T-O to 15 m (50 ft) 200 m (656 ft)
Landing from 15 m (50 ft) 450 m (1,476 ft)
Landing run 400 m (1,312 ft)
Range with max fuel 194 nm (360 km; 223 miles)
g limits +10/−10

MUDRY CAP 231 EX
TYPE: Single-seat competition aerobatic aircraft.
PROGRAMME: First flight 18 December 1991 (F-WZCI) with Barrett 194 kW (260 hp) engine.
CUSTOMERS: Being sold in parallel with CAP 231; six produced by April 1993.
COSTS: Price $220,000 (1992 depending on exchange rate) ready to fly in USA certificated for exhibition, with normal Mudry warranties; $155,000 if delivered without engine, propeller, paint and flight testing.
DESIGN FEATURES: Wing in carbonfibre produced by Extra (see Germany); similar to wing of Extra 260, but with different rear spar/fuselage attachment and different aileron assister 'spades'; bare wing weighs 90 kg (198 lb); aerofoil is modified symmetrical V16F; thickness/chord ratio 15 per cent at root, 12.5 per cent at tip; area greater than CAP 230, but CL_{MAX} lower.
FLYING CONTROLS: All push-pull rods with forces equal in all axes; stick movement reduced to 18 cm (7 in) in both axes; ailerons cover most of trailing-edge and assisted by 'spades'; elevator geared tab; two footrests on rudder pedals for aerobatics and cross-country flying.
STRUCTURE: Steel tube fuselage; carbonfibre wing.
LANDING GEAR: Glassfibre bow main landing gear, with Cleveland hydraulic brakes and parking brake; mainwheel tyres 5.00-5; improved tailwheel.
POWER PLANT: One 224 kW (300 hp) Textron Lycoming AEIO-540 flat-six driving Mühlbauer MTV9 BC/200-15 propeller; 300 hp is continuous power; electric starting. One 60 litre (15.8 US gallon; 13.2 Imp gallon) ferry tank in each wing; one 67 litre (17.7 US gallon; 14.7 Imp gallon) aerobatic tank in fuselage; electric contents gauges for wing tanks, sight tube for fuselage tank; average cruising fuel consumption at 145 knots (270 km/h; 167 mph) is 45 litres (11.9 US gallons; 9.9 Imp gallons)/h.
ACCOMMODATION: Pilot has custom seat clipped into cockpit, elbow supports for high g loads; throttle quadrant moved rearwards; rudder pedals adjustable over 15 cm (5½ in).
SYSTEMS: Battery and connectors located to simplify engine starting by jump leads.
DIMENSIONS, EXTERNAL: As for CAP 231 except:
Wing span 7.60 m (24 ft 11¼ in)
Wing aspect ratio 5.86
WEIGHTS AND LOADINGS: As for CAP 231 except:
Weight empty 603 kg (1,329 lb)
PERFORMANCE: As for CAP 231 except:
Max rate of climb at S/L 1,050 m (3,445 ft)/min
Rate of roll:
at 161 knots (300 km/h; 186 mph) 330°/s
at 176 knots (327 km/h; 203 mph) 360°/s

REIMS AVIATION
REIMS AVIATION SA
Aérodrome de Reims Prunay, BP 2745, F-51062 Reims Cedex
Telephone: 33 26 48 46 46
Fax: 33 26 49 13 60
Telex: REMAVIA 830754 F
PRESIDENT DIRECTOR-GENERAL: Jean-Paul Pellissier
MARKETING MANAGER: Philip Evans
EXTERNAL RELATIONS: Christian Jousset

Originally Société Nouvelle des Avions Max Holste, founded 1956. Reims Aviation licenced to manufacture Cessna aircraft for sale in Europe, Africa and Asia, but stopped making small piston-engined types when Cessna did; Reims had built 6,335 aircraft of all types by 1 January 1992. Reims developed twin-turboprop F 406 Caravan II (see below); programme continues, but Cessna sold its 49 per cent share in Reims Aviation to Compagnie Française Chauffour Investissement (CFCI) in early 1989. Reims Aviation supervised in early 1993 by Comité Interministériel de Restructuration Industrielle (CIRI); was then seeking investment from new risk-sharing partners.

Reims makes components for Dassault Falcons, ATR 42/72 and Airbus A300/A310; partnership work on A330/A340, but subcontracts for Dassault declined by 75 per cent in 1992 and for Aerospatiale by 20 per cent. Other activity is maintenance of general aviation aircraft. Workforce was 470 in 1992; office and factory floor space covers 28,288 m² (304,490 sq ft).

REIMS F 406 CARAVAN II
TYPE: Twin-turboprop unpressurised light business and utility transport.
PROGRAMME: Announced mid-1982; first flight (F-WZLT) 22 September 1983; French certification 21 December 1984, FAA later; first flight production F 406 (F-ZBEO) 20 April 1985.
CURRENT VERSIONS: **F 406 Caravan II:** Passenger, freight (with underbelly cargo pod), target towing and coastal patrol version, last-named with underfuselage pod containing Terma SLAR and SAT IR linescanner.
Vigilant: Version for Scottish Fisheries Protection Agency with new belly radome containing GEC-Marconi Seaspray 2000 radar.
CUSTOMERS: Total 69 of the initial production of 85 ordered by January 1993; 67 delivered. French Customs Service received four fitted with Sextant Nadir in Gemini navigation system and Bendix/King 1500 radar in belly radome; French Army uses two for target towing; largest fleet is 28 flying with Aviation Lease Holland BV as freighters; Scottish Fisheries has two Vigilants for fisheries patrol; Zimbabwe purchased four Caravan IIs for rural development programme.
COSTS: Standard aircraft $1,590,000 (1992); Zimbabwe contract for four aircraft reportedly $8 million.
DESIGN FEATURES: Extrapolated from Cessna Conquest airframe; wing section NACA 23018 at root and 23012 at tip; dihedral 3° 30′ on centre-section, 4° 55′ on outer panels; twist −3°; incidence 2° at root; fin offset 1° to port; tailplane dihedral 9°; cabin not pressurised.
FLYING CONTROLS: Conventional, with trim tabs in elevators, port aileron and rudder; hydraulically operated Fowler flaps.
STRUCTURE: Conventional light metal with three-spar fail-safe wing centre-section to SFAR 41C; two-spar outer wings.
LANDING GEAR: Hydraulically retractable tricycle type with single wheel on each unit. Main units retract inward into wing, nosewheel rearward. Emergency extension by means of a 138 bar (2,000 lb/sq in) rechargeable nitrogen bottle. Cessna oleo-pneumatic shock absorbers. Main units of articulated (trailing link) type. Single-disc hydraulic brakes. Parking brake.
POWER PLANT: Two Pratt & Whitney Canada PT6A-112 turboprops (each 373 kW; 500 shp), each driving a McCauley 9910535-2 three-blade reversible-pitch and automatically feathering metal propeller. Fuel capacity 1,823 litres (481 US gallons; 401 Imp gallons).

Reims F 406 Vigilant maritime surveillance aircraft of Scottish Fisheries

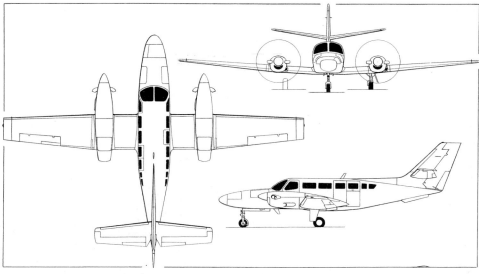

Reims F 406 Caravan II light business and utility transport *(Dennis Punnett)*

ACCOMMODATION: Crew of two and up to 12 passengers, in pairs facing forward, with centre aisle, except at rear of cabin in 12/14-seat versions. Alternative basic configurations for six VIP passengers in reclining seats in business version, and for operation in mixed passenger/freight role. Business version has partition between cabin and flight deck, and toilet on starboard side at rear. Split main door immediately aft of wing, on port side, with built-in airstair in downward hinged lower portion. Optional cargo door forward of this door to provide single large opening. Overwing emergency exit on each side. Passenger seats removable for cargo carrying, or for conversion to ambulance, air photography, maritime surveillance and other specialised roles. Baggage compartments in nose, with three doors, at

Reims F 406 Caravan II target tug of the French Army

rear of cabin and in rear of each engine nacelle. Ventral cargo pod optional.

SYSTEMS: Freon air-conditioning system of 17,500 BTU capacity, plus engine bleed air and electric boost heating. Electrical system includes 28V 250A starter/generator on each engine and 39Ah nickel-cadmium battery. Hydraulic system, pressure 120 bars (1,750 lb/sq in), for operation of landing gear. Separate hydraulic system for brakes. Goodrich pneumatic de-icing of wings and tail unit, and electric windscreen de-icing, optional.

AVIONICS: Standard avionics comprise Bendix/King Silver Crown equipment consisting of dual nav/com, ADF and marker beacon receiver. Bendix/King Gold Crown avionics and autopilot optional. Provision for equipment to FAR Pt 135A standards, including dual controls and instrumentation for co-pilot, IFR com/nav, Bendix/King RDS 82 weather radar and additional emergency exit.

EQUIPMENT: Optional cargo interior includes heavy duty sidewalls, utility floorboards, cabin floodlighting and cargo restraint nets.

DIMENSIONS, EXTERNAL:

Wing span	15.08 m (49 ft 5½ in)
Wing aspect ratio	9.7
Length overall	11.89 m (39 ft 0¼ in)
Height overall	4.01 m (13 ft 2 in)
Tailplane span	5.87 m (19 ft 3 in)
Wheel track	4.28 m (14 ft 0½ in)
Wheelbase	3.81 m (12 ft 5⅞ in)
Propeller diameter	2.36 m (7 ft 9 in)
Cabin door: Height	1.27 m (4 ft 2 in)
Width	0.58 m (1 ft 10¾ in)
Cargo double door (optional): Total width	1.24 m (4 ft 1 in)

DIMENSIONS, INTERNAL:

Cabin (incl flight deck): Length	5.71 m (18 ft 8¾ in)
Max width	1.42 m (4 ft 8 in)
Max height	1.31 m (4 ft 3¼ in)
Min height (at rear)	1.21 m (3 ft 11½ in)
Width of aisle	0.29 m (11½ in)
Volume	8.64 m³ (305 cu ft)
Baggage compartment (nose): Length	2.00 m (6 ft 6¾ in)
Volume	0.74 m³ (26.0 cu ft)
Nacelle lockers: Length	1.55 m (5 ft 1¼ in)
Width	0.73 m (2 ft 4¾ in)
Baggage volume: total, internal	2.22 m³ (78.5 cu ft)
incl cargo pod	3.52 m³ (124.3 cu ft)

AREAS:

Wings, gross	23.48 m² (252.75 sq ft)
Ailerons (total)	1.36 m² (14.64 sq ft)
Trailing-edge flaps	3.98 m² (42.84 sq ft)
Fin	4.05 m² (43.59 sq ft)
Rudder, incl tab	1.50 m² (16.15 sq ft)
Tailplane	5.81 m² (62.54 sq ft)
Elevators, incl tabs	1.66 m² (17.87 sq ft)

WEIGHTS AND LOADINGS:

Weight empty, equipped	2,460 kg (5,423 lb)
Max payload	1,563 kg (3,446 lb)
Max fuel	1,444 kg (3,183 lb)
Max ramp weight	4,502 kg (9,925 lb)
Max T-O and landing weight	4,468 kg (9,850 lb)
Max zero-fuel weight	3,856 kg (8,500 lb)
Max wing loading	190.3 kg/m² (38.97 lb/sq ft)
Max power loading	5.99 kg/kW (9.85 lb/shp)

PERFORMANCE:

Max operating Mach No. (MMO)	0.52
Max operating speed (VMO)	229 knots (424 km/h; 263 mph) IAS
Max cruising speed	246 knots (455 km/h; 283 mph)
Econ cruising speed	200 knots (370 km/h; 230 mph)
Stalling speed: clean	94 knots (174 km/h; 108 mph) IAS
wheels and flaps down	81 knots (150 km/h; 93 mph) IAS
Max rate of climb at S/L	564 m (1,850 ft)/min
Rate of climb at S/L, OEI	121 m (397 ft)/min
Service ceiling	9,145 m (30,000 ft)
Service ceiling, OEI	4,935 m (16,200 ft)
T-O run	526 m (1,725 ft)
T-O to 15 m (50 ft)	803 m (2,635 ft)
Landing from 15 m (50 ft), without reverse pitch	674 m (2,212 ft)
Range with max fuel, at max cruising speed, 45 min reserves	1,153 nm (2,135 km; 1,327 miles)

ROBIN

AVIONS PIERRE ROBIN

1 route de Troyes, Darois, F-21121 Fontaine-les-Dijon Cedex
Telephone: 33 80 44 20 50
Fax: 33 80 35 60 80
Telex: 350 818 ROBIN F
PRESIDENT DIRECTOR GENERAL: Georges Megrelis
DOMESTIC SALES MANAGER: Michel Pelletier
PUBLIC RELATIONS: Jacques Bigenwald

Formed October 1957 as Centre Est Aéronautique; name changed to Avions Pierre Robin 1969; acquired July 1988 by Compagnie Française Chauffour Investissement (CFCI) and incorporated into Aéronautique Service group with Robin SA (after-sales support company of Dijon Val-Suzon) and SN Centrair sailplane manufacturer; Pierre Robin left company 1990. Total of 3,220 aircraft produced by January 1993; 74 aircraft delivered in 1992, including 69 DR 400s and three R 3000s. Workforce 140 in 1992; factory area 11,500 m² (123,785 sq ft).

ROBIN 200

TYPE: Two-seat light trainer.

PROGRAMME: New production version of Robin HR 200/120B, which had first flown in 1971 and was built through much of 1970s (see 1977-78 *Jane's*); incorporates minor modifications.

CUSTOMERS: Approx 10 ordered by early 1993.

COSTS: FFr420,550 with Package 1 avionics; FFr497,785 with Package 2.

DESIGN FEATURES: As for HR 200/120B but with new instrument panel, adjustable seats, new engine cowlings and propeller spinner, and improved anti-corrosion treatment. Wing section NACA 64A515 (mod); dihedral 6° 18′ from roots; incidence 6°; no sweepback.

FLYING CONTROLS: Cable-actuated plain Frise-type ailerons, one-piece all-moving tailplane with trim and anti-servo tabs, and rudder; electrically actuated slotted trailing-edge flaps.

STRUCTURE: All-metal; aluminium alloy stressed skin and ribs.

LANDING GEAR: Non-retractable tricycle type, with single wheel and tyre of similar size on each unit; nosewheel leg offset to starboard, steered by rudder bar; streamline leg and wheel fairings; damped tailskid; hydraulic disc brakes and parking brake.

POWER PLANT: One 88 kW (118 hp) Textron Lycoming O-235-L2A flat-four engine, driving a two-blade fixed-pitch propeller. Fuel capacity 120 litres (31.7 US gallons; 26.4 Imp gallons). Auxiliary tanks optional.

ACCOMMODATION: Pilot and passenger side by side under forward sliding jettisonable canopy with anti-glare tint; dual stick controls; dual left hand throttles; dual toe brakes.

SYSTEMS: Cabin ventilated and heated, with windscreen defrosting standard. Electrical system includes 12V 32Ah battery, 12V 50A alternator and starter.

AVIONICS: *Package 1:* Magnetic compass, rate of climb indicator, Bendix/King KY 97 nav/com, dual PTT switches, SPA 400 intercom, flight hour recorder, landing light, navigation lights and strobe light. *Package 2:* Also has gyro-horizon and directional gyro with engine-driven vacuum pump, electric turn and bank indicator, Bendix/King KX 155-38 (in place of KY 97) nav/com, KI 208 VOR indicator, Bendix/King KT 76A transponder, ACK 30 Alti-encoder, and panel lighting.

DIMENSIONS, EXTERNAL:

Wing span	8.33 m (27 ft 4 in)
Wing chord, constant	1.50 m (4 ft 11 in)
Wing aspect ratio	5.55
Length overall	6.64 m (21 ft 9½ in)
Height overall	1.94 m (6 ft 4½ in)
Tailplane span	2.64 m (8 ft 8 in)
Wheel track	2.88 m (9 ft 5½ in)
Wheelbase	1.465 m (4 ft 9½ in)

AREAS:

Wings, gross	12.50 m² (134.5 sq ft)
Ailerons, total	1.06 m² (11.41 sq ft)
Trailing-edge flaps, total	1.34 m² (14.42 sq ft)
Elevators, incl tabs	2.03 m² (21.85 sq ft)

WEIGHTS AND LOADINGS:

Weight empty	525 kg (1,157 lb)
Max T-O weight	780 kg (1,719 lb)
Max wing loading	62.4 kg/m² (12.78 lb/sq ft)
Max power loading	8.86 kg/kW (14.57 lb/hp)

PERFORMANCE:

Cruising speed, 75% power	121 knots (225 km/h; 140 mph)
Stalling speed	50 knots (92 km/h; 57 mph)
Max rate of climb at S/L	234 m (768 ft)/min
Range	566 nm (1,050 km; 652 miles)
Endurance	4 h 35 min

ROBIN DR 400/100 CADET

TYPE: Two-seat training and touring aircraft.

PROGRAMME: Deliveries began 1987; production ended. Further details and an illustration in 1992-93 *Jane's*.

CUSTOMERS: Two delivered in 1991 and one in 1992, making total of 13.

ROBIN DR 400 DAUPHIN

TYPE: Three/four-seat light training and touring aircraft.

PROGRAMME: First flight original DR 400 Petit Prince 15 May 1972; French and UK certification 1977; DR 400 Dauphin introduced 1979; improvements introduced 1988.

CURRENT VERSIONS: **DR 400/120 Dauphin 2+2:** Production version with 83.5 kW (112 hp) engine, to carry two adults and two children.

DR 400/140B Dauphin 4: Full four-seater with 119 kW (160 hp) engine.

Robin 200 two-seat trainer (88 kW; 118 hp Textron Lycoming O-235-L2A engine)

FRANCE: AIRCRAFT—ROBIN

CUSTOMERS: 1992 production 24 Dauphin 2+2 and 17 Dauphin 4.

DESIGN FEATURES: Classic Jodel design; wing section NACA 23013.5 modified with leading-edge droop; centre panels parallel chord, slight twist; outer panels tapered with dihedral 14°; twist −6°.

FLYING CONTROLS: Slab tailplane with trimmable anti-balance tab; interchangeable ailerons; plain flaps; manually operated airbrake under main spar outboard of each main undercarriage leg.

STRUCTURE: All-wood; single box spar with ribs threaded over box; plywood covered leading-edge box; fabric covering elsewhere; fuselage plywood covered; flaps all-metal and interchangeable.

LANDING GEAR: Non-retractable tricycle type, with oleo-pneumatic shock absorbers and hydraulically actuated disc brakes. All three wheels and tyres are size 380 × 150 mm, pressure 1.57 bars (22.8 lb/sq in) on nose unit, 1.77 bars (25.6 lb/sq in) on main units. Nosewheel steerable via rudder bar. Fairings over all three legs and wheels. Tailskid with damper. Toe brakes and parking brake.

POWER PLANT: *Dauphin 2+2:* One 83.5 kW (112 hp) Textron Lycoming O-235-L2A flat-four engine, driving a Sensenich 72 CKS 6-0-56 two-blade fixed-pitch metal propeller, or Hoffmann two-blade fixed-pitch wooden propeller. *Dauphin 4:* One Textron Lycoming O-320-D flat-four engine developing 104.4 kW (140 hp) at 2,300 rpm and 119 kW (160 hp) at 2,700 rpm. Fuel tank in fuselage, usable capacity 100 litres (26.4 US gallons; 22 Imp gallons); optional 50 litre (13.2 US gallon; 11 Imp gallon) auxiliary tank. Oil capacity 5.7 litres (1.5 US gallons; 1.25 Imp gallons).

ACCOMMODATION: Enclosed cabin, with seats for three or four persons. Max weight of 154 kg (340 lb) on front pair and 136 kg (300 lb), including baggage, at rear in Dauphin 2+2. Additional 55 kg (121 lb) of disposable load in Dauphin 4. Access via forward sliding jettisonable transparent canopy. Dual controls standard. Cabin heated and ventilated. Baggage compartment with internal access.

EQUIPMENT: Standard equipment includes a 12V 50A alternator, 12V 32Ah battery, electric starter, audible stall warning and windscreen de-icing. Radio, blind-flying equipment, and navigation, landing and anti-collision lights, to customer's requirements.

Robin DR 400/120 Dauphin 2+2 light aircraft *(John Cook)*

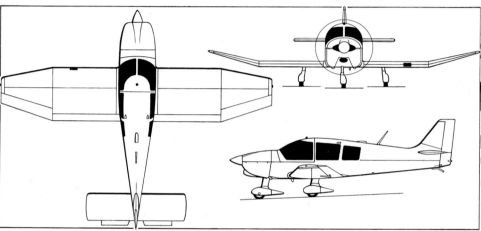

Robin DR 400/120 Dauphin 2+2 *(Jane's/Mike Keep)*

DIMENSIONS, EXTERNAL:
Wing span	8.72 m (28 ft 7¼ in)
Wing chord:	
centre-section, constant	1.71 m (5 ft 7½ in)
at tip	0.90 m (3 ft 0 in)
Wing aspect ratio	5.59
Length overall	6.96 m (22 ft 10 in)
Height overall	2.23 m (7 ft 3¾ in)
Tailplane span	3.20 m (10 ft 6 in)
Wheel track	2.60 m (8 ft 6¼ in)
Wheelbase	5.20 m (17 ft 0¾ in)
Propeller diameter	1.78 m (5 ft 10 in)

DIMENSIONS, INTERNAL:
Cabin: Length	1.62 m (5 ft 3¾ in)
Max width	1.10 m (3 ft 7¼ in)
Max height	1.23 m (4 ft 0½ in)
Baggage volume	0.39 m³ (13.75 cu ft)

AREAS:
Wings, gross	13.60 m² (146.39 sq ft)
Ailerons, total	1.15 m² (12.38 sq ft)
Flaps, total	0.70 m² (7.53 sq ft)
Fin	0.61 m² (6.57 sq ft)
Rudder	0.63 m² (6.78 sq ft)
Horizontal tail surfaces, total	2.88 m² (31.00 sq ft)

WEIGHTS AND LOADINGS:
Weight empty, equipped: 2+2	535 kg (1,179 lb)
4	580 kg (1,279 lb)
Max baggage: 2+2, 4	40 kg (88 lb)
Max T-O and landing weight: 2+2	900 kg (1,984 lb)
4	1,000 kg (2,205 lb)
Max wing loading: 2+2	66.2 kg/m² (13.56 lb/sq ft)
4	73.5 kg/m² (15.05 lb/sq ft)
Max power loading: 2+2	10.78 kg/kW (17.71 lb/hp)
4	8.38 kg/kW (13.78 lb/hp)

PERFORMANCE (at max T-O weight):
Never-exceed speed (V$_{NE}$): 2+2, 4	166 knots (308 km/h; 191 mph)
Max level speed at S/L: 2+2	130 knots (241 km/h; 150 mph)
4	143 knots (265 km/h; 165 mph)
Max cruising speed: 2+2	116 knots (215 km/h; 133 mph)
4	117 knots (216 km/h; 134 mph)
Stalling speed, flaps down:	
2+2	45 knots (82 km/h; 51 mph)
4	47 knots (87 km/h; 54 mph)
Max rate of climb at S/L: 2+2	183 m (600 ft)/min
4	264 m (865 ft)/min
Service ceiling: 2+2	3,660 m (12,000 ft)
4	4,265 m (14,000 ft)
T-O run: 2+2	235 m (771 ft)
4	245 m (804 ft)
T-O to 15 m (50 ft): 2+2	535 m (1,755 ft)
4	485 m (1,591 ft)
Landing from 15 m (50 ft): 2+2	460 m (1,510 ft)
4	470 m (1,542 ft)
Landing run: 2+2	200 m (656 ft)
4	220 m (722 ft)
Range with standard fuel at max cruising speed, no reserves: 2+2, 4	464 nm (860 km; 534 miles)

ROBIN DR 400/160 MAJOR

TYPE: Four-seat light aircraft.

PROGRAMME: First flight of original DR 400 Chevalier 29 June 1972; certificated France and UK same year; Major introduced 1980.

CUSTOMERS: Total 124 delivered by January 1993, including three in 1992.

DESIGN FEATURES: Main differences from Dauphin (see above) are external baggage compartment door on port side and extended wingroot leading-edges to house additional fuel tanks.

Differences from Dauphin listed below:

POWER PLANT: One 119 kW (160 hp) Textron Lycoming O-320-D flat-four engine, driving a Sensenich two-blade metal fixed-pitch propeller. Fuel tank in fuselage, capacity 110 litres (29 US gallons; 24 Imp gallons), and two tanks in wingroot leading-edges, giving total capacity of 190 litres (50 US gallons; 41.75 Imp gallons), of which 182 litres (48 US gallons; 40 Imp gallons) are usable. Provision for auxiliary tank, raising total capacity to 250 litres (66 US gallons; 55 Imp gallons). Oil capacity 7.5 litres (2 US gallons; 1.6 Imp gallons).

ACCOMMODATION: Seating for four persons, on adjustable front seats (max load 154 kg; 340 lb total) and rear bench seat (max load 154 kg; 340 lb total). Forward sliding transparent canopy. Up to 40 kg (88 lb) of baggage can be stowed aft of rear seats when four occupants are carried.

DIMENSIONS, EXTERNAL:
Wing aspect ratio	5.35
Propeller diameter	1.83 m (6 ft 0 in)
Baggage door: Height	0.47 m (1 ft 6½ in)
Width	0.55 m (1 ft 9½ in)

AREAS:
Wings, gross	14.20 m² (152.8 sq ft)

WEIGHTS AND LOADINGS:
Weight empty, equipped	570 kg (1,257 lb)
Max T-O and landing weight	1,050 kg (2,315 lb)
Max wing loading	74.2 kg/m² (15.20 lb/sq ft)
Max power loading	8.82 kg/kW (14.47 lb/hp)

PERFORMANCE (at max T-O weight):
Never-exceed speed (V$_{NE}$)	166 knots (308 km/h; 191 mph)
Max level speed at S/L	146 knots (271 km/h; 168 mph)
Max cruising speed (75% power) at 2,440 m (8,000 ft)	132 knots (245 km/h; 152 mph)
Econ cruising speed (65% power) at 3,200 m (10,500 ft)	130 knots (241 km/h; 150 mph)
Stalling speed: flaps up	56 knots (103 km/h; 64 mph)
flaps down	50 knots (93 km/h; 58 mph)
Max rate of climb at S/L	255 m (836 ft)/min
Service ceiling	4,115 m (13,500 ft)
T-O run	295 m (968 ft)
T-O to 15 m (50 ft)	590 m (1,936 ft)
Landing from 15 m (50 ft)	545 m (1,788 ft)
Landing run	250 m (820 ft)
Range with standard fuel at econ cruising speed, no reserves	825 nm (1,530 km; 950 miles)

ROBIN DR 400/180 REGENT

TYPE: Four/five-seat light aircraft.

PROGRAMME: First flight 27 March 1972; certificated 10 May 1972.

CUSTOMERS: Total 293 delivered by January 1993, including 15 in 1992.

DESIGN FEATURES: Typical Jodel design; two-seat rear bench.

Differences from DR 400/160 listed below:

POWER PLANT: One 134 kW (180 hp) Textron Lycoming O-360-A flat-four engine. Fuel tankage unchanged.

ACCOMMODATION: Basically as for DR 400/160. Baggage capacity 60 kg (132 lb).

DIMENSIONS, EXTERNAL:
Propeller diameter	1.93 m (6 ft 4 in)

WEIGHTS AND LOADINGS:
Weight empty, equipped	600 kg (1,322 lb)
Max T-O and landing weight	1,100 kg (2,425 lb)
Max wing loading	77.7 kg/m² (15.91 lb/sq ft)
Max power loading	8.21 kg/kW (13.47 lb/hp)

PERFORMANCE (at max T-O weight):
Max level speed at S/L	150 knots (278 km/h; 173 mph)
Max cruising speed (75% power) at 2,285 m (7,500 ft)	140 knots (260 km/h; 162 mph)
Econ cruising speed (60% power) at 3,660 m (12,000 ft)	132 knots (245 km/h; 152 mph)
Stalling speed: flaps up	57 knots (105 km/h; 65 mph)
flaps down	52 knots (95 km/h; 59 mph)
Max rate of climb at S/L	252 m (825 ft)/min
Service ceiling	4,720 m (15,475 ft)
T-O run	315 m (1,035 ft)
T-O to 15 m (50 ft)	610 m (2,000 ft)
Landing from 15 m (50 ft)	530 m (1,740 ft)
Landing run	249 m (817 ft)
Range with standard fuel at 65% power, no reserves	783 nm (1,450 km; 900 miles)

ROBIN DR 400 REMO 180R

TYPE: Glider tug.
PROGRAMME: First flight and certification 1972 as DR 400/180R (Remorqueur); flown 1985 with Porsche PFM 3200 engine as DR 400RP or Remo 212; became first Porsche powered aircraft to be certificated; Remo 212 production ceased 1990 after 29 had been built (details in 1991-92 *Jane's*).
CUSTOMERS: Total 293 Remorqueurs delivered by January 1993, including nine Remo 180Rs in 1992.
DESIGN FEATURES: Same as Régent except no external baggage door and baggage compartment covered in transparent Plexiglas to maximise rearward view; towing hook under tail; Dauphin wing without extended wingroot leading-edges.

Differences from DR 400/180 listed below:

POWER PLANT: One 134 kW (180 hp) Textron Lycoming O-360-A flat-four engine, driving (for glider towing) a Sensenich 76 EM 8S5 058 or Hoffmann HO-27-HM-180/138 two-blade propeller. For touring, a Sensenich 76 EM 8S5 064 propeller of same diameter is fitted.

AREAS:
Wings, gross   13.60 m² (146.39 sq ft)
WEIGHTS AND LOADINGS:
Weight empty, equipped   560 kg (1,234 lb)
Max T-O and landing weight   1,000 kg (2,205 lb)
Max wing loading   73.5 kg/m² (15.05 lb/sq ft)
Max power loading   7.46 kg/kW (12.25 lb/hp)
PERFORMANCE (glider tug, at max T-O weight):
Max level speed   146 knots (270 km/h; 168 mph)
Cruising speed (70% power) at 2,440 m (8,000 ft)
                          124 knots (230 km/h; 143 mph)
Stalling speed, flaps down   47 knots (87 km/h; 54 mph)
Max rate of climb at S/L   336 m (1,100 ft)/min
Max rate of climb at S/L, towing two-seat sailplane
                                    210 m (690 ft)/min
Service ceiling   6,100 m (20,000 ft)
T-O to 15 m (50 ft), towing single-seat sailplane
                                      375 m (1,230 ft)
Landing from 15 m (50 ft)   470 m (1,542 ft)
Range at econ cruising speed, with auxiliary fuel, no reserves   647 nm (1,200 km; 745 miles)

ROBIN DR 400/200R

TYPE: Glider tug, based on DR 400 Remo 180R.
PROGRAMME: Delivered from 1993.
DESIGN FEATURES: More powerful version of DR 400 Remo 180R, with 149 kW (200 hp) Textron Lycoming IO-360 engine driving a constant-speed propeller.

ROBIN R 3000 SERIES

TYPE: Four-seat light aircraft.
PROGRAMME: Development began 1978; two prototype R 3140s flew 1980 and 1981, second with compound-taper wing; marketing of R 3000 assigned to Socata from 1 September 1983 to 1 February 1988; manufacture of Series 120 ended 1987; developing retractable landing gear version with 149 kW (200 hp) Textron Lycoming IO-360.
CURRENT VERSIONS: **R 3000/140:** Formerly R 3140E, certificated 13 October 1983.
R 3000/160: Replaced Series 120; received US FAA certification December 1992.
R 3000 ATAL: Shown at Paris Air Show 1989; Aerospatiale ATAL pod weighing 8.6 kg (19 lb) containing steerable black and white, colour or LLTV camera and able to downlink picture to mobile ground stations from distance of 2.7 to 27 nm (5 to 50 km; 3.1 to 31 miles) or to fixed ground stations from 65 to 97 nm (120 to 180 km; 75 to 112 miles).
Peugeot Renault Volvo (PRV): First flight of R 3140 powered by PRV engine, 1983; engineering assistance from Renault; considered by Robin as 'flying test bench' for PRV power plant; current work on three-litre engine, with belt drive reduction replaced by 0.442:1 gearbox; dual electronic ignition and electronic fuel injection; rated at 134 kW (180 hp); France Aéromoteur (see Aero Engines section) formed to produce an aviation certificated PRV engine; JAR 'E'/FAR Pt 23 certification had been expected late 1992.
CUSTOMERS: Total 64 (all versions) delivered by January 1993.
DESIGN FEATURES: All-metal airframe with 'flat' version of Jodel wing; original intention that one single airframe should accept all different engines; seven proposed versions not produced.
Wing section NACA 43013.5 on constant chord inner panels and 43010.5 on tapered outer panels; dihedral 6°; incidence 3°; T tail.
FLYING CONTROLS: Fixed tailplane, full span elevator balance/trim tabs; electrically actuated slotted flaps.
STRUCTURE: All light alloy except glassfibre engine cowlings; single-spar wing.
LANDING GEAR: Non-retractable tricycle type. Nosewheel, steerable via rudder pedals, is self-centring and locks automatically after take-off. Robin long-stroke low pressure oleo-pneumatic shock absorbers. Mainwheel tyres size 380 × 150-6. Nosewheel tyre size 5.00-5. Cleveland disc brakes. Streamline polyester fairings on all three legs and wheels. Hydraulic disc brakes. Parking brake.

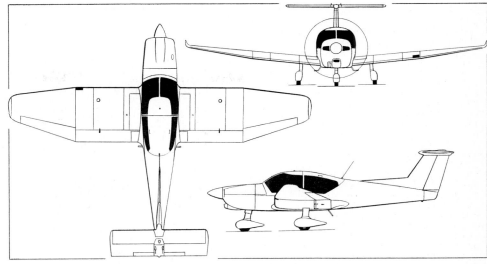

Robin R 3000/140 (O-320-D2A engine) *(Dennis Punnett)*

Robin R 3000/160 four-seat light aircraft *(John Cook)*

POWER PLANT: *R 3000/140:* One 119.3 kW (160 hp) Textron Lycoming O-320-D2A flat-four engine, driving a Sensenich 74DMS5-2-64 two-blade fixed-pitch metal propeller. Two integral fuel tanks in wing leading-edges, with total capacity of 160 litres (42.25 US gallons; 35.2 Imp gallons) standard, or 200 litres (52.8 US gallons; 44 Imp gallons) optional. Oil capacity 7.5 litres (2 US gallons; 1.6 Imp gallons). *R 3000/160:* One 134 kW (180 hp) Textron Lycoming O-320-A flat-four engine. Standard fuel capacity 225 litres (59.4 US gallons; 49.5 Imp gallons).
ACCOMMODATION: Four seats in pairs in enclosed cabin, with dual controls and brakes. Adjustable front seats, with inertia reel safety belts. Removable rear seats, with belts. Carpeted floor. Forward sliding jettisonable and tinted transparent canopy, with safety lock, accessible from both sides. Automatically retracting step on each side. Baggage capacity 40 kg (88 lb). Cabin heated and ventilated. Windscreen demister.
SYSTEMS: Electrical system includes 12V 60A alternator and 12V 32Ah battery.
AVIONICS: Three standards of optional avionics and equipment available. Series I includes horizon and directional gyros with vacuum pump, type 9100 electric turn co-ordinator, rate of climb indicator, C 2400 magnetic compass (exchange for standard C 2300), position lights and two beacons, anti-collision light and instrument panel lighting. Series II adds to Series I either Becker AR 2009/25 720-channel VHF, with NR 2029 VOR/LOC receiver and indicator; or Bendix/King KX 155/08 nav/com with audio and KI 203 VOR indicator. Series III adds to Series II either a Becker ATC 2000 transponder and type 2079 ADF; or Bendix/King KT 76A transponder and KR 87 digital ADF.
EQUIPMENT: Optional mission equipment includes agricultural spraygear with underwing spraybars and chemical tank, capacity 350 litres (92.5 US gallons; 77 Imp gallons). Equipment forming part of avionics packages can be found under Avionics.

DIMENSIONS, EXTERNAL:
Wing span   9.81 m (32 ft 2¼ in)
Wing chord: at root   1.72 m (5 ft 7¾ in)
         at tip   0.655 m (2 ft 1¾ in)
Wing aspect ratio   6.6
Length overall   7.51 m (24 ft 7¾ in)
Height overall   2.66 m (8 ft 8¾ in)
Tailplane span   3.20 m (10 ft 6 in)
Wheel track   2.64 m (8 ft 8 in)
Wheelbase   1.74 m (5 ft 8½ in)
Propeller diameter   1.83 m (6 ft 0 in)
Propeller ground clearance   0.30 m (11¾ in)
DIMENSIONS, INTERNAL:
Cabin: Length   2.70 m (8 ft 10¼ in)
       Max width   1.14 m (3 ft 8¾ in)
       Max height   1.20 m (3 ft 11¼ in)
       Floor area   2.60 m² (28.0 sq ft)
       Volume (incl baggage space) 2.4 m³ (84.75 cu ft)
Baggage space   0.43 m³ (15.2 cu ft)
AREAS:
Wings, gross   14.47 m² (155.75 sq ft)
Ailerons (total)   1.32 m² (14.21 sq ft)
Trailing-edge flaps (total)   2.02 m² (21.74 sq ft)
Vertical tail surfaces (total) 1.30 m² (14.00 sq ft)
Horizontal tail surfaces (total) 2.44 m² (26.26 sq ft)
WEIGHTS AND LOADINGS:
Weight empty: 140   600 kg (1,323 lb)
              160   650 kg (1,433 lb)
Max T-O and landing weight: 140   1,050 kg (2,315 lb)
                            160   1,150 kg (2,535 lb)
Max wing loading: 140   72.6 kg/m² (14.86 lb/sq ft)
                  160   79.5 kg/m² (16.28 lb/sq ft)
Max power loading: 140   10.10 kg/kW (16.54 lb/hp)
                   160   8.58 kg/kW (14.08 lb/hp)
PERFORMANCE (at max T-O weight):
Max level speed at S/L:
140   135 knots (250 km/h; 155 mph)
160   146 knots (270 km/h; 168 mph)
Max cruising speed (75% power) at optimum altitude:
140   130 knots (240 km/h; 149 mph)
160   138 knots (255 km/h; 158 mph)
Econ cruising speed (65% power):
140   119 knots (220 km/h; 136 mph)
160   128 knots (238 km/h; 148 mph)
Stalling speed, flaps down:
140   47 knots (87 km/h; 54 mph)
160   49 knots (91 km/h; 57 mph)
Max rate of climb at S/L: 140   258 m (846 ft)/min
                          160   267 m (875 ft)/min

FRANCE: AIRCRAFT—ROBIN/SOCATA

Service ceiling: 140	4,265 m (14,000 ft)
160	4,570 m (15,000 ft)
T-O run: 140	280 m (920 ft)
160	310 m (1,017 ft)
T-O to 15 m (50 ft): 140	525 m (1,725 ft)
160	565 m (1,854 ft)
Landing from 15 m (50 ft): 140	490 m (1,610 ft)
160	540 m (1,772 ft)
Landing run: 140	190 m (625 ft)
160	210 m (690 ft)
Range with max standard fuel, no reserves:	
75% power: 140	605 nm (1,120 km; 696 miles)
160	804 nm (1,490 km; 925 miles)
65% power: 140	640 nm (1,185 km; 736 miles)
160	868 nm (1,610 km; 1,000 miles)
Range with max optional fuel, no reserves:	
75% power: 140	756 nm (1,400 km; 870 miles)
65% power: 140	799 nm (1,480 km; 919 miles)

ROBIN X4

TYPE: Experimental four-seat light aircraft.
PROGRAMME: First flight (F-WKQX) 25 February 1991; exhibited at 1991 Paris Air Show, but no figures released.
DESIGN FEATURES: Objective to exploit comparative wind tunnel evaluation of new aerodynamics, including laminar sections, compared with DR 400; also to examine composite and mixed metal/composite structures; initial modest 86.5 kW (116 hp) Textron Lycoming engine provides severest test of aircraft performance capability; wing is wooden to facilitate rapid modification; fuselage composites.

Robin X4 experimental four-seat light aircraft (*John Cook*)

SOCATA
SOCIETE DE CONSTRUCTION D'AVIONS DE TOURISME ET D'AFFAIRES
(Subsidiary of Aerospatiale)

12 rue Pasteur, F-92150 Suresnes
Telephone: 33 (1) 45 06 37 60
Fax: 33 (1) 46 97 35 90
Telex: SOCATAS 520 828 F
WORKS AND AFTER-SALES SERVICE: Aérodrome de Tarbes-Ossun-Lourdes, BP 930, F-65009 Tarbes Cedex
Telephone: 33 62 41 73 00
Fax: 33 62 41 73 55
PRESIDENT AND DIRECTOR GENERAL:
  Jean-Marc de Raffin-Dourny
COMMERCIAL DIRECTOR: Alain Aubry
TECHNICAL DIRECTOR: Jean-Louis Rabilloud
MANAGER, PROMOTION AND COMMUNICATION:
  France Beaumont

Formed 1966 as a subsidiary of Aerospatiale responsible for light aircraft; sales of TB series, including turboprop TBM 700, totalled over 1,500 by late 1992, of which majority delivered; also makes components for Airbus A300/320/330/340, Lockheed C-130, ATR 42, Dassault Falcon, and Eurocopter Super Puma, Dauphin and Ecureuil; overhauls Morane-Saulnier MS 760 Paris. Covered floor area 57,000 m² (613,452 sq ft).

SOCATA TB 31 OMEGA

TYPE: Turboprop basic trainer.
PROGRAMME: Developed as private venture. First flight (F-WOMG) 30 April 1989; conversion of original Epsilon prototype previously re-engined with Turbomeca TP 319 Arrius.
CUSTOMERS: Excluded from rewritten US Air Force/Navy JPATS competition late 1992; still being promoted for export.
DESIGN FEATURES: Additional features include Martin-Baker Mk 15FC ejection seats and revised canopy; 60 per cent commonality with Epsilon, but wider manoeuvre envelope and greater fatigue tolerance; landing lights moved to undercarriage legs.
FLYING CONTROLS: Conventional mechanical; spring tabs in ailerons and elevators, ground adjustable tab on rudder; electric single-slotted flaps.
STRUCTURE: Light alloy stressed skin; elevators and rudder polyester fabric covered; corrosion protection to MIL-C-81773 and 83286.
POWER PLANT: One 360 kW (483 shp) Turbomeca TP 319 Arrius 1A2 turboprop, with FADEC, derated to 268 kW (360 shp) and fitted with a hydromechanical Hartzell propeller turning at 2,377 rpm. Fuel capacity in wing leading-edges of 278 litres (73.3 US gallons; 61.2 Imp gallons). Provision for two minutes of inverted flying.
ACCOMMODATION: One-piece canopy of Poly 76, hinged to starboard and including transparent separator between cockpits. MDC at junction of canopy frame for manual emergency evacuation. Martin-Baker Mk 15FC through-canopy ejection seats with zero altitude, 60 knot (111 km/h; 69 mph) capability.
SYSTEMS: Self-contained 4 h oxygen supply; Freon-based cockpit air-cooling.
AVIONICS: IFR equipped; CRT display of radio/navigation data (as on Portuguese Epsilons).

Socata TB 31 Oméga trainer (Turbomeca TP 319 Arrius 1A2 turboprop)

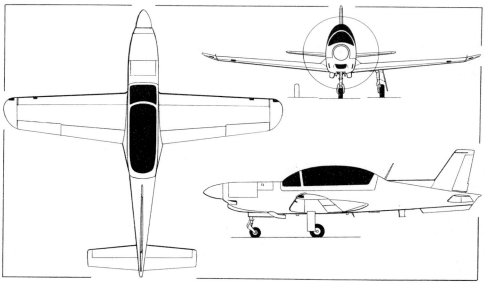

Socata TB 31 Oméga tandem-seat basic trainer (*Dennis Punnett*)

EQUIPMENT: Alkan E105-E200 armament selection indicator and E105-C02 control panel in front cockpit.
ARMAMENT: Armed versions available; four underwing hardpoints, as on Epsilon.

DIMENSIONS, EXTERNAL:
Wing span	7.92 m (25 ft 11¾ in)
Wing aspect ratio	6.97
Length overall	7.81 m (25 ft 7½ in)
Height overall	2.68 m (8 ft 9½ in)
Wheel track	2.30 m (7 ft 6½ in)
Wheelbase	1.80 m (5 ft 10¾ in)

AREAS:
Wings, gross	9.00 m² (96.9 sq ft)

WEIGHTS AND LOADINGS (approximate):
Weight empty, equipped	860 kg (1,896 lb)
Fuel weight	222 kg (489 lb)
Max T-O and landing weight	1,450 kg (3,197 lb)
Max wing loading	161.1 kg/m² (33.0 lb/sq ft)
Max power loading	4.03 kg/kW (6.62 lb/shp)

PERFORMANCE (calculated):
Never-exceed speed (V_{NE})
  321 knots (595 km/h; 370 mph) CAS

Max level speed:
  at 4,875 m (16,000 ft) 280 knots (519 km/h; 322 mph)
  at 9,145 m (30,000 ft) 210 knots (389 km/h; 242 mph)
Max cruising speed at 3,050 m (10,000 ft)
                          234 knots (434 km/h; 269 mph)
Econ cruising speed (75% power)
                          191 knots (354 km/h; 220 mph)
Stalling speed, power off, 25° flap, landing gear up or
  down   68 knots (126 km/h; 79 mph)
Max rate of climb at S/L 640 m (2,100 ft)/min
Service ceiling   9,145 m (30,000 ft)
T-O to 15 m (50 ft)   570 m (1,870 ft)
Range at 75% power at 6,100 m (20,000 ft), 20 min
  reserves   706 nm (1,308 km; 813 miles)
g limits   +7/−3.5

SOCATA TB 9 TAMPICO CLUB, TB 10 TOBAGO and TB 200 TOBAGO XL

TYPE: Four/five-seat light aircraft.
PROGRAMME: Design launched 1975; first flight of original TB 10 (F-WZJP), powered by 119 kW (160 hp) Textron Lycoming O-320, 23 February 1977; second prototype powered by 134 kW (180 hp) Textron Lycoming.
CURRENT VERSIONS: **TB 9 Tampico Club:** Four-seater with 119 kW (160 hp) Textron Lycoming O-320-D2A and Sensenich fixed-pitch propeller; superseded Tampico FP and CS in 1989 (see 1988-89 *Jane's*); fuel capacity 158 litres (41.75 US gallons; 34.75 Imp gallons); non-retractable landing gear; first flight 9 March 1979; French certification 27 September 1979.
  TB 10 Tobago: Four/five-seater with 134 kW (180 hp) Textron Lycoming O-360-A1AD and non-retractable landing gear; French certification 26 April 1979; FAA certification 27 November 1985. *Detailed description applies to this version, except where indicated.*
  TB 200 Tobago XL: Four/five-seater with 149 kW (200 hp) Textron Lycoming IO-360-A1B6; otherwise generally as TB 10; first flight 27 March 1991; French certification 30 October 1991.
CUSTOMERS: Total 236 Tampico Clubs ordered by 31 March 1992, of which 197 delivered; 68 delivered in 1992 and 87 ordered; Italian Aero Club ordered 47 plus another 63, for allocation to 90 Italian flying clubs from January 1991; 14 ordered 1991 by Embry-Riddle Aeronautical University; among recent orders are two IFR equipped for professional pilot training with Aeroavia based at Tires, Portugal.
  Total 536 TB 10 Tobagos ordered by 31 March 1992, of which 504 delivered; 55 delivered in 1992 and 57 ordered; eight used by SFACT for flying training for French air traffic controllers; 14 Tobago XL delivered in 1992 and 12 ordered.
DESIGN FEATURES: Wing section RA 16.3C3; thickness/chord ratio 16 per cent; dihedral 4° 30′.
FLYING CONTROLS: Slab tailplane with anti-balance tab; ground adjustable tabs on ailerons and rudder; strakes on lower edges of fuselage just aft of wing control turbulence under rear fuselage; electrically actuated flaps.
STRUCTURE: Conventional light alloy; single-spar wing; GFRP tips and engine cowlings.
LANDING GEAR: Non-retractable tricycle type, with steerable nosewheel. Oleo-pneumatic shock absorber in all three units. Mainwheel tyres size 6.00-6, 6-ply rating, pressure 2.3 bars (33 lb/sq in). Glassfibre wheel fairings on all three units. Hydraulic disc brakes. Parking brake.
POWER PLANT: One 134 kW (180 hp) Textron Lycoming O-360-A1AD flat-four engine, driving a Hartzell two-blade constant-speed propeller. Two integral fuel tanks in wing leading-edges; total capacity 210 litres (55.5 US gallons; 46 Imp gallons), of which 204 litres (54 US gallons; 45 Imp gallons) are usable. Oil capacity 7.5 litres (2 US gallons; 1.6 Imp gallons).
ACCOMMODATION: Four/five seats in enclosed cabin, with dual controls. Adjustable front seats with inertia reel seat belts. Removable rear bench seat with safety belts. Sharply inclined low-drag windscreen. Access via upward hinged window/doors of glassfibre. Baggage compartment aft of cabin, with external door on port side. Cabin carpeted, soundproofed, heated and ventilated. Windscreen defrosting standard.
SYSTEMS: Electrical system includes 12V 60A alternator and 12V 32A battery. Hydraulic system for brakes only.
AVIONICS: Bendix/King Silver Crown avionics to customer's specification. Current aircraft are equipped without extra charge with a basic nav pack that includes a rate of climb indicator, electric turn and bank indicator, horizontal and directional gyro, true airspeed indicator, EGT and outside air temperature indicators.
EQUIPMENT: Includes armrests for all seats, map pockets, anti-glare visors, stall warning indicator, tiedown fittings and towbar, landing and navigation lights, four individual cabin lights and instrument panel lighting.

DIMENSIONS, EXTERNAL (TB 9 and TB 10):
  Wing span   9.76 m (32 ft 0¼ in)
  Wing chord, constant   1.22 m (4 ft 0 in)
  Wing aspect ratio   8.0
  Length overall: TB 10   7.63 m (25 ft 0½ in)
    TB 200   7.70 m (25 ft 3 in)
  Height overall   3.20 m (10 ft 6 in)
  Tailplane span   3.20 m (10 ft 6 in)
  Wheelbase   1.96 m (6 ft 5 in)
  Propeller diameter   1.88 m (6 ft 2 in)

Socata TB 9 Tampico Club training aircraft of Institut Aéronautique Amaury de la Grange (*John Cook*)

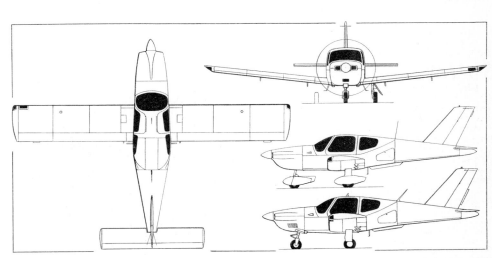

Socata TB 21 Trinidad TC, with additional side view (centre) of TB 10 Tobago (*Dennis Punnett*)

Propeller ground clearance   0.10 m (4 in)
Cabin doors (each): Width   0.90 m (2 ft 11½ in)
  Height   0.76 m (2 ft 6 in)
Baggage door: Width   0.64 m (2 ft 1¼ in)
  Max height   0.44 m (1 ft 5¼ in)

DIMENSIONS, INTERNAL (TB 9 and TB 10):
Cabin: Length:
  firewall to rear bulkhead 2.53 m (8 ft 3½ in)
  panel to rear bulkhead   2.00 m (6 ft 6¾ in)
Max width: at rear seats   1.28 m (4 ft 2¼ in)
  at front seats   1.15 m (3 ft 9¼ in)
Max height, floor to roof   1.12 m (3 ft 8 in)

AREAS (TB 9 and TB 10):
Wings, gross   11.90 m² (128.1 sq ft)
Ailerons (total)   0.91 m² (9.80 sq ft)
Trailing-edge flaps (total) 3.72 m² (40.04 sq ft)
Fin   0.88 m² (9.47 sq ft)
Rudder   0.63 m² (6.78 sq ft)
Horizontal tail surfaces (total) 2.56 m² (27.56 sq ft)

WEIGHTS AND LOADINGS:
Weight empty, with unusable fuel and oil:
  TB 9   655 kg (1,444 lb)
  TB 10   700 kg (1,543 lb)
  TB 200   715 kg (1,576 lb)
Baggage: TB 10   45 kg (100 lb)
  TB 200   65 kg (143 lb)
Max T-O weight: TB 9   1,058 kg (2,332 lb)
  TB 10, TB 200   1,150 kg (2,535 lb)
Max wing loading: TB 9   88.91 kg/m² (18.21 lb/sq ft)
  TB 10, TB 200   96.64 kg/m² (19.79 lb/sq ft)
Max power loading: TB 9   8.87 kg/kW (14.58 lb/hp)
  TB 10   8.57 kg/kW (14.08 lb/hp)
  TB 200   7.72 kg/kW (12.68 lb/hp)

PERFORMANCE (at max T-O weight):
Max level speed: TB 9   122 knots (226 km/h; 140 mph)
  TB 10   133 knots (247 km/h; 153 mph)
  TB 200   140 knots (259 km/h; 161 mph)
Max cruising speed (75% power):
  TB 9   107 knots (198 km/h; 123 mph)
  TB 10   127 knots (235 km/h; 146 mph)
  TB 200   130 knots (240 km/h; 149 mph)
Econ cruising speed (65% power):
  TB 9   100 knots (185 km/h; 115 mph)
  TB 10   117 knots (217 km/h; 135 mph)
  TB 200   121 knots (224 km/h; 139 mph)
Stalling speed:
  flaps up: TB 9   58 knots (107 km/h; 67 mph)
    TB 10   61 knots (112 km/h; 70 mph)
  flaps down: TB 9   48 knots (89 km/h; 56 mph)
    TB 10   52 knots (97 km/h; 60 mph)
    TB 200   53 knots (98 km/h; 61 mph)
Max rate of climb at S/L: TB 9 229 m (750 ft)/min
  TB 10   240 m (790 ft)/min
  TB 200   305 m (1,000 ft)/min
Service ceiling: TB 9   3,810 m (12,500 ft)
  TB 10, TB 200   3,960 m (13,000 ft)
T-O run: TB 9   340 m (1,116 ft)
  TB 10   325 m (1,067 ft)
T-O to 15 m (50 ft): TB 9   520 m (1,706 ft)
  TB 10   505 m (1,657 ft)
  TB 200   460 m (1,510 ft)
Landing from 15 m (50 ft): TB 9 420 m (1,378 ft)
  TB 10   425 m (1,395 ft)
  TB 200   450 m (1,476 ft)
Landing run: TB 9   195 m (640 ft)
  TB 10   190 m (623 ft)
Range with max standard fuel, allowances for T-O, climb, econ power cruise and descent, 45 min reserves:
  TB 9   450 nm (834 km; 518 miles)
  TB 10   653 nm (1,210 km; 752 miles)
  TB 200   590 nm (1,093 km; 678 miles)

SOCATA TB 20 and 21 TRINIDAD TC

TYPE: Four/five-seat touring and IFR training aircraft.
PROGRAMME: First flight TB 20 (F-WDBA) 14 November 1980; French certification 18 December 1981; FAA certification 27 January 1984; first delivery (F-WDBB) 23 March 1982; first flight TB 21, 24 August 1984; French certification 23 May 1985; FAA certification 5 March 1986.

CURRENT VERSIONS: **TB 20 Trinidad**: Basic version with 186 kW (250 hp) Textron Lycoming IO-540-C4D5D.
TB 21 Trinidad TC: Turbocharged version with 186 kW (250 hp) Textron Lycoming TIO-540-AB1AD.
CUSTOMERS: Total orders 472 TB 20s and 59 TB 21s by 31 March 1992; 458 TB 20s and 56 TB 21s delivered by then; 24 TB 20 and two TB 21 delivered in 1992 and 27 TB 20 and five TB 21 ordered; orders for civilian pilot training include 42 for French SFACT and others for IAAG and CIPRA; others in Australia, India and Tunisia; delivery of 28 to China completed October 1989; among recent orders is one TB 20 (plus options) for professional pilot training with Aeroavia based at Tires, Portugal.
DESIGN FEATURES: Mainly as for Tobago; dihedral 6° 30′.
FLYING CONTROLS: As Tobago, but rudder trim and flap preselector added.
STRUCTURE: Largely as Tobago.
LANDING GEAR: Hydraulically retractable tricycle type, with single wheel on each unit. Free fall emergency extension. Steerable nosewheel retracts rearward. Main units retract inward into fuselage. Hydraulic disc brakes. Parking brake.
POWER PLANT: One Textron Lycoming flat-six engine, as described in variant listings, driving a Hartzell HC-C2YK-1BF/F8477-4 two-blade metal propeller. Fuel tanks in wings, total usable capacity 326 litres (86 US gallons; 71.75 Imp gallons). Oil capacity 12.6 litres (3.3 US gallons; 2.8 Imp gallons).
SYSTEMS: Self-contained electrohydraulic system for landing gear actuation. Eros oxygen system is standard in TB 21; de-icing.
AVIONICS: Include PA three-axis autopilot with altitude preselect, HSI, RMI, R/Nav and Stormscope.
EQUIPMENT: In addition to basic nav pack described in Tampico/Tobago entry, current aircraft have as standard equipment a heated pitot, emergency static vent, cylinder head temperature gauge, emergency lighting systems, tinted windows and storm window.

DIMENSIONS, EXTERNAL: As for TB 10, except:
Length overall	7.71 m (25 ft 3½ in)
Height overall	2.85 m (9 ft 4¼ in)
Tailplane span	3.64 m (11 ft 11¼ in)
Wheelbase	1.91 m (6 ft 3¼ in)
Propeller diameter	2.03 m (6 ft 8 in)

AREAS: As for TB 10, except:
Horizontal tail surfaces (total)	3.06 m² (32.94 sq ft)

WEIGHTS AND LOADINGS:
Weight empty: TB 20	800 kg (1,763 lb)
TB 21	844 kg (1,861 lb)
Max baggage: TB 20, TB 21	65 kg (143 lb)
Max T-O weight: TB 20, TB 21	1,400 kg (3,086 lb)
Max wing loading:	
TB 20, TB 21	117.6 kg/m² (24.10 lb/sq ft)
Max power loading:	
TB 20, TB 21	7.51 kg/kW (12.35 lb/hp)

PERFORMANCE (at max T-O weight):
Max level speed:	
TB 20	167 knots (310 km/h; 192 mph)
TB 21 at 4,575 m (15,000 ft)	200 knots (370 km/h; 230 mph)
Max cruising speed (75% power) at 2,440 m (8,000 ft):	
TB 20	164 knots (303 km/h; 188 mph)
Best power cruising speed (75% power) at 7,620 m (25,000 ft): TB 21	187 knots (347 km/h; 215 mph)
Econ cruising speed (65% power):	
TB 20 at 3,660 m (12,000 ft)	160 knots (296 km/h; 184 mph)
TB 21 at 7,620 m (25,000 ft)	170 knots (315 km/h; 195 mph)
Stalling speed: flaps up:	
TB 20	64 knots (118 km/h; 74 mph)
TB 21	66 knots (121 km/h; 75 mph)
flaps and wheels down:	
TB 20	54 knots (99 km/h; 62 mph)
TB 21	55 knots (101 km/h; 63 mph)
Rate of climb: TB 20 at S/L	384 m (1,260 ft)/min
TB 21 at 610 m (2,000 ft)	332 m (1,090 ft)/min
TB 21 at 5,180 m (17,000 ft)	244 m (800 ft)/min
Service ceiling: TB 20	6,100 m (20,000 ft)
TB 21	7,620 m (25,000 ft)
T-O run: TB 20	295 m (968 ft)
TB 21	330 m (1,083 ft)
T-O to 15 m (50 ft): TB 20	479 m (1,572 ft)
TB 21	540 m (1,772 ft)
Landing from 15 m (50 ft): TB 20	530 m (1,739 ft)
TB 21	540 m (1,772 ft)
Landing run: TB 20	230 m (755 ft)

Range with max fuel, allowances for T-O, climb, cruise at best econ setting and descent, 45 min reserves:
TB 20 at 75% power at 2,135 m (7,000 ft)	885 nm (1,640 km; 1,019 miles)
TB 20 at 65% power at 3,050 m (10,000 ft)	964 nm (1,786 km; 1,110 miles)

Range with max fuel, no reserves:
TB 21 at 75% power at 7,620 m (25,000 ft)	890 nm (1,648 km; 1,024 miles)
TB 21 at 65% power at 7,620 m (25,000 ft)	1,030 nm (1,907 km; 1,185 miles)
Max ferry range at 6,100 m (20,000 ft):	
TB 20	1,158 nm (2,145 km; 1,332 miles)

Socata TB 21 Trinidad TC, the 1,000th TB series production aircraft

SOCATA TBM 700

TYPE: Six/seven-passenger pressurised business and multirole aircraft.
PROGRAMME: Launched jointly on 12 June 1987 by Socata (France) and Mooney (USA), with one-third of development costs funded by French government loan; three prototypes built: first flights 14 July 1988 (F-WTBM), 3 August 1989 (F-WKPG) and 11 October 1989 (F-WKDL); French certification received 31 January 1990; FAR Pt 23 type approval awarded 28 August 1990; first delivery 21 December 1990; Mooney withdrew from programme May 1991; Socata still negotiating over Mooney's 30 per cent share in mid-1992; TBM 700 offered to SFACT as part of a new pilot training system; studies into stretched 8/9- or 8/10-seat version began 1992.
CURRENT VERSIONS: Socata now offering military medevac, target towing, ECM, freight and vertical photography versions.
CUSTOMERS: Total 30 delivered during 1992; total 85 ordered. French Air Force received first of six for liaison duties with Groupe Aérien d'Entrainement et de Liaison (GAEL) on 27 May 1992.
COSTS: Standard aircraft $1.476 million (1992).
DESIGN FEATURES: Wing of Aerospatiale RA 16-43 root section with 6° 30′ dihedral from roots; twin strakes under rear fuselage; sweptback fin (with dorsal fin) and mass balanced rudder; non-swept tailplane with mass balanced elevators.
FLYING CONTROLS: Mechanical (pushrod/cable) controls, with electrically actuated trim tabs in port aileron, rudder and each elevator; scaled-down 'ATR' single-slotted Fowler flaps, also electrically actuated, along 71 per cent of each wing trailing-edge; slotted spoiler forward of each flap at outer end, linked mechanically to aileron; yaw damper.
STRUCTURE: Mainly of light alloy and steel except for control surfaces, flaps, most of tailplane and fin of Nomex honeycomb bonded to metal sheet; wing leading-edges and landing gear doors GFRP/CFRP; tailcone and wingtips GFRP; two-spar torsion box forms integral fuel tank in each wing.
LANDING GEAR: Hydraulically retractable tricycle type, with emergency manual operation. Inward retracting main units of trailing-link type; rearward retracting steerable nosewheel (±28°). Parker hydraulic disc brakes. Min ground turning radius (based on nosewheel) 23.98 m (78 ft 8 in).
POWER PLANT: One 522 kW (700 shp) Pratt & Whitney Canada PT6A-64 turboprop, driving a Hartzell four-blade constant-speed fully feathering reversible-pitch metal propeller. Fuel in integral tank in each wing, combined usable capacity 1,080 litres (285 US gallons; 237.5 Imp gallons). Gravity filling point in top of each tank. Oil capacity 12 litres (3.175 US gallons; 2.6 Imp gallons).
ACCOMMODATION: Adjustable seats for one or two pilots at front. Dual controls standard. Four seats in club layout aft of these, with centre aisle, or five seats in high-density layout. Optional toilet at rear of cabin. Upward/downward opening split door on port side aft of wing, with integral airstairs in lower half; overwing emergency exit on starboard side. Individual emergency oxygen mask for each passenger. Pressurised baggage compartment at rear of cabin, with internal access only; additional unpressurised compartment in nose, between engine and firewall, with external access via door on port side.
SYSTEMS: Engine bleed air pressurisation (to 0.43 bar; 6.2 lb/sq in) and air-cycle (optionally Freon) air-conditioning. Hydraulic system for landing gear only. Electrical system powered by two 28V 200A engine driven starter/generators (one main, one standby) and a 28V 25Ah nickel-cadmium battery. Pneumatic rubber-boot de-icing

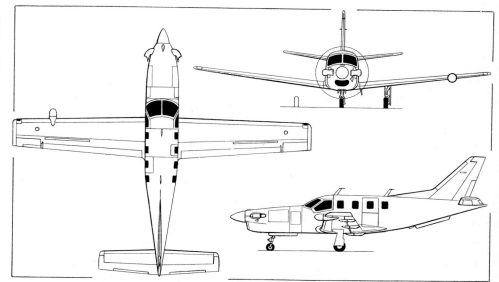

Socata TBM 700 single-turboprop transport *(Dennis Punnett)*

of wing/tailplane/fin leading-edges. Propeller blades anti-iced electrically, engine inlets by bleed air. Electric anti-icing and hot air demisting of windscreen.

AVIONICS: Bendix/King Silver Crown digital IFR package, including KY 196 VHF com transceiver; KX 165 VHF2/NAV2 with KI 206 indicator; KMA 24H audio panel with interphone; KNS 80 R/Nav with KI 525A indicator; KR 21 marker beacon receiver; KR 87 ADF with KA 44B combined loop/sense antenna and KI 229 RMI; KEA 130A encoding altimeter; KT 79 transponder; and KFC 275 flight director/autopilot with KAS 297 altitude preselect/alerter. Heated pitot and stall warning sensor heads. Wide range of options, including EFIS, GPS and weather radar.

DIMENSIONS, EXTERNAL:
Wing span	12.16 m (39 ft 10¾ in)
Wing chord, mean aerodynamic	1.51 m (4 ft 11½ in)
Wing aspect ratio	8.21
Length overall	10.43 m (34 ft 2½ in)
Height overall	3.99 m (13 ft 1 in)
Elevator span	4.88 m (16 ft 0 in)
Propeller diameter	2.31 m (7 ft 7 in)

DIMENSIONS, INTERNAL:
Cabin: Length (between pressure bulkheads)
4.56 m (14 ft 11½ in)
Max width 1.24 m (4 ft 0¾ in)
Max height 1.24 m (4 ft 0¾ in)
Volume 6.50 m³ (229.5 cu ft)
Baggage compartment volume:
front 0.25 m³ (8.83 cu ft)
rear 0.90 m³ (31.8 cu ft)

AREAS:
Wings, gross 18.00 m² (193.75 sq ft)
Vertical tail surfaces (total) 2.56 m² (27.55 sq ft)
Horizontal tail surfaces (total) 4.76 m² (51.24 sq ft)

WEIGHTS AND LOADINGS:
Weight empty, equipped 1,826 kg (4,025 lb)
Fuel weight (usable) 866 kg (1,910 lb)
Baggage: front 20 kg (44 lb)
rear 60 kg (132 lb)
Max T-O and landing weight 2,991 kg (6,595 lb)
Max ramp weight 3,003 kg (6,620 lb)
Max wing loading 166.2 kg/m² (34.04 lb/sq ft)
Max power loading 5.73 kg/kW (9.42 lb/shp)

Second prototype TBM 700 six/seven-passenger turboprop pressurised business aircraft

PERFORMANCE (A at AUW of 2,500 kg; 5,511 lb, B at max T-O weight):
Max cruising speed at 7,620 m (25,000 ft):
A 300 knots (555 km/h; 345 mph)
Max cruising speed at 9,145 m (30,000 ft):
A 250 knots (463 km/h; 288 mph)
Stalling speed, flaps and landing gear down
61 knots (113 km/h; 71 mph)
Max rate of climb at S/L: A 702 m (2,303 ft)/min
Certificated ceiling 9,150 m (30,000 ft)

Range: B
with max payload, 45 min reserves:
at max speed 1,001 nm (1,855 km; 1,152 miles)
at long-range cruising speed
1,261 nm (2,337 km; 1,452 miles)
with max fuel, no reserves:
at max speed 1,281 nm (2,374 km; 1,475 miles)
at long-range cruising speed
1,611 nm (2,985 km; 1,855 miles)
g limits: B +3.8/−1.5

SOGEPA
SOCIETE DE GESTION DE PARTICIPATIONS AERONAUTIQUES

PRESIDENT: Louis Gallois (Aerospatiale)
VICE-PRESIDENT: Serge Dassault (Dassault Aviation)

Under inter-company agreement signed on 17 September 1992 and published in December, state-owned holding company SOGEPA has acquired part holdings in both Aerospatiale and Dassault Aviation as a means of linking the two companies more closely. SOGEPA was formed in 1977, but remained dormant until 1992.

In February 1993, Mr Levi, Director of French space agency CNES, and Henri Martre, Chairman of aerospace industry association GIFAS, were appointed government representatives on the SOGEPA board. M Michot, Assistant Director General of Aerospatiale, and Charles Edelstenne, Dassault Aviation Vice-President for Economics and Finance, were appointed administrators of SOGEPA. Serge Dassault was appointed SOGEPA representative on Aerospatiale board; M Michot and M Renon, President of engine manufacturer SNECMA, joined Dassault Aviation board.

A strategic committee was appointed in SOGEPA, chaired for first year by Bruno Revellin-Falcoz (Dassault Vice-President Technical, Research and Co-operation). Aerospatiale is represented by M Michot, Jacques Plenier (Director of Aircraft Group), and M Delaye (Director of Space and Defence Group). Dassault Aviation is represented by Bruno Revellin-Falcoz, Charles Edelstenne and Michel Herchin (Vice-President Industrial and Social Affairs). A three-year joint research programme has been presented by both companies to Ministry of Defence armament directorate (DGA).

SOGEPA has taken 35 per cent share of Dassault Aviation capital and 20 per cent of Aerospatiale. French bank Credit Lyonnais bought 20 per cent share in Aerospatiale, thereby injecting new capital, in January 1993. Membership of SOGEPA is open to partners from other, and particularly European, countries.

GERMANY

DASA
DEUTSCHE AEROSPACE AG

PO Box 801109, D-8000 Munich 80
Telephone: 49 (89) 607-0
Fax: 49 (89) 607-26481
Telex: 5287-0 DASAM D
Teletext: 89720+DASAM D

CHAIRMAN, DEUTSCHE AEROSPACE SUPERVISORY BOARD:
Edzard Reuter
PRESIDENT OF BOARD OF MANAGEMENT AND CEO:
Jürgen E. Schrempp
MEMBERS OF THE BOARD OF MANAGEMENT:
Dr Manfred Bischoff (Finance and Control)
Hubert Dunkler (President Propulsion Systems Group and Chairman Board of Management MTU München GmbH)
Werner Heinzmann (President Space Systems Group, President Defence and Civil Systems Group, Chairman Board of Management Dornier GmbH)
Dr Hartwig Knitter (Personnel)
Hartmut Mehdorn (President Aircraft Group, President and CEO Board of Management Deutsche Aerospace Airbus GmbH, Hamburg)

Deutsche Aerospace, formed 19 May 1989, is the aircraft, defence, space and propulsion systems arm of the Daimler-Benz group; integrates Dornier, Messerschmitt-Bölkow-Blohm (MBB), MTU Motoren- und Turbinen-Union München and Telefunken SystemTechnik (TST); major reorganisation completed on 30 September 1992 included merger of TST (Ulm) with MBB (Ottobrunn) and transfer of Deutsche Aerospace business to former MBB, now trading as Deutsche Aerospace AG; former Deutsche Aerospace changed name to Daimler-Benz Luft- und Raumfahrt Holding AG; simultaneously German government transferred its 20 per cent share in Deutsche Airbus (Hamburg) to Deutsche Aerospace with retrospective effect from 1 January 1992; Hamburg company operating as Deutsche Aerospace Airbus, a subsidiary of Deutsche Aerospace, since 1 October 1992.

Workforce 81,872 at 31 December 1992; revenues DM17.3 billion in 1992 with Deutsche Airbus consolidated. Company organised into groups covering Aircraft, Space Systems, Propulsion Systems and Defence and Civil Systems; these groups have 12 divisions; all groups aim at international strategic co-operation.

Efforts to become an international competitor and equal partner in international programmes have included MTU general co-operation agreement with Pratt & Whitney, formation of Eurocopter with Aerospatiale, and senior membership of Airbus consortium through Deutsche Aerospace Airbus.

Deutsche Aerospace negotiated throughout 1992 to acquire 51 per cent of Netherlands company Fokker; on 27 April 1993, DASA acquired 51 per cent of a Fokker holding company shared with the Netherlands government; new capital also injected; DASA will acquire remaining government holding in three years; Fokker aircraft line to continue and new Fokker 70 (which see) being rapidly developed during 1993; DASA Regioliner, now known as Future Advanced Small Airliner, still offered if customers come forward.

Former East German Elbe Flugzeugwerke incorporated into Deutsche Aerospace Airbus and Flugzeugwerke Ludwig Felde into DASA Propulsion Group during 1991.

Aircraft Group consists of Military Aircraft, Regional Aircraft and Helicopter Divisions; activities include Airbus family, Tornado, Alpha Jet, Dornier regional airliners and helicopters. Participation in Airbus consortium is 37.9 per cent.

Space Systems Group consists of Satellites and Applications Systems, Space Transportation Systems and Propulsion, and Orbital Infrastructure Divisions; Group produces satellites for environmental and weather observation, reconnaissance and verification; international activities include Ariane 4 and 5, follow-on to Hermès spacecraft and Columbus space station; group is developing concepts for new applications and potential commercialisation.

Defence and Civil Systems Group consists of Dynamic Systems, Energy and Systems Technology, Command and Information Systems, Radar and Radio Systems Divisions; products include anti-tank, anti-ship and anti-aircraft missiles, dispenser systems, drones and training systems.

Propulsion Systems Group consists of Aircraft and Land/Marine Propulsion Applications Divisions. Aircraft propulsion projects include repair and overhaul of large civil engines and military engines, and joint development with Rolls-Royce, IAE, Pratt & Whitney and Turbomeca of such engines as RB199, EJ200, V2500, PW300, RTM 322 and MTR 390.

Aircraft Group divisions and their products and activities of Space Transportation Systems and Propulsion Division are described below.

HELICOPTER DIVISION

All former MBB/Bölkow helicopter activity now grouped under DASA banner, but called Eurocopter Deutschland, part of Eurocopter SA; see under this name in International section.

DEUTSCHE AEROSPACE AIRBUS GmbH

PO Box 950109, Kreetslag 10, D-2103 Hamburg 95
Telephone: 49 (40) 7437 0
Fax: 49 (40) 743 4422
Telex: 21950-0 DA D
WORKS: Hamburg, Bremen, Einswarden, Varel, Stade, Lemwerder, Munich, Laupheim, Speyer, Dresden
PRESIDENT AND CEO: Hartmut Mehdorn
CHAIRMAN, SUPERVISORY BOARD: Jürgen E. Schrempp
DIRECTOR, MARKETING AND SUPPORT: Horst Emker
DIRECTOR, ENGINEERING DESIGN AND TECHNOLOGY: Jürgen Thomas
DIRECTOR, PUBLIC RELATIONS: Peter Kirch

Deutsche Aerospace Airbus develops and manufactures about one-third of Airbus airliner family and about one-quarter of Fokker 100. Through the organisation Airspares and acting on behalf of Airbus Industrie, Deutsche Aerospace Airbus responsible for provision of spare parts worldwide for Airbus fleet. In Aircraft Service Centre (ASC) in Lemwerder, Transall is just one of many aircraft maintained and overhauled for operational service until 2010.

Deutsche Aerospace Airbus (22,000 employees in 10 locations) produces fuselage sections and vertical tails of A300-600, A310, A320, A330 and A340. Company is main cabin furnishing centre for Airbus family (except for A330/A340) and fits all movable wing parts to wing torsion boxes produced by British Aerospace. A321, a stretched version of A320, is assembled by Deutsche Aerospace Airbus. Aft fuselage for Fokker 100 is manufactured by Elbe Flugzeugwerke GmbH in Dresden.

AIRBUS SUPER TRANSPORTER

Together with Aerospatiale, Deutsche Aerospace Airbus has established Special Aircraft Transport International Company (SATIC). The German-French joint venture is responsible for manufacture of Super Guppy successor Airbus Super Transporter (AST), based on Airbus A300-600; see under SATIC in International section.

ULTRA HIGH CAPACITY AIRCRAFT (UHCA)

Deutsche Aerospace Airbus is working intensively in overall framework of Airbus Industrie consortium on study for an Airbus Very Large Commercial Transport (VLCT) for more than 600 passengers. Study currently performed by company under designation A2000 is German Airbus partner's contribution. DASA also involved in UHCA studies outside Airbus with Boeing, Aerospatiale, BAe and CASA. See on page 126 in International section.

EUROFLAG

In Euroflag group, company is co-operating with Aerospatiale, British Aerospace, CASA and Alenia on a successor for Transall and Hercules aircraft; see under Euroflag in International section.

SUPERSONIC COMMERCIAL TRANSPORT (SCT)

An international study group is working on a new supersonic airliner. Participants in study group on Supersonic Commercial Transport (SCT) include Aerospatiale (France), Alenia (Italy), Boeing Commercial Airplane Group (USA), British Aerospace (United Kingdom), Deutsche Aerospace Airbus (Germany), Japanese Aircraft Industries (Japan), McDonnell Douglas (USA) and Tupolev (Russia); see under SCTICSG in International section.

CRYOPLANE

Deutsche Aerospace Airbus leading a group of companies that include Tupolev and Kuznetsov in Russia to examine development of airliner fuelled with liquid hydrogen. Joint feasibility study in Germany and Russia was completed in early 1993.

DORNIER LUFTFAHRT GmbH (Subsidiary of Dornier GmbH – part of Deutsche Aerospace AG)

HEADQUARTERS: Dornier Airfield, D-8031 Wessling
Telephone: 49 (8153) 30-0
Fax: 49 (8153) 302901
Telex: 526412 DORW D

Dornier GmbH, formerly Dornier-Metallbauten, formed 1922 by late Professor Claude Dornier; has operated as a GmbH since 22 December 1972. Daimler-Benz AG acquired majority holding (65.5 per cent) in Dornier GmbH in 1985, but had reduced this to 57.55 per cent by 1 January 1989, when new three-group Dornier company structure came into being with Silvius Dornier (21.22 per cent) and Claudius Dornier heirs (21.22 per cent) as other shareholders. Daimler-Benz shareholding since assumed by Deutsche Aerospace AG.

Former Dornier System GmbH activities transferred to parent Dornier GmbH, based at Friedrichshafen. All of Dornier's aviation activities now undertaken by Dornier Luftfahrt GmbH (formerly Dornier Reparaturwerft GmbH) at Oberpfaffenhofen, which is a wholly owned subsidiary of Dornier GmbH. Dornier Medizintechnik GmbH of Munich also a wholly owned subsidiary of Dornier GmbH. Dornier Luftfahrt now basis of DASA Regional Aircraft Division (see below).

Dornier Luftfahrt is responsible for industrial support of Luftwaffe Alpha Jets and is in multi-nation Eurofighter 2000 (see International section); subcontractor to Deutsche Aerospace Airbus GmbH; supports the 18 NATO Boeing E-3A Sentry AWACS and NATO Trainer/Cargo Aircraft (TCA) (see 1989-90 *Jane's*); depot level maintenance on E-3As done by International Aerospace Management Company in Venice, Italy; assisted in design of Argentine IA 63 Pampa trainer; MoU on a future integrated military Pilot Training System (PTS-2000) (see International section) signed with Aermacchi April 1989; MBB joined early 1991; latest activities include new transonic laminar flow wing, and hypersonic aircraft technologies; technical and logistic servicing of German Navy Breguet Br 1150 Atlantic 1 and contributes to Atlantique 2; life extension modifications to 168 Bell UH-1D utility helicopters; service centre for business and general aviation aircraft. DASA/Dornier share in Advanced Amphibious Aircraft, already reduced from 50 to 5 per cent, reduced to zero in first quarter 1993.

REGIONAL AIRCRAFT DIVISION

PO Box 3, D-8031 Wessling
Telephone: 49 (8153) 30-0
Fax: 49 (8153) 302007
DIVISIONAL CHIEF EXECUTIVE: Hansjörg Kränzle
PRESS AND INFORMATION: Peter Kirch

DORNIER 228-212

TYPE: Twin-turboprop STOL light transport.
PROGRAMME: First flight of Dornier 228-100 prototype (D-IFNS) 28 March 1981; first flight 228-200 prototype (D-ICDO) 9 May 1981; British CAA certification 17 April 1984, FAR Pt 23 and Appendix A Pt 135 11 May 1984, Australian 11 October 1985; planned production for 1993, 15 aircraft.
CURRENT VERSIONS: For earlier production versions, see 1991-92 and earlier *Jane's*.

228-212: Current production version in Europe; certificated August 1989; introduced to increase payload on short routes; innovations include increased engine power, stronger landing gear, carbon brakes, improved anti-skid, two strakes under rear fuselage, increased max speed with flaps extended (VFE), modified flying controls and new avionics. *Detailed description applies to this version.*

228-202: Indian production version; licence production agreement with Hindustan Aeronautics signed 29 November 1983; five complete 228-201s delivered to regional airline Vayudoot 1984-85; three 228-101s delivered to Indian Coast Guard 1986-87; first complete sets of components delivered early 1985; first flight of HAL-assembled 228, 31 January 1986; see Indian section for further details.

228 Troop: Carries 17, 20 or 22 fully equipped troops; adaptable for paratrooping; fold-up seats at cabin sides, lightweight toilet, roller door, military nav/com and loadmaster intercom.

228 Paratroop: Carries 16, 19 or 21 persons, plus jumpmaster; no toilet.

228 Ambulance: Six stretchers in pairs and nine sitting patients/attendants; optional small galley, toilet, refrigerator, oxygen system and cabin intercom.

228 Cargo: As 228-212, but with all superfluous equipment removed; modified to US FARs; max payload 2,340 kg (5,159 lb) in 16.34 m³ (577 cu ft) cabin; large double door; six cargo nets at 140 cm (55 in) intervals secured to aluminium frames and seat rails; reinforced cabin floor, smoke detectors and glassfibre panels on side walls.

228 Maritime patrol, Maritime pollution surveillance and **Photogrammetry/geo-survey:** Described separately.

CUSTOMERS: Orders totalled 221 (all versions) by end 1992; 208 delivered, including 10 in 1992; 17 new orders in 1992.

DESIGN FEATURES: Special Dornier wing with Do A-5 supercritical aerofoil; 8° leading-edge sweep on outer wing panels; raked tips; no dihedral or anhedral; German certification accepted in Bhutan, Canada, India, Japan, Malaysia, Nigeria, Norway, Sweden and Taiwan.
FLYING CONTROLS: Mechanically actuated; variable incidence tailplane with actuator switch on aileron wheel; horn balanced elevators; rudder trim tab; single-slotted Fowler flaps augmented by drooping ailerons; two strakes under rear fuselage for low-speed stability.
STRUCTURE: Two-spar wing box; mainly light alloy structure, but with CFRP wingtips and tips of tailplane and elevators; GFRP nosecone, tips of rudder and fin; Kevlar landing gear fairings and in part of wing ribs; hybrid composites in fin leading-edge; fuselage unpressurised, built in five sections.
LANDING GEAR: Retractable tricycle type, with single mainwheels and twin-wheel nose unit; main units retract inward into fuselage fairings; hydraulically steerable nosewheels retract forward; Goodyear wheels and tyres, size 8.50-10 on mainwheels (10 ply rating); size 6.00-6, 6 ply rating, on nosewheels; low pressure tyres optional; Goodyear carbon brakes on mainwheels.
POWER PLANT: Two 578.7 kW (776 shp) Garrett TPE331-5-252D turboprops, each driving a Hartzell HC-B4TN-5ML/LT10574 four-blade constant-speed fully feathering reversible-pitch metal propeller. Primary wing box forms integral fuel tank with total usable capacity 2,386 litres (630 US gallons; 525 Imp gallons); oil capacity per engine 5.9 litres (1.56 US gallons; 1.30 Imp gallons).
ACCOMMODATION: Crew of one or two; pilots' seats adjustable fore and aft; two-abreast seating with central aisle; max capacity 19; flight deck door on port side; combined two-section passenger and freight door, with integral steps, on port side of cabin at rear; one emergency exit on port side

Dornier 228-212 of Flexair

of cabin, two on starboard side; baggage compartment at rear of cabin, accessible externally and from cabin; capacity 210 kg (463 lb). Enlarged baggage door optional; additional baggage space in fuselage nose, with separate access; capacity 120 kg (265 lb); modular units using seat rails for rapid changes of role.

SYSTEMS: Entire accommodation heated and ventilated; air-conditioning system optional; heating by engine bleed air. Hydraulic system, pressure 207 bars (3,000 lb/ sq in), for landing gear, brakes and nosewheel steering; handpump for emergency landing gear extension. Primary 28V DC electrical system, supplied by two 28V 250A engine driven starter/generators and two 24V 25Ah nickel-cadmium batteries; two 350VA inverters supply 115/26V 400Hz AC system. Air intake anti-icing standard; de-icing system optional for wing and tail unit leading-edges, windscreen and propellers.

AVIONICS: IFR instrumentation standard. Autopilot optional, to permit single-pilot IFR operation. Standard avionics include dual Bendix/King KY 196 VHF com, dual KN 53 VOR/ILS and KN 72 VOR/LOC converters; KMR 675 marker beacon receiver, dual or single KR 87 ADF and KT 76A transponder; dual or single Aeronetics 7137 RMI; dual or single DME; two Honeywell GH14B gyro horizons; two Bendix/King KPI 552 HSIs; dual ASIs; dual altimeters; dual ADIs; dual VSIs; Becker audio selector and intercom. Weather radar optional.

EQUIPMENT: Standard equipment includes complete internal and external lighting, hand fire extinguisher, first aid kit, gust control locks and tiedown kit.

DIMENSIONS, EXTERNAL:
Wing span	16.97 m (55 ft 8 in)
Wing aspect ratio	9.0
Length overall	16.56 m (54 ft 4 in)
Height overall	4.86 m (15 ft 11½ in)
Tailplane span	6.45 m (21 ft 2 in)
Wheel track	3.30 m (10 ft 10 in)
Wheelbase	6.29 m (20 ft 7½ in)
Propeller diameter	2.69 m (8 ft 10 in)
Propeller ground clearance	1.08 m (3 ft 6½ in)
Passenger door (port, rear): Height	1.34 m (4 ft 4¾ in)
Width	0.64 m (2 ft 1¼ in)
Height to sill	0.91 m (2 ft 11¾ in)
Freight door (port, rear): Height	1.34 m (4 ft 4¾ in)
Width, incl passenger door	1.28 m (4 ft 2½ in)
Emergency exits (each): Height	0.66 m (2 ft 2 in)
Width	0.48 m (1 ft 7 in)
Baggage door (nose): Height	0.50 m (1 ft 7½ in)
Width	1.20 m (3 ft 11¼ in)
Standard baggage door (rear):	
Height	0.90 m (2 ft 11½ in)
Width	0.53 m (1 ft 9 in)

DIMENSIONS, INTERNAL:
Cabin, excl flight deck and rear baggage compartment:
Length	7.08 m (23 ft 2¾ in)
Max width	1.346 m (4 ft 5 in)
Max height	1.55 m (5 ft 1 in)
Floor area	9.56 m² (102.9 sq ft)
Volume	14.70 m³ (519.1 cu ft)
Rear baggage compartment volume	2.60 m³ (91.8 cu ft)
Nose baggage compartment volume	0.89 m³ (31.4 cu ft)

AREAS:
Wings, gross	32.00 m² (344.3 sq ft)
Ailerons (total)	2.708 m² (29.15 sq ft)
Trailing-edge flaps (total)	5.872 m² (63.21 sq ft)
Fin, incl dorsal fin	4.50 m² (48.44 sq ft)
Rudder, incl tab	1.50 m² (16.15 sq ft)
Horizontal tail surfaces (total)	8.33 m² (89.66 sq ft)

WEIGHTS AND LOADINGS:
Weight empty, standard	3,258 kg (7,183 lb)
Operating weight empty	3,739 kg (8,243 lb)
Max payload	2,201 kg (4,852 lb)
Max fuel weight	1,885 kg (4,155 lb)
Max ramp weight	6,430 kg (14,175 lb)
*Max T-O weight	6,400 kg (14,110 lb)
Max landing weight	6,100 kg (13,448 lb)
Max wing loading	200.0 kg/m² (40.96 lb/sq ft)
Max power loading	5.53 kg/kW (9.09 lb/shp)

*Increasable to 6,600 kg (14,550 lb) in special cases

PERFORMANCE (at max T-O weight, S/L, ISA, except where indicated):
Never-exceed speed (VNE)	255 knots (472 km/h; 293 mph) IAS
Max operating speed (VMO)	223 knots (413 km/h; 256 mph) IAS
Max cruising speed:	
at S/L	222 knots (411 km/h; 255 mph)
at 3,050 m (10,000 ft)	234 knots (434 km/h; 269 mph)

Cruising speed at 4,575 m (15,000 ft), average cruise weight of 5,300 kg (11,684 lb)
220 knots (408 km/h; 253 mph)
Econ cruising speed	180 knots (333 km/h; 207 mph)
Stalling speed:	
flaps up	73 knots (136 km/h; 84 mph) IAS
flaps down	69 knots (128 km/h; 80 mph) IAS
Max rate of climb	570 m (1,870 ft)/min
Rate of climb, OEI	134 m (440 ft)/min

Service ceiling, 30.5 m (100 ft)/min rate of climb
8,535 m (28,000 ft)

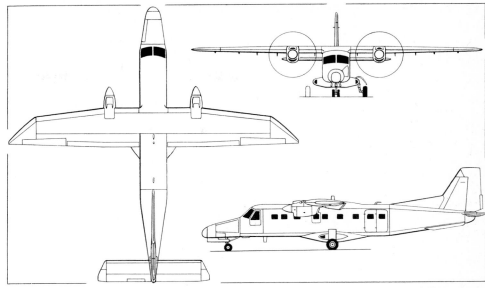

Deutsche Aerospace Dornier 228-212 (two Garrett TPE331-5-252D turboprops) *(Dennis Punnett)*

Service ceiling, OEI, 15 m (50 ft)/min rate of climb
3,960 m (13,000 ft)
T-O run	671 m (2,200 ft)
T-O to 10.7 m (35 ft)	793 m (2,605 ft)
Accelerate/stop distance, with anti-skid	710 m (2,330 ft)
Landing from 15 m (50 ft) at max landing weight	402 m (1,320 ft)

Range at 3,050 m (10,000 ft) with 19 passengers, reserves for 50 nm (93 km; 57 mile) diversion, 45 min hold and 5% fuel remaining:
at max cruising speed   560 nm (1,038 km; 645 miles)
at max range speed   630 nm (1,167 km; 725 miles)
Range with 775 kg (1,708 lb) payload, conditions as above:
at max cruising speed
1,160 nm (2,150 km; 1,335 miles)
at max range speed   1,320 nm (2,446 km; 1,520 miles)

DORNIER 228 MARITIME PATROL

PROGRAMME: Developed for Indian Coast Guard, Royal Thai Navy and others for maritime and fisheries patrol and border patrol.

DESIGN FEATURES: Modifications to standard 228 include major corrosion protection, radome beneath fuselage, four wing hardpoints for searchlight, Micronair spraypod and other equipment, roller door for dropping survival equipment and chute in rear cabin for dropping smoke markers and flares.

POWER PLANT: As for standard 228, with optional auxiliary fuel tanks to increase fuel capacity to 2,886 litres (762 US gallons; 635 Imp gallons).

ACCOMMODATION: Pilot and co-pilot with full dual controls and instruments as standard; co-pilot operates optional searchlight; two bubble observation windows in front of cabin (180° view) and photography window on port side which can be opened in flight. Console for radar operator on port side of cabin incorporating radar display, digital navigation display and intercom controls. Rest area on starboard side of rear cabin with optional folding table, galley or refrigerator and toilet.

AVIONICS: Standard items as detailed for 228-212. Optional exchange of standard avionics to Collins Pro Line II. Com/nav equipment comprises Collins AN/ARC-182 VHF/UHF transceiver, Collins DF-301E VHF/UHF direction finder, Collins radio altimeter ALT-55B, low altitude warning system, weather radar, Global/Wulfsberg GNS-500-5 nav system with search pattern mode, and optional GPS and Loran C. Bendix/King RDR-1500B maritime surveillance radar with 360° scan in underfuselage radome. Optional Litton AN/APS-504(V)5, THORN EMI Super Searcher and Eaton AN/APS-128. Day/night mission equipment includes Honeywell forward-looking infra-red system (FLIR), stabilised long-range observation system (SLOS), night vision goggles, Spectrolab 80 million candlepower searchlight, markers and flares, loudhailer and Nikon hand-held camera with data annotation.

DORNIER 228 MARITIME POLLUTION SURVEILLANCE

PROGRAMME: Developed for Netherlands Coast Guard (Kustwacht) and Finnish Frontier Guard.

DESIGN FEATURES: Modifications for pollution control as for Maritime Patrol version, but with large floor cutout to carry an IR/UV scanner plus photographic, television and IR cameras.

POWER PLANT: As for Maritime Patrol version.

ACCOMMODATION: Crew as for Maritime Patrol version. Main operator workstation on port side front of cabin, with all necessary controls and displays for remote sensing equipment; equipment rack for all the different computers, video recorders, power supplies and hard copy units on port side in mid-cabin; observer workstation with similar controls and displays as in operator console, is at rear of cabin, in front of toilet compartment. Two bubble observation windows port and starboard in front of cabin, each bulged to

Dornier 228-212 maritime pollution surveillance aircraft for the Netherlands Coast Guard

98 GERMANY: AIRCRAFT—DASA

give 180° view; two optical photo windows port and starboard in rear cabin can be opened in flight; rest area on starboard side of mid-cabin has seats and folding table.

AVIONICS: As for Maritime Patrol version, but also Racal R-Nav II navigation management system with search pattern mode, Decca, GPS, and Bendix/King RDR 1400C weather and search radar with 18 in antenna.

EQUIPMENT: Primary surveillance sensors from Terma Elektronik AS, Denmark, include Terma side-looking airborne radar (SLAR) with underfuselage antenna, and Daedalus ABS 3500 bi-spectral IR/UV scanner. To secure evidence of pollution, 228 equipped with downward-looking colour television, Nikon photographic and IR cameras and, for photographic documentation, Nikon hand-held camera interfacing with navigation system.

DORNIER 228 PHOTOGRAMMETRY/GEO-SURVEY

PROGRAMME: Besides being used in a purely photogrammetric version, 228 can serve wide variety of users as working platform in earth sciences field. The 228 allows good access to sensors and other equipment both in flight and for maintenance.

DESIGN FEATURES: Modifications to standard 228 as for Maritime Patrol version, but production model includes large sliding hatch in cabin floor for sensor installation. Wing hardpoints support different sensors, and various antennae are mounted on fuselage and tail.

POWER PLANT: As for Maritime Patrol version.

ACCOMMODATION: Crew dependent on mission, but basically as described for Maritime Patrol version.

AVIONICS: As for Maritime Patrol version.

OPTIONAL EQUIPMENT: Photogrammetry version has aerial survey cameras installed in floor cutout, Wild or Zeiss navigation telescope, operator station, flight track camera, intercom system, toilet modified as darkroom, rest area with folding table, and small galley or refrigerator.

Geo-survey version has VLF electromagnetometer mounted in nose thimble, magnetometer in tail sting, VLF or proton magnetometer installed in wingtips, electromagnetic reflection system mounted on wing hardpoints, and gamma-ray detector in lower fuselage. Aerial survey camera installed in floor cutout in rear fuselage.

DORNIER 328

TYPE: Twin-turboprop pressurised high-speed regional transport.

PROGRAMME: Development relaunched 3 August 1988; first flight (D-CHIC) 6 December 1991; first flights of second aircraft (D-CATI) 4 June 1992 and third and final development aircraft (D-COOL) 20 October 1992; first flight of first production aircraft (D-CITI) 23 January 1993. All six port propeller blades separated from a 328 on 14 December 1992 as a result of a controllability demonstration required for certification and two blades penetrated cabin; test flying continued; problems with overweight, cruising speed and range overcome by January 1993; European 19-country JAA 25 certification expected August 1993 with FAA certification shortly after; first delivery, to Air Engiadina, expected September 1993; eight 328s to be produced in 1993.

CURRENT VERSIONS: **328:** Standard version, *as detailed.*
328 S Stretch: See separate entry.

CUSTOMERS: Initial customers included Horizon Air of USA (35 firm + 25 optioned) and Air Engiadina of Switzerland (one); total of 45 firm orders and 29 options by June 1993. Potential market seen as more than 400 by year 2006.

DESIGN FEATURES: Combines basic TNT supercritical wing of Dornier 228 with new pressurised fuselage from NRT (Neue Rumpf Technologien) programme; internal volume designed to give passengers more seat width than in a Boeing 727 or 737 and stand-up headroom in aisle; target 78 dB noise level for 75 per cent of passengers.

FLYING CONTROLS: As for 228, but with new flaps and optional lateral control (one) and ground (two) spoilers ahead of each aileron; trim tab in each elevator and rudder.

STRUCTURE: Wing mainly light alloy structure, with aluminium-lithium alloy for most of wing skins, stringers and precision forgings; lithium alloy also used for fuselage longerons, stringers, window frames and skin panels; entire rear fuselage and tail surfaces of CFRP, except dorsal fin made of Kevlar/CFRP sandwich and aluminium alloy tailplane leading-edge; Kevlar/CFRP sandwich also used for wing trailing-edge structure, nosecone, tailcone and for long wing/fuselage fairing housing system components outside pressure hull; cabin doors of superplastic formed aluminium alloy; engine nacelles of superplastic formed titanium and carbon composite.

Daewoo Heavy Industries of South Korea manufactures fuselage panels, which are assembled into complete fuselages by risk-sharing partner Aermacchi in Italy which also manufactures flight deck structure; engine nacelles and doors by Westland Aerostructures; wings, rear fuselage and tail surfaces by Deutsche Aerospace Airbus at Neuaubing; final assembly at Oberpfaffenhofen. Daewoo share represents about 20 per cent of manufacturing time.

LANDING GEAR: ERAM (with SHL of Israel) retractable tricycle type, with twin Bendix wheels on each unit; nose unit retracts forward, main units into long Kevlar/CFRP sandwich unpressurised fairings on fuselage sides; tyre pressures 4.4 bars (64 lb/sq in) on nose unit, 8.0 bars (116 lb/sq in) on main units; Bendix brakes.

POWER PLANT: Two Pratt & Whitney Canada PW119B turboprops, each rated at 1,380 kW (1,850 shp) for normal take-off and 1,625 kW (2,180 shp) for short-field take-off, each driving a Hartzell six-blade composites propeller with electronic synchrophasing. Improved performance kit optional; all fuel in wing tanks, total capacity 4,268 litres (1,133 US gallons; 939 Imp gallons).

ACCOMMODATION: Flight crew of two and cabin attendant(s). Main cabin seats 30-33 passengers, three-abreast at 79 cm (31 in) or 76 cm (30 in) pitch, with single aisle; galley to rear of passenger seats; toilet at rear of cabin; large baggage hold between passenger cabin and rear pressure bulkhead, with external access via baggage door in port side; crew/passenger airstair door at front on port side, with Type III emergency exit opposite; Type III emergency exit on port side at rear of cabin, with service door Type II exit at rear on starboard side.

SYSTEMS: Air-conditioning and pressurisation systems standard (max differential 0.47 bar; 6.75 lb/sq in). Hydraulic and two independent AC/DC electrical systems housed in main landing gear fairings. APU optional.

AVIONICS: Standard fit includes Honeywell Primus 2000 with integrated five-tube EFIS, AFCS, FMS and EICAS (electronic flight instrument system, automatic flight control system, flight management system, and electronic indication, caution and advisory system); digital air data computer; AHRS with advanced fibre-optic laser gyros; radar altimeter; Primus 650 weather radar; and Primus II digital radio system. Standard options include Honeywell traffic

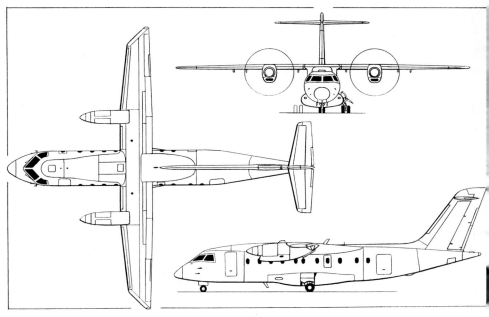

Dornier 328 (two P&WC PW119B turboprops) 30/33-passenger transport (*Dennis Punnett*)

Dornier 328 of lead customer Horizon Air

Dornier 328 flight deck with Honeywell Primus 2000 flight deck avionics including five tubes for flight instruments, systems and EICAS functions

alert and collision avoidance system (TCAS), GPS, MLS, Primus 870 colour weather radar and Laseref II laser inertial reference system. Horizon Air fitting head-up guidance system allowing landings in Cat. IIIa.

DIMENSIONS, EXTERNAL:

Wing span	20.98 m (68 ft 10 in)
Length overall	21.22 m (69 ft 7½ in)
Fuselage: Length	20.92 m (68 ft 7¾ in)
Max width	2.415 m (7 ft 11 in)
Max depth	2.426 m (7 ft 11½ in)
Height overall	7.239 m (23 ft 9 in)
Elevator span	6.70 m (21 ft 11¾ in)
Wheel track (c/l of shock struts)	3.22 m (10 ft 6¾ in)
Wheelbase	7.422 m (24 ft 4¼ in)
Propeller diameter	3.50 m (11 ft 5¾ in)
Propeller fuselage clearance	0.735 m (2 ft 5 in)
Passenger door (fwd, port): Height	1.70 m (5 ft 7 in)
Width	0.70 m (2 ft 3½ in)
Service door (rear, stbd): Height	1.25 m (4 ft 1¼ in)
Width	0.51 m (1 ft 8 in)
Baggage door (rear, port): Height	1.40 m (4 ft 7 in)
Width	0.92 m (3 ft 0¼ in)

DIMENSIONS, INTERNAL:

Cabin, excl flight deck:	
Length	10.32 m (33 ft 10¼ in)
Max width	2.18 m (7 ft 2 in)
Width at floor	1.83 m (6 ft 0 in)
Max height in aisle	1.89 m (6 ft 2½ in)
Baggage hold volume	6.30 m³ (222.5 cu ft)

WEIGHTS AND LOADINGS:

Max payload	3,450 kg (7,605 lb)
Max baggage load	750 kg (1,653 lb)
Max T-O weight	13,640 kg (30,071 lb)
Max zero-fuel weight	12,260 kg (27,029 lb)
Max landing weight	13,230 kg (29,167 lb)
Max power loading:	
normal T-O	4.94 kg/kW (8.13 lb/shp)
short-field T-O	4.20 kg/kW (6.90 lb/shp)

PERFORMANCE (estimated):

Max cruising speed	335 knots (620 km/h; 388 mph)
Design cruising altitude: normal	7,620 m (25,000 ft)
optional	9,450 m (31,000 ft)
Required T-O runway length	1,100 m (3,610 ft)
Landing field length	1,010 m (3,315 ft)
Range at max cruising speed with 30 passengers, with allowance for 100 nm (185 km; 115 mile) diversion and 45 min hold:	
at 7,620 m (25,000 ft)	729 nm (1,352 km; 840 miles)
at 9,450 m (31,000 ft)	840 nm (1,556 km; 967 miles)

DORNIER STRETCHED 328 S

TYPE: High-speed, 50-passenger regional turboprop transport.

PROGRAMME: Announced late 1990; launch expected late 1993; in service 1997.

DESIGN FEATURES: Full commonality with basic 328 retained in rear fuselage, tail surfaces, nose landing gear, cockpit, avionics and instrumentation; 328 fuselage stretched by plugs fore and aft of wing and centre-fuselage reinforced; wing extended by plugs outboard of engine nacelles and reinforced for higher operating weights; lateral control/ground spoilers standard. Passenger capacity to be decided between 48 and 50.

POWER PLANT: Engines not selected by January 1993, but Allison GMA or P&WC 300 series under consideration.

ACCOMMODATION: Two pilots; 50 passengers at 81 cm (32 in) pitch with 0.23 m³ (8.0 cu ft) aft baggage volume per passenger; cabin noise level 76 dB(A); toilet in forward cabin; emergency exit and service door to starboard; passenger door and baggage door to port.

SYSTEMS: Mainly as 328, but with new ECS.

AVIONICS: As for 328.

Dimensions, weights and performance depend on engine choice.

Model of Dornier Stretched 328 S for 50 passengers

MILITARY AIRCRAFT DIVISION

PO Box 801160, D-8000 Munich 80
Telephone: 49 (89) 607-0
Fax: 49 (89) 607 28740
HEAD OF STRATEGIC BUSINESS UNIT: Klaus-Jürgen Wolfert

Major military aircraft activities of Group include aircraft armament and airborne reconnaissance systems, disarmament verification systems (the former PRISMA system and BICES), simulation and training systems, and research into advanced aircraft systems, materials and manufacturing technologies. Division also makes Airbus subassemblies.

PANAVIA TORNADO

Division is prime contractor for German share of Tornado and assembled German Air Force and Navy aircraft. It also developed ECR version. Full details in International section.

EUROFIGHTER 2000

Division is prime contractor for German share of development of Eurofighter 2000 (formerly EFA). Full details under Eurofighter in International section.

X-31A

DASA (MBB) shared development, manufacture and testing of two NASA X-31A research aircraft (see under Rockwell International/DASA in International section).

PTS-2000

This multi-nation military pilot training system is briefly described under Aermacchi/DASA in International section.

RANGER 2000

Military Aircraft Division co-operated with Rockwell International to develop Fan Ranger (now Ranger 2000) turbofan-powered pilot trainer to compete for US Air Force/Navy JPATS programme. First flight at Manching, Germany, 15 January 1993. For details see under Rockwell International/DASA in International section.

DASA F-4F ICE PROGRAMME

Under German Defence Ministry programme known as ICE (improved combat effectiveness), 110 Luftwaffe F-4F Phantom IIs, primarily those of fighter wings JG 71 and JG 74, to be upgraded for lookdown/shootdown capability against multiple targets. Programme, for which DASA Military Aircraft Division is prime contractor, initiated late 1983 and reached end of definition phase two years later; entered full scale development phase December 1986; two development aircraft flying since 1990. First flight of fully modified F-4F ICE 2 May 1990. AMRAAM firing trials from October 1991 to September 1992 at US Navy Pacific Weapon Test Range at Point Mugu. Production ICE conversion started July 1991; first delivery April 1992; 24 delivered by June 1993.

Features include replacement of existing Westinghouse AN/APQ-120 radar with all-digital multi-mode Hughes AN/APG-65, built under licence in Germany by Telefunken SystemTechnik (now DASA Defence and Civil Systems Group), fitting up to four Hughes AIM-120 (AMRAAM) air-to-air missiles, new TST radar control console, optimisation (by Hughes) of cockpit display, installation of new Litef digital fire control computer, Honeywell H-423 laser inertial platform, GEC-Marconi Avionics CPU-143/A digital air data computer, new IFF system, Frazer-Nash AMRAAM launchers, MIL-STD-1553B digital databus with Advanced Operational Flight Program software, and improved resistance to electronic jamming and other countermeasures.

Further 40 Luftwaffe F-4Fs, serving in fighter-bomber role with JBGs 35 and 36, were to undergo partial update (databus, INS and ADC only, initially), with option of full ICE installation later.

German Air Force F-4F with AN/APG-65 radar in nose and carrying four AMRAAM and four AIM-9L Sidewinder missiles

SPACE TRANSPORTATION SYSTEMS AND PROPULSION DIVISION

Deutsche Aerospace AG, PO Box 881109, D-8000 Munich 80
Telephone: 49 (89) 607 34235
Fax: 49 (89) 607 34239
WORKS: Bremen and Ottobrunn

Besides producing major airframe and propulsion assemblies for Ariane 4 and 5, this division is responsible for DASA participation in new space transportation systems. The division was a member of the Hermes European spacecraft programme (see International section).

SÄNGER

Sänger two-stage completely reusable winged vehicle, with manned or unmanned second stage, is Germany's proposal for future space transportation system able to take off from and land in Europe.

First stage is to be air-breathing liquid hydrogen-propelled turboramjet hypersonic vehicle with cruise capability; second stage a reusable rocket-propelled vehicle for both manned missions and unmanned cargo transport and retrieval. Sänger selected by German Federal Ministry for Research and Technology as reference concept for national hypersonics technology programme.

Artist's impression of the Sänger fully reusable space vehicle being studied by the Space Transportation Systems and Propulsion Division of Deutsche Aerospace

EXTRA
EXTRA-FLUGZEUGBAU GmbH
Flugplatz Dinslaken Schwarze Heide, D-4224 Hünxe
Telephone: 49 (2858) 6851
Fax: 49 (2858) 7124
MANAGING DIRECTOR: Walter Extra
SALES MANAGER: Frank Versteegh, Hoflaan 1, 6824 BN Arnhem, Netherlands
Telephone: 31 (85) 514545
Fax: 31 (85) 514878

EXTRA 300
TYPE: Tandem two-seat aerobatic aircraft.
PROGRAMME: Design began January 1987; first flight (D-EAEW) 6 May 1988; LBA certification 16 May 1990; certificated FAA FAR Pt 23 Amendment 33 Normal and Aerobatic categories for USA and Europe; production started October 1988; production of 20 planned for 1993, selectable between 300 and 300S.
CURRENT VERSIONS: **Extra 300:** Production aircraft. Sometimes called **EA 300**. *Details apply to this version.*
Extra 300 S: Single-seat version of 300 with same power plant; wing shortened by 50 cm (19½ in) and more powerful ailerons; first flight 4 March 1992; certificated March 1992; FAA and French certification applied for early 1993.
CUSTOMERS: Six Extra 300s delivered to Chilean Air Force 1989-90; more than 40 delivered by end 1992.
COSTS: German price of both 300 and 300 S, DM295,000 without tax. Noise reduction system adds DM8,000.
DESIGN FEATURES: Designed for unlimited competition aerobatics; tapered, square-tipped wing with 4° leading-edge sweep; aerofoil symmetrical MA-15 at root, MA-12 at tip; no twist or dihedral.
FLYING CONTROLS: Rod and cable operated conventional; nearly full span ailerons; fixed tailplane; trim tab in starboard elevator.
STRUCTURE: Fuselage (excluding tail surfaces) of steel tube frame with part aluminium, part fabric covering; wing spars of carbon composites and shells of carbon composite sandwich; tail surfaces of carbon spars and glassfibre shells.
LANDING GEAR: Fixed cantilever composite arch main gear with single polyamid-faired wheels; Cleveland brakes; leaf-sprung steerable tailwheel.
POWER PLANT: One Textron Lycoming AEIO-540-L1B5 flat-six engine giving 224 kW (300 hp), driving Mühlbauer MTV-9-B-C/C 200-15 three-blade constant-speed propeller (MTV-14 optional). With either standard or four-blade propeller and Gomolzig Type 3 silencer, both Extra 300s meet German and US noise limits. Usable fuel capacity 38 litres (10 US gallons; 8.4 Imp gallons) AvGas 100LL in fuselage tank and 60 litres (15.9 US gallons; 13.2 Imp gallons) in each wing tank. Max 15 litres (4 US gallons; 3.3 Imp gallons) oil in Normal category, 11.4 litres (3 US gallons; 2.5 Imp gallons) for aerobatics.
ACCOMMODATION: Pilot and co-pilot/passenger in tandem under single-piece canopy opening to starboard; additional transparencies in lower sides of cockpit.
SYSTEMS: 12V generator and battery.
AVIONICS: Becker AR 3201 VHF radio standard.

DIMENSIONS, EXTERNAL:
Wing span: 300	8.00 m (26 ft 3 in)
300 S	7.50 m (24 ft 7¼ in)
Wing chord:	
at root: 300, 300 S	1.85 m (6 ft 0¾ in)
at tip: 300	0.83 m (2 ft 8¾ in)
300 S	0.93 m (3 ft 0½ in)
Wing aspect ratio: 300	5.98
300 S	5.39
Length overall: 300	7.12 m (23 ft 4¼ in)
300 S	6.65 m (21 ft 9¾ in)
Height overall: 300, 300 S	2.62 m (8 ft 7¼ in)
Tailplane span: 300	3.20 m (10 ft 6 in)
Wheelbase: 300	1.80 m (5 ft 11 in)
Propeller diameter: MTV-9	2.00 m (6 ft 6¾ in)
MTV-14	1.90 m (6 ft 2¾ in)

AREAS:
Wings, gross: 300	10.70 m² (115.17 sq ft)
300 S	10.44 m² (112.38 sq ft)

WEIGHTS AND LOADINGS:
Weight empty: 300	630 kg (1,389 lb)
Max T-O weight:	
300, aerobatic, and 300 S	820 kg (1,808 lb)
300, aerobatic, 2-seat	870 kg (1,918 lb)
300, normal	950 kg (2,094 lb)
Max wing loading:	
300, solo	76.5 kg/m² (15.67 lb/sq ft)
300, normal	88.8 kg/m² (18.19 lb/sq ft)
300 S	91.0 kg/m² (18.64 lb/sq ft)
Max power loading:	
300, solo	3.72 kg/kW (6.12 lb/hp)
300, normal	4.24 kg/kW (6.98 lb/hp)
300 S	3.66 kg/kW (6.02 lb/hp)

PERFORMANCE:
Never-exceed speed (V_{NE})	220 knots (407 km/h; 253 mph)
Max level speed	185 knots (343 km/h; 213 mph)
Max manoeuvring speed	158 knots (293 km/h; 182 mph)
Stalling speed	55 knots (102 km/h; 64 mph)
Max rate of climb at S/L	1,005 m (3,300 ft)/min
T-O to 15 m (50 ft)	approx 248 m (814 ft)
Landing from 15 m (50 ft)	approx 548 m (1,798 ft)
Range, 45 min reserves	526 nm (974 km; 605 miles)
g limits: Solo aerobatic	+10/−10
Two-seat aerobatic	+8/−8
Normal	+6/−3

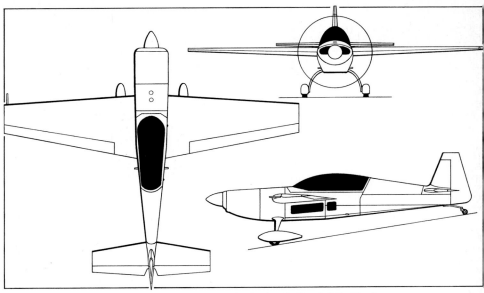
Extra 300 competition aerobatic aircraft *(Jane's/Mike Keep)*

Two Extra 300s in formation

Artist's impression of Extra 400 six-seat touring aircraft

EXTRA 400
TYPE: Six-seat, high performance touring aircraft.
PROGRAMME: Announced February 1993; first flight planned for 1994; will be certificated to JAR 23.
DESIGN FEATURES: High wing, T tail layout with single piston engine; spacious cabin; gust response system for smooth flight; low external noise.
STRUCTURE: Composites structural and non-structural parts.
LANDING GEAR: Retractable.

FFT
FFT GESELLSCHAFT FÜR FLUGZEUG- UND FASERVERBUND-TECHNOLOGIE mbH

Company ceased trading 31 December 1992. Speed Canard and Eurotrainer programmes could be revived by another company. Both types fully described in 1992-93 *Jane's*.

GROB
BURKHART GROB LUFT- UND RAUMFAHRT GmbH & Co KG
(Associated with Grob-Werke GmbH & Co KG)

Am Flugplatz, D-8939 Mattsies
Telephone: 49 (8268) 998-0
Fax: 49 (8268) 998-124
PRESIDENT AND CEO: Dr hc Dipl-Ing Burkhart Grob
VICE-PRESIDENT: Dipl-Ing Klaus-Harald Fischer
TECHNICAL DIRECTOR: Dipl-Ing Roland Fischer
MARKETING MANAGER: Konrad Lewald

Company founded 1971; has built more than 3,500 aircraft; name changed from Burkhart Grob Flugzeugbau to Burkhart Grob Luft- und Raumfahrt GmbH with light and heavy sections; light section produces G 103C Twin III Acro two-seat sailplane, G 103 Twin III SL with retractable engine and, on demand, the G 109B motor glider powered by Grob's own 67 kW (90 hp) type 2500 engine; heavy section deals with Egrett (see International section), the Strato C2 described below and with space activities in support of Weltraum-Institut Berlin.

German Luftwaffe cancelled LAPAS reconnaissance system based on Grob G-520 Egrett II in February 1993.

GROB G 115

TYPE: Two-seat light aircraft.
PROGRAMME: First flight November 1985; first flight of second prototype Spring 1986 with taller fin and rudder and relocated tailplane; LBA certification to FAR Pt 23 on 31 March 1987; British certification May 1988; later gained full public transport certification and German spinning clearance; 110 G 115 and 115A sold by end 1992 when production stopped; power plants and equipment changed and designations changed late 1992 for 1993 product line.
CURRENT VERSIONS: **G 115B**: Original G 115A upgraded from 85.8 kW (115 hp) to 119 kW (160 hp); retrofit of earlier aircraft possible; larger engine meets German noise limits.
G 115C: Newly designated 1993 model with 119 kW (160 hp) Textron Lycoming O-320 engine; fuel in wings, modified tail and many other improvements; first flight 26 January 1993; certification expected mid-1993; first deliveries expected September 1993; four pre-production aircraft produced in early 1993.
G 115D: Fully aerobatic version of G 115C powered by 134 kW (180 hp) Textron Lycoming AEIO-360; also due for first delivery third quarter of 1993; intended as professional trainer and glider tug.
COSTS: G 115C with fixed-pitch propeller price DM180,000 without tax; G 115D price DM235,000.
DESIGN FEATURES: Low-wing monoplane; wing section Eppler E 696; dihedral 5°; incidence 2°.
FLYING CONTROLS: Conventional; dual controls, aileron wheels in G 115C and sticks in G 115D; trim tab on port elevator; electrically operated flaps.
STRUCTURE: GFRP airframe.
LANDING GEAR: Non-retractable tricycle type, with wheel fairings; steerable nosewheel, size 5.00-5; mainwheels size 6.00-5; cantilever spring suspension; hydraulic toe-operated brakes; parking brake.
POWER PLANT: *G 115C:* One 119 kW (160 hp) Textron Lycoming O-320-D1A, driving two-blade fixed-pitch propeller. *G 115D:* One 134 kW (180 hp) Textron Lycoming AEIO-360-B1F, driving two-blade constant-speed propeller. Both power plants have exhaust silencer, electric starter and oil cooler. Fuel capacity 140 litre (37 US gallons; 30.8 Imp gallons) in two wing tanks with main mechanical and electric booster fuel pumps.
ACCOMMODATION: Two seats side by side under one-piece rearward sliding framed canopy, with dual controls; baggage space behind seats, with restraining net; heating; four-point harness in G 115C, five-point in G 115D.
SYSTEMS: Electrical; 24V alternator and battery.
AVIONICS: To customer's requirements. Instrument panel will accommodate full IFR instrumentation.
DIMENSIONS, EXTERNAL:
Wing span	10.00 m (32 ft 9½ in)
Wing aspect ratio	8.19
Length overall	7.44 m (24 ft 5 in)
Height overall	2.75 m (9 ft 0¼ in)
Wheel track	1.61 m (5 ft 3½ in)
Wheelbase	2.50 m (8 ft 2½ in)

AREAS:
Wings, gross	12.21 m² (131.43 sq ft)

WEIGHTS AND LOADINGS (Normal category):
Basic weight empty: C	650 kg (1,433 lb)
D	660 kg (1,455 lb)
Max T-O and landing weight: C, D	920 kg (2,028 lb)
Max wing loading: C, D	75.3 kg/m² (15.4 lb/sq ft)
Max power loading: C	7.73 kg/kW (12.68 lb/hp)
D	6.86 kg/kW (11.27 lb/hp)

PERFORMANCE:
Never-exceed speed (VNE):	
C	171 knots (317 km/h; 196 mph)
D	184 knots (341 km/h; 211 mph)
Max level speed at S/L, constant-speed propeller:	
C	135 knots (250 km/h; 155 mph)
D	146 knots (270 km/h; 168 mph)
Cruising speed, 75% power, constant-speed propeller:	
C	130 knots (240 km/h; 149 mph)
D	135 knots (250 km/h; 155 mph)
Stalling speed, flaps down:	
C, D	46 knots (85 km/h; 53 mph)
Max rate of climb at S/L: C	387 m (1,270 ft)/min
D	457 m (1,500 ft)/min
Service ceiling: C, D	4,875 m (16,000 ft)
T-O run: C	240 m (788 ft)
D	210 m (689 ft)
Landing run: C, D	180 m (591 ft)
Range, no reserves: C	650 nm (1,204 km; 748 miles)
D	520 nm (963 km; 598 miles)
Endurance: C	6 h 20 min
D	5 h 10 min
g limits: C	+4.4/−1.8
D	+6/−3

Grob G 115 club trainers

GROB G 115T

TYPE: Military and commercial pilot trainer.
PROGRAMME: Design began 1991; originally aimed at US Air Force Enhanced Flight Screener competition, but withdrawn in early 1992 because of shortage of time; first flight (D-EMGT) 11 June 1992; aircraft designed to conform to all relevant US MIL handling and layout standards and specifications for pilot trainers, but civil certification also expected 1994.
CUSTOMERS: None announced by early 1993.
COSTS: DM293,000 without tax with 175 kW (235 hp) engine; DM395,000 without tax with 194 kW (260 hp) engine.
DESIGN FEATURES: Low power loading, high wing loading, medium aspect ratio, smooth, accurate external form, low structure weight and low-drag laminar aerofoil give good aerobatic performance at full gross weight; cockpit and control layout dimensioned for two occupants in full military equipment.
FLYING CONTROLS: Control columns with sticktop electric elevator trim, adjustable rudder pedals and duplicated toe-operated brakes; handling meets MIL specifications; control surfaces operated by pushrods for low friction; electrically operated Fowler flaps.
STRUCTURE: All-composites structure laid up by hand without autoclaving for ease of repair in field.
LANDING GEAR: Retractable tricycle type with electro-hydraulic actuation; oleo dampers; steerable nosewheel; foot-powered hydraulic disc brakes; mainwheels 6.00-5; nosewheel 5.00-5.
POWER PLANT: One 194 kW (260 hp) Textron Lycoming AEIO-540-X86 flat-six engine driving a constant-speed three-blade propeller; four-blade propeller optional;

Grob G 115T commercial and military trainer

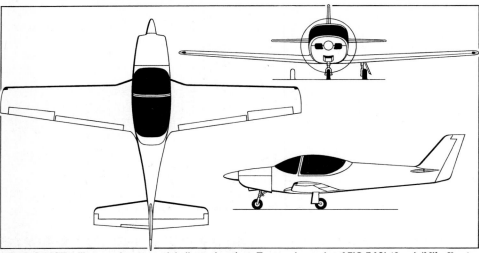

Grob G 115T military and commercial pilot trainer (one Textron Lycoming AEIO-540) *(Jane's/Mike Keep)*

exhaust silencer to keep noise below 68 dBA; Christen all-attitude oil system; EGT and cylinder head temperature gauges. Optional 175 kW (235 hp) engine and fixed-pitch propeller. Fuel capacity 280 litres (74 US gallons; 61.6 Imp gallons) located in wings; mechanical fuel pump with electric booster.

ACCOMMODATION: Two pilots side by side with room for crash helmets and parachutes; rudder pedals and seats adjustable; instrument panel space for two full blind flying panels and airways radio; five-point harness; full transparent sliding canopy; cabin heater and windscreen demister standard.

SYSTEMS: Electrical system 24V DC with 25Ah nickel-cadmium battery and 70VA alternator; full night lighting with integral instrument lighting; heated pitot and alternate static source.

AVIONICS: Standard aircraft has only basic flight instrumentation; full IFR equipment optional.

DIMENSIONS, EXTERNAL:
Wing span	10.00 m (32 ft 9¾ in)
Wing aspect ratio	7.63
Length overall	8.85 m (28 ft 0¾ in)
Height overall	2.57 m (8 ft 5¼ in)
Tailplane span	3.80 m (12 ft 5½ in)

DIMENSIONS, INTERNAL:
Cabin: Width	1.25 m (4 ft 1¼ in)
Height, from floor	1.20 m (3 ft 11¼ in)

AREAS:
Wings, gross	13.1 m² (141.0 sq ft)
Vertical tail surfaces (total)	1.44 m² (15.50 sq ft)
Horizontal tail surfaces (total)	3.04 m² (32.72 sq ft)

WEIGHTS AND LOADINGS:
Manufacturer's weight empty, according to equipment	850-890 kg (1,874-1,962 lb)
Useful load	400 kg (882 lb)
Max T-O weight	1,300 kg (2,866 lb)
Max wing loading	99.24 kg/m² (20.33 lb/sq ft)
Max power loading: 260 hp	6.70 kg/kW (11.02 lb/hp)
235 hp	7.42 kg/kW (12.20 lb/hp)

PERFORMANCE:
Never-exceed speed (V_NE)	209 knots (387 km/h; 240 mph)
Max cruising speed, 75% power, at 2,440 m (8,000 ft)	159 knots (295 km/h; 183 mph)
Stalling speed, flaps and gear down	54 knots (100 km/h; 62 mph)
Max rate of climb at S/L	426 m (1,400 ft)/min
Service ceiling	5,490 m (18,000 ft)
T-O run	507 m (1,663 ft)
T-O to 15 m (50 ft)	806 m (2,644 ft)
Landing from 15 m (50 ft)	450 m (1,477 ft)
Landing run	220 m (722 ft)
Range, no reserves	870 nm (1,612 km; 1,001 miles)
Endurance	3 h 30 min
g limits at max T-O weight	+6/−3

GROB GF 200

TYPE: Four-seat touring aircraft.
PROGRAMME: Announced early 1988; first flight of unpressurised prototype (D-EFKH) 26 November 1991; originally planned Porsche engine replaced by Textron Lycoming TIO-540; certification and deliveries planned for 1994.
CURRENT VERSIONS: **Unpressurised**: First batch of production GF 200s powered by piston engines and unpressurised.
Pressurised: Later aircraft will have turboprop engines; not selected by early 1993. Pressurised; cabin pressurisation tests under way early 1993.
COSTS: Standard aircraft, no IFR, DM445,000 without tax.
DESIGN FEATURES: Advanced profile wing with upswept tips; centre-mounted engine drives rear-mounted propeller through carbon composite extension shaft; swept fins and rudders above and below tailcone; T tail.

FLYING CONTROLS: Variable incidence tailplane and elevators; rudders in upper and lower fins; small strakes on both fins; strakes at wingroots; electrically operated area-extending Fowler flaps.
STRUCTURE: All-composites airframe.
LANDING GEAR: Tricycle retracting; main gear has levered suspension and hydraulically actuated wheel brakes; main legs retract inward; nose leg carries taxying light and retracts forward.
POWER PLANT: One 201 kW (270 hp) Textron Lycoming TIO-540-AF1A turbocharged intercooled flat-six engine; prototype flew with three-blade constant-speed Mühlbauer pusher propeller driven by Grob composites shaft; five-blade propeller of 1.80 m (5 ft 10¾ in) diameter considered; standard fuel tankage 350 litres (92.5 US gallons; 77 Imp gallons).
ACCOMMODATION: Pilot plus three passengers in pressurised cabin, after first unpressurised batch; access on port side between seat rows through split upward and downward opening door with integral airstairs.
SYSTEMS: 28V 70A electrical system, also powers hydraulic landing gear retraction.

DIMENSIONS, EXTERNAL:
Wing span	11.00 m (36 ft 0 in)
Wing aspect ratio	9.66
Length overall	8.50 m (27 ft 11 in)
Height overall	3.20 m (10 ft 6 in)

AREAS:
Wings, gross: flaps retracted	12.53 m² (134.9 sq ft)
flaps extended	14.09 m² (151.7 sq ft)

WEIGHTS AND LOADINGS:
Payload	580 kg (1,278 lb)
Max fuel weight	252 kg (555 lb)
Max T-O weight	1,600 kg (3,527 lb)
Max wing loading	127.69 kg/m² (26.1 lb/sq ft)
Max power loading	7.95 kg/kW (13.06 lb/hp)

PERFORMANCE:
Cruising speed: 75% power, at 6,100 m (20,000 ft)	202 knots (375 km/h; 233 mph)
100% power, at 6,100 m (20,000 ft)	227 knots (420 km/h; 261 mph)
Max rate of climb: at S/L	372 m (1,220 ft)/min
at 6,100 m (20,000 ft)	300 m (985 ft)/min
T-O run	370 m (1,214 ft)
T-O to 15 m (50 ft)	770 m (2,527 ft)
Max range at 6,100 m (20,000 ft), 45 min reserves	1,090 nm (2,020 km; 1,255 miles)

Unpressurised prototype of Grob GF 200 (Textron Lycoming TIO-540)

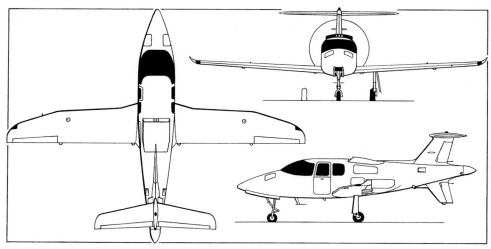

Grob GF 200 all-composites four-seat touring aircraft *(Jane's/Mike Keep)*

GROB STRATO 2C

TYPE: High altitude, long endurance stratospheric and climatic research aircraft.

PROGRAMME: Development begun under contract from German Ministry of Research and Technology's DLR research establishment at Oberpfaffenhofen April 1992; tooling began October 1992; engine installation changed from tractor to pusher configuration Spring 1993 (see Addenda); power plant tests to begin October 1993; first flight November 1994; delivery August 1995. Industrieanlagen-Betriebsgesellschaft participated in configuration development and will monitor and test the power plant.

CURRENT VERSIONS: **2C**: Basic observation prototype *as described*.

CUSTOMERS: German DLR.

COSTS: Reported contract value for one aircraft DM75 million.

DESIGN FEATURES: Can fly above 95 per cent of all atmosphere molecules; range sufficient to observe North or South Poles for eight hours operating from neighbouring continents; high aspect ratio wing (22) with laminar aerofoil and low wing loading, combined with very accurate and smooth surface contour, give efficiency at altitudes up to 24,000 m (78,740 ft) at speeds below Mach 0.53; low torsional moment of aerofoil simplifies wing structure; accuracy and finish of surfaces increase airspeed and range by 20 to 30 per cent.

FLYING CONTROLS: Conventional control surfaces and flaps.

STRUCTURE: Use of GFRP, CFRP and aramids reduces structure density from the normal 7.8 kg/dm^3 (0.282 lb/cu in) of metal to 1.7 kg/dm^3 (0.061 lb/cu in). Airframe laid up in half-shells in moulds, bonded together.

LANDING GEAR: Retractable tricycle type.

POWER PLANT: Compound power plant based on two 300 kW (402 hp) Teledyne Continental GT-550 liquid-cooled flat-six piston engines, driving variable-pitch five-blade propellers through reduction gearboxes; three-scroll turbocharger with 2.5:1 pressure ratio driven by exhaust of each engine; after driving turbocharger, exhaust gases recompressed by second scroll and ducted to separate two-stage turbocharger based on core of P&WC PW124 turboprop, which then feeds compressed fresh intake air to intake scroll of GT-550 turbocharger and thence to engine; total pressure ratio of second turbocharger 18:1, and air throughput at S/L 10 kg/s; automatic wastegate maintains engine power constant up to maximum altitude; to ensure heat dissipation at extreme altitudes, two intercoolers set in charge airstream of external turbocharger and third intercooler located between engine turbocharger and cylinders; two further radiators handle engine oil cooling and engine coolant (see accompanying drawing).

ACCOMMODATION: Two pilots side by side and two scientists at consoles facing starboard; toilet, galley and rest facilities provided for 40-hour sorties without need for pressure suits; scientific payloads inside pressurised compartment; crew door to port, aft of pilot.

SYSTEMS: Cabin air-conditioning and pressurisation provided by cooled compressed air drawn from engine turbochargers; oil-cooled engine-driven generator provides aircraft electrical power.

DIMENSIONS, EXTERNAL:
Wing span	56.50 m (185 ft 4½ in)
Wing aspect ratio	22
Fuselage: Length	22.40 m (73 ft 6 in)
Max diameter	2.00 m (6 ft 6¼ in)
Propeller diameter	6.00 m (19 ft 8¼ in)

DIMENSIONS, INTERNAL:
Pressure cabin: Length	5.50 m (18 ft 0½ in)

AREAS:
Wings, gross	145.0 m^2 (1,560.8 sq ft)

WEIGHTS AND LOADINGS:
Manufacturer's weight empty	5,800 kg (12,786 lb)
Max T-O weight (approx)	11,700 kg (25,794 lb)
Max wing loading	80.69 kg/m^2 (16.53 lb/sq ft)

PERFORMANCE:
Cruising speed: at 24,000 m (78,740 ft), design mission
  280 knots (520 km/h; 323 mph)
at 18,000 m (59,050 ft), long endurance mission
  186 knots (345 km/h; 214 mph)
Time to 26,000 m (85,300 ft), at 6,210 kg (13,690 lb) T-O weight
  10 min
Design mission altitude 24,000 m (78,740 ft)
Max operating altitude 26,000 m (85,300 ft)
Radius, design mission with 800 kg (1,763 lb) payload and 3,360 kg (7,407 lb) fuel, incl 8 h at 24,000 m (78,740 ft)
  1,888 nm (3,500 km; 2,174 miles)
Range, long endurance mission with 1,000 kg (2,204 lb) payload and 4,900 kg (10,802 lb) fuel, incl 48 h at 18,000 m (59,050 ft)
  9,766 nm (18,100 km; 11,245 miles)

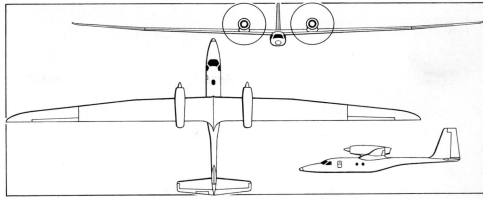

Grob Strato 2C high altitude long endurance surveillance aircraft *(Jane's/Mike Keep)*

Artist's impression of Grob Strato 2C surveillance aircraft in original tractor configuration

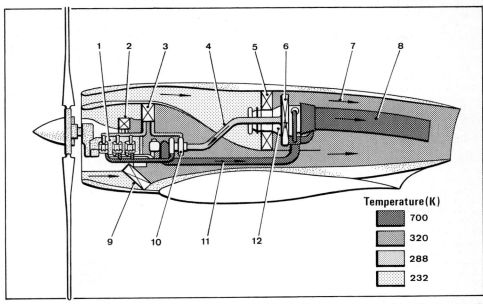

Grob Strato 2C power plant. (*1*) Teledyne Continental GT-550 liquid-cooled piston engine, (*2*) oil cooler, (*3*) intercooler stage 3, (*4*) piston engine charge air, (*5*) intercooler stage 1, (*6*) intercooler stage 2, (*7*) waste air from heat exchangers, (*8*) engine and compressor exhaust, (*9*) engine radiator, (*10*) turbocharger stage 3, (*11*) engine exhaust to compressor, (*12*) turbocharger stages 1 and 2. (Tractor configuration changed to pusher installation Spring 1993.)

HOFFMANN
WOLF HOFFMANN FLUGZEUGBAU KG

This company ceased trading in late 1992. Hoffmann H 40 programme belonged to Rhein-Flugzeugbau, which recovered prototype (D-EIOF); see under Rhein-Flugzeugbau.

RFB
RHEIN-FLUGZEUGBAU GmbH
(Owned by ABS International)

Flugplatz (PO Box 408), D-4050 Mönchengladbach 1
Telephone: 49 (2161) 6820
Fax: 49 (2161) 682200
Telex: 852 506
OWNER: Albert Blum
PRESIDENT: Hartmut Stiegler
CHIEF DESIGNER: Christoph Fischer
MARKETING MANAGER: Hans-Jörg Brandt

RFB sold by MBB to ABS International in 1990. Workforce is 250 and annual sales DM50 million.

RFB specialises in development and manufacture of GFRP wings and fuselages, components and assemblies of light alloy, steel and GFRP for aircraft in quantity production by other German companies, and spare parts and ground equipment.

Under contract to German government, RFB services military aircraft, and provides target towing flights and other services with special aircraft. It has Luftfahrt-Bundesamt (LBA) approval for aircraft development, manufacture, maintenance and overhaul; operates a factory certificated service centre for Beech, Mitsubishi, Partenavia and Piper aircraft, Bendix/King, Becker and Collins avionics, and P&WC PT6 engines; also airline components, parachutes, liferafts, lifejackets, evacuation slides for L-1011, 757/767/737 and MD-11 airliners; repair and overhaul of composite structural components.

Work on specialist ground effect vehicles on lines of its earlier Lippisch-type X113 and X114 has been continued with improved patented Power Augmentation System.

RFB has been engaged for many years in developing specialised applications for ducted propellers, one of which led to Fantrainer.

Prototype of Hoffmann H 40 two-seat all-composites trainer

RFB FANTRAINER SERIES
RFB continues to market Fantrainer 400 and 600, but no new sales reported in 1991 or subsequently. New, more powerful version being considered. Details and performance in 1990-91 *Jane's*. Royal Thai Air Force acquired 31 Fantrainer 400s and 16 Fantrainer 600s between 1982 and 1988.

ROCKWELL/DASA RANGER 2000
RFB became main subcontractor to DASA and Rockwell International in design of turbofan powered Ranger 2000 (which see in International section) intended for US Air Force/Navy JPATS. RFB designed and made all parts and tooling for two prototypes, delivering first set in July and second in October 1992. Parts included nose section, canopy frames, control system, aft fuselage including T tail, engine installation and airbrakes.

HOFFMANN H 40
TYPE: Two-seat basic trainer.
PROGRAMME: Design begun 1986; first flights of two prototypes 28 August 1988 and 4 April 1991; airframe under static test. When Hoffmann company ceased trading in late 1992, original owner Rhein-Flugzeugbau recovered design and prototype, but had not decided by April 1993 whether to continue development.
COSTS: About DM145,000 (1992).
DESIGN FEATURES: High-lift Wortmann FX-63-137 section; thickness/chord ratio 13.7 per cent; twist −1.5°; leading-edge sweptforward 3°; dihedral 4°; horn balanced ailerons; turned-down wingtips.
FLYING CONTROLS: Fixed T tailplane with one-piece horn balanced elevator and central trim tab; conventional mechanical actuation with aileron wheel control column; electrically actuated split flaps, set forward from trailing-edge, occupy 20 per cent of wing chord.
STRUCTURE: All GFRP except for CFRP wing spar caps; monocoque fuselage with integral fin; console between seats extends to rear fuselage, forming central keel through cabin; single-spar wing.
LANDING GEAR: Non-retractable tricycle landing gear with spring steel main legs and gas spring nose leg; single Cleveland 40-78 wheel on each leg; mainwheels have Goodyear 6.00-5 tyres, nosewheel 5.00-5; Cleveland 30-9 brakes; nosewheel steering range ±40°.

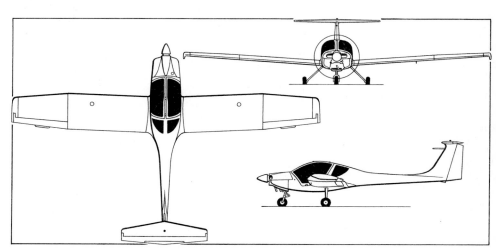
Hoffmann H 40 all-composites light aircraft (one Textron Lycoming O-235) (*Jane's*/Mike Keep)

POWER PLANT: One 86.5 kW (116 hp) Textron Lycoming O-235 flat-four engine driving three-blade constant-speed Mühlbauer MTV-F/160-03 propeller. Four aluminium fuel cells, two in each wing, containing total 100 litres (26.4 US gallons; 22 Imp gallons).
ACCOMMODATION: Side by side seats with central console; large gull-wing doors open upwards on centreline; baggage space behind seats.
AVIONICS: Becker AR 3021 VHF com radio; Becker ATC 2000 transponder.
DIMENSIONS, EXTERNAL:
Wing span	10.76 m (35 ft 3½ in)
Wing chord: at root	1.34 m (4 ft 4¾ in)
at tip	0.94 m (3 ft 1 in)
Wing aspect ratio	8.0
Length overall	7.26 m (24 ft 11¾ in)
Height overall	2.39 m (7 ft 10 in)
Tailplane span	3.00 m (9 ft 10 in)
Wheel track	2.22 m (7 ft 3¼ in)
Propeller diameter	1.60 m (5 ft 2¾ in)
Propeller ground clearance	0.25 m (9¾ in)

AREAS:
Wings, gross	13.62 m² (145.3 sq ft)
Trailing-edge flaps (total)	1.07 m² (11.5 sq ft)

WEIGHTS AND LOADINGS:
Operating weight empty	590 kg (1,300 lb)
Max payload	190 kg (419 lb)
Max T-O weight	850 kg (1,874 lb)
Max wing loading	62.4 kg/m² (12.78 lb/sq ft)
Max power loading	7.3 kg/kW (16.1 lb/hp)

PERFORMANCE (at max T-O weight):
Never-exceed speed (VNE)	134 knots (250 km/h; 155 mph)
Max level speed	113 knots (210 km/h; 130 mph)
Econ cruising speed	97 knots (180 km/h; 111 mph)
Stalling speed, power off, flaps up	46 knots (85 km/h; 53 mph)
Max rate of climb at S/L	213 m (700 ft)/min
Service ceiling	4,500 m (14,760 ft)
T-O to 15 m (50 ft)	350 m (1,148 ft)
Landing from 15 m (50 ft)	250 m (820 ft)
Range with max fuel	367 nm (680 km; 422 miles)

RUSCHMEYER
RUSCHMEYER LUFTFAHRTTECHNIK GmbH

Flugplatz, D-4520 Melle 1
Telephone: 49 (5422) 9493-0
Fax: 49 (5422) 9493-99
GENERAL MANAGER: Horst Ruschmeyer

Development of definitive production version of R 90-230 RG early 1992 being followed by batch production. By 1995, Ruschmeyer plans introduction of rest of family, as detailed below.

RUSCHMEYER R 90-230 RG
TYPE: Four-seat touring aircraft.
PROGRAMME: Development of original MF-85 started 1985; first prototype MF-85P-RG (V0001) flew with Porsche; changed to R 90-230 RG using flat rated Textron Lycoming IO-540 after cessation of Porsche production in March 1990; first flight of prototype V001 (D-EEHE) 8 August 1988; first flight of V002 (D-EERO) 25 September 1990; V003 (D-EERH) first flight 12 February 1992; LBA certification June 1992; FAA certification to follow. Three delivered 1992 and batch of 22 planned for 1993.

CURRENT VERSIONS: **R 90-230 RG**: Current production aircraft; *description applies to this version*.
R 90-230 FG: Non-retractable landing gear version of RG; 175 kW (235 hp) Textron Lycoming O-540-J, derated to 2,300 rpm and about 164 kW (220 hp); first flight planned for June 1993; certification expected end 1993.
R 90-180 FG: Non-retractable landing gear version; 149 kW (200 hp) Textron Lycoming IO-360, flat rated at 134 kW (180 hp); much modified structure; developed concurrently with R 90-230 FG; first flight expected mid-1994; first delivery early 1995.

R 90-250T RG: Presently intended top model of family; turbocharged 186 kW (250 hp) piston engine; intended cruising speed 234 knots (435 km/h; 270 mph) at 7,620 m (25,000 ft).

COSTS: Price of first production version DM342,000 (1992) plus VAT without avionics and optional items.

DESIGN FEATURES: Objectives were high performance by low-drag all-composites airframe, use of derated engine requiring less cooling, reduced noise by lower engine rpm, special silencer and matched four-blade constant-speed propeller, consequently reduced fuel consumption and emissions; noise level under German regulations demonstrated 66 dBA, 8 dBA below limit; also 10.2 dBA below limit during climb after take-off.

FLYING CONTROLS: Conventional, all push/pull rod operated to minimise friction; dual controls with sticks standard; fixed tailplane; trim tab on port elevator; stall strips on inboard wing leading-edge; flaps operated by self-contained electrohydraulic system.

STRUCTURE: All-composites moulded airframe with 70 per cent fewer parts than equivalent metal structure. Ruschmeyer qualified BASF Palatal A430 resin fibre composite material for aviation; advantages are durability at up to 72°C ambient, improved strength, field repair without special tools, negligible allergic risks during manufacture, and material is recyclable. Airframe certificated for 18,000 flying hours.

LANDING GEAR: Retractable tricycle; trailing-link levered mainwheel suspension retracts inward and covered by mechanically linked doors; foot-powered hydraulic disc brakes; nosewheel retracts rearward, also covered by doors.

POWER PLANT: One Textron Lycoming IO-540-C4D5 flat-six engine, derated to 171.5 kW (230 hp) at 2,400 rpm, with exhaust silencer and special induction and cooling system; Mühlbauer MTV-14-B four-blade composites constant-speed propeller with 7.5 cm (3 in) spacer. Standard fuel 250 litres (66 US gallons; 55 Imp gallons); usable fuel 236 litres (62.3 US gallons; 52 Imp gallons) in inner wings. Fuel 100/100LL Avgas. Oil capacity 11.4 litres (3 US gallons; 2.5 Imp gallons).

ACCOMMODATION: Four-seat interior with ergonomically designed seats; upward opening gull wing doors; heating and ventilation system with electric blower; baggage compartment accessible from outside.

SYSTEMS: Generator and battery; electrohydraulic actuation for flaps.

AVIONICS: Optional; room for full IFR system and autopilot.

DIMENSIONS, EXTERNAL:
Wing span	9.50 m (31 ft 2 in)
Length overall	7.93 m (26 ft 0¼ in)
Propeller diameter	1.88 m (6 ft 2 in)

DIMENSIONS, INTERNAL:
Cabin: Length	2.86 m (9 ft 4½ in)
Width	1.14 m (3 ft 9 in)
Height	1.24 m (4 ft 0¾ in)
Baggage compartment volume	0.80 m³ (28.25 cu ft)

AREAS:
Wings, gross	12.94 m² (139.3 sq ft)

WEIGHTS AND LOADINGS:
Standard weight empty	898 kg (1,980 lb)
Max useful load	452 kg (996 lb)
Max fuel weight: standard	173 kg (381 lb)
long range	274 kg (604 lb)
Max T-O and landing weight	1,350 kg (2,976 lb)
Max wing loading	104.3 kg/m² (21.36 lb/sq ft)
Max power loading	7.87 kg/kW (12.94 lb/hp)

Ruschmeyer R 90-230 RG (flat rated Textron Lycoming IO-540)

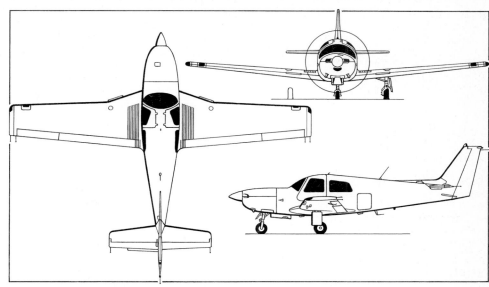

Ruschmeyer R 90-230 RG touring aircraft (*Jane's/Mike Keep*)

PERFORMANCE:
Max level speed at 3,050 m (10,000 ft)	174 knots (322 km/h; 200 mph)
Stalling speed: clean	67 knots (124 km/h; 77 mph)
landing configuration	58 knots (108 km/h; 67 mph)
Max rate of climb at S/L	347 m (1,140 ft)/min
Service ceiling	4,895 m (16,060 ft)
T-O run	260 m (853 ft)
T-O to 15 m (50 ft)	520 m (1,706 ft)
Landing from 15 m (50 ft)	480 m (1,575 ft)
Landing run	325 m (1,067 ft)

Range:
75% power, 165 knots (306 km/h; 190 mph) at 2,440 m (8,000 ft) 744 nm (1,378 km; 856 miles)
38% power, 114 knots (211 km/h; 131 mph) 1,004 nm (1,860 km; 1,156 miles)
45% power, 120 knots (222 km/h; 138 mph) at 915 m (3,000 ft) 1,479 nm (2,740 km; 1,703 miles)

GREECE

HAI
HELLENIC AEROSPACE INDUSTRY LTD
Athens Tower, Messogion Avenue 2-4, GR-115 27 Athens
Telephone: 30 (1) 77 99 678 or 77 99 622
Fax: 30 (1) 77 97 670
Telex: 221951 HAI GR
WORKS: TANAGRA, PO Box 23, GR-320 09 Schimatari, Viotia
Telephone: 30 (262) 5 20 00
Fax: 30 (262) 88 38 714
Telex: 299372 HAI GR
CHAIRMAN: P. Velissaropoulos
MANAGING DIRECTOR: Dan Lange
COMMERCIAL DIRECTOR: Dimitris Sarlis
MANAGER, ADVERTISING AND CUSTOMER RELATIONS:
 Thomas Nestor

Owned 87 per cent by Greek government and 13 per cent by Hellenic Industrial Bank, but to be privatised 1993 by sale of up to 49 per cent to two or three foreign investors. Aircraft Division, Engine Division, Electronics/Avionics Division and Manufacturing Division based in Tanagra works occupying 180 ha (445 acre) site with 150,000 m² (1,614,585 sq ft) covered floor space and 1992 workforce of 3,150; civil and military manufacturing and repair/maintenance for 22 countries in Middle East, Africa, Europe and North America, including full support Mirage F1, Atar 9K-50, F-4 Phantom, A-7 Corsair, Lockheed C-130 depot level maintenance, J79 for USAFE and UK RAF. Major manufacturing includes F-16 rear fuselages and intakes, Airbus A300 and A310 door frames, assembly and test of SNECMA M53.

HAI depot level maintenance on a Mirage F1-C

INDIA

ADA
AERONAUTICAL DEVELOPMENT AGENCY
PO Box 1718, Vimanapura Post Office, Bangalore 560 017
Telephone: 91 (812) 562060
Fax: 91 (812) 569445
Telex: 0845 8114 ADA IN
LCA PROGRAMME DIRECTOR: Dr K. Harinarayana
SENIOR MANAGER, INFORMATION TECHNOLOGY:
T. N. Prakash

LIGHT COMBAT AIRCRAFT (LCA)
TYPE: All-weather air superiority fighter and light close air support aircraft.
PROGRAMME: Development approved by Indian government 1983 as MiG-21/Ajeet replacement; project definition begun Spring 1987, completed late 1988; basic design finalised 1990; prototype construction by HAL started mid-1991; first flight expected mid-1996; two prototypes each to be powered by F404-GE-F2J3 afterburning turbofan; indigenous 83.4 kN (18,740 lb st) GTX-35VS Kaveri being developed for production aircraft; naval version also planned.
CUSTOMERS: Indian Air Force.
COSTS: Two prototypes funded with Rs15,000 million ($872 million) (1990).
DESIGN FEATURES: Advanced materials planned to achieve minimum weight; full quadruplex fly-by-wire control to confer agility with wide range of external stores; shoulder-mounted delta wings with compound sweep on leading-edges; large twist from inboard to outboard leading-edges; fixed-geometry intakes.
FLYING CONTROLS: Two-segment trailing-edge elevons and three-section leading-edge slats; vortex-shedding inboard leading-edges with inboard slots to form vortices over wingroot and fin; quadruplex fly-by-wire AFCS; HOTAS controls.
STRUCTURE: Materials to include aluminium-lithium alloy, carbonfibre composites and titanium alloys; CFRP wing, fin and rudder.
LANDING GEAR: Hydraulically retractable tricycle type. Single mainwheels; twin-wheel nose unit.
POWER PLANT: One 80.5 kN (18,100 lb st) General Electric F404-F2J3 afterburning turbofan in prototypes; Indian GTRE GTX-35VS Kaveri turbofan, with FADEC, under development for production aircraft. In-flight refuelling probe on starboard side of front fuselage. Provision for up to five 800 litre (211 US gallon; 176 Imp gallon) external fuel tanks.
ACCOMMODATION: Pilot only, on zero/zero ejection seat. Development will include two-seat training version.
SYSTEMS: Hydraulic system for powered flying controls, brakes and landing gear; electrical system for fly-by-wire and avionics power supply; environmental control system.
AVIONICS: V/UHF and UHF com radios and data link; INS, Tacan and radio altimeter; Hindustan Aeronautics multi-mode radar; IFF transponder/interrogator; FLIR; ECM (RWR, self-protection jammer and chaff/flare dispenser). Avionics architecture centred round powerful mission computer with three MIL-STD-1553B databuses. Cockpit displays include collimated HUD with holographic combiner, and left and right hand multi-function colour CRTs, and are NVG compatible.
ARMAMENT: Internally mounted GSh-23 twin-barrel 23 mm gun, with 220 rounds. Seven external stores stations (three under each wing and one under fuselage) for wide range of short/medium-range missiles and other ordnance.

DIMENSIONS, EXTERNAL:
Wing span 8.20 m (26 ft 10¾ in)
Wing aspect ratio 1.8
Length overall 13.20 m (43 ft 3¾ in)
Height overall 4.40 m (14 ft 5¼ in)
AREAS:
Wings, gross approx 37.50 m² (403.6 sq ft)
WEIGHTS AND LOADINGS:
Weight empty approx 5,500 kg (12,125 lb)
Max external stores load more than 4,000 kg (8,818 lb)
T-O weight (clean) 8,500 kg (18,740 lb)
Wing loading (clean) approx 226.7 kg/m² (46.4 lb/sq ft)
Power loading (prototypes, clean)
105.6 kg/kN (1.03 lb/lb st)
PERFORMANCE (estimated):
Max level speed at altitude Mach 1.6
Service ceiling above 15,240 m (50,000 ft)
g limits +9/−3.5

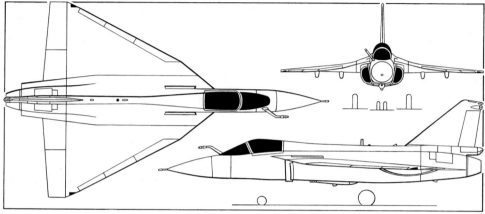

Provisional drawing of the ADA Light Combat Aircraft *(Jane's/Mike Keep)*

BHEL
BHARAT HEAVY ELECTRICALS LTD
BHEL House, Sirifort, New Delhi
Telephone: 91 (11) 6493437
CHAIRMAN: A. Gavisiddappa
WORKS: Heavy Electrical Equipment Plant, Ranipur, Hardwar 249403, Uttar Pradesh
Telephone: 91 (133) 426080 and 426457
Fax: 91 (133) 426462
Telex: 0599-206, 207 and 212
EXECUTIVE DIRECTOR: B. R. Gulati
PROJECT MANAGER, AVIATION: H. W. Bhatnagar

BHEL approved for manufacture of light aircraft by Indian Directorate General of Civil Aviation in May 1991 and by Indian Ministry of Industry in March 1991.

BHEL LT-1 SWATI
TYPE: Two-seat civil light trainer; intended for flying training, sport flying, surveillance, photography, short-hauls etc.
PROGRAMME: Designed and developed by Technical Centre of Directorate General of Civil Aviation, India; first flight 17 November 1990; technical certificate issued 1991 by Indian Institute of Technology, Kanpur; type certification awarded January 1992.
CURRENT VERSIONS: **Prototype:** Initially test flown (VT-XIV) for certification with 97 kW (130 hp) Rolls-Royce Continental O-240-A engine and 'taildragger' landing gear; details in 1992-93 *Jane's*.
Production version: Wooden wings, tricycle landing gear and O-235-N2C engine; first flight (VT-XSR) 22 November 1992; undergoing test certification early 1993.
COSTS: Approx Rs3.5 million (1992).
DESIGN FEATURES: Designed to FAR Pt 23 (Utility and Normal categories), capable of such manoeuvres as stall turns, barrel rolls, loops and spins, lazy eights and chandelles. Wing section NASA GA(W)-1 (constant); dihedral 4°; incidence 2°.
FLYING CONTROLS: Rod and cable controls for all three axes; plain flaps; fixed incidence tailplane.
STRUCTURE: Wings constant chord with single box spar of Himalayan spruce and ply covered leading-edges; ailerons and flaps similar; fuselage of chromoly steel tube with metal/GFRP engine cowling and skin on forward fuselage; fabric covering aft; GFRP fairings; tail surfaces light metal.
LANDING GEAR: Non-retractable bungee-sprung mainwheels, on side Vs and half-axles, with dual hydraulic brakes. Non-retractable steerable nosewheel.
POWER PLANT: One 86.5 kW (116 hp) Textron Lycoming O-235-N2C flat-four engine, driving a McCauley 1A103TCM-695 two-blade fixed-pitch metal propeller. Fuel capacity 90 litres (23.8 US gallons; 19.8 Imp gallons).
ACCOMMODATION: Two seats, with adjustable safety harnesses, side by side under one-piece cockpit canopy.
AVIONICS: VFR instrumentation standard. Provision for radio.

DIMENSIONS, EXTERNAL:
Wing span 9.20 m (30 ft 2¼ in)
Wing chord, constant 1.30 m (4 ft 3¼ in)
Wing aspect ratio 7.08
Length overall (flying attitude) 7.21 m (23 ft 8 in)
Height overall (flying attitude) 2.78 m (9 ft 1½ in)
Tailplane span 2.60 m (8 ft 6½ in)
Wheel track 2.00 m (6 ft 6¾ in)
Wheelbase 1.725 m (5 ft 8 in)
Propeller diameter 1.83 m (6 ft 0 in)
AREAS:
Wings, gross 11.96 m² (128.7 sq ft)
Ailerons (total) 1.10 m² (11.84 sq ft)
Trailing-edge flaps (total) 1.04 m² (11.19 sq ft)
Fin 0.75 m² (8.07 sq ft)
Rudder 0.62 m² (6.67 sq ft)
Tailplane 0.93 m² (10.01 sq ft)
Elevators (total, incl tab) 0.94 m² (10.12 sq ft)

First production BHEL LT-1 Swati two-seat light trainer

WEIGHTS AND LOADINGS (U: Utility, N: Normal category):
Weight empty, equipped (typical):
U, N 524 kg (1,155 lb)
Max T-O weight: U 730 kg (1,609 lb)
N 770 kg (1,698 lb)
Max wing loading: U 61.04 kg/m² (12.50 lb/sq ft)
N 64.38 kg/m² (13.19 lb/sq ft)
Max power loading: U 8.44 kg/kW (13.87 lb/hp)
N 8.90 kg/kW (14.63 lb/hp)
PERFORMANCE (at max T-O weight):
Never-exceed speed (V_{NE}):
U, N 135 knots (250 km/h; 155 mph)
Max level speed: U 109 knots (201 km/h; 125 mph)
Cruising speed (75% power):
U 96 knots (177 km/h; 110 mph)
Stalling speed 44 knots (81 km/h; 50 mph)
Max rate of climb at S/L: N 231 m (759 ft)/min
T-O run: N 240 m (788 ft)
Landing run: N 220 m (722 ft)
Range with max fuel: U, N 273 nm (507 km; 315 miles)
Max endurance: U, N 3 h 30 min
g limits: U +4.4/−1.76
N +3.8/−1.52

HAL
HINDUSTAN AERONAUTICS LIMITED
CORPORATE OFFICE: PO Box 5150, 15/1 Cubbon Road, Bangalore 560 001
Telephone: 91 (812) 266901
Fax: 91 (812) 268758 and 577533
Telex: 845 2266 HAL IN
CHAIRMAN: R. N. Sharma
DIRECTOR, CORPORATE PLANNING: S. N. Sachindran

Formed 1 October 1964; has 12 manufacturing divisions at seven locations (Bangalore, Nasik, Koraput, Hyderabad, Kanpur, Lucknow and Korwa), plus Design Complex; total workforce approx 40,600. Hyderabad Division manufactures avionics for all aircraft produced by HAL, plus air route surveillance and precision approach radars; Lucknow Division produces landing gears and other accessories under licence from manufacturers in UK, France and Russia; Korwa manufactures inertial navigation and nav/attack systems.

DESIGN COMPLEX
Post Bag 1789, Bangalore 560 017
Telephone: 91 (812) 565201 and 561020
Fax: 91 (812) 565188
Telex: 845 8083
GENERAL MANAGER: Dr C. R. Ramanujachar

Comprises Aircraft, Engine and Helicopter Design Bureaux; earlier designs have included HT-2, Pushpak, Krishak, Basant, HF-24 Marut, Kiran, HPT-32 and HTT-34. Most recent design is ALH, first flown 1992; effort also being given to design and development of LCA (see ADA entry in this section).

HAL ADVANCED LIGHT HELICOPTER (ALH)
TYPE: Multi-role twin-turbine light helicopter.
PROGRAMME: Agreement signed with MBB July 1984 to support design, development and production; design started November 1984; ground test vehicle runs began April 1991; four flying prototypes (two basic, one air force/army, one naval); first prototype (Z 3182) rolled out 29 June 1992, first flight 20 August and 'official' first flight 30 August 1992; not then fitted with ARIS; second prototype, equipped with ARIS, made its first flight 18 April 1993.
CURRENT VERSIONS: **Air force/army:** Skid gear, crashworthy fuel tanks, bulletproof supply tanks, IR and flame suppression; night attack capability; roles to include attack and SAR.
Naval: Retractable tricycle gear, harpoon decklock, foldable tailboom, pressure refuelling; fairings on fuselage sides to house mainwheels, flotation gear and batteries.
Civil: Roles to include passenger and utility transport, commuter/offshore executive, rescue/emergency medical service and law enforcement.
LAH (Light Attack Helicopter): Projected gunship version with tandem two-seat cockpit, gun turret, missile aiming system, weapon pylons and tailwheel landing gear.
CUSTOMERS: Indian government requirement for 250 or more for armed forces and Coast Guard to replace Chetaks/Cheetahs, but no firm contracts yet placed.
COSTS: Unit price of basic aircraft approx $4 million (1992).
DESIGN FEATURES: Four-blade main rotor with advanced aerofoils and sweptback tips; DASA FEL (fibre elastomer) rotor head, with blades held between pair of cruciform CFRP starplates; manual blade folding and rotor brake standard; integrated drive system transmission; four-blade bearingless crossbeam tail rotor on starboard side of fin; fixed tailplane; sweptback endplate fins offset to port; vibration damping by ARIS (anti-resonance isolation system), comprising four isolator elements between main gearbox and fuselage. Folding tailboom on naval variant.
FLYING CONTROLS: Integrated dynamic management by four-axis AFCS (actuators have manual as well as AFCS input); constant-speed rpm control, assisted by collective anticipator (part of FADEC and stability augmentation system acting through AFCS).
STRUCTURE: Main and tail rotor blades glassfibre/carbonfibre; Kevlar nosecone, crew/passenger doors, cowling, upper rear tailboom and most of tail unit; carbonfibre lower rear tailboom and fin centre panels; Kevlar/carbonfibre cockpit section; light alloy sandwich centre cabin and forward portion of tailboom.
LANDING GEAR: Non-retractable metal skid gear standard for air force/army version. Hydraulically retractable tricycle type on naval variant, with twin nosewheels and single mainwheels, latter retracting into fairings on fuselage sides which also house flotation gear and batteries; rearward retracting nose unit; harpoon decklock system. Spring skid under rear of tailboom on all versions, to protect tail rotor.
POWER PLANT: Two Turbomeca TM 333-2B turboshafts, with full authority digital electronic control (FADEC), rated at 746 kW (1,000 shp) for T-O, 788 kW (1,057 shp) max contingency and 663 kW (889 shp) max continuous. Transmission input from both engines combined through spiral bevel gears to collector gear on stub-shaft. ARIS system gives 6° of freedom damping. Power take-off from main and auxiliary gearboxes for transmission driven accessories. Pressure refuelling in naval version.
ACCOMMODATION: Flight crew of two, on crashworthy seats in military/naval versions. Main cabin seats 10 persons as standard, 14 in high density configuration. Crew door and rearward sliding passenger door on each side; clamshell cargo doors at rear of main cabin.
SYSTEMS: Electrical system standard. Hydraulic landing gear actuation on naval variant. Oxygen system.
AVIONICS: V/UHF, HF/SSB and standby UHF com radio, plus intercom. SFIM four-axis AFCS. Doppler navigation system, TAS system, ADF, radio altimeter, heading reference and IFF standard. Other avionics can include Omega nav system and weather radar.
EQUIPMENT: Depending on mission, can include two stretchers, external rescue hoist and 1,500 kg (3,307 lb) capacity cargo sling.
ARMAMENT: Cabin-side pylons for two torpedoes/depth charges or four anti-ship missiles on naval variant; on army/air force variant, these can be fitted with two anti-tank guided missiles, two pods of 68 mm rockets or two

Two views of the first prototype HAL Advanced Light Helicopter

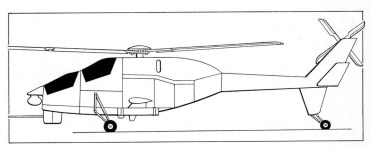

Provisional drawing of HAL's proposed light attack helicopter *(Jane's/Mike Keep)*

108 INDIA: AIRCRAFT—HAL

air-to-air missiles. Army/air force variant can also be equipped with ventral 20 mm gun turret or sling for carriage of land mines.

DIMENSIONS, EXTERNAL:
Main rotor diameter	13.20 m (43 ft 3¼ in)
Tail rotor diameter	2.55 m (8 ft 4½ in)
Length:	
overall, both rotors turning	15.87 m (52 ft 0¾ in)
fuselage	12.89 m (42 ft 3¾ in)
Height: overall, tail rotor turning	4.91 m (16 ft 1¼ in)
to top of rotor head	3.76 m (12 ft 4 in)
Tail unit span (over fins)	3.19 m (10 ft 5½ in)
Wheel track (naval version)	2.80 m (9 ft 2¼ in)
Skid track (army/air force version)	2.60 m (8 ft 6½ in)

DIMENSIONS, INTERNAL:
Cabin volume	7.33 m³ (258.9 cu ft)
Cargo compartment volume	2.16 m³ (76.28 cu ft)

AREAS:
Main rotor disc	136.85 m² (1,473.0 sq ft)
Tail rotor disc	5.11 m² (55.0 sq ft)

WEIGHTS AND LOADINGS (A: army/air force version, B: naval variant):
Weight empty, equipped: A, B	2,500 kg (5,511 lb)
Max fuel weight: A	1,040 kg (2,293 lb)
B	1,100 kg (2,425 lb)
Max sling load: A	1,000 kg (2,205 lb)
B	1,500 kg (3,307 lb)
Max T-O weight: A	4,000 kg (8,818 lb)
B	5,000 kg (11,023 lb)
Max disc loading: A	29.23 kg/m² (5.99 lb/sq ft)
B	36.54 kg/m² (7.48 lb/sq ft)

PERFORMANCE (estimated, at S/L, ISA):
Never-exceed speed (V_{NE})	164 knots (305 km/h; 189 mph)
Max level speed	151 knots (280 km/h; 174 mph)
Max cruising speed	132 knots (245 km/h; 152 mph)
Max rate of climb	540 m (1,770 ft)/min
Service ceiling	6,000 m (19,685 ft)
Hovering ceiling IGE	above 3,000 m (9,850 ft)

Range: with max fuel, 20 min reserves 431 nm (800 km; 497 miles)
with 700 kg (1,543 lb) payload 216 nm (400 km; 249 miles)
Endurance, 20 min reserves 3 h 48 min

Naval version of the HAL Advanced Light Helicopter, with additional side view (centre) of air force/army variant *(Jane's/Mike Keep)*

BANGALORE COMPLEX

Post Bag 1785, Bangalore 560 017
Telephone: 91 (812) 565201 and 561020
Telex: 845 2234
MANAGING DIRECTOR: Dr C. G. Krishnadas Nair

Aircraft Division
Post Bag 1788, Bangalore 560 017
Telephone: 91 (812) 570469
Fax: 91 (812) 561956
Telex: 845 2234
GENERAL MANAGER: B. Haridas

Helicopter Division
Post Bag 1790, Bangalore 560 017
Telephone: 91 (812) 572951
Fax: 91 (812) 560096
Telex: 845 2764
GENERAL MANAGER: U. D. Paradkar

Bangalore Complex includes Aircraft Division, Helicopter Division, Aerospace Division, Engine Division, Overhaul Division, Services Division, Foundry and Forge Division and Flight Operations. Programmes include Jaguar International and Adour engine, prototype of ADA LCA (which see), contract to supply up to 150 sets of BAe Jetstream ATP tailplanes (signed May 1987), and general repair and overhaul.

HAL (SEPECAT) JAGUAR INTERNATIONAL
Indian Air Force name: Shamsher (Assault Sword)

TYPE: Single-seat tactical attack aircraft and two-seat operational trainer.

PROGRAMME: Forty Jaguar Internationals with Adour Mk 804 engines delivered from UK; 45 more with Adour Mk 811s assembled in India; further 31 being manufactured under licence in India (first delivery early 1988); further 15 (ordered 1988) were cancelled 1989.

CURRENT VERSIONS: **Jaguar B:** Two-seat trainer version.
Jaguar S: Standard single-seat attack version; *description applies to this model, except where indicated.*
Anti-shipping Jaguar: Jaguars of IAF No. 6 Squadron, assigned to anti-shipping role, have Thomson-CSF Agave radar, Smiths Industries DARIN nav/attack system and Sea Eagle air-to-surface missiles; first modified aircraft delivered January 1986; eight delivered by end of 1992; two more to follow.

CUSTOMERS: Indian Air Force (116), comprising 101 single-seat and 15 two-seat combat-capable trainers. Basic strike version equips Nos. 5, 14, 16 and 27 Squadrons; anti-shipping version part-equips No. 6 Squadron.

DESIGN FEATURES: As British Jaguars (see International section of 1985-86 *Jane's*).

POWER PLANT: Two HAL built Rolls-Royce Turbomeca Adour Mk 811 turbofans, each rated at 24.6 kN (5,520 lb st) dry and 37.4 kN (8,400 lb st) with afterburning. Fixed geometry air intake on each side of fuselage aft of cockpit. Fuel in six tanks, one in each wing and four in fuselage. Total internal fuel capacity 4,200 litres (1,110 US gallons; 924 Imp gallons). Armour protection for critical fuel

Jaguar International of No. 14 ('Bulls') Squadron, Indian Air Force, with overwing Magic AAMs *(Peter Steinemann)*

One of the anti-shipping Jaguars of No. 6 Squadron, Indian Air Force, with nose-mounted Agave radar

system components. Provision for carrying three auxiliary drop tanks, each of 1,200 litres (317 US gallons; 264 Imp gallons) capacity, on fuselage and inboard wing pylons. Provision for in-flight refuelling, with retractable probe forward of cockpit on starboard side.

ACCOMMODATION (single-seater): Enclosed cockpit for pilot, with rearward hinged canopy and Martin-Baker IN9B zero/zero ejection seat. Windscreen bulletproof against 7.5 mm rifle fire.

ACCOMMODATION (trainer): Crew of two in tandem on Martin-Baker IN9B Mk II zero/zero ejection seats. Individual rearward hinged canopies. Rear seat 38 cm (15 in) higher than front seat. Bulletproof windscreen, as in single-seat version.

SYSTEMS: As detailed in 1985-86 *Jane's*.

AVIONICS: Include Smiths Industries DARIN (display attack and ranging inertial navigation) nav/attack system, incorporating SAGEM Uliss 82 INS, GEC-Marconi HUDWAC (head-up display and weapon aiming computer) and GEC-Marconi COMED 2045 (combined map and electronic display). HAL IFF-400 AM.

ARMAMENT: Two 30 mm Aden cannon in lower fuselage aft of cockpit in single-seater; single Aden gun on port side in two-seater. One stores attachment on fuselage centreline and two under each wing. Centreline and inboard wing points can each carry up to 1,134 kg (2,500 lb) of weapons, outboard underwing points up to 567 kg (1,250 lb) each. Typical alternative loads include one air-to-surface missile and two 1,200 litre (317 US gallon; 264 Imp gallon) drop tanks; eight 1,000 lb bombs; various combinations of free-fall, laser guided, retarded or cluster bombs; overwing Matra R.550 Magic missiles; air-to-surface rockets; or a reconnaissance camera pack.

DIMENSIONS, EXTERNAL:	
Wing span	8.69 m (28 ft 6 in)
Length overall, incl probe:	
single-seat	16.83 m (55 ft 2½ in)
two-seat	17.53 m (57 ft 6¼ in)
Height overall	4.89 m (16 ft 0½ in)
Wheel track	2.41 m (7 ft 11 in)
Wheelbase	5.69 m (18 ft 8 in)
AREAS:	
Wings, gross	24.18 m² (260.27 sq ft)
WEIGHTS AND LOADINGS:	
Typical weight empty	7,000 kg (15,432 lb)
Max external stores load (incl overwing)	4,763 kg (10,500 lb)
Normal T-O weight (single-seater, with full internal fuel and ammunition for built-in cannon)	10,954 kg (24,149 lb)
Max T-O weight with external stores	15,700 kg (34,612 lb)
Max wing loading	649.3 kg/m² (133 lb/sq ft)
Max power loading	209.9 kg/kN (2.06 lb/lb st)
PERFORMANCE:	
Max level speed at S/L	Mach 1.1 (729 knots; 1,350 km/h; 840 mph)
Max level speed at 11,000 m (36,000 ft)	Mach 1.6 (917 knots; 1,699 km/h; 1,056 mph)
Landing speed	115 knots (213 km/h; 132 mph)
T-O run: clean	565 m (1,855 ft)
with four 1,000 lb bombs	880 m (2,890 ft)
with eight 1,000 lb bombs	1,250 m (4,100 ft)
T-O to 15 m (50 ft) with typical tactical load	940 m (3,085 ft)
Landing from 15 m (50 ft) with typical tactical load	785 m (2,575 ft)
Landing run:	
normal weight, with brake-chute	470 m (1,540 ft)
normal weight, without brake-chute	680 m (2,230 ft)
overload weight, with brake-chute	670 m (2,200 ft)
Typical attack radius, internal fuel only:	
hi-lo-hi	460 nm (852 km; 530 miles)
lo-lo-lo	290 nm (537 km; 334 miles)
Typical attack radius with external fuel:	
hi-lo-hi	760 nm (1,408 km; 875 miles)
lo-lo-lo	495 nm (917 km; 570 miles)
Ferry range with external fuel	1,902 nm (3,524 km; 2,190 miles)
g limits	+8.6 (+12 ultimate)

HAL (AEROSPATIALE) SA 315B LAMA

Indian name: Cheetah

TYPE: Turbine powered general purpose helicopter.

PROGRAMME: Design begun for Indian armed forces late 1968; first flight 17 March 1969; French certification 30 September 1970; FAA certification 25 February 1972; production by HAL for Indian Army started 1972 and continuing.

CUSTOMERS: Total 407 French built Lamas delivered (production ended); Helibras (which see) assembled some in Brazil (designation Gavião); total of 197 delivered by HAL by 31 March 1991 (latest figure supplied).

DESIGN FEATURES: Three-blade main and tail rotors; articulated main rotor hub; hydraulic drag dampers and inter-blade restraint cables; rotor brake standard; blades can be folded; main rotor rpm 353; tail rotor rpm 2,001; main blade section NACA 63A (constant chord).

FLYING CONTROLS: Fully powered hydraulic with adjustable friction damper on stick in place of cyclic trim; fixed-shaft engine runs at constant speed with transmission; power adjustment by means of fuel flow variation; fixed tailplane; freewheel for autorotation.

Indian Army HAL Cheetah (licence built SA 315B Lama)

STRUCTURE: Light metal cabin structure; steel tube truss centre-section and tailboom; main rotor blades have aluminium alloy spar and skin wrapped round Moltoprene block filling; stainless steel leading-edge strip; tail rotor blades hollow aluminium alloy sheet aerofoil with leading-edge protective strip.

LANDING GEAR: Skid type, with removable wheels for ground manoeuvring. Pneumatic floats for normal operation from water, and emergency flotation gear, inflatable in the air, are available.

POWER PLANT: One 640 kW (858 shp) HAL built Turbomeca Artouste IIIB turboshaft, derated to 404 kW (542 shp). Fuel tank in fuselage centre-section, with capacity of 575 litres (152 US gallons; 126.5 Imp gallons), of which 573 litres (151.5 US gallons; 126 Imp gallons) are usable. Oil capacity 7 litres (1.85 US gallons; 1.55 Imp gallons).

ACCOMMODATION: Glazed cabin seats pilot and co-pilot or passenger side by side in front and three passengers behind. Jettisonable door on each side. Provision for external sling for loads of up to 1,000 kg (2,204 lb). Can be equipped for rescue (hoist capacity 160 kg; 352 lb), liaison, observation, training, agricultural, photographic and other duties. As an ambulance, can accommodate two stretchers and a medical attendant. Cabin heating optional.

SYSTEMS: Single hydraulic system. Electrical system includes engine starter/generator, 36Ah battery and external power socket. Oxygen system optional.

DIMENSIONS, EXTERNAL:	
Main rotor diameter	11.02 m (36 ft 1¾ in)
Tail rotor diameter	1.91 m (6 ft 3¼ in)
Distance between rotor centres	6.435 m (21 ft 1½ in)
Main rotor blade chord (constant)	0.35 m (13.8 in)
Length:	
overall, both rotors turning	12.91 m (42 ft 4¼ in)
fuselage	10.23 m (33 ft 6¾ in)
Height overall	3.09 m (10 ft 1¾ in)
Skid track	2.38 m (7 ft 9¾ in)
DIMENSIONS, INTERNAL:	
Cabin: Length	2.10 m (6 ft 10½ in)
Max width	1.40 m (4 ft 7 in)
Max height	1.28 m (4 ft 2¼ in)
Volume	3.10 m³ (109.5 cu ft)
AREAS:	
Main rotor disc	95.38 m² (1,026.7 sq ft)
Tail rotor disc	2.87 m² (30.84 sq ft)
WEIGHTS AND LOADINGS:	
Weight empty	995 kg (2,193 lb)
Max T-O weight: normal	1,750 kg (3,858 lb)
with externally slung cargo	1,850 kg (4,078 lb)
Max disc loading: normal	18.35 kg/m² (3.76 lb/sq ft)
with externally slung cargo	19.40 kg/m² (3.97 lb/sq ft)
PERFORMANCE (at max normal T-O weight at S/L):	
Never-exceed speed (VNE)	113 knots (210 km/h; 130 mph)
Max cruising speed	103 knots (192 km/h; 119 mph)
Max rate of climb	330 m (1,080 ft)/min
Service ceiling	6,400 m (21,000 ft)
Range with max fuel	296 nm (550 km; 341 miles)
Endurance	3 h 30 min

HAL (AEROSPATIALE) SA 316B ALOUETTE III

Indian name: Chetak

TYPE: Turbine powered general purpose helicopter.

PROGRAMME: First flight (original French Alouette III) 28 February 1959; remains in production only in India.

CUSTOMERS: More than 1,450 produced in France; ICA Brasov (now IAR) produced 200 in Romania; HAL output had totalled 322 by 31 March 1991 (latest figure supplied).

DESIGN FEATURES: Three-blade main and tail rotors; articulated main rotor hub has hydraulic drag dampers plus inter-blade restraint cables; constant chord blades can be folded; rotor brake standard.

FLYING CONTROLS: Hydraulic fully powered controls; cyclic trimming by means of adjustable friction damper on stick; fixed-shaft engine runs at constant speed with

Civil registered HAL Chetak (licence built Alouette III)

transmission; power adjustment by means of fuel flow variation; fixed tailplane with endplate fins.
STRUCTURE: Steel tube centre-fuselage covered with fairings; semi-monocoque tailboom; all-metal blades.
LANDING GEAR: Non-retractable tricycle type, manufactured under Messier-Bugatti licence. Hydraulic shock absorption. Nosewheel is fully castoring. Provision for skis or emergency pontoon landing gear.
POWER PLANT: One 649 kW (870 shp) HAL built Turbomeca Artouste IIIB turboshaft, derated to 410 kW (550 shp) for max continuous operation. Fuel in single tank in fuselage centre-section, with capacity of 575 litres (152 US gallons; 126.5 Imp gallons), of which 573 litres (151 US gallons; 126 Imp gallons) are usable.
ACCOMMODATION: Normal accommodation for pilot and six persons, with three seats in front and four-person folding seat at rear of cabin. Two baggage holds in centre-section, on each side of welded structure and enclosed by centre-section fairings. Provision for carrying two stretchers athwartships at rear of cabin, and two other persons, in addition to pilot. All passenger seats removable to enable aircraft to be used for freight carrying. Can also be adapted for cropspraying or aerial survey roles. Provision for external sling for loads of up to 750 kg (1,650 lb). One forward opening door on each side, immediately in front of two rearward sliding doors. Dual controls and cabin heating optional.
ARMAMENT: In assault role, can be equipped with wide range of weapons. A 7.62 mm machine gun (with 1,000 rds) can be mounted athwartships on tripod behind pilot's seat, firing to starboard, either through small window in sliding door or through open doorway with door locked open. Rear seat is removed to allow gun mounting to be installed. In this configuration, max accommodation is for pilot, copilot, gunner and one passenger, although normally only pilot and gunner carried. Alternatively, a 20 mm cannon (with 480 rds) can be carried on open turret-type mounting on port side of cabin. For this installation all seats except that of pilot are removed, as is port side cabin door, and crew consists of pilot and gunner. Instead of these guns, aircraft can be equipped with two or four wire-guided missiles on external jettisonable launching rails, gyro-stabilised sight, or 68 mm rocket pods.

DIMENSIONS, EXTERNAL:
Main rotor diameter	11.02 m (36 ft 1¾ in)
Main rotor blade chord (each)	0.35 m (13.8 in)
Tail rotor diameter	1.912 m (6 ft 3¼ in)
Spraybar span (agricultural version)	10.00 m (32 ft 9¾ in)
Length: overall, rotors turning	12.84 m (42 ft 1½ in)
fuselage, tail rotor turning	10.17 m (33 ft 4½ in)
Width overall, blades folded	2.60 m (8 ft 6¼ in)
Height to top of rotor head	2.97 m (9 ft 9 in)
Wheel track	2.602 m (8 ft 6½ in)

AREAS:
Main rotor disc	95.38 m² (1,026.6 sq ft)
Tail rotor disc	2.87 m² (30.9 sq ft)

WEIGHTS AND LOADINGS:
Weight empty, standard	1,230 kg (2,711 lb)
Max T-O weight	2,200 kg (4,850 lb)
Max disc loading	23.07 kg/m² (4.72 lb/sq ft)

PERFORMANCE (standard version at max T-O weight):
Never-exceed speed (V_{NE}) at S/L	113 knots (210 km/h; 130 mph)
Max cruising speed at S/L	100 knots (185 km/h; 115 mph)
Max rate of climb at S/L	260 m (850 ft)/min
Service ceiling	3,250 m (10,675 ft)
Hovering ceiling: IGE	2,850 m (9,350 ft)
OGE	1,500 m (4,920 ft)
Range with max fuel at S/L	257 nm (477 km; 296 miles)
Endurance	3 h

ACCESSORIES COMPLEX
PO Box 5150, 15/1 Cubbon Road, Bangalore 560 001
Telephone: 91 (812) 266901
Fax: 91 (812) 268758 and 577533
Telex: 845 2266 HAL IN
MANAGING DIRECTOR: H. C. Cholay

Previously comprised Korwa and Hyderabad Divisions (avionics) and Lucknow Divison (other accessories); now also includes Kanpur Division.

KANPUR DIVISION
Post Bag 225, Kanpur 208 008
Telephone: 91 (512) 42488 and 44021/5
Fax: 91 (512) 42344
Telex: 325 243 HALK IN
GENERAL MANAGER: G. S. Singhal

HAL (HAWKER SIDDELEY) HS 748
Development of airborne surveillance, warning and control (ASWAC) version of Kanpur built HS 748s started 1985; testbed flown 5 November 1990 with empty 4.8 m (15 ft 9 in) diameter rotating rotodome produced by DASA (MBB), which would provide technical assistance in any eventual production; programme run by Indian Defence Research and Development Organisation; avionics designed by Indian Electronics and Radar Development Establishment; manufacture by BEL (Bharat Electronics Ltd); foreign participation in radar design expected.

HAL (DORNIER) 228
TYPE: Multi-role civil/military turboprop twin.
PROGRAMME: Contract for licence manufacture of up to 150 Dornier 228s over 10 years signed 29 November 1983; some aircraft supplied complete from Germany; first flight Kanpur built aircraft 31 January 1986.
CURRENT VERSIONS: **Regional Airliner:** Five 228-201s delivered to Vayudoot March 1986.
Maritime Surveillance: Thirty-three 228-101s for Indian Coast Guard (after three from Germany); deliveries started 1987; in service with Nos. 744 and 750 Squadrons at Daman and Madras; for coastal, environmental and antismuggling patrol; 360° THORN EMI Marec 2 search radar under fuselage (being replaced in 15 aircraft by Super Marec ordered in 1987), Litton Omega navigation system, Swedish Space Corporation infra-red/ultra-violet linescanner for pollution detection, search and rescue liferafts, one million candlepower searchlight, side-mounted loudhailer, marine markers, and provision for two Micronair underwing spraypods to combat oil spills; sliding main cabin door for airdropping six- or 10-man liferafts. Normal crew two pilots, radar operator and observer. Optional armament includes two 7.62 mm twin-gun machine-gun pods or underwing air-to-surface missiles.
Anti-ship: Indian Navy plans to acquire 27 specially equipped Dornier 228-202s with anti-ship missiles and Super Marec radar.
Utility transport: Indian Air Force acquiring 43 228-202s; deliveries started 1987; replacing Otters, Devons and C-47s of Nos. 41 and 59 Squadrons; carry 21 field-equipped troops on inward-facing folding seats; large roller door at rear, port side.
Executive/Air taxi: Various configurations including six- or 10-seat executive or 15-passenger air taxi, with cabin attendant, and galley/wardrobe/toilet; built-in APU for air-conditioning and lighting in flight or on ground.
CUSTOMERS: See above; total 36 delivered by December 1991 (latest figure supplied); one export order from Mauritius 1990.

HAL HPT-32
TYPE: Two-seat primary and basic trainer and multi-role aircraft.
PROGRAMME: First flight (X2157) 6 January 1977; first flight of first production aircraft (third built) 31 July 1981; production started late 1987 or early 1988; domestic DGCA civil certification received 25 November 1991.
CUSTOMERS: Indian Air Force and Navy initial orders for 88 all delivered; follow-on batch of 32 due for delivery from 1993.
DESIGN FEATURES: Roles include instrument, aerobatic and night flying training, glider or target towing, weapon training and light strike, SAR, supply dropping and reconnaissance, as well as primary and basic training. Tapered straight wing; section NACA 64A₁-212; dihedral 5° from roots; incidence 2° 30′ at roots.
FLYING CONTROLS: Conventional mechanical; one-piece elevator has geared tab in port half and trim tab in starboard half; rudder trim tab; geared tab and ground adjustable tab in each aileron; plain flaps.
STRUCTURE: Fail-safe all-metal stressed skin structure.
LANDING GEAR: Non-retractable tricycle type, with HAL oleo-pneumatic shock absorber in each unit. Dunlop UK single mainwheels and nosewheel. Dunlop UK mainwheel tyres, size 446 × 151 × 166 mm, pressure 3.10 bars (45 lb/sq in); Dunlop India nosewheel tyre, size 361 × 126 × 127 mm, pressure 2.41 bars (35 lb/sq in). Dunlop UK air-cooled hydraulic disc brakes on mainwheels. Min ground turning radius 6.50 m (21 ft 4 in).
POWER PLANT: One 194 kW (260 hp) Textron Lycoming AEIO-540-D4B5 flat-six engine, driving a Hartzell two-blade constant-speed metal propeller. Total of 220 litres (58.1 US gallons; 48.4 Imp gallons) of fuel in four flexible tanks (two in each wing), plus a 9 litre (2.4 US gallon; 2 Imp gallon) metal collector tank in fuselage. Total fuel capacity 229 litres (60.5 US gallons; 50.4 Imp gallons). Overwing refuelling points. Oil capacity 13.6 litres (3.6 US gallons; 3 Imp gallons).
ACCOMMODATION: Side by side seats for two persons under rearward sliding jettisonable framed canopy. Seats adjustable in height by 127 mm (5 in). Full dual controls, and adjustable rudder pedals, for instructor and pupil. Cockpit ventilated.
SYSTEMS: Hydraulic system for brakes only. Electrical system (28V DC earth return type) powered by 70A alternator, with SAFT 24V nickel-cadmium standby battery. No air-conditioning, pneumatic, de-icing or oxygen systems.
AVIONICS: HAL (Hyderabad Divn) COM-150 main UHF and COM-104A standby VHF com; directional gyro. Blind-flying instrumentation.

DIMENSIONS, EXTERNAL:
Wing span	9.50 m (31 ft 2 in)
Wing chord: at root	2.24 m (7 ft 4¼ in)
at tip	0.92 m (3 ft 0¼ in)
Wing aspect ratio	6.0
Length overall	7.72 m (25 ft 4 in)
Fuselage: Max width	1.25 m (4 ft 1¼ in)
Height overall	2.88 m (9 ft 5½ in)
Tailplane span	3.60 m (11 ft 9¾ in)
Wheel track	3.45 m (11 ft 4 in)
Wheelbase	2.10 m (6 ft 10¾ in)
Propeller diameter	2.03 m (6 ft 8 in)
Propeller ground clearance (static)	0.23 m (9 in)

AREAS:
Wings, gross	15.00 m² (161.5 sq ft)
Ailerons (total)	1.04 m² (11.19 sq ft)
Trailing-edge flaps (total)	1.82 m² (19.59 sq ft)

HAL built Dornier 228-101 of the Indian Coast Guard

Vertical tail surfaces (above fuselage reference line)
　　　　　　　　　　　　　　　2.06 m² (22.17 sq ft)
Rudder (aft of hinge line), incl tabs 0.869 m² (9.35 sq ft)
Tailplane　　　　　　　　　　　3.024 m² (32.55 sq ft)
Elevator (aft of hinge line), incl tabs
　　　　　　　　　　　　　　　1.34 m² (14.42 sq ft)
WEIGHTS AND LOADINGS:
Basic weight empty　　　　　　　　890 kg (1,962 lb)
Fuel and oil (guaranteed minimum)　164 kg (361 lb)
Max T-O and landing weight　　　1,250 kg (2,756 lb)
Max wing loading　　　83.33 kg/m² (17.07 lb/sq ft)
Max power loading　　　6.44 kg/kW (10.60 lb/hp)
PERFORMANCE (at max T-O weight, ISA):
Never-exceed speed (V$_{NE}$) (structural)
　　　　　　　　　　　240 knots (445 km/h; 276 mph)
Max level speed at S/L
　　　　　　　　　143 knots (265 km/h; 164 mph) IAS
Max cruising speed at 3,050 m (10,000 ft)
　　　　　　　　　　　115 knots (213 km/h; 132 mph)
Econ cruising speed　　95 knots (176 km/h; 109 mph)
Stalling speed, 20° flap, engine idling
　　　　　　　　　　　　60 knots (110 km/h; 69 mph)
Max rate of climb at S/L　　　335 m (1,100 ft)/min
Service ceiling　　　　　　　　5,500 m (18,045 ft)
T-O run　　　　　　　　　　　　345 m (1,132 ft)
T-O to 15 m (50 ft)　　　　　　　545 m (1,788 ft)
Landing from 15 m (50 ft)　　　　487 m (1,598 ft)
Landing run　　　　　　　　　　220 m (720 ft)
Range at 3,050 m (10,000 ft) at econ cruise power
　　　　　　　　　　　401 nm (744 km; 462 miles)
g limits　　　　　　　　　　　　　　　　　+6/−3

HAL HPT-32 two-seat basic trainers of the Indian Air Force *(Peter Steinemann)*

MiG COMPLEX
PO Box 5150, 15/1 Cubbon Road, Bangalore 560 001
Telephone: 91 (812) 266901
Fax: 91 (812) 268758 and 577533
Telex: 845 2266 HAL IN
MANAGING DIRECTOR: H. K. L. Anand

Complex originally contained Nasik, Koraput and Hyderabad Divisions of HAL, respectively building airframes, engines and avionics of MiG-21 under licence; Indian MiG-21 production phased out 1986-87 as production of MiG-27M increased; Hyderabad Division now part of Accessories Complex.

NASIK DIVISION
Ojhar Township Post Office, Nasik 422 207, Maharashtra
Telephone: 91 (253) 78117 and 77901/10
Fax: 91 (253) 77907
Telex: 075 241
GENERAL MANAGER: N. R. Mohanty

HAL (MIKOYAN) MiG-21
Reported October 1992 that Indian Air Force considering upgrade of approx 100 MiG-21bis to offset delays in LCA programme. Probable improvements would include more powerful engine (GE F404 preferred), improved avionics (including AN/APG-66 radar), greater firepower and strengthened airframe. Estimated programme cost Rs400 million ($1.33 million).

HAL (MIKOYAN) MiG-27M
Indian name: Bahadur (Valiant)
NATO reporting name: Flogger-J
TYPE: Variable-sweep attack fighter.
PROGRAMME: HAL began licence assembly of MiG-27M 1984; first aircraft rolled out October 1984; Indian components thought to be included from 1988; entered IAF service 11 January 1986; plan for mid-life update and continuation of production to more than 200.
CURRENT VERSIONS: **MiG-27M**: Current aircraft, as Russian 'Flogger-J'.
Mid-life update: Will include Smiths/SAGEM DARIN nav/attack system as in IAF Jaguars.
CUSTOMERS: MiG-27M joined MiG-23s and -27s supplied direct from USSR; approx 100 so far completed; original 165 to be increased to over 200. Currently equips six of planned eight IAF squadrons (Nos. 2, 9, 18, 22, 31 and 222).
DESIGN FEATURES: As MiG-27M.
POWER PLANT: One Soyuz (Tumansky) R-29B-300 turbojet, rated at 81.39 kN (18,298 lb st) dry and 112.78 kN (25,353 lb st) with afterburning.
AVIONICS: Include PrNK-23M integrated nav/attack system and active/passive ECM systems.
ARMAMENT: One ventrally mounted six-barrel 30 mm cannon, and seven external hardpoints for up to 3,000 kg (6,614 lb)

HAL built Indian Air Force MiG-27M attack fighter of No. 9 ('Wolfpack') Squadron *(Peter Steinemann)*

of stores which can include 500 kg bombs, 57 mm S-24 rockets, two 'Kerry' air-to-surface or four R-60 air-to-air missiles.
WEIGHTS AND LOADINGS:
Max T-O weight　　　　　　　　18,000 kg (39,685 lb)
PERFORMANCE:
T-O run at S/L　　　　　　　　　　　800 m (2,625 ft)
Combat radius (low level)　210 nm (390 km; 242 miles)
Ferry range　　　　　　1,349 nm (2,500 km; 1,553 miles)

NAL
NATIONAL AERONAUTICAL LABORATORY
PO Bag 1779, Kodihalli, Bangalore 560 017
Telephone: 91 (812) 573351/4
Telex: 279 NAL IN
DIRECTOR: Prof R. Narasimha

LIGHT TRANSPORT AIRCRAFT (LTA)
TYPE: Twin-turboprop transport.
PROGRAMME: Contract for design and prototype construction of 14-seat general purpose transport; three prototypes planned; configuration still undergoing changes late 1992; possible Russian (Myasishchev) participation to share costs of Rs860 million ($27.5 million) programme.

TAAL
TANEJA AEROSPACE AND AVIATION LTD
Hosur, near Bangalore

Agreement signed Spring 1992 with AerCosmos of Milan to assemble Partenavia P.68 and Viator light twins in India at initial rate of five per year; $9 million programme launched 1992 to build new factory at Hosur, with first Indian assembled aircraft due for completion late 1993.

Taneja also reported to be participating in NAL project to develop indigenous two-seat trainer.

INDONESIA

CRSS
P. T. CIPTA RESTU SARANA SVAHA
R. E. Martadinata, Sekupang, Pulau Batam
Telephone: 62 (778) 321 758
Fax: 62 (778) 321 467
PRESIDENT: R. Siregar

Acquired assets (including AA200 prototype and AA300 moulds and plugs) of former Aerodis America Inc (see pages 336-7 of 1991-92 *Jane's*) in November 1991; intended continuing work on AA200 but has temporarily suspended AA300/AA330 programmes.

CRSS AA200 ORION
TYPE: Four-seat piston engined light aircraft.
PROGRAMME: David B. Thurston design, started 1982 for Aerodis America Inc; prototype construction began July 1989; first flight 7 April 1991; meets or exceeds FAR Pt 23 Amendment 34; claimed to be unrelated to French Grinvalds G-802 Orion homebuilt (Sport Aircraft section, 1988-89 *Jane's*); work being continued by CRSS with objective of certification in a European country, possibly Germany.
DESIGN FEATURES: Piston engine mounted aft of cabin, driving pusher propeller at tail through extension shaft; wing section NASA GA40U-A215 at root and GA37U-A212 at tip; dihedral 3°; incidence 1° 30′ at root.
FLYING CONTROLS: Frise-type ailerons, with differential deflection and actuated by pushrods; one-piece elevator carries electrically actuated trim tab at centre; rudder in ventral as well as in main fin; electrically actuated slotted flaps, maximum deflection 30°; dorsal and ventral fins; fixed tailplane.
STRUCTURE: Damage-tolerant two-spar wing with D-section leading-edge torsion box and of graphite epoxy skins on Nomex core and graphite spars; ailerons and flaps of Rohacell foam/glassfibre; wings have quick-detach fittings in channels across the fuselage; damage-tolerant fuselage of graphite/glassfibre/Nomex honeycomb/Rohacell foam, made in two halves and bonded at centreline; stringers and seven bulkheads; roll-over protection over cabin; T tail of honeycomb core with glassfibre skins.
LANDING GEAR: Electrohydraulically retractable tricycle type with steerable nosewheel. Mainwheels retract inward into wing, nosewheel forward. Oleo-strut shock absorbers in all units. Cleveland mainwheels, size 6.00-6, pressure 3.45 bars (50 lb/sq in). Cleveland nosewheel, size 5.00-5, pressure 2.76 bars (40 lb/sq in). Cleveland disc brakes.
POWER PLANT: One 134 kW (180 hp) Textron Lycoming O-360-A4A flat-four engine, mounted to rear of cabin and driving a three-blade constant-speed composite MT pusher propeller via a graphite/epoxy driveshaft and flexidyne coupler; 149 kW (200 hp) Textron Lycoming IO-360-A1B6 engine optional. Fuel contained in integral wing leading-edge tanks, maximum capacity 227 litres (60 US gallons; 50 Imp gallons). Fuel filler cap in upper surface of each wing. Oil capacity 7.6 litres (2 US gallons; 1.67 Imp gallons).
ACCOMMODATION: Enclosed cabin with four individual seats. Baggage compartment aft of seats. Single door on port side of fuselage, horizontally divided, upper half opening upwards while lower portion contains integral entrance step. Emergency exit via forward cabin window on starboard side. Cabin is stressed to withstand 20g impact. Inertia-reel shoulder harnesses and seatbelts, cabin soundproofing, heating and ventilation, standard.
SYSTEMS: Electrical system supplied by 28V 60A alternator.
AVIONICS: IFR avionics to customer choice.
EQUIPMENT: Dual controls (sticks) optional.
DIMENSIONS, EXTERNAL:
Wing span 9.13 m (29 ft 11½ in)
Wing chord: at root 1.60 m (5 ft 3 in)
    at tip 1.12 m (3 ft 8 in)
Wing aspect ratio 6.55
Length overall 7.75 m (25 ft 5 in)
Height overall 2.62 m (8 ft 7 in)
Tailplane span 3.30 m (10 ft 10 in)
Wheel track 3.23 m (10 ft 7 in)
Wheelbase 2.86 m (9 ft 4½ in)
Propeller diameter 1.88 m (6 ft 2 in)
Propeller ground clearance 0.53 m (1 ft 8¾ in)
Passenger door: Height 1.00 m (3 ft 3¼ in)
    Width 1.00 m (3 ft 3¼ in)
    Height to sill 0.65 m (2 ft 1½ in)
DIMENSIONS, INTERNAL:
Cabin: Length 2.48 m (8 ft 1½ in)
    Max width 1.07 m (3 ft 6 in)
    Max height 1.14 m (3 ft 8¾ in)
    Floor area 2.46 m² (26.43 sq ft)
    Volume 2.62 m³ (92.5 cu ft)
AREAS:
Wings, gross 12.73 m² (137.0 sq ft)
Ailerons (total) 0.99 m² (10.67 sq ft)
Trailing-edge flaps (total) 1.68 m² (18.07 sq ft)
Vertical tail surfaces (total) 1.55 m² (16.7 sq ft)
Tailplane 3.14 m² (33.8 sq ft)
Elevator, incl tab 1.21 m² (13.0 sq ft)
WEIGHTS AND LOADINGS:
Weight empty 635 kg (1,400 lb)
Max fuel weight 163 kg (360 lb)
Max T-O weight 1,134 kg (2,500 lb)
Max wing loading 89.1 kg/m² (18.25 lb/sq ft)
Max power loading 8.46 kg/kW (13.89 lb/hp)
PERFORMANCE (estimated, at max T-O weight, ISA):
Never-exceed speed (V_{NE})
    190 knots (352 km/h; 218 mph)
Max level speed 180 knots (334 km/h; 207 mph)
Max cruising speed at 2,285 m (7,500 ft)
    165 knots (306 km/h; 190 mph)
Stalling speed, landing gear and flaps down
    54 knots (100 km/h; 62 mph)
Max rate of climb at S/L 311 m (1,020 ft)/min
Service ceiling 4,420 m (14,500 ft)
Range with max fuel 780 nm (1,445 km; 898 miles)

CRSS AA300 RIGEL and AA330 THETA
TYPE: Two-seat military or civil jet trainer (AA300) and single-seat tactical aircraft (AA330).
PROGRAMME: Two Rigel prototypes began construction in 1991 (one with each engine option); Theta design started September 1988. Moulds and plugs of AA300 now in Indonesia, but work on both models suspended for time being; work on AA300 prototype to be resumed after certification of AA200.
DESIGN FEATURES: Based on Orion airframe; detachable rear fuselage to facilitate engine change and servicing; both aircraft stressed for aerobatics; stores hardpoints on fuselage centreline and under each wing.
FLYING CONTROLS: Ventral fin and rudder replaced by small ventral strake; otherwise as AA200.
LANDING GEAR: Mainwheel size 18.5 × 5.5. Tyre pressures: mainwheels 5.86 bars (85 lb/sq in); nosewheel 2.07 bars (30 lb/sq in). Maximum nosewheel steering angle ±30°.
POWER PLANT: One 8.45 kN (1,900 lb st) Garrett TFE109-3 or Williams FJ44 turbofan, derated to 7.11 kN (1,600 lb st). Maximum internal fuel capacity 757 litres (200 US gallons; 166.5 Imp gallons). Refuelling point in each wingtip.
ACCOMMODATION: Two in tandem in pressurised and air-conditioned cockpit in Rigel, one in Theta, on Martin-Baker Mk 15 lightweight ejection seats. Upward opening bubble canopy, electrohydraulically actuated. Headroom raised 10 cm (4 in). Windscreen arch tilted 6 cm (2.4 in) forward to improve rear-seat view. Sliding canopy optional on Theta.
SYSTEMS: 28V 300A electrical system. Pressurisation system, maximum differential 0.45 bar (6.5 lb/sq in). Oxygen system standard.
DIMENSIONS, EXTERNAL:
Wing span 8.85 m (29 ft 0½ in)
Wing aspect ratio 6.39
Length overall 7.90 m (25 ft 11 in)
Height overall 2.64 m (8 ft 8 in)
Tailplane span 3.30 m (10 ft 10 in)
Wheel track 2.95 m (9 ft 8 in)
Wheelbase 3.16 m (10 ft 4½ in)
DIMENSIONS, INTERNAL:
Cockpit (Rigel): Length 2.81 m (9 ft 2½ in)
    Max width 0.78 m (2 ft 6½ in)
    Max height 1.12 m (3 ft 8 in)
    Floor area 2.00 m² (21.5 sq ft)
AREAS:
Wings, gross 12.26 m² (132.0 sq ft)
Fin 1.05 m² (11.26 sq ft)
Rudder 0.365 m² (3.93 sq ft)
WEIGHTS AND LOADINGS:
Weight empty 1,030 kg (2,270 lb)
Max T-O weight 1,860 kg (4,100 lb)
Max wing loading 151.65 kg/m² (31.06 lb/sq ft)
Max power loading 220.12 kg/kN (2.16 lb/lb st)
PERFORMANCE (estimated, at max T-O weight, ISA):
Max level speed at 9,150 m (30,000 ft)
    370 knots (686 km/h; 426 mph)
Max cruising speed at 9,150 m (30,000 ft)
    335 knots (621 km/h; 386 mph)
Econ cruising speed at 9,150 m (30,000 ft)
    220 knots (408 km/h; 253 mph)
Stalling speed, landing gear and flaps down
    68 knots (126 km/h; 79 mph)
Max rate of climb at S/L 1,067 m (3,500 ft)/min
T-O to 15 m (50 ft) 314 m (1,030 ft)
Landing from 15 m (50 ft) 752 m (2,466 ft)
Range with max internal fuel, 45 min reserves
    1,050 nm (1,945 km; 1,209 miles)
g limits +9/−6

First prototype of CRSS AA200 Orion (Textron Lycoming O-360-A4A)

IPTN
INDUSTRI PESAWAT TERBANG NUSANTARA (Nusantara Aircraft Industries Ltd)
PO Box 1562, Jalan Pajajaran 154, Bandung 40174
Telephone: 62 (22) 633900 and 632145
Fax: 62 (22) 631696 and 633912
Telex: 28295 IPTN BD IA
HEAD OFFICE: 14th Floor, Bumi Daya Plaza Building, Jalan Imam Bonjol 61, Jakarta 10310
Telephone: 62 (21) 322395, 322232 or 322247
Fax: 62 (21) 3100081
Telex: 61246 IPTN JKT IA
PRESIDENT DIRECTOR: Prof Dr-Ing B. J. Habibie
CHIEF ENGINEER: Ir Djoko Sartono
PRESS RELATIONS MANAGER: Soleh Affandi

Originally formed by Indonesian government as PT Industri Pesawat Terbang Nurtanio (Nurtanio Aircraft Industry Ltd) 23 August 1976 to centralise all aerospace facilities in one company; capital formed by combining assets of Pertamina Advanced Technology and Aeronautical Division and Nurtanio Aircraft Industry (LIPNUR: see 1977-78 *Jane's*); present name adopted late 1985; weapon system division in Menang Tasikmalaya, West Java, develops and produces weaponry for IPTN military aircraft. Total of approx 300 aircraft and helicopters delivered by early 1992, of which nearly 20 per cent exported.

IPTN developed and manufactures CN-235 with CASA (see Airtech in International section); produces CASA NC-212 Aviocar, NBO-105 and NAS-332 Super Puma under licence; produces components for Boeing 737 and 767, Fokker 100, Lockheed Fort Worth F-16 and P&W engines; 1,393 m² (15,000 sq ft) power plant centre can maintain, overhaul and repair Allison 250, P&WC PT6A, Textron Lycoming LTS101, Turbomeca Makila, Garrett TPE331 and General Electric CT7. Agreement mid-1991 with BAe to collaborate on production of Hawks for Indonesian Air Force.

Workforce 15,500 in 1990 (latest figure received); factory area 69 ha (170.5 acres) including 600,000 m² (6,458,340 sq ft) covered area.

AIRTECH (CASA/IPTN) CN-235

See details of this joint programme under Airtech in International section.

IPTN N-250-100

TYPE: Twin-turboprop, pressurised 64/68-passenger regional transport.

PROGRAMME: First metal cut August 1992; rollout planned November 1994; first flight 1995; certification (FAR/JAR 25) 1996; four prototypes being built.

CURRENT VERSIONS: **N-250**: First prototype only; *description applies to this aircraft*.

N-250-100: Production version; *see Addenda for details*.

CUSTOMERS: Letters of intent from Merpati (65), Bouraq (62), Sempati (six) and FFV Aerotech (24): total 157.

DESIGN FEATURES: First fully indigenously designed transport; compared with CN-235, has larger fuselage cross-section, longer cabin and no rear-loading ramp; high wing has parallel chord centre-section and slightly swept outer wings; large fairing from wingroot to base of fin; T tail.

FLYING CONTROLS: Lucas/Liebherr fly-by-wire control of flaps, ailerons, spoilers, elevators and rudder; backup system for ailerons and elevators. Fixed tailplane; horn balanced elevators have two tabs each side, short-span ailerons have two tabs each and are supported by two spoiler panels each side; long-span fixed-vane double-slotted flaps.

STRUCTURE: Includes composites sandwich for flaps, spoilers, ailerons, elevators, rudder, wing/fin/tailplane tips and leading-edges, wingroot fairings, dorsal fin, tailcone, cowlings, radome and landing gear doors.

LANDING GEAR: Messier-Bugatti retractable tricycle type, with twin wheels on each unit. Main units retract into fairings on fuselage sides.

POWER PLANT: Two 2,386 kW (3,200 shp) Allison GMA 2100C turboprops (2,622 kW; 3,517 shp with APR), each driving a Dowty 110AF ARAD/A R384 six-blade propeller. Engines will have Lucas Aerospace FADEC.

ACCOMMODATION: Cabin seating (four-abreast with central aisle) for 50 tourist class passengers at 81 cm (32 in) pitch or 54 passengers at 76 cm (30 in) pitch. One or two cabin attendants. Storage compartments at front of cabin on port side and at rear on starboard side; galley and toilet at rear. Passenger door at front on port side, and service door at rear on starboard side. Type III emergency exits at front (starboard) and rear (port). Large baggage compartment aft of main cabin, with door on port side. Additional bulk storage in underfloor compartment, also with external access.

SYSTEMS: Hamilton Standard R90-3WR environmental control system. Flight Refuelling fuel management system.

AVIONICS: Rockwell Collins five-CRT Pro Line IV system with EICAS standard, plus digital air data and attitude/heading reference and solid-state weather radar. Options include Collins TWR-850 turbulence weather radar, TCAS-94, FMS-4050 flight management system, and sixth CRT display.

DIMENSIONS, EXTERNAL:
Wing span	28.00 m (91 ft 10¼ in)
Wing aspect ratio	12.06
Length overall	26.30 m (86 ft 3½ in)
Fuselage: Length	25.25 m (82 ft 10 in)
Max diameter	2.90 m (9 ft 6¼ in)
Height overall	8.37 m (27 ft 5½ in)
Tailplane span	9.40 m (30 ft 10 in)
Propeller diameter	3.81 m (12 ft 6 in)
Passenger door (fwd, port): Height	1.75 m (5 ft 9 in)
Width	0.75 m (2 ft 5½ in)
Service door (rear, stbd): Height	1.75 m (5 ft 9 in)
Width	0.75 m (2 ft 5½ in)
Baggage doors:	
Height: fwd	0.94 m (3 ft 1 in)
rear	1.25 m (4 ft 1¼ in)
Width: fwd	0.85 m (2 ft 9½ in)
rear	1.20 m (3 ft 11¼ in)

DIMENSIONS, INTERNAL:
Cabin: Length	13.23 m (43 ft 5 in)
Max width	2.70 m (8 ft 10¼ in)
Width at floor	2.41 m (7 ft 10¾ in)
Max height	1.925 m (6 ft 3¾ in)
Main baggage compartment	11.05 m³ (390.2 cu ft)
Underfloor bulk storage	0.60 m³ (21.2 cu ft)

AREAS:
Wings, gross	65.00 m² (699.7 sq ft)
Vertical tail surfaces (total)	15.52 m² (167.06 sq ft)
Horizontal tail surfaces (total)	17.35 m² (186.75 sq ft)

WEIGHTS AND LOADINGS (design):
Typical operating weight empty	13,665 kg (30,126 lb)
Max fuel weight	4,200 kg (9,259 lb)
Max payload	6,000 kg (13,227 lb)
Max ramp weight	22,100 kg (48,722 lb)
Max T-O weight	22,000 kg (48,502 lb)
Max landing weight	21,800 kg (48,061 lb)
Max zero-fuel weight	19,665 kg (43,354 lb)
Max wing loading	338.5 kg/m² (69.32 lb/sq ft)
Max power loading: with APR	4.61 kg/kW (7.58 lb/shp)
without APR	4.20 kg/kW (6.90 lb/shp)

PERFORMANCE (estimated):
Max cruising speed at 6,100 m (20,000 ft), ISA	330 knots (611 km/h; 380 mph)
Econ cruising speed	300 knots (556 km/h; 345 mph)
Max rate of climb at S/L	600 m (1,968 ft)/min
Rate of climb at S/L, OEI	240 m (787 ft)/min
Service ceiling	7,620 m (25,000 ft)
Service ceiling, OEI	5,180 m (17,000 ft)
T-O and landing field length	1,220 m (4,000 ft)
Max range with 50 passengers:	
basic	800 nm (1,482 km; 921 miles)
optional	1,100 nm (2,038 km; 1,266 miles)

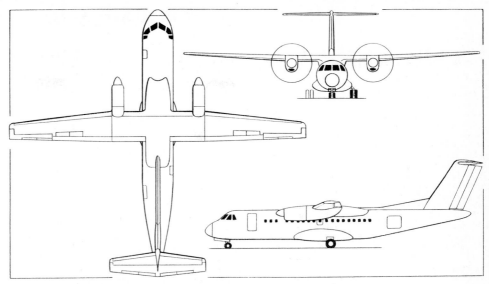

IPTN N-250 prototype twin-turboprop regional transport *(Dennis Punnett)*

IPTN N-2130

TYPE: Twin-turbofan regional airliner.

PROGRAMME: Projected 130-passenger aircraft with transonic performance and fly-by-wire controls; in design study stage 1992; possible launch in 1994 for first flight 2000.

IPTN (CASA) NC-212-200 AVIOCAR

Manufactured under licence in Indonesia as NC-212 since 1976; IPTN produced 29 NC-212-100 before switching to NC-212-200 (66 completed by mid-1993). Roles include civil passenger and cargo (including quick-change VIP), LAPES (low altitude parachute extraction system) airdropping, military transport, SAR, maritime patrol, medevac, photographic, survey and rainmaking.

IPTN (EUROCOPTER) NBO-105

Manufactured under licence from MBB (now Eurocopter) as NBO-105 since 1976; only rotors and transmission now supplied from Germany; stretched **NBO-105S** built from 101st aircraft onwards; **MPDS** (multi-purpose delivery system) can carry 50 mm to 81 mm unguided rockets, single or twin 0.30 in or 0.50 in machine-gun pods; and reconnaissance or FLIR pods. Also available in **FAC** (forward air control) version. Total 103 delivered by mid-1993 and continuing; capacity 2½ per month.

Customers for NBO-105 include Royal Jordanian Air Force, Indonesian Army, Navy (four), Police (six), Pelita Air Service (33), Indonesian Forestry Department, Indonesian Immigration Department, Indonesian Search and Rescue Agency, Gudang Garam, Gunung Madu and the Indonesian Civil Aviation Training Centre.

IPTN (EUROCOPTER) NAS-332 SUPER PUMA

Assembly of 11 AS 330J Pumas begun 1981; switched to AS 332C and L Super Puma in early 1983; first NAS-332 for Pelita Air Service rolled out 22 April 1983; 15 delivered by mid-1993, including four as commando and general purpose transports for Indonesian Navy, four for Pelita Air Service, and one VIP version delivered to Malaysian Ministry of Finance; available configurations include **ASW/MA** (anti-submarine warfare/maritime attack). Three NAS-332Ls sold to Iranian Oil Co 1992; two L1s delivered in 1993 for Indonesian Presidential Flight.

IPTN (BELL) NBELL-412

Licence agreement for Bell 412 (see Canadian section) signed November 1982; covers 100 helicopters; first flight April 1986; NBell-412 now 40 per cent Indonesian manufactured; deliveries totalled 17 by mid-1993, including Indonesian Army (four), Navy (four), civil operators (four). FN Herstal **EMA** (external mounting assembly) for 7.62 and 12.7 mm gun pods and 70 mm rocket pods, already certificated for Canadian and Italian Agusta built Bell 412s, qualified for NBell-412 and fitted to several helicopters; **CAS** (close air support) configuration also available. Production at standstill early 1992.

IPTN (LOCKHEED) HERCULES CONVERSION

TYPE: Modified military cargo transport (to civil passenger configuration).

PROGRAMME: Conversion of two L-100-30 Hercules; first (PK-MLT) undertaken by LACI with IPTN assistance, second (PK-MLS) by IPTN alone; STC obtained April 1990; redelivered August 1991 and April 1992.

CUSTOMERS: Merpati Nusantara Airlines (two).

DESIGN FEATURES: Airframe strengthened (incl thicker wing ribs) for 12,700 kg (28,000 lb) increase in max T-O weight; seating for 97 passengers installed; cabin headroom increased by 17.5 cm (6.9 in); intercom, interphone, smoke detectors and emergency lighting installed; ice warning system reinstated; hydraulic and environmental control systems modified.

Search and rescue NBO-105 of the Indonesian Air Force

INTERNATIONAL PROGRAMMES

AAA
ADVANCED AMPHIBIOUS AIRCRAFT
c/o Alenia, Rome, Italy
PARTICIPATING COMPANIES:
  Alenia: see under Italy
  HAI: see under Greece
  Per Udsen (Denmark)
  SDPR (former Yugoslavia)

Deutsche Aerospace reports it is no longer active in this programme; its share had already been reduced to five per cent. Little progress reported in 1992. Latest organisation reported joining was Yugoslav Directorate of Supply and Procurement and Soko which, prior to the war in Bosnia and Croatia, expected to have a 15 per cent share covering manufacture of rear fuselage and tail surfaces and some landing gear hydraulic components.

AAA
TYPE: Twin-turboprop multi-purpose amphibian; primary roles firefighting, search and rescue, maritime and anti-pollution surveillance and transport.
PROGRAMME: Definition phase by Alenia and Dornier started 1 April 1988; research and technology phase led by Alenia from 1 June 1990 to 30 September 1992; HAI (Greece) and Per Udsen (Denmark) became partners; Soko and SDPR (former Yugoslavia) and OGMA (Portugal) are associates. First flight planned for April 1995; certification end 1997.
CUSTOMERS: Estimated market for up to 300 such aircraft by year 2010.
DESIGN FEATURES: High-lift aerodynamics; low-drag hull; scoop filling of water tanks; rough-sea capability; should land in sea state 4 with 37 passengers.
FLYING CONTROLS: Mission-tailored flight management system.
STRUCTURE: Extensive use of corrosion-proof composites.
LANDING GEAR: Retractable tricycle type, with twin wheels on each unit; mainwheels retract into large sponsons on fuselage sides.
POWER PLANT: Two 2,013 kW (2,700 shp) class turboprops; four-blade propellers with advanced aerofoil section and composite blades.

Artist's impression of Advanced Amphibious Aircraft

DIMENSIONS, EXTERNAL (provisional):
  Wing span   33.67 m (110 ft 5½ in)
  Length of fuselage   22.00 m (72 ft 2 in)
  Height overall   9.73 m (30 ft 11 in)
AREAS:
  Wings, gross   98.00 m² (1,054.9 sq ft)
WEIGHTS AND LOADINGS (estimated):
  Typical operating weight empty 12,200 kg (26,896 lb)
  Max disposable payload   8,000 kg (17,637 lb)
  Max T-O weight: from water   19,400 kg (42,770 lb)
                  from land   23,000 kg (50,705 lb)
PERFORMANCE (estimated, at 23,000 kg; 50,705 lb max T-O weight except where indicated):
  Max cruising speed at 3,050 m (10,000 ft), ISA, at 90% max T-O weight   230 knots (426 km/h; 264 mph)
  Stalling speed, flaps down, AUW of 18,000 kg (39,683 lb)
                                60 knots (102 km/h; 69 mph) EAS
  Max rate of climb at S/L, ISA   518 m (1,700 ft)/min
  T-O run from land at S/L, ISA   840 m (2,756 ft)
  FAR Pt 25 balanced T-O field length 1,300 m (4,265 ft)
  Typical range with 4,000 kg (8,818 lb) payload
                                1,133 nm (2,100 km; 1,305 miles)
  Ferry range   2,347 nm (4,350 km; 2,703 miles)

AERMACCHI/DEUTSCHE AEROSPACE
PARTICIPATING COMPANIES:
  Aermacchi: see under Italy
  Deutsche Aerospace: see under Germany

AERMACCHI/DEUTSCHE AEROSPACE PTS-2000
TYPE: Tandem-seat advanced jet trainer.
PROGRAMME: Launched April 1989 by Aermacchi and Dornier, but German involvement now also includes Deutsche Aerospace Military Aircraft Division as result of DASA restructuring; joint studies continue of aircraft, market and complete training system; further companies may join programme.
CURRENT VERSIONS: **AT-2000**: Aircraft element of PTS-2000 training system.
CUSTOMERS: Target customers are advanced Western world air forces in next decade.
DESIGN FEATURES: No configuration details released by June 1993, but aircraft being used as baseline would be in 5,600-6,200 kg (12,400-13,800 lb) weight range and have 1:1 thrust/weight ratio, full fly-by-wire controls and modern cockpits; intended as component of complete military pilot training system including also simulator and computer-based instruction.

AIRBUS
AIRBUS INDUSTRIE
1 Rond Point Maurice Bellonte, F-31707 Blagnac Cedex, France
Telephone: 33 61 93 33 33
Fax: 33 61 93 37 92
Telex: AIRBU 530526 F
PARIS OFFICE: 12bis avenue Bosquet, F-75007 Paris, France
Telephone: 33 (1) 45 51 40 95
MANAGING DIRECTOR: Jean Pierson
COO: Volker von Tein
SENIOR VICE-PRESIDENT, COMMERCIAL: Stuart Iddles
GENERAL MANAGER, PRESS AND INFORMATION SERVICES: Barbara Kracht
AIRFRAME PRIME CONTRACTORS:
  Aerospatiale: see under France
  Deutsche Aerospace Airbus: see under Germany
  British Aerospace: see under UK
  CASA: see under Spain
  Alenia (in A321): see under Italy

Airbus Industrie set up December 1970 as Groupement d'Intérêt Economique to manage development, manufacture, marketing and support of A300; this management now extends to A300-600, A310, A319, A320, A321, A340/A330 and UHCA.

Airbus Industrie responsible for all work on these programmes by partner companies and has approximately 2,000 employees, including workers at its spare parts centre in Hamburg and its US subsidiary; approximately 35,000 more are directly employed on Airbus work within its five partners. Aerospatiale has 37.9 per cent interest in Airbus Industrie, Deutsche Aerospace Airbus 37.9 per cent, British Aerospace 20 per cent, CASA 4.2 per cent. Fokker is an associate in A300 and A310 and Belairbus (Belgian consortium) in A310, A320 and A330/A340. Alenia of Italy manufactures front fuselage plug for A321.

Airbus Industrie deliveries declining from 163 in 1991 to 157 in 1992 and 150 in 1993, but possibly increasing to 170 in 1995. Cancellations included 74 for Northwest Airlines in 1992. New orders totalled 136 in 1992. First A340 delivered to Air France in March 1993 was 1,000th Airbus airliner. Total order backlog at end 1992, not including subsequent changes, was 836.

Subsidiaries include Aeroformation and Airbus Industrie of North America.

TOTALS OF AIRBUS AIRLINERS
At 31 May 1993

	A319	A320	A321	A300	A310	A330	A340	Totals
Firm orders	6	657	153	481	265	130	113	1,805
Delivered	—	400	—	395	231	—	6	1,302
Operating	—	397	—	387	230	—	6	1,020

AIRBUS A300-600
TYPE: Large-capacity wide-bodied medium/long-range commercial airliner.
PROGRAMME: Launched May 1969; initial variants were A300B1 (first flight 28 October 1972, service entry May 1974: see 1971-72 *Jane's* for details), A300B2 and A300B4 (248 built: see 1984-85 and previous editions); A300-600 go-ahead December 1980; first flight (F-WZLR) 8 July 1983; certificated (with JT9D-7R4H1 engines) 9 March 1984; first delivery (to Saudia) 26 March 1984. Improved version with CF6-80C2 engines and other changes (see Current Versions) made first flight 20 March 1985; French certification for Cat. IIIb take-offs and landings 26 March 1985; first delivery of improved version (to Thai Airways) 26 September 1985. Extended range A300-600R (then known as -600ER) made first flight 9 December 1987, receiving European and FAA certification 10 and 28 March 1988 respectively, deliveries (to American Airlines) beginning 21 April 1988.
CURRENT VERSIONS: **A300-600**: Advanced version of A300B4-200; major production version since early 1984. Passenger and freight capacity increased by using rear fuselage of A310 with pressure bulkhead moved aft; wings have simple Fowler flaps and increased trailing-edge camber; forward facing two-person flight deck with CRT displays; new digital avionics; new braking control system; new APU; simplified systems; weight saving by use of composites for some secondary structural components; payload/range performance and fuel economy improved by comprehensive drag clean-up.

Improved version: Introduced 1985. Has CF6-80C2 or PW4000 as engine options, carbon brakes, wingtip fences and 'New World' flight deck; basic equipment of aircraft delivered from late 1991 further improved by incorporating standard options. *Detailed description applies to this version except where indicated.*

A300-600R: Extended range version of A300-600, differing mainly in having fuel trim tank in tailplane and higher MTOGW.

A300-600 Convertible: Convertible passenger/cargo version, described separately.

A300-600 Freighter: Non-passenger version, described separately.

Airbus Super Transporter: A300-600R conversion as Super Guppy replacement; see under SATIC later in this section.
CUSTOMERS: See table above.
DESIGN FEATURES: Mid-mounted wings with 10.5 per cent

Airbus A300-600R twin-turbofan extended range airliner in the insignia of Olympic Airlines

thickness/chord ratio, 28° sweepback at quarter-chord, and (since 1985) tip fences; circular-section pressurised fuselage; all-swept tail unit.

FLYING CONTROLS: Each wing has three-segment, two-position (T-O/landing) leading-edge slats (no cutout over engine pylon), Krueger flap at leading-edge wingroot, three cambered tabless flaps on trailing-edge, all-speed aileron between inboard flap and outer pair, and seven spoilers forward of flaps; flaps occupy 84 per cent of trailing-edge, increasing wing chord by 25 per cent when fully extended; ailerons deflect 9° 2′ downward automatically when flaps are deployed; all 14 spoilers used as lift dumpers: outboard 10 for roll control and inboard 10 as airbrakes; variable incidence tailplane. Ailerons/elevators/rudder fully powered by hydraulic servos (three per surface), controlled mechanically; secondary surfaces (spoilers/flaps/slats) fully powered hydraulically with electrical control, tailplane by two independent hydraulic motors electrically controlled with additional mechanical input; preselection of spoiler/lift dump lever permits automatic extension of lift dumpers on touchdown; flaps and slats have similar drive mechanisms, each powered by twin motors driving ball screwjacks on each surface with built-in protection against asymmetric operation.

STRUCTURE: Two-spar main wing box, integral with fuselage and incorporating fail-safe principles; third spar across inboard sections; semi-monocoque fuselage (frames and open Z-section stringers), with integrally machined skin panels in high-stress areas; primary structure is of high-strength, damage-tolerant aluminium alloy, with steel or titanium for some critical fuselage components, honeycomb panels or selected glassfibre laminates for secondary structures; metal slats, flaps and ailerons. Composites include AFRP for flap track fairings, rear wing/body fairings, cooling air inlet fairings and radome; GFRP for wing upper surface panels above mainwheel bays, fin leading/trailing-edges, fin-tip, fin/fuselage fairings, tailplane trailing-edges, elevator leading-edges, tailplane and elevator tips and elevator actuator access panel; carbon-reinforced GFRP for elevators and rudder; CFRP for spoilers, outer flap deflector doors and fin box; all CFRP moving surfaces have aluminium or titanium trailing-edges. Nosewheel doors and mainwheel leg fairing doors also of CFRP. Nose gear is structurally identical to that of B2/B4/A310; main gear is generally reinforced, with a new-design hinge arm and a new pitch damper hydraulic and electrical installation. Nacelles have CFRP cowling panels and are subcontracted to Rohr (California); pylon fairings are of AFRP. Aerospatiale builds nose (incl flight deck), lower centre-fuselage, four inboard spoilers, wing/body fairings and engine pylons; Deutsche Aerospace Airbus builds forward fuselage (flight deck to wing box), upper centre-fuselage, rear fuselage (incl tailcone), vertical tail, 10 outboard spoilers and some cabin doors, and also equips wings, installs interiors and seats; BAe designed wings and builds wing box; CASA manufactures horizontal tail, port and starboard forward passenger doors and mainwheel/nosewheel doors; Fokker produces wingtips, ailerons, flaps, slats and main gear leg fairings; large, fully equipped and inspected airframe sections shipped by Super Guppy to Aerospatiale at Toulouse for final assembly and painting, aircraft then being flown to Hamburg for outfitting and returned to Toulouse for final customer acceptance.

LANDING GEAR: Hydraulically retractable tricycle type, of Messier-Bugatti design, with Messier-Bugatti/Liebherr/ Dowty shock absorbers and wheels standard. Twin-wheel nose unit retracts forward, main units inward into fuselage. Free-fall extension. Each four-wheel main unit comprises two tandem mounted bogies, interchangeable left with right. Standard bogie size is 927 × 1,397 mm (36½ × 55 in); wider bogie of 978 × 1,524 mm (38½ × 60 in) is optional. Mainwheel tyres size 49 × 17-20 (standard) or 49 × 19-20 (wide bogie), with respective pressures of 12.4 and 11.1 bars (180 and 161 lb/sq in). Nosewheel tyres size 40 × 14-16, pressure 9.4 bars (136 lb/sq in). Steering angles 65°/95°. Messier-Bugatti/Liebherr/Dowty hydraulic disc brakes standard on all mainwheels. Normal braking powered by 'green' hydraulic system, controlled electrically through two master valves and monitored by a brake system control box to provide anti-skid protection. Standby braking (powered automatically by 'yellow' hydraulic system if normal 'green' system supply fails) controlled through a dual metering valve; anti-skid protection is ensured through same box as normal system, with emergency pressure supplied to brakes by accumulators charged from 'yellow' system. Automatic braking system optional. Duplex anti-skid units fitted, with a third standby hydraulic supply for wheel brakes. Bendix or Goodrich wheels and brakes available optionally. Min ground turning radius (effective, aft CG) 22.00 m (72 ft 2¼ in) about nosewheel, 34.75 m (114 ft 0 in) about wingtips.

POWER PLANT: Two turbofans in underwing pods. A300-600 was launched with 249 kN (56,000 lb st) Pratt & Whitney JT9D-7R4H1 and currently available with 249 kN (56,000 lb st) Pratt & Whitney PW4156 or 262.4 kN (59,000 lb st) General Electric CF6-80C2A1. A300-600R is offered with 273.6 kN (61,500 lb st) CF6-80C2A5 or 258 kN (58,000 lb st) PW4158. CF6-80C2A5 and PW4158 also available as options on A300-600. Fuel in two integral tanks in each wing, and fifth integral tank in wing centre-section, giving standard usable capacity of 62,000 litres (16,379 US gallons; 13,638 Imp gallons). Additional 6,150 litre (1,625 US gallon; 1,353 Imp gallon) fuel/trim tank in tailplane (-600R only) increases this total to 68,150 litres (18,004 US gallons; 14,991 Imp gallons). Optional extra fuel cell in aft cargo hold can increase total to 75,350 litres (19,906 US gallons; 16,575 Imp gallons) in -600R. Two standard refuelling points beneath starboard wing; similar pair optional under port wing.

ACCOMMODATION: Crew of two on flight deck, plus two observers' seats. Passenger seating in main cabin in six, seven, eight or nine-abreast layout with two aisles; typical mixed class layout has 266 seats (26 first class and 240 economy), six/eight abreast at 96/86 cm (40/32 in) seat pitch with two galleys and two toilets forward, one galley and two toilets in mid-cabin, and one galley and two toilets at rear; typical economy class layout for 289 passengers eight-abreast at 86 cm (34 in) pitch. Max capacity (subject to certification) 375 passengers. Closed overhead baggage lockers on each side (total capacity 10.48 m³; 370 cu ft) and in double-sided central 'super-bin' installation (total capacity 14.50 m³; 512 cu ft), giving 0.03 to 0.09 m³ (1.2 to 3.2 cu ft) per passenger in typical economy layout. Two outward parallel-opening Type A plug type passenger doors ahead of wing on each side, and one on each side at rear. Type I emergency exit on each side aft of wing. Underfloor baggage/cargo holds fore and aft of wings, with doors on starboard side; forward hold can accommodate twelve LD3 containers, or four 2.24 × 3.17 m (88 × 125 in) or, optionally, 2.43 × 3.17 m (96 × 125 in) pallets, or engine modules; rear hold can accommodate ten LD3 containers; additional bulk loading of freight provided for in an extreme rear compartment with usable volume of 17.3 m³ (611 cu ft); alternatively, rear hold can carry eleven LD3 containers, with bulk cargo capacity reduced to 9.0 m³ (318 cu ft); bulk cargo compartment can be used to transport livestock. Entire accommodation is pressurised, including freight, baggage and avionics compartments.

SYSTEMS: Air supply for air-conditioning system taken from

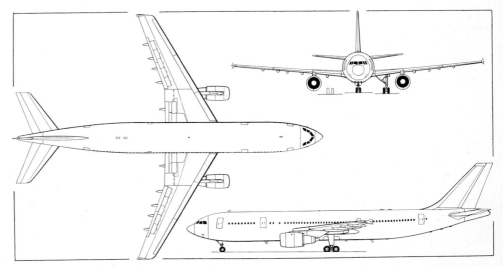

Airbus A300-600R wide-bodied medium-range airliner (GE CF6-80C2 turbofans) *(Dennis Punnett)*

engine bleed and/or APU via two high pressure points; conditioned air can also be supplied direct to cabin by two low pressure ground connections; ram air inlet for fresh air ventilation when packs not in use. Pressure control system (max differential 0.574 bar; 8.32 lb/sq in) consists of two identical, independent, automatic systems (one active, one standby); automatic switchover from one to other after each flight and in case of active system failure; in each system, pressure controlled by two electric outflow valves, function depending on pre-programmed cabin pressure altitude and rate of change of cabin pressure, aircraft altitude, and pre-selected landing airfield elevation. Automatic pre-pressurisation of cabin before take-off, to prevent noticeable pressure fluctuation during take-off. Hydraulic system comprises three fully independent circuits, operating simultaneously; each system includes reservoir of direct air/fluid contact type, pressurised at 3.5 bars (51 lb/sq in); fire resistant phosphate ester type fluid; nominal output flow 136 litres (35.9 US gallons; 30 Imp gallons)/min delivered at 207 bars (3,000 lb/sq in) pressure; 'blue' and 'yellow' systems have one pump each, 'green' system has two pumps. The three circuits provide triplex power for primary flying controls; if any circuit fails, full control of aircraft is retained without any necessity for action by crew. All three circuits supply ailerons, rudder and elevators; 'blue' circuit additionally supplies spoiler 7, spoiler/airbrake 4, airbrake 1, yaw damper and slats; 'green' circuit additionally supplies spoiler 6, flaps, Krueger flaps, slats, landing gear, wheel brakes, steering, tailplane trim, artificial feel, and roll/pitch/yaw autopilot; 'yellow' circuit additionally supplies spoiler 5, spoiler/airbrake 3, airbrake 2, flaps, wheel brakes, cargo doors, artificial feel, yaw damper, tailplane trim, and roll/pitch/yaw autopilot. Ram air turbine driven pump provides standby hydraulic power should both engines become inoperative. Main electrical power supplied under normal flight conditions by two integrated drive generators, one on each engine; third (auxiliary) generator, driven by APU, can replace either of main generators, having same electromagnetic components but not constant-speed drive; each generator rated at 90kVA, with overload ratings of 112.5kVA for 5 min and 150kVA for 5 s; APU generator driven at constant speed through gearbox. Three unregulated transformer-rectifier units (TRUs) supply 28V DC power. Three 25Ah nickel-cadmium batteries used for emergency supply and APU starting; emergency electrical power taken from main aircraft batteries and emergency static inverter, providing single-phase 115V 400Hz output for flight instruments, navigation, communications and lighting when power not available from normal sources. Hot air anti-icing of engines, engine air intakes, and outer segments of leading-edge slats; electrical heating for anti-icing flight deck front windscreens, demisting flight deck side windows, and for sensors, pitot probes and static ports, and waste water drain masts. Garrett GTCP 331-250F APU in tailcone, exhausting upward; installation incorporates APU noise attenuation. Self-contained fire protection system, and firewall panels protect main structure from an APU fire. APU provides bleed air to pneumatic system, and drives auxiliary AC generator during ground and in-flight operation; APU drives 90kVA oilspray-cooled generator, and supplies bleed air for main engine start or air-conditioning system. For current deliveries of A300-600, APU has improved relight capability, with starting capability throughout flight envelope. Modular box system provides passenger oxygen to all installation areas. For new A300-600s and -600Rs, two optional modifications offered for compliance with full extended-range operations (EROPS) requirements: hydraulically driven fourth generator and increased cargo hold fire suppression capability. EROPS kit qualified for aircraft with CF6-80C2 and JT9D-7R series engines, and since mid-1988 for those with PW4000 series.

AVIONICS: Standard communications radios include two VHF, two HF, one Selcal, interphone and passenger address systems, groundcrew call system, and voice recorder. Radio navigation avionics include two DME interrogators, two VOR receivers, two ATC transponders, one ADF, two marker beacon receivers, two ILS receivers, weather radar, and two radio altimeters. Full provisions for second weather radar and GPWS; third VHF; structural provision for such future systems as a Mode S ATC transponder. Two Honeywell digital air data computers standard. Most other avionics are to customer's requirements, only those related to blind landing system (Bendix/King or Collins ILS and Collins or TRT radio altimeter) being selected and supplied by manufacturer. Six identical and interchangeable CRT electronic displays (four EFIS and two ECAM: electronic flight instrument system and electronic centralised aircraft monitor), plus digitised electromechanical instruments with liquid crystal displays; basic digital AFCS comprises dual flight control computers (FCCs) for flight director and autopilot functions (for Cat. III automatic landing), single thrust control computer (TCC) for speed and thrust control, and two flight augmentation computers (FACs) to provide yaw damping, electric pitch trim, and flight envelope monitoring and protection; options include second FCC (for Cat. III automatic landing); second TCC; two flight management computers (FMCs) and two control display units for full flight management system. Basic aircraft also fitted with ARINC 717 data recording system, comprising digital flight data acquisition unit, digital flight data recorder and three-axis linear accelerometer; optional additional level of windshear protection is available.

DIMENSIONS, EXTERNAL:
Wing span 44.84 m (147 ft 1 in)
Wing aspect ratio 7.7
Length overall 54.08 m (177 ft 5 in)
Fuselage: Length 53.30 m (174 ft 10½ in)
  Max diameter 5.64 m (18 ft 6 in)
Height overall 16.53 m (54 ft 3 in)
Tailplane span 16.26 m (53 ft 4 in)
Wheel track 9.60 m (31 ft 6 in)
Wheelbase (c/l of shock absorbers) 18.60 m (61 ft 0 in)
Passengers doors (each): Height 1.93 m (6 ft 4 in)
  Width 1.07 m (3 ft 6 in)
  Height to sill: forward 4.60 m (15 ft 1 in)
    centre 4.80 m (15 ft 9 in)
    rear 5.50 m (18 ft 0½ in)
Emergency exits (each): Height 1.60 m (5 ft 3 in)
  Width 0.61 m (2 ft 0 in)
  Height to sill 4.87 m (15 ft 10 in)
Underfloor cargo door (forward):
  Height 1.71 m (5 ft 7½ in)
  Width 2.69 m (8 ft 10 in)
  Height to sill 3.07 m (10 ft 1 in)
Underfloor cargo door (rear):
  Height 1.71 m (5 ft 7½ in)
  Width 1.81 m (5 ft 11¼ in)
  Height to sill 3.41 m (11 ft 2¼ in)
Underfloor cargo door (extreme rear):
  Height (projected) 0.95 m (3 ft 1 in)
  Width 0.95 m (3 ft 1 in)
  Height to sill 3.56 m (11 ft 8 in)

DIMENSIONS, INTERNAL:
Cabin, excl flight deck:
  Length 40.21 m (131 ft 11 in)
  Max width 5.28 m (17 ft 4 in)
  Max height 2.54 m (8 ft 4 in)
Underfloor cargo hold:
  Length: forward 10.60 m (34 ft 9¼ in)
    rear 7.95 m (26 ft 1 in)
    extreme rear 3.40 m (11 ft 2 in)
  Max height 1.76 m (5 ft 9 in)
  Max width 4.20 m (13 ft 9¼ in)
Underfloor cargo hold volume:
  forward 75.1 m³ (2,652 cu ft)
  rear 55.0 m³ (1,942 cu ft)
  extreme rear 17.3 m³ (611 cu ft)

AREAS:
Wings, gross 260.0 m² (2,798.6 sq ft)
Leading-edge slats (total) 30.30 m² (326.15 sq ft)
Krueger flaps (total) 1.115 m² (12.00 sq ft)
Trailing-edge flaps (total) 47.30 m² (509.13 sq ft)
All-speed ailerons (total) 7.06 m² (75.99 sq ft)
Spoilers (total) 5.396 m² (58.08 sq ft)
Airbrakes (total) 12.59 m² (135.52 sq ft)
Fin 45.20 m² (486.53 sq ft)
Rudder 13.57 m² (146.07 sq ft)
Horizontal tail surfaces (total) 64.0 m² (688.89 sq ft)

*WEIGHTS AND LOADINGS (A: CF6-80C2A1/A5 engines, B: PW4156/4158 engines, both in 267-seat configuration):
Manufacturer's weight empty:
  A (600) 79,210 kg (174,628 lb)
  A (600R) 79,403 kg (175,053 lb)
  B (600) 79,151 kg (174,498 lb)
  B (600R) 79,318 kg (174,866 lb)
Operating weight empty:
  A (600) 89,672 kg (197,693 lb)
  A (600R) 89,813 kg (198,003 lb)
  B (600) 89,655 kg (197,655 lb)
  B (600R) 89,141 kg (196,522 lb)
Max payload (structural): A (600) 40,328 kg (88,908 lb)
  A (600R) 40,187 kg (88,597 lb)
  B (600) 40,345 kg (88,946 lb)
  B (600R) 40,261 kg (88,760 lb)
Underfloor cargo capacity (A and B):
  containerised 68,400 kg (150,795 lb)
  bulk 2,766 kg (6,100 lb)
Max usable fuel:
  600: standard 49,786 kg (109,760 lb)
  600R: standard 54,721 kg (120,640 lb)
    with optional cargo hold tank 58,618 kg (129,230 lb)
Max T-O weight (A and B):
  600 165,000 kg (363,765 lb)
  600R (standard) 170,500 kg (375,885 lb)
  600R (option) 171,700 kg (378,535 lb)
Max ramp weight (A and B):
  600 165,900 kg (365,745 lb)
  600R (standard) 171,400 kg (377,870 lb)
  600R (option) 172,600 kg (380,520 lb)
Max landing weight (A and B):
  600 138,000 kg (304,240 lb)
  600R (standard) 140,000 kg (308,645 lb)
Max zero-fuel weight (A and B):
  600, 600R (standard) 130,000 kg (286,600 lb)
Max wing loading: 600 635 kg/m² (130.0 lb/sq ft)
  600R (standard) 656 kg/m² (134.4 lb/sq ft)

* Production aircraft from late 1991 onward. See 1989-90 and previous editions for earlier versions

PERFORMANCE (at max T-O weight except where indicated: A and B as for Weights and Loadings):
Max operating speed (V_{MO}) from S/L to 8,140 m (26,700 ft) 335 knots (621 km/h; 386 mph) CAS
Max operating Mach number (M_{MO}) above 8,140 m (26,700 ft) 0.82
Max cruising speed at 7,620 m (25,000 ft) 480 knots (890 km/h; 553 mph)
Max cruising speed at 9,150 m (30,000 ft) Mach 0.82 (484 knots; 897 km/h; 557 mph)
Typical long-range cruising speed at 9,450 m (31,000 ft) Mach 0.80 (472 knots; 875 km/h; 543 mph)
Approach speed: 600 135 knots (249 km/h; 155 mph)
  600R 136 knots (251 km/h; 156 mph)
Max operating altitude 12,200 m (40,000 ft)
Runway ACN for flexible runway, category B:
  standard bogie & tyres: 600 56
    600R 59
    600R (option) 60
  optional bogie & tyres: 600 52
    600R 55
    600R (option) 56
T-O field length at S/L, ISA + 15°C:
  600: A 2,280 m (7,480 ft)
    B 2,190 m (7,185 ft)
  600R: A (C2A5 engines) 2,290 m (7,520 ft)
    B (PW4158 engines) 2,240 m (7,350 ft)
Landing field length: 600 1,536 m (5,040 ft)
  600R 1,555 m (5,100 ft)
Range (1991 and subsequent deliveries) at typical airline OWE with 267 passengers and baggage, reserves for 200 nm (370 km; 230 miles):
  600: A, B 3,600 nm (6,670 km; 4,145 miles)
  600R (standard fuel):
    A 4,000 nm (7,410 km; 4,600 miles)
    B 4,100 nm (7,600 km; 4,720 miles)
  600R (MTOGW option and additional fuel):
    A 4,150 nm (7,690 km; 4,780 miles)
    B 4,170 nm (7,725 km; 4,800 miles)

OPERATIONAL NOISE LEVELS (A300-600R, ICAO Annex 16, Chapter 3):
T-O: A 91.1 EPNdB (96.3 limit)
  B 92.2 EPNdB (96.3 limit)
Sideline: A 98.6 EPNdB (99.9 limit)
  B 97.7 EPNdB (99.9 limit)
Approach: A 99.8 EPNdB (103.3 limit)
  B 101.7 EPNdB (103.3 limit)

AIRBUS A300-600 CONVERTIBLE and A300-600 FREIGHTER

TYPE: Specialised versions of A300-600.
CURRENT VERSIONS: **Convertible:** For all-passenger or mixed passenger/cargo configuration. Typical options include accommodation (in mainly eight-abreast seating) for up to 375 passengers (subject to certification) on the upper deck; or 145 passengers (seven/eight abreast) plus six 2.44 × 3.17 m (96 × 125 in) pallets; or 83 passengers plus nine 96 × 125 in pallets; up to twenty 2.24 × 3.17 m (88 × 125 in) pallets; or five 88 × 125 in plus nine 96 × 125 in pallets.

Freighter: For freighting only; no passenger systems provided; various systems options give airlines ability to adapt basic aircraft to specific freight requirements; Airbus offers conversion with port-side forward freight door.
STRUCTURE: Generally similar to A300-600. Main differences are large forward upper deck cargo door, reinforced cabin floor, smoke detection system in main cabin; upper deck cargo door is on opposite side to door of forward underfloor hold, enabling loading or unloading to be carried out simultaneously at all positions.
CUSTOMERS: Federal Express became A300-600 Freighter launch customer June 1991 with order for 25; commitments for 50 more.
POWER PLANT: Options as for A300-600R.
DIMENSIONS, EXTERNAL: As A300-600R, plus:
Upper deck cargo door (fwd, port):
  Height (projected) 2.57 m (8 ft 5¼ in)
  Width 3.58 m (11 ft 9 in)
  Height to sill 4.91 m (16 ft 1 in)
DIMENSIONS, INTERNAL:
Cabin upper deck usable for cargo:
  Length 33.45 m (109 ft 9 in)
  Min height 2.01 m (6 ft 7 in)
  Max height:
    ceiling trim panels in place 2.22 m (7 ft 3½ in)
    without ceiling trim panels 2.44 m (8 ft 0 in)
  Volume 192-203 m³ (6,780-7,169 cu ft)
WEIGHTS AND LOADINGS (basic Convertible. A: with CF6-80C2A5 engines, B: with PW4158 engines):
Manufacturer's weight empty:
  A, passenger mode 81,900 kg (180,558 lb)
  B, passenger mode 81,820 kg (180,382 lb)
  A, freight mode 81,640 kg (179,985 lb)
  B, freight mode 81,560 kg (179,809 lb)
Operating weight empty:
  A, passenger mode 92,160 kg (203,178 lb)
  B, passenger mode 92,100 kg (203,045 lb)
  A, freight mode 83,470 kg (184,020 lb)
  B, freight mode 83,410 kg (183,887 lb)
Max payload (structural):
  A, passenger mode 37,840 kg (83,423 lb)

B, passenger mode	37,900 kg (83,555 lb)
A, freight mode	46,530 kg (102,581 lb)
B, freight mode	46,590 kg (102,713 lb)
Max T-O weight: A, B	170,500 kg (375,900 lb)
Max landing weight: A, B	140,000 kg (308,650 lb)
Max zero-fuel weight: A, B	130,000 kg (286,600 lb)

WEIGHTS AND LOADINGS (basic Freighter variant of -600R):
Manufacturer's weight empty:
- A   77,947 kg (171,840 lb)
- B   77,874 kg (171,680 lb)

Operating weight empty:
- A   78,854 kg (173,840 lb)
- B   78,781 kg (173,680 lb)

Max payload (structural):
- A, range mode   51,146 kg (112,760 lb)
- B, range mode   51,220 kg (112,920 lb)
- A, payload mode   54,946 kg (121,130 lb)
- B, payload mode   55,017 kg (121,290 lb)

Max T-O weight: A, B
- range mode   170,500 kg (375,900 lb)
- payload mode   165,100 kg (363,980 lb)

Max landing weight: A, B
- range mode   140,000 kg (308,650 lb)
- payload mode   140,600 kg (309,970 lb)

Max zero-fuel weight: A, B
- range mode   130,000 kg (286,600 lb)
- payload mode   133,800 kg (294,980 lb)

PERFORMANCE:
Range with max (structural) payload, allowances for 30 min hold at 460 m (1,500 ft) and 200 nm (370 km; 230 mile) diversion:
- A, B, range mode   2,650 nm (4,908 km; 3,050 miles)
- A, B, payload mode   1,900 nm (3,520 km; 2,186 miles)

AIRBUS SUPER AIRCRAFT TRANSPORTER
See SATIC entry later in this section.

AIRBUS A310
Canadian Forces designation: CC-150 Polaris

TYPE: Large-capacity wide-bodied medium/extended-range airliner.

PROGRAMME: Launched July 1978; first flight (F-WZLH) 3 April 1982; initial French/German certification 11 March 1983; first deliveries (Lufthansa and Swissair) 29 March 1983, entering service 12 and 21 April respectively; JAR Cat. IIIA certification (France/Germany) September 1983; UK certification January 1984; JAR Cat. IIIB November 1984; FAA type approval early 1985. First flight of extended-range A310-300 8 July 1985 (certificated with JT9D-7R4E engines 5 December 1985, delivered to launch customer Swissair 17 December); wingtip fences introduced as standard on A310-200 from Spring 1986 (first delivery: Thai Airways, 7 May); certification/delivery of A310-300 with CF6-80C2 engines April 1986, with PW4152s June 1987; version with ACTs (additional centre tanks) certificated November 1987 (first customer Wardair of Canada). Russian State Aviation Register certification October 1991 (first Western-built aircraft to achieve this status).

CURRENT VERSIONS: **A310-200**: Basic passenger version, *to which detailed description mainly applies*.
  A310-200C: Convertible version of A310-200; first delivery (Martinair) 29 November 1984.
  A310-200F: Freighter version.
  A310-300: Extended-range passenger version; second member of Airbus family to introduce delta shaped wingtip fences as standard. Extra range provided by increased basic max T-O weight (150,000 kg; 330,695 lb) and greater fuel capacity (higher max T-O weights optional); standard extra fuel capacity is in tailplane, allowing in-flight CG control for improved fuel efficiency; for extra long range, one or two ACTs (additional centre tanks) can be installed in part of cargo hold.

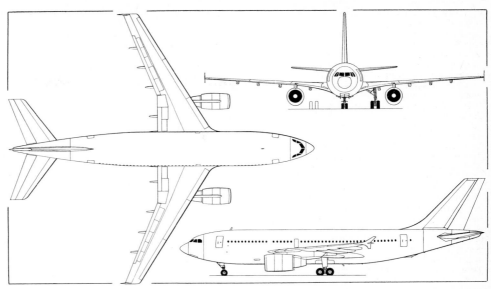

Airbus A310 medium/extended-range airliner *(Dennis Punnett)*

CUSTOMERS: See table on page 114.

DESIGN FEATURES: Retains same fuselage cross-section as A300, but with cabin 11 frames shorter and overall fuselage 13 frames shorter than A300B2/B4-100 and -200; new advanced-technology wings of reduced span and area; new and smaller horizontal tail surfaces; common pylons able to support all types of GE and PW engines offered; advanced digital two-man cockpit; landing gear modified to cater for size and weight changes. Wings have 28° sweepback at quarter-chord, root incidence 5° 3′, dihedral 11° 8′ (inboard) and 4° 3′ (outboard) at trailing-edge, and thickness/chord ratios of 15.2 (root), 11.8 (at trailing-edge kink), and 10.8 per cent (tip).

FLYING CONTROLS: Wing leading-edge movable surfaces as for A300-600; trailing-edges each have single Fowler flap outboard, vaned Fowler flap inboard, with all-speed aileron between; all 14 spoilers used as lift dumpers, inner eight also as airbrakes; two independent computer systems with different software provide redundancy and operational safety. Tail control surfaces as for A300-600.

STRUCTURE: Mainly of high-strength aluminium alloy except for outer shrouds (structure in place of low-speed ailerons), spoilers, wing leading-edge lower access panels and outer deflector doors, nosewheel doors, mainwheel leg fairing doors, engine cowling panels, elevators and fin box (CFRP); flap track fairings, flap access doors, rear wing/body fairings, pylon fairings, nose radome, cooling air inlet fairings and tailplane trailing-edges (AFRP); wing leading-edge top panels, panel aft of rear spar, upper surface skin panels above mainwheel bays, forward wing/body fairings, glideslope antenna cover, fin leading/trailing-edges, fin and tailplane tips (GFRP); and rudder (CFRP/GFRP). Wing box is two-spar multi-rib metal structure, with top and bottom load-carrying skins. Undertail bumper beneath rear fuselage, to protect structure against excessive nose-up attitude during T-O and landing. Manufacturing breakdown differs in detail from that of A300-600: Aerospatiale builds nose section (incl flight deck), lower centre-fuselage and wing box, rear wing/body fairings, engine pylons and airbrakes, and is responsible for final assembly; Deutsche Aerospace Airbus builds forward fuselage, upper centre-fuselage, rear fuselage and associated doors, tailcone, fin and rudder, flaps and spoilers, and fits control surfaces and equipment to main wing structure produced by BAe; CASA's contribution includes horizontal tail surfaces, nose-gear and mainwheel doors, and forward passenger doors; Fokker manufactures main landing gear leg doors, wingtips, all-speed ailerons and flap track fairings; wing leading-edge slats and forward wing/fuselage fairings produced by Belairbus.

LANDING GEAR: Hydraulically retractable tricycle type. Twin-wheel steerable nose unit (steering angle 65°/95°) as for A300. Main gear by Messier-Bugatti, each bogie comprising two tandem mounted twin-wheel units. Retraction as for A300-600. Standard tyre sizes: main, 46 × 16-20, pressure 11.2 bars (163 lb/sq in); nose, 40 × 14-16, pressure 9.0 bars (131 lb/sq in). Two options for low-pressure tyres on main units: (1) size 49 × 17-20, pressure 9.8 bars (143 lb/sq in); (2) size 49 × 19-20, pressure 8.9 bars (129 lb/sq in). Messier-Bugatti brakes and anti-skid units standard; Bendix type optional on A310-200. Carbon brakes standard since 1986. Min ground turning radius (effective, aft CG) 18.75 m (61 ft 6 in) about nosewheel, 33.00 m (108 ft 3¼ in) about wingtips.

POWER PLANT: Launched with two 213.5 kN (48,000 lb st) Pratt & Whitney JT9D-7R4D1 or 222.4 kN (50,000 lb st) General Electric CF6-80A3 turbofans; currently available with 238 kN (53,500 lb st) CF6-80C2A2, or 231.2 kN (52,000 lb st) Pratt & Whitney PW4152. Available from late 1991 with 262.4 kN (59,000 lb st) CF6-80C2A8 or 249.1 kN (56,000 lb st) PW4156A. Total usable fuel capacity 55,000 litres (14,530 US gallons; 12,098 Imp gallons) in A310-200. Increased to 61,100 litres (16,141 US gallons; 13,440 Imp gallons) in A310-300 by additional fuel in tailplane trim tank. Further 7,200 litres (1,902 US gallons; 1,584 Imp gallons) can be carried in each additional centre tank (ACT) in forward part of aft cargo hold. Two refuelling points, one beneath each wing outboard of engine.

ACCOMMODATION: Crew of two on flight deck; provision for third and fourth crew seats. Cabin, with six/seven/eight and nine-abreast seating, normally for 210 to 250 passengers, although certificated for up to 280; typical two-class layout for 220 passengers (20 first class, six-abreast at 101.6 cm; 40 in seat pitch, plus 200 economy class mainly eight-abreast at 81.3 cm; 32 in pitch); max capacity for 280 passengers nine-abreast in high-density configuration at pitch of 76 cm (30 in). Standard layout has two galleys and toilet at forward end of cabin, plus two galleys and four toilets at rear; depending on customer requirements, second toilet can be added forward, and toilets and galleys can be located at forward end at class divider position. Overhead baggage stowage as for A300-600, rising to 0.09 m³ (3.2 cu ft) per passenger in typical economy layout. Four passenger doors, one forward and one aft on each side; oversize Type I emergency exit over wing on each side. Underfloor baggage/cargo holds fore and aft of wings, each with door on starboard side; forward hold accommodates eight LD3 containers or three 2.24 × 3.17 m (88 × 125 in) standard or three 2.44 × 3.17 m (96 × 125 in) optional pallets; rear hold accommodates six LD3 containers, with optional seventh LD3 or LD1 position; LD3 containers can be carried two abreast, and/or standard pallets installed crosswise.

SYSTEMS: Garrett GTCP 331-250 APU. Air-conditioning system, powered by compressed air from engines, APU, or a ground supply unit, comprises two separate packs; air is distributed to flight deck, three separate cabin zones, electrical and electronic equipment, avionics bay and bulk cargo compartment; ventilation of forward cargo compartments optional. Pressurisation system has max normal differential of 0.57 bar (8.25 lb/sq in). Air supply for wing ice protection, engine starting and thrust reverser system bled from various stages of engine compressors, or supplied by APU or ground supply unit. Hydraulic system (three fully independent circuits operating at 207 bars; 3,000 lb/sq in; details as described for A300-600). Electrical system,

Airbus A310-300 extended-range twin-turbofan airliner in the livery of Saeta

118 INTERNATIONAL: AIRCRAFT—AIRBUS

similar to that of A300-600, consists of a three-phase 115/200V 400Hz constant frequency AC system and a 28V DC system; two 90kVA engine driven brushless generators for normal single-channel operation, with automatic transfer of busbars in the event of a generator failure; each has overload rating of 135kVA for 5 min and 180kVA for 5 s; third (identical) AC generator, directly driven at constant speed by APU, can be used during ground operations, and also in flight to compensate for loss of one or both engine driven generators; current production A310s have APU with improved relight capability, which can be started and operated throughout the flight envelope. Any one generator can provide sufficient power to operate all equipment and systems necessary for indefinite period of safe flight; DC power is generated via three 150A transformer-rectifiers; three nickel-cadmium batteries supplied. Flight crew oxygen system fed from rechargeable pressure bottle of 2,166 litres (76.5 cu ft) capacity; standard options are second 76.5 cu ft bottle, a 3,256 litre (115 cu ft) bottle, and an external filling connection; emergency oxygen sets for passengers and cabin attendants. Anti-icing of outer wing leading-edge slats and engine air intakes by hot air bled from engines; and of pitot probes, static ports and plates, and sensors, by electric heating. For current production A310s, an EROPS (extended-range operations) modification kit, as for the A300-600, is available.

AVIONICS: As described for A300-600.

DIMENSIONS, EXTERNAL:
Wing span	43.89 m (144 ft 0 in)
Wing chord: at root	8.38 m (27 ft 6 in)
at tip	2.18 m (7 ft 1¾ in)
Wing aspect ratio	8.8
Length overall	46.66 m (153 ft 1 in)
Fuselage: Length	45.13 m (148 ft 0¾ in)
Max diameter	5.64 m (18 ft 6 in)
Height overall	15.80 m (51 ft 10 in)
Tailplane span	16.26 m (53 ft 4¼ in)
Wheel track	9.60 m (31 ft 6 in)
Wheelbase (c/l of shock absorbers)	15.21 m (49 ft 10¾ in)
Passenger door (forward, port): Height	1.93 m (6 ft 4 in)
Width	1.07 m (3 ft 6 in)
Height to sill at OWE	4.54 m (14 ft 10¾ in)
Passenger door (rear, port): Height	1.93 m (6 ft 4 in)
Width	1.07 m (3 ft 6 in)
Height to sill at OWE	4.85 m (15 ft 11 in)
Servicing doors (forward and rear, stbd)	as corresponding passenger doors
Upper deck cargo door (A310C/F)	as A300-600
Emergency exits (overwing, port and stbd, each):	
Height	1.39 m (4 ft 6¾ in)
Width	0.67 m (2 ft 2½ in)
Underfloor cargo door (forward):	
Height	1.71 m (5 ft 7½ in)
Width	2.69 m (8 ft 10 in)
Height to sill at OWE	2.611 m (8 ft 6¾ in)
Underfloor cargo door (rear):	
Height	1.71 m (5 ft 7½ in)
Width	1.81 m (5 ft 11¼ in)
Height to sill at OWE	2.72 m (8 ft 11 in)
Underfloor cargo door (aft bulk hold):	
Height	0.95 m (3 ft 1½ in)
Width	0.95 m (3 ft 1½ in)
Height to sill at OWE	2.751 m (9 ft 0¼ in)

DIMENSIONS, INTERNAL:
Cabin, excl flight deck: Length	33.24 m (109 ft 0¾ in)
Max width	5.28 m (17 ft 4 in)
Max height	2.33 m (7 ft 7¾ in)
Volume	210.0 m³ (7,416.1 cu ft)
Forward cargo hold: Length	7.63 m (25 ft 0 in)
Max width	4.18 m (13 ft 8½ in)
Height	1.71 m (5 ft 7¼ in)
Volume	50.3 m³ (1,776.3 cu ft)
Rear cargo hold: Length	5.033 m (16 ft 6¼ in)
Max width	4.17 m (13 ft 8¼ in)
Height	1.67 m (5 ft 5¾ in)
Volume	34.5 m³ (1,218.4 cu ft)
Aft bulk hold: Volume	17.3 m³ (610.9 cu ft)
Total overall cargo volume	102.1 m³ (3,605.6 cu ft)

AREAS:
Wings, gross	219 m² (2,357.3 sq ft)
Leading-edge slats (total)	28.54 m² (307.20 sq ft)
Trailing-edge flaps (total)	36.68 m² (394.82 sq ft)
Ailerons (total)	6.86 m² (73.84 sq ft)
Spoilers (total)	7.36 m² (79.22 sq ft)
Airbrakes (total)	6.16 m² (66.31 sq ft)
Fin	45.20 m² (486.53 sq ft)
Rudder	13.57 m² (146.07 sq ft)
Tailplane	44.80 m² (482.22 sq ft)
Elevators (total)	19.20 m² (206.67 sq ft)

WEIGHTS AND LOADINGS (218-seat configuration. A: CF6-80C2A2 engines, B: PW4152s, C: CF6-80C2A8s, D: PW4156As):
Manufacturer's weight empty:		
200:	A	71,660 kg (157,983 lb)
	B	71,601 kg (157,853 lb)
300:	A	71,840 kg (158,380 lb)
	B	71,781 kg (158,250 lb)
Operating weight empty: 200:	A	80,142 kg (176,683 lb)
	B	80,125 kg (176,645 lb)
300:	A	80,344 kg (177,128 lb)
	B	80,296 kg (177,022 lb)
	C	80,329 kg (177,095 lb)
	D	80,255 kg (176,932 lb)
Max payload: 200:	A	32,860 kg (72,443 lb)
	B	32,880 kg (72,490 lb)
300:	A	32,656 kg (71,995 lb)
	B	32,704 kg (72,100 lb)
*Max usable fuel: 200		44,100 kg (97,224 lb)
300		49,039 kg (108,112 lb)
Max T-O weight: 200		142,000 kg (313,055 lb)
300		150,000 kg (330,695 lb)
options (300)		153,000 kg (337,305 lb)
		or 157,000 kg (346,125 lb)
		or 164,000 kg (361,560 lb)
Max landing weight: 200, 300		123,000 kg (271,170 lb)
options (200 and 300)		124,000 kg (273,375 lb)
Max zero-fuel weight: 200, 300		113,000 kg (249,120 lb)
options (200 and 300)		114,000 kg (251,330 lb)

* optional additional tank in aft cargo hold adds 5,779 kg (12,740 lb) of fuel and increases OWE/reduces max payload by 692 kg (1,525 lb). Two additional tanks add 11,560 kg (25,485 lb) of fuel and increase OWE/reduce max payload by 1,320 kg (2,910 lb)

PERFORMANCE (at basic max T-O weight except where indicated; engines as under Weights and Loadings):
Typical long-range cruising speed at 9,450-12,500 m (31,000-41,000 ft): A, B, C, D Mach 0.80
Approach speed at max landing weight:
A, B, C, D	135 knots (250 km/h; 155 mph)

T-O field length at S/L, ISA + 15°C:
200: A	1,860 m (6,100 ft)
B	1,799 m (5,900 ft)
300: A (at 150 tonne MTOGW)	2,408 m (7,900 ft)
B (at 150 tonne MTOGW)	2,225 m (7,300 ft)
C (at 164 tonne MTOGW)	2,560 m (8,400 ft)
D (at 164 tonne MTOGW)	2,400 m (7,875 ft)

Landing field length at S/L, at max landing weight (200 and 300):
A	1,479 m (4,850 ft)
B	1,555 m (5,100 ft)

Runway ACN for flexible runway, category B:
standard tyres: 200	43
300	49
optional tyres: 200	41
300	47

Range (1991 and subsequent deliveries) at typical airline OWE with 218 passengers and baggage, international reserves for 200 nm (370 km; 230 mile) diversion:
200: A, B	3,670 nm (6,800 km; 4,225 miles)
300 at basic MTOGW:	
A	4,310 nm (7,982 km; 4,960 miles)
B	4,300 nm (7,968 km; 4,951 miles)
300 at 164 tonne MTOGW with ACTs:	
C, D	5,170 nm (9,580 km; 5,953 miles)

OPERATIONAL NOISE LEVELS (ICAO Annex 16, Chapter 3):
T-O: 200: A	89.6 EPNdB (95.3 limit)
300: A	91.2 EPNdB (95.6 limit)
Sideline: 200: A	96.4 EPNdB (99.2 limit)
300: A	96.3 EPNdB (99.4 limit)
Approach: 200, 300: A	98.6 EPNdB (102.9 limit)

AIRBUS A320

TYPE: Twin-turbofan short/medium-range airliner.
PROGRAMME: Launched 2 March 1984; four-aircraft development programme (first flight 22 February 1987 by F-WWAI); JAR (UK/French/German/Dutch) certification of A320-100 with CFM56-5 engines, for two-crew operation, awarded 26 February 1988; first deliveries (Air France and British Airways) 28 and 31 March 1988 respectively; JAR certification of A320-200 with CFM56-5s received 8 November 1988, followed by FAA type approval for both models 15 December 1988; certification with V2500 engines (first flown 28 July 1988) received 20 April (JAR) and 6 July 1989 (FAA), deliveries with this power plant (to Adria Airways) beginning 18 May 1989.

CURRENT VERSIONS: **A320**: Initial version (21 ordered); details in 1987-88 *Jane's*. Superseded by A320-200.

A320-200: Now called simply **A320**. Standard version from third quarter 1988; differs from A320-100 in having wingtip fences, wing centre-section fuel tank and higher max T-O weights. *Detailed description applies to this version.*

A320 research: A320 used for riblet research 1989-91; riblet programme continued with in-service trial in A340-200 and -300 starting second half 1993; laminar-flow fin with porous skin to be flown on A320 prototype during 1993.

A321-100: Stretched version of A320-200; described separately.

A319: Shortened version of A320; described separately.

CUSTOMERS: See table on page 114.

DESIGN FEATURES: First subsonic commercial aircraft with fly-by-wire control throughout normal flight envelope, sidestick controllers instead of control columns, composites for major primary structures, and centralised maintenance system; advanced technology wings with 25° sweepback at quarter-chord, 5° 6′ 36″ dihedral, and incorporate gust load alleviation system, plus experience from A310 and significant commonality with other Airbus Industrie aircraft where cost-effective; 6° tailplane dihedral.

FLYING CONTROLS: A320 is first subsonic commercial aircraft equipped for fly-by-wire (FBW) control throughout entire normal flight regime, and first to have sidestick controller (one for each pilot) instead of control column and hand wheel. Thomson-CSF/SFENA digital FBW system features five main computers and operates, via electrical signalling and hydraulic jacks, all primary and secondary flight controls; pilot's pitch and roll commands applied through sidestick controller via two different types of computer; these have redundant architecture to provide safety levels at least as high as those of mechanical systems they replace; system incorporates flight envelope protection features to a degree that cannot be achieved with conventional mechanical control systems, and its computers will not allow aircraft's structural and aerodynamic limitations to be exceeded; even if pilot holds sidestick fully forward, it is impossible to go beyond aircraft's maximum operating speed (V_{MO}) for more than a few seconds; if pilot holds sidestick fully back, aircraft is controlled to an 'alpha floor' angle of attack, a safe airspeed above stall and throttles automatically opened to ensure positive climb. Nor is it possible to exceed g limits while manoeuvring.

Fly-by-wire system controls, ailerons, elevators, spoilers, flaps and leading-edge slats; rudder movement and tailplane trim connected to FBW system, but also signalled mechanically when used to provide final backup pitch and yaw control, which suffices for basic instrument flying. Each wing has five-segment leading-edge slats (one inboard, four outboard of engine pylon), two-segment Fowler trailing-edge flaps, and five-segment spoilers forward of outboard flap; all 10 spoilers used as lift dumpers, outer four also for roll and inner six as airbrakes; slat and flap controls by Liebherr and Lucas.

STRUCTURE: Generally similar to that of A310, but with AFRP for fuselage belly fairing skins; GFRP for fin leading-edge and fin/fuselage fairing; CFRP for wing fixed leading/

Airbus A320 twin-turbofan short/medium-range airliner of Monarch Airlines

trailing-edge bottom access panels and deflectors, trailing-edge flaps and flap track fairings, spoilers, ailerons, fin (except leading-edge), rudder, tailplane, elevators, nosewheel/mainwheel doors, and main-gear leg fairing doors. Aerospatiale builds entire front fuselage (forward of wing leading-edge), cabin rear doors, nosewheel doors, centre wing box and engine pylons, and is responsible for final assembly; centre and rear fuselage, tailcone, wing flaps, fin, rudder, and commercial furnishing undertaken by Deutsche Aerospace Airbus; British Aerospace builds main wings, including ailerons, spoilers and wingtips, and main landing gear leg fairings; Belairbus produces leading-edge slats; CASA responsible for tailplane, elevators, mainwheel doors, and sheet metal work for parts of rear fuselage.

LANDING GEAR: Hydraulically retractable tricycle type, with twin wheels and oleo-pneumatic shock absorber on each unit (four-wheel main-gear bogies, for low-strength runways, optional); Dowty main units retract inward into wing/body fairing; steerable Messier-Bugatti nose unit retracts forward; nosewheel steering angle ±75° (effective turning angle ±70°). Radial tyres standard, size 45 × 16-R20 on main gear and 30 × 8.8-R15 on nose gear; optional tyres for main gear are 49 × 17R20 or 49 × 19R20 radials, or 46 × 16-20 or 49 × 19-20 crossplies, and for nose gear 30 × 8.8-15 crossplies; tyres for main-gear bogie option are either 915 × 300R16 radials or 36 × 11-16 crossplies. Carbon brakes standard. Min width of pavement for 180° turn 23.1 m (75 ft 9½ in).

POWER PLANT: Two 104.5-111.2 kN (23,500-25,000 lb st) class CFM International CFM56-5A1 turbofans for first aircraft delivery in Spring 1988, with 111.2 kN (25,000 lb st) IAE V2500-A1 engines available for aircraft delivered from May 1989 and CFM56-5A3 from November 1990 (117.9 kN; 26,500 lb st); 117.9 kN (26,500 lb st) CFM56-5B4 and 111.2 kN (25,000 lb st) IAE V2525-A5 available from 1994. Nacelles by Rohr Industries; thrust reversers by Hispano-Suiza on CFM56 engines, by IAE for V2500s. Dual-channel FADEC (full authority digital engine control) system standard on each engine. For A320-200, standard fuel capacity in wing and wing centre-section tanks is 23,859 litres (6,303 US gallons; 5,248 Imp gallons); for A320-100, standard fuel capacity without centre-section tank is 15,843 litres (4,185 US gallons; 3,485 Imp gallons).

ACCOMMODATION: Standard crew of two on flight deck, with one (optionally two) forward facing folding seats for additional crew members; seats for four cabin attendants. Single-aisle main cabin has seating for up to 179 passengers, depending upon layout, with locations at front and rear of cabin for galley(s) and toilet(s); typical two-class layout has 12 seats four-abreast at 91.5 cm (36 in) pitch in 'super first' and 138 six-abreast at 81 cm (32 in) pitch economy class; alternative 152 six-abreast seats (84 business + 68 economy) at 86 and 78 cm (34 and 31 in) pitch respectively; single class economy layout could offer 164 seats at 81 cm (32 in) pitch, or up to 179 in high-density configuration. Compared with existing single-aisle aircraft, fuselage cross-section is significantly increased, permitting use of wider triple seats to provide higher standards of passenger comfort; five-abreast business class seating provides standard equal to that offered as first class on major competitive aircraft. In addition, wider aisle permits quicker turnrounds. Overhead stowage space superior to that available on existing aircraft of similar capacity, and provides ample carry-on baggage space; best use of under-seat space for baggage is provided by improved seat design and optimised positioning of seat rails. Passenger doors at front and rear of cabin on port side, forward one having optional integral airstairs; service door opposite each of these on starboard side. Two overwing emergency exits each side. Fuselage double-bubble cross-section provides increased baggage/cargo hold volume and working height, and ability to carry containers derived from standard interline LD3 type. As base is same as that of LD3, all existing wide-body aircraft and ground handling equipment can accept these containers without modification. Forward and rear underfloor baggage/cargo holds, plus overhead lockers; with 164 seats, overhead stowage space per seat is 0.056 m³ (2.0 cu ft). Mechanised cargo loading system will allow up to seven LD3-46 containers to be carried in freight holds (three forward and four aft).

SYSTEMS: Liebherr/ABG-Semca air-conditioning, Hamilton Standard/Nord-Micro pressurisation, hydraulic, Sundstrand electrical system, and new and more efficient Garrett APU. Primary electrical system powered by two Sundstrand 90kVA constant frequency generators, providing 115/200V three-phase AC at 400Hz; third generator of same type, directly driven at constant speed by APU, can be used during ground operations and, if required, during flight.

AVIONICS: Fully equipped digital avionics fit, to ARINC 700 series specification, including advanced digital automatic flight control and flight management systems; AFCS integrates functions of SFENA autopilot and Honeywell FMS; each pilot has two Thomson-CSF/VDO electronic flight instrumentation system (EFIS) displays (primary flight display and a navigation display); primary flight display is first on an airliner to incorporate speed, altitude and heading. Between these two pairs of displays are two Thomson-CSF/VDO electronic centralised aircraft monitor (ECAM) displays unique to Airbus Industrie and developed from the ECAM systems on the A310 and A300-600; upper display incorporates engine performance and warning, lower display carries warning and system synoptic diagrams. Honeywell air data and inertial reference system.

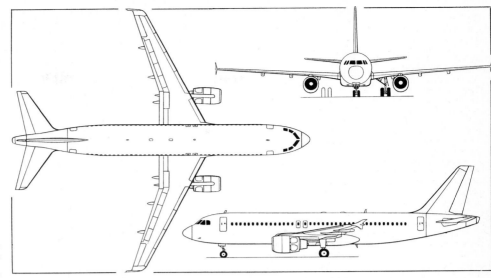

Airbus A320 twin-turbofan single-aisle 150/179-seat airliner *(Dennis Punnett)*

DIMENSIONS, EXTERNAL:
Wing span	33.91 m (111 ft 3 in)
Wing aspect ratio	9.4
Length overall	37.57 m (123 ft 3 in)
Fuselage: Max width	3.95 m (12 ft 11½ in)
Max depth	4.14 m (13 ft 7 in)
Height overall	11.80 m (38 ft 8½ in)
Tailplane span	12.45 m (40 ft 10 in)
Wheel track (c/l of shock struts)	7.59 m (24 ft 11 in)
Wheelbase	12.63 m (41 ft 5 in)
Passenger doors (port, forward and rear), each:	
Height	1.85 m (6 ft 1 in)
Width	0.81 m (2 ft 8 in)
Height to sill	3.415 m (11 ft 2½ in)
Service doors (stbd, forward and rear), each	as corresponding passenger doors
Overwing emergency exits (two port and two stbd), each:	
Height	1.02 m (3 ft 4¼ in)
Width	0.51 m (1 ft 8 in)
Underfloor baggage/cargo hold doors (stbd, forward and rear), each: Height	1.249 m (4 ft 1¼ in)
Width	1.82 m (5 ft 11½ in)

DIMENSIONS, INTERNAL:
Cabin, excl flight deck: Length	27.38 m (89 ft 10 in)
Max width	3.696 m (12 ft 1½ in)
Max height	2.22 m (7 ft 4 in)
Baggage/cargo hold volume:	
front	13.28 m³ (469 cu ft)
rear	25.48 m³ (900 cu ft)

AREAS:
Wings, gross	122.4 m² (1,317.5 sq ft)
Leading-edge slats (total)	12.64 m² (136.1 sq ft)
Trailing-edge flaps (total)	21.10 m² (227.1 sq ft)
Ailerons (total)	2.74 m² (29.49 sq ft)
Spoilers (total)	8.64 m² (93.00 sq ft)
Airbrakes (total)	2.35 m² (25.30 sq ft)
Vertical tail surfaces (total)	21.5 m² (231.4 sq ft)
Horizontal tail surfaces (total)	31.0 m² (333.7 sq ft)

WEIGHTS AND LOADINGS (Typical 150-passenger configuration. A: CFM56-5A1 engines, B: V2500-A1s):
Operating weight empty: A	41,583 kg (91,675 lb)
B	41,870 kg (92,307 lb)
Max payload: A	19,417 kg (42,807 lb)
B	19,220 kg (42,373 lb)
Max fuel	19,159 kg (42,238 lb)
Max T-O weight: standard	73,500 kg (162,040 lb)
option	77,000 kg (169,756 lb)
Max landing weight	64,500 kg (142,195 lb)
Max zero-fuel weight	60,500 kg (133,380 lb)
Max wing loading:	
standard	600.5 kg/m² (123.0 lb/sq ft)
option	629.08 kg/m² (128.8 lb/sq ft)

PERFORMANCE (at max T-O weight except where indicated; engines A and B as for Weights and Loadings, C: CFM56-5A3, D: 77,000 kg; 169,756 lb MTOW):
T-O distance at S/L, ISA + 15°C:	
A	2,336 m (7,665 ft)
B	2,294 m (7,526 ft)
C	2,042 m (6,700 ft)
D	2,286 m (7,500 ft)
Landing distance at max landing weight:	
A, C	1,470 m (4,823 ft)
B	1,442 m (4,730 ft)
Runway ACN (flexible runway, category B):	
twin-wheel, standard 45 × 16R20 tyres	41
four-wheel bogie option, 36 × 11-16 Type VII or 900 × 300-R16	22

Range with 150 passengers and baggage in two-class layout, FAR domestic reserves and 200 nm (370 km; 230 mile) diversion:

Airbus A320 flight deck with sidestick controllers outboard and 6-screen EFIS

120 INTERNATIONAL: AIRCRAFT—AIRBUS

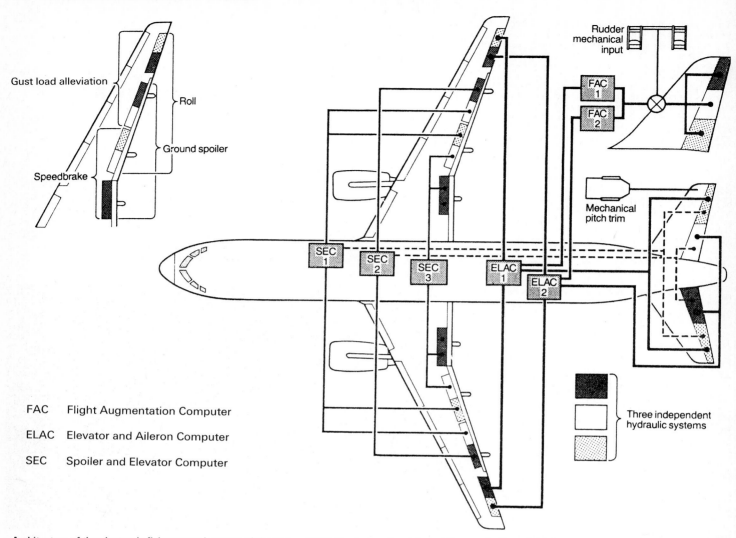

Architecture of the electronic flying control system of the Airbus A320. The fly-by-wire slats and flaps are independently controlled and not part of the primary system. Inputs to ELACs and SECs come from the Air Data Inertial Reference System, the Sidesticks and the Flight Guidance Computer. See flying controls heading in A320 entry

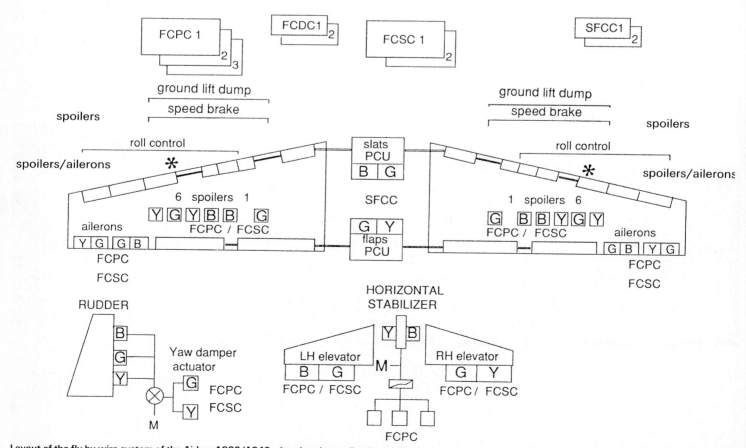

Layout of the fly-by-wire system of the Airbus A330/A340, showing the application of the three hydraulic systems, Blue, Yellow and Green, to the various surfaces and the various computers, three Flight Control Primary Computers, two Flight Control Data Concentrator computers, two Flight Control Secondary Computers and two Slat and Flap Control Computers. In the diagram M stands for direct mechanical control and PCU for powered control unit. See flying controls heading in A340 entry

standard: A, C		2,800 nm (5,190 km; 3,222 miles)
B		2,850 nm (5,280 km; 3,280 miles)
option: A, C		2,950 nm (5,465 km; 3,395 miles)
B		3,000 nm (5,559 km; 3,452 miles)

OPERATIONAL NOISE LEVELS (ICAO Annex 16, Chapter 3):
T-O: A	88.0 EPNdB (91.5 limit)
B	86.6 EPNdB (91.5 limit)
C	86.5 EPNdB (91.5 limit)
Sideline: A	94.4 EPNdB (96.8 limit)
B	92.8 EPNdB (96.8 limit)
C	94.8 EPNdB (96.8 limit)
Approach: A	96.2 EPNdB (100.5 limit)
B	96.6 EPNdB (100.5 limit)
C	96.0 EPNdB (100.5 limit)

AIRBUS A319

TYPE: Short-fuselage A320.
PROGRAMME: Airbus Board officially authorised start of sales 22 May 1992; full development to begin as soon as viable number of commitments received; target in-service date Spring 1996.
CUSTOMERS: ILFC first customer (six); Alitalia and Swissair reported to be possible customers in Spring 1993; 450 sales predicted.
COSTS: Total development cost estimated at $275 million, entirely financed by Airbus Industrie.
DESIGN FEATURES: Seven fuselage frames shorter than A320, giving overall length of 33.8 m (110 ft 11 in), but otherwise little changed. Seats 124 passengers in typical two-class layout, compared with 150 in A320 and 186 in A321; range of 2,150 nm (3,984 km; 2,472 miles) claimed as longest in this category of airliner.
FLYING CONTROLS: Same flight deck and flying control system as A320.
STRUCTURE: As A320. Aircraft to be assembled in Germany by Deutsche Aerospace Airbus, which already assembles the stretched A321; partner workshares rearranged to maintain workshare balance between France and Germany.
POWER PLANT: Two CFMI CFM56-5A or IAE V2500A5 turbofans, each giving 98-104 kN (22,000-23,500 lb st). Max fuel capacity 23,860 litres (6,300 US gallons; 5,248 Imp gallons).

WEIGHTS AND LOADINGS:
Typical operating weight empty	39,884 kg (87,930 lb)
Max payload	17,100 kg (37,698 lb)
Max T-O weight	64,000 kg or 68,000 kg (141,094 lb or 149,912 lb)
Max landing weight	61,000 kg (134,480 lb)

AIRBUS A321

TYPE: Twin-turbofan short/medium-range airliner.
PROGRAMME: Announced 22 May and launched 24 November 1989 as stretched version of A320; four development aircraft planned; first flight with V2530 lead engine 11 March 1993 (F-WWIA), second aircraft with alternative engine due three months later; initial certification planned for December 1993; service entry (Lufthansa) January 1994, with alternative engine model (Alitalia) three months later.
CUSTOMERS: See table on page 114.
DESIGN FEATURES: Stretched version of A320, with 4.27 m (14 ft 0 in) fuselage plug immediately forward of wing and 2.67 m (8 ft 9 in) plug immediately aft; overwing emergency exits replaced by two exits in each plug section; pairs of wing fuel tanks unified and system simplified; other changes include local structural reinforcement of existing assemblies, slightly extended wing trailing-edge with double-slotted flaps, uprated landing gear and higher T-O weights.

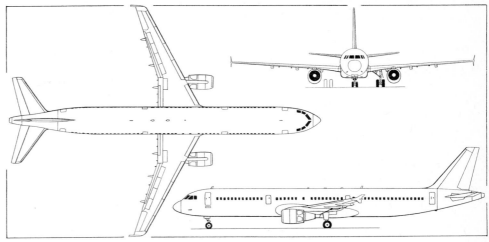

A321 stretched development of the Airbus A320 *(Dennis Punnett)*

First flight of Airbus A321 from Hamburg on 11 March 1993

STRUCTURE: As for A320 except for airframe changes noted under Design Features; front fuselage plug by Alenia, rear one by BAe; final assembly and outfitting by Deutsche Aerospace Airbus at Hamburg.
LANDING GEAR: Uprated, with 22 in wheel rims, 1,270 × 455R22 mainwheel tyres and increased energy brakes; wheels and brakes by Aircraft Braking Systems.
POWER PLANT: Offered initially with two CFM56-5B1 or V2530-A5 turbofans, both rated at 133.4 kN (30,000 lb st); V2530 is lead engine, but each type will be flight tested on development aircraft. CFM56-5B2 engines of 137.9 kN (31,000 lb st) will be available optionally. Fuel capacity 23,700 litres (6,261 US gallons; 5,213 Imp gallons).
ACCOMMODATION: Typically offers 24 per cent more seats and 40 per cent more hold volume than A320-200; examples are 186 passengers in two-class layout (16 first class at 91 cm; 36 in seat pitch and 170 economy class at 81 cm; 32 in), or 200 passengers in all-economy configuration. Each fuselage plug incorporates one pair of emergency exits, replacing single overwing pair of A320.
SYSTEMS: Choice of Garrett GTCP 36-300 or APIC APS 3200 APU; full commonality with A320 installation.

DIMENSIONS, EXTERNAL:
Wing span	34.09 m (111 ft 10 in)
Length overall	44.51 m (146 ft 0 in)
Height overall	11.81 m (38 ft 9 in)
Passenger and service doors (port/stbd, forward and rear)	as for A320
Emergency exits (forward stbd and rear port/stbd, each):	
Height	1.52 m (5 ft 0 in)
Width	0.76 m (2 ft 6 in)
Emergency exit (forward port, usable also as passenger door):	
Height	1.85 m (6 ft 1 in)
Width	0.76 m (2 ft 6 in)

DIMENSIONS, INTERNAL:
Cabin, excl flight deck: Length	34.39 m (112 ft 10 in)
Baggage/cargo hold volume:	
front	23.02 m³ (813 cu ft)
rear	29.02 m³ (1,025 cu ft)

WEIGHTS AND LOADINGS (Typical 186-passenger layout. A: CFM56-5B1, B:V2530-A5, C: CFM56-5B2):
Operating weight empty: A, C	47,512 kg (104,746 lb)
B	47,626 kg (104,997 lb)
Max payload: A	21,988 kg (48,475 lb)
B	21,874 kg (48,224 lb)
Max fuel: A, B, C	19,031 kg (41,956 lb)
Max T-O weight	83,000 kg (182,984 lb)
Max landing weight	73,500 kg (162,040 lb)
Max zero-fuel weight	69,500 kg (153,221 lb)
Max wing loading	671.6 kg/m² (137.5 lb/sq ft)

PERFORMANCE (estimated):
T-O distance at max T-O weight, S/L, ISA +15°C:	
A	2,339 m (7,674 ft)
B	2,328 m (7,638 ft)
C	2,276 m (7,468 ft)
Landing distance at max landing weight:	
A, B, C	1,587 m (5,208 ft)
Runway ACN (flexible runway, category B):	
standard	48
Range with 186 passengers and baggage at typical airline OWE, FAR domestic reserves and 200 nm (370 km; 230 mile) diversion:	
A, C	2,300 nm (4,260 km; 2,648 miles)
B	2,350 nm (4,352 km; 2,706 miles)

Artist's impression of 'destretched' Airbus A319

AIRBUS A321 structural layout

Plugs in the A321 fuselage can be identified by the additional doors fore and aft of the wing. Structure otherwise also represents the A320 and the shortened A319. Engines on the wing are CFM56-5B1. Additional engine is IAE V2530-A5.
(Drawing by Robert Roux for Revue Aerospatiale)

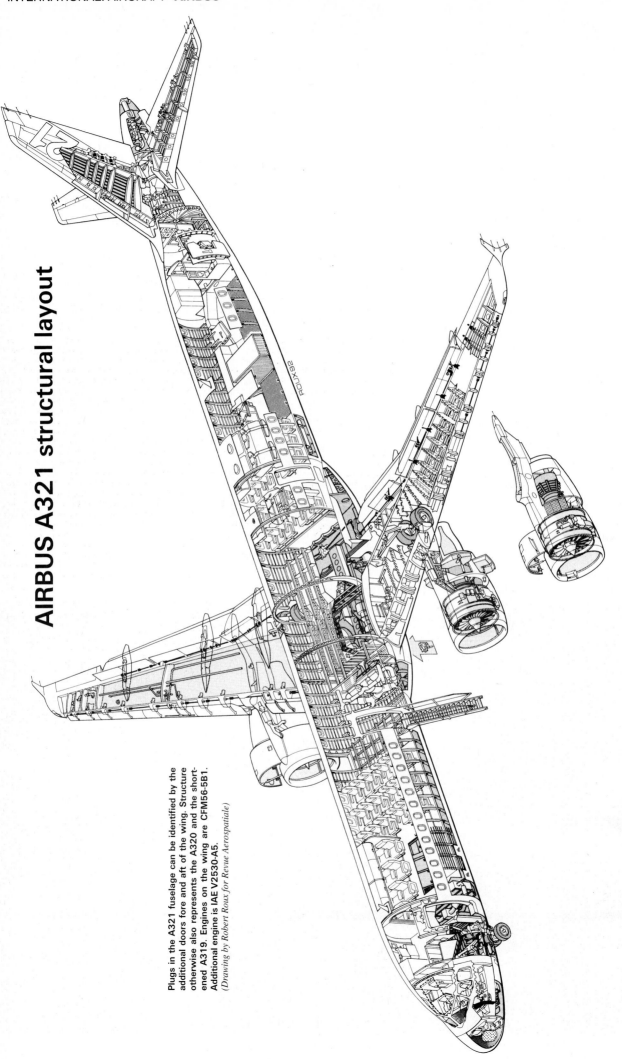

Airbus A321 cutaway drawing, showing the two different engines initially available

OPERATIONAL NOISE LEVELS (ICAO Annex 16, Chapter 3, estimated):
T-O: A 87.4 EPNdB (92.1 limit)
   B 87.3 EPNdB (92.1 limit)
Sideline: A 94.5 EPNdB (97.2 limit)
   B 95.9 EPNdB (97.2 limit)
Approach: A 97.0 EPNdB (100.9 limit)
   B 96.9 EPNdB (100.9 limit)

AIRBUS A340

TYPE: Large-capacity wide-bodied medium/long-range airliner.
PROGRAMME: Launched 5 June 1987 as combined programme with A330, differing mainly in number of engines and in engine-related systems; first six aircraft are four A340-300s and two -200s; first flight 25 October 1991 (A340-300); A340-200 and -300 certificated simultaneously by 18 European joint airworthiness authorities (JAA) on 22 December 1992; A340-300 and -200 entered service with Air France and Lufthansa March 1993; both received FAA certification 27 May 1993.
CURRENT VERSIONS: **A340-300**: Four-engined long-range version, carrying up to 375 passengers (standard) or 440 (optional) and powered initially by CFM56-5C2 turbofans.
Longer-range version of A340-300, previously referred to as 340-300X, able to carry typical load of 295 passengers over distances of 7,200 nm (13,340 km; 8,290 miles); powered by 151.2 kN (34,000 lb st) CFM56-5C4 turbofans; max T-O weight 271,000 kg (597,450 lb). Stronger landing gear, riblets and engine refinements probably to be added to achieve range. Option for 1996 delivery.
A340-200: Longer-range version of A340-300, with same initial power plant and shorter fuselage; exit-limited seating capacity 303 or 263 in three classes; 26 LD3s or nine standard pallets under floor; first flight (F-WWBA) 1 April 1992; entered service March 1993.
CUSTOMERS: See table on page 114.
DESIGN FEATURES: Capitalise on commonality with A330 (identical wing/cockpit/tail unit and same basic fuselage) to create aircraft for different markets, and also have much in common (eg, existing Airbus wide-body fuselage cross-sections, A310/A300-600 fin, advanced versions of A320 cockpit and systems) with existing Airbus range; FAA has approved cross-qualification for A320, A321, A330 and A340. New design wing (by BAe), approx 40 per cent larger than that of A300-600, has 30° sweepback and winglets. One strake on each engine nacelle and airflow smoothing bulge on inboard face of each outer engine pylon improve performance.
FLYING CONTROLS: In A340/A330 electronic flight control system (EFCS), roll axis is controlled by two individual outboard ailerons and five outboard spoiler panels on each wing; pitch axis control is by trimmable tailplane and separate left and right elevators; tailplane can also be mechanically controlled from flight deck, but fly-by-wire com-

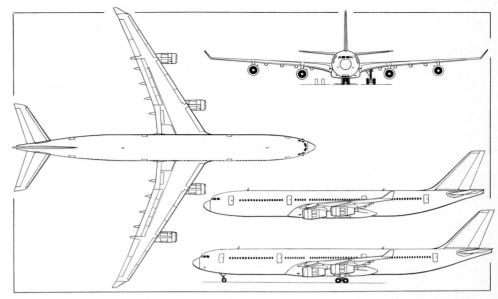

Airbus A340-300 four-turbofan long-range airliner, with additional side view (centre) of A340-200 *(Dennis Punnett)*

puter inputs are superimposed; single rudder is directly linked to rudder pedals, with dual yaw damping inputs superimposed. High-lift devices consist of full-span slats, flaps and aileron droop; speed braking and lift dumping by raising all six spoilers on each wing and raising all ailerons. Slats and flaps controlled outside main fly-by-wire complex by duplicated slat and flap control computers (SFCC). See diagram on page 120.

Control surface actuation by three hydraulic systems (green, yellow, blue); two powered control units (PCU) at each aileron and elevator are controlled either by primary or secondary flight control computers; single actuators at spoiler panels controlled by primary or secondary flight control computers; dual PCUs for fly-by-wire tailplane trimming, and for centrally located flap and slat actuators; three PCUs at rudder.

Fly-by-wire computers include three flight control primary computers (FCPC) and two flight control secondary computers (FCSC); each computer has two processors with different software; primary and secondary computers have different architecture and hardware; power supplies and signalling lanes are segregated; system provides stall protection, overspeed protection and manoeuvre protection as in the A320, but the A340/330 computer arrangement maintains the protections for longer in the face of failures of sensors and inputs; FCPC and FCSC all operate continuously and provide comparator function to active channels, but only one in active control at any one control surface; reconfiguration logic can provide alternative control after failures; different normal and alternative control laws apply fly-by-wire basically as a g demand in pitch and rate demand in roll, plus complex manoeuvre limitations; if all three inertial systems fail (removing attitude information), system reverts to direct mode in which control surface angle is directly related to sidestick position; ultimate control mode is direct control of rudder and tailplane angle from rudder pedals and manual trim wheel, which is sufficient for accurate basic instrument flight.

Pilots have sidestick controllers and normal rudder pedals; EFIS instrumentation consists of duplicated primary flight displays (PFD), navigation displays (ND) and electronic centralised aircraft monitors (ECAM); three display management computers with separate EFIS and ECAM channels, can each control all six displays in their four possible formats; flight path control by duplicated flight management and guidance and envelope computers (FMGEC); they control every phase of flight including course, attitude, engine thrust and flight planning using information from Global Positioning System and inertial systems; point of no return and fuel dumping calculations

Airbus A340 at time of delivery to Lufthansa in February 1993

Airbus A340 in take-off configuration showing three-leg main landing gear and flap/slat geometry

Flight decks of the A330 and A340 are almost identical and very similar to A320. The FAA has approved cross-qualification of pilots

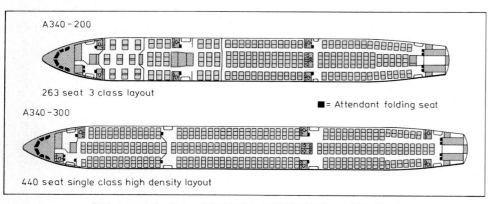

Cabin layouts for Airbus A340-300 and A340-200 (*Jane's*/Mike Keep)

made automatically for long-range flights in A340/330. Control system data is collected for maintenance purposes by two flight control data concentrator (FCDC) computers.

In normal flight, bank angle limited to 33° hands-off (autopilot control) and 67° with full stick displacement; airspeeds limited to 305 knots (565 km/h; 350 mph) and Mach 0.82 under automatic flight control; if stick is held fully forward, the nose is automatically raised when airspeed reaches V_{MO} + 15 knots (28 km/h; 17 mph); if nose is raised equivalent protections apply; 'alpha max' (13° clean and 19° with flap), slightly below max lift coefficient, is the greatest achievable with sidestick; 'alpha floor' is the angle of attack beyond which throttles progressively open to go-around power and airspeed is finally held steady just above stall, even if stick is continuously held back; alpha protection applied at 'normal' and 'hard' modes according to alpha rate; below protection speed (V_{PROT}) of 142 knots (263 km/h; 163 mph) automatic and manual trimming stops; outer ailerons remain centred at over 200 knots (371 km/h; 230 mph); if a spoiler panel fails, the symmetrical opposite panel stops operating; if rudder yaw damping fails, pairs of spoiler panels are used instead; minimum-speed marker on PFD adjusts to changing aircraft configuration, airbrake and pitch rate; fuel automatically transferred between wing and tailplane tanks to minimise trim drag when cruising above 7,620 m (25,000 ft).

Engines controlled by setting throttle levers to marks on quadrant, such as climb (CLB), max continuous and flexible take-off (Flex T-O); digital engine control makes detailed settings appropriate to altitude and temperature.

During landing and take-off, nosewheel steering, by rudder pedals, automatically disengages above 100 knots (185 km/h; 115 mph); demands for more than 4°/s nose-up pitch restrained near ground; max airspeeds for flap and slat signalled on airspeed scale; trim automatically cancels effect of flap, landing gear and airspeed; ailerons droop with full flap selection and deflect 25° up with spoilers in lift dump after touchdown; thrust reverser failure automatically countered by cancelling symmetrical opposite reverser; voice warning demands throttle closure at 6 m (20 ft) during landing flare; autothrottle disengaged when throttles closed at touchdown; on touch-and-go landing, trim automatically reset for take-off when Flex T-O power selected; engine failure in flight automatically compensated, with wings held level, slight heading drift and spoiler panels sucked down to avoid unnecessary drag.

STRUCTURE: A330 and A340 wings almost identical except latter strengthened in area of outboard engine pylon with appropriate modification of leading-edge slats 4 and 5; main three-spar wing box and leading/trailing-edge ribs and fittings of aluminium alloy, with Al-Li for some secondary structures; steel or titanium slat supports; approx 13 per cent (by weight) of wings is of CFRP, GFRP or AFRP, including outer flaps and flap track fairings, ailerons, spoilers, leading/trailing-edge fixed surface panels and (possibly) winglets; common fuselage for all three initial versions, except in overall length (A340-300 and A330 same size and longest, A340-200 eight frames shorter); construction generally similar to that of A310 and A300-600 except centre-section to accept new wing; tail unit (common to all versions) utilises same CFRP fin as A300-600 and A310; new tailplane incorporates trim fuel tank and has CFRP outer main boxes bridged by aluminium alloy centre-section. Work sharing along lines similar to those for A310 and A300-600, with percentages similar to those held in consortium. Aerospatiale thus responsible for cockpit, engine pylons, part of centre-fuselage, and final assembly and outfitting at Toulouse; British Aerospace (with Textron Aerostructures, USA, as subcontractor) for wings; Deutsche Aerospace Airbus for most of fuselage, fin and interior; CASA for tailplane; Belairbus Belgium for leading-edge slats and slat tracks.

LANDING GEAR: Main (four-wheel bogie) and twin-wheel nose units identical on all versions. A340 has additional twin-wheel auxiliary unit on fuselage centreline amidships. Goodyear tyres on all units. Landing gear and surrounding structure to be reinforced for longer range A340-300.

POWER PLANT: Four 138.8 kN (31,200 lb st) CFM56-5C2 turbofans initially; performance improvement programme (PIP) modifications to combustors and turbine cooling expected to improve fuel consumption by 1.5 per cent at entry into service and 1.0 per cent later. Max fuel capacity (-200 and -300) 135,000 litres (35,664 US gallons; 29,697 Imp gallons).

ACCOMMODATION: Crew of two on flight deck (all versions); flight deck can be supplied with humidified air. Passenger seating typically six-abreast in first class, seven-abreast in business class and eight-abreast in economy (nine-abreast optional), all with twin aisles; two-class configurations seat 335 passengers in A330-300 and A340-300, and 303 passengers in A340-200; more typically, a three-class layout seats 295 in A340-300 and 263 in the A340-200. Underfloor cargo holds house up to 32 LD3 containers or 11 standard 2.24 × 3.17 m (88 × 125 in) pallets in A340-300 and A330, and 26 LD3s or 9 pallets in A340-200; both front and rear cargo holds have doors wide enough to accept 2.44 × 3.17 m (96 × 125 in) pallets; all models have a 19.68 m³ (695 cu ft) bulk cargo hold aft of the rear cargo hold.

DIMENSIONS, EXTERNAL:
Wing span (all versions)	60.30 m (197 ft 10 in)
Wing aspect ratio (all versions)	10.0
Length overall:	
A340-200	59.39 m (194 ft 10 in)
A330-300, A340-300	63.65 m (208 ft 10 in)
Fuselage: Max diameter (all versions)	5.64 m (18 ft 6 in)
Height overall (all versions)	16.74 m (54 ft 11 in)
Wheel track (all versions)	10.49 m (34 ft 5 in)

AREAS:
Wings, gross (all versions)	363.1 m² (3,908.4 sq ft)

WEIGHTS AND LOADINGS:
Typical airline operating weight empty:	
A340-200: standard	122,769 kg (270,659 lb)
longer range version	125,709 kg (277,141 lb)
A340-300: standard	126,481 kg (278,843 lb)
longer range version	129,414 kg (285,309 lb)
Max payload:	
A340-200: standard	46,231 kg (101,922 lb)
longer range version	47,291 kg (104,259 lb)
A340-300: standard	47,519 kg (104,761 lb)
longer range version	48,586 kg (107,114 lb)
Max T-O weight:	
A340-200, -300: standard	257,000 kg (566,588 lb)
longer range version	271,000 kg (597,670 lb)
Max landing weight: A340-200	185,000 kg (407,855 lb)
A340-300	190,000 kg (418,878 lb)
Max zero-fuel weight: A340-200	173,000 kg (381,400 lb)
A340-300	178,000 kg (392,423 lb)

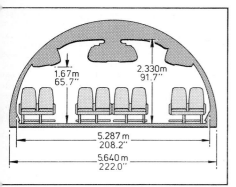

Airbus A340 passenger cabin cross-section
(Jane's/Mike Keep)

PERFORMANCE (estimated, definitions as for Weights and Loadings):
Max operating speed (M$_{MO}$) Mach 0.86
Typical operating speed Mach 0.82
Range at typical OWE, with allowances for 200 nm (370 km; 230 mile) diversion and international reserves:
A340-200:
   standard 7,350 nm (13,600 km; 8,450 miles)
   with additional fuel
      7,750 nm (14,400 km; 8,900 miles)
A340-300:
   standard 6,650 nm (12,300 km; 7,650 miles)
   longer range version
      7,200 nm (13,300 km; 8,290 miles)
OPERATIONAL NOISE LEVELS (estimated):
T-O, fly-over 96.0 EPNdB
Sideline 94.4 EPNdB
Approach 97.8 EPNdB

AIRBUS A330

TYPE: Wide-body, medium/long-range twin-engined airliner.
PROGRAMME: Developed simultaneously with four-engined A340; launched 5 June 1987; first flight (F-WWKA) 2 November 1992; certification with initial GE CF6-80E1 engines expected October 1993; first delivery (Air Inter) expected November 1993; certification with P&W PW4164 and transatlantic clearance expected May 1994; first delivery with R-R Trent 768 end 1994.
   All main structural and systems information common to both A330 and A340, listed in A340 entry. Differences in A330 given here.
CURRENT VERSIONS: **A330-300**: Seating capacity as for A340-300; 375 passengers standard, 440 passengers optional.

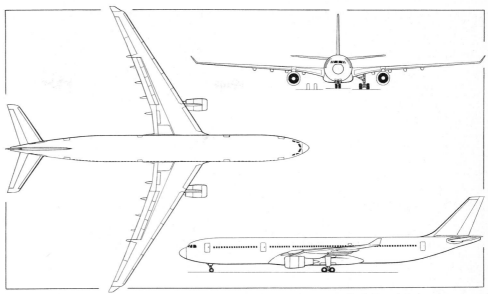

Airbus A330 twin-turbofan airliner, developed in parallel with the A340 *(Dennis Punnett)*

Airbus A330 completes first take-off 2 November 1992

First Airbus A330 in flight late 1992

Longer-range version: Previously A330-300X; higher gross weight to allow typical 335 passengers to be carried 5,300 nm (9,820 km; 6,100 miles), compared with about 4,700 nm (8,710 km; 5,412 miles) for basic A330-300; max T-O weight 230,000 kg (507,050 lb); optional for 1996 delivery.

POWER PLANT: First deliveries with two 300.3 kN (67,500 lb st) GE CF6-80E1A2 turbofans; alternative engines, using common pylon and mount, P&W PW4164 or R-R Trent 768. Longer-range version powered by PW4174 or Trent 775. Fuel capacity 93,500 litres (24,700 US gallons; 20,568 Imp gallons).

WEIGHTS AND LOADINGS (A330-300 basic versions; A: with CF6-80E1A2, B: with PW4164, C: with Trent 768. A330-300 longer-range version; D: with PW4174, E: with Trent 775):

Typical airline operating weight empty:
A 119,745 kg (263,993 lb)
B 120,311 kg (265,240 lb)
C 119,472 kg (263,391 lb)
D 123,487 kg (272,242 lb)
E 123,207 kg (271,625 lb)
Max payload: A 44,255 kg (97,565 lb)
B 43,689 kg (96,318 lb)
C 44,528 kg (98,167 lb)
D 49,513 kg (109,157 lb)
E 49,793 kg (109,775 lb)
Max T-O weight: A, B, C 212,000 kg (467,380 lb)
D, E 223,000 kg (491,630 lb)
Max landing weight: A, B, C 174,000 kg (383,605 lb)
D, E 185,000 kg (407,855 lb)

Max zero-fuel weight: A, B, C 164,000 kg (361,560 lb)
D, E 173,000 kg (381,400 lb)
PERFORMANCE:
Range at typical OWE, with allowances for 200 nm (370 km; 230 mile) diversion and international reserves
A330-300 with 335 passengers and baggage:
A 4,500 nm (8,339 km; 5,181 miles)
B, C 4,550 nm (8,432 km; 5,239 miles)
C (from 1996) 4,640 nm (8,598 km; 5,343 miles)
A330-300 at higher gross weight:
B, C 5,400 nm (10,000 km; 6,200 miles)
OPERATIONAL NOISE LEVELS (A330-300, estimated):
T-O, flyover 92.2 EPNdB
Sideline 96.8 EPNdB
Approach 97.8 EPNdB

AIRBUS/BOEING/DOUGLAS/AEROSPATIALE/DEUTSCHE AEROSPACE/JAPAN

VERY LARGE COMMERCIAL TRANSPORT/ULTRA-HIGH CAPACITY AIRCRAFT

TYPE: Airliners for 500 to over 1,000 passengers.

PROGRAMME: Principal airliner manufacturers, individually and together, studying high capacity airliners, largely aimed at Asia Pacific customers. Objective to achieve operating costs 15 to 20 per cent lower than those of Boeing 747-400. High development cost and market estimated at only 500 to 700 aircraft cannot support more than one or two types, so international project very likely.

CURRENT VERSIONS: **International VLCT study:** Boeing, Aerospatiale, British Aerospace, DASA, CASA and possibly Douglas signed MoU 26 January 1993 for one-year study of Very Large Capacity Transport to carry 500 to 800 passengers over ranges between 7,014 and 10,252 nm (13,000 and 19,000 km; 8,078 to 11,806 miles); further meeting planned for mid-1993 to review progress and decide future direction.

Japan: Boeing, Airbus and Deutsche Aerospace Airbus reported individually talking to Fuji, Mitsubishi and Kawasaki during 1992 and early 1993 about possible future participation.

Boeing/DASA: Boeing reported talking separately to DASA about VLCT prospects during 1992.

Boeing 747-500 stretches: In 1992, Boeing revealed two possible stretches of the 747-400; one lengthens present fuselage by two 3.5 m (11 ft 6 in) plugs, fore and aft of the wing, the forward plug also stretching the upper deck, increasing capacity by 84 passengers; increased gross and empty weights require reinforced wing of 747-400F freighter already being built during 1993. Other stretch continues upper deck to fin and accommodates 160 more passengers. Both aircraft suffer 20 to 30 per cent range reduction relative to 747-400. First stages of stretch being incorporated in Boeing 747-400 from mid-1990s.

Boeing NLA: New large aircraft; also called 747-X, possibly Boeing 787; potential fresh design approach to VLCT; market study confirmed April 1992; broad three-deck fuselage with 12-abreast seating on main deck and three-abreast LD3 containers in under-floor cargo hold; passenger capacity from 612 in three-class to maximum 800 in stretched version; gross weight 545,000 kg (1,201,500 lb); range similar to that of 747-400; folding outer wings; power plant four engines of type already under development for big twins.

Airbus UHCA: Reported possible launch date 1997 and designation A350; development cost ECU10 billion. Consortium studies envisage aircraft using more advanced materials, better pylon-wing integration, laminar flow and advanced electronic controls; power plant could be four of the engines of existing big twins, GE CF6-80E or GE90, Rolls-Royce Trent or P&W PW4000; airframe size goal is span and length not much bigger than 747; various cylindrical, oval and double-bubble fuselage cross-sections considered; Airbus consulted 10 major airlines and asked Airbus member companies to propose designs.

Aerospatiale ASX-500: First of two-aircraft family study for UHCA; 500 to 600 passengers; range 7,014 nm (13,000 km; 8,078 miles); fuselage length 65 m (213 ft 3 in).

Aerospatiale ASX-600: 600 to 800 passengers; range as above; fuselage length 80 m (262 ft 5½ in); possible combi version with half main deck used for freight.

DASA A-2000: Deutsche Aerospace Airbus showed model of possible D-2000 Ultra High Capacity Aircraft at Paris Air Show 1991. Further studies during 1992, now identified as A-2000, envisaged family of aircraft; basic aircraft would have either three-class 600-passenger layout with 74,000 kg (163,140 lb) payload of which 30 per cent would be freight, or an 800-passenger layout with 76,000 kg (167,550 lb) payload and no freight; larger aircraft would have three-class 800-passenger layout and 99,000 kg (218,257 lb) payload including 30 per cent freight; largest would be a 1,050-passenger layout with 185,000 kg (407,855 lb) payload or a 1,050-passenger layout with 100,000 kg (220,460 lb) payload and no freight; various fuselage layouts, wing geometries and structural features also investigated.

Douglas MD-12: Almost launched during 1992, but proposed association with Taiwan aircraft industry did not materialise. Douglas estimated development cost at $5.2 billion. MD-12 about same size as 747-400, designed to carry 430 passengers for 8,000 nm (14,825 km; 9,200 miles) or 511 passengers for 7,200 nm (13,340 km; 8,290 miles); cargo version would have carried 141,522 kg (312,000 lb) for 4,400 nm (8,154 km; 5,060 miles).

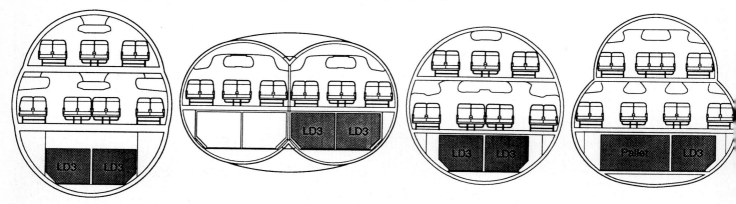

Four possible fuselage cross-sections for the proposed Airbus UHCA 600-passenger transport, showing business class seating

AIRTECH
AIRCRAFT TECHNOLOGY INDUSTRIES

PRESIDENT: Prof Dr-Ing B. J. Habibie (IPTN)
VICE-PRESIDENT: Javier Alvarez Vara (CASA)
PARTICIPATING COMPANIES:
  CASA: see under Spain
  IPTN: see under Indonesia

Airtech formed by CASA and IPTN to develop CN-235 twin-turboprop transport; design and production work shared 50-50.

AIRTECH (CASA/IPTN) CN-235

TYPE: Twin-turboprop civil and military transport.

PROGRAMME: Preliminary design began January 1980, prototype construction May 1981; one prototype completed in each country, with simultaneous rollouts 10 September 1983; first flights 11 November 1983 (by CASA's ECT-100) and 30 December 1983 (IPTN's PK-XNC); Spanish and Indonesian certification 20 June 1986; first flight of production aircraft 19 August 1986; FAA type approval (FAR Pts 25 and 121) 3 December 1986; deliveries began 15 December 1986 from IPTN line and 4 February 1987 from CASA; entered service (with Merpati Nusantara Airlines) 1 March 1988; licence assembly agreement with TAI for 50 of 52 ordered (see Turkish section), to lead eventually to local manufacture, announced January 1990.

CURRENT VERSIONS: **CN-235 Series 10:** Initial production version (15 built by each company), with CT7-7A engines; described in 1986-87 and earlier Jane's.

CN-235 Series 100: Generally as Series 10, but CT7-9C engines in new composite nacelles; replaced Series 10 in 1988 from 31st production aircraft. *Detailed description applies to this version except where indicated.*

CN-235 Series 200: Structural reinforcements to cater for higher operating weights, aerodynamic improvements to wing leading-edges and rudder, reduced field length requirements and much-increased range with max payload; certificated by FAA March 1992.

CN-235 M: Military transport version.

CN-235 MPA: Maritime patrol version, developed by IPTN; lengthened nose housing FLIR and ESM; second CN-235 prototype PK-XNC serving as testbed, with Litton APS-504(V)5 radar, GEC-Marconi MRT-S (multi-role turret system) FLIR and Marconi Defence Systems Sky Guardian ESM. Provision for quick-change configuration for general transport, communications or other duties.

CN-235 QC: Quick-change cargo/passenger version; certificated by Spanish DGAC May 1992.

CUSTOMERS: Total orders 194 (34 civil, 160 military and other) by November 1992, of which 28 civil and 160 military/other then delivered.

Military and other non-commercial customers are Botswana (Defence Force two), Brunei (Royal Air Wing three), Chile (Army four), Ecuador (Army one, Navy one) France (Air Force eight), Gabon (Air Force one), Indonesia (Air Force/Navy 24), Ireland (Air Corps three), South Korea (Air Force 12, for 1993-94 delivery), Morocco (Air Force seven), Oman (Royal Police three), Panama (National Guard one), Papua New Guinea (Defence Force two), Saudi Arabia (Air Force four, incl two VIP transports), Spain (Air Force 26, incl two Series 10 VIP) Turkey (Air Force 52, of which first two delivered January 1992) and United Arab Emirates (Abu Dhabi, one VIP and six troop/paratroop transports, for delivery from early 1993).

Commercial customers include Merpati Nusantara (15) Asahi Airlines (three), Binter Canarias (four), Binter Mediterráneo (five) and other Spanish regional carriers (six).

DESIGN FEATURES: Optimised for short-haul operations, enabling it to fly four 860 nm (1,593 km; 990 mile) stage lengths (with reserves) before refuelling and to operate from paved runways or unprepared strips; high mounted NACA 65_3-218 aerofoil wing with no-dihedral/constant chord centre-section; tapered outer panels have 3° dihedral and 3° 51′ 36″ sweepback at quarter-chord; pressurised fuselage (incl baggage compartment) of flattened circular cross-section, with upswept rear end incorporating cargo ramp/door; sweptback fin (with dorsal fin) and rudder; low-set non-swept fixed incidence tailplane and elevators; two small ventral fins; vortex generators on rudder and elevator leading-edges.

FLYING CONTROLS: Ailerons, elevators and rudder statically and dynamically balanced, with mechanical actuation (duplicated for ailerons); mechanical servo tab and electric trim tab in each aileron, rudder and starboard elevator, trim tab only in port elevator; single-slotted inboard and outboard trailing-edge flaps (each pair interchangeable port/ starboard), actuated hydraulically by Dowty irreversible jacks.

STRUCTURE: Conventional semi-monocoque, mainly of aluminium alloys with chemically milled skins; composites (mainly glassfibre or glassfibre/Nomex honeycomb sandwich, with some carbonfibre and Kevlar) for leading/trailing-edges of wing/tail moving surfaces, wing/fuselage and main landing gear fairings, wing/fin/tailplane tips, engine nacelles, ventral fins and nose radome. Propeller blades are of glassfibre, with metal spar and urethane foam core. CASA builds wing centre-section, inboard flaps, forward and centre fuselage, engine nacelles; IPTN builds outer wings, outboard flaps, ailerons, rear fuselage and tail unit; both manufacturers use numerical control machinery extensively. Final assembly line in each country. Part of tail unit built by ENAER Chile under subcontract from CASA. TAI (Turkey) to assemble under licence initially, progressing gradually to local manufacture of 50 aircraft for Turkish Air Force.

LANDING GEAR: Messier-Bugatti retractable tricycle type with levered suspension, suitable for operation from semi-prepared runways. Electrically controlled hydraulic extension/retraction, with mechanical backup for emergency extension. Oleo-pneumatic shock absorber in each unit. Each main unit comprises two wheels in tandem, retracting rearward into fairing on side of fuselage. Mainwheels semi-exposed when retracted. Single steerable nosewheel (±48°) retracts forward into unpressurised bay under flight deck. Dunlop 28 × 9.00-12 (12 ply rating) tubeless mainwheel tyres standard, pressure 5.17 bars (75 lb/sq in) on civil version, 5.58 bars (81 lb/sq in) on military version; low pressure mainwheel tyres optional, size 11.00-12/10, pressure 3.45 bars (50 lb/sq in). Dunlop 24 × 7.7 (10/12 ply rating) tubeless nosewheel tyre, pressure 5.65 bars (82 lb/sq in) on civil version, 6.07 bars (88 lb/sq in) on military version. Dunlop hydraulic differential disc brakes; Dunlop anti-skid units on main gear. Chilean Army aircraft used in Antarctic have wheel/ski gear. Min ground turning radius 9.50 m (31 ft 2 in) about nosewheel, 18.98 m (62 ft 3¼ in) about wingtip.

POWER PLANT: Two General Electric CT7-9C turboprops, each flat rated at 1,305 kW (1,750 shp) (S/L, to 41°C) for take-off and 1,394.5 kW (1,870 shp) up to 31°C with automatic power reserve. Hamilton Standard 14-RF21 four-blade constant-speed propellers, with full feathering and reverse-pitch capability. Fuel in two 1,042 litre (275 US gallon; 229 Imp gallon) integral main tanks in wing centre-section and two 1,592 litre (421 US gallon; 350 Imp gallon) integral outer-wing auxiliary tanks; total fuel capacity 5,264 litres (1,392 US gallons; 1,158 Imp gallons), of which 5,128 litres (1,355 US gallons; 1,128 Imp gallons) are usable. Single pressure refuelling point in starboard main landing gear fairing; gravity filling point in top of each tank. Propeller braking permits No. 2 engine to be used as on-ground APU. Oil capacity 13.97 litres (3.69 US gallons; 3.07 Imp gallons).

ACCOMMODATION: Crew of two on flight deck, plus cabin attendant (civil version) or third crew member (military version). Accommodation in commuter version for up to 44 passengers in four-abreast seating, at 76 cm (30 in) pitch, with 22 seats each side of central aisle. Toilet, galley and overhead luggage bins standard. Pressurised baggage compartment at rear of cabin, aft of movable bulkhead; additional stowage in rear ramp area and in overhead lockers. Can also be equipped as mixed passenger/cargo combi (eg, 19 passengers and two LD3 containers), or for all-cargo operation, with roller loading system, carrying four standard LD3 containers, five LD2s, or two 2.24 × 3.18 m (88 × 125 in) and one 2.24 × 2.03 m (88 × 80 in) pallets; or for military duties, carrying up to 48 troops or 46 paratroops. Other options include layouts for aeromedical (24 stretchers and four medical attendants), ASW/maritime patrol (with 360° search radar and Exocet missiles or Mk 46 torpedoes), electronic warfare, geophysical survey or aerial photographic duties. Main passenger door, outward and forward opening with integral stairs, aft of wing on port side, serving also as a Type I emergency exit. Type III emergency exit facing this door on starboard side. Crew/service downward opening door (forward, starboard) has built-in stairs, and serves also as a Type I emergency exit, or as passenger door in combi version; second Type III exit opposite this door on port side. Wide ventral door/cargo ramp in underside of upswept rear fuselage, for loading of bulky cargo. Accommodation fully air-conditioned and pressurised.

SYSTEMS: Hamilton Standard air-conditioning system, using engine compressor bleed air. AiResearch electropneumatic pressurisation system (max differential 0.25 bar; 3.6 lb/sq in) giving cabin environment of 2,440 m (8,000 ft) up to operating altitude of 5,485 m (18,000 ft). Hydraulic system, operating at nominal pressure of 207 bars (3,000 lb/sq in), comprises two engine driven, variable displacement axial electric pumps, a self-pressurising standby mechanical pump, and a modular unit incorporating connectors, filters and valves; system is employed for actuation of wing flaps, landing gear extension/retraction, wheel brakes, emergency and parking brakes, nosewheel steering, cargo

Spanish built Airtech CN-235 of the Irish Air Corps

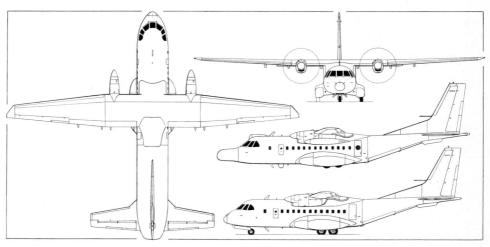

Airtech (CASA/IPTN) CN-235 twin-turboprop multi-purpose transport, with additional side view (centre) of CN-235 MPA *(Dennis Punnett)*

Airtech CN-235 twin-turboprop civil transport operated by Binter Canarias

Paradropping from a CN-235 M military transport

ramp and door, and propeller braking. Accumulator for backup braking system. 28V DC primary electrical system powered by two 400A Auxilec engine driven starter/generators, with two 24V 37Ah nickel-cadmium batteries for engine starting and 30 min (minimum) emergency power for essential services. Constant frequency single-phase AC power (115/26V) provided at 400Hz by three 600VA static inverters (two for normal operation plus one standby); two three-phase engine driven alternators for 115/200V variable frequency AC power. Fixed oxygen installation for crew of three (single cylinder at 124 bars; 1,800 lb/sq in pressure); three portable units and individual masks for passengers. Pneumatic boot anti-icing of wing (outboard of engine nacelles), fin and tailplane leading-edges. Electric anti-icing of propellers, engine air intakes, flight deck windscreen, pitot tubes and angle of attack indicators. No APU: starboard engine, with propeller braking, can be used to fulfil this function. Hand type fire extinguishers on flight deck (one) and in passenger cabin (two); smoke detector in baggage compartment. Engine fire detection and extinguishing system.

AVIONICS: Standard avionics include two Collins VHF-22B com radios, one Avtech DADS crew interphone, one Collins TDR-90 ATC transponder, two Collins VIR-32 VOR/ILS/marker beacon receivers, one Collins DME-42, one Collins ADF-60A, one Collins WXR-300 weather radar, two Collins 332D-11T vertical gyros, two Collins MCS-65 directional gyros, two Collins ADI-85A, two Collins HSI-85, two Collins RMI-36, one Collins APS-65 autopilot/flight director, one Collins ALT-55B radio altimeter, one Fairchild/Teledyne flight data recorder, one Fairchild A-100A cockpit voice recorder, one Avtech PACIS PA system, two Collins 345A-7 rate of turn sensors, one SFENA H-301 APM standby attitude director indicator, one Dorne & Margolin ELT 8-1 emergency locator transmitter, and one Sundstrand Mk II GPWS. Collins EFIS-85B five-tube CRT system standard. Options include second TDR-90, DME-42 and ADF-60A; plus Collins HF-230 com radio, Collins RNS-325 radar nav, Litton LTN-72R inertial nav or Global GNS-500A Omega navigation system.

EQUIPMENT: Navigation lights, anti-collision strobe lights, 600W landing light in front end of each main landing gear fairing, taxi lights, ice inspection lights, emergency door lights, flight deck and flight deck emergency lights, cabin and baggage compartment lights, individual passenger reading lights, and instrument panel white lighting, all standard.

ARMAMENT (military version): Three attachment points under each wing. Weapons can include McDonnell Douglas Harpoon anti-ship missiles; Indonesian MPA version can be fitted with two Mk 46 torpedoes or AM 39 Exocet anti-shipping missiles.

DIMENSIONS, EXTERNAL:
Wing span 25.81 m (84 ft 8 in)
Wing chord: at root 3.00 m (9 ft 10 in)
at tip 1.20 m (3 ft 11¼ in)
Wing aspect ratio 11.27
Length overall 21.40 m (70 ft 2½ in)
Fuselage: Max width 2.90 m (9 ft 6 in)
Max depth 2.615 m (8 ft 7 in)
Height overall 8.177 m (26 ft 10 in)
Tailplane span 10.60 m (34 ft 9¼ in)
Wheel track (c/l of mainwheels) 3.90 m (12 ft 9½ in)
Wheelbase 6.919 m (22 ft 8½ in)
Propeller diameter 3.35 m (11 ft 0 in)
Propeller ground clearance 1.66 m (5 ft 5¼ in)
Distance between propeller centres 7.00 m (22 ft 11½ in)
Passenger door (port, rear) and service door (stbd, fwd):
Height 1.70 m (5 ft 7 in)
Width 0.73 m (2 ft 4¾ in)
Height to sill 1.22 m (4 ft 0 in)
Paratroop doors (port and stbd, rear, each):
Height 1.75 m (5 ft 9 in)
Width 0.90 m (2 ft 11½ in)
Height to sill 1.22 m (4 ft 0 in)
Ventral upper door (rear): Length 2.366 m (7 ft 9¼ in)
Width 2.349 m (7 ft 8½ in)
Height to sill 1.22 m (4 ft 0 in)
Ventral ramp/door (rear): Length 3.042 m (9 ft 11¾ in)
Width 2.349 m (7 ft 8½ in)
Height to sill 1.22 m (4 ft 0 in)
Type III emergency exits (port, fwd, and stbd, rear):
Height 0.91 m (3 ft 0 in)
Width 0.51 m (1 ft 8 in)

DIMENSIONS, INTERNAL:
Cabin, excl flight deck: Length 9.65 m (31 ft 8 in)
Max width 2.70 m (8 ft 10½ in)
Width at floor 2.366 m (7 ft 9 in)
Max height 1.88 m (6 ft 2 in)
Floor area 22.822 m² (245.65 sq ft)
Volume 43.24 m³ (1,527.0 cu ft)
Baggage compartment volume:
ramp 5.30 m³ (187.2 cu ft)
overhead bins 1.68 m³ (59.3 cu ft)

AREAS:
Wings, gross 59.10 m² (636.1 sq ft)
Ailerons (total, incl tabs) 3.07 m² (33.06 sq ft)
Trailing-edge flaps (total) 10.87 m² (117.0 sq ft)
Fin, incl dorsal fin 11.38 m² (122.49 sq ft)
Rudder, incl tabs 3.32 m² (35.74 sq ft)
Tailplane 25.40 m² (273.4 sq ft)
Elevators (total, incl tabs) 4.25 m² (45.75 sq ft)

WEIGHTS AND LOADINGS:
Operating weight empty:
passenger version 9,800 kg (21,605 lb)
cargo and military versions 8,800 kg (19,400 lb)
Max fuel 4,230 kg (9,325 lb)
Max payload: passenger version:
Srs 100 4,000 kg (8,818 lb
Srs 200 4,300 kg (9,480 lb
cargo and military versions 6,000 kg (13,227 lb
Max weapon load (CN-235 M) 3,500 kg (7,716 lb
Max T-O weight: civil:
Srs 100 15,100 kg (33,289 lb
Srs 200 15,800 kg (34,833 lb
military 16,500 kg (36,376 lb
Max landing weight: civil:
Srs 100 14,900 kg (32,849 lb
Srs 200 15,600 kg (34,392 lb
military 16,500 kg (36,376 lb
Max zero-fuel weight: civil 14,100 kg (31,085 lb
military 14,800 kg (32,628 lb
Cabin floor loading:
cargo and military versions 1,504 kg/m² (308.0 lb/sq f
Max wing loading: civil:
Srs 100 255.5 kg/m² (52.36 lb/sq f
Srs 200 267.3 kg/m² (54.75 lb/sq f
military 279.2 kg/m² (57.18 lb/sq f
Max power loading (without APR): civil:
Srs 100 5.78 kg/kW (9.51 lb/shp
Srs 200 6.05 kg/kW (9.95 lb/shp
military 6.32 kg/kW (10.39 lb/shp

PERFORMANCE (civil versions at max T-O weight, ISA, except where indicated):
Max operating speed at S/L 240 knots (445 km/h; 276 mph) IA
Max cruising speed at 4,575 m (15,000 ft) 248 knots (460 km/h; 286 mph
Stalling speed at S/L:
flaps up 100 knots (186 km/h; 116 mph) IA
flaps down 84 knots (156 km/h; 97 mph) IA
Max rate of climb at S/L 465 m (1,527 ft)/mi
Rate of climb at S/L, OEI 128 m (420 ft)/mi
Service ceiling 7,620 m (25,000 f
Service ceiling, OEI 4,500 m (14,775 f
T-O run: Srs 100 1,217 m (3,993 f
Srs 200 1,051 m (3,450 f
T-O balanced field length at S/L:
Srs 100 1,406 m (4,615 f
Srs 200 1,275 m (4,185 f
Srs 200 at AUW of 14,646 kg (32,290 lb) 1,139 m (3,737 f
Landing from 15 m (50 ft) at S/L:
Srs 100 1,276 m (4,187 f
Srs 200 670 m (2,200 f
Range at 5,485 m (18,000 ft), reserves for 87 nm (161 km 100 mile) diversion and 45 min hold:
Srs 100 with max payload 450 nm (834 km; 518 miles
Srs 200 with max payload 957 nm (1,773 km; 1,102 miles
Srs 100 with max fuel 2,110 nm (3,910 km; 2,429 miles
Srs 200 with max fuel 1,974 nm (3,658 km; 2,273 miles

PERFORMANCE (CN-235 M at max T-O weight, ISA, except where indicated):
As for civil versions except:
Max rate of climb at S/L 579 m (1,900 ft)/mi
Rate of climb at S/L, OEI 156 m (512 ft)/mi
Service ceiling 8,110 m (26,600 f
Service ceiling, OEI 4,800 m (15,750 f
T-O to 15 m (50 ft) (CFL): Srs 100 1,290 m (4,235 f
Srs 200 1,165 m (3,825 f
Landing from 15 m (50 ft): Srs 100 772 m (2,530 f
Srs 200 652 m (2,140 f
Landing run, with propeller reversal:
Srs 100 398 m (1,306 f
Srs 200 400 m (1,313 f
Range at 6,100 m (20,000 ft), long-range cruising speed reserves for 45 min hold:
Srs 100 with max payload 810 nm (1,501 km; 932 miles
Srs 200 with max payload 825 nm (1,528 km; 950 miles
Srs 100 with 3,550 kg (7,826 lb) payload 2,350 nm (4,355 km; 2,706 miles
Srs 200 with 3,550 kg (7,826 lb) payload 2,400 nm (4,447 km; 2,763 miles

OPERATIONAL NOISE LEVELS (civil versions):
T-O 84.0 EPNdB
Approach 87.0 EPNdB
Sideline 86.0 EPNdB

AMX

AMX INTERNATIONAL LTD

PRESIDENT: Dott Ing Giovanni Gazzaniga (Alenia)

VICE-PRESIDENTS:
P. Vergnano (Alenia)
F. Grandi (Aermacchi)
R. Pesce (Embraer)

PARTICIPATING COMPANIES:
Alenia: see under Italy
Aermacchi: see under Italy
Embraer: see under Brazil

AMX

Brazilian Air Force designation: A-1

TYPE: Single-seat close air support, battlefield interdiction and reconnaissance aircraft, with secondary capability for offensive counter-air.

PROGRAMME: Resulted from 1977 Italian Air Force specification for small tactical fighter-bomber (see 1987-88 and previous *Jane's* for early background); original Aeritalia/Aermacchi partnership joined by Embraer July 1980; seven single-seat prototypes built (first flight 15 May 1984: further details in 1987-88 and earlier editions); production of first 30 (Italy 21, Brazil nine), and design of two-seater began mid-1986; first production aircraft rolled out at Turin 29 March 1988, making first flight 11 May; second contract (Italy 59, Brazil 25, incl six and three two-seater respectively) placed 1988; deliveries to Italian Air Force (six for Reparto Sperimentale di Volo at Pratica di Mare began April 1989; production A-1 for Brazilian Air Force (s/n 5500) made first flight 12 August 1989, deliveries (two to 1° Esquadrão of 16° Grupo de Aviação at Santa Cruz following from 17 October 1989; in-flight refuelling test programme completed (by Embraer) August-Septembe

1989; first flight by first (of three) two-seat AMX-Ts 14 March 1990, followed by second on 16 July; first flight of Embraer two-seater (s/n 5650) 14 August 1991; third production lot (Italy 56, Brazil 22, incl 18 and seven two-seaters) authorised early 1992; first production two-seater for Brazilian Air Force (serial number 5650) delivered 7 May 1992; Embraer deliveries in 1992 comprised seven single-seat and first two two-seat AMXs, plus 15 shipsets to Italian production line. Series production for Italy, Brazil and prospective export customers expected to continue through the 1990s.

CURRENT VERSIONS: **Single-seater:** Intended to replace G91R/Y and F-104G/S in Italian Air Force (eight squadrons) and EMB-326GB Xavante in Brazilian Air Force for close support/interdiction/reconnaissance, sharing counter-air duties with IDS Tornado (Italy) and F-5E/Mirage III (Brazil); in service with 2°, 3° and 51° Stormi (Italy) and 16° Grupo (Brazil); Brazilian Air Force aircraft differ primarily in avionics and weapon delivery systems, and have two 30 mm guns instead of Italian version's single multi-barrel 20 mm weapon. *Detailed description applies to single-seater except where indicated.*

Two-seater: Second cockpit accommodated by removing forward fuselage fuel tank and relocating environmental control system; dual controls, canopy, integration of rear cockpit GEC-Marconi HUD monitor, and oxygen systems, designed/redesigned by Embraer; intended both as **AMX-T** operational trainer and, suitably equipped, for such roles as EW, reconnaissance and maritime attack; three prototypes (two Italian, one Brazilian).

CUSTOMERS: Total of 192 (136 Italy/56 Brazil, incl 26 and 11 two-seaters) ordered by January 1993, of which more than 100 (Italy 70+/Brazil 16+) delivered; stated requirements are 266 single-seaters (Italy 187/Brazil 79) and 66 two-seaters (Italy 51/Brazil 15).

DESIGN FEATURES: Intended for high-subsonic/very low altitude day/night missions, in poor visibility and, if necessary, from poorly equipped or partially damaged runways; wing sweepback 31° on leading-edges, 27° 30' at quarter-chord; thickness/chord ratio 12 per cent.

FLYING CONTROLS: Hydraulically actuated ailerons and elevators; leading-edge slats and Fowler double-slotted trailing-edge flaps (each two-segment on each wing) actuated electrohydraulically; pair of hydraulically actuated spoilers forward of each flap pair, deployable separately in inboard and outboard pairs; fly-by-wire control of spoilers, rudder and variable incidence tailplane by Alenia/GEC Avionics flight control computer; ailerons, elevators, rudder have manual reversion for fly-home capability even with both hydraulic systems inoperative; spoilers serve also as airbrakes/lift dumpers.

STRUCTURE: Mainly aluminium alloy except for carbonfibre fin and elevators; shoulder-mounted wings, each with three-point attachment to fuselage, have three-spar torsion box with integrally stiffened skins; oval-section semi-monocoque fuselage, with rear portion (incl tailplane) detachable for engine access; work split gives programme leader Alenia 46.7 per cent (centre-fuselage, nose radome, tail surfaces, ailerons and spoilers); Aermacchi has 23.6 per cent (forward fuselage incl gun and avionics integration, canopy, tailcone) and Embraer 29.7 per cent (air intakes, wings, leading-edge slats, flaps, wing pylons, external fuel tanks and reconnaissance pallets); single-sourced production, with final assembly lines in Italy and Brazil.

LANDING GEAR: Hydraulically retractable tricycle type, of Messier-Bugatti levered suspension design, built in Italy by Magnaghi (nose unit) and in France by ERAM (main units). Single wheel and oleo-pneumatic shock absorber on each unit. Nose unit retracts forward; main units retract forward and inward, turning through approx 90° to lie almost flat in underside of engine air intake trunks. Nosewheel hydraulically steerable (±60°), self-centring, and fitted with anti-shimmy device. Mainwheel tyres size 670 × 210-12, pressure 9.65 bars (140 lb/sq in); nosewheel tyre size 18 × 5.5-8, pressure 10.7 bars (155 lb/sq in). Hydraulic brakes and fully modulated anti-skid system. No brake-chute. Runway arrester hook. Min ground turning radius 7.53 m (24 ft 8½ in).

POWER PLANT: One 49.1 kN (11,030 lb st) Rolls-Royce Spey Mk 807 non-afterburning turbofan, built under licence in Italy by Fiat, Piaggio and Alfa Romeo Avio, in association with Companhia Eletro-Mecânica (CELMA) in Brazil. Self-sealing, compartmented, rubber fuselage bag tanks and two integral wing tanks with combined capacity of 3,500 litres (924.6 US gallons; 770 Imp gallons). Auxiliary fuel tanks of up to 1,100 litres (290 US gallons; 242 Imp gallons) capacity can be carried on each inboard underwing pylon, and up to 580 litres (153 US gallons; 128 Imp gallons) on each outboard pylon. Single-point pressure or gravity refuelling of internal and external tanks. In-flight refuelling capability (probe and drogue system).

ACCOMMODATION: Pilot only, on Martin-Baker Mk 10L zero/zero ejection seat; 18° downward view over nose. One-piece wraparound windscreen; one-piece hinged canopy, opening sideways to starboard. Cockpit pressurised and air-conditioned. Tandem two-seat combat trainer/special missions version also in production.

SYSTEMS: Microtecnica environmental control system (ECS) provides air-conditioning of cockpit, avionics and reconnaissance pallets, cockpit pressurisation, air intake and

Single-seat AMXs of the 51° Stormo of the Italian Air Force

First production Brazilian AMX two-seat trainer; tail code SC indicates No. 1 Squadron's base at Santa Cruz

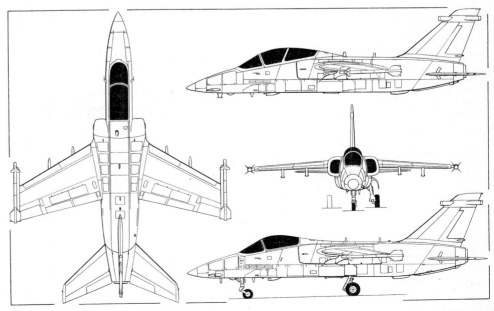

Alenia/Aermacchi/Embraer AMX, in production for the air forces of Italy and Brazil; upper side view shows two-seater (*Dennis Punnett*)

130 INTERNATIONAL: AIRCRAFT—AMX

AMX of Italy's Reparto Sperimentale di Volo, showing flaps, slats, pylons and port-side gun

AMX with Exocet anti-ship missiles

inlet guide vane anti-icing, windscreen demisting, and anti-g systems. Duplicated redundant hydraulic systems, driven by engine gearbox, operate at pressure of 207 bars (3,000 lb/sq in) for actuation of primary flight control system, flaps, spoilers, landing gear, wheel brakes, anti-skid system, nosewheel steering and gun operation. Primary electrical system AC power (115/200V at fixed frequency of 400Hz) supplied by two 30kVA IDG generators, with two transformer-rectifier units for conversion to 28V DC; 36Ah nickel-cadmium battery for emergency use, to provide power for essential systems in the event of primary and secondary electrical system failure. Aeroeletrônica (Brazil) external power control unit. Fiat FA 150 Argo APU for engine starting. APU driven electrical generator for ground operation. Liquid oxygen system.

AVIONICS: Six main subsystems: (1) UHF and VHF com, and IFF; (2) navigation (Litton Italia inertial system, with standby AHRS and Tacan, for Italian Air Force; VOR/ILS for Brazil); (3) Alenia computer based weapon aiming and delivery, incorporating radar and Alenia stores management system; (4) digital data displays (OMI/Alenia head-up, Alenia multi-function head-down, and weapons/nav selector); (5) data processing, with Microtecnica air data computer; (6) Elettronica active and passive ECM, including fin mounted radar warning receiver. Pointer ranging radar in Italian AMXs is I-band set modified from Elta (Israel) EL/M-2001B and built in Italy by FIAR; Brazilian aircraft have Tecnasa/SMA SCP-01 radar. GEC-Marconi MED 2067 video monitor display in rear cockpit of two-seater, for use as HUD monitor by instructor/navigator. Modular design and space provisions within aircraft permit retrofitting of alternative avionics, FLIR and provisions for use of night vision goggles if and when required. All avionics/equipment packages pallet mounted and positioned for rapid access.

EQUIPMENT: For reconnaissance missions, any one of three interchangeable Aeroeletrônica (Brazil) pallet mounted photographic systems can be carried, installed internally in forward fuselage; external infra-red/electro-optical pod can be carried on centreline pylon. Each system is fully compatible with aircraft, and does not affect operational capability; aircraft can therefore carry out reconnaissance missions without effect upon normal nav/attack and self-defence capabilities. Camera bay in lower starboard side of fuselage, forward of mainwheel bay.

ARMAMENT: One M61A1 multi-barrel 20 mm cannon, with 350 rds, in port side of lower forward fuselage of aircraft for Italian Air Force (one 30 mm DEFA 554 cannon on each side in aircraft for Brazilian Air Force). Single stores attachment point on fuselage centreline, plus two attachments under each wing, and wingtip rails for two AIM-9L Sidewinder or similar infra-red air-to-air missiles (MAA-1 Piranha on Brazilian aircraft). Fuselage and inboard underwing points each stressed for loads of up to 907 kg (2,000 lb); outboard underwing points stressed for 454 kg (1,000 lb) each. Triple carriers can be fitted to inboard underwing pylons, twin carriers to all five stations. Total external stores load 3,800 kg (8,377 lb). Attack weapons can include free-fall or retarded Mk 82/83/84 bombs, laser guided bombs, cluster bombs, air-to-surface missiles (including area denial, anti-radiation and anti-shipping weapons), electro-optical precision guided munitions, and rocket launchers. Exocet firing trials conducted 1991.

DIMENSIONS, EXTERNAL:
Wing span:	
excl wingtip missiles and rails	8.874 m (29 ft 1½ in)
over missiles	9.97 m (32 ft 8½ in)
Wing aspect ratio	3.7
Wing taper ratio	0.
Length: overall	13.23 m (43 ft 5 in)
fuselage	12.55 m (41 ft 2 in)
Height overall	4.55 m (14 ft 11¼ in)
Tailplane span	5.20 m (17 ft 0¾ in)
Wheel track	2.15 m (7 ft 0¾ in)
Wheelbase	4.70 m (15 ft 5 in)

AREAS:
Wings, gross	21.00 m² (226.04 sq ft)
Ailerons (total)	0.88 m² (9.47 sq ft)
Trailing-edge flaps (total)	3.86 m² (41.55 sq ft)
Leading-edge slats (total)	2.066 m² (22.24 sq ft)
Spoilers (total)	1.30 m² (13.99 sq ft)
Fin (exposed)	4.265 m² (45.91 sq ft)
Rudder	0.833 m² (8.97 sq ft)
Tailplane (total exposed)	5.10 m² (54.90 sq ft)
Elevators (total)	1.00 m² (10.76 sq ft)

WEIGHTS AND LOADINGS (all versions):
Operational weight empty	6,700 kg (14,771 lb)
Max fuel weight: internal	2,762 kg (6,089 lb)
external	1,760 kg (3,880 lb)
Max external stores load	3,800 kg (8,377 lb)
T-O weight (clean)	9,600 kg (21,164 lb)
Typical mission T-O weight	10,750 kg (23,700 lb)
Max T-O weight	13,000 kg (28,660 lb)
Normal landing weight	7,000 kg (15,432 lb)
Combat wing loading (clean)	457.1 kg/m² (93.62 lb/sq ft)
Max wing loading	619.05 kg/m² (126.79 lb/sq ft)

Structural cutaway of the Italo-Brazilian AMX

Max power loading	265.14 kg/kN (2.60 lb/lb st)

PERFORMANCE (A at typical mission weight of 10,750 kg; 23,700 lb with 907 kg; 2,000 lb of external stores, B at max T-O weight with 2,721 kg; 6,000 lb of external stores, ISA in both cases):

Max level speed	Mach 0.86
Max rate of climb at S/L	3,124 m (10,250 ft)/min
Service ceiling	13,000 m (42,650 ft)
T-O run at S/L: A	631 m (2,070 ft)
B	982 m (3,220 ft)
T-O to 15 m (50 ft) at S/L: B	1,442 m (4,730 ft)
Landing from 15 m (50 ft) at S/L: B	753 m (2,470 ft)
Landing run at S/L	464 m (1,520 ft)

Attack radius, allowances for 5 min combat over target and 10% fuel reserves:

lo-lo-lo: A	300 nm (556 km; 345 miles)
B	285 nm (528 km; 328 miles)
hi-lo-hi: A	480 nm (889 km; 553 miles)
B	500 nm (926 km; 576 miles)

Ferry range with two 1,000 litre (264 US gallon; 220 Imp gallon) drop tanks, 10% reserves

	1,800 nm (3,336 km; 2,073 miles)
g limits	+7.33/−3

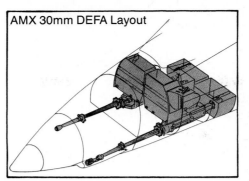
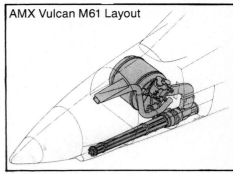

Brazilian (left) and Italian AMX gun installations

ATR

AVIONS DE TRANSPORT REGIONAL

1 Allée Pierre Nadot, F-31712 Blagnac Cedex, France
Telephone: 33 61 93 11 11
Fax: 33 61 30 07 40
Telex: 533 984 F/GIE ATR
CHAIRMAN OF THE BOARD (alternate):
  Louis Gallois (Aerospatiale)
  Fausto Cereti (Alenia)
PRESIDENT AND CEO:
  Henri-Paul Puel (Aerospatiale)
SENIOR VICE-PRESIDENT, SALES AND MARKETING:
  P. Revelli-Beaumont
SENIOR VICE-PRESIDENT, NEW PROJECTS:
  Philippe Lebouc
VICE-PRESIDENT, SALES ENGINEERING: Alain Fontaine
COMMUNICATIONS MANAGER: Mlle E. Broge
PARTICIPATING COMPANIES:
  Aerospatiale: see under France
  Alenia: see under Italy

First Aerospatiale/Aeritalia agreement July 1980; Groupement d'Intéret Economique (50:50 joint management company) formally established 5 February 1982 to develop ATR series of transport aircraft (Avions de Transport Regional). Discussions on future co-ordination with DASA/Fokker regional airliner operations under way April 1993.

ATR delivered 61 ATR 42/72 in 1991, 51 in 1992. Production to be stabilised at 55 to 60 a year from 1993 instead of rising to 96 by 1995; orders for 25 cancelled in 1992; new orders for 20 taken in 1992; total 315 delivered by end December 1992; total 505 (388 firm, 117 options) ordered by same date.

ATR marketing and support office opened in Washington 15 July 1986; ATR office in Singapore opened 18 November 1988; ATR Training Centre opened 1 July 1989.

ATR 42

TYPE: Twin-turboprop regional transport.
PROGRAMME: Joint launch by Aerospatiale and Aeritalia (now Alenia) in October 1981, following June 1981 selection of P&WC PW120 turboprop as basic power plant; first flights of two prototypes 16 August 1984 (F-WEGA) and 31 October 1984 (F-WEGB); first flight production aircraft 30 April 1985; simultaneous certification to JAR 25 by France and Italy 24 September 1985, followed by USA (FAR 25) 25 October 1985, Germany 12 February 1988, UK 31 October 1989; deliveries began 3 December 1985 (fourth aircraft, to Air Littoral); into service 9 December 1985.
CURRENT VERSIONS: **ATR 42-300:** Initial production version, with higher MTOGW and better payload/range than prototypes. *Description applies to this version, except where indicated.*
  ATR 42-320: Identical to 42-300 except for optional PW121 engines and improved hot/high performance; OWE increased/payload decreased by 5 kg (11 lb).
  ATR 42-500: Suggested version of ATR 42 powered by P&WC PW127, as ATR 72-210, for hot and high performance; better climb, and cruising speed over 300 knots (556 km/h; 345 mph).
  ATR 42 Cargo: Quick change (1 h) interior to hold nine containers with 4,000 kg (8,818 lb) payload.
  ATR 42 F: Military/paramilitary freighter with modified interior, reinforced cabin floor, flight-openable portside cargo/airdrop door; can carry 3,800 kg (8,377 lb) of cargo or 42 passengers over 1,250 nm (2,316 km; 1,439 miles). One delivered to Gabon (Presidential Guard) 1989.
  ATR Calibration: Navaid calibration version.
CUSTOMERS: Total 273 firm orders and 29 options by 31 December 1992; 238 delivered by that date.
DESIGN FEATURES: Designed to JAR 25/FAR 25; wing section Aerospatiale RA-XXX-43 (NACA 43 series derivative); thickness/chord ratio 18 per cent at root, 13 per cent at tip; constant chord, no-dihedral centre-section with 2° incidence at root; outer panels 3° 6′ sweepback at quarter-chord and 2° 30′ dihedral. Fuselage (incl baggage/cargo compartments) pressurised. Sweptback vertical and non-swept horizontal (T) tail surfaces.
FLYING CONTROLS: Mechanically actuated; lateral control by ailerons and single spoiler surface ahead of each outer flap; ailerons each have electrically actuated trim tab; fixed incidence tailplane; horn balanced rudder and elevators, each with electrically actuated trim tab; two-segment double-slotted flaps on offset hinges with Ratier-Figeac hydraulic actuators.
STRUCTURE: Two-spar fail-safe wings, mainly of aluminium alloy, with leading-edges of Kevlar/Nomex sandwich; wing top skin panels aft of rear spar are of Kevlar/Nomex with carbon reinforcement; flaps and ailerons have aluminium ribs and spars, with skins of carbonfibre/Nomex and carbon/epoxy respectively; fuselage is fail-safe stressed skin, mainly of light alloy except for Kevlar/Nomex sandwich nosecone, tailcone, wing/body fairings, nosewheel doors and main landing gear fairings; fin (attached to rearmost fuselage frame) and tailplane mainly of aluminium alloy; CFRP/Nomex sandwich rudder and elevators; dorsal fin of Kevlar/Nomex and GFRP/Nomex sandwich; engine cowlings of CFRP/Nomex and Kevlar/Nomex sandwich, reinforced with CFRP in nose and underside; propeller blades have metal spars and GFRP/polyurethane skins. Aerospatiale responsible for design and construction of wings, flight deck/cabin layout, installation of power plant/flight controls/electrical and de-icing systems, and final assembly/flight testing of civil passenger versions; Alenia builds fuselage/tail unit, installs landing gear/hydraulic system/air-conditioning/pressurisation systems. ATR 42/72 manufactured at St Nazaire and Nantes (France), Pomigliano d'Arco and Capodichino (Italy) and assembled in Toulouse.
LANDING GEAR: Hydraulically retractable tricycle type, of Messier-Bugatti/Magnaghi/Nardi trailing-arm design,

Aerospatiale/Alenia ATR 42 (rear) and 72 short-haul transports in the insignia of American Eagle, USA

132 INTERNATIONAL: AIRCRAFT—ATR

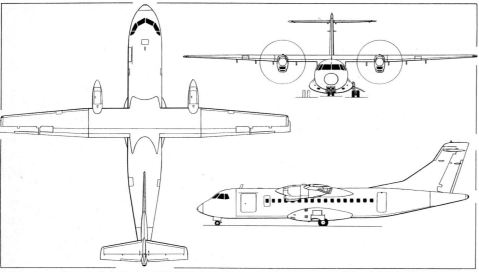

Aerospatiale/Alenia ATR 42 twin-turboprop regional transport (Dennis Punnett)

Internal layout of ATR 42 with baggage space forward as well as aft and airstairs at rear door

with twin wheels and oleo-pneumatic shock absorber on each unit. Nose unit retracts forward, main units inward into fuselage and large underfuselage fairing. Goodyear multi-disc mainwheel brakes and Hydro-Aire anti-skid units. Goodyear mainwheels and tubeless tyres, size 32 × 8.8-10PR, pressure 7.17 bars (104 lb/sq in). Low pressure tyres optional. Goodyear nosewheels and tubeless tyres, size 450 × 190-5TL, pressure 4.14 bars (60 lb/sq in). Min ground turning radius 17.08 m (56 ft 0½ in).

POWER PLANT: Two Pratt & Whitney Canada PW120 turboprops in 42-300, each flat rated at 1,342 kW (1,800 shp) for normal operation and 1,492 kW (2,000 shp) OEI and driving a 3.96 m (12 ft 11½ in) diameter Hamilton Standard 14SF four-blade constant-speed fully feathering and reversible-pitch propeller. Power plant for ATR 42-320 is two PW121s, each flat rated at 1,417 kW (1,900 shp) for normal operation and 1,567 kW (2,100 shp) OEI. Fuel in two integral tanks in spar box, total capacity 5,700 litres (1,506 US gallons; 1,254 Imp gallons). Single pressure refuelling point in starboard wing leading-edge. Gravity refuelling points in wing upper surface. Oil capacity 40 litres (10.6 US gallons; 8.8 Imp gallons).

ACCOMMODATION: Crew of two on flight deck; folding seat for observer. Seating for 42 passengers at 81 cm (32 in) pitch; or 46, 48 or 50 passengers at 76 cm (30 in) pitch, in four-abreast layout with central aisle. Passenger door, with integral steps, at rear of cabin on port side. Main baggage/cargo compartment between flight deck and passenger cabin, with access from inside cabin and separate loading door on port side; toilet, galley, wardrobe, and seat for cabin attendant, at rear of passenger cabin, with service door on starboard side; rear baggage/cargo compartment aft of passenger cabin; additional baggage space provided by overhead bins and underseat stowage. Entire accommodation, including flight deck and baggage/cargo compartments, pressurised and air-conditioned. Emergency exit via rear passenger and service doors, and by window exits on each side at front of cabin.

SYSTEMS: AiResearch air-conditioning and Softair pressurisation systems, utilising engine bleed air. Pressurisation system (nominal differential 0.41 bar; 6.0 lb/sq in) provides cabin altitude of 2,040 m (6,695 ft) at flight altitudes of up to 7,620 m (25,000 ft). Two independent hydraulic systems, each at pressure of 207 bars (3,000 lb/sq in), driven by electrically operated Abex pump and separated by interconnecting valve controlled from flight deck; system flow rate 7.9 litres (2.09 US gallons; 1.74 Imp gallons)/min; one system actuates wing flaps, spoilers, propeller brake, emergency wheel braking and nosewheel steering, and second system for landing gear and normal braking. Kléber-Colombes pneumatic system for de-icing of outer wing leading-edges, tailplane leading-edges and engine air intakes; optional de-icing of inner wing leading-edge and fin for severe conditions; noses of aileron and elevator horns have full-time electric anti-icing. Main electrical system is 28V DC, supplied by two Auxilec 12kW engine driven starter/generators and two nickel-cadmium batteries (43Ah and 15Ah), with two solid state static inverters for 115/26V single-phase AC supply; 115/200V three-phase supply from two 20kVA frequency-wild engine driven alternators for anti-icing of windscreen, flight deck side windows, stall warning and airspeed indicator pitots, propeller blades and control surface horns. Eros/Puritan oxygen system. Instead of APU, starboard propeller braked and engine run to give DC and 400Hz power, air-conditioning and hydraulic pressure.

AVIONICS: Bendix/King Gold Crown III com/nav equipment standard, Collins Pro Line II optional. Other standard avionics include Honeywell DFZ-600 AFCS, Honeywell P-800 weather radar, dual Honeywell AZ-800 digital air data computers and dual AH-600 AHRS with ASCB (avionics standard communication bus), GPWS, radio altimeter, and digital flight data recorder. EDZ-820 electronic flight instrumentation system with FMS/RNAV optional. Standard avionics package includes two VHF, two VOR/ILS/marker beacon receivers, radio compass, radio altimeter, DME, ATC transponder, cockpit voice recorder, intercom, PA system, and equipment to FAR Pt 121.

DIMENSIONS, EXTERNAL:
Wing span	24.57 m (80 ft 7½ in)
Wing chord: at root	2.57 m (8 ft 5¼ in)
at tip	1.41 m (4 ft 7½ in)
Wing aspect ratio	11.08
Length overall	22.67 m (74 ft 4½ in)
Fuselage: Max width	2.865 m (9 ft 4½ in)
Height overall	7.586 m (24 ft 10¾ in)
Elevator span	7.31 m (23 ft 11¾ in)
Wheel track (c/l of shock struts)	4.10 m (13 ft 5½ in)
Wheelbase	8.78 m (28 ft 9¾ in)
Propeller diameter	3.96 m (13 ft 0 in)
Distance between propeller centres	8.10 m (26 ft 7 in)
Propeller fuselage clearance	0.82 m (2 ft 8¼ in)
Propeller ground clearance	1.10 m (3 ft 7¼ in)
Passenger door (rear, port): Height	1.75 m (5 ft 9 in)
Width	0.75 m (2 ft 5½ in)
Height to sill (at OWE)	1.375 m (4 ft 6¼ in)
Service door (rear, stbd): Height	1.22 m (4 ft 0 in)
Width	0.61 m (2 ft 0 in)
Height to sill	1.375 m (4 ft 6¼ in)
Cargo/baggage door (fwd, port):	
Height	1.53 m (5 ft 0¼ in)
Width	1.275 m (4 ft 2¼ in)
Height to sill (at OWE)	1.15 m (3 ft 9¼ in)
Emergency exits (fwd, each): Height	0.91 m (3 ft 0 in)
Width	0.51 m (1 ft 8 in)
Crew emergency hatch (flight deck roof):	
Length	0.51 m (1 ft 8 in)
Width	0.483 m (1 ft 7 in)

DIMENSIONS, INTERNAL:
Cabin: Length (excl flight deck, incl toilet and baggage compartments)	14.66 m (48 ft 1¼ in)
Max width	2.57 m (8 ft 5¼ in)
Max width at floor	2.263 m (7 ft 5⅛ in)
Max height	1.91 m (6 ft 3¼ in)
Floor area	31.0 m² (333.7 sq ft)
Volume	58.0 m³ (2,048.25 cu ft)
Baggage/cargo compartment volume:	
front (42-46 passengers)	6.0 m³ (211.9 cu ft)
front (48 passengers)	4.8 m³ (169.5 cu ft)
front (50 passengers)	3.6 m³ (127.1 cu ft)
rear	4.8 m³ (169.5 cu ft)
overhead bins	1.5 m³ (53.0 cu ft)

AREAS:
Wings, gross	54.50 m² (586.6 sq ft)
Ailerons (total)	3.12 m² (33.58 sq ft)
Flaps (total)	11.00 m² (118.40 sq ft)
Spoilers (total)	1.12 m² (12.06 sq ft)
Fin, excl dorsal fin	12.48 m² (134.33 sq ft)
Rudder, incl tab	4.00 m² (43.05 sq ft)
Tailplane	11.73 m² (126.26 sq ft)
Elevators (total, incl tabs)	3.92 m² (42.19 sq ft)

WEIGHTS AND LOADINGS:
Operating weight empty (incl FAR 121 equipment):	
42-300	10,285 kg (22,674 lb)
42-320	10,290 kg (22,685 lb)
Max fuel weight	4,500 kg (9,920 lb)
Max payload: 42-300	4,915 kg (10,835 lb)
42-320	4,910 kg (10,824 lb)
Max T-O weight	16,700 kg (36,817 lb)
Max ramp weight	16,720 kg (36,860 lb)
Max zero-fuel weight	15,200 kg (33,510 lb)
Max landing weight	16,400 kg (36,156 lb)
Max wing loading	306.4 kg/m² (62.79 lb/sq ft)
Max power loading: 42-300	6.22 kg/kW (10.23 lb/shp)
42-320	5.90 kg/kW (9.69 lb/shp)

PERFORMANCE (42-300 at max T-O weight, to FAR Pt 25, incl Amendment 42, ISA, except where indicated):
Never-exceed speed (V_{NE})	Mach 0.55 (250 knots; 463 km/h; 287 mph CAS)
Max cruising speed at 5,180 m (17,000 ft), AUW of 16,200 kg (35,715 lb)	265 knots (490 km/h; 305 mph)
Econ cruising speed at 7,620 m (25,000 ft)	243 knots (450 km/h; 279 mph)
Stalling speed: flaps up	104 knots (193 km/h; 120 mph)
30° flap	81 knots (151 km/h; 94 mph)
Max rate of climb at S/L, AUW of 15,000 kg (33,069 lb)	640 m (2,100 ft)/min
Rate of climb at S/L, OEI, AUW as above	191 m (625 ft)/min
Max operating altitude	7,620 m (25,000 ft)
Service ceiling, OEI, at 97% of MTOGW and ISA + 10°C	2,315 m (7,595 ft)
T-O balanced field length:	
at S/L, ISA	1,090 m (3,576 ft)
at 915 m (3,000 ft), ISA + 10°C	1,300 m (4,265 ft)
Landing field length at S/L at max landing weight	1,030 m (3,380 ft)
Runway LCN at max T-O weight:	

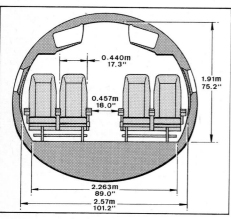

Fuselage cross-section of ATR-42/72
(Jane's/Mike Keep)

rigid pavement, 200 cm radius of relative stiffness:
standard tyres 19
low pressure tyres 16
76 cm flexible pavement, standard tyres 20
83 cm flexible pavement, low pressure tyres 16
Max range with 46 passengers, reserves for 87 nm (161 km; 100 mile) diversion and 45 min hold
1,050 nm (1,946 km; 1,209 miles)
Range with max fuel, reserves as above:
max cruising speed 2,420 nm (4,481 km; 2,785 miles)
long-range cruising speed
2,720 nm (5,040 km; 3,131 miles)
PERFORMANCE (42-320, conditions as above): As for 42-300 except:
Max cruising speed at 5,180 m (17,000 ft), AUW of 16,200 kg (35,715 lb)
269 knots (498 km/h; 310 mph)
Service ceiling, OEI, at 97% of max T-O weight, ISA + 10°C
3,140 m (10,300 ft)
T-O balanced field length:
at S/L, ISA 1,040 m (3,412 ft)
at 915 m (3,000 ft), ISA + 10°C 1,235 m (4,052 ft)
OPERATIONAL NOISE LEVELS:
T-O 82.8 EPNdB
Approach 96.7 EPNdB
Sideline 83.7 EPNdB

ATR 52C

TYPE: Rear-loading military/civil cargo version of ATR 72-210.
PROGRAMME: Announced April 1992; at end 1992 India, South Africa and Romania thought most likely risk-sharing partners to manufacture new rear fuselage and bring in 20 or so orders.
CUSTOMERS: Market estimated at 400 aircraft.
COSTS: Development cost estimated at $180 million in mid-1992.
DESIGN FEATURES: ATR 72-210 wings, fuselage and tail surfaces except for rear-loading feature and 3.176 m (10 ft 5 in) shorter fuselage; location of shortening depends on weight of rear loading feature.
FLYING CONTROLS: As ATR 72.
STRUCTURE: As ATR 72; rear loading ramp may be petal or ramp type; strengthened cargo floor.
LANDING GEAR: Same as ATR 72.
POWER PLANT: Two P&WC PW127; normal take-off power 1,849 kW (2,480 shp); OEI power 2,058 kW (2,760 shp); driving Hamilton Standard/Ratier-Figeac 247F composite-blade propellers. Fuel system as ATR 72-210.
ACCOMMODATION: Two-man flight deck with roof escape hatch; main cargo floor, excluding small service and seating area at front and loading ramp at rear, can accommodate three 2.24 m × 3.18 m (88 in × 125 in) pallets or five LD3 containers end-to-end along cabin centreline or three 2.24 m × 2.74 m (88 in × 108 in) pallets and one LD3; cargo restraint net at forward end of cabin; four roller tracks down main cargo floor; forward crew door to port and passenger door at rear to port; four cabin windows each side; front starboard unit is Type III emergency exit; medevac version has three tiers of four stretchers down each side wall (24 total), plus toilet, separate washbasin, two personnel seats and folding table at front of cabin; passenger layout has 50 seats four abreast at 76 cm (30 in) pitch, plus attendant folding seat and toilet.
SYSTEMS: As in ATR 72-210.
AVIONICS: As in ATR 72-210.
DIMENSIONS, EXTERNAL:
Wing span 27.05 m (88 ft 9 in)
Length overall 23.99 m (78 ft 8½ in)
Height overall 8.29 m (27 ft 2¼ in)
Wheel track 4.10 m (13 ft 5½ in)
Wheelbase 9.63 m (31 ft 7 in)
Tailplane span 7.31 m (23 ft 11¾ in)
Propeller diameter 3.96 m (12 ft 11¾ in)
Crew door: Height 1.75 m (5 ft 8¾ in)
Width 0.82 m (2 ft 8¼ in)
Passenger door: Height 1.75 m (5 ft 8¾ in)
Width 0.90 m (2 ft 11½ in)
DIMENSIONS, INTERNAL:
Cabin: Overall length, excl ramp
13.50 m (44 ft 3½ in)
Cargo floor length, excl ramp
12.33 m (40 ft 5½ in)

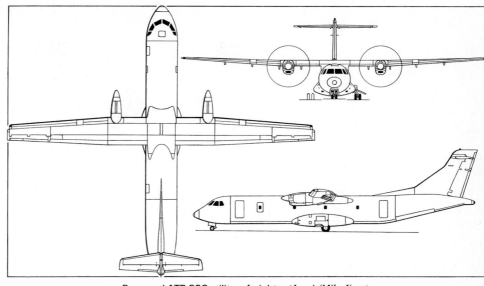

Proposed ATR 52C military freighter *(Jane's/Mike Keep)*

WEIGHTS AND LOADINGS:
Operating weight empty 11,600 kg (25,573 lb)
Max payload 7,500 kg (16,534 lb)
Max fuel weight 5,000 kg (11,023 lb)
Mission equipment allowance 500 kg (1,102 lb)
Max T-O weight 20,000 kg (44,092 lb)
Max zero-fuel weight 17,600 kg (38,800 lb)
Max landing weight 20,000 kg (44,092 lb)
PERFORMANCE (estimated):
Balanced T-O field length 1,050 m (3,444 ft)
T-O to 15 m (50 ft) 730 m (2,394 ft)
Range with 5,000 kg (11,023 lb) payload
798 nm (1,480 km; 919 miles)

ATR 72

TYPE: Twin-turboprop regional transport.
PROGRAMME: Announced at 1985 Paris Air Show; launched 15 January 1986; three development aircraft built: first flights 27 October 1988 (F-WWEY), 20 December 1988 and April 1989; French and US certification 25 September and 15 November 1989 respectively; deliveries, to Kar Air of Finland, began 27 October 1989.
CURRENT VERSIONS: **ATR 72-200**: Current production version. *Description applies to this version except where indicated.*
ATR 72-210: Improved hot/high version with PW127 engines rated at 1,849 kW (2,480 shp) and Hamilton Standard 247F propellers with composites blades on steel hubs; ATPCS power 2,059 kW (2,760 shp); carries 17 to 19 more passengers than standard ATR 72 in WAT-limited conditions; French and US certification 21 December 1992, German on 22 February 1993; first delivery December 1992; 60 ordered by American airlines, eight by Eurowings.
ATR 52C: Rear-loading cargo version; described separately.
CUSTOMERS: Total 203 ATR 72s ordered (115 firm + 88 options) by 31 December 1992; total 77 delivered at that date.
DESIGN FEATURES: Stretched version of ATR 42 (which see) with more power, more fuel, greater wing span/area, and longer fuselage for up to 74 passengers.
FLYING CONTROLS: As for ATR 42 but vortex generators ahead of ailerons and aileron horn balances shielded by wingtip extensions.
STRUCTURE: Generally as for ATR 42, but new wings outboard of engine nacelles have CFRP front and rear spars, self-stiffening CFRP skin panels and light alloy ribs, resulting in weight saving of 120 kg (265 lb); sweepback on outer panels 2° 18' at quarter-chord.
LANDING GEAR: Improved main units, with Dunlop wheels (tyres size 34 × 10R-16, pressure 7.86 bars; 114 lb/sq in) and structural carbon brakes; nose gear as ATR 42. Min ground turning radius 19.76 m (64 ft 10 in).
POWER PLANT: Two Pratt & Whitney Canada PW124B turboprops, each rated at 1,611 kW (2,160 shp) for normal take-off and 1,790 kW (2,400 shp) with ATPCS; Hamilton Standard 14SF-11 four-blade propellers; each new outer wing spar box forms additional 637 litre (168 US gallon; 140 Imp gallon) fuel tank; pressure refuelling point in starboard main landing gear fairing.
ACCOMMODATION: As ATR 42 but seating for 64, 66, 70 or (high density) 74 passengers, at respective seat pitches of 81/79/76/76 cm (32/31/30/30 in), plus second cabin attendant's seat. Single baggage compartment at rear of cabin; one or two at front, depending on seating layout and type of port forward door fitted. This can be a passenger or cargo door, with a service door opposite on starboard side. Service door on each side at rear, that on port side replaced by a passenger door when cargo door is fitted at front. Two additional emergency exits (one each side); both rear doors also serve as emergency exits. All doors are of plug type. Increased-capacity air-conditioning system.
AVIONICS: Flight deck equipment and layout generally as for ATR 42. Additions/improvements include engine monitoring mini-aids, and fuel repeater on refuelling panel.
DIMENSIONS, EXTERNAL: As ATR 42 except:
Wing span 27.05 m (88 ft 9 in)
Wing chord at tip 1.59 m (5 ft 2½ in)
Wing aspect ratio 12.0
Length overall 27.166 m (89 ft 1½ in)
Height overall 7.65 m (25 ft 1¼ in)

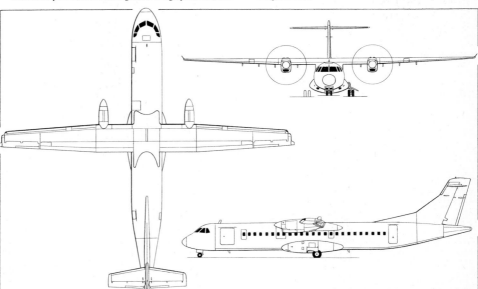

ATR 72-200 (two Pratt & Whitney Canada PW124B turboprops) *(Dennis Punnett)*

Model showing a rear loading ramp for the ATR 52C

134 INTERNATIONAL: AIRCRAFT—ATR/AVIOTECHNICA

ATR 72-200 of TAT (France), demonstrating its 13-container freight capacity

Wheelbase	10.77 m (35 ft 4 in)
Passenger door (fwd, port): Height	1.75 m (5 ft 9 in)
Width	0.82 m (2 ft 8¼ in)
Height to sill	1.12 m (3 ft 8 in)
Alternative cargo door (fwd, port):	
Height	1.53 m (5 ft 0¼ in)
Width	1.275 m (4 ft 2½ in)
Height to sill	1.12 m (3 ft 8 in)

DIMENSIONS, INTERNAL:
Cabin:
Length (excl flight deck, incl toilet and baggage compartments) 19.21 m (63 ft 0¼ in)
Cross-section as for ATR 42
Floor area 41.7 m² (449 sq ft)
Volume 76.0 m³ (2,684 cu ft)
Baggage/cargo compartment volume (with front passenger door):
front (64-66 passengers) 3.9 m³ (137.7 cu ft)
front (66 passengers with front cargo door) 5.8 m³ (204.8 cu ft)
front (74 passengers) 1.6 m³ (56.5 cu ft)
rear 4.8 m³ (169.5 cu ft)
AREAS: As ATR 42 except:
Wings, gross 61.0 m² (656.6 sq ft)
Ailerons (total) 3.75 m² (40.36 sq ft)
Flaps (total) 12.28 m² (132.18 sq ft)
Spoilers (total) 1.34 m² (14.42 sq ft)

WEIGHTS AND LOADINGS:
Operating weight empty 12,500 kg (27,558 lb)
Max fuel weight 5,000 kg (11,023 lb)
Max payload 7,200 kg (15,873 lb)
Max T-O weight 21,500 kg (47,400 lb)
Max ramp weight 21,530 kg (47,465 lb)
Max zero-fuel weight 19,700 kg (43,430 lb)
Max landing weight 21,350 kg (47,068 lb)
Max wing loading 352.5 kg/m² (72.20 lb/sq ft)
Max power loading 6.01 kg/kW (9.88 lb/shp)
PERFORMANCE (at max T-O weight except where indicated):
Max cruising speed at 4,575 m (15,000 ft)
 284 knots (526 km/h; 327 mph)
Econ cruising speed at 7,010 m (23,000 ft), at 95% MTOW
 248 knots (460 km/h; 286 mph)
Max operating altitude 7,620 m (25,000 ft)
Service ceiling, OEI, at 97% MTOW and ISA + 10°C
 2,680 m (8,800 ft)
T-O balanced field length:
 at S/L, ISA 1,408 m (4,620 ft)
 at 915 m (3,000 ft), ISA + 10°C, at 97% MTOW
 1,570 m (5,150 ft)
Landing field length at S/L, ISA 1,210 m (3,970 ft)
Still air range (ISA), reserves for 87 nm (161 km; 100 mile) diversion and 45 min continued cruise:
 max optional payload 645 nm (1,195 km; 742 miles)
 66 passengers 1,200 nm (2,222 km; 1,380 miles)

 max fuel and zero payload
 2,100 nm (3,889 km; 2,416 miles)
PERFORMANCE (ATR 72-210): Mainly as ATR 72-200 except:
Max cruising speed 280 knots (519 km/h; 322 mph)
Gross ceiling, OEI, ISA + 10°C, 97% MTOW
 4,330 m (14,200 ft)
Balanced T-O field length, ISA, S/L, MTOW
 1,205 m (3,953 ft)
Balanced T-O field length, ISA + 10°C, 915 m (3,000 ft), TOW 20,500 kg (45,195 lb) 1,280 m (4,200 ft)
Landing field length, ISA, S/L, MLW
 1,043 m (3,420 ft)

ATR 82

TYPE: Stretched ATR 72.
PROGRAMME: Technical and market studies begun in 1992 are expected to be completed by end 1993; risk-sharing partners being sought; in service three years after launch.
DESIGN FEATURES: Existing fuselage strengthened and stretched to accommodate 78 passengers at 81 cm (32 in) pitch with equal or increased baggage volume; optional second toilet; wing enlarged and strengthened.
COSTS: Estimated development cost $500 to 600 million.
POWER PLANT: Choice of turbofan or turboprop, both designed to confer high performance; power plant selection expected during 1993.

AVIOTECHNICA
AVIOTECHNICA LTD
BULGARIA: PO Box 423, Trakia, 4000 Plovdiv
Telephone: 539 (32) 83 10 40
Fax: 539 (32) 83 10 80
PRESIDENT: D. Dobrev
RUSSIA: PO Box 32, K-45, 103045 Moscow

'Aviation Technique' founded 1983 by Bulgarian company VMZ-Sopot as research, development and manufacturing subsidiary, including plant and Institute producing UAVs for countries in former Warsaw Pact. Company's primary objective was to manufacture tail surfaces, engine nacelles and landing gear for Ilyushin Il-114 airliner and components for Mi-17 and Mi-34 helicopters. Due to delays in Il-114 production, Aviotechnica Ltd created 1991 as new joint company with 51 per cent share held by Bulgarian companies VMZ-Sopot, Metalchim and Metalchim Trading, and 49 per cent by Russian companies Interavia and LMZ (Luhovitskii Mashinostroitelnii Zavod, manufacturer in MiG-29 programme).

First joint product of Aviotechnica Ltd is SL-90 Leshii (I-1) lightplane.

AVIOTECHNICA SL-90 LESHII (I-1) (SPRITE)

TYPE: Two/three-seat sport, touring, patrol, cartographic and agricultural light aircraft.
PROGRAMME: Construction started by Interavia (responsible for engineering documentation) in 1988; first three prototypes built by Interavia and Myasishchev EMZ (Experimentalnii Mashinostroitelnii Zavod) in Russia, with flight tests carried out at Mikhail Gromov Flight Test Institute near Moscow (first flight February 1990); certification is in accordance with FAR Pt 23 and Russian light aircraft standards; pre-series batch of 10 under construction early 1993 at LMZ (responsible for jigs, engineering documentation and component manufacture), with final assembly at both LMZ and Aviotechnica; first flight of first series aircraft February 1993 in Bulgaria; planned production programme involves 350 aircraft by LMZ and Aviotechnica.
CURRENT VERSIONS: To be available with choice of one standard or three optional engines (see Power Plant).
DESIGN FEATURES: Braced high-wing monoplane with extensively glazed cabin, extended rear fuselage, braced tailplane and sweptback vertical tail. Constant chord wing with TsAGI R-III-15 aerofoil section, 3° sweepforward, 1° dihedral and 2° incidence.
FLYING CONTROLS: Conventional coupled mechanical; tab-operated port aileron and starboard elevator; ground adjustable tab on rudder. Plain flaps.
STRUCTURE: Mixed-construction wing (duralumin frame with

duralumin skin on forward portion, cotton fabric covering aft); tailplane similar; all-metal (aluminium alloy) stressed skin fuselage. All-metal wing and tailplane to be introduced on later aircraft.

LANDING GEAR: Non-retractable, with cantilever self-sprung steel mainwheel legs and controllable/lockable tailwheel. Mainwheels size 400 × 150 mm, tailwheel 200 × 80 mm. Mainwheel brakes.

POWER PLANT: One 82 kW (110 hp) M-3 three-cylinder four-stroke radial engine standard, with Interavia VM-3 two-blade constant-speed wooden propeller. Fuel capacity (two wing tanks) 140 litres (37 US gallons; 30.8 Imp gallons). Alternatives (already flight tested) are RPD-25 rotary engine (more than 89.5 kW; 120 hp), Textron Lycoming O-235 (85.75 kW; 115 hp) and Textron Lycoming O-320 (112 kW; 150 hp).

ACCOMMODATION: Two seats side by side. Space for more than 80 kg (176 lb) of baggage, or third person, aft of front seats. Door on each side for pilot/passenger; separate door for baggage compartment.

EQUIPMENT: Russian or other, to customer's requirements.

DIMENSIONS, EXTERNAL:
Wing span	11.623 m (38 ft 1¾ in)
Wing chord, constant	1.262 m (4 ft 1½ in)
Wing aspect ratio	9.33
Length overall	6.40 m (21 ft 0 in)
Height overall	2.10 m (6 ft 10¾ in)
Tailplane span	3.30 m (10 ft 10 in)
Wheel track	2.30 m (7 ft 6½ in)
Wheelbase	3.63 m (11 ft 11 in)
Propeller diameter	1.85 m (6 ft 0¾ in)
Cabin doors, each: Height	0.95 m (3 ft 1½ in)
Max width	1.15 m (3 ft 9¼ in)
Baggage door: Max height	0.83 m (2 ft 8¾ in)
Max width	0.80 m (2 ft 7½ in)

DIMENSIONS, INTERNAL:
Cabin: Max width	1.15 m (3 ft 9¼ in)

AREAS:
Wings, gross	14.481 m² (155.9 sq ft)
Ailerons (total)	1.449 m² (15.60 sq ft)
Trailing-edge flaps (total)	1.542 m² (16.60 sq ft)
Fin	0.97 m² (10.44 sq ft)
Rudder	0.87 m² (9.36 sq ft)
Tailplane	1.44 m² (15.50 sq ft)
Elevators (total)	1.46 m² (15.72 sq ft)

WEIGHTS AND LOADINGS (M-3 engine):
Weight empty	510 kg (1,124 lb)
Fuel weight	110 kg (242.5 lb)
Typical mission T-O weight	780 kg (1,720 lb)
Max T-O weight	840 kg (1,852 lb)
Max wing loading	58.01 kg/m² (11.88 lb/sq ft)
Max power loading	10.24 kg/kW (16.83 lb/hp)

PERFORMANCE (at max T-O weight, M-3 engine, except where indicated):
Never-exceed speed (V_{NE})	121 knots (225 km/h; 140 mph)
Max level speed	103 knots (190 km/h; 118 mph)
Max cruising speed	86 knots (160 km/h; 99 mph)
Stalling speed	43 knots (80 km/h; 50 mph)
Max rate of climb at S/L	180 m (590 ft)/min
Service ceiling	4,000 m (13,125 ft)
T-O and landing run (grass)	200 m (657 ft)
g limits (metal wings/tailplane)	+6.3/−4

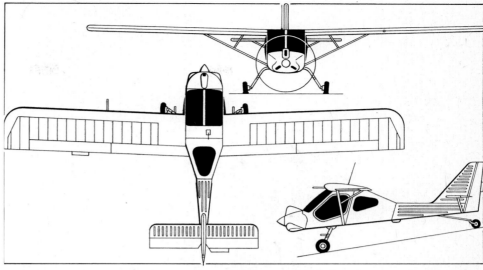

Aviotechnica SL-90 Leshii light aircraft (*Jane's/Mike Keep*)

Aviotechnica/Interavia SL-90 Leshii (I-1) two/three-seat light aircraft

AVRO
AVRO INTERNATIONAL AEROSPACE LTD (Division of British Aerospace Regional Aircraft)

Woodford Aerodrome, Chester Road, Stockport, Cheshire SK7 1QR, UK
Telephone: 44 (61) 439 5051
Fax: 44 (61) 955 4570
PUBLIC RELATIONS: Howard Borrington
PARTICIPATING COMPANIES:
British Aerospace: see under UK
Taiwan Aerospace Corporation: see under Taiwan

Formal signing of agreement between British Aerospace (BAe) and Taiwan Aerospace Corporation (TAC) for formation of Avro International Aerospace took place 19 January 1993. Both participants have 50 per cent share in the new company, which takes over former Regional Jetliner activities of BAe. Headquarters is at Woodford in UK, with major representative offices in Taiwan and Washington DC in USA. TAC paid BAe about £120 million for 50 per cent equity, with additional US$25 million to follow on delivery of first Taiwan assembled RJ.

AVRO INTERNATIONAL AEROSPACE REGIONAL JETLINER (RJ) SERIES

TYPE: Four turbofan short-range transport.
PROGRAMME: Developed from BAe 146 (see UK section), announced June 1992; first production RJ (RJ85 for Crossair, G-CROS) flew at Woodford 27 November 1992, delivered 2 April 1993; final assembly of RJs continuing at Woodford, with further line being established at Taichung in Taiwan; first Taichung assembled RJ anticipated for completion in early 1994. Introduced Jetstart and Jetkey airline support programmes, incorporating extensive maintenance cost guarantees; company guarantees to match twin-engined aircraft maintenance costs.

Avro International Aerospace RJ70, RJ85 and RJ100 representatives

136 INTERNATIONAL: AIRCRAFT—AVRO

First delivery of an RJ was this RJ85 to Crossair in April 1993

CURRENT VERSIONS: **RJ70**: Shortest fuselage version, accommodating 70-94 passengers; derated 27.27 kN (6,130 lb st) LF 507 engines. Deliveries scheduled to begin third quarter of 1993.

RJ85: Lengthened version for typically 85-112 passengers; 31.14 kN (7,000 lb st) LF 507 engines. Deliveries began 2 April 1993.

RJ100: Longest current model, for 100-128 passengers; engines as for RJ85.

RJ115: Same fuselage dimensions as RJ100 but featuring mid-cabin exits, increased capacity air-conditioning packs and higher design weights and fuel capacity as standard; 116 passengers in typical six-abreast configuration; same engines as RJ100.

RJX: Proposed twin-engined development for service entry in late 1990s; formerly announced as BAe 146 NRA (New Regional Airliner); programme to be reviewed by new company in light of current market projections, power plant options and opportunities arising from participation of a new partner; reportedly 120 passengers.

QT Quiet Trader: Freighter versions of RJ85 and RJ100.

CUSTOMERS: Thirty-three firm orders by April 1993, to Business Express (20 RJ70s), City Air Scandinavia (five RJ85s), Crossair (four RJ85s plus eight options), and Meridiana (four RJ85s plus options).

DESIGN FEATURES: Low operating noise levels; ability to operate from short or semi-prepared airstrips with minimal ground facilities; improved 'Spaceliner' interior; new LF 507 FADEC controlled engines with increased power and reduced temperatures; new digital flight deck with Cat. IIIa all-weather landing capability; and increased design weights.

High lift aerofoil section; thickness/chord ratio 15.3 per cent at root, 12.2 per cent at tip; anhedral 3° at trailing-edge; incidence 3° 6′ at root, 0° at tip; sweepback 15° at quarter-chord.

FLYING CONTROLS: Mechanically actuated balanced ailerons, with trim and servo tabs; manually operated balanced elevators, with trim and servo tabs; powered rudder. Single-section hydraulically actuated tabbed Fowler flaps, spanning 78 per cent of trailing-edges, with Dowty actuators; hydraulically operated roll spoiler outboard of three lift dumpers on each wing; no leading-edge lift devices; petal airbrakes form tailcone when closed. Honeywell automatic flight control/flight guidance system.

STRUCTURE: All-metal; fail-safe wings with machined skins, integrally machined spars and ribs; fail-safe fuselage with chemically etched skins; strengthened centre-section developed initially for RJ100 (will be standard on all RJ70s); nose free of stringers; remainder of fuselage has top hat stringers bonded to skins above keel area; Z section stringers wet assembled with bonding agent and riveted to skin in keel area; T tail with chemically etched skins bonded to top hat section stringers; fixed incidence tailplane.

Final assembly at Woodford; second line being set up at Taichung in Taiwan. Under risk-sharing arrangements, Textron Aerostructures (USA) manufactures wings.

LANDING GEAR: Hydraulically retractable tricycle type of Dowty design, with twin Dunlop wheels on each unit. Main units retract inward into fairings on fuselage sides; steerable (±70°) nose unit retracts forward. Oleo-pneumatic shock absorbers with wheels mounted on trailing axle. Simple telescopic nosewheel strut. Mainwheel tyres size 12.50-16 Type III, pressure (RJ70) 8.41 bars (122 lb/sq in). Nosewheel tyres size 7.50-10 Type III, pressure (RJ70) 7.79 bars (113 lb/sq in). Low pressure tyres optional. Dunlop multi-disc carbon brakes operated by duplicated hydraulic systems. Anti-skid units in both primary and secondary brake systems. Min ground turning radius about nosewheels: RJ70, 11.53 m (37 ft 10 in); RJ85, 12.55 m (41 ft 2 in); RJ100, 13.97 m (45 ft 10 in).

POWER PLANT: Four Textron Lycoming LF 507 turbofans, each rated at 31.14 kN (7,000 lb st), installed in underwing pylon pods on RJ85, RJ100 and RJ115. RJ70 has derated LF 507 engines as standard, providing 27.27 kN (6,130 lb st). No reverse thrust. Fuel in two integral wing tanks and integral centre-section tank (latter with vented and drained sealing diaphragm above passenger cabin); combined usable capacity 11,728 litres (3,098 US gallons; 2,580 Imp gallons). Optional auxiliary tanks in wingroot fairings, combined capacity 1,173 litres (310 US gallons; 258 Imp gallons), giving total capacity of 12,901 litres (3,408 US gallons; 2,838 Imp gallons). Single-point pressure refuelling, with coupling situated in starboard wing outboard of outer engine.

ACCOMMODATION: Crew of two pilots on flight deck, and two or three cabin staff. Standard observer's seat. RJ70 accommodates 70 passengers five-abreast at 84 cm (33 in) pitch and up to 94 six-abreast at 74 cm (29 in) pitch; RJ85 maximum capacity for 112 passengers six-abreast at 74 cm (29 in) pitch; RJ100 standard accommodation for 100 passengers five-abreast at 79 cm (31 in) pitch, and maximum 128 passengers. One outward opening passenger door forward and one aft on port side of cabin. Built-in airstairs optional. Servicing doors, one forward and one aft on starboard side of cabin. Freight and baggage holds under cabin floor. All accommodation pressurised and air-conditioned.

SYSTEMS: Normalair-Garrett cabin air-conditioning and pressurisation system, using engine bleed air. Electro-pneumatic pressurisation control with discharge valves at fore and aft of cabin. Max differential 0.47 bar (6.75 lb/sq in), giving 2,440 m (8,000 ft) equivalent altitude at 9,450 m (31,000 ft). Hydraulic system, duplicated for essential services, for landing gear, flaps, rudder, roll and lift spoilers, airbrakes, nosewheel steering, brakes and auxiliary fuel pumps; pressure 207 bars (3,000 lb/sq in). Electrical system powered by two 40kVA integrated-drive alternators to feed 115/200V 3-phase 400Hz primary systems. 28V DC power supplied by transformer-rectifier in each channel. Hydraulically powered emergency electrical power unit. Garrett GTCP 36-150 APU for ground and air usable air-conditioning and electrical power generation. Chemical oxygen system. Stall warning and identification system, comprising stick shaker (warning) and stick force (identification) elements, providing soft and hard corrective stick forces at the approach of stall conditions. Hot air de-icing of wing and tailplane leading-edges; windscreen electric anti-icing and demisting standard; rain repellent system optional.

AVIONICS: Honeywell digital flight guidance system incorporates fail-passive Cat. IIIa autopilot, auto-throttle, yaw damper and windshear detection and protection. Standard ARINC interface with radio nav systems allows choice of radio equipment. Basic avionics include dual VHF com audio system, passenger address system, cockpit voice recorder, marker beacon receiver, weather radar, dual radio altimeters, ground proximity warning system, dual DME, dual ATC transponders, dual VHF nav and dual ADF. Flight deck warning system. Optional avionics include third VHF com, Selcal, tape reproducer, TCAS, and single or dual HF com.

DIMENSIONS, EXTERNAL:
* Wing span: all versions, excl static dischargers

	26.21 m (86 ft 0 in)
Wing aspect ratio: all versions except RJX	8.97
Wing chord: at root	2.75 m (9 ft 0 in)
at tip	0.91 m (3 ft 0 in)
**Length overall: RJ70	26.20 m (85 ft 11½ in)
RJ85	28.60 m (93 ft 10 in)
RJ100, RJ115	30.99 m (101 ft 8¼ in)
Height overall: RJ70	8.61 m (28 ft 3 in)
RJ85	8.59 m (28 ft 2 in)
Fuselage max diameter	3.56 m (11 ft 8 in)
Tailplane span	11.09 m (36 ft 5 in)
Wheel track	4.72 m (15 ft 6 in)
Wheelbase: RJ70	10.09 m (33 ft 1½ in)
RJ85	11.20 m (36 ft 9 in)
RJ100, RJ115	12.52 m (41 ft 1 in)
Passenger doors (port, fwd and rear):	
Height	1.83 m (6 ft 0 in)
Width	0.85 m (2 ft 9½ in)
Height to sill: fwd	1.88 m (6 ft 2 in)
rear	1.98 m (6 ft 6 in)
Servicing doors (stbd, fwd and rear):	
Height	1.47 m (4 ft 10 in)
Width	0.85 m (2 ft 9½ in)
Height to sill: fwd	1.88 m (6 ft 2 in)
rear	1.98 m (6 ft 6 in)
Underfloor freight hold door (stbd, fwd):	
Height	1.09 m (3 ft 7 in)
Width	1.35 m (4 ft 5 in)
Height to sill	0.78 m (2 ft 7 in)
Underfloor freight hold door (stbd, rear):	
Height	1.04 m (3 ft 5 in)
Width	0.91 m (3 ft 0 in)
Height to sill	0.90 m (2 ft 11½ in)
Freight door (freighter versions):	
Height	1.93 m (6 ft 4 in)
Width: RJ70	2.92 m (9 ft 7 in)
RJ85, RJ100	3.33 m (10 ft 11 in)
Height to sill	1.93 m (6 ft 4 in)

* *Static discharger extends 6.3 cm (2½ in) from each wingtip*
** *Static dischargers on elevator extend length of all series by 18.4 cm (7¼ in)*

DIMENSIONS, INTERNAL:

Cabin (excl flight deck; incl galley and toilets):	
Length: RJ70	15.42 m (50 ft 7 in)
RJ85	17.81 m (58 ft 5 in)
RJ100, RJ115	20.20 m (66 ft 3¼ in)
Max width	3.42 m (11 ft 2½ in)
Max height	2.02 m (6 ft 7½ in)
Freight cabin (RJ85-QT):	
Cargo floor: Length	16.08 m (52 ft 9 in)
Width	3.23 m (10 ft 7 in)
Volume: Pallets/igloos	60.3 m³ (2,145 cu ft)
LD3 containers	42.66 m³ (1,422 cu ft)
Baggage/freight holds, underfloor:	
RJ70	13.7 m³ (479 cu ft)
RJ85	18.3 m³ (645 cu ft)
RJ100, RJ115	22.99 m³ (812 cu ft)

AREAS:

Wings, gross: all versions	77.30 m² (832.0 sq ft)
Ailerons (total)	3.62 m² (39.0 sq ft)
Trailing-edge flaps (total)	19.51 m² (210.0 sq ft)
Spoilers (total)	10.03 m² (108.0 sq ft)
Fin	15.51 m² (167.0 sq ft)
Rudder	5.30 m² (57.0 sq ft)
Tailplane	15.61 m² (168.0 sq ft)
Elevators, incl tabs (total)	10.03 m² (108.0 sq ft)

WEIGHTS AND LOADINGS:

Operating weight empty: RJ70	23,781 kg (52,430 lb)
RJ85	24,377 kg (53,742 lb)
RJ85-QT	23,045 kg (50,805 lb)
RJ100	25,362 kg (55,915 lb)
RJ100-QT	23,706 kg (52,263 lb)
RJ115	26,156 kg (57,665 lb)
Max payload: RJ70	8,650 kg (19,070 lb)

RJ85	11,457 kg (25,258 lb)
RJ85-QT	12,789 kg (28,195 lb)
RJ100	12,059 kg (26,585 lb)
RJ100-QT	13,715 kg (30,237 lb)
RJ115	11,265 kg (24,835 lb)
Max fuel weight: all series except RJ115:	
standard	9,362 kg (20,640 lb)
optional	10,298 kg (22,704 lb)
RJ115	10,298 kg (22,704 lb)
Max T-O weight: RJ70	38,102 kg (84,000 lb)
RJ85	42,184 kg (93,000 lb)
RJ100	44,225 kg (97,500 lb)
RJ115	46,040 kg (101,500 lb)
Max ramp weight: RJ70	38,329 kg (84,500 lb)
RJ85	42,410 kg (93,500 lb)
RJ100	44,452 kg (98,000 lb)
RJ115	46,266 kg (102,000 lb)
Max zero-fuel weight: RJ70	32,432 kg (71,500 lb)
RJ85	35,834 kg (79,000 lb)
RJ100, RJ115	37,421 kg (82,500 lb)
Max landing weight: RJ70	37,875 kg (83,500 lb)
RJ85	38,555 kg (85,000 lb)
RJ100, RJ115	40,143 kg (88,500 lb)
Max wing loading: RJ70	493.1 kg/m² (101.0 lb/sq ft)
RJ85	545.9 kg/m² (111.8 lb/sq ft)
RJ100	572.2 kg/m² (117.2 lb/sq ft)
RJ115	595.7 kg/m² (122.0 lb/sq ft)
Max power loading: RJ70	349.2 kg/kN (3.43 lb/lb st)
RJ85	338.6 kg/kN (3.32 lb/lb st)
RJ100	355.0 kg/kN (3.48 lb/lb st)
RJ115	369.5 kg/kN (3.63 lb/lb st)

PERFORMANCE (at max standard T-O weight, except where indicated):
Max operating Mach number (M_{MO}): all versions 0.73
Max operating speed (V_{MO}):
RJ70, RJ85 300 knots (555 km/h; 345 mph) IAS
RJ100, RJ115 305 knots (565 km/h; 351 mph) IAS
Cruising speed at 8,840 m (29,000 ft) for 300 nm (556 km; 345 mile) sector:
RJ70: high speed 432 knots (801 km/h; 498 mph)
long range 356 knots (660 km/h; 410 mph)
RJ85: high speed 432 knots (801 km/h; 498 mph)
long range 364 knots (674 km/h; 419 mph)
RJ100, RJ115: high speed
 432 knots (801 km/h; 498 mph)
long range 371 knots (687 km/h; 427 mph)
Stalling speed, 30° flap:
RJ70 97 knots (179 km/h; 111 mph) EAS
RJ85 101 knots (187 km/h; 116 mph) EAS
RJ100 104 knots (192 km/h; 119 mph) EAS
RJ115 106 knots (197 km/h; 122 mph) EAS
Stalling speed, 33° flap, at max landing weight:
RJ70, RJ85 93 knots (172 km/h; 107 mph) EAS
RJ100, RJ115 95 knots (176 km/h; 109 mph) EAS
T-O to 10.7 m (35 ft), S/L, ISA: RJ70 1,469 m (4,820 ft)
RJ85 1,564 m (5,130 ft)
RJ100 1,692 m (5,550 ft)
RJ115 1,835 m (6,020 ft)
FAR landing distance from 15 m (50 ft), S/L, ISA, at max landing weight: RJ70 1,173 m (3,850 ft)
RJ85 1,195 m (3,920 ft)
RJ100, RJ115 1,274 m (4,180 ft)
Range with standard fuel:
RJ70 1,952 nm (3,619 km; 2,249 miles)
RJ85 1,782 nm (3,302 km; 2,052 miles)
RJ100 1,671 nm (3,098 km; 1,925 miles)
RJ115 1,794 nm (3,325 km; 2,066 miles)
Range with max payload:
RJ70 1,104 nm (2,047 km; 1,272 miles)
RJ85 1,148 nm (2,127 km; 1,322 miles)
RJ100 1,165 nm (2,159 km; 1,342 miles)
RJ115 1,469 nm (2,723 km; 1,692 miles)
OPERATIONAL NOISE LEVELS (FAR Pt 36-12, estimated):
T-O: RJ70 81.8 EPNdB
RJ85 83.3 EPNdB
RJ100 84.9 EPNdB
RJ115 86.3 EPNdB
Approach: RJ70 97.5 EPNdB
RJ85 97.6 EPNdB
RJ100, RJ115 97.6 EPNdB
Sideline: RJ70 87.3 EPNdB
RJ85 88.1 EPNdB
RJ100 87.8 EPNdB
RJ115 87.6 EPNdB

DASSAULT/DORNIER

PARTICIPATING COMPANIES:
Dassault Aviation: see under France
Dornier: see under Deutsche Aerospace Military Aircraft Division, Germany

DASSAULT/DORNIER ALPHA JET

TYPE: Tandem-seat jet basic, low-altitude and advanced trainer and close support/battlefield reconnaissance aircraft.
PROGRAMME: Started 1970, initially for air forces of France and Germany; four prototypes (first flight 26 October 1973: see 1978-79 and earlier *Jane's*); production ended, but any version except Lancier (1989-90 *Jane's*) remains available. Last detailed description in 1990-91 *Jane's*, variants as listed in 1992-93 edition.

CUSTOMERS: Total 504 (excluding prototypes) built for Belgium (33), Cameroon (seven), Egypt (30 MS1 and 15 MS2, of which 26 and 11 assembled locally by AOI), France (176), Germany (175, of which 168 remain), Ivory Coast (seven), Morocco (24), Nigeria (24), Qatar (six) and Togo (seven). Planned German upgrade (see 1991-92 *Jane's*) cancelled; Luftwaffe now phasing out all but 35, with remaining aircraft awaiting government approval for resale.

DE CHEVIGNY/WILSON

c/o Aéro Club de France, 6 rue Galilée, F-75016 Paris, France

DE CHEVIGNY/WILSON EXPLORER

TYPE: Habitable exploration amphibian.
PROGRAMME: Design commissioned by Hubert de Chevigny from Avid Flyer designer Dean Wilson late 1987; construction began October 1988; first flight April 1991; operates in FAA Utility Experimental category. Flew for six months around Alaska, British Columbia and the Rockies in 1991, making travel/adventure programmes; extensively damaged in take-off accident 1991; second Explorer being built to fly in late 1993.
COSTS: $1 million (1991) without special living arrangements and equipment. Initial development financed by Bail Aviation, F-69003 Lyon, France.
DESIGN FEATURES: Aircraft designed to give protection from all climates and house five people in full comfort with space heating, kitchen and beds. Large rear cargo door in second Explorer replaces swing tail for loading bulky freight; hatch in windscreen for mooring and loading long objects. Full facilities for filming and observation including video cabling to camera mounts on outer wings; wing designed to double as working platform; hatch/window in floor for downward observation into water, equipment dropping and filming; ceiling hatch, openable in flight, gives access to top wing surface. Wing aerofoil Gottingen 387. Empty weight to be reduced by 317 kg (700 lb).
FLYING CONTROLS: Simple mechanical controls with separate flaps and ailerons replacing flaperons of first Explorer; four lateral control spoiler panels usable as lift dumpers.
STRUCTURE: All structure designed to be easy to repair in field. Fuselage shell, tailplane and fin made of welded steel tube covered with double layer of Dacron Ceconite; outer skin of boat hull of S-glass epoxy. Wing spars of Oregon pine; wooden ribs; all parts dipped in epoxy; wings skinned with plywood faced with several diagonal layers of S-glass epoxy, six coats of nitrate dope and six coats of butyrate doped with aluminium powder.
LANDING GEAR: Flying-boat hull with retractable water rudder under tail; tricycle retractable undercarriage for land operation; skis can be mounted on wheels and cover main undercarriage bays when wheels are retracted.
POWER PLANT: Two 175.4 kW (235 hp) Textron Lycoming O-540 flat-six engines, running on Mogas, each driving a two-blade fixed-pitch propeller. One 68.3 litre (18 US gallon; 14.98 Imp gallon) tank in each wing and a 586.6 litre (155 US gallon; 129 Imp gallon) main belly tank.
ACCOMMODATION: Custom interiors designed for living in all climates.

de Chevigny/Wilson Explorer amphibian (two Textron Lycoming O-540)

DIMENSIONS, EXTERNAL:
Wing span 19.81 m (65 ft 0 in)
Wing chord, constant 2.44 m (8 ft 0 in)
Length overall 12.19 m (40 ft 0 in)
Height overall 4.47 m (14 ft 8 in)
Fuselage: Max width 3.05 m (10 ft 0 in)
Width over sponsons 5.03 m (16 ft 6 in)
Tailplane span 6.10 m (20 ft 0 in)
Area of wing platform
 19.81 m (65 ft 0 in) × 0.99 m (3 ft 3 in)
DIMENSIONS, INTERNAL:
Cabin:
Length from pilot's seat to rear fixed bed
 6.55 m (21 ft 6 in)
Max width 3.00 m (9 ft 10 in)
Headroom (constant) 1.90 m (6 ft 2¾ in)
Floor area 27.0 m² (290.6 sq ft)
AREAS:
Wings, gross 46.92 m² (505.0 sq ft)
WEIGHTS AND LOADINGS:
Weight empty 1,950 kg (4,300 lb)
Max T-O weight 3,402 kg (7,500 lb)
Max wing loading 72 kg/m² (14.85 lb/sq ft)
Max power loading 9.69 kg/kW (15.95 lb/hp)
PERFORMANCE (preliminary observations):
Cruising speed, 55% power, 2,200 rpm
 86 knots (160 km/h; 99 mph)
Stalling speed 43 knots (79 km/h; 49 mph)

DEUTSCHE AEROSPACE AIRBUS/ TUPOLEV

PARTICIPATING COMPANIES:
Deutsche Aerospace Airbus: see under Germany
Tupolev: see under Russia

DEUTSCHE AEROSPACE AIRBUS/ TUPOLEV CRYOPLANE

Deutsche Aerospace Airbus co-operating in German-Russian effort started in 1990 to investigate environmentally compatible airliner powered by cryogenic hydrogen or natural gas instead of kerosene. Two-year feasibility study by Deutsche Aerospace Airbus, completed in September 1992, concluded that liquid hydrogen is safer than natural gas, kinder to environment and more readily available over long term.

Participants in next stage of Cryoplane project, under leadership of Deutsche Aerospace Airbus, include Tupolev, KKBM (formerly Kuznetsov), Deutsche Aerospace, Linde, MAN Technologie, Messer-Griesheim, Uhde, Garrett, Liebherr, Deutsche Lufthansa, Berlin Airport Corporation and Hamburg's Max Planck Meteorological Institute. Some years ago, Tupolev tested a Tu-155 with one engine fuelled by hydrogen (see 1990-91 *Jane's*).

Demonstration phase, possibly based on modified Airbus A310, could start in about 2000. Liquid hydrogen requires four times as much tankage volume as kerosene, and best location for such tankage is faired into upper part of fuselage, as shown in artist's impression. Such a system could be in service by 2010.

Artist's impression of the Deutsche Aerospace Airbus Cryoplane based on an Airbus A310 experimentally fuelled with cryogenic hydrogen

DOUGLAS/CATIC

PARTICIPATING COMPANIES:
Douglas Aircraft: see under USA (McDonnell Douglas)
CATIC: see under China

DOUGLAS/CATIC MD-95

Now regarded as a Douglas project. See under Douglas Aircraft in McDonnell Douglas entry in US section.

EGRETT

PARTICIPATING COMPANIES:
E-Systems Inc, Greenville Division, PO Box 6056, Greenville, Texas 75403-6056, USA
Telephone: 1 (903) 457 6462
Fax: 1 (903) 457 7674
DIRECTOR OF BUSINESS DEVELOPMENT, SURVEILLANCE AND VERIFICATION SYSTEMS: Elizabeth King
Grob: see under Germany
Garrett: see USA in Engines section
Egrett name derived from those of original three companies collaborating in its development.

G 520 EGRETT II and STRATO 1

TYPE: Multi-purpose high-altitude surveillance and relay aircraft. See also Grob Strato 2C in German section.
PROGRAMME: Begun 1986 and revealed April 1987 as joint programme by E-Systems' Greenville Division (programme leader/systems integrator), Grob (airframe design and construction) and Garrett (engine); first flight (proof of concept Egrett I) 24 June 1987; five others since flown (see Current Versions, total over 1,000 flying hours by March 1993); US-German venture Telos GmbH (E-Systems, Grob, MBB and Elekluft) formed June 1990 for in-country logistics and maintenance support for Egrett/D-500 associated programmes; second D-500 outfitted for sigint during 1991 (demonstration of imaging radar); two-seat trainer version flown April 1993.
CURRENT VERSIONS: **D-FGEI/N14ES:** Egrett I proof of concept (POC) aircraft, as described in 1991-92 and earlier *Jane's*; no longer flying.

D-FGEE: D-500 'pre-prototype' for Egrett II, differing from Egrett I in having 5.00 m (16 ft 4¾ in) greater wing span (to house additional antennae), pressurised cockpit, retractable main landing gear (to avoid masking underfuselage DF antenna), modified rear fuselage, and higher gross and payload weights; first flight 20 April 1989; certificated by LBA 22 March 1991 and FAA 13 September 1991.

D-FGEO: D-500 'integration prototype' for basic systems evaluation; first flight 9 September 1990; generally similar to D-FGEE; total flying hours of D-FGEI and D-FGEO more than 350 by January 1992.

D-FDEM (G 520): Egrett II demonstrator for German Air Force requirement; detachable wingtips and bulged payload modules beneath wingroots; first flight 19 January 1991; FAR Pt 23 certification (reciprocal with German LBA) December 1991. *Detailed description applies specifically to D-FDEM unless otherwise stated.*

D-FGRO Strato 1 (G 520): Grob-owned commercial demonstrator, fitted with winglets; pressurised cockpit (max differential 0.41 bar; 6.0 lb/sq in); first flight 5 June 1991.

D-FDST (G 520T): Trainer version: E-Systems awarded $27.4 million FMS contract August 1991 for two-seat trainer, to be built and flight tested by Grob; first flight 21 April 1993.
CUSTOMERS: US Congress notified March 1992 of German government intention to order nine single-seat D-500s (sic) to fulfil Luftwaffe LAPAS 1 requirement at estimated cost of $795 million including one ground station, spares and support material. Programme survived two defence budget

Egrett II military demonstrator

reviews in late 1992 but cancelled 3 February 1993 following further budget reductions. Believed intended to be completed by Egrett II demonstrator brought up to LAPAS 1 standard, plus G 520T trainer, making 11 aircraft total.

DESIGN FEATURES: Capacious fuselage and very high aspect ratio wings for HALE (high altitude, long endurance) performance; capable of manned or unmanned operation; bays in rear fuselage can accept various payloads to customer's requirements, in modular packages facilitating installation/servicing/removal; large fuselage bays provide easy access to payloads.

Laminar-flow wing section (modified Eppler E 580); leading-edge sweepback 4°; thickness/chord ratio 6.25 per cent; elliptical tip section; dihedral 3°; incidence 2° at root; 0° twist.

FLYING CONTROLS: Primary surfaces manual/mechanical (pushrods), with cable-controlled elevator and rudder trim; split flaps on wing inboard trailing-edges.

STRUCTURE: Single box-spar in wing; load-bearing skins of honeycomb core construction throughout; airframe built largely of glassfibre, carbonfibre and Kevlar reinforced composites, for low radio/radar reflectivity; propeller blades also of composites.

LANDING GEAR: Electrohydraulically retractable tricycle type, with single wheel and oleo-pneumatic shock absorber on each unit. Nose unit retracts rearward; main units forward into underwing pods. Mainwheel tyres size 17.5 × 5.75 in, pressure 12.41 bars (180 lb/sq in); nosewheel tyre size 18 × 4.4 in, pressure 6.62 bars (96 lb/sq in). Air-cooled hydraulic disc brakes. Nosewheel steerable ±55°. Min ground turning radius about nosewheel 8.50 m (27 ft 11 in).

POWER PLANT: One Garrett TPE331-14F-801L turboprop, flat rated at 559 kW (750 shp), driving a Hartzell HC-E4P-5/E11990K four-blade constant-speed, fully feathering, reversible-pitch propeller. Integral fuel tank occupying almost whole of each wing leading-edge, combined capacity 1,317 litres (348 US gallons; 290 Imp gallons); gravity fuelling point in upper surface of each wing near tip. (D-FGRO fuel capacity 1,117 litres; 295 US gallons; 246 Imp gallons standard, 1,382 litres; 365 US gallons; 304 Imp gallons optional.) Oil capacity 11.4 litres (3 US gallons; 2.5 Imp gallons).

ACCOMMODATION: Single-seat cockpit. Canopy opens sideways to starboard. Three avionics/equipment bays in each side of upper rear fuselage, plus three-bay ventral pod, all pressurised and with external access only. Cockpit is heated, ventilated and air-conditioned; D-FGRO is pressurised, D-FDEM is not.

SYSTEMS: AiResearch air cycle environmental control system uses engine bleed air for cockpit heating, ventilation and air-conditioning, and to pressurise avionics/equipment bays at 0.41 bar (5.9 lb/sq in). Electrically driven hydraulic system for landing gear actuation and mainwheel brakes only. Electrical power provided by 250A engine driven starter/generator and two (incl one standby) 24V 19Ah lead-acid batteries, with 10kVA generator and 200V inverter for 115V three-phase AC at 400Hz; 24V 19Ah battery. Liquid oxygen system, capacity 10 litres (610 cu in). Pneumatic boot de-icing of wing and tailplane leading-edges; electrically heated propeller blades and pitot tubes; bleed air heated engine air inlet.

AVIONICS: Commercial Bendix/King IFR system standard, comprising one each VHF com, UHF com, VHF nav, Tacan, marker beacon receiver, ATC transponder with IFF capability, ADF, KFC 325 flight director and autopilot. (N. B. PRISMA teaming agreement, described in previous editions, is no longer in existence.)

DIMENSIONS, EXTERNAL:
Wing span: FDEM	31.50 m (103 ft 4¼ in)
FGRO	33.00 m (108 ft 3¼ in)
Wing chord at root	1.90 m (6 ft 2¾ in)
Wing aspect ratio: FDEM	25.57
FGRO	27.45
Length overall	12.10 m (39 ft 8½ in)
Fuselage: Max width	1.04 m (3 ft 5 in)
Max depth	1.93 m (6 ft 4 in)
Height overall	5.68 m (18 ft 7¾ in)
Tailplane span	6.40 m (21 ft 0 in)
Wheel track: FDEM	4.57 m (15 ft 0 in)
FGRO	4.68 m (15 ft 4¼ in)
Wheelbase	3.71 m (12 ft 2 in)
Propeller diameter	3.05 m (10 ft 0 in)
Propeller ground clearance	0.39 m (1 ft 3¼ in)

DIMENSIONS, INTERNAL:
Cockpit, excl avionics bays:
Length	1.685 m (5 ft 6¼ in)
Max width	0.862 m (2 ft 10 in)
Max height	1.265 m (4 ft 1¾ in)
Floor area	1.45 m² (15.6 sq ft)
Volume	1.47 m³ (51.9 cu ft)

Avionics/equipment bays:
Combined volume	6.37 m³ (225 cu ft)

AREAS:
Wings, gross: FDEM	38.80 m² (417.6 sq ft)
FGRO	39.68 m² (427.1 sq ft)
Ailerons (total)	1.51 m² (16.25 sq ft)
Trailing-edge flaps (total)	2.82 m² (30.35 sq ft)
Fin	5.67 m² (61.03 sq ft)
Rudder	3.25 m² (34.98 sq ft)
Tailplane	7.50 m² (80.73 sq ft)
Elevators (total)	2.40 m² (25.83 sq ft)

Egrett D-FGRO Strato 1, fifth of the series, at the Paris Air Show in June 1991 (*Brian M. Service*)

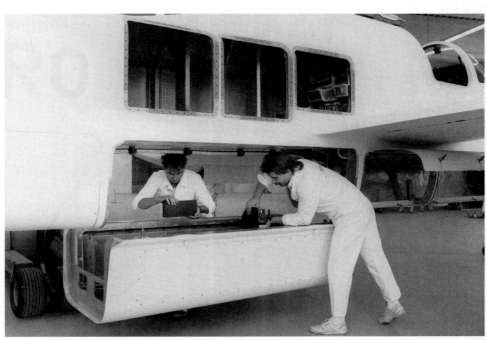

Egrett's deep fuselage has three upper bays each side for standard payload items; the lower pod, also three-bay, can be modified to accept payloads of variable shape or size. All bays are pressurised

Clockwise from top left: Egrett II removable equipment module (REM), synthetic aperture radar, long-range radar and EO/IR surveillance sensor

140 INTERNATIONAL: AIRCRAFT—EGRETT/EHI

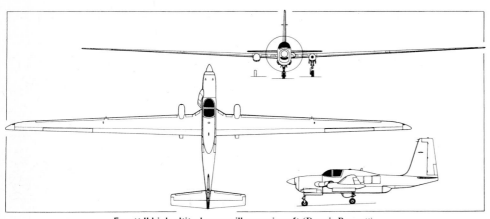

Egrett II high altitude surveillance aircraft *(Dennis Punnett)*

WEIGHTS AND LOADINGS:
Weight empty	3,063 kg (6,754 lb)
Max payload	998 kg (2,200 lb)
Max fuel load	1,055 kg (2,325 lb)
Max T-O weight	4,700 kg (10,362 lb)
Max landing weight	4,465 kg (9,843 lb)
Max zero-fuel weight	4,240 kg (9,347 lb)
Max wing loading: FDEM	121.13 kg/m² (24.81 lb/sq ft)
FGRO	118.46 kg/m² (24.26 lb/sq ft)
Max power loading	8.41 kg/kW (13.82 lb/shp)

PERFORMANCE (at max T-O weight):
Max operating speed up to 9,750 m (32,000 ft), and max cruising speed	153 knots (284 km/h; 176 mph)
Econ cruising speed	75 knots (139 km/h; 86 mph)
Stalling speed, engine idling:	
flaps up	66 knots (123 km/h; 76 mph)
flaps down	60 knots (111 km/h; 69 mph)
Max rate of climb at S/L	427 m (1,400 ft)/min
Service ceiling	16,000 m (52,495 ft)
T-O run	457 m (1,500 ft)
T-O to 15 m (50 ft)	839 m (2,750 ft)
Landing from 15 m (50 ft)	884 m (2,900 ft)
Landing run	482 m (1,580 ft)
Range with max payload, max range power at 13,715 m (45,000 ft), reserves for 45 min at 3,050 m (10,000 ft) at max range power	724 nm (1,341 km; 833 miles)
Range with max fuel	2,475 nm (4,586 km; 2,850 miles)
Max endurance	13 h
g limits (FGRO): flaps up	+3.28/−1.31
flaps down	+2

EVER

TYPE: Optionally piloted surveillance aircraft (Endurance Vehicle for Extended Reconnaissance).

PROGRAMME: Projected follow-on derivative of Egrett II, for piloted or unmanned operation; no timetable yet announced.

DESIGN FEATURES: Longer fuselage and greater wing span than Egrett II, and internal fuel increase sufficient for 24 h endurance or 4,000 nm (7,413 km; 4,606 mile) range; hose/drogue in-flight refuelling permitting endurance (as UAV) of up to 30 days; removable equipment modules for payload and its environmental control system (max payload 680 kg; 1,500 lb). Applications include coastal, battlefield or other wide-area surveillance, drug interdiction, naval battle group AEW, special forces support, verification, nuclear disaster monitoring and remote sensing.

EHI
EH INDUSTRIES LIMITED

PARTICIPATING COMPANIES:
Agusta: see under Italy
Westland: see under UK

CO-CHAIRMEN:
Alan Jones
Amedeo Caporaletti

MANAGING DIRECTOR: E. Striano

EH Industries formed June 1980 by Westland Helicopters and Agusta (50 per cent each) to undertake joint development of new anti-submarine warfare helicopter for Royal Navy and Italian Navy within an integrated programme under which naval, army and civil variants now being produced. Programme handled on behalf of both governments by British Ministry of Defence; Westland has design leadership for commercial version, Agusta for rear loading military/utility version; naval version being developed jointly for UK and Italian navies and export. IBM/Westland selected to manage Royal Navy programme in 1991; Canadian programme launched October 1992.

EH INDUSTRIES EH 101
Royal Navy name: Merlin
Canadian Forces designations: CH-148 Petrel and CH-149 Chimo

TYPE: Three-engined multi-role helicopter.

PROGRAMME: Stems from Westland WG 34 (see UK section of 1979-80 *Jane's*), selected by UK MoD as Sea King replacement late Summer 1978; broadly similar requirement by Italian Navy led to 1980 joint venture with Agusta; subsequent market research confirmed compatibility of basic EH 101 design with commercial payload/range and tactical transport/logistics requirements, resulting in decision to develop naval, civil and military variants based on common airframe. Nine-month project definition phase approved by UK/Italian governments 12 June 1981; full programme go-ahead announced 25 January 1984; design and development contract signed 7 March 1984; selected by Canadian government August 1987; Italian built iron bird ground test airframe followed by nine pre-production aircraft (PP1-9: see Current Versions); RTM 322 engines selected for RN Merlin June 1990; T700-GE-T6A engines selected for Italian Navy version September 1990; 4th UK/Italian government MoU signed 30 September 1991 (starting industrialisation phase); UK MoD contract for 44 Merlins 9 October 1991; Canadian order for 50 (35 CH-148 and 15 CH-149) announced 24 July 1992; British and Italian civil certification planned for late 1993; service entry for naval version due early 1996.

CURRENT VERSIONS (general): **Naval:** Primary roles ASW, ASV, anti-ship surveillance/tracking, amphibious operations and SAR; other roles may include AEW, vertrep and ECM (deception, jamming and missile seduction); designed for fully autonomous all-weather operation from land bases, large and small vessels (incl merchant ships) and oil rigs, and specifically from a 3,500 tonne frigate, with dimensions tailored to frigate hangar size. Capabilities include frigate launch and recovery in sea states 5-6 with ship on any heading and wind speed (from any direction) up to 50 knots (93 km/h; 57 mph); endurance and carrying capacity needed to meet expanding maritime tactical requirements of 21st century, including ability to operate distantly for up to 5 hours with state of the art equipment

PP5, the Royal Navy's Merlin dedicated pre-production aircraft *(Peter J. Cooper)*

and weapons. See under Italian Navy, Merlin and CH-148 headings below.

Heliliner: Commercial version (see below).

Utility: Military and civil versions with rear-loading facility (see CH-149 and Utility headings below).

CURRENT VERSIONS (specific): **PP1:** Westland built first pre-production aircraft; first flight (ZF641) 9 October 1987; completed official test centre review 1991; over 375 hours flown by September 1992, at which time being refitted with T700-GE-T6A engines and 3,878 kW (5,200 shp) main gearbox.

PP2: Agusta built second pre-production aircraft; first flight 26 November 1987; deck trials aboard Italian Navy *N. Gereale* and *Maestrale* July 1990; was to lead work on achieving new max T-O weight of 14,288 kg (31,500 lb), and had been fitted with ACSR (see Design Features), but lost in crash in Italy January 1993 during noise suppression trials.

PP3: Westland built; first flight (G-EHIL) 30 September 1988; first civil-configured aircraft; has conducted rotor vibration trials and flights with 'soft link' engine mounting struts; scheduled for icing trials early 1993; continuing programme also includes optimisation of ACSR, and weapon carriage and development trials.

PP4: Westland built; first flight (ZF644) 15 June 1989; overwater navigation equipment trials and AFCS development; to fly with RTM 322 engines February 1993, initiating two-year programme of ground running and 160 flight hours.

PP5: Westland built dedicated Merlin development aircraft; first flight (ZF649) 24 October 1989; Type 23 frigate interface trials in HMS *Norfolk* (incl decklock, recovery handling, weapons handling, tail and blade folding) August 1991; seagoing trials in HMS *Iron Duke* (incl overwater sonobuoy release) completed December 1992.

PP6: Agusta built dedicated Italian Navy development aircraft; first flight late April 1989; seagoing trials in *Giuseppe Garibaldi* and *Andrea Doria* completed mid-October 1991; rotor downwash trials on behalf of Canadian Forces to assess acceptability for SAR role.

PP7: Agusta built military utility development aircraft; first flight 18 December 1989; low-speed handling trials and tail rotor performance assessment 1992; fitted with ACSR 1992.

PP8: Westland built; first flight (G-OIOI) 24 April 1990; evaluations of ADF and DME, area nav, electronic instrumentation, civil AFCS and communications equipment; fitted with ACSR.

PP9: Agusta built final development aircraft; first flight (I-LIOI) 16 January 1991; represents civil utility version; extensive flight controls survey involving both ground and in-flight trials.

Civil utility: For commercial operators requiring rear-loading facility; represented by PP9.

Heliliner: Commercial variant; main certification programme being flown by PP3, with PP8 as demonstrator; intended to offer 550 nm (1,019 km; 633 mile) range, with full IFR reserves, carrying 30 passengers and baggage; flight crew of two, provision for cabin attendant, stand-up headroom, airline style seating, overhead baggage stowage, full environmental control, passenger entertainment, toilet and galley. Category A VTO performance, capable of offshore/oil rig operations or scheduled flights into city centres at high all-up weights under more rigorous future civil operating rules; rear-loading ramp optional.

Italian Navy: Development aircraft PP4 (basic) and PP6; will operate from both shore bases and aircraft/helicopter carriers.

Merlin: Royal Navy version, for which PP4 (basic) and PP5 are development aircraft; will operate from Type 23 frigates, 'Invincible' class carriers, RFAs and other ships, and land bases.

Military utility: Tactical or logistic transport variant (represented by PP7) with rear-loading ramp; able to airlift six tons or up to 30 combat equipped troops.

CH-148 Petrel: Designation of Canadian NSA (new shipborne aircraft) ASW version to replace CH-124A Sea Kings; T700-GE-T6C engines; crew of four (two pilots, navigator, sensor operator); 23 to be based on east coast (Nova Scotia) and 12 on west coast (British Columbia), operating from 12 patrol frigates, four 'Tribal' class destroyers and two auxiliaries. Agusta built airframes to be assembled in Canada by IMP Group; Paramax prime contractor for mission systems and integration; deliveries to begin late 1997.

CH-149 Chimo: Designation of Canadian NSH (new SAR helicopter) search and rescue version to replace 13 Boeing CH-113A Labradors; -T6C engines, as in CH-148; crew of five (two pilots, navigator, two SAR specialists), capacity for eight stretchers and 10 sitting casualties; based on military utility version; will operate from land bases. Airframes by Westland, assembly by IMP Group; Paramax prime contractor for mission systems and integration; deliveries to begin 1998.

CUSTOMERS: Royal Navy 44 Merlin ordered; Italian Navy expected to order 16 with options on further eight; Canadian Forces 35 CH-148 and 15 CH-149 ordered. Royal Air Force requires 25 of military utility version initially, to provide air mobility for British Army; long-range utility version for logistic and tactical support role also under consideration early 1993 by US Marine Corps, Netherlands and Italy.

COSTS: Royal Navy £1.5 billion (1991) for 44 Merlin; Canada C$4.4 billion (1992) for 50 CH-148/149.

DESIGN FEATURES: Three-engine power margin, fail-safe/damage-tolerant airframe and rotating components, high system redundancy, onboard monitoring of engines/transmission/avionics/utility systems; airframe/power plant/rotor and transmission systems/flight controls/utility systems common to all variants; five-blade main rotor with multiple load path hub and elastomeric bearings; blades of advanced aerofoil section with BERP-derived high-speed tips; four-blade teetering tail rotor; transmission has 45 min run-dry capacity; fuselage in four main modules (front and centre ones common to all variants, modified rear fuselage and slimmer tailboom on military/utility variant to accommodate rear-loading ramp); automatic power folding of main rotor blades and tail rotor pylon on naval variant, with emergency manual backup (tail section folds forward/downward, stowing starboard half of tailplane beneath rear fuselage). New active vibration cancelling system ACSR (active control of structural response) reduces vibration by 80 per cent at blade passing frequency.

FLYING CONTROLS: Front drive directly into main gearbox from all three engines, with all gears straddle mounted for greater rigidity; dual redundant digital AFCS.

STRUCTURE: Rotor head of composites surrounding a metal core; composite blades; fuselage mainly aluminium alloy, with bonded honeycomb main panels; composites for such complex shapes as forward fuselage, upper cowling panels, tail-fin, tailplane and windscreen. Engine air intakes of Kevlar reinforced with aero-web honeycomb. Westland responsible for front fuselage (incl flight deck and cabin)

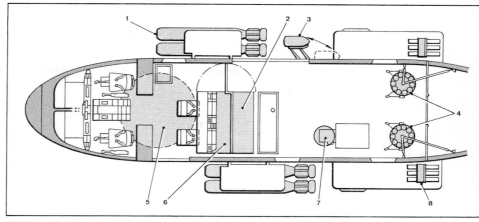

EH 101 typical ASW layout: (1) up to four homing torpedoes; (2) sensor processing; (3) retractable rescue hoist; (4) sonobuoy dispensers; (5) 360° radar; (6) mission console; (7) active dipping sonar; (8) light stores carriers *(Jane's/Mike Keep)*

PP8 pre-production Heliliner 30-seat passenger helicopter

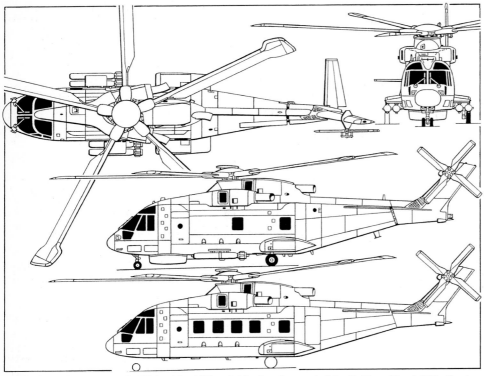

Three-view of the naval EH 101, with additional side view (bottom) of the Heliliner *(Jane's/Mike Keep)*

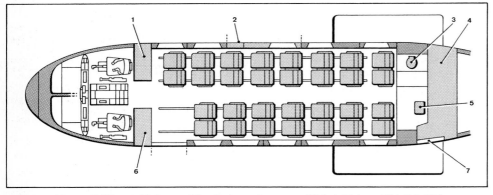

Typical 30-seat Heliliner cabin layout: (1) avionics and stowage; (2) freight door; (3) toilet; (4) baggage compartment; (5) cabin attendant's seat; (6) avionics; (7) baggage door *(Jane's/Mike Keep)*

Utility EH 101 can load up to 30 combat equipped troops via the rear ramp/door in less than 2 minutes

and main rotor blades, Agusta for rear fuselage, tail assembly, rotor head and transmission, hydraulic system and part of electrical system; single-sourced series production, with final assembly lines in Italy and UK.

LANDING GEAR: Hydraulically retractable tricycle type, with single mainwheels and steerable twin-wheel nose unit, designed and manufactured by AP Precision Hydraulics in association with Officine Meccaniche Aeronautiche. Main units retract into fairings on sides of fuselage. Goodrich wheels, tyres and brakes: main units have size 8.50-10 wheels with 24 × 7.7 tyres, unladen pressure 6.96 bars (101 lb/sq in); nosewheels have size 19.5 × 6.75 tyres, unladen pressure 8.83 bars (128 lb/sq in). Twin-mainwheel gear optional for all variants. FPT Industries emergency flotation bags.

POWER PLANT: Three Rolls-Royce Turbomeca RTM 322 turboshafts in Royal Navy Merlin (max contingency and intermediate contingency ratings 1,724 kW; 2,312 shp and 1,566 kW; 2,100 shp respectively); General Electric T700-GE-T6A turboshafts in Italian naval variant, rated at 1,278 kW (1,714 shp) max contingency, 1,254 kW (1,682 shp) intermediate and 1,071 kW (1,437 shp) max continuous at S/L, ISA (-T6C variant of this engine for Canadian aircraft). Engines for Italian naval variant will be assembled by Alfa Romeo Avio and Fiat. Commercial and utility variants powered by three General Electric CT7-6 turboshafts (CT7-6A in PP3) with ratings of 1,432 kW (1,920 shp) max and intermediate contingency, 1,230 kW (1,649 shp) max continuous. Transmission rating in commercial variant 3,878 kW (5,200 shp) max T-O and max continuous, 2,863 kW (3,840 shp) max continuous with OEI. Computerised fuel management system. Pressure refuelling point on starboard side; three gravity positions on port side.

ACCOMMODATION: One or two pilots on flight deck (naval version will be capable of single-pilot operation, commercial variant will be certificated for two-pilot operation). ASW version will normally also carry observer and acoustic systems operator. Martin-Baker crew seats in naval version, able to withstand 10.7 m (35 ft)/s impact. Socea or Ipeco crew seats in commercial variant. Commercial version able to accommodate 30 passengers four abreast at approx seat pitch of 76 cm (30 in), plus cabin attendant, with toilet, galley and baggage facilities (including overhead bins). Military variant can accommodate up to 30 (seated) or 45 (non-seated) combat equipped troops, 16 stretchers plus a medical team, palleted internal loads, or can carry externally slung loads of up to 5,443 kg (12,000 lb). Main passenger door/emergency exit at front on port side with additional emergency exits on starboard side and on each side of cabin at rear, above main landing gear sponson. Large sliding door at mid-cabin position on starboard side, with inset emergency exit. Commercial variant has baggage bay aft of cabin, with external access via door on port side. Cargo loading ramp/door at rear of cabin on military and utility versions. Cabin floor loading 976 kg/m² (200 lb/sq ft) on PP1.

SYSTEMS: Hamilton Standard/Microtecnica environmental control systems. Dual redundant integrated hydraulic system, pressurised by three Vickers pumps each supplying fluid at 207 bars (3,000 lb/sq in) nominal working pressure, with flow rates of 55, 59 and 60 litres (14.5, 15.6 and 15.9 US gallons; 12.1, 13.0 and 13.2 Imp gallons)/min respectively. Hydraulic system reservoirs are of the piston load pressurised type, with a nominal pressure of 0.97 bar (14 lb/sq in). Primary electrical system is 115/200V three-phase AC, powered by two Lucas brushless, oilspray-cooled 45kVA generators (90kVA if Lucas Spraymat blade ice protection system fitted), with one driven by main gearbox and the other by accessory gearbox, plus a third separately driven standby alternator. APU for main engine air-starting, and to provide electric power, plus air for ECS without running main engines or using external power supplies. Lucas Spraymat electric de-icing of main/tail blades standard on naval variant, optional on others; Dunlop electric anti-icing of engine air intakes. Fire detection and suppression systems by Graviner and Walter Kidde respectively. BAJ Ltd four-float emergency flotation system.

AVIONICS: Integrated avionics systems of naval and military variants based on two MIL-STD-1553B multiplex data buses that link basic aircraft management, avionics and mission systems. Integrated avionics system of commercial variant based on ARINC 429 data transfer bus. Smiths Industries/OMI SEP 20 dual redundant digital AFCS is standard, providing fail-operational autostabilisation and four-axis autopilot modes (auto hover, auto transitions to/from hover standard on naval variants, optional on commercial and military variants).

On naval variant, main processing element of management system is a dual redundant aircraft management computer which carries out navigation, control and display management, performance computation and health and usage monitoring of principal systems (engines, drive systems, avionics and utilities); it also controls basic bus. ASW version will have 360° search radar (GEC-Marconi Blue Kestrel in Merlin, Texas Instruments APS-137 derivative in CH-148) in chin radome, plus dipping sonar (Ferranti-Thomson/Sintra folding lightweight acoustic system for helicopters, FLASH, in Merlin), advanced sonobuoy processing equipment and Racal Orange Reaper ESM. GEC-Marconi AQS-903 acoustic processing system, NGL mission recorder and sonobuoy/flare dispenser and Chelton sonobuoy homers, in Merlin. ASST (anti-ship surveillance and tracking) version will carry equipment for tactical surveillance and OTH (over the horizon) targeting, to locate and relay to a co-operating frigate the position of a target vessel, and for midcourse guidance of frigate's missiles. On missions involving patrol of an exclusive economic zone it can also, with suitable radar, monitor every hour all surface contacts within area of 77,700 km² (30,000 sq miles); can patrol an EEZ 400 × 200 nm (740 × 370 km; 460 × 230 miles) twice in one sortie; and can effect boarding and inspection of surface vessels during fishing protection and anti-smuggling missions. AFCS sensors on naval variant include British Aerospace LINS 300 ring laser gyro inertial reference unit (IRU) and Litton Italia LISA-4000 strapdown AHRS; IRU also provides self-contained navigation, with Racal Doppler 91E velocity sensor; GEC-Plessey GPS receiver selected for Royal Navy variant, Euronav GPS for Italian Navy aircraft. Other avionics on naval variants include GEC-Plessey/Elettronica PA 5015 I-band radar altimeters, GEC-Marconi low airspeed sensing and air data system, THORN EMI pilot's mission display unit, Alenia/Racal cabin mission display unit, and Alenia/GEC-Marconi aircraft management computer. Elmer HF (two), and GEC-Marconi UHF (one) and AD 3400 V/UHF (two) com radios.

On civil variant, a Canadian Marconi CMA-900 flight management system will include a colour CRT display with graphics and alphanumeric capability; fuel flow, fuel quantity and specific range computations; tuning of nav/com radios; interfaces with electronic instrument systems; two-dimensional multi-sensor navigation; and built-in navigational database with update service. AFCS sensors on commercial variant include two Litton Italia LISA-4000 strapdown AHRS. Advanced flight deck incorporates standard Smiths Industries/OMI electronic instrument system (EIS) providing colour flight instrument, navigation and power systems displays. Standard avionics include Penny and Giles air data system, Racal intercom system, optional Collins or Bendix/King communications and navigation systems, optional Honeywell or Bendix/King weather radar.

EQUIPMENT: ASW variants will have two sonobuoy dispensers, external rescue hoist and Fairey Hydraulics (Merlin) or Indal HHRSD (CH-148) decklock.
ARMAMENT (naval and military utility versions): Naval version able to carry up to four homing torpedoes (probably Marconi Sting Ray on Merlin) or other weapons. ASV version designed to carry air-to-surface missiles and other weapons, for use as appropriate, from strikes against major units using sea-skimming anti-ship missiles to small-arms deterrence of smugglers. Armament optional on military utility versions.

DIMENSIONS, EXTERNAL:
Main rotor diameter 18.59 m (61 ft 0 in)
Tail rotor diameter 4.01 m (13 ft 2 in)
Length:
 overall, both rotors turning 22.81 m (74 ft 10 in)
 main rotor and tail pylon folded (naval variant) 16.00 m (52 ft 6 in)
Width: excl main rotor 4.52 m (14 ft 10 in)
 main rotor and tail pylon folded (naval variant) 5.49 m (18 ft 0 in)
Height: overall, both rotors turning 6.65 m (21 ft 10 in)
 main rotor and tail pylon folded (naval variant) 5.21 m (17 ft 1 in)

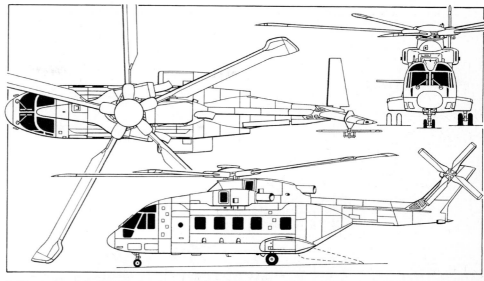

Utility version of the EH 101, showing modified rear fuselage with rear loading ramp/door *(Jane's/Mike Keep)*

Passenger door (fwd, port):
 Height 1.70 m (5 ft 7 in)
 Width 0.91 m (3 ft 0 in)
Sliding cargo door (mid-cabin, stbd):
 Height 1.55 m (5 ft 1 in)
 Width 1.83 m (6 ft 0 in)
Baggage compartment door (rear, port, Heliliner):
 Height 1.63 m (5 ft 4 in)
 Width 0.79 m (2 ft 7 in)
Rear-loading ramp/door (rear, military/utility variant):
 Height 1.80 m (5 ft 11 in)
 Width 2.11 m (6 ft 11 in)
DIMENSIONS, INTERNAL:
Cabin:
 Length: naval variant 7.09 m (23 ft 3 in)
  commercial/utility variant 6.50 m (21 ft 4 in)
 Max width 2.49 m (8 ft 2 in)
 Width at floor 2.26 m (7 ft 5 in)
 Max height 1.83 m (6 ft 0 in)
 Volume: naval variant 29.0 m³ (1,024 cu ft)
  Heliliner 27.5 m³ (970 cu ft)
Baggage compartment volume (Heliliner):
 3.82 m³ (135 cu ft)
AREAS:
Main rotor disc 271.51 m² (2,922.5 sq ft)
Tail rotor disc 12.65 m² (136.2 sq ft)
WEIGHTS AND LOADINGS (A: naval variant, B: Heliliner, C: military/utility variant):
Basic weight empty (estimated): A 7,121 kg (15,700 lb)
 B 6,967 kg (15,360 lb)
 C 7,284 kg (16,060 lb)
Operating weight empty (estimated):
 A 9,298 kg (20,500 lb)
 B (IFR, offshore equipped) 8,933 kg (19,695 lb)
 C 9,000 kg (19,840 lb)

Max fuel weight (four internal tanks, total):
 A 3,438 kg (7,580 lb)
 B, C 3,370 kg (7,430 lb)
Max fuel weight with optional auxiliary tank:
 A 4,298 kg (9,475 lb)
 B, C 4,213 kg (9,288 lb)
Disposable load/payload:
 A (four torpedoes) 960 kg (2,116 lb)
 B (30 passengers plus baggage) 2,721 kg (6,000 lb)
 C (30 combat equipped troops) 3,900 kg (8,598 lb)
Max T-O weight: A 13,000 kg (28,660 lb)
 or 13,530 kg (29,830 lb)
 B, C 14,288 kg (31,500 lb)
PERFORMANCE:
Never-exceed speed (VNE) at S/L, ISA
 167 knots (309 km/h; 192 mph) IAS
Average cruising speed 150 knots (278 km/h; 173 mph)
Best range cruising speed
 140 knots (259 km/h; 161 mph)
Best endurance speed 90 knots (167 km/h; 104 mph)
Range (B):
 standard fuel, offshore IFR equipped, with reserves, 30 passengers 500 nm (926 km; 575 miles)
 auxiliary fuel, offshore IFR equipped, with reserves, 30 passengers 625 nm (1,158 km; 720 miles)
 with zero T-O distance (Category A rules)
 300 nm (556 km; 345 miles)
Ferry range:
 B (standard fuel, IFR equipped, with reserves)
 950 nm (1,760 km; 1,094 miles)
 C (standard fuel plus internal auxiliary tanks)
 1,130 nm (2,094 km; 1,301 miles)

EMBRAER/FMA

PARTICIPATING COMPANIES:
Embraer: see under Brazil
FMA: see under Argentina

EMBRAER/FMA CBA-123 VECTOR

TYPE: Twin-turboprop regional and corporate transport.
PROGRAMME: Begun (as EMB-123) as Embraer private venture mid-1985; first details released April 1986, three months after initial collaboration agreement with FMA (see Addenda to 1986-87 *Jane's*); redesignated CBA-123 (Co-operation Brazil-Argentina) on becoming fully international joint programme (then with 70/30 per cent sharing) 21 May 1987; work sharing defined March 1988; first metal cut March 1989; three flying prototypes (two in Brazil, one in Argentina): first flights 18 July 1990 (PT-ZVE, c/n 801), 15 March 1991 (PT-ZVB, c/n 802) in Brazil; Argentine prototype (c/n 803) not yet flown (June 1993); certification to be to Transport Category of FAR/JAR Pt 25 and FAR Pt 36, ICAO Annex 16. Programme continued at slow pace in 1992, basically involving flying by both Embraer built prototypes, which by June 1993 had logged 975 hours' flying; decision on whether programme will proceed dependent on confirmation of proposed orders for 40 and 20, for use respectively by Brazilian and Argentine air forces, but hopes fading.
CURRENT VERSIONS: **Commuter** version standard, **Executive** version optional (second prototype reconfigured to Executive interior 1992).

CUSTOMERS: Total of 113 options held at end of 1992 from customers in Brazil, Argentina and nine other countries.
COSTS: Development programme budgeted at US$300 million in September 1990, including tooling for series production; unit price then quoted at US$4.8 million, but price reduction programme started in November 1990.
DESIGN FEATURES: Approx 60 per cent commonality with Embraer Brasilia (which see) but with shorter fuselage, new supercritical wing and rear-mounted engines; low-mounted wing has Embraer aerofoil sections (root EA 160316/tip EA 160313), 3° dihedral, 2° root incidence (tip 0°) and 4° 8′ sweepback at quarter-chord; fuselage waisted at rear; sweptback T tail; undertail bumper to protect propeller blades during T-O rotation.
FLYING CONTROLS: Mechanically actuated mass balanced ailerons and horn balanced elevators; hydraulically actuated tandem dual rudders; three-segment double-slotted Fowler flaps, actuated by electrically driven screwjacks; electrically signalled, hydraulically actuated two-segment ground spoilers/lift dumpers forward of each inboard flap; electrical (fly-by-wire) trimming of ailerons, rudders and variable incidence tailplane; single integrated flight director/autopilot AFCS standard, dual AFCS optional; stick pusher/shaker stall warning system.
STRUCTURE: Basic damage-tolerant structure of riveted 7000 series and 2024T-3 aluminium alloys, with extensive use of chemical milling; composites used for nose radome (glassfibre); wing/fin/tailplane leading-edges, wing/tailplane tips, wing/body fairing, dorsal fin, fin/tailplane bullet and engine pylon trailing-edges (Kevlar reinforced glass-

fibre); ailerons, elevators, rudders, flaps, spoilers, tailcone, landing gear doors, cowlings and engine pylon spars and skin (carbonfibre); and underseat panels (carbonfibre/ Nomex); pylons have steel mounts and titanium leading-edges; one-piece, three-spar wing; pressurised fuselage. FMA currently has approx 20 per cent of manufacture, comprising front portion of centre-fuselage, sections 1 and 2 of rear fuselage, tailcone and undertail bumper, tail unit (except rudders) and engine pylons, with Embraer building remainder; assembly line planned in each country.
LANDING GEAR: Hydraulically retractable tricycle type, with twin wheels and oleo-pneumatic shock absorber on each unit. Mainwheels retract inward into wing/underfuselage fairing; nose unit retracts forward. Emergency free-fall extension. Hydraulically actuated nosewheel steering (±50°). Mainwheel tyres size 22 × 6.75-10 (10 ply rating), pressure 7.24 bars (105 lb/sq in); nosewheel tyres size 16 × 4.4 (6 ply rating), pressure 5.17 bars (75 lb/sq in). Hydraulic carbon disc brakes with anti-skid provision. Backup braking system for use if one hydraulic system fails. Mechanical emergency and parking brake systems. High flotation tyres optional. Min ground turning radius 10.04 m (32 ft 11 in).
POWER PLANT: Two Garrett TPF351-20 or -20A turboprops, each derated to 969 kW (1,300 shp) and driving a Hartzell HC-E6A-5 contra-rotating, slow-turning constant-speed pusher propeller with reversible pitch, autofeathering, synchrophasing and six scimitar blades. Engines are pylon mounted in carbonfibre nacelles at rear of fuselage, and have FADEC control and a cruise/climb rating of 746 kW

146 INTERNATIONAL: AIRCRAFT—EUROCOPTER

EUROCOPTER INTERNATIONAL PACIFIC
PO Box 51, Bankstown, NSW 2200, Australia
Telephone: 61 (2) 794 9900
Fax: 61 (2) 791 0195

Sales and support company for Eurocopter products and programmes.

EUROCOPTER TIGER GmbH
Gustav Heinemann Ring 135 (PO Box 830356), D-8000 Munich 83, Germany
Telephone: 49 (89) 638250-0
Fax: 49 (89) 638250-50
CHIEF EXECUTIVE OFFICERS:
Bernard Darrieus
Ingo Jaschke

Company formed 18 September 1985 to manage development and manufacture of Tiger/Tigre/Gerfaut battlefield helicopter for French and German armies (see details later in this section); because it is working on a single government contract, Eurocopter Tiger is not a full member of Eurocopter; executive authority for programme is DFHB (Deutsch-Französisches Hubschrauberbüro) in Koblenz; procurement agency is German government BWB (Bundesamt für Wehrtechnik und Beschaffung).

EUROCOPTER/KAWASAKI BK 117
Jointly developed by MBB and Kawasaki; now being manufactured in BK 117 B-2 and C-1 versions. See under Eurocopter/Kawasaki later in this section.

EUROCOPTER/CATIC/SA EC 120
Former P120L; full development launched in October 1991. See under this project name later in this section.

EUROCOPTER/MIL/KAZAN/KLIMOV Mi-38
Agreement for co-operative development, production and marketing of 30-passenger Mi-38 helicopter signed 18 December 1992. See later in this section.

NH INDUSTRIES NH 90
Joint European medium military helicopter programme; described under NH Industries later in this section.

EUROFAR
European joint tilt-rotor programme; described under this designation later in this section.

AEROSPATIALE SA 315B LAMA
Now manufactured only by Hindustan Aeronautics in India (which see). Eurocopter sold 12 Lamas to Pakistan in 1992. In USA, Rocky Mountain Helicopters (RMH) has flown a Lama re-engined with a P&WC PT6B-35F; conversion by RMH subsidiary AeroProducts International; conversion reduces maintenance, fuel consumption, weight and noise.

AEROSPATIALE AS 330 PUMA
Sole source of AS 330 Puma is now IAR SA in Romania (which see). IAR now developing 2000 version.

EUROCOPTER AS 332 SUPER PUMA and AS 532 COUGAR
TYPE: Twin-turbine multi-role helicopter.
PROGRAMME: Early history and versions listed in 1985-86 and earlier *Jane's*; first flight AS 332 Super Puma (F-WZJA) 13 September 1978; six prototypes; deliveries began mid-1981; present version powered by Turbomeca Makila 1A1 introduced 1986; military versions renamed Cougar in 1990; first AS 332L (stretched fuselage) certificated to French IFR Cat. II 7 July 1983 and delivered to Lufttransport of Norway; certificated for flight into known icing 29 June 1983; FAA Cat. II certification with SFIM CDV 85 P4 four-axis AFCS and FAR Pt 25 Appendix C known icing clearance.
CURRENT VERSIONS: Designation suffixes: U military unarmed utility, A armed, S anti-ship/submarine, M naval SAR/surveillance, C military court (short) fuselage, L long fuselage, military or civil.
AS 532UC Cougar: Military short fuselage unarmed utility; seats up to 21 troops and two crew; cabin floor reinforced for 1,500 kg/m² (307 lb/sq ft).
AS 532UL Cougar: Military unarmed long fuselage version; cabin lengthened by 0.76 m (2 ft 6 in); extra fuel and two large windows in forward cabin plug; carries up to 25 troops and two crew.
AS 532AC Cougar: Armed AS 532UC.
AS 532AL Cougar: Armed AS 532UL.
AS 532MC Cougar: Naval short version equipped for SAR and surveillance roles.
AS 532SC Cougar: Naval short version with ASW/ASV equipment; folding tail rotor pylon, main rotor blades and deck harpoon device.
AS 332L1 Super Puma: Standard civil version with long fuselage and airline interior for 20 passengers; UK CAA IFR certification 21 April 1992; one delivered to British Inetrnational Helicopters.
AS 332L Tiger: Bristow Helicopters proprietary passenger interior with hard floor and foldable seats to allow for heavy equipment; externally accessible rear baggage compartment, automatically jettisonable cabin door and large-capacity liferafts; standard IFR avionics, flight management system, North Sea navaids, ice protection on blades, tailplane and main rotor droop stops.
AS 532 Horizon: Battlefield surveillance radar helicopter; flight trials of small Orphée radar under an AS 330B Puma started 1986; French Army wanted 20 AS 532 Cougar Mk II fitted with radar, dedicated ECM and data link (Orchidée programme) to be delivered in 1986; first flight of full-scale AS 532/Orchidée June 1990; programme cancelled for budgetary reasons August 1990; prototype without data link flew 24 missions in Gulf War February 1991 in Operation Horus; Horizon programme (Hélicoptère d'Observation Radar et d'Investigation sur Zone) with more effective operational concept and reduced development costs replaced Orchidée; development contract awarded to Eurocopter October 1992 for two AS 532UL Horizons with same radar capabilities and ECM as Orchidée, but in AS 532UL Mk I Cougar, with standard ECM and longer mission endurance; first flight Horizon with full radar 8 December 1992; delivery of two AS 532UL Horizons to French Army planned April 1994 for troop trials with full operational capability except for extended computer to be added later; French Army currently intends to procure total six AS 532UL Horizons.
AS 532U2/A2 Cougar Mk II: See separate entry.
AS 332L2 Super Puma Mk II: See separate entry.
CUSTOMERS: Total 405 AS 332/532s ordered from 38 countries (67 per cent of them military Cougars) by January 1992; total 73 AS 332L1 Super Pumas in civil offshore oil industry support operations by end 1991; French military orders include five for French Air Force (three for nuclear test facilities in Pacific and two for government VIP flying); French Army (ALAT) is replacing AS 330 Pumas with AS 532 Cougars; first 22 delivered to Force d'Action Rapide between late 1988 and end 1991; 10 Super Puma/Cougars ordered in 1992.

Export customers include Abu Dhabi (eight including two VIP), Brazil (16 including six AS 532SCs), Cameroon (one), Chile (army, two; navy, four AS 532SCs), China (six), Ecuador (eight including six army), Finland (three for border police, Gabon (one, Presidential Guard), Germany (three, border police), Indonesia (built under

French Army AS 532UL Cougar fitted with Horizon System (*Brian M. Service*)

licence); Japan (three, army/VIP), Jordan (eight), South Korea (three army/VIP, one AS 332L1), Mexico (two VIP), Nepal (two, Royal Flight), Nigeria (two), Oman (two, Royal Flight), Panama (one VIP), Saudi Arabia (12 AS 532SCs), Singapore (22), Spain (10 SAR HD.21s, two VIP HT.21s, 18 army tactical transport HT.21s), Sweden (10 SAR), Switzerland (15 Cougars), Togo (one), Venezuela (eight), Zaïre (one VIP). Bristow Helicopters ordered 31 19-passenger AS 332Ls for offshore oil support in special version called Tiger (see Current Versions).
DESIGN FEATURES: Four-blade main rotor turning clockwise seen from above; five-blade tail rotor on starboard side of tailboom; engines mounted above cabin have rear drive into main transmission at 23,840 rpm; main rotor turns at 265 rpm, tail rotor at 1,278 rpm; various lateral sponsons available housing partly retracted main landing gear and combinations of additional fuel or pop-out floats; optional air-conditioning system housed in casing on port forward flank of cabin; all civil versions certificated for IFR category A and B to FAR Pt 29.
FLYING CONTROLS: Dual fully powered hydraulic controls with full-time autostabilisation and yaw damping; machine remains flyable with autostabiliser switched off; cyclic trimming by stick friction adjustment; inverted slot on tailplane to maintain attitude-holding effect at low climb speeds; large ventral fin; saucer fairing on rotor head to smooth wake of hub; four-axis SFIM 155 autopilot standard.
STRUCTURE: Conventional light alloy airframe with some titanium; some crashworthy features; main rotor blades of GFRP with CFRP stiffening and Moltoprene filler; elastomeric drag dampers.

Some AS 332/532s built under licence by IPTN in Indonesia, some assembled by CASA in Spain and some equipped by F+W in Switzerland.
LANDING GEAR: Retractable tricycle high energy absorbing design by Messier-Bugatti; all units retract rearward hydraulically, mainwheels into sponsons on sides of fuselage; dual-chamber oleo-pneumatic shock absorbers; optional 'kneeling' capability for main units; twin-wheel self-centring nose unit, tyre size 466 × 176, pressure 7.0 bars (102 lb/sq in); single wheel on each main unit with tyre size 615 × 225-10 or 640 × 230-10, pressure 9.0 bars (130 lb/sq in); hydraulic differential disc brakes, con-

Eurocopter AS 332L1 Super Puma water bombing with 2,400 litre (634 US gallon; 527 Imp gallon) water tank with self-powered refill pipe

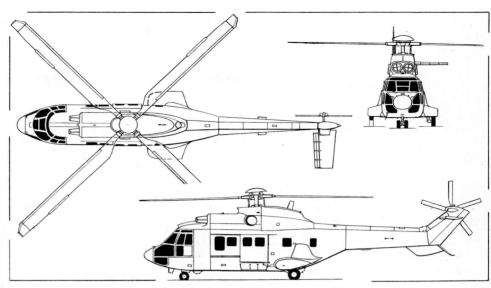

Eurocopter AS 532UL Cougar (long fuselage military unarmed utility) *(Dennis Punnett)*

Eurocopter AS 532UL Cougar of French Army with jeep on sling *(Alexandre Paringaux)*

trolled by foot pedals; lever operated parking brake; emergency pop-out flotation units can be mounted on main landing gear fairings and forward fuselage.

POWER PLANT: Two Turbomeca Makila 1A1 turboshafts, each with max contingency rating of 1,400 kW (1,877 shp) and max continuous rating of 1,184 kW (1,588 shp). Air intakes protected by a grille against ingestion of ice, snow and foreign objects; Centrisep multi-purpose intake optional for flight into sandy areas; AS 532UC/AC have five flexible fuel tanks under cabin floor, with total usable capacity of 1,497 litres (395 US gallons; 329 Imp gallons); AS 532SC has total basic capacity of 2,141 litres (565 US gallons; 471 Imp gallons); AS 332L1/532UL/532AL have a basic fuel system of six flexible tanks with total capacity of 2,020 litres (533 US gallons; 444 Imp gallons) in the 332 and 2,003 litres (529 US gallons; 440 Imp gallons) in the 532; provision for additional 1,900 litres (502 US gallons; 418 Imp gallons) in four auxiliary ferry tanks installed in cabin; two external auxiliary tanks with total capacity of 650 litres (172 US gallons; 143 Imp gallons) standard on AS 532SC, optional on other versions; for long range missions (mainly offshore) in AS 332L1, a special internal auxiliary tank can be fitted in cargo sling well, in addition to the two external tanks, to raise total usable fuel capacity to 2,994 litres (791 US gallons; 658 Imp gallons); refuelling point on starboard side of cabin; fuel system designed to avoid leakage following a crash; self-sealing tanks standard on military versions, optional on other versions; other options include a fuel dumping system and pressure refuelling.

ACCOMMODATION: One pilot (VFR) or two pilots side by side (IFR) on flight deck, with jump seat for third crew member or paratroop dispatcher; provision for composite light alloy/Kevlar armour for crew protection on military models; door on each side of flight deck and internal doorway connecting flight deck to cabin; dual controls, co-pilot instrumentation and crashworthy flight deck and cabin floors; max accommodation for 21 passengers in AS 532UC, 24 in AS 332L1 and 25 in AS 532UL; interiors available for VIP, air ambulance with six stretchers and eleven seated casualties/attendants, or nine stretchers and three seated; strengthened floor for cargo carrying, with lashing points; jettisonable sliding door on each side of main cabin or port side door with built-in steps and starboard side double door in VIP or airline configurations; removable panel on underside of fuselage, at rear of main cabin, for longer loads; removable door with integral steps for access to baggage racks optional; hatch in floor below main rotor contains hook for slung loads up to 4,500 kg (9,920 lb) on internally mounted cargo sling; cabin and flight deck heated, ventilated and soundproofed; demisting, de-icing, washers and wipers for pilots' windscreens.

SYSTEMS: Two independent hydraulic systems, supplied by self-regulating pumps driven by main gearbox. Each system supplies one set of powered flying controls; left-hand system also supplies autopilot, landing gear, rotor brake and wheel brakes; hydraulically actuated systems can be operated on ground from main gearbox (when a special disconnect system is installed to permit running of port engine with rotors stationary), or by external power through ground power receptacle. Emergency landing gear lowering by hand pump. Three-phase 200V AC electrical power supplied by two 20kVA 400Hz alternators, driven by port side intermediate shaft from main gearbox and available on ground under same conditions as hydraulic ancillary systems; two 28.5V DC transformer-rectifiers; main battery used for self-starting and emergency power in flight.

AVIONICS: Optional communications equipment includes VHF, UHF, tactical HF and HF/SSB radio and intercom. Navigational equipment includes ADF, radio altimeter, VLF Omega, Decca navigator and flight log, Doppler and VOR/ILS; SFIM 155 autopilot, with provision for coupling to self-contained navigation and landing systems; full IFR instrumentation optional; offshore models have nose-mounted radar. The search and rescue version has nose-mounted Bendix/King RDR 1400 or Honeywell Primus 500 search radar, Doppler, and Sextant Avionique Nadir or Decca self-contained navigation system, including navigation computer with SAR patterns, polar indicator, roller map display, hover indicator, route mileage indicator and ground speed and drift indicator. Nadir Mk 2 in French Army version; SFIM CDV 155 autopilot coupler contains automatic nav track including search patterns, transitions and hover; multi-function video display shows radar and route images, SAR patterns and hover indication;

148 INTERNATIONAL: Aircraft—EUROCOPTER

AS 332L1 Super Puma of Tokyo Police with air-conditioning pack between port-side flight deck and cabin windows

naval ASW and ASV helicopters can have nose-mounted Thomson-CSF Varan radar, linked to a tactical table in the cabin, and an Alcatel/Thomson-Sintra HS 312 sonar station at rear of cabin. Swedish Hkp 10s have Bendix/King 1500 radar with integrated FLIR, Racal RAMS flight management system including GEC-Marconi AHRS, Decca Doppler and GPS.

EQUIPMENT: Optional fixed or retractable rescue hoist (capacity 275 kg; 606 lb) starboard side; equipment for naval missions can include sonar, MAD and sonobuoys.

ARMAMENT (optional): Typical alternatives for army/air force missions are one 20 mm gun, two 7.62 mm machine guns, or two pods each containing twenty-two 68 mm or nineteen 2.75 in rockets. Armament for naval missions includes two AM 39 Exocet missiles or two lightweight torpedoes.

DIMENSIONS, EXTERNAL:
Main rotor diameter	15.60 m (51 ft 2¼ in)
Tail rotor diameter	3.05 m (10 ft 0 in)
Main rotor blade chord	0.60 m (1 ft 11½ in)
Length: overall, rotors turning	18.70 m (61 ft 4¼ in)
fuselage, incl tail rotor:	
AS 532UC/AC/SC	15.53 m (50 ft 11½ in)
AS 332L1/532AL/UL	16.29 m (53 ft 5½ in)
Width, blades folded:	
AS 532UC/AC/AL/UL/332L1	3.79 m (12 ft 5¼ in)
AS 532SC	4.04 m (13 ft 3 in)
Height: overall	4.92 m (16 ft 1¾ in)
blades and tail pylon folded:	
AS 532UC/AC/SC	4.80 m (15 ft 9 in)
to top of rotor head	4.60 m (15 ft 1 in)
Wheel track	3.00 m (9 ft 10 in)
Wheelbase: AS 532UC/AC/SC	4.49 m (14 ft 8¾ in)
AS 332L1/532AL/UL	5.28 m (17 ft 4 in)
Passenger cabin doors, each:	
Height	1.35 m (4 ft 5 in)
Width	1.30 m (4 ft 3¼ in)
Floor hatch, rear of cabin:	
Length	0.98 m (3 ft 2¾ in)
Width	0.70 m (2 ft 3½ in)

DIMENSIONS, INTERNAL:
Cabin: Length: AS 532UC/AC/SC	6.05 m (19 ft 10½ in)
AS 332L1/532AL/UL	6.81 m (22 ft 4 in)
Max width	1.80 m (5 ft 10¾ in)
Max height	1.55 m (5 ft 1 in)
Floor area: AS 532UC/AC/SC	7.80 m² (84 sq ft)
AS 332L1/532AL/UL	9.18 m² (98.8 sq ft)
Usable volume: AS 532UC/AC/SC	11.40 m³ (403 cu ft)
AS 332L1/532AL/UL	13.30 m³ (469.5 cu ft)

AREAS:
Main rotor disc	191.1 m² (2,057.4 sq ft)
Tail rotor disc	7.31 m² (78.64 sq ft)

WEIGHTS AND LOADINGS:
Weight empty (standard aircraft):	
AS 532UC/AC	4,330 kg (9,546 lb)
AS 532SC	4,500 kg (9,920 lb)
AS 332L1/532AL/UL	4,460 kg (9,832 lb)
Max T-O weight:	
AS 532UC/AC/AL/UL/SC, internal load	9,000 kg (19,841 lb)
AS 332L1, internal load	8,600 kg (18,960 lb)
all versions, with slung load	9,350 kg (20,615 lb)

PERFORMANCE (at max T-O weight):
Never-exceed speed (V_NE)	150 knots (278 km/h; 172 mph)
Cruising speed at S/L:	
AS 532UC/AC/AL/UL	141 knots (262 km/h; 163 mph)
AS 532SC	130 knots (240 km/h; 149 mph)
AS 332L1	144 knots (266 km/h; 165 mph)
Max rate of climb at S/L:	
AS 532UC/AC/AL/UL	420 m (1,378 ft)/min
AS 532SC	372 m (1,220 ft)/min
AS 332L1	486 m (1,594 ft)/min
Service ceiling: AS 332L1	4,600 m (15,090 ft)
AS 532UC/AC/SC/AL/UL	4,100 m (13,450 ft)
Hovering ceiling IGE:	
AS 532AC/UC/SC	2,700 m (8,860 ft)
AS 532AL/UL	2,800 m (9,185 ft)
AS 332L1	3,100 m (10,170 ft)
Hovering ceiling OGE:	
AS 532AC/UC/SC	1,600 m (5,250 ft)
AS 532AL/UL	1,650 m (5,415 ft)
AS 332L1	2,300 m (7,545 ft)
Range at S/L, standard tanks, no reserves:	
AS 532UC/AC	334 nm (618 km; 384 miles)
AS 532SC/332L1	470 nm (870 km; 540 miles)
AS 532AL/UL	455 nm (842 km; 523 miles)
Range at S/L with external (2 × 338 litre) and auxiliary (320 litre) tanks, no reserves:	
AS 532AL/UL	672 nm (1,245 km; 773 miles)

EUROCOPTER AS 332L2 SUPER PUMA Mk II and AS 532 COUGAR Mk II

TYPE: Twin-turbine medium helicopter.

PROGRAMME: First flight of development vehicle 6 February 1987; see delayed certification dates below; progressively replacing AS 332L1 and 532U/A Cougar on production line.

CURRENT VERSIONS: **AS 332L2 Super Puma Mk II**: Current production civil transport; French certification 2 April 1992; UK BCAR 29 certification 16 November 1992 includes hydraulically powered standby electrics for two hours' operation after complete main generator failure, crash resistant fuel tanks and health and usage monitoring system (HUMS).

AS 332L2 Super Puma Mk II VIP: New eight-passenger arrangement with attendant and fully equipped galley and toilet; two four-seat lounges; fine materials and fittings throughout; range with eight passengers and attendant, 635 nm (1,176 km; 730 miles).

AS 532U2 Cougar Mk II: Unarmed military utility transport; longest member of Cougar family, carries 29 troops and two-man crew.

AS 532A2 Cougar Mk II: Armed army/navy/air force helicopter.

CUSTOMERS: Launch civil customer Bristow Helicopters ordered 20 in June 1989 for North Sea offshore support; first delivery expected early 1993; Helikopter Service ordered 12 mid-1992; first delivery due August 1993.

DESIGN FEATURES: Composites plug in rear cabin area accommodates one extra row of seats and moves tail rotor rearwards; Spheriflex four-blade main rotor with longer blades having parabolic tips; four-blade Spheriflex tail rotor; standard enlarged composites sponsons can contain fuel, liferafts, air-conditioning and pop-out floats.

FLYING CONTROLS: Fully powered hydraulically actuated with SFIM 165 four-axis digital AFCS and coupler.

STRUCTURE: Largely as previous AS 332/532.

LANDING GEAR: Retractable tricycle high-impact absorbing landing gear; hydraulic retraction rearward, single mainwheels partially into sponsons and twin nosewheels into fuselage; self-centring nosewheels, mainwheel brakes.

POWER PLANT: Two Turbomeca Makila 1A2 rear drive free turbines; maximum emergency power each (OEI 30 seconds), 1,573 kW (2,109 shp); intermediate emergency power (OEI 2 min) 1,467 kW (1,967 shp); take-off power 1,376 kW (1,845 shp); max continuous power 1,236 kW (1,657 shp); protective intake grilles standard; optional momentum separators for dusty conditions. Transmission ratings: twin-engined max 2,410 kW (3,229 hp), max from single engine 1,666 kW (2,232 hp), max transient 20 seconds 2,651 kW (3,552 hp), max continuous 1,555 kW (2,084 hp).

Standard fuel tankage under cabin floor 2,020 litres (535 US gallons; 444 Imp gallons); auxiliary tankage includes 324 litres (85 US gallons; 71.3 Imp gallons) in cargo hook well; sponson tanks each holding 325 litres (86 US gallons; 71.5 Imp gallons); 600 litres (159 US gallons; 132 Imp gallons) in cabin fuel tank; one to five internal ferry tanks each holding 475 litres (126 US gallons; 104.5 Imp gallons). Optional crashproof tankage is standard 1,919 litres (507 US gallons; 422 Imp gallons); hook well tank 320 litres (84 US gallons; 70.4 Imp gallons); 325 litres (86 US gallons; 71.5 Imp gallons) in each sponson tank; cabin and internal ferry tanks not specially crashproofed; fuel dumping and pressure refuelling optional.

ACCOMMODATION: Single pilot in DGAC Category B civil operation; single pilot plus licensed crewman in Category A VFR; two pilots in IFR; military operation one pilot in VFR, two in IFR; civil transport 24 passengers and attendant in airline interior for 220 nm (408 km; 253 miles) or 19 passengers at 81 cm (32 in) seat pitch for 350 nm (648 km; 403 miles); military capacity, chief of squad plus 28 troops or chief of squad plus 18 troops in crashworthy seats; VIP version for eight to 14 passengers plus attendant; ambulance version for doctor, nine stretchers and four casualties plus attendant.

SYSTEMS: Simplified electrical system.

AVIONICS: EFIS flight instrumentation; civil or military IFR systems; choice of military com/nav and nose-mounted radars.

EQUIPMENT: Pop-out floats, rescue winch, external sling.

ARMAMENT: Armament choice as for AS 532A Cougar except for Exocet.

DIMENSIONS, EXTERNAL:
Main rotor diameter	16.20 m (53 ft 1½ in)
Tail rotor diameter	3.15 m (10 ft 4 in)

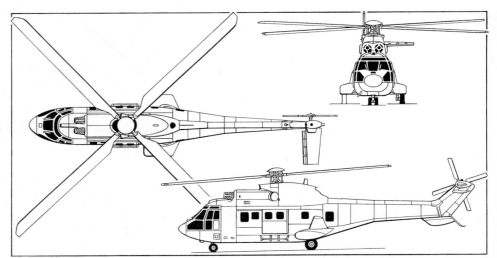

Eurocopter AS 332L2 Super Puma Mk II with Spheriflex main and tail rotor heads *(Jane's/Mike Keep)*

Length overall: rotors turning	19.50 m (63 ft 11 in)
main rotor folded	16.79 m (55 ft 0½ in)
Width: over sponsons	3.38 m (11 ft 1 in)
overall, main rotor folded	3.86 m (12 ft 8 in)
Height: overall, tail rotor turning	4.97 m (16 ft 4 in)
to top of rotor head	4.60 m (15 ft 1 in)
Tailplane span	2.17 m (7 ft 1½ in)
Wheel track	3.00 m (9 ft 10 in)
Wheelbase	5.28 m (17 ft 4 in)
Sliding cabin doors, each:	
Height	1.35 m (4 ft 5 in)
Width	1.30 m (4 ft 3¼ in)
Floor hatch: Length	0.98 m (3 ft 2¾ in)
Width	0.70 m (2 ft 3½ in)

DIMENSIONS, INTERNAL:
Cabin: Max length	7.87 m (25 ft 10 in)
Floor length	6.15 m (20 ft 2¼ in)
Max width	1.80 m (5 ft 10¾ in)
Max height	1.45 m (4 ft 9 in)

AREAS:
Main rotor disc	206.0 m² (2,217.4 sq ft)
Tail rotor disc	7.79 m² (83.88 sq ft)

WEIGHTS AND LOADINGS:
Manufacturer's weight empty: L2	4,660 kg (10,274 lb)
U2	4,760 kg (10,493 lb)
Useful load: L2	4,490 kg (9,899 lb)
U2	4,990 kg (11,000 lb)
Standard fuel weight: L2	1,596 kg (3,519 lb)
U2, crashworthy tanks	1,548 kg (3,412 lb)
Max normal T-O weight: L2	9,150 kg (20,172 lb)
U2	9,750 kg (21,495 lb)
Max slung load: L2/U2	4,500 kg (9,920 lb)
Max flight weight with slung load:	
L2/U2	10,000 kg (22,046 lb)
Max disc loading, normal T-O weight:	
L2	44.4 kg/m² (9.10 lb/sq ft)
U2	47.3 kg/m² (9.69 lb/sq ft)
Max 2-engine transmission power loading:	
L2	3.79 kg/kW (6.24 lb/shp)
U2	4.04 kg/kW (6.66 lb/shp)

PERFORMANCE:
Never-exceed speed (V_{NE}):	
L2/U2	177 knots (327 km/h; 203 mph)
Fast cruising speed: L2	150 knots (277 km/h; 172 mph)
U2	147 knots (273 km/h; 170 mph)
Econ cruising speed: L2	133 knots (247 km/h; 154 mph)
U2	131 knots (242 km/h; 150 mph)
Rate of climb at 70 knots (130 km/h; 81 mph) at S/L:	
L2	441 m (1,447 ft)/min
U2	384 m (1,260 ft)/min
Service ceiling (45.7 m; 150 ft/min climb), ISA:	
L2	5,180 m (16,995 ft)
U2	4,100 m (13,451 ft)
Hovering ceiling IGE, normal T-O weight, ISA, T-O power:	
L2	2,900 m (9,414 ft)
U2	2,540 m (8,333 ft)
Hovering ceiling OGE, normal T-O weight, ISA, T-O power:	
L2	2,250 m (7,382 ft)
U2	1,900 m (6,234 ft)
Hovering ceiling OGE, slung load weight, ISA, T-O power:	
L2/U2	380 m (1,247 ft)
Range, no reserves, standard fuel, econ cruise:	
L2	460 nm (851 km; 529 miles)
U2	430 nm (796 km; 494 miles)
Range, no reserves, max fuel, econ cruise:	
L2	805 nm (1,491 km; 927 miles)
U2	635 nm (1,176 km; 730 miles)
Endurance, standard fuel, at 70 knots (130 km/h; 81 mph)	
L2	4.54 h
U2	4.20 h

EUROCOPTER SA 342 GAZELLE

TYPE: Five-seat light utility helicopter.

PROGRAMME: First flight 7 April 1967, powered by Astazou III fixed-shaft turbine; earlier versions detailed in 1979-80, 1984-85 and 1991-92 *Jane's*.

CURRENT VERSIONS: **SA 341F:** French Army machines fitted with Giat elevating side-mounted 20 mm gun and Sextant Avionique sight elevating automatically with gun.

SA 342M: In low rate production for French Army ALAT attrition and fleet renewal; three produced in 1991.

SA 342M ATAM: 30 French Army machines each retrofitted with four Matra air-to-air Mistrals (ATAM) and T2000 sight.

SA 342M Viviane: Up to 70 French Army SA 342 normally fitted with daytime HOT sight being retrofitted with Viviane stabilised direct view/IR/laser roof-mounted sight to allow night firing of HOT missiles.

CUSTOMERS: French Army ALAT operating 170 SA 341L1 and 188 SA 342M in early 1992; total (all variants and customers) 1,254 delivered and 1,116 operating on 30 June 1991, excluding those produced by Westland in UK and Soko in former Yugoslavia; total 628 SA 341 delivered and 542 in service on 30 June 1991; total 626 SA 342 delivered and 574 in service on 30 June 1991.

POWER PLANT: One 640 kW (858 shp) Turbomeca Astazou fixed-shaft turboshaft with optional momentum separator

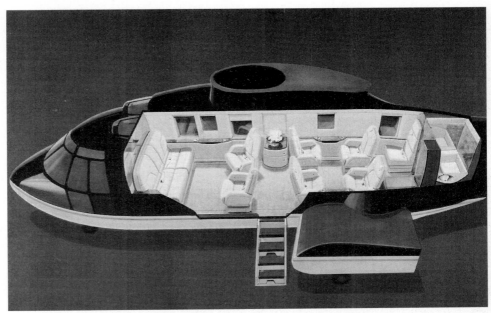

VIP interior of the Eurocopter AS 332L2 Super Puma Mk II showing eight passenger seats plus attendant and toilet

Eurocopter AS 332L2 Super Puma Mk II before delivery to Bristow Helicopters

Flight deck of Eurocopter AS 332L2 Super Puma Mk II showing EFIS flight instruments backed by integrated digital avionics, flight management system and duplex four-axis autopilot

150 INTERNATIONAL: AIRCRAFT—EUROCOPTER

Viviane all-weather sight for Euromissile HOT anti-tank missiles on an SA 342M Gazelle

on intake and upturned exhaust pipe; engine and rotor directly coupled and turn at constant speed, power controlled by fuel input. Usable fuel 545 litres (144 US gallons; 120 Imp gallons).

ACCOMMODATION: Pilot with gunsight on right; weapon aimer/observer on left; three seats in rear if equipment allows.
AVIONICS: ALAT instrument panel; VHF/UHF/FM communications radio and homing; SFIM PA 85G autopilot; Sextant Avionique Nadir/Decca Doppler 80 self-contained navigation system; night flying equipment.

DIMENSIONS, EXTERNAL:
Main rotor diameter	10.50 m (34 ft 5½ in)
Length overall, rotors turning	11.97 m (39 ft 3⅝ in)
Height overall	3.19 m (10 ft 5½ in)

WEIGHTS AND LOADINGS:
Max T-O weight: 342M	2,100 kg (4,630 lb)
Max disc loading	21.94 kg/m² (4.49 lb/sq ft)

PERFORMANCE:
Max cruising speed at S/L
140 knots (260 km/h; 161 mph)

EUROCOPTER AS 350 ECUREUIL/ASTAR and AS 550 FENNEC

Brazilian Air Force designations: CH-50 and TH-50 Esquilo
Brazilian Army designation: HA-1 Esquilo
Brazilian Navy designation: UH-12 Esquilo

TYPE: Five/six-seat light general purpose helicopter.
PROGRAMME: First flight (F-WVKH) powered by Textron Lycoming LTS101 turboshaft 27 June 1974; first flight second prototype (F-WVKI) powered by Turbomeca Arriel 14 February 1975; first production version AS 350B powered by 478 kW (641 shp) Arriel 1B certificated 27 October 1977; LTS101-powered AStar sold only in USA.
CURRENT VERSIONS: **AS 350BA Ecureuil:** Powered by 478 kW (641 shp) Turbomeca Arriel 1B and fitted with large main rotor blades of AS 350B2 (see below); AS 350B can be upgraded to AS 350BA in field; replaced AS 350B during 1992; French VFR certification 1991; UK and US certifications expected late 1992.

AS 350B2 Ecureuil: Powered by 546 kW (732 shp) Arriel 1D1 and with transmission uprated to 440 kW (590 shp); wide-chord new section main and tail rotor blades originally developed for AS 355 twin; certificated 26 April 1989; known as **SuperStar** in North America.

AS 350B3 Ecureuil: New variant under study; main innovation is 612 kW (820 shp) Arriel 2.

AS 350D AStar: Version of 350B for North American market; LTS101 engine.

AS 350 Firefighter: Conair system (see Canadian section) able to pick up water load in 30 seconds while hovering over water; demonstrated 1986.

AS 550 Fennec: Military version of AS 350B2 powered by 546 kW (732 shp) Arriel 1D1; standard features include taller landing gear, sliding doors, extended instrument panel; NVG-compatible cockpit; airframe reinforced for armament; provision for armoured seats; individual versions include utility **AS 550U2**, armed **AS 550A2**, missile-armed **AS 550C2**, unarmed naval **AS 550M2** and armed naval **AS 550S2**.

CUSTOMERS: Total 1,713 AS 350s of all types ordered from 55 countries by 31 May 1992; total 1,535 delivered; 102 Ecureuil/Fennec ordered during 1992.
Military customers include Singapore armed forces (six); Australia (RAAF 18 for training; RAN six utility); Danish Army (12 AS 550C2 with ESCO HeliTOW system ordered 1987, delivered 1990); French Army ALAT (needs up to 100 to replace Alouette IIs). Built under licence by Helibras in Brazil (which see).

DESIGN FEATURES: Starflex bearingless glassfibre main rotor head; all versions now have lifting section composite main rotor blades.
FLYING CONTROLS: Single fully powered controls with accumulators to delay manual control following hydraulic failure; cyclic trim by adjustable stick friction; inverted aerofoil tailplane to adjust pitch attitude in climb, cruise and descent; saucer fairing on rotor head to smooth wake; swept fins above and below tail. Autostabiliser and autopilot optional.
STRUCTURE: Main rotor head and much of airframe of glassfibre and aramids; main rotor blades automatically manufactured in composites; self-sealing composites fuel tank in military versions.
LANDING GEAR: Steel tube skid type. Taller version standard on military aircraft. Emergency flotation gear optional.
POWER PLANT: AS 350BA powered by one Turbomeca 478 kW (641 shp) Arriel 1B; AS 350B2/550 powered by one 546 kW (732 shp) Turbomeca Arriel 1D1; AS 350D powered by Textron Lycoming LTS101 sold only in USA. Plastics fuel tank (self-sealing on AS 550) with capacity of 540 litres (142.6 US gallons; 119 Imp gallons).
ACCOMMODATION: Two individual bucket seats at front of cabin and four-place bench standard; optional ambulance layout; large forward hinged door on each side of versions for civil use; optional sliding door at rear of cabin on port side; (sliding doors standard on military version); baggage compartment aft of cabin, with full-width upward hinged door on starboard side; top of baggage compartment reinforced to provide work platform on each side.
SYSTEMS: Hydraulic system includes four single-body servo units, operating at 40 bars (570 lb/sq in) pressure, and accumulators to delay reversion to manual control; electrical system includes a 4.5kW engine driven starter/generator, a 24V 16Ah nickel-cadmium battery and a ground power receptacle; cabin air-conditioning system optional.
AVIONICS: Optional com/nav radio equipment includes VHF/AM, ICS, VOR/ILS, marker beacon indicator, ADF, HF/SSB, transponder and DME; choice of SFIM PA 85T31, Honeywell HelCis or Collins APS-841H autopilot.
EQUIPMENT: Includes 907 kg (2,000 lb) cargo sling (1,160 kg; 2,557 lb for AS 350B2/550), a 135 kg (297 lb) electric hoist, a TV camera for aerial filming, and a 735 litre (194 US gallon; 161 Imp gallon) Simplex agricultural spraytank and boom system.
ARMAMENT (AS 550): Provision for wide range of weapons, including 20 mm Giat M621 gun, FN Herstal TMP twin 7.62 mm machine-gun pods, Thomson-Brandt 68.12 launchers for twelve 68 mm rockets, Forges de Zeebrugge launchers for seven 2.75 in rockets, and ESCO HeliTOW anti-tank missile system.

DIMENSIONS, EXTERNAL:
Main rotor diameter	10.69 m (35 ft 0¾ in)
Main rotor blade chord: AS 350BA/D	0.30 m (11.8 in)
AS 350B2/550	0.35 m (13.8 in)
Tail rotor diameter	1.86 m (6 ft 1¼ in)
Tail rotor blade chord: AS 350BA/D	0.185 m (7.28 in)
AS 350B2/550	0.205 m (8.07 in)
Length: overall, rotors turning	12.94 m (42 ft 5½ in)
fuselage	10.93 m (35 ft 10½ in)
Width: fuselage	1.80 m (5 ft 10¾ in)
overall, blades folded (ie, horizontal stabiliser span)	2.53 m (8 ft 3¾ in)
Height overall: AS 350BA/B2/D	3.14 m (10 ft 3½ in)
AS 550	3.34 m (10 ft 11½ in)
Skid track: AS 350BA/B2/D	2.17 m (7 ft 1½ in)
AS 550	2.28 m (7 ft 5¾ in)
Cabin doors (civil versions, standard, each):	
Height	1.15 m (3 ft 9¼ in)
Width	1.10 m (3 ft 7¼ in)

DIMENSIONS, INTERNAL:
Cabin: Length	2.42 m (7 ft 11¼ in)
Width at rear	1.65 m (5 ft 5 in)
Height	1.35 m (4 ft 5 in)
Baggage compartment volume	1.00 m³ (35.31 cu ft)

AREAS:
Main rotor disc	89.75 m² (966.1 sq ft)
Tail rotor disc	2.72 m² (29.25 sq ft)

WEIGHTS AND LOADINGS:
Weight empty: AS 350BA	1,146 kg (2,325 lb)
AS 350B2	1,153 kg (2,542 lb)
AS 350D	1,123 kg (2,475 lb)
AS 550	1,220 kg (2,689 lb)
Max T-O weight: AS 350BA	2,100 kg (4,630 lb)
AS 350B2/550	2,250 kg (4,960 lb)
AS 350D	1,950 kg (4,300 lb)
Max weight with slung load:	
AS 350BA	2,250 kg (4,960 lb)
AS 350D	2,100 kg (4,630 lb)

PERFORMANCE (AS 350BA/D at normal T-O weight, AS 350B2/550 at 2,200 kg; 4,850 lb):
Never-exceed speed (V_{NE}) at S/L:
AS 350BA/B2/550   155 knots (287 km/h; 178 mph)

Eurocopter AS 350B2 Ecureuil

AS 350D	147 knots (272 km/h; 169 mph)

Max cruising speed at S/L:
AS 350BA	126 knots (234 km/h; 145 mph)
AS 350B2/550	133 knots (246 km/h; 153 mph)
AS 350D	124 knots (230 km/h; 143 mph)

Max rate of climb at S/L:
AS 350BA	456 m (1,500 ft)/min
AS 350B2/550	534 m (1,750 ft)/min
Service ceiling: AS 350BA/B2/550	4,800 m (15,750 ft)
Hovering ceiling OGE: AS 350BA	1,980 m (6,500 ft)
AS 350B2/550	2,550 m (8,350 ft)

Range with max fuel at recommended cruising speed, no reserves:
AS 350BA	394 nm (730 km; 453 miles)
AS 350B2/550	360 nm (666 km; 414 miles)

EUROCOPTER AS 355 ECUREUIL 2 TWIN-STAR and AS 555 FENNEC

Brazilian Air Force designations: CH-55 and VH-55 Esquilo
Brazilian Navy designation: UH-12B Esquilo

TYPE: Twin-turbine light general purpose helicopter.
PROGRAMME: First flight of first prototype (F-WZLA) 28 September 1979; details of early production AS 355E/F versions in 1984-85 *Jane's*; AS 355F superseded by AS 355F1 in January 1984; AS 355F2 certificated 10 December 1985; AS 355N powered by two Turbomeca TM 319 Arrius certificated 1989; deliveries of this version began early 1992.
CURRENT VERSIONS: All current production AS 355 Ecureuil 2/555 Fennec are powered by two Turbomeca TM 319 Arrius turboshafts.
  AS 355N Ecureuil 2: Current civil production version, adaptable for passengers, cargo, police, ambulance, sling loading and other missions; known as **TwinStar** in USA.
  AS 555UN Fennec: Military utility version and French Army ALAT IFR pilot trainer.
  AS 555AN Fennec: Armed version; later machines adapted for centreline-mounted 20 mm gun and pylon-mounted rockets.
  AS 555CN Fennec: Missile-armed version under development in 1992.
  AS 555MN Fennec: Naval unarmed version. Can carry a 360° chin-mounted radar.
  AS 555SN Fennec: Armed naval version for ASW and over-the-horizon targeting operating from ships from 600 t upwards; armament includes two lightweight homing torpedoes or the cannon, missiles and rockets of land-based versions; avionics include Bendix/King RDR-1500 360° chin-mounted radar, Sextant Avionique Mk 3 MAD beneath tailboom, Sextant Nadir Mk 10 navigation system, Dassault Electronique RDN 85 Doppler with hover guidance and SFIM 85 T31 three-axis autopilot.
CUSTOMERS: Total 524 AS 355/555 ordered by mid-1992 including 75 powered by TM 319 Arrius. French Air Force ordered 52 Fennecs, first eight as AS 355F1s powered by Allison 250; flown by 67e Escadre d'Hélicoptères at Villacoublay for communications and, with side-mounted Giat M61 20 mm gun pod, by ETOM 68 in Guyana; delivery of remaining 44 (AS 555AN powered by Arrius) began 19 January 1990; from 24th onwards, provision for centrally mounted 20 mm cannon and T-100 sight, plus Matra Mistral missiles; French Army ordered four AS 555UNs for IFR training (delivery from February 1992); six more on option; Brazilian Air Force has 13 AS 555s, 11 with armament designated CH-55 and two VIP transports designated VH-55; Brazilian Navy acquired 11 UH-12Bs; Brazilian AS 555s assembled by Helibras in Brazil (which see).
DESIGN FEATURES: Starflex main rotor; two engine shafts drive into combiner gearbox containing freewheels; otherwise substantially as AS 350.
FLYING CONTROLS: Full dual powered flying controls without manual reversion; trim by adjustable stick friction; inverted aerofoil tailplane; swept fins above and below tailboom.
STRUCTURE: Light metal tailboom and central fuselage structure; thermoformed plastics for cabin structure.
LANDING GEAR: As for AS 350B2/550.
POWER PLANT: Two Turbomeca TM 319 1M Arrius turboshafts, each rated at 340 kW (456 shp) for take-off and 295 kW (395 shp) max continuous with full authority digital engine control (FADEC); this allows automatic sequenced starting of both engines, automatic top temperature and torque limiting and pre-selection of lower limits for practice OEI operation. Two integral fuel tanks, with total usable capacity of 730 litres (193 US gallons; 160 Imp gallons), in body structure.
ACCOMMODATION: As for AS 350B2, except sliding doors optional on both sides (standard on military aircraft); three baggage holds with external doors.
SYSTEMS: As for AS 350B2/550, except two hydraulic pumps and electric generators.
AVIONICS: Options include a second VHF/AM and radio altimeter; provision for IFR system, SFIM 85 T31 three-axis autopilot and CDV 85 T3 nav coupler.
EQUIPMENT: Casualty installations and winch available. See also current version descriptions.
ARMAMENT (AS 555): Optional alternative weapons include Thomson-Brandt or Forges de Zeebrugge rocket packs, Matra or FN Herstal machine-gun pods, a Giat M621

Armed Eurocopter AS 550C2 single-engined Fennec with ESCO HeliTOW system

Eurocopter AS 355N of Electricité de France with full single-engine hover capability holding men working on live high tension cables (further helicopter for photography only)

Eurocopter AS 555UN Fennec military utility helicopter (two Turbomeca TM 319 Arrius)

152 INTERNATIONAL: Aircraft—EUROCOPTER

Eurocopter AS 555AN Fennec twin-turbine armed helicopter with side view (top) of single-engined AS 350BA Ecureuil and scrap view (bottom) of AS 355N civil twin *(Dennis Punnett)*

20 mm gun, and HOT or TOW anti-tank missiles. Naval version carries two homing torpedoes in ASW role, or SAR winch.

DIMENSIONS, EXTERNAL: As for AS 350B2/550
DIMENSIONS, INTERNAL: As for AS 350B2/550
WEIGHTS AND LOADINGS:
Weight empty: 355N	1,382 kg (3,046 lb)
Max sling load: 555N	1,134 kg (2,500 lb)
Max T-O weight:	
555N, internal load	2,540 kg (5,600 lb)
555N, max sling load	2,600 kg (5,732 lb)

PERFORMANCE (AS 555N at max T-O weight, ISA):
Never-exceed speed (V_{NE})	150 knots (278 km/h; 172 mph)
Max cruising speed at S/L	121 knots (225 km/h; 140 mph)
Max rate of climb at S/L	408 m (1,340 ft)/min
Service ceiling	4,000 m (13,125 ft)
Hovering ceiling: IGE	2,600 m (8,530 ft)
OGE	1,550 m (5,085 ft)
Radius, SAR, two survivors	70 nm (129 km; 80.5 miles)
Range with max fuel at S/L, no reserves	389 nm (722 km; 448 miles)

Endurance, no reserves:
two torpedoes	1 h
one torpedo, or cannon or rocket pods	2 h 20 min
cannon plus rockets	1 h 50 min

EUROCOPTER AS 365N2 DAUPHIN 2

TYPE: Twin-turbine commercial general purpose helicopter.
PROGRAMME: Certificated for VFR in France November 1989; 500th Dauphin (all versions) delivered in November 1991 was 83rd AS 365N2.
CURRENT VERSIONS: **AS 365N2 Dauphin 2**: Current production model. *Details below refer to this version.*
  AS 366G1 (HH-65A Dolphin): US Coast Guard version (99 built, plus two ex-trials aircraft bought for evaluation by Israel); trial installation of 895 kW (1,200 shp) LHTEC T800 turboshaft ordered February 1990, but abandoned November 1991 when LTS101 performance promised to improve; re-engining with French Arriels also declined. For description see 1991-92 and earlier *Jane's*.
  AS 565UA/AA/CA Panther: Army versions; described separately.
  AS 565MA/SA Panther: Naval versions; described separately.
  Ambulance/EMS: Flight crew of two plus two or four stretchers along cabin sides loaded through rear doors; up to four seats can replace stretchers on one side; doctor sits in middle with equipment; rear doors open through 180° instead of sliding.
  Offshore support: Pilot plus up to 13 passengers, autopilot and navigation systems, pop-out floats.
CUSTOMERS: Total 557 AS 365/366 ordered for civil and military use from 44 countries by 1 January 1992; about 475 delivered; totals include 50 produced as Harbin Z-9/9A in China and 101 in HH-65A US Coast Guard versions; Chinese production continuing as Z-9A-100; 17 Dauphin/Panther ordered during 1992.
DESIGN FEATURES: Starflex main rotor hub; 11-blade Fenestron; main rotor blades have quick disconnect pins for manual folding; ONERA OA 212 (thickness/chord ratio 12 per cent) at root to OA 207 (thickness/chord ratio 7 per cent) at tip; adjustment tab near tip; leading-edge of tip swept at 45°; main rotor rpm 350; Fenestron rpm 3,665; rotor brake standard.
FLYING CONTROLS: Hydraulic dual fully powered; cyclic trim by adjustable friction damper; fixed tailplane, with endplate fins offset 10° to port; IFR systems available.
STRUCTURE: Mainly light alloy; machined frames fore and aft of transmission support; main rotor blades have CFRP spar and skins with Nomex honeycomb filling; Fenestron duct and fin of CFRP and Nomex/Rohacell sandwich; nose and power plant fairings of GFRP/Nomex sandwich; centre and rear fuselage assemblies, flight deck floor, roof, walls and bottom skin of fuel tank bays of light alloy/Nomex sandwich.
LANDING GEAR: Hydraulically retractable tricycle type; twin-wheel self-centring nose unit retracts rearward; single wheel on each rearward retracting main unit; all three units have oleo-pneumatic shock absorber; mainwheel tyres size 15 × 6.00, pressure 8.6 bars (125 lb/sq in); nosewheel tyres size 5.00-4, pressure 5.5 bars (80 lb/sq in); hydraulic disc brakes.
POWER PLANT: Two Turbomeca Arriel 1C2 turboshafts, each rated at 551 kW (739 shp) for T-O and 471 kW (631 shp) max continuous, side by side with stainless steel firewall between them. Standard fuel in four tanks under cabin floor and fifth tank in bottom of centre-fuselage; total capacity 1,135 litres (300 US gallons; 249.5 Imp gallons); provision for auxiliary tank in baggage compartment, with capacity of 180 litres (47.5 US gallons; 39.5 Imp gallons); or ferry tank in place of rear seats in cabin, capacity 475 litres (125.5 US gallons; 104.5 Imp gallons); refuelling point above landing gear door on port side. Oil capacity 14 litres (3.7 US gallons; 3 Imp gallons).
ACCOMMODATION: Standard accommodation for pilot and co-pilot or passenger in front, and two rows of four seats to rear; high density seating for one pilot and 13 passengers; VIP configurations for four to six persons in addition to pilot; three forward opening doors on each side; freight hold aft of cabin rear bulkhead, with door on starboard side; cabin heated and ventilated.
SYSTEMS: Air-conditioning system optional. Duplicated hydraulic system, pressure 60 bars (870 lb/sq in). Electrical system includes two 4.8kW starter/generators, one 24V 27Ah battery and two 250VA 115V 400Hz inverters.
AVIONICS: Two-pilot IFR instrument panel and SFIM 155 duplex autopilot standard. Optional avionics include SFIM CDV 85 autopilot coupler, EFIS, VHF and HF com/nav, VOR, ILS, ADF, transponder, DME, radar and self-contained nav system.
EQUIPMENT: Includes 1,600 kg (3,525 lb) capacity cargo sling, and 275 kg (606 lb) capacity hoist with 90 m (295 ft) cable length.
DIMENSIONS, EXTERNAL:
Main rotor diameter	11.94 m (39 ft 2 in)
Diameter of Fenestron	1.10 m (3 ft 7 5/16 in)

Eurocopter AS 365N2 Dauphin 2 of Japanese Hankyu Airlines

Eurocopter AS 365N2 Dauphin of Maryland State Police, with radar and FLIR in nose, searchlight and hoist

Main rotor blade chord: basic	0.385 m (1 ft 3¼ in)
outboard of tab	0.405 m (1 ft 4 in)
Length: overall, rotor turning	13.68 m (44 ft 10⅝ in)
fuselage	11.63 m (38 ft 1⅞ in)
Width, rotor blades folded	3.21 m (10 ft 6½ in)
Height: to top of rotor head	3.52 m (11 ft 6½ in)
overall (tip of fin)	3.98 m (13 ft 0¾ in)
Wheel track	1.90 m (6 ft 2¾ in)
Wheelbase	3.61 m (11 ft 10¼ in)
Main cabin door (fwd, each side):	
Height	1.16 m (3 ft 9½ in)
Width	1.14 m (3 ft 9 in)
Main cabin door (rear, each side):	
Height	1.16 m (3 ft 9½ in)
Width	0.87 m (2 ft 10¼ in)
Baggage compartment door (stbd):	
Height	0.51 m (1 ft 8 in)
Width	0.73 m (2 ft 4¾ in)
DIMENSIONS, INTERNAL:	
Cabin: Length	2.30 m (7 ft 6½ in)
Max width	1.92 m (6 ft 3½ in)
Max height	1.40 m (4 ft 7 in)
Floor area	4.20 m² (45.20 sq ft)
Volume	5.00 m³ (176 cu ft)
Baggage compartment volume	1.00 m³ (35.3 cu ft)
AREAS:	
Main rotor disc	111.9 m² (1,204.5 sq ft)
Fenestron disc	0.95 m² (10.23 sq ft)
WEIGHTS AND LOADINGS:	
Weight empty	2,239 kg (4,936 lb)
Max T-O weight:	
internal or external load	4,250 kg (9,370 lb)
PERFORMANCE (at max T-O weight):	
Never-exceed speed (V_{NE}) at S/L	160 knots (296 km/h; 184 mph)
Max cruising speed at S/L	154 knots (285 km/h; 177 mph)
Econ cruising speed at S/L	140 knots (260 km/h; 161 mph)
Max rate of climb at S/L	420 m (1,379 ft)/min
Service ceiling	4,300 m (14,100 ft)
Hovering ceiling: IGE	2,550 m (8,365 ft)
OGE	1,800 m (5,905 ft)
Max range with standard fuel at S/L	484 nm (897 km; 557 miles)
Endurance with standard fuel	4 h

EUROCOPTER AS 565 PANTHER (ARMY/AIR FORCE)

Brazilian Army designation: HM-1

TYPE: Unarmed and armed army/air force versions of AS 365N2 Dauphin 2.

PROGRAMME: First flight of AS 365M Panther (F-WZJV) 29 February 1984; first shown in production form 30 April 1986; armament integration and firing trials completed late 1986; first flight of second, improved prototype AS 365K (F-ZVLO) April 1987.

CURRENT VERSIONS: **AS 565UA:** Utility version; high-speed assault transport for eight to 10 troops with two crewmen over radius of 215 nm (400 km; 248 miles); other roles include reconnaissance, aerial command post, electronic warfare, target designation, search and rescue, four-stretcher casevac and slung loads up to 1,600 kg (3,525 lb).

AS 565AA: Armed version; fuselage outriggers can carry two packs of 22 Thomson-Brandt 68 mm rockets, two launchers for 19 Forges de Zeebrugge 2.75 in rockets, two Giat M621 20 mm gun pods with 180 rounds each, or four two-round packs of Matra Mistral air-to-air missiles.

AS 565CA: Anti-tank version; armed with HOT missiles and roof-mounted sight.

AS 565 Panther 800: Version fitted with two LHTEC T800s and IBM integrated avionics offered to US Army as UH-1H successor by Eurocopter and Vought Aircraft; first hover 24 March 1992, first flight 12 June.

CUSTOMERS: First order 36 AS 565AAs for Brazilian 1 BAvEx (Army Aviation Branch) at Taubaté delivered from 1989.

DESIGN FEATURES: As AS 365N2, but with greater use of composite materials and greater emphasis on survivability in combat areas; radar and IR signatures reduced by composites and special paints; noise signature low; powered control servos and engine controls armoured; cable cutters; run-dry transmission; crew seats tolerate 20g; entire basic airframe designed to withstand vertical impact at 7 m (23 ft)/s at max T-O weight.

FLYING CONTROLS: As naval AS 565, but with special fly-through and tactical modes in SFIM 155 autopilot.

POWER PLANT: Two Turbomeca Arriel 1M1 turboshafts with FADEC giving automatic sequenced two-engine starting and selectable top temperature limiting to suit mission; ratings (each) are contingency 584 kW (783 shp), take-off 558 kW (749 shp) and continuous 487 kW (653 shp). Fuel system withstands 14 m (46 ft)/s crash and tanks self-seal.

AVIONICS: Military communications, Sextant Avionique self-contained navigation system; optional Thomson-CSF TMV 011 Sherloc RWR, IR jammer and chaff/flare dispenser.

DIMENSIONS, EXTERNAL: As for AS 365N2, except:
Length of fuselage	12.11 m (39 ft 8¾ in)
Height overall	3.99 m (13 ft 1 in)

WEIGHTS AND LOADINGS:
Weight empty	2,193 kg (4,835 lb)

First Eurocopter AS 565AA Panther for Brazilian Army

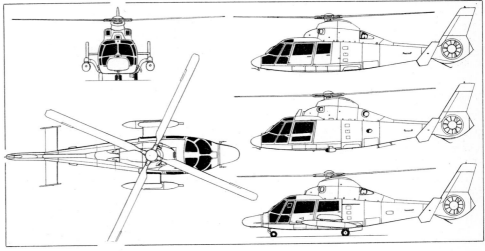

Eurocopter AS 565AA Panther with side views (centre) of HH-65A Dolphin and (top) AS 365N2 Dauphin
(Dennis Punnett)

Vought Aircraft/Eurocopter Panther 800 powered by two LHTEC T800 turboshafts

154 INTERNATIONAL: AIRCRAFT—EUROCOPTER

Max sling load	1,600 kg (3,527 lb)
Max T-O weight, internal or external load	4,250 kg (9,369 lb)

PERFORMANCE (at average mission weight of 4,000 kg; 8,818 lb):

Never-exceed speed (V$_{NE}$)	160 knots (296 km/h; 184 mph)
Max cruising speed at S/L	150 knots (278 km/h; 173 mph)
Max rate of climb at S/L	420 m (1,378 ft)/min
Hovering ceiling: IGE	2,600 m (8,531 ft)
OGE	1,850 m (6,070 ft)
Range with max standard fuel at S/L	472 nm (875 km; 544 miles)

EUROCOPTER AS 565 PANTHER (NAVY)

TYPE: Naval derivative of AS 365N2 Dauphin 2.

PROGRAMME: Launch order placed by Saudi Arabia 13 October 1980; first flight of modified AS 365N (c/n 5100) as prototype 22 February 1982; first flight production AS 365F (c/n 6014) equipped as SAR helicopter on 2 July 1982.

CURRENT VERSIONS: **AS 565MA Panther:** Unarmed naval search and sea surveillance version; rescue hoist, sea search radar, self-contained navigation, automatic hover transition; searchlight and deck-landing harpoon optional.

AS 565SA Panther: Armed ASW and anti-ship version; ASW version has Sextant Avionique MAD or Thomson-Sintra ASM HS 312 sonar and two homing torpedoes; four side-mounted AS.15TT radar-guided missiles for anti-ship role; SAR versions have nose-mounted Omera ORB 32 search radar; anti-ship versions have chin-mounted, roll-stabilised Agrion 15 radar.

CUSTOMERS: Saudi Arabia ordered four SAR/surveillance AS 565MAs and 20 anti-ship AS 565SAs with AS.15TT; Ireland ordered five for SAR with Bendix/King RDR 1500 radar, SFIM 155 autopilot, Sextant ONS 200A long-range navigation, Nadir Mk II nav computer, Dassault Electronique Cina B Doppler and five-screen EFIS; these can carry light weapons; French Navy ordered three AS 565F (MA) in 1988 for plane guard with carriers *Clémenceau* and *Foch*; 15 more AS 565MAs planned; eventual requirement 40; Chile ordered four AS 565MAs with Thomson-CSF TRWR (tactical radar and warning receiver), Varan radar and Exocet missile or Murène torpedo.

DESIGN FEATURES: Updated version of AS 365F shipborne Panther with enlarged 11-blade Fenestron for control during out-of-wind overwater hover; extended nose to house radar and additional avionics.

POWER PLANT: Two Turbomeca Arriel 1M1 turboshafts, each giving 558 kW (749 shp) for take-off and 487 kW (653 shp) continuously; standard fuel capacity 1,135 litres (300 US gallons; 250 Imp gallons); optional 180 litre (47.5 US gallons; 39.6 Imp gallons) auxiliary tank.

FLYING CONTROLS: As AS 365N2, with SFIM 155 autopilot.

ACCOMMODATION: Two-man crew; cabin can hold 10 passengers.

DIMENSIONS, EXTERNAL: As for AS 565 Panther (army/air force), except:

Width over missiles (565SA)	4.20 m (13 ft 9½ in)

WEIGHTS AND LOADINGS:

Weight empty	2,240 kg (4,938 lb)
Max sling load	1,600 kg (3,527 lb)
Max T-O weight, internal or external load	4,250 kg (9,370 lb)

PERFORMANCE (at average mission weight of 4,000 kg; 8,818 lb):

Never-exceed speed (V$_{NE}$)	160 knots (296 km/h; 184 mph)
Max cruising speed at S/L	148 knots (274 km/h; 170 mph)
Max rate of climb at S/L	420 m (1,380 ft)/min
Hovering ceiling: IGE	2,600 m (8,530 ft)
OGE	1,860 m (6,100 ft)
Range with max standard fuel at S/L	472 nm (875 km; 544 miles)

Radius of action: anti-shipping, with four missiles, 120 knots (222 km/h; 138 mph) cruising speed at 915 m (3,000 ft), ISA−20°C, 30 min reserves

	135 nm (250 km; 155 miles)
with two missiles	150 nm (278 km; 173 miles)
SAR, ISA+20°C, 30 min reserves, carrying six survivors	130 nm (241 km; 150 miles)

EUROCOPTER DAUPHIN X 380, AS 365X DGV and FBW

TYPE: Three experimental Dauphins testing future technology.

PROGRAMME: First flight modified Dauphin X 380 DTP (Développement Technique Probatoire) (F-WDFK) 20 March 1989; first flight fly-by-wire Dauphin FBW (c/n 6001/F-WZJJ) 6 April 1989; first flight high-speed AS 365X Dauphin Grande Vitesse (DGV) March 1991; set up Class E1e speed record of 200 knots (371 km; 230 mph) over 3 km triangular course 19 November 1991; also beat Class E1 world speed record.

DESIGN FEATURES: DTP and DGV have five-blade Spheriflex rotor head with integrated hub/mast of filament-wound CFRP mounted by single bearing in transmission casing; composites blades with drooped tapered tips; smaller Fenestron and aerodynamic rudder in fin; hub fully faired and blended with special low-drag engine/transmission fairing.

FBW aircraft designed to investigate complex attitude control laws; trials with sidestick controller began April 1990.

Eurocopter Dauphin AS 365X DGV record-breaking experimental helicopter

POWER PLANT: DTP powered by two Turbomeca Arriel 1C1 turboshafts; DGV powered by two special Arriel 1X with short-term capability for 660 kW (884 shp).

EUROCOPTER BO 105

TYPE: Five/six-seat twin-turbine light helicopter.

PROGRAMME: First flight of prototype 16 February 1967; production of 100 BO 105 M (VBH) and 212 BO 105 P (PAH-1) (see later entry and 1985-86 *Jane's*) completed 1984.

CURRENT VERSIONS: **BO 105 CB:** Basic production model since 1975.

BO 105 CBS: Optional model with cabin stretched 0.25 m (10 in) in rear seat area; additional window aft of rear door; FAA certification to SFAR Pt 29-4 early 1983; current model is BO 105 CBS-4. *Details apply to this model, except where indicated.*

EC Super Five: Uprated alternative to BO 105; described separately.

BO 105 LS: Hot and high variant powered by two Allison 250-C28C each rated at 410 kW (550 shp) for 2.5 min; produced exclusively by Eurocopter Canada (which see under Canada).

CUSTOMERS: Total more than 1,300 BO 105 of all models delivered to 40 countries by January 1992; these include Mexican Navy (12), Spanish Army (70 including 18 armed reconnaissance, 14 observation and 28 anti-tank), Swedish Army (20 delivered by September 1988), Swedish Air Force (four BO 105 CBS for IFR search and rescue).

DESIGN FEATURES: Four-blade main rotor with rigid titanium hub; only articulation is roller bearings for blade pitch change; all-composite blades of NACA 23012 section with drooped leading-edge and reflexed trailing-edge; 8° linear twist; pendulum vibration damper near each blade root; main rotor brake standard; two blades can be folded; two-blade semi-rigid tail rotor; main rotor rpm 424, tail rotor rpm 2,220.

FLYING CONTROLS: Fully powered controls through hydraulic actuation pack mounted on transmission casing; very high control power and aerobatic capability, including claimed sustained inverted 1g flight; IFR certification without autostabiliser.

STRUCTURE: Light alloy structure with GFRP main and tail rotor blades and top fairing. Main production source is Eurocopter Deutschland factory at Donauwörth, but Spanish models assembled by CASA, Indonesian models manufactured and assembled by IPTN and BO 105 LS exclusively manufactured by Eurocopter Canada (formerly MBB Helicopter Canada) in Ontario (which see).

LANDING GEAR: Skid type, with cross-tubes designed for energy absorption by plastic deformation in event of heavy landing. Inflatable emergency floats can be attached to skids.

POWER PLANT: Two 313 kW (420 shp) Allison 250-C20B turboshafts, each with a max continuous rating of 298 kW (400 shp). Bladder fuel tanks under cabin floor, usable capacity 570 litres (150.6 US gallons; 125.3 Imp gallons); fuelling point on port side of cabin; optional auxiliary tanks in freight compartment; oil capacity: engine 12 litres (3.2 US gallons; 2.6 Imp gallons), gearbox 11.6 litres (3.06 US gallons; 2.55 Imp gallons).

ACCOMMODATION: Pilot and co-pilot or passenger on individual longitudinally adjustable front seats with lap and shoulder harnesses; optional dual controls. Bench seat at rear for three persons, removable for cargo and stretcher carrying; EMS versions available; entire rear fuselage aft of seats

French Navy Eurocopter AS 565MA Panther plane guard helicopter

and under power plant available as freight and baggage space, with access through two clamshell doors at rear; two standard stretchers side by side in ambulance role; one forward opening hinged and jettisonable door and one sliding door on each side of cabin; ram air and electrical ventilation system; heating system optional.

SYSTEMS: Tandem fully redundant hydraulic system, pressure 103.5 bars (1,500 lb/sq in), for powered main rotor controls; system flow rate 6.2 litres (1.64 US gallons, 1.36 Imp gallons)/min; bootstrap/fluid reservoir, pressurised at 1.7 bars (25 lb/sq in); electrical system powered by two 150A 28V DC starter/generators and a 24V 25Ah nickel-cadmium battery; external power socket.

AVIONICS: Wide variety of avionics available including weather radar, Doppler and GPS navigation, SAS and autopilot.

EQUIPMENT: Standard equipment includes heated pitot, tie-down rings in cargo compartment, cabin and cargo compartment dome lights, position lights and collision warning lights; options include dual controls, heating system, windscreen wiper, rescue winch, landing light, searchlight, externally mounted loudspeaker, fuel dump valve, external load hook, settling protectors, snow skids, and manual main rotor blade folding.

DIMENSIONS, EXTERNAL:
Main rotor diameter	9.84 m (32 ft 3½ in)
Tail rotor diameter	1.90 m (6 ft 2¾ in)
Main rotor blade chord	0.27 m (10⅝ in)
Tail rotor blade chord	0.18 m (7 in)
Distance between rotor centres	5.95 m (19 ft 6¼ in)
Length: incl main and tail rotors	11.86 m (38 ft 11 in)
excl rotors: CB	8.56 m (28 ft 1 in)
CBS	8.81 m (28 ft 11 in)
fuselage pod: CB	4.30 m (14 ft 1 in)
CBS	4.55 m (14 ft 11 in)
Height to top of main rotor head	3.02 m (9 ft 11 in)
Width over skids: unladen	2.53 m (8 ft 3½ in)
laden	2.58 m (8 ft 5½ in)
Rear loading doors: Height	0.64 m (2 ft 1 in)
Width	1.40 m (4 ft 7 in)

DIMENSIONS, INTERNAL:
Cabin, incl cargo compartment:	
Max width	1.40 m (4 ft 7 in)
Max height	1.25 m (4 ft 1 in)
Volume	4.80 m³ (169 cu ft)
Cargo compartment: Length	1.85 m (6 ft 0¾ in)
Max width	1.20 m (3 ft 11¼ in)
Max height	0.57 m (1 ft 10½ in)
Floor area	2.25 m² (24.2 sq ft)
Volume	1.30 m³ (45.9 cu ft)

AREAS:
Main rotor disc	76.05 m² (818.6 sq ft)
Tail rotor disc	2.835 m² (30.5 sq ft)

WEIGHTS AND LOADINGS:
Weight empty, basic: CB	1,277 kg (2,815 lb)
CBS	1,301 kg (2,868 lb)
Standard fuel (usable)	456 kg (1,005 lb)
Max fuel, incl auxiliary tanks	776 kg (1,710 lb)
Max T-O weight	2,500 kg (5,511 lb)
Max disc loading	32.9 kg/m² (6.74 lb/sq ft)

PERFORMANCE (at max T-O weight):
Never-exceed speed (V_{NE}) at S/L	131 knots (242 km/h; 150 mph)
Max cruising speed at S/L	131 knots (242 km/h; 150 mph)
Best range speed at S/L	110 knots (204 km/h; 127 mph)
Max rate of climb at S/L, max continuous power	444 m (1,457 ft)/min
Vertical rate of climb at S/L, T-O power	90 m (295 ft)/min
Max operating altitude	3,050 m (10,000 ft)
Hovering ceiling, T-O power: IGE	1,525 m (5,000 ft)
OGE	457 m (1,500 ft)
Range with standard fuel and max payload, no reserves:	
at S/L	300 nm (555 km; 345 miles)
at 1,525 m (5,000 ft)	321 nm (596 km; 370 miles)
Ferry range with auxiliary tanks, no reserves:	
at S/L	519 nm (961 km; 597 miles)
at 1,525 m (5,000 ft)	550 nm (1,020 km; 634 miles)
Endurance with standard fuel and max payload, no reserves: at S/L	3 h 24 min

EUROCOPTER EC SUPER FIVE

TYPE: Alternative high-performance version of BO 105 CBS-4.

PROGRAMME: Derived from German Army PAH-1 upgrade programme; announced at Heli-Expo February 1993; certification expected Autumn 1993; deliveries by end 1993.

DESIGN FEATURES: New main rotor blades have parallel-chord DM-H4 aerofoil to 0.8 radius and tapered DM-H3 to tip; rotor lift increased 150 kg (330 lb); airframe vibration reduced to less than 0.1 g; improved stability.

Improvements include Cat. A T-O weight in ISA + 20°C increased by 130 kg (287 lb); OGE hover ceiling at max T-O weight in ISA + 20°C increased by 500 m (1,640 ft); power required to hover OGE at S/L in ISA + 20°C reduced by 26 kW (35 hp); max weight in OGE hover, OEI with 2.5 min power in ISA + 20°C, increased by 240 kg (529 lb); service ceiling OEI at max weight increased by 900 m (2,953 ft).

Eurocopter EC Super Five, to be introduced as alternative to BO 105 in late 1993

FLYING CONTROLS: As BO 105 CBS-4; dual controls optional.

POWER PLANT: As BO 105 CBS-4, but with scavenge oil filter increasing oil change interval to 200 hours; one-handed engine starting arrangement; transmission 30 min rating increased 10 per cent and 2.5 min rating up 15 per cent. Fuel consumption reduced by up to three per cent.

ACCOMMODATION: As BO 105 CBS-4.

SYSTEMS: As BO 105 CBS-4, but with improved hydraulic system.

EQUIPMENT: Windscreen wiper on pilot's side; 250W retractable landing light in nose.

PERFORMANCE (where different from BO 105 CBS-4: at average mission weight 2,300 kg; 5,070 lb):
Max cruising speed at S/L	131 knots (243 km/h; 151 mph)
Rate of climb, max continuous power at S/L	570 m (1,870 ft)/min
Service ceiling, OEI, 30 min power	1,700 m (5,577 ft)
Hovering ceiling, T-O power: IGE	3,200 m (10,500 ft)
OGE	2,430 m (7,970 ft)
Range at S/L	310 nm (574 km; 357 miles)

EUROCOPTER BO 105/PAH-1/VBH/BSH-1

TYPE: Military variants of BO 105.

PROGRAMME: Production of 100 BO 105 M/VBH (Verbindungs und Beobachtungs Hubschrauber) scouts and 212 BO 105 P/PAH-1 (Panzer Abwehr Hubschrauber) anti-tank helicopters completed 1984 (see 1985-86 *Jane's*); fitting of new rotor blades, improved oil cooling and intakes to 209 remaining PAH-1s approved September 1990; of these, 155 to have HOT 2 missile system with lightweight launchers and digital avionics and 54 were planned to become BSH (Begleitschutz Hubschrauber).

CURRENT VERSIONS: **PAH-1 Phase I (PAH-1A1):** Current retrofit from original PAH-1 (see above) for improved day fighting capability; first in service Summer 1991; programme to be completed mid-1994.

BSH: Begleitschutz Hubschrauber escort helicopter; conversion of 54 from PAH-1s cancelled early 1993.

PAH-1 Phase II: Added night firing capability with roof-mounted sight cancelled in budget adjustments January 1993.

CUSTOMERS: German Army, as above; Swedish Army adopted ESCO HeliTOW system.

COSTS: PAH-1 Phase II cost reported as DM630 million in January 1992.

DESIGN FEATURES: New rotor blades and lighter HOT system increase useful load by 180 kg (397 lb).

Data substantially as for BO 105 CB except:

WEIGHTS AND LOADINGS:
Empty weight, without fuel, missiles or crew	1,688 kg (3,721 lb)
German anti-tank mission weight	2,380 kg (5,247 lb)

PERFORMANCE:
Hovering ceiling, OGE, at mission weight	2,100 m (6,890 ft)

Eurocopter BO 105 M, designated Hkp 9 by the Swedish Army *(Peter J. Cooper)*

EUROCOPTER EC 135

TYPE: Former BO 108; five/seven-seat twin-turbine light helicopter.

PROGRAMME: First flight of technology prototype (D-HBOX) powered by two Allison 250-C20R turboshafts 15 October 1988; new all-composite bearingless tail rotor tested during 1990; Eurocopter announced in January 1991 BO 108 was to succeed BO 105; first flight of second prototype (D-HBEC) powered by two Turbomeca TM 319-1B Arrius, 5 June 1991; production main and tail rotors flight tested during 1992 in preparation for certification programme; two more prototypes, one powered by Arrius and the other by P&WC PW206Bs, to fly mid-1994; design revised late 1992 to increase max seating to seven; Advanced Fenestron adopted; VFR certification expected mid-1996; IFR certification mid-1997; deliveries to start mid-1996.

COSTS: Operating cost 25 per cent lower than BO 105; development programme funded by Eurocopter Deutschland and Eurocopter Canada, suppliers and German Ministries of Economics and Research and Technology.

DESIGN FEATURES: Designed to FAR Pt 27 including Category A and European JARs; four-blade FVW bearingless main rotor, as adopted for US Army RAH-66 armed scout helicopter; composites blades mounted on controlled flexibility composites arms giving flap, lag and pitch-change freedom; control demands transmitted from rods to root of blade by rigid CFRP pitch case; main rotor blades have DM-H3 and -4 aerofoils with non-linear twist and tapered transonic tips; main rotor axis tilted forward 5°; airframe drag 30 per cent lower than BO 105 by clean and compact external shape; cabin height retained by shallow transmission; vibration reduced by ARIS mounting between transmission and fuselage; all dynamically loaded components to have 3,000 h MTBR or be maintained on-condition.

Second prototype has EFIS-based IFR system; fuselage stretched 15 cm (5.9 in) and interior cabin width extended by 10 cm (3.9 in); main rotor diameter extended to 10.20 m (33 ft 5½ in); tail rotor replaced in 1992 by 10-blade Advanced Fenestron with blades positioned to minimise noise; fan blade tip speed 185 m (607 ft)/s; max T-O weight increased to 2,500 kg (5,511 lb).

FLYING CONTROLS: Conventional hydraulic fully powered controls with integrated electrical SAS servos; objective is single-pilot IFR with cost-effective stability augmentation.

STRUCTURE: Airframe mainly Kevlar/CFRP sandwich composites, except aluminium alloy sidewalls, pod lower module and cabin floor, tailboom and around cargo area; some titanium in engine bay; composite tailplane.

LANDING GEAR: Skid type, inclined rearward by 1°.

POWER PLANT: First prototype has two 335.5 kW (450 shp) class Allison 250-C20R-3 turboshafts, mounted side by side aft of the main rotor; second prototype has two 360 kW (480 shp) Turbomeca Arrius 1B (TM 319-1B) with full authority digital control; alternative engines will be 342 kW (458 shp) Pratt & Whitney Canada PW206Bs; two separate fan powered oil cooling systems. Fuel in underfloor tanks.

ACCOMMODATION: Pilot, plus four or six passengers on crashproof seats; forward hinged doors for crew; sliding doors for passengers; rear passenger headroom may be increased by further reducing depth of transmission. Rear of pod clamshell doors for bulky items/cargo; flights permissible with clamshell doors removed. Unobstructed cabin interior.

SYSTEMS: Redundant electrical supply systems to FAR Pt 27 standards. Fully redundant dual hydraulic systems.

AVIONICS: IFR capability; provisions for integrated weather radar.

DIMENSIONS, EXTERNAL:
Main rotor diameter	10.20 m (33 ft 5½ in)
Fenestron diameter	1.00 m (3 ft 3¼ in)
Length: overall, rotor turning	11.988 m (39 ft 4 in)
fuselage	10.024 m (32 ft 10½ in)
Height to top of main rotor head	3.234 m (10 ft 7¼ in)
Width of fuselage	1.56 m (5 ft 1½ in)
Width over skids	2.10 m (6 ft 10½ in)

DIMENSIONS, INTERNAL:
Cabin volume (approx)	5.00 m³ (176.6 cu ft)

AREAS:
Main rotor disc	81.71 m² (879.5 sq ft)
Fenestron disc	2.84 m² (30.52 sq ft)

WEIGHTS AND LOADINGS:
Weight empty	1,300 kg (2,866 lb)
Useful load	1,200 kg (2,646 lb)
Max T-O weight	2,500 kg (5,511 lb)

PERFORMANCE (at max T-O weight, at 1,500 m; 4,920 ft):
Max cruising speed	approx 146 knots (270 km/h; 168 mph)
Econ cruising speed	129 knots (240 km/h; 149 mph)
Max rate of climb	582 m (1,910 ft)/min
Hovering ceiling: IGE	4,500 m (14,775 ft)
OGE, ISA	3,700 m (12,150 ft)
OGE, ISA +20°C	2,500 m (8,200 ft)
Range with max fuel	461 nm (855 km; 531 miles)
Max endurance, no reserves	4 h 10 min

Second prototype BO 108, showing installation of two Turbomeca TM 319 Arrius, before redesign and redesignation as EC 135

Artist's impression of definitive Eurocopter EC 135 (formerly BO 108)

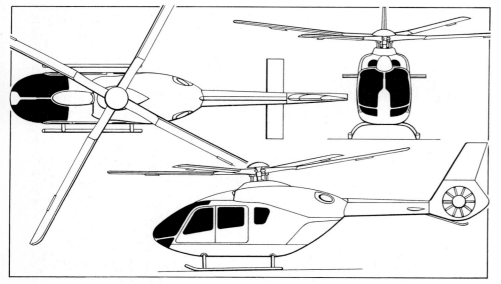

Eurocopter EC 135 seven-seat twin-engined helicopter (*Jane's/Mike Keep*)

EUROCOPTER TIGER/TIGRE/GERFAUT

TYPE: Twin-engined anti-tank and ground support helicopter.

PROGRAMME: Original 1984 French/German MoU to develop common new anti-tank helicopter amended 13 November 1987; FSD approved 8 December 1987; main development contract awarded 30 November 1989, when name Tiger (Germany)/Tigre (France) adopted; five development aircraft planned, including three unarmed aerodynamic prototypes to be used also for core avionics testing (PT1, 2 and 3), one (PT4) in HAP configuration and one (PT5) as PAH-2/HAC prototype; PT1 rolled out 4 February 1991; first flight 27 April 1991.

After 137 hours flying by late 1992 (successively fitted with aerodynamic mockups of mast-mounted and roof-mounted sights, nose-mounted gun and weapon containers), PT1 had demonstrated 170 knots (315 km/h; 195 mph), 3,962 m (13,000 ft), 2.7g and 5,700 kg

(12,566 lb) gross weight; PT2 in Gerfaut configuration and with full core avionics system rolled out at Ottobrunn 9 November 1992; first flight 22 April 1993; PT3 to fly October 1993; PT2 and 3 are avionics testbeds, but will be fitted with Gerfaut and Tiger armament (May 1995 and February 1997 respectively); PT4 (full Gerfaut) and PT5 (full Tiger) expected to fly October 1994 and March 1995; these schedules under pressure from possible budget changes; PT5 may be completed as UHU escort helicopter (see Current Versions).

Joint team at Marignane to flight test basic helicopter, update avionics during trials, and test HAP variant; similar team at Ottobrunn to qualify basic avionics, Euromep mission equipment package, and integrate weapons system.

CURRENT VERSIONS: **HAP Gerfaut**: Hélicoptère d'Appui et de Protection; escort and fire support version for French Army, for delivery from 1997; armed with 30 mm Giat AM-30781 automatic cannon in undernose turret, with 150-450 rds ammunition; four Matra Mistral air-to-air missiles and two pods each with twenty-two 68 mm unguided SNEB rockets delivering armour-piercing darts, mounted on stub-wings, or 12-round rocket pod instead of each pair of Mistrals, making total of 68 rockets; roof mounted TV, FLIR, laser rangefinder and direct view optics sensors; night vision goggles integrated in helmets.

PAH-2 Tiger: Panzer Abwehr Hubschrauber, 2nd generation; variant of common anti-tank version for German Army, for delivery from 1999; underwing pylons for up to eight HOT 2 or four HOT and four Trigat long-range anti-tank missiles (inboard) and four Stinger 2 air-to-air self-defence missiles (outboard); mast mounted TV/FLIR/laser rangefinder sighting system for gunner; nose-mounted FLIR night vision device for piloting.

UHU: Unterstützungs Hubschrauber; German MoD was considering in late 1992 reducing the number of Tiger anti-tank helicopters or broadening their capability to include close support or purchasing part of fleet in a specialised close support version called UHU.

HAC Tigre: Hélicoptère Anti-Char; anti-tank variant for French Army, for delivery from 1998; wing pylons for up to eight HOT 2 or Trigat missiles (or four HOT 2 and four Trigat) inboard, four Matra Mistral air-to-air missiles outboard; mast-mounted sight and pilot FLIR system similar to that of PAH-2.

CUSTOMERS: Estimated 427 required initially (France 75 HAP and 140 HAC, Germany 212 PAH-2); DM2 billion cut from Tiger programme January 1993; escort capability or version may be substituted for some PAH-2s; also under consideration by Spain; Eurocopter teamed with BAe Defence for UK Army competition for 91 to be ordered May 1995; further exports of 300 Tigers thought possible.

COSTS: Tiger/Gerfaut current development cost reported DM2.2 billion.

DESIGN FEATURES: FEL (fibre elastomer) main rotor designed for simplicity, manoeuvrability and damage tolerance; has infinite life except for inspection of elastomeric elements at more than 2,500 h intervals; hub consists of titanium centrepiece (including duct for mast-mounted sight) with composites starplates bolted above and below; flap and lead/lag motions of blades allowed by elastic bending of neck region and pitch change by elastic part of elastomeric bearings; lead/lag damping by solid-state visco-elastic damper struts faired into trailing-edge of each blade root; equivalent flapping hinge offset of 10.5 per cent gives high control power; whole main rotor contains 24 parts, not including standard bolts and bushings; blades have new DMH series aerofoils and replaceable parabolic anhedral tips for noise suppression; three-blade Spheriflex tail rotor has composite blades with OA series aerofoil and fork roots; built-in ram air engine exhaust suppressors.

FLYING CONTROLS: All versions have pilot in front, gunner in rear with full dual controls; both crew members can perform all tasks and weapon operation except anti-tank missile firing; each crew member has two colour multi-function displays and one central control and display unit; autopilot is part of basic avionics system (see under Avionics heading); fully powered hydraulic flying controls by SAMM/Liebherr; Labinal/Electrométal servo trim.

STRUCTURE: 80 per cent CFRP, block and sandwich and Kevlar sandwich; six per cent titanium and 11 per cent aluminium; airframe structure protected against lightning and EMP by embedded copper/bronze grid and copper bonding foil; stub-wings of aluminium spars with CFRP ribs and skins; titanium engine deck may be replaced by GFRP; airframe tolerates crash impacts at 10.5 m/s (32.8 ft/s) and meets MIL-STD-1290 crashworthiness standards; titanium main rotor hub centrepiece and tail rotor Spheriflex integral hub/mast; blade spars filament-wound; GFRP, CFRP skins and subsidiary spars and foam filling. Eurocopter France responsible for transmission, tail rotor, centre-fuselage (incl engine installation), aerodynamics, fuel and electrical systems, weight control, maintainability, reliability and survivability; Eurocopter Deutschland for main rotor, flight control and hydraulic systems, front and rear fuselage (incl cockpits), prototype assembly, flight characteristics and performance, stress and vibration testing and simulation.

LANDING GEAR: Non-retractable tailwheel type, with single wheel on each unit. Designed to absorb impacts of up to 6 m (20 ft)/s. Main gear by Messier-Bugatti, tail gear by Liebherr Aerotechnik.

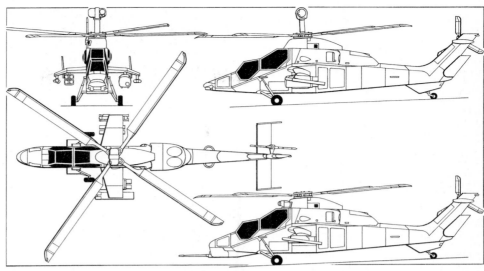

Eurocopter PAH-2 Tiger/HAC Tigre anti-tank helicopter, with additional side view (bottom) of the HAP Gerfaut escort and support version (*Jane's/Mike Keep*)

Eurocopter Tiger PT1 with dummy Osiris mast-mounted sight, Trigat and Mistral missiles on pylons and chin-mounted PVS sight

Eurocopter Tiger prototype PT1 with dummy roof-mounted STRIX sight, 30 mm gun in nose and rockets and Mistral missiles on pylons

POWER PLANT: Two MTU/Rolls-Royce/Turbomeca MTR 390 turboshafts, mounted side by side above centre fuselage (engine first flown in Panther testbed 14 February 1991); power ratings are max T-O 958 kW (1,285 shp), super emergency 1,160 kW (1,556 shp), max continuous 873 kW (1,171 shp); self-sealing crashworthy fuel tanks, with explosion suppression; total capacity 1,360 litres (359 US gallons; 299 Imp gallons).

ACCOMMODATION: Crew of two in tandem, with pilot in front and weapons system operator at rear; armoured, impact-absorbing seats; stepped cockpits, with flat-plate transparencies.

SYSTEMS: Redundant hydraulic, electrical and fuel systems.

AVIONICS: *Basic or core avionics* common to all three versions include bus/display system, com radio (French and German systems vary), autonomous nav system and radio/Doppler navaids, electronic countermeasures (including laser warning) and automatic flight control system, all connected to and controlled through redundant MIL-STD-1553B data highway. Navigation system, by Sextant Avionique, Teldix and DASA, is fully redundant; system contains two Sextant PIXYZ three-axis ring laser gyro units, two air data computers, two magnetic sensors, one Teldix/Canadian Marconi CMA 2012 Doppler radar, a radio altimeter and GPS; laser units occupy 3 MCU volume (3/8 ATR) each and weigh 5.5 kg (12.1 lb); these sensors also provide signals for flight control, information display and guidance; integrated duplex AFCS by Sextant and Nord Micro; AFCS computers produced by Sextant Avionique, VDO-Luft and Litef; colour liquid crystal flight displays showing symbology and imagery (two per cockpit for flight and weapon/systems information) by Sextant and VDO-Luft; each crewman has central control/display unit for inputting all radio, electronic systems and navigation selections; digital map display system by Dornier and VDO-Luft; engine and systems data are fed into the data-bus for in-flight indication and subsequent maintenance analysis.

Euromep (European mission equipment package) includes Aerospatiale/SATEL/THORN EMI/Eltro pilot vision subsystem (PVS), air-to-air subsystem (Stinger or Mistral), mast-mounted sight and missile subsystem and Euromep management system all connected to separate MIL-STD-1553B data highway; PVS has 40° field of view thermal imaging sensor steered by helmet position detector giving both crewmen day/night/bad weather vision, flight symbology and air-to-air aiming in helmet-mounted display; mast-mounted sight, gunner sight electronics and gunner's head-in target acquisition display and ATGW 3 subsystem connected by separate data highway; HOT 2 missile system also available. GEC-Marconi Avionics developing Knighthelm integrated helmet-mounted display for symbology and imagery for German Tigers.

Gerfaut combat support mission equipment package includes chin-mounted 30 mm gun, four Mistrals or rocket pods, SFIM/TRT STRIX gyro-stabilised roof-mounted sight above rear cockpit, helmet-mounted sights for both crewmen, Sextant Avionique HUD for pilot, armament control panel and fire control computer; roof-mounted sight includes direct-view optics with folding sight tube, television and IR channels and laser ranger/designator.

ARMAMENT: As listed under Current Versions.

DIMENSIONS, EXTERNAL:
Main rotor diameter	13.00 m (42 ft 7¾ in)
Tail rotor diameter	2.70 m (8 ft 10¼ in)
Length of fuselage	14.00 m (45 ft 11¼ in)
Height to top of rotor head	3.81 m (12 ft 6 in)
Height to top of tail rotor disc	4.32 m (14 ft 2 in)
Wheel track	2.40 m (7 ft 10½ in)
Wheelbase	7.65 m (25 ft 1 in)

AREAS:
Main rotor disc	132.7 m² (1,428.7 sq ft)
Tail rotor disc	5.72 m² (61.63 sq ft)

WEIGHTS AND LOADINGS:
Basic weight empty	3,300 kg (7,275 lb)
Mission T-O weight	5,300-5,800 kg (11,685-12,787 lb)
Max overload T-O weight	6,000 kg (13,227 lb)
Main rotor disc loading (max mission T-O weight)	43.70 kg/m² (8.95 lb/sq ft)
Power loading (max mission T-O weight, T-O power)	3.0 kg/kW (4.97 lb/hp)

PERFORMANCE (estimated, at AUW of 5,400 kg; 11,905 lb):
Cruising speed	135-151 knots (250-280 km/h; 155-174 mph)
Max rate of climb at S/L	more than 600 m (1,970 ft)/min
Hovering ceiling OGE	more than 2,000 m (6,560 ft)
Endurance, incl 20 min reserves	2 h 50 min

EUROCOPTER/CATIC/SA

PARTICIPATING COMPANIES:
Eurocopter: see this section
CATIC: see under China
Singapore Aerospace: see under Singapore

EC 120

TYPE: Formerly P120L; five-seat light helicopter.

PROGRAMME: Definition phase of original P120L launched 15 February 1990 as partnership of Eurocopter (programme leader, 61 per cent), CATIC (24 per cent) and Singapore Aerospace (15 per cent); redesigned with 500 kg (1,102 lb) lower gross weight and new engine and rotor and redesignated EC 120 January 1993; first flight expected early 1995; first deliveries in second half 1997. Eurocopter responsible for rotor system/transmission/final assembly/flight test/certification, CATIC (through HAMC, which see) for fuselage, landing gear and fuel system, and SA for tailboom, fin structure and doors; final assembly in France or Germany.

CUSTOMERS: Estimated sales about 900 in first decade and 1,600 to 2,000 total.

COSTS: Reported unit price FFr3.6 million.

DESIGN FEATURES: Three-blade main rotor on Spheriflex hub integrated with main shaft and transmission; two-stage reduction gear; eight-blade Fenestron.

STRUCTURE: Composites for main and tail rotor blades; Spheriflex head and shaft made as single composites assembly; metal centre fuselage; composites landing gear.

POWER PLANT: Not selected by early 1993, but could be Turbomeca TM 319 Arrius 1B1 or P&WC PW206B free turbine; power required 373 kW (500 shp). Fuel capacity 400 litres (106 US gallons; 88 Imp gallons).

ACCOMMODATION: Pilot and four passengers.

DIMENSIONS, EXTERNAL:
Main rotor diameter	10.20 m (33 ft 5½ in)
Main rotor blade chord	0.26 m (10¼ in)
Fenestron diameter	0.75 m (2 ft 5½ in)
Fenestron blade chord	0.058 m (2¼ in)
Length overall, blades folded	11.54 m (37 ft 10¼ in)
Width, including tailplane	2.40 m (7 ft 10½ in)
Height overall	3.27 m (10 ft 8¾ in)
Skid track	1.80 m (5 ft 10¾ in)

DIMENSIONS, INTERNAL:
Cabin: Length	1.80 m (5 ft 10¾ in)
Max width	1.50 m (4 ft 11 in)
Max height	1.31 m (4 ft 3½ in)

AREAS:
Main rotor disc	81.72 m² (879.3 sq ft)
Fenestron disc	0.44 m² (4.73 sq ft)

WEIGHTS AND LOADINGS:
Standard empty weight	850 kg (1,874 lb)
Max sling load	750 kg (1,653 lb)
Max useful load	700 kg (1,543 lb)
Max T-O weight	1,550 kg (3,417 lb)

PERFORMANCE:
Max cruising speed	130 knots (240 km/h; 149 mph)
Hovering ceiling OGE: ISA	2,400 m (7,875 ft)
ISA + 20°C	1,200 m (3,940 ft)

Artist's impression of EC 120 (redesigned P120L) light helicopter under development by Eurocopter, CATIC and Singapore Aerospace

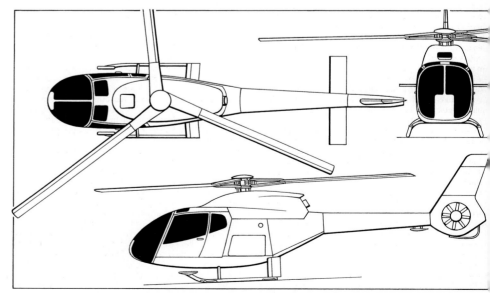

Eurocopter EC 120 single-engined five-seat light helicopter (*Jane's*/Mike Keep)

EUROCOPTER/KAWASAKI

PARTICIPATING COMPANIES:
Eurocopter: see this section
Kawasaki: see under Japan

EUROCOPTER/KAWASAKI BK 117

TYPE: Twin-turboshaft multi-purpose helicopter.
PROGRAMME: Developed jointly under agreement of 25 February 1977; four prototypes, first flight 13 June 1979; one pre-production aircraft, first flight 6 March 1981; first flights of production aircraft 24 December 1981 (JQ1001 in Japan) and 23 April 1982 (in Germany); certificated in Germany and Japan 9 and 17 December 1982 respectively, followed by US FAA 29 March 1983 (FAR Pt 29, Categories A and B, incl Amendments 29-1 to 29-16); deliveries began early 1983. See 1991-92 and previous editions for earlier A series and B-1.
CURRENT VERSIONS: **BK 117 B-2:** Production model from 1992 (with BK 117 C-1 below); certificated January 1992; max T-O weight increased to 3,350 kg (7,385 lb); payload increased by 150 kg (330 lb). *Details below apply to this version.*

BK 117 C-1: German version with new cockpit and Turbomeca Arriel 1E engines; first flight (F-WMBB) 6 April 1990; FAA certification received 7 December 1992. Performance similar to that of BK 117 B-2; payload increased 150 kg (331 lb); better hot and high performance; first deliveries end 1992.

NBK-117: Designation of aircraft licensed to be built by IPTN (see Indonesian section, 1991-92 *Jane's*) under November 1982 agreement with MBB. Only four built.

All-composites testbed: One aircraft built by MBB for 3½-year German MoD research programme; 80 per cent of airframe in CFRP and 20 per cent in AFRP (Kevlar). First flight 27 April 1989; flight test programme completed July 1989.

BK 117 P5: Fly-by-wire control system first flown in Japan (JQ0003 P5) 2 October 1992; system has triplex sensors of control input and flight, triplex computing and triple powered control unit valves but simplex pitch linkage; full authority signals automatically set tail rotor for control demand, airspeed, altitude and other factors. System is aimed at future Japanese OH-X observation helicopter.
CUSTOMERS: Total of 323 delivered by MBB/Eurocopter by 1 June 1992; Kawasaki total was 90 by 1 June 1992. Kawasaki agreed 1990 to supply CKD kits (about 30 over five-year period) for local assembly in South Korea by Hyundai Precision Industry; five kits delivered by end 1992, with four more due to follow by April 1993. BK 117 delivered by Kawasaki to Sendai Fire Department November 1992 has GPS/digital map system indicating helicopter's three-dimensional position, flight direction and drift on map in colour liquid crystal display; all terrain higher than helicopter automatically shown red.
DESIGN FEATURES: System Bölkow four-blade main rotor head, almost identical to that of BO 105; main rotor blades similar to but larger than those of BO 105, with NACA 23012/23010 (modified) section; optional two-blade folding. Two-blade semi-rigid tail rotor with MBB-S102E performance/noise optimised blade section; rotor rpm 383 (main), 2,169 (tail).
FLYING CONTROLS: Equipped as standard for single-pilot VFR operation; dual controls and dual VFR instrumentation optional; rotor brake and yaw CSAS standard on German built models, optional on Kawasaki aircraft; options include IFR instrumentation, two-axis (pitch/roll) CSAS and Honeywell SPZ-7100 dual digital AFCS. Mast moment indicator discourages excessive cyclic control inputs.
STRUCTURE: Main rotor has one-piece titanium hub with pitch-change bearings; fail-safe GFRP blades with stainless steel anti-erosion strip. Tail rotor, mounted on port side of central fin, has GFRP blades of high impact resistance. Main fuselage pod and tailboom are aluminium alloy with single-curvature sheets and (on fuselage) bonded aluminium sandwich panels; secondary fuselage components are compound curvature shells of Kevlar sandwich. Engine deck, to which tailboom is integrally attached, forms cargo compartment roof and is of titanium adjacent to engine bays. Detachable tailcone carries main fin/tail rotor support, and horizontal stabiliser with offset endplate fins. MBB responsible for rotor systems, tailboom, tail unit, skid landing gear, hydraulic system, engine firewall and cowlings, powered controls and systems integration; Kawasaki for fuselage, transmission, fuel and electrical systems, and standard equipment. Components single-sourced and exchanged for separate assembly lines at Donauwörth and Gifu; some components and accessories interchangeable with those of BO 105 (which see), from which hydraulic powered control system is also adapted.
LANDING GEAR: Non-retractable tubular skid type, of aluminium construction. Skids are detachable from crosstubes. Ground handling wheels standard. Emergency flotation gear, settling protectors and snow skids available optionally.
POWER PLANT: BK 117 B-2 has two Textron Lycoming LTS 101-750B-1 turboshafts, each rated at 442 kW (592 shp) for 30 min for take-off and 410 kW (550 shp) max continuous power. BK 117 C-1 has two Turbomeca Arriel 1E turboshafts each rated at 528 kW (708 shp) for take-off,

Eurocopter/Kawasaki BK 117 B-2 (two Textron Lycoming LTS 101) equipped for rescue

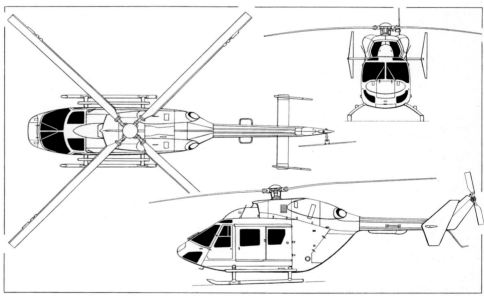

Eurocopter/Kawasaki BK 117 B-2 twin-turboshaft multi-purpose helicopter *(Dennis Punnett)*

516 kW (692 shp) max continuous and 560 kW (750 shp) for 2.5 min OEI.

Kawasaki KB 03 main transmission rated at 736 kW (986 shp) for twin-engine T-O, 632 kW (848 shp) max continuous; for single-engine operation 442 kW (592 shp) for 2½ min, 404 kW (542 shp) for 30 min and 368 kW (493 shp) max continuous.

Fuel in four flexible bladder tanks (forward and aft main tanks, with two supply tanks between), in compartments under cabin floor. Two independent fuel feed systems for engines and common main fuel tank. Total standard fuel capacity 697 litres (184 US gallons; 153 Imp gallons); a 200 litre (53 US gallon; 44 Imp gallon) auxiliary tank raises total capacity to 897 litres (237 US gallons; 197 Imp gallons).
ACCOMMODATION: Pilot and up to six (executive version), seven (Eurocopter standard version) or nine passengers (Kawasaki standard version). High-density layouts available for up to 10 passengers in addition to pilot. Level floor throughout cockpit, cabin and cargo compartment. Jettisonable forward hinged door on each side of cockpit, pilot's door having an openable window. Jettisonable rearward sliding passenger door on each side of cabin, lockable in open position. Fixed steps on each side. Two hinged, clamshell doors at rear of cabin, providing straight-in access to cargo compartment. Rear cabin window on each side. Aircraft can be equipped for offshore, medical evacuation (one or two stretchers side by side and up to six attendants), firefighting, search and rescue, law enforcement, cargo transport or other operations.
SYSTEMS: Ram air and electrical ventilation system. Fully redundant tandem hydraulic boost system (one operating and one standby), pressure 103.5 bars (1,500 lb/sq in), for flight controls. System flow rate 8.1 litres (2.14 US gallons; 1.78 Imp gallons)/min. Bootstrap/oil reservoir, pressure 1.7 bars (25 lb/sq in). Main DC electrical power from two 150A 28V starter/generators (one on each engine) and a 24V 25Ah nickel-cadmium battery. AC power provided by inverter; second AC inverter optional; emergency busbar provides direct battery power to essential services, external DC power receptacle.
AVIONICS: Basic instrumentation for single-pilot VFR operation includes airspeed indicator with electrically heated pitot tube and static ports, encoding altimeter, instantaneous vertical speed indicator, 4 in artificial horizon, 3 in standby artificial horizon, gyro magnetic heading system, HSI, magnetic compass, ambient air thermometer, and clock. (The 4 in and 3 in artificial horizons and HSI are optional on Kawasaki built aircraft.) Com/nav and other avionics to customer's requirements, including VHF-AM/FM, UHF and HF radios and full IFR packages with long-range navaid and multi-mode radar.
EQUIPMENT: Standard basic equipment includes rotor brake (optional on Kawasaki aircraft), annunciator panel, master caution light, rotor rpm/engine fail warning control unit, fuel quantity indicator and low level sensor, outside air temperature indicator, engine and transmission oil pressure and temperature indicators, two exhaust temperature indicators, dual torque indicator, triple tachometer, two N1 tachometers, full internal and external lighting, ground handling wheels, pilot's and co-pilot's windscreen wipers, floor covering, interior panelling and sound insulation, ashtrays, map/document case, tiedown rings in cabin and cargo compartment, engine compartment fire warning

160 INTERNATIONAL: AIRCRAFT—EUROCOPTER/KAWASAKI / EUROFAR

indicator, engine fire extinguishing system, portable fire extinguisher, first aid kit, and single colour exterior paint scheme. Optional equipment includes high-density seating arrangement, bleed air heating system, long-range fuel tank, emergency flotation gear, settling protectors, snow skids, main rotor blade folding kit, dual pilot operation kit, stretcher installation, external cargo hook, rescue hoist, SX 16 remotely controlled searchlight, external loudspeaker, sand filter and kits for rescue, law enforcement and VIP transport.

DIMENSIONS, EXTERNAL:
Main rotor diameter	11.00 m (36 ft 1 in)
Tail rotor diameter	1.956 m (6 ft 5 in)
Main rotor blade chord	0.32 m (1 ft 0½ in)
Length: overall, both rotors turning	13.00 m (42 ft 8 in)
fuselage, tail rotor blades vertical	9.91 m (32 ft 6¼ in)
Fuselage: Max width	1.60 m (5 ft 3 in)
Height: overall, both rotors turning	3.85 m (12 ft 7½ in)
to top of main rotor head	3.36 m (11 ft 0¼ in)
Tailplane span (over endplate fins)	2.70 m (8 ft 10¼ in)
Tail rotor ground clearance	1.90 m (6 ft 2¾ in)
Width over skids	2.50 m (8 ft 2½ in)

DIMENSIONS, INTERNAL:
Combined cabin and cargo compartment:
Max length	3.02 m (9 ft 11 in)
Width: max	1.49 m (4 ft 10½ in)
min	1.21 m (3 ft 11½ in)
Height: max	1.28 m (4 ft 2½ in)
min	0.99 m (3 ft 3 in)
Useful floor area	3.70 m² (39.83 sq ft)
Volume	5.00 m³ (176.6 cu ft)

AREAS:
Main rotor blades (each)	1.76 m² (18.94 sq ft)
Tail rotor blades (each)	0.0975 m² (1.05 sq ft)
Main rotor disc	95.03 m² (1,022.9 sq ft)
Tail rotor disc	3.00 m² (32.24 sq ft)

WEIGHTS AND LOADINGS:
Basic weight empty	1,727 kg (3,807 lb)
Fuel: standard usable	558 kg (1,230 lb)
incl auxiliary tank	718 kg (1,583 lb)
Max T-O weight, internal and external payload	3,350 kg (7,385 lb)
Max disc loading	35.25 kg/m² (7.22 lb/sq ft)
Max power loading	4.55 kg/kW (7.48 lb/shp)

PERFORMANCE (ISA; A at gross weight of 3,000 kg; 6,614 lb, B at 3,200 kg; 7,055 lb, C at 3,350 kg; 7,385 lb):
Never-exceed speed (VNE) at S/L:
A, B, C	150 knots (278 km/h; 172 mph)

Max cruising speed at S/L:
A	135 knots (250 km/h; 155 mph)
B	134 knots (248 km/h; 154 mph)
C	133 knots (247 km/h; 153 mph)

Max forward rate of climb at S/L:
A	660 m (2,165 ft)/min
B	582 m (1,910 ft)/min
C	540 m (1,770 ft)/min

Max certificated operating altitude:
A	4,575 m (15,000 ft)
B, C	3,050 m (10,000 ft)

Service ceiling, OEI, 46 m (150 ft)/min climb:
A	2,440 m (8,000 ft)
B	1,770 m (5,800 ft)
C, 30 m (110 ft)/min climb	1,280 m (4,200 ft)

Hovering ceiling IGE (zero wind):
A	3,565 m (11,700 ft)
B	2,925 m (9,600 ft)
C	2,500 m (8,200 ft)

Hovering ceiling IGE (17 knot; 32 km/h; 20 mph cross wind):
A	2,195 m (7,200 ft)
B	1,495 m (4,900 ft)
C	1,040 m (3,400 ft)

Hovering ceiling OGE:
A	2,955 m (9,700 ft)
B	2,285 m (7,500 ft)
C	1,280 m (4,200 ft)

Range at S/L with standard fuel, no reserves
292 nm (541 km; 335 miles)

EUROCOPTER/MIL/KAZAN/KLIMOV Mi-38

TYPE: Medium transport helicopter.
PROGRAMME: Agreement signed in Moscow 18 December 1992 between Eurocopter, Mil Moscow Helicopter Plant, Kazan Helicopter Production Plant (KVPO) and Klimov Corporation covering development and production of 30-passenger Mi-38 helicopter. Joint venture company to be formed to co-ordinate activities and arrange various certifications. Eurocopter plans to invest about $100 million in programme. First deliveries expected 1999.

Mil (see Russian section for Mi-38 description) will handle development of helicopter; Eurocopter will lead flight deck, avionic system and passenger accommodation and will be responsible for preparation for export from Russia; Klimov will be in charge of engine development, industrialisation and production; Kazan will manufacture helicopter and prepare it for domestic market.

EUROFAR
EUROPEAN FUTURE ADVANCED ROTORCRAFT

PARTICIPATING COMPANIES:
Aerospatiale: see under France
Eurocopter France: see this section
Eurocopter Deutschland: see this section
Westland: see under UK

EUROFAR

TYPE: Twin-turboshaft tilt-rotor transport.
PROGRAMME: Three year Phase 1 feasibility study completed end of 1991, including definition of a baseline aircraft; five-year Phase 2 now under way includes preparation for demonstrator. CASA and Agusta withdrew from Eurofar Phase 2. First flight end of decade; flight testing completed last quarter 2002; production start 2004; first flight production aircraft 2006; certification possibly 2009. Shares in programme are Aerospatiale/Eurocopter France 46 per cent, Eurocopter Deutschland 22 per cent and Westland 32 per cent.
CUSTOMERS: 93 per cent of sales expected in oil industry support, 7 per cent in regional airlines; geographic distribution, 40 per cent in USA, 24 per cent in Europe, 21 per cent in Asia and 15 per cent elsewhere.
COSTS: Phase 2 expenditure ECU9 million, 50 per cent from governments through Eureka. Phase 1 cost ECU32 million.
DESIGN FEATURES: 30-passenger airliner; high forward-swept wing with partially tilting nacelles (stationary engines); four-blade rotors; T-tail; cylindrical pressurised fuselage with APU; tricycle landing gear.
FLYING CONTROLS: Quadruplex fly-by-light electronic controls; automatic transition control.
STRUCTURE: CFRP/GFRP fuselage, wing and tail.
POWER PLANT: Two 3,200 kW (4,290 shp) max continuous power class turboshafts (modified PW300 foreseen).
ACCOMMODATION: 30 passengers, three-abreast seating with overhead stowage, toilet, galley. Two-pilot crew.

DIMENSIONS, EXTERNAL:
Wing span between rotor centres	14.66 m (48 ft 1¼ in)
Rotor diameter, each	11.21 m (36 ft 9¼ in)
Length overall	20.41 m (66 ft 11½ in)
Fuselage: Length	19.40 m (63 ft 7¾ in)
Diameter	2.48 m (8 ft 1½ in)
Height overall	6.645 m (21 ft 9½ in)

DIMENSIONS, INTERNAL:
Max height in aisle	1.83 m (6 ft 0 in)
Max width at shoulder level seated	2.23 m (7 ft 3¾ in)
Width of aisle	0.46 m (1 ft 6 in)
Seat width between armrests	0.43 m (1 ft 5 in)
Seat pitch	0.83 m (2 ft 8¾ in)

WEIGHTS AND LOADINGS:
Max vertical T-O weight (Category A)	13,650 kg (30,093 lb)
Power/weight ratio (nominal)	2.13 kg/kW (6.57 lb/shp)

PERFORMANCE:
Cruising speed	335 knots (621 km/h; 385 mph)
Rate of climb	660 m (2,165 ft)/min
Hovering ceiling, both engines, OGE	3,050 m (10,000 ft)
Ceiling, OEI	1,250 m (4,100 ft)
Range	600 nm (1,112 km; 690 miles)

Eurofar civil tilt-rotor aircraft for the next century, as depicted by CATIA computer-aided design system

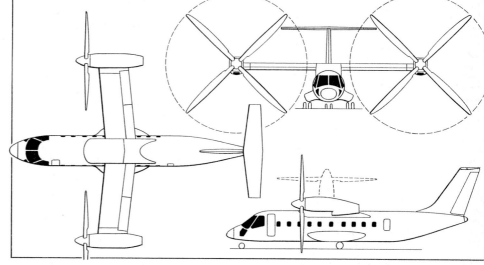

Possible configuration of Eurofar European tilt-rotor for the next century *(Jane's/Mike Keep)*

EUROFIGHTER
EUROFIGHTER JAGDFLUGZEUG GmbH
Arabellastrasse 16 (PO Box 860366), D-8000 Munich 81, Germany
Telephone: 49 (89) 92803-1
Fax: 49 (89) 92803-443
Telex: 5213908 or 5213744
CHAIRMAN: Dr Ing R. Mannu
MANAGING DIRECTOR: John Vincent
MARKETING: A. French
PUBLICITY EXECUTIVE: Ursula Kruse

Eurofighter GmbH formed to manage EFA (European Fighter Aircraft) programme June 1986, followed shortly after by Eurojet Turbo GmbH to manage engine programme; NEFMA (NATO European Fighter Management Agency) supervises EFA programme.

Merger planned early 1992 of Eurofighter and Panavia (which see), but not completed.

EUROFIGHTER 2000 (EFA)
TYPE: Single-seat, highly agile STOL-capable fighter, optimised for air defence/air superiority; secondary capability for ground attack.
PROGRAMME: Outline staff target for common combat aircraft issued December 1983 by air chiefs of staff of France, Germany, Italy, Spain and UK; initial feasibility study launched July 1984; France withdrew July 1985, shareholdings then being readjusted to 33 per cent each to UK and Germany, 21 per cent Italy and 13 per cent Spain; project definition phase completed September 1986; definitive ESR-D (European Staff Requirement–Development) issued September 1987, giving military requirements in greater detail; definition refinement and risk reduction stage completed December 1987; main engine and weapons system development contracts signed 23 November 1988.

Programme halted 1992 by German demands for substantial cost reduction and studies of alternative proposals, which submitted in October 1992; Italy and Spain froze EFA work mid-October. Seven possible alternative configurations for New EFA (NEFA) offered to Germany, being permutations of single (three types) or twin engines; canards; and cranked wing. Only two of seven cheaper than EFA — both inferior to developments of MiG-29 and Su-27. Defence ministers' conference of 10 December 1992 re-launched aircraft as Eurofighter 2000, delaying service entry by three years, to 2000, and allowing Germany to incorporate off-shelf avionics (probably AN/APG-65 radar), lower standard of defensive aids and other deletions to effect 30 per cent price cut. Production commitment due by Italy, Spain and UK in 1995 and by Germany in 1996; German service entry planned in 2002, but programme under-funding in 1993 possibly to result in further delay. Intended four production lines likely to be reduced.

Planned eight development aircraft (no prototypes apart from BAe EAP — see 1991-92 and earlier *Jane's*) reduced to seven (DA1-7) early 1991 coincident with 11 per cent cut in intended flight test programme to 4,500 hours. DA1 (DASA-built at Ottobrunn; airframe No. 01; Luftwaffe serial number 9829) by road to Manching 11 May 1992, for first flight (still awaited June 1993), which is to be followed by some 20 sorties before transfer to Warton for handling and envelope expansion trials wearing UK serial number ZH586; DA2 (BAe at Warton; airframe No. 02; ZH588) first engine run 30 August 1992, and assigned to envelope expansion; DA3 (Alenia at Turin/Caselle; airframe No. 04) first with EJ200 power plants for engine trials (originally scheduled from March 1993) and gun/weapon release trials; DA4 (BAe Warton; airframe No. 03; ZH590) first two-seat and first with full avionics, including ECR 90 radar, IRST and DASS; DA5 (DASA) avionics and weapons trials; DA6 (CASA at Seville) second two-seat; and DA7 (Alenia) all originally to have flown before end 1994. Further eight ground testing part-airframes. Defensive aids subsystem contract awarded to Euro-DASS 13 March 1992, but Germany and Spain declined to participate; will develop own equivalent systems. ECR 90 radar first flew in nose of modified BAe One-Eleven testbed (ZE433) at Bedford, 8 January 1993.
CURRENT VERSIONS: **Single-seater:** Standard version.
Two-seater: Combat-capable conversion trainer.
CUSTOMERS: Originally declared requirements for 765 (UK and Germany 250 each; Italy 165 and Spain 100); believed reduced by 1991 to some 630: Germany, 175 for eight squadrons; Italy, 130 including five squadrons of 14 each, plus OCU; Spain, 72; and UK unchanged. All require some two-seat EFAs; export orders also anticipated.
COSTS: £25-26.5 million, UK, 1992 unit cost; DM127 million, Germany early 1992, 10-year system price; reduced to DM89 million by late 1992 economies.
DESIGN FEATURES: Collaborative design by BAe, DASA, Alenia and CASA, incorporating some design and technology (incl low detectability) from BAe EAP programme; low wing, low aspect ratio tail-less delta with 53° leading-edge sweepback; underfuselage box with side by side engine air intakes, each with fixed upper wedge/ramp and vari-cowl (variable position lower cowl lip) with Dowty actuators.

GEC-Marconi ECR 90 radar for Eurofighter 2000 installed in BAe One-Eleven testbed

Internal arrangement of wing structure, airbrake, avionics bay and refuelling boom of the Eurofighter 2000

FLYING CONTROLS: Two-segment automatic slats on wing leading-edges, inboard and outboard flaperons on trailing-edges; all-moving foreplanes below windscreen; rudder; hydraulically actuated airbrake aft of canopy, forming part of dorsal spine; Liebherr primary flight control actuators. Full-authority quadruplex ACT (active control technology) digital fly-by-wire flight control system (team leader DASA) combines with mission adaptive configuring and aircraft's instability in pitch to provide required 'carefree' handling, gust alleviation and high sustained manoeuvrability throughout flight envelope; pitch and roll control via foreplane/flaperon ACT to provide artificial longitudinal stability; yaw control via rudder; no manual reversion. STANAG 3838 NATO standard databus.
STRUCTURE: Fuselage, wings (incl inboard flaperons), fin and rudder mainly of CFC (carbonfibre composites) except for foreplanes, outboard flaperons and exhaust nozzles (titanium); nose radome and fin-tip (GFRP); leading-edge slats, wingtip pods, fin leading-edge, rudder trailing-edge and major fairings (aluminium-lithium alloy); and canopy surround (magnesium alloy); manufacture includes such advanced techniques as superplastic forming and diffusion bonding; CASA-led joint structures team. BAe responsible for front fuselage, foreplanes, starboard leading-edge slats and flaperons; DASA the centre-fuselage, fin and rudder; Alenia the port wing, incl all movable surfaces; Alenia/CASA the rear fuselage; and CASA/BAe the starboard wing; no duplication of tooling; final assembly line at each manufacturer's facility now in doubt.

Most subsystems developed by multi-nation teams; in following description, for clarity, team leaders only are named:
LANDING GEAR: Dowty Aerospace retractable tricycle type. Single-wheel main units retract inward into fuselage, steerable nosewheel unit forward.
POWER PLANT: First two development aircraft each powered

162 INTERNATIONAL: AIRCRAFT—EUROFIGHTER/EUROFLAG

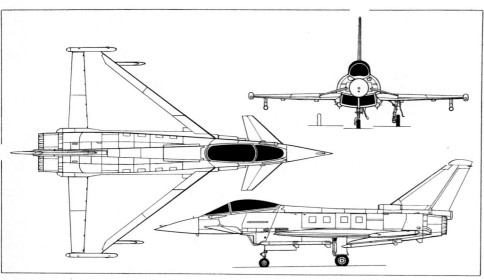

Eurofighter 2000 (European Fighter Aircraft), initially powered by two Turbo-Union RB199-122s and later by two Eurojet EJ200s *(Dennis Punnett)*

Mockup of RAF Eurofighter 2000 in air defence configuration *(Paul Jackson)*

Eurofighter 2000 DA2 prototype assembled by British Aerospace at Warton

by two Turbo-Union RB199-122 afterburning turbofan (each more than 71.2 kN; 16,000 lb st). DA3-DA7, an production aircraft, will have two Eurojet EJ200 advance technology turbofans (each of approx 60 kN; 13,490 lb dry and 90 kN; 20,250 lb nominal thrust with afterburn ing), mounted side by side in rear fuselage with ventra intakes. Staged EJ200 improvements available (but n funded) to 103 kN (23,155 lb st) and 117 kN (26,300 lb st DASA digital engine control system. Lucas Aerospace fue management system. Internal fuel capacity classified. Pro vision for in-flight refuelling and up to three external fue tanks: two 1,000 litre (264 US gallon; 220 Imp gallon) an one 1,500 litre (396 US gallon; 330 Imp gallon) carrie simultaneously.

ACCOMMODATION: Pilot only, on Martin-Baker Mk 16A zero zero ejection seat. Smiths Industries glareshields.

SYSTEMS: Normalair-Garrett environmental control system Magnaghi hydraulic system. Lucas Aerospace electrica system, with GEC-Marconi/Bendix variable speed con stant frequency generator and GEC-Marconi transforme rectifier units. Alenia-led utilities control system (UCS controlled by microcomputer. Garrett APU for engin starting, systems running and NBC filtering. Microturb UK air turbine starter motor.

AVIONICS: BAe has overall team leadership for avionic development and integration. Primary sensor will be GEC Marconi ECR 90 multi-mode pulse Doppler radar; second ary is Eurofirst PIRATE (Passive Infra-Red Airborn Tracking Equipment); advanced integrated defensive aid subsystem (DASS) includes RWR and active jamming po at each wingtip plus laser warning receiver, missil approach warning and towed decoys (Germany require only RWR and MAW; only Spain and UK to have LWR Rohde & Schwarz Saturn VHF/UHF communications. A avionics, flight control and utilities control systems will b integrated through databus highways with appropriat redundancy levels, using fibre optics and microprocessors Special attention given to reducing pilot workload. Ne cockpit techniques simplify flying aircraft safely and effec tively to limits of flight envelope while monitoring an managing aircraft and its operational systems, and detec ing/identifying/attacking desired targets while remainin safe from enemy defences. This achieved through high lev el of system integration and automation, including HOTA controls; GEC-Marconi wide-angle HUD able to display in addition to other symbology, FLIR pictures from senso pod-mounted externally to left of cockpit; helmet mounte sight (HMS), with direct voice input (DVI) for appropriat functions; and three Smiths Industries multi-functio head-down (MFHD) colour CRT displays. Other cockpi instrumentation includes Computing Devices video an voice recorder, GEC-Marconi (Elmer) crash surviva memory unit, and Teldix cockpit interface unit.

ARMAMENT: Interceptor will have internally mounted 27 mr Mauser gun on starboard side, plus mix of medium-rang AIM-120 AMRAAM or Aspide and short-range air-to-ai missiles carried externally; four AIM-120s carried i underfuselage troughs. Short-range missiles carried on M Aviation underwing ejector release units. EFA will, necessary, be able to carry considerable load of air-to-ai weapons. Total of 13 external stores stations: five (incl on wet) under fuselage and four (incl one wet) under eac wing.

DIMENSIONS, EXTERNAL:
Wing span	10.50 m (34 ft 5½ in
Wing aspect ratio	2.20
Length overall	14.50 m (47 ft 7 in
Height overall	approx 4.00 m (13 ft 1½ in

AREAS:
Wings, gross	50.0 m² (538.2 sq ft
Foreplanes	2.40 m² (25.83 sq ft

WEIGHTS AND LOADINGS (approx):
Weight empty	9,750 kg (21,495 lb
Internal fuel load	4,000 kg (8,818 lb
External stores load (weapons and/or fuel)	6,500 kg (14,330 lb
Max T-O weight	21,000 kg (46,297 lb

PERFORMANCE (design):
Max level speed Mach 2.0
T-O and landing distance with full internal fuel, tw AIM-120s and two dogfight missiles, ISA + 15°C
 500 m (1,640 ft
Combat radius
 250-300 nm (463-556 km; 288-345 miles
g limits with full internal fuel and two AIM-120s
 +9/−3

EUROFLAG

EUROFLAG srl

c/o Alenia, Via Vitorchino 81, I-00198 Rome, Italy
Telephone: 39 (6) 333 9046
Fax: 39 (6) 333 9047
GENERAL MANAGER: P. Felici
PARTICIPATING COMPANIES:
 Aerospatiale: see under France
 Alenia: see under Italy
 British Aerospace: see under UK
 CASA: see under Spain
 Deutsche Aerospace Airbus: see under Germany

Euroflag srl formed 17 June 1991, with headquarters in Alenia head office in Rome, to manage European FLA development; Aerospatiale, Alenia, British Aerospace, CASA and Deutsche Aerospace Airbus have equal shares in Euroflag srl; MoUs established 1992 with Flabel (SABCA and SONACA) of Belgium, OGMA of Portugal and TUSAŞ Aerospace Industries of Turkey to allow integrated participation in FLA programme; these companies may join Euroflag srl; BAe participation by private investment.

EUROFLAG FUTURE LARGE AIRCRAFT

TYPE: Long-range military tactical transport.
PROGRAMME: Original FIMA programme (see 1989-90 *Jane's*) replaced April 1989 by five-nation industry MoU to develop new technology transport to replace C-130 Hercules and C-160 Transall; Independent European Programme Group (IEPG) defined Outline European Staf Target (OEST) during 1991; Western European Union report in Autumn 1991 concluded Euroflag FLA should form core of future European military transport capability to support Rapid Reaction Corps; national armament direc-

tors of Belgium, France, Germany, Italy, Portugal, Spain and Turkey affirmed support for 12-month pre-feasibility study completed by Euroflag in late 1992; UK Ministry of Defence stayed aloof, but retained observer status; feasibility programme expected to start 1993; if full-scale development phase starts in 1996, first flight could follow in 2000 and first deliveries in 2003; initial NATO fleet size estimated at 300.

CURRENT VERSIONS: Primarily for personnel/cargo transport; derivatives may include air-refuelling tanker, surveillance/reconnaissance, long-range maritime patrol and AEW. Three-point tanker carrying up to 40,000 kg (88,185 lb) of transferable fuel over 200 nm (370 km; 230 miles) radius could take off from 1,525 m (5,000 ft) runway.

CUSTOMERS: Air forces of Belgium, France, Germany, Italy, Portugal, Spain and Turkey probable; UK possible; exports expected; overall market estimated at 700-1,000 aircraft.

DESIGN FEATURES: Currently seen as high-wing, T-tailed aircraft with rough-field landing gear and much larger cabin/hold floor area and cross-section than C-130/C-160, permitting high payload factors with low-density cargo, vehicles or mixed passenger/cargo loads, and much greater payload/range; will have flight refuelling; operable all-weather and night. Moving 375 tonnes (826,725 lb) of cargo over 3,777 nm (7,000 km; 4,350 miles) in one lift would require 32 FLAs compared with 58 C-130Js.

STRUCTURE: Modern design/manufacturing techniques expected to afford major reductions in maintenance manhour requirements and increases in aircraft availability/survivability.

POWER PLANT: Four advanced turbofans of 80.1 kN (18,000 lb st) each. Fuel capacity 52,500 litres (13,870 US gallons; 11,548 Imp gallons).

ACCOMMODATION: Two-man flight deck; typical loads include MLRS, Bradley AFV, Super Puma, two PAH-2 Tigers, up to 126 paratroops or 62 troops and eight 2.74 × 2.24 m (108 × 88 in) pallets.

DIMENSIONS, EXTERNAL (provisional):
Wing span 42.70 m (140 ft 1 in)
Length overall 40.30 m (132 ft 2½ in)
Height overall 13.30 m (43 ft 7½ in)
Fuselage: Max width 5.07 m (16 ft 7½ in)
Max depth 4.69 m (15 ft 4½ in)

DIMENSIONS, INTERNAL:
Cabin, excl ramp: Length 17.07 m (56 ft 0 in)
Floor width, continuous 4.00 m (13 ft 1½ in)
Floor to ceiling height, continuous 3.55 m (11 ft 7¾ in)
Ramp length 4.74 m (15 ft 6½ in)
Floor area 87.4 m² (940.8 sq ft)
Volume 300.0 m³ (10,595 cu ft)

AREAS:
Wings, gross 180.0 m² (1,937.5 sq ft)

WEIGHTS AND LOADINGS:
Max payload 25,000 kg (55,115 lb)
Max T-O weight 111,000 kg (244,710 lb)
Estimated power loading 346.44 kg/kN (3.40 lb/lb st)

PERFORMANCE (preliminary):
Cruising Mach number greater than Mach 0.7
Cruising TAS 415 knots (769 km/h; 478 mph)
Range: with 20,000 kg (44,092 lb) payload
up to 3,000 nm (5,560 km; 3,455 miles)
with 12,000 kg (26,455 lb) payload
approx 3,777 nm (7,000 km; 4,350 miles)

Artist's impression of the Euroflag Future Large Aircraft as currently envisaged

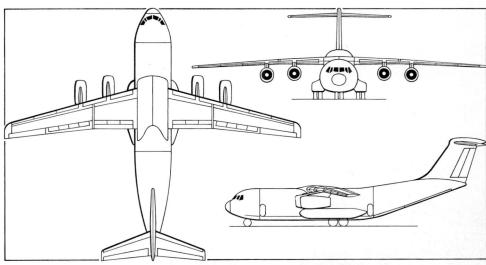

Euroflag Future Large Aircraft military transport (Jane's/Mike Keep)

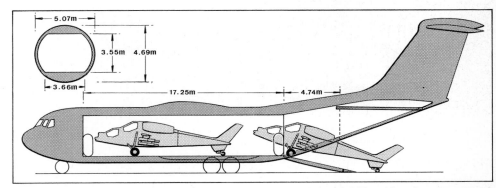

Euroflag Future Large Aircraft freight cabin compared with two Agusta A 129s with main rotor removed (Jane's/Mike Keep)

EURO-HERMESPACE

EURO-HERMESPACE SA
Avenue Yves Brunaud, F-31770 Colomiers, France

European Space Agency (ESA) ministerial level conference in Granada, on 9-10 November 1992, decided to abandon Hermès reusable space shuttle project and to begin studies for new spaceplane in co-operation with Russia. Details and an artist's impression of Hermès appeared in 1992-93 *Jane's*.

EUROPATROL

PARTICIPATING COMPANIES:
Alenia: see under Italy
British Aerospace: see under UK
Dassault Aviation: see under France
Deutsche Aerospace: see under Germany
CASA: see under Spain
Fokker: see under Netherlands

Grouping formed by above manufacturers in late Summer 1992 to move towards development of a European maritime patrol aircraft and mission system for the next century. Initial objectives are to work towards a common requirement, approach governments and encourage Independent European Programme Group to formulate an Outline European Staff Target. Existing, updated or new aircraft would be considered and the mission system would be suitable for retrofitting in existing aircraft.

KATRAN

DIRECTORS:
Victor D'Jamirze (Australia)
A. Rahman (Singapore)

KATRAN

TYPE: Four/ten-passenger twin-turbofan business jet.
PROGRAMME: Announced at Asian Aerospace show, February 1992; venture capital for launch then being sought in Canada, Japan, Singapore and USA. Nothing further heard during past year.

DESIGN FEATURES: Mid-mounted rear wing, canards and T tail.

STRUCTURE: All-composites.
Following data are provisional:

POWER PLANT: Two Garrett/General Electric CFE738 turbofans, each rated at 24.9 kN (5,600 lb st).

DIMENSIONS, EXTERNAL:
Wing span 17.56 m (57 ft 7¼ in)
Length overall 20.42 m (67 ft 0 in)
Fuselage: Max diameter 2.04 m (6 ft 8½ in)
Height overall 5.15 m (16 ft 10¾ in)

WEIGHTS AND LOADINGS:
Max-T-O weight 16,284 kg (35,900 lb)

PERFORMANCE (estimated):
Max cruising speed at 13,715 m (45,000 ft)
Mach 0.91 (518 knots; 960 km/h; 596 mph)
Econ cruising speed 480 knots (889 km/h; 553 mph)
Service ceiling 16,765 m (55,000 ft)
IFR range with 450 kg (992 lb) payload:
at max cruising speed
8,000 nm (14,825 km; 9,212 miles)
at econ cruising speed
9,700 nm (17,976 km; 11,170 miles)

MCDONNELL DOUGLAS/BAe

PARTICIPATING COMPANIES:
McDonnell Douglas: see under USA
British Aerospace: see under UK
VICE-PRESIDENT AND GENERAL MANAGER, AV-8:
Patrick J. Finneran Jr
VICE-PRESIDENT AND GENERAL MANAGER, T45TS:
Robert H. Soucy Jr

MCDONNELL DOUGLAS/BRITISH AEROSPACE HARRIER II

US Marine Corps designations: AV-8B and TAV-8B
RAF designations: Harrier GR. Mk 5, 5A and 7, and T. Mk 10
Spanish Navy designation: VA.2 Matador II

TYPE: Single-seat V/STOL close support, battlefield interdiction, night attack and reconnaissance aircraft.
PROGRAMME: Early background given in several previous editions; present collaborative programme began with two YAV-8B (converted AV-8A) aerodynamic prototypes (first flights 9 November 1978 and 19 February 1979); followed by four FSD aircraft (first flight 5 November 1981); first 12 pilot production AV-8Bs ordered FY 1982 (first flight 29 August 1983), deliveries to USMC beginning 12 January 1984; development programme for night attack version announced November 1984; first flights of RAF GR. Mk 5 development aircraft 30 April (ZD318) and 31 July 1985 (ZD319); first USMC operational AV-8B squadron (VMA-331) achieved IOC August 1985; first flight of two-seat TAV-8B (No. 162747) 21 October 1986; first flight of night attack AV-8B prototype (162966) 26 June 1987; first GR. Mk 5 for RAF (ZD324) handed over 1 July 1987; TAV-8B deliveries (to VMAT-203) began August

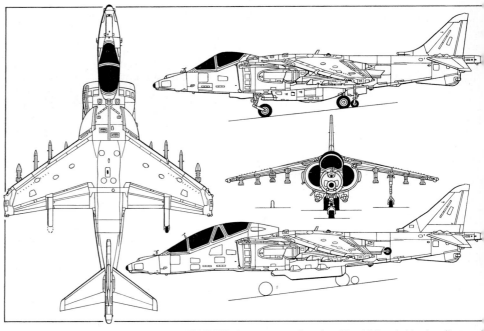

McDonnell Douglas/BAe Harrier GR. Mk 7 V/STOL close support aircraft, with additional side view (bottom) of T. Mk 10 *(Jane's/Mike Keep)*

McDonnell Douglas/BAe Harrier GR. Mk 5s of the Royal Air Force

Close-up of the nose of a night attack Harrier GR. Mk 7

1987; EAV-8B deliveries to Spain 1987-88; production contract for new-build GR. Mk 7s placed April 1988; first flight of Pegasus 11-61 power plant (ZD402) 10 June 1989; first production night attack AV-8B (163853) delivered to VMA-214 on 15 September 1989; first flight of RAF GR. Mk 7 (development aircraft, converted from GR. Mk 5) 29 November 1989; 27 GR. Mk 7s ordered April 1988 (later increased to 34); first flight of production Mk 7 (ZG471) May 1990; production contract for T. Mk 10 placed February 1990; see also separate entry for Harrier II Plus.

CURRENT VERSIONS: **AV-8B Harrier II:** US Marine Corps single-seat close support version. Night attack avionics (including FLIR bulge ahead of windscreen) from 167th AV-8B (163853) onwards (see Avionics paragraph), plus (from No. 182, 163874, and TAV-8B No. 16, 164120, in December 1990) uprated F402-RR-408 (Pegasus 11-61) engine.

AV-8B Harrier II Plus: Radar equipped night attack version; described separately.

TAV-8B Harrier II: US Marine Corps two-seat operational trainer, with longer forward fuselage and 0.43 m (1 ft 5 in) taller vertical tail than AV-8B; two cockpits in tandem; two underwing stores stations only; BAe major subcontractor for this version.

EAV-8B: Manufacturer's designation for Spanish Navy **VA.2 Matador II** single-seat export version.

Harrier GR. Mk 5: Royal Air Force single-seat battlefield air interdiction/close air support version. Two additional underwing stations, for Sidewinder missile carriage. Being converted to Mk 7.

Harrier GR. Mk 5A: Interim designation for 19 GR. Mk 5s prior to upgrade to full GR. Mk 7 standard.

Harrier GR. Mk 7: Royal Air Force single-seat night attack version, based on GR. Mk 5. New production and conversions.

Harrier T. Mk 10: Royal Air Force operational trainer for GR. Mk 7, based on TAV-8B airframe with eight underwing pylons, FLIR and night vision equipment of GR. Mk 7, but no ARBS.

CUSTOMERS: US Marine Corps ordered total of 280 by FY 1991, including four FSDs (ordered FY 1979), 24 TAV-8Bs and 24 Harrier II Plus; target procurement of 300 AV-8Bs and 28 TAV-8Bs unlikely to be achieved but six ordered June 1992 to replace 1991 Gulf War losses and to be built in II Plus configuration. VMAT-203 received first AV-8B (161573) 12 January 1984 and first TAV-8B (162963) August 1987; also operated by VMA-211 (re-equipped 1990, second Night Attack squadron), VMA-214 (first with Night Attack version; initial aircraft, 163853, delivered 15 September 1989), VMA-223 (1 October 1987), VMA-231 (September 1985), VMA-311 (1989; third Night Attack unit, 1992), VMA-331 (commissioned 30 January 1985 as first operational unit but disbanded 30 September 1992), VMA-513 (January 1987; fourth Night Attack unit, 1992) and VMA-542 (1986). Requirement to modify earlier aircraft to Plus standard with new fuselage shells and power plant upgraded to -408 series, between FY 1994 and FY 2000.

Royal Air Force ordered total of 109 by 1990, comprising two FSDs, 41 GR. Mk 5s, 19 GR. Mk 5As, 34 GR. Mk 7s and 13 T. Mk 10s (last mentioned ordered March 1990 for delivery from 1994). First service flight of Mk 5 (ZD324) with No. 233 OCU, 30 March 1988; OCU redesignated No. 20 (Reserve) Squadron on 1 September 1992; deliveries to No. 1 Squadron from 23 November 1988 and unit re-declared to NATO on 2 November 1989; No. 1 re-equipped to GR. Mk 7 (rebuilt Mk 5As) and flew first sortie (ZD434) 2 June 1992; to No. 3 Squadron from 17 March 1989 (ZD401); Mk 5A delivered mainly to storage from 21 August 1990 (ZD432); first Mk 7 delivery to A&AEE, Boscombe Down, 5 June 1990 (ZG472); to Strike/Attack Operational Evaluation Unit (SAOEU) from 17 August 1990 (ZG473)—unit flew RAF's first NVG Harrier mission, 11 December 1990 and first NVG/GPS live bomb drop 19 February 1992; to No. 4 Squadron from 12 September 1990 (ZG473); re-equipped No. 3 Squadron from 30 November 1990 (ZG479). Single-seat deliveries completed (ZG862) 2 June 1992. Contract for BAe upgrade of 58 (less any attrition) Mks 5/5A to Mk 7 awarded 2 November 1990; first ex-Mk 5 (ZD380) re-delivered to RAF 21 December 1990; first ex-Mk 5A (ZD430) on 9 April 1991. INS modified to FIN 1075G (incorporating GPS) on one SAOEU aircraft and six of No. 1 Squadron; first squadron flight (ZD437) 19 November 1992.

Spanish Navy received 12 EAV-8Bs for Novena Escuadrilla (Eslla 009) between 6 October 1987 and September 1988; operational aboard *Principe de Asturias*; single TAV-8B ordered March 1992; eight new-build Harrier II Plus ordered; remaining 11 EAV-8Bs to be converted to Plus standard; procurement 21, including single TAV-8B.

Italian Navy ordered a total of 18; two AV-8Bs ordered May 1989 and 16 Harrier II Plus, of which first three ordered late 1991 (to come from USMC batch); option on eight; first delivery (two TAV-8Bs, MM55032-55033, to *Giuseppe Garibaldi*) 23 August 1991.

Total firm orders 429, of which 364 delivered by February 1993 from MDC (268) and BAe (96).

Italian Navy TAV-8B Harrier II. Note single underwing pylon and strakes in place of gun pods

HARRIER II and II PLUS ORDERS

Customer		Single seat	Trainer	Remarks
US Marines	FY 1979 order	4		Prototypes
	FY 1982 order	12		Pilot production
	FY 1983 order	21		
	FY 1984 order	26	1	
	FY 1985 order	30	2	
	FY 1986 order	40	6	
	FY 1987 order	36	6	
	FY 1988 order	22	2	
	FY 1989 order	20	4	Incl 1 Plus, and 2 trainers to Italy
	FY 1990 order	22	2	Incl 6 Plus
	FY 1991 order	21	3	Incl 21 Plus, of which 3 to Italy
	FY 1992 order	6		All Plus
Sub-totals		**260**	**26**	
Royal Air Force	Development	2		Converted to Mk 7
	GR. Mk 5	41		Converted to Mk 7
	GR. Mk 5A	19		Converted to Mk 7
	GR. Mk 7	34		
	T. Mk 10		13	
Sub-totals		**96***	**13***	
Spanish Navy		20≠	1	Incl 8 Plus
Italian Navy		13†		All Plus; excludes 3 AV- and 2 TAV- from USMC. Option on 8
Grand totals		**389**	**40**	Options 8

Notes: Plus refers only to single-seat aircraft
*Assembled in UK
†Assembled in Italy
≠Eight Plus assembled in Spain

COSTS: £200 million (1990 contract) for 13 T. Mk 10 aircraft. $3,000 million (estimated) for rebuild of 114 AV-8Bs to Plus configuration. TAV-8B, $25 million (Spain 1992).

DESIGN FEATURES: Differences compared with Harrier GR. Mk 3/AV-8A (see under BAe in UK section of 1989-90 *Jane's*) include bigger wing and longer fuselage; use of graphite epoxy (carbonfibre) composite materials for wings and parts of fuselage and tail unit; adoption of supercritical wing section; addition of LIDS (lift improvement devices: strakes to replace gun/ammunition pods when armament not carried, plus retractable fence panel forward of pods) to augment lift for vertical take-off; larger wing trailing-edge flaps and drooped ailerons; redesigned forward fuselage and cockpit; redesigned engine air intakes to provide more VTO/STO thrust and more efficient cruise; two additional wing stores stations; wing outriggers relocated at mid-span to provide better ground manoeuvring capability; leading-edge root extensions (LERX) to enhance instantaneous turn rate and air combat capability; landing gear strengthened to cater for higher operating weights and greater external stores loads. Wing span and area increased by approx 20 per cent and 14.5 per cent respectively compared with GR. Mk 3/AV-8A; leading-edge sweep reduced by 10°; thickness/chord ratios 11.5 per cent (root)/7.5 per cent (tip); marked anhedral on wings and variable incidence tailplane. Increased size ('100 per cent') LERX from 79th UK production aircraft (ZG506); being retrofitted.

FLYING CONTROLS: Hydraulic actuation (by Fairey irreversible jacks) of drooping ailerons and slab tailplane; rudder actuated mechanically; single-slotted trailing-edge flaps with slot closure doors; manoeuvring at airspeeds below wingborne flight by jet reaction control valves in nose and tailcone and at each wingtip and by thrust vectoring; LIDS 'box' traps air cushion bounced off ground by engine exhaust in VTOL modes, providing enough extra lift to enable aircraft to take off vertically at a gross weight equal to its max hovering gross weight; large forward hinged airbrake beneath fuselage aft of rear main landing gear bay.

STRUCTURE: One-piece wing (incl main multi-spar torsion box, ribs and skins), ailerons, flaps, LERX, outrigger pods and fairings, forward part of fuselage, LIDS, tailplane and rudder, are manufactured mainly from graphite epoxy (carbonfibre) and other composites; centre and rear fuselage, wing leading-edges (reinforced against bird strikes on RAF aircraft), wingtips, tailplane leading-edges and tips, and fin, are of aluminium alloy; titanium used for front and rear underfuselage heatshields and small area forward of windscreen. McDonnell Douglas/BAe work split is 60/40 for AV-8B and EAV-8B, 50/50 for RAF aircraft. McDonnell Douglas builds entire wing, front and forward centre-fuselage (incl nosecone, air intakes, heatshields, engine access doors and forward fuel tanks) and underfuselage fences/strakes, for all aircraft, plus tailplanes for USMC and Spanish aircraft, and assembles all USMC/Spanish fuselages; BAe builds rear centre and rear fuselage (incl blast and heatshields, centre and rear fuel tanks, dorsal air intakes and tail bullets), fins and rudders, and the complete jet reaction control system, for all aircraft, plus tailplanes for RAF aircraft, and assembles all RAF fuselages; final assembly is by McDonnell Douglas for USMC/Italy/Spain, BAe for RAF.

LANDING GEAR: Retractable bicycle type of Dowty design, permitting operation from rough unprepared surfaces of very low CBR (California Bearing Ratio). Hydraulic actuation, with nitrogen bottle for emergency extension. Single steerable nosewheel retracts forward, twin coupled mainwheels rearward, into fuselage. Small outrigger units, at approx mid span between flaps and ailerons, retract rearward into streamline pods. Telescopic oleo-pneumatic main and outrigger gear; levered suspension nosewheel leg. Dunlop wheels, tyres, multi-disc carbon brakes and anti-skid system. Mainwheel tyres (size 26.0 × 7.75-13.00) and nosewheel tyre (size 26.0 × 8.75-11) all have pressure of 8.62 bars (125 lb/sq in). Outrigger tyres are size 13.5 × 6.00-4.00, pressure 10.34 bars (150 lb/sq in). McDonnell Douglas responsible for entire landing gear system.

POWER PLANT: One 105.87 kN (23,800 lb st) Rolls-Royce F402-RR-408 (Pegasus 11-61) vectored thrust turbofan in AV-8B (95.42 kN; 21,450 lb st F402-RR-406A/Pegasus 11-21 in aircraft delivered before December 1990); one 95.63 kN (21,500 lb st) Pegasus Mk 105 in Harrier GR. Mk 5/7; Mk 152-42 in EAV-8B. Redundant digital engine control system (DECS), with mechanical backup, standard from March 1987. Zero-scarf front nozzles. Air intakes have an elliptical lip shape, leading-edges reinforced against bird strikes, and a single row of auxiliary intake doors. Access to engine accessories through top of fuselage, immediately ahead of wing. Integral fuel tanks in wings, usable total 2,746 litres (725 US gallons; 604 Imp gallons) plus four fuselage tanks: front and rear, 609 litres (161 US gallons; 134 Imp gallons) each and left centre and right centre, 177 litres (47 US gallons; 39 Imp gallons) each. Internal fuel 4,319 litres (1,140 US gallons; 950 Imp gallons) usable; 4,410 litres (1,164 US gallons; 970 Imp gallons) total in single-seat versions; 4,150 litres (1,096 US gallons; 913 Imp gallons) total in two-seat versions. Water injection tank with capacity of 225 kg (495 lb). Retractable bolt-on in-flight refuelling probe optional. Each of four inner underwing stations capable of carrying a 1,135 litre (300 US gallon; 250 Imp gallon) auxiliary fuel tank; total internal and external fuel 8,956 litres (2,365 US gallons; 1,970 Imp gallons).

ACCOMMODATION: Pilot only, on zero/zero ejection seat (UPC/Stencel for USMC, Martin-Baker for RAF), in pressurised, heated and air-conditioned cockpit. AV-8B cockpit raised approx 30.5 cm (12 in) by comparison with AV-8A/YAV-8B, with redesigned one-piece wraparound windscreen (thicker on RAF aircraft than on those for USMC) and rearward sliding bubble canopy, to improve all-round field of view. Windscreen de-icing. Windscreens and canopies for all aircraft manufactured by McDonnell Douglas.

SYSTEMS: No. 1 hydraulic system has flow rate of 43 litres (11.4 US gallons; 9.5 Imp gallons)/min; flow rate of No. 2 system is 26.5 litres (7.0 US gallons; 5.8 Imp gallons)/min. Reservoirs nitrogen pressurised at 2.76-5.52 bars (40-80 lb/sq in). Other systems include Westinghouse variable speed constant frequency (VSCF) solid state electrical system, Lucas Mk 4 gas turbine starter/APU, Clifton Precision onboard oxygen generating system (OBOGS), and Graviner Firewire fire detection system. Dorsal airscoop at

166 INTERNATIONAL: AIRCRAFT—McDONNELL DOUGLAS/BAe

Cockpit of Harrier GR. Mk 7 displaying moving map (left) and FLIR (right) imagery on CRTs *(Paul Jackson)*

base of fin for avionics bay cooling system.

AVIONICS: Include dual Collins RT-1250A/ARC U/VHF com (GEC Avionics AD3500 ECM-resistant U/VHF-AM/FM in RAF GR. Mk 7 aircraft; military designation ARI 23447 but ARC-182/ARI 23387 in GR. Mk 5), R-1379B/ARA-63 all-weather landing receiver (AV-8B only), RT-1159A/ARN-118 Tacan (ARI 23368 for RAF), RT-1015A/APN-194(V) radar altimeter (ARI 23388 for RAF GR. Mk 5 but RT1042A/ARI 23388 in GR. Mk 7), Honeywell CV-3736/A com/nav/identification data converter, Bendix/King RT-1157/APX-100 IFF (Cossor IFF 4760 Mk 12/ARI 23389 transponder for RAF), Litton AN/ASN-130A inertial navigation system (replaced by GEC-Marconi FIN 1075 or 1075G with RAF), AiResearch CP-1471/A digital air data computer, Smiths Industries SU-128/A dual combining glass HUD and CP-1450/A display computer, IP-1318/A CRT Kaiser digital display indicator, and (RAF only) GEC-Marconi moving map display. Litton AN/ALR-67(V)2 fore/aft looking RWR (AV-8B only), UK MoD AN/ARR-51 FLIR receiver, Goodyear AN/ALE-39 flare/chaff dispenser (upper and lower rear fuselage; current two dispensers to be increased to six) (Tracor AN/ALE-40 in RAF aircraft). Primary weapon delivery sensor system for AV-8B and GR. Mks 5/7 is Hughes Aircraft AN/ASB-19(V)2 or (V)3 Angle Rate Bombing Set, mounted in nose and comprising a dual-mode (TV and laser) target seeker/tracker. ARBS functions in conjunction with Unisys CP-1429/AYK-14(V) mission computer (built as Computing Devices ACCS 2000 for RAF), Smiths Industries AN/AYQ-13 stores management system, display computer, HUD and digital display indicator. Flight controls that interface with reaction control system provided by Honeywell AN/ASW-46(V)2 stability augmentation and attitude hold system, currently being updated to high AOA capable configuration. RAF aircraft have an accident data recorder. Night attack versions equipped with GEC-Marconi nose-mounted FLIR, Smiths Industries wide-angle HUD/HDD, digital colour moving map display (Honeywell for USMC, GEC-Marconi for RAF) and pilot's NVGs (variant of GEC-Marconi Nite-Op) with compatible cockpit lighting. Provision for Sanders AN/ALQ-164 defensive ECM pod on centreline pylon. GR. Mks 5/7 have Marconi Defence Systems ARI 23333 Zeus internal ECM system comprising advanced RWR and multi-mode jammer with Northrop RF transmitter; and Plessey PVS 2000 pulse Doppler missile approach warning (MAW) radar, mounted in tailboom, which automatically activates appropriate countermeasures upon detecting approach of enemy missiles; optical missile approach warner for retrofit to USMC aircraft. Vinten VICON 18 Srs 403 long-range reconnaissance pod evaluated for RAF 1991 and adopted, 1993, together with VICON 57 multi-sensor pod. Carriage of Harrier GR. Mk 3's optical recce pod also possible. RAF defensive aids include Bofors BOL 304 chaff dispenser in rear of Sidewinder launch rails (Phimat pod on port outer wing pylon pending BOL availability).

EQUIPMENT: Backup mechanical instrumentation includes ASI, altimeter, AOA indicator, attitude indicator, cabin pressure altitude indicator, clock, flap position indicator, HSI, standby compass, turn and slip indicator, and vertical speed indicator. Anti-collision, approach, formation, in-flight refuelling, landing gear position, auxiliary exterior lights, and console, instrument panel and other internal lighting.

ARMAMENT: Two underfuselage packs, mounting on port side a five-barrel 25 mm cannon based on General Electric GAU-12/U, and 300 round container on starboard side, in AV-8B; or (RAF) two 25 mm Royal Ordnance Factories cannon with 100 rds/gun (derived from 30 mm Aden). Single 454 kg (1,000 lb) stores mount on fuselage centreline, between gun packs. Three stores stations under each wing on AV-8B, stressed for loads of up to 907 kg (2,000 lb) inboard, 454 kg (1,000 lb) on intermediate stations, and 286 kg (630 lb) outboard. Four inner wing stations are wet, permitting carriage of auxiliary fuel tanks; reduced manoeuvring limits apply when tanks mounted on intermediate stations. RAF aircraft and new production Harrier II Plus have additional underwing station, for Sidewinder air-to-air missile, ahead of each outrigger wheel fairing. Typical weapons include two or four AIM-9L Sidewinder, Magic or AGM-65E Maverick missiles, or up to six Sidewinders; up to sixteen 540 lb free-fall or retarded general purpose bombs, 12 BL 755 or similar cluster bombs, 1,000 lb free-fall or retarded bombs, ten Paveway laser guided bombs, eight fire bombs, 10 Matra 155 rocket pods (each with eighteen 68 mm SNEB rockets), or (in addition to underfuselage gun packs) two underwing gun pods. ML Aviation BRU-36/A bomb release units standard on all versions. RAF armoury expanded from early 1993 with CRV-7 rocket pods and CBU-87 cluster bombs. TAV-8B can carry six Mk 76 practice bombs or two LAU-68 rocket launchers for weapons training.

DIMENSIONS, EXTERNAL:
Wing span	9.25 m (30 ft 4 in)
Wing aspect ratio	4.0
Length overall (flying attitude):	
AV-8B	14.12 m (46 ft 4 in)
TAV-8B	15.32 m (50 ft 3 in)
GR. Mks 5/7	14.36 m (47 ft 1½ in)
T. Mk 10	15.79 m (51 ft 9½ in)
Height overall	3.55 m (11 ft 7¾ in)
Tailplane span	4.24 m (13 ft 11 in)
Outrigger wheel track	5.18 m (17 ft 0 in)

AREAS:
Wings, excl LERX, gross	21.37 m² (230.0 sq ft)
LERX (total): Pegasus 11-21	0.81 m² (8.7 sq ft)
Pegasus 11-61	1.24 m² (13.4 sq ft)
'100 per cent'	1.39 m² (15.0 sq ft)
Ailerons (total)	1.15 m² (12.4 sq ft)
Trailing-edge flaps (total)	2.88 m² (31.0 sq ft)
Ventral fixed strakes (total)	0.51 m² (5.5 sq ft)

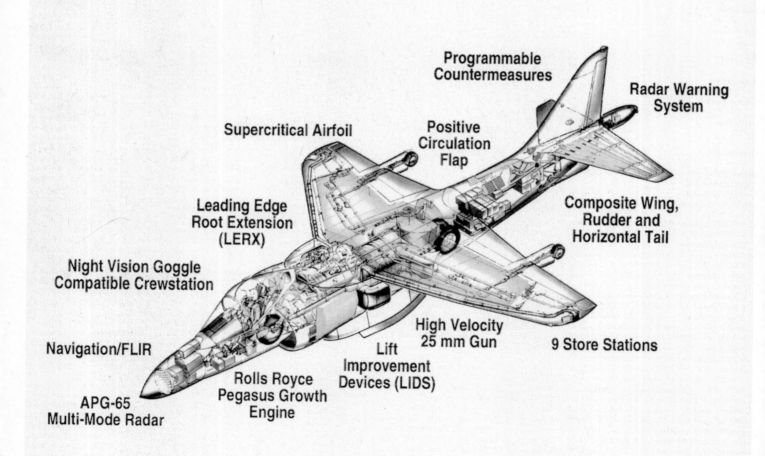

Cutaway of Harrier II Plus, showing radar installation and other features

Ventral retractable fence (LIDs)	0.24 m² (2.6 sq ft)
Ventral airbrake	0.42 m² (4.5 sq ft)
Fin	2.47 m² (26.6 sq ft)
Rudder, excl tab	0.49 m² (5.3 sq ft)
Tailplane	4.51 m² (48.5 sq ft)

WEIGHTS AND LOADINGS (single-seaters, except where indicated):
Operating weight empty (incl pilot and unused fuel):
AV-8B	6,336 kg (13,968 lb)
GR. Mk 7	7,050 kg (15,542 lb)
TAV-8B	6,451 kg (14,223 lb)
Max fuel: internal only*	3,519 kg (7,759 lb)
internal and external*	7,180 kg (15,829 lb)
Max external stores: Pegasus 11-61	6,003 kg (13,235 lb)
Pegasus 11-21/Mk 105†	4,899 kg (10,800 lb)

Max useful load (incl fuel, stores, weapons, ammunition, and water injection for engine):
VTO	approx 3,062 kg (6,750 lb)
STO	more than 7,710 kg (17,000 lb)
Basic flight design gross weight for 7g operation	10,410 kg (22,950 lb)

Max T-O weight:
435 m (1,427 ft) STO	14,061 kg (31,000 lb)
S/L VTO, ISA:	
AV-8B/Pegasus 11-61	9,342 kg (20,595 lb)
GR. Mk 7	8,700 kg (19,180 lb)
S/L VTO, 32°C	8,142 kg (17,950 lb)
Design max landing weight	11,340 kg (25,000 lb)
Max vertical landing weight	9,043 kg (19,937 lb)

* 205 kg (453 lb) less in TAV-8B
† throughout full manoeuvring envelope

PERFORMANCE:
Max Mach number in level flight:
at S/L	0.87 (575 knots; 1,065 km/h; 661 mph)
at altitude	0.98

STOL T-O run at max T-O weight:
ISA	435 m (1,427 ft)
32°C	518 m (1,700 ft)

Operational radius with external loads shown:
short T-O (366 m; 1,200 ft), twelve Mk 82 Snakeye bombs, internal fuel, 1 h loiter
90 nm (167 km; 103 miles)
hi-lo-hi, short T-O (366 m; 1,200 ft), seven Mk 82 Snakeye bombs, two 300 US gallon external fuel tanks, no loiter 594 nm (1,101 km; 684 miles)
deck launch intercept mission, two AIM-9 missiles and two external fuel tanks
627 nm (1,162 km; 722 miles)
Unrefuelled ferry range, with four 300 US gallon external tanks:
tanks retained	1,638 nm (3,035 km; 1,886 miles)
tanks dropped	1,965 nm (3,641 km; 2,263 miles)
Combat air patrol endurance at 100 nm (185 km; 115 miles) from base	3 h
g limits	+8/−3

MCDONNELL DOUGLAS/BRITISH AEROSPACE HARRIER II PLUS

TYPE: Enhanced capability derivative of AV-8B.
PROGRAMME: Intention to develop radar-equipped version of AV-8B announced as McDonnell Douglas/BAe private venture June 1987; radar integration efforts (with Hughes Aircraft) started 1988; tri-national MoU (USA/Italy/Spain) of 28 September 1990 approved joint funding to develop and integrate AN/APG-65 radar; US Navy contract of 3 December 1990 authorised development of prototype and completion to Harrier II Plus standard of 24 AV-8Bs ordered in FYs 1990-91; production MoU signed by USA/Italy/Spain March-December 1992; prototype (AV-8B No. 205, 164129) first flight 22 September 1992; deliveries to USMC due to start in April 1993 and finish November 1994 (beginning No. 233, 164542). Harrier consortium adds Hughes (for radar), Alenia (Italy) and CASA (Spain); Italy and Spain each have 15 per cent share.
CUSTOMERS: US Marine Corps (30 including six replacements for AV-8Bs lost in 1991 Gulf War); orders from Italian Navy (16, with option for eight more) and Spanish Navy (eight ordered March 1993, for delivery from 1996; plus 11 conversions). Initial three Italian aircraft diverted from USMC contract; balance of 13 authorised November 1992 (plus option on eight more) for final assembly by Alenia in Italy. USMC considering remanufacture of earlier AV-8Bs to Harrier II Plus standard. Refer also to AV-8B entry.
COSTS: $181.5 million (December 1990 contract) for development and production of 24 for USMC.
DESIGN FEATURES: Generally as for night attack AV-8B with F402-RR-408 engine, plus Hughes Aircraft AN/APG-65 multi-mode pulse Doppler radar (extending nose by 0.43 m; 1 ft 5 in), FLIR, and future weapon capability to include AMRAAM, Sparrow, Sea Eagle and Harpoon; enlarged LERX. RAF-type (eight pylon) wing. Fatigue life of 6,000 hours; improved ECM.
DIMENSIONS, EXTERNAL: As AV-8B except:
Length overall 14.55 m (47 ft 9 in)
WEIGHTS AND LOADINGS:
Operating weight empty	6,740 kg (14,860 lb)
Max T-O weight	14,061 kg (31,000 lb)

PERFORMANCE (estimated, with 137 m; 450 ft short T-O deck run, 6.5° ski-jump, 20 knot; 37 km/h; 23 mph wind over deck, air temperature 32°C, optimum cruise conditions, incl reserves for landing):
Anti-shipping combat radius with two Harpoons, two Sidewinders and two 1,136 litre (300 US gallon; 250 Imp gallon) drop tanks
609 nm (1,128 km; 701 miles)
Combat air patrol (incl 2 min combat) with four AMRAAM and two 300 US gallon tanks: time on station:
at 100 nm (185 km; 115 mile) radius	2 h 42 min
at 200 nm (370 km; 230 mile) radius	2 h 6 min

Sea surveillance combat radius (incl 50 nm; 92 km; 57 mile dash at S/L) with two Sidewinders and two 300 US gallon tanks 608 nm (1,127 km; 700 miles)

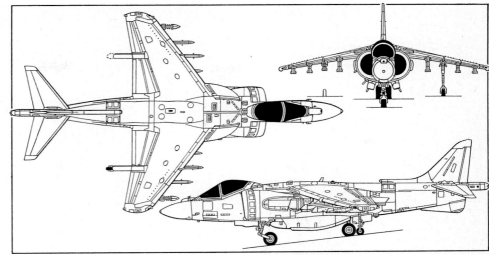

McDonnell Douglas/BAe Harrier II Plus radar-equipped version of AV-8B (*Jane's/Mike Keep*)

Prototype Harrier II Plus on its initial flight, 22 September 1992

McDonnell Douglas artist's impression of possible STOVL Strike Fighter technology demonstrator

MCDONNELL DOUGLAS/BRITISH AEROSPACE 'HARRIER III' AND SSF

McDonnell Douglas and British Aerospace set up in 1990 a joint study group to determine next stage of advanced short take-off/vertical landing development. Although actual definition still some time away, initial parameters suggest larger, folding wing of 9.75 m (32 ft) span, made from carbonfibre composites; longer fuselage and refined aerodynamics; developed Pegasus 11-61 engine; and EFA-generation avionics, including advanced radar. Harrier III might be aimed initially at replacing current Sea Harriers, with increased range and payload and in-service date of 2000-2005, but could also replace present generation of AV-8Bs.

In late 1992 BAe/McDonnell Douglas among five contenders in ARPA (US Advanced Research Projects Agency) contest for STOVL Strike Fighter (SSF) as possible F/A-18 Hornet replacement; consortium awarded $1.9 million payment of projected $27.7 million contract, March 1993 for further studies in competition with Lockheed; projected aircraft uses GE YF120 variable cycle engine with proportion of gas ducted forward to drive lifting fan for short take-offs; full-scale wind tunnel test planned for 1995 at NASA Ames.

McDONNELL DOUGLAS/BRITISH AEROSPACE T-45A GOSHAWK

TYPE: Two-seat intermediate and advanced jet trainer.
PROGRAMME: Selected 18 November 1981 (from five other candidates) as winner of US Navy VTXTS (now T45TS) competition for undergraduate jet pilot trainer to replace T-2C Buckeye and TA-4J Skyhawk; original plan was for initial 54 'dry' (land-based) T-45Bs followed by 253 carrier-capable 'wet' T-45As; B model eliminated in FY 1984 in favour of 300 'all-wet' T-45As; FSD phase began October 1984; construction of two prototypes by Douglas began February 1986; funding approved 16 May 1986 for first three production lots (incl 60 T-45As and 15 flight simulators during FYs 1988-90); Lot 1 production contract (12 aircraft) awarded 26 January 1988; FSD prototypes made first flights 16 April (162787) and November 1988 (162788); original planned date for first deliveries (October 1989) delayed by further airframe and power plant changes requested by US Navy; announced 19 December 1989 that entire T45TS programme to be transferred to McDonnell Aircraft Co at St Louis; modified FSD prototypes made first flights September and October 1990; two Douglas production aircraft (163599 and '600) delivered to NATC Patuxent River, Maryland, on 10 October and 15 November 1990; first carrier landing (162787 on USS *Kennedy*) 4 December 1991; first McAir production aircraft (163601) flew at St Louis 16 December 1991 and handed over to USN 23 January 1992. Digital/'glass' cockpit under development as 'Cockpit 21'; prototype to fly 1994; planned production line introduction at 73rd aircraft, 1996.
CURRENT VERSIONS: **T-45A Goshawk:** Based on BAe Hawk 60 series (see UK section), with airframe/power plant/avionics changes necessary to meet USN specification. Introduction expected to meet USN training requirements with 42 per cent fewer aircraft than at present, 25 per cent fewer flight hours, and 46 per cent fewer personnel.
CUSTOMERS: US Navy: two FSD prototypes and 268 production aircraft required, of which 48 contracted by January 1993 and 10 delivered by February 1993; complete T45TS programme also involves 24 (originally 32) flight simulators (built by Hughes Training Inc); 34 (originally 49) computer aided instructional devices, three training integration system mainframes, 200 terminals, plus academic materials and contractor operated logistic support. Five training squadrons (VT) to equip: VT-21, 22 and 23 of Training Wing 2 at Kingsville, Texas, in 1992-96; VT-7 and 19 of TW-1 at Meridian, Mississippi, 1996-99. VT-21 operational 27 June 1992; first student class begun October 1992. Primary sea platform is training carrier, USS *Forrestal*.

US Navy Procurement

FY 1988	Lot 1	12 aircraft
FY 1989	Lot 2	24 aircraft
FY 1992	Lot 3	12 aircraft
FY 1993	Lot 4	12 aircraft

DESIGN FEATURES: Generally as for two-seat BAe Hawk, with some redesign and strengthening for carrier operation, incl new main and nose landing gear and provision of nose tow launch bar and arrester hook.
FLYING CONTROLS: Differences from two-seat BAe Hawk include electrically actuated/hydraulically operated full-span wing leading-edge slats (operation limited to landing configuration); aileron/rudder interconnect; two fuselage-side airbrakes instead of single one under fuselage and associated autotrim system for horizontal stabiliser when brakes deployed; GEC-Marconi yaw damper computer and addition of 'smurf' (side mounted unit root fin), a small curved surface forward of each tailplane leading-edge root, to eliminate pitch-down during low-speed manoeuvres; Dowty actuators for slats and airbrakes.
STRUCTURE: Redesigned (incl deeper and longer forward fuselage) and strengthened to accommodate new landing gear and withstand carrier operation; twin airbrakes of composites material; fin height increased by 15.2 cm (6 in) and single ventral fin added; rudder modified; tailplane span increased by 10.2 cm (4 in); wingtips squared off; underfuselage arrester hook, deployable 20° to each side of longitudinal axis. BAe (principal subcontractor) builds wings, centre and rear fuselage, fin, tailplane, windscreen, canopy and flying controls.
LANDING GEAR: Wide-track hydraulically retractable tricycle type, stressed for vertical velocities of 7.28 m (23.9 ft)/s. Single wheel and long-stroke oleo (increased from 33 cm; 13 in of standard Hawk to 63.5 cm; 25 in) on each main unit; twin-wheel steerable nose unit with 40.6 cm (16 in) stroke. Articulated main gear, by AP Precision Hydraulics, is of levered suspension (trailing arm) type with a folding side-stay. Cleveland Pneumatic nose gear, with Sterer digital dual gain steering system (high gain for carrier deck operations). Nose gear has catapult launch bar and hold-

'Everything down' flypast by prototype T-45A Goshawk showing nose towbar, slats, flaps, airbrakes and hook

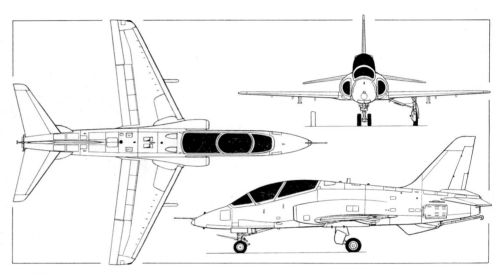

McDonnell Douglas/BAe T-45A Goshawk tandem-seat basic and advanced trainer (*Dennis Punnett*)

McDonnell Douglas/BAe T-45A Goshawk first carrier landing, 4 December 1991

back devices. Main units retract inward into wing, forward of front spar; nose unit retracts forward. All wheel doors sequenced to close after gear lowering; inboard mainwheel doors bulged to accommodate larger trailing arm and tyres. Gear emergency lowering by free fall. Goodrich wheels, tyres and brakes. Mainwheel tyres size 24 × 7.7-10; nosewheels have size 19 × 5.25-10 tyres. Tyre pressure (all units) 22.40 bars (325 lb/sq in) for carrier operation; reduced for land operation. Hydraulic multi-disc mainwheel brakes with Dunlop adaptive anti-skid system.
POWER PLANT: One 26.00 kN (5,845 lb st) Rolls-Royce Turbomeca F405-RR-401 (navalised Adour Mk 871) non-afterburning turbofan. (FSD aircraft powered by a 24.24 kN; 5,450 lb st F405-RR-400L, equivalent to Adour Mk 861-49.) Garrett F124 being assessed as alternative power plant under US Congressional directive. Air intakes and engine starting as described for BAe Hawk. Fuel system similar to BAe Hawk, but with revision for carrier operation. Total internal capacity of 1,768 litres (467 US gallons; 380 Imp gallons). Provision for carrying one 591 litre (156 US gallon; 130 Imp gallon) drop tank on each underwing pylon.
ACCOMMODATION: Similar to BAe Hawk, except that ejection seats are of Martin-Baker Mk 14 NACES (Navy aircrew common ejection seat) zero/zero rocket assisted type.
SYSTEMS: Air-conditioning and pressurisation systems, using engine bleed air. Duplicated hydraulic systems, each 207 bars (3,000 lb/sq in), for actuation of control jacks, slats, flaps, airbrakes, landing gear, arrester hook and anti-skid

wheel brakes. No. 1 system has flow rate of 36.4 litres (9.6 US gallons; 8.0 Imp gallons)/min, No. 2 system a rate of 22.7 litres (6.0 US gallons; 5.0 Imp gallons)/min. Reservoirs nitrogen pressurised at 2.75-5.5 bars (40-80 lb/sq in). Hydraulic accumulator for emergency operation of wheel brakes. Pop-up Dowty Aerospace ram air turbine in upper rear fuselage provides emergency hydraulic power for flying controls in event of engine or No. 2 pump failure. No pneumatic system. DC electrical power from single brushless generator, with two static inverters to provide AC power and two batteries for standby power. Onboard oxygen generating system (OBOGS).

AVIONICS: Avionics and cockpit displays optimised for carrier-compatible operations. AN/ARN-182 UHF/VHF com radios and AN/ARN-144 VOR/ILS by Collins, Honeywell AN/APN-194 radio altimeter, Bendix/King APX-100 IFF, Sierra AN/ARN-136A Tacan, US Navy AN/USN-2 standard attitude and heading reference system (SAHRS), Smiths Industries Mini-HUD (front cockpit), Racal Acoustics avionics/com management system, GEC-Marconi yaw damper computer, Electrodynamics airborne data recorder and Teledyne caution/warning system. Digital avionics system from 1996 onwards: two 127×127 mm (5×5 in) monochrome multi-function screens in both cockpits, Navstar GPS and MIL-STD-1553B databus.

ARMAMENT: No built-in armament, but weapons delivery capability for advanced training is incorporated. Single pylon under each wing for carriage of practice multiple bomb rack, rocket pods or auxiliary fuel tank. Provision also for carrying single stores pod on fuselage centreline. CAI Industries gunsight in rear cockpit.

*DIMENSIONS, EXTERNAL:
Wing span	9.39 m (30 ft 9¾ in)
Wing chord: at root	2.87 m (9 ft 5 in)
at tip	0.89 m (2 ft 11 in)
Wing aspect ratio	4.99
Length: overall, incl nose probe	11.99 m (39 ft 4 in)
fuselage	10.89 m (35 ft 9 in)
Height overall	4.08 m (13 ft 4¾ in)
Tailplane span	4.59 m (15 ft 0¾ in)
Wheel track (c/l of shock struts)	3.90 m (12 ft 9½ in)
Wheelbase	4.31 m (14 ft 1¾ in)

* before USN modifications

AREAS:
Wings, gross	17.66 m² (190.1 sq ft)
Ailerons (total)	1.05 m² (11.30 sq ft)
Trailing-edge flaps (total)	2.50 m² (26.91 sq ft)
Airbrakes (total)	0.88 m² (9.47 sq ft)
Fin	2.61 m² (28.1 sq ft)
Rudder, incl tab	0.58 m² (6.24 sq ft)
Tailplane	4.43 m² (47.64 sq ft)

WEIGHTS AND LOADINGS:
Weight empty	4,619 kg (10,184 lb)
Internal fuel	1,440 kg (3,176 lb)
Max T-O weight	6,363 kg (14,028 lb)

PERFORMANCE (at max T-O weight):
Design limit diving speed at 1,000 m (3,280 ft)	575 knots (1,065 km/h; 662 mph)
Max true Mach number in dive	1.04
Max level speed at 2,440 m (8,000 ft)	543 knots (1,006 km/h; 625 mph)
Max level Mach number at 9,150 m (30,000 ft)	0.84
Max rate of climb at S/L	2,440 m (8,000 ft)/min
Time to 9,150 m (30,000 ft), clean	7 min 40 s
Service ceiling	12,190 m (40,000 ft)
T-O to 15 m (50 ft)	1,100 m (3,610 ft)
Landing from 15 m (50 ft)	1,009 m (3,310 ft)
Ferry range, internal fuel only (20 min reserves)	826 nm (1,532 km; 952 miles)
g limits	+7.33/-3

NAMC/PAC

PARTICIPATING COMPANIES:
Nanchang Aircraft Manufacturing Co: see under China
Pakistan Aeronautical Complex: see under Pakistan

NAMC/PAC K-8 KARAKORUM 8

TYPE: Tandem-seat jet basic trainer and light ground attack aircraft.

PROGRAMME: Launched publicly (as L-8) by NAMC at 1987 Paris Air Show as proposed export aircraft to be developed jointly with international partner; with Pakistan as that partner (25 per cent share), detail design began July 1987, with aircraft redesignated K-8 and named after mountain range forming part of China/Pakistan border. Construction started January 1989; three flying prototypes: 001 (first flight 21 November 1990), 003 (first flight 18 October 1991) and 004; nearly 500 flights made by January 1993; 002 is static test aircraft. Initial batch of 15 now in production (first flight late 1992); PAC to begin manufacturing tailplanes/elevators from ninth of these.

CUSTOMERS: Pakistan Air Force; PLA Air Force.

DESIGN FEATURES: Tapered, non-swept low wings, with NACA 64A-114 root and NACA 64A-412 tip sections; 2° incidence at root, 3° dihedral from roots; sweptback vertical/non-swept horizontal tail surfaces; intended for full basic flying training plus parts of primary and advanced syllabi, but capable also of light ground attack missions.

FLYING CONTROLS: Mechanically actuated primary control surfaces; variable incidence tailplane; trim tab in rudder and port elevator; two-position Fowler flaps, and split airbrake under each side of rear fuselage, hydraulically actuated; ailerons have hydraulic boost.

STRUCTURE: All-metal damage-tolerant main structure; ailerons of honeycomb, fin and rudder of composites. PAC will build horizontal tail and some other components, up to approx 25 per cent of airframe.

LANDING GEAR: Retractable tricycle type, with single wheel and oleo-pneumatic shock absorber on each unit. Main units retract inward into underside of fuselage; nosewheel, which has hydraulic steering, retracts forward. Mainwheel tyres size 561 × 169 mm, pressure 6.9 bars (100 lb/sq in). Chinese hydraulic disc brakes. Anti-skid units.

POWER PLANT: One 16.01 kN (3,600 lb st) Garrett TFE731-2A-2A turbofan, with FADEC, mounted in rear fuselage, with intake and splitter plate on each side of fuselage; to be licence built in China by SMPMC (Zhuzhou). Fuel in two flexible tanks in fuselage and one integral tank in wing centre-section, combined capacity 1,000 litres (264 US gallons; 220 Imp gallons); single refuelling point in fuselage. Provision for carrying one 250 litre (66 US gallon; 55 Imp gallon) drop tank on inboard pylon under each wing.

ACCOMMODATION: Instructor and pupil in tandem, on Martin-Baker CN10LW zero/zero ejection seats; rear seat elevated 28 cm (11 in). One-piece wraparound windscreen; canopy opens sideways to starboard. Cockpits pressurised and air-conditioned.

SYSTEMS: AiResearch air-conditioning and pressurisation system, with max differential of 0.27 bar (3.91 lb/sq in). Hydraulic system, pressure 207 bars (3,000 lb/sq in), for operation of landing gear extension/retraction, wing flaps, airbrakes, aileron boost, nosewheel steering and wheel brakes. Flow rate 15 litres (3.96 US gallons; 3.30 Imp gallons)/min, with air pressurised reservoir, plus emergency backup hydraulic system. Abex AP09V-8-01 pump. Electrical systems 28.5V DC (primary) and 24V DC (auxiliary), with 115/26V single-phase AC and 36V three-phase AC available, both at 400Hz. Gaseous oxygen system for occupants. Demisting of cockpit transparencies.

AVIONICS: Bendix/King avionics, including UHF/VHF and Tacan, in first two prototypes. Collins EFIS-86 system selected for first 100 aircraft, incorporating CRT primary flight and navigation displays for each crew member plus

K-8 004, the third flying prototype, at Asian Aerospace, Singapore, in February 1992 *(Paul Beaver)*

First prototype of the Sino-Pakistan NAMC/PAC K-8

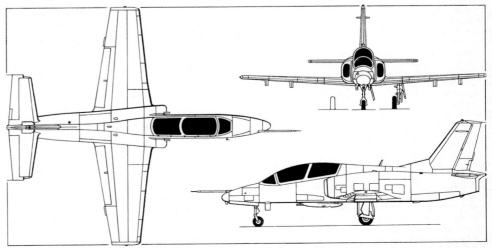
NAMC/PAC K-8 Karakorum 8 jet trainer and light attack aircraft *(Jane's/Mike Keep)*

dual display processing units and selector panels for tandem operation. Collins AN/ARC-186 VHF and Magnavox AN/ARC-164 UHF com radios. Interfaces to Collins EFIS-86T system include KNR 634A VOR/ILS with marker beacon receiver, ADF, KTU-709 Tacan, Type 265 radio altimeter, WL-7 radio compass, AHRS and air data computer. Standby flight instruments include ASI, rate of climb indicator, barometric altimeter, emergency horizon and standby compass. Blind-flying instrumentation standard.

ARMAMENT (optional): One 23 mm gun pod under centre-fuselage; self-computing optical gunsight in cockpit, plus gun camera. Two external stores points under each wing, capable of carrying gun pods, rocket pods, bombs, missiles, auxiliary fuel tanks (inboard pair only), or a reconnaissance pod. Each stores station can carry up to 250 kg (551 lb).

DIMENSIONS, EXTERNAL:
Wing span	9.63 m (31 ft 7¼ in)
Wing aspect ratio	5.45
Length overall, incl nose pitot	11.60 m (38 ft 0¾ in)
Height overall	4.21 m (13 ft 9¾ in)
Wheel track	2.43 m (7 ft 11¾ in)
Wheelbase	4.38 m (14 ft 4½ in)

AREAS:
Wings, gross	17.02 m² (183.2 sq ft)
Ailerons (total, incl tab)	1.096 m² (11.80 sq ft)
Trailing-edge flaps (total)	2.69 m² (28.95 sq ft)
Fin	1.976 m² (21.27 sq ft)
Rudder, incl tab	1.06 m² (11.41 sq ft)
Tailplane	2.716 m² (29.23 sq ft)
Elevators (total, incl tab)	1.084 m² (11.67 sq ft)

WEIGHTS AND LOADINGS:
Weight empty, equipped	2,687 kg (5,924 lb)
Max fuel: internal	780 kg (1,720 lb)
external (2 drop tanks)	390 kg (860 lb)
Max external stores load	950 kg (2,094 lb)
T-O weight clean	3,630 kg (8,003 lb)
Max T-O weight with external stores	4,330 kg (9,546 lb)
Max wing loading	254.4 kg/m² (52.11 lb/sq ft)
Max power loading	270.4 kg/kN (2.65 lb/lb st)

PERFORMANCE (at clean T-O weight):
Never-exceed speed (VNE)	512 knots (950 km/h; 590 mph) IAS
Max level speed at S/L	432 knots (800 km/h; 497 mph)
Approach speed	108 knots (200 km/h; 124 mph)
Unstick speed	100 knots (185 km/h; 115 mph)
Touchdown speed, 35° flap	89 knots (165 km/h; 103 mph)
Max rate of climb at S/L	1,620 m (5,315 ft)/min
Service ceiling	13,000 m (42,650 ft)
T-O run	410 m (1,345 ft)
Landing run	512 m (1,680 ft)
Range:	
max internal fuel	755 nm (1,400 km; 870 miles)
max internal/external fuel	1,214 nm (2,250 km; 1,398 miles)
Endurance: max internal fuel	3 h
max internal/external fuel	4 h 25 min
g limits	+7.33/−3

NH INDUSTRIES
NH INDUSTRIES sarl
Le Quatuor, Batiment C, 42 Route de Galice, F-13082 Aix en Provence, France
Telephone: 33 42 95 97 00
Fax: 33 42 95 97 49
GENERAL MANAGER: Jean-Pierre Barthélemy
PARTICIPATING COMPANIES:
  Agusta: see under Italy
  Eurocopter Deutschland: see this section
  Eurocopter France: see this section
  Fokker Aircraft: see under Netherlands

NH Industries established 1992 to manage design and development of NH 90; shares are Agusta 26.9 per cent, Eurocopter Deutschland 24.0 per cent, Eurocopter France 42.4 per cent, Fokker 6.7 per cent; NH Industries is prime contractor for design and development, industrialisation and production, logistic support, marketing and sales.

Joint agency NAHEMA (NATO Helicopter Management Agency) formed February 1992 by the four governments, within NATO framework, to manage programme; NAHEMA located alongside NH Industries in Aix en Provence.

NH INDUSTRIES NH 90
TYPE: Multi-role naval (NFH) and tactical transport (TTH) medium helicopter.
PROGRAMME: Initial studies by NIAG SG14 in 1983-84; September 1985 MoU between defence ministers of France, UK, Germany, Italy and Netherlands led to 14-month feasibility/pre-definition study for new naval/military NH 90 (NATO helicopter for the 1990s); initial design phase approved December 1986; second MoU in September 1987 led to pre-definition phase and completion of weapons system definition in 1988; UK withdrew from programme April 1987; German workshare reduced early 1990, Italian participation renegotiated later 1990; French and German launch decision 26 April 1990; design and development contract signed 1 September 1992; five prototypes and iron bird planned, with prototype basic vehicle produced in France, TTH in Germany and NFH in Italy; first flight P01 expected end 1995 and initial deliveries by 1999.
CURRENT VERSIONS: **NFH** (NATO Frigate Helicopter): Naval version, primarily for autonomous ASW and ASVW; additional applications include vertrep, SAR, transport and anti-air warfare support; designed for all-weather/severe ship motion environment; fully integrated mission system for crew of three (optionally four); ECM, anti-radar and IR protection systems.
  TTH (Tactical Transport Helicopter): Land-based army/air force version, primarily for tactical transport, airmobile operations and SAR; additional applications include tactical support, special EW, VIP transport and training; defensive weapons suite; rear-loading ramp/door to be provided for French Army to accommodate light armoured anti-tank missile vehicle; high manoeuvrability and survivability for NOE operation near front line; radar, FLIR and NVG compatibility integrated with high-precision navigation system for day and night; low radar signature, ECM, anti-radar and IR self-protection.
CUSTOMERS: Estimated total requirements (early 1992) for armed forces of participating countries are 726 (France 220, Italy 214, Germany 272, Netherlands 20), of which 182 are NFH and 544 TTH.
COSTS: Reported development cost $812 million. Eurocopter financing 18.45 per cent of cost; Eurocopter France pays 27.6 per cent of French share and Eurocopter Germany 22.5 per cent of German share.
DESIGN FEATURES: Titanium Spheriflex main rotor hub with elastomeric bearings; four composite blades; main rotor has French QA3 aerofoils with 3½ per cent apparent hinge offset, thickness/chord ratio tapering from 12 to 9 per cent, and parabolic tapered tips; higher harmonic control to damp vibration; main rotor turns at 256.6 rpm with 219 m (719 ft)/s tip speed; Spheriflex tail rotor with four blades of similar design and construction; tail rotor turns at 1,235.4 rpm with 207 m (679 ft)/s tip speed; automatic folding of main rotor blades and tail pylon in NFH; overall design will aim for low vulnerability/detectability, reduced maintenance requirements, and day/night operability within temperature range of −40°C to +50°C.
FLYING CONTROLS: Quadruplex fly-by-wire controls with sidestick controllers and pedals eliminates cross coupling between control axes.
STRUCTURE: All-composites fuselage with low radar signature; fail-safe design of structure, rotating parts and systems for high safety levels; rotor blades have two-box foam-filled spars with honeycomb rear section skinned GFRP/CFRP coated with radar absorbing paint and with nickel leading-edge; electric de-icing mats built-in.
LANDING GEAR: Retractable crashworthy tricycle gear with twin-wheel nose unit and single-wheel main units.
POWER PLANT: Two engines; required power from each engine during normal operation at 1,000 m (3,280 ft) in ISA + 15°C is 1,360 kW (1,824 shp) for 30 min, 1,250 kW (1,676 shp) max continuous, OEI max contingency (2½ min) 1,484 kW (1,990 shp), emergency at S/L ISA, 1,942 kW (2,604 shp). Choice of Rolls-Royce Turbomeca (with Piaggio and MTU), RTM 322-01/9 or General Electric (with Alfa Romeo) T700-T6E turboshafts; engine control by FADEC; engine RFP issued April 1992. Transmission rating 2,300 kW (3,084 shp) with both types of engine, 1,850 kW (2,481 shp) for 30 s OEI; main gearbox can run dry for 30 min; fuel system has crash-resistant self-sealing cells.
ACCOMMODATION: Minimum crew one pilot VFR and IFR; NFH crew pilot, co-pilot/TACCO on flight deck and one system operator in cabin; optionally, two pilots on flight deck and one TACCO, one SENSO in cabin; TTH crew two pilots or one pilot and one crewman on flight deck and

Mockup of NH 90 international military and naval helicopter

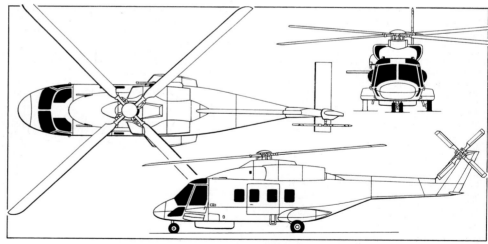
NH 90 NATO helicopter for the 1990s (*Jane's/Mike Keep*)

up to 20 equipped troops or one two-tonne tactical vehicle in cabin.

SYSTEMS: Full redundancy for all vital systems; hydraulic system has two mechanically driven and one electric pump; electrical system has two batteries and four AC generators feeding DC buses through transformer rectifiers; APU for electrical engine starting, environmental control on ground and emergency use on flight; fire detection and suppression in engine bays, APU and cabin; emergency flotation gear.

AVIONICS: Core avionic system based on dual MIL-STD-1553B data highways; several redundant computers allow integration of aircraft and mission equipment; automatic monitoring and diagnostic system for maintainability; mission and maintenance data presented to crew on redundant colour multi-function displays, automatically or on demand; integrated management of communication, identification and navigation systems, and of 360° radar, low-frequency dipping sonar, sonobuoys, AFCS, OTHT, MAD, EW, IR/radar decoy system, helmet-mounted display and missile launch detector; each mission system has its own STANAG 3838B redundant databus.

ARMAMENT: TTH can have area suppression and self-defence armament; NFH to carry air-to-surface missiles and anti-submarine torpedoes; air-to-air missiles optional.

DIMENSIONS, EXTERNAL:
Main rotor diameter 16.30 m (53 ft 5½ in)
Tail rotor diameter 3.20 m (10 ft 6 in)
Main rotor blade chord 0.65 m (2 ft 1½ in)
Tail rotor blade chord 0.32 m (1 ft 0½ in)
Length: overall, rotors turning 19.60 m (64 ft 3¾ in)
 fuselage, tail rotor turning 16.11 m (52 ft 10¼ in)
 folded (NFH) 13.50 m (44 ft 3½ in)
Height: folded (NFH) 4.10 m (13 ft 5½ in)
 overall, tail rotor turning 4.22 m (13 ft 10¼ in)
Width: over mainwheel fairings 3.69 m (12 ft 1¼ in)
 folded 3.80 m (12 ft 5½ in)
Wheel track 3.20 m (10 ft 6 in)
Wheelbase 6.083 m (19 ft 11½ in)
DIMENSIONS, INTERNAL:
Cabin: Length, excl rear ramp 4.00 m (13 ft 1½ in)
 Max width 2.00 m (6 ft 6¾ in)
 Max height 1.58 m (5 ft 2¼ in)
 Max volume 18.0 m³ (635.7 cu ft)
AREAS:
Main rotor disc 208.67 m² (2,246.1 sq ft)
Tail rotor disc 8.04 m² (86.57 sq ft)
WEIGHTS AND LOADINGS:
Weight empty, equipped: TTH 5,504 kg (12,134 lb)
 NFH 6,428 kg (14,171 lb)

Standard fuel (usable) 1,902 kg (4,193 lb)
Max payload (both) more than 2,000 kg (4,409 lb)
Max T-O weight: TTH 8,700 kg (19,180 lb)
 NFH 9,100 kg (20,062 lb)
Max disc loading: TTH 41.7 kg/m² (8.54 lb/sq ft)
 NFH 43.6 kg/m² (8.93 lb/sq ft)
Max power loading: TTH 3.78 kg/kW (6.22 lb/shp)
 NFH 3.96 kg/kW (6.50 lb/shp)
PERFORMANCE (estimated, at appropriate max T-O weight, ISA):
Dash speed at S/L:
 TTH 162 knots (300 km/h; 186 mph)
Max cruising speed at S/L:
 TTH 156 knots (290 km/h; 180 mph)
Normal cruising speed at S/L:
 TTH 140 knots (259 km/h; 161 mph)
Service ceiling: TTH 4,250 m (13,945 ft)
Absolute ceiling: TTH 6,000 m (19,685 ft)
Hovering ceiling: IGE: TTH 3,600 m (11,810 ft)
 OGE: TTH 3,000 m (9,850 ft)
Ferry range: TTH 650 nm (1,204 km; 748 miles)
Time on station, 60 nm (111 km; 69 miles) from base, 30 min reserves: NFH 3 h
Max endurance at 75 knots (140 km/h; 87 mph):
 NFH 5 h 5 min

PANAVIA
PANAVIA AIRCRAFT GmbH
Arabellastrasse 16 (PO Box 860629), D-8000 Munich 81, Germany
Telephone: 49 (89) 9217 238/9
Fax: 49 (89) 9217 903
Telex: 05 29 825
MANAGING DIRECTOR: Oskar Friedrich
PUBLIC RELATIONS: Ralf Wolf
PARTICIPATING COMPANIES:
 Alenia: see under Italy
 British Aerospace: see under UK
 Deutsche Aerospace: see under Germany

Panavia formed 26 March 1969 as industrial prime contractor to design, develop and produce an all-weather MRCA (multi-role combat aircraft) for air forces of UK, Germany (incl Navy), Italy and Netherlands; Netherlands withdrew July 1969, shareholdings then being readjusted to UK and Germany 42.5 per cent each, Italy 15 per cent. Tornado programme, one of largest European industrial ventures yet undertaken, is guided and monitored on behalf of the three governments by NAMMO (NATO MRCA Management Organisation), whose executive agency NAMMA (formed 15 December 1968) is co-located with Panavia; Tornado production involves three major versions: IDS (interdictor/strike), ECR (electronic combat and reconnaissance) and ADV (air defence variant).

PANAVIA TORNADO IDS
RAF designations: GR. Mks 1, 1A and 4

TYPE: All-weather close air support/battlefield interdiction, interdiction/counter-air strike, naval strike and reconnaissance aircraft.

PROGRAMME: Six-government feasibility study (originally involving Belgium and Canada) initiated 17 July 1968; project definition began 1 May 1969; development phase started 22 July 1970; structural design completed August 1972; first flight 14 August 1974 by first of nine prototypes (P01-09: see 1978-79 *Jane's* for details); Tornado name adopted September 1974; German procurement approved 19 May 1976; production programme initiated 29 July 1976 by three-government MoU for 809 aircraft in six batches (640 IDS, 165 ADV, plus four pre-series aircraft brought up to IDS production standard); first flight 5 February 1977 by first of six pre-series Tornados (PP11-16: see 1980-81 and earlier editions); Italian production approved 8 March 1977; first flights by production IDS in UK and Germany 10 and 27 July 1979 respectively; deliveries for operational conversion training (to Tri-national Tornado Training Establishment at RAF Cottesmore) began 1 July 1980; first flight by Italian production aircraft 25 September 1981; operational squadron deliveries began 1982 to RAF (6 January), Germany (Navy, 2 July) and Italy (27 August); first export delivery (to Saudi Arabia) March 1986; batch 7 contract (57 IDS, 35 ECR and 32 ADV) awarded 10 June 1986; contract to develop mid-life update for RAF GR. Mk 1 awarded 16 March 1989; 26 GR. Mk 4s (part of proposed batch 8) cancelled 18 June 1990. Final Italian aircraft (MM7088) delivered 12 July 1989; final German IDS (4622) 31 May 1989; final ECR and last German-built Tornado (4657) handed over 19 December 1991 and delivered 28 January 1992; final RAF GR. Mk 1 (ZG794) delivered 19 November 1992. New reconnaissance system sought by 1994 for 40 ex-naval aircraft being transferred to German Air Force.

CURRENT VERSIONS: **ECR:** Electronic combat and reconnaissance version, utilising IDS airframe; described separately.

 German Air Force IDS: Equips eight squadrons: two each with JBGs 31 (Nörvenich), 33 (Büchel) and 34 (Memmingen), one with JBG 32 (Lechfeld), all NATO-assigned, plus one with weapons training unit JBG 38 (Jever); others at TTTE in UK and WTD 61 at Manching; AG 51 to re-form at Schleswig/Jagel with 40 ex-Navy Tornados in reconnaissance role; all German Tornados cleared for in-flight refuelling by USAF KC-10A and KC-135 tankers; mid-life upgrade (MLU) programme under development for German and Italian Tornados, to confer more accurate navigation for blind attacks, improved capability for sortie generation, increased range and target acquisition capability, reduced penetration altitude, improved ESM, better threat suppression and greater reliability/maintainability. MLU confirmed, January 1993, as including FLIR, GPS and other avionics improvements; Apache modular stand-off weapon available from 1996. Principal weapons B61, MW-1, AIM-9L, AGM-65, AGM-88.

 German Navy IDS: Equips four operational squadrons: two (NATO-assigned) with Marinefliegergeschwader 1 (Jagel) and one with MFG 2 (Eggebeck) for strike missions against sea and coastal targets, plus one (1/MFG 2) for reconnaissance using MBB/Alenia multi-sensor pod; mid-life update under development (see preceding paragraph). Squadrons to be reduced to two (total 65 aircraft) with transfer of 40 Tornados from MFG 1 to air force. Principal weapons carried are BL 755, AIM-9L, AGM-88, Kormoran 1 and (from 1995) 2.

 Italian Air Force IDS: In service with three operational squadrons: Gruppi 154 (6° Stormo at Brescia-Ghedi), 155 (50° Stormo, Piacenza) and the Kormoran anti-shipping Gruppo 156 (36° Stormo, Gioia del Colle), plus the Reparto Sperimentale di Volo at Pratica di Mare and TTTE in UK; mid-life update under development (see German Air Force paragraph for details). All delivered. Principal weapons carried are B61, MW-1, AIM-9L, AGM-65D, AGM-88, Kormoran 1.

 Royal Air Force GR. Mk 1: UK IDS version, meeting Air Staff Requirement 392, equipping four NATO-assigned squadrons with RAF Germany (Nos. IX, 14, 17 and 31 at Brüggen) and two SACEUR-declared squadrons in UK (Nos. 27 and 617 at Marham), plus UK training units (TTTE at Cottesmore, TWCU at Lossiemouth), Strike/Attack Operational Evaluation Unit (Boscombe Down), A&AEE, and DRA Bedford (including ZA326 for REVISE stand-off missile trials). Squadron deliveries (to No. IX) began 6 January 1982; modification for tactical nuclear weapon carriage began 1984; first combat use (Gulf War) 17 January 1991; to be redesignated GR. Mk 4 (which see) after receiving mid-life update. Interim Phase 2 modifications introduced 1990 include RAM on forward-facing surfaces, Have Quick 2 secure radios, filters for NVG use, Navstar GPS and provision for 2,250 litre (594 US gallon; 495 Imp gallon) drop tanks. Mk XII Mode 4 IFF for 1991 Gulf War deployment. Principal weapons carried are WE177B, 1,000 lb bomb, JP 233, BL 755, AIM-9L, ALARM, Paveway II. No. IX Squadron operational with ALARM, 1 January 1993 (but weapon first used two years previously). Nos. 27 (re-numbered 12) and 617 Squadrons to convert to Sea Eagle missile (see GR. Mk 1B).

 Royal Air Force GR. Mk 1A: UK day/night all-weather tactical reconnaissance version, equipping No. 2 Squadron (Marham) and No. 13 (Honington); one development aircraft (ZA402, first flight 11 July 1985) followed by 15 others also converted from GR. Mk 1 (delivered from 3 April 1987) and 14 new-production Mk 1As (delivered from 13 October 1989 to 5 December 1990). Retains air-to-surface role except for deletion of guns; identifiable by small underbelly blister fairing (immediately behind laser rangefinder pod) and transparent side panels for BAe SLIR (sideways looking infra-red) system and Vinten Linescan 4000 surveillance system; has Computing Devices Co signal processing and video recording system (first video-based tac/recce system with replay facility) offering capability for future real-time reconnaissance data relay; first operational mission in Gulf War, night of 18/19 January 1991.

 Royal Air Force GR. Mk 1B: Retrofit by BAe at Warton of 24 GR. Mk 1s for maritime attack; IOC 1994 with four BAe Sea Eagle (or two, plus two drop tanks); second-

RAF Tornado GR. Mk 1s with CBLS 200 practice-bomb carriers under fuselage (*Paul Jackson*)

172   INTERNATIONAL: AIRCRAFT—PANAVIA

German Navy Tornado IDS with SFC 28-300 buddy refuelling pod *(Paul Jackson)*

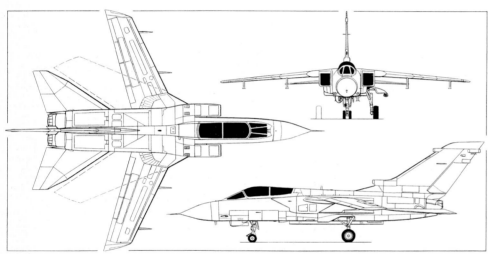

Panavia Tornado IDS multi-role combat aircraft *(Dennis Punnett)*

phase modification to permit up to five Sea Eagles (or four, plus one tank). First aircraft is ZA407.

Royal Air Force GR. Mk 4: Designation to be applied to GR. Mk 1s after receiving MLU (mid-life update) in satisfaction of Air Staff Requirement 417; modifications include new Marconi Defence Systems EW suite; updated weapon control system, advanced video recording system with ground replay facility, and computer loading system; new Ferranti HUD with computer generated symbology, TF display 3D ground profile, and digital storage/image generating system for moving map display; Smiths colour CRT head-down display. Three development aircraft undergoing modification in early 1993; prototype is pre-series aircraft P15/XZ631; ZD708 and ZG773 to follow. Planned terrain-referenced navigation cancelled.

Royal Saudi Air Force IDS: Equips two squadrons (Nos. 7 and 66), each including six configured for reconnaissance; deliveries March 1986 to October 1987 (first 20); further 28 (completing initial order) between May 1989 and 25 November 1991. Second batch of 48 ordered January 1993; deliveries from 1996. Principal weapons carried are JP 233, 1,000 lb bomb, AIM-9L, ALARM; Sea Eagle on order.

Tornado 2000: Proposed successor to RAF GR. Mk 4; would be single-seat, optimised for low-level/high-speed/long-range penetration and able to carry standoff weapons; longer fuselage than current IDS, containing increased fuel tankage; faceted nose section and pitot intakes to minimise radar signature.

CUSTOMERS: Total of 780 IDS/ECR (plus four refurbished pre-series IDS) ordered for Germany (359: Air Force 157 strike incl two pre-series, 35 ECR and 55 dual control, Navy 112 strike incl 12 dual), Italy (100: Air Force 88 strike incl one pre-series, plus 12 dual), Saudi Arabia (96: Air Force 28 attack, six recce, 14 dual and 48 unknown) and UK (229: Air Force 164 strike, 14 recce, and 51 dual incl one pre-series). Total of 732 IDS delivered by 31 December 1992.

COSTS: $7,500 million programme for 48 aircraft, RSAF, 1993.

DESIGN FEATURES: Continuously variable geometry shoulder wings, with leading-edge sweep angles from 25° (minimum) to 67° (maximum) on movable portions (limited to 63° when 2,250 litre drop tanks carried), 60° on fixed inboard portions; modest overall dimensions; high wing loading to minimise low-altitude gust response; swivelling wing pylons to retain stores alignment with fuselage.

FLYING CONTROLS: Full span double-slotted fixed-vane flaperons (four segments per side), all-moving tailplane (tailerons) and inset rudder, all actuated by electrically controlled tandem hydraulic jacks; full span wing leading-edge slats (three segments each side); two upper-surface spoilers/lift dumpers forward of each central pair of flaperons; tailerons operate together for pitch control and differentially for roll control; spoilers provide augmented roll control at unswept and intermediate wing positions at low speed; Krueger flap on leading-edge of each wing glove box; door type airbrake each side on top of rear fuselage; wing sweep hydraulically powered via ballscrew actuators (aircraft can land safely with wings fully swept if sweep mechanism fails); triple-redundant CSAS (command stability augmentation system), APFD (autopilot/flight director) and TFE (terrain following E-scope), as detailed under Avionics.

STRUCTURE: Basically all-metal (mostly aluminium alloy with integrally stiffened skins, titanium alloy for wing carry-through box and pivot attachments); FRP for nosecone, dielectric panels and interface between fixed and movable portions of wings; Teflon plated wing pivot bearings; elastic seal between outer wings and fuselage sides; nosecone hinges sideways to starboard for access to radar antennae; slice of fuselage immediately aft of nosecone also hinges to starboard for access to forward avionics bay and rear of radars; passive ECM antenna fairing near top of fin; ram air intake for heat exchanger at base of fin. Alenia builds entire outer wings (incl moving surfaces), with Microtecnica as prime contractor for sweep system; BAe (Warton) builds front and rear fuselage portions (incl engine installation) and entire tail unit; DASA is prime contractor for centre-fuselage (incl engine intake ducts, wing centre-section box, pivot mechanism, and interface with outer wings); radar-transparent nosecone by Telefunken System-Technik.

The following details apply to the basic IDS production version; subsystem details are listed by team leader only, for the sake of clarity.

LANDING GEAR: Hydraulically retractable tricycle type, with forward retracting twin-wheel steerable nose unit. Single-wheel main units retract forward and upward into centre section of fuselage. Emergency extension system, using nitrogen gas pressure. Development and manufacture of complete landing gear and associated hydraulics headed by Dowty (UK). Dunlop aluminium alloy wheels, hydraulic multi-disc brakes and low-pressure tyres (to permit operation from soft, semi-prepared surfaces) and Goodyear anti-skid units. Mainwheel tyres size 30 × 11.50-14.5, Type VIII (24 or 26 ply); nosewheel tyres size 18 × 5.5, Type VIII (12 ply). Runway arrester hook beneath rear of fuselage.

POWER PLANT: Two Turbo-Union RB199-34R turbofans, fitted with bucket type thrust reversers and installed in rear fuselage with downward opening doors for servicing and engine change. Mk 101 engines of early production aircraft nominally rated at 38.7 kN (8,700 lb st) dry and 66.0 kN (14,840 lb st) with afterburning (uninstalled); RAF aircraft have engines downrated to 37.7 kN (8,475 lb st) in squadron service (37.0 kN; 8,320 lb st dry for TTTE) to extend service life. Mk 103 engines, introduced in May 1983 (engine number 761), dry rated nominally at 40.5 kN (9,100 lb st) uninstalled (38.5 kN; 8,650 lb st for RAF) and provide 71.5 kN (16,075 lb st) with afterburning. RAF ordered 100 modification kits in 1983 to upgrade Mk 101 engined aircraft to Mk 103 standard. All internal fuel in multi-cell Uniroyal self-sealing integral fuselage tanks and/or wing box tanks, all fitted with press-in fuel sampling and water drain plugs, and all refuelled from a single-point NATO connector. Wing tanks total 955 litres (252 US gallons; 210 Imp gallons); fuselage tanks total 4,887 litres (1,290 US gallons; 1,075 Imp gallons). Additional 551 litre (145.5 US gallon; 121 Imp gallon) tank in fin (on RAF aircraft only). Detachable and retractable in-flight refuelling probe can be mounted on starboard side of fuselage, adjacent to cockpit. Provision for one or two drop tanks to be carried on underfuselage shoulder pylons (1,500 litres; 396 US gallons; 330 Imp gallons) and single tanks on inboard underwing pylons (1,500 or 2,250 litres; 396 or 594 US gallons; 330 or 495 Imp gallons). Some German Navy, Italian Air Force and (from 1991) RAF aircraft adapted to carry a Sargent-Fletcher Type 28-300 1,135 litre (300 US gallon; 250 Imp gallon) buddy type hose/drogue refuelling pod. Dowty afterburning fuel control system.

ACCOMMODATION: Crew of two on tandem Martin-Baker Mk 10A zero/zero ejection seats under Kopperschmidt one-piece, rear-hinged, upward opening canopy. Flat centre armoured windscreen panel and curved side panels, built by Lucas Aerospace, incorporate Sierracote electrically conductive heating film for windscreen anti-icing and demisting. Canopy (and windscreen in emergency) demisted by engine bleed air. Windscreen, hinged at front, can be opened forward and upward, allowing access to back of pilot's instrument panel. Seats provide safe escape at zero altitude and at speeds from zero up to 630 knots (1,166 km/h; 725 mph) IAS.

SYSTEMS: Cockpit air-conditioned and pressurised (max differential 0.36 bar; 5.25 lb/sq in) by Normalair-Garrett conventional air cycle system (with bootstrap cold air unit) using engine bleed air with ram air precooler, Marston intercooler and Teddington temperature control system. Nordmicro air intake control system, and Dowty engine intake ramp control actuators. Two independent hydraulic systems, each of 276 bars (4,000 lb/sq in pressure), are supplied from two separate, independently driven Vickers pumps, each mounted on an engine accessory gearbox. Each system supplied from separate bootstrap type reservoir. Systems provide fully duplicated power for primary flight control system, tailerons, rudder, flaps, slats, wing sweep, pitch Q-feel system, and refuelling probe. Port system also supplies power for Krueger flaps, inboard spoilers, port air intake ramps, canopy, and wheel brakes; starboard system for airbrakes, outboard spoilers, starboard air intake ramps, landing gear, nosewheel steering, and radar stabilisation and scanning. Main system includes Dowty accumulators and Teves power pack. Fairey Hydraulics system for actuation of spoilers, rudder and taileron control. Provision for reversion to single-engine drive of both systems, via mechanical cross-connection between two engine auxiliary gearboxes, in event of single engine failure. In event of double engine flameout, emergency pump in No. 1 system has sufficient duration for re-entry into engine cold relight boundary. Flying control circuits protected from loss of fluid due to leaks in other circuits by isolating valves which shut off utility circuits if reservoir contents drop below predetermined safety level. Electrical system consists of a 115/200V AC three-phase 400Hz constant frequency subsystem and 28V DC subsystem. Power generated by two Rotax automatically controlled oil-cooled brushless AC generators integrated with a constant speed drive unit and driven by engines via KHD accessory gearbox. Normally, each engine drives its own accessory gearbox, but provision also made for either engine to drive opposite gearbox through cross-drive system. In event of generator failure, remaining unit can supply total aircraft load. Both gearboxes and generators can be driven by APU when aircraft is on ground. Generators supply two main AC busbars and an AC essential busbar. DC power provided from two fan-cooled transformer/rectifier units (power being derived from main AC system), these feeding power to two main DC busbars, one essential DC busbar and a battery busbar. Either TRU can supply total aircraft DC load. Fifth DC busbar provided for maintenance purposes only. Rechargeable nickel-cadmium battery provides power for basic flightline servicing and starting APU; in event of main electrical system or double TRU failure, it is connected automatically to essential services busbar to supply essential electrical loads. Normalair-Garrett demand type oxygen system, using 10 litre (2.6 US gallon; 2.2 Imp gallon) lox converter. Emergency oxygen system on each seat. GEC Avionics flow metering system. Eichweber fuel gauging system and Flight Refuelling flexible couplings. Graviner fire detection and extinguishing systems. Rotax contactors. Smiths engine speed and temperature indicators. Telefunken SystemTechnik intake de-icing system.

AVIONICS: Communications equipment includes GEC-Plessey PTR 1721 (UK and Italy) or Rohde und Schwarz (Germany) UHF/VHF transceiver; Telefunken SystemTechnik UHF/ADF (UK and Germany only); SIT emergency UHF with Rohde und Schwarz switch; BAe HF/SSB aerial tuning unit; Rohde und Schwarz (UK and Germany) or Montedel (Italy) HF/SSB radio; Ultra communications control system; GEC-Marconi central suppression unit (CSU); Leigh voice recorder; Chelton UHF communications and landing system aerials.

Primary self-contained nav/attack system includes a European built Texas Instruments multi-mode forward looking, terrain following ground mapping radar; GEC-Marconi FIN 1010 three-axis digital inertial navigation system (DINS) and combined radar/map display; Decca Type 72 Doppler radar system, with Kalman filtering of Doppler and inertial inputs for extreme navigational accuracy; Microtecnica air data computer; Litef Spirit 3 central digital computer (64K initially, but progressively upgraded to 256K); Alenia radio/radar altimeter (to be replaced on RAF aircraft by GEC-Marconi AD1990 covert radar altimeter); Smiths electronic HUD with Davall camera; GEC-Marconi nose-mounted laser rangefinder and marked target seeker; GEC-Marconi TV tabular display; Astronautics (USA) bearing, distance and heading indicator and contour map display. Defensive equipment includes Siemens (Germany) or Cossor SSR-3100 (UK and Saudi Arabia) IFF transponder; and Elettronica ARI 23284 RWR (being replaced in GR. Mk 1 from 1987 by Marconi Defence Systems Hermes RHWR). Production batches 6 and 7 (556th IDS onwards) incorporate MIL-STD-1553B databus, upgraded radar warning equipment and active ECM, improved missile control unit, and integration of HARM anti-radar missile.

Flight control system includes GEC-Marconi triplex command stability augmentation system (CSAS), incorporating fly-by-wire and autostabilisation; GEC-Marconi autopilot and flight director (APFD), using two self-monitoring digital computers; GEC-Marconi triplex transducer unit (TTU), with analog computing and sensor channels; GEC-Marconi terrain following E-scope (TFE); Fairey quadruplex electrohydraulic actuator; and Microtecnica air data set. APFD provides preselected attitude, heading or barometric height hold, heading and track acquisition, and Mach number or airspeed hold with autothrottle. Flight director operates in parallel with, and can be used as backup for, autopilot, as a duplex digital system with extensive range of modes. Automatic approach, terrain following and radio height-holding modes also available. Other instrumentation includes Smiths HSI, VSI and standby altimeter; Lital standby AHRS; SEL (with Setac) or (in UK aircraft) GEC-Marconi AD2770 (without Setac) Tacan; Cossor CILS 75/76 ILS; Bodenseewerk attitude director indicator; Dornier flight data recorder. Marconi Sky Shadow (jamming/deception) and BOZ 101 (Germany), 102 (Italy) or 107 (UK) chaff/flare ECM pods. Elta (Israel)/Telefunken SystemTechnik Cerberus II, III or (from late 1994) IV jammer pods on German and Italian aircraft. GEC-Marconi TIALD (thermal imaging airborne laser designator) night/adverse visibility pods for RAF Nos. 31 and 617 Squadron Tornados. (Can also carry similar Thomson-CSF CLDP pod.) Various terrain reference navigation systems also developed, including BAe Terprom, GEC-Marconi Spartan and Penetrate, as are night vision systems incorporating FLIR and NVGs. German Navy and Italian Air Force Tornados can carry DASA/Alenia multi-sensor reconnaissance pod on centre-line pylon. RAF GR. Mk 1A fitted with infra-red cameras in ammunition bay. Dornier onboard life monitoring system (OLMOS) retrofitted to all German Air Force Tornados 1991-92.

ARMAMENT: Fixed armament comprises two 27 mm IWKA-Mauser cannon, one in each side of lower forward fuselage, with 180 rds/gun. Other armament varies according to version, with emphasis on ability to carry wide range of advanced weapons. GEC-Marconi stores management system; Sandall Mace 355 and 762 mm (14 and 30 in) ejector release units standard on UK Tornados; German and Italian aircraft use multiple weapon carriage system (MWCS) ejector release units. ML Aviation CBLS 200 practice bomb carriers also standard. Battlefield interdiction version capable of carrying weapons for hard or soft targets. Weapons carried on seven fuselage and wing hardpoints: one centreline pylon fitted with single ejection release unit (ERU), two fuselage shoulder pylons each with three ERUs, and, under each wing, one inboard and one outboard pylon each with single ERU. Among weapons carried (see also Current Versions) are B61 and British WE177B nuclear bombs; Sidewinder air-to-air, and ALARM or HARM anti-radiation missiles; JP 233 low-altitude airfield attack munition dispenser; Paveway II laser guided bomb; Maverick, Sea Eagle and Kormoran air-to-surface missiles; napalm; BL755 cluster bombs (277 kg; 611 lb Mk 1 or 264 kg; 582 lb Mk 2); MW-1 munitions dispenser; 1,000 lb bombs; smart or retarded bombs; BLU-1B 750 lb fire bombs; Matra 250 kg ballistic and retarded bombs; Lepus flare bombs; LAU-51A and LR-25 rocket launchers. Matra Apache stand-off weapons dispenser from 1996.

Tornado GR. Mk 1A reconnaissance variant in the markings of No. 13 Squadron, RAF *(Paul Jackson)*

Saudi Arabia ordered 48 more Tornados in 1993, when early-production RSAF IDS aircraft (illustrated) were undergoing upgrading by BAe at Warton

DIMENSIONS, EXTERNAL:
Wing span: fully spread	13.91 m (45 ft 7½ in)
fully swept	8.60 m (28 ft 2½ in)
Length overall	16.72 m (54 ft 10¼ in)
Height overall	5.95 m (19 ft 6¼ in)
Tailplane span	6.80 m (22 ft 3½ in)
Wheel track	3.10 m (10 ft 2 in)
Wheelbase	6.20 m (20 ft 4 in)

AREAS:
Wings, gross (to fuselage c/l, 25° sweepback)	26.60 m² (286.3 sq ft)

WEIGHTS AND LOADINGS:
Basic weight empty	approx 13,890 kg (30,620 lb)
Weight empty, equipped	14,091 kg (31,065 lb)
Fuel:	
internal: wing/fuselage tanks	4,650 kg (10,251 lb)
fin tank (RAF only)	440 kg (970 lb)
drop tanks (each): 1,500 litre	1,197 kg (2,640 lb)
2,250 litre	1,796 kg (3,960 lb)
Nominal max external stores load	more than 9,000 kg (19,840 lb)
Max T-O weight:	
clean, full internal fuel	20,411 kg (45,000 lb)
with external stores	approx 27,950 kg (61,620 lb)

PERFORMANCE:
Max Mach number in level flight at altitude, clean	2.2
Max level speed, clean	above 800 knots (1,480 km/h; 920 mph) IAS
Max level speed with external stores	Mach 0.92 (600 knots; 1,112 km/h; 691 mph)
Landing speed	approx 115 knots (213 km/h; 132 mph)
Time to 9,150 m (30,000 ft) from brake release	less than 2 min
Automatic terrain following	down to 61 m (200 ft)
Required runway length	less than 900 m (2,950 ft)
Landing run	370 m (1,215 ft)
Max 360° rapid roll clearance with full lateral control	4g
Radius of action with heavy weapons load, hi-lo-lo-hi	750 nm (1,390 km; 863 miles)
Ferry range	approx 2,100 nm (3,890 km; 2,420 miles)
g limit	+7.5

PANAVIA TORNADO ECR

TYPE: Electronic combat and reconnaissance version of Tornado IDS.

PROGRAMME: Selected by German Luftwaffe to supplement existing in-service tactical reconnaissance aircraft such as Wild Weasel F-4G Phantom; 35 included in batch 7 production contract signed 10 June 1986; two development aircraft (s/n 9803 and 9879) converted from IDS (first

174　INTERNATIONAL: AIRCRAFT—PANAVIA

Tornado IT ECR (electronic combat and reconnaissance) aircraft of the Italian Air Force, carrying two underfuselage HARM anti-radiation missiles

flight 18 August 1988); first production aircraft (s/n 4623) made first flight 26 October 1989; deliveries (to 2/JBG 38 initially, later to 2/JBG 32) began 21 May 1990; ended with s/n 4657, 28 January 1992, but not operational until April 1993 when emitter location system became available. Italian prototype (conversion of IDS MM7079) rolled out 19 March 1992; first flight 20 July 1992.

CURRENT VERSIONS: **ECR:** Standoff reconnaissance and border control, reconnaissance via image-forming and electronic means, electronic support, and employment of anti-radar guided missiles. Italian equivalent designated **IT ECR**.

CUSTOMERS: Germany (Air Force 35 new-build); Italy converting 16 IDS to IT ECR for 102° Gruppo of 6° Stormo at Ghedi.

POWER PLANT (German ECR only): Mk 105 version of RB199 engine, providing approx 10 per cent more thrust than Mk 103.

AVIONICS: Include Texas Instruments ELS (emitter location system) initially fitted in five German trials aircraft in 1992; Honeywell/Sondertechnik infra-red linescan; Zeiss PAMIR-N (Passive Modular Infra-Red) system; onboard processing/storing/transmission systems for reconnaissance data; advanced tactical displays for pilot and weapons officer. IT ECR has video recording of data (dry silver film in German variant) and different radar warning system.

ARMAMENT: Both internal cannon deleted; external load stations can be used for ECR or fighter-bomber missions, or a combination of both; normally configured to carry two HARM missiles, two AIM-9L Sidewinders, active ECM pod, chaff/flare dispenser pod and two 1,500 litre (396 US gallon; 330 Imp gallon) underwing fuel tanks.

PANAVIA TORNADO ADV
RAF designations: Tornado F. Mks 2, 2A and 3

TYPE: All-weather air defence interceptor, air superiority fighter and combat patrol aircraft.

PROGRAMME: Feasibility studies for ADV (air defence variant) for UK, begun in 1968, given impetus by MoD Air Staff Requirement 395 of 1971 for interceptor with advanced radar and Sky Flash air-to-air missiles; full scale development authorised 4 March 1976; three prototypes (first flight 27 October 1979) included in production batch 1; first flight by F. Mk 2 production aircraft 5 March 1984; last F. Mk 2 delivered 9 October 1985; first flight by F. Mk 3 made 20 November 1985; first export order (by Saudi Arabia) placed 26 September 1985. Production ended 1993.

CURRENT VERSIONS: **Royal Air Force F. Mk 2:** Designation of first 18 (batch 4) production ADVs, with RB199 Mk 103 engines (further details in 1989-90 and earlier *Jane's*); currently in store except for one each with BAe at Warton, A&AEE at Boscombe Down, DRA at Farnborough. DRA aircraft (ZD902; dual control) first flown 18 August 1992 as TIARA — Tornado Integrated Avionics Research Aircraft; to receive Blue Vixen radar (Sea Harrier FRS. Mk 2 programme), GEC-Marconi IRST system, Marconi Defence Systems RHWR and JTIDS Class 2. Front cockpit configured as single-seat fighter, with three multi-function displays, helmet-mounted sight and, possibly voice-activated functions.

Royal Air Force F. Mk 2A: Designation to be applied to F. Mk 2s after being upgraded largely to F. Mk 3 standard except for retention of RB 103 engines. Intended upgrade in doubt; 14 aircraft remain in storage.

Royal Air Force F. Mk 3: Current definitive production version (batches 5-7), delivered from 28 July 1986 (to No. 229 OCU/65 Squadron at Coningsby, which became No. 56 [Reserve] Squadron on 1 July 1992); now equips seven UK air defence squadrons (Nos. 5 and 29 at Coningsby, Nos. 11, 23 and 25 at Leeming and Nos. 43 and 111 at Leuchars) although reduction to six is possible; primary missions are air defence of UK, protection of NATO's northern and western approaches, and long-range air defence of UK maritime forces; main differences from F. Mk 2 are uprated (Mk 104) engines, automatic wing sweep and manoeuvring systems, and improved avionics (see Flying Controls, Structure, Power Plant and Avionics paragraphs). Stage 1 update (new-build from Block 13, plus retrofits), from 1989, introduced HOTAS-type 'combat stick' for pilot, type 'AA' radar upgrade, improvements to Hermes RHWR, 5 per cent combat boost switch for power plants, and chaff/flare dispensers beneath rear fuselage. All except chaff/flare systems implemented by early 1992, although auto wing sweep disconnected. Stage 2G radar upgrade under trial in 1992 for installation in mid-1990s.

Royal Saudi Air Force ADV: Equips Nos. 29 and 34 Squadrons at Dhahran; deliveries began (to No. 29) 20 March 1989; ended 8 October 1990.

CUSTOMERS: Total 197 ordered for UK (RAF 173 incl 52 dual control) and Saudi Arabia (Air Force 24 incl six dual); RAF total includes eight dual control aircraft transferred from cancelled Omani order. Final delivery (ZH559) Spring 1993.

DESIGN FEATURES: Structural changes reduce drag, especially at supersonic speed, compared with IDS version, and longer fuselage provides more space for avionics and additional 10 per cent internal fuel.

FLYING CONTROLS: Similar to IDS, but with AWS (automatic wing sweep), AMDS (automatic manoeuvre device system) and SPILS (spin prevention and incidence limiting system); AWS allows scheduling of four different sweep angles (25° at speeds up to Mach 0.73, 45° from there up to Mach 0.88, 58° up to Mach 0.95 and 67° above Mach 0.95), enabling specific excess power at transonic speeds and turning capability at subsonic speeds to be maximised; buffet-free handling can be maintained, to limits defined by SPILS, by using AMDS, which schedules with wing incidence to deploy either flaperons and slats at 25° sweep angle or slats-only at 45° (beyond 45°, both flaperons and slats are scheduled in). AWS provisions embodied in late production aircraft but not activated (is reactive system; air combat requires wing sweep selection in advance of manoeuvre). Fly-by-wire CSAS/APFD system modified for increased roll rate and reduced pitch stick forces.

STRUCTURE: Generally as IDS version except: fuselage lengthened forward of front cockpit to accommodate longer radome, and aft of rear cockpit to allow Sky Flash missiles to be carried in two tandem pairs; CG shift compensated by extending fixed inboard portions of wings to increase chord and give 67° leading-edge sweep angle; Krueger flaps deleted; afterburner nozzles extended by 360 mm (14 in) on F. Mk 3, requiring modification to adjacent contours of rudder and tailerons; port internal gun deleted; wing/tailplane/fin leading-edges of some F. Mk 3s

Royal Air Force Tornado F. Mk 3 of No. 11 Squadron *(Paul Jackson)*

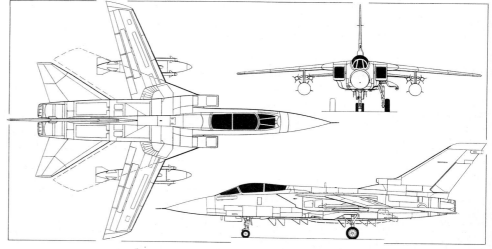

Panavia Tornado F. Mk 3 all-weather air defence interceptor *(Dennis Punnett)*

coated with radar absorbent material (RAM) for early 1991 Gulf operations; inner faces of engine air intake ducts similarly treated. Fatigue diminution measures for future embodiment include 80 kg (176 lb) wingtip weights and wing fuel auto hold fuel management system which empties fuselage tanks first–latter installed in ZH552 of No. 56 Squadron in 1993.

LANDING GEAR: As IDS version, but nosewheel steering augmented to minimise 'wander' on landing.

POWER PLANT: Two Turbo-Union RB199-34R Mk 104 turbofans, each with uninstalled rating of 40.5 kN (9,100 lb st) dry and 73.5 kN (16,520 lb st) with afterburning. Lucas digital engine control. Internal fuel capacity (incl fin tank) 7,143 litres (1,886 US gallons; 1,571 Imp gallons). Internally mounted, fully retractable in-flight refuelling probe in port side of nose, adjacent to cockpit. Provision for drop tanks of 1,500 or 2,250 litres (396 or 594 US gallons; 330 or 495 Imp gallons) capacity to be carried on shoulder pylons and underwing pylons.

ACCOMMODATION: As for IDS version.

SYSTEMS: Generally as described for IDS version, with addition of radar-dedicated cold air unit to cool Foxhunter radar, and pop-up ram air turbine to assist recovery in event of engine flameout at high altitude in zoom climb.

AVIONICS: Among those in IDS retained in ADV are communications equipment (GEC-Plessey VHF/UHF transceiver, SIT emergency UHF, Rohde und Schwarz HF/SSB, Ultra communications control system and Epsylon cockpit voice recorder); GEC-Marconi triplex fly-by-wire CSAS and APFD; Litef Spirit 3 central digital computer (capacity 224K) and data transmission system; Smiths electronic head-up and navigator's radar display; GEC-Marconi FIN 1010 INS (to which is added a second 1010 to monitor HUD); GEC-Marconi Tacan; Cossor ILS; and Cossor IFF transponder. Those deleted include Texas Instruments nose radar, Decca 72 Doppler radar with terrain following, GEC-Marconi laser rangefinder and marked target seeker, and Lital standby AHRS.

Nose-mounted Marconi Defence Systems AI Mk 24 Foxhunter multi-mode track-while-scan pulse Doppler radar with FMICW (frequency modulated interrupted continuous wave), with which is integrated new Cossor IFF-3500 interrogator and radar signal processor to suppress ground clutter. This system is intended to enable aircraft to detect targets more than 100 nm (185 km; 115 miles) away, and to track several targets simultaneously. A ground mapping mode for navigation backup is available. GEC-Marconi is subcontractor for Foxhunter transmitter and aerial scanning mechanism. 'AA' standard Foxhunter, meeting original RAF specification, installed from early 1989 (beginning ZE862 and first RSAF aircraft; ie, Block 13); earlier radars modified from Type 'W' and Type 'Z'. New data processor, being introduced during mid-1990s, offers final Foxhunter standard considerably more capable than earlier versions of this radar, in particular more automation to improve close combat capability. Modification kits will bring radars already in service up to new 'AB' standard. Pilot's HDD is added, GEC-Marconi displayed data video recorder (DDVR) replaces navigator's wet-film display recorder, and Marconi Defence Systems Hermes modular RHWR added. HUD/HDD are on front instrument panel only, radar control and data link presentations on rear panel only; both panels have weapon control and RHWR displays. GEC-Marconi FH 31A AC driven 3 in horizon gyro in rear cockpit, in addition to providing attitude display for navigator, feeds pitch and roll signals to other avionics systems in aircraft in certain modes. Lucas digital electronic engine control unit (DECU 500). ESM (electronic surveillance measures) and ECCM standard; installation of JTIDS data link due from 1993, initially in RAF Block 15/16 aircraft; ADV can contribute significantly to transfer of vital information over entire tactical area and can, if necessary, partially fulfil roles of both AEW and ground based radar. Smiths Industries/Computing Devices Company missile management system (MMS), which also controls tank jettison, has provision for pilot override, optimised for visual attack. Studies being undertaken for 1553B multiplex digital databus associated with AMRAAM and Sidewinder replacement.

Two Tracor AN/ALE-40(V) chaff/flare dispensers beneath rear fuselage of RSAF aircraft; initially on some RAF Mk 3s, but replaced early 1991 by Vinten VICON 78 Srs 210 flare dispensers; VICON 78 Srs 400 to be standard on all RAF Mk 3s; Bofors Phimat chaff dispenser on starboard outer Sidewinder pylon (RAF, optional); future installation of Bofors BOL 304 dispenser in rear of Sidewinder launch rails.

ARMAMENT: Fixed armament of one 27 mm IWKA-Mauser cannon in starboard side of lower forward fuselage. Four BAe Sky Flash semi-active radar homing medium-range air-to-air missiles semi-recessed under centre-fuselage, carried on internally mounted Frazer-Nash launchers; two European built AIM-9L Sidewinder infra-red homing short-range air-to-air missiles on each inboard underwing station (outboard stations not used on RAF ADVs). Sky Flash missiles, each fitted with MSDS monopulse seeker head, can engage targets at high altitude or down to 75 m (250 ft), in face of heavy ECM, and at standoff ranges of more than 25 nm (46 km; 29 miles). Release system permits missile to be fired over Tornado's full flight envelope. For future, ADV will be able to carry, instead of Sky Flash and Sidewinder, up to six Hughes AIM-120 AMRAAM or BAe Active Sky Flash medium-range and four new-generation short-range air-to-air missiles.

DIMENSIONS, EXTERNAL: As for IDS version, except:
Length overall 18.68 m (61 ft 3½ in)

WEIGHTS AND LOADINGS (approx):
Operational weight empty 14,500 kg (31,970 lb)
Fuel:
   internal: wing/fuselage tanks 5,250 kg (11,574 lb)
   fin tank as for IDS version
   drop tanks as for IDS version
Max external fuel 5,806 kg (12,800 lb)
Nominal max external stores load 8,500 kg (18,740 lb)
Max T-O weight 27,986 kg (61,700 lb)

PERFORMANCE:
Max Mach number in level flight at altitude, clean 2.2
Max level speed, clean 800 knots (1,480 km/h; 920 mph) IAS
Rotation speed, depending on AUW 145-160 knots (269-297 km/h; 167-184 mph)
Normal touchdown speed 115 knots (213 km/h; 132 mph)
Demonstrated roll rate at 750 knots (1,390 km/h; 864 mph) and up to 4 g 180°/s
Operational ceiling approx 21,335 m (70,000 ft)
T-O run:
   with normal weapon and fuel load 760 m (2,500 ft)
   ferry configuration (four 1,500 litre drop tanks and full weapon load) approx 1,525 m (5,000 ft)
T-O to 15 m (50 ft) under 915 m (3,000 ft)
Landing from 15 m (50 ft) approx 610 m (2,000 ft)
Landing run with thrust reversal 370 m (1,215 ft)
Intercept radius:
   supersonic more than 300 nm (556 km; 345 miles)
   subsonic more than 1,000 nm (1,853 km; 1,151 miles)
Endurance
   2 h combat air patrol at 300-400 nm (555-740 km; 345-460 miles) from base, incl time for interception and 10 min combat

TORNADO PRODUCTION

		IDS									ADV					
		RAF		GAF		GN		IAF		RSAF		RAF		RSAF			Cum
Batch	Block	S	T	S	T	S	T	S	T	S	T	S	T	S	T	Total	total
1	1	8	8	3	11							2	1			33	33
1	2	3	4	–	3											10	43
2	3	8	7	7	3	1	–	2	2							30	73
2	4	18	4	8	5			3	2							40	113
2	5	13	5			11	5	5	1							40	153
3	6	25	7			32	–	9	4							77	230
3	7	35	1	24	12			14	1							87	317
4	8	23	6	27	4			13	–			–	6			79	396
4	9	22	2	29	4			14	–			10	2			83	479
5	10	2	–	29	–	–	5	11	2	11	3	12	6			81	560
5	11			2	–	32	–	16	–	3	3	22	12			90	650
5	12					2	–									2	652
6	12			21	5	9	2					39	7			83	735
6	13			5	8	13	–					12	10	18	6	72	807
7	14	14[1]	–	18[2]	–					10	4	7	–			53	860
7	15	7	6	17[2]	–					10[3]	4	17	–			61	921
7	16											–	8			8	929
		178	50	190	55	100	12	87	12	34	14	121	52	18	6		
		228		245		112		99		48		173		24			
		732										197					
Prototype		3	1	3	–			2	–							9	
Pre-series		1	1[4]	2[4]	1			1[4]	–							6	
		234		251		112		102		48		173		24			
		747										197				944	
On order										48 (RSAF)						48	
		795										197				992	

Notes: [1] New-build GR. Mk 1A
[2] Tornado ECR
[3] Including six reconnaissance (equivalent of GR. Mk 1A)
[4] Refurbished to production standard
RAF–Royal Air Force
GAF–German Air Force
GN–German Navy
IAF–Italian Air Force
RSAF–Royal Saudi Air Force
S–single control
T–dual control

PHÖNIX-AVIATECHNICA — see Aviotechnica Ltd in this section and Phoenix-Aviatechnica in Russian section

PROMAVIA/MiG

PARTICIPATING COMPANIES:
Promavia: see under Belgium
MiG: see under Russia

PROMAVIA/MiG ATTA 3000

TYPE: Projected tandem-seat derivative of Jet Squalus, for full advanced training syllabus and/or tactical use.

PROGRAMME: Announced by Promavia 1989; may be proposed to USAF and USN for JPATS requirement, though no US partner yet announced; also being promoted internationally, together with Jet Squalus, as military training package offering complete primary/basic/advanced capability to take trainees through to advanced combat types such as F-16, F/A-18 and Tornado.

Contract signed 18 July 1992 between Promavia SA and A. I. Mikoyan Aviation Scientific Production Complex MiG provides for design, development, construction and flight testing of two prototypes, first of which planned to fly in Russia mid-1993 and second in Canada by end of 1993; third airframe, for fatigue testing, also to be assembled in Russia. Power plant, ejection seats and avionics to be supplied mainly by US manufacturers.

DESIGN FEATURES: Would have two stepped zero/zero seats in tandem and EFIS avionics; proposed in both single- and twin-engined form.

POWER PLANT: As Jet Squalus. Internal fuel capacity 757 litres (200 US gallons; 166.5 Imp gallons).

ARMAMENT: Could be armed in weapons training/tactical role with 7.62, 12.7 or 20 mm gun pods, 70 mm rocket launchers, infra-red air-to-air missiles, or bombs of up to Mk 82 size.

DIMENSIONS, EXTERNAL:
Wing span 9.20 m (30 ft 2¼ in)
Wing aspect ratio 6.13
Length overall 9.96 m (32 ft 8 in)
Height overall 3.60 m (11 ft 9¾ in)

AREAS:
Wings, gross 13.81 m² (148.65 sq ft)

WEIGHTS AND LOADINGS (estimated):
Weight empty, equipped 1,769 kg (3,900 lb)
Max T-O weight 3,265 kg (7,198 lb)
Max landing weight 2,520 kg (5,555 lb)

Max wing loading	236.4 kg/m² (48.42 lb/sq ft)
Max power loading:	
TFE109-1	552.25 kg/kN (5.41 lb/lb st)
FJ44	386.57 kg/kN (3.79 lb/lb st)

PERFORMANCE (estimated, FJ44 engine):

Max level speed	485 knots (899 km/h; 558 mph)
Stalling speed, flaps down	72 knots (134 km/h; 83 mph)
Max rate of climb at S/L	2,165 m (7,100 ft)/min
Service ceiling	12,200 m (40,000 ft)
T-O run	244 m (800 ft)
Landing run	366 m (1,200 ft)
Ferry range	950 nm (1,760 km; 1,094 miles)

Promavia/MiG ATTA 3000 tandem-seat trainer project *(Jane's/Mike Keep)*

ROCKWELL INTERNATIONAL/ DEUTSCHE AEROSPACE

PARTICIPATING COMPANIES:
Rockwell International: see under USA
Deutsche Aerospace: see under Germany

Known also by programme title EFM (Enhanced Fighter Maneuverability), X-31A is first US 'X' series experimental aircraft developed jointly with another country, and was one of first NATO co-operative efforts part-funded under Nunn-Quayle R&D initiative. ARPA, acting through US Naval Air Systems Command, is working with German Ministry of Defence to manage development programme. NASA and US Air Force joined programme in January 1992 and flight testing moved to Dryden research centre at Edwards AFB.

ROCKWELL INTERNATIONAL/DEUTSCHE AEROSPACE X-31A EFM

TYPE: Single-seat combat manoeuvrability research aircraft.
PROGRAMME: Evolved from work begun at MBB (now Deutsche Aerospace Military Aircraft Division) in 1977; joined by Rockwell 1983; feasibility study began November 1984, followed by US/German MoU May 1986 and start of one-year Phase 2 (vehicle preliminary design) September 1986; two prototypes funded August 1988 and assembled by Rockwell under 22-month Phase 3; first prototype (BuAer No. 164584) rolled out 1 March 1990, making first flight 11 October 1990; first flight of second prototype (164585) 19 January 1991; first aircraft made first flight with thrust vectoring paddles installed 14 February 1991; post-stall testing started November 1991 and 52° angle of attack reached by end 1991, after total 108 flights; International Test Organisation formed when testing moved to Dryden in January 1992.

Phase 4 high angle of attack (AOA) tests started June 1992; final target 70° AOA with 45° bank reached 18 September; first ever 360° rolls at 70° AOA performed 6 November; post-stall programme completed March 1993 and followed by tactical utility trials with military pilots at Naval Air Test Center Patuxent River; first one X-31A with post-stall controls against the other X-31A with post-stall controls disabled; then dissimilar combat against operational aircraft. Completion of flight trials due late 1993.
COSTS: Programme costs shared USA 75/Germany 25 per cent.
DESIGN FEATURES: Low mounted cranked delta wings with Rockwell transonic aerofoil section (thickness/chord ratio 5.5 per cent), incorporating camber and twist; no dihedral or anhedral; incidence 0°; sweepback at quarter-chord 48° 6' inboard, 36° 36' outboard; swept-back foreplanes, fin and rudder; no horizontal tail surfaces. Design integrates several technologies to expand manoeuvring flight envelope, including vectored thrust, integrated control systems and pilot assistance; enhanced manoeuvrability could yield significant exchange ratio advantages in future close-in fighter combat, and X-31A is intended to break so-called stall barrier by allowing close-in aerial combat beyond normal stall angles of attack; design also expected to enable extremely rapid target acquisition and fuselage pointing for future low-speed, transonic and supersonic engagements; earlier programmes such as Rockwell HiMAT RPV and MBB's TKF-90 contributed much useful data to X-31A design and development. Rockwell primarily responsible for configuration, aerodynamics and construction, DASA for control systems and thrust vectoring design, plus some major components and subassemblies (incl wings).

FLYING CONTROLS: Inboard and outboard trailing-edge flaperons, two-segment leading-edge flaps, all-moving active foreplanes, and rudder; door type airbrake each side

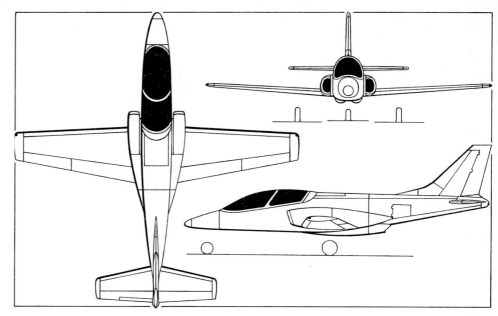

Rockwell International/Deutsche Aerospace X-31A EFM enhanced fighter manoeuvrability research aircraft, showing the three thrust deflector panels at the tail

Rockwell International/Deutsche Aerospace X-31A flying at an angle of attack of 70°

of rear fuselage; AlliedSignal (Electrodynamics and AiResearch Divisions) electrical signalling system for control surfaces; Honeywell flight control computers; AlliedSignal Bendix (modified V-22) rudder and foreplane actuators. Pitch and roll stability and control by flaperons, pitch and yaw by thrust vectoring, pitch (up to 70° angle of attack) by foreplanes and engine intake control lip; leading-edge flaps also scheduled for high AOA stability and control, and for conventional performance; three thrust vectoring paddles attached to rear of nozzles can deflect engine exhaust through approx 10° for yaw control, and can also act as additional airbrakes for rapid deceleration.

STRUCTURE: Wings have aluminium spars and ribs, graphite epoxy (carbonfibre) upper and lower skins; aluminium flaps, fin and rudder, honeycomb ailerons and foreplanes, all with graphite epoxy skins; fuselage mostly has conventional bulkheads and stringers of aluminium: forward panels are honeycomb with graphite epoxy skin, mid-fuselage has aluminium skin, rear 0.76 m (2 ft 6 in) has titanium bulkheads and skin; nose radome is glassfibre.

LANDING GEAR: Menasco landing gear adapted from F-16; hydraulically retractable tricycle type, main units retracting forward into fuselage, nose unit rearward. Main units have Goodrich (Cessna Citation III) wheels and brakes and Vought Aircraft A-7D tyres (pressure 15.51 bars; 225 lb/sq in). Syndex tail braking parachute.

POWER PLANT: One 71.17 kN (16,000 lb st class with afterburning) General Electric F404-GE-400 turbofan. Single fuel tank in fuselage, with gravity feed filler just aft of canopy. Single ventral air intake, with movable lower lip.

ACCOMMODATION: Pilot only, on Martin-Baker SJU-5/6 ejection seat in pressurised, heated and air-conditioned cockpit. Windscreen and rear-hinged, upward opening canopy from McDonnell Douglas F/A-18 Hornet. General Electric's Aerospace Business Group assisted in cockpit development.

SYSTEMS: Include a Sundstrand electrical power generator, AlliedSignal Garrett (modified F-16 hydrazine system) emergency power unit to provide 4½ minutes of electrical hydraulic power, and Garrett (hydrazine powered) emergency air start system from Northrop F-20.

DIMENSIONS, EXTERNAL:
Wing span	7.26 m (23 ft 10 in)
Wing aspect ratio	2.51
Foreplane span	2.64 m (8 ft 8 in)
Length overall: incl nose probe	14.85 m (48 ft 8½ in)
excl probe	13.21 m (43 ft 4 in)
fuselage, excl probe	12.39 m (40 ft 8 in)
Height overall	4.44 m (14 ft 7 in)
Wheel track	2.29 m (7 ft 6⅓ in)
Wheelbase	3.51 m (11 ft 6⅓ in)

AREAS:
Wings, gross	21.02 m² (226.3 sq ft)
Foreplanes (total)	2.19 m² (23.60 sq ft)
Ailerons (total)	1.29 m² (13.88 sq ft)
Trailing-edge flaps (total)	1.73 m² (18.66 sq ft)
Leading-edge flaps: inboard (total)	0.60 m² (6.42 sq ft)
outboard (total)	0.77 m² (8.28 sq ft)
Fin, incl dorsal fin	2.68 m² (28.87 sq ft)
Rudder	0.81 m² (8.68 sq ft)

WEIGHTS AND LOADINGS:
Weight empty, equipped	5,175 kg (11,410 lb)
Fuel weight	1,876 kg (4,136 lb)
Normal flying weight	6,622 kg (14,600 lb)
Max T-O weight	7,228 kg (15,935 lb)

PERFORMANCE (estimated, at max T-O weight):
Never-exceed (VNE) and max level speed:	
S/L to 8,535 m (28,000 ft)	1,485 knots (2,752 km/h; 1,710 mph)
8,535-12,200 m (28,000-40,000 ft)	Mach 1.3
Max rate of climb at S/L	13,106 m (43,000 ft)/min
Max operating altitude	12,200 m (40,000 ft)
T-O run	457 m (1,500 ft)
T-O to 15 m (50 ft)	823 m (2,700 ft)
Landing from 15 m (50 ft)	1,128 m (3,700 ft)
Landing run	823 m (2,700 ft)
Design g limits	+9/−4

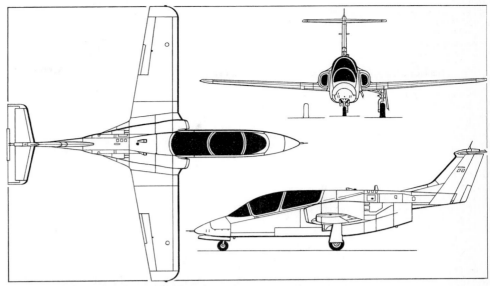

Rockwell International/DASA Ranger 2000 (one P&WC JT15D-5C turbofan) *(Jane's/Mike Keep)*

First take-off of Ranger 2000 from DASA base at Manching 15 January 1993

ROCKWELL INTERNATIONAL/DASA RANGER 2000

TYPE: Military jet trainer, initially aimed at US JPATS programme (see US Air Force in US section).

PROGRAMME: International programme launched (as FanRanger) May 1991; bird strike and ejection trials completed first half 1992; final assembly of first of two prototypes started June 1992; first flight (D-FANA) at Manching 15 January 1993; delivery of prototypes with German certification to Rockwell International North American Aircraft Division expected June 1993 in time for fly-off and JPATS selection early 1994. AAI signed as ground training partner before aircraft and ground elements were separated in early 1993. Production Rangers would be assembled in USA by Rockwell North American Aircraft.

CUSTOMERS: Potentially US Navy and Air Force for JPATS basic trainer programme; total potential market seen as 1,500 to 2,000 over next 20 years.

COSTS: Production programme costs to be shared 50:50 between Rockwell and DASA; prototype development shared 60 per cent Rockwell, 40 per cent DASA; DASA investment $25.4 million.

FLYING CONTROLS: Conventional manual with fixed high-mounted tailplane and central elevator trim tab; servo tabs in ailerons.

STRUCTURE: Cabin section, wings, centre fuselage and engine nacelle of CFRP/GFRP, produced by DASA Augsburg and Rhein Fluzeugbau (RFB) (which see under Germany); metal rear fuselage and tail by RFB. Structure evolved from that of RFB Fantrainer with cabin section based on single structural keel beam.

LANDING GEAR: Retractable tricycle.

POWER PLANT: One 14.19 kN (3,190 lb st) P&WC JT15D-5C turbofan; same engine as US Air Force Beech T-1A Jayhawk Tanker Transport Training System (TTTS) and SIAI-Marchetti S.211A. Maximum internal fuel 860 litres (227 US gallons; 189 Imp gallons).

ACCOMMODATION: Two pilots in tandem on UPCO Stencel zero/zero ejection seats; rear seat raised; birdproof windscreen; cockpit pressurised and air-conditioned.

AVIONICS: Rockwell Collins four-tube EFIS flight instruments in each cockpit.

DIMENSIONS, EXTERNAL:
Wing span	10.46 m (34 ft 4 in)
Wing aspect ratio	7.40
Length overall	7.85 m (25 ft 9¼ in)
Height overall	3.91 m (12 ft 10 in)

AREAS:
Wings, gross	15.55 m² (167.4 sq ft)

WEIGHTS AND LOADINGS: Claimed to be commercial secret

PERFORMANCE:
Max airspeed at S/L	329 knots (610 km/h; 379 mph)
Max true airspeed at 9,145 m (30,000 ft)	392 knots (726 km/h; 451 mph)
Approach speed, landing configuration	107 knots (198 km/h; 123 mph)
Operational ceiling	10,670 m (35,000 ft)
Range with max internal fuel	971 nm (1,800 km; 1,118 miles)

RUSJET

Announced early 1992 that one South African and two Russian companies were collaborating in development of a small utility transport, known as LCPT (light cargo passenger turboprop), but project subsequently abandoned. For details, see 1992-93 *Jane's*.

SATIC
SPECIAL AIRCRAFT TRANSPORT INTERNATIONAL COMPANY (Joint subsidiary of Deutsche Aerospace Airbus and Aerospatiale)

SATIC A300-600ST SUPER TRANSPORTER
TYPE: Replacement for Airbus Super Guppy assembly transporters.
PROGRAMME: Announced December 1990; Latécoère chosen as lead contractor for programme in May 1992 with overall integration responsibility and to design cargo hold, floor, pressure bulkhead and rear side doors; SOGERMA-SOCEA to make upward hinging nose door, side panels and airtight floor above pressurised flight deck; Hurel-Dubois will produce nose landing gear fittings and structure of transverse pressure bulkhead aft of flight deck; Socata to produce rear lower fuselage lobe; CASA (Spain) will produce forward upper fuselage panels and Elbe Flugzeugwerke (Deutsche Aerospace Airbus, formerly East German) will make the redesigned tail section; first subassemblies to be delivered to Latécoère plant near Toulouse May 1993; first nose section to be delivered August 1993; first flight September 1994; in service 1995; remaining three delivered by 1998.
CUSTOMERS: Airbus Industrie requires four Super Transporters to transport large Airbus subassemblies between group factories.
DESIGN FEATURES: Based on new A300-600 airframe, with enlarged unpressurised upper fuselage, accessed via new upward-hinging door above flight deck; new tail unit with endplate fins; pressurised flight deck below main deck floor level permits roll-on/roll off loading of main hold; total payload 50,000 kg (110,230 lb).

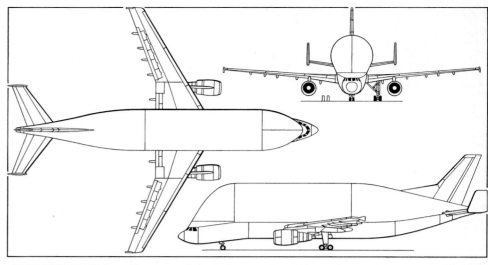

SATIC A300-600ST Super Transporter being built to transport Airbus airframe sections (*Jane's/Mike Keep*)

SATIC Super Transporter based on Airbus A300-600 airframe

SCTICSG
SUPERSONIC COMMERCIAL TRANSPORT INTERNATIONAL CO-OPERATION STUDY GROUP

Various follow-on SST programmes have been discussed and research continues. Five-year joint study for Concorde successor announced by British Aerospace and Aerospatiale 9 May 1990, combining individual earlier efforts on AST (Advanced Supersonic Transport) and ATSF (Avion de Transport Supersonique Futur) respectively under the joint designation Alliance (Aerospatiale denied existence of Alliance early 1993) followed on 19 May 1990 by formation of five-company study group (including Deutsche Aerospace Airbus, Boeing and McDonnell Douglas) to work for about one year examining environmental implications, market potential, certification basis, and benefits of global co-operation.

Five-man Supersonic Transport Development Council established 1991 by Society of Japanese Aerospace Companies to act as liaison committee for possible Japanese participation; announced at Paris Air Show June 1991 that Japanese aircraft industries (Society of Japanese Aerospace Companies) and Alenia of Italy had joined technical and marketing study activities; Tupolev of Russia now also participating. In March 1991, Boeing and Deutsche Aerospace Airbus signed agreement in principle to pursue their joint studies in selected areas outside scope of those being examined by SCTICSG.

Aerospatiale/BAe research favours retention of ogival delta wing with leading-edge sweep varying from approx 70° inboard to 50° at tip, optimised for both subsonic (overland) and supersonic cruise; redesigned nose (possibly with larger 'droop-snoot'); four new fuel-efficient variable-cycle engines in separate underwing nacelles; fly-by-wire flying controls; 250-300 passenger seating capacity; performance parameters to include speed of Mach 2.05-2.4 and max range of 6,475 nm (12,000 km; 7,455 miles).

Artist's impression of 'general supersonic transport configuration' by Boeing

SEPECAT
SOCIETE EUROPEENNE DE PRODUCTION DE L'AVION E. C. A. T.

PARTICIPATING COMPANIES:
  British Aerospace: see under UK
  Dassault Aviation: see under France
PRESIDENT: J. P. Weston (BAe)
VICE-PRESIDENT: R. Dubost (Dassault)
PUBLIC RELATIONS: D. Kamiya (BAe)

Anglo-French company formed May 1966 by Breguet Aviation and British Aircraft Corporation to design and produce Jaguar strike fighter/trainer; production now in India only (see HAL entry).

SEPECAT JAGUAR INTERNATIONAL

TYPE: Single-seat tactical support aircraft and two-seat operational or advanced trainer.
PROGRAMME: Initiated 17 May 1965 by Defence Ministries of UK and France, originally to build 200 **Jaguars** each (with Mk 102 Adour turbofans) for Royal Air Force and Armée de l'Air. Production completed 1982, but RAF aircraft upgraded (1978-84) with more powerful Adour Mk 104s and later with GEC Ferranti FIN 1064 INS (see 1989-90 *Jane's*). One T. Mk 2A fitted for trials with GEC-Marconi ATLANTIC (airborne targeting low altitude navigation, thermal imaging and cueing) underwing podded FLIR system; weapons listed for RAF aircraft in 1990-91 *Jane's* now also include CBU-87 cluster bombs and 19-round LAU-5003B/A pods of 70 mm CRV7 rockets. Other modifications to RAF Jaguars taking part in 1991 Gulf War included RWR upgraded to Sky Guardian standard; Vinten VICON 18 Srs 600 LOROP pod fitted; second ARI 23315/4 UHF radio (with modified antenna) installed; engine control amplifier modified for increased thrust and provision for overwing AAMs (as Jaguar International).
CURRENT VERSIONS: **Jaguar International:** Export version (first flight 19 August 1976); European production of this ended in 1985, but manufacture of final 46 continues by HAL in India (which see for detailed description).
CUSTOMERS: Total of 403 European built Jaguars produced for RAF (203) and Armée de l'Air (200), plus 139 Jaguar Internationals for air forces of Ecuador (12), India (85, including 45 in CKD form for local assembly), Nigeria (18) and Oman (24).

Following data refer to Royal Air Force Jaguar GR. Mk 1A and augment previous editions of Jane's:

POWER PLANT: Two Rolls-Royce Turbomeca Adour Mk 104 turbofans, each rated at 23.4 kN (5,270 lb st) dry and 35.1 kN (7,900 lb st) with afterburning; optional modification (TGT increased from 700 to 725°C) for high ambient temperature operation raises outputs to 23.8 kN (5,350 lb st) and 36.0 kN (8,100 lb st). Usable internal fuel 4,183 litres (1,104 US gallons; 920 Imp gallons) in four tanks: front fuselage 909 litres (240 US gallons; 200 Imp gallons); centre fuselage 1,173 litres (310 US gallons; 258 Imp gallons); rear fuselage 1,146 litres (302 US gallons; 252 Imp gallons) and wing 955 litres (252 US gallons; 210 Imp gallons). Three drop tanks, each with 1,186 litres (313 US gallons; 261 Imp gallons) usable fuel.
AVIONICS: ARI 23315/4 V/UHF radio (Magnavox AN/ARC-164 UHF and GEC AD120 VHF/AM); ARI 23159 standby VHF; ARI 23181 HF; ARI 23427/1 (Vinten 1200-04) voice recorder—replaced by HUD video; ARI 23231/3 (GEC-Marconi) laser ranger and marked target seeker; ARI 23205/4 (AN/APN-91) Tacan; ARI 23232 (AN/APN-198) radar altimeter; ARI 23134/3 SSR/IFF; ARI 18223 RWR (upgraded to Marconi Sky Guardian).

WEIGHTS AND LOADINGS:
Empty: GR. Mk 1 (1973)	7,390 kg (16,292 lb)
GR. Mk 1A (1992)	7,700 kg (16,976 lb)
Max fuel weight: internal (usable)	3,337 kg (7,357 lb)
external (usable)	2,844 kg (6,270 lb)
Max T-O weight	15,500 kg (34,172 lb)

PERFORMANCE:
Max level speed, clean:	
up to 6,100 m (20,000 ft)	Mach 1.25 (625 knots; 1,159 km/h; 720 mph)
above 6,100 m (20,000 ft)	Mach 1.4 (600 knots; 1,112 km/h; 691 mph)
Service ceiling	13,720 m (45,000 ft)
g limits	+8/−3

RAF Jaguar GR. Mk 1As with reconnaissance pods and AIM-9L Sidewinder missiles beginning a mission over northern Iraq *(Paul Jackson)*

SOKO/AVIOANE

PARTICIPATING COMPANIES:
  Soko: see under Bosnia-Hercegovina
  Avioane: see under Romania

SOKO J-22 ORAO (EAGLE) and AVIOANE IAR-93

TYPE: Single-seat close support, ground attack and tactical reconnaissance aircraft, with secondary capability as low level interceptor. Combat capable two-seat versions used also for advanced flying and weapon training.
PROGRAMME: Joint design by Yugoslav and Romanian engineers, started 1970 under original project name Yurom, to meet requirements of both air forces; two single-seat prototypes started in each country 1972, making simultaneous first flights 31 October 1974; first flight in each country of a two-seat prototype 29 January 1977; each manufacturer then built 15 pre-production aircraft (first flights 1978); series production began in Romania (IAv Craiova, now Avioane) 1979, in Yugoslavia (Soko) 1980; Yugoslav production stopped by damage and dismantling of Mostar factory in Bosnia-Hercegovina in 1992.
CURRENT VERSIONS: **IAR-93A:** Romanian single- and two-seat versions with non-afterburning Viper Mk 632-41 turbojets; first flight 1981; production completed.
  IAR-93B: Romanian single- and two-seat versions with afterburning Viper Mk 633-47 turbojets; first flight 1985. Avionics of two-seaters being upgraded early 1993.
  IJ-22 and INJ-22 Orao: Yugoslav non-afterburning pre-series aircraft for tactical reconnaissance (INJ-22 also a conversion trainer); total 15 single-seat IJ-22 and two-seat INJ-22 built.
  NJ-22 Orao: Production two-seat tactical reconnaissance version, some with/some without afterburning engines; first flight 18 July 1986; total 35 ordered (all delivered). Improved version planned.
  J-22 Orao: Production single-seat attack version, some with/some without afterburning engines; first flight 20 October 1983; in production. Plans to upgrade with modern avionics including integration of radar and inertial nav/attack system via new databus; intake and wing leading-edge de-icing.
CUSTOMERS: Romanian Air Force (26 single-seat and 10 two-seat IAR-93A, all delivered, and 165 IAR-93B); former Yugoslav Air Force (15 IJ/INJ-22, 35 NJ-22, all delivered; 165 J-22 ordered, of which 74 delivered by early 1992).
DESIGN FEATURES: Wings of NACA 65A-008 (modified) section, shoulder-mounted with 0° incidence and 3° 30′ anhedral from roots; sweepback 35° at quarter-chord and approx 43° on outer leading-edges; incidence 0°; inboard leading-edges extended forward (except on prototypes/pre-production) at 65° sweepback; small boundary layer fence on upper surface of each outer panel; fuselage has hydraulically actuated door type perforated airbrake underneath on each side forward of mainwheel bay, dorsal spine fairing, pen-nib fairing above exhaust nozzles, and detachable rear portion for access to engine bays; all-sweptback tail surfaces, plus ventral strake beneath rear fuselage each side.
FLYING CONTROLS: Internally balanced plain ailerons, single-slotted trailing-edge flaps, two-segment leading-edge slats (three-segment on prototype/pre-production aircraft), low-set all-moving tailplane and inset rudder, all hydraulically actuated (except leading-edge slats: electrohydraulic), with PPT servo-actuators for primary control surfaces; no aileron or rudder tabs.
STRUCTURE: Conventionally built, almost entirely of aluminium alloy except for honeycomb rudder and tailplane on later production aircraft; two-spar wings with ribs, stringers and partially machined skin; wing spar box forms integral fuel tanks on production aircraft; pre-series aircraft have rubber fuel cells, forward of which are sandwich panels.
LANDING GEAR: Hydraulically retractable tricycle type of Messier-Bugatti design, with single-wheel hydraulically steerable nose unit and twin-wheel main units. All units retract forward into fuselage. Two-stage PPT hydro/nitrogen shock absorber in each unit. Mainwheels and tubeless tyres size 615 × 225 mm; pressure 4.5 bars (65.3 lb/sq in). Nosewheel and tubeless tyre size 450 × 190 mm, pressure 4.0 bars (58.0 lb/sq in), on all afterburning versions. Hydraulic disc brakes on each mainwheel unit, and electronic anti-skid system. Min ground turning radius 7.00 m (22 ft 11½ in). Bullet fairing at base of rudder contains hydraulically deployed 4.2 m (13 ft 9½ in) diameter braking parachute.
POWER PLANT (non-afterburning versions): Two 17.79 kN (4,000 lb st) Turbomecanica/Orao (licence built Rolls-Royce) Viper Mk 632-41 turbojets, mounted side by side in rear fuselage; air intake on each side of fuselage, below cockpit canopy. Pre-series aircraft have seven fuselage tanks and two collector tanks, with combined capacity of 2,480 litres (655 US gallons; 545.5 Imp gallons), and two 235 litre (62 US gallon; 51.75 Imp gallon) rubber fuel cells in wings, giving total internal fuel capacity of 2,950 litres (779 US gallons; 649 Imp gallons). J-22 and NJ-22 have five fuselage tanks, two collectors and two integral wing tanks for total internal capacity of 3,120 litres (824 US gal-

Single-seat interceptor version of the Romanian Air Force's IAR-93B, with underwing air-to-air missiles

INTERNATIONAL: Aircraft—SOKO/AVIOANE

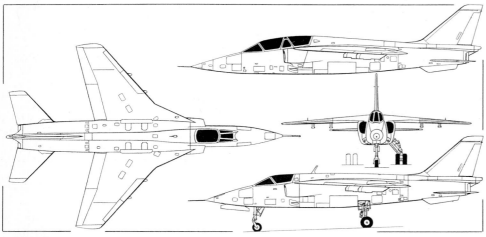

Single-seat J-22 Orao close support/ground attack aircraft, with additional side view (top) of two-seat NJ-22
(Dennis Punnett)

Soko built NJ-22 Orao tandem-seat operational trainer of the former Yugoslav Air Force

lons; 686 Imp gallons). Provision for carrying three 500 litre (132 US gallon; 110 Imp gallon) auxiliary fuel tanks, one on underfuselage stores attachment and one inboard under each wing. Pressure refuelling point in fuselage below starboard air intake; gravity refuelling points in fuselage near starboard wing trailing-edge and in each external tank. Oil capacity 6.25 litres (1.65 US gallons; 1.37 Imp gallons).

POWER PLANT (afterburning versions): Two Turbomecanica/Orao (licence built Rolls-Royce) Viper Mk 633-47 turbojets, each rated at 17.79 kN (4,000 lb st) dry and 22.24 kN (5,000 lb st) with afterburning. Production aircraft have five fuselage and two collector tanks, plus two integral wing tanks, giving total internal capacity of 3,120 litres (824 US gallons; 686 Imp gallons). Drop tanks, refuelling points and oil capacity as for non-afterburning versions.

ACCOMMODATION: Single-seat or tandem two-seat cockpit(s), with Martin-Baker zero/zero seat for each occupant (RU10J in IAR-93, Y-10LB in Orao), capable of ejection through canopy. Canopy of single-seat IAR-93A and J-22 hinged at rear and actuated electrically to open upward; single-seat IAR-93B, and all two-seaters, have manually operated canopies opening sideways to starboard. All accommodation pressurised, heated and air-conditioned. Dual controls in two-seat versions.

SYSTEMS: Bootstrap type environmental control system for cockpit pressurisation (max differential 0.214 bar; 3.1 lb/sq in), air-conditioning, and windscreen de-icing/demisting. Two Prva Petoletka independent hydraulic systems, each of 207 bars (3,000 lb/sq in) pressure, flow rate 48 litres (12.7 US gallons; 10.6 Imp gallons)/s, for actuation of leading-edge slats, trailing-edge flaps, ailerons, tailplane, rudder, airbrakes, landing gear extension/retraction, mainwheel brakes, nosewheel steering, brake-chute and afterburner nozzles. Backup system for landing gear, wheel brakes and primary flight controls. No pneumatic system. Main electrical system is 28V DC, supplied by two Lucas BC-0107 9kW engine driven starter/generators through two voltage regulators and a switching system, and a 24V 36Ah nickel-cadmium battery; four 300VA static inverters for 115V AC power at 400Hz. High pressure (150 bars; 2,175 lb/sq in) gaseous oxygen system for crew. Planned retrofit of air intake and wing leading-edge de-icing.

AVIONICS: Standard avionics include VHF/UHF air-to-air and air-to-ground com radio (20W transmission power); gyro unit (Honeywell SGP500 twin-gyro platform in Orao), radio altimeter, ADF, radio compass and marker beacon receiver; IFF (Romanian aircraft only); and GEC-Marconi three-axis stability augmentation system, incorporating a basic bank/attitude hold autopilot and emergency wings-level facility. Oraos also have Collins VIR-30 VOR/ILS and DME-40; and Iskra SO-1 RWR. Reconnaissance pod (1-18GHz VRT10 radio surveillance pod, photographic/jamming pod or photographic/IR pod on IJ/INJ-22), or P10-65-13 passive jamming pod (Orao) on underfuselage station. Chaff and IR decoy launch pods (up to three per aircraft) can also be carried.

EQUIPMENT: Landing light under nose, forward of nosewheel bay; taxying light on nosewheel shock strut.

ARMAMENT (IAR-93A/B): Two 23 mm GSh-23L twin-barrel cannon in lower front fuselage, below engine air intakes, with 200 rds/gun. Gun camera and GEC-Marconi D282 gyro gunsight. Five external stores stations, of which inboard underwing pair and fuselage centreline station are each stressed for loads up to 500 kg (1,102 lb); outboard underwing stations stressed for up to 300 kg (661 lb) each, giving max external stores load of 1,500 kg (3,307 lb). Typical weapon loads can include two or three 500 kg bombs; four or five 250 kg bombs; four multiple carriers each with three 100 kg or 50 kg bombs; two such multiple carriers plus two L-57-16MD launchers each with sixteen 57 mm rockets; four L-57-16MD launchers; four launchers each with two 122 mm, one 128 mm or one 240 mm rocket (122 and 240 mm not used on Oraos); a GSh-23L cannon pod with four L-57-16MD rocket launchers; four 160 kg KPT-150 or similar munition dispensers; or four L-57-32 launchers each with thirty-two 57 mm rockets. Some IAR-93Bs equipped to carry up to eight air-to-air missiles, on twin launch rails, on four underwing stations. Centreline and inboard underwing points can carry drop tanks.

ARMAMENT (Orao): Guns, gun camera, drop tanks and centreline reconnaissance pod as for IAR-93. All four wing stations stressed for 500 kg (1,102 lb), and fuselage station for 800 kg (1,763 lb), giving max external stores capacity of 2,800 kg (6,173 lb). Typical weapon loads include five 50 kg, 100 kg, 250 kg or 500 kg bombs; four multiple carriers for totals of twelve 50 or 100 kg or eight 250 kg bombs; four FLAB-350 napalm bombs (each 360 kg; 794 lb); five BL755 bomblet dispensers, or eight on four multiple carriers; sixteen BRZ-127 5 in HVAR rockets; four pods of L-57-16MD or L-128-04 (4 × 128 mm) rockets, or eight pods on multiple carriers; five 500 kg AM-500 sea mines; or two launch rails for AGM-65B Maverick or Yugoslav developed Grom air-to-surface missiles. The 100 kg and 250 kg bombs can be parachute retarded.

DIMENSIONS, EXTERNAL:
Wing span	9.30 m (30 ft 6¼ in)
Wing chord: at root	4.20 m (13 ft 9⅜ in)
at tip	1.40 m (4 ft 7⅛ in)
Wing aspect ratio	3.33
Length overall, incl probe:	
single-seater	14.90 m (48 ft 10⅝ in)
two-seater	15.38 m (50 ft 5½ in)
Length of fuselage:	
single-seater	13.02 m (42 ft 8½ in)
two-seater	14.44 m (47 ft 4½ in)
Fuselage: Max width	1.62 m (5 ft 3¾ in)
Height overall	4.52 m (14 ft 10 in)
Tailplane span	4.59 m (15 ft 0¾ in)
Wheel track (c/l of shock struts)	2.50 m (8 ft 2½ in)
Wheelbase: single-seater	5.40 m (17 ft 8½ in)
two-seater	5.88 m (19 ft 3½ in)

AREAS:
Wings, gross	26.00 m² (279.86 sq ft)
Ailerons (total)	2.39 m² (25.73 sq ft)
Trailing-edge flaps (total)	3.33 m² (35.84 sq ft)
Leading-edge slats (total)	2.22 m² (23.90 sq ft)
Fin	3.55 m² (38.21 sq ft)
Rudder	0.88 m² (9.47 sq ft)
Tailplane	7.31 m² (78.68 sq ft)

WEIGHTS AND LOADINGS:
Weight empty, equipped: IAR-93A	6,150 kg (13,558 lb)
IAR-93B	5,750 kg (12,676 lb)
J-22	5,500 kg (12,125 lb)
IJ-22	5,755 kg (12,687 lb)
Max internal fuel: IAR-93A	2,457 kg (5,416 lb)
IAR-93B	2,400 kg (5,291 lb)
J-22	2,430 kg (5,357 lb)
IJ-22	2,360 kg (5,203 lb)
Max external stores load: IAR-93A	1,500 kg (3,307 lb)
IAR-93B	2,500 kg (5,511 lb)
J-22, IJ-22	2,800 kg (6,173 lb)
Normal T-O weight clean: IAR-93A	8,826 kg (19,458 lb)
IAR-93B	8,400 kg (18,519 lb)
J-22	8,170 kg (18,012 lb)
IJ-22 (with recce pod)	8,500 kg (18,739 lb)
Max T-O weight: IAR-93A	10,326 kg (22,765 lb)
IAR-93B	10,900 kg (24,030 lb)
J-22	11,080 kg (24,427 lb)
IJ-22	9,500 kg (20,944 lb)
Max landing weight: IAR-93A	8,826 kg (19,458 lb)
IAR-93B	9,360 kg (20,635 lb)
J-22	9,500 kg (20,944 lb)
IJ-22	8,600 kg (18,960 lb)
Max wing loading: IAR-93A	397.1 kg/m² (81.3 lb/sq ft)
IAR-93B	419.2 kg/m² (85.9 lb/sq ft)
J-22	426.1 kg/m² (87.3 lb/sq ft)
IJ-22	365.4 kg/m² (74.8 lb/sq ft)
Max power loading: IAR-93A	289.8 kg/kN (2.84 lb/lb st)
IAR-93B	245.2 kg/kN (2.40 lb/lb st)
J-22	249.2 kg/kN (2.44 lb/lb st)
IJ-22	213.7 kg/kN (2.09 lb/lb st)

PERFORMANCE (IAR-93A: at max T-O weight, IAR-93B: at 8,400 kg; 18,519 lb T-O weight, J-22: at clean T-O weight with 50% internal fuel, IJ-22: at T-O weight of 8,500 kg; 18,739 lb, except where indicated):
Max level speed at S/L:	
IAR-93A	577 knots (1,070 km/h; 665 mph)
IAR-93B	586 knots (1,086 km/h; 675 mph)
J-22	610 knots (1,130 km/h; 702 mph)
IJ-22	566 knots (1,050 km/h; 652 mph)
Max level speed at altitude:	
J-22 at 11,000 m (36,100 ft)	Mach 0.96
IJ-22 at 8,000 m (26,250 ft)	Mach 0.93
Max cruising speed:	
IAR-93A at 7,000 m (22,965 ft)	394 knots (730 km/h; 453 mph)
IAR-93B at 5,000 m (15,240 ft)	587 knots (1,089 km/h; 676 mph)
J-22, IJ-22 at 11,000 m (36,100 ft)	Mach 0.7
Stalling speed at S/L:	
IAR-93A	130 knots (241 km/h; 150 mph)
IAR-93B	148 knots (274 km/h; 171 mph)
J-22, IJ-22, gear and flaps down	100 knots (185 km/h; 115 mph)
Max rate of climb at S/L:	
IAR-93A	2,040 m (6,693 ft)/min
IAR-93B	3,900 m (12,800 ft)/min
J-22	5,340 m (17,520 ft)/min
IJ-22	2,280 m (7,480 ft)/min
Service ceiling: IAR-93A	10,500 m (34,450 ft)
IAR-93B	13,600 m (44,625 ft)
J-22	15,000 m (49,210 ft)
IJ-22	13,500 m (44,290 ft)
Time to 6,000 m (19,685 ft): J-22	1 min 20 s
IJ-22	3 min 12 s
T-O run: IAR-93A	1,500 m (4,921 ft)
IAR-93B	800 m (2,625 ft)

J-22 at 9,443 kg (20,818 lb) with four BL755s	880 m (2,888 ft)
IJ-22	1,200 m (3,937 ft)
T-O to 15 m (50 ft): IAR-93A	1,600 m (5,249 ft)
IAR-93B	1,150 m (3,775 ft)
J-22	1,255 m (4,118 ft)
IJ-22	1,800 m (5,906 ft)
Landing from 15 m (50 ft): IAR-93A	1,650 m (5,413 ft)
IAR-93B	1,520 m (4,987 ft)
IAR-93B with brake-chute	990 m (3,250 ft)
J-22	1,295 m (4,249 ft)
IJ-22	1,400 m (4,594 ft)
Landing run: IAR-93A	720 m (2,362 ft)
IAR-93B	1,050 m (3,445 ft)
IAR-93B with brake-chute	690 m (2,265 ft)
J-22	755 m (2,477 ft)
IJ-22	850 m (2,789 ft)

Landing run with brake-chute:
IAR-93A, IAR-93B	670 m (2,200 ft)
J-22	530 m (1,739 ft)
IJ-22	600 m (1,969 ft)

Tactical radius, IAR-93A, IAR-93B:
lo-lo-lo with four rocket launchers, 5 min over target
140 nm (260 km; 161 miles)
hi-hi-hi patrol with three 500 litre drop tanks, 45 min over target
205 nm (380 km; 236 miles)
lo-lo-hi with two rocket launchers, six 100 kg bombs and one 500 litre drop tank, 10 min over target
243 nm (450 km; 280 miles)
hi-hi-hi with four 250 kg bombs and one 500 litre drop tank, 5 min over target
286 nm (530 km; 329 miles)

Tactical radius, J-22:
hi-lo-hi with four BL755s and 1,500 litre centreline drop tank, 2 min reheat
282 nm (522 km; 324 miles)
hi-lo-hi with four 500 kg air mines and 1,500 litre centreline drop tank, 1.2 min reheat
248 nm (460 km; 286 miles)
hi-lo-hi with eight 250 kg bombs and one 500 litre drop tank, 2 min reheat
200 nm (370 km; 300 miles)

Ferry range:
IAR-93A, IAR-93B, with three 500 litre drop tanks
1,025 nm (1,900 km; 1,180 miles)
J-22, with two 500 litre drop tanks, at 6,000 m (19,685 ft)
712 nm (1,320 km; 820 miles)

g limits: IAR-93A, IAR-93B, J-22, IJ-22   +8/−4.2

SUKHOI/GULFSTREAM

SUKHOI/GULFSTREAM S21-G SSBJ (SUPERSONIC BUSINESS JET)

Gulfstream Aerospace has terminated its involvement in this supersonic business jet project. Sukhoi (see Russian section) believed still to be continuing its design work.

IRAN

SEYEDO SHOHADA

DEFENCE INDUSTRIES, SEYEDO SHOHADA PROJECT
Km 5, Qom Road, Kashan
Telephone: 98 (31) 20459 or 23606
DESIGNER/BUILDER: Akbar Akhundzadeh

ZAFAR 300

TYPE: Two-seat gunship and agricultural helicopter, converted from Bell 206A.
PROGRAMME: Design began 20 March 1987; construction began 21 April 1988; first flight 31 January 1989; flown 100 hours by end 1990; further modifications then under way, but no news since then. Details and photograph in 1992-93 *Jane's*.

IRAQ

IAF

IRAQI AIR FORCE
Ministry of Defence, Bab Al-Muadam, Baghdad

BAGHDAD 1

AEW version of Iraqi Il-76MD ('Candid-B') converted by Iraqi Air Force; first shown at Iraqi defence exhibition 1989; locally built Thomson-CSF Tigre surveillance radar, normally trailer-mounted, installed inverted under tail of Il-76; signal processing modified to reduce ground clutter; radio and radar ESM added; navigation system modified; claimed to have well over 180° scan; transmission to ground by voice or real-time data link; Iraqi-designed generator; can track and identify targets at 189 nm (350 km; 217 miles); four radar operators; reported used during Iran-Iraq war. Illustrated in 1991-92 and earlier *Jane's*.

ADNAN 1

Improved AEW aircraft based on Il-76MD ('Candid-B') with back-to-back radar antennae in rotating aerial housing (rotodome); two large strakes under rear fuselage maintain directional stability and nose-down pitching moment at low airspeed; rotodome diameter estimated at 9 m (29 ft 6 in) and

Adnan 1 airborne early warning and control aircraft, modified from an Iraqi Air Force Il-76

about 4 m (13 ft 1 in) above fuselage; detection range reported as "few hundred kilometres".

Baghdad television reported start of testing in December 1990; at least three Adnans reported operational at start of Gulf War Desert Storm on 17 January 1991; one put out of action at Al Taqaddum 23 January 1991; two others were flown to Iran late January.

Tanker version of Il-76 shown on Iraqi television 20 January 1991 with single hose/drogue pod at base of rear loading ramp.

ISRAEL

IAI

ISRAEL AIRCRAFT INDUSTRIES LTD
Ben-Gurion International Airport, Israel 70100
Telephone: 972 (3) 935 3111 and 935 8111
Fax: 972 (3) 935 3131 and 971 2290
Telex: 381014, 381033 and 381002 ISRAV IL
PRESIDENT AND CEO: M. Keret
CORPORATE EXECUTIVE VICE-PRESIDENTS:
  A. Ostrinsky
  Dr M. Dvir
VICE-PRESIDENT, MARKETING: D. Onn
DIRECTOR OF CORPORATE COMMUNICATIONS: D. Suslik

Founded 1953 as Bedek Aviation; name changed to Israel Aircraft Industries 1 April 1967; number of divisions reduced from five to four February 1988 (Aircraft, Electronics, Technologies and Bedek Aviation).

Covered floor space 680,000 m² (7.32 million sq ft) in July 1988; total workforce 17,300 in early 1993 (but being reduced by 1,500); approved as repair station and maintenance organisation by Israel Civil Aviation Administration, US Federal Aviation Administration, UK Civil Aviation Authority and Israeli Air Force.

Products include aircraft of own design; in-house produced airframe systems and avionics; service, upgrading and retrofit packages for civil and military aircraft and helicopters (see Bedek Aviation Division); other activities include space technology, missile and ordnance development, and seaborne and ground equipment.

Aircraft Division
Follows this entry

Bedek Aviation Division
Follows Aircraft Division entry

Electronics Division
PO Box 105, Yahud Industrial Zone, Israel 56000
Telephone: 972 (3) 531 5555 and 531 4021
Fax: 972 (3) 536 5205 and 536 3975
Telex: 341450 MBT IL
GENERAL MANAGER: M. Ortasse

IAI's largest Division; covered floor area 150,000 m² (1,614,585 sq ft); workforce 6,700 in 1988. Plants include Elta Electronic Industries (wholly owned subsidiary of IAI), MBT Systems and Space Technology, Tamam Precision Instruments Industries and MLM System Engineering and Integration. Division products include electronics and electro-optical systems and components, space technologies (including SDI environment) and wide range of civil and military hardware and software products and services. Tamam producing laser night targeting system (NTS) for AH-1S attack helicopters of Israeli Air Force (40) and AH-1W SuperCobras of US Marine Corps (43 to date), to replace M-65 TOW sighting system; first redelivery to Israeli Air Force 17 December 1992, first to USMC due January 1993.

Technologies Division
Follows Bedek Aviation Division entry

ISRAEL: AIRCRAFT—IAI

IAI Kfir C7 multi-mission combat aircraft, with mixed weapon load of bombs, air-to-air missiles and drop tanks

AIRCRAFT DIVISION
Ben-Gurion International Airport, Israel 70100
Telephone: 972 (3) 934 4136 and 971 1471
Fax: 972 (3) 972 1266
Telex: 381014 and 381033 ISRAV IL
GENERAL MANAGER: S. Alkon

Established February 1988. Five autonomous plants: Lahav, military aircraft; Matan, civil aircraft; Malat, unmanned air vehicles; Malkam, manufacturing; Tashan, engineering and testing. Military aircraft plant operating third Lavi prototype as advanced combat technology demonstrator. Civil work includes production of Astra, support for IAI Arava and Westwind and development of future civil airframes. Manufacturing plant produces structural components for foreign and domestic customers. Engineering services include analysis, design, development, integration and testing of platforms and systems for domestic and foreign military and civil customers.

IAI SHAHAL (BIBLICAL LION)
IAI funded design study of small, low-cost lightweight fighter with high agility and armed with smart weapons.

IAI TD (TECHNOLOGY DEMONSTRATOR)
TYPE: Two-seat technology demonstrator.
PROGRAMME: Original Lavi programme go-ahead February 1980; full-scale development started October 1982; first flights of single-seat prototypes 31 December 1986 (B-1) and 30 March 1987 (B-2); programme terminated by Israeli government 30 August 1987, but IAI continues flying B-3 two-seat third prototype as technology demonstrator (TD) for Israeli avionics and future systems; first flight of TD, 25 September 1989. Description in 1991-92 and earlier *Jane's*.

IAI KFIR (LION CUB)
TYPE: Single-seat interceptor and attack fighter.
PROGRAMME: C7 first delivered to Israeli Air Force Summer 1983; now being phased out and offered for export with Atar 9K-50 and GE F404 as alternatives to standard J79. First flight with Atar 9K-50 believed made Spring 1991. Described in 1991-92 and earlier *Jane's*, and in *Jane's Civil and Military Aircraft Upgrades*.
CUSTOMERS: Colombia received 13 Kfir C7s in 1989; Philippines (16 C7s and two TC7s) and Taiwan (34 C7s and six TC7s) have agreed to purchase subject to US (engine) approval (given March 1992) and domestic funding.

IAI NAMMER (TIGER)
See Bedek Aviation Division entry.

IAI 1125 ASTRA
TYPE: Twin-turbofan business transport.
PROGRAMME: Construction of two prototypes started April 1982; first flight (4X-WIN, c/n 4001) 19 March 1984; first flight 4X-WIA (c/n 4002) August 1984; third airframe for static and fatigue testing; first flight production Astra (4X-CUA) 20 March 1985; FAR Pts 25 and 36 certification 29 August 1985; first delivery 30 June 1986.
CURRENT VERSIONS: **Original Astra:** Initial production version: see 1992-93 and earlier *Jane's*.
Astra SP: Announced at NBAA convention October 1989; new interior; upgraded avionics with Collins digital autopilot and EFIS; aerodynamic refinements for high altitude performance; range with NBAA reserves extended by 63 nm (117 km; 72 miles). *Detailed description applies to this version.*
Astra IV: Improved version in design stage early 1993; described separately.
CUSTOMERS: Total 66 civil Astras delivered by February 1993, of which majority to US customers.
COSTS (Astra SP): US$7,537,200 (1992).
DESIGN FEATURES: Wing section high efficiency IAI Sigma 2; leading-edge sweep 34° inboard, 25° outboard; trailing-edge sweep on outer panels.
FLYING CONTROLS: Control surfaces operated by pushrods and hydraulically powered; tailplane incidence controlled by three motors running together to protect against runaway or elevator disconnect; ailerons can be separated in case of jam; spoiler/lift dumper panels ahead of Fowler flaps; dual actuated rudder trim tab; flaps interconnected with leading-edge slats, both electrically actuated.
STRUCTURE: One-piece, two-spar wing with machined ribs and skin panels, attached by four main and five secondary frames; wing/fuselage fairings, elevator and fin tips and tailcone of GFRP; ailerons, spoilers, inboard leading-edges and wingtips of Kevlar and Nomex honeycomb; nose avionics bay door and nosewheel doors of Kevlar; Kevlar reinforced nacelle doors and panels; chemically milled fuselage skins; some titanium fittings; heated windscreens of laminated polycarbonate with external glass layer to resist scratching.
LANDING GEAR: SHL hydraulically retractable tricycle type, with oleo-pneumatic shock absorber and twin wheels on each unit. Trailing-link main units retract inward, nosewheels forward. Tyre sizes 23 × 7 in (main), 16 × 4.4 in (nose). Hydraulic extension, retraction and nosewheel steering; hydraulic multi-disc anti-skid mainwheel brakes. Compressed nitrogen cylinder provides additional power source for emergency extension.
POWER PLANT: Two 16.46 kN (3,700 lb st) Garrett TFE731-3A-200G turbofans, with Grumman hydraulically actuated target type thrust reversers, pod-mounted in Grumman nacelle on each side of rear fuselage. Standard fuel in integral tank in wing centre-section, two outer-wing tanks, and upper and lower tanks in centre-fuselage (combined usable capacity 4,910 litres; 1,297 US gallons; 1,080 Imp gallons). Additional fuel can be carried in 378.5 litre (100 US gallon; 83.3 Imp gallon) removable auxiliary tank in forward area of baggage compartment. Single pressure refuelling point in lower starboard side of fuselage aft of wing, or single gravity point in upper fuselage, allow refuelling of all tanks from one position. Fuel sequencing automatic.
ACCOMMODATION: Crew of two on flight deck. Dual controls standard. Sliding door between flight deck and cabin. Standard accommodation in pressurised cabin for six persons, two in forward facing seats at front and four in club layout; galley (port or starboard) at front of cabin, coat closet forward (stbd), toilet at rear. All six seats individually adjustable fore and aft, laterally, and can be swivelled or reclined; all fitted with armrests and headrests. Two wall mounted foldaway tables between club seat pairs. Coat closet houses stereo tape deck. Maximum accommodation for nine passengers. Plug type airstair door at front on port side; emergency exit over wing on each side. Heated baggage compartment aft of passenger cabin, with external access. Service compartment in rear fuselage houses aircraft batteries (or optional APU), electrical relay boxes, inverters and miscellaneous equipment. Cabin soundproofing improved compared with Westwind 2.
SYSTEMS: AiResearch environmental control system, using engine bleed air, with normal pressure differential of 0.615 bar (8.9 lb/sq in). Garrett GTCP36-150(W) APU available optionally. Two independent hydraulic systems, each at pressure of 207 bars (3,000 lb/sq in). Primary system operated by two engine driven pumps for actuation of brakes, anti-skid, landing gear, nosewheel steering, spoilers/lift dumpers and ailerons. Backup system, operated by electrically driven pump, provides power for emergency/parking brake, ailerons and thrust reversers. Electrical system comprises two 300A 28V DC engine driven starter/generators, with two 1kVA single-phase solid state inverters operating in unison to supply single-phase 115V AC power at 400Hz and 26V AC power for aircraft instruments. Two 24V nickel-cadmium batteries for engine starting and to permit operation of essential flight instruments and emergency equipment. 28V DC external power receptacle standard. Pneumatic de-icing of wing leading-edge slats and tailplane leading-edges; thermal anti-icing of engine intakes. Oxygen system for crew (pressure demand) and passengers (drop-down masks) supplied by single 1.35 m³ (48 cu ft) cylinder. Two-bottle freon type engine fire extinguishing system standard.
AVIONICS: Standard avionics suite comprises Collins EFIS-86C five-tube electronic flight instrumentation system, dual FCS-80 flight director systems, dual VHF-22A com, dual VIR-32 nav, APS-85 autopilot, ADS-85 air data system, AHS-85 AHRS, VNI-80D vertical nav system, provisions for GNS-1000, GNS-X or UNS 1A flight management system, dual DME-42, ADF-60A, dual RMI-36, dual C-14 compass systems, dual TDR-90 transponders, ALT-50A radio altimeter, Baker dual audio systems, and WXT-250A colour weather radar.
EQUIPMENT: Standard equipment includes electric windscreen wipers, electric (warm air) windscreen demisting, cockpit and cabin fire extinguishers, axe, first aid kit, wing ice inspection lights, landing light in each wingroot, taxying light inboard of each mainwheel door, navigation and strobe lights at wingtips and tailcone, rotating beacons under fuselage and on top of fin, and wing/tailplane static wicks.
DIMENSIONS, EXTERNAL:

Wing span	16.05 m (52 ft 8 in)
Wing aspect ratio	8.8
Length overall	16.94 m (55 ft 7 in)

IAI 1125 Astra SP twin-turbofan business transport (*Peter J. Cooper*)

IAI ASTRA IV

TYPE: Twin-turbofan business/commuter transport.
PROGRAMME: Design finalised late 1992 in anticipation of 1993 launch; this would enable October 1994 first flight, December 1995 certification and mid-1996 delivery start; co-production with Yakovlev (see Russian section) being discussed.
CURRENT VERSIONS: Seen as eight-passenger **business/executive** (standard model), with possible 19-passenger quick-change interior for **regional transport** operators.
DESIGN FEATURES: Same wing as Astra SP, but new wide-body fuselage, longer and with more headroom.
POWER PLANT: Twin 48.93 kN (11,000 lb) thrust class rear-mounted turbofans; Allison GMA 3007, Garrett/General Electric CFE738 and Pratt & Whitney Canada PW305 being considered.

DIMENSIONS, EXTERNAL:
Wing span	17.48 m (57 ft 4 in)
Length overall	19.56 m (64 ft 2 in)
Height overall	6.30 m (20 ft 8 in)
Tailplane span	6.40 m (21 ft 0 in)
Wheel track (c/l of shock struts)	3.30 m (10 ft 10 in)

DIMENSIONS, INTERNAL:
Cabin: Length, excl flight deck	7.39 m (24 ft 3 in)
Max width	2.18 m (7 ft 2 in)
Max height	1.91 m (6 ft 3 in)
Baggage compartment volume	3.68 m³ (130.0 cu ft)

WEIGHTS (estimated):
Max fuel weight	6,214 kg (13,700 lb)
Max payload	1,905 kg (4,200 lb)
Max ramp weight	14,810 kg (32,650 lb)
Max T-O weight	14,742 kg (32,500 lb)
Max landing weight	12,474 kg (27,500 lb)
Max zero-fuel weight	10,206 kg (22,500 lb)

PERFORMANCE (design):
Max operating Mach number (M_{MO})	0.855
Max operating speed (V_{MO})	360 knots (667 km/h; 414 mph) IAS
Max cruising speed	476 knots (882 km/h; 548 mph)
Long-range cruising speed	430 knots (797 km/h; 495 mph)
Max operating altitude	13,715 m (45,000 ft)
Range with 4 passengers, NBAA IFR reserves	3,650 nm (6,764 km; 4,203 miles)

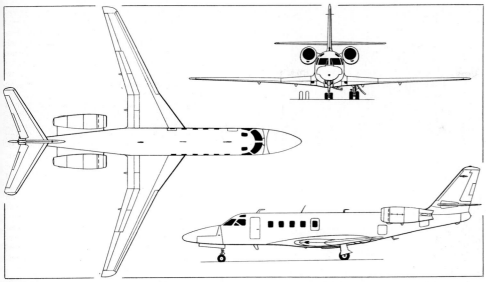

IAI 1125 Astra business transport (two Garrett TFE731-3A-200G turbofans) *(Dennis Punnett)*

Fuselage: Max width	1.57 m (5 ft 2 in)
Max depth	1.905 m (6 ft 3 in)
Height overall	5.54 m (18 ft 2 in)
Tailplane span	6.40 m (21 ft 0 in)
Wheel track (c/l of shock struts)	2.77 m (9 ft 1 in)
Wheelbase	7.34 m (24 ft 1 in)
Passenger door (fwd, port): Height	1.37 m (4 ft 6 in)
Width	0.66 m (2 ft 2 in)
Overwing emergency exits (each):	
Height	0.69 m (2 ft 3 in)
Width	0.48 m (1 ft 7 in)

DIMENSIONS, INTERNAL:
Cabin: Length: incl flight deck	6.86 m (22 ft 6 in)
excl flight deck	5.23 m (17 ft 2 in)
Max width	1.50 m (4 ft 11 in)
Max height	1.70 m (5 ft 7 in)
Baggage compartment volume	1.56 m³ (55 cu ft)

AREAS:
Wings, gross	29.40 m² (316.6 sq ft)

WEIGHTS AND LOADINGS (A: without, B: with, long-range fuel tank):
Basic operating weight empty	5,999 kg (13,225 lb)
Max usable fuel: A	3,942 kg (8,692 lb)
B	4,248 kg (9,365 lb)
Fuel with max payload: A, B	3,470 kg (7,650 lb)
Max payload: A, B	1,259 kg (2,775 lb)
Max ramp weight: A, B	10,727 kg (23,650 lb)
Max T-O weight: A, B	10,659 kg (23,500 lb)
Max landing weight: A, B	9,389 kg (20,700 lb)
Max zero-fuel weight: A, B	7,257 kg (16,000 lb)
Max wing loading	362.4 kg/m² (74.23 lb/sq ft)
Max power loading	323.8 kg/kN (3.18 lb/lb st)

PERFORMANCE (at max T-O weight ISA except where indicated):
Max cruising speed at 10,670 m (35,000 ft), AUW of 7,257 kg (16,000 lb)	463 knots (858 km/h; 533 mph)
Max operating speed (V_{MO}/M_{MO}):	
S/L to 7,620 m (25,000 ft)	363 knots (673 km/h; 418 mph) IAS
above 7,620 m (25,000 ft)	Mach 0.855
Stalling speed at max landing weight:	
flaps and gear up	111 knots (206 km/h; 128 mph)
flaps and gear down	92 knots (171 km/h; 106 mph)
Max rate of climb at S/L	1,112 m (3,650 ft)/min
Rate of climb at S/L, OEI	335 m (1,100 ft)/min
Max certificated altitude	13,715 m (45,000 ft)
Service ceiling, OEI	5,790 m (19,000 ft)

FAR 25 balanced field length:
S/L, ISA	1,600 m (5,250 ft)
S/L, ISA + 27°C	976 m (3,200 ft)
FAR 25 landing field length at S/L at max landing weight	829 m (2,720 ft)
Range with 4 passengers, 45 min VFR reserves	2,814 nm (5,215 km; 3,241 miles)

OPERATIONAL NOISE LEVELS (FAR 36 at max T-O weight, estimated):
T-O: normal	89.9 EPNdB
with thrust cutback	84.1 EPNdB
Approach	89.8 EPNdB
Sideline	89.7 EPNdB

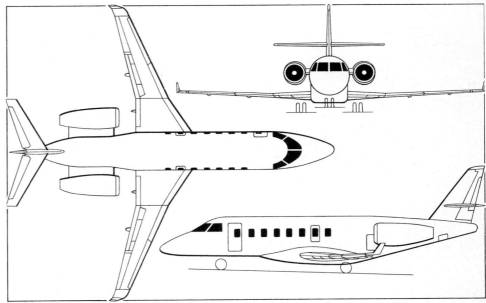

Preliminary drawing of the IAI Astra IV twin-turbofan transport *(Jane's/Mike Keep)*

BEDEK AVIATION DIVISION

Ben-Gurion International Airport, Israel 70100
Telephone: 972 (3) 935 3111 and 971 1240
Fax: 972 (3) 971 2298
Telex: 381014 and 381033 ISRAV IL
GENERAL MANAGER: A. Raz (Brig-Gen Retd)

Internationally approved as single-site civil and military airframe, power plant, systems and accessory service and upgrading centre; plants are Matam (aircraft services and infrastructure), Shaham (aircraft maintenance and upgrading), Masham (engine maintenance) and Mashav (components maintenance). Workforce about 3,000; covered floor space 110,000 m² (1,184,030 sq ft).

Current programmes (actual or available) include Phantom 2000, Super Phantom, modification of large transport aircraft to passenger/cargo, tanker, AEW and reconnaissance configurations. More than 25 types of aircraft being handled, including Boeing 707/727/737/747/767, McDonnell Douglas DC-8/-9/-10 and Lockheed C-130; combat aircraft include A-4 Skyhawk, F-4 Phantom, F-5, F-15 Eagle, F-16 Fighting Falcon, various MiG fighters and Mirage III/5; 30 types of civil and military engine include JT3D, JT8D, JT9D, F100, J79, Atar 9C/9K-50, TFE731, T56, PT6, Allison 250, T53 and T64; more than 6,000 types of accessory and instruments serviced; Bedek provides total technical support and holds warranty and/or approved service centre appointments for domestic and foreign regularity agencies, air arms and manufacturers; approving agencies include Israeli CAA, Israeli Air Force, US military, US FAA, UK CAA and German LBA.

IAI COMBAT AIRCRAFT UPGRADING

Offers modular modernisation packages, including digital weapon delivery and navigation system (WDNS) with multiple air-to-air and air-to-ground weapon delivery modes and navigation with up to 40 waypoints; stores management and release system (SMRS) integrated in WDNS for dumb and smart weapons; active and passive ECM, including podded or internal jammers, radar warning and flare/chaff dispensers; customised new digital ASDI, HSI and ADC; airframe and/or engine modifications and refits. More extensive details can be found in *Jane's Civil and Military Aircraft Upgrades*.

Typical programmes include:

Mirage III/5: Includes Kfir-type foreplanes and strengthened landing gear reducing T-O run or allowing higher T-O weight. Additional fuel tank can be installed aft of cockpit; extended nose for additional avionics such as control and stability augmentation systems. Radar warning with omni-directional threat analysis and WDNS-391 inertial weapon delivery and navigation system with head-up operation in all air-to-air and air-to-ground modes. Martin-Baker Mk 10 ejection seat; missiles or ECM pods on wingtip stations; two or four additional external stores stations; flare/chaff dispensers under rear fuselage.

Mirage III/5 Option 1: Raising thrust by 30 per cent and improving sfc by 20 per cent by fitting General Electric/Flygmotor F404/RM12 giving 55.6 kN (12,500 lb st) dry and 80.7 kN (18,140 lb st) with afterburning; aircraft empty weight reduced by 453.5 kg (1,000 lb), internal fuel increased by 544 kg (1,200 lb); max T-O weight increased by 2,721 kg (6,000 lb).

Mirage III/5 Option 2: Elta EL/M-2011 or M-2032 lightweight coherent pulse Doppler fire control radar; air target information presented on HUD; improved air-to-ground ranging; extensive built-in test and calibration; adaptability to other avionics; growth potential through fully software controlled LRUs and MIL-STD-1553B databus.

Nammer (Tiger): Mirage III/5 upgrade with Atar 9K-50 engine, longer fuselage, canards, dog-tooth wing leading-edges, improved avionics/weapons capability. Further details in 1992-93 and earlier *Jane's,* but report of Spring 1991 first flight now thought to be erroneous.

Skyhawk: Modifications (already applied to Israeli Air Force A-4s) include life extension, complete rewiring, dual-disc mainwheel brakes, nosewheel steering, wing spoilers, two additional underwing hardpoints, engine tailpipe extension, brake parachute in fairing under tail. Wingroot cannon 30 mm; modern WDNS; extended nose compartment and saddleback hump for more and lighter avionics; flare/chaff dispensers under tail. Dual control two-seat modification also offered.

Phantom 2000: Israeli Air Force (IAF) programme to extend life beyond 2000, also available for export; first flight Phantom 2000 prototype 11 August 1987; first redelivery to IAF 9 April 1989; first operational use 5 February 1991. Prototypes completed by IAF; airframe changes to production aircraft by IAI; redeliveries exceeded 20 by mid-1991 and then continuing at average two per month.

Modifications include reinforced skins and fuel cells in fuselage and wings; complete rewiring; dual MIL-STD-1553B databuses; replaced and rerouted hydraulic lines; number of avionics boxes reduced; built-in test features added; small strakes added above intake flanks to improve stability and manoeuvrability; cockpit comfort and instrument layout modernised.

New IAF avionics installation includes Norden/UTC multi-mode high resolution radar (deliveries delayed but due to begin April 1992), El-Op (Kaiser licence) wide angle diffractive-optics HUD, Elbit multi-function electronic displays for both crewmen, digital WDNS with HOTAS (hands on throttle and stick), Orbit integrated com and com/nav systems, improved ECM and self-protection systems; core of system is derivative of ACE-3 currently fitted in IAF F-16C/Ds; overall avionics integrator is Elbit Computers Ltd.

Super Phantom: Prototype refitted with P&W PW1120s test-flown 1986-87, giving excellent performance improvements (see 1989-90 *Jane's*), but hopes for IAF retrofit effectively ended with Lavi cancellation. Modification on offer for export; includes canards and conformal ventral fuel tank.

F-5 Plus: Shaham package. Avionics improvements include HUD, two multi-function displays and HOTAS, modern pulse Doppler radar, video camera and recorder, optional helmet-mounted display, improved weapon delivery and ECM, new weapons, and maintainability improvements. Other options include modern ejection seat, secure communications, flight refuelling and podded or nose-mounted reconnaissance equipment.

Chilean Air Force upgrade of 12 F-5Es and two F-5Fs, ordered early 1990, includes Elta EL/M-2032B multi-mode pulse Doppler fire-control radar, improved WDNS, INS, El-Op HUD, two head-down CRT displays, HOTAS, Aeronautics (Israel) air data computer, modular mission and display processor and multiple databus, and full EW suite (passive RWR, jammer and chaff/flare dispenser). Nose modifications to accommodate larger radar antenna include removal of port M39 20 mm cannon and relocation of some subsystems. Refit reportedly includes provision of Python air-to-air missiles. First two (one F-5E, one F-5F) under flight test early 1993; remainder to be converted by ENAER Chile under IAI supervision.

Public debut of the IAI/Elta Phalcon 707 AEW aircraft at the Paris Air Show, June 1993 *(Photo Link)*

F-15: Bedek involved in upgrading of F-15 structures, maintainability and various utility and mission systems.

MiG-21 and MiG-23: Similar to upgrades of Western types; added to range 1991/92. Contract reported late 1992 to upgrade 100 MiG-21s of Romanian Air Force, with work carried out by Aerostar (which see) under IAI supervision. *(see Addenda for further details).*

IAI S-2UP TRACKER

Several upgrades offered by Shaham for Grumman S-2 Tracker, including re-engining with Garrett TPE331 turboprops, modern search radar, electro-optical systems, EW, ESM and ECM and new flight systems. Argentine Navy ('prototype' and five customer kits) is first customer. *See also Addenda.*

IAI LARGE TRANSPORT CONVERSIONS

Described in more detail in *Jane's Civil and Military Aircraft Upgrades.* Conversions include the following:

Boeing 707/720: Numerous 707s and 720s converted to cargo, sigint, hose or boom refuelling or other configurations (see earlier *Jane's*); **sigint/tanker** with wingtip refuelling pods and Elta EL/L-8300 sigint system illustrated in 1987-88 *Jane's.*

Elta Electronics **AEW Phalcon** with EL/2075-derived solid-state L-band radar with six conformal phased-array antennae (two on each side of fuselage, one in enlarged nose and one under tail); Phalcon also includes advanced monopulse IFF, wide-range ESM and comint data processing systems. Launch customer Chile (one); first flight (4X-JYT) 12 May 1993.

Tanker modifications include reinforced outer wings, supports for additional fuel tanks where applicable, fuselage reinforcement for boom support point or tail hose/reel unit, additional hydraulic system to power pumps and boom or tail hose/reel, adapted fuel system, external lighting and director lights, refuelling controls, boom operator station with 3D electro-optical viewing system and customer-specified avionics. Combi tankers with boom and two wingtip hose/reel units delivered to IAF; four kits produced for Royal Australian Air Force.

WEIGHTS AND LOADINGS (707-320C tanker, approx):

Operational weight empty	65,770 kg (145,000 lb)
*Internal fuel weight	72,575 kg (160,000 lb)
**Additional fuel weight	up to 13,605 kg (30,000 lb)
Tanker T-O weight	151,950 kg (335,000 lb)

**90,300 litres (23,855 US gallons; 19,863 Imp gallons)*
***17,034 litres (4,500 US gallons; 3,747 Imp gallons)*

Boeing 747-100 and -200 Freighter: Bedek Boeing 747-100 freighter prototype received FAA STC late 1992. Changes include 3.05 × 3.40 m (10 ft 0 in × 11 ft 2 in) upward-opening freight door to port aft of wing, local fuselage reinforcement, strengthened cabin floor, fully powered ball mat/roller cargo handling system, cargo restraint, bulkhead between cargo and passenger cabins and interior modifications for selected passenger/cargo combinations. Choices include (1) all-cargo with 29 main-deck standard pallets or containers, (2) Combi with passengers at front and 7-13 pallets aft and (3) all-passenger with custom interior. Similar conversions of 747-200 and provision for non-standard available.

British operator Electra Aviation ordered 10 all-cargo 747-100 conversions late 1990 for delivery starting in mid-1992; other customers include Lufthansa (six), also for all-cargo.

WEIGHTS AND LOADINGS (747-100 Combi, estimated):

Operational weight empty	148,325 kg (327,000 lb)
Max payload	98,883 kg (218,000 lb)
Max T-O weight	334,750 kg (738,000 lb)
Max landing weight	265,350 kg (585,000 lb)
Max zero-fuel weight	247,435 kg (545,500 lb)

Lockheed C-130/L-100 Hercules: Several successful conversions already of C-130 series to in-flight refuelling tanker and sigint platform, with appropriate airframe modifications and avionics refits. Operational configurations currently being offered for any C-130B to C-130H variant, or L-100 commercial counterparts, include: (1) probe and drogue tanker, with transfer fuel in 11,356 litre (3,000 US gallon; 2,498 Imp gallon) cargo compartment tank plus two underwing fuel pods; (2) maritime patrol and ASW, with surveillance, acoustic, MAD, armament or stores management systems, and operator stations; (3) C^3I and EW, with comint, elint, communications and EW systems to customer's requirements; (4) search and rescue, with rescue kit, flare storage/launcher and operator station on logistic pallet installed on rear loading ramp; (5) emergency assistance, with insulated cabin mounted on logistic pallet for ambulance or flying hospital missions or in firefighting configuration with up to 11,356 litres (3,000 US gallons; 2,498 Imp gallons) of water and retardant in pallet mounted tanks in cargo hold; (6) VIP, 65-seat passenger or passenger/cargo combi transport, with full airliner type seating, galley and toilet facilities, pallet-mounted in air-conditioned environment.

WEIGHTS AND LOADINGS (C-130H tanker, approx):

Operational weight empty	35,380 kg (78,000 lb)
*Internal fuel weight	29,030 kg (64,000 lb)
**Additional fuel weight	10,885 kg (24,000 lb)
Tanker T-O weight	75,295 kg (166,000 lb)
Max overload T-O weight	79,380 kg (175,000 lb)

**36,643 litres (9,860 US gallons; 8,060 Imp gallons)*
***13,627 litres (3,600 US gallons; 2,997 Imp gallons)*

Cockpit upgrades: Bedek offers to convert three-man to two-man flight decks, including advanced monitoring and control systems and engine indication and crew alerting (EICAS). Candidate aircraft include Boeing 727 and McDonnell Douglas DC-10.

TECHNOLOGIES DIVISION

PO Box 190, Lod Industrial Zone, Lod 71101
Telephone: 972 (8) 239111 and 223050
Fax: 972 (8) 222792
Telex: 381520 SHLD IL
GENERAL MANAGER: D. Arzi

Division runs four plants: SHL (Servo Hydraulics Lod), Ramta Structures and Systems, MATA Helicopters and Golan Industries.

SHL designs and produces hydraulic system components, flight control servos, landing gears and brake systems, air-actuated chucks, miniature gears, clutches and brakes. Products fitted in Kfir, Arava, Westwind, Astra, IDF Black Hawks. Manufacturing approvals include Boeing, Dornier, Lockheed Fort Worth and General Electric.

Ramta makes metal and advanced composites structures for F-4, F-16, E-2C, Kfir, Westwind and Astra, as well as ground vehicles and patrol boats.

MATA repairs, modifies and remanufactures helicopter structures and components and produces equipment and systems for helicopters. Recent contracts include upgrade of two Agusta-Bell 212s for Venezuelan Navy. Its subsidiary Golan designs and produces aircraft crew and passenger seats (including crashworthy troop seats for Bell/Boeing V-22 Osprey), aircraft wheels and cockpit controls. Current contract for components and seats for McDonnell Douglas Explorer helicopter.

IAI YASUR (HELICOPTER) 2000

Led by MATA with Elbit as avionics integrator; upgrading 30 IAF CH-53D heavy-lift helicopters (go-ahead given June 1990); first flight of first Yasur 2000, 4 June 1992; redeliveries began 2 February 1993. Modifications include life extension of airframe beyond 2000, new mission computer, moving map display, two multi-function displays, new autopilot, external fuel tanks, flight refuelling boom, rescue hoist, crashworthy seats, cockpit armour, improved landing lights, internal batteries, electric pump for reloading APP accumulator and new APP clutch.

ITALY

AERONAUTICA MACCHI

AERONAUTICA MACCHI SpA

Via P. Foresio, I-21040 Venegono Superiore
Telephone: 39 (331) 865912
Fax: 39 (331) 865910
CHAIRMAN: Dott Fabrizio Foresio

Original Macchi company founded 1913 in Varese, and produced famous line of high-speed flying-boats and seaplanes. Aeronautica Macchi became holding company with all operating activities in wholly owned company Aermacchi SpA; group includes, besides Aermacchi SpA, SICAMB (airframe and equipment manufacturing, including licence production of Martin-Baker ejection seats), OMG (precision machining), and Logic (electronic equipment). A 25 per cent holding in Aeronautica Macchi was acquired by Aeritalia, now Alenia, in 1983.

Transfer of SIAI-Marchetti fixed-wing interests of Agusta to Aermacchi, proposed during 1992, did not take place.

AERMACCHI

AERMACCHI SpA (Subsidiary of Aeronautica Macchi SpA)
Via P. Foresio, PO Box 101, I-21040 Venegono Superiore
Telephone: 39 (331) 813111
Fax: 39 (331) 827595
CHAIRMAN: Dott Fabrizio Foresio
MANAGING DIRECTOR: Dott Ing Giorgio Brazzelli
GENERAL MANAGERS:
  Dott Ing Bruno Cussigh
  Dott Ing Romano Antichi
TECHNICAL DIRECTOR: Dott Ing Massimo Lucchesini
COMMERCIAL MANAGER: Dott Cesare Cozzi
PUBLIC RELATIONS MANAGER: Franca Grandi

Aermacchi is aircraft manufacturing company of Aeronautica Macchi group; plants at Venegono airfield occupy total area of 274,000 m² (2,949,310 sq ft), including 52,000 m² (559,720 sq ft) covered space; flight test centre has covered space of 5,100 m² (54,900 sq ft) in total area of 28,000 m² (301,390 sq ft); total workforce at end of 1992 was 1,908.

Aermacchi active in aerospace ground equipment; also has important roles in AMX, Tornado, Ariane and Eurofighter 2000 programmes.

AERMACCHI/DASA PTS-2000

Aermacchi and Dornier signed MoU in April 1989 for future integrated military pilot training system (PTS-2000) to meet pilot training requirements of year 2000; DASA Military Aircraft Division joined group early 1991. See International section.

AMX

Aermacchi teamed with Alenia and Embraer in AMX combat aircraft programme (see International section) for Italian and Brazilian air forces.

DORNIER 328

Memorandum for co-operation on Dornier 328 programme (see German section) signed April 1989; Aermacchi designed and is building forward fuselage and assembles main fuselage using barrel sections produced by South Korean Daewoo Heavy Industries.

AERMACCHI MB-339A

TYPE: Two-seat basic and advanced trainer and attack aircraft.
PROGRAMME: First flights of two prototypes (MM588) 12 August 1976 and (MM589) 20 May 1977; first flight of production MB-339A 20 July 1978; initial 51 (first aircraft delivered 8 August 1979) included batch used as radio calibration aircraft at Pratica di Mare and 20 PANs for Frecce Tricolori aerobatic team (in service 27 April 1982); last of 101 IAF aircraft delivered 1987; development continues.
CURRENT VERSIONS: **MB-339A:** Current standard military basic/advanced trainer (full details in 1990-91 *Jane's*). *Description applies to this version except where indicated.*
MB-339 PAN: Special version for Italian Air Force national aerobatic team Pattuglia Acrobatica Nazionale; smoke generator added and no tip tanks.
MB-339AM: Special anti-ship version armed with OTO Melara Marte Mk 2A missile; avionics, equivalent to MB-399C, include new inertial navigator, Doppler radar, navigation and attack computers, head-up display and multi-function display. Prototype converted from MB-339A.
MB-339B: Powered by 19.57 kN (4,400 lb st) Viper Mk 680-43; larger tip tanks. Prototype only; now being used in MB-339AM development programme.
Radiomisure: Small batch of radio calibration aircraft produced for Italian Air Force.
T-Bird II: Modified for US JPATS competition; described separately.
CUSTOMERS: Italian Air Force 101; Argentine Navy (10 delivered 1980); Peruvian Air Force (16 delivered 1981-82); Royal Malaysian Air Force (13 delivered 1983-84); Dubai (two delivered 1984, three in 1987, two in 1992); Nigeria (12 delivered 1985); Ghana Air Force (two delivered 1987). Possible further order from Dubai.
DESIGN FEATURES: For full details, see 1990-91 *Jane's*; all IAF trainers camouflaged for use as emergency close air support.
POWER PLANT: One Italian-made Rolls-Royce Viper Mk 632-43 turbojet, rated at 17.8 kN (4,000 lb st). Fuel in two-cell rubber fuselage tank, capacity 781 litres (206 US gallons; 172 Imp gallons), and two integral tip tanks, combined capacity 632 litres (167 US gallons; 139 Imp gallons) or 1,000 litres (264 US gallons; 220 Imp gallons) with larger tip tanks; bigger tip tanks also available for MB-339A. Total internal capacity 1,413 litres (373 US gallons; 311 Imp gallons) usable or 1,781 litres (470.5 US gallons; 392 Imp gallons) with larger tip tanks. Single-point pressure refuelling receptacle in port side of fuselage, below wing trailing-edge. Gravity refuelling points on top of fuselage and each tip tank. Provision for two drop tanks, each of 325 litres (86 US gallons; 71.5 Imp gallons) usable capacity, on centre underwing stations. In-flight refuelling now available optionally.

Italian Air Force Aermacchi MB-339B (one Rolls-Royce Viper Mk 680-43)

AVIONICS: Conventional military IFR avionics and instruments; EFIS cockpits now being offered.
EQUIPMENT: Provision for towing type A-6B (1.83 × 9.14 m; 6 × 30 ft) and A-4 aerial banner targets; tow attachment point on inner surface of ventral airbrake. External stores can include photographic pod with four 70 mm Vinten cameras; or a single underwing Elettronica ECM pod.
ARMAMENT: Up to 2,040 kg (4,500 lb) of external stores can be carried on six underwing hardpoints, inner four of which are stressed for loads of up to 454 kg (1,000 lb) each and outer two for up to 340 kg (750 lb) each. Integration of Marte Mk 2A anti-ship missiles began 1991; first launch 17 June 1992. Provisions on two inner stations for two Macchi gun pods, each containing either a 30 mm DEFA 553 cannon with 120 rds, or a 12.7 mm AN/M-3 machine gun with 350 rds. Other typical loads can include two Matra 550 Magic or AIM-9 Sidewinder air-to-air missiles on two outer stations; four 1,000 lb or six 750 lb bombs; six SUU-11A/A 7.62 mm Minigun pods with 1,500 rds/pod; six Matra 155 launchers, each for eighteen 68 mm rockets; six Matra F-2 practice launchers, each for six 68 mm rockets; six LAU-68/A or LAU-32G launchers, each for seven 2.75 in rockets; six Aerea AL-25-50 or AL-18-50 launchers, each with twenty-five or eighteen 50 mm rockets respectively; six Aerea AL-12-80 launchers, each with twelve 81 mm rockets; four LAU-10/A launchers, each with four 5 in Zuni rockets; four Thomson-Brandt 100-4 launchers, each with four 100 mm Thomson-Brandt rockets; six Aerea BRD bomb/rocket dispensers; six Aermacchi 11B29-003 bomb/flare dispensers; six Thomson-Brandt 14-3-M2 adaptors, each with six 100 mm anti-runway bombs or 120 mm tactical support bombs. Provision for Alenia 8.105.924 fixed reflector sight or Saab RGS 2 gyroscopic gunsight; a gunsight can also be installed in rear cockpit, to enable instructor to evaluate manoeuvres performed by student pilot. All gunsights can be equipped with fully automatic Teledyne TSC 116-2 gun camera.

DIMENSIONS, EXTERNAL:
Wing span over tip tanks	10.858 m (35 ft 7½ in)
Wing aspect ratio	6.1
Length overall	10.972 m (36 ft 0 in)
Height overall	3.994 m (13 ft 1¼ in)

WEIGHTS AND LOADINGS:
Weight empty, equipped	3,125 kg (6,889 lb)
Fuel load (internal, usable)	1,100 kg (2,425 lb)
T-O weight, clean	4,400 kg (9,700 lb)
Max T-O weight with external stores	5,895 kg (13,000 lb)

PERFORMANCE (at clean T-O weight, ISA):
Mach/IAS limit	Mach 0.85 (500 knots; 926 km/h; 575 mph)
Max level speed at S/L	485 knots (898 km/h; 558 mph) IAS
Stalling speed	80 knots (149 km/h; 93 mph)
Max rate of climb at S/L	2,010 m (6,595 ft)/min

Trial firing of OTO Melara Marte Mk 2A anti-ship missile from an MB-339

ITALY: AIRCRAFT—AERMACCHI

Lockheed/Aermacchi/Rolls-Royce T-Bird II trainer proposed for US Air Force/Navy JPATS programme

LOCKHEED/AERMACCHI/ROLLS-ROYCE T-BIRD II

In October 1989, Aermacchi and Lockheed signed co-operation agreement to compete in Joint Primary Aircraft Training System (JPATS) for US Air Force and Navy, using T-Bird II, a 'missionised' version of the MB-339A; if successful, Lockheed would be prime contractor and system integrator and have manufacturing licence. Rolls-Royce joined team in September 1990. Demonstrator delivered to Lockheed's factory at Marietta, Georgia, on 20 May 1992. Power plant is Rolls-Royce RB582 of 17.79 kN (4,000 lb st) with improved maintainability.

AERMACCHI MB-339C

TYPE: Two-seat advanced fighter lead-in trainer and attack aircraft.
PROGRAMME: Development begun 1982/83; first flight (I-AMDA) 17 December 1985.
CUSTOMERS: Royal New Zealand Air Force ordered 18 in May 1990; deliveries to be completed in 1993.
DESIGN FEATURES: Designed to MIL-A-8860A for 10,000 h service life. Wing section NACA 64A-114 (mod) at centreline, 64A-212 (mod) at tip; quarter-chord sweepback 8° 29'.
FLYING CONTROLS: Power assisted ailerons with servo tabs to assist manual reversion; fixed tailplane and manual elevators and rudder; electrically controlled servo tab for control assistance and trimming on elevator; hydraulically actuated single-slotted flaps; two ventral strakes under tail; electro-hydraulically actuated airbrake panel under forward fuselage; wing fence ahead of aileron inboard edge; both pilots have HUD; rear pilot elevated sufficiently to be able to aim guns and air-to-surface weapons and fly visual approaches (see nav/attack system under Avionics heading).
STRUCTURE: All-metal; stressed-skin wings with main and auxiliary spars and spanwise stringers; bolted to fuselage; tip tanks permanently attached; rear fuselage detachable by four bolts for engine access.
LANDING GEAR: Hydraulically retractable tricycle type with oleo-pneumatic shock absorbers, suitable for operation from semi-prepared runways. Hydraulically steerable nosewheel retracts forward; main units retract outward into wings. Low pressure mainwheel tubeless tyres size 545 × 175-10 (14 ply rating); nosewheel tubeless tyre size 380 × 150-4 (6 ply rating). Emergency extension system. Hydraulic disc brakes with anti-skid system. Min ground turning radius 8.45 m (27 ft 9 in).
POWER PLANT: One Rolls-Royce Viper Mk 680-43 turbojet, rated at 19.57 kN (4,400 lb st). Fuel in two-cell rubber fuselage tank, capacity 781 litres (206 US gallons; 172 Imp gallons), and two wingtip tanks with combined capacity of 1,000 litres (264 US gallons; 220 Imp gallons). Total internal usable capacity 1,781 litres (470.5 US gallons; 392 Imp gallons). Single-point pressure refuelling point in port side of fuselage, below wing trailing-edge. Gravity refuelling points on top of fuselage and each tip tank. Provision for two drop tanks, each of 325 litres (86 US gallons; 71.5 Imp gallons) usable capacity, on centre underwing stations.
ACCOMMODATION: Crew of two in tandem, on Martin-Baker IT10LK zero/zero ejection seats in pressurised cockpit. Rear seat elevated 32.5 cm (1 ft 1 in). Rearview mirror for each occupant. Two-piece moulded transparent canopy, opening sideways to starboard.
SYSTEMS: Pressurisation system max differential 0.24 bar (3.5 lb/sq in); cockpit designed for 40,000 pressurisation cycles. Bootstrap type air-conditioning system, also providing air for windscreen and canopy demisting. Hydraulic system, pressure 172.5 bars (2,500 lb/sq in), for actuation of flaps, aileron servos, airbrake, landing gear, wheel brakes and nosewheel steering. Backup system for wheel brakes and emergency extension of landing gear. Main electrical DC power from one 28V 9kW engine driven starter/generator and one 28V 6kW secondary generator. Two 24V 22Ah nickel-cadmium batteries for engine starting. Fixed frequency 115/26V AC power from two 600VA single phase static inverters. External power receptacle. Low pressure demand oxygen system, operating at 28 bars (400 lb/sq in). Anti-icing system for engine air intakes.
AVIONICS: Typical avionics installation includes Collins AN/ARN-118(V), type RT-1159/A Tacan or Bendix/King KDM 706A DME; Collins 51RV-4B VOR/ILS and MKI-3 marker beacon receiver; Collins ADF-60A ADF/L; Collins DF 301E VHF/UHF ADF (optional); GEC-Marconi AD-660 Doppler velocity sensor integrated with Litton LR-80 inertial platform; GEC-Marconi 620K navigation computer; Kaiser Sabre head-up display/weapons aiming computer; Alenia CRT multifunction display; Alenia/Honeywell HG7505 radar altimeter; FIAR P 0702 laser rangefinder; ELT-156 radar warning system; Logic stores management system; Bendix/King AN/APX-100 IFF; Astronautics AN/ARU-50/A attitude director indicator; Astronautics AN/AQU-13 HSI; HOTAS (hands on throttle and stick) controls. Single underwing Elettronica/ELT-555 ECM pod combined with a flare/chaff dispenser, onboard RWR and indicators.
EQUIPMENT: Includes Fairchild Weston video camera; photographic pod with four 70 mm Vinten cameras; Tracor AN/ALE-40 chaff/flare dispenser.

Aermacchi MB-339C of the Royal New Zealand Air Force

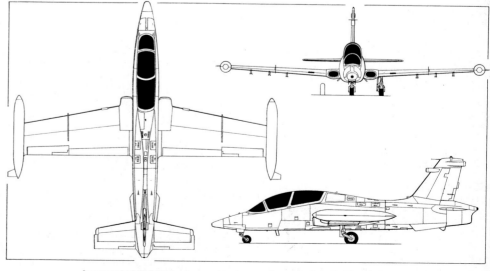

Aermacchi MB-339C advanced trainer and attack aircraft (*Dennis Punnett*)

ARMAMENT: Up to 1,814 kg (4,000 lb) of external stores on six underwing hardpoints. Four inner hardpoints each stressed for up to 454 kg (1,000 lb) load, and two outer hardpoints each for up to 340 kg (750 lb) load. RNZAF aircraft fitted for AIM-9 Sidewinder and AGM-65 Maverick. Provision on two inner stations for installation of two Macchi gun pods, each containing either a 30 mm DEFA 553 cannon with 120 rounds or a 12.7 mm AN/M-3 machine gun with 350 rounds. Other typical loads can include two Matra 550 Magic or AIM-9 Sidewinder air-to-air missiles on two outer stations; six general purpose or cluster bombs of appropriate weights; six AN/SUU-11A/A 7.62 mm Mini-gun pods, each with 1,500 rounds; six Matra 155 launchers, each for eighteen 68 mm rockets; six AN/LAU-68/A or AN/LAU-32G launchers, each for seven 2.75 in rockets; six Aerea AL-25-50 or AL-18-50 launchers, each with twenty-five or eighteen 50 mm rockets respectively; six Aerea AL-18-80 launchers, each with twelve 81 mm rockets; four AN/LAU-10/A launchers, each with four 5 in Zuni rockets; four Thomson-Brandt 100-4 launchers, each with four 100 mm Thomson-Brandt rockets; six Bristol Aerospace LAU-5002 launchers for CRV-7 high velocity rockets; six Aerea BRD bomb/rocket dispensers; six Aermacchi 11B29-003 bomb/flare dispensers; six Thomson-Brandt 14-3-M2 adapters, each with six BAP 100 anti-runway bombs or BAT 120 tactical support bombs.

DIMENSIONS, EXTERNAL:
Wing span over tip tanks	11.22 m (36 ft 9¾ in)
Wing aspect ratio	6.1
Length overall	11.24 m (36 ft 10½ in)
Height overall	3.994 m (13 ft 1¼ in)
Elevator span	4.08 m (13 ft 4½ in)
Wheel track	2.483 m (8 ft 1¾ in)
Wheelbase	4.369 m (14 ft 4 in)

AREAS:
Wings, gross	19.30 m^2 (207.7 sq ft)
Ailerons (total)	1.328 m^2 (14.29 sq ft)
Trailing-edge flaps (total)	2.21 m^2 (23.79 sq ft)
Airbrake	0.68 m^2 (7.32 sq ft)
Fin	2.37 m^2 (25.51 sq ft)
Rudder, incl tab	0.61 m^2 (6.57 sq ft)
Tailplane	3.38 m^2 (36.38 sq ft)
Elevators (total incl tabs)	0.979 m^2 (10.54 sq ft)

WEIGHTS AND LOADINGS:
Weight empty, equipped	3,310 kg (7,297 lb)
Fuel load (internal, usable)	1,388 kg (3,060 lb)
T-O weight, clean	4,884 kg (10,767 lb)
Max T-O weight with external stores	6,350 kg (14,000 lb)
Max wing loading	329.0 kg/m^2 (67.39 lb/sq ft)
Max power loading	324.7 kg/kN (3.18 lb/lb st)

PERFORMANCE (at trainer clean T-O weight, ISA, except where indicated):
Max level speed at S/L	486 knots (900 km/h; 558 mph)
Max level speed at 9,150 m (30,000 ft)	Mach 0.77 (441 knots; 815 km/h; 508 mph)
Max speed for landing gear extension	175 knots (324 km/h; 202 mph)
T-O speed	100 knots (185 km/h; 115 mph)
Approach speed over 15 m (50 ft) obstacle	98 knots (182 km/h; 113 mph)
Stalling speed	80 knots (149 km/h; 93 mph)
Max rate of climb at S/L	2,160 m (7,085 ft)/min
Time to 9,150 m (30,000 ft)	6.7 min
Service ceiling (30.5 m; 100 ft/min rate of climb)	14,240 m (46,700 ft)
T-O run at S/L	490 m (1,608 ft)
Landing run at S/L	455 m (1,493 ft)
Max ferry range with two underwing drop tanks, 10% reserves	1,100 nm (2,038 km; 1,266 miles)
Max endurance with drop tanks	3 h 50 min
g limit	7.33

AGUSTA

AGUSTA SpA (called G. F. Gestioni Industriali since 1 January 1993)
Via Giovanni Agusta 520, I-21017 Cascina Costa di Samarate (VA)
Telephone: 39 (331) 229111
Fax: 39 (331) 222595
Telex: 332569 AGUCA I
OFFICES:
Via Caldera 21, I-20153 Milan
Telephone: 39 (2) 452751
Fax: 39 (2) 48204701
Telex: 333280 AGUMI I
Via Abruzzi 11, I-00187 Rome
Telephone: 39 (6) 49801
Fax: 39 (6) 6799944
Telex: 614398 AGURO I
CHAIRMAN: Gen Basilio Cottone
CEO: Amedeo Caporaletti
INTERNATIONAL CO-OPERATION: Ing Luigi Passini
MARKETING AND SALES: Dott Enrico Guerra
PLANNING: Dott Roberto Leverone
RESEARCH AND DEVELOPMENT BUSINESS UNIT:
Ing Bruno Lovera

Formed in 1977; the Agusta group (see 1980-81 *Jane's*) completely reorganised from 1 January 1981 under new holding company Agusta SpA; became part of Italian public holding company EFIM, employing nearly 10,000 people in 12 factories in various parts of Italy.

As part of recovery of liquidated state-owned EFIM, Agusta group, including OMI, OTO Melara, Breda Meccanica Bresciana, Galileo and SMA, was transferred to Finmeccanica from 1 January 1993 to 30 June 1993 for peppercorn rent of LIt1. Decision was then to be taken on disposal of Agusta group and its constituents, probably by sale to other company or companies. Incorporation of Agusta as helicopter group of Alenia seemed unlikely.

Various domestic activities of Agusta are grouped under location of works. Familiar names of Costruzioni Aeronautiche Giovanni Agusta, SIAI-Marchetti, Caproni Vizzola Costruzioni Aeronautiche and BredaNardi Costruzioni Aeronautiche no longer used; in addition, Elicotteri Meridionali is domestic affiliate and Agusta has or had international affiliates in South Korea and Turkey. The various works and affiliates are as follows:

DOMESTIC WORKS
Benevento Works
(ex-FOMB—Fonderie e Officine Meccaniche di Benevento SpA)
Contrada Ponte Valentino — S.S.90bis, I-82100 Benevento
Telephone: 39 (824) 53440/53441/53447
Fax: 39 (824) 53418
Telex: 710667
Specialises in aircraft co-production and overhaul of helicopters and multi-engined aircraft.

Brindisi Works
(ex-IAM—Industrie Aeronautiche Meridionali SpA)
Contrada Santa Teresa Pinti, I-72100 Brindisi
Telephone: 39 (831) 8911
Fax: 39 (831) 452659
Telex: 813360

Cascina Costa Works
(ex-Costruzioni Aeronautiche Giovanni Agusta SpA)
Via Giovanni Agusta 520, I-21017 Cascina Costa di Samarate (VA)
Telephone: 39 (331) 229111
Fax: 39 (331) 222595
Telex: 332569 AGUCA I

Monteprandone Works
(ex-BredaNardi Costruzioni Aeronautiche SpA)
Casella Postale 108, San Benedetto del Trento (Ascoli Piceno), Monteprandone (AP)
Telephone: 39 (735) 801721
Fax: 39 (735) 701927
Telex: 560165 BRENAR I

Rome Works
(ex-Ottico Meccanica Italiana SpA)
Via della Vasca Navale 79/81, I-00146 Rome
Telephone: 39 (6) 55421
Fax: 39 (6) 5593753
Telex: 610137

Sesto Calende Works
(ex-SIAI-Marchetti SpA)
Via Indipendenza 2, I-21018 Sesto Calende (VA)
Telephone: 39 (331) 924421
Fax: 39 (331) 922525
Telex: 332601 SI AIA VI
Planned to be transferred to Aermacchi.

Somma Lombarda Works
(ex-Caproni Vizzola Costruzioni Aeronautiche SpA)
Via Per Tornavento 15, I-21019 Somma Lombarda
Telephone: 39 (331) 230826
Fax: 39 (331) 230622
Telex: 332554 CAVIZ I

Tradate Works
(ex-Agusta Sistemi SpA)

DOMESTIC AFFILIATES
Elicotteri Meridionali SpA
Via G. Agusta 1, Frosinone

S. E. I.—Servizi Elicotteristici Italiani
Via della Vasca Navale 79/81, I-00146 Rome
Telephone: 39 (6) 49801

Italcompositi SpA
Via Pomarico, Pisticci, MT

INTERNATIONAL AFFILIATES
Agusta Aerospace Corporation
2655 Interplex Drive, Travose, Philadelphia 19047, USA
Telephone: 1 (215) 281 1400
Fax: 1 (215) 281 0440
Telex: 6851181
CHAIRMAN AND CEO: Ing Giuseppe Orsi

Agusta Aerospace Services SA
Keiberg Park, Excelsiorlanne 23, Zaventem B-1930, Belgium
Telephone: 32 (2) 648585/6485515
Telex: 63349
GENERAL MANAGER: Dott Riccardo Baldini

Agusta Aviation Far East Pte Ltd
11-03 United Square, 101 Thomson Road, 1130 Singapore
Telephone: 65 273 3100
Telex: 36126
GENERAL MANAGER: Dott Fulvio Maurogiovanni

EH Industries Ltd
500 Chiswick High Road, London W4 5RG, UK
Telephone: 44 (81) 995 8221
Fax: 44 (81) 995 5207/5990

EH Industries Inc
141 Laurier Avenue West, 6th Floor, Ottawa, Ontario K1P 5J3, Canada
Telephone: 1 (613) 563 2180
Fax: 1 (613) 233 0399

Monacair SAM
Héliport de Fontvieille, Principality of Monaco, MC-98000

OMI Corporation of America
1319 Powhatan Street, Alexandria, Virginia 22314, USA
Telephone: 1 (703) 549 9191
Telex: 899141
CHAIRMAN: James F. Byrnes

CASCINA COSTA WORKS

Original Agusta company established 1907 by Giovanni Agusta; acquired licence for Bell Helicopter Model 47 in 1952; first flight of first Agusta example 22 May 1954; some other Bell models still in production; also produced various versions of Sikorsky S-61 under licence and is partner with Westland in EH 101 (see under EHI in International section); participates in Eurofar tilt-rotor and NH 90 programmes (see International). Own designs include A 109 multi-role helicopter, A 129 anti-tank and projected A 139 battlefield utility transport.

AGUSTA A 109C

TYPE: Twin-turbine light transport helicopter.
PROGRAMME: First flight 4 August 1970; deliveries of A 109A started early 1976; single-pilot IFR certification 20 January 1977; deliveries of uprated A 109 Mk II began September 1981; A 109C certificated 1989.
CURRENT VERSIONS: **A 109C**: Certificated in USA by Agusta Aerospace Corporation in early 1989; transmission uprated from 552 kW (740 shp) to 589 kW (790 shp); 'wide body' cabin; new composites main rotor blades; Wortmann aerofoil on tail rotor; strengthened landing gear; max T-O weight raised to 2,720 kg (5,996 lb), affording 109 kg (240 lb) increase in payload. Other civil/public service roles include law enforcement (max useful load 1,130 kg; 2,491 lb) and coastal patrol with 360° radar (max useful load 1,105 kg; 2,436 lb).

A 109CM: Military version of civil A 109C powered by two 335.6 kW (450 hp) Allison 250-C20R; can have sliding doors and fixed landing gear; ventral fin removed; first customer Belgian Army with 18 scout versions and 28 anti-tank versions designated **109BA**; ordered 1988; first delivery, from offset supplier Sabca, February 1992; first 109BA with equipped weight reduced to originally specified 1,944 kg (4,286 lb) delivered 25 November 1992; max T-O weight increased to 2,850 kg (6,283 lb); this was 11th of 46 A 109BAs; composites sliding doors, relocated batteries and custom Collins/Alcatel Bell avionics; scouts have roof-mounted Saab Helios stabilised observation sight; anti-tank system has roof-mounted Saab/ESCO HeliTOW 2 sight and Hughes TOW-2A missiles on lateral pylons; firing trials completed in Sardinia late 1992. Other roles for 109CM include electronic warfare, command and control, medevac, shipborne ASV/ASW and UAV launching.

A 109EOA: Italian Army scout version powered by Allison 250-C20R, 24 delivered in 1988 as Elicottero d'Osservazione Avanzata to Aviazione Leggera dell'Esercito; fitted with sliding doors, roof-mounted SFIM M334-25 daytime sight with CILAS laser ranger, variety of armament options, fixed landing gear, crashworthy fuel tanks and ECM. Max flight weight with slung load 2,850 kg (6,283 lb); performance generally as coastal patrol A 109C.

Max: Medevac configuration certificated in USA by Agusta Aerospace Corporation early 1989; large upward-opening bulged doors and fairings give 3.96 m^3 (140 cu ft) cabin volume and allow for two stretchers across main cabin and two sitting attendants/patients; engineered by Custom Aircraft Completions of Teterboro, New Jersey.
CUSTOMERS: Total more than 500 all versions sold; order for 65 A 109s from Asian Helicopter Corporation in 1989 believed to be largest single civil helicopter order until then; others ordered by Turkish Ministry of Interior; total 35 A 109s delivered in 1992.
DESIGN FEATURES: Fully articulated four-blade metal main rotor hub with tension/torsion blade attachment; delta-hinged two-blade tail rotor; manual blade folding and rotor brake optional. Main blade section NACA 23011 with

188 ITALY: AIRCRAFT—AGUSTA

Monaco-registered Agusta A 109C (Kenneth Munson)

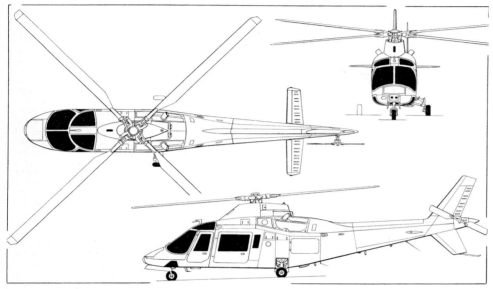

Agusta A 109C civil helicopter (two Allison 250-C20R turboshafts) (Dennis Punnett)

Agusta A 109CM/BA for Belgian Army Aéromobilité programme

Agusta A 109EOA of the Italian Army with roof-mounted sight, grenade launcher pods and antennae for HF, Doppler, signals homing, radar warning and jamming and communications

drooped leading-edge; thickness/chord ratios 11.3 per cent at root, 6 per cent at tip.
FLYING CONTROLS: Fully powered hydraulic; IFR system with autopilot available.
STRUCTURE: Prior to introduction of composites, main and tail rotor blades were bonded aluminium alloy with Nomex core; airframe conventional all-metal; Mk II introduced self-damping engine mounts, redesigned tailboom, removable floor in baggage compartment and systems improvements.
LANDING GEAR: Retractable tricycle type, with oleo-pneumatic shock absorber in each unit. Single mainwheels and self-centring nosewheel castoring ±45°. Hydraulic retraction, nosewheel forward, mainwheels upward into fuselage. Hydraulic emergency extension and locking. Magnaghi disc brakes on mainwheels. All tyres are Kléber-Colombes tubeless of same size (650 × 6) and pressure (5.9 bars; 85 lb/sq in). Tailskid under ventral fin. Emergency pop-out flotation gear and fixed snow skis optional.
POWER PLANT: Two Allison 250-C20R turboshafts, each rated at 335 kW (450 shp) for 5 min for T-O and 283 kW (380 shp) max continuous; flat rated at 258 kW (346 shp) for twin-engine operation; engines mounted side by side in upper rear fuselage and separated from passenger cabin and from each other by firewalls. Transmission ratings 589 kW (790 shp) for take-off and 567 kW (760 shp) for max continuous twin-engined operation, with max contingency rating of 607 kW (814 shp) for 6 s. Rating for single-engined operation is 336 kW (450 shp) for take-off (5 min limit). Two bladder fuel tanks in lower rear fuselage, combined capacity 560 litres (148 US gallons; 123 Imp gallons), of which 550 litres (145.3 US gallons; 121 Imp gallons) are usable. Refuelling point in each side of fuselage, near top of each tank. Oil capacity 7.7 litres (2.0 US gallons; 1.7 Imp gallons) for each engine and 12 litres (3.2 US gallons; 2.6 Imp gallons) for transmission. Provision for internal auxiliary tank containing up to 170 litres (44.9 US gallons; 37.4 Imp gallons) of fuel.

DIMENSIONS, EXTERNAL:
Main rotor diameter	11.00 m (36 ft 1 in)
Tail rotor diameter	2.03 m (6 ft 8 in)
Length overall, rotors turning	13.05 m (42 ft 9¾ in)
Fuselage: Length	10.706 m (35 ft 1½ in)
Height over tail fin	3.30 m (10 ft 10 in)
Tailplane span	2.88 m (9 ft 5½ in)
Width over mainwheels	2.45 m (8 ft 0½ in)
Wheelbase	3.535 m (11 ft 7¼ in)
Passenger doors (each): Height	1.06 m (3 ft 5¾ in)
Width	1.15 m (3 ft 9¼ in)
Height to sill	0.65 m (2 ft 1½ in)
Baggage door (port, rear): Height	0.51 m (1 ft 8 in)
Width	1.00 m (3 ft 3¼ in)

DIMENSIONS, INTERNAL:
Cabin, excl flight deck: Length	1.63 m (5 ft 4¼ in)
Max width	1.32 m (4 ft 4 in)
Max height	1.28 m (4 ft 2½ in)
Volume	2.82 m³ (100 cu ft)
Baggage compartment volume	0.52 m³ (18.4 cu ft)

AREAS:
Main rotor blades (each)	1.84 m² (19.8 sq ft)
Tail rotor blades (each)	0.203 m² (2.185 sq ft)
Main rotor disc	95.03 m² (1022.9 sq ft)
Tail rotor disc	3.24 m² (34.87 sq ft)

WEIGHTS AND LOADINGS:
Basic weight empty, equipped	1,590 kg (3,503 lb)
Max external slung load	907 kg (2,000 lb)
Max baggage	150 kg (331 lb)
Max certificated T-O weight	2,720 kg (5,997 lb)
Max disc loading	28.6 kg/m² (5.86 lb/sq ft)
Max power loading	4.61 kg/kW (7.59 lb/shp)

PERFORMANCE: (A: civil EMS and law enforcement versions, B: coastal patrol version):
Never-exceed speed (VNE):	
A, B	168 knots (311 km/h; 193 mph)
Max cruising speed:	
A	154 knots (285 km/h; 177 mph)
B	152 knots (281 km/h; 175 mph)
Max rate of climb at S/L: A	522 m (1,710 ft)/min
B	516 m (1,700 ft)/min
Rate of climb at S/L, OEI:	
A	108 m (354 ft)/min
B	96 m (315 ft)/min
Service ceiling: A, B	4,572 m (15,000 ft)
Service ceiling, OEI: A, B	2,134 m (7,000 ft)
Hovering ceiling IGE: A	3,350 m (10,990 ft)
B	3,200 m (10,500 ft)
Hovering ceiling OGE: A	2,438 m (9,283 ft)
B	2,286 m (7,500 ft)
Range with max standard fuel, no reserves, best speed and height: A	420 nm (778 km; 483 miles)
B	340 nm (630 km; 391 miles)
Endurance with max fuel, no reserves, best speed and height: A	4 h 20 min
B	3 h 50 min

AGUSTA A 109K

TYPE: Military and special civil utility helicopter.
PROGRAMME: First flight April 1983; first flight of production representative second aircraft March 1984.
CURRENT VERSIONS: **A 109KM**: Military version; roles include anti-tank/scout, escort, command and control,

utility, ECM and SAR/medevac; fixed landing gear; sliding side doors.

A 109KN: Shipboard version with equivalent roles to A 109KM, including anti-ship, over-the-horizon surveillance and targeting and vertical replenishment.

A 109K2: Special civil rescue version first sold to Swiss REGA non-profit rescue service; REGA equipment includes steerable searchlight, winch, Canadian Marconi CMA-900 FMS with GPS, Elbit moving map display, single-pilot IFR system; K2 delivered to Eastern Idaho Regional Medical Center equipped like Swiss REGA helicopters.

CUSTOMERS: Swiss REGA mountain rescue service ordered 15 A 109K2s; first aircraft delivered December 1991; first export to USA early 1993 to Eastern Idaho Regional Medical Center.

COSTS: Price of 15 REGA A 109K2s with spares, logistics and training approx $70 million (1991).

DESIGN FEATURES: Composites main rotor blades and hub with elastomeric bearings; special coating on blades; slightly smaller tail rotor with Wortmann aerofoil and stainless steel skins; optional rotor brake.

Data below for military and civil versions and dimensions different from A 109C.

LANDING GEAR: Non-retractable tricycle type, giving increased clearance between fuselage and ground. Changes restricted to replacement of nose leg actuator by fixed strut, and replacement of each main leg actuator by fixed strut and V support frame.

POWER PLANT: Two Turbomeca Arriel 1K1 turboshafts, each rated at 575 kW (771 shp) for 2.5 min, 550 kW (737 shp) for take-off (30 min) and 471 kW (632 shp) max continuous power. Engine particle separator optional. Main transmission uprated to 671 kW (900 shp) for take-off and max continuous twin-engined operation; single-engine emergency rating is 477 kW (640 shp). Standard usable fuel capacity 550 litres (145 US gallons; 121 Imp gallons), with optional 150 litre (39.6 US gallon; 33 Imp gallon) auxiliary tanks. Optional closed circuit refuelling system and optional 200 litre (53 US gallon; 44 Imp gallon) ferry tanks in cabin. Self-sealing fuel tanks optional. Independent fuel and oil system for each engine.

AVIONICS: Radar and laser warning systems, FLIR, gyrostabilised sight, night vision goggles and chaff/flare dispenser are among options.

EQUIPMENT: Options include rescue hoist, searchlight and cargo platform.

ARMAMENT (optional): Total of four stores attachments, two on each side of cabin, on outriggers. Typical loads include two 7.62 mm or 12.7 mm gun pods, 70 mm or 80 mm rocket launchers, or up to eight TOW anti-armour missiles (with roof mounted sight), Stinger air-to-air missiles, UAVs, plus 7.62 or 12.7 mm side-firing gun in cabin.

DIMENSIONS, EXTERNAL:
Tail rotor diameter	2.00 m (6 ft 6¾ in)
Length of fuselage	11.106 m (36 ft 5¼ in)

AREAS:
Tail rotor disc	3.143 m² (33.83 sq ft)

WEIGHTS AND LOADINGS:
Weight empty: KM/KN, K2	1,622 kg (3,576 lb)
Max T-O weight: KM/KN	2,850 kg (6,283 lb)
K2	2,720 kg (5,997 lb)
Max flight weight with slung load:	
KM/KN, K2	3,000 kg (6,614 lb)
Max disc loading: KM/KN	30.0 kg/m² (6.14 lb/sq ft)
K2	28.6 kg/m² (5.86 lb/sq ft)
Max power loading: KM/KN	4.24 kg/kW (6.98 lb/shp)
K2	4.05 kg/kW (6.66 lb/shp)

PERFORMANCE (at max T-O weight except where indicated):
Never-exceed speed (V_{NE})	150 knots (278 km/h; 172 mph)
Max cruising speed at S/L, clean:	
KM/KN	142 knots (263 km/h; 163 mph)
K2	144 knots (266 km/h; 166 mph)
Max rate of climb at S/L: KM/KN	618 m (2,020 ft)/min
K2	678 m (2,220 ft)/min
Rate of climb at S/L, OEI: KM/KN	168 m (560 ft)/min
K2	198 m (660 ft)/min
Service ceiling: KM/KN, K2	6,100 m (20,000 ft)
Service ceiling, OEI: KM/KN	3,050 m (10,000 ft)
K2	3,230 m (10,600 ft)
Hovering ceiling IGE: KM/KN	5,030 m (16,500 ft)
K2	5,425 m (17,800 ft)
Hovering ceiling OGE: KM/KN	4,023 m (13,200 ft)
K2	4,450 m (14,600 ft)
Max range, best height and speed:	
KM/KN, K2	394 nm (730 km; 453 miles)

AGUSTA A 129 MANGUSTA (MONGOOSE)

TYPE: Light anti-tank and scout helicopter.

PROGRAMME: Italian Army specification issued 1972; A 129 given go-ahead March 1978; final form settled 1980; detail design completed 30 November 1982; first flights of five development aircraft 11 September 1983, 1 July and 5 October 1984, 27 May 1985 and 1 March 1986; first delivery October 1990; night flight proving under way 1992.

CURRENT VERSIONS: **Anti-tank:** Italian Army version; also demonstrated for export (*described below*).

Shipborne: Proposed maritime anti-ship version; no orders received by mid-1993.

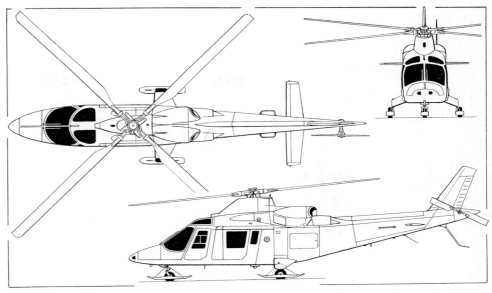

Agusta A 109K2 civil rescue and utility helicopter (two Turbomeca Arriel 1K1 turboshafts) *(Dennis Punnett)*

T800 power: Export version; first flight of prototype powered by two Allison/Garrett LHTEC T800-LHT-800 turboshafts October 1988; demonstrated in Gulf during 1990; T800 gives between 20 and 40 per cent more power than Gems; in potential growth versions, transmission uprated to 910 kW (1,220 hp) and gross weight increased to 4,400 kg (9,700 lb).

CUSTOMERS: First five of 60 for Italian Army anti-tank squadrons delivered October 1990 after delay of more than a year to allow fitting of Saab/ESCO HeliTOW system with nose-mounted sight; 10 more to higher equipment standard delivered by end 1992; second batch of 30 expected, some convertible scouts with mast-mounted sights, but these may be cancelled and 20 of the first 60 may have to be convertible; first A 129 with Honeywell/OMI helicopter IR navigation system (HIRNS) and helmet display delivered early 1993; A 129 proposed for UK Army anti-tank helicopter competition to be decided late 1994. Initial Dutch Army order for 20 shelved.

DESIGN FEATURES: Fully articulated four-blade main rotor with blades retained by single elastomeric bearing and restrained by hydraulic drag damper and mechanical droop stop; each blade has CFRP and Nomex main spar, Nomex honeycomb leading- and trailing-edges, composites skins, stainless steel leading-edge abrasion strip and frangible tip; control linkage runs inside driveshaft to reduce radar signature, avoid icing and improve ballistic tolerance; blades tolerant to 12.7 mm hits, possibly also 23 mm; delta-hinge two-blade tail rotor with broad-chord blades for ballistic tolerance.

Main transmission has independent oil cooling system; intermediate and tail rotor gearboxes grease lubricated; all designed for at least 30 min run dry; accessory gearbox can be run independently on ground by No. 1 engine engaged by pilot-operated clutch.

FLYING CONTROLS: Full-time dual electronic flight controls, with full manual reversion, provide automatic heading hold, autohover, autopilot modes and autostabiliser modes, all selectable by pilot; gunner in front seat has cyclic sidearm controller, normal collective lever and pedals and has full access to AFCS; electrical inputs from AFCS integrated with hydraulic powered control units; fly-by-wire standby system under development.

STRUCTURE: Composite materials account for 45 per cent of fuselage weight (less engines) and 16.1 per cent total empty weight; material used for fuselage panels, nosecone, tailboom, tail rotor pylon, engine nacelles, canopy frame and maintenance panels; 70 per cent of airframe surface is composites; bulkhead in nose and A frame running up through fuselage to rotor pylon protect crew against rollover; overall infra-red suppressing paint; airframe meets MIL-STD-1290 crashworthiness covering vertical velocity changes of 11.2 m (36.75 ft)/s and longitudinal changes of 13.1 m (43 ft)/s.

LANDING GEAR: Non-retractable tailwheel type, with single wheel on each unit. Two-stage hydraulic shock strut in each main unit designed to withstand normal loads and hard landings at descent rates in excess of 10 m (32.8 ft)/s.

POWER PLANT: Two Rolls-Royce Gem 2 Mk 1004D turboshafts, each with a max continuous rating of 615 kW (825 shp) for normal twin-engined operation; intermediate contingency rating of 657 kW (881 shp) for 1 h; max contingency rating of 704 kW (944 shp) for 2½ min; and emergency rating (S/L, ISA) of 759 kW (1,018 shp) for 20 s. Transmission rating is 969 kW (1,300 shp) (two engines), 704 kW (944 shp) for single-engined operation, with emergency rating of 759 kW (1,018 shp); power input into transmission is at 27,000 rpm from the Gem and 23,000 rpm in opposite direction from T800. Production Gems licence built in Italy by Piaggio. Fireproof engine compartment, with engines widely spaced to improve survivability from enemy fire. Two separate fuel systems, with cross-feed capability; interchangeable self-sealing and crash resistant tanks, self-sealing lines, and digital fuel feed control. Tanks can be foam-filled for fire protection. Single-point pressure refuelling. Infra-red exhaust suppression system and low engine noise levels. Separate independent lubrication oil cooling system for each engine. Provision

Prototype Agusta A 129 attack helicopter powered by LHTEC T800 turboshafts, flying over desert during a sales demonstration

for auxiliary (self-ferry) fuel tanks on inboard underwing stations.

ACCOMMODATION: Pilot and co-pilot/gunner in separate cockpits in tandem. Elevated rear (pilot's) cockpit. External crew field of view exceeds MIL-STD-850B. Each cockpit has a flat plate low-glint canopy with upward hinged door panels on starboard side, blow-out port side panel for exit in emergency, and Martin-Baker crashworthy seat with sliding side panels of composites armour. Landing gear design and crashworthy seats reduce impact from 50g to 20g in crash.

SYSTEMS: Hydraulic system includes three main circuits dedicated to flight controls and two independent circuits for rotor and wheel braking. Main system operates at pressure of 207 bars (3,000 lb/sq in) and is fed by three independent power groups, two integrated and driven mechanically by the main transmission, the third integrated and driven by the tail rotor gearbox. Tandem actuators are provided for main and tail rotor flight controls. Hydraulic system flow rate 23.6 litres (6.2 US gallons; 5.2 Imp gallons)/min in each main group. Spring type reservoirs, pressurised at 0.39 bar (5.6 lb/sq in).

AVIONICS: All main functions are handled and monitored by fully integrated digital multiplex system (IMS), which controls navigation, flight management, weapon control, autopilot, monitoring of transmission and engine condition, fuel/hydraulic/electrical systems, caution and warning systems. IMS is managed by two Harris central computers, each capable of operating independently, backed by two interface units which pick up outputs from sensors and avionic equipment and transfer them, via redundant MIL-STD-1553B databuses, to main computers for real-time processing. Processed information is presented to pilot and co-pilot/gunner on separate graphic/alphanumeric head-down multi-function displays (MFDs) with standard multi-function keyboards for easy access to information, including area navigation using up to 100 waypoints, weapons status and selection, radio tuning and mode selection, caution and warning, and display of aircraft performance; conventional instruments and dials are provided as backup. IMS computer can store up to 100 pre-set frequencies for HF, VHF and UHF radio management. Navigation is controlled by navigation computer of IMS coupled to Doppler radar and radar altimeter with low airspeed indicator, normally used for rocket aiming, providing backup velocity data when the Doppler is beyond limits. Synthetic map presentation of waypoints, target areas and dangerous areas is shown on pilot's or co-pilot's MFD. A Litton strap-down inertial reference for both flight control and navigation is being substituted for the present gyro system.

AFCS provides either three-axis stabilisation or full attitude and heading hold, automatic hover, downward transition to hover or holds for altitude, heading and airspeed or groundspeed and automatic track following.

A 129 has full day/night operational capability, with equipment designed to give both crew members a view outside helicopter irrespective of light conditions; pilot's night vision system (HIRNS: helicopter infra-red night system) allows nap-of-the-earth (NOE) flight by night, a picture of world outside being generated by Honeywell mini-FLIR sensor mounted on a Ferranti steerable platform at nose of aircraft and projected to both crewmen through the monocle of the Honeywell integrated helmet and display sighting system (IHADSS), to which it is slaved by helmet position sensors. Symbology containing information required for flight is superimposed onto image, giving a true head-up reference. HeliTOW sight gives the co-pilot/gunner direct view optics and FLIR, plus laser for ranging. A 129 also has provision for mast mounted sight (MMS). Active and passive self-protection systems (ECCM and ECM) standard on Italian Army A 129; Have Quick frequency-hopping radio will follow; onboard nav/weapon system can connect directly or by data link with Italian CATRIN C³I combat information system. Passive electronic warfare systems include radar jammer and radar and laser warning receivers, chaff/flare dispensers and IR jammer; cockpits coloured to allow use of night vision goggles as supplementary night flying aid.

ARMAMENT: Four underwing attachments stressed for loads of up to 300 kg (661 lb) each; all stations incorporate articulation which allows pylon to be elevated 2° and depressed 10° to increase missile launch envelope; they are aligned with aircraft automatically, with no need for boresighting. Initial armament of up to eight thermal tracking TOW 2 or 2A wire guided anti-tank missiles (two, three or four in carriers suspended from each wingtip station), with Saab/ESCO HeliTOW aiming system; with these can be carried, on inboard stations, either two 7.62, 12.7 or 20 mm gun pods, or two launchers each for seven air-to-surface rockets. For general attack missions, rocket launchers can be carried on all four stations (two nineteen-tube plus two seven-tube); Italian Army has specified SNIA-BPD 81 mm and 70 mm rockets. Alternatively, A 129 can carry six Hellfire anti-tank missiles (three beneath each wingtip); eight HOT missiles; AIM-9L Sidewinder, Matra Mistral, Javelin or Stinger air-to-air missiles for aerial combat; two gun pods plus two nineteen-tube rocket launchers; or grenade launchers.

Lucas 0.50 in self-contained gun turret qualified, but not used by the Italian Army. A 12.7 mm turret has also been fired and 20 mm or 12.7 mm Gatling turrets have been investigated. Optional upgrades offered for export include an autotracking sight, a laser designator for Hellfire and an MMS for scouting.

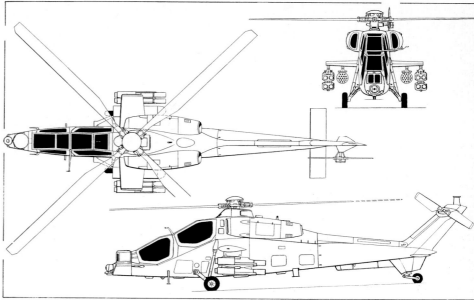

Agusta A 129 light anti-tank, attack and advanced scout helicopter (Dennis Punnett)

DIMENSIONS, EXTERNAL:
Main rotor diameter	11.90 m (39 ft 0½ in)
Tail rotor diameter	2.24 m (7 ft 4¼ in)
Wing span	3.20 m (10 ft 6 in)
Width over TOW pods	3.60 m (11 ft 9¾ in)
Length overall, both rotors turning	14.29 m (46 ft 10½ in)
Fuselage: Length	12.275 m (40 ft 3¼ in)
Max width	0.95 m (3 ft 1½ in)
Height:	
over tail fin, tail rotor horizontal	2.65 m (8 ft 8¼ in)
tail rotor turning	3.315 m (10 ft 10½ in)
to top of rotor head	3.35 m (11 ft 0 in)
Tailplane span	3.00 m (9 ft 10 in)
Wheel track	2.20 m (7 ft 3½ in)
Wheelbase	6.955 m (22 ft 9¾ in)

AREAS:
Main rotor disc	111.2 m² (1,196.95 sq ft)
Tail rotor disc	3.94 m² (42.42 sq ft)

WEIGHTS AND LOADINGS:
Weight empty, equipped	2,529 kg (5,575 lb)
Max internal fuel load	750 kg (1,653 lb)
Max external weapons load	1,200 kg (2,645 lb)
Max T-O weight	4,100 kg (9,039 lb)
Max disc loading	33.3 kg/m² (6.8 lb/sq ft)
Max power loading	4.23 kg/kW (6.95 lb/shp)

PERFORMANCE:
At mission T-O weight of 3,700 kg (8,157 lb), at 2,000 m (6,560 ft), ISA+20°C, except where indicated, A 129 is designed to meet following performance requirements:
Dash speed	170 knots (315 km/h; 196 mph)
Max level speed at S/L	140 knots (259 km/h; 161 mph)
Max rate of climb at S/L	655 m (2,150 ft)/min
Hovering ceiling: IGE	3,750 m (12,300 ft)
OGE	3,015 m (9,900 ft)

Basic 2 h 30 min mission profile with 8 TOW and 20 min fuel reserves:
Fly 54 nm (100 km; 62 miles) to battle area, mainly in NOE mode, 90 min loiter (incl 45 min hovering), and return to base
Max endurance, no reserves	3 h
g limits	+3.5/–0.5

AGUSTA A 139 UTILITY

Agusta aims to create a family of new helicopters, with a civil/military utility version combining the dynamics, systems and integrated avionics of A 129 with completely new cabin-type fuselage; in original form, designated Light Battlefield Helicopter, A 139 could carry weapons and have T800 engines. The cabin could accommodate eight to ten passengers or six stretchers and two attendants; slung loads of 2,000 kg (4,409 lb) could be carried. No progress reported by May 1993.

AGUSTA-BELL 206B-3 JETRANGER III

JetRanger has been manufactured under licence from Bell since end of 1967; deliveries began 1972 of Agusta-Bell 206B JetRanger II, and of JetRanger III at end of 1978. Approx 1,000 JetRangers built by Agusta; production continues. See Canadian section for JetRanger III description.

AGUSTA-BELL 212

Agusta-Bell 212 is twin-engined utility transport helicopter generally similar to Bell Model 212 Twin Two-Twelve described in Canadian section; small numbers still being produced.

AGUSTA-BELL 212ASW

Extensively modified version of AB 212, intended primarily for anti-submarine search, classification and attack missions, and for attacks on surface vessels, while operating

Agusta-Bell 412 Griffon of the Italian Army (Paul Jackson)

AGUSTA—AIRCRAFT: ITALY 191

om decks of small ships, but suitable also for search and rescue, electronic warfare, fire support, troop transport, liaison and utility roles; more than 100 in service with Italian and other navies; customers have included navies of Greece (12, including some in electronic warfare configuration), Iraq (five), Turkey (12 in both ASW and ASV configurations) and Venezuela (six).

Occasional machines being produced. Details appeared in the 1990-91 *Jane's*.

AGUSTA-BELL 412 HP and GRIFFON

TYPE: Multi-purpose medium helicopter.
PROGRAMME: First flight of Bell 412 SP August 1979; deliveries started January 1981; Agusta licence production of civil version started 1981; first flight of military Griffon August 1982; deliveries began January 1983; Bell 412 HP (see Canadian section) certificated 29 June 1990.
CURRENT VERSIONS: **Griffon**: Military derivatives developed for direct fire support, scouting, assault transport, equipment transport, SAR and maritime surveillance.
CUSTOMERS: Those in Italy include Army (18), Carabinieri (20) and Special Civil Protection fleet (four, with two to come), Coast Guard (ultimately 24), national fire service (four) and national forest service (three); others include Zimbabwe Air Force (10), Ugandan Army (two), and Finnish coastguard (two); Royal Netherlands Air Force ordered three for search and rescue and ambulance in October 1992, cost about LIt30 billion; total 25 AB 212/412 delivered during 1992.
DESIGN FEATURES: Griffon has reinforced impact-absorbing landing gear, selective armour protection and differences noted below.
POWER PLANT: One 1,342 kW (1,800 shp) Pratt & Whitney Canada PT6T-3BE Twin Pac (single-engine ratings 764 kW; 1,025 shp for 2½ min and 723 kW; 970 shp for 30 min). IR emission reduction devices optional. Fuel capacity 1,249 litres (330 US gallons; 275 Imp gallons). Single-point refuelling. Two 76 or 341 litre (20 or 90 US gallon; 16.7 or 75 Imp gallon) auxiliary fuel tanks optional; single-point refuelling.
ACCOMMODATION: One or two pilots on flight deck, on energy-absorbing, armour protected seats. Fourteen crash-attenuating troop seats in main cabin in personnel transport roles, six patients and two medical attendants in ambulance version, or up to 1,814 kg (4,000 lb) of cargo or other equipment. Space for 181 kg (400 lb) of baggage in tailboom. Total of 51 fittings in cabin floor for attachment of seats, stretchers, internal hoist or other special equipment.
SYSTEMS: Generally as for Bell 212/412.
ARMAMENT: Wide variety of external weapon options for Griffon includes swivelling turret for 12.7 mm gun, two 25 mm Oerlikon cannon, four or eight TOW anti-tank missiles, two launchers each with nineteen 2.75 in SNORA or twelve 81 mm rockets, 12.7 mm machine guns (in pods or door mounted), four air-to-air or air defence suppression missiles, or, for attacking surface vessels, four Sea Skua or similar air-to-surface missiles.
WEIGHTS AND LOADINGS:
Weight empty, equipped (standard configuration)
   2,950 kg (6,505 lb)
Max T-O weight   5,400 kg (11,905 lb)
PERFORMANCE (at max T-O weight, ISA):
Never-exceed speed (VNE) at S/L
   140 knots (259 km/h; 161 mph)
Cruising speed: at S/L 122 knots (226 km/h; 140 mph)
 at 1,500 m (4,920 ft) 125 knots (232 km/h; 144 mph)
 at 3,000 m (9,840 ft) 123 knots (228 km/h; 142 mph)
Max rate of climb at S/L 438 m (1,437 ft)/min
Rate of climb at S/L, OEI 168 m (551 ft)/min
Service ceiling, 30.5 m (100 ft)/min climb rate
   5,180 m (17,000 ft)
Service ceiling, OEI, 30.5 m (100 ft)/min climb rate
   2,320 m (7,610 ft)

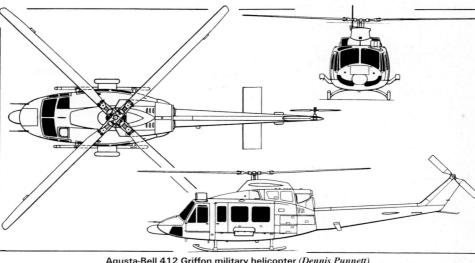

Agusta-Bell 412 Griffon military helicopter (*Dennis Punnett*)

Hovering ceiling: IGE 3,110 m (10,200 ft)
 OGE 1,585 m (5,200 ft)
Range with max standard fuel at appropriate cruising speed (see above), no reserves:
 at S/L 354 nm (656 km; 407 miles)
 at 1,500 m (4,920 ft) 402 nm (745 km; 463 miles)
 at 3,000 m (9,840 ft) 434 nm (804 km; 500 miles)
Max endurance: at S/L 3 h 36 min
 at 1,500 m (4,920 ft) 4 h 12 min

AGUSTA-SIKORSKY AS-61 and ASH-3H

Agusta licence manufacture of Sikorsky S-61S and S-61R in various civil and military forms started 1967; deliveries of anti-submarine ASH-3Ds to Italian Navy began 1969. Agusta is exclusive repair and overhaul agent for Europe and the Mediterranean basin. Production has ceased, but could be restarted in 36 months. Models most recently in production were utility **AS-61**, **SH-3D/TS** (Trasporto Speciale) VIP transport, and upgraded **SH-3H** multi-role naval version.

AGUSTA-SIKORSKY AS-61R (HH-3F) PELICAN

Agusta began production of this multi-purpose search and rescue helicopter in 1974; deliveries began 1976; production line reopened to make two for national civil protection service (SNPC) and 13 rescue helicopters for Italian Air Force; these have new radar, Loran, FLIR and navigation computer, all of which will be retrofitted in remaining 19 of original production batch; nine redelivered during 1992. See further details in 1991-92 *Jane's*.

Late production Agusta-Sikorsky HH-3F (AS-61R) of the Italian Air Force, with chaff/flare dispensers, in Mogadishu in 1993 (*Paul Jackson*)

MONTEPRANDONE WORKS

Formerly BredaNardi, continues to produce NH-500E under licence for Italian government central helicopter pilot training school; licensed to manufacture, and sell in central Europe, McDonnell Douglas 530F and 520N. Group also sells NH-300C imported from Schweizer. Descriptions of 500/530 series can be found under McDonnell Douglas Helicopters heading and of 300C under Schweizer heading in US section of this edition.

SESTO CALENDE WORKS

MAIN WORKS: Vergiate (Varese)
AIRFIELD AND WORKS: Sesto Calende (Varese) and Malpensa.

Founded in 1915, SIAI-Marchetti produced wide range of military and civil landplanes and flying-boats up to end of Second World War. Now known simply as Sesto Calende works of Agusta, current products include piston, turboprop and turbofan powered trainers; since 1970s has been engaged in co-production of licence built Boeing CH-47C, Bell 204/205/212/412, and Sikorsky S-61A, SH-3D/H and HH-3F helicopters.

On 6 October 1988 a memorandum of intent signed with Grumman Aircraft Systems (a division of Grumman Corporation), to offer version of S.211 for USAF/USN's Joint Primary Aircraft Training System (JPATS) requirement (see US section).

Sesto Calende works undertakes overhaul and repair of various types of aircraft (notably C-130 Hercules, DHC-5 Buffalo and Cessna Citation II); participates in national or multi-national programmes, producing parts for Alenia G222, Panavia Tornado, AMX, Airbus A310 and Atlantique 2.

Works at Sesto Calende, Vergiate and Malpensa total 1,370,267 m² (14,749,416 sq ft) in area, of which 119,494 m² (1,286,221 sq ft) are covered.

Transfer of SIAI-Marchetti from Agusta to Aermacchi to form Italian trainer aircraft entity agreed, but not finalised by early 1993.

SIAI-MARCHETTI SF.260

TYPE: Military basic trainer.
PROGRAMME: Originated as F.250 designed by Stelio Frati and made by Aviamilano (see 1965-66 *Jane's*); civil SF.260A and B detailed in 1980-81 *Jane's*; military training SF.260C detailed in 1985-86 *Jane's*; claimed still in production as military trainer.
CURRENT VERSIONS: **SF.260D**: Improved and updated civil version, replacing SF.260C, with aerodynamic and structural improvements developed for military SF.260M. Certificated by RAI on 14 December 1985 and by FAA in October 1986.

SF.260M: Two/three-seat military trainer, developed from civil SF.260A and first flown on 10 October 1970.

Introduced a number of important structural and aerodynamic improvements, many of which were subsequently applied to later models. Meets requirements for basic flying training; instrument flying; aerobatics, including deliberate spinning; night flying; navigation flying; and formation flying. Detailed customer list in 1984-85 and earlier editions of *Jane's. Main description applies to this version except where indicated.*

SF.260W Warrior: Trainer/tactical support version of SF.260M; first flight (I-SJAV) in May 1972. Two or four underwing pylons, for up to 300 kg (661 lb) of external stores, and cockpit stores selection panel. Able to undertake a wide variety of roles, including low-level strike; forward air control; forward air support; armed reconnaissance; and liaison. Also meets same requirements as SF.260M for use as a trainer. Customers as listed in 1984-85 and earlier *Jane's*. One aircraft (described in 1980-81 and earlier editions) completed as **SF.260SW Sea Warrior** surveillance/SAR/supply version.

SF.260TP: Turboprop powered development. Described separately.

CUSTOMERS: Total more than 860 produced; Turkish Air Force ordered 40 SF.260Ds for military training with an agreement for co-production by the TAI aerospace manufacturing group; 10 delivered complete; first Turkish-built aircraft completed April 1992.

DESIGN FEATURES: Wing section NACA 64_1-212 (modified) at root, $64_1$1-210 (modified) at tip; dihedral 6.5° 20′ from roots on military versions, 5° on SF.260D.

FLYING CONTROLS: Differential Frise ailerons, each with servo tab; electrically operated slotted flaps; controls operated by rods and cables.

STRUCTURE: All-metal stressed skin structure; wing skin formed of butt-jointed panels flush riveted; single main spar and auxiliary rear spar; press-formed ribs; wings bolted together on centreline and attached to fuselage by six bolts.

LANDING GEAR: Electrically retractable tricycle type, with manual emergency actuation. Inward retracting main gear, of trailing arm type, and rearward retracting nose unit, each embodying Magnaghi oleo-pneumatic shock absorber (type 2/22028 in main units). Each welded steel tube main leg is hinged to the main and rear spars. Nose unit is of leg and fork type, with coaxial shock absorber and torque strut. Cleveland P/N 3080A mainwheels, with size 6.00-6 tube and tyre (6-ply rating), pressure 2.45 bars (35.5 lb/sq in). Cleveland P/N 40-77A nosewheel, with size 5.00-5 tube and tyre (6-ply rating), pressure 1.96 bars (28.4 lb/sq in). Cleveland P/N 3000-500 independent hydraulic single-disc brake on each mainwheel; parking brake. Nosewheel steering (±20°) operated directly by rudder pedals.

POWER PLANT: One 194 kW (260 hp) Textron Lycoming O-540-E4A5 flat-six engine, driving a Hartzell HC-C2YK-1BF/8477-8R two-blade constant-speed metal propeller. AEIO-540-D4A5 engine available optionally. Fuel in two light alloy tanks in wings, capacity of each 49.5 litres (13.1 US gallons; 10.9 Imp gallons); and two permanent wingtip tanks, capacity of each 72 litres (19 US gallons; 15.85 Imp gallons). Total internal fuel capacity 243 litres (64.2 US gallons; 53.3 Imp gallons), of which 235 litres (62.1 US gallons; 51.7 Imp gallons) are usable. Individual refuelling point on top of each tank. In addition, SF.260W may be fitted with two 80 litre (21.1 US gallon; 17.5 Imp gallon) auxiliary tanks on underwing pylons. Oil capacity (all models) 11.4 litres (3 US gallons; 2.5 Imp gallons).

ACCOMMODATION (SF.260M; W similar): Side by side front seats (for instructor and pupil in SF.260M), with third seat centrally at rear. Front seats individually adjustable fore and aft, with forward folding backs and provision for back type parachute packs. Dual controls standard. All three seats equipped with lap belts and shoulder harnesses. Baggage compartment aft of rear seat. Upper portion of canopy tinted. Emergency canopy release handle for each front seat occupant. Steel tube windscreen frame for protection in the event of an overturn.

SYSTEMS (SF.260M; other models generally similar): Hydraulic system for mainwheel brakes only. No pneumatic system. 24V DC electrical system of single-conductor negative earth type, including 70A Prestolite engine-mounted alternator/rectifier and 24V 24Ah Varley battery, for engine starting, flap and landing gear actuation, fuel booster pumps, electronics and lighting. Sealed battery compartment in rear of fuselage on port side. Connection of an external power source automatically disconnects the battery. Heating system for carburettor air intake. Emergency electrical system for extending landing gear if normal electrical actuation fails; provision for mechanical extension in the event of total electrical failure. Cabin heating, and windscreen de-icing and demisting, by heat exchanger using engine exhaust air. Additional manually controlled warm air outlets for general cabin heating. Oxygen system optional.

AVIONICS (SF.260M; W generally similar): Basic instrumentation to customer's requirements. Blind-flying instrumentation and communications equipment optional: typical selection includes dual Collins 20B VHF com; Collins VIR-31A VHF nav; Collins ADF-60A; Collins TDR-90 ATC transponder; Collins PN-101 compass; ID-90-000 RMI; and Gemelli AG04-1 intercom. Instrument panel can be slid rearward to provide access to rear of instruments.

EQUIPMENT: Military equipment to customer's requirements. External stores can include one or two reconnaissance pods with two 70 mm automatic cameras, or two supply containers. Landing light in nose, below spinner.

ARMAMENT (SF.260W): Two or four underwing hardpoints, able to carry external stores on NATO standard pylons up to a maximum of 300 kg (661 lb) when flown as a single-seater. Typical alternative loads can include one or two SIAI gun pods, each with one or two 7.62 mm FN machine-guns and 500 rds; two Aerea AL-8-70 launchers each with eight 2.75 in rockets; two LAU-32 launchers each with seven 2.75 in rockets; two Aerea AL-18-50 launchers each with eighteen 2 in rockets; two Aerea AL-8-68 launchers each with eight 68 mm rockets; two Aerea AL-6-80 launchers each with six 81 mm rockets; two LUU-2/B parachute flares; two SAMP EU 32 125 kg general purpose bombs or EU 13 120 kg fragmentation bombs; two SAMP EU 70 50 kg general purpose bombs; Mk 76 11 kg practice bombs; two cartridge throwers for 70 mm multi-purpose cartridges, F 725 flares or F 130 smoke cartridges. One or two photo-reconnaissance pods with two 70 mm automatic cameras; two supply containers.

DIMENSIONS, EXTERNAL:
Wing span over tip tanks 8.35 m (27 ft 4¾ in)
Wing chord: at root 1.60 m (5 ft 3 in)
mean aerodynamic 1.325 m (4 ft 4¼ in)
at tip 0.784 m (2 ft 6⅞ in)
Wing aspect ratio (excl tip tanks) 6.3
Wing taper ratio 2.2
Length overall 7.10 m (23 ft 3½ in)
Fuselage: Max width 1.10 m (3 ft 7¼ in)
Max depth 1.042 m (3 ft 5 in)
Height overall 2.41 m (7 ft 11 in)
Elevator span 3.01 m (9 ft 10½ in)
Wheel track 2.274 m (7 ft 5½ in)
Wheelbase 1.66 m (5 ft 5¼ in)
Propeller diameter 1.93 m (6 ft 4 in)
Propeller ground clearance 0.32 m (1 ft 0½ in)

DIMENSIONS, INTERNAL:
Cabin: Length 1.66 m (5 ft 5¼ in)
Max width 1.00 m (3 ft 3¼ in)
Height (seat cushion to canopy) 0.98 m (3 ft 2½ in)
Volume 1.50 m³ (53 cu ft)
Baggage compartment volume 0.18 m³ (6.36 cu ft)

AREAS:
Wings, gross 10.10 m² (108.70 sq ft)
Ailerons (total, incl tabs) 0.762 m² (8.20 sq ft)
Trailing-edge flaps (total) 1.18 m² (12.70 sq ft)
Fin 0.76 m² (8.18 sq ft)
Dorsal fin 0.16 m² (1.72 sq ft)
Rudder, incl tab 0.60 m² (6.46 sq ft)
Tailplane 1.46 m² (15.70 sq ft)
Elevator, incl tab 0.96 m² (10.30 sq ft)

WEIGHTS AND LOADINGS:
Manufacturer's basic weight empty:
M 755 kg (1,664 lb)
W 770 kg (1,697 lb)
Weight empty, equipped: D 755 kg (1,664 lb)
M 815 kg (1,797 lb)
W 830 kg (1,830 lb)
Fuel:
in-wing and wingtip tanks (all versions) 169 kg (372.5 lb)
underwing tanks (W only) 114 kg (251.5 lb)
Typical mission weights:
M, trainer (clean) 1,140 kg (2,513 lb)
W, two 47 kg (103.5 lb) machine-gun pods and full internal fuel 1,163 kg (2,564 lb)
W, one Alkan 500B cartridge thrower, one two-camera reconnaissance pod and full internal fuel 1,182 kg (2,605 lb)
W, trainer with 94 kg (207 lb) external stores 1,249 kg (2,753 lb)
W, self-ferry with two 80 litre (21.1 US gallon; 17.5 Imp gallon) underwing tanks 1,285 kg (2,833 lb)
W, two 125 kg bombs and 150 kg (331 lb) internal fuel 1,300 kg (2,866 lb)
W, two AL-8-70 rocket launchers and 160 kg (353 lb) internal fuel 1,300 kg (2,866 lb)
Max T-O weight: D, M, Aerobatic 1,100 kg (2,425 lb)
D, Utility 1,100 kg (2,425 lb)
M, Utility 1,200 kg (2,645 lb)
W, max permitted 1,300 kg (2,866 lb)
Max wing loading: D 109 kg/m² (22.4 lb/sq ft)
M 119 kg/m² (24.4 lb/sq ft)
W 129 kg/m² (26.4 lb/sq ft)
Max power loading: D 5.68 kg/kW (9.33 lb/hp)
M 6.19 kg/kW (10.17 lb/hp)
W 6.70 kg/kW (11.01 lb/hp)

PERFORMANCE (D at AUW of 1,102 kg; 2,430 lb, M at AUW of 1,200 kg; 2,645 lb, W at 1,300 kg; 2,866 lb, except where indicated):
Never-exceed speed (VNE):
D, M 235 knots (436 km/h; 271 mph)
Max level speed at S/L:
D 187 knots (347 km/h; 215 mph)
M 180 knots (333 km/h; 207 mph)
W 165 knots (305 km/h; 190 mph)
Max cruising speed (75% power):
D at 3,050 m (10,000 ft) 178 knots (330 km/h; 205 mpl
M at 1,500 m (4,925 ft) 162 knots (300 km/h; 186 mpl
W at 1,500 m (4,925 ft) 152 knots (281 km/h; 175 mpl
Stalling speed, flaps and landing gear up:
M 74 knots (137 km/h; 86 mpl
W 88 knots (163 km/h; 102 mpl
Stalling speed, flaps and landing gear down:
D 60 knots (111 km/h; 70 mpl
M 68 knots (126 km/h; 79 mpl
W 72 knots (134 km/h; 83 mpl
Max rate of climb at S/L: D 546 m (1,791 ft)/mi
M 457 m (1,500 ft)/mi
W 381 m (1,250 ft)/mi
Time to 1,500 m (4,925 ft): M 4 mi
W 6 min 20
Time to 2,300 m (7,550 ft): M 6 min 50
W 10 min 20
Time to 3,000 m (9,850 ft): M 10 mi
W 18 min 40
Service ceiling: D 5,790 m (19,000 f
M 4,665 m (15,300 f
W 4,480 m (14,700 f
T-O run at S/L: D 480 m (1,575 f
M 384 m (1,260 f
T-O to 15 m (50 ft) at S/L: M 606 m (1,988 f
W 825 m (2,707 f
Landing from 15 m (50 ft) at S/L: D 445 m (1,460 f
M 539 m (1,768 f
W 645 m (2,116 f
Landing run at S/L: D, M 345 m (1,132 f
Operational radius:
W, 6 h 25 min single-seat armed patrol mission a 1,163 kg (2,564 lb) AUW, incl 5 h 35 min over oper-ating area, 20 kg (44 lb) fuel reserves 50 nm (92 km; 57 mile
W, 3 h 38 min single-seat strike mission, incl two 5 mi loiters over separate en-route target areas, 20 k (44 lb) fuel reserves 250 nm (463 km; 287 mile
W, 4 h 54 min single-seat strike mission, incl 5 min ov target area, 20 kg (44 lb) fuel reserves 300 nm (556 km; 345 mile
W, 4 h 30 min single-seat photo-reconnaissance missio at 1,182 kg (2,605 lb) AUW, incl three 1 h loiters ov separate en-route operating areas, 20 kg (44 lb) fu reserves 150 nm (278 km; 172 mile
W, 6 h 3 min two-seat self-ferry mission with two 8 litre (21.1 US gallon; 17.5 Imp gallon) underwin tanks, at 1,285 kg (2,833 lb) AUW, 30 kg (66 lb) fu reserves 926 nm (1,716 km; 1,066 mile
Range with max fuel:
D (two-seat) 805 nm (1,490 km; 925 mile
M (two-seat) 890 nm (1,650 km; 1,025 mile
g limits (M):
at max Aerobatic T-O weight +6/–
at max Utility T-O weight without external load +4.4/–2.

SIAI-MARCHETTI SF.260TP

TYPE: Turboprop powered version of SF.260M/W.

PROGRAMME: First flight July 1980; airframe virtuall unchanged aft of firewall except for inset rudder trim ta and automatic fuel feed system. More than 60 SF.260TF ordered by military customers; Philippine Air Forc ordered 16 in February 1992.

SF.260M/W description applies also to TP, except in th following details:

POWER PLANT: One Allison 250-B17D turboprop, flat rated a 261 kW (350 shp) and driving a Hartzell HC-B3TF-7A T10173-25R three-blade constant-speed fully featherin and reversible-pitch propeller. Fuel capacity as fo SF.260M/W; automatic fuel feed system. Oil capacity litres (1.8 US gallons; 1.5 Imp gallons).

DIMENSIONS, EXTERNAL:
Length overall 7.40 m (24 ft 3¼ in

WEIGHTS AND LOADINGS:
Weight empty, equipped 750 kg (1,654 lb
Max power loading: trainer 4.60 kg/kW (7.56 lb/shp
Warrior 4.98 kg/kW (8.19 lb/shp

PERFORMANCE (at trainer Utility T-O weight of 1,200 kg 2,645 lb, ISA):
Never-exceed speed (VNE) 236 knots (437 km/h; 271 mph
Max level speed at 3,050 m (10,000 ft) 228 knots (422 km/h; 262 mph
Max cruising speed at 2,440 m (8,000 ft) 216 knots (400 km/h; 248 mph
Econ cruising speed at 4,575 m (15,000 ft) 170 knots (315 km/h; 195 mph
Stalling speed at S/L, flaps down, power off 68 knots (126 km/h; 79 mph
Max rate of climb at S/L 660 m (2,065 ft)/mi
Service ceiling 7,500 m (24,600 ft
T-O run 298 m (978 ft
T-O to 15 m (50 ft) 467 m (1,532 ft
Landing from 15 m (50 ft) 533 m (1,749 ft
Landing run, without reverse pitch 307 m (1,007 ft
Range at 4,575 m (15,000 ft) with max fuel, 30 mi reserves 512 nm (949 km; 589 miles

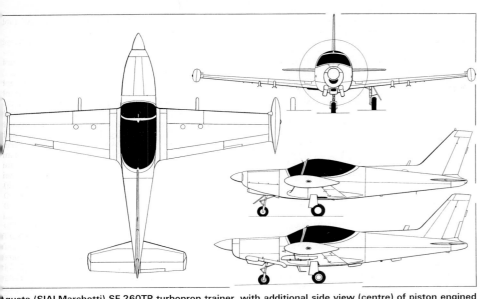

Agusta (SIAI-Marchetti) SF.260TP turboprop trainer, with additional side view (centre) of piston engined SF.260M *(Dennis Punnett)*

ACCOMMODATION: Two pilots in tandem, stepped up 28 cm (11 in) in rear, on Martin-Baker Mk 10 lightweight ejection seats. Blast screen between seats. Pressurised and air-conditioned cockpit under one-piece framed canopy opening sideways to starboard.

SYSTEMS: ECS for cockpit pressurisation and air-conditioning, using engine bleed air for heating, freon vapour for cooling. Max pressure differential 0.24 bar (3.5 lb/sq in). Hydraulic system, pressure 207 bars (3,000 lb/sq in), for actuation of airbrake, landing gear, freon compressor and aileron boost, and independent actuation of wheel brakes. Primary electrical system is 28V DC, using engine driven starter/generator; nickel-cadmium battery; two static inverters supply AC power for instruments and avionics. External power receptacle in port side of lower fuselage aft of wing. Demand type main oxygen system, at 124 bars (1,800 lb/sq in) pressure, sufficient to supply both occupants for 4 hours, plus bottles for emergency oxygen supply.

AVIONICS: Standard avionics fit includes two V/UHF com, ADF, VOR/ILS and DME or Tacan, IFF or ATC, ICS, AHRS, HSI and AI. Provision for dual gyro stabilised gunsight system with miniaturised video recording or film camera. Additional provisions for R/Nav, radar altimeter, Doppler radar, head-up display, radar warning system and ECM.

EQUIPMENT: Inboard wing stations can carry two photo-reconnaissance pods, each with four cameras and infra-red linescan.

SIAI-MARCHETTI S.211

TYPE: Two-seat turbofan trainer and light attack aircraft.

PROGRAMME: First revealed June 1977; first flight (I-SITF) 10 April 1981; deliveries began November 1984.

CURRENT VERSIONS: **S.211**: Standard production version for all customers so far. *Description applies to this version.*

S.211A: Uprated version with more powerful 14.19 kN (3,190 lb st) P&WC JT15D-5C; one flying in USA for demonstrations, one in Italy for development; wing has new aerofoil; higher powered engine gives shorter field length, higher rate of climb, better sustained turn rate and higher max speed; max airspeed at 7,620 m (25,000 ft) 410 knots (760 km/h; 472 mph); landing gear reinforced for 3.96 m (13 ft)/s sink rate; empty weight 2,020 kg (4,453 lb); T-O weight for training 2,900 kg (6,393 lb); max T-O weight 3,500 kg (7,716 lb). First flight (I-PATS) September 1992.

JPATS: S.211A entered in US Air Force/Navy JPATS training system competition in association with Grumman Aircraft Systems (which see); first S.211A delivered to Grumman 1992; second, with EFIS, early 1993; workshare 70:30 with Grumman manufacturing front fuselage and assembling aircraft.

CUSTOMERS: Philippines (18 S.211 ordered in 1988); Singapore (30); 14 of Philippine aircraft assembled by Philippine Aerospace Development Corporation (which see); first aircraft delivered September 1989; completed end 1991; first six Singapore aircraft delivered as kits; remainder manufactured by SAMCO subsidiary of Singapore Aerospace (SA, which see). Philippines option for further 18 not taken up by early 1993.

DESIGN FEATURES: Shoulder wing with NASA GAW-1 aerofoil; sweepback 15° 30′ at quarter-chord; thickness/chord ratio 15 per cent at root, 13 per cent at tip; anhedral 2° from root; twist −3° 17′.

FLYING CONTROLS: Variable incidence tailplane; servo tab in manually operated, horn balanced elevator; ailerons powered by single hydraulic actuator in fuselage with electrically actuated trim bias in aileron linkage; trim tab in rudder; electrically actuated Fowler flaps; airbrake panel under centre-fuselage.

STRUCTURE: Two-spar metal wing forming integral tank in torsion box; wings bolted to fuselage; upper and lower skins formed in single sheets; 60 per cent of external surfaces of fuselage in composites.

LANDING GEAR: Hydraulically retractable tricycle type, of Messier-Bugatti/Magnaghi design. Oleo-pneumatic shock absorber in each unit. All units retract forward into fuselage (main units turning through 90° to lie flat in undersides of engine air intake trunks). Nosewheel steerable ±18°. Mainwheels size 6.50-8; nosewheel size 5.00-5 with water deflecting tyre. Designed for sink rate of 4 m (13 ft)/s. Wheel brakes actuated hydraulically, independently of main hydraulic system. Provision for emergency free-fall extension.

POWER PLANT: One 11.13 kN (2,500 lb st) Pratt & Whitney Canada JT15D-4C turbofan, with electronic fuel control, mounted in rear of fuselage; lateral intake each side of fuselage, with splitter plate. Fuel in 650 litre (171.5 US gallon; 143 Imp gallon) integral wing tank and 150 litre (39.5 US gallon; 33 Imp gallon) fuselage tank; total capacity 800 litres (211 US gallons; 176 Imp gallons). Single gravity refuelling point in top surface of starboard wing. Electric fuel pump for engine starting and emergency use. Fuel and oil systems permit inverted flight. Provision for two 270 litre (71.3 US gallon; 59.4 Imp gallon) drop tanks on inboard underwing stores points. Oil capacity 10 kg (22 lb).

Agusta (SIAI-Marchetti) S.211A in the guise of the Grumman-Agusta bid for the USAF/USN JPATS order

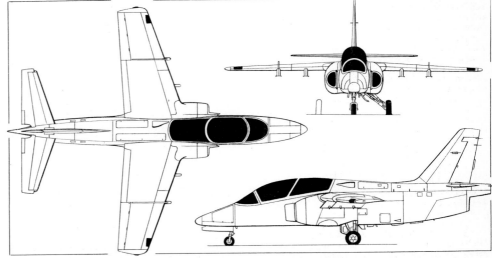

Agusta (SIAI-Marchetti) S.211 basic trainer and light attack aircraft *(Dennis Punnett)*

ARMAMENT: Four underwing hardpoints, stressed for loads of up to 330 kg (727.5 lb) inboard, 165 kg (364 lb) outboard; max external load 660 kg (1,455 lb). Typical loads can include four single- or twin-gun 7.62 mm machine-gun pods, four 12.7 mm gun pods, or (inboard only) two 20 mm gun pods; four AL-18-50 (18 × 50 mm), Matra F2 (6 × 68 mm), LAU-32 (7 × 2.75 in), or AL-6-80 (6 × 81 mm) rocket launchers, or (inboard only) two Matra 155 (18 × 68 mm), SNORA RWK-020 (12 × 81 mm) or 100 mm rocket launchers; four bombs or practice bombs of up to 150 kg size, or (inboard only) two bombs or napalm containers of up to 300 kg; or four 74 mm cartridge throwers. Ferranti ISIS D-211 optical weapon aiming system optional.

DIMENSIONS, EXTERNAL:
Wing span	8.43 m (27 ft 8 in)
Wing chord: at root	2.151 m (7 ft 0¾ in)
mean aerodynamic	1.646 m (5 ft 4¾ in)
at tip	1.00 m (3 ft 3¼ in)
Wing aspect ratio	5.1
Length overall	9.31 m (30 ft 6½ in)
Height overall	3.80 m (12 ft 5½ in)
Tailplane span	3.96 m (13 ft 0 in)
Wheel track	2.29 m (7 ft 6 in)
Wheelbase	4.02 m (13 ft 2¼ in)

AREAS:
Wings, gross	12.60 m² (135.63 sq ft)
Airbrake	0.42 m² (4.52 sq ft)
Vertical tail surfaces (total)	2.01 m² (21.64 sq ft)
Horizontal tail surfaces (total)	3.378 m² (36.36 sq ft)

WEIGHTS AND LOADINGS:
Weight empty, equipped	1,850 kg (4,078 lb)
Max usable fuel: internal	622 kg (1,371 lb)
external	390 kg (860 lb)
Max T-O weight: trainer, clean	2,750 kg (6,063 lb)
armed version	3,150 kg (6,944 lb)
Max wing loading:	
trainer, clean	218.25 kg/m² (44.70 lb/sq ft)
armed version	250.00 kg/m² (51.20 lb/sq ft)
Max power loading:	
trainer, clean	247.4 kg/kN (2.42 lb/lb st)
armed version	283.4 kg/kN (2.78 lb/lb st)

PERFORMANCE (at T-O weight of 2,500 kg; 5,511 lb except where indicated):
Never-exceed speed (V_{NE})	Mach 0.80 (400 knots; 740 km/h; 460 mph EAS)
Max cruising speed at 7,620 m (25,000 ft)	360 knots (667 km/h; 414 mph)
Rotation speed	90 knots (167 km/h; 104 mph)
Stalling speed, flaps down	74 knots (138 km/h; 86 mph)
Max rate of climb at S/L	1,280 m (4,200 ft)/min
Time to 6,100 m (20,000 ft)	6 min 12 s
Service ceiling	12,200 m (40,000 ft)
T-O run (S/L, ISA)	390 m (1,280 ft)
T-O to 15 m (50 ft)	512 m (1,680 ft)
Landing from 15 m (50 ft)	705 m (2,313 ft)
Landing run (S/L, ISA)	361 m (1,185 ft)
Min air turning radius at S/L	less than 305 m (1,000 ft)

Typical attack radius with four rocket launchers, AUW of 3,150 kg (6,944 lb):
hi-lo-hi, out and back at 265 knots (491 km/h; 305 mph) at 9,150 m (30,000 ft), 2 h 50 min mission (incl 5 mi over target), 60 kg (132 lb) of fuel remaining 300 nm (556 km; 345 miles)
lo-lo-lo, out and back at 250 knots (463 km/h; 288 mph at less than 305 m (1,000 ft), 1 h 5 min mission (incl min over target), 60 kg (132 lb) of fuel remaining 12 nm (231 km; 144 miles)
Max range on internal fuel, 30 min reserves
900 nm (1,668 km; 1,036 miles
Ferry range (AUW of 3,150 kg; 6,944 lb, max internal external fuel) at 270 knots (500 km/h; 311 mph) a 9,150 m (30,000 ft), 90 kg (198 lb) of fuel remaining
1,340 nm (2,483 km; 1,543 miles
Endurance, 30 min reserves 3 h 50 mi
Sustained g limit at 4,575 m (15,000 ft) 3.
g limits: +6/−3 clea
 +5/−2.5 with external store

SIAI-MARCHETTI SF.600TP CANGURO (KANGAROO)
First flight F.600 Canguro (I-CANG), built by General Avia and then powered by 261 kW (350 hp) Textron Lycoming TIO-540-J flat-six piston engines, 30 December 1978 (se 1979-80 *Jane's*); current standard power plant is two 335 kW (450 shp) Allison 250-B17F turboprops. About 15 produce in Italy; Philippine Aerospace Development Corporatio (PADC) may produce Canguro under licence. For technica details see 1991-92 *Jane's*.

ELICOTTERI MERIDIONALI SpA
Formed with assistance from Agusta October 1967; remains separate commercial entity affiliated to Agusta. In 1968 EM acquired rights to co-production, marketing and servicing of Boeing CH-47C Chinook transport helicopter for customers in Italy and certain foreign countries; Italian production of CH-47C airframe by Agusta's Sesto Calende works.

Works of EM occupy total area of more than 300,000 m² (3,229,170 sq ft); participates in manufacturing of Agusta A 109 and A 129, Agusta-Bell 212/412, and Agusta-Sikorsky ASH-3H helicopters; has facilities for overhaul, repair and field assistance; designated overhaul organisation for all types of Italian Army helicopter, and is distributor in Italy for Allison 250 turboshaft engines.

EM (BOEING) CH-47C CHINOOK
Italian manufacture of CH-47C began in Spring 1970 for Italian Army Aviation; later customers included Egypt (15), Greece (10), Libya (20), Morocco (nine), Iran (68 of 95 originally ordered) and US Army (11).

Ten **CH-47C Plus** produced for operation by Italian Army on behalf of Civil Protection Agency; upgrading programme begun to fit earlier aircraft with new Textron Lycoming T55-L-712E engines, composite rotor blades and more advanced transmission system; max T-O weight increased to 22,680 kg (50,000 lb).

Agusta developed, jointly with Hosp Ital SpA (a division of Cogefar) of Milan, an **ESFC** (emergency surgery flying centre) version of Chinook for use as mobile hospital (details in 1983-84 *Jane's*); one delivered to Italian Army in 1987; six more CH-47Cs ordered for Italian Army Light Aviation unit at Castelnuovo di Porto specialising in disaster relief and firefighting.

Italian Army's entire fleet of Chinooks being overhauled at rate of three a year by EM; first of 23 then-operational aircraft redelivered March 1986; roles include firefighting using 5,000 litre (1,321 US gallon; 1,100 Imp gallon) metal tank total four CH-47s redelivered during 1992.

Italian built Boeing CH-47C Plus with T55-L-712E engines on relief operations in Mogadishu *(Paul Jackson)*

ALENIA
Via E. Petrolini 2, I-00197 Rome
Telephone: 39 (6) 807781
Fax: 39 (6) 8072215 or 8075184
Telex: 611395 Alenia I
CHAIRMAN: Fausto Cereti
MANAGER DIRECTOR: Enrico Gemelli
GENERAL MANAGER, OPERATIONS: Raffaele Esposito
GENERAL MANAGER, FINANCE AND CONTROL:
Paolo Micheletta
MARKETING AND SALES: Nicolas Zalonis
NEW MARKET DEVELOPMENT: Umberto di Nardo
VICE-PRESIDENT, PUBLIC RELATIONS AND PRESS:
Fabio Dani

Alenia resulted from merger in December 1991 of Aeritalia and Selenia, both members of state-owned IRI industrial group; new company organised into four sectors: aeronautics, space, military systems and civil systems.

On 8 October 1992, Alenia reorganised to form central operational departments, avoiding duplication between sectors; industrial activities organised into areas called aeronautics, aero engines, space, defence and control systems, naval systems and commercial systems; areas subdivided into divisions; thus, aeronautics area consists of Defence Aircraft Division, Aerostructures Division, Commuter Division and Transport Aircraft Overhaul and Modification Division; aero engines area consists of Alfa Romeo Avio; total 20 divisions in all areas.

In further major reorganisation, the boards of Finmeccanica, Alenia, Ansaldo and Elsag Bailey, all members of the state-owned IRI industrial group, proposed to merge into Finmeccanica; latter already owned 88.8 per cent of Alenia, 60 per cent of Ansaldo and 77.7 per cent of Elsag Bailey; shareholders approved merger 18 February 1993; merger was to be initiated 15 March 1993. Exact nomenclature and organisation still unknown. Ultimate objective is privatisation.

Agusta, OMI, OTO Melara, Breda Meccanica Bresciana, Galileo and SMA, members of the liquidated state-owned industrial group EFIM, transferred to Finmeccanica from 1 January 1993 to 30 June 1993, pending decision on disposal of Agusta group and its constituents (see Agusta entry for details). Incorporation of Agusta as helicopter group of Alenia seemed unlikely.

Alenia had 29,000 employees and 42 factories in early 1993; 1992 sales totalled LIt5,000 billion.

ADVANCED AMPHIBIOUS AIRCRAFT
See International section for state of programme; Deutsche Aerospace withdrew its remaining participation early 1993.

AIRBUS A321
In January 1992 Alenia delivered first fuselage section to Aerospatiale at St Nazaire for further integration with other aft fuselage frames (see International section).

AMX
Alenia has largest share (46.5 per cent) in AMX attack aircraft, also involving Aermacchi and Embraer. Details in International section.

ATR 42/72
Alenia is equal partner with Aerospatiale of France in these regional transport aircraft; details under ATR in International section.

BOEING 707TT
Alenia's Naples/Capodichino plant converted four Boeing 707-320Cs from passenger to TT (Tanker/Transport) for Italian Air Force; included full refurbishing of airframe and interior, fitting of flight refuelling equipment and related systems; first flight of modified aircraft Autumn 1991; first delivery early 1992.

DASSAULT FALCON 2000
Since September 1990, Alenia has been risk-sharing partner in design, development and production of Falcon 2000; with help of some Italian airframe subcontractors, Alenia manufactures rear fuselage, including nacelles, and tail unit.

DOUGLAS DC-8 and BOEING 727
Alenia subsidiary Dee Howard (see in USA section) is modifying entire United Parcels Service fleet of DC-8-70s and 727s; work includes digital avionics and 'dark cockpit' equipment. Aeronavali Venezia also converted DC-8s into freighters between 1982 and late 1980s.

DOUGLAS MD-11 and DC-10F
Alenia subsidiary Aeronavali Venezia currently modifying MD-11s and DC-10s from passenger to all-cargo; work includes fitting of large side cargo door; first modified MD-11 delivered to Federal Express end 1992; first DC-10-10F delivered also to Federal Express end 1992.

EUROFIGHTER 2000
Alenia has 21 per cent share in development; details in International section.

EUROFLAG
Alenia is member of Euroflag Future Large Aircraft Group (see International section); company head office based with Alenia in Rome.

NAMC A-5M
Under joint programme with CATIC, Alenia has modernised avionics of Nanchang Q-5 II ('Fantan') attack aircraft (see under NAMC in Chinese section). Upgrading includes all-weather autonomous and radio-assisted navigation, air and ground communications, IFF, weapon aiming for ground attack and air-to-air self-defence modes and passive ECM; avionics system designed around central computers and dual redundant MIL-STD-1553B databus. Three equipped aircraft (one since lost) have flown; development continues.

PANAVIA TORNADO
Alenia has 15 per cent participation in manufacturing programme for Tornado (see International section), being responsible for radomes and movable wings, including control surfaces. All 100 aircraft for Italian Air Force delivered.

ALENIA G222
US Air Force designation: C-27A Spartan
TYPE: Pressurised twin-turboprop tactical transport.
PROGRAMME: First flight 18 July 1970; early history in 1987-88 *Jane's*.
CURRENT VERSIONS: **G222**: Standard military transport. *Detailed description applies to this version.*
G222RM Radiomisure: Navaid calibration version (see 1989-90 *Jane's*).
G222SAA Sistema Aeronautico Antincendio: Firefighting version (see 1989-90 *Jane's*).
G222T: Powered by Rolls-Royce Tyne turboprops; 20, including two in VIP configuration, sold to Libya (see 1986-87 *Jane's*).
G222VS Versione Speciale: Electronic warfare version (see 1989-90 *Jane's*).
C-27A Spartan (G222-710): US Air Force transport for Panama and South America; fitted with mission equipment by Chrysler Technologies Airborne Systems (see CTAS in US section); US Air Force ordered total 10 in August 1990 and February 1991; eight options. See also different weights and performance of USAF C-27A under Chrysler Technologies entry.
CUSTOMERS: From April 1978, Italian Air Force received 46 G222s, including 30 standard transports, 10 G222SAAs, four G222RMs and two G222VSs. Italian Ministry for Civil Defence received five for rapid intervention squadron equipped for firefighting, aeromedical evacuation and airlift. Export customers included Argentine Army (three), Dubai Air Force (one), Libyan Air Force (20 G222T), Nigerian Air Force (five), Somali Air Force (two), Venezuelan Army (two) and Air Force (six); reported orders from Congo, Guatemala and Yemen removed from sales list by Alenia in 1992. US Air Force ordered 10 C-27A (see above).
DESIGN FEATURES: Full description in 1987-88 *Jane's*.
STRUCTURE: Subcontractors include Aermacchi (outer wings), Piaggio (wing centre-section), Agusta (tail unit), CIRSEA (landing gear) and Aeronavali Venezia (airframe components).
POWER PLANT: Two Fiat built General Electric T64-GE-P4D turboprops, each flat rated at 2,535 kW (3,400 shp) at ISA + 25°C and driving a Hamilton Standard 63E60-27 three-blade variable- and reversible-pitch propeller. Fuel in integral tanks; two in outer wings, combined capacity 6,800 litres (1,796 US gallons; 1,495 Imp gallons); two centre-section tanks, combined capacity 5,200 litres (1,374 US gallons; 1,143 Imp gallons); cross-feed provision to either engine. Total overall fuel capacity 12,000 litres (3,170 US gallons; 2,638 Imp gallons).
ACCOMMODATION: Two-pilot crew on flight deck with third seat; provision for loadmaster or jumpmaster when required. Standard troop transport version has 34 foldaway sidewall seats and 12 stowable seats for 46 fully equipped troops. Paratroop transport version can carry up to 40 fully equipped paratroops, and is fitted with 32 sidewall seats, plus eight stowable seats, door jump platforms and static lines. Cargo transport version can accept standard pallets of up to 2.24 m (88 in) wide, and can carry up to 9,600 kg (21,164 lb) of freight. Hydraulically operated rear loading ramp and upward opening door in underside of upswept rear fuselage, which can be opened in flight for airdrop operations. In cargo version, five pallets of up to 1,000 kg (2,205 lb) each can be airdropped from rear opening, or single pallet of up to 5,000 kg (11,023 lb). Paratroop jumps can be made either from this opening or from rear side doors. Entire accommodation pressurised and air-conditioned.

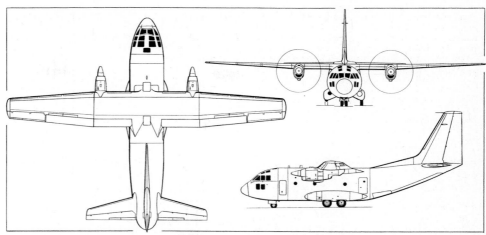

Alenia G222 (C-27A Spartan) twin-turboprop general purpose military transport *(Dennis Punnett)*

Italian Air Force Alenia G222 in new camouflage paintwork in Mogadishu *(Paul Jackson)*

DIMENSIONS, EXTERNAL:
Wing span	28.70 m (94 ft 2 in)
Wing aspect ratio	10.0
Length overall	22.70 m (74 ft 5½ in)
Height overall	10.57 m (34 ft 8¼ in)
Fuselage: Max diameter	3.55 m (11 ft 7¾ in)
Rear-loading ramp/door: Width	2.45 m (8 ft 0½ in)
Height	2.25 m (7 ft 4½ in)

DIMENSIONS, INTERNAL:
Main cabin: Length	8.58 m (28 ft 1¾ in)
Width	2.45 m (8 ft 0½ in)
Height	2.25 m (7 ft 4½ in)
Floor area: excl ramp	21.00 m² (226.0 sq ft)
incl ramp	25.68 m² (276.4 sq ft)
Volume	58.0 m³ (2,048 cu ft)

WEIGHTS AND LOADINGS:
Operating weight empty	15,700 kg (34,610 lb)
Max payload (cargo)	9,000 kg (19,840 lb)
Max fuel load	9,400 kg (20,725 lb)
Max T-O weight	28,000 kg (61,730 lb)
Max landing weight	26,500 kg (58,420 lb)
Max zero-fuel weight	24,700 kg (54,454 lb)
Max cargo floor loading	1,500 kg/m² (307.2 lb/sq ft)
Max wing loading	341.5 kg/m² (69.9 lb/sq ft)
Max power loading	5.52 kg/kW (9.08 lb/shp)

PERFORMANCE (at max T-O weight except where indicated):
Max level speed at 4,575 m (15,000 ft)	263 knots (487 km/h; 303 mph)
Long-range cruising speed at 6,000 m (19,680 ft)	236 knots (437 km/h; 272 mph)
Airdrop speed (paratroops or cargo)	110-140 knots (204-259 km/h; 127-161 mph) IAS
Stalling speed, flaps down	92 knots (171 km/h; 106 mph)
Time to 4,500 m (14,760 ft)	16 min
Max rate of climb at S/L	381 m (1,250 ft)/min
Rate of climb at S/L, OEI	101 m (330 ft)/min
Service ceiling	7,835 m (25,700 ft)
Service ceiling, OEI	3,660 m (12,000 ft)
T-O run	686 m (2,250 ft)
Landing run at max landing weight	872 m (2,860 ft)
Range with max payload, at optimum cruising speed and height	680 nm (1,260 km; 783 miles)
Ferry range with max fuel	2,530 nm (4,688 km; 2,913 miles)
g limit	+2.5

AVIOLIGHT

AVIOLIGHT SRL
PRESIDENT: Prof Ing Luigi Pascale
Subsidiary of Tecnam; held rights to Partenavia single-engined aircraft; no longer in business early 1993; aircraft operations being continued by Tecnam, which see.

General Avia Pinguinos: left to right, F.22/B (160 hp), F.22/R (160 hp) and F.22/A (116 hp)

GENERAL AVIA
GENERAL AVIA COSTRUZIONI AERONAUTICHE SRL
HEAD OFFICE: Via U. Comandini 38, I-00173 Rome
Telephone: 39 (6) 72 31 651
Fax: 39 (6) 72 34 536
WORKS: Via Trieste 22-24, I-20096 Pioltello, Milan
Telephone: 39 (2) 92 66 774 and 92 16 12 86
Fax: 39 (2) 92 16 03 95
MANAGING DIRECTORS:
 Dott Ing Silvio Angelucci
 Maurizio Ruggiero
TECHNICAL DIRECTOR: Dott Ing Stelio Frati
MARKETING: Cmdte Alessandro Ghisleni
PUBLIC RELATIONS: Carla Bielli
QUALITY CONTROL: Dott Ing Giovanni Ballarin

Dott Ing Stelio Frati is well known for many successful light aircraft which, as a freelance designer, he has developed since 1950; these include F.8 Falco, F.15 Picchio, the F.250 (now manufactured by Agusta as the SF.260) and F.20 Pegaso (see 1981-82 *Jane's*). General Avia developed and built prototype of Canguro transport. In 1984 started construction of F1300 Jet Squalus, financed by Promavia of Belgium (which see); Promavia bought out whole Jet Squalus programme.

GENERAL AVIA F.22/A, F.22/B and F.22/C PINGUINO (PENGUIN)
TYPE: Side by side two-seat primary trainer.
PROGRAMME: First flight of prototype F.22/A (I-GEAD) from Orio al Serio (Bergamo) 13 June 1989; certification achieved May 1993; certification of 119 kW (160 hp) F.22/B expected Summer 1993; deliveries from batch production to start late 1993.
CURRENT VERSIONS: **F.22/A:** Basic version powered by 86.5 kW (116 hp) engine.
 F.22/B: Powered by 119 kW (160 hp) engine.
 F.22/C: See Addenda.
 Retractable: All three Pinguinos to be offered with optional retractable landing gear.
DESIGN FEATURES: Classic Frati design.
FLYING CONTROLS: Conventional, with fixed tailplane and elevator trim tab; electrically actuated flaps.
STRUCTURE: All-metal single-spar wing built as one piece; all-metal fuselage.
LANDING GEAR: First aircraft have non-retractable tricycle type, with steerable nosewheel; oleo shock absorbers; faired main legs. Retractable landing gear optional.
POWER PLANT: F.22/A powered by one 86.5 kW (116 hp) Textron Lycoming O-235-N2C flat-four engine, driving a two-blade wooden propeller. F.22/B powered by 119 kW (160 hp) O-320-D2A flown in second prototype; electric starter. Fuel capacity 105 litres (28 US gallons; 23 Imp gallons) in F.22/A and 135 litres (36 US gallons; 30 Imp gallons) in F.22/B; optional long-range tank.
ACCOMMODATION: Two seats side by side: sliding canopy.
AVIONICS: Bendix/King Silver Crown nav/com, ADF, audio console and transponder.

DIMENSIONS, EXTERNAL:
Wing span	8.50 m (27 ft 10¾ in)
Wing chord: at root	1.589 m (5 ft 2½ in)
at tip	0.876 m (2 ft 10½ in)
Wing aspect ratio	6.7
Length overall	7.40 m (24 ft 3¼ in)
Height overall	2.84 m (9 ft 3¾ in)
Tailplane span	3.00 m (9 ft 10 in)
Wheel track	2.90 m (9 ft 6¼ in)
Wheelbase	1.86 m (6 ft 1¼ in)
Propeller diameter	1.78 m (5 ft 10 in)

AREAS:
Wings, gross	10.82 m² (116.25 sq ft)
Fin	0.738 m² (7.94 sq ft)
Rudder	0.505 m² (5.44 sq ft)
Tailplane	1.24 m² (13.35 sq ft)
Elevator, incl tab	1.02 m² (10.98 sq ft)

WEIGHTS AND LOADINGS:
Weight empty, equipped: F.22/A	510 kg (1,124 lb)
F.22/B	530 kg (1,168 lb)
Max T-O weight:	
Aerobatic: F.22/A	750 kg (1,653 lb)
F.22/B	800 kg (1,763 lb)
Utility: F.22/A	780 kg (1,719 lb)
F.22/B	850 kg (1,873 lb)
Max wing loading:	
Aerobatic: F.22/A	69.32 kg/m² (14.20 lb/sq ft)
F.22/B	73.94 kg/m² (15.14 lb/sq ft)
Utility: F.22/A	72.08 kg/m² (14.82 lb/sq ft)
F.22/B	78.55 kg/m² (16.13 lb/sq ft)
Max power loading:	
Aerobatic: F.22/A	8.67 kg/kW (14.25 lb/hp)
F.22/B	6.7 kg/kW (10.74 lb/hp)
Utility: F.22/A	9.0 kg/kW (14.8 lb/hp)
F.22/B	7.14 kg/kW (11.7 lb/hp)

PERFORMANCE:
Max level speed: F.22/A	128 knots (237 km/h; 147 mph)
F.22/B	145 knots (269 km/h; 166 mph)
Cruising speed, 75% power:	
F.22/A	120 knots (220 km/h; 138 mph)
F.22/B	130 knots (241 km/h; 149 mph)
Stalling speed, flaps down:	
F.22/A, F.22/B	48 knots (89 km/h; 55 mph)
Max rate of climb at S/L: F.22/A	213 m (700 ft)/min
F.22/B	335 m (1,100 ft)/min
Service ceiling: F.22/A	4,100 m (13,450 ft)
F.22/B	5,030 m (16,500 ft)
T-O run: F.22/A	295 m (968 ft)
F.22/B	240 m (787 ft)
Landing run: F.22/A	160 m (525 ft)
F.22/B	170 m (557 ft)
Max range, standard tank:	
F.22/A	485 nm (899 km; 557 miles)
F.22/B	595 nm (1,102 km; 684 miles)
Max range, long-range tank:	
F.22/A	595 nm (1,102 km; 684 miles)
F.22/B	730 nm (1,352 km; 839 miles)

GENERAL AVIA F.22/R PINGUINO-SPRINT
TYPE: Two-seat light club aircraft.
PROGRAMME: First flight (I-GEAE) 16 November 1990; certification expected Summer 1993.
DESIGN FEATURES: Airframe largely as F.22 Pinguino, but powered by 119 kW (160 hp) Textron Lycoming O-320-D1A driving constant-speed propeller and with retractable landing gear.

WEIGHTS AND LOADINGS:
Weight empty, incl radio	575 kg (1,268 lb)
Max T-O weight: Aerobatic	750 kg (1,653 lb)
Utility	850 kg (1,873 lb)

General Avia F.22/R Pinguino-Sprint

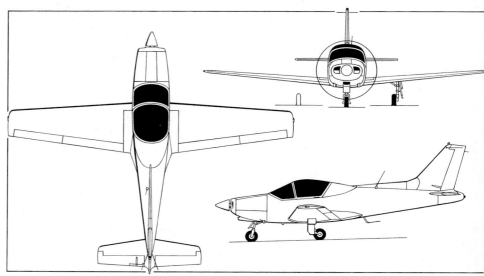

General Avia F.22/R Pinguino-Sprint (*Jane's*/Mike Keep)

Max wing loading:	
Aerobatic	69.32 kg/m² (14.20 lb/sq ft)
Utility	78.56 kg/m² (16.09 lb/sq ft)
Max power loading:	
Aerobatic	6.29 kg/kW (10.33 lb/hp)
Utility	7.13 kg/kW (11.71 lb/hp)

PERFORMANCE (at 800 kg; 1,764 lb AUW):
Max level speed at S/L	164 knots (305 km/h; 189 mph)
Stalling speed, flaps down	53 knots (98 km/h; 61 mph)
Max rate of climb at S/L	420 m (1,378 ft)/min
Service ceiling	5,650 m (18,535 ft)
T-O run	200 m (657 ft)
Landing run	230 m (755 ft)
Max range	700 nm (1,300 km; 805 miles)

GENERAL AVIA F.220 AIRONE (HERON)

TYPE: Light four-seat touring aircraft.
PROGRAMME: First flight expected September 1993.
DESIGN FEATURES: Extrapolation of F.22/R Pinguino-Sprint.
FLYING CONTROLS: As Pinguinos, dual controls, but with aileron wheels.
STRUCTURE: Metal stressed-skin structure analogous to that of F.22/R Pinguino-Sprint.
LANDING GEAR: Retractable tricycle; steerable nosewheel; electromechanical actuation.
POWER PLANT: One 149 kW (200 hp) Textron Lycoming IO-360-A1A flat-four engine, driving a Hartzell constant-speed propeller. Fuel capacity 245 litres (64.7 US gallons; 53.9 Imp gallons) in integral wing tanks.
ACCOMMODATION: Four seats under hard roof; door on each side.
AVIONICS: Bendix/King Silver Crown IFR with dual nav/com and GPS.

DIMENSIONS, EXTERNAL:
Wing span	9.70 m (31 ft 10 in)
Wing aspect ratio	8.0

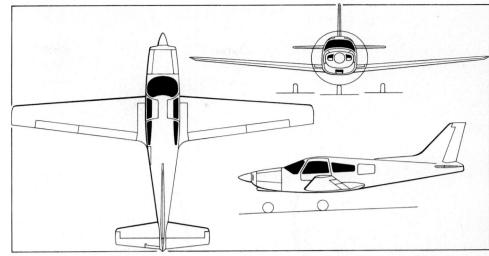

General Avia F.220 Airone four-seater (one Textron Lycoming IO-360) *(Jane's/Mike Keep)*

Length overall	9.40 m (30 ft 10 in)

AREAS:
Wings, gross	11.76 m² (126.6 sq ft)
Horizontal tail surfaces (total)	2.26 m² (24.32 sq ft)
Vertical tail surfaces (total)	1.24 m² (13.35 sq ft)

WEIGHTS AND LOADINGS:
Weight empty, with radio	700 kg (1,543 lb)
Max T-O weight	1,150 kg (2,535 lb)
Max wing loading	97.79 kg/m² (20.03 lb/sq ft)
Max power loading	7.72 kg/kW (12.68 lb/hp)

PERFORMANCE:
Max level speed	185 knots (343 km/h; 213 mph)
Cruising speed, 75% power at 2,440 m (8,000 ft)	175 knots (324 km/h; 201 mph)
Stalling speed	57 knots (106 km/h; 66 mph)
Max rate of climb at S/L	427 m (1,400 ft)/min
Service ceiling	5,945 m (19,500 ft)
T-O run	259 m (850 ft)
Landing run	244 m (800 ft)
Max range at 2,440 m (8,000 ft)	1,000 nm (1,853 km; 1,151 miles)
Endurance	6 h 0 min

PARTENAVIA

PARTENAVIA COSTRUZIONI AERONAUTICHE srl (An AerCosmos company)

HEAD OFFICE: 24 Corso Vittorio Emanuele, I-20122 Milan
Telephone: 39 (2) 76001847
Fax: 39 (2) 783147
FACTORY: Via Giovanni Pascoli 7, I-80026 Casoria, Naples
Telephone: 39 (81) 584 2831
PRESIDENT AND CEO: Dott Luciano Zanotti
GENERAL MANAGER: Dott Ing Ciro d'Amato
MARKETING AND COMMUNICATIONS MANAGER:
Dott Barbara Contini Champier

Company founded 1957 by Prof Ing Luigi Pascale; under control of Aeritalia/Alenia from July 1981 to end 1992; taken over by AerCosmos, an Aerospace Engineering Ltd company, March 1993.

From 1991, factory beside Naples Capodichino airport only produced subassemblies for various Partenavia light twins, some of which were assembled by Piaggio. AerCosmos is relaunching production of four Partenavia light twins: basic P.68C light piston twin, P.68TC turbocharged version, P.68 Observer 2 with transparent nose, and 11-seat turboprop-powered AP.68TP 600 Viator.

See also new projects in Addenda.

PARTENAVIA P.68C, P.68TC and P.68 OBSERVER 2

TYPE: Seven-seat light twin.
PROGRAMME: Production of P.68C started 1978 and P.68TC 1980; output slowed down in late 1980s; being restarted 1993.
CURRENT VERSIONS: **P.68C**: Basic version.
P.68TC: As P.68C, but with turbosupercharged engines for better hot and high performance.
P.68 Observer 2: For use by government and specialised services for patrol, surveillance and search; largely transparent nose section with lowered, compact instrument panel; can carry variety of electro-optical sensors; slightly different equipment from other versions.
CUSTOMERS: Italian police and other government services. Discussions completed during May 1993 with Indian manufacturer Taneja Aerospace (which see), to produce P.68s under licence near Bangalore.
DESIGN FEATURES: High wing with NACA 63-3515 aerofoil section, and Hoerner tips; dihedral 1°, incidence 1° 3′.
FLYING CONTROLS: Conventional rod and cable actuated, with slab tailplane and anti-balance tab acting also as trim tab; trim tab in rudder; electrically operated single-slotted flaps.
STRUCTURE: Light alloy stressed skin fuselage with frames and longerons; stressed skin two-spar torsion box wing; metal stressed-skin tailplane and fin; fuselage/wing fairings mainly GFRP.
LANDING GEAR: Non-retractable, with spring steel main legs; oleo suspension for nosewheel, steered from rudder pedals; mainwheels Cleveland 40-142 with Pirelli 8-ply 6.00-6 or 7.00-6 tyres; nosewheel Cleveland 40-77B with Goodyear six-ply 5.00-5 or 6.00-6 tyre; Cleveland type 30-61 foot-powered hydraulic disc brakes; streamlined wheel fairings optional. P.68 Observer 2 has larger mainwheel tyres as standard. Min ground turning radius 5.70 m (18 ft 8 in).
POWER PLANT: *P.68C:* Two 149 kW (200 hp) Textron Lycoming IO-360-A1B6 flat-four engines, each driving a Hartzell HC-C2YK-2C two-blade constant-speed propeller. *P.68TC:* Two 156.6 kW (210 hp) Textron Lycoming TIO-360-C1A6D; same propellers as P.68C. Fuel capacity 269 litres (71 US gallons; 59 Imp gallons) in integral tank in each wing, of which 260 litres (68.7 US gallons; 57 Imp gallons) usable; over-wing gravity refuelling. Oil capacity 7.5 litres (1.98 US gallons; 1.65 Imp gallons) for each engine. P.68 Observer 2 has unsupercharged engines and optional extra tanks giving total 150 litres (39.6 US gallons; 33 Imp gallons) more usable fuel.
ACCOMMODATION: One or two pilots and five or six passengers; cabin has two forward-facing seats in middle and three-seat rear bench; club seating optional; baggage door at rear and pilot door at front on starboard side; passenger door to port in centre cabin; max baggage weight allowance 181 kg (399 lb), accessible from inside cabin. P.68 Observer 2 has no front starboard door for pilots.
SYSTEMS: Two 24V 70Ah alternators (100Ah in P.68 Observer 2) and one 24V 17Ah battery; Goodrich pneumatic de-icing boots optional; air-conditioning optional.
AVIONICS: Choice of VFR or full IFR Bendix/King Silver Crown avionics with KFC 150 autopilot. P.68 Observer 2 can also carry FLIR, weather radar, ATAL video surveillance pod with data downlink and side-looking radar (SLAR).

DIMENSIONS, EXTERNAL (C: P.68C, TC: P.68TC, O: P.68 Observer 2):
Wing span	12.00 m (39 ft 4½ in)
Wing chord, constant	1.55 m (5 ft 1 in)
Wing aspect ratio	7.74
Length overall: C, TC	9.55 m (31 ft 4 in)
O, normal	9.35 m (30 ft 8 in)
O, with ATAL pod in nose	9.55 m (31 ft 4 in)
Height overall	3.40 m (11 ft 1¾ in)
Tailplane span	3.90 m (12 ft 9½ in)
Wheel track	2.40 m (7 ft 10½ in)
Wheelbase	3.80 m (12 ft 5½ in)
Propeller diameter: all versions	1.83 m (6 ft 0 in)

Partenavia/AerCosmos P.68C (two Textron Lycoming IO-360)

ITALY: AIRCRAFT—PARTENAVIA

Partenavia/AerCosmos P.68 Observer 2 with fully transparent nose for observation flying

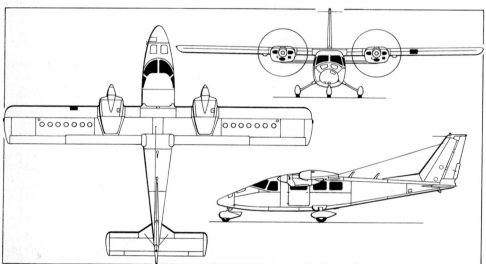

Partenavia/AerCosmos P.68C *(Jane's/Mike Keep)*

Propeller ground clearance	0.77 m (2 ft 6¼ in)
DIMENSIONS, INTERNAL:	
Cabin: Length	3.58 m (11 ft 9 in)
Max width	1.16 m (3 ft 9½ in)
Max height	1.20 m (3 ft 11¼ in)
Baggage compartment volume	0.56 m³ (19.78 cu ft)
AREAS:	
Wings, gross	18.60 m² (200.2 sq ft)
Ailerons (total)	1.79 m² (19.27 sq ft)
Trailing-edge flaps (total)	2.37 m² (25.51 sq ft)
Fin	1.59 m² (17.11 sq ft)
Rudder, incl tab	0.44 m² (4.74 sq ft)
Tailplane, incl tab	4.41 m² (47.47 sq ft)
WEIGHTS AND LOADINGS:	
Weight empty equipped: C, TC	1,300 kg (2,866 lb)
Max T-O weight: C, TC	1,990 kg (4,387 lb)
O	2,084 kg (4,594 lb)
Max ramp weight: O	2,100 kg (4,630 lb)
Max zero-fuel weight: O	1,890 kg (4,167 lb)
Max landing weight: all	1,980 kg (4,365 lb)
Max wing loading: C, TC	106.99 kg/m² (21.92 lb/sq ft)
O	112.04 kg/m² (22.94 lb/sq ft)
Max power loading: C	6.68 kg/kW (10.97 lb/hp)
TC	6.36 kg/kW (10.45 lb/hp)
O	6.99 kg/kW (11.49 lb/hp)
PERFORMANCE:	
Never-exceed speed (V_{NE}):	
C, TC	193 knots (358 km/h; 222 mph)
O	193.5 knots (359 km/h; 223 mph)
Max level speed:	
C at S/L	174 knots (322 km/h; 200 mph)
TC at 3,660 m (12,000 ft)	190 knots (352 km/h; 219 mph)
O at S/L	173 knots (321 km/h; 199 mph)
Max cruising speed, 75% power:	
C at 2,285 m (7,500 ft)	166 knots (308 km/h; 191 mph)
TC at 4,265 m (14,000 ft)	175 knots (324 km/h; 201 mph)
O at 2,135 m (7,000 ft)	165 knots (306 km/h; 190 mph)
Cruising speed, 55% power:	
C at 3,050 m (10,000 ft)	150 knots (278 km/h; 173 mph)
TC at 4,265 m (14,000 ft)	152 knots (282 km/h; 175 mph)
O at 3,660 m (12,000 ft)	150 knots (278 km/h; 173 mph)
Stalling speed, power off:	
flaps up: C, TC	70 knots (129 km/h; 80 mph)
O	68 knots (126 km/h; 79 mph)
flaps down: TC, O	57 knots (106 km/h; 66 mph)
C	58 knots (108 km/h; 67 mph)
Max rate of climb at S/L: C	457 m (1,500 ft)/min
TC	472 m (1,550 ft)/min
O	378 m (1,240 ft)/min
Max rate of climb, OEI: C	82 m (270 ft)/min
TC	88 m (290 ft)/min
O	64 m (210 ft)/min
Service ceiling: C	5,850 m (19,200 ft)
TC	8,230 m (27,000 ft)
O	5,850 m (19,200 ft)
Service ceiling, OEI: C	2,100 m (6,900 ft)
TC	4,420 m (14,500 ft)
O	1,770 m (5,800 ft)
T-O run: C, TC	230 m (755 ft)
O	241 m (791 ft)
T-O distance to 15 m (50 ft): C	396 m (1,300 ft)
TC	385 m (1,264 ft)
O	400 m (1,313 ft)
Landing distance from 15 m (50 ft):	
C, TC	488 m (1,601 ft)
O	600 m (1,970 ft)
Landing run: C, TC	215 m (706 ft)
O	221 m (725 ft)
Range with max payload: TC	300 nm (556 km; 345 miles)
O with max fuel	590 nm (1,093 km; 679 miles)
Range with max fuel:	
C	1,210 nm (2,242 km; 1,393 miles)
TC	1,040 nm (1,927 km; 1,197 miles)
O	1,525 nm (2,826 km; 1,756 miles)

PARTENAVIA AP.68TP 600 VIATOR

TYPE: Eleven-seat multi-purpose twin-turboprop.
PROGRAMME: First flight 29 March 1985; production started 1989.
CUSTOMERS: Initial customers were government services.
DESIGN FEATURES: Cantilever, untapered high wing of same span and form as P.68C; aerofoil section NACA 63-3515, with Hoerner wingtip; dihedral 1°, incidence 1° 3′.
FLYING CONTROLS: Rod and cable actuated primary surfaces; fixed tailplane; elevator has vortex generators under leading-edge; down spring in elevator circuit; stall strips on wing leading-edges; trim tabs for elevator, rudder and ailerons; electrically actuated single-slotted flaps.
STRUCTURE: All-metal stressed skin fuselage; two-spar torsion box wing; metal control surfaces.
LANDING GEAR: Retractable tricycle; main gear retracts hydraulically inward into fuselage fairings; nosewheel forward; Cleveland mainwheels size 40-163EA; Cleveland 40-778 nosewheel; McCreary 6.50-8 mainwheel tyres and 6.00-6 on nosewheel; Cleveland powered hydraulic disc brakes. Min ground turning radius 5.45 m (17 ft 11 in).
POWER PLANT: Two Allison 250-B17C turboprops, each flat rated at 244.6 kW (328 shp), driving a Hartzell HC-B3TF-7A three-blade constant-speed reversible propeller. One integral fuel tank in each wing and a 38 litre (10 US gallon; 8.35 Imp gallon) tank in each engine nacelle, giving total capacity of 840 litres (222 US gallons; 184.8 Imp gallons). Oil capacity 5.7 litres (1.5 US gallons; 1.25 Imp gallons) each engine.

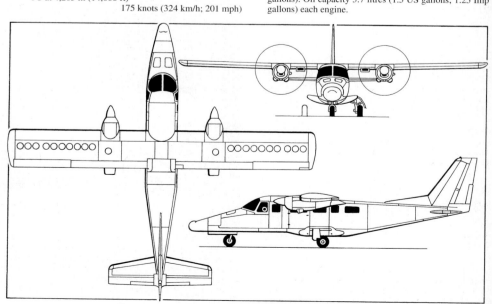

Partenavia/AerCosmos AP.68TP Viator *(Jane's/Mike Keep)*

PARTENAVIA/PIAGGIO—AIRCRAFT: ITALY 199

Partenavia/AerCosmos AP.68TP Viator

ACCOMMODATION: One or two pilots, nine or ten passengers; two doors to port, one for pilot and one for passengers; two doors to starboard, one for co-pilot and one for baggage; baggage compartment variable by using part of cabin; baggage accessible in flight.
SYSTEMS: Two 150Ah 28V DC starter/generators; one 24V 29Ah battery; hydraulics for brakes and landing gear retraction, pressurised by electric pump; electric anti-icing for engine intake and propellers, and pneumatic de-icing boots, standard.
AVIONICS: VFR radio and instruments standard; full IFR with weather radar and Bendix/King Silver Crown radios optional; observation equipment optional.

DIMENSIONS, EXTERNAL:
Wing span 12.00 m (39 ft 4½ in)
Wing chord, constant 1.55 m (5 ft 1 in)
Wing aspect ratio 7.74
Length overall 11.27 m (36 ft 11½ in)
Height overall 3.64 m (11 ft 11 in)
Tailplane span 4.01 m (13 ft 1¾ in)
Wheel track 2.167 m (7 ft 1¼ in)
Wheelbase 3.51 m (11 ft 6¼ in)
Propeller diameter 2.03 m (6 ft 8 in)
Propeller ground clearance 0.725 m (2 ft 4½ in)
Port doors (each): Height 1.01 m (3 ft 3¾ in)
  Width 0.80 m (2 ft 7½ in)
  Height to sill 0.79 m (2 ft 7 in)
Starboard doors (each): Height 0.92 m (3 ft 0¼ in)
  Width 1.10 m (3 ft 7¼ in)
  Height to sill 0.79 m (2 ft 7 in)

DIMENSIONS, INTERNAL:
Cabin, excl cockpit and baggage compartment:
  Length 5.29 m (17 ft 4¼ in)
  Max width 1.13 m (3 ft 8½ in)
  Max height 1.26 m (4 ft 1½ in)
  Floor area 5.75 m² (61.89 sq ft)
Baggage/cargo compartment volume 0.65 m³ (22.95 cu ft)

AREAS:
Wings, gross 18.60 m² (200.2 sq ft)
Ailerons (total) 1.76 m² (18.94 sq ft)
Trailing-edge flaps (total) 2.42 m² (26.05 sq ft)
Horizontal tail surfaces (total) 5.06 m² (54.47 sq ft)
Vertical tail surfaces (total) 4.54 m² (48.87 sq ft)

WEIGHTS AND LOADINGS:
Operating weight empty 1,680 kg (3,704 lb)
Max payload 870 kg (1,918 lb)
Max fuel weight 675 kg (1,488 lb)
Max T-O weight 3,000 kg (6,614 lb)
Max ramp weight 3,025 kg (6,669 lb)
Max zero-fuel weight 2,550 kg (5,622 lb)
Max landing weight 2,850 kg (6,283 lb)
Max wing loading 153.23 kg/m² (31.38 lb/sq ft)
Max power loading 6.13 kg/kW (10.08 lb/shp)

PERFORMANCE:
Max level speed at S/L 200 knots (370 km/h; 230 mph)
Max cruising speed at 3,050 m (10,000 ft)
  214 knots (397 km/h; 247 mph)
Econ cruising speed at 3,050 m (10,000 ft)
  170 knots (315 km/h; 196 mph)
Stalling speed, power off:
  flaps up 75 knots (139 km/h; 87 mph)
  flaps down 65 knots (120 km/h; 75 mph)
Max rate of climb at S/L 503 m (1,650 ft)/min
Rate of climb at S/L, OEI 59 m (194 ft)/min
Service ceiling 7,925 m (26,000 ft)
Service ceiling, OEI 3,475 m (11,400 ft)
T-O run 400 m (1,313 ft)
T-O to 15 m (50 ft) 600 m (1,970 ft)
Landing from 15 m (50 ft) 700 m (2,297 ft)
Landing run 320 m (1,050 ft)
Range: with max payload 530 nm (982 km; 610 miles)
  with max fuel 860 nm (1,594 km; 990 miles)

PIAGGIO
INDUSTRIE AERONAUTICHE E MECCANICHE RINALDO PIAGGIO SpA
Via Cibrario 4, I-16154 Genova Sestri, Genoa
Telephone: 39 (10) 6481-1
Fax: 39 (10) 6520160
Telex: 270695 AERPIA I
WORKS: Genova Sestri, Finale Ligure (SV), and Wichita, Kansas, USA
BRANCH OFFICE: Via A. Gramsci 34, I-00197 Rome
PRESIDENT: Dott Rinaldo Piaggio
DIRECTOR GENERAL AND CEO: Ing Roberto Mannu
DIRECTOR OF OPERATIONS: Ing R. Ludovico Ughi
MARKETING AND SALES: Enzo Traini

Aircraft production begun at Genoa Sestri 1916 and later extended to Finale Ligure; present company formed 29 February 1964; covered floor area Sestri and Finale Ligure 120,000 m² (1,291,670 sq ft); workforce 1,600. Piaggio Aviation Inc with 10,000 m² (107,640 sq ft) factory in Wichita and head office in Dover, Delaware, founded 9 September 1987; now produces P.180 Avanti forward fuselages; Duncan Aviation formed Duncan Piaggio Aircraft with Piaggio in October 1990; original plan to complete green P.180s for American market not put into effect, but Duncan now manufacturing fuselages in Wichita (workforce approx 60) for Genoa assembly line.

Alenia acquired 31 per cent holding in Piaggio in 1988 (adjusted to 24.5 per cent in 1992, but raised to 30.9 per cent in early 1993 following further restructuring and reorganisation).

P.180 Avanti and P.166-DL3SEM in production; Piaggio also produces engines (see Aero Engines section) and shelters, and manufactures subassemblies of Alenia G222, Panavia Tornado, AMX and Dassault Falcon 2000.

PIAGGIO P.180 AVANTI
TYPE: Twin-turboprop high-speed corporate transport.
PROGRAMME: Launched 1982; Gates Learjet became partner in 1983, but withdrew for economic reasons on 13 January 1986; all existing Learjet P.180 tooling and first three forward fuselages transferred to Piaggio; first flights of two prototypes I-PJAV 23 September 1986 and I-PJAR 14 May 1987; first Italian certification 7 March 1990; first flight full production P.180 30 May 1990; Italian and US certification 2 October 1990; first customer delivery 30 September 1990; French certification March 1993.
CURRENT VERSIONS: Increased gross weight giving higher payload/range decided 1991 (see figures) and early aircraft retrofitted with minor modifications to allow new weights; weights increased again in 1992, as shown.

Piaggio P.180 Avanti (two P&WC PT6A-66 turboprops)

CUSTOMERS: Total 17 delivered by March 1993 to Italian Defence Ministry (three) and corporate and airline customers in USA, Canada and Europe. Six ordered by Air Entreprise of France, for delivery by end 1994; second aircraft ordered by Alpi Eagles of Italy. Production of 10 planned for 1993.
COSTS: 1992 sales price $4.35 million in USA and Europe.
DESIGN FEATURES: Three-surface control with foreplane and T tail to allow unobstructed cabin with maximum headroom

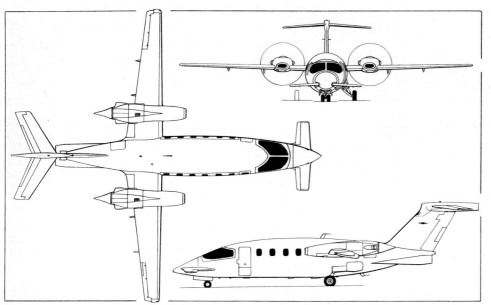

Piaggio P.180 Avanti corporate transport *(Dennis Punnett)*

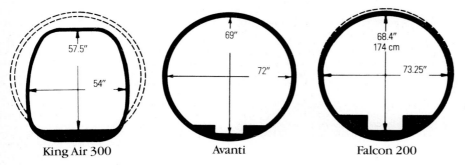

Comparision of cabin cross-sections of P.180 Avanti, centre, King Air 300, left, and Falcon 200

to be placed forward of mid-mounted wing carry-through structure; pusher turboprops aft of cabin and wing reduce cabin noise and propeller vortices on wing; mid-wing avoids root bulges of low-set wings and spar does not pass through cabin; lift from foreplane allows horizontal tail to act as lifting surface and thereby reduce required wing area.

Wing section Piaggio PE 1491 G (mod) at root, PE 1332 G at tip; thickness/chord ratio 13 per cent; dihedral 2°; no sweep; foreplane aerofoil Piaggio PE 1300 GN4 unswept; 5° anhedral on foreplane and tailplane.

FLYING CONTROLS: Variable-incidence swept tailplane; tab in each elevator; electrically actuated trim tab in starboard aileron; trim tab in rudder; two 'delta fin' strakes under tail; electrically actuated outboard and inboard flaps on wing synchronised with flaps in foreplanes; dual control circuits; heated stall warning system.

STRUCTURE: Fuselage precision stretch-formed in large seamless sections and inner structure matched to precise outer contour; CFRP in high stress areas and Kevlar/epoxy elsewhere used for tail unit, engine nacelles, foreplanes, wing outboard flaps, nosecone, tailcone and landing gear doors; wing main spar integral with rear pressure bulkhead and landing gear attachments.

Composites parts manufactured by Sikorsky and Edo; wings and tail section produced by Piaggio in Genoa and forward fuselages by Piaggio Aviation in Wichita; final assembly in Genoa.

LANDING GEAR: Dowty Aerospace hydraulically retractable tricycle type, with single-wheel main units and steerable, twin-wheel nose unit. Main units retract rearward into sides of fuselage; nose unit retracts forward. Dowty hydraulic shock absorbers. Tyre sizes 6.50-10 (main) and 5.00-4 (nose). Multi-disc carbon brakes.

POWER PLANT: Two 1,107 kW (1,485 shp) Pratt & Whitney Canada PT6A-66 turboprops, flat rated at 634 kW (850 shp), each mounted above wing in all-composite nacelle and driving a Hartzell five-blade constant-speed fully feathering reversible-pitch pusher propeller; propellers handed to counter-rotate. Fuel in two fuselage tanks totalling 680 litres (180 US gallons; 149.5 Imp gallons) and two 460 litre (121.5 US gallon; 101 Imp gallon) wing tanks; total fuel capacity 1,600 litres (423 US gallons; 352 Imp gallons). Optional single pressure refuelling point in lower centre-fuselage. Gravity refuelling point in upper part of rudder.

ACCOMMODATION: Crew of one or two on flight deck; certificated for single pilot operation. Seating in main cabin for up to nine passengers, with galley, fully enclosed toilet and coat storage area; choice of nine-passenger high-density or five-seat VIP cabins. Club passenger seats are armchair type, which can be reclined, tracked and swivelled, and locked at any angle. Foldaway tables can be extended between facing club seats. Two-piece wraparound electrically heated windscreen. Rectangular cabin windows, including one emergency exit at front on starboard side. Indirect lighting behind each window ring, plus individual overhead lights. Airstair door at front on port side. Baggage compartment aft of rear pressure bulkhead, with door immediately aft of wing on port side. Entire cabin area pressurised and air-conditioned.

SYSTEMS: AiResearch bleed air ECS, with max pressure differential of 0.62 bar (9.0 lb/sq in). Single hydraulic system driven by electric motor, with handpump for emergency backup, for landing gear, brakes and steering. Electrical system powered by two starter/generators and 25V 38Ah nickel-cadmium battery; 0.62 m³ (22 cu ft) oxygen system. Hot air anti-icing of main wing outer leading-edges; electric anti-icing for foreplane and windscreen; rubber boot for engine intakes, with dynamic particle separator; propeller blades de-iced by engine exhaust; tail de-icing.

AVIONICS: Standard Collins EFIS (three CRTs), Collins VHF com/nav equipment, Collins WXR-840 weather radar, Collins dual transponder TDR-90, Collins single DME and ADF system, Collins radio altimeter, Collins primary and secondary compass systems, Aeronetics dual RMI, Collins APS-65A digital autopilot with yaw damper, J. E. T. dual vertical gyro, and Collins electronic air data system.

DIMENSIONS, EXTERNAL:
Wing span	14.03 m (46 ft 0½ in)
Foreplane span	3.38 m (11 ft 1 in)
Wing chord: at root	1.82 m (5 ft 11¾ in)
at tip	0.62 m (2 ft 0½ in)
Foreplane chord: at root	0.79 m (2 ft 7 in)
at tip	0.55 m (1 ft 9⅔ in)
Wing aspect ratio	12.30
Foreplane aspect ratio	5.05
Length overall	14.41 m (47 ft 3½ in)
Fuselage: Length	12.53 m (41 ft 1¼ in)
Max width	1.95 m (6 ft 4¾ in)
Height overall	3.94 m (12 ft 11 in)
Tailplane span	4.25 m (13 ft 11½ in)
Wheel track	2.84 m (9 ft 4 in)
Wheelbase	5.79 m (19 ft 0 in)
Propeller diameter	2.16 m (7 ft 1 in)
Propeller ground clearance	0.80 m (2 ft 7½ in)
Distance between propeller centres	4.13 m (13 ft 6½ in)
Passenger door (fwd, port): Height	1.35 m (4 ft 5 in)
Width	0.61 m (2 ft 0 in)
Height to sill	0.58 m (1 ft 10¾ in)
Baggage door (rear, port): Height	0.60 m (1 ft 11¾ in)
Width	0.70 m (2 ft 3½ in)
Height to sill	1.38 m (4 ft 6½ in)
Emergency exit (stbd): Height	0.67 m (2 ft 2¼ in)
Width	0.48 m (1 ft 7 in)

DIMENSIONS, INTERNAL:
Passenger cabin: Length	4.45 m (14 ft 7¼ in)
Max width	1.85 m (6 ft 0¾ in)
Max height	1.75 m (5 ft 9 in)
Volume	10.62 m³ (375.0 cu ft)
Baggage compartment: Floor length	1.70 m (2 ft 3½ in)
Max length	2.10 m (6 ft 10¾ in)
Volume	1.25 m³ (44.14 cu ft)

AREAS:
Wings, gross	16.00 m² (172.22 sq ft)
Ailerons (total, incl tab)	0.66 m² (7.10 sq ft)
Trailing-edge flaps (total)	1.60 m² (17.23 sq ft)
Foreplane	2.25 m² (24.22 sq ft)
Foreplane flaps (total)	0.58 m² (6.30 sq ft)
Fin	4.73 m² (50.91 sq ft)
Rudder, incl tab	1.05 m² (11.30 sq ft)
Tailplane	3.83 m² (41.23 sq ft)
Elevators (total, incl tabs)	1.24 m² (13.35 sq ft)

WEIGHTS AND LOADINGS:
Weight empty, equipped	3,402 kg (7,500 lb)

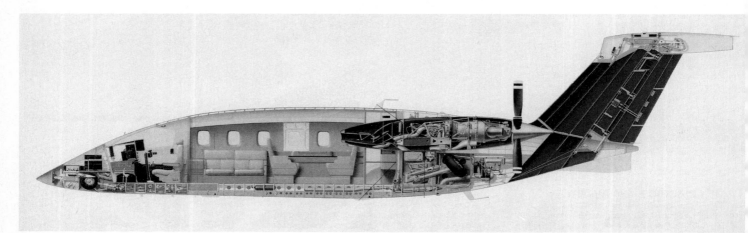

Internal arrangement of Piaggio P.180 Avanti. Black engine nacelle and fin indicate CFRP structure

Operating weight empty, one pilot	3,479 kg (7,670 lb)
Max usable fuel load	1,193 kg (2,630 lb)
Max payload	907 kg (2,000 lb)
Payload with max fuel	567 kg (1,250 lb)
Max T-O weight	5,239 kg (11,550 lb)
Max ramp weight	5,262 kg (11,600 lb)
Max landing weight	4,965 kg (10,945 lb)
Max zero-fuel weight	4,309 kg (9,500 lb)
Max wing loading	327.43 kg/m² (67.07 lb/sq ft)
Max power loading	4.13 kg/kW (6.79 lb/shp)

PERFORMANCE:
Max operating Mach number	0.67
Max operating speed	260 knots (482 km/h; 299 mph) IAS
Max level speed at 8,625 m (28,300 ft)	395 knots (732 km/h; 455 mph)
Stalling speed at max landing weight:	
flaps up	109 knots (202 km/h; 125 mph)
flaps down	94 knots (174 km/h; 108 mph)
Max rate of climb at S/L	899 m (2,950 ft)/min
Rate of climb at S/L, OEI	230 m (755 ft)/min
Service ceiling	12,500 m (41,000 ft)
Service ceiling, OEI	7,620 m (25,000 ft)
T-O to 15 m (50 ft) ISA, S/L at max T-O weight	869 m (2,850 ft)
Landing from 15 m (50 ft) ISA, S/L at max landing weight	872 m (2,860 ft)
Range at 11,890 m (39,000 ft), one pilot and six passengers:	
NBAA IFR reserves	1,400 nm (2,594 km; 1,611 miles)
VFR reserves	1,700 nm (3,150 km; 1,956 miles)

Piaggio P.166-DL3SEM used by the Italian Ministry of Merchant Marine

PIAGGIO P.166-DL3SEM

TYPE: Twin-turboprop multi-role utility aircraft.
PROGRAMME: P.166DL2 described in 1978-79 *Jane's* and earlier versions in previous issues; first flight of current version 3 July 1976; Italian and US certification 1978; production continuing on demand in 1993.
CURRENT VERSIONS: **P.166-DL3SEM**: Current production reconnaissance version able to carry chin-mounted radar and underwing FLIR sensor; training, transport, medical and special patrol/observation versions offered.
CUSTOMERS: Total 31 produced by February 1992; batch of 12 for Italian Ministry of Merchant Marine Capitanerie completed in mid-1990; 10 ordered for Italian Guardia di Finanza customs service for offshore patrol; seven delivered in 1992.
DESIGN FEATURES: Full details in 1990-91 *Jane's*.
POWER PLANT: Two Textron Lycoming LTP 101-700 turboprops, each flat rated at 447.5 kW (600 shp) and driving a Hartzell HC-B3DL/LT10282-9.5 three-blade constant-speed fully feathering metal pusher propeller. Fuel in two 212 litre (56 US gallon; 46.5 Imp gallon) outer-wing main tanks, two 323 litre (85.3 US gallon; 71 Imp gallon) wing-tip tanks, and a 116 litre (30.6 US gallon; 25.5 Imp gallon) fuselage collector tank; total standard internal fuel capacity 1,186 litres (313.3 US gallons; 260.9 Imp gallons). Auxiliary fuel system available optionally, comprising a 232 litre (61.3 US gallon; 51 Imp gallon) fuselage tank, transfer pump and controls; with this installed, total usable fuel capacity is increased to 1,418 litres (374.6 US gallons; 312 Imp gallons). Gravity refuelling points in each main tank and tip tank. Provision for two 177 or 284 litre (46.8 or 75 US gallon; 39 or 62.5 Imp gallon) underwing drop tanks. Air intakes and propeller blades de-iced by engine exhaust.
ACCOMMODATION: Crew of two on raised flight deck, with dual controls. Aft of flight deck, accommodation consists of a passenger cabin, utility compartment and baggage compartment. Access to flight deck via passenger/cargo double door on port side, forward of wing, or via individual crew door on each side of flight deck. External access to baggage compartment via port side door aft of wing. Passenger cabin extends from rear of flight deck to bulkhead at wing main spar; fitting of passenger carrying, cargo or other interiors facilitated by two continuous rails on cabin floor, permitting considerable flexibility in standard or customised interior layouts. Standard seating for eight passengers, with individual lighting, ventilation and oxygen controls. Flight deck can be separated from passenger cabin by a screen. Door in bulkhead at rear of cabin provides access to utility compartment, in which can be fitted a toilet, bar, or mission equipment for various roles. Entire accommodation heated, ventilated and soundproofed. Emergency exit forward of wing on starboard side. Windscreen hot-air demisting standard. Windscreen wipers, washers and methanol spray de-icing optional.

DIMENSIONS, EXTERNAL:
Wing span over tip tanks	14.69 m (48 ft 2½ in)
Length overall	11.88 m (39 ft 0 in)
Height overall	5.00 m (16 ft 5 in)
Cabin door: Height	1.38 m (4 ft 6 in)
Width	1.28 m (4 ft 2 in)

WEIGHTS AND LOADINGS:
Weight empty, equipped	2,688 kg (5,926 lb)
Max fuel	1,036 kg (2,284 lb)
Max payload	1,092 kg (2,407 lb)
Max T-O weight	4,300 kg (9,480 lb)
Max ramp weight	4,320 kg (9,524 lb)
Max zero-fuel weight	3,800 kg (8,377 lb)
Max landing weight	4,085 kg (9,006 lb)
Max wing loading	161.9 kg/m² (33.16 lb/sq ft)
Max power loading	4.81 kg/kW (7.90 lb/shp)

PERFORMANCE (at max T-O weight except where indicated):
Never-exceed speed (V_{NE})	220 knots (407 km/h; 253 mph) CAS
Max level and max cruising speed at 3,050 m (10,000 ft)	215 knots (400 km/h; 248 mph)
Range, VFR:	
with max payload	750 nm (1,390 km; 863 miles)
with max fuel	1,150 nm (2,131 km; 1,324 miles)

SIVEL

SIVEL srl
Via Nazionale 39, I-23030 Chiuro (SO)
Telephone: 39 (342) 483122/483124
Fax: 39 (342) 482742
INFORMATION AND SALES: SIB Trade srl
Telephone: 39 (30) 3700644
Fax: 39 (30) 305692
MANAGING DIRECTOR: Mario Balzarini

SIVEL SD27

TYPE: Two-seat very light trainer and club aircraft.
PROGRAMME: Prototype flying 1992; certification to JAR/VLA expected second half 1993; deliveries to begin immediately afterwards.
CUSTOMERS: None announced by March 1993.
COSTS: First 20 aircraft offered at LIt90 million including electric constant-speed propeller and Bendix/King digital nav/com radio; radio and constant-speed propeller optional thereafter.
DESIGN FEATURES: New design combining efficient aerodynamics and lightweight structure able to perform all manoeuvres required for club training within the limits of JAR VLA certification.
FLYING CONTROLS: Conventional, with fixed tailplane; swept fin and rudder; flaps.
STRUCTURE: Basic all-metal structure.
LANDING GEAR: Mainwheels carried on cantilever spring struts; swivelling nosewheel on tubular spring leg; steering by differential braking.
POWER PLANT: One 59.7 kW (80 hp) Rotax 912A four-cylinder four-stroke, driving fixed-pitch two-blade wooden propeller or optional MT two-blade constant-speed propeller. Fuel in 80 litre (21.1 US gallon; 17.6 Imp gallon) tank in fuselage.

DIMENSIONS, EXTERNAL:
Wing span	10.00 m (32 ft 9½ in)
Wing aspect ratio	8.0

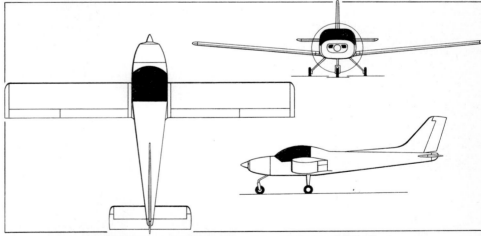

Sivel SD27 very light aircraft (one Rotax 912A) *(Jane's/Mike Keep)*

Length overall	6.57 m (21 ft 6½ in)
Height overall	2.76 m (9 ft 0½ in)
Propeller diameter	1.70 m (5 ft 7 in)

DIMENSIONS, INTERNAL:
Cabin: Max width	1.10 m (3 ft 7¼ in)

AREAS:
Wings, gross	12.50 m² (134.5 sq ft)

WEIGHTS AND LOADINGS:
Empty weight	370 kg (816 lb)
Max T-O weight	620 kg (1,367 lb)
Max wing loading	49.6 kg/m² (10.16 lb/sq ft)
Max power loading	10.39 kg/kW (17.09 lb/hp)

PERFORMANCE:
Never-exceed speed (V_{NE})	132 knots (245 km/h; 152 mph)
Max level speed	105 knots (195 km/h; 121 mph)
Max cruising speed, 75% power, at 2,440 m (8,000 ft)	95 knots (176 km/h; 109 mph)
Stalling speed, power off:	
flaps up	45 knots (83 km/h; 52 mph)
flaps down	41 knots (76 km/h; 48 mph)
Max rate of climb at S/L at 55 knots (102 km/h; 63 mph)	282 m (925 ft)/min
Service ceiling	4,000 m (13,125 ft)
T-O run	150 m (493 ft)
T-O to 15 m (50 ft)	300 m (984 ft)
Landing from 15 m (50 ft)	130 m (427 ft)
Landing run	100 m (328 ft)
Max range	539 nm (1,000 km; 621 miles)
Endurance	6 h

TECNAM
TECNAM srl
1a Traversa Via G. Pascoli, I-80026 Casoria (Naples)
Telephone: 39 (81) 758 3210, 758 8751 and 758 8854
Fax: 39 (81) 758 4528
PRESIDENT: Dott Ing Luigi Pascale Langer
MANAGING DIRECTOR: Dott Giovanni Pascale Langer

Company founded 1986, after Pascale brothers were released from the original Pascale company, Partenavia, which had been placed under control of Alenia in 1981. Tecnam manufactures tailplane and other components of ATR 42/72, fuselage panels for MD-80 (temporarily stopped in early 1993), fuselages of former Partenavia (now Partenavia/AerCosmos) P.68, parts of Agusta A 109 and wing of Meteor Mirach 200 UAV. Workshops qualified to NATO AQA04. Now developing new P92 Echo JAR VLA light aircraft described below.

TECNAM P92 ECHO
TYPE: Very light two-seat trainer/club aircraft.
PROGRAMME: Prototype under construction 1992; first flight expected Summer 1993.
DESIGN FEATURES: Objectives of lightness, simplicity and good aerodynamics dictated braced high wing with metal torsion box and composite leading-edge; aerofoil chosen for good performance at low Reynolds number; untapered with 1° dihedral; underside of fuselage mostly flat for ease of kit assembly; large one-piece windscreen, inward-tapered inboard wing leading-edges, large windows in doors and rear-view window give 360° view.
FLYING CONTROLS: Full dual controls and two throttles; differential Frise ailerons; slab tailplane with anti-balance/trim tab; electrically actuated flaps cover half trailing-edge; panel space for blind-flying instruments and com/nav radio.
STRUCTURE: Each wing easily removed and replaced without adjustment; ailerons of metal structure with Dacron covering; steel tube cabin section contains strong frame carrying wing attachment, landing gear and seats; metal sheet skin panels also provide diagonal bracing; stressed skin rear fuselage; tailplane of metal structure with Dacron covering aft of spar; rudder all-metal with GFRP tip fairing; tailplane halves based on tubular spars which can be unpinned and quickly removed from central tube for transport; engine cowling consists of GFRP lower shell and upper shell partly metal, both removable by undoing four quick latches to reveal whole power plant. Structural testing to proof load completed.

LANDING GEAR: Non-retractable tricycle type; main legs steel alloy self-sprung; hand-powered hydraulic disc brakes operated together from single lever in cockpit; main leg attachments accessible outside fuselage; all tyres 5.00-5; levered suspension nosewheel with rubber-in-compression spring; nosewheel steered from rudder pedals; designed for rough ground operation.
POWER PLANT: One 47.7 kW (64 hp) Rotax 582 two-cylinder water-cooled two-stroke engine with Type B (1:2.58) reduction gear driving GSC P.4160 three-blade wooden propeller; radiator under engine. Fuel contained in tanks in tapered portion of each wingroot; fuel cocks located at tanks so that wing can be removed without draining tanks.
ACCOMMODATION: Door each side; side-by-side seats with four-point harness; baggage space behind seats.
SYSTEMS: 100W 12V alternator and battery.
AVIONICS: Capacity for nav/com radio.
DIMENSIONS, EXTERNAL:

Wing span	9.60 m (31 ft 6 in)
Wing chord, constant	1.40 m (4 ft 7 in)
Wing aspect ratio	6.98
Length overall	6.30 m (20 ft 8 in)
Height overall	1.125 m (3 ft 8¼ in)
Tailplane span	2.90 m (9 ft 6 in)
Wheel track	1.80 m (5 ft 10¾ in)
Wheelbase	1.60 m (5 ft 3 in)
Propeller diameter	1.65 m (5 ft 5 in)

AREAS:

Wings, gross	13.20 m² (142.1 sq ft)
Tailplane	1.972 m² (21.23 sq ft)

WEIGHTS AND LOADINGS:

Basic weight empty	260 kg (573 lb)
Max T-O weight	450 kg (992 lb)
Max wing loading	34.1 kg/m² (6.98 lb/sq ft)
Max power loading	9.37 kg/kW (15.5 lb/hp)

PERFORMANCE (with 14° pitch propeller):

Max level speed at S/L	100 knots (185 km/h; 115 mph)
Cruising speed at 2,350 propeller rpm (65% power)	86 knots (160 km/h; 99 mph)
Cruising speed at 2,100 propeller rpm (55% power)	73 knots (135 km/h; 84 mph)
Stalling speed: flaps up	37 knots (67 km/h; 42 mph)
flaps down	34 knots (63 km/h; 39 mph)
Max rate of climb at S/L at 59 knots (110 km/h; 68 mph)	300 m (985 ft)/min
T-O run	90 m (296 ft)
T-O to 15 m (50 ft)	250 m (821 ft)
Endurance	2 h 30 min

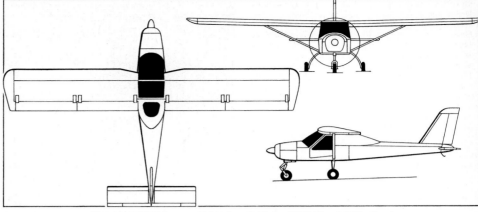

Tecnam P92 Echo very light aircraft (Rotax 582) *(Jane's/Mike Keep)*

TERZI
TERZI AERODINE
Piazzale A. Baiamonti 1, I-20154 Milan
Telephone: 39 (2) 3360 9080
Fax: 39 (2) 3360 7996

Company offers Terzi T30 Katana specialised aerobatic aircraft, in kit form, and Stiletto.

TERZI T-9 STILETTO
TYPE: Two-seat very light sporting aircraft.
PROGRAMME: First flights (I-STIL) December 1990 (with Limbach engine), November 1991 (with Rotax engine); JAR VLA certification being sought early 1993.
DESIGN FEATURES: Wing section Wortmann FX67-K-150/17 (constant chord); dihedral 2° 30'; twist 0°; wings easily detachable at centreline joint; fin and rudder have NACA 64A-008 section. Aircraft designed to be easily dismantled for repair and storage.
FLYING CONTROLS: Slab tailplane with Wortmann FX-71-L-150/20 aerofoil; manually operated three-position flaps.
STRUCTURE: Light alloy single-spar wing with auxiliary spars spigoted to fuselage; riveted skin panels; steel tube forward fuselage with removable GFRP shell; stressed skin metal tailboom.
LANDING GEAR: Non-retractable tricycle type with GFRP cantilever legs mounted on fuselage frame; hydraulic disc brakes controlled from rudder pedals; fully castoring nosewheel.
POWER PLANT: One 59 kW (79 hp) Rotax 912A flat-four driving an MT Elcoprop two-blade wooden propeller. The 912A is certificated to JAR 22. Fuel in 80 litre (21.1 US gallon; 17.6 Imp gallon) tank behind seats.
ACCOMMODATION: Two individually adjustable seats; upward opening canopy; baggage shelf behind seats.
SYSTEMS: 12/14V electrical system with alternator and battery.
AVIONICS: Panel space for VHF nav/com, ADF and transponder.
DIMENSIONS, EXTERNAL:

Wing span	10.26 m (33 ft 8 in)
Wing aspect ratio	8.56
Length overall	6.85 m (22 ft 5¾ in)
Height overall	2.30 m (7 ft 6½ in)
Wheel track	2.01 m (6 ft 7¼ in)

AREAS:

Wings, gross	12.30 m² (132.4 sq ft)

WEIGHTS AND LOADINGS:

Weight empty	380 kg (838 lb)
Max T-O weight	650 kg (1,433 lb)

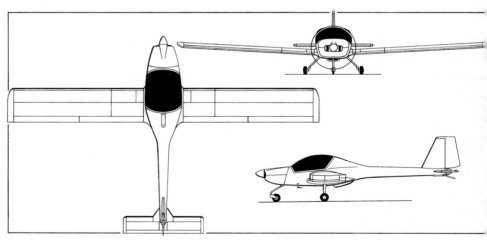

Terzi Aerodine T-9 Stiletto very light aircraft *(Jane's/Mike Keep)*

Terzi Aerodine T-9 Stiletto (one Rotax 912A)

Max wing loading	52.84 kg/m² (10.82 lb/sq ft)
Max power loading	11.09 kg/kW (18.22 lb/hp)

PERFORMANCE:
Max level speed	108 knots (200 km/h; 124 mph)
Cruising speed, 60% power	87 knots (162 km/h; 100 mph)
Stalling speed, flaps down	40 knots (74 km/h; 46 mph)
Max rate of climb at S/L	244 m (800 ft)/min
T-O to 15 m (50 ft)	190 m (624 ft)
Landing from 15 m (50 ft)	175 m (575 ft)
g limits	+3.8/−2.2

TERZI T30 KATANA

TYPE: Single-seat competition aerobatic aircraft.
PROGRAMME: Designed by Pietro Terzi; first flight 16 January 1991. Available as kit or fully built and certificated to FAR 21 in Experimental category; *data below refer to E version*.
DESIGN FEATURES: Metal and composites low-wing monoplane; non-retractable tailwheel landing gear.
STRUCTURE: Aluminium alloy wings (detachable) and tail unit; welded steel tube fuselage with removable composite shells of PVC core with Kevlar and carbon cloth facing.
POWER PLANT: Prototype flew with 224 kW (300 hp) Textron Lycoming O-540, but definitive E version has 298 kW (400 hp) Textron Lycoming IO-720A, with Christen inverted operating kit, driving Mühlbauer four-blade constant-speed propeller. Aerobatic fuel tank mounted in fuselage and two cruising tanks in wing leading-edges: total capacity 195 litres (51.5 US gallons; 42.9 Imp gallons).

DIMENSIONS, EXTERNAL:
Wing span	7.77 m (25 ft 6 in)
Length overall	6.70 m (21 ft 11¾ in)
Height overall	2.57 m (8 ft 5¼ in)

AREAS:
Wings, net	10.60 m² (114.1 sq ft)

Terzi T30 Katana single-seat aerobatic competition aircraft

WEIGHTS AND LOADINGS (Aerobatic):
Max T-O weight	880 kg (1,940 lb)
Max wing loading	83.02 kg/m² (17.00 lb/sq ft)
Max power loading	2.95 kg/kW (4.85 lb/hp)

PERFORMANCE:
Max level speed	205 knots (380 km/h; 236 mph)
Stalling speed	53 knots (98 km/h; 61 mph)
Max rate of climb at S/L	1,380 m (4,527 ft)/min
g limits	±12

JAPAN

CAC
COMMERCIAL AIRPLANE COMPANY

Toranomon Daiichi Building, 2-3, Toranomon 1-chome, Minato-ku, Tokyo 105
Telephone: 81 (3) 3503 3221
Fax: 81 (3) 3508 2418
Telex: 222 2863
PRESIDENT: Masahiko Iwata
GENERAL MANAGER, BUSINESS DEPARTMENT: I. Yamada

CAC took over (from CTDC) Japan's share of Boeing 767 manufacture in July 1982; Kawasaki builds fuselage forward main barrel section and main wing ribs, Mitsubishi the rear main barrel section, Fuji the wing/body fairings; Japan Aircraft Manufacturing Company (Nippi) and ShinMaywa are subcontractors.

FUJI
FUJI HEAVY INDUSTRIES LTD
(Fuji Jukogyo Kabushiki Kaisha)

Subaru Building, 7-2, 1-chome, Nishi-shinjuku Shinjuku-ku, Tokyo 160
Telephone: 81 (3) 3347 2525
Fax: 81 (3) 3347 2588
Telex: 232 2268 FUJI J
PRESIDENT: Isamu Kawai

Aerospace Division
GENERAL MANAGERS:
Yasuyuki Kogure (Managing Director)
Kisaburo Wani (Commercial Business)
Satoshi Idei (Marketing and Sales, Defence Programmes)

Utsunomiya Manufacturing Division
1-11, Yonan 1-chome, Utsunomiya, Tochigi 320
Telephone: 81 (286) 58 1111
DIRECTOR AND GENERAL MANAGER: Tsutou Ono
DEPUTY GENERAL MANAGER: Masaomi Kadoya

Established 15 July 1953 as successor to Nakajima. Utsunomiya Manufacturing Division (aircraft and rolling stock plants) occupies 60 ha (148.3 acre) site, including 188,490 m² (2,028,887 sq ft) floor area; employed 3,000 people in April 1992.

In addition to T-5, Fuji producing Bell UH-1H and AH-1S HueyCobra helicopters; wing main assemblies for JMSDF Lockheed P-3C Orions (see Kawasaki entry); main landing gear doors and some titanium airframe parts for Japanese built McDonnell Douglas F-15Js (see Mitsubishi); and wings, tailplanes and canopies for Kawasaki T-4 (which see). Commercial aircraft components produced are spoilers, inboard and outboard ailerons for Boeing 747; outboard flaps for Boeing 757; wing/body fairings and main landing gear doors for Boeing 767 and 777, plus centre wing box for 777; outboard ailerons for McDonnell Douglas MD-11; rudders and elevators for Fokker 50.

Fuji participating in Japanese SST and hypersonic transport programmes, and has important composites department; Fuji and Kawasaki investigating configuration of NAL fan-lift VTOL airliner (which see).

FUJI T-5

TYPE: Two/four-seat primary trainer; turboprop version (KM-2Kai) of KM-2.
PROGRAMME: Fuji refitted KM-2 with Allison 250-B17D turboprop in 1984; first flown 28 June 1984 as KM-2D; JCAB certification (Aerobatic and Utility categories) gained 14 February 1985; ordered March 1987 as replacement for 31 JMSDF KM-2s; first flight of production KM-2Kai 27 April 1988; deliveries began 30 August 1988.
CURRENT VERSIONS: **T-5:** For JMSDF; Aerobatic and Utility applications; *description applies to this version*.
CUSTOMERS: Total of 29 T-5s ordered by JMSDF by March 1993, of which 15 delivered; three more approved in FY 1993 budget.
DESIGN FEATURES: Turboprop engine; sliding canopy; modernised cockpit; visibility, payload and cockpit volume increased.
Wing section NACA 23016.5 at root, NACA 23012 at tip; no sweep; dihedral 6°; incidence 4° at root, 1° at tip.
FLYING CONTROLS: Mechanically actuated plain ailerons, balanced elevators and rudder; single-slotted wing flaps; aileron anti-servo tabs, port tab controllable for trim; controllable tabs in elevators; rudder anti-servo tab.

First production Fuji KM-2Kai (T-5) for the JMSDF

STRUCTURE: All-metal light alloy.
LANDING GEAR: Electrically retractable tricycle type, with emergency manual control. Oleo-pneumatic shock absorber in each unit. Main units retract inward into wings, nose unit rearward into fuselage. Single Parker wheel and Goodyear tyre on each main unit, size 6.50-8 (6-ply); Goodyear nosewheel and tyre, size 5.00-5 (4-ply). Nose unit steerable ±16°. Parker single-disc hydraulic brakes. Min ground turning radius 7.38 m (24 ft 2½ in).
POWER PLANT: One Allison 250-B17D turboprop, flat rated at 261 kW (350 shp), driving a Hartzell HC-B3TP-7A/T10173-18 three-blade constant-speed fully feathering propeller. Two bladder type fuel tanks in each wing, one of 94.6 litres (25 US gallons; 20.8 Imp gallons) capacity and one of 87 litres (23 US gallons; 19.2 Imp gallons); total capacity 363 litres (96 US gallons; 80 Imp gallons). Gravity refuelling point in top surface of each wing. Oil capacity 9.5 litres (2.5 US gallons; 2.1 Imp gallons).
ACCOMMODATION: Enclosed cabin seating two persons side by side, with dual controls (Aerobatic version), or four persons in pairs in Utility version. Rearward sliding canopy. Accommodation heated and ventilated.
SYSTEMS: Hydraulic system for brakes only. Electrical system includes 30V 150A starter/generator, two 160VA static inverters, and 24Ah battery for engine starting and emergency.
AVIONICS: Standard items include UHF and VHF radio, intercom, ADF, Tacan, SIF and IFR training hood.

DIMENSIONS, EXTERNAL:
Wing span	10.04 m (32 ft 11¼ in)
Wing chord: at root	2.13 m (6 ft 11¾ in)
at tip	1.07 m (3 ft 6¼ in)
Wing aspect ratio	6.11
Length overall	8.44 m (27 ft 8¼ in)
Height overall	2.96 m (9 ft 8½ in)
Elevator span	3.71 m (12 ft 2 in)
Wheel track	2.92 m (9 ft 7 in)
Wheelbase	2.27 m (7 ft 5½ in)
Propeller diameter	2.12 m (6 ft 11½ in)
Propeller ground clearance	0.37 m (1 ft 2½ in)

DIMENSIONS, INTERNAL:
Cabin: Length	2.90 m (9 ft 6¼ in)
Max width	1.27 m (4 ft 2 in)
Max height	1.33 m (4 ft 4½ in)

AREAS:
Wings, gross	16.50 m² (177.6 sq ft)
Ailerons (total, incl tabs)	1.09 m² (11.73 sq ft)
Trailing-edge flaps (total)	1.98 m² (21.31 sq ft)
Fin, incl dorsal fin	1.28 m² (13.78 sq ft)
Rudder, incl tab	0.66 m² (7.10 sq ft)
Tailplane	3.46 m² (37.24 sq ft)
Elevators (total, incl tabs)	1.39 m² (14.96 sq ft)

WEIGHTS AND LOADINGS (A: Aerobatic, U: Utility):
Weight empty: A, U	1,082 kg (2,385 lb)
Max fuel weight: A, U	644 kg (1,420 lb)
Max T-O weight: A	1,585 kg (3,494 lb)
U	1,805 kg (3,979 lb)
Max wing loading: A	96.06 kg/m² (19.67 lb/sq ft)
U	109.39 kg/m² (22.40 lb/sq ft)
Max power loading: A	6.07 kg/kW (9.98 lb/shp)
U	6.92 kg/kW (11.37 lb/shp)

PERFORMANCE (at max Aerobatic T-O weight except where indicated):
Never-exceed speed (VNE)	223 knots (413 km/h; 256 mph) EAS
Max level speed at 2,440 m (8,000 ft)	193 knots (357 km/h; 222 mph)
Econ cruising speed at 2,440 m (8,000 ft)	155 knots (287 km/h; 178 mph)
Stalling speed, flaps and landing gear down, power off	56 knots (104 km/h; 65 mph)
Max rate of climb at S/L	518 m (1,700 ft)/min
Service ceiling	7,620 m (25,000 ft)
T-O run	302 m (990 ft)
T-O to 15 m (50 ft)	430 m (1,410 ft)
Landing from 15 m (50 ft)	515 m (1,690 ft)
Landing run	174 m (570 ft)
Range with max payload (Utility version), MIL-C-5011A reserves	510 nm (945 km; 587 miles)

FUJI (BAe) 125
JASDF designation: U-125

TYPE: Twin-turbofan aircraft for navaid calibration and SAR.
PROGRAMME: BAe 125 Corporate 800 (see UK section) selected under JASDF H-X programme to replace Mitsubishi MU-2J and MU-2E in navaid calibration and SAR roles respectively; Fuji outfitting aircraft to JASDF specification; first U-125 delivered to JASDF 18 December 1992.
CURRENT VERSIONS: **U-125**: For navaid flight check role, to replace MU-2J.
U-125A: Search and rescue version, to replace MU-2E; 360° search radar, FLIR, airdroppable marker flares and rescue equipment.
CUSTOMERS: JASDF (three U-125 and three U-125A).

FUJI-BELL UH-1H and ADVANCED 205B
TYPE: Single-engined general purpose helicopters.
PROGRAMME: Fuji (now only production source) manufactures Bell UH-1H under sublicence from Mitsui and Co, Bell's Japanese licensee; earlier production programme (see 1989-90 and previous *Jane's*) involved 34 Bell 204Bs and 22 Bell 204B-2s. First flight of Fuji built UH-1H, 17 July 1973. Joint Fuji-Bell upgrade of 205 produced Advanced 205B (formerly 205A-1); first flight of prototype (N19AL) in Texas 23 April 1988; demonstrated to US Army that year and in Japan, Far East and Southeast Asia 1989.
CURRENT VERSIONS: **UH-1H**: Standard production model for JGSDF; *described in detail below*.
Advanced 205B: Upgrade of Bell 205, with UH-1N/212-type tapered rotor blades, Textron Lycoming T53-L-703 engine, 212-type transmission rated at 962 kW (1,290 shp); none yet ordered.
CUSTOMERS: Total of 133 UH-1Hs ordered for JGSDF by March 1992. Further 16 approved for FY 1993.
DESIGN FEATURES: Same airframe and dynamic components as Bell UH-1H but has tractor tail rotor and Kawasaki built engine; two-blade teetering main rotor; interchangeable metal blades being replaced by new composite blades; stabilising bar above and at right angles to main rotor blades; underslung feathering axis head; two-blade tail rotor; main rotor rpm 294-324. See Current Versions for Advanced 205B.
FLYING CONTROLS: Duplicated hydraulic fully powered controls with adjustable centring spring on cyclic stick for trimming; small synchronised elevator on rear fuselage, connected to cyclic control to increase allowable CG travel.
STRUCTURE: All-metal, except for new main rotor blades (see Design Features); honeycomb tail rotor blades.
LANDING GEAR: Tubular skid type. Lock-on ground handling wheels and inflatable nylon float bags available.
POWER PLANT: One 1,044 kW (1,400 shp) Kawasaki built Textron Lycoming T53-K-13B turboshaft, mounted aft of transmission on top of fuselage and enclosed in cowlings.

Artist's impression of the BAe 125-800 in JASDF U-125A configuration

Transmission rating 820 kW (1,100 shp). Five interconnected rubber fuel cells, total capacity 844 litres (223 US gallons; 186 Imp gallons), of which 799 litres (211 US gallons; 176 Imp gallons) are usable. Overload fuel capacity of 1,935 litres (511 US gallons; 425 Imp gallons) usable, obtained by installation of kit comprising two 568 litre (150 US gallon; 125 Imp gallon) internal auxiliary fuel tanks interconnected with basic fuel system.
ACCOMMODATION: Pilot and 11-14 troops, or six stretchers and medical attendant, or 1,759 kg (3,880 lb) of freight. Crew doors open forward and are jettisonable. Two doors each side of cargo compartment; front door hinged to open forward and is removable, rear door slides aft. Forced air ventilation system.
AVIONICS: FM, UHF, VHF radios, IFF transponder, Gyromatic compass system, DF set, VOR receiver and intercom standard.
EQUIPMENT: Standard equipment includes bleed air heater and defroster, comprehensive range of engine and flight instruments, power plant fire detection system, 30V 300A DC starter/generator, navigation, landing and anti-collision lights, controllable searchlight, hydraulically boosted controls. Optional equipment includes external cargo hook, auxiliary fuel tanks, rescue hoist, 150,000 BTU muff heater.

DIMENSIONS, EXTERNAL:
Main rotor diameter	14.63 m (48 ft 0 in)
Tail rotor diameter	2.59 m (8 ft 6 in)
Main rotor blade chord	0.53 m (1 ft 9 in)
Tail rotor blade chord	2.56 m (8 ft 4¾ in)
Length:	
overall (main rotor fore and aft)	17.62 m (57 ft 9⅝ in)
fuselage	12.77 m (41 ft 10¾ in)
Height:	
overall, tail rotor turning (excl fin tip antenna)	4.41 m (14 ft 5½ in)
to top of main rotor head	3.60 m (11 ft 9¾ in)

Fuji-Bell UH-1H in JGSDF camouflage

Tailplane span	2.84 m (9 ft 4 in)
Width over skids	2.91 m (9 ft 6½ in)

DIMENSIONS, INTERNAL:
Cabin: Max width	2.34 m (7 ft 8 in)
Max height	1.25 m (4 ft 1¼ in)
Volume (excl flight deck)	approx 6.23 m³ (220 cu ft)

AREAS:
Main rotor disc	168.11 m² (1,809.56 sq ft)
Tail rotor disc	5.27 m² (56.7 sq ft)

WEIGHTS AND LOADINGS:
Weight empty, equipped	2,390 kg (5,270 lb)
Max T-O and landing weight	4,309 kg (9,500 lb)
Max disc loading	25.63 kg/m² (5.25 lb/sq ft)
Max power loading	5.26 kg/kW (8.64 lb/shp)

PERFORMANCE (at max T-O weight):
Max level and max cruising speed	110 knots (204 km/h; 127 mph)
Max rate of climb at S/L	488 m (1,600 ft)/min
Service ceiling	3,840 m (12,600 ft)
Hovering ceiling: IGE	4,145 m (13,600 ft)
OGE	335 m (1,100 ft)
Range at S/L	252 nm (467 km; 290 miles)

FUJI-BELL AH-1S

TYPE: Fuji-built Bell anti-armour helicopter.
PROGRAMME: Fuji selected FY 1982 as prime contractor for licence manufacturing of AH-1S for JGSDF; Kawasaki building T53 engine; first flight 2 July 1984.
CURRENT VERSIONS: **Fuji-Bell AH-1S**: Based on US Army AH-1F.
CUSTOMERS: JGSDF has ordered 83 of 88 planned, following operational evaluation of two Bell AH-1Es (bought 1977-78) later upgraded to F standard; 70 delivered by March 1993; to equip five anti-tank squadrons, with surplus for attrition and training; first four squadrons based at Obihiro on Hokkaido, Hachinohe, Metabaru and Kisarazu; also serve with training school at Akeno.
DESIGN FEATURES: Corresponds to US Army AH-1F; cockpits to be adapted for NVGs and integrated nav/com control and display panel.

Fuji-built AH-1S HueyCobra for the JGSDF

ISHIDA
ISHIDA AEROSPACE RESEARCH INC
Alliance Airport, 2301 Horizon Drive, Fort Worth, Texas 76177, USA
Telephone: 1 (817) 837 8000
Fax: 1 (817) 837 8020
PUBLIC RELATIONS: Ellen Forbes
PARENT COMPANY:
The Ishida Corporation, 41 Myoken-cho, Showa-ku, Nagoya 466
  Telephone: 81 (52) 833 8167
  Fax: 81 (52) 834 1546
  CHAIRMAN AND CEO: Taiichi Ishida
MANAGEMENT COMPANY:
TW-68 Industries Inc, 88 Howard Street, Room 1904, San Francisco, California 94105, USA
  Telephone: 1 (415) 495 3492
  Fax: 1 (415) 495 3798
  EXECUTIVE VICE-PRESIDENT: Dr J. David Kocurek

Ishida Foundation, sponsor of TW-68 programme, founded Ishida Aerospace Research Inc in Texas 1990 to carry through engineering design and prototype fabrication to beginning of flight testing. DMAV (Dual Mode Air Vehicle) is its consultant.

Construction of 2,694 m² (29,000 sq ft) development centre at Alliance Airport began 11 June 1990; occupied September 1991; IAR workforce approx 50 by September 1992.

Ishida Corporation funded and directed development of TW-68 in USA, and markets Swearingen SJ30 in Asia and Australia; *but see also Addenda.*

ISHIDA TW-68

TYPE: Four-engined tilt-wing light transport.
PROGRAMME: Under development by Ishida Aerospace Research, with DMAV (see US section) as consultant; signed MoU with Pratt & Whitney Canada 3 October 1990 for supply of PT6A engines; tiltable version of PT6A-67 (820 to over 1,119 kW; 1,100 to over 1,500 shp) later selected; Lucas Western Inc contracted 1991 for preliminary design and eventual development of combiner gearbox; revised engine selection (1,500 shp class PT6B-67R announced early 1993); first flight and certification anticipated March 1996 and 1999 respectively.
CURRENT VERSIONS: **Initial passenger version**: Pressurised, with cabin differential over 0.34 bar (5 lb/sq in) to allow cruising at 7,620 m (25,000 ft). Nine to 14 passengers or freight. *Details refer to this version.*
  Stretched version: Stretched 19-passenger pressurised derivative anticipated.
  Utility: Unpressurised derivative for offshore oil support, search and rescue and other roles.
COSTS: Estimated aircraft cost $6 million-plus (1992).
DESIGN FEATURES: Tilting wing for horizontal and vertical flight modes; conventional aeroplane fuselage; conventional T tailplane, with elevator; boat-tail tailcone contains

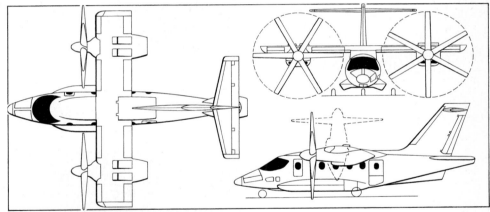
Ishida TW-68 in latest configuration, with modified tail pitch attitude control system (*Jane's/Mike Keep*)

Artist's impression of the TW-68 in city-centre passenger transport operation

The Lucas Western combiner gearbox for the TW-68

horizontal fan for pitch attitude control in hover; four turboshaft engines in two pairs which, together with combiner gearboxes and cross shafting, should achieve full Category A performance OGE with OEI; engines flat rated; combiner gearboxes rated at 1,493 kW (2,000 shp) to each propeller; six-blade propellers. Tilt-wing system claimed to offer lower weight and complexity than other V/STOL configurations; streamlines wing with propeller downwash during hover rather than leaving it athwart downwash as in tilt-rotor, which normally reduces rotor lift by some 10 per cent; downwash of propellers, which are smaller than rotors of corresponding tilt-rotor, is more intense but should still leave calm area between two columns of air to facilitate picking up loads or survivors.

FLYING CONTROLS: Plain hydraulic boosted controls; autostabilisation for hover (no fly-by-wire); mechanical mixer box to alter function of control surfaces and devices progressively during transition; in hover, double-slotted flaperons operate differentially for yaw control, a hydraulically driven fan in boat-tail controls pitch attitude, and lateral control is by varying propeller power output; in forward flight, flaperons act as ailerons, elevator on tailplane replaces fan (which is stopped), and normal rudder becomes effective; no cyclic pitch control.

POWER PLANT: Four P&WC PT6B-67R turboshafts, two in each nacelle, adapted for tilting operation and totalling between about 2,985 kW (4,000 shp) and 4,478 kW (6,000 shp), each pair driving a Dowty six-blade slow-turning propeller (800 rpm for VTOL, 680 rpm in cruise). See under Weights and Loadings for fuel details.

ACCOMMODATION: Flight crew of one or two. Cabin seating for nine (corporate), 14 (commuter) or a maximum of 16 passengers, or equivalent freight or other payload.

DIMENSIONS, EXTERNAL:
Wing span	12.44 m (40 ft 9½ in)
Total span (incl propellers)	13.53 m (44 ft 4¾ in)
Wing chord, constant	2.16 m (7 ft 1 in)
Wing aspect ratio	5.37
Length overall	15.42 m (50 ft 7¼ in)
Fuselage: Max width	1.78 m (5 ft 10 in)
Height overall	5.55 m (18 ft 2½ in)
Propeller diameter	5.46 m (17 ft 10¾ in)

DIMENSIONS, INTERNAL:
Cabin: Length	5.59 m (18 ft 4 in)
Max width	1.85 m (6 ft 1 in)
Max height	1.76 m (5 ft 9½ in)
Volume	14.55 m³ (514 cu ft)
Baggage volume	3.57 m³ (126 cu ft)

AREAS:
Wings, gross	28.80 m² (310.0 sq ft)

WEIGHTS AND LOADINGS:
Weight empty, equipped	5,618 kg (12,386 lb)
Max fuel weight	2,097 kg (4,624 lb)
Payload with max fuel	903 kg (1,990 lb)
Max T-O weight	8,618 kg (19,000 lb)
Max wing loading	299.2 kg/m² (61.29 lb/sq ft)

PERFORMANCE (estimated):
Max level speed	350 knots (648 km/h; 403 mph)
Max cruising speed at 4,875 m (16,000 ft)	310 knots (574 km/h; 357 mph)
Best range speed	250 knots (463 km/h; 288 mph)
Average rate of climb to 7,620 m (25,000 ft)	847m (2,778 ft)/min
Max certificated ceiling	8,840 m (29,000 ft)
Hovering ceiling IGE:	
normal	1,495 m (4,900 ft)
OEI	1,035 m (3,400 ft)
Max range:	
14 passengers	760 nm (1,408 km; 875 miles)
9 passengers	920 nm (1,705 km; 1,059 miles)

JADC
JAPAN AIRCRAFT DEVELOPMENT CORPORATION
Toranomon Daiichi Building, 2-3, Toranomon 1-chome, Minato-ku, Tokyo 105
Telephone: 81 (3) 3503 3225
Fax: 81 (3) 3504 0368
Telex: 222 2863 JADC J
CHAIRMAN: Hiroshi Ohba

JADC is co-ordinating body for Japan's 20 per cent share in Boeing 777; participation includes design, testing, manufacturing and sales financing; 230 staff assigned to Boeing by end of 1991 for design and other activities.

JADC also co-ordinates work on YSX airliner, for which Project Office opened July 1989; feasibility study conducted in FYs 1989 and 1990 followed by change of status in FY 1991 from government commissioned survey project to government supported programme involving whole Japanese aircraft industry. From 15 August 1991, JADC also commissioned to undertake supersonic transport development survey programme work previously handled by SJAC.

YSX
TYPE: Proposed regional airliner.
PROGRAMME: Entered pre-development phase 1991 after two-year feasibility study; configuration design, programme planning and finding international partners expected to take several years, leading to in-service date of late 1990s.
COSTS: Estimated $1 billion for development; Japanese government contributed $7 million for development early 1993.
DESIGN FEATURES: To compete with turboprop rivals, but powered by twin turbofans; provisionally rear-engined, but underwing configuration possible; low wing with 15° leading-edge sweep and aspect ratio of 10; circular-section fuselage with 3.15 m (10 ft 4 in) max internal width.
POWER PLANT: Two 53.38 kN (12,000 lb st) class turbofans.

Model of possible future Japanese supersonic transport *(Mark Lambert)*

ACCOMMODATION: Five-abreast seating for 75 passengers.
WEIGHTS AND LOADINGS:
Max T-O weight	approx 30,750 kg (67,790 lb)

PERFORMANCE (estimated):
Max cruising speed	Mach 0.76
Range with max fuel	more than 1,000 nm (1,850 km; 1,150 miles)

SST-X001
Model of possible Japanese supersonic transport design displayed February 1992 at Asian Aerospace exhibition, Singapore; Japanese studies (for eventual international collaboration) favour Mach 2.0-2.5 design of approx 400,000 kg (881,850 lb) gross weight, with 90 m (295 ft) long fuselage, and max range of 6,500 nm (12,045 km; 7,485 miles). Service entry target date 2005-2010.

KAWASAKI
KAWASAKI JUKOGYO KABUSHIKI KAISHA (Kawasaki Heavy Industries Ltd)
1-18 Nakamachi-Dori, 2-chome, Chuo-ku, Kobe
TOKYO AND AEROSPACE GROUP OFFICE: World Trade Center Building, 4-1, Hamamatsu-cho 2-chome, Minato-ku, Tokyo 105
Telephone: 81 (3) 3435 2111
Fax: 81 (3) 3436 3037
Telex: 242-4371 KAWAJU J
PRESIDENT: Hiroshi Ohba
Aerospace Group
MANAGING DIRECTOR AND GROUP SENIOR GENERAL MANAGER: Setsuo Futatsugi
WORKS: Gifu, Nagoya 1 and 2, Akashi, Seishin and Harima

Kawasaki Aircraft Co built many US aircraft under licence from 1955; amalgamated with Kawasaki Dockyard Co and Kawasaki Rolling Stock Mfg Co to form Kawasaki Heavy Industries Ltd 1 April 1969; Aerospace Group employs some 4,700 people; Kawasaki has 25 per cent holding in Nippi (which see).

Kawasaki is currently prime contractor on T-4 programme and for OH-X new small observation helicopter; co-developer and co-producer, with Eurocopter, of BK 117 helicopter (see International section); manufactures McDonnell Douglas MD 500 helicopters under licence (315 delivered by March 1992); is prime contractor for Japanese licence production of Lockheed P-3C/Update II and III Orions for JMSDF and Boeing CH-47 Chinooks for Japanese armed forces. Subcontract work includes rear fuselages, wings and tail units for Mitsubishi-built McDonnell Douglas F-15 Eagles; and forward and centre-fuselage panels and wing ribs for Boeing 767. Nominated as prime contractor for maintenance and support of JASDF Grumman E-2C Hawkeyes and Lockheed C-130 Hercules. Undertaking feasibility study for JDA for future C-1/C-130 transport replacement.

Kawasaki also extensively involved in satellites and launch vehicles, and HOPE orbiting spaceplane to operate with NASA Space Station; is member of International Aero Engines consortium and produces Textron Lycoming T53 and T55 engines under licence (see Aero Engines section); overhauls engines; and builds hangars, docks, passenger bridges and similar airport equipment.

KAWASAKI (LOCKHEED) P-3C ORION

TYPE: Land-based maritime patrol and ASW aircraft.

PROGRAMME: Kawasaki is prime contractor for JMSDF Orions; first of four P-3Cs assembled by Kawasaki from US-supplied knocked-down components flown 17 March 1982 and delivered 26 May to Fleet Squadron 51 at Atsugi Air Base; production continues.

CURRENT VERSIONS: **P-3C/Update II and III**: Versions built under licence in Japan; Japan Defence Agency plans to modernise current Update II configuration to Update III (see Lockheed) from 1996.

EP-3: Two Orions, ordered FYs 1987 and 1988, equipped for electronic surveillance; first flight (9171) October 1990; delivered March and November 1991; third ordered in FY 1992; fourth approved in FY 1993 and fifth planned in FY 1994 or 1995; NEC and Mitsubishi Electric low and high frequency detector systems.

UP-3C: One flying testbed for JMSDF; ordered in FY 1991.

UP-3D: Two ECM trainers for JMSDF; to be ordered during FYs 1994-95.

Proposed Variants: Ocean surveillance, military transport and systems testbed variants proposed to Japan Defence Agency.

CUSTOMERS: 109 to be purchased by JMSDF, of which 99 ordered by FY 1991, one more P-3C and one EP-3 approved for FY 1993; first three (US built) P-3Cs handed over to JMSDF in April 1981; 88 Kawasaki built P-3Cs delivered by 31 March 1993; two squadrons of 10 aircraft each based at Atsugi, two at Hachinohe, two at Kanoya, one at Shimofusa, one at Iwakuni and one at Naha.

STRUCTURE: Kawasaki builds centre-fuselages and is responsible for final assembly and flight testing; Fuji, Mitsubishi, Nippi and ShinMaywa participate in production of airframe; IHI manufactures Allison T56-IHI-14 engines.

KAWASAKI T-4

TYPE: Tandem two-seat intermediate jet trainer and liaison aircraft, replacing Lockheed T-33A and Fuji T-1A/B.

PROGRAMME: Kawasaki named prime contractor 4 September 1981 by Japan Defence Agency; T-4 based on Kawasaki KA-851 design, by engineering team led by Kohki Isozaki; basic design studies completed October 1982; funding approved in FYs 1983 and 1984 for four flying prototypes; prototype construction began April 1984; first flight of first XT-4 (56-5601) 29 July 1985; all four prototypes delivered between December 1985 and July 1986, preceded by static and fatigue test aircraft; production began FY 1986; first flight of production T-4, 28 June 1988; production deliveries started 20 September 1988; Fuji and Mitsubishi each have 30 per cent share in production programme.

CURRENT VERSIONS: **T-4**: Standard version, *as detailed*.

Enhanced version: Proposed to Japan Defence Agency as possible replacement for Mitsubishi T-2.

CUSTOMERS: JASDF requires about 200 for pilot training, liaison and other duties; total of 135 ordered by 31 March 1993; 95 then delivered; used by Flying Training Squadrons of 1st Air Wing at Hamamatsu, near Tokyo, and of eight other Wings in Japan. Further nine T-4s approved in FY 1993 budget.

DESIGN FEATURES: High subsonic manoeuvrability; ability to carry external loads under wings and fuselage; anhedral mid-mounted wings, with extended chord outer panels giving dog-tooth leading-edges; tandem stepped cockpits with dual controls; baggage compartment in centre-fuselage for liaison role.

Supercritical wing section; thickness/chord ratio 10.3 per cent at root, 7.3 per cent at tip; anhedral 7° from roots; incidence 0°; sweepback at quarter-chord 27° 30′.

FLYING CONTROLS: Hydraulically actuated controls; plain hinged ailerons with Teijin powered actuators; all-moving tailplane and rudder use Mitsubishi servo actuators; double-slotted trailing-edge flaps; no tabs; airbrake on each side of rear fuselage.

STRUCTURE: Aluminium alloy wings, with slow crack growth characteristics; CFRP ailerons, fin, rudder and airbrakes; aluminium alloy flaps with AFRP trailing-edges; aluminium alloy tailplane with CFRP trailing-edge; aluminium alloy fuselage with slow crack growth characteristics, and minimum use of titanium in critical areas. Kawasaki builds forward fuselage and is responsible for final assembly and flight testing; Fuji builds rear fuselage, wings and tail unit; Mitsubishi builds centre-fuselage and engine air intakes.

LANDING GEAR: Hydraulically retractable tricycle type, with Sumitomo oleo-pneumatic shock absorber in each unit. Single-wheel main units retract forward and inward; steerable nosewheel retracts forward. Bendix (Kayaba) mainwheels, tyre size 22 × 5.5-13.8, pressure 19.31 bars (280 lb/sq in); Bendix (Kayaba) nosewheel, tyre size 18 × 4.4-11.6, pressure 12.76 bars (185 lb/sq in). Bendix (Kayaba) carbon brakes and Hydro-Aire (Sumitomo) anti-skid units on mainwheels. Min ground turning radius 9.45 m (31 ft 0 in).

POWER PLANT: Two 16.28 kN (3,660 lb st) Ishikawajima-Harima F3-IHI-30 turbofans, mounted side by side in centre-fuselage. Internal fuel in two 401.25 litre (106 US gallon; 88.3 Imp gallon) wing tanks and two Japanese-built Goodyear rubber bag tanks in fuselage, one of 776 litres (205 US gallons; 170.7 Imp gallons) and one of 662.5 litres (175 US gallons; 145.7 Imp gallons). Total internal capacity 2,241 litres (592 US gallons; 493 Imp gallons). Single pressure refuelling point in outer wall of port engine air intake. Provision to carry one 454 litre (120 US gallon; 100 Imp gallon) ShinMaywa drop tank on each underwing pylon. Oil capacity 5 litres (1.3 US gallons; 1.1 Imp gallons).

ACCOMMODATION: Crew of two in tandem in pressurised and air-conditioned cockpit with wraparound windscreen and one-piece sideways (to starboard) opening canopy. Dual controls standard; rear (instructor's) seat elevated 27 cm (10.6 in). UPC (Stencel) SHIS-3J ejection seats and Teledyne McCormick Selph canopy severance system, licence built by Daicel Chemical Industries. Baggage compartment in centre of fuselage, with external access via door on port side.

SYSTEMS: Shimadzu bootstrap type air-conditioning and pressurisation system (max differential 0.28 bar; 4.0 lb/sq in). Two independent hydraulic systems (one each for flight controls and utilities), each operating at 207 bars (3,000 lb/sq in) and each with separate air/fluid reservoir pressurised at 3.45 bars (50 lb/sq in). Flow rate of each hydraulic system 45 litres (12 US gallons; 10 Imp gallons)/min. No pneumatic system. Electrical system powered by two 9kW Shinko engine driven starter/generators. Tokyo Aircraft Instruments onboard oxygen generating system.

AVIONICS: Mitsubishi Electric J/ARC-53 UHF com, Nagano JRC J/AIC-103 intercom, Nippon Electric J/ARN-66 Tacan, Toyo Communication (Teledyne Electronics) J/APX-106 SIF, Japan Aviation Electronics (Honeywell) J/ASN-3 AHRS, Tokyo Keiki (Honeywell) J/ASK-1 air data computer, Shimadzu (Kaiser) J/AVQ-1 HUD, and Tokyo Aircraft Instrument J/ASH-3 VGH recorder.

EQUIPMENT: Two Nippi pylons under each wing for carriage of drop tanks (see Power Plant); one Nippi pylon under fuselage, on which can be carried target towing equipment, ECM/chaff dispenser or air sampling pod.

ARMAMENT: No built-in armament.

Lockheed EP-3 Orion electronic surveillance aircraft for the JMSDF, modified by Kawasaki

Kawasaki built P-3C of the JMSDF's Fleet Air Wing 5, based at Naha *(Shojiro Ootake)*

JAPAN: AIRCRAFT—KAWASAKI

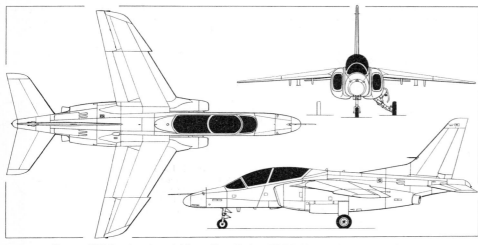

Kawasaki T-4 trainer (two Ishikawajima-Harima F3-IHI-30 turbofans) *(Dennis Punnett)*

DIMENSIONS, EXTERNAL:
Wing span	9.94 m (32 ft 7½ in)
Wing chord: at root	3.11 m (10 ft 2½ in)
at tip	1.12 m (3 ft 8 in)
Wing aspect ratio	4.7
Length: overall	13.00 m (42 ft 8 in)
fuselage	11.96 m (39 ft 3 in)
Height overall	4.60 m (15 ft 1¼ in)
Tailplane span	4.40 m (14 ft 5¼ in)
Wheel track	3.20 m (10 ft 6 in)
Wheelbase	5.10 m (16 ft 9 in)

DIMENSIONS, INTERNAL:
Cockpit: Length	3.20 m (10 ft 6 in)
Max width	0.69 m (2 ft 3 in)
Max height	1.40 m (4 ft 7¼ in)

AREAS:
Wings, gross	21.00 m² (226.05 sq ft)
Ailerons (total)	1.51 m² (16.25 sq ft)
Trailing-edge flaps (total)	2.93 m² (31.54 sq ft)
Fin	3.78 m² (40.69 sq ft)
Rudder	0.91 m² (9.80 sq ft)
Tailplane	6.04 m² (65.02 sq ft)

WEIGHTS AND LOADINGS:
Weight empty	3,790 kg (8,356 lb)
T-O weight, clean	5,690 kg (12,544 lb)
Max design T-O weight	7,500 kg (16,535 lb)
Max wing loading	357.1 kg/m² (73.15 lb/sq ft)
Max power loading	230.34 kg/kN (2.26 lb/lb st)

PERFORMANCE (in clean configuration. A: at weight of 4,850 kg; 10,692 lb with 50% fuel, B: at T-O weight of 5,690 kg; 12,544 lb):
Max level speed (A): at height	Mach 0.9
at S/L	560 knots (1,038 km/h; 645 mph)
Cruising speed: B	Mach 0.75
Stalling speed: A	90 knots (167 km/h; 104 mph)
Max rate of climb at S/L: B	3,050 m (10,000 ft)/min
Service ceiling: B	15,240 m (50,000 ft)
T-O run, 35°C: B	610 m (2,000 ft)
Landing run: A	640 m (2,100 ft)
Range (B) at Mach 0.75 cruising speed:	
internal fuel only	700 nm (1,297 km; 806 miles)
with two 120 US gallon drop tanks	900 nm (1,668 km; 1,036 miles)
g limits	+7.33/−3

Kawasaki T-4 twin-turbofan intermediate trainer of the Japan Air Self-Defence Force *(Shojiro Ootake)*

KAWASAKI (BOEING) CH-47 CHINOOK
JASDF/JGSDF designation: CH-47J

TYPE: Twin-engined tandem-rotor helicopter.
PROGRAMME: FY 1984 defence budget approved purchase of three Boeing CH-47s, two for JGSDF and one for JASDF; first two built in USA and delivered Spring 1986, third delivered CKD for assembly in Japan; Kawasaki granted manufacturing licence for Japanese services' Chinooks; first CH-47Js delivered late 1986.
CURRENT VERSIONS: **CH-47J**: Generally similar to US CH-47D.
CUSTOMERS: JGSDF has eventual requirement for 42 and JASDF for 16; 34 and 15 ordered respectively by 31 March 1993, of which 23 and 12 delivered. Further one for JASDF and two for JGSDF approved in FY 1993 budget.

Kawasaki-built Boeing CH-47J Chinook in JASDF camouflage *(Shojiro Ootake)*

KAWASAKI (MCDONNELL DOUGLAS) MD 500D
JGSDF/JMSDF designation: OH-6D

TYPE: Light helicopter.
PROGRAMME: First flight of initial Kawasaki licence-built Hughes (now McDonnell Douglas) Helicopters 369D (500D) 2 December 1977; JCAB Normal category certification awarded 20 April 1978. Replacement now being sought under provisional designation OH-X (which see).
CUSTOMERS: Nine delivered for civil operation by March 1993; JGSDF ordered 140, of which 126 delivered by end of March 1992; nine delivered to JMSDF by 31 March 1993 for training role. FY 1993 budget approved 13 for JGSDF.

KAWASAKI OH-X

TYPE: Scout/observation light helicopter to replace OH-6D.
PROGRAMME: Japan Defence Agency (JDA) awarded ¥2.7 billion ($22.5 million) in FY 1992 to cover basic design phase; RFPs issued by Technical Research & Development Institute (TRDI) Spring 1992; Kawasaki selected as prime contractor (60 per cent of programme) September 1992, with Fuji and Mitsubishi (20 per cent each) as partners; Observation Helicopter Engineering Team (OHET) formed by these three companies began preliminary design phase 1 October 1992; programme to include seven prototypes (four flying, three for ground test), with first flight planned for 1996 and first deliveries 1999.
CUSTOMERS: Japan Ground Self-Defence Force requirement for 150-200.
COSTS: Total development programme estimated at ¥80 billion ($666 million) (1992); Kawasaki contract (to December 1993) valued at ¥2,513 million ($21 million).

Kawasaki (McDonnell Douglas) OH-6D light helicopter of the JGSDF *(Shojiro Ootake)*

DESIGN FEATURES: Kawasaki bearingless and ballistic-tolerant rotors and transmission system; tandem seating.
STRUCTURE: Rotor blades and most of airframe manufactured from composites; centre-fuselage by Mitsubishi, tail unit/canopy/stub-wings by Fuji, rest by Kawasaki.
POWER PLANT: Twin 597 kW (800 shp) class XTSI-1 turboshafts of Japanese design, under development since 1991 by Mitsubishi (prime contractor) with Kawasaki and IHI as subcontractors. Prototype engine scheduled for completion by March 1993. Possibility of off-the-shelf alternative engines not ruled out.
ACCOMMODATION: Crew of two in tandem.
AVIONICS: Integrated cockpits with LCDs, TV targeting sight, FLIR, laser rangefinder.
ARMAMENT: Two lightweight, short-range air-to-air missiles on stub-wings.

MITSUBISHI

MITSUBISHI JUKOGYO KABUSHIKI KAISHA (Mitsubishi Heavy Industries Ltd)
5-1, Marunouchi 2-chome, Chiyoda-ku, Tokyo 100
Telephone: 81 (3) 3212 3111
Fax: 81 (3) 3212 9852
Telex: J22443
NAGOYA AIRCRAFT WORKS: 10, Oye-cho, Minato-ku, Nagoya 455
PRESIDENT: Kentaro Aikawa
EXECUTIVE VICE-PRESIDENTS:
 Nobuichi Tsuruoka
 Yu Tashiro
 Takaaki Yamada (Managing Director, and General Manager of Aircraft and Special Vehicle Headquarters)
GENERAL MANAGER, AIRCRAFT DEPARTMENT:
 Ichiro Ogawa

Present Komaki South plant built 1952; Nagoya facility currently accommodates Aerospace Systems Works and Guidance & Propulsion Systems Works, with combined floor area of 552,463 m² (5,946,666 sq ft).

Developed MU-2, MU-300, T-2 supersonic trainer and close support F-1 for JASDF. Built 167 HSS-2/2A/2B and 18 S-61A helicopters under Sikorsky licence (last aircraft delivered 2 March 1990). Is currently prime contractor for Japanese F-4E upgrades, F-15J, FS-X and SH/UH-60J helicopters. Subcontract work includes forward and rear fuselages for Kawasaki JMSDF Orions; Boeing 767 rear passenger cabin sections (partly subcontracted by Mitsubishi to ShinMaywa); and McDonnell Douglas MD-11 tailcones. Participating in JDA feasibility study for new transport aircraft to replace C-1 and C-130.

Aero engine activities detailed in Engine section; also produces rocket engines and participates in H-I and H-II launchers and Japanese Experimental Module for US Space Station and for HOPE orbiting spaceplane.

Japan Defence Agency has authorised service introduction of AAM-3 missile developed with NEC seeker and proximity fuse and Komatsu warhead; will replace AIM-9s on F-15Js and F-4EJs.

MITSUBISHI F-4EJKai and RF-4EJ

TYPE: Tactical fighter and interceptor (F-4EJKai) and reconnaissance-fighter (RF-4EJ).
PROGRAMME: Mitsubishi prime contractor in major equipment and weapon system update for JASDF F-4s; prototype F-4EJKai first flight 17 July 1984, delivered 13 December 1984. For details see *Jane's Civil and Military Aircraft Upgrades*.
CURRENT VERSIONS: **F-4EJKai**: Avionics and weapons upgrade of F-4EJ; some expected to be allocated to close air support instead of interceptor duties.
 RF-4EJ: 17 F-4EJs being converted to this model, joining 14 existing aircraft. Upgraded avionics. Mitsubishi in charge of programme.
CUSTOMERS: Current plans to convert 100 of JASDF's 125 remaining F-4EJs to F-4EJKai configuration and 17 to RF-4EJ; initial batch of 80 F-4EJKai conversions under way, of which 59 completed by January 1993. Four F-4EJKai and five RF-4EJ approved in FY 1993 budget.
DESIGN FEATURES: F-4EJKai has Westinghouse J/APG-66J radar; other advanced avionics include Litton J/ASN-4 INS, HUD and J/APR-4Kai RWR; lookdown/shootdown capability with AIM-7E/F Sparrows or AIM-9P/L Sidewinders; can carry two ASM-1 anti-shipping missiles. RF-4E upgrades receiving Texas Instruments AN/APQ-172 forward looking radars and Melco-developed variant of Thomson-CSF Astac elint pod.

MITSUBISHI (McDONNELL DOUGLAS) F-15J EAGLE

TYPE: Air superiority fighter.
PROGRAMME: Two US built F-15Js followed by eight assembled in Japan from US supplied CKD kits; first flight of CKD Eagle 26 August 1981, delivered 11 December; production to continue into late 1990s.
CURRENT VERSIONS: **F-15J**: Single-seater; having engine upgrade (see Design Features).
 F-15DJ: Two-seat combat-capable trainer (all US built).
CUSTOMERS: Japan Defence Agency plans to procure 223 Eagles for JASDF, including 14 (two F-15Js and 12 F-15DJs) built in USA; 196 F-15Js funded by January 1993, of which 172 then delivered; four more approved for FY 1993.

First JASDF squadron was No. 202 (5th Air Wing) at Nyutabaru, activated December 1982; other units are Nos. 201 and 203 Squadrons (2nd Air Wing) at Chitose, Hokkaido, Nos. 204 and 305 (7th Air Wing) at Hyakuri, No. 303 (6th Air Wing) at Komatsu, and No. 304 (8th Air Wing) and an 'aggressor' unit at Tsuiki.
DESIGN FEATURES: Japanese designed and manufactured ECM and radar warning systems. Starting FY 1991, F-15J's F100-PW-100 engines being upgraded to -220E standard, offering additional 13.34 kN (3,000 lb) thrust, and digital electronic engine control (DEEC) added.
STRUCTURE: Mitsubishi building forward and centre-fuselages, and responsible for final assembly and flight testing; subcontractors are Fuji (landing gear doors), Kawasaki (wings and tail assembly), Nippi (pylons and missile launchers), ShinMaywa (drop tanks), Sumitomo (landing gear), and IHI (engines).
AVIONICS: RT-1360A UHF/VHF transceiver, data link communications set, J/ALQ-8 ECM, J/APR-4A RWR and ALE-45 (J) chaff/flare dispenser.

Mitsubishi F-15J Eagle air superiority fighters of the JASDF

MITSUBISHI FS-X

TYPE: Single-seat support fighter.
PROGRAMME: Modified F-16C selected as Japan's FS-X replacement for Mitsubishi F-1 in October 1987; Mitsubishi appointed prime contractor November 1988; initial contracts awarded for airframe design March 1989 and prototype active phased-array radar February 1990; General Electric F110-GE-129 Improved Performance Engine selected December 1990. Programme delayed by questions of development sharing with General Dynamics (now Lockheed Fort Worth) and technology transfer to Japan, but agreed at Japan 60 per cent/USA 40 per cent cost sharing; first subcontracts to GD let February 1990 for design and development of rear fuselage, wing, leading-edge flaps, avionics and computer based test equipment. Active phased-array radar (Mitsubishi), EW (ECM/ESM), mission computer and IRS being developed using Japanese domestic technology.

Full scale wooden mockup of the Mitsubishi FS-X fighter derived from the F-16C

210   JAPAN: AIRCRAFT—MITSUBISHI/NAL

Preliminary drawing of the Mitsubishi FS-X, with additional side view (top) of two-seat TFS-X *(Jane's/Mike Keep)*

Worth providing rear fuselage, leading-edge flaps, avionics systems and some test equipment.
POWER PLANT: One General Electric F110-GE-129 turbofan (129.0 kN; 29,000 lb st with afterburning).
AVIONICS: Include Yokogawa LCD multi-function display and Shimadzu holographic display; active phased-array radar; mission computer; inertial reference system; integrated EW system.
ARMAMENT: One internal M61A1 Vulcan 20 mm multi-barrel gun. Five external stores stations (one on centreline and two on each wing); Frazer-Nash common rail launchers, to be built and installed by Nippi, will be configured initially for Sparrow medium-range air-to-air missiles; armament eventually expected to include Mitsubishi AAM-3 air-to-air and XASM-2 anti-shipping missiles.
*DIMENSIONS, EXTERNAL (approx):
  Wing span   11.00 m (36 ft 1 in)
  Length overall   16.00 m (52 ft 6 in)
  Height overall   5.00 m (16 ft 4¾ in)
*WEIGHTS AND LOADINGS (estimated):
  Weight empty   9,600 kg (21,164 lb)
  Max external stores load   nearly 9,000 kg (19,840 lb)
  Max T-O weight   22,000 kg (48,500 lb)
PERFORMANCE:
  Max level speed at height   approx Mach 2.0
*See also Addenda

Japan totally responsible for FS-X programme, including all funding; subcontractors include Kawasaki and Fuji (see Structure). Programme to involve four flying prototypes (two single-seat, two tandem two-seat) and two for static test; construction due to begin second quarter 1993; first flight planned for third quarter 1995.
CURRENT VERSIONS: **FS-X:** Single-seat support fighter.
   TFS-X: Combat-capable two-seater.
CUSTOMERS: JASDF (sole user) has requirement for approx 72 to replace F-1, plus others for additional roles, to overall total of up to 130.
COSTS: Total JDA expenditure since 1988 ¥202 billion ($1.68 billion), including ¥75.7 billion ($575 million) in FY 1992 to include first prototype and radar development; further ¥96.5 billion ($804 million) in FY 1993 provides for three more flying prototypes and two for ground test. First two Mitsubishi contracts to GD totalled $280.5 million; follow-on contract to Lockheed Fort Worth (5 February 1993) valued at $74.2 million.
DESIGN FEATURES: New co-cured composite wing of Japanese design, with greater span, root chord and area than that of F-16; tapered trailing-edge; slightly longer radome and forward fuselage to house new radar and other mission avionics; longer mid-fuselage and shorter jetpipes; increased-span tailplane; addition of brake-chute; adoption of increased performance engine.
FLYING CONTROLS: Initially planned vertical canards now deleted; CCV functions will be achieved by digital fly-by-wire system, developed jointly by Japan Aviation Electronics and Bendix/King. Based on earlier Mitsubishi work with T-2 CCV testbed. Available modes to include control augmentation, relaxed static stability, manoeuvre load control, decoupled yaw and manoeuvre enhancement.
STRUCTURE: All-composites co-cured wing; fuselage, tail and other structures also to use advanced materials and structure technology, including radar-absorbent material. Other Japanese airframe companies involved include Fuji (tail unit) and Kawasaki (fuselage mid-section). Lockheed Fort

MITSUBISHI (SIKORSKY) SH-60J
TYPE: Anti-submarine helicopter.
PROGRAMME: Detail design of S-70B-3 version of Sikorsky SH-60B Seahawk (see US section), to meet JMSDF requirements, started August 1983; Japanese avionics and equipment integrated by Technical Research and Development Institute of JDA. First flight of first of two XSH-60J prototypes, based on imported airframes, 31 August 1987; evaluation by 51st Air Development Squadron of JMSDF at Atsugi completed early 1991; first production SH-60J flown 10 May 1991, delivered 26 August.
CUSTOMERS: JMSDF requirement for nearly 100; 47 ordered by December 1992, of which 21 delivered; four more approved for FY 1993; half of force will be land based.
POWER PLANT: T700-401C engines manufactured by IHI.
ACCOMMODATION: Crew of three plus five systems operators/observers.
AVIONICS: Japanese items include search radar, sonar, MAD, ESM, HCDS, ring laser gyro AHRS, data link, tactical data processing, automatic flight management system.
EQUIPMENT: RAST, sonobuoys, rescue hoist and cargo sling.

MITSUBISHI (SIKORSKY) UH-60J
TYPE: Combat search and rescue helicopter.
PROGRAMME: Detail design to Japanese requirements started April 1988; one US built aircraft imported, followed by two CKD kits for licence assembly (first flight February 1991, delivered to JASDF 28 February 1991); second kit aircraft first flew March 1991, delivered to JASDF 29 March 1991). Remainder being built in Japan.
CURRENT VERSIONS: **UH-60J:** Standard military version for JASDF and JMSDF; can fly one-hour search at 250 nm (463 km; 288 miles) from base.
   Civil version: Under discussion with Sikorsky late 1991/early 1992. No recent news.
CUSTOMERS: Excluding CKD aircraft, 18 ordered by December 1992 (JASDF 10, JMSDF eight), of which six completed (JASDF three, JMSDF three); one more for JASDF and two for JMSDF approved for FY 1993. JASDF has total requirement for 46, JMSDF for 18.
POWER PLANT: T700-401C in JMSDF aircraft, T700-701A in those for JASDF, both manufactured by IHI.
ACCOMMODATION: Crew of four (JMSDF) or five (JASDF) plus up to 12 other persons.
AVIONICS: FLIR and nose-mounted weather radar.
EQUIPMENT: Rescue hoist, ESSS and cargo sling.

Mitsubishi (Sikorsky) UH-60J combat search and rescue helicopter of the JASDF

NAL
NATIONAL AEROSPACE LABORATORY
7-44-1 Jindaijihigashi-machi, Chofu City, Tokyo 182
Telephone: 81 (422) 47 5911
DIRECTOR-GENERAL: Dr Kazuyuki Takeuchi
DEPUTY DIRECTOR-GENERAL: Kazuaki Takashima
   NAL is government establishment responsible for research and development in aeronautical and space sciences and technologies.

NAL SPACEPLANE STUDIES
TYPE: Spaceplane studies.
PROGRAMME: NAL conducting research programme for possible future Japanese spaceplane; tunnel testing of alternative configurations began 1982; Dornier 228-200 received 1988 for simulation of final approach and landing phases of eventual spaceplane. Extensive system study on single-stage-to-orbit (SSTO) aircraft carried out with aid of computational fluid dynamics on supercomputers; research on hypersonic air-breathing engines and lightweight high-temperature structures being conducted. Construction of ramjet/scramjet engine test facility initiated April 1989 as

Artist's impression of the NAL spaceplane which it is intended to develop as an international project

three-year plan, by partly remodelling existing high altitude rocket test facility at Kakuda, to test scramjets at Mach 4 to 8. Composite structure test facilities completed 1992 for test and evaluation of advanced materials and structures. Hypersonic wind tunnel test section being enlarged from 0.50 m (1 ft 7¾ in) to 1.20 m (3 ft 11¼ in) as part of three-year programme. Test vehicle should fly during 1990s, leading to operational vehicle as Japanese national goal, but with foreign co-operation. About half of NAL's research workers on aero-spaceplane technology.
COSTS: Expenditure about $5 million for technology research in 1992, plus about $12 million for facility construction.
DESIGN FEATURES: Dornier 228-200 equipped with INS, GPS and special workstation from which flight characteristics can be changed and controlled by computers.
POWER PLANT: LACE (liquefied air cycle engine) for low speeds and in vacuum (operation as a rocket engine in a vacuum); scramjet for high speed within Earth atmosphere.
ACCOMMODATION: Capacity for 10 persons.
DIMENSIONS, EXTERNAL:
Length overall approx 94.0 m (308 ft)
WEIGHTS AND LOADINGS:
Fuselage weight approx 110,000 kg (242,500 lb)
Max T-O weight approx 350,000 kg (771,615 lb)

NAL VTOL AIRLINER STUDIES

TYPE: Medium-range lift/cruise fan 100-passenger VTOL airliner.
PROGRAMME: Study by NAL, Fuji and Ishikawajima; intended to fly 2010.
DESIGN FEATURES: Three core engines to drive either six lift fans beside centre-fuselage in aerodynamic wingroot sections or two propulsion fans at tail; rear-mounted swept-back wings with large winglets; sweptback foreplanes; airscoops above rear fuselage.
FLYING CONTROLS: Pitch-axis control by reaction jets or small fans; rudders in winglets.

Artist's impression of the NAL fan-lift VTOL medium-range airliner

PERFORMANCE:
Cruising speed Mach 0.8
Range 1,350 nm (2,500 km; 1,550 miles)

NIPPI
NIHON HIKOKI KABUSHIKI KAISHA (Japan Aircraft Manufacturing Co Ltd)
HEAD OFFICE AND YOKOHAMA PLANT: 3175 Showa-machi, Kanazawa-ku, Yokohama 236
Telephone: 81 (45) 773 5111
Fax: 81 (45) 771 1253
Telex: (3822) 267 NIPPI J
PRESIDENT: Teruaki Yamada
EXECUTIVE VICE-PRESIDENT: Kanji Sonoda
ATSUGI PLANT: 28 Soyagi 2-chome, Yamato City, Kanagawa Prefecture 242
Telephone: 81 (462) 60 2111
Fax: 81 (462) 64 4945
Telex: 3872 308 NIPPI A J

Yokohama plant, including head office, opened 1935 and now has 71,609 m² (770,800 sq ft) floor area and employs about 1,170 persons; manufactures wing in-spar ribs for Boeing 767/777, elevators for Boeing 757, underwing 'Y-barrels' for McDonnell Douglas MD-11, components and assemblies for Kawasaki-built P-3C Orion (engine nacelles) and T-4 (pylons) and Mitsubishi-built F-15J (pylons and launchers), major dynamic components for Kawasaki CH-47J, V2500 engine ducts, shelters for Mitsubishi-built Patriot ground-to-air missiles, body structures for Japanese satellites, tail units for Japanese built rocket vehicles, and targets for Japan Defence Agency.
Atsugi plant has 42,485 m² (457,300 sq ft) floor area and employs about 600 persons; chiefly engaged in overhaul, repair and maintenance of various types of aeroplanes and helicopters, including those of JDA and Maritime Safety Agency, and carrier based aircraft of US Navy. Kawasaki has 25 per cent holding in Nippi.

NIPPI (NAMC) YS-11E(EL)
TYPE: Elint aircraft, converted from NAMC YS-11C transport.
PROGRAMME: First flight 12 September 1991; delivery was due December 1991.
CUSTOMERS: Japan Air Self-Defence Force (one); JASDF also has two earlier conversions as YS-11E ECM trainers.
DESIGN FEATURES: Main differences from standard YS-11 are replacement of Rolls-Royce Dart turboprops with 2,605 kW (3,493 ehp) GE T64-IHI-10Js and Hamilton Standard propellers, plus installation of NEC J/ALQ-7 ECM system.

SHINMAYWA
SHINMAYWA INDUSTRIES LTD
Nippon Building, 6-2, Otemachi 2-chome, Chiyoda-ku, Tokyo 100
Telephone: 81 (3) 3245 6611
Fax: 81 (3) 3245 6616
Telex: 222 2431 SMIC T J
PRESIDENT: Shinji Tamagawa
EXECUTIVE VICE-PRESIDENT: Shiko Saikawa
HEAD OFFICE: 5-25, Kosone-Cho 1-chome, Nishinomiya 663
Telephone: 81 (798) 47 0331
Fax: 81 (798) 41 0755
Telex: 5644493
Aircraft Division
EXECUTIVE MANAGING DIRECTOR AND GENERAL MANAGER:
 Yukio Koya
BOARD DIRECTOR AND DEPUTY GENERAL MANAGER:
 Junpei Matsuo
DIRECTOR AND GENERAL MANAGER:
 Takaoki Kiho (Konan Plant)
DIRECTOR AND GENERAL MANAGER:
 Yasushi Ikeda (Sales, Head Office)
GENERAL MANAGERS (SALES):
 Yushi Tanaka (International)
 Junsei Nagai (Domestic)
WORKS: Konan and Tokushima

Former Kawanishi Aircraft Company, became Shin Meiwa in 1949 and renamed ShinMaywa Industries in June 1992; major overhaul centre for Japanese and US military and commercial aircraft. Principal activities are production of US-1A for JMSDF, and overhaul work on amphibians. Manufactures external drop tanks for Mitsubishi-built F-15Js and Kawasaki T-4s; nose and tail cones, ailerons and trailing-edge flaps for Kawasaki-built P-3C Orions; tailplanes for Mitsubishi-built SH-60Js; internal cargo handling system for Kawasaki-built CH-47J; wing and tail engine pylons for McDonnell Douglas MD-11; thrust reverser doors for McDonnell Douglas MD-80, under subcontract to Rohr Inc; fixed trailing-edges for Boeing 757/767, under subcontract to Vought; and other components for Boeing 767, under subcontract to Mitsubishi; designated to design and manufacture wing/body fairings for Boeing 777. In 1991, modified five Learjet 36As into U-36A naval fleet training support aircraft for JMSDF; in-service centre for U-36A and Fairchild aircraft.

Has been studying and looking for partners to develop Amphibious Air Transport System, which is 30/50-passenger airliner powered by two wing-mounted turbofans with upper surface blowing; range would vary from 500 nm (926 km; 575 miles) with full payload to 1,200 nm (2,224 km; 1,380 miles) with full fuel; take-off distance 1,000 m (3,280 ft) on water and 800 m (2,624 ft) on soft ground; cruising speed between 300 and 360 knots (556-667 km/h; 345-414 mph).

SHINMAYWA US-1A
TYPE: Four-turboprop STOL search and rescue amphibian.
PROGRAMME: First flown 16 October 1974; first delivery (as US-1) 5 March 1975 (see 1985-86 *Jane's*); all now have T64-IHI-10J engines as US-1As.
CURRENT VERSIONS: **US-1A**: SAR amphibian, developed from PS-1 ASW flying-boat; manufacturer's designation **SS-2A**. *Data apply to this version.*
 Firefighting amphibian: PS-1 modified in 1976 to firefighting configuration; 7,348 kg (16,200 lb) capacity water tank in centre-fuselage aft of step. Since then, US-1A modified experimentally, with more than 13,608 kg (30,000 lb) tank capacity; tank system developed by Conair of Canada.
CUSTOMERS: 13 US-1/1As delivered; eight in service with No. 71 SAR Squadron, JMSDF, at Iwakuni and Atsugi bases. Fourteenth aircraft being built in 1992; fifteenth approved in FY 1993 budget.
DESIGN FEATURES: Boundary layer control system and extensive flaps for propeller slipstream deflection for very low landing and take-off speeds; low-speed control and stability enhanced by blowing rudder, flaps and elevators, and by use of automatic flight control system (see Flying Controls). Fuselage high length/beam ratio; V shaped single-step planing bottom, with curved spray suppression strakes along sides of nose and spray suppressor slots in lower fuselage sides aft of inboard propeller line; double-deck interior. Large dorsal fin.
FLYING CONTROLS: Automatic flight control system controlling elevators, rudder and outboard flaps. Hydraulically powered ailerons, elevators with tabs and rudder, all with 'feel' trim. High-lift devices include outboard leading-edge slats over 17 per cent of wing span and large outer and inner blown trailing-edge flaps deflecting 60° and 80° respectively; outboard flaps can be linked with ailerons; inboard flaps, elevators and rudder blown by BLC system. Two spoilers in front of outer flaps on each wing. Inverted slats on tailplane leading-edge.
STRUCTURE: All-metal; two-spar wing box.
LANDING GEAR: Flying-boat hull, plus hydraulically retractable Sumitomo tricycle landing gear with twin wheels on all units. Steerable nose unit. Oleo-pneumatic shock absorbers. Main units, which retract rearward into fairings on hull sides, have size 40 × 14-22 (type VII) tyres, pressure 7.79 bars (113 lb/sq in). Nosewheel tyres size 25 × 6.75-18 (type VII), pressure 20.69 bars (300 lb/sq in). Three-rotor hydraulic disc brakes. No anti-skid units. Min ground turning radius 18.80 m (61 ft 8¼ in) towed, 21.20 m (69 ft 6¾ in) self-powered.
POWER PLANT: Four 2,605 kW (3,493 ehp) Ishikawajima-built General Electric T64-IHI-10J turboprops, each driving a Sumitomo-built Hamilton Standard 63E60-27 three-blade constant-speed reversible-pitch propeller. Fuel in five wing tanks, with total usable capacity of 11,640 litres (3,075 US gallons; 2,560.5 Imp gallons) and two fuselage tanks (10,849 litres; 2,866 US gallons; 2,386.5 Imp gallons); total usable capacity 22,489 litres (5,941 US gallons; 4,947 Imp gallons). Pressure refuelling point on port side, near bow hatch. Oil capacity 152 litres (40.2 US gallons; 33.4 Imp gallons). Aircraft can be refuelled on open sea, either from surface vessel or from another US-1A with detachable at-sea refuelling equipment.
ACCOMMODATION: Crew of three on flight deck (pilot, co-pilot and flight engineer), plus navigator/radio operator's seat in main cabin. Latter can accommodate up to 20 seated survivors or 12 stretchers, one auxiliary seat and two observers' seats. Sliding rescue door on port side of fuselage, aft of wing.
SYSTEMS: Cabin air-conditioning system. Two independent hydraulic systems, each 207 bars (3,000 lb/sq in). No. 1 system actuates ailerons, outboard flaps, spoilers, elevators, rudder and control surface 'feel'; No.2 system actuates ailerons, inboard and outboard flaps, wing leading-edge slats, elevators, rudder, landing gear extension/

ShinMaywa US-1A of the JMSDF (four Ishikawajima/GE T64-IHI-10J turboprops)

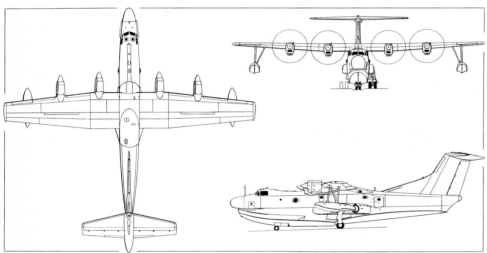

ShinMaywa US-1A ocean-going search and rescue amphibian (*Dennis Punnett*)

retraction and lock/unlock, nosewheel steering, mainwheel brakes and windscreen wipers. Emergency system, also of 207 bars (3,000 lb/sq in), driven by 24V DC motor, for actuation of inboard flaps, landing gear extension/retraction and lock/unlock, and mainwheel brakes. Oxygen system for all crew and stretcher stations. Garrett GTCP85-131J APU provides power for starting main engines and shaft power for 40kVA emergency AC generator. BLC system includes a C-2 compressor, driven by a 1,014 kW (1,360 shp) Ishikawajima-built General Electric T58-IHI-10-M2 gas turbine, housed in upper centre portion of fuselage, which delivers compressed air at 14 kg (30.9 lb)/s and pressure of 1.86 bars (27 lb/sq in) for ducting to inner and outer flaps, rudder and elevators. Electrical system includes 115/200V three-phase 400Hz constant frequency AC and three transformer-rectifiers to provide 28V DC. Two 40kVA AC generators, driven by Nos. 2 and 3 main engines. Emergency 40kVA AC generator driven by APU. 24V emergency DC power from two 34Ah nickel-cadmium batteries. De-icing of wing and tailplane leading-edges. Fire detection and extinguishing systems standard.

AVIONICS: HIC-3 interphone, HRC-107 HF, N-CU-58/HRC antenna coupler, HGC-102 teletypewriter, HRC-106 radio, HRC-110 radio, HRN-101 ADF, AN/ARA-50 UHF/DF, HRN-105B Tacan, HRN-115-1 GPS nav system, HRN-106 ILS marker beacon receiver, AN/APN-171 (N2) radio altimeter, HPN-101B wave height meter, AN/APC-187C Doppler radar, AN/AYK-2 navigation computer, A/A24G-9 TAS transmitter, N-PT-3 dead reckoning plotting board, HRA-5 nav display, AN/APS-115-1 search radar, AN/APX-68N IFF transponder, RRC-15 emergency transmitter and N-ID-66/HRN BDHI.

EQUIPMENT: Marker launcher, 10 marine markers, six green markers, two droppable message cylinders, 10 float lights, pyrotechnic pistol, parachute flares, two flare storage boxes, binoculars, two rescue equipment kits, two droppable liferaft containers, rescue equipment launcher, lifeline pistol, lifeline, three lifebuoys, loudspeaker, hoist unit, rescue platform, lifeboat with outboard motor, camera, and 12 stretchers. Sea anchor in nose compartment. Stretchers can be replaced by troop seats.

DIMENSIONS, EXTERNAL:
Wing span	33.15 m (108 ft 9 in)
Wing chord: at root	5.00 m (16 ft 4¾ in)
at tip	2.39 m (7 ft 10 in)
Wing aspect ratio	8.1
Length overall	33.46 m (109 ft 9¼ in)
Height overall	9.95 m (32 ft 7¾ in)
Tailplane span	12.36 m (40 ft 8½ in)
Wheel track	3.56 m (11 ft 8¼ in)
Wheelbase	8.33 m (27 ft 4 in)
Propeller diameter	4.42 m (14 ft 6 in)
Rescue hatch (port side, rear fuselage):	
Height	1.58 m (5 ft 2¼ in)
Width	1.46 m (4 ft 9½ in)

AREAS:
Wings, gross	135.82 m² (1,462.0 sq ft)
Ailerons (total)	6.40 m² (68.90 sq ft)
Inner flaps (total)	9.40 m² (101.18 sq ft)
Outer flaps (total)	14.20 m² (152.85 sq ft)
Leading-edge slats (total)	2.64 m² (28.42 sq ft)
Spoilers (total)	2.10 m² (22.60 sq ft)
Fin	17.56 m² (189.0 sq ft)
Dorsal fin	6.32 m² (68.03 sq ft)
Rudder	7.01 m² (75.50 sq ft)
Tailplane	23.04 m² (248.0 sq ft)
Elevators, incl tab	8.78 m² (94.50 sq ft)

WEIGHTS AND LOADINGS (search and rescue):
Manufacturer's weight empty	23,300 kg (51,367 lb)
Weight empty, equipped	25,500 kg (56,218 lb)
Usable fuel: JP-4	17,518 kg (38,620 lb)
JP-5	18,397 kg (40,560 lb)
Max oversea operating weight	36,000 kg (79,365 lb)
Max T-O weight: from water	43,000 kg (94,800 lb)
from land	45,000 kg (99,200 lb)
Max wing loading	331.4 kg/m² (67.9 lb/sq ft)
Max power loading	4.32 kg/kW (7.10 lb/ehp)

PERFORMANCE (search and rescue, land T-O. A: at 36,000 kg; 79,365 lb weight, B: at 43,000 kg; 94,800 lb, C: at max T-O weight):
Max level speed: C	276 knots (511 km/h; 318 mph)
Max level speed at 3,050 m (10,000 ft):	
A	282 knots (522 km/h; 325 mph)
Cruising speed at 3,050 m (10,000 ft):	
C	230 knots (426 km/h; 265 mph)
Max rate of climb at S/L: A	713 m (2,340 ft)/min
C	488 m (1,600 ft)/min
Service ceiling: A	8,655 m (28,400 ft)
C	7,195 m (23,600 ft)
T-O to 15 m (50 ft) on land, 30° flap, BLC on (ISA):	
C	655 m (2,150 ft)
T-O distance on water, 40° flap, BLC on (ISA):	
B	555 m (1,820 ft)
Landing from 15 m (50 ft) on land, AUW of 36,000 kg (79,365 lb), 40° flap, BLC on, with reverse pitch (ISA):	
A	810 m (2,655 ft)
Landing distance on water, AUW of 36,000 kg (79,365 lb), 60° flap, BLC on (ISA):	
A	220 m (722 ft)
Runway LCN requirement: B	42
Max range at 230 knots (426 km/h; 265 mph) at 3,050 m (10,000 ft)	2,060 nm (3,817 km; 2,372 miles)

KOREA, SOUTH

BHK
BELL HELICOPTER KOREA INC
8-10, Hyorim-ri, Sapkyo-eup, Yesan-kun, Chungcheong Nam-do
Telephone: 82 (458) 37 1991/5
Fax: 82 (458) 37 1996

REPRESENTATIVE DIRECTOR: Jin W. Song
SEOUL OFFICE: Hanjong (Bell Korea) Building, 1460-10 Suhcho-Dong, Suhcho-Ku, Seoul 137 070
Telephone: 82 (2) 597 1927/29
Fax: 82 (2) 588 6091
Telex: (787) K 25622 KORBELL

Joint venture corporation between Bell Helicopter Textron and Korea Technologies Corporation (KTC), a subsidiary of United Industries International. Since April 1988, licensed to inspect, overhaul and carry out depot level maintenance on Bell helicopters, train air and ground crews, and manufacture parts for subassemblies of Bell 212 and 412 HP made by Samsung Aerospace and Daewoo.

DHI
DAEWOO HEAVY INDUSTRIES CO LTD
6 Manseog-dong, Dong-gu, Inchon
Telephone: 82 (32) 760 1114
Fax: 82 (32) 762 1546
Telex: DHILTD K 28473
PRESIDENT: Kyung-Hoon Lee
Aerospace Division
SEOUL OFFICE: Daewoo Center Building (20th Floor), 541, 5-ga, Namdaemun-ro, Jung-gu, Seoul 100 714
Telephone: 82 (2) 779 1669
Fax: 82 (2) 756 2679
Telex: K 23301
GENERAL MANAGER: K. W. Park
CHANGWON PLANT: 3 Block, Sungju Complex, Changwon
Telephone: 82 (551) 80 6501
Fax: 82 (551) 85 2383
EXECUTIVE MANAGING DIRECTOR: Hyo-Sang Cho
DIRECTORS, SALES OFFICE:
  Du-Il Kim
  Young-Min Kim
DIRECTOR, PRODUCTION CONTROL: Kwang-Jin Lee
DIRECTOR, PROJECT MANAGEMENT: Duck-Joo Ra

DHI established October 1976 as major member of Daewoo Group, although roots go back to founding of Chosun Machine Works in 1937. Aerospace Division at Changwon established 1984, began aircraft component manufacture 1985; currently occupies 38.5 ha (95.1 acre) site, incl 43,187 m² (464,860 sq ft) of floor space, with 1992 workforce of 800. Capabilities include fuselage and wing construction, manufacture of helicopter structures and dynamic components, final assembly of indigenous trainer, UAV design and manufacture, and satellite platform manufacture.

DHI is prime contractor for Korean Indigenous Trainer and Agricultural Remote Control Helicopter (Korint and ARCH: see below), and for Korean Light Scout Helicopter (KLH). Contenders in latter programme include Agusta A 109CM and Eurocopter BO 105 CB; winner due to be selected by mid-1993 (unless delayed again), with deliveries from 1995 to 1999. KLH requirement now reportedly reduced from original 100-plus to 86 (50 firm and 36 options).

Other current work includes approx 100 centre-fuselage assemblies for Korean Fighter Programme (see SSA entry), wing and fuselage components for Boeing 747, complete fuselage shells for Dornier 328 (first delivery 13 January 1991, total 400 due by 2003), rotor hub assemblies for Bell 212/412 HP, airframe of Lynx helicopter, and wing construction for BAe Hawk Mk 67 and Lockheed P-3C Orion. Also developing UAV known as Doyosae.

First prototype of the Daewoo KTX-1 Korint two-seat primary trainer

DAEWOO KTX-1 KORINT
TYPE: Turboprop powered tandem-seat primary trainer.
PROGRAMME: Built to design of government agency; five prototypes (three flying, two for ground test), first of which rolled out November 1991 and made first flight December 1991; second completed by mid-1992; full scale development due to continue until 1997, series production due to start 1998. Fuselage built by Samsung (front) and Korean Air (centre and rear), wings and tail unit by Daewoo.
DESIGN FEATURES: See accompanying illustration.
LANDING GEAR: Retractable tricycle type; single wheel on each unit.
POWER PLANT: One 410 kW (550 shp) Pratt & Whitney Canada PT6A-25A turboprop in first prototype; three-blade propeller. More powerful engine expected in subsequent prototypes.

DAEWOO AGRICULTURAL REMOTE CONTROL HELICOPTER (ARCH)
TYPE: Agricultural helicopter.
PROGRAMME: First prototype due for delivery February 1993; development to continue until 1996; deliveries will be to Korean government.
DESIGN FEATURES: Based on Kamov design (Ka-37); coaxial two-blade rotors; skid landing gear; ventral pod for chemicals.
POWER PLANT: One rotary engine.
DIMENSIONS, EXTERNAL:
  Rotor diameter (each)   4.80 m (15 ft 9 in)
  Length overall   2.60 m (8 ft 6½ in)
  Max width   1.20 m (3 ft 11¼ in)
  Height overall   1.60 m (5 ft 3 in)
WEIGHTS AND LOADINGS:
  Max payload   50 kg (110 lb)
PERFORMANCE (estimated):
  Max level speed   39 knots (72 km/h; 45 mph)
  Normal operating speed   10 knots (18 km/h; 11 mph)
  Normal operating height   1-2 m (3.3-6.6 ft)
  Operating radius   500 m (1,640 ft)
  Endurance   40-60 min

KA
KOREAN AIR
KAL Building, CPO Box 864, 41-3 Seosomun-Dong, Chung-Ku, Seoul
Telephone: 82 (2) 751 7114
Telex: KALHO K 27526
CHAIRMAN AND CEO: C. H. Cho
PRESIDENT: Y. H. Cho
Aerospace Division
Marine Center Building 22FL, 118-2-ka, Namdaemun-Ro, Chung-Ku, Seoul
Telephone: 82 (2) 726 6240
Fax: 82 (2) 756 7929
Telex: KALHO K 27526 (SELDBKE)
EXECUTIVE MANAGING VICE-PRESIDENT: Y. T. Shim
MANAGING VICE-PRESIDENT, MARKETING AND SALES:
  J. K. Lee
Korean Institute of Aeronautical Technology (KIAT)
Address/telephone/fax/telex details as for Aerospace Division
VICE-PRESIDENT: S. Y. Yoo

Aerospace Division of Korean Air established 1976 to manufacture and develop aircraft; at November 1992, occupied 64.75 ha (160 acre) site at Kim Hae, including floor area of 130,065 m² (1.4 million sq ft); workforce then about 2,200. Has overhauled RoKAF aircraft since 1978; programmed depot maintenance of US military aircraft in Pacific area began 1979, including structural repair of F-4s, systems modifications for F-16s, MSIP upgrading of F-15s and overhaul of C-130s. Began production in 1981 of first domestically manufactured fighter (Northrop F-5E/F), completing deliveries to Korean Air Force 1986. Since April 1988 has delivered wingtip extensions and flap track fairings for Boeing 747 and fuselage components for McDonnell Douglas MD-11.

Since 1990, KA has manufactured UH-60P helicopters under licence from Sikorsky (essentially same as current US Army UH-60L, with T700-GE-701C engines and 2,535 kW; 3,400 shp transmission, upgraded with added avionics).

Korean Institute of Aeronautical Technology (KIAT), established as division of KA in 1978, has grown to become a major Korean aerospace industry R&D centre.

As part of Korean industry development programme from 1988, KA has designed and developed light aircraft (Chang-Gong 91); co-developed (with McDonnell Douglas) the MD 520MK military helicopter derived from the MD 500; and domestically co-developed the KTX-1 primary trainer (see Daewoo entry) for the RoKAF.

Korean Air CHK-91 Chang-Gong 91 four/five-seat light aircraft

KA CHK-91 CHANG-GONG 91 (BLUE SKY 91)
TYPE: Four/five-seat light cabin monoplane.
PROGRAMME: Partly funded by Hankook fibre company and Samsun industrial company; design began 21 June 1988; construction of prototype started 20 December 1990; first flight 22 November 1991; static test aircraft also completed and second flying prototype completed at end of November 1992; domestic Ministry of Transportation certification in Utility category expected by end of June 1993.
COSTS: Programme cost $5 million (1991).
DESIGN FEATURES: Designed for S/L speed range of 51-135 knots (95-250 km/h; 59-155 mph); conventional low-wing, fixed-gear cabin monoplane. Wing section NACA 63²-415; dihedral 6° from roots; incidence 2°; twist 3°; sweepback 2° 15' on leading-edge, 0° 52' 48" at quarter-chord.
FLYING CONTROLS: Manual/mechanical; piano hinged Frise ailerons with fixed tabs, all-moving tailplane with large central geared tab, balanced rudder; single-slotted trailing-edge flaps.
STRUCTURE: Aluminium alloy single-spar wings; light alloy fuselage; graphite/epoxy tail unit, wheel speed fairings, doors, engine cowling.
LANDING GEAR: Non-retractable tricycle type, with single wheel, oleo-pneumatic shock absorber and speed fairing on each unit. Cleveland 40-86B mainwheels and 40-77B nosewheel, former with Cleveland 30-55 disc brakes. Tyre sizes 6.00-6 (main) and 5.00-5 (nose), both 6-ply; pressures 3.45 bars (50 lb/sq in) and 2.90 bars (42 lb/sq in) respectively. Nosewheel steerable ±25°. Min ground turning radius 9.45 m (31 ft 0 in).
POWER PLANT: One Textron Lycoming IO-360-A1B6 flat-four engine (149 kW; 200 hp at 2,700 rpm), driving a Hartzell HC-C2YK-IBF/F7666A-2 two-blade constant-speed propeller. Two integral fuel tanks in wings each holding 106 litres (28 US gallons; 23.3 Imp gallons). Total fuel capacity 212 litres (56 US gallons; 46.6 Imp gallons). Refuelling points in wing upper surfaces. Oil sump capacity 7.6 litres (2 US gallons; 1.7 Imp gallons).
ACCOMMODATION: Side by side seats in front for pilot and one passenger. Second pair of seats behind these, to rear of

KOREA, SOUTH/MALAYSIA: AIRCRAFT—KA/AIROD

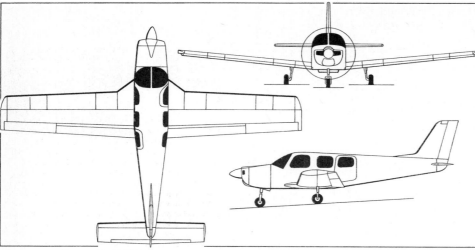

Korean Air CHK-91 Chang-Gong 91 (149 kW; 200 hp IO-360) *(Janes/Mike Keep)*

which is a fifth (child's) seat. Space for 45.4 kg (100 lb) of baggage behind rearmost seat. Front-hinged, outward opening door at front on starboard side, over wing, and at rear (aft of wing) on port side. Entire accommodation ventilated.

SYSTEMS: 14V DC electrical system powered by engine driven alternator. No air-conditioning, oxygen, hydraulic or pneumatic systems.

AVIONICS: Bendix/King KX 155 nav/com, KN 62A DME, KR 87 ADF, KT 79 transponder, KR 21 marker beacon receiver, KHA 24H audio control and KI 525A pictorial nav indicator.

DIMENSIONS, EXTERNAL:
Wing span	10.20 m (33 ft 5½ in)
Wing chord: at root	1.66 m (5 ft 5½ in)
at tip	1.21 m (3 ft 11½ in)
Wing aspect ratio	7.44
Length: overall	7.74 m (25 ft 4¾ in)
fuselage	7.00 m (22 ft 11½ in)
Height overall	2.70 m (8 ft 10¼ in)
Tailplane span	3.69 m (12 ft 1¼ in)
Wheel track	2.59 m (8 ft 6 in)
Wheelbase	1.71 m (5 ft 7¼ in)
Propeller diameter	1.88 m (6 ft 2 in)
Propeller ground clearance	0.34 m (12.1 in)
Passenger door (fwd, stbd):	
Height	0.97 m (3 ft 2 in)
Width	0.91 m (3 ft 0 in)
Height to sill	0.91 m (3 ft 0 in)
Passenger door (rear, port):	
Height	0.97 m (3 ft 2 in)
Width	0.79 m (2 ft 7¼ in)
Height to sill	0.70 m (2 ft 3½ in)

DIMENSIONS, INTERNAL:
Cabin: Max length	3.05 m (10 ft 0 in)
Max width	1.17 m (3 ft 10 in)
Max height	1.86 m (6 ft 1¼ in)
Floor area	3.27 m² (35.2 sq ft)
Volume	2.14 m³ (75.7 cu ft)
Baggage compartment volume	0.34 m³ (12.0 cu ft)

AREAS:
Wings, gross	14.86 m² (160.0 sq ft)
Ailerons (total)	1.18 m² (12.67 sq ft)
Trailing-edge flaps (total)	2.00 m² (21.55 sq ft)
Fin, incl dorsal fin	0.81 m² (8.70 sq ft)
Rudder, incl tab	0.52 m² (5.60 sq ft)
Tailplane, incl tab	2.60 m² (28.00 sq ft)

WEIGHTS AND LOADINGS:
Weight empty, equipped	839 kg (1,850 lb)
Max fuel weight	159 kg (350 lb)
Max T-O and landing weight	1,225 kg (2,700 lb)
Max wing loading	82.39 kg/m² (16.87 lb/sq ft)
Max power loading	8.22 kg/kW (13.50 lb/hp)

PERFORMANCE (at max T-O weight):
Max level speed at S/L	135 knots (250 km/h; 155 mph)
Max cruising speed at 1,525 m (5,000 ft)	131 knots (243 km/h; 151 mph)
Econ cruising speed (65% power) at 1,525 m (5,000 ft)	116 knots (215 km/h; 134 mph)
Stalling speed at S/L, 65% power at 2,350 rpm:	
0° flap	63 knots (117 km/h; 73 mph)
20° flap	57 knots (106 km/h; 66 mph)
40° flap	51 knots (95 km/h; 59 mph)
Max rate of climb at S/L	298 m (978 ft)/min
Service ceiling	5,180 m (17,000 ft)
T-O run at S/L	199 m (620 ft)
T-O to 15 m (50 ft)	372 m (1,220 ft)
Landing from 15 m (50 ft)	442 m (1,450 ft)
Landing run	262 m (860 ft)
Range at 1,525 m (5,000 ft), max fuel, 65% power at 2,350 rpm, no reserves	762 nm (1,412 km; 877 miles)

KA (McDONNELL DOUGLAS) 500E, 520N and 530F

Korean Air manufactured 307 Model 500D and 500MD helicopters under licence from Hughes/McDonnell Douglas between 1976 and September 1988. Under new offset agreement signed 1984 is now providing McDonnell Douglas Helicopter Co with major fuselage assemblies (approx 80 a year) for 500E, 520N and 530F (see US section), these being made into specific model required by Intertech in USA. Modified military version of 520, designated **520MK Black Tiger**, was co-developed by KA and MDHC in 1989.

SSA
SAMSUNG AEROSPACE
15/16th Floor, Dongbang Main Building, PO Box 9762, Seoul
Telephone: 82 (2) 751 8850
Telex: 22933
MARKETING MANAGER: Han-sung Kim

Prime contractor in licence manufacture of F-16s for RoKAF; also produces wing fixed trailing-edges for Boeing 757/767 and horizontal tail surfaces for Dash 8. Samsung and Bell Helicopter Korea are approved by South Korean government to produce medium helicopters in Korea; agreement with Bell calls for SSA to produce major fuselage and tailboom assemblies for Bell 212 and 412 HP; initial production began May 1988. BHK and SSA have facilities to build complete helicopters at SSA's manufacturing centres in Changwon and Sacheon; plans called for 93 per cent local manufacture by early 1990s.

SAMSUNG (LOCKHEED FORT WORTH) F-16
TYPE: Multi-role fighter.
PROGRAMME: Announcement to co-produce 120 F-16C/Ds with improved weapon delivery systems made by South Korean government 28 March 1991; programme known as Korean Fighter Programme (KFP); Samsung is main Korean contractor, with Daewoo and Hanjin as main subcontractors; original plan to acquire 12 aircraft and their engines off the shelf, assemble 36 from kits and manufacture remaining 72 locally.
COSTS: $5.2 billion for 120-aircraft programme.
POWER PLANT: One Pratt & Whitney F100-PW-229 turbofan.

SAMSUNG KTX-2
TYPE: Jet basic trainer and light attack aircraft.
PROGRAMME: Long-term programme for indigenous replacement for RoKAF T-33As and T-37Cs; Samsung prime contractor, with team including Daewoo and Korean Air. No other details yet known.

MALAYSIA

AIROD
AIROD SENDIRIAN BERHAD (Subsidiary of Aerospace Industries Malaysia)
Subang International Airport, Pejabat Pos Kampung Tunku, Locked Bag 4004, 47309 Petaling Jaya
Telephone: 60 (3) 746 5112
Fax: 60 (3) 746 1395
Telex: MA 37910
MANAGING DIRECTOR AND CEO: Dato Ariff Awang
SENIOR MARKETING DIRECTOR: Dato Wan Majid
PUBLIC RELATIONS DIRECTOR: Col (Ret'd) M. A. Theseira

Airod formed 1985 as joint venture of Aerospace Industries Malaysia and Lockheed Aircraft Services International, USA.

SIKORSKY S-61A NURI UPGRADE
TYPE: Multi-purpose military helicopter.
PROGRAMME: March 1991 contract to upgrade 34 aircraft between 1991 and end of 1993 with modernised navigation and other avionics; first example redelivered late 1991; 21st handed over 16 February 1993.
CUSTOMERS: Royal Malaysian Air Force (34).
COSTS: Contract value $50 million.
DESIGN FEATURES: To enhance operational capability and reduce crew workload.
AVIONICS: Include new nose-mounted weather radar, Doppler navigation system and radar altimeter.

Installation of the nose radar during S-61 Nuri avionics upgrade

NETHERLANDS

FOKKER
NV KONINKLIJKE NEDERLANDSE VLIEGTUIGFABRIEK FOKKER

CORPORATE CENTRE: PO Box 12222, NL-1100 AE Amsterdam-Zuidoost
Telephone: 31 (20) 6056666
Fax: 31 (20) 6057015
Telex: 11526 FMHS NL
CHAIRMAN: Erik Jan Nederkoorn
OPERATING COMPANIES:
Fokker Aircraft BV, PO Box 12222, NL-1100 AE Amsterdam-Zuidoost
Telephone: 31 (20) 6056666
Fax: 31 (20) 6057015
Telex: 11526 FMHS NL
EXECUTIVE VICE-PRESIDENT, MARKETING AND SALES: Edo van den Assem
Fokker Aircraft Services BV, PO Box 3, NL-4630 AA Hoogerheide
Fokker Space and Systems BV, PO Box 12222, NL-1100 AE Amsterdam-Zuidoost
Fokker Special Products BV, PO Box 59, NL-7900 AB Hoogeveen
Aircraft Financing and Trading BV, PO Box 12222, NL-1100 AE Amsterdam-Zuidoost
Avio-Diepen BV, PO Box 5952, NL-2280 HZ Rijswijk

Royal Netherlands Aircraft Factories NV Fokker founded by Anthony Fokker 1919; forms main aircraft industry in Netherlands, employing some 13,000 people in early 1992; under new company structure effective from 1 January 1987, Fokker comprises a corporate centre and six operating companies. Main products are Fokker 50, 70 and 100 airliners and their derivatives. Fokker Defence Marketing handles all defence activities.

Provisional agreement reached in October 1992 for Deutsche Aerospace (DASA) to acquire 51 per cent holding in Fokker; final plan, agreed in February 1993, is for two-stage acquisition by DASA of Dutch government's 31.8 per cent holding in Fokker (first stage immediately, second stage in 1996); DASA to establish new company, Fokker Holding, owning 51 per cent of NV Fokker, with remaining 49 per cent retained by Dutch government for three years; Fokker/DASA agreement signed 27 April 1993; provision for Aerospatiale and Alenia to join Fokker Holding in future.

Some 6,000 employed at Schiphol plant, Amsterdam, in Fokker 50 and 100 assembly and test flying facilities, design offices, spare parts stores, R&D department, numerically controlled milling department, electronics division, space integration and test facilities, and computer facilities.

Drechsteden plant employs about 2,200; most engaged on detail production and component assembly for Fokker 50 and 100 and F-16.

Ypenburg employs 800 in installation of F-16 centre-fuselages and construction of composites components for Fokker 50 and 100 and Westland Lynx (radomes and fairings); new 30,000 m² (322,917 sq ft) composites and bonding plant began operating May 1990.

Fokker Aircraft Services BV situated at Woensdrecht, with about 900 workforce; specialises in maintenance, overhaul, repair and modification of civil and military aircraft. ELMO plant (Fokker Aircraft), also at Woensdrecht, employs about 750; produces electrical and electronic systems and cable harnesses.

Hoogeveen (Fokker Special Products BV) engaged in industrial products activities, such as licence programmes, shelters, missile launchers, pylons, fuel tanks and thermoplastics components; workforce about 600.

As part of European F-16 programme, Fokker was responsible for component manufacture and assembly of F-16s for Netherlands (213, delivered between June 1979 and 27 February 1992), Norway (72) and Denmark (12); it continues to produce F-16 centre-fuselages, wing moving surfaces, main landing gear doors and legs, tailplanes, rudders and fin leading-edges for USAF and export assembly lines of Lockheed Fort Worth.

FOKKER 50

TYPE: Twin-turboprop short-haul transport.
PROGRAMME: Follow-on development of F27 Friendship, announced 24 November 1983; more than 80 per cent of components new or modified; two prototypes used modified F27 fuselages; first flight of first prototype (PH-OSO) 28 December 1985; first flight of first production Fokker 50 (PH-DMO) 13 February 1987; JAR 25 certification by Dutch RLD 15 May 1987; first delivery (to Lufthansa Cityline) 7 August 1987; FAA type approval (FAR Pt 25) 16 February 1989. Offered in range with seating capacities from 46 to 68 and four-door or three-door configuration (see Accommodation), all with maximum commonality.
CURRENT VERSIONS: **Series 100:** Baseline model for up to 58 passengers; PW125B turboprops; available as four-door **50-100** or three-door **50-120**. *Detailed description applies to Series 100, except where indicated.*
Series 300: High performance version of Series 100; same seating capacity but more powerful PW127B turboprops, providing superior performance from short runways, airfields characterised by obstacles, and airfields in hot and high environments; available as four-door **50-300** or three-door **50-320**; first delivery, to launch customer Avianca of Colombia, due early 1993.
Series 400: Planned stretched version with up to 68 seats in 2.40 m (7 ft 10½ in) longer fuselage; PW127B turboprops; four-door version designated **50-400**, three-door version **50-420**.
Special mission versions: Various configurations: described separately.
Utility: Available for specific government and military requirements; configurable to wide variety of missions including passenger or cargo transport, paratrooping and medevac.
CUSTOMERS: Firm orders for 182 by 1 June 1993, of which over 150 delivered; 26 scheduled to be built in 1993.
Customers include Aer Lingus Commuter (six), Aircraft Financing & Trading (one), Ansett Express (eight), Austrian Airlines (eight), Avianca (six), Crossair (five), Icelandair (four), KLM Cityhopper (10), Kenya Airways (three), Lufthansa CityLine (31), Luxair (four), Maersk Air (eight), Malaysia Airlines (11), Nakanihon Airline Service (two), Norwegian Air Shuttle (three), Pelangi Air (two), Philippine Airlines (10), Rio-Sul (three), Royal Thai Police (one), SAS Commuter (22), SEEA (two), Sudan Airways (two), Taiwan government (three), Tanzania government (one), and VLM (two), plus 24 undisclosed.
DESIGN FEATURES: Based on F27 proven airframe but with significant design and structural changes, allied to more efficient and fuel-efficient new technology engines in redesigned nacelles, driving specially designed six-blade propellers; 12 per cent higher cruising speed; carbon/aramid/glassfibre components in areas of wings, tailplane, fin, radome, engine nacelles and propellers; 'Foklet' horn balance at each wingtip to increase lateral stability at low airspeeds; passenger door relocated at front; greater passenger comfort and convenience, with more windows, new-design interior with extensive noise reduction; all-new cockpit, with EFIS; advanced digital avionics; twin-wheel nose gear; latest technology systems; pneumatic system replaced by hydraulic; improved airport handling.

Wing section NACA 64_4-421 (modified) at root, 64_4-415 (modified) at tip; unswept; dihedral 2° 30′; washout 2° on outer wings; incidence 3° 30′.
FLYING CONTROLS: Mechanically (cable) actuated ailerons, with inboard spring tab and outboard geared tab (starboard geared tab acting also as electrically actuated trim tab); mechanically interconnected elevators, with starboard trim tab; rudder has trim tab, geared tab and horn balance; hydraulically actuated, mechanically interconnected single-slotted trailing-edge flaps with electrical backup.
STRUCTURE: Primary structure is all-metal riveted and metal-bonded stressed skin; detachable AFRP wing leading-edges; composite wing trailing-edge skins supported by composite or metal ribs; bonded skin/stringer ailerons have composites leading-edges; metal flaps; fin and fixed incidence tailplane have metal primary structure; wingtip 'Foklets' are of metal reinforced composites; composites also for nosecone, fairings, nosewheel doors, access doors, cabin floor, engine air intakes and nacelle cowlings, tail unit leading-edges and part of dorsal fin. Subcontractors include Dassault (centre and rear fuselage), Fuji (rudder and elevators), Deutsche Aerospace Airbus (wing trailing-edge and control surfaces, tailcone and dorsal fin), Sabca (outer wing skins and wingtips), and Dowty (propellers and landing gear).
LANDING GEAR: Dowty retractable tricycle type, with twin wheels on each unit. Main units attached to wings, retracting rearward hydraulically into rear extension of engine nacelle; nose unit retracts forward. Long-stroke oleo-pneumatic shock absorber in each unit (single-stage on nose unit, double-acting on main units). Goodyear wheels and tyres on all units. Standard mainwheel tyres size 34 × 10.75-R16, nosewheel tyres 24 × 7.7-10. Goodyear hydraulic brakes, incorporating anti-skid system. Hydraulic nosewheel steering (±73°); free-castoring angle of ±130° available for towing. Min ground turning radius 18.07 m (59 ft 3½ in).
POWER PLANT: Series 100 powered by two Pratt & Whitney Canada PW125B turboprops, each flat rated at 1,864 kW (2,500 shp) up to 30°C ambient at S/L. Series 300 and 400 powered by P&WC PW127Bs flat rated at 2,050 kW (2,750 shp) up to 30.8°C for T-O. All have Dowty

Fokker 50 Series 100 twin-turboprop transport (two Pratt & Whitney Canada PW125B engines) of Rio-Sul (Brazil)

216 NETHERLANDS: AIRCRAFT—FOKKER

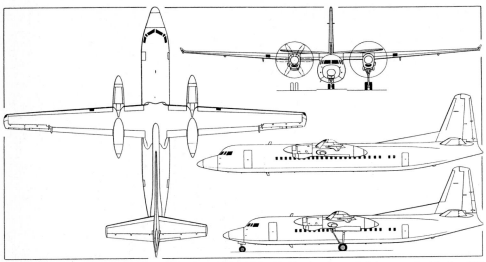

Fokker 50 twin-turboprop short-haul transport, with additional side view (upper) of the stretched Series 400 *(Dennis Punnett)*

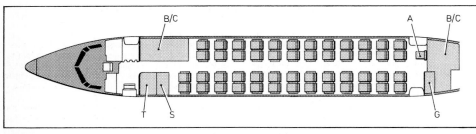

Fokker 50 standard configuration (50 seats at 81 cm; 32 in pitch). A: attendant seat, B: baggage, C: cargo, G: galley, S: stowage, T: toilet *(Jane's/Mike Keep)*

Aerospace propellers, designed specifically for Fokker 50, with six all-composite blades and Beta control; precise propeller rpm control, plus phase synchronisation of ±2°, reduces aircraft noise. Digital blade matching system eliminates all propeller-induced vibration in cabin. Fuel in two integral tanks located between two spars of central wing box outboard of nacelles, with total standard capacity of 5,136 litres (1,357 US gallons; 1,130 Imp gallons). Single-point pressure refuelling; overwing gravity points.

ACCOMMODATION: Crew of two and observer seat on flight deck; one or two cabin attendants, depending on configuration. Dedicated door concept (every ground handling activity has its own door) offers a four-door configuration, three-door configuration, and three-door with optional large cargo door. All configurations have a four-abreast layout and 46 cm (18 in) aisle width. Passenger door, with electrically actuated integral airstairs, at front on port side. Four-door Fokker 50-100/300/400 have forward and rear baggage/cargo compartments and rear galley, each serviced/loaded via a dedicated adjacent door; as Type III emergency exits are not required, layout is flexible from 46 seats at 86 cm (34 in) pitch to 68 seats in 400) at 76 cm (30 in); stowage area and lavatory at front of cabin.

Three-door Fokker 50-120/320/420 feature reconfigured cabin allowing number of seats or cargo volume to be increased without detriment to passenger comfort. In this configuration, port rear galley service door is deleted and two Type III emergency exits introduced, galley is at front of cabin and baggage/cargo compartment at rear; cargo wall can be placed at one of four set positions, allowing number of seats and cargo volume to be adjusted. This configuration allows up to 68 seats at 76 cm (30 in) pitch. Various other major options and configurations available for 100/120, 300/320 and 400/420, including roll-on/roll-off containers, heavy duty floor, convertible configuration and VIP interiors. Large cargo door at rear on port side offered as option on 50-120/320/420, allowing loading of palletised cargo and containers.

Entire accommodation pressurised and air-conditioned. Windscreens anti-iced electrically, cabin windows and flight deck side windows demisted by hot air.

SYSTEMS: Hamilton Standard air-conditioning system. AiResearch digital cabin pressure control system. Max pressure differential 0.38 bar (5.46 lb/sq in). Hydraulic system, operating at 207 bars (3,000 lb/sq in) pressure via two engine driven Abex pumps, for landing gear actuation, brakes, nosewheel steering and flap drive. Pneumatic de-icing of wing, fin and tailplane leading-edges, using engine bleed air. Engine air intakes, propeller blades and spinners de-iced electrically. Primary electrical system powered by Sundstrand 30/40kVA integrated drive generator mounted on propeller gearbox of each engine, supplying 115/200V three-phase AC at 400Hz, with two 300A transformer-rectifiers and two 43Ah nickel-cadmium batteries for 28V DC power. Auxiliary Power International Corporation (APIC) APS 1000 APU optional. Configured for ground use, APU (combined with Sundstrand 115V 20kVA oil-cooled generator, accessory controls and provisions) provides AC self-sufficiency during turnaround cycles, servicing and maintenance. APU also provides air supply for flight deck and cabin air-conditioning and is fitted to fireproof bulkhead aft of wheel bay in starboard nacelle.

AVIONICS: Dual Honeywell EDZ-806 electronic flight instrument system (EFIS) with CRT displays for primary flight and navigation information, and space provisions for central multifunction display. Standard avionics include Honeywell SPZ-600 AFCS with Cat. I landing (Cat. II optional); Honeywell FZ-500 dual flight director systems; Bendix/King dual Series III VHF com, single Series III ADF and DME (latter including frequency hold facility), and Series III ATC transponder; Honeywell Primus P-650 weather radar with dual presentation on EFIS; dual Bendix/King Series III VHF nav with VOR/ILS/marker beacon receiver; TRT AHV-530A (ARINC 552A) radio altimeter with dual presentation on EFIS; dual Litton LTR 81.01 AHRS; Sundstrand Mk II GPWS (ARINC 549); Honeywell AZ-800 air data computer; Fairchild A100 (ARINC 557) cockpit voice recorder; Collins 346-2B (ARINC 560) PA system; Sundstrand 980-4100 DXUS (ARINC 573) flight data recorder, incl underwater locator beacon; and Teledyne Model 70-275 flight data acquisition unit. Full provisions for Cat. II landing on AFCS, single Collins 628T-2A HF com to ARINC 559A2 and second ADF; space provisions for second DME, second ATC transponder, VLF-Omega or Bendix/King KNS 660 nav system, and Dorne & Margolin ELT.

DIMENSIONS, EXTERNAL (Series 100 and 300):
Wing span	29.00 m (95 ft 1¾ in)
Wing chord: at root	3.464 m (11 ft 4½ in)
at tip	1.40 m (4 ft 7 in)
Wing aspect ratio	12.0
Length overall	25.247 m (82 ft 10 in)
Fuselage: Max width	2.70 m (8 ft 10¼ in)
Height overall (static)	8.317 m (27 ft 3½ in)
Tailplane span	9.746 m (31 ft 11¾ in)
Wheel track	7.20 m (23 ft 7½ in)
Wheelbase	9.70 m (31 ft 10 in)
Propeller diameter	3.66 m (12 ft 0 in)
Propeller ground clearance	1.162 m (3 ft 9¾ in)
Propeller fuselage clearance	0.593 m (1 ft 11¼ in)
Passenger door (fwd, port): Height	1.78 m (5 ft 10 in)
Width	0.76 m (2 ft 6 in)
Service door (rear, port) and baggage door (fwd, stbd):	
each: Height	1.27 m (4 ft 2 in)
Width	0.61 m (2 ft 0 in)
Standard cargo door (rear, stbd):	
Height	1.27 m (4 ft 2 in)
Width	0.86 m (2 ft 9¾ in)
Optional large cargo door (rear, port):	
Height	1.65 m (5 ft 5 in)
Width	1.30 m (4 ft 3¼ in)

DIMENSIONS, EXTERNAL (Series 400): As Series 100/300 except:
Length overall	27.69 m (90 ft 10¼ in)
Height overall (static)	8.34 m (27 ft 4½ in)
Wheelbase	11.23 m (36 ft 10¼ in)

DIMENSIONS, INTERNAL (Series 100 and 300):
Cabin, excl flight deck: Length	15.96 m (52 ft 4 in)
Width at floor	2.11 m (6 ft 11 in)
Max width	2.50 m (8 ft 2½ in)
Max height	1.96 m (6 ft 5¼ in)
Floor area (excl toilet)	30.2 m² (325.0 sq ft)
Baggage/cargo volume (standard commuter version):	
main compartments	7.38 m³ (260.6 cu ft)
wardrobe compartment	0.82 m³ (29.0 cu ft)
overhead bins	2.22 m³ (78.4 cu ft)

DIMENSIONS, INTERNAL: (Series 400): As Series 100/300 except:
Cabin, excl flight deck: Length	18.36 m (60 ft 3 in)
Floor area (excl toilet)	34.8 m² (374.6 sq ft)
Baggage/cargo volume (standard commuter version):	
main compartments	8.23 m³ (290.6 cu ft)
wardrobe compartment	0.77 m³ (27.2 cu ft)
overhead bins	2.75 m³ (97.1 cu ft)

AREAS (Series 100, 300 and 400):
Wings, gross	70.0 m² (753.5 sq ft)
Ailerons (total)	3.66 m² (39.40 sq ft)
Trailing-edge flaps (total)	17.15 m² (184.60 sq ft)
Fin, incl dorsal fin	17.60 m² (189.44 sq ft)
Rudder, incl tab	3.17 m² (34.12 sq ft)
Tailplane	16.00 m² (172.22 sq ft)
Elevators (total, incl tab)	3.17 m² (34.12 sq ft)

WEIGHTS AND LOADINGS (A: Series 100, B: Series 300, C: Series 400):
Typical operating weight empty:	
A, B	12,520 kg (27,602 lb)
C	13,423 kg (29,592 lb)
Max fuel load: A, B, C	4,123 kg (9,090 lb)
Max payload: A, B	6,080 kg (13,404 lb)
C	7,177 kg (15,823 lb)
Max ramp weight: A, B, standard	19,990 kg (44,070 lb)
A, B, optional	20,865 kg (46,000 lb)
C	22,295 kg (49,152 lb)
Max T-O weight: A, B, standard	19,950 kg (43,980 lb)
A, B, optional	20,820 kg (45,900 lb)
C	22,250 kg (49,052 lb)
Max landing weight:	
A, B, standard	19,500 kg (42,990 lb)
A, B, optional	19,730 kg (43,500 lb)
C	21,500 kg (47,400 lb)
Max zero-fuel weight: A, B	18,600 kg (41,000 lb)
C	20,600 kg (45,415 lb)
Max wing loading:	
A, B, standard	285.0 kg/m² (58.37 lb/sq ft)
A, B, optional	297.4 kg/m² (60.92 lb/sq ft)
C	317.9 kg/m² (65.11 lb/sq ft)
Max power loading:	
A, standard	5.35 kg/kW (8.80 lb/shp)
A, optional	5.59 kg/kW (9.18 lb/shp)
B, standard	4.87 kg/kW (8.00 lb/shp)
B, optional	5.08 kg/kW (8.35 lb/shp)
C	5.43 kg/kW (8.92 lb/shp)

PERFORMANCE (Series 100):
Max operating Mach number	0.507
Typical cruising speed	282 knots (522 km/h; 325 mph)
Typical climb speed	170 knots (315 km/h; 196 mph) CAS
Typical descent speed	227 knots (420 km/h; 261 mph) CAS
Max operating altitude	7,620 m (25,000 ft)
Service ceiling, OEI, AUW of 17,770 kg (39,176 lb) ISA	4,300 m (14,100 ft)
Runway LCN (51 cm; 20 in flexible pavement), 34 × 10.75-R16 tyres at 5.86 bars (85 lb/sq in):	
AUW of 19,050 kg (42,000 lb)	16.9
AUW of 20,820 kg (45,900 lb)	18.4
T-O field length for typical mission T-O weight at S/L, ISA, 15° flap	890 m (2,920 ft)
Landing field length for typical mission landing weight at S/L, ISA, 35° flap	1,020 m (3,347 ft)
Range with 50 passengers and baggage, reserves for 45 min continued cruise at long-range schedule* and 87 nm (161 km; 100 mile) diversion:	
at standard MTOW:	
high-speed* procedure	1,109 nm (2,055 km; 1,277 miles)
min fuel procedure	1,216 nm (2,253 km; 1,400 miles)
at optional MTOW:	
high-speed* procedure	1,533 nm (2,841 km; 1,765 miles)
min fuel procedure	1,665 nm (3,085 km; 1,917 miles)

* *relevant speed and altitude details not supplied*

PERFORMANCE (Series 300 and 400, estimated):
Max operating Mach number: 300, 400	0.507
Typical cruising speed:	
300	284 knots (526 km/h; 327 mph)
400	274 knots (508 km/h; 316 mph)
Typical climb speed:	
300, 400	170 knots (315 km/h; 196 mph)

Fokker 50 Maritime Enforcer Mk 2 demonstrator, with underwing homing torpedoes and fuselage-mounted anti-ship missiles

Typical descent speed:
300, 400 227 knots (421 km/h; 261 mph)
Max operating altitude: 300, 400 7,620 m (25,000 ft)
Service ceiling OEI, ISA:
300 at 17,830 kg (39,308 lb) AUW 4,970 m (16,300 ft)
400 at 19,865 kg (43,794 lb) AUW 3,935 m (12,900 ft)
T-O field length for typical mission T-O weight at S/L, ISA, 15° flap: 300 850 m (2,790 ft)
400 1,135 m (3,725 ft)
Landing field length for typical mission landing weight at S/L, ISA, 35° flap: 300 1,015 m (3,330 ft)
400 1,125 m (3,691 ft)
Range with 50 passengers and baggage, reserves for 45 min continued cruise at long-range schedule and 87 nm (161 km; 100 mile) diversion:
at standard max T-O weight:
300, high-speed procedure
  1,097 nm (2,033 km; 1,263 miles)
400, high-speed procedure
  1,092 nm (2,023 km; 1,257 miles)
300, min fuel procedure
  1,186 nm (2,198 km; 1,365 miles)
400, min fuel procedure
  1,177 nm (2,181 km; 1,355 miles)
at optional max T-O weight:
300, high-speed procedure
  1,521 nm (2,818 km; 1,751 miles)
300, min fuel procedure
  1,628 nm (3,017 km; 1,874 miles)
OPERATIONAL NOISE LEVELS:
T-O 81.0 EPNdB
Approach 96.7 EPNdB
Sideline 85.0 EPNdB

FOKKER 50 SPECIAL MISSION VERSIONS

TYPE: Twin-turboprop multi-purpose aircraft.
CURRENT VERSIONS: **Black Crow Mk 2**: Communications/electronic intelligence version, with mission suite centred on ARCOSystems AR-7000 sigint system; applications include search, interception, direction finding, locating, analysis and recording of radar signals and communications transmissions from airborne, shipborne and land-based sources; max endurance 14 hours.

Kingbird Mk 2: Airborne early warning and airborne command/control version, with phased-array surveillance radar, IFF interrogator and ESM among standard sensors; can search for and track multiple airborne intruders (incl cruise missiles) and monitor seaborne and land-based targets; typical on-station time of 8 hours at 300 nm (556 km; 345 miles) from base.

Maritime Mk 2: Basic unarmed coastguard/maritime patrol version; duties can include coastal surveillance, search and rescue, and environmental control; airframe heavily treated with anti-corrosion measures: crew of up to six. Texas Instruments AN/APS-134 search radar in ventral radome, plus IR detection system for night viewing; can simultaneously detect and track up to 20 surface targets. *Details apply to this version, except where indicated.*

Maritime Enforcer Mk 2: Similar to Maritime Mk 2 but equipped for armed surveillance, anti-submarine and anti-ship warfare; enhanced systems (see under Avionics) and provisions for external stores (armament chosen and installed by operator); typical mission endurance of 10 hours. Second Fokker 50 converted into prototype/demonstrator (PH-LXX, first flight 10 December 1992), with US/British/Canadian equipment. In production; first deliveries expected early 1994.

Sentinel Mk 2: Border surveillance and reconnaissance version, capable of more than 2,000 nm (3,706 km; 2,303 miles) per mission; Motorola AN/APS-135(V) SLAR or Texas Instruments AN/APS-134(V)7 synthetic aperture radar (latter also with ship detection capability), plus pod-mounted E-O imaging system.

Troopship Mk 3: Multi-purpose military transport version; fuselage lengthened by 0.90 m (2 ft 11½ in) at front to incorporate starboard side 3.05 × 1.78 m (10 ft 0 in × 5 ft 10 in) cargo door; QC interior has capacity for 54 passengers, 50 armed/equipped troops, paratroops, 24 NATO stretchers plus sitting patients/medical attendants, or 5,000 kg (11,023 lb) of cargo; range with max cargo more than 1,400 nm (2,595 km; 1,612 miles).

CUSTOMERS: Maritime Enforcer Mk 2 selected by Republic of Singapore Air Force March 1991 (believed four ordered, plus four on option); one Black Crow sold to unnamed customer. RNethAF discussing possible Troopship requirement early 1993; this could be standard F50 size or stretched version (approx 60 seats).

Details generally as Fokker 50 Series 100 except as follows:

LANDING GEAR: Mainwheel tyre pressures 5.52 bars (80 lb/sq in) standard, 4.50 bars (65 lb/sq in) for low pressure tyres. Nosewheel tyre pressure increased to 3.80 bars (55 lb/sq in).

POWER PLANT: Additional centre-wing fuel tank of 2,310 litres (610 US gallons; 508 Imp gallons) capacity, and two 938 litre (248 US gallon; 206.5 Imp gallon) tanks on underwing pylons, giving overall total fuel capacity of 9,322 litres (2,463 US gallons; 2,051 Imp gallons).

ACCOMMODATION: Crew of two, with folding seat for third member if required. Main cabin of Maritime Mk 2 fitted out as tactical compartment (for two to four operators), containing advanced avionics, galley, toilet and crew rest area. Enforcer accommodates crew of eight including two pilots; tactical co-ordinator (TACCO) responsible for off-airways navigation and overall efforts of mission crew; acoustic sensor operator (ASO) to handle active and passive sonobuoys, acoustic receivers and processor display system; non-acoustic sensor operator (NASO) controlling search radar and electronic surveillance subsystem; and two observers. Bubble windows for observers at front of main cabin. Rear cabin door openable in flight.

SYSTEMS: Oxygen system includes individual supply for each tactical crew member. Methyl bromide fire extinguishing system, with flame detectors.

AVIONICS (Maritime Enforcer Mk 2): Com/nav equipment comprises two Collins AN/ARC-217(V) HF transceivers and HF com, three Collins AN/ARC-210 VHF/UHF transceivers, interphone, crew address system, two Litton LTN-92 inertial GPS navigation systems, two Honeywell AZ-800 air data computers, Bendix/King DFS-43 radio compass, two TRT AHV-530 radio altimeters, Collins DF-301E VHF/UHF direction finder, two Bendix/King VNS-41 VOR/ILS/marker beacon receivers, Honeywell EDZ-806 EFIS, Teledyne AN/APX-109 IFF transponder/interrogator, Honeywell SPZ-600 AFCS, Honeywell P-650 weather radar, Sundstrand Mk VII low altitude warning system, and Collins Tacan. Texas Instruments AN/APS-134(V)7 360° search radar in ventral radome. Central tactical computer and display system, radar detection and display system, on-top position indicator/receiver, and CDC AN/UYS-503 sonobuoy processing system with Alliant M101E acoustic signal recorder and Flightline AN/ARR-502 sonobuoy signal receiver. Litton AN/ALR-85(V)3 electronic surveillance and monitoring equipment to detect radar transmissions, which can be classified and recorded and their bearings transferred to

Artist's impression of the Fokker 50 Troopship Mk 3 with large cargo door

tactical display. Alliant AN/ASH-34 mission recorder. GEC-Marconi VOO-1069 TICM II infra-red detection system (IRDS). CAE AN/ASQ-504(V) MAD. Rockwell TE-237P-6 data link with available ground or shipborne systems can be provided.

EQUIPMENT: NGL and Alkan 8025 active/passive sonobuoy launchers; 60 sonobuoys in rear of cabin. Nobeltech BOP-300 chaff/flare dispenser. Auxiliary fuel tanks can be carried on central underwing pylons.

ARMAMENT (Maritime Enforcer Mk 2): Fokker installs Alkan T930 stores management system and provisions for armament; weapon mix and purchase is up to customer. Two 907 kg (2,000 lb) stores attachments on fuselage and three under each wing (capacities 295 kg; 650 lb inboard, 680 kg; 1,500 lb in centre, and 113 kg; 250 lb outboard). Typical ASW armament can include two or four Mk 44, Mk 46, Sting Ray or A244/S torpedoes and/or depth bombs. For anti-shipping warfare, two AM 39 Exocet, AGM-65F Maverick, AGM-84A Harpoon, Sea Skua, Sea Eagle or similar air-to-surface missiles can be carried.

WEIGHTS AND LOADINGS (A: Maritime Mk 2, B: Maritime Enforcer Mk 2):

Operating weight empty: A	13,314 kg (29,352 lb)
B (typical)	14,796 kg (32,620 lb)
Max fuel (incl pylon tanks): A	7,257 kg (16,000 lb)
B	7,511 kg (16,560 lb)
Normal T-O weight: A, B	20,820 kg (45,900 lb)
Max T-O weight: A, B	21,545 kg (47,500 lb)
Emergency overload T-O weight:	
B	22,680 kg (50,000 lb)
Max landing weight: A, B	19,730 kg (43,500 lb)
Max zero-fuel weight: A, B	18,144 kg (40,000 lb)
Max wing loading: A, B	291.6 kg/m² (59.75 lb/sq ft)
Max power loading: A, B	6.39 kg/kW (10.5 lb/shp)

PERFORMANCE (at normal T-O weight except where indicated):

Normal cruising speed	259 knots (480 km/h; 298 mph)
Typical search speed at 610 m (2,000 ft)	
	150 knots (277 km/h; 172 mph)
Service ceiling: A, B	7,620 m (25,000 ft)
Service ceiling, OEI: A	3,565 m (11,700 ft)
Runway LCN (42 per cent tyre deflection) at 15,875 kg (35,000 lb) AUW: A:	
rigid pavement, L 76.2 cm (30 in)	10.4
flexible pavement, h 25.4 cm (10 in)	11.4
flexible pavement, h 12.7 cm (5 in)	9.0
Runway LCN (42 per cent tyre deflection) at 20,410 kg (45,000 lb) AUW: A:	
rigid pavement, L 76.2 cm (30 in)	16.0
flexible pavement, h 25.4 cm (10 in)	14.8
flexible pavement, h 12.7 cm (5 in)	12.0
Runway CBR, unpaved soil, h 25.4 cm (10 in), 3,000 passes: A:	
AUW of 15,875 kg (35,000 lb)	6.2%
AUW of 20,410 kg (45,000 lb)	7.8%
T-O run at S/L, AUW of 21,320 kg (47,000 lb):	
ISA	1,525 m (5,000 ft)
ISA + 20°C	1,700 m (5,575 ft)

Landing distance (unfactored, ISA at S/L), landing weight of 18,990 kg (41,866 lb) 762 m (2,500 ft)
Max radius of action with 1,814 kg (4,000 lb) mission load 1,200 nm (2,224 km; 1,382 miles)
Max range, no reserves 3,680 nm (6,820 km; 4,237 miles)
Max endurance, reserves for 30 min hold, 5% fuel remaining 14 h 42 min

FOKKER 100

TYPE: Twin-turbofan short/medium-haul airliner.

PROGRAMME: Announced simultaneously with Fokker 50; derived from F28 Mk 4000, which it superseded in production; built in collaboration with Deutsche Aerospace Airbus and Shorts; first flight of first prototype (PH-MKH) 30 November 1986; second prototype (PH-MKC) joined flight test and certification programme 25 February 1987. Complies with FAR Pt 36 Stage 3 noise requirements; Dutch RLD certification to JAR Pt 25 received 20 November 1987, followed by Cat. IIIB autoland certificate June 1988. First flight of Swissair aircraft 30 December 1987, delivered 29 February 1988; FAA type approval granted 30 May 1989; certification of version with higher rated Tay Mk 650 engines (first flown on PH-MKH on 8 June 1988) received 1 July 1989; first delivery of Tay Mk 650 version same day to USAir. Output planned to rise to 60 a year by early 1993; new main landing gear to be introduced, and large upward opening cargo doors and forward opening passenger door to become standard, by end of 1993.

CURRENT VERSIONS: **Fokker 100**: Standard airliner; *description applies to this version.*
ELFIN: Testbed (European Laminar Flow INvestigation), fitted with full-chord laminar flow glassfibre glove around mid-section of starboard wing; first flight (PH-MKC) 12 November 1991; programme successfully completed September 1992.
Fokker 100QC: Quick-change version, manufactured as standard 100 by Fokker and modified to QC specification by a subcontractor; 20-minute changeover by three-person ground crew claimed. Modifications include large (3.40 × 1.93 m; 11 ft 2 in × 6 ft 4 in) cargo door at front on port side; 11 seat pallets (interchangeable with cargo containers). Capacity in all-cargo role for five LD9/LD7 containers plus one half-size container, or up to 11 LD3 containers. Max structural payload 11,500 kg (25,353 lb), range with typical 10,000 kg (22,046 lb) cargo load estimated at more than 1,600 nm (2,965 km; 1,842 miles). All-passenger version seats 88 with smaller overhead bins and additional side-bins.
Fokker Executive Jet 100: VIP/corporate shuttle version; optional belly fuel tanks for extended range; customised interior.
Fokker 70: Shortened version: described separately.
Fokker 130: Stretched version: described separately.

CUSTOMERS: Firm orders totalled 245 Fokker 100s by 1 June 1993, of which 160 delivered; 114 more then on option. Customers (incl 52 leases) include Air Ivoire (two), Air Littoral (six), Air UK (five), American Airlines (75), Aviacsa (four), Bangkok Airways (one), Berline (two), British Airways (three), China Eastern Airlines (10), Corse Méditerranée (two), Deutsche BA (one), Garuda Indonesia (12), Iran Air (six), Ivory Coast government (one), Korean Air (12), Mexicana (10), Pelita Air Service (one), Portugalia (five), Sempati Air (seven), Swissair (10), TABA (two), TAM Brazil (nine), TAT European Airlines (18), USAir (40) and GPA Fokker 100 (31), plus nine undisclosed.

DESIGN FEATURES: Compared to F28 Mk 4000, Fokker 100 has stretched fuselage, extended and redesigned wings, Rolls-Royce Tay Mk 620 or 650 turbofans, completely new CRT and digital ARINC 700 flight deck, standard Cat. IIIa, lowest OWE/seat in its class, new cabin interior, and extensively modernised systems. Extensive use of composites.
Major options include intermediate (44,452 kg; 98,000 lb) and high (45,813 kg; 101,000 lb) max T-O weights, latter from 1993; higher thrust Tay Mk 650 engines; Cat. IIIb autoland; higher capacity air-conditioning system; downward opening passenger door with integral stairs; and moving belt loading system. Polished outer skin available as customer requirement.
Fokker designed transonic wing sections, offering substantially improved aerodynamic efficiency, especially at high speed; thickness/chord ratio up to 12.3 per cent on inner panels, 9.6 per cent at tip; dihedral 2° 30′; sweepback at quarter-chord 17° 27′.

FLYING CONTROLS: Hydraulically actuated; irreversible ailerons (in third mode, both ailerons driven manually via servo tabs), boosted elevators with manual backup, and rudder with manual third mode; variable incidence tailplane (third mode is electric operation); double-slotted Fowler flaps with electrical alternative extension; five panel lift dumpers in front of flaps on each wing; airbrakes form rear end of fuselage.

STRUCTURE: Light alloy, hot-bonded for 45,000-cycle crack-free life and 90,000-cycle economic repair life; fail-safe except for CFRP ailerons and flaps, AFRP wing/fuselage fairing panels, honeycomb sandwich/multiple spar fin, AFRP dorsal fin, CFRP rudder, and CFRP/GFRP with Nomex core quickly detachable sandwich floor panels. Light metal clamshell type airbrakes at end of fuselage. Nacelles manufactured from composite material. Deutsche Airbus builds large fuselage sections and tail section, Shorts the wings; Grumman is subcontractor for engine nacelles and thrust reversers; main landing gear by Dowty until end of 1993, Menasco thereafter.

LANDING GEAR: Hydraulically retractable tricycle type, with twin wheels on each unit. Main units retract inward into wing/body fairing; Dowty nosewheel retracts forward. Shock absorber in each unit. Goodyear tyres, size H40 × 14-19 on main units (pressure 9.38 bars; 136 lb/sq in), size 24 × 7.7-10 (pressure 6.21 bars; 90 lb/sq in) on nose unit. Loral multiple-disc carbon brakes, with anti-skid system. Steerable nose unit (effective angle ±76°).

POWER PLANT: Two 61.6 kN (13,850 lb st) Rolls-Royce Tay Mk 620 turbofans standard, fitted with thrust reversers and pylon mounted on sides of rear fuselage. Option of 67.2 kN (15,100 lb st) Tay Mk 650 turbofans. Fuel in 4,820 litres (1,273 US gallon; 1,060 Imp gallon) main tank in each wing as standard. In 1993, at same time as 101,000 lb max T-O weight option, an integral centre wing tank with capacity of 3,725 litres (984 US gallons; 819 Imp gallons) becomes standard, raising total capacity to 13,365 litres (3,531 US gallons; 2,940 Imp gallons). Refuelling point under starboard wing, near wing/fuselage belly fairing. Oil capacity (two engines) 41 kg (90 lb).

ACCOMMODATION: Crew of two on flight deck; three cabin attendants. Standard accommodation for 107 passengers in five-abreast seating (3+2) at 81 cm (32 in) pitch. Optional layouts include 12 first class seats (four-abreast) at 91 cm (36 in) pitch plus 85 economy class (five-abreast) at 32 in; 55 business class at 86 cm (34 in) plus 50 economy class, all five-abreast; or 122 tourist class passengers at 74 cm (29 in) pitch. Reduced galley and stowage space in 122-seat layout. Standard layout includes two galleys, two toilets, two wardrobes, two other stowage/wardrobe compartments, and overhead bins. Outward and forward opening passenger door at front of cabin on port side; outward opening service/emergency door opposite on starboard side. Optional downward opening passenger door with integral stairs. Two overwing emergency exits (inward opening plug type) each side. Two underfloor baggage/cargo holds (one forward of wing, one aft), with three identical upward opening cargo doors on starboard side. Option for moving belt loading system. Entire accommodation pressurised and air-conditioned.

SYSTEMS: AiResearch air-conditioning and pressurisation system (max differential 0.52 bar; 7.45 lb/sq in). Two fully independent hydraulic systems for actuation of flight controls, landing gear, brakes and nosewheel steering. AiResearch pneumatic system. Sundstrand integrated drive generator electrical supply system (now with proximity switches instead of microswitches, for higher reliability). Oxygen system for flight crew and passengers. AiResearch thermal anti-icing system for wings and tail unit. Electric anti-icing of flight deck windows, pitot tubes, static vents, angle of attack vanes and ice detector probe. Garrett GTCP36-150RR APU, with digital control, standard; can be operated at up to 10,670 m (35,000 ft).

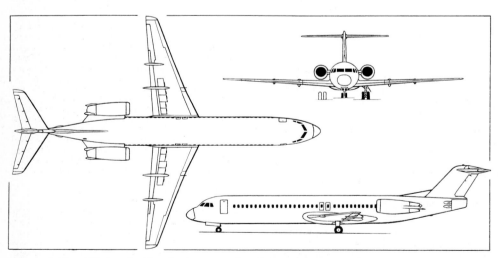

Fokker 100 short/medium-haul transport (two Rolls-Royce Tay turbofans) *(Dennis Punnett)*

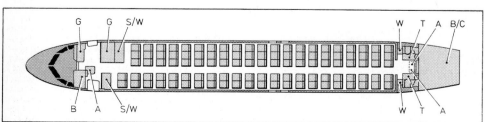

Fokker 100 standard configuration (107 seats at 81 cm; 32 in pitch). A: attendant seat, B: baggage, C: cargo, G: galley, S: stowage, T: toilet, W: wardrobe *(Jane's/Mike Keep)*

Fokker 100 twin-turbofan short/medium-range transport in the insignia of Air UK

AVIONICS: Standard avionics include Collins EFIS electronic display systems: primary flight display (PFD) and navigation display (ND) for each pilot, and multi-function display system (MFDS), consisting of two CRTs on centre flight instrument display panel. 'Dark cockpit philosophy' emphasised in every system. EFIS and MFDS display units identical in size. Collins digital aircraft flight control and augmentation system (AFCAS) for Cat. IIIa automatic landing plus dual-channel full flight regime autothrottle system. Dual flight management system (ARINC 702) by Honeywell, triple AHRS (ARINC 705), dual digital air data systems (ARINC 706), dual radio altimeters (ARINC 707), weather radar (ARINC 708, on ND), dual DME (ARINC 709), dual ILS (ARINC 710), dual VOR with marker beacon receiver (ARINC 711), single ADF (ARINC 712), PA system (ARINC 715), dual VHF com (ARINC 716), digital flight data acquisition unit (ARINC 717), flight data recorder (ARINC 717), ATC transponder (ARINC 718), GPWS (ARINC 723), flight warning computer system (ARINC 726) with full flight envelope protection, cockpit voice recorder (ARINC 557), and digitally controlled audio management system (ARINC 736).

Optional avionics include Selcal (ARINC 714), third VHF com (ARINC 716), aircraft condition monitoring system (ARINC 717), second ATC (ARINC 718), single or dual HF com (ARINC 719), ACARS (ARINC 724), TCAS (ARINC 735), windshear, Cat. IIIb autoland capability, data link, music reproducer, and noise abatement profile (certificated November 1990) as optional addition to AFCS.

DIMENSIONS, EXTERNAL:
Wing span	28.08 m (92 ft 1½ in)
Wing chord: at root	5.28 m (17 ft 4 in)
at tip	1.26 m (4 ft 1½ in)
Wing aspect ratio	8.4
Length overall	35.53 m (116 ft 6¾ in)
Fuselage: Length	32.50 m (106 ft 7½ in)
Max diameter	3.30 m (10 ft 10 in)
Height overall	8.50 m (27 ft 10½ in)
Tailplane span	10.04 m (32 ft 11¼ in)
Wheel track (c/l of shock struts)	5.04 m (16 ft 6½ in)
Wheelbase	14.01 m (45 ft 1½ in)
Passenger door (fwd, port): Height	1.82 m (5 ft 11¾ in)
Width	0.78 m (2 ft 6¾ in)
Service door (fwd, stbd): Height	1.30 m (4 ft 3¼ in)
Width	0.63 m (2 ft 0¾ in)
Cargo compartment doors (fwd and rear, stbd):	
Height (each)	1.43 m (4 ft 8¼ in)
Width (each)	1.44 m (4 ft 8¾ in)
Height to sill (MTOW):	
fwd hold, fwd door	1.20 m (3 ft 11¼ in)
fwd hold, rear door	1.27 m (4 ft 2 in)
aft hold door	1.36 m (4 ft 5½ in)
Overwing emergency exits (four):	
Height (each)	0.91 m (3 ft 0 in)
Width (each)	0.51 m (1 ft 8 in)

DIMENSIONS, INTERNAL:
Cabin, excl flight deck: Length	21.19 m (69 ft 6¼ in)
Max length of seating area	18.80 m (61 ft 8¼ in)
Max width	3.10 m (10 ft 2 in)
Width at floor	2.89 m (9 ft 5¾ in)
Max height	2.01 m (6 ft 7¼ in)
Max floor area	58.48 m² (629.5 sq ft)
Max volume	107.58 m³ (3,799 cu ft)
Overhead stowage bins (total)	5.15 m³ (181.9 cu ft)
Additional baggage space (total)	3.00 m³ (105.9 cu ft)
Underfloor compartment volume:	
fwd	9.48 m³ (334.8 cu ft)
aft	7.24 m³ (255.7 cu ft)

AREAS:
Wings, gross	93.50 m² (1,006.4 sq ft)
Ailerons (total)	3.53 m² (38.0 sq ft)
Trailing-edge flaps (total)	17.08 m² (183.85 sq ft)
Lift dumpers (total)	5.30 m² (57.05 sq ft)
Rudder	2.30 m² (24.76 sq ft)
Elevators (total)	3.96 m² (42.63 sq ft)
Airbrakes (total)	3.62 m² (38.97 sq ft)

WEIGHTS AND LOADINGS (A: standard weights and Tay 620, B: intermediate gross weight and Tay 650, C: 1993 high gross weight and Tay 650):
Typical operating weight empty:	
A	24,593 kg (54,218 lb)
B	24,727 kg (54,513 lb)
C	24,747 kg (54,557 lb)
Max payload (weight-limited): A	11,108 kg (24,489 lb)
B	12,013 kg (26,484 lb)
C	11,993 kg (26,440 lb)
Max ramp weight: A	43,320 kg (95,504 lb)
B	44,680 kg (98,500 lb)
C	46,040 kg (101,500 lb)
Max T-O weight: A	43,090 kg (95,000 lb)
B	44,450 kg (98,000 lb)
C	45,810 kg (101,000 lb)
Max landing weight: A	38,780 kg (85,500 lb)
B, C	39,915 kg (88,000 lb)
Max zero-fuel weight: A	35,830 kg (78,991 lb)
B, C	36,740 kg (81,000 lb)
Max wing loading: A	460.8 kg/m² (94.39 lb/sq ft)
B	475.4 kg/m² (97.37 lb/sq ft)
C	489.9 kg/m² (100.35 lb/sq ft)
Max power loading: A	350.0 kg/kN (3.43 lb/lb st)
B	330.7 kg/kN (3.25 lb/lb st)
C	340.8 kg/kN (3.34 lb/lb st)

PERFORMANCE (A, B and C as for Weights and Loadings):
Max operating Mach number: A, B, C	0.77
Max operating speed at 7,775 m (25,500 ft), ISA:	
A, B, C	462 knots (856 km/h; 532 mph)
Approach speed at max landing weight:	
A	128 knots (237 km/h; 147 mph)
B, C	130 knots (241 km/h; 150 mph)
Service ceiling: A, B, C	10,670 m (35,000 ft)
FAR T-O field length at S/L, ISA, at max T-O weight:	
A	1,855 m (6,086 ft)
B	1,720 m (5,643 ft)
C	1,825 m (5,988 ft)
FAR landing field length at S/L, ISA, at max landing weight: A	1,320 m (4,330 ft)
B, C	1,350 m (4,420 ft)
Range with 107 passengers and baggage:	
A	1,290 nm (2,390 km; 1,485 miles)
B	1,550 nm (2,872 km; 1,784 miles)
C	1,680 nm (3,113 km; 1,934 miles)

OPERATIONAL NOISE LEVELS (A, B and C as for Weights and Loadings):
T-O (flyover): A	83.4 EPNdB
B	81.8 EPNdB
C	82.7 EPNdB
T-O (sideline): A	89.3 EPNdB
B	91.7 EPNdB
C	91.6 EPNdB
Approach: A	93.1 EPNdB
B	93.0 EPNdB
C	93.0 EPNdB

FOKKER 70

TYPE: Twin-turbofan short/medium-haul airliner.
PROGRAMME: Authorisation to proceed given November 1992; fuselage structure derived from Fokker 100 by removing two fuselage plugs (one forward and one aft of wing); built on same production line as Fokker 100; built in collaboration with Deutsche Aerospace (fuselage sections) and Shorts (wing); modification of second Fokker 100 prototype (PH-MKC) into Fokker 70 configuration started 9 October 1992; first flight 4 April 1993; first delivery due late 1994.
CURRENT VERSIONS: **Fokker 70**: Standard airliner; *description applies to this version.*
Fokker 70A: Dedicated US version for regional operators, with additional main deck cargo hold limiting seating capacity to 70 seats; MZFW and MLW 1,361 kg (3,000 lb) lower than standard weights.
Fokker Executive Jet 70: VIP/shuttle version; extended range available optionally with belly tanks; interior to be custom built.
DESIGN FEATURES: Fuselage shortened by 4.62 m (15 ft 2 in) compared to Fokker 100; one pair of overwing emergency exits removed; Rolls-Royce Tay Mk 620 turbofans; downward opening passenger door with integral stairs; digital ARINC 700 CRT flight deck; Cat. II capability; considerable use of composite materials. Major options include intermediate (38,100 kg; 84,000 lb) and high (39,915 kg; 88,000 lb) max take-off weights, forward opening passenger door, Cat. IIIa autoland capability, integral centre wing tank, second rear toilet, second forward toilet. Polished outer skin available as customer requirement.

Fokker 70 features Fokker 100's transonic wing, offering substantially improved aerodynamic efficiency, especially at high speed; thickness/chord ratio up to 12.3 per cent on inner panels, 9.6 per cent at tip; dihedral 2° 30'; sweepback at quarter-chord 17° 27'.
FLYING CONTROLS: As for Fokker 100.
STRUCTURE: As for Fokker 100.
LANDING GEAR: Hydraulically retractable tricycle type, with twin wheels on each unit. Main units, by Menasco, retract inward into wing/body fairing; nose unit, by Dowty, retracts forward. Shock absorber in each unit. Tyres size H40 × 14-19 on main units (pressure 9.1 bars; 132 lb/sq in), 24 × 7.7-10 on nose unit (pressure 6.34 bars; 92 lb/sq in). Loral multiple-disc carbon brakes, with anti-skid system. Steerable nose unit (effective angle ±76°).
POWER PLANT: As described for Fokker 100, but without option for Tay Mk 650 engines.
ACCOMMODATION: Crew of two on flight deck; two cabin attendants. Standard accommodation for 79 passengers in five-abreast seating (3+2) at 78.5/81 cm (31/32 in) pitch. Standard layout includes one galley (forward, starboard), one toilet (aft, port), three wardrobes (one forward, starboard, and two aft, port and starboard), and two stowages (one forward, port, and one aft, starboard), plus overhead bins. Outward and downward opening passenger door with integral stairs at front of cabin on port side. Outward and forward opening service/emergency door opposite on starboard side. Optional passenger door outward and forward opening. One overwing emergency exit (inward opening

220 NETHERLANDS: AIRCRAFT—FOKKER

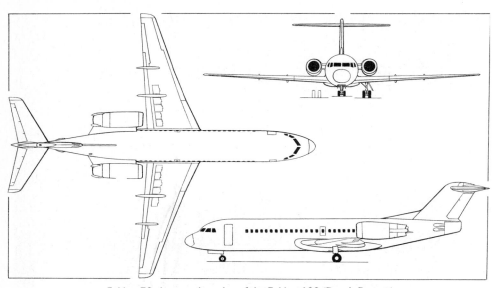

Fokker 70 shortened version of the Fokker 100 *(Dennis Punnett)*

plug type) on each side. Two underfloor baggage/cargo holds (one forward of wing, one aft), with two upward opening doors on starboard side.

SYSTEMS: As described for Fokker 100.

AVIONICS: Standard avionics include Collins EFIS electronic display systems: primary flight display (PFD) and navigation display (ND) for each pilot, and multi-function display system (MFDS), consisting of two CRTs on centre flight instrument display panel. 'Dark cockpit philosophy' emphasised in every system. EFIS and MFDS display units identical in size. Collins digital aircraft flight control and augmentation system (AFCAS) plus dual-channel full flight regime autothrottle system, Cat. II autopilot capability, dual AHRS + YRS (ARINC 705), dual digital air data systems (ARINC 706), dual radio altimeters (ARINC 707), weather radar (ARINC 708, on ND), dual DME (ARINC 709), dual ILS (ARINC 710), dual VOR with marker beacon receiver (ARINC 711), single ADF (ARINC 712), PA system (ARINC 715), dual VHF com (ARINC 716), flight data recorder (ARINC 717), digital flight data acquisition unit (ARINC 717), ATC transponder (ARINC 718), GPWS (ARINC 723), flight warning computer system (ARINC 726) with full flight envelope protection, digital control audio management system (ARINC 736), cockpit voice recorder (ARINC 557).

Optional avionics include dual Honeywell flight management system (ARINC 702), Selcal (ARINC 714), aircraft condition monitoring system (ARINC 717), second ATC (ARINC 718), single or dual HF com (ARINC 719), ACARS (ARINC 724), TCAS (ARINC 735), windshear, third VHF com, Cat. IIIa autoland system, single LNAV, data link, music reproducer, third radio altimeter, noise abatement profile as optional addition to AFCS.

DIMENSIONS, EXTERNAL: As Fokker 100 except:
Length overall 30.91 m (101 ft 4¾ in)
Fuselage: Length 27.88 m (91 ft 5½ in)
Wheelbase 11.54 m (37 ft 10½ in)
Passenger door (fwd, port): Height 1.91 m (6 ft 3¼ in)
Width 0.86 m (2 ft 9¾ in)
Cargo compartment door (fwd):
Height 1.43 m (4 ft 8¼ in)
Width 1.44 m (4 ft 8¾ in)
Cargo compartment door (aft):
Height 1.22 m (4 ft 0 in)
Width 0.97 m (3 ft 2¼ in)

DIMENSIONS, INTERNAL: As Fokker 100 except:
Cabin, excl flight deck: Length 16.57 m (54 ft 4¼ in)
Max length of seating area 13.31 m (43 ft 8 in)
Max floor area 45.07 m² (485.1 sq ft)
Max volume 84.02 m³ (2,967.1 cu ft)
Overhead stowage bins (total) 3.77 m³ (133.1 cu ft)
Additional baggage space (total) 5.03 m³ (177.6 cu ft)
Fwd cargo hold 8.11 m³ (286.4 cu ft)
Aft cargo hold 4.64 m³ (163.9 cu ft)

AREAS: As Fokker 100

WEIGHTS AND LOADINGS (A: standard weights and fuel, B: intermediate gross weight and optional fuel, C: high gross weight and optional fuel):
Typical operating weight empty:
A, B, C 22,673 kg (49,985 lb)
Max payload (weight-limited): A 9,302 kg (20,507 lb)
B 9,982 kg (22,006 lb)
C 10,890 kg (24,008 lb)
Max ramp weight: A 36,965 kg (81,493 lb)
B 38,325 kg (84,492 lb)
C 40,140 kg (88,493 lb)

Max T-O weight: A 36,740 kg (80,997 lb)
B 38,100 kg (83,996 lb)
C 39,915 kg (87,997 lb)
Max landing weight: A (normal) 34,020 kg (75,000 lb)
A (optional), B (normal) 35,830 kg (78,991 lb)
B (optional), C 36,740 kg (80,997 lb)
Max zero-fuel weight: A 31,975 kg (70,492 lb)
B 32,655 kg (71,991 lb)
C 33,365 kg (73,557 lb)
Max wing loading: A 392.9 kg/m² (80.47 lb/sq ft)
B 407.5 kg/m² (83.46 lb/sq ft)
C 426.9 kg/m² (87.44 lb/sq ft)
Max power loading: A 298.2 kg/kN (2.92 lb/lb st)
B 309.3 kg/kN (3.03 lb/lb st)
C 324.0 kg/kN (3.18 lb/lb st)

PERFORMANCE (estimated; A, B and C as in Weights and Loadings):
Max operating Mach number 0.7
Max operating speed at 7,770 m (25,500 ft), ISA:
A, B, C 462 knots (856 km/h; 532 mph)
Approach speed at max landing weight:
A 118 knots (219 km/h; 136 mph)
B 121 knots (224 km/h; 139 mph)
C 122 knots (226 km/h; 140 mph)
Service ceiling: A, B, C 10,670 m (35,000 ft)
T-O field length at S/L, ISA, at max T-O weight:
A 1,391 m (4,564 ft)
B 1,469 m (4,820 ft)
C 1,573 m (5,161 ft)
Landing field length at S/L, ISA, at max landing weight:
A 1,208 m (3,963 ft)
B 1,251 m (4,105 ft)
C 1,274 m (4,180 ft)
Range with 79 passengers and baggage:
A 1,080 nm (2,001 km; 1,243 miles)
B 1,415 nm (2,622 km; 1,629 miles)
C 1,840 nm (3,410 km; 2,119 miles)

OPERATIONAL NOISE LEVELS (A, estimated):
T-O (flyover) 78 EPNdB
T-O (sideline) 90 EPNdB
Approach 90 EPNdB

FOKKER 130

TYPE: Twin-turbofan airliner; proposed stretch derivative of Fokker 100.

PROGRAMME: Wind tunnel testing started June 1991; wing configuration freeze scheduled for second half of 1993; first deliveries 1997.

DESIGN FEATURES: Fuselage lengthened by 6.22 m (20 ft 5 in); 1.50 m (4 ft 11 in) wingroot plugs.

POWER PLANT: To be decided.

ACCOMMODATION: 137 passengers at 81 cm (32 in) pitch; 128 passengers in mixed class configuration (12 seats at 91 cm; 36 in pitch and 116 at 81 cm; 32 in).

DIMENSIONS, EXTERNAL:
Wing span 31.09 m (102 ft 0 in)
Length overall 41.75 m (136 ft 11¾ in)
Fuselage: Length 38.80 m (127 ft 3½ in)
Max diameter 3.30 m (10 ft 10 in)

First prototype Fokker 70 twin-turbofan short/medium-haul airliner

Height overall	8.74 m (28 ft 8 in)
Tailplane span	10.04 m (32 ft 11 in)
Wheel track	5.15 m (16 ft 10¾ in)
Wheelbase	19.10 m (62 ft 8 in)
Passenger door (fwd, port): Height	1.98 m (6 ft 6 in)
Width	0.96 m (3 ft 1¾ in)
Service door (fwd, stbd): Height	1.38 m (4 ft 6¼ in)
Width	0.72 m (2 ft 4¼ in)
Service door (aft, port): Height	1.53 m (5 ft 0¼ in)
Width	0.73 m (2 ft 4¾ in)

WEIGHTS AND LOADINGS (estimated):
Operating weight empty	31,555 kg (69,566 lb)
Max payload (weight-limited)	16,525 kg (34,431 lb)
Max ramp weight: standard	55,565 kg (122,500 lb)
optional	59,195 kg (130,502 lb)
Max T-O weight: standard	55,340 kg (122,000 lb)
optional	58,965 kg (129,995 lb)
Max landing weight	51,255 kg (113,000 lb)
Max zero-fuel weight	48,080 kg (106,000 lb)

PERFORMANCE (estimated):
Max operating Mach number	0.77
Max operating speed at 7,770 m (25,500 ft), ISA	462 knots (856 km/h; 532 mph)
Service ceiling	10,670 m (35,000 ft)
T-O field length at S/L, ISA, at max T-O weight	1,829 m (6,000 ft)
Landing field length at S/L, ISA, at max landing weight	1,433 m (4,700 ft)
Range with 137 passengers and baggage:	
standard	1,700 nm (3,150 km; 1,958 miles)
optional	2,365 nm (4,383 km; 2,723 miles)

OPERATIONAL NOISE LEVELS: Comply with FAR Pt 36 Stage 3, ICAO Annex 16 Chapter 3

Model of the 128/137-passenger twin-turbofan Fokker 130

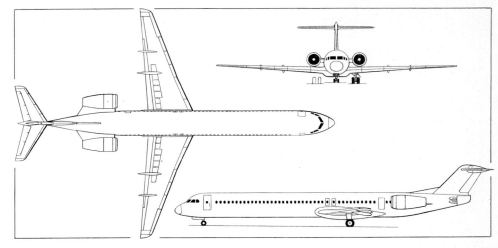

Fokker 130 proposed stretched derivative of the Fokker 100 (Dennis Punnett)

NEW ZEALAND

PAC
PACIFIC AEROSPACE CORPORATION LIMITED
Private Bag HN 3027, Hamilton Airport, Hamilton
Telephone: 64 (7) 843 6144 and 6069
Fax: 64 (7) 843 6134
Telex: NZ 21242 PACORP
CHIEF EXECUTIVE: Gordon Pennefather
SALES MANAGERS:
 Bob Geer
 Alan Thoresen

Former New Zealand Aerospace Industries Ltd reconstituted 1 July 1982 as PAC; maintains spares support for Airtrainer CT4A/CT4B, Fletcher FU24-954 series and Cresco 08-600; continuing to market FU24-954, Cresco 08-600 and variants of CT4 Airtrainer; manufactured installation kits for RNZAF's Project Kahu A-4K Skyhawk upgrade (see 1991-92 *Jane's* and 1993-94 *Jane's Civil and Military Aircraft Upgrades*). Shareholding divided between ASTA of Australia (75.1 per cent) and Lockheed Corporation of USA (24.9 per cent); will handle subcontract work from ASTA for Airbus and Boeing.

PAC FLETCHER FU24-954
TYPE: Agricultural and general purpose aircraft.
PROGRAMME: First flight of US-built FU24 prototype July 1954; first flight of first US production aircraft five months later; type certificate granted 22 July 1955; all manufacturing and sales rights transferred to New Zealand 1964; factory refurbishment/upgrade programme offered by PAC for earlier FU24 series aircraft.
CURRENT VERSIONS: **FU24-954**: Current standard model.
CUSTOMERS: Total of 297 FU24 series aircraft built by mid-1992, including 110 for New Zealand customers, 40 for Australia and 13 for Pakistan; other export deliveries made to Malaysia, Sudan, Syria (five), Thailand, Turkey, Uruguay and Venezuela.
DESIGN FEATURES: Primary configuration as agricultural aircraft; with appropriate hopper base, can also be used for firefighting; as utility aircraft, hopper is removed and new hopper top and floor are installed. Suitable for cargo/freight or passenger-carrying, or as aerial ambulance.
Wing section NACA 4415 (constant); 8° dihedral on outer wing panels; incidence 2°.
FLYING CONTROLS: Conventional mechanical; horn balanced ailerons (single row of vortex generators forward of each); all-moving slab tailplane with full span anti-servo/trim tab; ground adjustable tab on rudder; single-slotted flaps.
STRUCTURE: Conventional light alloy; two-spar wing; cockpit area stressed for $25g$ impact.
LANDING GEAR: Non-retractable tricycle type; steerable nosewheel. Fletcher air-oil shock struts. Cleveland wheels and hydraulic disc brakes on main units; Goodyear tyres, size 8.50-6 (6 ply); pressures vary from 0.76 to 2.07 bars (11-30 lb/sq in).
POWER PLANT: One 298 kW (400 hp) Textron Lycoming IO-720-A1A or A1B flat-eight engine, driving a Hartzell HC-C3YR-1R/847SR three-blade constant-speed metal propeller. Fuel tanks in wing leading-edges; total usable capacity 254 litres (67 US gallons; 55.8 Imp gallons) normal, 481 litres (127 US gallons; 105.75 Imp gallons) with optional long-range tanks.
ACCOMMODATION: Two-seat cockpit with sliding canopy and roll-over protection. Rear compartment, with freight door at rear to port, holds six passengers, equivalent freight or a 1,211 litre (320 US gallon; 266 Imp gallon) liquid or 1,066 kg (2,350 lb) dry hopper. Transland or Micronair application equipment. Options include dual controls, heating, full instruments, radio, dual mainwheels and wheel/leg fairings.

DIMENSIONS, EXTERNAL:
Wing span	12.81 m (42 ft 0 in)
Length overall	9.70 m (31 ft 10 in)
Height overall	2.84 m (9 ft 4 in)
Tailplane span	4.22 m (13 ft 10 in)
Wheel track	3.71 m (12 ft 2 in)
Wheelbase	2.28 m (7 ft 6 in)
Propeller diameter	2.18 m (7 ft 2 in)
Passenger/cargo door (port, rear):	
Height	0.97 m (3 ft 2 in)
Width	0.94 m (3 ft 1 in)

DIMENSIONS, INTERNAL:
Cabin: Length	3.18 m (10 ft 5 in)
Max width	1.22 m (4 ft 0 in)
Max height	1.27 m (4 ft 2 in)
Floor area	3.87 m² (41.7 sq ft)
Volume aft of hopper	3.37 m³ (119.0 cu ft)
Hopper volume	1.22 m³ (43.0 cu ft)

AREAS:
Wings, gross	27.31 m² (294.0 sq ft)
Vertical tail surfaces (total)	1.90 m² (20.50 sq ft)
Horizontal tail surfaces (total)	3.91 m² (42.10 sq ft)

WEIGHTS AND LOADINGS:
Weight empty, equipped	1,188 kg (2,620 lb)
Max disposable load (Agricultural)	1,275 kg (2,810 lb)
Normal max T-O weight	2,204 kg (4,860 lb)
Max Agricultural T-O weight	2,463 kg (5,430 lb)
Max cabin floor loading	1,884.6 kg/m² (386 lb/sq ft)
Wing loading at Normal max T-O weight	80.71 kg/m² (16.53 lb/sq ft)
Power loading at Normal max T-O weight	7.39 kg/kW (12.15 lb/hp)

PERFORMANCE (at Normal max T-O weight):
Never-exceed speed (V_{NE})	143 knots (265 km/h; 165 mph)
Max level speed at S/L	126 knots (233 km/h; 145 mph)
Max cruising speed (75% power)	113 knots (209 km/h; 130 mph)
Operating speed for spraying (75% power)	90-115 knots (167-212 km/h; 104-132 mph)
Stalling speed: flaps up	55 knots (102 km/h; 64 mph)
flaps down	49 knots (91 km/h; 57 mph)
Max rate of climb at S/L	264 m (805 ft)/min
Service ceiling	4,875 m (16,000 ft)
T-O run	244 m (800 ft)
T-O to 15 m (50 ft)	500 m (1,640 ft)
Landing from 15 m (50 ft)	390 m (1,280 ft)
Landing run	207 m (680 ft)
Swath width (agricultural models):	
oily	23 m (75 ft)
aqueous	21.3-24.4 m (70-80 ft)
dust	7.6-15.2 m (25-50 ft)
Range with max normal fuel, 45 min reserves	383 nm (709 km; 441 miles)

NEW ZEALAND: AIRCRAFT—PAC

Loading the hopper of a Pacific Aerospace Corporation Cresco 08-600

PAC CRESCO 08-600

TYPE: Turboprop powered agricultural and general purpose aircraft.

PROGRAMME: Design began 1977; first flight of prototype (ZK-LTP) 28 February 1979; first flight of production aircraft early 1980; entered service January 1982. Marketing and support continuing for agricultural/utility and counter-insurgency versions; first flight of PT6A-34AG version (ZK-TMN, c/n 10) 18 November 1992.

CURRENT VERSIONS: **Cresco 08-600:** Standard version, with Textron Lycoming LTP 101 turboprop. *Detailed description applies to this version except where indicated.*

Cresco 08-600-34AG: With higher powered PT6A-34AG engine; launched 1992 (first delivery 23 December).

CUSTOMERS: Ten completed by January 1993, including six for domestic customers and one agricultural and two utility configured aircraft for Bangladesh, with passenger conversion kits supplied.

DESIGN FEATURES: Turboprop development of FU24; many parts interchangeable with FU24-954; larger hopper (capacity 1,779 litres; 470 US gallons; 391.4 Imp gallons).

POWER PLANT: One Textron Lycoming LTP 101-700A-1A turboprop standard, flat rated at 447 kW (599 shp) and driving a Hartzell HC-B3TN-3D/T10282 three-blade constant-speed fully feathering reversible-pitch metal propeller. Alternative choice of 559 kW (750 shp) Pratt & Whitney Canada PT6A-34AG engine available optionally, with propeller Beta control. Four fuel tanks in wing centre-section, total capacity 545.5 litres (144 US gallons; 120 Imp gallons). Two refuelling points in upper surface of each wing. Oil capacity 5.5 litres (1.4 US gallons; 1.2 Imp gallons). Chin mounted engine air intake, fitted with Centrisep filter panel.

DIMENSIONS, EXTERNAL:
Wing span	12.81 m (42 ft 0 in)
Wing chord, constant	2.13 m (7 ft 0 in)
Wing aspect ratio	6.0
Length overall	11.07 m (36 ft 4 in)
Height overall	3.63 m (11 ft 10¾ in)
Tailplane span	4.62 m (15 ft 2 in)
Wheelbase	2.77 m (9 ft 1¼ in)
Propeller diameter	2.59 m (8 ft 6 in)
Propeller ground clearance	0.305 m (12 in)
Cargo door (port): Height	0.99 m (3 ft 3 in)
Width	0.94 m (3 ft 1 in)
Height to sill	0.91 m (3 ft 0 in)

DIMENSIONS, INTERNAL:
Passenger/cargo compartment (aft of hopper):
Length	3.05 m (10 ft 0 in)
Max width	0.91 m (3 ft 0 in)
Max height	1.22 m (4 ft 0 in)
Floor area	2.79 m² (30.0 sq ft)
Volume	3.40 m³ (120.0 cu ft)
Hopper volume	1.77 m³ (62.5 cu ft)

AREAS:
Wings, gross	27.31 m² (294.0 sq ft)
Ailerons (total)	1.97 m² (21.18 sq ft)
Trailing-edge flaps (total)	3.05 m² (32.87 sq ft)
Fin	2.16 m² (23.23 sq ft)
Rudder	0.63 m² (6.77 sq ft)
Tailplane	5.09 m² (54.74 sq ft)
Elevator, incl tab	2.32 m² (25.00 sq ft)

WEIGHTS AND LOADINGS (A: standard aircraft with LTP 101-700A-1A engine, B: with PT6A-34AG):
Weight empty, equipped: A	1,270 kg (2,800 lb)
B	1,247 kg (2,750 lb)
Max fuel: A, B	435 kg (960 lb)

Max disposable load (fuel + hopper):
A, Normal	1,578 kg (3,480 lb)
A, Agricultural	1,828 kg (4,030 lb)
B, Normal	1,524 kg (3,360 lb)
B, Agricultural	2,393 kg (5,276 lb)
Max T-O weight: A, Normal	2,925 kg (6,450 lb)
A, Agricultural	3,175 kg (7,000 lb)
B, Normal	2,925 kg (6,450 lb)
B, Agricultural	3,794 kg (8,366 lb)
Max landing weight: A, B	2,925 kg (6,450 lb)

Wing loading at Normal max T-O weight:
A, B	107.12 kg/m² (21.94 lb/sq ft)

Wing loading at Agricultural max T-O weight:
A	116.25 kg/m² (23.81 lb/sq ft)
B	138.91 kg/m² (28.45 lb/sq ft)

Power loading at Normal max T-O weight:
A	6.55 kg/kW (10.77 lb/shp)
B	5.23 kg/kW (8.60 lb/shp)

Power loading at Agricultural max T-O weight:
A	7.11 kg/kW (11.69 lb/shp)
B	6.78 kg/kW (11.15 lb/shp)

PERFORMANCE (at max Normal T-O weight except where indicated, A and B as above):

Never-exceed speed (VNE):
A, B	177 knots (328 km/h; 204 mph)

Max level speed at S/L:
A	148 knots (274 km/h; 170 mph)
B	152 knots (282 km/h; 175 mph)

Max cruising speed (75% power) at 305 m (1,000 ft):
A	133 knots (246 km/h; 153 mph)
B	155 knots (287 km/h; 178 mph)

Stalling speed, flaps down, power off:
A at 2,767 kg (6,100 lb) AUW	52 knots (97 km/h; 60 mph)
B at max normal T-O weight	55 knots (102 km/h; 64 mph)

Max rate of climb at S/L: A	379 m (1,245 ft)/min
B	457 m (1,500 ft)/min
Absolute ceiling: A	5,485 m (18,000 ft)
B	11,580 m (38,000 ft)
T-O run: A	323 m (1,058 ft)
B	244 m (800 ft)
T-O to 15 m (50 ft): A	436 m (1,430 ft)
B	397 m (1,300 ft)
Landing from 15 m (50 ft): A, B	500 m (1,640 ft)
Landing run: B	275 m (900 ft)

Range with max fuel, no reserves:
A at 3,175 kg (7,000 lb) MTOGW	460 nm (852 km; 529 miles)
B at 2,925 kg (6,450 lb) MTOGW	467 nm (865 km; 537 miles)

ZK-TMN, the tenth Cresco built, is the first with optional PT6A-34AG engine

PAC AIRTRAINER CT4

TYPE: Two/three-seat aerobatic basic trainer.

PROGRAMME: New Zealand redesign of Australian Victa Airtourer; first flight 23 February 1972; total 94 (plus two prototypes) built before production ended 1977; production line reopened 1990 to build civil CT4Bs for Ansett Flying College (first two delivered June 1991); first flight of CT4C turboprop prototype (ZK-FXM, converted RNZAF CT4B) 21 January 1991. Original prototype was demonstrated in USA 1992 as prelude to EFS candidature.

CURRENT VERSIONS: **CT4B:** New-build version with 157 kW (210 hp) Teledyne Continental IO-360-HB9 flat-six engine.

CT4C: Turboprop version, with 313 kW (420 shp) Allison 250-B17D in lengthened nose.

CT4CR: As CT4C but with retractable landing gear; development deferred. Details in 1992-93 *Jane's*.

CT4E: Developed version of CT4B with more powerful engine (Textron Lycoming AEIO-540-L1B5: 224 kW; 300 hp at 2,600 rpm) and three-blade Hartzell constant-speed metal propeller; wing mounted slightly farther forward than on CT4B; first flight 14 December 1991; NZ certification (FAR Pt 23 Amendment 36) 8 May 1992.

CUSTOMERS: Total of 114 CT4s (all versions) produced by January 1993, comprising 63 CT4As (Royal Australian Air Force 37, Royal Thai Air Force 24, Royal Thai Police one, NZ Warbirds one), 50 CT4Bs (RAAF 13, RNZAF 19, RTAF six, BAe/Ansett Flying College 12) and one CT4E; 18 of Royal Thai Air Force's original CT4As undergoing fatigue life extension of wings, performed by RTAF personnel with PAC assistance.

COSTS: BAe/Ansett contract reportedly valued at NZ$3.5 million ($2.05 million).

DESIGN FEATURES: Detachable wingtips to allow fitting of optional wingtip fuel tanks. Wing section NACA 23012 (modified) at root, NACA 4412 (modified) at tip; dihedral

6° 45′ at chord line; incidence 3° at root; twist 3°; root chord increased by forward sweep of inboard leading-edges; small fence on outer leading-edges.

FLYING CONTROLS: Aerodynamically and mass balanced bottom-hinged ailerons, statically balanced one-piece elevator with rod and mechanical linkage, rudder with rod and cable linkage; ground adjustable tab on rudder, electric trim control for elevator and rudder; electrically actuated single-slotted flaps.

STRUCTURE: Light alloy stressed skin except for Kevlar/GFRP wingtips and engine cowling; ailerons and flaps have fluted skins.

LANDING GEAR: Tricycle type, non-retractable on CT4B, C and E, with cantilever spring steel main legs; steerable (±25°) nosewheel carried on telescopic strut and oleo shock absorber. Main units have Dunlop Australia wheels and tubeless tyres size 15 × 6 in, pressure 1.65 bars (24 lb/sq in); nosewheel has tubeless tyre size 14 × 4.5 in, pressure 1.24 bars (18 lb/sq in). Single-disc toe operated hydraulic brakes, with hand operated parking lock. Min ground turning radius 2.90 m (9 ft 6 in). Landing gear designed to shear before any excess impact loading is transmitted to wing, to minimise structural damage in event of crash landing.

POWER PLANT: See under Current Versions. CT4B has total standard fuel capacity of 204.5 litres (54 US gallons; 45 Imp gallons); wingtip tanks, each of 77 litres (20.5 US gallons; 17 Imp gallons) capacity, available optionally. CT4E has internal capacity (two integral wing tanks) of 182 litres (48 US gallons; 40 Imp gallons). Gravity fuelling point in each wing. Oil capacity (CT4E) 11.4 litres (3 US gallons; 2.5 Imp gallons).

ACCOMMODATION: Two seats side by side under hinged, fully transparent Perspex canopy. Space to rear for optional third seat or 52 kg (115 lb) of baggage or equipment (77 kg; 170 lb in CT4E). Dual controls standard.

DIMENSIONS, EXTERNAL (all versions, except where indicated):
Wing span	7.92 m (26 ft 0 in)
Wing chord: at root	2.18 m (7 ft 2 in)
at tip	0.94 m (3 ft 1 in)
Wing aspect ratio	5.25
Length overall: B	7.06 m (23 ft 2 in)
C	7.14 m (23 ft 5 in)
E	7.26 m (23 ft 9¾ in)
Height overall	2.59 m (8 ft 6 in)
Fuselage: Max width	1.12 m (3 ft 8 in)
Max depth	1.40 m (4 ft 7¼ in)
Tailplane span	3.61 m (11 ft 10 in)
Wheel track	2.97 m (9 ft 9 in)
Wheelbase	1.71 m (5 ft 7⅜ in)

DIMENSIONS, INTERNAL (all versions):
Cabin: Length	2.74 m (9 ft 0 in)
Max width	1.08 m (3 ft 6½ in)
Max height	1.35 m (4 ft 5 in)

AREAS (all versions):
Wings, gross	11.98 m² (129.0 sq ft)
Ailerons (total)	1.07 m² (11.56 sq ft)
Flaps (total)	2.10 m² (22.60 sq ft)
Fin	0.78 m² (8.44 sq ft)
Rudder, incl tab	0.58 m² (6.26 sq ft)
Tailplane	1.46 m² (15.71 sq ft)
Elevator	1.56 m² (16.74 sq ft)

WEIGHTS AND LOADINGS:
Weight empty, equipped: E	780 kg (1,720 lb)
Max fuel weight: E	149 kg (328 lb)
Max T-O weight: B, C	1,202 kg (2,650 lb)
E	1,179 kg (2,600 lb)
Max wing loading: B, C	100.3 kg/m² (20.54 lb/sq ft)
E	98.38 kg/m² (20.15 lb/sq ft)
Max power loading: B	7.68 kg/kW (12.62 lb/hp)
C	8.06 kg/kW (13.25 lb/shp)
E	5.26 kg/kW (8.67 lb/hp)

PERFORMANCE (B at AUW of 1,088 kg; 2,400 lb, C at 1,111 kg; 2,450 lb, E at 1,179 kg; 2,600 lb):
Never-exceed speed (V_{NE}):	
E	230 knots (426 km/h; 264 mph)
Max level speed at S/L:	
B	144 knots (267 km/h; 166 mph)
C	205 knots (380 km/h; 236 mph)
E	163 knots (302 km/h; 188 mph)
Max level speed at 3,050 m (10,000 ft):	
C	208 knots (385 km/h; 239 mph)
Max level speed at 6,100 m (20,000 ft):	
C	195 knots (361 km/h; 224 mph)
Cruising speed, 75% power:	
B	140 knots (259 km/h; 161 mph)
E at 2,590 m (8,500 ft)	158 knots (293 km/h; 182 mph)
Stalling speed at S/L:	
flaps up: C	57 knots (106 km/h; 66 mph)
flaps down: B, C	44 knots (82 km/h; 51 mph)
E	46 knots (86 km/h; 53 mph)
Max rate of climb at S/L, ISA:	
B	381 m (1,250 ft)/min
C	843 m (2,765 ft)/min
E	580 m (1,900 ft)/min
Service ceiling: B	4,420 m (14,500 ft)
E	7,925 m (26,000 ft)
T-O run at S/L, ISA: B	224 m (733 ft)
C	117 m (384 ft)
E	183 m (600 ft)
T-O to 15 m (50 ft) at S/L, ISA: C	206 m (675 ft)
E	275 m (900 ft)
Landing from 15 m (50 ft): E	244 m (800 ft)
Landing run: B	156 m (510 ft)
E	168 m (550 ft)
Range with max fuel (75% power), ISA, no reserves, at S/L: B	600 nm (1,112 km; 691 miles)
C	464 nm (860 km; 534 miles)
Range with max fuel at 1,830 m (6,000 ft), 45% power, no reserves: E	530 nm (982 km; 610 miles)

PAC Airtrainer CT4B for BAe/Ansett Flying College of Australia

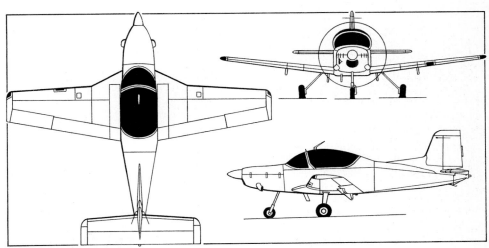

PAC Airtrainer CT4C turboprop conversion of the CT4B (*Jane's/Mike Keep*)

Prototype of the PAC CT4E Airtrainer, with more powerful (224 kW; 300 hp) flat-six engine

NIGERIA

AIEP
AERONAUTICAL INDUSTRIAL ENGINEERING AND PROJECT MANAGEMENT COMPANY LTD

General Aviation Service Centre, PO Box 5662, Old Kaduna Airport
Telephone: 234 (62) 217573
Fax: 234 (62) 217325
Telex: 71327 AIEP NG
MANAGING DIRECTOR: Klaus Gloege

AIEP has technical partnership agreement with Dornier Luftfahrt of Germany for Air Beetle programme, enabling it to draw upon German company's expertise.

AIEP AIR BEETLE

TYPE: Two-seat military trainer, extensively modified from Van's RV-6A homebuilt (see US section).
PROGRAMME: Three prototypes built, construction starting 1988, ending October 1991; first flight 1989; all three flown (1,000 hours total by end of 1991); first aircraft (672 hours) then used for static testing, third undergoing evaluation and testing by Nigerian Air Force early 1992 prior to acceptance as replacement for BAe Bulldog trainer; second aircraft continuing flying to test systems/component changes. Series production expected to start February 1992, but no confirmation of this received.
CUSTOMERS: Nigerian Air Force to be initial customer; interest from other countries; to be marketed outside Nigeria once production starts.
DESIGN FEATURES: Conventional fixed-gear, low-wing lightplane; aim was for fully aerobatic piston-powered military/civil primary trainer able to use either mogas or avgas fuel; constant chord wings (modified NACA 23013.5 section) with 3° dihedral and 1° incidence; no sweep or twist.
FLYING CONTROLS: Manual/mechanical elevators, rudder and differential ailerons, all with electric trim; plain trailing-edge flaps.
STRUCTURE: Conventional all-metal (Alclad aluminium alloy); two-spar wings and tail unit.
LANDING GEAR: Non-retractable tricycle type, with cantilever mainwheel and nosewheel legs of carbon reinforced GFRP and spring steel respectively. Single Cleveland wheel on each unit, with McCreary (main) and Lamb (nose) tyres; tyre pressure 2.76 bars (40 lb/sq in) on all units. Cleveland single-disc mainwheel brakes. All wheels have GFRP speed fairing. Min ground turning radius 1.77 m (5 ft 9½ in).
POWER PLANT: One Textron Lycoming O-360-A1A flat-four engine (134 kW; 180 hp at 2,700 rpm), driving a Hartzell two-blade constant-speed metal propeller. Integral fuel tank in each wing, combined capacity 140 litres (37 US gallons; 30.8 Imp gallons); gravity fuelling point in top of each tank. Oil capacity 7.6 litres (2 US gallons; 1.7 Imp gallons).

Third prototype AIEP Air Beetle primary trainer in Nigerian Air Force markings

ACCOMMODATION: Two seats side by side under rearward sliding canopy; baggage compartment aft of seats.
SYSTEMS: DC electrical system, supplied by 28V alternator and 24V battery.
AVIONICS: Bendix/King KX 155 com transceiver; conventional IFR instrumentation (different manufacturers' equipment currently being evaluated); dual controls standard.

DIMENSIONS, EXTERNAL:
Wing span	7.01 m (23 ft 0 in)
Length overall	6.15 m (20 ft 2¼ in)
Fuselage: Max width	1.04 m (3 ft 5 in)
Height overall	2.30 m (7 ft 6½ in)
Tailplane span	2.54 m (8 ft 4 in)
Propeller diameter	1.85 m (6 ft 0¾ in)
Propeller ground clearance	0.26 m (10¼ in)

AREAS:
Wings, gross	10.20 m² (109.8 sq ft)
Ailerons (total)	0.71 m² (7.64 sq ft)
Trailing-edge flaps (total)	0.848 m² (9.13 sq ft)
Fin	0.38 m² (4.09 sq ft)
Rudder	0.41 m² (4.41 sq ft)
Tailplane	1.97 m² (21.20 sq ft)
Elevators (total)	2.10 m² (22.60 sq ft)

WEIGHTS AND LOADINGS:
Weight empty, equipped	476 kg (1,050 lb)
Max fuel weight	151 kg (333 lb)
Max T-O and landing weight	816 kg (1,800 lb)
Max wing loading	80.02 kg/m² (16.39 lb/sq ft)
Max power loading	6.09 kg/kW (10.0 lb/hp)

PERFORMANCE (at max T-O weight):
Never-exceed speed (V_{NE})	175 knots (324 km/h; 201 mph)
Max level speed at S/L	150 knots (278 km/h; 173 mph)
Max cruising speed, 75% power at 3,050 m (10,000 ft)	155 knots (287 km/h; 178 mph)
Econ cruising speed, 55% power at 2,440 m (8,000 ft)	135 knots (250 km/h; 155 mph)
Stalling speed: flaps up	52 knots (97 km/h; 60 mph)
flaps down	50 knots (93 km/h; 58 mph)
Max rate of climb at S/L	610 m (2,000 ft)/min
Service ceiling	6,100 m (20,000 ft)
T-O run	200 m (656 ft)
T-O to 15 m (50 ft)	300 m (985 ft)
Landing from 15 m (50 ft)	270 m (886 ft)
Landing run	205 m (673 ft)
Range with max fuel, allowances for T-O and 45 min reserves at 45% power	590 nm (1,093 km; 679 miles)

NORWAY

LUNDS TEKNISKE
LUNDS TEKNISKE

Vikaveien 2, N-8600 Mo
Telephone: 47 (87) 52 100
Fax: 47 (87) 55 065
OWNER/MANAGER: Arne Lund

LUNDS TEKNISKE SILHOUETTE

TYPE: Single-seat sport/recreational homebuilt and motor glider; designed and tested to FAR Pt 23 standards.
PROGRAMME: Rights to Silhouette homebuilt purchased from US Silhouette Aircraft Inc; Lunds Tekniske currently offers kits; original prototype first flew in USA on 3 July 1984.
CUSTOMERS: Forty-two kits sold by January 1992, when last update received; 16 then assembled.
COSTS: Kit: US$13,000 in sport version, including engine, propeller, spinner, hydraulics, brakes and instruments; US$13,775 in motor glider form, including above components.
DESIGN FEATURES: Very slim streamline composites airframe; high aspect ratio wings; provision for bolt-on wingtip extensions for use as motor glider; non-retractable tricycle landing gear.
FLYING CONTROLS: Half span ailerons; flaps to be available; centreline dive brake optional.
STRUCTURE: Wings have glassfibre spars, pre-cut Styrofoam cores, glassfibre/epoxy skins and pre-moulded wingtips; Styrofoam/glassfibre ailerons. Fuselage of glassfibre and Nomex honeycomb, pre-moulded in halves with integral tail fin, around plywood/Styrofoam/glassfibre bulkheads. Tail surfaces of foam and glassfibre.
POWER PLANT: One 30 kW (40 hp) Rotax 447 or 37.3 kW (50 hp) Rotax 503. Fuel capacity with standard wing 45 litres (12 US gallons; 10 Imp gallons).

DIMENSIONS, EXTERNAL:
Wing span: standard	9.75 m (32 ft 0 in)
with optional extensions	12.50 m (41 ft 0 in)
Wing aspect ratio: standard	13.46
with extensions	18.69
Length overall	5.87 m (19 ft 3 in)
Height overall	2.03 m (6 ft 8 in)
Propeller diameter	1.47 m (4 ft 10 in)

AREAS:
Wings, gross: standard	7.06 m² (76.0 sq ft)
with extensions	8.36 m² (90.0 sq ft)

WEIGHTS AND LOADINGS:
Weight empty, standard wings	262 kg (578 lb)
Max T-O weight	374 kg (824 lb)
Max wing loading:	
standard wings	52.97 kg/m² (10.85 lb/sq ft)
with extensions	44.74 kg/m² (9.16 lb/sq ft)

PERFORMANCE (standard wings):
Max level speed	122 knots (225 km/h; 140 mph)
Cruising speed	104 knots (193 km/h; 120 mph)
Stalling speed	45 knots (84 km/h; 52 mph)
Max rate of climb at S/L	244 m (800 ft)/min
Service ceiling	4,265 m (14,000 ft)
T-O run	244 m (800 ft)
Landing run	183 m (601 ft)
Range with max fuel, no reserves	390 nm (724 km; 450 miles)
g limits	+4.2/−2

Long-wing Lunds Tekniske Silhouette motor glider (foreground) and version with standard wings (rear) *(Howard Levy)*

PAKISTAN

PAC
PAKISTAN AERONAUTICAL COMPLEX
Kamra, District Attock
WORKS: F-6 Rebuild Factory; Mirage Rebuild Factory; Kamra Avionics and Radar Factory; Aircraft Manufacturing Factory (all at Kamra)
Telephone: 92 (51) 580260/5
Fax: 92 (51) 584162
Telex: 5601 PAC KAMRA PK
DIRECTOR GENERAL: Air Vice-Marshal M. Yusuf Khan
MANAGING DIRECTORS:
  Air Cdre Abdul Wahid (AMF)
  Air Cdre Imtiaz Rasul Khan (F-6RF)
  Air Cdre Azhar Hussain (MRF)
  Air Cdre Muhammad Irshad (KARF)

Pakistan Aeronautical Complex is organ of Pakistan Ministry of Defence; consists of four factories, as follows:

Aircraft Manufacturing Factory (AMF) came into operation mid-1981, as licence production centre for Saab Safari/Supporter (Pakistani name Mushshak); major facilities include equipment for all Mushshak GFRP component manufacture; 1992 workforce approx 1,000. Collaborating with NAMC in China in developing Karakorum 8 jet trainer.

F-6 Rebuild Factory, or F-6RF, established 1980 primarily for overhauling Pakistan's Shenyang F-6s and their accessories; authorised to manufacture about 7,000 spares items for FT-5, F-6 and FT-6, and produces 1,140 litre (301 US gallon; 250 Imp gallon) F-6 auxiliary fuel tanks; production of 500 and 800 litre (132 and 211 US gallon; 110 and 176 Imp gallon) supersonic drop tanks for F-7P began mid-1991; 1992 workforce approx 2,000.

F-6RF possesses modern technical facilities for various engineering processes such as surface treatment, heat treatment, forging, casting, non-destructive testing, and machine tools required to manufacture items from raw materials; overhauls Pakistan Air Force FT-5s, FFT-6s and A-5Cs and was preparing in 1992 to assume responsibility for rebuild and overhaul of Chinese F/FT-7s in PAF service.

Mirage Rebuild Factory (MRF) began operating 1978; has site area of over 81 ha (200.15 acres) and nearly 2,000 engineers and technicians; can accomplish complete overhaul of Mirage III/5, Atar 9C engine, and all associated aircraft components and engine accessories; current overhaul capacity 8-10 aircraft and over 50 engines each year; can overhaul/rebuild third country Mirage III/5s, engines, components and accessories; overhauling United Arab Emirates Air Force Mirages since 1988; overhaul of 42 Dassault/Commonwealth Mirage IIIOAs and eight IIIDs, bought from Australia, transferred from PAF Air Logistics Depot at Sharea Faisal to MRF; first IIID received January 1991.

Facility being upgraded to undertake increased life core (ILC) modification, overhaul and upgrade of Pratt & Whitney F100-PW-220E turbofans and F-16 engine jet fuel starters, and will soon have limited capability to service and overhaul F100 engine accessories.

Kamra Avionics and Radar Factory (KARF) recently established to rebuild PAF radars and avionics; employs 18 engineers and 135 technicians. At present (1992) rebuilding Siemens 45-E mobile pulse Doppler radar, complex components and modules of MPDR-45, and Siemens and associated power generators. Has modern, environmentally controlled electronics workshops and test equipment including automated Pegamat tester; by early 1992 had rebuilt nine radars, 27 generators and over 6,000 components; intends to rebuild MPDR-60 and MPDR-90 radars and Siemens Control and Reporting Centre and to assemble radar line replaceable units; is investigating international market for co-production of avionics, including radar warning receivers and airborne radios and radars.

NAMC/PAC K-8 KARAKORUM 8
Joint venture (with China) turbofan powered military trainer: see International section. Some components to be manufactured by AMF.

PAC (AMF) MUSHSHAK (PROFICIENT) and SHAHBAZ (FALCON)
TYPE: Two/three-seat training and observation light aircraft.
PROGRAMME: 15 Mushshaks supplied complete from Sweden; 92 then assembled from CKD kits at Risalpur between 1975 and 1982; completely indigenous production followed, with 120 delivered by December 1991; 58 rebuilt at Kamra 1980-91; production rate approx 2½ per month in mid-1992; engines, instruments, electrical equipment and radios imported, but most other items manufactured locally. Shahbaz (first flight July 1987) received US FAR Pt 23 certification 1989.
CURRENT VERSIONS: **Mushshak:** Standard production version. *Description applies to this version.*
  Shahbaz: Similar to Mushshak; 156.6 kW (210 hp) Teledyne Continental TSIO-360-MB turbocharged engine; four completed by September 1992.
CUSTOMERS: Total of 227 Mushshaks acquired/produced by December 1991, 50 for PAF and rest for Army.
DESIGN FEATURES: Based on Swedish Saab Safari/Supporter; armament option; higher-powered Shahbaz has take-off run at 1,402 m (4,600 ft) ASL in +41°C (106°F) reduced by 25 to 30 per cent to 290 m (590 ft).

Wing thickness/chord ratio 10 per cent; dihedral 1° 30'; incidence 2° 48'; sweepforward 5° from roots.
FLYING CONTROLS: Mass balanced ailerons with servo tab in starboard unit, rudder with trim tab, and one-piece mass balanced tailplane with large anti-servo and trimming tab; electrically actuated plain sealed flaps.
STRUCTURE: All-metal, except for GFRP tailcone, engine cowling panels, wing strut/landing gear attachment fairings and fin tip.
LANDING GEAR: Non-retractable tricycle type. Cantilever composite spring main legs. Goodyear 6.00-6 mainwheels and 5.00-5 steerable nosewheel. Cleveland disc brakes on main units.
POWER PLANT: One 149 kW (200 hp) Textron Lycoming IO-360-A1B6 flat-four engine, driving a Hartzell HC-C2YK-4F/FC7666A-2 two-blade constant-speed metal propeller. Two integral wing fuel tanks, total capacity 190 litres (50.2 US gallons; 41.8 Imp gallons). Oil capacity 7.5 litres (2.0 US gallons; 1.6 Imp gallons). From 10-20 s inverted flight (limited by oil system) permitted.
ACCOMMODATION: Side by side adjustable seats, with provision for back type or seat type parachutes, for two persons beneath fully transparent upward hinged canopy. Dual controls standard. Space aft of seats for 100 kg (220 lb) of baggage (with external access on port side) or, optionally, a rearward facing third seat. Upward hinged door, with window, beneath wing on port side. Cabin heated and ventilated.
SYSTEMS: 28V 50A DC electrical system.
AVIONICS: Provision for full blind-flying instrumentation and radio.
ARMAMENT: Provision for six underwing attachment points, inner two stressed to carry up to 150 kg (330 lb) each and outer four up to 100 kg (220 lb) each. Possible armament loads include two 7.62 mm or 5.56 mm machine-gun pods, two pods each with seven 75 mm or 2.75 in air-to-surface

AMF Mushshaks (Saab Safari/Supporters) built under licence by PAC

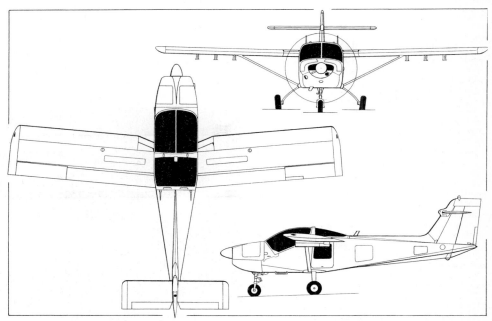

Pakistan Aeronautical Complex Mushshak two/three-seat light aircraft (*Dennis Punnett*)

rockets, four pods each with seven 68 mm rockets, eighteen 75 mm rockets, or six Bofors Bantam wire guided anti-tank missiles.

EQUIPMENT: Options include ULV cropspraying kit, target towing kit, or underwing supply/relief containers.

DIMENSIONS, EXTERNAL:
Wing span	8.85 m (29 ft 0½ in)
Wing chord (outer panels, constant)	1.36 m (4 ft 5½ in)
Length overall	7.00 m (22 ft 11½ in)
Height overall	2.60 m (8 ft 6½ in)
Tailplane span	2.80 m (9 ft 2¼ in)
Wheel track	2.20 m (7 ft 2½ in)
Wheelbase	1.59 m (5 ft 2¾ in)
Propeller diameter	1.88 m (6 ft 2 in)
Cabin door (port): Height	0.78 m (2 ft 6¾ in)
Width	0.52 m (1 ft 8½ in)

DIMENSIONS, INTERNAL:
Cabin: Max width	1.10 m (3 ft 7¼ in)
Max height (from seat cushion)	1.00 m (3 ft 3¼ in)

AREAS:
Wings, gross	11.90 m² (128.1 sq ft)
Ailerons (total)	0.98 m² (10.55 sq ft)
Flaps (total)	1.55 m² (16.68 sq ft)
Fin	0.77 m² (8.29 sq ft)
Rudder, incl tab	0.73 m² (7.86 sq ft)
Tailplane, incl tab	2.10 m² (22.6 sq ft)

WEIGHTS AND LOADINGS (A: Aerobatic, U: Utility, N: Normal category):
Weight empty, equipped	646 kg (1,424 lb)
Max external stores load	300 kg (661 lb)
Max T-O weight: A	900 kg (1,984 lb)
U	1,000 kg (2,205 lb)
N	1,200 kg (2,645 lb)
Max wing loading: A	75.63 kg/m² (15.49 lb/sq ft)
U	84.03 kg/m² (17.21 lb/sq ft)
N	100.84 kg/m² (20.65 lb/sq ft)
Max power loading: A	6.04 kg/kW (9.92 lb/hp)
U	6.71 kg/kW (11.02 lb/hp)
N	8.05 kg/kW (13.23 lb/hp)

PERFORMANCE (at max T-O weight, Utility category):
Never-exceed speed (V_NE)	197 knots (365 km/h; 227 mph)
Max level speed at S/L	128 knots (238 km/h; 148 mph)
Cruising speed at S/L	113 knots (210 km/h; 130 mph)
Stalling speed, power off:	
flaps up	60 knots (111 km/h; 69 mph)
flaps down	54 knots (100 km/h; 63 mph)
Max rate of climb at S/L	312 m (1,024 ft)/min
Time to 1,830 m (6,000 ft)	7 min 30 s
Service ceiling	4,800 m (15,750 ft)
T-O run	150 m (493 ft)
T-O to 15 m (50 ft)	305 m (1,000 ft)
Landing from 15 m (50 ft)	350 m (1,149 ft)
Landing run	140 m (460 ft)
Max endurance (65% power) at S/L, 10% reserves	5 h 10 min
g limits: A	+6/−3
U	+5.4/−2.7
N	+4.8/−2.4

PERU

INDAER PERU
INDUSTRIA AERONAUTICA DEL PERU SA
Chinchon 1070, Lima 27
Telephone: 51 (14) 42 0355
Fax: 51 (14) 42 4953
Telex: 20125 PE MINAR

PRESIDENT: Alfredo Arrisueno
GENERAL MANAGER: Fernando Carulla

Indaer Peru is nation's sole aircraft manufacturer; abandoned original plans to assemble/build Aermacchi MB-339As; chose instead programme for general aviation aircraft to provide alternative to both microlights and more expensive light aircraft. Agreements with Aero Boero of Argentina (to produce AB 115 and AB 180) and Pilatus of Switzerland (production of PC-6 Turbo-Porter) not implemented up to early 1993.

No news received since January 1990 of IAP-001/002/003 Chuspi/Urpi lightplane programme; for all available details and illustrations, see 1992-93 and earlier *Jane's*.

PHILIPPINES

ACT — *see PAI*

AIPI
AEROTECH INDUSTRIES PHILIPPINES INC
PADC Hangar 5, Domestic Airport Road, NAIA, 1300 Pasay City, Metro Manila
Telephone: 63 (2) 833 0630 and 832 2741
Fax: 63 (2) 832 3588
Telex: 62561 AERO PN

AIPI is member of Aerotech Group and affiliate of Aerotech Holding BV, Netherlands; represents wide range of European, Far East and North American producers of aerospace, defence and other products, and is local area manufacturing, repair and service activity centre for Aerotech Group.

Current programmes and activities include design and manufacture of 12.7 mm ventral gun pod for Philippine Air Force Agusta (SIAI-Marchetti) S.211s; supply of components in composites; distribution of spares for various types of aircraft; aircraft overhaul; sales, marketing, distribution and representation of aviation and defence products.

PACI
PHILIPPINE AIRCRAFT COMPANY INC
RPMCI Hangar, Manila Domestic Airport, PO Box 7633, Airport Airmail Exchange, 1300 Pasay City, Metro Manila
Telephone: 63 (2) 832 2777 and 832 3375
Fax: 63 (2) 833 0605
Telex: 66621 WPAC PN
PRESIDENT: Brig Gen Rudolfo G. Hautea (Ret'd)
CHAIRMAN AND CEO: Rolando P. Moscardon

PACI builds and markets SkyStar (formerly Denney) Kitfox (see US section) as Skyfox in Western Pacific area, and supports these; also manufactures parts and components for other US companies.

PACI SKYFOX

TYPE: Two-seat light aircraft.

PROGRAMME: PACI agreement with SkyStar Corporation (then Denney Aerocraft) of USA (see US section) announced November 1987; local production, as Skyfox, began early 1988; first two certificated and had flown over 140 hours by late 1990. First prototype, with Rotax 532 engine, fitted with Louisiana Agrilite spraygear; Mk II type certification due by end of 1993; production certification to start six months later.

CURRENT VERSIONS: **Skyfox I:** Initial version, with 47.7 kW (64 hp) Rotax 532 single-ignition engine.

Skyfox II: Improved version, with Rotax 582 engine, dual ignition, dual brakes, cabin doors with extended transparencies for enhanced field of view, and venturi driven artificial horizon. Future improvement modifications to wings and tail planned. Certification anticipated by end of 1993. *Description applies to this version.*

CUSTOMERS: Including prototypes, total of seven (three Skyfox I, four Skyfox II) completed by March 1993, with two more Skyfox II then nearing completion. Skyfox II prototype evaluated 1990-91 by Philippine Integrated Police Force (PIPF).

DESIGN FEATURES: Airframe description generally as for Kitfox in US section.

POWER PLANT: One 47.7 kW (64 hp) Rotax Bombardier 582 UL two-cylinder two-stroke dual-ignition engine, with electric starter, driving a fixed-pitch, three-blade wooden propeller. Fuel capacity (two wing tanks) 45.5 litres (12 US gallons; 10 Imp gallons). Gravity refuelling.

DIMENSIONS, EXTERNAL:
Wing span	9.55 m (31 ft 4 in)
Wing chord, constant	1.07 m (3 ft 6 in)
Wing aspect ratio	7.65
Length: overall	5.41 m (17 ft 9 in)
wings folded	6.40 m (21 ft 0 in)
Width, wings folded	2.39 m (7 ft 10 in)
Height overall	2.29 m (7 ft 6 in)
Wheel track	1.45 m (4 ft 9 in)
Wheelbase	3.86 m (12 ft 8 in)
Propeller ground clearance	0.56 m (1 ft 10 in)

AREAS:
Wings, gross (incl flaperons)	11.92 m² (128.3 sq ft)
Tailplane	1.46 m² (15.72 sq ft)

WEIGHTS AND LOADINGS:
Weight empty	230.5 kg (508 lb)
Max T-O weight	431 kg (950 lb)
Max zero-fuel weight	398 kg (878 lb)
Max wing loading	36.15 kg/m² (7.40 lb/sq ft)
Max power loading	9.03 kg/kW (14.84 lb/hp)

PERFORMANCE (A: single-seat at 363 kg; 800 lb AUW; B: two-seat at 431 kg; 950 lb):

PACI Skyfox II (Rotax 582 engine) with ventral cargo pod (*Anglo Philippine Aviation*)

Max level speed at S/L:
A, B 86 knots (160 km/h; 100 mph)
Max cruising speed (75% power) at S/L:
A 78 knots (145 km/h; 90 mph)
B 74 knots (137 km/h; 85 mph)
Stalling speed, power off, flaps up:
A 28 knots (52 km/h; 32 mph)
B 33 knots (62 km/h; 38 mph)
Stalling speed, power on, flaps down:
A 22 knots (41 km/h; 25 mph)
B 27 knots (49 km/h; 30 mph)
Max rate of climb at S/L: A 183 m (600 ft)/min
B 153 m (500 ft)/min
Service ceiling (approx): A 2,895 m (9,500 ft)
B 2,745 m (9,000 ft)
T-O run: A 83 m (270 ft)
B 118 m (385 ft)
T-O to 15 m (50 ft): A 150 m (490 ft)
B 196 m (641 ft)
Landing from 15 m (50 ft): A 199 m (650 ft)
B 249 m (816 ft)
Landing run: A 51 m (166 ft)
B 97 m (316 ft)
Range at 65% power: A 252 nm (467 km; 290 miles)
B 221 nm (410 km; 255 miles)

PADC
PHILIPPINE AEROSPACE DEVELOPMENT CORPORATION
PO Box 7395, Domestic Post Office, Lock Box 1300, Domestic Road, Pasay City, Metro Manila
Telephone: 63 (2) 832 2741/50
Fax: 63 (2) 832 2568
Telex: 66019 PADC PN
CHAIRMAN: Jesus B. Garcia Jr
PRESIDENT: Prudencio M. Reyes Jr
EXECUTIVE VICE-PRESIDENT: Antonio S. Duarte
SUBSIDIARIES:
Philippine Helicopter Services Inc (PHSI)
Maintenance and overhaul centre for in-country BO 105 helicopters, including all inspections from 50 to 2,500 hours; overhaul and repair of MDHC (Hughes) helicopter rotor blades; capability to repair and overhaul power steering units, hydraulic pumps and other components of MAN commercial vehicles; overhaul and repair of Zahnradfabrik Friedrichshafen AG products.
Philippines East Asia Cargo Airlines Inc (PEAC)
Formed 1990 (as Air Philippines Corporation) as joint venture with Transnational Transport Ltd (TNT) of Australia to undertake international air freight services.

PADC established 1973 as government arm for development of Philippine aviation industry; is now an attached agency of Department of Transportation and Communication (DOTC) and has technical workforce of about 200. Main activities are aircraft manufacturing and assembly; maintenance engineering; aircraft and spare parts sales; service centres for Eurocopter BO 105 helicopter and Pilatus Britten-Norman Islander. Recently completed programmes included licence assembly of 67 Islanders (including 22 for Philippine Air Force) and 44 BO 105s.
Assembly of initial 18 S.211 jet trainers under subcontract to Agusta of Italy completed 1991; kits for further six S.211s

PADC assembled S.211 with Aerotech 12.7 mm ventral gun pod *(Anglo Philippine Aviation)*

due for delivery to PADC in 1993; first two (of 18) shipsets of SF.260 TP components from Italy received by March 1993, with further four sets to follow later in year; also contracted to produce 40 vertical fins for SF.260 (10 delivered by March 1993); total work package with Agusta involves 1 million man-hours over period 1989-1996; may eventually produce SF.600 Canguro in Philippines (negotiations continuing).
PADC has maintenance/repair/overhaul centre for Allison 250 series turbine engines and for Textron Lycoming and Teledyne Continental piston engines of up to 298 kW (400 hp). Its Maintenance and Engineering Dept is appointed as Allison AMOC (authorised maintenance and overhaul centre) by Hawker Pacific of Australia (regional distributor for Allison) and undertakes FMS work on 250-C30 engines for Sikorsky helicopters and 250-B17 turboprops and propellers for Philippine Air Force (PAF) Nomads.
PADC is Eurocopter International's Philippine agent for government and selected private sales, and may become service centre for in-country Aerospatiale helicopters. Potential programmes under negotiation in early 1993 included service centre status for Manila-based Beech aircraft; co-production of IAI Kfir if selected by Philippine Air Force; and assistance to Northrop in maintenance and overhaul of PAF F-5 aircraft.

PAI
PACIFIC AERONAUTICAL INC
5 First Avenue, Mactan Export Processing Zone, Lapu-Lapu City 6015
Telephone: 63 (32) 400 283/4/5
Fax: 63 (32) 400 386
PAI (formerly Aviation Composite Technology: see 1992-93 and earlier *Jane's*) has completed and sold one Lancair 320 (RP-X328) to a domestic private owner; Lancair production (see description in US section) continuing in 1993. Former Apache 1 programme derived from Lancair design (see 1991-92 *Jane's*) has apparently been discontinued.

SEFA
SEFA ASIA INC
PH Box 1059, Philcom Building, 8755 Paseo de Roxas, Makati, Metro Manila
Telephone: 63 (2) 802 3173
Fax: 63 (2) 842 2776
PRESIDENT: Manuel Fernandez
WORKS: People's Technology Complex, Carmona, Cavite
Telephone: 63 (90) 201 4272
SEFA (Société d'Etudes et de Fabrications Aéronautiques) has designed and is building the Sea Bird two-seat amphibian at its Cavite factory.

SEFA SEA BIRD
TYPE: Two-seat light amphibian.
PROGRAMME: Prototype first flight expected November 1993.
DESIGN FEATURES: Low mounted gull wing, with retractable stabilising float at each tip; engine pylon-mounted above fuselage; sweptback fin and rudder and non-swept T tailplane; landing gear can be extended under water to facilitate beaching.
FLYING CONTROLS: Primary control surfaces conventional mechanical; one-piece balanced elevator. No flaps.
STRUCTURE: All-composites (GFRP/epoxy sandwich with Kevlar and CFRP reinforcement).
LANDING GEAR: Hydraulically retractable mainwheels (rearward) and tailwheel.
POWER PLANT: One 59.7 kW (80 hp) Rotax 912 four-cylinder two-stroke engine, with dual ignition, pylon-mounted above fuselage and driving a three-blade pusher propeller. Automotive fuel.
ACCOMMODATION: Two seats side by side.
DIMENSIONS, EXTERNAL:
Wing span 12.00 m (39 ft 4½ in)
Wing aspect ratio 8.23
Length overall 6.20 m (20 ft 4 in)
AREAS:
Wings, gross 17.50 m² (188.4 sq ft)

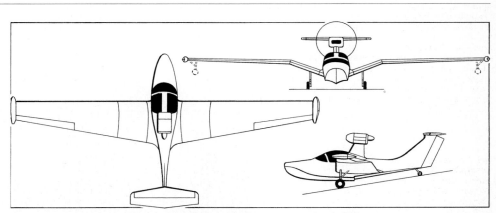
SEFA Asia Sea Bird two-seat light amphibian *(Jane's/Mike Keep)*

WEIGHTS AND LOADINGS:
Weight empty 200 kg (441 lb)
Max T-O weight 450 kg (992 lb)
Max wing loading 25.71 kg/m² (125.5 lb/sq ft)
Max power loading 7.55 kg/kW (12.40 lb/hp)
PERFORMANCE (estimated):
Max level speed 92 knots (170 km/h; 105 mph)
Cruising speed 78 knots (145 km/h; 90 mph)
Stalling speed 33 knots (60 km/h; 38 mph)

UDRD
UNIVERSAL DYNAMICS RESEARCH AND DEVELOPMENT
Penthouse, Zeta II Building, 191 Salcedo Street, Legaspi Village, Makati, Metro Manila
Telephone: 63 (2) 878 997 and 876 493
Telex: 66497 DEFCO PN or 23708 DEFCON PH
PRESIDENT AND GENERAL MANAGER:
Capt Panfilo V. Villaruel

UDRD DEFIANT 500 and 1000
TYPE: Tandem two-seat turboprop trainer.
PROGRAMME: First flight of Defiant 300 with 224 kW (300 hp) Textron Lycoming IO-540-K1B5 engine February 1988; flew twice but abandoned because of problems obtaining local hardwood for manufacture; donated 9 March 1993 to Philippine State College of Aeronautics. UDRD decided in 1990 to produce all-metal turboprop version as Defiant 500, but funds for development and production still being sought in 1993.
CURRENT VERSIONS: **Defiant 500:** Turboprop trainer; see 1992-93 *Jane's* for description and illustration.
Defiant 1000: Proposed single-seat counter-insurgency version; 1,342 kW (1,800 shp) turboprop.

POLAND

PEZETEL
PEZETEL FOREIGN TRADE ENTERPRISE LTD
Aleja Stanów Zjednoczonych 61, PL-04-028 Warsaw 50
Telephone: 48 (22) 108001
Fax: 48 (22) 132356 and 132835
Telex: 814651
GENERAL MANAGER: Jerzy Krezlewicz, MA
MARKETING MANAGER: Włodzimierz Skrzypiec
MANAGER OF AVIATION DEPARTMENT: Kazimierz Niepsuj
MANAGER OF PUBLICITY DEPARTMENT:
   Wojciech Kowalczyk, MA

Panstwowe Zakłady Lotnicze (State Aviation Works) formed by industrial syndicate in 1928 from existing factories to produce aircraft for domestic use and export; until 1981 aviation industry organised under ZPLS-PZL (Aircraft and Engine Industry Union) control; activities came under Bureau of Ministers control 1982 (see earlier *Jane's*). Aviation and diesel engine industry encompassed 19 factories and other establishments in early 1992, with combined workforce of approx 50,000; currently undergoing further reorganisation following democratisation and dissolution of USSR, with aviation industry grouped from 1991 under council of factories headed by Pezetel. Consortium formed by PZL Mielec/Swidnik/Okecie (aircraft), PZL Rzeszów (aero engines) and PZL Hydral-Wroclaw (hydraulics) concerns. Consortium being assisted by Polish Industrial Development Agency in attempt to raise new investment capital. Eventual plan is to privatise industry.

Polish aviation industry has relied substantially on aircraft engines (see Engines section) and equipment (military, propellers, and ground equipment for agricultural aircraft and helicopters) of its own design, as well as on co-operation and co-production with manufacturers from East and West; also undertakes component manufacture for Russian Il-86 and Il-96 airliners. Disappearance of former Soviet market leaves Polish Air Force as only major military customer. Pezetel handles export sales of Polish aviation products.

IL
INSTYTUT LOTNICTWA (Aviation Institute)
Al. Krakowska 110/114, PL-02-256 Warsaw-Okecie
Telephone: 48 (22) 460993 and 460171
Fax: 48 (22) 464232
Telex: 813537
GENERAL MANAGER: Roman Czerwiński, MScEng
CHIEF CONSULTANT FOR SCIENTIFIC AND TECHNICAL
  CO-OPERATION: Jerzy Grzegorzewski, MScEng

Founded 1926; directly subordinate to Ministry of Heavy and Machine Building Industry and responsible for most research and development work in Polish aviation industry; conducts scientific research, including investigation of problems associated with low-speed and high-speed aerodynamics, static and fatigue tests, development and testing of aero engines, flight instruments, space science instrumentation, and other equipment, flight tests, and materials technology; also responsible for construction of aircraft and aero engines.

PZL I-22 IRYDA (IRIDIUM)
Now in production and service and described under WSK-PZL Mielec heading.

PZL PIRANIA
TYPE: Two-seat ground attack aircraft.
PROGRAMME: Design began late 1991; no recent indication of status.
CUSTOMERS: Proposed as possible alternative to PZL-230F Skorpion (see under PZL Warszawa-Okecie) for Polish Air Force.
DESIGN FEATURES: Low-wing configuration, with two rear-mounted engines.
POWER PLANT: Two Polish K-15 turbofans or Motorlet M 601 T turboprops, pylon mounted on sides of rear fuselage (latter driving pusher propellers). Provision for drop fuel tanks.
ACCOMMODATION: Two seats in tandem.
AVIONICS: To include lookdown/shootdown radar.
ARMAMENT: Underfuselage 12.7 mm or 20 mm gun(s); eight underwing stations for missiles, munition dispensers, rockets or free-fall bombs.
PERFORMANCE (estimated. A: K-15 powered, B: M 601 T):
  Cruising speed: A   421 knots (780 km/h; 485 mph)
               B   302 knots (560 km/h; 348 mph)
  Stalling speed, flaps down:
    A, B   87 knots (160 km/h; 100 mph)
  Combat radius, depending on weapons load
    162-324 nm (300-600 km; 186-373 miles)
  Ferry range with drop tanks
    2,158 nm (4,000 km; 2,485 miles)

WSK-PZL MIELEC
WYTWÓRNIA SPRZETU KOMUNIKACYJNEGO-PZL MIELEC (Transport Equipment Manufacturing Centre, Mielec)
ul. Ludowego Wojska Polskiego 3, PL-39-300 Mielec
Telephone: 48 (196) 7010
Fax: 48 (14) 214785 or 48 (196) 7451
Telex: 0632293 C WSK PL
GENERAL MANAGER: Stanislaw Zmuda
SALES AND MARKETING DIRECTOR:
  Tadeusz Witek
MARKETING MANAGER: Janusz Soboń

WSK-PZL Mielec is largest and best equipped aircraft factory in Poland; founded 1938; had produced over 15,000 aircraft by 1 January 1993. OBR SK Mielec (Osrodek Badawczo Rozwojowy Sprzetu Komunikacyjnego — Transport Equipment Research and Development Centre) responsible for R&D.

In addition to production of aircraft of its own design and ex-Soviet types, manufactures components for Ilyushin Il-86 (including fins, tailplanes, engine pylons, wing slats and flaps) and Il-96; began manufacture and subassembly March 1991 of components for Socata TB series light aircraft (see French section), with nine TB 10 Tobagos completed by end of that year (output planned to rise to 13 aircraft in 1992 and 30 in 1993).

PZL MIELEC I-22 IRYDA (IRIDIUM)
TYPE: Two-seat basic and advanced jet trainer, reconnaissance and close support aircraft.
PROGRAMME: Launched 1977, leading to 1980 Polish Ministry of National Defence 'Iskra-22' requirement for combat-capable jet trainer; designed by OBR SK/IL/PZL Warszawa team led initially by Dr Eng Alfred Baron and from 1987 by Dr Eng W. Gnarowski; construction began 1982; static test aircraft completed 1983; five flying prototypes, all with PZL-5 turbojets, of which first one (c/n 1ANP01-02) made first flight 3 March 1985 but lost in crash 31 January 1987; first flights of remaining four 26 June 1988 (SP-PWB, c/n 1ANP01-03), 13 May 1989 (SP-PWC, c/n 1ANP01-04), 22 October 1989 (SP-PWD, c/n AN001-01) and 4 July 1991 (SP-PWE, c/n AN001-02); fatigue test aircraft completed 1990; factory and state tests (1,040 flights totalling 900 hours) completed by end March 1992, followed by award of state test certificate 14 April 1992.

First order (for nine pre-production aircraft) announced during Poznan air show September 1991; first of these (serial number 103) made first flight 5 May 1992, handed over to Polish Air Force (with second aircraft 105) 24 October, entering service 27 November 1992; fourth prototype (SP-PWD) refitted with K-15 turbojets, making new 'first' flight 22 December 1992 as demonstrator for M-92; next four pre-series aircraft due for delivery by end of 1993.
CURRENT VERSIONS: **I-22:** Basic two-seat training version with PZL-5 engines. *Detailed description applies to this version except where indicated.*
  M-92: As I-22, but with 14.71 kN (3,307 lb st) Instytut Lotnictwa K-15 turbojets.
  M-93: Possible combat trainer version, with strengthened wings, enhanced weapon-carrying capability, provision for Martin-Baker Mk 10L zero/zero ejection seats, upgraded avionics and new engines (14.12 kN; 3,175 lb SNECMA Larzac 04-C20, 14.86 kN; 3,340 lb st Rolls Royce Viper 535, 14.19 kN; 3,190 lb st P&WC JT15D-5 and 17.65 kN; 3,968 lb st Polish D-18A being considered).
  M-95: Proposed two-seat reconnaissance and close support version; described separately.
  M-97S: Proposed single-seat ground attack version; described separately.
  M-97MS: Proposed single-seat fighter/ground attack version; described separately.
CUSTOMERS: Polish Air Force (requirement for 50 or more, of which two delivered in 1992 and four more due to follow in 1993).
COSTS: Development programme Zł1.5 billion (1991); I-22 approx $3.5 million (1993).
DESIGN FEATURES: Intended to replace TS-11 Iskra and LiM-with Polish Air Force; covers full spectrum of pilot, navigation, air combat, reconnaissance and ground attack training, with day/night and bad weather capability; able to operate from unprepared airstrips and carry variety of ordnance; airframe designed to be tolerant of battle damage, capable of quick and inexpensive repair, and stressed to standard that permits later use of more powerful engines and carriage of greater weapons load without jeopardising permissible load factors; service life calculated on basis of 2,500 flying hours or 10,000 take-offs and landings.

Laminar flow aerofoil section (NACA 64A-010 at root, 64A-210 at tip), with 0° incidence and 1° 43′ 48″ geometric twist; sweepback 18° on leading-edges, 14° 27′ 36″ at quarter-chord; no sweep on trailing-edges; anhedral from roots 3° on wings, 6° on tailplane.
FLYING CONTROLS: Mechanically (rod) actuated mass balanced, differentially operable ailerons (with hydraulic boost and manual reversion), elevators and rudder; ground adjustable tab on rudder and each aileron; hydraulically actuated single-slotted trailing-edge flaps (25° for T-O and

'Green' TB-9 Tampico manufactured by PZL Mielec for Socata of France *(R. J. Malachowski)*

landing, 62° max), with emergency pneumatic operation; variable incidence tailplane (0°/−7.5°) and twin airbrakes in upper rear fuselage also actuated hydraulically.

STRUCTURE: All-metal light alloy (aluminium/magnesium/steel) stressed skin, with limited use of glassfibre (eg, fuselage/fin fairing); two-spar wing, built as one unit with centre and inboard portions forming integral fuel tanks. Titanium heatshields in engine bays. Elevator trailing-edges of honeycomb sandwich from third pre-production aircraft onward.

LANDING GEAR: Retractable tricycle type, with oleo-pneumatic shock absorber, single wheel and tubeless tyre on each unit. Hydraulic extension and retraction: nose unit retracts forward, main units forward and upward into engine nacelles. Auxiliary pneumatic system for emergency extension. Mainwheel tyres size 670 × 210 mm, pressure 7.0 bars (102 lb/sq in); nosewheel tyre 430 × 170 mm, pressure 5.78 bars (84 lb/sq in). Nosewheel steerable ±45°. Hydraulic disc brakes on mainwheels; auxiliary mainwheel parking brake serves also as emergency brake. Max rate of descent 220 m (722 ft)/min at AUW of 5,650 kg (12,456 lb). SH 21U-1 brake-chute (area 15 m²; 161.5 sq ft, drag coefficient 0.45) in fuselage tailcone. Small tail bumper under rear of fuselage.

POWER PLANT: Two 10.79 kN (2,425 lb st) PZL-5 (formerly SO-3W22) non-afterburning turbojets, pod mounted on lower sides of centre-fuselage. Fuel in three integral wing tanks (combined capacity 1,090 litres; 288 US gallons; 240 Imp gallons) and two rubber tanks and two header tanks in fuselage (combined capacity 1,340 litres; 354 US gallons; 295 Imp gallons), to give total internal capacity of 2,430 litres (642 US gallons; 535 Imp gallons). Provision for one 380 litre (100 US gallon; 83.6 Imp gallon) auxiliary tank under each wing. Fuel system permits up to 30 s of inverted flight. Single-point pressure refuelling (at front of port engine nacelle), plus four gravity filling points (two in upper fuselage, one on each wing). Oil capacity 10 litres (2.6 US gallons; 2.2 Imp gallons) per engine.

ACCOMMODATION: Pressurised, heated and air-conditioned cockpit, with tandem seating for pupil (in front) and instructor; rear seat elevated 400 mm (15¾ in). For solo flying, pilot occupies front seat. Back-type parachute, oxygen bottle and emergency pack for both occupants. Individual framed canopies, each hinged at rear and opening upward pneumatically. Rearview mirror in front cockpit. VS1-BRI/P rocket assisted ejection seats, fitted with canopy breakers, can be operated at zero altitude and at speeds down to 81 knots (150 km/h; 94 mph). Dual controls standard; front cockpit equipped for IFR flying. Windscreen anti-iced by electric heating, supplemented by alcohol spray. Remaining transparencies anti-iced and demisted by hot engine bleed air.

SYSTEMS: Cockpits pressurised (max differential 3.5 bars; 50.8 lb/sq in) and air-conditioned by engine bleed air. Air from air-conditioning system also used to pressurise crew's g suits. Main hydraulic system, pressure 210 bars (3,045 lb/sq in), actuates landing gear extension and retraction, wing flaps, airbrakes, tailplane incidence, brake-chute deployment, differential braking of mainwheels, and nosewheel steering. Auxiliary hydraulic system for aileron control boost. Pneumatic system comprises three separate circuits, each supplied by a nitrogen bottle pressurised at 150 bars (2,175 lb/sq in): one powers emergency extension of wing flaps for landing, one the emergency extension of landing gear; third is for canopy opening, closing and sealing, windscreen fluid anti-icing system, and hydraulic reservoir pressurisation. All three bottles charged simultaneously through a common nozzle. Electrical system, powered by two 9kW PR-9 DC starter/generators, supplies 115V single-phase AC (via two 1kVA LUN 2458-8 static converters) and 36V three-phase AC (via two 500VA PT-500C converters), both at 400Hz; two 24V 25Ah 20NKBN-25 batteries provide DC power in event of a double failure. Each AC voltage supplied by one main converter and one standby, latter automatically assuming full load if a main converter fails. Crew oxygen system, capacity 9 litres (549 cu in). Engine intakes anti-iced by engine bleed air; fire detection and extinguishing system (two freon bottles in rear fuselage). Electronic control system for gun firing and weapon release.

AVIONICS: Bays in nose and under floor of rear cockpit include Unimor RS6113 VHF/UHF multi-channel com radio; ARK-15M ADF; IL RW-5 radio altimeter; IL ORS-2M marker beacon receiver (from third pre-series aircraft onward); SOD-57M ILS; SRO-2 IFF; and SPO-10 radar warning system. Blind-flying instrumentation. Flight data recorder in dorsal fin fillet.

ARMAMENT: One 23 mm GSz-23L twin-barrel gun in ventral pack, with up to 200 rds in fuselage (normal load 50 rds on training missions); ASP-PFD-I22 gyro gunsight, S-13-100 nose mounted gun camera and (optionally) SSz-45-1-100-05 firing effect monitor camera. Four UBP-I-22 underwing multiple stores carriers normally, each stressed for load of up to 500 kg (1,102 lb) (but max external stores load 1,100 kg; 2,425 lb); on these can be carried various bombs of 50, 100, 250 or 500 kg size (alternative MBD2-67U carriers for 50 and 100 kg bombs); from 32 to 128 57 mm S-5 unguided rockets in Mars-2, Mars-4, UB-16-57U or UB-32A-1 launchers; or R-60 air-to-air homing missiles on APU60IM launch rails. Alternative underwing loads can include UPK-23-250 23 mm gun pods, Zeus-1

I-22 second prototype SP-PWB, displaying range of bombs, rockets and other stores that can be carried by the Iryda (*R. J. Malachowski*)

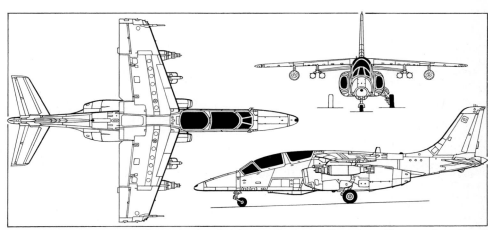

PZL I-22 Iryda tandem-seat advanced trainer in initial production configuration with external armament (*Jane's/Mike Keep*)

Serial number 103, handed over to the Polish Air Force in October 1992, is one of the first two pre-production Irydas (*R. J. Malachowski*)

230   POLAND: AIRCRAFT—WSK-PZL MIELEC

SP-PWD, the fourth I-22 prototype, re-engined with IL K-15 turbojets which significantly improve performance *(R. J. Malachowski)*

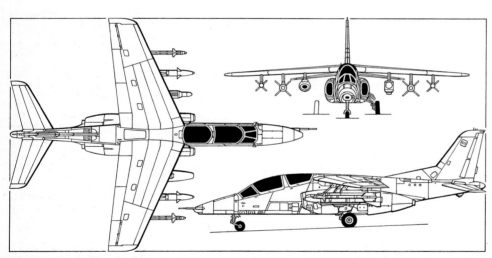

The proposed M-95 reconnaissance and close support derivative of the I-22, with sweptback wings, revised nose/tail contours and other modifications *(Jane's/Mike Keep)*

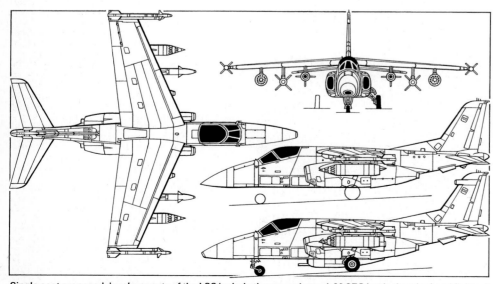

Single-seat proposed developments of the I-22 include the ground attack M-97S (main drawing) and fighter/ground attack M-97MS (upper side view) *(Jane's/Mike Keep)*

7.62 mm gun pods, a ZR-8MB bomblet dispenser, Saturn camera pods, 80 mm S-8 rockets in B-8M launchers, drop tanks or other weapons.

DIMENSIONS, EXTERNAL:
Wing span	9.60 m (31 ft 6 in)
Wing chord: at root	2.92 m (9 ft 7 in)
at tip	1.25 m (4 ft 1¼ in)
Wing aspect ratio	4.6
Length overall	13.22 m (43 ft 4½ in)
Height overall	4.30 m (14 ft 1¼ in)
Tailplane span	4.20 m (13 ft 9¼ in)
Wheel track	2.71 m (8 ft 10¾ in)
Wheelbase	4.91 m (16 ft 1¼ in)

AREAS:
Wings, gross	19.92 m² (214.4 sq ft)
Ailerons (total)	1.362 m² (14.66 sq ft)
Trailing-edge flaps (total)	3.22 m² (34.66 sq ft)
Airbrakes (total)	0.476 m² (5.12 sq ft)
Fin	2.705 m² (29.12 sq ft)
Rudder, incl tab	0.975 m² (10.49 sq ft)
Tailplane	3.57 m² (38.43 sq ft)
Elevators (total)	1.26 m² (13.56 sq ft)

WEIGHTS AND LOADINGS (I-22):
Weight empty, equipped: clean	4,560 kg (10,053 lb)
Max internal fuel	1,890 kg (4,167 lb)
Max external stores load	1,100 kg (2,425 lb)
T-O weight with fuselage fuel only	5,770 kg (12,720 lb)
Max T-O weight: clean	6,620 kg (14,594 lb)
with external stores	6,900 kg (15,212 lb)
Max landing weight	6,600 kg (14,550 lb)
Max wing loading: clean	332.3 kg/m² (68.06 lb/sq ft)
with external stores	346.4 kg/m² (70.95 lb/sq ft)
Max power loading: clean	306.77 kg/kN (3.01 lb/lb st)
with external stores	319.8 kg/kN (3.14 lb/lb st)

WEIGHTS AND LOADINGS (M-92):
Weight empty, equipped	4,740 kg (10,450 lb)
Max T-O weight	7,300 kg (16,093 lb)
Max landing weight	as for I-22
Max wing loading	366.5 kg/m² (75.07 lb/sq ft)
Max power loading	248.13 kg/kN (2.43 lb/lb st)

WEIGHTS AND LOADINGS (M-93, estimated):
Weight empty, equipped, depending upon power plant	4,300-4,550 kg (9,480-10,031 lb)
Max fuel weight: internal	1,895 kg (4,178 lb)
external	595 kg (1,312 lb)
Max external stores load	2,000 kg (4,409 lb)
Max T-O weight:	
clean, depending upon power plant	6,360-6,620 kg (14,021-14,594 lb)
with external stores	7,600 kg (16,755 lb)
Max landing weight	6,600 kg (14,550 lb)
Max wing loading:	
clean, depending upon power plant	319.3-332.3 kg/m² (65.40-68.06 lb/sq ft)
with external stores	381.5 kg/m² (78.14 lb/sq ft)
Max power loading:	
clean, Larzac engines	225.21 kg/kN (2.21 lb/lb st)
clean, Viper engines	222.24 kg/kN (2.18 lb/lb st)
with external stores, Larzac engines	269.12 kg/kN (2.64 lb/lb st)
with external stores, Viper engines	255.72 kg/kN (2.51 lb/lb st)

PERFORMANCE (I-22 Iryda. A: clean aircraft at 5,850 kg; 12,897 lb T-O weight, B: with 820 kg; 1,808 lb of external stores comprising four 50 kg bombs and two 16-round 57 mm rocket launchers for T-O weight of 6,600 kg; 14,550 lb, except where indicated):
Max limiting Mach number: A	0.80
B	0.70
Max permissible diving speed:	
A	512 knots (950 km/h; 590 mph)
B	445 knots (826 km/h; 513 mph)
Max level speed at S/L:	
A	424 knots (785 km/h; 488 mph)
B	369 knots (684 km/h; 425 mph)
Max level speed at height:	
A at 5,000 m (16,400 ft)	450 knots (835 km/h; 519 mph)
B at 4,000 m (13,125 ft)	374 knots (694 km/h; 431 mph)
Stalling speed, flaps down:	
A	110 knots (203 km/h; 126 mph)
B	118 knots (218 km/h; 136 mph)
Max rate of climb at S/L: A	1,500 m (4,925 ft)/min
B	1,008 m (3,305 ft)/min
Rate of climb at S/L, OEI: A	300 m (985 ft)/min
Time to 5,000 m (16,400 ft): A	4 min 24 s
B	6 min 48 s
Service ceiling: A	12,000 m (39,375 ft)
B	9,500 m (31,175 ft)
Service ceiling, OEI: A	5,000 m (16,400 ft)
T-O run: A	730 m (2,395 ft)
B	1,250 m (4,100 ft)
T-O to 15 m (50 ft): A	1,200 m (3,940 ft)
B	2,200 m (7,220 ft)
Landing from 15 m (50 ft):	
A with brake-chute	1,340 m (4,400 ft)
A without brake-chute	2,160 m (7,090 ft)
B with brake-chute	1,570 m (5,150 ft)
B without brake-chute	2,240 m (7,350 ft)

Landing run: A with brake-chute 640 m (2,100 ft)
A without brake-chute 1,400 m (4,595 ft)
B with brake-chute 760 m (2,495 ft)
B without brake-chute 1,570 m (5,150 ft)
Combat radius at 500 m (1,640 ft):
B 108 nm (200 km; 124 miles)
Range at 5,000 m (16,400 ft):
A at 6,650 kg (14,660 lb) T-O weight
620 nm (1,150 km; 714 miles)
B at 6,900 kg (15,211 lb) T-O weight
326 nm (605 km; 376 miles)
Endurance at 5,000 m (16,400 ft), weights as for ranges
above: A 2 h 33 min
B 1 h 32 min
g limits:
clean, AUW below 5,865 kg (12,930 lb) +8/−4
clean, AUW above 5,865 kg (12,930 lb) +7.3/−4
with external stores +6/−3
PERFORMANCE (M-92 Iryda):
Max level speed at 5,000 m (16,400 ft), AUW of 6,200 kg
(13,668 lb) 498 knots (923 km/h; 574 mph)
Max cruising speed at 5,000 m (16,400 ft), AUW as above
459 knots (850 km/h; 528 mph)
Econ cruising speed at 5,000 m (16,400 ft), AUW as above
313 knots (580 km/h; 360 mph)
Stalling speed at AUW of 5,800 kg (12,786 lb):
flaps up 130 knots (240 km/h; 149 mph)
flaps down 110 knots (204 km/h; 127 mph)
Max rate of climb at S/L at max landing weight
2,640 m (8,660 ft)/min
Service ceiling 12,000 m (39,375 ft)
T-O run at AUW of 5,600 kg (12,345 lb)
500 m (1,640 ft)
T-O to 15 m (50 ft) at AUW of 5,600 kg (12,345 lb)
850 m (2,790 ft)
Landing from 15 m (50 ft), with brake-chute, at AUW of
5,000 kg (11,023 lb) 1,375 m (4,515 ft)
Landing run, with brake-chute, at AUW of 5,000 kg
(11,023 lb) 750 m (2,460 ft)
Range:
at max landing weight at 11,000 m (36,100 ft), max
internal fuel 593 nm (1,100 km; 683 miles)
at max T-O weight at 6,000 m (19,675 ft), max external
stores 388 nm (720 km; 447 miles)
PERFORMANCE (M-93 Iryda, estimated. A: with Larzac
engines, B: with Vipers, both at AUW of 5,600 kg;
12,345 lb except where indicated):
Max limiting Mach number: A, B 0.83
Max level speed at S/L: A 475 knots (880 km/h; 547 mph)
B 513 knots (950 km/h; 590 mph)
Max level speed at 5,000 m (16,400 ft):
A 475 knots (880 km/h; 547 mph)
B 515 knots (955 km/h; 593 mph)
Stalling speed, flaps down:
A, B 100 knots (185 km/h; 115 mph)
Max rate of climb at S/L: A 2,310 m (7,580 ft)/min
B 2,670 m (8,760 ft)/min
Rate of climb at S/L, OEI: A 600 m (1,970 ft)/min
B 780 m (2,560 ft)/min
Time to 10,000 m (32,800 ft): A 10 min
B 6 min
Service ceiling: A 11,900 m (39,050 ft)
B 13,000 m (42,650 ft)
Service ceiling, OEI: A 6,000 m (19,675 ft)
B 9,500 m (31,175 ft)
T-O run, AUW of 6,300 kg (13,890 lb): A 820 m (2,690 ft)
B 745 m (2,445 ft)
T-O to 15 m (50 ft), AUW as above: A 1,280 m (4,200 ft)
B 1,140 m (3,740 ft)
Landing from 15 m (50 ft) at max landing weight:
A with brake-chute 1,490 m (4,890 ft)
A without brake-chute 1,790 m (5,875 ft)
B with brake-chute 1,495 m (4,905 ft)
B without brake-chute 1,795 m (5,890 ft)
Landing run at max landing weight:
A with brake-chute 725 m (2,380 ft)
A without brake-chute 1,025 m (3,365 ft)
B with brake-chute 730 m (2,395 ft)
B without brake-chute 1,030 m (3,380 ft)
Combat radius at 500 m (1,640 ft) at AUW of 7,500 kg
(16,534 lb) incl 700 kg (1,543 lb) of external stores, 12
min fuel reserves: A 192 nm (357 km; 222 miles)
B 127 nm (236 km; 146 miles)

PZL MIELEC M-95 and M-97 IRYDA

TYPE: Proposed derivatives of M-93 Iryda.
PROGRAMME: Announced 1992.
CURRENT VERSIONS: **M-95:** Two-seater for reconnaissance and close support. New, slightly sweptback wing of supercritical section and increased span/area; modified nose and tail contours (see drawing); Larzac 04P20 (Polish designation of 04-C20) or Viper 535 engines as in M-93, or their derivatives, or 16.35 kN; 3,675 lb st General Electric J85-GE-J1 turbojets; internal 30 mm gun instead of ventral 23 mm; eight underwing stations for wider assortment of ordnance or other stores including air-to-air, air-to-surface or anti-ship missiles, conventional or guided bombs; more comprehensive avionics. Structure strengthened for higher operating weights.

M-97S: Single-seat ground attack version. Modifications as for M-95, plus necessary reconfiguration in cockpit area. Illustrated in accompanying drawing.
M-97MS: Single-seat fighter/ground attack version. Generally as M-97S except for equipment and armament.
DIMENSIONS, EXTERNAL:
Wing span: M-95, M-97 10.83 m (35 ft 6½ in)
Length overall: M-95 13.80 m (45 ft 3¼ in)
M-97 12.90 m (42 ft 4 in)
Height overall: M-95, M-97 4.30 m (14 ft 1¼ in)
AREAS:
Wings, gross: M-95, M-97 23.0 m² (247.6 sq ft)
WEIGHTS AND LOADINGS*:
Max external stores load:
M-95 2,100-3,000 kg (4,630-6,614 lb)
M-97 2,500-3,500 kg (5,511-7,716 lb)
Max T-O weight:
M-95 8,000-8,500 kg (17,637-18,739 lb)
M-97 8,500-9,000 kg (18,739-19,841 lb)
PERFORMANCE* (M-95, M-97, estimated):
Max level speed
502-593 knots (930-1,100 km/h; 578-684 mph)
Max rate of climb
2,400-4,200 m (7,875-13,780 ft)/min
T-O to 15 m (50 ft) 570-820 m (1,870-2,690 ft)
* *Depending upon power plant*

PZL MIELEC TS-11R ISKRA

TYPE: Coastal reconnaissance modification of Iskra-Bis DF.
PROGRAMME: Original Iskra production totalled four prototypes and 423 production aircraft (31 Iskra 100, 45 Iskra 100-Bis A, 134 Iskra 100-Bis B, five Iskra 200 ART-Bis C and 208 Iskra 200 SB-Bis DF). TS-11R is replacement for SBLiM-2As (modified MiG-15UTIs) in Polish Naval Air Force, which were retired in early 1991; prototype (c/n 3H1917) converted at Mielec June 1991, delivered at end of year to 7th Special Air Regiment of PNAF at Siemirowice for evaluation; five more delivered subsequently.
CUSTOMERS: Polish Naval Air Force (six).
DESIGN FEATURES: Modified by installing weather radar in reshaped nose behind dielectric nosecone and replacing dual controls in rear cockpit with radar VDU and artificial horizon.
AVIONICS: Bendix/King RDS-81 weather radar in nose; GPS navigation system; three (Russian) AFA-39 optical cameras.

PZL MIELEC (ANTONOV) An-28

NATO reporting name: Cash
TYPE: Twin-turboprop short-range transport.
PROGRAMME: Developed by Antonov in former USSR for service on Aeroflot's shortest routes, particularly those operated by An-2s into places relatively inaccessible to other fixed-wing aircraft; official Soviet flight testing completed 1972; first pre-production An-28 (SSSR-19723) originally retained same engines as prototype, but re-registered SSSR-19753 April 1975 when flown with current engines; production assigned to PZL Mielec 1978; temporary Soviet NLGS-2 type certificate awarded 4 October 1978 to second Soviet-built pre-production aircraft. Polish manufacture started with initial batch of 15; first flight of Polish An-28 (SSSR-28800) 22 July 1984; version received full Soviet type certificate 7 February 1986.
CURRENT VERSIONS: **An-28:** Standard Polish production version since 1984, but now suspended. *Detailed description applies to this version except where indicated.*
An-28B1R Bryza: Search and rescue version for use by Polish agencies for SAR missions in Baltic; improved avionics, GPS receiver, Doppler navigation, Polish designed SRN-441XA search radar in ventral radome, two 100 kg illumination bombs, rescue dinghy, stretchers, automatic data link and other specialist equipment. Prototype (SP-PDC) completed 1992 and delivered to Polish Air Force 1993.
An-28B1T: Adapted for airdrop operations, with clamshell rear doors replaced by single door sliding forward under fuselage, similar to that of An-26 (which see). First use may be to transport and drop firefighting teams. Prototype (SP-PDE) converted 1992.
AN-28PT: Westernised version, generally as standard An-28 but with Polish engines and three-blade propellers replaced by PT6A-65B turboprops and five-blade Hartzell propellers, and Bendix/King nav/com, weather radar and other avionics. Prototype conversion began early 1991; first flight expected April/May 1993; certification trials due for completion 1993 (FAR Pt 23 Amendment 34 and complying with Russian NLGS-2), leading to first deliveries shortly after. Mielec hoped to deliver up to 10 of this version by end of 1993 and begin full scale new production 1994.
AEW version: In design study phase; probable dorsally mounted SLAR pod, following configuration of Swedish Air Force Saab 340 AEW.
Stretched version: Development of An-28PT; same power plant, 24-26 passengers, higher operating weights, ski landing gear, provision for containerised cargo; Antonov bureau assisting in programme.
CUSTOMERS: Total of 187 ordered by 1 January 1992, of which 173 then delivered. Polish Air Force requirement for eight An-28B1Rs (two ordered by March 1993) for search and rescue.
DESIGN FEATURES: Suited to carrying passengers, cargo and mail, scientific expeditions, geological survey, forest fire patrol, air ambulance or rescue operations, and parachute training. Braced wings; will not stall because of action of automatic slats; if an engine fails, patented upper surface spoiler forward of aileron on opposite wing is opened automatically, resulting in wing bearing dead engine dropping only 12° in 5 s instead of 30° without spoiler; patented fixed tailplane slat improves handling during high angle of attack climbout; under icing conditions, if normal anti-icing system fails, ice collects on slat rather than tailplane, to retain controllability; short stub-wing extends from lower fuselage to carry main landing gear and support wing bracing struts, curving forward and downward at front to serve as mudguards; underside of rear fuselage upswept, incorporating clamshell doors for passenger/cargo loading; twin fins and rudders, mounted on inverted-aerofoil, no-dihedral fixed incidence tailplane.
Wing section TsAGI R-II-14; thickness/chord ratio 14 per cent; constant chord, non-swept, no-dihedral centre-section, with 4° incidence; tapered outer panels with 2° dihedral, negative incidence and 2° sweepback at quarter-chord.
FLYING CONTROLS: Unpowered single-slotted mass and aerodynamically balanced ailerons (port aileron has trim tab), designed to droop with large, hydraulically actuated, two-segment double-slotted flaps; elevators with electrically actuated trim tabs; twin rudders each with electrically actuated trim tab; automatic leading-edge slats over full span of wing outer panels; slab type spoiler forward of each aileron and each outer flap segment at 75 per cent chord; fixed slat under full span of tailplane leading-edge.
STRUCTURE: Mostly metal; duralumin ailerons with fabric covering; duralumin slats with CFRP skins; CFRP spoilers and trim tabs. Air intakes lined with epoxy laminate.
LANDING GEAR: Non-retractable tricycle type, with single Soviet-built wheel and PZL oleo-pneumatic shock absorber on each unit. Main units, mounted on small stub-wings, have wide tread balloon tyres of Russian manufacture, size 720 × 320 mm, pressure 3.5 bars (51 lb/sq in). Steerable (±50°) and self-centring nosewheel, with size 595 × 185 × 280 mm Stomil (Poland) tyre, pressure 3.5 bars (51 lb/sq in). Russian multi-disc hydraulic brakes on main units, and Russian inertial anti-skid units. Min ground turning radius 16.00 m (52 ft 6 in). Ski gear under development.
POWER PLANT: Two 716 kW (960 shp) WSK-PZL Rzeszów PZL-10S (TWD-10B) turboprops in standard An-28, each driving an AW-24AN three-blade automatic propeller with full feathering and reversible-pitch capability. An-28PT (from late 1993) has two 820 kW (1,100 shp) Pratt & Whitney Canada PT6A-65B turboprops and Hartzell HC-B5MP-3M10876ASK five-blade propellers. Two centre-section and two outer-wing integral fuel tanks in wing spar boxes, with total capacity of 1,960 litres (518 US gallons; 431 Imp gallons). Refuelling point on each tank. Oil capacity 16 litres (4.2 US gallons; 3.5 Imp gallons) per engine.
ACCOMMODATION: Pilot and co-pilot on flight deck, which has bulged side windows and electric anti-icing for wind-

Prototype TS-11R Iskra conversion *(Piotr Butowski)*

232   POLAND: AIRCRAFT—WSK-PZL MIELEC

Polish-built Antonov An-28 short-range transport (two PZL Rzeszów PZL-10S turboprops)

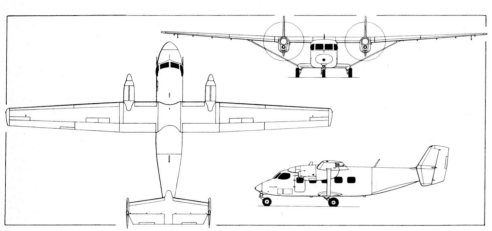

Antonov An-28 short-range transport, produced in Poland by WSK-PZL Mielec *(Dennis Punnett)*

An-28B1R prototype, with ventral SAR radar bulge

screens, and is separated from main cabin by bulkhead with connecting door. Dual controls standard. Jettisonable emergency door at front on each side. Standard cabin layout of passenger version has seats for 17 people, with six single seats on port side, one single seat and five double seats on starboard side of aisle, at 72 cm (28 in) pitch. Aisle width 34.5 cm (13.5 in). Five passenger windows each side of cabin. Seats fold back against walls when aircraft is operated as a freighter or in mixed passenger/cargo role, seat attachments providing cargo tiedown points. Hoist of 500 kg (1,102 lb) capacity able to deposit cargo in forward part of cabin. Entire cabin heated, ventilated and soundproofed. Outward/downward opening clamshell double door, under upswept rear fuselage, for passenger and cargo loading. Emergency exit at rear of cabin on each side.

SYSTEMS: No air-conditioning, pressurisation or pneumatic systems. Hydraulic system, pressure 150 bars (2,175 lb/sq in), for flap and spoiler actuation, mainwheel brakes and nosewheel steering, with emergency backup system for spoiler extension and mainwheel braking. Primary electrical system is three-phase AC, with two engine driven alternators providing 200/115V power for heating systems, engine vibration monitoring, fuel pump, radio, recorders and instrument lights. Transformer-rectifiers on this system provide 36V AC power for pressure gauges, artificial horizon, navigation and recording equipment, and 27V DC for control systems and signalling, internal and external lighting, firefighting system, propeller pitch control and feathering, radio, and engine starting and monitoring systems. In emergency, single-phase 115V AC can be provided by PO-250A converter, 36V AC by a static inverter and 27V DC by two 25Ah 20NKBN-25 batteries. Thermal (engine bleed air) anti-icing of outer-wing, fin and tailplane leading-edges and engine air intakes. Electric anti-icing of flight deck windscreens, propellers, spinners and pitot heads. Oxygen system (for crew plus two passengers) optional. No APU.

AVIONICS: Standard avionics include Baklan-5 (USSR) VHF com radio, R-855UM (USSR) VHF emergency locator transmitter, ARK-15 radio compass, MRP-66 marker beacon receiver, RW-5 or A-037 radio altimeter, Grebien-1 navigation unit, BUR-1-2A flight recorder, and SGU-6 intercom. Blind-flying instrumentation standard.

Alternative Bendix/King suite comprises KY 196 and KX 165 VHF/AM and HF com radio, KMA 24H-70/71 audio selector panel and intercom and dual emergency VHF; VOR/ILS (dual), DME, ADF, marker beacon receiver, RMI and dual KCS 55A pictorial nav system; KNS 81 digital area nav and GC 381A radar graphics computer; RDS 81 digital weather radar; KT 79 radar altimeter transponder; and (from late 1993) KFC 275 or KFC 325 AFCS.

DIMENSIONS, EXTERNAL:
Wing span	22.063 m (72 ft 4½ in)
Wing chord: at root	2.20 m (7 ft 2½ in)
mean aerodynamic	1.886 m (6 ft 2¼ in)
at tip	1.10 m (3 ft 7¼ in)
Wing aspect ratio	12.25
Length overall	13.10 m (42 ft 11¾ in)
Fuselage: Length	12.68 m (41 ft 7¼ in)
Max width	1.90 m (6 ft 2¾ in)
Max depth	2.14 m (7 ft 0¼ in)
Height overall	4.90 m (16 ft 1 in)
Tailplane span	5.14 m (16 ft 10¼ in)
Wheel track	3.405 m (11 ft 2 in)
Wheelbase	4.354 m (14 ft 3½ in)
Propeller diameter: AW-24AN	2.80 m (9 ft 2¼ in)
Hartzell	2.82 m (9 ft 3 in)
Propeller ground clearance:	
AW-24AN	1.25 m (4 ft 1¼ in)
Distance between propeller centres	5.20 m (17 ft 0¾ in)
Rear clamshell doors: Length	2.40 m (7 ft 10½ in)
Total width: at top	1.00 m (3 ft 3¼ in)
at sill	1.40 m (4 ft 7 in)
Emergency exits (rear, each): Height	0.91 m (3 ft 0 in)
Width	0.51 m (1 ft 8 in)

DIMENSIONS, INTERNAL:
Cabin, excl flight deck: Length	5.26 m (17 ft 3 in)
Max width	1.74 m (5 ft 8½ in)
Max height	1.60 m (5 ft 3 in)
Floor area	approx 7.5 m² (80.73 sq ft)
Volume	approx 14.0 m³ (494.4 cu ft)

AREAS:
Wings, gross	39.72 m² (427.5 sq ft)
Ailerons (total)	4.33 m² (46.61 sq ft)
Trailing-edge flaps (total)	7.986 m² (85.96 sq ft)
Spoilers (total)	1.667 m² (17.94 sq ft)

Foldaway seating in the An-28 cabin

The prototype An-28B1T, showing the slide-under door which replaces the normal clamshell pair
(R. J. Malachowski)

Fins (total)	10.00 m² (107.64 sq ft)
Rudders (total, incl tabs)	4.00 m² (43.06 sq ft)
Tailplane	8.85 m² (95.26 sq ft)
Elevators (total, incl tabs)	2.56 m² (27.56 sq ft)

WEIGHTS AND LOADINGS:
Weight empty, equipped	3,900 kg (8,598 lb)
Max fuel load	1,529 kg (3,371 lb)
Max payload	1,750 kg (3,858 lb)
Fuel with max payload	650 kg (1,433 lb)
Payload with max fuel	871 kg (1,920 lb)
Max T-O and landing weight	6,500 kg (14,330 lb)
Max zero-fuel weight	5,884 kg (12,972 lb)
Normal wing loading	153.5 kg/m² (31.5 lb/sq ft)
Max power loading:	
PZL-10S	4.64 kg/kW (7.62 lb/shp)
PT6A-65B	3.96 kg/kW (6.51 lb/shp)

PERFORMANCE (standard An-28 at max T-O weight):
Never-exceed speed (V_{NE})	210 knots (390 km/h; 242 mph)
Max level and max cruising speed at 3,000 m (9,850 ft)	189 knots (350 km/h; 217 mph)
Econ cruising speed at 3,000 m (9,850 ft)	181 knots (335 km/h; 208 mph)
Lift-off speed	73 knots (135 km/h; 84 mph)
Approach speed	70 knots (130 km/h; 81 mph)
Landing speed, flaps down	76 knots (140 km/h; 87 mph)
Max rate of climb at S/L	500 m (1,640 ft)/min
Rate of climb at S/L, OEI	210 m (689 ft)/min
Service ceiling	above 6,000 m (19,685 ft)
T-O run	265 m (870 ft)
T-O to 10.7 m (35 ft)	410 m (1,345 ft)
Landing from 15 m (50 ft)	315 m (1,035 ft)
Landing run	170 m (558 ft)
Range:	
max payload, no reserves	302 nm (560 km; 348 miles)
max fuel and 1,000 kg (2,205 lb) payload, 30 min reserves	736 nm (1,365 km; 848 miles)
g limit	+3

PZL MIELEC M-18 DROMADER (DROMEDARY)

TYPE: Agricultural and firefighting aircraft.
PROGRAMME: Designed to meet requirements of FAR Pt 23; prototype first flights 27 August and 2 October 1976; M-18 awarded Polish type certificate 27 September 1978; 10 pre-production aircraft built, of which eight used for operational trials; now certificated in Australia, Brazil, Canada, Czech Republic, France, Germany, Poland, USA and Yugoslavia; series production began 1979.
CURRENT VERSIONS: **M-18:** Initial single-seat agricultural version (see 1988-89 *Jane's*); production ended 1984 but available to order.
M-18A: Two-seat agricultural version, for operators requiring to transport mechanic/loader to improvised airstrip; production began 1984, following Polish supplementary type certification 14 February 1984; FAA type certificate for M-18 extended to M-18A September 1987; in production. *Detailed description applies to M-18A except where indicated.*
M-18AS: Two-seat training version, with smaller hopper to make space for instructor's cockpit aft of pilot; rear cockpit installation readily interchangeable with that of M-18A; first flown 21 March 1988; in production (five built by 1 January 1992).
M-18B: Modernised version with more powerful Polish K-9 engine; expected to be announced 1993.
T45 Turbine Dromader: Turboprop version; 895 kW (1,200 shp) Pratt & Whitney Canada PT6A-45AG with Hartzell propeller; developed by James Mills in co-operation with Melex USA Inc (see Melex entry).
Firefighter: Prototype first flown 11 November 1978; **amphibious water bomber floatplane** variant under consideration.
CUSTOMERS: Total of 610 built (all versions) by 1 January 1992, 90 per cent for export; sold to operators in Australia, Brazil, Bulgaria, Canada, Chile, China, Cuba, Czech Republic, Germany, Greece (30 for firefighting), Hungary, Iran, Morocco, Nicaragua, Poland, Portugal, Spain, Swaziland, Trinidad, Turkey, USA, Venezuela and Yugoslavia.
DESIGN FEATURES: Emphasis on crew safety; all parts exposed to chemical contact treated with polyurethane or epoxy enamels, or manufactured of stainless steel; detachable fuselage side panels for airframe inspection and cleaning; braced tailplane.
Wing sections NACA 4416 at root, NACA 4412 at end of centre-section and on outer panels; incidence 3°.
FLYING CONTROLS: Mechanically actuated, mass and aerodynamically balanced slotted ailerons with trim tabs, using pushrods; aerodynamically and mass balanced rudder and elevators with trim tabs, actuated by cables and pushrods respectively; hydraulically actuated two-section trailing-edge slotted flaps.
STRUCTURE: All-metal; stainless steel capped duralumin wing spar; fuselage mainframe of helium-arc welded chromoly steel tube, oiled internally against corrosion; duralumin fuselage side panels and stainless steel bottom covering; corrugated tail unit skins.
LANDING GEAR: Non-retractable tailwheel type. Oleo-pneumatic shock absorber in each unit. Main units have tyres size 800 × 260 mm, and are fitted with hydraulic disc brakes, parking brake and wire cutters. Fully castoring tailwheel, lockable for take-off and landing, with size 380 × 150 mm tyre.
POWER PLANT: One PZL Kalisz ASz-62IR nine-cylinder radial air-cooled supercharged engine (721 kW; 967 hp at 2,200 rpm), driving a PZL Warszawa AW-2-30 four-blade constant-speed aluminium propeller. Integral fuel tank in each outer wing panel, combined usable capacity 400 or 712 litres (105.7 or 188 US gallons; 88 or 156.6 Imp gallons). Gravity feed header tank in fuselage.
ACCOMMODATION: Single adjustable seat in fully enclosed, sealed and ventilated cockpit stressed to withstand 40g impact. Additional cabin located behind cockpit and separated from it by a wall. Latter equipped with rigid seat with protective padding and safety belt, port-side jettisonable door, windows (port and starboard), fire extinguisher, and ventilation valve. Communication with pilot provided via window in dividing wall, and by intercom. In M-18AS, standard hopper is replaced by smaller one, permitting installation of bolt-on instructor's cabin. Second cockpits of M-18A and M-18AS quickly interchangeable. Glass-fibre cockpit roof and rear fairing, latter with additional small window each side. Rear cockpit of M-18AS has more extensive glazing. Adjustable shoulder type safety harness. Adjustable rudder pedals. Quick-opening door on each side of front cockpit; port door jettisonable.
SYSTEMS: Hydraulic system, pressure 98-137 bars (1,421-1,987 lb/sq in), for flap actuation, disc brakes and dispersal system. Electrical system powered by 28.5V 100A generator, with 24V 25Ah nickel-cadmium battery and over-voltage protection relay.
AVIONICS: RS6102 (Polish built), Bendix/King KX 175B or KY 195B com transceiver, KI 201C nav receiver, VOR-OBS indicator, gyro compass, radio compass and stall warning.
EQUIPMENT: Glassfibre epoxy hopper, with stainless steel tube bracing, forward of cockpit; capacity (M-18A) 2,500 litres (660 US gallons; 550 Imp gallons) of liquid or 1,350 kg (2,976 lb) of dry chemical (1,850 kg; 4,078 lb under CAM 8 conditions). Smaller hopper in M-18AS. Deflector cable from cabin roof to fin. M-18 variants can be fitted optionally with several different types of agricultural and firefighting systems, as follows: spray system with 54/96 nozzles on spraybooms; dusting system with standard, large or extra large spreader; atomising system with six atomisers; water bombing installation; and fire bombing installation with foaming agents. Aerial application roles can include seeding, fertilising, weed or pest control, defoliation, forest and bush firefighting, and patrol flights. Special wingtip lights permit agricultural flights at night, and aircraft can operate in both temperate and tropical climates. Landing lights, taxi light and night working light optional. Navigation lights, cockpit light, instrument panel lights and two rotating beacons standard. Built-in jacking and tiedown points in wings and rear fuselage; towing lugs on main landing gear. Cockpit fire extinguisher and first aid kit.

DIMENSIONS, EXTERNAL:
Wing span	17.70 m (58 ft 0¾ in)
Wing chord, constant	2.286 m (7 ft 6 in)
Wing aspect ratio	7.8
Length overall	9.47 m (31 ft 1 in)
Height: over tail fin	3.70 m (12 ft 1¾ in)
overall (flying attitude)	4.60 m (15 ft 1 in)

234 POLAND: AIRCRAFT—WSK-PZL MIELEC

M-18A standard version of the Dromader agricultural aircraft

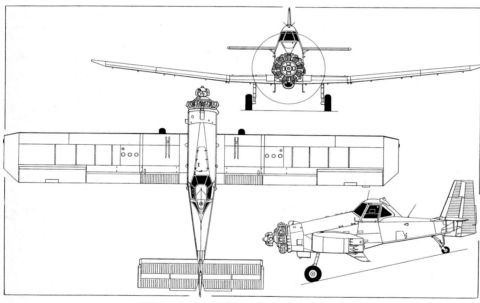

PZL Mielec M-18A Dromader *(Dennis Punnett)*

The M-18AS version of the Dromader has an extra cockpit at the rear for an instructor

Tailplane span	5.60 m (18 ft 4½ in)
Wheel track	3.48 m (11 ft 5 in)
Propeller diameter	3.30 m (10 ft 10 in)
Propeller ground clearance (tail up)	0.23 m (9 in)

DIMENSIONS, INTERNAL:
Hopper volume	2.50 m³ (88.3 cu ft)

AREAS:
Wings, gross	40.00 m² (430.5 sq ft)
Ailerons (total)	3.84 m² (41.33 sq ft)
Trailing-edge flaps (total)	5.69 m² (61.25 sq ft)
Vertical tail surfaces (total)	2.65 m² (28.5 sq ft)
Horizontal tail surfaces (total)	6.50 m² (70.0 sq ft)

WEIGHTS AND LOADINGS (M-18A):
Basic weight empty	2,710 kg (5,975 lb)
Weight empty, equipped	2,750-2,860 kg (6,063-6,305 lb)
Payload: FAR 23	1,050-1,350 kg (2,315-2,976 lb)
CAM 8	1,550-1,850 kg (3,417-4,078 lb)
Max T-O weight: FAR 23	4,200 kg (9,259 lb)
Restricted	4,700 kg (10,362 lb)
CAM 8 (max)	5,300 kg (11,684 lb)
Max landing weight	4,200 kg (9,259 lb)
Max wing loading (FAR 23)	105.0 kg/m² (21.51 lb/sq ft)
Max power loading (FAR 23)	5.83 kg/kW (9.58 lb/hp)

PERFORMANCE (M-18A at 4,200 kg; 9,259 lb T-O weight, ISA. A: without agricultural equipment, B: with spreader equipment):
Never-exceed speed (V_{NE})	
A	151 knots (280 km/h; 174 mph)
Max level speed: A	138 knots (256 km/h; 159 mph)
B	128 knots (237 km/h; 147 mph)
Cruising speed at S/L:	
A	110 knots (205 km/h; 127 mph)
B	102 knots (190 km/h; 118 mph)
Normal operating speed:	
A	124 knots (230 km/h; 143 mph)
B	92 knots (170 km/h; 106 mph)
Stalling speed, power off, flaps up:	
A, B	65 knots (119 km/h; 74 mph)
Stalling speed, power off, flaps down:	
A, B	59 knots (109 km/h; 68 mph)
Max rate of climb at S/L: A	414 m (1,360 ft)/min
B	340 m (1,115 ft)/min
Service ceiling: A	6,500 m (21,325 ft)
T-O run: A	180-200 m (590-656 ft)
B	210-245 m (689-805 ft)
Landing run: A, B	260-300 m (853-984 ft)
Max range, no reserves:	
A, 400 litres (105.7 US gallons; 88 Imp gallons) fuel	291 nm (540 km; 335 miles)
A, 712 litres (188 US gallons; 156.6 Imp gallons) fuel	523 nm (970 km; 602 miles)
g limits: FAR 23	+3.4/−1.4
CAM 8	+3/−1.2

PZL MIELEC M-20 MEWA (GULL)

TYPE: Six/seven-seat executive transport, liaison, survey and ambulance aircraft.

PROGRAMME: Developed from PA-34-200T Seneca II light twin under 1977 agreement with Piper Aircraft Corporation, USA; nine Piper kits supplied to Pezetel 1978-80; adapted to accept PZL-F (Polish Franklin) engines 1978 and made first flight (SP-PKA) 25 July 1979; first four completed as M-20 00 and next five as M-20 01 (first flight, by SP-PKE, 22 September 1982); Polish certification of 00/01 22 September 1983; fifth M-20 01 converted as M-20 02 prototype, making first flight in this form 10 October 1985, but non-availability of production PZL-F engines resulted in switch to Teledyne Continental power plant in current M-20 03 model, which made first flight (SP-DMA) 13 October 1988 and received Polish certification 12 December same year; 03 since certificated by Germany (2 October 1991) and Australia (3 March 1992).

CURRENT VERSIONS: See 1987-88 and earlier *Jane's* for **M-20 00/01/02**; no longer in production.
 M-20 03: Current production version, with Polish built airframe and Teledyne Continental engines. Marketed as **Gemini** for export. *Detailed description applies to this version.*
 M-20 04: Under development for first flight in 1994 or 1995; strengthened wing with new main spar; max T-O weight increase to 2,156 kg (4,753 lb), Bendix/King avionics and 28V electrical system.

CUSTOMERS: Four M-20 00 and five M-20 01 built 1979-80 and 1983-84 respectively (one 01 converted to M-20 02); 12 M-20 03 completed by early 1993, at which time orders for up to 60 more, including customers in USA.

COSTS: M-20 03 $353,000 (1993).

DESIGN FEATURES: Can be operated from concrete runways or grass strips. Wing section NACA 65_2-415 (constant chord); 7° dihedral from roots; 2° incidence; leading-edges sweptforward at root.

FLYING CONTROLS: Frise differential ailerons, aerodynamically and mass balanced rudder with anti-servo tab, and slab type all-moving tailplane with trim tab; single-slotted trailing-edge flaps.

STRUCTURE: Safe-life aluminium alloy.

LANDING GEAR: Electrohydraulically retractable tricycle type, with single wheel and oleo-pneumatic shock strut on each unit. Mainwheels retract inward into wings, nosewheel forward. Size 6.00-6 wheels on all three units (McCreary main, Air Hawk nose), tyre pressures 3.79 bars (55 lb/sq in) on main units, 2.76 bars (40 lb/sq in) on nose unit. Nosewheel steerable ±27°. Emergency gravity extension. Cleveland disc brakes; parking brake. Min ground turning radius about nosewheel 5.00 m (16 ft 5 in).

POWER PLANT: Two 164 kW (220 hp) Teledyne Continental TSIO/LTSIO-360-KB turbocharged, counter-rotating flat-six engines, each driving a Hartzell BHC-2CYF-2CK three-blade constant-speed propeller. Two 92 litre (24.3 US gallon; 20.3 Imp gallon) fuel tanks in each wing leading-edge; total standard fuel capacity 368 litres (97.2 US gallons; 81 Imp gallons). Optional auxiliary tank in each leading-edge can increase this total to 480 litres (127 US gallons; 105.5 Imp gallons). Gravity fuelling points in top of each wing. Oil capacity 7.6 litres (2 US gallons; 1.7 Imp gallons) per engine.

ACCOMMODATION: Passenger version seats one or two pilots plus five or four passengers, with optional seventh seat. Forward opening door at front (starboard) and rear (port). Space for 68 kg (150 lb) of baggage in nose and 91 kg (200 lb) aft of rear seats. Ambulance version can carry one stretcher patient, two medical attendants and one other person in addition to pilot. Stretcher rack replaces right hand centre seat and, like seat, can be quickly and easily removed. Rack has special guides which can be connected to door threshold to facilitate stretcher loading; they can be folded back when stretcher is on board and locked. Hooks in cabin ceiling for suspending transfusion set; oxygen installation for patient. Doctor's seat (centre, left) has earphone and microphone, enabling him to contact ground for assistance; nurse's seat at rear. Modified electrical system permits incubator to be installed.

SYSTEMS: Two independent hydraulic systems, one operating at 154 bars (2,233 lb/sq in) for landing gear extension/retraction and one at 103.5 bars (1,500 lb/sq in) for wheel braking. Electrical system powered by two 24V 55A alternators and a 24V 25Ah battery. Pneumatic wing, tailplane and fin leading-edge anti-icing, and electric de-icing of propeller blades, optional.

AVIONICS: Multi-channel VOR/LOC radio and blind-flying instrumentation standard. DME, marker beacon receiver, radio compass and three-axis autopilot optional. Polish ARL 1601 ADF, CG 121 slaved gyro, RS 6102 VHF com transceiver, MRP-66 marker transceiver and SSA-1 audio control panel. Bendix/King Silver Crown package optional.

DIMENSIONS, EXTERNAL:	
Wing span	11.86 m (38 ft 11 in)
Wing chord: at root	1.88 m (6 ft 2 in)
at tip	1.60 m (5 ft 3 in)
Wing aspect ratio	7.3
Length overall	8.72 m (28 ft 7¼ in)
Height overall	3.02 m (9 ft 11 in)
Tailplane span	4.13 m (13 ft 6½ in)
Wheel track	3.37 m (11 ft 0¾ in)
Wheelbase	2.13 m (7 ft 0 in)
Propeller diameter	1.93 m (6 ft 4 in)
DIMENSIONS, INTERNAL:	
Cabin: Length	3.17 m (10 ft 4¾ in)
Max width	1.24 m (4 ft 0¾ in)
Max height	1.07 m (3 ft 6¼ in)
Volume	5.53 m³ (195.3 cu ft)
AREAS:	
Wings, gross	19.18 m² (206.5 sq ft)
Ailerons (total)	1.17 m² (12.59 sq ft)
Trailing-edge flaps (total)	1.94 m² (20.88 sq ft)
Fin	1.96 m² (21.10 sq ft)
Rudder	0.89 m² (9.58 sq ft)
Tailplane, incl tab	3.60 m² (38.75 sq ft)
WEIGHTS AND LOADINGS:	
Weight empty (standard)	1,320 kg (2,910 lb)
Max T-O weight	2,070 kg (4,563 lb)
Max landing weight	1,970 kg (4,343 lb)
Max zero-fuel weight	1,810 kg (3,990 lb)
Max wing loading	107.9 kg/m² (22.10 lb/sq ft)
Max power loading	6.86 kg/kW (11.13 lb/hp)
PERFORMANCE (at max T-O weight):	
Never-exceed speed (VNE)	194 knots (360 km/h; 223 mph)
Max level speed at 4,500 m (14,765 ft)	194 knots (360 km/h; 223 mph)
Max cruising speed at 7,560 m (24,800 ft)	173 knots (320 km/h; 199 mph)
Econ cruising speed (45% power) at 7,560 m (24,800 ft)	168 knots (311 km/h; 193 mph)
Stalling speed: flaps up	67 knots (124 km/h; 77 mph)
flaps down	61 knots (112 km/h; 70 mph)
Max rate of climb at S/L	456 m (1,496 ft)/min
Rate of climb at S/L, OEI	135 m (443 ft)/min
Service ceiling	7,620 m (25,000 ft)
Service ceiling, OEI	2,375 m (7,800 ft)
T-O run	400 m (1,313 ft)
T-O to 15 m (50 ft)	444 m (1,457 ft)
Landing from 15 m (50 ft)	715 m (2,346 ft)
Landing run	600 m (1,969 ft)
Range, 45 min reserves:	
with max standard fuel	669 nm (1,240 km; 770 miles)
with max standard and auxiliary fuel	989 nm (1,833 km; 1,139 miles)

Ambulance version of the PZL Mielec M-20 Mewa, a Polish version of the Piper PA-34-200T Seneca II *(R. J. Malachowski)*

PZL MIELEC M-26 ISKIERKA (LITTLE SPARK)

TYPE: Tandem two-seat primary training aircraft, for civil pilot training and military pilot selection.

PROGRAMME: Designed to FAR Pt 23; two versions being developed; first flight of first prototype (SP-PIA) with PZL-F engine 15 July 1986; first flight of Textron Lycoming-engined prototype (SP-PIB) 24 June 1987; flight testing completed; Polish certification obtained 26 October 1991. No production start by early 1993.

CURRENT VERSIONS: **M-26 00:** PZL-F engine and Polish avionics.

M-26 01: Textron Lycoming AEIO-540 engine and Bendix/King avionics.

DESIGN FEATURES: Selected parts and assemblies of M-20 Mewa used in design of wings, tail unit, landing gear, power plant, and electrical and power systems. Wing section NACA 65_2-415 (constant chord); 7° dihedral from roots; 2° incidence; leading-edges sweptforward at root. Fixed incidence tailplane.

FLYING CONTROLS: Frise ailerons, balanced rudder, and elevators with starboard trim tab; single-slotted trailing-edge flaps.

STRUCTURE: Safe-life aluminium alloy.

LANDING GEAR: Retractable tricycle type, actuated hydraulically, with single wheel and oleo strut on each unit. Mainwheels retract inward into wings, nosewheel rearward. Size 6.00-6 wheels on all three units; tyre pressures 3.43 bars (50 lb/sq in) on main units, 2.16 bars (31 lb/sq in) on nose unit. PZL Hydral hydraulic disc brakes on mainwheels. Parking brake.

POWER PLANT (M-26 00): One 153 kW (205 hp) PZL-F 6A-350CA flat-six engine, driving a PZL Warszawa-Okecie US 142 three-blade constant-speed propeller, or a two-blade Hartzell BHC-C2YF-2CKUF constant-speed propeller. One 92 litre (24.3 US gallon; 20.2 Imp gallon) fuel tank in each wing leading-edge, plus a 9 litre (2.4 US gallon; 2.0 Imp gallon) fuselage tank, to give total capacity of 193 litres (51 US gallons; 42.4 Imp gallons). Gravity fuelling point in top of each wing tank. Oil capacity 10 litres (2.6 US gallons; 2.2 Imp gallons).

POWER PLANT (M-26 01): One 224 kW (300 hp) Textron Lycoming AEIO-540-L1B5D flat-six engine, driving a Hoffmann HO-V123K-V/200AH-10 three-blade constant-speed propeller. Second tank in each wing. Total fuel capacity 377 litres (99.6 US gallons; 82.8 Imp gallons). Gravity fuelling point in top of each outer wing tank. Oil capacity 15 litres (4.0 US gallons; 3.3 Imp gallons).

ACCOMMODATION: Tandem seats for pupil (in front) and instructor, under framed canopy which opens sideways to starboard. Rear seat elevated. Baggage compartment aft of rear seat. Both cockpits heated and ventilated.

SYSTEMS: Two independent hydraulic systems, one operating at 154 bars (2,233 lb/sq in) for landing gear extension/retraction and one at 103 bars (1,494 lb/sq in) for wheel braking. DC electrical power supplied by 24V alternator (50A in M-26 00, 100A in M-26 01) and 25Ah battery.

AVIONICS: Polish ARL 1601 ADF, CG 121 slaved gyro, RS 6102 VHF com transceiver, SSA-1 audio control panel, and ORS-2M marker beacon receiver standard. Bendix/King avionics optional.

EQUIPMENT: Landing light in port wing leading-edge.

DIMENSIONS, EXTERNAL:	
Wing span	8.60 m (28 ft 2½ in)
Wing chord: at root	1.88 m (6 ft 2 in)
at tip	1.60 m (5 ft 3 in)
Wing aspect ratio	5.3
Length overall	8.30 m (27 ft 2¾ in)
Height overall	2.96 m (9 ft 8½ in)
Tailplane span	3.80 m (12 ft 5½ in)
Wheel track	2.93 m (9 ft 7¼ in)
Wheelbase	1.93 m (6 ft 4 in)
Propeller diameter	1.90 m (6 ft 2¾ in)
DIMENSIONS, INTERNAL:	
Cockpits: Total length	2.91 m (9 ft 6½ in)
Max width	0.88 m (2 ft 10½ in)
Max height	1.30 m (4 ft 3¼ in)
AREAS:	
Wings, gross	14.00 m² (150.7 sq ft)
Ailerons (total)	1.17 m² (12.59 sq ft)
Trailing-edge flaps (total)	1.06 m² (11.41 sq ft)
Fin	1.96 m² (21.10 sq ft)

Second prototype PZL Mielec M-26 Iskierka (224 kW; 300 hp AEIO-540 engine) *(Press Office Sturzenegger)*

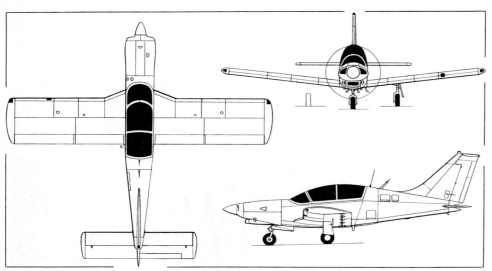

PZL Mielec M-26 Iskierka tandem two-seat primary trainer *(Dennis Punnett)*

Rudder	0.89 m² (9.58 sq ft)
Tailplane	3.30 m² (35.52 sq ft)
Elevators (total, incl tab)	1.15 m² (12.38 sq ft)

WEIGHTS AND LOADINGS (M-26 01):
Weight empty	940 kg (2,072 lb)
Max fuel weight	271 kg (597 lb)
Max T-O and landing weight	1,400 kg (3,086 lb)
Max wing loading	100.0 kg/m² (20.5 lb/sq ft)
Max power loading	6.26 kg/kW (10.29 lb/hp)

PERFORMANCE (M-26 01 at max T-O weight except where indicated):
Never-exceed speed (V_{NE})	215 knots (400 km/h; 248 mph)
Max level speed at S/L	178 knots (330 km/h; 205 mph)
Stalling speed, flaps down	60 knots (110 km/h; 69 mph)
Max rate of climb at S/L	420 m (1,378 ft)/min
T-O to 15 m (50 ft)	570 m (1,870 ft)
Landing from 15 m (50 ft)	685 m (2,248 ft)
Range with max fuel, 30 min reserves	874 nm (1,620 km; 1,006 miles)
g limits	+7/−3.5 at 1,100 kg (2,425 lb) AUW
	+4/−1.72 at max T-O weight

PZL MIELEC (ANTONOV) An-2 ANTEK
NATO reporting name: Colt

TYPE: Single-engined general purpose biplane.
PROGRAMME: First flight of An-2 prototype, designed to specification of USSR Ministry of Agriculture and Forestry, 31 August 1947; went into production in USSR 1948, with 746 kW (1,000 hp) ASh-62 engine; over 5,000 built in USSR by 1960; licence rights granted to China, where first Y-5 completed December 1957 and limited production continues (see SAP heading in that section); since 1960, apart from few dozen Soviet-built An-2Ms (see 1971-72 *Jane's*), continued production primarily by PZL Mielec; original licence agreement provided for An-2T and An-2R versions; first flight of Polish An-2 on 23 October 1960.
CURRENT VERSIONS: **An-2P:** Passenger version; 12 adult passengers and two children; improved cabin layout and comfort, better soundproofing, new propeller and spinner, weight saving instruments and equipment compared with Soviet model; entered production 1968. *Detailed description applies to this version.*
An-2R: Agricultural version; 1,350 kg (2,976 lb) capacity GFRP reinforced epoxy resin hopper or 1,400 litre (370 US gallon; 308 Imp gallon) capacity tank for dry or liquid chemicals; 7,782 built by Mielec up to end of 1991.
An-2T: General purpose transport version; 12 passengers plus baggage or 1,500 kg (3,306 lb) of cargo.
An-2TP: Passenger version; similar to An-2T but with higher cabin standard.
Further details of Geofiz, LW, PK, P-Photo, PR, S and TD specialised versions in 1983-84 and earlier *Jane's*.
CUSTOMERS: Over 11,950 An-2s built by Mielec for domestic use and export to USSR/CIS (10,440), Bulgaria, former Czechoslovakia, Egypt, France, former East Germany, Hungary, Iraq, North Korea, Mongolia, Netherlands, Nicaragua, Romania, Sudan, Tunisia, Turkey, the UK, Venezuela and former Yugoslavia; in 1989, AICSA of Colombia assembled two An-2s from kits supplied via Pezetel.
DESIGN FEATURES: Unequal span single-bay biplane; braced wings and tail; fuselage circular section forward, rectangular in cabin section, oval in tail section; fin integral with rear fuselage. RPS wing section, thickness/chord ratio 14 per cent (constant); dihedral, both wings, approx 2° 48′.
FLYING CONTROLS: Dual controls and blind-flying instrumentation standard. Mechanically actuated differential ailerons, elevators and rudder, using cables and push/pull rods; electric trim tab in port aileron, rudder and port elevator; full span automatic leading-edge slots on upper wings; electrically actuated slotted trailing-edge flaps on both wings.
STRUCTURE: All-metal, with fabric covering on wings aft of main spar and on tailplane.
LANDING GEAR: Non-retractable split axle type, with long stroke oleo-pneumatic shock absorbers. Mainwheel tyres size 800 × 260 mm, pressure 2.25 bars (32.7 lb/sq in). Pneumatic shoe brakes on main units. Fully castoring and self-centring PZL Krosno tailwheel, size 470 × 210, with electropneumatic lock. For rough field operation shock absorbers can be charged from compressed air cylinder installed in rear fuselage. Interchangeable ski landing gear available optionally.
POWER PLANT: One 746 kW (1,000 hp) PZL Kalisz ASz-62IR nine-cylinder radial air-cooled engine, driving an AW-2 four-blade variable-pitch metal propeller. Six fuel tanks in upper wings, with total capacity of 1,200 litres (317 US gallons; 264 Imp gallons). Fuel consumption 120-170 litres (31.7-45 US gallons; 26.4-37.4 Imp gallons)/h. Oil capacity 120 litres (31.7 US gallons; 26.4 Imp gallons).
ACCOMMODATION: Crew of two on flight deck, with access via passenger cabin. Standard accommodation for 12 passengers, in four rows of three with single aisle. Two foldable seats for children in aisle between first and second rows, and infant's cradle at front on starboard side. Toilet at rear on starboard side. Overhead racks for up to 160 kg (352 lb) of baggage, with space for coats and additional 40 kg (88 lb) of baggage between rear pair of seats and toilet. Emergency exit on starboard side at rear. Walls of cabin lined with glass-wool mats and inner facing of plywood to reduce internal noise level. Cabin floor carpeted. Cabin heating and starboard windscreen de-icing by engine bleed air; port and centre windscreens electrically de-iced. Cabin ventilation by ram air intakes on underside of top wings.
SYSTEMS: Compressed air cylinder, of 8 litres (0.28 cu ft) capacity, for pneumatic charging of shock absorbers and operation of tailwheel lock at 49 bars (711 lb/sq in) pressure and operation of mainwheel brakes at 9.80 bars (142 lb/sq in). Contents of cylinder maintained by AK-50 P engine driven compressor, with AD-50 automatic relief device to prevent overpressure. DC electrical system supplied with basic 27V power (and 36V or 115V where required) by engine driven generator and storage battery. CO_2 fire extinguishing system with automatic fire detector.
AVIONICS: R-842 HF and RS-6102 or Baklan-5 VHF lightweight radio transceivers, A-037 radio altimeter (RW-UM before April 1989), ARK-9 radio compass, MRP-56P marker beacon receiver, GIK-1 gyro compass, GPK-48 gyroscopic direction indicator and SPU-7 intercom.

DIMENSIONS, EXTERNAL:
Wing span: upper	18.18 m (59 ft 7¾ in)
lower	14.24 m (46 ft 8½ in)
Wing chord (constant): upper	2.45 m (8 ft 0½ in)
lower	2.00 m (6 ft 6¾ in)
Wing aspect ratio: upper	7.6
lower	7.1
Wing gap	2.17 m (7 ft 1½ in)
Length overall: tail up	12.74 m (41 ft 9½ in)
tail down	12.40 m (40 ft 8¼ in)
Height overall: tail up	6.10 m (20 ft 0 in)
tail down	4.01 m (13 ft 2 in)
Tailplane span	7.20 m (23 ft 7½ in)
Wheel track	3.36 m (11 ft 0¼ in)
Wheelbase	8.19 m (26 ft 10½ in)
Propeller diameter	3.60 m (11 ft 9¾ in)
Propeller ground clearance	0.69 m (2 ft 3¼ in)
Cargo door (port): Mean height	1.55 m (5 ft 1 in)
Mean width	1.39 m (4 ft 6¾ in)
Emergency exit (stbd, rear): Height	0.65 m (2 ft 1½ in)
Width	0.51 m (1 ft 8 in)

DIMENSIONS, INTERNAL:
Cargo compartment: Length	4.10 m (13 ft 5½ in)
Max width	1.60 m (5 ft 3 in)
Max height	1.80 m (5 ft 11 in)

AREAS:
Wings, gross: upper	43.54 m² (468.7 sq ft)
lower	27.98 m² (301.2 sq ft)
Ailerons (total)	5.90 m² (63.5 sq ft)
Trailing-edge flaps (total)	9.60 m² (103 sq ft)
Fin	3.20 m² (34.4 sq ft)
Rudder, incl tab	2.65 m² (28.52 sq ft)
Tailplane	7.56 m² (81.4 sq ft)
Elevators (total, incl tab)	4.72 m² (50.81 sq ft)

WEIGHTS AND LOADINGS:
Weight empty	3,450 kg (7,605 lb)
Max fuel weight	900 kg (1,984 lb)
Max T-O weight	5,500 kg (12,125 lb)
Max landing weight	5,250 kg (11,574 lb)
Max zero-fuel weight	4,800 kg (10,582 lb)
Max wing loading	76.82 kg/m² (15.7 lb/sq ft)
Max power loading	7.38 kg/kW (12.13 lb/hp)

PERFORMANCE (at AUW of 5,250 kg; 11,574 lb):
Max level speed at 1,750 m (5,740 ft)	139 knots (258 km/h; 160 mph)
Econ cruising speed	100 knots (185 km/h; 115 mph)
Min flying speed	49 knots (90 km/h; 56 mph)
T-O speed	43 knots (80 km/h; 50 mph)
Landing speed	46 knots (85 km/h; 53 mph)
Max rate of climb at S/L	210 m (689 ft)/min
Service ceiling	4,400 m (14,425 ft)
Time to 4,400 m (14,425 ft)	30 min
T-O run: hard runway	150 m (492 ft)
grass	170 m (558 ft)
T-O to 15 m (50 ft): hard runway	475 m (1,558 ft)
grass	495 m (1,624 ft)
Landing from 15 m (50 ft): hard runway	427 m (1,401 ft)
grass	432 m (1,417 ft)
Landing run: hard runway	170 m (558 ft)
grass	185 m (607 ft)
Range at 1,000 m (3,280 ft) with 500 kg (1,102 lb) payload	485 nm (900 km; 560 miles)

Standard Mielec-built An-2P passenger transport biplane *(R. J. Malachowski)*

PZL SWIDNIK SA
ZYGMUNTA PULAWSKIEGO-PZL SWIDNIK
(Zygmunt Pulawski Transport Equipment Manufacturing Centre, Swidnik)

ul. Przodowników Pracy 1, 21-045 Swidnik k/Lublina
Telephone: 48 (81) 12061, 12071, 13061 and 13071
Fax: 48 (81) 12178
Telex: 0642301 WSK PL
GENERAL MANAGER: Mieczyslaw Majewski, MScEng
DIRECTOR OF RESEARCH AND DEVELOPMENT:
   Ryszard Kochanowski, MScEng
SALES MANAGER: Ryszard Cukierman, MScEng

Swidnik factory established 1951; engaged initially in manufacturing components for LiM-1 (MiG-15) jet fighter; began licence production of Soviet Mi-1 helicopter in 1955, building some 1,700 as SM-1s, followed by 450 Swidnik developed SM-2s; design office formed at factory to work on variants/developments of SM-1 and original projects such as SM-4 Latka.

Swidnik works named after famous pre-war PZL designer, Zygmunt Pulawski, September 1957; currently employs about 10,000 persons; production concentrates on W-3 Sokól, developments of Soviet designed Mi-2, manufacture of wing and tailplane slats for An-28, and components for Russian Ilyushin Il-86 and French Eurocopter helicopters. Polish government announced intention to privatise Swidnik factory 22 February 1991 (effective January 1992), initially as state-owned limited company.

Plans announced early 1993 to set up subassembly factory for Sokól in Halifax, Nova Scotia, to be known as PZL Canada, to assemble W-3 for North/South American and Pacific Rim markets, possibly with Canadian or US engines and avionics.

PZL SWIDNIK (MIL) Mi-2
NATO reporting name: Hoplite

TYPE: Twin-turbine general purpose light helicopter.
PROGRAMME: Designed in USSR by Mikhail L. Mil and first flown September 1961; January 1964 agreement assigned further development, production and marketing exclusively to Polish industry; first flight of Polish example 4 November 1965; series production began 1965. Has undergone continuous development and upgrading, with versions for new applications developed to meet specific customers' requirements (see list in 1984-85 and earlier *Jane's*). Production at standstill early 1992, but reported continuing on limited basis March 1993.
CURRENT VERSIONS: **Civil Mi-2:** Recent versions include examples for convertible passenger/cargo transport; ambulance and rescue (**Mi-2R**); agricultural dusting and spraying (conventional or ULV); freighter with external sling and electric hoist; training; and aerial photography (able to carry photographic, photogrammetric, thermal imaging or TV cameras for oblique or vertical pictures). *Details below refer to basic civil Mi-2, except where indicated.*
Mi-2B: Different electrical system and more modern navigational aids; manufactured in same versions (except agricultural) as basic Mi-2, and has same flight performance; empty equipped weights 2,300 kg (5,070 lb) for passenger version, 2,293 kg (5,055 lb) for cargo version; T-O weight unchanged; no rotor blade de-icing. Production total not large.
Mi-2RM: Naval version.
Mi-2T: Military transport version.
Mi-2URN: Combat support/armed reconnaissance version; as Mi-2US but with two Mars 2 launchers (each 16 S-5 57 mm unguided rockets) instead of pylon mounted gun pods; PKV gunsight in cockpit for aiming all weapons; in service from 1973.
Mi-2URP: Anti-tank version; cabin-side outriggers for four 9M14M Malyutka (AT-3 'Sagger') wire guided missiles; four additional missiles in cargo compartment; in service from 1976; later version can carry four 9M32 Strela 2 missiles. Illustrated in 1988-89 *Jane's*.
Mi-2US: Gunship version; 23 mm NS-23KM cannon on port side of fuselage, two 7.62 mm gun pods on each side pylon, two other 7.62 mm PK type pintle mounted machine guns in rear of cabin.
CUSTOMERS: Over 5,250 built for civil and military operators, majority exported; military operators include air forces of Bulgaria, former Czechoslovakia, Germany, Hungary, Iraq, North Korea, Libya, Poland (approx 130 Mi-2CH, T, URN, URP, US and other variants, mostly with No. 49 and No. 56 Combat Helicopter Regiments and No. 47 Helicopter Training Regiment), Russia and Syria; civil operators in European and various developing countries, including agricultural models in Bulgaria, Czech Republic, Egypt, Iraq, Poland and Russia.
DESIGN FEATURES: Three-blade main rotor with hydraulic blade vibration dampers; flapping, drag and pitch hinges on each blade; anti-flutter weights on leading-edges, balancing plates on trailing-edges. Coil spring counterbalance in main and tail rotor systems; pitch change centrifugal loads on tail rotor carried by ribbon type steel torsion elements. Blades do not fold; rotor brake fitted. Main rotor blade section NACA 230-12M. Main rotor shaft driven via gearbox on each engine; three-stage WR-2 main gearbox, intermediate gearbox and tail rotor gearbox; main rotor/engine rpm ratio 1:24.6, tail rotor/engine rpm ratio 1:4.16;

PZL Swidnik (Mil) Mi-2T helicopter of the Polish Air Force (*Press Office Sturzenegger*)

main gearbox provides drive for auxiliary systems and take-off for rotor brake; freewheel units permit disengagement of failed engine and autorotation.
FLYING CONTROLS: Hydraulic system for cyclic and collective pitch control boosters; variable incidence horizontal stabiliser, controlled by collective pitch lever.
STRUCTURE: Main and tail rotor blades have extruded duralumin spar with bonded honeycomb trailing-edge pockets; pod and boom fuselage of sheet duralumin, bonded and spot welded or riveted to longerons and frames, in three main assemblies (nose including cockpit, central section and tailboom); steel alloy main load bearing joints.
LANDING GEAR: Non-retractable tricycle type, plus tailskid. Twin-wheel nose unit. Single wheel on each main unit. Oleo-pneumatic shock absorbers in all units, including tailskid. Main shock absorbers designed to cope with both normal operating loads and possible ground resonance. Mainwheel tyres size 600 × 180, pressure 4.41 bars (64 lb/sq in). Nosewheel tyres size 400 × 125, pressure 3.45 bars (50 lb/sq in). Pneumatic brakes on mainwheels. Metal ski landing gear optional.
POWER PLANT: Two 298 kW (400 shp) Polish built Isotov GTD-350 turboshafts, mounted side by side above cabin. Fuel in single rubber tank, capacity 600 litres (158.5 US gallons; 131 Imp gallons), under cabin floor. Provision for carrying 238 litre (63 US gallon; 52.4 Imp gallon) external tank on each side of cabin. Refuelling point in starboard side of fuselage. Oil capacity 25 litres (6.6 US gallons; 5.4 Imp gallons).
ACCOMMODATION: Normal accommodation for one pilot on flight deck (port side). Seats for up to eight passengers in air-conditioned cabin, comprising back to back bench seats for three persons each, with two optional extra starboard side seats at rear, one behind the other. All passenger seats removable for carrying up to 700 kg (1,543 lb) of internal freight. Access to cabin via forward hinged doors on each side at front of cabin and aft on port side. Pilot's sliding window jettisonable in emergency. Ambulance version has accommodation for four stretchers and medical attendant, or two stretchers and two sitting casualties. Side by side seats and dual controls in pilot training version. Cabin heating, ventilation and air-conditioning standard.
SYSTEMS: Cabin heating, by engine bleed air, and ventilation; heat exchangers warm atmospheric air for ventilation system during cold weather. Hydraulic system, pressure 65 bars (940 lb/sq in), for cyclic and collective pitch control boosters. Hydraulic fluid flow rate 7.5 litres (1.98 US gallons; 1.65 Imp gallons)/min. Vented reservoir, with gravity feed. Pneumatic system, pressure 49 bars (710 lb/sq in), for mainwheel brakes. AC electrical system, with two STG-3 3kW engine driven starter/generators and 208V 16kVA three-phase alternator. 24V DC system, with two 28Ah lead-acid batteries. Main and tail rotor blades de-iced electrically; engine air intake de-icing by engine bleed air. Electric de-icing of windscreen.
AVIONICS: Standard items include two transceivers (MF/HF), gyro compass, radio compass, radio altimeter, intercom system and blind-flying panel. Nose and tail warning radar fitted to some military versions.
EQUIPMENT: Agricultural version carries hopper on each side of fuselage (total capacity 1,000 litres; 264 US gallons; 220 Imp gallons of liquid or 750 kg; 1,650 lb of dry chemical) and either a spraybar to rear of cabin on each side or distributor for dry chemicals under each hopper. Swath width covered by spraying version is 40-45 m (130-150 ft). For search and rescue, electric hoist, capacity 120 kg (264 lb), is fitted. In freight role an underfuselage hook can be fitted for suspended loads of up to 800 kg (1,763 lb). Polish press has illustrated version equipped for laying smokescreens. Electrically operated wiper for pilot's windscreen. Freon fire extinguishing system, for engine bays and main gearbox compartment, can be actuated automatically or manually.
ARMAMENT: See Current Versions.
DIMENSIONS, EXTERNAL:

Main rotor diameter	14.50 m (47 ft 6⅞ in)
Main rotor blade chord (constant, each)	0.40 m (1 ft 3¾ in)
Tail rotor diameter	2.70 m (8 ft 10¼ in)
Length: overall, rotors turning	17.42 m (57 ft 2 in)
fuselage	11.40 m (37 ft 4¾ in)
Height to top of rotor head	3.75 m (12 ft 3½ in)
Stabiliser span	1.85 m (6 ft 0¾ in)
Wheel track	3.05 m (10 ft 0 in)
Wheelbase	2.71 m (8 ft 10¾ in)
Tail rotor ground clearance	1.59 m (5 ft 2¾ in)
Cabin door (port, rear): Height	1.065 m (3 ft 5¾ in)
Width	1.115 m (3 ft 8 in)
Cabin door (stbd, front): Height	1.11 m (3 ft 7¾ in)
Width	0.75 m (2 ft 5½ in)
Cabin door (port, front): Height	1.11 m (3 ft 7¾ in)
Width	0.78 m (2 ft 6¾ in)

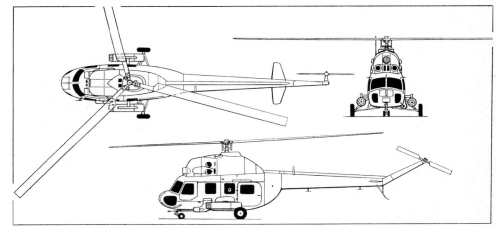

Mi-2 twin-turboshaft helicopter in URN armed configuration (*Dennis Punnett*)

238 POLAND: AIRCRAFT—PZL SWIDNIK

Civil Mi-2 demonstrating its turning capability (*R. J. Malachowski*)

DIMENSIONS, INTERNAL:
  Cabin:
    Length: incl flight deck   4.07 m (13 ft 4¼ in)
            excl flight deck   2.27 m (7 ft 5½ in)
    Mean width   1.20 m (3 ft 11¼ in)
    Mean height   1.40 m (4 ft 7 in)
AREAS:
  Main rotor blades (each)   2.40 m² (25.83 sq ft)
  Tail rotor blades (each)   0.22 m² (2.37 sq ft)
  Main rotor disc   166.4 m² (1,791.11 sq ft)
  Tail rotor disc   5.73 m² (61.68 sq ft)
  Horizontal stabiliser   0.70 m² (7.53 sq ft)
WEIGHTS AND LOADINGS:
  Weight empty, equipped:
    passenger version   2,402 kg (5,295 lb)
    cargo version   2,372 kg (5,229 lb)
    ambulance version   2,410 kg (5,313 lb)
    agricultural version   2,372 kg (5,229 lb)
  Basic operating weight empty:
    single-pilot versions   2,365 kg (5,213 lb)
    dual control version   2,424 kg (5,344 lb)
  Max payload, excl pilot, oil and fuel   800 kg (1,763 lb)
  Normal T-O weight (and max T-O weight of agricultural version)   3,550 kg (7,826 lb)
  Max T-O weight (special versions)   3,700 kg (8,157 lb)
  Max disc loading   22.4 kg/m² (4.6 lb/sq ft)
PERFORMANCE (at 3,550 kg; 7,826 lb T-O weight):
  Never-exceed speed (V_{NE}) at 500 m (1,640 ft):
    agricultural version   84 knots (155 km/h; 96 mph)
    other versions   113 knots (210 km/h; 130 mph)
  Max cruising speed at 500 m (1,640 ft):
    agricultural version (without agricultural equipment)
            102 knots (190 km/h; 118 mph)
    other versions   108 knots (200 km/h; 124 mph)
  Max level speed with agricultural equipment
            84 knots (155 km/h; 96 mph)
  Econ cruising speed at 500 m (1,640 ft):
    for max range   102 knots (190 km/h; 118 mph)
    for max endurance   54 knots (100 km/h; 62 mph)
  Max rate of climb at S/L   270 m (885 ft)/min
  Time to 1,000 m (3,280 ft)   5 min 30 s
  Time to 4,000 m (13,125 ft)   26 min
  Service ceiling   4,000 m (13,125 ft)
  Hovering ceiling: IGE   approx 2,000 m (6,560 ft)
                 OGE   approx 1,000 m (3,280 ft)
  Range at 500 m (1,640 ft):
    max payload, 5% fuel reserves
            91 nm (170 km; 105 miles)
    max internal fuel, no reserves
            237 nm (440 km; 273 miles)
    max internal and auxiliary fuel, 30 min reserves
            313 nm (580 km; 360 miles)
    max internal and auxiliary fuel, no reserves
            430 nm (797 km; 495 miles)
  Endurance at 500 m (1,640 ft), no reserves:
    max internal fuel   2 h 45 min
    max internal and auxiliary fuel   5 h
  Endurance (agricultural version), 5% reserves:
    spraying   40 min
    dusting   50 min

PZL SWIDNIK KANIA/KITTY HAWK

TYPE: Twin-turboshaft multi-purpose light helicopter.
PROGRAMME: Developed in collaboration with Allison in USA; two prototypes produced by converting Mi-2 airframes; first flight of first prototype (SP-PSA) 3 June 1979; Polish supplementary type certificate to Mi-2 on 1 October 1981; full type certificate as FAR Pt 29 (Transport Category B) day and night SVFR multi-purpose utility helicopter with Category A engine isolation on 21 February 1986 as considerably improved Kania Model 1. Full description in 1990-91 and earlier *Jane's*. Not currently in production, but being offered as Mi-2 replacement or upgrade.
CURRENT VERSIONS: **Kania Model 1**: Intended for passenger transport (with standard, executive or customised interiors), cargo transport (internal or slung load), agricultural (LV and ULV spraying/spreading/dusting), medevac, training, rescue, and aerial surveillance configurations.
CUSTOMERS: Four prototypes and six or seven production aircraft built.

PZL SWIDNIK W-3 SOKÓL (FALCON)

TYPE: Twin-turboshaft medium weight multi-purpose helicopter.
PROGRAMME: Developed in second half of 1970s; first flight of first of five prototypes 16 November 1979, and used in subsequent wide ranging tiedown tests; remaining prototypes completed to embody changes resulting from tests; manufacturer's flight trials resumed 6 May 1982 with second prototype (SP-PSB); certification trials with two other helicopters carried out in wide range of operating conditions, including heavy icing and extreme temperatures of −60°C and +50°C; certification to Russian NLGW regulations received by early 1993; production started 1985 and continuing at nine a year in 1993; US FAA and German certification hoped for by mid-1993.
CURRENT VERSIONS: **W-3 Sokól**: Standard civil and military version. *Detailed description applies to this model except where indicated.*
  W-3L Traszka (Newt): Proposed stretched version (1.20 m; 3 ft 11¼ in longer), seating 14 troops; engines uprated to 746 kW (1,000 shp). Development suspended.
  W-3RM Anakonda: Offshore search and rescue version; watertight cabin, six inflatable flotation bags, additional window in lower part of each flight deck door.
  W-3 Huzar (Hussar): Armed version with undernose 20 mm gun, cabin-side weapon outriggers, roof-mounted sight with TV and FLIR cameras; Mars-2 16-round rocket launchers and Grot anti-tank missiles standard; ZR-8 rocket launchers, Gad launchers for 9M32M Strzala anti-aircraft missiles, laser rangefinder and helmet sight optional.
  W-3U-1 Alligator: Anti-submarine version; due in service 1997.
  W-3EW: Planned electronic warfare version.
  W-3MS: Provisional designation for proposed battlefield attack version, with tandem cockpits and fuselage-mounted weapon pylons.
CUSTOMERS: 48 of first 50 delivered by mid-1991, including three for Polish Air Force, 12 to Myanmar and 20 of 35 ordered by Aeroflot; further batch of 20 then under construction. Remainder of Aeroflot order cancelled 1991; W-3RM evaluated against Mi-14 'Haze' by Polish Naval Air Regiment 1989-91; six being built for Polish Navy, of which one in service and two more imminent early 1993. Huzar ordered 1993 by unnamed customer.
DESIGN FEATURES: All-new Polish design; larger than Mi-2 and Kania. Four-blade fully articulated main rotor and three-blade tail rotor; main rotor has pendular Salomon type vibration absorber for smooth flight and low vibration levels; main rotor blades have tapered tips; rotor brake fitted; transmission driven via main rotor, intermediate and tail rotor gearboxes. Tail fin integral with tailboom; horizontal stabiliser not interconnected with main rotor control system.
FLYING CONTROLS: Three hydraulic boosters for longitudinal, lateral and collective pitch control of main rotor; one booster for tail rotor control.
STRUCTURE: Rotor blades (main and tail) and single-spar horizontal stabiliser of laminated GFRP impregnated with epoxy resin; tail rotor driveshaft of duralumin tube with splined couplings; duralumin fuselage; GFRP fin trailing-edge.
LANDING GEAR: Non-retractable tricycle type, plus tailskid beneath tailboom. Twin-wheel castoring nose unit; single wheel on each main unit. Oleo-pneumatic shock absorber in each unit. Mainwheel tyres size 500 × 250 mm; nosewheel tyres size 400 × 150 mm. Pneumatic disc brakes on mainwheels. Metal ski landing gear optional. Six inflatable flotation bags on Anakonda.
POWER PLANT: Two WSK-PZL Rzeszów TWD-10W turboshafts (Polish-made Mars TVD-10), each with rating of 671 kW (900 shp) for T-O and emergency ratings of 746 kW (1,000 shp) and 858 kW (1,150 shp) for 30 min and 2½ min OEI respectively. Particle separators on engine intakes, and inlet de-icing, standard. Power plant equipped with advanced electronic fuel control system for maintaining rotor speed at pilot-selected value amounting to ±5 per cent of normal rpm, and also for torque sharing as well

PZL Swidnik Kania twin-turboshaft derivative of the Mi-2

PZL SWIDNIK—AIRCRAFT: POLAND 239

W-3RM Anakonda maritime SAR version in Polish Naval Air Force markings *(Mark Lambert)*

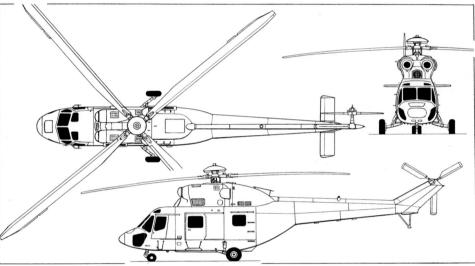
PZL Swidnik W-3 Sokól twin-turboshaft helicopter *(Dennis Punnett)*

Prototype of armed version of the Sokól *(R. J. Malachowski)*

as for supervising engine limits during start-up and normal or OEI operation. Engines and main rotor gearbox mounted on bed frame, eliminating drive misalignment due to deformations of fuselage. Bladder fuel tanks beneath cabin floor, with combined capacity of 1,700 litres (449 US gallons; 374 Imp gallons). Auxiliary tank, capacity 1,100 litres (290.5 US gallons; 242 Imp gallons), optional.

ACCOMMODATION: Pilot (port side), and co-pilot or flight engineer, side by side on flight deck, on adjustable seats with safety belts. Dual controls and dual flight instrumentation optional. Accommodation for 12 passengers in main cabin or up to eight survivors plus two-person rescue crew and doctor in Anakonda SAR version. Seats removable for carriage of internal cargo. Ambulance version can carry four stretcher cases and medical attendant. Baggage space at rear of cabin. Door with bulged window on each side of flight deck; large sliding door for passenger and/or cargo loading on port side at forward end of cabin; second sliding door at rear of cabin on starboard side. Optically flat windscreens, improving view and enabling wipers to sweep a large area. Accommodation soundproofed, heated (by engine bleed air) and ventilated.

SYSTEMS: Two independent hydraulic systems, working pressure 90 bars (1,300 lb/sq in), for controlling main and tail rotors, unlocking collective pitch control lever, and feeding damper of directional steering system. Flow rate 11 litres (2.9 US gallons; 2.4 Imp gallons)/min in each system. Vented gravity feed reservoir, at atmospheric pressure. Pneumatic system for actuating hydraulic mainwheel brakes. Electrical system providing both AC and DC power. Electric anti-icing of rotor blades. Fire detection/extinguishing system. Air-conditioning and oxygen systems optional. Neutral gas system optional, for inhibiting fuel vapour explosion.

AVIONICS: Standard IFR nav/com avionics permit adverse weather operation by day or night. Weather radar optional. Stability augmentation system standard. Chrom (NATO 'Pin Head') IFF transponder and modified Syrena RWR in military version. Additional avionics in Anakonda include AP Decca navigator, SA-813 radar and SPOR search and detection system.

EQUIPMENT: Cargo version equipped with 2,100 kg (4,630 lb) capacity external hook and 150 kg (331 lb) capacity rescue hoist. W-3RM has 267 kg (589 lb) capacity electric hoist, stretchers, two-person rescue basket, rescue belts, liferafts for six people, rope ladder, portable oxygen equipment, electric blankets and vacuum flasks, various types of buoy (light, smoke and radio) and marker, binoculars, flare pistol and searchlights.

ARMAMENT: Polish Air Force aircraft can be fitted with GSh-23 twin-barrel cannon pack on lower port side of fuselage, plus cabin outriggers for four AT-6 (NATO 'Spiral') anti-tank missiles and two 12-round launchers for 80 mm air-to-surface unguided rockets.

DIMENSIONS, EXTERNAL:
Main rotor diameter	15.70 m (51 ft 6 in)
Tail rotor diameter	3.03 m (9 ft 11¼ in)
Length: overall, rotors turning	18.85 m (61 ft 10⅛ in)
fuselage	14.21 m (46 ft 7½ in)
Height to top of rotor head	4.12 m (13 ft 6¼ in)
Stabiliser span	3.45 m (11 ft 3¾ in)
Wheel track	3.40 m (11 ft 2 in)
Wheelbase	3.55 m (11 ft 7¾ in)
Passenger/cargo doors:	
Height (each):	1.20 m (3 ft 11¼ in)
Width: port	0.95 m (3 ft 1½ in)
starboard	1.25 m (4 ft 1¼ in)

DIMENSIONS, INTERNAL:
Cabin: Length	3.20 m (10 ft 6 in)
Max width	1.55 m (5 ft 1 in)
Max height	1.40 m (4 ft 7 in)

AREAS:
Main rotor disc	193.6 m² (2,083.8 sq ft)
Tail rotor disc	7.21 m² (77.6 sq ft)

WEIGHTS AND LOADINGS:
Minimum basic weight empty	3,300 kg (7,275 lb)
Basic operating weight empty (multi-purpose versions)	3,630 kg (8,002 lb)
Max payload, internal or external	2,100 kg (4,630 lb)
Normal T-O weight	6,100 kg (13,448 lb)
Max T-O weight	6,400 kg (14,110 lb)
Max disc loading	33.06 kg/m² (6.77 lb/sq ft)

PERFORMANCE (standard W-3 at normal T-O weight at 500 m; 1,640 ft, ISA, except where indicated):
Never-exceed speed (V_{NE})	145 knots (270 km/h; 167 mph)
Max level speed	138 knots (255 km/h; 158 mph)
Max cruising speed	127 knots (235 km/h; 146 mph)
Econ cruising speed	119 knots (220 km/h; 137 mph)
Max rate of climb at S/L	510 m (1,673 ft)/min
Rate of climb at S/L, OEI:	
at 30 min rating	96 m (315 ft)/min
at 2½ min emergency rating	186 m (610 ft)/min
Service ceiling:	
at normal T-O weight	5,100 m (16,725 ft)
at T-O weight below normal	up to 6,000 m (19,680 ft)
Service ceiling, OEI:	
at 30 min rating	1,800 m (5,905 ft)

240 POLAND: AIRCRAFT—PZL SWIDNIK/PZL WARSZAWA-OKECIE

W-3 Sokół multi-purpose helicopter, now PZL Swidnik's major helicopter programme *(Peter J. Cooper)*

Artist's impression of the projected stretched W-3L

PZL SWIDNIK SW-4

TYPE: Four/five-seat, single-engined multi-purpose light helicopter.

PROGRAMME: Development began 1985, originally based on use of Polish GTD-350 turboshaft; full-scale mockup completed 1987 (see description and illustrations in 1991-92 and earlier *Jane's*); major redesign undertaken 1989-90, now using Allison 250 engine in more streamlined fuselage with modified tail unit; possibility that prototype construction could begin 1993.

CURRENT VERSIONS: Intended applications include passenger and cargo transport, training and agricultural use.

DESIGN FEATURES: Three-blade main rotor; arrowhead tail fin on port side, with two-blade tail rotor to starboard; narrow tailplane with small endplate fins; skid landing gear.

POWER PLANT: One 335 kW (450 shp) Allison 250-C20R turboshaft. Transmission rating 335 kW (450 shp) for T-O, 283 kW (380 shp) max continuous. Standard fuel capacity 450 litres (119 US gallons; 99 Imp gallons).

DIMENSIONS, EXTERNAL:
Main rotor diameter	9.00 m (29 ft 6⅓ in)
Tail rotor diameter	1.50 m (4 ft 11 in)
Length:	
overall, both rotors turning	10.50 m (34 ft 5½ in)
fuselage	8.30 m (27 ft 2¾ in)
Height overall	2.60 m (8 ft 6½ in)
Skid track	1.80 m (5 ft 11 in)

DIMENSIONS, INTERNAL:
Cabin: Length	2.34 m (7 ft 8¼ in)
Max width	1.48 m (4 ft 10¼ in)
Max height	1.32 m (4 ft 4 in)

AREAS:
Main rotor disc	63.62 m² (684.8 sq ft)
Tail rotor disc	1.77 m² (19.05 sq ft)

WEIGHTS AND LOADINGS:
Weight empty	751 kg (1,655 lb)
Normal T-O weight	1,512 kg (3,333 lb)
Max T-O weight	1,700 kg (3,748 lb)
Max disc loading	26.72 kg/m² (5.47 lb/sq ft)
Max power loading	5.07 kg/kW (8.33 lb/shp)

PERFORMANCE (estimated, at normal T-O weight, ISA):
Never-exceed speed (V_{NE})	151 knots (280 km/h; 174 mph)
Max level speed at 500 m (1,640 ft)	132 knots (245 km/h; 152 mph)
Max cruising speed, 70% power at 500 m (1,640 ft)	127 knots (235 km/h; 146 mph)
Service ceiling	6,500 m (21,325 ft)
Hovering ceiling: IGE	3,500 m (11,480 ft)
OGE	2,900 m (9,515 ft)
Range: standard fuel, no reserves	323 nm (600 km; 373 miles)
with auxiliary fuel tank	539 nm (1,000 km; 621 miles)

at 2½ min emergency rating
	approx 2,300 m (7,545 ft)
Hovering ceiling: IGE	3,000 m (9,845 ft)
OGE	2,100 m (6,890 ft)
Range:	
standard fuel, 5% reserves	367 nm (680 km; 422 miles)
standard fuel, no reserves	386 nm (715 km; 444 miles)
with auxiliary fuel, 5% reserves	626 nm (1,160 km; 721 miles)
with auxiliary fuel, no reserves	661 nm (1,225 km; 761 miles)
Endurance:	
standard fuel, 5% reserves	3 h 50 min
standard fuel, no reserves	4 h 5 min
with auxiliary fuel, 5% reserves	6 h 41 min
with auxiliary fuel, no reserves	7 h 5 min

PERFORMANCE (W-3RM at max T-O weight of 6,400 kg; 14,110 lb at 500 m; 1,640 ft except where indicated):
Max level speed	124 knots (230 km/h; 143 mph)
Cruising speed: max continuous power	118 knots (218 km/h; 135 mph)
econ cruise power	111 knots (206 km/h; 128 mph)
Max rate of climb at S/L	492 m (1,615 ft)/min
Rate of climb at S/L, OEI	60 m (197 ft)/min
Service ceiling	4,650 m (15,250 ft)
Hovering ceiling: IGE	2,500 m (8,200 ft)
OGE	660 m (2,165 ft)
Range, 30 min reserves:	
standard fuel	334 nm (620 km; 385 miles)
with auxiliary fuel	574 nm (1,065 km; 662 miles)

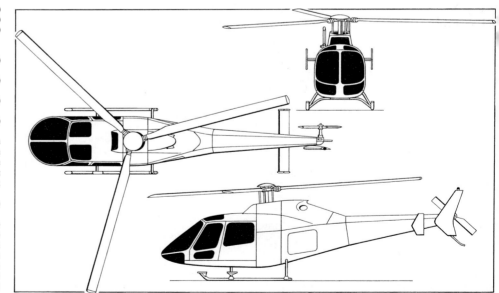

PZL Swidnik SW-4 in latest configuration, with Allison 250-C20R turboshaft *(Jane's/Mike Keep)*

PZL WARSZAWA-OKECIE
PANSTWOWE ZAKŁADY LOTNICZE WARSZAWA-OKECIE (State Aviation Works, Warsaw-Okecie)
Aleja Krakowska 110/114, PL-00-971 Warsaw
Telephone: 48 (22) 460903
Fax: 48 (22) 462701
Telex: 817735
GENERAL MANAGER: Ryszard Leja, MSc
SALES MANAGER: Zbigniew Pesko, MScEc

Okecie factory founded in 1928; responsible for light aircraft development and production, and for design and manufacture of associated agricultural equipment for its own aircraft and those built at other Polish factories; has produced over 3,600 aircraft since 1945.

Javelin Aerospace Ltd of 50 Dawson Street, Dublin 2, Ireland *(Telephone:* 353 (1) 6796434; *Fax:* 353 (1) 6796327) is worldwide distributor (with some exceptions) for Wilga, Flaming, Turbo Javelin (Turbo-Kruk), Koliber and Turbo Orlik. Melex USA Inc (see US section) markets Wilga in various versions.

Light Aircraft and Agricultural Equipment Pilot Plant
DIRECTOR: Marian Kolodynski, BScEng
GENERAL DESIGNER: Andrzej Frydrychewicz, MScEng

Main function of this PZL Warszawa-Okecie plant is to develop and perform research tasks, build and flight test prototypes.

PZL-104 WILGA (ORIOLE) 35 and 80

TYPE: Single-engined general purpose light aircraft.
PROGRAMME: First flight of prototype Wilga 1, 24 April 1962 (see 1968-69 *Jane's*); first flight of improved Wilga 35, 28 July 1967; production of Wilga 35 and 32 began 1968; both received Polish type certificate 31 March 1969 (Wilga 32 described in 1974-75 *Jane's*; see 1975-76 edition for Lipnur Gelatik Indonesian modified Wilga 32); first flight of Wilga 80, 30 May 1979.
CURRENT VERSIONS: **Wilga 35 and 80**: Currently built versions; meet requirements of British BCAR and US FAR Pt 23 respectively; 80 has carburettor air intake further aft; *details below apply to these versions, except where indicated.*
  Wilga 35A and 80A: Aeroclub versions, with glider towing hook.
  Wilga 35H and 80H: With Airtech (Canada) LAP-3000 floats.
  Wilga 35M: Multi-purpose variant of Wilga 35A under study in 1990; extended operational range; 260.5 kW (360 hp) M-14P radial engine, driving (on prototype) W530TA-D35 two-blade constant-speed propeller or (on any production version) a PZL-144 propeller; max T-O and landing weight unchanged.
  Wilga 35R and 80R: Agricultural versions.
  Wilga 80-550: Version for Western markets, with 224 kW (300 hp) Teledyne Continental IO-550 flat-six engine; see under Melex heading in US section.
CUSTOMERS: Total of 931 (all versions) sold by 1 January 1992 to customers in Australia, Austria, Belgium, Bulgaria, Canada, Cuba, former Czechoslovakia, Denmark, Egypt, Finland, France, Germany, Hungary, Indonesia, Italy, North Korea, New Zealand, Poland, Romania, South Africa, Spain, Sweden, Switzerland, Turkey, UK, USA, former USSR (more than 380), Venezuela and former Yugoslavia.
COSTS: $85,900 for radial-engined version (1992) and approx $98,000 for Melex version with Continental engine.
DESIGN FEATURES: Used for wide variety of general aviation and flying club duties; high-mounted cantilever wings; braced tail unit; tall landing gear legs. Wing section NACA 2415; dihedral 1°.
FLYING CONTROLS: Aerodynamically and mass balanced slotted ailerons can be drooped to supplement flaps during landing; tab on starboard aileron; aerodynamically, horn and mass balanced one-piece elevator and rudder; trim tab in centre of elevator; manually operated slotted flaps; fixed slat on wing leading-edge along full span.
STRUCTURE: All-metal, with beaded skins; single-spar wings, with leading-edge torsion box; fuselage in two portions, forward incorporating main wing spar carry-through structure; rear section is tailcone; cabin floor of metal sandwich, with paper honeycomb core, covered with foam rubber; aluminium tailplane bracing strut.
LANDING GEAR: Non-retractable tailwheel type. Semi-cantilever main legs, of rocker type, have oleo-pneumatic shock absorbers. Low-pressure tyres size 500 × 200 mm on mainwheels. Hydraulic brakes. Steerable tailwheel, tyre size 255 × 110 mm, carried on rocker frame with oleo-pneumatic shock absorber. Metal ski landing gear and Airtech Canada LAP-3000 twin-float landing gear optional.
POWER PLANT: One 194 kW (260 hp) PZL AI-14RA nine-cylinder supercharged radial air-cooled engine (AI-14RA-KAF in Wilga 80), driving a PZL US-122000 two-blade constant-speed wooden propeller. Two removable fuel tanks in each wing, with total capacity of 195 litres (51.5 US gallons; 43 Imp gallons). Refuelling point on each side of fuselage, at junction with wing. For longer-range operation, additional 90 litre (23.8 US gallon; 19.8 Imp gallon) fuel tank can be installed in place of rear pair of seats. Oil capacity 16 litres (4.2 US gallons; 3.5 Imp gallons).
ACCOMMODATION: Passenger version accommodates pilot and three passengers, in pairs, with adjustable front seats. Baggage compartment aft of seats, capacity 35 kg (77 lb). Rear seats can be replaced by additional fuel tank for longer-range operation. Upward opening door on each side of cabin, jettisonable in emergency. In parachute training version, starboard door is replaced by two tubular uprights with central connecting strap, and starboard front seat is rearward facing. Jumps are facilitated by step on starboard side and by parachute hitch. Controllable towing hook can be attached to tail landing gear permitting Wilga, in this role, to tow single glider of up to 650 kg (1,433 lb) weight or two or three gliders with combined weight of 1,125 kg (2,480 lb).
SYSTEMS: Hydraulic system pressure 39 bars (570 lb/sq in). Engine started pneumatically by compressed air stored in 7 litre (0.25 cu ft) capacity bottle at pressure of 49 bars (710 lb/sq in) charged by engine. Electrical system powered by DC generator and 24V 10Ah battery.
AVIONICS: VHF transceiver and blind-flying instrumentation standard; RS-6102 (of Polish design), R-860 II, R860 IIM, Bendix/King KY 195 or other radio. ARL-1601 VHF, ARK-9, Bendix/King KR 85 or AV-200 ADF, GB-1 gyro compass, K2-715 airspeed and altitude recorder, optional.
EQUIPMENT: Sun visors, exhaust silencer and windscreen wiper optional.

PZL-104 Wilga 35A *(R. J. Malachowski)*

DIMENSIONS, EXTERNAL:
Wing span: 35	11.12 m (36 ft 5¾ in)
80	11.13 m (36 ft 6¼ in)
Wing chord, constant	1.40 m (4 ft 7¼ in)
Wing aspect ratio	8.0
Length overall: 35	8.10 m (26 ft 6¾ in)
80	8.03 m (26 ft 4¼ in)
Height overall	2.96 m (9 ft 8½ in)
Tailplane span	3.70 m (12 ft 1¾ in)
Wheel track	2.75 m (9 ft 0¼ in)
Wheelbase	6.70 m (21 ft 11¾ in)
Propeller diameter	2.65 m (8 ft 8 in)
Passenger doors (each): Height	1.00 m (3 ft 3¼ in)
Width	1.50 m (4 ft 11 in)

DIMENSIONS, INTERNAL:
Cabin: Length	2.20 m (7 ft 2½ in)
Max width	1.20 m (3 ft 10 in)
Max height	1.50 m (4 ft 11 in)
Floor area	2.20 m² (23.8 sq ft)
Volume	2.40 m³ (85 cu ft)
Baggage compartment	0.50 m³ (17.5 cu ft)

AREAS:
Wings, gross	15.50 m² (166.8 sq ft)
Ailerons (total)	1.57 m² (16.90 sq ft)
Trailing-edge flaps (total)	1.97 m² (21.20 sq ft)
Fin	0.97 m² (10.44 sq ft)
Rudder	0.92 m² (9.90 sq ft)
Tailplane	3.16 m² (34.01 sq ft)
Elevator, incl tab	1.92 m² (20.67 sq ft)

WEIGHTS AND LOADINGS (Wilga 35A and 80):
Weight empty, equipped	870 kg (1,918 lb)
Max T-O and landing weight	1,300 kg (2,866 lb)
Max wing loading	83.9 kg/m² (17.18 lb/sq ft)
Max power loading	6.70 kg/kW (11.02 lb/hp)

PERFORMANCE (Wilga 35A, at max T-O weight):
Never-exceed speed (V_{NE})	150 knots (279 km/h; 173 mph)
Max level speed	105 knots (194 km/h; 120 mph)
Cruising speed (75% power)	85 knots (157 km/h; 97 mph)
Cruising speed for max range	74 knots (137 km/h; 85 mph)
Stalling speed: flaps up	35 knots (65 km/h; 41 mph)
flaps down	30 knots (56 km/h; 35 mph)
Max rate of climb at S/L	276 m (905 ft)/min
Time to 1,000 m (3,280 ft)	3 min
Service ceiling	4,040 m (13,250 ft)
T-O run (grass)	121 m (397 ft)
Landing run	106 m (348 ft)
Range with max fuel, 30 min reserves	275 nm (510 km; 317 miles)

PZL-105 FLAMING (FLAMINGO)

TYPE: Single-engined general purpose light aircraft (passenger or cargo transport, sport and aero club flying, glider towing, parachute training, air ambulance, patrol, geophysical survey, agricultural and other uses).
PROGRAMME: Originally referred to as Wilga 88, but entirely new design, developed as successor to Wilga 35/80 series; static test airframe and three flying prototypes being built; first prototype (SP-PRC), with M-14P engine, made first flight 19 December 1989 and was expected to gain FAR Pt 23 (Amendments 1-28) certification in Normal category 1993 as PZL-105M; second aircraft (SP-PRD), with IO-720 engine as prototype for PZL-105L, made first flight in Summer 1991; flight trials toward certification continuing Spring 1993.
CURRENT VERSIONS: **PZL-105L**: Textron Lycoming IO-720 flat-six engine.
  PZL-105M: VMKB (Vedeneyev) M-14P radial engine.

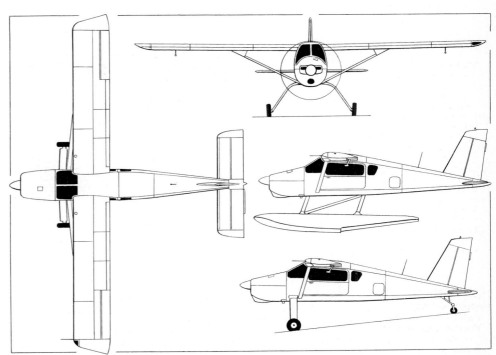

PZL Warszawa-Okecie PZL-105L (Textron Lycoming IO-720 flat-six engine), with additional side view of optional floatplane version *(Dennis Punnett)*

242 POLAND: AIRCRAFT—PZL WARSZAWA-OKECIE

Prototype PZL Warszawa-Okecie PZL-105M six-seat general utility aircraft

DESIGN FEATURES: Braced high wings (single strut each side); cantilever tail unit; STOL characteristics combined with better speed, range and payload than Wilga; choice of power plants and landing gears; operable from unprepared airstrips. Wing section (constant) similar to NASA GA(W)-1; dihedral 1°; incidence 4°; no twist.

FLYING CONTROLS: Mechanically actuated (by push/pull rods and cables) single-slotted flaperons, rudder and elevators; fixed tab on port flaperon, ground adjustable tab on rudder, electrically actuated trim tab on port elevator; rudder and elevators horn balanced; electrically operated single-slotted Fowler trailing-edge flaps.

STRUCTURE: All-metal except for glassfibre wing/rudder/elevator tips and spinner; two-spar wing box; entire structure protected against weather, corrosion and abrasion and finished in polyurethane paint.

LANDING GEAR: Non-retractable type, with single mainwheels on glassfibre spring cantilever legs and steerable tailwheel with oleo-pneumatic shock absorption. Tyre sizes 500 × 200 mm (main) and 250 × 125 mm (tail), with pressures of 3.5 bars (50.75 lb/sq in) and 2.5 bars (36.25 lb/sq in) respectively. Single-disc hydraulic brake on each mainwheel. Optional gear includes floats, skis (with snow brakes) and wheel/skis.

POWER PLANT: One 268.5 kW (360 hp) VMKB (Vedeneyev) M-14P nine-cylinder air-cooled radial engine in PZL-105M, driving a W-530TA-D35 two-blade constant-speed wooden propeller; PZL-105L has a 298 kW (400 hp) Textron Lycoming IO-720-A1B flat-eight engine and Hartzell three-blade constant-speed metal propeller; spinner is glassfibre on PZL-105M, metal on PZL-105L. Fuel in integral tanks in wings, total capacity 270 litres (71.3 US gallons; 59.4 Imp gallons). Gravity fuelling point in upper surface of each wing. Oil capacity (PZL-105M) 25 litres (6.6 US gallons; 5.5 Imp gallons).

ACCOMMODATION: Fully enclosed, heated and ventilated cabin, with seats for up to six persons including pilot (four individual seats in two rows and two-place rear bench seat). All passenger seats and bench can be quickly removed to provide space for up to 450 kg (992 lb) of cargo. Cabin has flat floor, flush with door sill. Large door each side, with upward opening top half; lower halves open downward and incorporate steps. Doors are non-structural. Dual controls optional. Cabin heated and ventilated.

SYSTEMS: Hydraulic system for brakes only. Electrical system (28V DC) supplied by GSR 3000/M engine driven generator in PZL-105M (24V 70A alternator in PZL-105L) and 24V 15Ah nickel-cadmium battery. PZL-105M has pneumatic system (pressure 50 bars; 725 lb/sq in) for engine starting.

AVIONICS: Initial choice includes RS-6102 radio, ARL-1601 VHF and GB-1 gyro compass. Provision for Becker avionics includes VHF transceiver, nav receiver, ADF, gyro compass, marker beacon receiver, ATC transponder, RMI/HSI and audio panel. Navigation/anti-collision/cabin/landing lights standard.

DIMENSIONS, EXTERNAL:
Wing span	12.97 m (42 ft 6½ in)
Wing chord, constant	1.35 m (4 ft 5¼ in)
Wing aspect ratio	9.95
Length overall: M	8.70 m (28 ft 6½ in)
L	8.67 m (28 ft 5¼ in)
Height overall	2.87 m (9 ft 5 in)
Wheel track	3.10 m (10 ft 2 in)
Wheelbase	6.10 m (20 ft 0¼ in)
Propeller diameter: M	2.40 m (7 ft 10½ in)
L	2.18 m (7 ft 1¾ in)
Cabin doors (two, each): Height	1.00 m (3 ft 3¼ in)
Width	1.60 m (5 ft 3 in)

DIMENSIONS, INTERNAL:
Cabin: Length	2.80 m (9 ft 2¼ in)
Max width	1.19 m (3 ft 11 in)
Max height	1.20 m (3 ft 11¼ in)

AREAS:
Wings, gross	16.90 m² (181.9 sq ft)

WEIGHTS AND LOADINGS:
Standard weight empty, equipped: M	1,130 kg (2,491 lb)
L	1,150 kg (2,535 lb)
Max T-O weight: M, L	1,850 kg (4,078 lb)
Max wing loading: M, L	109.5 kg/m² (22.43 lb/sq ft)
Max power loading: M	6.89 kg/kW (11.33 lb/hp)
L	6.21 kg/kW (10.20 lb/hp)

PERFORMANCE (at max T-O weight):
Never-exceed speed (VNE)	165 knots (306 km/h; 190 mph)
Max level speed: M	132 knots (245 km/h; 152 mph)
*L	140 knots (260 km/h; 162 mph)
Max cruising speed: M	111 knots (205 km/h; 127 mph)
*L	127 knots (235 km/h; 146 mph)
Econ cruising speed: M	101 knots (188 km/h; 117 mph)
*L	117 knots (216 km/h; 134 mph)
Stalling speed, flaps down, engine idling:	
M, L	55 knots (102 km/h; 64 mph)
Max rate of climb at S/L: M	330 m (1,085 ft)/min
*L	336 m (1,100 ft)/min
Service ceiling: M	4,140 m (13,575 ft)
*L	5,070 m (16,625 ft)
T-O run: *M	176 m (578 ft)
*L	189 m (620 ft)
T-O to 15 m (50 ft): *M	338 m (1,109 ft)
*L	351 m (1,152 ft)
Landing from 15 m (50 ft): *M	326 m (1,070 ft)
*L	330 m (1,083 ft)
Landing run: *M	153 m (502 ft)
*L	157 m (515 ft)
Range with max fuel:	
at max cruising speed:	
M	563 nm (1,044 km; 648 miles)
*L	460 nm (853 km; 530 miles)
at econ cruising speed:	
M	585 nm (1,085 km; 674 miles)
*L	529 nm (981 km; 609 miles)
g limits: M, L	+3.8/−1.52

* estimated

PZL-106B KRUK (RAVEN)

TYPE: Single-engined agricultural aircraft.

PROGRAMME: Original PZL-106 designed early 1972; first flight of first prototype (SP-PAS), with IO-720 engine, 17 April 1973; five further prototypes, latter four with PZL-3S engine; superseded by current B series from early 1980s. Production at standstill early 1992, but due to resume 1993 with new option of larger hopper (see Equipment paragraph).

CURRENT VERSIONS: Details of **PZL-106A**, **AR** and **AT** in 1985-86 and earlier *Jane's*; **PZL-106AS** in 1991-92 and earlier editions.

PZL-106B: Three prototypes of improved series; first flown (SP-PKW) 15 May 1981; redesigned wings with new aerofoil section, increased span and area, trailing-edge flaps, and shortened V bracing struts; weights and performance in 1985-86 and earlier *Jane's*.

PZL-106BR: Version with geared PZL-3SR engine; first flown 8 July 1983; tested with wingtip vanes (three at each tip); total 64 built by 1 January 1990. *Detailed description applies to BR and BS.*

PZL-106BS: Uprated version, with PZL (Shvetsov) ASz-62IR engine, for Restricted category operation, with higher max T-O weight and increased chemical load; prototype first flown 8 March 1982; 15 built by 1 January 1990.

PZL-106BT: Turboprop version (marketed in West as **Turbo Javelin**), described separately.

CUSTOMERS: Manufactured for member countries of CMEA (Council for Mutual Economic Aid); total of over 246 (all versions) built by 1 January 1990, including 144 PZL-106As produced 1976-81. Sales in 1990 and 1991 not notified; 12 sold in 1992; orders and options held for more than 50 (mostly Turbo-Kruks, and mainly from Africa and Latin America) in early 1993.

DESIGN FEATURES: Structure corrosion resistant, and additionally protected by external finish of polyurethane enamel; strut braced wings with upward cambered wingtips and braced tailplane; production aircraft have low-mounted tailplanes and greater chemical load than prototypes in larger hopper.
Wing section NACA 2415; dihedral 4° from roots; incidence 6° 6'; sweepback 1° at quarter-chord.

FLYING CONTROLS: Slotted ailerons, with ground adjustable tabs on first 10 aircraft, electrically actuated trim tab in port aileron thereafter; mass balanced elevators with port trim tab; rudder with automatic tab; full span four-segment fixed leading-edge slats on each wing; trailing-edge flaps.

STRUCTURE: Duralumin wing structure, with metal and polyester fabric covering and GFRP wingtips; duralumin

Second Flaming, with IO-720 engine as prototype for PZL-105L

PZL WARSZAWA-OKECIE—AIRCRAFT: POLAND 243

PZL-106BR Kruk (PZL-3SR radial engine) *(R. J. Malachowski)*

ailerons and trailing-edge flaps with polyester fabric covering; GFRP/foam core sandwich slats; duralumin V bracing struts. Welded steel tube fuselage, with polyurethane enamel coating and quickly removable light alloy and GFRP panels; structure can be pressure tested for cracks. Duralumin tail unit; fixed surfaces metal covered, control surfaces polyester fabric covered.

LANDING GEAR: Non-retractable tailwheel type, with oleo-pneumatic shock absorber in each unit. Mainwheels, with low-pressure tyres size 800 × 260 mm, each carried on side V and half-axle. Mainwheel tyre pressure 2.0 bars (29 lb/sq in). Pneumatically operated hydraulic disc brakes on mainwheels. Parking brake. Steerable tailwheel, with tubeless tyre size 350 × 135 mm, pressure 2.5 bars (36.25 lb/sq in).

POWER PLANT: *(PZL-106BR):* One 448 kW (600 hp) PZL-3SR seven-cylinder radial air-cooled geared and supercharged engine, driving a PZL US-133000 four-blade constant-speed metal propeller. *(PZL-106BS):* One 746 kW (1,000 hp) PZL (Shvetsov) ASz-62IR nine-cylinder radial air-cooled engine and AW-2-30 propeller. Fuel in two integral wing tanks, total capacity 560 litres (148 US gallons; 123 Imp gallons); can be increased to total of 950 litres (251 US gallons; 209 Imp gallons) by using hopper as auxiliary fuel tank. Gravity refuelling point on each wing; semi-pressurised refuelling point on starboard side of fuselage. Oil capacity 54 litres (14.3 US gallons; 11.9 Imp gallons) max in BR, 67 litres (17.7 US gallons; 14.7 Imp gallons) in BS. Carburettor air filter fitted.

ACCOMMODATION: Single vertically adjustable seat in enclosed, ventilated and heated cockpit with steel tube overturn structure. Provision for instructor's cockpit with basic dual controls, forward of main cockpit and offset to starboard, for training of pilots in agricultural duties. Optional rearward facing second seat (for mechanic) to rear. Jettisonable window/door on each side of cabin. Pilot's seat and seat belt designed to resist 40*g* impact. Cockpit air-conditioning optional.

SYSTEMS: Pneumatic system, rated at 49 bars (710 lb/sq in), for brakes and agricultural equipment. Electrical power, from 3kW 27.5V DC generator and 24V 15Ah battery, for engine starting, pneumatic system control, aircraft lights, instruments, transceiver and semi-pressurised refuelling.

AVIONICS: VHF com transceiver standard; 720-channel UHF transceiver optional.

EQUIPMENT: Easily removable non-corroding (GFRP) hopper/tank, forward of cockpit, can carry more than 1,000 kg (2,205 lb) (see Weights and Loadings) of dry or liquid chemical, and has max capacity of 1,400 litres (370 US gallons; 308 Imp gallons). Enlarged hopper (capacity 1,500 litres; 397 US gallons; 330 Imp gallons) optional. Turnround time, with full load of chemical, is approx 28 s. Hopper quick-dump system can release 1,000 kg of chemical in 5 s or less. Pneumatically operated intake for loading dry chemicals optional. Distribution system for liquid chemical (jets or atomisers) powered by fan driven centrifugal pump. Dispersal system, with positive on/off action for dry chemicals, gives effective swath widths of 30-35 m (100-115 ft). For ferry purposes, hopper can be used to carry additional fuel instead of chemical. When converted into two-seat trainer (see Accommodation), standard hopper can be replaced easily by container with reduced capacity tank for liquid chemical. Steel cable cutter on windscreen and each mainwheel leg; steel deflector cable runs from top of windscreen cable cutter to tip of fin. Windscreen washer and wiper standard. Other equipment includes clock, rearview mirror, second (mechanic's) seat (optional), landing light, anti-collision light, and night working lights (optional).

DIMENSIONS, EXTERNAL:
Wing span	14.90 m (48 ft 10½ in)
Wing chord, constant	2.16 m (7 ft 1 in)
Wing aspect ratio	6.9
Length overall: BR	9.25 m (30 ft 4½ in)
BS	9.34 m (30 ft 7¾ in)
Height overall	3.32 m (10 ft 10¾ in)
Tailplane span	5.77 m (18 ft 11¼ in)
Wheel track	3.10 m (10 ft 2¼ in)
Wheelbase	7.41 m (24 ft 3¾ in)
Propeller diameter: BR	3.10 m (10 ft 2 in)
BS	3.30 m (10 ft 10 in)
Propeller ground clearance (tail up)	0.39 m (1 ft 3¼ in)
Crew doors (each): Height	0.91 m (2 ft 11¾ in)
Width	1.06 m (3 ft 5¾ in)
Baggage door: Height	0.70 m (2 ft 3½ in)
Width	0.60 m (1 ft 11¾ in)

DIMENSIONS, INTERNAL:
Cabin: Length	1.37 m (4 ft 6 in)
Max width	1.25 m (4 ft 1¼ in)
Max height	1.30 m (4 ft 3¼ in)
Floor area	1.12 m² (12.05 sq ft)
Rear cockpit/baggage compartment:	
Length	1.40 m (4 ft 7 in)
Width	1.00 m (3 ft 3¼ in)
Depth	0.60 m (1 ft 11¾ in)

AREAS:
Wings, gross	31.69 m² (341.1 sq ft)
Ailerons (total)	4.34 m² (46.72 sq ft)
Trailing-edge flaps (total)	4.44 m² (47.79 sq ft)
Leading-edge slats (total)	4.25 m² (45.75 sq ft)
Fin	1.26 m² (13.56 sq ft)
Rudder, incl tab	1.62 m² (17.44 sq ft)
Tailplane	3.34 m² (35.95 sq ft)
Elevators, incl tab	4.22 m² (45.42 sq ft)

WEIGHTS AND LOADINGS:
Weight empty, equipped: BR	1,790 kg (3,946 lb)
BS	2,080 kg (4,585 lb)
Max chemical payload: BR	1,300 kg (2,866 lb)
BS	1,150 kg (2,535 lb)
Max T-O and landing weight:	
BR, BS	3,000 kg (6,614 lb)
BR (Restricted category)	3,450 kg (7,606 lb)
BS (Restricted category)	3,500 kg (7,716 lb)
Max wing loading (Restricted category):	
BR	108.86 kg/m² (22.30 lb/sq ft)
BS	110.44 kg/m² (22.62 lb/sq ft)
Max power loading (Restricted category):	
BR	7.70 kg/kW (12.68 lb/hp)
BS	4.69 kg/kW (7.72 lb/hp)

PERFORMANCE (at max T-O weight):
Never-exceed speed (V_{NE}):	
BR, BS	145 knots (270 km/h; 167 mph)
Max level speed at S/L:	
BR, BS	116 knots (215 km/h; 134 mph)
Operating speed with max chemical load:	
BR	81-86 knots (150-160 km/h; 93-99 mph)
BS	86 knots (160 km/h; 99 mph)
Stalling speed at S/L:	
BR, BS	54 knots (100 km/h; 62 mph)
Max rate of climb at S/L (with agricultural equipment):	
BR	228 m (748 ft)/min
BS	372 m (1,220 ft)/min
T-O run (with agricultural equipment):	
BR	250 m (820 ft)
BS	120 m (394 ft)
Landing run (with agricultural equipment):	
BR, BS	200 m (656 ft)
Range with max standard fuel:	
BR, BS	485 nm (900 km; 559 miles)

PZL-106BT TURBO-KRUK

TYPE: Turboprop version of Kruk agricultural aircraft.
PROGRAMME: First flight of prototype (SP-PAA) 18 September 1985.
CURRENT VERSIONS: **Turbo-Kruk:** Standard version; marketed in West as **Turbo Javelin**.
CUSTOMERS: 21 production aircraft built by 1 January 1992; operated in Egypt, Germany and Poland. Further orders and options early 1993 from Latin America and Africa (see main Kruk entry).
DESIGN FEATURES: Compared with Kruk, has turboprop engine, 6° sweepback at quarter-chord, 6° dihedral, taller fin and larger chemical load.
Description as for piston engined Kruk except as follows:
POWER PLANT: One 544 kW (730 shp) Motorlet M 601 D turboprop; Avia V 508 D three-blade propeller.

DIMENSIONS, EXTERNAL:
Wing span	15.00 m (49 ft 2½ in)
Wing aspect ratio	7.1
Length overall	10.24 m (33 ft 7¼ in)
Height overall	3.82 m (12 ft 6½ in)
Propeller diameter	2.50 m (8 ft 2½ in)

AREAS:
Fin	1.82 m² (19.59 sq ft)

WEIGHTS AND LOADINGS:
Weight empty, equipped	1,680 kg (3,704 lb)
Max chemical payload	1,300 kg (2,866 lb)

PZL-106BT Turbo-Kruk equipped with forest firefighting kit enabling it to drop a 1,500 litre water 'bomb' in 1.5 seconds *(R. J. Malachowski)*

244 POLAND: AIRCRAFT—PZL WARSZAWA-OKECIE

PZL-106BT Turbo-Kruk with conventional spraybar agricultural kit

Max T-O weight	3,500 kg (7,716 lb)
Max landing weight	3,000 kg (6,614 lb)
Max wing loading	110.44 kg/m² (22.62 lb/sq ft)
Max power loading	6.07 kg/kW (9.97 lb/shp)

PERFORMANCE (at max T-O weight):
Never-exceed speed (VNE)
    145 knots (270 km/h; 167 mph)
Max level speed at S/L:
  without agricultural equipment
    135 knots (250 km/h; 155 mph)
  with agricultural equipment
    116 knots (215 km/h; 134 mph)
Operating speed with max chemical load
    81-92 knots (150-170 km/h; 93-106 mph)
Stalling speed at S/L   49 knots (90 km/h; 56 mph)
Max rate of climb at S/L (with agricultural equipment)
    360 m (1,180 ft)/min
T-O run (with agricultural equipment)   230 m (755 ft)
Landing run (with agricultural equipment)
    130 m (427 ft)
Range with max standard fuel
    485 nm (900 km; 559 miles)

PZL-107 KACZOR (DRAKE)

TYPE: Agricultural aircraft.
PROGRAMME: Various design configurations explored during second half of 1980s (see Design Features for latest 1990 form); intended for development during 1990s and production from year 2000, but no funding since 1990.
DESIGN FEATURES: Intended for full range of agricultural operations (including simultaneous sowing and protection) and forestry services (including firefighting); three hopper/tanks (two for fluids, one for dry chemicals) grouped around CG. Joined-wing (back-staggered sesquiplane) lifting system comprises low mounted, short-span front wing, connected by endplates to mid span of mid mounted main wing to form rigid frame; main wing (except for new tips) is that of PZL-105 Flaming, and is additionally supported by single bracing strut each side; fuselage aft of cabin flattened to form vertical fin, with rudder attached; anhedral horizontal surfaces attached atop these in T configuration.

Design intended to eliminate undesirable downwash effects on distribution of chemicals, and to offer optimum protection of cabin area in event of a crash. Estimated coverage 140 ha (346 acres)/h with dry chemicals, 200 ha (494.2 acres)/h with liquids. Average fuel consumption during agricultural operations 140 litres (37 US gallons; 30.8 Imp gallons)/h.

PZL-110 KOLIBER (HUMMING-BIRD)

TYPE: Two/four-seat training and multi-purpose light aircraft; licence built and uprated version of Socata Rallye 100 ST.
PROGRAMME: First PZL-110, modified to receive 86.5 kW (116 hp) PZL-F (Franklin) engine, first flown 18 April 1978; first flight of Koliber 150 prototype (SP-PHA) 27 September 1988; Polish type certificate for 150 awarded January 1989.
CURRENT VERSIONS: **Koliber:** See 1991-92 and earlier *Jane's* for details of initial **Series I/II/III** with 93.2 kW (125 hp) PZL-Franklin engines; production (total of 40 built) ended December 1989 and included 675 sets of spare fuselage/wing/aileron/flap/control surface assemblies, some of them exported to France.
Koliber 150: Current production version, *as described below*; higher powered Textron Lycoming engine and detail improvements. Currently (end 1992) certificated in Denmark, Germany, Norway, Poland, Sweden and UK; French certification expected 1994.
Koliber 235 (PZL-111): More powerful version; described separately.
CUSTOMERS: Koliber 150s sold to Denmark, Germany, Netherlands, Sweden and UK.
COSTS: Koliber 150 approx $60,000 (1993).
DESIGN FEATURES: Wing section NACA 63A-416 (modified); dihedral 7° 7′ 30″; incidence 4°.
FLYING CONTROLS: Mechanically actuated, aerodynamically and mass balanced ailerons with ground adjustable tabs; aerodynamically and mass balanced elevator with control tab, rudder with ground adjustable tab; full span automatic leading-edge slats; electrically actuated Fowler flaps.
STRUCTURE: All-metal; flaps, elevators and rudder have corrugated skins; wing torsion box and trailing-edge segments electrically spot-welded.
LANDING GEAR: Non-retractable tricycle type, with leg fairings and oleo-pneumatic shock absorption. Castoring nosewheel, size 330 × 130 mm; mainwheels size 380 × 150 mm. Tyre pressures 1.4 and 1.8 bars (20.3 and 26.1 lb/sq in) respectively. Hydraulic disc brakes.
POWER PLANT: One 112 kW (150 hp) Textron Lycoming O-320-E2A flat-four engine, driving a Sensenich 74DM6-054, -056 or -058 two-blade constant-speed metal propeller. Fuel in two metal tanks in wings, with total capacity of 105 litres (27.7 US gallons; 23.1 Imp gallons). Refuelling points above wings. Oil capacity 6 litres (1.6 US gallons; 1.3 Imp gallons).
ACCOMMODATION: Two side by side seats, plus two-person bench seat at rear, under large rearward sliding canopy. Dual controls. Heating and ventilation standard.
SYSTEMS: 12V electrical system, with 50A alternator and 30Ah battery.
AVIONICS: Bendix/King KX 155 or Narco Mk 12D UHF transceiver, ADF, VOR, electrically powered gyro attitude indicator, turn and bank indicator, and directional gyro.
EQUIPMENT: For training role, includes pupil's window blinds for instrument training, front seat backrests suitable for use with back type parachutes, safety belts, and accelerometers.

DIMENSIONS, EXTERNAL:
Wing span	9.75 m (31 ft 11¾ in)
Wing chord, constant	1.30 m (4 ft 3 in)
Wing aspect ratio	7.5
Length overall	7.37 m (24 ft 2¼ in)
Height overall	2.80 m (9 ft 2¼ in)
Tailplane span	3.67 m (12 ft 0½ in)
Wheel track	2.01 m (6 ft 7¼ in)
Wheelbase	1.71 m (5 ft 7¼ in)
Propeller diameter	1.78 m (5 ft 10 in)

AREAS:
Wings, gross	12.68 m² (136.5 sq ft)
Ailerons (total)	1.56 m² (16.79 sq ft)
Trailing-edge flaps (total)	2.40 m² (25.83 sq ft)
Vertical tail surfaces (total)	1.74 m² (18.73 sq ft)
Horizontal tail surfaces (total)	3.48 m² (37.50 sq ft)

WEIGHTS AND LOADINGS (U: Utility, N: Normal category):
Weight empty, equipped	548 kg (1,208 lb)
Max T-O weight: U	770 kg (1,697 lb)
N	850 kg (1,874 lb)
Max wing loading: U	60.72 kg/m² (12.44 lb/sq ft)
N	67.03 kg/m² (13.73 lb/sq ft)
Max power loading: U	8.90 kg/kW (14.63 lb/hp)
N	7.60 kg/kW (12.49 lb/hp)

PERFORMANCE (at max T-O weight):
Never-exceed speed (VNE):	
U	145 knots (270 km/h; 167 mph)
N	134 knots (250 km/h; 155 mph)
Max level speed at S/L:	
U, N	108 knots (200 km/h; 124 mph)
Max cruising speed at S/L:	
U, N	92 knots (170 km/h; 106 mph)

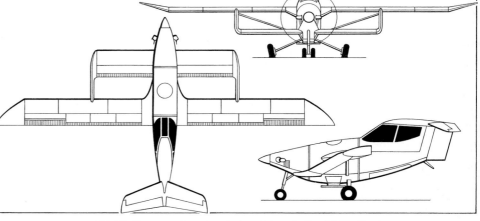

PZL-107 Kaczor projected agricultural aircraft (*Jane's/Mike Keep*)

PZL-110 Koliber 150, Polish-built version of the Socata Rallye (*R. J. Malachowski*)

Econ cruising speed:	
U, N	75 knots (140 km/h; 87 mph)
Stalling speed:	
U, N, flaps up	50 knots (92 km/h; 58 mph)
U, N, flaps down	45 knots (82 km/h; 51 mph)
Max rate of climb at S/L: U	264 m (866 ft)/min
N	216 m (708 ft)/min
Service ceiling: U, N	3,700 m (12,140 ft)
T-O run at S/L: U	140 m (459 ft)
N	167 m (548 ft)
T-O to 15 m (50 ft) at S/L: U	340 m (1,116 ft)
N	397 m (1,303 ft)
Landing from 15 m (50 ft) at S/L: U	290 m (952 ft)
N	320 m (1,050 ft)
Landing run at S/L: U	125 m (411 ft)
N	138 m (453 ft)
Range at 500 m (1,640 ft) with max fuel, no reserves	
	324 nm (600 km; 373 miles)
g limits	+4.4/−1.76 (ultimate)

PZL-111 KOLIBER 235

TYPE: Higher-powered version of Koliber 150.
PROGRAMME: Construction of first prototype began August 1990; static testing being completed March 1993.
CURRENT VERSIONS: **Koliber 235:** With 175.3 kW (235 hp) Lycoming engine. First prototype is of this version; first flight expected 1993.
Koliber 180: Alternative version with 134 kW (180 hp) Lycoming.
Koliber 160: Second alternative: 119 kW (160 hp) Lycoming engine.
DESIGN FEATURES: Designed at PZL Warszawa-Okecie Pilot Plant (project engineer Ryszard Kaczkowski); approx 40 per cent commonality with PZL-110 series. Intended for basic, navigation and general training; glider towing; sport flying; touring; executive flights; and special missions.
POWER PLANT: One 175.3 kW (235 hp) Textron Lycoming O-540-B4B5 flat-six engine; Hartzell HC-C2YK-13F/F8468A4 two-blade metal propeller. Fuel capacity 170 litres (45 US gallons; 37.4 Imp gallons).
SYSTEMS: 12V electrical system with 18Ah battery.
DIMENSIONS, EXTERNAL: As Koliber 150 except:

Length overall	7.32 m (24 ft 0¼ in)
Propeller diameter	2.03 m (6 ft 8 in)

WEIGHTS AND LOADINGS (U: Utility, N: Normal category):

Weight empty	650 kg (1,433 lb)
Max T-O weight: U	1,000 kg (2,204 lb)
N	1,150 kg (2,535 lb)
Max wing loading: U	77.52 kg/m² (15.88 lb/sq ft)
N	89.15 kg/m² (18.26 lb/sq ft)
Max power loading: U	5.70 kg/kW (9.38 lb/hp)
N	6.56 kg/kW (10.79 lb/hp)

PERFORMANCE (estimated):

Max level speed	130 knots (240 km/h; 149 mph)
Econ cruising speed	108 knots (200 km/h; 124 mph)
Stalling speed, flaps down	44 knots (80 km/h; 50 mph)
Max rate of climb at S/L	360 m (1,180 ft)/min
Service ceiling	4,500 m (14,775 ft)
Range with max fuel	539 nm (1,000 km; 621 miles)

PZL-126 MRÓWKA (ANT)

TYPE: Single-seat light agricultural and ecological support aircraft.
PROGRAMME: Design of aircraft, and innovatory new airborne spraying system, initiated late 1970s; propeller, rear fuselage and parts of landing gear built by students of factory's training college; preliminary design completed late 1982 and initial detail design work in second quarter of 1983; project then delayed for some years (see 1990-91 *Jane's*); first flight of first prototype (SP-PMA) 20 April 1990.
Agrolot foundation created July 1990 to finance specialist groups building aircraft at Warszawa-Okecie.
CURRENT VERSIONS: **PZL-126:** First prototype. *Detailed description applies to this aircraft.*
PZL-126P: Second prototype, with extended span (7.50 m; 24 ft 7¼ in) wings; owing to non-availability of series production PZL-Franklin engines, will have a 74.5 kW (100 hp) Teledyne Continental O-200-A engine and McCauley 1A100MCM-6950 propeller (diameter 1.75 m; 5 ft 9 in). Not yet flown (early 1993).
DESIGN FEATURES: Very small size, with constant chord wings and upswept rear fuselage; economical for use on farms and smallholdings, and considered suitable also for patrol and liaison missions (eg for detecting/controlling forest fires, identification of diseased vegetation, and monitoring areas of polluted land and water); new airborne spraying system in form of vaned and pressurised wingtip pods of chemical, attached by quick-fastening locks to facilitate rapid replacement of empty pods by full ones (these replaced by extended wingtips on first prototype, increasing span by 1.00 m; 3 ft 3¼ in); special retractable device for spreading biological agents installed in fuselage underside aft of cockpit. Meets requirements of FAR Pt 23 (USA) and BCAR Section K (UK); can be dismantled quickly for long-distance transportation in cabin of An-2, or towed on own landing gear by light all-terrain vehicle; optional taildragger landing gear; optional wing with full span flaps, airbrakes and spoilers on PZL-126P. Wing section NASA GA(W)-1.

First prototype PZL-126 Mrówka experimental agricultural aircraft, without wingtip pods *(R. J. Malachowski)*

FLYING CONTROLS: Trailing-edge flaperons outboard and single-slotted flaps inboard; alternative wings to be tested (see Design Features); one-piece elevator with starboard trim tab; rudder with ground adjustable tab.
STRUCTURE: Mostly metal but with some wing and fuselage components of GFRP/epoxy.
LANDING GEAR: Non-retractable tricycle gear standard. Cantilever self-sprung mainwheel legs; shock absorber in nosewheel unit. Size 350 × 135 mm tyres on all three wheels, pressure 1.23 bars (17.8 lb/sq in) on main units and 0.78 bar (11.3 lb/sq in) on nose unit. Mainwheels fitted with differential hydraulic disc brakes. Tailwheel configuration optional.
POWER PLANT: One 44.7 kW (60 hp) PZL-F 2A-120-C1 flat-twin engine, driving a two-blade fixed-pitch wooden propeller. (Interchangeable propellers for agricultural flying or patrol mission.) Integral fuel tanks in wing torsion box. Fuel capacity 70 litres (18.5 US gallons; 15.4 Imp gallons).
ACCOMMODATION: Single adjustable seat under one-piece moulded canopy, opening sideways to starboard. Seat and canopy taken from SZD-51 Junior sailplane.
SYSTEMS: Hydraulic system for mainwheel brakes only; 12V DC electrical system.
AVIONICS: VFR instrumentation standard, plus 720-channel UHF com and 10-channel radio telephone.
EQUIPMENT: Dedicated system for spraying with low volume liquid chemicals (pyrethroids) consists of 25 litre (6.6 US gallon; 5.5 Imp gallon) pod at each wingtip. Spraying controlled electrically by push-button on throttle lever and effected by dispersing liquid under pressure via atomiser at rear of each pod. Area of 25 ha (61.8 acres) can be covered with one pair of full pods. Biological agents, such as eggs of Trichogramma wasp, carried in capsules in paper tape wound on reel housed in lower fuselage behind cockpit and extended through openable hatch in floor. One spreader holds 3 kg (6.6 lb) package of eggs, on four reels, and at drop rate of four capsules every 50 m (164 ft) can cover area of 800 ha (1,977 acres) on single loading. Like spray system, spreader's actuation is electrical, by means of push-button on throttle lever. Other equipment can include cameras and first aid appliances.
DIMENSIONS, EXTERNAL:

Wing span: excl pods	6.00 m (19 ft 8¼ in)
incl pods	6.30 m (20 ft 8 in)
Wing chord, constant	0.75 m (2 ft 5½ in)
Length overall	4.66 m (15 ft 3½ in)
Height overall	2.53 m (8 ft 3½ in)
Tailplane span	2.00 m (6 ft 6¾ in)
Wheel track	1.94 m (6 ft 4½ in)
Wheelbase	0.92 m (3 ft 0¼ in)
Propeller diameter	1.40 m (4 ft 7 in)
Propeller ground clearance	0.18 m (7 in)

AREAS:

Wings, gross: excl pods	4.50 m² (48.44 sq ft)
incl pods	5.12 m² (55.11 sq ft)
Flaps (total)	0.572 m² (6.16 sq ft)
Flaperons (total)	0.414 m² (4.46 sq ft)
Vertical tail surfaces (total)	0.68 m² (7.32 sq ft)
Horizontal tail surfaces (total)	1.02 m² (10.98 sq ft)

WEIGHTS AND LOADINGS:

Max T-O and landing weight	420 kg (926 lb)
Max wing loading:	
without pods (prototype)	93.33 kg/m² (19.12 lb/sq ft)
with pods	82.03 kg/m² (16.80 lb/sq ft)
Max power loading	9.39 kg/kW (15.43 lb/hp)

PERFORMANCE:

Operating speed	75 knots (140 km/h; 87 mph)

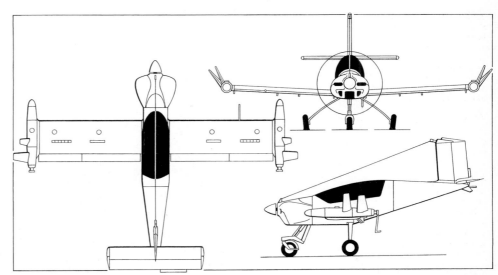

PZL-126 Mrówka experimental light agricultural aircraft *(Jane's/Mike Keep)*

POLAND: AIRCRAFT—PZL WARSZAWA-OKECIE

PZL-130TM Turbo-Orlik prototype (c/n 005), one of the first two aircraft delivered to the Polish Air Force *(R. J. Malachowski)*

PZL-130T TURBO-ORLIK

TYPE: Tandem-seat basic and advanced turboprop trainer.
PROGRAMME: Development of piston engined Orlik (Spotted Eaglet) discontinued 1990 due to non-availability of Russian M-14Pm engine and inadequate power of Polish alternative (full description of Orlik in 1989-90 *Jane's*). Development of turboprop derivative began 1985 with collaboration of Airtech Canada (no longer involved in programme), by refitting third of six Orlik prototypes (SP-PCC, c/n 004) with P&WC PT6A-25A; this aircraft, re-registered SP-RCC, made first flight 13 July 1986 and received provisional type certification to FAR Pt 23 on 23 January 1987, but lost in crash later same month.

Test programme continued with two further Turbo-Orliks: c/n 007 (first flight 12 January 1989) and c/n 008 (SP-WCA, first flight early 1990), former with 560 kW (750 shp) Motorlet M 601 E engine and latter with 410 kW (550 shp) PT6A-25A; these prototypes designated **PZL-130TM** and **PZL-130TP** respectively.

Modifications resulting from Polish Air Force trials and overseas demonstration flights (Israel and South Africa) included provision for ejection seats, use of M 601 T for Polish Air Force aircraft and PT6A-62 for export, extensive nav/com avionics, and strengthened landing gear to cater for increased MTOGW of 2,700 kg (5,952 lb); c/n 010 is static test aircraft for TC version, and c/n 011 first flying TC prototype. First deliveries, to Polish Air Force Academy at Deblin, were two PZL-130TMs (c/n 005 and 006) in October 1992, followed by two PZL-130TBs (c/n 009 and 012) December 1992; c/n 017 in final assembly March 1993.
CURRENT VERSIONS: **PZL-130TB**: For Polish Air Force and potential export to East European or other countries accustomed to Russian/Soviet equipment. Differences from TM prototype include greater wing span, increased wing incidence (to lower nose in flight), double-slotted flaps, redesigned ventral fin, ejection seats, revised canopy, cockpit layout closely resembling that of Sukhoi Su-22, more powerful brakes, steerable nosewheel, six (instead of four) underwing hardpoints, and higher max T-O weight. Standard engine is M 601 E, but fully aerobatic M 601 T available as option. First example (c/n 009) rolled out May 1991, making first flight 18 September 1991. *Detailed description applies to TB and TC, except where otherwise indicated.*

PZL-130TC: Export version, with 708 kW (950 shp) P&WC PT6A-62 engine; Bendix/King avionics, Martin-Baker ejection seats and Hamilton Standard air-conditioning system; c/n 011 completed as prototype/demonstrator.

PZL-130TD: Alternative export version; same equipment as TC but lower powered 559 kW (750 shp) PT6A-25C engine.

PZL-130TE: 'Economy' export version, with more limited equipment, no ejection seats and 410 kW (550 shp) PT6A-25A engine.
CUSTOMERS: Polish Air Force (two PZL-130TM and eight PZL-130TB ordered); preliminary order also placed for 40 more, but not yet certain whether these will be TBs or one of the variants with PT6A engines. Has been evaluated by Israel and South Africa; interest also shown by Hungary and some Latin American countries.
DESIGN FEATURES: Two seats in tandem, with ejection seats in most versions; modular installation of instruments and displays; wings have detachable leading-edges and raked tips; shallow underfin beneath tailcone; fixed incidence tailplane; unspecified modifications to horizontal tail surfaces reported early 1992.

Wing section NACA 64_2215 (modified); 6° (±3°) twist; 5° dihedral from roots.
FLYING CONTROLS: Frise differential ailerons, elevators and rudder all aerodynamically and mass balanced and actuated mechanically (ailerons by pushrods and torque tube, elevators by rods and cables, rudder by cables); electrically actuated trim tabs on port aileron, port elevator and rudder. Three-position double-slotted trailing-edge flaps, also actuated electrically.
STRUCTURE: All-metal, except for GFRP/epoxy wingtips. One-piece light alloy multi-spar wing box, stiffened by riveted omega formers, forms integral fuel tanks; wing trailing-edges and fuselage skin panels stiffened by L formers, electrically spot-welded; fin integral with fuselage.
LANDING GEAR: Hydraulically retractable tricycle type, all three units retracting into fuselage (mainwheels inward, steerable nosewheel rearward). PZL Warszawa-Okecie oleo-pneumatic shock absorber in each unit. All three wheels same size, with 500 × 200 mm tubeless tyres. Differential hydraulic multi-disc brakes; parking brakes for main and nosewheels. No anti-skid system.
POWER PLANT: One 560 kW (750 shp) Motorlet M 601 E turboprop in PZL-130TB, driving a three-blade propeller; PZL-130TC has a Pratt & Whitney Canada PT6A-62, flat rated to 708 kW (950 shp), and Hartzell HC-D4N-2A/D9512A four-blade variable-pitch fully feathering propeller. See Current Versions for engines of TD and TE. Fuel in four integral wing tanks, total usable capacity 540 litres (142.7 US gallons; 118.8 Imp gallons). Overwing refuelling point for each tank. Fuel and oil systems permit up to 30 s of inverted flight. Provision for two 340 litre (90 US gallon; 74.8 Imp gallon) underwing drop tanks.
ACCOMMODATION: Tandem seating for pupil (in front) and instructor under one-piece canopy which opens sideways to starboard. Rear seat elevated 65 mm (2.6 in). Polish LFK-K1 ejection seats in PZL-130TB (zero height/70-323 knots; 130-600 km/h; 81-373 mph); Martin-Baker CH15A, McDonnell Douglas Minipac II or similar seats optional in TC and TD. Baggage space aft of rear seat. Full dual controls standard. Cockpit heating and ventilation; canopy demisting.
SYSTEMS: Cockpit ventilation, cooling, heating and canopy demisting provided by integral system utilising air cycle unit and engine bleed air. Engine driven hydraulic pump provides power for landing gear actuation and wheel brakes, with hydraulic accumulator for emergency braking and pneumatic (nitrogen bottle) power for emergency landing gear extension. Lear Siegler 6kW starter/generator and two 24V 15Ah nickel-cadmium batteries provide power for 27.5V DC electrical system, with solid-state inverter for 115/26V AC power; ground power receptacle fitted. Diluter demand oxygen system, capacity 2.716 litres (166 cu in).
AVIONICS: Single or dual VHF, UHF and/or HF com systems, navigation systems and instrumentation to customer's specification. Standard fit in PZL-130TC includes Bendix/King KTR 908 VHF com, KNR 634 nav with glideslope, KPJ 525 HSI, KDF 806 ADF, KDM 706 DME, KTU 709 Tacan, KXP 756 transponder and Omega long-range nav system. Avionics in PZL-130TB similar but of East European manufacture.
ARMAMENT: Three hardpoints under each wing, inboard and centre ones stressed for loads of 160 kg (353 lb) each and outboard stations for 80 kg (176 lb) each. Typical external loads for Polish Air Force PZL-130TB include free-fall bombs of up to 100 kg, Tejsy and Mokrzycko bomb canisters, Zeus gun pods each with two 7.62 mm machine guns, launchers for 57 mm or 80 mm rockets, or Strela IR air-to-air missiles; plus S-17 gunsight and S-13 gun camera. Export versions can be equipped with Alkan or FN Herstal pylons and armament to customer's requirements.

DIMENSIONS, EXTERNAL:
Wing span	9.00 m (29 ft 6¼ in)
Wing aspect ratio	6.23
Length overall	9.00 m (29 ft 6¼ in)
Fuselage: Max width	0.90 m (2 ft 11½ in)
Height overall	3.53 m (11 ft 7 in)
Wheel track	3.10 m (10 ft 2 in)
Wheelbase	2.90 m (9 ft 6¼ in)
Propeller diameter: TC	2.40 m (7 ft 10½ in)

AREAS:
Wings, gross	13.00 m² (139.93 sq ft)

WEIGHTS AND LOADINGS:
Weight empty: TB	1,600 kg (3,527 lb)
TC	1,450 kg (3,197 lb)
TD	1,380 kg (3,042 lb)
TE	1,300 kg (2,866 lb)
Max external stores load: all	800 kg (1,764 lb)
Max T-O weight:	
Aerobatic: TB, TC, TD	2,000 kg (4,409 lb)
TE	1,900 kg (4,189 lb)
Utility: TB, TC, TD, TE	2,700 kg (5,952 lb)
Max wing loading:	
Aerobatic: TB, TC, TD	153.8 kg/m² (31.50 lb/sq ft)
TE	146.1 kg/m² (29.92 lb/sq ft)
Utility: TB, TC, TD, TE	207.7 kg/m² (42.54 lb/sq ft)
Max power loading:	
Aerobatic: TB, TD	3.58 kg/kW (5.88 lb/shp)
TC	2.82 kg/kW (4.64 lb/shp)
TE	4.63 kg/kW (7.62 lb/shp)
Utility: TB, TD	4.83 kg/kW (7.94 lb/shp)
TC	3.81 kg/kW (6.27 lb/shp)
TE	6.58 kg/kW (10.82 lb/shp)

PERFORMANCE (estimated, at max Aerobatic T-O weight, clean configuration, except where indicated):
Max level speed at S/L:	
TB, TD	245 knots (454 km/h; 282 mph)
TC	274 knots (508 km/h; 316 mph)
TE	232 knots (430 km/h; 267 mph)
Max level speed at 6,000 m (19,685 ft):	
TB, TD	270 knots (501 km/h; 311 mph)
TC	302 knots (560 km/h; 348 mph)

PZL Warszawa-Okecie PZL-130TC Turbo-Orlik tandem-seat trainer (PT6A-62 turboprop) *(Jane's/Mike Keep)*

Full size mockup of the PZL-230F Skorpion as unveiled in early 1993 *(Grzegorz L. Holdanowicz)*

Three-quarter rear view of the PZL-230F mockup with 30 mm GAU-8 underfuselage gun pod

TE	251 knots (465 km/h; 289 mph)
Max rate of climb at S/L:	
TB, TD	798 m (2,620 ft)/min
TC	1,236 m (4,055 ft)/min
TE	708 m (2,320 ft)/min
Service ceiling: TC	10,060 m (33,000 ft)
T-O run at S/L: TB, TD	222 m (729 ft)
TC	172 m (565 ft)
TE	260 m (853 ft)
T-O to 15 m (50 ft): TC	266 m (873 ft)
Landing run: TC	184 m (604 ft)
Range with max fuel: TB	523 nm (970 km; 602 miles)
TC	501 nm (930 km; 577 miles)
TD	620 nm (1,150 km; 714 miles)
TE	680 nm (1,260 km; 783 miles)
Range, clean configuration, no reserves, at 299 knots (555 km/h; 345 mph):	
TC	593 nm (1,100 km; 683 miles)
Range with two 340 litre (90 US gallon; 74.8 Imp gallon) drop tanks, no reserves, at 277 knots (514 km/h; 319 mph) cruising speed:	
TC	1,241 nm (2,300 km; 1,429 miles)
g limits: Aerobatic	+6/−3
Utility	+4.4/−1.76

PZL-230F SKORPION

TYPE: Single-seat small agile battlefield attack (SABA) aircraft.
PROGRAMME: Preliminary PZL-230 design revealed late 1990 with twin-turbofan power plant; changed to twin pusher PT6A-67A turboprops 1991 (see 1991-92 *Jane's*); further redesign (PZL-230F) reverts to turbofans and slimmer profile. Possible first flight 1996. *All details provisional.*
CUSTOMERS: To meet Polish Air Force requirement; interest expressed for up to 50, to enter service in 2000.
DESIGN FEATURES: Double-delta main wings, outer portions of which all-moving, working with close-coupled all-moving canards for simultaneous pitch and roll control as well as conventional T-O/landing function; claimed to make possible flight at 50° angle of attack and ability to make 180° turn in only 5 s; inward canted tail-fins. Intended for battlefield close support close to FLOT, able to use unprepared airstrips.
FLYING CONTROLS: Electronic fly-by-wire, probably of Western design.
STRUCTURE: Mainly composites.
LANDING GEAR: Retractable tricycle type.
POWER PLANT: Twin rear mounted turbofans (23.13 kN; 5,200 lb st class PW305 proposed).
ACCOMMODATION: Pilot only, under rear-hinged, upward opening canopy; Martin-Baker Mk 10L zero/zero ejection seat, inclined at 34° from perpendicular. Armoured cockpit to protect pilot against rounds of up to 12.7 mm calibre.
ARMAMENT: Internally mounted 25 mm General Electric GAU-12/U five-barrel gun, with 220 rds, plus a 30 mm GAU-8 gun in ventral pod.

DIMENSIONS, EXTERNAL:
Wing span	9.00 m (29 ft 6½ in)
Length overall	9.30 m (30 ft 6¼ in)

WEIGHTS AND LOADINGS:
Max external stores load	4,000 kg (8,818 lb)
Max T-O weight	10,000 kg (22,046 lb)

PERFORMANCE:
Max level speed	539 knots (1,000 km/h; 621 mph)
Combat radius	162 nm (300 km; 186 miles)
g limit	+9

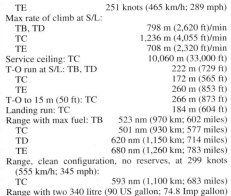
PZL-230F Skorpion single-seat agile battlefield attack aircraft *(Jane's/Mike Keep)*

PORTUGAL

OGMA
OFICINAS GERAIS DE MATERIAL AERONÁUTICO (General Aeronautical Material Workshops)
P-2615 Alverca
Telephone: 351 (1) 9581000
Fax: 351 (1) 9581288
Telex: 14479 OGMA P
DIRECTOR: Maj-Gen Adriano de Aldeia Portela
DEPUTY DIRECTOR AND PRODUCTION MANAGER:
 Col A. Nogueira Pinto
COMMERCIAL DIRECTOR: Lt Col M. Renato Oliveira
PUBLIC RELATIONS: Jorge M. F. Pires

Founded 1918; is department of Portuguese Air Force responsible for depot level maintenance and repair of its aircraft, avionics, engines, ground communications and radar equipment, and can undertake similar work for civil or military national or foreign customers; has total covered area of 116,000 m² (1,248,612 sq ft) and approx 2,700 workforce.

Under 1959 contract, undertakes IRAN, refurbishing and rehabilitation, periodic inspection and emergency maintenance and crash repair of US Air Force and US Navy aircraft. Manufactured main and tail rotor structures for French-built SA 315B Lamas and components for other Aerospatiale helicopters; ten Portuguese Air Force AS 330C Pumas upgraded and re-engined with Turbomeca Makila turboshafts in 1991. Manufactures important structural components for Dornier 228.

Engine repair and maintenance facility, with 28,000 m² (301,390 sq ft) covered area, overhauls military and commercial turbojets/turbofans (up to 146.8 kN; 33,000 lb), and turboprops/turboshafts (up to 5,667 kW; 7,600 shp); has two fully computerised test cells; is equipped with plasma spray, two vacuum furnaces, complete cleaning and electroplating facilities, non-destructive testing, shot-peening and other specific equipment; besides work for Portuguese Air Force, also overhauls under contract Artouste III and Turmo IV turboshafts for Turbomeca, and, as maintenance/overhaul centre for Allison, T56 engines and gearboxes for USAF and other customers.

Performs major maintenance on C-130/L-100 Hercules as a Lockheed Service Center, and on Alouette III, Puma and Ecureuil helicopters as a Eurocopter France Station-Service; is also line service centre for Dassault Falcon 20/50 and Garrett TFE731 engines.

Avionics Division has 6,400 m² (68,900 sq ft) premises, equipped to maintain new generation avionics, communications systems, test equipment and calibration laboratories; is licensed by Litton Systems of Canada to carry out level 2 and 2A maintenance on LTN-72 INS equipment.

ROMANIA

IAROM
IAROM SA
GENERAL MANAGER: D. Gozia

New holding company for Romanian aircraft industry, replacing former CNIAR (1991-92 and earlier *Jane's*). Aircraft and aero engine companies remain essentially government owned for time being, though not necessarily government funded (government owns 35 per cent, companies themselves 35 per cent, remaining 30 per cent available for Romanian private ownership through voucher system, though regulations planned to allow for foreign investment and attraction of foreign management skills). Following overthrow of Ceausescu regime, industry continues to operate, but at very reduced rate; is looking for opportunities to work for or with Western companies and projects.

Export sales of Romanian aerospace products handled (though no longer exclusively) by:
Tehnoimportexport SA
2 Doamnei Street (PO Box 110), Bucharest
Telephone: 40 (1) 156724
Telex: 10254 TEHIE R
GENERAL DIRECTOR: Dipl Eng M. Bortes
DEPUTY DIRECTOR (AIRCRAFT DIVISION):
 Dipl Ec G. Popescu
CONTRACT MANAGER: Dipl Ec E. M. Preda
Centrul de Incercari in Zbor SA
1 Aeroportului Street, R-1100 Craiova
Telephone: 40 (941) 24557

Official ground and flight test establishment for military and civil prototypes and upgraded aircraft.

INAv

INSTITUTUL DE AVIATIE SA (Aviation Institute SA) (formerly INCREST)
44A Ficusului Boulevard, R-70544 Bucharest
Telephone: 40 (1) 6137259
Fax: 40 (1) 3655980
Telex: 11142

INAv is one of several institutes formed from former INCREST, whose policy and planning functions were assumed in 1991 by holding company Orcas SA. INCREST designed Romanian portion of IAR-93 and IAR-99 Şoim. INAv also designed AG-6 agricultural biplane (see Aerostar entry) and IAR-705 transport (see under Avioane).

INAv IAR-503A

TYPE: Tandem two-seat turboprop trainer.
PROGRAMME: Being designed 1991, based on P&WC PT6A-25C engine; no other details known.

AEROSTAR

AEROSTAR SA
BACAU Grup Industrial Aeronautic (Aeronautical Industrial Group)
9 Condorilor Street, R-5500 Bacau
Telephone: 40 (93) 130070
Fax: 40 (93) 170513
Telex: 21339 AERO R
GENERAL DIRECTOR: Dipl Eng Ioan Galusca
PRODUCTION MANAGER: Dipl Eng Alexandru Stanciu
TECHNICAL MANAGER: Dipl Eng Simion Ungurasu
COMMERCIAL MANAGER: Dipl Eng Vasile Manciu

Factory founded 1953 as URA (later IRAv, then IAv Bacau), originally as repair centre for Romanian Air Force Yak-17/23, MiG-15/17/19/21 and Il-28/H-5 front-line military aircraft and Aero L-29/L-39 jet trainers; this work still continuing for later of these types; built first Romanian prototype of IAR-93 between 1972 and 1974. Five other specialised work sections: (1) landing gears, hydraulic and pneumatic equipment; (2) special production; (3) light aircraft; (4) engines and reduction gears; (5) avionics. Factories have production buildings floor area totalling 50,396 m² (542,457 sq ft) plus further 31,618 m² (340,333 sq ft) of auxiliary buildings; workforce in early 1992 totalled 8,385, including 1,974 technical personnel.

Landing gears and/or hydraulic/pneumatic equipment have been produced for IAR-316B (Alouette), IAR-330 (Puma) and Ka-126 helicopters; IAR-93, IAR-99 and Rombac 1-11 jet aircraft; and Iak-52 and IAR-822/823/827 and AG-6 light aircraft. Engines include former Soviet M-14P for Iak-52, M-14V26 for Ka-26 and RU-19A-300 APU for An-24; reduction gears include R-26 and VR-126. Avionics factory produces radio altimeters, radio compasses, marker beacon receivers, IFF and other radio/radar items.

AEROSTAR (YAKOVLEV) Iak-52

TYPE: Tandem two-seat piston engined primary trainer (licence version of former Soviet Yak-52).
PROGRAMME: Design in USSR began 1975; series production assigned to IAv Bacau (now Aerostar) under Comecon programme (Council for Mutual Economic Assistance), following Romanian-USSR inter-governmental agreement of 1976; construction began 1977 and first Romanian prototype made first flight May 1978; series production began 1979; first deliveries (to DOSAAF, USSR) 1979; military certification for Romanian Air Force 1985 and first deliveries 1986; production continuing in 1993 but slowed from 150 per year at peak to fewer than 100; new Westernised Condor version now proposed (see Current Versions).
CURRENT VERSIONS: **Iak-52:** Romanian designation of standard Yak-52. *Detailed description applies to this version except where indicated.*

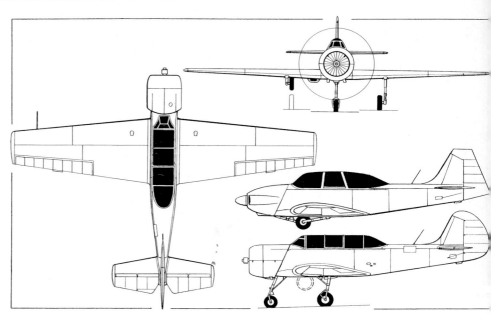

Aerostar (Yakovlev) Iak-52 tandem two-seat primary trainer, with additional side view (centre) of the Condor Lycoming-engined version *(Dennis Punnett)*

Condor: Proposed Westernised version, under development for FAR Pt 23 certification; 224 kW (300 hp) Textron Lycoming AEIO-540 flat-six engine and two-blade Hartzell propeller; redesigned rudder and canopy. Prototype under construction 1992.
CUSTOMERS: More than 1,700 produced by early 1993, mainly for former USSR and Romanian Air Force.
COSTS: $105,000 (1993).
DESIGN FEATURES: Tandem-cockpit variant of Yak-50, with unchanged span and length, but with semi-retractable landing gear to reduce damage in wheels-up landing; straight-tapered wings; 'glasshouse' canopy. Wing section Clark YN with thickness/chord ratios of 14.5 per cent at root and 9 per cent at tip; dihedral 2° from roots; incidence 2° at root; no sweep.
FLYING CONTROLS: Mechanical actuation of mass balanced slotted ailerons (by rods), mass balanced elevators (rods/cables) and horn balanced rudder (cables); manually operated trim tab in port elevator, ground adjustable tab on rudder and each aileron. Ailerons deflect 22° up/16° down, elevators 25° up/down, rudder 27° left/right. Pneumatically actuated trailing-edge split flaps (max deflection 45°).
STRUCTURE: All-metal (D-16 and AK-6 aluminium alloy) stressed skin, with 30 HGSA steel reinforcement in high-stress areas, except for fabric covered primary control surfaces; single-spar wings.
LANDING GEAR: Semi-retractable tricycle type, with single wheel on each unit. Pneumatic actuation, nosewheel retracting rearward, main units forward. All three wheels remain fully exposed to airflow, against undersurface of fuselage and wings respectively, to offer greater safety in event of wheels-up emergency landing. Aerostar oleo-pneumatic shock absorbers. Mainwheel tyres size 500 × 150 mm; nosewheel tyre size 400 × 150 mm. Tyre pressure (all units) 3.0 bars (43 lb/sq in). Pneumatic mainwheel brakes, operated differentially from pedals. Min ground turning radius 6.22 m (20 ft 5 in). Non-retractable plastic coated duralumin skis, with shock struts, can be fitted in place of wheels for Winter operations.
POWER PLANT: One 268 kW (360 hp) Aerostar built VMKB (Vedeneyev) M-14P nine-cylinder air-cooled radial, driving a V-530TA-D35 two-blade constant-speed wooden propeller. Adjustable louvres in front of cowling to regulate cooling. Two-part cowling, split on horizontal centreline. Two aluminium alloy fuel tanks, in wingroots forward of spar, each with capacity of 61 litres (16.1 US gallons; 13.5 Imp gallons). Collector tank in fuselage of 5.5 litres (1.45 US gallons; 1.25 Imp gallons) capacity supplies engine during inverted flight. Total internal fuel capacity 122 litres (32.2 US gallons; 27 Imp gallons). Gravity fuelling point in upper surface of each wing. Oil capacity 20 litres (5.3 US gallons; 4.4 Imp gallons).
ACCOMMODATION: Tandem seats for pupil (at front) and instructor under long 'glasshouse' canopy, with separate rearward sliding hood over each seat. Dual controls standard. Seats and rudder pedals adjustable. Heating and ventilation standard. Optional 0.20 m³ (7.06 cu ft) baggage compartment, accessible from rear seat.
SYSTEMS: No hydraulic system. Independent main and emergency pneumatic systems, pressure 50 bars (725 lb/sq in), for flap and landing gear actuation, engine starting and brake control. Pneumatic systems supplied by 11 litre (main) and 3 litre (emergency) compressed air bottles (671 and 183 cu in), mounted behind rear seat and recharged in flight by an AK-50T engine driven compressor. Electrical system (27V DC) supplied by 3kW engine driven generator and (in port wing) 12V 23Ah ASAM battery; two static inverters in fuselage for 36V AC power at 400Hz. Heated pitot tube and stall speed sensor. Oxygen system available optionally.
AVIONICS: Balkan 5 VHF radio, SPU-9 intercom, ARK-15M automatic radio compass and GMK-1A gyro compass (all of CIS origin).
DIMENSIONS, EXTERNAL:

Wing span	9.30 m (30 ft 6¼ in)
Wing chord: at root	1.997 m (6 ft 6¾ in)
mean	1.64 m (5 ft 4½ in)

Aerostar (Yakovlev) Iak-52 primary trainer of the Romanian Air Force

AEROSTAR—AIRCRAFT: ROMANIA 249

AEROSTAR AG-6

TYPE: Single-seat agricultural biplane.
PROGRAMME: Designed at INCREST (now INAv, which see) in mid-1980s; prototype construction by IAv Bacau (now Aerostar) began 1986; first flight (YR-BGX) 12 January 1989; flight programme for FAR Pt 23 certification continuing. Intended for agricultural and forestry work; production decision awaited following suspension of Ka-126 programme.
DESIGN FEATURES: Braced biplane with near-equal-span wings; single-seat reinforced cockpit; hopper forward of cockpit.
FLYING CONTROLS: Conventional primary control surfaces, all mechanically actuated; ailerons, in synchronised pairs, on upper and lower wings; electrically operated trim tab in starboard elevator, fixed tab on rudder. No flaps.
STRUCTURE: Welded chromoly steel tube fuselage; two-spar wings, fin and tailplane. Duralumin covering on forward fuselage (attached by quick-release fastenings), wings, fin and tailplane; rear fuselage and duralumin framed ailerons/elevators/rudder are fabric covered. N type interplane struts; glassfibre wingtip fairings. Anti-corrosion protection; strengthened cockpit frame.
LANDING GEAR: Non-retractable mainwheels and tailwheel. Semi-cantilever main gear legs of rocker type with oleo-pneumatic shock absorbers. Cleveland 40-90B mainwheels with 30-23A hydraulic brakes and 7.00-6 (6-ply rating) Goodyear tyres. Tost 4.00-4 steerable tailwheel.
POWER PLANT: One 268.5 kW (360 hp) VMKB (Vedeneyev) M-14P nine-cylinder air-cooled radial engine, driving (on prototype) a V-530TA-D35 two-blade variable-pitch propeller; Hoffmann three-blade variable-pitch propeller intended for production aircraft. Cooling louvres in front of engine cowling. Fuel tanks in upper wing centre-section, total usable capacity 200 litres (52.8 US gallons; 44 Imp gallons).
ACCOMMODATION: Single seat in fully enclosed, heated and ventilated cockpit aft of wings. Downward opening door on port side.

Prototype Aerostar Condor development of the Iak-52, showing new rudder shape; modified canopy (see drawing) not fitted in this view

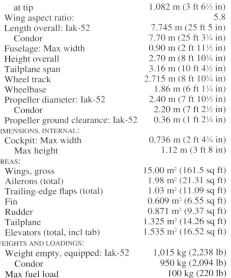

at tip	1.082 m (3 ft 6½ in)
Wing aspect ratio:	5.8
Length overall: Iak-52	7.745 m (25 ft 5 in)
Condor	7.70 m (25 ft 3¼ in)
Fuselage: Max width	0.90 m (2 ft 11½ in)
Height overall	2.70 m (8 ft 10¼ in)
Tailplane span	3.16 m (10 ft 4½ in)
Wheel track	2.715 m (8 ft 10¾ in)
Wheelbase	1.86 m (6 ft 1¼ in)
Propeller diameter: Iak-52	2.40 m (7 ft 10½ in)
Condor	2.20 m (7 ft 2½ in)
Propeller ground clearance: Iak-52	0.36 m (1 ft 2¼ in)

DIMENSIONS, INTERNAL:
Cockpit: Max width	0.736 m (2 ft 4¾ in)
Max height	1.12 m (3 ft 8 in)

AREAS:
Wings, gross	15.00 m² (161.5 sq ft)
Ailerons (total)	1.98 m² (21.31 sq ft)
Trailing-edge flaps (total)	1.03 m² (11.09 sq ft)
Fin	0.609 m² (6.55 sq ft)
Rudder	0.871 m² (9.37 sq ft)
Tailplane	1.325 m² (14.26 sq ft)
Elevators (total, incl tab)	1.535 m² (16.52 sq ft)

WEIGHTS AND LOADINGS:
Weight empty, equipped: Iak-52	1,015 kg (2,238 lb)
Condor	950 kg (2,094 lb)
Max fuel load	100 kg (220 lb)
Max T-O weight: Iak-52	1,305 kg (2,877 lb)
Condor	1,250 kg (2,756 lb)
Max wing loading: Iak-52	87.0 kg/m² (17.82 lb/sq ft)
Condor	83.33 kg/m² (17.07 lb/sq ft)
Max power loading: Iak-52	4.86 kg/kW (7.99 lb/hp)
Condor	5.58 kg/kW (9.18 lb/hp)

PERFORMANCE (Iak-52 at max T-O weight):
Never-exceed speed (V_NE)	194 knots (360 km/h; 223 mph)
Max level speed: at S/L	154 knots (285 km/h; 177 mph)
at 1,000 m (3,280 ft)	145 knots (270 km/h; 167 mph)
Econ cruising speed at 1,000 m (3,280 ft)	103 knots (190 km/h; 118 mph)
Landing speed	60 knots (110 km/h; 69 mph)
Stalling speed, flaps down, engine idling	46-49 knots (85-90 km/h; 53-56 mph)
Max rate of climb at S/L	420 m (1,378 ft)/min
Service ceiling	4,000 m (13,125 ft)
Time to 4,000 m (13,125 ft)	15 min
T-O run	170 m (558 ft)
T-O to 15 m (50 ft)	200 m (656 ft)
Landing from 15 m (50 ft)	350 m (1,149 ft)
Landing run	300 m (985 ft)
Range at 500 m (1,640 ft), max fuel, 20 min reserves	296 nm (550 km; 341 miles)
g limits	+7/−5

PERFORMANCE (Condor, estimated):
Max level speed	189 knots (350 km/h; 217 mph)
Max rate of climb at S/L	600 m (1,970 ft)/min
Service ceiling	as Iak-52
T-O and landing run	as Iak-52
Range with max fuel	647 nm (1,200 km; 745 miles)

Prototype Aerostar AG-6 single-seat agricultural biplane

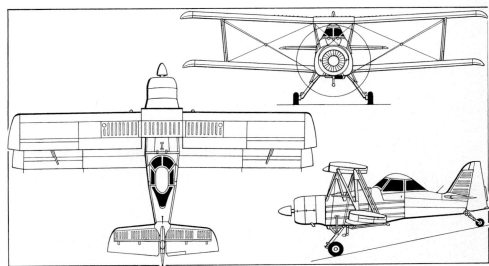

Aerostar AG-6 (VMKB/Vedeneyev M-14P engine) *(Jane's/Mike Keep)*

250 ROMANIA: AIRCRAFT—AEROSTAR/AVIOANE

SYSTEMS: Hydraulic system for brakes only. Pneumatic system for engine starting and driving agricultural equipment. Electrical system (28V DC) supplied by engine driven alternator and silver-zinc battery.
AVIONICS: Bendix/King KY 92 UHF com radio.
EQUIPMENT: Glassfibre hopper for dry or liquid chemical (capacity 1,050 litres; 277.4 US gallons; 231 Imp gallons) forward of cockpit. Dispersal gear under lower wings (for liquid) or under fuselage (for dry chemical).

DIMENSIONS, EXTERNAL:
Wing span: upper	10.56 m (34 ft 7¾ in)
lower	10.26 m (33 ft 8 in)
Wing chord (mean aerodynamic, each)	1.30 m (4 ft 3¼ in)
Length overall	7.45 m (24 ft 5¼ in)
Height overall	3.415 m (11 ft 2½ in)
Tailplane span	3.74 m (12 ft 3¼ in)
Wheel track	2.502 m (8 ft 2½ in)
Wheelbase	4.988 m (16 ft 4½ in)
Propeller diameter: Hoffmann	2.40 m (7 ft 10½ in)

DIMENSIONS, INTERNAL:
Hopper volume	1.05 m³ (37.08 cu ft)

AREAS:
Wings, gross	26.00 m² (279.86 sq ft)
Vertical tail surfaces (total)	1.56 m² (16.79 sq ft)
Horizontal tail surfaces (total)	3.54 m² (38.10 sq ft)

WEIGHTS AND LOADINGS:
Weight empty, equipped	1,020 kg (2,249 lb)
Max T-O weight: FAR Pt 23	1,500 kg (3,307 lb)
Restricted category	1,900 kg (4,189 lb)
Max wing loading: FAR Pt 23	57.7 kg/m² (11.82 lb/sq ft)
Restricted category	73.1 kg/m² (14.97 lb/sq ft)
Max power loading: FAR Pt 23	5.59 kg/kW (9.18 lb/hp)
Restricted category	7.08 kg/kW (11.63 lb/hp)

PERFORMANCE (at max FAR Pt 23 T-O weight, spraying configuration, except where indicated):
Never-exceed speed (VNE)	118 knots (220 km/h; 136 mph)
Max level speed	102 knots (190 km/h; 118 mph)
Cruising speed	94 knots (175 km/h; 109 mph)
Working speed	81-94 knots (150-175 km/h; 93-109 mph)
Approach speed	69 knots (127.5 km/h; 79 mph)
Stalling speed, power off	60 knots (110 km/h; 69 mph)
Max rate of climb at S/L	210 m (690 ft)/min
Service ceiling: FAR Pt 23	4,000 m (13,125 ft)
Restricted category	3,400 m (11,155 ft)
T-O run	230 m (755 ft)
Landing run	200 m (657 ft)
Range with max fuel, 30 min reserves	280 nm (520 km; 323 miles)

AVIOANE

AVIOANE SA (formerly IAv Craiova)
1 Aeroportului Street, R-1100 Craiova
Telephone: 40 (941) 24170
Fax: 40 (941) 24382
Telex: 41290 COCOR R
MANAGING DIRECTOR: Dipl Eng Grigore Leoveanu

Founded 1 February 1972 as IAv Craiova, changing to present name 29 March 1991; a major domestic and international aviation enterprise, producing wide range of products and services for both military and civil aviation. Activities include aircraft and equipment design and manufacture, repair and overhaul, life cycle management and integrated logistics support. Principal military programmes are IAR-93 and IAR-99; has now begun construction of IAR-705 twin-turboprop regional transport.

IAR-93

For details see under Soko/Avioane entry in International section.

AVIOANE IAR-99 ŞOIM (HAWK)

TYPE: Advanced jet trainer and light ground attack aircraft.
PROGRAMME: Existence revealed at 1983 Paris Air Show; designed by INAv (Institutul de Aviatie) at Bucharest; three prototypes built at Craiova, incl one for structural testing; first flight 21 December 1985.
CURRENT VERSIONS: **IAR-99**: Standard version for Romanian Air Force; *detailed description applies to this version.*
IAR-109 Swift: Upgraded version; described separately.
CUSTOMERS: Initial 20 delivered to Romanian Air Force from 1987; further 30 reportedly on order, of which six delivered by mid-1991.
DESIGN FEATURES: Tandem cockpits (rear seat elevated); straight wings. Wing section NACA 64$_1$A-214 (modified) at centreline; 64$_1$A-212 (modified) at tip; dihedral 3° from roots; quarter-chord sweepback 6° 35′; incidence 1° at root.
FLYING CONTROLS: Statically balanced ailerons hydraulically actuated, with manual reversion; horn balanced elevators and statically balanced rudder actuated mechanically by push/pull rods; servo tab in port aileron, trim tabs in rudder and each elevator, all operated electrically; ailerons deflect 15° up/15° down, elevators 20° up/10° down, rudder 25° to left and right. Hydraulically actuated single-slotted flaps, deflecting 20° for T-O and 40° for landing, retract gradually when airspeed reaches 162 knots (300 km/h; 186 mph); twin hydraulically actuated airbrakes under rear fuselage.
STRUCTURE: All-metal; aluminium honeycomb ailerons/elevators/rudder; semi-monocoque fuselage includes honeycomb panels for fuel tank compartments; machined wing skin panels form integral fuel tanks.
LANDING GEAR: Retractable tricycle type, with single wheel and oleo-pneumatic shock absorber on each unit. Mainwheels retract inward, non-steerable nosewheel forward, all being fully enclosed by doors when retracted. Landing light in port wingroot leading-edge. Mainwheels fitted with tubeless tyres, size 552 × 164-10, pressure 7.5 bars (108.8 lb/sq in), and hydraulic disc brakes with anti-skid system. Nosewheel has tubeless tyre size 445 × 150-6, pressure 4.0 bars (58.0 lb/sq in).
POWER PLANT: One Turbomecanica Romanian-built Rolls-Royce Viper Mk 632-41M turbojet, rated at 17.79 kN (4,000 lb st). Fuel in two flexible bag tanks in centre-fuselage, capacity 900 litres (238 US gallons; 198 Imp gallons), and four integral tanks between wing spars, combined capacity 470 litres (124 US gallons; 103 Imp gallons). Total internal fuel capacity 1,370 litres (362 US gallons; 301 Imp gallons). Gravity refuelling point on top of fuselage. Provision for two drop tanks, each of 225 litres (59.5 US gallons; 49.5 Imp gallons) capacity, on inboard underwing stations. Max internal/external fuel capacity 1,820 litres (481 US gallons; 400 Imp gallons).
ACCOMMODATION: Crew of two in tandem, on zero/zero ejection seats in pressurised and air-conditioned cockpit. Rear seat elevated 35 cm (13.8 in). Dual controls standard. One-piece canopy with internal screen (trainer), or (in ground attack version) individual canopies, all opening sideways to starboard.
SYSTEMS: Engine compressor bleed air for pressurisation, air-conditioning, anti-g suit and windscreen anti-icing system, and to pressurise fuel tanks. Hydraulic system, operating at pressure of 206 bars (2,990 lb/sq in), for actuation of landing gear and doors, flaps, airbrakes, ailerons and main wheel brakes. Emergency hydraulic system for operation of landing gear, doors, flaps and wheel brakes. Main electrical system, supplied by 9kW 28V DC starter/generator with 28.5V 36Ah nickel-cadmium battery, ensures operation of main systems, in case of emergency, and engine starting. Two 750VA static inverters supply two secondary AC networks: 115V/400Hz and 26V/400Hz. Oxygen system for two crew for 2 h 30 min.
AVIONICS: Standard avionics include VHF/UHF com radio, intercom, radio altimeter, marker beacon receiver, SRR-2 IFF, ADF, gyro platform and flight recorder.
ARMAMENT: Removable ventral gun pod containing 23 mm GSh-23 gun with 200 rds. Gun/rocket firing and weapon release controls, including electrically controlled AA-1F gyroscopic gunsight and AFCT-1 gun camera, in front cockpit only. Four underwing hardpoints stressed for loads of 250 kg (551 lb) each. Typical underwing stores can include four 250 kg bombs; four triple carriers each for three 50 kg bombs (or two 100 kg and one 50 kg); four L 16-57 launchers each containing sixteen 57 mm air-to-surface rockets; four L 32-42 launchers each containing thirty-two 42 mm air-to-surface rockets; infra-red air-to-air missiles (inner pylons only); two twin-7.62 mm machine-gun pods with 800 rds/pod (inboard pylons only); and auxiliary fuel tanks (see under Power Plant) on inboard pylons.

DIMENSIONS, EXTERNAL:
Wing span	9.85 m (32 ft 3¾ in)
Wing chord: at root	2.305 m (7 ft 6¾ in)
mean	1.963 m (6 ft 5¼ in)
at tip	1.30 m (4 ft 3¼ in)
Wing aspect ratio	5.19
Length overall	11.009 m (36 ft 1½ in)
Height overall	3.898 m (12 ft 9½ in)
Elevator span	4.12 m (13 ft 6¼ in)
Wheel track	2.686 m (8 ft 9¾ in)
Wheelbase	4.378 m (14 ft 4½ in)

AREAS:
Wings, gross	18.71 m² (201.4 sq ft)
Ailerons (total)	1.56 m² (16.79 sq ft)
Flaps (total)	2.54 m² (27.34 sq ft)
Fin, incl dorsal fin	1.919 m² (20.66 sq ft)
Rudder	0.629 m² (6.77 sq ft)
Tailplane	3.123 m² (33.62 sq ft)
Elevators (total)	1.248 m² (13.43 sq ft)

WEIGHTS AND LOADINGS:
Weight empty, equipped	3,200 kg (7,055 lb)
Max fuel weight: internal	1,100 kg (2,425 lb)
external	350 kg (772 lb)

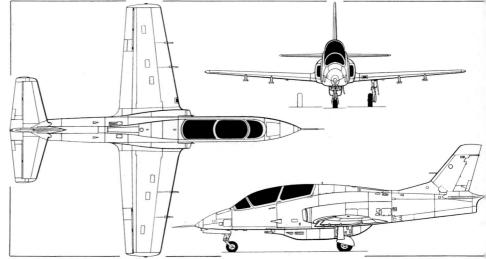

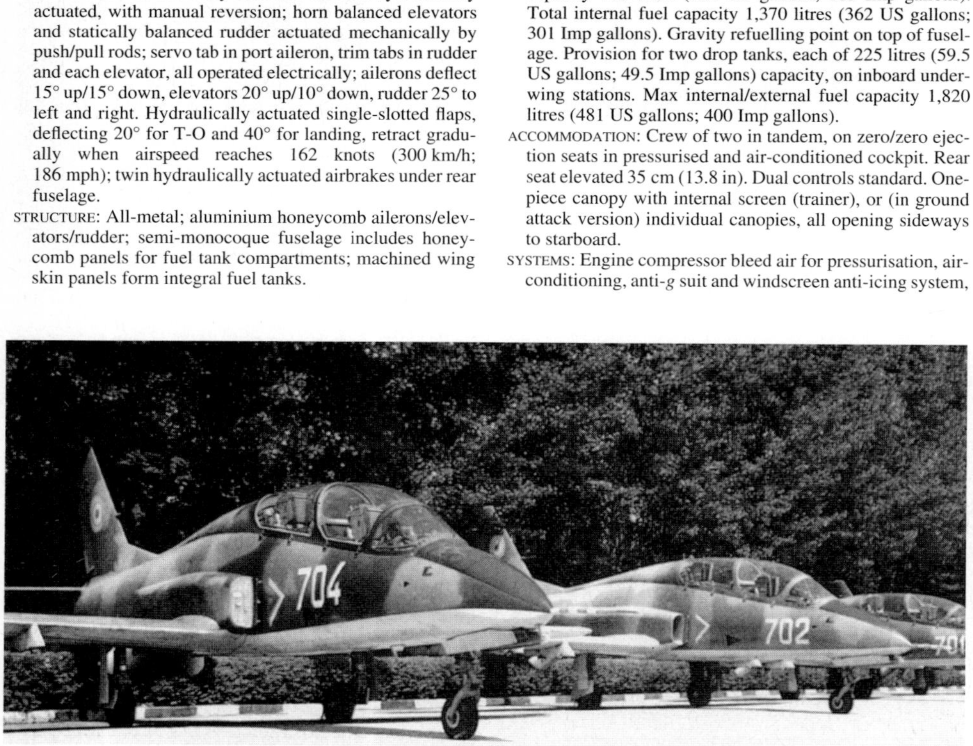

Romania's advanced jet trainer, the Avioane IAR-99 Şoim *(Dennis Punnett)*

Avioane IAR-99 Şoim tandem-seat jet trainers of the Romanian Air Force

Max T-O weight: trainer	4,400 kg (9,700 lb)
ground attack	5,560 kg (12,258 lb)
Max wing loading: trainer	235.2 kg/m² (48.17 lb/sq ft)
ground attack	297.2 kg/m² (60.86 lb/sq ft)
Max power loading: trainer	247.5 kg/kN (2.42 lb/lb st)
ground attack	312.7 kg/kN (3.06 lb/lb st)

PERFORMANCE:
Max Mach number	0.76
Max level speed at S/L: trainer	467 knots (865 km/h; 537 mph)
Max rate of climb at S/L	2,100 m (6,890 ft)/min
Service ceiling	12,900 m (42,325 ft)
Min air turning radius	330 m (1,083 ft)
T-O run: trainer	450 m (1,477 ft)
ground attack	960 m (3,150 ft)
T-O to 15 m (50 ft): trainer	750 m (2,461 ft)
ground attack	1,350 m (4,430 ft)
Landing from 15 m (50 ft): trainer	740 m (2,428 ft)
ground attack	870 m (2,855 ft)
Landing run: trainer	550 m (1,805 ft)
ground attack	600 m (1,969 ft)

Typical combat radius (1 pilot, ventral gun, internal fuel only):
lo-lo-hi, four 16-round rocket pods, AUW 5,000 kg (11,023 lb)	189 nm (350 km; 217 miles)
hi-lo-hi, two 16-round rocket pods, two 50 kg and four 100 kg bombs, AUW 5,280 kg (11,640 lb)	186 nm (345 km; 214 miles)
hi-hi-hi, four 250 kg bombs, AUW 5,480 kg (12,081 lb)	208 nm (385 km; 239 miles)

Max range with internal fuel:
trainer	593 nm (1,100 km; 683 miles)
ground attack	522 nm (967 km; 601 miles)
Max endurance with internal fuel: trainer	2 h 40 min
ground attack	1 h 46 min
g limits	+7/−3.6

AVIOANE IAR-109 SWIFT

TYPE: All-through jet trainer and close support aircraft (upgraded IAR-99).

PROGRAMME: Announced Autumn 1992, at which time development aircraft (modified IAR-99, serial number 712) under test.

CURRENT VERSIONS: **IAR-109T**: Tandem-seat trainer.
IAR-109TF: Combat trainer and close support/COIN version.

DESIGN FEATURES: Same airframe and power plant as IAR-99; upgraded and expanded avionics and enhanced weapons delivery capability.

ACCOMMODATION: Martin-Baker Mk 10L lightweight zero/zero ejection seats.

AVIONICS: Multiplex 1553B integrated databus system in IAR-109TF with high capacity mission display processor, HUD, up-front control panel, cockpit TV centre and video tape recorder, ring laser gyro INS, digital ADC, radio altimeter, EFIS (ADI and HSI), laser rangefinder, HOTAS, IFF transponder, and customised communications and radio navigation equipment. Full ILS package in both versions.

ARMAMENT: Underwing pylons adaptable for Eastern or Western weapons, of types generally similar to those listed for IAR-99, plus capability for infra-red air-to-air missiles, precision guided munitions and larger (300 litre; 79 US gallon; 66 Imp gallon) drop tanks.

WEIGHTS AND LOADINGS: As for IAR-99 except:
Max T-O weight: trainer	4,800 kg (10,582 lb)

PERFORMANCE: As for IAR-99 except:
Unstick speed: T	108 knots (200 km/h; 124 mph)
Approach speed: T	127 knots (235 km/h; 146 mph)
Landing speed: T	103 knots (190 km/h; 118 mph)
Stalling speed: T	93 knots (172 km/h; 107 mph)
Time to 6,100 m (20,000 ft): T	5 min
T-O run: T	500 m (1,640 ft)
Landing from 15 m (50 ft): T	900 m (2,953 ft)
Landing run: T	600 m (1,969 ft)

Typical combat radius with c/l gun, two rocket pods, two 225 litre drop tanks and max internal fuel, 10% reserves:
TF, lo-lo-lo	151 nm (280 km; 174 miles)
TF, hi-lo-hi	178 nm (330 km; 205 miles)

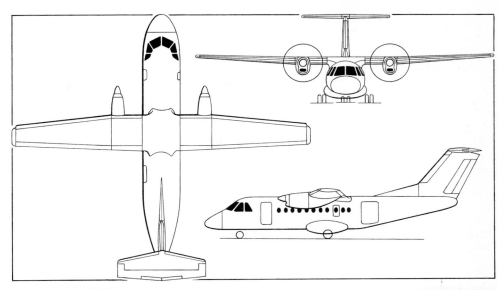

Preliminary drawing (*Jane's/Mike Keep*) and model of the IAR-705-30 twin-turboprop regional transport

AVIOANE IAR-705

TYPE: Project for twin-turboprop regional transport.

PROGRAMME: Design and feasibility studies, and fuselage mockup, completed by INAv by mid-1991. Initial production expected for Romanian Air Force to replace about 50 An-24s; Avioane stated to have started development by Autumn 1992.

CURRENT VERSIONS: **IAR-705-30**: Baseline version, to seat 30 passengers at 76 cm (30 in) pitch; PW118 engines.
IAR-705-50: 50-passenger derivative, with stretched fuselage and engines farther out on increased-span wing; PW123 engines for increased power.

DESIGN FEATURES: High-wing, T tail, cargo door, pressurised fuselage, wing de-icing. Conforms to FAR Pt 25.

FLYING CONTROLS: Conventional three-axis, plus wing flaps and spoilers.

LANDING GEAR: Fully retractable tricycle type, with twin wheels on each unit. Hydraulic actuation, mainwheels retracting into fuselage-side fairings; steerable nose unit. Hydraulic mainwheel brakes.

POWER PLANT: Two 1,342 kW (1,800 shp) Pratt & Whitney Canada PW118 turboprops in 705-30; 1,775 kW (2,380 shp) PW123s in 705-50. Hamilton Standard 14RF propellers in -30, 14SF in -50. See under Weights and Loadings for fuel details.

ACCOMMODATION: Two-person flight crew, plus optional third seat for observer. Four-abreast seating, with central aisle, for 30 or 50 passengers according to model. Passenger door at front on port side, with integral steps; service door opposite this on starboard side. Emergency exit under wing on each side. Large baggage/cargo door at rear on port side. Entire accommodation pressurised and air-conditioned.

SYSTEMS: Environmental control system, using engine bleed air, to provide cabin pressure equivalent to 2,400 m (7,875 ft) up to operating altitude of 7,620 m (25,000 ft). Hydraulic main system, pressure 207 bars (3,000 lb/sq in), for flaps, spoilers, landing gear extension/retraction, nosewheel steering and mainwheel braking; emergency backup system for gear and flap extension and mainwheel brakes. Electrical power supplied by two 28V 400A DC starter/generators, with static inverters for single-phase AC (115V

Development prototype for the IAR-109 Swift

and 26V) at 400Hz. Two 15/22kVA AC generators for propeller, spinner and windscreen de-icing; two 24V 45Ah batteries for engine starting, ground support equipment and emergency use. Pneumatic boot de-icing of wing leading-edges.

AVIONICS: Standard FAR 121 avionics to include VHF com radio, VOR/ILS/marker beacon receiver, ADF, radio altimeter, DME, ATC transponder, weather radar, air data computer, crew interphone, CVR, FDR and central warning system.

DIMENSIONS, EXTERNAL:
Wing span: 705-30	22.40 m (73 ft 6 in)
705-50	26.40 m (86 ft 7½ in)
Length overall: 705-30	20.76 m (68 ft 1¼ in)
705-50	24.76 m (81 ft 2¾ in)
Height overall: 705-30, 705-50	6.55 m (21 ft 6 in)
Tailplane span: 705-30, 705-50	6.40 m (21 ft 0 in)
Wheel track (c/l of main units):	
705-30, 705-50	3.80 m (12 ft 5½ in)

AREAS:
Wings, gross: 705-30	47.7 m² (513.5 sq ft)
705-50	58.0 m² (625.0 sq ft)

WEIGHTS AND LOADINGS:
Operating weight empty: 705-30	8,100 kg (17,857 lb)
705-50	11,000 kg (24,251 lb)
Max fuel load: 705-30	2,400 kg (5,291 lb)
705-50	4,000 kg (8,818 lb)
Standard fuel load: 705-30	1,600 kg (3,527 lb)
705-50	2,250 kg (4,960 lb)
Max payload: 705-30	2,800 kg (6,173 lb)
705-50	4,750 kg (10,472 lb)
Max T-O weight: 705-30	12,500 kg (27,558 lb)
705-50	18,000 kg (39,683 lb)
Max landing weight: 705-30	12,300 kg (27,117 lb)
705-50	17,800 kg (39,242 lb)
Max zero-fuel weight: 705-30	10,900 kg (24,030 lb)
705-50	15,750 kg (34,723 lb)
Max wing loading: 705-30	262 kg/m² (53.7 lb/sq ft)
705-50	310 kg/m² (63.6 lb/sq ft)
Max power loading: 705-30	4.66 kg/kW (7.65 lb/shp)
705-50	5.07 kg/kW (8.34 lb/shp)

PERFORMANCE (estimated):
Max cruising speed at 4,500 m (14,760 ft):
705-30	253 knots (470 km/h; 292 mph)
705-50	270 knots (500 km/h; 310 mph)

Max range, ISA, reserves for 200 nm (370 km; 230 mile diversion and 45 min hold:
705-30, 705-50	647 nm (1,200 km; 745 miles)

IAR

IAR SA (formerly ICA)
1 Aeroportului Street, PO Box 198, R-2200 Brasov
Telephone: 40 (921) 50014
Fax: 40 (921) 51304
Telex: 61266 ICAER R
MANAGING DIRECTOR: Dipl Eng N. Banea

Factory, created 1968, continues work begun in 1926 by IAR-Brasov and undertaken 1950-59 as URMV-3 Brasov; occupies 1.6 ha (3.95 acre) site and had 1993 workforce of 3,000. Currently manufactures Romanian designed light aircraft and (under licence from Eurocopter France) Puma helicopters; IS-28/29 series sailplanes/motor gliders (see 1992-93 and earlier *Jane's*); aircraft components and equipment.

IAR erroneously reported in 1992 to be continuing production of IAR-823 (1988-89 *Jane's*) and IAR-316B (licence Aerospatiale Alouette III). Facts are that IAR-823 production ended 1983 (80 built: 68 for domestic customers and 12 for export); IAR-316B production totalled 200 (108 military, 82 civil), of which 125 domestic and 75 exported, and ceased in 1987.

IAR-46

TYPE: Two-seat Utility category very light aircraft.
PROGRAMME: Definition phase and marketing studies started early 1991; detail design began late 1991; development phase initiated mid-1992; first flight scheduled for May 1993; certification to JAR VLA expected December 1993; deliveries, depending on market response, could begin second half 1994.
DESIGN FEATURES: High aspect ratio low-wing monoplane with raked tips and T tail. Dihedral 2° from flat centre-section; incidence 4° at root; no twist.
FLYING CONTROLS: Conventional mechanical; trim tab in each elevator. Plain trailing-edge flaps.
STRUCTURE: All-metal except fabric covering on elevators and rudder and GFRP for non-stressed fairings.
LANDING GEAR: Semi-retractable single mainwheels with rubber in compression shock absorption and toe operated brakes; non-retractable tailwheel.
POWER PLANT: One 59 kW (79 hp) Rotax 912A flat-four engine, driving a Mühlbauer MTV-1. A/170-08 two-blade variable-pitch propeller with composite blades. Fuel in single tank in fuselage, capacity 78 litres (20.6 US gallons; 17.2 Imp gallons).
ACCOMMODATION: Two adjustable seats side by side; dual controls standard. Fixed windscreen and rearward sliding canopy. Baggage compartment aft of seats. Cockpit ventilated; heating optional.
AVIONICS: Standard VFR instrumentation to JAR 22.1303 and JAR 22.1305. Options include anti-collision and position lights, horizon and directional gyros, and turn co-ordinator.

DIMENSIONS, EXTERNAL:
Wing span	11.42 m (37 ft 5½ in)
Wing chord: at root	1.40 m (4 ft 7 in)
at tip	0.929 m (3 ft 0½ in)
Wing aspect ratio	9.40
Length overall	7.85 m (25 ft 9 in)
Height overall	2.15 m (7 ft 0½ in)
Propeller diameter	1.70 m (5 ft 7 in)

DIMENSIONS, INTERNAL:
Cockpit: Length	1.54 m (5 ft 0½ in)
Max width	1.04 m (3 ft 5 in)
Max height (from seat cushion)	0.85 m (2 ft 9½ in)

AREAS:
Wings, gross	13.87 m² (149.30 sq ft)
Ailerons (total)	0.82 m² (8.83 sq ft)
Trailing-edge flaps (total)	1.36 m² (14.64 sq ft)
Vertical tail surfaces (total)	1.50 m² (16.15 sq ft)
Horizontal tail surfaces (total)	2.73 m² (29.39 sq ft)

WEIGHTS AND LOADINGS:
Weight empty	500-530 kg (1,102-1,168 lb)
Max T-O weight	750 kg (1,653 lb)
Max wing loading	54.07 kg/m² (11.08 lb/sq ft)
Max power loading	12.74 kg/kW (20.93 lb/hp)

PERFORMANCE (estimated, at max T-O weight at S/L, ISA):
Never-exceed speed (V$_{NE}$)
151 knots (280 km/h; 174 mph)

Max level speed	110 knots (204 km/h; 127 mph)
Max cruising speed	97 knots (180 km/h; 112 mph)
Econ cruising speed	89 knots (165 km/h; 102 mph)
Speed for best rate of climb	54 knots (100 km/h; 62 mph)
Stalling speed, flaps up	43 knots (78 km/h; 49 mph)
Max rate of climb	604 m (1,982 ft)/min
Service ceiling	5,000 m (16,400 ft)
T-O run	235 m (771 ft)
T-O to 15 m (50 ft)	450 m (1,477 ft)
Landing run	100 m (328 ft)
Max range, no reserves	459 nm (850 km; 528 miles)

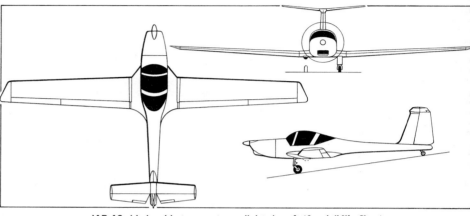

IAR-46 side by side two-seat very light aircraft (*Jane's/Mike Keep*)

IAR IAR-828

TYPE: Single-seat agricultural aircraft.
PROGRAMME: Originally known as IAR-827TP (IAR-827 prototype YR-MGA refitted with P&WC PT6A-15AG turboprop); first flight 7 September 1981; testing towards certification continuing 1991, following unspecified modifications, but programme subsequently cancelled. Description and illustration in 1992-93 *Jane's*.

IAR (EUROCOPTER FRANCE) IAR-330L PUMA

TYPE: Twin-turboshaft medium transport helicopter.
PROGRAMME: Agreement for licence production of AS 330 Puma in Romania concluded 1977; 100 initially covered, since expanded; current production rate about 10 a year; Romania sole producer of Puma; see IAR entry in 1991-92 *Jane's* for details of earlier French and other production.

CURRENT VERSIONS: **Civil:** For oil platform, police security and patrol, casevac, aerial photography, drug interdiction, power/pipeline patrol, surveillance and other duties.
Coastguard: Customs & Excise patrol, search and rescue, sea survey and emergency missions.
Military: Artillery observation, troop transport, medevac, cargo/sling transport, border security, anti-armour attack and air-to-air defence.
Naval: Anti-submarine, air-to-surface attack, reconnaissance, personnel transport and air ambulance.
VIP/Executive: 8/15-seat VIP transport or airborne command post.
Puma 2000: Upgraded version; described separately.

CUSTOMERS: 165 built by mid-1991; most for Romanian Air Force but 60-70 exported to Pakistan, South Africa (50), Sudan and others. No updated total received for 1993, but IAR reports Romanian Air Force order for six and export order (customer not named) for 10 second-hand aircraft plus option for 10 new-built Pumas.

DESIGN FEATURES: Four-blade main rotor, with fully articulated hub and integral hydraulically actuated rotor brake (on main gearbox, stopping rotor 15 s after engine shutdown); blade cuffs, with horns, connected by link-rods to swashplate articulated by three hydraulic twin-cylinder servo-control units; each blade attached to sleeve by two quick-disconnect pins for rapid folding. Starboard five-blade tail rotor with flapping hinges. Mechanical shaft and gear drive; five-stage main gearbox, atop cabin behind engines, has two separate inputs from engines; first stage drives (from each engine) intermediate shaft directly driving alternator and ventilation fan, and indirectly two hydraulic pumps; at second stage, action of both units

IAR-330L Puma in coastguard configuration, with emergency flotation bags (*Peter J. Cooper*)

synchronises on single main driveshaft by freewheeling spur gears — if one or both engines stops, gears rotated by remaining turbine or autorotating rotor, maintaining drive to ancillary systems. Drive to tail rotor via shafting and intermediate angle gearbox, terminating at right-angle tail rotor gearbox. Turbine output 23,000 rpm, main rotor shaft 265 rpm; tail rotor shaft 1,278 rpm. Engine installation outside main fuselage shell; port horizontal stabiliser at tail.

FLYING CONTROLS: Fully powered dual hydraulic with full-time yaw damper operating series hydraulic servo in control runs; autopilot, with provision for coupling to self-contained navigation and microwave landing systems (see Avionics).

TRUCTURE: Moulded main blades have GFRP roving spar, GFRP/CFRP fabric skins and Moltoprene/honeycomb filler, with stainless steel covered leading-edges; titanium leading-edge shielding of heating mat if optional blade de-icing fitted on either rotor. Tail rotor blades metal. All-metal airframe structure; local use of titanium alloy under engines.

ANDING GEAR: Messier-Bugatti semi-retractable tricycle type, with twin wheels on each unit. Main units retract upward hydraulically into fairings on sides of fuselage; self-centring nose unit retracts rearward. When landing gear is down, nosewheel jack is extended and mainwheel jacks are telescoped. Dual-chamber oleo-pneumatic shock absorbers. All tyres same size (7.00-6), of tubeless type, pressure 6.0 bars (85 lb/sq in) on all units. Hydraulic differential disc brakes, controlled by foot pedals. Lever-operated parking brake. Emergency pop-out flotation units can be mounted on rear landing gear fairings and forward fuselage.

OWER PLANT: Two Turbomecanica Romanian-built Turbomeca Turmo IVCA turboshafts, each rated at 1,184 kW (1,588 shp) (S/L, ISA) for max emergency power, 1,114 kW (1,494 shp) for T-O, 1,029 kW (1,380 shp) intermediate emergency power and 941 kW (1,262 shp) max continuous. Intake anti-icing. Engines mounted side by side above cabin forward of main rotor and separated by firewall. They are coupled to main rotor transmission box, with shaft drive to tail rotor, and form a completely independent system from fuel tanks up to main gearbox inputs. Fuel in four flexible tanks and one auxiliary tank beneath cargo compartment floor, total capacity 1,544 litres (408 US gallons; 339.5 Imp gallons). External auxiliary tanks (two, each 350 litres; 92.5 US gallons; 77 Imp gallons capacity) available. For long-range missions (mainly off-shore) one or two special internal tanks (each 215 litres; 56.8 US gallons; 47.25 Imp gallons) can be fitted in cabin. Each engine supplied normally by two interconnected primary tanks, lower halves of which have self-sealing walls for protection against small-calibre projectiles. Refuelling point on starboard side of main cabin. Oil capacity 22 litres (5.8 US gallons; 4.8 Imp gallons) for engines, 25.5 litres (6.7 US gallons; 5.6 Imp gallons) for transmission.

ACCOMMODATION: Crew of one (VFR) or two (IFR) side by side on anti-crash seats on flight deck, with jump-seat for third crew member if required. Door on each side of flight deck. Internal doorway connects flight deck to cabin, with folding seat in doorway for extra crew member or cargo supervisor. Dual controls standard. Accommodation in main cabin for 16 individually equipped troops, six stretchers and seven seated patients, or equivalent freight; number of troops can be increased to 20 in high-density version. Alternative 17/20-seat civil passenger or 8/15-seat executive interiors (incl toilet) available. Strengthened floor for cargo-carrying, with lashing points. Jettisonable sliding door on each side of main cabin. Removable panel on underside of fuselage, at rear of main cabin, permits longer loads to be accommodated and also serves as emergency exit. Hatch in floor below centreline of main rotor for carrying loads of up to 3,200 kg (7,055 lb) on internally mounted cargo sling. Fixed or retractable rescue hoist (capacity 275 kg; 606 lb) can be mounted externally on starboard side of fuselage. Cabin and flight deck heated, ventilated and soundproofed. Demisting, de-icing, washers and wipers for pilots' windscreens. Air-conditioning system optional.

SYSTEMS: Two independent hydraulic systems, each 172 bars (2,500 lb/sq in), supplied by self-regulating pumps driven by main gearbox. Each system supplies one set of servo unit chambers, left hand system also supplying autopilot, landing gear, rotor brake and wheel brakes. Freewheels in main gearbox ensure that both systems remain in operation, for supplying servo-controls, if engines are stopped in flight. Other hydraulically actuated systems can be operated on ground from main gearbox, or through ground power receptacle. Independent auxiliary system, fed through handpump, can be used in emergency to lower landing gear and pressurise accumulator for parking brake on ground. Three-phase 200V AC electrical power supplied by 20kVA 400Hz alternator, driven by port side intermediate shaft from main gearbox and available on ground under same conditions as hydraulic ancillary systems. Second 20kVA alternator optional. 28.5V 10kW DC power provided from AC system by two transformer-rectifiers. Main aircraft battery used for self-starting and emergency power in flight. For latter purpose, emergency 400VA inverter can supply essential navigation equipment from battery, permitting at least 20 min continued flight in event of main power failure. Optional electric de-icing of main and tail rotor blades, with heating mat protected by titanium shielding on leading-edge of each blade. De-icing of engines and engine air intakes by warm air bled from compressor. Anti-snow shield for Winter operations.

AVIONICS: Optional VHF, UHF, tactical HF and HF/SSB com radio and intercom, radio compass, radio altimeter, GPS, VLF Omega, Decca navigator and flight log, Doppler, and VOR/ILS with glidepath. Autopilot, with provision for coupling to self-contained navigation and microwave landing systems. Full IFR instrumentation optional. Search and rescue version has nose-mounted Bendix/King RDR 1400 or Honeywell Primus 40 or 50 search radar, Doppler, and Decca self-contained navigation system, including navigation computer, polar indicator, roller-map display, hover indicator, route mileage indicator and ground speed and drift indicator. Roof-mounted sight for missiles, laser/radar warning system, chaff/flare dispensers and smoke launchers optional in armed version.

ARMAMENT: Armed version equipped with two forward firing 23 mm guns in streamline pods attached to lower sides of fuselage at front; steel tube carriers attached to sides of main cabin can carry two or four unguided air-to-ground rocket pods (16 or 32 × 57 mm or 2 × 122 mm), plus two wire-guided or four laser-guided anti-tank missiles, or two or four infra-red air-to-air missiles, anti-infantry mines or anti-submarine mines; 12.7 mm machine-gun pintle mounted in each cabin doorway. Alternative loads on cabin outriggers can include two or four 7.62 mm GMP-2 machine-gun pods (550 rds/pod) or four 100 kg bombs.

DIMENSIONS, EXTERNAL:

Main rotor diameter	15.08 m (49 ft 5¾ in)
Tail rotor diameter	3.04 m (9 ft 11½ in)
Distance between rotor centres	9.20 m (30 ft 2¼ in)
Main rotor blade chord	0.60 m (1 ft 11½ in)
Tail rotor ground clearance	2.10 m (6 ft 10¾ in)
Length: overall, both rotors turning	18.22 m (59 ft 9¼ in)
fuselage	14.06 m (46 ft 1½ in)
main blades folded, tail rotor turning	14.82 m (48 ft 7½ in)
Height:	
overall, tail rotor turning	5.14 m (16 ft 10½ in)
to top of rotor head	4.54 m (14 ft 10¾ in)
Width: blades folded	3.62 m (11 ft 10½ in)
over wheel fairings	3.00 m (9 ft 10 in)
Wheel track	2.38 m (7 ft 10¾ in)
Wheelbase	4.045 m (13 ft 3 in)
Tail rotor ground clearance	2.10 m (6 ft 10¾ in)
Passenger cabin doors, each:	
Height	1.35 m (4 ft 5 in)
Width	1.35 m (4 ft 5 in)
Height to sill	1.00 m (3 ft 3¼ in)
Floor hatch, rear of cabin:	
Length	0.98 m (3 ft 2¾ in)
Width	0.70 m (2 ft 3½ in)

DIMENSIONS, INTERNAL:

Cabin: Length	6.05 m (19 ft 10 in)
Max width	1.80 m (5 ft 10¾ in)
Max height	1.55 m (5 ft 1 in)
Floor area	7.80 m² (84 sq ft)
Usable volume	11.40 m³ (403 cu ft)

AREAS:

Main rotor blades (each)	4.00 m² (43.06 sq ft)
Tail rotor blades (each)	0.28 m² (3.01 sq ft)
Main rotor disc	176.7 m² (1,902.1 sq ft)
Tail rotor disc	7.26 m² (78.13 sq ft)
Horizontal stabiliser	1.34 m² (14.42 sq ft)

WEIGHTS AND LOADINGS:

Weight empty, standard aircraft	3,615 kg (7,970 lb)
Max fuel load: standard internal	1,220 kg (2,690 lb)
external auxiliary tanks	552 kg (1,217 lb)
Max cargo sling load	3,200 kg (7,055 lb)
Max T-O and landing weight	7,400 kg (16,315 lb)
Max disc loading	41.88 kg/m² (8.58 lb/sq ft)

PERFORMANCE (A: at 6,000 kg; 13,230 lb AUW, B: at max T-O weight):

Never-exceed speed (VNE):	
A	158 knots (294 km/h; 182 mph)
B	142 knots (263 km/h; 163 mph)
Max cruising speed: A	146 knots (271 km/h; 168 mph)
B	139 knots (258 km/h; 160 mph)
Normal cruising speed: B	134 knots (248 km/h; 154 mph)
Max rate of climb at S/L: A	552 m (1,810 ft)/min
B	366 m (1,200 ft)/min
Service ceiling (30 m; 100 ft/min rate of climb):	
A	6,000 m (19,680 ft)
B	4,800 m (15,750 ft)
Hovering ceiling IGE: A, ISA	4,400 m (14,435 ft)
A, ISA +20°C	3,700 m (12,135 ft)
B, ISA	2,300 m (7,545 ft)
B, ISA +20°C	1,600 m (5,250 ft)
Hovering ceiling OGE: A, ISA	4,250 m (13,940 ft)
A, ISA +20°C	3,600 m (11,810 ft)
B, ISA	1,700 m (5,575 ft)
B, ISA +20°C	1,050 m (3,445 ft)
Max range at normal cruising speed, no reserves:	
A	309 nm (572 km; 355 miles)
B	297 nm (550 km; 341 miles)
Max endurance at 70 knots (130 km/h; 81 mph), no reserves	3 h 9 min

IAR IAR-330 PUMA 2000

TYPE: Upgraded version of current Puma.

PROGRAMME: Design being finalised late 1992; prototype rollout expected 1993.

CURRENT VERSIONS: Upgrade to be offered on **330J** and **330L**.

DESIGN FEATURES: More modern and higher powered engines (type not specified); EFIS flight deck and GPS navigation; obstacle avoidance database; advanced night vision system. Elbit of Israel prime contractor for new avionics.

ACCOMMODATION: 23 seats or high density and greater cargo loads.

AVIONICS: Standard aircraft will include HOCAS (hands off cyclic and stick), helmet mounted HUD and GPS navigation. Options include El-Op multi-sensor stabilised integrated system (MSIS) with nose mounted TV/FLIR cameras, digital image processing and communications system, engine performance indicator and radar/laser warning systems. New EFIS flight deck will use 1553B databus technology and versatile central onboard computer, plus comprehensive mission management software including laser rangefinder/target designator and weapon delivery system. Night vision system comprises helmet mounted NVGs for both pilots, supplemented by HUD. Multi-function displays (MFDs) to include five different vector or raster moving maps, ILS landing information, and video images of battlefield from TV/FLIR cameras or pre-recorded tapes. Downlink option for MFD imagery.

ARMAMENT: 20 mm guns and Hellfire or Swingfire missiles, or other ordnance of customer's choice.

IAR (KAMOV) Ka-126

NATO reporting name: Hoodlum-B

Responsibility for this programme has reverted to Kamov in Russia, which see, although IAR states that Romanian participation could be resumed ''in the foreseeable future''.

ROMAERO

ROMAERO SA (formerly IAv Bucuresti)

44 Ficusului Boulevard, Sector 1 (PO Box 18), R-70544 Bucharest
Telephone: 40 (1) 3335082
Fax: 40 (1) 3794456
Telex: 11691 IAvB R
MANAGING DIRECTOR: Dipl Eng Eugeniu Smirnov
TECHNICAL DIRECTOR: Dipl Eng Dan Gozia
COMMERCIAL DIRECTOR: Dipl Eng Anatolie Merling

Established 1951 and operated successively under several names (see 1991-92 *Jane's* for details); became commercial company under present name 20 November 1990; manufactures Rombac 1-11 and Islander; repairs and overhauls various large and small aircraft; agent and repair centre for Textron Lycoming engines; manufactures aircraft equipment.

ROMAERO ROMBAC 1-11 (BAC/BAe ONE-ELEVEN)

TYPE: Twin-turbofan short/medium-range transport.

PROGRAMME: Romaero is Romanian prime contractor for licence manufacture of 20 BAC/BAe One-Elevens (see UK section of 1974-75 and 1981-82 *Jane's*), known as Rombac 1-11; corresponding programme covers Romanian production of Rolls-Royce Spey 512-14DW engines. One BAe-built Srs 487 freighter and two Srs 525/1s delivered 1981-82; industrial transfer to Romania completed 1986; Romanian assembled Srs 560 first flight 18 September and certification November 1982 (entered service with Tarom January 1983).

CURRENT VERSIONS: **Series 495:** Combines standard fuselage and accommodation of British built Srs 400 with wings and power plant of Srs 560 and modified landing gear (using low-pressure tyres to permit operation from secondary low-strength runways with poorer grade surfaces); 10th 1-11, long delayed but expected to fly mid-1993, is first convertible passenger/cargo **Model 497** (equivalent to BAC/BAe Srs 475 but redesigned by Romaero and approved by BAe, using components from Srs 487 and 560). *Description applies to Series 495, except where indicated.*

Series 560 (Model 561RC): Stretched version, derived from British Series 300/400; lengthened fuselage (2.54 m; 8 ft 4 in forward of wings, 1.57 m; 5 ft 2 in aft); accommodates up to 109 passengers; wing extensions increase span by 1.52 m (5 ft); main landing gear strengthened; heavier wing planks to cater for increased AUW. See Landing Gear, Accommodation and tabulated data for other differences.

Romanian assembled Rombac 1-11-561RC of Tarom (two Rolls-Royce Spey Mk 512-14DW turbofans)

Series 2560: Basically as Series 560, seating 96-115 passengers, but re-engined with Rolls-Royce Tay 650 turbofans to meet current noise and environmental pollution requirements; first example originally due for delivery at end of 1991, but present status of this version uncertain due to lack of funding; may fly late 1993.

CUSTOMERS: Nine Model 561RCs completed by July 1990, of which seven delivered to Tarom (incl one leased to Ryanair) and one to Romavia. Reported options for Srs 2560 from Tarom (five) and Romavia (three), but unclear whether these are new-build or 561RC conversions. Kiwi International Airlines (USA) announced agreement early 1993 to purchase 11 Series 2560s (plus five on option) if Tay programme goes ahead.

DESIGN FEATURES: Modified NACA cambered wing section; thickness/chord ratio 12.5 per cent at root, 11 per cent at tip; dihedral 2°; incidence 2° 30′; sweepback 20° at quarter-chord.

FLYING CONTROLS: Autopilot and provision for automatic throttle control (see Avionics). Manually actuated ailerons, with servo tabs (port tab used for trimming). Hydraulically actuated variable incidence T tailplane (controlled through duplicated hydraulic units), elevators and rudder (using tandem jacks), Fowler flaps, spoiler/airbrakes on upper surface of wings, and lift dumpers (inboard of spoilers).

STRUCTURE: Mostly all-metal fail-safe, mainly of copper based aluminium alloy. Three-shear-web wing torsion box with integrally machined skin/stringer panels; Redux bonded light alloy honeycomb ailerons; Srs 495 flaps have GFRP coating; fuselage uses continuous frames and stringers.

LANDING GEAR: Retractable tricycle type, with twin wheels on each unit. Hydraulic retraction, nose unit forward, main units inward. Oleo-pneumatic shock absorbers. Hydraulic nosewheel steering. Wheels have tubeless tyres, 5-plate heavy duty hydraulic disc brakes, and anti-skid units. Mainwheel tyres size 40 × 12 on Srs 560, pressure 11.03 bars (160 lb/sq in); size 44 × 16 on Srs 495, pressure 5.72 bars (83 lb/sq in). Nosewheel tyres size 24 × 7.25 on Srs 560, pressure 7.58 bars (110 lb/sq in); size 24 × 7.7 on Srs 495, pressure 7.24 bars (105 lb/sq in). Min ground turning radius (nosewheel to outer wingtip) 15.24 m (50 ft 0 in) for Srs 495, 17.07 m (56 ft 0 in) for Srs 560.

POWER PLANT: Two Rolls-Royce Spey Mk 512-14DW turbofans, each rated at 55.8 kN (12,550 lb st), pod-mounted on sides of rear fuselage. Fuel in integral wing tanks with usable capacity of 10,160 litres (2,684 US gallons; 2,235 Imp gallons) and centre-section tank of 3,968 litres (1,048 US gallons; 873 Imp gallons) usable capacity; total usable fuel 14,129 litres (3,732 US gallons; 3,108 Imp gallons). Executive versions can be fitted with auxiliary fuel tanks of up to 5,791 litres (1,530 US gallons; 1,274 Imp gallons) usable capacity. Pressure refuelling point in fuselage forward of wing on starboard side. Provision for gravity refuelling. Oil capacity (total engine oil) 13.66 litres (3.6 US gallons; 3 Imp gallons) per engine. Engine hush kits standard.

ACCOMMODATION (Srs 495): Crew of two on flight deck and up to 89 passengers in main cabin. Single class or mixed class layout, with movable divider bulkhead to permit any first/tourist ratio. Typical mixed class layout has 16 first class (four abreast) and 49 tourist (five abreast) seats. Galley units normally at front on starboard side. Coat space available on port side aft of flight deck. Ventral entrance with hydraulically operated airstair. Forward passenger door on port side incorporates optional power operated airstair. Galley service door forward on starboard side. Overwing emergency exit on each side. Two baggage and freight holds under floor, fore and aft of wings, with doors on starboard side. Upward opening forward freight door available at customer's option. Entire accommodation air-conditioned.

ACCOMMODATION (Srs 560): Crew of two on flight deck and up to 109 passengers in main cabin. Two overwing emergency exits on each side. One toilet on each side of cabin at rear. Otherwise generally similar to Srs 495.

SYSTEMS: Fully duplicated air-conditioning and pressurisation systems. Max pressure differential 0.52 bar (7.5 lb/sq in). Thermal (engine bleed air) de-icing of wing, fin and tailplane leading-edges. Hydraulic system, pressure 207 bars (3,000 lb/sq in), operates flaps, spoilers, rudder, elevators, tailplane, landing gear, brakes, nosewheel steering, ventral and forward airstairs and windscreen wipers. No pneumatic system. Electrical system utilises two 30kVA AC generators, driven by constant speed drive and starter units, plus similar generator mounted on APU and shaft driven. Gas turbine APU in tailcone to provide ground electric power, air-conditioning and engine starting, also some system checkout capability. APU run during take-off to eliminate performance penalty of bleeding engine air for cabin air-conditioning.

AVIONICS: Communications and navigation avionics generally to customers' requirements. Typical installation includes dual VHF com to ARINC 546, dual VHF nav to ARINC 547A, including glideslope receivers, marker beacon receiver, flight/service interphone system, ADF, ATC transponder to ARINC 532D, DME, weather radar. Compass system and flight director system (dual) also installed. Autopilot system. Provision on Srs 560 for additional equipment, including automatic throttle control, for low weather minima operation.

DIMENSIONS, EXTERNAL:
Wing span	28.50 m (93 ft 6 in)
Wing chord: at root	5.00 m (16 ft 5 in)
at tip	1.61 m (5 ft 5 in)
Wing aspect ratio	8.5
Length:	
overall: Srs 495	28.50 m (93 ft 6 in)
Srs 560	32.61 m (107 ft 0 in)
fuselage: Srs 495	25.55 m (83 ft 10 in)
Srs 560	29.67 m (97 ft 4 in)
Height overall	7.47 m (24 ft 6 in)
Tailplane span	8.99 m (29 ft 6 in)
Wheel track	4.34 m (14 ft 3 in)
Wheelbase: Srs 495	10.08 m (33 ft 1 in)
Srs 560	12.62 m (41 ft 5 in)
Passenger door (fwd, port): Height	1.73 m (5 ft 8 in)
Width	0.84 m (2 ft 9 in)
Height to sill	2.08 m (6 ft 10 in)
Ventral entrance, bulkhead door:	
Height	1.83 m (6 ft 0 in)
Width	0.66 m (2 ft 2 in)
Height to sill	2.08 m (6 ft 10 in)
Freight door (fwd, starboard):	
Height (projected)	0.79 m (2 ft 7 in)
Width	0.91 m (3 ft 0 in)
Height to sill	1.04 m (3 ft 5 in)
Freight door (rear, starboard):	
Height (projected)	0.71 m (2 ft 4 in)
Width	0.91 m (3 ft 0 in)
Height to sill	1.17 m (3 ft 10 in)
Freight door, main deck (optional, fwd, Srs 495):	
Height	1.85 m (6 ft 1 in)
Width	3.05 m (10 ft 0 in)
Galley service door (fwd, starboard):	
Height (projected)	1.22 m (4 ft 0 in)
Width	0.69 m (2 ft 3 in)
Height to sill	2.08 m (6 ft 10 in)

DIMENSIONS, INTERNAL (Srs 495):
Cabin, excl flight deck: Length	17.32 m (56 ft 10 in)
Max width	3.15 m (10 ft 4 in)
Max height	1.98 m (6 ft 6 in)
Floor area	approx 47.4 m² (510 sq ft)
Freight hold: fwd	10.02 m³ (354 cu ft)
rear	4.42 m³ (156 cu ft)

DIMENSIONS, INTERNAL (Srs 560):
Cabin, excl flight deck: Length	21.44 m (70 ft 4 in)
Total floor area	approx 59.5 m² (640 sq ft)
Freight holds (total volume)	19.45 m³ (687 cu ft)

AREAS (Srs 495, 560):
Wings, gross	95.78 m² (1,031.0 sq ft)
Ailerons (total)	2.86 m² (30.8 sq ft)
Flaps (total)	16.26 m² (175.0 sq ft)
Spoilers (total)	2.30 m² (24.8 sq ft)
Fin	7.86 m² (84.6 sq ft)
Rudder, incl tab	3.05 m² (32.8 sq ft)
Tailplane	17.43 m² (187.6 sq ft)
Elevators, incl tab	6.54 m² (70.4 sq ft)

WEIGHTS AND LOADINGS:
Operating weight empty, typical:	
Srs 495 (89 seats)	23,286 kg (51,339 lb)
Srs 560 (109 seats)	25,267 kg (55,704 lb)
Max payload, typical: Srs 495	10,733 kg (23,661 lb)
Srs 560	11,474 kg (25,296 lb)
Max T-O weight:	
Srs 495: standard	41,730 kg (92,000 lb)
optional	44,680 kg (98,500 lb)
Srs 560: standard	45,200 kg (99,650 lb)
optional	47,400 kg (104,500 lb)
Max ramp weight:	
Srs 495: standard	41,955 kg (92,500 lb)
optional	44,905 kg (99,000 lb)
Srs 560: standard	45,450 kg (100,200 lb)
optional	47,625 kg (105,000 lb)
Max landing weight:	
Srs 495: standard	38,100 kg (84,000 lb)
optional	39,465 kg (87,000 lb)
Srs 560	39,465 kg (87,000 lb)
Max zero-fuel weight:	
Srs 495: standard	33,110 kg (73,000 lb)
optional	34,020 kg (75,000 lb)
Srs 560	36,740 kg (81,000 lb)
Max wing loading: Srs 495	466.3 kg/m² (95.5 lb/sq ft)
Srs 560	495.1 kg/m² (101.4 lb/sq ft)
Max power loading: Srs 495	400.2 kg/kN (3.92 lb/lb st)
Srs 560	424.5 kg/kN (4.16 lb/lb st)

PERFORMANCE (at standard max T-O weights):
Design diving speed (S/L)	410 knots (760 km/h; 472 mph) EAS
Max level and max cruising speed at 6,400 m (21,000 ft)	470 knots (870 km/h; 541 mph)
Econ cruising speed at 10,670 m (35,000 ft)	410 knots (760 km/h; 472 mph)
Stalling speed (landing flap setting, at standard max landing weight):	
Srs 495	98 knots (182 km/h; 113 mph) EAS
Srs 560	100 knots (186 km/h; 115 mph) EAS
Rate of climb at S/L at 300 knots (555 km/h; 345 mph) EAS: Srs 495	786 m (2,580 ft)/min
Srs 560	722 m (2,370 ft)/min
Max cruising altitude	10,670 m (35,000 ft)
Runway LCN, rigid pavement (l 30): Srs 495	32
Srs 560	53
T-O run at S/L, ISA: Srs 495	1,676 m (5,500 ft)
Srs 560	1,981 m (6,500 ft)
Balanced T-O to 10.7 m (35 ft) at S/L, ISA:	
Srs 495	1,798 m (5,900 ft)
Srs 560	2,225 m (7,300 ft)
Landing distance (BCAR) at S/L, ISA, at standard max landing weight: Srs 495	1,440 m (4,725 ft)
Srs 560	1,455 m (4,775 ft)
Max still air range, ISA, with reserves for 200 nm (370 km; 230 mile) diversion and 45 min hold:	
Srs 495	1,933 nm (3,582 km; 2,226 miles)
Srs 560	1,897 nm (3,515 km; 2,184 miles)
Still air range with typical capacity payload, ISA, reserves as above:	
Srs 495 at 44,680 kg (98,500 lb)	1,454 nm (2,694 km; 1,674 miles)
Srs 560 at 47,400 kg (104,500 lb)	1,327 nm (2,459 km; 1,528 miles)
Srs 495 executive aircraft with additional 5,602 litres (1,479 US gallons; 1,232 Imp gallons) fuel and 10 passengers	2,875 nm (5,325 km; 3,308 miles)

ROMAERO (PILATUS BRITTEN-NORMAN) ISLANDER

TYPE: Twin-engined feederliner.
PROGRAMME: First flight of Romanian built Islander (built by former IRMA) 4 August 1969.
CUSTOMERS: Initial commitment to build 215 completed 1976; over 450 now delivered to Pilatus Britten-Norman.

RUSSIA

AST

AVIASPETSTRANS CONSORTIUM
Zhukovsky 5, Moscow Region
Telephone: 7 (095) 556 59 93
Fax: 7 (095) 292 65 11
Telex: 411700 AST
DIRECTOR GENERAL: Valentin A. Korchagin

Consortium formed 1989 by Russia's Arctic and Antarctic Research Institute, Scientific Research Institute for Civil Aviation, Gasprom gas enterprises, Institute of Oceangeology Engineering Centre, Peoples of the North Foundation, Myasishchev Design Bureau and Promstroybank. These have a common interest in air transport infrastructure in remote regions of the north, Siberia and Far East. Aviaspetstrans will market services such as monitoring systems and navigation aids as well as aircraft. First aircraft project, headed by Chief Designer/Director General V. A. Korchagin, is amphibious twin-turboshaft Yamal, capable of autonomous operation in regions with minimal transport infrastructure.

AVIASPETSTRANS YAMAL
TYPE: Twin-turboshaft cargo/passenger multi-purpose amphibian.
PROGRAMME: Work performed by 48 scientific and research institutes and aviation design bureaux under Gosaviaregistr supervision, consistent with FAR requirements. Development delegated to Myasishchev OKB. Full-scale mockup exhibited at MosAeroshow '92; prototypes to be manufactured in Moscow.
DESIGN FEATURES: Basically conventional small amphibian. Unique power plant; two turboshafts side by side on top decking over wing centre-section, driving single pusher propeller mounted behind tail unit, through helicopter-like combining reduction gearbox; position of engine air intakes protects them from water and foreign object ingestion. Flying-boat hull and two fixed underwing stabilising floats. All between-flights servicing of systems and equipment possible from inside fuselage, through access panels in galley ceiling, protecting engineer from low temperature, precipitation, mosquitoes and effects of waves on water. Similar access practicable to remedy in-flight failure.

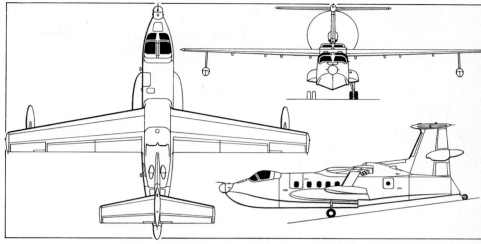

Aviaspetstrans Yamal twin-turboshaft multi-purpose amphibian (*Jane's/Mike Keep*)

LANDING GEAR: Retractable tailwheel type; twin wheels on each main unit; all units retract upward, main units into sponsons each side of hull. Optional nosewheel gear under development.
POWER PLANT: Two RKBM RD600S (TVD-1500) turboshafts; six-blade propeller, dimensions and aerodynamics similar to those of Il-114 propellers. Fuel tank between spars in each wing.
ACCOMMODATION: Provision for passengers, freight, stretchers and medical equipment, cameras, equipment for variety of commercial and military duties.
AVIONICS: Radar in nose thimble.
EQUIPMENT: Optional equipment being developed for ecological monitoring, forest protection and firefighting, ice and fish reconnaissance, 200-mile economic zone patrol, and air/sea rescue.
DIMENSIONS, EXTERNAL:
  Wing span   21.40 m (70 ft 2½ in)
  Length overall   16.825 m (55 ft 2½ in)
  Height overall   5.367 m (17 ft 7½ in)
DIMENSIONS, INTERNAL:
  Passenger cabin: Volume   23.4 m³ (826 cu ft)
AREAS:
  Wings, gross   51.9 m² (558.7 sq ft)
WEIGHTS AND LOADINGS:
  Max payload   2,000 kg (4,410 lb)
PERFORMANCE (estimated, at max T-O weight):
  Max cruising speed   226 knots (420 km/h; 261 mph)
  T-O run: on land   225 m (740 ft)
    on water   230 m (755 ft)
  Range: with max payload   270 nm (500 km; 310 miles)
    with 500 kg (1,100 lb) payload
       more than 1,350 nm (2,500 km; 1,553 miles)

BERIEV

TAGANROG AVIATSIONNYI NAUCHNO-TEKHNICHESKIY KOMPLEKS IMIENI G. M. BERIEVA (TANTK) (Taganrog Aviation Scientific-Technical Complex named after G. M. Beriev) (TASTC)
Instrumentalniy Toupic 347927, Taganrog
Telephone: 7 (86344) 49839, 49901, 49964
Fax: 7 (86344) 41454
PRESIDENT AND GENERAL DESIGNER: Gennady S. Panatov

This OKB founded by Georgy Mikhailovich Beriev (1902-1979) in 1932; except during Second World War, 1942-45, it has been based at Taganrog, in northeast corner of Sea of Azov; since 1948 has been centre for all Russian seaplane development. In 1990 was redesignated as shown.

BERIEV A-40 ALBATROSS
NATO reporting name: Mermaid
TYPE: Twin-turbofan maritime patrol amphibian.
PROGRAMME: Conceived as military amphibian to carry extensive avionics and operational systems in primary anti-submarine warfare form; design started 1983; prototype construction began 1985; first flight December 1986; construction of production aircraft started 1987; mentioned Spring 1988 by Rear Adm William O. Studeman, then US director of naval intelligence, as seaplane with provisional Western designation 'Tag-D' (implying fourth unidentified type photographed by US reconnaissance satellite over Taganrog), for possible ASW/surveillance/minelaying role; identified as A-40 Albatross, designed by Alexei K. Konstantinov "for search and rescue", when prototype flown over Tushino Airport during Aviation Day display, 20 August 1989; feature in *Krasnaya Zvezda*, 6 August 1989, stated that in this role A-40 will be confined to SAR missions near coast, and next task confronting designers was to produce similar aircraft capable of operating anywhere in Pacific; A-40 prototype has set 14 records, lifting payload of 10,000 kg to 13,281 m (43,573 ft); second prototype has flown; US Department of Defense stated in 1991 *Military Forces in Transition* document that acceptance trials of A-40 by Russian Naval Aviation had started and that it will replace Il-38 and Be-12, though not on one-for-one basis.
CURRENT VERSIONS: **A-40:** ASW/surveillance/minelaying version.
  Be-40P Passenger: Projected transport for 105 passengers, five abreast; aisle between three-seat (port) and two-seat units; two toilets at rear (port); other facilities fore and aft of passengers; three flight crew plus cabin staff. Range with max payload 2,160 nm (4,000 km; 2,485 miles).
  Be-40PT Cargo-passenger: Projected transport with max payload of 10,000 kg (22,045 lb); 37 or 70 passengers, five abreast at front of cabin, freight to rear. Range with max payload 2,265 nm (4,200 km; 2,610 miles).
  Be-42: Search and rescue version; see separate entry.
CUSTOMERS: Initial order 20 A-40s for CIS Naval Aviation.
DESIGN FEATURES: Largest amphibian yet built. Swept wings of moderate aspect ratio, with high-lift devices; single-step hull of high length to beam ratio, with what Russian press describes as "the world's first development of a variable-rise bottom, providing a considerable improvement in stability and controllability in the water, as well as a reduction in g loads when landing and taking off at sea"; small wedge-shape boxes aft of step aid 'unsticking' from water in wave heights up to 2.2 m (7 ft 2½ in); all-swept T tail; high-mounted engines protected from spray by strakes on each side of nose and by wings; large underwing pod each side of hull, faired into wingroot; wing leading-edge sweep 23° 13'; supercritical wing sections, thickness/chord ratio 14.5% to 11.3%; incidence 3° 23' at root; wing twist 4° 30'; no dihedral or anhedral; large dorsal fin.
FLYING CONTROLS: Entire span of each wing trailing-edge occupied by aileron and two-section area-increasing double-slotted flaps; full-span leading-edge slats; outer spoilers assist ailerons; variable-incidence tailplane; conventional rudder and elevators. Powered controls, with spring feel and electric trim.
STRUCTURE: All-metal semi-monocoque boat type fuselage, with heavy gauge double-chine planing bottom forward of step; conventional two-spar wings; honeycomb-core

This photograph shows the unique double-chine 'variable rise bottom' of the Beriev A-40 Albatross (*J. M. G. Gradidge*)

RUSSIA: AIRCRAFT—BERIEV

Beriev A-40 Albatross amphibian (two Aviadvigatel D-30KPV turbofans) *(Martin Fricke)*

sandwich panels and composites used widely; water rudder at rear of hull.

LANDING GEAR: Hydraulically retractable tricycle type; twin-wheel nose unit retracts rearward; main four-wheel bogies retract rearward into large underwing pods, rotating round pivot to stow as tandem twin-wheels on each side; oleo-nitrogen shock absorbers; nosewheel tyres size 840 × 290 mm, pressure 6.9-7.35 bars (100-106 lb/sq in); main-wheel tyres size 1,030 × 350 mm, pressure 9.3-9.8 bars (135-142 lb/sq in); multi-disc brakes on mainwheels, with inertial anti-skid units. Ground turning radius 19.25 m (63 ft 2 in). Nosewheel steering angle ±55°.

POWER PLANT: Two Aviadvigatel D-30KPV turbofans, pylon-mounted above wingroot pods, with outward toed exhaust efflux; each 117.7 kN (26,455 lb st). RKBM RD-60K booster turbojet, rated at 24.5 kN (5,510 lb st), in fairing on each turbofan pylon, slightly aft and slightly inboard of D-30KPV nozzle, with vertically split eyelid jetpipe closure at rear. D-30KPVs to be replaced later by two 147.1 kN (33,070 lb st) turbofans. Fuel tanks in wing torsion box, capacity 35,100 litres (9,272 US gallons; 7,721 Imp gallons). Total oil capacity 94.5 litres (25 US gallons; 20.75 Imp gallons). Flight refuelling probe above nose.

ACCOMMODATION: Crew of eight: two pilots, flight engineer, radio operator, navigator/observer and three observers. Door on each side to rear of flight deck, port door outward opening, starboard door inward opening.

SYSTEMS: Flight deck and crew quarters air-conditioned and pressurised by engine bleed and APU. Four hydraulic systems at 207 bars (3,000 lb/sq in); 190 litres (50 US gallons; 42 Imp gallons) of AMG-10 fluid; flow rate 55 litres (14.5 US gallons; 12 Imp gallons)/min. Pneumatic system pressure 207 bars (3,000 lb/sq in); capacity of bottles 95 litres (3.35 cu ft). Three-phase 115/220V 400Hz AC electrical system; single-phase 115V 400Hz AC system; 27V DC system; supplied by two engine driven 60kVA AC generators and three static inverters; three batteries. Gaseous oxygen bottle, pressure 147 bars (2,135 lb/sq in). Provision for de-icing tail unit, slats, engine air intakes and windscreen. TA-12 APU, operable up to 7,000 m (23,000 ft).

AVIONICS: Inertial flight and nav system; standard com radio and IFF; navigation, search, surveillance and wave height measurement radar; conventional instrumentation.

EQUIPMENT: Slim container, probably for ESM, above each wingtip float pylon; bombing equipment; optical/TV sight.

ARMAMENT: Stores bay in bottom of hull, aft of step.

DIMENSIONS, EXTERNAL:
Wing span	41.62 m (136 ft 6½ in)
Wing chord: at root	7.282 m (23 ft 10¾ in)
at tip	2.24 m (7 ft 4¼ in)
Wing aspect ratio	8.6
Length overall, incl nose probe	43.839 m (143 ft 10 in)
Length of fuselage	38.915 m (127 ft 8 in)
Max diameter of fuselage	3.50 m (11 ft 6 in)
Height overall	11.066 m (36 ft 3¾ in)
Tailplane span	11.87 m (38 ft 11½ in)
Wheel track, c/l of oleos	4.96 m (16 ft 3¼ in)
Wheelbase	14.835 m (48 ft 8 in)
Crew doors: Height	1.10 m (3 ft 7¼ in)
Width	0.70 m (2 ft 3½ in)
Stores bay: Length	6.50 m (21 ft 4 in)
Width	1.76 m (5 ft 9¼ in)
Emergency exits: Height	1.10 m (3 ft 7¼ in)
Width	0.70 m (2 ft 3½ in)

DIMENSIONS, INTERNAL:
Cabin, excl flight deck:	
Length: pressurised	7.90 m (25 ft 11 in)
unpressurised	13.9 m (45 ft 7¼ in)
Max width	3.25 m (10 ft 8 in)
Max height	2.10 m (6 ft 10½ in)
Floor area, pressurised	38.6 m² (415.5 sq ft)
Unpressurised cabin volume	58.0 m³ (2,048 cu ft)

AREAS:
Wings, gross	200.0 m² (2,152.8 sq ft)
Ailerons (total)	6.142 m² (66.11 sq ft)
Flaps (total)	36.55 m² (393.43 sq ft)
Slats (total)	23.00 m² (247.58 sq ft)
Spoilers (total)	6.25 m² (67.28 sq ft)
Fin	20.99 m² (225.94 sq ft)
Rudder	8.76 m² (94.30 sq ft)
Tailplane	28.086 m² (302.32 sq ft)
Elevators (total)	7.104 m² (76.47 sq ft)

WEIGHTS AND LOADINGS:
Max payload	6,500 kg (14,330 lb)
Max fuel weight	35,000 kg (77,160 lb)
Max T-O and ramp weight	86,000 kg (189,595 lb)
Max landing weight: on land	73,000 kg (160,935 lb)
on water	85,000 kg (187,390 lb)
Max wing loading	430 kg/m² (88.07 lb/sq ft)
Max power loading	365.3 kg/kN (3.58 lb/lb st)

PERFORMANCE (at max T-O weight):
Max Mach No. in level flight	0.7
Never-exceed speed (V_NE)	350 knots (650 km/h; 404 mph) EAS
Max level speed at 6,000 m (19,700 ft)	410 knots (760 km/h; 472 mph)
Max cruising speed at 6,000 m (19,700 ft)	388 knots (720 km/h; 447 mph)
Stalling speed: flaps up	146 knots (270 km/h; 168 mph)
flaps down	99 knots (182 km/h; 113 mph)
Rate of climb at S/L, OEI	1,800 m (5,900 ft)/min
Service ceiling	9,700 m (31,825 ft)
T-O run	1,000 m (3,280 ft)
T-O to 15 m (50 ft)	1,100 m (3,610 ft)
Landing from 15 m (50 ft)	1,450 m (4,760 ft)

Prototype Beriev A-40 Albatross ASW/surveillance/minelaying amphibian, taking off from Sea of Azov

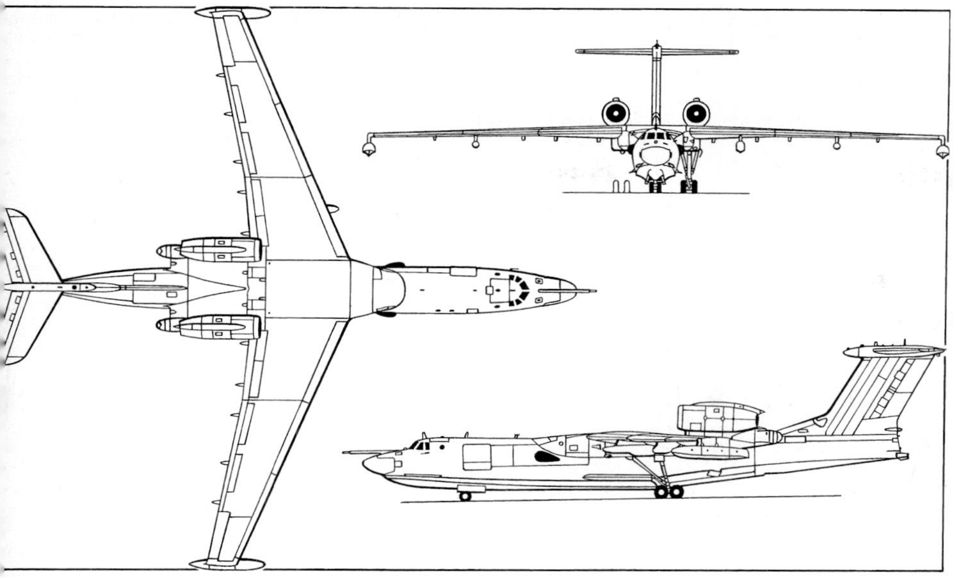

Beriev Be-42 twin-turbofan search and rescue amphibian *(Jane's/Mike Keep)*

Landing run 900 m (2,955 ft)
Range:
with max payload 2,212 nm (4,100 km; 2,547 miles)
with max fuel 2,967 nm (5,500 km; 3,417 miles)
PERATIONAL NOISE LEVELS:
T-O 104 EPNdB
Climb 99 EPNdB
Landing 102 EPNdB

BERIEV Be-42

ATO reporting name: Mermaid

YPE: Twin-turbofan search and rescue amphibian.

ROGRAMME: Design started 1988.

ESIGN FEATURES: As A-40, but specialised SAR equipment and no booster turbojets. No wingtip ESM containers.

CCOMMODATION: Nine crew: two pilots, flight engineer, radio operator, navigator, flight technician, senior medical attendant, two assistants. Provision for 54 survivors, who enter via side hatches, aided by mechanised ramps. Two outward opening doors on port side; two inward opening doors starboard.

QUIPMENT: Flares. Power boats, two LPS-6 liferafts; onboard equipment to combat hypothermia for 20 survivors, AT-2 transfusion equipment, DI-C-0.4 defibrillator, EK-1T-03M2 electrocardiograph, other resuscitation and surgical equipment and medicines. Electro-optical sensors and searchlights to detect shipwreck survivors by day or night.

MENSIONS, EXTERNAL:
Forward cabin door, each side:
Height 1.80 m (5 ft 10¾ in)
Width 1.95 m (6 ft 4¾ in)
Rear cabin door, each side:
Height 0.70 m (2 ft 3½ in)
Width 1.45 m (4 ft 9 in)

MENSIONS, INTERNAL:
Cabin volume, unpressurised 69.0 m³ (2,436 cu ft)
Underfloor hold volume 6.3 m³ (222 cu ft)

EIGHTS AND LOADINGS:
Max payload 5,000 kg (11,025 lb)

RFORMANCE (estimated, at max T-O weight):
Max level speed at 6,000 m (19,700 ft)
458 knots (850 km/h; 528 mph)
Max cruising speed at 5,000 m (16,400 ft)
431 knots (800 km/h; 497 mph)
Max rate of climb at S/L 840 m (2,755 ft)/min
Rate of climb at S/L, OEI 210 m (690 ft)/min
Service ceiling 10,000 m (32,800 ft)
Service ceiling, OEI 4,100 m (13,450 ft)
Range with max payload (54 survivors)
1,888 nm (3,500 km; 2,175 miles)

BERIEV Be-200

YPE: Twin-turbofan multi-role amphibian; primary mission firefighting.

OGRAMME: Details announced, and model displayed, at 1991 Paris Air Show; full-scale mockup under construction mid-1991; two prototypes to be built in partnership with ILTA Trade Finance SA of Geneva, Switzerland; first flight scheduled 1994, deliveries 1997.

JRRENT VERSIONS: **Firefighting:** Tanks under cabin floor, capacity 12 m³ (423 cu ft) water; tanks in cabin for 1.2 m³ (42.3 cu ft) liquid chemicals; seats along sidewalls of cabin; 12 tonnes of water scooped from seas with waves up to 1.2 m (4 ft). Fully fuelled, Be-200 can drop total 320,000 kg (705,465 lb) of water in successive flights when airfield-reservoir distance is 108 nm (200 km; 125 miles) and reservoir-fire zone distance is 5.4 nm (10 km; 6.2 miles); or 140,000 kg (308,640 lb) when distances are respectively 108 nm (200 km; 125 miles) and 27 nm (50 km; 31 miles).

Passenger: Two flight crew, two cabin attendants, and 64 tourist class passengers four-abreast in pairs, with centre aisle, at seat pitch of 75 cm (29.5 in).

Cargo: Payload 8,000 kg (17,635 lb) in unobstructed cabin 17.0 m (55 ft 9 in) long, 2.6 m (8 ft 6 in) wide and 1.9 m (6 ft 3 in) high. Estimated range 595 nm (1,100 km; 685 miles) with 7,000 kg (15,430 lb) payload.

Ambulance: Two flight crew, seven medical personnel, 30 stretchers in three tiers.

Search and rescue: Sensors and searchlights for detecting survivors, and onboard medical equipment.

DESIGN FEATURES: Scaled-down version of A-40. Details generally similar; underwing stabilising floats moved inboard from tips, winglets added, twin-wheel main landing gear units, and no booster turbojets. Supercritical wing sections, thickness/chord ratio 16% to 11.5%.

FLYING CONTROLS: Area-increasing single-slotted flaps.

LANDING GEAR: Twin-wheel main units, tyre size 950 × 300 mm, pressure 9.8-10.3 bars (142-150 lb/sq in); nosewheel tyre size 620 × 180 mm, pressure 7.35-7.85 bars (106-114 lb/sq in). Ground turning radius 17.4 m (57 ft 1 in). Nosewheel steering angle ±45°.

POWER PLANT: Two ZMKB Progress D-436T turbofans, each 73.6 kN (16,550 lb st). Much reduced fuel capacity compared with A-40. Total oil capacity 22 litres (5.8 US gallons; 4.85 Imp gallons).

ACCOMMODATION: Two flight crew; up to 64 tourist class passengers, or 10 to 32 first class and business class passengers at up to 102 cm (40 in) seat pitch, with provision for galley, toilet and baggage stowage. Two outward opening doors each side, cargo door and emergency exits; forward and rear freight/baggage holds.

SYSTEMS: Three hydraulic systems at 207 bars (3,000 lb/sq in); 150 litres (39.6 US gallons; 33 Imp gallons) of MGJ-5U fluid; flow rate 70 litres (18.5 US gallons; 15.4 Imp gallons)/min. Capacity of pneumatic system bottles 29 litres (1.02 cu ft). Electrical system, de-icing and APU as A-40.

AVIONICS: MN-85 weather radar; INS; EFIS displays.

DIMENSIONS, EXTERNAL:
Wing span 31.88 m (104 ft 7¼ in)
Wing chord: at root 5.578 m (18 ft 3½ in)
at tip 1.716 m (5 ft 7½ in)
Length overall 32.049 m (105 ft 1¾ in)
Length of fuselage 29.18 m (95 ft 9 in)
Max diameter of fuselage 2.86 m (9 ft 4½ in)
Height overall 8.90 m (29 ft 2½ in)
Tailplane span 10.114 m (33 ft 2¼ in)
Wheel track 4.30 m (14 ft 1¼ in)
Wheelbase 11.143 m (36 ft 6½ in)
Width over stabilising floats 25.60 m (84 ft 0 in)
Passenger doors (each): Height 1.70 m (5 ft 7 in)
Width 0.90 m (2 ft 11½ in)
Cargo door: Height 1.80 m (5 ft 10¾ in)
Width 2.00 m (6 ft 6¾ in)
Emergency exits: Height 1.70 m (5 ft 7 in)
Width 0.90 m (2 ft 11½ in)

DIMENSIONS, INTERNAL:
Cabin, excl flight deck: Length 17.00 m (55 ft 9 in)
Max width 2.60 m (8 ft 6¼ in)
Max height 1.90 m (6 ft 2¾ in)
Floor area 39.0 m² (420 sq ft)
Volume: forward baggage hold 8.8 m³ (310 cu ft)
rear baggage hold 4.5 m³ (159 cu ft)
main cabin and freight/baggage holds, total, cargo configuration 84 m³ (2,966 cu ft)

AREAS:
Wings, gross 117.44 m² (1,264.2 sq ft)
Ailerons (total) 3.56 m² (38.32 sq ft)
Flaps (total) 20.43 m² (219.91 sq ft)
Slats (total) 12.61 m² (135.74 sq ft)
Spoilers (total) 4.59 m² (49.41 sq ft)
Fin 12.60 m² (135.63 sq ft)
Rudder 4.60 m² (49.52 sq ft)
Tailplane 17.96 m² (193.33 sq ft)
Elevators (total) 6.96 m² (74.92 sq ft)

WEIGHTS AND LOADINGS:
Max fuel weight 12,260 kg (27,025 lb)
Max T-O and ramp weight 36,000 kg (79,365 lb)
Max airborne weight (after water scooping)
43,000 kg (94,800 lb)
Max landing weight, land or water 35,000 kg (77,160 lb)
Max wing loading 306.5 kg/m² (62.78 lb/sq ft)
Max power loading 244.5 kg/kN (2.40 lb/lb st)

PERFORMANCE (estimated, at max T-O weight):
Max Mach No. in level flight 0.69
Never-exceed speed (VNE)
329 knots (610 km/h; 379 mph) EAS
Max level speed at 7,000 m (22,965 ft)
388 knots (720 km/h; 447 mph)
Max cruising speed at 8,000 m (26,250 ft)
377 knots (700 km/h; 435 mph)
Stalling speed: flaps up 116 knots (215 km/h; 134 mph)
flaps down 84 knots (155 km/h; 97 mph)
Max rate of climb at S/L 840 m (2,755 ft)/min
Rate of climb at S/L, OEI 168 m (550 ft)/min
Service ceiling 11,000 m (36,090 ft)
Service ceiling, OEI 5,500 m (18,045 ft)
T-O to 15 m (50 ft): on land 600 m (1,970 ft)
on water 1,000 m (3,280 ft)

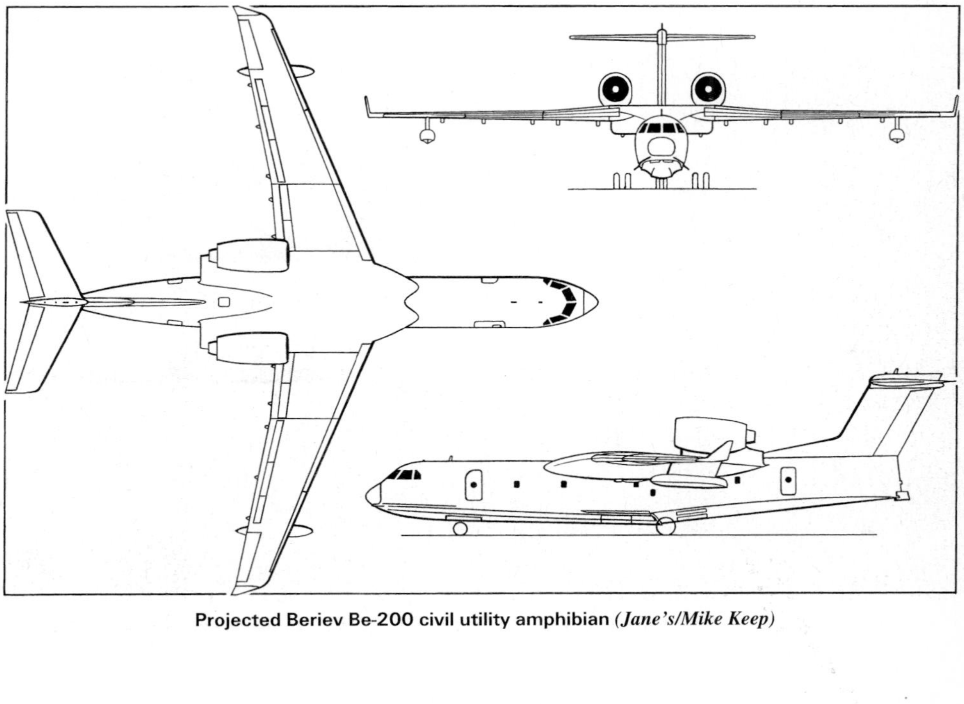

Projected Beriev Be-200 civil utility amphibian *(Jane's/Mike Keep)*

258 RUSSIA: AIRCRAFT—BERIEV

Landing from 15 m (50 ft): on land	1,050 m (3,445 ft)
on water	1,100 m (3,610 ft)
Water scooping distance to 15 m (50 ft)	1,450 m (4,760 ft)
Range with 4,000 kg (8,818 lb) payload	1,133 nm (2,100 km; 1,305 miles)
with max fuel	2,158 nm (4,000 km; 2,485 miles)

OPERATIONAL NOISE LEVELS:
T-O	96 EPNdB
Climb	90 EPNdB
Landing	98 EPNdB

BERIEV Be-103

TYPE: Twin-engined light business amphibian.
PROGRAMME: Design started 1992; model exhibited and initial data released at MosAeroshow '92; prototype construction began 1993; first flight scheduled for first half of 1994.
DESIGN FEATURES: Low-wing monoplane, with water-displacing wings of moderate sweep with large wingroot extensions; two-step boat hull; no stabilising floats; engine pylon-mounted on each side of rear fuselage, aft of wings; sweptback fin and rudder; tailplane mid-set on fin. Wing leading-edge sweep 22°; wing section NACA 2412M; dihedral 5° 3′ on outer wings; incidence 1°; aspect ratio 6.46.
FLYING CONTROLS: Manually operated; slab tailplane and ailerons by rods, rudder by cables; electric trim; spring feel i tailplane control; no flaps.
STRUCTURE: All-metal semi-monocoque boat-type fuselage all-metal single-spar wings.
LANDING GEAR: Pneumatically retractable tricycle type; singl wheel on each unit; mainwheels retract inward into win centre-section; nosewheel retracts forward; oleo-nitroge shock absorbers; brakes on mainwheels; nosewheel tyr size 400 × 150 mm, pressure 3 bars (43 lb/sq in); mair wheel tyres size 500 × 150 mm, pressure 3 bars (43 lb sq in); disc brakes; self-centring nosewheel steerable ±35°
POWER PLANT: Two 133 kW (178 hp) Bakanov M-17 pisto engines, each driving an AV-103 three-blade variable pitch tractor propeller, with optional reversible pitch; fue tank in each wing, total capacity 450 litres (119 US gal lons; 99 Imp gallons).
ACCOMMODATION: Pilot and five passengers in pairs; upwar opening door on each side, hinged on centreline; baggag freight compartment aft of cabin. Optional ambulance, al cargo, patrol, ecological monitoring use.
SYSTEMS: Interior heated and ventilated. Pneumatic system bottle capacity 6 litres (365 cu in) of air, pressure 49 bar (711 lb/sq in). 27V DC electrical system and three-phas 36V 400Hz AC, supplied by 3kW DC generators, rectifier and 25Ah battery.
AVIONICS: Digital flight control and nav equipment, VHF con radio, emergency locator beacon, intercom; position, angl rate, linear acceleration and angle of attack sensors; con ventional instrumentation.

DIMENSIONS, EXTERNAL:
Wing span	12.72 m (41 ft 9 in
Wing chord: at root	6.214 m (20 ft 4¾ ir
at tip	0.832 m (2 ft 9 in
Wing aspect ratio	6.4
Length overall	10.65 m (34 ft 11½ ir
Length of fuselage	9.96 m (32 ft 8 in
Height overall	3.90 m (12 ft 9½ ir
Tailplane span	3.90 m (12 ft 9½ in
Wheel track	2.91 m (9 ft 6½ in
Wheelbase	4.145 m (13 ft 7¼ in
Propeller diameter	1.60 m (5 ft 3 in
Distance between propeller centres	3.00 m (9 ft 10¼ in

DIMENSIONS, INTERNAL:
Cabin: Length	3.00 m (9 ft 10¼ in
Max width	1.25 m (4 ft 1¼ in
Max height	1.23 m (4 ft 0½ ir
Floor area	3.0 m² (32.3 sq ft
Volume	4.15 m³ (146.5 cu ft

AREAS:
Wings, gross	25.10 m² (270.2 sq ft
Ailerons, total	0.804 m² (8.65 sq ft
Fin	2.86 m² (30.8 sq ft
Rudder	1.54 m² (16.6 sq ft
Tailplane, total	3.68 m² (39.6 sq ft

WEIGHTS AND LOADINGS:
Weight empty	1,210 kg (2,668 lb
Max payload, incl fuel	550 kg (1,212 lb
Max fuel	320 kg (705 lb
Max T-O and landing weight	1,760 kg (3,880 lb
Max zero-fuel weight	1,675 kg (3,692 lb
Max wing loading	70.12 kg/m² (14.36 lb/sq ft
Max power loading	6.62 kg/kW (10.90 lb/hp

PERFORMANCE (estimated at max T-O weight):
Never-exceed speed (VNE)	178 knots (330 km/h; 205 mph
Max level speed at 2,000 m (6,560 ft)	156 knots (290 km/h; 180 mph
Max cruising speed	148 knots (275 km/h; 171 mph

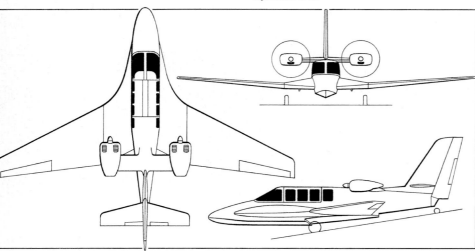

Beriev Be-103 twin-engined light multi-purpose amphibian *(Jane's/Mike Keep)*

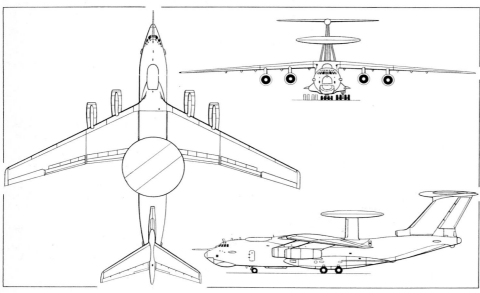

Beriev A-50 AEW&C version of Ilyushin Il-76, known to NATO as 'Mainstay' *(Dennis Punnett)*

Beriev A-50 'Mainstay', the AEW&C version of the Ilyushin Il-76 *(Swedish Air Force, via FLYGvapenNYTT)*

Nose of A-50 is much modified by comparison with that of Ilyushin Il-76 transport *(Linda Jackson)*

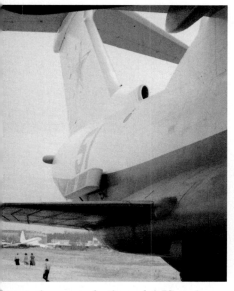

Flare pack on rear fuselage of A-50, and aerodynamic surface on side of landing gear fairing *(Mark Lambert)*

Stalling speed	54 knots (100 km/h; 62 mph)
Max rate of climb at S/L	390 m (1,280 ft)/min
Rate of climb at S/L, OEI	90 m (295 ft)/min
T-O run: on land	215 m (705 ft)
on water	390 m (1,280 ft)
T-O to 15 m (50 ft): on land	340 m (1,115 ft)
on water	500 m (1,640 ft)
Landing from 15 m (50 ft): on land	550 m (1,805 ft)
on water	590 m (1,935 ft)
Landing run: on land	190 m (625 ft)
on water	350 m (1,150 ft)
Range: with max payload	270 nm (500 km; 310 miles)
with max fuel	1,400 nm (2,600 km; 1,615 miles)

BERIEV A-50
NATO reporting name: Mainstay

TYPE: Four-turbofan airborne early warning and control aircraft.

PROGRAMME: Development from Ilyushin Il-76 transport began in 1970s to replace Tu-126s of APVO; production began early 1980s; maintained at rate of five a year until 1990 when only two delivered, causing problems for fighter force; service entry 1984; two operated round-the-clock over Black Sea during 1991 Gulf War, monitoring USAF flights from Turkey to Iraq and watching for possible stray US cruise missiles heading for CIS territory. Continued development by Beriev OKB.

CUSTOMERS: About 25 operational in 1992, primarily with MiG-29, MiG-31 and Su-27 counter-air fighters of home defence force and tactical air forces.

DESIGN FEATURES: Derivative of Il-76 (which see), with conventionally located rotating 'saucer' radome (diameter 9 m; 29 ft 6 in), lengthened fuselage forward of wings, flight refuelling noseprobe, satellite nav/com antenna above fuselage forward of wing, new IFF and comprehensive ECM; normal nose glazing around navigator's station replaced by non-transparent fairings; intake for avionics cooling at front of dorsal fin; no rear gun turret; small aerodynamic surface on each outer landing gear fairing, 'shadowing' circular form of rotating radome aft of wing trailing-edge.

FLYING CONTROLS: As for Il-76.
STRUCTURE: As for Il-76, except as noted under Design Features.
LANDING GEAR: As for Il-76.
POWER PLANT: As for Il-76. In-flight refuelling difficult because of severe buffeting induced by rotating radome in tanker's slipstream.
ACCOMMODATION: Normal crew of 15. No onboard toilet.
AVIONICS: Considered capable of detecting and tracking aircraft and cruise missiles flying at low altitude over land and water, and of helping to direct fighter operations over combat areas as well as enhancing air surveillance and defence of CIS. Colour CRT displays for radar observers; RWR; satellite data link to ground stations.
EQUIPMENT: Flare pack on each side of rear fuselage; wingtip countermeasures pods under development.
PERFORMANCE: Normally operates on figure-of-eight course at 10,000 m (33,000 ft), with 54 nm (100 km; 62 miles) between centres of the two orbits.

ILYUSHIN

AVIATSIONNYI KOMPLEKS IMIENI S. V. ILYUSHINA (Aviation Complex named after S. V. Ilyushin)

125319 Moscow
Telephone: 7 (095) 943 83 25
Fax: 7 (095) 212 21 320
Telex: 411956 Sokol
GENERAL DESIGNER: Genrikh V. Novozhilov
CHIEF DESIGNER: I. Ya. Katyrev
FOREIGN ECONOMIC RELATIONS DEPARTMENT:
  V. A. Belyakov

The Ilyushin OKB is named after Sergei Vladimirovich Ilyushin, who died 9 February 1977, aged 82. OKB was founded 1933. About 60,000 aircraft of Ilyushin design have been built.

ILYUSHIN Il-20
NATO reporting name: Coot-A

TYPE: Military elint/reconnaissance variant of Il-18 four-turboprop airliner.
PROGRAMME: First observed 1978; small number operational.
DESIGN FEATURES: Il-18 airframe basically unchanged; under-fuselage container, approx 10.25 m (33 ft 7½ in) long and 1.15 m (3 ft 9 in) deep, assumed to house side-looking radar; container, approx 4.4 m (14 ft 5 in) long and 0.88 m (2 ft 10½ in) deep, on each side of forward fuselage contains door over camera or other sensor; antennae and blisters include eight on undersurface of centre and rear

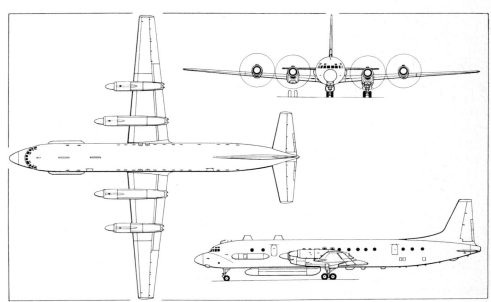

Ilyushin Il-20 (NATO 'Coot-A') elint/reconnaissance development of the Il-18 airliner *(Dennis Punnett)*

260   RUSSIA: AIRCRAFT—ILYUSHIN

Ilyushin Il-20 (NATO 'Coot-A') electronic intelligence (elint) and reconnaissance aircraft *(Swedish Air Force, via FLYGvapenNYTT)*

fuselage, with two large plates projecting above forward fuselage.

Detailed description of Il-18 airliner in 1979-80 and earlier editions of *Jane's*; *following abbreviated details of Il-18D indicate likely features retained by Il-20:*

POWER PLANT: Four 3,126 kW (4,190 ehp) ZMKB Progress/Ivchenko AI-20M turboprops, each with AV-68I four-blade reversible-pitch propeller. Ten flexible fuel tanks in inboard panel of each wing and integral tank in outboard panel; total capacity 23,700 litres (6,261 US gallons; 5,213 Imp gallons). Some Il-18 airliners have additional bag tanks in centre-section, giving total capacity of 30,000 litres (7,925 US gallons; 6,600 Imp gallons).

DIMENSIONS, EXTERNAL:
Wing span	37.42 m (122 ft 9¼ in)
Wing chord: at root	5.61 m (18 ft 5 in)
at tip	1.87 m (6 ft 2 in)
Wing aspect ratio	10.0
Length overall	35.9 m (117 ft 9 in)
Height overall	10.17 m (33 ft 4 in)
Tailplane span	11.80 m (38 ft 8½ in)
Wheel track	9.00 m (29 ft 6 in)
Wheelbase	12.78 m (41 ft 10 in)
Propeller diameter	4.50 m (14 ft 9 in)
Cabin doors (each): Height	1.40 m (4 ft 7 in)
Width	0.76 m (2 ft 6 in)
Height to sill	2.90 m (9 ft 6 in)

DIMENSIONS, INTERNAL:
Flight deck: Volume	9.36 m³ (330 cu ft)
Cabin, excl flight deck:	
Length	approx 24.0 m (79 ft 0 in)
Max width	3.23 m (10 ft 7 in)
Max height	2.00 m (6 ft 6¾ in)
Volume	238 m³ (8,405 cu ft)

AREAS:
Wings, gross	140.0 m² (1,506.9 sq ft)

WEIGHTS AND LOADINGS (Il-18D airliner):
Max payload	13,500 kg (29,750 lb)
Max T-O weight	64,000 kg (141,100 lb)
Max wing loading	457.1 kg/m² (93.6 lb/sq ft)
Max power loading	5.12 kg/kW (8.42 lb/ehp)

PERFORMANCE (Il-18D airliner, at max T-O weight):
Max cruising speed	364 knots (675 km/h; 419 mph)
Econ cruising speed	337 knots (625 km/h; 388 mph)
Operating height	8,000-10,000 m (26,250-32,800 ft)
T-O run	1,300 m (4,265 ft)
Landing run	850 m (2,790 ft)
Range, 1 h reserves:	
with max fuel	3,508 nm (6,500 km; 4,040 miles)
with max payload	1,997 nm (3,700 km; 2,300 miles)

ILYUSHIN Il-22
NATO reporting name: Coot-B

Many Il-22 airborne command post adaptations of the Il-18 transport are operational with CIS air forces. It would be logical to expect variety of external fairings and antennae, differing from one aircraft to another, depending on its specific duties. Il-22 shown in an accompanying illustration was identified by bullet fairing at fin tip, long and shallow container under front fuselage, and many small blade antennae above and below fuselage.

Ilyushin Il-22 (NATO 'Coot-B') airborne command post conversion of Il-18 airliner *(Piotr Butowski)*

ILYUSHIN Il-38
NATO reporting name: May

TYPE: Intermediate-range shore-based four-turboprop maritime patrol aircraft.
PROGRAMME: Development of Il-18 airliner, first reported 1970; about 59 serve currently with CIS Naval Aviation; first export order, from India, for five in 1975.
CUSTOMERS: CIS Naval Aviation; Indian Navy (INAS 315 at Dabolim, Goa).
DESIGN FEATURES: Basic Il-18 airframe, with lengthened fuselage, and wings moved forward to cater for effect of role equipment and stores on CG position; few cabin windows; large undernose radome; MAD tail sting; wing dihedral 3° from roots; mean thickness/chord ratio 14 per cent.
FLYING CONTROLS: Flying controls cable actuated; mass and aerodynamically balanced ailerons with electric trim tabs; hydraulically assisted elevators and rudder, each with electric trim tab; additional rudder spring tab; hydraulically actuated double-slotted wing trailing-edge flaps.
STRUCTURE: All-metal; three-spar wing centre-section, two spars in outer wings; circular-section fail-safe semi-monocoque fuselage, with rip-stop doublers around window cutouts, door frames and more heavily loaded skin panels.
LANDING GEAR: Retractable tricycle type, strengthened by comparison with Il-18. Hydraulic actuation. Four-wheel bogie main units, with 930 × 305 mm tyres and hydraulic brakes. Steerable (±45°) twin-wheel nose unit, with 700 × 250 mm tyres. Hydraulic brakes. Pneumatic emergency braking.

POWER PLANT: Four ZMKB Progress/Ivchenko AI-20M turboprops, each 3,126 kW (4,190 ehp), with AV-68I four-blade reversible-pitch metal propellers. Multiple bag fuel tanks in centre-section and in inboard panel of each wing and integral tank in outboard panel; total capacity 30,000 litres (7,925 US gallons; 6,600 Imp gallons). Pressure fuelling through four international standard connections in inner nacelles. Provision for overwing fuelling. Oil capacity 58.5 litres (15.45 US gallons; 12.85 Imp gallons) per engine. Engines started electrically.
ACCOMMODATION: Pilot and co-pilot side by side on flight deck, with dual controls; flight engineer to rear. Number of operational crew believed to be nine, but unconfirmed. Flight deck separated from main cabin by pressure bulkhead to reduce hazards following decompression of either. Main cabin has few windows and contains search equipment, electronic equipment and crew stations. Door on starboard side at rear of cabin (location of Il-18 service door).
SYSTEMS: Cabin max pressure differential 0.49 bar (7.1 lb/sq in). Eight engine driven generators for 28V DC and 115V 400Hz AC supply. Hydraulic system, pressure 207 bars (3,000 lb/sq in), for landing gear retraction, nosewheel steering, brakes, elevator and rudder actuators, flaps, weapon bay doors and radar antennae. Electrothermal de-icing for wings and tail unit.
AVIONICS: Navigation/weather radar in nose. Search radar (NATO 'Wet Eye') in undernose radome. MAD tail sting. Automatic navigation equipment, radio compasses and radio altimeter probably similar to those of Il-18.

Ilyushin Il-38 (four ZMKB Progress/Ivchenko AI-20M turboprops) over the Mediterranean, escorted by an RAF Hawk in 63 Squadron markings *(Royal Air Force)*

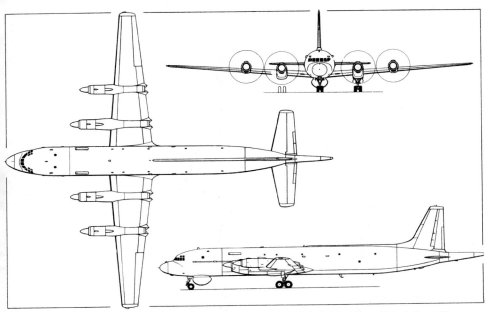

Ilyushin Il-38 anti-submarine/maritime patrol derivative of the Il-18 airliner *(Dennis Punnett)*

ARMAMENT: Two weapons/stores bays forward and aft of wing carry-through structure on most aircraft, for attack weapons and sonobuoys.

DIMENSIONS, EXTERNAL:
As listed under Il-20 entry, except:
Length overall   39.60 m (129 ft 10 in)

WEIGHTS AND LOADINGS:
Weight empty   36,000 kg (79,367 lb)
Max T-O weight   63,500 kg (140,000 lb)
Max wing loading   453.6 kg/m² (92.9 lb/sq ft)
Max power loading   5.08 kg/kW (8.35 lb/ehp)

PERFORMANCE:
Max level speed at 6,400 m (21,000 ft)
     390 knots (722 km/h; 448 mph)
Max cruising speed at 8,230 m (27,000 ft)
     330 knots (611 km/h; 380 mph)
Patrol speed at 600 m (2,000 ft)
     216 knots (400 km/h; 248 mph)
Min flying speed   103 knots (190 km/h; 118 mph)
T-O run   1,300 m (4,265 ft)
Landing run with propeller reversal   850 m (2,790 ft)
Range with max fuel   3,887 nm (7,200 km; 4,473 miles)
Patrol endurance with max fuel   12 h

ILYUSHIN Il-62M/MK

NATO reporting name: Classic

Small-scale production of this four-turbofan airliner continues. Details and an illustration last appeared in the 1989-90 *Jane's*.

ILYUSHIN Il-76

NATO reporting name: Candid
Indian Air Force name: Gajaraj

TYPE: Four-turbofan medium/long-range transport.
PROGRAMME: Design began late 1960s, led by G. V. Novozhilov, to replace turboprop An-12; prototype (SSSR-86712) flew 25 March 1971; official 1974 film showed development squadron of Il-76s, with twin-gun rear turrets, as vehicles for airborne troops; series production began 1975, exceeded 700 by early 1992; delivery of more than 50 a year continues, from Chkalov Plant, Tashkent, Uzbekistan.
CURRENT VERSIONS: **Il-76** ('Candid-A'): Initial basic production version.
Il-76T ('Candid-A'): Developed version; additional fuel in wing centre-section, above fuselage; heavier payload; no armament. *Description applies to this version.*
Il-76M ('Candid-B'): As Il-76T but military; up to 140 troops or 125 paratroops carried as alternative to freight; rear gun turret (not always fitted on export aircraft) containing two 23 mm twin-barrel GSh-23L guns; small ECM fairings (optional on export aircraft) between centre windows at front of navigator's compartment, on each side of front fuselage, and each side of rear fuselage; packs of ninety-six 50 mm IRCM flares on landing gear fairings and/or on sides of rear fuselage of aircraft operating into combat areas.
Il-76TD ('Candid-A'): Unarmed; generally as Il-76T but with Aviadvigatel D-30KP-2 turbofans, maintaining full power to ISA +23°C against ISA +15°C for earlier models; max T-O weight and payload increased; 10,000 kg

Model of Il-76MF, with cargo hold lengthened by 6.6 m (21 ft 8 in) *(Mark Lambert)*

262 RUSSIA: AIRCRAFT—ILYUSHIN

Ilyushin Il-76TD freight transport (four Aviadvigatel D-30KP-2 turbofans) *(Paul Duffy)*

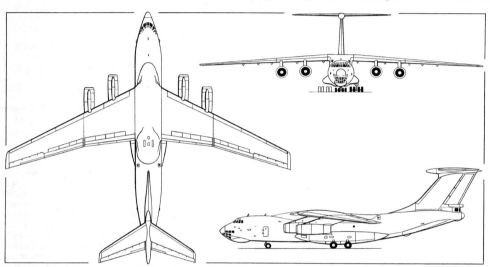

Ilyushin Il-76MD four-turbofan military transport *(Dennis Punnett)*

Ilyushin Il-76DMP, equipped for water-bombing use *(Piotr Butowski)*

(22,046 lb) additional fuel increases max fuel range by 648 nm (1,200 km; 745 miles); first identified when SSSR-76467 passed through Shannon Airport, Ireland, November 1982; fully operational July 1983; one specially equipped with seats, soundproofing, buffet kitchen, toilet and working facilities, to carry members of Antarctic expeditions between Maputo, Mozambique, and Molodozhnaya Station, Antarctica (proving flight February 1986 with 94 passengers, 14,000 kg; 30,865 lb of scientific equipment, cargo and baggage containers).

Under development is **new version** of Il-76TD **with CFM56 turbofans**, each rated at 138.8 kN (31,200 lb st); range increased 20-30 per cent; fuel burn decreased; noise reduced to comply with ICAO Ch 3 Appendix 16.

Il-76MF: Stretched version with four Aviadvigatel PS-90AN turbofans, each 156.9 kN (35,275 lb st). Cargo hold lengthened 6.6 m (21 ft 8 in) by two plugs, fore and aft of wings; max payload 52,000 kg (114,640 lb); range, with reserves, with 40,000 kg (88,185 lb) payload 2,805 nm (5,200 km; 3,230 miles). First flight scheduled for 1994. To be built in Tashkent for civil and military use.

Il-76MD ('Candid-B'): Military version; generally as Il-76M but with improvements of Il-76TD.

Il-76MDK: Adaptation of Il-76MD to enable cosmonauts to experience several tens of seconds of weightlessness during training.

Il-76DMP: Firefighting conversion of Il-76 demonstrated first in 1990; up to 44 tonnes of water/fire retardant in two cylindrical tanks in hold; discharge, replenishment and draining systems; drop zone aiming devices; up to 384 meteorological cartridges in dispensers for weather modification; able to water-bomb an area of 500 × 100 m (1,650 × 330 ft) in 6 s, or to carry, and parachute when required, 40 firefighters; all airborne fire equipment (weight 5,000 kg; 11,025 lb) can be installed in standard Il-76, or removed, in four hours; tank replenishment time 10-15 min; discharge time 6-7 s, with option of successive discharge of tanks to cover 600 × 80 m (1,970 × 260 ft); airspeed during discharge 130-215 knots (240-400 km/h; 150-248 mph).

Il-76LL: Engine testbed conversion, carrying gas turbine of up to 245 kN (55,100 lb st), including turboprops, in place of normal port inner D-30KP; provisions for five test engineers; four Il-76LLs are available, on commercial contract basis, from Gromov Flight Research Institute; engines tested by 1991 include NK-86, PS-90A and D-18T turbofans and D-236 propfan.

Specialised variants and developments of Il-76 include the AEW&C **A-50** known to NATO as 'Mainstay' (described separately under Beriev OKB entry); the **'Be-976'**, with rotating radome like A-50 and wingtip avionics pods but with nose glazing retained, used as radar picket at Zhukovsky to observe flight tests; the **Il-78** (NATO 'Midas') flight refuelling tanker (described separately); the AEW&C **Adnan 1** and single-point flight refuelling tanker modified for Iraq. India is reportedly considering conversion of standard Il-76s for AEW duties.

CUSTOMERS: First-line units of CIS Transport Aviation force (VTA) (more than 500 Il-76/76M/76MD); air forces of Algeria, Czech Republic, India (24, given name Gajaraj), Iraq, Poland; commercial operators include Aeroflot and its successors (more than 120, including Il-76Ts and Ms, forming military reserve), HeavyLift Cargo Airlines (UK), Iraqi Airways (about 30 Il-76Ts and Ms, operated for military), Jamahiriya Libyan Arab Airlines (21 Il-76Ts and Ms), Syrianair (two Il-76Ts, two Ms); Il-76Ms of airlines have no guns in turret; first of at least two Il-76MDs delivered for Cubana had no tail turret.

DESIGN FEATURES: Late 1960s requirement was to carry 40 tonnes of freight 2,700 nm (5,000 km; 3,100 miles) in less than six hours, with ability to operate from short unprepared airstrips, in the most difficult weather conditions experienced in Siberia, the north of the Soviet Union and the Far East, while much simpler to service and able to fly much faster than An-12; wings mounted above fuselage to leave interior unobstructed; rear loading ramp/door; unique landing gear, with two large external fairings for each main gear. Wing anhedral constant outboard of centre-section; sweepback 25° at quarter-chord; thickness/chord ratio 13 per cent at root, 10 per cent at tip; all tail surfaces sweptback.

FLYING CONTROLS: Hydraulically boosted; manual operation possible in emergency; mass balanced ailerons, with balance/trim tabs; two-section triple-slotted trailing-edge flaps over approx 75 per cent of each semi-span; eight upper-surface spoilers forward of flaps on each wing, four on each inner and outer panel; leading-edge slats over almost entire span, two on each inner panel, three on each outer panel; variable incidence T tailplane; elevators and rudder aerodynamically balanced, each with tab.

STRUCTURE: All-metal; five-piece wing of multi-spar fail-safe construction, centre-section integral with fuselage; basically circular-section semi-monocoque fail-safe fuselage; underside of upswept rear fuselage made up of two outward hinged clamshell doors, upward hinged panel between doors, and downward hinged ramp; all tail surfaces sweptback.

LANDING GEAR: Hydraulically retractable tricycle type. Steerable nose unit has two pairs of wheels, side by side, with central oleo. Main gear on each side has two units in tandem, each unit with four wheels on single axle. Low-pressure tyres size 1,300 × 480 mm on mainwheels, 1,100 × 330 mm on nosewheels. Nosewheels retract forward. Main units retract inward into two large ventral fairings under fuselage, with additional large fairing on each side of lower fuselage over actuating gear. During retraction mainwheel axles rotate around leg, so that wheels stow with axles parallel to fuselage axis (ie: wheels remain vertical but at 90° to direction of flight). All doors on wheel wells close when gear is down, to prevent fouling of legs by snow, ice, mud, etc. Oleo-pneumatic shock absorbers. Tyre pressure can be varied in flight from 2.5 to 5 bars

(36-73 lb/sq in) to suit different landing strip conditions. Hydraulic brakes on mainwheels.

POWER PLANT: Four Aviadvigatel D-30KP turbofans, each 117.7 kN (26,455 lb st), in individual underwing pods. Each pod on large forward-inclined pylon and fitted with clamshell thrust reverser. Integral fuel tanks between spars of inner and outer wing panels. Total fuel capacity 109,480 litres (28,922 US gallons; 24,083 Imp gallons).

ACCOMMODATION: Crew of seven, including two freight handlers. Conventional side by side seating for pilot and co-pilot on spacious flight deck. Station for navigator below flight deck in glazed nose. Crew door on port side of nose. Forward hinged main cabin door on each side of fuselage forward of wing. Two windows on each side of hold serve as emergency exits. Hold has reinforced floor of titanium alloys, with folding roller conveyors, and is loaded via rear ramp. Entire accommodation pressurised. Advanced mechanical handling systems for containerised and other freight, which can include standard ISO containers, each 12 m (39 ft 4½ in) long, building machinery, heavy crawlers and mobile cranes. Typical loads include six containers measuring either 2.99 × 2.44 × 2.44 m (9 ft 9¾ in × 8 ft × 8 ft) or 2.99 × 2.44 × 1.90 m (9 ft 9¾ in × 8 ft × 6 ft 2¾ in) and with loaded weights of 5,670 kg (12,500 lb) or 5,000 kg (11,025 lb) respectively; or twelve containers measuring 1.46 × 2.44 × 1.90 m (4 ft 9¼ in × 8 ft × 6 ft 2¾ in) and each weighing 2,500 kg (5,511 lb) loaded; or six pallets measuring 2.99 × 2.44 m (9 ft 9¾ in × 8 ft) and each weighing 5,670 kg (12,500 lb); or twelve pallets measuring 1.46 × 2.44 m (4 ft 9¼ in × 8 ft) and each weighing 2,500 kg (5,511 lb). Folding seats along sidewalls in central portion of hold. Quick configuration changes made by use of modules, each able to accommodate 36 passengers in four-abreast seating, litter patients and medical attendants, or cargo. Three such modules can be carried, each approx 6.10 m (20 ft) long, 2.44 m (8 ft) wide and 2.44 m (8 ft) high; loaded through rear doors by two overhead travelling cranes, and secured to cabin floor with cargo restraints. Two winches at front of hold, each with capacity of 3,000 kg (6,615 lb). Cranes embody total of four hoists, each with capacity of 2,500 kg (5,511 lb). Ramp can be used as additional hoist, with capacity of up to 30,000 kg (66,140 lb) to facilitate loading of large vehicles and those with caterpillar tracks. Pilot's and co-pilot's windscreens can each be fitted with two wipers, top and bottom.

SYSTEMS: Flight deck only, or entire interior, can be pressurised; max differential 0.50 bar (7.25 lb/sq in). Hydraulic system includes servo motors and motors to drive flaps, slats, landing gear and its doors, ramp, rear fuselage clamshell doors and load hoists. Flying control boosters supplied by electric pumps independent of central hydraulic supply. Electrical system includes engine driven generators, auxiliary generators driven by APU, DC converters and batteries. It powers pumps for flying control system boosters, radio and avionics, and lighting systems.

AVIONICS: Full equipment for all-weather operation by day and night, including computer for automatic flight control and automatic landing approach. Weather radar in nose; navigation and ground mapping radar in undernose radome.

EQUIPMENT: APU in port side landing gear fairing for engine starting and to supply all aircraft systems on ground, making aircraft independent of ground facilities.

Radar picket conversion of Il-76, known as 'Be-976' *(Piotr Butowski)*

DIMENSIONS, EXTERNAL:
Wing span 50.50 m (165 ft 8 in)
Wing aspect ratio 8.5
Length overall 46.59 m (152 ft 10¼ in)
Fuselage: Max diameter 4.80 m (15 ft 9 in)
Height overall 14.76 m (48 ft 5 in)
Rear loading aperture: Width 3.40 m (11 ft 1¾ in)
Height 3.45 m (11 ft 4 in)

DIMENSIONS, INTERNAL:
Cabin: Length: excl ramp 20.00 m (65 ft 7½ in)
incl ramp 24.50 m (80 ft 4½ in)
Width 3.40 m (11 ft 1¾ in)
Height 3.46 m (11 ft 4¼ in)
Volume 235.3 m³ (8,310 cu ft)

AREAS:
Wings, gross 300.0 m² (3,229.2 sq ft)

WEIGHTS AND LOADINGS:
Max payload: T 40,000 kg (88,185 lb)
TD 50,000 kg (110,230 lb)
Max fuel: T 84,840 kg (187,037 lb)
Max T-O weight: T 170,000 kg (374,785 lb)
TD 190,000 kg (418,875 lb)
Max landing weight: TD 151,500 kg (333,995 lb)
Permissible axle load (vehicles):
T 7,500-11,000 kg (16,535-24,250 lb)
Permissible floor loading:
T 1,450-3,100 kg/m² (297-635 lb/sq ft)
Max wing loading: T 566.7 kg/m² (116.05 lb/sq ft)
TD 633.3 kg/m² (129.72 lb/sq ft)
Max power loading: T 361.1 kg/kN (3.54 lb/lb st)
TD 403.6 kg/kN (3.95 lb/lb st)

PERFORMANCE:
Max level speed: T, TD 459 knots (850 km/h; 528 mph)
Cruising speed:
T, TD 405-432 knots (750-800 km/h; 466-497 mph)
T-O speed: T 114 knots (210 km/h; 131 mph)
Approach and landing speed:
T 119-130 knots (220-240 km/h; 137-149 mph)

Normal cruising height:
T, TD 9,000-12,000 m (29,500-39,370 ft)
Absolute ceiling: T approx 15,500 m (50,850 ft)
T-O run: T 850 m (2,790 ft)
TD 1,700 m (5,580 ft)
Landing run: T 450 m (1,475 ft)
TD 900-1,000 m (2,950-3,280 ft)
Range with max payload:
TD 1,970 nm (3,650 km; 2,265 miles)
Nominal range with 40,000 kg (88,185 lb) payload:
T 2,700 nm (5,000 km; 3,100 miles)
Max range, with reserves:
T 3,617 nm (6,700 km; 4,163 miles)
Range with 20,000 kg (44,090 lb) payload:
TD 3,940 nm (7,300 km; 4,535 miles)

ILYUSHIN Il-76 COMMAND POST

TYPE: Airborne command post version of Il-76MD transport.
PROGRAMME: Two examples (SSSR-76450 and -76451) photographed at Zhukovsky Flight Research Centre 1992.
DESIGN FEATURES: Large canoe-shaped fairing above fuselage forward of wings; five small antennae above centre-section; other small antennae, and air intake scoops, under front fuselage and at rear of main landing gear fairings; long and shallow fairing forward of dorsal fin on each side at top of fuselage; large downward inclined flat plate antenna on each side under tailcone; long pod-mounted probe on pylon under each outer wing; nose glazing around navigator's compartment deleted and flight deck rear side windows covered; downward facing exhaust near end of port landing gear fairing; partially retracted basket-drogue of what appears to be a VLF trailing wire aerial under rear fuselage.

Ilyushin Il-76 command post aircraft at Zhukovsky Flight Research Centre *(Sebastian Zacharias)*

ILYUSHIN Il-78M

NATO reporting name: Midas

TYPE: Four-turbofan probe-and-drogue flight refuelling tanker.

PROGRAMME: Development began in late 1970s, to replace modified Myasishchev 3MS2 and 3MN2 (NATO 'Bison') used previously in this role; first operational Il-78 with single refuelling pod entered service 1987, supporting tactical and strategic combat aircraft; fully developed Il-78 is three-point tanker.

CURRENT VERSIONS: **Il-78:** Initial version with two cylindrical fuel storage tanks inside cargo hold; convertible into transport by removal of tanks; in addition to tanks in hold, fuel can be transferred from standard tanks in wing torsion box. Able also to refuel aircraft on ground, using conventional hoses.

Il-78M: Current standard; non-convertible, with third storage tank in hold, increasing available transfer fuel by 20,000 kg (44,090 lb); all three tanks fixed; lower structure weight; increased max T-O weight.

CUSTOMERS: Twelve operational with 230th Air Tanker Regiment in early 1992.

DESIGN FEATURES: Development of basic Il-76MD; three-point tanker, with UPAZ-1A refuelling system, utilising refuelling pods of same type under outer wings and on port side of rear fuselage; rear turret retained as flight refuelling observation station, without guns; rear-facing radar rangefinder built into bottom of upswept rear fuselage.

POWER PLANT: Four Aviadvigatel D-30KP-2 turbofans; each 117.7 kN (26,455 lb st).

ACCOMMODATION: Crew of seven.

AVIONICS: Kupol navigation system and RSBN short-range nav system to permit all-weather day/night mutual detection and approach by receiver aircraft from distances up to 160 nm (300 km; 185 miles). Systems control automatically the convergence and warn of too-close approach. Refuelling permitted only in direct visibility.

DIMENSIONS, EXTERNAL: Same as Il-76, except:
Distance of refuelling pods from c/l:
  Underwing pods 16.40 m (53 ft 9¾ in)
  Rear fuselage pod 3.00 m (9 ft 10¼ in)

WEIGHTS AND LOADINGS:
Weight empty: 78 98,000 kg (216,050 lb)
Fuel load:
  Wing tanks: 78 90,000 kg (198,412 lb)
  Fuselage tanks: 78 28,000 kg (61,728 lb)
Max T-O weight:
  On concrete runway: 78 190,000 kg (418,875 lb)
  78M 210,000 kg (462,965 lb)
  On runway with bearing strength less than 6 kg/cm² (85 lb/sq in): 78, 78M 157,000 kg (346,120 lb)
Max landing weight: 78 151,500 kg (333,995 lb)

PERFORMANCE:
Nominal cruising speed:
  78 405 knots (750 km/h; 466 mph)
Refuelling speed:
  78 232-318 knots (430-590 km/h; 267-366 mph)
Refuelling height: 78 2,000-9,000 m (6,560-29,525 ft)
Max T-O run: 78M 2,080 m (6,825 ft)
Refuelling radius:
  with 60,000-65,000 kg (132,275-143,300 lb) transfer fuel:
    78 540 nm (1,000 km; 620 miles)
  with 32,000-36,000 kg (70,545-79,365 lb) transfer fuel:
    78 1,350 nm (2,500 km; 1,553 miles)

Air-to-air refuelling pods under port wing and on rear fuselage of Il-78M *(Peter J. Cooper)*

Ilyushin Il-78M (NATO 'Midas') three-point refuelling tanker *(Piotr Butowski)*

Il-78M flight refuelling tanker with drogues streamed, in formation with Tu-95MS and escorting MiG-29 fighters *(Piotr Butowski)*

ILYUSHIN Il-86

NATO reporting name: Camber

TYPE: Four-turbofan medium-range wide-bodied passenger transport.

PROGRAMME: Construction of two prototypes started 1974; first flight 22 December 1976 from Ilyushin OKB headquarters at old Moscow Central Airport, Khodinka, from 1,820 m (5,970 ft) runway, to official flight test centre, by prototype SSSR-86000; first production Il-86 (SSSR-86002) flew at Voronezh assembly plant 24 October 1977; first delivery (SSSR-86004) to Aeroflot 24 September 1979; scheduled services began 26 December 1980; first international service, Moscow-East Berlin, 3 July 1981; 97 built by April 1993; subject to funding, delivery of CFM56-5 turbofans for late production aircraft will begin 1994.

CUSTOMERS: Aeroflot and its successors only.

DESIGN FEATURES: Conventional low/mid swept wing; two-deck fuselage, intended to be entered via lower-deck doors, into stowage compartments for coats and hand baggage, and up stairways to passenger deck (deletion of lower-deck airstairs and internal stairways optional); additional centreline bogie between main landing gear units; dihedral from roots on wings and tailplane; wing sweepback 35° at quarter-chord; all tail surfaces swept.

FLYING CONTROLS: Hydraulic actuation, without manual reversion for primary surfaces; aileron and two-section double-slotted flaps occupy entire trailing-edge of each wing; multi-section upper-surface spoilers and airbrakes forward of each flap section; full-span leading-edge slats; variable incidence tailplane; rudder and elevators each two-section.

STRUCTURE: All-metal; inner wings three-spar, outer panels two-spar; shallow fence above wing in line with each engine pylon; circular-section semi-monocoque fuselage; floors of both decks of honeycomb and carbonfibre reinforced plastics. Flaps, wing slats, engine pylons, tailplane and fin manufactured by PZL Mielec, Poland.

LANDING GEAR: Retractable four-unit type. Forward retracting steerable twin-wheel nose unit; three four-wheel bogie main units. Two of latter retract inward into wingroot fairings; third is mounted centrally under fuselage, slightly forward of the others, and retracts forward. (Main landing gear made at Kuybyshev.) Mainwheel tyres size 1,300 × 480 mm; nosewheel tyres size 1,120 × 450 mm.

POWER PLANT: Initially, four KKBM (Kuznetsov) NK-86 turbofans, each 127.5 kN (28,660 lb st), on pylons forward of wing leading-edges. Combined thrust reversers/noise attenuators. CFM56-5 turbofans intended for some aircraft. Integral fuel tanks in wings, capacity 114,000 litres (30,116 US gallons; 25,077 Imp gallons).

ACCOMMODATION: Two pilots and flight engineer, with provision for navigator. Flight engineer's seat normally faces to starboard, aft of co-pilot, but can pivot to central forward-facing position to enable engineer to operate throttles. Upper deck, on which all seats are located, divided into three separate cabins by wardrobes, a serving area connected by elevator to lower deck galley, and cabin staff accommodation, with eight toilets at front (two) and rear (six) of aircraft. Unusually large windows, indirect lighting in walls and in ceiling panels, and enclosed baggage lockers at top of side walls. Preponderance of metal and natural fibre materials rather than plastics throughout cabins to enhance safety in an emergency. Up to 350 passengers in basic nine-abreast seating throughout, with two aisles, each 55 cm (21.6 in) wide. Mixed class layout for 28 passengers six-abreast in front cabin, and 206 passengers eight-abreast in other two cabins. Three airstair doors (made in Kharkov) hinge down from port side of lower deck; one is forward of wing, others aft. Four further doors at upper deck level on each side, for emergency use (using dual inflatable escape slides) and for use at airports where utilisation of high level boarding steps or bridges preferred. Coats and hand baggage stowed on lower deck before passengers climb one of three fixed stairways to main deck. (Optional deletion of lower deck airstair doors and stairways reduces operating weight empty by 3,000 kg; 6,610 lb and permits installation of 25 more seats on upper deck.) Cargo holds on lower deck accommodate heavy or registered baggage and freight in 8 standard LD3 containers, or 16 LD3 containers if some carry-on baggage racks omitted. Access via upward hinged doors forward of starboard wingroot leading-edge and at side of rear hold. Containers can be loaded and unloaded by self-propelled truck with built-in roller conveyor. Films can be shown in

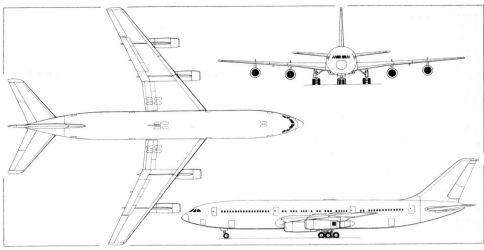

Ilyushin Il-86 four-turbofan wide-bodied passenger transport (*Dennis Punnett*)

Model of Il-86 with CFM56-5 turbofans (*Air Portraits*)

Ilyushin Il-86 wide-bodied transport (four KKBM NK-86 turbofans) (*Anton Wettstein*)

flight, and there is choice of 12 tape recorded audio programmes. A bar-buffet can be provided on lower deck in place of baggage and freight accommodation in forward vestibule.

SYSTEMS: Four self-contained hydraulic systems, each operated by one of engines, for actuation of flying control surfaces, tailplane variable incidence, spoilers, airbrakes, slats, flaps, landing gear, nosewheel steering, wheel brakes, anti-skid system, and upper level doors when passenger gangways used. All hot pipelines of air-conditioning system, and all fuel supply lines, outside pressure cell. Primary 200/155V 400Hz AC electrical system powered by four 40kVA engine driven generators. Secondary 36V three-phase AC and 27V DC systems. Five accumulators and static transformer. Smoke detection sensors in baggage, freight and equipment stowage areas. Pulse generating de-icing system consuming 500 times less energy than conventional hot air or electrical system. APU in tailcone.

AVIONICS: All avionics equipment within pressurised part of fuselage. Flight control and nav systems provide for automatic climb to selected height, control of rate of climb and automatic descent, and automatic landing in ICAO Cat. IIIA conditions. Pre-programmable Doppler nav system with readout display screen, on which microfilmed maps can be projected. Position of aircraft indicated by cursor, driven by computer. Nav system updated automatically by inputs from VOR or VOR/DME radio beacons.

DIMENSIONS, EXTERNAL:
Wing span	48.06 m (157 ft 8¼ in)
Length overall	59.54 m (195 ft 4 in)
Fuselage: Length	56.10 m (184 ft 0¾ in)
Max diameter	6.08 m (19 ft 11½ in)
Height overall	15.81 m (51 ft 10½ in)
Tailplane span	20.57 m (67 ft 6 in)
Wheel track (c/l of outer shock struts)	11.15 m (36 ft 7 in)
Wheelbase	21.34 m (70 ft 0 in)

DIMENSIONS, INTERNAL:
Main cabins: Height	2.61 m (8 ft 7 in)
Max width	approx 5.70 m (18 ft 8½ in)

AREAS:
Wings, gross	320.0 m² (3,444 sq ft)

WEIGHTS AND LOADINGS (NK-86 engines):
Max payload	42,000 kg (92,600 lb)
Max fuel	88,350 kg (194,775 lb)
Max T-O weight (dependent on size and type of runway)	190,000-208,000 kg (418,875-458,560 lb)
Max landing weight	175,000 kg (385,800 lb)
Max wing loading	650.0 kg/m² (133.1 lb/sq ft)
Max power loading	407.9 kg/kN (4.0 lb/lb st)

PERFORMANCE (designed, with NK-86 engines):
Normal cruising speed at 9,000-11,000 m (30,000-36,000 ft)
  486-512 knots (900-950 km/h; 559-590 mph)
Approach speed
  130-141 knots (240-260 km/h; 149-162 mph)
Field length for T-O and landing
  2,300-2,600 m (7,550-8,530 ft)
* Range: with 40,000 kg (88,185 lb) payload
  1,944 nm (3,600 km; 2,235 miles)
* with max fuel 2,480 nm (4,600 km; 2,858 miles)
* *Reports suggest that these design ranges are not being achieved. The former East German airline Interflug quoted a max range of 1,350 nm (2,500 km; 1,550 miles) in its sales literature*

ILYUSHIN Il-86 COMMAND POST

TYPE: Airborne command post version of Il-86 transport.
PROGRAMME: First observed at Zhukovsky Flight Research Centre 1992; four seen to be completed at that time (SSSR-86146-49).
DESIGN FEATURES: Large boat-shaped fairing above fuselage forward of wings; large pod with ram air intake under each inner wing; long and shallow dished fairing forward of fin root; strake antenna under rear fuselage; small fin-like component on port side lower fuselage, carrying what appears to be drogue for VLF trailing wire aerial. SSSR-86146 and -86147 have large blade aerials above centre and rear fuselage and under forward fuselage.

ILYUSHIN Il-96-300

TYPE: Four-turbofan wide-bodied passenger transport.
PROGRAMME: First of five prototypes (SSSR-96000) flew at Khodinka 28 September 1988, second on 28 November 1989; two used for static and fatigue testing; areas of commonality with Il-86 permitted planned test programme to be reduced to 750 flights totalling 1,200 hours; route proving trials by first production aircraft (SSSR-96005) conducted late 1991; nine flying 1992; certification received 29 December 1992; CIS airlines to receive 70 by 1995.
CUSTOMERS: Aeroflot and its successors (more than 100); CSA of Czech Republic (intent to buy).
DESIGN FEATURES: Superficial resemblance to Il-86, but new design, with different engines to overcome performance deficiencies of original Il-86; new structural materials and state of the art technology intended to provide life of 60,000 hours and 12,000 landings; no lower-deck passenger entry; winglets standard; wing and tailplane dihedral from roots; supercritical wings, with 30° sweep at quarter-chord; sweepback at quarter-chord 37° 30′ on tailplane, 45° on fin. Current development aiming at range of 6,475 nm (12,000 km; 7,450 miles) with 300 passengers.
FLYING CONTROLS: Triplex fly-by-wire, with manual reversion; each wing trailing-edge occupied by, from root, double-slotted inboard flap, small inboard aileron, two-section single-slotted flaps, and outboard aileron used only as gust damper and to smooth out buffeting; seven-section full-span leading-edge slats on each wing; three airbrakes forward of each inboard trailing-edge flap; six spoilers forward of outer flaps; inboard pair supplement ailerons,

Ilyushin Il-86 airborne command post *(Sebastian Zacharias)*

Ilyushin Il-86 airborne command posts. Note blade antennae above and below fuselage of rear aircraft, and apparent drogue for VLF trailing wire aerial on both aircraft *(Sebastian Zacharias)*

Flight deck of Ilyushin Il-96-300 four-turbofan wide-bodied airliner *(Air Portraits)*

Ilyushin Il-96-300 four-turbofan wide-bodied passenger transport *(Piotr Butowski)*

others operate as airbrakes and supplementary ailerons; variable incidence tailplane; two-section rudder and elevators, without tabs.

STRUCTURE: Basically all-metal, including new high-purity aluminium alloy, with composites flaps, main-deck floors and underfloor holds of honeycomb and CFRP; inner wings three-spar, outer panels two-spar; each wing has seven machined skin panels, three top surface, four bottom, with integral stiffeners; circular-section semi-monocoque fuselage; leading- and trailing-edges of fin and tailplane of composites. Some components manufactured by PZL Mielec, Poland.

LANDING GEAR: Retractable four-unit type. Forward retracting steerable twin-wheel nose unit; three four-wheel bogie main units. Two of latter retract inward into wingroot/fuselage fairings; third is mounted centrally under fuselage, to rear of others, and retracts forward after the bogie has itself pivoted upward 20°. Oleo-pneumatic shock absorbers. Nosewheel tubeless tyres size 1,260 × 460 mm; mainwheel tubeless tyres size 1,300 × 480 mm. Tyre pressure (all) 11.65 bars (169 lb/sq in).

POWER PLANT: Four Aviadvigatel PS-90A turbofans, each 156.9 kN (35,275 lb st), on pylons forward of wing leading-edges. Thrust reversal standard. Integral fuel tanks in wings and fuselage centre-section, total capacity 148,260 litres (39,166 US gallons; 32,613 Imp gallons).

ACCOMMODATION: Pilot, co-pilot and flight engineer; two seats for supplementary crew or observer. Ten or 12 cabin staff. Basic all-tourist configuration has two cabins for 66 and 234 passengers respectively, nine-abreast at 87 cm (34.25 in) seat pitch, separated by buffet counter, video stowage and two lifts from galley on lower deck. Two aisles, each 55 cm (21.65 in) wide. Two toilets and wardrobe at front, six more toilets, a rack for cabin staff's belongings and seats for cabin staff at rear. Seats recline, and are provided with individual tables, ventilation, earphones and attendant call button. Indirect lighting is standard. 235-seat mixed class version has front cabin for 22 first class passengers, six-abreast in pairs, at 102 cm (40 in) seat pitch and with aisles 75.5 cm (29.7 in) wide; centre cabin with 40 business class seats, eight-abreast at 90 cm (35.4 in) seat pitch and with aisles 56.5 cm (22.25 in) wide; and rear cabin for 173 tourist class passengers, basically nine-abreast at 87 cm (34.25 in) seat pitch, with aisle width of 55 cm (21.65 in). Unlike Il-86, passenger cabin is entered through three doors on port side of upper deck, at front and rear and forward of the wings. Opposite each door, on starboard side, is emergency exit door. Lower deck houses front cargo compartment for six ABK-1.5 (LD3) containers or igloo pallets, central compartment aft of wing for ten ABK-1.5 containers or pallets, and tapering compartment for general cargo at rear. Three doors on starboard side provide separate access to each compartment. Galley and lifts are between front cargo compartment and wing, with separate door aft of front cargo compartment door.

SYSTEMS: Four independent hydraulic systems, using fire- and explosion-proof fluid, at pressure of 207 bars (3,000 lb/sq in). APU in tailcone.

AVIONICS: On the flight deck conventional standby instruments are retained, but primary flight information is presented on dual twin-screen colour CRTs, fed by triplex INS, a satellite-based and Omega navigation system and other sensors. Triplex flight control and flight management systems, together with a head-up display, permit fully automatic en route control and operations in ICAO Cat. IIIA minima. Duplex engine and systems monitoring and failure warning systems feed in-flight information to both the flight engineer's station and monitors on the ground. Autothrottle is based on IAS, without angle of attack protection. Another electronic system provides real-time automatic weight and CG situation data.

DIMENSIONS, EXTERNAL:
Wing span: excl winglets	57.66 m (189 ft 2 in)
over winglets	60.105 m (197 ft 2½ in)
Wing aspect ratio	9.5
Length overall	55.35 m (181 ft 7¼ in)
Fuselage: Length	51.15 m (167 ft 9¾ in)
Max diameter	6.08 m (19 ft 11½ in)
Height overall	17.57 m (57 ft 7¾ in)
Tailplane span	20.57 m (67 ft 6 in)
Wheel track	10.40 m (34 ft 1½ in)
Wheelbase	20.065 m (65 ft 10 in)
Passenger doors (three): Height	1.83 m (6 ft 0 in)
Width	1.07 m (3 ft 6 in)
Height to sill: Nos. 1 and 2	4.54 m (14 ft 10¾ in)
No. 3	4.80 m (15 ft 9 in)
Emergency exit doors (three):	
Height	1.825 m (5 ft 11¾ in)
Width	1.07 m (3 ft 6 in)
Cargo compartment doors (front and centre):	
Height	1.825 m (5 ft 11¾ in)
Width	1.78 m (5 ft 10 in)
Height to sill: front	2.34 m (7 ft 8¼ in)
centre	2.48 m (8 ft 1¾ in)
Cargo compartment door (rear):	
Height	1.38 m (4 ft 6¼ in)
Width	0.972 m (3 ft 2¼ in)
Height to sill	2.74 m (9 ft 0 in)
Galley door: Height	1.20 m (3 ft 11¼ in)
Width	0.80 m (2 ft 7½ in)

DIMENSIONS, INTERNAL:
Cabins, excl flight deck:	
Height	2.60 m (8 ft 6¼ in)
Max width	approx 5.70 m (18 ft 8½ in)
Volume	350 m³ (12,360 cu ft)
Cargo hold volume: front	37.10 m³ (1,310 cu ft)
centre	63.80 m³ (2,253 cu ft)
rear	15.00 m³ (530 cu ft)

AREAS:
Wings, gross	391.6 m² (4,215 sq ft)
Vertical tail surfaces (total)	61.0 m² (656.6 sq ft)
Horizontal tail surfaces (total)	96.5 m² (1,038.75 sq ft)

WEIGHTS AND LOADINGS:
Basic operating weight	117,000 kg (257,940 lb)
Max payload	40,000 kg (88,185 lb)
Max fuel	114,902 kg (253,311 lb)
Max T-O weight	216,000 kg (476,200 lb)
Max landing weight	175,000 kg (385,810 lb)
Max zero-fuel weight	157,000 kg (346,120 lb)
Max wing loading	551.6 kg/m² (113.0 lb/sq ft)
Max power loading	344.2 kg/kN (3.37 lb/lb st)

PERFORMANCE (estimated):
Normal cruising speed at 10,100-12,100 m (33,135-39,700 ft)	459-486 knots (850-900 km/h; 528-559 mph)
Approach speed	140 knots (260 km/h; 162 mph)
Balanced T-O runway length	2,600 m (8,530 ft)
Balanced landing runway length	1,980 m (6,500 ft)
Range, with UASA reserves:	
with max payload	4,050 nm (7,500 km; 4,660 miles)
with 30,000 kg (66,140 lb) payload	4,860 nm (9,000 km; 5,590 miles)
with 15,000 kg (33,070 lb) payload	5,940 nm (11,000 km; 6,835 miles)

OPERATIONAL NOISE LEVELS:
Il-96-300 is designed to conform with ICAO Chapter 3 Annex 16 noise requirements

ILYUSHIN Il-96M and Il-96T

TYPE: Four-turbofan wide-bodied passenger or freight transport.

PROGRAMME: Projected initially as Il-96-350; designation changed to Il-96M in 1990, when model exhibited at Moscow Aerospace '90; then at initial design stage; Pratt & Whitney supplying 10 PW2337 engines for two certification aircraft; conversion of Il-96-300 prototype (RA-96000) to **Il-96MO** prototype began early 1992; rolled out at Moscow City Airport 30 March 1993; first flight 6 April 1993; planned certification to FAR 25 and ICAO Annex 16 noise levels in 1995; deliveries to begin 1996.

CURRENT VERSIONS: **Il-96M**: Basic version, as described and illustrated; scheduled to fly 1995.

Il-96MK: Projected development with ducted engines rated at 175-195 kN (38,000-43,000 lb st), with 17:1 or 18:1 bypass ratio.

Il-96T: Freighter; cargo door 3.60 m × 2.60 m (11 ft 9¾ in × 8 ft 6¼ in) forward of wing on port side; to carry standard international containers and pallets; max payload 92,000 kg (202,820 lb).

CUSTOMERS: Letters of intent from Russia International Airlines, Uzbekistan Airways and Far East Aviation for 30 aircraft by April 1993.

DESIGN FEATURES: Basically as Il-96-300; lengthened fuselage of unchanged cross-section, permitting smaller tail fin; wings identical.

POWER PLANT: Four Pratt & Whitney PW2337 turbofans, each 164.6 kN (37,000 lb st); nacelles supplied by Rohr Industries, USA.

ACCOMMODATION: Two flight crew; three passenger arrangements proposed: (1) 18 first class passengers at 152.4 cm (60 in) seat pitch, with 0.77 m (2 ft 6¼ in) aisle; 44 business class at 91.5 cm (36 in) pitch, with 0.625 m (2 ft 0½ in) aisle; 250 tourist class at 86.4 cm (34 in) pitch, with 0.55 m (1 ft 9½ in) aisle. (2) 85 business class and 250 tourist class. (3) three tourist class cabins for 124, 162 and 89

Prototype of increased-capacity Ilyushin Il-96M four-turbofan wide-bodied transport

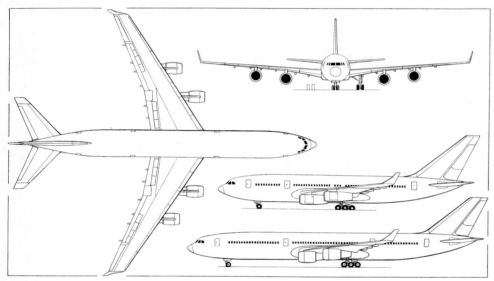

Ilyushin Il-96M powered by four P&W PW2337s, with additional side view (top) of the Il-96-300, powered by four Aviadvigatel PS-90As *(Dennis Punnett)*

Balanced landing runway length 2,250 m (7,385 ft)
Range (30,000 kg; 66,138 lb payload, international rules)
6,195 nm (11,482 km; 7,136 miles)

ILYUSHIN Il-102

TYPE: Two-seat armoured close support aircraft.
PROGRAMME: First flight of first prototype 25 September 1982; revealed in public at MosAeroshow '92; offered in versions tailored to customers' requirements.
DESIGN FEATURES: Low-wing monoplane, with engines in ducts at wingroots; pilot seated high in armoured cockpit; gunner above mid-fuselage, with swivelling turret at tail; underwing fairing pods for mainwheels; three internal bays for individual bombs in each wing outboard of wheel pods.
LANDING GEAR: Retractable tricycle type; rearward retracting nosewheel; forward retracting twin-wheel main units.
POWER PLANT: Two Klimov/Sarkisov RD-33E turbofans; each 51 kN (11,465 lb st).
ARMAMENT: Bomb bays inside wings; two underfuselage pylons side by side, three under each wing outboard of bomb bays; max combat load 7,200 kg (15,875 lb) of bombs, rocket packs, R-60 (NATO AA-8 'Aphid') and R-73 (AA-11 'Archer') self-defence air-to-air missiles, guns and auxiliary fuel tanks; twin-barrel 23 mm GSh-23 gun in rear turret for air-to-air and air-to-ground use; AO-17A gun mounted under fuselage.
DIMENSIONS, EXTERNAL:
Wing span 16.90 m (55 ft 5½ in)
Wing aspect ratio 6.0

passengers. Eight emergency exits. Underfloor hold for 32 standard LD3 containers.
AVIONICS: Rockwell Collins digital avionics, with six-CRT EFIS, EICAS, triple-redundant flight management system to ARINC 700 specifications, three inertial platforms with laser gyros, GPWS, windshear detection radar, GPS and Glonass satellite nav receivers, SAT-900 satcom, Collins Srs 700 com/nav radios, Litton Flagship LTN-101 inertial reference centre, Ball Airlink conformal antennae. Designed for Cat. IIIb fully automatic landings.

DIMENSIONS, EXTERNAL:
Wing span: excl winglets 57.66 m (189 ft 2 in)
over winglets 60.105 m (197 ft 2½ in)
Length overall 63.94 m (209 ft 9 in)
Fuselage: Length 60.50 m (198 ft 6 in)
Max diameter 6.08 m (19 ft 11½ in)
Height overall 15.72 m (51 ft 7 in)
Wheel track 10.40 m (34 ft 1½ in)
DIMENSIONS, INTERNAL:
Cabin: Length 49.13 m (161 ft 2¼ in)
WEIGHTS AND LOADINGS:
Operating weight empty 132,400 kg (291,887 lb)
Max payload 58,000 kg (127,866 lb)
Max T-O weight 270,000 kg (595,238 lb)
Max power loading 410.0 kg/kN (4.02 lb/lb st)
PERFORMANCE (estimated):
Max Mach no. (M_{MO}) 0.86
Normal cruising speed at 9,000-12,000 m (29,500-39,370 ft)
459-469 knots (850-870 km/h; 528-540 mph)
Balanced T-O runway length 3,350 m (11,000 ft)

Ilyushin Il-102 twin-turbofan close support aircraft at MosAeroshow '92

Rear view of Il-102, showing tail gun turret *(Mark Lambert)*

Starboard wing of Il-102, with internal bomb bays open *(Mark Lambert)*

Length overall	17.75 m (58 ft 3 in)
Height overall	5.08 m (16 ft 8 in)
AREAS:	
Wings, gross	47.6 m² (512.4 sq ft)
WEIGHTS AND LOADINGS:	
Weight empty	13,000 kg (28,660 lb)
Max T-O weight	22,000 kg (48,500 lb)
PERFORMANCE:	
Max level speed	512 knots (950 km/h; 590 mph)
Min control speed (V$_{MC}$)	135 knots (250 km/h; 155 mph)
T-O and landing run	300 m (985 ft)
Combat radius	215-270 nm (400-500 km; 248-310 miles)

ILYUSHIN Il-103

TYPE: Two/five-seat light aircraft for primary training and general aviation.
PROGRAMME: Exhibited in model form at Moscow Aerospace '90; programme go-ahead 1990; first flight scheduled 1993; deliveries to begin 1994, after FAR Pt 23 certification.
DESIGN FEATURES: Conventional low-wing monoplane, with non-retractable tricycle landing gear, originally to meet DOSAAF requirement for 500 military/civil pilot trainers; slight wing dihedral from roots; upswept wingtip trailing-edges; swept vertical tail surfaces.
FLYING CONTROLS: Conventional ailerons, elevators, rudder and trailing-edge flaps; trim tab in elevator trailing-edge.
STRUCTURE: Initially all-metal; use of composites and advanced materials possible in later versions.
POWER PLANT: One 157 kW (210 hp) Teledyne Continental IO-360-ES flat-six engine; alternative engines available to meet customers' requirements. Fuel in wings.
ACCOMMODATION: Two seats side by side at front of cabin; bench seat for two or three at rear; space for 220 kg (485 lb) freight with rear bench seat removed; two gull-wing window/doors, hinged on centreline, at front of canopy.
AVIONICS: Unspecified Western equipment.

DIMENSIONS, EXTERNAL:	
Wing span	10.60 m (34 ft 9 in)
Length overall	7.95 m (26 ft 1 in)
Height overall	3.00 m (9 ft 10 in)
Tailplane span	3.60 m (11 ft 9¾ in)
Wheel track	2.00 m (6 ft 6¾ in)
DIMENSIONS, INTERNAL:	
Cabin: Max width	1.36 m (4 ft 5½ in)
Max height	1.22 m (4 ft 0 in)
Baggage space	0.2 m³ (7 cu ft)
WEIGHTS AND LOADINGS (A: training, B: utility):	
Weight empty: A	720 kg (1,587 lb)
B	765 kg (1,686 lb)
Payload: A	180 kg (397 lb)
B	440 kg (970 lb)
Max fuel: A, B	150 kg (330 lb)
Max T-O weight: A	965 kg (2,127 lb)
B	1,310 kg (2,888 lb)
Max power loading: A	6.15 kg/kW (10.13 lb/hp)
B	8.34 kg/kW (13.75 lb/hp)
PERFORMANCE (estimated):	
Max level speed at 3,000 m (9,850 ft):	
A	143 knots (265 km/h; 165 mph)
B	140 knots (260 km/h; 161 mph)
Cruising speed at 3,000 m (9,850 ft):	
A	127 knots (235 km/h; 146 mph)
B	121 knots (225 km/h; 140 mph)
Stalling speed, flaps up:	
A	53 knots (98 km/h; 61 mph)
B	60 knots (110 km/h; 69 mph)
Stalling speed, flaps down:	
A	46 knots (85 km/h; 53 mph)
B	52 knots (95 km/h; 59 mph)
Max rate of climb at S/L: A	480 m (1,575 ft)/min
B	330 m (1,080 ft)/min
T-O run: A	160 m (525 ft)
T-O to 15 m (50 ft): B	465 m (1,525 ft)
Landing from 15 m (50 ft): B	500 m (1,640 ft)
Landing run: A	165 m (542 ft)
Max range with three passengers:	
B	669 nm (1,240 km; 770 miles)
Max endurance: A	2 h
g limits: A	+6/−3
B	+4.4/−2.2
Radius of turn at 130 knots (240 km/h; 149 mph):	
A	85 m (280 ft)
Time of full turn at 130 knots (240 km/h; 149 mph):	
A	8 s
Rate of roll: A	84°/s

ILYUSHIN Il-108

TYPE: Projected twin-turbofan business or third-level airline transport.
PROGRAMME: First shown in model form 1990; no decision to build prototype by Spring 1993; current work involves market analysis, search for investors, and more precise definition of design parameters using different engine options. *Following data are provisional:*
DESIGN FEATURES: Conventional low-wing/T-tail configuration, with rear-mounted engines; wings and tail unit swept; winglets added late 1990; retractable tricycle landing gear with twin wheels on each unit. Manual flying controls.
POWER PLANT: Two ZMKB/PS DV-2 turbofans, each 21.58 kN (4,852 lb st); other engines being considered.
ACCOMMODATION: Proposed in two versions: (1) two crew and nine passengers, with facilities for work or relaxation on long flights. (2) 15 passengers, three-abreast with 40 cm (15.75 in) aisle; both versions with toilet, wardrobes, front and rear baggage compartments.

DIMENSIONS, EXTERNAL:	
Wing span	15.00 m (49 ft 2½ in)
Length overall	15.85 m (52 ft 0 in)
Fuselage: Length	14.00 m (45 ft 11¼ in)
Max diameter	2.35 m (7 ft 8½ in)

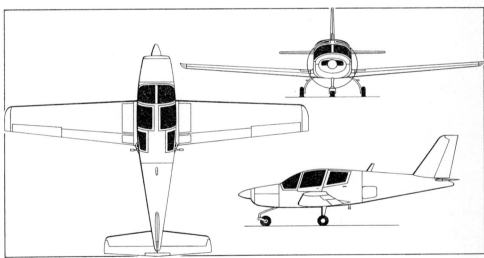
Ilyushin Il-103 two/five-seat multi-purpose light aircraft *(Jane's/Mike Keep)*

Full-scale mockup of Ilyushin Il-103 at MosAeroshow '92 *(Mark Lambert)*

Model of projected Ilyushin Il-108 business/regional transport *(Piotr Butowski)*

Height overall	5.50 m (18 ft 0½ in)
Tailplane span	5.29 m (17 ft 4¼ in)
Wheel track	3.00 m (9 ft 10¼ in)
WEIGHTS AND LOADINGS:	
Max payload	1,500 kg (3,306 lb)
Max T-O weight	14,300 kg (31,525 lb)
Max power loading	331.3 kg/kN (3.25 lb/lb st)
PERFORMANCE (estimated):	
Econ cruising speed at 12,000 m (39,370 ft)	
	432 knots (800 km/h; 497 mph)
Balanced T-O runway length	1,800 m (5,905 ft)
Range at econ cruising speed:	
with max payload	2,428 nm (4,500 km; 2,796 miles)
with 15 passengers and max fuel	
	2,615 nm (4,850 km; 3,010 miles)
with nine passengers	
	3,235 nm (6,000 km; 3,725 miles)

ILYUSHIN Il-114

TYPE: Twin-turboprop short-range passenger and freight transport.

PROGRAMME: Design finalised 1986, as replacement for aircraft in An-24 class; prototype (SSSR-54000) first flew at Zhukovsky flight test centre 29 March 1990; two more flying prototypes, two for static tests; in production at Tashkent and by the Moscow Aircraft Production Group (MAPO); first production aircraft flew 7 August 1992; certification and first deliveries planned for 1993.

CURRENT VERSIONS: **Il-114**: *As described in detail.*

Il-114M: With TV7M-117 turboprops, increased max T-O weight and payload of 7,000 kg (15,430 lb).

CUSTOMERS: Aeroflot and its successors (350).

DESIGN FEATURES: Conventional low-wing monoplane; only fin and rudder swept; slight dihedral on wing centre-section, much increased on outer panels; operation from unpaved runways practical.

FLYING CONTROLS: Manual actuation; each wing trailing-edge occupied entirely by aileron, with servo and trim tabs, and hydraulically actuated double-slotted trailing-edge flaps, inboard and outboard of engine nacelle; two airbrakes (inboard) and spoiler (outboard) forward of flaps; spoilers supplement ailerons differentially in event of engine failure during take-off; trim and servo tabs in rudder, trim tab in each elevator.

STRUCTURE: Approx 10 per cent of airframe by weight made of composites; two-spar wings; removable leading-edge on outer panels; circular-section aluminium alloy semi-monocoque fuselage built as five subassemblies; metal tail unit (CFRP tailplane and fin boxes planned for later aircraft).

LANDING GEAR: Retractable tricycle type, with twin wheels on each unit. All retract forward hydraulically; emergency extension by gravity. Oleo-pneumatic shock absorbers. Tyres size 620 × 80 mm on nosewheels, 880 × 305 mm on mainwheels. Nosewheels steerable ±55°. Disc brakes on mainwheels. All wheel doors remain closed except during retraction or extension of landing gear.

POWER PLANT: Two 1,839 kW (2,466 shp) Klimov TV7-117 turboprops, each driving a low-noise six-blade SV-34 CFRP propeller. Integral fuel tanks in wings, capacity 8,360 litres (2,208 US gallons; 1,839 Imp gallons). APU in tailcone.

ACCOMMODATION: Flight crew of two, plus stewardess. Emergency exit window each side of flight deck. Four-abreast seats for 64 passengers in main cabin, at 75 cm (29.5 in) seat pitch, with central aisle 45 cm (17.72 in) wide. Provision for rearrangement of interior for increased seating, and lengthening of fuselage for 70-75 passengers. Airstair door at front of cabin, further door at rear; both on port side, opening outward. Galley, cloakroom and toilet at rear; emergency escape slide by service door on starboard side. Type III emergency exit over each wing. Service doors at front and rear of cabin on starboard side. Baggage compartments forward of cabin on starboard side and to rear of cabin, plus overhead baggage racks. Optional carry-on baggage shelves in lobby by main door at front.

SYSTEMS: Dual redundant pressurisation and air-conditioning system using bleed air from both engines. Two independent hydraulic systems, pressure 207 bars (3,000 lb/sq in), for landing gear actuation, wheel brakes, nosewheel steering and flaps. Three-phase 115/220V 400Hz AC electrical system powered by 40kW alternator on each engine. Secondary 24V DC system. Wing and tail unit leading-edges de-iced electrically. Electrothermal anti-icing system for propeller blades and windscreen. Engine air intakes de-iced by hot air.

AVIONICS: Digital avionics for automatic or manual control by day or night, including automatic approach and landing in limiting weather conditions (ICAO Cat. I and II). Two colour CRTs for each pilot for flight and navigation information. Centrally mounted CRT for engine and systems data.

DIMENSIONS, EXTERNAL:

Wing span	30.00 m (98 ft 5¼ in)
Length overall	26.87 m (88 ft 2 in)
Diameter of fuselage	2.86 m (9 ft 4½ in)
Height overall	9.32 m (30 ft 7 in)
Tailplane span	11.10 m (36 ft 5 in)
Wheel track	8.40 m (27 ft 6½ in)
Wheelbase	9.13 m (29 ft 11½ in)
Propeller diameter	3.60 m (11 ft 9¾ in)
Propeller ground clearance	0.50 m (1 ft 7¾ in)
Propeller fuselage clearance	0.97 m (3 ft 2¼ in)
Passenger doors (each): Height	1.70 m (5 ft 7 in)
Width	0.90 m (2 ft 11¼ in)
Service door (front): Height	1.30 m (4 ft 3¼ in)
Width	0.96 m (3 ft 1¾ in)
Service door (rear): Height	1.38 m (4 ft 6¼ in)
Width	0.61 m (2 ft 0 in)

Prototype Ilyushin Il-114 twin-turboprop airliner at 1991 Paris Air Show *(Brian M. Service)*

ILYUSHIN/KAMOV—AIRCRAFT: RUSSIA

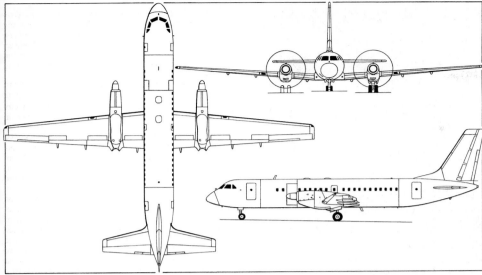

Ilyushin Il-114 short-range passenger and freight transport (*Jane's/Mike Keep*)

Emergency exit (each): Height	0.91 m (2 ft 11¾ in)
Width	0.51 m (1 ft 8 in)
DIMENSIONS, INTERNAL:	
Length between pressure bulkheads	22.24 m (72 ft 11½ in)
Cabin: Max height	1.92 m (6 ft 3½ in)
WEIGHTS AND LOADINGS:	
Operating weight empty	15,000 kg (33,070 lb)
Max payload	6,500 kg (14,330 lb)
Max fuel	6,500 kg (14,330 lb)
Max T-O weight	22,700 kg (50,045 lb)
Max power loading	6.085 kg/kW (10.01 lb/shp)
PERFORMANCE (estimated):	
Nominal cruising speed	270 knots (500 km/h; 310 mph)
Approach speed	100 knots (185 km/h; 115 mph)
Optimum cruising height	8,100 m (26,575 ft)
T-O run: paved	1,200 m (3,940 ft)
unpaved	1,400 m (4,600 ft)
Landing run: paved or unpaved	1,300 m (4,265 ft)
Range, with reserves:	
with 64 passengers	540 nm (1,000 km; 621 miles)
with 1,500 kg (3,300 lb) payload	2,590 nm (4,800 km; 2,980 miles)

KAMOV

VERTOLETNYI NAUCHNO-TEKHNICHES-KIY KOMPLEKS IMENI N. I. KAMOVA (VNTK) (Helicopter Scientific and Technology Complex named after N. I. Kamov)
March 8th Street, Lubertsy 140007, Moscow Region
Telephone: 7 (095) 700 32 04, 7 (095) 171 37 43
Fax: 7 (095) 700 30 71
Telex: 206112 Kamov
GENERAL DESIGNER: Sergei Victorovich Mikheyev, PhD
DEPUTY GENERAL DESIGNER: Veniamin Kasjanikov
CHIEF DESIGNERS:
  Juri Sokovikov
  Vyacheslav Krygin
DEPUTY CHIEF DESIGNER: Evgeny Pak

Formed in 1947, this OKB continues work of Prof Dr Ing Nikolai Ilyich Kamov, a leading designer of rotating wing aircraft from late 1920s, who died on 24 November 1973, aged 71; all Kamov helicopters in current service have coaxial contra-rotating rotors; Ka-62, under development, has single main rotor.

KAMOV Ka-25

NATO reporting name: Hormone

TYPE: Twin-turbine multi-purpose military helicopter.
PROGRAMME: Prototype flew 1961; shown in Soviet Aviation Day flypast, Tushino Airport, Moscow, July 1961, carrying two dummy air-to-surface missiles (ASMs not fitted to production aircraft); about 460 built 1966-75, of which 100 remain operational with CIS Navy.
CURRENT VERSIONS: Four major variants in service:
  Ka-25PL (protivolodochny) ('Hormone-A'): Ship-based anti-submarine helicopter, operated from frigates, cruisers, helicopter carriers *Moskva* and *Leningrad* and carrier/cruisers of 'Kiev' class; equipped with surveillance radar and dipping sonar; armed with one torpedo. *Detailed description applies to Ka-25PL.*
  Ka-25Ts (tseleukazatel) ('Hormone-B'): Special electronics variant, providing over-the-horizon target acquisition for ship-launched cruise missiles including SS-N-3B (NATO 'Shaddock') from 'Kresta I' cruisers, SS-N-12 ('Sandbox') from 'Kiev' class and 'Slava' class cruisers, SS-N-19 ('Shipwreck') from battle cruisers *Kirov* and *Frunze*, and SS-N-22 ('Sunburn') from 'Sovremenny' class destroyers. 'Kiev' and 'Kirov' class ships each carry three 'Hormone-Bs', other classes one; larger undernose radome (NATO 'Big Bulge') than Ka-25PL, with spherical undersurface; cylindrical radome under rear cabin for missile mid-course guidance and helicopter-to-ship data link; when radar operates, all landing gear wheels can retract upward to minimise interference to emissions; cylindrical fuel container each side of lower fuselage.
  Ka-25BShZ: Equipped to tow minesweeping gear; has been used to clear mines in Red Sea.
  Ka-25PS ('Hormone-C'): Search and rescue version, with special role equipment, including hoist.
CUSTOMERS: CIS Naval Aviation, India, Syria, Vietnam and Yugoslavia.
DESIGN FEATURES: Folding three-blade coaxial rotors, requiring no tail rotor, and triple tail fins ensure compact stowed overall dimensions on board ship; engines above cabin and external mounting of operational equipment and auxiliary fuel leaves interior uncluttered.
FLYING CONTROLS, STRUCTURE, LANDING GEAR: See previous editions.
POWER PLANT: Two 662 kW (888 shp) Mars GTD-3F turboshafts, side by side above cabin forward of rotor driveshaft, on early aircraft. Later aircraft have 735 kW (986 shp) GTD-3M turboshafts. Independent fuel supply to each engine. Provision for external fuel tank each side of cabin. Single-point fuelling.
ACCOMMODATION: Pilot and co-pilot side by side on flight deck, with rearward sliding door each side. Rearward sliding door to rear of main landing gear on port side. Up to 12 folding seats for passengers optional.
AVIONICS: Equipment available for all versions includes autopilot, radio, nav system, radio compass, lighting for all-weather operation by day or night, and hoist above cabin door. IFF antennae (NATO 'Odd Rods') above nose and alongside central tail fin. Dipping sonar in compartment at rear of main cabin, immediately forward of tailboom, and search radar (NATO 'Short Horn') in flat-bottom undernose radome (diameter 1.25 m; 4 ft 1 in) on Ka-25PL, which can have canister of sonobuoys mounted externally aft of starboard main landing gear. Most aircraft have cylindrical housing for ESM above tailboom, shallow blister fairing to rear of cylindrical housing and similar housing under rear of cabin for data link.
ARMAMENT: Doors under fuselage of some aircraft enclose weapons bay for 450 mm (18 in) ASW torpedo.

Kamov Ka-25PL ('Hormone-A') anti-submarine helicopter of the CIS Naval Aviation (*Piotr Butowski*)

DIMENSIONS, EXTERNAL:	
Rotor diameter (each)	15.74 m (51 ft 7¾ in)
Length of fuselage	9.75 m (32 ft 0 in)
Height to top of rotor head	5.37 m (17 ft 7½ in)
Width over tail-fins	3.76 m (12 ft 4 in)
Wheel track: front	1.41 m (4 ft 7½ in)
rear	3.52 m (11 ft 6½ in)
Cabin door: Height	1.10 m (3 ft 7¼ in)
Width	1.20 m (3 ft 11¼ in)
DIMENSIONS, INTERNAL:	
Cabin, excl flight deck:	
Length	3.95 m (12 ft 11½ in)
Max width	1.50 m (4 ft 11 in)
Max height	1.25 m (4 ft 1¼ in)
AREAS:	
Rotor disc (each)	194.6 m² (2,095 sq ft)
WEIGHTS AND LOADINGS:	
Weight empty	4,765 kg (10,505 lb)
Max T-O weight	7,200 kg (15,873 lb)
PERFORMANCE:	
Max level speed	113 knots (209 km/h; 130 mph)
Normal cruising speed	104 knots (193 km/h; 120 mph)
Service ceiling	3,350 m (11,000 ft)
Range, with reserves:	
with standard fuel	217 nm (400 km; 250 miles)
with external tanks	351 nm (650 km; 405 miles)

KAMOV Ka-26

NATO reporting name: Hoodlum-A

TYPE: Light twin-engined multi-purpose helicopter.
PROGRAMME: Prototype first flew 1965; production aircraft entered agricultural service in Soviet Union 1967; 850 built for many civil and military roles.
CURRENT VERSIONS: *Abbreviated details follow*; full description appeared last in 1987-88 edition; turbine-powered Ka-126 ('Hoodlum-B'), described separately.
CUSTOMERS: Delivered for civilian use in 15 countries; military operators include air forces of Bulgaria and Hungary.

272 RUSSIA: AIRCRAFT—KAMOV

Chemical spraying version of Kamov Ka-26 twin-engined helicopter *(Press Office Sturzenegger)*

DESIGN FEATURES: Airframe comprises backbone structure carrying flight deck, coaxial contra-rotating rotors, landing gear, engine pods and twin tailbooms; space aft of flight deck, between main landing gear units and under rotor transmission, can accommodate interchangeable modules for passenger/freight transport, air ambulance, aerial survey, forest firefighting, mineral prospecting, pipeline and power transmission line construction, search and rescue, and (of primary importance) agricultural equipment.

POWER PLANT: Two 239 kW (320 hp) Vedeneyev M-14V-26 air-cooled radial piston engines, in pods on short stub-wings at top of fuselage.

ACCOMMODATION: Fully enclosed cabin, with door each side, normally for operation by single pilot; second seat and dual controls optional. Cabin warmed and demisted by air from combustion heater, which also heats passenger compartment when fitted. Air filter on nose of agricultural version, which has chemical hopper (capacity 900 kg; 1,985 lb) and dust spreader or spraybars in module space, on aircraft's centre of gravity. This equipment quickly removable and can be replaced by cargo/passenger pod for four or six persons, with provision for seventh passenger beside pilot; or two stretcher patients, two seated casualties and medical attendant in ambulance role. Alternatively, Ka-26 can be operated with open platform for hauling freight or hook for slinging bulky loads at end of cable or in cargo net.

DIMENSIONS, EXTERNAL:
Rotor diameter (each)	13.00 m (42 ft 7¾ in)
Vertical separation between rotors	1.17 m (3 ft 10 in)
Length of fuselage	7.75 m (25 ft 5 in)
Height overall	4.05 m (13 ft 3½ in)
Width: over engine pods	3.64 m (11 ft 11½ in)
over agricultural spraybars	11.20 m (36 ft 9 in)
Tailplane span	4.60 m (15 ft 1 in)
Wheel track: mainwheels	2.42 m (7 ft 11½ in)
nosewheels	0.90 m (2 ft 11½ in)
Wheelbase	3.48 m (11 ft 5 in)
Passenger pod door: Height	1.40 m (4 ft 7 in)
Width	1.25 m (4 ft 1¼ in)

DIMENSIONS, INTERNAL:
Passenger pod:
Length, floor level	1.83 m (6 ft 0 in)
Width, floor level	1.25 m (4 ft 1¼ in)
Headroom	1.40 m (4 ft 7 in)

AREAS:
Rotor disc (each)	132.7 m² (1,430 sq ft)

WEIGHTS AND LOADINGS:
Operating weight empty: stripped	1,950 kg (4,300 lb)
cargo/platform	2,085 kg (4,597 lb)
cargo/hook	2,050 kg (4,519 lb)
passenger	2,100 kg (4,630 lb)
agricultural	2,216 kg (4,885 lb)
Fuel weight: transport	360 kg (794 lb)
other versions	100 kg (220 lb)
Payload: transport	900 kg (1,985 lb)
agricultural duster	1,065 kg (2,348 lb)
agricultural sprayer	900 kg (1,985 lb)
with cargo platform	1,065 kg (2,348 lb)
flying crane	1,100 kg (2,425 lb)
Max T-O weight: all versions	3,250 kg (7,165 lb)

PERFORMANCE (at max T-O weight):
Max level speed	91 knots (170 km/h; 105 mph)
Max cruising speed	73 knots (135 km/h; 84 mph)
Econ cruising speed	49-59 knots (90-110 km/h; 56-68 mph)
Agricultural operating speed range	16-62 knots (30-115 km/h; 19-71 mph)
Service ceiling	3,000 m (9,840 ft)
Service ceiling, OEI	500 m (1,640 ft)
Hovering ceiling at AUW of 3,000 kg (6,615 lb):	
IGE	1,300 m (4,265 ft)
OGE	800 m (2,625 ft)
Range with 7 passengers, 30 min fuel reserves	242 nm (450 km; 280 miles)
Max range with auxiliary tanks	400 nm (740 km; 460 miles)
Endurance at econ cruising speed	3 h 42 min

KAMOV Ka-27 and Ka-28
NATO reporting names: Helix-A and D

TYPE: Twin-turbine multi-purpose military helicopter.

PROGRAMME: Design started 1969 to overcome inability of Ka-25 to operate dipping sonar at night and in adverse weather; first flight of prototype December 1974; first open reference in US Department of Defense's 1981 *Soviet Military Power* document, which stated that "Hormone variant" helicopters could be carried in telescoping hangar on 'Sovremenny' class of guided missile destroyers, for ASW missions; photographs of two on stern platform of *Udaloy*, first of new class of ASW guided missile destroyers, taken by Western pilots during Baltic exercises, September 1981; at least 16 observed on 'Kiev' class carrier/cruiser *Novorossiysk* during ship's maiden deployment 1983, as stage in continuous replacement of Ka-25s with Ka-27s; will equip *Admiral of the Fleet Kuznetsov* and any other large Russian carriers.

CURRENT VERSIONS: **Ka-27PL** ('Helix-A'): Basic ASW helicopter with three crew; operational since 1982; normally operated in pairs, one tracking hostile submarine, other dropping depth charges. CIS Naval Aviation has more than 100.

Ka-27PS ('Helix-D'): Search and rescue and plane guard helicopter, first seen on *Novorossiysk*; as Ka-27PL, but some operational equipment deleted; external fuel tank each side of cabin, as civil Ka-32; winch beside port cabin door.

Ka-28 ('Helix-A'): Export version of Ka-27PL, said by Yugoslavia to have 1,618 kW (2,170 shp) TV3-117BK turboshafts and 3,680 kg (8,113 lb) of fuel in 12 tanks. Installation of four Kh-35 active radar homing anti-ship missiles under development.

Ka-29 ('Helix-B'): Described separately.

Ka-32 (civil 'Helix-C'): Described separately; general description (which see) applies also to Ka-27 and Ka-28.

CUSTOMERS: CIS Naval Aviation; India (13 Ka-28, incl three for training) and Yugoslavia (Ka-28).

DESIGN FEATURES: Basic configuration very like Ka-25, but longer and more capacious fuselage pod, no central tail fin and different undernose radome; similar overall dimensions with rotors folded enable Ka-27 to stow in shipboard hangars and use deck lifts built for Ka-25.

Kamov Ka-28 anti-submarine helicopter (NATO 'Helix-A') in Yugoslav service *(Press Office Sturzenegger)*

Kamov Ka-27PS search and rescue helicopter (NATO 'Helix-D') *(Piotr Butowski)*

Type 2A42 gun on port outrigger of Ka-29TB

Ka-29TB (NATO 'Helix-B') combat transport helicopter *(Linda Jackson)*

ACCOMMODATION: Crew of three: pilot, tactical co-ordinator, ASW systems operator.
AVIONICS: Undernose 360° search radar; IFF (NATO 'Odd Rods'); directional ESM radomes above rear of engine bay fairing and at tailcone tip; Doppler box under tailboom; radar warning receivers on nose and above tailplane; dipping sonar behind clamshell doors at rear of fuselage pod; MAD; otherwise as Ka-32.
EQUIPMENT: Infra-red jammer (NATO 'Hot Brick') at rear of engine bay fairing; station keeping light between ESM radome and jammer; chaff/flare dispensers; colour coded identification flares; sonobuoys stowed internally; otherwise as Ka-32.
ARMAMENT: Ventral weapons bay for torpedoes, depth charges, other stores.
DIMENSIONS, EXTERNAL: As Ka-32
PERFORMANCE (Yugoslav Ka-28):
  Max level speed at 10,700 kg (23,590 lb) AUW
    135 knots (250 km/h; 155 mph)
  Max cruising speed
    124-129 knots (230-240 km/h; 143-149 mph)
  Max rate of climb at S/L   750 m (2,460 ft)/min
  Radius of action against submarine cruising at up to 40 knots (75 km/h; 47 mph) at depth of 500 m (1,640 ft)
    108 nm (200 km; 124 miles)

KAMOV Ka-29
NATO reporting name: Helix-B

TYPE: Twin-turbine assault transport and electronic warfare helicopter.
PROGRAMME: Entered service with Northern and Pacific Fleets 1985; photographed on board assault ship *Ivan Rogov* in Mediterranean 1987, thought to be Ka-27B and given NATO reporting name 'Helix-B'; identified as Ka-29TB (Transportno Boyevoya – combat transport) at Frunze (Khodinka) air show, Moscow, August 1989; EW version completed initial shipboard trials on aircraft carrier *Admiral of the Fleet Kuznetsov* (then *Tbilisi*) 1990; both versions expected to equip this ship.
CURRENT VERSIONS: **Ka-29TB** ('Helix-B'): Development of Ka-27 for transport and close support of seaborne assault troops. *Detailed description generally as for Ka-32 except as under.*
  Ka-29RLD: Early warning version; two examples seen on *Admiral of the Fleet Kuznetsov* by early 1992; basic airframe of Ka-29TB, but many equipment changes; shallow pannier extends full length of underfuselage; two large panniers each side of cabin, fore and aft of main landing gear on helicopter numbered 032 (forward panniers only on 031); starboard airstair-type cabin door, aft of flight deck, divided horizontally into upward and downward opening sections, with box fairing in place of window; hatch window deleted above starboard rear pannier; APU repositioned above rear of engine bay fairing, with slot type air intake at front of housing, displacing usual ESM and IR jamming pods; tailcone extended by conical, probably dielectric, fairing; no visible gun door on nose, or armour plating on flight deck and engine bay; no stores pylons or outriggers; unidentified structure at rear of fuselage.
CUSTOMERS: CIS Naval Aviation (more than 30 Ka-29TBs).
POWER PLANT: Two Klimov TV3-117V turboshafts, each 1,633 kW (2,190 shp). Engines started by APU.
ACCOMMODATION: Wider flight deck than Ka-27 for two crew; three flat-plate windscreen glazings instead of two-piece curved transparency; main cabin port-side door, aft of landing gear, divided horizontally into upward and down-

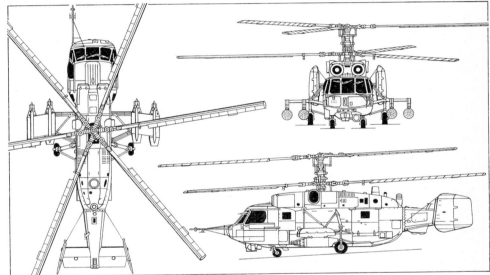

Kamov Ka-29TB combat transport helicopter *(Jane's/Mike Keep)*

Kamov Ka-29RLD early warning helicopter on *Admiral of the Fleet Kuznetsov (Tass)*

ward opening sections, to facilitate rapid exit of up to 16 assault troops; four stretcher patients and six seated casualties in ambulance role; flight deck and engine bay heavily armoured.

AVIONICS: Undernose sensor pods appear similar to electro-optics and RF missile guidance pods of Mil Mi-24V; ESM 'flower pot' above rear of engine bay fairing, forward of infra-red jammer (NATO 'Hot Brick'); station keeping light between ESM and jammer; Doppler box under tailboom; radar warning receivers and IFF ('Slap Shot').

ARMAMENT: Four-barrel Gatling type 7.62 mm machine gun, with 1,800 rds, flexibly mounted behind downward-articulated door on starboard side of nose; four pylons on outriggers, for two four-round clusters of 9M114 Shturm (AT-6 'Spiral') air-to-surface missiles and two 57 mm or 80 mm rocket packs. Alternative loads include four rocket packs, two pods each containing a 23 mm gun and 250 rds, or two ZB-500 auxiliary fuel tanks. Provision for 30 mm Type 2A42 gun above port outrigger, with ammunition feed from cabin.

DIMENSIONS, EXTERNAL:
Rotor diameter (each) 15.90 m (52 ft 2 in)
Blade length, aerofoil section (each)
  5.45 m (17 ft 10½ in)
Blade chord 0.48 m (1 ft 7 in)
Vertical separation of rotors 1.40 m (4 ft 7 in)
Length overall, excl noseprobe and rotors
  11.30 m (37 ft 1 in)
Height overall 5.40 m (17 ft 8½ in)
Width: between centrelines of outboard pylons
  5.65 m (18 ft 6½ in)
over tail fins and centred rudders
  3.65 m (12 ft 0 in)
flight deck 2.20 m (7 ft 2 in)
Mainwheel track 3.50 m (11 ft 6 in)
Nosewheel track 1.40 m (4 ft 7 in)
Wheelbase 3.00 m (9 ft 10 in)

AREAS:
Rotor disc (each) 198.5 m² (2,138 sq ft)

WEIGHTS AND LOADINGS:
Weight empty 5,520 kg (12,170 lb)
Max load: internal 2,000 kg (4,409 lb)
  external 4,000 kg (8,818 lb)
Normal T-O weight 11,000 kg (24,250 lb)
Max T-O weight 12,600 kg (27,775 lb)

PERFORMANCE:
Max level speed at S/L 151 knots (280 km/h; 174 mph)
Nominal cruising speed 127 knots (235 km/h; 146 mph)
Service ceiling 4,300 m (14,100 ft)
Combat radius, with 6-8 attack runs over target
  54 nm (100 km; 62 miles)
Range: max standard fuel 248 nm (460 km; 285 miles)
  ferry 400 nm (740 km; 460 miles)

KAMOV Ka-32
NATO reporting name: Helix-C

TYPE: Twin-turbine civil utility helicopter.

PROGRAMME: Development of Ka-27/32 began 1969; first flight of common prototype December 1974; first Ka-32 (SSSR-04173) exhibited Minsk Airport late 1981, during fourth CMEA scientific/technical conference on use of aircraft in national economy, carrying slung truck in flying display; prototype of utility version shown at Paris Air Show June 1985; in production.

CURRENT VERSIONS: **Ka-32T** ('Helix-C'): Utility transport, ambulance and flying crane; limited avionics; for carriage of internal or external freight, and passengers, along airways and over local routes, including support of offshore drilling rigs.
Ka-32S ('Helix-C'): Maritime version; more comprehensive avionics, including undernose radar, for IFR operation from icebreakers in adverse weather and over terrain devoid of landmarks; 300 kg (661 lb) load hoist standard; duties include ice patrol, guidance of ships through icefields, unloading and loading ships (up to 30 tonnes an hour, 360 tonnes a day), maritime search and rescue.
Ka-32K: Flying crane with retractable gondola for second pilot under cabin. Model shown 1990. Operational testing completed 1992.

CUSTOMERS: Aeroflot and its successors (about 150 being delivered).

DESIGN FEATURES: Conceived as completely autonomous 'compact truck', to stow in much the same space as Ka-25 with rotors folded, despite greater power and capability, and to operate independently of ground support equipment; special attention paid to ease of handling with single pilot; overall dimensions minimised by use of coaxial rotors, requiring no tail rotor, and twin fins on short tailboom; upper rotor turns clockwise, lower rotor anti-clockwise; rotor mast tilted forward 3°; twin turbines and APU above cabin, leaving interior uncluttered; lower fuselage sealed for flotation.

FLYING CONTROLS: Dual hydraulically powered flight control systems, without manual reversion; spring stick trim; yaw control by differential collective pitch applied through rudder pedals; mix in collective system maintains constant total rotor thrust during turns, to reduce pilot workload when landing on pitching deck, and to simplify transition to hover and landing; twin rudders intended mainly to improve control in autorotation, but also effective in co-ordinating turns; helicopter not designed for negative g loading; flight can be maintained on one engine at max T-O weight.

Kamov Ka-32K flying crane, with retractable pilot gondola extended under rear of cabin

STRUCTURE: Titanium and composites used extensively, with particular emphasis on corrosion resistance; fully articulated three-blade coaxial contra-rotating rotors have all-composites blades with carbonfibre and glassfibre main spars, pockets (13 per blade) of Kevlar type material, and filler similar to Nomex; blades have non-symmetrical aerofoil section; each has ground adjustable tab; each lower blade carries adjustable vibration damper, comprising two dependent weights, on root section, with further vibration dampers in fuselage; tip light on each upper blade; blades fold manually outboard of all control mechanisms, to folded width within track of main landing gear; rotor hub is 50 per cent titanium/50 per cent steel; rotor brake standard; all-metal fuselage; composites tailcone; fixed incidence tailplane, elevators, fins and rudders have aluminium alloy structure, composite skins; fins toe inward approx 25°; fixed leading-edge slat on each fin prevents airflow over fin stalling in crosswinds or at high yaw angles.

LANDING GEAR: Four-wheel type. Oleo-pneumatic shock absorbers. Castoring nosewheels. Mainwheel tyres size 600 × 180 mm. Nosewheel tyres size 400 × 150 mm. Skis optional.

POWER PLANT: Two 1,633 kW (2,190 shp) Klimov TV3-117V turboshafts, with automatic synchronisation system, side by side above cabin, forward of rotor driveshaft. Main gearbox brake standard. Oil cooler fan aft of gearbox. Cowlings hinge downward as maintenance platforms. All standard fuel in tanks under cabin floor and inside container each side of centre-fuselage. Provision for auxiliary tanks in cabin. Single-point refuelling behind small forward hinged door on port side, where bottom of tailboom meets rear of cabin.

ACCOMMODATION: Pilot and navigator side by side on air-conditioned flight deck, in adjustable seats. Rearward sliding jettisonable door with blister window each side. Seat behind navigator, on starboard side, for observer, loadmaster or rescue hoist operator. Alcohol windscreen anti-icing. Direct access to cabin from flight deck. Heated and ventilated main cabin can accommodate freight or 16 passengers, on three folding seats at rear, six along port sidewall and seven along starboard sidewall. Lifejackets under seats. Fittings to carry stretchers. No provisions for toilet or galley. Pyramid structure can be fitted on floor beneath rotor driveshaft to prevent swinging of external cargo sling loads. Rearward sliding door aft of main landing gear on port side, with steps below. Emergency exit door opposite. Hatch to avionics compartment on port side of tailboom.

SYSTEMS: Dual hydraulically powered flight control systems without manual reversion. Spring stick trim. Electro-thermal de-icing of entire profiled portion of each blade switches on automatically when helicopter enters icing conditions. Hot air engine intake anti-icing. APU in rear of engine bay fairing on starboard side, for engine starting and to power all essential hydraulic and electrical services on ground, eliminating need for GPU.

AVIONICS: Include electromechanical flight director controlled from autopilot panel, Doppler hover indicator, two HSI and air data computer. Autopilot can provide automatic approach and hover at height of 25 m (82 ft) over landing area, on predetermined course, using Doppler. Radar altimeter. Doppler box under tailboom.

EQUIPMENT: Doors at rear of fuel tank bay provide access to small compartment for auxiliary fuel, or liferafts which eject during descent in emergency, by command from flight deck. Container each side of fuselage, under external fuel containers, for emergency flotation bags, deployed by water contact. Optional rescue hoist, capacity 300 kg (661 lb), between top of door opening and landing gear.

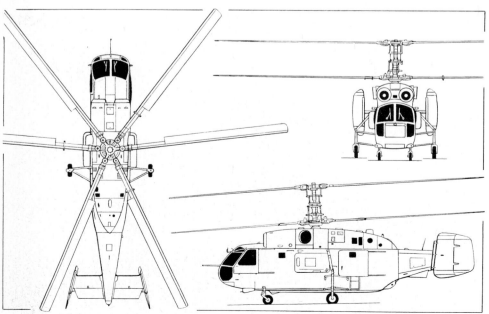

Kamov Ka-32T ('Helix-C') utility helicopter (two Klimov TV3-117V turboshafts) (*Dennis Punnett*)

KAMOV—AIRCRAFT: RUSSIA 275

Kamov Ka-32T ('Helix-C') on operational lease to Heliswiss *(Graham Dinsdale)*

STRUCTURE: Composite materials constitute 35 per cent by weight of structure, including rotors. Approx 350 kg (770 lb) of armour protect pilot and critical airframe parts; canopy and windscreen panels are heavy bulletproof glass.
LANDING GEAR: Retractable tricycle type; twin-wheel nose unit and single mainwheels all semi-exposed when up; all wheels retract rearward.
POWER PLANT: Two 1,633 kW (2,190 shp) Klimov TV3-117VK turboshafts.
ACCOMMODATION: Double-wall steel armoured cockpit, able to protect pilot from hits by 20 mm and 23 mm gunfire over ranges as close as 100 m (330 ft). Specially designed K-37 ejection system for safe ejection at any altitude; following explosive separation of rotor blades and cockpit roof, pilot is extracted from cockpit by large rocket; alternatively, he can jettison doors and stores before rolling out of cockpit sideways.
SYSTEMS: All systems configured for operational deployment away from base for at least two weeks without need for maintenance ground equipment; refuelling, avionics and weapon servicing performed from ground level.
AVIONICS: Four computers to meet navigation, mission control and display demands. To reduce pilot workload and introduce a degree of low-observability, target location and designation is intended to be done by other aircraft; equipment behind windows in nose includes laser marked target seeker and rangefinder; other equipment includes FLIR pod, TV, HUD and cockpit CRT; pilot has MiG-29 type helmet sight; when pilot has target centred on HUD, he pushes button to lock sighting and autopilot into one unit.
ARMAMENT: Up to eighty 80 mm S-8 air-to-surface rockets in four underwing packs; or up to 16 Vikhr (AT-9) tube-launched laser-guided air-to-surface missiles with range of 8-10 km (5-6.2 miles) capable of penetrating 900 mm of reactive armour; single-barrel 30 mm 2A42 gun on

Optional external load sling, with automatic release and integral load weighing and stabilisation systems.

DIMENSIONS, EXTERNAL:
Rotor diameter (each)	15.90 m (52 ft 2 in)
Length overall: excl rotors	11.30 m (37 ft 1 in)
rotors folded	12.25 m (40 ft 2¼ in)
Width, rotors folded	4.00 m (13 ft 1½ in)
Height to top of rotor head	5.40 m (17 ft 8½ in)
Wheel track: mainwheels	3.50 m (11 ft 6 in)
nosewheels	1.40 m (4 ft 7 in)
Wheelbase	3.02 m (9 ft 11 in)
Cabin door: Height	approx 1.20 m (3 ft 11¼ in)
Width	approx 1.20 m (3 ft 11¼ in)

DIMENSIONS, INTERNAL:
Cabin: Length	4.52 m (14 ft 10 in)
Max width	1.30 m (4 ft 3 in)
Max height	1.32 m (4 ft 4 in)

AREAS:
Rotor disc (each)	198.5 m² (2,138 sq ft)

WEIGHTS AND LOADINGS:
Weight empty	6,500 kg (14,330 lb)
Max payload: internal	4,000 kg (8,818 lb)
external	5,000 kg (11,023 lb)
Normal T-O weight	11,000 kg (24,250 lb)
Max flight weight with slung load	12,600 kg (27,775 lb)

PERFORMANCE (at AUW of 11,000 kg; 24,250 lb):
Max level speed	135 knots (250 km/h; 155 mph)
Max cruising speed	124 knots (230 km/h; 143 mph)
Service ceiling	5,000 m (16,400 ft)
Hovering ceiling OGE	3,500 m (11,480 ft)
Range with max fuel	432 nm (800 km; 497 miles)
Endurance with max fuel	4 h 30 min

KAMOV Ka-50 WEREWOLF
NATO reporting name: Hokum

TYPE: Twin-turbine close support helicopter.
PROGRAMME: Project design completed December 1977; first prototype flew 27 July 1982; first Western report Summer 1984; first photograph published in US Department of Defense's *Soviet Military Power* document 1989; series production has been under way for three years at Arsenyer after competitive evaluation with Mi-28; production Ka-50 exhibited at 1992 Farnborough Air Show; entering service 1993.
CURRENT VERSIONS: **Hokum-A:** Basic single-seat close support helicopter.
Hokum-B: Tandem two-seat trainer, under development.
CUSTOMERS: Initial production for Russian Land Forces Army Aviation.
DESIGN FEATURES: World's first single-seat close-support helicopter. Coaxial, contra-rotating and widely separated three-blade rotors, with swept blade tips; small fuselage cross-section, with nose sensors; flat-screen cockpit, heavily armour protected, with rearview mirror above windscreen; small sweptback tail fin, with inset rudder and large tab; high-set tailplane on rear fuselage, with endplate auxiliary fins; retractable landing gear; mid-set unswept wings, carrying ECM pods at tips; four underwing weapon pylons; engines above wingroots, with prominent exhaust heat suppressors; high agility for fast, low-flying, close-range attack role; partially dismantled can be air-ferried in Il-76 freighter.

Kamov Ka-50 Werewolf ('Hokum') close support helicopter *(Piotr Butowski)*

Kamov Ka-50 Werewolf ('Hokum') close support helicopter

276 RUSSIA: AIRCRAFT—KAMOV

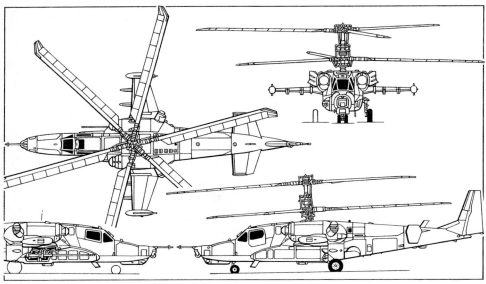

Basic single-seat combat version of Kamov Ka-50 Werewolf ('Hokum') with scrap view of gun installation on starboard side (*Jane's/Mike Keep*)

LANDING GEAR: Retractable tailwheel type; single mainwheels retract inward and upward into bottom of fuselage; twin tailwheels retract into tailboom; shock absorber in each unit. Optional inflatable pontoons for emergency use on water.

POWER PLANT (basic Ka-62): Two RKBM Rybinsk RD-600 turboshafts, each 956 kW (1,282 shp); fuel tanks under floor, capacity 1,450 litres (383 US gallons; 319 Imp gallons). APU in centre-fuselage.

ACCOMMODATION: Crew of one or two; optional bulkhead divider between flight deck and cabin; up to 16 passengers in four rows; forward hinged door each side of flight deck; large forward-sliding door and small rearward-hinged door each side of cabin; baggage hold to rear of cabin. VIP configuration to be available, typically with five seats, refreshment bar and toilet.

SYSTEMS: Interior heated and air-conditioned.

EQUIPMENT: Stretchers, hoist above port cabin door, cargo tie-downs, and other items as necessary for variety of roles, including transport of slung freight; air ambulance/operating theatre; search and rescue; patrol of highways, forests, electric power lines, gas and oil pipelines; survey of ice areas; surveillance of territorial waters, economic areas and fisheries; mineral prospecting; and servicing of offshore gas and oil rigs.

DIMENSIONS, EXTERNAL:
Main rotor diameter	13.50 m (44 ft 3½ in)
Length overall, rotors turning	15.75 m (51 ft 8 in)
Fuselage length	13.25 m (43 ft 5¾ in)
Height overall	4.10 m (13 ft 5½ in)
Width: over endplate fins	3.00 m (9 ft 10¼ in)
over mainwheels	2.50 m (8 ft 2½ in)
Wheelbase	4.725 m (15 ft 6 in)

AREAS:
Main rotor disc	143.1 m² (1,540 sq ft)

WEIGHTS AND LOADINGS:
Max payload:	
internal: 62, 62R, 62M	2,000 kg (4,409 lb)
external: 62, 62R, 62M	2,500 kg (5,510 lb)
Normal T-O weight: 62R	6,000 kg (13,225 lb)
Max T-O weight:	
62, 62R, 62M, internal load	6,250 kg (13,775 lb)
62, 62R, 62M, external load	6,500 kg (14,330 lb)

Cockpit of Kamov Ka-50 Werewolf (*Brian M. Service*)

starboard side of fuselage, with 500 rds, can be depressed up to 30° and traversed 5-6° hydraulically and is kept on target in azimuth by tracker which turns helicopter on its axis; two ammunition boxes in centre-fuselage. Provision for alternative weapons, including 23 mm gun pods, Kh-25MP (AS-12 'Kegler') anti-radiation missiles, air-to-air missiles or dispenser weapons.

Ka-62R: To be certificated to Western standards, for sale outside CIS; two 1,566 kW (2,100 shp) Rolls-Royce Turbomeca RTM322 turboshafts.

Ka-62M: As Ka-62R, but with two 1,212 kW (1,625 shp) General Electric T700/CT7-2D turboshafts.

DESIGN FEATURES: Configuration resembles such Western types as Eurocopter Dauphin, with single main rotor and fan-in-fin similar to latter's Fenestron; retractable tailwheel landing gear.

STRUCTURE: Between 50 and 55 per cent of structure made of composites, including four blades of main rotor; fuselage sides, doors, floor and roof, tailboom, fin, vertical stabilisers, and fan blades of carbon reinforced Kevlar.

DIMENSIONS, EXTERNAL:
Rotor diameter (each)	14.5 m (47 ft 7 in)
Length overall, rotors turning	16.0 m (52 ft 6 in)
Height overall	4.93 m (16 ft 2 in)

AREAS:
Rotor disc (each)	165.13 m² (1,777.4 sq ft)

WEIGHTS AND LOADINGS:
Normal T-O weight	9,800 kg (21,605 lb)
Max T-O weight	10,800 kg (23,810 lb)

PERFORMANCE:
Max speed:	
in shallow dive	189 knots (350 km/h; 217 mph)
in level flight	167 knots (310 km/h; 193 mph)
Vertical rate of climb at 2,500 m (8,200 ft)	600 m (1,970 ft)/min
Hovering ceiling OGE	4,000 m (13,125 ft)
Combat radius (estimated)	135 nm (250 km; 155 miles)
Endurance	4 h
g limit	+3

KAMOV Ka-62

TYPE: Medium-sized twin-turbine multi-purpose helicopter.
PROGRAMME: Construction of prototype Ka-62 (then known as V-62) began early 1990; Ka-62R variant announced September 1990.
CURRENT VERSIONS: **Ka-62:** Basic model for domestic market, *as described in detail*; first flight planned for 1993, certification 1996.

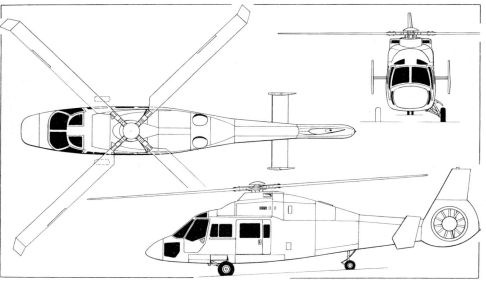

Kamov Ka-62 twin-turbine 16-passenger helicopter (*Dennis Punnett*)

Kamov Ka-62 utility helicopter (two RKBM Rybinsk RD-600 turboshafts) (*Mark Lambert*)

Max disc loading:
 62, 62R, 62M   44.08 kg/m² (9.03 lb/sq ft)
PERFORMANCE (estimated, at max T-O weight with internal load):
Max level speed:
 62, 62R, 62M   162 knots (300 km/h; 186 mph)
Cruising speed: 62   140 knots (260 km/h; 161 mph)
 62R, 62M   143 knots (265 km/h; 165 mph)
Max rate of climb at S/L: 62   702 m (2,300 ft)/min
 62R, 62M   750 m (2,460 ft)/min
Rate of climb at S/L, OEI: 62 72 m (236 ft)/min
 62R, 62M   108 m (354 ft)/min
Service ceiling: 62, 62M   above 5,000 m (16,400 ft)
 62R   above 6,000 m (19,685 ft)
Hovering ceiling OGE: 62   2,500 m (8,200 ft)
 62R   3,600 m (11,810 ft)
 62M   2,800 m (9,185 ft)
Range with max standard fuel at 500 m (1,640 ft), no reserves: 62   391 nm (725 km; 450 miles)
 62R   353 nm (655 km; 407 miles)
 62M   359 nm (665 km; 413 miles)
Range with auxiliary fuel at 500 m (1,640 ft), no reserves:
 62   605 nm (1,120 km; 695 miles)
 62R   545 nm (1,010 km; 627 miles)
 62M   553 nm (1,025 km; 637 miles)

KAMOV Ka-126
NATO reporting name: Hoodlum-B

TYPE: General purpose light helicopter; turboshaft version of Ka-26.

PROGRAMME: Development began 1984; early mockup had two small turboshafts above cabin; single turboshaft adopted subsequently; ground test vehicle completed early 1986; first flight of prototype (SSSR-01963) later 1986; first flight of first of four Soviet-built pre-production Ka-126s 19 October 1988; production by IAR SA of Romania planned; first flight of Romanian-built Ka-126 made 31 December 1988; first flight of Romanian production aircraft 14 February 1989; responsibility for programme now reverted to Kamov; FAA certification (VFR) scheduled 1994.

CUSTOMERS: Seven of initial series of 12 ordered from IAR delivered to Kamov by Spring 1992.

DESIGN FEATURES: Compared with Ka-26 has turboshaft engine, significant increase in payload/endurance/range, and greater year-round utilisation; updated equipment includes new low-volume spraygear, Hungarian-designed pellet dispensing system, low-cost nav/com radio.

Contra-rotating coaxial three-blade rotors; hinge rotor head with 'rake' type blade attachment, to be succeeded eventually by hingeless head made of titanium and composites. Ka-26 blades of initial series to be succeeded by GFRP and CFRP blades with twin-contour spar, load-carrying rear section and electrothermal anti-icing; rotor brake standard; non-folding blades. Three-stage gearbox with planetary gear trains, of alloy steel and aluminium casting, flange mounted with four load-carrying bolts. Accessories include cooling fan, hydraulic pump and AC generator; engine input 6,000 rpm.

FLYING CONTROLS: Mechanical with irreversible hydraulic actuators. Automatic rotor constant-speed control; conventional four-channel control (longitudinal, lateral, cyclic and differential pitch). Two endplate fins and rudders, toed inward 15°; non-controllable horizontal stabiliser.

STRUCTURE: Airframe materials primarily aluminium alloys, steel alloys and composite sandwich panels of GFRP with honeycomb filler.

LANDING GEAR: Non-retractable four-wheel type. Main units, at rear, carried by stub-wings. All four units embody oleo-pneumatic shock absorber. Forward wheels of castoring type, self-centring, no brakes. Rear wheels have pneumatic brakes. Mainwheel tyres size 595 × 185 mm, pressure 2.45 bars (35.5 lb/sq in); forward wheel tyres size 300 × 125 mm, pressure 3.43 bars (50 lb/sq in). Skis optional. Provision for large inflatable pontoons, across front of aircraft forward of front wheels and under each mainwheel.

POWER PLANT: One 522 kW (700 shp) Mars (Omsk) TV-O-100 turboshaft, installed centrally in streamline fairing above cabin. Electronic-hydraulic automatic two-channel control system, with manual control in case of electronic governor failure. Front driveshaft with plate coupling to gearbox. Fuel in two forward and one aft tank, total capacity 800 litres (211.3 US gallons; 176 Imp gallons). Provision for two external tanks, on sides of fuselage, total capacity 320 litres (84.5 US gallons; 70.4 Imp gallons). Single-point main tank refuelling, on port side of aft tank.

ACCOMMODATION: Fully enclosed cabin, with rearward sliding door each side; normal operation by single pilot; second seat and dual controls optional. Cabin ventilated, and warmed and demisted by air from combustion heater, which also heats passenger cabin when fitted. Air filter on nose of agricultural version. Space aft of cabin, between main landing gear legs and under transmission, can accommodate variety of interchangeable payloads. Cargo/passenger pod accommodates four or six persons on folding sidewall seats, with provision for seventh passenger beside pilot; two clamshell doors at rear of pod, emergency exit each side and hatch in floor. Ambulance pod accommodates two stretcher patients, two seated casualties and medical attendant. For agricultural work, chemical hopper

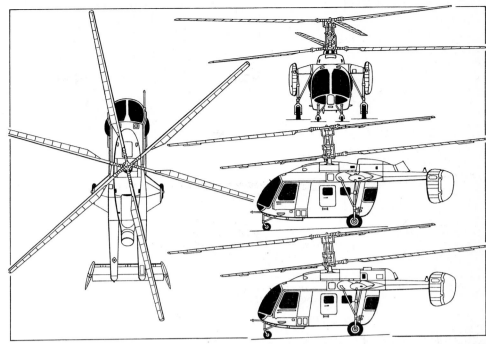
Kamov Ka-126 turbine powered development of Ka-26, with additional side elevation (bottom) of Ka-128 *(Jane's/Mike Keep)*

Kamov Ka-126 (522 kW; 700 shp Mars TV-O-100 turboshaft) *(John Fricker)*

Kamov Ka-126 general purpose light helicopter (Mars TV-O-100 turboshaft) *(Piotr Butowski)*

(capacity 1,000 litres; 264 US gallons; 220 Imp gallons) and dust spreader or spraybars are fitted in this position, on aircraft's CG. Aircraft can also be operated with either an open platform for hauling freight or hook for slinging loads at end of a cable or in a cargo net.

SYSTEMS: Single hydraulic system, with manual override, for control actuators. Main electrical system 27V 3kW DC, with backup 40Ah battery; secondary system 36/115V AC with two static inverters; 115/200V AC system with 16kVA generator (6kVA to power agricultural equipment and rotor anti-icing). Electrothermal rotor blade de-icing; hot air engine air intake anti-icing; alcohol windscreen anti-icing; electrically heated pitot. Pneumatic system for mainwheel brakes, tyre inflation, agricultural equipment control, pressure 39-49 bars (570-710 lb/sq in). Oxygen system optional.

AVIONICS: Include radar altimeter, ADF, main and backup artificial horizons, VHF and HF radio, intercom, and emergency locator beacon.

DIMENSIONS, EXTERNAL:
Rotor diameter (each)	13.00 m (42 ft 7¾ in)
Rotor blade chord	0.25 m (9.84 in)
Length overall, rotors turning	13.00 m (42 ft 7¾ in)
Length of fuselage	7.775 m (25 ft 6 in)
Width, excl rotors	3.224 m (10 ft 7 in)
Height to top of rotor head	4.155 m (13 ft 7½ in)
Wheel track: mainwheels	2.56 m (8 ft 4¾ in)
front wheels	0.90 m (2 ft 11½ in)
Wheelbase	3.479 m (11 ft 5 in)
Crew cabin doors (each): Height	1.06 m (3 ft 5¾ in)
Width	0.83 m (2 ft 8½ in)
Height to sill	0.60 m (1 ft 11¾ in)
Passenger pod doorway: Height	1.40 m (4 ft 7 in)
Width	1.40 m (4 ft 7 in)
Emergency exits (each): Height	0.86 m (2 ft 9¾ in)
Width	0.66 m (2 ft 2 in)
Floor hatch: Length	0.74 m (2 ft 5⅛ in)
Width	0.54 m (1 ft 9¼ in)

DIMENSIONS, INTERNAL:
Passenger pod: Length	2.04 m (6 ft 8¼ in)
Max width, at floor	1.21 m (3 ft 11½ in)
Height	1.38 m (4 ft 6¼ in)
Volume	2.47 m³ (87.2 cu ft)

WEIGHTS AND LOADINGS:
Weight empty	1,915 kg (4,222 lb)
Max slung payload	1,000 kg (2,205 lb)
Max internal fuel	640 kg (1,411 lb)
Auxiliary fuel	256 kg (564 lb)
Max T-O weight	3,250 kg (7,165 lb)

PERFORMANCE:
Never-exceed speed (V_{NE})	97 knots (180 km/h; 112 mph)
Econ cruising speed at S/L	86 knots (160 km/h; 99 mph)
Max rate of climb at S/L	396 m (1,300 ft)/min
Vertical rate of climb at S/L	15 m (49 ft)/min
Service ceiling	3,850 m (12,630 ft)
Hovering ceiling IGE: ISA	970 m (3,180 ft)
ISA + 20°C	100 m (320 ft)
Hovering ceiling OGE: ISA	80 m (260 ft)
Range: with max payload	136 nm (253 km; 157 miles)
with max internal fuel	384 nm (713 km; 443 miles)
with auxiliary fuel, no reserves	547 nm (1,015 km; 630 miles)
Max endurance, internal fuel, no reserves	5 h 36 min

KAMOV Ka-226

TYPE: Twin-turbine utility and agricultural helicopter.

PROGRAMME: Announced at 1990 Helicopter Association International convention, Dallas, USA; prototype to fly 1993; delivery of production Allison engines scheduled to begin 1994; FAA certification in 1995 to be sought; production will be in Russia.

DESIGN FEATURES: Refined development of Ka-26/126; changes to shape of nose, twin tail fins and rudders, and passenger pod; passenger cabin has much larger windows and remains interchangeable with variety of payload modules including agricultural systems with hopper capacity of 1,000 litres (264 US gallons; 220 Imp gallons); new rotor system, interchangeable with standard coaxial system, will become available later.

LANDING GEAR: Non-retractable four-wheel type. Main units at rear, carried by stub-wings. All units embody oleo-pneumatic shock absorber. Forward units of castoring type, without brakes. Rear wheels have pneumatic brakes.

POWER PLANT: Two 313 kW (420 shp) Allison 250-C20B turboshafts, side by side aft of rotor mast, in same position as single turboshaft of Ka-126, with individual driveshafts to rotor gearbox. Transmission rating 616 kW (840 shp). Standard fuel capacity 750 litres (198 US gallons; 165 Imp gallons), in tanks above and forward of payload module area. Provision for two external tanks, on sides of fuselage, total capacity 320 litres (84.5 US gallons; 70.4 Imp gallons).

ACCOMMODATION: Cabin module has two three-place rearward and forward facing bench seats; baggage compartment behind rear wall. Seventh passenger beside pilot on flight deck.

AVIONICS: Cockpit instrumentation and avionics to customer's choice, including Bendix/King equipment for IFR flight.

EQUIPMENT: Specially equipped payload modules available for variety of roles.

DIMENSIONS, EXTERNAL:
Rotor diameter (each)	13.00 m (42 ft 7¾ in)
Length overall, excl rotors	8.10 m (26 ft 7 in)
Width over stub-wings	3.22 m (10 ft 6¾ in)
Height to top of rotor head	4.15 m (13 ft 7½ in)
Wheel track: nosewheels	0.90 m (2 ft 11½ in)
mainwheels	2.56 m (8 ft 4¾ in)
Wheelbase	3.48 m (11 ft 5 in)
Passenger pod: Length	2.04 m (6 ft 8¼ in)
Width	1.28 m (4 ft 2¼ in)
Depth	1.40 m (4 ft 7¼ in)

AREAS:
Rotor disc (each)	132.7 m² (1,430 sq ft)

WEIGHTS AND LOADINGS:
Weight empty	1,952 kg (4,304 lb)
Max payload	1,300 kg (2,865 lb)
Max internal fuel	600 kg (1,322 lb)
Auxiliary fuel	256 kg (564 lb)
Normal T-O weight	3,100 kg (6,835 lb)
Max T-O weight	3,400 kg (7,500 lb)

PERFORMANCE (estimated):
Never-exceed speed (V_{NE})	110 knots (205 km/h; 127 mph)
Econ cruising speed	100 knots (185 km/h; 115 mph)
Max rate of climb at S/L	540 m (1,770 ft)/min
Vertical rate of climb at S/L	168 m (550 ft)/min
Service ceiling	5,050 m (16,565 ft)
Hovering ceiling IGE: ISA	2,020 m (6,625 ft)
ISA + 20°C	1,200 m (3,935 ft)
Hovering ceiling OGE: ISA	1,280 m (4,200 ft)
ISA + 20°C	600 m (1,970 ft)
Range: with max payload	20 nm (37 km; 23 miles)
with max internal fuel	325 nm (602 km; 374 miles)
with auxiliary fuel, no reserves	470 nm (873 km; 542 miles)
Max endurance, internal fuel, no reserves	4 h 38 min

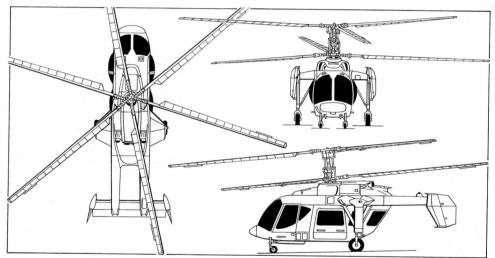

Kamov Ka-226, a refined development of the Ka-126 utility helicopter (*Jane's/Mike Keep*)

KAMOV Ka-128

TYPE: General purpose light helicopter; differs from Ka-126 only in power plant and addition of an intermediate gearbox.

PROGRAMME: Development began 1992; first flight scheduled 1993, with FAA certification (VFR) 1994.

POWER PLANT: One 531 kW (712 shp) Turbomeca Arriel 1D1 turboshaft, with similarly rated transmission.

WEIGHTS AND LOADINGS:
Weight empty	1,820 kg (4,013 lb)
Max slung payload	1,000 kg (2,205 lb)
Max internal fuel	640 kg (1,411 lb)
Auxiliary fuel	256 kg (564 lb)
Max T-O weight	3,250 kg (7,165 lb)

PERFORMANCE (estimated):
Never-exceed speed (V_{NE})	105 knots (195 km/h; 121 mph)
Econ cruising speed at S/L	86 knots (160 km/h; 99 mph)
Max rate of climb at S/L	468 m (1,535 ft)/min
Vertical rate of climb at S/L	90 m (295 ft)/min
Service ceiling	5,200 m (17,060 ft)
Hovering ceiling IGE: ISA	1,600 m (5,250 ft)
ISA + 20°C	850 m (2,785 ft)
Hovering ceiling OGE: ISA	750 m (2,460 ft)
Range: with max payload	198 nm (367 km; 228 miles)
with max internal fuel	383 nm (710 km; 441 miles)
with auxiliary fuel, no reserves	570 nm (1,057 km; 656 miles)
Max endurance, internal fuel, no reserves	5 h 39 min

Full-scale mockup of Kamov Ka-226 at MosAeroshow '92

MiG

AVIATSIONNYI NAUCHNO-PROIZVODST-VENNIY KOMPLEKS - ANPK MiG IMIENI A. I. MIKOYANA (A. I. Mikoyan Aviation Scientific-Production Complex 'MiG' - ANPK MiG)

6 Leningradskoye Shosse, 125299 Moscow
Telephone: 7 (095) 158 82 46
Fax: 7 (095) 943 00 27
Telex: 411 700 MIGA No. 006022
GENERAL DESIGNER: Rostislav Apollosovich Belyakov

Colonel-General Artem Ivanovich Mikoyan, who died on 9 December 1970 aged 65, was head of OKB responsible for MiG series of fighter aircraft from 1939. With Mikhail Iosifovich Guryevich (1893-1976), a mathematician, he collaborated in design of first really effective Soviet jet fighter, MiG-15, which began to enter squadron service in numbers in 1949. Subsequent designs by the OKB have resulted in more MiGs entering worldwide service than any other fighter 'family' since Second World War.

In addition to aircraft described in this entry, MiG OKB is manufacturing horizontal tail surfaces for Dassault Falcon business jets.

MIKOYAN MiG-23

NATO reporting names: Flogger-A, B, C, E, F, G, H and K
TYPE: Single-seat variable-geometry air combat fighter and two-seat operational trainer.
PROGRAMME: Development began 1964; 23-11/1 prototype first flew 10 June 1967 and was displayed during Aviation Day flypast, Domodyedovo Airport, Moscow, 9 July 1967; pre-series aircraft delivered to Soviet air forces 1970; initial series production interceptors delivered 1973; with MiG-27, superseded MiG-21 as primary equipment Soviet tactical air forces and APVO home defence interceptor force; production in USSR ended mid-1980s, but continues in India; replacement of early variants with MiG-29s and Su-27s continues.
CURRENT VERSIONS: Early versions last described in 1990-91 edition.

MiG-23M ('Flogger-B'): Most-produced production version; first flown June 1972; single-seat air combat fighter; first aircraft of former Soviet Union with demonstrated ability to track and engage targets flying below its own altitude; Soyuz/Khachaturov R-29-300 turbojet, rated at 122.5 kN (27,540 lb st) with afterburning; no wing leading-edge flaps initially (retrofitted later); Sapfir-23D-Sh J-band radar (NATO 'High Lark'); Sirena-3 radar warning system; Doppler; TP-23 infra-red search/track pod under cockpit; standard in Soviet air forces from about 1975.

MiG-23MF ('Flogger-B'): Export version of MiG-23M, in service with non-Soviet Warsaw Pact air forces from 1978.

MiG-23UB ('Flogger-C'): Tandem two-seat operational training/combat version; Tumansky R-27F2M-300 turbojet, rated at 98 kN (22,045 lb st) with afterburning; individual canopy over each seat; rear seat raised, with retractable periscopic sight; deepened dorsal spine fairing aft of rear canopy. First flown May 1969; in production 1970-78.

MiG-23MS ('Flogger-E'): Export version of MiG-23M with R-27F2M-300 engine; equipped to lower standard; smaller radar ('Jay Bird', search range 15 nm; 29 km; 18 miles, tracking range 10 nm; 19 km; 12 miles) in shorter nose radome; no infra-red sensor or Doppler; armed with R-3S (K-13T; NATO AA-2 'Atoll') or R-60 (K-60; NATO AA-8 'Aphid') air-to-air missiles and GSh-23 gun.

MiG-23B ('Flogger-F'): Single-seat light attack aircraft based on MiG-23S interceptor airframe; forward fuselage redesigned; instead of ogival radome, nose sharply tapered in side elevation, housing PrNK Sokol-23S nav/attack system; twin-barrel 23 mm GSh-23L gun retained in bottom of centre-fuselage; armour on sides of cockpit; wider, low-pressure tyres; Lyulka AL-21F-300 turbojet, rated at 11.27 kN (25,350 lb st) with afterburning; fuel tanks designed to fill with neutral gas as fuel level drops, to prevent explosion after impact; active and passive ECM; six attachments under fuselage and wings for wide range of weapons; project started 1969; first flight 20 August 1970; 24 built; developed as MiG-23BN/BM/BK and MiG-27 series.

MiG-23BN ('Flogger-F'): As MiG-23B except for Soyuz/Khachaturov R-29B-300 turbojet, rated at 11.27 kN (25,350 lb st) with afterburning, and Sokol-23N nav/attack system.

MiG-23BM ('Flogger-F'): As MiG-23BN except for PrNK-23 nav/attack system slaved to a computer.

MiG-23BK: Further equipment changes. NATO reporting name 'Flogger-H' identifies aircraft with small fairing for radar warning receiver each side of bottom fuselage, forward of nosewheel doors. Iraqi aircraft have Dassault-type fixed flight refuelling probe forward of windscreen.

MiG-23ML ('Flogger-G'): Much redesigned and lightened version (L of designation for *logkiy*; light) built in series 1976-81; basically as MiG-23M, but Soyuz/Khachaturov R-35-300 turbojet; rear fuselage fuel tank deleted; much smaller dorsal fin; modified nosewheel leg; Sapfir-23ML lighter weight radar; new undernose pod for

MiG-23MF ('Flogger-B') of Indian Air Force's No. 224 Squadron *(Peter Steinemann)*

MiG-23UB two-seat trainer of Czech Air Force *(P. R. Foster)*

TP-23M IRST; new missiles. *Detailed description applies to MiG-23ML.*

MiG-23P ('Flogger-G'): Modified version of MiG-23ML; digital nav system computer guides aircraft under automatic control from the ground and informs pilot when to engage afterburning and to fire his missiles and gun.

MiG-23MLD ('Flogger-K'): Mid-life update of MiG-23ML (D of designation for *dorabotannyy*; modified); identified by dogtooth notch at junction of each wing glove leading-edge and intake trunk; system introduced to extend and retract leading-edge flaps automatically when wing sweep passes 33° (system disengaged and flaps retracted when speed exceeds 485 knots; 900 km/h; 560 mph and wings at 72° sweep); new IFF antenna forward of windscreen; R-73A (NATO AA-11 'Archer') close-range air-to-air missiles on fuselage pylons; pivoting pylons under outer wings; radar warning receivers and chaff/flare dispensers added; built-in simulation system enables pilot to train for weapon firing and air-to-surface missile guidance without use of gun or missiles.
CUSTOMERS: Much reduced numbers (from around 1,800) serve with CIS tactical and air defence units, and Naval Aviation; exported to Afghanistan, Algeria ('Flogger-E/F'), Angola ('C/E'), Bulgaria ('B/C/H'), Cuba ('C/E/F'), Czech Republic ('B/C/G/H'), Egypt ('C/F'), Ethiopia ('F'), former East Germany ('B/C/G'), Hungary ('B/C'), India ('B/C/H'), Iraq ('E/H'), North Korea ('E'), Libya ('C/E/F'), Poland ('B/C/H'), Romania ('B/C'), Syria ('F/G'), Vietnam ('F'), Yemen ('F').
DESIGN FEATURES: Shoulder-wing variable-geometry configuration; sweep variable manually in flight or on ground to 16°, 45° or 72° (values given in manuals and on pilot's panel; true values 18° 40′, 47° 40′ and 74° 40′ respectively); two hydraulic wing sweep motors driven separately by main and control booster systems; if one system fails, wing sweep system remains effective at 50 per cent normal angular velocity; rear fuselage detachable between wing and tailplane for engine servicing; lower portion of large ventral fin hinged to fold to starboard when landing gear extended, for ground clearance; leading-edge sweepback 72° on fixed wing panels, 57° on horizontal tail surfaces, 65° on fin.
FLYING CONTROLS: Hydraulically actuated; full-span single-slotted trailing-edge flaps, each in three sections; outboard sections operable independently when wings fully swept; no ailerons; two-section upper surface spoilers/lift dumpers, forward of mid and inner flap sections each side, operate differentially in conjunction with horizontal tail surfaces (except when disengaged at 72° sweep), and collectively for improved runway adherence and braking after touchdown; leading-edge flap on outboard two-thirds of each main (variable geometry) panel, coupled to trailing-edge flaps; all-moving horizontal tail surfaces operate differentially and symmetrically for aileron and elevator function respectively; ground adjustable tab on each horizontal surface; rudder actuated by hydraulic booster with spring artificial feel; four door type airbrakes, two on each side of rear fuselage, above and below horizontal tail surface.
STRUCTURE: All-metal; two main spars and auxiliary centre spar in each wing; extended chord (dogtooth) on outer panels visible when wings swept; fixed triangular inboard wing panels; welded steel pivot box carry-through structure; basically circular section semi-monocoque fuselage, flattened each side of cockpit; lateral air intake trunks blend into circular rear fuselage; splitter plate, with boundary layer bleeds, forms inboard face of each intake; two

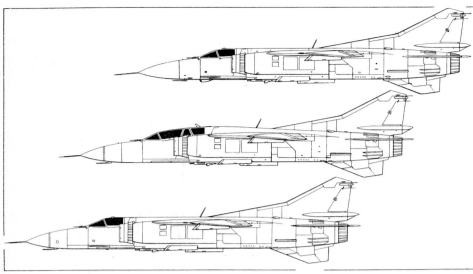

Top to bottom: Side views of the 'Flogger-B', 'Flogger-C' and 'Flogger-E' variants of the MiG-23 series *(Dennis Punnett)*

Max wing loading: spread	476.6 kg/m² (97.6 lb/sq ft)
swept	521.0 kg/m² (106.7 lb/sq ft)
Max power loading	139.6 kg/kN (1.37 lb/lb st)

PERFORMANCE:
Max level speed: at height, 72° sweep
　　　　Mach 2.35 (1,350 knots; 2,500 km/h; 1,553 mph)
　at height, 16° sweep
　　　　Mach 0.88 (507 knots; 940 km/h; 584 mph)
　at S/L, 72° sweep
　　　　Mach 1.10 (728 knots; 1,350 km/h; 838 mph)
Max rate of climb at S/L　　　14,400 m (47,250 ft)/min
Service ceiling　　　　　　　18,500 m (60,700 ft)
T-O run　　　　　　　　　　500 m (1,640 ft)
Landing run　　　　　　　　750 m (2,460 ft)
Combat radius:
　with six air-to-air missiles
　　　　　　　　620 nm (1,150 km; 715 miles)
　with 2,000 kg (4,410 lb) of bombs
　　　　　　　　378 nm (700 km; 435 miles)
Range:
　with max internal fuel
　　　　　　　1,050 nm (1,950 km; 1,210 miles)
　with three external tanks
　　　　　　　1,520 nm (2,820 km; 1,750 miles)
g limit: below Mach 0.85　　　　　　+8.5
　　　　above Mach 0.85　　　　　　+7.5

rectangular auxiliary intake doors in each trunk, under inboard wing leading-edge, are sucked open to increase intake area at take-off and low airspeeds; pressure relief vents under rear fuselage; fin and forward portion of horizontal surfaces conventional light alloy structures; rudder and rear of horizontal surfaces have honeycomb core.

LANDING GEAR: Hydraulically retractable tricycle type; single wheel on each main unit and steerable twin-wheel nose unit; mainwheel tyres size 830 × 300; nosewheel tyres size 520 × 125; main units retract inward into rear of air intake trunks; main fairings to enclose these units attached to legs; small inboard fairing for each wheel bay hinged to fuselage belly. Nose unit, with mudguard over each wheel, retracts rearward. Mainwheel disc brakes and anti-skid units. Brake parachute, area 21 m² (226 sq ft), in cylindrical fairing at base of rudder with split conic doors.

POWER PLANT: One Soyuz/Khachaturov R-35-300 turbojet, rated at up to 127.5 kN (28,660 lb st) with max afterburning. Water injection system, capacity 28 litres (7.4 US gallons; 6.15 Imp gallons). Three fuel tanks in fuselage, aft of cockpit, and six in wings; internal fuel capacity 4,250 litres (1,122 US gallons; 935 Imp gallons). Variable geometry air intakes and variable nozzle. Provision for jettisonable external fuel tank, capacity 800 litres (211 US gallons; 176 Imp gallons), on underfuselage centreline; two more under fixed wing panels. Two additional external tanks of same capacity may be carried on non-swivelling pylons under outer wings for ferry flights, with wings in fully forward position. Attachment for assisted take-off rocket each side of fuselage aft of landing gear.

ACCOMMODATION: Pilot only, on zero/zero ejection seat in air-conditioned and pressurised cockpit, under small hydraulically actuated rearward hinged canopy. Bullet-proof windscreen.

AVIONICS: Modernised SAU-23AM automatic flight control system, coupled to Polyot short-range navigation and flight system. Sapfir-23ML J-band multi-mode radar (NATO 'High Lark 2': search range 38 nm; 70 km; 43 miles, tracking range 29 nm; 55 km; 34 miles) behind dielectric nosecone; no radar scope; instead, picture is projected onto head-up display. RSBN-6S short-range radio nav system; ILS, with antennae (NATO 'Swift Rod') under radome and at tip of fin trailing-edge; suppressed UHF antennae form tip of fin and forward fixed portion of ventral fin; yaw vane above fuselage aft of radome; angle of attack sensor on port side. SRO-2 (NATO 'Odd Rods') IFF antenna immediately forward of windscreen. TP-23M undernose infra-red sensor pod, Sirena-3 radar warning system, and Doppler equipment standard on CIS version. Sirena-3 antennae in horns at inboard leading-edge of each outer wing and below ILS antenna on fin.

EQUIPMENT: ASP-17ML gunsight; small electrically heated rearview mirror on top of canopy; retractable landing/taxying light under each engine air intake.

ARMAMENT: One 23 mm GSh-23L twin-barrel gun in fuselage belly pack; large flash eliminator around muzzles; 200 rds. Two pylons in tandem under centre-fuselage, one under each engine air intake duct, and one under each fixed inboard wing panel, for radar guided R-23R (K-23R; NATO AA-7 'Apex'), infra-red R-23T (K-23T; AA-7 'Apex') and/or infra-red R-60T (AA-8 'Aphid') air-to-air missiles, B-8 packs of twenty 80 mm S-8 air-to-surface rockets, UB-32-57 packs of thirty-two 57 mm S-5 rockets, S-24 240 mm rockets, bombs, container weapons, UPK-23-250 pods containing a GSh-23L gun, various sensor and equipment pods or other external stores. Use of twin launchers under air intake ducts permits carriage of four R-60 missiles, plus two R-23 on underwing pylons.

DIMENSIONS, EXTERNAL:
Wing span: fully spread	13.965 m (45 ft 10 in)
fully swept	7.779 m (25 ft 6¼ in)
Length overall: incl nose probe	16.71 m (54 ft 10 in)
Fuselage length: nosecone tip to jetpipe nozzle	15.65 m (51 ft 4¼ in)
Height overall	4.82 m (15 ft 9¾ in)
Wheel track	2.658 m (8 ft 8¾ in)
Wheelbase	5.772 m (18 ft 11¼ in)

AREAS:
Wings, gross: spread	37.35 m² (402.0 sq ft)
swept	34.16 m² (367.7 sq ft)

WEIGHTS AND LOADINGS:
Weight empty	10,200 kg (22,485 lb)
Max external weapon load	3,000 kg (6,615 lb)
T-O weight	14,700-17,800 kg (32,405-39,250 lb)

MIKOYAN MiG-27

NATO reporting names: Flogger-D and J
Indian Air Force name: Bahadur (Valiant)

TYPE: Single-seat variable-geometry ground attack aircraft.

PROGRAMME: Based on MiG-23B airframe, with enhanced power plant and nav/attack system for specific role; entered service second half of 1970s; production in USSR completed mid-1980s, continues in India.

CURRENT VERSIONS: **MiG-27** ('Flogger-D'): Initial version for former Soviet tactical air forces; as MiG-23BM but fixed engine air intakes with shorter inner wall 80 mm (3 in) from side of fuselage for boundary layer bleed; two-position afterburner nozzles; modifications to wing sweep and flap extension controls to cope with changes of CG on weapon carriage and release, and for take-off and landing. *MiG-23 data (which see) apply also to MiG-27, except as noted below.*

MiG-27K ('Flogger-D'): Improved version, based on MiG-23BK; PrNK-23K nav/attack system, providing automatic flight control, gun firing and weapons release; new PMS mode for release of weapons and gun firing during manoeuvres, and PKS mode for automatic navigation to target and attack on preprogrammed co-ordinates at night or in cloud; new 30 mm six-barrel GSh-6-30 gun; SUV weapon selector and fire control system; active ECM jammer; laser rangefinder and marked target tracker behind small sloping window in nose, permitting use of laser guided missiles; blister fairing under nose of later aircraft, with windows, providing added rearward designation capability for laser guided weapon delivery; IFF blade antenna under each side of nose aft of designator fairing; RSBN ILS antenna port side of nose; no bullet shaped antennae above wingroot glove pylons; external cockpit armour deleted on all but early production aircraft; wingroot leading-edge extensions.

MiG-27D ('Flogger-J'): Further improved; identified 1981; delivered in successively upgraded versions; PrNK-23M nav/attack system, permitting use of new stores including three-camera reconnaissance pod and 23 mm SPPU-22 gun pods, each containing twin-barrel gun that can be depressed to attack ground targets, with 260 rds. Final model has wider and deeper nose, with lip at top over larger and less sloping window for more advanced Klen laser rangefinder.

MiG-27M: Version of MiG-27D built in India under licence by HAL as Bahadur (see Indian section). Export **MiG-27L** similar.

CUSTOMERS: CIS tactical air forces and Naval Aviation.

POWER PLANT: One Soyuz/Khachaturov R-29B-300 turbojet, rated at 78.40 kN (17,625 lb st) dry and 112.7 kN (25,335 lb st) with max afterburning; fixed air intakes and two-position (on/off) afterburner nozzle consistent with

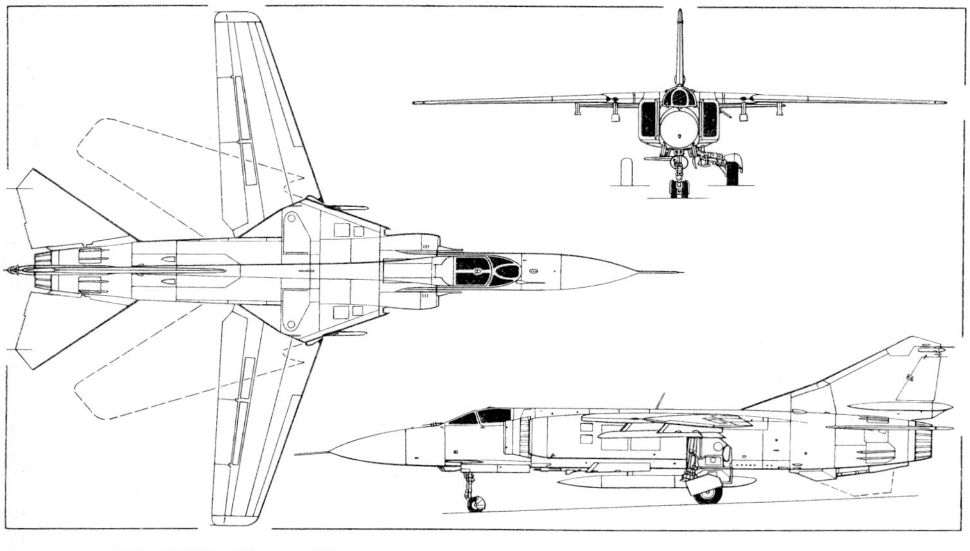

MiG-23MLD ('Flogger-K') single-seat variable-geometry interceptor *(Dennis Punnett)*

Nose of late production MiG-27D 'Flogger-J' *(Linda Jackson)*

MiG—AIRCRAFT: RUSSIA 281

MiG-27D ('Flogger-J') showing the restyled nose, absence of bullet shape antennae on the wing gloves, addition of wingroot leading-edge extensions, and depressed barrel of gun in port SPPU-22 underwing pod

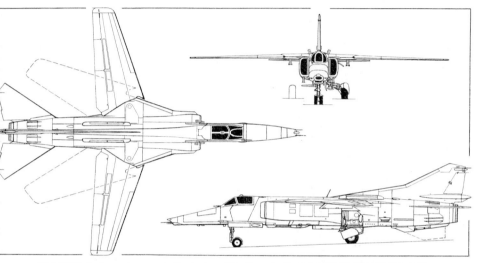

MiG-27M single-seat ground attack aircraft built under licence in India *(Dennis Punnett)*

MiG-27K ('Flogger-D') armed with Kh-31 air-to-surface missile *(Piotr Butowski)*

primary requirement of transonic speed at low altitude; internal fuel capacity 5,400 litres (1,426 US gallons; 1,188 Imp gallons); provision for up to three 790 litre (209 US gallon; 174 Imp gallon) external tanks.

AVIONICS (basic MiG-27): PrNK-23 nav/attack system; SAU-1 automatic flight control system; INS; SPS-141 IR jammer; RI-65 16-item vocal warning system; SUA-1 angle of attack indicator; SG-1 radar warning system; SO-69 transponder; SRO-1P IFF; RV-5R/RV-10 radio altimeters; Fone telemetry system; bullet-shape antenna above each glove pylon associated with missile guidance.

EQUIPMENT: Flare dispenser, in form of short fence, on fuselage each side of dorsal fin.

ARMAMENT: One 23 mm GSh-23L twin-barrel gun in fuselage belly pack (30 mm GSh-6-30 in MiG-27K/D/M, with 260 rds); bomb/JATO rack each side of rear fuselage; five other pylons for external stores; 4,000 kg (8,818 lb) of weapons, including tactical nuclear bombs, R-3S (K-13T; NATO AA-2D 'Atoll-D') and R-13M air-to-air missiles, Kh-23 (NATO AS-7 'Kerry') radio command guided air-to-surface missiles (laser guided Kh-29, AS-14 'Kedge' on MiG-27D), 240 mm S-24 rockets, UB-32 or UB-16 packs of 57 mm rockets, twenty-two 50 or 100 kg, nine 250 kg or eight 500 kg bombs, napalm containers.

DIMENSIONS, EXTERNAL: As MiG-23, plus:
  Length overall   17.076 m (56 ft 0¼ in)

Height overall	5.00 m (16 ft 5 in)
Tailplane span	5.75 m (18 ft 10¼ in)

AREAS:
Horizontal tail surfaces	6.88 m² (74.06 sq ft)

WEIGHTS AND LOADINGS (A: MiG-27, K: MiG-27K):
Weight empty: A	11,908 kg (26,252 lb)
Max internal fuel: A	4,560 kg (10,053 lb)
Normal T-O weight: A	18,100 kg (39,905 lb)
Max T-O weight from unprepared surface:	
K	18,107 kg (39,920 lb)
Max T-O weight: A	20,300 kg (44,750 lb)
K	20,670 kg (45,570 lb)
Max landing weight: K	17,000 kg (37,475 lb)
Max wing loading:	
A, spread	544.7 kg/m² (111.3 lb/sq ft)
A, swept	594.3 kg/m² (121.7 lb/sq ft)
K, spread	553.4 kg/m² (113.4 lb/sq ft)
K, swept	605.1 kg/m² (123.9 lb/sq ft)
Max power loading: A	180.0 kg/kN (1.76 lb/lb st)
K	183.25 kg/kN (1.80 lb/lb st)

PERFORMANCE (basic MiG-27):
Max level speed: at 8,000 m (26,250 ft)	
	Mach 1.7 (1,017 knots; 1,885 km/h; 1,170 mph)
at S/L, clean	
	Mach 1.1 (728 knots; 1,350 km/h; 839 mph)
Landing speed	140-146 knots (260-270 km/h; 162-168 mph)
Max rate of climb at S/L	12,000 m (39,370 ft)/min
Service ceiling	14,000 m (45,900 ft)
T-O run	950 m (3,120 ft)
Landing run: with brake-chute	900 m (2,950 ft)
without brake-chute	1,300 m (4,265 ft)
Combat radius (lo-lo-lo), with 7% reserves:	
with two Kh-29 missiles	121 nm (225 km; 140 miles)
with two Kh-29 missiles and three 790 litre external tanks	291 nm (540 km; 335 miles)
g limit	+7.0

MIKOYAN MiG-25
NATO reporting name: Foxbat

TYPE: Single-seat interceptor, reconnaissance aircraft and two-seat conversion trainer.

PROGRAMME: Design started 1959 as Ye-155P supersonic high-altitude interceptor to counter all potential threats, from low-flying cruise missiles to A-11 (SR-71A reconnaissance aircraft) under US development; programme launched officially February 1962; Ye-155R reconnaissance version designed and built 1961-62; Ye-155R-1 first to fly 6 March 1964; Ye-155P-1 interceptor prototype flew 9 September 1964; early history in previous editions of *Jane's*; production, as MiG-25/25R series, completed mid-1980s.

CURRENT VERSIONS: **MiG-25P** ('Foxbat-A'): Single-seat interceptor derived from Ye-155P-1 prototype; two R-15B-300 turbojets, each 100.1 kN (22,500 lb st) with afterburning and with 150 h service life; Smertch-A (NATO 'Fox Fire') radar, search range 54 nm (100 km; 62 miles), tracking range 27 nm (50 km; 31 miles); CIS aircraft all converted to MiG-25PDS.

MiG-25RB ('Foxbat-B'): Single-seat high-altitude reconnaissance-bomber, derived from Ye-155R-1 prototype; production began as MiG-25R, for reconnaissance only, in 1969; bombing capability added to redesignated RB in 1970; no guns or air-to-air missiles; R-15BD-300 turbojets; any one of three interchangeable photographic/elint modules, with five camera windows and flush dielectric panels, carried aft of small dielectric nosecap, instead of interceptor's Smertch radar; slightly reduced wing span; wing leading-edge sweep constant 41° from root to tip; first aircraft produced in former USSR with INS updated by Doppler; specially developed automatic bombing system makes possible all-weather day/night precision attacks at supersonic speed from heights above 20,000 m (65,600 ft) against targets with known geographic co-ordinates, carrying four 500 kg bombs under wings, two under fuselage; SRS-4A/B elint equipment; fuel tank in each fin, providing additional 700 litres (185 US gallons; 154 Imp gallons) capacity; provision for 5,300 litre (1,400 US gallon; 1,165 Imp gallon) underbelly tank; able to fly long distances at cruising speed of Mach 2.35, max speed of Mach 2.83 with full bomb load.

MiG-25RBV and **RBT** ('Foxbat-B'): As MiG-25RB, with different equipment, including SRS-9 elint on RBV. Produced 1978-82.

MiG-25PU ('Foxbat-C'): Training version of MiG-25P; redesigned nose section containing separate cockpit with individual canopy for instructor, forward of standard cockpit and at lower level; gun firing and weapon release simulation standard; some systems modified and updated, permitting simulation of failures; no radar in nose; no combat capability; first rollout 1968; max speed limited to Mach 2.65.

MiG-25RU ('Foxbat-C'): Training version of MiG-25R; identical to MiG-25PU except for absence of combat simulation system; no reconnaissance sensors; first rollout 1972.

MiG-25RBK ('Foxbat-D'): Produced simultaneously with RB series from 1972; reconnaissance modules contain different elint and other avionics and no cameras; bombing capability retained.

282 RUSSIA: AIRCRAFT—MiG

MiG-25R ('Foxbat-B') of the Indian Air Force *(Peter Steinemann)*

MiG-25PU ('Foxbat-C') of the Indian Air Force *(Peter Steinemann)*

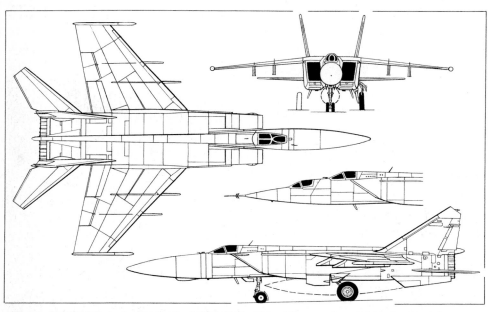

Mikoyan MiG-25PDS single-seat fighter (NATO 'Foxbat-E'), with scrap view of front fuselage of two-seat MiG-25PU ('Foxbat-C') *(Dennis Punnett)*

MiG-25RBS ('Foxbat-D'): As MiG-25RBK but with different sensors; followed RBK in production 1975-82; all RBSs upgraded to **MiG-25RBSh** during servicing from 1981. Other equipment changes produced **MiG-25RBF** in 1981.

MiG-25PD ('Foxbat-E'): Development of MiG-25P produced 1978-82; uprated R-15BD-300 turbojets with life of 1,000 h; Sapfir-25 radar and IRST, giving look-down/shootdown capability comparable with MiG-23M; two R-40 (K-40; NATO AA-6 'Acrid') and four R-60 (K-60; NATO AA-8 'Aphid') missiles; provision for same underbelly tank as on MiG-25R series.

MiG-25PDS ('Foxbat-E'): As MiG-25PD but converted from MiG-25P, from 1979. Front fuselage lengthened by 250 mm (10 in) to accommodate flight refuelling equipment on some aircraft. *Detailed description applies to MiG-25PDS, except where indicated.*

MiG-25BM ('Foxbat-F'): Defence suppression aircraft derived from MiG-25RB; development started 1972; produced 1982-85; ECM in place of reconnaissance module in lengthened nose, with dielectric panel each side; small blister each side at rear of radome; dielectric panel at front of each outboard weapon pylon; underbelly auxiliary fuel tank as MiG-25R series; four Kh-58 (NATO AS-11 'Kilter') anti-radiation missiles underwing to attack surface-to-air missile radars over standoff ranges.

CUSTOMERS: More than 300 with CIS air defence and tactical air forces; Algeria ('Foxbat-A/B'); India ('Foxbat-B/C'); Iraq ('Foxbat-A'); Libya ('Foxbat-A/B/D/E'); Syria ('Foxbat-A/B').

DESIGN FEATURES: With MiG-31 derivative, is fastest combat aircraft yet identified in squadron service; original role demanded high-speed high-altitude capability and weapon system for attack over considerable range; high swept wings with anti-flutter body (max diameter 30 cm; 11.8 in at each tip; slim front fuselage, with ogival nosecone blended into rectangular air intake trunks with wedge intakes; inner wall of intakes curved at top and not parallel with outer wall; hinged panel forms lower intake lip enabling area to be varied electronically; fuselage under surface dished between engines; all-swept tail surfaces twin 11° outward canted fins and twin outward canted ventral fins, all with large flush antennae; wing anhedral 5° from roots; leading-edge sweepback 42° 30′ inboard, 41° outboard of each outer missile pylon; sweepback at quarter-chord 32°; sweepback on tailplane 50°, fins 60°. Two shallow upper surface fences on each wing, in line with weapon pylons.

FLYING CONTROLS: Aileron at centre of each wing trailing-edge; plain flap on inboard 37 per cent; all-moving horizontal tail surfaces able to operate differentially at high speeds; inset rudders; no tabs; airbrakes above and below jetpipes at rear of fuselage.

STRUCTURE: Airframe 80 per cent tempered steel, 8 per cent titanium in areas subject to extreme heating, such as wing and tail unit leading-edges, 11 per cent D19 heat-resistant aluminium alloy, by weight; two main wing spars forming torsion box, auxiliary front spar and two auxiliary rear spars; 14 primary fuselage frames, many intermediate frames and stringers.

LANDING GEAR: Retractable tricycle type. Single wheel, with high pressure tyre of 1.30 m (51.2 in) diameter, on each forward retracting main unit; wheel stows vertically between air intake duct and outer skin of each trunk. Twin wheel forward retracting nose unit. Retractable sprung tail skid on each ventral fin. Twin circular (60 m²; 645 sq ft) or cruciform (50 m²; 538 sq ft) brake-chutes in fairing above and between jet nozzles.

POWER PLANT: Two Soyuz/Tumansky R-15BD-300 single-shaft turbojets, each rated at 110 kN (24,700 lb st) with afterburning, in compartment of silver-coated steel. Water/methanol injection standard. Fuel in two welded structural tanks occupying 70 per cent of volume of fuselage between cockpit and engine bay, in saddle tanks around intake ducts, and in integral tank in each wing, filling almost entire volume inboard of outer fence; total capacity 17,660 litres (4,665 US gallons; 3,885 Imp gallons); provision for 5,300 litre (1,400 US gallon; 1,165 Imp gallon) underbelly tank.

ACCOMMODATION: Pilot only, on KM-1 zero-height/70 knot (130 km/h; 81 mph) ejection seat similar to that fitted to some versions of MiG-21. Canopy hinged to open sideways, to starboard.

SYSTEMS: Electronic fuel control system.

AVIONICS: Sapfir-25 fire control radar in nose, search range 54 nm (100 km; 62 miles), tracking range 40 nm (75 km; 46 miles), forward of avionics compartment housing navigation radar; K-10T radar scope; infra-red search/track sensor pod under front fuselage. SO-63 transponder, SRZO-2 (NATO 'Odd Rods') IFF, and SOD-57M ATC SIF, with antennae in starboard fin tip; Sirena-3 360° radar warning system with receivers in centre of each wingtip anti-flutter body and starboard fin tip; SAU-155 automatic flight control system; unidentified ECCM, decoys and jammers. RSB-70/RPS HF, R-832M VHF-UHF; R-831 UHF communications equipment; RSBN-6S short-range nav; SP-50 (NATO 'Swift Rod') ILS; MRP-56P marker beacon receiver; RV-UM or RV-4 radio altimeter, and ARK-10 radio compass.

Rear view of MiG-25BM shows underfuselage auxiliary fuel tank

Kh-58 'Kilter' missiles under wing of MiG-25BM ('Foxbat-F')

Lengthened nose distinguishes the MiG-25PDS

QUIPMENT: Retractable landing light under front of each intake trunk. Backup optical weapon sight.

RMAMENT: No gun. Air-to-air missiles on four underwing attachments; originally one radar guided R-40R (K-40R; NATO AA-6 'Acrid') and one infra-red R-40T (K-40T; NATO AA-6 'Acrid') under each wing; alternatively, one R-40 and two R-60s under each wing and, later, one R-23 (K-23; NATO AA-7 'Apex') and two R-73A (NATO AA-11 'Archer') or R-60T (NATO AA-8 'Aphid') under each wing.

IMENSIONS, EXTERNAL:

Wing span: MiG-25P	14.015 m (45 ft 11¾ in)
MiG-25RB	13.418 m (44 ft 0¼ in)
Wing aspect ratio: MiG-25P	3.4
Length overall	23.82 m (78 ft 1¾ in)
Length of fuselage	19.40 m (63 ft 7¾ in)
Height overall	6.10 m (20 ft 0¼ in)
Wheel track	3.85 m (12 ft 7½ in)
Wheelbase	5.14 m (16 ft 10½ in)

REAS:

Wings, gross: MiG-25P	61.40 m^2 (660.9 sq ft)
Vertical tail surfaces (total)	16.00 m^2 (172.2 sq ft)
Horizontal tail surfaces (total)	9.81 m^2 (105.6 sq ft)

EIGHTS AND LOADINGS:

Max internal fuel: P	14,570 kg (32,120 lb)
R series	15,245 kg (33,609 lb)
Max fuel with underbelly tank: P	18,940 kg (41,755 lb)
Take-off weight:	
P, clean, max internal fuel	34,920 kg (76,985 lb)
P, four R-40 missiles, max internal fuel	36,720 kg (80,950 lb)
R series, normal	37,000 kg (81,570 lb)
R series, max	41,200 kg (90,830 lb)
Max wing loading: P	598 kg/m^2 (122.5 lb/sq ft)
R series	671 kg/m^2 (137.4 lb/sq ft)
Max power loading: P	166.9 kg/kN (1.64 lb/lb st)
R series	187.3 kg/kN (1.84 lb/lb st)

ERFORMANCE:

Max permitted Mach No. at height: P, R series	2.83
Max level speed: at 13,000 m (42,650 ft):	
P, R series	1,620 knots (3,000 km/h; 1,865 mph)
at S/L: P, R series	Mach 0.98 (647 knots; 1,200 km/h; 745 mph)
T-O speed: P	195 knots (360 km/h; 224 mph)
Landing speed: P	157 knots (290 km/h; 180 mph)
Time to 20,000 m (65,600 ft) at Mach 2.35: P	8.9 min
Time to 19,000 m (62,335 ft): R series, clean	6.6 min
R series, with 2,000 kg (4,410 lb) of bombs	8.2 min
Service ceiling: P	20,700 m (67,900 ft)
R series, clean	21,000 m (68,900 ft)
T-O run: P	1,250 m (4,100 ft)
Landing run, with brake-chute: P	800 m (2,625 ft)
Range with max internal fuel:	
P, supersonic	675 nm (1,250 km; 776 miles)
P, subsonic	933 nm (1,730 km; 1,075 miles)
R series, supersonic	882 nm (1,635 km; 1,015 miles)
R series, subsonic	1,006 nm (1,865 km; 1,158 miles)
Range with 5,300 litre external tank:	
R series, supersonic	1,150 nm (2,130 km; 1,323 miles)
R series, subsonic	1,295 nm (2,400 km; 1,491 miles)
Endurance: P	2 h 5 min
g limit: P, supersonic	+4.5

MIKOYAN MiG-29

NATO reporting name: Fulcrum
Indian Air Force name: Baaz (Eagle)

TYPE: All-weather single-seat counter-air fighter, with attack capability, and two-seat combat trainer.

PROGRAMME: Technical assignment (operational requirement) issued 1972, to replace MiG-21, MiG-23, Su-15 and Su-17; initial order placed simultaneously; detail design began 1974; first of 19 prototypes (9-01 to 9-19) flew 6 October 1977; photographed by US satellite, Ramenskoye flight test centre, November 1977 and given interim Western designation 'Ram-L'; second prototype flew June 1978; second and fourth prototypes lost through engine failures; after major design changes (see previous editions of *Jane's*) production began 1982, deliveries to Frontal Aviation 1984; operational early 1985; first detailed Western study possible after visit of demonstration team to Finland July 1986; production of basic MiG-29 combat aircraft by Moscow Aircraft Production Group (MAPO), and of MiG-29UB combat trainers at Nizhny Novgorod, for CIS air forces completed, but manufacture for export continues and new versions under development.

CURRENT VERSIONS: **MiG-29** ('Fulcrum-A'): Land-based single-seat dual-role fighter, identified in three forms: (1) Initial production with two ventral tail fins like Su-27. (2) Ventral fins deleted; dorsal fins able to be extended forward in form of removable overwing 'fences' containing IRCM flare dispensers. (3) As second variant, with extended-chord rudders. Not piped to carry underwing drop tanks. Some upgraded to 'Fulcrum-C' standard, but not redesignated MiG-29S. *Detailed description refers to this version.*

MiG-29UB ('Fulcrum-B'): Combat trainer; second seat forward of normal cockpit, under continuous canopy, with periscope for rear occupant; underwing stores pylons retained.

MiG-29S ('Fulcrum-C'): As 'Fulcrum-A' variant 3, but deeply curved top to fuselage aft of cockpit, containing avionics; internal fuel capacity increased by 75 litres (20 US gallons; 17 Imp gallons); first version with two 1,150 litre (304 US gallon; 253 Imp gallon) underwing drop tanks. Max total fuel capacity 8,240 litres (2,177 US gallons; 1,813 Imp gallons), for max range of 1,565 nm (2,900 km; 1,800 miles). N019M radar, able to engage two targets simultaneously. Able to carry R-77 air-to-air missiles or up to 4,000 kg (8,820 lb) of bombs. Produced concurrently and operational in same units as 'Fulcrum-A'. Max T-O weight 19,700 kg (43,430 lb).

MiG-29SE ('Fulcrum-C'): Export version of MiG-29S; downgraded N019ME radar.

MiG-29TVK: Modified 'Fulcrum-A' for carrier training; arrester hook fitted; standard landing gear; non-folding wings.

MiG-29K (K for korabelnyy; ship-based) ('Fulcrum-D'): Maritime version, used for ski-jump take-off and deck landing trials on carrier *Admiral of the Fleet Kuznetsov* (formerly *Tbilisi*), beginning 1 November 1989; two, converted from early 'Fulcrum-As'; upward folding outer wing panels with bulged tips, probably for ESM; sharp-edge slightly raised LERX; enlarged dorsal spine; increased-chord horizontal tail surfaces, with dogtooth leading-edge; 86.3 kN (19,400 lb st) RD-33K turbofans; new N010 radar in radome with single-curvature profile;

MiG-29UB ('Fulcrum-B') two-seat combat trainer of Polish Air Force *(Press Office Sturzenegger)*

284　RUSSIA: AIRCRAFT—MiG

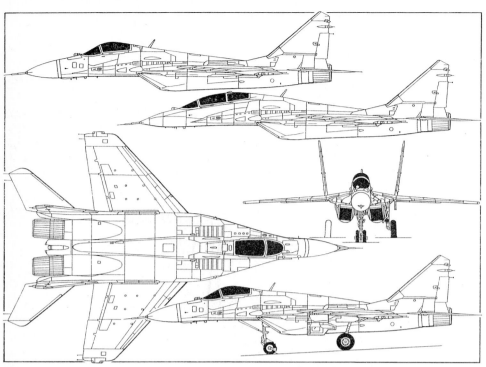

Three-view drawing of MiG-29S 'Fulcrum-C' fighter. Additional side views at top show late-model 'Fulcrum-A' and two-seat 'Fulcrum-B' *(Jane's/Mike Keep)*

MiG-29TVK modified from 'Fulcrum-A' for carrier training *(R. J. Malachowski)*

MiG-29 ('Fulcrum-A') fighters of Nos. 47 and 28 Squadrons, Indian Air Force *(Peter Steinemann)*

eight underwing hardpoints; original FOD doors in air intakes replaced by lighter retractable grids, permitting deletion of overwing louvres and internal ducting in lightweight aluminium-lithium alloy centre-section, providing increased fuel tankage (2,550 litres; 674 US gallons; 561 Imp gallons in centre-section); flight refuelling capability strengthened landing gear; arrester hook; single large over-fuselage airbrake; no APU airscoop on rear fuselage; different IRST. Exhibited at Machulishche airfield, Minsk, February 1992, with typical anti-ship armament of four Kh-31P/A (AS-17 'Krypton') air-to-surface missiles and four R-73A (AA-11 'Archer') air-to-air missiles. Reportedly, not selected for operational use on *Admiral of the Fleet Kuznetsov*.

MiG-29KVP: Prototype 9-18, fitted with arrester hook and tested at land base as demonstrator for MiG-29K.

MiG-29M ('Fulcrum-E'): Advanced version with quadruplex fly-by-wire controls and 'glass' cockpit with two multi-function CRTs (not push-button, but HOTAS); first of six prototypes flown late 1989; first exhibited at Machulishche airfield, February 1992; same new basic airframe and RD-33K turbofans as MiG-29K, but without wing folding and arrester hook; new terrain following and ground mapping radar in more tapered radome; new IRST; enlarged engine air intakes, welded aluminium/lithium front fuselage, welded steel behind; nose lengthened by approx 20 cm (7½ in); longer canopy; wider and longer dorsal spine, terminating in spade-like structure that extends beyond jet nozzles; more rounded wingtip trailing edge; new centre-section without louvres, as MiG-29K; wing leading-edge vortex generators to increase attainable angle of attack; modifications to extend aft centre of gravity limit for relaxed stability; claimed more comfortable to fly, with better manoeuvrability, improved cruise efficiency; eight underwing hardpoints; for 4,500 kg (9,920 lb) stores, including four laser-guided Kh-25ML (AS-10 'Karen') or Kh-29L (AS-14 'Kedge'), radar homing Kh-31P/A (AS-17 'Krypton') or TV guided Kh-29T (AS-14 'Kedge') air-to-surface missiles, eight R-77 AMRAAM class air-to-air missiles, or 500kg TV-guided bombs. *(Further details in Addenda.)*

A **'fifth generation' version** with multi-axis thrust vectoring engine nozzles has been test flown at Zhukovsky flight research centre. Another MiG-29 has been flight tested with fibre optics.

CUSTOMERS: CIS air forces and Naval Aviation (more than 600); Bulgaria; Cuba; Czech Republic and Slovakia (18 single-seat, two two-seat); Germany (20 single-seat, four two-seat delivered to former East Germany); India (65 single-seat, five two-seat); Iran (14 plus others acquired from Iraq); Iraq (35 single-seat, six two-seat delivered originally); North Korea; Poland; Romania (12 single-seat, two two-seat); Syria; Yugoslavia (14 single-seat, two two-seat).

DESIGN FEATURES: All-swept low-wing configuration, with wide ogival wing leading-edge root extensions (LERX) lift-generating fuselage, twin tail fins carried on booms outboard of widely spaced engines with wedge intakes; doors in intakes, actuated by extension and compression of nosewheel leg, prevent ingestion of foreign objects during take-off and landing; gap between roof of each intake and skin of wingroot extension for boundary layer bleed; fire control and mission computers link radar with laser rangefinder and infra-red search/track sensor, in conjunction with helmet mounted target designator; radar able to track 10 targets simultaneously; targets can be approached and engaged without emission of detectable radar or radio signals; sustained turn rate much improved over earlier Soviet fighters; thrust-to-weight ratio better than one; allowable angles of attack at least 70 per cent higher than previous fighters; difficult to get into stable flat spin, reluctant to enter normal spin, recovers as soon as controls released; wing leading-edge sweepback 73° 30′ on LERX, 42° on outer panels; anhedral approx 2°; tail fins canted outward 6°; leading-edge sweep 47° 50′ on fins, approx 50° on horizontal surfaces. Design flying life 2,500 h.

FLYING CONTROLS: Mechanical controls, hydraulically powered, with autopilot and rate dampers; computer controlled leading-edge manoeuvring flaps over full wing span except tips, and plain trailing-edge flaps; inset ailerons, each with manually adjustable trim tab; inset rudders and all-moving (collectively and differentially) horizontal tail surfaces, without tabs; interconnect allows rudders to augment roll rate; mechanical yaw stability augmentation system; hydraulically actuated forward hinged airbrakes above and below rear fuselage between jetpipes.

STRUCTURE: Approx seven per cent of airframe, by weight, of composites; remainder metal, including aluminium-lithium alloys; trailing-edge wing flaps, ailerons and vertical tail surfaces of carbonfibre honeycomb; approx 65 per cent of horizontal tail surfaces aluminium alloy, remainder carbonfibre; semi-monocoque all-metal fuselage, sharply tapered and downswept aft of flat-sided cockpit area, with ogival dielectric nosecone; small vortex generator each side of nose helps to overcome early tendency to aileron reversal at angles of attack above 25°; tail surfaces carried on slim booms alongside engine nacelles.

LANDING GEAR: Retractable tricycle type, made by Hydromash, with single wheel on each main unit and twin nosewheels. Mainwheels retract forward into wingroots, turning through 90° to lie flat above leg; nosewheels, on trailing-link oleo, retract rearward between engine air intakes. Hydraulic retraction and extension, with mechanical emergency release. Nosewheels steerable ±8° for taxying, T-O and landings, ±30° for slow speed manoeuvring in confined areas (selector in cockpit). Mainwheel tyres size 840 × 290 mm; nosewheel tyres size 570 × 140 mm. Pneumatic steel brakes. Mudguard to rear of nosewheels. Container for cruciform brake-chute in centre of boat-tail between engine nozzles.

POWER PLANT: Two Klimov/Sarkisov RD-33 turbofans, each 49.4 kN (11,110 lb st) dry and 54.9-81.4 kN (12,345-18,300 lb st) with afterburning. Engine ducts canted at approx 9°, with wedge intakes, sweptback at approx 35°, under wingroot leading-edge extensions. Multi-segment ramp system, including top-hinged forward door (containing a very large number of small holes) inside each intake that closes the duct while aircraft is taking off or landing, to prevent ingestion of foreign objects, ice or snow. Air is then fed to each engine through louvres in top of wingroot leading-edge extension and perforations in duct closure door. Four integral fuel tanks in inboard portion of each wing and in fuselage between wings; total capacity 4,365 litres (1,153 US gallons; 960 Imp gallons). Attachment for 1,500 litre (396 US gallon; 330 Imp gallon) non-conformal external fuel tank under fuselage, between ducts. Single-point pressure refuelling through receptacle in port wheel well. Overwing receptacles for manual fuelling. Airscoop for GTDE-117 turboshaft APU, rated at 73 kW (98 eshp) for engine starting, above rear fuselage on port side; exhaust passes through underbelly fuel tank when fitted.

ACCOMMODATION: Pilot only, on 10° inclined K-36DM zero/zero ejection seat, under rearward hinged transparent blister canopy in high-set cockpit. Sharply inclined one-piece curved windscreen. Three internal mirrors provide rearward view.

SYSTEMS: Hydraulic system pressure 207 bars (3,000 lb/sq in), with 80 litres (21 US gallons; 17.5 Imp gallons) fluid.

AVIONICS: RP-29 (N019 Sapfir-29) coherent pulse Doppler lookdown/shootdown engagement radar (NATO 'Slot Back'; search range 54 nm; 100 km); 62 miles, tracking range 38 nm; 70 km; 43 miles), target tracking limits 60° up, 38° down, 67° each side, collimated with laser rangefinder; infra-red search/track sensor (fighter detection range 8 nm; 15 km; 9.25 miles) forward of windscreen (protected by removable fairing on non-operational flights); R-862 com radio; ARK-19 DF; inertial navigation system; SRO-2 (NATO 'Odd Rods') IFF transponder and SRZ-15 interrogator; Sirena-3 360° radar warning system, with sensors on wingroot extensions, wingtips and port fin. Two SO-69 ECM antennae under conformal dielectric fairings in leading-edge of each wingroot extension; head-up display; and helmet-mounted target designation system for off-axis aiming of air-to-air missiles.

EQUIPMENT: BVP-30-26M chaff/flare dispenser, with thirty 26 mm cartridges.

ARMAMENT: Six close-range R-60MK (NATO AA-8 'Aphid') infra-red air-to-air missiles, or four R-60MK and two medium-range radar guided R-27R1 (AA-10A 'Alamo-A'), on three pylons under each wing; alternative air combat weapons include R-73E (AA-11 'Archer') close-range infra-red missiles. Able to carry FAB-250 bombs, KMGU-2 submunitions dispensers, 3B-500 napalm tanks, and 80 mm, 130 mm and 240 mm rockets in attack role. One 30 mm GSh-301 gun in port wingroot leading-edge extension, with 150 rds.

DIMENSIONS, EXTERNAL:
Wing span	11.36 m (37 ft 3¼ in)
Wing chord: at c/l	5.60 m (18 ft 4½ in)
at tip	1.27 m (4 ft 2 in)
Wing aspect ratio	3.5
Length overall: incl noseprobe	17.32 m (56 ft 10 in)
excl noseprobe	16.28 m (53 ft 5 in)
Length of fuselage, excl noseprobe	14.875 m (48 ft 9¾ in)
Height overall	4.73 m (15 ft 6¼ in)
Tailplane span	7.78 m (25 ft 6¼ in)

MiG-29K ('Fulcrum-D') on the deck of the carrier *Admiral of the Fleet Kuznetsov*

More deeply curved spine identifies this MiG-29S ('Fulcrum-C') *(Neville M. Beckett)*

Rear view shows single over-fuselage airbrake and arrester hook of MiG-29K *(Mark Lambert)*

Kh-31 ('Krypton') ASMs and R-73A ('Archer') AAMs on MiG-29K *(Linda Jackson)*

286　RUSSIA: AIRCRAFT—MiG

MiG-29M, armed with eight R-77 AMRAAM class missiles *(Brian M. Service)*

Wheel track	3.09 m (10 ft 1¾ i
Wheelbase	3.645 m (11 ft 11½ i
AREAS:	
Wings, gross	38.0 m² (409.0 sq
WEIGHTS AND LOADINGS:	
Operating weight empty	10,900 kg (24,030 l
Max weapon load	3,000 kg (6,615 l
Max fuel load	4,640 kg (10,230 l
Normal T-O weight (interceptor)	15,240 kg (33,600 l
Max T-O weight	18,500 kg (40,785 l
Max wing loading	486.8 kg/m² (99.7 lb/sq
Max power loading	113.6 kg/kN (1.11 lb/lb s
PERFORMANCE:	
Max level speed: at height	
	Mach 2.3 (1,320 knots; 2,445 km/h; 1,520 m
at S/L	Mach 1.06 (700 knots; 1,300 km/h; 805 mp
T-O speed	119 knots (220 km/h; 137 m
Acceleration:	
325-595 knots (600-1,100 km/h; 373-683 mph)	13.5
595-700 knots (1,100-1,300 km/h; 683-805 mph)	8.7
Approach speed	140 knots (260 km/h; 162 mp
Landing speed	127 knots (235 km/h; 146 mp
Max rate of climb at S/L	19,800 m (65,000 ft)/m
Service ceiling	17,000 m (55,775 f
T-O run	250 m (820 f
Landing run, with brake-chute	600 m (1,970 f
Range:	
with max internal fuel	810 nm (1,500 km; 932 mile
with underbelly auxiliary tank	
	1,133 nm (2,100 km; 1,305 mile
g limits: above Mach 0.85	+
below Mach 0.85	+

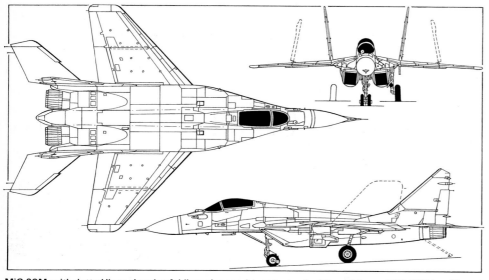

MiG-29M, with dotted lines showing folding wings and arrester hook of the shipborne MiG-29K. Land-based aircraft have lighter nosewheel leg and splash guard *(Jane's/Mike Keep)*

MIKOYAN MiG-31
NATO reporting name: Foxhound

TYPE: Two-seat twin-engined strategic interceptor.
PROGRAMME: First flown, as Ye-155MP (original MiG-25MP), 16 September 1975; first reported reliabl by Lt Viktor Belyenko, Soviet defector to Japan i 'Foxbat-A', September 1976; production started 197 NATO reporting name 'Foxhound' made publi mid-1982; delivery, to replace MiG-23s and Su-15s, bega by early 1983; production continues at Nizhny Novgoro
CURRENT VERSIONS: **MiG-31** ('Foxhound-A'): Two-seat, al weather, all-altitude interceptor, able to be guided auto matically, and to engage targets, under ground contro *Detailed description applies to this version.*
MiG-31M ('Foxhound-B'): Improved intercepto small side windows only for rear cockpit; wider dorsa spine; more rounded wingtips, with flush dielectric areas a front and rear; larger, curved fin root extensions; modifie and extended wingroot leading-edge extensions; retrac able flight refuelling probe transferred to starboard side four new-type underwing pylons for R-77 AMRAAN class active radar-guided missiles. Prototype lost on August 1991; further examples flying 1992, at least on (057) with cylindrical wingtip ECM jammer pods carrying upper and lower winglets (see accompanying illustration
CUSTOMERS: CIS air forces (more than 160).
DESIGN FEATURES: Basic MiG-25 configuration retained, bu very different aircraft, strengthened to permit supersoni flight at low altitude; more powerful engines than MiG-25 major requirement increased range, not speed; advance

MiG-31 ('Foxhound-A') all-weather interceptor *(Piotr Butowski)*

digital avionics; Zaslon radar was first electronically scanned phased array type to enter service, enabling MiG-31 to track 10 targets and engage four simultaneously, including targets below and behind its own location; fuselage weapon mountings added; crew increased to two. Wing anhedral 4° from roots; sweepback approx 40° on leading-edge, 32° at quarter-chord, with small sharply swept wingroot extensions; all-swept tail surfaces, with twin outward canted fins and dihedral horizontal surfaces.

FLYING CONTROLS: Large-span ailerons and flaps; wing leading-edge slats in four sections on each wing; all-moving horizontal tail surfaces; inset rudders.

STRUCTURE: Airframe 50 per cent arc-welded nickel steel, 16 per cent titanium, 33 per cent light alloy; negligible composites, except for radome; three-spar wings; no wingtip fairings or mountings; small forward-hinged airbrake under front of each intake trunk; undersurface of centre-fuselage not dished between engine ducts like MiG-25; much enlarged air intakes; jet nozzles extended rearward; shallow fairing extends forward from base of each fin leading-edge; fence above each wing in line with stores pylon.

LANDING GEAR: Retractable tricycle type; offset tandem twin wheels on each main unit, retracting forward into air intake trunk, facilitate operation from unprepared ground and gravel; rearward retracting twin nosewheel unit with mudguard.

POWER PLANT: Two Aviadvigatel D-30F6 turbofans, each 151.9 kN (34,170 lb st) with afterburning; internal fuel capacity approx 20,250 litres (5,350 US gallons; 4,455 Imp gallons); provision for two underwing tanks, each 2,500 litres (660 US gallons; 550 Imp gallons); semi-retractable flight refuelling probe on port side of front fuselage.

ACCOMMODATION: Pilot and weapon systems operator in tandem under individual rearward-hinged canopies; rear canopy has only limited side glazing and blends into shallow dorsal spine fairing which extends to forward edge of jet nozzles.

AVIONICS: Zaslon electronically scanned phased array fire control radar (NATO 'Flash Dance') in nose, with search range of 108 nm (200 km; 124 miles) and tracking range of 65 nm (120 km; 75 miles); capable of tracking ten targets and attacking four simultaneously; retractable infra-red search/track sensor under cockpit; radar warning receivers.

EQUIPMENT: Active infra-red and electronic countermeasures.

ARMAMENT: Four R-33 (NATO AA-9 'Amos') semi-active radar homing long-range air-to-air missiles in pairs under fuselage; two R-40T (AA-6 'Acrid') medium-range infra-red missiles on inner underwing pylons; four infra-red R-60 (K-60, AA-8 'Aphid') air-to-air missiles on outer underwing pylons, in pairs. Front pair of AA-9s is semi-recessed in fuselage, rear pair on short pylons. GSh-6-23 six-barrel Gatling-type 23 mm gun inside fairing on starboard side of lower fuselage, adjacent to main landing gear, with 260 rds.

DIMENSIONS, EXTERNAL:
Wing span	13.464 m (44 ft 2 in)
Wing aspect ratio	2.94
Length overall	22.688 m (74 ft 5¼ in)
Height overall	6.15 m (20 ft 2¼ in)

AREAS:
Wings, gross	61.6 m² (663.0 sq ft)

WEIGHTS AND LOADINGS:
Weight empty	21,825 kg (48,115 lb)
Internal fuel	16,350 kg (36,045 lb)
Max T-O weight:	
with max internal fuel	41,000 kg (90,390 lb)
with max internal fuel and two underwing tanks	46,200 kg (101,850 lb)
Max wing loading	750.0 kg/m² (153.6 lb/sq ft)
Max power loading	152.0 kg/kN (1.49 lb/lb st)

PERFORMANCE:
Max permitted Mach No. at height	Mach 2.83
Max level speed: at 17,500 m (57,400 ft)	1,620 knots (3,000 km/h; 1,865 mph)
at S/L	810 knots (1,500 km/h; 932 mph)
Max cruising speed at height	Mach 2.35
Econ cruising speed	Mach 0.85
Time to 10,000 m (32,800 ft)	7 min 54 s
Service ceiling	20,600 m (67,600 ft)
T-O run at max T-O weight	1,200 m (3,940 ft)
Landing run	800 m (2,625 ft)
Radius of action with max internal fuel and four R-33 missiles: at Mach 2.35	388 nm (720 km; 447 miles)
at Mach 0.85	647 nm (1,200 km; 745 miles)
at Mach 0.85 with two underwing tanks	755 nm (1,400 km; 870 miles)
at Mach 0.85 with two underwing tanks and one flight refuelling	1,185 nm (2,200 km; 1,365 miles)
Ferry range, max internal and external fuel, no missiles	1,780 nm (3,300 km; 2,050 miles)
Max endurance with underwing tanks:	
unrefuelled	3 h 36 min
refuelled in flight	6-7 h
g limit: supersonic	+5

MIKOYAN MiG-33

Reported designation of fully developed MiG-29M configuration for Russian air forces; claimed to offer five-fold increase in air-to-air performance over basic MiG-29. More powerful engines, possibly designated RD-37; new radar and IRST; fly-by-wire control.

MIKOYAN MiG-AT*

TYPE: Two-seat advanced jet trainer.

PROGRAMME: One of designs by five OKBs to meet official Russian requirement to replace Aero L-29 and L-39 Albatros.

COSTS: Estimated unit price of production MiG-AT $5.5 million.

* See Addenda for update

MiG-31M at Machulishche, with flight refuelling probe deployed, R-77 AMRAAM class missiles underwing, and ECM pods with winglets at wingtips

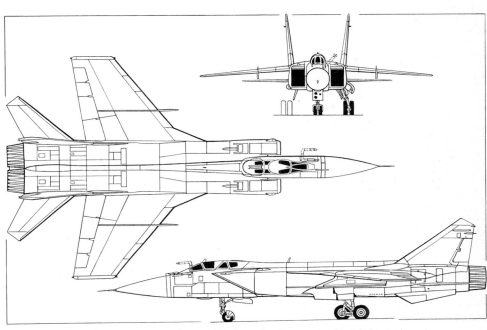
MiG-31 (NATO 'Foxhound-A') all-weather interceptor (Dennis Punnett)

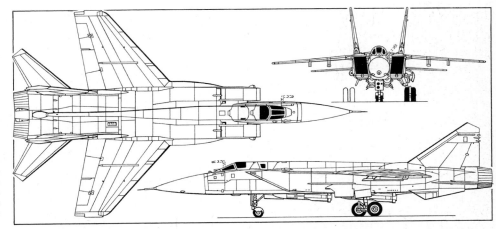
This three-view shows changes introduced on the MiG-31M ('Foxhound-B') (Jane's/Mike Keep)

DESIGN FEATURES: Conventional unswept low-wing monoplane; twin engines above wingroots; swept fin; ventral fin; unswept T tailplane; retractable tricycle landing gear. Designed for manoeuvrability comparable with front-line combat aircraft, and service life of 10,000 flying hours or 25 years, with 20,000-25,000 landings. See three-view for further details.

POWER PLANT: Two Klimov turbofans with 0.9 bypass ratio, each 14.7 kN (3,300 lb st).

ACCOMMODATION: Two crew in tandem under blister canopy; rear seat raised.

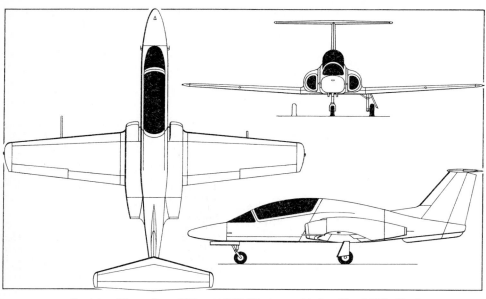

Provisional three-view of Mikoyan MiG-AT advanced trainer *(Jane's/Mike Keep)*

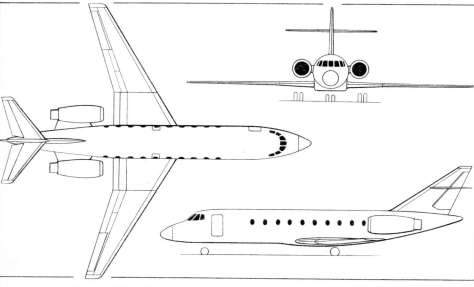

Mikoyan MiG 18-50 long-range airliner/business transport *(Dennis Punnett)*

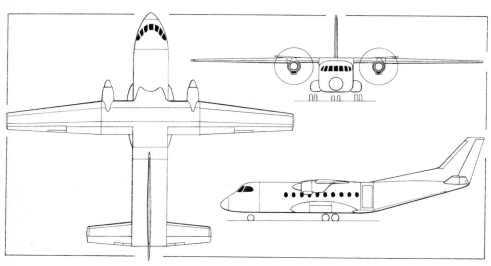

Mikoyan MiG SVB twin-turboprop pressurised passenger/freight transport *(Dennis Punnett)*

DIMENSIONS, EXTERNAL:	
Wing span	10.10 m (33 ft 1¾ in)
Wing aspect ratio	5.37
Length overall	11.00 m (36 ft 1 in)
AREAS:	
Wings, gross	19.0 m² (205 sq ft)
WEIGHTS AND LOADINGS:	
Max T-O weight	more than 5,000 kg (11,020 lb)
Max landing weight	4,000 kg (8,820 lb)

PERFORMANCE (estimated):	
Max level speed at S/L	540 knots (1,000 km/h; 621 mph)
T-O speed	97 knots (180 km/h; 112 mph)
Approach speed	114 knots (210 km/h; 131 mph)
Landing speed	92 knots (170 km/h; 106 mph)
Max rate of climb at S/L	4,800 m (15,750 ft)/min
Service ceiling	16,500 m (54,135 ft)
Range with max fuel	863 nm (1,600 km; 994 miles)
Sustained turn capability	8 g
g limits	+8/−3

MIKOYAN MiG 18-50

TYPE: Projected twin-turbofan long-range airliner/business transport.

PROGRAMME: Announced early 1990 as one of several designs by Mikoyan OKB under konversiya directives; at continuing development stage early 1992.

CURRENT VERSIONS: **Long-range airliner** for 50 passengers four-abreast in pairs at 75 cm (29.5 in) seat pitch, with 40 cm (15.75 in) centre aisle. Also **long-range 18-passenger business jet**. Both versions with forward and rear baggage compartments and aft toilet. Increase to 75/1 passengers being studied for airliner, in lengthened fuselage.

DESIGN FEATURES: Conventional all-swept low-wing transport with rear-mounted engines; digital navigation system automatic and manual all-weather, all-season, round-the clock flight and landing approach to ICAO Cat. II; operation to be practicable from 1,800 m (5,906 ft) CIS Class airfields.

LANDING GEAR: Retractable tricycle type, with twin wheels each unit.

POWER PLANT: Two Progress/Lotarev D-36 turbofans, each 63.74 kN (14,330 lb st).

DIMENSIONS, EXTERNAL:
Wing span 23.30 m (76 ft 5½ i
Wing aspect ratio 8.
Length overall 24.90 m (81 ft 8½ i
Height overall 7.55 m (24 ft 9¼ i
Fuselage diameter 2.80 m (9 ft 2¼ i

DIMENSIONS, INTERNAL:
Passenger cabin: Length 10.66 m (34 ft 11½ i
Width 2.60 m (8 ft 6¼ i
Height 1.90 m (6 ft 3 i

AREAS:
Wings, gross 64.00 m² (689.0 sq

WEIGHTS AND LOADINGS (A: 50 seats, B: 18 seats):
Weight empty, equipped: A, B 20,200 kg (44,530 l
Max payload: A 4,500 kg (9,920 l
B 1,620 kg (3,570 l
Max fuel: A, B 14,000 kg (30,865 l
Max T-O weight: A 39,000 kg (85,980 l
B 36,000 kg (79,365 l
Max wing loading: A 609.4 kg/m² (124.8 lb/sq
Max power loading: A 305.9 kg/kN (3.0 lb/lb s

PERFORMANCE (A: 50 seats, B: 18 seats; estimated):
Nominal cruising speed:
A, B 458 knots (850 km/h; 528 mp
Service ceiling: A 11,000 m (36,000
B 12,500 m (41,000
Balanced field length: A, B 1,800 m (5,906
Range, with 45 min reserves:
A 4,530 nm (8,400 km; 5,220 mile
B 5,395 nm (10,000 km; 6,215 mile

MIKOYAN MiG SVB

TYPE: Pressurised twin-turboprop passenger/freight transport

PROGRAMME: Announced early 1990 as Mikoyan konversiy project; at continuing study stage.

DESIGN FEATURES: Requested by Asian Republics for day night all-weather operation into short fields at heights up 4,000 m (13,125 ft) in hot (up to 40°C) mountainou regions; conventional high-wing configuration, wi sweptback fin and rudder; large fairings on sides of fuse age for retracted landing gear. Digital avionics standard.

LANDING GEAR: Retractable tricycle type; main bogies eac with two tandem pairs of wheels; twin-wheel nose unit.

POWER PLANT: Two 1,839 kW (2,466 shp) Klimov TV7-11 turboprops, each driving SV-34 six-blade low nois propeller.

ACCOMMODATION: Flight crew of two; two cargo handlers cabin attendants, as appropriate; rear loading ramp, integral ceiling hoist and cargo floor for freight operations optional seats for 50 passengers, five-abreast, seat pitc 78 cm (30.7 in); aisle width 40 cm (15.75 in). All accommodation fully pressurised.

DIMENSIONS, EXTERNAL:
Wing span 25.90 m (84 ft 11¾ in
Wing aspect ratio 10.8
Length overall 22.22 m (72 ft 11 in
Height overall 8.07 m (26 ft 5¾ in

DIMENSIONS, INTERNAL:
Cabin: Width 2.96 m (9 ft 8½ in
Max height 2.00 m (6 ft 6¾ in

AREAS:
Wings, gross 62.0 m² (667.4 sq ft

WEIGHTS AND LOADINGS:
Max freight payload 5,000 kg (11,023 lb
Max fuel 2,100 kg (4,630 lb
Max T-O weight 19,400 kg (42,770 lb
Max wing loading 312.9 kg/m² (64.09 lb/sq ft
Max power loading 5.21 kg/kW (8.55 lb/shp

PERFORMANCE (estimated):
Nominal cruising speed at 6,000 m (19,685 ft)
297 knots (550 km/h; 342 mph
Balanced field length at 2,100 m (6,900 ft), 30°C
1,800 m (5,905 ft
Range with 5,000 kg (11,025 lb) payload, 45 min reserves 810 nm (1,500 km; 932 miles

MIL

MOSKOVSKII VERTOLETNAY ZAVOD IMIENI M. L. MILYA (MVZ) (Moscow Helicopter Plant named after M. L. Mil)

Sokolnichyesky Val, 107113 Moscow
Telephone: 7 (095) 264 90 83
Fax: 7 (095) 264 47 62
Telex: 412144 MIL SU
GENERAL DESIGNER: Mark V. Vineberg
CHIEF DESIGNER: Alexei Ivanov

OKB founded 1947 by Mikhail Leontyevich Mil, who was involved with Soviet gyroplane and helicopter development from 1929 until his death on 31 January 1970, aged 60. His original Mi-1, first flown September 1948 and introduced into service 1951, was first series production helicopter built in former USSR. More than 25,000 helicopters of Mil design built by September 1992, representing 95 per cent of all helicopters in CIS.

MIL Mi-2 (V-2)
NATO reporting name: Hoplite

Built exclusively in Poland and described under Polish entry for WSK-PZL Swidnik.

MIL Mi-6 and Mi-22
NATO reporting name: Hook

TYPE: Twin-turbine heavy transport helicopter.
PROGRAMME: Joint military/civil requirement issued 1954; prototype flew 5 June 1957 as, by far, world's largest helicopter of that time; five built for development testing; initial pre-series of 30; more than 800 built for civil/military use, ending 1981; developments included Mi-10 and Mi-10K flying cranes; Mi-6 dynamic components used in duplicated form on V-12 (Mi-12) of 1967, which remains largest helicopter yet flown.
CURRENT VERSIONS: **Mi-6** ('Hook-A'): Basic transport. *Description applies to this version.* Full description last appeared in 1983-84 *Jane's*.

Mi-6 ('Hook-B'): Command support helicopter; large flat-bottom U-shape antenna under tailboom; X configuration blade antennae forward of horizontal stabilisers; large heat exchanger on starboard side of cabin; small cylindrical container aft of starboard rear cabin door.

Mi-22 ('Hook-C'): Developed command support version with large sweptback plate antenna above forward part of tailboom instead of 'Hook-B's' U-shape antenna; small antennae under fuselage; pole antenna on starboard main landing gear of some aircraft.

CUSTOMERS: CIS ground forces, primarily to haul guns, armour, vehicles, supplies, freight and troops in combat areas, but also in command support roles; air forces of Algeria, Iraq, Peru and Vietnam; Peruvian Army air force.
DESIGN FEATURES: Two small shoulder wings offload rotor by providing some 20 per cent of total lift in cruising flight. Removed when aircraft is operated as flying crane.
POWER PLANT: Two 4,045 kW (5,425 shp) Aviadvigatel/Soloviev D-25V (TV-2BM) turboshafts, mounted side by side above cabin, forward of main rotor shaft. Eleven internal fuel tanks, capacity 6,315 kg (13,922 lb); two external tanks, one on each side of cabin, capacity 3,490 kg (7,695 lb); provision for two ferry tanks inside cabin, capacity 3,490 kg (7,695 lb).
ACCOMMODATION: Crew of five: two pilots, navigator, flight engineer and radio operator. Four jettisonable doors and overhead hatch on flight deck. Electrothermal anti-icing system for glazing of flight deck and navigator's compartment. Equipped normally for cargo operation, with easily removable tip-up seats along side walls; when these seats are supplemented by additional seats in centre of cabin, 65-90 passengers can be carried, with cargo or baggage in aisles. Normal military seating for 70 combat equipped troops. As ambulance, 41 stretcher cases and two medical attendants on tip-up seats can be carried; one attendant's station, provided with intercom to flight deck; provision for portable oxygen installations for patients. Cabin floor stressed for loadings of 2,000 kg/m² (410 lb/sq ft); provision for cargo tiedown rings. Rear clamshell doors and ramps operated hydraulically. Standard equipment includes electric winch of 800 kg (1,765 lb) capacity and pulley block system. Central hatch in cabin floor for cargo sling for bulky loads. Three jettisonable doors, fore and aft of main landing gear on port side and aft of landing gear on starboard side.
AVIONICS: VHF and HF communications radio, intercom, radio altimeter, radio compass, three-channel autopilot, marker beacon receiver, directional gyro and full all-weather instrumentation.
ARMAMENT: Some military Mi-6s have 12.7 mm machine gun in nose.

DIMENSIONS, EXTERNAL:
Main rotor diameter	35.00 m (114 ft 10 in)
Tail rotor diameter	6.30 m (20 ft 8 in)
Length: overall, rotors turning	41.74 m (136 ft 11½ in)
fuselage, excl nose gun and tail rotor	33.18 m (108 ft 10½ in)
Height overall	9.86 m (32 ft 4 in)
Wing span	15.30 m (50 ft 2½ in)
Wheel track	7.50 m (24 ft 7¼ in)
Wheelbase	9.09 m (29 ft 9¾ in)
Rear loading doors: Height	2.70 m (8 ft 10¼ in)
Width	2.65 m (8 ft 8¼ in)
Passenger doors:	
Height: front door	1.70 m (5 ft 7 in)
rear doors	1.61 m (5 ft 3½ in)
Width	0.80 m (2 ft 7½ in)
Sill height: front door	1.40 m (4 ft 7¼ in)
rear doors	1.30 m (4 ft 3¼ in)
Central hatch in floor	1.44 m (4 ft 9 in) × 1.93 m (6 ft 4 in)

DIMENSIONS, INTERNAL:
Cabin: Length	12.00 m (39 ft 4½ in)
Max width	2.65 m (8 ft 8¼ in)
Max height: at front	2.01 m (6 ft 7 in)
at rear	2.50 m (8 ft 2½ in)
Cabin volume	80 m³ (2,825 cu ft)

AREAS:
Main rotor disc	962.1 m² (10,356 sq ft)
Tail rotor disc	31.17 m² (335.5 sq ft)

WEIGHTS AND LOADINGS:
Weight empty	27,240 kg (60,055 lb)
Max internal payload	12,000 kg (26,450 lb)
Max slung cargo	8,000 kg (17,637 lb)
Fuel load: internal	6,315 kg (13,922 lb)
with external tanks	9,805 kg (21,617 lb)
Max T-O weight with slung cargo at altitudes below 1,000 m (3,280 ft)	38,400 kg (84,657 lb)
Normal T-O weight	40,500 kg (89,285 lb)
Max T-O weight for VTO	42,500 kg (93,700 lb)
Max disc loading	44.17 kg/m² (9.05 lb/sq ft)

PERFORMANCE (at max T-O weight for VTO):
Max level speed	162 knots (300 km/h; 186 mph)
Max cruising speed	135 knots (250 km/h; 155 mph)
Service ceiling	4,500 m (14,750 ft)
Range with 8,000 kg (17,637 lb) payload	334 nm (620 km; 385 miles)
Range with external tanks and 4,500 kg (9,920 lb) payload	540 nm (1,000 km; 621 miles)
Max ferry range (tanks in cabin)	781 nm (1,450 km; 900 miles)

MIL Mi-8 (V-8)
NATO reporting name: Hip

TYPE: Twin-turbine multi-purpose helicopter.
PROGRAMME: Development began May 1960, to replace piston-engined Mi-4; first prototype, with single AI-24V turboshaft and four-blade main rotor, flew June 1961, given NATO reporting name 'Hip-A'; second prototype ('Hip-B'), with two production standard TV2-117 engines and five-blade main rotor, flew August 1962; more than

Mil Mi-6 heavy general purpose helicopter (two Aviadvigatel/Soloviev D-25V turboshafts) of the CIS air forces *(R. J. Malachowski)*

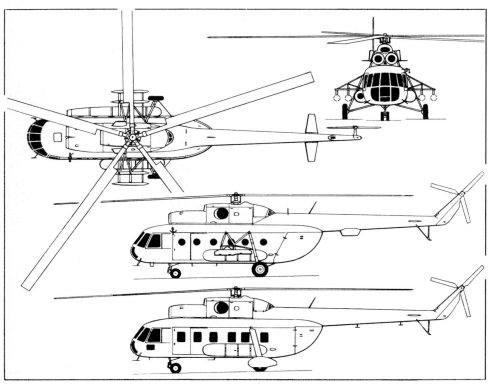

'Hip-C' military version of Mil Mi-8 twin-turbine helicopter, with additional side view (bottom) of commercial version *(Dennis Punnett)*

Mil Mi-8 ('Hip-C') military helicopter of the German Navy *(Peter J. Cooper)*

Model of Mil Mi-8TT with large external tanks for LPG fuel *(Mark Lambert)*

'Hip-D' has additional antennae as well as the canisters to which reference is made in the accompanying text. It has an airborne communications role

10,000 Mi-8s and uprated Mi-17s (which see) marke and delivered from Kazan and Ulan-Ude plants* for ci and military use; Mi-8 production completed; many co verted to Mi-17 standard.

* **Kazan Helicopter Production Association (KVPO)**
  10 let Oktyabrya 13/30, 420036 Kazan, Tatarostan
  Telephone: 7 (8432) 54 46 41 or 54 31 41
  Fax: 7 (8432) 54 52 52
  Telex: 224848 AGAT SU

* **Ulan-Ude Aviation Industrial Association**
  1 Khorinskaya Street, 670009 Ulan-Ude
  Telephone: 7 (830122) 3 06 38 or 4 30 31
  Fax: 7 (830122) 3 01 47
  Telex: 288110 AVIA

CURRENT VERSIONS: **Mi-8** ('Hip-C'): Civil passenger helico ter; standard seating for 28-32 persons in main cabin w large square windows. *Detailed description applies to t version, except where indicated.*

Mi-8T ('Hip-C'): Civil utility version; normal paylo internal or external freight, but 24 tip-up passenger se along cabin sidewalls optional; square cabin windows.

Mi-8TT: Modified TV2-117TG engines permit ope ation on liquefied petroleum gas (LPG) and kerosene. LP contained in large tanks, on each side of cabin, under lo pressure. Engines switch to kerosene for take-off and lan ing. Reduced harmful exhaust emissions in flight off anti-pollution benefits. Modification to operate on LP requires no special equipment and can be effected on i service Mi-8s at normal maintenance centre. Weigh unchanged. Large tanks reduce payload by 100-150 (220-330 lb) over comparable ranges, with little effect performance. First flight on LPG made 1987.

Mi-8 Salon ('Hip-C'): De luxe version of standa Mi-8; normally 11 passengers, on eight-place inwar facing couch on port side, two chairs and swivelling seat starboard side, with table; square windows; air-to-grou radio telephone and removable ventilation fans; compa ment for attendant, with buffet and crew wardrobe, fo ward of cabin; toilet (port) and passenger wardrobe (sta board) to each side of cabin rear entrance; alternati nine-passenger configuration; max T-O weight 10,400 (22,928 lb); range 205 nm (380 km; 236 miles) with min fuel reserve.

Military versions, with smaller circular cabin window are:

'Hip-C': Standard assault transport of CIS army suppo forces; twin-rack for stores each side, to carry 128 × 57 m rockets in four packs, or other weapons; more than 1,500 CIS army service; some uprated to Mi-17 standard **Mi-8MT** and **Mi-8MTB**, with port-side tail rotor.

'Hip-D': Airborne communications role; as 'Hip-C' b rectangular section canisters on outer stores racks; add antennae above and below forward part of tailboom.

'Hip-E': Development of 'Hip-C'; flexibly mounte 12.7 mm machine gun in nose; triple stores rack each sid to carry total 192 rockets in six packs, plus four M1 Skorpion (NATO AT-2 'Swatter') anti-tank missil (semi-automatic command to line of sight) on rails abo racks; about 250 in CIS ground forces; some uprated Mi-17 standard as **Mi-8MTBK**, with port-side tail rotor.

'Hip-F': Export 'Hip-E'; missiles changed to six 9M (NATO AT-3 'Saggers'; manual command to line sight).

'Hip-G': See separate entry on Mi-9.
'Hip-H': See separate entry on Mi-17.
'Hip-J': ECM version; additional small boxes each si of fuselage, fore and aft of main landing gear legs.

'Hip-K' (Mi-8PP): ECM communications jammer; re tangular container and array of six cruciform dipole ante nae each side of cabin; no Doppler box under tailboo heat exchangers under front fuselage; some uprated Mi-17 standard, with port-side tail rotor. See also Mi- 'Hip-K derivative'.

STOMERS: CIS ground forces (estimated 2,400 Mi-8/17s); CIS air forces; at least 40 other air forces; civil operators worldwide.

SIGN FEATURES: Conventional pod and boom configuration; five-blade main rotor, inclined forward 4° 30′ from vertical; interchangeable blades of basic NACA 230 section, solidity 0.0777; spar failure warning system; drag and flapping hinges a few inches apart; blades carried on machined spider; pendulum vibration damper; three-blade starboard tail rotor; transmission comprises type VR-8 two-stage planetary main reduction gearbox giving main rotor shaft/engine rpm ratio of 0.016:1, intermediate and tail rotor gearboxes, main rotor brake, and drives off main gearbox for tail rotor, fan, AC generator, hydraulic pumps and tachometer generators; tail rotor pylon forms small vertical stabiliser; horizontal stabiliser near end of tailboom; clamshell rear-loading freight doors.

YING CONTROLS: Mechanical system, with irreversible hydraulic boosters; main rotor collective pitch control linked to throttles.

RUCTURE: All-metal; main rotor blades each have extruded light alloy spar carrying root fitting, 21 honeycomb-filled trailing-edge pockets and blade tip; balance tab on each blade; each tail rotor blade made of spar and honeycomb filled trailing-edge; semi-monocoque fuselage.

NDING GEAR: Non-retractable tricycle type; steerable twin-wheel nose unit, locked in flight; single wheel on each main unit; oleo-pneumatic (gas) shock absorbers. Mainwheel tyres 865 × 280 mm; nosewheel tyres 595 × 185 mm. Pneumatic brakes on mainwheels; pneumatic system can also recharge tyres in the field, using air stored in main landing gear struts. Optional mainwheel fairings.

WER PLANT: Two 1,250 kW (1,677 shp) Klimov TV2-117A turboshafts (1,434 kW; 1,923 shp TV3-117MTs in Mi-8MT/MTB/MTBK). Main rotor speed governed automatically, with manual override. Single flexible internal fuel tank, capacity 445 litres (117.5 US gallons; 98 Imp gallons); two external tanks, each side of cabin, capacity 745 litres (197 US gallons; 164 Imp gallons) in port tank, 680 litres (179.5 US gallons; 149.5 Imp gallons) in starboard tank; total standard fuel capacity 1,870 litres (494 US gallons; 411.5 Imp gallons). Provision for one or two ferry tanks in cabin, raising max total capacity to 3,700 litres (977 US gallons; 814 Imp gallons). Fairing over starboard external tank houses optional cabin air-conditioning equipment at front. Engine cowling side panels form maintenance platforms when open, with access via hatch on flight deck. Total oil capacity 60 kg (132 lb).

COMMODATION: Two pilots side by side on flight deck, with provision for flight engineer's station; separated by curtain from main cabin. Windscreen de-icing standard. Basic passenger version furnished with 24-26 four-abreast track mounted tip-up seats at pitch of 72-75 cm (28-29.5 in), with centre aisle 32 cm (12.5 in) wide; removable bar, wardrobe and baggage compartment. Seats and bulkheads of basic version quickly removable for cargo carrying. Mi-8T and standard military versions have cargo tiedown rings in floor, winch of 150 kg (330 lb) capacity and pulley block system to facilitate loading of heavy freight, an external cargo sling system (capacity 3,000 kg; 6,614 lb), and 24 tip-up seats along side walls of cabin. All versions can be converted for air ambulance duties, with accommodation for 12 stretchers and tip-up seat for medical attendant. Large windows on each side of flight deck slide rearward. Sliding, jettisonable main passenger door at front of cabin on port side; electrically operated rescue hoist (capacity 150 kg; 330 lb) can be installed at this doorway. Rear of cabin made up of clamshell freight loading doors, which are smaller on commercial versions, with downward hinged passenger airstair door centrally at rear. Hook-on ramps used for vehicle loading.

STEMS: Standard heating system can be replaced by full air-conditioning system; heating of main cabin cut out when carrying refrigerated cargoes. Two independent hydraulic systems, each with own pump; operating pressure 44-64 bars (640-925 lb/sq in). DC electrical supply from two 27V 18kW starter/generators and six 28Ah storage batteries; AC supply for automatically controlled electrothermal de-icing system and some radio equipment supplied by 208/115/36/7.5V 400Hz generator, with 36V three-phase standby system. Engine air intake de-icing standard. Provision for oxygen system for crew and, in ambulance version, for patients. Freon fire extinguishing system in power plant bays and service fuel tank compartments, actuated automatically or manually. Two portable fire extinguishers in cabin.

IONICS: R-842 HF transceiver, frequency range 2 to 8 MHz and range up to 540 nm (1,000 km; 620 miles); R-860 VHF transceiver on 118 to 135.9 MHz effective up to 54 nm (100 km; 62 miles), intercom, radio telephone, ARK-9 automatic radio compass, RV-3 radio altimeter with 'dangerous height' warning, and four-axis autopilot to give yaw, roll and pitch stabilisation under any flight conditions, stabilisation of altitude in level flight or hover, and stabilisation of pre-set flying speed; Doppler radar box under tailboom.

UIPMENT: Instrumentation for all-weather flying by day and night: two gyro horizons, two airspeed indicators, two main rotor speed indicators, turn indicator, two altimeters, two rate of climb indicators, magnetic compass, radio altimeter, radio compass and astrocompass for Polar flying.

Mi-8PP ('Hip-K') communications jamming variant ***(P. R. Foster)***

Military versions can be fitted with external flight deck armour, infra-red suppressor above forward end of tailboom and flare dispensers above rear cabin window on each side.

ARMAMENT: See individual model descriptions of military versions.

DIMENSIONS, EXTERNAL:

Main rotor diameter	21.29 m (69 ft 10¼ in)
Tail rotor diameter	3.91 m (12 ft 9⅞ in)
Distance between rotor centres	12.65 m (41 ft 6 in)
Length: overall, rotors turning	25.24 m (82 ft 9¾ in)
fuselage, excl tail rotor	18.17 m (59 ft 7⅜ in)
Width of fuselage	2.50 m (8 ft 2½ in)
Height overall	5.65 m (18 ft 6½ in)
Wheel track	4.50 m (14 ft 9 in)
Wheelbase	4.26 m (13 ft 11¾ in)
Fwd passenger door: Height	1.41 m (4 ft 7¼ in)
Width	0.82 m (2 ft 8¼ in)
Rear passenger door: Height	1.70 m (5 ft 7 in)
Width	0.84 m (2 ft 9 in)
Rear cargo door: Height	1.82 m (5 ft 11½ in)
Width	2.34 m (7 ft 8¼ in)

DIMENSIONS, INTERNAL:

Passenger cabin: Length	6.36 m (20 ft 10¼ in)
Width	2.34 m (7 ft 8¼ in)
Height	1.80 m (5 ft 10¾ in)
Cargo hold (freighter):	
Length at floor	5.34 m (17 ft 6¼ in)
Width	2.34 m (7 ft 8¼ in)
Height	1.80 m (5 ft 10¾ in)
Volume	23 m³ (812 cu ft)

AREAS:

Main rotor disc	356 m² (3,832 sq ft)
Tail rotor disc	12.01 m² (129.2 sq ft)

WEIGHTS AND LOADINGS:

Weight empty:	
civil passenger version	6,799 kg (14,990 lb)
civil cargo version	6,624 kg (14,603 lb)
military versions (typical)	7,260 kg (16,007 lb)
Max payload: internal	4,000 kg (8,820 lb)
external	3,000 kg (6,614 lb)
Fuel: standard tanks	1,450 kg (3,197 lb)
with 2 auxiliary tanks	2,870 kg (6,327 lb)
Normal T-O weight	11,100 kg (24,470 lb)
T-O weight: with 28 passengers, each with 15 kg (33 lb) of baggage	11,570 kg (25,508 lb)
with 2,500 kg (5,510 lb) of slung cargo	11,428 kg (25,195 lb)
Max T-O weight for VTO	12,000 kg (26,455 lb)
Max disc loading	33.7 kg/m² (6.90 lb/sq ft)

PERFORMANCE (civil Mi-8T):

Max level speed at 1,000 m (3,280 ft):	
normal AUW	140 knots (260 km/h; 161 mph)
Max level speed at S/L:	
normal AUW	135 knots (250 km/h; 155 mph)
max AUW	124 knots (230 km/h; 142 mph)
with 2,500 kg (5,510 lb) of slung cargo	97 knots (180 km/h; 112 mph)
Max cruising speed:	
normal AUW	119 knots (220 km/h; 137 mph)
max AUW	97 knots (180 km/h; 112 mph)
Service ceiling: normal AUW	4,500 m (14,765 ft)
max AUW	4,000 m (13,125 ft)
Hovering ceiling at normal AUW:	
IGE	1,900 m (6,235 ft)
OGE	800 m (2,625 ft)
Ranges: cargo version at 1,000 m (3,280 ft), with standard fuel, 5% reserves	242 nm (450 km; 280 miles)
with 24 passengers at 1,000 m (3,280 ft), with 20 min fuel reserves	270 nm (500 km; 311 miles)
cargo version, with auxiliary fuel, 5% reserves	518 nm (960 km; 596 miles)

MIL Mi-9

NATO reporting name: Hip-G

Designation Mi-9 applies to airborne command post variant of Mi-8; 'hockey stick' antennae projecting from rear of cabin and from undersurface of tailboom, aft of Doppler radar box; rearward inclined short whip antenna above forward end of tailboom; strakes on fuselage undersurface.

MIL Mi-14

NATO reporting name: Haze

TYPE: Twin-turbine shore-based amphibious helicopter.

PROGRAMME: Development of Mi-8; first flew September 1969, under designation **V-14** and with Mi-8 power plant; changed to Mi-17 engines for production, which continues.

CURRENT VERSIONS: **Mi-14PL** ('Haze-A'): Basic ASW version; four crew; large undernose radome; OKA-2 retractable sonar in starboard rear of planing bottom, forward of

Mil Mi-14PL ('Haze-A') with MAD bird mounted in low position ***(Piotr Butowski)***

292 RUSSIA: AIRCRAFT—MIL

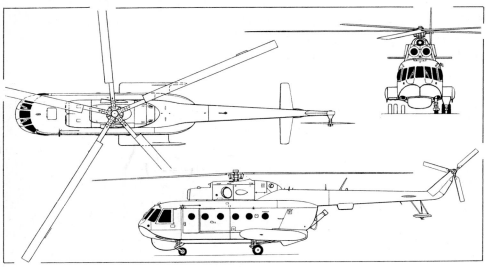

Mil Mi-14PL ASW helicopter (NATO 'Haze-A') *(Dennis Punnett)*

Mi-14PS search and rescue helicopter ('Haze-C') in Polish service *(Piotr Butowski)*

two probable sonobuoy or signal flare chutes; APM-60 towed magnetic anomaly detection (MAD) bird stowed against rear of fuselage pod (moved to lower position on some aircraft); weapons include torpedoes, bombs and depth charges in enclosed bay in bottom of hull; VAS-5M-3 liferaft (in all versions).

Mi-14PW: Polish designation of Mi-14PL.

Mi-14BT ('Haze-B'): Mine countermeasures version; fuselage strake, for hydraulic tubing, and air-conditioning pod on starboard side of cabin; no MAD; container for searchlight, to observe MCM gear during deployment and retrieval, under tailboom forward of Doppler box.

Mi-14PS ('Haze-C'): Search and rescue version, carrying ten 20-place liferafts; room for ten survivors in cabin, including two on stretchers; provision for towing many more survivors in liferafts; fuselage strake and air-conditioning pod as Mi-14BT; double-width sliding door at front of cabin on port side, with retractable rescue hoist able to lift up to three persons in basket; searchlight each side of nose and under tailboom; three crew.

CUSTOMERS: At least 230 delivered; most to CIS armed forces; Bulgaria (10), Cuba (14), the former East Germany (eight, incl Mi-14BT now converted for SAR), North Korea, Libya (12), Poland (12 Mi-14PW, five Mi-14PS), Romania (six), Syria (12), Yugoslavia.

DESIGN FEATURES: Developed from Mi-8; power plant and dynamic components as Mi-17; new features include boat hull, sponson carrying inflatable flotation bag each side at rear and small float under tailboom; fully retractable landing gear with two forward retracting single-wheel nose units and two rearward retracting twin-wheel main units.

AVIONICS (Mi-14PL): Type 12-M undernose radar, R-842-M HF transceiver, R-860 VHF transceiver, SBU-7 intercom, RW3 radio altimeter, ARK-9 and ARK-U2 ADFs, DISS-15 Doppler, Chrom Nikiel IFF, AP34-B autopilot/autohover system and SAU-14 autocontrol system.

DIMENSIONS, EXTERNAL:
Main rotor diameter	21.29 m (69 ft 10¼ in)
Main rotor blade chord	0.52 m (1 ft 8½ in)
Tail rotor diameter	3.91 m (12 ft 9⅞ in)
Length: overall, rotors turning	25.30 m (83 ft 0 in)
fuselage: Mi-14PL	18.38 m (60 ft 3½ in)
Mi-14PS	18.78 m (61 ft 7½ in)
Height overall	6.93 m (22 ft 9 in)
Wheelbase	4.13 m (13 ft 6½ in)

AREAS:
Main rotor disc	356 m² (3,832 sq ft)
Tail rotor disc	12.01 m² (129.2 sq ft)

WEIGHTS AND LOADINGS:
Weight empty: Mi-14PL	11,750 kg (25,900
Max T-O weight	14,000 kg (30,865
Max disc loading	39.3 kg/m² (8.05 lb/sq

PERFORMANCE:
Max level speed	124 knots (230 km/h; 143 mp
Max cruising speed	116 knots (215 km/h; 133 mp
Normal cruising speed	110 knots (205 km/h; 127 m
Service ceiling	3,500 m (11,500
Range with max fuel	612 nm (1,135 km; 705 mile
Endurance with max fuel	5 h 56 m

MIL Mi-17 and Mi-171
NATO reporting names: Hip-H and K derivative

TYPE: Twin-turbine multi-purpose helicopter.

PROGRAMME: First displayed at 1981 Paris Air Show; su cessor to Mi-8 for civil use and export; exports began (Cuba) 1983; production continues at Kazan and Ulan-U plants, from where Mi-17 series is marketed (see Mi entry for addresses).

CURRENT VERSIONS: **Mi-17** ('Hip-H'): Mid-life update of Mi with more powerful turboshafts, giving overall perform ance improvement, particularly hover ceiling; in CIS mi tary service, many existing Mi-8s uprated to Mi-17 sta dard (see Mi-8 entry). *Detailed description applies to bas Mi-17, except where indicated.*

Mi-17P ('Hip-K derivative'): ECM communicatio jammer; two observed in Hungarian service in 199 antenna array much more advanced than that of Mi ('Hip-K'); large 32-element array, resembling vertica segmented panel, aft of main landing gear each side; fo element array to rear on tailboom each side; large rado each side of cabin, below jet nozzle; triangular container place of rear cabin window each side; six heat exchange under front fuselage.

Mi-171 ('Hip-H'): First displayed 1989 Paris A Show; more powerful TV3-117VM turboshafts, ea 1,545 kW (2,070 shp); improved rates of climb and hov ceilings; other weights and performance genera unchanged; example displayed (SSSR-95043) had bee produced for Ministry of Health of former USSR as flyi hospital equipped to highest practicable standards for rel tively small helicopter; interior, with equipment develope in Hungary, had provision for three stretchers, operati table, extensive surgical and medical equipment, acco modation for doctor/surgeon and three nursing attendan In production and marketed by Ulan-Ude Aviation Indu trial Association.

Mil Mi-17M, latest production version of Mi-17 series from Kazan *(Peter J. Cooper)*

Electronic warfare version of Mi-17 in Czech Air Force service *(S. May)*

Mi-17M ('Hip-H'): Current production version from Kazan Helicopter Production Association; TV3-117VM engines as Mi-171; optional nose radar and flotation gear (as illustrated).

A **further military variant**, presumably with electronic warfare role, first seen in Czech and Slovak Air Force service at Dobrany-Line air base, near Plzen, 1991; each of two examples had a tandem pair of large cylindrical containers mounted each side of cabin; assumed that containers made of dielectric material and contain receivers to locate and analyse hostile electronic emissions; each of two operator's stations in main cabin has large screens, computer-type keyboards and oscilloscope; several blade antennae project from tailboom.

CUSTOMERS: Many operational side by side with Mi-8s in CIS armed forces; Angola, Cuba (16), Czech Republic, Hungary, India, North Korea, Nicaragua, Papua New Guinea, Peru, Poland, Slovakia.

DESIGN FEATURES: Distinguished from basic Mi-8 by port-side tail rotor; shorter engine nacelles, with air intakes extending forward only to mid-point of door on port side at front of cabin; small orifice each side forward of jetpipe; correct rotor speed maintained automatically by system that also synchronises output of the two engines.

POWER PLANT (basic Mi-17): Two 1,434 kW (1,923 shp) Klimov TV3-117MT turboshafts; should one engine stop, output of the other increased automatically to contingency rating of 1,637 kW (2,195 shp), enabling flight to continue; APU for pneumatic engine starting; deflectors on engine air intakes prevent ingestion of sand, dust and foreign objects.

ACCOMMODATION: Configuration and payloads generally as Mi-8; but civilian Mi-17 promoted as essentially a cargo carrying helicopter, with secondary passenger transport role.

SYSTEMS (Mi-171): AC electrical supply from two 40kW three-phase 115/220V 400Hz GT40/P-48V generators.

AVIONICS (Mi-171): Baklan-20 and Yadro-1G1 com radio; ARK-15M radio compass, ARK-UD radio compass, DISS-32-90 Doppler; AGK-77 and AGR-74V automatic horizons; BKK-18 attitude monitor; ZPU-24 course selector; A-037 radio altimeter; A-723 long-range nav; 8A-813 weather radar.

EQUIPMENT: AI-9V APU for engine starting; options as for Mi-8, plus, on military versions, external cockpit armour, ASO-2 chaff/flare dispenser under tailboom and IR jammer (NATO 'Hot Brick') at forward end of tailboom.

ARMAMENT: Options as for Mi-8, plus 23 mm GSh-23 gun packs.

DIMENSIONS, EXTERNAL: As for Mi-8, except:
Distance between rotor centres	12.661 m (41 ft 6½ in)
Length: overall, rotors turning	25.352 m (83 ft 2 in)
fuselage, excl tail rotor	18.424 m (60 ft 5⅜ in)
Height to top of main rotor head	4.755 m (15 ft 7¼ in)
Wheel track	4.510 m (14 ft 9½ in)
Wheelbase	4.281 m (14 ft 0½ in)

DIMENSIONS, INTERNAL: As for Mi-8
AREAS: As for Mi-8
WEIGHTS AND LOADINGS:
Weight empty, equipped: Mi-17	7,100 kg (15,653 lb)
Mi-171	7,055 kg (15,555 lb)
Internal fuel: Mi-171	2,027 kg (4,469 lb)
Internal fuel plus one aux tank:	
Mi-171	2,737 kg (6,034 lb)
Internal fuel plus two aux tanks:	
Mi-171	3,447 kg (7,600 lb)
Max payload:	
internal: Mi-17, Mi-171	4,000 kg (8,820 lb)
external, on sling:	
Mi-17, Mi-171	3,000 kg (6,614 lb)
Normal T-O weight: Mi-17, Mi-171	11,100 kg (24,470 lb)
Max T-O weight: Mi-17, Mi-171	13,000 kg (28,660 lb)
Max disc loading:	
Mi-17, Mi-171	36.5 kg/m² (7.48 lb/sq ft)

PERFORMANCE:
Max level speed:	
Mi-17, max AUW	135 knots (250 km/h; 155 mph)
Mi-171, normal AUW	135 knots (250 km/h; 155 mph)
Mi-171, max AUW	124 knots (230 km/h; 143 mph)
Max cruising speed:	
Mi-17, max AUW	129 knots (240 km/h; 149 mph)
Mi-171, normal AUW	124 knots (230 km/h; 143 mph)
Mi-171, max AUW	113 knots (210 km/h; 130 mph)
Service ceiling: Mi-17, normal AUW	5,000 m (16,400 ft)
Mi-17, max AUW	3,600 m (11,800 ft)
Mi-171, normal AUW	5,700 m (18,700 ft)
Mi-171, max AUW	4,500 m (14,760 ft)
Hovering ceiling OGE:	
Mi-17, max AUW	1,760 m (5,775 ft)
Mi-171, normal AUW	3,980 m (13,055 ft)
Mi-171, max AUW	1,700 m (5,575 ft)
Range with max standard fuel, 5% reserves:	
Mi-17, normal AUW	267 nm (495 km; 307 miles)
Mi-17, max AUW	251 nm (465 km; 289 miles)
Range at 500 m (1,640 ft), max AUW, 30 min reserves:	
Mi-171, internal fuel	307 nm (570 km; 354 miles)
Mi-171, internal fuel plus one aux tank	440 nm (815 km; 506 miles)
Mi-171, internal fuel plus two aux tanks	575 nm (1,065 km; 661 miles)

Mil Mi-17 ('Hip-H') in Indian Air Force service *(Peter Steinemann)*

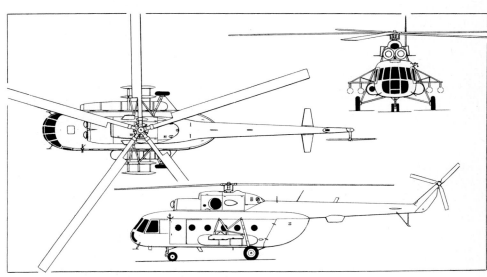

Mi-17 military general purpose helicopter, with external stores carriers *(Dennis Punnett)*

Mi-17P ECM jammer ('Hip-K derivative') of the Hungarian armed forces *(Flight/Sixma)*

MIL Mi-22

See Mi-6 entry earlier in this section.

MIL Mi-24

NATO reporting name: Hind

TYPE: Twin-turbine gunship helicopter, with transport capability.

PROGRAMME: Development began second half of 1960s, as first fire support helicopter in former USSR, with accommodation for eight armed troops; first reported in West 1972; photographs became available 1974, when two units of approx squadron strength based in East Germany; reconfiguration of front fuselage changed primary role to gunship; new version first observed 1977; used operationally in Chad, Nicaragua, Ceylon (Sri Lanka), Angola, Afghanistan and Iran/Iraq war, when at least one Iranian F-4 Phantom II destroyed by AT-6 (NATO 'Spiral') anti-tank missile from Mi-24; low-rate production continues, only for export.

CURRENT VERSIONS: **Mi-24A** ('Hind-A, B and C'): Early versions with pilot and co-pilot/gunner in tandem under large-area continuous glazing; large flight deck; last described in 1989-90 *Jane's*.

Mi-24D ('Hind-D'): First gunship version, observed 1977; basically as late model 'Hind-A' with TV3-117 engines and port-side tail rotor, but entire front fuselage redesigned above floor forward of engine air intakes; heavily armoured separate cockpits for weapon operator and pilot in tandem; flight mechanic optional, in main cabin; transport capability retained; undernose JakB-12.7 four-barrel 12.7 mm machine gun in turret, slaved to adjacent electro-optical sighting pod, for air-to-air and air-to-surface use; Falanga (Phalanx) anti-tank missile system; nosewheel leg extended to increase ground clearance of sensor pods; nosewheels semi-exposed when retracted. **Mi-24DU** training version has no gun turret. (See also Mi-25.) *Detailed description applies to Mi-24D, except where indicated.*

Mi-24V ('Hind-E'): As Mi-24D, but modified wingtip launchers and four underwing pylons; weapons include up to twelve 9M114 (NATO AT-6 'Spiral') radio guided tube-launched anti-tank missiles in pairs in Shturm (Attack) missile system; enlarged undernose missile guidance pod on port side, with fixed searchlight to rear; R-60 (K-60; NATO AA-8 'Aphid') air-to-air missiles optional on underwing pylons; pilot's HUD replaces former reflector gunsight. (See also Mi-35.)

Mi-24VP: Variant of Mi-24V with twin-barrel 23 mm GSh-23 gun, with about 300 rds, in place of four-barrel 12.7 mm gun in nose; photographed 1992; small production series built.

Mi-24P ('Hind-F'): First shown in service in 1982 photographs; P of designation refers to pushka cannon; as Mi-24V, but nose gun turret replaced by GSh-30-2 twin-barrel 30 mm gun (with 750 rds) in semi-cylindrical pack on starboard side of nose; bottom of nose smoothly faired above and forward of sensors.

Mi-24R ('Hind-G1'): Identified at Chernobyl after April 1986 accident at nuclear power station; no undernose electro-optical and RF missile guidance pods; instead of wingtip weapon mounts, has 'clutching hand' mechanisms on lengthened pylons, to obtain six soil samples per sortie, for NBC (nuclear/biological/chemical) warfare analysis; lozenge shape housing with exhaust pipe of air filtering system under port side of cabin; bubble window on starboard side of main cabin; small rearward-firing marker flare pack on tailskid; deployed individually throughout CIS ground forces.

Mi-24K ('Hind-G2'): As Mi-24R, but with large camera in cabin, lens on starboard side; believed for reconnaissance and artillery spotting. No target designator pod under nose; upward hinging cover for IR sensor.

Mi-24 Ecological Survey Version: Seen 1991 with large flat sensor 'tongue' projecting from nose in place of gun turret; large rectangular sensor pod on outer starboard underwing pylon; unidentified modification replaces rear cabin window on starboard side.

Mi-25: Export Mi-24D, including those for Afghanistan, Cuba and India.

Mi-35: Export Mi-24V.

Mi-35P: Export Mi-24P.

The Mil OKB has given details of an **upgraded Mi-24/35** designed to meet the latest air mobility requirements of the Russian Army. Features include Mi-28 main and tail rotors and transmission; a 23 mm nose gun like that of the Mi-24VP; and 9K114-9 Attacka laser beam-riding developments of the tube-launched AT-6 'Spiral' anti-tank missile.

CUSTOMERS: More than 2,300 produced at Arsenyev and Rostov; about 1,250 in CIS army service, most with helicopter attack regiments of Mi-8/17s and Mi-24s; air forces of Afghanistan, Algeria, Angola, Bulgaria, Cuba, Czech Republic, the former East Germany, Hungary, India, Iraq, Libya, Mozambique, North Korea, Nicaragua, Peru, Poland, Vietnam, Yemen.

DESIGN FEATURES: Typical helicopter gunship configuration, with stepped tandem seating for two crew and heavy weapon load on stub-wings; fuselage unusually wide for role, due to requirement for carrying eight troops; dynamic components and power plant originally as Mi-8, but soon upgraded to Mi-17-type power plant and port-side tail rotor. Main rotor blade section NACA 230, thickness/chord ratio 11-12 per cent; tail rotor blade section NACA 230M; stub-wing anhedral 12°, incidence 19°; wings contribute approx 25 per cent of lift in cruising flight; fin offset 3°.

STRUCTURE: Five-blade constant-chord main rotor; forged and machined steel head, with conventional flapping, drag and pitch change articulation; each blade has aluminium alloy spar, skin and honeycomb core; spars nitrogen pressurised for crack detection; hydraulic lead/lag dampers; balance tab on each blade; aluminium alloy three-blade tail rotor; main rotor brake; all-metal semi-monocoque fuselage pod and boom; 5 mm hardened steel integral side armour on front fuselage; all-metal shoulder wings with no movable surfaces; swept fin/tail rotor mounting; variable incidence horizontal stabiliser.

LANDING GEAR: Tricycle type; rearward retracting steerable twin-wheel nose unit; single-wheel main units with oleo-pneumatic shock absorbers and low pressure tyres, size 720 × 320 mm on mainwheels, 480 × 200 mm on nosewheels. Main units retract rearward and inward into aft end

Mi-24R ('Hind-G1'), with 'clutching hand' wingtip fittings for NBC warfare *(Lutz Freundt)*

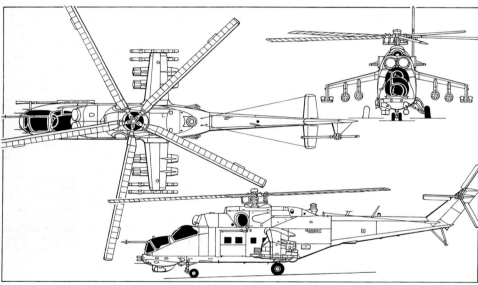

Mil Mi-24P gunship, known to NATO as 'Hind-F' *(Jane's/Mike Keep)*

Mil Mi-24P ('Hind-F') helicopter gunship with twin-barrel 30 mm gun *(Piotr Butowski)*

of fuselage pod, turning through 90° to stow almost vertically, discwise to longitudinal axis of fuselage, under prominent blister fairings. Tubular tripod skid assembly, with shock strut, protects tail rotor in tail-down take-off or landing.

POWER PLANT: Two Klimov TV3-117 turboshafts, each with max rating of 1,633 kW (2,190 shp), side by side above cabin, with output shafts driving rearward to main rotor shaft through combining gearbox. 5 mm hardened steel armour protection for engines. Main fuel tank in fuselage to rear of cabin, with bag tanks behind main gearbox. Internal fuel capacity 1,500 kg (3,307 lb); can be supplemented by 1,000 kg (2,205 lb) auxiliary tank in cabin (Mi-24D); provision for carrying (instead of auxiliary tank) up to four external tanks, each 500 litres (132 US gallons; 110 Imp gallons), on two inner pylons under each wing. Optional deflectors and separators for foreign objects and dust in air intakes; and infra-red suppression exhaust mixer boxes over exhaust ducts.

ACCOMMODATION: Pilot (at rear) and weapon operator on armoured seats in tandem cockpits under individual canopies; dual flying controls, with retractable pedals in front cockpit; if required, flight mechanic on jump-seat in cabin, with narrow passage between flight deck and cabin. Front canopy hinged to open sideways to starboard; footstep under starboard side of fuselage for access to pilot's rearward hinged door; rear seat raised to give pilot unobstructed forward view; anti-fragment shield between cockpits. Main cabin can accommodate eight persons on folding seats, or four stretchers; at front of cabin on each side is a door, divided horizontally into two sections hinged to open upward and downward respectively, with integral step on lower portion. Optically flat bulletproof glass windscreen, with wiper, for each crew member.

SYSTEMS: Cockpits and cabin heated and ventilated. Dual electrical system, with three generators, giving 36, 115 and 208V AC at 400Hz, and 27V DC. Retractable landing/taxying light under nose; navigation lights; anti-collision light above tailboom. Stability augmentation system. Electrothermal de-icing system for main and tail rotor blades. AI-9V APU mounted transversely inside fairing aft of rotor head.

AVIONICS: Include VHF and UHF radio, autopilot, ARK-15M radio compass, ARK-U2 radio compass with R-852 receiver, RV-5 radio altimeter, blind-flying instrumentation, and ADF navigation system with Doppler-fed mechanical map display. Air data sensor boom forward of top starboard corner of bulletproof windscreen at extreme nose. Undernose pods for electro-optics (starboard) and Raduga-F missile guidance (port). Many small antennae and blisters, including SRO-2 Khrom (NATO 'Odd Rods') IFF transponder and Sirena-3M radar warning antennae on each side of front fuselage and on trailing-edge of tail rotor pylon. Infra-red jammer (Jspanka microwave pulse lamp) in 'flower pot' container above forward end of tailboom.

EQUIPMENT: Gun camera on port wingtip. Colour-coded identification flare system. ASO-2V flare dispensers under tailboom forward of tailskid assembly initially; later triple racks (total of 192 flares) on sides of centre-fuselage.

ARMAMENT: One remotely controlled YakB-12.7 four-barrel Gatling type 12.7 mm machine gun, with 1,470 rds, in VSPU-24 undernose turret with field of fire 60° to each side, 20° up, 60° down; gun slaved to KPS-53AV undernose sighting system with reflector sight in front cockpit; four 9M17P Skorpion (NATO AT-2 'Swatter') anti-tank missiles on 2P32M twin rails under endplate pylons at wingtips; four underwing pylons for UB-32 rocket pods (each thirty-two S-5 type 57 mm rockets), B-8V-20 pods each containing twenty 80 mm S-8 rockets, UPK-23-250 pods each containing a GSh-23L twin-barrel 23 mm gun, GUV pods each containing either one four-barrel 12.7 mm YakB-12.7 machine gun with 750 rds and two four-barrel 7.62 mm 9-A-622 machine guns with total 1,100 rds or a 30 mm grenade launcher, up to 1,500 kg (3,300 lb) of conventional bombs, mine dispensers, or other stores. Helicopter can be landed to install reload weapons carried in cabin. PKV reflector gunsight for pilot. Provisions for firing AKMS guns from cabin windows.

DIMENSIONS, EXTERNAL (Mi-24P):
Main rotor diameter	17.30 m (56 ft 9¼ in)
Main rotor blade chord	0.58 m (1 ft 10¾ in)
Tail rotor diameter	3.908 m (12 ft 10 in)
Wing span	6.536 m (21 ft 5½ in)
Width of fuselage	1.70 m (5 ft 7 in)
Length overall:	
excl rotors and gun	17.506 m (57 ft 5¼ in)
rotors turning	21.35 m (70 ft 0½ in)
Height: to top of rotor head	3.97 m (13 ft 0½ in)
overall: rotors turning	6.50 m (21 ft 4 in)
Span of horizontal stabiliser	3.27 m (10 ft 9 in)
Wheel track	3.03 m (9 ft 11½ in)
Wheelbase	4.39 m (14 ft 5 in)

DIMENSIONS, INTERNAL:
Main cabin: Length	2.825 m (9 ft 3¼ in)
Width	1.46 m (4 ft 9½ in)
Height	1.20 m (3 ft 11¼ in)

AREAS:
Main rotor disc	235.06 m² (2,530.2 sq ft)
Tail rotor disc	11.99 m² (129.1 sq ft)

WEIGHTS AND LOADINGS (Mi-24P):
Weight empty	8,200 kg (18,078 lb)
Max external stores	2,400 kg (5,290 lb)
Normal T-O weight	11,200 kg (24,690 lb)
Max T-O weight	12,000 kg (26,455 lb)
Max disc loading	51.1 kg/m² (10.5 lb/sq ft)

PERFORMANCE (Mi-24P):
Max level speed	180 knots (335 km/h; 208 mph)
Cruising speed	145 knots (270 km/h; 168 mph)
Econ cruising speed	117 knots (217 km/h; 135 mph)
Max rate of climb at S/L	750 m (2,460 ft)/min
Service ceiling	4,500 m (14,750 ft)
Hovering ceiling OGE	1,500 m (4,920 ft)
Combat radius:	
with max military load	86 nm (160 km; 99 miles)
with two external fuel tanks	121 nm (224 km; 139 miles)
with four external fuel tanks	155 nm (288 km; 179 miles)
Range:	
standard internal fuel	270 nm (500 km; 310 miles)
with auxiliary tanks	540 nm (1,000 km; 620 miles)
Max endurance	4 h

MIL Mi-25
NATO reporting name: Hind-D

Some export variants of Mi-24D, including those for Angola, India and Peru, are designated Mi-25; change presumably signifies different equipment standards.

MIL Mi-26
NATO reporting name: Halo

TYPE: Twin-turbine heavy transport helicopter.

PROGRAMME: Development started early 1970s; aim was payload capability 1.5-2 times greater than that of any previous production helicopter; first prototype flew 14 December 1977; one of several prototype or pre-production Mi-26s (SSSR-06141) displayed at 1981 Paris Air Show; in-field evaluation, probably with military development squadron, began early 1983; fully operational 1985; export deliveries started (to India) June 1986; production continues, with **marketing by Rostov Helicopter Manufacturing Association, 5 Novatorov Street, 344038 Rostov-on-Don, Russia.**

Mil Mi-24VP with twin-barrel 23 mm gun in nose turret *(Piotr Butowski)*

CURRENT VERSIONS: **Mi-26** basic military transport; **Mi-26T** basic freight transport; **Mi-26P** seating 63 passengers; **Mi-26MS** medical evacuation version; **Mi-26A** with PNK-90 nav systems; **Mi-26M** planned upgrade; **Mi-26TZ** projected tanker version; 1990 edition of US Department of Defense's *Soviet Military Power* stated "New variants of 'Halo' are likely in the early 1990s to begin to replace 'Hooks' specialised for command support". Further information in Addenda. *Detailed description applies to Mi-26, except where indicated.*

CUSTOMERS: CIS armed forces (more than 60), India (10).

DESIGN FEATURES: Largest production helicopter; empty weight comparable to that of Mi-6 and, as specified, is approx 50 per cent of max T-O weight; weight saved by in-house design of main gearbox providing multiple torque paths, GFRP tail rotor blades, titanium main and tail rotor heads, main rotor blades of mixed metal and GFRP, use of aluminium-lithium alloys in airframe; conventional pod and boom configuration, but first successful use of eight-blade main rotor, of smaller diameter than Mi-6 rotor; payload and cargo hold size similar to those of Lockheed C-130 Hercules; auxiliary wings not required; rear-loading ramp/doors; main rotor rpm 132.

FLYING CONTROLS: Hydraulically powered cyclic and collective pitch controls actuated by small parallel jacks, with redundant autopilot and stability augmentation system inputs.

STRUCTURE: Eight-blade constant-chord main rotor; flapping and drag hinges, droop stops and hydraulic drag dampers; no elastomeric bearings or hinges; each blade has one-piece tubular steel spar and 26 GFRP aerofoil shape full-chord pockets, honeycomb filled, with ribs and stiffeners and non-removable titanium leading-edge abrasion strip; blades have moderate twist, taper in thickness toward tip, and are attached to small forged titanium head of unconventional design; each has ground adjustable trailing-edge tab; five-blade constant-chord tail rotor, starboard side, has GFRP blades, forged titanium head; conventional transmission, with tail rotor shaft inside cabin roof; all-metal riveted semi-monocoque fuselage with clamshell rear doors; flattened tailboom undersurface; engine bay of titanium for fire protection; all-metal tail surfaces; swept vertical stabiliser/tail rotor support profiled to produce

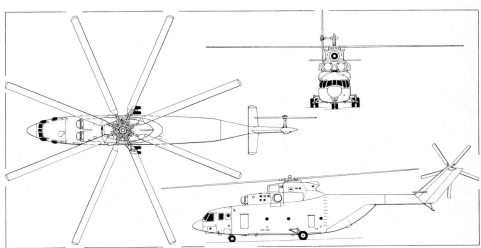

Mil Mi-26, first helicopter to operate successfully with an eight-blade main rotor *(Dennis Punnett)*

Mil Mi-26 heavy lift helicopter (two ZMKB Progress D-136 turboshafts) in Indian Air Force service *(Peter Steinemann)*

sideways lift; ground adjustable variable incidence horizontal stabiliser.

LANDING GEAR: Non-retractable tricycle type; twin wheels on each unit; steerable nosewheels; mainwheel tyres size 1,120 × 450 mm. Retractable tailskid at end of tailboom to permit unrestricted approach to rear cargo doors. Length of main legs adjusted hydraulically to facilitate loading through rear doors and to permit landing on varying surfaces. Device on main gear indicates take-off weight to flight engineer at lift-off, on panel on shelf to rear of his seat.

POWER PLANT: Two 8,267 kW (11,086 shp) ZMKB Progress D-136 free-turbine turboshafts, side by side above cabin, forward of main rotor driveshaft. Air intakes fitted with particle separators to prevent foreign object ingestion, and have both electrical and bleed air anti-icing systems. Above and behind is central oil cooler intake. VR-26 fan-cooled main transmission, rated at 14,710 kW (19,725 shp), with air intake above rear of engine cowlings. System for synchronising output of engines and maintaining constant rotor rpm; if one engine fails, output of other is increased to max power automatically. Independent fuel system for each engine; fuel in eight underfloor rubber tanks, feeding into two header tanks above engines, which permit gravity feed for a period in emergencies; max standard internal fuel capacity 12,000 litres (3,170 US gallons; 2,640 Imp gallons); provision for four auxiliary tanks. Two large panels on each side of main rotor mast fairing, aft of engine exhaust outlet, hinge downward as work platforms.

ACCOMMODATION: Crew of five on flight deck: pilot (on port side) and co-pilot side by side, navigator's seat between pilots, and seats for flight engineer and loadmaster to rear. Four-seat passenger compartment aft of flight deck. Loads in hold include two airborne infantry combat vehicles and a standard 20,000 kg (44,090 lb) ISO container; about 20 tip-up seats along each side wall of hold; max military seating for 80 combat equipped troops. (Mi-26MS carries 60 litter patients and 4/5 attendants.) Heated windscreen, with wipers; four large blistered side windows on flight deck; forward pair swing open slightly outward and rearward. Downward hinged doors, with integral airstairs, at front of hold on port side, and each side of hold aft of main landing gear units. Hold loaded via downward hinged lower door, with integral folding ramp, and two clamshell upper doors forming rear wall of hold when closed; doors opened and closed hydraulically, with backup handpump for emergency use. Two electric hoists on overhead rails, each with capacity of 2,500 kg (5,511 lb), enable loads to be transported along cabin; winch for hauling loads, capacity 500 kg (1,100 lb); roller conveyor in floor and load lashing points throughout hold. Flight deck and hold fully air-conditioned.

SYSTEMS: Two hydraulic systems, operating pressure 207 bars (3,000 lb/sq in). 28V DC electrical system. APU under flight deck, with intake louvres (forming fuselage skin when closed) and exhaust on starboard side, for engine starting and to supply hydraulic, electrical and air-conditioning systems on ground. Electrically heated leading-edge of main and tail rotor blades for anti-icing. Only flight deck pressurised.

AVIONICS: All items necessary for day and night operations in all weathers are standard, including integrated flight/nav system and automatic flight control system, weather radar in the hinged (to starboard) nosecone, Doppler, map display, HSI, and automatic hover system. Com/nav equipment generally similar to that of Yak-42.

EQUIPMENT: Hatch for load sling in bottom of fuselage, in line with main rotor shaft; sling cable attached to internal winching gear. Closed circuit TV cameras to observe slung payloads. Operational equipment on military version includes infra-red jammers and suppressors, infra-red decoy dispensers and colour-coded identification flare system.

ARMAMENT: None.

DIMENSIONS, EXTERNAL:
Main rotor diameter	32.00 m (105 ft 0 in)
Tail rotor diameter	7.61 m (24 ft 11½ in)
Length: overall, rotors turning	40.025 m (131 ft 3¾ in)
fuselage, excl tail rotor	33.727 m (110 ft 8 in)
Height to top of rotor head	8.145 m (26 ft 8¾ in)
Width over mainwheels	8.15 m (26 ft 9 in)
Wheel track	7.17 m (23 ft 6¼ in)
Wheelbase	8.95 m (29 ft 4½ in)

DIMENSIONS, INTERNAL:
Freight hold:	
Length: ramp trailed	15.00 m (49 ft 2½ in)
excl ramp	12.00 m (39 ft 4¼ in)
Width	3.20 m (10 ft 6 in)
Height	2.95-3.17 m (9 ft 8 in to 10 ft 4¾ in)
Volume	121 m³ (4,273 cu ft)

AREAS:
Main rotor disc	804.25 m² (8,657 sq ft)
Tail rotor disc	45.48 m² (489.6 sq ft)

WEIGHTS AND LOADINGS:
Weight empty	28,200 kg (62,170 lb)
Max payload, internal or external	20,000 kg (44,090 lb)
Normal T-O weight	49,600 kg (109,350 lb)
Max T-O weight	56,000 kg (123,450 lb)
Max disc loading	69.6 kg/m² (14.26 lb/sq ft)
Max transmission power loading	3.81 kg/kW (6.26 lb/shp)

PERFORMANCE:
Max level speed	159 knots (295 km/h; 183 mph)
Normal cruising speed	137 knots (255 km/h; 158 mph)
Service ceiling	4,600 m (15,100 ft)
Hovering ceiling OGE, ISA	1,800 m (5,900 ft)
Range: with max internal fuel at max T-O weight, 5% reserves	432 nm (800 km; 497 miles)
with four aux tanks	1,036 nm (1,920 km; 1,190 miles)

MIL Mi-28

NATO reporting name: Havoc

TYPE: Twin-turbine all-weather combat helicopter.

PROGRAMME: Design started 1980; first prototype flew 10 November 1982; development 90 per cent complete by June 1989, when third prototype demonstrated at Paris Air Show; certification completed Autumn 1991; production to begin at Rostov 1994.

CURRENT VERSIONS: Members of Mil OKB stated that versions under development for naval amphibious assault support, night attack (Mi-28N) and air-to-air missions.

DESIGN FEATURES: Conventional gunship configuration, with two crew in stepped cockpits; original three-blade tail rotor superseded by 'scissors' type comprising two independent two-blade rotors set as narrow X (35°/145°) on same shaft, known in USA as Δ^3 (delta 3) type; resulting flapping freedom relieves flight loads; agility enhanced by doubling hinge offset of main rotor blades compared with Mi-24; survivability emphasised; crew compartments protected by titanium and ceramic armour and armoured glass transparencies; vital structural elements shielded by less vital; single hit will not knock out both engines; vital units and parts are redundant and widely separated; multiple self-sealing fuel tanks in centre fuselage enclosed in composites second skin, outside metal fuselage skin; no explosion, fire or fuel leakage results if tanks hit by bullet or shell fragment; energy-absorbing seats and landing gear protect crew in crash landing at descent rate of 12 m (40 ft)/s; crew doors are rearward hinged, to open quickly and remain open in emergency; parachutes are mandatory for CIS military helicopter aircrew; if Mi-28 crew had to parachute, emergency system would jettison doors, blast away stub-wings, and inflate bladder beneath each door sill; as crew jumped, they would bounce off bladders and clear main landing gear; no provision for rotor separation; port-side door, aft of wing, provides access to avionics compartment large enough to permit combat rescue of two or three persons on ground, although it lacks windows, heating and ventilation; handcrank, inserted into end of each stub-wing, enables stores of up to 500 kg (1,100 lb) to be winched on to pylons without hoists or ground equipment; current 30 mm gun is identical with that of CIS Army ground vehicles and uses same ammunition; jamming averted by attaching twin ammunition boxes to sides of gun mounting, so that they turn, elevate and depress with gun; main rotor shaft has 5° forward tilt, providing tail rotor clearance; transmission capable of running without oil for 20-30 minutes; main rotor rpm 242; with main rotor blades and wings removed, helicopter is air transportable in An-22 or Il-76 freighter.

FLYING CONTROLS: Hydraulically powered mechanical type; horizontal stabiliser linked to collective; controls for pilot only.

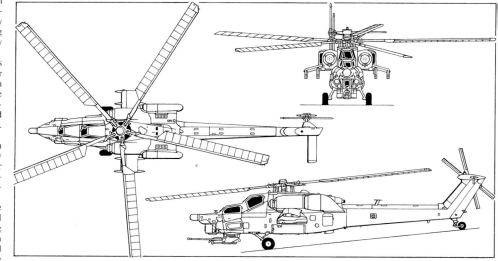

Mil Mi-28 combat helicopter (two Klimov TV3-117 turboshafts) *(Jane's/Mike Keep)*

MIL—AIRCRAFT: RUSSIA 297

Main rotor blade chord	0.67 m (2 ft 2⅜ in)
Tail rotor diameter	3.84 m (12 ft 7¼ in)
Tail rotor blade chord	0.24 m (9½ in)
Length overall, excl rotors	16.85 m (55 ft 3½ in)
Fuselage max width	1.85 m (6 ft 1 in)
Width over stub-wings	4.87 m (16 ft 0 in)
Height to top of rotor head	3.82 m (12 ft 6½ in)
Height overall	4.81 m (15 ft 9½ in)
Wheel track	2.29 m (7 ft 6¼ in)
Wheelbase	11.00 m (36 ft 1 in)
AREAS:	
Main rotor disc	232.3 m² (2,501 sq ft)
Tail rotor disc	11.58 m² (124.7 sq ft)
WEIGHTS AND LOADINGS:	
Weight empty	7,000 kg (15,432 lb)
Normal T-O weight	10,400 kg (22,928 lb)
Max T-O weight	11,500 kg (25,353 lb)
Max disc loading	49.5 kg/m² (10.14 lb/sq ft)
PERFORMANCE:	
Max level speed	162 knots (300 km/h; 186 mph)
Max cruising speed	143 knots (265 km/h; 164 mph)
Service ceiling	5,800 m (19,025 ft)
Hovering ceiling OGE	3,500 m (11,480 ft)
Range, max standard fuel	248 nm (460 km; 285 miles)
Ferry range	593 mm (1,100 km; 683 miles)
Endurance with max fuel	2 h
g limits	+3/−0.5

Mil Mi-28 ('Havoc') combat helicopter *(Mark Lambert)*

STRUCTURE: Five-blade main rotor; blades have very cambered high-lift section and sweptback tip leading-edge; full-span upswept tab on trailing-edge of each blade; structure comprises numerically controlled, spirally wound glassfibre D-spar, blade pockets of Kevlar-like material with Nomex-like honeycomb core, and titanium erosion strip on leading-edge; each blade has single elastomeric root bearing, mechanical droop stop and hydraulic drag damper; four-blade GFRP tail rotor with elastomeric bearings for flapping; rotor brake lever on starboard side of cockpit; strong and simple machined titanium main rotor head with elastomeric bearings, requiring no lubrication; power output shafts from engines drive main gearbox from each side; tail rotor gearbox, at base of tail pylon, driven by aluminium alloy shaft inside composite duct on top of tailboom; sweptback mid-mounted wings have light alloy primary box structure, leading- and trailing-edges of composites; no wing movable surfaces; provision for countermeasures pod on each wingtip, housing chaff/flare dispensers and sensors, probably RWR; light alloy semi-monocoque fuselage, with titanium armour around cockpits and vulnerable areas; composites access door aft of wing on port side; swept fin has light alloy primary box structure, composites leading- and trailing-edges; cooling air intake at base of fin leading-edge, exhaust at top of trailing-edge; two-position composites horizontal stabiliser.

LANDING GEAR: Non-retractable tailwheel type, with single wheel on each unit; mainwheel tyres size 720 × 320 mm, pressure 5.4 bars (78 lb/sq in); castoring tailwheel with tyre size 480 × 200 mm.

POWER PLANT: Two Klimov TV3-117VMA turboshafts, each 1,633 kW (2,190 shp), in pod above each wingroot; three jetpipes inside downward deflected composites nozzle fairing on each side of third prototype shown in Paris 1989; upward deflecting type also tested. Ivchenko AI-9V APU in rear of main pylon structure supplies compressed air for engine starting and to drive small turbine for pre-flight ground checks. Deflectors for dust and foreign objects forward of air intakes, which are de-iced by engine bleed air. Internal fuel capacity approx 1,900 litres (502 US gallons; 418 Imp gallons); provision for four external fuel tanks on underwing pylons.

ACCOMMODATION: Navigator/gunner in front cockpit; pilot behind, on elevated seat; transverse armoured bulkhead between; flat non-glint tinted transparencies of armoured glass; navigator/gunner's door on port side, pilot's door on starboard side.

SYSTEMS: Cockpits air-conditioned and pressurised by engine bleed air. Duplicated hydraulic systems, pressure 152 bars (2,200 lb/sq in). 208V AC electrical system supplied by two generators on accessory section of main gearbox, ensuring continued supply during autorotation. Low airspeed system standard, giving speed and drift via main rotor blade-tip pitot tubes at −50 to +70 km/h (−27 to +38 knots; −31 to +43 mph) in forward flight, and ±70 km/h (±38 knots; ±43 mph) in sideways flight. Main and tail rotor blades electrically de-iced.

AVIONICS: Conventional IFR instrumentation, with autostabilisation, autohover, and hover/heading hold lock in attack mode; standard UHF/VHF nav/com; pilot has HUD and centrally mounted CRT for basic TV (and, later, FLIR/LLTV); radio for missile guidance in nose radome; radar warning receivers; small IFF fairing each side of nose and tail. Daylight optical weapons sight and laser ranger/designator in double-glazed nose turret above gun, with which it rotates through ±110°; wiper on outer glass protects inner optically flat panel. Cylindrical container each side of this turret for FLIR and low light level TV night vision systems (not fitted to aircraft in Paris, 1989). Aircraft designed for use with night vision goggles.

EQUIPMENT: Two slots, one above the other on port side of tailboom, for colour-coded identification flares. Three pairs of rectangular formation-keeping lights in top of tailboom; further pair in top of main rotor pylon fairing. Infra-red suppressors, radar and laser warning receivers standard; optional countermeasures pod on each wingtip, housing chaff/flare dispensers and sensors.

ARMAMENT: One NPPU-28 30 mm turret mounted gun (with 250 rds) at nose, able to rotate ±100°, elevate 13° and depress 40°; maximum rate of fire 900 rds/min air-to-air and air-to-ground. (New specially designed gun under development.) Two pylons under each stub-wing, each with capacity of 480 kg (1058 lb), typically for total of sixteen 9M114 (NATO AT-6 'Spiral') radio guided tube-launched anti-tank missiles and two UB-20 pods of twenty 80 mm C-8 or 130 mm C-13 rockets. Gun fired and guided weapons launched normally only from front cockpit; unguided rockets fired from both cockpits. (When fixed, gun can be fired also from rear cockpit.)

DIMENSIONS, EXTERNAL:
Main rotor diameter   17.20 m (56 ft 5 in)

Wingtip countermeasures pod on Mi-28
(Nick Cook/Jane's Defence Weekly)

MIL Mi-34

NATO reporting name: Hermit

TYPE: Lightweight two/four-seat multi-purpose helicopter.

PROGRAMME: First flight 1986; two prototypes and structure test airframe completed by mid-1987, when exhibited for first time at Paris Air Show; first helicopter built in former USSR to perform normal loop and roll; series production scheduled to begin 1993.

CURRENT VERSIONS: **Mi-34**: Single-engined version; *as described below.*
Mi-34 VAZ: Twin-engined development; described separately.

DESIGN FEATURES: Intended primarily for training and international competition flying; Mil claims Mi-34 is aerobatic; conventional pod and boom configuration; piston engine of same basic type as that in widely used Yakovlev fixed-wing training aircraft and Kamov Ka-26 helicopters; suitable also for light utility, mail delivery, observation and liaison duties, and border patrol.

FLYING CONTROLS: Mechanical, with no hydraulic boost.

STRUCTURE: Semi-articulated four-blade main rotor with flapping and cyclic pitch hinges, but natural flexing in lead/lag plane; blades of GFRP with CFRP reinforcement, attached by flexible steel straps to head like that of McDonnell Douglas MD 500; two-blade tail rotor of similar composites construction, on starboard side; riveted light alloy fuselage; sweptback tail fin with small unswept T tailplane.

LANDING GEAR: Conventional non-retractable skids on arched support tubes; small tailskid to protect tail rotor.

POWER PLANT: One 239 kW (320 hp) M-14V-26 nine-cylinder radial air-cooled engine mounted sideways in centre-fuselage. Fuel system for inverted flight.

ACCOMMODATION: Normally one or two pilots, side by side, in enclosed cabin, with optional dual controls. Rear of cabin contains low bench seat, available for two passengers and offering flat floor for cargo carrying. Forward hinged door on each side of flight deck and on each side of rear cabin.

AVIONICS: VHF radio, ADF, radio altimeter, magnetically slaved compass system incorporating radio magnetic indicator.

EQUIPMENT: Gyro horizon.

DIMENSIONS, EXTERNAL:
Main rotor diameter   10.00 m (32 ft 9¾ in)

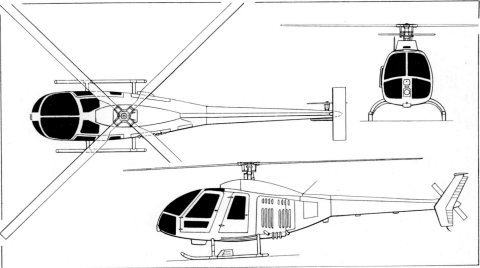

Mil Mi-34 two/four-seat training and competition helicopter *(Dennis Punnett)*

Mil Mi-34 light multi-purpose helicopter (M-14V-26 engine) *(Brian M. Service)*

Main rotor blade chord	0.22 m (8⅔ in)
Tail rotor diameter	1.48 m (4 ft 10¼ in)
Tail rotor blade chord	0.16 m (6⅓ in)
Length of fuselage	8.71 m (28 ft 7 in)
Width of fuselage	1.42 m (4 ft 8 in)
Skid track	2.24 m (7 ft 4¼ in)
AREAS:	
Main rotor disc	78.5 m² (845 sq ft)
Tail rotor disc	1.72 m² (18.52 sq ft)
WEIGHTS AND LOADINGS:	
Fuel weight	120 kg (265 lb)
T-O weight, training and aerobatic missions:	
Aerobatic	1,100 kg (2,425 lb)
Normal	1,280 kg (2,822 lb)
Max	1,350 kg (2,976 lb)
Max disc loading	17.2 kg/m² (3.52 lb/sq ft)
PERFORMANCE (at normal T-O weight, except where indicated):	
Max level speed	113 knots (210 km/h; 130 mph)
Max cruising speed	97 knots (180 km/h; 112 mph)
Normal cruising speed	86 knots (160 km/h; 99 mph)
Max speed rearward at AUW of 1,020 kg (2,249 lb)	
	70 knots (130 km/h; 80 mph)
Service ceiling	4,500 m (14,765 ft)
Hovering ceiling	1,500 m (4,920 ft)
Range at T-O weight of 1,250 kg (2,755 lb):	
with 165 kg (363 lb) payload	
	97 nm (180 km; 112 miles)
with 140 kg (309 lb) payload	
	194 nm (360 km; 224 miles)
g limits at AUW of 1,020 kg (2,249 lb) and speeds of 27-81 knots (50-150 km/h; 31-93 mph) for short periods	+2.5/−0.5

MIL Mi-34 VAZ

TYPE: Lightweight two/four-seat twin-engined helicopter.
PROGRAMME: Now in development; powered by two rotary engines, produced by VAZ motorcar works in Togliattigrad; prototype exhibited at MosAeroshow '92; first flight expected 1993.
DESIGN FEATURES: Basically as Mi-34 but completely new rotor head based on carbonfibre star plate, anchoring pitch cases that are integral with main blade section (sleeve over starplate arm); droop stops under blade root; no drag dampers; starplate clamped between upper and lower metal forgings attached to main shaft; pitch-change fittings also metal forgings. Resultant large apparent hinge offset gives high control response. Main rotor tip speed 205 m (672 ft)/s.
POWER PLANT: Two VAZ-430 twin-chamber rotary engines, each 169 kW (227 hp) for T-O, 198.5 kW (266 hp) contingency. Internal fuel capacity 245 litres (64.7 US gallons; 54 Imp gallons); aux fuel capacity 245 litres (64.7 US gallons; 54 Imp gallons). Engines burn Mogas and have power:weight ratio of 0.5 kg (1.1 lb)/hp.
ACCOMMODATION: As Mi-34, plus optional stretcher for EMS duties.
SYSTEMS: Cabin heated; de-icing system for main rotor blades and cabin windows.
AVIONICS: Navigation equipment for flying in adverse weather optional.
DIMENSIONS, EXTERNAL: As Mi-34
DIMENSIONS, INTERNAL:

Cabin: Length	2.40 m (7 ft 10½ in)
Width	1.60 m (5 ft 3 in)
Height	1.30 m (4 ft 3¼ in)
WEIGHTS AND LOADINGS:	
Max T-O weight	1,960 kg (4,320 lb)
PERFORMANCE (estimated):	
Max level speed	113 knots (210 km/h; 130 mph)
Nominal cruising speed	100 knots (185 km/h; 115 mph)
Range:	
with 400 kg (882 lb) payload, 30 min reserves	
	162 nm (300 km; 186 miles)
with max internal fuel, no reserves	
	259 nm (480 km; 298 miles)
with max internal and aux fuel, no reserves	
	528 nm (980 km; 609 miles)

MIL Mi-35

Mi-35 is export version of Mi-24V; **Mi-35P** is export version of Mi-24P.

MIL Mi-38

TYPE: Twin-turbine multi-role medium-range helicopter.
PROGRAMME: Model shown at 1989 Paris Air Show, when aircraft at mockup stage; first flight scheduled 1992-93; production planned to begin in Kazan 1996. Under December 1992 agreement, Eurocopter will integrate cockpit, avionics and passenger systems, and will adapt Mi-38 for international market.
COSTS: Civil export price about $30 million.

DESIGN FEATURES: Conventional pod and boom configuration; power plant above cabin; six-blade main rotor with considerable non-linear twist and swept tips; two independent two-blade tail rotors, set as narrow X on same shaft; portside door at front of cabin; clamshell rear loading doors and ramp; hatch in cabin floor, under main rotor driveshaft, for tactical/emergency cargo airdrop and for cargo sling attachment; optional windows for survey cameras in place of hatch; sweptback fin/tail rotor mounting; small horizontal stabiliser; Mi-38 planned as Mi-8/17 series replacement; designed to FAR Pt 29 standards, for day/night operation over temperature range −60°C to +50°C; Western engines optional.
FLYING CONTROLS: Fly-by-wire, with manual backup.
STRUCTURE: Composites main and tail rotors, with elastomeric bearings; main rotor has hydraulic drag dampers; single lubrication point, at driveshaft; fuselage mainly composites.
LANDING GEAR: Non-retractable tricycle type; single wheel on each main unit; twin nosewheels; low-pressure tyres; optional pontoons for emergency use in overwater missions.
POWER PLANT: Two Klimov TV7-117V turboshafts, each flat rated at 1,728 kW (2,318 shp) for T-O; single-engine rating 2,610 kW (3,500 shp) and transmission rated for same power. Power plant above cabin, to rear of main reduction gear; air intakes and filters in sides of cowling. Bag fuel tanks beneath floor of main cabin; provision for external auxiliary fuel tanks. Liquid petroleum gas fuel planned as alternative to aviation kerosene.
ACCOMMODATION: Crew of two on flight deck, separated from main cabin by compartment for majority of avionics; single-pilot operation possible for cargo missions. Lightweight seats for 30 passengers as alternative to unobstructed hold for 5,000 kg (11,020 lb) freight. Ambulance and air survey versions planned. Provision for hoist over port-side door, remotely controlled hydraulically actuated rear cargo ramp, powered hoist on overhead rails in cabin, and roller conveyor system in cabin floor and ramp.
SYSTEMS: Air-conditioning by compressor bleed air, or APU on ground, maintains temperature of not more than 25°C on flight deck in outside temperature of 40°C, and not less than 15°C on flight deck and in main cabin in outside temperature of −50°C. Three independent hydraulic systems; any one able to maintain control of helicopter in emergency. Electrical system has three independent AC generators, two batteries, and transformer/rectifiers for DC supply; electric rotor blade de-icing. Independent fuel system for each engine, with automatic cross-feed; forward part of cowling houses VD-100 APU, hydraulic, air-conditioning, electrical and other system components.
AVIONICS: Five colour CRTs on flight deck for use in flight and by servicing personnel on ground. Pre-set flight control system allows full autopilot, autohover and automatic landing. Equipment monitoring, failure warning and damage control system. Avionics controlled by large central computer, linked also to automatic navigation system with Doppler, ILS, satellite navigation system, weather/navigation radar (range 54 nm; 100 km; 62 miles), autostabilisation system and automatic radio compass. Closed-circuit TV for monitoring cargo loading and slung loads. Options include low-cost electromechanical instrumentation based on that of Mi-8, sensors for weighing and CG positioning of cargo in cabin, and for checking weight of slung loads.
DIMENSIONS, EXTERNAL:

Main rotor diameter	21.10 m (69 ft 2¾ in)
Tail rotor diameter	3.84 m (12 ft 7¼ in)
Length overall, excl rotors	19.70 m (64 ft 7½ in)
Height to top of rotor head	5.13 m (16 ft 10 in)
Stabiliser span	4.20 m (13 ft 9½ in)
Wheel track, c/l of oleos	3.30 m (10 ft 10 in)

Prototype of twin-engined Mil Mi-34 VAZ light helicopter *(Mark Lambert)*

Cockpit of prototype Mil Mi-34 VAZ *(Mark Lambert)*

Full-scale mockup of Mil Mi-38 *(Mark Lambert)*

Mockup cockpit of Mil Mi-38 *(R. J. Malachowski)*

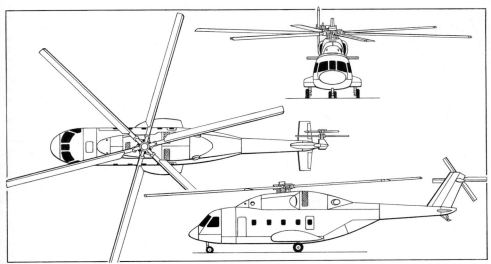
Mil Mi-38 medium-range helicopter to replace the Mi-8/17 *(Jane's/Mike Keep)*

Wheelbase	6.61 m (21 ft 8¼ in)
Forward freight door: Width	1.50 m (4 ft 11 in)
Height	1.70 m (5 ft 7 in)
Floor hatch: Length	1.15 m (3 ft 9¼ in)
Width	0.75 m (2 ft 5½ in)

DIMENSIONS, INTERNAL:
Main cabin: Length to ramp	6.80 m (22 ft 3½ in)
Length, incl fwd part of tailboom	10.70 m (35 ft 1¼ in)
Max width	2.36 m (7 ft 9 in)
Width at floor	2.20 m (7 ft 2½ in)
Height: centre	1.80 m (5 ft 11 in)
rear	1.85 m (6 ft 1 in)

AREAS:
Main rotor disc	349.67 m² (3,763.8 sq ft)
Tail rotor disc	11.58 m² (124.7 sq ft)

WEIGHTS AND LOADINGS (provisional):
Max payload, internal or external	5,000 kg (11,020 lb)
Normal T-O weight	13,460 kg (29,672 lb)
Max T-O weight	14,500 kg (31,965 lb)
Max disc loading	41.4 kg/m² (8.49 lb/sq ft)
Max transmission power loading	5.56 kg/kW (9.13 lb/shp)

PERFORMANCE (estimated, at max T-O weight):
Max level speed	148 knots (275 km/h; 171 mph)
Cruising speed	135 knots (250 km/h; 155 mph)
Service ceiling	6,500 m (21,325 ft)
Hovering ceiling OGE	2,500 m (8,200 ft)

Range, 30 min reserves:
with 5,000 kg (11,020 lb) payload
    175 nm (325 km; 202 miles)
with 4,500 kg (9,920 lb) payload (30 passengers and baggage)
    286 nm (530 km; 329 miles)
with 3,500 kg (7,715 lb) payload and standard fuel
    430 nm (800 km; 497 miles)
with 1,800 kg (3,965 lb) payload and auxiliary fuel
    700 nm (1,300 km; 808 miles)

MIL Mi-40

TYPE: Twin-turboshaft infantry combat helicopter.
PROGRAMME: Project announced Summer 1992.
DESIGN FEATURES: Development of Mi-24 concept, to deliver 8-10 troops to battlefield in armoured cabin behind flight deck and to provide fire support. Dynamic system based on that of Mi-28; for other details see three-view.
LANDING GEAR: Non-retractable tailwheel type; single wheel on each unit.
ARMAMENT: Undernose gun turret; up to eight tube-launched air-to-surface missiles and two B-8V20 rocket packs under stub-wings. Basic missions specify four missiles, gun and eight or ten troops; and eight missiles, gun and rocket pods, with or without ten troops.

DIMENSIONS, EXTERNAL:
Main rotor diameter	17.20 m (56 ft 5 in)
Tail rotor diameter	3.84 m (12 ft 7¼ in)
Length overall, excl rotors	16.60 m (54 ft 5¾ in)
Height to top of main rotor equipment pod	4.40 m (14 ft 5¼ in)
Width over wingtip pods	5.20 m (17 ft 0¾ in)

WEIGHTS AND LOADINGS:
Weight empty	7,675 kg (16,920 lb)
Max internal fuel	1,170 kg (2,580 lb)
Troops and weapons	1,098-1,800 kg (2,420-3,968 lb)
T-O weight, according to mission	10,372-11,402 kg (22,865-25,137 lb)
Max T-O weight	11,500 kg (25,352 lb)

PERFORMANCE (estimated):
Max level speed	159 knots (295 km/h; 183 mph)
Nominal cruising speed	140 knots (260 km/h; 162 mph)
Service ceiling	5,800 m (19,025 ft)
Hovering ceiling OGE	3,600 m (11,810 ft)

Range with max internal fuel, 5% reserve
    248 nm (460 km; 285 miles)

MIL Mi-46T

TYPE: Twin-turboshaft passenger/freight transport helicopter.
PROGRAMME: Project announced Summer 1992.
DESIGN FEATURES: Replacement for Mi-6, with T-O weight just over half that of Mi-26. General configuration similar to Mi-26; seven-blade main rotor, five-blade tail rotor; non-retractable tricycle landing gear with twin wheels on each unit; few windows in hold; rear loading ramp and doors; engines above hold, forward of main rotor driveshaft.
POWER PLANT: Two unspecified 5,965 kW (8,000 shp) turboshafts.
ACCOMMODATION: Equipped normally to carry freight.

WEIGHTS AND LOADINGS:
Weight empty	16,200 kg (35,715 lb)
Max T-O weight	30,000 kg (66,137 lb)

PERFORMANCE (estimated):
Nominal cruising speed	145 knots (270 km/h; 168 mph)

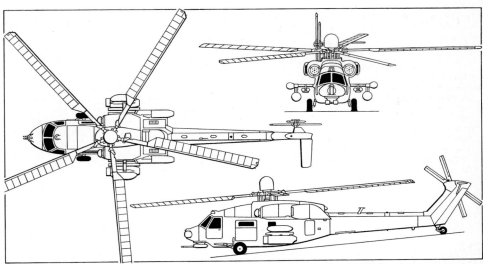
Mil Mi-40 twin-turboshaft infantry combat helicopter *(Jane's/Mike Keep)*

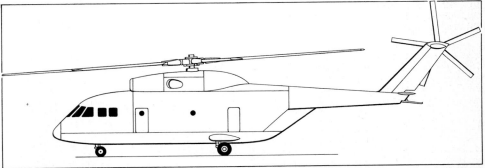
Provisional drawing of Mil Mi-46T twin-turboshaft passenger/freight transport helicopter *(Jane's/Mike Keep)*

300 RUSSIA: AIRCRAFT—MIL/MOLNIYA

Artist's impression of Mil Mi-46K flying crane counterpart of Mi-46T (*Jane's/Mike Keep*)

Hovering ceiling at max T-O weight 2,300 m (7,545 ft)
Range with 10,000 kg (22,045 lb) payload at max T-O
weight, 30 min reserve 215 nm (400 km; 248 miles)

MIL Mi-46K

TYPE: Twin-turboshaft flying crane helicopter.
PROGRAMME: Project announced Summer 1992.
DESIGN FEATURES: Replacement for Mi-10K, of typical large flying crane configuration, utilising fuselage nose, flight deck, power plant and dynamic components of Mi-46T; flat bottomed, shallow centre and rear fuselage, with short stub-wings and long tripod-braced mainwheel units; glazed gondola for second pilot, facing rearward behind front fuselage pod, to control helicopter during loading and unloading. Sling cable directly under main rotor driveshaft for payload.
POWER PLANT: Two unspecified 5,965 kW (8,000 shp) turboshafts.
WEIGHTS AND LOADINGS:
Weight empty 19,700 kg (43,430 lb)
Max T-O weight 36,500 kg (80,467 lb)
PERFORMANCE (estimated):
Hovering ceiling at max T-O weight 2,300 m (7,545 ft)
Range with 11,000 kg (24,250 lb) payload at max T-O
weight, 30 min reserves 215 nm (400 km; 248 miles)

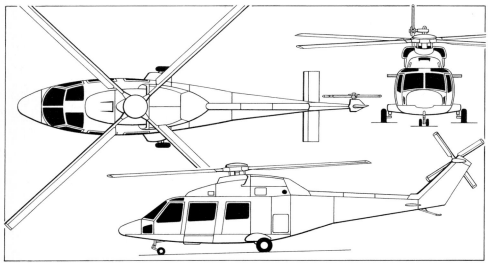

Mil Mi-54 twin-turbine commercial utility helicopter (*Jane's/Mike Keep*)

MIL Mi-54

TYPE: Twin-turbine commercial utility helicopter.
PROGRAMME: Project announced Summer 1992; model displayed at MosAeroshow '92; single-engined version for Asian market also being studied.
DESIGN FEATURES: Configuration shown in accompanying illustrations; four-blade main and tail rotors; high excess power will permit operation as Cat. A helicopter from sites of limited size; intended for passenger/cargo, oil rig support, ambulance and executive duties.
LANDING GEAR: Non-retractable tricycle type; single wheel on each main unit, carried on short sponsons; twin nosewheels; tailskid.
POWER PLANT: Two unspecified turboshafts mounted side by side above cabin.
ACCOMMODATION: Normal seating for 10-12 passengers.
DIMENSIONS, EXTERNAL:
Main rotor diameter 13.50 m (44 ft 3½ in)
Tail rotor diameter 2.60 m (8 ft 6½ in)
Length overall, excl rotors 13.20 m (43 ft 3¾ in)
Height to top of rotor head 3.55 m (11 ft 7¾ in)
Wheel track 3.00 m (9 ft 10¼ in)
Wheelbase 3.90 m (12 ft 9½ in)
WEIGHTS AND LOADINGS:
Payload 1,000-1,300 kg (2,205-2,865 lb)
Max T-O weight 4,000 kg (8,820 lb)
PERFORMANCE (estimated):
Max level speed 151 knots (280 km/h; 174 mph)
Nominal cruising speed 140 knots (260 km/h; 161 mph)
Hovering ceiling OGE 2,000 m (6,560 ft)
Range, 30 min reserves:
with max internal fuel 431 nm (800 km; 497 miles)
with 1,200 kg (2,645 lb) payload
242 nm (450 km; 280 miles)

Model of Mil Mi-54 displayed at MosAeroshow '92 (*Mark Lambert*)

MOLNIYA
MOLNIYA SCIENTIFIC AND INDUSTRIAL ENTERPRISE

Novoposelkovaya 4, Moscow 123459
Telephone: 7 (095) 493 33 35
Fax: 7 (095) 493 43 88
DIRECTOR AND GENERAL DESIGNER: Gleb Lozino-Lozinsky

To compensate for reduced funding for Buran space shuttle orbiter programme, in which it was much involved, Molniya has projected a series of civil aircraft. First to fly was six-seat Molniya-1. Others, at various design stages, include **Molniya-100**, an 18-seat twin-engined airliner; **Molniya-300** twin-engined business jet; and six-engined **Heracles** freighter, which would be world's heaviest aeroplane, with max T-O weight of 900,000 kg (1,984,125 lb) and payload of up to 450,000 kg (992,060 lb) suspended under its inverted gull-wing centre-section. Payloads could include a shuttle orbiter or large rocket stages.

MOLNIYA-1

TYPE: Single-engined six-seat light aircraft.
PROGRAMME: Announced at MosAeroshow '92; international commercial association to market and operate Molniya-1 being established with Russian government support. Prototype flew 17 December 1992; initial production series of 20 being built 1993.
DESIGN FEATURES: Low-wing monoplane; foreplanes mid-mounted on nose; wings at rear of fuselage pod, carrying twin tailbooms with tailplane bridging tips of sweptback vertical tail surfaces; non-retractable tricycle landing gear, with single wheel on each unit; fairing over each wheel; engine mounted at rear of fuselage pod, driving three-blade pusher propeller. Folding wings intended to permit driving on roads and transportation in container-hangar measuring 4 × 8.2 × 3 m (13 ft 1½ in × 26 ft 11 in × 9 ft 10 in).

Designed to permit flying by pilot of average ability after 8 to 16 hours' instruction.
POWER PLANT: One unspecified unsupercharged piston engine.
EQUIPMENT: Optional extras include radio, fax, autopilot, satellite navigation system, digital instruments, navaids, air-conditioning, skis and floats.
ACCOMMODATION: Six persons in pairs; seats quickly removable for freight carrying.
DIMENSIONS, EXTERNAL:
Wing span 8.50 m (27 ft 10¾ in)
Length overall 7.86 m (25 ft 9½ in)
Width, wings folded 3.60 m (11 ft 9¾ in)
WEIGHTS AND LOADINGS:
Max payload 505 kg (1,115 lb)
Max fuel 180 kg (395 lb)
Max T-O weight 1,580 kg (3,480 lb)

PERFORMANCE (estimated):
  Max level speed   221 knots (410 km/h; 255 mph)
  Nominal cruising speed   194 knots (360 km/h; 224 mph)
  Min flying speed   54 knots (100 km/h; 62 mph)
  Landing speed   60 knots (110 km/h; 69 mph)
  T-O run   260 m (855 ft)
  Range with max payload at 145 knots (270 km/h; 168 mph)   645 nm (1,200 km; 745 miles)

Molniya-1 six-seat single-engined light aircraft *(Jane's/Mike Keep)*

MYASISHCHEV
MYASISHCHEV DESIGN BUREAU
140160 Zhukovsky, Moscow Region
Telephone: 7 (095) 272 60 41
Fax: 7 (095) 556 55 83
GENERAL DESIGNER: V. Novikov
CHIEF DESIGNER: Leonid Sokolov

The Myasishchev OKB was founded in 1951, under the leadership of Professor Vladimir Mikhailovich Myasishchev, who died on 14 October 1978. In the 1950s it developed and built the M-4 and 3 M (NATO 'Bison') subsonic strategic bombers and M-50 ('Bounder') supersonic strategic missile carrier. In the 1970s and '80s the bureau was engaged in development of multi-purpose subsonic high-altitude aircraft, and in building the VM-T Atlant aircraft to transport sections of the Energia rocket launch vehicle and airframe of the Buran space shuttle orbiter. In 1981 the design bureau was named after Prof V. M. Myasishchev.

MYASISHCHEV M-17 and M-55
NATO reporting name: Mystic
TYPE: Single-seat high-altitude reconnaissance and research aircraft.
PROGRAMME: M-17 ('Mystic-A') prototype observed at Ramenskoye flight test centre 1982, given provisional US designation 'Ram-M'; two prototypes, plus two M-17R (M-55, 'Mystic-B') prototypes and two pre-production 'Bs', completed by early 1992; publicised as civilian research aircraft; 25 international records set by 'Mystic-A' 28 March-14 May 1990, in classes Cli/j, for speed, climb and height.
CURRENT VERSIONS: **M-17 Stratosfera** ('Mystic-A'): First two aircraft, each with single RKBM Rybinsk RD-36-51V turbojet, rated at 68.6 kN (15,430 lb st) in non-afterburning form; aircraft SSSR-17401 used in 1989-90 for record-flying and atmospheric research flights, with a joint Chilean-Russian programme to investigate the Antarctic ozone layer scheduled in 1992; aircraft SSSR-17103 in outdoor display of historic aircraft, Monino.

M-55 Geofizika ('Mystic-B'): Basically as Stratosfera, but with two Aviadvigatel PS-30-V12 engines, each 49 kN (11,025 lb st), side by side at rear of fuselage pod; lengthened and more capacious nose; raised cockpit; small underfuselage radome forward of nosewheels; able to loiter for 4 h 12 min at 20,000 m (65,600 ft) with 1,500 kg (3,305 lb) of sensors, or 5 h at 17,000 m (55,775 ft); example in officially released photographs (SSSR-01552) has Aeroflot insignia and said to be for environmental research; smaller wing span than 'Mystic-A'. Designed initially for reconnaissance, as suggested by military designation of **M-17R**.

M-55: Variant with conventional fuselage in design stage; sweptback conventional tail unit, and twin engines on sides of fuselage pod beneath wingroots.
DESIGN FEATURES: High aspect ratio high-wing configuration, with central fuselage pod and twin tailbooms with bridging T-tailplane; side air ducts for engine(s), mounted at rear of fuselage pod; constant wing anhedral from roots; long and shallow ventral strake under port tailboom; assumed to have similar powered sailplane performance to US Lockheed U-2 series; large compartment in lower fuselage for cameras and other sensors.
FLYING CONTROLS: Aileron, with tab, and three-section flaps over most of each wing trailing-edge; elevators and twin rudders, with tabs.
LANDING GEAR: Retractable tricycle type; twin wheels on each unit; all units retract rearward, mainwheels into tailbooms.
POWER PLANT: See Current Versions.
ACCOMMODATION: Single seat under rearward hinged canopy.

Myasishchev M-17 ('Mystic-A') single-engined high-altitude atmospheric research aircraft *(Tass)*

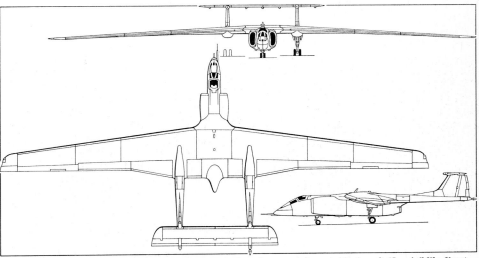
Myasishchev M-17 ('Mystic-A') reconnaissance and environmental research aircraft *(Jane's/Mike Keep)*

Myasishchev M-55/M-17R ('Mystic-B'), a twin-engined version of the M-17 *(Tass)*

DIMENSIONS, EXTERNAL (estimated. A: 'Mystic-A', B: 'Mystic-B'):
Wing span: A		40.7 m (133 ft 6½ in)
B		37.5 m (123 ft 0½ in)
Length overall: A		21.2 m (69 ft 6½ in)
B		24.00 m (78 ft 9 in)
Height overall: A		5.25 m (17 ft 3 in)
B		4.8 m (15 ft 9 in)
Wheel track: A		6.65 m (21 ft 10 in)
Wheelbase: A		5.6 m (18 ft 4½ in)

WEIGHTS AND LOADINGS:
Payload: B	1,500 kg (3,307 lb)
Max T-O weight: A, record flights	19,950 kg (44,000 lb)
Power loading: A, record flights	290.8 kg/kN (2.85 lb/lb st)

PERFORMANCE:
Max level speed at height:	
B	377-405 knots (700-750 km/h; 435-466 mph)
Service ceiling: B	20,000 m (65,600 ft)
Landing run: B	1,750 m (5,745 ft)
Max endurance at 17,000 m (55,775 ft): A	2 h 30 min
B	6 h 30 min

The twin-turbofan power plant of the Myasishchev M-55/M-17R is shown clearly in this photograph *(Piotr Butowski)*

Mockup of Myasishchev Gzhel turboprop powered light aircraft *(Mark Lambert)*

MYASISHCHEV M-101 GZHEL

TYPE: Six/seven-passenger turboprop light aircraft.
PROGRAMME: First shown in model form at Moscow Aerospace '90 exhibition; developed full-scale mockup exhibited at MosAeroshow '92; under development at Gorki; scheduled to fly 1993.
DESIGN FEATURES: Conventional low-wing monoplane; swept-back vertical tail surfaces.
LANDING GEAR: Retractable tricycle type; single wheel on each unit; nosewheel retracts rearward, mainwheels inward into wingroots.
POWER PLANT: One 560 kW (751 shp) Motorlet M 601E turboprop, driving three-blade propeller.
ACCOMMODATION: Pilot and six or seven passengers, in pairs in enclosed cabin; door to flight deck on port side; large double door for passengers and freight loading aft of wing on port side; emergency exit on starboard side.

WEIGHTS AND LOADINGS:
Max payload	700 kg (1,543 lb)
Max T-O weight	2,500 kg (5,510 lb)

PERFORMANCE (estimated):
Max cruising speed	243 knots (450 km/h; 280 mph)
Landing speed	65 knots (120 km/h; 75 mph)
Service ceiling	7,600 m (24,935 ft)
Runway required	500 m (1,640 ft)
Range with max fuel	1,400 nm (2,600 km; 1,615 miles)

MYASISHCHEV M-103 SKIF

TYPE: Four/five-seat light aircraft.
PROGRAMME: Shown in model form at Moscow Aerospace '90 exhibition; prototype flying 1992; further work deferred until 1995.
DESIGN FEATURES: Conventional cantilever high-wing monoplane; pod and boom fuselage; conventional tail unit with sweptback vertical surfaces.
LANDING GEAR: Non-retractable tricycle type, with single wheel on each unit; cantilever spring main legs.
POWER PLANT: One 221 kW (296 hp) M-16 piston engine, driving three-blade propeller.
ACCOMMODATION: Enclosed cabin for four or five persons.

MYASISHCHEV DELPHIN

TYPE: Projected twin-turboprop 9/14-seat business aircraft.
PROGRAMME: At design study stage.
DESIGN FEATURES: High aspect ratio low-mounted wings, with slight sweepback; pressurised circular section fuselage; T tailplane on sweptback fin; engines pylon-mounted each side of rear fuselage, with pusher propellers.
LANDING GEAR: Retractable tricycle type.
ACCOMMODATION: Nine to 14 persons in enclosed cabin; door at front on port side; overwing emergency exit starboard side.

WEIGHTS AND LOADINGS:
Max payload	1,300 kg (2,865 lb)
Max T-O weight	5,700 kg (12,565 lb)

PERFORMANCE (estimated):
Max cruising speed	296 knots (550 km/h; 341 mph)
Range with nine passengers	1,080 nm (2,000 km; 1,240 miles)

MYASISHCHEV/AVIASPETSTRANS YAMAL

See Aviaspetstrans entry in this section.

Model of Myasishchev Skif four/five-seat light aircraft *(Piotr Butowski)*

Artist's impression of Myasishchev Delphin business aircraft

PHOENIX-AVIATECHNICA

JOINT STOCK COMPANY 'PHOENIX-AVIATECHNICA'

Ulansky per 11A-34, 36, 101000 Moscow
POSTAL ADDRESS: K-45, PO Box 32, 103045 Moscow
Telephone: 7 (095) 207 18 61
Fax: 7 (095) 207 80 34
PRESIDENT: Alexander Koshelev

This company is developing the LKhS-4 light four-seat utility aircraft for short-range commuter and taxi services throughout Russia.

PHOENIX-AVIATECHNICA LKhS-4

TYPE: Twin-engined light utility aircraft.

PROGRAMME: Design started 1989; prototype construction began 1992, for 1993 first flight; two flying prototypes, one for static test; production scheduled for 1994.

COSTS: Programme cost 100 million roubles (1992); target cost production aircraft 10 million roubles.

DESIGN FEATURES: Designed to Russian AP 23 and FAR 23 standards. Equal-span unswept biplane of high aspect ratio, with pod and boom fuselage, constant-chord wing and tail surfaces, twin pusher engines. Wings fold back for transportation and storage. All three tail surfaces, including elevators and rudder, interchangeable. Wing thickness/chord ratio 15.5 per cent; dihedral 3°; no incidence or twist.

FLYING CONTROLS: Conventional mechanical; full-span flaps on lower wings; fixed ventral tail fin.

STRUCTURE: Aluminium alloy primary structure, with corrugated skin; some GFRP composite panels; no stringers. Two-spar wings with four ribs in each; central main beam and four frames in fuselage, with monocoque tailboom; two spars and three ribs in each fixed tail section.

LANDING GEAR: Non-retractable tricycle type; single wheel on each unit; suitable for operation from unprepared fields; cantilever spring steel legs; wheels and tyres from factories in Kámensk-Uralsky and Yaroslavl respectively; all three tyres size 400 × 150 mm on 200 mm wheels; tyre pressure 1.52 bars (22 lb/sq in); disc brakes; min ground turning radius 5.0 m (16 ft 5 in).

POWER PLANT: Two 59.7 kW (80 hp) Volszhsky BA3-21083 piston engines, each driving a two-blade Stupino metal pusher propeller. Two fuel tanks in fuselage, each 50 litres (13.2 US gallons; 11 Imp gallons) capacity. Oil capacity 6 litres (1.6 US gallons; 1.3 Imp gallons).

ACCOMMODATION: Pilot and three passengers, in pairs in fully enclosed cabin; automobile-type door each side; emergency exit through hinged windscreen. Hold for hand baggage in rear fuselage, accessible from cabin. Interior heated and ventilated.

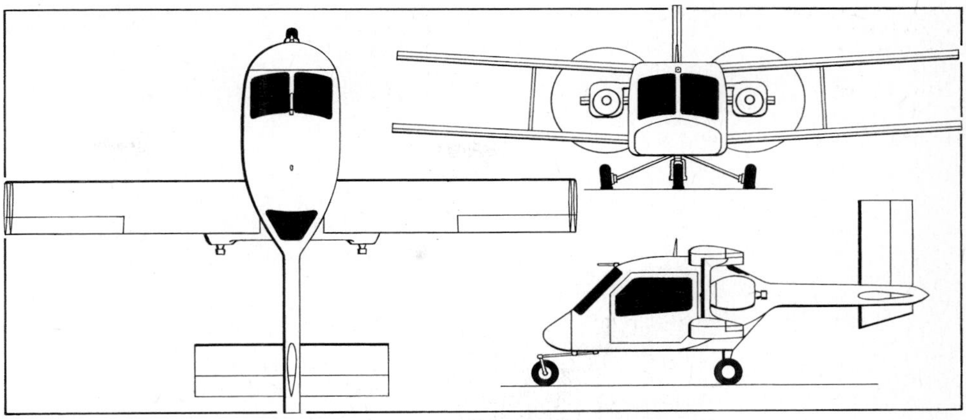

Phoenix-Aviatechnica LKhS-4 twin-engined utility aircraft *(Jane's/Mike Keep)*

SYSTEMS: Hydraulic system for brakes. DC electrical system with two 100W 27V generators and two 27V 23Ah batteries. De-icing system for wing and tail unit leading-edges and windscreen.

AVIONICS: BRI3 ultra short wave radio, range 27 nm (50 km; 31 miles). Instrumentation and navigation equipment to FAR 23 for VFR.

DIMENSIONS, EXTERNAL:

Wing span	8.20 m (26 ft 10¾ in)
Wing chord, constant	0.80 m (2 ft 7½ in)
Length overall	5.00 m (16 ft 5 in)
Width, wings folded	3.00 m (9 ft 10¼ in)
Height overall	2.80 m (9 ft 2¼ in)
Tailplane span	2.80 m (9 ft 2¼ in)
Wheel track	2.04 m (6 ft 8½ in)
Wheelbase	2.50 m (8 ft 2½ in)
Propeller diameter (each)	1.60 m (5 ft 3 in)
Propeller/ground clearance	0.25 m (9¾ in)
Distance between propeller centres	2.05 m (6 ft 9 in)
Doors (each): Height	1.00 m (3 ft 3¼ in)
Max width	1.20 m (3 ft 11¼ in)
Height to sill	0.60 m (1 ft 11¾ in)
Emergency exit (windscreen):	
Height	0.80 m (2 ft 7½ in)
Width	0.70 m (2 ft 3½ in)

DIMENSIONS, INTERNAL:

Cabin: Length	1.80 m (5 ft 10¾ in)
Max width	1.40 m (4 ft 7 in)
Max height	1.30 m (4 ft 3 in)
Floor area	2.50 m² (26.9 sq ft)
Volume	3.30 m³ (116.5 cu ft)
Baggage hold: Volume	0.50 m³ (17.6 cu ft)

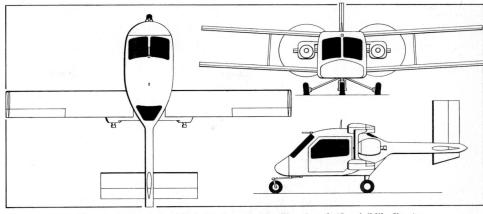

AREAS:

Wings, gross	13.00 m² (139.94 sq ft)
Ailerons (total)	0.77 m² (8.29 sq ft)
Trailing-edge flaps (total)	1.54 m² (16.58 sq ft)
Fin, incl ventral fin	1.03 m² (11.09 sq ft)
Rudder	0.41 m² (4.41 sq ft)
Tailplane	1.42 m² (15.29 sq ft)
Elevators (total)	0.82 m² (8.82 sq ft)

WEIGHTS AND LOADINGS:

Weight empty	550 kg (1,213 lb)
Max fuel	80 kg (176 lb)
Max T-O and landing weight	950 kg (2,095 lb)
Max wing loading	73.08 kg/m² (14.97 lb/sq ft)
Max power loading	7.96 kg/kW (13.1 lb/hp)

PERFORMANCE (estimated at max T-O weight):

Never-exceed speed (VNE)	140 knots (260 km/h; 161 mph)
Max level speed at 1,000 m (3,280 ft)	135 knots (250 km/h; 155 mph)
Max and econ cruising speed at 500 m (1,640 ft)	108 knots (200 km/h; 124 mph)
Stalling speed, flaps down	44 knots (80 km/h; 50 mph)
Max rate of climb at S/L	390 m (1,280 ft)/min
Rate of climb at S/L, OEI	102 m (335 ft)/min
Service ceiling	4,000 m (13,125 ft)
Service ceiling, OEI	3,000 m (9,840 ft)
T-O run	200 m (660 ft)
T-O to 15 m (50 ft)	400 m (1,320 ft)
Landing from 15 m (50 ft)	400 m (1,320 ft)
Landing run	200 m (660 ft)
Range with max fuel and max payload	270 nm (500 km; 310 miles)

ROKS-AERO

ROKS-AERO CORPORATION

Novy Arbat 36, Office 1333, 121205 Moscow
Telephone: 7 (095) 290 75 46 and 75 50
Fax: 7 (095) 203 29 70
PRESIDENT AND GENERAL DESIGNER: Evgeny P. Grunin
INTERNATIONAL PROGRAMMES AND SALES: Vitaly V. Shikera
DEPUTY GENERAL DESIGNER: Nikolai I. Frolov

Known initially as ROS-Aeroprogress, ROKS-Aero Corporation was founded in 1990, to design and manufacture utility, commuter, amphibian, aerobatic, agricultural, firefighting, training and attack aircraft, WIG (wing in ground effect) vehicles, replicas and other airborne vehicles. It is a member of the Business Aviation Association, which includes the Moscow Aviation Production Association, Yakovlev's 'Skorost' factory, the Myasishchev experimental plant, and aviation works in Komsomolsk-on-Amur, Smolensk, Novosibirsk, Ulan-Ude and Luchovitsy. Its divisions are:

Utility Aircraft Division
DIRECTOR: Gennadiy A. Kolokolnikov
CHIEF DESIGNER: Anatoliy Y. Grishcechkin

Amphibian and Special Aircraft Division
DIRECTOR: Pavel G. Tkatchenco
CHIEF DESIGNER: Mikhail S. Remizov

Trainer and Aerobatics Division
DIRECTOR: Sergei V. Ivanov
CHIEF DESIGNER: Vladimir P. Lapshin

Business and Touring Aircraft Division
DIRECTOR: Vitaly G. Shelmanov
CHIEF DESIGNER: Leonard A. Tarasevich

Kitplane and Ultralight Aircraft Division
DIRECTOR: Anatoly A. Lastovkin

Replica Division
DIRECTOR: Mikhail V. Korenkov

ROKS-AERO T-101 GRATCH

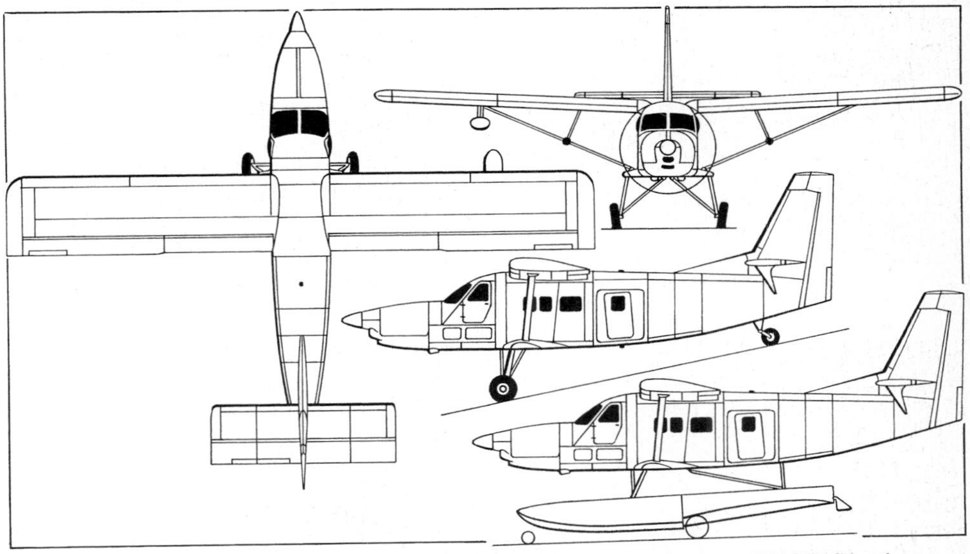

ROKS-Aero T-101 Gratch turboprop utility aircraft, with additional side view of T-101V floatplane *(Jane's/Mike Keep)*

TYPE: Turboprop powered general utility aircraft.

PROGRAMME: Design started by Utility Aircraft Division in September 1991, as monoplane successor to Antonov An-2/3 biplane; construction of first of five prototypes started April 1992; first flight scheduled April 1993; construction of production aircraft initiated January 1993; first flight of production aircraft scheduled May 1993, with certification to AP 23 and FAR 23 in March 1994, following first customer delivery in June 1993.

CURRENT VERSIONS: **T-101**: Basic passenger/cargo transport.
Detailed description applies specifically to T-101.

T-101V: As basic T-101 but with amphibious float landing gear. Described separately.

T-101P: Firefighting version, on non-amphibious floats.

T-101L: As basic T-101 but with ski landing gear.

T-101Skh: Redesigned agricultural aircraft; fuselage and tail unit as basic T-101; strut braced low wings with considerable dihedral; each mainwheel on strut braced oleo; spraybars under wings; new cockpit with large flat windscreen and large side windows.

T-101S: Military version of basic T-101; small sweptback winglets; two stores pylons under each wing; added small stub-wings, each with a weapon pylon and wingtip mount for a gun pod or other store.

Further developments, with tricycle landing gear, increased wing span, more powerful engines and other changes, will be designated **T-102**, **T-103** and **T-104**.

CUSTOMERS: Not yet announced; orders totalled 250 in early 1993.

DESIGN FEATURES: Single-turboprop aircraft for Normal category passenger/cargo transportation and utility

Full-scale mockup of ROKS-Aero T-101V Gratch floatplane

applications; high-mounted braced wing, non-retractable landing gear and unpressurised cabin; STOL capable, with wide CG range; large passenger/freight door. Unswept, constant chord wings; P-11-14 wing section, dihedral 3°, incidence 3° constant; sweptback fin and rudder with large dorsal fin; braced constant chord tailplane and elevators.

FLYING CONTROLS: Conventional mechanical control (rods and cables). Aileron deflection 30° up, 14° down; elevator deflection 42° up, 22.5° down; rudder deflection ±28°. Trim tabs on port aileron, port elevator and rudder. Electrically actuated single-section slotted trailing-edge flap and two-section automatic leading-edge slats on each wing; flap deflection 25° for take-off, 40° for landing.

STRUCTURE: All-metal (aluminium alloy and high tensile steel) structure; two-spar wings, metal skinned with integral stringers, and ribs; two-spar fin and tailplane; semi-monocoque fuselage, with frames, stringers and stressed skin.

LANDING GEAR: Non-retractable tailwheel type with single wheel on each unit. Main legs of tripod type, with oleo-pneumatic shock absorption; KT-135D mainwheels, with tyres size 720 × 320 mm, pressure 3.43 bars (50 lb/sq in); K-392 tailwheel, with tyre size 380 × 200 mm, pressure 3.43 bars (50 lb/sq in); hydraulic brakes and anti-skid units on mainwheels. Skis and floats optional.

POWER PLANT: One 754 kW (1,011 shp) Mars (Omsk) TVD-10B turboprop, driving AV-24AN three-blade constant-speed propeller with reverse pitch and full feathering. Three fuel tanks in each wing, each 200 litres (52.8 US gallons; 44 Imp gallons); total fuel capacity 1,200 litres (317 US gallons; 264 Imp gallons). Oil tank capacity 30 litres (7.9 US gallons; 6.6 Imp gallons).

ACCOMMODATION: Crew of one or two and nine passengers or equivalent freight. Forward opening door each side of flight deck; large upward opening freight door aft of wing on port side, with integral inward opening passenger door; door between flight deck and cabin; starboard emergency exit. Cabin ventilated and heated by engine bleed air.

SYSTEMS: Hydraulic system for brakes; max flow 4 litres (1.05 US gallons; 0.88 Imp gallons)/min, at 147 bars (2,135 lb/sq in). Three-phase 120/208V 400Hz AC electrical system, supplied by 6kVA 200A BU6BK brushless alternator, with emergency DC power supply and 24V 40Ah nickel-cadmium battery. Electric de-icing of propeller blades and spinner; engine air intake de-iced by heated oil.

AVIONICS: Radio com/nav equipment for VFR and IFR operation by day and night in all terrains, including KC-MC compact compass system, AGB-96 gyro horizon, PNP-72-14 nav, air data system with digital airspeed and altitude indication, GROM satellite nav, VBM-1PB standby altimeter, A-723 long-range radio nav, GRAN low-altitude radio altimeter, ARK-M ADF, A-611 marker beacon receiver, AP-93 autopilot, R-855A emergency locator beacon and ARB-NK emergency radio buoy.

DIMENSIONS, EXTERNAL:
Wing span	18.18 m (59 ft 8 in)
Wing chord: at root	2.40 m (7 ft 10½ in)
at tip	2.45 m (8 ft 0½ in)
Wing aspect ratio	7.58
Length overall	15.042 m (49 ft 4¼ in)
Fuselage: Max width	1.80 m (5 ft 10⅞ in)
Max height	2.52 m (8 ft 3¼ in)
Height overall	6.67 m (21 ft 10½ in)
Tailplane span	5.80 m (19 ft 0½ in)
Wheel track	3.24 m (10 ft 7½ in)
Wheelbase	8.30 m (27 ft 3 in)
Propeller diameter	2.80 m (9 ft 2¼ in)
Propeller ground clearance	0.97 m (3 ft 2¼ in)
Flight deck door (each): Height	1.20 m (3 ft 11¼ in)
Width at top	0.425 m (1 ft 4¾ in)
Width at bottom	0.67 m (2 ft 2¼ in)
Passenger door: Height	1.42 m (4 ft 8 in)
Width	0.81 m (2 ft 7⅞ in)
Freight door: Height	1.53 m (5 ft 0¼ in)
Width	1.46 m (4 ft 9½ in)
Emergency exit: Height	0.535 m (1 ft 9 in)
Width	0.87 m (2 ft 10¼ in)

DIMENSIONS, INTERNAL:
Cabin: Length	4.20 m (13 ft 9¼ in)
Max width	1.60 m (5 ft 3 in)
Max height	1.80 m (5 ft 10⅞ in)
Floor area	6.72 m² (72.3 sq ft)
Volume	12.10 m³ (427.3 cu ft)
Flight deck/cabin door: Height	1.345 m (4 ft 5 in)
Width	0.51 m (1 ft 8 in)

AREAS:
Wings, gross	43.63 m² (469.65 sq ft)
Ailerons (total)	5.66 m² (60.93 sq ft)
Trailing-edge flaps (total)	4.02 m² (43.27 sq ft)
Leading-edge slats (total)	5.98 m² (64.37 sq ft)
Fin, incl dorsal fin	5.455 m² (58.72 sq ft)
Rudder, incl tab	2.955 m² (31.81 sq ft)
Tailplane	6.34 m² (68.25 sq ft)
Elevator, incl tab	4.10 m² (44.13 sq ft)

WEIGHTS AND LOADINGS:
Weight empty, equipped	3,330 kg (7,342 lb)
Max payload	1,600 kg (3,527 lb)
Max fuel	950 kg (2,095 lb)
Max T-O and landing weight	5,500 kg (12,125 lb)
Max wing loading	126 kg/m² (25.8 lb/sq ft)
Max power loading	7.30 kg/kW (12.00 lb/shp)

PERFORMANCE (estimated):
Max level speed at 3,000 m (9,840 ft)	164 knots (305 km/h; 190 mph)
Max cruising speed at 3,000 m (9,840 ft)	161 knots (298 km/h; 185 mph)
Econ cruising speed at 3,000 m (9,840 ft)	127 knots (235 km/h; 146 mph)
Service ceiling	3,600 m (11,800 ft)
T-O run	454 m (1,490 ft)
T-O to 15 m (50 ft)	615 m (2,020 ft)
Landing from 15 m (50 ft)	370 m (1,215 ft)
Range with max fuel	712 nm (1,320 km; 820 miles)

ROKS-AERO T-101V GRATCH

TYPE: Floatplane version of T-101.
PROGRAMME: Design started January 1991; full-scale mockup exhibited at MosAeroshow '92; prototype construction scheduled to begin May 1993.
DESIGN FEATURES: As for T-101, but with amphibious float landing gear.
FLYING CONTROLS: As for T-101.
STRUCTURE: As for T-101, but strengthened at float landing gear attachment points.
LANDING GEAR: Twin-float amphibious type, strut-mounted to fuselage; nosewheel and mainwheel on each float, retracting rearward into float; mainwheel tyre size 510 × 150 mm, nosewheel tyre size 325 × 145 mm; no shock absorbers or brakes; hydraulic retraction; min ground turning radius 7.0 m (23 ft). Water rudder, towing point and tiedown fitting on each float.
POWER PLANT: One 1,029 kW (1,380 shp) Mars (Omsk) TVD-20 turboprop, driving AV-17 three-blade constant-speed propeller with reverse pitch. Six fuel tanks in wings, capacity 1,240 litres (327 US gallons; 272 Imp gallons). Oil capacity 8 litres (2.1 US gallons; 1.75 Imp gallons).
ACCOMMODATION: Two crew and 12 passengers; doors as for T-101.

DIMENSIONS, EXTERNAL:
Wing span	18.50 m (60 ft 8½ in)
Wing chord (constant)	2.40 m (7 ft 10½ in)
Wing aspect ratio	8.14
Length overall	15.225 m (49 ft 11½ in)
Fuselage: Length	15.042 m (49 ft 4¼ in)
Max width	1.80 m (5 ft 10⅞ in)
Max height	2.52 m (8 ft 3¼ in)
Height overall	6.766 m (22 ft 2½ in)
Tailplane span	5.80 m (19 ft 0½ in)
Wheel track	3.70 m (12 ft 1¾ in)
Wheelbase	4.896 m (16 ft 0¾ in)
Propeller diameter	3.60 m (11 ft 9¾ in)
Propeller water clearance	1.00 m (3 ft 3½ in)

DIMENSIONS, INTERNAL: As for T-101
AREAS: As for T-101, except:
Wings, gross	42.02 m² (452.3 sq ft)

WEIGHTS AND LOADINGS:
Weight empty, equipped	3,700 kg (8,157 lb)
Max payload	1,500 kg (3,307 lb)
Max fuel	920 kg (2,028 lb)
Max T-O weight	5,715 kg (12,600 lb)
Max wing loading	136 kg/m² (27.85 lb/sq ft)
Max power loading	5.55 kg/kW (9.13 lb/shp)

PERFORMANCE (estimated):
Max level speed at 4,000 m (13,125 ft)	151 knots (280 km/h; 174 mph)
Econ cruising speed at 3,000 m (9,840 ft)	130 knots (240 km/h; 149 mph)
Min stalling speed: flaps up	89 knots (165 km/h; 103 mph)
Max rate of climb at S/L	360 m (1,180 ft)/min
T-O run	350 m (1,150 ft)
T-O to 15 m (50 ft)	550 m (1,805 ft)
Landing from 15 m (50 ft)	500 m (1,640 ft)
Landing run	300 m (985 ft)
Range, 30 min reserves:	
with max payload	324 nm (600 km; 373 miles)
with max fuel	647 nm (1,200 km; 745 miles)

ROKS-AERO T-106

TYPE: Twin-engined development of T-101 Gratch.
PROGRAMME: At initial design stage 1993.
DESIGN FEATURES: Basic configuration and many components similar to T-101 series; nose faired, with provision for radar and equipment or baggage; wings braced to short

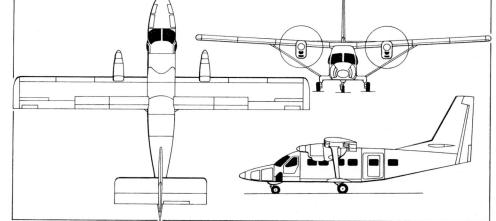

ROKS-Aero T-106 twin-turboprop multi-purpose aircraft (Jane's/Mike Keep)

stub-wings carrying mainwheels of landing gear; engines under inner wings. Intended for passenger/freight transport, ambulance, offshore patrol of 200-mile economic zones, aerial photography, geological survey, agricultural and other general duties.

FLYING CONTROLS: Mechanical control. High-lift wings, with slotted aileron, two-segment double-slotted flaps and three-section leading-edge slats on each wing; spoiler forward of each aileron; trim tab in each aileron, each half of tailplane, and rudder.

LANDING GEAR: Non-retractable tricycle type; single wheel on each trailing-link leg.

POWER PLANT: Two Mars (Omsk) TVD-10B or Pratt & Whitney Canada PT6A turboprops.

ACCOMMODATION: Two crew on flight deck; door on each side and in rear bulkhead to cabin, as T-101; large passenger/freight door on port side of main cabin. Mixed passenger/freight configurations planned.

DIMENSIONS, EXTERNAL:
Wing span	19.90 m (65 ft 3½ in)
Length overall	15.56 m (51 ft 0¾ in)
Height overall	6.42 m (21 ft 0¾ in)
Wheel track	3.366 m (11 ft 0½ in)
Wheelbase	4.53 m (14 ft 10¼ in)
Propeller diameter (three-blade)	2.80 m (9 ft 2¼ in)
Distance between propeller centres	5.70 m (18 ft 8½ in)

DIMENSIONS, INTERNAL:
Cabin: Length	5.75 m (18 ft 10¼ in)
Max width	1.65 m (5 ft 5 in)
Max height	1.85 m (6 ft 0¾ in)
Volume	15.00 m³ (530 cu ft)

WEIGHTS AND LOADINGS:
Max payload	2,000 kg (4,409 lb)
Max fuel	920 kg (2,028 lb)
Max T-O weight	5,950 kg (13,117 lb)

PERFORMANCE (estimated):
Max level speed	189 knots (350 km/h; 217 mph)
Service ceiling	4,000 m (13,125 ft)
Range: with max payload	189 nm (350 km; 217 miles)
with max fuel	647 nm (1,200 km; 745 miles)

ROKS-AERO T-401 SOKOL

TYPE: Single-engined light multi-purpose aircraft.

CURRENT VERSIONS: Designed for passenger/freight transport, ambulance, offshore patrol, primary training and agricultural applications.

PROGRAMME: Project design started 1990; detail design under way 1993; construction of prototype scheduled to begin January 1994.

CUSTOMERS: Initial requirement 200 aircraft.

DESIGN FEATURES: Conventional cantilever high-wing monoplane, with sweptback vertical tail surfaces; non-retractable tricycle landing gear, with fairing over each wheel; wing section MS (1)-3M, dihedral 1° 30′, incidence 2°. Small auxiliary fin on each side of tailplane on floatplane version.

FLYING CONTROLS: Conventional three-axis mechanical control. Trim tabs in port elevator and rudder. Large flap and aileron occupy full span of each wing trailing-edge.

STRUCTURE: All-metal. Conventional two-spar wing; square-section semi-monocoque fuselage.

LANDING GEAR: Non-retractable tricycle type; single wheel on each unit; mainwheel tyres size 500 × 150 mm; nosewheel tyre size 400 × 150 mm. Min ground turning radius 4 m (13 ft 1½ in). Max steering angle of nosewheel ±40°. Optional floats and skis.

POWER PLANT: One 265 kW (355 hp) M-14PR nine-cylinder air-cooled radial engine; alternatives include a Textron Lycoming piston engine; V530TA-D35 three-blade constant-speed propeller. Fuel capacity 350 litres (92.5 US gallons; 77 Imp gallons).

ACCOMMODATION: Six persons in pairs, including one or two pilots in basic version. All passenger seats removable for freight carrying. Ambulance version carries one stretcher patient, one or two medical attendants, and medical equipment including optionally An-8 narcosis apparatus, Ki-4 oxygen supply or Volga-MT automatic respirator, Ambu manual breathing apparatus, Salut electrocardiographic equipment and EKSN electrocardiostimulator. Provision for photographic and video equipment, PZS-68 public address system and various work stations.

DIMENSIONS, EXTERNAL:
Wing span	13.66 m (44 ft 9¾ in)
Wing chord: at centreline	1.882 m (6 ft 2 in)
at wingtip	1.046 m (3 ft 5¼ in)
Wing aspect ratio	9.33
Length overall	8.90 m (29 ft 2½ in)
Fuselage: Max width	1.50 m (4 ft 11 in)
Max height	1.65 m (5 ft 5 in)
Height overall	4.386 m (14 ft 4¾ in)
Tailplane span	5.10 m (16 ft 9 in)
Wheel track	3.30 m (10 ft 10 in)
Wheelbase	2.67 m (8 ft 9¼ in)
Propeller diameter	2.40 m (7 ft 10½ in)
Propeller ground clearance	0.338 m (1 ft 1¼ in)
Flight deck door (each side):	
Height	1.10 m (3 ft 7¼ in)
Width at top	0.47 m (1 ft 6½ in)
Width at bottom	0.78 m (2 ft 6½ in)
Cabin door: Height	1.18 m (3 ft 10½ in)
Width	1.02 m (3 ft 4 in)

DIMENSIONS, INTERNAL:
Cabin, incl flight deck:	
Length	4.41 m (14 ft 5½ in)
Max width	1.34 m (4 ft 4¾ in)
Max height	1.25 m (4 ft 1¼ in)
Floor area	4.94 m² (53.2 sq ft)
Volume	6.175 m³ (218 cu ft)
Baggage space, passenger version	1.0 m³ (35 cu ft)
Volume of hold, freighter	4.76 m³ (168 cu ft)

AREAS:
Wings, gross	20.00 m² (215.3 sq ft)
Ailerons (total)	1.142 m² (12.29 sq ft)
Trailing-edge flaps (total)	4.345 m² (46.77 sq ft)
Fin, incl dorsal fin	2.33 m² (25.08 sq ft)
Rudder	1.24 m² (13.35 sq ft)
Tailplane	2.48 m² (26.68 sq ft)
Elevators (total)	1.856 m² (19.98 sq ft)

WEIGHTS AND LOADINGS:
Weight empty, equipped	1,430 kg (3,153 lb)
Max payload	450 kg (992 lb)
Max fuel	320 kg (705 lb)
Max T-O and landing weight	2,030 kg (4,475 lb)
Max wing loading	101.5 kg/m² (20.50 lb/sq ft)
Max power loading	7.66 kg/kW (12.6 lb/hp)

PERFORMANCE (estimated):
Max level speed at 3,000 m (9,840 ft)	156 knots (290 km/h; 180 mph)
Max cruising speed at 3,000 m (9,840 ft)	145 knots (270 km/h; 167 mph)
Econ cruising speed at 3,000 m (9,840 ft)	116 knots (215 km/h; 134 mph)
Stalling speed, flaps down	68 knots (125 km/h; 78 mph)
Max rate of climb at S/L	260 m (855 ft)/min
Service ceiling	4,000 m (13,125 ft)
T-O run	255 m (837 ft)
T-O to 15 m (50 ft)	430 m (1,410 ft)
Landing from 15 m (50 ft)	545 m (1,790 ft)
Landing run	233 m (765 ft)
Range: with max payload, 30 min reserves	420 nm (780 km; 485 miles)
with max fuel	880 nm (1,630 km; 1,012 miles)

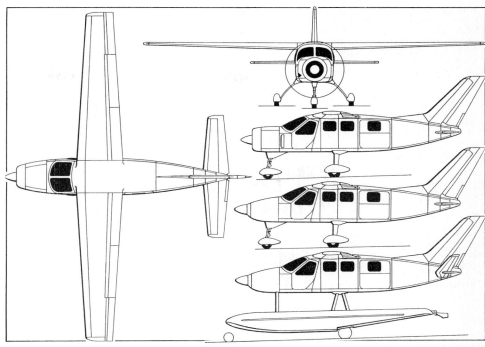

ROKS-Aero T-401 Sokol light multi-purpose aircraft, with additional side views showing versions with Textron Lycoming engine and float landing gear (*Jane's/Mike Keep*)

ROKS-AERO T-407

TYPE: Single-engined light multi-purpose aircraft.

PROGRAMME: Under design early 1993.

DESIGN FEATURES: Conventional strut braced high-wing monoplane of extremely simple design; constant chord wings and tailplane; unswept vertical tail surfaces; square section fuselage. Wing section P-II, dihedral 1° 30′, incidence 2°.

FLYING CONTROLS: Conventional mechanical control. Slotted aileron and slotted flap on full span of each wing trailing-edge. Horn balanced elevators and rudder. Trim tab in starboard tailplane.

STRUCTURE: All-metal, with steel-tube engine mounting and rear fuselage truss structure.

LANDING GEAR: Non-retractable tricycle type; single wheel on each unit. Each mainwheel on tripod mounting, with oleo-pneumatic shock absorber. Mainwheel tyres size 500 × 150 mm; nosewheel tyre size 400 × 150 mm.

POWER PLANT: One 265 kW (355 hp) M-14P nine-cylinder air-cooled radial piston engine; V530TA-D35 three-blade propeller. Fuel tanks in inner wings, capacity 380 litres (100 US gallons; 83.5 Imp gallons).

ACCOMMODATION: Basic seating for one or two persons on flight deck; up to five in cabin on two rearward facing seats

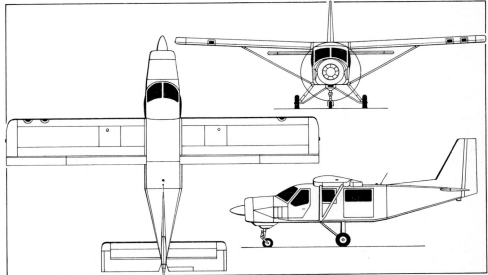

ROKS-Aero T-407 (M-14P radial piston engine) (*Jane's/Mike Keep*)

and rear bench seat. Baggage space behind rear seats. Alternative cabin configurations for freight carrying, ambulance, agricultural, offshore patrol and other utility duties. Forward opening door each side of flight deck. Large horizontally divided two-piece door at rear of cabin, port side.

DIMENSIONS, EXTERNAL:
Wing span	12.12 m (39 ft 9¼ in)
Wing chord (constant)	2.00 m (6 ft 6¾ in)
Wing aspect ratio	5.84
Length overall	10.00 m (32 ft 9¾ in)
Height overall	4.40 m (14 ft 5¼ in)
Wheel track	2.72 m (8 ft 11 in)
Wheelbase	3.05 m (10 ft 0 in)

DIMENSIONS, INTERNAL:
Cabin, excl flight deck:
Length	2.60 m (8 ft 6¼ in)
Max width	1.34 m (4 ft 4¾ in)
Max height	1.30 m (4 ft 3 in)
Volume	4.50 m³ (159 cu ft)

AREAS:
Wings, gross	25.14 m² (270.6 sq ft)
Ailerons (total)	2.43 m² (26.16 sq ft)
Trailing-edge flaps (total)	3.10 m² (33.37 sq ft)

WEIGHTS AND LOADINGS:
Weight empty	1,250 kg (2,755 lb)
Max fuel	300 kg (661 lb)
Fuel with max payload	150 kg (331 lb)
Max payload	600 kg (1,322 lb)
Payload with max fuel	450 kg (992 lb)
Max T-O weight	2,080 kg (4,585 lb)
Max wing loading	82.74 kg/m² (16.94 lb/sq ft)
Max power loading	7.85 kg/kW (12.91 lb/hp)

PERFORMANCE (estimated):
Max cruising speed	116 knots (215 km/h; 133 mph)
Nominal cruising height	2,000 m (6,560 ft)
T-O run	480 m (1,575 ft)
T-O to 15 m (50 ft)	600 m (1,970 ft)
Landing from 15 m (50 ft)	410 m (1,345 ft)
Landing run	290 m (952 ft)
Range, 45 min reserves:	
with max payload	410 nm (760 km; 472 miles)
with max fuel	960 nm (1,780 km; 1,105 miles)

ROKS-AERO T-433 FLAMINGO

TYPE: Single-engined light multi-purpose amphibian.
PROGRAMME: Design started May 1991; construction of first prototype scheduled to begin July 1993.
CURRENT VERSIONS: Passenger/cargo transport; civil variants for economic zone patrol and fish survey, search and rescue, ecology monitoring and training; military patrol and reconnaissance.
DESIGN FEATURES: Conventional light amphibian, with boat hull, mid-mounted wings; tailplane mounted high on sweptback fin; wide-track retractable tricycle landing gear; stabilising float under each outer wing; engine pod on pylon above centre-fuselage; unpressurised cabin. Wing section P301, dihedral 4°, incidence 3° at root, 0° at tip.
FLYING CONTROLS: Conventional mechanical control, via rods. Trim tab in each aileron, port elevator and rudder. Single-slotted Fowler type flap in two sections on each wing.
STRUCTURE: All-metal (aluminium alloy and high-tensile steel). Two-spar wing with ribs, stringers and metal skin; semi-monocoque fuselage with single-step planing bottom.
LANDING GEAR: Retractable tricycle type; single wheel on each unit; hydraulic retraction, mainwheels inward into wing-roots, nosewheel forward into wheelbay in hull; oleo-pneumatic shock absorber in each unit; mainwheel tyres size 500 × 150 mm, nosewheel tyre size 400 × 150 mm; pneumatic brakes on mainwheels. Min ground turning radius 8.0 m (26 ft 3 in). Nosewheel steering angle ±35°.
POWER PLANT: One 265 kW (355 hp) M-14P nine-cylinder air-cooled radial piston engine; V530 two-blade constant-speed propeller. Two wing fuel tanks, total capacity 300 litres (79 US gallons; 66 Imp gallons); refuelling points above wings. Oil capacity 30 litres (7.9 US gallons; 6.6 Imp gallons).
ACCOMMODATION: Pilot and passenger on individual front seats; three-place rear bench seat; passenger seats removable for other duties. Baggage hold behind rear seats. Upward opening canopy door each side. Cabin heated by engine bleed air and ventilated.

DIMENSIONS, EXTERNAL:
Wing span	14.20 m (46 ft 7 in)
Wing chord: at root	1.891 m (6 ft 2½ in)
at tip	1.022 m (3 ft 4¼ in)
Wing aspect ratio	9.75
Length overall	10.62 m (34 ft 10¼ in)
Fuselage: Max width	1.60 m (5 ft 3 in)
Max height	1.65 m (5 ft 5 in)
Height overall	3.93 m (12 ft 10¾ in)
Tailplane span	4.96 m (16 ft 3¼ in)
Wheel track	5.25 m (17 ft 2¾ in)
Wheelbase	3.99 m (13 ft 1 in)
Propeller diameter	2.40 m (7 ft 10½ in)
Canopy doors (each): Height	0.65 m (2 ft 1½ in)
Width at top	1.30 m (4 ft 3 in)
Width at bottom	1.00 m (3 ft 3¼ in)

DIMENSIONS, INTERNAL:
Cabin: Length	2.50 m (8 ft 2½ in)
Max width	1.50 m (4 ft 11 in)
Max height	1.20 m (3 ft 11¼ in)
Floor area	3.60 m² (38.75 sq ft)
Volume	4.10 m³ (145 cu ft)
Baggage hold: Volume	0.60 m³ (21 cu ft)
Freight volume, seats removed	2.50 m³ (88 cu ft)

AREAS:
Wings, gross	20.68 m² (222.6 sq ft)
Ailerons (total)	1.044 m² (11.24 sq ft)
Trailing-edge flaps (total)	4.36 m² (46.93 sq ft)
Fin, incl dorsal fin	2.67 m² (28.74 sq ft)
Rudder	1.56 m² (16.79 sq ft)
Tailplane	2.49 m² (26.80 sq ft)
Elevators (total)	1.65 m² (17.76 sq ft)

WEIGHTS AND LOADINGS:
Weight empty, equipped	1,470 kg (3,240 lb)
Max payload	370 kg (815 lb)
Max fuel	220 kg (485 lb)
Max T-O and landing weight	2,050 kg (4,520 lb)
Max wing loading	99.13 kg/m² (20.31 lb/sq ft)
Max power loading	7.74 kg/kW (12.73 lb/hp)

PERFORMANCE (estimated, at max T-O weight):
Max level speed at S/L	132 knots (245 km/h; 152 mph)
Max cruising speed at 3,000 m (9,840 ft)	
	124 knots (230 km/h; 143 mph)
Econ cruising speed at 3,000 m (9,840 ft)	
	97 knots (180 km/h; 112 mph)
Stalling speed, flaps up at S/L	
	62 knots (115 km/h; 72 mph)
Max rate of climb at S/L	255 m (835 ft)/min
Service ceiling	5,900 m (19,350 ft)
T-O run: on land	205 m (675 ft)
on water	300 m (985 ft)
T-O to 15 m (50 ft): on land	365 m (1,200 ft)
on water	500 m (1,640 ft)
Landing from 15 m (50 ft): on land	445 m (1,460 ft)
on water	475 m (1,560 ft)
Landing run: on land	225 m (740 ft)
on water	270 m (885 ft)
Range with max payload, 30 min reserves:	
at 3,000 m (9,840 ft)	340 nm (630 km; 390 miles)
at S/L	237 nm (440 km; 273 miles)
Range with max fuel, 30 min reserves:	
at 3,000 m (9,840 ft)	377 nm (700 km; 435 miles)
at S/L	485 nm (900 km; 560 miles)
Max range, no reserves:	
at 3,000 m (9,840 ft)	500 nm (930 km; 578 miles)
at S/L	595 nm (1,100 km; 685 miles)

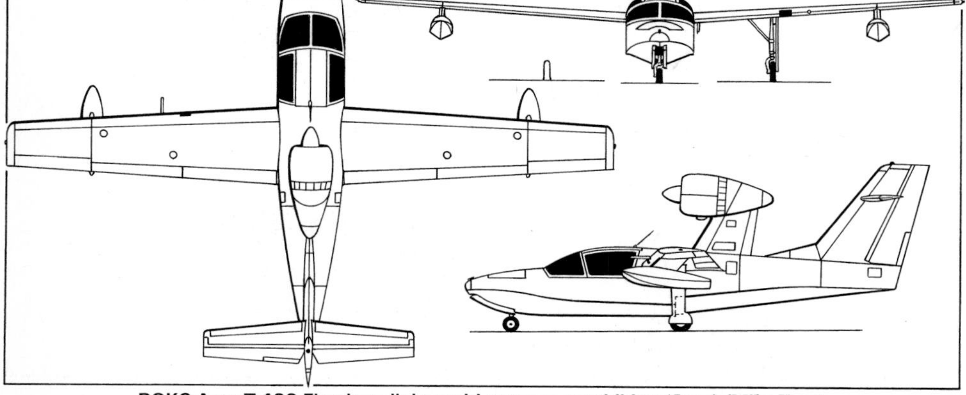

ROKS-AERO T-501

TYPE: Tandem two-seat turboprop basic trainer.
PROGRAMME: Details released and model displayed March 1992; construction of prototype started by Mikoyan in April 1992, under $22 million contract from RAP; contract also covers second prototype and static test airframe; prototypes scheduled to fly March and June 1993 respectively; CIS air forces envisage possible need for 800 T-501s.
DESIGN FEATURES: Conventional low-wing monoplane; unswept wings without dihedral or anhedral; sweptback vertical tail surfaces; unswept horizontal tail surfaces; retractable tricycle landing gear; conventional raised rear cockpit; for other details see accompanying three-view drawing.
POWER PLANT: One 754 kW (1,010 shp) Mars (Omsk) TVD-10B turboprop. Provision for drop tank under each wing.
ARMAMENT: Provision for underwing armament.

DIMENSIONS, EXTERNAL:
Wing span	11.00 m (36 ft 1 in)
Wing aspect ratio	7.33
Length overall	9.66 m (31 ft 8½ in)

AREAS:
Wings, gross	16.5 m² (177.6 sq ft)

WEIGHTS AND LOADINGS:
Max fuel weight	500 kg (1,102 lb)
Max external load	500 kg (1,102 lb)
Max T-O weight	2,670 kg (5,886 lb)
Max wing loading	161.8 kg/m² (33.14 lb/sq ft)
Max power loading	3.50 kg/kW (5.74 lb/shp)

PERFORMANCE (estimated):
Max level speed	
	285-307 knots (530-570 km/h; 330-354 mph)
T-O speed	73 knots (135 km/h; 84 mph)
Landing speed	65 knots (120 km/h; 75 mph)
Max rate of climb at S/L	1,260 m (4,135 ft)/min
T-O run	160 m (525 ft)
Landing run	190 m (625 ft)

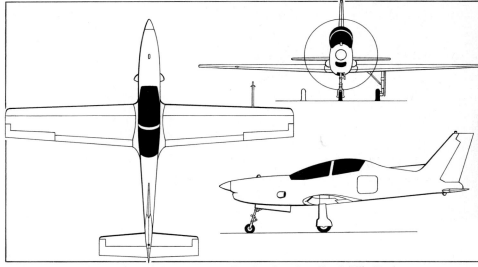

ROKS-Aero T-501 turboprop military basic trainer (Jane's/Mike Keep)

ROKS-Aero T-433 Flamingo light multi-purpose amphibian (Jane's/Mike Keep)

Model of ROKS-Aero T-501 with underwing armament *(Piotr Butowski)*

Min turning radius	90 m (295 ft)
Min 360° turn time	11 s
Max range with external fuel:	
at high altitude	970 nm (1,800 km; 1,115 miles)
at low altitude	540 nm (1,000 km; 620 miles)

ROKS-AERO T-602 OREL (EAGLE)

TYPE: Twin-engined light business aircraft.
DESIGN FEATURES: Conventional low-wing monoplane, with twin engines wing-mounted; large two-section passenger/freight door aft of wing on port side; swept vertical tail surfaces; unswept horizontal tail surfaces on fuselage tailcone; unswept wings with slight dihedral; for other details see accompanying three-view drawing.
LANDING GEAR: Retractable tricycle type; single wheel on each main unit; rearward-retracting twin nosewheels.
POWER PLANT: Two unspecified 265 kW (355 hp) engines, each driving a three-blade propeller.
ACCOMMODATION: One or two pilots; passengers or freight in main cabin.

DIMENSIONS, EXTERNAL:
Wing span	13.66 m (44 ft 9¾ in)
Length overall	12.12 m (39 ft 9 in)

WEIGHTS AND LOADINGS:
Max fuel weight	720 kg (1,585 lb)
Max payload	800 kg (1,764 lb)
Max T-O weight	3,200 kg (7,055 lb)

PERFORMANCE (estimated):
Max level speed	188 knots (350 km/h; 217 mph)
Max cruising speed	172 knots (320 km/h; 199 mph)
T-O run	380 m (1,247 ft)
Landing run	475 m (1,560 ft)
Range: with max payload	852 nm (1,580 km; 980 miles)
with auxiliary fuel	1,208 nm (2,240 km; 1,390 miles)
Endurance: with max payload	7 h
with max fuel and 1,000 kg (2,205 lb) payload	10 h

ROKS-AERO T-610 VOYAGE

TYPE: Single-turboprop civil and military multi-purpose aircraft.
PROGRAMME: Design started November 1991; construction of first prototype began October 1992; first flight scheduled August 1993.
CURRENT VERSIONS: Suitable for civil passenger/freight transport, forest patrol, aerial photography and ambulance missions; military transport and airdropping of paratroops, equipment and supplies.
DESIGN FEATURES: Conventional strut braced high-wing monoplane; sweptback vertical tail surfaces; non-retractable landing gear; unpressurised accommodation; large freight door. STOL capable. Wide CG range. Wing section P301, dihedral 3°, incidence 1° 30′ at root, −1° 30′ at tip.
FLYING CONTROLS: Conventional mechanical control, via cables and rods; ailerons supplemented by small single-section upper-surface spoilers forward of outer flap sections; long-span two-section Fowler type flap on each wing; trim tab in each aileron, port elevator and rudder.
STRUCTURE: Two-spar wings, with single bracing strut each side; fuel tanks in wingroots; semi-monocoque fuselage. Basic structure aluminium alloy, with some high-tensile steel.
LANDING GEAR: Non-retractable tricycle landing gear, with single wheel on each leg; mainwheels carried on cantilever steel leaf-springs; nosewheel on leaf-spring, with oleo damper; mainwheel tyres size 700 × 200 mm; nosewheel tyre size 480 × 200 mm; hydraulic brakes on mainwheels. Floats and skis optional. Min ground turning radius 6.5 m (21 ft 4 in). Nosewheel steering angle ±40°.
POWER PLANT: One 560 kW (751 shp) Motorlet M 601E turboprop in each prototype; one 529 kW (710 shp) Mars (Omsk) TV-O-100, or one 610 kW (818 shp) Saturn AL-34, or one P&WC PT6A-114, in production aircraft; V510 five-blade constant-speed propeller with reverse thrust. Wing fuel tanks, total capacity 1,300 litres (343 US gallons; 286 Imp gallons).
ACCOMMODATION: Eleven seats, arranged 2-3-3-2-1; optional twelfth seat; front three passenger seats rearward-facing. Horizontally divided two-piece passenger door on port side at rear of cabin; airstairs in lower section; large rearward sliding cargo door on starboard side; forward hinged door each side of flight deck. Baggage/freight compartments forward of flight deck and at rear of cabin. Interior heated by engine bleed air and ventilated.

DIMENSIONS, EXTERNAL:
Wing span	16.16 m (53 ft 0¼ in)
Wing chord: at root	2.228 m (7 ft 3¾ in)
at tip	1.238 m (4 ft 0¾ in)
Wing aspect ratio	9.33
Length overall	12.08 m (39 ft 7½ in)
Fuselage: Max width	1.72 m (5 ft 7¾ in)
Max height	1.70 m (5 ft 7 in)
Height overall	4.50 m (14 ft 9 in)
Tailplane span	6.52 m (21 ft 4¾ in)
Wheel track	3.64 m (11 ft 11½ in)
Wheelbase	3.48 m (11 ft 5 in)
Propeller diameter	2.30 m (7 ft 6½ in)
Propeller ground clearance	0.40 m (1 ft 3¾ in)
Crew doors (each): Height	1.20 m (3 ft 11¼ in)
Width at top	0.52 m (1 ft 8½ in)
Width at bottom	0.84 m (2 ft 9 in)
Cabin door, port: Height	1.20 m (3 ft 11¼ in)
Width	0.64 m (2 ft 1¼ in)
Cargo door, starboard: Height	1.20 m (3 ft 11¼ in)
Width	1.30 m (4 ft 3¼ in)

DIMENSIONS, INTERNAL:
Cabin, incl flight deck and toilet:	
Length	5.00 m (16 ft 4¾ in)
Max width	1.60 m (5 ft 3 in)
Max height	1.40 m (4 ft 7 in)
Floor area	7.00 m² (75.35 sq ft)
Volume	9.80 m³ (346 cu ft)
Baggage compartment volume:	
Nose	0.50 m³ (17.65 cu ft)
Rear	0.70 m³ (24.72 cu ft)
Freight volume, seats removed	6.80 m³ (240 cu ft)

AREAS:
Wings, gross	28.00 m² (301.4 sq ft)
Ailerons (total)	1.213 m² (13.06 sq ft)
Trailing-edge flaps (total)	6.38 m² (68.68 sq ft)
Spoilers (total)	0.303 m² (3.26 sq ft)
Fin, incl dorsal fin	2.388 m² (25.71 sq ft)
Rudder	1.312 m² (14.12 sq ft)
Tailplane	4.28 m² (46.07 sq ft)
Elevators (total)	2.80 m² (30.14 sq ft)

WEIGHTS AND LOADINGS:
Weight empty, equipped	1,955 kg (4,310 lb)
Max payload	1,100 kg (2,425 lb)
Max fuel	1,000 kg (2,205 lb)
Max T-O weight	3,850 kg (8,488 lb)
Max landing weight	3,540 kg (7,805 lb)
Max wing loading	137.5 kg/m² (28.16 lb/sq ft)
Max power loading (M 601E engine)	6.88 kg/kW (11.30 lb/shp)

PERFORMANCE (estimated):
Stalling speed, flaps up, at 3,000 m (9,840 ft)	81 knots (150 km/h; 93 mph)
Max rate of climb at S/L	336 m (1,102 ft)/min
Service ceiling	10,000 m (32,800 ft)
T-O run	278 m (912 ft)
T-O to 15 m (50 ft)	530 m (1,740 ft)
Landing from 15 m (50 ft):	
with reverse thrust	256 m (840 ft)
without reverse thrust	826 m (2,710 ft)
Range with max payload, 30 min reserves:	
at 3,000 m (9,840 ft)	647 nm (1,200 km; 745 miles)
at 6,000 m (19,685 ft)	863 nm (1,600 km; 995 miles)
Range with max fuel, 30 min reserves:	
at 3,000 m (9,840 ft)	998 nm (1,850 km; 1,150 miles)
at 6,000 m (19,685 ft)	1,257 nm (2,330 km; 1,447 miles)
Max range, no reserve:	
at 3,000 m (9,840 ft)	1,250 nm (2,320 km; 1,440 miles)
at 6,000 m (19,685 ft)	1,619 nm (3,000 km; 1,864 miles)

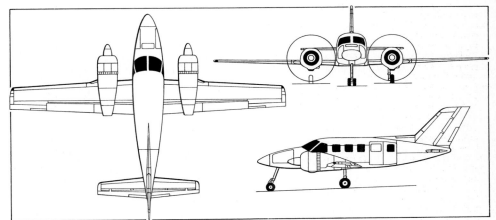

ROKS-Aero T-602 Orel light business transport *(Jane's/Mike Keep)*

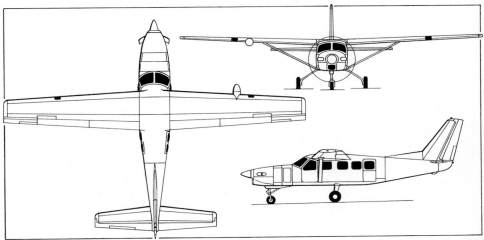

ROKS-Aero T-610 Voyage turboprop multi-purpose aircraft *(Jane's/Mike Keep)*

SUKHOI
SUKHOI DESIGN BUREAU AVIATION SCIENTIFIC-INDUSTRIAL COMPLEX

23A Polikarpov Street, Moscow 125284
Telephone: 7 (095) 945 65 25
Fax: 7 (095) 200 42 43
Telex: 414716 SUHOI SU
GENERAL DESIGNER AND CEO: Mikhail Petrovich Seemonov
FIRST DEPUTY GENERAL DESIGNER: Mikhail A. Pogosian
DIRECTOR, MANUFACTURER: Vladimir N. Avramenko
DEPUTY GENERAL DESIGNERS:
 Alexander F. Barkovsky
 Nikolai F. Nikitin
 Aleksei I. Knishev
 Boris V. Rakitin (Sport Aviation Projects)
 Vladimir M. Korchagin (Avionics)
 Alexander I. Blinov (Strength Problems)

OKB named for Pavel Osipovich Sukhoi, who headed it from 1939 until his death in September 1975. It remains one of two primary Russian centres for development of fighter and attack aircraft, and is widening its activities to include civilian aircraft, under konversiya programme, in some cases with Western partners.

Sukhoi also active in design and construction of large ground effect vehicles, including 100/150-passenger A.90.150 Ekranoplan described in *Jane's High-Speed Marine Craft 1993-94.*

Sukhoi Su-22M-4 ('Fitter-K') of the Afghan Air Force *(Peter Steinemann)*

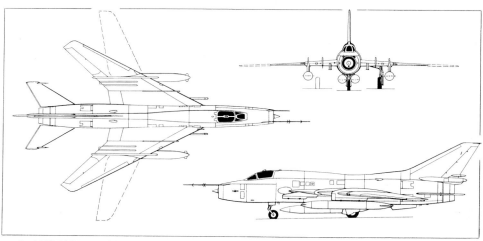

Su-17M-4 ('Fitter-K'), final single-seat version of the variable geometry 'Fitter' series *(Dennis Punnett)*

SUKHOI Su-17, Su-20 and Su-22
NATO reporting names: Fitter-C, D, E, F, G, H, J and K

TYPE: Single-seat variable geometry ground attack fighter, reconnaissance aircraft and two-seat combat trainer.

PROGRAMME: Prototype S-22I or Su-7IG (Izmenyaemaya Geometriya; variable geometry) was minimal conversion of fixed-wing Su-7 (NATO 'Fitter-A'); only 4.2 m (13 ft 9 in) of each wing pivoted, outboard of large fence and deepened inboard glove panel; first flew 2 August 1966; shown at Aviation Day display July 1967; given NATO reporting name 'Fitter-B'; two squadrons of **Su-17** 'improved Fitter-Bs' in Soviet air forces 1972; AL-21F-3 engine then replaced AL-7 in major Soviet air force production versions, beginning with 'Fitter-C'. Production ended 1991.

CURRENT VERSIONS: **Su-17M** (S-32M, 'Fitter-C'): Single-seat attack aircraft; AL-21F-3 engine; eight stores pylons; additional wing fence on each glove panel; curved dorsal fin; operational with CIS air forces and Naval Aviation since 1971 in relatively small numbers. *Detailed description applies to basic Su-17, except where indicated otherwise.*

Su-17R: Reconnaissance version of Su-17M.

Su-17M-2/M-2D (S-32M2, 'Fitter-D'): Generally as Su-17M, but forward fuselage lengthened by 0.38 m (15 in) and drooped 3° to improve pilot's view while keeping intake face vertical; added undernose Doppler navigation radar pod; Klen laser rangefinder in intake centrebody.

Su-17UM-3 ('Fitter-G'): Two-seat trainer version of Su-17M-3 with combat capability; drooped forward fuselage and deepened spine like Su-17UM-2D; taller vertical tail surfaces; removable ventral fin; starboard wingroot gun only; laser rangefinder in intake centrebody.

Su-17M-3 (S-52, 'Fitter-H'): Improved single-seater; same deepened spine and tail modifications as Su-17UM-3; Doppler navigation radar internally in deepened undersurface of nose; gun in each wingroot; launcher for R-60 (AA-8 'Aphid') air-to-air missile between each pair of underwing pylons; approx 165 'Fitter-H/Ks' equipped for tactical reconnaissance carry, typically, centreline sensor pod, active ECM pod under port wing glove, two underwing fuel tanks.

Su-17M-4 (S-54, 'Fitter-K'): Single-seat version, identified 1984; cooling air intake at front of dorsal fin; otherwise as Su-17M-3. Max weapon load 4,250 kg (9,370 lb), including nuclear weapons, bombs, rocket pods, S-25 tube-launched rockets with 325 mm head, 23 mm SPPU-22 gun pods, two R-3 or R-13M (AA-2 'Atoll'), R-60 (AA-8 'Aphid') or R-73A (AA-11 'Archer') air-to-air missiles, Kh-23 (AS-7 'Kerry') or Kh-25ML (AS-10 'Karen') air-to-surface missiles, or a reconnaissance pod. When four SPPU-22 gun pods are fitted, with downward attack capability, the two underfuselage pods can be arranged to fire rearward. Chaff/flare and decoy dispensers standard.

Su-20 (S-32MK, 'Fitter-C'): Export version of Su-17M.

Su-20R: Reconnaissance version of Su-20.

All Su-17s and Su-20s have AL-21F-3 engine; some Su-22 export aircraft have Tumansky R-29BS-300 (112.8 kN; 25,350 lb st with afterburning) in more bulged rear fuselage, with rearranged small external air intakes on rear fuselage and shorter plain metal shroud terminating fuselage, as follows:

Su-22U ('Fitter-E'): Tandem two-seat trainer developed from Su-17M-2, with Tumansky engine; no Doppler pod; deepened dorsal spine fairing for additional fuel tankage; port wingroot gun deleted.

Su-22 ('Fitter-F'): Export Su-17M-2; modified undernose electronics pod, R-29 engine; gun in each wingroot; weapons include R-3 (AA-2 'Atoll') air-to-air missiles; aircraft supplied to Peru had Sirena-2 limited coverage radar warning system and virtually no navigation aids; some basic US-supplied avionics retrofitted.

Su-22UM-3K ('Fitter-G'): Export Su-17UM-3; AL-21F-3 or R-29B engine.

Su-22M-3 ('Fitter-J'): As Su-17M-3 but R-29 engine; internal fuel tankage 6,270 litres (1,656 US gallons; 1,379 Imp gallons); more angular dorsal fin; AA-2 ('Atoll') air-to-air missiles.

Su-22M-4 ('Fitter-K'): As Su-17M-4; AL-21F-3 engine.

CUSTOMERS: CIS air forces (many of 1,060 Su-17s of all versions deployed in late 1980s since put into storage, assigned to training schools and passed to Naval Aviation to supplement 75 deployed originally for anti-shipping strike and amphibious support in Baltic Sea and Pacific areas); air forces of Afghanistan (Su-22M-4), Algeria (Su-20), Angola (Su-22), Czech Republic (Su-20/22M-4), the former East Germany (Su-22/22M-4), Egypt (Su-20), Hungary (Su-22), Iraq (Su-20), North Korea (Su-22), Libya (Su-22), Peru (Su-22), Poland (Su-20/22M-4), Slovakia (Su-22M-4), Syria (Su-22), Vietnam (Su-22), Yemen (Su-22).

DESIGN FEATURES: Modest amount of variable geometry added to original fixed-wing Su-7 permitted doubled external load from strips little more than half as long, and 30 per cent greater combat radius; progressive refinements led to very effective final versions. Conventional mid-wing all-swept monoplane, except for variable geometry outer wings with manually selected positions of 30°, 45°, 63°; wide-span fixed centre-section glove panels; basically circular fuselage with dorsal spine; ram intake with variable shock-cone centrebody; pitot on port side of nose, transducer to provide pitch and yaw data for fire control computer starboard; anti-flutter bodies near tailplane tips.

ECM pod under wing of Su-22M-4 ('Fitter-K') *(Letectvi + Kosmonautika/ Václav Jukl)*

Polish Air Force Su-22M-4 ('Fitter-K') armed with forward and rearward firing SPPU-22 gun pods, rocket pods and R-60 'Aphid' air-to-air missiles *(Piotr Butowski)*

FLYING CONTROLS: Slotted ailerons operable at all times; slotted trailing-edge flap on each variable geometry wing panel operable only when wings spread; area-increasing flap on each centre-section glove panel; full-span leading-edge slats on variable geometry wing panels; top and bottom door-type airbrakes each side of rear fuselage, forward of tailplane; all-moving horizontal tail surfaces; conventional rudder; no tabs.

STRUCTURE: All-metal; semi-monocoque fuselage; large main wing fence on each side, at junction of fixed and movable panels, square-cut at front, with attachment for external store; shorter fence above glove panel each side.

LANDING GEAR: Retractable tricycle type, with single wheel on each unit; nosewheel retracts forward, requiring blistered door to enclose it; main units retract inward into centre-section. Container for single cruciform brake-chute between base of rudder and tailpipe.

POWER PLANT: One Saturn/Lyulka AL-21F-3 turbojet, 76.5 kN (17,200 lb st) dry and 110.3 kN (24,800 lb st) with afterburning. Fuel capacity increased to 4,550 litres (1,202 US gallons; 1,000 Imp gallons) by added tankage in dorsal spine fairing; provision for up to four 800 litre (211 US gallon; 176 Imp gallon) PTB-800 drop tanks on outboard wing pylons and under fuselage; when underfuselage tanks carried, only the two inboard wing pylons may be used for ordnance, to total 1,000 kg (2,204 lb). Two solid propellant rocket units attached optionally to rear fuselage to shorten T-O run.

ACCOMMODATION: Pilot only, on ejection seat, under rearward hinged transparent canopy; rearview mirror above canopy.

AVIONICS: SRD-5M (NATO 'High Fix') I-band ranging radar in intake centrebody; ASP-5ND fire control system; HUD standard; Sirena-3 radar warning system providing 360° coverage, with antennae in slim cylindrical housing above brake-chute container and in each centre-section leading-edge, between fences; SRO-2M (NATO 'Odd Rods') IFF transponder; SOD-57M ATC/SIF, with transponder housing beneath brake-chute container; SP-50 ILS, RSB-70 HF and RSIU-5/R-831 UHF/VHF.

EQUIPMENT: Front and rear pairs of ASO chaff/flare dispensers to each side of dorsal spine fairing; KDS decoy dispensers in forward section of spine fairing.

ARMAMENT: Two 30 mm NR-30 guns, each with 80 rds, in wingroot leading-edges; nine weapon pylons (one on centreline, two tandem pairs under fuselage, one under each centre-section leading-edge, one under each main wing fence) for more than 3,175 kg (7,000 lb) of bombs, including nuclear weapons, rocket pods, 23 mm gun pods and guided missiles such as air-to-surface Kh-23 (NATO AS-7 'Kerry'), AS-9 ('Kyle') and Kh-25ML (AS-10 'Karen').

DIMENSIONS, EXTERNAL:
Wing span: fully spread	13.80 m (45 ft 3 in)
fully swept	10.00 m (32 ft 10 in)
Wing aspect ratio: fully spread	4.8
fully swept	2.7
Length overall, incl probes	18.75 m (61 ft 6¼ in)
Fuselage length	15.40 m (50 ft 6¼ in)
Height overall	5.00 m (16 ft 5 in)

AREAS (estimated):
Wings, gross: fully spread	40.0 m² (430.0 sq ft)
fully swept	37.0 m² (398.0 sq ft)

WEIGHTS AND LOADINGS (Su-17M-4):
Max external stores	4,250 kg (9,370 lb)
Normal T-O weight	16,400 kg (36,155 lb)
Max T-O weight	19,500 kg (42,990 lb)
Max wing loading: spread	487.5 kg/m² (100 lb/sq ft)
swept	527 kg/m² (108 lb/sq ft)
Max power loading	177.3 kg/kN (1.74 lb/lb st)

PERFORMANCE (Su-17M-4):
Max level speed: at height	Mach 2.09
at S/L	Mach 1.14
	(755 knots; 1,400 km/h; 870 mph)
Service ceiling	15,200 m (49,865 ft)
T-O run	900 m (2,955 ft)
Landing run	950 m (3,120 ft)
Range with max fuel:	
at high altitude	1,240 nm (2,300 km; 1,430 miles)
at low altitude	755 nm (1,400 km; 870 miles)

SUKHOI Su-24
NATO reporting name: Fencer

TYPE: Two-seat variable geometry 'frontal bomber', reconnaissance and EW aircraft.

PROGRAMME: Design started 1964 under Yevgeniy S. Felsner, Pavel Sukhoi's successor, to replace Il-28 and Yak-28 attack aircraft; T-6-1 prototype, first flown June 1967 and now at Monino, had fixed delta wings with downswept tips, and four Koliesov auxiliary booster motors mounted vertically in fuselage for improved take-off performance; T-6-2IG variable geometry prototype chosen for production; first flight January 1970; by 1981, delivery rate 60-70 a year; production of Su-24M/MR/MP continues.

CURRENT VERSIONS: **Su-24** ('Fencer-A'): Had rectangular rear fuselage box enclosing jet nozzles; few early aircraft only; deployed with trials unit 1974.

Su-24 ('Fencer-B'): First operational version, 1976; deeply dished bottom skin to rear fuselage between jet nozzles; larger brake-chute housing at base of rudder.

Su-24 ('Fencer-C'): Introduced 1981; important avionics changes; multiple nose fitting instead of former simple probe; triangular fairing for RWR on side of each engine air intake, forward of fixed wingroot, and each side of fin tip; chord of fin leading-edge extended forward, except at tip, giving kinked profile.

Su-24M ('Fencer-D'): Major attack version; first flew

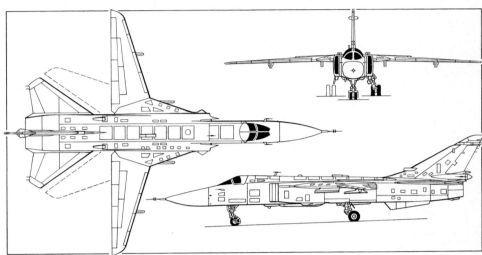

Sukhoi Su-24M ('Fencer-D') variable geometry attack aircraft *(Jane's/Mike Keep)*

Sukhoi Su-24M ('Fencer-D') of Russian Air Force *(R. J. Malachowski)*

Fully equipped Sukhoi Su-24MR ('Fencer-E') maritime reconnaissance aircraft *(Swedish Air Force, via FLYGvapenNYTT)*

Cockpit of Sukhoi Su-24MR ('Fencer-E') *(Brian M. Service)*

Height overall	6.19 m (20 ft 3¾ in)
Wheel track	3.31 m (10 ft 10½ in)
Wheelbase	8.51 m (27 ft 11 in)

AREAS:
Wings, gross: 16° sweep	55.17 m² (593.8 sq ft)
69° sweep	51.02 m² (549.2 sq ft)

WEIGHTS AND LOADINGS (Su-24MK):
Weight empty, equipped	19,000 kg (41,885 lb)
Max internal fuel	9,764 kg (21,525 lb)
Max external stores	8,000 kg (17,635 lb)
Normal T-O weight	36,000 kg (79,365 lb)
Max T-O weight	39,700 kg (87,520 lb)
Max wing loading	945.2 kg/m² (193.6 lb/sq ft)
Max power loading	180.5 kg/kN (1.77 lb/lb st)

PERFORMANCE (Su-24MK):
Max level speed, clean: at height	Mach 1.35
at S/L	Mach 1.08 (712 knots; 1,320 km/h; 820 mph)
Service ceiling	17,500 m (57,400 ft)
T-O run	1,300 m (4,265 ft)
Landing run	950 m (3,120 ft)
Combat radius:	
lo-lo-lo	over 174 nm (322 km; 200 miles)
lo-lo-hi with 2,500 kg (5,500 lb) of weapons	515 nm (950 km; 590 miles)
hi-lo-hi, with 3,000 kg (6,615 lb) of weapons and two external tanks	565 nm (1,050 km; 650 miles)
g limit	+6.5

1977; entered service 1983; believed to have terrain-following radar instead of earlier terrain-avoidance system; added flight refuelling capability, with centrally mounted retractable probe forward of windscreen; nose approx 0.75 m (2 ft 6 in) longer to accommodate new avionics bay; large overwing fences with integral extended wingroot glove pylons when carrying Kh-29 (NATO AS-14 'Kedge') missiles; undernose antennae deleted; laser ranger/designator housing aft of nosewheel bay; single long noseprobe; export version is **Su-24MK**.

Su-24MR ('Fencer-E'): Reconnaissance version of Su-24M used by tactical and naval air forces; internal equipment includes Shtik side-looking airborne multi-mission radar in nose, Zima IR reconnaissance system, Aist-M TV reconnaissance system, panoramic and oblique cameras in ventral fairing. A Shpil-2M laser pod can be carried on centreline, with a Tangazh elint pod or Efir-1M radiation detector pod on starboard underwing swivelling pylon and two R-60 air-to-air missiles under port wing. Data can be transmitted to ground by data link. No over-wing fences; shorter nose radome, with flush dielectric side panels on nose; domed centre-fuselage air intake for heat exchanger; 'hockey stick' antenna at bottom of fuselage under each engine air intake nose section; provision for two 3,000 litre (792 US gallon; 660 Imp gallon) underwing auxiliary fuel tanks; flight refuelling and air-to-surface missile capabilities retained; tactical air force units include two squadrons on Chinese border; deliveries to Baltic fleet, replacing Tu-16s, began Summer 1985.

Su-24MP ('Fencer-F'): Electronic warfare/jamming/sigint version to replace Yak-28 'Brewer-E'; dielectric nose panels differ from those of Su-24MR; added small fairing under nose; no underside electro-optics; 'hockey stick' antennae as on Su-24MR; centreline EW pod.

CUSTOMERS: More than 900 delivered from Komsomolsk factory; about 25 per cent form primary strike components of CIS Air Armies; reassignment of other Air Army Su-24s increased capability of Military Districts/Groups of Forces and Naval Aviation forces; often replacing Su-17s and operating with 'Wild Weasel' MiG-25BMs; others with air forces of Iraq (most of 24 now in Iran, which may order more), Libya (15), Syria (12 ordered).

DESIGN FEATURES: Variable geometry shoulder wing; slight anhedral from roots; triangular fixed glove box; three-position (16°, 45°, 69°) pivoted outer panels, each with pivoting stores pylon; slab-sided rectangular section fuselage; integral engine air intake trunks, each with splitter plate and outer lip inclined slightly downward; chord of lower part of tail fin extended forward, giving kinked leading-edge; leading-edge sweepback 59° 30′ on fin, 30° on inset rudder, 50° on horizontal tail surfaces; basic operational task, as designated 'frontal bomber', to deliver wide range of air-to-surface missiles for defence suppression, with some hard target kill potential; specially developed long-range navigation system and electro-optical weapons systems make possible penetration of hostile airspace at night or in adverse weather with great precision, to deliver ordnance within 55 m (180 ft) of target.

FLYING CONTROLS: Full-span leading-edge slats, drooping aileron and two-section double-slotted trailing-edge flaps on each outer wing panel; differential spoilers forward of flaps for roll control at low speeds and use as lift dumpers on landing; airbrake under each side of centre-fuselage; inset rudder; all-moving horizontal tail surfaces operate collectively for pitch control, differentially for roll control, assisted by wing spoilers except when wings fully swept.

STRUCTURE: All-metal; semi-monocoque fuselage; two slightly splayed ventral fins.

LANDING GEAR: Retractable tricycle type, with twin wheels on each unit; main units retract forward and inward into air intake duct fairings; nose unit retracts rearward. Trailing-link main units; mainwheel tyres size 950 × 300 mm, pressure 11.75 bars (170 lb/sq in); nosewheel tyres size 660 × 200 mm; mudguard on nosewheels; cruciform brake-chute.

POWER PLANT: Two Saturn/Lyulka AL-21F-3A turbojets, each 109.8 kN (24,690 lb st) with afterburning; fixed engine air intakes (hence lower max speed now confirmed). Internal fuel capacity 11,700 litres (3,090 US gallons; 2,574 Imp gallons) can be supplemented by four 1,250 litre (330 US gallon; 275 Imp gallon) external tanks on underbelly and glove pylons. Probe-and-drogue flight refuelling capability, including operation as buddy tanker.

ACCOMMODATION: Crew of two (pilot and weapon systems officer) side by side on K-36 ejection seats; cockpit width 1.65 m (5 ft 5 in); jettisonable canopy, hinged to open upward and rearward in two panels, split on centreline.

AVIONICS: Large radar in nose; laser ranger/designator under front fuselage; radar warning receivers on sides of engine air intakes and tail fin; LO-82 Mak-UL missile warning receiver above centre-fuselage; ram air intakes for heat exchangers above centre-fuselage and at base of fin; Geran-F active jamming system.

ARMAMENT (Su-24M): Nine pylons under fuselage, each wingroot glove and outer wings for guided and unguided air-to-surface weapons, including TN-1000 and TN-1200 nuclear weapons, up to four TV or laser guided bombs, missiles such as Kh-23 (NATO AS-7 'Kerry'), Kh-25ML (AS-10 'Karen'), Kh-58 (AS-11 'Kilter'), Kh-25MP (AS-12 'Kegler'), Kh-59 (AS-13 'Kingpost'), Kh-29 (AS-14 'Kedge') and Kh-31A/P (AS-17 'Krypton'), rockets of 57 mm to 370 mm calibre, bombs (typically 38 × 100 kg FAB-100), 23 mm gun pods or external fuel tanks; two R-60 (AA-8 'Aphid') air-to-air missiles can be carried for self-defence. No internal weapons bay. One GSh-6-23M six-barrel 23 mm Gatling type gun inside fairing on starboard side of fuselage undersurface; fairing for recording camera on other side.

DIMENSIONS, EXTERNAL:
Wing span: 16° sweep	17.63 m (57 ft 10 in)
69° sweep	10.36 m (34 ft 0 in)
Wing aspect ratio: 16° sweep	5.63
69° sweep	2.10
Length overall, incl probe	24.53 m (80 ft 5¾ in)

SUKHOI Su-25 and Su-28
NATO reporting name: Frogfoot

TYPE: Single-seat close support aircraft and two-seat trainer.
PROGRAMME: Development began 1968; prototype, known as T-8-1, flew 22 February 1975, with two 25.5 kN (5,732 lb st) non-afterburning versions of Tumansky RD-9 turbojets and underbelly twin-barrel AO-17A 30 mm gun in fairing; in eventual developed form, second prototype, T-8-2, had more powerful non-afterburning versions of R-13, designated R-95Sh, wingtip avionics/speed brake pods, underwing weapon pylons, and internal gun; observed by satellite at Ramenskoye flight test centre 1977, given provisional US designation 'Ram-J'; entered production 1978 with R-95 turbojets; trials unit, followed by squadron of 12, sent to Afghanistan for co-ordinated low-level close support of Soviet ground forces in mountain terrain, with Mi-24 helicopter gunships; fully operational 1984; attack versions built initially at Tbilisi, Georgia; production at Tbilisi ended 1989 (approx 330 built); production now at Ulan-Ude; all production for CIS completed 1991-92; see separate entry on Su-25T/Su-34.

CURRENT VERSIONS: **Su-25** ('Frogfoot-A'): Single-seat close support aircraft. Export version **Su-25K** (kommercheskiy; commercial). *Detailed description applies basically to late production Su-25K.*

Su-25UB ('Frogfoot-B'): Tandem two-seat operational conversion and weapons trainer; first photographs Spring 1989; rear seat raised considerably, giving hump-back appearance; separate hinged portion of continuous framed canopy over each cockpit; taller tail fin, increasing overall height to 5.20 m (17 ft 0¾ in); new IFF blade antenna forward of windscreen instead of SRO-2 (NATO 'Odd Rods'); weapons pylons and gun retained. Export version **Su-25UBK**.

Su-25UT ('Frogfoot-B'): As Su-25UB, but without weapons; prototype first flew 6 August 1985; demonstrated 1989 Paris Air Show as **Su-28**; overall length 15.36 m (50 ft 4¾ in); few only.

Su-25UTG (G for gak; hook) ('Frogfoot-B'): As Su-25UT, with added arrester hook under tail; used initially for deck landing training on dummy flight deck

Sukhoi Su-25 ('Frogfoot-A') in lightly armed training configuration *(Piotr Butowski)*

marked on runway at Saki naval airfield, Ukraine; on 1 November 1989 was third aircraft to land for trials on carrier *Admiral of the Fleet Kuznetsov*, after Su-27K and MiG-29K; ten built in 1989-90; five in Ukrainian use at Saki; one lost; four at Severomorsk, Kola Peninsula, for service individually on *Kuznetsov*; to be supplemented by Su-25UBPs.

Su-25UBP: Ten standard Su-25UBs being converted to Su-25UBP (Palubnyi: shipborne) in 1993 for service on *Admiral of the Fleet Kuznetsov*.

Su-25T: See separate entry.

Su-25BM (BM for buksir misheney; target towing aircraft): As Su-25 attack aircraft, with added underwing pylons for rocket propelled targets released for missile training by fighter pilots.

CUSTOMERS: CIS (then Soviet) air forces lost 23 in Afghanistan, and passed some to Naval Aviation 1990; exports to Afghanistan, Bulgaria, Czechoslovakia, Hungary and Iraq (45).

DESIGN FEATURES: Shoulder-mounted wings; approx 20° sweepback; anhedral from roots; extended chord leading-edge dogtooth on outer 50 per cent each wing; wingtip pods each split at rear to form airbrakes that project above and below pod when extended; retractable landing light in base of each pod, outboard of small glareshield and aft of dielectric nosecap for ECM; semi-monocoque fuselage, with 24 mm (0.94 in) welded titanium armoured cockpit; pitot on port side of nose, transducer to provide data for fire control computer on starboard side; conventional tail unit; variable incidence tailplane, with slight dihedral.

Emphasis on survivability led to features accounting for 7.5 per cent of normal T-O weight, including armoured cockpit; pushrods instead of cables to actuate flying control surfaces (duplicated for elevators); damage-resistant main load-bearing members; widely separated engines in stainless steel bays; fuel tanks filled with reticulated foam for protection against explosion.

Maintenance system packaged into four pods for carriage on underwing pylons; covers onboard systems checks, environmental protection, ground electrical power supply for engine starting and other needs, and pressure refuelling from all likely sources of supply in front-line areas; engines can operate on any fuel likely to be found in combat area, including MT petrol and diesel oil.

FLYING CONTROLS: Hydraulically actuated ailerons, with manual backup; multiple tabs in each aileron; double-slotted two-section wing trailing-edge flaps; full-span leading-edge slats, two segments per wing; manually operated elevators and two-section inset rudder; upper rudder section operated through sensor vanes and transducers on nose probe and automatic electromechanical yaw damping system; tabs in lower rudder segment and each elevator.

STRUCTURE: All-metal; three-spar wings; semi-monocoque slab-sided fuselage.

LANDING GEAR: Hydraulically retractable tricycle type; mainwheels retract to lie horizontally in bottom of engine air intake trunks. Single wheel with low-pressure tyre on each levered suspension unit, with oleo-pneumatic shock absorber; mudguard on forward retracting steerable nosewheel, which is offset to port; mainwheel tyres size 840 × 360 mm, pressure 9.3 bars (135 lb/sq in); nosewheel tyre size 660 × 200 mm, pressure 7.35 bars (106 lb/sq in); brakes on mainwheels. Twin cruciform brake-chutes housed in tailcone.

POWER PLANT: Two Soyuz/Tumansky R-195 turbojets in long nacelles at wingroots, each 44.18 kN (9,921 lb st); 5 mm thick armour firewall between engines; current upgraded R-195 turbojets have pipe-like fitment at end of tailcone, from which air is expelled to lower exhaust temperature and so reduce infra-red signature; non-waisted under surface to rear cowlings, which have additional small airscoops (as three-view). Fuel tanks in fuselage between cockpit and wing front spar, and between rear spar and fin leading-edge, and in wing centre-section; provision for four PTB-1500 external fuel tanks on underwing pylons.

ACCOMMODATION: Single K-36L zero/zero ejection seat under sideways hinged (to starboard) canopy, with small rear-view mirror on top; flat bulletproof windscreen. Folding ladder for access to cockpit built into port side of fuselage.

SYSTEMS: 28V DC electrical system, supplied by two engine driven generators.

AVIONICS: Laser rangefinder and target designator under flat sloping window in nose. SRO-1P (NATO 'Odd Rods') or (later) SRO-2 IFF transponder, with antennae forward of windscreen and under tail. Sirena-3 radar warning system antenna above fuselage tailcone.

EQUIPMENT: ASO-2V chaff/flare dispensers (total of 256 flares) can be carried above root of tailplane and above rear of engine ducts. Strike camera in top of nosecone.

ARMAMENT: One twin-barrel AO-17A 30 mm gun with rate of fire of 3,000 rds/min in bottom of front fuselage on port side, with 250 rds (sufficient for a one second burst during each of five attacks). Eight large pylons under wings for 4,400 kg (9,700 lb) of air-to-ground weapons, including UB-32A rocket pods (each 32 × 57 mm S-5), B-8M1 rocket pods (each 20 × 80 mm S-8), 240 mm S-24 and 330 mm S-25 guided rockets, Kh-23 (NATO AS-7 'Kerry'), Kh-25 (AS-10 'Karen') and Kh-29 (AS-14 'Kedge') air-to-surface missiles, laser guided rocket-boosted 350 kg, 490 kg and 670 kg bombs, 500 kg incendiary, anti-personnel and chemical cluster bombs, and SPPU-22 pods each containing a 23 mm GSh-23 gun with twin barrels that can pivot downward for attacking ground targets, and 260 rds. Two small outboard pylons for R-3S (K-13T; NATO AA-2D 'Atoll') or R-60 (AA-8 'Aphid') air-to-air self-defence missiles.

Sukhoi Su-25UB ('Frogfoot-B') operational conversion and weapons trainer *(Lutz Freundt)*

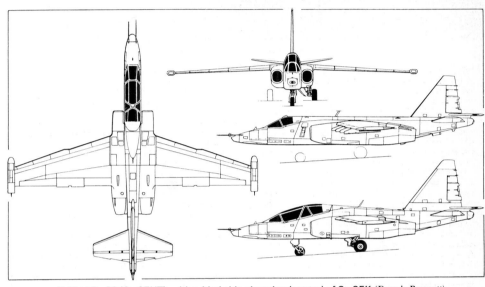

Sukhoi Su-28 (Su-25UT), with added side elevation (centre) of Su-25K *(Dennis Punnett)*

DIMENSIONS, EXTERNAL (Su-25K):
Wing span	14.36 m (47 ft 1½ in)
Wing aspect ratio	6.12
Length overall	15.53 m (50 ft 11½ in)
Height overall	4.80 m (15 ft 9 in)

AREAS:
Wings, gross	33.7 m² (362.75 sq ft)

WEIGHTS AND LOADINGS:
Weight empty	9,500 kg (20,950 lb)
Max T-O weight	14,600-17,600 kg (32,187-38,800 lb)
Max landing weight	13,300 kg (29,320 lb)
Max wing loading	522.2 kg/m² (107.0 lb/sq ft)
Max power loading	199.2 kg/kW (1.96 lb/lb st)

PERFORMANCE:
Max level speed at S/L	Mach 0.8
	(526 knots; 975 km/h; 606 mph)
Max attack speed, airbrakes open	
	372 knots (690 km/h; 428 mph)
Landing speed (typical)	108 knots (200 km/h; 124 mph)
Service ceiling: clean	7,000 m (22,965 ft)
with max weapons	5,000 m (16,400 ft)
T-O run: typical	600 m (1,970 ft)
with max weapon load from unpaved surface	
	under 1,200 m (3,935 ft)
Landing run: normal	600 m (1,970 ft)
with brake-chutes	400 m (1,312 ft)
Range with 4,400 kg (9,700 lb) weapon load and two external tanks:	
at S/L	405 nm (750 km; 466 miles)
at height	675 nm (1,250 km; 776 miles)
g limits: with 1,500 kg (3,306 lb) of weapons	+6.5
with 4,400 kg (9,700 lb) of weapons	+5.2

SUKHOI Su-25T/Su-34

TYPE: Specialised single-seat anti-tank development of Su-25.

PROGRAMME: First of three **Su-25T** (for anti-tank) development aircraft, flown August 1984, was converted Su-25UB airframe with humped rear cockpit faired over; internal space used to house new avionics and extra tonne of fuel; initial batch of ten built 1990-91 for air force acceptance testing, completed 1993; redesignated **Su-25TM** after equipment changes; first exhibited at Dubai '91 air show as export **Su-25TK**; delivery of eight to Russian Air Force 1993, under designation **Su-34**.

DESIGN FEATURES: Embodies lessons learned during war in Afghanistan; new nav system makes possible flights to and from combat areas under largely automatic control; equipment in widened nose includes TV activated some 5 nm (10 km; 6 miles) from target; subsequent target tracking, weapon selection and release automatic; wingtip countermeasures pods; gun transferred to underbelly position, on starboard side of farther-offset nosewheel.

AVIONICS: Voskhod nav/attack system and Schkval electro-optical system for precision attacks on armour; larger window in nose for TV, laser rangefinder and target designator of improved capability, all using same stabilised mirror, 23× magnification lens and cockpit CRT; HUD; nav system has two digital computers and inertial platform; radar

Cockpit of Sukhoi Su-25 *(Brian M. Service)*

Fully armed Sukhoi Su-25T with underfuselage sensor pack *(Piotr Butowski)*

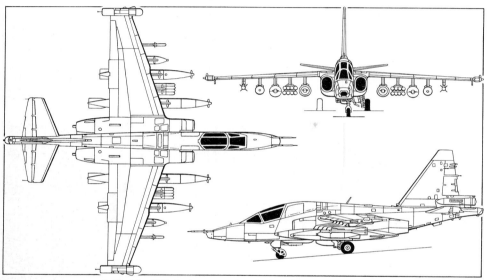

Sukhoi Su-25T single-seat anti-tank aircraft *(Jane's/Mike Keep)*

warning/emitter location system; Kinzhal (dagger) radar; Khod (motion) centreline IR pack replaces Merkuri LLLTV pack of Su-25T.

EQUIPMENT: Chaff/flare dispensers in top of fuselage tailcone and in large cylindrical housing at base of rudder that also contains IR jammer optimised against Stinger and Redeye missile frequencies.

ARMAMENT: One twin-barrel 30 mm gun; ten external stores attachments; two eight-round underwing clusters of Vikhr (AT-9) tube-launched primary attack missiles able to penetrate 900 mm of reactive armour; other weapons include laser-guided Kh-25ML (AS-10 'Karen') and Kh-29L (AS-14 'Kedge'), rocket/ramjet Kh-31A/P (AS-17 'Krypton') and anti-radiation Kh-58 (AS-11 'Kilter') air-to-surface missiles, KAB-500 laser-guided bombs, S-25L laser-guided rockets and R-60 (AA-8 'Aphid') air-to-air missiles.

DIMENSIONS, EXTERNAL:
Wing span	14.52 m (47 ft 7¾ in)
Length overall	15.33 m (50 ft 3½ in)
Height overall	5.20 m (17 ft 0¾ in)

WEIGHTS AND LOADINGS:
Max internal fuel	3,840 kg (8,465 lb)
Max combat load	4,360 kg (9,612 lb)
Max T-O weight	19,500 kg (42,990 lb)

PERFORMANCE:
Max level speed at S/L	512 knots (950 km/h; 590 mph)
Service ceiling	10,000 m (32,800 ft)
T-O and landing run: normal	600 m (1,970 ft)
unpaved runway	700 m (2,300 ft)
Combat radius with 2,000 kg (4,410 lb) weapon load:	
at S/L	215 nm (400 km; 248 miles)
at height	378 nm (700 km; 435 miles)
Ferry range	1,350 nm (2,500 km; 1,550 miles)
g limits	+6.5

SUKHOI Su-26M

TYPE: Single-seat aerobatic competition aircraft.
PROGRAMME: Su-26 prototype first flew June 1984; took part in World Aerobatic Championships, Hungary, August 1984 (details in 1985-86 *Jane's*); modified Su-26Ms, identified by sharp-cornered (rather than rounded) rudder and reduced fuselage side glazing, gained both men's and women's team prizes 1986 Championships, UK; further refined **Su-26M** shown 1989 Paris Air Show; pilots of former USSR had won 61 gold medals in competitions when **Su-26MX** (X for export) appeared at Farnborough Air Show 1990; in production.
CUSTOMERS: Marketed worldwide, including USA and UK.
DESIGN FEATURES: Typical aerobatic competition aircraft; mid-wing of specially developed symmetrical section, variable along span, slightly concave in region of ailerons to increase their effectiveness; leading-edge somewhat sharper than usual to improve responsiveness to control surface movement; thickness/chord ratio 18 per cent root, 12 per cent at tip; no dihedral, incidence or sweep quarter-chord.

FLYING CONTROLS: Mechanical actuation, ailerons and elevators by pushrods, rudder by cables; each aileron has ground adjustable tab on trailing-edge and two suspended triangular balance tabs; no flaps; horn balanced rudder and elevators, each with ground adjustable tab.

STRUCTURE: Composites comprise more than 50 per cent airframe weight; one-piece wing, covered with honeycomb composite panels; foam filled front box spar with CFRP booms and wound glassfibre webs; channel section rear spar of CFRP; titanium truss ribs; plain ailerons have CFRP box spar, GFRP skin and foam filling; fuselage has basic welded truss structure of VNS-2 high strength stainless steel tubing; lower nose section of truss removable for wing detachment; quickly removable honeycomb composite skin panels; light alloy engine cowlings; integral fin and tailplane construction same as wings; rudder and elevator construction same as ailerons.

LANDING GEAR: Non-retractable tailwheel type; arched cantilever mainwheel legs of titanium alloy; mainwheels size 400 × 150 mm, with hydraulic disc brakes; steerable tailwheel, on titanium spring, connected to rudder.

POWER PLANT: One 265 kW (355 hp) Vedeneyev M-14P nine cylinder radial engine; three-blade Gerd Mühlbauer variable-pitch metal propeller; optional V-530TA-D3 two-blade variable-pitch propeller. Steel tube engine mounting. Fuel tank in fuselage forward of front spar capacity 63 litres (16.6 US gallons; 13.8 Imp gallons); tank in each wing leading-edge, total 200 litres (53 US gallons; 44 Imp gallons), for ferry flights; oil capacity 22.6 litres (6 US gallons; 5 Imp gallons); fuel and oil systems adapted for inverted flight; pneumatic engine starting system.

ACCOMMODATION: One-piece pilot's seat of GFRP, inclined at 45° and designed for use with PLP-60 backpack parachute; rearward hinged jettisonable canopy; safety harness anchored to fuselage structure.

SYSTEMS: Electrical system 24/28V, with 3kW generator, batteries and external supply socket.

AVIONICS: Briz VHF radio.

DIMENSIONS, EXTERNAL:
Wing span	7.80 m (25 ft 7 in)
Wing chord: at root	1.95 m (6 ft 4¾ in)
at tip	1.10 m (3 ft 7¼ in)
Wing aspect ratio	5.0
Length overall	6.845 m (22 ft 5½ in)
Height overall	2.78 m (9 ft 1½ in)
Tailplane span	2.95 m (9 ft 8¼ in)
Wheel track	2.20 m (7 ft 2½ in)
Wheelbase	5.05 m (16 ft 6¾ in)
Propeller diameter	2.40 m (7 ft 10½ in)

AREAS:
Wings, gross	11.80 m² (127.0 sq ft)
Ailerons (total)	1.18 m² (12.70 sq ft)
Fin	0.34 m² (3.66 sq ft)
Rudder	0.89 m² (9.58 sq ft)
Tailplane	1.10 m² (11.84 sq ft)
Elevators (total)	1.53 m² (16.47 sq ft)

WEIGHTS AND LOADINGS:
Weight empty	705 kg (1,554 lb)
Max T-O weight	1,000 kg (2,205 lb)
Max wing loading	84.75 kg/m² (17.36 lb/sq ft)
Max power loading	3.73 kg/kW (6.13 lb/hp)

PERFORMANCE:
Never-exceed speed (V_NE)	243 knots (450 km/h; 280 mph)
Max level speed at S/L	167 knots (310 km/h; 192 mph)
Normal cruising speed	140 knots (260 km/h; 161 mph)
T-O speed	65 knots (120 km/h; 75 mph)
Landing speed	62 knots (115 km/h; 72 mph)

Sukhoi Su-26MX aerobatic competition aircraft *(P. Gaillard)*

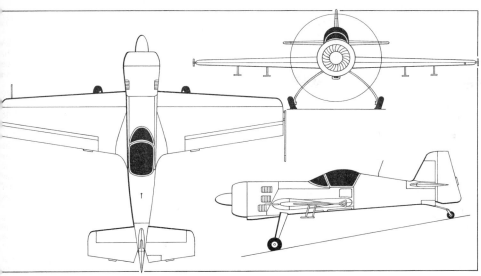

Sukhoi Su-26M single-seat aerobatic aircraft *(Jane's/Mike Keep)*

Stalling speed	60 knots (110 km/h; 69 mph)
Max rate of climb at S/L	1,080 m (3,540 ft)/min
Service ceiling	4,000 m (13,125 ft)
Rate of roll	more than 360°/s
T-O run	160 m (525 ft)
Landing run	250 m (820 ft)
Ferry range at 1,000 m (3,280 ft)	
	432 nm (800 km; 497 miles)
g limits:	+12/−10 (operating)
	+23 (ultimate)

SUKHOI Su-27

NATO reporting name: Flanker

TYPE: Single-seat all-weather counter-air fighter and single/two-seat ground attack aircraft; two-seat combat trainer.

PROGRAMME: Development began 1969 under leadership of Pavel Sukhoi; T-10-1 prototype (first of 15), built under Mikhail Seemonov's supervision, was flown 20 May 1977 by Vladimir Ilyushin; development was not easy; two pilots lost their lives before major airframe redesign resulted in production configuration; entered service 1984; production centred in plant at Komsomolsk, Khabarovsk Territory; ground attack role observed in 1991; new versions being developed.

CURRENT VERSIONS: **Su-27** ('Flanker-A'): Prototypes, with curved wingtips, rearward retracting nosewheel, tail fins mounted centrally above engine housings; first observed Ramenskoye by US satellite, given provisional designation 'Ram-J'.

Su-27 ('Flanker-B'): Single-seat land-based production version; square wingtips, carrying air-to-air missile launchers in air combat role, cylindrical ECM jammer pods (each approx 4 m; 13 ft 1½ in long, like those of Beriev A-40) in ground attack configuration; tail fins outboard of engine housings; extended tailcone; forward retracting nosewheel; first flown 20 April 1981; standard CIS air forces equipment. *Detailed description applies to this version, except where indicated.*

Su-27UB ('Flanker-C'): Tandem two-seat trainer version of 'Flanker-B' with full combat capability; taller fin; overall height 6.357 m (20 ft 10¼ in).

Su-27K (K for *korabelnyy*; ship-based) ('Flanker-D'): Counter-air/attack version for ramp-assisted operation from navy carriers, first mentioned as 'Flanker-B variant 2' by Rear Adm William O. Studeman, USN, Spring 1988; basically similar to production Su-27 but with movable foreplanes, first tested on experimental Su-27 known as T-10-24; foreplanes operate at all speeds and only collectively, not differentially (7° up/70° down); made first conventional (non-V/STOL) landing by Soviet aircraft on ship, the *Admiral of the Fleet Kuznetsov* (formerly *Tbilisi*), 1 November 1989; hydraulically folding outer wing (through 135°) and outer tailplane panels; folded width 7.4 m (24 ft 3½ in); strengthened landing gear with twin-wheel nose unit; added arrester hook; flight refuelling capability; long tailcone of land-based versions shortened to prevent tailscrapes during take-off and landing; IRST with wider angle of view; provision for auxiliary fuel tank or buddy flight refuelling pack on centreline pylon; T-O run 120 m (395 ft) on carrier with 14° ramp; carrier approach speed 130 knots (240 km/h; 150 mph). (See Su-33.)

Su-27KU (Su-27IB; Istrebitel Bombardirovschik: fighter-bomber): A side by side two-seat Su-27 variant was first seen in a *Tass* photograph showing this aircraft approaching (but not necessarily landing on) the carrier *Admiral of the Fleet Kuznetsov*; described as deck landing trainer, but no wing folding or deck arrester hook; foreplanes and twin nosewheels like Su-27K; completely new and wider front fuselage, with wing extensions taken forward as chines to tip of nose; shallow nose, without radar, reminiscent of US Lockheed SR-71 and source of Russian nickname 'Platypus'; deep fairing behind wide curved canopy, offering space for increased fuel, avionics, etc; no IRST forward of windscreen; louvres on engine air intake ducts reconfigured; nosewheel leg moved forward to retract rearward; four nosewheel doors replace usual single door; no ventral fins; K-36 ejection seats staggered and splayed to separate pilots after ejection; 30 mm gun retained. This, or similar, aircraft exhibited to CIS leaders at Machulishche airfield, Minsk, February 1992, under designation Su-27IB, with simulated attack armament on 10 external stores pylons (under each intake duct, on each wingtip, three under each wing); rocket/ramjet powered Kh-31A/P (NATO AS-17 'Krypton') air-to-surface missiles under ducts, R-73A (AA-11 'Archer') air-to-air missiles on wingtips; a 500 kg laser-guided bomb inboard, TV/laser-guided Kh-29 (AS-14 'Kedge') on central pylon and R-77 AMRAAM type air-to-air missile outboard under each wing; retractable flight refuelling probe beneath port windscreen; internal ECM. Reported in production to replace Mig-27, Su-22 and some Su-24s.

Su-27PU: Two-seat long-range interceptor, developed from Su-27UB, with new radar and avionics, flight refuelling probe, and buddy refuelling capability. Full combat capability of single-seat version. Able to operate with four standard Su-27s; only Su-27PU has radar operating, enabling it to assign targets to other fighters by radio data link. Prototype (T-10PU), built 1986, was first Su-27 used for testing UPAZ-A Sakhalin flight refuelling system; modified subsequently to full Su-27PU operational standard; series production at Irkutsk. Because of Russian Air Force funding problems, first two production Su-27PUs sold to aerobatic team at Zhukovsky Flight Research Institute.

P-42: A specially prepared Su-27, set 27 official world records, including climb to 12,000 m (39,370 ft) in 55.542 seconds; some records are in FAI category for STOL aircraft.

Su-30: See Addenda.

Su-35: See separate entry.

CUSTOMERS: More than 200 delivered to CIS home defence interceptor force and fighter components of Legnica and Vinnitsa air armies; China received first eight of 22 ordered in 1991.

DESIGN FEATURES: Developed to replace Yak-28P, Su-15 and Tu-28P/128 interceptors in APVO, and to escort Su-24 deep-penetration strike missions; exceptional range on internal fuel made flight refuelling unnecessary until

Sukhoi Su-27 ('Flanker-B') in ground attack configuration *(AviaData)*

Sukhoi Su-27UB ('Flanker-C') of 'Russian Knights' display team *(Paul Jackson)*

314 RUSSIA: AIRCRAFT—SUKHOI

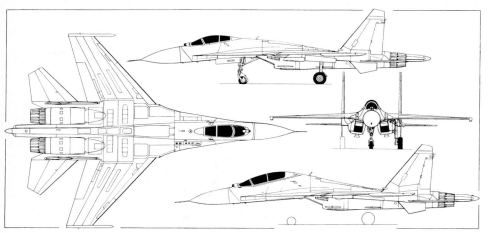

Sukhoi Su-27, with added side elevation (bottom) of two-seat Su-27UB *(Dennis Punnett)*

Sukhoi Su-27K, with full armament and wings folded *(R. J. Malachowski)*

Sukhoi Su-27KU (Su-27IB) fighter-bomber at Machulishche airfield *(Tass)*

Su-24s received probes; external fuel tanks still not considered necessary; all-swept mid-wing configuration, with long curved wing leading-edge root extensions, lift generating fuselage, twin tail fins and widely spaced engines with wedge intakes; rear-hinged doors in intake hinge up to prevent ingestion of foreign objects during take-off and landing; integrated fire control system with pilot's helmet mounted target designator; exceptional high-Alpha performance; basic wing leading-edge sweepback 42°; anhedral approx 2° 30′.

FLYING CONTROLS: Four-channel analog fly-by-wire, no mechanical backup; inherently unstable; no ailerons; full-span leading-edge flaps and plain half-span inboard flaperons controlled manually for take-off and landing, computer controlled in flight; differential/collective tailerons operate in conjunction with flaperons and rudders for pitch and roll control; flight control system limits g loading to +9 and normally limits angle of attack to 30-35°; angle of attack limiter can be overruled manually for certain flight manoeuvres; large door-type airbrake in top of centre-fuselage.

STRUCTURE: All-metal, with many titanium components but no composites; comparatively conventional two-spar wings; basically circular section semi-monocoque fuselage, sloping down sharply aft of canopy; cockpit high-set behind drooped nose; large ogival dielectric nosecone; long rectangular blast panel forward of gun on starboard side, above wingroot extension; uncanted vertical tail surfaces on narrow decks outboard of engine housings; fin extensions beneath decks form parallel, widely separated ventral fins.

LANDING GEAR: Retractable tricycle type, made by Hydromash, with single wheel on each unit; mainwheels retract forward into wingroots; steerable nosewheel, with mudguard, also retracts forward; mainwheel tyres 1030 × 350 mm, pressure 12.25-15,7 bars (178-227 lb/sq in) nosewheel tyre 680 × 260 mm, pressure 9.3 bars (135 lb/sq in); brake-chute housed in fuselage tailcone.

POWER PLANT: Two Saturn/Lyulka AL-31F turbofans, each 122.6 kN (27,557 lb st) with afterburning. Large auxiliary air intake louvres in bottom of each three-ramp engine duct near primary wedge intake; two rows of small vertical louvres in each side/wall of wedge, and others in top face; fine-grille screen hinges up from bottom of each duct to shield engine from foreign object ingestion during take-off and landing.

ACCOMMODATION: Pilot only, on K-36MD zero/zero ejection seat, under large rearward opening transparent blister canopy, with low sill.

SYSTEMS: Two independent hydraulic systems, pressure 275 bars (4,000 lb/sq in), for nosewheel steering, landing gear and wing flaps.

AVIONICS: Track-while-scan coherent pulse Doppler lookdown/shootdown radar (antenna diameter approx 1.0 m; 3 ft 4 in) with reported search range of 130 nm (240 km; 150 miles) and tracking range of 100 nm (185 km; 115 miles); infra-red search/track (IRST) sensor in transparent housing forward of windscreen; Sirena-3 360° radar warning receivers, outboard of each bottom air intake lip and at tail. Integrated fire control system enables radar, IRST and laser rangefinder to be slaved to pilot's helmet mounted target designator and displayed on wide-angle HUD; autopilot able to restore aircraft to right-side-up level flight from any attitude when 'panic button' depressed.

EQUIPMENT: Three banks of chaff/flare dispensers in bottom of long tailcone extension.

ARMAMENT: One 30 mm GSh-301 gun in starboard wingroot extension, with 149 rds. Up to 10 air-to-air missiles in air combat role, on tandem pylons under fuselage between engine ducts, beneath each duct, under each centre-wing and outer-wing, and at each wingtip. Typically, two short-burn semi-active radar homing R-27R (NATO AA-10A 'Alamo-A') in tandem under fuselage; two short-burn infra-red homing R-27T (AA-10B 'Alamo-B') missiles on centre-wing pylons; and long-burn semi-active radar homing R-27ER (AA-10C 'Alamo-C') or infra-red R-27ET (AA-10D 'Alamo-D') beneath each engine duct. The four outer pylons carry either R-73A (AA-11 'Archer') or R-60 (AA-8 'Aphid') close-range infra-red missiles. R-33 (AA-9 'Amos') missiles optional in place of AA-10s. Five-round packs of 130 mm rockets, or larger rocket pods, under wings in ground attack role.

DIMENSIONS, EXTERNAL ('Flanker-B'):
Wing span	14.70 m (48 ft 2¾ in)
Length overall, excl nose probe	21.935 m (71 ft 11½ in)
Height overall	5.932 m (19 ft 5½ in)
Fuselage: Max width	1.50 m (4 ft 11 in)
Tailplane span	9.90 m (32 ft 6 in)
Distance between fin tips	4.30 m (14 ft 1¼ in)
Wheel track	4.33 m (14 ft 2½ in)
Wheelbase	5.88 m (19 ft 3½ in)

WEIGHTS AND LOADINGS (B: 'Flanker-B', C: 'Flanker-C'):
Max T-O weight: B	22,000-30,000 kg (48,500-66,135 lb)
C	22,500 kg (49,600 lb)
Max power loading: B	121.1 kg/kN (1.20 lb/lb st)
C	90.8 kg/kN (0.90 lb/lb st)

PERFORMANCE:
Max level speed:
at height: B, C   Mach 2.35
         (1,345 knots; 2,500 km/h; 1,550 mph)
at S/L: B, C   Mach 1.1
         (725 knots; 1,345 km/h; 835 mph)

Rate of roll	approx 270°/s
Service ceiling: B, C	18,000 m (59,055 ft)
T-O run: B	500 m (1,640 ft)
C	550 m (1,805 ft)
Landing run: B	600 m (1,970 ft)
C	650 m (2,135 ft)
Combat radius: B	810 nm (1,500 km; 930 miles)
Range with max fuel:	
B	over 2,160 nm (4,000 km; 2,485 miles)
C	1,620 nm (3,000 km; 1,865 miles)
g limit (operational): B, C	+9

SUKHOI Su-29

TYPE: Tandem two-seat training/single-seat aerobatic aircraft.
PROGRAMME: Announced at Moscow Aerospace '90, when development was at advanced stage; prototype first flew 1991; shipments to Pompano Air Center, Florida, USA, began Spring 1992, with 18 of 24 deliveries for 1992 already sold.
DESIGN FEATURES: Two-seat development of Su-26MX; wing span and overall length increased; inclined seats; continuous jettisonable rearward-hinged transparent canopy over both cockpits; wheel fairings optional; dual controls standard; tail unit unchanged.
POWER PLANT: One 265 kW (355 hp) Vedeneyev M-14P nine-cylinder radial engine; three-blade MT propeller.

DIMENSIONS, EXTERNAL:
Wing span	8.20 m (26 ft 10¾ in)
Wing aspect ratio	5.49
Length overall	7.32 m (24 ft 0¼ in)
Height overall	2.87 m (9 ft 5 in)
Propeller diameter	2.54 m (8 ft 4 in)

AREAS:
Wings, gross	12.24 m² (131.75 sq ft)

WEIGHTS AND LOADINGS (A: two persons, B: pilot only):
Weight empty: A	760 kg (1,675 lb)
Max T-O weight: A	1,100 kg (2,425 lb)
B	850 kg (1,874 lb)
Max wing loading: A	89.87 kg/m² (18.41 lb/sq ft)
B	69.44 kg/m² (14.22 lb/sq ft)
Max power loading: A	4.10 kg/kW (6.74 lb/hp)
B	3.17 kg/kW (5.21 lb/hp)

PERFORMANCE:
Never-exceed speed (V_{NE}):	
B	243 knots (450 km/h; 280 mph)
Max level speed: A	183 knots (340 km/h; 211 mph)
Nominal cruising speed: B	159 knots (295 km/h; 183 mph)
Stalling speed: A, B	59 knots (110 km/h; 68 mph)
Max rate of roll: B	more than 360°/s
Max rate of climb at S/L: A	960 m (3,150 ft)/min
B	1,080 m (3,543 ft)/min
Service ceiling: B	4,000 m (13,125 ft)
T-O and landing run: A	120 m (395 ft)
Max ferry range: B	520 nm (965 km; 600 miles)
g limits: B	+11/−9 (competition)
A	+9/−7 (training)
A, B	+23 (ultimate)

SUKHOI Su-31T/U

TYPE: Single-seat aerobatic competition and training aircraft.
PROGRAMME: Announced 1992; prototype demonstrated at 1992 Farnborough Air Show.
CURRENT VERSIONS: **Su-31T**: Basic version; non-retractable landing gear; *described in detail.*
Su-31U: As Su-31T but retractable landing gear; scheduled to fly 1993.
DESIGN FEATURES: Basically single-seat version of Su-29 with uprated engine; 35° inclination of seat enables pilot to employ repeatedly a g load of +12/−10, giving advantages in controlling aircraft within limited flying area and to perform very complicated manoeuvres.
STRUCTURE: More than 70 per cent composites by weight; centre fuselage is welded truss of high-strength stainless steel tube, with detachable skin panels of honeycomb filled composite sandwich; rear fuselage is semi-monocoque of composites with honeycomb filler; one-piece two-spar wing is of carbonfibre and organic composites, covered with honeycomb sandwich skin; tail unit is all-composite.
POWER PLANT: One 294 kW (395 hp) Vedeneyev M-14PF nine-cylinder radial engine. Basic fuel capacity 70 litres (18.5 US gallons; 15.5 Imp gallons); provision for 140 litre (37 US gallon; 31 Imp gallon) drop tank for ferrying, or two tanks in wings, total capacity 200 litres (53 US gallons; 44 Imp gallons).

DIMENSIONS, EXTERNAL (Su-31T):
Wing span	8.20 m (26 ft 10¾ in)
Length overall	6.83 m (22 ft 5 in)
Height overall	2.74 m (9 ft 0 in)

WEIGHTS AND LOADINGS (Su-31T):
Weight empty	670 kg (1,477 lb)
Max T-O weight	780 kg (1,720 lb)

PERFORMANCE (Su-31T at max T-O weight):
Max never-exceed speed (V_{NE})	243 knots (450 km/h; 280 mph)
Max level speed	183 knots (340 km/h; 211 mph)
T-O speed	60 knots (110 km/h; 69 mph)

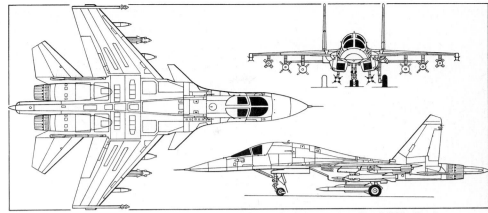
Three-view drawing of Sukhoi Su-27KU (Su-27IB) fighter-bomber *(Jane's/Mike Keep)*

Sukhoi Su-27PU two-seat long-range interceptor *(Piotr Butowski)*

Sukhoi Su-29 two-seat specialised aerobatic aircraft *(Paul Jackson)*

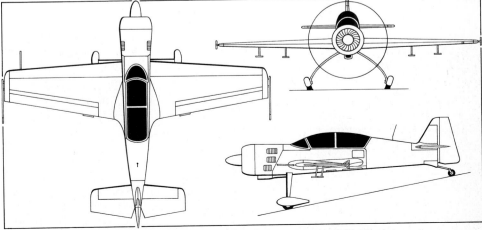
Sukhoi Su-29 training and aerobatic aircraft *(Jane's/Mike Keep)*

Sukhoi Su-31T single-seat aerobatic competition and training aircraft *(Peter J. Cooper)*

Landing speed	57 knots (105 km/h; 66 mph)
Rate of climb at S/L	1,080 m (3,540 ft)/min
Service ceiling	4,000 m (13,125 ft)
Rate of roll	400°/s
T-O run	140 m (460 ft)
Landing run	250 m (820 ft)
Max ferry range	400-645 nm (740-1,200 km; 460-745 miles)
g limits	+12/−10

SUKHOI Su-33

TYPE: Carrier-based development of Su-27K (which see).
PROGRAMME: Initial batch of about 20 being completed at Komsomolsk-on-Amur in 1993, for operation on *Admiral of the Fleet Kuznetsov*, primarily for air defence. Generally as Su-27K, except as follows:
POWER PLANT: Two Saturn/Lyulka AL-35F turbofans; each reported 133 kN (29,900 lb st) with afterburning.
AVIONICS: Include nav systems specialised for use over sea.
ARMAMENT: Basically as Su-27, plus ability to carry Kh-31 (AS-17 'Krypton') air-to-surface missiles underwing and 4,500 kg (9,920 lb) Kh-41 ASM-MSS anti-ship missile, known as Mosquito, on centreline.

Sukhoi Su-35 advanced development of the Su-27 series *(Neville M. Beckett)*

SUKHOI Su-35

TYPE: Single-seat all-weather counter-air fighter and ground attack aircraft.
PROGRAMME: Prototype first flown May 1985; exhibited at 1992 Farnborough Air Show; in final stages of flight testing early 1993; scheduled entry into Russian Air Force service mid-1990s, for effective service until 2015-2020.
DESIGN FEATURES: Advanced development of Su-27; airframe basically same as that of Su-33, with foreplanes, but without specifically shipboard features such as folding wings and arrester hook; airframe, power plant, avionics and armament all upgraded compared with Su-27; digital fly-by-wire controls; claimed to be first series-built fighter with static instability and tandem 'triplane' layout, with foreplanes; taller twin tail fins; reprofiled front fuselage; enlarged tailcone for rearward-facing radar; thrust vectoring nozzles (±15°) to be offered for later use.
STRUCTURE: Composites used for components such as leading-edge flaps, nosewheel door.
POWER PLANT: Two Saturn/Lyulka AL-35F turbofans; each reported 133 kN (29,900 lb st) with afterburning.
AVIONICS: New multi-mode radar able to acquire airborne targets at ranges up to 216 nm (400 km; 250 miles), surface targets up to 108 nm (200 km; 125 miles); simultaneous tracking of more than 15 air targets and engagement of six; low-altitude terrain following/avoidance; rearward-facing radar will enable 'over the shoulder' firing of radar-guided air-to-air missiles; IRST moved to starboard; all combat flight phases computerised; enhanced ECM, including wingtip jammer pods. Shown at Farnborough with GEC Ferranti TIALD (thermal imaging airborne laser designator) night/adverse visibility pod fitted for possible future use.
ARMAMENT: One 30 mm GSh-30 gun in starboard wingroot extension. Mountings for up to 14 stores, including R-27 (AA-10 'Alamo-A/B/C/D'), R-40 (AA-6 'Acrid'), R-60 (AA-8 'Aphid'), R-73A (AA-11 'Archer') and R-77 (AMRAAM class) air-to-air missiles, Kh-25ML (AS-10 'Karen'), Kh-25MP (AS-12 'Kegler'), Kh-29 (AS-14 'Kedge') and Kh-31 (AS-17 'Krypton') air-to-surface missiles, KAB-500 bombs and rocket packs. Max weapon load 8,000 kg (17,635 lb).

DIMENSIONS, EXTERNAL:
Wing span, over ECM pods	15.00 m (49 ft 2½ in)
Length overall	22.00 m (72 ft 2¼ in)
Height overall	6.00 m (19 ft 8 in)

PERFORMANCE:
Max level speed: at height	Mach 2.35 (1,350 knots; 2,500 km/h; 1,555 mph)
at S/L	Mach 1.14 (755 knots; 1,400 km/h; 870 mph)
Service ceiling	18,000 m (59,055 ft)
Balanced runway length	1,200 m (3,940 ft)
Range: with max internal fuel	more than 2,160 nm (4,000 km; 2,485 miles)
with flight refuelling	more than 3,500 nm (6,500 km; 4,040 miles)

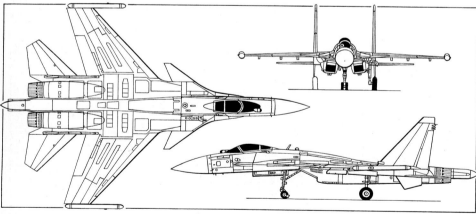
Sukhoi Su-35 single-seat counter-air and ground attack fighter *(Jane's/Mike Keep)*

SUKHOI S-21

TYPE: Supersonic business aircraft.
PROGRAMME: Projected originally as joint programme by Sukhoi and Gulfstream Aerospace of USA; separate studies for aircraft of this type started by Sukhoi 1987, Gulfstream 1988, and recorded subsequently in International section of *Jane's*; first draft of detailed specification of current redesign submitted by Sukhoi October 1991; Gulfstream withdrew from programme 1992; first flight, expected originally 1994, now deferred until late 1990s to await new technology and international working arrangements.
DESIGN FEATURES: Twin-engined low/mid-wing aircraft; wings have compound sweep, with highly swept inner panels; oval-section fuselage; all-moving swept foreplanes; low-set tailplane.

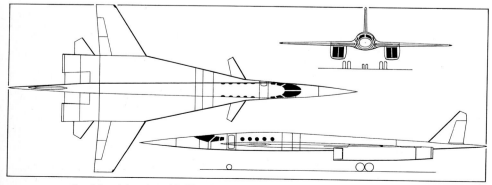
Provisional drawing of Sukhoi S-21 supersonic business jet *(Jane's/Mike Keep)*

FLYING CONTROLS: Digital fly-by-wire AFCS.
LANDING GEAR: Retractable tricycle type with main four-wheel bogies, twin nosewheels. Main units retract upward into engine air intake trunks, nose unit forward.
POWER PLANT: Prototype(s) may be powered by Saturn/Lyulka AL-31F augmented turbofans. Production aircraft intended to have two Rolls-Royce/Saturn engines.
ACCOMMODATION: Crew of two and six to 10 passengers.
DIMENSIONS, EXTERNAL:
Wing span 19.90 m (65 ft 3½ in)
Wing aspect ratio 3.1
Length overall 40.50 m (132 ft 10½ in)
Height overall 7.40 m (24 ft 3⅓ in)
DIMENSIONS, INTERNAL:
Passenger cabin: Length 8.50 m (27 ft 10½ in)
Max width 1.80 m (5 ft 10¾ in)
Max height 1.85 m (6 ft 0¾ in)
WEIGHTS AND LOADINGS (estimated):
Max T-O weight 48,000 kg (105,820 lb)
PERFORMANCE (estimated):
Max cruising speed Mach 2.0
Cruising altitude 15,500-19,500 m (50,850-63,975 ft)
T-O and landing run 1,980 m (6,500 ft)
Range, NBAA IFR reserves
 4,060 nm (7,410 km; 4,605 miles)

SUKHOI Su-37

This project for a multi-role combat aircraft has been cancelled.

SUKHOI S-51

TYPE: Medium-size long-range supersonic transport.
PROGRAMME: At design study stage; initial letter of designation refers to Sukhoi General Designer Mikhail Seemonov.
DESIGN FEATURES: General similarity to smaller S-21; rear-mounted sweptback low wings, with very long curved root extensions, carrying four engine ducts and tailplanes on wide rearward extension of inboard semi-span; small winglets; sweptback foreplanes; small diameter, long fuselage, with flush windscreen in conical nose; swept vertical tail surfaces. Intended to fly non-stop from Moscow to any airport in Europe and Asian continent, or one-stop to any world capital; range similar when flying supersonic or high subsonic, permitting operation over highly populated areas.
LANDING GEAR: Retractable tricycle type; twin nosewheels; main bogies each two pairs of wheels in tandem.
POWER PLANT: Variable-cycle type.
ACCOMMODATION: Crew of three; from 51 to 68 passengers.
DIMENSIONS, EXTERNAL:
Wing span 27.12 m (89 ft 0 in)
Length overall 50.70 m (166 ft 4 in)
Height overall 9.56 m (31 ft 4 in)
WEIGHTS AND LOADINGS:
Max T-O weight 90,000 kg (198,410 lb)
PERFORMANCE (estimated):
Max supersonic speed Mach 2.0
 (1,150 knots; 2,125 km/h; 1,320 mph)
Max subsonic speed Mach 0.9
 (540 knots; 1,000 km/h; 620 mph)
Service ceiling 17,000-19,000 m (55,775-62,335 ft)
Max range: subsonic or supersonic
 4,965 nm (9,200 km; 5,715 miles)

SUKHOI S-54

TYPE: Two-seat advanced jet trainer and light combat aircraft.
PROGRAMME: One of designs by five OKBs to meet official Russian requirement to replace Aero L-29 and L-39 Albatros; configuration refined 1992; avionics and systems units being tested on modified Su-25 and Su-27.
DESIGN FEATURES: Described by Sukhoi as scaled-down development of Su-27, with speed, operating altitude and manoeuvrability commensurate with combat aircraft; unconventional all-swept mid-wing configuration; twin fins mounted at wing trailing-edges; twin engine air intakes under wingroots; retractable tricycle landing gear. See accompanying three-view for further details.
FLYING CONTROLS: Fly-by-wire as Su-27; pre-programmable to make aircraft easier or harder to fly, dependent on pupil's ability; optional 'panic button' to return aircraft to straight and level flight from any attitude, and push-button spin recovery; optional playback system to record student's every move in flight.
POWER PLANT: One Soyuz/Tumansky R-195FS turbojet, modified from Su-25 power plant; optional alternatives could include F404, RB199 and PD33, with minimal airframe modification.
ACCOMMODATION: Two crew in tandem under blister canopy; rear seat raised.
AVIONICS: Avionics and cockpit interior same as for current and advanced tactical aircraft.
ARMAMENT: Compatible with air-to-air and air-to-surface guided weapons.
DIMENSIONS, EXTERNAL:
Wing span 9.08 m (29 ft 9½ in)
Length overall 12.30 m (40 ft 4¼ in)
Height overall 4.47 m (14 ft 8 in)

Model of one version of the Sukhoi S-21 supersonic transport (*Piotr Butowski*)

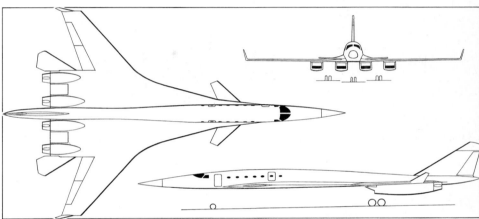

Sukhoi S-51 four-engined supersonic passenger airliner (*Jane's/Mike Keep*)

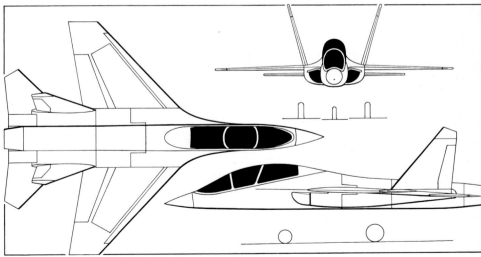

Projected Sukhoi S-54 tandem two-seat trainer derivative of Su-27 (*Jane's/Mike Keep*)

Tailplane span approx 5.26 m (17 ft 3 in)
Wheel track approx 2.45 m (8 ft 0½ in)
Wheelbase approx 3.32 m (10 ft 10¾ in)
WEIGHTS AND LOADINGS: Not available
PERFORMANCE (estimated):
Max level speed: at height
 Mach 1.55 (890 knots; 1,650 km/h; 1,025 mph)
 at S/L Mach 0.98 (645 knots; 1,200 km/h; 745 mph)
T-O speed 98 knots (180 km/h; 112 mph)
Landing speed 92 knots (170 km/h; 106 mph)
Service ceiling 18,000 m (59,050 ft)
T-O run 360 m (1,180 ft)
Landing run 500 m (1,640 ft)
Range with max fuel:
 at S/L 440 nm (820 km; 510 miles)
 at height 1,080 nm (2,000 km; 1,240 miles)
g limits +9/-3

SUKHOI S-80

TYPE: Twin-turboprop multi-purpose STOL transport.
PROGRAMME: First, largest and most advanced design by Sukhoi under konversiya programme of former Soviet industry. Work began 1989 by Sukhoi-Europe/Asia joint stock

318 RUSSIA: AIRCRAFT—SUKHOI

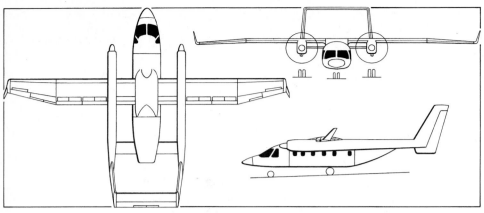

Sukhoi S-80 twin-turboprop multi-purpose STOL transport (*Jane's*/Mike Keep)

Model of Sukhoi S-80 in current redesigned configuration

company, with founder members Sukhoi Design Bureau ASIC, Komsomolsk-on-Amur Aviation Production Association, Rybinsk Motor Engineering Design Bureau, Ramenskoye Instrument Engineering Design Bureau, and Instrument Engineering R&D Institute. Manufacture of cargo/passenger version of S-80 well advanced at largest Komsomolsk-on-Amur plant early 1993; first flight of prototype scheduled late 1993; series production to begin 1995.

CURRENT VERSIONS: **S-80P**: Basic cargo/passenger version; up to 24 passengers or 2,500 kg (5,510 lb) of freight.
  S-80DT: Military light assault transport.
  S-80GR: Adapted to carry geological exploration equipment.
  S-80M: Ambulance for 10 stretcher patients and attendant.
  S-80PT: Patrol/transport version.
  S-80R: Fishery surveillance version.

CUSTOMERS: Planned to produce approx 1,800-2,000 S-80s for Russia and former USSR states by 2005.

DESIGN FEATURES: Basically conventional high-wing, podded fuselage, twin-boom, rear-loading configuration, but with short tandem-wing surfaces between each tailboom and rear fuselage; unswept wings with no dihedral or anhedral on large-span inner panels, anhedral and small sweptback winglets on outer panels; sweptback vertical tail surfaces, toed slightly inward, with bridging horizontal surfaces. Systems, accessories and components of Su-25, Su-27 and Su-35 aircraft embodied in S-80. Automatic built-in systems testing.

FLYING CONTROLS: Basic three-axis control; three-view suggests complex wing leading-edge and trailing-edge surfaces to ensure STOL capability.

STRUCTURE: No details, but said to employ "advanced and high technology".

LANDING GEAR: Retractable tricycle type; twin wheels on each unit; main units retract into tailbooms.

POWER PLANT: Two 1,104 kW (1,480 shp) Rybinsk TVD-1500S turboprops, each driving a six-blade propeller. APU for autonomous operation at remote, unprepared sites.

ACCOMMODATION: Two crew side by side on flight deck; up to 24 passengers or 10 stretcher patients in main cabin, or equivalent freight or role equipment. Typical S-80P carries seven passengers in single seats on port side of cabin and 12 in double seats on starboard side, with baggage space, toilet, wardrobe or galley at front, as specified by customer. Typical freighter has a row of seats behind flight deck and unobstructed main hold for a small vehicle or cargo. Door at centre of cabin on port side; rear loading doors and ramp.

AVIONICS: New airborne radar; advanced flight deck equipment.

EQUIPMENT: Options include equipment for air photography.

DIMENSIONS, EXTERNAL:
  Wing span   32.16 m (105 ft 6¼ in)
  Length overall   16.68 m (54 ft 9 in)
  Height overall   5.58 m (18 ft 3¾ in)
DIMENSIONS, INTERNAL:
  Cabin: Length   6.30 m (20 ft 8 in)
    Max width   2.13 m (7 ft 0 in)
    Max height   1.90 m (6 ft 2¾ in)
WEIGHTS AND LOADINGS:
  Max payload   2,500 kg (5,510 lb)
  Max T-O weight   8,200 kg (18,075 lb)
PERFORMANCE (estimated):
  Max level speed   259 knots (480 km/h; 298 mph)
  T-O run   350 m (1,150 ft)
  Landing run   180 m (590 ft)

SUKHOI S-84

TYPE: Six-seat touring, transport and general purpose aircraft.
PROGRAMME: Development started October 1992.
CURRENT VERSIONS: Projected for passenger carrying, training, ambulance and patrol missions.
DESIGN FEATURES: Cantilever high-wing monoplane; streamline fuselage, with flush canopy, upswept at rear to carry unswept tailplane with twin endplate fins and rudders. Pusher propeller pod-mounted on tailplane trailing-edge.
FLYING CONTROLS: Conventional three-axis.
LANDING GEAR: Retractable tricycle type; single wheel on each unit.
POWER PLANT: One Teledyne Continental TSIOL-550-A piston engine, buried in centre-fuselage behind wings.
ACCOMMODATION: Pilot and passenger on individual front seats; four de luxe seats in pairs to rear. Upward hinged canopy panel each side, above downward hinged airstair door. Alternative configurations according to role.
DIMENSIONS, EXTERNAL:
  Wing span   12.39 m (40 ft 8 in)
  Length overall   10.94 m (35 ft 10¾ in)
  Height overall   3.87 m (12 ft 8½ in)
PERFORMANCE (estimated):
  Nominal cruising speed   162 knots (300 km/h; 186 mph)
  T-O run at max T-O weight 290 m (950 ft)
  Landing run   160 m (525 ft)
  Range with max fuel   1,106 nm (2,050 km; 1,275 miles)

SUKHOI S-86

TYPE: Seven-seat twin-turboprop multi-purpose transport.
PROGRAMME: Completely redesigned successor to original S-86 project described in 1991-92 *Jane's*; details released at Dubai '91 air show.
DESIGN FEATURES: Unconventional high-wing configuration; sweptforward wings of high aspect ratio at rear of fuselage, with centre-section of much increased chord extended rearward to carry widely separated twin tail units; small foreplanes; mainwheels of landing gear retract into large sponsons; wingtip pods with winglets; power plant drives contra-rotating propellers between tail fins.
FLYING CONTROLS: Aileron and two-section trailing-edge flap on each wing; twin rudders and elevators on tail unit; movable trailing-edge control surfaces on foreplanes.
LANDING GEAR: Retractable tricycle type; single wheel on each unit.
POWER PLANT: Two unspecified 404 kW (542 shp) turboprops driving pusher contra-rotating propellers through combining gearbox; flush engine air intake each side of rear fuselage.
ACCOMMODATION: Pressurised cabin; three pairs of seats, second and third rows club-style with table between; seventh seat to rear, opposite washroom and toilet on starboard side; baggage compartment on port side aft of rear seat; door each side of flight deck; passenger door in centre of cabin on port side; alternative cabin arrangements for cargo, medical and patrol missions.
SYSTEMS: Pressurisation system gives cabin altitude 2,400 m (7,875 ft) at normal cruising height.
DIMENSIONS, EXTERNAL:
  Wing span   15.87 m (52 ft 0¾ in)
  Wing aspect ratio   11.9
  Length overall   11.57 m (37 ft 11½ in)
  Height overall   3.90 m (12 ft 9½ in)

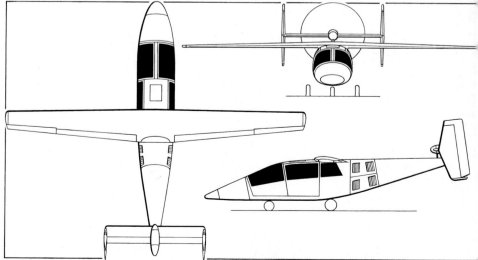

Sukhoi S-84 six-seat touring and transport aircraft (*Jane's*/Mike Keep)

DIMENSIONS, INTERNAL:
 Cabin: Length   6.00 m (19 ft 8¼ in)
        Width   1.64 m (5 ft 4½ in)
        Height   1.50 m (4 ft 11 in)
        Volume   11.20 m³ (395 cu ft)
AREAS:
 Wings, gross   21.03 m² (226.4 sq ft)
WEIGHTS AND LOADINGS:
 Operating weight empty   2,900 kg (6,395 lb)
 Max fuel   1,350 kg (2,976 lb)
 Max payload   540 kg (1,190 lb)
 Max T-O weight   4,500 kg (9,920 lb)
 Max wing loading   214.0 kg/m² (43.8 lb/sq ft)
 Max power loading   11.0 kg/kW (18.0 lb/shp)
PERFORMANCE (estimated):
 Max level speed   323 knots (600 km/h; 373 mph)
 Service ceiling   10,500 m (34,450 ft)
 T-O to, and landing from, 15 m (50 ft)   500 m (1,640 ft)
 Range with max fuel   1,885 nm (3,500 km; 2,175 miles)

SUKHOI S-99

This project, described in 1992-93 edition, has been cancelled and replaced with the S-84.

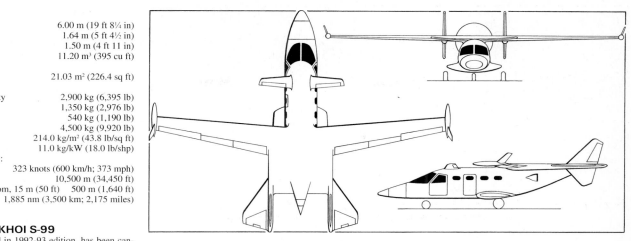

Sukhoi S-86 twin-turboprop multi-purpose transport (*Jane's/Mike Keep*)

TUPOLEV
AVIATSIONNYI NAUCHNO-TEKHNICHES-KIY KOMPLEKS IMIENI A. N. TUPOLEVA (Aviation Scientific-Technical Complex named after A. N. Tupolev) (ANTK)

Naberejnaia Akademika Tupoleva, Moscow 111250
Telephone: 7 (095) 267 25 08
DEPUTY CHIEF OF OKB: Andrei I. Kandalov
CHIEF DESIGNERS:
 Lev Aronovich Lanovski (Commercial Aircraft)
 Dmitry S. Markov
 L. L. Selyakov
 Alexander S. Shengardt

Andrei Nikolayevich Tupolev was leading figure in Central Aero-Hydrodynamic Institute (TsAGI) in Moscow from when it was founded, in 1929, until his death on 23 December 1972, aged 84. Bureau that bears his name concentrates primarily on large military and civil aircraft.

TUPOLEV Tu-16
NATO reporting name: Badger

TYPE: Twin-jet medium bomber, maritime reconnaissance/attack and electronic warfare aircraft.
PROGRAMME: Prototype flown by N. S. Rybka, under OKB designation Tu-88, on 27 April 1952; achieved disappointing max speed of 510 knots (945 km/h; 587 mph) and was overweight; Andrei Tupolev delayed production until second prototype flew in 1953 with uprated AM-3A turbojets and 5,500 kg (12,125 lb) weight reduction; max speed increased to 535 knots (992 km/h; 616 mph) at cost of max IAS of only 378 knots (700 km/h; 435 mph) at low altitude instead of originally required Mach 0.9; deliveries began 1954; nine Tu-16s took part in May Day 1954 flypast over Moscow, 54 in Aviation Day flypast 1955; approx 2,000 produced in many versions; manufacture ended late 1950s; late versions all converted bombers; all but earliest have uprated AM-3M (RD-3M) engines.

CURRENT VERSIONS: **Tu-16A** ('Badger-A'): 1954; original strategic bomber version, with conventional or nuclear free-fall weapons; glazed nose, with small undernose radome; defensive armament of seven 23 mm AM-23 guns; licence production, as **H-6**, by Xian Aircraft Company of China ended in late 1980s.
 Tu-16KS ('Badger-B'): 1954; first version with air-to-surface missiles (two KS-1; AS-1 'Kennel' turbojet powered ASMs underwing); now retired.
 Tu-16Z: 1955; experimental flight refuelling tanker, using unique wingtip-to-wingtip transfer technique (see production Tu-16N).
 Tu-16T: 1959; torpedo bomber.
 Tu-16S Korvet: 1965; Tu-16T adapted for search and rescue duties; large radio-controlled rescue boat under fuselage; continues in use.
 Tu-16K-10 ('Badger-C'): 1958; anti-shipping version; first seen 1961 Soviet Aviation Day display; Mikoyan K-10S (NATO AS-2 'Kipper') turbojet powered missile in underbelly recess; wide nose radome (NATO 'Puff Ball') instead of glazing and nose gun of Tu-16A; no provision for free-fall bombs; about 100 supplied to Northern, Baltic, Black Sea and Pacific Fleet shore bases.
 Tu-16R ('Badger-D'): Maritime/electronic reconnaissance version, introduced early 1960s; nose as Tu-16K-10 but larger undernose radome; three elint radomes in tandem under weapons bay; large number of cameras in weapons bay. Tu-16Rs carry crew of eight to ten.
 Tu-16 ('Badger-E'): Photographic/electronic reconnaissance version; as Tu-16A but cameras in weapons bay; two additional radomes under fuselage, larger one aft.
 Tu-16R ('Badger-F'): Similar to 'Badger-E' but elint pod on pylon under each wing; late versions have various small radomes under centre-fuselage.
 Tu-16K-11/16 ('Badger-G'): 1959; as Tu-16KS but with two KSR-11 (K-11) or KSR-2 (K-16; AS-5 'Kelt') underwing rocket-powered missiles, which can be carried over range greater than 1,735 nm (3,220 km; 2,000 miles). Free-fall bombing capability retained; delivered mainly to Naval anti-shipping squadrons.
 Tu-16K-26 ('Badger-G Mod'): Modified to carry KSR-5 (K-26; AS-6 'Kingfish') Mach 3 missile, with nuclear or conventional warhead, under each wing; large radome, presumably associated with missile operation, under centre-fuselage, replacing chin radome; external device on glazed nose may help to ensure correct attitude of Tu-16K-26 during missile launch; delivered to Northern, Black Sea and Pacific Fleets.
 Tu-16K-10-26 ('Badger-C Mod'): 1962; modified to carry two KSR-5 (K-26; AS-6 'Kingfish') missiles underwing, in addition to K-10S capability.
 Tu-16N: 1963; as Tu-16A but equipped as flight refuelling tanker, using wingtip-to-wingtip technique to refuel other Tu-16s, or probe-and-drogue system to refuel Tu-22s; added tankage in bomb bay.
 Tu-16P (Postanovchik Pomiekh: jammer) ('Badger-H'): stand-off or escort ECM aircraft; primary function chaff dispensing to protect missile carrying strike force; two teardrop radomes, fore and aft of weapons bay, house passive receivers to identify enemy radar signals and establish length of chaff strips to be dispensed; dispensers with total capacity of up to 9,075 kg (20,000 lb) of chaff located in weapons bay, with three chutes in doors; hatch aft of weapons bay; two blade antennae aft of bay (that on port side with V twin blades); glazed nose and chin radome.
 Tu-16PP ('Badger-J'): ECM version specialised for active jamming in all frequencies; modified from standard bomber; some equipment in canoe shape radome protruding from weapons bay, surrounded by heat exchangers and exhaust ports; anti-radar noise jammers operate in A to I bands inclusive; glazed nose as Tu-16A; some aircraft (as illustrated) have large flat-plate antennae at wingtips.
 Tu-16R ('Badger-K'): Electronic reconnaissance variant; nose as Tu-16A; two teardrop radomes, inside and forward of weapons bay (closer together than on 'Badger-H'); four small pods on centreline in front of rear radome; chaff dispenser aft of weapons bay.
 Tu-16 ('Badger-L'): Naval electronic warfare variant, like 'Badger-G' but with equipment of the kind fitted to the Tu-95 'Bear-G', including an ECM nose thimble, small pods on centre fuselage forward of engine air intakes, and 'solid' extended tailcone containing special equipment instead of tail gun position; sometimes has pod on pylon under each wing.
 Tu-16LL: Testbed for AL-7F-1, VD-7, D-36 and other engines, and equipment.
 Tu-16 Tsiklon: 1979; meteorological laboratory.
 Tu-104G: Redesignation of Tu-16s used by Aeroflot for urgent mail transportation.

CUSTOMERS: Few of 2,000 built for air forces of former USSR and Naval Aviation remain operational; air armies continue to deploy up to 20 Tu-16N tankers, 90 ECM and 15 Tu-16R versions, in absence of Tu-22Ms equipped for these roles; Naval Aviation has a few attack models (mostly 'Badger-G'), 70 Tu-16Ns and up to 80 reconnaissance/ECM variants; air forces of Egypt, Indonesia (retired), Iraq.

DESIGN FEATURES: All-swept high mid-wing configuration; heavy engine nacelles form root fairings; wing section PR-1-10S-9, with 15.7 per cent thickness/chord ratio, on inboard panels; SR-11-12 section, of 12 per cent thickness/chord, outboard; anhedral 3° from roots; incidence 1°; sweepback 41° inboard, 37° outboard at leading-edges, 35° at quarter-chord; 42° sweepback at quarter-chord on tail fin and tailplane; tailplane incidence −1° 30′.

'Badger-J' ECM jamming version of Tu-16PP (*Swedish Air Force*)

RUSSIA: AIRCRAFT—TUPOLEV

Tupolev Tu-16R ('Badger-D') maritime/electronic reconnaissance aircraft *(Royal Navy)*

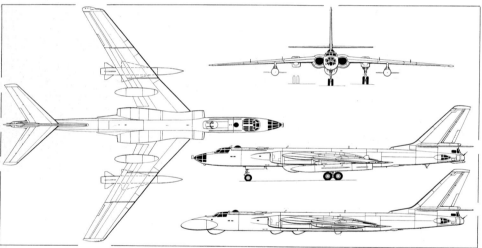

Tupolev Tu-16K-26 'Badger-G Mod', with additional side view (bottom) of Tu-16R 'Badger-D' *(Dennis Punnett)*

FLYING CONTROLS: Conventional, with aileron wheel and rudder pedals; hydraulically boosted, with large trim tabs in each aileron, elevators and rudder; except in region of landing gear nacelles, entire wing trailing-edges comprise aerodynamically balanced Frise/TsAGI ailerons and electrically operated outboard and inboard sections of TsAGI (modified Fowler) flaps.

STRUCTURE: All-metal; two-spar wings made in centre-section (integral with fuselage), inner and outer sections; nacelle for landing gear at junction of inner and outer wing panels each side; circular semi-monocoque fuselage in five sections; nose houses navigator's pressure cabin with double-glazed nose panels in magnesium alloy frame, pilot's pressure cabin, forward gunner's cabin and radar equipment; second and fourth sections house fuel tanks, with weapon compartment between them; tail section contains pressure cabin for radio operator/gunner and rear gunner.

LANDING GEAR: Hydraulically retractable tricycle type; twin-wheel nose unit retracts rearward; main four-wheel bogies retract into housings projecting beyond the wing trailing-edge; mainwheel tyres 1,100 × 330 mm; nosewheel tyres 900 × 275 mm; anti-skid brakes on mainwheels.

POWER PLANT: Early Tu-16s have two Mikulin AM-3A turbojets, each 85.21 kN (19,155 lb st) at S/L; later aircraft have RD-3M-500 (AM-3M) turbojets, each 93.05 kN (20,920 lb st). Engines semi-recessed into sides of fuselage, giving unplanned area ruling; divided air intake ducts: main duct passes through wing torque box between spars; secondary duct passes under wing to feed into primary airflow in front of engine. Engines separated from wings and fuselage by firewalls; jetpipes inclined outward 3° to shield fuselage from effects of exhaust gases. Fuel in 27 wing and fuselage tanks, total capacity 43,800 litres (11,570 US gallons; 9,635 Imp gallons); provision for underwing auxiliary fuel tanks and for flight refuelling. Tu-16 tankers trail hose from starboard wingtip; receiving equipment is in port wingtip extension.

ACCOMMODATION: Normal crew of six on ejection seats, with two pilots side by side; navigator/bombardier, on seat with armoured sides and base, in glazed nose of all versions except 'Badger-C and D'; manned tail position plus lateral observation blisters in rear fuselage under tailplane. Entry via two hatches in bottom of fuselage, in front and rear structural sections.

AVIONICS: PBR-4 Rubin 1 mapping radar; SP-50M for IFR to ICAO Cat. I standard; RSDN Chaika Loran; NavSat receiver; AP-6E autopilot, NAS-1 nav including DISS Trassa Doppler; R-807 and R-808 HF; RSIU-3M UHF; IFF, two ARK-15 ADFs, SPU-10 intercom; RV-5 and RV-18 radar altimeters; Sirena-2 radar warning receivers.

EQUIPMENT: Differs according to role; all Tu-16s can carry cameras in small section of fuselage, ahead of weapons bay, for reconnaissance.

ARMAMENT: PV-23 integrated armament firing control system; forward dorsal and rear ventral barbettes each contain two 23 mm AM-23 guns; two more in tail position controlled by an automatic gun ranging radar set; seventh, fixed, gun on starboard side of nose of versions with no glazing. Bomb load of up to 9,000 kg (19,800 lb) in weapons bay 6.5 m (21 ft) long in standard bomber, under control of navigator; normal bomb load 3,000 kg (6,600 lb). Naval versions can carry air-to-surface winged standoff missiles.

DIMENSIONS, EXTERNAL ('Badger-G'):
Wing span 32.99 m (108 ft 3 in)
Wing mean aerodynamic chord 5.021 m (16 ft 5¾ in)
Wing aspect ratio 6.6
Length overall 34.80 m (114 ft 2 in)
Height overall 10.36 m (34 ft 0 in)
Basic diameter of fuselage 2.50 m (8 ft 2½ in)
Tailplane span 11.75 m (38 ft 6½ in)
Wheel track 9.775 m (32 ft 0¾ in)
Wheelbase 10.91 m (35 ft 9½ in)

AREAS:
Wings, gross 164.65 m² (1,772.3 sq ft)
Ailerons (total) 14.77 m² (159.0 sq ft)
Flaps (total) 25.17 m² (270.9 sq ft)
Vertical tail surfaces 23.30 m² (250.8 sq ft)
Horizontal tail surfaces 34.45 m² (370.8 sq ft)

WEIGHTS AND LOADINGS ('Badger-G'):
Weight empty, equipped 37,200 kg (82,000 lb)
Weight of max fuel 34,360 kg (75,750 lb)
Normal T-O weight 75,000 kg (165,350 lb)
Max landing weight 50,000 kg (110,230 lb)
Max wing loading 455.5 kg/m² (93.3 lb/sq ft)
Max power loading 403.0 kg/kN (3.95 lb/lb st)

PERFORMANCE ('Badger-G', at max T-O weight):
Max level speed at 6,000 m (19,700 ft) 566 knots (1,050 km/h; 652 mph)
Service ceiling 15,000 m (49,200 ft)
Range with 3,000 kg (6,600 lb) bomb load 3,108 nm (5,760 km; 3,580 miles)

TUPOLEV Tu-95 and Tu-142
NATO reporting name: Bear

TYPE: Four-turboprop long-range bomber and maritime reconnaissance aircraft.

PROGRAMME: **Tu-95/1** prototype, with four 8,950 kW (12,000 ehp) Kuznetsov 2TV-2F turboprops, flew 1952, was destroyed during testing. **Tu-95/2** with 8,950 kW (12,000 ehp) TV-12 turboprops, flew 1955. Seven **Tu-95** (NATO 'Bear-A') took part in 1955 Aviation Day flypast; operational with strategic attack force 1956; **Tu-95M** ('Bear-A') was modernised production version; experimental **Tu-95K** of 1956 airdropped the MiG-19 SM-2 aircraft equipped to test features of the Kh-20 missile system; production **Tu-95K-20** ('Bear-B') of 1959 was armed with a Kh-20 (AS-3 'Kangaroo') air-to-surface missile; **Tu-95KD** of 1961 was similar to Tu-95K-20, with added flight refuelling noseprobe. 'Bear' series remained in almost continuous, latterly small scale, production for 38 years, ending 1992. Variants included **Tu-96** of 1956, high-altitude high-speed bomber development with NK-1

TUPOLEV—AIRCRAFT: RUSSIA 321

Tupolev Tu-142M ('Bear-F' Mod 4) at MosAeroshow '92 *(Linda Jackson)*

The Tu-95K-22 'Bear-G' is a reconfigured 'Bear-B' or 'C' equipped to carry Kh-22 'Kitchen' missiles *(UK Ministry of Defence)*

engines, built but not flown; **Tu-116/114D**, with Tu-95 airframe adapted as civil aircraft; **Tu-119**, a Tu-95M converted but not flown as testbed for a nuclear engine; and production **Tu-126** (NATO 'Moss') AEW&C aircraft (see 1990-91 *Jane's*).

CURRENT VERSIONS: **Tu-95RTs** ('Bear-D'): Maritime reconnaissance aircraft, first identified 1967; glazed nose; undernose radar (NATO 'Short Horn'); large underbelly radome for I band surface search radar ('Big Bulge'); elint blister fairing each side of rear fuselage; nose refuelling probe; variety of blisters and antennae, including streamlined fairing on each tailplane tip. Defensive armament comprises three pairs of 23 mm NR-23 guns in remotely controlled rear dorsal and ventral barbettes and manned tail turret; two glazed blisters on rear fuselage, under tailplane, used for sighting by gunner controlling all these guns; dorsal and ventral barbettes can also be controlled from station aft of flight deck. Housing for I band tail warning radar ('Box Tail') above tail turret is larger than on previous variants; no offensive weapons; tasks include pinpointing of maritime targets for missile launch crews on ships and aircraft that are themselves too distant for precise missile aiming and guidance; about 15 operational, probably converted from 'Bear-A' strategic bombers. A 'Bear-D' was the first Tu-95 seen, 1978, with faired tailcone housing special equipment instead of normal tail turret and radome (similar tail on 'Bear-G').

Tu-95MR ('Bear-E'): Strategic reconnaissance conversion of Tu-95M ('Bear-A'); armament, refuelling probe and rear fuselage elint fairings as Tu-95RTs; six camera windows in weapons bay, in pairs in line with wing flaps; seventh window to rear on starboard side. Few only.

Tu-142 ('Bear-F'): Anti-submarine version; prototype flew 1968; first of extensively redesigned Tu-142 series; highly cambered wings; double-slotted flaps; longer fuselage forward of wings; rudder of increased chord; space in longer pressure cabin for improved galley and relief crew on long missions. Deployed initially by Naval Aviation 1970; re-entered production mid-1980s. Initial 'Bear-Fs' had 12-wheel main landing gear bogies, retracting into enlarged and lengthened fairings aft of inboard engine nacelles, and undernose radar; main underfuselage J band radar housing considerably farther forward than on 'Bear-D' and smaller; no large blister fairings under and on sides of rear fuselage; nosewheel doors bulged, suggesting larger or low-pressure tyres; two stores bays for sonobuoys, torpedoes, nuclear or conventional depth charges in rear fuselage, one replacing usual rear ventral gun turret and leaving tail turret as sole defensive gun position. Later 'Bear-Fs' identified as follows:

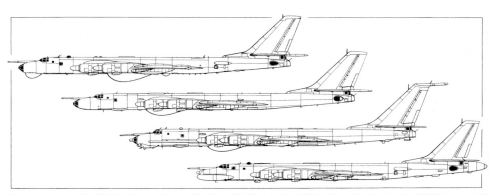

Top to bottom: Tu-95RTs 'Bear-D', Tu-142 'Bear-F' Mod 1, Tu-142M 'Bear-F' Mod 4, Tu-95K-22 'Bear-G' *(Dennis Punnett)*

Mod 1: Reverted to standard size nacelles and standard four-wheel main landing gear bogies; chin mounted J band radar deleted; fewer protrusions.

Mod 2 (**Tu-142M**): First flown 1975; nose lengthened by 23 cm (9 in); roof of flight deck raised; angle of flight refuelling probe lowered by 4°; inertial navigation system standard.

Mod 3: MAD boom added to fin tip; fairings at tailplane tips deleted; rear stores bay lengthened and narrowed.

Mod 4: Chin radar reinstated; self-protection ECM thimble radome on nose; other fairings added; observation blister each side of rear fuselage deleted; entered service 1985; further deliveries 1991.

All versions of Tu-142M can carry eight Kh-35 active

322   RUSSIA: AIRCRAFT—TUPOLEV

Tupolev Tu-142MR ('Bear-J') maintains communications between CIS national command authorities and the Navy's nuclear submarines, over long range. The ventral pod for the VLF trailing-wire antenna is under the centre-fuselage in the weapon bay area *(G. Jacobs/Jane's Intelligence Review)*

Tu-95K-22 ('Bear-G') has two large pylons under the wingroots on which to carry Kh-22 ('Kitchen') missiles *(UK Ministry of Defence)*

Deep flight deck glazing and new undernose antennae are features of the Tu-95MS ('Bear-H') cruise missile carrier *(Piotr Butowski)*

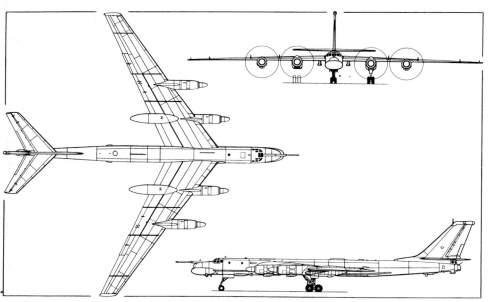
Tupolev Tu-95MS strategic bomber, known to NATO as 'Bear-H' *(Dennis Punnett)*

radar homing anti-ship missiles in underwing pairs; IOC is scheduled for 1994.

Tu-95K-22 ('Bear-G'): Bomber and elint conversion of 'Bear-B/C'; two Kh-22 (AS-4 'Kitchen') air-to-surface missiles, on pylon under each wingroot; new large undernose radome ('Down Beat'); ECM thimble under flight refuelling probe; streamlined ECM pod each side at bottom of both centre and rear fuselage; 'solid' tailcone containing special equipment, as on some 'Bear-Ds'; ventral gun turret sole defensive armament.

Tu-95M-5: Missile carrier, with two KSR-5 (AS-6 'Kingfish') missiles, flown 1972.

Tu-95M-55: Flown 1976; carrier for unidentified missile.

Tu-95MS ('Bear-H'): Late production version; crew of seven; based on Tu-142 airframe but fuselage same length as Tu-95. Initial **Tu-95MS6** version carries six RK-55 (AS-15A 'Kent') long-range cruise missiles on an internal rotary launcher; the **Tu-95MS16** carries two more under each wingroot, and a cluster of three between each pair of engines, for a total of 16. Built at Kuybyshev; achieved IOC 1984; larger and deeper radome built into nose; small fin tip fairing; no elint blister fairings on sides of rear fuselage; ventral gun turret deleted; some aircraft have single

twin-barrel 23 mm gun instead of usual pair in tail turret; active electronic jammer; RWR and missile warning receivers; chaff/flare dispensers.

Tu-142MR ('Bear-J'): Identified 1986; Soviet counterpart of US Navy E-6A and EC-130Q Tacamo, with VLF communications avionics to maintain on-station/all-ocean link between national command authorities and nuclear missile armed submarines under most operating conditions; large ventral pod for VLF trailing wire antenna several kilometres long, under centre-fuselage in weapons bay area; undernose fairing as 'Bear-F' Mod 4; fin-tip pod with trailing-edge of kind on some 'Bear-Hs'; satcom dome aft of flight deck canopy; modified 'Bear-F' airframe. With Northern and Pacific Fleets.

Tu-95U: Conversion of Tu-95M for training.

CUSTOMERS: CIS strategic bomber force (151, most Tu-95K-22 and Tu-95MS; 89 in Russia, 40 in Kazakhstan, 22 in Ukraine); Naval Aviation (15 'Bear-D', 60 'Bear-F' mostly Mod 3/4, few 'Bear-J'); Indian Navy (10 'Bear-F' Mod 3).

DESIGN FEATURES: Unique large, high-performance, four-turboprop combat aircraft, able to carry largest air-launched missiles and outsize radars; all-swept mid-wing configuration; fuselage same diameter as US Boeing B-29/Soviet Tu-4, with similar crawlway linking crew compartments fore and aft of weapons bay; main landing gear retracts into wing trailing-edge nacelles; contraprops with high tip speeds; wing section SR-5S, thickness/chord ratio 12.5 per cent at root; slight anhedral; sweepback at quarter chord 37° on inner wings, 35° outer panels.

FLYING CONTROLS: All flying control surfaces hydraulically boosted; three-segment aileron and two-segment area-increasing flap each wing; trim tab in each inboard aileron segment; upper-surface spoiler forward of each inboard aileron; adjustable tailplane incidence; trim tab in rudder and each elevator.

STRUCTURE: All-metal; four spars in each inner wing, three outboard; three boundary layer fences above each wing; circular section semi-monocoque fuselage containing three pressurised compartments; tail gunner's compartment not accessible from others.

LANDING GEAR: Hydraulically retractable tricycle type; main units consist of four-wheel bogies, with tyres approx

Tupolev Tu-22 photographed from an investigating interceptor of the Swedish Air Force *(via FLYGvapenNYTT)*

1.50 m (5 ft) diameter and hydraulic internal expanding brakes; twin-wheel steerable nose unit; all units retract rearward, main units into nacelles built on to wing trailing-edge; retractable tail bumper consisting of two small wheels; braking parachute may be used to reduce landing run.

POWER PLANT: Four KKBM Kuznetsov NK-12MV turboprops, each 11,033 kW (14,795 ehp); eight-blade contra-rotating reversible-pitch Type AV-60N propellers. Fuel in wing tanks, normal capacity 95,000 litres (25,100 US gallons; 20,900 Imp gallons); flight refuelling probe above nose of most current operational aircraft.

ACCOMMODATION: See notes applicable to individual versions and under Design Features.

SYSTEMS: Thermal anti-icing of wing and tailplane leading-edges.

AVIONICS ('Bear-D'): Large I-band radar (NATO 'Big Bulge') in blister fairing under centre-fuselage, for reconnaissance and to provide data on potential targets for anti-shipping aircraft or surface vessels; in latter mode, PPI presentation is data linked to missile launch station. Four-PRF range J-band circular and sector scan navigation radar (NATO 'Short Horn'); I-band tail warning radar (originally NATO 'Bee Hind'; later 'Box Tail') in housing at base of rudder; SRO-2 IFF (NATO 'Odd Rods'), A-321 ADF, A-322Z Doppler radar, A-325Z/321B Tacan/DME and ILS.

ARMAMENT: See notes applicable to individual versions.

EQUIPMENT ('Bear-D'): Two remotely controlled chaff/flare dispensers.

DIMENSIONS, EXTERNAL ('Bear-F'):
Wing span	51.10 m (167 ft 8 in)
Wing aspect ratio	8.39
Length overall	49.50 m (162 ft 5 in)
Height overall	12.12 m (39 ft 9 in)

AREAS:
Wings, gross	311.1 m² (3,349 sq ft)

WEIGHTS AND LOADINGS (A: 'Bear-F' Mod 3, B: 'Bear-H'):
Weight empty: B	120,000 kg (264,550 lb)
Max fuel: A	87,000 kg (191,800 lb)
Max T-O weight: A	185,000 kg (407,850 lb)
B	188,000 kg (414,470 lb)
Max wing loading: A	594.7 kg/m² (121.8 lb/sq ft)
B	604.3 kg/m² (123.8 lb/sq ft)
Max power loading: A	4.19 kg/kW (6.89 lb/shp)

PERFORMANCE:
Max level speed at 7,620 m (25,000 ft):
B	440 knots (815 km/h; 506 mph)
Nominal cruising speed	384 knots (711 km/h; 442 mph)
Service ceiling	12,000 m (39,370 ft)
Combat radius with 11,340 kg (25,000 lb) payload	3,455 nm (6,400 km; 3,975 miles)

TUPOLEV Tu-22

NATO reporting name: Blinder

TYPE: Twin-jet supersonic bomber and maritime reconnaissance aircraft.

PROGRAMME: First shown publicly 1961, when 10 took part in Aviation Day flypast, Moscow; one had Kh-22 (NATO AS-4 'Kitchen') air-to-surface missile semi-recessed in weapons bay; 22 shown in 1967 display at Domodyedovo, all with refuelling probes, most with AS-4s; about 250 manufactured.

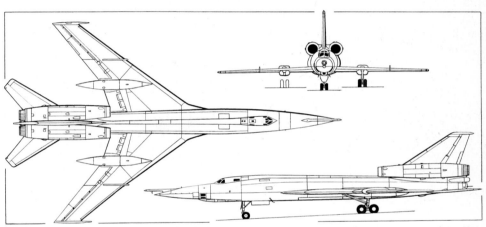
Tupolev Tu-22 twin-jet supersonic bomber ('Blinder-A') *(Dennis Punnett)*

CURRENT VERSIONS: **Tu-22** ('Blinder-A'): Basic reconnaissance bomber with supersonic dash capability; fuselage weapons bay for free-fall nuclear and conventional bombs; entered limited service.

Tu-22K ('Blinder-B'): As 'Blinder-A' but weapons bay doors redesigned to carry Kh-22 (AS-4 'Kitchen') air-to-surface missile semi-recessed; larger radar and partially retractable flight refuelling probe on nose.

Tu-22R ('Blinder-C'): Maritime reconnaissance version; initially with six camera windows in weapons bay doors; flight refuelling probe; modified version illustrated has a reconnaissance pack, approx 5.5 m long and 1.0 m wide (18 ft × 3 ft 3 in), and possible ESM installation just forward of wing on each side. The tail gun is removed; a 2.4 m (7 ft 10½ in) tail extension is fitted, as on some other Tu-22s.

Tu-22U ('Blinder-D'): Training version; raised cockpit for instructor aft of standard flight deck, with stepped-up canopy.

Tu-22P ('Blinder-E'): Electronic warfare version, with equipment in weapons bay, modifications to nosecone, additional dielectric panels, etc.

CUSTOMERS: CIS air forces have approx 75 alongside Tu-16s in Smolensk and Irkutsk air armies, most for ECM jamming and reconnaissance; Naval Aviation has approx 30 bombers and 20 maritime reconnaissance/ECM variants, which were based mainly in southern Ukraine and Estonia to protect sea approaches; air forces of Libya and Iraq each have a few.

DESIGN FEATURES: All-swept mid-wing configuration; engine nacelles mounted above rear fuselage, each side of fin, with translating intake lips; main landing gear units retract into pods on wing trailing-edge; small wingtip pods; wing section modified TsAGI SR-5S; constant slight anhedral from roots; leading-edge sweepback 52° inboard of fence on each wing, 45° outboard, with small acutely swept segment at root; fuselage basically circular section, with area ruling at wingroots.

FLYING CONTROLS: Hydraulically powered control surfaces; two-section ailerons, with tab in each inboard section; tracked plain flaps inboard and outboard of each landing gear pod; all-moving horizontal tail surfaces at bottom of fuselage; aerodynamically balanced rudder, with inset tab.

STRUCTURE: All-metal; semi-monocoque fuselage.

LANDING GEAR: Retractable tricycle type; wide track four-wheel bogie main units retract rearward into pods built on to wing trailing-edges; oleo-pneumatic shock absorbers; main legs designed to swing rearward for additional cushioning during taxying and landing on rough runways; twin-wheel nose unit retracts rearward; small retractable skid to protect rear fuselage in tail-down landing or take-off; twin brake-chutes standard.

POWER PLANT: Initially, two Koliesov (now RKBM) VD-7M turbojets, each 156.9 kN (35,275 lb st) with afterburning, in nacelles above rear fuselage, on each side of tail fin; lip of each intake is in the form of a ring which can be translated forward by jacks for take-off; air entering ram intake is then supplemented by air ingested through annular slot between ring and main body of nacelle. Later series have RD-7M-2 turbojets, each 161.8 kN (36,375 lb st) with afterburning. Jetpipes have convergent-divergent nozzle inside outer fairing. Semi-retractable flight refuelling probe on nose of 'Blinder-B/C', with triangular guard underneath to prevent drogue damaging nosecone. Provision for JATO rockets.

ACCOMMODATION: Crew of three in tandem; row of windows in bottom of fuselage, aft of nose radome, at navigator/systems operator's station; pilot has upward ejection seat, other crew members have downward ejection seats.

AVIONICS: Rubin nav/attack radar in nose; PRS-3 or PRS-4 tail warning and gun fire control radar ('Bee Hind') at base of rudder. DISS Doppler, ARK-11 ADF, RV-25 radar altimeter, TP-1A TV gunsight, PSB-11 bomb sight.

EQUIPMENT: Chaff/flare countermeasures dispensers and bombing assessment cameras carried in rear of wheel pods of some aircraft.

ARMAMENT: Weapons bay in centre-fuselage, with double-fold doors on 'Blinder-A', for 24 FAB-500, one FAB-9000 or other weapons; special doors with panels shaped to accommodate recessed Kh-22 (AS-4 'Kitchen') missile on

324 RUSSIA: AIRCRAFT—**TUPOLEV**

Variant of Tupolev Tu-22R ('Blinder-C') reconnaissance aircraft (foreground) and Tu-22U ('Blinder-D') training version *(Duncan Cubitt)*

'Blinder-B'. Single 23 mm NR-23 gun in radar directed tail turret.
DIMENSIONS, EXTERNAL:
Wing span	23.50 m (77 ft 1¼ in)
Wing aspect ratio	3.41
Length overall	42.60 m (139 ft 9 in)
Height overall	10.00 m (32 ft 9¾ in)

AREAS:
Wings, gross	162.0 m² (1,744 sq ft)

WEIGHTS AND LOADINGS:
Max fuel load	42,500 kg (93,695 lb)
Max weapon load	12,000 kg (26,455 lb)
Normal T-O weight	85,000 kg (187,390 lb)
Max T-O weight	92,000 kg (202,820 lb)
Max T-O weight with four JATO rockets	94,000 kg (207,230 lb)
Normal landing weight	60,000 kg (132,275 lb)
Max wing loading (without JATO)	568.0 kg/m² (116.3 lb/sq ft)
Max power loading (without JATO)	284.3 kg/kN (2.79 lb/lb st)

PERFORMANCE (RD-7M-2 engines):
Max level speed at 12,200 m (40,000 ft)	Mach 1.52 (870 knots; 1,610 km/h; 1,000 mph)
Service ceiling, supersonic	13,300 m (43,635 ft)
T-O run	2,250 m (7,385 ft)
Landing run	2,170 m (7,120 ft)
Landing run with brake-chutes	1,650 m (5,415 ft)
Radius of action	700-1,185 nm (1,300-2,200 km; 807-1,365 miles)
Range with max fuel	2,640 nm (4,900 km; 3,045 miles)
Ferry range	3,050 nm (5,650 km; 3,510 miles)

TUPOLEV Tu-22M
NATO reporting name: Backfire

TYPE: Twin-engined variable geometry medium bomber and maritime reconnaissance/attack aircraft.

PROGRAMME: NATO revealed existence of a Soviet variable geometry bomber programme Autumn 1969; prototype observed July 1970 on ground near Kazan plant; confirmed subsequently as twin-engined design by Tupolev OKB; at least two prototypes built, with first flight estimated 1971; up to 12 pre-production models by early 1973, for development testing, weapons trials and evaluation; production has averaged 30 a year at S. P. Gorbunov manufacturing complex, Kazan, Western Russia; first displayed in public in West at 1992 Farnborough Air Show.

CURRENT VERSIONS: **Tu-22M-2** ('Backfire-B'): First series production version; differs from original Tu-22M ('Backfire-A') in having increased wing span, and wing trailing-edge pods eliminated except for shallow underwing fairings, no longer protruding beyond trailing-edge; slightly inclined lateral air intakes, with large splitter plates; seen usually with optional flight refuelling nose probe removed and its housing replaced by long fairing. Power plant of two Kuznetsov/KKBM NK-22 turbofans (each 215.75 kN; 48,500 lb st) same as Tu-22M. Initial armament was normally one Kh-22 (NATO AS-4 'Kitchen') air-to-surface missile semi-recessed under fuselage; current aircraft also have rack for a Kh-22 under each fixed wing centre-section panel, for a maximum three 'Kitchens', with optional external stores racks under engine air intake trunks; two GSh-23 twin-barrel 23 mm guns in tail mounting with barrels side by side horizontally

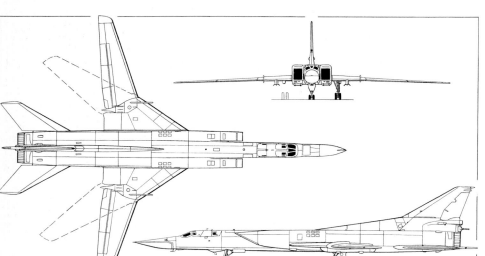

Tupolev Tu-22M-3 (NATO 'Backfire-C') bomber and maritime reconnaissance/attack aircraft *(Dennis Punnett)*

Tupolev Tu-22M-2 ('Backfire-B') with wings spread, photographed from an interceptor of the Swedish Air Force *(via FLYGvapenNYTT)*

Tupolev Tu-22M-3 ('Backfire-C') bombers carrying two versions of Kh-22 (AS-4 'Kitchen') missile underwing *(Piotr Butowski)*

initially beneath ogival radome, now with drum-shape radome of larger diameter.

Tu-22M-3 ('Backfire-C'): Advanced long-range bomber and maritime version; first known deployment with Black Sea fleet air force 1985; new engines of approx 25 per cent higher rating; wedge type engine air intakes; upturned nosecone; no visible flight refuelling probe; rotary launcher in weapons bay for six Kh-15P (AS-16 'Kickback') short-range attack missiles; provision for four more underwing as alternative to standard two Kh-22 (AS-4 'Kitchens'); single GSh-23 twin-barrel 23 mm gun, with barrels superimposed, in aerodynamically improved tail mounting, beneath large drum-shape radome. *Detailed description applies specifically to Tu-22M-3.*

CUSTOMERS: Smolensk and Irkutsk air armies have more than 200 to attack deep theatre targets; CIS Naval Aviation's four Fleet air forces have more than 160.

DESIGN FEATURES: Capable of performing nuclear strike, conventional attack and anti-ship missions; low-level penetration features ensure better survivability than for earlier Tupolev bombers; deployment of Kh-15P (AS-16 'Kickback') short-range attack missiles in Tu-22Ms has increased significantly their weapon carrying capability. Low/mid-wing configuration; large-span fixed centre-section and two variable-geometry outer wing panels (20°, 30°, 65° sweepback); no anhedral or dihedral; leading-edge fence towards tip of centre-section each side; basically circular fuselage forward of wings, with ogival dielectric nosecone; centre-fuselage faired into rectangular section air intake trunks, each with large splitter plate and assumed to embody complex variable geometry ramps: no external area ruling of trunks; all-swept tail surfaces, with large dorsal fin.

FLYING CONTROLS: Full-span leading-edge slat, aileron and three-section slotted trailing-edge flaps aft of spoilers/lift dumpers on each outer wing panel; all-moving horizontal tail surfaces; inset rudder.

LANDING GEAR: Retractable tricycle type; rearward-retracting twin nosewheels; each mainwheel bogie comprises three pairs of wheels in tandem, with varying distances between each pair (narrowest track for front pair, widest track for centre pair); mainwheel tyres size 1,030 × 350 mm, pressure 11.75 bars (170 lb/sq in); nosewheel tyres size 1,000 × 280 mm, pressure 9.32 bars (135 lb/sq in); bogies pivot inward from vestigial fairing under centre-section on each side into bottom of fuselage. Brake-chute housed inside large door under rear fuselage.

POWER PLANT: Two Kuznetsov/KKBM NK-25 turbofans, side by side in rear fuselage, each 245.2 kN (55,115 lb st) with afterburning. Provision for JATO rockets.

ACCOMMODATION: Pilot and co-pilot side by side, under upward opening gull-wing doors hinged on centreline; two crew members further aft, as indicated by position of windows between flight deck and air intakes. Ejection seats for all four crew members.

AVIONICS: Large missile targeting and navigation radar (NATO 'Down Beat') inside dielectric nosecone; radar ('Fan Tail') for tail turret, above guns. Fairing with flat glazed front panel under front fuselage, for video camera to provide visual assistance for weapon aiming from high altitude. Very advanced ECM and ECCM; infra-red missile approach warning sensor above fuselage aft of cockpit.

Tupolev Tu-22M-3 with wings fully spread

Tupolev Tu-22M-3 with wings fully swept

ARMAMENT: Max offensive weapon load three Kh-22 (NATO AS-4 'Kitchen') air-to-surface missiles, one semi-recessed under centre-fuselage, one under fixed centre-section panel of each wing; or 24,000 kg (52,910 lb) of conventional bombs or mines, half carried internally and half on racks under wings and engine air intake trunks. Internal bombs can be replaced by rotary launcher for six Kh-15P (AS-16 'Kickback') short-range attack missiles, with four more underwing as alternative to Kh-22s. Normal weapon load is single Kh-22 or 12,000 kg (26,455 lb) of bombs. Typical loads two FAB-3000, eight FAB-1500, 42 FAB-500 or 69 FAB-250 or -100 bombs (figures indicate weight in kg), or eight 1,500 kg or eighteen 500 kg mines. One GSh-23 twin-barrel 23 mm gun, with barrels superimposed, in radar directed tail mounting.

DIMENSIONS, EXTERNAL (Tu-22M-3):
Wing span: fully spread 34.28 m (112 ft 5¾ in)
  fully swept 23.30 m (76 ft 5½ in)
Wing aspect ratio: fully spread 6.40
  fully swept 3.09
Length overall 42.46 m (139 ft 3¾ in)
Height overall 11.05 m (36 ft 3 in)

AREAS:
  Wings, gross: 20° sweep   183.58 m² (1,976.1 sq ft)
               65° sweep   175.8 m² (1,892.4 sq ft)
WEIGHTS AND LOADINGS (Tu-22M-3):
  Max weapon load   24,000 kg (52,910 lb)
  Fuel load   approx 50,000 kg (110,230 lb)
  Max T-O weight   124,000 kg (273,370 lb)
  Max T-O weight with JATO   126,400 kg (278,660 lb)
  Normal landing weight   78,000 kg (171,955 lb)
  Max landing weight   88,000 kg (194,000 lb)
  Max wing loading (without JATO):
    20° sweep   675.45 kg/m² (138.34 lb/sq ft)
    65° sweep   705.35 kg/m² (144.45 lb/sq ft)
  Max power loading (without JATO)
                                253 kg/kN (2.48 lb/lb st)
PERFORMANCE (Tu-22M-3):
  Max level speed: at high altitude
                  Mach 1.88 (1,080 knots; 2,000 km/h; 1,242 mph)
    at low altitude   Mach 0.9
  Nominal cruising speed   485 knots (900 km/h; 560 mph)
  T-O speed   200 knots (370 km/h; 230 mph)
  Normal landing speed   154 knots (285 km/h; 177 mph)
  Service ceiling   13,300 m (43,635 ft)
  T-O run   2,000-2,100 m (6,560-6,890 ft)
  Normal landing run   1,200-1,300 m (3,940-4,265 ft)
  Unrefuelled combat radius
                810-1,187 nm (1,500-2,200 km; 932-1,365 miles)
  g limit   +2.5

A pointed dielectric nosecone, like that of Tu-160, distinguishes the Tu-134UBL pilot trainer *(Piotr Butowsk*

TUPOLEV Tu-160
NATO reporting name: Blackjack

TYPE: Four-engined variable geometry long-range strategic bomber.

PROGRAMME: Designed as Aircraft 70 under leadership of V. I. Bliznuk; prototype observed by satellite at Ramenskoye flight test centre 25 November 1981 (photograph in 1982-83 *Jane's*), first flew 19 December 1981; US Defense Secretary Frank Carlucci invited to inspect twelfth aircraft built, at Kubinka air base, near Moscow, 2 August 1988; deliveries to 184th Regiment, Priluki air base, Ukraine, began May 1987; production at Kazan airframe plant ended 1992.

CUSTOMERS: Approx 20 deployed in two squadrons at Priluki.

DESIGN FEATURES: Intended for high altitude standoff role carrying ALCMs and for defence suppression, using short-range attack missiles similar to US Air Force SRAMs, along path of bomber making low altitude penetration to attack primary targets with free-fall nuclear bombs or missiles; this implies capability of subsonic cruise/supersonic dash at almost Mach 2 at 18,300 m (60,000 ft) and transonic flight at low altitude. About 20 per cent longer than USAF B-1B, with greater unrefuelled combat radius and much higher max speed; low-mounted variable geometry wings, with very long and sharply swept fixed root panel; small diameter circular fuselage; horizontal tail surfaces mounted high on fin, upper portion of which is pivoted one-piece all-moving surface; large dorsal fin; engines mounted as widely separated pairs in underwing ducts, each with central horizontal V wedge intakes and jetpipes extending well beyond wing centre-section trailing-edge; manually selected outer wing sweepback 20°, 35° and 65°; when wings fully swept, inboard portion of each trailing-edge flap hinges upward and extends above wing as large fence; unswept tail fin; sweptback horizontal surfaces, with conical fairing for brake-chute aft of intersection.

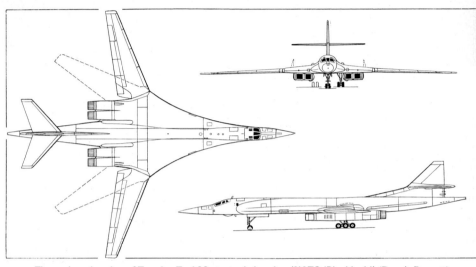

Three-view drawing of Tupolev Tu-160 strategic bomber (NATO 'Blackjack') *(Dennis Punnett)*

FLYING CONTROLS: Fly-by-wire. Full-span leading-edge flaps, long-span double-slotted trailing-edge flap and inset drooping aileron on each wing; all-moving vertical and horizontal one-piece tail surfaces.

STRUCTURE: Slim and shallow fuselage blended with wingroots and shaped for maximum hostile radar signal deflection.

LANDING GEAR: Twin nosewheels retract rearward; main gear comprises two bogies, each with three pairs of wheels; retraction very like that on Tu-154 airliner; as each leg pivots rearward, bogie rotates through 90° around axis of centre pair of wheels, to lie parallel with retracted leg; gear retracts into thickest part of wing, between fuselage a inboard engine on each side; so track relatively sma Nosewheel tyres size 1,080 × 400 mm; mainwheel tyr size 1,260 × 425 mm.

POWER PLANT: Four Samara/Trud NK-321 turbofans, each 2 kN (55,115 lb st) with afterburning. In-flight refuelli probe retracts into top of nose.

ACCOMMODATION: Four crew members in pairs, on individu K-36 ejection seats; one window each side of flight de can be moved inward and rearward for ventilation ground; flying controls use fighter type sticks rather tha yokes or wheels; crew enter via nosewheel bay.

Tupolev Tu-160 strategic bomber, known to NATO as 'Blackjack' *(Piotr Butowski)*

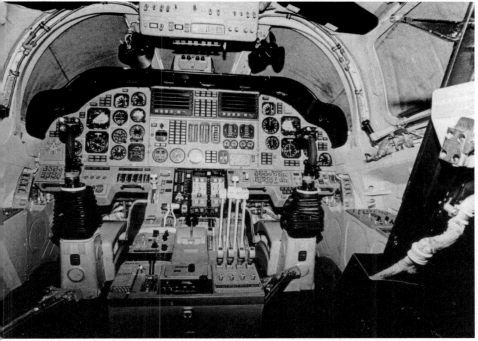

Cockpit of Tupolev Tu-160 supersonic strategic bomber

AVIONICS: Radar in slightly upturned dielectric nosecone claimed to provide terrain following capability; fairing with flat glazed front panel, under forward fuselage, for video camera to provide visual assistance for weapon aiming; astro-inertial nav with map display; no head-up display or CRTs; active jamming self-defence system.

ARMAMENT: No guns. Internal stowage for up to 16,330 kg (36,000 lb) of free-fall bombs, short-range attack missiles or ALCMs; a rotary launcher can be installed in each of two 12.80 m (42 ft) long weapon bays, carrying 12 Kh-15P (AS-16 'Kickback') SRAMs or six ALCMs, currently RK-55s (AS-15 'Kent'), intended to be superseded by supersonic AS-19s ('Koala').

DIMENSIONS, EXTERNAL:
Wing span: fully spread (20°)	55.70 m (182 ft 9 in)
35° sweep	50.70 m (166 ft 4 in)
fully swept (65°)	35.60 m (116 ft 9¾ in)
Length overall	54.10 m (177 ft 6 in)
Height overall	13.10 m (43 ft 0 in)
Tailplane span	13.25 m (43 ft 5¾ in)
Wheel track	5.40 m (17 ft 8½ in)
Wheelbase	17.88 m (58 ft 8 in)

WEIGHTS AND LOADINGS:
Max weapon load	16,330 kg (36,000 lb)
Max T-O weight	275,000 kg (606,260 lb)
Max power loading	280 kg/kN (2.75 lb/lb st)

PERFORMANCE:
Max level speed at high altitude
 Mach 1.88 (1,080 knots; 2,000 km/h; 1,240 mph)

Econ cruising speed	432 knots (800 km/h; 497 mph)
Service ceiling	18,300 m (60,000 ft)

Max unrefuelled range
 6,475 nm (12,000 km; 7,455 miles)

TUPOLEV Tu-134UBL and Tu-134BSh

TYPE: Conversions of Tu-134B airliners (NATO 'Crusty') for training pilots and navigators of CIS strategic aviation.

PROGRAMME: Conversions produced by Kharkov factory; first deliveries of Tu-134UBL to 184th Regiment, flying Tu-160, Priluki air base, early 1991.

CURRENT VERSIONS: **Tu-134UBL** (uchebno-boevoi dla lotchikov: trainer for pilots): New nose, containing Tu-160 radar, increases overall length nearly 5 m to 41.92 m (137 ft 6½ in); aerodynamic characteristics almost unchanged at angles of attack up to 30° (20° at high speed); deterioration of longitudinal and lateral stability at greater angles of attack requires reduced max T-O weight of 44,250 kg (97,550 lb) from Tu-134B's 47,600 kg (104,935 lb); max speed at 10,000 m (32,800 ft) reduced slightly to 464 knots (860 km/h; 534 mph).

Tu-134BSh (bombardirovochno-shturmanskiy: bomber navigational): As Tu-134UBL but Tu-22M radar in nose for fourth year training of student navigators at military school, Tambov; consoles in cabin with bombsight displays for 12 pupils; underwing racks for training bombs.

TUPOLEV Tu-154M and Tu-154S
NATO reporting name: Careless

TYPE: Three-turbofan medium-range transport.

PROGRAMME: Basic Tu-154 announced Spring 1966 to replace Tu-104, Il-18 and An-10 on Aeroflot medium/long stages up to 3,240 nm (6,000 km; 3,725 miles); first of six prototype/pre-production models flew 4 October 1968; regular services began 9 February 1972; more than 600 Tu-154s and Tu-154As, Bs and B-2s with uprated turbofans and other refinements delivered, over 500 to Aeroflot (last described 1985-86 *Jane's*); prototype Tu-154M (SSSR-85317), converted from standard Tu-154B-2 (see 1990-91 *Jane's*), first flew 1982; first two production aircraft delivered to Aeroflot from Kuybyshev plant 27 December 1984; nearly 1,000 built by September 1992; production continues.

CURRENT VERSIONS: **Tu-154M**: Basic airliner with alternative standard configurations for up to 180 passengers; executive version available; with all passenger seats removed can carry light freight. *Detailed description applies to this version.*

Tu-154M2: Modernised version; two Perm/Soloviev PS-90A turbofans, consuming 62 per cent as much fuel per passenger as Tu-154M; area navigation system. Intended to fly 1995. Life 20,000 h or 15,000 cycles.

Tu-154S: Specialised freight version, announced Autumn 1982; offered primarily as Tu-154B conversion; unobstructed main cabin cargo volume 72 m³ (2,542 cu ft); freight door 2.80 m (9 ft 2¼ in) wide and 1.87 m (6 ft 1½ in) high in port side of cabin, forward of wing, with ball mat inside and roller tracks full length of cabin floor; typical load nine standard international pallets 2.24 × 2.74 m (88 × 108 in) plus additional freight in standard underfloor baggage holds, volume 38 m³ (1,341 cu ft); nominal range 1,565 nm (2,900 km; 1,800 miles) with 20,000 kg (44,100 lb) cargo.

CUSTOMERS: Approx 75 Tu-154Ms operational with Aeroflot and its successors; Air Great Wall, China (two), Balkan Bulgarian Airlines (six); CAAC; China Northwest Airlines (10), China Southwest Airlines (five), China United Airlines (nine), Cubana (four), LOT, Poland (eight, plus three on order), Syrianair (three) and Xinjiang Airlines, China (five).

DESIGN FEATURES: Conventional all-swept low-wing configuration; two podded turbofans on sides of rear fuselage, third in extreme rear fuselage with intake at base of fin; nacelle to house retracted main landing gear on trailing-edge of each wing; wing sweep 35° at quarter-chord; anhedral on outer panels; geometric twist along span; circular section fuselage; sweepback at quarter-chord 40° on T tailplane; leading-edge sweep 45° on fin.

FLYING CONTROLS: Hydraulically actuated ailerons, triple-slotted flaps, four-section spoilers forward of flaps on each wing; electrically actuated slats on outer 80 per cent each wing leading-edge; tab in each aileron; electrically actuated variable incidence tailplane; rudder and elevators hydraulically actuated by irreversible servo controls; tab in each elevator.

STRUCTURE: All-metal; riveted three-spar wings, centre spar extending to just outboard of inner edge of aileron; semi-monocoque fail-safe fuselage; rudder and elevators of honeycomb sandwich construction.

LANDING GEAR: Hydraulically retractable tricycle type; main bogies, each three pairs of wheels in tandem, retract rearward into fairings on wing trailing-edge; rearward retract-

Tupolev Tu-160 ('Blackjack') bomber with wings fully swept and flap-end 'fences' visible above wingroots *(Paul Duffy)*

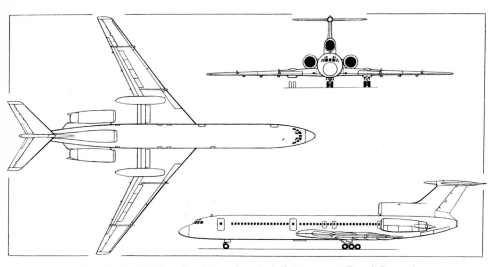

Tupolev Tu-154M medium-range three-turbofan transport *(Dennis Punnett)*

ing anti-shimmy twin-wheel nose unit, steerable through ±63°; disc brakes and anti-skid units on mainwheels.

POWER PLANT: Three Aviadvigatel D-30KU-154-II turbofans, each 104 kN (23,380 lb st), in pod each side of rear fuselage and inside extreme rear of fuselage; two lateral engines have clamshell thrust reversers. Integral fuel tanks in wings: four tanks in centre-section and two in outer wings; for reasons of trim, all fuel fed to collector tank in centre-section and thence to engines; single-point refuelling.

ACCOMMODATION: Crew of three; two pilots and flight engineer, with provisions for navigator and five cabin staff. Two passenger cabins, separated by service compartments; alternative configurations for 180 economy class passengers, 164 tourist class with hot meal service, or 154 tourist/economy plus separate first class cabin seating eight to 24 persons; mainly six-abreast seating with centre aisle; washable non-flammable materials used for all interior furnishing. Fully enclosed baggage containers. Toilet, galley and wardrobe to customer's requirements. Executive and light cargo configurations available. Passenger doors forward of front cabin and between cabins on port side, with emergency and service doors opposite; all four doors open outward; six emergency exits: two overwing and one immediately forward of engine nacelle each side. Two pressurised baggage holds under floor of cabin, with two inward opening doors; smaller unpressurised hold under rear of cabin.

SYSTEMS: Air-conditioning pressure differential 0.58 bar (8.4 lb/sq in). Three independent hydraulic systems, working pressure 207 bars (3,000 lb/sq in), powered by engine driven pumps; Nos. 2 and 3 systems each have additional electric backup pump; systems actuate landing gear retraction and extension, nosewheel steering, and operation of ailerons, rudder, elevators, flaps and spoilers. Three-phase 200/115V 400Hz AC electrical system supplied by three 40kVA alternators; additional 36V 400Hz AC and 27V DC systems and four storage batteries; TA-92 APU in rear fuselage. Hot air anti-icing of wing, fin and tailplane leading-edges, and engine air intakes; wing slats heated electrically. Engine fire extinguishing system in each nacelle; smoke detectors in baggage holds.

AVIONICS: Avionics meet ICAO standards for Cat. II weather minima and include updated navigation system with triplex INS; automatic flight control system operates throughout flight except during take-off to 400 m (1,312 ft) and landing from 30 m (100 ft); automatic go-round and automatic speed control provided by autothrottle down to 10 m (33 ft) on landing. Weather radar, transponder, Doppler, dual HF and VHF com and emergency VHF, cockpit voice recorder and GPWS standard.

DIMENSIONS, EXTERNAL:
Wing span	37.55 m (123 ft 2½ in)
Wing aspect ratio	7.0
Length overall	47.90 m (157 ft 1¾ in)
Height overall	11.40 m (37 ft 4¾ in)
Diameter of fuselage	3.80 m (12 ft 5½ in)
Tailplane span	13.40 m (43 ft 11½ in)
Wheel track	11.50 m (37 ft 9 in)
Wheelbase	18.92 m (62 ft 1 in)
Passenger doors (each): Height	1.73 m (5 ft 7 in)
Width	0.80 m (2 ft 7½ in)
Height to sill	3.10 m (10 ft 2 in)
Servicing door: Height	1.28 m (4 ft 2½ in)
Width	0.61 m (2 ft 0 in)
Emergency door: Height	1.28 m (4 ft 2½ in)
Width	0.64 m (2 ft 1¼ in)
Emergency exits (each): Height	0.90 m (2 ft 11½ in)
Width	0.48 m (1 ft 7 in)
Main baggage hold doors (each):	
Height	1.20 m (3 ft 11¼ in)
Width	1.35 m (4 ft 5 in)
Height to sill	1.80 m (5 ft 11 in)
Rear (unpressurised) hold:	
Height	0.90 m (2 ft 11½ in)
Width	1.10 m (3 ft 7¼ in)
Height to sill	2.20 m (7 ft 2½ in)

DIMENSIONS, INTERNAL:
Cabin: Width	3.58 m (11 ft 9 in)
Height	2.02 m (6 ft 7½ in)
Volume	163.2 m³ (5,763 cu ft)
Main baggage holds: front	21.5 m³ (759 cu ft)
rear	16.5 m³ (582 cu ft)
Rear underfloor hold	5.0 m³ (176 cu ft)

AREAS:
Wings, gross	201.45 m² (2,169 sq ft)
Horizontal tail surfaces (total)	42.20 m² (454.24 sq ft)

WEIGHTS AND LOADINGS:
Basic operating weight empty	55,300 kg (121,915 lb)
Max payload	18,000 kg (39,680 lb)
Max fuel	39,750 kg (87,633 lb)
Max T-O weight	100,000 kg (220,460 lb)
Max landing weight	80,000 kg (176,366 lb)
Max zero-fuel weight	74,000 kg (163,140 lb)
Max wing loading	496.4 kg/m² (101.6 lb/sq ft)
Max power loading	320.5 kg/kN (3.14 lb/lb st)

PERFORMANCE:
Max cruising speed	513 knots (950 km/h; 590 mph)
Max cruising height	11,900 m (39,000 ft)
Balanced field length for T-O and landing	2,500 m (8,200 ft)

Range:
with max payload 2,100 nm (3,900 km; 2,425 miles)
with 12,000 kg (26,455 lb) payload
    2,805 nm (5,200 km; 3,230 miles)
with max fuel and 5,450 kg (12,015 lb) payload
    3,563 nm (6,600 km; 4,100 miles)

TUPOLEV Tu-204

TYPE: Twin-turbofan medium-range airliner.

PROGRAMME: Development to replace Tu-154 announced 1983; preliminary details available Spring 1985; programme finalised 1986; first prototype (SSSR-64001) with PS-90AT engines, flown 2 January 1989 by Tupolev chief test pilot A. Talavkine; three more prototypes followed, plus two for structural and fatigue testing; second version, with RB211-535E4-B engines, flew 14 August 1992 and was demonstrated at Farnborough Air Show in following month; version with PW2240 engines expected at 1993 Paris Air Show; production by Aviastar joint stock company at Ulyanovsk plant scheduled at 80 to 90 a year to meet initial requirement for up to 500 for Aeroflot and its successors.

CURRENT VERSIONS: Three versions projected originally, with identical dimensions and internal arrangements; details of 93,500 kg (206,125 lb) **Tu-204** and 99,500 kg (219,355 lb) **Tu-204-100** in 1992-93 edition; production now centred on Tu-204-200 series, with further increase in max T-O weight and choice of Russian, British or US power plant, as follows:

Tu-204-210: With 156.9 kN (35,275 lb st) Aviadvigatel PS-90AT turbofans; additional fuel compared with Tu-204-100, carried in wing centre-section and adjacent baggage hold. Expected to be main production version for traditional Russian markets and domestic use.

Tu-204-220: Initially with 191.7 kN (43,100 lb st) Rolls-Royce RB211-535E4 turbofans in prototype (SSSR-64004), RB211-535F5s in production aircraft **funded and marketed by British Russian Aviation Corporation (BRAVIA), Moor House, London Wall, London EC2Y 5ET, England (Tel: 071 382 8279; Fax 071 382 8271)**. JAR certification scheduled May 1994; entry into service July 1994.

Tu-204-230: With 185.5 kN (41,700 lb st) Pratt & Whitney PW2240 turbofans.

DESIGN FEATURES: Conventional low/mid-wing configuration with all surfaces sweptback, and winglets; wing dihedral from roots; sweepback 28°; supercritical section; thickness/chord ratio 14 per cent at root, 9-10 per cent at tip; negative twist; semi-monocoque oval section pressurised fuselage; torsion box of fin forms integral fuel tank, used for automatic trimming of CG in flight; design life 45,000 flights or 60,000 flight hours.

FLYING CONTROLS: Triplex digital fly-by-wire, with triplex analog backup; conventional 'Y' control yokes selected after evaluation of alternative sidestick on Tu-154 testbed; inset aileron outboard of two-section double-slotted flap on each wing trailing-edge; two-section upper surface air

Tupolev Tu-154M medium-range airliner (three Aviadvigatel D-30KU-154-II turbofans) in service with CAAC of China *(Peter J. Bish)*

brake forward of each centre-section flap; five-section spoiler forward of each outer flap; four-section leading-edge slat over full span of each wing; conventional rudder and elevators; no tabs.

STRUCTURE: Approx 18 per cent of airframe by weight of composites; three-piece two-spar wing, with metal structure, part composite skin; carbonfibre skin on spoilers, airbrakes and flaps; glassfibre wingroot fairings; all-metal fuselage, utilising aluminium-lithium and titanium; nose radome and some access panels of composites; extensive use of composites in tail unit, particularly for leading-edges of fixed surfaces and for rudder and elevators.

LANDING GEAR: Hydraulically retractable tricycle type; electrohydraulically steerable twin-wheel nose unit (±10° via rudder pedals; ±70° by electric steering control) retracts forward; four-wheel bogie main units retract inward into wing/fuselage fairings. Carbon disc brakes, electrically controlled. Tyre size 1,070 × 390 mm on main-wheels, 840 × 290 mm on nosewheels.

POWER PLANT: Two turbofans (see notes on versions), underwing in composite cowlings. Fuel in six integral tanks in wings, in centre-section and adjacent baggage hold, and in tail fin, total capacity 40,730 litres (10,760 US gallons; 8,960 Imp gallons); torsion box of fin forms integral fuel tank for automatic trimming of CG in flight, as well as for optional standard use.

ACCOMMODATION: Can be operated by pilot and co-pilot, but Aeroflot specified requirement for a flight engineer; provision for fourth seat for instructor or observer. Three basic single-aisle passenger arrangements: (1) 190 seats, with 12 seats four-abreast in first class cabin at front at pitch of 99 cm (39 in), 35 business class seats six-abreast at pitch of 96 cm (38 in) in centre cabin, and 143 tourist class seats six-abreast at pitch of 81 cm (32 in) at rear; (2) 196 seats, with 12 seats four-abreast at pitch of 99 cm (39 in) in first class cabin at front and 184 six-abreast tourist class seats at pitch of 81 cm (32 in) at rear; (3) 214 seats, all six-abreast at tourist class pitch of 81 cm (32 in). All configurations have buffet/galley and toilet immediately aft of flight deck, and two more toilets, large buffet/galley and service compartment at rear of passenger accommodation; overhead stowage for hand baggage. Other layouts optional, with increased galley and toilet provisions. Passenger doors at front and rear of cabin on port side; service doors opposite. Type I emergency exit doors fore and aft of wing each side; inflatable slide for emergency use at each of eight doors. Two underfloor baggage/freight holds: forward hold accommodates five type 2AK-0.7 or 2AK-0.4 containers; rear hold accommodates seven containers. Fully automatic container loading system; manual backup.

SYSTEMS: Triplex fly-by-wire digital control system, with triplex analog backup. Three independent hydraulic systems, pressure 207 bars (3,000 lb/sq in). Ailerons, elevators, rudder, spoilers and airbrakes operated by all three systems; flaps, leading-edge slats, brakes and nosewheel steering operated by two systems; landing gear retraction and extension by all three systems. Electrical power supplied by two 200/115V 400Hz AC generators and a 27V DC system. Type TA-12-60 APU in tailcone.

AVIONICS: To ICAO Cat. III approach standards. Avionics of Russian design, or integrated by Sextant Avionique or Honeywell, with standard nav/com equipment by Collins or Bendix optional. EFIS equipment comprises two colour CRTs for flight and navigation information for each pilot, plus two central CRTs for engine and systems data; other

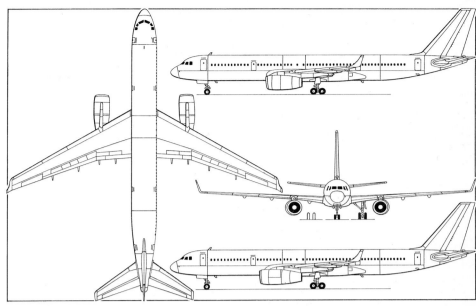

Tupolev Tu-204-210 medium-range transport (two Aviadvigatel PS-90AT turbofans), with additional side view (top) of Tu-204-220 *(Jane's/Mike Keep)*

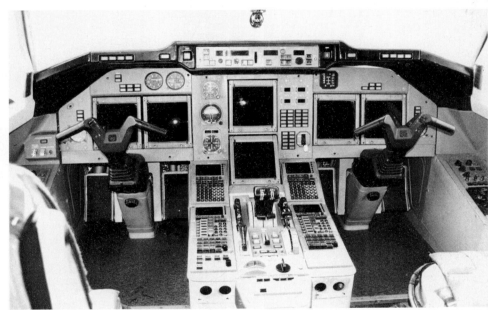

Cockpit of Tupolev Tu-204-220 twin-turbofan airliner *(Peter J. Cooper)*

Tupolev Tu-204-220 with Rolls-Royce RB211-535E4 turbofans at 1992 Farnborough Air Show *(Brian M. Service)*

equipment includes triplex automatic flight control, automatic approach and landing system, permitting operation in ICAO Cat. IIIA minima, VHF and HF radio, intercom, VOR, DME, inertial navigation system and satellite navigation system.

DIMENSIONS, EXTERNAL:
Wing span	42.00 m (137 ft 9½ in)
Wing aspect ratio	9.67
Length overall	46.22 m (151 ft 7¾ in)
Fuselage cross-section	3.80 m × 4.10 m (12 ft 5½ in × 13 ft 5½ in)
Height overall	13.88 m (45 ft 6½ in)
Tailplane span	15.00 m (49 ft 2½ in)
Wheel track	7.82 m (25 ft 8 in)
Wheelbase	17.00 m (55 ft 9¼ in)
Passenger doors (each): Height	1.85 m (6 ft 0¾ in)
Width	0.84 m (2 ft 9 in)
Service doors (each): Height	1.60 m (5 ft 3 in)
Width	0.65 m (2 ft 1½ in)
Emergency exit doors (each): Height	1.442 m (4 ft 8¾ in)
Width	0.61 m (2 ft 0 in)
Baggage holds: Height to sill	2.71 m (8 ft 10¾ in)

DIMENSIONS, INTERNAL:
Cabin, excl flight deck: Length	30.18 m (99 ft 0 in)
Max width	3.57 m (11 ft 8½ in)
Max height	2.28 m (7 ft 6 in)
Fwd cargo hold: Height	1.162 m (3 ft 9¾ in)
Volume	11.00 m³ (388 cu ft)
Rear cargo hold: Height	1.162 m (3 ft 9¾ in)
Volume	15.4 m³ (544 cu ft)
Bulk volume	5.4 m³ (190 cu ft)

AREAS:
Wings, gross	182.4 m² (1,963.4 sq ft)

WEIGHTS AND LOADINGS (Tu-204-220):
Operational weight empty	59,000 kg (130,070 lb)
Max payload:	
space limited (196 seats)	19,565 kg (43,132 lb)
weight limited	25,200 kg (55,555 lb)
Max baggage/freight: fwd hold	3,625 kg (7,990 lb)
rear hold	5,075 kg (11,190 lb)
Max fuel	32,700 kg (72,090 lb)
Max ramp weight	111,750 kg (246,360 lb)
Max T-O weight	110,755 kg (244,170 lb)
Max landing weight	89,500 kg (197,310 lb)
Max zero-fuel weight	84,200 kg (185,625 lb)
Max wing loading	607.2 kg/m² (124.4 lb/sq ft)
Max power loading	288.9 kg/kN (2.83 lb/lb st)

PERFORMANCE (estimated, basic Tu-204 with PS-90AT engines, at max T-O weight):
Cruising speed at 10,650-12,200 m (34,950-40,000 ft)	437-458 knots (810-850 km/h; 503-528 mph)
T-O speed	145 knots (269 km/h; 167 mph)
Approach speed	132 knots (245 km/h; 152 mph)
Time to cruising height after T-O	22-25 min
T-O run	1,230 m (4,035 ft)
T-O run, OEI	2,030 m (6,660 ft)
Balanced T-O field length (30°C)	2,500 m (8,200 ft)
Balanced landing field length	2,130 m (6,990 ft)
Landing run	850 m (2,800 ft)
Range at Mach 0.78 at 11,000 m (36,100 ft) with max fuel and payload of 19,000 kg (41,887 lb), representing 196 passengers	2,077 nm (3,850 km; 2,392 miles)

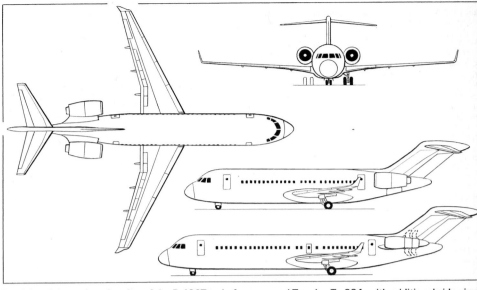

Provisional three-view drawing of the D-436T turbofan-powered Tupolev Tu-334, with additional side view (bottom) of stretched version with propfan engines (*Dennis Punnett*)

Model of Tu-334 medium-range passenger transport with BR 700 turbofans (*Stephane Guilbaud*)

TUPOLEV Tu-334

TYPE: Twin-turbofan or twin-propfan medium-range airliner.
PROGRAMME: Under development as replacement for Tu-134s on some Aeroflot routes; first flight of turbofan version scheduled 1993, for service 1994; propfan version at early design stage.
CURRENT VERSIONS: **Turbofan version:** For 86-102 passengers.
Propfan version: Basically similar; lengthened fuselage for 104-137 passengers; slightly reduced wing span and overall height; two rear-mounted pusher propfans, probably ZMKB Progress D-227s, each 78.5-88.25 kN (17,635-19,840 lb st).
DESIGN FEATURES: To meet Aeroflot's urgent requirements, Tu-334's wings have much in common with those of Tu-204, and its fuselage is shortened version of that of Tu-204, with identical flight deck; configuration is all-swept low/mid-wing, with rear-mounted engines and T tail; wings have supercritical section, with winglets; circular section semi-monocoque fuselage; wings have dihedral from roots, tailplane anhedral.
FLYING CONTROLS: Almost certainly fly-by-wire; aileron, two-section trailing-edge flap, two airbrakes forward of inner flap and four spoilers forward of outer flap on each wing; four-section leading-edge slat over full span each wing; inner flaps probably double-slotted, outer flaps single-slotted; conventional rudder and elevators; no tabs.
LANDING GEAR: Retractable tricycle type; twin-wheels on each unit; main units retract inward into wing/fuselage fairings; trailing-link mainwheel legs.
POWER PLANT: Two ZMKB Progress D-436T turbofans, each rated at 73.5 kN (16,535 lb st) (alternatively, two BMW Rolls-Royce BR 700 series turbofans, in initial version. Two propfans in later version.
ACCOMMODATION (turbofan): Crew of two or three on flight deck; provision for fourth seat for instructor or observer. Three basic single-aisle passenger arrangements: (1) 86 seats, with eight seats four-abreast in first class cabin at front, at pitch of 102 cm (40 in) and with 73 cm (28.75 in) aisle, 12 business class seats six-abreast at 96 cm (38 in) pitch and with 47 cm (18.5 in) aisle in centre cabin, and 66 tourist class seats six-abreast at 81 cm (32 in) pitch and with 47 cm (18.5 in) aisle at rear; (2) 92 seats, with eight first class and 84 tourist class at same pitches as 86-seater; (3) 102 seats, all tourist, at 81 cm (32 in) pitch. All configurations have buffet/galley and toilet immediately behind flight deck, a further buffet/galley, toilet and service compartment at rear; overhead stowage for hand baggage. Passenger doors at front and rear of cabin on port side; service doors opposite. Underfloor baggage/freight holds; doors on starboard side.
ACCOMMODATION (propfan): Flight deck unchanged. Four basic single-aisle passenger arrangements: (1) 104 seats, with eight first class, 12 business and 84 tourist at same seat pitches as in turbofan version; (2) 116 seats; with eight first class and 108 tourist; (3) 126 all-tourist; (4) 137 economy class at 75 cm (29.5 in) seat pitch. Facilities as turboprop version, plus emergency exit over wing each side.

DIMENSIONS, EXTERNAL (T: turbofan, P: propfan):
Wing span: T	29.8 m (97 ft 9 in)
P	29.1 m (95 ft 5¾ in)
Length overall: T	33.0 m (108 ft 3 in)
P	36.9 m (121 ft 0¾ in)
Height overall: T	8.8 m (28 ft 10½ in)
P	8.4 m (27 ft 6½ in)

DIMENSIONS, INTERNAL:
Cabin: Height: above aisle	2.155 m (7 ft 0¾ in)
beneath hand baggage racks	1.70 m (5 ft 7 in)
Floor width	3.50 m (11 ft 5¾ in)

AREAS:
Wings, gross: T	83.2 m² (895.6 sq ft)

WEIGHTS AND LOADINGS:
Max payload: T	11,000 kg (24,250 lb)
P	13,500 kg (29,760 lb)
Max T-O weight: T	43,600 kg (96,120 lb)
P	47,000 kg (103,615 lb)
Max wing loading: T	524.0 kg/m² (107.3 lb/sq ft)
Max power loading: T	296.6 kg/kN (2.91 lb/lb st)

PERFORMANCE (estimated):
Nominal cruising speed:	
T	431-442 knots (800-820 km/h; 497-510 mph)
P	431 knots (800 km/h; 497 mph)
Balanced runway length at 30°C:	
T, P	2,200-2,300 m (7,220-7,545 ft)
Range:	
T, with 9,251 kg (20,395 lb) payload (102 passengers)	1,620 nm (3,000 km; 1,865 miles)
P, with 11,430 kg (25,198 lb) payload (126 passengers)	1,860 nm (3,450 km; 2,143 miles)

YAKOVLEV

MOSKOVSKII MASHINOSTROITELNYY ZAVOD 'SKOROST' IMIENI A. S. YAKOVLEVA (Moscow Machine-building Factory 'Speed' named after A. S. Yakovlev) **(MMZ)**

68 Leningradsky Prospekt, Moscow 125315
Telephone: 7 (095) 157 17 34
Fax: 7 (095) 157 47 26
CHAIRMAN AND GENERAL DESIGNER: Alexander N. Dondukov
FIRST DEPUTY GENERAL DESIGNER: V. G. Dmitriev
FIRST DEPUTY GENERAL DESIGNER AND EXECUTIVE DIRECTOR:
  O. F. Demchenko
DEPUTY GENERAL DESIGNER (MARKETING AND CONVERSION):
  A. I. Gurtovoy
DEPUTY GENERAL DESIGNERS AND PROGRAMME MANAGERS:
  V. G. Dmitriev (Yak-46)
  S. A. Yakovlev (Yak-40TL and Yak-48)
  V. A. Mitjkin (Yak-112)
CHIEF DESIGNERS AND PROGRAMME MANAGERS:
  A. G. Rachimbaiev (Yak-42)
  Ju. I. Jankjevich (Yak-58, UAV)
  N. N. Dolzhenkov (Yak-UTS)
  K. F. Popovich (Yak-141)
  D. K. Drach (Yak-52, Yak-54, Yak-55M)
CHIEF OF MARKETING AND SALES DEPARTMENT: A. S. Ivanov
CHIEF OF EXTERIOR DEPARTMENT: E. M. Tarasov

Founder of this OKB, Alexander Sergeyevich Yakovlev, died on 22 August 1989, aged 83. He was one of most versatile Soviet designers. Products of his OKB ranged from transonic long-range fighters to Yak-24 tandem-rotor helicopter, an operational V/STOL carrier-based fighter and a variety of training, competition aerobatic and transport aircraft.

Association known as Skorost has been established by Yakovlev OKB and a number of aircraft and engine factories in CIS, including Saratov and Smolensk airframe plants and ZMKB Progress/Zaporozhye engine plants, to co-operate in production of civil aircraft. It was negotiating collaborative agreement with Hyundai of South Korea in 1992.

Yakovlev Yak-38M (NATO 'Forger-A') single-seat naval V/STOL combat aircraft *(Brian M. Service)*

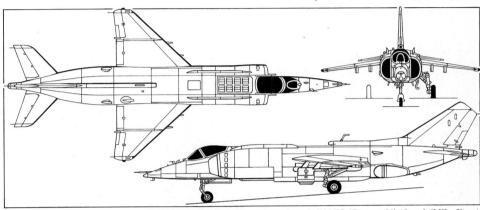

Yakovlev Yak-38M single-seat V/STOL carrier-based combat aircraft (NATO 'Forger-A') *(Jane's/Mike Keep)*

YAKOVLEV Yak-36M/Yak-38
NATO reporting name: Forger

TYPE: Single-seat ship-based V/STOL combat aircraft and two-seat trainer.
PROGRAMME: Prototype **Yak-36M** flew 1971; production began 1975 as world's second operational V/STOL jet combat aircraft after UK Harrier; deployed with Navy development squadron on *Kiev*, first of class of four 40,000 ton carrier/cruisers to put to sea 1976; subsequently 12 production Yak-38s on this ship and on each of sister ships, *Minsk*, *Novorossiysk* and *Admiral of the Fleet Gorshkov* (then *Baku*), later, on 65,000 ton aircraft carrier *Admiral of the Fleet Kuznetsov*; experimental operation from specially configured ro-ro ships reported. In process of retirement from service.
CURRENT VERSIONS: **Yak-38** ('Forger-A'): Basic single-seat combat aircraft; primary roles reconnaissance, strikes against small ships and fleet defence against shadowing maritime reconnaissance aircraft. Full description in 1991-92 and earlier *Jane's*; abbreviated description below.

Yak-38M ('Forger-A'): Developed single-seater; Soyuz/Tumansky R-28V-300 turbojet (65.7 kN; 14,770 lb st); each liftjet uprated to 31.9 kN (7,165 lb st); areas of engine air intakes and liftjet compartment correspondingly increased. Operational since 1984; one of first Yak-38Ms instrumented to check new features shown at 1992 Farnborough Air Show.

Yak-38U ('Forger-B'): Two-seat trainer; second cockpit forward of normal cockpit, with ejection seat at lower level, under continuous transparent canopy; to compensate for longer nose, plug inserted in fuselage aft of wing, lengthening constant-section portion without modification of tapering rear fuselage; no ranging radar or weapon pylons; two on each 'Kiev' class ship.
CUSTOMERS: CIS Naval Aviation (more than 230).
POWER PLANT: Primary power plant is Soyuz/Tumansky R-27V-300 turbojet (59.8 kN; 13,450 lb st), in centre-fuselage and exhausting through single pair of hydraulically actuated vectoring side nozzles aft of wings; no afterburner. Two RKBM Rybinsk RD-36-35FVR liftjets (each 30.4 kN; 6,835 lb st) in tandem immediately aft of cockpit, inclined 13° from vertical, exhausting downward and rearward, and used also to adjust pitch and trim. Fuel tanks in fuselage, forward and aft of main engine; drop tanks, each estimated 600 litres (158 US gallons; 132 Imp gallons), can be carried underwing.
ACCOMMODATION: Pilot only, on zero/zero ejection seat, under sideways hinged (to starboard) transparent canopy; armoured glass windscreen. Electronic system ejects pilot automatically if aircraft height and descent rate sensed to indicate emergency.
AVIONICS: Ranging radar in nose; SRO-2 (NATO 'Odd Rods') IFF transponder, with antennae forward of windscreen; other avionics in rear fuselage.
ARMAMENT: No installed armament. Two pylons under fixed panel of each wing for total 2,000 kg (4,410 lb) external stores, including gun pods each containing 23 mm twin-barrel GSh-23 cannon, rocket packs, bombs up to 500 kg each, Kh-23 (NATO AS-7 'Kerry') short-range air-to-surface missiles, armour-piercing anti-ship missiles, R-60 (AA-8 'Aphid') air-to-air missiles and auxiliary fuel tanks.

DIMENSIONS, EXTERNAL (estimated):
Wing span	7.32 m (24 ft 0 in)
Wing aspect ratio	2.90
Width, wings folded	4.88 m (16 ft 0 in)
Length overall: 'Forger-A'	15.50 m (50 ft 10¼ in)
'Forger-B'	17.68 m (58 ft 0 in)
Height overall	4.37 m (14 ft 4 in)
Tailplane span	3.81 m (12 ft 6 in)
Wheel track	2.90 m (9 ft 6 in)
Wheelbase	5.50 m (18 ft 0 in)

AREAS (estimated):
Wings, gross	18.5 m² (199 sq ft)

WEIGHTS AND LOADINGS (estimated):
Basic operating weight, incl pilot(s):	
'Forger-A'	7,485 kg (16,500 lb)
'Forger-B'	8,390 kg (18,500 lb)
Max T-O weight	11,700 kg (25,795 lb)
Max wing loading:	
'Forger-A'	632.4 kg/m² (129.6 lb/sq ft)
Max power loading:	
'Forger-A'	91.4 kg/kN (0.90 lb/lb st)

PERFORMANCE ('Forger-A', estimated, at max T-O weight):
Max level speed at height	Mach 0.95 (545 knots; 1,009 km/h; 627 mph)
Max level speed at S/L	Mach 0.8 (528 knots; 978 km/h; 608 mph)
Max rate of climb at S/L	4,500 m (14,750 ft)/min
Service ceiling	12,000 m (39,375 ft)
Combat radius:	
with air-to-air missiles and external tanks, 75 min on station	100 nm (185 km; 115 miles)
with max weapons, lo-lo-lo	130 nm (240 km; 150 miles)
with max weapons, hi-lo-hi	200 nm (370 km; 230 miles)

SKOROST/YAKOVLEV Yak-40TL

TYPE: Twin-turbofan conversion of Yak-40 three-engined transport.
PROGRAMME: Basic Yak-40 was last described fully in 1978-79 *Jane's*; Yak-40TL conversion announced 1991, now available.
DESIGN FEATURES: To increase fuel efficiency and reduce operating costs, three 14.7 kN (3,300 lb st) Ivchenko AI-25 turbofans replaced with two Textron Lycoming LF 507-1N turbofans, each 31.14 kN (7,000 lb st), on each side of rear fuselage; air intake for former central engine removed and rear fuselage rebuilt with dorsal fin and tailcone (see accompanying three-view drawing); 10 per cent improvement in cruising speed; hot and high performance improved; updated avionics to customer's choice; 27/32-passenger accommodation unchanged.

WEIGHTS AND LOADINGS:
Basic operating weight	10,865 kg (23,950 lb)

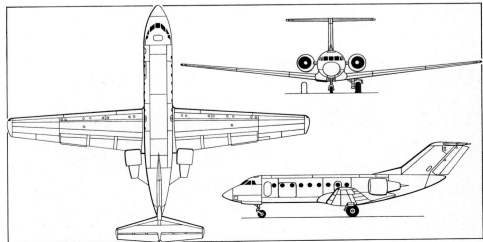

Skorost/Yakovlev Yak-40TL twin-turbofan conversion of Yak-40 transport *(Jane's/Mike Keep)*

Normal payload	2,720 kg (6,000 lb)
Max fuel	4,000 kg (8,820 lb)
Max T-O weight	16,575 kg (36,540 lb)
Max wing loading	236.8 kg/m² (48.5 lb/sq ft)
Max power loading	266.1 kg/kN (2.61 lb/lb st)

PERFORMANCE (estimated):
 Nominal cruising speed at 8,000 m (26,250 ft)
  297 knots (550 km/h; 342 mph)
 Balanced runway length 1,400 m (4,600 ft)
 Range at nominal cruising speed, 30 min fuel reserves:
  with normal payload 674 nm (1,250 km; 775 miles)
  with max fuel 1,050 nm (1,950 km; 1,210 miles)

YAKOVLEV Yak-42

NATO reporting name: Clobber

TYPE: Three-turbofan short/medium-range passenger transport.

PROGRAMME: Three prototypes ordered initially; first prototype (SSSR-1974) flew 7 March 1975, with 11° wing sweepback and furnished in 100-seat local service form, with carry-on baggage and coat stowage fore and aft of cabin; second prototype (SSSR-1975) had 23° sweepback and more cabin windows, representative of 120-seat version with three more rows of seats and no carry-on baggage areas; third prototype (SSSR-1976, later SSSR-42303) introduced small refinements (described in previous editions of *Jane's*); flight testing proved 23° wing superior; first series of production aircraft, built at Smolensk to replace some Aeroflot Tu-134s, generally similar to SSSR-42303 as exhibited 1977 Paris Air Show; changes for production included substitution of four-wheel main landing gear bogies for twin-wheel units on prototypes; scheduled service on Aeroflot's Moscow-Krasnodar route began late 1980.

CURRENT VERSIONS: **Yak-42**: Basic standard, *described in detail*; Aeroflot had 72 in Autumn 1990.
 Yak-42D: Increased fuel; range increased to 1,185 nm (2,200 km; 1,365 miles) with 120 passengers; about 20 delivered by early 1991; production 15 a year, to continue five more years.
 Yak-42E-LL: Propfan testbed; starboard D-36 turbofan replaced by ZMKB Progress D-236 geared propfan (8,090 kW; 10,850 shp), driving SV-36 tractor contraprops with eight forward blades and six rear blades; propeller diameter 4.2 m (13 ft 9½ in). First flown 15 March 1991; exhibited 1991 Paris Air Show.
 Yak-42F: Basically standard Yak-42 adapted for electro-optical research; equipment in very large cylindrical pod under inboard panel of each wing. Unknown number used by Aeroflot for Earth resources and environmental survey.
 Yak-42M: See separate entry.

CUSTOMERS: 100 built by Autumn 1990, mostly for Aeroflot and its successors; two for Cubana, six for China General Aviation Corporation.

DESIGN FEATURES: Basic design objectives simple construction, reliability in operation, economy, and ability to operate in remote areas with widely differing climatic conditions; design is in accordance with CIS civil airworthiness standards and US FAR 25; engines conform with international smoke and noise limitations, and aircraft is intended to operate in ambient temperatures from −50°C to +50°C; APU for engine starting and services removes need for ground equipment.
 Conventional all-swept low-wing configuration; two podded turbofans on sides of rear fuselage, third in extreme rear fuselage with intake at base of fin; no wing dihedral or anhedral; sweepback 23° at quarter-chord; basically circular fuselage, blending into oval rear section; T tail.

FLYING CONTROLS: Hydraulically actuated; two-section ailerons, each with servo tab on inner section, trim tab on outer section; two-section single-slotted trailing-edge flap, three-section spoiler forward of outer flap, full-span leading-edge flap on each wing; one-piece variable incidence tailplane (from 1° up to 12° down); trim tab in each elevator, trim tab and spring servo tab in rudder.

STRUCTURE: All-metal; two-spar torsion box wing structure; riveted, bonded and welded semi-monocoque fuselage.

LANDING GEAR: Hydraulically retractable tricycle heavy-duty type, made by Hydromash; four-wheel bogie main units retract inward into flattened fuselage undersurface; twin nosewheels retract forward; hydraulic backup for extension only; emergency extension by gravity; oleo-nitrogen shock absorbers; steerable nose unit of levered suspension type; low pressure tyres; 930 × 305 mm on nosewheels; hydraulic disc brakes on mainwheels; nosewheel brakes to stop wheel rotation after take-off.

POWER PLANT: Three ZMKB Progress D-36 three-shaft turbofans, each 63.74 kN (14,330 lb st); centre engine, inside rear fuselage, has S-duct air intake; outboard engines in pod on each side of rear fuselage; no thrust reversers. Integral fuel tanks between spars in wings, capacity approx 23,175 litres (6,120 US gallons; 5,100 Imp gallons).

ACCOMMODATION: Crew of two side by side, with provision for flight engineer; two or three cabin attendants. Single passenger cabin, with 120 seats six-abreast, at pitch of 75 cm (29.5 in), with centre aisle 45 cm (17.7 in) wide, in high-density configuration. Alternative 104-passenger (96 tourist, eight first class) local service configuration, with carry-on baggage and coat stowage compartments fore and aft of cabin. Main airstair door hinges down from undersurface of rear fuselage; second door forward of cabin on port side, with integral airstairs; service door opposite. Galley and crew coat stowage between flight deck and front vestibule; passenger coat stowage and toilet between vestibule and cabin; second coat stowage and toilet at rear of cabin. Two underfloor holds for cargo, mail and baggage in nets or standard containers, loaded through door on starboard side, forward of wing; chain-drive handling system for containers built into floor; forward hold for six containers, each 2.2 m³ (77.7 cu ft); rear hold takes two similar containers. Provision for convertible passenger/cargo interior, with enlarged loading door on port side of front fuselage. Two emergency exits overwing each side. All passenger and crew accommodation pressurised, air-conditioned, and furnished with non-flammable materials.

SYSTEMS: APU standard, for engine starting, and for power and air-conditioning supply on ground and, if necessary, in flight.

AVIONICS: Flight and navigation equipment for operation by day and night under adverse weather conditions, with landings on concrete or unpaved runways in ICAO Category II

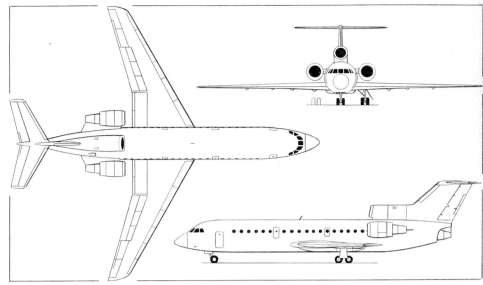

Yakovlev Yak-42 three-turbofan short/medium-range passenger transport *(Dennis Punnett)*

Yakovlev Yak-42D short/medium-range transport *(Paul R. Duffy)*

Yakovlev Yak-42E-LL testbed for a ZMKB Progress D-236 geared propfan in starboard nacelle *(Brian M. Service)*

Yakovlev Yak-42F fitted with electro-optical pods. Forward half of pod shell rotates to uncover windows *(Air Portraits)*

weather minima down to 30 m (100 ft) visibility at 300 m (985 ft). Type SAU-42 automatic flight control system and automatic navigation system standard.

DIMENSIONS, EXTERNAL:
Wing span	34.88 m (114 ft 5¼ in)
Wing aspect ratio	8.11
Length overall	36.38 m (119 ft 4¼ in)
Fuselage diameter	3.80 m (12 ft 5½ in)
Height overall	9.83 m (32 ft 3 in)
Tailplane span	10.80 m (35 ft 5 in)
Wheel track	5.63 m (18 ft 5¾ in)
Wheelbase	14.78 m (48 ft 6 in)
Passenger door (fwd):	
Height	1.50 m (4 ft 11 in)
Width	0.83 m (2 ft 8½ in)
Passenger entrance (rear): Height	1.78 m (5 ft 10 in)
Width	0.81 m (2 ft 7¾ in)
Cargo door (convertible version):	
Height	2.025 m (6 ft 7¾ in)
Width	3.23 m (10 ft 7 in)
Baggage/cargo hold door: Height	1.35 m (4 ft 5 in)
Width	1.145 m (3 ft 9 in)
Height to sill	1.45 m (4 ft 9 in)

DIMENSIONS, INTERNAL:
Cabin: Length	19.89 m (65 ft 3 in)
Max width	3.60 m (11 ft 9¾ in)
Max height	2.08 m (6 ft 9¾ in)
Baggage compartment volume (100-seater):	
fwd	19.8 m³ (700 cu ft)
rear	9.5 m³ (335 cu ft)

AREAS:
Wings, gross	150.0 m² (1,615 sq ft)
Vertical tail surfaces (total)	23.29 m² (250.7 sq ft)
Horizontal tail surfaces (total)	27.60 m² (297.1 sq ft)

WEIGHTS AND LOADINGS:
Weight empty, equipped:	
104 seats	34,500 kg (76,058 lb)
120 seats	34,515 kg (76,092 lb)
Max payload	13,000 kg (28,660 lb)
Max fuel	18,500 kg (40,785 lb)
Max T-O weight	57,000 kg (125,660 lb)
Max landing weight	51,000 kg (112,433 lb)
Max wing loading	380.0 kg/m² (77.8 lb/sq ft)
Max power loading	298.1 kg/kN (2.92 lb/lb st)

PERFORMANCE:
Max cruising speed at 7,620 m (25,000 ft)	
	437 knots (810 km/h; 503 mph)
Econ cruising speed	405 knots (750 km/h; 466 mph)
T-O speed	119 knots (220 km/h; 137 mph) IAS
Approach speed	114 knots (210 km/h; 131 mph) IAS
Max cruising height	9,600 m (31,500 ft)
T-O balanced field length	2,200 m (7,220 ft)
Landing from 15 m (50 ft)	1,100 m (3,610 ft)
Range at econ cruising speed, with 3,000 kg (6,615 lb) fuel reserves:	
with max payload	745 nm (1,380 km; 857 miles)
with 120 passengers (10,800 kg; 23,810 lb payload)	
	1,080 nm (2,000 km; 1,243 miles)
with 104 passengers (9,360 kg; 20,635 lb payload)	
	1,240 nm (2,300 km; 1,430 miles)
with max fuel and 42 passengers	
	2,210 nm (4,100 km; 2,545 miles)

YAKOVLEV Yak-42M

TYPE: Three-turbofan short/medium-range passenger transport.
PROGRAMME: Development began 1987; prototype construction started 1991; production to begin 1995, at Saratov.
CURRENT VERSIONS: Three single-class configurations planned, for 150, 156 and 168 passengers; two mixed-class for 110 and 122 passengers.
CUSTOMERS: Aeroflot and its successors are considering purchase of 200.

DESIGN FEATURES: Much more than a stretched Yak-42, only fuselage cross-section is not new; wings have supercritical section devised in collaboration with TsAGI, with sweepback of 25°, increased span and higher aspect ratio; double-slotted trailing-edge flaps, and winglets; together with new turbofans (two underwing and one, like DC-10, at base of tail fin), this gives increased performance; fly-by-wire flight controls and five-CRT EFIS are standard; avionics upgraded for ICAO Cat. IIIa minima.
FLYING CONTROLS: Fly-by-wire.
LANDING GEAR: Hydraulically retractable tricycle type; four-wheel bogie main units; twin nosewheels.
POWER PLANT: Three ZMKB Progress D-436M turbofans, each rated at 73.5 kN (16,535 lb st) and fitted with thrust reverser.
ACCOMMODATION: Crew of two side by side on flight deck, with provision for flight engineer if required; basic accommodation for 156 tourist class passengers, six-abreast at 78 cm (30.7 in) seat pitch, with centre aisle; toilet and buffet/galley immediately aft of flight deck; two toilets at rear of cabin; alternative single class arrangements for 150 passengers at 81 cm (32 in) seat pitch, and 168 passengers; two mixed class configurations, for eight business and 114 tourist, or eight first class, 30 business and 72 tourist; main airstair door hinges down from undersurface of rear fuselage; second door forward of cabin on port side, with airstairs; service door opposite; emergency exit doors forward and at rear of cabin each side; two overwing emergency exit windows each side.
AVIONICS: TsPNK-42M digital flight control and navigation system and five-CRT EFIS standard.

Model of three-turbofan Yakovlev Yak-42M

DIMENSIONS, EXTERNAL:
Wing span	36.25 m (118 ft 11¼ in)
Wing aspect ratio	10.95
Length overall	38.00 m (124 ft 8 in)

AREAS:
Wings, gross	120.0 m² (1,292 sq ft)

WEIGHTS AND LOADINGS:
Weight empty, equipped	37,400 kg (82,450 lb)
Payload: with max fuel (156 passengers)	
	9,600 kg (21,165 lb)
normal	14,820 kg (32,672 lb)
max	16,500 kg (36,375 lb)
Max T-O weight	63,600 kg (140,210 lb)
Max wing loading	530 kg/m² (108.5 lb/sq ft)
Max power loading	288.4 kg/kN (2.83 lb/lb st)

PERFORMANCE (estimated):
Max cruising speed	448 knots (830 km/h; 515 mph)
Econ cruising speed	431 knots (800 km/h; 497 mph)
Normal cruising height	11,100 m (36,400 ft)
Balanced runway length at 30°C	2,200 m (7,220 ft)
Range, with reserves:	
with max payload	1,050 nm (1,950 km; 1,210 miles)
with normal payload	1,350 nm (2,500 km; 1,550 miles)
with max fuel	2,428 nm (4,500 km; 2,795 miles)

YAKOVLEV Yak-46 (Propfan)

TYPE: Twin-propfan short/medium-range passenger transport.
PROGRAMME: Design study and model shown at Moscow Aerospace '90 air show; prototype could fly 1995, if given go-ahead.
CURRENT VERSIONS: Four versions projected: single class with 168 tourist seats; mixed class for 12 first class and 114 tourist passengers; convertible passenger/freighter; casualty evacuation version.
DESIGN FEATURES: Airframe identical with Yak-42M, except for new rear fuselage and tail unit, with two side-mounted pusher propfans on pylons.
POWER PLANT: Two ZMKB Progress D-27 propfans, each rated at 109.8 kN (24,690 lb st), driving contraprops, one with eight blades, other with six blades.
ACCOMMODATION: Generally similar to comparable versions of Yak-42M, except for deletion of rear underfuselage airstairs.
DIMENSIONS, EXTERNAL: As Yak-42M, except:
Length overall	41.00 m (134 ft 6 in)
Propeller diameter (each)	3.80 m (12 ft 5¾ in)

Projected Yakovlev Yak-46 twin-propfan airliner *(Jacques Marmain)*

30 passengers	2,850 kg (6,285 lb)
max	3,600 kg (7,935 lb)
Max fuel	7,400 kg (16,315 lb)
Max ramp weight	18,300 kg (40,350 lb)
Max T-O weight	18,200 kg (40,125 lb)
Normal landing weight	13,500 kg (29,760 lb)
Max landing weight	15,000 kg (33,070 lb)
Max power loading	292.4 kg/kN (2.87 lb/lb st)

PERFORMANCE (estimated):

Max cruising speed	458 knots (850 km/h; 528 mph)
Econ cruising speed at 12,500 m (41,000 ft)	
	431 knots (800 km/h; 497 mph)
T-O speed	132 knots (245 km/h; 153 mph)
Approach speed	130 knots (240 km/h; 150 mph)
Balanced runway length, ISA	1,630 m (5,350 ft)
Normal landing run: dry runway	840 m (2,755 ft)
wet runway	1,620 m (5,320 ft)
Range with 30 passengers	
	2,466 nm (4,570 km; 2,840 miles)
Range with max fuel, 8 passengers, with reserves	
	3,505 nm (6,500 km; 4,040 miles)

YAKOVLEV Yak-52

Tandem two-seat piston engined primary trainer produced by Aerostar SA in Romania (which see).

YAKOVLEV Yak-54

TYPE: Tandem two-seat sporting and aerobatic training aircraft.
PROGRAMME: Announced 1992; prototype at 1993 Paris Air Show, scheduled to fly before year-end; five to be built at Saratov in 1993.
DESIGN FEATURES: Conventional mid-wing configuration; symmetrical section; no dihedral, anhedral or incidence; almost full-span ailerons, elevators and rudder all horn balanced; each aileron has large suspended balance tab; non-retractable tailwheel type landing gear, with cantilever main legs and small wheels. Designed on basis of systems and units of Yak-55M.
STRUCTURE: All-metal; two-spar wings; semi-monocoque fuselage; conventional tail unit; titanium spring main landing gear legs.

WEIGHTS AND LOADINGS:

Weight empty, equipped	37,300 kg (82,230 lb)
Max payload	17,500 kg (38,580 lb)
Max T-O weight	61,300 kg (135,140 lb)
Max wing loading	510.8 kg/m² (104.6 lb/sq ft)
Max power loading	279.1 kg/kN (2.74 lb/lb st)

PERFORMANCE (estimated):

Nominal cruising speed	448 knots (830 km/h; 515 mph)
Nominal cruising height	11,100 m (36,400 ft)
Balanced runway length, ISA	2,100 m (6,890 ft)
Range:	
with max payload	971 nm (1,800 km; 1,118 miles)
with normal payload	1,888 nm (3,500 km; 2,175 miles)

YAKOVLEV Yak-46 (Turbofan)

TYPE: Twin-turbofan short/medium-range passenger transport.
PROGRAMME: Design study announced early 1991.
CURRENT VERSIONS: As for Yak-46 (Propfan).
DESIGN FEATURES: Despite having same designation, this aircraft is entirely different from Yak-46 (Propfan) except for same basic fuselage and accommodation; use of underwing advanced turbofans has necessitated moving wings forward; in turn, this has permitted use of conventional swept tail unit, with tailplane mounted on tailcone; fly-by-wire flight controls and advanced avionics as Yak-46 (Propfan).
POWER PLANT: Two Samara/SSPE/Trud high bypass turbofans, each rated at 107.9 kN (24,250 lb st).

DIMENSIONS, EXTERNAL: As Yak-42M, except:

Length overall	38.80 m (127 ft 3½ in)

WEIGHTS AND LOADINGS:

Weight empty, equipped	34,840 kg (76,808 lb)
Max payload	17,500 kg (38,580 lb)
Max T-O weight	60,200 kg (132,715 lb)
Max wing loading	501.7 kg/m² (102.7 lb/sq ft)
Max power loading	278.7 kg/kN (2.74 lb/lb st)

PERFORMANCE (estimated):

Nominal cruising speed	
	448-458 knots (830-850 km/h; 515-528 mph)
Nominal cruising height	11,100 m (36,400 ft)
Balanced runway length, ISA	2,100 m (6,890 ft)
Range:	
with max payload	1,187 nm (2,200 km; 1,367 miles)
with normal payload	1,860 nm (3,450 km; 2,143 miles)

YAKOVLEV Yak-48

TYPE: Projected 8/10-passenger twin-turbofan business aircraft and 30-passenger regional transport.
PROGRAMME: Design greatly refined since original concept displayed in model form at Moscow Aerospace '90; Western power plant and avionics now standard.
DESIGN FEATURES: All-swept low-wing configuration with compound wing leading-edge sweepback, rear podded turbofans of US manufacture and winglets; tailplane near tip of tail fin; tricycle landing gear with twin wheels on each unit; advanced avionics and equipment.
POWER PLANT: Two Textron Lycoming LF 507-1N turbofans, each 31.14 kN (7,000 lb st); APU in rear fuselage.
ACCOMMODATION: Two crew side by side on flight deck; door from flight deck to lobby with main passenger door on port side, service door opposite, and buffet; wide-body cabin for two pairs of club-style armchairs with tables between, four-place sofa on starboard side, two more facing seats with table on port side; toilet and wardrobe to rear; large baggage space aft, with door beneath port engine pod; overwing emergency exit each side. Regional jet layout has 30 seats four-abreast in pairs with centre aisle; coat space at front of cabin, buffet and toilet at rear; unchanged baggage compartment. No airframe modification or structural strengthening required for role change.
AVIONICS: Integrated digital flight control and navigation systems, with CRTs, for flight to ICAO Cat. II minima.

DIMENSIONS, EXTERNAL:

Wing span	21.60 m (70 ft 10½ in)
Length overall	21.40 m (70 ft 2½ in)
Height overall	6.00 m (19 ft 8¼ in)
Wheel track	3.20 m (10 ft 6 in)
Wheelbase	7.50 m (24 ft 7¼ in)

WEIGHTS AND LOADINGS:

Operating weight empty, incl crew	9,840 kg (21,695 lb)
Payload: 8 passengers	960 kg (2,116 lb)

Yakovlev Yak-48, a projected 8/10-passenger business aircraft *(Jane's/Mike Keep)*

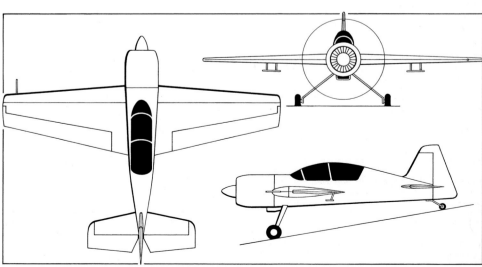

Yakovlev Yak-54 tandem two-seat sporting aircraft *(Jane's/Mike Keep)*

POWER PLANT: One 265 kW (355 hp) Vedeneyev M-14P nine-cylinder air-cooled radial engine; three-blade variable-pitch propeller.
ACCOMMODATION: Two seats in tandem under continuous transparent canopy.
DIMENSIONS, EXTERNAL:
Wing span   8.16 m (26 ft 9¼ in)
Length overall   6.91 m (22 ft 8 in)
AREAS:
Wings, gross   12.89 m² (138.75 sq ft)
WEIGHTS AND LOADINGS:
Max T-O weight: one pilot   850 kg (1,874 lb)
        two occupants   990 kg (2,182 lb)
Max wing loading   76.8 kg/m² (15.73 lb/sq ft)
Max power loading   3.74 kg/kW (6.15 lb/hp)
PERFORMANCE (estimated):
Max level speed   243 knots (450 km/h; 280 mph)
Stalling speed   60 knots (110 km/h; 69 mph)
Rate of roll   345°/s
Rate of climb at S/L   900 m (2,950 ft)/min
Ferry range   377 nm (700 km; 435 miles)
g limits   +9/–7

Yakovlev Yak-55M single-seat aerobatic competition aircraft

YAKOVLEV Yak-55M

TYPE: Single-seat competition aerobatic aircraft.
PROGRAMME: Original Yak-55 made unanticipated appearance at 11th World Aerobatic Championships, Spitzerberg, Austria, August 1982; considerable subsequent refinement, including early change to stronger, tapered wings of thinner section for 1984 Championships; current Yak-55M, with further refinements, accompanied Soviet team to 1989 European Aerobatic Championships in Hungary; series production began 1991.
DESIGN FEATURES: Mid-wing configuration; symmetrical section, no dihedral, anhedral or incidence; almost full-span ailerons, elevators and rudder all horn balanced, with ground adjustable tab; each aileron also has large suspended balance tab; non-retractable tailwheel type landing gear, with bowed cantilever main legs and small wheels.
STRUCTURE: All-metal; two-spar wings; semi-monocoque fuselage; conventional tail unit; titanium spring main landing gear legs; rearward sliding canopy.
POWER PLANT: One 265 kW (355 hp) Vedeneyev M-14P nine-cylinder air-cooled radial engine; two-blade controllable-pitch propeller; wing fuel tanks, capacity 120 litres (31.5 US gallons; 26 Imp gallons).
DIMENSIONS, EXTERNAL:
Wing span   8.10 m (26 ft 6¾ in)
Wing aspect ratio   5.13
Length overall   7.00 m (22 ft 11½ in)
Height overall   2.80 m (9 ft 2¼ in)
Tailplane span   3.15 m (10 ft 4 in)
Wheel track   2.24 m (7 ft 4¼ in)
AREAS:
Wings, gross   12.8 m² (137.8 sq ft)
WEIGHTS AND LOADINGS:
Max T-O weight   840 kg (1,852 lb)
Max wing loading   65.6 kg/m² (13.44 lb/sq ft)
Max power loading   3.13 kg/kW (5.14 lb/hp)
PERFORMANCE:
Max level speed   243 knots (450 km/h; 280 mph)
Stalling speed   57-60 knots (105-110 km/h; 66-69 mph)
Max rate of climb at S/L   930 m (3,050 ft)/min
Rate of roll   345°/s
g limits   +9/–6

YAKOVLEV Yak-56

Programme for this two-seat aerobatic aircraft has been cancelled.

YAKOVLEV Yak-58

TYPE: Six-seat light multi-purpose aircraft.
PROGRAMME: Shown in model form at Moscow Aerospace '90; full-scale mockup exhibited February 1991; OKB claims receipt of 250 letters of intent to purchase. Could fly in 1993.
DESIGN FEATURES: Constant chord unswept wing with dihedral from roots and cambered tips; fuselage pod mounted above wing, with annular duct at rear to house air-cooled pusher engine; short twin booms carry sweptback and slightly toed-in tail fins and bridging horizontal tail surface.
FLYING CONTROLS: Mechanical control system for ailerons, twin rudders and elevator; pneumatically operated trailing-edge flaps.
LANDING GEAR: Pneumatically actuated retractable tricycle type; single wheel and low-pressure tyre on each trailing-link unit, for operation from unprepared strips; mainwheels retract inward, nosewheel forward.
POWER PLANT: One 265 kW (355 hp) Vedeneyev M-14PT nine-cylinder radial engine enclosed in annular duct; three-blade variable-pitch pusher propeller. Location reduces noise in cabin.
ACCOMMODATION: Six persons in pairs in enclosed cabin; large sliding door on starboard side, facilitating freight loading when passenger seats removed, or despatch of parachutists. Planned uses include business, taxi and ambulance transport; surveillance of forests, high tension cables, oilfields and fisheries; mail and freight operation.

Full-scale mockup of Yakovlev Yak-58 six-seat business aircraft

SYSTEMS: Two independent pneumatic systems for flaps and landing gear actuation and engine starting.
DIMENSIONS, EXTERNAL:
Wing span   12.70 m (41 ft 8 in)
Wing aspect ratio   8.06
Length overall   8.55 m (28 ft 0½ in)
Height overall   3.16 m (10 ft 4½ in)
AREAS:
Wings, gross   20.0 m² (215.3 sq ft)
WEIGHTS AND LOADINGS:
Weight empty   1,270 kg (2,800 lb)
Max payload   450 kg (992 lb)
Max T-O weight   2,100 kg (4,630 lb)
Max wing loading   105.0 kg/m² (21.5 lb/sq ft)
Max power loading   7.84 kg/kW (12.9 lb/hp)
PERFORMANCE (estimated):
Max level speed   162 knots (300 km/h; 186 mph)
Max cruising speed   153 knots (285 km/h; 177 mph)
Landing speed   68 knots (125 km/h; 78 mph)
Service ceiling   4,000 m (13,125 ft)
T-O run   610 m (2,000 ft)
Landing run   600 m (1,970 ft)
Range with max payload, 45 min reserves
                                more than 540 nm (1,000 km; 620 miles)

Prototype Yakovlev Yak-112 four-seat multi-purpose lightplane *(Vladimir Shcheglov)*

YAKOVLEV Yak-112

TYPE: Four-seat multi-purpose light aircraft.
PROGRAMME: Winner of 1988 official design competition for two-seat primary trainer for CIS aero clubs; developed subsequently to current four-seat configuration; model shown at Moscow Aerospace '90 and full-scale mockup at 1991 Paris Air Show; first flight of prototype 20 October 1992; production scheduled to begin at Saratov and Irkutsk 1993.
CURRENT VERSIONS: Intended to carry passengers, light cargo and mail, for pilot training, glider towing, ambulance duties, pipeline and cable patrol, fisheries surveillance and agricultural use.
CUSTOMERS: Order total 500 mid-1993; exports primarily to India and South America.
DESIGN FEATURES: Conventional high-wing configuration; constant-chord unswept wing, dihedral 1°; cambered wingtips; single bracing strut each side; pod and boom fuselage, with heavily glazed cabin offering exceptional all-round view; swept tail fin; composites used extensively.
LANDING GEAR: Non-retractable tricycle type; single wheel, with low-pressure tyre, on each unit; cantilever spring mainwheel legs; floats and skis optional.
POWER PLANT: One 157 kW (210 hp) six-cylinder Teledyne Continental IO-360-ES initially; four-cylinder Textron

Yakovlev Yak-141 prototype V/STOL combat aircraft

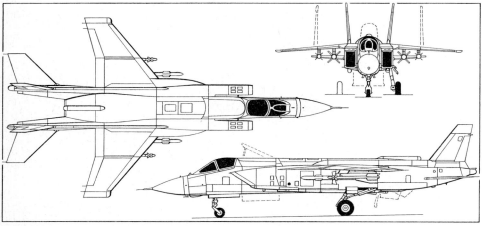
Yakovlev Yak-141 ('Freestyle') single-seat carrier-based V/STOL combat aircraft (*Jane's/Mike Keep*)

Yakovlev Yak-141 hovering, with main jet nozzle vectored 95° (*Charles Bickers*)

Lycoming IO-360-A1B6 piston engine or Rybinsk DN-200 opposed piston diesel engine of same power are future alternatives; Hartzell two-blade two-position propeller.
ACCOMMODATION: Four persons in pairs in enclosed cabin.
AVIONICS: Bendix/King avionics.
DIMENSIONS, EXTERNAL:
Wing span   10.60 m (34 ft 9½ in)
Wing aspect ratio   6.63
Length overall   6.96 m (22 ft 10 in)
Height overall   2.90 m (9 ft 6 in)
AREAS:
Wings, gross   16.96 m² (182.6 sq ft)
WEIGHTS AND LOADINGS:
Weight empty   775 kg (1,709 lb)
Max payload   270 kg (595 lb)
Max T-O weight   1,290 kg (2,844 lb)
Max wing loading   76.06 kg/m² (15.58 lb/sq ft)
Max power loading   8.22 kg/kW (13.54 lb/hp)
PERFORMANCE:
Max cruising speed   113 knots (210 km/h; 130 mph)
Econ cruising speed   102 knots (190 km/h; 118 mph)
Landing speed   68 knots (125 km/h; 78 mph)
Service ceiling   4,000 m (13,125 ft)
T-O and landing run   500 m (1,640 ft)
Range, 45 min reserves:
  with max payload   458 nm (850 km; 528 miles)
  with max fuel   647 nm (1,200 km; 745 miles)

YAKOVLEV Yak-141
NATO reporting name: Freestyle

TYPE: Single-seat carrier-based V/STOL multi-purpose combat aircraft.

PROGRAMME: Authentic details first released by Yakovlev OKB at 1991 Paris Air Show; project design started 1975; first flight of prototype March 1989; first vertical take-off 29 December 1989; two prototypes (Nos. 75 and 77) accumulated 200 flying hours by June 1991; two others built for structural, systems and engine testing; intended originally to replace Yak-38 for air defence of 'Kiev' class carrier/cruisers, with secondary attack capability; 12 international records homologated by FAI, including time-to-height with payload in VTOL category; flight tests planned to continue until 1995 but programme stopped due to termination of Defence Ministry funding; No. 77 prototype crashed when landing on carrier/cruiser *Admiral Gorshkov* on 5 October 1991; first prototype demonstrated at 1992 Farnborough Air Show; Yakovlev OKB continuing development in refined land-based and naval combat aircraft forms.

DESIGN FEATURES: Multi-engine lift/thrust configuration as Yak-38, but twin fins widely separated on flat-sided tailbooms, extending well beyond nozzle of propulsion engine; inner surface of each tailboom protected by curved titanium heatshield; rectangular wedge engine air intake each side of fuselage; recirculation of jet efflux restricted by large door that hinges down forward of vectored main nozzle, and smaller doors between this and liftjets; shallow 'fence' forward of each fin root, as on MiG-29 but longer, probably housing chaff/flare dispensers; bulged wingtips for 'puffer-jet' stability control system.

FLYING CONTROLS: Triplex full-authority digital fly-by-wire control of aerodynamic surfaces and 'puffer-jets', with inputs from inertial and area nav systems, via nav computer, and from air data computer system, with provision for satellite navigation; all-moving horizontal tail surfaces; wing leading-edge flaps.

STRUCTURE: Extensive use of aluminium/lithium; 26 per cent by weight CFRP, including wing flaps, slats, leading- and trailing-edges, and tail surfaces; sweptback wings fold upward at mid-span for stowage; wing leading-edge extension (also CFRP) on side of each intake duct, forward of wingroot.

LANDING GEAR: Retractable tricycle type with single wheel on each unit; nosewheel retracts rearward; mainwheels, on trailing-link legs, retract forward into engine ducts. Mainwheel tyre size 880 × 230 mm; nosewheel tyre size 500 × 150 mm. Brake-chute housing on centreline above jet nozzle.

POWER PLANT: Primary power plant is Soyuz RD-79V-300 turbofan, 88.25 kN (19,840 lb st) dry, 152.0 kN (34,170 lb st) with afterburning; door beneath nozzle allows it to be vectored 65° downward for short take-off, 95° downward and forward for vertical landing. RD-79 lift thrust is approx 80 per cent of cruise rating. Two RKBM Rybinsk RRD-41 liftjets, each 40.2 kN (9,040 lb st), inclined at 10° from vertical immediately aft of cockpit in installation similar to Yak-38 liftjets, able to vector rearward to 24° from vertical for STOL, and 2° forward of vertical for braking; puffer-jet stability controls at wingtips and nose (at tail of first prototype only); computerised engine control system. Conformal centreline 2,000 litre (528 US gallon; 440 Imp gallon) external fuel tank.

ACCOMMODATION: Pilot only, on Zvezda K-36 zero/zero ejection seat under blister canopy; flat bulletproof windscreen; automatic ejection system for pilot in emergency during vertical and transition flight modes. Tandem two-seat trainer at mockup stage.

AVIONICS: Manual or automatic flight control from take-off to landing, day and night, in all weathers; multi-mode fire-control radar similar to that of MiG-29, with slightly smaller antenna, providing information to HUD and multifunction displays via computer that also receives input from IFF and stores management systems, with optional laser/TV target designator and helmet-mounted display.

ARMAMENT: One 30 mm gun, 120 rds; four underwing hardpoints for R-27 (NATO AA-10 'Alamo'), R-73 (AA-11 'Archer') or R-77 air-to-air missiles, anti-ship missiles, Kh-31A/P (AS-17 'Krypton') or Kh-25 (AS-12 'Kegler') air-to-surface missiles, 500 kg bombs, rockets or 23 mm gun pods.

DIMENSIONS, EXTERNAL:
Wing span   10.10 m (33 ft 1¾ in)
Wing aspect ratio   3.22
Width, wings folded   5.90 m (19 ft 4¼ in)
Length overall   18.30 m (60 ft 0 in)
Height overall   5.00 m (16 ft 5 in)
AREAS:
Wings, gross   31.70 m² (341.2 sq ft)
WEIGHTS AND LOADINGS:
Weight empty   11,650 kg (25,685 lb)
Max fuel: internal   4,400 kg (9,700 lb)
  external   1,750 kg (3,858 lb)
Max external weapons load, STOL   2,600 kg (5,730 lb)
Max T-O weight: VTOL   15,800 kg (34,830 lb)
  STOL   19,500 kg (42,990 lb)
PERFORMANCE:
Max level speed: at height
       Mach 1.7 (970 knots; 1,800 km/h; 1,118 mph)
  at S/L   Mach 1.02 (672 knots; 1,250 km/h; 775 mph)

Service ceiling	15,000 m (49,200 ft)
STOL T-O distance	30-100 m (100-328 ft)
STOL landing distance	240 m (790 ft)
Combat radius, STOL, with 2,000 kg (4,410 lb) weapons	372 nm (690 km; 428 miles)
Range, STOL, with 1,000 kg (2,205 lb) weapons:	
at S/L	545 nm (1,010 km; 627 miles)
at 10,000-12,000 m (32,800-39,370 ft)	1,133 nm (2,100 km; 1,305 miles)
Loiter time, 54 nm (100 km; 62 miles) from base	1 h 30 min
g limit, 50% fuel	+7

YAKOVLEV Yak-144

TYPE: Twin-turboprop airborne early warning and control aircraft.

PROGRAMME: Selected for naval service after abandonment of Antonov An-72 derivative known to NATO as 'Madcap'; Admiral Vladimir Chernavin, CinC of CIS Navy, told *Jane's Defence Weekly* in March 1992 that it is a "radio-electronic support aircraft... equivalent of the US Hawkeye and will be used for observation, target indication and homing"; two models, one with folded wings, displayed on scale model of aircraft carrier *Admiral of the Fleet Kuznetsov* seen by visitors to MosAeroshow '92; programme terminated late 1993.

DESIGN FEATURES: Configuration almost identical with Grumman E-2 Hawkeye; high-wing monoplane; rotodome over rear fuselage; dihedral tailplane with toed-in twin endplate fins and rudders; turboprop engine under each inner wing; wings fold upward and inward from point outboard of each engine.

YAKOVLEV Yak-UTS*

TYPE: Two-seat advanced jet trainer.

PROGRAMME: One of designs by five OKBs to meet official Russian requirement to replace Aero L-29 and L-39 Albatros; first flight scheduled for late 1994/early 1995; navalised version proposed for carrier training.

DESIGN FEATURES: Unconventional all-swept mid-wing monoplane, except for straight wing trailing-edge; no dihedral or anhedral; winglets; full-span leading-edge slats permit flight at angles of attack up to 32°; low-mounted tailplane; engines in ducts under wingroots, beneath LERX extending almost to nose; retractable tricycle landing gear; mainwheels retract into engine ducts. Design service life 15,000 flying hours. See accompanying three-view for further details.

POWER PLANT: Two ZKBM Progress AI-25T31M turbofans; each 16.6 kN (3,730 lb st), in prototypes; Klimov/SNECMA turbofans, each 19.2 kN (4,320 lb st), in production aircraft.

ACCOMMODATION: Two crew in tandem under four-piece blister canopy; rear seat raised.

DIMENSIONS, EXTERNAL:
Wing span	11.25 m (36 ft 11 in)
Length overall	12.40 m (40 ft 8¼ in)
Height overall	4.60 m (15 ft 1 in)

WEIGHTS AND LOADINGS:
Max fuel: internal	1,800 kg (3,968 lb)
with external tank	2,500 kg (5,511 lb)
Max T-O weight	5,500 kg (12,125 lb)
Max power loading	165.7 kg/kN (1.63 lb/lb st)

* See Yak-130 in Addenda

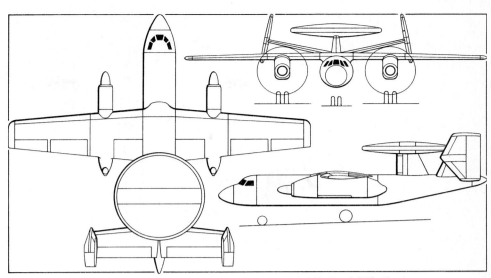

Provisional three-view drawing of Yakovlev Yak-144 *(Jane's/Mike Keep)*

Model of shipboard version of Yak-UTS with wings folded *(Piotr Butowski)*

Models of Yakovlev Yak-144 AEW&C aircraft on *Admiral of the Fleet Kuznetsov (Mark Lambert)*

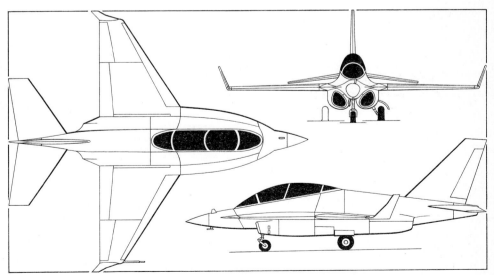

Three-view of projected Yakovlev Yak-UTS advanced trainer *(Jane's/Mike Keep)*

PERFORMANCE (estimated):
Max level speed:	
at height	over 485 knots (900 km/h; 560 mph)
at S/L	460 knots (850 km/h; 530 mph)
T-O speed	97 knots (180 km/h; 112 mph)
Landing speed	92 knots (170 km/h; 106 mph)
T-O run	250-330 m (820-1,085 ft)
Landing run	425-520 m (1,395-1,705 ft)
Max ferry range with conformal tank	1,350 nm (2,500 km; 1,550 miles)
g limits	+8/-3

SINGAPORE

SA
SINGAPORE AEROSPACE LTD
540 Airport Road, Paya Lebar, Singapore 1953
Telephone: 65 287 1111
Fax: 65 280 9713
Telex: RS 55851 (or 51158) SAERO
CHAIRMAN AND CEO: Lim Hok San
PRESIDENT: Quek Poh Huat
VICE-PRESIDENT MARKETING: Michael Ng

Formed early 1982 as government-owned Singapore Aircraft Industries Ltd, controlled by Ministry of Defence Singapore Technology Holding Company Pte Ltd; renamed Singapore Aerospace April 1989; total 1991 workforce (latest update received) 2,936.

Major programmes include refurbishing and A-4 to TA-4 conversion for RSAF and other air forces; overhaul, repair and refurbishing of many types including C-130 Hercules, F-5E/F Tiger II, Hunter, Strikemaster and several models of Bell and Eurocopter helicopters. Work started at Seletar; new 15,000 m² (161,460 sq ft) factory at Paya Lebar opened October 1983.

Assembly of 30 SIAI-Marchetti S.211s (see under Agusta in Italian section) begun for RSAF 1985; 17 of 22 Super Pumas for RSAF assembled by SA; company is partner in EC 120 helicopter programme (see International section); to manufacture 100 shipsets of nosewheel doors worth $12 million for Boeing 777 (option for another 100), and 150 sets of passenger doors for Fokker 100.

Six subsidiaries are as follows:

Singapore Aerospace Engineering Pte Ltd,
Seletar West Camp, Singapore 2879
Telephone: 65 481 5955
Fax: 65 482 0245
Telex: RS 25507 SAMAIR
VICE-PRESIDENT/GENERAL MANAGER: Bob Tan

Maintenance, modification and repair of civil and military aircraft and helicopters; authorised service centre for Beechcraft, Bell Helicopter Textron and Eurocopter France (Super Puma).

Singapore Aerospace Systems Pte Ltd,
505A Airport Road, Paya Lebar, Singapore 1953
Telephone: 65 287 2222
Fax: 65 284 4414
Telex: RS 55851 SAERO
VICE-PRESIDENT/GENERAL MANAGER: Foo Hee Liat

Maintenance, overhaul and repair of civil and military aircraft components and equipment; authorised service centre for Aerospatiale, AlliedSignal (Bendix/King), Astronautics, ECE, GEC-Marconi, Hydraulic Research Textron, J. E. T., Kollsman, Lucas Aerospace, Revue Thommen, Rockwell Collins, Sextant Avionique, Superflexit, Teledyne, Thomson-CSF and Thomson-TRT Défense.

Singapore Aerospace Engines Pte Ltd,
501 Airport Road, Paya Lebar, Singapore 1953
Telephone: 65 285 1111
Fax: 65 282 3010
Telex: RS 33268 SAENG
VICE-PRESIDENT/GENERAL MANAGER: Chong Kok Pan

Overhaul and repair of Pratt & Whitney JT8D and JT15D, Rolls-Royce Avon Mk 207, Wright J65, Allison T56/501, General Electric J85 and Textron Lycoming T53 gas-turbine engines.

Singapore Aerospace Manufacturing Pte Ltd,
503 Airport Road, Paya Lebar, Singapore 1953
Telephone: 65 284 6255
Fax: 65 288 0965 and 284 2704
Telex: RS 38216 SAMPL
VICE-PRESIDENT/GENERAL MANAGER: Goh Chin Khee

Manufacture of aircraft structure and aero engine components, external stores and composites structures.

Singapore Aerospace Supplies Pte Ltd,
540 Airport Road, Paya Lebar, Singapore 1953
Telephone: 65 287 2033
Fax: 65 284 1167 and 280 6179
Telex: RS 51158 SASUPP
VICE-PRESIDENT/GENERAL MANAGER: Donald Seow

Supplies wide range of parts and components for civil and military aircraft; material support specialist for Singapore Aerospace.

Singapore Aviation Services Company Pte Ltd,
540 Airport Road, Paya Lebar, Singapore 1953
Telephone: 65 280 0817
Fax: 65 382 1509
VICE-PRESIDENT/GENERAL MANAGER: Chiang Woon Seng

Specialises in commercial aircraft engineering, especially Section 41 modification of Boeing 747, heavy maintenance and structural modification, widebody interior conversions and refurbishing, ageing aircraft modification, avionics upgrade and modification, painting, finishing and corrosion control programme.

Super Skyhawks of the 'Black Knights' display team of the Republic of Singapore Air Force

EUROCOPTER/CATIC/SA EC 120
Five-seat light helicopter (SA programme share 15 per cent). Details in International section.

SA (MCDONNELL DOUGLAS) A-4S SUPER SKYHAWK
TYPE: Single-seat attack aircraft.
PROGRAMME: First two McDonnell Douglas A-4S Skyhawks fitted with General Electric non-afterburning F404-GE-100D in 1986; Phase I (engine) conversion of 52 launched 1987; Phase II (avionics) upgrade started in 1991. See also *Jane's Civil and Military Aircraft Upgrades 1993-94*.
CURRENT VERSIONS: **A-4S-1**: First upgrade, re-engined only: see Power Plant.
A-4SU: Second upgrade stage: see Avionics.
CUSTOMERS: RSAF received 52 A-4S-1 and TA-4S Phase I aircraft (first squadron declared operational March 1992); Phase II being introduced; further airframes and engines reported being acquired.
DESIGN FEATURES: F404 engine increases dash speed by 15 per cent, initial climb by 35 per cent and acceleration by 40 per cent; turn rate and take-off also improved (for further details, see 1989-90 *Jane's*). Wing section NACA 0008 to 0005; quarter-chord sweep 33° 12'.
FLYING CONTROLS: Powered controls; electrically variable incidence tailplane; automatic slats; airbrakes in fuselage sides; lift-dumping spoilers above split flaps.
POWER PLANT: One 48.04 kN (10,800 lb st) General Electric F404-GE-100D non-afterburning turbofan with gas turbine starting. Internal fuel in integral wing tanks (combined capacity 2,142.5 litres; 566 US gallons; 471.3 Imp gallons) and self-sealing centre-fuselage tank (870.5 litres; 230 US gallons; 191.5 Imp gallons); total capacity 3,013 litres (796 US gallons; 662.8 Imp gallons). One 1,136 or 1,514 litre (300 or 400 US gallon; 250 or 333 Imp gallon) centreline drop tank; 568 or 1,136 litre (150 or 300 US gallon; 125 or 250 Imp gallon) drop tank on each inboard wing pylon. Single-point fuelling receptacle in rear fuselage; all tanks allow pressure or gravity fuelling; optional centreline buddy refuelling store contains 1,136 litres (300 US gallons; 250 Imp gallons), hydraulic hose/drogue unit with 18.3 m (60 ft) hose and ram-air driven hydraulic pump (only in-wing and underwing fuel is transferable; transfer rate about 681 litres; 180 US gallons; 150 Imp gallons/min).
ACCOMMODATION: McDonnell Douglas Escapac ejection seat(s); individual rear-hinged upward opening canopy (ies); bullet-resistant windscreen; pressurised and air-conditioned.
AVIONICS: Phase II avionics comprise GEC-Marconi (formerly GEC Ferranti) 4150 HUD, new HDD, Litton LN-93 laser inertial navigation system and Maverick missile interface.

ARMAMENT: Two 20 mm Mk 12 guns in wingroots; centreline and four underwing pylons for bombs, rockets, air-to-surface missiles and gun pods; missiles and tanks on inner pylons only.

DIMENSIONS, EXTERNAL:
Wing span	8.38 m (27 ft 6 in)
Wing aspect ratio	2.91
Length overall	12.72 m (41 ft 8⅝ in)
Height overall	4.57 m (14 ft 11⅞ in)
Wheel track	2.37 m (7 ft 9½ in)
Wheelbase	3.64 m (11 ft 11⅛ in)

AREAS:
Wings, gross	24.14 m² (259.82 sq ft)

WEIGHTS AND LOADINGS:
Operating weight empty	4,649 kg (10,250 lb)
Max fuel weight: internal	2,364 kg (5,213 lb)
external (one 400 and two 300 US gallon tanks)	2,961 kg (6,529 lb)
Max T-O weight	10,206 kg (22,500 lb)
Max landing weight	7,257 kg (16,000 lb)
Max zero-fuel weight	7,841 kg (17,287 lb)
Max wing loading	422.8 kg/m² (86.6 lb/sq ft)
Max power loading	212.6 kg/kN (2.08 lb/lb st)
Thrust/weight ratio	0.48

PERFORMANCE (at max T-O weight except where indicated):
Never-exceed speed (V_{NE}) at S/L	628 knots (1,163 km/h; 723 mph)
Max level speed at S/L	609 knots (1,128 km/h; 701 mph)
Max cruising speed at 9,150 m (30,000 ft)	445 knots (825 km/h; 512 mph)
Econ cruising speed at 10,670 m (35,000 ft)	424 knots (786 km/h; 488 mph)
Stalling speed at S/L	133 knots (247 km/h; 154 mph)
Max rate of climb at S/L	3,326 m (10,913 ft)/min
Combat ceiling	12,200 m (40,000 ft)
T-O run	1,220 m (4,000 ft)
T-O to 15 m (50 ft)	1,768 m (5,800 ft)
Landing from 15 m (50 ft) at max landing weight	1,590 m (5,215 ft)
Landing run at max landing weight	1,372 m (4,500 ft)

Range, 113 kg (250 lb) fuel reserves:
with max payload	625 nm (1,158 km; 720 miles)
with max internal/external fuel	2,046 nm (3,791 km; 2,356 miles)

SA (NORTHROP) F-5E/RF-5E CONVERSION
Three-year programme started in 1990 for Singapore Aerospace, with Northrop support, to convert eight of 28 RSAF F-5E Tiger IIs to RF-5E TigerEye; new nose, pallet-mounted cameras/sensors, FIAR Grifo F/X Plus radar, AiResearch air data computer and updated nav/com (see also *Jane's Civil and Military Aircraft Upgrades 1993-94*). RF-5E last described in 1986-87 *Jane's*.

SOUTH AFRICA

AEROTEK
AERONAUTICAL SYSTEMS TECHNOLOGY (Division of Council for Scientific and Industrial Research)
PO Box 395, Pretoria 0001
Telephone: 27 (12) 841 2227
Fax: 27 (12) 86 8803
SENIOR DESIGN MANAGER: Eric Sparrow
PROGRAMME MANAGER: Dr A. J. Vermeulen

Designed NGT turboprop trainer (described in Atlas entry); now developing all-composites Hummingbird observation aircraft.

AEROTEK HUMMINGBIRD
TYPE: Two-seat observation light aircraft.
PROGRAMME: Prototype in final assembly December 1992; first flight early 1993; photo in Addenda.
CUSTOMERS: Aimed at civil/police patrol and surveillance market as alternative to helicopters.

AEROTEK/ATLAS—AIRCRAFT: SOUTH AFRICA

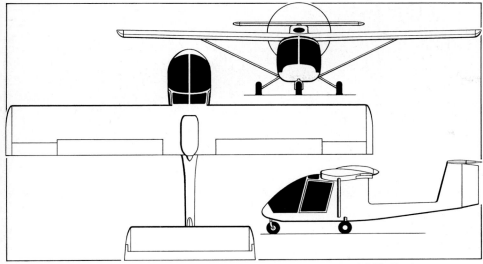

Aerotek Hummingbird two-seat observation light aircraft *(Jane's/Mike Keep)*

DESIGN FEATURES: Strut braced high-wing monoplane with T tail; pod and boom fuselage; helicopter-like view from cockpit. Tailplane/elevator/rudder removable for transportation/storage.
  Wing section NASA GA(W)-1 (constant); dihedral 2°; washout 1°; no sweepback.
FLYING CONTROLS: Conventional mechanical (mainly cables) for primary surfaces. Plain ailerons of 25 per cent chord; Fowler flaps of 30 per cent chord.
STRUCTURE: All-composites (GFRP and Nomex). Single wing strut each side. Fin integral with fuselage. Tailplane and elevator each of one-piece construction.
LANDING GEAR: Non-retractable tricycle type, with single wheel and size 5.00-5 tyre on each unit; nosewheel steerable. Cantilever self-sprung mainwheel legs and independent mainwheel brakes.
POWER PLANT: One 74.6 kW (100 hp) Alvis NR 642-GF-90 rotary engine, reduced to 2,550 rpm through integral gearbox and driving a two-blade, two-position pusher propeller. Fuel tank in each wing, combined capacity 80 litres (21.1 US gallons; 17.6 Imp gallons). Rotax 914 engine optional.
ACCOMMODATION: Two seats side by side in fully enclosed and extensively glazed cabin; second seat removable; provision for third seat behind front pair in emergency. Forward opening window/door each side.
AVIONICS: VFR instrumentation to JAR VLA standard.

DIMENSIONS, EXTERNAL:
Wing span	10.90 m (35 ft 9¼ in)
Wing aspect ratio	7.27
Length overall	6.50 m (21 ft 4 in)
Height overall	2.10 m (6 ft 10¾ in)
Elevator span	4.00 m (13 ft 1½ in)

DIMENSIONS, INTERNAL:
Cockpit: Max width	1.27 m (4 ft 2 in)

AREAS:
Wings, gross	16.35 m² (176.0 sq ft)
Vertical tail surfaces (total)	1.60 m² (17.22 sq ft)
Horizontal tail surfaces (total)	4.00 m² (43.06 sq ft)

WEIGHTS AND LOADINGS:
Weight empty	360 kg (794 lb)
Max T-O weight	650 kg (1,433 lb)
Max wing loading	39.75 kg/m² (8.14 lb/sq ft)
Max power loading	8.71 kg/kW (14.33 lb/hp)

PERFORMANCE (estimated):
Max level speed at S/L	105 knots (195 km/h; 121 mph)
Max cruising speed at S/L	95 knots (176 km/h; 109 mph)
Loiter speed, 10° flap	41 knots (76 km/h; 47 mph)
Stalling speed, flaps down	30 knots (56 km/h; 35 mph)
Max rate of climb at S/L	220 m (725 ft)/min
Service ceiling	4,575 m (15,000 ft)
T-O and landing run	less than 100 m (328 ft)
Endurance at max cruising speed	4 h 30 min

ATLAS
ATLAS AVIATION (PTY) LIMITED
(A Division of Denel (Pty) Ltd)

PO Box 11, Kempton Park 1620, Transvaal
Telephone: 27 (11) 927 4116
Fax: 27 (11) 973 5353
Telex: 742403 BONAERO
CEO: J. J. Eksteen
SIMERA MARKETING: PO Box 117, 1620 Kempton Park, Transvaal
  Telephone: 27 (11) 927 4117
  Fax: 27 (11) 973 5353
PUBLIC RELATIONS: Sam J. Basch

Atlas Aircraft Corporation (see 1991-92 and earlier *Jane's*) founded 1964; incorporated into Armscor Group 1969; good manufacturing, design and development facilities for airframes, engines, missiles and avionics; developed Cheetah from Mirage III, Rooivalk from Puma, Oryx (Super Puma) from Puma, V3B and V3C dogfight missiles, and many weapons installations. Restructuring of Armscor on 1 April 1992 created Denel as self-sufficient commercial industrial group, in which Atlas Aviation is military aircraft manufacturing branch of Simera, which is starting up civil aircraft work; other Armscor aerospace divisions now in Denel Aerospace Group with Simera; Atlas workforce over 5,000 in 1992.

Prototype all-composites NGT turboprop trainer developed by Atlas and Aerotek

ATLAS/AEROTEK NGT TURBOPROP TRAINER

TYPE: Tandem two-seat turboprop trainer.
PROGRAMME: Design started 1986 by government research agency Aerotek (which see: Division of Council for Scientific and Industrial Research), initially as demonstrator for composites manufacturing technology (Project Ovid); first flight of Atlas-built prototype 29 April 1991; proposed later that year as replacement for South African Air Force T-6G Harvards; could still be in contention if proposed purchase of Pilatus PC-7 fails to materialise.
CUSTOMERS: SAAF requirement for 60 of chosen trainer.
DESIGN FEATURES: Objectives were pilot comfort, good view, high performance, aerobatics and simplicity to meet NATO trainer specification. Straight tapered wing, slightly sweptback vertical tail, stepped tandem cockpits; fully aerobatic to FAR Pt 23+ standards.
STRUCTURE: All-composites (Kevlar/GFRP) airframe; one-piece wing with honeycomb box main spar, auxiliary spar and integral fuel tanks; made in large moulds.
LANDING GEAR: Retractable tricycle type adapted from existing aircraft.
POWER PLANT: One 559 kW (750 shp) Pratt & Whitney Canada PT6A-34A turboprop.
ACCOMMODATION: Two ejection seats in tandem, with rear seat elevated; one-piece canopy hinging to starboard.

DIMENSIONS, EXTERNAL:
Wing span	10.90 m (35 ft 9¼ in)
Wing aspect ratio	6.60
Length overall	10.00 m (32 ft 9¼ in)

AREAS:
Wings, gross	18.00 m² (193.75 sq ft)

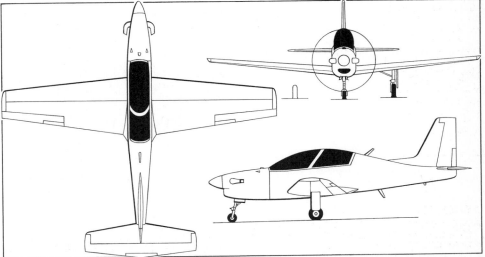

Atlas/Aerotek NGT trainer powered by P&WC PT6A-34A turboprop *(Jane's/Mike Keep)*

340 SOUTH AFRICA: AIRCRAFT—ATLAS

WEIGHTS AND LOADINGS:
  Max T-O weight: Aerobatic, clean   2,000 kg (4,409 lb)
    with external stores   2,750 kg (6,062 lb)
PERFORMANCE (at clean Aerobatic max T-O weight, ISA):
  Max level speed at 3,660 m (12,000 ft)
                                       260 knots (482 km/h; 299 mph)
  Max cruising speed at 3,660 m (12,000 ft)
                                       240 knots (445 km/h; 276 mph)
  Stalling speed, clean, power off
                                       70 knots (130 km/h; 81 mph)
  Max rate of climb at S/L more than 915 m (3,000 ft)/min
  Service ceiling   10,975 m (36,000 ft)
  T-O to 15 m (50 ft)   500 m (1,640 ft)
  Landing from 15 m (50 ft)   650 m (2,133 ft)
  Max range, 45 min reserves
                                       890 nm (1,650 km; 1,025 miles)
  Max endurance, 45 min reserves   4 h 30 min
  g limit at 200 knots (370 km/h; 230 mph)   +7/−4
  Max sustained g at 4,575 m (15,000 ft)   +2.5
  Roll rate at 240 knots (445 km/h; 276 mph)   129°/s

Single-seat Cheetah EZ developed by Atlas from the Mirage III

ATLAS CHEETAH

TYPE: Single-seat fighter and reconnaissance aircraft and two-seat operational trainer.
PROGRAMME: Revealed July 1985; first Cheetah DZ, converted from SAAF two-seat Mirage III-D2Z serial 845, rolled out July 1986; operational Summer 1987.
CURRENT VERSIONS: **Cheetah DZ:** First conversions were two-seaters, originally III-BZ, DZ and D2Z. Two-seater more elaborately equipped, probably to act as pathfinder for single-seaters.
  Cheetah EZ and R2: Conversions of up to 15 Mirage III-EZ, four III-RZ and four III-R2Z single-seaters; no R2 conversions yet confirmed.
  Cheetah ACW: Upgraded version, revealed April 1992 (ACW: advanced combat wing); fixed and drooped wing leading-edges, kinked outboard of dog-tooth, reduce supersonic drag, increase sustained turn rate by 14 per cent and allow increased in-wing fuel tankage; max T-O weight 600 kg (1,323 lb) higher than Cheetah EZ. Prototype converted from Mirage III-R2Z Serial 855; has reportedly been flown down to 80 knots (148 km/h; 92 mph) and at 33° angle of attack.
CUSTOMERS: Single-seaters operated by No. 5 Squadron at Louis Trichardt base; two-seaters with No. 89 Combat Training School also being absorbed by this unit following closure of Pietersburg base.
DESIGN FEATURES: Two-seater has longer nose containing multi-mode radar, RWR antennae, avionics cooling system and external cable ducts from nose to centre-fuselage equipment bay; like Israeli Kfirs, Cheetah has dog-tooth outboard leading-edges and fixed foreplanes; ordinary fences replace slot fences; single-seater has only ranging radar in nose. Nose-probe higher and offset to port on R2. About 50 per cent of airframe renewed.
FLYING CONTROLS: As Mirage III, but with small strakes beside nose to direct yaw-correcting vortices against fin at high angles of attack.
POWER PLANT: One SNECMA Atar 9K-50 turbojet (70.6 kN; 15,873 lb st with afterburning). Bolt-on flight refuelling probe on starboard side of cockpit.
AVIONICS: EZ has ranging radar in nose, RWR antennae in nose and fin trailing-edge, apparently chaff/flare dispenser in former rocket motor fairing, radar altimeter, Elbit HUD and nav/attack system (perhaps including inertial system) and possibly South African-made helmet-mounted sight. DZ two-seater has larger lookdown search and track radar in nose with large cooling system at base of pitot head.
ARMAMENT: Normal Mirage III 30 mm DEFA gun in fuselage underside; Armscor V3B Kukri or V3C Darter dogfight missiles; possibly Matra R.550 semi-active air-to-air missiles; AS.30 air-to-surface missiles, possibly with designator pod. Cheetah seen loaded with eight 500 lb bombs, two underwing tanks and two V3B missiles.

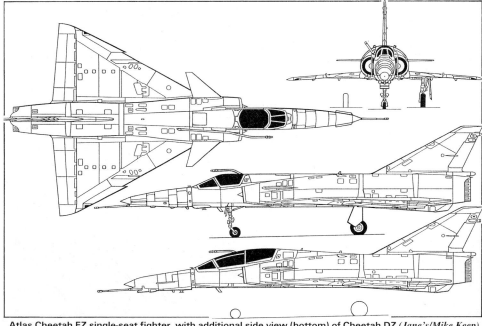

Atlas Cheetah EZ single-seat fighter, with additional side view (bottom) of Cheetah DZ (*Jane's/Mike Keep*)

Cheetah DZ two-seat combat trainer

ATLAS CSH-2 ROOIVALK (RED KESTREL)

TYPE: Combat support helicopter (gunship development of French AS 330 Puma).
PROGRAMME: Gunship development began with two South African Air Force contracts of March 1981; first outcome was single XH-1 Alpha cockpit/gun turret testbed (modified Alouette III) which first flew 1985 (see 1987-88 *Jane's*); followed 1986 by two XTP-1s (modified Pumas: see 1989-90 *Jane's*) to evaluate more powerful engines, locally produced components in composites, electro-optics and air-launched weapons.
  Design of Rooivalk **XDM** (experimental development model) first prototype (originally designated XH-2, later CSH-2) began late 1984 to meet South African Air Force user requirement specification, leading to public rollout 15 January and first flight 11 February 1990; ending of war in Namibia and subsequent defence cuts removed SAAF need for the aircraft, and programme temporarily halted later in 1990, but restarted as mainly company funded venture with reported export interest and discussions with possible development partners 1990-91; **ADM** (advanced

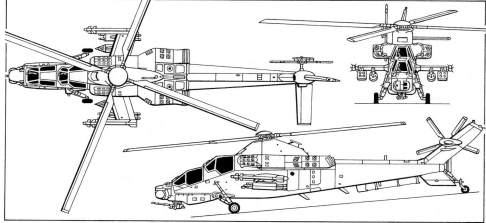

Atlas Rooivalk ADM twin-turbine combat support helicopter (*Jane's/Mike Keep*)

ATLAS—AIRCRAFT: SOUTH AFRICA 341

Atlas CSH-2 Rooivalk XDM first prototype, armed with V3B Kukri air-to-air missiles, pods of 68 mm unguided rockets, enclosed four-round 80 mm anti-tank missile pods and a chin mounted 20 mm GA-1 gun

Close-up of Rooivalk prototype with alternative XC-30 turret for 30 mm DEFA gun

Rooivalk ADM (nearest camera) and XDM prototypes. ADM has HF rail aerial under rear fuselage, with UHF blade antennae further forward and swept; large (VHF?) blade on tail rotor pylon; fin extension above tail rotor; chaff/flare box on rear fuselage; Pacer low-airspeed sensor above rotor head; and HUD in rear cockpit. Engine filter and IR exhaust suppressor can be seen on XDM, but are not yet fitted to ADM

demonstration model) second prototype, with fully integrated avionics and full weapons system, made first flight during second quarter of 1992; approx 250 hours flown by XDM and 50 hours by ADM by late 1992; SAAF funding expired at end of FY 1992, but alternative funding (reportedly from South African Army) enabling programme to continue through FYs 1993 and 1994.

CUSTOMERS: Original South African Air Force requirement for 40-50 now unlikely, but recent interest shown by South African Army, China and Pakistan.

DESIGN FEATURES: Based on reverse engineering of AS 330 Puma; engines moved aft to clear stepped tandem cockpits; composites for rotor blades and some fuselage structure; rear drive taken inboard and forward to modified transmission; nose-mounted target acquisition turret and chin mounted gun; sweptback stub-wings for weapon carriage; upper part of fin modified in 1991 to have tapering chord and increased sweepback.

FLYING CONTROLS: Duplex AFCS; HOCAS (hands on collective and stick).

LANDING GEAR: Puma tricycle landing gear replaced by forward mounted main gear with two-stage high-absorption main legs and tailwheel at base of lower fin. Gear designed to withstand landing impact of up to 6 m (20 ft)/s.

POWER PLANT: Two Topaz (locally upgraded Turbomeca Turmo IV) turboshafts, with rear drive turned to drive forward into rear of transmission; no heat shielding on exhaust; particle separators on intakes. Transmission has 40 min run-dry capability at reduced power settings. Fuel tankage (self-sealing) in fuselage, under stub-wings. Provision for drop tank on each outboard underwing station.

ACCOMMODATION: Pilot (front) and co-pilot/gunner in stepped, tandem cockpits with crashworthy seats and armour protection. All transparencies flat-plate or single-curvature. Dual flight controls.

SYSTEMS: Dual redundant hydraulic and electrical systems.

AVIONICS: Nose-mounted, gyro-stabilised turret contains target detection and tracking system (TDATS) incorporating a FLIR with automatic guidance and tracking, TV camera and laser rangefinder; Doppler-based navigation system with moving-map displays. Twin weapon aiming computers. Dual redundant mission computers and MIL-STD-1553B databuses. Both crewmen have helmet sights, head-up displays and three multi-function CRT displays. Provision for IR jammers or other ECM equipment. Cockpit instruments possibly compatible with NVGs. Avionics prime contractor is ATE (Advanced Technologies and Engineering), supported by 15 subcontractors.

ARMAMENT: Kentron GA-1 Rattler 20 mm gun in TC-20 steerable chin turret; cleared also for 30 mm weapon. One wingtip and two underwing stores stations each side. Eighteen-round launchers for 68 mm unguided rockets and/or four-round launchers for ZT-3 Swift laser guided anti-tank missiles on underwing pylons; V3B Kukri or V3C Darter air-to-air missile at each wingtip. Kentron is prime contractor for weapons integration.

DIMENSIONS, EXTERNAL:
Main rotor diameter	15.08 m (49 ft 5¾ in)
Tail rotor diameter	3.04 m (9 ft 11¾ in)
Length: fuselage	16.65 m (54 ft 7½ in)
overall, rotors turning	19.00 m (62 ft 4 in)
Height overall	4.58 m (15 ft 0¼ in)

AREAS:
Main rotor disc	178.6 m² (1,922.5 sq ft)
Tail rotor disc	7.26 m² (78.15 sq ft)

WEIGHTS AND LOADINGS:
Weight empty	5,270 kg (11,618 lb)
Typical mission T-O weight	7,500 kg (16,535 lb)
Max T-O weight	9,400 kg (20,723 lb)
Max disc loading	52.6 kg/m² (10.77 lb/sq ft)

PERFORMANCE (at 7,500 kg; 16,535 lb combat weight, except where indicated. A: ISA at S/L, B: ISA + 27°C at 1,525 m; 5,000 ft):

Never-exceed speed (V_NE):
  A, B 167 knots (309 km/h; 192 mph)
Max cruising speed:
  A, B 154 knots (285 km/h; 177 mph)
Max rate of climb: A 744 m (2,440 ft)/min
  B 457 m (1,500 ft)/min
Max vertical rate of climb: A 686 m (2,250 ft)/min
  B 177 m (580 ft)/min
Service ceiling: A 6,100 m (20,000 ft)
  B 5,150 m (16,900 ft)
Hovering ceiling IGE: A 4,360 m (14,300 ft)
  B 3,110 m (10,200 ft)
Hovering ceiling OGE: A 3,445 m (11,300 ft)
  B 2,410 m (7,900 ft)
Range with max internal fuel, no reserves:
  A 380 nm (704 km; 437 miles)
  B 507 nm (940 km; 584 miles)
Range at max T-O weight with external fuel:
  A 650 nm (1,205 km; 748 miles)
  B 720 nm (1,335 km; 829 miles)
Endurance with max internal fuel, no reserves:
  A 3 h 36 min
  B 4 h 55 min
Endurance at max T-O weight with external fuel:
  A 6 h 52 min
  B 7 h 22 min
g limits +2.6/−0.5

ATLAS (EUROCOPTER FRANCE) AS 330 PUMA CONVERSIONS

Two conversions of Puma (of which SAAF received 66) exist, in addition to XTP-1 described in 1989-90 *Jane's*:

Oryx (previously known as Gemsbok): AS 330 airframe refitted with Makila power plant, uprated transmission similar to that of Rooivalk, cockpit displays redesigned for single-pilot operation, ventral fin and tailplane of AS 532 Cougar, plus nose radome; programme started 1984; entered service from 1988 as replacement for SAAF SA 321 Super Frelons.

Gunship: Four Atlas proposals offered as possible quick-change, lower cost alternative to Rooivalk, some of which already evaluated by SAAF. Strap-on beams aft of cabin sliding doors, each with two pylons and wingtip missile rail; Atlas armoured seats; still able to transport 12 troops; Grinaker Avitronics integrated airborne communications system. **Option 1:** With Kentron TC-20 ventral gun turret and helmet sight. **Option 2:** With four launchers for 68 mm rockets for area suppression. **Option 3:** As Option 2, but Kentron nose-mounted HSOS (helicopter stabilised optronic sight) added; integration on development aircraft scheduled for late 1994. **Option 4:** Full anti-tank version with ZT-3 Swift laser-guided missiles.

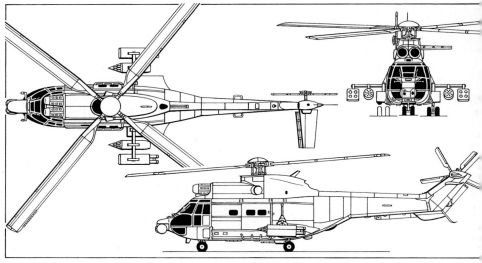

Atlas gunship modification of the AS 330 Puma *(Jane's/Mike Keep)*

Opposite: Atlas Puma gunship with display of full weapon carrying potential

Below left: Mockup of Puma gunship's nose-mounted HSOS for aiming ZT-3 anti-tank missiles. Below right: Strap-on nature of weapons system (ie no carry-through to obstruct cabin interior) enables helicopter to maintain its primary 12-troop transport capability; armament mix shows 68 mm rocket launchers and four-round anti-tank missile launch boxes

CELAIR
CELAIR (PTY) LTD

Company has ceased aircraft manufacture and is no longer in operation. Details of six-seat Eagle 300 and GA-1 Celstar sailplane can be found in 1992-93 *Jane's*.

PROFESSIONAL AVIATION
PROFESSIONAL AVIATION SERVICES (PTY) LTD

Terminal Building, Lanseria Airport, PO Box 3171, Randburg, Johannesburg 2125
Telephone: 27 (11) 659 2860
Fax: 27 (11) 659 1336/1331
Telex: 4-21865 SA
DIRECTORS:
  R. C. H. Garbett
  C. T. Garbett

Formed 1979; buys, sells, charters, leases and manages many types of fixed-wing aircraft and helicopters; maintains and supports aircraft; undertakes turboprop conversions of DC-3/C-47s.

PROFESSIONAL AVIATION (DOUGLAS) JET PROP DC-3 AMI

TYPE: Stretched C-47 with turboprop retrofit.
PROGRAMME: Began 1991; includes conversion of approx 30 C-47s in current SAAF fleet; first aircraft for redelivery (to No. 35 Squadron at D. F. Malan Airport, Cape Town) in 1992; four converted by September 1992.
CUSTOMERS: SAAF (approx 30); plus four civil conversions.
COSTS: Approx SAR5 million ($1.7 million) per aircraft (1992).
DESIGN FEATURES: All lifed items zero-lifed; fuselage lengthened by 1.02 m (3 ft 4 in) forward of wing centre-section; fully overhauled main and tail landing gear (with new disc brakes and tyres), wiring and hydraulic lines; one-piece windscreens; new instrument panel; new electrical system; turboprop engines and five-blade propellers; upgraded avionics; optional auxiliary fuel system; redesigned interior for cargo and/or passengers. Substantial payload/range gains. Conversion takes approx six weeks.
POWER PLANT: Two 1,062 kW (1,424 shp) Pratt & Whitney Canada PT6A-65AR turboprops; Hartzell five-blade metal propellers with autofeathering, reverse pitch and Beta control. Option for outer-wing auxiliary fuel tanks increasing capacity by 1,514 or 3,028 litres (400 or 800 US gallons; 333 or 666 Imp gallons).
ACCOMMODATION: Interior volume increased by 35 per cent, permitting layouts for 34-40 passengers, passenger/cargo mix or all-cargo; quick-change from cargo to passenger interior in 60 min; galley optional. Cargo door at rear on port side; cargo liners and floor tracks installed.
SYSTEMS: New 24V electrical system with two 250A starter/generators. Full fire detection and extinguishing systems. Propeller and engine inlet de-icing.
AVIONICS: Bendix/King package including VLF/Omega, HF, GPS, GPWS, flight data recorder and cockpit voice recorder; coupled three-axis autopilots, certificated for Cat. III operation.

DIMENSIONS, EXTERNAL: As C-47 except:
  Length overall   20.68 m (67 ft 10¼ in)
  Propeller diameter   2.82 m (9 ft 3 in)
  Cargo door: Height   1.50 m (4 ft 11 in)
    Width   1.50 m (4 ft 11 in)
WEIGHTS AND LOADINGS:
  Basic weight empty   7,257 kg (16,000 lb)
  Max payload: standard 4,000 kg (8,818 lb)
    optional   4,500 kg (9,921 lb)
  Max T-O weight   12,202 kg (26,900 lb)
PERFORMANCE:
  Max cruising speed   185 knots (343 km/h; 213 mph)
  Max rate of climb at S/L   305 m (1,000 ft)/min
  Service ceiling, OEI   4,265 m (14,000 ft)
  Min runway requirement 1,000 m (3,280 ft)
  Min landing distance   300 m (984 ft)
  Range with max payload 350 nm (648 km; 403 miles)
  Endurance with max payload   3 h
  Max IFR range at 1,220 m (4,000 ft) with 2,000 kg (4,409 lb) payload and max auxiliary fuel, reserves for 1 h diversion and 45 min hold
                         2,000 nm (3,706 km; 2,303 miles)
  Max endurance, conditions as above   14 h

Professional Aviation Jet Prop DC-3 (two P&WC PT6A-65AR turboprops)

SPAIN

AISA
AERONAUTICA INDUSTRIAL SA

Carretera del Aeroclub s/n, PO Box 27.094, E-28044 Madrid
Telephone: 34 (1) 396 3247
Fax: 34 (1) 208 3958
PRESIDENT: J. A. Pérez-Nievas
GENERAL MANAGER: Carlos Herraiz
DESIGN MANAGER: Rafael Moreno

Founded 1923; adopted present name 1934; produced several liaison, training and sporting aircraft for Spanish Air Force and aero clubs. Division of CESELSA Group since 1986; has manufactured 1,011 aircraft of 35 different types and overhauled more than 1,500. Cuatros Vientos factory has 10,272 m² (110,567 sq ft) covered area; 1993 workforce 160.

Repairs and overhauls Beech aircraft, Bell 47/204/205/206/212s, McDonnell Douglas/Hughes 500s and Boeing CH-47s for Spanish armed forces and civil operators; repairs and overhauls in-flight refuelling systems under licence from Sargent-Fletcher of California, USA; produces landing gear shock absorbers and actuators for Dassault Mirage F1, 2000 and Falcon series and Dassault/Dornier Alpha Jet, and hydraulic components for Airbus; produces structural components for CASA C-212 Aviocar; makes helicopter structures and funnels for dipping sonars; involved in Eurofighter EFA; participating in conversion of Spanish Boeing CH-47Cs to CH-47D; dedicated to aircraft upgrading and modification; currently (early 1993) in development of EW programmes for Spanish Air Force.

CASA
CONSTRUCCIONES AERONAUTICAS SA

Avenida de Aragón 404 (PO Box 193), E-28022 Madrid
Telephone: 34 (1) 585 7000
Fax: 34 (1) 585 7666/7
Telex: 41696 and 41726 CASA E
WORKS: Getafe, Illescas, Tablada, San Pablo, San Fernando, Puerto Real and Cádiz
PRESIDENT AND CHAIRMAN OF THE BOARD:
  Javier Alvarez Vara
PUBLIC RELATIONS AND PRESS MANAGER:
  José de Sanmillán

Founded 1923; owned 97.42 per cent by state holding company INI, 2.56 per cent by DASA (Germany) and 0.02 per cent by others; INI ready to reduce its holding in favour of other European interests.

CASA formed Industria de Turbopropulsores (ITP) in 1989 to make parts for and assemble turbojets, including EJ200 for Eurofighter, F404 for F/A-18, possibly R-R Pegasus for Spanish Matadors (Harriers) and eventually for CASA AX; Compania Espanola de Sistemas Aeronauticas (CESA) formed with 40 per cent holding by Lucas Aerospace; CASA space division greatly extended; largely auto-

Spanish Navy SH-3D Sea King modified by CASA to AEW configuration (*Press Office Sturzenegger*)

SPAIN: AIRCRAFT—CASA

mated composites assembly plant opened at Illescas, near Toledo, January 1991.

Largest current programme is 4.2 per cent share in Airbus; makes horizontal tail, landing gear doors, wing ribs and skins, leading/trailing-edges and passenger doors for A300/310/320/330/340, many in composites. Second largest programme is share in EFA with membership of avionics, control systems, flight management and structure teams; shares starboard wing with BAe and rear fuselage with Alenia; CASA responsible for integration and software creation for communications system. Own programmes include C-212 Aviocar and C-101 Aviojet; CN-235 (shared with IPTN through Airtech); CASA 3000 high-speed regional airliner; looking for partner for AX fighter/trainer.

Contract to design, stress, test and manufacture whole wing of Saab 2000 won October 1989; designed and is making tailplane of McDonnell Douglas MD-11; makes outer flaps for Boeing 757, tail components for Spanish Sikorsky S-70s; member of European Future Large Aircraft Group Euroflag (see International section); extending life of 23 Northrop SF-5B trainers.

CASA made centre-fuselages for Spanish Mirage F1s and components for Spanish F/A-18s; assembled 40 ENAER T-35C Pilláns (E.26 Tamiz) for Spanish Air Force, 57 BO 105s for Spanish Army, 19 more for Spanish government agencies and two BK 117s for ICONA; assembled 12 of 18 AS 532B₁ Cougars for Spanish Army and installed THORN EMI Searchwater AEW radar in three Spanish Navy Sikorsky SH-3Ds.

CASA's seven factories have 335,000 m² (3,605,900 sq ft) covered floorspace; 1993 workforce 9,200.

AIRTECH (CASA-IPTN) CN-235

Details of this twin-turboprop transport appear in International section.

CASA AX

Government-funded pre-feasibility and feasibility studies under way since 1988 for this advanced trainer and ground attack aircraft to succeed Spanish F-5As and Bs from about 2000; current study due to end in 1993; could use non-afterburning versions of EJ200 or F404 turbofans.

CASA 3000

TYPE: Twin-turboprop regional airliner.
PROGRAMME: Market studies completed September 1991; definition phase completed May 1993; CASA expecting to launch in 1993 if risk-sharing partner found; first flight forecast for March 1996, JAA certification May 1997 and service entry June 1997.
COSTS: Project launch cost estimated at $700 million; Spanish Ministry of Industry to provide Pta32.9 billion ($328 million) over period 1992-97.
DESIGN FEATURES: Conventional twin-engined low-wing configuration; new carbonfibre wing and new-design fuselage; wing and tailplane non-swept but with marked dihedral; sweptback vertical tail. Designed for Cat. II landings (Cat. IIIa optional) and flight into known icing conditions.
FLYING CONTROLS: Ailerons operated mechanically, all other control surfaces electrohydraulically. One aileron and two spoilers on each wing, latter operable symmetrically on ground to act as airbrakes; two independent elevators, with artificial feel (Q-feel) and stall warning (stick shaker and stick pusher) capability; two (vertically split) rudder surfaces operated electrically (fly-by-wire). Two-segment hydraulically operated single-slotted flaps on each wing, max deflection 40°.

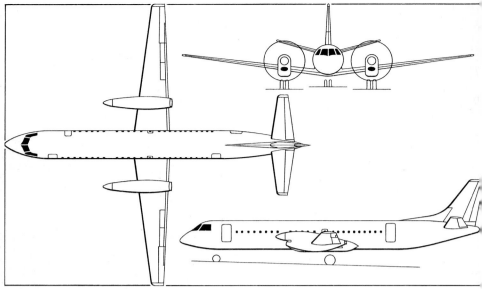

Preliminary drawing of the CASA 3000 twin-turboprop regional airliner (*Jane's*/Mike Keep)

Artist's impression of CASA's new 68/78-seat CASA 3000

STRUCTURE: Mainly of aluminium alloys, with extensive metal to metal bonding; composites for radome, movable control surfaces, fairings, trailing-edges, cabin flooring, and possibly other structural components. Primary structure designed for 60,000 h service life.
LANDING GEAR: Retractable tricycle type, with twin wheels on each unit. All units retract forward. Designed for service life of 75,000 landings, following safe-life criteria.
POWER PLANT: Two turboprops, each rated at 3,691 kW (4,950 shp) for T-O (3,989 kW; 5,350 shp with APR), each driving an 880 rpm six-blade propeller. FADEC engine control system. Fuel capacity 6,460 litres (1,706 US gallons; 1,42 Imp gallons).
ACCOMMODATION: Flight crew of pilot, co-pilot and observer Basic cabin arranged for 72 passengers in mainly four abreast seating at 81 cm (32 in) pitch, with galley and three stowage areas at front and toilet at rear, plus one attendant seat at front and one at rear. Baggage compartment aft of passenger area. Alternative layouts for 78 passengers at 76 cm (30 in) pitch (high density), 68 at 32 in pitch, or 68 in mixed class (22 at 86 cm; 34 in and 46 at 32 in). Passenger door at front and rear on port side; service door at front and baggage compartment door at rear on starboard side. Type III overwing emergency exit on each side. Entire accommodation pressurised and air-conditioned.
SYSTEMS: Cabin pressure differential of 0.48 bar (7.0 lb/sq in) providing 2,285 m (7,500 ft) environment at altitude of 9,450 m (31,000 ft).
AVIONICS: Navigation control and sensors, central radio tuning units, aircraft systems monitoring sensors and flight management system connected through digital buses to two avionics computers. These computers provide output to two primary flight parameter displays, two multi-function displays, two EICAS displays and Cat. II (optionally Cat. IIIa) autopilot. Options to include GPS, ACARS (ARINC communication and reporting system) and MLS

DIMENSIONS, EXTERNAL:
Wing span	27.67 m (90 ft 9½ in)
Length overall	29.70 m (97 ft 5¼ in)
Height overall	8.68 m (28 ft 5¾ in)
Wheel track (c/l of shock struts)	9.292 m (30 ft 6 in)
Wheelbase	10.428 m (34 ft 2½ in)
Propeller fuselage clearance	1.16 m (3 ft 9¾ in)

DIMENSIONS, INTERNAL:
Cabin: Max width	2.64 m (8 ft 8 in)
Width at floor	2.27 m (7 ft 5½ in)
Max height	1.97 m (6 ft 5½ in)

CASA 3000 passenger cabin mockup

Baggage compartment volume 14.8 m³ (522.7 cu ft)
Carry-on baggage volume 9.7-10.8 m³ (342.6-381.4 cu ft)
WEIGHTS AND LOADINGS:
  Operating weight empty 17,200 kg (37,919 lb)
  Max payload 7,500 kg (16,534 lb)
  Max T-O weight 28,300 kg (62,390 lb)
  Max landing weight 27,800 kg (61,288 lb)
  Max zero-fuel weight 24,700 kg (54,454 lb)
  Max power loading (with APR) 3.55 kg/kN (5.83 lb/shp)
PERFORMANCE (estimated):
  Max cruising speed 350 knots (648 km/h; 403 mph)
  FAR 25 T-O balanced field length at max T-O weight:
    ISA at S/L 1,500 m (4,920 ft)
    ISA at 610 m (2,000 ft) 1,625 m (5,330 ft)
    ISA + 15°C at S/L 1,600 m (5,250 ft)
    ISA + 15°C at 610 m (2,000 ft) 1,750 m (5,740 ft)
  Range at max cruising speed:
    with max payload 757 nm (1,402 km; 871 miles)
    with 78 passengers and baggage
      790 nm (1,464 km; 909 miles)
    with 72 passengers and baggage
      1,000 nm (1,853 km; 1,151 miles)

CASA C-212 SERIES 300 AVIOCAR

TYPE: Twin-turboprop general purpose transport.
PROGRAMME: See 1991-92 and previous editions for early history; Series 100 last described in 1981-82 *Jane's* and Series 200 in 1987-88 edition; Series 300 certificated to FAR Pts 25, 121 and 135 in December 1987.
CURRENT VERSIONS: **Series 300 Airliner**: Standard seating for 26 passengers, or 24 if toilet included.
  Series 300 Utility: Standard seating for 23 passengers, or 21 if toilet included, with max capacity for 26.
  Series 300M: Military troop/cargo/general purpose transport.
  Series 300ASW: Anti-submarine version; described separately.
  Series 300DE: Elint/ECM version; described separately.
  Series 300MP: Maritime patrol version; described separately.
  Series 300P: As standard Series 300 but powered by 820 kW (1,100 shp) P&WC PT6A-65 turboprops for improved hot and high performance; Spanish certification 1989; no announced orders up to Spring 1993.
CUSTOMERS: Total 436 of all versions (208 civil/228 military) sold by February 1993, of which more than 400 delivered, incl 164 Series 100 and approx 240 Series 200. Over 30 Series 300 then ordered, incl 14 Series 300M for air forces of Bolivia (one), Colombia (two), France (five), Lesotho (three) and Panama (three), plus nine Series 300MPs (see separate entry) and at least six civil examples. Military customers for Series 100/200 listed in 1991-92 and earlier editions.
DESIGN FEATURES: Wing section NACA 65_3-218; no dihedral; incidence 2° 30′; swept winglets canted upwards at 45°; meets FAR Pt 36 noise limits.
FLYING CONTROLS: Mechanical; fixed incidence tailplane; trim tab in port aileron, trim and geared tabs in rudder and each elevator; double-slotted flaps.
STRUCTURE: All-metal light alloy fail-safe structure; unpressurised; two-spar tailplane and fin. Wing centre-section, forward and rear passenger doors and dorsal fin manufactured by AISA.
LANDING GEAR: Non-retractable tricycle type, with single mainwheels and single steerable nosewheel. CASA oleo-pneumatic shock absorbers. Goodyear wheels and tyres, main units size 11.00-12 Type III (10-ply rating), nose unit size 24-7.7 Type VII (8-ply rating). Tyre pressure 3.86 bars (56 lb/sq in) on main units, 4.0 bars (58 lb/sq in) on nose unit. Goodyear hydraulic disc brakes on mainwheels. No brake cooling. Anti-skid system optional.
POWER PLANT: Two Garrett TPE331-10R-513C turboprops, each flat rated at 671 kW (900 shp) and equipped with automatic power reserve (APR) system providing 690 kW (925 shp) in event of one engine failing during take-off. Dowty Aerospace R-334/4-82-F/13 four-blade constant-speed fully feathering reversible-pitch propellers. Fuel in four integral wing tanks, with total capacity of 2,040 litres (539 US gallons; 449 Imp gallons), of which 2,000 litres (528 US gallons; 440 Imp gallons) usable. Gravity refuelling point above each tank. Single pressure refuelling point in starboard wing leading-edge. Additional fuel can be carried in one 1,000 litre or two 750 litre (264 or 198 US gallon; 220 or 165 Imp gallon) optional ferry tanks inside cabin, and/or two 500 litre (132 US gallon; 110 Imp gallon) auxiliary underwing tanks. Oil capacity 4.5 litres (1.2 US gallons; 1.0 Imp gallon) per engine.
ACCOMMODATION: Crew of two on flight deck; cabin attendant in civil version. For troop transport role, main cabin can be fitted with 25 inward facing seats along cabin walls, to accommodate 24 paratroops with instructor/jumpmaster; or seats for 25 fully equipped troops. As ambulance, cabin is normally equipped to carry 12 stretcher patients and four medical attendants. As freighter, up to 2,700 kg (5,952 lb) of cargo can be carried in main cabin, including two LD1, LD727/DC-8 or three LD3 containers, or light vehicles. Cargo system, certificated to FAR Pt 25, includes roller loading/unloading system and 9g barrier net. Photographic

CASA C-212-300M Aviocar of the Panamanian Air Force

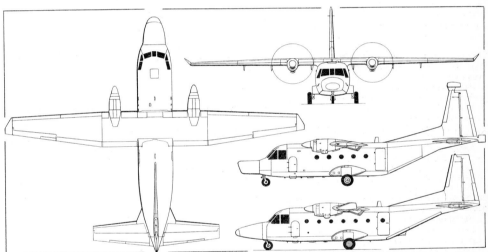
CASA C-212 Series 300 Aviocar, with additional side view of C-212DE (ECM) version *(Dennis Punnett)*

Rear view of the C-212 Aviocar showing the in-flight-openable rear ramp/door

version equipped with two Wild RC-20/30 vertical cameras and darkroom. Navigation training version has individual desks/consoles for instructor and five pupils, in two rows, with appropriate instrument installations. Civil passenger transport version has standard seating for up to 26 in mainly three-abreast layout at 72 cm (28.5 in) pitch, with provision for quick change to all-cargo or mixed passenger/cargo interior. Toilet, galley and 400 kg (882 lb) capacity baggage compartment standard, plus additional 150 kg (330 lb) in nose bay. VIP transport version can be furnished to customer's requirements. Forward and outward opening door on port side immediately aft of flight deck; forward/outward opening passenger door on port side aft of wing; inward opening emergency exit opposite each door on starboard side. Additional emergency exit in roof of forward main cabin. Two-section underfuselage loading ramp/door aft of main cabin is openable in flight for discharge of paratroops or cargo, and can be fitted with optional external wheels for door protection during ground manoeuvring. Interior of rear loading door can be used for additional baggage stowage in civil version. Entire accommodation heated and ventilated; air-conditioning optional.

SYSTEMS: Freon cycle or (on special mission versions) engine bleed air air-conditioning system optional. Hydraulic system, operating at service pressure of 138 bars (2,000 lb/sq in), provides power via electric pump to actuate mainwheel brakes, flaps, nosewheel steering and rear cargo ramp/door. Handpump for standby hydraulic power in case of electrical failure or other emergency. Electrical system supplied by two 9kW starter/generators, three batteries and three static converters. Pneumatic boot and engine bleed air de-icing of wing and tail unit leading-edges; electric de-icing of propellers and windscreens. Oxygen system for crew (incl cabin attendant); two portable oxygen cylinders for passenger supply. Engine and cabin fire protection systems.

AVIONICS: Standard avionics include Collins VHF com, VOR/ILS, ADF, DME, ATC transponder, radio altimeter, intercom (with Gables control) and PA system; Honeywell directional gyro and AFCS; and Bendix/King weather radar. Blind-flying instrumentation standard. Optional avionics include second Collins ADF and transponder; Collins HF and UHF com; Global Omega nav; Dorne & Margolin marker beacon receiver; and Fairchild flight data and cockpit voice recorders.

ARMAMENT (military versions, optional): Two machine-gun pods or two rocket launchers, or one launcher and one gun pod, on hardpoints on fuselage sides (capacity 250 kg; 551 lb each).

DIMENSIONS, EXTERNAL:
Wing span 20.28 m (66 ft 6½ in)
Wing chord: at root 2.50 m (8 ft 2½ in)
at tip 1.25 m (4 ft 1¼ in)
Wing aspect ratio 10.0
Length overall 16.15 m (52 ft 11¾ in)
Fuselage: Max width 2.30 m (7 ft 6½ in)
Height overall 6.60 m (21 ft 7¾ in)
Tailplane span 8.40 m (27 ft 6¾ in)
Wheel track 3.10 m (10 ft 2 in)
Wheelbase 5.46 m (17 ft 11 in)
Propeller diameter 2.79 m (9 ft 2 in)
Propeller ground clearance (min) 1.27 m (4 ft 2 in)
Distance between propeller centres 5.27 m (17 ft 3¾ in)
Passenger door (port, rear):
Max height 1.58 m (5 ft 2¼ in)
Max width 0.70 m (2 ft 3½ in)
Crew and servicing door (port, fwd):
Max height 1.10 m (3 ft 7¼ in)
Max width 0.58 m (1 ft 10¾ in)
Rear loading door: Max length 3.66 m (12 ft 0 in)
Max width 1.70 m (5 ft 7 in)
Max height 1.80 m (5 ft 11 in)
Emergency exit (stbd, fwd): Height 1.10 m (3 ft 7¼ in)
Width 0.58 m (1 ft 10¾ in)
Emergency exit (stbd, rear): Height 0.94 m (3 ft 1 in)
Width 0.55 m (1 ft 9¾ in)

DIMENSIONS, INTERNAL:
Cabin (excl flight deck and rear loading door):
Length: passenger version 7.22 m (23 ft 8¼ in)
cargo/military 6.50 m (21 ft 4 in)
Max width 2.10 m (6 ft 10¾ in)
Max height 1.80 m (5 ft 11 in)
Floor area: passenger 13.51 m² (145.4 sq ft)
cargo/military 12.21 m² (131.4 sq ft)
Volume: passenger 23.7 m³ (837 cu ft)
cargo/military 22.0 m³ (777 cu ft)
Cabin: volume incl flight deck and rear loading door 30.4 m³ (1,073.6 cu ft)
Baggage compartment volume 3.6 m³ (127 cu ft)

AREAS:
Wings, gross 41.0 m² (441.33 sq ft)
Ailerons (total, incl tab) 7.50 m² (80.73 sq ft)
Trailing-edge flaps (total) 14.92 m² (160.60 sq ft)
Fin, incl dorsal fin 4.22 m² (45.42 sq ft)
Rudder, incl tab 2.05 m² (22.07 sq ft)
Tailplane 9.01 m² (96.98 sq ft)
Elevators (total, incl tabs) 3.56 m² (38.32 sq ft)

WEIGHTS AND LOADINGS:
Manufacturer's weight empty 3,780 kg (8,333 lb)
Weight empty, equipped (cargo) 4,400 kg (9,700 lb)
Max payload: cargo 2,700 kg (5,952 lb)
military 2,820 kg (6,217 lb)
Max fuel: standard 1,600 kg (3,527 lb)
with underwing auxiliary tanks 2,400 kg (5,291 lb)
Max T-O weight: standard 7,700 kg (16,975 lb)
military version 8,000 kg (17,637 lb)
Max ramp weight 7,750 kg (17,085 lb)
Max landing weight 7,450 kg (16,424 lb)
Max zero-fuel weight 7,100 kg (15,653 lb)
Max cabin floor loading 732 kg/m² (150 lb/sq ft)
Max wing loading: standard 187.8 kg/m² (38.46 lb/sq ft)
military version 195.1 kg/m² (39.96 lb/sq ft)
Max power loading: standard 5.74 kg/kW (9.43 lb/shp)
military version 5.96 kg/kW (9.80 lb/shp)

PERFORMANCE (at max T-O weight. A: passenger version, B: freighter, C: military version at 8,000 kg; 17,637 lb MTOGW):
Max operating speed (VMO):
A, B, C 200 knots (370 km/h; 230 mph)
Max cruising speed at 3,050 m (10,000 ft):
A, B, C 191 knots (354 km/h; 220 mph)
Econ cruising speed at 3,050 m (10,000 ft):
A, B, C 162 knots (300 km/h; 186 mph)
Stalling speed in T-O configuration:
A, B, C 78 knots (145 km/h; 90 mph)
Max rate of climb at S/L: A, B, C 497 m (1,630 ft)/min
Rate of climb at S/L, OEI: A, B, C 95 m (312 ft)/min
Service ceiling: A, B, C 7,925 m (26,000 ft)
Service ceiling, OEI: A, B, C 3,380 m (11,100 f
FAR T-O distance: A, B 817 m (2,680 f
FAR landing distance: A, B 866 m (2,840 f
MIL-7700C T-O distance to 15 m (50 ft):
C 610 m (2,000 f
MIL-7700C landing distance from 15 m (50 ft):
C 462 m (1,516 f
MIL-7700C landing run: C 285 m (935 f
Required runway length for STOL operation:
C 384 m (1,260 f
Range (civil operation, IFR reserves):
with 25 passengers, at max cruising speed 237 nm (440 km; 273 miles
with 1,713 kg (3,776 lb) payload 773 nm (1,433 km; 890 miles
Range (military operation):
with max payload 450 nm (835 km; 519 miles
with max standard fuel and 2,120 kg (4,674 lb) payloa 907 nm (1,682 km; 1,045 miles
with max standard and auxiliary fuel and 1,192 k (2,628 lb) payload 1,446 nm (2,680 km; 1,665 miles

CASA C-212 SERIES 300 AVIOCAR (SPECIAL MISSION VERSIONS)

TYPE: Specialised adaptations of standard C-212 Series 300.

CURRENT VERSIONS: **ASW**: Anti-submarine version; 360° sca radar under fuselage; ESM.

DE: Elint/ECM version: see Equipment paragraph development began 1981.

MP: Maritime patrol version; nose- or belly-mounte search radar in enlarged radome; FLIR; additiona antennae.

CUSTOMERS: Deliveries of versions based on Series 100/20 listed in 1991-92 and earlier *Jane's*; customers for Serie 300-based variants include Argentine Prefectura Nava (five), Portuguese Air Force (two MP) and Spanish Minis try of Finance/Customs Service (two); Angolan Air Forc has requirement for six MP.

POWER PLANT: As C-212 Series 300. Two 500 litre (132 U gallon; 110 Imp gallon) underwing auxiliary fuel tanks.

ACCOMMODATION: Flight crew of two; four systems operator in ASW version; radar operator and two observer station in MP version.

AVIONICS (ASW): IFR nav/com includes single HF and UH and two VHF radios, ICS, autopilot, flight director UHF/DF and VLF/Omega. Underfuselage radar with 360 scan, MAD, IFF/SIF, OTPI, tactical and sonobuoy pro cessing systems and ESM. Radar console, weapon control and intervalometer added to flight deck; nav/com boxes i rack aft of pilot; starboard rack behind pilot contains radar sonobuoy, MAD and ESM boxes; three consoles on star board side of cabin: (1) radar control and display, ESM an intercom; (2) tactical display, MAD recorder and intercom (3) sonobuoy controls, acoustic controls and displays.

AVIONICS (MP): IFR nav/com as for ASW version. Nose mounted or belly-mounted AN/APS-128 100kW searc radar with 360° scan; FLIR. Radar repeater console an searchlight controls added to flight deck; nav/com (inc UHF/DF and VLF/Omega) and radar in rack behind pilot radar operator console on starboard side of cabin; observe stations at rear.

CASA C-212-300MP maritime patrol aircraft of the Prefectura Naval Argentina

Maritime patrol Series 300 Aviocar leased to the US Coast Guard in 1991

EQUIPMENT (MP): Sonobuoy and smoke marker launchers; searchlight.

EQUIPMENT (DE): Automatic signal interception, classification and identification; jamming emitters; map for plotting location and characteristics of hostile radars.

ARMAMENT (ASW): Can include torpedoes such as Mk 46 and Sting Ray; air-to-surface missiles such as AS15TT and Sea Skua; unguided air-to-surface rockets.

CASA C-101 AVIOJET

Spanish Air Force designation: **E.25 Mirlo (Blackbird)**
Chilean Air Force designations: **T-36 and A-36 Halcón (Hawk)**

TYPE: Tandem two-seat basic and advanced trainer and attack aircraft.

PROGRAMME: First flight 27 June 1977; four prototypes; first flight Spanish production aircraft 1979; nav/attack system modernisation of DD completed 1992.

CURRENT VERSIONS: **C-101EB**: Trainer for Spanish Air Force; powered by 15.57 kN (3,500 lb st) Garrett TFE731-2-2J; see 1983-84 *Jane's*.

C-101BB: Armed export version, powered by Garrett 16.46 kN (3,700 lb st) TFE731-3-1J; Chilean version is **C-101BB-02** and Honduras **C-101BB-03**.

C-101CC: Light attack version, powered by 19.13 kN (4,300 lb st) Garrett TFE731-5-1J with military power reserve (MPR) rating of 20.91 kN (4,700 lb st); first flight 16 November 1983; two prototypes; designations **CC-02** for Chile, **CC-04** for Jordan. *Detailed description applies to this version, except where indicated.*

C-101DD: Enhanced training and attack version; first flight 20 May 1985; avionics include GEC-Marconi HUD, weapon aiming computer and inertial platform; engine as for C-101CC; no orders announced by Spring 1993.

CUSTOMERS: Spanish Air Force 88 EB (E.25 Mirlo) delivered from 17 March 1980, later increased to 92; serving with Spanish Academia General del Aire, Grupo de Escuelas and Patrulla Aguila display team; Chilean Air Force 14 BB-02 (**T-36 Halcón**) plus 23 CC-02 (**A-36 Halcón**) of which 22 assembled and part manufactured in Chile by ENAER; Honduras four BB-03 with custom avionics; 16 CC-04 delivered to Royal Jordanian Air Force 1987-88; Jordan and Chile hold options for eight more.

DESIGN FEATURES: MBB (Germany) and Northrop (USA) supported design; internal fuel allows autonomous ferry flight Spain to Canary Islands. Wing section Norcasa 15 symmetrical; thickness/chord ratio 15 per cent; quarter-chord sweep 1° 53'; dihedral 5°; incidence 1°.

FLYING CONTROLS: Rod-operated elevator and rudder; hydraulically powered ailerons with electrically actuated trimmable centring spring and manual reversion; ground adjustable tab on port aileron; electrically actuated variable incidence tailplane; electric trim tab in rudder; hydraulically actuated slotted flaps; airbrake panel under centre-fuselage; ventral fins under jetpipe of armed versions.

STRUCTURE: Conventional light alloy with three-spar, fail-safe wing; six-bolt attachment to fuselage; titanium flap tracks; flaps and ailerons of GFRP/honeycomb sandwich; rudder, elevators and airbrake of aluminium honeycomb. All tail surfaces manufactured by AISA.

LANDING GEAR: Hydraulically retractable tricycle type, with single wheel and oleo-pneumatic shock absorber on each unit. Forward retracting Dowty Aerospace nose unit, with non-steerable nosewheel and chined tubeless tyre size 457 × 146 (18 × 5.75-8). Inward retracting mainwheels with tubeless tyres size 622 × 216 (24.5 × 8.5-10) and hydraulically actuated multi-disc brakes.

POWER PLANT: One Garrett TFE731 non-afterburning turbofan (see variants for details), with lateral intake on each side of fuselage abreast of second cockpit. Fuel in one 1,155 litre (305 US gallon; 254 Imp gallon) fuselage bag tank, one 575 litre (152 US gallon; 126.5 Imp gallon) integral tank in wing centre-section, and two outer wing integral tanks, each of 342 litres (90.4 US gallons; 75.25 Imp gallons). Total internal fuel capacity 1,730 litres (457 US gallons; 380.5 Imp gallons) normal, 2,414 litres (637.8 US gallons; 531 Imp gallons) maximum, of which 1,667 litres (440 US gallons; 367 Imp gallons) and 2,337 litres (617 US gallons; 514 Imp gallons), respectively, are usable. Fuel system permits up to 30 s of inverted flight. Pressure refuelling point beneath port air intake; gravity fuelling point for each tank. No provision for external fuel tanks. Oil capacity 8.5 litres (2.2 US gallons; 1.8 Imp gallons).

ACCOMMODATION: Crew of two in tandem, on Martin-Baker Mk 10L zero/zero ejection seats, under individual canopies which open sideways to starboard and are separated by internal screen. Rear (instructor's) seat elevated 32.5 cm (12¾ in). Cockpit pressurised and air-conditioned by engine bleed air. Dual controls standard.

SYSTEMS: Hamilton Standard three-wheel bootstrap type air-conditioning and pressurisation system, differential 0.28 bar (4.07 lb/sq in), using engine bleed air. Single hydraulic system, pressure 207 bars (3,000 lb/sq in), for landing gear, ailerons, flaps, airbrake, anti-skid units and wheel brakes. Backup system comprising compressed nitrogen bottle for landing gear extension and accumulator for aileron boosters and emergency braking. Pneumatic system for air-conditioning, pressurisation and canopy seal. Electrical system includes 28V 9kW DC starter/generator, two 700VA static inverters for 115/26V single phase AC power, and two 24V 23Ah nickel-cadmium batteries for emergency DC power and engine starting. High pressure gaseous oxygen system.

AVIONICS: C-101EB and BB as listed in earlier *Jane's*. Standard C-101CC equipped with Magnavox AN/ARC-164 UHF com, Collins AN/ARC-186 VHF com, Collins VIR-31A VOR/ILS, Collins DME-40, Collins ADF-60, Andrea AN/AIC-18 interphone, Teledyne/CASA AN/APX-101 IFF/SIF, Dorne and Margolin DMELT 8.1 ELT, Honeywell STARS IVC flight director, Lear Siegler LSI 6000D gyro platform, ADI-500C, RD-550A HSI, Avimo RGS2 gunsight (front and rear cockpits), and CASA SCAR-81 armament control system. Wide range of alternative avionics available for export versions, including Collins AN/ARN-118 Tacan, General Instrument AN/ALR-66 radar warning receiver and Vinten Vicon 78 chaff/flare dispenser.

C-101DD specific equipment includes GEC-Marconi FD 4513 head-up display and weapon aiming computer, Litton LN-39 inertial platform, Alenia mission computer, all linked by MIL-1553 digital bus, FIAR P 0702 laser ranger, HOTAS controls, Collins AN/ARC-182(V) UHF/VHF-AM/FM com, Collins AN/ARC-186 VHF-AM/FM com, Collins VIR-130A VOR/ILS, Alenia radar altimeter, Ferranti video camera and rear seat monitor.

ARMAMENT: Large bay below rear cockpit suitable for quick-change packages, including 30 mm DEFA 553 cannon pod with 130 rds, twin 12.7 mm Browning M3 machine-gun pod with 220 rds/gun, reconnaissance camera, ECM package or laser designator. Six underwing hardpoints, capacities 500 kg (1,102 lb) inboard, 375 kg (827 lb) centre and 250 kg (551 lb) outboard; total external stores load 2,250 kg (4,960 lb). Typical armament can include one 30 mm cannon with up to 130 rds, or two 12.7 mm guns, in fuselage; and four LAU-10 pods of 5 in rockets, six 250 kg BR250 bombs, four LAU-3/A rocket launchers, four 125 kg BR125 bombs and two LAU-3/A launchers, two AGM-65 Maverick missiles, or two AIM-9L Sidewinders or Matra Magics.

DIMENSIONS, EXTERNAL:
Wing span	10.60 m (34 ft 9⅜ in)
Wing chord: at c/l	2.36 m (7 ft 9 in)
at tip	1.41 m (4 ft 7½ in)
Wing aspect ratio	5.6
Length overall	12.50 m (41 ft 0 in)
Height overall	4.25 m (13 ft 11¼ in)
Tailplane span	4.32 m (14 ft 2 in)
Wheel track (c/l of shock struts)	3.18 m (10 ft 5¼ in)
Wheelbase	4.77 m (15 ft 7¾ in)

AREAS:
Wings, gross	20.00 m² (215.3 sq ft)
Ailerons (total)	1.18 m² (12.70 sq ft)
Trailing-edge flaps (total)	2.50 m² (26.91 sq ft)
Fin	2.10 m² (22.60 sq ft)
Rudder	1.10 m² (11.84 sq ft)
Tailplane	3.44 m² (37.03 sq ft)
Elevators	1.00 m² (10.76 sq ft)

WEIGHTS AND LOADINGS:
Weight empty, equipped	3,500 kg (7,716 lb)
Max fuel weight: usable	1,822 kg (4,017 lb)
total	1,882 kg (4,149 lb)
Max external stores load	2,250 kg (4,960 lb)

E.25 Mirlo (CASA C-101EB) of the Spanish Air Force

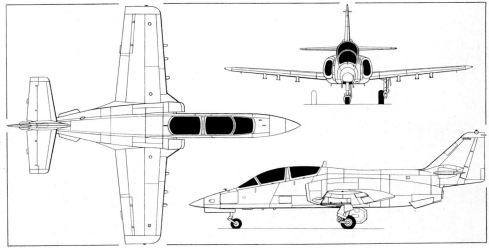

CASA C-101CC Aviojet light attack aircraft (*Dennis Punnett*)

Spanish Air Force C-101EB in standard trainer finish (Press Office Sturzenegger)

```
DD (with MPR)                    240.7 kg/kN (2.36 lb/lb st
  ground attack: BB              340.8 kg/kN (3.34 lb/lb st
  CC, DD (with MPR)              301.0 kg/kN (2.95 lb/lb st
```
PERFORMANCE (C-101BB at 4,400 kg; 9,700 lb AUW C-101CC and DD at 4,350 kg; 9,590 lb, CC in training configuration with 50% normal fuel, except where indicated):
```
Max limiting Mach No. (all)                               0.8
Never-exceed speed (VNE) (all)
                         450 knots (834 km/h; 518 mph) IAS
Max level speed at S/L:
  BB                       373 knots (691 km/h; 430 mph
  CC, DD                   423 knots (784 km/h; 487 mph
Max level speed at height:
  BB at 7,620 m (25,000 ft)
                           430 knots (797 km/h; 495 mph
  CC and DD at 6,100 m (20,000 ft)
                           443 knots (821 km/h; 510 mph
  CC and DD at 4,575 m (15,000 ft) with MPR
                           450 knots (834 km/h; 518 mph
Econ cruising speed at 9,145 m (30,000 ft) (all)
                    Mach 0.56 (330 knots); 612 km/h; 380 mph
Unstick speed (all)      110 knots (204 km/h; 127 mph
Touchdown speed (all)     92 knots (170 km/h; 106 mph
Stalling speed (all):
  flaps up                 99 knots (183 km/h; 114 mph) IAS
  flaps down               88 knots (164 km/h; 102 mph) IAS
Max rate of climb at S/L: BB   1,152 m (3,780 ft)/min
  CC and DD (normal)           1,517 m (4,975 ft)/min
  CC and DD (with MPR)         1,939 m (6,360 ft)/min
Time to 7,620 m (25,000 ft): BB          8 min 30 s
  CC, DD (at 4,500 kg; 9,921 lb)         6 min 30 s
Service ceiling: BB              12,200 m (40,000 ft
  CC, DD                         13,410 m (44,000 ft
T-O run: BB                            630 m (2,065 ft
  CC, DD (at 4,500 kg; 9,921 lb)       560 m (1,835 ft
T-O to 15 m (50 ft): BB                850 m (2,790 ft
  CC, DD (at 4,500 kg; 9,921 lb)       750 m (2,460 ft
Landing from 15 m (50 ft) (all, at 4,000 kg; 8,818 lb)
                                       800 m (2,625 ft
Landing run (all, at 4,000 kg; 8,818 lb)
                                       480 m (1,575 ft
Typical interdiction radius (lo-lo-lo) with four 250 kg
  bombs and 30 mm gun:
  CC and DD, 5 min over target, 30 min reserves
                                260 nm (482 km; 299 miles)
Typical close air support radius (lo-lo-lo):
  CC and DD with four 19 × 2.75 in rocket launchers and
    30 mm gun, 30 min loiter over battle area, 5 min over
    target, 30 min reserves
                                250 nm (463 km; 287 miles)
  CC and DD with two Maverick missiles and 30 mm gun
    8 min over target, 30 min reserves
                                325 nm (602 km; 374 miles)
Typical ECM mission radius:
  BB and CC, 3 h 15 min loiter over target, 30 min
    reserves                    330 nm (611 km; 380 miles)
Typical photo-reconnaissance radius (hi-lo-lo):
  BB and CC, 30 min reserves
                                520 nm (964 km; 599 miles)
Armed patrol mission, no underwing stores, 100 nm
  (185 km; 115 mile) transit from base to patrol area:
  BB, CC and DD with one 30 mm or two 12.7 mm guns,
    45 min reserves
   3 h 30 min at 200 knots (370 km/h; 230 mph) at S/L
Ferry range (all), 30 min reserves
                              2,000 nm (3,706 km; 2,303 miles)
Typical training mission endurance (all):
  two 1 h 10 min general handling missions, incl aero-
    batics, with 20 min reserves after second mission
Max endurance (all)                                    7 h
g limits (all):
  at 4,900 kg (10,802 lb) AUW                      +7.5/−3.9
  at 6,300 kg (13,890 lb) AUW                      +5.5/−2
```

CASA C-101CC, demonstrator for Jordanian CC-04

```
T-O weight:                                 Wing loading:
  trainer, clean: BB, CC  5,000 kg (11,023 lb)   trainer, clean: BB, CC  250.0 kg/m² (51.20 lb/sq ft)
  DD                      5,030 kg (11,089 lb)   DD                      251.5 kg/m² (51.51 lb/sq ft)
  ground attack: BB       5,600 kg (12,345 lb)   ground attack: BB       280.0 kg/m² (57.35 lb/sq ft)
  CC, DD                  6,300 kg (13,890 lb)   CC, DD                  315.0 kg/m² (64.52 lb/sq ft)
Max landing weight:                         Power loading:
  3.66 m (12 ft)/s sink rate  4,700 kg (10,361 lb)   trainer, clean: BB      303.7 kg/kN (2.98 lb/lb st)
  3.05 m (10 ft)/s sink rate  5,800 kg (12,787 lb)   CC (with MPR)           239.3 kg/kN (2.35 lb/lb st)
```

SWEDEN

IG JAS
INDUSTRIGRUPPEN JAS AB
CHAIRMAN: Lars V. Kylberg (Saab-Scania)
PRESIDENT AND CEO: Hans Ahlinder (Saab-Scania)
PUBLIC RELATIONS: Jan Ahlgren

JAS Industry Group formed 1981 to represent Saab-Scania, Ericsson, Volvo Flygmotor and FFV in JAS 39 Gripen programme; acts as contractor for Försvarets Materielverk (Defence Materiel Administration, FMV) and co-ordinates JAS 39 programme.

JAS 39 GRIPEN (GRIFFIN)
TYPE: Single-seat all-weather, all-altitude fighter, attack and reconnaissance aircraft.

PROGRAMME: Funded definition and development began June 1980; initial proposals submitted 3 June 1981; government approved programme 6 May 1982; initial FMV development contract 30 June 1982 for five prototypes and 30 production aircraft, with options for next 110; overall go-ahead confirmed Spring 1983; first test runs of RM12 engine January 1985; Gripen HUD first flown in Viggen testbed February 1987; study for two-seat JAS 39B authorised July 1989. First of five single-seat prototypes (39-1) rolled out 26 April 1987; made first flight 9 December 1988 but lost in landing accident after fly-by-wire problem 2 February 1989; subsequent first flights 4 May 1990 (39-2), 20 December 1990 (39-4), 25 March 1991 (39-3) and 23 October 1991 (39-5); total of 825 flights by 15 February 1993; modified Viggen (37-51) retired at end of 1991 after assisting with avionics trials (nearly 250 flights); second production batch (110 aircraft) approved 3 June 1992; first production Gripen (39101) made first flight 10 September 1992 and joined test programme in lieu of 39-1; flight test results in many cases better than specification due to lower zero drag; flight test programme to continue until 1995.

First production aircraft for Swedish Air Force (39102) made first flight 4 March 1993 and was handed over to FMV 8 June 1993; first 30 aircraft due for delivery 1993-96, next 110 in 1996-2001; first unit will be F7 Wing at Såtenäs. First flight of JAS 39B due 1996, delivery 1998.

CURRENT VERSIONS: **JAS 39A**: Standard single-seater.
JAS 39B: Two-seater, under development; 0.50 m (1 ft 7¾ in) fuselage plug and lengthened cockpit canopy. Primary roles conversion and tactical training. Avionics essentially as for JAS 39A except no HUD in rear cockpit; instead, HUD image from front seat can be presented on flight data display in rear cockpit. Redesigned environmental control system.
JAS 39C and D: Potential improved Swedish versions of A and B with enhanced data handling capability.
JAS 39X: Potential future export version, to upgraded standard of C/D.

CUSTOMERS: Eventual Swedish Air Force requirement approx 300, to equip 16 squadrons; first 30 ordered with prototypes and full scale development 30 June 1982; next 110 will include 14 JAS 39Bs.

COSTS: Planned cost of SEK25.7 billion in 1982, increased to SEK48.5 billion 1991 by inflation; FMV has reported total cost increase of SEK9.3 billion for period 1982-2001; SEK19 billion spent by December 1992, including SEK12 billion to IG JAS. Up to SEK300 million approved late

Prototype 39-2 with canards 'snowploughed' and airbrakes fully deployed

A Gripen prototype displays the range of weapons that can be carried by this multi-role aircraft

WS 39 munitions dispenser on the inboard wing station of prototype 39-5

1991 for JAS 39B development. Flyaway price $25-30 million (1992).

DESIGN FEATURES: Intended to replace AJ/SH/SF/JA/AJS versions of Saab Viggen, in that order, and remaining J 35 Drakens; partners are Saab Military Aircraft, Ericsson Radar Electronics, Volvo Flygmotor and FFV Aerotech; to operate from 800 m (2,625 ft) Swedish V90 road strips; simplified maintenance and quick turnround with conscript groundcrew.

Delta wing with squared tips for missile rails has 45° leading-edge sweepback; foreplane leading-edge sweep 43°.

FLYING CONTROLS: Lear Astronics triplex fly-by-wire system (two-channel digital with Moog electrically signalled servo valves on powered control units, plus single-channel analog FBW standby); Saab Combitech aircraft motion sensors and throttle actuator; mini-stick and HOTAS controls.

Leading-edge with dog-tooth and automatic flaps on Lucas Aerospace 'geared hinge' rotary actuators; three elevon surfaces at each trailing-edge; outboard functions as elevon, middle as elevon and flap, inboard as stabiliser and flap; individual movable foreplanes for low-speed flight and manoeuvre, which also 'snowplough' for aerodynamic braking after landing; airbrake each side of tail.

STRUCTURE: First 3½ carbonfibre wing sets produced by BAe; all subsequent carbonfibre parts (30 per cent of airframe) made by Saab, including wing boxes, foreplanes, fin and all major doors and hatches.

LANDING GEAR: AP Precision Hydraulics retractable tricycle gear, single mainwheels retracting hydraulically forward into fuselage; steerable twin-wheel nose unit retracts rearward. Goodyear wheels, tyres, carbon disc brakes and anti-skid units. Nosewheel braking. Entire gear designed for high rate of sink.

POWER PLANT: One General Electric/Volvo Flygmotor RM12 (F404-GE-400) turbofan, rated initially at approx 54 kN (12,140 lb st) dry and 80.5 kN (18,100 lb st) with afterburning. Near-rectangular intakes, each with splitter plate. Fuel in self-sealing main tank and collector tank in fuselage. Active control of CG location provided by Intertechnique fuel management system.

ACCOMMODATION: Pilot only, on Martin-Baker S10LS zero/zero ejection seat. Hinged canopy (opening sideways to port) and one-piece windscreen by Lucas Aerospace.

SYSTEMS: BAe environmental control system for cockpit airconditioning, pressurisation and avionics cooling. Hughes-Treitler heat exchanger. Two main Dowty hydraulic systems and one auxiliary system, with Abex pumps. Sundstrand main electrical power generating system (40kVA constant speed, constant frequency at 400Hz) comprises an integrated drive generator, generator control unit and current transformer assembly. Lucas Aerospace auxiliary and emergency power system, comprising gearbox-mounted turbine, hydraulic pump and 10kVA AC generator, to provide emergency electric and hydraulic power in event of engine or main generator failure. In emergency role, turbine is driven by engine bleed or APU air; if this is not available the stored energy mode, using pressurised oxygen and methanol, is selected automatically. Microturbo TGA 15 APU and DA 15 air turbine starter for engine starting, cooling air and standby electric power.

AVIONICS: Bofors Aerotronics AMR 345 VHF/UHF-AM/FM com transceiver. Honeywell laser inertial navigation system. Ericsson EP-17 electronic display system incorporates Hughes Aircraft wide-angle HUD, using advanced diffraction optics, to combine symbology and video images. GEC-Marconi FD 5040 video camera for weapon aiming; three Ericsson head-down CRT displays; minimum of conventional analog instruments, for backup only. Left hand (flight data) head-down display normally replaces all conventional flight instruments. Central display shows computer generated map of area surrounding aircraft with tactical information superimposed. Right hand CRT is a multi-sensor display showing information on targets acquired by radar, FLIR and weapon sensors. Ericsson SDS 80 central computing system (D80E 32-byte computer, Pascal/D80 high order language and programming support environment); three MIL-STD-1553 databuses, one of which links flight data, navigation, flight control, engine control and main systems. BAe three-axis strapdown gyro-magnetic unit provides standby attitude and heading information.

Ericsson/GEC-Marconi Avionics PS-05/A multi-mode pulse Doppler target search and acquisition (lookdown/shootdown) radar (weight 156 kg; 344 lb). For fighter missions, this system provides fast target acquisition at long range; search and multi-target track-while-scan; quick scanning and lock-on at short ranges; and automatic fire control for missiles and cannon. In attack and reconnaissance roles its operating functions are search against sea and ground targets; mapping, with normal and high resolution; and navigation. FLIR pod, carried externally under starboard air intake trunk, forward of wing leading-edge, is for attack and reconnaissance missions at night, providing heat picture of target on right hand head-down CRT. Radar warning equipment and countermeasures.

ARMAMENT: Internally mounted 27 mm Mauser BK27 automatic cannon in fuselage and infra-red dogfight missiles on wingtips. Five other external hardpoints (two under each wing and one on centreline) for short and medium range air-to-air missiles such as RB71 (Sky Flash), RB74 (AIM-9L Sidewinder) or AMRAAM; air-to-surface missiles such as RB75 Maverick; anti-shipping missiles such as Saab RB 15F; conventional or retarded bombs; air-to-sur-

Aircraft 39102, in the markings of F7 Wing, is the first production aircraft to be delivered to the Swedish Air Force

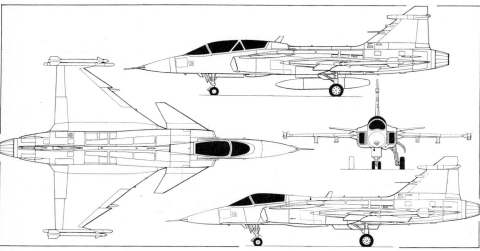

JAS 39A Gripen multi-role combat aircraft for the Swedish Air Force, with additional side view (top) of two-seat JAS 39B (Dennis Punnett)

face rockets; or external fuel tanks. DASA (MBB) DW 39 submunitions dispenser under development. Swedish Defence Materiel Administration (FMV) has proposed developing TSA (Tungt Styrt Attackvapen; heavy guided attack weapon) glide bomb, with warhead weighing several hundred kg, for attacking large targets such as bridges. Series of mission pods to be developed.

DIMENSIONS, EXTERNAL (approx):
Wing span	8.40 m (27 ft 6¾ in)
Length overall: JAS 39A	14.10 m (46 ft 3 in)
JAS 39B	14.60 m (47 ft 10¾ in)
Height overall	4.50 m (14 ft 9 in)
Wheel track	2.40 m (7 ft 10½ in)
Wheelbase	5.75 m (18 ft 10½ in)

WEIGHTS AND LOADINGS:
Operating weight empty	6,622 kg (14,600 lb)
Internal fuel weight	2,268 kg (5,000 lb)
Design T-O weight, clean	approx 8,000 kg (17,635 lb)
Max T-O weight with external stores	12,473 kg (27,500 lb)

PERFORMANCE:
Max level speed	supersonic at all altitudes
T-O and landing strip length	approx 800 m (2,625 ft)
g limit	+

Cockpit of the JAS 39A

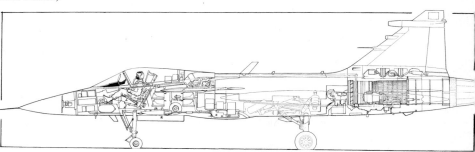

Inboard profile of the JAS 39A Gripen

MFI
MALMÖ FORSKNINGS & INNOVATIONS AB
Stenyxegatan 32, S-213 76 Malmö
Telephone: 46 (49) 21 99 50
Fax: 46 (40) 21 81 11
CHIEF DESIGNER: Håkan Langebro

MFI has manufactured various designs by Björn Andreasson, including BA-12 Slandan (see Microlights section, 1992-93 *Jane's*), BA-14 and earlier MFI-9. A recent move, and preoccupation with manufacture of new floor panels for Swedish Navy ASW helicopters, has slowed development of BA-14B, but company hoped to begin flight testing early 1993. MFI has also developed MFI-11 (new version of MFI-9B) and designed Stiletto light twin.

MFI BA-14B
TYPE: Two/four-seat trainer and utility aircraft.
PROGRAMME: Construction of BA-14 Starling (first flight 25 August 1988), designed by Björn Andreasson, began 1987; became joint venture project between MFI and FFV Aerotech, but later redesigned by MFI to FAR Pt 23A; resumption of flight testing anticipated in early 1993.
DESIGN FEATURES: Strut braced, no-dihedral high wing with approx 3° forward sweep; sweptback vertical tail with dorsal fin; upswept rear fuselage with baggage/freight door amidships.
FLYING CONTROLS: Ailerons, mass balanced rudder, and all-moving tailplane with large inset tab; mechanical actuation; plain flaps.
STRUCTURE: Mostly composites: GFRP wing with CFRP reinforced spar caps; fuselage GFRP/CFRP semi-monocoque. Wing carry-through structure of welded steel tube.
LANDING GEAR: Non-retractable tricycle type, with mainwheels on arched GFRP leaf spring and nosewheel on shock strut. Mainwheel brakes. Twin-float gear optional.
POWER PLANT: One 119 kW (160 hp) Textron Lycoming IO-320 flat-four engine, driving a Hoffmann HO-V72A variable-pitch wood/GFRP propeller. Fuel capacity 80 litres (21 US gallons; 17.6 Imp gallons).
ACCOMMODATION: Two seats side by side under large one-piece transparent canopy. Provision for two further seats. Baggage compartment aft of seats.

DIMENSIONS, EXTERNAL:
Wing span	9.00 m (29 ft 6½ in)
Wing aspect ratio	7.71
Length overall	7.20 m (23 ft 7½ in)
Height overall	2.90 m (9 ft 6¼ in)
Propeller diameter	1.85 m (6 ft 0¾ in)

AREAS:
Wings, gross	10.50 m² (113.02 sq ft)

WEIGHTS AND LOADINGS:
Weight empty	615 kg (1,356 lb)
Max T-O weight	900 kg (1,984 lb)
Max wing loading	85.71 kg/m² (17.55 lb/sq ft)
Max power loading	7.55 kg/kW (12.40 lb/hp)

PERFORMANCE:
Max level speed at S/L	124 knots (230 km/h; 143 mph)
Max cruising speed at S/L	113 knots (210 km/h; 130 mph)
Econ cruising speed at S/L	108 knots (200 km/h; 124 mph)
Stalling speed, power off	46 knots (84 km/h; 53 mph)
Max rate of climb at S/L	335 m (1,100 ft)/min
T-O and landing run	250 m (821 ft)
g limits	+4.4/-2.2

MFI BA-14B training and utility aircraft (*Jane's*/Mike Keep)

MFI MFI-11
TYPE: Two-seat light aircraft.
PROGRAMME: Developed 1991-92; to be produced under licence by German consortium, probably starting in 1993 or 1994.
CURRENT VERSIONS: Choice of **landplane** or **floatplane** versions; latter has increased wing area.
DESIGN FEATURES: Generally similar to BA-14B, but with increased wing and tailplane span, new engine and cowling, wider fuselage panels of glassfibre, and new interior.
LANDING GEAR: Non-retractable tricycle type; twin-float gear optional.
POWER PLANT: One Textron Lycoming O-235-H2C flat-four engine (85.75 kW; 115 hp at 2,800 rpm). Standard fuel capacity 80 litres (21.1 US gallons; 17.6 Imp gallons).
ACCOMMODATION: Two seats side by side.

DIMENSIONS, EXTERNAL (L: landplane; F: floatplane):
Wing span: L, F	7.83 m (25 ft 8¼ in)
Wing aspect ratio: L	6.71
F	6.02

Length overall: L	6.33 m (20 ft 9¼ in)	
F	6.68 m (21 ft 11 in)	
Height overall: L	1.98 m (6 ft 6 in)	
F	2.66 m (8 ft 8¾ in)	

AREAS:
Wings, gross: L	9.14 m² (98.38 sq ft)	
F	10.18 m² (109.58 sq ft)	

WEIGHTS AND LOADINGS:
Weight empty: L	425 kg (937 lb)	
F	480 kg (1,058 lb)	
Max payload: L	205 kg (452 lb)	
F	220 kg (485 lb)	
Max T-O weight: L	630 kg (1,389 lb)	
F	700 kg (1,543 lb)	
Max wing loading: L	68.93 kg/m² (14.12 lb/sq ft)	
F	68.76 kg/m² (14.08 lb/sq ft)	
Max power loading: L	7.35 kg/kW (12.08 lb/hp)	
F	8.16 kg/kW (13.42 lb/hp)	

PERFORMANCE (estimated, at max T-O weight, S/L, ISA):
Max level speed: L	130 knots (240 km/h; 149 mph)	
F	103 knots (190 km/h; 118 mph)	
Max cruising speed: L	116 knots (215 km/h; 134 mph)	
F	92 knots (170 km/h; 106 mph)	
Stalling speed: L	43 knots (80 km/h; 50 mph)	
F	49 knots (90 km/h; 56 mph)	
Max rate of climb: L	456 m (1,496 ft)/min	
F	229 m (751 ft)/min	

Range with max fuel, 45 min reserves:
L at 75% power	331 nm (614 km; 381 miles)	
F at 75% power	288 nm (534 km; 332 miles)	
L at 65% power	418 nm (775 km; 481 miles)	
F at 65% power	360 nm (668 km; 415 miles)	

Max endurance, 45 min reserves:
L at 75% power	2 h 54 min	
F at 75% power	3 h 6 min	
L at 65% power	4 h 0 min	
F at 65% power	4 h 12 min	
g limits	+3.8/−1.5 (Normal category)	

Floatplane version of the two-seat MFI-11 (*Jane's/Mike Keep*)

MFI STILETTO

TYPE: Twin-engined light aircraft.
PROGRAMME: Under study late 1992; probably to be marketed first in kit form, but licensed under FAR/JAR Pt 23 later.
DESIGN FEATURES: See accompanying three-view drawing.
LANDING GEAR: Retractable tricycle type.
POWER PLANT: Two Zoche Aero diesel engines (ZO 01A of 112 kW; 150 hp or ZO 02A of 224 kW; 300 hp, both at 2,500 rpm). Fuel capacity 300 litres (79.25 US gallons; 66 Imp gallons).
ACCOMMODATION: Four seats in fully enclosed cabin. Space aft of seats for 100 kg (220 lb) of baggage.

DIMENSIONS, EXTERNAL:
Wing span	9.36 m (30 ft 8½ in)
Wing aspect ratio	6.99
Length overall	8.18 m (26 ft 10 in)
Height overall	2.56 m (7 ft 5 in)
Tailplane span	3.32 m (10 ft 10¾ in)

AREAS:
Wings, gross	12.53 m² (134.9 sq ft)
Vertical tail surfaces (total)	1.90 m² (20.45 sq ft)
Horizontal tail surfaces (total)	2.68 m² (28.85 sq ft)

WEIGHTS AND LOADINGS:
Weight empty	750 kg (1,653 lb)
Max payload	436 kg (961 lb)
Max T-O weight	1,400 kg (3,086 lb)
Max wing loading	111.7 kg/m² (22.88 lb/sq ft)
Max power loading (ZO 01A)	6.25 kg/kW (10.29 lb/hp)

PERFORMANCE (estimated, with 150 hp engines, ISA):
Max level speed: at S/L	186 knots (344 km/h; 214 mph)	
at 2,750 m (9,000 ft)	260 knots (481 km/h; 299 mph)	
Cruising speed:		
75% power at S/L	166 knots (307 km/h; 191 mph)	
75% power at 2,750 m (9,000 ft)	232 knots (430 km/h; 267 mph)	
65% power at S/L	152 knots (282 km/h; 175 mph)	
65% power at 2,750 m (9,000 ft)	213 knots (395 km/h; 245 mph)	
Stalling speed: at S/L	54 knots (100 km/h; 62 mph)	
at 2,750 m (9,000 ft)	63 knots (116 km/h; 72 mph)	
Range with max fuel, 65% power at 2,750 m (9,000 ft)	1,519 nm (2,815 km; 1,749 miles)	

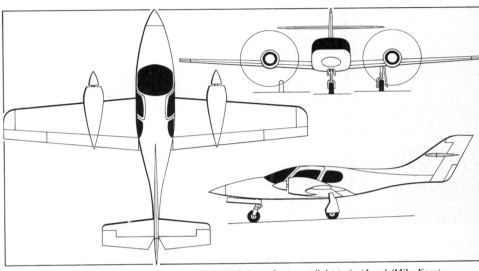

General appearance of the projected MFI Stiletto four-seat light twin (*Jane's/Mike Keep*)

SAAB-SCANIA

SAAB-SCANIA AB
S-581 88 Linköping
Telephone: 46 (13) 18 00 00
Fax: 46 (13) 18 54 27
PRESIDENT AND CEO: Lars V. Kylberg
Saab Military Aircraft (Division of Saab-Scania AB)
GENERAL MANAGER: Hans Ahlinder
VICE-PRESIDENTS:
 Billy Fredriksson (Development)
 Tomy Ivarsson (Strategic Planning)
 Kurt Ahlborg (Marketing and Sales)
PUBLIC RELATIONS: Jan Ahlgren
WORKS: Linköping
Saab Aircraft AB (Subsidiary of Saab-Scania AB)
S-581 88 Linköping
Telephone: 46 (13) 18 20 00
Fax: 46 (13) 18 38 33
PRESIDENT: Christer Skogsborg
VICE-PRESIDENTS:
 Johan Öster (Product Programmes)
 Jeffrey Marsh (Marketing and Sales)
PUBLIC RELATIONS DIRECTOR: Rolf Erichs
WORKS: Linköping, Malmö and Ödeshög
MARKETING SUBSIDIARIES:
 Saab Aircraft International Ltd, Leworth House, 14-16 Sheet Street, Windsor, Berkshire SL4 1BG, UK
 Telephone: 44 (753) 859991
 Fax: 44 (753) 858884
 Telex: 847815 SFIWIN G
 PRESIDENT: Jeffrey Marsh
 VICE-PRESIDENT, SALES AND MARKETING: Philip Male
 PUBLIC RELATIONS: Mike Savage
 Saab Aircraft of America Inc, Loudoun Tech Center, 21300 Ridgetop Circle, Sterling, Virginia 22170, USA
 Telephone: 1 (703) 406 7200
 Fax: 1 (703) 406 7272
 PRESIDENT: Ove Dahlén
 PUBLIC RELATIONS: Ron Sherman
Saab Service Partner AB (Subsidiary of Saab-Scania AB)
PRESIDENT: Stellan Eklof

Original Svenska Aeroplan AB founded at Trollhättan 1937 to make military aircraft; amalgamated 1939 with Aircraft Division of Svenska Järnvägsverkstädernas rolling stock factory at Linköping; renamed Saab Aktiebolag May 1965; merged with Scania-Vabis 1968 to combine automotive interests; Malmö Flygindustri acquired 1968. More than 4,000 military and commercial aircraft delivered since 1940; has held dealership for McDonnell Douglas (formerly Schweizer/Hughes) helicopters in Scandinavia and Finland since 1962.

Total Saab-Scania workforce (four business areas) of 30,000, of whom 6,500 in Saab Aircraft at December 1992 including 5,700 at Linköping. Saab Aircraft Division (see 1991-92 *Jane's*) reorganised from 31 March 1992 into Saab Military Aircraft (producing JAS 39 Gripen), Saab Aircraft AB (producing Saab 340B and 2000) and Saab Service Partner AB. (Last named provides plant, operations, personnel, computer support, airfield, transportation and other service activities for the other two.) New Saab 340 building at Linköping, opened July 1982, extended in 1986 and 1991 to cater for increased Saab 340 production and Saab 2000 assembly; now exceeds 60,000 m² (645,834 sq ft).

SAAB MILITARY AIRCRAFT

Development and production of JAS 39 Gripen; extensive type servicing of Saab 37 Viggen in Sweden and Saab 35 Draken in Austria, Denmark, Finland and Sweden. Last of 66 upgraded J 35J Drakens for Swedish Air Force (see 1991-92 *Jane's*) delivered August 1991.

JAS 39 GRIPEN
Described under IG JAS heading in this section.

SAAB AJS 37 VIGGEN (THUNDERBOLT)

TYPE: Multi-role interceptor, attack and reconnaissance fighter.
PROGRAMME: Last production Viggen (JA 37) delivered 29 June 1990; AJS 37 conversion announced May 1991; converting 115 AJ, SF and SH 37s into interchangeable defence, attack and reconnaissance fighters; to be operational late 1993.
CUSTOMERS: Total 329 Viggens produced for Swedish Air Force, comprising 110 AJ 37, 26 SF 37, 26 SH 37, 18 Sk 37

and 149 JA 37 (see detailed description of JA 37 in 1990-91 *Jane's* and shortened description in 1991-92 edition).
COSTS: AJS 37 programme SEK300 million (1991) ($48.3 million).

DESIGN FEATURES: Combines all three missions in one aircraft.
AVIONICS: New computer-based threat analysis/mission planning system; multi-processors and databus to increase computing power and improve communications; updated ECM.

ARMAMENT: Wider range of weapons to include BK/DWS and six types of missile including RB05, RB15F, RB2 24J and RB74; same reconnaissance pod as JAS 39 Gripe

SAAB AIRCRAFT AB

Principal programmes are Saab 340B and Saab 2000 regional airliners. Production of combined total of 40, including 10 Saab 2000s, planned for 1993.

SAAB 340B
Swedish Air Force designation: Tp100

TYPE: Twin-turboprop regional and business transport.
PROGRAMME: Go-ahead for joint Saab-Fairchild 340 programme given September 1980; Saab took over programme November 1985; Fairchild continued as subcontractor until 1987, when designation changed to Saab 340; first flight (SE-ISF) 25 January 1983; first flight of first production 340A (fourth aircraft, SE-E04) 5 March 1984; Swedish certification 15 May 1984, followed by ten European civil aviation directorates and US FAA sharing FAR/JAR certification June 1984; first operator Crossair; first corporate 340A delivered November 1985; 340B announced late 1987, certificated 3 July 1989; 200th Saab 340 delivered 14 August 1990, 300th on 20 May 1992.
CURRENT VERSIONS: **340A:** Initial production version delivered from June 1984 to September 1989, when 340B introduced; 340A engine power increased from 1,215 kW (1,630 shp) to 1,294 kW (1,735 shp) and propeller enlarged mid-1985; earlier aircraft retrofitted; improved cabin by Metair, meeting 1990 fire regulations, introduced mid-1988. Total of 159 built, incl three prototypes; full description in 1989-90 *Jane's*.
340B: Hot and high version, first flight April 1989; replaced 340A from c/n 160 onwards; first delivery (Crossair) September 1989; powered by GE CT7-9B with APR; improved payload/range, higher weights and increased tailplane span. *Detailed description applies to this version.*
340 AEW: Saab 340B with dorsal Ericsson Erieye S-band side-looking airborne reconnaissance radar; prototype to be delivered to Swedish Air Force (which has options on five more) during first quarter of 1995.
Future 340: Planning started on 340 with 2000 features giving higher cruising speed, lower interior noise and new avionics.
CUSTOMERS: Total 398 firm orders for 340A (159) and B (239) by early May 1993, of which 326 delivered.
Customers for 340A were Aigle Azur (one), Air Nelson (six), Brit Air (six), Business Air (five), Business Express (17), Chautauqua Airlines (two), Comair (19), Crossair (eight), Deutsche BA (nine), Finnaviation (five), Formosa Airlines (three), Kendell Airlines (six), LAPA (two), Metro Flight (16), Northwest Airlink (26), Skyways (eight), Swedair (13), TAN (one) and Tatra Air (two), plus four corporate sales.
Saab 340B ordered by Aer Lingus Commuter (four), AMR Eagle (90), Business Express (56), Chautauqua Airlines (four), China Southern Airlines (four), Crossair (15), Finnaviation (one), Formosa Airlines (two), Golden Air (four), Hazelton Airlines (four), Japan Air Commuter (eight), KLM Cityhopper (12), Metro Flight (10), Northwest Airlink (13), Regional Airlines (five) and Swedish government (two), plus five unannounced.
COSTS: Standard civil aircraft $9 million (1992).
DESIGN FEATURES: Wing section NASA MS(1)-0313; tapered planform with quarter-chord sweep 3° 36′; thickness/chord ratio 16 per cent at root, 12 per cent at tip; dihedral 7° from root; incidence 2° at root; swept fin; dihedral tailplane; pressurised cabin with circular cross-section.

Artist's impression of the Saab 340 AEW surveillance version with Ericsson dorsal radar

FLYING CONTROLS: Mechanical controls; fixed tailplane; trimmable ground tab in each aileron and elevator, trimmab spring tab in rudder; all tabs actuated electromechanically small strakes under rear fuselage; hydraulically actuate single-slotted flaps. Collins APS-85 autopilot system.
STRUCTURE: Fail-safe mainly all-metal structure; two-spa wing; stringers and skins of 2024/7075 aluminium; cab built in three sections; all doors of aluminium honeycom double-shell GFRP radome; CFRP sandwich cabin floo flaps have aluminium spars, honeycomb panels faced wi aluminium sheet and Kevlar leading- and trailing-edge ailerons, rudder and elevators have Kevlar skins and GFR leading-edges; fin integral with fuselage; tailplane and f contain aluminium honeycomb; mainwheel doors Kevla sandwich. Propeller blades are moulded glassfibre/poly urethane foam/carbonfibre.
LANDING GEAR: Retractable tricycle type, of AP Precisio Hydraulics design and manufacture, with twin Goodyea wheels and oleo-pneumatic shock absorber on each uni

Saab 340B regional transport for delivery to Ireland's Aer Lingus Commuter

SAAB—AIRCRAFT: SWEDEN

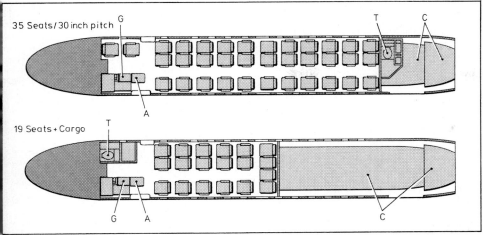

Saab 340B standard 35-passenger layout (top) and 19-passenger/cargo combi. A: attendant's seat, C: cargo, G: galley, T: toilet *(Jane's/Mike Keep)*

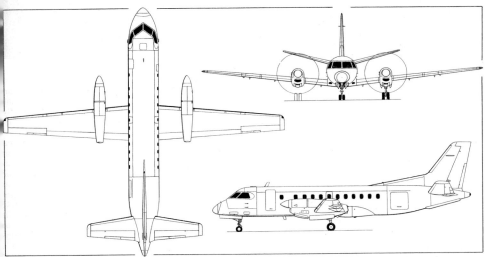

Saab 340B regional airliner (two General Electric CT7-9B turboprops) *(Dennis Punnett)*

Hydraulic actuation. All units retract forward, main units into engine nacelles. Hydraulically steerable nose unit (±60°), with shimmy damper. Mainwheel tyres size 24 × 7.7-10, pressure 8.27 bars (120 lb/sq in); nosewheel tyres size 17.5 × 6.25-6, pressure 4.0 bars (58 lb/sq in). Independent Goodyear carbon hydraulic disc brakes on main units, with Hydro Aire anti-skid control. Min ground turning radius 8.90 m (29 ft 2½ in).

POWER PLANT: Two General Electric CT7-9B turboprops, each rated at 1,305 kW (1,750 shp) for normal T-O and 1,394 kW (1,870 shp) with automatic power reserve. Dowty (Hamilton Standard optional) four-blade slow-turning constant-speed propellers, with full autofeathering and reverse pitch capability. Fuel in two integral tanks in each wing; total capacity 3,220 litres (850.5 US gallons; 708 Imp gallons). Single-point pressure refuelling inlet in starboard outer wing leading-edge. Overwing gravity refuelling point in each wing. Oil capacity 13.8 litres (3.65 US gallons; 3.04 Imp gallons).

ACCOMMODATION: Two pilots and provision for observer on flight deck; attendant's seat (forward, port) in passenger cabin. Main cabin accommodates up to 37 passengers (35 standard), in 12 rows of three, with aisle, and rearward facing seat(s) on starboard side at front. One or both rearward facing seats can be replaced by an optional galley module and/or baggage/wardrobe module. Seat pitch 76 cm (30 in). Standard provision for galley, wardrobe or storage module on port side at front of cabin, regardless of installations on starboard side. Toilet at front or rear of cabin. In former case, QC operation (conversion from passenger to freight interior or vice versa) is possible. Also available is a VIP-to-airliner convertible, as well as a fixed-installation combi with 19 passengers and 1,500 kg (3,307 lb) of cargo. Passenger door (plug type) at front of cabin on port side, with separate forward stowing airstair. Type II emergency exit opposite this on starboard side, and Type III over wing on each side. Crew escape hatch in flight deck roof. Baggage space under each passenger seat; overhead stowage bins. Main baggage/cargo compartment aft of passenger cabin (from which it is accessible), with large plug type door on port side. Entire accommodation pressurised, including baggage compartment.

SYSTEMS: Hamilton Standard ECS (max pressure differential 0.48 bar; 7.0 lb/sq in) maintains S/L cabin environment up to altitude of 3,660 m (12,000 ft) and 1,525 m (5,000 ft) environment up to 7,620 m (25,000 ft). Single on-demand hydraulic system, operating at up to 207 bars (3,000 lb/sq in), for actuation of landing gear, wheel and propeller braking, nosewheel steering and wing flaps. System is powered by single 28V DC electric motor driven pump, rated delivery 9.5 litres (2.5 US gallons; 2.1 Imp gallons)/min. Self-

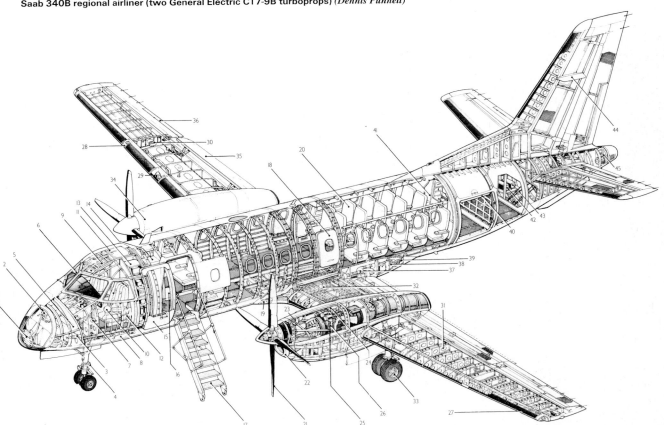

Saab 340B: (1) Weather radar; (2) Radar transceiver; (3) Nose gear actuating rod; (4) Nosewheel steering cam; (5) Hydraulic bay; (6) Windscreen; (7) Angle of incidence sensor; (8) Electrical system access; (9) Pilot's overhead control panel; (10) Ground communication port; (11) Flight deck escape hatch; (12) ASI antennae; (13) Main avionics bay; (14) Office; (15) Passenger door; (16) Cabin attendant's seat; (17) Airstair; (18) Overwing emergency exit; (19) Landing light; (20) Three-abreast passenger seating; (21) Composite-blade propeller; (22) Electrically anti-iced engine air intake; (23) Engine driven AC generator; (24) DC starter/generator; (25) General Electric CT7-9B turboprop; (26) Engine accessory gearbox; (27) Pneumatic de-icing boot; (28) Airflow detector; (29) Pressure refuelling point; (30) Gravity refuelling point; (31) Wing outboard fuel tank; (32) Wing inboard fuel tank; (33) Carbon disc brakes; (34) Propeller brake (optional); (35) Single-slotted flap; (36) Aileron tabs; (37) Air-conditioning unit; (38) Main battery (one of two); (39) Power supply distribution box; (40) Cargo door; (41) Toilet; (42) Flight data recorder; (43) Cockpit voice recorder; (44) VOR/LOC antenna; (45) Discharge valves

354 SWEDEN: AIRCRAFT—SAAB

pressurising main reservoir with 5.08 litres (0.18 cu ft) capacity, operating at pressure of 1.79-2.69 bars (26-39 lb/sq in). Hydraulic backup via four accumulators and pilot operated handpump, working via 2.5 litre (0.09 cu ft) emergency reservoir. Electrical power supplied by two 28V 400A DC engine driven starter/generators, each connected to separate busbar. Variable frequency 115/200V for heating circuits provided by two 26kVA AC generators; single-phase 115V and 26V AC at 400Hz for avionics provided by static inverters. Two 43Ah nickel-cadmium batteries for ground power and engine starting; standby 5Ah lead-acid battery for emergency use. External power receptacle. Engine bleed air for air-conditioning, pressurisation and pneumatic boot de-icing of wing and tail unit leading-edges. Oxygen system (11.2 litre; 2.96 US gallon; 2.46 Imp gallon cylinder operating at 127.5 bars; 1,850 lb/sq in pressure) includes portable and first aid units. Plug-in connections for oxygen masks. Flight deck windows have electric anti-icing and electrically driven windscreen wipers. Cox & Co electric anti-icing for engine air intakes, propellers, AOA and OAT sensors and pitot heads. Demisting by means of air-conditioning system. Kidde engine fire detection system. No APU.

AVIONICS: Standard avionics include two Collins VHF com, Collins PA system and Telephonics cabin interphone; two electronic flight display systems; Collins APS-85 AFCS; colour weather radar; GPWS; LAS flight data recorder; Fairchild cockpit voice recorder; and standby gyro horizon indicator, magnetic compass and VOR/ILS indicator.

DIMENSIONS, EXTERNAL:
Wing span	21.44 m (70 ft 4 in)
Wing aspect ratio	11.0
Length overall	19.73 m (64 ft 8¾ in)
Fuselage: Max diameter	2.31 m (7 ft 7 in)
Height overall	6.97 m (22 ft 10½ in)
Tailplane span	9.24 m (30 ft 3¾ in)
Wheel track	6.71 m (22 ft 0 in)
Wheelbase	7.14 m (23 ft 5 in)
Propeller diameter	3.35 m (11 ft 0 in)
Propeller ground clearance	0.51 m (1 ft 8 in)
Passenger door: Height	1.60 m (5 ft 3 in)
Width	0.69 m (2 ft 3¼ in)
Height to sill	1.63 m (5 ft 4 in)
Cargo door: Height	1.30 m (4 ft 3 in)
Width	1.35 m (4 ft 5 in)
Height to sill	1.66 m (5 ft 5½ in)
Emergency exit (fwd, stbd): Height	1.32 m (4 ft 4 in)
Width	0.51 m (1 ft 8 in)
Emergency exits (overwing, each):	
Height	0.91 m (3 ft 0 in)
Width	0.51 m (1 ft 8 in)

DIMENSIONS, INTERNAL:
Cabin, excl flight deck, toilet and galley:
Length	10.39 m (34 ft 1 in)
Max width	2.16 m (7 ft 1 in)
Width at floor	1.70 m (5 ft 7 in)
Max height	1.83 m (6 ft 0 in)
Volume	33.4 m³ (1,179.5 cu ft)
Baggage/cargo compartment volume:	
with rear toilet	6.8 m³ (240.0 cu ft)
with fwd toilet	8.3 m³ (293.1 cu ft)

AREAS:
Wings, gross	41.81 m² (450.0 sq ft)
Ailerons (total)	2.12 m² (22.84 sq ft)
Trailing-edge flaps (total)	8.07 m² (86.84 sq ft)
Fin (incl dorsal fin)	7.77 m² (83.64 sq ft)
Rudder (incl tab)	2.76 m² (29.71 sq ft)
Tailplane	11.28 m² (121.42 sq ft)
Elevators (total, incl tabs)	3.29 m² (35.40 sq ft)

WEIGHTS AND LOADINGS:
Operating weight empty	8,140 kg (17,945 lb)
Max payload (weight limited)	3,880 kg (8,554 lb)
Max fuel load	2,581 kg (5,690 lb)
Max ramp weight	13,210 kg (29,123 lb)
Max T-O weight	13,155 kg (29,000 lb)
Max landing weight	12,930 kg (28,505 lb)
Max zero-fuel weight	12,020 kg (26,500 lb)
Max wing loading	314.6 kg/m² (64.44 lb/sq ft)
Max power loading (with APR)	4.72 kg/kW (7.75 lb/shp)

PERFORMANCE (at max T-O weight, ISA, except where indicated):
Max operating speed (VMO)	250 knots (463 km/h; 288 mph)
Max operating Mach No. (MMO)	0.5
Max cruising speed:	
at 4,575 m (15,000 ft)	282 knots (522 km/h; 325 mph)
at 6,100 m (20,000 ft)	280 knots (519 km/h; 322 mph)
Best range cruising speed at 7,620 m (25,000 ft)	252 knots (467 km/h; 290 mph)
Stalling speed: flaps up	106 knots (197 km/h; 123 mph)
T-O flap	95 knots (176 km/h; 110 mph)
approach flap	92 knots (171 km/h; 106 mph)
landing flap	88 knots (164 km/h; 102 mph)
Max rate of climb at S/L	610 m (2,000 ft)/min
Rate of climb at S/L, OEI	160 m (525 ft)/min
Service ceiling: standard	7,620 m (25,000 ft)
optional	9,450 m (31,000 ft)
Service ceiling, OEI (1.1% net gradient at 95% MTOW)	3,810 m (12,500 ft)
T-O to 15 m (50 ft) at S/L:	
JAR	1,290 m (4,233 ft)
FAR	1,322 m (4,338 ft)
T-O to 15 m (50 ft) at 1,525 m (5,000 ft):	
JAR (low flap setting)	1,835 m (6,021 ft)
FAR	1,657 m (5,437 ft)
Landing from 15 m (50 ft) at max landing weight at S/L:	
JAR	1,035 m (3,396 ft)
FAR	1,065 m (3,495 ft)
Landing from 15 m (50 ft) at max landing weight at 1,525 m (5,000 ft): JAR	1,165 m (3,823 ft)
FAR	1,200 m (3,937 ft)
Runway LCN: flexible pavement	8
rigid pavement	10

Range with 35 passengers and baggage, reserves for 45 min hold at 1,525 m (5,000 ft) and 100 nm (185 km; 115 mile) diversion:
at max cruising speed	805 nm (1,491 km; 927 miles)
at long-range cruising speed	935 nm (1,732 km; 1,076 miles)

Range with 37 passengers, reserves as above:
at max cruising speed	695 nm (1,288 km; 800 miles)
at long-range cruising speed	795 nm (1,473 km; 915 miles)

OPERATIONAL NOISE LEVELS (FAR Pt 36, Appendix C):
T-O	78.6 EPNdB
Sideline	85.9 EPNdB
Approach	91.6 EPNdB

SAAB 2000

TYPE: Twin-turboprop high-speed regional transport.

PROGRAMME: Definition started Autumn 1988; launched 15 December 1988 with Crossair commitment for 25 firm and 25 on option; formal go-ahead May 1989; Allison GMA 2100 selected as power plant July 1989; first metal cut February 1990; major subcontractor items delivered 1991 by Westland (rear fuselage, March), Valmet (tail unit, July), CASA (wing, August) and Allison (engines, August); first aircraft (SE-001) rolled out 14 December 1991 and made first flight 26 March 1992. Three aircraft in 14-month certification programme, of which c/n 002 made first flight 3 July 1992, followed by 003 on 28 August; 003 first to production standard; static and fatigue test airframes also completed; first flight of production aircraft (SE-004) 17 March 1993; certification due third quarter 1993, initially to JAR 25 (Amendment 13) and FAR 25 (Amendment 70), but will be extended later to include Cat. IIIa operation; deliveries due to start with c/n 004 to Crossair September 1993.

COSTS: Swedish government lending Saab between $163 million and $187 million (1989) for development between 1989 and 1994, to be repaid from sales from 31st aircraft until 2009. Finnish Flygplansfabriken investing $8.4 million (1989) and ADCO $5.3 million (1989) in Valmet tail unit subcontract; total Valmet agreement valued at $69.8 million (1989); Westland rear fuselage production contract worth £40 million (1990).

CUSTOMERS: Firm orders for 46 by April 1993 included Air Marshall Islands (two), Crossair (20), Salenia (five) and Deutsche BA (five), plus conditional order from Northwest Airlink/Express Airlines (10); options at same date totalled 148 including Air Marshall Islands (two), AMR Eagle (50), Brit Air (four), Business Express (10), Comair (20), Crossair (25), Deutsche BA (five), Kendell Airlines (two), Salenia (five) and Skywest (20).

DESIGN FEATURES: Objective to combine jet speeds with turboprop economy, aiming at 360 knots (667 km/h; 414 mph) cruising speed, climb to 6,100 m (20,000 ft) in 10 minutes and cruising altitudes between 5,485 m and 9,450 m (18,000-31,000 ft). CAD/CAM designed throughout; same fuselage cross-section as 340B, but longer; same wing section, but span stretched 15 per cent to give 33 per cent more area; engines farther outboard.

FLYING CONTROLS: Rod and cable control linkages for aileron and elevators, with electrically actuated trim tab in each aileron and each elevator plus spring tab in each elevator; electrically signalled, hydraulically powered rudder, with dual Dowty actuators; hydraulically actuated single slotted flaps with offset hinges.

STRUCTURE: Wing and fuselage primary structures of metal bonded aluminium alloy, as in Saab 340B, with honeycomb sandwich fin, tailplane and doors; two-spar wings, fin and tailplane; composites for ailerons (CFRP/Nomex), flaps (CFRP skins), wing/body fairings (Kevlar/Nomex), nosecone (GFRP/Nomex), rudder and elevator (GFRP leading-edges and CFRP skins), dorsal fin, propeller blades, and cabin floor (carbonfibre sandwich); super plastic formed/diffusion bonded titanium nacelle firefloors.

Major subcontractors are CASA (wing, including design, stressing and testing), Westland Helicopters (rear fuselage and, with Hispano-Suiza, engine cowlings), Valmet (fin, rudder, tailplane and elevators) and Fischer Advanced Composites (dorsal fin).

LANDING GEAR: AP Precision Hydraulics retractable tricycle type, with twin wheels and oleo-pneumatic shock absorbers on each unit; all units retract forward. ABS wheels and carbon brakes; Goodyear tyres; Hydro Aire anti-skid system; Ozone Inc nosewheel steering. Min ground turning radius 18.85 m (61 ft 10 in).

POWER PLANT: Two Allison GMA 2100A turboprops, each flat rated at 3,076 kW (4,125 shp) with APR (S/L, ISA, at 1,100 rpm); Lucas Aerospace full dual-channel FADEC engine control (single-lever control of engine and propellers). Dowty Aerospace R381 slow-turning, constant speed propellers with six swept blades, full autofeathering and reverse pitch; propellers 950 rpm in cruise; blades of both propellers held in phase at all times. Fuel in two integral tanks in each outer wing, total usable capacity 5,18 litres (1,370 US gallons; 1,140 Imp gallons). Single pressure refuelling point in starboard outer wing panel; over wing gravity refuelling point in each wing.

ACCOMMODATION: Flight crew of three or four, including cabin attendant(s). Standard AIM Aviation (UK) 'European' cabin has 50 seats three-abreast at 81 cm (32 in) pitch with single aisle, but cabin length can be extended for additional galley and wardrobe space by moving rear bulkhead aft into baggage space, permitting seating to be increased to 58 at 76 cm (30 in) pitch. Main baggage compartment aft of passenger cabin, with door on port side; provision for additional, smaller baggage area at front of cabin on starboard side. Passenger airstair door at front on port side; service/emergency door at rear on starboard side; overwing Type III emergency exit each side. Entire accommodation pressurised and air-conditioned.

SYSTEMS: Hamilton Standard Recircair ECS, using engine bleed air; max pressure differential 0.48 bar (7.0 lb/sq in). Hydraulic system, with Abex pumps, for landing gear, flap and rudder actuation. Goodrich pneumatic-boot de-icing system for wing/fin/tailplane leading-edges and engine intakes. Electrical system has two 45kVA variable-frequency engine driven generators for three-phase 115/200V AC power and three 28V DC batteries. Electric anti-icing of propeller blades; Swedlow electrically heated windscreen panels. Sundstrand APU for engine starting and ECS. Scott oxygen system. Pacific Scientific fire detection and Kidde-Graviner fire extinguishing systems.

AVIONICS: Standard suite based on Collins Pro Line IV package with six CRT displays and integrated avionics processing system (IAPS); also include Collins WXR-84

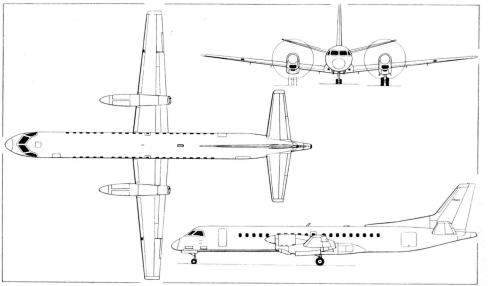

Saab 2000 short/medium-range 50/58-passenger regional transport (Dennis Punnett)

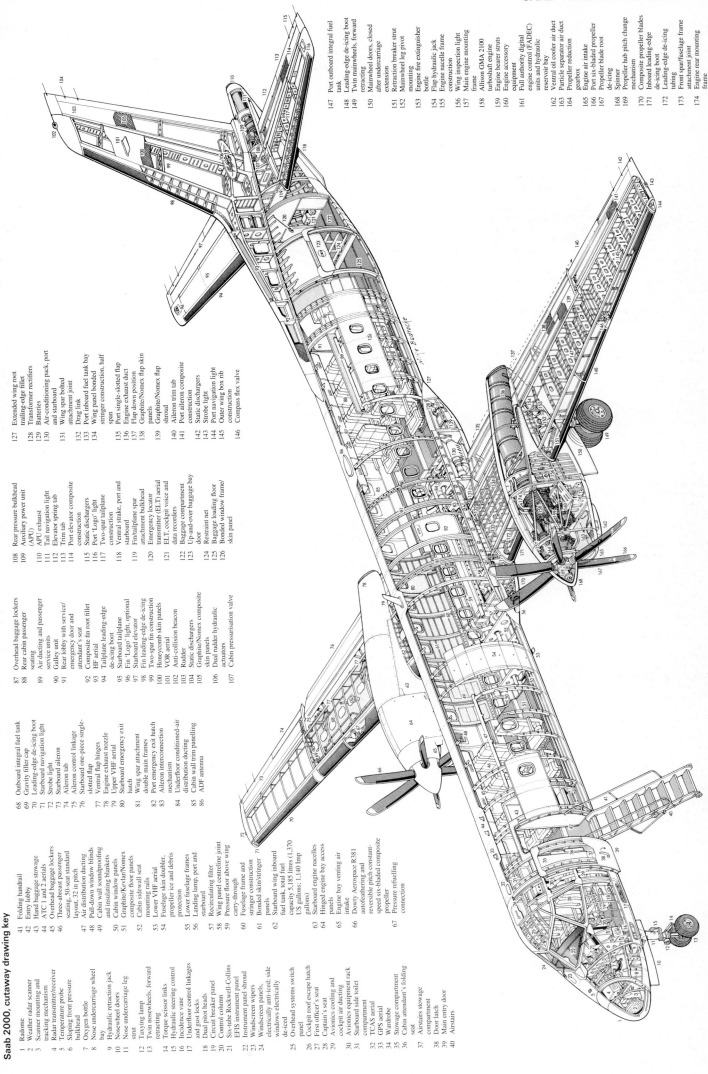

Saab 2000, cutaway drawing key

1. Radome
2. Weather radar scanner
3. Scanner mounting and tracking mechanism
4. Radar transmitter/receiver
5. Temperature probe
6. Sloping front pressure bulkhead
7. Oxygen bottle
8. Nose undercarriage wheel bay
9. Hydraulic retraction jack
10. Nosewheel doors
11. Nose undercarriage leg strut
12. Taxying lamp
13. Twin nosewheels, forward retracting
14. Torque scissor links
15. Hydraulic steering control
16. Incidence vane
17. Underfloor control linkages and gust locks
18. Dual pitot heads
19. Circuit breaker panel
20. Control column
21. Six-tube Rockwell-Collins EFIS instrument panel
22. Instrument panel shroud
23. Windscreen wipers
24. Windscreen panels, electrically anti-iced; side windows electrically de-iced
25. Overhead systems switch panel
26. Cockpit roof escape hatch
27. First officer's seat
28. Captain's seat
29. Avionics cooling and cockpit air ducting
30. Avionics equipment rack
31. Starboard side toilet compartment
32. TCAS aerial
33. GPS aerial
34. Wardrobe
35. Stowage compartment
36. Cabin attendant's folding seat
37. Airstairs stowage compartment
38. Door latch
39. Main entry door
40. Airstairs
41. Folding handrail
42. Entry lobby
43. Hand baggage stowage
44. ATC 1 and 2 aerials
45. Overhead baggage lockers
46. Three-abreast passenger seating, 50-seat standard layout, 32 in pitch
47. Air distribution ducting
48. Pull-down window blinds
49. Cabin wall soundproofing and insulating blankets
50. Cabin window panels
51. Graphite/Kevlar/Nomex composite floor panels
52. Cabin sidewall seat mounting rails
53. Lower VHF aerial
54. Fuselage skin doubler, propeller ice and debris protection
55. Lower fuselage frames
56. Landing lamp, port and starboard
57. Recirculating filter
58. Wing panel centreline joint
59. Pressure floor above wing carry-through
60. Fuselage frame and stringer construction
61. Bonded skin/stringer panels
62. Starboard wing inboard fuel tank, total fuel capacity 5,185 litres (1,370 US gallons; 1,140 Imp gallons)
63. Hinged engine bay access panels
64. Starboard engine nacelles
65. Engine bay venting air intake
66. Dowty Aerospace R381 autofeathering and reversible pitch constant-speed six-bladed composite propeller
67. Pressure refuelling connection
68. Outboard integral fuel tank
69. Gravity filler cap
70. Leading-edge de-icing boot
71. Starboard navigation light
72. Strobe light
73. Starboard aileron
74. Aileron tab
75. Aileron control linkage
76. Starboard one-piece single-slotted flap
77. Ventral flap hinges
78. Engine exhaust nozzle
79. Upper VHF aerial
80. Starboard emergency exit hatch
81. Wing spar attachment double main frames
82. Port emergency exit hatch
83. Aileron interconnection mechanism
84. Underfloor conditioned-air distribution ducting
85. Cabin wall trim panelling
86. ADF antenna
87. Overhead baggage lockers
88. Rear cabin passenger seating
89. Air ducting and passenger service units
90. Galley unit
91. Rear lobby with service/ emergency door and attendant's seat
92. Composite tin root fillet
93. HF aerial
94. Tailplane leading-edge de-icing boot
95. Starboard tailplane
96. Starboard elevator
97. Fin leading-edge de-icing boot
98. Two-spar fin construction
99. Honeycomb skin panels
100. Anti-collision beacon
101. VOR aerial
102. Emergency locator transmitter (ELT) aerial
103. Rudder
104. Static dischargers
105. Graphite/Nomex composite skin panels
106. Dual rudder hydraulic actuators
107. Cabin pressurisation valve
108. Rear pressure bulkhead
109. Auxiliary power unit (APU)
110. APU exhaust
111. Tail navigation light
112. Elevator spring tab
113. Trim tab
114. Port elevator composite construction
115. Static dischargers
116. Port 'Logo' light
117. Two-spar tailplane construction
118. Ventral strake, port and starboard
119. Fin/tailplane spar attachment bulkhead
120. Two-spar fin construction
121. ELT, cockpit voice and data recorders
122. Baggage compartment
123. Up-and-over baggage bay door
124. Restraint net
125. Baggage loading floor
126. Bonded window frame/skin panel
127. Extended wing root trailing-edge fillet
128. Transformer rectifiers
129. Batteries
130. Air-conditioning pack, port and starboard
131. Wing spar bolted attachment joint
132. Drag link
133. Port inboard fuel tank bay
134. Wing panel bonded stringer construction, half span
135. Port single-slotted flap
136. Engine exhaust duct
137. Flap down position
138. Graphite/Nomex flap shroud
139. Graphite/Nomex flap panels
140. Aileron trim tab
141. Port aileron composite construction
142. Static dischargers
143. Strobe light
144. Port navigation light
145. Outer wing box rib construction
146. Compass flex valve
147. Port outboard integral fuel tank
148. Leading-edge de-icing boot
149. Twin mainwheels, forward retracting
150. Mainwheel doors, closed after undercarriage extension
151. Retraction breaker strut
152. Mainwheel leg pivot mounting
153. Engine fire extinguisher bottle
154. Flap hydraulic jack
155. Engine nacelle frame construction
156. Wing inspection light
157. Main engine mounting frame
158. Allison GMA 2100 turboshaft engine
159. Engine bearer struts
160. Engine accessory equipment
161. Full authority digital engine control (FADEC) units and hydraulic reservoir bay
162. Ventral oil cooler air duct
163. Particle separator air duct
164. Propeller reduction gearbox
165. Engine air intake
166. Port six-bladed propeller
167. Propeller blade root de-icing
168. Spinner
169. Propeller hub pitch change mechanism
170. Composite propeller blades
171. Inboard leading-edge de-icing boot
172. Leading-edge de-icing tubing
173. Front spar/fuselage frame attachment joint
174. Engine rear mounting frame

SWEDEN: AIRCRAFT—SAAB

The third prototype Saab 2000, which made its first flight in August 1992

Saab 2000 first prototype in the livery of launch customer Crossair

solid state weather radar, EICAS, com/nav/pulse and radio tuning units, digital air data system and AHRS. Optional avionics include TCAS, flight management system and turbulence weather radar.

DIMENSIONS, EXTERNAL:
Wing span	24.76 m (81 ft 2¾ in)
Wing aspect ratio	11.0
Length overall	27.03 m (88 ft 8¼ in)
Fuselage: Max diameter	2.31 m (7 ft 7 in)
Height overall	7.73 m (25 ft 4 in)
Tailplane span	10.36 m (34 ft 0 in)
Wheel track	8.23 m (27 ft 0 in)
Wheelbase	10.97 m (36 ft 0 in)
Propeller diameter	3.81 m (12 ft 6 in)
Propeller ground clearance	0.46 m (1 ft 6 in)
Passenger door: Height	1.60 m (5 ft 3 in)
Width	0.69 m (2 ft 3 in)
Height to sill	1.81 m (5 ft 11¼ in)
Baggage door (rear, port):	
Height	1.30 m (4 ft 3 in)
Width	1.35 m (4 ft 5 in)
Height to sill	1.85 m (6 ft 0¾ in)
Service/emergency door (rear, stbd):	
Height	1.22 m (4 ft 0 in)
Width	0.61 m (2 ft 0 in)
Emergency exits (overwing, each):	
Height	0.91 m (3 ft 0 in)
Width	0.51 m (1 ft 8 in)

DIMENSIONS, INTERNAL:
Cabin, excl flight deck, toilet and galley:
Length	17.25 m (56 ft 7¼ in)
Max width	2.16 m (7 ft 1 in)
Width at floor	1.70 m (5 ft 7 in)
Max height	1.83 m (6 ft 0 in)
Volume	52.7 m³ (1,860.0 cu ft)
Baggage/cargo compartment:	
Volume	10.2 m³ (360.0 cu ft)

AREAS:
Wings, gross	55.74 m² (600.0 sq ft)
Vertical tail surfaces (total)	13.01 m² (140.0 sq ft)
Horizontal tail surfaces (total)	18.35 m² (197.5 sq ft)

WEIGHTS AND LOADINGS:
Operating weight empty	13,500 kg (29,762 lb)
Max payload (weight limited)	5,900 kg (13,007 lb)
Max fuel load	4,165 kg (9,182 lb)
Max ramp weight	22,200 kg (48,942 lb)
Max T-O weight	22,000 kg (48,500 lb)
Max landing weight	21,500 kg (47,400 lb)
Max zero-fuel weight	19,400 kg (42,770 lb)
Max wing loading	394.7 kg/m² (80.84 lb/sq ft)
Max power loading	3.58 kg/kW (5.88 lb/shp)

PERFORMANCE (at max T-O weight ISA, except where indicated):
Max operating speed (V_{MO}):
 below 3,050 m (10,000 ft)
  250 knots (463 km/h; 288 mph) IAS
 above 3,050 m (10,000 ft)
  270 knots (500 km/h; 311 mph) IAS
Max operating Mach No. (M_{MO}) 0.62
Max cruising speed:
 at 7,620 m (25,000 ft) 366 knots (678 km/h; 421 mph)
 at 9,450 m (31,000 ft) 353 knots (653 km/h; 406 mph)
Long-range cruising speed at 9,450 m (31,000 ft)
  300 knots (556 km/h; 345 mph)
Max rate of climb at S/L 725 m (2,380 ft)/min
Rate of climb at S/L, OEI 183 m (600 ft)/min
Time to 6,100 m (20,000 ft) 10 min
Service ceiling 9,450 m (31,000 ft)
Service ceiling, OEI, at 95% of MTOW:
 ISA 6,645 m (21,800 ft)
 ISA + 10°C 5,975 m (19,600 ft)
T-O to 15 m (50 ft): at S/L 1,360 m (4,465 ft)
 at 1,525 m (5,000 ft) 1,680 m (5,515 ft)
Landing from 15 m (50 ft) at max landing weight:
 at S/L 1,250 m (4,105 ft)
 at 1,525 m (5,000 ft) 1,390 m (4,560 ft)
Runway LCN (paved runways) max 15
Range at 9,450 m (31,000 ft) with 50 passengers and baggage, reserves for 45 min hold at 1,525 m (5,000 ft) and 100 nm (185 km; 115 mile) diversion:
 at max cruising speed
  1,255 nm (2,326 km; 1,445 miles)
 at long-range cruising speed
  1,425 nm (2,641 km; 1,641 miles)

OPERATIONAL NOISE LEVELS (estimated):
T-O	89.0 EPNdB
Sideline	94.0 EPNdB
Approach	98.0 EPNdB

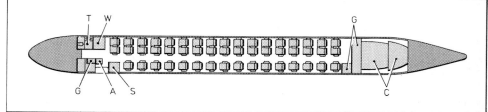

Saab 2000 standard 50-passenger layout. A: attendant's seat, C: baggage/cargo, G: galley, S: stowage, T: toilet, W: wardrobe (Jane's/Mike Keep)

SWITZERLAND

DÄTWYLER — see MDB Flugtechnik

FFA
FFA FLUGZEUGWERKE ALTENRHEIN AG
CH-9423 Altenrhein
Telephone: 41 (71) 43 51 11
Fax: 41 (71) 43 53 30
Telex: 882906 FFA CH
PRESIDENT: Charles Brönimann
CHIEF ENGINEER: A. Gasser
VICE-PRESIDENT, MARKETING: J. E. Watts-Phillips

Originated as Swiss branch of German Dornier company, becoming all-Swiss in 1948; acquired by Justus Dornier group in Zurich January 1987, taking present name 1 June 1987. Apart from Bravo, activities include producing spares for Swiss-built Northrop F-5E/F, overhaul and maintenance for Swiss Air Force and general aviation, and subcontract work for various foreign aircraft manufacturers. Current workforce about 300, of whom about 200 employed on aviation-related activities.

FFA AS 202/18A BRAVO
TYPE: Two/three-seat trainer and sporting aircraft.
PROGRAMME: Inherited from SIAI-Marchetti of Italy; first flight by AS 202/15 Swiss prototype (HB-HEA) 7 March 1969; 34 of this version built (see 1981-82 *Jane's*), plus one prototype AS 202/26A (1985-86 edition); AS 202/18A made first flight (HB-HEY) 22 August 1974, certificated by Swiss 12 December 1975 and FAA on 17 December 1976.
CURRENT VERSIONS: **AS 202/18A:** Two/three-seat aerobatic version; subtypes include **A2** (higher max T-O/landing weight, extended canopy, electric trim), **A3** (as A2 but mechanical trim, 24V electrics) and **A4** (as A2 but with British CAA approved special instrumentation); A4s of BAe Flying College named **Wren**. Full description in 1991-92 and earlier editions: following is shortened version.
CUSTOMERS: Total of 180 AS 202/18As delivered by end of 1989 (see 1987-88 *Jane's* for customer list); no further orders or deliveries since then, but aircraft remains available.
POWER PLANT: One 134 kW (180 hp) Textron Lycoming AEIO-360-B1F flat-four engine, driving a Hartzell HC-C2YK-1BF/F7666A-2 two-blade constant-speed propeller; Hoffmann three-blade propeller optional. Two wing leading-edge rubber fuel tanks with total capacity of 170 litres (44.9 US gallons; 37.4 Imp gallons). Refuelling point above each wing. Starboard tank has additional flexible fuel intake for aerobatics. Christen 801 fully aerobatic oil system, capacity 7.6 litres (2 US gallons; 1.6 Imp gallons).
ACCOMMODATION: Seats for two persons side by side in Aerobatic versions, under rearward sliding jettisonable transparent canopy. Space at rear in Utility versions for third seat or 100 kg (220 lb) of baggage. Dual controls, cabin ventilation and heating standard.
AVIONICS: Provision for VHF radio, VOR, ADF, Nav-O-Matic 200A autopilot, blind-flying instrumentation or other special equipment at customer's option.

DIMENSIONS, EXTERNAL:
Wing span	9.75 m (31 ft 11¾ in)
Wing aspect ratio	6.5
Length overall	7.50 m (24 ft 7¼ in)
Height overall	2.81 m (9 ft 2¾ in)
Tailplane span	3.67 m (12 ft 0½ in)
Wheel track	2.25 m (7 ft 4½ in)
Wheelbase	1.78 m (5 ft 10 in)
Propeller diameter	1.88 m (6 ft 2 in)

AREAS:
Wings, gross	13.86 m² (149.2 sq ft)

WEIGHTS AND LOADINGS:
Weight empty, equipped	710 kg (1,565 lb)
Max useful load (incl fuel): Aerobatic	177 kg (390 lb)
Utility	248 kg (546 lb)
Max T-O and landing weight:	
Aerobatic: A/A1	950 kg (2,094 lb)
A2/A3	980 kg (2,160 lb)
A4	1,010 kg (2,226 lb)
Utility: current 18A models	1,080 kg (2,380 lb)
Max wing loading: Utility	75.8 kg/m² (15.52 lb/sq ft)
Max power loading: Utility	7.84 kg/kW (12.86 lb/hp)

PERFORMANCE (Utility category at max T-O weight):
Never-exceed speed (V_{NE})	173 knots (320 km/h; 199 mph)
Max level speed at S/L	130 knots (241 km/h; 150 mph)
Max cruising speed (75% power) at 2,440 m (8,000 ft)	122 knots (226 km/h; 141 mph)
Econ cruising speed (55% power) at 3,050 m (10,000 ft)	109 knots (203 km/h; 126 mph)
Stalling speed, engine idling:	
flaps up	62 knots (115 km/h; 71 mph)
flaps down	49 knots (90 km/h; 56 mph)
Max rate of climb at S/L	244 m (800 ft)/min
Service ceiling	5,180 m (17,000 ft)
T-O run at S/L	215 m (705 ft)
T-O to 15 m (50 ft) at S/L	415 m (1,360 ft)
Landing from 15 m (50 ft)	465 m (1,525 ft)
Landing run	210 m (690 ft)
Range with max fuel, no reserves	615 nm (1,140 km; 707 miles)
Max endurance	5 h 30 min
g limits	+6/−3

FFA AS 202/18A4 Bravo (Wren) of the BAe Flying College (Peter J. Cooper)

MDB
MDB FLUGTECHNIK AG
Flugplatz, CH-3368 Bleienbach-Langenthal
Telephone: 41 (63) 28 31 11
Fax: 41 (63) 23 24 29
Telex: 982626 MDC CH
PRESIDENT: Max Dätwyler
MANAGER, AIRCRAFT MARKETING: Ulrich Wenger

Main activities of MDC Max Dätwyler AG (see 1991-92 *Jane's*) are in rotogravure, metal processing and other non-aerospace areas. All aircraft-related activities (repair and modification of light aircraft, and subcontract work for Swiss aircraft industry) concentrated during 1991 into new and separate MDB Flugtechnik.

MDB MD3 SWISS-TRAINER
TYPE: Two-seat trainer.
PROGRAMME: Originated in late 1960s, but much redesigned later with view to maximising common-module interchangeability; first flight of MD3-160 (HB-HOH) 12 August 1983; first flight second prototype (HB-HOJ) 1990; Swiss federal certification to FAR Pt 23 awarded 22 January 1991; first flight of first pre-production aircraft (HB-HNA) July 1992; FAA certification received 2 September 1992; full scale series production to be established outside Switzerland.
CURRENT VERSIONS: **MD3-116:** Primary trainer, powered by 86.5 kW (116 hp) Textron Lycoming O-235-N2A flat-four engine; first prototype currently fitted with this engine and successfully completed performance and handling test flights Autumn 1991.
MD3-160: Aerobatic trainer and glider tug, with more powerful Textron Lycoming O-320-D2A engine; description applies to this version.
MD3-160A: Fully aerobatic version, with AEIO-320-D2B fuel-injection engine and modified fuel system; second pre-production aircraft is to this version.
DESIGN FEATURES: Components interchangeable with each other, and left with right, are: (1) ailerons, flaps, elevators and rudder; (2) wing leading-edges; (3) wing inboard panels; (4) wing outboard panels; (5) wingtips; (6) tailplane halves and fin; (7) elevator and rudder tips; (8) aileron, elevator and rudder tabs.
Wing section NACA 64₂-15414 (modified); dihedral 5° 30'; incidence 2°.
FLYING CONTROLS: Mechanical; single-slotted ailerons; all primary surfaces mass balanced; trim tabs in port aileron, each elevator and rudder; balance tab in starboard aileron; electrically operated flaps.
STRUCTURE: All-metal except for glassfibre fairings and cowling; designed for easy construction under licence, even in countries with no developed aircraft industry; rear fuselage detachable.
LANDING GEAR: Non-retractable tricycle type with steerable nosewheel (30° left, 44° right). Main-gear legs are cantilever steel struts, descending at 45° from fuselage main bulkhead. Nose gear fitted with oleo-pneumatic shock absorber. Cleveland 6.00-6 mainwheels and 5.00-5 nosewheel. Tyre pressure 2.41 bars (35 lb/sq in) on all units. Independent Cleveland hydraulic disc brake on each mainwheel. Speed fairings on all three wheels.
POWER PLANT: One Textron Lycoming O-320-D2A flat-four engine (119 kW; 160 hp at 2,700 rpm), driving a McCauley 1C172/AGM7462 two-blade fixed-pitch metal propeller. Exhaust system extends full length under fuselage to extreme rear of tailcone, exhaust gases being emitted through narrow slot running along pipe. (Aircraft meets noise requirements with short or long exhaust.) Integral fuel tank in each wing: total capacity 148 litres (39 US gallons; 32.6 Imp gallons). Refuelling point in top of each tank. Oil capacity 7.6 litres (2 US gallons; 1.7 Imp gallons).
ACCOMMODATION: Side by side adjustable seats for pilot and one passenger. Five-point fixed seat belts. Forward sliding canopy. Space behind seats for 50 kg (110 lb) of baggage. Dual controls, cabin ventilation and heating standard.
SYSTEMS: Hydraulic system for brakes only. 28V 60A engine driven alternator and 24V 30Ah battery provide electrical power for engine starting, lighting, instruments and com/nav equipment.
AVIONICS: Provision for VHF radio, VOR, ADF, GPS, transponder or other items at customer's option.
EQUIPMENT: Equipment for glider towing optional.

DIMENSIONS, EXTERNAL:
Wing span	10.00 m (32 ft 9¾ in)
Wing chord, constant	1.50 m (4 ft 11 in)
Wing aspect ratio	6.67
Length overall	7.10 m (23 ft 3½ in)
Height overall	2.92 m (9 ft 7 in)
Tailplane span	3.00 m (9 ft 10 in)
Wheel track	2.05 m (6 ft 8¾ in)
Wheelbase	1.56 m (5 ft 1½ in)
Propeller diameter	1.88 m (6 ft 2 in)

DIMENSIONS, INTERNAL:
Cabin, from firewall to rear bulkhead:	
Length	1.30 m (4 ft 3¼ in)
Max width	1.12 m (3 ft 8 in)
Max height	1.08 m (3 ft 6½ in)

358 SWITZERLAND: AIRCRAFT—MDB/PILATUS

First pre-production MD3-160 Swiss-Trainer, for delivery to a US civil flying training school

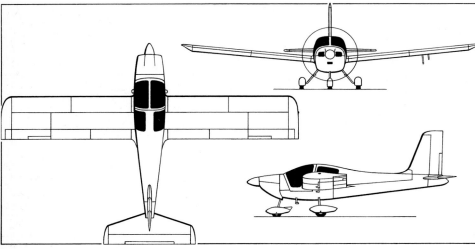

MDB MD3-160 Swiss-Trainer (*Jane's/Mike Keep*)

AREAS:	
Wings, gross	15.00 m² (161.5 sq ft)
Ailerons (total)	1.13 m² (12.16 sq ft)
Trailing-edge flaps (total)	1.96 m² (21.10 sq ft)
Fin	0.89 m² (9.58 sq ft)
Rudder	0.51 m² (5.49 sq ft)
Tailplane	1.71 m² (18.41 sq ft)
Elevators (total)	1.04 m² (11.19 sq ft)

WEIGHTS AND LOADINGS (A: Aerobatic, U: Utility):
Weight empty	640 kg (1,411 lb)
Max T-O weight: A	840 kg (1,852 lb)
U	920 kg (2,028 lb)
Max landing weight: A	840 kg (1,852 lb)
U	891 kg (1,964 lb)
Max wing loading: A	56.0 kg/m² (11.47 lb/sq ft)
U	61.3 kg/m² (12.56 lb/sq ft)
Max power loading: A	7.06 kg/kW (11.58 lb/hp)
U	7.73 kg/kW (12.68 lb/hp)

PERFORMANCE (A: Aerobatic at 840 kg; 1,852 lb AUW; U: Utility category at 920 kg; 2,028 lb):
Max permissible diving speed (V_D):	
A	196 knots (363 km/h; 226 mph)
U	174 knots (323 km/h; 201 mph)
Never-exceed speed (V_{NE}):	
A	174 knots (323 km/h; 201 mph)
U	158 knots (293 km/h; 182 mph)
Max structural cruising speed:	
A	141 knots (261 km/h; 162 mph)
U	126 knots (233 km/h; 145 mph)
Max cruising speed (75% power) at 1,525 m (5,000 ft):	
A	138 knots (256 km/h; 159 mph)
U	123 knots (229 km/h; 142 mph)
Max manoeuvring speed:	
A	121 knots (224 km/h; 139 mph)
U	110 knots (204 km/h; 127 mph)
Max speed for flap extension:	
A, U	90 knots (167 km/h; 104 mph)
Stalling speed, power off:	
flaps up	56 knots (104 km/h; 65 mph)
flaps down	46 knots (85 km/h; 53 mph)
Max rate of climb at S/L: A	336 m (1,102 ft)/min
Max rate of climb (75% power) towing 365 kg (805 lb) sailplane	104 m (341 ft)/min
T-O run: A	138 m (453 ft)
U	165 m (541 ft)
T-O to 15 m (50 ft): A	274 m (899 ft)
U	338 m (1,109 ft)
Landing from 15 m (50 ft) at max landing weight:	
A, U	309 m (1,011 ft)
Landing run at max landing weight:	
A, U	135 m (443 ft)
Range with max fuel, no reserves	588 nm (1,090 km; 677 miles)
g limits: A	+6/−3
U	+4.4/−2.2

PILATUS
PILATUS FLUGZEUGWERKE AG
CH-6370 Stans
Telephone: 41 (41) 63 61 11
Fax: 41 (41) 61 33 51
Telex: 866202 PIL CH
CHAIRMAN: Dr Ernst Thomke
SENIOR VICE-PRESIDENTS:
 Oscar J. Schwenk (Operations)
 Oscar Bründler (Finance and Controlling)

Formed December 1939; now part of Oerlikon-Bührle Group; purchased assets of Britten-Norman (Bembridge) Ltd of UK 24 January 1979, which now operates as Pilatus Britten-Norman Ltd.

Current products include PC-6 Turbo-Porter, PC-7 Turbo-Trainer, PC-9 and PC-12; makes components for BAe Jetstream 41 as risk-sharing partner; overhauls Swiss Air Force Hunters; has proposed modified PC-9 with Beech Aircraft for US Air Force/Navy JPATS trainer programme.

PILATUS PC-6 TURBO-PORTER
US Army designation: UV-20A Chiricahua
TYPE: Single-turboprop multi-purpose STOL utility transport.
PROGRAMME: First flight piston-engined prototype 4 May 1959 (see contemporary *Jane's*); current PC-6/B2-H4 introduced mid-1985. Production being increased from seven to 25 per year in 1992.
CURRENT VERSIONS: **Early versions:** PC-6/A, A1, A2, B, B1, B2 and C2-H2 with various turboprops (see 1974-75 *Jane's*); about 40 produced as agricultural aircraft with spray or dusting gear.
 PC-6/B2-H4: Gross weight for FAR Pt 23 passenger carrying increased by 600 kg (1,323 lb), giving up to 570 kg (1,257 lb) greater payload; changes include turned-up wingtips, enlarged dorsal fin, uprated mainwheel shock absorbers, new tailwheel assembly and slight airframe reinforcement. H4 changes can be retrofitted to B2-H2. *Description applies to B2-H4.*
CUSTOMERS: Total of over 500 (all versions) produced, including licence manufacture in USA; in service in more than 50 countries; military operators include Angola, Argentina, Austria, Chad, Dubai, Ecuador, France, Iran, Myanmar, Peru, Sudan, Switzerland, Thailand and US Army.
COSTS: French Army order (1992, five aircraft) FFr30 million ($5.5 million).
DESIGN FEATURES: Wing section NACA 64-514 (constant) with span-increasing wingtip fairings; dihedral 1°; incidence 2°.
FLYING CONTROLS: Mechanical primary surfaces; Flettner tabs on elevator; geared aileron tabs; electrically actuated double-slotted flaps; electrically actuated variable incidence tailplane.
STRUCTURE: All-metal; single-strut wing bracing.
LANDING GEAR: Non-retractable tailwheel type. Oleo shock absorbers of Pilatus design in all units. Steerable/lockable tailwheel. Goodyear Type II mainwheels and GA 284 tyres size 24 × 7 or 7.50 × 10 (pressure 2.21 bars; 32 lb/sq in); oversize Goodyear Type III wheels and tyres optional, size 11.0 × 12, pressure 0.88 bar (12.8 lb/sq in). Goodyear tailwheel with size 5.00-4 tyre. Goodyear disc brakes. Pilatus wheel/ski gear optional.
POWER PLANT: One 507 kW (680 shp) Pratt & Whitney Canada PT6A-27 turboprop (flat rated at 410 kW; 550 shp at S/L), driving a Hartzell HC-B3TN-3D/T-10178 C or CH, or T10173 C or CH constant-speed fully feathering reversible-pitch propeller with Beta mode control. Standard fuel in integral wing tanks, usable capacity 644 litres (170 US gallons; 142 Imp gallons). Two underwing auxiliary tanks, each of 245 litres (65 US gallons; 54 Imp gallons), available optionally. Oil capacity 12.5 litres (3.3 US gallons; 2.75 Imp gallons).
ACCOMMODATION: Cabin has pilot's seat forward on port side, with one passenger seat alongside, and is normally fitted with six quickly removable seats, in pairs, to rear of these for additional passengers. Up to 11 persons, including pilot, can be carried in 2-3-3-3 high density layout; or up to ten parachutists; or two stretchers plus three attendants in ambulance configuration. Floor level, flush with door sill, with seat rails. Forward opening door beside each front seat. Large rearward sliding door on starboard side of main cabin. Port side sliding door optional. Double doors, without central pillar, on port side. Hatch in floor 0.58 × 0.90 m (1 ft 10¾ in × 2 ft 11½ in), openable from inside cabin, for aerial camera or supply dropping. Hatch in cabin rear wall 0.50 × 0.80 m (1 ft 7 in × 2 ft 7 in) permits stowage of six passenger seats or accommodation of freight items up to 5.0 m (16 ft 5 in) in length. Walls lined with lightweight soundproofing and heat insulation material. Adjustable heating and ventilation systems. Dual controls optional.
SYSTEMS: Cabin heated by engine bleed air. Scott 8,500 oxygen system optional. 200A 30V starter/generator and 24V 34Ah (optionally 40Ah) nickel-cadmium battery.
EQUIPMENT: Generally to customer's requirements, but can include stretchers for ambulance role, aerial photography and survey gear, agricultural equipment, or 800 litre (211 US gallon; 176 Imp gallon) water tank in cabin, with quick release system, for firefighting role. Stainless steel 1,330 litre (351.5 US gallon; 292.5 Imp gallon) agricultural tank, installed behind front seats, can also be used in firebombing role. With 46- or 62-nozzle underwing spraybooms, aircraft can spray a 45 m (148 ft) swath; optional ULV system (four to six atomisers or two to six Micronairs) can increase swath width to 400 m (1,310 ft).

DIMENSIONS, EXTERNAL:
Wing span	15.87 m (52 ft 0¾ in)
Wing chord, constant	1.90 m (6 ft 3 in)
Wing aspect ratio	8.4
Length overall	11.00 m (36 ft 1 in)
Height overall (tail down)	3.20 m (10 ft 6 in)
Elevator span	5.12 m (16 ft 9½ in)
Wheel track	3.00 m (9 ft 10 in)
Wheelbase	7.87 m (25 ft 10 in)
Propeller diameter	2.56 m (8 ft 5 in)
Cabin double door (port) and sliding door (starboard):	
Max height	1.04 m (3 ft 5 in)
Width	1.58 m (5 ft 2¼ in)

DIMENSIONS, INTERNAL:
Cabin, from back of pilot's seat to rear wall:	
Length	2.30 m (7 ft 6½ in)
Max width	1.16 m (3 ft 10 ft 6 in)
Max height (at front)	1.28 m (4 ft 2½ in)
Height at rear wall	1.18 m (3 ft 10½ in)
Floor area	2.67 m² (28.6 sq ft)
Volume	3.28 m³ (107 cu ft)

PILATUS—AIRCRAFT: SWITZERLAND

Pilatus PC-6/B2-H4 Turbo-Porter STOL utility aircraft

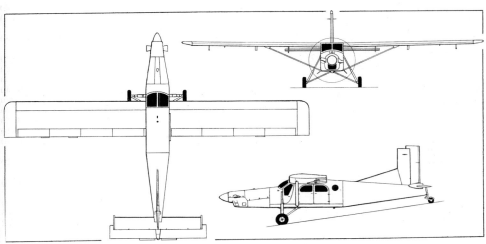

Pilatus PC-6/B2-H4 Turbo-Porter (*Dennis Punnett*)

AREAS:
Wings, gross 30.15 m² (324.5 sq ft)
Ailerons (total) 3.83 m² (41.2 sq ft)
Flaps (total) 3.76 m² (40.5 sq ft)
Fin 1.70 m² (18.3 sq ft)
Rudder, incl tab 0.96 m² (10.3 sq ft)
Tailplane 4.03 m² (43.4 sq ft)
Elevator, incl tab 2.11 m² (22.7 sq ft)
WEIGHTS AND LOADINGS:
Weight empty, equipped 1,270 kg (2,800 lb)
Max fuel weight: internal 508 kg (1,120 lb)
  underwing 392 kg (864 lb)
Max payload:
  with reduced internal fuel 1,130 kg (2,491 lb)
  with max internal fuel 1,062 kg (2,341 lb)
  with max internal and underwing fuel
 571 kg (1,259 lb)
Max T-O weight, Normal (CAR 3):
  wheels (standard) 2,800 kg (6,173 lb)
  skis 2,600 kg (5,732 lb)
Max landing weight: wheels 2,660 kg (5,864 lb)
  skis 2,600 kg (5,732 lb)
Max cabin floor loading 488 kg/m² (100 lb/sq ft)
Max wing loading (Normal):
  wheels 92.87 kg/m² (19.03 lb/sq ft)
  skis 86.23 kg/m² (17.67 lb/sq ft)
Max power loading (Normal):
  wheels 6.83 kg/kW (11.22 lb/shp)
  skis 5.13 kg/kW (8.43 lb/shp)
PERFORMANCE (at max T-O weight, ISA, Normal category):
Never-exceed speed (V_{NE})
 151 knots (280 km/h; 174 mph) IAS
Econ cruising speed at 3,050 m (10,000 ft)
 115 knots (213 km/h; 132 mph)
Stalling speed, power off, flaps down
 52 knots (96 km/h; 60 mph)
Max rate of climb at S/L 287 m (941 ft)/min
Max operating altitude 7,620 m (25,000 ft)
T-O run at S/L 197 m (646 ft)
Landing run at S/L 127 m (417 ft)
Max range at 115 knots (213 km/h; 132 mph) at 3,050 m (10,000 ft), no reserves:

with max payload 394 nm (730 km; 453 miles)
with max internal fuel 500 nm (926 km; 576 miles)
with max internal and underwing fuel
 870 nm (1,612 km; 1,002 miles)
g limits +3.72/−1.5

PILATUS PC-7 TURBO-TRAINER
Swiss Air Force designation: PC-7/CH

TYPE: Two-seat turboprop trainer.
PROGRAMME: First flight production PC-7 18 August 1978; Swiss federal civil aerobatic category certification 5 December 1978; first deliveries December 1978; Swiss utility category 6 April 1979; French DGAC certification 16 May 1983; US FAR Pt 23 certification 12 August 1983; meets selected group of US military trainer category specifications.
CUSTOMERS: Total deliveries over 430 by early 1993; customers (see also 1992-93 *Jane's*) include air forces of Abu Dhabi, Angola, Austria, Bolivia, Chad, Chile, France (for CEV), Guatemala, Iran, Iraq, Malaysia, Mexico, Myanmar, Netherlands, Switzerland, Uruguay and three undisclosed countries; others include CIPRA of France and three US private owners; French Patrouille ECCO formation display team operates four with smoke generators.
DESIGN FEATURES: Constant chord centre-section and tapered outer sections with 1° quarter-chord sweepback; wing section NACA 64_2A-415 at root and 64_1A-612 at tip; dihedral 7° on outer panels.
FLYING CONTROLS: Elevators and rudder cable operated; mass balanced ailerons pushrod operated; trim tab in starboard half of elevator; anti-servo tab in rudder; trim tab in port aileron; strakes under rear fuselage and at tailplane leading-edges; dorsal fin; electrically operated split flaps extending under fuselage.
STRUCTURE: All-metal with some GFRP fairings on wings and fuselage; one-piece single-spar wing with auxiliary spar, ribs and stringer-reinforced skin; tail surface structure similar to wing.
LANDING GEAR: Electrically actuated retractable tricycle type, with emergency manual extension. Mainwheels retract inward, nosewheel rearward. Oleo-pneumatic shock absorber in each unit. Castoring nosewheel, with shimmy dampers. Goodrich mainwheels and tyres, size 6.50-8, pressure 4.5 bars (65 lb/sq in). Goodrich nosewheel and tyre, size 6.00-6, pressure 2.75 bars (40 lb/sq in). No mainwheel doors. Goodrich hydraulic disc brakes on mainwheels. Parking brake.
POWER PLANT: One 485 kW (650 shp) Pratt & Whitney Canada PT6A-25A turboprop, flat rated at 410 kW (550 shp at S/L), driving a Hartzell HC-B3TN-2/T10173C-8 three-blade constant-speed fully feathering propeller. Fuel in integral tanks in outer wing leading-edges, total usable capacity 474 litres (125 US gallons; 104 Imp gallons). Overwing refuelling point on each tank. Engine oil system permits up to 30 s of inverted flight. Provision for two 152 or 240 litre (40 or 63.5 US gallon; 33.5 or 52.75 Imp gallon) underwing drop tanks. Oil capacity 16 litres (4.2 US gallons; 3.5 Imp gallons).
ACCOMMODATION: Adjustable seats for two persons in tandem (instructor at rear), beneath rearward sliding jettisonable Plexiglas canopy. Martin-Baker Mk CH 15A lightweight ejection seats, available optionally, offer safe escape for both occupants at speeds between 60 knots on runway and 300 knots in the air (111-556 km/h; 69-345 mph), and at altitudes up to 6,700 m (22,000 ft). Dual controls standard. Cockpits ventilated and heated by engine bleed air, which can also be used for windscreen de-icing. Space for 25 kg (55 lb) of baggage aft of seats, with external access.
SYSTEMS: Freon air-conditioning and oxygen systems standard. Hydraulic system for mainwheel brakes only. No pneumatic system. 28V DC operational electrical system, incorporating Lear Siegler 30V 200A starter/generator and Marathon 36Ah or 42Ah nickel-cadmium battery; two static inverters for AC power supply. Ground power receptacle in port side of rear fuselage. Goodrich propeller electric de-icing system optional.
AVIONICS: Basic flight and navigation instrumentation in both cockpits, except for magnetic compass (front cockpit only). Additional nav and com equipment to customer's requirements. Other optional equipment includes hood to screen pupil from rear cockpit during IFR training.
EQUIPMENT: Landing/taxying light standard on each main-wheel leg. Swiss Air Force PC-7s to receive (1994) Ericsson A100 Erijammer ECM training pod.

DIMENSIONS, EXTERNAL:
Wing span 10.40 m (34 ft 1 in)
Wing chord: mean aerodynamic 1.64 m (5 ft 5 in)
  mean geometric 1.60 m (5 ft 3 in)
Wing aspect ratio 6.5
Length overall 9.78 m (32 ft 1 in)
Height overall 3.21 m (10 ft 6 in)
Tailplane span 3.40 m (11 ft 2 in)
Wheel track 2.54 m (8 ft 4 in)
Wheelbase 2.32 m (7 ft 7 in)
Propeller diameter 2.36 m (7 ft 9 in)
AREAS:
Wings, gross 16.60 m² (179.0 sq ft)
Ailerons (total) 1.621 m² (17.45 sq ft)
Trailing-edge flaps (total) 2.035 m² (21.90 sq ft)
Fin, incl dorsal fin 1.062 m² (11.43 sq ft)
Rudder, incl tab 0.959 m² (10.32 sq ft)
Tailplane 1.783 m² (19.19 sq ft)
Elevators, incl tab 1.395 m² (15.02 sq ft)
WEIGHTS AND LOADINGS (A: Aerobatic, U: Utility category):
Basic weight empty 1,330 kg (2,932 lb)
Max T-O weight: A 1,900 kg (4,188 lb)
  U 2,700 kg (5,952 lb)
Max ramp weight: U 2,711 kg (5,976 lb)
Max landing weight:
  A (military specification) 1,804 kg (3,977 lb)
  A (FAR Pt 23) 1,900 kg (4,188 lb)
  U 2,565 kg (5,655 lb)
Max zero-fuel weight 1,664 kg (3,668 lb)
Max wing loading: A 114.5 kg/m² (23.44 lb/sq ft)
  U 162.7 kg/m² (33.31 lb/sq ft)
Max power loading: A 4.63 kg/kW (7.61 lb/shp)
  U 6.59 kg/kW (10.82 lb/shp)
PERFORMANCE (at max T-O weight, ISA, except where indicated):
Never-exceed speed (V_{NE}):
  A, U 270 knots (500 km/h; 310 mph) EAS
Max operating speed:
  A, U 270 knots (500 km/h; 310 mph) EAS
Max cruising speed at 6,100 m (20,000 ft):
  A 222 knots (412 km/h; 256 mph)
  U 196 knots (364 km/h; 226 mph)
Econ cruising speed at 6,100 m (20,000 ft):
  A 171 knots (317 km/h; 197 mph)
  U 165 knots (305 km/h; 190 mph)
Manoeuvring speed:
  A 175 knots (325 km/h; 202 mph) EAS
  U 181 knots (335 km/h; 208 mph) EAS
Max speed with flaps and landing gear down:
  A, U 135 knots (250 km/h; 155 mph) EAS
Stalling speed, flaps and landing gear up, power off:
  A 71 knots (131 km/h; 82 mph) EAS
  U 83 knots (154 km/h; 96 mph) EAS
Stalling speed, flaps and landing gear down, power off:
  A 64 knots (119 km/h; 74 mph) EAS
  U 74 knots (138 km/h; 86 mph) EAS
Max rate of climb at S/L: A 655 m (2,150 ft)/min
  U 393 m (1,290 ft)/min

360   SWITZERLAND: AIRCRAFT—PILATUS

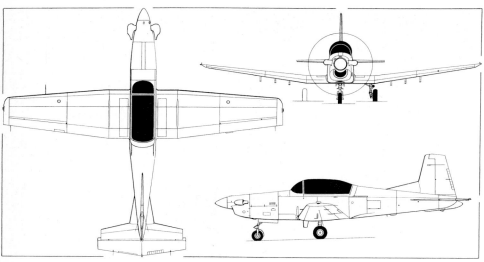

Pilatus PC-7 Turbo-Trainer (Pratt & Whitney Canada PT6A-25A turboprop) *(Dennis Punnett)*

Royal Netherlands Air Force PC-7 Turbo-Trainer on a pre-delivery flight over the Swiss Alps

Luftwaffe by civilian company; increased fuel for 3 h 20 min mission; two Southwest RM-24 winches on inboard pylons with targets stowed aft of winch; TAS 06 acoustic scoring system.

PC-9 Mk II: Modified for US Air Force/Navy JPATS trainer programme; described separately.

CUSTOMERS: Total sales more than 180 to 11 customers including Germany (10) and air forces of Australia (67), Myanmar (four), Saudi Arabia (30) and Thailand (20); Royal Saudi Air Force order 26 September 1985 (first delivery 15 December 1985); Royal Australian Air Force order 10 July 1986; two delivered complete in 1987, then six kits and 11 sets of components; remainder built under licence by Hawker de Havilland and ASTA (which see); first RAAF delivery 14 December 1987; delivery of target tugs to Germany started September 1990; first of three as chase planes for US Army delivered Spring 1991.

DESIGN FEATURES: Meets FAR Pt 23 (Amendments 1 to 28) and special Swiss federal civil conditions for both aerobatic and utility categories; complies with selected parts of US military training specifications; approx 9 per cent structural commonality with PC-7; parallel chord wing centre-section; tapered outer panels with quarter-chord sweepback 1°. Wing section PIL15M825 at root, PIL12M850 at tip; dihedral 7° from centre-section; incidence 1° at root; twist −2°.

FLYING CONTROLS: Cable operated elevator and rudder; mass balanced, electrically actuated trim tab in starboard half of elevator; electrically actuated trim/anti-balance tab in rudder controlled from rocker switch on power control lever; ailerons operated by pushrods; electrically operated lateral trim; hydraulically operated split flaps; hydraulically operated airbrake panel under centre-fuselage.

STRUCTURE: All-metal with some GFRP wing/fuselage fairings; one-piece wing with auxiliary spar, ribs and stringer-reinforced skin.

LANDING GEAR: Retractable tricycle type, with hydraulic actuation in both normal and emergency modes. Mainwheels retract inward into wing centre-section, nosewheel rearward; all units enclosed by doors when retracted. Oleo-pneumatic shock absorber in each leg. Hydraulically actuated nosewheel steering. Goodrich wheels and tyres, with Goodrich multi-piston hydraulic disc brakes on mainwheels. RAAF version has low pressure tyres for grass field operation. Parking brake.

POWER PLANT: One 857 kW (1,150 shp) Pratt & Whitney Canada PT6A-62 turboprop, flat rated at 708 kW (950 shp), driving a Hartzell HC-D4N-ZA/09512A four-blade constant-speed fully feathering propeller. Single lever engine control. Fuel in two integral tanks in wing leading-edges, total usable capacity 535 litres (141.3 US gallons; 117.7 Imp gallons). Overwing refuelling point on each side. Fuel system includes 12 litre (3.2 US gallon; 2.6 Imp gallon) aerobatics tank in fuselage, forward of front cockpit, which permits up to 60 s of inverted flight. Provision for two 154 or 248 litre (40.7 or 65.5 US gallon; 33.9 or 54.5 Imp gallon) drop tanks on centre underwing attachment points. Total oil capacity 16 litres (4.2 US gallons; 3.5 Imp gallons).

ACCOMMODATION: Two Martin-Baker Mk CH 11A adjustable ejection seats, each with integrated personal survival pack

Time to 5,000 m (16,400 ft): A	9 min
U	17 min
Max operating altitude	7,620 m (25,000 ft)
Service ceiling: A	10,060 m (33,000 ft)
U	7,925 m (26,000 ft)
T-O run at S/L: A	240 m (787 ft)
U	780 m (2,560 ft)
T-O to 15 m (50 ft) at S/L: A	400 m (1,312 ft)
U	1,180 m (3,870 ft)
Landing from 15 m (50 ft) at S/L at max landing weight:	
A	510 m (1,675 ft)
U	800 m (2,625 ft)
Landing run at S/L at max landing weight:	
A	295 m (968 ft)
U	505 m (1,655 ft)
Max range at cruise power at 5,000 m (16,400 ft), 5% fuel plus 20 min reserves:	
A	647 nm (1,200 km; 745 miles)
U	1,420 nm (2,630 km; 1,634 miles)
Endurance at 6,100 m (20,000 ft), with reserves:	
A, at max speed	3 h
A, for max range	4 h 22 min
U, at max speed	2 h 36 min
U, for max range	3 h 45 min
g limits: A	+6/−3
U	+4.5/−2.25

PILATUS PC-9

TYPE: Two-seat turboprop trainer.
PROGRAMME: Design began May 1982; aerodynamic elements tested on PC-7 1982-83; first flights by two pre-production PC-9s: HB-HPA on 7 May 1984 and HB-HPB on 20 July 1984; aerobatic certification 19 September 1985.
CURRENT VERSIONS: **Standard PC-9:** *Description applies to this version.*
PC-9/A: Australian version.
PC-9B: German target towing version, operated for

Pilatus PC-9/As of the Royal Australian Air Force's 'Roulettes' display team

and fighter-standard pilot equipment. Stepped tandem arrangement with rear seat elevated 15 cm (6.3 in). Seats operable, through canopy, at zero height and speeds down to 60 knots (112 km/h; 70 mph). Anti-g system optional. One-piece acrylic Perspex windscreen; one-piece framed canopy, incorporating rollover bar, opens sideways to starboard. Dual controls standard. Cockpit heating, cooling, ventilation and canopy demisting standard. Space for 25 kg (55 lb) of baggage aft of seats, with external access.

SYSTEMS: AiResearch ECS, using air cycle and engine bleed air, for cockpit heating/ventilation and canopy demisting. Fairey Systems hydraulic system, pressure 207 bars (3,000 lb/sq in), for actuation of landing gear, mainwheel doors, nosewheel steering, flaps and airbrake; system max flow rate 18.8 litres (4.97 US gallons; 4.14 Imp gallons)/min. Bootstrap oil/oil reservoir, pressurised at 3.45-207 bars (50-3,000 lb/sq in). Oil/nitrogen accumulator, also charged to 207 bars (3,000 lb/sq in), provides emergency hydraulic power for flaps and landing gear. Primary electrical system (28V DC operational, 24V nominal) powered by Lear Siegler 30V 200A starter/generator and 24V 40Ah battery; two static inverters supply 115/26V AC power at 400Hz. Ground power receptacle. Electric anti-icing of pitot tube, static ports and AOA transmitter standard; electric de-icing of propeller blades optional. Diluter demand oxygen system, selected and controlled individually from panel in each cockpit.

AVIONICS: Both cockpits fully instrumented to customer specifications, with Logic computer operated integrated engine and systems data display. Single or dual VHF, UHF and/or HF com radios to customer's requirements. Audio integrating system controls audio services from com, nav and interphone systems. Customer-specified equipment provides flight environmental, attitude and direction data, and ground-transmitted position determining information. Optional equipment includes Bendix/King CRT displays (electronic ADI and HSI, standard on PC-9/A), J. E. T. head-up displays, encoding altimeter and ELT.

EQUIPMENT: Retractable 250W landing/taxying light in each main landing gear leg bay.

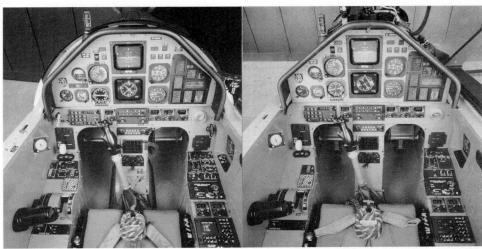
Front (left) and rear cockpits of the PC-9, showing EFIS and LCD engine displays

DIMENSIONS, EXTERNAL:
Wing span	10.124 m (33 ft 2½ in)
Wing chord: mean aerodynamic	1.65 m (5 ft 5 in)
mean geometric	1.61 m (5 ft 3½ in)
Wing aspect ratio	6.3
Length overall	10.175 m (33 ft 4¾ in)
Height overall	3.26 m (10 ft 8⅓ in)
Wheel track	2.54 m (8 ft 4 in)
Propeller diameter	2.44 m (8 ft 0 in)

AREAS:
Wings, gross	16.29 m² (175.3 sq ft)
Ailerons (total)	1.57 m² (16.90 sq ft)
Trailing-edge flaps (total)	1.77 m² (19.05 sq ft)
Airbrake	0.30 m² (3.23 sq ft)
Fin	0.86 m² (9.26 sq ft)
Rudder, incl tab	0.90 m² (9.69 sq ft)
Tailplane	1.80 m² (19.38 sq ft)
Elevator, incl tab	1.60 m² (17.22 sq ft)

WEIGHTS AND LOADINGS (A: Aerobatic, U: Utility):
Basic weight empty	1,685 kg (3,715 lb)
Max T-O weight: A	2,250 kg (4,960 lb)
U	3,200 kg (7,055 lb)
Max ramp weight: A	2,260 kg (4,982 lb)
U	3,210 kg (7,077 lb)
Max landing weight: A	2,250 kg (4,960 lb)
U	3,100 kg (6,834 lb)
Max zero-fuel weight: A	1,900 kg (4,188 lb)
Max wing loading: A	138.1 kg/m² (28.3 lb/sq ft)
U	196.4 kg/m² (40.2 lb/sq ft)
Max power loading: A	3.18 kg/kW (5.22 lb/shp)
U	4.52 kg/kW (7.42 lb/shp)

PERFORMANCE (at appropriate max T-O weight, ISA, propeller speed 2,000 rpm):
Max permissible diving speed (V_D): A, U
   Mach 0.73 (360 knots; 667 km/h; 414 mph EAS)
Max operating speed: A, U
   Mach 0.68 (320 knots; 593 km/h; 368 mph EAS)
Max level speed:
   A at S/L   270 knots (500 km/h; 311 mph)
   A at 6,100 m (20,000 ft)
      300 knots (556 km/h; 345 mph)
Max manoeuvring speed:
   A   210 knots (389 km/h; 242 mph) EAS
   U   200 knots (370 km/h; 230 mph) EAS
Max speed with flaps and/or landing gear down:
   A and U   150 knots (278 km/h; 172 mph) EAS
Stalling speed, engine idling:
   A, flaps and landing gear up
      79 knots (147 km/h; 91 mph) EAS
   U, flaps and landing gear up
      93 knots (172 km/h; 107 mph) EAS
   A, flaps and landing gear down
      70 knots (130 km/h; 81 mph) EAS
   U, flaps and landing gear down
      86 knots (159 km/h; 99 mph) EAS
Max rate of climb at S/L: A   1,247 m (4,090 ft)/min
Time to 4,575 m (15,000 ft): A   4 min 30 s
Max operating altitude   7,620 m (25,000 ft)
Service ceiling   11,580 m (38,000 ft)
T-O run at S/L: A   227 m (745 ft)
T-O to 15 m (50 ft) at S/L: A   440 m (1,444 ft)

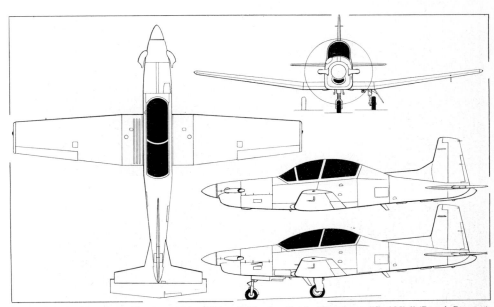
Pilatus PC-9 basic/advanced trainer, with additional side view (top) of Beech-modified Mk II (Dennis Punnett)

First production prototype (nearest camera) and engineering testbed for the Beech-modified PC-9 Mk II

Landing from 15 m (50 ft) at S/L: A   530 m (1,739 ft)
Landing run at S/L:
   A (normal braking action)   417 m (1,368 ft)
Max range at cruise power at 7,620 m (25,000 ft), 5% fuel plus 20 min reserves   887 nm (1,642 km; 1,020 miles)
Endurance (typical mission power settings)
    2 sorties of 1 h duration plus 20 min reserves
g limits: A   +7/−3.5
        U   +4.5/−2.25

PILATUS/BEECHCRAFT PC-9 Mk II

TYPE: Modified PC-9 for US JPATS competition.
PROGRAMME: Proposed for US Air Force/Navy JPATS trainer programme following August 1990 agreement with Beech Aircraft Corporation; two PC-9s supplied by Pilatus as demonstrators, one of which (N26BA) converted as engineering testbed, making first flight with new engine and canopy September 1992. Beech also manufactured two production prototypes (first flight 23 December 1992 by N8284M; second to fly Summer 1993) for operational evaluation 1993. British Aerospace Defence is team member for ground based training systems.
DESIGN FEATURES: Approx 70 per cent redesign including strengthened fuselage, pressurised cockpits with larger and more birdstrike resistant canopy, more powerful engine, increased fuel capacity and single-point pressure fuelling, new digital avionics.
STRUCTURE: Generally as for standard PC-9, but modified dorsal fin; possible use of honeycomb panels in fuselage; empty weight increased by approx 317.5 kg (700 lb).
POWER PLANT: One 1,274 kW (1,708 shp) Pratt & Whitney Canada PT6A-68 turboprop, flat rated at 932 kW (1,250 shp). Wing tank fuel capacity increased to 700 litres (185 US gallons; 154 Imp gallons).
ACCOMMODATION: Martin-Baker Mk 16 zero/zero ejection seats; rear seat position modified to improve instructor's view. Cockpits pressurised (max differential 0.24 bar;

362 SWITZERLAND: AIRCRAFT—PILATUS

3.5 lb/sq in). Canopy deeper, strengthened with extra frame at front, and made from more impact resistant acrylic.

AVIONICS: All-digital: Bendix/King suite to be evaluated in one production prototype and Collins in the other. Provision required for GPS navigation, microwave landing system, collision avoidance and HUD.

PILATUS PC-12

TYPE: Single-turboprop pressurised utility/business transport.
PROGRAMME: Announced at NBAA Show October 1989; first flight P.01 (HB-FOA) 31 May 1991; announced October 1991 that parent company Oerlikon-Bührle would fund remaining development cost to make PC-12 a single-company programme; more than 290 hours in 295 flights by November 1992; first flight of second prototype (HB-FOB) 28 May 1993; Swiss certification by late 1993 to be followed by US certification to FAR Pt 23 Amendment 40 (covering FAR Pt 135 commercial and Pt 91 general operations); first production PC-12s planned to be delivered in mid-1994.

CURRENT VERSIONS: **PC-12**: Nine passengers under FAR Pt 23, six in business layout; 1.12 m³ (39.5 cu ft) baggage compartment. *Detailed description applies to standard PC-12 except where indicated.*
PC-12F: Freighter with 9.24 m³ (326.3 cu ft) usable cargo capacity.
Combi: Four passengers plus 6.0 m³ (211.9 cu ft) freight.
Military: Transport, parachute training, ambulance and other missions; high-density seating for 14.

CUSTOMERS: Expected share of 640 out of total market for 870 over 15 years, 60 per cent of it in USA and Canada; total of 28 orders and options by early 1993 from Australian Royal Flying Doctor Service (four), Japan, Thailand, Europe, and North and South America.

COSTS: Initial price $1.85 million (1993). Company funding approved for first 35 production aircraft.

DESIGN FEATURES: Claimed 100 knots (185 km/h; 115 mph) faster than Cessna Caravan I and longer range than Beech Super King Air B200; can fly three 200 nm sectors in six-hour flight; to be approved for single/two-pilot VFR/IFR operation into known icing.

Wing sections (modified NASA-GA(W)-1 series), LS(1)-0417-MOD at root and LS(1)-0313 at tip, help to reduce tailplane loads; latter's T mounting reduces trim changes with power and protects against lorry strikes; CG range 25-46 per cent of MAC. Modifications following early flight trials include introduction of winglets, increased wing span, addition of tailplane/fin bullet fairing and enlarged dorsal fin and ventral strake.

FLYING CONTROLS: Mechanical, by push/pull rods and cables; servo tab in each aileron; electrically actuated Flettner tab in rudder; short span mass balanced ailerons; electrically actuated Fowler flaps cover 67 per cent of wing trailing-edge; electrically actuated (dual redundant) variable incidence tailplane.

STRUCTURE: All-metal basic structure; composites for ventral and dorsal fins (Kevlar/honeycomb sandwich), wingtips (glassfibre), fairings, engine cowling (glassfibre/honeycomb sandwich) and interior trim; titanium firewall; two-spar wing with integral fuel tankage; airframe to be proved for 20,000 hour life.

Swiss Federal Aircraft Factory makes rudder, tailplane and elevator. Second production location being sought, probably in USA.

LANDING GEAR: Nardi hydraulically retractable tricycle type, with single wheel on each unit; nosewheel steerable ±60°. Suitable for operation from grass strips. Goodrich low-pressure mainwheel tyres (size 8.50-10); Goodrich nosewheel tyre, size 17.5 × 6.25-6. Propeller ground clearance maintained with nose leg compressed and nosewheel tyre flat. Main gear retracts inward into wings, nose gear rearward under flight deck. Min ground turning radius about nosewheel 4.12 m (13 ft 6¼ in).

POWER PLANT: One 1,197 kW (1,605 shp) Pratt & Whitney Canada PT6A-67B turboprop, flat rated to 895 kW (1,200 shp) for T-O and 746 kW (1,000 shp) for climb and cruise. Hartzell HC-E4A-3D/E10477K constant-speed reversible-pitch four-blade aluminium propeller, turning at 1,700 rpm. Fuel tanks in wings, total capacity 1,540 litres (407 US gallons; 339 Imp gallons), of which 1,522 litres (402 US gallons; 335 Imp gallons) are usable. Gravity fuelling point in top of each wing. Oil capacity 11 litres (2.9 US gallons; 2.4 Imp gallons).

ACCOMMODATION: Two-seat flight deck: aircraft to be approved for single pilot, but dual controls and second flight instrument panel optional. Limit of nine passengers under FAR Pt 23, or business layout for six, both with toilet. Airstair crew/passenger door at front, upward opening cargo door at rear, both on port side; Type III emergency exit above wing on starboard side.

SYSTEMS: Normalair Garrett ECS, max pressure differential 0.4 bar (5.8 lb/sq in). Vickers Systems (Germany) hydraulic system, pressure 207 bars (3,000 lb/sq in), for landing gear actuation. Lucas electrical power system (28V DC) supplied by two engine driven starter/generators (300A main and 100A standby) and 24V 40Ah nickel-cadmium battery, with two 150VA static inverters for 115/26V AC power. Pneumatic boot de-icing of wing and tailplane leading-edges; Goodrich electric de-icing of propeller blades; electric heating for windscreen; bleed air de-icing of engine air intake.

AVIONICS: Bendix/King KFC 325 AFCS with EFS 40 4 in EFIS standard, plus KY 196A VHF com, KX 155 com/nav, KMA 24H audio control panel and intercom, KN 62A DME, KR 87 digital ADF with KI 229 RMI, KAS 297C altitude/vertical speed preselector, KR 21 marker beacon receiver, KT 71 transponder, KEA 130A encoding altimeter and Narco ELT-910 emergency locator transmitter. R/Nav, Bendix/King 20582VP weather radar, GPS, HF and co-pilot EFIS panel optional.

DIMENSIONS, EXTERNAL:
Wing span	16.08 m (52 ft 9 in)
Wing aspect ratio	10.02
Length overall	14.38 m (47 ft 2¼ in)
Height overall	4.26 m (13 ft 11¾ in)
Elevator span	5.15 m (16 ft 10¾ in)
Wheel track	4.53 m (14 ft 10½ in)
Wheelbase	3.54 m (11 ft 7½ in)
Propeller diameter	2.67 m (8 ft 9 in)
Propeller ground clearance	0.32 m (1 ft 0½ in)
Passenger door: Height	1.35 m (4 ft 5¼ in)
Width	0.64 m (2 ft 1¼ in)
Cargo door: Height	1.32 m (4 ft 4 in)
Width	1.40 m (4 ft 7 in)
Emergency exit: Height	0.69 m (2 ft 3¼ in)
Width	0.48 m (1 ft 7 in)

DIMENSIONS, INTERNAL:
Cabin: Length, excl flight deck	5.16 m (16 ft 11 in)
Max width	1.53 m (5 ft 0¼ in)
Width at floor	1.30 m (4 ft 3¼ in)
Max height	1.45 m (4 ft 9 in)
Usable cargo volume (PC-12F)	9.24 m³ (326.3 cu ft)
Baggage compartment (PC-12)	1.12 m³ (39.5 cu ft)

AREAS:
Wings, gross	25.81 m² (277.8 sq ft)

WEIGHTS AND LOADINGS:
Weight empty: PC-12F	2,183 kg (4,813 lb)
PC-12, standard	2,386 kg (5,260 lb)
Max payload: PC-12F	1,400 kg (3,086 lb)
PC-12	1,197 kg (2,639 lb)
Max ramp weight	4,020 kg (8,862 lb)
Max T-O weight	4,000 kg (8,818 lb)

Rollout photograph of the Pilatus PC-12 with original wing

PC-12 single-turboprop pressurised utility and business transport prototype in its current form

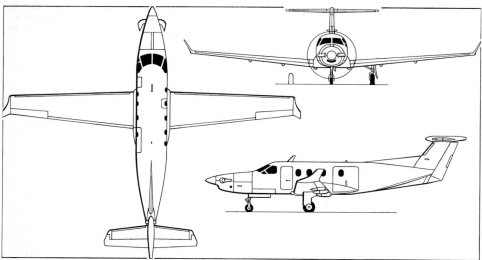

Pilatus PC-12 pressurised light utility and business transport (*Jane's*/Mike Keep)

Max landing weight	3,800 kg (8,377 lb)	Max cruising speed at 6,100 m (20,000 ft)		T-O to 15 m (50 ft)	555 m (1,821 ft)
Max zero-fuel weight	3,700 kg (8,157 lb)		268 knots (496 km/h; 309 mph)	Landing from 15 m (50 ft)	560 m (1,838 ft)
Max wing loading	155.0 kg/m² (31.75 lb/sq ft)	Stalling speed:		Landing run	425 m (1,395 ft)
Max power loading	4.47 kg/kW (7.35 lb/shp)	flaps and gear up	85 knots (158 km/h; 98 mph)	Max IFR range at 232 knots (430 km/h; 267 mph) at	
PERFORMANCE:		flaps and gear down	61 knots (113 km/h; 71 mph)	7,620 m (25,000 ft), 45 min reserves	
Max permissible diving speed		Max rate of climb at S/L	622 m (2,040 ft)/min		1,600 nm (2,965 km; 1,842 miles)
	280 knots (518 km/h; 322 mph) CAS	Max operating altitude	9,145 m (30,000 ft)	g limits: flaps up	+3.4/−1.36
Max operating speed		Service ceiling	10,670 m (35,000 ft)	flaps down	+2.0
	240 knots (444 km/h; 276 mph) CAS	T-O run	310 m (1,018 ft)		

SWISS FEDERAL AIRCRAFT FACTORY (F+W)
EIDGENÖSSISCHES FLUGZEUGWERK— FABRIQUE FEDERALE D'AVIONS— FABBRICA FEDERALE DI AEROPLANI

CH-6032 Emmen
Telephone: 41 (41) 59 41 11
Fax: 41 (41) 55 25 88
Telex: 868 505 FWE CH
MANAGING DIRECTOR: Hansjürg Kobelt

Swiss government official aircraft and missile research, development, production, maintenance and modification establishment; Emmen factory has 140,000 m² (1,506,946 sq ft) floor space; workforce about 800.

R&D activities include aerodynamics and flight mechanics department with four wind tunnels for speeds up to Mach 4-5 and test cells for piston and jet engines with and without afterburners, all with digital data acquisition and processing; wind tunnel work done for outside clients, including subsonic testing for Hermès spacecraft, work for surface transport designers and building industry; structural and systems engineering department deals with aircraft, helicopters and space hardware, with specialities in fatigue analysis and testing of entire aircraft structures; electronics and missile systems department deals with all system aspects of aircraft and helicopter avionics and missiles; fourth R&D department is for prototype fabrication, flight test, instrumentation and system environmental testing.

Production department can handle mechanical, sheet metal and composites and electronic, electrical, electro-mechanical and electro-optical subassemblies. Current and recent programmes include fitting foreplanes to Swiss Air Force Mirage IIIs; producing slats for McDonnell Douglas MD-80 series and wingtips for Airbus A320; assembly (completed) of 19 of 20 BAe Hawk Mk 66s and integration of 12 Super Pumas for Swiss Air Force; development, manufacture and delivery of Rafale drop tanks to Dassault Aviation. F+W is general contractor for McDonnell Douglas Dragon anti-tank missile; TOW missile programme begun 1986; Stinger missile programme begun 1989; co-fabrication with Contraves of all shrouds for Ariane and Titan space launchers.

Proprietary products include low-level dispenser bombing system, acoustic systems for failure and flight envelope warning, all-electronic linear angle of attack and g indicators, scoring indicators for air-to-air and air-to-ground shooting with microcomputer-based ground station, multi-component strain-gauge balances for forces from few hundred grammes to several tonnes, water separators for air-conditioning systems, POHWARO water rockets (see 1977-78 *Jane's*). F+W took part in Farner KZD 85 target drone development; subsequently developed and delivered Ranger UAVs with hydraulic launcher and ground support (see *Jane's Battlefield Surveillance Systems*), of which a reconnaissance and surveillance unit is now in operation.

Dassault Mirage IIIS, upgraded by F+W for the Swiss Air Force *(Paul Jackson)*

F+W MIRAGE IMPROVEMENT PROGRAMME

Main improvements authorised by Swiss government in 1985 include adding fixed foreplanes and small strakes beside nose to 30 Mirage IIIS, 18 IIIRS, two IIIBS and two IIIDS.

Other changes include new audible warning and visual angle of attack indicators to warn of flight envelope limits, replacing Martin-Baker Mk 4 ejection seats with Mk 6 zero/ zero, adding IR and passive/active ECM, more powerful VHF radio, wing refurbishing, fittings for two 500 litre (132 US gallon; 110 Imp gallon) underwing drop tanks and 730 litre (193 US gallon; 160.5 Imp gallon) centreline tank, improved blast deflectors to allow gun firing at high angles of attack, and new camouflage paint scheme. Work to continue through 1990s. See also *Jane's Civil and Military Aircraft Upgrades*.

TAIWAN

AIDC
AERO INDUSTRY DEVELOPMENT CENTER

PO Box 90008-10, Taichung 40722
Telephone: 886 (4) 252 3051/2
Fax: 886 (4) 256 2295
OTHER WORKS: Kangshan
DIRECTOR GENERAL: Dr Hsi-chun M. Hua
VICE-DIRECTOR GENERAL: Dr Shih-sen Wang

Established 1969 to succeed Bureau of Aircraft Industry (BAI), formed 1946, which moved to Taiwan in 1948; now subsidiary of government Chung Shan Institute of Science and Technology (CSIST); 1993 workforce approx 6,000.

Produced 118 Bell UH-1H (Model 205) helicopters under licence 1969-76 for Chinese Nationalist Army; built PL-1A prototype (based on Pazmany PL-1) and 55 PL-1B Chien-Shou primary trainers for Republic of China Air Force between 1968 and 1974 (see 1975-76 *Jane's*); built under licence 248 Northrop F-5E Tiger IIs and 36 two-seat F-5Fs between 1974 and 1986 (see 1986-87 edition).

Designed and produced T-CH-1 Chung-Hsing turboprop basic trainer (see 1981-82 *Jane's*); developed and produced AT-3A/B Tzu-Chung twin-turbofan advanced trainer; has begun production of IDF Ching-Kuo fighter.

AIDC CHING-KUO

TYPE: Single-seat air superiority fighter.
PROGRAMME: IDF (Indigenous Defensive Fighter) programme initiated May 1982 after US refusal to allow purchase of Northrop F-20 Tigershark or GD F-16; US development assistance received from General Dynamics), engine (Garrett), radar (Westinghouse) and various subsystems; design frozen 1985; named after former Taiwan president in 1988. Four flying prototypes built (three single-seat and one two-seat), with first flights 28 May 1989 (77-8001), 27 September 1989 (78-8002), 10 January 1990 (78-8003) and 10 July 1990 (two-seater, 79-8004), of which '002 lost 12 July 1991 following vibration during transonic acceleration; modified intakes on '003; total prototype flying hours 760 by end of November 1992 (1,169 flights). Pre-production batch of 10 started October 1990; four of these handed over to RoCAF 31 December 1992; six more due to be delivered by end of 1993; Sky Sword I missile launch demonstrated late 1992; first 60 production aircraft authorised; first production deliveries to RoCAF due 1994; peak production rate of four per month in 1996.
CUSTOMERS: Republic of China Air Force required 256, including 40-50 two-seaters, by 1999, to replace F/TF-104Gs and F-5E/Fs, but virtually halved following recent purchase of F-16s.
COSTS: Reported programme cost of $10 billion (1991) for 256 aircraft; flyaway bare aircraft $30 million ($40 million with training and spares).
DESIGN FEATURES: Collective project name An Hsiang (Safe Flight); individual programme names for airframe (Ying Yang: Soaring Eagle), engine (Yun Han: Cloud Man), avionics (Tien Lei: Sky Thunder) and main missile armament (Tien Chien: Sky Sword).
Slightly swept blended wing/body with large leading-edge strakes and tip-mounted missiles; fixed intakes; transonic area rule; 8,000 hour airframe design life.
FLYING CONTROLS: Lear Astronics digital fly-by-wire control system. Sidestick controller. Near-full-span flaperons and large all-moving tailerons controlled jointly with leading-edge flaps by fly-by-wire system; strakes designed to shed large vortices over wing at high angles of attack.
STRUCTURE: All-metal initially, but composites may be introduced (tailplane already changed to composites).
LANDING GEAR: Retractable tricycle type of Menasco design, with single wheel and oleo-pneumatic shock absorber on each unit. Nose unit retracts forward, main units inward/ upward into engine air intake trunks.
POWER PLANT: Two ITEC (Garrett/AIDC) TFE1042-70 (F125) turbofans initially, side by side in rear fuselage, each developing 26.80 kN (6,025 lb st) dry and 42.08 kN (9,460 lb st) with afterburning and fitted with FADEC. Elliptical air intakes, with splitter plates, mounted low on centre-fuselage beneath wingroots. Internal fuel capacity approx 2,517 litres (665 US gallons; 554 Imp gallons). Plans for improved performance engine (see earlier editions) shelved 1992.
ACCOMMODATION: Pilot only, on Martin-Baker Mk 12 zero/ zero ejection seat. One-piece bubble canopy, hinged at rear and opening upward. Two-seater canopy opens sideways to port. Cockpit(s) pressurised and air-conditioned.
SYSTEMS: AiResearch ECS. Westinghouse variable-speed constant-frequency electrical power generating system.
AVIONICS: Golden Dragon 53 (GD-53) multi-mode pulse Doppler radar, modified version of GE Aerospace AN/ APG-67 (V) incorporating also some elements of Westinghouse AN/APG-66, has range of approx 81 nm (150 km; 93 miles), capability for air and sea search, and lookdown/shootdown capability. Honeywell H423 inertial navigation system. Bendix/King cockpit displays (two multi-function and one head-up). (No second HUD in two-seater.) Digital databus and data processor.
ARMAMENT: One 20 mm M61A Vulcan cannon in port side of fuselage, beneath extended wingroot leading-edge. Photo-Sonics gun camera. Six attachment points for external stores: two under fuselage, one under each wing and one at each wingtip. First prototype at rollout shown with four Sky Sword I short-range infra-red homing air-to-air missiles (two underwing and two at wingtips). Other combi-

Fourth prototype of the AIDC Ching-Kuo indigenous defensive fighter

on each side of fuselage. Inclined ram air intakes, each with splitter plate, abreast of rear cockpit. Engine starting by onboard battery or ground power. All fuel carried in fuselage, in two equal-size rubber impregnated nylon bladder tanks, with combined capacity of 1,630 litres (430.6 US gallons; 358.5 Imp gallons). Two independent fuel systems, one for each engine, with crossfeed to allow fuel from either or both systems to be fed to either or both engines. Pressure fuelling point forward of, and below, port air intake for internal and external tanks. A 568 litre (150 US gallon; 125 Imp gallon) auxiliary drop tank can be carried on each inboard underwing pylon. Oil capacity 5.7 litres (1.5 US gallons; 1.25 Imp gallons) total, 1.9 litres (0.5 US gallon; 0.42 Imp gallon) usable. Fire warning and extinguishing systems for each engine bay.

ACCOMMODATION: Crew of two in tandem on zero/zero ejection (through canopy) seats, under individual manually operated canopies which open sideways to starboard. Crew separated by internal windscreen. Independent miniature detonation cord (MDC) system to break each canopy for ground and in-flight emergency egress. MDC can be operated from outside cockpit on ground. Rear seat elevated 30 cm (12 in). Dual controls standard.

SYSTEMS: AiResearch bootstrap air cycle ECS, for cockpit air-conditioning and pressurisation (max differential 0.34 bar; 5 lb/sq in), canopy seal, demisting, and pressurisation of g suits, hydraulic reservoirs and external fuel tanks. Two independent hydraulic systems, pressure 207 bars (3,000 lb/sq in), with engine driven pumps (flow rate 34.4 litres; 9.09 US gallons; 7.57 Imp gallons/min). Air type reservoir, pressurised at 2.41 bars (35 lb/sq in). Flight control hydraulic system provides power only for operation of primary flying control surfaces. Utility system serves primary flying control surfaces, landing gear, landing gear doors, airbrakes, wheel brakes, nosewheel steering and stability augmentation system. Primary electrical power supplied by two 28V 12kW DC starter/generators, one on each engine. One 40Ah nickel-cadmium battery for engine starting. Two static inverters supply AC power at 400Hz. External DC power socket on starboard side of centre fuselage. Hydraulic and electrical systems can be sustained by either engine. Liquid oxygen system, capacity 5 litres (1.3 US gallons; 1.1 Imp gallons), for crew.

AVIONICS: Most radio and nav equipment located in large avionics bays in forward fuselage. Standard avionics include UHF com, intercom, IFF/SIF, Tacan, panel mounted VOR/ILS/marker beacon indicator, AHRS and angle of attack system, plus full blind-flying instrumentation. Wide range of optional avionics available.

EQUIPMENT: Can be equipped with A/A37U-15TTS aerial target system, carried on centreline and outboard pylons.

ARMAMENT (AT-3B): Manually adjustable gunsight and camera in forward cockpit. Large weapons bay beneath rear cockpit can house variety of stores, including quick-change semi-recessed machine-gun packs. Disposable weapons can be carried on centreline pylon (stressed for 907 kg; 2,000 lb load), two inboard underwing pylons (each 635 kg; 1,400 lb and capable of accepting triple ejector racks), two outboard underwing pylons (each 272 kg; 600 lb), and wingtip launch rails (each of 91 kg; 200 lb capacity), subject to max external stores load of 2,721 kg (6,000 lb). Weapons that can be carried include GP, SE, cluster and fire bombs; SUU-25A/A, -25C/A and -25E/A flare dispensers; LAU-3/A, -3A/A, -3B/A, -10/A, -10A/A, -60/A, -68A/A and -68B/A rocket launchers; wingtip infrared air-to-air missiles; and rocket pods, practice bombs and bomb or rocket training dispensers.

DIMENSIONS, EXTERNAL:
Wing span 10.46 m (34 ft 3¼ in)
Wing chord: at root 2.80 m (9 ft 2¼ in)
    at tip 1.40 m (4 ft 7 in)

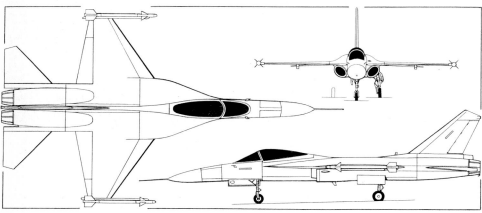
AIDC Ching-Kuo fighter (two Garrett TFE1042-70 turbofans) *(Dennis Punnett)*

nations may include two medium-range Sky Sword II radar homing air-to-air missiles under fuselage in addition to four Sky Sword Is; or three Hsiung Feng II (Male Bee II) sea-skimming anti-shipping missiles (under wings and fuselage) plus two wingtip Sky Sword Is. In attack role, underwing and underfuselage hardpoints could be occupied by Maverick (or similar) missiles, single or cluster bombs, or rocket pods.

DIMENSIONS, EXTERNAL (estimated):
Wing span: incl missile rails 8.53 m (28 ft 0 in)
    over wingtip missiles 9.00 m (29 ft 6 in)
Length overall: excl nose probe 13.26 m (43 ft 6 in)
    incl nose probe 14.48 m (47 ft 6 in)
WEIGHTS AND LOADINGS (estimated):
Internal fuel weight 1,950 kg (4,300 lb)
Max T-O weight 9,072 kg (20,000 lb) class
PERFORMANCE (estimated):
Max level speed approx Mach 1.7
Max rate of climb at S/L 15,240 m (50,000 ft)/min
Service ceiling 16,760 m (55,000 ft)
g limit +6.5

AIDC AT-3 TZU-CHUNG

TYPE: Tandem two-seat twin-turbofan military trainer and close support aircraft.

PROGRAMME: Development contract placed July 1975; first flights of two prototypes (0801 and 0802) 16 September 1980 and 30 October 1981; production of order for 60 started in March 1982; first flight of production aircraft (0803) 6 February 1984; deliveries began March 1984, completed by early 1990, but wingroot leading-edge improvements flight tested on prototypes in early 1992.

CURRENT VERSIONS: **AT-3A**: Standard trainer; *detail description applies to this version, except where indicated.*
AT-3B: Close air support aircraft; two prototype conversions flown 1989 and 20 AT-3As converted.

CUSTOMERS: RoCAF (60 AT-3A/B).

DESIGN FEATURES: Supercritical wing section; quarter-chord sweepback 7° 20'; thickness/chord ratio 10 per cent; dihedral 0° 46'; incidence 1° 30'.

FLYING CONTROLS: Hydraulic fully powered slab tailplane, rudder and sealed-gap ailerons; yaw damper; electro-hydraulically controlled single-slotted flaps; two airbrake panels forward of mainwheel wells.

STRUCTURE: One-piece multi-spar metal wing with thick machined skins forming torsion box; fin integral with rear fuselage; three-part fuselage; steel, magnesium and graphite/epoxy components in fuselage; airbrake panels of laminated graphite composites; ailerons have metal honeycomb core.

LANDING GEAR: Hydraulically retractable tricycle type, with single wheel on each unit. Main units retract inward into fuselage, nosewheel forward. Oleo-pneumatic shock absorber in each unit. Two-position extending nose leg increases static angle of attack by 3° 30', to reduce T-O run, and is shortened automatically during retraction. Emergency extension by gravity. Mainwheels and tyres size 24 × 8.00-13, pressure 8.96 bars (130 lb/sq in). Hydraulically steerable nose unit, with wheel and tyre size 18 × 6.50-8, pressure 5.51 bars (80 lb/sq in). All-metal multi-disc brakes.

POWER PLANT: Two Garrett TFE731-2-2L non-afterburning turbofans (each 15.57 kN; 3,500 lb st), installed in nacelle

Camouflaged AT-3A of the Republic of China Air Force. Twenty AT-3As have been converted as AT-3B close air support aircraft

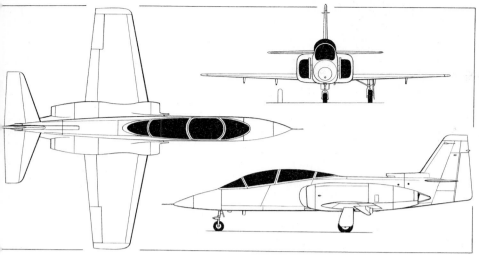

AIDC AT-3A tandem-seat jet trainer *(Dennis Punnett)*

Wing aspect ratio	5.0	AREAS:		
Length overall, incl nose probe	12.90 m (42 ft 4 in)	Wings, gross	21.93 m² (236.05 sq ft)	
Height overall	4.36 m (14 ft 3¼ in)	Ailerons (total)	1.33 m² (14.32 sq ft)	
Tailplane span	4.83 m (15 ft 10¼ in)	Trailing-edge flaps (total)	2.53 m² (27.23 sq ft)	
Wheel track	3.96 m (13 ft 0 in)	Fin	3.45 m² (37.14 sq ft)	
Wheelbase	5.49 m (18 ft 0 in)	Rudder	1.15 m² (12.38 sq ft)	
		Tailplane	5.02 m² (54.04 sq ft)	

WEIGHTS AND LOADINGS:
Weight empty, equipped 3,855 kg (8,500 lb)
Max fuel: internal 1,270 kg (2,800 lb)
    external 884 kg (1,950 lb)
Max external stores load 2,721 kg (6,000 lb)
Normal T-O weight:
  trainer, clean 5,216 kg (11,500 lb)
Max T-O weight with external stores
    7,938 kg (17,500 lb)
Max landing weight 7,360 kg (16,225 lb)
Max wing loading 362 kg/m² (74.14 lb/sq ft)
Max power loading 254.9 kg/kN (2.5 lb/lb st)
PERFORMANCE (at max T-O weight):
Max limiting Mach No. 1.05
Max level speed:
  at S/L 485 knots (898 km/h; 558 mph)
  at 11,000 m (36,000 ft)
    Mach 0.85 (488 knots; 904 km/h; 562 mph)
Max cruising speed at 11,000 m (36,000 ft)
    Mach 0.83 (476 knots; 882 km/h; 548 mph)
Stalling speed:
  flaps and landing gear up
    100 knots (185 km/h; 115 mph)
  flaps and landing gear down
    90 knots (167 km/h; 104 mph)
Max rate of climb at S/L 3,078 m (10,100 ft)/min
Service ceiling 14,625 m (48,000 ft)
T-O run 458 m (1,500 ft)
T-O to 15 m (50 ft) 671 m (2,200 ft)
Landing from 15 m (50 ft) 945 m (3,100 ft)
Landing run 671 m (2,200 ft)
Range with max internal fuel
    1,230 nm (2,279 km; 1,416 miles)
Endurance with max internal fuel 3 h 12 min

TAC
TAIWAN AEROSPACE CORPORATION

Room 2901, 333 Keelung Road, Sector 1, Taipei
Telephone: 886 (2) 345 0030
Fax: 886 (2) 757 6451
CHAIRMAN: Earle Ho
PRESIDENT: Dr Denny Ko
PUBLIC AFFAIRS MANAGER: Suzanne Wu
WORKS: Taichung and future subassembly plant for British Aerospace RJ series

Established 27 September 1991 as foundation for civil aircraft industry; 29 per cent owned by Taiwan government; start-up capital $200-250 million; government investment could rise to 49 per cent; aims to develop national manufacturing capability for aircraft, engines, avionics and materials by 2000.

MoU signed with British Aerospace Corporation (BAe, which see) 23 September 1992 for possible acquisition of 50 per cent share in Regional Jet (RJ) commercial business, involving up to £120 million investment; followed 19 January 1993 by agreement to form joint venture company Avro International Aerospace Ltd (see International section) to produce new RJ family in both Taiwan and UK and to support BAe 146s already delivered. First Taiwan assembled RJ due to be completed 1994. Agreement also provided for continuation during 1993 of feasibility study for possible 120-passenger RJX derivative.

TURKEY

AYMET
AYMET AEROSPACE INDUSTRIES

Esenboga International Airport, Ankara

AYMET (BELL) 206L-4 LONGRANGER IV

Signed $25 million agreement with Bell Helicopter Textron early 1993 for assembly of LongRanger IV (see Canadian section) in Turkey; 24-aircraft programme, of which Aymet to outfit 10 'green' aircraft from Bell, followed by local assembly of next 14. New factory to be built at Esenboga Airport for the purpose.

TAI
TUSAŞ AEROSPACE INDUSTRIES INC
(TUSAŞ Havacilik ve Uzay Sanayi A. Ş.)

PO Box 18, TR-06692 Kavaklidere, Ankara
Telephone: 90 (4) 298 1800
Fax: 90 (4) 298 1408/1425
Telex: 44640 TAIA TR
MANAGING DIRECTOR: J. R. Jones
DIRECTOR OF PROGRAMMES: Kaya Ergenç

Formed 15 May 1984; shareholders (by percentage) are TAI (49), Turkish Armed Forces Foundation (1.9), Turkish Air League (0.1), Lockheed Fort Worth (42) and General Electric (7); TAI aims to produce modern weapon systems, aircraft and helicopters, modernise aircraft, conduct R&D and develop and manufacture weapon systems.

R&D activities include computer studies such as turbulence models and jets in a cross-flow, undertaken with General Dynamics for NASA, NATO-AGARD, METU and UDI; TAI developing UAV and participating in Future Large Aircraft programme (see Euroflag in International section).

TAI site of 230 ha (568.3 acres) includes 100,000 m² (1,076,390 sq ft) covered floor area; 62,500 m² (672,687 sq ft) main assembly building; factory contains facilities and high technology machinery including large chemical processing and milling plant and advanced computer capability, to manufacture modern aircraft; workforce 2,138 at end of 1991 (latest figure received); TAI had then trained 908 Turkish employees at Mürted training centre.

TAI (AIRTECH) CN-235

TAI assembling and producing parts for 50 of 52 Airtech CN-235s (see International section) ordered by Turkey 1991. First flight by Turkish assembled CN-235 made 24 September 1992; three delivered by March 1993 (first one 13 November 1992).

TAI (AGUSTA) SF.260D

TAI assembling 10 SIAI-Marchetti SF.260Ds (from Agusta kits) for Turkish Air Force, to be followed by manufacture of further 24; first four delivered 1991.

TAI (LOCKHEED FORT WORTH) F-16C/D FIGHTING FALCON

TAI under contract to Lockheed Fort Worth to co-produce 152 of 160 F-16C/D for Turkish Air Force under Peace Onyx programme and to manufacture rear and centre-fuselages and wings for US Air Force; first F-16C delivered to Turkish Air Force 30 November 1987; 82 delivered and 15 in assembly or flight test by end 1991 (latest figure received). Agreement signed early 1992 for further 40 F-16C/Ds, plus long-lead items for 40 more.

TAI (SIKORSKY) UH-60L BLACK HAWK

TAI to co-produce last 50 of 95 UH-60Ls ordered under $1.1 billion contract of 8 December 1992; production to start August 1994 and run for four years. Further orders may follow.

TAI-assembled F-16C Fighting Falcons over the TUSAŞ factory

UKRAINE

ANTONOV
ANTONOV DESIGN BUREAU
1 Tupolev Street, Kiev 252062
Telephone: 7 (044) 442 61 24
Fax: 7 (044) 449 99 96
Telex: 131309 OZON SU
GENERAL DESIGNER: Pyotr Vasilyevich Balabuyev

Antonov OKB was founded in 1946 by Oleg Konstantinovich Antonov, who died 4 April 1984, aged 78. In current production are An-32 at Kiev, An-72/74 at Kharkov, An-124 at Ulyanovsk. Small An-2 and An-28 are built by PZL Mielec, Poland, plus small batch production of An-2 (as Y-5B) in China. More than 22,000 aircraft of Antonov design have been built; more than 1,500 have been exported, to 42 countries.

ANTONOV An-2
NATO reporting name: Colt
TYPE: Single-engined general purpose biplane.
PROGRAMME: Prototype flew as SKh-1 on 31 August 1947. More than 5,000 An-2s were built at Kiev, ending in mid-1960s after limited manufacture of specialised agricultural An-2M. Production transferred to PZL Mielec, Poland, from where more than 11,950 delivered since 1960. China acquired licence and has built **Yunshuji-5 (Y-5)** versions 1957 to date. (See SAP, China, and WSK-PZL Mielec, Poland.)

ANTONOV An-12
NATO reporting name: Cub
TYPE: Four-turboprop transport and electronic warfare aircraft; described in previous editions.
PROGRAMME: Prototype flew 1958, with Kuznetsov NK-4 turboprops, as rear-loading development of An-10 airliner; more than 900 built with AI-20K engines for military and civil use, ending in USSR in 1973; standard medium-range paratroop and cargo transport of Soviet Military Transport Aviation (VTA) from 1959; replacement with Il-76 began 1974, and fewer than 100 believed still in military service; Shaanxi Aircraft Company, China, manufactures redesigned **Yunshuji-8 (Y-8)** transport version and derivatives (see SAC, China).

ANTONOV An-24
NATO reporting name: Coke
Production of An-24 twin-turboprop short-haul transport in Ukraine ended in 1978, after about 1,100 had been delivered. Versions known as **Y7-100** and **-200** (which see) continue in production at Xian in China.

ANTONOV An-26
NATO reporting name: Curl
TYPE: Twin-turboprop pressurised short-haul transport.
PROGRAMME: First exhibited 1969 Paris Air Show; more than 1,000 built before superseded by An-32; derivative **Y7H-500** built by Xian Aircraft Company (see XAC, China).
CURRENT VERSIONS: **An-26** ('Curl-A'): Original version; electrically/manually operated conveyor flush with cabin floor for freight handling.
An-26B ('Curl-A'): Improved version, announced 1981, to carry three standard freight pallets, each 2.44 m (8 ft) long, 1.46 m (4 ft 9½ in) wide and 1.60 m (5 ft 3 in) high, with total weight of 5,500 kg (12,125 lb). Improved freight handling equipment.
'Curl-B': Signals intelligence (sigint) version; many short blade antennae mounted above and below fuselage.
CUSTOMERS: Military An-26s assigned to air commands in CIS regiments and squadrons; exported to at least 27 air forces; Angolan and Mozambique aircraft have bomb racks. Aeroflot and its successors have more than 200, available as military reserve. Civil operators include Aero Caribbean (Cuba), Air Mongol, Ariana Afghan Airlines, CAAC (China), Cubana, Syrianair, Tarom (Romania) and Trans Arabian.
DESIGN FEATURES: Generally similar to earlier An-24RT specialised freighter, with auxiliary turbojet; more powerful turboprops and redesigned 'beaver-tail' rear fuselage. Oleg Antonov's special loading ramp forms underside of rear fuselage when retracted, slides forward under rear of cabin to facilitate direct loading and when airdropping cargo. Wing anhedral 2° on outer panels; incidence 3°; sweepback on outer panels 6° 50′ at quarter-chord, 9° 41′ on leading-edge; swept vertical and horizontal tail; tailplane dihedral 9°.
FLYING CONTROLS: Mechanical controls; mass balanced servo compensated ailerons with electrically operated glassfibre trim tabs; manual tab in each elevator; electrical trim/servo tab in rudder; hydraulically actuated tracked and slotted TsAGI flaps, single-slotted on centre-section, double-slotted outboard of nacelles.
STRUCTURE: Conventional light alloy; two-spar wing, built in centre, two inner and two detachable outer sections, with skin attached by electric spot welding; bonded/welded semi-monocoque fuselage in front, centre and rear portions, with 'bimetal' (duralumin-titanium) bottom skin for protection during operation from unpaved airfields; blister on each side of fuselage forward of rear ramp carries track to enable ramp to slide forward; large dorsal fin; ventral strake each side of ramp.
LANDING GEAR: Hydraulically retractable tricycle type; twin wheels on each unit. Emergency extension by gravity. All units retract forward. Shock absorbers of oleo-nitrogen type on main units; nitrogen-pneumatic type on nose unit. Mainwheel tyres size 1,050 × 400 mm, pressure 5.9 bars (85 lb/sq in). Nosewheel tyres size 700 × 250 mm, pressure 3.9 bars (57 lb/sq in). Hydraulic disc brakes and anti-skid units on mainwheels. Nosewheels steered hydraulically ±45° while taxying and controllable ±10° during take-off and landing. Min ground turning radius 22.3 m (73 ft 2 in).
POWER PLANT: Two 2,074 kW (2,780 ehp) ZMKB Progress AI-24VT turboprops; four-blade constant-speed fully feathering propellers. One 7.85 kN (1,765 lb st) RU 19A-300 auxiliary turbojet in starboard nacelle for use, as required, during take-off, climb and level flight, and for self-contained starting of main engines. Two independent but interconnected fuel systems, with 5,500 kg (12,125 lb) of fuel in integral tanks in inner wings and ten bag tanks in centre-section. Pressure refuelling socket in starboard engine nacelle. Gravity fuelling point above each tank area. Carbon dioxide inert gas system to create fireproof condition inside fuel tanks.
ACCOMMODATION: Crew of five (pilot, co-pilot, radio operator, flight engineer and navigator); station at rear of cabin on starboard side for loading supervisor or load dispatcher. Optional domed observation window for navigator on port side of flight deck. Toilet on port side aft of flight deck; crew door, small galley and oxygen bottle stowage on starboard side. Emergency escape hatch in door immediately aft of flight deck. Large downward hinged rear ramp/door hinged to an anchorage mounted on tracks running forward under blister fairings. Ramp/door slides forward under fuselage for direct loading on to cabin floor or for air dropping freight. When doing so, its rear is supported by the pivoted swinging arm on each side which raises and lowers door in alternative fixed-hinge mode. Door can be locked in any intermediate position. Electrically powered mobile winch, capacity 2,000 kg (4,409 lb), hoists crate through rear entrance and runs on rail in cabin ceiling to position payload. Electrically and manually operated conveyor, capacity 4,500 kg (9,920 lb), flush with cabin floor of original An-26, facilitates loading and airdropping freight. An-26B has removable rollgangs, mechanism for moving pallets inside hold, and moorings, enabling two men to load and unload three pallets in 30 min. Rollgang can be stowed against sides of cabin. Both versions accommodate motor vehicles, including GAZ-69 and UAZ-469 military vehicles, or cargo up to 1.50 m (59 in) high by 2.10 m (82.6 in) wide. Height of rear edge of cargo door surround above cabin floor is 1.50 m (4 ft 11 in). Cabin pressurised and air-conditioned; optional tip-up seat

Antonov An-26 of the German Air Force *(Paul Jackson)*

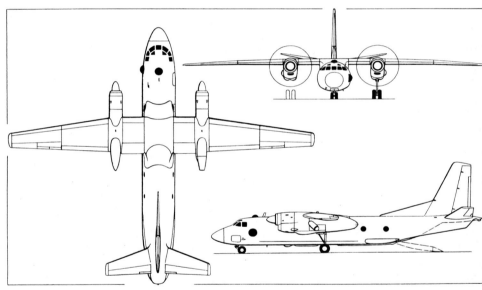

Antonov An-26 twin-turboprop short-haul transport *(Dennis Punnett)*

along each wall for total of 38 to 40 persons. Conversion to troop transport, or to ambulance for 24 stretcher patients and medical attendant, takes 20 to 30 min in the field.

SYSTEMS: Air-conditioning hot air tapped from 10th compressor stage of each engine, with heat exchanger and turbocooler in each nacelle. Cabin pressure differential 0.29 bar (4.27 lb/sq in). Main and emergency hydraulic systems, pressure 151.7 bars (2,200 lb/sq in), for landing gear retraction, nosewheel steering, flaps, brakes, windscreen wipers, propeller feathering and operation of cargo ramp and emergency escape doors. Handpump to operate doors only and to build up pressure in main system. Electrical system includes two 27V DC starter/generators on engines, standby generator on auxiliary turbojet, and three storage batteries for emergency use. Two engine driven alternators provide 115V 400Hz single-phase AC supply, with standby inverter. Basic source of 36V 400Hz three-phase AC supply is two inverters, with standby transformer. Permanent oxygen system for pilot, installed equipment for other crew members and three portable bottles for personnel in cargo hold. Bleed air thermal de-icing system for wing and tail unit leading-edges. Electric windscreen de-icing. Electric de-icing system for propeller blades and hubs; hot air system for engine air intakes.

AVIONICS: Two VHF transceivers, HF, intercom, two ADF, radio altimeter, glidepath receiver, glideslope receiver, marker beacon receiver, weather/navigation radar, directional gyro and flight recorder standard. Optional flight director system, astrocompass and autopilot.

EQUIPMENT: Parachute static line attachments and retraction devices, tiedowns, jack to support ramp sill, flight deck curtains, sun visors and windscreen wipers standard. Optional OPB-1R sight for pinpoint dropping of freight, medical equipment, and liquid heating system. Provision for chaff/flare dispensers pylon-mounted on each side of lower fuselage below wings (seen on Afghan aircraft), and for water-bombing equipment.

ARMAMENT: Provision for bomb rack on fuselage below each wingroot trailing-edge.

DIMENSIONS, EXTERNAL:
Wing span	29.20 m (95 ft 9½ in)
Wing aspect ratio	11.7
Length overall	23.80 m (78 ft 1 in)
Height overall	8.575 m (28 ft 1½ in)
Width of fuselage	2.90 m (9 ft 6 in)
Depth of fuselage	2.50 m (8 ft 2½ in)
Tailplane span	9.973 m (32 ft 8¾ in)
Wheel track (c/l shock struts)	7.90 m (25 ft 11 in)
Wheelbase	7.651 m (25 ft 1¼ in)
Propeller diameter	3.90 m (12 ft 9½ in)
Propeller ground clearance	1.227 m (4 ft 0¼ in)
Crew door (stbd, front): Height	1.40 m (4 ft 7 in)
Width	0.60 m (1 ft 11¾ in)
Height to sill	1.47 m (4 ft 9¾ in)
Loading hatch (rear): Length	3.15 m (10 ft 4 in)
Width at front	2.40 m (7 ft 10½ in)
Width at rear	2.00 m (6 ft 6¾ in)
Height to sill	1.47 m (4 ft 9¾ in)
Height to top edge of hatchway	3.014 m (9 ft 10¾ in)
Emergency exit (in floor at front):	
Length	1.02 m (3 ft 4¼ in)
Width	0.70 m (2 ft 3½ in)
Emergency exit (top): Diameter	0.65 m (2 ft 1½ in)
Emergency exits (one each side of hold):	
Height	0.60 m (1 ft 11¾ in)
Width	0.50 m (1 ft 7½ in)

DIMENSIONS, INTERNAL:
Cargo hold: Length of floor	11.50 m (37 ft 8¾ in)
Width of floor	2.78 m (9 ft 1½ in)
Max height	1.91 m (6 ft 3 in)

AREAS:
Wings, gross	74.98 m² (807.1 sq ft)
Vertical tail surfaces (total, incl dorsal fin)	15.85 m² (170.61 sq ft)
Horizontal tail surfaces (total)	19.83 m² (213.45 sq ft)

WEIGHTS AND LOADINGS:
Weight empty	14,750 kg (32,518 lb)
Normal payload	4,500 kg (9,920 lb)
Max payload	5,500 kg (12,125 lb)
Normal T-O and landing weight	23,000 kg (50,706 lb)
Max T-O and landing weight	24,000 kg (52,911 lb)
Max wing loading	320.1 kg/m² (65.6 lb/sq ft)
Max power loading	5.79 kg/kW (9.52 lb/ehp)

PERFORMANCE (at normal T-O weight):
Cruising speed at 6,000 m (19,685 ft)	235 knots (435 km/h; 270 mph)
T-O speed	116 knots (215 km/h; 134 mph)
Landing speed	102 knots (190 km/h; 118 mph)
Max rate of climb at S/L	480 m (1,575 ft)/min
Service ceiling	7,500 m (24,600 ft)
T-O run, on concrete	870 m (2,855 ft)
T-O to 15 m (50 ft)	1,240 m (4,068 ft)
Landing from 15 m (50 ft)	1,740 m (5,709 ft)
Landing run, on concrete	650 m (2,135 ft)
Range, no reserves:	
with max payload	669 nm (1,240 km; 770 miles)
with max fuel	1,434 nm (2,660 km; 1,652 miles)

ANTONOV An-28

NATO reporting name: **Cash**

An-28 manufactured in WSK-PZL Mielec works in Poland (see Polish section).

Sigint version of An-26 known to NATO as 'Curl-B' *(Piotr Butowski)*

Firefighting version of Antonov An-26 dumping water *(Piotr Butowski)*

Starboard water tank and flare pack of An-26 water-bomber *(Linda Jackson)*

ANTONOV An-32

NATO reporting name: **Cline**
Indian Air Force name: **Sutlej**

TYPE: Twin-turboprop short/medium-range transport.

PROGRAMME: Prototype (SSSR-83966) first exhibited 1977 Paris Air Show; export deliveries to India began 1984; current production, 40 a year, largely for CIS armed forces.

CURRENT VERSIONS: Offered initially with choice of two AI-20 turboprops; all production **An-32**s have AI-20D Series 5 engines, as described in detail. Specialised versions available for firefighting, fisheries surveillance, agricultural and air ambulance use, last-named complete with operating theatre. **An-32B** first seen early 1993 at Nairobi; reported to be upgraded with approx 149 kW (200 shp) additional power from each turboprop to provide increase of 500 kg (1,100 lb) max payload. Firefighting **An-32P** (protivopozharny) has a tank on each side of fuselage; total water capacity 8,000 kg (17,635 lb); able to drop 30 smoke-jumpers; provision for flare packs to induce atmospheric precipitation artificially over fire; max T-O weight 29,000 kg (63,930 lb); cruising speed 215 knots (400 km/h; 248 mph); operating speed 124-130 knots (230-240 km/h; 143-149 mph); radius of action 80 nm (150 km; 93 miles). An-32P has not been built in series.

Antonov An-32 short/medium-range transport of No. 48 Squadron, Indian Air Force *(Peter Steinemann)*

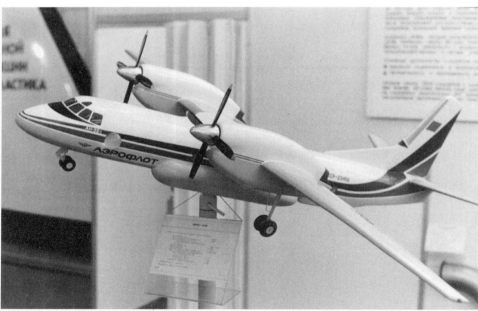
Model of Antonov An-32P firefighting aircraft, with side-mounted water tanks (photograph in Addenda) *(Piotr Butowski)*

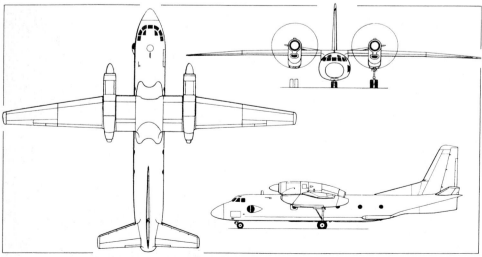
Antonov An-32 transport powered by two ZMKB Progress AI-20D turboprops *(Dennis Punnett)*

CUSTOMERS: CIS air forces and government agencies (about 100 with Department of Agriculture); Afghanistan, India (123, named Sutlej after a Punjabi river), and Peru (15); reported customers include Cape Verde, Nicaragua, São Tome and Principe, and Tanzania.

DESIGN FEATURES: Development of An-26, with triple-slotted trailing-edge flaps outboard of engines, automatic leading edge slats, enlarged ventral fins and full-span slotted tailplane; improved landing gear retraction, de-icing and air conditioning, electrical system and engine starting; large increase in power compared with An-26 improves take-off performance, service ceiling and payload under hot and high conditions; overwing location of engines reduces possibility of stone or debris ingestion, but requires nacelles of considerable depth to house underwing landing gear; operation possible from unpaved strips at airfields 4,000-4,500 m (13,125-14,750 ft) above sea level in ambient temperature of ISA + 25°C; APU helps to ensure independence of ground servicing equipment, including onboard engine starting at these altitudes.

FLYING CONTROLS: As An-26 except for high-lift wings (see Design Features).

STRUCTURE: Generally as An-26.

LANDING GEAR: Hydraulically retractable tricycle type, basically as An-26. All shock absorbers of oleo-nitrogen type. Tyre sizes and pressures unchanged.

POWER PLANT: Two ZMKB Progress AI-20D Series 5 turboprops, each 3,760 kW (5,042 ehp); four-blade constant speed reversible-pitch propellers; TG-16M APU in rear of starboard landing gear fairing.

ACCOMMODATION: Crew of three (pilot, co-pilot and navigator), with provision for flight engineer. Rear loading hatch and forward-sliding ramp/door, as An-26, plus winch and hoist, capacity 3,000 kg (6,615 lb), for freight handling. Cargo or vehicles can be airdropped by parachute, including extraction of large loads by drag parachute, with aid of removable roller conveyors and guide rails on floor of hold. Payloads include 12 freight pallets, 50 passengers or 42 parachutists and a jumpmaster on rows of tip-up seats along each cabin wall, or 24 stretcher patients and up to three medical personnel.

SYSTEMS: Accommodation fully pressurised and air conditioned. Systems basically as An-26 but generally improved.

AVIONICS: Basically as An-26.

EQUIPMENT: Basically as An-26.

ARMAMENT: Provision for four bomb racks, two on each side of fuselage below wings (fitted to aircraft for Peru).

DIMENSIONS, EXTERNAL: As for An-26, except:
Length overall	23.78 m (78 ft 0¼ in)
Height overall	8.75 m (28 ft 8½ in)
Tailplane span	10.23 m (33 ft 6¾ in)
Propeller diameter	4.70 m (15 ft 5 in)
Propeller ground clearance	1.55 m (5 ft 1 in)

DIMENSIONS, INTERNAL:
Cargo hold: Length	15.68 m (51 ft 5¼ in)
Max width	2.78 m (9 ft 1¼ in)
Max height	1.84 m (6 ft 0½ in)
Volume	66.0 m³ (2,330 cu ft)

AREAS:
Wings, gross	74.98 m² (807.1 sq ft)

Ailerons (total)	6.12 m² (65.88 sq ft)
Flaps (total)	15.00 m² (161.46 sq ft)
Vertical tail surfaces (total, incl dorsal fin)	
	17.22 m² (185.36 sq ft)
Horizontal tail surfaces (total)	20.30 m² (218.5 sq ft)

WEIGHTS AND LOADINGS:
Weight empty	16,800 kg (37,038 lb)
Weight empty, equipped	17,308 kg (38,158 lb)
Max payload	6,700 kg (14,770 lb)
Max fuel	5,445 kg (12,004 lb)
Max fuel with max payload	2,267 kg (4,998 lb)
Max ramp weight	27,250 kg (60,075 lb)
Max T-O weight	27,000 kg (59,525 lb)
Max landing weight	25,000 kg (55,115 lb)
Max wing loading	360.1 kg/m² (73.75 lb/sq ft)
Max power loading	3.59 kg/kW (5.90 lb/ehp)

PERFORMANCE:
Max cruising speed	286 knots (530 km/h; 329 mph)
Econ cruising speed	254 knots (470 km/h; 292 mph)
Landing speed	100 knots (185 km/h; 115 mph)
Optimum cruising height	8,000 m (26,250 ft)
Service ceiling	9,400 m (30,840 ft)
Service ceiling, OEI	4,800 m (15,750 ft)
T-O run on concrete	760 m (2,495 ft)
T-O to 15 m (50 ft)	1,200 m (3,940 ft)
Landing run	470 m (1,542 ft)
Range: with max payload	647 nm (1,200 km; 745 miles)
with max fuel	1,360 nm (2,520 km; 1,565 miles)

Antonov An-32B twin-turboprop transport, photographed at Nairobi *(Paul Jackson)*

ANTONOV An-38

TYPE: Twin-turboprop light multi-purpose regional airliner.
PROGRAMME: Details announced, and model displayed, at 1991 Paris Air Show; first prototype to fly 1993; deliveries scheduled to begin 1994.
CURRENT VERSIONS: To be available as passenger and cargo transport, and for aerial photography, survey, forest patrol, VIP transport, ambulance, fishery and ice patrol duties.
DESIGN FEATURES: Developed from PZL Mielec (Antonov) An-28 (see Polish section). New high-efficiency engines; lengthened passenger cabin; optional weather radar and automatic flight control system; improved sound and vibration insulation; wheel or ski landing gear; rear cargo door and cargo handling system; operating temperatures from −50°C to +45°C, including 'hot and high' conditions.
POWER PLANT: Two RKBM/Rybinsk TVD-1500 or Garrett TPE331-14GR turboprops, each rated at 1,104 kW (1,480 shp); AV-36 quiet reversible-pitch propellers.
ACCOMMODATION: Two crew side by side on flight deck; passenger cabin equipped normally with 26 seats; 27 seats optional; seats and baggage compartment can be folded quickly against cabin wall to provide clear space for 2,500 kg (5,510 lb) of freight. Cabin door with airstairs on port side, with service door opposite; emergency exit each side. Cargo door under upswept rear fuselage slides forward under cabin for direct loading/unloading of freight.

DIMENSIONS, EXTERNAL:
Wing span	22.063 m (72 ft 4½ in)
Length overall	15.67 m (51 ft 5 in)
Height overall	4.30 m (14 ft 1¼ in)
Cargo door: Length	2.20 m (7 ft 2½ in)
Width	1.40 m (4 ft 7 in)

WEIGHTS AND LOADINGS:
Max payload	2,500 kg (5,510 lb)
Max T-O weight	8,150 kg (17,965 lb)

PERFORMANCE (estimated):
Nominal cruising speed	188-205 knots (350-380 km/h; 217-236 mph)
T-O	350 m (1,150 ft)
T-O to 10.7 m (35 ft)	530 m (1,740 ft)
Landing from 15 m (50 ft)	450 m (1,477 ft)
Landing run	270 m (886 ft)
Balanced field length	650-750 m (2,135-2,460 ft)
Range, 45 min fuel reserves:	
with 26 passengers	323 nm (600 km; 372 miles)
with 17 passengers	782 nm (1,450 km; 901 miles)
max, with 9 passengers	1,187 nm (2,200 km; 1,367 miles)

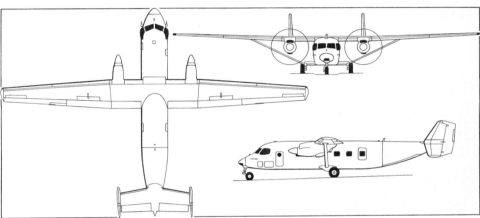

Antonov An-38, stretched version of the An-28 *(Jane's/Mike Keep)*

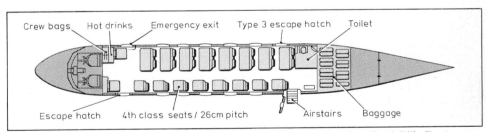

Antonov An-38 cabin arranged for 25 passengers and a cabin attendant *(Jane's/Mike Keep)*

ANTONOV An-70T and An-77

TYPE: Four-propfan medium-size wide-body transport.
PROGRAMME: Development to replace some An-12s remaining in air force service announced by *Izvestia* 20 December 1988; at 1991 Paris Air Show Antonov OKB reported prototype under assembly at Kiev, with planned demonstration at 1993 show; funding passed from former Soviet Air Force to Ukrainian government in past year; preliminary details released and model displayed at Moscow Aero Engine and Industry Show April 1992; D-27 propfan scheduled to fly in Il-76 testbed mid-1992; **An-70T** prototype first flight planned for late 1993. **An-77** variant proposed to UK Ministry of Defence as C-130 replacement.
DESIGN FEATURES: Slightly larger than projected international Euroflag transport, much smaller than US McDonnell Douglas C-17A; conventional high-wing configuration, with wings and tail surfaces slightly sweptback; wing anhedral from roots; loading ramp/doors under upswept rear fuselage with adjustable sill height and built-in cargo handling system; horizontal tail surfaces on rear fuselage; propfans mounted conventionally on wing leading-edge.

Model of An-70T pressurised propfan-powered transport *(Stephane Guilbaud)*

370 UKRAINE: AIRCRAFT—ANTONOV

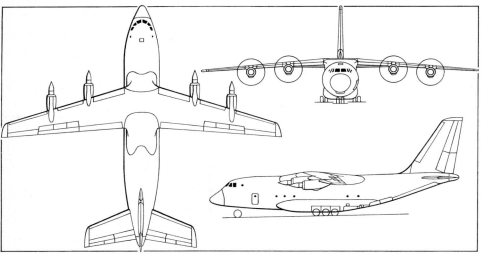

Antonov An-70T transport (four ZMKB Progress D-27 propfans) *(Jane's/Mike Keep)*

Design life 20,000 cycles and 45,000 flying hours in 25 years. Operable 3,500 hours per year, with seven to eight man-hours of maintenance per flying hour. Cost-effective with only 200 flying hours per month.

STRUCTURE: Approx 28 per cent of airframe, by weight, made of composites.

LANDING GEAR: Twin-wheel nose unit; each main unit has three pairs of wheels in tandem, retracting into large fairing on side of cabin; can operate from unpaved surfaces of bearing ratio 6 kg/cm² (85.3 lb/sq in).

POWER PLANT: Four ZMKB Progress D-27 propfans, each 10,290 kW (13,800 shp). Stupino SV-27 contra-rotating propellers, each with eight blades in front and six at rear. Reversible-pitch blades of scimitar form, with electric anti-icing.

ACCOMMODATION: Three flight crew (two pilots and flight engineer) plus loadmaster; provision for converting cockpit for two crew operation; seats in forward fuselage for two cargo attendants; freight loaded via rear ramp; freight can be carried on PA-5.6 rigid pallets, PA-3, PA-4 and PA-6.8 flexible pallets, in UAK-2.5, UAK-5 and UAK-10 containers; unpackaged freight, wheeled and tracked vehicles, food and perishables can be carried; crew door at front of cabin on port side; cargo hold pressurised and air-conditioned.

SYSTEMS: Aircraft systems automated to simplify operation and decrease probability of crew errors. Onboard systems testing.

AVIONICS: State-of-the-art flight and navigation systems for operation in adverse weather conditions and landing in ICAO Cat. II conditions.

DIMENSIONS, EXTERNAL:
Wing span 44.06 m (144 ft 6¼ in)
Length overall 40.25 m (132 ft 0¾ in)
Height overall 16.10 m (52 ft 10 in)
Propeller diameter 4.50 m (14 ft 9 in)
Rear loading aperture: Height 4.10 m (13 ft 5½ in)
Width 4.00 m (13 ft 1½ in)

DIMENSIONS, INTERNAL:
Cargo hold: Floor length 18.60 m (61 ft 0¼ in)
Floor length, incl ramp 22.40 m (73 ft 6 in)
Floor width 4.00 m (13 ft 1½ in)
Height 4.10 m (13 ft 5½ in)

WEIGHTS AND LOADINGS:
Max payload (incl 5,000 kg; 11,025 lb on ramp) 30,000 kg (66,135 lb)
Normal T-O weight 112,000 kg (246,900 lb)
Max T-O weight 123,000 kg (271,165 lb)
Max power loading 2.99 kg/kW (4.91 lb/shp)

PERFORMANCE (estimated):
Nominal cruising speed 405 knots (750 km/h; 466 mph)
Cruising height 8,600-9,600 m (28,200-31,500 ft)
T-O run: normal T-O weight 1,500 m (4,920 ft)
max T-O weight 1,800 m (5,905 ft)
Runway length at max T-O weight:
for T-O 1,720 m (5,645 ft)
for landing 1,950 m (6,400 ft)
Landing run 1,900 m (6,235 ft)
Range with 20,000 kg (44,090 lb) payload:
at normal T-O weight 2,885 nm (5,350 km; 3,325 miles)
at max T-O weight 3,910 nm (7,250 km; 4,505 miles)
Range with 30,000 kg (66,135 lb) payload:
at normal T-O weight 1,670 nm (3,100 km; 1,925 miles)
at max T-O weight 2,985 nm (5,530 km; 3,435 miles)

ANTONOV An-72 and An-74
NATO reporting name: Coaler

TYPE: Twin-turbofan light STOL transport.

PROGRAMME: First of two prototype An-72s, built at Kiev, flew 22 December 1977; after eight pre-series aircraft manufacture transferred to Kharkov; An-74 polar transport announced February 1984; An-72P maritime patrol version demonstrated 1992; production of An-72/74 variants currently 20 a year.

CURRENT VERSIONS: **An-72A** ('Coaler-C'): Light STOL transport for military and civil use; extended wings, lengthened fuselage and other changes compared with An-72 ('Coaler-A') prototypes; crew of two or three.

An-72AT ('Coaler-C'): Cargo version of An-72A, able to carry international standard containers.

An-72S ('Coaler-C'): Three-compartment executive transport; toilets, wardrobe and galley for hot and cold meals at front; centre compartment has small table, three-place sofa, wardrobe and baggage space starboard, table, two pivoting armchairs and intercom port, with six armchairs optional in place of sofa and wardrobe; 12 pairs of armchairs in rear compartment. Provision for adapting to carry light vehicle, freight, 38 persons in seats along side walls and in centre row, or eight stretcher patients.

An-72P: Maritime patrol version, described separately.

An-74 ('Coaler-B'): For all-weather operation in Arctic/Antarctic, to assist in setting up scientific stations on ice floes, airdrop supplies to motorised expeditions and observe changes in icefields; flight crew of five; wing, tail unit and engine air intake de-icing; advanced navigation aids, including inertial navigation system; provision for wheel/ski landing gear; much increased fuel capacity; airframe identical with An-72A except for two blister windows at rear of flight deck and front of cabin on port side and larger nose radome.

An-74A ('Coaler-B'): As An-72A, but with avionics and larger radome of An-74.

An-74T ('Coaler-B'): Cargo version of An-74A; payload 10,000 kg (22,045 lb); loading winch; roller conveyors in floor.

An-74TK ('Coaler-B'): Convertible cargo/passenger aircraft with 52 folding passenger seats or all-cargo or all-passenger or combi layouts. Built-in loading equipment.

An-74T-100 ('Coaler-B'): As An-74T, with navigator station.

An-74TK-100: As An-74TK, with navigator station.

An-74P-100 ('Coaler-B'): Executive transport version of An-74A; three-compartment cabin for work, rest and motorcar; full entertainment, galley and air-ground communication; designation -100 indicates provision of navigator station. Max T-O weight 43,800 kg (96,560 lb).

CUSTOMERS: More than 150 An-72/74s built by September 1992, for military use.

DESIGN FEATURES: Primary role as STOL replacement for turboprop An-26, with emphasis on freight carrying; ejection of exhaust efflux over upper wing surface and down over large multi-slotted flaps gives considerable increase in lift; special ramp/door as An-26; low-pressure tyres and multi-wheel landing gear for operation from unprepared strips, ice or snow; high-set engines avoid foreign object

Antonov An-74 ('Coaler-B') twin-turbofan STOL transport *(Jacques Marmain)*

ANTONOV—AIRCRAFT: UKRAINE 371

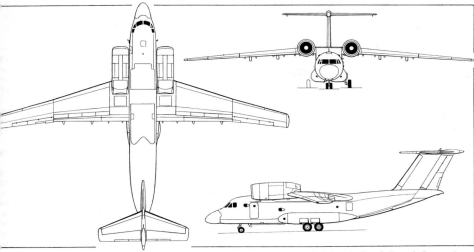

Antonov An-74 ('Coaler-B') STOL transport (two ZMKB Progress D-36 turbofans) *(Dennis Punnett)*

ingestion; wing leading-edge sweepback 17°; anhedral approx 10° on outer wings; normal T-O flap setting 25-30°, max deflection 60°; sweptback fin and rudder.

FLYING CONTROLS: Power actuated ailerons, with two tabs in port aileron, one starboard; double-hinged rudder, with tab in lower portion of two-section aft panel; during normal flight only lower rear rudder segment is used; both rear segments used in low-speed flight; forward segment is actuated automatically to offset thrust asymmetry; horn balanced and mechanically actuated, aerodynamically balanced elevators, each with two tabs; hydraulically actuated full-span wing leading-edge flaps outboard of nacelles; trailing-edge flaps double-slotted in exhaust efflux, triple-slotted between nacelles and outer wings; four-section spoilers forward of triple-slotted flaps; two outer sections on each side raised before landing, remainder opened automatically on touchdown by sensors actuated by weight on main landing gear; inverted leading-edge slat on tailplane linked to wing flaps.

STRUCTURE: All-metal; multi-spar wings mounted above fuselage; wing skin, spoilers and flaps of titanium aft of engine nacelles; circular semi-monocoque fuselage, with rear ramp/door; tapered fairing forward of T-tail fin/tailplane junction, blending into ogival rear fairing.

LANDING GEAR: Hydraulically retractable tricycle type, primarily of titanium. Rearward retracting steerable twin-wheel nose unit. Each main unit comprises two trailing-arm legs in tandem, each with a single wheel, retracting inward through 90° so that wheels lie horizontally in bottom of fairings, outside fuselage pressure cell. Oleo-pneumatic shock absorber in each unit. Low pressure tyres, size 720 × 310 mm on nosewheels, 1,050 × 400 mm on mainwheels. Hydraulic disc brakes. Telescopic strut hinges downward, from rear of each side fairing, to support fuselage during direct loading of hold with ramp/door under fuselage.

POWER PLANT: Two ZMKB Progress D-36 high bypass ratio turbofans, each 63.74 kN (14,330 lb st). Integral fuel tanks between spars of outer wings. Thrust reversers standard.

ACCOMMODATION: Pilot and co-pilot/navigator side by side on flight deck of An-72, plus flight engineer, with provision for fourth person; pilot, co-pilot, navigator, radio operator and flight engineer in An-74. Heated windows. Two windscreen wipers. Flight deck and cabin pressurised and air-conditioned. Main cabin designed primarily for freight, including four UAK-2.5 containers or four PAV-2.5 pallets each weighing 2,500 kg (5,511 lb); An-72 has folding seats along side walls and removable central seats for 68 passengers. It can carry 57 parachutists, and has provision for 24 stretcher patients, 12 seated casualties and an attendant in ambulance configuration. An-74 can carry eight mission staff in combi role, in two rows of seats, with tables, and with two bunks, one on each side of cabin aft of seats. Bulged observation windows on port side for navigator and hydrologist. Provision for wardrobe and galley. Movable bulkhead between passenger and freight compartments, with provision for 1,500 kg (3,307 lb) of freight in rear compartment. Downward hinged and forward sliding rear ramp/door for loading trucks and tracked vehicles, and for direct loading of hold from trucks, as described under An-26 entry. It is openable in flight, enabling freight loads of up to 7,500 kg (16,535 lb), with a maximum of 2,500 kg (5,511 lb) per individual item, to be airdropped by parachute extraction system. Removable mobile winch, capacity 2,500 kg (5,511 lb), assists loading of containers up to 1.90 × 2.44 × 1.46 m (6 ft 3 in × 8 ft × 4 ft 9½ in) in size, pallets 1.90 × 2.42 × 1.46 m (6 ft 3 in × 7 ft 11 in × 4 ft 9½ in) in size, and other bulky items. Cargo straps and nets stowed in lockers on each side of hold when not in use. Provision for roller conveyors in floor. Main crew and passenger door at front of cabin on port side. Emergency exit and servicing door at rear of cabin on starboard side.

SYSTEMS: Air-conditioning system for comfort to altitude of 10,000 m (32,800 ft), with independent temperature control in flight deck and main cabin areas. Used to refrigerate main cabin when perishable goods carried. Hydraulic system for landing gear, flaps and ramp. Electrical system powers auxiliary systems, flight deck equipment, lighting and mobile hoist. Thermal de-icing system for leading-edges of wings and tail unit (including tailplane slat), engine air intakes and cockpit windows. Provision for APU in starboard landing gear fairing. This can be used to heat cabin; under cold ambient conditions, servicing personnel can gain access to major electric, hydraulic and air-conditioning components without stepping outside.

AVIONICS: HF com, VHF com/nav, ADF. Navigation/weather radar in nose. Doppler-based automatic navigation system, linked to onboard computer, is pre-programmed before take-off on push-button panel to right of map display. Failure warning panels above windscreen display red lights for critical failures, yellow lights for non-critical failures, to minimise time spent on monitoring instruments and equipment. 'Odd Rods' IFF standard. An-74 has enhanced avionics, including INS.

DIMENSIONS, EXTERNAL:
Wing span	31.89 m (104 ft 7½ in)
Wing aspect ratio	10.31
Length overall (An-72)	28.07 m (92 ft 1¼ in)
Fuselage: Max diameter	3.10 m (10 ft 2 in)
Height overall	8.65 m (28 ft 4½ in)
Wheel track	4.15 m (13 ft 7½ in)
Wheelbase	8.12 m (26 ft 7¾ in)
Rear loading door: Length	7.10 m (23 ft 3½ in)
Width	2.40 m (7 ft 10½ in)

DIMENSIONS, INTERNAL:
Cabin: Length	10.50 m (34 ft 5¼ in)
Width at floor level	2.15 m (7 ft 0½ in)
Height	2.20 m (7 ft 2½ in)

AREAS:
Wings, gross	98.62 m² (1,062 sq ft)

WEIGHTS AND LOADINGS:
Weight empty	19,050 kg (42,000 lb)
Max fuel	12,950 kg (28,550 lb)
Max payload: normal	10,000 kg (22,045 lb)
Max T-O weight:	
from 1,800 m (5,905 ft) runway	34,500 kg (76,060 lb)
from 1,500 m (4,920 ft) runway	33,000 kg (72,750 lb)
from 1,000 m (3,280 ft) runway	27,500 kg (60,625 lb)
Max wing loading	349.8 kg/m² (71.62 lb/sq ft)
Max power loading	270.6 kg/kN (2.65 lb/lb st)

PERFORMANCE (An-72A. A: at T-O weight of 33,000 kg; 72,750 lb, B: at T-O weight of 27,500 kg; 60,625 lb on 1,000 m; 3,280 ft unprepared runway):
Max level speed at 10,000 m (32,800 ft):	
A	380 knots (705 km/h; 438 mph)
Cruising speed at 10,000 m (32,800 ft):	
A, B	297-324 knots (550-600 km/h; 342-373 mph)
Approach speed: A	97 knots (180 km/h; 112 mph)
Service ceiling: A	10,700 m (35,100 ft)
B	11,800 m (38,715 ft)
Service ceiling, OEI: A	5,100 m (16,730 ft)
B	6,800 m (22,300 ft)
T-O run: A	930 m (3,052 ft)
B	620 m (2,035 ft)
T-O to 10.7 m (35 ft): A	1,170 m (3,840 ft)
B	830 m (2,725 ft)
Landing run: A	465 m (1,525 ft)
B	420 m (1,380 ft)
Range, with 45 min reserves:	
A with max payload	430 nm (800 km; 497 miles)
A with 7,500 kg (16,535 lb) payload	1,080 nm (2,000 km; 1,240 miles)
A with max fuel	2,590 nm (4,800 km; 2,980 miles)
B with 5,000 kg (11,020 lb) payload	430 nm (800 km; 497 miles)
B with max fuel	1,760 nm (3,250 km; 2,020 miles)

PERFORMANCE (An-74):
Generally as for An-72A, except:
Range, with 2 h reserves:	
with max payload	620 nm (1,150 km; 715 miles)
with 5,000 kg (11,020 lb) payload	1,726 nm (3,200 km; 1,988 miles)
with 1,500 kg (3,307 lb) payload	2,860 nm (5,300 km; 3,293 miles)

ANTONOV An-72P
NATO reporting name: Coaler

TYPE: Twin-turbofan STOL maritime patrol aircraft.

PROGRAMME: First seen 1992; displayed at 1992 Farnborough Air Show; in production.

CUSTOMERS: Initial order 20 for Russian military use.

DESIGN FEATURES: Basically identical to An-72 transport; intended for armed surveillance of coastal areas within 200 nm (370 km; 230 miles) of shore, day and night in all weathers.

ACCOMMODATION: For maritime missions, flight crew supplemented by navigator and radio operator, stationed by bulged windows at rear of flight deck on port side and immediately forward of wing leading-edge on starboard side respectively. Provision in main cabin for 40 persons on sidewall and centreline removable seats, including ramp-mounted seats; or 22 fully equipped paratroops; or 16 stretcher patients and medical attendant; or up to 5,000 kg (11,025 lb) of ammunition, equipment or vehicles, with seats stowed.

AVIONICS: Permit automated navigation at all stages of flight, precise fixing of co-ordinates, speed and heading of surface ships, air-to-air and air-to-surface communication with aircraft, ships and coastguard to support missions. TV scanning system in port main landing gear fairing.

EQUIPMENT: Oblique camera on port side opposite radio operator's station; daylight and night mapping cameras in tail-cone; SFP-2A flares for night use.

ARMAMENT: One 23 mm gun pod forward of starboard main landing gear fairing; UB-32M rocket pack under each wing; four 100 kg (220 lb) bombs in roof of hold, above

Antonov An-72P after touchdown, with thrust reverser doors open and flaps extended *(Brian M. Service)*

rear loading hatch, with ramp slid forward under cabin to make release practicable.
Other data generally as for An-72 transport, except:

WEIGHTS AND LOADINGS:
Mission load 650 kg (1,433 lb)
Max T-O weight 32,000 kg (70,545 lb)
PERFORMANCE:
Patrol speed at 500-1,000 m (1,640-3,280 ft)
162-189 knots (300-350 km/h; 186-217 mph)
Service ceiling 10,100 m (33,135 ft)
Field requirement 1,400 m (4,600 ft)
Max endurance 5 h 18 min

ANTONOV An-124
NATO reporting name: Condor

TYPE: Long-range heavy-lift four-turbofan freight transport.
PROGRAMME: Prototype (SSSR-680125) first flew 26 December 1982; second prototype (SSSR-82002 *Ruslan*, named after giant hero of Russian folklore immortalised by Pushkin) exhibited 1985 Paris Air Show; lifted payload of 171,219 kg (377,473 lb) to 10,750 m (35,269 ft) on 26 July 1985, exceeding by 53 per cent C-5A Galaxy's record for payload lifted to 2,000 m and setting 20 more records; entered service January 1986, transporting units of US/Canadian Euclid 154 tonne dumper truck for Yakut diamond miners; set closed circuit distance record 6-7 May 1987 by flying 10,880.625 nm (20,150.921 km; 12,521.201 miles) in 25 h 30 min; deliveries to VTA, to replace An-22, began 1987; in September 1990, during Gulf crisis, an An-124 carried 451 Bangladeshi refugees from Amman to Dacca, after being fitted with chemical toilets, a 570 litre (150 US gallon; 125 Imp gallon) drinking water tank and foam rubber cabin lining in lieu of seats; civil type certificate An-124-100 granted by AviaRegistr of Interstate Aviation Committee of CIS on 30 December 1992. Two records set by Air Foyle with An-124s: heaviest commercial air cargo movement of three 43-tonne transformers and ancillary equipment, totalling 143 tonnes, from Barcelona, Spain, to Noumea, New Caledonia; and single heaviest cargo item comprising 120-tonne Liebherr crane from Krivoy Rog, Ukraine, to Berlin, Germany; production continues, at Ulyanovsk and Kiev; four built in 1992, 12 planned for delivery 1993.

CUSTOMERS: In early 1993 deliveries totalled more than 30 including prototypes: majority to VTA; first international commercial operator was Air Foyle of Luton, England, which wet leases three An-124s from Antonov OKB, with option on two more; five available to HeavyLift (UK) / VolgaDnepr (Russia), for charter operations (two based at Stansted, England).

DESIGN FEATURES: World's largest production aircraft; configuration similar to Lockheed C-5 Galaxy, except for low-mounted tailplane; upward hinged visor type nose and rear fuselage ramp/door for simultaneous front and rear loading/unloading; 100 per cent fly-by-wire control system; titanium floor throughout constant-section main hold, which is lightly pressurised, with a fully pressurised cabin for passengers above; landing gear for operation from unprepared fields, hard packed snow and ice covered swampland; steerable nose- and mainwheels permit turns on 45 m (148 ft) wide runway. Supercritical wings, with anhedral; sweepback approx 35° on inboard leading-edge, 32° outboard; all tail surfaces sweptback.

FLYING CONTROLS: Fly-by-wire, with all surfaces hydraulically actuated; two-section ailerons, three-section single-slotted Fowler flaps and six-section full-span leading-edge flaps on each wing; small slot in outer part of two inner flap sections each side to optimise aerodynamics; eight spoilers on each wing, forward of trailing-edge flaps; no wing fences, vortex generators or tabs; hydraulic flutter dampers on ailerons; rudder and each elevator in two sections, without tabs but with hydraulic flutter dampers; fixed incidence tailplane; control runs (and other services) channelled along fuselage roof.

STRUCTURE: Basically conventional light alloy, but 5,500 kg (12,125 lb) of composites make up more than 1,500 m² (16,150 sq ft) of surface area, giving weight saving of more than 2,000 kg (4,410 lb); each wing has one-piece root-to-tip upper surface extruded skin panel, strip of carbonfibre skin panels on undersurface forward of control surfaces, and glassfibre tip; front and rear of each flap guide fairing of glassfibre, centre portion of carbonfibre; central frames of semi-monocoque fuselage each comprise four large forgings; fairings over intersection of fuselage double-bubble lobes in line with wing, from rear of flight deck to plane of fin leading-edge, primarily of glassfibre, with central, and lower underwing, portions of carbonfibre; other glassfibre components include tailplane tips, nosecone, tailcone and most bottom skin panels forming blister underfairing between main landing gear legs; carbonfibre components include strips of skin panels forward of each tail control surface, nose and main landing gear doors, some service doors, and clamshell doors aft of rear loading ramp.

LANDING GEAR: Hydraulically retractable nosewheel type, made by Hydromash, with 24 wheels. Two independent forward retracting and steerable twin-wheel nose units, side by side. Each main gear comprises five independent inward retracting twin-wheel units; front two units on each side steerable. Each mainwheel bogie enclosed by separate upper and lower doors when retracted. Nosewheel doors and lower mainwheel doors close when gear extended. All wheel doors of carbonfibre. Main gear bogies retracted individually for repair or wheel change. Mainwheel tyres size 1,270 × 510 mm. Nosewheel tyres size 1,120 × 450 mm. Aircraft can 'kneel', by retracting nosewheels and settling on two extendable 'feet', giving floor of hold a 3.5° slope to assist loading and unloading. Rear of cargo hold lowered by compressing main gear oleos. Carbon brakes normally toe operated, via rudder pedals. For severe braking, pedals depressed by toes and heels.

POWER PLANT: Four ZMKB Progress D-18T turbofans, each 229.5 kN (51,590 lb st); thrust reversers standard. Engine cowlings of glassfibre; pylons have carbonfibre skin at rear

Antonov An-72P STOL maritime patrol aircraft *(Mark Lambert)*

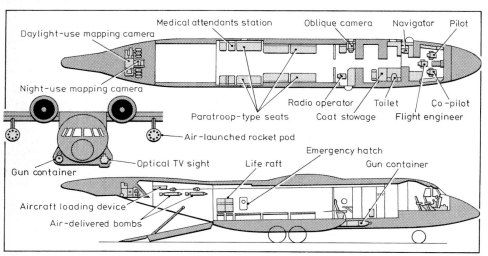

Disposition of equipment and armament in the military Antonov An-72P. The ramp moves down and forward to lie underneath the fuselage when bombs are dropped from the cargo gantry *(Jane's/Mike Keep)*

Bombs in roof of hold of Antonov An-72P *(Neville M. Beckett)*

Cockpit of An-72P maritime patrol aircraft *(Peter J. Cooper)*

end. All fuel in ten integral tanks in wings, total capacity 348,740 litres (92,128 US gallons; 76,714 Imp gallons).

ACCOMMODATION: Crew and passenger accommodation on upper deck; freight and/or vehicles on lower deck. Flight crew of six, in pairs, on flight deck, with place for loadmaster in lobby area (10-12 cargo handlers and servicing staff carried on commercial flights). Pilot and co-pilot on fully adjustable seats, which rotate for improved access. Two flight engineers, on wall-facing seats on starboard side, have complete control of master fuel cocks, detailed systems instruments, and digital integrated data system with CRT monitor. Behind pilot are navigator and communications specialist, on wall-facing seats. Between flight deck and wing carry-through structure, on port side, are toilets, washing facilities, galley, equipment compartment, and two cabins for up to six relief crew, with table and facing bench seats convertible into bunks. Aft of wing carry-through is passenger cabin for 88 persons. Hatches in upper deck provide access to wing and tail unit for maintenance when workstands not available. Flight deck and passenger cabin each accessible from cargo hold by hydraulically folding ladder, operated automatically with manual override. Rearward sliding and jettisonable window each side of flight deck. Primary access to flight deck via airstair door, with ladder extension, forward of wing on port side. Smaller door forward of this and slightly higher. Door from main hold aft of wing on starboard side. Upper deck doors at rear of flight deck on starboard side and at rear of passenger cabin on each side. Emergency exit from upper deck aft of wing on each side. Hydraulically operated visor type upward hinged nose takes 7 min to open fully, with simultaneous extension of folding nose loading ramp. When open, nose is steadied by reinforcing arms against wind gusts. No hydraulic, electrical or other system lines broken when nose is open. Radar wiring passes through hollow tube in hinge. Hydraulically operated rear loading doors take 3 min to open, with simultaneous extension of three-part folding ramp. This can be locked in intermediate position for direct loading from truck. Aft of ramp, centre panel of fuselage undersurface hinges upward; clamshell door to each side opens downward. Completely unobstructed lower deck freight hold has titanium floor, attached 'mobilely' to lower fuselage structure to accommodate changes of temperature, with rollgangs and retractable attachments for cargo tiedowns. Narrow catwalk along each sidewall facilitates access to, and mobility past, loaded freight. Payloads include largest CIS main battle tanks, complete missile systems, 12 standard ISO containers, oil well equipment and earth movers; HeavyLift/VolgaDnepr aircraft transport Airbus wings in Europe. No personnel carried normally on lower deck in flight, because of low pressurisation. Two electric travelling cranes in roof of hold, each with two lifting points, offer total lifting capacity of 20,000 kg (44,100 lb). Two winches each pull a 3,000 kg (6,614 lb) load.

SYSTEMS: Entire interior of aircraft is pressurised and air-conditioned. Max pressure differential 0.55 bar (7.8 lb/sq in) on upper deck, 0.25 bar (3.55 lb/sq in) on lower deck. Four independent hydraulic systems. Quadruple redundant fly-by-wire flight control system, with mechanical emergency fifth channel to hydraulic control servos. Special secondary bus electrical system. Landing lights under nose and at front of each main landing gear fairing. APU in rear of each landing gear fairing for engine starting, can be operated in the air or on the ground to open loading doors for airdrop from rear or normal ground loading/unloading, as well as for supplying electrical, hydraulic and air-conditioning systems. Bleed air anti-icing of wing leading-edges. Electro-impulse de-icing of fin and tailplane leading-edges.

AVIONICS: Conventional flight deck equipment, including automatic flight control system panel at top of glareshield, weather radar screen and moving map display forward of throttle and thrust reverse levers on centre console. No electronic flight displays. Dual attitude indicator/flight director and HSIs, and vertical tape engine instruments. Two dielectric areas of nose visor enclose forward-looking weather radar and downward-looking ground mapping/navigation radar. Hemispherical dielectric fairing above centre fuselage for satellite navigation receiver. Quadruple INS, plus Loran and Omega.

EQUIPMENT: Small two-face mirror, of V form, enables pilots to adjust their seating position until their eyes are reflected in the appropriate mirror, which ensures optimum field of view from flight deck.

DIMENSIONS, EXTERNAL:
Wing span 73.30 m (240 ft 5¼ in)
Wing aspect ratio 8.55

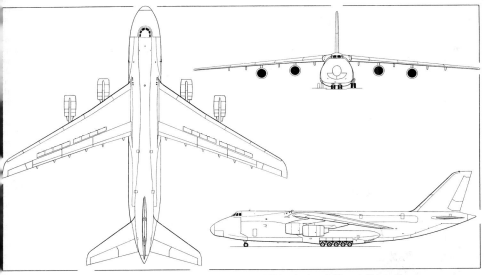
Antonov An-124 (four ZMKB Progress D-18T turbofans) *(Dennis Punnett)*

One of six 31.5-tonne excavators carried to Kuwait from Stansted Airport by Air Foyle Antonov An-124

Antonov An-124 heavy freight transport in the insignia of HeavyLift/VolgaDnepr *(Mike Perry)*

374 UKRAINE: AIRCRAFT—ANTONOV

Length overall	69.10 m (226 ft 8½ in)
Height overall	20.78 m (68 ft 2¼ in)
DIMENSIONS, INTERNAL:	
Cargo hold: Length	36.0 m (118 ft 1¼ in)
Max width	6.4 m (21 ft 0 in)
Max height	4.4 m (14 ft 5¼ in)
Volume	1,000 m³ (35,315 cu ft)
AREAS:	
Wings, gross	628.0 m² (6,760.0 sq ft)
WEIGHTS AND LOADINGS:	
Operating weight empty	175,000 kg (385,800 lb)
Max payload	150,000 kg (330,693 lb)
Max T-O weight	405,000 kg (892,872 lb)
Max zero-fuel weight	325,000 kg (716,500 lb)
Max wing loading	644.9 kg/m² (132.1 lb/sq ft)
Max power loading	441.2 kg/kN (4.32 lb/lb st)
PERFORMANCE:	
Max cruising speed	467 knots (865 km/h; 537 mph)
Normal cruising speed at 10,000-12,000 m (32,800-39,370 ft)	432-459 knots (800-850 km/h; 497-528 mph)
Approach speed	124-140 knots (230-260 km/h; 143-162 mph)
T-O balanced field length at max T-O weight	3,000 m (9,850 ft)
T-O run	2,520 m (8,270 ft)
Landing run at max landing weight	900 m (2,955 ft)
Range:	
with max payload	2,430 nm (4,500 km; 2,795 miles)
with max fuel	8,900 nm (16,500 km; 10,250 miles)
OPERATIONAL NOISE LEVELS:	
Stated to meet ICAO requirements	

ANTONOV An-180

TYPE: Twin-propfan medium-range passenger transport.
PROGRAMME: First details, and model, at 1991 Paris Air Show; wholly internal programme at that time; configuration changed, with engines on tailplane tips, 1992; prototype intended for display at 1995 Paris Show.
DESIGN FEATURES: Unique configuration, with propfans at tips of dihedral tailplane; low-mounted sweptback wings, circular-section fuselage and all-sweptback tail surfaces. Designed to equal, or exceed, highest Western standards, burning less fuel per passenger-km per engine than equivalent Western aircraft; equipped for automated navigation at all flight stages, and for ICAO Cat. IIIa automatic landings. Airframe design life 60,000 flying hours.
POWER PLANT: Two ZMKB Progress D-27 tractor propfans, each rated at 10,305 kW (13,820 shp), mounted on tips of tailplane.
ACCOMMODATION: Two or three crew; single-class seating for 162-174 passengers or mixed class seating for 144-156, in pairs, six-abreast, with two aisles; seven LD3 containers underfloor; four seats for cabin attendants by passenger doors (port side) and service doors (starboard); one toilet at front, two at rear. Available in mixed passenger/freight and all-freight versions with side loading door. Emergency exit over each wing.

DIMENSIONS, EXTERNAL:	
Wing span	35.80 m (117 ft 5½ in)
Length overall	41.70 m (136 ft 9¾ in)
Height overall	10.90 m (35 ft 9 in)
WEIGHTS AND LOADINGS:	
Max payload	18,000 kg (39,680 lb)
Max T-O weight	67,500 kg (148,810 lb)
PERFORMANCE (estimated):	
Nominal cruising speed	432 knots (800 km/h; 497 mph)
Nominal cruising height	9,100-10,100 m (29,850-33,135 ft)
Nominal field length	2,200-2,600 m (7,220-8,530 ft)
Range, with reserves, from 2,600 m (8,530 ft) field:	
with max payload	1,780 nm (3,300 km; 2,050 miles)
with 162 passengers	2,615 nm (4,850 km; 3,010 miles)
with 144 passengers	3,020 nm (5,600 km; 3,480 miles)
with max fuel	4,045 nm (7,500 km; 4,660 miles)

ANTONOV An-218

TYPE: Twin-turbofan passenger transport.
PROGRAMME: Announced at 1991 Paris Air Show; first flight scheduled for 1995; certification and service entry planned for 1996-97.
CURRENT VERSIONS: **An-218-100**: Basic version, *as described in detail*.
An-218-200: Extended range; two D-18TR (modified D-18TM) or PW4060, CF6-80C2B6, or RB211-524H4 engines; additional fuel tanks in cargo compartments; enhanced cabin comfort for 220 passengers; range 5,720 nm (10,600 km; 6,585 miles).
An-218-300: Long-range version; fuselage shortened by 11.5 m (37 ft 8¾ in) by removal of plugs; additional fuel in cargo compartments; 195 passengers; range 6,045 nm (11,200 km; 6,960 miles).
An-218-400: Under study; 400 passengers.
DESIGN FEATURES: Generally similar to Airbus A330, but slightly smaller and lighter in weight, with less fuel.
POWER PLANT: Two ZMKB Progress D-18TM turbofans, each 245 kN (55,115 lb st), in underwing pods.
ACCOMMODATION: Two pilots on modern flight deck with electronic displays; three basic cabin configurations: (1) 294 passengers: 18 first class six-abreast at 102 cm (40 in) pitch, 74 business class at 87 cm (34.25 in) pitch, and 202 econ class at 81 cm (32 in) pitch. (2) 316 passengers: 38 first class and 278 econ class. (3) 350 passengers seven/eight-abreast, all econ class. 24 freight/baggage containers in underfloor compartments. Four cabin doors on port side; service doors opposite.

DIMENSIONS, INTERNAL:	
Cargo compartments:	
Volume: front	81.0 m³ (2,860 cu ft)
rear	57.5 m³ (2,030 cu ft)
Baggage compartment volume for outsize cargoes	42.0 m³ (1,483 cu ft)
WEIGHTS AND LOADINGS:	
Max payload	42,000 kg (92,600 lb)
PERFORMANCE (estimated):	
Nominal cruising speed	458-470 knots (850-870 km/h; 528-540 mph)
Required runway length, 30°C	3,200 m (10,500 ft)
Range, with reserves: with 350 passengers	
	3,400-4,045 nm (6,300-7,500 km; 3,915-4,660 miles)
with 292 passengers	
	3,885-4,800 nm (7,200-8,900 km; 4,475-5,530 miles)
Fuel efficiency	18 g/passenger-km

ANTONOV An-225 MRIYA (DREAM)
NATO reporting name: Cossack

TYPE: Six-turbofan heavy transport for internal/external payloads.
PROGRAMME: Design studies began mid-1985; prototype (SSSR-480182) made 75 min first flight "from 1,000 m (3,280 ft) runway" on 21 December 1988, three weeks after unveiling at Kiev; total of 106 records set on 22 March 1989 during 3½ h flight: taking off at 508,200 kg (1,120,370 lb), with 156,300 kg (344,576 lb) payload, flew 2,000 km closed circuit at 438.75 knots (813.09 km/h; 505.24 mph), with max altitude of 12,340 m (40,485 ft) *en route*; first flight carrying *Buran* orbiter on back made 13 May 1989 from Baikonur. No more An-225s completed by early 1993, but second aircraft under unhurried assembly; possibility of building aircraft for commercial use under consideration, as well as various modifications.
DESIGN FEATURES: First aircraft built to fly at gross weight exceeding one million pounds; designed to replace Myasishchev VM-T Atlant as external load-carrier for space orbiters, components of Energiya rocket launch vehicles and other outsize loads; based on An-124, with extended wings and lengthened fuselage to permit 50 per cent increase in max T-O weight and payload; basic cabin cross-section and visor type nose door unchanged; rear loading ramp/door deleted and rear fuselage reconfigured with twin fins and rudders on dihedral tailplane to avoid airflow problems when carrying piggyback loads; main landing gear uprated from five to seven pairs of wheels on each side; six engines instead of four, of same type; basically standard An-124 wings attached to new centre-section; anhedral on outer wings only; sweepback 35° on inboard half-span, 32° outboard; all tail surfaces sweptback.
FLYING CONTROLS: Fly-by-wire, with all surfaces hydraulically actuated; each wing has two-section aileron, three-section single-slotted Fowler flaps on outer panels and

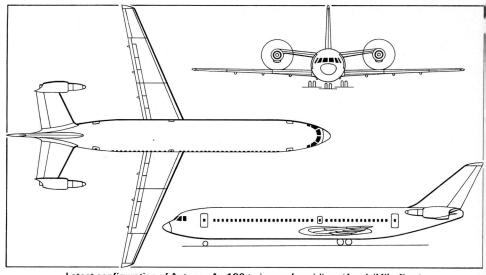

Latest configuration of Antonov An-180 twin-propfan airliner (*Jane's/Mike Keep*)

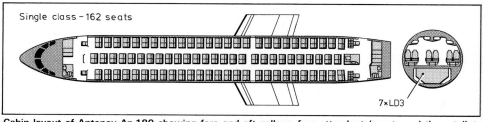

Cabin layout of Antonov An-180 showing fore and aft galleys, four attendants' seats and three toilets (*Jane's/Mike Keep*)

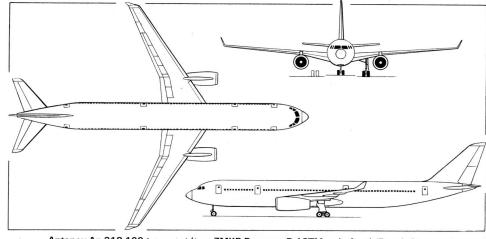

Antonov An-218-100 transport (two ZMKB Progress D-18TM turbofans) (*Dennis Punnett*)

ANTONOV—AIRCRAFT: UKRAINE

World's largest aeroplane, the Antonov An-225 Mriya heavy transport *(Piotr Butowski)*

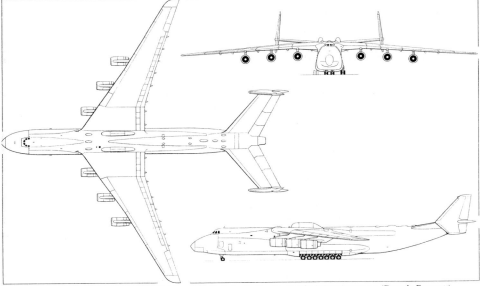

Antonov An-225 Mriya (NATO 'Cossack') six-turbofan heavy freight transport *(Dennis Punnett)*

single section on centre-section, and six-section leading-edge flaps on outer panels only; eight airbrakes (inboard) and eight spoilers (outboard) on each wing upper surface and centre-section forward of flaps; no wing fences, vortex generators or tabs; two-section rudder and three-section elevator on each side; control runs (and other services) channelled along fuselage roof.

STRUCTURE: Generally as for An-124.

LANDING GEAR: Hydraulically retractable nosewheel type, made by Hydromash. Two independent forward retracting and steerable twin-wheel nose units, side by side. Each main gear comprises seven independent inward retracting twin-wheel units, with tyres size 1,270 × 510 mm. Rear four pairs of wheels on each side steerable. Each main-wheel bogie enclosed by separate upper and lower doors when retracted. Nosewheel doors and lower mainwheel doors close when gear extended. Aircraft can 'kneel', by retracting nosewheels and settling on two extendable 'feet', giving floor a slope to assist loading and unloading. Min turning radius about nosewheels 60.0 m (197 ft).

POWER PLANT: Six ZMKB Progress D-18T turbofans, each 229.5 kN (51,590 lb st) and fitted with thrust reverser. Engine cowlings of glassfibre. All fuel in integral tanks in wings, including additional tanks in new centre-section. Max capacity estimated at well over 300,000 kg (661,375 lb).

ACCOMMODATION: Flight crew of six, in pairs, on flight deck, with place for loadmaster in lobby area. Pilot and co-pilot on fully adjustable seats that rotate for improved access. Two flight engineers, on wall-facing seats on starboard side. Navigator and communications specialist behind pilot, on wall-facing seats. Rest area for relief crew slightly larger than that of An-124. Cabin for 60-70 persons above hold aft of wing carry-through. Primary access to flight deck via airstair door, with ladder extension, forward of wing on port side. Door to main hold aft of wing on starboard side. Hydraulically operated visor type upward hinged nose takes 7 min to open fully, with simultaneous extension of folding nose loading ramp. Completely unobstructed lower deck freight hold, 43.0 m (141 ft) long, has titanium floor, attached 'mobilely' to lower fuselage structure to accommodate changes of temperature, with rollgangs and retractable attachments for cargo tiedowns. Interior can be heated with warm air from perforated tube above floor on each side of hold. Internal loads can include vehicles, ground test and field maintenance equipment required by external loads. Two longitudinal mounting beams for external loads above wing centre-section. Small blister fairings forward of beams and forward of tailplane cover load attachments. Under consideration is a scheme to use an eight-engined version of the An-225 as a launcher for future space vehicles like the British Interim Hotol or space combat aircraft.

AVIONICS: Generally similar to An-124, with automatic flight control system and moving map display; no electronic flight displays. Two dielectric areas of nose visor enclose forward-looking weather radar and downward-looking ground mapping/navigation radar. Quadruple INS, plus Loran and Omega.

DIMENSIONS, EXTERNAL:
Wing span	88.40 m (290 ft 0 in)
Wing aspect ratio	8.64
Length overall	84.00 m (275 ft 7 in)
Height overall	18.20 m (59 ft 8½ in)

Antonov An-225 carrying space shuttle *Buran*

Tailplane span	32.65 m (107 ft 1½ in)	
Wheel track	8.84 m (29 ft 0 in)	
Wheelbase, from main landing gear mid-point	29.10 m (95 ft 5½ in)	

DIMENSIONS, INTERNAL:
Cargo hold: Length of floor	43.0 m (141 ft 0 in)
Max width	6.4 m (21 ft 0 in)
Max height	4.4 m (14 ft 5¼ in)

AREAS:
Wings, gross	905.0 m² (9,741 sq ft)
Horizontal tail surfaces, total	265.5 m² (2,858 sq ft)
Vertical tail surfaces, total	141.7 m² (1,525 sq ft)

WEIGHTS AND LOADINGS:
Max payload, internal or external	250,000 kg (551,150 lb)
Max T-O weight	600,000 kg (1,322,750 lb)
Max wing loading	663 kg/m² (135.8 lb/sq ft)
Max power loading	435.7 kg/kN (4.27 lb/lb st)

PERFORMANCE (estimated):
Cruising speed	432-458 knots (800-850 km/h; 497-528 mph)
T-O balanced field length	3,500 m (11,485 ft)
Range: with 200,000 kg (440,900 lb) internal payload	2,425 nm (4,500 km; 2,795 miles)
with 100,000 kg (220,450 lb) internal payload	5,180 nm (9,600 km; 5,965 miles)
with max fuel	8,310 nm (15,400 km; 9,570 miles)

UNITED KINGDOM

AVIATION SCOTLAND
AVIATION SCOTLAND LTD
Fraser House, 205 Bath Street, Glasgow G2 4HZ
FACTORY: 9 Bertram Street, Burnbank, Hamilton ML3 0QX
Telephone: 44 (698) 820933
Fax: 44 (698) 829089
CHAIRMAN: Douglas Fraser
MANAGING DIRECTOR AND CEO: David Jordan
CHIEF ENGINEER: David Cummings

Formed to resume production of and further develop former Island Aircraft ARV-1 Super2; design and manufacturing rights purchased late 1991; manufacturing plant and machinery transferred to freehold premises in Burnbank.

AVIATION SCOTLAND ARV-1 SUPER2
TYPE: Two-seat light cabin monoplane.
PROGRAMME: Original prototype Super2 (G-OARV) first flew 11 March 1985, piloted by Mr Hugh Kendall; two further trials/demonstrator aircraft in programme which resulted in BCAR Section K certification July 1986; production halted 1987 but resumed 1988 for short period only. Jigs transferred to Hamilton late 1991; relaunched November 1991 at Copex '91 UK show at Sandown; overseas relaunch at Copex Caribbean and Latin American show in Miami February 1992; CAA A1 approval received October 1992; first Aviation Scotland aircraft flew November 1992 with AE 75 engine manufactured by Mid-West Aero Engines; MWAE rotary engine, in testbed Super2 G-BMWG, flew January 1993 for engine development purposes; Super2 currently marketed only in original form; company received CAA certification to carry out modifications in March 1993.
CURRENT VERSIONS: Basic original version available, with AE 75 engine.
CUSTOMERS: 18 built by 1988; no later information.
COSTS: £35,000 (1993) plus VAT.
DESIGN FEATURES: Design objective low initial cost and very low maintenance cost; Mid-West AE 75 three-cylinder liquid-cooled engine with aluminium radiator in recessed duct in rear fuselage; superplastically formed aluminium alloy pressings to save weight. Designed to BCAR Section K and FAR Pt 23; also intended to be suitable for home building in kit form under auspices of Popular Flying Association. Wing section NACA 2415 (modified); 5° 6' forward sweep.
FLYING CONTROLS: Mass balanced ailerons operated by torque tube through leading-edge of manually operated plain flaps; single-spar mass balanced rudder and elevator with trim tab.
STRUCTURE: Single-spar aluminium alloy wing, cold bonded and flush riveted. Single uncompromised main bulkhead carries wings, bracing struts, controls, seats, fuel tank and main landing gear; conventional double beam forward of main bulkhead carries firewall, nose landing gear and engine; front fuselage skinning in four panels; rear fuselage of aluminium alloy construction with single curvature skinning. Tail unit conventional aluminium alloy structure with three-spar fixed surfaces.
LANDING GEAR: Non-retractable tricycle type. Cantilever main legs of tapered steel leaf spring. All three wheels size 3.50 × 6.0, with tyres size 13 × 4.00-6, pressure 1.72 bars (25 lb/sq in). Hydraulic disc brakes. Nose leg of leading link design, with rubber in tension springing, damped by adjustable Spax gas filled shock absorber.
POWER PLANT: One 57.4 kW (77 hp) Mid-West Aero Engines AE 75 three-cylinder, 750 cc liquid-cooled inline two-stroke engine, driving a two-blade fixed-pitch Hoffmann propeller through 2.7:1 reduction gearing. Electronically variable ignition timing. Serck aluminium radiator in recessed duct in rear fuselage. Normal fuel capacity 50 litres (13.2 US gallons; 11 Imp gallons) in single fuselage tank. Optional fuel capacity 68 litres (18 US gallons; 15 Imp gallons). Refuelling point in top of fuselage.
ACCOMMODATION: Enclosed cabin seating two side by side; shoulder location of wing with forward sweep, low panel line and close cowling of small engine combine to provide optimum view from cabin, particularly in turns. Seats adjustable for height, and fold to reveal baggage compartment; additional storage space under seats. Rearward hinged canopy is one-piece Perspex moulding with GFRP frame. Dual controls standard.
SYSTEMS: Wheel disc brakes operated hydraulically. Electrical system includes 11A or 15A alternator.
AVIONICS: Radio equipment to customers' requirements.
EQUIPMENT: Standard equipment includes three basic flight instruments plus engine instruments. Optional vacuum instruments driven by dual venturis mounted under front fuselage.

Aviation Scotland ARV-1 Super2 two-seat monoplane

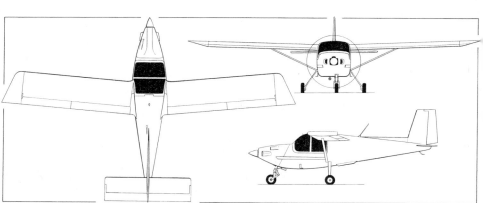

Aviation Scotland ARV-1 Super2 *(Dennis Punnett)*

DIMENSIONS, EXTERNAL:
Wing span	8.69 m (28 ft 6 in)
Width, wings folded	2.54 m (8 ft 4 in)
Wing aspect ratio	8.8
Length overall	5.49 m (18 ft 0 in)
Height overall	2.31 m (7 ft 7 in)
Tailplane span	2.54 m (8 ft 4 in)
Wheel track	1.83 m (6 ft 0 in)
Wheelbase	1.74 m (5 ft 8½ in)
Propeller diameter	1.60 m (5 ft 3 in)
Propeller ground clearance	0.23 m (9 in)

DIMENSIONS, INTERNAL:
Cabin: Length	1.27 m (4 ft 2 in)
Max width	0.99 m (3 ft 3 in)
Max height	1.09 m (3 ft 7 in)

AREAS:
Wings, gross	8.59 m² (92.5 sq ft)
Ailerons (total)	0.60 m² (6.5 sq ft)
Trailing-edge flaps (total)	0.89 m² (9.6 sq ft)
Fin	0.59 m² (6.4 sq ft)
Rudder	0.26 m² (2.8 sq ft)

Tailplane	1.23 m² (13.2 sq ft)	PERFORMANCE (at max T-O weight):		flaps down	48 knots (88 km/h; 55 mph)
Elevators, incl tab	0.55 m² (5.9 sq ft)	Never-exceed speed (V_NE)		Rate of climb at S/L	189 m (620 ft)/min
WEIGHTS AND LOADINGS:			126 knots (232 km/h; 149 mph)	T-O to 15 m (50 ft)	712 m (2,336 ft)
Weight empty, equipped	313 kg (690 lb)	Max level speed	97 knots (180 km/h; 112 mph)	Landing from 15 m (50 ft)	470 m (1,540 ft)
Normal fuel weight	37 kg (80 lb)	Max cruising speed	90 knots (167 km/h; 104 mph)	Range: normal fuel	270 nm (500 km; 311 miles)
Max T-O and landing weight	499 kg (1,100 lb)	Econ cruising speed	80 knots (148 km/h; 92 mph)	optional fuel	370 nm (685 km; 426 miles)
Max wing loading	58.35 kg/m² (11.95 lb/sq ft)	Stalling speed, power off:			
Max power loading	9.20 kg/kW (15.13 lb/hp)	flaps up	54 knots (99 km/h; 62 mph)		

BAe
BRITISH AEROSPACE plc
HEADQUARTERS: Warwick House, PO Box 87, Farnborough Aerospace Centre, Farnborough, Hampshire GU14 6YU
Telephone: 44 (252) 373232
Fax: 44 (252) 383000
Telex: 859598 BAEFBR G
CHAIRMAN: John C. Cahill
CHIEF EXECUTIVE: Richard Evans, CBE
DIRECTOR, PUBLIC AFFAIRS: Ian Woodward

British Aircraft Corporation (Holdings) Ltd, Hawker Siddeley Aviation Ltd, Hawker Siddeley Dynamics Ltd and Scottish Aviation Ltd nationalised and merged as British Aerospace 1977; became private sector public limited company January 1981; residual HM Government shareholding sold May 1985; responsibility for business operations devolved 1989 to divisions and subsidiaries, currently employing 105,000. Main subsidiaries comprise British Aerospace Airbus Ltd, British Aerospace Defence Ltd (incorporating Military Aircraft, Systems & Services, Dynamics and Royal Ordnance divisions), Jetstream Holdings Ltd, Arlington Property Holdings Ltd, Arlington Securities plc, British Aerospace Space Systems Ltd, and Rover Group Holdings plc; overseas subsidiaries are British Aerospace Australia Ltd, Steinheil Optronik GmbH, Arkansas Aerospace Inc, British Aerospace Inc, British Scandinavian Aviation AB, and Ballast Nedam BV. Corporate Jets Ltd sold to Raytheon 1993.

Associated companies (details in International section) are SEPECAT, Panavia Aircraft GmbH, Eurofighter Jagdflugzeug GmbH, and Avro International Aerospace Ltd. BAe and Taiwan Aerospace Corporation (TAC) each have 50 per cent holding in the new joint venture Avro International Aerospace Ltd, which incorporates BAe's Regional Jet business. Other interests include British Aerospace (Holdings) Inc (formed to group aerospace, defence and automotive interests in North America), 20 per cent partnership in Airbus Industrie, and partnership in Euromissile Dynamics Group (EMDG).

BRITISH AEROSPACE AIRBUS LTD
FILTON: Filton House, Bristol BS99 7AR
Telephone: 44 (272) 693831
Fax: 44 (272) 362828
CHESTER: Broughton, near Chester, Clwyd CH4 0DR
Telephone: 44 (244) 520444
Fax: 44 (244) 535839
CHAIRMAN: R. M. McKinlay
MANAGING DIRECTOR: C. V. Geoghegan
ENGINEERING DIRECTOR: S. Swadeling
DIRECTOR, AVIATION SERVICES: A. C. Duke
COMPANY PUBLIC AFFAIRS MANAGER: H. Berry

Twenty per cent partner in Airbus Industrie. Responsible for design, development and production of wings for the Airbus series of airliners, as well as overall design and supply of fuel systems. For most Airbus models, also undertakes design and supply of landing gear and, for A321, design and manufacture of one fuselage section.

BRITISH AEROSPACE CORPORATE JETS LTD
HATFIELD: 3 Bishop Square, St Albans Road West, Hatfield, Hertfordshire AL10 9NE
Telephone: 44 (707) 262345
Fax: 44 (707) 264057
CHAIRMAN: Robert L. Kirk
PRESIDENT AND CEO: W. W. Boisture
MANAGING DIRECTOR, OPERATIONS: Tom Nicholson
ENGINEERING DIRECTOR: R. D. Laugher
SENIOR VICE-PRESIDENT, COMMERCIAL MARKETING AND SALES: W. G. Wilson
PROJECT DIRECTOR: M. C. Steer

Responsible for the design, development, production, marketing and support of BAe 125 and BAe 146 Statesman. Sold to Raytheon Inc June 1993 for £250 million ($386 million).

BAe 125-700-II
Update of BAe 125-700 airframes by Arkansas Aerospace Inc, Little Rock, announced October 1990. Refurbished flight deck; updated avionics; new cabin interior. Details in 1991-92 *Jane's* and *Jane's Civil and Military Aircraft Upgrades*.

BAe 125 CORPORATE 800
US Air Force designation: C-29A
JASDF designation: U-125
TYPE: Twin-turbofan business transport.
PROGRAMME: First flight of prototype (G-BKTF) 26 May 1983 (max operating altitude achieved during the 3 h 8 min flight); type certificate gained 4 May 1984 and Public Transport Category C of A on 30 May 1984; FAA certification 7 June 1984; Russian certification Spring 1993.
CURRENT VERSIONS: **800A:** For North American market.
800B: General version outside North America.
P.134/P.135: See separate entry.
CUSTOMERS: By January 1993, 240 sold (840 BAe 125s of all versions ordered from 41 countries; 12 Corporate 800s delivered in 1992. First of six C-29As for USAF delivered on 24 April 1990, equipped with LTV Sierra Research Division inspection equipment for the combat flight inspection and navigation (C-FIN) mission, replacing CT-39A and C-140A calibration fleet with 1866th FCS at Scott AFB; in September 1991, control of these six aircraft transferred to FAA, Oklahoma City. BAe 125 selected under JASDF H-X programme for navaid calibration and SAR roles (see Fuji entry under Japan); first of three U-125s for former role handed over 18 December 1992; delivery of other two 1993 and 1994 respectively. JASDF has further requirement for 27 BAe 125 Corporate 800s in U-125A search and rescue role, to be purchased on fiscal basis (three ordered in FY 1992) for delivery 1995-2003; 360° scan radar and FLIR sensor to be fitted; to be capable of dropping marker flares and rescue equipment.
DESIGN FEATURES: Advanced version of BAe 125; other civil/military roles include communications, air ambulance, airways inspection and crew training; improvements compared with Series 700 include curved windscreen, sequenced nosewheel doors, extended fin leading-edge, larger ventral fuel tank, and increased wing span which reduces induced drag, improves aerodynamic efficiency and carries extra fuel; outboard 3.05 m (10 ft) of each wing redesigned. TFE731-5R-1H turbofans improve airfield/climb performance and increase maximum speed and range; redesigned interior, with increased headroom by relocating oxygen dropout units to sidewall panels, and 12.2 cm (4.8 in) extra width at shoulder level by sculpturing sidewall panels around fuselage frames; flight deck incorporates five-tube Collins EFIS-85, with centrally mounted multi-function display showing flight plans and checklists.

BAe wing sections; thickness/chord ratio 14 per cent at root, 8.35 per cent at tip; dihedral 2°; incidence 2° 5′ 42″ at root, −3° 5′ 49″ at tip; sweepback 20° at quarter-chord; small fairings on tailplane undersurface eliminate turbulence around elevator hinge cutouts.

FLYING CONTROLS: Manually operated mass balanced ailerons, elevators and rudder, each with geared tab; port aileron tab trimmed manually via screwjack. Hydraulically actuated four-position double-slotted flaps; mechanically operated hydraulic cutout prevents asymmetric flap operation; upper and lower airbrakes, with interconnected controls to prevent asymmetric operation, form part of flap shrouds and provide lift dumping for landing. Fixed incidence tailplane.
STRUCTURE: All-metal. One-piece wings, dished to pass under fuselage and attached by four vertical links, side link and drag spigot; two-spar fail-safe wings, with partial centre spar of approx two-thirds span, to form integral fuel tankage; single-piece skins on each upper and lower wing semi-spans; detachable leading-edges; fail-safe fuselage

BAe 125 Corporate 800 business transport

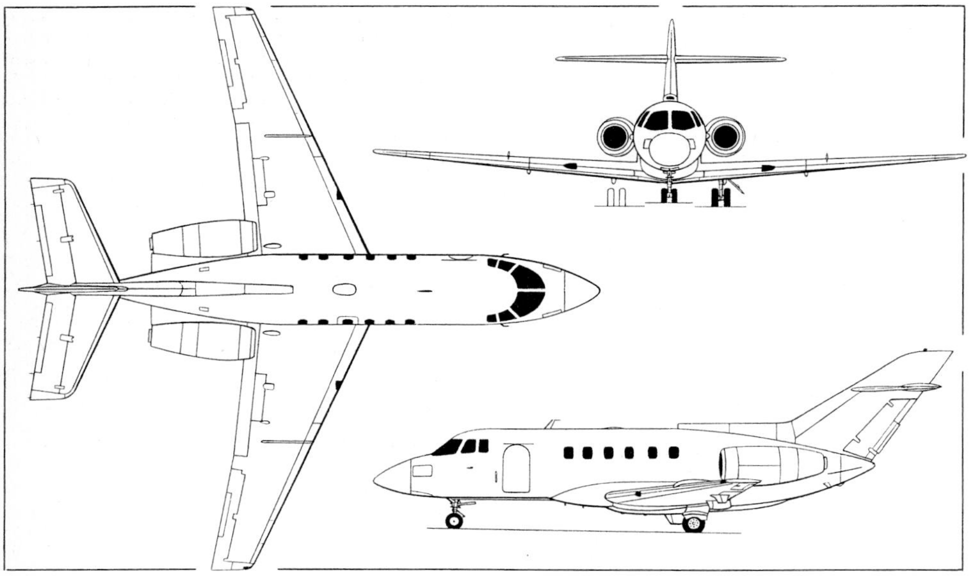

BAe 125 Corporate 800 (two Garrett TFE731-5R-1H turbofans) ***(Dennis Punnett)***

structure of mainly circular cross-section, incorporating Redux bonding.

LANDING GEAR: Retractable tricycle type, with twin wheels on each unit. Hydraulic retraction: nosewheels forward, mainwheels inward into wings. Oleo-pneumatic shock absorbers. Fully castoring nose unit, steerable ±45°. Dunlop mainwheels and 12-ply tubeless tyres, size 23 × 7-12. Dunlop nosewheels and 6-ply tubeless tyres, size 18 × 4.25-10. Dunlop triple-disc hydraulic brakes with Maxaret anti-skid units on all mainwheels. Min ground turning radius about nosewheel 9.14 m (30 ft 0 in).

POWER PLANT: Two 19.13 kN (4,300 lb st) Garrett TFE731-5R-1H turbofans, mounted on sides of rear fuselage in pods designed and manufactured by Grumman Aerospace. Optional thrust reversers developed by Dee Howard Company. Integral fuel tanks in wings, with combined capacity of 4,820 litres (1,273 US gallons; 1,060 Imp gallons). Rear underfuselage tank of 854 litres (226 US gallons; 188 Imp gallons) capacity, giving total capacity of 5,674 litres (1,499 US gallons; 1,248 Imp gallons). Single pressure refuelling point at rear of ventral tank. Overwing refuelling point near each wingtip.

ACCOMMODATION: Crew of two on flight deck, which is fully soundproofed, insulated and air-conditioned. Dual controls standard. Seat for third crew member. Standard executive layout has forward and rear baggage compartments, forward galley, wardrobe and toilet at rear. Individual recessed lights, air louvres and oxygen masks above each passenger position. Cabin styling offers a choice of interchangeable furnishing units to suit individual requirements, with up to 14 seats. The wide seats swivel through 180°, are adjustable fore, aft and sideways, and can be reclined and used as bed. Primary configuration since mid-1987 is open plan layout, with traditional club four seating moved to rear of cabin, the partial bulkhead at front removed and galley relocated behind enlarged forward baggage area. Accommodation in forward cabin comprises two individual swivelling chairs and a three-place settee. Outward opening door at front on port side, with integral airstairs. Emergency exit over wing on starboard side. Optional heated baggage pannier in place of rear underfuselage fuel tank.

SYSTEMS: AiResearch air-conditioning and pressurisation system. Max cabin differential 0.59 bar (8.55 lb/sq in). Oxygen system standard, with dropout masks for passengers. Hydraulic system, pressure 186-207 bars (2,700-3,000 lb/sq in), for operation of landing gear, mainwheel doors, flaps, spoilers, nosewheel steering, mainwheel brakes and anti-skid units. Two accumulators, pressurised by engine bleed air, one for main system pressure, other providing emergency hydraulic power for wheel brakes in case of main system failure. Independent auxiliary system for lowering landing gear and flaps in event of main system failure. DC electrical system utilises two 30V 12kW engine driven starter/generators and two 24V 23Ah nickel-cadmium batteries. A 24V 4Ah battery provides separate power for standby instruments. AC electrical system includes two 1.25kVA static inverters, providing 115V 400Hz single-phase supplies, one 250VA standby static inverter for avionics, and two engine driven 208V 7.4kVA frequency-wild alternators for windscreen anti-icing. Ground power receptacle on starboard side at rear of fuselage for 28V external DC supply. Turbomach T-62T-40C8D-1 auxiliary power unit. TKS liquid system de-icing/anti-icing on leading-edges of wings and tailplane. Engine ice protection system supplied by engine bleed air. Graviner triple FD Firewire fire warning system and two BCF engine fire extinguishers. Stall warning system indicates approach to the stall, and an identification system operates a stick pusher to initiate a nose down pitching movement if the approach to the stall exceeds a predetermined rate.

AVIONICS: Digital avionics received FAA Cat. II certification in 1987. Standard Collins avionics include dual VHF-422A com transceivers, 628T-3 HF com transceiver, dual VIR-32 VHF nav receivers with marker beacon indicator, dual ADF-60B, dual AHS-85E attitude and heading reference systems, dual DME-42 DME, dual TDR-90 ATC transponders, WXR-300 weather radar, dual ADS-86 air data systems, APS-85 autopilot, dual EFIS-86E-4 electronic flight instrument systems and ALT-55B radio altimeter, plus Baker M1045 audio system, Global GNS-1000 flight management system, Pioneer KE-8300 stereo tape and FM/AM radio.

DIMENSIONS, EXTERNAL:	
Wing span	15.66 m (51 ft 4½ in)
Wing chord (mean)	2.29 m (7 ft 6¼ in)
Wing aspect ratio	7.06
Length overall	15.60 m (51 ft 2 in)
Height overall	5.36 m (17 ft 7 in)
Fuselage: Max diameter	1.93 m (6 ft 4 in)
Tailplane span	6.10 m (20 ft 0 in)
Wheel track (c/l of shock absorbers)	2.79 m (9 ft 2 in)
Wheelbase	6.41 m (21 ft 0½ in)
Passenger door (fwd, port):	
Height	1.30 m (4 ft 3 in)
Width	0.69 m (2 ft 3 in)
Height to sill	1.07 m (3 ft 6 in)
Emergency exit (overwing, stbd):	
Height	0.91 m (3 ft 0 in)
Width	0.51 m (1 ft 8 in)
DIMENSIONS, INTERNAL:	
Cabin (excl flight deck): Length	6.50 m (21 ft 4 in)
Max width	1.83 m (6 ft 0 in)
Max height	1.75 m (5 ft 9 in)
Floor area	5.11 m² (55.0 sq ft)
Volume	17.10 m³ (604.0 cu ft)
Baggage compartments:	
forward	0.74 m³ (26.0 cu ft)
rear	0.74 m³ (26.0 cu ft)
pannier (optional)	0.79 m³ (28.0 cu ft)
AREAS:	
Wings, gross	34.75 m² (374.0 sq ft)
Ailerons (total)	2.05 m² (22.1 sq ft)
Airbrakes: upper (total)	0.74 m² (8.0 sq ft)
lower (total)	0.46 m² (5.0 sq ft)
Trailing-edge flaps (total)	4.83 m² (52.0 sq ft)
Fin (excl dorsal fin)	6.43 m² (69.2 sq ft)
Rudder	1.32 m² (14.2 sq ft)
Horizontal tail surfaces (total)	9.29 m² (100.0 sq ft)
WEIGHTS AND LOADINGS:	
Basic weight empty	6,676 kg (14,720 lb)
Typical operating weight empty	6,858 kg (15,120 lb)
Max payload	1,306 kg (2,880 lb)
Max ramp weight	12,480 kg (27,520 lb)
Max T-O weight	12,430 kg (27,400 lb)
Max zero-fuel weight	8,164 kg (18,000 lb)
Max landing weight	10,590 kg (23,350 lb)
Max wing loading	357.69 kg/m² (73.26 lb/sq ft)
Max power loading	325.1 kg/kN (3.19 lb/lb st)
PERFORMANCE:	
Max limiting Mach number	0.87
Max level speed and max cruising speed at 8,840 m (29,000 ft)	456 knots (845 km/h; 525 mph)
Econ cruising speed at 11,900-13,100 m (39,000- 43,000 ft)	400 knots (741 km/h; 461 mph)
Stalling speed in landing configuration at typical landing weight	92 knots (170 km/h; 106 mph)
Max rate of climb at S/L	945 m (3,100 ft)/min
Time to 10,670 m (35,000 ft)	19 min
Service ceiling	13,100 m (43,000 ft)
T-O balanced field length at max T-O weight	1,713 m (5,620 ft)
Landing from 15 m (50 ft) at typical landing weight (6 passengers and baggage)	1,372 m (4,500 ft)
Range:	
with max payload	2,870 nm (5,318 km; 3,305 miles)
with max fuel, NBAA VFR reserves	3,000 nm (5,560 km; 3,454 miles)

BAe P.134/P.135

TYPE: Military versions of BAe 125 Corporate 800 business jet, currently under development.

CURRENT VERSIONS: **P.134:** Signals intelligence version to collect communications intelligence and electronic intelligence for airborne analysis or datalink to ground station for near-real-time assessment, or onboard storage for post-flight analysis; estimated 6.5 h endurance at 13,100 m (43,000 ft); max T-O weight 12,450 kg (27,447 lb).

P.135: To have underbelly synthetic aperture radar, to look through smoke or weather; to be able to reconnoitre 18,500 nm (34,285 km; 21,300 mile) swath in 6.5 h mission at 13,100 m (43,000 ft).

BAe 125 CORPORATE 1000

TYPE: Twin-turbofan business transport.

PROGRAMME: Development initiated 1988 as BAe 125 Series 900, with larger cabin and increased range; substantial structural and systems changes, and modifications to meet latest FAR and JAR certification requirements; launched as BAe 1000 October 1989; first flight (G-EXLR) 16 June 1990 and of second aircraft 26 November 1990; 800 hour flight development programme with three aircraft achieved certification on 21 October 1991, followed by FAA certification on 31 October and initial customer deliveries in December.

CUSTOMERS: Launch customers were J. C. Bamford Excavators Ltd (two) and The Yeates Group (one) in UK, United Technologies in USA (three, plus one for Pratt & Whitney Canada subsidiary); Aravco Ltd in UK ordered two for business jet charter; orders totalled 27 by January 1993 (18 delivered in 1992).

DESIGN FEATURES: Intercontinental development of BAe 125 Corporate 800 to offer 27 per cent increase in range and 15 per cent improvement in field performance; 0.84 m (2 ft 9 in) fuselage stretch via plugs forward and aft of wings; additional cabin window each side; PW305 turbofans with thrust reversers standard; fuel tank in extended forward wing fairing, capacity 623 litres (164 US gallons; 137 Imp gallons); ventral fuel tank capacity increased by 150 litres (39.6 US gallons; 33 Imp gallons); new lightweight systems; restyled cabin interior with increased headroom; improved modular galley facilities; external baggage hatch; Honeywell SPZ8000 digital avionics; underwing vortillon.

BAe wing sections. Wing thickness/chord ratio 14 per cent at root, 8.35 per cent at tip; dihedral 2°; incidence 2° 5′ 42″ at root, −3° 5′ 49″ at tip; sweepback 20° at quarter-chord.

FLYING CONTROLS: Similar to BAe 125 Corporate 800. Split elevator circuit with normally locked break-out strut which splits automatically in event of control jam.

STRUCTURE: Similar to BAe 125 Corporate 800 but with upper fence on each wing replaced by underwing vortillon; intermediate fuselage frames fitted to meet latest FAR/JAR requirement; secondary rear pressure bulkhead forms forward bulkhead of rear baggage area (with baggage access door) allowing altitude certification to 13,100 m (43,000 ft) in USA; push-in/slide-up plug-type exterior baggage access door (63.5 × 35.5 cm; 25 × 14 in) on port side, interlocked to prevent engine start-up if open.

LANDING GEAR: Retractable tricycle type, with twin wheels on each unit. Hydraulic retraction, nosewheels forward, mainwheels inward into wings. Oleo-pneumatic shock absorbers. Fully castoring nose unit, steerable ±45°. Dunlop mainwheels and 12-ply tubeless tyres, size 23 × 7.12. Dunlop nosewheels and 6-ply tubeless tyres, size 18 × 4.25-10. Dunlop triple-disc hydraulic brakes with Maxaret anti-skid units on mainwheels.

POWER PLANT: Two 23.13 kN (5,200 lb st) Pratt & Whitney Canada PW305 turbofans on sides of rear fuselage, in Rohr Aerospace pods. Rohr target-type thrust reversers standard. Integral fuel tanks in wings, with combined capacity of 4,819 litres (1,273 US gallons; 1,060 Imp gallons). Rear underfuselage ventral tank of 1,023 litres (270 US gallons; 225 Imp gallons) capacity and forward underfuselage ventral tank forming wingroot fairing, capacity 623 litres (165 US gallons; 137 Imp gallons), giving total capacity of 6,464 litres (1,707 US gallons; 1,422 Imp gallons). Single pressure refuelling point at rear of aft ventral tank. Overwing refuelling point near each wingtip.

ACCOMMODATION: Standard executive layout has rear baggage compartment with forward wardrobe, forward galley comprising sink unit on port side, cooking and food stowage including fridge on starboard side, airliner style toilet at rear. Seats swivel through 360°. Standard seating for eight, with club four seating at the front of the cabin, a three-place settee on the port side, and single seat on the starboard side with an entertainment/bar unit behind, including video and compact disc player; max seating for 15. Otherwise as described for 125 Corporate 800.

BAe—AIRCRAFT: UK 379

BAe 125 Corporate 1000 twin-turbofan business aircraft

SYSTEMS: Hydraulic system also operates thrust reversers. Three accumulators, pressurised by engine bleed air, are for emergency hydraulic power for wheel brakes in case of main system failure. DC electrical system utilises two 30V 12kW (restricted to 9kW) engine driven starter/generators and two 26V 43Ah nickel-cadmium batteries. A 24V 5Ah battery provides separate power for standby instruments. The AC system is deleted on the BAe 1000, those instruments requiring AC having their own dedicated inverter. Two engine driven 208V 7.4kVA frequency-wild alternators for windscreen anti-icing can also provide DC power through a standby transformer rectifier unit in emergency. Otherwise as described for 125 Corporate 800.

AVIONICS: Standard Honeywell SPZ8000 avionics fit, includes dual EDZ-818 electronic flight instrument system; dual DFZ-800 automatic flight control system; dual SRZ-850 integrated radio/audio system; dual ADZ-810 air data system; dual Laseref III inertial reference system; DG-1086 global positioning system; dual FMZ-900 flight management system with colour display units; Primus 870 weather radar; AA-300 radio altimeter; dual HF com; Motorola N1335B Selcal, and LSZ-850 lightning detection system.

DIMENSIONS, EXTERNAL: As for BAe 125 Corporate 800 except:
Length overall	16.42 m (53 ft 10½ in)
Height overall	5.21 m (17 ft 1 in)
Wheelbase	6.91 m (22 ft 8 in)
Baggage door: Height	0.64 m (2 ft 1 in)
Width	0.36 m (1 ft 2 in)

DIMENSIONS, INTERNAL: As for BAe 125 Corporate 800 except:
Cabin (excl flight deck): Length	7.44 m (24 ft 5 in)
Floor area	5.85 m² (62.95 sq ft)
Volume	approx 19.26 m³ (680.0 cu ft)
Baggage compartments, volume:	
rear (main)	1.47 m³ (52.0 cu ft)
forward (wardrobe)	0.28 m³ (10.0 cu ft)
optional (wardrobe extension)	0.28 m³ (10.0 cu ft)

AREAS: As for BAe 125 Corporate 800

WEIGHTS AND LOADINGS:
Weight empty	7,629 kg (16,820 lb)
Typical operating weight empty	7,983 kg (17,600 lb)
Max payload	1,224 kg (2,700 lb)
Max T-O weight	14,060 kg (31,000 lb)
Max ramp weight	14,105 kg (31,100 lb)
Max zero-fuel weight	9,208 kg (20,300 lb)
Max landing weight	11,340 kg (25,000 lb)
Max wing loading	404.7 kg/m² (82.89 lb/sq ft)
Max power loading	304.1 kg/kN (2.98 lb/lb st)

PERFORMANCE (at max T-O weight, ISA, except where indicated):
Max limiting Mach number	0.87
Max level and max cruising speed at 8,840 m (29,000 ft)	468 knots (867 km/h; 539 mph)
Econ cruising speed at 11,890-13,100 m (39,000-43,000 ft)	402 knots (745 km/h; 463 mph)
Stalling speed in landing configuration at typical landing weight	89 knots (165 km/h; 103 mph)
Time to 10,670 m (35,000 ft)	22 min
Service ceiling	13,100 m (43,000 ft)
T-O balanced field length	1,830 m (6,000 ft)
Landing from 15 m (50 ft) at typical landing weight (6 passengers and baggage)	1,280 m (4,200 ft)
Range:	
with max payload	3,440 nm (6,375 km; 3,961 miles)
with max fuel, NBAA VFR reserves	3,635 nm (6,736 km; 4,185 miles)

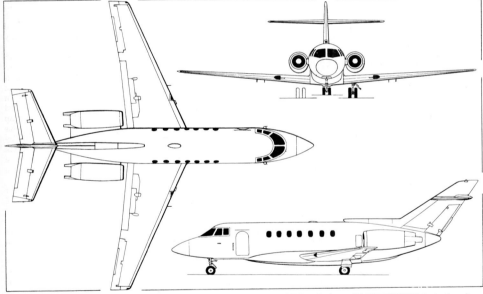

BAe 125 Corporate 1000 stretched development of BAe 125 family (two P&WC PW305 turbofans) *(Dennis Punnett)*

JETSTREAM HOLDINGS LTD
JETSTREAM AIRCRAFT LTD
Prestwick International Airport, Ayrshire KA9 2RW
Telephone: 44 (292) 79888
Fax: 44 (292) 79703
Telex: 77432
CHAIRMAN: M. J. Turner
MANAGING DIRECTOR: Allan MacDonald
PRESIDENT, MARKETING AND SALES: Stephen Smestad
DIRECTOR, PRODUCT ENGINEERING: William B. Black
DIRECTOR, TECHNICAL ENGINEERING: Gordon McConnell
PRODUCT PROMOTIONS MANAGER: Robert Gillies

Through its operating subsidiary, Jetstream Aircraft Ltd, this BAe company is responsible for design, manufacture, marketing and sales of Jetstream family of aircraft. Production and marketing of ATP being moved to Prestwick (from Woodford), following reorganisation of BAe's commercial aircraft business. Additions to range are Jetstream 61, a development of ongoing ATP, and proposed Jetstream 51 and 71 (see Addenda).

BAe JETSTREAM SUPER 31
TYPE: Commuter/executive light transport.
PROGRAMME: Jetstream 31 development began 5 December 1978; first flight of development aircraft (G-JSSD), converted from Jetstream 1, 28 March 1980; production go-ahead announced January 1981; first flight of production aircraft (G-TALL) 18 March 1982; certificated to BCAR Section D in UK 22 June 1982, SFAR 41C in USA 30 November 1982, by LBA in Germany July 1983, DoA in Australia February 1984, BCA in Sweden December 1984, RLD in Netherlands March 1985, DoT Canada 1987, Italy May 1987, Switzerland December 1987 and DGAC in France March 1993; customer deliveries began December 1982; 200th (first Super 31) handed over 3 October 1988; 300th to Pan Am Express at Farnborough 6 September 1990.

Jetstream Super 31 announced at 1987 Paris Air Show; certificated in FAA's 19-seat Commuter category (formulated under FAR Pts 23-24) 30 September 1988; also certificated by CAA 6 September 1988 under International Public Transport Category of BCAR Section D; also certificated in Canada February 1989, Australia January 1990, Switzerland April 1990 and Japan April 1991; Jetstream 31 and Super 31 fleet completed 2.83 million flying hours by 31 October 1992.

CURRENT VERSIONS: **Airliner:** 18/19 passengers. Up to 680 nm (1,260 km; 783 miles) stage length, without refuelling, with 18 passengers, baggage and full IFR reserves. Quick-change (QC) facility allows conversion from 18-seat to 12/10-seat layouts in under 2 h; optional underfuselage baggage pod, 4.62 m (15 ft 2 in) external and 3.21 m (10 ft 6½ in) internal length, provides stowage for extra six cases and additional soft luggage; penalties of pod are 3-4 knot (6-7.5 km/h; 3.5-4.5 mph) reduction in TAS at cruise and 1.5 per cent reduction in specific air range; water methanol injection system developed for hot and high conditions; Super 31 cleared for use from unsealed runways.

Corporate: Executive version for 8/10 passengers; up to 1,050 nm (1,945 km; 1,208 miles) range with nine passengers, baggage and full IFR reserves; typical interior has six reclining/swivelling chairs, three-place divan, hot/cold meal galley, cocktail cabinet, wardrobe and washroom/toilet.

Executive Shuttle: Company and business charter version, with typically 1,050 nm (1,945 km; 1,208 mile) range with 12 passengers and full IFR reserves.

Special Role: Specialist applications version, suited to military communications, casualty evacuation, multi-engine training, cargo carrying, airfield calibration, resources survey and protection. Jetstream **31EZ** exclusive economic zone patrol version has 360° scan underbelly radar, increased fuel, observation windows and searchlight. Two Jetstream 31s, specially equipped with Tornado IDS avionics, delivered to Royal Saudi Air Force in 1987 for navigator training.

CUSTOMERS: Total of 378 firm orders for Jetstream 31s (220) and Super 31s (158) placed by 31 December 1992; 375 delivered by 19 April 1993; first Super 31 order announced 13 October 1987 by Wings West of USA (15). AMR Eagle received 15 Super 31s, bringing its fleet to 75. See Current Versions for RSAF navigation trainers.

DESIGN FEATURES: Current Super 31 has improvements in performance and passenger comfort, and wider flexibility for regional and interline operations; benefits derive from

380 UK: AIRCRAFT—BAe

BAe Jetstream Super 31 light transport of AMR Eagle, with underfuselage baggage pod

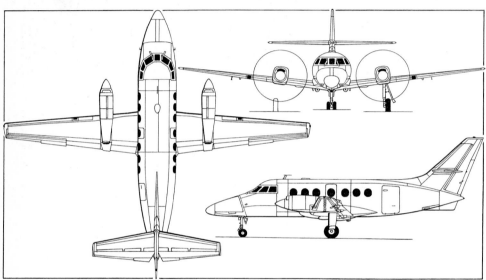

BAe Jetstream Super 31 twin-turboprop commuter/executive light transport (*Jane's/Mike Keep*)

introduction of more powerful TPE331-12 turboprops, flat rated at 760 kW (1,020 shp) at temperatures up to 25°C, giving 8 per cent improvement at up to 20°C and up to 18 per cent power improvement at high altitude; max T-O weight increased by 400 kg (882 lb); zero-fuel weight increased by 200 kg (441 lb); internal changes include sidewash lighting, recontoured furnishing panels giving 76 mm (3 in) increased width at head height, increased floor width and sunken aisle width, and reduced noise/vibration window panels.

Wing section NACA 63A418 at root, NACA 63A412 at tip; dihedral 7° from roots; incidence 2° at root, 0° at tip; sweepback 0° 34′ at quarter-chord; nose and tail sections unpressurised.

FLYING CONTROLS: Manually operated Frise ailerons, elevators and rudder, each with trim tab; hydraulically operated double-slotted flaps; no slats or leading-edge flaps; fixed incidence tailplane. AFCS optional (see Avionics).

STRUCTURE: Aluminium alloy construction; fail-safe wing structure of front, main and rear spars, with chordwise ribs; wing skins chemically etched and reinforced with bonded spanwise stringers; semi-monocoque fail-safe fuselage, with chemically milled skin panels.

LANDING GEAR: Retractable tricycle type, with nosewheel steering (±45°). Hydraulic retraction, mainwheels inward into wings, twin nosewheels forward. British Aerospace oleo-pneumatic shock absorbers in all units. Dunlop wheels and tyres: mainwheel tyres size 28 × 9.00-12, pressure 5.72 bars (83 lb/sq in); nosewheel tyres size 6.00-6, pressure 2.90 bars (42 lb/sq in). Anti-skid units. Min ground turning radius about nosewheel 6.50 m (21 ft 4 in).

POWER PLANT: Two 760 kW (1,020 shp) Garrett TPE331-12UAR turboprops, each driving a McCauley four-blade variable- and reversible-pitch fully feathering metal propeller. Fuel in integral tank in each wing, total usable capacity 1,850 litres (489 US gallons; 407 Imp gallons). Refuelling point on top of each outer wing. Water methanol injection optional.

ACCOMMODATION: Two seats side by side on flight deck, with provision for dual controls, though aircraft can be approved (subject to local regulations) for single-pilot operation. Main cabin can be furnished in commuter layout for up to 19 passengers at 76/79 cm (30/31 in) pitch, or with executive interior for 8/10 passengers, but optional layouts are available, including a QC (quick change) option enabling an operator to change from a commuter layout to 12-seat executive configuration in around 1¼ hours. Downward opening passenger door, with integral airstairs, at rear of cabin on port side. Emergency exit over wing on each side. Baggage compartment in rear of cabin, aft of main door. Entire accommodation pressurised, heated, ventilated and air-conditioned. Toilet; galley and bar optional. Jetstream completion is by Field Aircraft at East Midlands Airport, Castle Donington, UK.

SYSTEMS: Air-conditioning system with cabin pressurisation at max differential of 0.38 bar (5.5 lb/sq in), providing a 2,440 m (8,000 ft) cabin altitude at 7,620 m (25,000 ft). Single hydraulic system, pressure 138 bars (2,000 lb/sq in), with two engine driven pumps, each capable of supplying 20.7 litres (5.46 US gallons; 4.55 Imp gallons)/min. One pump is capable of supplying all hydraulic systems. Combined air/oil reservoir, pressurised to 1.24 bars (18 lb/sq in), for main and emergency supply, for actuation of flaps, landing gear, brakes and nosewheel steering. Goodrich rubber boot de-icing for wing and tail leading-edges.

AVIONICS: Standard Airliner avionics by Collins include dual VHF-22A com transceivers with CTL-22 controls; dual VIR-32 nav receivers with VOR/ILS/marker receivers and CTL-32 controls; dual RMI-36 radio magnetic indicators; ADF-60A with CTL-62 controller; DME-42 with IND-42C indicator; TDR-90 transponder with CTL-92 controller; WXR-270 colour weather radar with IND-270 display unit; dual Honeywell C-14 compass systems; dual Racal B692 communications control and public address system and six cabin loudspeakers. Optional Enhanced Specification avionics package adds Honeywell SPZ-500 automatic flight control system; dual Honeywell RD-550 horizontal situation indicators (replacing RD-450s); Collins DCP-270 information display control panel; Honeywell AL-245 altitude alert controller; dual Honeywell VG-14A vertical gyro subsystems, and toilet loudspeaker, to above. Standard flight deck equipment includes dual Honeywell GH-14 gyro horizons; dual Honeywell RD-450 horizontal situation indicators; dual Kollsman MAAS 39948 airspeed indicators, A.32069 vertical speed indicators and B.45152 encoding altimeters; dual Davtron M811B digital clocks/stopwatches; Aerospace Optics 84-207-1 flight director mode indicator, dual Aerospace Optics 99-230 marker beacon lights; SFENA H.301 standby artificial horizon, and Sangamo Weston S149-1-169 OAT gauge. Enhanced Specification adds or substitutes Honeywell AD-550C attitude director indicator, BA-141 encoding altimeter, VS-200 vertical speed indicator and PC-500 autopilot controller to above.

DIMENSIONS, EXTERNAL:
Wing span	15.85 m (52 ft 0 in)
Wing chord: at root	2.19 m (7 ft 2½ in)
at tip	0.80 m (2 ft 7¼ in)
Wing aspect ratio	9.95
Length: overall	14.37 m (47 ft 1½ in)
fuselage	13.40 m (43 ft 11½ in)
Height overall	5.38 m (17 ft 8 in)
Fuselage: Max diameter	1.98 m (6 ft 6 in)
Tailplane span	6.60 m (21 ft 8 in)
Wheel track	5.94 m (19 ft 6 in)
Wheelbase	4.60 m (15 ft 1 in)
Propeller diameter	2.69 m (8 ft 10 in)
Passenger door: Height	1.42 m (4 ft 8 in)
Width	0.86 m (2 ft 10 in)
Emergency exits:	
Starboard side: Height	0.91 m (3 ft 0 in)
Width	0.56 m (1 ft 10 in)
Port side: Height	0.71 m (2 ft 4 in)
Width	0.48 m (1 ft 7 in)

DIMENSIONS, INTERNAL:
Cabin, excl flight deck: Length	7.39 m (24 ft 3 in)
Max width	1.85 m (6 ft 1 in)
Max height	1.80 m (5 ft 11 in)
Floor area	8.35 m² (90 sq ft)
Volume (trimmed aircraft)	16.99 m³ (600 cu ft)
Baggage compartment volume:	
Airliner	2.13-2.74 m³ (75.2-96.7 cu ft)
Corporate	1.34-1.48 m³ (47.2-52.2 cu ft)
Baggage pod (optional)	1.39 m³ (49.0 cu ft)

AREAS:
Wings, gross	25.20 m² (271.3 sq ft)
Ailerons, aft of hinge line (total)	1.52 m² (16.4 sq ft)
Trailing-edge flaps (total)	3.25 m² (35.0 sq ft)
Vertical tail surfaces (total)	7.72 m² (83.1 sq ft)
Horizontal tail surfaces (total)	7.80 m² (84.0 sq ft)

WEIGHTS AND LOADINGS:
Operating weight empty	4,578 kg (10,092 lb)
Max fuel	1,372 kg (3,024 lb)
Max payload	1,805 kg (3,980 lb)
Baggage pod weight empty	59 kg (130 lb)
Max capacity of baggage pod	197 kg (435 lb)
Max T-O weight	7,350 kg (16,204 lb)
Max ramp weight	7,400 kg (16,314 lb)
Max landing weight	7,080 kg (15,609 lb)
Max zero-fuel weight	6,500 kg (14,330 lb)
Max wing loading	291.6 kg/m² (59.72 lb/sq ft)
Max power loading	4.84 kg/kW (7.94 lb/shp)

PERFORMANCE (at max T-O weight, except where stated):
Max cruising speed at 4,575 m (15,000 ft)	264 knots (489 km/h; 304 mph)
Econ cruising speed at 7,620 m (25,000 ft)	244 knots (452 km/h; 281 mph)
Stalling speed, flaps down	86 knots (159 km/h; 99 mph)
Max rate of climb at S/L	683 m (2,240 ft)/min
Rate of climb at S/L, OEI	137 m (450 ft)/min
Certificated ceiling	7,620 m (25,000 ft)
Service ceiling, OEI	3,660 m (12,000 ft)
T-O field length: BCAR Section D	1,570 m (5,151 ft)
FAR 23 (CC)	1,569 m (5,148 ft)
Landing field length, at max landing weight:	
BCAR Section D	1,220 m (4,003 ft)
FAR 23 (CC)	1,364 m (4,475 ft)
Accelerate/stop distance:	
SFAR 41C	1,362 m (4,470 ft)
Range with IFR reserves:	
19 passengers	643 nm (1,192 km; 741 miles)
18 passengers, flight attendant and galley	600 nm (1,111 km; 690 miles)

BAe JETSTREAM 41

TYPE: Commuter airliner.

PROGRAMME: Development announced 24 May 1989; full-scale cabin mockup displayed at 1989 Paris Air Show. Risk-sharing partners Field Aircraft Ltd of UK (electrical and avionics looms, interior furnishings) and Pilatus Flugzeugwerke AG of Switzerland (manufacture of tail assemblies and ailerons); Gulfstream Aerospace Technologies of Oklahoma, USA, to build 200 wing sets for 1990-2000 delivery; ML Slingsby Group to produce large composite components. Rollout 27 March 1991; first flight (G-GCJL) 25 September 1991; second aircraft (G-PJRT) flew February 1992, third (G-OXLI) 27 March 1992 and fourth (G-JMAC) 8 July 1992. Total of 1,790 hours flown during four-aircraft certification programme. JAA certification awarded 23 November 1992; first deliveries 25 November 1992 to Loganair and Manx Airlines. FAA certification (FAR Pt 25 to Amendment 71) April 1993. Production rate of 33 per year by end of 1993, increasing to 45 per year subsequently.

In April 1993, Jetstream Aircraft announced payload/range and performance increases for Jetstream 41s delivered from early 1994; max T-O and landing weights increase by 454 kg (1,000 lb), max zero-fuel weight increases by 317 kg (700 lb), and max IFR range with 29 passengers at 100 kg (220 lb) each increases to more than 730 nm (1,352 km; 840 miles). In addition the flat-rating of the engines increases to 1,230 kW (1,650 shp), and flapless take-off capability is introduced.

CURRENT VERSIONS: **Passenger:** Standard version. *Detailed description applies to this version, with weights and performances for 1993 and previously delivered aircraft.*

QC/Combi: Proposed; Combi may be offered with 18 airliner or 10 executive shuttle seats, three containers loaded through rear baggage door. Incorporation of large freight door in parallel section of rear fuselage being evaluated, sized to accommodate standard containers; QC seat-rail mounted guide rollers and ballmats for 9g restrained rigid containers.

CUSTOMERS: By 31 December 1992 firm orders placed by Manx Airlines (five), Loganair (three), Sun-Air (two), Atlantic Coast Airlines (17), total 27, of which seven delivered (to Manx and Loganair) by June 1993. Further commitments totalled 92. First two delivered to ACA also by June 1993, with balance 1993-95.

DESIGN FEATURES: Fuselage cross-section as Jetstream Super 31 but stretched 4.88 m (16 ft 0 in); forward plug 2.51 m (8 ft 3 in) with airstair door, rear plug 2.36 m (7 ft 9 in) with Type III emergency exit; large inward opening rear baggage door to 4.81 m³ (170 cu ft) hold; wing mounted below fuselage to permit clear cabin aisle; rearward extended fairing encloses 1.35 m³ (47.5 cu ft) baggage hold. Wing span increased, Super 31 wing extended inboard from nacelles; revised ailerons, improved flaps, increased in span and chord; increased fuel capacity in wing, with pressure refuelling; lift spoilers inboard of nacelles. New nacelles for TPE331-14GR/HR turboprops with increased propeller/fuselage clearance; forward retracting main landing gear with twin wheels; increased chord rudder, increased tailplane area. V windscreen; duplicated control runs; floor mounted control columns.

FLYING CONTROLS: Similar to Super 31.
STRUCTURE: Similar to Super 31.
LANDING GEAR: Dunlop wheels and tyres: twin wheels on each main leg, size 22 × 6.75, tyre pressure 8.27 bars (120 lb/sq in); twin nosewheels, size 17.5 × 6.25, tyre pressure 2.9 bars (42 lb/sq in). Dunlop three-rotor steel brakes; anti-skid standard. Steerable nosewheels (±45°). Min ground turning radius, based on nosewheel, 10.35 m (33 ft 1½ in).
POWER PLANT: Two 1,118 kW (1,500 shp) Garrett TPE331-14GR/HR flat rated turboprops, handed, each driving a McCauley five-blade constant-speed feathering metal propeller. Usable fuel capacity 3,305 litres (873 US gallons; 727 Imp gallons).
ACCOMMODATION: Flight crew of two; one attendant seated centrally at rear of cabin; and up to 29 passengers seated in ten double seats on right of cabin and nine singles on left at min seat pitch of 76.2 cm (30 in). Externally serviced toilet at rear on port side. Standard light galley at rear on port side. Passenger door with integral airstair at front of cabin on port side. Emergency exit over wing on each side and at rear of cabin on starboard side. Stowage for carry-on baggage in forward cabin on starboard side. Main baggage hold at rear of cabin, with external door; additional stowage in ventral wingroot fairing.
SYSTEMS: Normalair-Garrett air-conditioning system with cabin pressurisation at max differential of 0.39 bar (5.7 lb/sq in), providing 2,440 m (8,000 ft) cabin altitude at 7,925 m (26,000 ft) or 2,225 m (7,300 ft) at American max operating altitude of 7,620 m (25,000 ft). Hydraulic system, pressure 138 bars (2,000 lb/sq in), with variable delivery pump on each engine, one pump capable of supplying all services; system operates landing gear, nosewheel steering, flaps, spoilers and wheelbrakes. Emergency system operated by handpump under flight deck floor for flaps and landing gear. Electrical power by dual channel 28V DC system, provided by starter/generator on each engine, rated at 550A continuous, and two 27Ah batteries; 26V 400Hz AC system powered by two 1.25kVA static inverters for left and right hand instrument and navigation systems, either inverter capable of supplying 115V AC to flight data recorder. Electrical de-icing for propellers, pneumatic boot de-icing for wing, tailplane and fin leading-edges. Electric anti-icing for windscreen, pitot/static heads, airflow sensors and elevator horns. Engine intake bleed air anti-icing. Two-bottle fire extinguishing system in nacelles.
AVIONICS: Honeywell EDZ-805 four-tube EFIS and Primus II radar standard. Dual nav/com with VOR/ILS/markers; ADF; DME; dual transponders; flight director; and weather radar standard. Five-tube EFIS and autopilot optional.

DIMENSIONS, EXTERNAL:
Wing span	18.29 m (60 ft 0 in)
Wing chord: at root	2.70 m (8 ft 10½ in)
at tip	0.89 m (2 ft 11 in)
Wing aspect ratio	10.26
Length overall	19.25 m (63 ft 2 in)
Height overall	5.74 m (18 ft 10 in)
Tailplane span	6.68 m (21 ft 11 in)

BAe Jetstream 41 twin-turboprop commuter airliner flying with Loganair

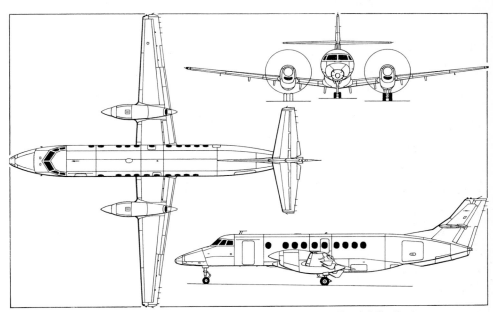

The 29-passenger BAe Jetstream 41 commuter airliner (*Jane's/Mike Keep*)

Wheel track	6.10 m (20 ft 0 in)
Wheelbase	7.32 m (24 ft 0 in)
Propeller diameter	2.90 m (9 ft 6 in)
Propeller ground clearance	0.43 m (1 ft 5 in)
Passenger door: Height	1.42 m (4 ft 8 in)
Width	0.74 m (2 ft 5 in)
Height to sill	1.24 m (4 ft 1 in)
Baggage door: Height	1.22 m (4 ft 0 in)
Width	1.35 m (4 ft 5 in)
Height to sill	1.68 m (5 ft 6 in)
Emergency exits, overwing: Height	0.99 m (3 ft 3 in)
Width	0.51 m (1 ft 8 in)
Emergency exit, rear: Height	1.22 m (4 ft 0 in)
Width	0.51 m (1 ft 8 in)

DIMENSIONS, INTERNAL:
Cabin: Length	9.55 m (31 ft 4 in)
Max width	1.85 m (6 ft 1 in)
Max height	1.78 m (5 ft 10 in)
Volume	29.45 m³ (1,040 cu ft)
Baggage compartment volume:	
rear	4.81 m³ (170 cu ft)
ventral wingroot fairing	1.35 m³ (47.5 cu ft)

AREAS:
Wings, gross	32.59 m² (350.8 sq ft)
Ailerons (total)	1.94 m² (20.90 sq ft)
Trailing-edge flaps (total)	5.28 m² (56.80 sq ft)
Spoilers (total)	0.91 m² (9.78 sq ft)
Rudder, incl tab	2.03 m² (21.87 sq ft)
Tailplane	8.58 m² (92.35 sq ft)
Elevators, incl tabs	2.43 m² (26.18 sq ft)

WEIGHTS AND LOADINGS:
Weight empty	6,350 kg (14,000 lb)
Max fuel weight	2,703 kg (5,960 lb)
Max payload	3,039 kg (6,700 lb)
Max T-O weight	10,433 kg (23,000 lb)
Max ramp weight	10,472 kg (23,110 lb)
Max zero-fuel weight	9,389 kg (20,700 lb)
Max landing weight	10,115 kg (22,300 lb)
Max wing loading	320.29 kg/m² (65.6 lb/sq ft)
Max power loading	4.67 kg/kW (7.67 lb/shp)

*PERFORMANCE (at max T-O weight, ISA, except where indicated):
Never-exceed speed (V_{NE}):	
to 5,300 m (17,400 ft)	315 knots (583 km/h; 362 mph) CAS
above 5,300 m (17,400 ft)	Mach 0.65
Max level and max cruising speed at 6,100 m (20,000 ft)	295 knots (547 km/h; 340 mph)
Econ cruising speed at 6,100 m (20,000 ft)	260 knots (482 km/h; 299 mph)
Max rate of climb at S/L	670 m (2,200 ft)/min
Service ceiling	7,925 m (26,000 ft)
Service ceiling, OEI	4,575 m (15,000 ft)
T-O run	1,523 m (4,997 ft)
Landing run	1,250 m (4,101 ft)
Range with 29 passengers, IFR reserves	681 nm (1,263 km; 785 miles)

* See also Programme paragraph for 1994 weight and performance improvements

382   UK: AIRCRAFT—BAe

BAe Jetstream ATP of Biman Bangladesh Airlines

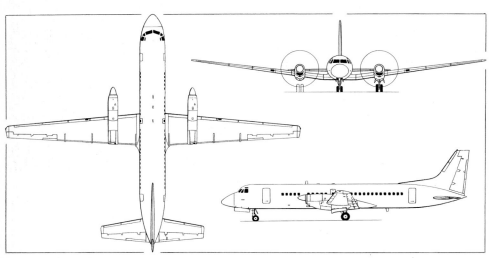

Three-view drawing of twin-turboprop BAe Jetstream ATP *(Dennis Punnett)*

BAe JETSTREAM ATP

TYPE: Twin-turboprop regional airliner.
PROGRAMME: Announced 1 March 1984 to succeed Super 748; first flight of prototype (G-MATP) 6 August 1986; first production aircraft (G-BMYM) flew 20 February 1987; JAR 25 certification March 1988; FAR 25 certification August 1988; first revenue service by British Midland Airways 9 May 1988. Fleet had completed 203,175 flying hours and 273,548 landings by 31 December 1992; ATP demonstrator departed UK April 1993 on first leg of six week sales tour of Middle East and Asia-Pacific regions.
CURRENT VERSIONS: **ATP**: Standard airliner; *description applies to this version (1994 production)*.
Jetstream 61: Improved development; described separately.
CUSTOMERS: Launch customer Airlines of Britain (British Midland, Loganair and Manx Airlines) for eight, later increased to 13. Other customers comprise Air Wisconsin (14); Biman Bangladesh (two); British Airways (14); LAR of Portugal (three); SATA of Portugal (three); Merpati Nusantara (five); THT (four). Total of 59 ordered by 31 December 1992, of which 51 delivered; remaining eight are for Air Wisconsin (four), Manx (three) and THT (one).
DESIGN FEATURES: Sill height of forward passenger door allows use of jetways at regional airports. Take-off and landing sound measurements have shown ATP to be 2-7 EPNdB quieter than any other airliner in class. Digital avionics.
FLYING CONTROLS: Ailerons (horn balanced), elevators and rudder actuated mechanically; geared tab in each aileron, trim tab in each elevator, trim tab and spring tab in rudder. Fowler trailing-edge flaps. Dual AFCS.
STRUCTURE: All-metal. Two-spar fail-safe wings; spars do not intrude into passenger cabin. Circular section fail-safe fuselage. Slightly swept vertical and non-swept horizontal tail surfaces.
LANDING GEAR: Retractable tricycle type, of Dowty design, with twin-wheel main units and twin-wheel steerable ($\pm 47°$) nose unit. All units retract forward, main units into bottom of engine nacelles. Oleo-pneumatic shock absorbers. Mainwheels fitted with 34×12-14 tubeless tyres. Nosewheels fitted with 22×6.75-10 tubeless tyres. Mainwheels have fusible plugs operating at 199°C. All wheels have 'roll on rim' capability. Dunlop carbon brakes and Maxaret anti-skid units on mainwheels. Inner and outer brakes on each leg supplied from three independent hydraulic systems via engine driven pump or standby DC pump. Min ground turning radius about nosewheel 9.75 m (32 ft 0 in).
POWER PLANT: Two 2,051 kW (2,750 shp) Pratt & Whitney Canada PW127D turboprops. Max continuous rating 1,781 kW (2,388 shp). BAe/Hamilton Standard slow-turning propellers, each having six blades of advanced aerodynamic profile and lightweight composite construction in a one-piece steel hub. Max thrust in reverse pitch 2,948 kg (6,500 lb). Fuel in two integral wing tanks, with combined usable capacity of 6,364 litres (1,681 US gallons; 1,400 Imp gallons). Single pressure refuelling point under starboard outer wing. Max fuel transfer rate 636 litres (168 US gallons; 140 Imp gallons)/min.
ACCOMMODATION: Crew of two on flight deck; two cab attendants. Main cabin has standard pressurised accomm dation for 64-68 passengers, at seat pitch of 79 cm (31 i in four-abreast layout with central aisle. Alternative la outs provide 60 to 72 seats and passenger/freight con configurations. Galley at rear of cabin on starboard si toilet forward on port side. Separate passenger doors front (with airstairs) and rear of cabin on port side. Co partment for carry-on baggage on port side of cabin, f ward of front row of seats. Two baggage/freight compa ments, one forward on starboard side and one aft of ma cabin, both with external access. Overhead lockers abc passenger seats. Forward cabin bulkhead can be moved seat rails to permit flexibility for multi-sector or mix passenger/cargo operations.
SYSTEMS: Hamilton Standard ECS with twin packs offeri sub-zero delivery temperature capability. Automatic pre urisation system, giving altitude equivalent to 2,440 (8,000 ft) at 7,620 m (25,000 ft). Pressure differential 0. bar (5.5 lb/sq in). Each engine drives an Abex varia delivery hydraulic pump providing hydraulic power a regulated pressure of 169 bars (2,450 lb/sq in) for landi gear actuation, nosewheel steering, brakes and airstai Auxiliary hydraulic power is supplied from a separate L pump and reservoir for emergency operation of the landi gear and brakes. The system also provides hydraulic pre ure for servicing when the engines are not running. Ma system has a flow rate of 41 litres (11 US gallons; 9 I gallons)/min controlled to 169 bars (2,450 lb/sq i emergency system has a flow rate of 2.25 litres (0.6 US g lon; 0.5 Imp gallon)/min controlled to 172.4 b. (2,500 lb/sq in). Air/oil reservoirs pressurised to 1.25 b (18 lb/sq in). Electrical power provided by Lucas 200V 45kVA variable frequency alternator mounted on ea engine. 28V DC subsystem from either two TRUs or t 36Ah nickel-cadmium batteries. Second subsystem p vides 1.5kVA 200/115V constant frequency power fr two static inverters. Optional Garrett GTCP36-150J AI for air-conditioning on the ground, and electrical power battery charging, engine starting assist and other ta Pneumatic boot de-icing of wings (outboard of eng nacelles), fin and tailplane leading-edges.
AVIONICS: Digital avionics system using ARINC 429 d transmission, Smiths SDS-201 four-tube EFIS, Bend King avionics. Twin VHF com, twin VHF nav, scanni DME with additional frequency under R/Nav contr ADF, ATC transponder, CVR, FDR and digital GPW Bendix/King RDS-86 colour weather radar, with check facility, can display weather on EFIS nav display. Built test and recording facility. Dual AFCS, each with Litt LTR 81-01 AHRS and Smiths digital DADS, for Cat. ILS capability. Options include second DME, seco ADF, second transponder, R/Nav, MLS and single HF.
DIMENSIONS, EXTERNAL:
Wing span   30.63 m (100 ft 6
Length overall   26.00 m (85 ft 4
Height overall   7.59 m (24 ft 11
Wheel track   8.46 m (27 ft 9

Wheelbase	9.70 m (31 ft 9¾ in)
Propeller diameter	4.19 m (13 ft 9 in)
Propeller fuselage clearance	0.80 m (2 ft 7½ in)
Passenger doors (each): Height	1.73 m (5 ft 8 in)
Width	0.71 m (2 ft 4 in)
Height to sill: fwd door	1.96 m (6 ft 5 in)
rear door	1.71 m (5 ft 7½ in)

DIMENSIONS, INTERNAL:

Cabin: Length	19.20 m (63 ft 0 in)
Max width	2.50 m (8 ft 2⅝ in)
Max height	1.93 m (6 ft 4 in)
Volume	75.1 m³ (2,652 cu ft)

Baggage/freight compartment volume (68 seats):

forward hold	3.62 m³ (128 cu ft)
rear hold	5.10 m³ (180 cu ft)
carry-on stowage	2.01 m³ (71 cu ft)
overhead lockers (max)	3.02 m³ (106.6 cu ft)

AREAS:

Wings, gross   78.32 m² (843.0 sq ft)

WEIGHTS AND LOADINGS:

Operating weight empty	14,242 kg (31,400 lb)
Max fuel	5,080 kg (11,200 lb)
Max payload	7,167 kg (15,800 lb)
Max T-O weight	23,678 kg (52,200 lb)
Max landing weight	23,133 kg (51,000 lb)
Max zero-fuel weight	21,410 kg (47,200 lb)
Max wing loading	302.3 kg/m² (61.92 lb/sq ft)
Max power loading	5.78 kg/kW (9.49 lb/shp)

PERFORMANCE:

Cruising speed for 150 nm (278 km; 173 mile) sector, ISA:
high speed at 3,960 m (13,000 ft)
    271 knots (502 km/h; 312 mph)
econ cruise at 5,485 m (18,000 ft)
    236 knots (437 km/h; 272 mph)

T-O field length:
at max T-O weight   1,345 m (4,410 ft)
for 200 nm (370 km; 230 mile) sector, 68 passengers, reserves for 100 nm (185 km; 115 mile) diversion, plus 45 min hold at 3,050 m (10,000 ft)
    1,052 m (3,450 ft)

Landing field length at max landing weight
    1,164 m (3,818 ft)

Range, with reserves for 100 nm (185 km; 115 mile) diversion and 45 min hold at 3,050 m (10,000 ft):
with max payload   619 nm (1,147 km; 713 miles)
with 68 passengers (6,477 kg; 14,280 lb)
    939 nm (1,741 km; 1,082 miles)
Ferry range   2,320 nm (4,299 km; 2,671 miles)

BAe JETSTREAM 61

TYPE: Seventy-seat development of ATP.
PROGRAMME: Announced 26 April 1993, for delivery from first quarter of 1994; further improvements/development anticipated.
DESIGN FEATURES: Improved Pratt & Whitney Canada PW127D engines, of 2,051 kW (2,750 shp) each, combine with increased design weights to offer significant payload/range improvements and better hot and high and T-O field length performances; passenger loads in hot and high conditions increased by up to 11 compared to ATP. Revised 70-seat standard layout at 79 cm (31 in) pitch; plans include 'new-look' spacious interior modelled on Jetstream 41, and package of operating cost and maintainability improvements; aircraft said to offer lowest seat/mile cost of any new turboprop.

WEIGHTS AND LOADINGS:

Max payload	7,452 kg (16,430 lb)
Max T-O weight	23,678 kg (52,200 lb)
Max zero-fuel weight	21,772 kg (48,000 lb)
Max landing weight	23,133 kg (51,000 lb)

PERFORMANCE (estimated, with 70 passengers at 100 kg; 220 lb each, IFR reserves):

Max cruising speed	270 knots (500 km/h; 311 mph)
T-O run at MTOW, ISA, S/L	1,329 m (4,360 ft)
Range: max	637 nm (1,180 km; 733 miles)
sectors, two unrefuelled (each)	
	275 nm (509 km; 316 miles)

from 1,220 m (4,000 ft) airfield, ISA + 20°C
    200 nm (370 km; 230 miles)

BRITISH AEROSPACE REGIONAL AIRCRAFT LTD

BAe ASSET MANAGEMENT ORGANISATION

Bishop Square, Hatfield, Hertfordshire AL10 9NE
Telephone: 44 (707) 262345
Fax: 44 (707) 264057

This organisation is dealing with BAe 146 fleet leasing and sales. Final BAe 146s were built for China 1993. (See also Regional Jets under Avro International in International section.)

BAe 146

RAF designation: BAe 146 CC. Mk 2
TYPE: Four-turbofan short-range transport.
PROGRAMME: Development with government support announced by former Hawker Siddeley Aviation August 1973, as HS 146; programme halted after few months because of UK economic problems, but research and design continued on limited basis; continued funding by BAe allowed manufacture of assembly jigs, system test rigs, and continuing design and wind tunnel testing; BAe Board's decision to give programme full go-ahead as private venture approved by government 10 July 1978.
BAe 146 Series 100 prototype (G-SSSH) first flew 3 September 1981; type certificate gained 4 February 1983 and Transport category CAA certificate 20 May 1983. First Series 200 (G-WISC) flew 1 August 1982; type certificate gained 4 February 1983. Aerodynamic prototype of Series 300 (G-LUXE) flew 1 May 1987 (conversion of Series 100 G-SSSH); first production Series 300 flew June 1988 and certificated 6 September. 146-QC Convertible announced August 1988 and prototype displayed at 1989 Paris Air Show.
By 1990, construction of 40 aircraft a year was target, aided by second final assembly line at BAe Woodford, near Manchester; first Woodford assembled BAe 146 (c/n 2106) flew 16 May 1988. In December 1989, Field Aircraft Ltd of East Midlands Airport appointed as approved maintenance, overhaul and completion centre. August 1990, RJ Series announced (see Avro in International section) as successor to BAe 146; first flight of 146-300 with RJ's LF 507 turbofans 2 March 1991.
CURRENT VERSIONS: **Series 100:** Designed to operate from short or semi-prepared airstrips with minimal ground facilities; normal seating for 82-94. Rollout 20 May 1981. *Detailed description applies to standard Series 100/200/300, except where indicated.*
Series 200: Seating for 82-112; fuselage lengthened 2.39 m (7 ft 10 in) by five frame pitches; increased maximum T-O weight and zero-fuel weight; underfloor cargo volume increased by 35 per cent.
Series 300: Development of Series 100; increased length by inserting 2.46 m (8 ft 1 in) forward fuselage plug and 2.34 m (7 ft 8 in) rear plug; accommodates 103 passengers five-abreast at 79 cm (31 in) pitch, with wardrobe and galley space; high-density seating for 128 passengers six-abreast at 74 cm (29 in) pitch, subject to incorporation of Type III emergency exits in centre-fuselage. Thicker centre-fuselage skin permits 44,225 kg (97,500 lb) T-O weight; first few have 43,090 kg (95,000 lb) max T-O weight, 37,648 kg (83,000 lb) max landing weight and 35,153 kg (77,500 lb) max zero-fuel weight. Small number fitted with LF 507-1H engines and with some structural strengthening for design weight increases.
146-QT Quiet Trader: Freighter versions of all Series; cabin volume allows 146-200 freighter to carry six standard 2.74 × 2.24 m (108 × 88 in) pallets, with space for extra half pallet, or up to nine standard LD3 containers; minor modifications to standard floor allow maximum payload of 11,827 kg (26,075 lb), and floor stressing permits maximum individual pallet load of 2,721 kg (6,000 lb); additional capacity of 300-QT enables payloads of up to 12,490 kg (27,535 lb). First freighter conversion (a Series 200) undertaken by Hayes International Corporation (now Pemco Aeroplex) in USA, and entered service 5 May 1987 with TNT International Aviation Services (see Customers).
146-QC Convertible: Quick-change convertible passenger/freight version of Series 200-QT and 300-QT; 200-QC Convertible has 10,039 kg (22,132 lb) gross payload in freighter configuration and range with standard tankage of 1,045 nm (1,936 km; 1,203 miles); in passenger layout, 85 (five-abreast) can be carried up to 1,289 nm (2,388 km; 1,484 miles). Comparable data for 300-QC are 10,877 kg (24,002 lb) up to 1,087 nm (2,014 km; 1,251 miles); and 96 passengers up to 1,233 nm (2,285 km; 1,420 miles).
Statesman: Executive versions of all Series; cabin area allows flexibility of interior design; staterooms, staff

BAe 146 Series 200 in the insignia of Air UK

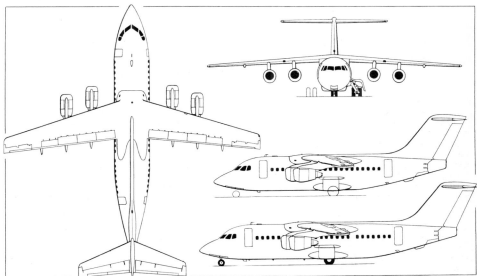

BAe 146 Series 200, with additional side view (centre) of Series 100 *(Dennis Punnett)*

384   UK: AIRCRAFT—BAe

BAe 146 Series 300 four-turbofan transport in the colours of Ansett New Zealand

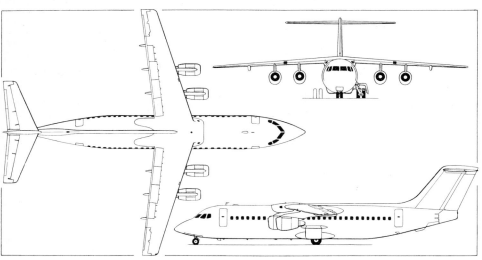

BAe 146 Series 300, the second stretched version of this regional transport *(Dennis Punnett)*

quarters, additional galley and wardrobe space can be provided. BAe Corporate Jets (which see) was responsible for this version.

CUSTOMERS: First delivery to Dan-Air (Series 100) 21 May 1983; scheduled operations began 27 May. First production Series 200 delivered to Air Wisconsin June 1983, beginning scheduled services 27 June, following FAA certification of Series 100 and 200. First production Series 300 delivered to Air Wisconsin 28 December 1988. In June 1987 TNT and BAe (see 146-QT under Current Versions) reached agreements for long-term commitment to purchase 72 146-QTs over five years; number for TNT freight network and remainder for sale or lease through Ansett Worldwide Aviation Services. Firm orders by TNT increased to 23 by March 1990, when 11 Series 200-QTs delivered to TNT in Europe and two in Australia for Ansett Air Freight. First of 10 TNT Series 300-QTs delivered to Malmo Aviation of Sweden at end of 1989. Prototype 146-QC Convertible sold to Ansett New Zealand in December 1989 after two months of service trials. First Statesman delivered in April 1986 as first of two BAe CC. Mk 2s for The Queen's Flight of RAF; third aircraft ordered 4 October 1989 and delivered in December 1990.

By December 1992 orders totalled 34 Series 100s, 91 Series 200s, 57 Series 300s, 13 Series 200-QTs, four Series 200-QCs and 10 Series 300-QTs; total 209, of which 205 had then been delivered. Production finished early 1993. Fleet then flown 1.6 million hours in 1.76 million flights with 54 operators.

Series 100s went to Abu Dhabi Royal Flight, Air Botswana, Air China, Air UK, Air Wisconsin, Australian Airlink, China Eastern, China Northwest, Dan-Air, Druk Air, Manx Airlines, PT National Air Charter, Safair, Thai Airways International and The Queen's Flight (RAF); Series 200 to Air Atlantic, Air BC, Air Nova, Air UK, Air Wisconsin, Ansett WA, Ansett New Zealand, Atlantic Airways, Aviacsa, Business Express, Crossair, Delta Air Transport, EPC Contiflug, LAN-Chile, Loganair, Meridiana, Pelita Air Services, Sal Air, USAir, WestAir and Zimbabwe; Series 200-QC to Air Jet, Ansett New Zealand, British Air Ferries and National Jet Systems; Series 200-QT to TNT Aviation and Ansett Air Freight; Series 300 to Air UK, Air Wisconsin, Ansett New Zealand, Crossair, Dan-Air, East-West Airlines, Makung Airlines, Meridiana and Thai Airways International; and Series 300-QT to TNT.

DESIGN FEATURES: Low operating noise levels; ability to operate from short or semi-prepared airstrips with minimal ground facilities. BAe high lift aerofoil section: thickne chord ratio 15.3 per cent adjacent to fuselage, 12.2 per c at tip; anhedral 3° at trailing-edge; incidence 3° 6′ at fu age side, 0° at tip; sweepback 15° at quarter-chord.

FLYING CONTROLS: Mechanically actuated balanced ailero with trim and servo tabs; manually operated balanced e ators, with trim and servo tabs; powered rudder. Sin section hydraulically actuated tabbed Fowler flaps, sp ning 78 per cent of trailing-edges, with Dowty actuat hydraulically operated roll spoiler outboard of three dumpers on each wing; no leading-edge lift devices; p airbrakes form tailcone when closed. Smiths automa flight control/flight guidance system.

STRUCTURE: All-metal; fail-safe wings with machined sk integrally machined spars and ribs; fail-safe fuselage w chemically etched skins; strengthened centre-sect developed initially for Series 300 became standard on later Series 100s; flight deck and tailcone areas free stringers; remainder of fuselage has top hat string bonded to skins above keel area; Z section stringers assembled with bonding agent and riveted to skin in k area; T tail with chemically etched skins bonded to top section stringers; fixed incidence tailplane.

LANDING GEAR: Hydraulically retractable tricycle type, Dowty design, with twin Dunlop wheels on each u Main units retract inward into fairings on fuselage sid steerable (±70°) nose unit retracts forward. Oleo-pn matic shock absorbers with wheels mounted on trail axle. Simple telescopic nosewheel strut. Mainwheel ty size 12.50-16 Type III, pressure (Series 100) 8.42 b (122 lb/sq in). Nosewheel tyres size 7.50-10 Type pressure (Series 100) 7.80 bars (113 lb/sq in). Low pre ure tyres optional. Dunlop multi-disc carbon brakes op ated by duplicated hydraulic systems. Anti-skid units both primary and secondary brake systems. Min grou turning radius about nosewheels: Series 100, 11.53 (37 ft 10 in); Series 200, 12.55 m (41 ft 2 in); Series 3 13.97 m (45 ft 10 in).

POWER PLANT: Four Textron Lycoming ALF 502R-5 tur fans, each rated at 31.0 kN (6970 lb st), installed in py mounted underwing pods. No reverse thrust. Fuel in t integral wing tanks and integral centre-section tank (latter with a vented and drained sealing diaphragm abc passenger cabin), having a combined usable capacity 11,728 litres (3,098 US gallons; 2,580 Imp gallon Optional auxiliary tanks in wingroot fairings, with co bined capacity of 1,173 litres (310 US gallons; 258 I gallons), giving total capacity of 12,901 litres (3,408 gallons; 2,838 Imp gallons). Single-point pressure refue ing, with coupling situated in starboard wing outboard outer engine.

ACCOMMODATION: Crew of two pilots on flight deck, and tw or three cabin staff. Optional observer's seat. Series 1 has accommodation in main cabin for 82 passengers with six-abreast seating at 84 cm (33 in) pitch, and up to 94 seats six-abreast at 74 cm (29 in) pitch. Series 200 has max capacity for 112 passengers with six-abreast seating at 74 cm (29 in) pitch. Series 300 has standard accommodation for 103 passengers, with five-abreast seating at 79 cm (31 in) pitch, and max seating for 128 passengers. All seating layouts have two toilets and a forward galley as standard. One outward opening passenger door forward and one aft on port side of cabin. Built-in airstairs optional. Servicing doors, one forward and one aft, on starboard side of cabin. Freight and baggage holds under cabin floor. All accommodation pressurised and air-conditioned. Windscreen electrical anti-icing and demisting standard; rain repellent system optional.

STEMS: BAe/Normalair-Garrett cabin air-conditioning and pressurisation system, using engine bleed air. Electropneumatic pressurisation control with discharge valves at front and rear of cabin. Max differential 0.47 bar (6.75 lb/sq in), giving 2,440 m (8,000 ft) equivalent altitude at 9,450 m (31,000 ft). Hydraulic system, duplicated for essential services, for landing gear, flaps, rudder, roll and lift spoilers, airbrakes, nosewheel steering, brakes and auxiliary fuel pumps; pressure 207 bars (3,000 lb/sq in). Electrical system powered by two 40kVA integrated-drive alternators to feed 115/200V 3-phase 400Hz primary systems. 28V DC power supplied by transformer-rectifier in each channel. Hydraulically powered emergency electrical power unit. Garrett GTCP 36-150 APU (36-100 in early aircraft) for ground air-conditioning and electrical power generation. High pressure gaseous oxygen system, pressure 124 bars (1,800 lb/sq in). Stall warning and identification system, comprising stick shaker (warning) and stick force (identification) elements, providing soft and hard corrective stick forces at the approach of stall conditions. Series 100: stick force soft with flaps up; Series 200: force soft above 185 knots (343 km/h; 213 mph) regardless of configuration. Hot air de-icing of wing and tailplane leading-edges.

IONICS: Smiths SEP 10 automatic flight control and flight guidance system incorporates a simplex Cat. I autopilot with a flight director display and separate attitude reference for each pilot. Addition of extra equipment and wiring permits coupled approaches to Cat. II minima. Standard ARINC interface with radio nav system allows choice of radio equipment. Basic avionics include dual VHF com, audio system, passenger address system, cockpit voice recorder, dual compass systems, dual ADIs with separate attitude reference driven by single computer, marker beacon receiver, weather radar, dual radio altimeters, ground proximity warning system, dual DME, dual ATC transponders, dual VHF nav and dual ADF. Dowty-UEL flight deck warning system. Optional avionics include third VHF com, area navigation system, Selcal, tape reproducer, and single or dual HF com. Honeywell EFIS with Smiths LED engine instrument display incorporated as standard from Summer 1990, although electromechanical instruments continue to be available as customer option.

MENSIONS, EXTERNAL:

Wing span: all versions, excl static dischargers	26.21 m (86 ft 0 in)

Note: Static discharger extends 6.3 cm (2½ in) from each wingtip

Wing aspect ratio: all versions	8.97
Wing chord: at root	2.75 m (9 ft 0 in)
at tip	0.91 m (3 ft 0 in)
Length overall: Series 100	26.20 m (85 ft 11½ in)
Series 200	28.60 m (93 ft 10 in)
Series 300	30.99 m (101 ft 8¼ in)

Note: Static dischargers on elevator extend length of all series by 18.4 cm (7¼ in)

Height overall: Series 100	8.61 m (28 ft 3 in)
Series 200	8.59 m (28 ft 2 in)
Fuselage max diameter	3.56 m (11 ft 8 in)
Tailplane span	11.09 m (36 ft 5 in)
Wheel track	4.72 m (15 ft 6 in)
Wheelbase: Series 100	10.09 m (33 ft 1½ in)
Series 200	11.20 m (36 ft 9 in)
Series 300	12.52 m (41 ft 1 in)
Passenger doors (port, fwd and rear):	
Height	1.83 m (6 ft 0 in)
Width	0.85 m (2 ft 9½ in)
Height to sill: fwd	1.88 m (6 ft 2 in)
rear	1.98 m (6 ft 6 in)
Servicing doors (stbd, fwd and rear):	
Height	1.47 m (4 ft 10 in)
Width	0.85 m (2 ft 9½ in)
Height to sill: fwd	1.88 m (6 ft 2 in)
rear	1.98 m (6 ft 6 in)
Underfloor freight hold door (stbd, fwd):	
Height	1.09 m (3 ft 7 in)
Width	1.35 m (4 ft 5 in)
Height to sill	0.78 m (2 ft 7 in)
Underfloor freight hold door (stbd, rear):	
Height	1.04 m (3 ft 5 in)
Width	0.91 m (3 ft 0 in)
Height to sill	0.90 m (2 ft 11½ in)
Freight door (Freighter versions):	
Height	1.93 m (6 ft 4 in)
Width: Series 100	2.92 m (9 ft 7 in)
Series 200	3.33 m (10 ft 11 in)
Height to sill	1.93 m (6 ft 4 in)

DIMENSIONS, INTERNAL:

Cabin (excl flight deck, incl galley and toilets):	
Length: Series 100	15.42 m (50 ft 7 in)
Series 200	17.81 m (58 ft 5 in)
Series 300	20.20 m (66 ft 3¼ in)
Max width	3.42 m (11 ft 2½ in)
Max height	2.02 m (6 ft 7½ in)
Freight cabin, Series 200-QT:	
Cargo floor: Length	16.08 m (52 ft 9 in)
Width	3.23 m (10 ft 7 in)
Volume: pallets/igloos	60.3 m³ (2,145 cu ft)
LD3 containers	42.66 m³ (1,422 cu ft)
Baggage/freight holds, underfloor:	
Series 100	13.7 m³ (479 cu ft)
Series 200	18.3 m³ (645 cu ft)
Series 300	22.99 m³ (812 cu ft)

AREAS:

Wings, gross: all versions	77.30 m² (832.0 sq ft)
Ailerons (total)	3.62 m² (39.0 sq ft)
Trailing-edge flaps (total)	19.51 m² (210.0 sq ft)
Spoilers (total)	10.03 m² (108.0 sq ft)
Fin	15.51 m² (167.0 sq ft)
Rudder	5.30 m² (57.0 sq ft)
Tailplane	15.61 m² (168.0 sq ft)
Elevators, incl tabs	10.03 m² (108.0 sq ft)

WEIGHTS AND LOADINGS (see also under Current Versions):

Operating weight empty:	
Series 100	23,336 kg (51,447 lb)
Series 200	23,897 kg (52,684 lb)
Series 200-QT	22,545 kg (49,704 lb)
Series 300	24,835 kg (54,752 lb)
Series 300-QT	23,189 kg (51,125 lb)
Max payload: Series 100	7,735 kg (17,053 lb)
Series 200	10,122 kg (22,316 lb)
Series 200-QT	11,474 kg (25,296 lb)
Series 300	10,771 kg (23,748 lb)
Series 300-QT	12,644 kg (27,875 lb)
Max fuel weight:	
All series: standard	9,362 kg (20,640 lb)
optional	10,298 kg (22,704 lb)
Max T-O weight: Series 100	38,102 kg (84,000 lb)
Series 200	42,184 kg (93,000 lb)
Series 300	44,225 kg (97,500 lb)
Max ramp weight: Series 100	38,329 kg (84,500 lb)
Series 200	42,410 kg (93,500 lb)
Series 300	44,452 kg (98,000 lb)
Max zero-fuel weight: Series 100	31,071 kg (68,500 lb)
Series 200	34,019 kg (75,000 lb)
Series 300	35,607 kg (78,500 lb)
Max landing weight: Series 100	35,153 kg (77,500 lb)
Series 200	36,741 kg (81,000 lb)
Series 300	38,328 kg (84,500 lb)
Max wing loading:	
Series 100	493.0 kg/m² (101.0 lb/sq ft)
Series 200	545.7 kg/m² (111.8 lb/sq ft)
Series 300	572.2 kg/m² (117.2 lb/sq ft)
Max power loading:	
Series 100, standard	307.3 kg/kN (3.01 lb/lb st)
Series 200	340.2 kg/kN (3.34 lb/lb st)
Series 300	358.5 kg/kN (3.52 lb/lb st)

PERFORMANCE (at max standard T-O weight, except where indicated):

Max operating Mach No. (MMO):	
all versions	0.73
Max operating speed (VMO):	
Series 100	300 knots (555 km/h; 345 mph) CAS
Series 200	295 knots (546 km/h; 339 mph) CAS
Series 300	305 knots (565 km/h; 351 mph) IAS
Cruising speed at 8,840 m (29,000 ft) for 300 nm (556 km; 345 mile) sector:	
Series 100/200:	
high-speed	432 knots (801 km/h; 498 mph)
long-range	361 knots (669 km/h; 416 mph)
Series 300:	
high-speed	432 knots (801 km/h; 498 mph)
long-range	377 knots (699 km/h; 434 mph)
Stalling speed, 30° flap:	
Series 100	97 knots (180 km/h; 112 mph) EAS
Series 200, 300	102 knots (189 km/h; 118 mph) EAS
Stalling speed, 33° flap, at max landing weight:	
Series 100	89 knots (165 km/h; 103 mph) EAS
Series 200, 300	92 knots (170 km/h; 106 mph) EAS
T-O to 10.7 m (35 ft), S/L, ISA:	
Series 100	1,219 m (4,000 ft)
Series 200, 300	1,509 m (4,950 ft)
FAR landing distance from 15 m (50 ft), S/L, ISA, at max landing weight: Series 100	1,067 m (3,500 ft)
Series 200	1,103 m (3,620 ft)
Series 300	1,228 m (4,030 ft)
Range with standard fuel:	
Series 100	1,620 nm (3,002 km; 1,865 miles)
Series 200	1,570 nm (2,909 km; 1,808 miles)
Series 300	1,520 nm (2,817 km; 1,750 miles)
Range with max payload:	
Series 100	880 nm (1,631 km; 1,013 miles)
Series 200	1,130 nm (2,094 km; 1,301 miles)
Series 200-QT	1,150 nm (2,131 km; 1,324 miles)
Series 300	1,040 nm (1,927 km; 1,197 miles)

OPERATIONAL NOISE LEVELS (FAR Pt 36-12, certificated):

T-O: Series 100	81.8 EPNdB
Series 200	85.2 EPNdB
Series 300	86.5 EPNdB
Approach: Series 100	95.6 EPNdB
Series 200	95.8 EPNdB
Series 300	95.6 EPNdB
Sideline: Series 100	87.7 EPNdB
Series 200	87.3 EPNdB
Series 300	86.7 EPNdB

RITISH AEROSPACE DEFENCE LTD

BAe Defence formed 1 January 1992 with four divisions, lilitary Aircraft and Dynamics (see below), Systems and rvices, and Royal Ordnance.

ilitary Aircraft Division

ARTON: Warton Aerodrome, Preston, Lancashire PR4 1AX
lephone: 44 (772) 633333
ax: 44 (772) 634724
MLESBURY: Samlesbury, Balderstone, Lancashire BB2 7LF
lephone: 44 (25) 481 2371
ax: 44 (25) 481 3623
JNSFOLD: Dunsfold Aerodrome, Godalming, Surrey GU8 4BS
lephone: 44 (483) 272121
ax: 44 (483) 200341
OUGH: Brough, North Humberside HU15 1EQ
lephone: 44 (482) 667121
ax: 44 (482) 666625
AIRMAN: J. P. Weston
ANAGING DIRECTOR: A. H. Baxter
CHNICAL DIRECTOR: D. Gardner
ECUTIVE VICE-PRESIDENT, MARKETING AND SALES: J. M. Wooding
RECTOR OF PUBLIC AFFAIRS, DEFENCE COMPANY: H. Colver
BLIC RELATIONS MANAGER, WARTON: D. Kamiya
BLIC RELATIONS MANAGER, DUNSFOLD: J. S. Godden
BLIC RELATIONS MANAGER, BROUGH: N. Dean

BAe Military Aircraft headquarters transferred to Warton 1989; main activities include development of Eurofighter 2000 (EFA) with Germany, Italy and Spain; design, development, production and support of Panavia Tornado, with Deutsche Aerospace and Alenia; design, development, production and support of V/STOL Harrier, Sea Harrier and, with McDonnell Douglas, AV-8B/GR. Mk 5/7 Harrier II; design, development, production and support of Hawk and, with McDonnell Douglas, T-45A Goshawk. Also provides product support for earlier aircraft still in use and contributes to BAe 146/Avro International RJ and Airbus programmes; offers defence support services and overseas and specialist training facilities, including technical and management courses.

Dynamics Division

STEVENAGE: Six Hills Way, Stevenage, Hertfordshire SG1 2DA
Telephone: 44 (438) 312422
Fax: 44 (438) 753377
LOSTOCK: Lostock Lane, Bolton, Lancashire BL6 4BR
Telephone: 44 (204) 696555
Fax: 44 (204) 693908
CHAIRMAN: J. P. Weston
MANAGING DIRECTOR: D. A. Laybourn
ENGINEERING DIRECTOR: N. Wallwork
HEAD OF PUBLIC AFFAIRS: Charles Carr
PUBLIC RELATIONS MANAGER: D. Dodds

Derived from British Aerospace (Dynamics) Ltd, which was formed on 1 January 1989, incorporating former Air Weapons Division (Hatfield and Lostock), Army Weapons Division (Stevenage), and Naval and Electronic Systems Division (Bristol, Bracknell, Plymouth and Weymouth). Details of missiles and other weapon systems can be found in appropriate *Jane's* publications.

BAe HAWK (TWO-SEAT VERSIONS)

RAF designations: Hawk T. Mks 1 and 1A
US Navy designation: T-45A Goshawk

TYPE: Two-seat basic and advanced jet trainer, with air defence and ground attack roles.

PROGRAMME: Early history of HS P1182 Hawk in 1989-90 *Jane's*; first generation Hawk remains in production and marketed with advanced 100 Series and single-seat 200 Series (detailed separately) to meet customers' requirements; Hawk design leadership transferred from Kingston to Brough 1988, and final assembly and flight test from Dunsfold to Warton 1989; Hawk 50 Series main exports made December 1980 to October 1985; Hawk 100 enhanced ground attack export model announced mid-1982; first flight of 100 Series aerodynamic prototype (G-HAWK/ZA101 converted as Mk 100 demonstrator) 21 October 1987; trials of wingtip Sidewinder rails started at Warton in April 1990. Warton assembly line officially opened 24 October 1991. RAF Hawk re-wing programme began 1989; initial 85 aircraft being re-worked by BAe;

BAe Hawk T. Mk 1 of No. 74 (Reserve) Squadron in experimental black high-conspicuity colour scheme *(Paul Jackson)*

BAe Hawk T. Mk 1 target facilities aircraft of No. 100 Squadron, RAF *(Paul Jackson)*

second batch of 59 for delivery over three-year period beginning September 1994.

CURRENT VERSIONS: **Hawk T. Mk 1:** Basic two-seater for RAF flying and weapon training; 23.13 kN (5,200 lb st) Adour 151-01 (-02 in Red Arrows aircraft) non-afterburning turbofan; two dry underwing hardpoints; underbelly 30 mm gun pack; three-position flaps; simple weapon sight in some aircraft of Nos. 4 and 7 FTSs; unarmed versions at Central Flying School. Following basic Tucano stage, future RAF fast-jet pilots undertake 100 hours of advanced flying, weapons and tactical training with either No. 4 or No. 7 FTS; Hawks introduced to navigator training syllabus at No. 6 FTS, Finningley; first delivery 10 September 1992. No. 100 Squadron received 15 Mks 1/1A from September 1991, replacing Canberras in target-towing role.

Hawk T. Mk 1A: Contract January 1983 to wire 89 Hawks (including Red Arrows) for AIM-9L Sidewinder on each inboard wing pylon and optional activation of previously unused outer wing hardpoints; last conversion redelivered 30 May 1986; 72 NATO-declared, for point defence and participation in RAF's Mixed Fighter Force, to accompany radar equipped Tornado ADVs on medium-range air defence sorties.

Hawk 50 Series: Initial export version with 23.13 kN (5,200 lb st) Adour 851 turbofan; max operating weight increased by 30 per cent, disposable load by 70 per cent, range by 30 per cent; revised tailcone shape to improve directional stability at high speed; larger nose equipment bay; four wing pylons, all configured for single or twin store carriage; each pylon cleared for 515 kg (1,135 lb) load; wet inboard pylons for 455 litre (122 US gallon; 100 Imp gallon) fuel tanks; improved cockpit, with angle of attack indication, fully aerobatic twin gyro AHRS and new weapon control panel; optional brake-chute; suitable for day VMC ground attack and armed reconnaissance with camera/sensor pod.

T-45A Goshawk: US Navy version (see McDonnell Douglas/BAe in International section).

Hawk 60 Series: Development of 50 Series with 25.35 kN (5,700 lb st) Adour 861 turbofan; leading-edge devices and four-position flaps to improve lift capability; low-friction nose leg, strengthened wheels and tyres, and adaptive anti-skid system; 591 or 864 litre (156 or 228 US gallon; 130 or 190 Imp gallon) drop tanks; provision for Sidewinder or Magic AAMs; max operating weight increased by further 17 per cent over 50 Series, disposable load by 33 per cent and range by 30 per cent; improved field performance, acceleration, rate of climb and turn rate. Recent orders comprise seven (designated Mk 51A) for Finland, 20 Mk 67s ('long-nosed' version with nosewheel steering) ordered by South Korea and five improved Mk 60As for Zimbabwe. Abu Dhabi upgrading 15 surviving Mk 63s to Mk 63A from 1991, incorporating Adour Mk 871 and new combat wing (four pylons and wingtip AAM rails); first two rebuilds at Brough; remainder at Al Dhafra. *Description applies to Hawk 60 Series, except where otherwise specified.*

Hawk 100 Series: Enhanced ground attack development of 60 Series, announced mid-1982, to exploit Hawk's stores carrying capability; two-seater, with perhaps pilot only on combat missions; 26.0 kN (5,845 lb st) Adour Mk 871 turbofan; new combat wing incorporating fixed leading-edge droop for increased lift and manoeuvrability from Mach 0.3 to 0.7; seven stores pylons; full-width flap vanes; manually selected combat flaps; detail changes to wing dressing; structural provision for wingtip missile pylons; MIL-STD-1553B databus; advanced Smiths Industries HUD/WAC and new air data sensor package with optional laser ranging and FLIR in extended nose; improved weapons management system allowing pre-selection in flight and display of weapon status; manual or automatic weapon release; passive radar warning; HOTAS controls; full colour multi-purpose CRT display in each cockpit; provision for ECM pod. Demonstrator ZA101 (see Programme). Production prototype Mk 102D (ZJ100) flown 29 February 1992. Orders from Abu Dhabi (placed 1989), Indonesia (signed June 1993), Malaysia (signed 10 December 19 and Oman (signed 30 July 1990).

Hawk 200 Series: Single-seat multi-role vers (described separately).

CUSTOMERS: See table. Additional prospects under negot tion include: Abu Dhabi, small quantity of Mk 63 required; Australia, MoU with ASTA March 1991 licenced production if Hawk 100/200 chosen for 30-50 a craft RAAF trainer requirement; Brunei, 16 Hawk 10 India, provisional selection of Hawk 60, July 1992, co prising 94 for IAF and 11 for Navy; Indonesia, coll orative production agreement with IPTN June 1991, anticipation of orders eventually totalling 144 aircraft; P lippines, commitment announced August 1991 for Hawks; and Saudi Arabia, requirement for 60 Hawk 20

DESIGN FEATURES: Fully aerobatic two-seat advanced trainer, adaptable for ground attack and air defence; des capable of other optional roles, with wing improveme on developed Series to enhance combat efficiency; sin non-afterburning engine; elevated rear cockpit to enha forward view; underwing hardpoints; wingtip AAM ra

Wing thickness/chord ratio 10.9 per cent at root, 9 cent at tip; dihedral 2°; sweepback 26° on leading-ed 21° 30′ at quarter-chord. Anhedral tailplane.

FLYING CONTROLS: Ailerons and one-piece all-moving ta plane actuated hydraulically by tandem actuators; rud mechanically operated, with electrically actuated trim t Hydraulically actuated double-slotted flaps, outbo 300 mm (12 in) of flap vanes normally deleted; small fe on each wing leading-edge; 100 and 200 Series use spec combat wing with full width flap vanes (refer Hawk 2 entry); large airbrake under rear fuselage, aft of wings; t small ventral fins; smurfs (refer Hawk 200 entry) on Series. Hydraulic yaw damper on 100 Series rudder.

STRUCTURE: Aluminium alloy; one-piece wing, w machined torsion box of two main spars, auxiliary sp ribs and skins with integral stringers; most of box for integral fuel tank; honeycomb-filled ailerons; composi wing fences; frames and stringers fuselage; swept tail s faces. Wing attached to fuselage by six bolts.

LANDING GEAR: Wide track hydraulically retractable tricy type, with single wheel on each unit. AP Precisi Hydraulics oleos and jacks. Main units retract inward i wing, ahead of front spar; castoring (optionally pow steered) nosewheel retracts forward. Dunlop mainwhee brakes and tyres size 6.50-10, pressure 9.86 bars (143 lb in). Nosewheel and tyre size 4.4-16, pressure 8.27 b (120 lb/sq in). Tail bumper fairing under rear fusela Anti-skid wheel brakes. Tail braking parachute, diame 2.64 m (8 ft 8 in), on Mks 52/53 and all 60 Series aircra

POWER PLANT: One Rolls-Royce Turbomeca Adour non-af burning turbofan, as described under Current Versio Adour Mk 861A for Switzerland assembled locally by S zer Bros. Air intake on each side of fuselage, forward wing leading-edge. Engine starting by Microturbo integ gas turbine starter. Fuel in one fuselage bag tank of 8 litres (220 US gallons; 183 Imp gallons) capacity and in gral wing tank of 823 litres (217 US gallons; 181 Imp g lons); total fuel capacity 1,655 litres (437 US gallons; 3 Imp gallons). Pressure refuelling point near front of p engine air intake trunk; gravity point on top of fusela Provision for carrying one 455, 591 or 864 litre (120, or 228 US gallon; 100, 130 or 190 Imp gallon) drop tank each inboard underwing pylon, according to Series.

ACCOMMODATION: Crew of two in tandem under one-pie fully transparent acrylic canopy, opening sideways to st board. Fixed front windscreen able to withstand a 0.9 (1.98 lb) bird at 454 knots (841 km/h; 523 mph). Improv front windscreen fitted retrospectively to RAF Hawks, to withstand a 1 kg (2.2 lb) bird at 528 knots (978 km 607 mph). Separate internal screen in front of rear cock Rear seat elevated. Martin-Baker Mk 10LH zero/z rocket assisted ejection seats, with MDC (miniature det ating cord) system to break canopy before seats eje MDC can also be operated from outside the cockpit ground rescue. Dual controls standard. Entire accomm dation pressurised, heated and air-conditioned.

SYSTEMS: BAe cockpit air-conditioning and pressurisat systems, using engine bleed air. Two hydraulic syster flow rate: System 1, 36.4 litres (9.6 US gallons; 8 Imp g lons)/min; System 2, 22.7 litres (6 US gallons; 5 Imp g lons)/min. Systems pressure 207 bars (3,000 lb/sq in). S tem 1 for actuation of control packs, flaps, airbrake, landi gear and anti-skid wheel brakes. Compressed nitrog accumulators provide emergency power for flaps and la ing gear at pressure of 2.75 to 5.5 bars (40 to 80 lb/sq i System 2 dedicated to powering flying controls. Hydrau accumulator for emergency operation of wheel brak Pop-up Dowty ram air turbine in upper rear fuselage p vides emergency hydraulic power for flying controls event of engine or No. 2 pump failure. No pneumatic s tem. DC electrical power from single 124kW brushle generator, with two static inverters to provide AC pow and two batteries for standby power. Gaseous oxygen s tem for crew.

AVIONICS: The RAF standard of flight instruments inclu GEC-Marconi gyros and inverter, two Honeywell RAI-in remote attitude indicators and magnetic detector u and Smiths-Newmark compass system. Radio and na gation equipment includes Sylvania UHF and VHF, C sor CAT.7,000 Tacan, Cossor ILS with CILS.75/76 loc

BAe's original two-seat Hawk demonstrator, modified to the aerodynamic shape of the Hawk 100 Series and fitted with wingtip Sidewinders

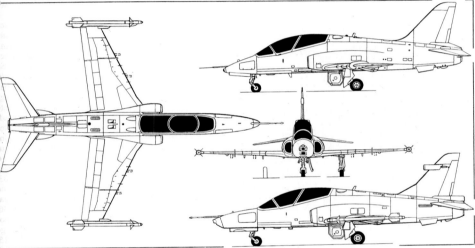

British Aerospace Hawk 100 with wingtip Sidewinders and additional side view (top) of Hawk 60 Series
(Dennis Punnett)

British Aerospace Hawk 100 Series demonstrator ZA101 (above) and pre-production Hawk 200

iser/glideslope receiver and marker receiver, and IFF/SSR (Cossor 2720 Mk 10A IFF in aircraft for Finland). Srs 60 avionics include Smiths-Newmark 6000-05 AHRS, Magnavox UHF, Collins V/UHF, Cossor IFF, Collins VOR/ILS and ADF, Smiths radar altimeter and Collins Tacan. Hawk 100 standard equipment includes BAe LINS300 IN, Smiths Industries 1500 Series HUD, HUD/WAC, and 3000 Series colour MPD, GEC data transfer system and Vinten colour video recording system; Collins AN/ARC-182 V/UHF, Magnavox AN/ARC-164 UHF, Cossor 4720 IFF, Collins AN/ARC-118 Tacan, Collins VIR-31A VOR/ILS and Smiths Industries 0103-KTX-1 radar altimeter, all integrated via a dual redundant MIL-STD-1553B databus. 100 Series attack sensors comprise RWR and optional FLIR and GEC-Marconi Type 105H laser rangefinder; chaff/flare dispenser at base of fin. GEC-Marconi F.195 weapon sight and camera recorder in each cockpit of about 90 RAF aircraft; GEC-Marconi ISIS 195 sight and Vinten video camera and recorder in all 50 and 60 Series aircraft. (Saab RGS2 sights in Finland Mk 51.)

ARMAMENT: Underfuselage centreline mounted 30 mm Aden Mk 4 cannon with 120 rounds (VKT 12.7 mm machine gun beneath Finnish aircraft), and two or four hardpoints underwing, according to Series. Provision for pylon in place of the ventral gun pack. In RAF training roles, normal max external load is about 680 kg (1,500 lb), but the uprated Hawk 60 and 100 Series are cleared for an external load of 3,000 kg (6,614 lb), or 500 kg (1,102 lb) at 8g. Typical weapon loadings on 60 Series include 30 mm or 12.7 mm centreline gun pod and four packs each containing eighteen 68 mm rockets; centreline reconnaissance pod and four packs each containing twelve 81 mm rockets; seven 1,000 lb free-fall or retarded bombs; four launchers each containing four 100 mm rockets; nine 250 lb or 250 kg bombs; thirty-six 80 lb runway denial or tactical attack bombs; five 600 lb cluster bombs; four Sidewinder/Magic air-to-air missiles; four CBLS 100/200 carriers each containing four practice bombs and four rockets; or two 592 litre (156 US gallon; 130 Imp gallon) drop tanks and two Maverick air-to-surface missiles. Vinten reconnaissance pod available for centre pylon. Similar options on 100 Series, plus wingtip air-to-air missiles.

DIMENSIONS, EXTERNAL:
Wing span: normal 9.39 m (30 ft 9¾ in)
  incl tip Sidewinders 9.94 m (32 ft 7⅜ in)
Wing chord: at root 2.65 m (8 ft 8¼ in)
  at tip 0.90 m (2 ft 11½ in)
Wing aspect ratio 5.28
Length overall: excl probe (incl tailplane):
  Mk 1, 50 and 60 Series 11.17 m (36 ft 7¾ in)
  100 Series 11.68 m (38 ft 4 in)
  incl probe: Mk 1, 50 and 60 Series
    11.86 m (38 ft 11 in)
  100 Series 12.42 m (40 ft 9 in)
Height overall:
  Mk 1, 50 and 60 Series 3.99 m (13 ft 1¼ in)
  100 Series 4.16 m (13 ft 8 in)
Tailplane span 4.39 m (14 ft 4¾ in)
Wheel track 3.47 m (11 ft 5 in)
Wheelbase 4.50 m (14 ft 9 in)

AREAS:
Wings, gross 16.69 m² (179.6 sq ft)
Ailerons (total):
  Mk 1, 50 and 60 Series 1.05 m² (11.30 sq ft)
  100 Series 0.97 m² (10.44 sq ft)
Trailing-edge flaps (total) 2.50 m² (26.91 sq ft)
Airbrake 0.53 m² (5.70 sq ft)
Fin: Mk 1, 50 and 60 Series 2.51 m² (27.02 sq ft)
  100 Series 2.61 m² (28.10 sq ft)
Rudder, incl tab 0.58 m² (6.24 sq ft)
Tailplane 4.33 m² (46.61 sq ft)

WEIGHTS AND LOADINGS:
Weight empty: 60 Series 4,012 kg (8,845 lb)
  100 Series 4,400 kg (9,700 lb)
Max weapon load (60, 100 Series) 3,000 kg (6,614 lb)
Max fuel weight:
  internal (usable) 1,304 kg (2,875 lb)
  external (usable) 932 kg (2,055 lb)
Max T-O weight: T. Mk 1 5,700 kg (12,566 lb)
  50 Series 7,350 kg (16,200 lb)
  60, 100 Series 9,100 kg (20,061 lb)
Max landing weight: T. Mk 1 4,649 kg (10,250 lb)
  60 Series 7,650 kg (16,865 lb)
Max wing loading:
  T. Mk 1 341.5 kg/m² (69.97 lb/sq ft)
  50 Series 440.4 kg/m² (90.2 lb/sq ft)
  60, 100 Series 545.2 kg/m² (111.7 lb/sq ft)
Max power loading:
  T. Mk 1 246.5 kg/kN (2.42 lb/lb st)
  50 Series 318.18 kg/kN (3.12 lb/lb st)
  60 Series 359.0 kg/kN (3.52 lb/lb st)
  100 Series 350.0 kg/kN (3.43 lb/lb st)

PERFORMANCE:
Never-exceed speed (V_{NE}), clean: at S/L
  Mach 0.87 (575 knots; 1,065 km/h; 661 mph EAS)
  at and above 5,180 m (17,000 ft)
  Mach 1.2 (575 knots; 1,065 km/h; 661 mph EAS)
Max level speed at S/L:
  50 Series 535 knots (990 km/h; 615 mph)
  60, 100 Series 549 knots (1,017 km/h; 632 mph)
Max level flight Mach number 0.88

BAe HAWK CUSTOMERS

Customer	Mark	Total	Deliveries	Squadrons
Abu Dhabi	63	16[8]	Oct 1984 - May 1985	FTS
	102[1 6 10]	18*	Apr 1993 -	On order
Dubai	61	8	Mar 1983 - Sep 1983	Ftr Sqdn
	61	1	Jun 1988	Ftr Sqdn
Finland	51	50[2]	Dec 1980 - Oct 1985	11, 21, 31, Koul LLv, TLLv, Koelentue
	51A	7*	1993	On order
Indonesia	53	20	Sep 1980 - Mar 1984	103
	100	12*		
	200	12*		
Kenya	52	12	Apr 1980 - Feb 1982	
Korea, South	67	20*	Sep 1992 -	
Kuwait	64	12	Nov 1985 - 1987	
Malaysia	108	10*	Jan 1994 -	On order
	208[6 9]	18*	Jul 1994 - Mar 1995	On order
Oman	103[6 10]	4*		On order
	203[9]	12*		On order
Saudi Arabia	65	30	Aug 1987 - Oct 1988	21, 37
Switzerland	66	20[3]	Nov 1989 - Nov 1991	55, 255
UK	T. Mk 1	176[7]	Nov 1976 - Feb 1982	4 FTS (74, 234 Sqdns), 7 FTS (19, 92 Sqdns), Red Arrows, 100 Sqdn, 6 FTS Requirement
USA	T-45A	270[4]		
Zimbabwe	60	8	Jul 1982 - Oct 1982	2
	60A	5*	Jun 1992 - Sep 1992	2
Demonstrator	60/102D	2[5]		
	200/200/200RDA	3[5]		
TOTAL		**746**		

NOTES
* Built at Brough, Hamble and Samlesbury and assembled at Warton; unless stated otherwise, remainder built at Kingston, assembled at Dunsfold (290) and Bitteswell (1)
[1] Laser nose
[2] 46 assembled by Valmet in Finland
[3] 19 assembled by F + W in Switzerland
[4] Production by McDonnell Douglas in USA
[5] One assembled at Warton
[6] Wingtip Sidewinders
[7] 89 converted to T. Mk 1A
[8] 15 converted to 63A
[9] Fixed refuelling probe
[10] Radar warning receiver

could be beyond target's radar envelope). Above configuration awaits flight clearance.

Two standards of equipment envisaged, depending on customer's mission requirements:

Day operation: Fit comprises gyro stabilised attack sight and AHRS, with radio aid navigation. Capabilities may be extended by adding INS, HUD/WAC; other options are HOTAS controls, laser rangefinder, IFF and RWR.

All-weather operation: Westinghouse AN/APG-66 advanced multi-mode radar for all-weather target acquisition and navigational fixing capabilities (Sea Eagle, Sky Flash and other weapons employable, subject to flight clearance.

CUSTOMERS: Oman (12 Mk 203 ordered July 1990), Indonesia (12 ordered June 1993), Malaysia (18 **Mk 208** ordered December 1990, for delivery from July 1994. Saudi Arabia signed MoU covering second batch of some 60 Hawks, substantial proportion **Mk 205** with APG-66 radar. All customers also ordered two-seat Hawks.

DESIGN FEATURES: Except for taller fin, fixed wing leading edge droop to enhance lift and manoeuvrability at Mach 0.3 to 0.7, manually selected combat flaps (less than ¼-flap setting) available below 350 knots (649 km/h; 403 mph) IAS to allow sustained 5g+ at 300 knots (556 km/h; 345 mph) at sea level, full-width flap vanes reinstated and detail modifications to wing dressing, Hawk 200 virtually identical to current production Hawk two-seater aft of cockpit, giving 80 per cent airframe commonality. Intended to take advantage of new miniaturised, low-cost avionics and intelligent weapons; Hawk 100 type avionics fit allows radar, but proposed alternative FLIR/laser rangefinder nose no longer on offer; intended integral cannon also deleted; all four underwing pylons capable of 907 kg (2,000 lb) load, within max 3,493 kg (7,700 lb) external load; optional wingtip rails make possible four Sidewinders or similar AAMs (inboard pylons not cleared for these missiles).

FLYING CONTROLS: See Hawk two-seater; smurfs (strake ahead of each half of tailplane to restore control authority at high angles of attack).

LANDING GEAR: Mainwheel tyres size 559 × 165-279, pressure 16.2 bars (235 lb/sq in). Nosewheel tyre size 457 × 146 203, pressure 7.24 bars (105 lb/sq in). Optional single Tornado-type nosewheel for increased T-O weight.

POWER PLANT: One Rolls-Royce Turbomeca Adour Mk 871 non-afterburning turbofan, with uninstalled rating of 26 kN (5,845 lb st). Optional fixed refuelling probe on starboard side of windscreen.

ACCOMMODATION: Pilot only, on Martin-Baker Type 10 zero/zero ejection seat, under side-hinged (to starboard) canopy.

SYSTEMS: 25kVA generator with DC transformer-rectifier. Fairey Hydraulics yaw control system added, comprising rudder actuator and servo control system, incorporating a autostabiliser computer. Lucas Aerospace artificial feel system. 12kVA APU for engine starting, ground running and emergency power.

```
Stalling speed, flaps down
                          96 knots (177 km/h; 110 mph)
Max rate of climb at S/L       3,600 m (11,800 ft)/min
Time to 9,145 m (30,000 ft), clean: 60 Series   6 min 12 s
    100 Series                                   7 min 30 s
Service ceiling: 60 Series        14,175 m (46,500 ft)
    100 Series                    13,565 m (44,500 ft)
T-O run: 60 Series                  710 m (2,330 ft)
    100 Series                      640 m (2,100 ft)
Landing run: 60 Series              550 m (1,800 ft)
    100 Series                      605 m (1,980 ft)
Combat radius:
    with 2,268 kg (5,000 lb) weapon load
                               538 nm (998 km; 620 miles)
    with 908 kg (2,000 lb) weapon load
                               781 nm (1,448 km; 900 miles)
Ferry range: clean       1,313 nm (2,433 km; 1,510 miles)
60 Series, with two 864 litre (228 US gallon; 190 Imp
    gallon) drop tanks
                   more than 1,600 nm (2,964 km; 1,842 miles)
100 Series, as above
                   more than 1,400 nm (2,594 km; 1,612 miles)
Endurance, 100 nm (185 km; 115 miles) from base
                                          approx 4 h
g limits                                       +8/-4
```

BAe HAWK 200 SERIES (SINGLE-SEAT VERSIONS)

TYPE: Single-seat multi-role combat aircraft.

PROGRAMME: Intention to build demonstrator (ZG200) announced 20 June 1984; first flight 19 May 1986 but lost 2 July in g-induced loss of consciousness (GLOC) accident; replaced by first pre-production Hawk 200 (ZH200), first flown 24 April 1987; third demonstrator Series 200RDA (ZJ201) with full avionics and systems, including Westinghouse AN/APG-66H radar, flown 13 February 1992.

CURRENT VERSIONS: Missions can include:

Airspace denial: Two Sidewinders (more can be carried) and (subject to flight clearance) two 864 litre (228 US gallon; 190 Imp gallon) drop tanks, enabling 3 hour loiter on station at low level and 100 nm (185 km; 115 miles) from base, or one hour on station 550 nm (1,018 km; 633 miles) from base, or max intercept radius 720 nm (1,333 km; 828 miles).

Close air support: Typically five 1,000 lb and four 500 lb bombs, precision delivered up to 104 nm (192 km; 120 miles) from base in lo-lo mission.

Battlefield interdiction: Typically 1,360 kg (3,000 lb) load on hi-lo-hi mission over 510 nm (945 km; 587 mile) radius.

Long-range photo reconnaissance: 1,723 nm (3,190 km; 1,982 mile) range, with two external tanks and pod containing cameras and infra-red linescan (rapid role change permits follow-on attack by same aircraft), or lo-lo day/night radius 510 nm (945 km; 586 miles).

Long-range deployment: 1,950 nm (3,610 km; 2,244 mile) ferry range, using two 864 litre (228 US gallon; 190 Imp gallon) and one 592 litre (156 US gallon; 130 Imp gallon) external tanks, unrefuelled and with 190 Imp gallon tanks retained (reserves allow 10 min over destination at 150 m; 500 ft). Larger tanks not yet flight-cleared.

Anti-shipping attack: Sea Eagle missile and two 864 litre (228 US gallon; 190 Imp gallon) tanks, enabling ship attack 666 nm (1,234 km; 767 miles) from base and return with 10 per cent fuel reserves (puts ships in wide area of North Atlantic within range of shore bases; weapon release

First flight of radar-equipped Hawk 200RDA, 13 February 1992

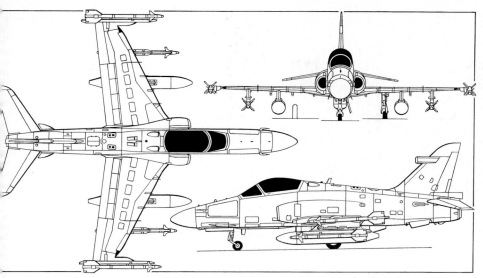

The single-seat British Aerospace Hawk 200 Series with nose mounted radar (*Jane's/Mike Keep*)

VIONICS: Similar to Hawk 100. GEC-Marconi multi-function display in cockpit; combined com/nav interface allows control of all functions from one panel; HOTAS controls optional. Marconi Defence Systems Sky Guardian 200 RWR in aircraft for Oman and Abu Dhabi Mk 102s. Smiths HUD optional. Chaff/flare dispenser (Vinten Vicon 78 Srs 300 or equivalent) at base of fin.

RMAMENT: None internally. All weapon pylons cleared for 8*g* manoeuvres with 500 kg (1,102 lb) loads.

IMENSIONS, EXTERNAL: As Hawk Series 100, except:
Length overall (no probe)	11.33 m (37 ft 2 in)
Wheelbase	3.56 m (11 ft 8 in)

WEIGHTS AND LOADINGS:
Basic weight empty	4,450 kg (9,810 lb)
Max fuel: internal (usable)	1,360 kg (3,000 lb)
external (usable)	932 kg (2,055 lb)
Max weapon load	3,000 kg (6,614 lb)
Max T-O weight	9,100 kg (20,061 lb)
Max wing loading	545.3 kg/m² (111.7 lb/sq ft)
Max power loading	350.04 kg/kN (3.43 lb/lb st)

ERFORMANCE (estimated; no external stores or role equipment unless stated):
Never-exceed speed (V_{NE}):
at S/L	Mach 0.87 (575 knots; 1,065 km/h; 661 mph EAS)
at and above 5,180 m (17,000 ft)	Mach 1.2 (575 knots; 1,065 km/h; 661 mph EAS)
Max level speed at S/L	549 knots (1,017 km/h; 632 mph)
Econ cruising speed at 12,500 m (41,000 ft)	430 knots (796 km/h; 495 mph)
Stalling speed, flaps down	96 knots (177 km/h; 110 mph) IAS
Max rate of climb at S/L	3,508 m (11,510 ft)/min
Time to 9,145 m (30,000 ft)	7 min 24 s
Service ceiling	13,720 m (45,000 ft)
Runway LCN: flexible pavement	15
rigid pavement	10
T-O run	630 m (2,070 ft)
Landing run	598 m (1,960 ft)
Ferry range (with two drop tanks)	more than 1,400 nm (2,594 km; 1,612 miles)
g limits	+8/−4

BAe HARRIER

RN designations: T. Mks 4N/8
Indian Navy designation: T. Mk 60

TYPE: V/STOL close support and reconnaissance aircraft with two-seat trainer derivative.

PROGRAMME: World's first operational fixed-wing V/STOL attack fighter; supplanted in production by AV-8B Harrier II/Harrier GR. Mks 5/7 (see International section: McDonnell Douglas/BAe), except limited manufacture as two-seat compatible trainer for Sea Harrier until 1992.

CURRENT VERSIONS: **Harrier T. Mk 4 and 4A and Harrier T. Mk 4N and 8:** Two-seat trainer versions; RAF flies Mk 4A and laser-nosed Mk 4; three Royal Navy two-seaters designated Harrier T. Mk 4N. Two-seat Harrier production for UK totalled 31, including prototypes; some RAF trainers transferred to RN; authorisation February 1992 for five naval Harrier trainers to receive cockpits representative of Sea Harrier FRS. Mk 2 (including FIN 1031B INS, but without radar), being redesignated T. Mk 8; first conversion, ZB605. *Technical details as for Sea Harrier, except as given below.*

Harrier T. Mk 60: Two-seat operational trainer for Indian Navy; T. Mk 4A configuration, but complete Sea Harrier avionics except Blue Fox radar; four ordered, delivered between March 1984 and January 1992.

Sea Harrier FRS. Mks 1, 2 and 51: Described separately.

AV-8B Harrier II, Harrier GR. Mk 5/7 and export variants: See under McDonnell Douglas/British Aerospace in International section.

CUSTOMERS: See table and previous editions of *Jane's*.
DESIGN FEATURES: See Programme and previous editions of *Jane's*.
LANDING GEAR: T. Mk 4 tyre pressures 6.90 bars (100 lb/sq in) on nose unit, 6.55 bars (95 lb/sq in) on main and outrigger units.
POWER PLANT: One Rolls-Royce Pegasus Mk 103 (or navalised Mk 104) vectored thrust turbofan (95.6 kN; 21,500 lb st).
ACCOMMODATION: Crew of two (Mk 4) on Martin-Baker Mk 9D zero/zero rocket ejection seats.
AVIONICS: GEC-Plessey U/VHF, Ultra standby UHF, GEC-Marconi AD 2770 Tacan and Cossor IFF, GEC-Marconi FE 541 inertial navigation and attack system (INAS), with Honeywell C2G compass, Smiths electronic HUD of flight information, and air data computer. GEC-Marconi Type 106 laser ranger and marked target seeker (LRMTS) and Marconi Defence Systems ARI.18223 RWR in most RAF Harriers.

DIMENSIONS, EXTERNAL: As for Sea Harrier except:
Wing span	7.70 m (25 ft 3 in)
Length overall:	
two-seat (laser nose)	17.50 m (57 ft 5 in)
Height overall: two-seat	4.17 m (13 ft 8 in)

WEIGHTS AND LOADINGS:
Weight empty (pilot/s plus four pylons; no guns):
T. Mk 4	6,693 kg (14,755 lb)
T. Mk 4A	6,568 kg (14,480 lb)
Internal fuel	2,295 kg (5,060 lb)
Max T-O weight: two-seat	11,880 kg (26,200 lb)
Max wing loading	636.0 kg/m² (130.3 lb/sq ft)
Max power loading	124.27 kg/kN (1.22 lb/lb st)

PERFORMANCE (single-seat):
Max level speed at S/L	635 knots (1,176 km/h; 730 mph)
Time to 12,200 m (40,000 ft) from vertical T-O	2 min 23 s
Service ceiling	15,600 m (51,200 ft)

Ferry range	1,850 nm (3,425 km; 2,129 miles)
g limits	+7.8/−4.2

BAe SEA HARRIER

RN designations: FRS. Mks 1 and 2
Indian Navy designation: FRS. Mk 51

TYPE: V/STOL fighter, reconnaissance and strike aircraft.
PROGRAMME: Development of P1184 Sea Harrier announced by British government 15 May 1975; first flight (XZ450) 20 August 1978; first delivery to Royal Navy (XZ451) 18 June 1979; first ship trials (HMS *Hermes*) November 1979.

Ski jump launching ramp (proposed by Lt Cdr D. R. Taylor, RN) take-off trials ashore 1977, and at sea from 30 October 1980; HMS *Invincible* and *Illustrious* first fitted with 7° ramps, HMS *Ark Royal* 12°; latter allows 1,135 kg (2,500 lb) increased load for same take-off run or 50-60 per cent shorter run at same weight; HMS *Invincible* recommissioned with 13° ramp 18 May 1989; HMS *Illustrious* began similar 2½ year re-work, May 1991.

UK MoD gave BAe project definition contract January 1985 for mid-life update of RN Sea Harrier FRS. Mk 1s; upgraded aircraft redesignated FRS. Mk 2; aerodynamic development FRS. Mk 2 converted at Dunsfold from Mk 1 ZA195; first flight 19 September 1988; first flight of second development aircraft (XZ439) 8 March 1989; contract for conversion of further 29 Mk 1s to Mk 2s signed by UK MoD 7 December 1988; modifications begun at Kingston October 1990, continuing at Dunsfold and Brough; redelivery from 2 April 1993, for later augmentation by newly built Mk 2s; additional five conversions under consideration.

New FRS. Mk 2 AAM launch rail first tested by live AIM-9L (from Mk 1) 2 November 1988; AMRAAM trials began in USA (using XZ439) with first launch on 29 March 1993, flying from Eglin AFB; airborne testing of Blue Vixen radar began in RAE One-Eleven (ZF433), completing 114 hour/121 sortie programme November 1987; development work transferred to RAE BAe 125 (XW930), first flown with A-version radar 26 August 1988; second 125-600B (ZF130) given full FRS. Mk 2 weapon system, including representative cockpit in co-pilot's position and Sidewinder acquisition round on underwing pylon (first flight at Woodford 20 May 1988; began development flying at Dunsfold December 1988; not fitted with B-version Blue Vixen radar until September 1989); first flight of B-version radar in Sea Harrier XZ439 24 May 1990. First Mk 2 deck landing, by ZA195 on HMS *Ark Royal*, 7 November 1990. Operational Evaluation Unit for Mk 2 formed at Boscombe Down, June 1993.

CURRENT VERSIONS: **FRS. Mk 1:** Initial Royal Navy version; Pegasus 104 engine; first used operationally during Falkland Islands campaign 1982, from HMS *Hermes* and *Invincible* (29 flew 2,376 sorties, destroying 22 enemy aircraft in air-to-air combat without loss; four lost in accidents and two to ground fire). Total 37 remained, June 1993, including Mk 2 conversions. *Following description applies to Sea Harrier FRS. Mk 1, except where indicated otherwise.*

FRS. Mk 51: Similar to Mk 1, for Indian Navy.

FRS. Mk 2: Differs externally from Mk 1 by less pointed nose radome; longer rear fuselage, resulting from 35 cm (1 ft 1¾ in) plug aft of wing trailing-edge; revisions of antennae and external stores. Internal changes include GEC-Marconi Blue Vixen pulse Doppler radar, offering all-weather lookdown/shootdown capability, with inherent track-while-scan, multiple target engagement, greatly increased missile launch range, enhanced surface target acquisition, and improved ECCM performance. Current weapons plus AIM-120 AMRAAM on Dowty/Frazer-Nash launch rails compatible with AIM-9L Sidewinder.

Improved systems built around MIL-STD-1553B databus, with dual redundant data highway, allowing computerised time sharing of information processed in databus control and interface unit.

TWO-SEAT HARRIER PRODUCTION

Airframe No.	Variant	Customer	First delivery	First aircraft	Total
1-2	T. Mk 2	RAF	24 Apr 1969*	XW174	2
3-11	T. Mk 2	RAF	21 May 1970	XW264-272	9
12	T. Mk 52	HSA	16 Sep 1971*	G-VTOL	1
13-15	T. Mk 2A	RAF	1 Oct 1971	XW925	3
16-17	T. Mk 2A	RAF	22 Aug 1973	XW933	2
18-25	Mk 54	USMC	1 Oct 1976	159378	8
26-28	T. Mk 4	RAF	8 Mar 1976	XZ145	3
29-30	Mk 56	Spain	25 Feb 1976	VAE.1-1	2
31	T. Mk 4	RN	1 May 1979	XZ445	1
32-35	T. Mk 4	RAF	10 Mar 1983	ZB600	4
36-38	T. Mk 4N	RN	21 Sep 1983	ZB604	3
39-40	T. Mk 60	IN	3 Feb 1984	IN651	2
41-44	T. Mk 4	RAF	Jun 1987	ZD990	4
45	T. Mk 60	IN	10 Apr 1990	IN653	1
46	T. Mk 60	IN	Jan 1992	IN654	1

* First flight

Note: Mks 2/2A converted to Mk 4 or 4A

BAe Sea Harrier FRS. Mk 2 aerodynamic prototype

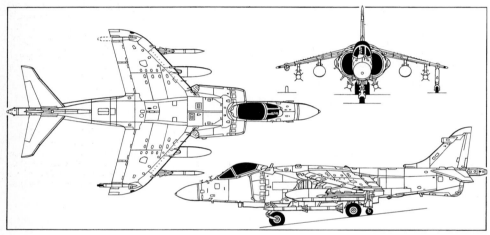

BAe Sea Harrier FRS. Mk 2 V/STOL fighter, reconnaissance and attack aircraft (*Jane's/Mike Keep*)

Redesign of cockpit includes new HUD and dual multi-purpose HDDs; proposed JTIDS terminal delayed following cancellation by US Navy of originally proposed equipment; JTIDS/Sea Harrier integration study (for retrofit) under consideration 1992. All time-critical weapon systems controls positioned on up-front control panel, or on throttle and stick (HOTAS).

Wingtip extensions of 20 cm (8 in) and 30 cm (1 ft 0 in) test-flown to enhance stability carrying AMRAAM, but proved unnecessary by 1990 trials.

CUSTOMERS: Royal Navy ordered three development aircraft plus batches of 21, 10, 14 and nine by September 1984; all built as Mk 1s, last completed June 1988; intent to order at least 10 Mk 2s revealed March 1990, but to be converted to firm order for 18-22 in 1993. Naval Intensive Flying Trials Unit (No. 700A Squadron) commissioned at RNAS Yeovilton 18 September 1979 and became normally shore based No. 899 HQ Squadron, April 1980; front line units Nos. 800 and 801 Squadrons, with eight aircraft each (previously five); non-radar T. Mk 4N two-seat trainers (three) received (see Harrier entry). First 'production' conversion to Mk 2 (XZ497) handed over 2 April 1993.

Six similar FRS. Mk 51s handed over to Indian Navy from January 1983; delivered (after pilot training in UK) between December 1983 and July 1984; used by No. 300 (White Tiger) Squadron from INS *Vikrant*; two T. Mk 60 two-seat trainers received (see Harrier entry); ten more FRS. Mk 51s and one T. Mk 60 ordered by Indian government November 1985; delivered December 1989 to September 1991; letter of intent for seven more FRS. Mk 51s and one T. Mk 60 issued September 1986, to equip INS *Viraat* (former HMS *Hermes*); third batch deliveries began September 1991 and completed April 1992. Total orders 80.

DESIGN FEATURES: Single-engined V/STOL system with four rotatable exhaust nozzles that can be set through 98.5° from fully aft position; short take-off made with nozzles initially fully aft, then turned partially downward for liftoff and continued forward acceleration; nozzles can be vectored at high speed to tighten turn radius or decelerate suddenly; control at less than wing-borne airspeed automatically transferred to puffer jets at wingtips, nose and tail, also enhancing combat manoeuvres.

Main differences from land-based Harriers include elimination of magnesium components, introduction of raised cockpit, revised operational avionics, and installation of multi-mode GEC-Marconi radar with air-to-air intercept and air-to-surface modes in redesigned nose that folds to port; Pegasus 104 turbofan of Mk 1 incorporates additional anti-corrosion features and generates more electrical power than land-based Pegasus 103. See Current Versions for FRS. Mk 2 features.

Wing section BAe (HS) design; thickness/chord ratio 10 per cent at root, 5 per cent at tip; anhedral 12°; incidence 1° 45′; and sweepback at quarter-chord 34°.

FLYING CONTROLS: Plain ailerons irreversibly operated by tandem hydraulic jacks; one-piece variable incidence tailplane, with 15° anhedral, irreversibly operated by tandem hydraulic jacks; manually operated rudder with trim tab; flaps; jet reaction control valve built into front of each outrigger wheel fairing and in nose and tailcone; large airbrake under fuselage; ventral fin under rear fuselage.

STRUCTURE: One-piece aluminium alloy three-spar safe-life wing with integrally machined skins (Brough built); entire wing unit removable to provide access to engine; revised inboard one-third of FRS. Mk 2 wing incorporates additional fence, kinked leading-edge, re-positioning of dog-tooth fillet closer to fuselage, and reduction of overwing vortex generators from 12 to 11; 67.3 cm (2 ft 2½ in) wing extensions available for ferrying; ailerons, flaps, rudder and tailplane trailing-edge of bonded aluminium alloy honeycomb construction; safe-life fuselage of frames and stringers, mainly aluminium alloy but with titanium skins at rear and some titanium adjacent to engine and other special areas; access to power plant through top of fuselage, ahead of wings; FRS. Mk 2 has deepened and stiffened nose structure, plus new rear fuselage, lengthened by 35 cm (1 ft 1¾ in); fin tip carries suppressed VHF aerial.

LANDING GEAR: Retractable bicycle type of Dowty Aerospace manufacture, permitting operation from rough unprepared surfaces of CBR as low as 3 to 5 per cent. Hydraulic actuation, with nitrogen bottle for emergency extension of landing gear. Single steerable nosewheel retracts forward, twin coupled mainwheels rearward, into fuselage. Small outrigger units retract rearward into fairings slightly inboard of wingtips. Nosewheel leg of levered suspension liquid spring type. Dowty telescopic oleo-pneumatic main and outrigger gear. Dunlop wheels and tyres, size 26.00 × 8.75-11 (nose unit), 27.00 × 7.74-13 (main units) and 13.50 × 6.4 (outriggers). Dunlop multi-disc brakes and Dunlop Hytrol adaptive anti-skid system.

POWER PLANT: One Rolls-Royce Pegasus Mk 104 or (FRS. Mk 2) Mk 106 vectored thrust turbofan (95.6 kN; 21,500 lb st) with four exhaust nozzles of the two-vane cascade type rotatable through 98.5° from fully aft position. Engine bleed air from HP compressor used for jet reaction control system and to power duplicated air motor for nozzle actuation. Low drag intake cowls each have eight automatic suction relief doors aft of leading-edge to improve intake efficiency by providing extra engine air at low forward or zero speeds. A 227 litre (60 US gallon; 50 Imp gallon) tank supplies demineralised water for thrust restoration in high ambient temperatures for STO, VTO and vertical landings Fuel in five integral tanks in fuselage and two in wings with total capacity of approx 2,865 litres (757 US gallons 630 Imp gallons). This can be supplemented by two 455 litre (120 US gallon; 100 Imp gallon) jettisonable combat tanks, or two 864 litre (228 US gallon; 190 Imp gallon tanks, or two 1,500 litre (396 US gallon; 330 Imp gallon ferry tanks on the inboard wing pylons. Ground refuelling point in port rear nozzle fairing. Provision for in-flight refuelling probe above the port intake cowl.

ACCOMMODATION: Pilot only, on Martin-Baker Mk 10H zero zero rocket ejection seat which operates through the miniature detonating cord equipped canopy of the pressurised heated and air-conditioned cockpit. Seat raised 28 cm (11 in) compared with Harrier. Manually operated rearward sliding canopy. Birdproof windscreen, with hydraulically actuated wiper. Windscreen washing system.

SYSTEMS: Three-axis limited authority autostabiliser for V/STOL flight. Pressurisation system of BAe design, with Normalair-Garrett and Delaney Gallay major components max pressure differential 0.24 bar (3.5 lb/sq in). Two hydraulic systems; flow rate: System 1, 36 litres (9.6 US gallons; 8 Imp gallons)/min; System 2, 23 litres (6 US gallons; 5 Imp gallons)/min. Systems, pressure 207 bar (3,000 lb/sq in), actuate Fairey flying control and general services and a retractable ram air turbine inside top of rear fuselage, driving a small hydraulic pump for emergency power. Turbine deleted from FRS. Mk 2. Hydraulic reservoirs nitrogen pressurised at 2.75 to 5.5 bars (40 to 80 lb/sq in). AC electrical system with transformer-rectifiers to provide required DC supply. Two 15kVA generators. Two 28V 25Ah batteries, one of which energises a 24V motor to start Lucas Mk 2 gas turbine starter/APU. This unit drives a 6kVA auxiliary alternator for ground readiness servicing and standby. Bootstrap cooling unit for equipment bay with intake at base of dorsal fin. Autopilot function on Fairey Hydraulics, giving throughput to aileron and tailplane power controls as well as to three-axis autostabs. British Oxygen liquid oxygen system of 4.5 litres (1.2 US gallons; 1 Imp gallon) capacity in Royal Navy aircraft Indian Navy has gaseous oxygen system.

AVIONICS: Nose mounted GEC-Marconi Blue Fox (Blue Vixen in FRS. Mk 2) multi-mode radar, with TV raster daylight viewing tube which conveys flight information, as well as radar data, to pilot. New and larger Smiths electronic HUD and 20,000 word digital weapon aiming computer. Autopilot, radar altimeter and Decca Doppler 72 radar. GEC-Marconi self-aligning attitude and heading reference platform and digital navigation computer. Radio navaids include UHF homing, GEC-Marconi AD 2770 Tacan with offset facility and I band transponder. Radio com by multi-channel GEC-Marconi PTR 377 U/VHF with VHF standby via D 403M transceiver. Passive electronic surveillance and warning of external radar illumination by receiver with forward and rear hemisphere antennae in fin and tailcone respectively. Intended 1994 retrofit of Mk XII IFF.

EQUIPMENT: Optically flat panel in nose, on port side, for F.95 oblique camera, which is carried as standard. A cockpit voice recorder with in-flight playback facility supplements the reconnaissance cameras, and facilitates rapid debriefing and mission evaluation.

ARMAMENT: No built-in armament. Combat load carried on four underwing and one underfuselage pylons, all with ML ejector release units. Inboard wing points and fuselage point stressed for loads up to 907 kg (2,000 lb) each, and outboard underwing pair for loads up to 295 kg (650 lb) each; two strake fairings under the fuselage can each be replaced by a 30 mm Aden gun pod and ammunition or, on FRS. Mk 2, by two AIM-120 AMRAAMs. Aircraft cleared for operations with maximum external load exceeding 2,270 kg (5,000 lb), and has flown with weapon load of 3,630 kg (8,000 lb). FRS. Mk 2 outboard pylons restressed to 454 kg (1,000 lb). Able to carry 30 mm guns, bombs, rockets and flares of UK and US designs. Alternative stores loads of RN Sea Harriers include a WE177 nuclear bomb; free-fall (1,030 lb) and parachute-retarded (1,120 lb) bombs; Lepus flares; and ML CBLS 100 carriers for Portsmouth Aviation 3 kg and 14 kg practice bombs. Four AIM-9 Sidewinder missiles carried on the outboard underwing pylons (Matra Magic instead of Sidewinder on Indian Navy aircraft); provision for two air-to-surface missiles of Sea Eagle or Harpoon type. FRS. Mk 2 accommodates up to four AIM-120 AMRAAMs, or two AIM-120s and four AIM-9L Sidewinders, on Frazer-Nash common rail launchers. BAe ALARM anti-radiation missile may replace AIM-120.

DIMENSIONS, EXTERNAL:
Wing span: normal	7.70 m (25 ft 3 in)
ferry	9.04 m (29 ft 8 in)
Length overall: FRS. Mk 1	14.50 m (47 ft 7 in)
FRS. Mk 2	14.17 m (46 ft 6 in)
Length overall, nose folded:	
FRS. Mk 1	12.73 m (41 ft 9 in)
FRS. Mk 2	13.16 m (43 ft 2 in)
Height overall	3.71 m (12 ft 2 in)
Tailplane span	4.24 m (13 ft 11 in)
Outrigger wheel track	6.76 m (22 ft 2 in)
Wheelbase, nosewheel to mainwheels	approx 3.45 m (11 ft 4 in)

AREAS:
Wings, gross	18.68 m² (201.1 sq ft)
Fin (excl ventral fin): two-seat	3.57 m² (38.4 sq ft)
Rudder, incl tab	0.49 m² (5.3 sq ft)
Tailplane	4.41 m² (47.5 sq ft)

WEIGHTS AND LOADINGS (FRS. Mk 1):
Operating weight empty	6,374 kg (14,052 lb)
Max fuel: internal	2,295 kg (5,060 lb)
external	2,404 kg (5,300 lb)
Max weapon load: STO	3,630 kg (8,000 lb)
VTO	2,270 kg (5,000 lb)
Max T-O weight	11,880 kg (26,200 lb)
Max wing loading	636.0 kg/m² (130.3 lb/sq ft)
Max power loading	124.27 kg/kN (1.22 lb/lb st)

PERFORMANCE (FRS. Mk 1):
Max Mach No. at high altitude	1.25
Max level speed at low altitude	above 640 knots (1,185 km/h; 736 mph) EAS
Typical cruising speed:	
high altitude, for well over 1 h on internal fuel	above Mach 0.8
low altitude	350-450 knots (650-833 km/h; 404-518 mph), with rapid acceleration to 600 knots (1,110 km/h; 690 mph)
STO run at max T-O weight, without ski-jump	approx 305 m (1,000 ft)
Time from alarm to 30 nm (55 km; 35 miles) combat area	under 6 min
High altitude intercept radius, with 3 min combat and reserves for VL	400 nm (750 km; 460 miles)
Attack radius	250 nm (463 km; 288 miles)
g limits	+7.8/−4.2

COMBAT PROFILES (FRS. Mk 2, from carrier fitted with a 12° ski-jump ramp, at ISA + 15°C and with a 20 knot; 37 km/h; 23 mph wind over the deck):

Combat air patrol: Up to 1½ hours on station at a radius of 100 nm (185 km; 115 miles), carrying four AMRAAMs, or two AMRAAMs and two 30 mm guns, plus two 864 litre (228 US gallon; 190 Imp gallon) combat drop tanks.

Reconnaissance: Low level cover of 130,000 nm² (446,465 km²; 172,380 sq miles) at a radius of 525 nm (970 km; 600 miles) from the carrier, with outward and return flights at medium/high level, carrying two 30 mm guns and two 864 litre (228 US gallon; 190 Imp gallon) combat drop tanks. Overall flight time 1 h 45 min.

Surface attack (hi-lo-hi): Radius of action to missile launch 200 nm (370 km; 230 miles), carrying two Sea Eagle missiles and two 30 mm guns. Take-off deck run for the above missions is 137 m, 107 m and 92 m (450 ft, 350 ft and 300 ft) respectively, with vertical landing.

Interception: A typical deck-launched interception could be performed against a Mach 0.9 target at a radius of 116 nm (215 km; 133 miles), or a Mach 1.3 target at 95 nm (175 km; 109 miles), after initial radar detection of the approaching target at a range of 230 nm (425 km; 265 miles), with the Sea Harrier at 2 min alert status, carrying two AMRAAM missiles.

EUROFIGHTER 2000

British Aerospace participating with companies from Germany, Italy and Spain (see International section).

F-111

In 1978, BAe's Filton plant began depot level maintenance of General Dynamics F-111 strike aircraft based in UK with USAF's 20th TFW (F-111E) and 48th TFW (F-111F). A second five-year contract, covering approximately 150 F-111s, was agreed with USAF 1 October 1988, this also including structural fatigue testing and avionics modifications as applied to US-based F-111s. Work terminated by withdrawals of strike-tasked F-111s from Europe. Final F-111F (73-0711) — also 300th F-111 overhaul at Filton — returned to USAF on 5 February 1991. Last F-111E departure (68-0076) 16 September 1992.

V-22 OSPREY

BAe and Bell/Boeing concluded MoU in 1987 covering examination by British company of latter's V-22 tilt-rotor aircraft in both military and civilian applications within European NATO area. In 1989, Aeritalia (now Alenia) and Dornier (now DASA) reached agreement with BAe to pursue prospects of V-22 tilt-rotor aircraft within Europe on collaborative tri-national basis, with support of Bell/Boeing.

Potential applications for a tilt-rotor aircraft in UK military service are foreseen by BAe as shore-based anti-submarine warfare (four Sting Ray torpedoes), ship-based AEW, commando assault (24 troops), air mobility and special forces (12 troops and extra fuel). Civilian uses include resource development (carrying 30+ oil/gas rig workers) and commuter transport.

HAC/PAH-2 TIGRE/TIGER

Eurocopter and BAe Defence agreed on 27 February 1992 to study single-source supply of helicopters to British MoD. First venture is Eurocopter Tiger armed with TRIGAT missiles. Later possibilities include a UK Wessex/Puma replacement.

E-3D SENTRY AEW. Mk 1

BAe selected by Boeing 1988 to perform installation and checkout (I&CO) duties on all except first of RAF's seven E-3Ds (assisted by Boeing team on first aircraft); responsible for reception of aircraft, installation and checking of some equipment, and maintenance in pre-delivery phase between arrival in UK and handover; most work undertaken at RAF Waddington; Sentry Training Squadron formed at Waddington 1 June 1990; second RAF Sentry (ZH102) delivered to BAe 4 July 1990, followed by third on 9 January 1991. (First E-3D arrived fully equipped 3 November 1990; to RAF 22 May 1991.) First handover to RAF following I&CO at Waddington was ZH102 on 24 March 1991; official acceptance 26 March. No. 8 Squadron converted from Shackleton AEW. Mk 2 (NATO's last piston-engined front-line aircraft) to Sentry on 1 July 1991. Seventh Sentry to UK 22 August 1991; handed over to RAF 12 May 1992 as JTIDS testbed; No. 8 Squadron declared to NATO 1 July 1992.

BAe (BAC/VICKERS) VC10 C. Mk 1(K), K. Mk 2, 3 and 4 TANKERS

Modification continues of RAF VC10s to tanker configuration. Details appear in *Jane's Civil and Military Aircraft Upgrades*. First C. Mk 1K conversion (XV101) flew at Hurn 11 June 1992; delivered to A&AEE at Boscombe Down 17 August 1992.

BRITISH AEROSPACE SPACE SYSTEMS LTD
Communications Satellites Division
STEVENAGE: Gunnels Wood Road, Stevenage, Hertfordshire SG1 2AS
Telephone: 44 (438) 313456
Fax: 44 (438) 736637

Telex: 82130/82197
Earth Observation and Science Division
BRISTOL: PO Box No. 5, Filton, Bristol BS12 7QW
Telephone: 44 (272) 693831
Fax: 44 (272) 363812
Telex: 449452

MANAGING DIRECTOR: David Hunt
Responsible for Skynet 4 UK military communications satellite, civil communications satellites, and pallets for US Space Shuttle (see *Jane's Spaceflight Directory*). Also undertook development study for Hotol space transport (see 1992-93 *Jane's*).

BAF
BRITISH AIR FERRIES
Viscount House, Southend Airport, Essex SS2 6YL
Telephone: 44 (702) 354435
Fax: 44 (702) 331914
Telex: 995687 and 995576

BAF (BAC/VICKERS) VISCOUNT LIFE EXTENSION

BAF is licensed by CAA to carry out life-extension modification programme on Viscount 800 airframes to requirements formulated by BAF/BAe, facilitating further 15 years' service or 75,000 flights. Details appear in *Jane's Civil and Military Aircraft Upgrades*.

CFM
COOK FLYING MACHINES (trading as CFM Metal-Fax Ltd)
Unit 2D, Eastlands Industrial Estate, Leiston, Suffolk IP16 4LL
Telephone: 44 (728) 832353/833076
Fax: 44 (728) 832944
Telex: 9877703 CHACOM G
MANAGING DIRECTOR: David G. Cook

While continuing to produce Shadow series of aircraft from its Suffolk works, CFM has licensed Laron Aviation Technologies of Portalis, New Mexico, to produce Shadow kits for North and South American and Canadian market, and was in 1992 negotiating for licensed production at a plant at Durban, Natal, for African, Australasian and Far Eastern markets.

CFM SHADOW SERIES C and C-D

TYPE: Tandem two-seat microlight; conforms to FAI and UK CAA requirements.
PROGRAMME: Shadow first flew as prototype in 1983; has been in continuous production since 1984; Type approval to BCAR CAP 482 Section S gained May 1985; current standard Shadow designated Series **C**, with **C-D** having dual controls, although **Series B/B-D** can still be bought. ULV cropspraying trials demonstrated its suitability in this role, carrying 64 litre (16.8 US gallon; 14 Imp gallon) chemical tank; multi-function surveillance fit for photography (Hasselblad 'lookdown' camera platform), video recording, linescan or thermal imagery, or closed-circuit TV microwave transmission to a command vehicle/station also available.

COSTS: (1992) Assembled: Series B, £15,410; Series B-D, £15,900; Series C, £16,370; Series C-D, £16,860. Kit: Series B, £9,625; Series B-D, £9,900; Series C, £10,425; Series C-D, £10,700. (All prices exclusive of VAT.)
DESIGN FEATURES: Wing design allows no 'defined stall' and will not permit spin. Overseas options include agricultural spraygear and other specialised equipment.
STRUCTURE: Wings of aluminium alloy and wood, with foam/glassfibre ribs, plywood covering on forward section and polyester fabric aft. Fuselage pod of Fibrelam; aluminium tube tailboom.
LANDING GEAR: Non-retractable tricycle landing gear, with brakes. Overseas options include floats.
POWER PLANT: One 38 kW (51 hp) Rotax 503.2V. Standard fuel capacity 23 litres (6 US gallons; 5 Imp gallons). Options include a 73 litre (19.2 US gallon; 16 Imp gallon) auxiliary fuel tank.

DIMENSIONS, EXTERNAL:
Wing span	10.03 m (32 ft 11 in)
Wing aspect ratio	6.69
Length overall	6.40 m (21 ft 0 in)
Height overall	1.75 m (5 ft 9 in)
Propeller diameter	1.30 m (4 ft 3 in)

AREAS:
Wings, gross	15.0 m² (162.0 sq ft)

WEIGHTS AND LOADINGS:
Weight empty	158 kg (349 lb)
Pilot weight range:	
front cockpit	54.5-100 kg (120-220 lb)
rear cockpit	0-100 kg (0-220 lb)
Max T-O weight	348 kg (767 lb)
Max wing loading	23.09 kg/m² (4.73 lb/sq ft)
Max power loading	9.16 kg/kW (15.04 lb/hp)

PERFORMANCE (at max T-O weight):
Max level speed	89 knots (164 km/h; 102 mph)
Econ cruising speed	65 knots (121 km/h; 75 mph)
Min flying speed	22 knots (41 km/h; 25 mph)
Max rate of climb at S/L:	
pilot only	335-366 m (1,100-1,200 ft)/min
dual	213-244 m (700-800 ft)/min
Service ceiling	7,620 m (25,000 ft)
T-O run from metalled surface	90 m (295 ft)
Landing run, with brakes	75 m (246 ft)
Range, standard fuel	139 nm (257 km; 160 miles)
Endurance, standard fuel	2 h
g limits	+6/−3 static, ultimate

UK: AIRCRAFT—CFM/CMC

CFM Streak Shadow

CFM STREAK SHADOW

TYPE: Tandem two-seat ultralight/homebuilt; dual controls standard.

PROGRAMME: Construction of prototype started 1987 and this first flew June 1988; first flight of production aircraft October 1988; 50 kits delivered by January 1992.

COSTS: (1992) Assembled: not available. Kits: Series S-A1, £11,750; Series S-A, £12,200. (All prices exclusive of VAT.)

DESIGN FEATURES: Derivative version of Shadow, designated a microlight in Europe and Experimental homebuilt in USA; has new wing design, light airframe weight and more powerful engine; fuel capacity similar to latest C version of Shadow; wing design continues to allow no 'defined stall' and is spin resistant.

FLYING CONTROLS: See Structure. Electric trim.

STRUCTURE: Similar to Shadow but with foam/glassfibre wings (CFM aerofoil section), and control surfaces with aluminium alloy ribs and polyester fabric covering (ailerons, flaps, rudder and elevators).

POWER PLANT: One 47.7 kW (64 hp) Rotax 582, or optional MWAE 622. Standard fuel capacity 54.6 litres (14.4 US gallons; 12 Imp gallons). Optional 73 litre (19.2 US gallon; 16 Imp gallon) auxiliary fuel tank.

DIMENSIONS, EXTERNAL:
Wing span 8.53 m (28 ft 0 in)
Wing aspect ratio 5.6
Length overall 6.40 m (21 ft 0 in)
Height overall 1.75 m (5 ft 9 in)
Propeller diameter 1.32 m (4 ft 4 in)

AREAS:
Wings, gross 13.01 m² (140.0 sq ft)

WEIGHTS AND LOADINGS:
Weight empty 176 kg (388 lb)
Max T-O weight 408 kg (900 lb)
Max wing loading 31.39 kg/m² (6.43 lb/sq ft)
Max power loading 8.55 kg/kW (14.06 lb/hp)

PERFORMANCE:
Max level speed 105 knots (195 km/h; 121 mph)
Cruising speed:
 80% power 87 knots (161 km/h; 100 mph)
 50% power 65 knots (121 km/h; 75 mph)
Min flying speed, engine idling
  28 knots (52 km/h; 32 mph)
Max rate of climb at S/L:
 pilot only 549 m (1,800 ft)/min
 dual 335-396 m (1,100-1,300 ft)/min
Service ceiling 9,150 m (30,000 ft)
Range 347 nm (643 km; 400 miles)
Endurance, standard fuel 4 h 30 min
g limits +6/−3

CFM SHADOW II

This specialised single-seat version of Shadow is marketed by Mission Technologies Inc (Mi-Tex) of Hondo, Texas, USA, as manned surveillance system aircraft for day and night use (two ordered by a NATO country in 1988, with six more on option). Uses one 44.7 kW (64 hp) Rotax 532 engine, and has wing span of 8.53 m (28 ft 0 in). Further details in *Jane's Battlefield Surveillance Systems*.

CMC

CHICHESTER-MILES CONSULTANTS LTD

West House, The Old Rectory, Ayot St Lawrence, Welwyn, Hertfordshire AL6 9BT

Telephone and Fax: 44 (438) 820341

CHAIRMAN: Ian Chichester-Miles

Ian Chichester-Miles, formerly Chief Research Engineer of BAe Hatfield, established Chichester-Miles Consultants to develop Leopard high performance light business jet.

CMC LEOPARD

TYPE: Four-seat light business aircraft.

PROGRAMME: Design began January 1981; mockup completed early 1982; detail design and construction of prototype by Designability Ltd of Dilton Marsh, Wiltshire, began July 1982 under CMC contract; first flight of unpressurised prototype (G-BKRL) 12 December 1988 at RAE Bedford; by December 1991 had made 50 flights investigating basic handling qualities at speeds up to 200 knots (371 km/h; 230 mph) IAS; new tailplane incorporating AS&T liquid anti-icing system on leading-edge subsequently installed on prototype prior to resumption of flight testing aimed at expanding airspeed, altitude and CG envelopes.

Design of second (pre-production) aircraft began April 1989. Features strengthened structure, new landing gear, pressurised cabin, de-icing, reprofiled nose for EFIS avionics, substitution of oleo-pneumatic main landing gear legs for current rubber-in-compression units, and powered by 3.11 kN (700 lb st) class Williams International FJX-1 turbofans.

DESIGN FEATURES: Streamline composites airframe; sweptback supercritical technology wings; sweptback tail unit; twin low-cost turbofans; pressurised and air-conditioned cabin; AS&T liquid de-icing and decontamination system on wing and tailplane leading-edges of production aircraft (see Programme); warm air de-icing of engine intake leading-edges. First prototype has lower-powered engines; lacks full pressurisation/air-conditioning system, de-icing, advanced avionics and instrumentation of second aircraft and planned production model.

ARA designed wing section and 3D profiles combining laminar flow and supercritical technology; thickness/chord ratio 14 per cent at root, 11 per cent at tip; wing sweepback at quarter-chord 25°.

FLYING CONTROLS: All-moving fin; two independent tailplane sections operated collectively for pitch control and differentially for roll control; no ailerons. Full-span electrically actuated trailing-edge plain flaps, with ±45° deflections for high drag landing and airbraking/lift dumping; no spoilers.

STRUCTURE: Two-spar wings, primarily of GFRP, with some carbonfibre reinforcement; carbonfibre flaps; fuselage built in three sections as unpressurised nose housing avionics and nosewheel gear, pressurised cabin (production aircraft), and unpressurised rear housing baggage bay, with fuel tanks below and equipment bays to rear; fuselage primarily GFRP with some carbonfibre reinforcement (fore

First prototype of CMC Leopard business jet (*B. J. Cunnington/CMC*)

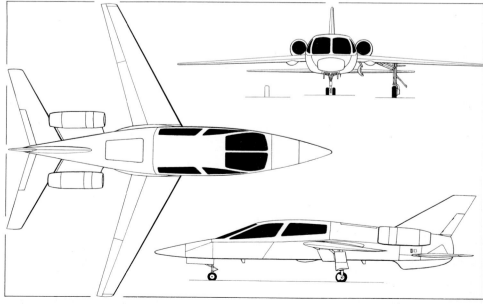

CMC Leopard (two Williams International FJX-1 turbofans) (*Dennis Punnett*)

and aft bulkheads, engine and tailplane axle frames moulded in); pressure cabin section divided approximately along aircraft horizontal datum, with upper section formed by electrically actuated upward opening canopy hinged at windscreen leading-edge; bonded-in acrylic side windows carry pressurisation tension; nose opens for access to avionics; light alloy engine nacelles, with stainless steel firewalls; composites fin and tailplane; fin sternpost projects to bottom of rear fuselage; low-set tailplane in two independent sections, each mounted on steel axle projecting from side of rear fuselage; carbonfibre tabs.

LANDING GEAR: Electrically retractable tricycle type, main units retracting inward into wingroot wells, nosewheels forward. Nose-well closure doors linked mechanically to landing gear units. Gravity extension assisted by bias springs and aerodynamic drag. Production aircraft main doors will be electrically actuated. Long-stroke shock absorber in each unit, using synthetic elastomers in compression. New, oleo-pneumatic main landing gear is being designed for production aircraft. Main units, each with single Cleveland wheel, size 5.00-5, have tyres size 11 × 4, pressure 4.82 bars (70 lb/sq in) on prototype, 11.56 bars (170 lb/sq in) on production aircraft. Unpowered steerable twin-wheel nose unit has wheels size 4.00-3 and tyres size 8.5 × 2.75 in, pressure 2.75 bars (40 lb/sq in) on prototype, 3.8 bars (55 lb/sq in) on production aircraft. Hydraulic disc brakes. Parking brake.

POWER PLANT: Prototype 001 has two Noel Penny Turbines NPT 301-3A turbojets each of nominal 1.33 kN (300 lb st) rating. Production aircraft will have two Williams International FJX-1 turbofans, each of about 3.11 kN (700 lb st). Each engine in nacelle, mounted on crossbeam located in rear fuselage. Fuel tanks in fuselage, below baggage bay. First prototype has total fuel capacity of 455 litres (120 US gallons; 100 Imp gallons). Production aircraft will have maximum capacity of 673 litres (178 US gallons; 148 Imp gallons). Refuelling point on upper surface of fuselage.

ACCOMMODATION: Cabin seats four, in two pairs, on semi-reclining (35°) seats beneath upward opening jettisonable canopy. Options include dual controls, and accommodation for pilot, stretcher and attendant in medevac role. Unpressurised baggage bay aft of cabin, capacity 63 kg (140 lb), with external door in upper surface of fuselage.

SYSTEMS (production aircraft): Air-conditioning and pressurisation (max differential 0.66 bar; 9.6 lb/sq in) by engine bleed air. Electrical system powered by dual engine driven 3kVA starter/generators. Hydraulic system for brakes only. De-icing.

AVIONICS (production aircraft): Full nav/com and storm-avoidance systems. Bendix/King avionics, mounted in nose bay, supply EADI and EHSI information to two CRTs in pilot's instrument panel, together with weather radar. CRT displays can be transferred in the failure mode. Reduced scale electromechanical standby flight instruments. All avionics systems fully integrated with digital autopilot.

DIMENSIONS, EXTERNAL:
Wing span	7.16 m (23 ft 6 in)
Wing chord: at root	1.14 m (3 ft 9 in)
at tip	0.36 m (1 ft 2 in)
Wing aspect ratio	8.78
Length overall	7.52 m (24 ft 8 in)
Height: overall	2.06 m (6 ft 9 in)
to canopy sill	0.76 m (2 ft 6 in)
Tailplane span	3.91 m (12 ft 10 in)
Wheel track	3.45 m (11 ft 4 in)
Wheelbase	3.20 m (10 ft 6 in)

DIMENSIONS, INTERNAL:
Cabin: Length	2.74 m (9 ft 0 in)
Max width	1.14 m (3 ft 9 in)
Max height	0.94 m (3 ft 1 in)
Baggage bay volume	0.40 m³ (14 cu ft)

AREAS:
Wings, gross	5.85 m² (62.9 sq ft)
Trailing-edge flaps (total)	1.24 m² (13.3 sq ft)
Fin	0.86 m² (9.3 sq ft)
Tailplane (incl tabs)	2.14 m² (23.0 sq ft)

WEIGHTS AND LOADINGS (A: first prototype, B: production aircraft, estimated):
Weight empty, equipped: A	862 kg (1,900 lb)
B	998 kg (2,200 lb)
Max fuel weight: A	367 kg (810 lb)
B	544 kg (1,200 lb)
Max T-O weight: A	1,156 kg (2,550 lb)
B	1,814 kg (4,000 lb)
Max zero-fuel weight: A	1,043 kg (2,300 lb)
B	1,361 kg (3,000 lb)
Max landing weight: A	1,156 kg (2,550 lb)
B	1,701 kg (3,750 lb)
Max wing loading: A	197.7 kg/m² (40.5 lb/sq ft)
B	310.3 kg/m² (63.6 lb/sq ft)
Max power loading: A	433.7 kg/kN (4.25 lb/lb st)
B	240.1 kg/kN (2.35 lb/lb st)

PERFORMANCE (production aircraft, estimated, ISA):
Never-exceed speed (VNE)	Mach 0.81 (300 knots; 556 km/h; 345 mph EAS)
Max level speed at 9,450 m (31,000 ft)	469 knots (869 km/h; 540 mph)
Max and econ cruising speed at 13,715 m (45,000 ft)	434 knots (804 km/h; 500 mph)
Stalling speed, full flap, at AUW of 1,497 kg (3,300 lb)	84 knots (156 km/h; 97 mph)
Max rate of climb at S/L	1,960 m (6,430 ft)/min
Rate of climb at S/L, OEI	631 m (2,070 ft)/min
Service ceiling	16,765 m (55,000 ft)
Service ceiling, OEI	9,145 m (30,000 ft)
T-O to 15 m (50 ft)	727 m (2,385 ft)
T-O balanced field length	838 m (2,750 ft)
Landing factored field length	854 m (2,800 ft)
Landing from 15 m (50 ft) at AUW of 1,497 kg (3,300 lb)	778 m (2,550 ft)
Range: max payload, with reserves	1,500 nm (2,775 km; 1,725 miles)
max fuel and reduced payload	1,915 nm (3,548 km; 2,205 miles)

CROPLEASE

CROPLEASE PLC

Vicarage House, 58-60 Kensington Church Street, London W8 4DB
Telephone: 44 (71) 376 0448
Fax: 44 (71) 376 0340
CHAIRMAN: Richard Cox-Johnson
MANAGING DIRECTOR: Andrew Mackinnon
CHIEF DESIGNER: Desmond Norman, CBE

Following Norman Aeroplane Company receivership October 1988, design, manufacture and marketing rights to Fieldmaster acquired by Andrew Mackinnon of Irish-based Croplease Ltd; Croplease plc formed April 1989 and acquired Fieldmaster rights and Croplease Ltd business.

CROPLEASE NAC 6 FIELDMASTER

TYPE: Two-seat agricultural, pollution control, or firefighting (Firemaster) aircraft.

PROGRAMME: Designed by Desmond Norman; financially supported by UK National Research Development Corpn; prototype (G-NRDC), built by Norman Aeroplane Co, first flown at Sandown (Isle of Wight) 17 December 1981; total 300 flight hours by March 1986 during tests, demonstrations and in-service spraying trials by an operator; first production Fieldmaster (G-NACL) flew 29 March 1987; certification received 27 April 1987; G-NACL flew 28 November 1989 with 917 kW (1,230 shp) PT6A-65AG turboprop driving five-blade Hartzell propeller, in Firemaster configuration; production of five Fieldmasters restarted using UTVA (Yugoslavia) manufactured components; to France Aviation at Cannes on permit for evaluation July-October 1990; returned for certification testing; CAA aerial work certification received late May 1991; two aircraft (G-NACM, G-NACN) to France Aviation 1 July 1991.

CURRENT VERSIONS: **Fieldmaster**: Agricultural version.
Fieldmaster 65: Optionally powered by PT6A-65AG.
Firemaster 65: Dedicated firefighting/water bombing variant (see Programme); firefighting modifications include additional 53 litre (14 US gallon; 11.6 Imp gallon) tank for fire retardant (mixed with water before release).

CUSTOMERS: Two production aircraft began season of fire patrol flights and water bombing sorties in Maritime Alps under France Aviation contract July 1987; turnaround time between sorties averaged 3 min; these two included in five Fieldmasters bought by Croplease Ltd of Shannon, Ireland.

DESIGN FEATURES: Structural titanium chemical hopper forming integral part of fuselage, outer surface contoured as part of fuselage skin; hopper capacity 2,032 kg (4,480 lb) of dry or 2,366 litres (625 US gallons; 520 Imp gallons) of liquid chemicals; power plant mounted on front of hopper; rear fuselage attached aft of hopper; wings attached directly to hopper sides; flaps embody liquid spray dispersal system (24 nozzles each wing; Micronair 12-nozzle system optional) discharging into flap downwash to ensure spray droplets achieve best crop penetration; strong rollover cockpit structure.

Croplease Firemaster 65 (Pratt & Whitney Canada PT6A-65AG) for French Alpes Maritimes region

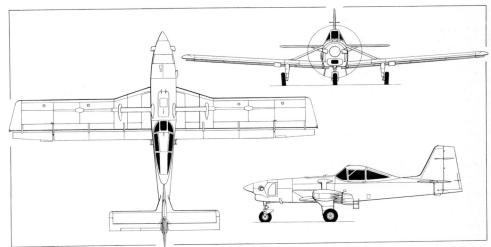

Croplease NAC 6 Fieldmaster agricultural aircraft (Pratt & Whitney Canada PT6A-34AG turboprop)
(Dennis Punnett)

NACA 23012 aerofoil section, modified on inner panels, where forward extension of leading-edge reduces thickness/chord ratio of 8.6 per cent at wingroot; dihedral 4° 15'; incidence 4° 30'.

FLYING CONTROLS: Auxiliary aerofoil ailerons; servo tab in starboard aileron, linked mechanically to rudder pedals, ensuring some bank with rudder movement; fixed incidence tailplane; trim tab in port elevator; servo tab in rudder controlled by stick movement, moved automatically with bank; electrically actuated wide span auxiliary aerofoil trailing-edge flaps, incorporating spray nozzle plumbing. Removable dual controls available.

STRUCTURE: Overwing streamline section wing bracing struts; all-metal wings with corrosion proofing; forward fuselage section comprises structural titanium hopper incorporating large door, vent system, inspection windows and light; light alloy semi-monocoque rear fuselage section (corrosion proofed: see Design Features); braced light alloy tail unit.

LANDING GEAR: Non-retractable tricycle type, with single wheel on each unit. Nosewheel has alternative steerable or castoring facility. Main units of levered suspension type. Nosewheel tyre size 7.00-8, pressure 3.45 bars (50 lb/sq in); mainwheels have tubed tyres, diameter 736 mm (29 in), pressure 3.79 bars (55 lb/sq in). Cleveland hydraulic disc brakes. Landing gear incorporates wire cutters.

POWER PLANT: One 559 kW (750 shp) Pratt & Whitney Canada PT6A-34AG turboprop, driving a Hartzell type HC-B3TN-3/T10282+4 three-blade fully feathering reversible-pitch metal propeller. Optional 917 kW (1,230 shp) P&WC PT6A-65AG driving five-blade propeller. Four integral fuel tanks, two per wing. Main tanks, each of 378.5 litres (100 US gallons; 83.3 Imp gallons), inboard; ferry tanks, each of 367 litres (97 US gallons; 80.7 Imp gallons), outboard. Total fuel capacity 1,491 litres (394 US gallons; 328 Imp gallons). Oil capacity 13 litres (3.5 US gallons; 2.9 Imp gallons). Engine air intake has a Centrisep filtration system.

ACCOMMODATION: Standard accommodation for pilot only, on fully adjustable seat in enclosed cockpit, with rollover protective structure. Rear trainee/observer seat optional. Dual controls optional, those for pupil easily removable. Crew safety helmets optional with headsets optional. Baggage space in fuselage. Sideways hinged door on each side. Birdproof armoured glass windscreen, two-speed windscreen wiper and windscreen wash system optional. Accommodation ventilated; air-conditioning and heating system optional. Wirecutters forward of windscreen, and cable deflecting wire from top of windscreen to tip of fin.

SYSTEMS: Electrical system includes 24V 200A starter/generator. Hydraulic system for brakes only. Central warning system standard.

AVIONICS: Intercom standard. Avionics, and IFR instrument package, to customer requirements.

EQUIPMENT: Standard equipment includes an external power socket. Optional equipment includes airframe and engine hour meter; instrument lighting, navigation lights, fin and wingtip strobe lights; two forward looking retractable work lights, each 765,000 candlepower; automatic flagman installation; firefighting dump door and water scoop; Transland gatebox, high volume spreader, quick disconnect flange kit, and side loading system; and Micronair installation, with flowmeter and rpm indicator.

DIMENSIONS, EXTERNAL (with PT6A-34AG):
Wing span 16.23 m (53 ft 3 in)
Wing chord (excl flaps): at root 2.01 m (6 ft 7¼ in)
  at tip 1.45 m (4 ft 9 in)
Wing aspect ratio 7.96
Length overall 11.02 m (36 ft 2 in)
Height overall 4.12 m (13 ft 6 in)
Wheel track 5.28 m (17 ft 4 in)
Wheelbase 3.35 m (11 ft 0 in)
Propeller diameter 2.69 m (8 ft 10 in)

DIMENSIONS, INTERNAL:
Hopper volume 2.36 m³ (83.0 cu ft)

AREAS:
Wings, gross 33.25 m² (358.0 sq ft)

WEIGHTS AND LOADINGS:
Standard weight empty 2,266 kg (4,995 lb)
Max T-O and landing weight:
  BCAR Aerial Work Category 3,855 kg (8,500 lb)
  UK CAA AN 90 4,535 kg (10,000 lb)
Max zero-fuel weight 3,855 kg (8,500 lb)
Max wing loading (AN 90) 136.37 kg/m² (27.93 lb/sq ft)
Max power loading (AN 90) 8.11 kg/kW (13.33 lb/shp)

PERFORMANCE (clean, with main landing gear fairings installed, at max Aerial Work Category T-O weight, S/L, ISA, except where indicated):
Never-exceed speed (V_{NE}) 172 knots (318 km/h; 198 mph)
Max level speed: at S/L 143 knots (265 km/h; 165 mph)
  at 1,830 m (6,000 ft) 147 knots (272 km/h; 169 mph)
Design manoeuvring speed 126 knots (233 km/h; 145 mph)
Stalling speed: flaps up 70 knots (129 km/h; 81 mph)
  30° flap 60 knots (111 km/h; 69 mph)
Max rate of climb at S/L 293 m (960 ft)/min
Service ceiling 5,550 m (18,200 ft)
T-O run 419 m (1,375 ft)
T-O to 15 m (50 ft) 625 m (2,050 ft)
Landing from 15 m (50 ft) 472 m (1,550 ft)
Landing run at typical landing weight of 2,720 kg (6,000 lb), with propeller reversal 152 m (500 ft)
Range at 3,050 m (10,000 ft) with two crew, 454 kg (1,000 lb) of equipment and max fuel, no reserves 1,000 nm (1,853 km; 1,150 miles)
g limits +3.4/−1.7

FLS

FLS AEROSPACE (LOVAUX) LTD
Bournemouth International Airport, Christchurch, Dorset BH23 6NW
Telephone: 44 (202) 570380
Fax: 44 (202) 580567
Telex: 41500, 41530
CHAIRMAN: Peter D. Purdy
MANAGING DIRECTOR: Brian Curry
CHIEF DESIGNER: Roger Hardy
HEAD OF SALES: Andrew Richardson

Wholly owned subsidiary of FLS Aerospace, division of F. L. Shmidt Group; specialises in deep maintenance and overhaul of military and civil aircraft, particularly of BAC One-Eleven, Shorts 360 and Boeing 737. Acquired assets of Brooklands Aerospace Group 27 July 1990, including Optica observation aircraft originally developed by Edgley Aircraft Ltd. Responsibility transferred from Brooklands' Old Sarum factory to Lovaux at Bournemouth for further engineering and operational improvements prior to certification. Early history, see Brooklands entry 1990-91 *Jane's*. Acquired SAH-1 from Orca Aircraft in October 1991, now renamed Sprint.

FLS OPTICA OA7-300

TYPE: Three-seat slow-flying observation and surveillance aircraft.

PROGRAMME: First flight 14 December 1979; production started 1983 but suspended after acquisition by Lovaux; aircraft redeveloped and prepared for FAA certification, which was awarded December 1991; production restarted at Bournemouth in 1992.

CURRENT VERSIONS: **Optica**: Three-seat observation aircraft, suited particularly for pipeline and powerline inspection, photography, crime prevention and pursuit, fire patrol, game spotting, Customs and Excise duties, pollution control, reconnaissance, narcotics control, disaster monitoring, survey, leisure and sightseeing, security escort, coastguard, search, accident control, police duties, film/TV and press reporting.

CUSTOMERS: Excluding prototype, demonstrator and nine aircraft destroyed before delivery: seven Opticas delivered by late 1990. Opticas now operating (in 1993) in Australia, continental Europe, Middle East, UK and USA.

DESIGN FEATURES: 'Insect eye' cabin with outward view approximately 270° in vertical plane and 340° in horizontal plane, combining all-round field-of-view of a helicopter with lower operating costs of fixed-wing aircraft; ducted propulsor power plant unit, offering very low vibration levels and exceptional quietness, both within cabin and from ground; claimed fuel consumption 16-24 per cent of comparable light turbine helicopter, and 31 per cent of operating costs (fixed and direct for 500 hours per year); low wing loading, pre-set inboard flaps and low stalling speed facilitate continuous low-speed en route flight; generous flap area provides good field performance from both hard and soft strips; stability increases at low speeds; stressed to BCAR Section K (non-aerobatic category) and FAR Pt 23 (Normal category).

Wing section NASA GA(W)-1; thickness/chord ratio of 17 per cent; dihedral 3° on outer panels; incidence 0°.

FLYING CONTROLS: Bottom hinged, mass balanced slotted ailerons outboard of outer flaps, operated by pushrods; balanced rudders; elevators with inset trailing-edge trim tab. Fowler trailing-edge flaps (29 per cent of total wing chord) inboard and outboard of tailbooms; electrically actuated outboard flaps can be set at angles up to 50° for landing; inboard flaps set permanently at 10°, giving effect of slotted wing, for continuous low speed flying; no spoilers or airbrakes; twin inward canted fins; fixed incidence tailplane.

STRUCTURE: Constant chord single-spar non-swept wings of L72 duralumin stressed skin construction; GFRP wingtips;

FLS Club Sprint trainer and Optica OA7-300 slow flying observation aircraft

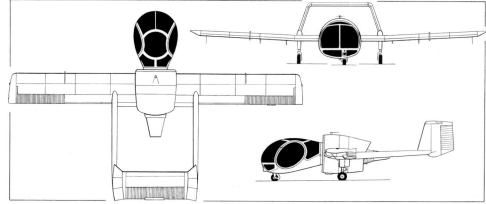

FLS Optica OA7-300 (Textron Lycoming IO-540 engine) *(Dennis Punnett)*

aluminium alloy cabin with ICI Perspex windows (optionally tinted), attached to fan shroud and rest of airframe by six stators of steel tube and aluminium alloy shear web construction; steel tube and aluminium alloy nose beam supports cabin floor; horizontal window frame member just above floor level, together with nosewheel box, designed to withstand 9*g* impact; two movable 7.5 kg (16.5 lb) ballast weights may be positioned on nose beam; landing lamp and taxi/standby landing lamp mounted in nose beam; aluminium alloy stressed skin tubular twin tailbooms; limited GFRP in non-load-bearing areas of tail unit, including fin/tailplane fillets; two movable 9 kg (20 lb) ballast weights may be positioned in fins. Propulsor pod attached to fan shroud with four Lord rubber mountings, and supported by four stators of steel channel and aluminium alloy shear web construction, with steel tube engine bearers. Some fairings of GFRP.

LANDING GEAR: Non-retractable tricycle type, with steerable nosewheel offset to port. Mainwheel legs embody rubber in compression shock absorption. Nosewheel shock absorption by bungee rubber in tension. Single wheel on each unit, tyre sizes 6.00-6 (main) and 5.00-5 (nose). Hydraulic disc brakes on mainwheels. Parking brake. Nosewheel mudguard of GFRP.

POWER PLANT: Ducted propulsor unit, with engine and fan forming a power pod separate from the main shroud. Five-blade Hoffmann fixed-pitch fan, driven by a 194 kW (260 hp) Textron Lycoming IO-540-V4A5D flat-six engine, mounted in a duct downstream of the fan. Fuel tank of 128 litres (33.8 US gallons; 28 Imp gallons) capacity in each wing leading-edge, immediately outboard of tailbooms and forward of wing spar. Tanks are of full wing section, but are designed not to be stressed by wing bending and torsion. Total usable fuel capacity 250 litres (66 US gallons; 55 Imp gallons). Refuelling point in upper surface of each wing. Oil capacity 7.6 litres (2.0 US gallons; 1.7 Imp gallons).

ACCOMMODATION: Cabin designed to accommodate up to three persons side by side on fore and aft adjustable seats, with either single- or two-pilot operation (left hand and centre seats). Dual controls standard. Baggage space aft of seats. Single elliptical door on each side, hinged at front and opening forward. Can be flown with doors removed. Cabin heated, by hot air from engine, and ventilated. A Janitrol combustion heater is offered as an extra.

SYSTEMS: Hydraulics for mainwheel brakes only. Electrical system (24V) includes engine driven alternator and storage battery for engine starting and actuation of flaps.

AVIONICS: Standard nav/com avionics by Bendix/King (Silver Crown). Alternative avionics could be provided.

EQUIPMENT: Special equipment which has been tested successfully includes FLIR, Barr & Stroud IR18 Mk II thermal imager and an air-to-ground video relay. Other equipment such as GEC-Marconi TICM II, searchlights and loudspeakers is being assessed.

DIMENSIONS, EXTERNAL:
Wing span 12.00 m (39 ft 4 in)
Wing chord: basic, constant 1.32 m (4 ft 4 in)
over 10° fixed flaps 1.45 m (4 ft 9 in)
Wing aspect ratio 9.1
Length overall 8.15 m (26 ft 9 in)
Height over fan shroud (excl aerial) 1.98 m (6 ft 6 in)
Diameter of fan shroud 1.68 m (5 ft 6 in)
Diameter of fan 1.22 m (4 ft 0 in)
Shroud ground clearance 0.25 m (10 in)
Height over tailplane 2.31 m (7 ft 7 in)
Tail unit span:
c/l of tailbooms 3.40 m (11 ft 2 in)
intersection fin chord 2.60 m (8 ft 6½ in)
Wheel track 3.40 m (11 ft 2 in)
Wheelbase 2.73 m (9 ft 0 in)
Doors (each): Long axis 1.35 m (4 ft 5 in)
Short axis 0.96 m (3 ft 1¾ in)
Height to sill 0.51 m (1 ft 8 in)

DIMENSIONS, INTERNAL:
Cabin: Length 2.44 m (8 ft 0 in)
Max width (to door Perspex) 1.68 m (5 ft 6 in)
Max height 1.35 m (4 ft 5 in)
Floor area 0.72 m^2 (7.75 sq ft)

AREAS:
Wings, gross 15.84 m^2 (170.5 sq ft)
Ailerons (total) 1.55 m^2 (16.68 sq ft)
Trailing-edge flaps:
inboard (total) 0.61 m^2 (6.57 sq ft)
outboard (total) 1.59 m^2 (17.12 sq ft)
Fins (total) 1.98 m^2 (21.31 sq ft)
Rudders (total) 1.10 m^2 (11.84 sq ft)
Tailplane 1.62 m^2 (17.44 sq ft)
Elevator, incl tab 1.26 m^2 (13.56 sq ft)

WEIGHTS AND LOADINGS:
Weight empty, equipped 948 kg (2,090 lb)
Max cabin load 231 kg (510 lb)
Crew (pilot/observer) 154 kg (340 lb)
Max T-O weight 1,315 kg (2,899 lb)
Max wing loading 83.0 kg/m^2 (17.0 lb/sq ft)
Max power loading 6.78 kg/kW (11.2 lb/hp)

PERFORMANCE (at max T-O weight, forward limit CG):
Never-exceed speed (VNE) 140 knots (259 km/h; 161 mph)
Max level speed 115 knots (213 km/h; 132 mph)
Cruising speed:
50% power 86 knots (159 km/h; 99 mph)
70% power 103 knots (191 km/h; 119 mph)
Loiter speed (40% power) 70 knots (130 km/h; 81 mph)
Stalling speed, outboard flaps up 58 knots (108 km/h; 67 mph)
Max rate of climb at S/L 247 m (810 ft)/min
Service ceiling 4,275 m (14,000 ft)
T-O run 331 m (1,084 ft)
T-O to 15 m (50 ft) 472 m (1,548 ft)
Landing from 15 m (50 ft) 555 m (1,820 ft)
Landing run 278 m (912 ft)
Range with max fuel (45 min reserves):
at 70% power 334 nm (619 km; 385 miles)
at 70 knots (130 km/h; 81 mph) 570 nm (1,056 km; 656 miles)
Endurance (45 min reserves):
at 70% power 3 h 40 min
at 70 knots (130 km/h; 81 mph) 8 h
g limits +3.8/–1.5

FLS SPRINT

TYPE: Two-seat aerobatic primary trainer.

PROGRAMME: Designed by Sydney A. Holloway as Trago Mills SAH-1; initiated October 1977; prototype (G-SAHI) flew 23 August 1983; UK certification obtained 12 December 1985. Orca Aircraft formed August 1988 to take over project but placed under administration in Summer 1989; design and manufacturing rights sold to Lovaux in October 1991. Production restarted.

CURRENT VERSIONS: **Club Sprint:** Current name of standard SAH-1 type. *Description applies to this version, except where indicated.*

Sprint 160: Enhanced version with 119 kW (160 hp) Textron Lycoming AEIO-320-D1B engine, driving Hoffman HO-V72 168 constant-speed propeller; unlimited inverted flying capability; flight test programme commenced March 1992.

DESIGN FEATURES: Wing section NACA 2413.6 (constant); dihedral 5° from roots; incidence 3° at root, 1° at tip.

FLYING CONTROLS: Slotted ailerons and trailing-edge single-slotted flaps; horn balanced elevators with full span trim tab in starboard elevator; horn balanced rudder.

STRUCTURE: Tapered, non-swept aluminium alloy wings with L65 spar booms, L72 sheet skins, stabilised with PVC foam. Aluminium alloy stressed skin fuselage with centre-section spars. Aluminium alloy cantilever tail unit, stabilised with PVC foam.

LANDING GEAR: Non-retractable tricycle type, with single wheel on each unit. Oleo-pneumatic shock absorber in nosewheel leg; spring steel main legs. Cleveland mainwheels and tyres size 6.00-6, pressure 1.24 bars (18.0 lb/sq in). Nosewheel and tyre size 5.00-5, pressure 1.03 bars (15 lb/sq in). Cleveland hydraulic brakes.

POWER PLANT: One 88 kW (118 hp) Textron Lycoming O-235-L2A flat-four engine, driving two-blade fixed-pitch propeller. Integral fuel tank in each wing leading-edge, total capacity 114 litres (30 US gallons; 25 Imp gallons). Refuelling point in upper surface of each wing. Oil capacity 5.7 litres (1.5 US gallons; 1.25 Imp gallons).

ACCOMMODATION: Two seats side by side under rearward sliding bubble canopy. Baggage space aft of seats. Cockpit heated and ventilated.

SYSTEM: 60A engine driven alternator.

EQUIPMENT: Blind-flying instrumentation standard.

AVIONICS: Radio to customer's specification.

DIMENSIONS, EXTERNAL:
Wing span: Club, 160 9.36 m (30 ft 8⅜ in)
Wing chord:
at root: Club, 160 1,515 m (4 ft 11⅔ in)
at tip: Club, 160 0.81 m (2 ft 8 in)
Wing aspect ratio: Club, 160 7.5
Length overall: Club, 160 6.67 m (21 ft 10¾ in)
Height overall: Club, 160 2.32 m (7 ft 7½ in)
Tailplane span: Club 2.74 m (9 ft 0 in)
160 3.05 m (10 ft 0 in)
Wheel track: Club 2.40 m (7 ft 10½ in)
160 2.29 m (7 ft 6 in)
Wheelbase: Club, 160 1.46 m (4 ft 9½ in)
Propeller diameter: Club, 160 1.68 m (5 ft 6 in)
Propeller ground clearance: Club 0.30 m (12 in)
160 0.33 m (13 in)

DIMENSIONS, INTERNAL:
Cockpit: Length 1.52 m (5 ft 0 in)
Max width 1.21 m (3 ft 11½ in)
Baggage space 0.4 m^3 (14.0 cu ft)

AREAS:
Wings, gross 11.15 m^2 (120.0 sq ft)
Ailerons (total) 0.89 m^2 (9.6 sq ft)
Trailing-edge flaps (total) 1.30 m^2 (14.0 sq ft)
Fin 0.96 m^2 (10.3 sq ft)
Rudder 0.63 m^2 (6.8 sq ft)
Tailplane 1.11 m^2 (12.0 sq ft)
Elevators, incl tab 0.93 m^2 (10.0 sq ft)

WEIGHTS AND LOADINGS:
Weight empty, equipped: Club 499 kg (1,100 lb)
160 592 kg (1,306 lb)
Max fuel load: Club, 160 85 kg (188 lb)
Max baggage weight (not permitted for aerobatic flight):
160 45.4 kg (100 lb)
Max T-O weight: Club 748 kg (1,650 lb)
160 871 kg (1,920 lb)
Max wing loading: Club 71.19 kg/m^2 (14.58 lb/sq ft)
160 78.12 kg/m^2 (16.00 lb/sq ft)
Max power loading: Club 8.50 kg/kW (13.98 lb/hp)
160 7.32 kg/kW (12.00 lb/hp)

PERFORMANCE (at max T-O weight):
Never-exceed speed (VNE):
Club 164 knots (304 km/h; 188 mph) EAS
160 182 knots (337 km/h; 209 mph) EAS
Max level speed at S/L:
Club 122 knots (226 km/h; 140 mph)
160 140 knots (259 km/h; 161 mph)
Max cruising speed, 75% power at S/L:
Club 110 knots (204 km/h; 127 mph)
160 126 knots (234 km/h; 145 mph)
Econ cruising speed, 50% power at S/L:
Club, 160 93 knots (172 km/h; 107 mph)
Stalling speed:
flaps up: Club 54 knots (100 km/h; 63 mph) EAS
160 57 knots (106 km/h; 66 mph) EAS
flaps down: Club 48 knots (89 km/h; 56 mph) EAS
160 52 knots (97 km/h; 60 mph) EAS
Max rate of climb at S/L: Club 279 m (915 ft)/min
160 393 m (1,290 ft)/min
Service ceiling: Club 5,000 m (16,400 ft)
160 6,645 m (21,800 ft)
T-O to 15 m (50 ft): Club 392 m (1,285 ft)
160 315 m (1,033 ft)
Landing from 15 m (50 ft): Club 290 m (953 ft)
160 315 m (1,032 ft)
Range with max fuel, 45 min reserves:
Club, at 78 knots (145 km/h; 90 mph) at 1,525 m (5,000 ft) 620 nm (1,149 km; 714 miles)
160, at 95 knots (176 km/h; 109 mph) at 3,050 m (10,000 ft) 500 nm (926 km; 575 miles)
g limits +6/–3

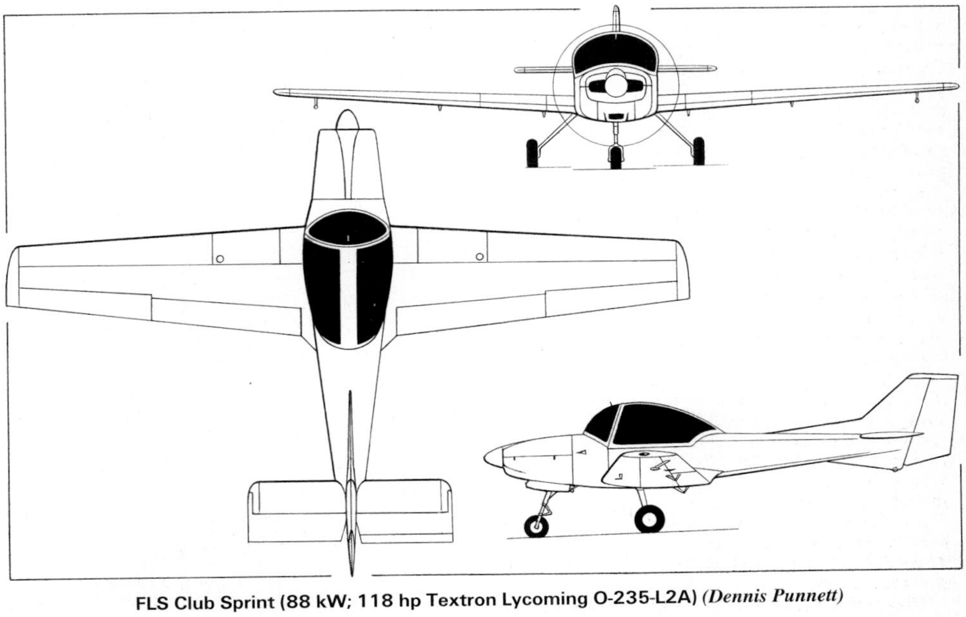

FLS Club Sprint (88 kW; 118 hp Textron Lycoming O-235-L2A) *(Dennis Punnett)*

NASH

NASH AIRCRAFT LTD (Subsidiary of Kinetrol Ltd)

Trading Estate, Farnham, Surrey GU9 9NU
Telephone: 44 (252) 733838
Fax: 44 (252) 713042
Telex: 858567

DIRECTORS:
A. R. B. Nash (Managing)
Roy G. Procter
R. C. Nash

NASH PETREL

TYPE: Two-seat light aircraft.
PROGRAMME: First flight of prototype (G-AXSF) 8 November 1980 with O-320-D2A engine (replaced 1982 by 134 kW; 180 hp Textron Lycoming O-360-A3A); new high tailplane fitted 1983, upswept wingtips 1984 (further details in 1990-91 and earlier *Jane's*). Pre-production batch of five Petrels started in about 1986; work towards certification proceeding, albeit slowly.
DESIGN FEATURES: Optimised for club/touring use, especially ease of handling and for glider towing.
FLYING CONTROLS: Manual operation of primary surfaces and NACA slotted flaps.
STRUCTURE: See 1990-91 *Jane's* for prototype; pre-production aircraft identical aerodynamically, but differ structurally.

Prototype of the Nash Petrel two-seat light aircraft *(R. J. Malachowski)*

PILATUS BRITTEN-NORMAN

PILATUS BRITTEN-NORMAN LTD (Subsidiary of Oerlikon-Bührle Holding Ltd)

Bembridge, Isle of Wight PO35 5PR
Telephone: 44 (983) 872511
Fax: 44 (983) 873246
Telex: 86277/86866
MANAGING DIRECTOR: A. Stansfeld
PRODUCTION DIRECTOR: E. Taylor
TECHNICAL DIRECTOR: R. Wilson
HEAD OF CUSTOMER SERVICES: Richard M. Blake
HEAD OF MARKETING: W. Stark

Pilatus Aircraft Ltd of Switzerland acquired Britten-Norman (Bembridge) Ltd 1979, including Isle of Wight facilities and former Fairey SA Islander/Trislander production hardware at Gosselies, Belgium.

Previous company history appeared in 1978-79 and earlier *Jane's*.

PILATUS BRITTEN-NORMAN BN2B ISLANDER

TYPE: Twin-engined feederline transport.
PROGRAMME: Prototype (G-ATCT) first flight 13 June 1965 with two 157 kW (210 hp) Rolls-Royce Continental IO-360-B engines and 13.72 m (45 ft) span wings; subsequently re-engined with Textron Lycoming O-540s and flown 17 December 1965; wing span also increased by 1.22 m (4 ft) to initial production standard; production prototype BN2 (G-ATWU) flown 20 August 1966; domestic C of A received 10 August 1967; FAA type certificate 19 December 1967; Romanian manufacture (see Romaero entry) began 1969.
CURRENT VERSIONS: **BN2 Islander:** Initial piston engined production model (see earlier *Jane's*).
BN2A Islander: Piston version built from 1 June 1969 (see 1977-78 *Jane's*).
BN2B Islander: Current standard piston engined version; higher max landing weight; improved interior design; available with two engine choices and optional wingtip fuel tanks as **BN2B-26** with O-540s and **BN2B-20** with IO-540s (BN2B-27 and -21 no longer available). Features include range of passenger seats and covers, more robust door locks, improved door seals and stainless steel sills, redesigned fresh air system to improve ventilation in hot and humid climates, smaller diameter propellers to decrease cabin noise, and redesigned flight deck and instrument panel. *Detailed description applies to this version.*

Series of modification kits available as standard or option for new production aircraft and can be fitted retrospectively to existing aircraft; extended nose, incorporating 0.62 m³ (22 cu ft) additional baggage space, introduced as option 1972; Jonas Aircraft developed Rajay turbocharging installation, optional for 194 kW (260 hp) engines as bolt-on unit.
BN2B Defender: Described separately (several versions).
BN2B Maritime Defender: Described separately.
BN2T Turbine Islander: Described separately.
MSSA: Described separately.
CUSTOMERS: By January 1993 deliveries of Islanders and Defenders totalled 1,160; recent customers include Fiji's Air Wakaya, Falkland Islands, Isles of Scilly Skybus, Central African Republic government, Berjaya Air Charter (Malaysia), Roc Aviation (Taiwan), and Zimbabwe Dept of Civil Aviation.
DESIGN FEATURES: STOL characteristics; three-blade propellers available on version of BN2B-26 giving quieter noise signature; NACA 23012 wing section; no dihedral; incidence 2°; no sweepback.
FLYING CONTROLS: Slotted ailerons, with starboard ground adjustable tab, operated by pushrods and cables; mass balanced elevator; rudder and elevator actuated by pushrods and cables; trim tabs in rudder and elevator; single-slotted flaps operated electrically; fixed incidence tailplane. Dual controls standard.
STRUCTURE: L72 aluminium-clad aluminium alloys; two-spar wing torsion box in one piece; flared-up wingtips; integral fuel tanks in wingtips optional; four-longeron fuselage of pressed frames and stringers; two-spar tail unit with pressed ribs.
LANDING GEAR: Non-retractable tricycle type, with twin wheels on each main unit and single steerable nosewheel. Cantilever main legs mounted aft of rear spar. All three legs fitted with oleo-pneumatic shock absorbers. All five wheels and tyres size 16 × 7-7, supplied by Goodyear. Tyre pressure: main 2.41 bars (35 lb/sq in); nose 2.00 bars (29 lb/sq in). Foot operated air-cooled Cleveland hydraulic brakes on main units. Parking brake. Wheel/ski gear available optionally. Min ground turning radius 9.45 m (31 ft 0 in).
POWER PLANT: Two Textron Lycoming flat-six engines, each driving a Hartzell HC-C2YK-2B or -2C two-blade constant-speed feathering metal propeller; optional three-blade Hartzell HC-C3YR-2UF/FC8468-8R for O-540-E4C5 engines. Propeller synchronisers optional. Standard power plant is the 194 kW (260 hp) O-540-E4C5, but the 224 kW (300 hp) IO-540-K1B5 can be fitted at customer's option. Optional Rajay turbocharging installation on 194 kW (260 hp) engines, to improve high altitude performance. Integral fuel tank between spars in each wing, outboard of engine. Total fuel capacity (standard) 518 litres (137 US gallons; 114 Imp gallons). Usable fuel 492 litres (130 US gallons; 108 Imp gallons). With optional fuel tanks in wingtips, total capacity is increased to 855 litres (226 US gallons; 188 Imp gallons). Additional pylon mounted underwing auxiliary tanks, each of 227 litres (60 US gallons; 50 Imp gallons) capacity, available optionally. Refuelling point in upper surface of wing above each internal tank. Total oil capacity 22.75 litres (6 US gallons; 5 Imp gallons).
ACCOMMODATION: Up to 10 persons, including pilot, on side by side front seats and four bench seats. No aisle. Seat backs fold forward. Access to all seats via three forward opening doors, forward of wing and at rear of cabin on port side and forward of wing on starboard side. Baggage compartment at rear of cabin, with port side loading door in standard versions. Exit in emergency by removing door windows. Special executive layouts available. Can be operated as freighter, carrying more than a ton of cargo; in this configuration passenger seats can be stored in rear baggage bay. In ambulance role, up to three stretchers and two attendants can be accommodated. Other layouts possible, including photographic and geophysical survey, parachutist transport or trainer (with accommodation for up to eight parachutists and dispatcher), firefighting, environmental protection and cropspraying.
SYSTEMS: Southwind cabin heater standard. 45,000 BTU Stewart Warner combustion unit, with circulating fan, provides hot air for distribution at floor level outlets and at windscreen demisting slots. Fresh air, boosted by propeller slipstream, is ducted to each seating position for on-ground ventilation. Electrical DC power, for instruments, lighting and radio, from two engine driven 24V 50A self-rectifying alternators and a controller to main busbar and circuit breaker assembly. Emergency busbar is supplied by a 24V 17Ah heavy duty lead-acid battery in the event of a twin alternator failure. Ground power receptacle provided. Optional electric de-icing of propellers and windscreen, and pneumatic de-icing of wing and tail unit leading-edges. Oxygen system available optionally for all versions.
AVIONICS: Standard items include blind-flying instrumentation, autopilot, and a wide range of VHF and HF communications and navigation equipment. Intercom, including second headset, and passenger address system are standard.
DIMENSIONS, EXTERNAL:
Wing span 14.94 m (49 ft 0 in)
Wing chord, constant 2.03 m (6 ft 8 in)
Wing aspect ratio 7.4

Pilatus Britten-Norman BN2B-26 of the Hampshire Police Authority's Aerial Support Unit *(Peter. J. Cooper)*

Length overall	10.86 m (35 ft 7¾ in)
Fuselage: Max width	1.21 m (3 ft 11½ in)
Max depth	1.46 m (4 ft 9¾ in)
Height overall	4.18 m (13 ft 8¾ in)
Tailplane span	4.67 m (15 ft 4 in)
Wheel track (c/l of shock absorbers)	3.61 m (11 ft 10 in)
Wheelbase	3.99 m (13 ft 1¼ in)
Propeller diameter	1.98 m (6 ft 6 in)
Cabin door (front, port):	
Height	1.10 m (3 ft 7½ in)
Width: top	0.64 m (2 ft 1¼ in)
Height to sill	0.59 m (1 ft 11¼ in)
Cabin door (front, starboard):	
Height	1.10 m (3 ft 7½ in)
Max width	0.86 m (2 ft 10 in)
Height to sill	0.57 m (1 ft 10½ in)
Cabin door (rear, port): Height	1.09 m (3 ft 7 in)
Width: top	0.635 m (2 ft 1 in)
bottom	1.19 m (3 ft 11 in)
Height to sill	0.52 m (1 ft 8½ in)
Baggage door (rear, port): Height	0.69 m (2 ft 3 in)

DIMENSIONS, INTERNAL:
Passenger cabin, aft of pilot's seat:
Length	3.05 m (10 ft 0 in)
Max width	1.09 m (3 ft 7 in)
Max height	1.27 m (4 ft 2 in)
Floor area	2.97 m² (32 sq ft)
Volume	3.68 m³ (130 cu ft)

Baggage space aft of passenger cabin
1.39 m³ (49 cu ft)

Freight capacity:
aft of pilot's seat, incl rear cabin baggage space
4.70 m³ (166 cu ft)
with four bench seats folded into rear cabin baggage space
3.68 m³ (130 cu ft)

AREAS:
Wings, gross	30.19 m² (325.0 sq ft)
Ailerons (total)	2.38 m² (25.6 sq ft)
Flaps (total)	3.62 m² (39.0 sq ft)
Fin	3.41 m² (36.64 sq ft)
Rudder, incl tab	1.60 m² (17.2 sq ft)
Tailplane	6.78 m² (73.0 sq ft)
Elevator, incl tabs	3.08 m² (33.16 sq ft)

WEIGHTS AND LOADINGS (A: 194 kW; 260 hp engines, B: 224 kW; 300 hp engines):
Weight empty, equipped (without avionics):
A	1,866 kg (4,114 lb)
B	1,925 kg (4,244 lb)
Max payload: A	929 kg (2,048 lb)
B	870 kg (1,918 lb)
Payload with max fuel: A	692 kg (1,526 lb)
B	633 kg (1,396 lb)
Max fuel weight: standard: A, B	354 kg (780 lb)
with optional tanks in wingtips: A, B	585 kg (1,290 lb)
Max T-O and landing weight: A, B	2,993 kg (6,600 lb)
Max zero-fuel weight (BCAR):	
A, B	2,855 kg (6,300 lb)
Max wing loading: A, B	99.1 kg/m² (20.3 lb/sq ft)
Max floor loading, without cargo panels:	
A, B	586 kg/m² (120 lb/sq ft)
Max power loading: A	7.71 kg/kW (12.7 lb/hp)
B	6.68 kg/kW (11.0 lb/hp)

PERFORMANCE (at max T-O weight. A and B as above):
Never-exceed speed (V_NE):
A, B 183 knots (339 km/h; 211 mph) IAS
Max level speed at S/L:
A	148 knots (274 km/h; 170 mph)
B	151 knots (280 km/h; 173 mph)

Max cruising speed (75% power) at 2,135 m (7,000 ft):
A	139 knots (257 km/h; 160 mph)
B	142 knots (264 km/h; 164 mph)

Cruising speed (67% power) at 2,750 m (9,000 ft):
A	134 knots (248 km/h; 154 mph)
B	137 knots (254 km/h; 158 mph)

Cruising speed (59% power) at 3,660 m (12,000 ft):
A	130 knots (241 km/h; 150 mph)
B	132 knots (245 km/h; 152 mph)

Stalling speed:
flaps up: A, B 50 knots (92 km/h; 57 mph) IAS
flaps down: A, B 40 knots (74 km/h; 46 mph) IAS
Max rate of climb at S/L: A 262 m (860 ft)/min
B 344 m (1,130 ft)/min
Rate of climb at S/L, OEI: A 44 m (145 ft)/min
B 60 m (198 ft)/min
Absolute ceiling: A 4,145 m (13,600 ft)
B 6,005 m (19,700 ft)
Service ceiling: A 3,445 m (11,300 ft)
B 5,240 m (17,200 ft)
Service ceiling, OEI: A 1,525 m (5,000 ft)
B 1,980 m (6,500 ft)
T-O run at S/L, zero wind, hard runway:
A 278 m (913 ft)
B 264 m (866 ft)
T-O run at 1,525 m (5,000 ft): A 396 m (1,299 ft)
B 372 m (1,221 ft)
T-O to 15 m (50 ft) at S/L, zero wind, hard runway:
A 371 m (1,218 ft)
B 352 m (1,155 ft)
T-O to 15 m (50 ft) at 1,525 m (5,000 ft):
A 528 m (1,732 ft)
B 496 m (1,628 ft)

Landing from 15 m (50 ft) at S/L, zero wind, hard runway:
A, B 299 m (980 ft)
Landing from 15 m (50 ft) at 1,525 m (5,000 ft):
A, B 357 m (1,170 ft)
Landing run at 1,525 m (5,000 ft): A, B 171 m (560 ft)
Landing run at S/L, zero wind, hard runway:
A, B 140 m (460 ft)
Range at 75% power at 2,135 m (7,000 ft):
A, standard fuel 622 nm (1,153 km; 717 miles)
A, with optional tanks
1,023 nm (1,896 km; 1,178 miles)
B, standard fuel 555 nm (1,028 km; 639 miles)
B, with optional tanks 920 nm (1,704 km; 1,059 miles)
Range at 67% power at 2,750 m (9,000 ft):
A, standard fuel 713 nm (1,322 km; 822 miles)
A, with optional tanks
1,159 nm (2,147 km; 1,334 miles)
B, standard fuel 577 nm (1,070 km; 665 miles)
B, with optional tanks 975 nm (1,807 km; 1,123 miles)
Range at 59% power at 3,660 m (12,000 ft):
A, standard fuel 755 nm (1,400 km; 870 miles)
A, with optional tanks
1,216 nm (2,253 km; 1,400 miles)
B, standard fuel 613 nm (1,136 km; 706 miles)
B, with optional tanks
1,061 nm (1,965 km; 1,221 miles)

PILATUS BRITTEN-NORMAN BN2T TURBINE ISLANDER

TYPE: Twin-turboprop development of piston engined Islander.
PROGRAMME: First flight of prototype (G-BPBN) 2 August 1980 with two Allison 250-B17C turboprops; British CAA certification received end of May 1981; first production aircraft delivered December 1981; FAR Pt 23 US type approval 15 July 1982; full icing clearance to FAR Pt 25 gained 23 July 1984.
CUSTOMERS: Forty-seven delivered by January 1993.
DESIGN FEATURES: Turboprops enable use of available low cost jet fuel instead of scarce and costly Avgas, and offer particularly low operating noise level; available for same range of applications as Islander, including military versions (described separately).
Description of BN2B Islander applies also to BN2T, except as follows:
POWER PLANT: Two 298 kW (400 shp) Allison 250-B17C turboprops, flat rated at 238.5 kW (320 shp), and each driving a Hartzell three-blade constant-speed fully feathering metal propeller. Usable fuel 814 litres (215 US gallons; 179 Imp gallons). Pylon mounted underwing tanks, each of 227 litres (60 US gallons; 50 Imp gallons) capacity, are available optionally for special purposes. Total oil capacity 5.7 litres (1.5 US gallons; 1.25 Imp gallons).
ACCOMMODATION: Generally as for BN2B. In ambulance role can accommodate, in addition to pilot, a single stretcher, one medical attendant and five seated occupants; or two stretchers, one attendant and three passengers; or three stretchers, two attendants and one passenger. Other possible layouts include photographic and geophysical survey; parachutist transport or trainer (with accommodation for up to eight parachutists and a dispatcher); and pest control or other agricultural spraying. Maritime Turbine Islander/Defender versions available for fishery protection, coastguard patrol, pollution survey, search and rescue, and similar applications. In-flight sliding parachute door optional.
AVIONICS: Radar, VLF/Omega nav system, radar altimeter, marine band and VHF transceivers.
EQUIPMENT: According to mission, can include fixed tail sting or towed bird magnetometer, spectrometer, or electromagnetic detection/analysis equipment (geophysical survey); one or two cameras, navigation sights and appropriate avionics (photographic survey); 188.7 litre (50 US gallon; 41.5 Imp gallon) Micronair underwing spraypods complete with pump and rotary atomiser (pest control/agricul-

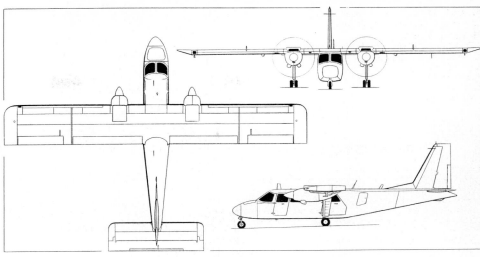

Pilatus Britten-Norman BN2T Turbine Islander *(Dennis Punnett)*

tural spraying versions); dinghies, survival equipment and special crew accommodation (maritime versions).
DIMENSIONS, EXTERNAL: As for BN2B, except:
Length overall: standard nose	10.86 m (35 ft 7¾ in)
weather radar nose	11.07 m (36 ft 3¾ in)
Propeller diameter	2.03 m (6 ft 8 in)

WEIGHTS AND LOADINGS:
Weight empty, equipped	1,832 kg (4,040 lb)
Payload with max fuel	608 kg (1,340 lb)
Max T-O weight	3,175 kg (7,000 lb)
Max landing weight	3,084 kg (6,800 lb)
Max zero-fuel weight	2,994 kg (6,600 lb)
Max wing loading	105.17 kg/m² (21.54 lb/sq ft)
Max power loading	5.33 kg/kW (8.75 lb/shp)

PERFORMANCE (standard Turbine Islander/Defender, at max T-O weight, ISA, except where indicated):
Max cruising speed:
at 3,050 m (10,000 ft) 170 knots (315 km/h; 196 mph)
at S/L 154 knots (285 km/h; 177 mph)
Cruising speed, 72% power:
at 3,050 m (10,000 ft) 150 knots (278 km/h; 173 mph)
at 1,525 m (5,000 ft) 143 knots (265 km/h; 165 mph)
Stalling speed, power off:
flaps up 52 knots (97 km/h; 60 mph) IAS
flaps down 45 knots (84 km/h; 52 mph) IAS
Max rate of climb at S/L 320 m (1,050 ft)/min
Rate of climb at S/L, OEI 66 m (215 ft)/min
Service ceiling over 7,620 m (25,000 ft)
Absolute ceiling, OEI over 3,050 m (10,000 ft)
T-O run 255 m (837 ft)
T-O to 15 m (50 ft) 381 m (1,250 ft)
Landing from 15 m (50 ft) 339 m (1,110 ft)
Landing run 231 m (757 ft)
Range (IFR) with max fuel, reserves for 45 min hold plus 10% 590 nm (1,093 km; 679 miles)
Range (VFR) with max fuel, no reserves
728 nm (1,349 km; 838 miles)

PILATUS BRITTEN-NORMAN DEFENDER

TYPE: Variants of the civil BN2B and BN2T, specially developed for military and para-military roles.
CUSTOMERS: Include Belgian Army, Belize Defence Force, Botswana Defence Force, Jamaica Defence Force, Mauritanian Air Force, Seychelles Defence Force and Surinam Air Force.
DESIGN FEATURES: Similar to those of civil Islander, retaining STOL capability, but with four underwing hardpoints for standard NATO pylons to attach fuel tanks, weapons and other stores; number of additional airframe options, including sliding door on rear left side which can be opened in flight are also offered. Concept of Defender is to provide a low-cost airframe which can be fitted with best available sensors to meet operational needs of customers. A number of specialist role variants are offered; Mauritius BN2T is for EEZ and fishery patrol, equipped with Bendix/King radar and FLIR for 24-hour availability.

PILATUS BRITTEN-NORMAN BATTLEFIELD SURVEILLANCE DEFENDER

TYPE: Battlefield surveillance aircraft (modified BN2T) developed from ASTOR.
PROGRAMME: ASTOR (Airborne STand-Off Radar) was acronym for British MoD programme to develop airborne surveillance radar to provide overall picture of battle area; intended to operate in conjunction with Phoenix battlefield reconnaissance UAV (to obtain closer look at target/area shown by ASTOR to require detailed examination) and/or Joint STARS surveillance aircraft.
Two platforms evaluated for ASTOR (BAC/English Electric Canberra the high-level option); more versatile BN2T Defender operated from medium heights; radar trials included GEC equipment in Canberra and GEC Ferranti (now GEC-Marconi) equipment in Defender

Pilatus Britten-Norman BN2T Maritime Defender of Mauritius National Coast Guard *(Peter. J. Cooper)*

Pilatus Britten-Norman Battlefield Surveillance Defender

(modified THORN EMI Searchwater in Canberra testbed since 1982).

Work on ASTOR (then CASTOR) Turbine Defender testbed (G-DLRA, c/n 2140; now ZG989) began 5 March 1984; converted aircraft first flight 12 May, with flat-bottomed undernose radome containing ballast only; new nose offered adequate performance and virtually unaltered handling characteristics; G-DLRA fitted with modified version of THORN EMI Skymaster radar in AEW Defender type nose radome 1988; tests as low-level ASTOR platform, flown at altitudes up to 3,050 m (10,000 ft), ended December 1990 (Canberra programme March 1991). Aircraft system now renamed.

DESIGN FEATURES: Nose radome considered best location for optimum lookdown capability for scanner; nosewheel leg lengthened by 30.5 cm (12 in) and Trislander main landing gear fitted to give necessary ground clearance. Tests involved new electrically driven antenna and additional signal processing techniques intended to provide required detection and azimuth resolution of slow moving targets against ground clutter background; data link trials 1990 evaluated system's interoperability with Joint STARS (see Avionics).

STRUCTURE: See Programme for changes to nose.

LANDING GEAR: Modified BN2A Mk III Trislander main landing gear. Longer nosewheel leg to provide adequate ground clearance for radome. Main landing gear tyre size 6.50 × 8, pressure 2.4 bars (35 lb/sq in).

AVIONICS: THORN EMI Skymaster multi-mode radar in nose, modified originally to meet UK General and Air Staff Requirement 3956, to provide primary intelligence information in immediate battle zone and beyond, while operating well within friendly territory. Has 360° scan, and offers wide area of coverage against moving and fixed targets. Associated transmitter, receiver and processing equipment housed in fuselage of Defender, which is flown and operated by two-man crew. Data acquired are processed and transmitted automatically, via airborne link, to one or more ground stations.

DIMENSIONS, EXTERNAL: As Turbine Islander/Defender except:
Length overall 12.37 m (40 ft 7¼ in)
WEIGHTS AND LOADINGS:
Design max T-O weight 3,630 kg (8,000 lb)

PILATUS BRITTEN-NORMAN ELINT DEFENDER

TYPE: Economically viable electronic warfare platform.
PROGRAMME: Jointly funded with Racal Radar Defence Systems Ltd; BN2T (G-DEMO) fitted with Racal Kestrel lightweight EW avionics suite as demonstrator 1989.
DESIGN FEATURES: Kestrel system provides ability to compile comprehensive library of electronic intelligence, constantly updated; offers total radar band coverage with instant onboard analysis and/or data recording for ground replay and analysis, and allows peacetime strategic intelligence role or tactical elint in active environment; flight crew plus equipment operator.

Underwing pods carry two outward-facing antenna units for Kestrel system; further two antennae opposite sides of tail fin; precision navigation system, such as Omega or an INS, standard; could remain on task at 87 nm (161 km; 100 miles) from base for four hours.

PILATUS BRITTEN-NORMAN MARITIME DEFENDER

TYPE: All-weather, day or night, maritime and coastal patrol aircraft for EEZ protection, fishery control, anti-smuggling and SAR roles.
CUSTOMERS: Falkland Islands Government Air Service, Cyprus Police, Pakistan Maritime Security Agency, Mauritius National Coast Guard, Morocco Gendarmerie, Indian Navy.
DESIGN FEATURES: Available with either piston or turboprop engines. Dimensions identical to those of basic aircraft but with four underwing hardpoints; two 227 litre (60 US gallon; 50 Imp gallon) fuel tanks, usually on inboard pods; strengthened floor; full length seat rails; rear port side sliding door; and nose-mounted sector scan radar.
AVIONICS: Pilot and co-pilot's panels incorporate modern GA instruments with Omega/GPS and integrated autopilot. Radar altimeter, HF, VHF, MF and UHF(FM) radios and all standard ATC navigation aids available. Primary sensor is radar with sector scan, but 360° scan available to customer's choice; thermal imaging systems and video and film cameras linked to navigation system.

PILATUS BRITTEN-NORMAN BORDER PATROL AND INTERNAL SECURITY DEFENDER

British Army designation: Islander AL. Mk 1
RAF designations: Islander CC. Mks 2/2A

TYPE: BN2B or BN2T fitted with sensors and weapons for border patrol, internal security and police duties.
PROGRAMME: Developed from aircraft bought by UK MoD for the British Army Air Corps in 1989. Additional aircraft subsequently delivered to Army Air Corps, Royal Ulster Constabulary and RAF. Piston-engined aircraft of lower specification used by Hampshire Police Authority.
ACCOMMODATION: Up to 10 troops or paratroops.
AVIONICS: Full ATC, tactical navigation aids and communications suite fitted, with OMEGA and GPS as primary navigation system. Thermal imaging equipment can be fitted on fuselage or underwing stations and can be complemented by range of vertical and oblique cameras mounted inside cabin.
ARMAMENT: Four underwing stations for wide assortment of weapons.

PILATUS BRITTEN-NORMAN DEFENDER 4000

TYPE: Enhanced variant of BN2T.
PROGRAMME: Marketing experience with BN2T revealed many military and government agencies expressed need for greater payload; Defender 4000 (BN2T-4) developed to meet this requirement.
DESIGN FEATURES: New wing of 16.15 m (53 ft) span based on Trislander wing and fitted to strengthened fuselage and landing gear; redesigned nose and tail unit. To maintain take-off and climb performance at higher weights, Allison 250-B17F engines fitted; 3,856 kg (8,500 lb) AUW, offering 100 per cent increase in payload.
POWER PLANT: Two 298 kW (400 shp) Allison 250-B17F turboprops.
AVIONICS: In keeping with modernised airframe, a new EFIS-based flight deck has also been designed for new variant.
PERFORMANCE: As for BN2T

PILATUS BRITTEN-NORMAN MSSA

TYPE: Multi-sensor surveillance aircraft.
PROGRAMME: Developed (as BN2T-4R) with Westinghouse Electric Corporation as low-cost, off-the-shelf system for surveillance, border and fisheries patrol, drug interdiction and special operations/low-intensity conflict applications. Demonstrator (G-MSSA, conversion of AEW Defender G-TEMI) rolled out at Baltimore on 10 September 1991, leased to US Navy Spring 1992 for trials at NATC Patuxent River.
DESIGN FEATURES: Based on Defender 4000.
AVIONICS: Integrated multi-sensor system includes AN/APG-66 radar with 360° rotating antenna; WF-360 FLIR and GPS receiver; dual UHF and VHF radios; LTN-92 ring laser gyro INS; real-time video data link; high-resolution multi-function displays.
CUSTOMERS: Turkish government announced intention to purchase unspecified number of MSSAs in 1993.

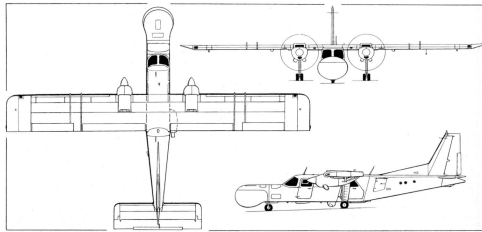

Pilatus Britten-Norman MSSA with Westinghouse Electric nose-mounted radar *(Dennis Punnett)*

SHORTS
SHORT BROTHERS PLC (Subsidiary of Bombardier Inc)

PO Box 241, Airport Road, Belfast BT3 9DZ, Northern Ireland
Telephone: 44 (232) 458444
Fax: 44 (232) 732974
Telex: 74688
OTHER WORKS: Newtownards, Castlereagh, Belfast (3), Dunmurry, Newtownabbey
LONDON OFFICE: 14 Queen Anne's Gate, London SW1H 9AA
Telephone: 44 (71) 222 4555
Fax: 44 (71) 976 8505
MARKETING SUBSIDIARY: Skytech Aviation Services Ltd, Monachus House, Hartley Wintney, nr. Farnborough, Hampshire RG27 8TP
Telephone: 44 (252) 84 4891
Fax: 44 (252) 84 3718
PRESIDENT: R. W. R. McNulty, CBE
EXECUTIVE VICE-PRESIDENT: A. F. C. Roberts, OBE
DIRECTOR, PUBLIC AFFAIRS: R. J. Gordon

After more than 40 years in UK government ownership, Shorts, oldest established aircraft manufacturer in the world, was acquired in October 1989 by Bombardier, a Canadian Corporation engaged in design, development, manufacturing and marketing activities in fields of transportation equipment, motorised consumer products, aerospace and defence. Shorts is European Group of Bombardier Aerospace, which also comprises North American Group (Canadair, de Havilland and Learjet); operates internationally from headquarters in Belfast, Northern Ireland, with subsidiary offices in London, Washington, Bahrain and Hong Kong; currently employs approximately 8,000 personnel worldwide on broad-based work programme in five operating units: Aircraft, Nacelle Systems, Defence Systems, Manufacturing and Fokker divisions. Major new investment by Shorts in plant, facilities and machinery has been carried out since 1989. Modern equipment, extended capacity, new systems, and updated working practices have contributed to establishing totally new standards of quality and productivity.

Two manufacturing centres, Newtownabbey and Dunmurry, cater for advanced composites programme; over 27,871 m² (300,000 sq ft) production floor area allocated to manufacture of components for Shorts aerostructure contracts with Boeing, Rolls-Royce, International Aero Engines, Fokker and British Aerospace.

Shorts has major partnership role in Canadair Regional Jet, responsible for designing and manufacturing 9.75 m (32 ft) long centre section of fuselage, including forward and aft fuselage extension plugs, as well as wing flaps, ailerons, spoilers and spoilerons, and design of vanes.

In September 1992 it was announced that Shorts will design and build complete fuselage and tail unit for new Learjet 45 mid-size business jet in Belfast.

Following intense international competition, Shorts Tucano was selected by UK Ministry of Defence as basic trainer for RAF. Deliveries of first of 130 aircraft commenced 1988; has also exported the aircraft. Shorts designed, developed and now produces Fokker 100 wing as risk sharing partner with Fokker of Netherlands and Deutsche Aerospace Airbus of Germany; 57 wing sets delivered 1991. Shorts also currently sole source supplier of high technology components for Boeing's 737, 747 and 757 programmes. In December 1992 Shorts was awarded three major follow-on contracts by Boeing, together worth over $125 million, for supply of all-composite rudders for Boeing 737, main landing gear doors for Boeing 747 and inboard trailing-edge wing flaps for Boeing 757. In 1991 Shorts was also awarded contract to supply nose landing gear doors for Boeing 777.

Shorts is source of high-technology aerostructures, and specialises in design and manufacture of engine nacelle systems and components of turbofan engines through its Nacelle Systems Division. The Division makes a significant contribution to many Rolls-Royce power plant programmes, and uses its resources in innovative use of new materials and technologies to improve product lines. Company is sole source supplier of nacelle components for RB211 engines used on Boeing 747, 757 and 767, and for design, development and manufacture of nacelles for Trent 775 for new generation of large wide-body aircraft such as A330. Shorts builds complete nacelles for Textron Lycoming LF 507 engines on all Avro International Aerospace Regional Jetliner variants and is responsible for design, development and manufacture of some 40 per cent of nacelle for International Aero Engines V2500 D series engine. Division also recently won contract to supply complete nacelle for General Electric CF34 used on Canadair Regional Jet and Challenger business jet, deliveries commencing late 1993. In November 1992 Shorts and Hurel-Dubois were selected by BMW Rolls-Royce GmbH to provide nacelles for BR700 range of aero engines. In December 1992 Shorts and Hurel-Dubois announced formation of International Nacelle Systems (INS), which will offer complete engine nacelle capability to aircraft and engine manufacturers worldwide. Through a separate entity, Société International de Nacelles (Toulouse) (SINT), the two companies are establishing aero podding facility at Toulouse, France.

Shorts is a world leader in close air defence systems (main products Starburst and Javelin). Starburst was in service with British Forces during Gulf War, when its reliability was stated by UK MoD to be "unparalleled". Defence Systems Division markets Shorland range of armoured vehicles, Skeet aerial target, and a new weapon simulation and training product known as S1 Multi Arms Trainer.

Short Brothers Air Services Limited operates aerial targets for UK MoD and overseas countries.

SHORTS 330 and 330-UTT

Orders and options for 330, 330-UTT and Sherpa totalled 136 by September 1992, when final C-23B was delivered. No longer in production. Full details plus photographs and three-view drawings can be found in 1992-93 *Jane's*.

SHORTS SHERPA
US Air Force and Army designations: C-23A
US Army National Guard designation: C-23B

Final C-23B (90-7016) was delivered to Puerto Rico STARC/US Army National Guard on 1 September 1992, ending Sherpa production. Details in 1992-93 *Jane's*, except that US Forestry Service received seven ex-US Air Force C-23As, US Army seven C-23As (ex-US Air Force), and a second civil registered Sherpa is also in service with government of Quebec Province, Canada.

SHORTS S312 TUCANO
RAF designation: Tucano T. Mk 1
TYPE: Two-seat turboprop basic trainer and (exports) COIN.
PROGRAMME: Co-operation agreement between Shorts and Embraer of Brazil to develop new version of EMB-312 Tucano, to meet or exceed all UK MoD Air Staff Target 412 requirements as Jet Provost replacement for RAF, announced May 1984; selected by UK government 21 March 1985, partly because least expensive of competitors; first flight of EMB-312 with required Garrett engine (PP-ZTC) in Brazil 14 February 1986; airfreighted to UK after 14.35 flight test hours, reassembled as G-14-007 (later G-BTUC); first flight of G-14-007 in UK 11 April 1986; first flight of production T. Mk 1 (ZF135) 30 December 1986; formal rollout 20 January 1987; ZF135 delivered to Aeroplane & Armament Experimental Establishment, Boscombe Down, 26 June 1987 for provisional type certificate trials; deliveries to Central Flying School, Scampton, began (with ZF138) 16 June 1988; flying by Tucano Course Design Team (CFS) started early August 1988; formal handover of first aircraft 1 September 1988; Shorts contracted January 1990 to modify first 50 completed aircraft with strengthened flying controls, revised com/nav equipment and structural changes to extend fatigue life, as introduced on production line. Shorts also responsible for logistic supplies; Airwork has maintenance contract at Church Fenton, Hunting Aircraft at Cranwell and (from 1 April 1993) Scampton, and Shorts at Finningley; five GEC-Marconi simulators delivered to bases from 1990. Weapon trials completed April 1991.

CUSTOMERS: Initially, 130 for RAF; student flying began at No. 7 FTS, Church Fenton, December 1989; first solo 11 January 1990; course completed 7 September 1990. Course Design Team disbanded November 1989 and CFS Tucano Squadron began working up to 15 aircraft; 30 aircraft to No. 3 FTS at RAF College, Cranwell, from (ZF144) 15 November 1990 onwards; No. 7 FTS closed April 1992, passing some Tucanos to No. 1 FTS at Linton-on-Ouse; navigator school, No. 6 FTS at Finningley, received 10 and flew first sortie on 19 May 1992; also intended for Refresher Flying Flight within No. 1 FTS; final RAF Tucano (ZF516) first flown 23 December 1992, delivered January 1993. RAF's baseline Tucano course 146 h 30 min for students with 30-60 hours previous piston-engine training; helicopter pilots receive 63½ Tucano hours, multi-engined aircraft pilots 140 hours.

Exports to Kenya (12 armed **T. Mk 51s** for weapons training), ordered 1988 and first flown ('811', temporarily ZH203) 11 October 1989; delivered 14 June 1990 to 4 June 1991. Kuwait (16 **T. Mk 52s** and support equipment, including simulator); first aircraft ('101', temporarily ZH506) flown 21 September 1990; last aircraft flown 20 June 1991; all stored in UK pending reconstruction of Kuwaiti air bases and formation of No. 19 Squadron. Total announced orders 158; all built, but further sales prospects being pursued in 1993; additional aircraft crashed on test

Shorts Tucano T. Mk 1 used by No. 6 FTS at Finningley for navigator training *(Paul Jackson)*

Shorts Tucano T. Mk 51 before delivery to Kenya for light attack duties

400   UK: AIRCRAFT—SHORTS

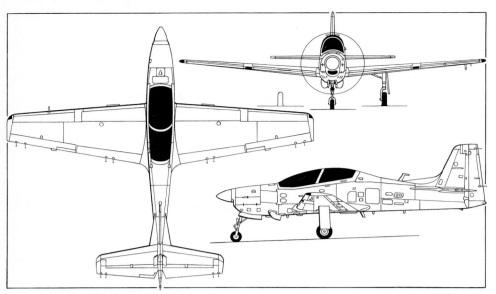

Shorts S312 Tucano basic trainer (Garrett TPE331-12B turboprop) *(Jane's/Mike Keep)*

Cockpits of the RAF Tucano T. Mk 1

Shorts Tucano T. Mk 52 in Kuwait Air Force markings

before Kenyan delivery and replaced; additional airframe destroyed before flight.

Main description of Tucano under Embraer heading in Brazilian section. Shorts Tucano T. Mk 1 for RAF differs in following respects:

DESIGN FEATURES: Based on EMB-312, but only 20 per cent commonality. Modifications include TPE331-12B engine offering higher speed (particularly at low level) and increased rate of climb; ventral airbrake to control descent speed; structural strengthening for increased manoeuvre loads and fatigue life (12,000 hours); new cockpit layout use of UK equipment; wing incidence 1° 13′; four wing strongpoints for armament training/light attack (for exports).

FLYING CONTROLS: Strengthened; hydraulically actuated ventral airbrake.

STRUCTURE: 7075-T73511 and 7075-T76, and 2024-T3 sheet aluminium alloy two-spar wing torsion box; wing leading edges strengthened against birdstrikes.

LANDING GEAR: Stressed for maximum sink rate of 4.0 m (13 ft)/s. Nosewheel unit supplied by Fairey Hydraulics, steerable ±20°. Dunlop wheels and tyres. Dunlop hydraulic single-disc brakes on mainwheels. Parking brake.

POWER PLANT: One 820 kW (1,100 shp) Garrett TPE331-12B-701A turboprop, driving a Hartzell HC-D4N-5C D9327K four-blade constant-speed fully feathering reversible-pitch propeller. Additional 38 kW (51 ehp) derived from exhaust. Provision for two external tanks with total capacity of 640 litres (169 US gallons; 142 Imp gallons). Oil capacity 9.0 litres (2.4 US gallons; 2.0 Imp gallons).

ACCOMMODATION: Instructor and pupil in tandem, on Martin-Baker 8LC Mk 1 lightweight ejection seats effective down to zero altitude and 70 knots (130 km/h; 81 mph) and with through-canopy capability. Lucas two-piece canopy strengthened to withstand 1 kg (2.2 lb) birdstrike at 240 knots (444 km/h; 276 mph) and incorporating miniature detonating cord. No pressurisation.

SYSTEMS: Cockpit air-conditioning by Normalair-Garrett with engine bleed air plus recirculated cockpit air through a regenerative turbofan system. Single hydraulic system pressure 217.5 bars (3,150 lb/sq in), for landing gear extension and retraction, landing gear doors and airbrake. Accumulator to lower landing gear in emergency; pressure 125 bars (1,810 lb/sq in). DC electrical power provided by a 28V 250A Lucas starter/generator and two 24Ah nickel-cadmium batteries. Static inverter (260VA) for 115V single phase AC power at 400Hz. Normalair-Garrett oxygen system from a single bottle, capacity 2,250 litres (80 cu ft at 124 bars (1,800 lb/sq in). Emergency oxygen bottle capacity 70 litres (2.5 cu ft), mounted on each ejection seat. Engine air intake de-iced by engine bleed air; propeller pitot head, static vents, and stall warning vane de-iced electrically.

AVIONICS: Standard avionics package includes Collins avionics control system, Bendix/King IFF, Collins DME, Collins VOR/ILS/MB, J. E. T. gyromagnetic compass, Collins VHF/UHF, Collins VHF and Racal CCS; navigation data presented on EHSI and RMI in each cockpit; avionics have dual control; aircraft integrated monitoring system fitted as standard. Optional ADF and Tacan; provision for optional stores management system and optical gunsights; optional armament package includes Avimo optical gunsight, Saab-Mekel gunsight camera, ML Aviation stores management system, Base Ten weapon control electronics and other equipment from Dowty and Guardian Electronics.

ARMAMENT: Optional provision on export variants for four underwing hardpoints: inboard points stressed to 290 kg (640 lb) and outboard points to 225 kg (496 lb). Loads can include SAMP Type 25 250 kg bombs, SAMP Type BL6 125 kg bombs, LAU-32 rocket launcher, Matra F2b rocket launcher, FN Herstal twin 12.7 mm machine-gun pod and FN Herstal heavy machine-gun pods, ML Aviation CBLS200 practice bomb carriers; ERU-119 ejector release units.

DIMENSIONS, EXTERNAL:
Wing span	11.28 m (37 ft 0 in)
Propeller diameter	2.39 m (7 ft 10 in)
Propeller ground clearance	0.32 m (12.6 in)

AREAS:
Fin, excl dorsal fin	2.08 m² (22.40 sq ft)
Rudder, incl tab	1.57 m² (16.9 sq ft)
Tailplane, incl fillets	4.57 m² (49.20 sq ft)

WEIGHTS AND LOADINGS (A: aerobatic configuration, B: full weapons configuration):
Basic weight empty (max): A	2,210 kg (4,872 lb)
B	2,232 kg (4,920 lb)
Max internal fuel: A, B	545 kg (1,202 lb)
Max T-O weight: A	2,935 kg (6,470 lb)
B	3,600 kg (7,937 lb)
Max landing weight: A	2,935 kg (6,470 lb)
B	3,600 kg (7,937 lb)
Max zero-fuel weight: A	2,318 kg (5,110 lb)
Max wing loading: A	151.3 kg/m² (31.0 lb/sq ft)
B	185.6 kg/m² (38.02 lb/sq ft)
Max power loading: A	3.58 kg/kW (5.88 lb/shp)
B	4.39 kg/kW (7.21 lb/shp)

PERFORMANCE (typical weight of 2,900 kg; 6,393 lb, except where indicated):
Never-exceed speed (V_{NE})	330 knots (611 km/h; 380 mph)
Max level and max continuous speeds:	
ISA at S/L	269 knots (498 km/h; 310 mph)

ISA at 3,050 m (10,000 ft)
277 knots (513 km/h; 319 mph)
Max speed, flaps and/or landing gear down
145 knots (269 km/h; 167 mph)
Stalling speed, power off at 2,700 kg (5,952 lb) AUW:
flaps and landing gear up
77 knots (141 km/h; 88 mph) EAS
flaps and landing gear down
70 knots (130 km/h; 81 mph) EAS

Max rate of climb at S/L 997 m (3,270 ft)/min
Service ceiling 10,365 m (34,000 ft)
T-O run 363 m (1,190 ft)
T-O to 15 m (50 ft) 590 m (1,930 ft)
Landing from 15 m (50 ft) 625 m (2,050 ft)
Landing run 360 m (1,180 ft)
Radius of action, 500 kg (1,102 lb) weapon load, 5 min over target 390 nm (723 km; 499 miles)
Range at 7,620 m (25,000 ft) with max fuel, 10% reserves:
internal fuel only 954 nm (1,767 km; 1,099 miles)
with external fuel 1,790 nm (3,317 km; 2,061 miles)
Endurance at econ cruising speed at 7,620 m (25,000 ft), 10% reserves 5 h 12 min
g limits:
aerobatic at 2,935 kg (6,470 lb) AUW +6.5/−3.3
with weapons, at 3,600 kg (7,937 lb) AUW +4.4/−2.2
with weapons, at 3,100 kg (6,834 lb) AUW, stores less than 185 kg (408 lb) each +6.0/−3.0

SLINGSBY

SLINGSBY AVIATION LIMITED (Subsidiary of ML Holdings plc)

Ings Lane, Kirkbymoorside, North Yorkshire YO6 6EZ
Telephone: 44 (751) 32474
Fax: 44 (751) 31173
Telex: 57597 SLINAV G
MANAGING DIRECTOR:
Eur Ing Dr David Holt, PhD, CEng
CHIEF DESIGNER: Barry Mellers, MSc
MARKETING DIRECTOR: John C. Dignan, MBIM
CUSTOMER SERVICE MANAGER: Malcolm Drinkell

Specialises in application of modern composite materials; formerly manufacturer of sailplanes but now concentrating on development and production of T67 Firefly series; recently completed 12 T67C model for Canadian Forces through Canadair. Currently producing 113 T67M260s for US Air Force. Fleet of T67M Mk II Firefly aircraft being supplied to Hunting Aircraft Ltd, which was awarded a contract to supply flying training to RAF on 8 January 1993. Approx 11,148 m² (120,000 sq ft) works area and 240 workforce early 1992.

Other activities include design and manufacture of hovercraft, large wind turbines and various marine structures and components; maintenance of type certificates, and product support, for former Airship Industries Skyship 500 and 600; design and manufacture of major components for Westinghouse Sentinel 1000 airship (which see) as main subcontractor; design and development of carbonfibre components for Bell/Boeing V-22 Osprey; design, development and production of ventral fairing/baggage bay, oil cooler ducts, and other composite components for the BAe Jetstream 41.

UK MoD work includes technical support and repair for RAF Air Cadet gliders/motor gliders, including full-scale dynamic fatigue test of Grob Viking.

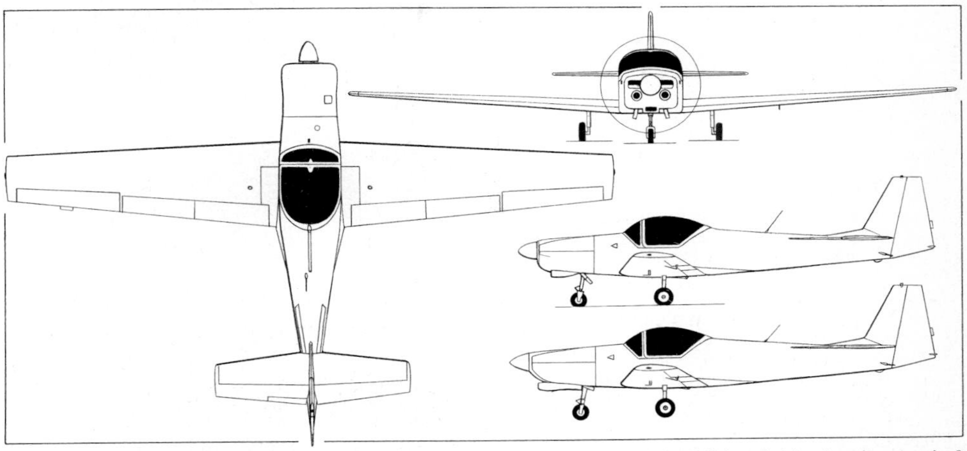

Slingsby T67M Mk II Firefly (Textron Lycoming AEIO-320-D1B engine) with additional side view (bottom) of T67M260 *(Dennis Punnett)*

SLINGSBY T67 FIREFLY

TYPE: Two-seat aerobatic, training and sporting aircraft.

PROGRAMME: Current composites constructed Firefly developed from wooden Slingsby T67A (licence built version of French Fournier RF6B - see 1982-83 *Jane's*); T67B gained CAA certification 18 September 1984; T67C was CAA certificated 15 December 1987.

CURRENT VERSIONS: **T67B**: Basic version; 86.5 kW (116 hp) Textron Lycoming O-235-N2A engine and two-blade fixed-pitch propeller; no longer produced.

T67C: Similar to T67B, but 119 kW (160 hp) Textron Lycoming O-320-D2A engine, metal fixed-pitch propeller, 24V 70A engine driven alternator and 24V 15Ah battery. Subvariants **T67C1** with normal fuselage fuel tank and one-piece canopy; **T67C2** with fuselage tank and two-piece canopy; **T67C3** with wing fuel tanks and two-piece canopy. *Detailed description applies to T67C.*

T67M: Military variants; described separately.

CUSTOMERS: Over 100 civil/military T67s delivered to customers in 12 countries by July 1993. Nine T67C3s purchased by Netherlands government Civil Aviation Flying School for KLM and Royal Netherlands Navy pilot training; 12 T67C3s for Canadian Department of National Defence for military primary flying training; plus other T67C variants for UK schools, including CSE and Bristow Helicopters.

DESIGN FEATURES: Wing section NACA 23015 at root, 23013 at tip; dihedral 3° 30′; incidence 3°.

FLYING CONTROLS: Manually operated mass balanced Frise-type ailerons, without tabs; mass balanced elevators with manually operated port trim tab (electric trim optional); rudder; trailing-edge fixed hinge flaps; spin strakes forward of tailplane roots.

STRUCTURE: GFRP; single-spar wings with double skin (corrugated inner skin bonded to plain outer skin) and conventional ribs in heavy load positions; conventional frame and top-hat stringer fuselage; stainless steel firewall between cockpit and engine; fixed incidence tailplane of similar construction to wings (built-in VOR antenna); fin incorporates VHF antenna.

LANDING GEAR: Non-retractable tricycle type. Oleo-pneumatic shock absorber in each unit. Steerable nosewheel. Mainwheel tyres size 6.00-6, pressure 1.4 bars (20 lb/sq in). Nosewheel tyre size 5.00-5, pressure 2.5 bars (37 lb/sq in). Hydraulic disc brakes. Parking brake.

POWER PLANT: One flat-four engine as described under Current Versions. Fuselage fuel tank, immediately aft of firewall, in T67C1 and T67C2, capacity 114 litres (30 US gallons; 25 Imp gallons). Refuelling point on fuselage upper surface, forward of windscreen. T67C3: wing fuel tanks as T67M. Oil capacity 4 litres (1.06 US gallons; 0.88 Imp gallon). Oil system permits short periods of inverted flight.

ACCOMMODATION: Two seats side by side, originally (T67C1) under one-piece transparent canopy, which swings upward and rearward for access to cockpit. T67C2/C3 have fixed windscreen and rearward hinged upward opening rear section. Dual controls standard. Adjustable rudder pedals. Cockpit heated and ventilated. Baggage space aft of seats.

SYSTEMS: Hydraulic system for brakes only. Vacuum system for blind-flying instrumentation. Electrical power supplied by 24V 70A engine driven alternator and 24V 15Ah battery.

AVIONICS: Standard avionics include artificial horizon and directional gyro, with vacuum system and vacuum gauge, electric turn co-ordinator, rate of climb indicator, recording tachometer, stall warning system, clock, outside air temperature gauge, accelerometer. Optional avionics, available to customer requirements, include equipment by Becker, Bendix/King and Narco, up to full IFR standard.

EQUIPMENT: Includes tiedown rings and towbar; cabin fire extinguisher, crash axe, heated pitot; instrument, landing, navigation and strobe lights. Optional equipment includes T67M-type blue tinted canopy, external power socket, and wingtip mounted smoke system.

DIMENSIONS, EXTERNAL:
Wing span 10.59 m (34 ft 9 in)
Wing chord: at root 1.53 m (5 ft 0¼ in)
at tip 0.83 m (2 ft 8¾ in)
Wing aspect ratio 8.9
Length overall 7.32 m (24 ft 0¼ in)
Height overall 2.36 m (7 ft 9 in)
Tailplane span 3.40 m (11 ft 1¾ in)
Wheel track 2.44 m (8 ft 0 in)
Wheelbase 1.50 m (4 ft 11 in)
Propeller diameter 1.88 m (6 ft 2 in)

DIMENSIONS, INTERNAL:
Cockpit: Length 2.05 m (6 ft 8¾ in)
Max width 1.08 m (3 ft 6½ in)
Max height 1.08 m (3 ft 6½ in)

AREAS:
Wings, gross 12.63 m² (136.0 sq ft)
Ailerons (total) 1.24 m² (13.35 sq ft)
Trailing-edge flaps (total) 1.74 m² (18.73 sq ft)
Fin 0.80 m² (8.61 sq ft)
Rudder 0.82 m² (8.8 sq ft)
Tailplane 1.65 m² (17.76 sq ft)
Elevators (incl tab) 0.99 m² (10.66 sq ft)

WEIGHTS AND LOADINGS:
Weight empty (basic): T67C2 649 kg (1,430 lb)
T67C3 658 kg (1,450 lb)
Max fuel: T67C1/C2 82 kg (181 lb)
T67C3 114 kg (252 lb)
Max baggage: T67C 30 kg (66 lb)
Max T-O, landing and aerobatic weights:
T67C2 907 kg (2,000 lb)
T67C3 952 kg (2,100 lb)
Max wing loading: T67C2 71.82 kg/m² (14.71 lb/sq ft)
T67C3 75.38 kg/m² (15.44 lb/sq ft)
Max power loading: T67C2 7.62 kg/kW (12.50 lb/hp)
T67C3 8.00 kg/kW (13.13 lb/hp)

PERFORMANCE (at max T-O weight):
Never-exceed speed (VNE) 180 knots (334 km/h; 207 mph)
Max level speed at S/L 127 knots (235 km/h; 146 mph)
Max cruising speed (75% power) at 2,440 m (8,000 ft) 125 knots (232 km/h; 144 mph)
Stalling speed, power off, flaps up:
T67C2 52 knots (97 km/h; 60 mph)
T67C3 53 knots (99 km/h; 61 mph)
Stalling speed, power off, flaps down 49 knots (91 km/h; 57 mph)
Max rate of climb at S/L: T67C2 274 m (900 ft)/min
Service ceiling 3,660 m (12,000 ft)
T-O run: T67C2 201 m (660 ft)
T-O to 15 m (50 ft): T67C2 442 m (1,450 ft)
Landing from 15 m (50 ft): T67C2 533 m (1,750 ft)
Landing run: T67C2 232 m (760 ft)
Range with max fuel (65% power at 2,440 m; 8,000 ft), allowances for T-O and climb, 45 min reserves at 45% power: T67C2 360 nm (666 km; 414 miles)
T67C3 516 nm (957 km; 595 miles)
Endurance at best econ setting to 'dry tanks':
T67C2 4 h 30 min
T67C3 7 h 20 min
g limits +6/−3

SLINGSBY T67M Mk II FIREFLY

TYPE: Two-seat military basic trainer.

PROGRAMME: To speed T67M programme, T67A G-BJNG modified to Firefly 160 standard (119 kW; 160 hp engine and constant-speed propeller) for tests including spinning trials at extreme CG limits; first flight of T67M Firefly 160 (G-BKAM) 5 December 1982; CAA certification 20 September 1983; designation changed to T67M Mk II as two-piece canopy introduced.

CUSTOMERS: Sold to Holland for military grading, Japan and UK for airline training, and Switzerland for aerobatic and general flying training. Being used by RAF for flying training from July 1993 (see main company introduction).

DESIGN FEATURES: T67M based on T67C, except as detailed.

STRUCTURE: Generally as T67C.

LANDING GEAR: Generally as T67C.

POWER PLANT: One 119 kW (160 hp) Textron Lycoming AEIO-320-D1B flat-four engine, driving a Hoffmann HO-V72 two-blade constant-speed composite propeller. Fuel and oil systems suitable for inverted flight. Fuel tanks in leading-edge of wings, capacity 159 litres (42 US gallons; 35 Imp gallons). Refuelling point in upper wing surface. Oil capacity 7.7 litres (2.0 US gallons; 1.675 Imp gallons).

ACCOMMODATION: As for T67C, except that current aircraft have blue tinted canopy with fixed windscreen and upward hinged, rearward opening rear section. Inertia reel lockable shoulder harness standard, air-conditioning optional.

AVIONICS: Avionics to customer requirements. Blind-flying instrumentation standard.

DIMENSIONS, EXTERNAL: As for T67C

AREAS: As for T67C

WEIGHTS AND LOADINGS:
Weight empty, equipped 658 kg (1,450 lb)
Max fuel weight 114 kg (252 lb)
Max T-O, aerobatic and landing weight 953 kg (2,100 lb)
Max wing loading 75.38 kg/m² (15.44 lb/sq ft)
Max power loading 8.00 kg/kW (13.12 lb/hp)

PERFORMANCE (at max T-O weight):
Never-exceed speed (VNE) 180 knots (334 km/h; 207 mph)
Max level speed at S/L 136 knots (252 km/h; 157 mph)
Max cruising speed, 75% power at 2,440 m (8,000 ft) 127 knots (235 km/h; 146 mph)

402 UK: AIRCRAFT—SLINGSBY/WESTLAND

Stalling speed, power off, flaps down
49 knots (91 km/h; 57 mph)
Max rate of climb at S/L 305 m (1,000 ft)/min
Service ceiling 4,575 m (15,000 ft)
T-O run 219 m (718 ft)
T-O to 15 m (50 ft) 445 m (1,458 ft)
Landing from 15 m (50 ft) 547 m (1,794 ft)
Landing run 258 m (847 ft)
Range with max fuel at 65% power, at 2,440 m (8,000 ft), allowances for T-O, climb and 45 min reserves at 45% power 548 nm (1,015 km; 631 miles)
Endurance, at best econ setting to 'dry tanks'
6 h 50 min
g limits at 884 kg (1,950 lb) AUW +6/−3

SLINGSBY T67M200 FIREFLY

TYPE: Two-seat military basic trainer.
PROGRAMME: First flight 16 May 1985; CAA certification 13 October 1985; representative airframe underwent long term fatigue test to simulate 75,000 flying hours.
CUSTOMERS: 29 used in five countries by January 1992; first customer Turkish Aviation Institute, Ankara (16 delivered from 1985); Dutch operator King Air (three T67M200s, plus one T67M Mk II) as screening trainers for prospective RNethAF pilots; Royal Hong Kong Auxiliary Air Force (four); Norwegian government's Flying Academy (six).
DESIGN FEATURES: Development of T67M; 149 kW (200 hp) Textron Lycoming AEIO-360-A1E engine; Hoffman HO-V123 three-blade variable-pitch composite propeller; fuel/oil systems for inverted flight.
DIMENSIONS, EXTERNAL: As for T67C/M except:
Propeller diameter 1.80 m (5 ft 11 in)
WEIGHTS AND LOADINGS:
Weight empty 700 kg (1,543 lb)
Max fuel 114 kg (252 lb)
Max baggage 30 kg (66 lb)
Max T-O weight: Utility 1,020 kg (2,250 lb)
Aerobatic 975 kg (2,150 lb)
Max landing weight 975 kg (2,150 lb)
Max wing loading 80.75 kg/m² (16.54 lb/sq ft)
Max power loading 6.85 kg/kW (11.25 lb/hp)
PERFORMANCE:
Never-exceed speed (V_{NE})
180 knots (334 km/h; 207 mph)
Max level speed at S/L 140 knots (259 km/h; 161 mph)
Max cruising speed (75% power at 2,440 m; 8,000 ft)
130 knots (240 km/h; 149 mph)
Stalling speed, power off, flaps down
51 knots (95 km/h; 59 mph)
Max rate of climb at S/L 350 m (1,150 ft)/min
T-O run 221 m (725 ft)
T-O to 15 m (50 ft) 409 m (1,340 ft)
Landing from 15 m (50 ft) 564 m (1,850 ft)
Landing run 265 m (870 ft)
Range with max fuel (65% power at 2,440 m; 8,000 ft), allowances for T-O and climb, 45 min reserves at 45% power 450 nm (833 km; 518 miles)
Endurance, at best econ setting to 'dry tanks'
6 h 20 min
g limits +6/−3

Pre-production Slingsby T67M260 Firefly military basic trainer (*John Dignan*)

SLINGSBY T67M260 FIREFLY

US Air Force designation: T-3A Firefly
TYPE: Two-seat military basic trainer.
PROGRAMME: Selected by USAF to meet EFS (Enhanced Flight Screener) requirement May 1992; Slingsby to manufacture airframe; Northrop Worldwide Aircraft Services to be responsible for final assembly, test and delivery, also to operate contractor logistic support activity. Prototype (G-BLUX) flown May 1991; operationally evaluated at Wright-Patterson AFB; hot and high trials at USAF Academy, Colorado Springs, in Summer 1991. Pre-production aircraft (G-EFSM) flown September 1992.
CUSTOMERS: US Air Force (38 initially, with two options to take total to 113), to replace Cessna T-41. Deliveries to start July 1993, finish October 1995.
COSTS: $11.7 million for first 38 aircraft (with support) ($34.7 million if all options taken up); total programme cost (incl contractor logistic support) $54.8 million.
DESIGN FEATURES: Development of T67M; 194 kW (260 hp) Textron Lycoming AEIO-540-D4A5 flat-six engine, driving Hoffman HO-V123K-X/180DT three-blade constant-speed composite propeller; cabin air-conditioning system added; higher max T-O and aerobatic weight, to allow 227 kg (500 lb) for two pilots and equipment plus full fuel load of 159 litres (42 US gallons; 35 Imp gallons).
Data as for T67C/M except:
DIMENSIONS, EXTERNAL:
Length overall 7.57 m (24 ft 10 in)
WEIGHTS AND LOADINGS (with air-conditioning installed):
Weight empty 807 kg (1,780 lb)
Max fuel 114 kg (252 lb)
Max T-O weight: Utility/Aerobatic 1,143 kg (2,520 lb)
Max wing loading 90.8 kg/m² (18.6 lb/sq ft)
Max power loading 5.89 kg/kW (9.69 lb/hp)
PERFORMANCE (estimated from testing and analysis):
Never-exceed speed (V_{NE})
195 knots (361 km/h; 224 mph)
Max level speed at S/L 152 knots (282 km/h; 175 mph)
Max cruising speed (75% power at 2,590 m; 8,500 ft)
150 knots (278 km/h; 173 mph)
Stalling speed, power off 59 knots (110 km/h; 68 mph)
Max rate of climb at S/L 480 m (1,575 ft)/min
T-O run, 18° flap 137 m (450 ft)
T-O to 15 m (50 ft), 18° flap 363 m (1,190 ft)
Landing from 15 m (50 ft), full flap 637 m (2,090 ft)
Landing run, full flap 296 m (970 ft)

WALLIS

WALLIS AUTOGYROS LTD

Reymerston Hall, Norfolk NR9 4QY
Telephone: 44 (362) 850418
MANAGING DIRECTOR: Wg Cdr K. H. Wallis, CEng, FRAeS, FRSA, RAF (Ret'd)

Full details and illustrations of autogyros produced by Wallis can be found in 1992-93 *Jane's*. Additional information is that WA-116/F (G-ATHM) now holds only four world records; WA-116/F/S (G-BLIK) now holds closed circuit speed and distance records previously held by G-ATHM in Class E-3/3a (thus holds 12 records); WA-116/X continues to test various engines, including Wallis-modified 1,800 cc Subaru in new fuselage; WA-118/M with supercharged Meteor Alfa 1 radial engine is still extant, as is WA-120/R-R; and WA-121/Mc (G-BAHH) was standing by in early 1993 to attempt time-to-climb to 6,000 m record and improve upon present 5,643 m (18,516 ft) altitude record.

WESTLAND

WESTLAND GROUP PLC

Yeovil, Somerset BA20 2YB
Telephone: 44 (935) 75222
Fax: 44 (935) 702131
Telex: 46277 WHL YEO G
LONDON OFFICE: 4 Carlton Gardens, Pall Mall, SW1Y 5AB
Telephone: 44 (71) 839 4061
CHAIRMAN: Sir Leslie Fletcher, DSC, FCA
DEPUTY CHAIRMAN: Alec Daly
CHIEF EXECUTIVE: Alan Jones, MA, FEng, FIProdE, FRAeS
PUBLIC RELATIONS DIRECTOR: Christopher Loney

Westland Aircraft Ltd (now Westland Group plc) formed July 1935, taking over aircraft branch of Petters Ltd (known previously as Westland Aircraft Works) that had designed/built aircraft since 1915; entered helicopter industry having acquired licence to build US Sikorsky S-51 as Dragonfly 1947; developed own Widgeon from Dragonfly; technical association with Sikorsky Division of United Technologies continued after decision to concentrate on helicopter design, development and construction.

Acquisition of Saunders-Roe Ltd 1959, Helicopter Division of Bristol Aircraft Ltd and Fairey Aviation Ltd 1960, and British Hovercraft Corporation's Aerospace Division 1983, plus subsequent restructuring into Divisions, detailed in 1989-90 *Jane's*; Divisions later consolidated into current limited liability companies, as Westland Helicopters Ltd, Westland Aerospace Ltd and Westland Technologies Ltd.

Financial reconstruction package approved February 1986, with United Technologies (USA) and Fiat (Italy) acquiring minority shareholdings; Fiat withdrew 1988; GKN acquired 22 per cent holding in Westland.

Westland Group has five subsidiary companies: Westland Aerospace, specialising in composites and metallic structures; Westland Engineering, producing transmission systems and rotor blades; Westland Helicopters (see below); Westland Industries, producing a wide range of components and subassemblies; and Westland Technologies, which includes Normalair-Garrett, a supplier of control systems. Westland Group employs 8,700 persons.

Current programmes include construction of composite engine cowlings for de Havilland DHC-8 Dash 8, BAe (Jetstream Aircraft) Jetstream 41 and Dornier 328; composite structures for Airbus, Boeing and McDonnell Douglas aircraft; Saab 340B engine nacelles; and Saab 2000 rear fuselage structure and engine nacelles.

EH Industries Ltd (see EHI in International section) is joint Westland/Agusta (Italy) management company developing EH 101 helicopter; collaboration with Agusta extended to include design, manufacture and marketing across joint product range; EHI Inc (USA) and EHI Canada are subsidiaries of EHI Ltd; Westland Group activities in USA and Central America represented by wholly owned subsidiary, Westland Inc.

WESTLAND HELICOPTERS LIMITED

Yeovil, Somerset BA20 2YB
Telephone: 44 (935) 75222
Fax: 44 (935) 704201
Telex: 46277 WHL YEO G
MANAGING DIRECTOR: Richard I. Case
MARKETING DIRECTOR: A. Lewis
HEAD OF HM GOVERNMENT BUSINESS: G. N. Cole
PUBLIC RELATIONS MANAGER: Vivien Davis

Sea King and Lynx in production; Westland and Agusta of Italy collaborate on EH 101 development and manufacture (see EHI in International section); agreement with United Technologies permits Sikorsky Black Hawk production as WS 70.

Under June 1989 agreement, Westland obtained co-production rights for McDonnell Douglas AH-64 Apache; if selected for British Army Air Corps, production of up to 150 envisaged (Longbow Apache version preferred); Westland to be prime contractor to UK MoD.

WESTLAND—AIRCRAFT: UK 403

EH 101
Westland and Agusta of Italy are partners in programme; see EHI in International section.

WESTLAND SEA KING
TYPE: Anti-submarine, search and rescue and airborne early warning helicopter.
PROGRAMME: Licence to develop/manufacture Sikorsky S-61 obtained 1959; developed initially for Royal Navy as advanced ASW helicopter with prolonged endurance; SAR, tactical troop transport, casualty evacuation, cargo carrying, long-range self-ferry secondary roles; ordered for RN 1967; first flight of production HAS. Mk 1 (XV642) 7 May 1969.
CURRENT VERSIONS: **Sea King HAR. Mk 3A:** SAR version for RAF; first flight 6 September 1977; 16 HAR. Mk 3s delivered by 1979, plus three in 1985; intention to order additional six officially announced 19 February 1992 and effected in October 1992. Designated HAR. Mk 3A, these latter to have Racal RNAV2 computer, Racal Doppler 91, BAe GM9 compass, Smiths-Newmark SN500 flight control system, Collins HF9000 HF radio, Cossor STR2000 series GPS, Motorola MX-1000(R) mountain rescue radio, THORN EMI ARI5955/2 search radar, Rockwell Collins AN/ARC-182 VHF/UHF, Bendix/King VOR/ILS and Bendix/King DME/ADF. Operated by No. 202 Squadron at Boulmer (HQ) and detachments at Boulmer, Brawdy, Manston, Lossiemouth and Leconfield, plus No. 78 Squadron on Falkland Islands. Revised UK organisation, by 1996, to be Nos. 22/202 Squadrons with detachments at Boulmer, Chivenor (from April 1994), Leconfield, Lossiemouth, Valley (from mid-1996) and Wattisham (from mid-1994); plus Sea King OCU at St Mawgan (from April 1993). Accommodation comprises two flight crew, air electronics/winch operator and loadmaster/winchman; up to six stretchers, or two stretchers and 11 seated survivors, or 19 persons; nav system of initial production Mk 3 includes Decca TANS F computer, accepting Mk 19 Decca nav receiver and Type 71 Doppler inputs; THORN EMI ARI5955 radar; No. 78 Squadron helicopters fitted with RWR and chaff/flare dispensers.

Sea King HAS. Mk 5: Updated ASW/SAR version for Royal Navy; 30 new aircraft handed over 2 October 1980 to July 1986; one HAS. Mk 1, 20 HAS. Mk 2s and 35 HAS. Mk 2As brought to same standard by 1987 at Fleet Air Arm workshops; four became HAR. Mk 5s and others HAS. Mk 6s (which see); nav/attack system utilises TANS G coupled to Decca 71 Doppler and THORN EMI Sea Searcher radar (in larger radome); Racal MIR-2 Orange Crop ESM, passive sonobuoy dropping equipment, and associated GEC-Marconi LAPADS acoustics processing and display equipment; four crew, with sonar operator also monitoring LAPADS as additional crew station; cabin enlarged by moving rear bulkhead 1.72 m (5 ft 7¾ in) aft; max T-O weight 9,525 kg (21,000 lb).

New equipment allows pinpoint of enemy submarine at greater range and attack with torpedoes; can monitor signals from own sonobuoys and those dropped by RAF Nimrod in joint search; can remain on station for long periods up to 87 nm (160 km; 100 miles) from ship.

Sea King HAS. Mk 6: Uprated RN ASW version; large blade aerial under starboard side of nose; five new aircraft ordered from October 1987 and delivered January-August 1990; 25 HAS. Mk 5s being retrofitted to standard at RN Fleetlands workshop using Westland-supplied kits; further batch of 44 kits followed; first flight of conversion (Mk 5 XZ581) 15 December 1987; first flight of new Mk 6 (ZG816) 7 December 1989; entered service (ZA136) with Intensive Flight Trials Unit within No. 824 Squadron (detached from Culdrose to Prestwick) 15 April 1988, squadron later disbanding; issued to 819 Sqn, Prestwick, April 1989; and Culdrose squadrons 810 (November 1989), 820 (January 1990), 826 (February 1990), 814 (October 1990) and 706 (June 1991). Mk 6 Operational Evaluation Unit at Boscombe Down is detachment of 819 Squadron following disbandment of 826 Squadron on 17 July 1993.

AQS-902G-DS enhanced sonar system (31 ordered from GEC-Marconi under 1987 contract, plus upgrade to standard of 112 previous AQS-902C sonobuoy processing systems), replacing Mk 5's analog computing element of Plessey 195 dipping sonar with digital processor (changing designation to GEC-Marconi 2069 : 44 ordered initially in June 1989 and delivered from August 1991), and presenting integrated information from sonobuoys and dipping sonar on single CRT display; sonar dunking depth increased from 75 m (245 ft) to about 213 m (700 ft); GEC-Plessey PTR 446 improved IFF; upgraded ESM to Orange Reaper standard; two GEC-Marconi AD3400 VHF/UHF secure speech radios; CAE Electronics internal AIMS (advanced integrated magnetic anomaly detection system) retrofit contract awarded to Westland December 1991 for 73 conversion kits (clearance trials completed 1989); 227-363 kg (500-800 lb) weight saving offers improved performance (equivalent to 30 min extra fuel).

Advanced Sea King: 1,092 kW (1,465 shp) Rolls-Royce Gnome H.1400-1T engines; uprated main gearbox with emergency lubrication and strengthened main lift frames; composite main and tail rotor blades; improved search radar; max AUW 9,752 kg (21,500 lb) for improved payload/range; through-life costs reduced. *Following details apply to Advanced Sea King and, in airframe and power train features, to new production of earlier versions.*

Sea King Mk 43B: Norwegian Air Force SAR helicopter; one (322, temporarily ZH566) delivered 28 July 1992, supplementing 11 Mk 43s received 1972-78. Nine survivors being upgraded to Mk 43B standard with additional Bendix/King RDR-1300C nose radar, Racal Doppler 91, RNAV2 and Mk 32 Decca, plus FLIR Inc 2000F FLIR; redelivery began, also on 28 July 1992, with 071.

CUSTOMERS: Royal Navy 56 HAS. Mk 1s, 21 HAS. Mk 2s, 30 HAS. Mk 5s and five HAS. Mk 6s delivered, but progressively modified; fleet in January 1992 comprised 10 AEW. Mk 2As, 38 HAS. Mk 5s, five HAR. Mk 5s, and 32 HAS. Mk 6s. Royal Air Force 25 HAR. Mk 3/3As, of which six on order.

Exports: Australia 12 Mk 50/50A, Belgium five Mk 48; Egypt six Mk 47; Germany 22 Mk 41; India 12 Mk 42, three Mk 42A and 20 Mk 42B; Norway 10 Mk 43, one Mk 43A and one Mk 43B; and Pakistan six Mk 45. Further 89 built as Commandos (which see) including 49 officially designated 'Sea King'; plus two lost on trials prior to delivery.

Total Sea King/Commando planned production: 326 (see table), of which 320 built by January 1993. (Royal Navy also received four Sikorsky S-61Ds for development 1966-67; one remains.) Further details of previous variants in earlier editions of *Jane's*; details of upgrades in *Jane's Civil and Military Aircraft Upgrades*.

DESIGN FEATURES: Based on Sikorsky S-61/SH-3 airframe and rotor system; Rolls-Royce Gnome turboshaft engines; specialised equipment to British requirements; composite

Westland Sea King HAR. Mk 3 of No. 78 Squadron, RAF, in grey tactical camouflage used in the Falkland Islands *(Paul Jackson)*

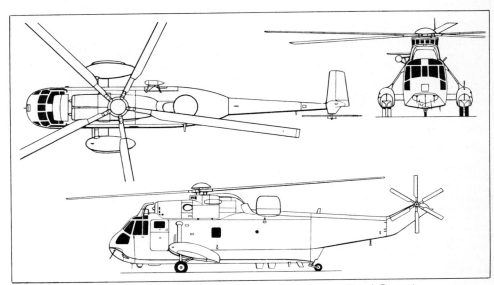
Westland Sea King HAS. Mk 5 anti-submarine helicopter *(Dennis Punnett)*

Westland Sea King HAS. Mk 6 conversion from Mk 5 (note blade aerial below forward fuselage) *(Royal Navy)*

404 UK: AIRCRAFT—WESTLAND

New-build (upper) and converted Westland Sea King Mk 43Bs of the Royal Norwegian Air Force

rotor blades; new six-blade tail rotor for increased capability in side wind; unbraced tail stabiliser; increased fuel capacity. Automatic main rotor blade folding and spreading is standard; for shipboard operation the tail pylon can also be folded.

FLYING CONTROLS: Mk 31 AFCS provides radio altitude displays for both pilots; artificial horizon displays; three-axis stabilisation in pilot controlled manoeuvres; attitude hold, heading hold and height hold in cruising flight; controlled transition manoeuvres to and from the hover; automatic height control and plan position control in the hover; and an auxiliary trim facility.

POWER PLANT: Two 1,238 kW (1,660 shp) (max contingency rating) Rolls-Royce Gnome H.1400-1T turboshafts, mounted side by side above cabin. Transmission rating 2,013 kW (2,700 shp). Fuel in six underfloor bag tanks, total capacity 3,714 litres (981 US gallons; 817 Imp gallons). Internal auxiliary tank, capacity 863 litres (228 US gallons; 190 Imp gallons), may be fitted for long-range ferry purposes. Pressure refuelling point on starboard side, two gravity points on port side. Flat plate debris guard for engine air intakes. Optional Centrisep air cleaner unit.

ACCOMMODATION: Crew of four in ASW role; accommodation for up to 22 survivors — or 18 if radar fitted — in SAR role; and up to 28 troops in utility role. Alternative layouts for nine stretchers and two attendants; or 15 VIPs. Two-section airstair door at front on port side, cargo door at rear on starboard side. Entire accommodation heated and ventilated. Cockpit doors and windows, and two windows each side of cabin, can be jettisoned in an emergency.

SYSTEMS: Three main hydraulic systems. Primary and auxiliary systems operate main rotor control. System pressure 103.5 bars (1,500 lb/sq in); flow rate 22.7 litres/min at 87.9 bars (6 US gallons; 5 Imp gallons/min at 1,275 lb/sq in). Unpressurised reservoir. Utility system for main landing gear, sonar and rescue winches, blade folding and rotor brake. System pressure 207 bars (3,000 lb/sq in); flow rate 41 litres/min at 186.2 bars (10.8 US gallons; 9 Imp gallons/min at 2,700 lb/sq in). Unpressurised reservoir. Electrical system includes two 20kVA 200V three-phase 400Hz engine driven generators, 26V single-phase AC supply fed from aircraft's 40Ah nickel-cadmium battery through an inverter, and DC power provided as secondary system from two 200A transformer-rectifier units.

AVIONICS (ASW models): Fully integrated all-weather hunter/killer weapon system, capable of operating independently of surface vessels. GEC-Marconi 2069, GEC-Plessey Type 195, Bendix/King AN/AQS-13B or Alcatel HS-312 dipping sonar, GEC-Marconi Doppler navigation system, THORN EMI Super Searcher radar in dorsal radome, transponder beneath rear fuselage, Honeywell AN/APN-171 radar altimeter, BAe GM9B Gyrosyn compass system, Smiths-Newmark Mk 31 AFCS. Observer/navigator has tactical display on which sonar contacts are integrated with search radar and navigational information. Radio equipment comprises Collins AN/ARC-182 UHF/VHF and homer, Ultra D 403M standby UHF, Collins 718U-5 HF radio, Racal B693 intercom, Telebrief system and IFF provisions. CAE Electronics AN/ASQ-504(V) internal MAD ordered for RN Sea Kings in 1987 and fitted from 1988 onwards. Ericsson Radar Electronics AN/ALQ-167 Yellow Veil modular jamming equipment installed internally in Mk 5 from about 1986.

AVIONICS (non-ASW models): Wide range of radio and navigation equipment installed, including VHF/UHF communications, VHF/UHF homing, radio compass, Doppler navigation system, radio altimeter, VOR/ILS, radar and transponder, of Collins, GEC-Plessey, Honeywell and GEC-Marconi manufacture. THORN EMI ARI5955 spine radar in SAR versions, plus Bendix/King RDR-1300C nose radar for Norwegian Mk 43B. Honeywell compass system and Smiths-Newmark AFCS.

EQUIPMENT: Two No. 4 marine markers, four No. 2 Mk 2 smoke floats, Ultra Electronics mini-sonobuoys, in ASW versions. Sea Kings equipped for search and rescue have Breeze BL 10300 variable speed hydraulic rescue hoist of 272 kg (600 lb) capacity mounted above starboard side cargo door. Second electric hoist optional.

ARMAMENT: Up to four Mk 46, Whitehead A244S or Sting Ray homing torpedoes; or four Mk 11 depth charges or one Clevite simulator; two BAe Sea Eagle or Aerospatiale Exocet anti-ship missiles. For secondary role a mounting is provided on the rear frame of the starboard door for a general purpose machine gun.

DIMENSIONS, EXTERNAL:
Main rotor diameter	18.90 m (62 ft 0 in)
Tail rotor diameter	3.16 m (10 ft 4 in)
Distance between rotor centres	11.10 m (36 ft 5 in)
Main rotor blade chord	0.46 m (1 ft 6¼ in)
Length:	
overall, rotors turning	22.15 m (72 ft 8 in)
main rotor folded	17.42 m (57 ft 2 in)
rotors and tail folded	14.40 m (47 ft 3 in)
Height: overall, rotors turning	5.13 m (16 ft 10 in)
rotors spread and stationary	4.85 m (15 ft 11 in)
to top of rotor head	4.72 m (15 ft 6 in)
Fuselage: Length	17.02 m (55 ft 10 in)
Max width	2.16 m (7 ft 1 in)
Width: overall, rotors folded	
with flotation bags	4.98 m (16 ft 4 in)
without flotation bags	4.77 m (15 ft 8 in)

SEA KING/COMMANDO PRODUCTION

Variant	Customer	Quantity	First flights	Operators
HAS. Mk 1	Fleet Air Arm	56	7 May 1969 - May 1972	Converted to Mk 2
HAS. Mk 2	Fleet Air Arm	21	18 Jun 1976 - 13 Jul 1977 (13)	Converted to Mk 5
			7 Feb 1979 - 24 Aug 1979 (8)	Converted to Mk 5
AEW. Mk 2A	Fleet Air Arm	-	Conversions	849 Sqdn
HAR. Mk 3	Royal Air Force	19	6 Sep 1977 - 1 Dec 1978 (15)	202 Sqdn
			14 Aug 1980 (1)	202 Sqdn
			22 May 1985 - 21 Aug 1985 (3)	202 Sqdn
HAR. Mk 3A	Royal Air Force	6		On order
HC. Mk 4*	Fleet Air Arm	40	26 Sep 1979 - 20 Sep 1982 (15)	845, 846, 707 Sqdns
			8 Dec 1983 - 9 Aug 1984 (8)	845, 846, 707 Sqdns
			10 Sep 1985 - 10 Dec 1985 (14)	845, 846, 707 Sqdns
			2 Jun 1990-18 Sep 1990 (3)	845, 846, 707 Sqdns
Mk 4*	MoD(PE)	1	10 Apr 1989	ETPS
Mk 4X*	MoD(PE)	2	19 Jan 1982 - 19 Nov 1982	DRA
HAS. Mk 5	Fleet Air Arm	30	26 Aug 1980 - 2 Sep 1982 (17)	
			9 Aug 1984 - 23 Apr 1985 (8)	
			Jan 1986 - Jul 1986 (5)	
HAR. Mk 5	Fleet Air Arm	-	Conversions	771 Sqdn
HAS. Mk 6	Fleet Air Arm	5	7 Dec 1989 - 27 Jun 1990	See text
Mk 41	German Navy	22	6 Mar 1972 - 21 Aug 1974 (21)	1/MFG 5
			Jul 1975 (1)	1/MFG 5
Mk 42	Indian Navy	12	1971 - 23 Jul 1971 (6)	INAS 330, 336
			17 Jul 1973 - 4 May 1974 (6)	INAS 330, 336
Mk 42A	Indian Navy	3	23 Nov 1979 - 12 Feb 1980	INAS 330, 336
Mk 42B	Indian Navy	20	17 May 1985 - 11 Apr 1990	INAS 339
Mk 42C*	Indian Navy	6	25 Sep 1986 - 5 Jan 1988	INAS 330, 336
Mk 43	Norwegian Air Force	10	19 May 1972 - 30 Sep 1972	Skv 330
Mk 43A	Norwegian Air Force	1	6 Jul 1978	Skv 330
Mk 43B	Norwegian Air Force	1	28 Jul 1992	Skv 330
Mk 45	Pakistani Navy	6	30 Aug 1974 - 10 Dec 1974	111 Sqdn
Mk 47	Egyptian Navy	6	11 Jul 1975 - 5 Nov 1975	(Alexandria)
Mk 48	Belgian Air Force	5	19 Dec 1975 - 1976	40 Sqdn
Mk 50	Australian Navy	10	30 Jun 1974 - Apr 1975	HS 817
Mk 50A	Australian Navy	2	7 Dec 1982 - 9 Feb 1983	HS 817
Mk 1*	Egyptian Air Force	5	12 Sep 1973 - 29 Sep 1973	
Mk 2*	Egyptian Air Force	17	16 Jan 1975 - 27 Jan 1976	
Mk 2A*	Qatari Air Force	3	9 Aug 1975 - 16 Mar 1976	8 Sqdn
Mk 2B*	Egyptian Air Force	2	13 Mar 1975 - 25 Jul 1975	
Mk 2C*	Qatari Air Force	1	9 Oct 1975	8 Sqdn
Mk 2E*	Egyptian Air Force	4	1 Sep 1978 - 4 Dec 1978	
Mk 3*	Qatari Air Force	8	14 Jun 1982 - 5 Oct 1983	9 Sqdn
Mk 41/42B	Destroyed pre-delivery	2		
Total		**326**		

* Commando variant

Wheel track (c/l of shock absorbers) 3.96 m (13 ft 0 in)
Wheelbase 7.14 m (23 ft 5 in)
Cabin door (port): Height 1.68 m (5 ft 6 in)
  Width 0.91 m (3 ft 0 in)
Cargo door (stbd): Height 1.52 m (5 ft 0 in)
  Width 1.73 m (5 ft 8 in)
  Height to sill 1.14 m (3 ft 9 in)
DIMENSIONS, INTERNAL:
Cabin: Length 7.59 m (24 ft 11 in)
  Max width 1.98 m (6 ft 6 in)
  Max height 1.92 m (6 ft 3½ in)
  Floor area (incl area occupied by radar, sonar etc)
    13.94 m² (150 sq ft)
  Volume 28.03 m³ (990 cu ft)
AREAS:
Main rotor disc 280.6 m² (3,020.3 sq ft)
Tail rotor disc 7.8 m² (83.9 sq ft)
WEIGHTS AND LOADINGS (A: anti-submarine, B: anti-surface vessel, C: airborne early warning, D: SAR, E: troop transport, F: external cargo, G: VIP):
Basic weight: with sponsons 5,393 kg (11,891 lb)
  without sponsons 5,373 kg (11,845 lb)
Weight empty, equipped (typical): A 7,428 kg (16,377 lb)
  B 7,570 kg (16,689 lb)
  C 7,776 kg (17,143 lb)
  D 6,241 kg (13,760 lb)
  E 5,712 kg (12,594 lb)
  F 5,686 kg (12,536 lb)
  G 7,220 kg (15,917 lb)
Max underslung or internal load 3,628 kg (8,000 lb)
Max T-O weight 9,752 kg (21,500 lb)
Max disc loading 34.75 kg/m² (7.12 lb/sq ft)
Max power loading 4.44 kg/kW (7.29 lb/shp)
PERFORMANCE (at max T-O weight, ISA):
Never-exceed speed (V$_{NE}$, British practice) at S/L
    122 knots (226 km/h; 140 mph)
Cruising speed at S/L 110 knots (204 km/h; 126 mph)
Max rate of climb at S/L 619 m (2,030 ft)/min
Max vertical rate of climb at S/L 246 m (808 ft)/min
Service ceiling, OEI 1,220 m (4,000 ft)
Max contingency ceiling (1 hour rating)
    1,067 m (3,500 ft)
Hovering ceiling: IGE 1,982 m (6,500 ft)
  OGE 1,433 m (4,700 ft)
Radius of action:
  A (2 h on station, incl three torpedoes)
    125 nm (231 km; 144 miles)
  B (2 h on station, incl two Sea Eagles)
    110 nm (204 km; 126 miles)
  C (2 h 24 min on station) 100 nm (185 km; 115 miles)
  D (picking up 20 survivors)
    220 nm (407 km; 253 miles)
  E (28 troops) range 300 nm (556 km; 345 miles)
  F (1,814 kg; 4,000 lb external load)
    225 nm (417 km; 259 miles)
  G 580 nm (1,075 km; 668 miles)
Range with max standard fuel, at 1,830 m (6,000 ft)
    800 nm (1,482 km; 921 miles)
Ferry range with max standard and auxiliary fuel, at 1,830 m (6,000 ft) 940 nm (1,742 km; 1,082 miles)
PERFORMANCE (at typical mid-mission weight):
Never-exceed speed (V$_{NE}$, British practice) at S/L
    146 knots (272 km/h; 169 mph)
Cruising speed at S/L 132 knots (245 km/h; 152 mph)

WESTLAND COMMANDO

TYPE: Twin-turboshaft tactical military helicopter.
PROGRAMME: First flight 12 September 1973; first flight HC. Mk 4 (ZA290) 26 September 1979; HC. Mk 4s delivered November 1979 to October 1990; HC. Mk 4 ZF115 (28th delivery) first Commando type completed with composites main rotor blades (first flight 14 November 1985); composite blades retrofitted to HC. Mk 4s.
CURRENT VERSIONS: See 1989-90 *Jane's* for Mk 1, Mk 2 and Mk 3 details.
  Sea King HC. Mk 4: Royal Navy utility Commando Mk 2; folding main rotor blades; folding tail pylon; non-retractable landing gear; 28 equipped troops or 2,720 kg (6,000 lb) cargo internally; 3,628 kg (8,000 lb) max slung load; parachuting/abseiling equipment; Decca TANS with chart display and Decca 71 Doppler nav system; 7.62 mm cabin machine gun; can operate in Arctic or tropics; serves with Nos. 707, 845 and 846 (Naval Air Commando) and 772 (SAR) Squadrons. Two **Mk 4Xs** with less operational equipment delivered to RAE (incl for Blue Kestrel radar trials) 1982-83; one **Mk 4** to Empire Test Pilots' School 3 May 1989. *Following data apply to late production aircraft.*
CUSTOMERS: See table; conversion undertaken in 1990-92 of Qatari Mk 2s to Mk 3 standard with new nav/com equipment). Six Indian Sea King Mk 42Cs and 39 RN Sea King HC. Mk 4s essentially Commandos (three of latter lost, all in Falklands conflict); four UK MoD Mk 4s. Total production 89; last delivery (HC. Mk 4 ZG822) 15 October 1990.
DESIGN FEATURES: Based on Sea King; optimised payload/range and endurance for tactical troop transport, logistic support and cargo transport, and casualty evacuation primary roles, or air-to-surface attack and SAR secondary roles; five-blade composite main rotor, attached to hub by multiple bolted joint; NACA 0012 blade section; rotor brake; optional automatic main rotor blade folding; five-blade composite tail rotor; twin input four-stage reduction main gearbox, with single bevel intermediate and tail gearboxes; main rotor/engine rpm ratio 93.43; tail rotor/engine rpm ratio 15.26; unpressurised; stub-wings instead of Sea King sponsons; non-retractable landing gear.
STRUCTURE: Light alloy stressed skin; optional tail pylon folding.
LANDING GEAR: Non-retractable tailwheel type, with twin-wheel main units. Oleo-pneumatic shock absorbers. Main-wheel tyres size 6.50-10, tailwheel tyre size 6.00-6.
POWER PLANT: As for current versions of Advanced Sea King.
ACCOMMODATION: Crew of two on flight deck. Seats along cabin sides, and single jump seat, for up to 28 troops. Over-load capacity 45 troops. Two-piece airstair door at front on port side, cargo door at rear on starboard side. Entire accommodation heated and ventilated. Cockpit doors and windows, and two windows each side of main cabin, are jettisonable in an emergency.
SYSTEMS: Primary and secondary hydraulic systems for flight controls. No pneumatic system. Electrical system includes two 20kVA alternators.
AVIONICS: Wide range of radio, radar and navigation equipment available to customer's requirements. Royal Navy Sea King HC. Mk 4s modified in late 1990 for participation in Gulf War with Racal Prophet Mk 1 RWRs, Honeywell AN/AAR-47 missile approach warning system, Loral AN/ALQ-157 IR jammers (detachable), chaff/flare dispensers; Racal RNS252 'Super TANS' navigation equipment linked with TNL8000 GPS; NVG compatible cockpit lighting.
EQUIPMENT: Cargo sling and rescue hoist optional.
ARMAMENT: Wide range of guns, missiles etc may be carried, to customer's requirements: typically, pintle-mounted gun (7.62 mm or 20 mm) in cabin doorway and machine-gun pod (0.50 in or 7.62 mm) on each side of forward fuselage with reflector sight for pilot. If fitted, sponsons may mount one rocket pod each (FZ M159C 2.75 in; Matra F4 68 mm; Thomson-Brandt 68-22 68 mm; Medusa 81 mm; or SNIA HL-12 80 mm).
DIMENSIONS, EXTERNAL: As Advanced Sea King except:
  Wheelbase 7.21 m (23 ft 8 in)
DIMENSIONS, INTERNAL: As Advanced Sea King (SAR version)
AREAS: As for Advanced Sea King, plus:
  Main rotor blades (each) 4.14 m² (44.54 sq ft)
  Tail rotor blades (each) 0.23 m² (2.46 sq ft)
  Tailplane 1.80 m² (19.40 sq ft)
WEIGHTS AND LOADINGS:
  Operating weight empty (troop transport, 2 crew, typical)
    5,620 kg (12,390 lb)
  Max underslung load 3,628 kg (8,000 lb)
  Max T-O weight 9,752 kg (21,500 lb)
PERFORMANCE (at max T-O weight): As given for Advanced Sea King, plus:
  Range with max payload (28 troops), reserves for 30 min standoff 214 nm (396 km; 246 miles)

WESTLAND LYNX

TYPE: Twin-engined multi-purpose helicopter.
PROGRAMME: Developed within Anglo-French helicopter agreement confirmed 2 April 1968; Westland given design leadership; first flight of first of 13 prototypes (XW835) 21 March 1971; first flight of fourth prototype (XW838) 9 March 1972, featuring production type monobloc rotor head; first flights of British Army Lynx prototype (XX153) 12 April 1972, French Navy prototype (XX904) 6 July 1973, production Lynx (RN HAS. Mk 2 XZ229) 20 February 1976; first RN operational unit (No. 702 Squadron) formed on completion of intensive flight trials December 1977; AH. Mk 5 first flew (ZE375) 23 February 1985; other development details and records in 1975-76 and subsequent *Jane's*. Production shared 70 per cent Westland, 30 per cent Aerospatiale; for details of G-LYNX's 1986

Westland Sea King HC. Mk 4 (Commando) of No. 707 Squadron, fitted with latest defensive aids package *(Paul Jackson)*

Westland Sea King Mk 4 of Empire Test Pilots' School recently modified with nose radar *(Paul Jackson)*

406 UK: AIRCRAFT—WESTLAND

Westland Lynx HAS. Mk 3CTS of No. 700L Squadron displaying large antenna beneath boom (Mk 3S feature) and nose flotation bags accompanying CTS conversion *(Peter J. Cooper)*

Westland Lynx HAS. Mk 3 ZD267 used for Mk 8 development with Sea Owl thermal imager and chin radome *(Peter J. Cooper)*

Westland Lynx AH. Mk 7 conversion from Mk 1 in Gulf War desert camouflage *(Peter J. Cooper)*

world helicopter absolute speed record, and Lynx AH. Mk 7 XZ170's 1989 agility trials, see 1990-91 *Jane's*.

CURRENT VERSIONS: **Lynx AH. Mk 1:** British Army general purpose and utility version; 113 built (106 remained January 1992 incl those converted to AH. Mk 7). Described in 1986-87 and earlier *Jane's*.

Lynx HAS. Mk 3: Second Royal Navy version for advanced shipborne anti-submarine and other duties; similar to Mk 2, with GEC-Marconi Seaspray search and tracking radar in modified nose; capable of anti-submarine classification and strike, air to surface vessel search and strike, SAR, reconnaissance, troop transport fire support, communications and fleet liaison, and vertrep; can carry Sea Skua; Mk 2's Gem 2 engines replaced by two 835 kW (1,120 shp) Gem 41-1 engines; 23 delivered March 1982 to April 1985; seven more in **HAS. Mk 3S** configuration (first flight, ZF557, 12 October 1987) delivered November 1987 to November 1988; additionally, ZD560 built in approx Mk 7 configuration, delivered to Empire Test Pilots' School April 1988; further 53 obtained through modification of all existing HAS. Mk 2s by 1989.

Lynx HAS. Mk 3ICE is Mk 3 lacking some operational equipment for general duties aboard Antarctic survey vessel, HMS *Endurance;* two converted. Main RN operators Nos. 815 and 829 Squadrons, combined on 26 March 1993 as No. 815 Squadron. 19 used by Armilla Patrol in Arabian Gulf modified to **HAS. Mk 3GM** (Gulf Mod), with better cooling, or **HAS. Mk 3S/GM**, also with Mk 3S modifications. Augmenting new-build Mk 3Ss, 36 modified by RN Aircraft Yard at Fleetlands from April 1989; Mk 3S is Phase 1 of Mk 8 conversion programme, involving GEC-Marconi AD 3400 secure speech radios (blade aerial beneath mid-point of tailboom) and upgraded ESM; programme continues, including Mk 3S/GM. Phase 2 is Lynx **HAS. Mk 3CTS**, adding RAMS 4000 central tactical system; prototype (XZ236 ex Mk 3) flew 25 January 1989; further six for RN trials (one ex Mk 3; five ex Mk 3S); deliveries to Operational Flight Trials Unit, Portland, from April 1989; unit became No. 700L Squadron 6 July 1990 with three Lynx; remaining three deployed to destroyer and frigates at sea from 3 December 1990 (HMS *Newcastle*); CTS service clearance granted August 1991; Mk 3CTS has flotation bag each side of nose. RN Lynx status January 1993: 21 Mk 3, 12 Mk 3GM, 31 Mk 3S, seven Mk 3CTS, five Mk 3S/GM, two Mk 3ICE and two Mk 8 (Excludes two Mk 3S sold to Portugal as Mk 95s.)

Lynx AH. Mk 7: Uprated British Army version, meeting GSR 3947; with improved systems, reversed-direction tail rotor with improved composite blades to reduce noise and enhance extended period hover at high weights; 4,876 kg (10,750 lb) AUW; 13 ordered, eight from Mk 1 contract (two cancelled); first flight (ZE376) 7 November 1985; 11th delivered July 1987.

RN workshops at Fleetlands converting Mk 1s to Mk 7s; first conversion (XZ641) redelivered 30 March 1988; box-type exhaust diffusers added from early 1989; programme continues; approximately 75 conversions by late 1992. Interim conversion is Lynx **AH. Mk 1GT** with uprated engines and rotors, but lacking Mk 7's improved electronic systems; first conversion (XZ195) 1991. GEC-Ferranti (now GEC-Marconi) AWARE-3 radar warning receiver selected 1989 for retrofit, designated ARI23491; Mk 1 XZ668 to Westland for trial installation 22 November 1991.

Lynx HAS. Mk 8: For RN; equivalent to export **Super Lynx** (see separate entry); passive identification system; 5,125 kg (11,300 lb) max T-O weight; improved (reversed-direction) tail rotor control; BERP composite main rotor blades; Racal RAMS 4000 central tactical system (CTS eases crew's workload by centrally processing sensor data and presents mission information on multi-function CRT display; 15 systems ordered 1987, 106 September 1989); original Seaspray Mk 1 radar re-positioned in new chin radome; GEC-Marconi Sea Owl thermal imager (×5 or ×30 magnifying system on gimballed mount, with elevation +20° to −30° and azimuth +120° to −120°; ordered October 1989) in former radar position; MIR-2 ESM updated; three Mk 3s used in development programme as tactical system (XZ236), dummy Sea Owl/chin radome (ZD267) and avionics (ZD266) testbeds; see Lynx Mk 3 for Phases 1 and 2 of Lynx Mk 8 programme.

Definitive Mk 8 (Phase 3) conversions begun 1992 with addition of Sea Owl, CAE internal MAD, further radar and navigation upgrades, (including RACAL RNS252 'Super TANS' with associated TNL8000 GPS), composites BERP main rotor blades and reversed-direction tail rotor. Conversion planned of 45 Mks 3/3S/3CTS to Mk 8; contract award to Westland, May 1992 for first seven conversions; initial deliveries due, Spring 1994; remainder for conversion by RN at Fleetlands; first Westland-produced conversion kit for RN to be delivered in February 1995; initial contract covers 18 kits.

Lynx AH. Mk 9: UK Army Air Corps equivalent of export **Battlefield Lynx** (see separate entry); tricycle wheel landing gear; max T-O weight 5,125 kg (11,300 lb); advanced technology composites main rotor blades; exhaust diffusers; no TOW capability; first flight of prototype (converted company demonstrator XZ170) 29 November 1989; 16 new aircraft (beginning ZG884, flown 20 July 1990) ordered for delivery from 1991, plus eight Mk 7 conversions (contract awarded November 1991); for

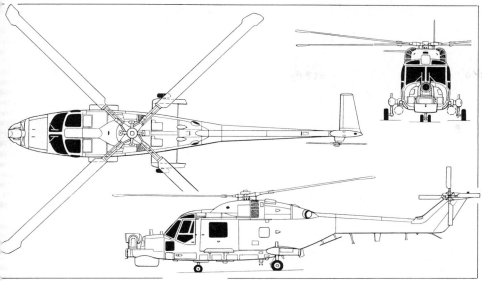

Westland Lynx HAS. Mk 8, based on the Super Lynx advanced export version *(Dennis Punnett)*

Westland Lynx AH. Mk 9 for the Army Air Corps

Westland Lynx Mk 90 (conversion of Mk 80) in Danish naval markings and fitted with FLIR Systems Inc Safire FLIR turret *(Paul Jackson)*

Nos. 672 and 664 Squadrons of 9 Regiment, Dishforth, to support 24th Airmobile Brigade; some outfitted as advanced command posts, remainder for tactical transport role. Deliveries to A&AEE, Boscombe Down from 22 May 1991 (ZG884); to No. 672 Squadron from 19 December 1991 (ZG889); No. 664 Squadron from June 1992; final aircraft (ZG923) flown 30 June 1992; first 'production' conversion (ZF538) to Westland for modifications 3 February 1992.

Other versions and operators where orders completed, see 1990-91 *Jane's*. Royal Netherlands Navy upgraded five UH-14As and eight SH-14Cs to **SH-14D** standard, with Alcatel dipping sonar, UHF radios, RWR, FLIR Systems Inc 2000HP FLIR, Trimble Type 2200 GPS, new radar altimeter, composites rotor blades and Mk 42 Gem power plants. Nine SH-14Bs, already with sonar, raised to SH-14D standards, but in interim SH-14Cs upgraded to SH-14B through deletion of MAD and addition of sonar. UH-14As are first full SH-14D conversions, from 1990; programme designated STAMOL (Standaardisatie en Modernisering Lynx); envisages standard fleet comprising 16 with sonar and six with provisions for sonar installation. Completed early 1993.

Denmark upgrading Mk 80s to Mk 90 with Gem Mk 42 power plant, Racal tactical data system, Racal MIR-2 passive radar detection system, composites blades and four-bag emergency flotation system; max T-O weight increased to 4,876 kg (10,750 lb).

CUSTOMERS: Production totalled 364 at June 1992 and January 1993 (incl two demonstrators but not 13 prototypes); further three awaiting construction; production details in table; interest expressed during 1992 in further procurement by Germany (12) and Brazil (seven).

COSTS: £3.5 million (1991) to upgrade eight Mk 7s to Mk 9s. £20 million (1992) to convert seven Mk 3s to Mk 8s.

Following description applies to military general purpose and naval versions with Gem 2 engines, except where indicated:

DESIGN FEATURES: Compact design suited to hunter-killer ASW and missile-armed anti-ship naval roles from frigates or larger ships (superseding ship-guided helicopters), armed/unarmed land roles with cabin large enough for squad, or other tasks; manually folding tail pylon on naval versions; single four-blade semi-rigid main rotor (foldable), each blade attached to main rotor hub by titanium root plates and flexible arm; rotor drives taken from front of engines into main gearbox mounted above cabin ahead of engines; in flight, accessory gears (at front of main gearbox) driven by one of two through shafts from first stage reduction gears; four-blade tail rotor, drive taken from main ring gear; single large window in each main cabin sliding door; provision for internally mounted armament, and for exterior universal flange mounting each side for other weapons/stores.

FLYING CONTROLS: Rotor head controls actuated by three identical tandem servojacks and powered by two independent hydraulic systems; control system incorporates simple stability augmentation system; each engine embodies independent control system providing full authority rotor speed governing, pilot control being limited to selection of desired rotor speed range; in event of one engine failure, system restores power up to single engine max contingency rating; main rotor can provide negative thrust to increase stability on deck after touchdown on naval versions; hydraulically operated rotor brake mounted on main gearbox; sweptback fin/tail rotor pylon, with starboard half-tailplane.

STRUCTURE: Conventional semi-monocoque pod and boom, mainly light alloy; glassfibre access panels, doors, fairings, pylon leading/trailing-edges, and bullet fairing over tail rotor gearbox; composites main rotor blades; main rotor hub and inboard flexible arm portions built as complete unit, as titanium monobloc forging; tail rotor blades have light alloy spar, stainless steel leading-edge sheath and rear section as for main blades.

LANDING GEAR (general purpose military version): Non-retractable tubular skid type. Provision for a pair of adjustable ground handling wheels on each skid. Flotation gear optional.

LANDING GEAR (naval versions): Non-retractable oleo-pneumatic tricycle type. Single-wheel main units, carried on sponsons, fixed at 27° toe-out for deck landing; can be manually turned into line and locked fore and aft for movement of aircraft into and out of ship's hangar. Twin-wheel nose unit steered hydraulically through 90° by the pilot to facilitate independent take-off into wind. Sprag brakes (wheel locks) fitted to each wheel prevent rotation on landing or inadvertent deck roll. These locks disengaged hydraulically and re-engage automatically in event of hydraulic failure. Max vertical descent 2.29 m (7½ ft)/s; with lateral drift 0.91 m (3 ft)/s for deck landing. Flotation gear, and hydraulically actuated harpoon deck lock securing system, optional.

POWER PLANT: Two Rolls-Royce Gem 2 turboshafts, each with max contingency rating of 671 kW (900 shp) in Lynx AH. 1, HAS. 2 and early export variants. Later versions have Gem 41-1 or 41-2 engines, each with max contingency rating of 835 kW (1,120 shp), or Gem 42-1 engines, each with max contingency rating of 846 kW (1,135 shp). Transmission rating 1,372 kW (1,840 shp). Engines of British and French Lynx in service converted to Mk 42 standard

LYNX PRODUCTION

Variant	Customer	Quantity	First flights	Operators
AH. Mk 1	Army Air Corps	113*	11 Feb 1977 - 24 Jan 1984	Note A
HAS. Mk 2	Fleet Air Arm	60	20 Feb 1976 - 26 May 1981	Converted to Mk 3
HAS. Mk 2(FN)	French Navy	26	4 May 1977 - 5 Sep 1979	31F, 34F, 20S
HAS. Mk 3	Fleet Air Arm	31	4 Jan 1982 - 21 Oct 1988	815, 702 Sqdns
HAS. Mk 4(FN)	French Navy	14	1 Apr 1982 - 26 Aug 1983	31F, 34F, 20S
Mk 5	MoD(PE)	3*	21 Nov 1984 - 23 Feb 1985	DRA
AH. Mk 6	Royal Marines		None built	-
AH. Mk 7	Army Air Corps	11*	23 Apr 1985 - 5 Jun 1987	Note A
HAS. Mk 8	Fleet Air Arm	-	Conversions	
AH. Mk 9	Army Air Corps	16*†	20 Jul 1990 - 21 Jun 1992	664, 672 Sqdns
Mk 21	Brazilian Navy	9	30 Sep 1977 - 14 Apr 1978	1° EHA
Mk 22	Egyptian Navy		None built	-
Mk 23	Argentine Navy	2	17 May 1978 - 23 Jun 1988	Withdrawn from service
Mk 24	Iraqi Army		None built	-
Mk 25[1]	Netherlands Navy	6	23 Aug 1976 - 16 Sep 1977	7 Sqdn
Mk 26	Iraqi Army (armed)		None built	-
Mk 27[2]	Netherlands Navy	10	6 Oct 1978 - 12 Nov 1979	860 Sqdn
Mk 28	Qatari Police	3*	2 Dec 1977 - 12 Apr 1978	Withdrawn from service
Mk 80	Danish Navy	8	3 Feb 1980 - 15 Sep 1981	Søvårnets Flyvetjåneste
Mk 81[3]	Netherlands Navy	8	9 Jul 1980 - 24 Mar 1981	860 Sqdn
Mk 82	Egyptian Army		None built	-
Mk 83	Saudi Army		None built	-
Mk 84	Qatari Army		None built	-
Mk 85	UAE Army		None built	-
Mk 86	Norwegian Coast Guard	6	23 Jan 1981 - 11 Sep 1981	Skv 337
Mk 87	Argentine Navy		Embargoed	-
Mk 88	German Navy	19	26 May 1981 - 10 Dec 1988	3/MFG 3
Mk 89	Nigerian Navy	3	29 Sep 1983 - 14 Mar 1984	101 Sqdn
Mk 90[4]	Danish Navy	1	19 Apr 1988[6]	Søvårnets Flyvetjåneste
Mk 95[5]	Portuguese Navy	3†		On order
Mk 99	South Korean Navy	12†	16 Nov 1989 - 14 May 1991	627 Sqdn
Total		**364**		
Lynx 3		1	14 Jun 1984	
Demonstrator		2	18 May 1979 - 29 Mar 1982	
Prototypes		13	21 Mar 1971 - 5 Mar 1975	
		380		

NOTES

Note A: Army Air Corps 651, 652, 653, 654, 655, 657, 659, 661, 662, 663, 665, 667, 669 and 671 Sqdns; 3 Commando Brigade Air Sqdn: B Flight

[1] Netherlands designation UH-14A
[2] Netherlands designation SH-14B
[3] Netherlands designation SH-14C
[4] Plus one conversion from demonstrator; others from Mk 80
[5] Plus two conversions from Mk 3
[6] Completion and first flight at Vaerløse, Denmark
* Army version
† Super/Battlefield Lynx

during regular overhauls from 1987 onwards. Danish, Netherlands and Norwegian Lynx similarly retrofitted. Usable fuel capacity 973 litres (257 US gallons; 214 Imp gallons) in five internal tanks. Optional 214 litres (56.4 US gallons; 47 Imp gallons) beneath bench seat in rear of cabin. For ferrying, two tanks each of 436 litres (115.3 US gallons; 96 Imp gallons) in cabin, replacing bench tank. Max usable fuel 1,845 litres (488 US gallons; 406 Imp gallons). Naval Lynx equipped with bench tank as standard, and ferry tanks for long-range surveillance, thus gross capacities: 1,200 litres (317 US gallons; 264 Imp gallons) standard; 1,862 litres (492 US gallons; 409.5 Imp gallons) maximum. Pressure or gravity refuelling. Engine oil tank capacity 6.8 litres (1.8 US gallons; 1.5 Imp gallons). Main rotor gearbox oil capacity 28 litres (7.4 US gallons; 6.2 Imp gallons).

ACCOMMODATION: Pilot and co-pilot or observer on side by side seats. Dual controls optional. Individual forward hinged cockpit door and large rearward sliding cabin door on each side; all four doors jettisonable. Cockpit accessible from cabin area. Maximum high density layout (military version) for one pilot and 10 armed troops or paratroops, on lightweight bench seats in soundproofed cabin. Alternative VIP layouts for four to seven passengers, with additional cabin soundproofing. Seats can be removed quickly to permit carriage of up to 907 kg (2,000 lb) of freight internally. Tiedown rings provided. In casualty evacuation role, with a crew of two, Lynx can accommodate up to six Alphin stretchers and a medical attendant. Both basic versions have secondary capability for search and rescue (up to nine survivors) and other roles.

SYSTEMS: Two independent hydraulic systems, pressure 141 bars (2,050 lb/sq in). Third hydraulic system provided in naval version when sonar equipment, MAD or hydraulic winch system installed. No pneumatic system. 28V DC electrical power supplied by two 6kW engine driven starter/generators and an alternator. External power sockets. 24V 23Ah (optionally 40Ah) nickel-cadmium battery fitted for essential services and emergency engine starting. 200V three-phase AC power available at 400Hz from two 15kVA transmission driven alternators. Optional cabin heating and ventilation system. Optional supplementary cockpit heating system. Electric anti-icing and demisting of windscreen, and electrically operated windscreen wipers, standard; windscreen washing system optional.

AVIONICS: All versions equipped as standard with navigation, cabin and cockpit lights; adjustable landing light under nose; and anti-collision beacon. Avionics common to all roles (general purpose and naval versions) include GEC-Marconi duplex three-axis automatic stabilisation equipment; BAe GM9 Gyrosyn compass system; Decca tactical air navigation system (TANS); Decca 71 Doppler, E2C standby compass; and Racal intercom. Optional role equipment for both versions includes GEC-Marconi Mk 34 AFCS; Collins VOR/ILS; DME; Collins AN/ARN-118 Tacan; I-band transponder (naval version only); GEC-Plessey PTR 446, Collins APX-72, Siemens STR 700/375 or Italtel APX-77 IFF; and vortex sand filter for engine air intakes. Additional units fitted in naval version, when sonar is installed, to provide automatic transition to hover and automatic Doppler hold in hover.

British Army Lynx equipped with TOW missiles have roof mounted Hughes sight manufactured under licence by British Aerospace. Roof sight upgraded with night vision capability in far infra-red waveband; first test firing of TOW with added GEC-Marconi thermal imager took place in October 1988. Sanders AN/ALQ-144 infra-red jammer installed beneath tailboom of some British Army Lynx from 1987 pending availability of exhaust diffusers. Requirement for RWR satisfied by 1989 selection of GEC-Marconi AWARE-3 (ARI23491) system. Optional equipment, according to role, can include lightweight sighting system with alternative target magnification, vertical and/or oblique cameras, flares for night operation, low light level TV, infra-red linescan, searchlight, and specialised communications equipment.

Naval Lynx has specialised equipment for its primary duties. Detection of submarines is by means of dipping sonars or magnetic anomaly detector. Dipping sonars operated by hydraulically powered winch and cable hover mode facilities within the AFCS. Racal MIR-2 Orange Crop passive radar detection system in RN Lynx; retrofitted to Danish Mk 90. CAE Electronics AN/ASQ-504(V) internal MAD ordered for RN Lynx in 1990. Tracor M-130 chaff/flare dispensers and Ericsson Radar Electronics AN/ALQ-167(V) D-J band anti-anti-ship missile jamming pods installed on RN Lynx patrolling Arabian Gulf, 1987. Two Loral Challenger IR jammers above cockpit of RN Lynx during 1991 Gulf War. GEC-Marconi ARI5979 Seaspray Mk 1 lightweight search and tracking radar, for detecting small surface targets in low visibility/high sea conditions. Matra AF 530 or APX-334 stabilised sight in French naval Lynx. Optional GEC Sandpiper FLIR on RN Lynx; FLIR Systems 2000HP specified for Netherlands SH-14D upgrade.

EQUIPMENT: For search and rescue, with three crew, both ve sions can have a waterproof floor and a 272 kg (600 lb capacity clip-on hydraulic hoist on starboard side of cabi Cable length 30 m (98 ft).

ARMAMENT: For armed escort, anti-tank or air-to-surfac strike missions, army version can be equipped with tw 20 mm cannon mounted externally so as to permit carriag also of anti-tank missiles or pintle-mounted 7.62 m machine-gun inside cabin. External pylon can be fitted c each side of cabin for variety of stores, including two Min gun or other self-contained gun pods; two rocket pods; up to eight Aerospatiale/MBB HOT, Rockwell Hellfir Hughes TOW, or similar air-to-surface missile Additional six or eight missiles carried in cabin. For AS role, armament includes two Mk 44, Mk 46 or Sting Ra homing torpedoes, one each on an external pylon on eac side of fuselage, and six marine markers; or two Mk 1 depth charges. Alternatively, up to four BAe Sea Sku semi-active homing missiles; on French Navy Lynx, fou AS.12 or similar wire guided missiles. Self-protection F HMP 0.50 in machine-gun pod optional on RN Lynx.

DIMENSIONS, EXTERNAL (A: military version, N: naval version

Main rotor diameter (A, N)	12.80 m (42 ft 0 i
Tail rotor diameter (A, N)	2.21 m (7 ft 3 ir
Length overall:	
A, N, both rotors turning	15.163 m (49 ft 9 ir
N, main rotor blades and tail folded	
	10.618 m (34 ft 10 ir
Width overall, main rotor blades folded:	
A	3.75 m (12 ft 3¾ ir
N	2.94 m (9 ft 7¾ ir
Height overall: both rotors stopped:	
A	3.504 m (11 ft 6 ir
N	3.48 m (11 ft 5 ir
main rotor blades and tail folded:	
N	3.20 m (10 ft 6 ir
Tailplane half-span	1.78 m (5 ft 10 ir
Skid track: A	2.032 m (6 ft 8 ir
Wheel track: N	2.778 m (9 ft 1.4 ir
Wheelbase: N	2.94 m (9 ft 7¾ ir

DIMENSIONS, INTERNAL:

Cabin, from back of pilots' seats:	
Min length	2.057 m (6 ft 9 ir
Max width	1.778 m (5 ft 10 ir
Max height	1.422 m (4 ft 8 ir
Floor area	3.72 m² (40.04 sq f
Volume	5.21 m³ (184 cu f
Cabin doorway: Width	1.37 m (4 ft 6 ir
Height	1.19 m (3 ft 11 ir

AREAS:

Main rotor disc	128.7 m² (1,385.4 sq f
Tail rotor disc	3.84 m² (41.28 sq f

WEIGHTS AND LOADINGS (A and N as above):

Manufacturer's empty weight: A	2,578 kg (5,683 lb
N	2,740 kg (6,040 lb
Manufacturer's basic weight: A	2,658 kg (5,860 lb
N	3,030 kg (6,680 lb
Operating weight empty, equipped:	
A, troop transport (pilot and 10 troops)	
	2,787 kg (6,144 lb
A, anti-tank strike (incl weapon pylons, firing equipment and sight)	3,072 kg (6,772 lb
A, search and rescue (crew of three)	
	2,963 kg (6,532 lb
N, anti-submarine strike	3,343 kg (7,370 lb
N, reconnaissance (crew of two)	3,277 kg (7,224 lb
N, anti-submarine classification and strike	
	3,472 kg (7,654 lb
N, air to surface vessel search and strike (crew of two and four Sea Skuas)	3,414 kg (7,526 lb
N, search and rescue (crew of three)	
	3,416 kg (7,531 lb
N, dunking sonar search and strike	
	3,650 kg (8,047 lb
Max T-O weight: A	4,535 kg (10,000 lb
N (Gem Mk 41)	4,763 kg (10,500 lb
N (Gem Mk 42)	4,876 kg (10,750 lb
Max disc loading: A	35.24 kg/m² (7.22 lb/sq ft
N (Gem Mk 41)	37.00 kg/m² (7.58 lb/sq ft
M (Gem Mk 42)	37.88 kg/m² (7.76 lb/sq ft
Max power loading: A	3.31 kg/kW (5.43 lb/shp
N (Gem Mk 41)	3.47 kg/kW (5.71 lb/shp
N (Gem Mk 42)	3.55 kg/kW (5.84 lb/shp

PERFORMANCE (at normal max T-O weight at S/L, ISA, excep where indicated, A and N as above):

Max continuous cruising speed:	
A	140 knots (259 km/h; 161 mph
N	125 knots (232 km/h; 144 mph
A (ISA + 20°C)	130 knots (241 km/h; 150 mph
N (ISA + 20°C)	114 knots (211 km/h; 131 mph
Speed for max endurance:	
A, N (ISA and ISA + 20°C)	
	70 knots (130 km/h; 81 mph
Max forward rate of climb: A	756 m (2,480 ft)/mir
N	661 m (2,170 ft)/mir
A (ISA + 20°C)	536 m (1,760 ft)/mir
N (ISA + 20°C)	469 m (1,540 ft)/mir
Max vertical rate of climb:	
A	472 m (1,550 ft)/mir
N	351 m (1,150 ft)/mir

A (ISA + 20°C)	390 m (1,280 ft)/min
N (ISA + 20°C)	244 m (800 ft)/min
Hovering ceiling OGE: A	3,230 m (10,600 ft)
N	2,575 m (8,450 ft)

Typical range, with reserves:
  A, troop transport   292 nm (540 km; 336 miles)
Radius of action, out and back at max sustained speed, allowances for T-O and landing, 30 min loiter in search area, 3 min hover for each survivor, and 10% fuel reserves at end of mission:
  N, search and rescue (crew of 3 and 2 survivors)
    115 nm (212 km; 132 miles)
  N, search and rescue (crew of 3 and 7 survivors)
    96 nm (178 km; 111 miles)
Time on station at 50 nm (93 km; 58 miles) radius, out and back at max sustained speed, with 2 torpedoes, smoke floats and marine markers, allowances for T-O and landing and 10% fuel reserves at end of mission:
  N, anti-submarine classification and strike, loiter speed on station   2 h
  N, anti-submarine strike, loiter on station   2 h 29 min
  N, dunking sonar search and strike, 50% loiter speed and 50% hover on station   1 h 5 min
Time on station at 50 nm (93 km; 58 miles) radius, out and back at max sustained speed, with crew of 2 and 4 Sea Skuas, allowances and reserves as above:
  N, air to surface vessel strike, en route radar search and loiter speed on station   1 h 36 min
Max range: A   340 nm (630 km; 392 miles)
  N   320 nm (593 km; 368 miles)
Max endurance: A   2 h 57 min
  N (ISA + 20°C)   2 h 50 min
Max ferry range with auxiliary cabin tanks:
  A   724 nm (1,342 km; 834 miles)
  N   565 nm (1,046 km; 650 miles)

WESTLAND SUPER LYNX and BATTLEFIELD LYNX

TYPE: Twin-engined multi-purpose export helicopters.
PROGRAMME: Battlefield Lynx mockup displayed at 1988 Farnborough air show (converted demonstrator G-LYNX), featuring wheeled landing gear, exhaust diffusers and provision for anti-helicopter missiles each side of fuselage; first flight of wheeled prototype (converted trials AH. Mk 7 XZ170) 29 November 1989; first flight of South Korean Super Lynx (90-0701, temporarily ZH219) 16 November 1989 (also first Lynx with Seaspray Mk 3); first flight of Portuguese Lynx (1001, temporarily ZH580, ex-RN ZF559) 27 March 1992.
CURRENT VERSIONS: **Super Lynx**: Upgraded export naval Lynx, approx equivalent to Lynx HAS. Mk 8 (see previous entry).
  Battlefield Lynx: Upgraded export army Lynx; approx equivalent to Lynx AH. Mk 9 (see previous entry). Demonstrator G-LYNX fitted with two 1,007 kW (1,350 shp) LHTEC T800 turboshafts as **Battlefield Lynx 800** private venture (LHTEC funding power plants and gearboxes, Westland providing airframe for full flight demonstration programme); first flight 25 September 1991; programme terminated early 1992 after 17 hours.
CUSTOMERS: Super Lynx ordered by South Korea 1988 (12 **Mk 99** with Racal Avionics Doppler 71/TANS N nav system, Seaspray Mk 3 360° radar, AN/AQS-18 dipping sonar and Sea Skua), handed over between 26 July 1990 and May 1991 for 'Sumner' and 'Gearing' class destroyers; Portugal ordered five (first two ex-Royal Navy modified airframes) Super Lynx **Mk 95** 1990 (plus three options) with Racal RNS252 'Super TANS' and Doppler 91 navigation systems and some US equipment including AN/AQS-18 sonar and Bendix/King RDR 1500 radar; deliveries due in 1993 for 'Vasco da Gama' class (MEKO 200) frigates.
DESIGN FEATURES: Upgraded export Lynx; max T-O weight 5,125 kg (11,300 lb); all-weather day/night capability; extended payload/range; advanced technology swept-tip (BERP) composite main rotor blades offering improved speed and aerodynamic efficiency and reduced vibration; Mk 7's dynamic improvements (reversed direction tail rotor for improved control, etc); non-retractable wheeled landing gear.
STRUCTURE: Composite main and tail rotor blades.
LANDING GEAR: Non-retractable tricycle type. Twin nose-wheels; single mainwheels. Oleo-pneumatic struts capable of absorbing 1.83 m (6 ft)/s descent rate.
POWER PLANT: Two Rolls-Royce Gem 42-1 turboshafts, each rated at 835 kW (1,120 shp). Transmission rating 1,372 kW (1,840 shp). Exhaust diffusers for infra-red suppression optional on Battlefield Lynx.
AVIONICS: Super Lynx has Seaspray Mk 3 or Bendix/King RDR 1500 360° scan radar in chin fairing. (UK Mk 8 has Seaspray Mk 1 re-packaged, plus GEC-Marconi Sea Owl thermal imaging equipment above nose.) Vinten Vicon 78 chaff dispenser; Vinten Vipa 1 reconnaissance pod; or Agiflite reconnaissance camera system. Optional Bendix/King AN/AQS-18 or Thomson-Sintra HS-312 sonars.
  Battlefield Lynx may be equipped with Goodyear AN/ALE-39 chaff/flare dispensers and (subject to development) Dalmo-Victor AN/APR-39 RWR. Secure speech radio. Decca Doppler 71 and TANS 9447 navigation; Honeywell/Smiths AN/APN-198 radar altimeter; Collins

First Westland Super Lynx Mk 95 for the Portuguese Navy

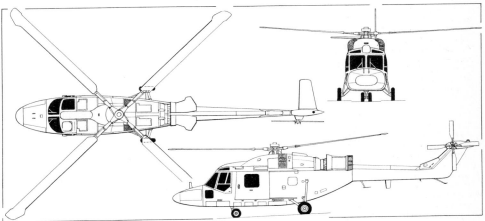

Westland Battlefield Lynx for export, equivalent to the British Army's Lynx AH. Mk 9 *(Dennis Punnett)*

First Westland Super Lynx Mk 99 in South Korean naval insignia *(Patrick H. F. Allen)*

206A ADF; Cossor 2720 IFF; Collins VIR 31A VOR/ILS. Sextant 250 sight for fixed armament. Vipa 1 and Agiflite cameras as Super Lynx.

ARMAMENT: Super Lynx as standard naval Lynx, including four Sea Skua or two Penguin anti-ship missiles; or four Stinger anti-helicopter missiles. Battlefield Lynx may carry two Giat 20 mm cannon pods; two FN Herstal pods with two 7.62 mm machine-guns each; or two M.159C pods containing nineteen 2.75 in rockets each. Eight TOW anti-tank missiles on modified mounting, with BAe sight for gunner. Optionally HOT or Hellfire missiles.

DIMENSIONS, EXTERNAL (A: Battlefield Lynx, N: Super Lynx):
Tail rotor diameter	2.36 m (7 ft 9 in)
Length overall:	
A, N, both rotors turning	15.24 m (50 ft 0 in)
A, rotors folded	13.24 m (43 ft 5¼ in)
N, main rotor blades and tail folded	10.85 m (35 ft 7¼ in)
Width overall, main rotor blades folded:	
A	3.02 m (9 ft 10⅞ in)
N	2.94 m (9 ft 7¾ in)
Height overall: tail rotor turning:	
A	3.73 m (12 ft 3 in)
N	3.67 m (12 ft 0½ in)
main rotor blades and tail folded:	
N	3.25 m (10 ft 8 in)
Tailplane half-span	1.32 m (4 ft 4 in)
Wheel track: A	2.80 m (9 ft 2¼ in)
Wheelbase: A, N	3.02 m (9 ft 11 in)

AREAS:
Tail rotor disc	4.37 m² (47.04 sq ft)

WEIGHTS AND LOADINGS:
Basic weight empty: A	3,178 kg (7,006 lb)
N	3,291 kg (7,255 lb)
Operating weight empty (including crew and appropriate armament):	
A, anti-tank (eight TOW)	3,949 kg (8,707 lb)
A, reconnaissance	3,444 kg (7,592 lb)
A, transport (unladen)	3,496 kg (7,707 lb)
N, anti-submarine warfare	4,207 kg (9,276 lb)
N, ASV (four Sea Skua)	4,252 kg (9,373 lb)
N, ASV (two Penguin)	4,461 kg (9,834 lb)
N, surveillance and targeting	3,597 kg (7,929 lb)
N, search and rescue	3,658 kg (8,064 lb)
Max underslung load	1,361 kg (3,000 lb)
Max T-O weight	5,125 kg (11,300 lb)
Max disc loading	39.82 kg/m² (8.16 lb/sq ft)
Max power loading	3.74 kg/kW (6.14 lb/shp)

PERFORMANCE:
Max continuous cruising speed:	
A	138 knots (256 km/h; 159 mph)
Radius of action:	
A, anti-tank, 2 h on station with four TOWs	25 nm (46 km; 29 miles)
N, anti-submarine, 2 h 20 min on station, dipping sonar and one torpedo	20 nm (37 km; 23 miles)
N, point attack with four Sea Skuas	148 nm (274 km; 170 miles)
N, surveillance, 4.1 h on station	75 nm (139 km; 86 miles)
Range:	
A, tactical transport	370 nm (685 km; 426 miles)

WESTLAND WS 70L

TYPE: Combat assault squad transport helicopter.
PROGRAMME: After full partnership agreement with United Technologies (USA), Westland received US State Department approval to produce Sikorsky Black Hawk as WS 70; demonstrator (ZG468) assembled from Sikorsky kit to US Army UH-60A battlefield transport standards; first flight April 1987 (used for training and market support).
CURRENT VERSIONS: **WS 70L**: Equivalent to US UH-60L (see Sikorsky entry in US section), except as detailed below.
CUSTOMERS: Saudi Arabia signed provisional agreement for 88 WS 70Ls July 1988 (part of larger equipment deal with UK Government); confirmation awaited 1993.
COSTS: Westland board assigned £3 million for demonstrator (ZG468).
POWER PLANT: Two 1,224 kW (1,641 shp) General Electric T700-GE-701C turboshafts.
AVIONICS (recommended): Communications equipment includes UHF and V/UHF and VHF/FM homing. Navigation equipment includes AN/ASN-43 compass, Doppler, plus customer specified equipment.

WEIGHTS AND LOADINGS:
Weight empty	4,964 kg (10,943 lb)
Max T-O weight	9,979 kg (22,000 lb)

PERFORMANCE:
Range with 1,814 kg (4,000 lb) payload	300 nm (556 km; 345 miles)
Ferry range	1,145 nm (2,122 km; 1,318 miles)

UNITED STATES OF AMERICA

AAC

AEROSTAR AIRCRAFT CORPORATION
3608 South Davison Boulevard, Spokane, Washington 99204-5799
Telephone: 1 (509) 455 8872
Fax: 1 (509) 838 0831
PRESIDENT: Steve Speer
VICE-PRESIDENT, MARKETING: James S. Christy

Subsidiary of Machen (which see) formed to develop jet-powered version of former Piper Aerostar. Company also extensively upgrades piston-engined Aerostars.

AAC AEROSTAR 3000

Pressurised Piper 601P, 602P and 700P re-engined with 8.45 kN (1,900 lb st) Williams FJ44 turbofans in underslung pods. Expected price $975,000 (1991); cruising speed 400 knots (740 km/h; 460 mph); range 1,500 nm (2,780 km; 1,725 miles). Conversion stopped; new-build Aerostar being developed with improved pressurisation and range, retaining published cruising speed.

Model of AAC Aerostar 3000 turbofan-powered light twin

AASI

ADVANCED AERODYNAMICS AND STRUCTURES INC
10703 Vanowen Street, North Hollywood, California 91605
Telephone: 1 (818) 753 1888
Fax: 1 (818) 753 8554
CHAIRMAN: Song-Gen Yeh
PRESIDENT: Dr Carl C. Chen
CEO AND GENERAL MANAGER: Darius Sharifzadeh
EXECUTIVE VICE-PRESIDENT: Bill Leeds
VICE-PRESIDENT, MANUFACTURING: David Tracy
DIRECTOR, MARKETING: Gene Comfort

AASI succeeded Aerodynamics & Structures Inc (ASI), formed by former airline pilot Darius Sharifzadeh to develop Jetcruzer. Factory being extended to 18,581 m² (200,000 sq ft) in readiness for production. Workforce 130.

AASI JETCRUZER 450 and 500P

TYPE: Six-seat turboprop business and utility transport.
PROGRAMME: Design began March 1983; aerodynamic design undertaken in UK 1984-85 by 'Sandy' Burns; layout prepared by Ladislao Pazmany; structural design by David Kent of Light Transport Design in UK; wind tunnel tests by University of San Diego; prototype construction started June 1988; first exhibited at NBAA show October 1988 with 313 kW (420 shp) Allison 250-C20S engine; first flight 11 January 1989 (N5369M) and had flown 150 hours by March 1991; pre-production prototype (N102JC) made first flight April 1991 and first flight in production form 13 September 1992; certification to FAR Pt 23 expected mid-1993; five then on line; aiming for single-engine FAA Part 135 public transport IFR certification.

CURRENT VERSIONS: **Jetcruzer 450**: Unpressurised model, *as detailed below*. Freight and air ambulance configurations available.
Jetcruzer 500P: Pressurised model with smaller windows.
Jetcruzer ML-1: UAV unmanned military model.
Jetcruzer ML-2: Piloted military model, costing $1.1 million.

CUSTOMERS: Total of 35 sold by December 1992; sales potential claimed to be 150 a year.
COSTS: $995,000 (1993).
DESIGN FEATURES: Canard safety; high speed; short field T-O/landing performance; all CAD; rear-mounted main wing using NACA 2412 aerofoil, with 20° sweepback at quarter-chord and 4° dihedral; tip-mounted fins; unswept foreplane using NASA LS 0417 (Mod) section. Compared to prototype, production Jetcruzer 500P has 25 cm (10 in) fuselage stretch, addition of fuel-carrying strakes at wingroots with 34° sweepback, leading-edge flaps on main wing, 43 cm (17 in) nose extension to allow forward retraction of nosewheel, and vertical tail surfaces canted 15° inboard to improve aileron control.
FLYING CONTROLS: Pushrod actuated; ailerons in main wing, elevators on foreplane, rudders in tip-mounted fins. Hydraulically actuated/mechanically controlled sealed leading-edge flaps on wings and foreplanes (30° droop). Autopilot optional.
STRUCTURE: Aluminium alloy wings, foreplane and vertical tail surfaces; monocoque fuselage of graphite composite Nomex honeycomb sandwich with embedded aluminium mesh.

AASI Jetcruzer 450 unpressurised six-seat business and utility aircraft (P&WC PT6A-27 turboprop engine)

LANDING GEAR: Retractable tricycle oleo-pneumatic type.
POWER PLANT: One 507 kW (680 shp) Pratt & Whitney Canada PT6A-27 turboprop, driving a Hartzell three-blade constant-speed pusher propeller at 2,200 rpm (with reverse and autofeathering). Fuel capacity 757 litres (200 US gallons; 166.5 Imp gallons).
ACCOMMODATION: Standard and club seating for six, including pilot, stressed to new FAA 26 *g* load limit. Three cabin windows on each side. Airstep gullwing door on each side.
SYSTEMS: Pressurisation (500P), air-conditioning, de-icing and anti-icing.
AVIONICS: EFIS. GPS. Dual Bendix/King KX 165 VHF com/nav with 65; KLN Loran nav; KR 87 ADF; KT 70 transponder; KN 63 DME; KMA 24H audio-control console or Collins system. Bendix/King KC 192 autopilot or Collins system optional. Bendix/King RD5-82UP colour radar or Collins system optional. Stormscope optional. Bendix/King KI 256 flight director (ADI) standard; ED 461 HSI standard.

DIMENSIONS, EXTERNAL:
Wing span	10.95 m (35 ft 11 in)
Wing aspect ratio	6.7
Foreplane span	5.77 m (18 ft 11 in)
Length overall	7.52 m (24 ft 8 in)
Fuselage length	7.49 m (24 ft 7 in)
Height overall	2.62 m (8 ft 7 in)
Propeller diameter	2.03 m (6 ft 8 in)
Propeller ground clearance	0.71 m (2 ft 4 in)
Cabin door (port): Height	1.09 m (3 ft 7 in)
Width	0.97 m (3 ft 2 in)
Cabin door (stbd): Height	1.09 m (3 ft 7 in)
Width	0.97 m (3 ft 2 in)

DIMENSIONS, INTERNAL:
Cabin: Length	3.40 m (11 ft 2 in)
Volume	4.15 m³ (146.5 cu ft)

AREAS:
Wings, gross	17.89 m² (192.6 sq ft)
Ailerons (total)	0.81 m² (8.75 sq ft)
Foreplane, incl elevators	3.71 m² (39.90 sq ft)
Elevators (total, incl tabs)	0.42 m² (4.53 sq ft)
Fins (total)	0.81 m² (8.75 sq ft)
Rudders (total)	0.33 m² (3.60 sq ft)

WEIGHTS AND LOADINGS:
Weight empty	1,111 kg (2,450 lb)
Max T-O weight	2,268 kg (5,000 lb)
Max wing loading	126.75 kg/m² (25.96 lb/sq ft)
Max power loading	4.47 kg/kW (7.35 lb/shp)

PERFORMANCE (at max T-O weight):
Max operating speed (V_{MO}) and max cruising speed at 6,700 m (22,000 ft)	262 knots (486 km/h; 302 mph)
Stalling speed:	
slats extended	57 knots (106 km/h; 66 mph)
slats retracted	69 knots (128 km/h; 80 mph)
Service ceiling	7,620 m (25,000 ft)
T-O to 15 m (50 ft)	436 m (1,430 ft)
Landing from 15 m (50 ft)	594 m (1,950 ft)
Max range at 3,050 m (10,000 ft)	1,344 nm (2,490 km; 1,547 miles)

AASI JETCRUZER 650

TYPE: 10/13-seat turboprop commuter and utility transport.
PROGRAMME: Construction of prototype and first production aircraft began December 1992; AASI states that certification under FAR Pt 23 expected September-December 1993, with initial deliveries January 1994.
CURRENT VERSIONS: **Jetcruzer 650**: *As detailed below*, for commuter, freight carrying and air ambulance uses.
ML-4: Military version.
CUSTOMERS: Seven ordered by 20 December 1992.
COSTS: $1.185 million unpressurised; $1.34 million pressurised.
DESIGN FEATURES: Enlarged version of Jetcruzer 450/500P, with greatly increased accommodation; 260 knot (483 km/h; 300 mph) cruising speed and short field performance; economical operation; low maintenance costs; low noise.
FLYING CONTROLS: As for Jetcruzer 450/500P.
STRUCTURE: As for Jetcruzer 450/500P.
POWER PLANT: One 691 ekW (927 ehp) Pratt & Whitney Canada PT6A-135A turboprop, driving a Hartzell three-blade pusher propeller. Fuel capacity 757 litres (200 US gallons; 166.5 Imp gallons) in wing tanks.
ACCOMMODATION: Thirteen persons (incl one or two pilots) in international models, 10 for US market, on standard or club seats. Two airstep gullwing doors. Air-conditioning and optionally pressurised.
SYSTEMS: Optional pressurisation; air-conditioning, de-icing and anti-icing.
AVIONICS: IFR, dual nav/com, EFIS, radar, GPS and Loran.

DIMENSIONS, EXTERNAL:
Wing span	11.33 m (37 ft 2 in)
Foreplane span	5.49 m (18 ft 0 in)
Fuselage: Length	10.36 m (34 ft 0 in)
Max depth	1.55 m (5 ft 1 in)
Max width	1.52 m (5 ft 0 in)
Height overall	2.44 m (8 ft 0 in)
Wheel track	3.66 m (12 ft 0 in)

DIMENSIONS, INTERNAL:
Cabin: Length	5.49 m (18 ft 0 in)
Max width	1.50 m (4 ft 11 in)
Max height	1.52 m (5 ft 0 in)

WEIGHTS AND LOADINGS:
Weight empty	1,588 kg (3,500 lb)
Max T-O weight	2,948 kg (6,500 lb)
Max power loading	4.27 kg/ekW (7.01 lb/ehp)

PERFORMANCE (estimated):
Max cruising speed at 6,700 m (22,000 ft)	260 knots (483 km/h; 300 mph)
Stalling speed, flaps down, power off	61 knots (113 km/h; 71 mph)
Max rate of climb at S/L	716 m (2,350 ft)/min
Service ceiling	9,150 m (30,000 ft)
T-O run	391 m (1,280 ft)
Landing run	412 m (1,352 ft)
Range	1,000 nm (1,853 km; 1,151 miles)

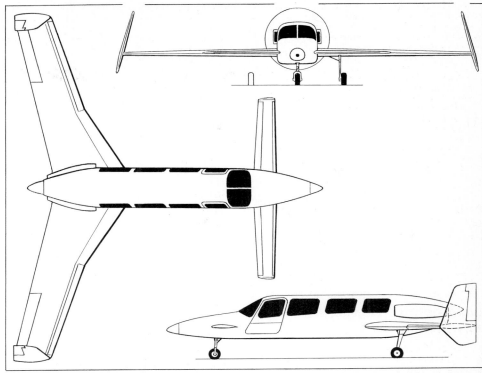
AASI Jetcruzer 650 commuter and utility transport (P&WC PT6A-135A turboprop engine) *(Dennis Punnett)*

Artist's impression of the AASI Stratocruzer 1250-ER intercontinental-range transport (two Rolls-Royce Williams FJ44 turbofans)

AASI STRATOCRUZER 1250-ER

TYPE: 10/13-seat commercial, corporate and utility twin-turbofan transport.
PROGRAMME: Design started September 1991; prototype expected to be exhibited at NBAA Convention September 1993; certification to FAR Pt 23 expected end of 1994.
CURRENT VERSIONS: **Stratocruzer 1250-ER**: Intercontinental civil version, *as detailed below*. Can be used for commercial operations, corporate transportation, freight carrying, medevac and other roles.
ML-5: Military version; envisaged SAR layout has two pilots plus three other crew and two radar scanner consoles.
CUSTOMERS: Four ordered at $3.2 million each.
DESIGN FEATURES: Intercontinental range; high speed; low cost and maintenance. Design based on Jetcruzer general configuration, but with swept foreplanes, extended inboard wing leading-edges, winglets, tailfin with high-mounted swept tailplane, and twin turbofans pod-mounted on pylons at fin root.
FLYING CONTROLS: Ailerons on main wings with trim tabs, elevators on tailplane, and rudder; foreplane elevator trim, wing leading-edge flaps, dual yaw dampers; mechanical trim system; autopilot.
STRUCTURE: As for Jetcruzer.
LANDING GEAR: Similar arrangement to Jetcruzer, but with dual wheels on each unit; hydraulically actuated; nosewheel steering; mainwheel disc brakes, anti-skid brake system, parking brake, emergency brake system and speed brake system.

412 USA: AIRCRAFT—AASI/AIR TRACTOR

POWER PLANT: Two 8.45 kN (1,900 lb st) Rolls-Royce Williams FJ44 turbojets; AASI requested 10 to 15 per cent thrust increase to improve engine-out performance. Four fuel tanks in fuselage and wings; fuel jettison system.

ACCOMMODATION: Thirteen persons (incl one or two pilots) on international models, 10 for US market. Alternative layouts are for two pilots and four executives plus rest room; two pilots plus two stretcher patients and three attendants for medevac role; or single pilot plus freight. Galley, toilet, standard or club seating are among customer options. Air-conditioning, heating, cabin temperature control, ventilation, and pressurisation. Two doors.

SYSTEMS: Cabin pressurisation, air-conditioning and heating, ram air ventilation, and oxygen with crew masks and passenger drop-down masks; electrical with two generators, dual 24V batteries, solid state voltage regulators, and two starters; hydraulic; de-icing and anti-icing, including wing and foreplane leading-edges, tailfin, winglets, engine inlets and windscreen; heated static ports and pitot tubes.

AVIONICS: Collins suite; typically EADI, EHSI, airspeed indicator, RMI, nav indicator, standby attitude indicator, altimeter, VSI, angle of attack indicator, clock, audio panel, EFIS, engine temperature and fan speed indicator, engine oil pressure and temperature indicators, fuel flow and quantity indicators, trim position indicators, annunciator panel, autopilot, colour weather radar, GPS, nav/com panel, DME, oxygen pressure indicator, hour meter, digital clocks, AC/DC voltmeters, wing temperature indicator, flap position indicator, magnetic compass, and warning systems.

DIMENSIONS, EXTERNAL:
Wing span	14.02 m (46 ft 0 in)
Length overall	10.97 m (36 ft 0 in)
Fuselage length	10.36 m (34 ft 0 in)
Diameter of fuselage	1.55 m (5 ft 1 in)
Height overall	4.04 m (13 ft 3 in)

WEIGHTS AND LOADINGS:
Weight empty	2,654 kg (5,850 lb)
Fuel weight	2,585 kg (5,700 lb)
Max T-O weight	5,670 kg (12,500 lb)
Max power loading	335.5 kg/kN (3.29 lb/lb st)

PERFORMANCE (estimated):
Max cruising speed at 11,275 m (37,000 ft)	419 knots (776 km/h; 482 mph)
Stalling speed, flaps down	81 knots (150 km/h; 94 mph)
Max rate of climb at S/L	1,115 m (3,650 ft)/min
Max cruising altitude	13,715 m (45,000 ft)
T-O run	1,250 m (4,100 ft)
Landing run	1,115 m (3,650 ft)
Range, with 20 min VFR reserves	3,200 nm (5,930 km; 3,684 miles)

AERO UNION
AERO UNION CORPORATION
PO Box 247, Municipal Airport, 100 Lockheed Avenue, Chico, California 95926
Telephone: 1 (916) 896 3000
Fax: 1 (916) 893 8585
PRESIDENT: Dale P. Newton
SECRETARY-TREASURER/GENERAL MANAGER: Victor E. Alvistur
DIRECTOR, SALES AND MARKETING: John Oswald
DIRECTOR, INTERNATIONAL MARKETING: John E. Gyarfas

Company established 1959 for aerial firefighting; manufactures tank systems for Douglas DC-3, DC-4, DC-6 and DC-7, Fairchild C-119, Grumman S-2, Lockheed C-130/L-100 Hercules, Lockheed P-3 Orion and Electra; also produces aircraft fuel tanks, airstairs, environmental control systems, cargo pallet roller systems, retardant delivery and aerial spraying systems, dorsal fins; sole manufacturer of Model 1080 Air Refuelling Store developed by Beechcraft an acquired by Aero Union in 1985. Details of all current programmes (see 1991-92 *Jane's*) now shown in *Jane's Civil and Military Aircraft Upgrades*.

AIR & SPACE
AIR & SPACE AMERICA INC
4460 Shemwell Lane, Paducah, Kentucky 42003
Telephone: 1 (502) 898 2403
Fax: 1 (502) 898 8691
PRESIDENT: Don Farrington

Manufacturing and marketing subsidiary of Farrington Aircraft Corporation.

AIR & SPACE 18A
TYPE: Two-seat autogyro.

PROGRAMME: Original aircraft (68 built) produced in mid-1960s (certificated 1965) until Air & Space Manufacturing Inc declared bankrupt in 1966. Farrington Aircraft announced in June 1990 that 30 of these had been modernised under seven STCs and that it intended to restart production in this improved form, starting deliveries September 1990; first new production aircraft (N902AS) granted Standard Airworthiness Certificate 2 April 1991 and sold; production temporarily halted in 1992 pending accumulation of 50 orders to support production launch; prototype flown with 149 kW (200 hp) IO-360 engine to be offered as optional engine; to be followed by max T-O weight increase to 907 kg (2,000 lb), which provide empty weight of 601 kg (1,325 lb), an anticipated useful load of 306 kg (675 lb), and possibility of three-seat interior. Proposed applications for 18A include patrol/surveillance, environmental monitoring, law enforcement, pipeline inspection and broadcasting, as well as private flying.

COSTS: Standard VFR-equipped $99,995 (May 1993).

DESIGN FEATURES: Three-blade rotor, fully enclosed tandem-seat cabin, non-retractable tricycle landing gear, three-fin tail unit (all-moving central and two fixed outer). Original aircraft described in 1967-68 *Jane's*; Farrington improvements include trimmable two-position collective pitch control, improved engine cowling, noise-reduced exhaust and strengthened nosewheel leg.

POWER PLANT: One 134 kW (180 hp) Textron Lycoming O-360 flat-four engine with Hartzell two-blade pusher propeller. Fuel capacity 113.5 litres (30 US gallons; 25 Imp gallons).

Air & Space 18A two-seat autogyro

DIMENSIONS, EXTERNAL:
Rotor diameter	10.67 m (35 ft 0 in)
Length, without rotor	6.06 m (19 ft 10½ in)
Height overall	2.95 m (9 ft 8 in)

DIMENSIONS, INTERNAL:
Cabin length	2.78 m (9 ft 1½ in)

AREAS:
Rotor disc	89.38 m² (962.1 sq ft)

WEIGHTS AND LOADINGS:
Weight empty	596 kg (1,315 lb)
Max baggage weight	45 kg (100 lb)
Max T-O weight	816.5 kg (1,800 lb)
Max disc loading	9.13 kg/m² (1.87 lb/sq ft)

PERFORMANCE:
Never-exceed speed (VNE)	95 knots (177 km/h; 110 mph)
Cruising speed (75% power) at 2,135 m (7,000 ft)	80 knots (148 km/h; 92 mph)
Min level flight speed	24 knots (44 km/h; 27 mph)
Range with reserves	173 nm (322 km; 200 miles)
Endurance with reserves	3

AIR TRACTOR
AIR TRACTOR INC
PO Box 485, Municipal Airport, Olney, Texas 76374
Telephone: 1 (817) 564 5616
Fax: 1 (817) 564 2348
Telex: 910 890 4792
PRESIDENT: Leland Snow

Air Tractor agricultural aircraft based on 35-year experience of Leland Snow, who produced Snow S-2 series, which later became Rockwell S-2R (see earlier *Jane's*); 1,100th Air Tractor delivered November 1991; seven models available in 1993, powered by various P&WC PT6A and R-1340 engines.

AIR TRACTOR AT-401 AIR TRACTOR
TYPE: Single-seat agricultural aircraft.

PROGRAMME: AT-401 developed 1986 from AT-301, with increased wing span and larger hopper; combined production of AT-401/AT-502 (which see) running at eight/nine a month in early 1993; total of 92 built 1992; similar planned for 1993. AT-401A version with Polish PZL-3S radial engine has been abandoned, with just one aircraft produced (see 1992-93 *Jane's*).

CUSTOMERS: By January 1993, 156 AT-401s delivered to Australia, Brazil, Canada, Colombia, Mexico, Spain and USA.

COSTS: Standard AT-401 $168,000; with customer-supplied engine $139,900.

DESIGN FEATURES: Wing aerofoil NACA 4415; dihedral 3° 30′; incidence 2°.

FLYING CONTROLS: Ailerons, elevators and rudders have boost tabs; ailerons droop 10° when electrically operated Fowler flaps deflected to maximum 26°.

STRUCTURE: Two-spar wing structure of 2024-T3 light alloy, with alloy steel lower spar cap; bonded doubler inside wing leading-edge to resist impact damage; glassfibre wingroot fairings and skin overlaps sealed against chemical ingress; wing ribs and skins zinc chromated before assembly; flaps and ailerons of light alloy. Fuselage of 4130N steel tube, oven stress relieved and oiled internally, with skin panels of 2024-T3 light alloy attached by Camloc fasteners for quick removal; rear fuselage lightly pressurised to prevent chemical ingress; cantilever fin and strut braced tailplane of light alloy, metal-skinned and sealed against chemical ingress.

LANDING GEAR: Non-retractable tailwheel type. Cantilever heavy duty E-4340 spring steel main gear, thickness 28.6 mm (1.125 in); flat spring suspension for castoring and lockable tailwheel. Cleveland mainwheels with tyre size 8.50-10 (8-ply), pressure 2.83 bars (41 lb/sq in). Tailwheel tyre size 5.00-5. Cleveland four-piston brakes with heavy duty discs. Optional 29.00-11 Cleveland mainwheels with six-piston brakes.

POWER PLANT: One remanufactured 447 kW (600 hp) Pratt & Whitney R-1340 air-cooled radial engine with speed ring cowling, driving a Hamilton Standard 12D40/6101A-12 propeller; optional propellers include a Pacific Propeller 22D40/AG200-2 Hydromatic two-blade constant-speed metal and Hydromatic 23D40 three-blade propeller. Air Tractor has designed and is producing new FAA approved replacement crankshaft for R-1340; other new replacement parts available include main and thrust bearings, master rod bearings and blower (impeller) bearings.

ACCOMMODATION: Single seat with nylon mesh cover in enclosed cabin which is sealed to prevent chemical ingress. Downward hinged window/door on each side. 'Line of sight' instrument layout, with swing-down lower instrument panel for ease of access for instrument maintenance. Baggage compartment in bottom of fuselage, aft of cabin, with door on port side. Cabin ventilation by 0.10 m (4 in) diameter airscoop.

SYSTEMS: 24V electrical system, supplied by 35A engine driven alternator; 60A alternator optional.

AVIONICS: Optional avionics include Bendix/King KX 15 nav/com and Narco ELT-10 emergency locator transmitter.

EQUIPMENT: Agricultural dispersal system comprises a 1,514 litre (400 US gallon; 333 Imp gallon) Derakane vinylester resin/glassfibre hopper mounted in forward fuselage with hopper window and instrument panel mounted hopper quantity gauge; 0.97 m (3 ft 2 in) wide Transland gatebox, Transland 5 cm (2 in) bottom loading valve; Agrinautic 6.4 cm (2½ in) spraypump with Transland on/off valve and two-blade wooden fan, and 41-nozzle stainless steel spray system with streamlined booms. Ground start receptacle and three-colour polyurethane paint finish standard Optional equipment includes night flying package comprising strobe and navigation lights; night working lights retractable 600W landing light in port wingtip, and ferry fuel system. Alternative agricultural equipment includes Transland 22358 extra high volume spreader, Transland 54401 NorCal Swathmaster, and 40 extra spray nozzles for high volume spraying.

DIMENSIONS, EXTERNAL:
Wing span	14.97 m (49 ft 1¼ in)
Wing chord, constant	1.83 m (6 ft 0 in)
Wing aspect ratio	8.20
Length overall	8.23 m (27 ft 0 in)
Height overall	2.59 m (8 ft 6 in)
Propeller diameter: standard	2.77 m (9 ft 1 in)
optional	2.59 m (8 ft 6 in)

AREAS:
Wings, gross	27.31 m² (294.0 sq ft)
Ailerons (total)	3.55 m² (38.2 sq ft)
Trailing-edge flaps (total)	3.75 m² (40.4 sq ft)
Fin	0.90 m² (9.7 sq ft)
Rudder	1.30 m² (14.0 sq ft)
Tailplane	2.42 m² (26.0 sq ft)
Elevators, incl tabs	2.36 m² (25.4 sq ft)

WEIGHTS AND LOADINGS:
Weight empty, spray equipped   1,875 kg (4,135 lb)
Max T-O weight   3,565 kg (7,860 lb)
Max landing weight   2,721 kg (6,000 lb)
Max wing loading   130.51 kg/m² (26.73 lb/sq ft)
Max power loading   7.97 kg/kW (13.1 lb/hp)
PERFORMANCE (at max T-O weight, ISA, except where indicated):
Max cruising speed at S/L, hopper empty
 135 knots (251 km/h; 156 mph)
Cruising speed at 1,220 m (4,000 ft)
 124 knots (230 km/h; 143 mph)
Typical working speed
 104-122 knots (193-225 km/h; 120-140 mph)
Stalling speed at 2,721 kg (6,000 lb):
 flaps up   64 knots (118 km/h; 73 mph)
 flaps down   53 knots (99 km/h; 61 mph)
Stalling speed as usually landed
 47 knots (87 km/h; 54 mph)
Max rate of climb at S/L:
 at max landing weight   335 m (1,100 ft)/min
 at max T-O weight   158 m (520 ft)/min
T-O run   402 m (1,318 ft)
Range, econ cruising speed at 2,440 m (8,000 ft), no reserves   547 nm (1,014 km; 630 miles)

AIR TRACTOR AT-400 and AT-402 TURBO AIR TRACTOR

TYPE: Turboprop agricultural aircraft with choice of wing spans.
PROGRAMME: Following on from AT-400, AT-402 first flight August 1988; certificated November 1988; first delivery late 1988.
CURRENT VERSIONS: **AT-400**: Basic version, with short-span wing.
 AT-402: Combines fuselage, tail surfaces and landing gear of AT-400 with longer wing of AT-401.
CUSTOMERS: Total of 86 AT-400s and 66 AT-402s delivered by 1 January 1993.
COSTS: Standard AT-402 $398,000; with customer-supplied PT6A-15AG, -27, -28, -34AG engine $179,500.
DESIGN FEATURES: Both variants have 1,514 litre (400 US gallon; 333 Imp gallon) hopper with 0.97 m (3 ft 2 in) wide gatebox; both variants powered by 507 kW (680 shp) P&WC PT6A-15AG, -27 or -28, either new or customer-furnished, driving Hartzell three-blade constant-speed reversible-pitch propeller. All models have steel alloy lower wing spar caps for unlimited fatigue life and reinforced leading-edge to prevent bird strike damage, size 29-11 high-flotation tyres and wheels as standard, 250A starter/generator and two 24V 21Ah batteries; standard fuel capacity 644 litres (170 US gallons; 142 Imp gallons); optional fuel tankage 818 litres (216 US gallons; 180 Imp gallons) or 886 litres (234 US gallons; 195 Imp gallons); optional equipment includes Transland extra high volume dispersal system.
DIMENSIONS, EXTERNAL:
Wing span: AT-400   13.75 m (45 ft 1¼ in)
 AT-402   14.97 m (49 ft 1¼ in)
WEIGHTS AND LOADINGS:
Weight empty, spray equipped   1,696 kg (3,739 lb)
Certificated gross weight (FAR 23)   2,721 kg (6,000 lb)
Typical operating weight (CAM 8)   3,538 kg (7,800 lb)
Max wing loading   141.1 kg/m² (28.9 lb/sq ft)
Max power loading   6.98 kg/kW (11.47 lb/shp)
PERFORMANCE (at max T-O weight except where indicated):
Max level speed at S/L:
 clean   174 knots (322 km/h; 200 mph)
 with dispersal equipment
 160 knots (298 km/h; 185 mph)
Cruising speed at 283.3 kW (380 shp) at 2,440 m (8,000 ft)
 142 knots (264 km/h; 164 mph)
Typical working speed
 113-126 knots (209-233 km/h; 130-145 mph)
Stalling speed at 2,721 kg (6,000 lb) AUW:
 flaps up   64 knots (118 km/h; 73 mph)
 flaps down   53 knots (99 km/h; 61 mph)
Stalling speed as usually landed
 46 knots (86 km/h; 53 mph)
Max rate of climb at S/L, dispersal equipment installed:
 AUW of 2,721 kg (6,000 lb)   495 m (1,625 ft)/min
 AUW of 3,565 kg (7,860 lb)   305 m (1,000 ft)/min
T-O run at AUW of 3,565 kg (7,860 lb)   247 m (810 ft)
Landing run as usually landed   122 m (400 ft)

AIR TRACTOR AT-501

No AT-501s sold during 1992; withdrawn from production line.

AIR TRACTOR AT-502

TYPE: Turboprop powered agricultural aircraft for low-density fertilisers.
PROGRAMME: First flight April 1987; certificated 23 June 1987; 41 AT-502/502As sold in 1992; combined production of AT-401/AT-502 at eight/nine a month in early 1993; reportedly the world's best selling turboprop agricultural aircraft.

Air Tractor AT-401 Air Tractor (Pratt & Whitney R-1340 radial)

Air Tractor AT-402 agricultural aircraft

CURRENT VERSIONS: **AT-502**: Standard version, *described below*.
 AT-502A: Similar to AT-502 but with used 820 kW (1,100 shp) PT6A-45R; slow turning (1,425 rpm) five-blade Hartzell propeller; enlarged vertical tail surfaces. For operation in mountainous/high density altitude conditions. Prototype first flight February 1992; certificated April 1992; 12 delivered by 1 January 1993.
CUSTOMERS: 140 AT-502s delivered by 1 January 1993.
COSTS: Standard AT-502 $408,000; with customer-supplied PT6A-15AG, -27, -28, -34AG engine $189,500.
DESIGN FEATURES: Larger chemical hopper than AT-400 to handle low-density nitrogen based fertilisers such as urea; safety glass centre windscreen with wiper. Differences from AT-400 include approx 1.52 m (5 ft 0 in) increase in wing span, alloy steel lower spar cap and bonded doubler on inside of wing leading-edge for increased resistance to impact damage, and glassfibre wingroot fairings; fuselage length increased by 0.56 m (1 ft 10 in), width increased by 0.25 m (10 in), and larger diameter tubular frame members used to cater for increased max T-O weight.
LANDING GEAR: Non-retractable tailwheel type. Heavy duty E-4340 spring steel main gear, thickness 37.2 mm (1.31 in); flat spring for castoring and lockable tailwheel. Cleveland mainwheels, tyre size 29.00-11, pressure 3.45 bars (50 lb/sq in); tailwheel tyre size 5.00-5; Cleveland six-piston brakes with heavy duty discs.
POWER PLANT: One 507 kW (680 shp) Pratt & Whitney Canada PT6A-15AG, PT6A-27 or PT6A-28, or 559 kW (750 shp) PT6A-34 or PT6A-34AG turboprop, driving a Hartzell HCB3TN-3D/T10282+4 three-blade metal propeller. Standard fuel capacity 644 litres (170 US gallons; 142 Imp gallons). Optional capacities 818 litres (216 US gallons; 180 Imp gallons) and 886 litres (234 US gallons; 195 Imp gallons).
ACCOMMODATION: As for AT-400, but with new quick-detachable instrument panel and removable fuselage skin panels for ease of maintenance.
SYSTEMS: Two 24V 42Ah batteries and 250A starter/generator.
AVIONICS: Optional avionics include Bendix/King KX 155 nav/com and KR 87 ADF, KY 196 com radio, KT 76A transponder and Narco ELT-10 emergency locator transmitter.
EQUIPMENT: Agricultural dispersal system comprises a 1,900 litre (502 US gallon; 418 Imp gallon) Derakane vinylester resin/glassfibre hopper mounted in forward fuselage with hopper window and instrument panel mounted hopper quantity gauge; 0.97 m (3 ft 2 in) wide Transland gatebox; Transland 6.4 cm (2½ in) bottom loading valve; Agrinautics 6.4 cm (2½ in) spraypump with Transland on/off valve and two-blade wooden fan, and 40-nozzle stainless steel spray system with streamlined booms. Optional dispersal equipment includes 7.6 cm (3 in) spray system with 119 spray nozzles, and automatic flagman. Standard equipment includes safety glass centre windscreen panel, ground start receptacle, three-colour polyurethane paint finish, strobe and navigation lights; windscreen washer and wiper; and

414   USA: AIRCRAFT—AIR TRACTOR

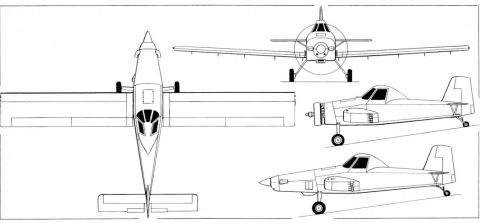

Air Tractor AT-502 agricultural aircraft with extra side view (top) of AT-401 *(Dennis Punnett)*

twin nose-mounted landing/taxi lights. Optional equipment includes night flying package, comprising night working lights, retractable 600W landing light in port wingtip; fuel flowmeter, fuel totaliser and ferry fuel system. Alternative agricultural equipment includes Transland 22356 extra high volume spreader, Transland 54401 NorCal Swathmaster, 41 extra spray nozzles for high volume spraying, and eight- or ten-unit Micronair Mini Atomiser unit.

DIMENSIONS, EXTERNAL:
Wing span	15.24 m (50 ft 0 in)
Wing chord, constant	1.83 m (6 ft 0 in)
Wing aspect ratio	8.33
Length overall	9.91 m (32 ft 6 in)
Height overall	2.99 m (9 ft 9½ in)
Wheel track	3.11 m (10 ft 2½ in)
Wheelbase	6.64 m (21 ft 9½ in)
Propeller diameter	2.69 m (8 ft 10 in)

AREAS:
Wings, gross	27.87 m² (300.0 sq ft)
Ailerons (total)	3.53 m² (38.0 sq ft)
Trailing-edge flaps (total)	3.75 m² (40.4 sq ft)
Fin	0.90 m² (9.7 sq ft)
Rudder	1.30 m² (14.0 sq ft)
Tailplane	2.41 m² (26.0 sq ft)
Elevators (total, incl tab)	2.44 m² (26.3 sq ft)

WEIGHTS AND LOADINGS:
Weight empty, spray equipped	1,870 kg (4,123 lb)
Max T-O weight (CAM 8)	4,173 kg (9,200 lb)
Max landing weight	3,629 kg (8,000 lb)
Max wing loading	138.36 kg/m² (28.33 lb/sq ft)
Max power loading	7.61 kg/kW (12.5 lb/shp)

PERFORMANCE (at max T-O weight, except where indicated, with spray equipment installed):
Never-exceed speed (VNE) and max level speed at S/L, hopper empty 156 knots (290 km/h; 180 mph)
Cruising speed at 2,440 m (8,000 ft), 283.3 kW (380 shp) 136 knots (253 km/h; 157 mph)
Typical working speed 104-126 knots (193-233 km/h; 120-145 mph)
Stalling speed at 3,629 kg (8,000 lb):
  flaps up 72 knots (134 km/h; 83 mph)
  flaps down 60 knots (111 km/h; 69 mph)
Stalling speed as usually landed 46 knots (86 km/h; 53 mph)
Max rate of climb at S/L, AUW of 3,629 kg (8,000 lb):
  with PT6A-15AG 311 m (1,020 ft)/min
  with PT6A-34AG 360 m (1,180 ft)/min
Max rate of climb at S/L:
  with PT6A-15AG 232 m (760 ft)/min
  with PT6A-34AG 282 m (925 ft)/min
T-O run at AUW of 3,629 kg (8,000 lb):
  with PT6A-15AG 244 m (800 ft)
  with PT6A-34AG 222 m (730 ft)
T-O run:
  with PT6A-15AG 356 m (1,170 ft)
  with PT6A-34AG 302 m (990 ft)
Range with max fuel 435 nm (805 km; 500 miles)

AIR TRACTOR AT-503/AT-503A

TYPE: Two-seat agricultural and training aircraft.
PROGRAMME: Design began September 1985 to meet US State Department requirement for anti-narcotics aerial application aircraft; first flight 25 April 1986; certificated 2 October 1986.
CURRENT VERSIONS: **AT-503**: Single aircraft (see Customers).
AT-503A: Production version, certificated 26 November 1990; AT-502 wings, tail unit, landing gear and engine installation. Prototype and first production aircraft with PT6A-34 turboprop engine rated at 559 kW (750 shp). Full dual controls; dry and liquid dispersal systems can be operated from either seat; empty weight 136 kg (300 lb) greater than otherwise comparable AT-502 due to extra weight of second cockpit and 68 kg (150 lb) ballast installed on engine mount ring. Specifications, external dimensions, areas, weights and performance identical to AT-502 except weight empty 2,032 kg (4,480 lb).
CUSTOMERS: Sole prototype AT-503 delivered to Spain, for fire suppression duties. Prototype AT-503A went to US customer in Spring 1991; first production aircraft to Brazil (Aviação Agricola), also in Spring 1991; second to US customer late 1991; none delivered in 1992.
DESIGN FEATURES: AT-503 became two-seat Air Tractor; missions include anti-drug enforcement, fire bombing, forest spraying in mountainous terrain, high-volume seeding or fertiliser application.
FLYING CONTROLS: As for AT-502.
STRUCTURE: Wings as for AT-400 except increased wing span and flaps and heavier gauge skins. Fuselage is 0.56 m (1 ft 10 in) longer, 0.25 m (10 in) wider and has larger diameter structure tubes.

AIR TRACTOR AT-802

TYPE: Two-seat agricultural and firefighting aircraft.
PROGRAMME: Design started July 1989; first flight of prototype (N802LS) 30 October 1990; second aircraft flew November 1991, with used PT6A-45R and configured as agricultural model with spraybooms, pump and Transland gate box. Production deliveries to start second quarter 1993.
CURRENT VERSIONS: **AT-802**: Two-seater powered by P&W PT6A-45R; certificated for gross weight of 6,804 (15,000 lb) 27 April 1993.
AT-802A: Single-seat version; third production aircraft in this configuration; can be powered by used PT6A-45 PT6A-65B or PT6A-65AG engine. First flight 6 July 199 FAA certification gained 17 December 1992; certificat for gross weight of 6,804 kg (15,000 lb) with PT6A-45 27 April 1993.
AT-802AF: Single-seat firefighting versio PT6A-67AF engine. FAA certification at 5,670 (12,500 lb) 17 December 1992; certificated at 7,257 (16,000 lb) 27 April 1993, giving useful load of 4,445 (9,800 lb).
COSTS: Standard AT-802A $860,000 with PT6A-67R; wi customer-supplied PT6A-45R, -65AG or -67R $295,00 Standard AT-802AF $960,000. *Data apply to this versi except where indicated.*
DESIGN FEATURES: Largest model built by company to da full dual controls for training; also designed for firefig ing; programmable logic computer with cockpit contr panel and digital display enables pilot to select covera level and opens hydraulically operated 'bomb bay' dr doors to prescribed width, closing them when select amount of retardant released. Wing aerofoil section NAC 4415; dihedral 3° 30′; incidence 2°.
FLYING CONTROLS: Boost tabs in each aileron, elevator a rudder; electrically operated Fowler trailing-edge fla deflect to maximum 30°.
STRUCTURE: Two-spar wing structure of 2024-T3 light all with alloy steel upper and lower spar caps and bonded do bler on inside of leading-edge for impact damage resi ance; ribs and skins zinc chromated before assembl glassfibre wingroot fairing and skin overlaps sealed agair chemical ingress; flaps and ailerons of light alloy. Fusela of welded 4130N steel tube, oven stress relieved and oil internally, with skin panels of 2024-T3 light alloy attach by Camloc fasteners for quick removal; rear fusela lightly pressurised to prevent chemical ingress. Cantilev fin and strut braced tailplane of light alloy, metal skinn and sealed against chemical ingress.
LANDING GEAR: Non-retractable tailwheel type. Cantilev heavy duty E-4340 spring steel main legs, thickne 39.6 mm (1.56 in); flat spring suspension for castoring a lockable tailwheel. Cleveland mainwheels with tyre si 11.00-12 (10-ply), pressure 4.14 bars (60 lb/sq in). Ta

Air Tractor AT-503A dual-control agricultural and training aircraft

Air Tractor AT-802 two-seat agricultural and firefighting aircraft

AIR TRACTOR/AMERICAN CHAMPION—AIRCRAFT: USA

Air Tractor AT-802A single-seat agricultural aircraft

wheel tyre size 6.00-6. Cleveland eight-piston brakes with heavy duty discs.
POWER PLANT: One Pratt & Whitney Canada PT6A-67R or -67AF turboprop, rated at 1,062 kW (1,424 shp) at 99°F, driving a Hartzell five-blade feathering and reversible-pitch constant-speed metal propeller. (Used engine options for AT-802A: see Current Versions.) Fuel in two integral wing tanks, total usable capacity 946 litres (250 US gallons; 208 Imp gallons); optional tanks to increase capacity to 1,363 litres (360 US gallons; 300 Imp gallons). Engine air is filtered through two large pleated paper industrial truck filters.
ACCOMMODATION: Two seats in tandem in enclosed cabin, which is sealed to prevent chemical ingress and protected with overturn structure. Four downward hinged doors, two on each side. Windscreen is safety-plate auto glass, with washer and wiper. Air-conditioning system standard.
SYSTEMS: Hydraulic system, pressure 207 bars (3,000 lb/sq in).
AVIONICS: Advanced nav/com, including Loran.

EQUIPMENT: One removable Derakane vinylester hopper forward of cockpit and 227 litre (60 US gallon; 50 Imp gallon) gate tank in ventral bulge, for agricultural chemical, fire retardant or water; total capacity 3,066 litres (810 US gallons; 674 Imp gallons).

DIMENSIONS, EXTERNAL:
Wing span	17.68 m (58 ft 0 in)
Wing chord, constant	2.07 m (6 ft 9½ in)
Wing aspect ratio	8.60
Length overall	11.07 m (36 ft 4 in)
Height overall	3.35 m (11 ft 0 in)
Wheel track	3.11 m (10 ft 2½ in)
Wheelbase	7.25 m (23 ft 9½ in)
Propeller diameter	2.92 m (9 ft 7 in)

AREAS:
Wings, gross	36.33 m² (391.0 sq ft)
Ailerons (total)	4.61 m² (49.6 sq ft)
Trailing-edge flaps (total)	5.54 m² (59.6 sq ft)
Fin	1.24 m² (13.4 sq ft)
Rudder	1.57 m² (16.9 sq ft)
Tailplane	3.44 m² (37.0 sq ft)
Elevators (total, incl tab)	3.00 m² (32.3 sq ft)

WEIGHTS AND LOADINGS:
Weight empty, equipped	2,858 kg (6,300 lb)
Max T-O and landing weight	7,257 kg (16,000 lb)
Max wing loading	199.8 kg/m² (40.92 lb/sq ft)
Max power loading	6.83 kg/kW (11.24 lb/shp)

PERFORMANCE (at max T-O weight except where indicated):
Max level speed at S/L	182 knots (338 km/h; 210 mph)
Max cruising speed at 1,675 m (5,500 ft)	169 knots (314 km/h; 195 mph)
Stalling speed, power off, flaps down, at max landing weight	79 knots (147 km/h; 91 mph)
Max rate of climb at S/L	244 m (800 ft)/min
Service ceiling	3,960 m (13,000 ft)
T-O run	671 m (2,200 ft)
Range with max fuel	434 nm (804 km; 500 miles)

ALLIEDSIGNAL

ALLIEDSIGNAL AVIATION SERVICES

Los Angeles International Airport, 6201 West Imperial Highway, Los Angeles, California 90045
Telephone: 1 (310) 568 3729
Fax: 1 (310) 568 3715
Telex: 181827 A/B AIRE AVI LSA
VICE-PRESIDENT, MARKETING & SALES: Richard A. Graser
MANAGER, MARKETING SERVICES: William E. Whittaker

Former Garrett General Aviation Services (see 1992-93 *Jane's*) was incorporated into Garrett Engine Division in 1992, but from 1 December that year organisation was renamed AlliedSignal Aviation Services.

ALLIEDSIGNAL 731 FALCON 20B RETROFIT PROGRAMME

TYPE: Engine retrofit for in-service Falcon 20s.
PROGRAMME: Announced May 1987; new engine pylons developed by Dassault Aviation in France (which see); first flight development Falcon 20F (F-WTFE) from Bordeaux-Mérignac 7 October 1988; first flight second aircraft (Falcon 20C F-WTFF) 26 January 1989; certificated USA and France March 1989.
CUSTOMERS: Total 61 Falcons of all models retrofitted by end January 1993, including first European conversion delivered by Europe Falcon Service on 6 July 1989.
COSTS: Standard aircraft $3.75 million.
DESIGN FEATURES: As listed in 1991-92 *Jane's*; only

AlliedSignal 731 Falcon 20B (TFE731-5BR turbofans)

TFE731-5BR engine currently used in retrofits, with resulting Falcon 20B designation. Retrofit undertaken by AlliedSignal at Los Angeles, California, or Springfield, Illinois; Falcon Jet Service Center at Little Rock, Arkansas; Europe Falcon Service, Le Bourget Airport, Paris, France; and TransAirCo, Geneva, Switzerland. Details now transferred to *Jane's Civil and Military Aircraft Upgrades*.

ALLIEDSIGNAL (CESSNA) CITATION 500 UPGRADE PROGRAMME

Reported that AlliedSignal is offering upgrade to Citation 500, including Bendix/King EFS-50 EFIS and digital AFCS. Certification tests were scheduled to begin in March 1993, allowing initial redeliveries of upgraded aircraft in Autumn that year. Cost of upgrade, to include single-pilot IFR operation, is $125,000.

AMERICAN CHAMPION

AMERICAN CHAMPION AIRCRAFT CORPORATION

PO Box 37, 32032 Washington Avenue, Highway D, Rochester, Wisconsin 53167
Telephone: 1 (414) 534 6315
Fax: 1 (414) 534 2395
PRESIDENT AND CEO: Jerry K. Mehlhaff
GENERAL MANAGER: Dale Gauger

ACAC offers new-build Citabria (now called Explorer), Super Decathlon and Scout, formerly marketed by Bellanca Aircraft Corporation (see 1979-80 *Jane's*) and then Champion Aircraft Company (see 1985-86 *Jane's*). In addition, ACAC offers its new metal spar wing for retrofit to existing GCBC aircraft.

ACAC EXPLORER

TYPE: Two-seat light cabin monoplane, suited to training and touring. Formerly Citabria.
PROGRAMME: Full details of Citabria (7GCBC) can be found in 1985-86 *Jane's*; now includes modified instrument panel, new window and door, and improved ventilation and heating system. New certification expected Autumn 1993.

COSTS: Standard aircraft $60,000 approx.
DESIGN FEATURES: Options include streamline wheel fairings and voltage regulator.

ACAC SUPER DECATHLON

TYPE: Two-seat light cabin monoplane and aerobatic competition aircraft.
PROGRAMME: FAA certification under FAR Pt 23, for both Normal and Aerobatic categories, granted 1970.
CURRENT VERSIONS: **Super Decathlon:** Constant-speed propeller.
Super Decathlon Fixed Pitch: Fixed-pitch propeller, new paint scheme; offers economy of Explorer with all aerobatic manoeuvres of Super Decathlon.
COSTS: Standard price $69,900.
CUSTOMERS: 80 Decathlon and Super Decathlon delivered since September 1990.
DESIGN FEATURES: Higher rated engine than Decathlon; cleared for limited inverted flight; choice of constant-speed or fixed-pitch propeller. Wing section NACA 1412 (modified); dihedral 1°; incidence 1° 30′.
FLYING CONTROLS: Ailerons, elevators with port trim tab, and rudder.
STRUCTURE: Wing has wooden spars and aluminium alloy ribs, Dacron covered; GFRP wingtips; Dacron covered ailerons; aluminium alloy (front) and steel tube (aft) bracing struts. Fuselage and tail unit are welded steel tube structures, Dacron covered.
LANDING GEAR: Non-retractable tailwheel type; cantilever spring steel main legs; mainwheel tyres size 17 × 6-6, pressure 1.66 bars (24 lb/sq in); tailwheel tyre size 8.30 × 2.50-2.80, pressure 2.07 bars (30 lb/sq in); Cleveland disc brakes; optional wheel fairings.
POWER PLANT: One 134 kW (180 hp) Textron Lycoming AEIO-360-H1A flat-four engine, driving a Hartzell constant-speed or fixed-pitch propeller. Fuel capacity 151.4 litres (40 US gallons; 33.3 Imp gallons). Oil capacity 7.5 litres (2 US gallons; 1.7 Imp gallons).
ACCOMMODATION: Tandem seats in enclosed cabin.

DIMENSIONS, EXTERNAL:
Wing span	9.75 m (32 ft 0 in)
Wing chord, constant	1.63 m (5 ft 4 in)
Wing aspect ratio	6.06
Length overall	6.98 m (22 ft 10¾ in)
Height overall	2.36 m (7 ft 9 in)
Tailplane span	3.10 m (10 ft 2¼ in)

AREAS:
Wings, gross	15.71 m² (169.1 sq ft)

WEIGHTS AND LOADINGS:
Weight empty	596 kg (1,315 lb)
Max T-O weight	816 kg (1,800 lb)

416 USA: AIRCRAFT—AMERICAN CHAMPION/ARCTIC

Max wing loading	51.75 kg/m² (10.6 lb/sq ft)
Max power loading	6.08 kg/kW (10.0 lb/hp)

PERFORMANCE:
Max level speed at 2,315 m (7,000 ft)	137 knots (254 km/h; 158 mph)
Cruising speed at 75% power	130 knots (241 km/h; 150 mph)
Stalling speed	47 knots (87 km/h; 54 mph)
Max rate of climb at S/L	375 m (1,230 ft)/min
Service ceiling	4,875 m (16,000 ft)
Range, no reserves:	
at 75% power	509 nm (944 km; 587 miles)
at 55% power	542 nm (1,006 km; 625 miles)
g limits	+6/−5

ACAC SCOUT

TYPE: Two-seat light utility aircraft.
PROGRAMME: Original 8GCBC version with 134 kW (180 hp) engine received type approval 1974.
CUSTOMERS: Three delivered between September 1990 and May 1993.
COSTS: $65,900 with 134 kW (180 hp) Textron Lycoming O-360-C2A engine and McCauley 1A200HFA8041 fixed-pitch propeller; $67,900 with O-360-C1A engine and 1.93 m (6 ft 4 in) Hartzell HC-C2YR-1BF/F7666A-O constant-speed propeller. Retrofit metal spar wing for existing 8GCBC, $13,800 with 136 litre (36 US gallon; 30 Imp gallon) wing tank (usable capacity), $14,800 with optional 265 litre (70 US gallon; 58 Imp gallon) tank (usable).
DESIGN FEATURES: Upgraded version of Bellanca/Champion Scout, with new lighter and stronger metal spar wing with 300 per cent less deflection than previous wooden wings; circuit breakers; modern avionics; revised interior; and high gloss weather resistant exterior finish. Can operate from short fields while towing glider or banner; low speed assists pipeline, border patrol, forestry and wildlife management roles. Options include wheel fairings, cropduster package, long-range fuel tank, glider tow assembly, and new voltage regulator.

Wing section NACA 4412; dihedral 1°; incidence 1°. Hoerner wingtips.
FLYING CONTROLS: Ailerons, elevators and rudder; 27° droop trailing-edge flaps.
STRUCTURE: All-metal, with Dacron covering.
LANDING GEAR: Non-retractable tailwheel type. Cantilever spring steel main gear with wheels and tyres size 8.50 × 6 (four-ply); heavy duty tailwheel; optional floats, skis or tundra tyres. Dual toe brakes; parking brake.
POWER PLANT: See Costs. Oil capacity 7.5 litres (2 US gallons; 1.7 Imp gallons).
ACCOMMODATION: Tandem seating for two; dual controls.
SYSTEMS: Heating and ventilation; vacuum pump system; electrical with starter, 60A alternator, ammeter, 12V Gel Cell battery, voltage regulator with overvolt protector, control panel with circuit breakers.
AVIONICS: Terra, Narco and Bendix/King suites available; include communications, intercom, emergency locator transmitter, and gyro panel groups; other instruments at customer's request.
DIMENSIONS, EXTERNAL (A: fixed-pitch propeller, B: constant-speed propeller):

Wing span	11.02 m (36 ft 2 in)
Wing chord, constant	1.52 m (5 ft 0 in)
Wing aspect ratio	7.27
Length overall: A	6.93 m (22 ft 9 in)
B	7.01 m (23 ft 0 in)
Height overall	2.64 m (8 ft 8 in)
Tailplane span	3.10 m (10 ft 2¼

AREAS:
Wings, gross	16.70 m² (180.0 sq

WEIGHTS AND LOADINGS (A and B as above):
Weight empty: A	597 kg (1,315
B	603 kg (1,330
Max T-O weight: A and B, Normal	975 kg (2,150
A, Restricted	1,179 kg (2,600
Max wing loading: Normal	58.3 kg/m² (11.94 lb/sq
B, Restricted	70.5 kg/m² (14.4 lb/sq
Max power loading: Normal	7.28 kg/kW (11.94 lb/
B, Restricted	8.80 kg/kW (14.44 lb/l

PERFORMANCE (A and B as above):
Max level speed at S/L:	
A, B	117 knots (217 km/h; 135 m
Cruising speed, 75% power:	
A	106 knots (196 km/h; 122 m
B	113 knots (209 km/h; 130 m
Stalling speed, flaps down:	
A, B	45 knots (84 km/h; 52 m
Max rate of climb at S/L: A	329 m (1,080 ft)/n
B	311 m (1,020 ft)/n
T-O run: A, B	156 m (510
Range, no reserves:	
A, at 75% power	333 nm (618 km; 384 mil
A, at 75% power, optional fuel	680 nm (1,260 km; 783 mil
A, at 55% power, optional fuel	779 nm (1,444 km; 897 mil
B, at 75% power	343 nm (636 km; 395 mil
B, at 75% power, optional fuel	700 nm (1,297 km; 806 mil
B, at 55% power, optional fuel	802 nm (1,487 km; 924 mil

AMERICAN EUROCOPTER
AMERICAN EUROCOPTER CORPORATION
(Subsidiary of Eurocopter SA)
2701 Forum Drive, Grand Prairie, Texas 75053-4005
Telephone: 1 (214) 641 1-0000
Fax: 1 (214) 641 3550
CHAIRMAN: Guy Essautier
CEO: David Smith

This company combines former MBB Helicopter of West Chester, Pennsylvania, and Aerospatiale Helicopters of Grand Prairie, Texas, both of which had modification and assembly facilities as well as sales activities. MBB site now concentrating on support. Aerospatiale Helicopters had agreement with LTV (now Vought) to propose AS 565 P ther to US Army as utility transport, and a LHTEC T8(powered Panther has been demonstrated. Two re-engin AS 350BA Ecureuils were demonstrated to US Army NTH (new training helicopter) programme, won by B TH-67 Creek in April 1993; subsequent Enstrom object overruled by US Army.

AMERICAN GENERAL
AMERICAN GENERAL AIRCRAFT CORPORATION
PO Box 5757, Greenville, Mississippi 38704
Telephone: 1 (601) 332 2422
Fax: 1 (601) 334 9950/9945
CHAIRMAN: Robert E. Crowley
PRESIDENT: James E. Cox
VICE-PRESIDENT, MARKETING: Robert R. Martin

Production and marketing rights for Gulfstream American AA-1 Lynx, AA-5A Cheetah, AA-5B Tiger and GA-7 Cougar (see 1979-80 *Jane's* and earlier) bought from Gulfstream Aerospace Corporation in June 1989. Trainer version of Tiger in production as AG-5B; agreement signed early 1991 for licence manufacture of GA-7 Cougar by Tbilisi Aircraft Manufacturing Association in former Soviet Dimitriov aircraft factory which produced Su-25s and MiG-21s; American General was to supply materials and maintain quality control; aircraft were to be assembled, furnished, painted and marketed in USA; operation in abeyance in mid-1993 for one or two years.

AMERICAN GENERAL AG-5B TIGER

TYPE: Four-seat private aircraft.
PROGRAMME: Modified/updated version of AA-5B (1979-80 *Jane's*); certificated 1990; first AG-5B delivered September 1990; 59 produced in 1992, of which 70 per cent exported; 96 planned for 1993.
CUSTOMERS: Include Florida Institute of Technology School of Aeronautics (up to 15).
DESIGN FEATURES: Changes from AA-5B include Sensenich propeller for 134 kW (180 hp) Textron Lycoming O-360-A4K flat-four engine; fuel capacity 199 litres (52.6 US gallons; 43.8 Imp gallons); CFRP engine cowling; four-point seat restraints; updated instrument pan control yoke and throttle quadrant; 24V electrical syste with circuit breakers; wingtip landing lights; recessed ar collision light in fin.

DIMENSIONS, EXTERNAL:
Wing span	9.60 m (31 ft 6
Length overall	6.71 m (22 ft 0

WEIGHTS AND LOADINGS:
Weight empty	595 kg (1,311
Max T-O weight	1,088 kg (2,400
Max power loading	8.1 kg/kW (13.3 lb/

PERFORMANCE:
Max cruising speed (75% power) at 2,590 m (8,500 ft)	143 knots (265 km/h; 165 m
Service ceiling	4,205 m (13,800
Range with max fuel (75% power)	550 nm (1,019 km; 633 mile

ARCTIC
ARCTIC AIRCRAFT COMPANY
PO Box 190141, Anchorage International Airport, Alaska 99519
Telephone: 1 (907) 243 1580
Fax: 1 (907) 562 2549
PRESIDENT: William A. Diehl

ARCTIC AIRCRAFT INTERSTATE S1B2 ARCTIC TERN

TYPE: Tandem two-seat sporting and general utility aircraft.
PROGRAMME: Updated and much improved version of Interstate S1A, first flown 1940. Built to CAR.04a (aerobatic) standard; certificated for operation on optional Edo floats 20 January 1981; no new Terns produced since 1990; company hopes to restart production early 1994.
COSTS: Price $69,900.
DESIGN FEATURES: Braced high mounted wings. Wing section NACA 23012.
FLYING CONTROLS: Ailerons, elevators, rudder and flaps.
STRUCTURE: Wings have Sitka spruce spars, light alloy ribs and Dacron covering. Hoerner glassfibre wingtips. Welded steel tube fuselage and tail unit, Dacron covered.
LANDING GEAR: Non-retractable tailwheel type, with brakes. Optional parking brake, floats or skis.
POWER PLANT: One 112 kW (150 hp) Textron Lycoming O-320 flat-four engine. Fuel capacity 151 litres (40 US gallons; 33.3 Imp gallons). Underbelly auxiliary fuel tank optional.

Arctic Aircraft Interstate S1B2 Arctic Tern

DIMENSIONS, EXTERNAL:
Wing span 11.18 m (36 ft 8 in)
Wing aspect ratio 7.22
Length overall: landplane 7.01 m (23 ft 0 in)
  seaplane 7.32 m (24 ft 0 in)
Height overall 2.13 m (7 ft 0 in)
AREAS:
Wings, gross 17.30 m² (186.2 sq ft)
WEIGHTS AND LOADINGS (L: landplane, F: with Edo 2000 floats):
Weight empty: L 487 kg (1,073 lb)
  F 521.5 kg (1,150 lb)
Baggage capacity 45 kg (100 lb)
Max T-O weight: L 862 kg (1,900 lb)
  F 965 kg (2,127 lb)
Max wing loading: L 49.8 kg/m² (10.2 lb/sq ft)
  F 55.8 kg/m² (11.42 lb/sq ft)
Max power loading: L 7.70 kg/kW (12.67 lb/hp)
  F 8.62 kg/kW (14.18 lb/hp)
PERFORMANCE:
Max cruising speed at S/L, 75% power:
  L 102 knots (188 km/h; 117 mph)
  F 91 knots (169 km/h; 105 mph)
Cruising speed, 65% power at optimum altitude:
  L 96 knots (178 km/h; 111 mph)
Stalling speed, flaps down:
  L, F 30 knots (55 km/h; 34 mph)
Max rate of climb at S/L, at max T-O weight:
  L 389 m (1,275 ft)/min
  F 305 m (1,000 ft)/min
Service ceiling: L, F 5,790 m (19,000 ft)
T-O run at max T-O weight: L 99 m (325 ft)
Landing from 15 m (50 ft): L 137 m (450 ft)
Range with max fuel, 45 min reserves:
  L, 75% power 479 nm (888 km; 552 miles)
  L, 65% power 566 nm (1,049 km; 652 miles)
Range with max fuel, no reserves:
  F 499 nm (925 km; 575 miles)

ARCTIC AIRCRAFT INTERSTATE S-4 PRIVATEER

TYPE: Four-seat utility aircraft.
PROGRAMME: Not certificated by May 1993.
COSTS: Assembled: $72,300.
CUSTOMERS: Four assembled aircraft and eight kits sold.

Arctic Aircraft Interstate S-4 Privateer

DESIGN FEATURES: Longer span version of Arctic Tern, with increased fuselage width to allow for four persons.
POWER PLANT: One 119 kW (160 hp) Textron Lycoming O-320-B2B flat-four engine. Fuel capacity 151 litres (40 US gallons; 33.3 Imp gallons). Optional underbelly auxiliary fuel tank.
DIMENSIONS, EXTERNAL: As for Arctic Tern, except:
  Wing span 11.48 m (37 ft 8 in)
  Wing aspect ratio 7.51
AREAS:
  Wings, gross 17.56 m² (189.0 sq ft)
WEIGHTS AND LOADINGS:
  Weight empty 521 kg (1,148 lb)
  Baggage capacity 45 kg (100 lb)
  Max T-O weight 1,021 kg (2,250 lb)
  Max wing loading 58.1 kg/m² (11.9 lb/sq ft)
  Max power loading 8.58 kg/kW (14.06 lb/hp)
PERFORMANCE:
  Cruising speed: 75% power at 1,065 m (3,500 ft)
    107 knots (198 km/h; 123 mph)
  65% power at 1,065 m (3,500 ft)
    103 knots (191 km/h; 119 mph)
  Stalling speed, flaps down 33 knots (61 km/h; 38 mph)
  Max rate of climb at S/L 317 m (1,040 ft)/min
  Service ceiling 4,575 m (15,000 ft)
  T-O to, and landing from, 15 m (50 ft) 153 m (500 ft)
  Range: at 75% power at 915 m (3,000 ft), 45 min reserves 600 nm (1,112 km; 691 miles)
    at 65% power, reserves as above
      650 nm (1,204 km; 748 miles)

AVIAT

AVIAT INC

Airport Box 1149, South Washington Street, Afton, Wyoming 83110
Telephone: 1 (307) 886 3151
Fax: 1 (307) 886 9674
CHAIRMAN: Malcolm T. White
PRESIDENT: Robert Merritt
VICE-PRESIDENT, MARKETING: Verdean G. Heiner

Pitts Aerobatics company plus manufacturing and marketing rights of Pitts Special aircraft acquired by Christen Industries November 1983; Pitts Aerobatics factory at Afton, Wyoming, became headquarters of Christen Industries, which in turn was acquired April 1991 by Aviat Inc of Delaware (wholly owned subsidiary of White International Ltd, Guernsey, Channel Islands). Aviat now owns production and type certificates for Christen range.

AVIAT A-1 HUSKY

TYPE: Two-seat light utility aircraft.
PROGRAMME: Computer-aided design started November 1985; first flight (N6070H) 1986; FAA certification 1987, including Edo 2000 floats and skis, glider/banner towing hook approved.
CUSTOMERS: Total 230 delivered by end of 1992, including two to US Dept of Interior and four to US Dept of Agriculture; also operated by US police agencies and in Kenya for wildlife protection patrols.
COSTS: $74,995 basic; $86,500 typically equipped.
DESIGN FEATURES: Missions include bush flying, border patrol, fish and wildlife protection and pipeline inspection. Wing has modified Clark Y US 35B section; drooped Plane Booster wingtips.
FLYING CONTROLS: Symmetrical section ailerons with spade-type mass balance; trim tabs in elevators; slotted flaps. Fixed tailplane; trim by adjustable bungee.
STRUCTURE: Wing has two aluminium spars, metal ribs and metal leading-edge, Dacron covering overall. Twin bracing struts each side of wings and wire and strut braced tail unit. Light alloy flaps and ailerons, with Dacron covering. Fuselage and tail have chrome molybdenum steel tube frames, covered in Dacron except for metal skin to rear fuselage.
LANDING GEAR: Non-retractable tailwheel type. Two faired side Vs and half-axles hinged to bottom of fuselage, with internal (under front seat) bungee cord shock absorption. Cleveland mainwheels, tyres size 8.00-6 as standard; 6.00-6 or 8.50-6 tyres optional. Oversize 24 × 10-6 tundra tyres optional. Cleveland mainwheel brakes. Steerable leaf-spring tailwheel. Wheel-replacement or wheel-retract skis and floats optional.
POWER PLANT: One 134 kW (180 hp) Textron Lycoming O-360-C1G flat-four engine, driving a Hartzell two-blade constant-speed metal propeller. Fuel in two metal tanks, one in each wing, total capacity 208 litres (52 US gallons; 45.75 Imp gallons), of which 189 litres (50 US gallons; 41.6 Imp gallons) are usable. Fuel filler point in upper surface of each wing, near root.
ACCOMMODATION: Enclosed cabin seating two in tandem, with dual controls. Downward hinged door on starboard side, with upward hinged window above. Skylight window in roof.
SYSTEMS: Electrical system includes lights and 60A alternator.
DIMENSIONS, EXTERNAL:
  Wing span 10.73 m (35 ft 2½ in)
  Wing aspect ratio 6.89

Aviat A-1 Husky two-seat utility aircraft

  Length overall 6.88 m (22 ft 7 in)
  Height overall 2.01 m (6 ft 7 in)
  Propeller diameter 1.93 m (6 ft 4 in)
AREAS:
  Wings, gross 16.72 m² (180.0 sq ft)
  Ailerons (total) 1.43 m² (15.4 sq ft)
  Trailing-edge flaps (total) 2.09 m² (22.5 sq ft)
  Fin 0.43 m² (4.66 sq ft)
  Rudder 0.62 m² (6.76 sq ft)
  Tailplane 1.48 m² (15.9 sq ft)
  Elevators, incl tabs 1.31 m² (14.1 sq ft)
WEIGHTS AND LOADINGS:
  Weight empty 540 kg (1,190 lb)
  Max T-O weight 816 kg (1,800 lb)
  Max baggage weight 22.7 kg (50 lb)
  Max wing loading 48.8 kg/m² (10.0 lb/sq ft)
  Max power loading 7.45 kg/kW (10.0 lb/hp)
PERFORMANCE:
  Never-exceed speed (V_{NE})
    132 knots (245 km/h; 152 mph)

418   USA: AIRCRAFT—AVIAT/AVID

Cruising speed: 75% power at 1,220 m (4,000 ft)	
	122 knots (226 km/h; 140 mph)
55% power	115 knots (212 km/h; 132 mph)
Stalling speed: flaps up	48 knots (80 km/h; 49 mph)
flaps down	37 knots (68 km/h; 42 mph)
Max rate of climb at S/L	457 m (1,500 ft)/min
Service ceiling	6,100 m (20,000 ft)
T-O run	46 m (150 ft)
T-O to 15 m (50 ft)	229 m (750 ft)
Landing from 15 m (50 ft)	427 m (1,400 ft)
Landing run, full flap	107 m (350 ft)
Range with max fuel, 75% power, 45 min reserves	
	550 nm (1,019 km; 633 miles)

AVIAT ACRO-HUSKY

Still in R&D; was expected to fly in 1993. Features shorter wing span with NACA 23012 section; 134 kW (180 hp) Textron Lycoming engine with full inverted-flight oil/fuel systems; symmetrical ailerons with spade-type balance; no flaps; g limits +6/−3.

PITTS S-1T SPECIAL

TYPE: Single-seat aerobatic biplane.
PROGRAMME: Original single-seat Pitts Special built and flown 1944; early factory-built S-1S described in 1987-88 *Jane's*. Production of current S-1T factory version started early 1981; FAA certification Autumn 1982.
CUSTOMERS: Total 60 delivered by January 1993, including one during 1992. Available to special order only.
COSTS: $91,995.
DESIGN FEATURES: New features include more powerful engine and wings moved forward 11.5 cm (4½ in) to compensate; symmetrical wing and aileron sections. Wing section M6; thickness/chord ratio 12 per cent; dihedral upper wings 0°, lower wings 3°; incidence upper wing 1° 30′, lower wings 0°; sweepback 6° 40′ upper wing only.
FLYING CONTROLS: Symmetrical ailerons on upper and lower wings; lower with spade-type aerodynamic balance; trim tab on each elevator; fixed tailplane; no flaps.
STRUCTURE: Fabric covered wooden wing and ailerons; single interplane strut and duplicated flying and landing wires; fabric covered steel tube fuselage with wooden stringers and aluminium top decking and side panels, remainder fabric covered; tail surfaces fabric covered steel tube.
LANDING GEAR: Non-retractable tailwheel type. Rubber cord shock absorption. Cleveland mainwheels with 6-ply tyres, size 5.00-5, pressure 2.07 bars (30 lb/sq in). Cleveland hydraulic disc brakes. Steerable tailwheel. Glassfibre fairing on mainwheels.
POWER PLANT: One 149 kW (200 hp) Textron Lycoming AEIO-360-A1E flat-four engine, driving a Hartzell two-blade constant-speed propeller. Fuel tank aft of firewall, capacity 75 litres (20 US gallons; 16.6 Imp gallons). Refuelling point on upper surface of fuselage, forward of windscreen. Oil capacity 7.5 litres (2 US gallons; 1.7 Imp gallons). Inverted fuel and oil systems standard.
ACCOMMODATION: Single seat. Sliding cockpit canopy standard.

DIMENSIONS, EXTERNAL:
Wing span, upper	5.28 m (17 ft 4 in)
Wing chord (constant, both)	0.91 m (3 ft 0 in)
Wing aspect ratio	5.8
Length overall	4.72 m (15 ft 6 in)
Height overall	1.91 m (6 ft 3 in)
Tailplane span	1.98 m (6 ft 6 in)
Propeller diameter	1.93 m (6 ft 4 in)

AREAS:
Wings, gross	9.15 m² (98.5 sq ft)

WEIGHTS AND LOADINGS:
Weight empty	376 kg (830 lb)
Max T-O weight	521 kg (1,150 lb)
Max wing loading	57.05 kg/m² (11.68 lb/sq ft)
Max power loading	3.50 kg/kW (5.75 lb/hp)

PERFORMANCE (at max T-O weight):
Never-exceed speed (VNE)	
	176 knots (326 km/h; 203 mph)
Max level speed at S/L	161 knots (298 km/h; 185 mph)
Max cruising speed at S/L	
	152 knots (282 km/h; 175 mph)
Stalling speed	56 knots (103 km/h; 64 mph)
Max rate of climb at S/L	853 m (2,800 ft)/min
Range with max fuel, 55% power, 30 min reserves	
	268 nm (497 km; 309 miles)
g limits	+6/−3.5

Aviat Pitts S-2B two-seat aerobatic aircraft

PITTS S-2B

TYPE: Two-seat aerobatic biplane; successor to S-2A.
PROGRAMME: Prototype completed September 1982; certificated in FAR Pt 23 aerobatic category Spring 1983; won first place Advanced Category of 1982 US Nationals with two occupants.
CUSTOMERS: Total 36 delivered during 1992.
COSTS: $111,800 basic.
DESIGN FEATURES: 194 kW (260 hp) engine and wings moved 15 cm (6 in) forward to compensate: more front cockpit space. Wing sections NACA 6400 series on upper wing, 00 series on lower wings. See 1987-88 and earlier *Jane's*.
FLYING CONTROLS: Ailerons on upper and lower wings with aerodynamic spade-type balances on lower ailerons; trim tab on each elevator; fixed tailplane; no flaps.
STRUCTURE: Wings generally as S-1T; fuselage 4130 steel tube with wooden stringers, aluminium top decking and side panels; remainder Dacron covered. Steel tube, metal skinned fixed tail surfaces; Dacron covered control surfaces.
LANDING GEAR: Non-retractable tailwheel type. Rubber cord shock absorption. Steerable tailwheel. Streamline fairings on mainwheels.
POWER PLANT: One 194 kW (260 hp) Textron Lycoming AEIO-540-D4A5 flat-six engine, driving a Hartzell two-blade constant-speed metal propeller. Fuel tank in fuselage, immediately aft of firewall, capacity 110 litres (29 US gallons; 24 Imp gallons). Refuelling point on fuselage upper surface forward of windscreen. Oil capacity 11.35 litres (3 US gallons; 2.5 Imp gallons). Inverted fuel and oil systems standard.
ACCOMMODATION: Two seats in tandem cockpits, with dual controls. Sideways opening one-piece canopy covers both cockpits. Space for 9.1 kg (20 lb) baggage aft of rear seat when flown in non-aerobatic category.
SYSTEMS: Electrical system powered by 12V 40A alternator and non-spill 12V battery.

DIMENSIONS, EXTERNAL:
Wing span: upper	6.10 m (20 ft 0 in)
lower	5.79 m (19 ft 0 in)
Wing chord (constant, both)	1.02 m (3 ft 4 in)
Length overall	5.71 m (18 ft 9 in)
Height overall	2.02 m (6 ft 7½ in)

AREAS:
Wings, gross	11.6 m² (125.0 sq ft)

WEIGHTS AND LOADINGS:
Weight empty	521 kg (1,150 lb)
Max T-O weight	737 kg (1,625 lb)

Max wing loading	63.55 kg/m² (13.0 lb/sq)
Max power loading	3.80 kg/kW (6.25 lb/h)

PERFORMANCE (at max T-O weight):
Never-exceed speed (VNE)	
	182 knots (338 km/h; 210 mp)
Max cruising speed	152 knots (282 km/h; 175 mp)
Stalling speed	52 knots (97 km/h; 60 mp)
Max rate of climb at S/L	823 m (2,700 ft)/m
Service ceiling	6,400 m (21,000
Range with max fuel, 55% power, 30 min reserves	
	277 nm (513 km; 319 mile

PITTS S-2S

TYPE: Single-seat version of S-2B aerobatic biplane.
PROGRAMME: First flight of prototype 9 December 1977; pr duction began late 1978; full certification June 1981.
CUSTOMERS: Total 34 delivered by January 1993. Available special order only.
COSTS: $105,470 basic.
DESIGN FEATURES: Forward fuselage shortened by 0.36 (14 in) to accommodate 194 kW (260 hp) Textron Lyco ing AEIO-540-D4A5 flat-six engine, driving Hartzell tw blade, constant-speed metal propeller; fuel capac increased to 132.5 litres (35 US gallons; 29.1 Imp gallon oil capacity as for S-2B.
FLYING CONTROLS: Generally as for S-2B.
STRUCTURE: Generally as for S-2B.

DIMENSIONS, EXTERNAL: As for S-2B except:
Length overall	5.28 m (17 ft 4

WEIGHTS AND LOADINGS:
Weight empty	499 kg (1,100
Max T-O weight	680 kg (1,500
Max wing loading	58.6 kg/m² (12.0 lb/sq
Max power loading	3.51 kg/kW (5.77 lb/h

PERFORMANCE (at max T-O weight):
Never-exceed speed (VNE)	
	176 knots (326 km/h; 203 mp
Max level speed at S/L	162 knots (301 km/h; 187 mp
Max cruising speed at S/L	
	152 knots (282 km/h; 175 mp
Stalling speed	51 knots (94 km/h; 58 mp
Max rate of climb at S/L	853 m (2,800 ft)/m
g limits	+6/−3

AVID

AVID AIRCRAFT INC

PO Box 728, 4823 Aviation Way, Caldwell, Idaho 83606
Telephone: 1 (208) 454 2600
Fax: 1 (208) 454 8608
GENERAL MANAGER: Jim Metzger
MANAGER, SALES: Gary Walton

Avid Aircraft currently markets Avid Flyer Mark IV, Magnum and Catalina amphibian; details of amphibian can be found in Private Aircraft section of 1992-93 *Jane's*.

AVID AIRCRAFT AVID FLYER MARK IV

TYPE: Side by side two-seat, dual-control homebuilt.
PROGRAMME: First flown 1983; takes from 200 to 400 working hours to assemble; available as single kit, or six separate kits to spread cost of purchase. Strongly influenced design of Indaer-Peru Chuspi light aircraft (see under Peru in 1992-93 *Jane's*).
CUSTOMERS: Approx 1,500 kits delivered by early 1992.
COSTS: Kit: $17,995.
DESIGN FEATURES: Strut braced wings; two forms available (interchangeable), as original **High Gross STOL** wi unique near full span auxiliary aerofoil flaperons, a shorter span **Aerobatic Speedwing** using new wing se tion; cruising and stalling speeds raised with Aerobat Speedwing fitted. Wings fold. Baggage compartment wi external door. Cabin heating.
FLYING CONTROLS: Flaperons (see Design Features), elevato with adjustable trim tab, and rudder.
STRUCTURE: Aluminium wing spars and plywood ribs, co ered with heat shrunk Dacron. Welded steel tube fuselag

rudder, tailplane and elevators, Dacron covered except for fuselage nose which has pre-moulded GFRP cowlings. Fin integral with fuselage.

LANDING GEAR: Non-retractable tricycle or tailwheel landing gear, with tundra tyres and brakes. Optional floats, skis and wheel-skis.

POWER PLANT: One 48.5 kW (65 hp) Rotax 582 two-stroke engine, driving a two- or three-blade propeller; fixed-pitch propeller for STOL, three-blade ground adjustable for Speedwing. Fuel capacity 53 litres (14 US gallons; 11.7 Imp gallons) for High Gross STOL and 68 litres (18 US gallons; 15 Imp gallons) for Aerobatic Speedwing. Similar capacity fuel tanks may be added in port wing.

AVIONICS: Standard are altimeter, airspeed indicator, compass, slip indicator, tachometer, water temperature gauge and dual EGT.

DIMENSIONS, EXTERNAL (A: STOL, B: Aerobatic Speedwing):
Wing span: A	9.11 m (29 ft 10½ in)
B	7.30 m (23 ft 11½ in)
Wing chord:	
A, B without flaperons	1.07 m (3 ft 6 in)
A, B with flaperons	1.30 m (4 ft 3 in)
Wing aspect ratio: A	7.03
B	5.50
Length overall	5.46 m (17 ft 11 in)
Height overall	1.80 m (5 ft 11 in)
Propeller diameter: A	1.88 m (6 ft 2 in)
	1.73 m (5 ft 8 in)

AREAS (A and B as above):
Wings, gross: A	11.38 m² (122.46 sq ft)
B	9.04 m² (97.31 sq ft)

WEIGHTS AND LOADINGS (A and B as above):
Weight empty: A	200-231 kg (440-510 lb)
B	231 kg (510 lb)
Baggage capacity	16 kg (35 lb)
Max T-O weight: A	522 kg (1,150 lb)
B	476 kg (1,050 lb)
Max wing loading: A	45.85 kg/m² (9.39 lb/sq ft)
B	52.68 kg/m² (10.79 lb/sq ft)
Max power loading: A	10.76 kg/kW (17.69 lb/hp)
B	9.81 kg/kW (16.15 lb/hp)

PERFORMANCE (A and B as above):
Never-exceed speed (V_{NE}):	
A	117 knots (217 km/h; 135 mph)
B	130 knots (241 km/h; 150 mph)
Max level speed at 1,525 m (5,000 ft):	
A	91 knots (169 km/h; 105 mph)
Max cruising speed: A	78 knots (145 km/h; 90 mph)
B	104 knots (193 km/h; 120 mph)
Stalling speed: A, flaps down, engine idling	
	32 knots (58 km/h; 36 mph)
B	40 knots (74 km/h; 46 mph)
Max rate of climb at S/L: A	305 m (1,000 ft)/min
B	259 m (850 ft)/min
Service ceiling: A, B	more than 3,810 m (12,500 ft)
T-O run: A	43 m (140 ft)
B	92 m (300 ft)
T-O to 15 m (50 ft): A	61 m (200 ft)
B	183 m (600 ft)
Landing run: A	61 m (200 ft)
B	183 m (600 ft)
Range, no reserves: A	295 nm (547 km; 340 miles)
B	491 nm (910 km; 566 miles)
g limits	+6/−3

Avid Aircraft Flyer Mark IV (Rotax 582 engine)

AVID AIRCRAFT MAGNUM

TYPE: Side by side two-seat, dual-control homebuilt.
COSTS: Kit: $18,995 without engine.
DESIGN FEATURES: Similar layout to Avid Flyer, but using Textron Lycoming engine and having baggage area large enough to fit optional jump seat for small adult or two children.
FLYING CONTROLS: Similar to Avid Flyer.
STRUCTURE: Similar to Avid Flyer, but with Ceconite covering.
LANDING GEAR: Non-retractable tailwheel type, with Cleveland wheels and brakes; wheel fairings.
POWER PLANT: One Textron Lycoming engine in 85.75-134 kW (115-180 hp) range, including O-235, O-320 and O-360. Fuel capacity 106 litres (28 US gallons; 23.3 Imp gallons).
ACCOMMODATION: Two seats side by side, plus optional jump seat for small adult or two children in 0.81 m³ (28.5 cu ft) baggage area.

DIMENSIONS, EXTERNAL:
Wing span	10.06 m (33 ft 0 in)
Wing chord	1.30 m (4 ft 3 in)
Wing aspect ratio	7.76
Length overall	6.40 m (21 ft 0 in)
Height overall	1.86 m (6 ft 1¼ in)

DIMENSIONS, INTERNAL:
Cabin: Width	1.12 m (3 ft 8 in)

AREAS:
Wings, gross	13.03 m² (140.25 sq ft)

WEIGHTS AND LOADINGS:
Weight empty	465 kg (1,025 lb)
Baggage capacity	68 kg (150 lb)
Max T-O weight	748 kg (1,650 lb)
Max wing loading	57.44 kg/m² (11.76 lb/sq ft)
Power loading (119 kW; 160 hp)	6.29 kg/kW (10.31 lb/hp)

PERFORMANCE (119 kW; 160 hp O-320 engine):
Never-exceed speed (V_{NE})	130 knots (241 km/h; 150 mph)
Cruising speed	113 knots (209 km/h; 130 mph)
Stalling speed	32 knots (58 km/h; 36 mph)
Max rate of climb at S/L	305 m (1,000 ft)/min
Service ceiling	more than 3,810 m (12,500 ft)
T-O run	approx 77 m (250 ft)
T-O to 15 m (50 ft)	99 m (325 ft)
Landing run	77 m (250 ft)

Avid Aircraft Magnum (Textron Lycoming engine)

AVTEK
AVTEK CORPORATION

4680 Calle Carga, Camarillo, California 93010
Telephone: 1 (805) 482 2700
Fax: 1 (805) 987 0068
PRESIDENT: Robert F. Adickes
SENIOR VICE-PRESIDENT, ENGINEERING: Niels Andersen
DIRECTOR OF MARKETING: Robert D. Honeycutt

Company founded 1982 to develop Avtek 400A; investors include Valmet Aviation (Finland), Air Rotor GmbH (Germany), Nomura Securities (Japan), Dow Chemical (USA) and Per Udsen Co Aircraft Industry A/S. In October 1991, Danish Ministry of Foreign Affairs arranged transfer of Avtek technology, production and marketing to EuroAvtek A/S in Denmark; this new company jointly owned by Avtek Corporation, Per Udsen and Danish government.

AVTEK 400A

TYPE: Six/10-seat all-composite, turboprop multi-role twin.
PROGRAMME: Design started March 1981; proof-of-concept aircraft N400AV (see 1985-86 *Jane's*) flew 17 September 1984; N400AV then fitted with P&WC PT6A-135M engines (PT6A-35s with counter-rotating gearboxes from PT6A-66); extensive changes made Spring 1985, including fuselage stretch 20 cm (8 in) forward and 76 cm (2 ft 6 in) aft of front pressure bulkhead, widened cabin, new outer wing and enlarged fuel tanks in forward-swept root extensions, foreplane with greater span and reduced chord, ventral strakes (known as delta fins), relocated main landing gear legs, and specially developed P&WC PT6A-3s mounted closer to wings. Pre-production second prototype expected to fly 1993; FAA certification planned for late 1994.

CURRENT VERSIONS: **400A**: Principal version, *as detailed below*.
Explorer: Valmet purchased option on 20 March 1985 to build Explorer derivative of Avtek 400A for maritime surveillance, liaison, coastal patrol, aerial survey, search and reconnaissance, and ESM, tailored to individual customer requirements. Fuel capacity increased by 208 litres (55 US gallons; 45.8 Imp gallons) and max T-O weight raised to 3,402 kg (7,500 lb).
419 Express: 19-passenger commuter version; PT6A-45 engines; to be certificated after FAR 23 gained by 400A. Length increased to 16.62 m (54 ft 5 in) and max T-O weight raised to 5,669 kg (12,499 lb).
CUSTOMERS: Avtek claims deposit payments received for 82 Avtek 400As.

USA: AIRCRAFT—AVTEK/AYRES

Artist's impression of Avtek 400A all-composite multi-role twin

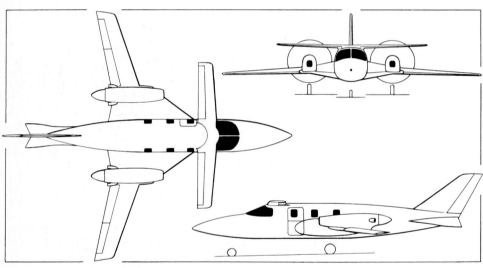

Avtek 400A six/10-seat twin-turboprop aircraft (Dennis Punnett)

COSTS: Standard 400A $1.75 million; Explorer $2 million plus sensors and systems; 419 Express $4.2 million.

DESIGN FEATURES: Initial design by Al W. Mooney, founder of Mooney Aircraft; refined by Niels Andersen, Ford Johnston and Irvin Culver; computer analysis of configuration and wind tunnel testing by NASA; materials research by Dow Chemical and Dr Leo Windecker; Dow Chemical basic patents on Windecker Eagle (first all-composites aircraft to receive civil certification) licensed to Avtek. Avtek 12 aerofoil section; anhedral 2° 30′ from roots; sweepback 50° inboard, 15° 30′ outboard; foreplane dihedral 1°; fuselage pressurised.

FLYING CONTROLS: Actuation by pushrods and cranks throughout; mass balanced elevators on foreplane; mass balanced rudder without trim tab; two-section ailerons, with inboard sections also electrically actuated as pitch-axis trim surfaces; no flaps.

STRUCTURE: All-composite structure, 72 per cent DuPont Kevlar and Nomex, 16 per cent graphite/carbonfibre, smaller quantities of R-glass, S-glass, aluminium and nickel fibres; wire mesh incorporated to protect against lightning.

LANDING GEAR: Hydraulically retractable tricycle type, main units retracting inward and nosewheel forward. Emergency extension system. Oleo-pneumatic shock absorber in each unit. Single wheel on each unit, mainwheels size 6.00-6 with Goodyear tyres size 17.5-6.25 × 7.5, pressure 7.0 bars (102 lb/sq in). Steerable nosewheel unit with wheel size 5.00-5 and Goodyear tyre size 14.2-5.5 × 6.5, pressure 5.0 bars (73 lb/sq in). Cleveland hydraulic disc brakes.

POWER PLANT: Two Pratt & Whitney Canada PT6A-3L/R turboprops derived from PT6A-135 and flat rated at 507 kW (680 shp), one mounted within nacelle above each wing. Hartzell four-blade constant-speed fully feathering reversible-pitch pusher propellers (metal blades on prototype, Kevlar on production version). Propellers are opposite rotating, with automatic synchrophasing, full Beta control reversing and autofeathering. In-flight start capability. Integral fuel tank in each wingroot leading-edge. Total fuel capacity 1,003 litres (265 US gallons; 221 Imp gallons), of which 984 litres (260 US gallons; 216.5 Imp gallons) are usable. Refuelling point in upper surface of each wing.

ACCOMMODATION: Pilot and five to nine passengers according to interior layout. Optional configurations include eight passenger Pullman or Salon; eight-passenger Lounge; six passenger Conference; Ambulance with stretcher, medical equipment and seats for three ambulatory patients or medical attendants; and Cargo with seats for pilot and one passenger. Two-section door on port side of cabin, with steps incorporated in lower half. Emergency exit on starboard side opposite cabin door. Baggage compartment at rear of cabin with internal access. Unpressurised baggage compartment in nose with external door on port side. Accommodation pressurised, air-conditioned, heated and ventilated.

SYSTEMS: AiResearch bleed air pressurisation system with max differential of 0.52 bar (7.6 lb/sq in), and air cycle air conditioning system. Electrically driven hydraulic pump provides pressure of 138 bars (2,000 lb/sq in) for landing gear actuation. Electrical system includes dual 28V 300A engine driven generators, dual 29Ah storage batteries and external power socket. Oxygen system of 1.39 m³ (49 cu ft) capacity, pressure 128 bars (1,850 lb/sq in), provides constant flow for passengers and demand flow for pilot. Anti-icing of windscreen by electrical system, of propellers by engine efflux; electrically heated pitot. Engine fire extinguishing system optional. Choice of wing and foreplane pneumatic, alcohol, electric or engine bleed de-icing. Dual anti-icing inlets for each engine.

AVIONICS: Wide range of optional avionics by Bendix/King, Collins and Honeywell, including EFIS and EICAS, and Honeywell colour weather radar. Full IFR instrumentation optional.

DIMENSIONS, EXTERNAL:
Wing span	10.67 m (35 ft 0 in)
Wing aspect ratio	8.5
Foreplane span	6.92 m (22 ft 8½ in)
Length: overall	11.99 m (39 ft 4 in)
fuselage	10.41 m (34 ft 2 in)
Height overall	3.47 m (11 ft 4¾ in)
Propeller diameter	1.93 m (6 ft 4 in)
Passenger door (port): Height	1.17 m (3 ft 10 in)
Width	0.76 m (2 ft 6 in)
Height to sill	0.76 m (2 ft 6 in)
Baggage door (port, nose): Height	0.51 m (1 ft 8 in)
Width	0.56 m (1 ft 10 in)
Height to sill	0.79 m (2 ft 7 in)
Emergency exit (stbd): Height	0.51 m (1 ft 8 in)
Width	0.67 m (2 ft 2½ in)

DIMENSIONS, INTERNAL:
Cabin: Length	3.14 m (10 ft 3½ in)
Max width	1.40 m (4 ft 7¼ in)
Max height	1.37 m (4 ft 6 in)
Baggage hold volume:	
nose	0.62 m³ (22.0 cu ft)
cabin	1.24 m³ (44.0 cu ft)

AREAS:
Wings, gross	13.40 m² (144.2 sq ft)
Foreplane, gross	4.52 m² (48.7 sq ft)
Elevators (total)	0.92 m² (9.9 sq ft)
Ailerons (total)	0.60 m² (6.5 sq ft)
Fin	1.16 m² (12.5 sq ft)
Rudder	0.90 m² (9.7 sq ft)

WEIGHTS AND LOADINGS:
Weight empty, equipped	1,714 kg (3,779 lb)
Max ramp weight	2,976 kg (6,560 lb)
Max T-O and landing weight	2,948 kg (6,500 lb)
Max wing loading	220.1 kg/m² (45.08 lb/sq ft)
Max power loading	2.91 kg/kW (4.78 lb/shp)

PERFORMANCE (estimated):
Max level speed at S/L	255 knots (473 km/h; 294 mph)
Max cruising speed:	
at 3,050 m (10,000 ft)	297 knots (550 km/h; 342 mph)
at 6,700 m (22,000 ft)	364 knots (675 km/h; 419 mph)
at 12,500 m (41,000 ft)	338 knots (626 km/h; 389 mph)
Stalling speed	83 knots (154 km/h; 96 mph)
Max rate of climb at S/L	1,411 m (4,630 ft)/min
Rate of climb at S/L, OEI	578 m (1,897 ft)/min
Service ceiling	12,950 m (42,500 ft)
Service ceiling, OEI	10,060 m (33,000 ft)
T-O to 15 m (50 ft)	463 m (1,520 ft)
Landing from 15 m (50 ft)	390 m (1,280 ft)
Range with max fuel:	
no reserves	2,276 nm (4,218 km; 2,621 miles)
NBAA IFR reserves	1,922 nm (3,562 km; 2,213 miles)

AYRES

AYRES CORPORATION

PO Box 3090, One Rockwell Avenue, Albany, Georgia 31708-5201
Telephone: 1 (912) 883 1440
Fax: 1 (912) 439 9790
Telex: 547629 AYRESPORT ABN
VICE-PRESIDENT, SALES: Daniel Lewis

Ayres Corporation bought manufacturing and world marketing rights to Thrush Commander-600 and -800 from Rockwell International General Aviation Division in November 1977.

AYRES THRUSH S2R-R1340

TYPE: Single/two-seat agricultural aircraft.
PROGRAMME: In production.
CURRENT VERSIONS: **Thrush S2R-R1340**: Basic version, powered by Pratt & Whitney R-1340 Wasp air-cooled radial engine and with one or two seats. *Description applies to this version.*

Thrush S2R-R1820: Powered by Wright R-1820 air-cooled radial (described separately).

Turbo-Thrush S2R: Basic S2R powered by Pratt & Whitney Canada PT6A-11, -15, -34 or -65AG; described separately.

CUSTOMERS: Operating in 70 countries.
DESIGN FEATURES: Cantilever wing with 3° 30′ dihedral, wingroots sealed against chemical entry; wing extension adding 2.79 m² (30 sq ft) standard; deflector cable from cockpit to tip of fin.
FLYING CONTROLS: Plain ailerons; servo tab in each elevator; electrically actuated flaps.
STRUCTURE: Two-spar light alloy wing with 4130 chrome molybdenum steel spar caps; welded chrome molybdenum steel tube fuselage structure covered with quickly removable light alloy skin panels; underfuselage skin of stainless steel; all-metal tail surfaces with strut braced tailplane.

metal ailerons and flaps.

LANDING GEAR: Non-retractable tailwheel type. Main units have rubber in compression shock absorption and 29 × 11.00-10 wheels with 10-ply tyres. Hydraulically operated disc brakes. Parking brakes. Wire cutters on main gear. Steerable, locking tailwheel, size 12.5 × 4.5 in.

POWER PLANT: One 447 kW (600 hp) Pratt & Whitney R-1340 Wasp nine-cylinder air-cooled radial engine, driving a Hamilton Standard 12D40/EAC AG-100-2 two-blade constant-speed metal propeller. Fuel contained in wing tanks with combined capacity of 401 litres (106 US gallons; 88.3 Imp gallons).

ACCOMMODATION: Single adjustable mesh seat in 'safety pod' sealed cockpit enclosure, with steel tube overturn structure. Tandem seating optional, with forward facing second seat. Dual controls optional with forward facing rear seat, for pilot training. Adjustable rudder pedals. Downward hinged door on each side. Tempered safety glass windscreen. Cockpit wire cutter. Dual inertia reel safety harness with optional second seat. Baggage compartment standard on single-seat aircraft. Windscreen wiper and washer.

SYSTEM: Electrical system powered by a 24V 50A alternator. Lightweight 24V 35Ah battery.

AVIONICS: To customer's requirements.

EQUIPMENT: GFRP hopper forward of cockpit can hold 1,514 litres (400 US gallons; 333 Imp gallons) of liquid or 1,487 kg (3,280 lb) of dry chemical. Hopper has a 0.33 m² (3.56 sq ft) lid, openable by two handles, and cockpit viewing window. Standard equipment includes Universal spray system with external 50 mm (2 in) stainless steel plumbing, 50 mm pump with wooden fan, Transland gate, 50 mm valve, quick-disconnect pump mount and strainer. Streamlined spraybooms with outlets for 68 nozzles. Micro-adjust valve control (spray) and calibrator (dry). A 63 mm (2.5 in) side loading system is installed on the port side. Stainless steel rudder cables. Navigation lights, instrument lights and two strobe lights. Optional equipment includes a rear cockpit to accommodate forward facing seat for passenger, or flying instructor if optional dual controls installed; space can be used alternatively for cargo. Other optional items are a Transland high-volume spreader; agitator installation; 10-unit AU5000 Micronair installation in lieu of standard booms and nozzles; Transland gatebox with stiffener casting; quick-disconnect flange and kit; night working lights including landing light and wingtip turn lights; cockpit fire extinguisher; and water bomber configuration.

Ayres Thrush S2R-R1340 (600 hp Pratt & Whitney R-1340 Wasp engine)

Ayres Thrush S2R-R1820 (1,200 hp Wright R-1820 engine)

Ayres Turbo-Thrush S2R-T34 with optional dual cockpit and pitot inlet

DIMENSIONS, EXTERNAL:
Wing span	14.48 m (47 ft 6 in)
Wing aspect ratio	6.33
Length overall (tail up)	8.95 m (29 ft 4½ in)
Height overall	2.79 m (9 ft 2 in)
Tailplane span	5.18 m (17 ft 0 in)
Wheel track	2.72 m (8 ft 11 in)
Propeller diameter	2.74 m (9 ft 0 in)

DIMENSIONS, INTERNAL:
Hopper volume	1.50 m³ (53.0 cu ft)

AREAS:
Wings, gross	33.13 m² (356.6 sq ft)

WEIGHTS AND LOADINGS:
Weight empty, equipped	1,678 kg (3,700 lb)
Max T-O weight: CAR 3	2,721 kg (6,000 lb)
CAM 8	3,130 kg (6,900 lb)
Max wing loading	103.0 kg/m² (21.1 lb/sq ft)
Max power loading	7.0 kg/kW (11.5 lb/hp)

PERFORMANCE (with spray equipment installed and at CAR 3 max T-O weight, except where indicated):
Max level speed	122 knots (225 km/h; 140 mph)
Max cruising speed, 70% power	108 knots (200 km/h; 124 mph)
Working speed, 70% power	91-100 knots (169-185 km/h; 105-115 mph)
Stalling speed: flaps up	58 knots (108 km/h; 67 mph)
flaps down	55 knots (101 km/h; 63 mph)
Stalling speed at normal landing weight:	
flaps up	47 knots (87 km/h; 54 mph)
flaps down	45 knots (84 km/h; 52 mph)
Max rate of climb at S/L	317 m (1,040 ft)/min
Service ceiling	4,575 m (15,000 ft)
T-O run	215 m (705 ft)
Landing run	139 m (455 ft)
Ferry range with max fuel at 70% power	350 nm (648 km; 403 miles)

AYRES THRUSH S2R-R1820/510

TYPE: Single/two-seat agricultural aircraft.

DESIGN FEATURES: As S2R-R1340 except for bigger hopper and more powerful engine.

Details as for S2R-R1340 except as follows:

POWER PLANT: One 895 kW (1,200 hp) Wright R-1820 Cyclone nine-cylinder air-cooled radial engine, driving a Hamilton Standard three-blade constant-speed metal propeller. Fuel system as for S2R-R1340, but total usable fuel capacity 863 litres (228 US gallons; 190 Imp gallons).

EQUIPMENT: Generally as for S2R-R1340, except that the chemical hopper is of 1,930 litres (510 US gallons; 425 Imp gallons) capacity.

DIMENSIONS, EXTERNAL: As for S2R-R1340, except:
Wing span	13.54 m (44 ft 5 in)
Wing aspect ratio	6.04
Length overall	9.60 m (31 ft 6 in)
Height overall	2.92 m (9 ft 7 in)
Wheel track	2.74 m (9 ft 0 in)

DIMENSIONS, INTERNAL:
Hopper volume	1.93 m³ (68.2 cu ft)

AREAS:
Wings, gross	30.34 m² (326.6 sq ft)

WEIGHTS AND LOADINGS:
Weight empty, equipped	2,263 kg (4,990 lb)
Typical operating weight (CAM 8)	4,536 kg (10,000 lb)
Max wing loading	149.5 kg/m² (30.62 lb/sq ft)
Max power loading	5.07 kg/kW (8.33 lb/hp)

PERFORMANCE (with spray equipment, at CAM 8 T-O weight, except where indicated):
Max level speed	138 knots (256 km/h; 159 mph)
Cruising speed, 50% power	135 knots (249 km/h; 155 mph)
Working speed, 30-50% power	87-130 knots (161-241 km/h; 100-150 mph)
Stalling speed: flaps up	61 knots (113 km/h; 70 mph)
flaps down	58 knots (107 km/h; 66 mph)
Stalling speed at normal landing weight:	
flaps up	52 knots (95 km/h; 59 mph)
flaps down	50 knots (92 km/h; 57 mph)
Max rate of climb at S/L	620 m (2,033 ft)/min
Service ceiling	8,535 m (28,000 ft)
T-O run	168 m (550 ft)
Landing run at normal landing weight	290 m (950 ft)
Ferry range at 40% power	582 nm (1,078 km; 670 miles)

AYRES TURBO-THRUSH S2R

TYPE: Single/two-seat turboprop agricultural and multi-role aircraft.

CURRENT VERSIONS: **S2R-T11**: 373 kW (500 shp) PT6A-11AG turboprop; standard 1,514 litre (400 US gallon; 333 Imp gallon) chemical hopper.
S2R-T15: 507 kW (680 shp) P&WC PT6A-15AG turboprop; standard or optional 1,930 litre (510 US gallon; 425 Imp gallon) hopper.

422 USA: AIRCRAFT—AYRES/BEDE

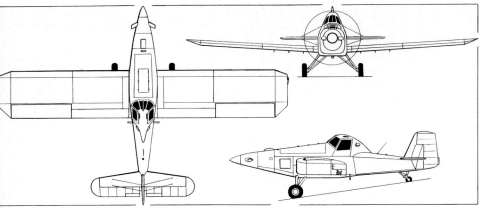

Ayres Turbo-Thrush S2R-T34 with optional 1,930 litre (510 US gallon; 425 Imp gallon) hopper *(Dennis Punnett)*

Ayres Turbo-Thrush S2R-G10 (Garrett TPE331-10 turboprop)

S2R-T34: 559 kW (750 shp) P&WC PT6A-34AG turboprop; standard or optional hoppers.
S2R-T65 NEDS: Narcotics Eradication Delivery System; 1,026 kW (1,376 shp) P&WC PT6A-65AG turboprop and 2.82 m (9 ft 3 in) five-blade propeller; 19 delivered to US State Department (see Customers). NEDS equipment includes Bendix/King VLF/Omega 660, ADF, VOR, HF and VHF avionics.
S2R-G6: Introduced 1992; 559 kW (750 shp) Garrett TPE331-6 turboprop; standard or optional hopper.
S2R-G10: First flown November 1992; CAM 8 certification; 701 kW (940 shp) Garrett TPE331-10 turboprop; two 435 litre (115 US gallon; 95.75 Imp gallon) fuel tanks, with 863 litres (228 US gallons; 190 Imp gallons) usable; hopper capacity 1,930 litres (510 US gallons; 425 Imp gallons); pilot, with dual cockpit option.
CUSTOMERS: US State Department ordered 19 S2R-T65 NEDS during 1983-88 for use by International Narcotics Matters Bureau.
DESIGN FEATURES: All but NEDS variant have Hartzell three-blade, constant-speed, feathering and reversing propellers; usable fuel capacity 863 litres (228 US gallons; 190 Imp gallons); claimed advantages include much improved take-off and climb, 454 kg (1,000 lb) higher payload because of lower engine weight, operation on aviation turbine fuel or diesel, 3,500 h TBO, quieter operation and ability to feather propeller without stopping engine while refuelling and reloading. Wing extensions (see S2R-R1340) optional on Turbo-Thrush.

DIMENSIONS, EXTERNAL: As for S2R-R1820 except:
Length overall 10.06 m (33 ft 0 in)
Height overall 2.79 m (9 ft 2 in)
AREAS: As for S2R-R1820
WEIGHTS AND LOADINGS (A: standard hopper and PT6A, B: optional hopper and PT6A, C: with TPE331-6, D: with TPE331-10):
Weight empty: A 1,633 kg (3,600 lb)
B 1,769 kg (3,900 lb)
C 1,905 kg (4,200 lb)
D 2,041 kg (4,500 lb)
Max T-O weight (CAR 3): A, B 2,721 kg (6,000 lb)
Typical operating weight (CAM 8):
A 3,719 kg (8,200 lb)
B 3,856 kg (8,500 lb)
C 4,400 kg (9,700 lb)
D 4,082 kg (9,000 lb)
CAM 8 wing loading: A 122.6 kg/m² (25.11 lb/sq ft)
B 127.1 kg/m² (26.03 lb/sq ft)
C 145.0 kg/m² (26.70 lb/sq ft)
D 134.5 kg/m² (27.56 lb/sq ft)
PERFORMANCE (A and B with PT6A-34AG engine at max T-weight except where indicated, C with TPE331-6, D: with TPE331-10):
Never-exceed speed (V_{NE}):
D 138 knots (256 km/h; 159 mph)
Max level speed with spray equipment:
A, B, D 138 knots (256 km/h; 159 mph)
Cruising speed:
A and B, 50% power 130 knots (241 km/h; 150 mph)
C, 55% power 130 knots (241 km/h; 150 mph)
Working speed, 30-50% power
82-130 knots (153-241 km/h; 95-150 mph)
Stalling speed: flaps up 61 knots (113 km/h; 70 mph)
flaps down 57 knots (106 km/h; 66 mph)
Stalling speed at normal landing weight:
flaps up 51 knots (95 km/h; 59 mph)
flaps down 50 knots (92 km/h; 57 mph)
Max rate of climb at S/L: A, B, C 530 m (1,740 ft)/min
D 762 m (2,500 ft)/min
Service ceiling: A, B, C 7,620 m (25,000 ft)
D 3,660 m (12,000 ft)
T-O run: A, B, D 183 m (600 ft)
C at 4,400 kg (9,700 lb) 366 m (1,200 ft)
Landing from 15 m (50 ft) 366 m (1,200 ft)
Landing run: A, B, C 152 m (500 ft)
D 244 m (800 ft)
Landing run with propeller reversal:
A, B, C 91 m (300 ft)
Range: D 500 nm (926 km; 575 miles)
Ferry range at 40% power
664 nm (1,231 km; 765 miles)

BASLER
BASLER TURBO CONVERSIONS INC
255 West 35th Avenue, PO Box 2305, Oshkosh, Wisconsin 54903-2305
Telephone: 1 (414) 236 7820
Fax: 1 (414) 236 0381
PRESIDENT: Thomas R. Weigt
VICE-PRESIDENT: Bob Clark

Company associated with Basler Flight Service Inc at same address; factory occupies 6,968 m² (75,000 sq ft); about 142 employees. Additional centres being set up in Taiwan and Poland. Performs conversions of Douglas DC-3 airframes, with Pratt & Whitney Canada PT6A-67R turboprops; 2.54 m (8 ft 4 in) fuselage stretch; centre and outer sections of wings strengthened; redesigned wingtips improve stall characteristics and low speed control response; lighter engines increase useful load. Max T-O and landing weight increased to 13,041 kg (28,750 lb). Fifteen completed (of 20 ordered) by January 1993. Detailed in 1991-92 *Jane's*; can now be found in *Jane's Civil and Military Aircraft Upgrades*.

Basler Turbo Conversions DC-3 *(Peter J. Cooper)*

BEDE
BEDE JET CORPORATION
Spirit of St Louis Airport, 18421 Edison Avenue, Chesterfield, Missouri 63005
Telephone: 1 (314) 537 2333
Fax: 1 (314) 536 2822
INFORMATION: James R. Bede

Details of current Bede BD-4 sporting and utility monoplane can be found in 1992-93 *Jane's*.

BEDE BD-10
TYPE: Tandem two-seat supersonic homebuilt.
PROGRAMME: Design began 1983; construction of prototype started 1989 and first flight 8 July 1992 (N2BJ); construction of production kits/aircraft began 1992; initial kit deliveries anticipated August 1993.
CUSTOMERS: 45 ordered by February 1993; then one built, 10 under construction.
COSTS: Kit: $221,000. Assembled aircraft: $495,000.
DESIGN FEATURES: Supersonic flight. Wings with leading-edge sweepback strakes and anhedral; twin fins.
FLYING CONTROLS: Non-boosted controls, using push-pull rods; sidestick controller for low speed flying and centre floor-mounted stick for high speed flight; dual controls; all-moving tailplane with anti-servo tab; twin rudders; double-slotted flaps.
STRUCTURE: 60 per cent metal, 40 per cent composites; aluminium alloy and aluminium honeycomb/aluminium or GFRP skin sandwich construction.
LANDING GEAR: Retractable tricycle type; electrical/mechanical actuation; all wheels retract aft; oleo-pneumatic shock absorbers in all units; Cleveland wheels and brakes.
POWER PLANT: One 13.12 kN (2,950 lb st) General Electric CJ-610 turbojet; alternative GE J85 engine. Eventually to be available with Williams Rolls-Royce FJ44. Fuel in five centre fuselage cells, total capacity 1,132 litres (299 US gallons; 249 Imp gallons). Single refuelling point in top of fuselage. Oil capacity 7.6 litres (2 US gallons; 1.7 Imp gallons).
ACCOMMODATION: Two persons in tandem cockpits.
SYSTEMS: No hydraulics. Electrically de-iced engine inlet. Cockpit heating, ventilation and air-conditioning.
AVIONICS: Weather radar; EFIS.
DIMENSIONS, EXTERNAL:
Wing span 6.55 m (21 ft 6 in)
Wing aspect ratio 4.7
Length overall 8.79 m (28 ft 10 in)
Height overall 2.46 m (8 ft 1 in)
AREAS:
Wings, gross 9.10 m² (98.0 sq ft)
WEIGHTS AND LOADINGS:
Weight empty 717 kg (1,580 lb)

BEDE/BEECH—AIRCRAFT: USA 423

Max T-O weight	1,878 kg (4,140 lb)
Max wing loading	206.3 kg/m² (42.24 lb/sq ft)
Max power loading (CJ-610)	143.2 kg/kN (1.40 lb/lb st)

PERFORMANCE (estimated):

Max level speed	Mach 1.4
Max cruising speed	Mach 0.91
Stalling speed	53-98 knots (97-181 km/h; 60-112 mph)
Max rate of climb at S/L	9,145 m (30,000 ft)/min
Service ceiling	13,715 m (45,000 ft)
T-O run at 1,270 kg (2,800 lb) AUW	183 m (600 ft)
Landing run at 953 kg (2,100 lb) AUW	458 m (1,500 ft)
g limits	+12/−9

Bede Jet Corporation BD-10 supersonic light aircraft

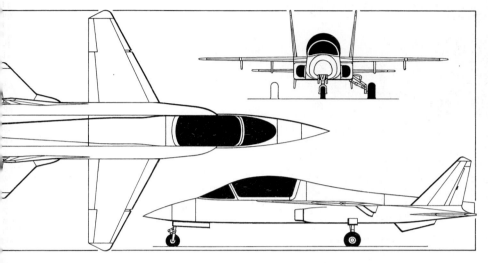

Bede Jet Corporation BD-10 supersonic light aircraft (*Jane's/Mike Keep*)

BEECH

BEECH AIRCRAFT CORPORATION
(Subsidiary of Raytheon Company)

709 East Central, Wichita, Kansas 67201-0085
Telephone: 1 (316) 676 7111
Fax: 1 (316) 676 8286
BRANCH DIVISION: Salina, Kansas
PRESIDENT: Jack Braly
EXECUTIVE VICE-PRESIDENT: Charles W. Dieker (Marketing)
VICE-PRESIDENT, AEROSPACE: Dr William A. Edgington
VICE-PRESIDENT, ENGINEERING: Robert D. Dickerson
VICE-PRESIDENT, AIRCRAFT DESIGN: Joseph F. Furnish
VICE-PRESIDENT, MANUFACTURING: Cecil Miller
VICE-PRESIDENT, MARKETING: Doug Mahin
VICE-PRESIDENT, COMMERCIAL MARKETING: C. Don Cary
DIRECTOR, CORPORATE AFFAIRS: James M. Gregory
DIRECTOR, AIRLINE SALES AND MARKETING: Mike Scheidt
MANAGER, PUBLIC RELATIONS: Michael S. Potts

Beech Aircraft Corporation founded 1932 by Mr and Mrs Beech; became wholly owned subsidiary of Raytheon 8 February 1980, but continues to operate separately building civil and military aircraft, missile targets, and components for aircraft (incl control surfaces for Boeing 737) and missiles; Salina division supplies all wings, non-metallic interior components, ventral fins, nosecones and tailcones used in Wichita production and builds major subassemblies for the Beechjet. Additional Wichita products include subcontracted composites and metal winglets and composites landing gear doors for McDonnell Douglas C-17.

Wholly owned subsidiaries include Beech Aerospace Services Inc (BASI) of Madison, Mississippi (worldwide logistic support for Army/Air Force/Navy C-12s, Army U-21s and Beech MQM-107 targets, and of US Navy T-34C and T-44 trainers in the United States); Beech Acceptance Corporation Inc (business aircraft retail financing and leasing); Travel Air Insurance Company Ltd, Bermuda (aircraft liability insurance); Beech Holdings Inc (marketing support to parent company). Wholly owned sales outlets of Beech Holdings Inc include Beechcraft East Inc, Farmingdale, New York, and Bedford, Massachusetts; Hangar One Inc, Atlanta Hartsfield, De-Kalb Peachtree and Fulton County airports in Georgia; Birmingham in Alabama; and Opa Locka, Orlando and Tampa in Florida; United Beechcraft Inc, Wichita, Kansas; Beechcraft West, Ontario, Van Nuys and Fresno, California; Hedrick Beechcraft in Houston, Dallas, Corpus Christi and San Antonio, Texas; Indiana Beechcraft Inc, Indianapolis, Indiana, the location of Beech Commuter Spare Parts Supply; Hartzog Aviation, Rockford, Illinois.

The first of two Beech constructed Pilatus PC-9 Mk IIs (N8284M), to take part in the US Air Force/Navy JPATS trainer programme, first flew on 23 December 1992 (see Switzerland).

Beech Aircraft has about 10,900 employees worldwide and occupies 371,612 m² (4,000,000 sq ft) of plant area at its two major facilities in Wichita (workforce 6,750) and Salina, Kansas (workforce 650).

Total production by Beech reached 50,342 at start of 1993. Deliveries of Beech aircraft during 1992 included 172 Bonanzas, 28 Barons, 86 King Airs, 34 Model 1900Cs, 31 Beechjets and seven Starship 1s.

Founder's widow Mrs Olive Ann Beech, Chairman Emeritus since 1982, died on 6 July 1993, aged 89.

BEECHCRAFT BONANZA F33A/C

TYPE: Four/five-seat single-engined executive aircraft.
PROGRAMME: First flight 14 September 1959; known as Debonair until 1967; 1985 model introduced cargo door, three-blade propeller, super soundproofing and full IFR avionics.
CURRENT VERSIONS: **F33A**: Similar to V35B Bonanza (1984-85 *Jane's*), but with conventional vertical fin and tailplane. Certificated in Utility category. *Description applies to F33A, except where indicated.*
F33C: Aerobatic version without air-conditioning, big cargo door and fifth seat; produced to special order.
CUSTOMERS: F33C sales include three civilian F33Cs in 1986-87, Imperial Iranian Air Force (16), Mexican Air Force (21), Mexican Navy (five), Netherlands Government Flying School (16), and Spanish Air Force/Air Ministry (74). F33A special order trainers include Lufthansa (three), Singapore International Airlines (four) and Airline Training Centre. Total 3,271 Bonanza 33s built by 1 January 1993, including 76 in 1992 (49 domestic and 27 international deliveries).
DESIGN FEATURES: Beech modified NACA 23016.5 wing section at root, modified 23012 at tip; dihedral 6°; incidence 4° at root and 1° at tip.
FLYING CONTROLS: Mechanical; electrically actuated trim tab in each elevator; ground adjustable tabs in ailerons and rudder; single-slotted, three-position light alloy flaps.
STRUCTURE: Conventional light alloy, with two-spar wing torsion box and stressed-skin tail surfaces. Reinforced fuselage, wings and tail on F33C.
LANDING GEAR: Electrically retractable tricycle type, with steerable nosewheel. Mainwheels retract inward into wings, nosewheel rearward. Beech oleo-pneumatic shock absorbers in all units. Cleveland mainwheels, size 6.00-6, and tyres, size 7.00-6, pressure 2.28-2.76 bars (33-40 lb/sq in). Cleveland nosewheel and tyre, size 5.00-5, pressure 2.76 bars (40 lb/sq in). Cleveland ring-disc hydraulic brakes. Parking brake. Magic Hand landing gear system optional.
POWER PLANT: One 212.5 kW (285 hp) Teledyne Continental IO-520-BB flat-six engine, driving a McCauley three-blade constant-speed metal propeller. Manually adjustable engine cowl flaps. Two standard fuel tanks in wing leading-edges, with total usable capacity of 280 litres (74 US gallons; 61.6 Imp gallons). Refuelling points above tanks. Oil capacity 11.5 litres (3 US gallons; 2.5 Imp gallons).
ACCOMMODATION: Enclosed cabin with four individual seats in pairs as standard, plus optional forward facing fifth seat (F33A only). Baggage compartment and hatshelf aft of seats. Passenger door and baggage compartment door on starboard side. Heater standard. Large cargo door, on starboard side of fuselage, standard on F33A. F33A has removable seat cushions, to accommodate parachutes, and a quick-release passenger door.
SYSTEMS: Optional 12,000 BTU refrigeration type air-conditioning system (F33A only) comprises evaporator located beneath pilot's seat, condenser on lower fuselage, and engine mounted compressor. Air outlets on centre console, with two-speed blower. Electrical system supplied by 28V 60A alternator, 24V 15.5Ah battery; a 100A alternator is available as an option, as is a standby generator. Hydraulic system for brakes only. Pneumatic system for instrument gyros and refrigeration type air-conditioning system optional. Oxygen system and electric propeller de-icing optional.
AVIONICS: Standard avionics include Bendix/King KX 155 760-channel com transceiver, 200-channel nav/glideslope receiver/converter with KI 209 VOR/ILS indicator, Bendix/King KY 155 760-channel com transceiver, 200-channel nav receiver/converter with KI 208 VOR/LOC indicator, KR 87 ADF with 227-00 indicator, KN 63 DME with KDI 572 indicator, DME hold and nav 1/nav 2 switching, KT 70 Mode S transponder, KEA 130A encoding altimeter, KMA 24-03 audio control/marker beacon receiver, microphone, headset, cabin speaker and static wicks. A wide range of optional avionics is available, including Bendix/King KFC 150 autopilot, yaw damper, KLN 88 Loran C and Foster WX-1000 Stormscope.
EQUIPMENT: Standard equipment includes LCD digital chronometer, EGT and OAT gauges, rate of climb indicator, turn co-ordinator, 3 in horizon and directional gyros, four fore and aft adjustable and reclining seats, armrests, headrests, single diagonal strap shoulder harness with inertia reel for all occupants, pilot's storm window, sun visors, ultra-violet-proof windscreen and windows, large cargo door (F33A only), emergency locator transmitter, stall warning device, alternate static source, heated pitot, rotating beacon, three-light strobe system, carpeted floor, super soundproofing, control wheel map lights, entrance door courtesy light, internally lit instruments, coat hooks, glove compartment, in-flight storage pockets, approach plate holder, utility shelf, cabin dome light, reading lights, instrument post lights, control wheel map light, electroluminescent

Beechcraft F33A Bonanza four/five-seat executive aircraft

sub-panel lighting, landing light, taxi light, full-flow oil filter, three-colour polyurethane exterior paint, external power socket and towbar. Optional equipment includes dual controls, co-pilot's wheel brakes, air-conditioning, fifth seat, fresh air vent blower and ground com switch. F33C has accelerometer, second boost pump with indicator light, non-baffled fuel cells and heavy-gauge rudder cables.

DIMENSIONS, EXTERNAL:
Wing span	10.21 m (33 ft 6 in)
Wing chord: at root	2.13 m (7 ft 0 in)
at tip	1.07 m (3 ft 6 in)
Wing aspect ratio	6.20
Length overall	8.13 m (26 ft 8 in)
Height overall	2.51 m (8 ft 3 in)
Tailplane span	3.71 m (12 ft 2 in)
Wheel track	2.92 m (9 ft 7 in)
Wheelbase	2.13 m (7 ft 0 in)
Propeller diameter	2.13 m (7 ft 0 in)
Passenger door: Height	0.91 m (3 ft 0 in)
Width	0.94 m (3 ft 1 in)
Baggage compartment door:	
Height	0.61 m (2 ft 0 in)
Width	0.99 m (3 ft 3 in)

DIMENSIONS, INTERNAL:
Cabin, aft of firewall: Length	3.07 m (10 ft 1 in)
Max width	1.07 m (3 ft 6 in)
Max height	1.27 m (4 ft 2 in)
Volume	3.31 m^3 (117.0 cu ft)
Baggage space	0.99 m^3 (35.0 cu ft)

AREAS:
Wings, gross	16.80 m^2 (181.0 sq ft)
Ailerons (total)	1.06 m^2 (11.4 sq ft)
Trailing-edge flaps (total)	1.98 m^2 (21.3 sq ft)
Fin	0.93 m^2 (10.0 sq ft)
Rudder, incl tab	0.52 m^2 (5.6 sq ft)
Tailplane	1.75 m^2 (18.82 sq ft)
Elevators, incl tabs	1.67 m^2 (18.0 sq ft)

WEIGHTS AND LOADINGS:
Weight empty	1,017 kg (2,242 lb)
Max T-O and landing weight	1,542 kg (3,400 lb)
Max T-O weight, F33C in Aerobatic category	1,270 kg (2,800 lb)
Max ramp weight	1,548 kg (3,412 lb)
Max wing loading	91.8 kg/m^2 (18.8 lb/sq ft)
Max power loading	7.26 kg/kW (11.93 lb/hp)

PERFORMANCE (at max T-O weight, except cruising speeds at mid-cruise weight):
Max level speed at S/L	182 knots (338 km/h; 209 mph)
Cruising speed: 75% power at 1,830 m (6,000 ft)	172 knots (319 km/h; 198 mph)
65% power at 3,050 m (10,000 ft)	160 knots (297 km/h; 184 mph)
55% power at 3,660 m (12,000 ft)	145 knots (269 km/h; 167 mph)
45% power at 2,440 m (8,000 ft)	136 knots (253 km/h; 157 mph)
Stalling speed, power off:	
flaps up	64 knots (118 km/h; 74 mph) IAS
30° flap	51 knots (94 km/h; 59 mph) IAS
Max rate of climb at S/L	353 m (1,157 ft)/min
Service ceiling	5,443 m (17,858 ft)
T-O run	305 m (1,000 ft)
T-O to 15 m (50 ft)	530 m (1,740 ft)
Landing from 15 m (50 ft)	396 m (1,300 ft)
Landing run	232 m (760 ft)

Range with max usable fuel, allowances for engine start, taxi, T-O, climb and 45 min reserves at 45% power:
75% power at 1,830 m (6,000 ft)	715 nm (1,325 km; 823 miles)
65% power at 3,050 m (10,000 ft)	810 nm (1,501 km; 932 miles)
55% power at 3,660 m (12,000 ft)	860 nm (1,593 km; 990 miles)
45% power at 2,440 m (8,000 ft)	889 nm (1,648 km; 1,023 miles)

BEECHCRAFT BONANZA A36

TYPE: Four/six-seat, single-engined business and utility aircraft.
PROGRAMME: Developed from Bonanza V35B, but with vertical fin; current A36 introduced 3 October 1983, succeeding model powered by 212.5 kW (285 hp) Continental IO-520-BB; certificated in FAA Utility category.
CURRENT VERSIONS: **Bonanza A36**: Standard version, *as described in detail*.
Bonanza A36AT (Airline Trainer): Certificated 1991 for use in Europe; modifications to propeller, engine and exhaust systems to reduce external noise; flyover sound level 71.8 dBA; Teledyne Continental IO-550-B, limited to 216 kW (290 hp) at 2,550 rpm, driving Hartzell three-blade constant-speed propeller; special exhaust silencers; additional engine cooling louvres; four seats.
CUSTOMERS: Include Saudi Arabian Airlines (four) in April 1985 for pilot training; Finnair Training Centre at Pori, Finland (three), in 1987; Japan Air Lines (five) in 1990; Lufthansa (12 A36ATs, to replace F33As) in 1991; Japan Air Lines (23); RLS Netherlands (eight A36ATs) in 1991; and All Nippon Airways (20, delivered 1992-94). Total 3,205 36 Bonanzas delivered by 1 January 1993, including 81 in 1992 (36 domestic and 45 international deliveries).
DESIGN FEATURES: Structure as for F33A, but fuselage 0.25 m (10 in) longer; cabin volume increased by 0.54 m^3 (18.9 cu ft); baggage volume increased by 0.28 m^3 (10 cu ft); large double freight doors starboard side aft of wing; current model has improved instrument panel controls, lighting and systems; options same as for F33A, plus instrument post lights or internal instrument lighting; courtesy lights for entrance and step; co-pilot's vertically adjusting seat; refrigeration-type air-conditioning.
FLYING CONTROLS: As for F33A. Dual controls standard.
STRUCTURE: As for F33A, with addition of two vortex generators on wings.
LANDING GEAR: As for F33A; new landing gear warning system introduced in 1989.
POWER PLANT: One 224 kW (300 hp) Teledyne Continental IO-550-B flat-six engine, driving a McCauley three-blade constant-speed propeller (Lufthansa aircraft derated to 212.5 kW; 285 hp and have smaller propeller). The engine is equipped with an altitude-compensating fuel pump which automatically leans and enriches the fuel/air mixture during climb and descent respectively. Fuel capacity as for F33A.
ACCOMMODATION: Enclosed cabin seating four to six persons on individual seats. Pilot's seat is vertically adjustable. Dual controls standard. Two rear removable seats and two folding seats permit rapid conversion to utility configuration. Optional club seating with rear facing third and fourth seats, executive writing desk, refreshment cabinet, headrests for third and fourth seats, reading lights and fresh air outlets for fifth and sixth seats. Double doors of bonded aluminium honeycomb construction on starboard side facilitate loading of cargo. As an air ambulance, one stretcher can be accommodated with ample room for medical attendant and/or other passengers. Extra windows provide improved view for passengers. Stowage for 181 k (400 lb) of baggage.
SYSTEMS: Electrical system as for F33A. Hydraulic system f brakes only. Pneumatic system for instrument gyros, ar refrigeration type air-conditioning system, optional.
AVIONICS: Standard avionics include Bendix/King KX 1 760-channel nav/com, with KI 208 VOR/LOC Omni co verter/indicator, but a wide range of optional avionics available, including Bendix/King KX 165 or KY 196 TSO/IFR transceiver, KAS 297. An optional ground com munication switch permits use of one com radio witho turning on the battery master switch.
EQUIPMENT: Optional equipment as detailed for F33 Bonanza.

DIMENSIONS, EXTERNAL: As for F33A except:
Length overall	8.38 m (27 ft 6 i
Height overall	2.62 m (8 ft 7 i
Wheelbase	2.39 m (7 ft 10¼ i
Propeller diameter: A36	2.03 m (6 ft 8 i
A36AT	1.93 m (6 ft 4 i
Rear passenger/cargo door: Height	0.89 m (2 ft 11 i
Width	1.14 m (3 ft 9 i

DIMENSIONS, INTERNAL:
Cabin, aft of firewall:	
Length, incl extended baggage compartment	3.84 m (12 ft 7 i
Max width	1.07 m (3 ft 6 i
Max height	1.27 m (4 ft 2 i
Volume	3.85 m^3 (135.9 cu

AREAS: As for F33A

WEIGHTS AND LOADINGS:
Weight empty, standard:	
A36, A36AT	1,041 kg (2,295 l
Max T-O weight: A36	1,655 kg (3,650 l
A36AT	1,633 kg (3,600 l
Max ramp weight: A36	1,661 kg (3,663 l
A36AT	1,639 kg (3,613 l
Max wing loading: A36	98.6 kg/m^2 (20.2 lb/sq
A36AT	97.16 kg/m^2 (19.9 lb/sq
Max power loading: A36	7.40 kg/kW (12.17 lb/h
A36AT	7.59 kg/kW (12.46 lb/h

PERFORMANCE (max speed at minimum weight; cruisi speeds at mid-cruise weight):
Max level speed	184 knots (340 km/h; 212 mp
Max cruising speed: 2,500 rpm at 1,830 m (6,000 ft)	176 knots (326 km/h; 202 mp
2,300 rpm at 2,440 m (8,000 ft)	167 knots (309 km/h; 192 mp
2,100 rpm at 1,830 m (6,000 ft)	160 knots (296 km/h; 184 mp
2,100 rpm at 3,050 m (10,000 ft)	153 knots (283 km/h; 176 mp
Stalling speed, power off:	
flaps up	68 knots (126 km/h; 78 mph) IA
30° flap	59 knots (109 km/h; 68 mph) IA
Max rate of climb at S/L	368 m (1,208 ft)/m
Service ceiling	3,640 m (18,500 f
T-O run: flaps up: A36	360 m (1,182 f
A36AT	450 m (1,476 f
12° flap: A36	296 m (971 f
A36AT	402 m (1,316 f
T-O to 15 m (50 ft):	
flaps up: A36	640 m (2,100 f
A36AT	662 m (2,170 f
12° flap: A36	583 m (1,913 f
A36AT	618 m (2,025 f
Landing from 15 m (50 ft)	442 m (1,450 f
Landing run	280 m (920 f

Range with max usable fuel, with allowances for engir start, taxi, T-O, climb and 45 min reserves at econ crui power: 2,500 rpm at 3,660 m (12,000 ft)
875 nm (1,621 km; 1,008 mile
2,300 rpm at 3,660 m (12,000 ft)
903 nm (1,672 km; 1,039 mile
2,100 rpm at 1,830 m (6,000 ft)
914 nm (1,694 km; 1,052 mile

BEECHCRAFT TURBO BONANZA B36TC

TYPE: Turbocharged version of A36 Bonanza.
PROGRAMME: Certificated as A36TC 7 December 1978; 27 A36TCs delivered; improved B36TC introduced 1982.
CUSTOMERS: Total of 531 delivered by 1 January 1993, inclu ing 15 in 1992 (14 domestic and one international).
DESIGN FEATURES: Compared with A36TC, B36TC has great span; wing section NACA 23010.5 at tip; 0° incidence tip; greater fuel capacity.
Data below summarise differences from A36TC:
POWER PLANT: One 223.7 kW (300 hp) Teledyne Continent TSIO-520-UB turbocharged flat-six engine, driving three-blade constant-speed metal propeller. Fixed engir cowl flaps. Two fuel tanks in each wing leading-edge, wi total usable capacity of 386 litres (102 US gallons; 85 Im gallons). Refuelling points above tanks. Oil capacity 11 litres (3 US gallons; 2.5 Imp gallons).
SYSTEMS: Air-conditioning optional.
AVIONICS: As for A36.
EQUIPMENT: As for A36, except EGT gauge not availabl Turbine inlet temperature gauge is standard.

Beechcraft B36TC Turbo Bonanza four/six-seat cabin monoplane

DIMENSIONS, EXTERNAL: As for A36, except:
Wing span	11.53 m (37 ft 10 in)
Wing chord at tip	0.91 m (3 ft 0 in)
Wing aspect ratio	7.61
Propeller diameter	1.98 m (6 ft 6 in)

DIMENSIONS, INTERNAL: As for A36
AREAS:
Wings, gross	17.47 m² (188.1 sq ft)

WEIGHTS AND LOADINGS:
Weight empty, standard	1,104 kg (2,433 lb)
Max T-O and landing weight	1,746 kg (3,850 lb)
Max ramp weight	1,753 kg (3,866 lb)
Max wing loading	100.1 kg/m² (20.5 lb/sq ft)
Max power loading	7.81 kg/kW (12.8 lb/hp)

PERFORMANCE (at max T-O weight, except speeds are at mid-cruise weight):
Max level speed at 6,700 m (22,000 ft)	213 knots (394 km/h; 245 mph)

Cruising speed at 7,620 m (25,000 ft):
79% power	200 knots (370 km/h; 230 mph)
75% power	195 knots (361 km/h; 224 mph)
69% power	188 knots (348 km/h; 216 mph)
56% power	173 knots (320 km/h; 199 mph)

Stalling speed, power off:
flaps up	65 knots (120 km/h; 75 mph) IAS
30° flap	57 knots (106 km/h; 66 mph) IAS
Max rate of climb at S/L	321 m (1,053 ft)/min
Service ceiling	over 7,620 m (25,000 ft)
T-O run, 15° flap	311 m (1,020 ft)
T-O to 15 m (50 ft), 15° flap	649 m (2,130 ft)
Landing run	298 m (976 ft)

Range with max fuel, allowances for engine start, taxi, T-O, cruise climb, descent, and 45 min reserves at 50% power:
79% power at 7,620 m (25,000 ft)	956 nm (1,770 km; 1,100 miles)
75% power at 7,620 m (25,000 ft)	984 nm (1,822 km; 1,132 miles)
69% power at 7,620 m (25,000 ft)	1,022 nm (1,892 km; 1,176 miles)
56% power at 6,100 m (20,000 ft)	1,092 nm (2,022 km; 1,256 miles)

BEECHCRAFT BARON 58

TYPE: Four/six-seat piston-twin business aircraft.
PROGRAMME: Developed from Baron D55; certificated in FAA Normal category 19 November 1969.
CUSTOMERS: Include All Nippon Airways (12 ordered 1992; six for delivery 1993 and six for 1994), Indonesian Civil Flying Academy, Java (four), Centre National de Formation Aviation Civile of M'Vengue, Gabon (three), US FAA staff pilot proficiency training programme (eight), Lufthansa pilot training (19), Air France (three), Japan Air Lines (three), Singapore Airlines (two), and Seneca College, Toronto, Canada (two), all for pilot training; 1992 deliveries totalled 28 (21 domestic and seven international). Total 2,310 Baron 58s, including 58Ps and 58TCs, delivered by 1 January 1993.
DESIGN FEATURES: Changes from Baron 55 include forward cabin extended by 0.254 m (10 in), wheelbase extended forward, double passenger/cargo doors starboard side, extended propeller hubs, redesigned nacelles for better cooling, and fourth window on each side.
Wing section NACA 23015.5 at root, 23010.5 at tip; dihedral 6°; incidence 4° at root and 0° at tip.
FLYING CONTROLS: Manually operated trim tabs in elevators, rudder and port aileron; electrically operated single-slotted flaps.
STRUCTURE: Light alloy with two-spar wing box; elevators have smooth magnesium alloy skins.
LANDING GEAR: Electrically retractable tricycle type. Main units retract inward into wings, nosewheel aft. Beech oleo-pneumatic shock absorbers in all units. Steerable nosewheel with shimmy damper. Cleveland wheels, with mainwheel tyres size 6.50-8, pressure 3.59-3.96 bars (52-56 lb/sq in). Nosewheel tyre size 5.00-5, pressure 3.79-4.14 bars (55-60 lb/sq in). Cleveland ring-disc hydraulic brakes. Heavy duty brakes optional. Parking brake. New warning system introduced in 1989.
POWER PLANT: Two 224 kW (300 hp) Teledyne Continental IO-550-C flat-six engines, each driving a McCauley three-blade constant-speed fully feathering metal propeller. The standard fuel system has a usable capacity of 514 litres (136 US gallons; 113 Imp gallons), with optional usable capacity of 628 litres (166 US gallons; 138 Imp gallons). Optional 'wet wingtip' installation also available, increasing usable capacity to 734 litres (194 US gallons; 161.5 Imp gallons).
ACCOMMODATION: Standard model has four individual seats in pairs in enclosed cabin, with door on starboard side. Single diagonal strap shoulder harness with inertia reel standard on all seats. Pilot's vertically adjusting seat is standard. Co-pilot's vertically adjusting seat, folding fifth and sixth seats, or club seating comprising folding fifth and sixth seats and aft facing third and fourth seats, are optional. Executive writing desk available as option with club seating. Baggage compartment in nose, capacity 136 kg (300 lb). Double passenger/cargo doors on starboard side of cabin provide access to space for 181 kg (400 lb) of baggage or cargo behind the third and fourth seats. Pilot's storm window. Openable windows adjacent to the third and fourth seats are used for ground ventilation and as emergency exits. Cabin heated and ventilated. Windscreen defrosting standard.
SYSTEMS: Cabin heated by Janitrol 50,000 BTU heater, which serves also for windscreen defrosting. Oxygen system of 1.41 m³ (49.8 cu ft) or 1.87 m³ (66 cu ft) capacity optional. Electrical system includes two 28V 60A engine driven alternators with alternator failure lights and two 12V 25Ah batteries. Two 100A alternators optional. Hydraulic system for brakes only. Pneumatic pressure system for air driven instruments, and optional wing and tail unit de-icing system. Oxygen system, cabin air-conditioning and windscreen electric anti-icing optional.
AVIONICS: Standard avionics include Bendix/King KX 155-09 760-channel com transceiver with audio amplifier, 200-channel nav receiver with KI 208 VOR/LOC converter/indicator, KR 87 ADF with KI 227-00 indicator, Bendix/King combined loop/sense antenna, microphone, headset, cabin speaker, nav and com antennae. Bendix/King weather radar and/or Stormscope optional. Optional avionics by Bendix/King and Collins.
EQUIPMENT: Standard equipment includes dual controls, blind-flying instruments, control wheel clock, outside air temperature gauge, sensitive altimeter, turn co-ordinator, pilot's storm window, sun visors, ultraviolet-proof windscreen and cabin windows, armrests, adjustable rudder

Beechcraft Baron 58 four/six-seat cabin monoplane

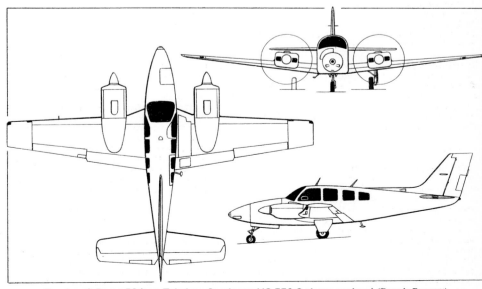

Beechcraft Baron 58 (two Teledyne Continental IO-550-C piston engines) *(Dennis Punnett)*

pedals (retractable on starboard side), emergency locator transmitter, heated pitot head, instrument panel floodlights, map light, lighted trim tab position indicator, step and entrance door courtesy lights, reading lights, navigation and position lights, steerable taxi light, dual landing lights, cabin carpeting and soundproofing, headrests, heated fuel vents, cabin dome light, door ajar warning light, nose baggage compartment light, heated fuel and stall warning vanes, external polyurethane paint finish, EGT and CHT gauges, synchroscope, engine winterisation kit, towbar and external power socket. Options include a true airspeed indicator, engine and flight hour recorders, instantaneous vertical speed indicator, alternate static source, internally illuminated instruments, digital beacon, strobe lights, electric windscreen anti-icing, wing ice detection light, static wicks, cabin club seating, executive writing desk, refreshment cabinet, cabin fire extinguisher, ventilation blower, super soundproofing, and approach plate holder.

DIMENSIONS, EXTERNAL:
Wing span	11.53 m (37 ft 10 in)
Wing chord: at root	2.13 m (7 ft 0 in)
at tip	0.90 m (2 ft 11.6 in)
Wing aspect ratio	7.19
Length overall	9.09 m (29 ft 10 in)
Height overall	2.97 m (9 ft 9 in)
Tailplane span	4.85 m (15 ft 11 in)
Wheel track	2.92 m (9 ft 7 in)
Wheelbase	2.72 m (8 ft 11 in)
Propeller diameter	1.98 m (6 ft 6 in)
Rear passenger/cargo doors:	
Max height	0.89 m (2 ft 11 in)
Width	1.14 m (3 ft 9 in)
Baggage door (fwd): Height	0.56 m (1 ft 10 in)
Width	0.64 m (2 ft 1 in)

DIMENSIONS, INTERNAL:
Cabin, incl rear baggage area:	
Length	3.84 m (12 ft 7 in)
Max width	1.07 m (3 ft 6 in)
Max height	1.27 m (4 ft 2 in)
Floor area	3.72 m^2 (40.0 sq ft)
Volume	3.85 m^3 (135.9 cu ft)
Baggage compartment: fwd	0.51 m^3 (18.0 cu ft)

AREAS:
Wings, gross	18.51 m^2 (199.2 sq ft)
Ailerons (total)	1.06 m^2 (11.40 sq ft)
Trailing-edge flaps (total)	1.98 m^2 (21.30 sq ft)
Fin	1.46 m^2 (15.67 sq ft)
Rudder, incl tab	0.81 m^2 (8.75 sq ft)
Tailplane	4.95 m^2 (53.30 sq ft)
Elevators, incl tabs	1.84 m^2 (19.80 sq ft)

WEIGHTS AND LOADINGS:
Weight empty	1,619 kg (3,570 lb)
Max T-O weight	2,495 kg (5,500 lb)
Max landing weight	2,449 kg (5,400 lb)
Max ramp weight	2,506 kg (5,524 lb)
Max wing loading	143.4 kg/m^2 (27.6 lb/sq ft)
Max power loading	5.60 kg/kW (9.2 lb/hp)

PERFORMANCE (at max T-O weight, except cruising speeds at average cruise weight):
Max level speed at S/L	208 knots (386 km/h; 239 mph)
Max cruising speed, 2,500 rpm at 1,525 m (5,000 ft)	203 knots (376 km/h; 234 mph)
Cruising speed, 2,500 rpm at 3,050 m (10,000 ft)	198 knots (367 km/h; 228 mph)
Econ cruising speed, 2,100 rpm at 3,660 m (12,000 ft)	163 knots (302 km/h; 188 mph)
Stalling speed, power off:	
flaps up	84 knots (156 km/h; 97 mph) IAS
flaps down	75 knots (139 km/h; 86 mph) IAS
Max rate of climb at S/L	529 m (1,735 ft)/min
Rate of climb at S/L, OEI	119 m (390 ft)/min
Service ceiling	6,306 m (20,688 ft)
Service ceiling, OEI	2,220 m (7,284 ft)
T-O run	427 m (1,400 ft)
T-O to 15 m (50 ft)	701 m (2,300 ft)
Landing from 15 m (50 ft)	747 m (2,450 ft)
Landing run	434 m (1,425 ft)

Range with 734 litres (194 US gallons; 161.5 Imp gallons) usable fuel, with allowances for engine start, taxi, T-O, climb and 45 min reserves at econ cruise power:
  max cruising speed (power/altitude settings as above)
    1,150 nm (2,130 km; 1,324 miles)
  cruising speed (power/altitude settings as above)
    1,411 nm (2,615 km; 1,625 miles)
  econ cruising speed (power/altitude settings as above)
    1,575 nm (2,919 km; 1,814 miles)

BEECHCRAFT KING AIR C90B

TYPE: Six/10-seat turboprop pressurised business twin.
PROGRAMME: Announced at NBAA show at Houston October 1991; superseded King Air 90, A90, B90, C90, C90-1, C90A; introduced four-blade McCauley propellers, special interior soundproofing, updated and redesigned interior, updated cockpit features and interior noise and vibration levels substantially reduced.
CUSTOMERS: King Air C90A customers include Japan Air Lines and Japan Civil Aviation College; Indira Gandhi Academy; University of North Dakota; United States Navy (T-44A); Japan Maritime Self Defence Force; Canadair (for Canadian Forces' flight training); and Spanish Air Ministry; 28 delivered during 1992 (three C90As, two domestic and one international; 25 C90Bs, 10 domestic and 15 international); total 1,317 commercial and 226 military King Air 90/A90/B90/C90/C90-1/C90A delivered by 1 January 1993.
DESIGN FEATURES: Wing section NACA 23014.1 (modified) at root, 23016.22 (modified) at outer end of centre-section, 23012 at tip; dihedral 7°; incidence 4° 48' at root, 0° at tip; tailplane 7° dihedral.
FLYING CONTROLS: Trim tabs on port aileron, in both elevators and rudders; single-slotted aluminium flaps.
STRUCTURE: Generally light alloy; magnesium ailerons.
LANDING GEAR: Hydraulically retractable tricycle type. Nosewheel retracts rearward, mainwheels forward into engine nacelles. Mainwheels protrude slightly beneath nacelles when retracted, for safety in a wheels-up emergency landing. Fully castoring steerable nosewheel with shimmy damper. Beech oleo-pneumatic shock absorbers. Goodrich mainwheels with tyres size 8.50-10, pressure 3.79 bars (55 lb/sq in). Goodrich nosewheel with tyre size 6.50-10, pressure 3.59 bars (52 lb/sq in). Goodrich heat-sink and air-cooled multi-disc hydraulic brakes. Parking brakes. Min ground turning radius 10.82 m (35 ft 6 in).
POWER PLANT: Two 410 kW (550 shp) Pratt & Whitney Canada PT6A-21 turboprops, each driving a McCauley four-blade constant-speed fully feathering propeller. Propeller auto ignition system, environmental fuel drain collection system, magnetic chip detector, automatic propeller feathering and propeller synchrophaser standard. Fuel in two tanks in engine nacelles, each with usable capacity of 231 litres (61 US gallons; 50.8 Imp gallons), and auxiliary bladder tanks in outer wings, each with capacity of 496 litres (131 US gallons; 109 Imp gallons). Total usable fuel capacity 1,454 litres (384 US gallons; 320 Imp gallons). Refuelling points in top of each engine nacelle and in wing leading-edge outboard of each nacelle. Oil capacity 13.2 litres (3.5 US gallons; 2.9 Imp gallons) per engine.
ACCOMMODATION: Two seats side by side in cockpit with dual controls standard. Normally, four reclining seats in main cabin, in pairs facing each other fore and aft. Standard furnishings include cabin forward partition, with fore and aft partition curtain and coat rack, hinged nose baggage compartment door, seat belts and inertia reel shoulder harness for all seats. Optional arrangements seat up to eight persons, with lateral tracking chairs, and refreshment cabinets. Baggage racks at rear of cabin on starboard side, with optional toilet on port side. Door on port side aft of wing, with built-in airstairs. Emergency exit on starboard side of cabin. Entire accommodation pressurised, heated and air-conditioned. Electrically heated windscreen, windscreen defroster and windscreen wipers standard.
SYSTEMS: Pressurisation by dual engine bleed air system with pressure differential of 0.34 bar (5.0 lb/sq in). Cabin heated by 45,000 BTU dual engine bleed air system and auxiliary electrical heating system. Hydraulic system for landing gear actuation. Electrical system includes two 28V 250A starter/generators, 24V 45Ah air-cooled nickel-cadmium battery with failure detector. Oxygen system, 0.62 m^3 (2 cu ft), 1.39 m^3 (49 cu ft) or 1.81 m^3 (64 cu ft) capacity optional. Vacuum system for flight instruments. Automatic pneumatic de-icing of wing/fin/tailplane leading-edge standard. Engine and propeller anti-icing systems standard. Engine fire detection and extinguishing system optional.
AVIONICS: Standard Collins Pro Line II avionics package includes APS-65 autopilot/flight director with ADI-84 FD and EFD-74 EHSI; dual VHF-22A transceivers with CTL-22 controls; dual VIR-32 VOR/LOC/GLS/MK receivers with CTL-32 controls; ADF-60A with CTL-6 control; DME-42 with IND-42 indicator; RMI-30; dual MCS-65 compass systems; WXR-270 colour weather radar; dual TDR-90 transponders; ALT-50A radio altimeter; dual DB systems Model 415 audio systems; pilot United 5506-S encoding altimeter; dual Flite-Tronic PC-250 inverters; dual 2 in electric turn and bank indicators; co-pilot's 3 in horizon indicator; co-pilot's HSI avionics master switch; edgelite radio panel; ELT; ground clearance switch to com 1; control wheel push-to-talk switches; dual hand microphones; dual Telex headsets; and sectional instrument panel. Optional avionics include Bendix/King Silver Crown and Gold Crown packages; Foster LNS-616B RNav/Loran C; and Foster WX-1000 Stormscope.
EQUIPMENT: Standard equipment includes dual blind-flying instrumentation with sensitive altimeters and dual instantaneous VSIs; standby magnetic compass; OAT gauge; LCD digital chronometer clock; vacuum gauge; de-icing pressure gauge; cabin rate of climb indicator; cabin altitude and differential pressure indicator; flight hour recorder; automatic solid state warning and annunciator panel; pilot and co-pilot's four-way adjustable seats with shoulder harness and lap belts; dual cockpit speakers; adjustable sun visors; map pocket; cigarette lighter; two ashtrays; fresh air outlets; coffee cup holders; oxygen outlets with overhead mounted diluter demand masks with microphones; pilot and co-pilot's approach plate/map cases; four fully adjustable cabin seats in club arrangement with removable headrests, shoulder harnesses, lap belts, adjustable reading lights, retractable inboard armrests; cabin floor carpeting; adjustable polarised sunshades; cabin coat cable with hangers; magazine rack; removable low profile toilet with shoulder harness, lap belt; baggage webbing; relief tube; aisle-facing storage seat with air, light, ashtray, water tank; ice chest drawer, removable bottle/decanter rack and padded partition; two cabinets, forward left and right sides of cabin, with ice chest, heated liquid container and storage drawers; two cabin tables; cabin fire extinguisher; 'No smoking — Fasten seat belt' sign with audible chime; internal corrosion proofing; urethane exterior paint; tow bar; propeller restraints; wing ice lights; dual landing lights; nosewheel taxi light; flush position lights; dual rotating beacons; dual map lights; indirect cabin lighting; two overhead cabin spotlights; entrance door light; and compartment lights; primary and secondary instrument

Beechcraft King Air C90B six/10-seat business aircraft

lighting systems; rheostat-controlled white cockpit lighting; entrance door step lights; wingtip recognition lights; wingtip and tail strobe lights; and vertical tail illumination lights. Optional equipment includes lateral tracking cabin seats; cabin door support cable; cabin partition sliding doors; central aisle carpet runner; electric flushing toilet; engine fire detection system; 1.81 m³ (64 cu ft) oxygen bottle; cockpit relief tube; co-pilot's control wheel digital chronometer; cabinet with three drawers and stereo tape deck storage; pilot-to-cabin paging with four stereo speakers; ADF audio to cabin paging; control wheel frequency transfer and memory advance switch; co-pilot's control wheel transponder ident switch; and passenger stereo headsets.

DIMENSIONS, EXTERNAL:
Wing span	15.32 m (50 ft 3 in)
Wing chord: at root	2.15 m (7 ft 0½ in)
at tip	1.07 m (3 ft 6 in)
Wing aspect ratio	8.59
Length overall	10.82 m (35 ft 6 in)
Height overall	4.34 m (14 ft 3 in)
Tailplane span	5.26 m (17 ft 3 in)
Wheel track	3.89 m (12 ft 9 in)
Wheelbase	3.73 m (12 ft 3 in)
Propeller diameter	2.29 m (7 ft 6 in)
Propeller ground clearance	0.34 m (1 ft 1½ in)
Passenger door: Height	1.30 m (4 ft 3½ in)
Width	0.69 m (2 ft 3 in)
Height to sill	1.22 m (4 ft 0 in)

DIMENSIONS, INTERNAL:
Total pressurised length	5.43 m (17 ft 10 in)
Cabin: Length	3.86 m (12 ft 8 in)
Max width	1.37 m (4 ft 6 in)
Max height	1.45 m (4 ft 9 in)
Floor area	6.50 m² (70.0 sq ft)
Volume	8.88 m³ (313.6 cu ft)
Baggage compartment, rear	1.51 m³ (53.5 cu ft)

AREAS:
Wings, gross	27.31 m² (293.94 sq ft)
Ailerons (total)	1.29 m² (13.90 sq ft)
Trailing-edge flaps (total)	2.72 m² (29.30 sq ft)
Fin	2.20 m² (23.67 sq ft)
Rudder, incl tab	1.30 m² (14.00 sq ft)
Tailplane	4.39 m² (47.25 sq ft)
Elevators, incl tabs (total)	1.66 m² (17.87 sq ft)

WEIGHTS AND LOADINGS:
Weight empty	3,005 kg (6,625 lb)
Max T-O weight	4,581 kg (10,100 lb)
Max ramp weight	4,608 kg (10,160 lb)
Max landing weight	4,354 kg (9,600 lb)
Max wing loading	167.7 kg/m² (34.4 lb/sq ft)
Max power loading	5.59 kg/kW (9.2 lb/shp)

PERFORMANCE (at max T-O weight except where indicated):
Max cruising speed at AUW of 3,855 kg (8,500 lb):
at 3,660 m (12,000 ft)	242 knots (448 km/h; 278 mph)
at 4,880 m (16,000 ft)	247 knots (457 km/h; 284 mph)
at 6,400 m (21,000 ft)	243 knots (450 km/h; 280 mph)

Stalling speed, power off:
wheels and flaps up
  88 knots (163 km/h; 101 mph) IAS
wheels and flaps down
  78 knots (144 km/h; 90 mph) IAS
Max rate of climb at S/L	610 m (2,003 ft)/min
Rate of climb at S/L, OEI	151 m (494 ft)/min
Service ceiling	8,810 m (28,900 ft)
Service ceiling, OEI	3,990 m (13,100 ft)
T-O run	620 m (2,033 ft)
T-O to 15 m (50 ft)	826 m (2,710 ft)
Accelerate/stop distance	1,113 m (3,650 ft)

Landing from 15 m (50 ft) at max landing weight, with propeller reversal 698 m (2,290 ft)
Landing run at max landing weight, with propeller reversal 384 m (1,260 ft)
Range with max fuel at max cruising speed, incl allowance for starting, taxi, take-off, climb, descent and 45 min reserves at max range power, ISA, at:
6,400 m (21,000 ft)	1,075 nm (1,992 km; 1,238 miles)
4,875 m (16,000 ft)	933 nm (1,729 km; 1,074 miles)
7,315 m (24,000 ft)	1,165 nm (2,159 km; 1,341 miles)

Max range at econ cruising power, allowances as above, at:
6,400 m (21,000 ft)	1,277 nm (2,366 km; 1,470 miles)
4,875 m (16,000 ft)	1,155 nm (2,140 km; 1,330 miles)
7,315 m (24,000 ft)	1,340 nm (2,483 km; 1,543 miles)

BEECHCRAFT SUPER KING AIR B200
Swedish Air Force designation: Tp 101

TYPE: Twin-turboprop pressurised passenger, cargo or business light transport.
PROGRAMME: Design of Super King Air 200 began October 1970; first flight (c/n BB1) 27 October 1972; certificated FAR Pt 23 plus icing requirements of FAR Pt 25, 14 December 1973; design of B200 (prototype c/n BB343) began March 1980; production started May 1980; FAA certification 13 February 1981; on sale March 1981.
CURRENT VERSIONS: **Super King Air B200**: Basic version; *detailed description applies to B200*.
Super King Air B200C: As B200 but with 1.32 × 1.32 m (4 ft 4 in × 4 ft 4 in) cargo door.

Beechcraft Super King Air B200 fourteen-passenger pressurised transport

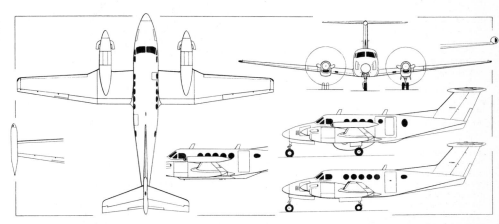

Beechcraft Super King Air B200 twin-turboprop transport, with additional side view of Maritime Patrol B200T (centre right); scrap views of wingtip tanks and centre-fuselage of photo survey aircraft for IGN
(Dennis Punnett)

Super King Air B200T: Standard provision for removable tip tanks, adding total 401 litres (106 US gallons; 88.25 Imp gallons), making total 2,460 litres (650 US gallons; 541 Imp gallons). Span without tip tanks 16.92 m (55 ft 6 in).
Super King Air B200CT: Combines tip tanks and cargo door as standard.
Maritime patrol B200T: Described separately.
Super King Air 300 and 300LW: Described separately.
Super King Air 350: Described separately.
CUSTOMERS: French Institut Géographique National acquired two B200T fitted with twin Wild RC-10 Superaviogon camera installations and Doppler navigation in February 1977; max endurance 10.3 h; high flotation landing gear; special French certification for max T-O weight 6,350 kg (14,000 lb) and max landing weight 6,123 kg (13,500 lb). Egyptian government acquired one Super King Air in 1978 for water, uranium and other natural resources exploration over Sinai and Egyptian deserts as follow-up to satellite surveys; fitted with remote sensing gear, specialised avionics and special cameras. Navaid checking versions of Super King Air used by Taiwan government (one) and Malaysian government (two). Special missions Super King Air delivered to Taiwan Ministry of Interior May 1979; Royal Hong Kong Auxiliary Air Force (two) 1986 and 1987; two Super King Air 200s operated by Swedish Air Force since 1989 as **Tp 101**. Total of 28 Super King Air B200s and one B200C delivered 1992. Total 1,533 commercial and private and 340 military (described separately) delivered to US armed forces plus those to foreign customers by 1 January 1993.
DESIGN FEATURES: Pratt & Whitney Canada PT6A-41 turboprops of Super King Air 200 replaced by 634 kW (850 shp) P&WC PT6A-42s for better cruise and altitude performance; max zero-fuel weight raised by 272 kg (600 lb); cabin pressure differential increased from 0.41 bar (6.0 lb/sq in) to 0.44 bar (6.5 lb/sq in). Wing aerofoil NACA 23018 to 23016.5 over inner wing, 23012 at tip; dihedral 6°; incidence 3° 48′ at root, −1° 7′ at tip; swept vertical and horizontal tail.
FLYING CONTROLS: Trim tabs in port aileron and both elevators; anti-servo tab in rudder; single-slotted trailing-edge flaps; fixed tailplane.
STRUCTURE: Two-spar light alloy wing; safe-life semi-monocoque fuselage.
LANDING GEAR: Hydraulically retractable tricycle type, with twin wheels on each main unit. Single wheel on steerable nose unit, with shimmy damper. Main units retract forward, nosewheel rearward. Beech oleo-pneumatic shock absorbers. Goodrich mainwheels and tyres size 18 × 5.5, pressure 7.25 bars (105 lb/sq in). Oversize and/or 10-ply mainwheel tyres optional. Goodrich nosewheel size 6.50 × 10, with tyre size 22 × 6.75-10, pressure 3.93 bars (57 lb/sq in). Goodrich hydraulic multiple-disc brakes. Parking brake.
POWER PLANT: Two 634 kW (850 shp) Pratt & Whitney Canada PT6A-42 turboprops, each driving a Hartzell three-blade constant-speed reversible-pitch metal propeller with autofeathering and synchrophasing. Bladder fuel cells in each wing, with main system capacity of 1,461 litres (386 US gallons; 321.5 Imp gallons) and auxiliary system capacity of 598 litres (158 US gallons; 131.5 Imp gallons). Total usable fuel capacity 2,059 litres (544 US gallons; 453 Imp gallons). Two refuelling points in upper surface of each wing. Wingtip tanks optional, providing an additional 401 litres (106 US gallons; 88.25 Imp gallons) and raising maximum usable capacity to 2,460 litres (650 US gallons; 541 Imp gallons). Oil capacity 29.5 litres (7.8 US gallons; 6.5 Imp gallons).
ACCOMMODATION: Pilot only, or crew of two side by side, on flight deck, with full dual controls and instruments as standard. Six cabin seats standard, each equipped with seat belts and inertia reel shoulder harness; alternative layouts for a maximum of 13 passengers in cabin and 14th beside pilot. Partition with sliding door between cabin and flight

deck, and partition at rear of cabin. Door at rear of cabin on port side, with integral airstair. Large cargo door optional. Inward opening emergency exit on starboard side over wing. Lavatory and stowage for up to 249 kg (550 lb) baggage in rear fuselage. Maintenance access door in rear fuselage; radio compartment access doors in nose. Cabin is air-conditioned and pressurised, with radiant heat panels to warm cabin before engine starting.

SYSTEMS: Cabin pressurisation by engine bleed air, with a maximum differential of 0.44 bar (6.5 lb/sq in). Cabin air-conditioner of 34,000 BTU capacity. Auxiliary cabin heating by radiant panels standard. Oxygen system for flight deck, and 0.62 m^3 (22 cu ft) oxygen system for cabin, with automatic drop-down face masks; standard system of 1.39 m^3 (49 cu ft); 1.81 m^3 (64 cu ft) or 2.15 m^3 (76 cu ft) optional. Dual vacuum system for instruments. Hydraulic system for landing gear retraction and extension, pressurised to 171-191 bars (2,475-2,775 lb/sq in). Separate hydraulic system for brakes. Electrical system has two 250A 28V starter/generators and a 24V 45Ah air-cooled nickel-cadmium battery with failure detector. AC power provided by dual 250VA inverters. Engine fire detection system standard; engine fire extinguishing system optional. Pneumatic de-icing of wings and tailplane standard. Anti-icing of engine air intakes by hot air from engine exhaust, electrothermal anti-icing for propellers.

AVIONICS: Standard Collins Pro Line II avionics and cockpit equipment generally as for King Air C90A except: dual DME-42, dual RMI-30; pilot's ALT-80A encoding altimeter; cockpit-to-cabin paging; dual maximum allowable airspeed indicators; flight director indicator; co-pilot's 24 hour clock; oxygen pressure indicator and blue-white cockpit lighting standard. Optional avionics include Collins Pro Line II with 5 in FCS; EFIS-85B(-14); Honeywell SPZ-4000 autopilot with 4 in or 5 in flight director systems or EDZ-605 three-tube EFIS; Bendix/King Gold Crown avionics packages; Bendix/King RDS-84VP colour weather radar; Fairchild 17M-700-274 flight data recorder; Fairchild A-100A cockpit voice recorder; Bendix/King KHF-950 or Collins HF-230 HF transceiver; Bendix/King CC-2024C or Collins DCP-270 radar checklist; Foster LNS-616B RNAV/Loran C; Bendix/King KNS-660 RNAV/VLF/Omega; Foster WX-1000+ Stormscope; and Wulfsberg Flitefone VII.

EQUIPMENT: Standard/optional equipment generally as for King Air C90A except fluorescent cabin lighting, one-place couch with storage drawers, flushing toilet (B200) or chemical toilet (B200C), cabin radiant heating, cockpit/cabin partition with sliding doors, and airstair door with hydraulic snubber and courtesy light, standard. FAR Pt 135 operational configuration includes cockpit fire extinguisher and 2.15 m^3 (76 cu ft) oxygen bottle as standard. A range of optional cabin seating and cabinetry configuration is available, including quick-removable fold-up seats.

DIMENSIONS, EXTERNAL:

Wing span	16.61 m (54 ft 6 in)
Wing chord: at root	2.18 m (7 ft 1¾ in)
at tip	0.90 m (2 ft 11⅝ in)
Wing aspect ratio	9.80
Length overall	13.34 m (43 ft 9 in)
Height overall	4.57 m (15 ft 0 in)
Tailplane span	5.61 m (18 ft 5 in)
Wheel track	5.23 m (17 ft 2 in)
Wheelbase	4.56 m (14 ft 11½ in)
Propeller diameter	2.50 m (8 ft 2½ in)
Propeller ground clearance	0.37 m (1 ft 2½ in)
Distance between propeller centres	5.23 m (17 ft 2 in)
Passenger door: Height	1.31 m (4 ft 3½ in)
Width	0.68 m (2 ft 2¾ in)
Height to sill	1.17 m (3 ft 10 in)
Cargo door (optional): Height	1.32 m (4 ft 4 in)
Width	1.24 m (4 ft 1 in)
Nose avionics service doors (port and stbd):	
Max height	0.57 m (1 ft 10½ in)
Width	0.63 m (2 ft 1 in)
Height to sill	1.37 m (4 ft 6 in)
Emergency exit (stbd): Height	0.66 m (2 ft 2 in)
Width	0.50 m (1 ft 7¾ in)

DIMENSIONS, INTERNAL:

Cabin (from forward to rear pressure bulkhead):	
Length	6.71 m (22 ft 0 in)
Max width	1.37 m (4 ft 6 in)
Max height	1.45 m (4 ft 9 in)
Floor area	7.80 m^2 (84 sq ft)
Volume	11.10 m^3 (392 cu ft)
Baggage hold, rear of cabin:	
Volume	1.51 m^3 (53.5 cu ft)

AREAS:

Wings, gross	28.15 m^2 (303.0 sq ft)
Ailerons (total)	1.67 m^2 (18.0 sq ft)
Trailing-edge flaps (total)	4.17 m^2 (44.9 sq ft)
Fin	3.46 m^2 (37.2 sq ft)
Rudder, incl tab	1.40 m^2 (15.1 sq ft)
Tailplane	4.52 m^2 (48.7 sq ft)
Elevators, incl tabs (total)	1.79 m^2 (19.3 sq ft)

WEIGHTS AND LOADINGS:

Weight empty	3,675 kg (8,102 lb)
Max fuel	1,653 kg (3,645 lb)
Max T-O and landing weight	5,670 kg (12,500 lb)
Max ramp weight	5,710 kg (12,590 lb)
Max zero-fuel weight	4,990 kg (11,000 lb)
Max wing loading	201.6 kg/m^2 (41.3 lb/sq ft)
Max power loading	4.47 kg/kW (7.35 lb/shp)

PERFORMANCE (at max T-O weight ISA, except where indicated):

Never-exceed speed (VNE)	259 knots (480 km/h; 298 mph) IAS
Max operating Mach No.	0.52
Max level speed at 7,620 m (25,000 ft), average cruise weight	294 knots (545 km/h; 339 mph)
Max cruising speed at 7,620 m (25,000 ft), average cruise weight	289 knots (536 km/h; 333 mph)
Econ cruising speed at 7,620 m (25,000 ft), average cruise weight, normal cruise power	282 knots (523 km/h; 325 mph)
Stalling speed:	
flaps up	99 knots (183 km/h; 114 mph) IAS
flaps down	75 knots (139 km/h; 86 mph) IAS
Max rate of climb at S/L	747 m (2,450 ft)/min
Rate of climb at S/L, OEI	226 m (740 ft)/min
Service ceiling	over 10,670 m (35,000 ft)
Service ceiling, OEI	6,675 m (21,900 ft)
T-O run, 40% flap	566 m (1,856 ft)
T-O to 15 m (50 ft), 40% flap	786 m (2,579 ft)
Landing from 15 m (50 ft):	
without propeller reversal	867 m (2,845 ft)
with propeller reversal	632 m (2,074 ft)
Landing run	536 m (1,760 ft)
Range with max fuel, allowances for start, taxi, climb, descent, and 45 min reserves at max range power, ISA:	
max cruise power at:	
5,485 m (18,000 ft)	1,190 nm (2,205 km; 1,370 miles)
8,230 m (27,000 ft)	1,550 nm (2,872 km; 1,785 miles)
9,450 m (31,000 ft)	1,750 nm (3,243 km; 2,015 miles)
10,670 m (35,000 ft)	1,965 nm (3,641 km; 2,263 miles)
econ cruise power at:	
5,485 m (18,000 ft)	1,517 nm (2,811 km; 1,747 miles)
8,230 m (27,000 ft)	1,860 nm (3,447 km; 2,142 miles)
9,450 m (31,000 ft)	1,974 nm (3,658 km; 2,273 miles)

BEECHCRAFT MARITIME PATROL B200T

TYPE: Maritime patrol or multi-mission aircraft.

PROGRAMME: Maritime Patrol 200T announced 9 April 1979; current version B200T for surface and sub-surface monitoring of exclusive economic zones, pollution detection, inspecting offshore installations, search and rescue; special missions include aerial photography, environmental and ecological research, airways and ground-based navaid checking, target towing, ambulance flying.

CUSTOMERS: Japan Maritime Safety Agency (17), Algerian Ministry of Defence (two), Peruvian Navy (five), Puerto Rico (one), Uruguayan Navy (one), Germany (one), Malaysia (two), France (two), University of Wyoming (one). No deliveries in 1992. Malaysian Air Force to take one in 1993 and three in 1994, to replace Lockheed C-130H-MPs with No. 4 Squadron at Subang.

DESIGN FEATURES: Modifications from standard Super King Air B200 include new outboard wings with provision for tip tanks, strengthened landing gear, two bubble observation windows at rear, hatch for dropping survival equipment, 360° radome under fuselage; standard avionics include VLF/Omega coupled to autopilot to allow programmed search patterns; integrated avionics, displays and controls; optional wingtip ESM antennae.

LANDING GEAR: Strengthened to cater for higher operating weights.

POWER PLANT: As for Super King Air B200, including removable wingtip tanks which increase maximum usable fuel capacity by 401 litres (106 US gallons; 88.25 Imp gallons) to a total of 2,460 litres (650 US gallons; 541 Imp gallons).

AVIONICS: Standard items as listed under Design Features. Optional avionics include ESM integrated with INS, VHF-FM com, HF and VHF com, Northrop Seehawk FLIR, LLLTV, sonobuoys and processor, OTPI, multi-spectral scanner, tactical navigation computer, and two alternative search radar systems, both with 360° scan and weather avoidance capability and integrated with INS.

DIMENSIONS, EXTERNAL: As for Super King Air B200, except:

Wing span over tip tanks	17.25 m (56 ft 7 in)
Wing aspect ratio	10.35

DIMENSIONS, INTERNAL: As for Super King Air B200, except:

Cabin: Length (excl flight deck)	5.08 m (16 ft 8 in)

WEIGHTS AND LOADINGS (N: Normal category, R: Restricted category):

Weight empty: N, R	3,744 kg (8,255 lb)
Max T-O weight: N	5,670 kg (12,500 lb)
R	6,350 kg (14,000 lb)
Max landing weight: N	5,670 kg (12,500 lb)
R	6,123 kg (13,500 lb)

PERFORMANCE (at max Normal T-O weight, except where indicated):

Max cruising speed, AUW of 4,990 kg (11,000 lb) at 4,265 m (14,000 ft)	265 knots (491 km/h; 305 mph)
Typical patrol speed	140 knots (259 km/h; 161 mph)
Range with max fuel, patrolling at 227 knots (420 km/h; 261 mph) at 825 m (2,700 ft), 45 min reserves	1,790 nm (3,317 km; 2,061 miles)
Typical endurance at 140 knots (259 km/h; 161 mph), at 610 m (2,000 ft), 45 min reserves	6 h 36 min
Max time on station, with wingtip fuel tanks	9 h

BEECHCRAFT SUPER KING AIR 200/B200 (US MILITARY VERSIONS)

US basic military designation: C-12

TYPE: Military versions of Super King Air 200/B200.

PROGRAMME: US Army procured first three Super King Airs designated RU-21Js in 1971; US Army ordered 60 military passenger-carrying Super King Airs designated C-12As between FYs 1973-77; worldwide deployment began July 1975.

CUSTOMERS: Total of 358 ordered by early 1993; 340 then delivered.

CURRENT VERSIONS: **C-12A**: Initial A200 version, powered by 559 kW (750 shp) P&WC PT6A-38 turboprops with Hartzell three-blade fully feathering reversible-pitch propellers; auxiliary tanks. Total 91 delivered (US Army 60; US Air Force 30; Greek Air Force one); entered service July 1975; details in 1980-81 *Jane's*. See C-12C and C-12E for C-12A re-engining.

UC-12B: US Navy/Marine Corps version (Model A200C) with 634 kW (850 shp) PT6A-41 turboprops, cargo door, high flotation landing gear. US Navy (49), US Marines (17), delivered by May 1982 for various base communications flights.

C-12C: As C-12A, but with PT6A-41 turboprops. Deliveries (US Army 14) complete; US Army C-12A fleet re-engined as C-12Cs; five civilianised 1989 for covert operations.

C-12D: Model A200CT. As US Army C-12C but cargo door, high flotation landing gear and provision for tip tanks. US Army (24), US Air Force (six); additional 21 built but converted to RC-12 before delivery to US Army (16) and Israel (five). Wing span (over tip tanks) 16.92 m (55 ft 6 in).

RC-12D Improved Guardrail V: Model A200CT. US Army special mission version; carries AN/USD-9 Improved Guardrail remote-controlled communications intercept and direction finding system with direct reporting to tactical commanders at corps level and below; aircraft survivability equipment (ASE) system, Carousel IV-E INS and Tacan system, radio data link, AN/ARW-83(V)5 airborne relay with antennae above and below wings; wingtip ECM pods; associated equipment includes AN/TSQ-105(V)4 integrated processing facility, AN/ARM-63(V)4 AGE flightline van and AN/TSC-87 tactical commander's terminal. System prime contractor ESL Inc. US Army had 13 RC-12D Improved Guardrail Vs converted from C-12Ds, with deliveries starting in Summer 1983; one to HQ Forces Command at Fort McPherson, Georgia, remainder to 1st Military Intelligence Battalion, Wiesbaden, Germany, and 2nd MIB at Stuttgart, Germany; German-based aircraft reassigned to US-based units in late 1991. Five further aircraft to Israel. Wing span (over ECM pods) 17.63 m (57 ft 10 in).

C-12E: Designation of 29 US Air Force C-12As retrofitted with PT6A-42 turboprops. Assigned to various embassies.

C-12F: Operational support aircraft (OSA), similar to Model B200C with PT6A-42 engines; payload choices include eight passengers, more than 1,043 kg (2,300 lb) freight, two litter patients plus attendants; cargo door standard. First delivery May 1984. US Air Force purchased 40 after initial five-year lease; US Army National Guard (26 ordered FYs 1985-87); Air National Guard (six ordered FY 1984).

UC-12F: US Navy equivalent of USAF C-12F with PT6A-42 turboprops. US Navy received first of 12 in 1986; two modified to **RC-12F** Range Surveillance Aircraft (RANSAC) for Pacific Missile Range Facilities, Barking Sands, Hawaii.

RC-12G: US Army special mission aircraft; similar to RC-12D but max T-O weight increased to 6,804 kg (15,000 lb). Mission equipment contractor Sanders Associates. Three delivered to Howard AFB, Panama CZ in 1985, after conversion from C-12D.

RC-12H Guardrail Common Sensor (Minus): US Army special mission aircraft, similar to RC-12D but max T-O weight increased to 6,804 kg (15,000 lb); system contractor ESL Inc. Six delivered in 1988 to 3rd Military Intelligence Battalion, Pyongtaek, South Korea.

C-12J: Variant of Beechcraft 1900C (see 1991–92 *Jane's*).

RC-12K Guardrail Common Sensor: Similar to RC-12D except 820 kW (1,100 shp) PT6A-67 turboprops and max T-O weight 7,257 kg (16,000 lb). US Army ordered nine in October 1985, of which eight replaced RC-12Ds in 1st MIB, May 1991; ninth retained by Beech for RC-12N conversion.

C-12L: Three RU-21Js (71-21058 to 21060) stripped of Guardrail equipment in 1979 for transport duties but not re-designated until mid-1980s.

UC-12M: US Navy designation of C-12F. Twelve delivered from 1987; two conversions to **RC-12M**

RANSAC (with additional rooftop antennae) ordered 1988 for Pacific Missile Test Center, Point Mugu.

RC-12N Guardrail Common Sensor: Generally similar to RC-12K, but equipped with dual EFIS and aircraft survivability equipment/avionics control system (ASE/ACS). ASE suite includes AN/APR-39 radar warning receiver, AN/APR-44 radar warning system, AN/ALQ-136, AN/ALQ-156 and AN/ALQ-162 countermeasure sets and M130 dispenser. Avionics suite includes AN/ARC-186 or AN/ARC-201 VHF-FM radio, AN/ARC-164 Have Quick II UHF-AM radio; AN/APX-100 IFF transponder; three KY-58 and one KIT-1A secure communications systems; Carousel IV INS; AN/ASN-149 GPS receiver. Prototype converted from ninth RC-12K; further 21 ordered (three in 1988, nine in 1989, three in 1991 and six in January 1992). Deliveries originally intended for 1992-94, but 12 now being converted for later delivery as RC-12P (which see).

RC-12P Guardrail Common Sensor: Beech is modifying late RC-12Ns to RC-12P configuration; RC-12P has identical avionics and power plant but different mission equipment, smaller, lighter wing pods and increased T-O weight of 7,484 kg (16,500 lb). First batch of seven being modified under subcontract from ESL Inc, for delivery late 1994 and 1995; second batch of five will be converted under direct contract from US Army and will be delivered between September 1995 and June 1996.

CUSTOMERS: See under individual variants above.

BEECHCRAFT SUPER KING AIR 300

Production certificate for Super King Air 300 expired 17 October 1991, due to temporary nature of SFAR 41C to which aircraft was originally certificated; 300LW (see next entry) developed for European market now sole 300 variant available to US and worldwide customers; 219 300s delivered, including four in 1991. Beech awarded $6.05 million contract by FAA early 1993 to install, in latter's 19-aircraft fleet of 300s, equipment to monitor engine and airframe wear; aircraft are used for national airways navigation calibration.

Details appeared in 1991-92 *Jane's*.

BEECHCRAFT SUPER KING AIR 300LW

TYPE: Lightweight version of Super King Air 300.
PROGRAMME: Announced September 1988 for European market; became sole 300 series aircraft on offer from October 1991, when production certificate for basic Super King Air 300 expired.
CURRENT VERSIONS: **Super King Air 300LW:** Basic version; special European certification max T-O weight of 5,670 kg (12,500 lb) to limit airways user fees; max ramp weight 5,715 kg (12,600 lb); otherwise similar to 300.
CUSTOMERS: 22 delivered by 1 January 1993, including one to domestic and five to international customers in 1992.
DESIGN FEATURES: Two PT6A-60A turboprops; 'pitot cowl' engine intakes; aerodynamically faired exhausts; wing leading-edges extended 12.7 cm (5 in) forward; propellers moved forward 13.2 cm (5.2 in); hydraulically actuated landing gear; interior equipment changes.
LANDING GEAR: Hydraulically retractable tricycle type. Goodrich mainwheels and tyres size 19 × 6.75-8, pressure 6.20 bars (90 lb/sq in) at max T-O weight. Goodrich nosewheel and tyre size 22 × 6.75-10, pressure 3.79-4.13 bars (55-60 lb/sq in).
POWER PLANT: Two 783 kW (1,050 shp) Pratt & Whitney Canada PT6A-60A turboprops, each driving a Hartzell four-blade constant-speed fully feathering reversible-pitch metal propeller. Bladder cells and integral tanks in each wing, with total capacity of 1,438 litres (380 US gallons; 316.5 Imp gallons); auxiliary tanks inboard of engine nacelles, capacity 601 litres (159 US gallons; 132.5 Imp gallons). Total fuel capacity 2,039 litres (539 US gallons; 449 Imp gallons). No provision for wingtip tanks. Oil capacity 30.2 litres (8 US gallons; 6.66 Imp gallons).
ACCOMMODATION: As for B200, except for additional emergency exit on port side of cabin, opposite starboard emergency exit and of the same dimensions. Pilot and co-pilot storm windows standard. Cabin features single-piece upper sidewall panels, indirect overhead lighting system with rheostat controls, stereo system with graphic equaliser and overhead speakers, larger executive tables incorporating magnetic game boards, seats with inflatable lumbar support adjustment, fore-and-aft, reclining and lateral tracking movement as standard. Crew seats have 2.5° or 5° tilt positions. Emergency exit lighting standard. Electric heating on ground standard. Optional radiant heat panels of B200 not available.
SYSTEMS: As for B200, except for automatic bleed air type heating and 22,000 BTU cooling system with high capacity ventilation system; 2.18 m³ (77 cu ft) oxygen system standard; hydraulic landing gear retraction and extension system; two 300A 28V starter/generators with triple bus electrical distribution system.
AVIONICS: Generally as for B200.
EQUIPMENT: Generally as for B200.
DIMENSIONS, EXTERNAL: As for B200 except:
Length overall 13.36 m (43 ft 10 in)
Height overall 4.37 m (14 ft 4 in)
Propeller diameter 2.67 m (8 ft 9 in)

Beechcraft RC-12N Guardrail Common Sensor of the US Army

Beechcraft C-12F operational support aircraft of US Air Force (*Peter J. Cooper*)

Propeller ground clearance 0.25 m (10 in)
Emergency exit (each side of cabin, above wing):
Height 0.66 m (2 ft 2 in)
Width 0.95 m (1 ft 7¾ in)
WEIGHTS AND LOADINGS:
Weight empty 3,878 kg (8,550 lb)
Max baggage weight 249 kg (550 lb)
Max T-O and landing weight 5,670 kg (12,500 lb)
Max ramp weight 5,715 kg (12,600 lb)
Max zero-fuel weight 5,216 kg (11,500 lb)
Max wing loading 201.6 kg/m² (41.3 lb/sq ft)
Max power loading 3.62 kg/kW (5.95 lb/shp)
PERFORMANCE:
Never-exceed speed (V_{NE})
 259 knots (480 km/h; 298 mph) IAS
Max operating Mach No. (M_{MO}) 0.58
Max level and cruising speed
 317 knots (587 km/h; 365 mph)
Econ cruising speed 309 knots (573 km/h; 356 mph)
Stalling speed: flaps up 95 knots (176 km/h; 110 mph)
 flaps down 78 knots (145 km/h; 90 mph)
Max rate of climb at S/L 999 m (3,277 ft)/min
Rate of climb at S/L, OEI 327 m (1,074 ft)/min
Max certificated ceiling 10,670 m (35,000 ft)
Service ceiling, OEI 7,882 m (25,855 ft)
T-O run, 40% flap, at T-O weight of 5,670 kg (12,500 lb)
 421 m (1,381 ft)
T-O to 15 m (50 ft), 40% flap, at T-O weight of 5,670 kg (12,500 lb) 514 m (1,686 ft)
Accelerate/stop distance, 40% flap 1,067 m (3,500 ft)
Landing from 15 m (50 ft) 784 m (2,570 ft)
Landing run, without propeller reversal 393 m (1,289 ft)
Range with max fuel, allowances for start, taxi, T-O, climb, descent and 45 min reserves at max range power:
max cruise power at:
5,485 m (18,000 ft)
 960 nm (1,779 km; 1,105 miles)
7,315 m (24,000 ft)
 1,156 nm (2,142 km; 1,331 miles)
8,535 m (28,000 ft)
 1,333 nm (2,470 km; 1,535 miles)
10,670 m (35,000 ft)
 1,744 nm (3,232 km; 2,008 miles)

max range power at:
5,485 m (18,000 ft)
 1,335 nm (2,474 km; 1,537 miles)
8,535 m (28,000 ft)
 1,780 nm (3,298 km; 2,049 miles)
10,670 m (35,000 ft)
 2,086 nm (3,865 km; 2,402 miles)

BEECHCRAFT SUPER KING AIR 350

TYPE: Eight/12-passenger turboprop business aircraft.
PROGRAMME: Replaced Super King Air 300 (1991-92 *Jane's*); first flight (N120SK) September 1988; introduced at NBAA Convention 1989; certificated to FAR Pt 23 (commuter category); first delivery 6 March 1990.
CURRENT VERSIONS: **Super King Air 350:** Basic version, *as described in detail*.
Super King Air 350C: Has 132 × 132 cm (52 × 52 in) freight door with built-in airstair passenger door.
RC-350 Guardian: Elint version, converted from 350 prototype 1991 by Beech Aircraft Corporation; mission avionics include Raytheon AN/ALQ-142 ESM, Watkins-Johnson 9195C communications interceptor, Honeywell laser INS, GPS receiver and Cubic secure digital datalink; can loiter on station at 10,670 m (35,000 ft) for more than 6 h; can locate/monitor radar emitters in 20MHz to 18GHz range, and intercept communications within 20-1,400MHz bandwidths. Wingtip pods house AN/ALQ-142 antennae, underfuselage bulge contains antenna for comint system.
CUSTOMERS: Total 88 Super King Air 350s and 350Cs delivered by 1 January 1993; first 350C delivery in 1990 to Rossing Uranium, Namibia; 12 domestic and one international Super King Air 350 deliveries in 1992, one international 350C in same period.
DESIGN FEATURES: Compared with Super King Air 300, fuselage stretched 0.86 m (2 ft 10 in) by plugs 0.37 m (1 ft 2½ in) forward of main spar and 0.49 m (1 ft 7½ in) aft; wing span increased by 0.46 m (1 ft 6 in) with NASA winglets 0.61 m (2 ft 0 in) high; two additional cabin windows each side; double club seating for eight passengers; optionally two more seats in rear of cabin, one passenger on toilet seat and one in co-pilot seat if operating single-crew, making maximum 12 passengers. Can depart with full payload and full tanks.

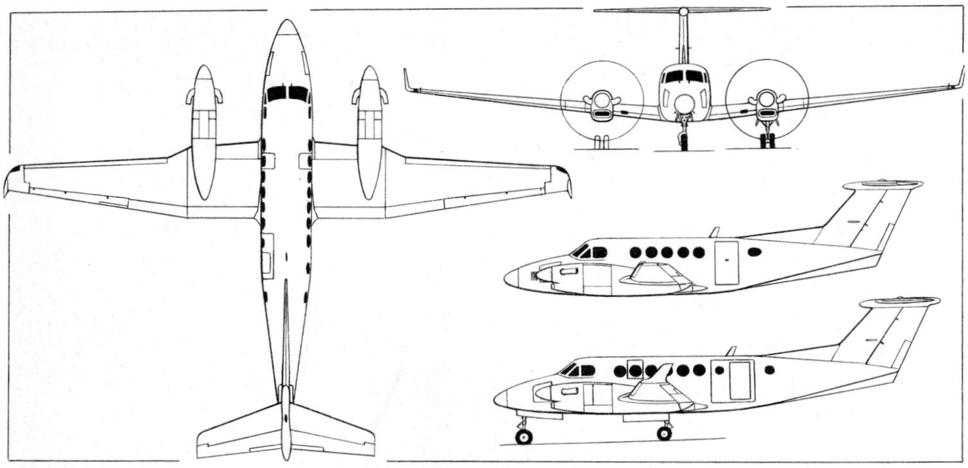

Beechcraft Super King Air 350, with additional side view (top) of Super King Air 300LW *(Dennis Punnett)*

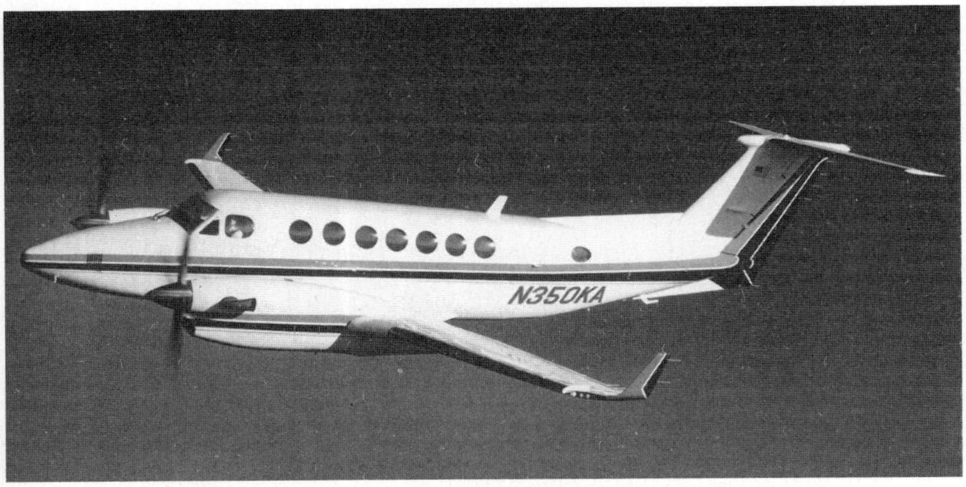

Beechcraft Super King Air 350 eight/12-passenger turboprop business aircraft

FLYING CONTROLS: Automatic cable tensioner in aileron circuit and larger elevator bobweight; larger rudder anti-servo tab; ailerons and rudder cleaned up.

POWER PLANT: As for 300LW.

DIMENSIONS, EXTERNAL: As for 300LW except:

Wing span over winglets	17.65 m (57 ft 11 in)
Wing aspect ratio	10.79
Length overall	14.22 m (46 ft 8 in)

DIMENSIONS, INTERNAL:

Cabin, excl cockpit: Length	5.94 m (19 ft 6 in)
Max width	1.37 m (4 ft 6 in)
Height	1.45 m (4 ft 9 in)

AREAS:

Wings, gross	28.80 m² (310.0 sq ft)

WEIGHTS AND LOADINGS:

Weight empty	4,096 kg (9,030 lb)
Max fuel weight	1,638 kg (3,611 lb)
Max T-O and landing weight	6,804 kg (15,000 lb)
Max ramp weight	6,849 kg (15,100 lb)
Max zero-fuel weight	5,670 kg (12,500 lb)
Max wing loading	236.3 kg/m² (48.4 lb/sq ft)
Max power loading	4.34 kg/kW (7.14 lb/shp)

PERFORMANCE:

Max level speed 315 knots (584 km/h; 363 mph)

Max cruising speed, AUW of 5,896 kg (13,000 lb) at:
- 7,315 m (24,000 ft) 311 knots (576 km/h; 358 mph)
- 10,670 m (35,000 ft) 290 knots (537 km/h; 334 mph)

Cruising speed, normal cruising power, AUW of 5,896 kg (13,000 lb) at:
- 7,315 m (24,000 ft) 301 knots (558 km/h; 347 mph)
- 10,670 m (35,000 ft) 281 knots (521 km/h; 324 mph)

Cruising speed, max range power, AUW of 5,896 kg (13,000 lb) at:
- 5,485 m (18,000 ft) 210 knots (389 km/h; 242 mph)
- 10,670 m (35,000 ft) 240 knots (445 km/h; 276 mph)

Stalling speed at max landing weight, flaps and wheels down 81 knots (150 km/h; 94 mph)

Max rate of climb at S/L, AUW of 6,804 kg (15,000 lb) 832 m (2,731 ft)/min

Rate of climb at S/L, OEI, AUW of 6,350 kg (14,000 lb) 236 m (775 ft)/min

Service ceiling above 10,670 m (35,000 ft)

Service ceiling, OEI, AUW of 6,804 kg (15,000 lb) 6,555 m (21,500 ft)

T-O balanced field length, AUW of 6,804 kg (15,000 lb) 1,139 m (3,737 ft)

Landing from 15 m (50 ft) at AUW of 6,804 kg (15,000 lb) 802 m (2,631 ft)

Landing run 408 m (1,338 ft)

Range with 2,040 litres (539 US gallons; 449 Imp gallons) usable fuel, allowances for start, T-O, climb and descent plus 45 min reserves:

max cruising power at:
- 5,485 m (18,000 ft) 1,067 nm (1,977 km; 1,228 miles)
- 7,315 m (24,000 ft) 1,252 nm (2,320 km; 1,441 miles)
- 8,535 m (28,000 ft) 1,407 nm (2,607 km; 1,620 miles)
- 10,670 m (35,000 ft) 1,724 nm (3,195 km; 1,985 miles)

normal cruising power, allowances as above:
- 5,485 m (18,000 ft) 1,085 nm (2,010 km; 1,249 miles)
- 7,315 m (24,000 ft) 1,308 nm (2,424 km; 1,506 miles)
- 8,535 m (28,000 ft) 1,474 nm (2,731 km; 1,697 miles)
- 10,670 m (35,000 ft) 1,771 nm (3,282 km; 2,039 miles)

max range power, allowances as above:
- 5,485 m (18,000 ft) 1,421 nm (2,633 km; 1,636 miles)
- 8,535 m (28,000 ft) 1,756 nm (3,254 km; 2,022 miles)
- 10,670 m (35,000 ft) 1,894 nm (3,510 km; 2,181 miles)

NBAA VFR range, 4 passengers, 30 min reserves 2,031 nm (3,763 km; 2,338 miles)

BEECHCRAFT 1900C AIRLINER and 1900 EXEC-LINER

No longer manufactured; 255 delivered from 1984. Two elint-configured 1900C-1s delivered to Egyptian Air Force in September 1992, joining earlier examples. See 1991-92 *Jane's* for details. Replaced by 1900D.

BEECHCRAFT 1900D

TYPE: Development of 1900C (1991-92 *Jane's*).

PROGRAMME: Announced at US Regional Airlines Association meeting 1989; prototype (N5584B) first flight 1 March 1990; certification to FAR Pt 23 Amendment 34 received March 1991; full certification with supplements received, and deliveries (to Mesa Airlines) began, November 1991; planned production four per month; replaced earlier 1900C in current product line.

CUSTOMERS: Launch customer United Express partner Mesa Airlines (58 ordered originally, followed by further 20 in December 1992 for 1993-95 delivery); Autec Range Services (two); Mark Air Express (10); Commutair (26). Total sales in 1992, 26.

DESIGN FEATURES: Flat floor with stand-up headroom; cabin volume increased by 28.5 per cent compared to 1900C; winglets add better hot and high performance; tailplane and fin swept; each tailplane carries small ventral fin (tailet) near tip; auxiliary horizontal fixed tail surface (stabilon) each side of rear fuselage improve centre of gravity range; twin ventral strakes improve directional stability and turbulence penetration; small horizontal vortex generator on fuselage ahead of wingroots.

Wing aerofoil NACA 23018 (modified) at root, 23012 (modified) at tip; dihedral 6°; incidence 3° 29′ at root, −1° 4′ at tip; no sweepback at quarter-chord.

FLYING CONTROLS: Mechanical, with automatic cable tensioner in aileron circuit; trim tabs in elevators, rudder and port aileron; primary and secondary controls routed separately to improve rotor burst protection; single-slotted trailing-edge flaps in two sections on each wing.

STRUCTURE: Wing has continuous main spar with fail-safe structure riveted and bonded; fuselage pressurised and mainly bonded.

LANDING GEAR: Hydraulically retractable tricycle type; main units retract forward and nose unit rearward; Beech oleo-pneumatic shock absorber in each unit. Twin Goodyear wheels on each main unit, size 6.50 × 10, with Goodyear tyres size 22 × 6.75-10 (tubeless 10-ply rating), pressure 6.07 bars (88 lb/sq in); Goodyear steerable nosewheel size 6.60 × 8, with Goodyear tyres size 22 × 6.75-8 (tubeless 10-ply rating), pressure 6.07 bars (88 lb/sq in). Multiple-disc hydraulic brakes. Optional Beech Hydro-Aire anti-skid units, power steering and brake de-icing. Ground turning radius based on wingtip clearance 12.55 m (41 ft 2 in).

POWER PLANT: Two Pratt & Whitney Canada PT6A-67D turboprops, each flat rated at 954 kW (1,279 shp), driving a Hartzell four-blade constant-speed fully feathering reversible-pitch composite propeller. Wet wing fuel storage with a total capacity of 2,528 litres (668 US gallons; 556 Imp gallons), of which 2,519 litres (665 US gallons; 554 Imp gallons) usable. Refuelling point in each wing leading-edge, inboard of engine nacelle. Oil capacity (total) 29.5 litres (7.8 US gallons; 6.5 Imp gallons).

ACCOMMODATION: Crew of one (FAR Pt 91) or two (FAR Pt 135) on flight deck, with standard accommodation in cabin of commuter version for 19 passengers in single airline standard seats on each side of centre aisle. Forward carry-on baggage lockers, underseat baggage stowage, rear baggage compartment. Forward and rear doors, incorporating airstairs, on port side. Upward hinged cargo door. Three emergency exits over wing (two on starboard side plus one on port side). Accommodation air-conditioned, heated, ventilated and pressurised. Executive version has 10/18-passenger cabin with forward and rear compartments, combination lavatory/passenger seat and two beverage bars at cabin compartment division. Club, double club and triple club seating optional. Customised interior to customer's choice.

SYSTEMS: Bleed air cabin heating and pressurisation, max differential 0.35 bar (5.1 lb/sq in). Air cycle and vapour cycle air-conditioning. Hydraulic system, pressure 207 bars (3,000 lb/sq in), for landing gear actuation. Electrical system includes two 300A engine driven starter/generators and one 22Ah nickel-cadmium battery. Constant flow oxygen system of 4.3 m³ (152 cu ft) capacity standard. Engine inlet screen anti-ice protection, exhaust heated engine inlet lips, fuel vent heating, electric propeller and windscreen de-icing systems standard. Brake de-icing optional. Pneumatic de-icing boots on wings, tailplane, tailets and stabilons.

AVIONICS: Four tube Collins EFIS-84, Collins Pro Line II digital technology radios, dual flight directors, cabin briefer, cockpit voice recorder, flight data recorder and identical pilot/co-pilot panels standard. Primary display system consists of multi-colour CRT displays, remote display processor unit and system control units. CRT displays provide conventional electronic attitude director indicator (EADI) and electronic horizontal situation indicator (EHSI) functions. Dual Loran optional.

DIMENSIONS, EXTERNAL:

Wing span over winglets	17.67 m (57 ft 11⅘ in)
Wing chord: at root	2.18 m (7 ft 1⅚ in)
at tip	0.91 m (2 ft 11⅚ in)
Wing aspect ratio	10.8
Length overall	17.63 m (57 ft 10 in)
Height overall	4.57 m (14 ft 11⅚ in)
Tailplane span	5.63 m (18 ft 5⅘ in)
Wheel track	5.23 m (17 ft 2 in)
Wheelbase	7.25 m (23 ft 9⅖ in)
Propeller diameter	2.78 m (9 ft 1½ in)
Propeller ground clearance	0.35 m (1 ft 1⅚ in)
Distance between propeller centres	5.23 m (17 ft 2 in)
Passenger door: Height	1.63 m (5 ft 4⅕ in)
Width	0.64 m (2 ft 1 3/10 in)
Cargo door: Height	1.45 m (4 ft 9 in)
Width	1.32 m (4 ft 4 in)
Emergency exits (each): Height	0.80 m (2 ft 7⅖ in)
Width	0.51 m (1 ft 8 in)

DIMENSIONS, INTERNAL:
  Cabin, incl flight deck and rear baggage compartment:
    Length   12.03 m (39 ft 5½ in)
    Max width   1.37 m (4 ft 6 in)
    Max height   1.80 m (5 ft 11 in)
    Floor area   15.28 m² (164.5 sq ft)
    Pressurised volume   26.0 m³ (918 cu ft)
  Volume of passenger cabin   18.12 m³ (640 cu ft)
  Baggage space, cabin: forward   0.48 m³ (17.0 cu ft)
                     underseat   0.91 m³ (32.0 cu ft)
                     rear   4.95 m³ (175 cu ft)
AREAS:
  Wings, gross   28.80 m² (310.0 sq ft)
  Ailerons (total)   1.67 m² (18.0 sq ft)
  Trailing-edge flaps (total)   4.17 m² (44.9 sq ft)
  Fin   4.86 m² (52.3 sq ft)
  Rudder (incl tab)   1.40 m² (15.1 sq ft)
  Tailets (total)   0.63 m² (6.8 sq ft)
  Tailplane   6.32 m² (68.0 sq ft)
  Elevator (incl tab)   1.79 m² (19.3 sq ft)
  Stabilons (total)   1.44 m² (15.5 sq ft)
WEIGHTS AND LOADINGS:
  Weight empty (typical)   4,785 kg (10,550 lb)
  Max fuel (usable)   2,022 kg (4,458 lb)
  Max baggage   939 kg (2,070 lb)
  Max ramp weight   7,738 kg (17,060 lb)
  Max T-O weight   7,688 kg (16,950 lb)
  Max landing weight   7,303 kg (16,100 lb)
  Max zero-fuel weight   6,804 kg (15,000 lb)
  Max wing loading   266.9 kg/m² (54.68 lb/sq ft)
  Max power loading   4.03 kg/kW (6.63 lb/shp)
PERFORMANCE (at max T-O weight except where indicated):
  Max cruising speed at AUW of 6,804 kg (15,000 lb):
    at 2,440 m (8,000 ft)   276 knots (511 km/h; 318 mph)
    at 4,875 m (16,000 ft)   288 knots (533 km/h; 331 mph)
    at 7,620 m (25,000 ft)   278 knots (515 km/h; 320 mph)
  T-O speed, T-O flap setting
                             105 knots (195 km/h; 121 mph) IAS
  Approach speed at max landing weight
                             116 knots (215 km/h; 133 mph)
  Stalling speed at max T-O weight:
    wheels and flaps up   101 knots (187 km/h; 116 mph)
    wheels down, T-O flap setting
                             90 knots (167 km/h; 104 mph)
  Stalling speed at max landing weight, wheels and flaps
    down   84 knots (156 km/h; 97 mph)
  Max rate of climb at S/L   800 m (2,625 ft)/min
  Rate of climb at S/L, OEI   206 m (675 ft)/min
  Service ceiling   10,058 m (33,000 ft)
  Service ceiling, OEI   5,181 m (17,000 ft)
  T-O field length, T-O flap setting   1,139 m (3,737 ft)
  Landing from 15 m (50 ft) at max landing weight
                                   823 m (2,700 ft)
  Range with 10 passengers, at long range cruise power, with
    allowances for starting, taxi, T-O, climb and descent,
    with reserves (45 min hold at 1,525 m; 5,000 ft)
                                   1,498 nm (2,776 km; 1,725 miles)

Beechcraft 1900D regional transport of USAir's Commutair feeder airline

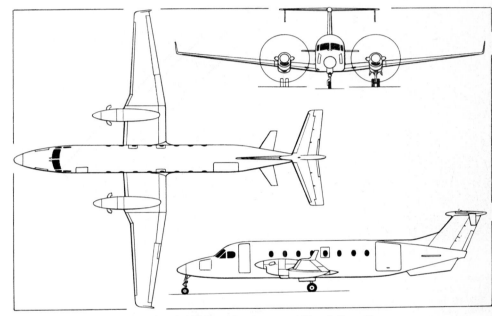

Beechcraft 1900D regional transport *(Dennis Punnett)*

BEECHCRAFT 2000 STARSHIP 1

TYPE: Eight/nine-passenger business turboprop.
PROGRAMME: Scaled Composites Inc (which see) flew first 85 per cent scale proof-of-concept demonstrator 29 August 1983; first flight of full-scale prototype (N2000S) at Wichita 15 February 1986, powered by Pratt & Whitney Canada PT6A-65A-4 engines while awaiting PT6A-67As; first flight second prototype (N3042S) 14 June 1986, fitted with Collins advanced integrated avionics intended for production aircraft; first flight of third prototype (N3234S; c/n NC-3) 5 January 1987, with furnished cabin and intended for function and reliability testing. N3234S appeared at Paris Air Show 1987; three other prototype airframes used for static, damage tolerance and pressure cycle testing.
  Basic certification granted 14 June 1988; full certification for two-crew operation gained December 1989, followed by FAA crew determination study; single-pilot certification requiring functioning autopilot and flight management system granted May 1990; first flight of first full production aircraft (N2000S; c/n NC-4) 25 April 1989; used for demonstration flights. Starship 2000A version announced at NBAA Convention October 1991; certificated by FAA April 1992.
CURRENT VERSIONS: **Starship 2000**: Initial production model.
  Starship 2000A: Improved version; new six-passenger interior with increased passenger/baggage space and roomier aft toilet; increases of 92 kg (202 lb) in max fuel capacity, 222 kg (490 lb) in max payload and 181 kg (400 lb) in max T-O weight; stall speed and T-O field length reduced; max range increased to 1,576 nm (2,920 km; 1,814 miles). First 2000A with public transport Part 135 certification operating from Las Vegas, Nevada, early 1993. *Detailed description applies to this version.*
CUSTOMERS: 20 in use by customers, including one in Europe, by January 1993; 14 delivered in 1992 (seven to domestic and seven to international customers).
DESIGN FEATURES: Pusher turboprops with propellers aft of wing; integral fuel tanks are separate structures forming root extensions bolted to main wing; almost cylindrical cabin section mounted far forward of engines and propellers to reduce cabin noise; engines mounted close together to reduce yaw moment under asymmetric thrust; wing has specially developed aerofoil; dihedral 1° 18′ 36″; incidence 2° at root; sweepback 24° 24′ at quarter-chord.
FLYING CONTROLS: Pitch axis control by elevators on foreplane and elevons on main wing; directional control and yaw stabilisation through wingtip fins, called tipsails, and rudders; electrically actuated foreplane moved from 30° cruise sweep setting to 4° forward sweep as centre of lift moves when wing-mounted Fowler trailing-edge flaps are extended; vortex generators ahead of elevators energise flow over control surface when foreplane is swept. Trim tabs in elevators, rudders and elevons (for lateral trim); four vortillons under leading-edge of each outer wing and trailing-edge fences between flaps and elevons control airflow near stall; fixed ventral fin also acts as tail bumper.
STRUCTURE: Wing is continuous tip to tip structure of Nomex honeycomb and graphite epoxy monocoque, semi-monocoque and honeycomb sandwich with spars bonded to upper and lower skin assemblies; aluminium used in such points as integrally machined landing gear attachments and tipsail mountings; fuselage structure similar to wings, manually laid-up and autoclaved; Bell Aerospace manufactures composites foreplanes.
LANDING GEAR: Retractable tricycle type, hydraulically operated with emergency extension. Main units retract inward, nose unit forward. Beech oleo-pneumatic shock absorbers. Twin Goodyear mainwheels with tyres size 19.5 × 6.75-10, 8-ply rated, pressure 6.55 bars (95 lb/sq in). Single Goodrich nosewheel with tyre size 19.55 × 6.50-8, 10-ply rated, pressure 4.48 bars (65 lb/sq in). Goodyear multi-disc antiskid brakes with carbon heat sink.
POWER PLANT: Two Pratt & Whitney Canada PT6A-67A turboprops, each flat rated at 895 kW (1,200 shp) and driving a McCauley five-blade fully feathering and reversible-pitch metal pusher propeller. Fuel, total usable capacity 2,139 litres (565 US gallons; 470 Imp gallons), contained in integral wing tanks with flush refuelling point in upper surface of each wing.
ACCOMMODATION: Certificated for single pilot operation, but provision for two crew with dual controls on four-way adjustable reclining seats on flight deck, separated from cabin by bulkhead with door. Standard seating for eight passengers in NC-4 to NC-20 production aircraft only, six in double club arrangement with centre aisle in reclining, lateral tracking, and fore and aft adjustable seats with retractable inboard armrests, and two on forward facing couch (not fitted from NC-21) at rear of cabin, each with shoulder harness, lap belt, adjustable headliner-mounted reading light and fresh air outlet, cup holder, ashtray. Four folding work tables with vinyl tops at forward and rear club seating positions. From NC-21, six-passenger cabin, with seventh passenger in co-pilot's position and possible eighth on bolted toilet seat. Fail-safe dual pane cabin windows with electrically powered Polaroid dimming. Forward baggage compartment with flushing toilet, relief tube and cabin privacy door. Rear baggage compartment, accessible in flight via door in rear partition. Cabin is pressurised, bleed air heated, with vapour cycle cooling and high capacity ventilation systems. Single airstair door at forward end of cabin on port side, with courtesy light. Emergency exit at rear end of cabin on starboard side, over wing.
SYSTEMS: Pressurisation system with max differential of 0.58 bar (8.4 lb/sq in) to provide a cabin altitude of 2,440 m (8,000 ft) at 12,500 m (41,000 ft). Freon vapour cycle cooling system. Engine bleed air provides pressurisation, heating and ventilation. 28V DC three-bus electrical system supplied by single air-cooled 34Ah battery and 300A 28V starter/generator mounted on each engine and connected in parallel. Oxygen cylinder capacity 2.18 m³ (77 cu ft) rated at 124 bars (1,800 lb/sq in), mounted in nose, provides passenger oxygen supply automatically via drop-down masks until cabin altitude reaches 4,115 m (13,500 ft). Quick-donning masks for crew. B. F. Goodrich fully automatic self-initiating ice detection and de-icing system with Silver Estane pneumatic boots on wing and foreplane leading-edges, and anti-icing systems for windscreen, engine air inlets, fuel vents, pitot static probes and stall warning sensor.
AVIONICS: Collins integrated avionics package comprising 12 colour and two monochrome CRT displays in 'all glass' cockpit. Pilot and co-pilot have duplicated instrument panels, each with two 15.2 × 17.8 cm (6 × 7 in) EFIS displays for primary flight and navigation functions and two 10.2 ×

First production Beechcraft Starship 1 all-composite twin-turboprop business aircraft

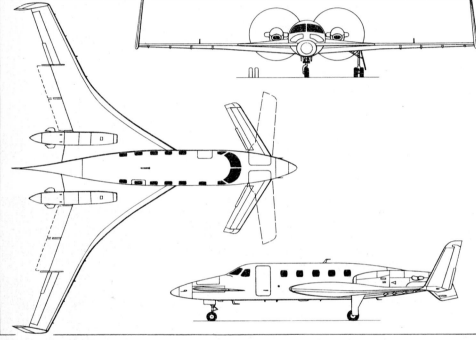

Beechcraft Starship 1 eight/nine-passenger business aircraft (Dennis Punnett)

10.2 cm (4 × 4 in) airspeed indicator and altitude/vertical speed indicator CRTs. Monochrome sensor display units (SDUs) provide heading and VOR, FMS and ADF bearings, and serve as secondary nav display. Dual control/display units (CDUs) control EFIS, weather radar, navigation radios and flight management functions. Engine indication caution and advisory system (EICAS) provides nearly 100 specific pieces of information in analog or digital form on a 15.2 × 17.8 cm (6 × 7 in) colour CRT display, with a priority message system to override extraneous information. Dual multi-function displays (MFDs) provide weather radar images from Collins TWR-850 Doppler turbulence weather radar system, maps, checklists and diagnostic and maintenance data, and serve as backup to EICAS. Two radio tuning units provide gas discharge tube alphanumerics for displaying navigation and transponder frequencies and codes, and can be used for display of engine parameters if EICAS fails. Dual flight management system (FMS) keyboards control all navigation frequencies, selected from onboard microdisc storage which is updated every 28 days. Standard avionics also include dual Collins com and nav receivers, dual transponders, ADF, DME, radio altimeter, dual compass system with strapdown attitude/heading reference system, and dual audio system with pilot/co-pilot interphone, aural warnings to flight deck, cabin speaker system, and emergency locator transmitter. Dual altitude awareness panels, a course heading panel, reversionary switching panels and standby electromagnetic airspeed indicator, gyro horizon and altimeter are provided. Information from sensors and data acquisition units located throughout the aircraft is available to all instruments through an ARINC 429 digital databus system. Teledyne stall warning system. Stick pusher limits max attainable angle of attack.

EQUIPMENT: Standard cockpit equipment includes dual audio speakers, dual hand-held microphones and crew headsets, avionics master switch, primary and secondary instrument lighting systems, electrically heated windscreen and cockpit side windows, hot air windscreen defroster, oxygen outlets and console mounted pressure/diluter demand crew oxygen masks with microphones, pilot and co-pilot map cases, and lighted control wheel approach plate holder. Upright refreshment cabinet with heated liquid container, cup dispenser, ice chest, four decanters, waste container and overboard drain on forward port side; low pyramid bar with ice and general stowage between each double club arrangement. Centre aisle stowage container. Low nap wool carpet, vinyl carpet protector, and No smoking and Fasten seat belt signs with audible chime, standard. Other equipment includes nosewheel bay microphone/earphone jack plug and ground clearance switch linked to com 1, external power receptacle, external oxygen filler ports and pressure gauge, dual heated pitot and static heads, heated stall warning system, wing ice-detection lights, dual wing landing lights, nosewheel-mounted taxi and landing lights, position lights, selectable high- and low-intensity anti-collision strobe lights, cabin fire extinguisher, towbar, pitot tube covers, static wicks, and control locks.

DIMENSIONS, EXTERNAL:
Wing span (reference)	16.60 m (54 ft 4¾ in)
Wing aspect ratio	10.53
Winglet height, each	2.45 m (8 ft 0½ in)
Foreplane span: sweptforward	7.82 m (25 ft 8 in)
sweptback	6.69 m (21 ft 11½ in)
Length overall	14.05 m (46 ft 1 in)
Fuselage: Length	13.67 m (44 ft 10 in)
Diameter (constant section)	1.78 m (5 ft 10 in)
Height overall	3.94 m (12 ft 11 in)
Wheel track	5.13 m (16 ft 10 in)
Wheelbase	6.86 m (22 ft 6 in)
Propeller diameter	2.64 m (8 ft 8 in)
Propeller ground clearance	0.84 m (2 ft 9 in)
Distance between propeller centres	3.07 m (10 ft 1 in)
Passenger door: Height	1.28 m (4 ft 2½ in)
Width	0.71 m (2 ft 4 in)
Emergency exit: Height	0.56 m (1 ft 10 in)
Width	0.66 m (2 ft 2 in)

DIMENSIONS, INTERNAL:
Cabin, excl flight deck: Length	5.08 m (16 ft 8 in)
Max width	1.68 m (5 ft 6 in)
Max height	1.61 m (5 ft 3½ in)
Floor area	5.94 m² (64.0 sq ft)
Volume (between pressure bulkheads)	13.08 m³ (462 cu ft)
Baggage holds: forward	0.55 m³ (19.5 cu ft)
rear	0.99 m³ (35.0 cu ft)

AREAS:
Wings, gross	26.09 m² (280.9 sq ft)
Elevons (total)	1.59 m² (17.1 sq ft)
Trailing-edge flaps (total)	4.78 m² (51.5 sq ft)
Ventral fin	1.25 m² (13.5 sq ft)
Foreplane (forward position)	5.67 m² (61.0 sq ft)
Elevators (total)	1.01 m² (10.9 sq ft)
Winglets (total)	4.92 m² (53.0 sq ft)
Rudders (total, incl tabs)	1.04 m² (11.2 sq ft)

WEIGHTS AND LOADINGS:
Weight empty, equipped	4,574 kg (10,085 lb)
Max baggage weight	311 kg (685 lb)
Max payload	1,141 kg (2,515 lb)
Max fuel weight	1,717 kg (3,786 lb)
Max T-O weight	6,758 kg (14,900 lb)
Max ramp weight	6,808 kg (15,010 lb)
Max zero-fuel weight	5,715 kg (12,600 lb)
Max landing weight	6,205 kg (13,680 lb)
Max wing loading	259.0 kg/m² (53.0 lb/sq ft)
Max power loading	3.78 kg/kW (6.21 lb/shp)

PERFORMANCE (at max T-O weight, ISA, except where indicated):
Max limiting Mach number	0.60
Max cruising speed:	
at 7,620 m (25,000 ft)	335 knots (621 km/h; 386 mph)
at 10,670 m (35,000 ft)	329 knots (610 km/h; 379 mph)
Econ cruising speed at 10,670 m (35,000 ft)	307 knots (569 km/h; 354 mph)
Stalling speed: flaps up	97 knots (180 km/h; 112 mph)
flaps down	92 knots (171 km/h; 106 mph)
Max rate of climb at S/L	838 m (2,748 ft)/min
Rate of climb at S/L, OEI	204 m (670 ft)/min
Max certificated altitude	12,500 m (41,000 ft)
Service ceiling	10,910 m (35,800 ft)
Service ceiling, OEI	5,485 m (18,000 ft)
T-O to 11 m (35 ft)	1,175 m (3,854 ft)
Landing from 15 m (50 ft)	729 m (2,390 ft)

Range at 10,670 m (35,000 ft), max usable fuel, with reserves:
at max cruise power	1,494 nm (2,768 km; 1,720 miles)
at econ cruise power	1,514 nm (2,805 km; 1,743 miles)
at max range power	1,576 nm (2,920 km; 1,814 miles)

BEECHCRAFT 400A BEECHJET

US Air Force designation: T-1A Jayhawk
JASDF designation: T-400

TYPE: Twin-turbofan business aircraft and military trainer.
PROGRAMME: Beech acquired rights to Diamond II from Mitsubishi Heavy Industries and Mitsubishi Aircraft International, December 1985; made improvements to aircraft and renamed it Beechjet 400. First Beech assembled Beechjet rolled out 19 May 1986. During 1989, Beech moved entire manufacturing operation to Wichita. Announced new Beechjet 400A November 1989, featuring certification to 13,715 m (45,000 ft), larger and more comfortable cabin, all Collins avionics with digital EFIS; customer deliveries began November 1990.

CURRENT VERSIONS: **Beechjet 400:** Initial production version (see earlier *Jane's*); superseded by 400A.

Beechjet 400A: Announced at 1989 NBAA show; production 400A first flight 22 September 1989; FAA certification received 20 June 1990; deliveries began November 1990. Also certificated by May 1993 in Australia, Canada, Germany, Italy and UK; French type approval expected later same year.

Beechjet 400T/T-1A Jayhawk: US Air Force selected McDonnell Douglas, Beechcraft and Quintron to supply Tanker Transport Training System (TTTS) on 21 February 1990, including original requirement for 211 Beechjet 400Ts (now expected to be 180) designated T-1A Jayhawks; represents missionised version of 400A, sharing many components and characteristics with commercial counterpart; differences include cabin-mounted avionics, increased air-conditioning capability, increased fuel capacity with single-point refuelling, and strengthened windscreen and leading-edges for low-level birdstrike protection. First aircraft (90-0400) delivered 17 January 1992; deliveries scheduled at approx three per month until 1997; Beech Plant IV at Wichita extended by 9,290 m² (100,000 sq ft) 1991 for Jayhawk production.

IOC for USAF Jayhawks January 1993, for Air Training Command Specialised Undergraduate Pilot Training (SUPT) programme at Reese AFB. Jayhawks to be based at Columbus AFB (Mississippi), Vance AFB (Oklahoma), and Randolph, Reese and Laughlin AFBs (Texas) for 14th,

71st, 12th, 64th and 47th FTWs respectively. T-1A will be used for training crews for KC-10, KC-135, C-5 and C-17.

Beechjet T-400: JASDF version, featuring thrust reversers, long-range inertial navigation and direction finding systems; interior changes.

CUSTOMERS: Total 64 Beechjet 400s and 29 400As delivered by 1 January 1992; US Air Force (113 T-1As so far ordered, of which 40 delivered by 10 June 1993); JASDF (three ordered in FY 1992 for TC-X trainer requirement; deliveries in first quarter of 1994).

COSTS: Jayhawk programme cost $1.3 billion; Beech contracts for 113 aircraft, $489 million. Civil 400A quoted at $4.3 million in October 1990.

DESIGN FEATURES: Swept wing has computer designed three-dimensional Mitsubishi MAC510 aerofoil; thickness/chord ratio 13.2 per cent at root, 11.3 per cent at tip; dihedral 2° 30′; incidence 3° at root, −3° 30′ at tip; sweepback 20° at quarter-chord. New features of Beechjet 400A include increased payload and certificated ceiling, greater cabin volume achieved by moving rear-fuselage fuel tank forward under floor (balanced by moving toilet to rear of cabin), improved soundproofing, and emergency door moved one window forward to facilitate forward club seating. Collins Pro Line 4 EFIS avionics standard.

T-1A Jayhawk features include student pilot in left seat, instructor on right and pupil/observer behind instructor; strengthened landing gear; more bird resistant windscreen and tail surfaces; fewer cabin windows; strengthened wing carry-through structure and engine attachment points to meet low-level flight stresses; rails for four passenger seats in cabin for personnel transport; avionics relocated from nose to rack in cabin to facilitate nose installation of air-conditioning; emergency door moved forward to position opposite main cabin door to allow straight-through egress; improved brakes; additional fuel tank; single-point pressure refuelling; Rockwell Collins five-tube EFIS; digital autopilot; turbulence-detection radar; central diagnostic and maintenance system; Tacan with air-to-air capability.

FLYING CONTROLS: Variable incidence tailplane and elevators for pitch axis; lateral control by small ailerons and almost full semi-span, narrow chord spoilers used also as airbrakes and lift dumpers; narrow chord Fowler type flaps, double-slotted inboard and single-slotted outboard, occupy most of trailing-edges and are hydraulically actuated; midspan leading-edge fences on wing; small horizontal strakes on fuselage at base of fin; small ventral fin.

STRUCTURE: Wings include integrally machined metal upper and lower skins joined to two box spars forming integral fuel tank; tailplane and fin similar. Wing, fuselage and tail unit certificated failsafe for unlimited life (with periodic inspections and maintenance).

LANDING GEAR: Retractable tricycle type, with single wheel and oleo-pneumatic shock absorber on each unit. Hydraulic actuation, controlled electrically. Emergency free-fall extension. Nosewheel, which is steerable by rudder pedals, retracts forward; mainwheels retract inward into fuselage. Goodyear wheels and tyres; Aircraft Braking Systems brakes.

POWER PLANT: Two Pratt & Whitney Canada JT15D-5 turbofans, each rated at 12.9 kN (2,900 lb st) for take-off. Rohr thrust reversers optional on 400A, but not fitted to T-1A. Total usable fuel capacity 2,775 litres (733 US gallons; 610.3 Imp gallons). One refuelling point in top of each wing, and one in rear fuselage for fuselage tank, capacity 1,158 litres (305.8 US gallons; 254.6 Imp gallons). Oil capacity 7.7 litres (2 US gallons; 1.7 Imp gallons).

ACCOMMODATION: Crew of two on flight deck of 400A; T-1A has seats for trainee pilot, co-pilot/instructor and observer. Standard double club layout of 400A seats eight passengers in pressurised cabin, with eight tracking, reclining seats in facing pairs, each with integral headrest and armrest and shoulder harness. Fold-out writing table between each pair of seats. Private flushing lavatory at rear with sliding doors and optional lighted vanity unit and hot water supply. With seat belts, this compartment can serve as an additional passenger seat. Interior options include substitution of carry-on baggage compartment, volume 0.34 m³ (12.0 cu ft), for one of the forward club seats, and hot and cold service refreshment centre with integral stereo entertainment system. Independent temperature control for flight deck and cabin heating systems standard. In-flight telephone optional. Rear baggage compartment with external access, capacity 204 kg (450 lb). Optional four passenger seats in main cabin of T-1A.

SYSTEMS: Pressurisation system, with normal differential of 0.63 bar (9.1 lb/sq in). Backup pressurisation system, using engine bleed air, for use in emergency. Hydraulic system, pressure 103.5 bars (1,500 lb/sq in), for actuation of flaps, landing gear and other services. Each variable volume output engine driven pump has a maximum flow rate of 14.76 litres (3.9 US gallons; 3.25 Imp gallons)/min, and one pump can actuate all hydraulic systems. Reservoirs, capacity 4.16 litres (1.1 US gallons; 0.9 Imp gallon), pressurised by filtered engine bleed air at 1.03 bars (15 lb/sq in). All systems are, wherever possible, of modular conception: for example, entire hydraulic installation can be removed as a single unit. Stick shaker as backup stall warning device.

AVIONICS: Standard avionics include pilot's integrated Collins Pro Line 4 EFIS featuring two-tube (optional three/four-tube) colour CRT primary flight display (PFD) and multifunction display (MFD) units mounted side by side, and control/display unit. PFD displays airspeed, altitude, verti-

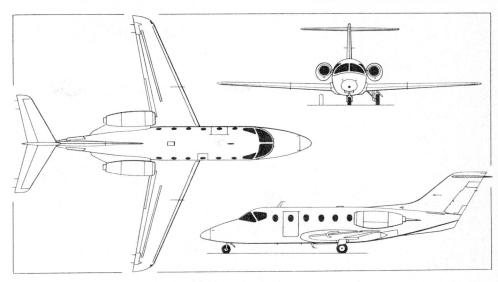

Beechcraft Beechjet 400A (two P&WC JT15D-5 turbofans) *(Dennis Punnett)*

Beechcraft Beechjet 400A twin-turbofan business aircraft

Beechcraft T-1A Jayhawk Tanker Transport Training System aircraft

cal speed, flight director, attitude and horizontal situation information, while MFD displays navigation, radar, map, checklist and fault annunciation information. Smaller, single or dual CRTs mounted on central console function as independent navigation sensor displays or backup displays for main CRTs. Co-pilot's PFD/MFD optional. EFIS installation features strapdown attitude/heading referencing system, electronic map navigation display, airspeed trend information and V-speeds on Mach airspeed display, and solid state Doppler turbulence detection radar.

DIMENSIONS, EXTERNAL:
Wing span	13.25 m (43 ft 6 in)
Wing aspect ratio	7.5
Length overall	14.75 m (48 ft 5 in)
Fuselage: Length	13.15 m (43 ft 2 in)
Max width	1.68 m (5 ft 6 in)
Max depth	1.85 m (6 ft 1 in)
Height overall	4.24 m (13 ft 11 in)
Tailplane span	5.00 m (16 ft 5 in)
Wheel track	2.84 m (9 ft 4 in)
Wheelbase	5.86 m (19 ft 3 in)
Crew/passenger door: Height	1.27 m (4 ft 2 in)
Width	0.71 m (2 ft 4 in)

DIMENSIONS, INTERNAL:
Cabin:	
Max length, incl flight deck	6.37 m (20 ft 11 in)
Length, excl flight deck	4.78 m (15 ft 8 in)
Max width	1.50 m (4 ft 11 in)
Max height	1.45 m (4 ft 9 in)
Volume: incl flight deck	11.32 m³ (400 cu ft)
excl flight deck	8.64 m³ (305 cu ft)
Baggage compartment volume	0.68 m³ (24.0 cu ft)

AREAS:
Wings, net	22.43 m² (241.4 sq ft)
Trailing-edge flaps (total)	4.22 m² (45.4 sq ft)
Spoilers (total)	0.57 m² (6.2 sq ft)
Fin, incl dorsal fin	5.91 m² (63.6 sq ft)
Rudder, incl yaw damper	0.99 m² (10.7 sq ft)
Tailplane	5.25 m² (56.5 sq ft)
Elevators, incl tab	1.55 m² (16.7 sq ft)

WEIGHTS AND LOADINGS:
Basic operating weight, incl crew, avionics and interior fittings	4,740 kg (10,450 lb)
Max fuel weight	2,228 kg (4,912 lb)
Max T-O weight	7,303 kg (16,100 lb)
Max ramp weight	7,393 kg (16,300 lb)
Max zero-fuel weight	5,896 kg (13,000 lb)
Max landing weight	7,121 kg (15,700 lb)
Max wing loading	325.6 kg/m² (66.69 lb/sq ft)
Max power loading	284.26 kg/kN (2.78 lb/lb st)

PERFORMANCE (at max T-O weight except where indicated):
Max limiting Mach number	0.78
Max level speed at 8,230 m (27,000 ft)	468 knots (867 km/h; 539 mph)
Typical cruising speed at 12,500 m (41,000 ft)	450 knots (834 km/h; 518 mph)
Long-range cruising speed at 12,500 m (41,000 ft)	392 knots (726 km/h; 451 mph)
Stalling speed, flaps down, idling power	93 knots (173 km/h; 107 mph) CAS
Max operating altitude: 400A	13,715 m (45,000 ft)
T-1A	12,500 m (41,000 ft)
FAA (FAR 25) T-O to 10.7 m (35 ft) at S/L, ISA	1,308 m (4,290 ft)
FAA landing distance from 15 m (50 ft) at S/L, ISA, max landing weight	1,072 m (3,514 ft)
Range with four passengers, max fuel, ISA, zero wind, with allowance for climb and descent, long-range cruise power, NBAA VFR reserves	1,900 nm (3,521 km; 2,187 miles)

BELL

BELL HELICOPTER TEXTRON INC
(Subsidiary of Textron Inc)

PO Box 482, Fort Worth, Texas 76101
Telephone: 1 (817) 280 8415
Fax: 1 (817) 280 8221
PRESIDENT: Webb F. Joiner
SENIOR VICE-PRESIDENT, OPERATIONS: Lloyd Shoppa
SENIOR VICE-PRESIDENT, MARKETING: Peter H. Parsinen
SENIOR VICE-PRESIDENT, PRODUCT SUPPORT AND US GOVERNMENT BUSINESS: John R. Murphey
VICE-PRESIDENT, RESEARCH AND ENGINEERING:
  Troy M. Gaffey
DIRECTOR, PUBLIC AFFAIRS AND ADVERTISING:
  Carl L. Harris

From 1970-81, Bell Helicopter Textron was unincorporated division of Textron Inc; became wholly owned subsidiary of Textron Inc from 3 January 1982. Bell Helicopter Canada (see Canada) formed at Montreal/Mirabel under contract with Canadian government October 1983; transfer to Mirabel completed January 1987 of Bell 206B JetRanger and 206L LongRanger production to make room for V-22 Osprey. Production of Bell 212/412 transferred mid-1988 and early 1989 respectively; Bell 230 programme also undertaken in Canada. Bell TH-206 JetRanger variant selected March 1993 (as TH-67 Creek) for US Army NTH requirement; further details under Bell in Canadian section and US Army in this section.

More than 32,000 Bell helicopters manufactured worldwide, including over 9,000 commercial models.

Bell helicopters built in USA detailed here. Those built in Canada listed under Canada; several models built under licence by Agusta in Italy and Fuji in Japan; Bell Helicopter Korea (BHK, which see) will co-produce helicopters with Bell Helicopter Textron in Republic of Korea; Bell Helicopter de Venezuela CA, joint venture with Maquinarias Mendoza CA and Aerotecnica SA, established early 1984 in Caracas for marketing and support; Bell Helicopter Asia (Pte) Ltd is wholly owned Singapore-based company for marketing and support in Southeast Asia.

BELL 205 UH-1HP HUEY II

TYPE: Proposed update of US Army UH-1H.
DESIGN FEATURES: More powerful T53 engine; Bell 212 main rotor system; uprated transmission; strengthened tailboom; tractor tail rotor. Rebuild offers 28 per cent additional T-O power, 275 per cent increase in hover ceiling, 12-15 per cent additional speed, and extended component operational lives reducing maintenance by 38 per cent and breaking even on conversion cost within five years. Empty weight increased by 159 kg (350 lb) and internal payload capacity by 288 kg (635 lb); gross weight 4,763 kg (10,500 lb).

Manufacture of original Bell 205/UH-1 Iroquois continues under licence in Japan by Fuji (which see). Details of other UH-1 modification programmes appear under Global, Tridair and UNC in this edition, and in *Jane's Civil and Military Aircraft Upgrades*.
POWER PLANT: One 1,342 kW (1,800 shp) Textron Lycoming T53-L-703 turboshaft. Transmission rating increased to 954.5 kW (1,280 shp).
PERFORMANCE:

Cruising speed	120 knots (222 km/h; 138 mph)
Max rate of climb at S/L	536 m (1,760 ft)/min
Vertical rate of climb at S/L	122 m (400 ft)/min
Hovering ceiling, OGE	1,625 m (5,325 ft)
Service ceiling	5,140 m (16,870 ft)

Bell AH-1F HueyCobra of the US Army

BELL 209 HUEYCOBRA (MODERNISED VERSIONS)

US Army designations: **AH-1E, AH-1F, TAH-1F, AH-1P, AH-1S** and **TH-1S**

TYPE: Two-seat close support and attack helicopter.
PROGRAMME: AH-1S first ordered as TOW-capable version of AH-1G in 1975; programme included conversion of earlier AH-1Gs and three-stage production of new aircraft with various degrees of upgrading; all versions designated AH-1S until March 1987, when new-build AH-1s allotted dormant UH-1 Iroquois suffixes AH-1P, AH-1E and AH-1F. Remains in production in Japan by Fuji (which see).
CURRENT VERSIONS: **AH-1S:** Formerly AH-1S(MOD); 92 AH-1Qs (early TOW-capable AH-1G) upgraded by 1979; 87 AH-1Qs upgraded in 1986-88 with Textron Lycoming T53-L-703 engines, Kaman rotor blades (see AH-1P) and TOW system, but retaining original curved canopies. Total of 377 conversions from AH-1G/Q with either round or flat glass canopies, including 15 in unarmed **TH-1S** Night Stalker configuration (converted from TH-1Gs) for training AH-64 crews to operate night vision system and integrated helmet and display sighting system (IHADSS); TH-1s used by 1-285 Avn, Arizona ArNG.

AH-1P: First batch of 100 new-production TOW Cobras (formerly called Production AH-1S), beginning with 76-22567, delivered 1977-78, two becoming AH-1F prototypes; improvements include flat-plate canopy, upturned exhaust, improved nap-of-the-earth (NOE) instrument panel, continental US (CONUS) navigation equipment, radar altimeter, improved communication radios, uprated engine and transmission, push/pull anti-torque control and, from 67th aircraft onwards, Kaman composite rotor blades with tapered tips.

AH-1E: Formerly Enhanced Cobra Armament System (ECAS) or Up-gun AH-1S; next 98 new-build aircraft, from 77-22673, with AH-1P improvements plus universal 20/30 mm gun turret (invariably fitted with long barrel 20 mm cannon); improved wing stores management system for 2.75 in rockets; automatic compensation for off-axis gun firing; 10kVA alternator for increased power. Delivered 1978-79.

AH-1F: Fully upgraded TOW version, previously designated Modernised AH-1S; 149 manufactured for US Army, beginning 78-23095, in 1979-86, including 50 transferred to Army National Guard; also 378 AH-1Gs and two AH-1Ps converted to full AH-1F standard between November 1979 and June 1982, including 41 as **TAH-1F** trainers; improvements of AH-1P and AH-1E added, plus new fire control system having laser rangefinder and tracker, ballistics computer, low airspeed sensor probe, Kaiser pilots' HUD, Doppler navigation system, IFF transponder, infra-red jammer above engine, hot metal and plume infra-red suppressor, closed-circuit refuelling, new secure voice communications, Kaman composite rotor blades.

Retrofits: Later modifications have included C-Nite FLIR for TOW sight fitted to 52 US Army AH-1Fs (reduced from planned 500) in South Korea from 1990. Air-to-Air Stinger (ATAS) and Cobra Fleet Life Extension (C-Flex), engine air filter, redesigned swashplate, M43 nuclear/biological/chemical mask, AN/AVR-2 laser warning and improved SCAS roll modifications. C-Flex items already completed include Nite Fix lighting, AH-1G-to-AH-1S upgrade and K-Flex driveshaft; remaining C-Flex work includes rotor improvements, improved TOW test set and radio upgrade. C-Nite also ordered for Pakistan, South Korea, Japan (licenced production) and US Army National Guard. Israel ordered Tamam night targeting system (NTS) for upgrade of TOW sight; first of 40 sets delivered 17 December 1992.
CUSTOMERS: See Current Versions for US Army orders. Japan received two AH-1Es and converted to AH-1F; AH-1F manufactured under licence by Fuji (which see) with 81 locally built funded to end of FY 1992. Israel (six AH-1E 30 AH-1F), Jordan (24 AH-1F), Pakistan (20 AH-1F) South Korea (70 AH-1F), Thailand (four AH-1F).
DESIGN FEATURES: Kaman composite blades, fitted from 67th AH-1P onwards, tolerate hits by 23 mm shells, have tungsten carbide bearing sleeves and outer 15 per cent of blade is tapered in chord and thickness; tailboom strengthened against 23 mm hits; airframe has infra-red suppressant paint finish.
POWER PLANT: One 1,342 kW (1,800 shp) Textron Lycoming T53-L-703 turboshaft. Transmission rated at 962 kW (1,290 shp) for take-off and 845 kW (1,134 shp) continuous. Closed circuit refuelling on AH-1F. Fuel capacity 980 litres (259 US gallons; 216 Imp gallons). Upward facing exhaust on AH-1E; IR suppression nozzle on AH-1F.
ACCOMMODATION: Flat-plate canopy introduced on AH-1F has seven planes of viewing surfaces, designed to minimise glint and reduce possibility of visual detection during nap-of-the-earth (NOE) flying; it also provides increased headroom for pilot. Improved instrument layout and lighting, compatible with use of night vision goggles. Improved, independently operating window/door ballistic jettison system to facilitate crew escape in emergency.
SYSTEMS: 10kVA 400Hz AC alternator with emergency bus added to electrical system. Dual hydraulic systems; pressure 103.5 bars (1,500 lb/sq in), maximum flow rate 22.7 litres (6 US gallons; 5 Imp gallons)/min; emergency (electrical pump) pressure 69 bars (1,000 lb/sq in). Open reservoir. Battery driven Abex standby pump, for use in event of main hydraulic system failure, can be used for collective pitch control and for boresighting turret and TOW missile system. Improved environmental control and fire detection systems.
AVIONICS: Standard lightweight avionics equipment (SLAE) includes AN/ARC-114 FM, AN/ARC-164 UHF/AM voice com, and E-Systems (Memcor Division) AN/ARC-115

Bell UH-1HP Huey II demonstrator (N16144)

VHF/AM voice com (compatible with KY-58 single-channel secure voice system). Other avionics include AN/ASN-128 Doppler nav system in AH-1F; HSI; VSI; radar altimeter; push/pull anti-torque controls for tail rotor; co-pilot's standby magnetic compass. C-Flex upgrade includes introduction of Magnavox AN/ARC-164(V) UHF/AM, Collins AN/ARC-186 VHF/AM-FM, ITT AN/ARC-201 (SINCGARS) VHF/FM, and LaBarge AN/ARN-89B D/F. AH-1F introduced fire control subsystem which includes Kaiser HUD for pilot, Teledyne Systems digital fire control computer for turreted weapon and underwing rockets, omnidirectional airspeed system to improve cannon and rocket accuracy, Hughes laser rangefinder (accurate over 10,000 m; 32,800 ft), and Rockwell AN/AAS-32 automatic airborne laser tracker. Other operational equipment includes Hughes LAAT stabilised sight (see 1987-88 *Jane's*), GEC Avionics M-143 air data subsystem, Bendix/King AN/APX-100 solid-state IFF transponder, Lockheed Sanders AN/ALQ-144 infra-red jammer (above engine), M-130 chaff/flare dispensers and AN/APR-39 RWR. Israeli aircraft being retrofitted with Rafael NTSF-65 thermal imaging sight.

ARMAMENT: M65 system with eight Hughes TOW missiles, disposed as two two-round clusters on each outboard underwing station. Inboard wing stations remain available for other stores, including retrofit of air-to-air version of GD Stinger SAM (ATAS), two per side. Beginning with first AH-1E, M28 (7.62/40 mm) turret in earlier HueyCobras replaced by new electrically powered General Electric universal turret, designed to accommodate either 20 mm or 30 mm weapon and improve standoff capability, although only 20 mm M197 three-barrel cannon (with 750 rds) mounted in this turret. Rate of fire 675 rds/min. Turret position is controlled by pilot or co-pilot/gunner through helmet sights, or by co-pilot using M65 TOW missile system's telescopic sight unit. Field of fire up to 110° to each side of aircraft, 20.5° upward and 50° downward. Also from first AH-1E, helicopter equipped with Baldwin Electronics M138 wing stores management subsystem, providing means to select and fire, singly or in groups, any one of five types of external 2.75 in rocket store. These mounted in launchers each containing from seven to 19 tubes, additional to TOW missile capability.

DIMENSIONS, EXTERNAL:
Main rotor diameter	13.41 m (44 ft 0 in)
Main rotor blade chord (from 67th AH-1P onward)	0.76 m (2 ft 6 in)
Tail rotor diameter	2.59 m (8 ft 6 in)
Tail rotor blade chord	0.305 m (1 ft 0 in)
Wing span	3.28 m (10 ft 9 in)
Length overall, rotors turning	16.18 m (53 ft 1 in)
Width: fuselage	0.99 m (3 ft 3 in)
over TOW pods	3.56 m (11 ft 8 in)
Height to top of rotor head	4.09 m (13 ft 5 in)
Elevator span	2.11 m (6 ft 11 in)
Width over skids	2.13 m (7 ft 0 in)

AREAS:
Main rotor disc	141.26 m² (1,520.23 sq ft)
Tail rotor disc	5.27 m² (56.75 sq ft)

WEIGHTS AND LOADINGS (AH-1S):
Operating weight empty	2,993 kg (6,598 lb)
Mission weight	4,524 kg (9,975 lb)
Max T-O and landing weight	4,535 kg (10,000 lb)
Max disc loading	32.10 kg/m² (6.58 lb/sq ft)
Max power loading	4.72 kg/kW (7.75 lb/shp)

PERFORMANCE (AH-1S at max T-O weight, ISA):
Never-exceed speed (V_{NE}) (TOW configuration)	170 knots (315 km/h; 195 mph)
Max level speed (TOW configuration)	123 knots (227 km/h; 141 mph)
Max rate of climb at S/L, normal rated power	494 m (1,620 ft)/min
Service ceiling, normal rated power	3,720 m (12,200 ft)
Hovering ceiling IGE	3,720 m (12,200 ft)
Range at S/L with max fuel, 8% reserves	274 nm (507 km; 315 miles)
g limits	+2.5/−0.5

BELL 209 IMPROVED SEACOBRA and SUPERCOBRA

US Navy/Marine Corps designations: AH-1J, AH-1T and AH-1W

TYPE: Two-seat, twin-engined close support and attack helicopter.

CURRENT VERSIONS: **AH-1J SeaCobra:** Initial twin-engined version for US Marine Corps (67) and Imperial Iranian Army Aviation (202); production ended February 1975 (see 1987-88 *Jane's*). Remains with HMA-773 squadron of USMC Reserve (HMA-775 converted to AH-1W, 1992).

AH-1T Improved SeaCobra: Improved AH-1J for US Marine Corps (see 1987-88 and earlier *Jane's*); total 57 built, but all 42 survivors converted to AH-1W by 1992.

AH-1W SuperCobra: Bell flew AH-1T powered by two GE T700-GE-700; first flight of proposed improved AH-1T+, including GE T700-GE-401 engines, 16 November 1983. Congress approved 44 AH-1W SuperCobras early 1984 for FYs 1985 and 1986; these, plus one composite maintenance trainer, delivered from 27 March 1986 to August 1988 to HML/A-169, 267, 367 and 369 at Camp Pendleton, California, and also co-located HMT-303 for crew training; total new-build planned, 190 including two Gulf War attrition replacements; 122 on order up to FY 1992, plus 12 requested for both FYs 1993 and 1994, including initial 14 of 36 for USMC Reserve delivered to HMA-775 at Camp Pendleton, California, from June 1992; 100th delivery 8 August 1991. Further 42 AH-1Ts upgraded to AH-1W for HML/A-167 and 269 at New River, North Carolina; last seven (from HML/A-269) completed in 1992. *Detailed description applies to AH-1W.*

Cobra Venom: See separate entry.

CUSTOMERS: US Marine Corps (see under Current Versions); Turkish Land Forces received five AH-1Ws in 1990; five more planned for 1993; all diverted from USMC contracts; Taiwan signed letter of offer and acceptance February 1992, for 18 plus 24 options; to receive new AH-1Ws in 1993.

COSTS: $10.7 million (1992) projected unit cost.

DESIGN FEATURES: Two-blade main rotor similar to that of Bell Model 214 with strengthened rotor head incorporating Lord Kinematics Lastoflex elastomeric and Teflon faced bearings. Blade aerofoil Wortmann FX-083 (modified);

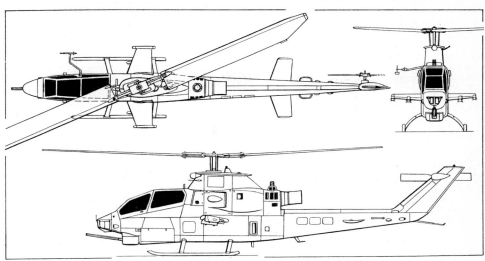

Bell AH-1F HueyCobra (Textron Lycoming T53-L-703 turboshaft) *(Dennis Punnett)*

Pakistan Army Bell AH-1F Cobra with TOW missiles

Bell AH-1W SuperCobra in USMC desert camouflage

436 USA: AIRCRAFT—BELL

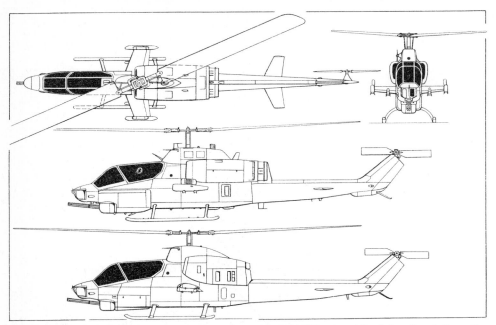

Bell AH-1W SuperCobra, with added side view (bottom) of AH-1T Improved SeaCobra *(Dennis Punnett)*

Turkish Land Forces Bell AH-1W SuperCobra (163923)

normal 311 rpm. Tail rotor also similar to that of Model 214 with greater diameter and blade chord; normal 1,460 rpm. Rotor brake standard. Stub-wings, NACA 0030 section at root; NACA 0024 at tip; incidence 14°; sweepback 14.7°.

Missions of AH-1W include anti-armour, escort, multiple-weapon fire support, including air-to-air with Sidewinder, armed reconnaissance, search and target acquisition. US AH-1Ws to be fitted with Kollsman laser Night Targeting System (NTS) based on Israeli Tamam NTSF-65 CLNAS; prototype conversion authorised December 1991; initial 25 sets built by Tamam and delivered from January 1993; further production jointly with Kollsman; total of 250 required by USMC. Testing of enhanced electronic warfare system began August 1989. Hughes AGM-65D Maverick ASM test-fired in August 1990; up to 12 Maverick-capable SuperCobras required.

STRUCTURE: Main rotor blades have aluminium spar and aluminium faced honeycomb aft of spar; tail rotor has aluminium honeycomb with stainless steel skin and leading-edge. Airframe conventional all-metal semi-monocoque.

LANDING GEAR: Non-retractable tubular skid type. Ground handling wheels optional.

POWER PLANT: Two General Electric T700-GE-401 turboshafts, each rated at 1,285 kW (1,723 shp). Transmission rating 1,515 kW (2,032 shp) for take-off; 1,286 kW (1,725 shp) continuous. Fuel (JP5) contained in two interconnected self-sealing rubber fuel cells in fuselage, with protection from damage by 0.50 in ballistic ammunition, total usable capacity 1,162 litres (306.8 US gallons; 255.7 Imp gallons). Gravity refuelling point in forward fuselage, pressure refuelling point in rear fuselage. Provision for carriage on underwing stores stations of two or four external fuel tanks each of 291 litres (77 US gallons; 64 Imp gallons) capacity; or two 378 litre (100 US gallon; 83 Imp gallon) tanks; or two 100 and two 77 US gallon tanks; large tanks on outboard pylons only. Oil capacity 19 litres (5 US gallons; 4.2 Imp gallons).

ACCOMMODATION: Crew of two in tandem, with co-pilot/gunner in front seat and pilot at rear. Cockpit is heated, ventilated and air-conditioned. Dual controls; lighting compatible with night vision goggles, and armour protection standard. Forward crew door on port side and rear crew door on starboard side, both upward opening.

SYSTEMS: Three independent hydraulic systems, pressure 207 bars (3,000 lb/sq in), for flight controls and other services. Electrical system comprises two 28V 400A DC generators, two 24V 34.5Ah batteries and three inverters: main 115V AC, 1kVA, single-phase at 400Hz, standby 115V AC, 750VA, three-phase at 400Hz and a dedicated 115V AC 365VA single-phase for AIM-9 missile system. AiResearch environmental control unit.

AVIONICS: Kaiser HUD compatible with PNVS-5 and ANVIS-6 night vision goggles. AN/ASN-75B compass set, AN/ARN-89B ADF, AN/APX-100(V) IFF transponder, AN/ARN-118 Tacan, AN/APN-154(V) radar beacon set, two AN/ARC-182(V) communication radios, AN/APN-194 radar altimeter, AN/APR-39(V) pulse radar signal detecting set, AN/APR-44(V) CW radar warning system, KY-58 TSEC secure voice set and AN/ALQ-144(V) IR countermeasures set. Improved countermeasures suite in USMC AH-1Ws replaces AN/APR-39 and AN/APR-44 by AN/APR-39(XE2) radar warning and adds AN/AVR-2 laser warner and AN/AAR-47 plume detecting set. Dual AN/ALE-39 chaff system with one MX-7721 dispenser mounted on top of each stub-wing. From January 1991, new-build AH-1Ws have Teledyne AN/APN-217 Doppler-based navigation system with Collins CDU-800 control/display unit and dual Collins ICU-800 processors. Tamam/Kollsman NTSF-65 thermal imaging sight to be retrofitted (within M-65 sighting system for TOW missiles) from 1993 (see Design Features); alternative McDonnell Douglas Electronic Systems NightHawk system also offered for export SuperCobras, following 1992-93 flight-testing.

ARMAMENT: Electrically operated General Electric undernose A/A49E-7(V4) turret housing an M197 three-barrel 20 mm gun. A 750-rd ammunition container is located in the fuselage directly aft of the turret; firing rate is 650 rds/min; a 16-round burst limiter is incorporated in the firing switch. Gun can be tracked 110° to each side, 18° upward, and 50° downward, but barrel length of 1.52 m (5 ft) makes it imperative that the M197 is centralised before wing stores are fired. Underwing attachments for up to four LAU-61A (19 tube), LAU-68A, LAU-68A/A, LAU-68B/A or LAU-69A (seven-tube) 2.75 in Hydra 70 rocket launcher pods; two CBU-55B fuel-air explosive weapons; four SUU-44/A flare dispensers; two M118 grenade dispensers; Mk 45 parachute flares; or two GPU-2A or SUU-11A/A Minigun pods. Provision for carrying totals of up to eight TOW missiles, eight AGM-114 Hellfire missiles, two AIM-9L Sidewinder or AGM-122A Sidearm missiles, on outboard underwing stores stations. Canadian Marconi TOW/Hellfire control system enables AH-1W to fire both TOW and Hellfire missiles on same mission. Hughes AGM-65D Maverick capability demonstrated 1990; under consideration for USMC. Addition planned of further two pylons (total six) with proportional increase in weapons capability.

DIMENSIONS, EXTERNAL:
Main rotor diameter	14.63 m (48 ft 0 in)
Main rotor blade chord	0.84 m (2 ft 9 in)
Tail rotor diameter	2.97 m (9 ft 9 in)
Tail rotor blade chord	0.305 m (1 ft 0 in)
Distance between rotor centres	8.89 m (29 ft 2 in)
Wing span	3.28 m (10 ft 9 in)
Wing aspect ratio	3.74
Length: overall, rotors turning	17.68 m (58 ft 0 in)
fuselage	13.87 m (45 ft 6 in)
Width overall	3.28 m (10 ft 9 in)
Height: to top of rotor head	4.11 m (13 ft 6 in)
overall	4.44 m (14 ft 7 in)
Ground clearance, main rotor, turning	2.74 m (9 ft 0 in)
Elevator span	2.11 m (6 ft 11 in)
Width over skids	2.24 m (7 ft 4 in)

AREAS:
Main rotor blades (each)	6.13 m² (66.0 sq ft)
Tail rotor blades (each)	0.45 m² (4.835 sq ft)
Main rotor disc	168.11 m² (1,809.56 sq ft)
Tail rotor disc	6.94 m² (74.70 sq ft)
Vertical fin	2.01 m² (21.70 sq ft)
Horizontal tail surfaces	1.41 m² (15.20 sq ft)

WEIGHTS AND LOADINGS:
Weight empty	4,634 kg (10,216 lb)
Mission fuel load (usable)	946 kg (2,086 lb)
Max useful load (fuel and disposable ordnance)	2,065 kg (4,552 lb)
Max T-O and landing weight	6,690 kg (14,750 lb)
Max disc loading	39.80 kg/m² (8.15 lb/sq ft)
Max power loading	4.42 kg/kW (7.26 lb/shp)

PERFORMANCE (at max T-O weight, ISA):
Never-exceed speed (V_{NE})	190 knots (352 km/h; 219 mph)
Max level speed at S/L	152 knots (282 km/h; 175 mph)
Max cruising speed	150 knots (278 km/h; 173 mph)
Rate of climb at S/L, OEI	244 m (800 ft)/min
Service ceiling	more than 4,270 m (14,000 ft)
Service ceiling, OEI	more than 3,660 m (12,000 ft)
Hovering ceiling: IGE	4,495 m (14,750 ft)
OGE	915 m (3,000 ft)
Range at S/L with standard fuel, no reserves	317 nm (587 km; 365 miles)

BELL AH-1W COBRA VENOM

Cobra Venom is a version of AH-1W being offered by Bell Helicopter with GEC-Marconi as prime contractor for British Army future attack helicopter requirement. Avionics will include night vision and new sensors for day/night all-weather operation, autonomous navigation and linked defensive aids, colour multi-function flat panel displays, integrated night vision helmets with night trackers and combined situation and digital map displays.

BELL 406 (AHIP)

US Army designations: OH-58D Kiowa and Kiowa Warrior

TYPE: Two-seat scout and attack helicopter.

PROGRAMME: Bell won US Army Helicopter Improvement Program (AHIP) 21 September 1981; first flight of OH-58D 6 October 1983; deliveries started December 1985; first based in Europe June 1987. Production running at 1992 minimum economic rate of three a month, compared with capacity for 12.

CURRENT VERSIONS: **Prime Chance:** Fifteen special armed OH-58Ds modified from September 1987 under Operation Prime Chance for use against Iranian high-speed boats in Gulf; delivery started after 98 days, in December 1987; firing clearance for Stinger, Hellfire, 0.50 in gun and seven-tube rocket pods completed in seven days; aircraft remained at sea in Gulf 1991.

Kiowa Warrior: Armed version, to which all planned OH-58Ds are being modified; integrated weapons pylons, uprated transmission and engine, lateral CG limits increased, raised gross weight, EMV protection of avionics bays, localised strengthening, RWR, IR jammer, video recorder, SINCGARS radios, laser warning receiver and tilting vertical fin; armament same as Prime Chance; integrated avionics and lightened structure. See illustration overleaf for stealth kit modification (details of kit in 1992-93 *Jane's*); converted aircraft have since reverted to standard.

Multi-Purpose Light Helicopter (MPLH): Further modification of Kiowa Warrior; features include squatting landing gear, quick-folding rotor blades, horizontal stabiliser and tilting fin to allow helicopter to be transported in cargo aircraft and flown to cover 10 minutes after unloading from C-130. Later additions include cargo hook for up to 907 kg (2,000 lb) slung load and fittings for external carriage of six outward-facing troop seats or four stretchers.

OH-58X Light Utility Variant: Contender for anticipated US Army requirement; fourth development OH-58D modified in 1992 with partial stealth features (including chisel nose); Kodak FLIR with 30° field-of-view; 907 kg (2,000 lb) capacity cargo hook; Allison T703-AD-700 power plant with 429 kW (575 shp) transmission; and Honeywell avionics including ring-laser INS with GPS and Integrated Helmet Display System. Avionics relocated to nose, freeing two passenger seats; third seat available if mast sight avionics removed; target empty weight 1,400 kg (3,085 lb).

CUSTOMERS: US Army: initial plan to modify 592 OH-58A to OH-58D reduced to 477; again reduced to 207, but Congressionally mandated re-orders increased total to 363 (excluding five prototypes) by 1993 from new goal of 507, as under:

FY	Lot	Qty	First aircraft
1983	I	16	83-24129
1985	II	44	85-24690
1986	III	39	86-8901
1987	IV	36	87-0725
1988	V	36	88-0285
1989	VI	36	89-0082
1990	VII	36	90-0346
1991	VIII	36	91-0536
1992	IX	36	
1992	X	12*	
1993	XI	36	

*Gulf War attrition replacements

New-build Kiowa Warrior from 202nd aircraft (89-0112), May 1991; initially to 'C' and 'D' Troops of 4-17 Aviation, Fort Bragg. First Warrior retrofit contract (28 helicopters) to Bell January 1992; further 75 in FY 1993 budget. Taiwan ordered 12 OH-58Ds in February 1992 and reserved 14 options; deliveries from July 1993. Main effort now is provision for armament and accompanying upgrades in new aircraft and retrofitting existing OH-58D to armed Kiowa Warrior configuration; 81 Kiowa Warriors to be raised to ultimate Multi-Purpose Light Helicopter (MPLH) standard for use by US Army quick reaction forces of XVIII (32 aircraft) and 82nd (49 aircraft) Airborne Divisions.

COSTS: $9.42 million (1990) programme unit cost. Flyaway $4.9 million (Kiowa) or $6.7 million (Kiowa Warrior). Retrofit to Warrior configuration $1.34 million (1992-93 average).

DESIGN FEATURES: Four-blade Bell soft in plane rotor with carbon composites yoke, elastomeric bearings and composites blades. Main rotor rpm 395; tail rotor rpm 2,381. McDonnell Douglas/Northrop mast-mounted sight containing TV and IR optics and laser designator/ranger; Honeywell integrated control of mission functions, navigation, communications, systems and maintenance functions based on large electronic primary displays for pilot and observer/gunner; hands-on cyclic and collective controls for all combat functions; automatic target hand-off system in some OH-58Ds operates air-to-air as well as air-to-ground using digital frequency hopping; system indicates location and armament state of other helicopters; some OH-58Ds have real-time video downlink capable of relaying to US Army Guardrail aircraft, to headquarters 22 nm (40 km; 25 miles) away or, via satellite, to remote locations.

Stage 1 of Multi-Stage Improvement Program (MSIP) includes fitting GPS receiver, improved Doppler, digital data loader and MIL-STD-1750 processors.

FLYING CONTROLS: Full-powered controls, including tail rotor, with four-way trim and trim release; stability and control augmentation system (SCAS) using AHRS gyro signals; automatic bob-up and return to hover mode; Doppler blind hover guidance mode; co-pilot/observer's cyclic stick can be disconnected from controls and locked centrally.

STRUCTURE: Basic OH-58 structure reinforced; armament cross-tube fixed above rear cabin floor; avionics occupy rear cabin area, baggage area and nose compartment.

LANDING GEAR: Light alloy tubular skids bolted to extruded cross-tubes.

POWER PLANT: One Allison 250-C30R (T703-AD-700) turboshaft (C30X with improved diffuser in Kiowa Warrior), with an intermediate power rating of 485 kW (650 shp) at S/L ISA. Transmission rating: Kiowa 339 kW (455 shp) continuous; Kiowa Warrior 410 kW (550 shp) continuous.

Bell OH-58D Kiowa Warrior with weapons and mast-mounted sight

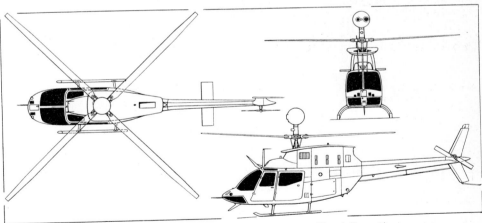

Bell OH-58D Kiowa scout and attack helicopter *(Dennis Punnett)*

One self-sealing crash resistant fuel cell, capacity 424 litres (112 US gallons; 93 Imp gallons) located aft of the cabin area. Refuelling point on starboard side of fuselage. Oil capacity 5.7 litres (1.5 US gallons; 1.2 Imp gallons).

ACCOMMODATION: Pilot and co-pilot/observer seated side by side. Door on each side of fuselage. Accommodation is heated and ventilated.

SYSTEMS: Single hydraulic system, pressure 69 bars (1,000 lb/sq in), for main and tail rotor controls and SCAS system. Maximum flow rate 11.36 litres (3 US gallons; 2.5 Imp gallons)/min. Open-type reservoir. Primary electrical power provided by 10kVA 400Hz three-phase 120/208V AC alternator with 200A 28V DC transformer-rectifier unit for secondary DC power. Backup power provided by 500VA 400Hz single-phase 115V AC solid state inverter and 200A 28V DC starter/generator.

AVIONICS: Multi-function displays for vertical and horizontal situation indication, mast-mounted sight day/night viewing and communications control, with selection via control column handgrip switches. Five com transceivers, data link and secure voice equipment. Plessey (PESC) AN/ASN-157 Doppler strapdown INS. Equipped for day/night VFR. Mast-mounted sight houses 12x magnification TV camera, auto-focusing IR thermal imaging sensor and laser rangefinder/designator, with automatic target tracking and in-flight automatic boresighting. Night vision goggles; AHRS; and airborne target handoff subsystem (ATHS). Germany-based OH-58Ds have real-time video downlink which can be relayed via Guardrail-capable aircraft. Phase 1 additions, introduced on production line in 1990 in preparation for Kiowa Warrior, include doubled computer capacity to 88K, added weapons selection/aiming and multi-target acquisition/track displays, IR jammer, second RWR and laser warning, video recorder, data transfer system, SINCGARS and Have Quick II radios, ANVIS display and symbology system and EMV hardening. Full production standard Kiowa Warrior avionics include AN/APR-44(V)3 radar warning receiver, AN/ALQ-144(V)1

438   USA: AIRCRAFT—BELL / BELL/BOEING

Stealth Kiowa Warrior (89-0090) with chisel nose, blade root cuffs, composites side door, tail rotor hub cover, modified engine fairing and radar-absorbent coatings

Quick folding system for OH-58D Multi-Purpose Light Helicopter

IR jammer, AN/AVR-2 laser detection system, AN/APR-39(V)1 or -39A(V)1 radar warning and AN/ARC-201 SINCGARS secure voice/data radio.
EQUIPMENT: M-43 NBC mask in Phase 1 and Warrior aircraft.
ARMAMENT: Four Stinger air-to-air or Hellfire air-to-surface missiles, or two seven-round 2.75 in rocket pods, or two Global Helicopter Technology CFD-5000 pods for 7.62 mm and 0.50 in machine guns, mounted on outrigger on cabin sides. IR jammer standard on armed version.

DIMENSIONS, EXTERNAL:
Main rotor diameter	10.67 m (35 ft 0 in)
Main rotor blade chord (mean)	0.24 m (9½ in)
Tail rotor diameter	1.65 m (5 ft 5 in)
Length: overall, rotors turning	12.85 m (42 ft 2 in)
fuselage (pitot to skid)	10.48 m (34 ft 4¾ in)
fin tilted for air transport	10.29 m (33 ft 9¼ in)
Width: rotors folded, clean	1.97 m (6 ft 5½ in)
armament fitted, pylons folded for air transport (MPLH)	2.39 m (7 ft 10 in)
Height: overall	3.93 m (12 ft 10⅝ in)
squatting and folded for air transport (MPLH)	2.73 m (8 ft 11⅝ in)
Horizontal stabiliser span	2.29 m (7 ft 6 in)
Skid track	1.88 m (6 ft 2 in)
Cabin doors (port and stbd, each):	
Height	1.04 m (3 ft 5 in)
Width	0.91 m (3 ft 0 in)
Height to sill	0.66 m (2 ft 2 in)

AREAS:
Main rotor blades (each)	1.38 m² (14.83 sq ft)
Tail rotor blades (each)	0.13 m² (1.43 sq ft)
Main rotor disc	89.37 m² (962.0 sq ft)
Tail rotor disc	2.14 m² (23.04 sq ft)
Horizontal stabiliser	1.11 m² (11.92 sq ft)
Fin	0.87 m² (9.33 sq ft)

WEIGHTS AND LOADINGS (K: OH-58D Kiowa, KW: armed Kiowa Warrior):
Weight empty: K	1,381 kg (3,045 lb)
KW	1,492 kg (3,289 lb)
Max external load: KW	907 kg (2,000 lb)
Max fuel weight: K, KW	341 kg (752 lb)
Max T-O and landing weight: K	2,041 kg (4,500 lb)
KW	2,495 kg (5,500 lb)
Max disc loading: K	22.84 kg/m² (4.68 lb/sq ft)
KW	27.91 kg/m² (5.72 lb/sq ft)
Max power loading: K	6.02 kg/kW (9.89 lb/shp)
KW	5.82 kg/kW (9.56 lb/shp)

PERFORMANCE (at max T-O weight, clean; K and KW as above):
Never-exceed speed (V_{NE})	130 knots (241 km/h; 149 mph)
Max level speed at 1,220 m (4,000 ft)	128 knots (237 km/h; 147 mph)
Max cruising speed: K	118 knots (219 km/h; 136 mph)
KW	112 knots (208 km/h; 129 mph)
Econ cruising speed at 1,220 m (4,000 ft)	110 knots (204 km/h; 127 mph)
Max rate of climb: at S/L, ISA	469 m (1,540 ft)/min
at 1,220 m (4,000 ft), 35°C (95°F)	over 366 m (1,200 ft)/min
Vertical rate of climb: at S/L, ISA	232 m (760 ft)/min
at 1,220 m (4,000 ft), 35°C (95°F)	over 152 m (500 ft)/min
Service ceiling	over 3,660 m (12,000 ft)
Hovering ceiling: IGE, ISA	over 3,660 m (12,000 ft)
OGE, ISA	3,415 m (11,200 ft)
OGE, 35°C (95°F)	1,735 m (5,700 ft)
Range: K and KW	250 nm (463 km; 288 miles)
Endurance: K and KW	2 h 24 min

BELL 406 CS COMBAT SCOUT
The Model 406 CS Combat Scout is no longer marketed; details and illustrations in 1992-93 *Jane's*.

BELL/BOEING V-22 OSPREY
Joint programme; described below under Bell/Boeing heading.

BELL/BOEING
BELL HELICOPTER TEXTRON and BOEING HELICOPTERS
PROGRAMME MANAGER: Colonel James H. Schaeffer, USMC

BELL/BOEING V-22 OSPREY
TYPE: Twin-engined tilt-rotor multi-mission aircraft.
PROGRAMME: Based on Bell/NASA XV-15 tilt-rotor; initiated as US Department of Defense Joint Services Advanced Vertical Lift Aircraft (JVX), run by US Army, FY 1982; programme transferred to US Navy January 1983; 24-month US Navy preliminary design contract 26 April 1983; aircraft named V-22 Osprey January 1985; seven-year full-scale development (FSD) began 2 May 1986 with order for six prototypes (Nos. 1, 3 and 6 by Bell; Nos. 2, 4 and 5 by Boeing) plus static test airframes.
  No. 1 (BuAer No. 163911) rolled out at Arlington, Texas, 23 May 1988; first flight 19 March 1989; achieved first transition from helicopter to aeroplane mode 14 September 1989. First flight No. 2 at Arlington 9 September and No. 4 at Wilmington, Delaware, 21 December 1989; first flight No. 3 on 9 May 1990 at Arlington; first flight No. 5 at Wilmington ended with non-fatal crash due to avionics mis-wiring, 11 June 1991; No. 6 construction suspended but re-started to assume some duties of No. 5. Test assignments include No. 1, flight envelope expansion and flight loads examination; No. 2, fly-by-wire development; No. 3, flight loads, vibration and acoustics and sea trials; No. 4, first with full avionics and ejection seats, initial shipboard compatibility and propulsion studies; No. 6, initial US Air Force mission equipment tests. Initial government trials, April 1990, involved 15 hours' flying by three military test pilots in No. 2. No. 4 lost on 94th sortie (103 hours) 21 July 1992.
  Sea trials aboard USS *Wasp*, 4-7 December 1990, involved No. 3 in landing and take-off tests and No. 4 in fit and function tests. By end 1990, trials included transition landings and take-offs, wing stall tests, single-engine tests and flights up to 349 knots (647 km/h; 402 mph). V-22 awarded National Aeronautic Association's Collier Trophy in 1990 for "greatest achievement in aeronautics in past year"; claimed to meet or exceed 32 multi-service mission requirements. Trials programme is 4,000 hours, of which 763 hours flown in 643 sorties up to July 1992 temporary grounding. Flying resumed in April 1993 with No. 3, followed by No. 2 shortly thereafter, these having been modified with improved firewalls, nacelle drains and pylon driveshaft heatshields as consequence of No. 4's crash.
  Original schedule called for delivery to US Marines late 1991/early 1992, US Air Force 1993 and US Navy 1995. Production funding cut FY 1990, but development contract maintained; some long-lead items for production aircraft funded FY 1991; Congress appropriated FY 1992 funding for three of additional six 'production representative' Engineering/Manufacturing Development (EMD) aircraft; second three planned for FY 1993 funding; however

programme revised in late 1992, and six to comprise four new and two rebuilt prototypes; operational testing probably deferred until these new aircraft available.

EMD Ospreys, to begin flying in 1997, have significant changes from earlier aircraft, including 1,000 kg (2,205 lb) reduction in empty weight; aluminium cockpit cage, replacing titanium, but with smaller windows to preserve structural strength; upgraded flight controls; enhanced engine and drive system; improved tail unit construction (built by Bell, in place of Grumman) including fibre placement aft fuselage; redesigned rotor system; absence of fin tuning weights; improved wing constructional techniques; redesigned wiring; and pyrotechnic escape hatches.

Marine Corps downgraded requirements for CH-46 replacement, 1992, making conventional helicopter solution possible; Corps requires 1998 IOC.

CURRENT VERSIONS: **MV-22A**: Basic US Marine Corps transport; original requirement for 552 (now 507) to replace CH-46 Sea Knight and CH-53 Sea Stallion. Three-man crew and 24 combat-equipped troops or cargo carried at 250 knots (463 km/h; 288 mph) over radius of 200 nm (370 km; 230 miles), with ability to hover mid-way at 915 m (3,000 ft) OGE at 33°C (91.4°F).

HV-22A: US Navy combat search and rescue (CSAR), special warfare and fleet logistics model to replace HH-3s. Original requirement for 50.

CV-22A: US Air Force long-range special missions aircraft. Original requirement for 80 reduced to 55; should carry 12 troops or 1,306 kg (2,880 lb) internal cargo over 520 nm (964 km; 599 mile) radius at 250 knots (463 km/h; 288 mph), with ability to hover OGE at 1,220 m (4,000 ft) at 35°C (95°F).

SV-22A: Tentative US Navy ASW version to replace S-3 Viking. Original requirement for up to 300. Required to deposit and pick up recoverable large ASW sensors and operate them from high altitude.

US Army: Original requirement for 231 V-22s, based on USMC transport, withdrawn. Documented requirement remains for V-22 in medevac, special operations and combat assault support roles.

CUSTOMERS: See above. Also European marketing co-operation with British Aerospace (which see), Dornier and Alenia. Japan Maritime Self Defence Force provisional commitment to fund two SAR Ospreys in 1994, plus two in 1995.

COSTS: FY 1991 budget assigned $238 million to continue R&D. FY 1992 budget earmarked $625 million to begin manufacture of three of planned six 'production representative' V-22s; also identified $165 million of FY 1989 production money to be converted to RDT&E funding. Estimated cost (1991) to complete full-scale development, $2,100 million. Unit cost, US Navy (1992 estimate) $5-12 million.

DESIGN FEATURES: Engines, transmission and proprotors tilt through 97° 30′ between forward flight and steepest approach gradient or tail-down hover; cross-shaft keeps both proprotors turning after engine loss; transmission rated at 3,408 kW (4,570 shp) normally and 4,415 kW (5,920 shp) with one engine inoperative; APU mounted in fuselage drives into cross-shaft for starting. Three-blade contra-rotating proprotors have special high-twist tapered format blades with elastomeric bearings and powered folding mechanisms; separate swashplates produce respectively yaw and fore-and-aft translation in hover and sideways flight in level attitude. Wing-fold sequence from helicopter mode involves power-folding of blades parallel to wing leading-edge, tilting engine nacelles down to horizontal and rotating entire wing/engine/proprotor group clockwise on stainless steel carousel to lie over fuselage.

FLYING CONTROLS: Three-lane fly-by-wire (Moog actuators) with automatic stabilisation, full autopilot and formation-flying modes. Automatic control of configuration change during transition and of transfer of control from aerodynamic surfaces to rotor-blade pitch changing; flaperons and ailerons droop during hover to reduce negative lift of wing.

Rotors have separate cyclic control swashplates for sideways flight and fore-and-aft control (symmetrical for forward and rearward flight and differential for yaw) in hover. Lateral attitude controlled in hover by differential rotor thrust, but lateral swashplate allows sideways flight in level attitude controlled by button on control column. Integrated electronic cockpit with six electronic display screens; helicopter-style control columns rather than aileron wheels, but left-handed power levers move forward for full power in opposite sense to helicopter collective lever.

STRUCTURE: Approx 59 per cent of airframe is composites and just 454 kg (1,000 lb) of empty weight is metal; main composites are Hercules IM-6 graphite/epoxy in wing and AS4 in fuselage and tail; nacelle cowlings and pylon supports are of GFRP. Fuselage mainly 'black aluminium' consisting of conventional stringers, frames and preformed skin, all in composites, assembled with metal fasteners. Wing box is high-strength, very stiff torsion box made up from one-piece upper and lower skins with moulded ribs and bonded stringers; two-segment graphite single-slotted flaperons with titanium fittings; three-segment detachable leading-edge of aluminium alloy with Nomex honeycomb core. Wing locking and unlocking with Lucas Aerospace actuators; fuselage sponsons contain landing gear, air-conditioning unit and fuel; tail unit of Hercules AS4 graphite/

Bell/Boeing V-22 prototype No. 2 during in-flight refuelling trials

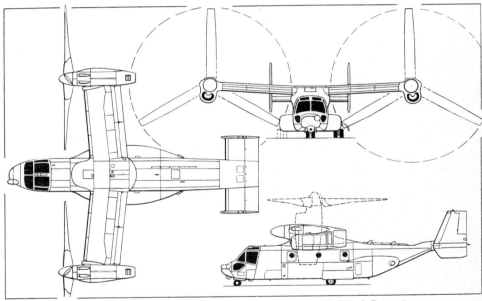

Bell/Boeing V-22 Osprey tilt-rotor multi-mission aircraft (*Dennis Punnett*)

Engine/proprotor ground test rig for the V-22 Osprey

USA: AIRCRAFT—BELL/BOEING / BOEING

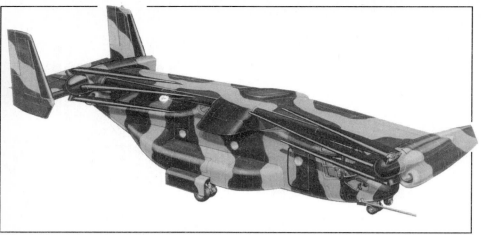

Diagram illustrating the V-22 with wing and rotor blades in the stowed position

epoxy, built by Grumman. Bell contributes wing, nacelles, proprotor and transmission systems and integrates engines. Boeing responsible for fuselage, tail unit, landing gear and fairings and integrates avionics.

LANDING GEAR: Dowty hydraulically actuated retractable tricycle type, with twin wheels and oleo-pneumatic shock absorbers on each unit, Menasco Canada steerable nose unit. Dowty Toronto two-stage shock absorption in main gear is designed for landing impacts of up to 3.66 m (12 ft)/s normal, 4.48 m (14.7 ft)/s maximum, and has been drop tested to 7.32 m (24 ft)/s. All units retract rearward, main gear into sponsons on lower sides of centre-fuselage. Manual and nitrogen pressurised standby systems for emergency extension. Parker Bertea wheels and multi-disc hydraulic carbon brakes.

POWER PLANT: Two Allison T406-AD-400 (501-M80C) turboshafts, each with T-O and intermediate rating of 4,586 kW (6,150 shp) and max continuous rating of 4,392 kW (5,890 shp), installed in Bell-built tilting nacelles at wingtips and driving a three-blade proprotor. Each nacelle has a Garrett infra-red emission suppressor at the rear. Air particle separator and Lucas inlet/spinner ice protection system for each engine. Lucas Aerospace FADEC for each engine, with analog electronic backup control. Pratt & Whitney originally named as second production source for engines, starting with production lot 5. Internal fuel (JP-5) in up to 13 crash resistant, self-sealing (nitrogen pressurised) cells: one 1,431 kg (3,155 lb) forward cell in each sponson, a 925 kg (2,040 lb) cell in rear of starboard sponson, four 227 kg (500 lb) auxiliary cells in each wing leading-edge, and a 306 kg (675 lb) engine feed cell outboard of the auxiliary tanks in each wing. Total capacity 7,627 litres (2,015 US gallons; 1,678 Imp gallons). (Not all versions have all tanks.) Pressure refuelling point in starboard sponson leading-edge; gravity point in upper surface of each wing. Simmonds fuel management system. Provision for a further 7,427 kg (16,374 lb) of fuel to be carried in four additional joint services auxiliary fuel tanks, each 2,279 litres (602 US gallons; 501 Imp gallons), in main cabin for self-deployment mission. In-flight refuelling probe in lower starboard side of forward fuselage.

ACCOMMODATION: Normal crew complement of pilot (in starboard seat), co-pilot and crew chief in USMC variant. Flight crew accommodated on Simula Inc crashworthy armoured seats capable of withstanding strikes from 0.30 in armour piercing ammunition, 30g forward and 14.5g vertical decelerations. Seats are manufactured from a boron carbide/polyethylene laminate. Flight deck has overhead and knee-level side transparencies in addition to large windscreen and main side windows, plus an overhead rearview mirror. Main window frame is of titanium. Main cabin has composites floor panels, and can accommodate up to 24 combat-equipped troops, on inward facing crashworthy foldaway seats, plus two gunners; up to 12 litters plus medical attendants; or an equivalent cargo load with energy absorbing tiedowns. Cargo handling provisions include a 907 kg (2,000 lb) capacity cargo winch and pulley system and removable roller rails. Main cabin door at front on starboard side, top portion of which opens upward and inward, lower portion (with built-in steps) downward and outward. Full width rear loading ramp/door in underside of rear fuselage, operated by Parker Bertea hydraulic actuators. Emergency exit windows on port side; escape hatch in fuselage roof aft of wing.

SYSTEMS: Environmental control system, utilising engine bleed air; control unit in rear of port main landing gear sponson. Three hydraulic systems (two independent main systems and one standby), all at operating pressure of 345 bars (5,000 lb/sq in), with Parker Bertea reservoirs. Electrical power supplied by two 40kVA constant frequency AC generators, two 50/80kVA variable frequency DC generators (one driven by APU), rectifiers, and a 15Ah battery. GE Aerospace triple redundant digital fly-by-wire flight control system, incorporating triple primary FCS (PFCS) and triple automatic FCS (AFCS) processors, and triple flight control computers (FCC) each linked to a MIL-STD-1553B databus; two PFCSs and one AFCS are fail-operational. FBW system signals hydraulic actuation of flaperons, elevator and rudders, controls aircraft transition between helicopter and aeroplane modes, and can be programmed for automatic management of airspeed, nacelle tilting and angle of attack. FCCs provide interfaces for swashplate, conversion actuator, flaperon, elevator, rudder and pylon primary actuators, flight deck central drive, force feel, and nosewheel steering. Dual 1750A processors for PFCS and single 1750A for AFCS incorporated in each FCC. Non-redundant standby analog computer (in development aircraft only) provides control of aircraft, including FADEC and pylon actuation, in the event of FBW system failure. Sundstrand Turbomach 261 kW (350 shp) APU, in rear portion of wing centre-section, provides power for mid-wing gearbox which, in turn, drives two electrical generators and an air compressor. Anti-icing of windscreens and engine air intakes; de-icing of proprotors and spinners. Clifton Precision combined oxygen (OBOGS) and nitrogen (OBIGGS) generating systems for cabin and fuel tank pressurisation respectively. Systron Donner pneumatic fire protection systems for engines, APU and wing dry bays.

AVIONICS: VHF/AM-FM, HF/SSB and (USAF only) UHF secure voice com; Tacan, VOR/ILS, AHRS, radar altimeter and digital map displays; IFF; Honeywell AN/AAR-47 missile warning system; radar/infra-red warning system; J. E. T. ADI-350W standby attitude indicator; Aydin Vector data acquisition and storage system. Major tactical sensors are a Hughes Aircraft AN/AAQ-16 FLIR detector in undernose fairing and (USAF and USN only) a Texas Instruments AN/APQ-174 terrain following multifunction radar in offset (to port) nose thimble, with two Allied Signal IP-1555 full colour multifunction displays. Two Control Data AN/AYK-14 mission computers with Boeing/IBM software. Pilots' night vision system and Honeywell integrated helmet display system.

EQUIPMENT: Chaff/flare dispensers. Provision for rescue hoist over forward (starboard) cabin door.

DIMENSIONS, EXTERNAL:
Rotor diameter, each	11.58 m (38 ft 0 in)
Rotor blade chord: at root	0.90 m (2 ft 11½ in)
at tip	0.56 m (1 ft 10 in)
Wing span: excl nacelles	14.02 m (46 ft 0 in)
incl nacelles	15.52 m (50 ft 11 in)
Wing chord, constant	2.54 m (8 ft 4 in)
Distance between proprotor centres	14.25 m (46 ft 9 in)
Width: overall, rotors turning	25.78 m (84 ft 7 in)
folded	5.61 m (18 ft 5 in)
Length: fuselage, excl probe	17.47 m (57 ft 4 in)
overall, wings stowed/blades folded	19.09 m (62 ft 7½ in)
Height: over tail fins	5.28 m (17 ft 4 in)
wings stowed/blades folded	5.51 m (18 ft 1 in)
overall, nacelles vertical	6.90 m (22 ft 7½ in)
Nacelle ground clearance, nacelles vertical	1.31 m (4 ft 3¾ in)
Proprotor ground clearance, nacelles vertical	6.35 m (20 ft 10 in)
Tail span, over fins	5.61 m (18 ft 5 in)
Wheel track (c/l of outer mainwheels)	4.62 m (15 ft 2 in)
Wheelbase	6.59 m (21 ft 7½ in)
Dorsal escape hatch: Length	1.02 m (3 ft 4 in)
Width	0.74 m (2 ft 5 in)

DIMENSIONS, INTERNAL:
Cabin: Length	7.37 m (24 ft 2 in)
Max width	1.80 m (5 ft 11 in)
Max height	1.83 m (6 ft 0 in)
Usable volume	24.3 m³ (858 cu ft)

AREAS:
Rotor discs, each	105.4 m² (1,134 sq ft)
Rotor blades (each)	12.15 m² (130.76 sq ft)
Wing, total incl flaperons and fuselage centre-section	35.49 m² (382.0 sq ft)
Flaperons, total	8.25 m² (88.8 sq ft)
Tailplane	8.22 m² (88.5 sq ft)
Elevators, total	4.79 m² (51.54 sq ft)
Fins (each)	10.81 m² (116.4 sq ft)
Rudders (each)	1.64 m² (17.6 sq ft)

WEIGHTS AND LOADINGS:
Weight empty, equipped	14,463 kg (31,886 lb)
Max fuel weight: standard	6,215 kg (13,700 lb)
with self-ferry cabin tanks	13,641 kg (30,074 lb)
Max internal payload (cargo)	9,072 kg (20,000 lb)
Cargo hook capacity: single	4,536 kg (10,000 lb)
two hooks (combined weight)	6,804 kg (15,000 lb)
Rescue hoist capacity	272 kg (600 lb)
Normal mission T-O weight: VTO	21,545 kg (47,500 lb)
STO	24,947 kg (55,000 lb)
Max STO weight for self-ferry	27,442 kg (60,500 lb)
Max floor loading (cargo)	1,464 kg/m² (300 lb/sq ft)
Max power loading	8.06 kg/kW (13.23 lb/shp)

PERFORMANCE (estimated):
Max cruising speed:	
at S/L, helicopter mode	100 knots (185 km/h; 115 mph)
at S/L, aeroplane mode	275 knots (509 km/h; 316 mph)
at optimum altitude, aeroplane mode	314 knots (582 km/h; 361 mph)
Max forward speed with max slung load	130 knots (241 km/h; 150 mph)
Service ceiling	7,925 m (26,000 ft)
T-O run at normal mission STO weight	less than 152 m (500 ft)

Range:
VTO at 21,146 kg (46,619 lb) gross weight, incl 5,443 kg (12,000 lb) payload
    1,200 nm (2,224 km; 1,382 miles)
STO at 24,947 kg (55,000 lb) gross weight, incl 9,072 kg (20,000 lb) payload
    1,800 nm (3,336 km; 2,073 miles)
STO at 27,442 kg (60,500 lb) self-ferry gross weight, no payload
    2,100 nm (3,892 km; 2,418 miles)

BOEING
THE BOEING COMPANY

PO Box 3707, Seattle, Washington 98124
Telephone: 1 (206) 655 2121
Fax: 1 (206) 655 1171
CHAIRMAN AND CEO: Frank A. Shrontz
PRESIDENT: Philip M. Condit
MANAGER, PUBLIC RELATIONS AND ADVERTISING: Harold Carr

Company founded July 1916. On 2 January 1990, Boeing Defense & Space Group formed to co-ordinate Aerospace & Electronics, Helicopters, Military Airplanes and Advanced Systems divisions of The Boeing Company. Simultaneously, former Military Airplanes divisions at Wichita reduced in size by transfer of some activities to Boeing Commercial Airplane Group.

Operating components of The Boeing Company include:
BOEING COMMERCIAL AIRPLANE GROUP
See next entry

BOEING DEFENSE & SPACE GROUP
Boeing Computer Services
Electronic Systems Division
Follows Commercial Airplane Group
Helicopters Division
Follows Electronic Systems Division
Military Airplanes Division
Follows Helicopters Division
Product Support Division
Follows Military Airplanes Division

BOEING COMMERCIAL AIRPLANE GROUP

PO Box 3707, Seattle, Washington 98124-2207
Telephone: 1 (206) 237 2121
Fax: 1 (206) 237 1706
Telex: 0650 329430 BOEING CO C
PRESIDENT: Dean D. Thornton

EXECUTIVE VICE-PRESIDENTS:
  Alan R. Mulally (General Manager, 777 Division)
  R. R. Albrecht
  Robert L. Dryden (Manufacturing)
  B. Gissing (Operations)
SENIOR VICE-PRESIDENT, ENGINEERING: Ben A. Cosgrove

SENIOR VICE-PRESIDENT, GOVERNMENT AND INTERNATIONAL AFFAIRS: L. W. Clarkson
VICE-PRESIDENT, COMMUNICATIONS: Gerald A. Hendin

Boeing Commercial Airplane Group, headquartered at Renton, near Seattle, reorganised into three divisions in 1983: Renton Division produced 707 (until 1991) and produces 737 nd 757; Everett Division produces 747 and 767 and will produce 777; Fabrication Division provides manufacturing for other divisions. Materiel Division, created 1984, covers purchasing, quality control and vendor supplies.

Output in 1992 was 218 737s, 61 747s, 99 757s and 63 767s. New orders covered 114 737s, 28 747s, 38 757s, 21 767s and 42 777s. Grand total 8,663 Boeing jet airliners ordered by end March 1993 and 7,264 delivered. Slowdown in airliner deliveries and reduction in defence orders will cause Boeing to reduce its total workforce by 28,000 by end 1994, of which about 19,000 in Seattle airliner production area; airliner deliveries expected to recover in about four years.

Boeing airliner sales reached $30,184 million in 1992, with net earnings calculated at $1,635 million before an accounting change (1991 figures were $29,314 million and $1,567 million respectively); after accounting change, net earnings $552 million.

BOEING 7J7/YXX

Agreement between Boeing and Japanese to develop 7J7/YXX signed 1984, but not under active development in early 1993.

BOEING 707

USAF designations: VC-137, E/KE-3, E-6A and E-8

First flight Boeing 367-80, original prototype of 707, 15 July 1954; developed version ordered in large numbers by US Air Force as KC-135 (Boeing 717); details of commercial 707 and 720 in 1980-81 and earlier *Jane's*; last commercial 707 was 707-320C for Moroccan Government delivered March 1982.

Last 707 to fly was final British Royal Air Force E-3D on 14 June 1991, the 1,009th airframe; 1,010th, an E-3, retained by Boeing as trials aircraft. Boeing Military Airplanes offers tanker/transport conversions of ex-airline 707s (which see); 707/720 conversions also offered by Israel Aircraft Industries (which see), Alenia in Italy (which see) and Comtran International Inc, USA (which see).

BOEING 727

First 727-100 rolled out 27 November 1962; first flight 9 February 1963; FAA certification 24 December 1963; into service with Eastern Air Lines 1 February 1964. 727-200 flew 27 July 1967; FAA certification 30 November 1967; into service with Northeast Airlines 14 December 1967. Last of 572 727-100s delivered October 1972 and of 1,260 727-200s September 1984. Current re-engining programmes by Valsan Partners and Dee Howard (which see). Boeing 727-100 last described in 1973-74 *Jane's* and 727-200 in 1983-84.

BOEING 737-100 and 737-200

First flight of 737-100, 9 April 1967; FAA certification 15 December 1967; 30 built. Superseded by 737-200; first flight 8 August 1967; added to 737-100 type certificate 21 December 1967; first delivery to United Air Lines 29 December 1967. Last of 1,114 Boeing 737-200s (and 30 737-100s) delivered August 1988; total includes 19 **T-43A** navigation trainers for US Air Force and three **Surveillers** for Indonesian Air Force. Details of early versions and developments in 1974-75 *Jane's*; 737-200 last described in *Jane's* 1990-91.

BOEING 737-300

TYPE: First of three 737 variants in production.
PROGRAMME: Production go-ahead March 1981; first flight 24 February 1984; certificated 14 November 1984; first delivery (to USAir) 28 November 1984; 737-300 for Ansett Worldwide (and subsequent lease to British Midland Airways) rolled out at Renton 19 February 1990 (as 1,833rd 737); 737 orders passed 3,000 when Southwest Airlines ordered 34 in third quarter 1992. 120-minute EROPS approved November 1986; approval withdrawn July 1989 due to concerns related to operation in heavy rain and hail; approval restored 14 September 1990. Former Soviet Commonwealth of Independent States Interstate Aviation Committee certificated Boeing 737 family with P&W or CFM engines 18 January 1993; first delivery for CIS registration (737-300 to National State Aviacompany Turkmenistan) 12 November 1992. Production rate for 737 series reduced from 21 to 14 a month in October 1992, and to 10 in October 1993; 2,500th 737 rolled out 16 June 1993.
CURRENT VERSIONS: **737-300**: Basic airliner, *as detailed below.*
  Executive: Typically for about 20 passengers, with conference room, bedroom, bathroom and full dining facilities. Three sold by 29 February 1992, including one to Royal Thai Air Force.
CUSTOMERS: See table for orders/deliveries by 31 March 1993; CAAC ordered 20 Boeing 737-300s for China Southern Airlines in April 1993.
DESIGN FEATURES: Fuselage stretched 2.64 m (8 ft 8 in) compared with 737-200, by 1.12 m (3 ft 8 in) plug forward of wing box and 1.52 m (5 ft 0 in) aft; underfloor freight volume increased by 5.47 m³ (193 cu ft); wing aerofoil modified by 4.4 per cent extension of leading-edge outboard of engines; new slats; new flap sections and track fairings aft of engines; additional lateral control spoilers outboard; each wingtip extended by 28 cm (11 in); increased dorsal fin area and tailplane span.
FLYING CONTROLS: All surfaces powered by two independent hydraulic systems with manual reversion for ailerons and elevator. Elevator servo tabs unlock on manual reversion. Rudder has standby hydraulic actuator and system. Three outboard powered overwing spoiler panels on each wing assist lateral control and also act as airbrakes. Variable incidence tailplane has two electric motors and manual standby.

Leading-edge Krueger flaps inboard and three sections of slats outboard of engines. Two airbrake/lift dumper panels on each wing, inboard and outboard of engines. Triple-slotted trailing-edge flaps inboard and outboard of engines.

FAA Category II landing minima system standard using SP-300 dual digital integrated flight director/autopilot; Category IIIA capability optional. Common pilot type ratings for 737-200, -300, -400 and -500.
STRUCTURE: Aluminium alloy dual-path fail-safe two-spar wing structure. Aluminium alloy two-spar tailplane. Graphite composite ailerons, elevators and rudder, latter built by Short Brothers (UK). Aluminium honeycomb spoiler/airbrake panels and trailing-edges of slats and flaps. Fuselage structure fail-safe aluminium. Some fins made by Xian Aircraft Co in China. Elevators, rudder and aileron contain graphite/Kevlar and CFRP; other unstressed components in GFRP and CFRP.
LANDING GEAR: Hydraulically retractable tricycle type, with Boeing oleo-pneumatic shock absorbers. Inward retracting main units have no doors, wheels forming wheel well seal; nose unit retracts forward. Free-fall emergency extension. Compared with 737-200, nose unit is repositioned downwards by 13 cm (5 in) and modified to ensure adequate ground clearance for larger engine nacelles. Twin nosewheels have tyres size 27 × 7.75. Main units have heavy duty twin wheels, H40 × 14.5-19 heavy duty tyres, and Bendix or Goodrich heavy duty wheel brakes as standard. Mainwheel tyre pressure 13.45-14.00 bars (195-203 lb/sq in). Nosewheel tyre pressure 11-45-11.85 bars

BOEING COMMERCIAL AND NON-COMMERCIAL ORDERS AND DELIVERIES
(up to 31 March 1993)

	Commercial		Non-Commercial		Total	
	Ordered	Delivered	Ordered	Delivered	Ordered	Delivered
707-E3			63	62	63	62
707-E3A			2	2	2	2
707-E3D			3	3	3	3
707-E6A			17	17	17	17
707-KE3			8	8	8	8
707-120	60	60	3	3	63	63
707-120B	78	78			78	78
707-220	5	5			5	5
707-320	69	69			69	69
707-320B	170	170	4	4	174	174
707-320C	306	306	31	31	337	337
707-420	37	37			37	37
720-000	64	64	1	1	65	65
720-000B	89	89			89	89
727-100	405	405	2	2	407	407
727-100C	164	164			164	164
727-200	1237	1237	8	8	1245	1245
727-200F	15	15			15	15
737-TBD	222	0			222	0
737-T43A			19	19	19	19
737-100	29	29	1	1	30	30
737-200	980	980	11	11	991	991
737-200C	101	101	3	3	104	104
737-300	990	781	3	3	993	784
737-400	423	291			423	291
737-500	272	227			272	227
747-E4A			3	3	3	3
747-E4B			1	1	1	1
747-SP	43	43	2	2	45	45
747-100	167	167			167	167
747-100B	9	9			9	9
747-100SR	29	29			29	29
747-200B	223	223	2	2	225	225
747-200C	13	13			13	13
747-200F	69	69	4	4	73	73
747-200M	78	78			78	78
747-300	55	55	1	1	56	56
747-300M	21	21			21	21
747-300SR	4	4			4	4
747-400	387	194	2	2	389	196
747-400D	13	11			13	11
747-400F	16	0			16	0
747-400M	45	34			45	34
757-200	760	498	1	1	761	499
757-200M	1	1			1	1
757-200PF	61	34			61	34
767-TBD	17	0			17	0
767-200	124	124			124	124
767-200ER	94	91			94	91
767-300	93	66			93	66
767-300ER	282	198			282	198
767-300F	30	0			30	0
777-200	118	0			118	0
Totals	8468	7070	195	194	8663	7264

NOTES
Boeing-owned aircraft not included above: one 727-100 E002 scrapped; one 747-100 RA001 Boeing Field; one 757-200 NA001 Boeing Field; one 767-200 VA001 Boeing Field (Airborne Optical Adjunct); remaining 707-E3 is temporarily retained at Boeing for test purposes. In May 1993, Continental ordered 50+50 737s, 25+25 757s, 12+18 767-300ERs and 5+5 777 B Markets.

Sub-types: TBD, variant to be decided; ER, extended range; SR, short range; F, freighter; SP (747), short body, long range; PF, package freighter; M, combi; D, domestic (Japan).

442 USA: AIRCRAFT—BOEING

Boeing 737-300 of British Midland Airways

(166-172 lb/sq in).

POWER PLANT: Basic aircraft has two CFM International CFM56-3C-1 turbofans rated at either 88.97 kN (20,000 lb st) or 97.86 kN (22,000 lb st), introduced 1988. Engines pylon-mounted forward of wings, and higher than those of 737-200; each has external strake on inboard side. Standard fuel capacity up to 20,104 litres (5,311 US gallons; 4,422 Imp gallons), with integral fuel cells in wing centre-section and integral wing tanks. Fuel options up to 23,830 litres (6,295 US gallons; 5,242 Imp gallons) with Rogerson tanks in underfloor cargo bays (from 1989). Single-point pressure refuelling under leading-edge of starboard wing.

ACCOMMODATION: Crew of two side by side on flight deck. Alternative cabin layouts seat from 128 to 149 passengers. Typical arrangements offer eight first class seats four-abreast at 96.5 cm (38 in) pitch and 120 tourist class seats six-abreast at 81 cm (32 in) in mixed class; and 141 or 149 all-tourist class at seat pitches of 81 cm (32 in) or 76 cm (30 in) respectively. One plug type door at each corner of cabin, with passenger doors on port side and service doors on starboard side. Airstair for forward cabin door optional. Overwing emergency exit on each side. One or two galleys and one lavatory forward, and one or two galleys and lavatories aft, depending on configuration. New lightweight interior, using advanced crushed core materials, providing total overhead baggage capacity of 6.80 m³ (240 cu ft), equivalent to 0.048 m³ (1.7 cu ft) per passenger. Underfloor freight holds forward and aft of wing, with doors on starboard side.

SYSTEMS: AiResearch bleed air control system for thermal anti-icing, air-conditioning and pressurisation systems; max differential 0.52 bar (7.5 lb/sq in); two functionally independent hydraulic systems with a third standby system, using fire resistant hydraulic fluid, for flying controls, flaps, slats, landing gear, nosewheel steering and brakes; pressure 207 bars (3,000 lb/sq in). No pneumatic system. Electrical supply since 1991 from two 50kVA variable speed constant frequency generators. AlliedSignal Garrett GTCP-5-129(C) APU (GTCP36-280 from 1988 and APS 2000 from 1991) for air supply and electrical power in flight and on ground as well as engine starting.

AVIONICS: Flight management computer provides lateral, vertical and time navigation using pilot-set waypoints, nav database or stored company routes; time navigation used to arrive at any fix at a preset time and shows required take-off time bracket to make good RTA; dual digital flight management computers to be delivered from March 1993; EFIS screens show map, flight plan, full or partial compass rose, weather and, optionally, integrated airspeed scale; electronic engine instrument system has coloured LED dials, with secondary panel, secondary engine and hydraulics indications; windshear alerting with recovery guidance in attitude indicator; full flight regime autothrottle; dual laser gyro inertial system; digital colour weather radar. These avionics common to 737-300, -400 and -500.

DIMENSIONS, EXTERNAL:
Wing span	28.88 m (94 ft 9 in)
Wing chord at root	4.71 m (15 ft 5.6 in)
Wing aspect ratio	7.91
Length overall	33.40 m (109 ft 7 in)
Height overall	11.13 m (36 ft 6 in)
Tailplane span	12.70 m (41 ft 8 in)
Wheel track	5.23 m (17 ft 2 in)
Wheelbase	12.45 m (40 ft 10 in)

Main passenger door (port, fwd):
Height	1.83 m (6 ft 0 in)
Width	0.86 m (2 ft 10 in)
Height to sill	2.62 m (8 ft 7 in)

Passenger door (port, rear): Height 1.83 m (6 ft 0 in)
Width	0.76 m (2 ft 6 in)
Width with airstair	0.86 m (2 ft 10 in)
Height to sill	2.74 m (9 ft 0 in)

Emergency exits (overwing, port and stbd, each):
Height	0.97 m (3 ft 2 in)
Width	0.51 m (1 ft 8 in)

Galley service door (stbd, fwd):
Height	1.65 m (5 ft 5 in)
Width	0.76 m (2 ft 6 in)
Height to sill	2.62 m (8 ft 7 in)

Service door (stbd, rear): Height 1.65 m (5 ft 5 in)
Width	0.76 m (2 ft 6 in)
Height to sill	2.74 m (9 ft 0 in)

Freight hold door (stbd, fwd): Height 1.22 m (4 ft 0 in)
Width	1.30 m (4 ft 3 in)
Height to sill	1.30 m (4 ft 3 in)

Freight hold door (stbd, rear): Height 1.22 m (4 ft 0 in)
Width	1.22 m (4 ft 0 in)
Height to sill	1.55 m (5 ft 1 in)

DIMENSIONS, INTERNAL:
Cabin, incl galley and toilet:
Length	23.52 m (77 ft 2 in)
Max width	3.45 m (11 ft 4 in)
Max height	2.13 m (7 ft 0 in)

Freight hold volume: basic 30.2 m³ (1,068 cu ft)
with max optional fuel 22.4 m³ (792 cu ft)

AREAS:
Wings, gross	105.4 m² (1,135.0 sq ft)
Ailerons (total)	2.49 m² (26.8 sq ft)
Trailing-edge flaps (total)	16.87 m² (181.6 sq ft)
Slats (total)	7.23 m² (77.8 sq ft)
Ground spoilers (total)	5.00 m² (53.8 sq ft)
Flight spoilers (total)	2.64 m² (28.4 sq ft)
Fin	23.13 m² (249.0 sq ft)
Rudder	5.22 m² (56.2 sq ft)
Tailplane	31.31 m² (337.0 sq ft)
Elevators, incl tabs (total)	6.55 m² (70.5 sq ft)

WEIGHTS AND LOADINGS (A: basic aircraft, B: long-range option):
Operating weight empty: A	31,895 kg (70,320 lb)
B	up to 32,459 kg (71,560 lb)
Max T-O weight: A	56,472 kg (124,500 lb)
B	up to 62,822 kg (138,500 lb)
Max ramp weight: A	56,699 kg (125,000 lb)
B	up to 63,050 kg (139,000 lb)
optional, with controlled CG	63,500 kg (140,000 lb)
Max zero-fuel weight: A	47,627 kg (105,000 lb)
B	up to 49,713 kg (109,600 lb)
Max landing weight: A	51,719 kg (114,000 lb)
B	up to 52,888 kg (116,600 lb)

PERFORMANCE: (A: at brake release weight of 56,472 kg; 124,500 lb, B: at optional BRW of 62,822 kg; 138,500 lb):
T-O field length, S/L, at 29°C (84°F):
A	2,027 m (6,650 ft)
B	2,749 m (9,020 ft)

Wet landing field length, 40° flap, at max landing weight
A, B	1,603 m (5,260 ft)

Still air range with 140 passengers, T-O at S/L:
A	2,850 nm (5,280 km; 3,275 miles)
B	3,400 nm (6,300 km; 3,910 miles)

OPERATIONAL NOISE LEVELS: 737-300, -400 and -500 only exceed FAR Pt 36, Stage 3/ICAO Annex 16 Chapter noise in approach with 40° flap setting

BOEING 737-400

TYPE: Stretched version of 737-300.

PROGRAMME: Announced June 1986; rolled out 26 January 1988; first flight 19 February 1988; certificated for up to 188 passengers 2 September 1988; first delivery (to Piedmont Airlines) 15 September 1988. High gross weight structure variant rolled out 23 December 1988; certificated by FAA and delivered to first customer 21 March 1989. EROPS approval granted 14 September 1990. CIS certification with CFM engines 18 January 1993, as for 737-300. Production rate of 737 being reduced in October 1993.

CURRENT VERSIONS: **Basic and long-range:** As described below.

High gross weight structure: Optional strengthened centre-fuselage, wing and landing gear, slats and Kruegers, increased fuel capacity by means of Rogerson tanks in aft cargo bay.

CUSTOMERS: See table for orders/deliveries at 31 March 1993.

DESIGN FEATURES: Incorporates all the new technology of 737-300. Fuselage has 1.83 m (6 ft 0 in) plug forward of wing and 1.22 m (4 ft 0 in) aft, totalling 3.05 m (10 ft); outer wings and landing gear strengthened for max landing weights from 54,885 to 56,245 kg (121,000 to 124,000 lb). Tail bumper standard on all 737-400s. At gross weights above 63,049 kg (139,000 lb) loading must be controlled to preserve CG. Avionics and systems as for 737-300 except for modified avionics software and improved environmental control system.

POWER PLANT: Two CFM56-3C-1 turbofans rated at 97.86 kN (22,000 lb st) or 104.5 kN (23,500 lb st). Basic fuel capacity 20,104 litres (5,311 US gallons; 4,422 Imp gallons); max long-range option fuel capacity 23,830 litres (6,295 US gallons; 5,242 Imp gallons).

DIMENSIONS, EXTERNAL: As for 737-300 except:
Length overall	36.45 m (119 ft 7 in)

DIMENSIONS, INTERNAL: As for 737-300 except:
Cabin, incl galley and toilet:
Length	27.18 m (89 ft 2 in)

WEIGHTS AND LOADINGS (A: basic aircraft, B: long-range option):
Operating weight empty: A	33,434 kg (73,710 lb)
B	34,271 kg (75,555 lb)
Max T-O weight: A	62,822 kg (138,500 lb)
B	68,039 kg (150,000 lb)
Max ramp weight: A	63,049 kg (139,000 lb)
B	68,265 kg (150,500 lb)
Max zero-fuel weight: A	51,256 kg (113,000 lb)
B	53,070 kg (117,000 lb)
Max landing weight: A	54,885 kg (121,000 lb)
B	56,245 kg (124,000 lb)

PERFORMANCE (A: at T-O weight of 62,822 kg; 138,500 lb, B: at optional T-O weight of 68,039 kg; 150,000 lb):
T-O field length, S/L, at 30°C: A   2,315 m (7,600 ft)
  B, with optional higher thrust engines
                                      2,500 m (8,200 ft)
Wet landing field length, 40° flap:
  A at 54,885 kg (121,000 lb) landing weight
                                        1,725 m (5,650 ft)
  B at 56,245 kg (124,000 lb) landing weight
                                        1,850 m (6,070 ft)
Range with 146 passengers, T-O at S/L:
  A   2,700 nm (5,000 km; 3,105 miles)
  B   3,200 nm (5,930 km; 3,680 miles)

BOEING 737-500

TYPE: Short-body version of 737-300, replacing 737-200.
PROGRAMME: Initially known as 737-1000; announced as 737-500 on 20 May 1987; first flight 20 June 1989; certificated 12 February 1990 after 375 hour test programme; first delivery (to Southwest Airlines) 28 February 1990; EROPS approval 14 September 1990. CIS certification with CFM engines 18 January 1993, as for 737-300 and -400. Production rate of 737 being reduced in October 1993.
CURRENT VERSIONS: Max T-O weights ranging from 52,390 to 60,554 kg (115,500 to 133,500 lb).
CUSTOMERS: Launch customers were Braathens SAFE of Norway (25 firm) and Southwest Airlines (20 firm and 20 optioned); see table on page 441 for orders/deliveries at 31 March 1993.
DESIGN FEATURES: Incorporates advanced technology of 737-300 and -400, but fuselage shortened. Engine thrust and fuel capacity options detailed below. New nosewheel tyres. Systems and avionics as for 737-300 with minor variations.
POWER PLANT: Two CFM International CFM56-3C-1 turbofans, rated at 88.97 kN (20,000 lb st) or derated to 82.29 kN (18,500 lb st) according to gross weight. Electronic power control allows fixed-throttle climb and limits fan speed and EGT overshoots. Basic fuel capacity 20,104 litres (5,311 US gallons; 4,422 Imp gallons); long-range option fuel capacity 23,830 litres (6,295 US gallons; 5,242 Imp gallons).
DIMENSIONS, EXTERNAL: As for 737-300 except:
  Length overall   31.01 m (101 ft 9 in)
WEIGHTS AND LOADINGS (A: basic aircraft, B: long-range option):
  Operating weight empty: A   30,953 kg (68,240 lb)
                            B   31,515 kg (69,480 lb)
  Max T-O weight: A   52,390 kg (115,500 lb)
                     B   60,554 kg (133,500 lb)
  Max ramp weight: A   52,617 kg (116,000 lb)
                    B   60,781 kg (134,000 lb)
  Max zero-fuel weight: A   46,493 kg (102,500 lb)
                        B   46,720 kg (103,000 lb)
  Max landing weight: A, B   49,895 kg (110,000 lb)
PERFORMANCE:
  Range with 108 passengers:
    A   1,700 nm (3,150 km; 1,955 miles)
    B   2,420 nm (4,485 km; 2,785 miles)

BOEING 737X

TYPE: Advanced version of 737.
PROGRAMME: Proposed configuration design completed first quarter 1993 and discussed with 22 existing 737 operators; deliveries could start in 1996-98. Lead customers in mid-1993 thought to be Finnair and SAS. *See also Addenda.*
DESIGN FEATURES: Main objectives are US trans-continental range, greater reliability to give lower maintenance and operating costs, and cruise speed increased to Mach 0.8; fuselage might be lengthened 4.05 m (13 ft 3½ in) from 737-400 and wing span increased 5 m (16 ft 4¼ in); passenger capacity could lie between 150 (737-400) and 200 (757), but 737X will also be offered in -300X, -400X and -500X versions with passenger capacity from 100 to 170 as well as the stretched -400X.
Wing area would be increased by 20 per cent by greater span and chord; aerofoil modified for cruising at Mach 0.78 to 0.80; additional fuel capacity for 600 to 800 nm (1,112 to 1,483 km; 690 to 920 miles); engine choice lay between CFMI CFM56-3XS and IAE V2500-B5 in mid-1993; engine objectives are greater thrust and lower noise and fuel consumption; flying control and flight deck systems to have maximum commonality with earlier 737s.

BOEING 747 (DISCONTINUED MODELS)

Programme announced 13 April 1966 (first-ever wide-body jet airliner), with Pan American order for 25; official programme launch 25 July 1966; first flight 9 February 1969; FAA certification 30 December 1969; first delivery (to Pan Am) 12 December 1969; first route service New York-London flown 22 January 1970. In May 1990 Boeing decided to market only the -400; last -200 (a -200F Freighter for Nippon Cargo Air Lines) delivered 19 November 1991.
For all variants prior to 747-400, see 1990-91 and earlier editions. Production variants, listed in table on earlier page, totalled 724 (205 -100, 45 SP, 393 -200 and 81 -300).

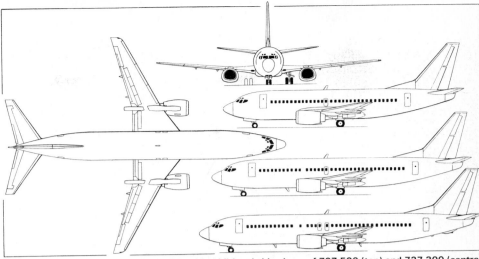

Three-view drawing of Boeing 737-400, with additional side views of 737-500 (top) and 737-300 (centre) *(Dennis Punnett)*

Boeing 737-500 of Maersk Air taking off from Boeing's Renton factory airfield

Nineteen Pan American 747s modified as passenger/cargo C-19As by Boeing Military Airplanes for Civil Reserve Air Fleet (see 1990-91 edition).
Major upgrade of 747-100 and -200B, including side cargo door, offered by Boeing to give main deck cargo capability. Alternatives include all-passenger or all-cargo Special Freighter; 6- or 12-pallet Combi; all-passenger, all-freight or mixed passenger/cargo Convertible. Main elements of modification include cargo door measuring 3.05 m (10 ft 0 in) high by 3.40 m (11 ft 2 in) wide, strengthened floor, powered or manual cargo handling systems. Optional increase in gross weight.
Original fuse pins, which allow engines to break away from wing in case of excessive loading, replaced by second-generation pins during 1993 following a number of engine separations, mainly from older 747s converted to freighters; third-generation pin and additional struts expected to be certificated by FAA in mid-1993.

BOEING 747-400

TYPE: 747 with extended capacity and range.
PROGRAMME: Announced May 1985; design go-ahead July 1985; rollout 26 January 1988; first flight 29 April 1988; certificated with P&W PW4056 10 January 1989; certificated with GE CF6-80C2B1F 8 May 1989; R-R RB211-524G 8 June 1989; R-R RB211-524H 11 May 1990. Since May 1990, -400 is the only 747 marketed. Production to be reduced from five to three a month in second quarter 1994.
CURRENT VERSIONS: **747-400:** Basic passenger version; standard and three optional gross weights (see below). *Detailed description applies to this version, except where indicated.*
747-400 Combi: Passenger/freight version; certificated 1 September 1989; max 266 three-class passengers with freight, 413 without; port side rear freight door; main deck limit seven pallets at 27,215 kg (60,000 lb); underfloor and fuel capacities as for passenger 747; 34 delivered by 30 April 1993. For all gross weights, max landing weight 285,763 kg (630,000 lb) and max zero-fuel weight 256,280 kg (565,000 lb). All three engine options available.
747-400F: All-freight version. See separate entry.
747-400 Domestic: Special high-density two-class 568-passenger version; certificated 10 October 1991; ordered by Japan Air Lines (six), All Nippon (six) and Japan Air System (one). Max T-O weight 272,155 kg (600,000 lb) but can be certificated to 394,625 kg (870,000 lb). Structurally reinforced; no winglets; lower engine thrust; five more upper deck windows; revised avionics software and cabin pressure schedule; brake cooling fans; five pallets, 14 LD-1 containers and bulk cargo under floor; GE or P&W engines.
747-400 Performance Improvement Package (PIP): Announced April 1993; first stage includes gross weight increase of 907 kg (2,000 lb) or more, longer chord dorsal fin made of CFRP, slower transfer of fuel from tailplane trim tank to prolong trim drag alleviation, and wing spoilers held down more tightly to reduce profile drag and leakage; this stage to be applied to production aircraft as soon as ready and to be retrofittable; first-stage PIP being flight tested in leased United Airlines 747-400 May 1993. Second PIP stage, available by mid-1996 if launched in 1993, would increase gross weight to 421,840 kg (930,000 lb) and range to 8,000 nm (14,825 km; 9,210 miles); required airframe reinforcement would include strengthened wing carry-through box and landing gear and thicker top and bottom skins over inboard portion of wing torsion boxes; fuel capacity of tailplane and wing increased; second stage structural changes would facilitate 6.0 m (19 ft 8¼ in) fuselage stretch with capacity for 80 more passengers planned for 747X, which could emerge in 1997; this 747X would have same range as 747-200 and -300.

444　USA: AIRCRAFT—BOEING

CUSTOMERS: Northwest Orient Airlines ordered 10 -400s with PW4000s and 420-passenger interior October 1985; first delivery 26 January 1989. Total orders for 747-400, 464 at 30 April 1993 to 24 operators; includes 13 -400 Domestics, 16 Freighters and 37 Combis; 246 delivered, including 12 Domestics and 34 Combis.

DESIGN FEATURES: Wing has special Boeing aerofoil and 3.66 m (12 ft 0 in) greater span than 747-300; sweepback at quarter-chord 37° 30′; thickness/chord ratio 13.44 per cent inboard, 7.8 per cent at mid-span, 8 per cent outboard; dihedral at rest 7°; incidence 2°; winglets, canted 22° outward and swept 60°, increase range by three per cent; upper deck extended rearwards by 7.11 m (23 ft 4 in).

FLYING CONTROLS: Generally as for 747-200; rudder travel increased to ±30° reduces V_{MCG} by 10 knots (18 km/h; 11 mph); six-screen EFIS and 600 fewer lights, gauges and switches than early 747s keeps two-crew workload equivalent to that in 737, 757 and 767.

STRUCTURE: Wing and tail surfaces are aluminium alloy dual-path fail-safe structures; advanced aluminium alloys in wing torsion box save 2,721 kg (6,000 lb); advanced aluminium honeycomb spoiler panels; CFRP winglets and main deck floor panels; advanced graphite/phenolic and Kevlar/graphite in cabin fittings and engine nacelles; frame/stringer/stressed skin fuselage with some bonding. Improved corrosion protection and further coverage with compound being introduced during 1993.

LANDING GEAR: Hydraulically retractable tricycle type; twin-wheel nose unit retracts forward; main gear consists of four four-wheel bogies; two, mounted side by side under fuselage at wing trailing-edge, retract forward; two, mounted under wings, retract inward; carbon disc brakes on all mainwheels, with individually controlled digital anti-skid units; mainwheel diameter increased to 56 cm (22 in); 125 cm (49 in) diameter low-profile tyres; new wheels save 816 kg (1,800 lb) weight.

POWER PLANT: Four 252.4 kN (56,750 lb st) Pratt & Whitney PW4056, 258 kN (57,900 lb st) General Electric CF6-80C2B1F, 258 kN (58,000 lb st) Rolls-Royce RB211-524G or 270 kN (60,600 lb st) Rolls-Royce RB211-524H turbofans.

Alternative engines, not yet certificated at 30 April 1993, are 266.9 kN (60,000 lb st) PW4060, 275.8 kN (62,000 lb st) PW4062, 273.6 kN (61,500 lb st) CF6-80C2B1F1 or CF6-80C2B7F.

Fuel capacity 204,333 litres (53,985 US gallons; 44,987 Imp gallons) with P&W and R-R engines; 203,500 litres (53,765 US gallons; 44,804 Imp gallons) with GE engines; at 377,842 kg (833,000 lb) and 394,625 kg (870,000 lb) T-O weights, fuel capacity including tailplane tank is 216,846 litres (57,285 US gallons; 47,737 Imp gallons) with P&W and R-R engines and 216,013 litres (57,065 US gallons; 47,554 Imp gallons) with GE engines; optional tailplane tank holds 12,492 litres (3,300 US gallons; 2,748 Imp gallons) transferable fuel (must be full for take-off at 394,625 kg; 870,000 lb gross weight).

ACCOMMODATION: Two-crew flight deck, with seats for two observers; two-bunk crew rest cabin accessible aft of flight deck. Optional overhead cabin crew rest compartments above rear of main deck cabin (four bunks, four seats; eight bunks, two seats; two bunks, two seats, five sleeper seats). Typical 421-seat three-class configuration accommodates 42 business class on upper deck; 24 first class in front cabin, 29 business class in middle cabin and 326 economy class in rear cabin on main deck. Max upper deck capacity 69 economy class. Centre overhead stowage bins 0.16 m³ (5.7 cu ft) volume per 1.02 m (40 in) long bin; outboard bins 0.45 m³ (15.9 cu ft) volume per 1.52 m (60 in) long bin; 0.083 m³ (2.95 cu ft) bin volume per passenger (three-class). Two modular upper deck toilets, 14 on main deck, relocatable and vacuum-drained into four waste tanks. Basic galley configuration one on upper deck, seven centreline and two sidewall on main deck; toilets and galleys can be quickly relocated if required fittings are installed; advanced integrated audio/video/announcement system.

Underfloor freight: forward compartment, five 2.44 m (96 in) × 3.18 m (125 in) pallets or 16 LD-1 containers; aft compartment, 14 LD-1 containers and 23.6 m³ (835 cu ft) bulk cargo or 16 LD-1 and 13.9 m³ (490 cu ft) bulk cargo; pallets and LD-1s can be interlined with Boeing 767.

SYSTEMS: P&WC PW901A APU with 180kVA generator.

DIMENSIONS, EXTERNAL:
Wing span	64.44 m (211 ft 5 in)
Wing span, fully fuelled	64.92 m (213 ft 0 in)
Length: overall	70.66 m (231 ft 10 in)
fuselage	68.63 m (225 ft 2 in)
Height overall	19.41 m (63 ft 8 in)
Tailplane span	22.17 m (72 ft 9 in)
Wheel track	11.00 m (36 ft 1 in)
Wheelbase	25.60 m (84 ft 0 in)
Passenger doors (ten, each): Height	1.93 m (6 ft 4 in)
Width	1.07 m (3 ft 6 in)
Height to sill	approx 4.88 m (16 ft 0 in)
Baggage door (front hold): Height	1.68 m (5 ft 6 in)
Width	2.64 m (8 ft 8 in)
Height to sill	approx 2.64 m (8 ft 8 in)
Baggage door (forward door, rear hold):	
Height	1.68 m (5 ft 6 in)
Width	2.64 m (8 ft 8 in)
Height to sill	approx 2.69 m (8 ft 10 in)
Bulk loading door (rear door, rear hold):	
Height	1.19 m (3 ft 11 in)
Width	1.12 m (3 ft 8 in)
Height to sill	approx 2.90 m (9 ft 6 in)
Freighter cargo door (port): Height	3.05 m (10 ft 0 in)
Width	3.40 m (11 ft 2 in)
Height to sill	4.87 m (16 ft 0 in)

WEIGHTS AND LOADINGS (letters denote engine installation as follows: P: PW4056, C: CF6-80C2B1F, R: RB211-524G/H):
Operating weight empty: P	180,985 kg (399,000 lb)
P at max optional T-O weight	181,484 kg (400,100 lb)
C	181,030 kg (399,100 lb)
C at max optional T-O weight	181,529 kg (400,200 lb)
R	182,255 kg (401,800 lb)
R at max optional T-O weight	182,754 kg (402,900 lb)

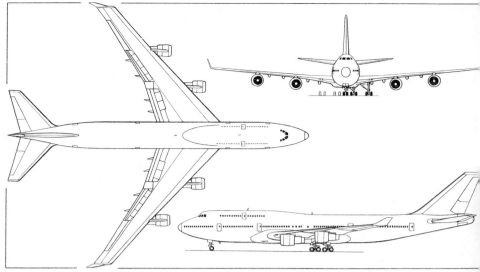

Boeing 747-400 advanced long-range airliner (General Electric CF6-80C2 engines) *(Dennis Punnett)*

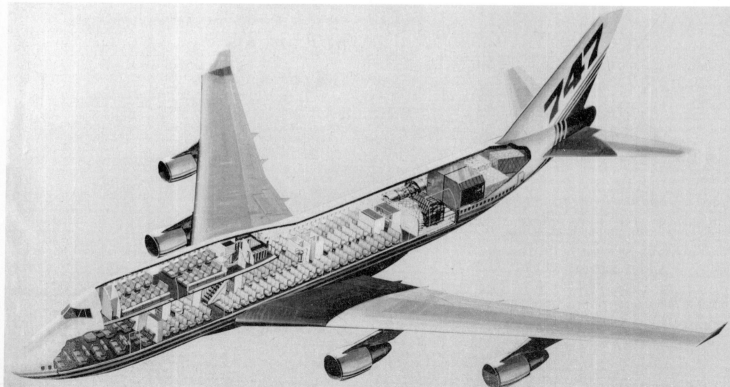

Internal arrangement of Boeing 747-400 Combi with 266 passengers in three classes and seven freight pallets in the rear cargo hold. Standard underfloor capacity remains

Max T-O weight: P, C, R 362,875 kg (800,000 lb)
 or 385,555 kg (850,000 lb)
 or 394,625 kg (870,000 lb)
Max ramp weight: P, C, R 364,235 kg (803,000 lb)
 or 386,915 kg (853,000 lb)
 or 395,986 kg (873,000 lb)
Max zero-fuel weight: P, C, R 242,670 kg (535,000 lb)
Max landing weight: at standard max T-O weight:
 P, C, R 260,360 kg (574,000 lb)
 at alternate max T-O weights:
 P, C, R 285,765 kg (630,000 lb)
RFORMANCE (engines as designated under Weights and Loadings):
Approach speed at basic landing weight:
 P, C, R 146 knots (270 km/h; 168 mph)

Approach speed at highest optional landing weight:
 P, C, R 153 knots (284 km/h; 176 mph)
Initial cruise altitude at highest optional T-O weight:
 P, C, R 10,030 m (32,900 ft)
FAR T-O field length at S/L, ISA, at highest optional T-O weight: P, C 3,322 m (10,900 ft)
 R 3,352 m (11,000 ft)
FAR landing field length at max landing weight of 285,765 kg (630,000 lb): P, C, R 2,072 m (6,800 ft)
Design range, typical international rules, 420 three-class passengers, at highest optional T-O weight:
 P 7,165 nm (13,278 km; 8,239 miles)
 C 7,230 nm (13,398 km; 8,314 miles)
 R 7,100 nm (13,157 km; 8,165 miles)

BOEING 747-400F

TYPE: All-freight version of 747-400.
PROGRAMME: Structurally complete February 1993; first flight (N6005C) 7 May 1993; certification expected Autumn 1993.
CUSTOMERS: Six customers (Air France, Asiana, Cargolux, Cathay Pacific, KLM and Singapore Airlines) ordered 16 aircraft by May 1993.
DESIGN FEATURES: 747-200F fuselage (short upper deck) with additional changes combined with stronger and larger 747-400 wing; strengthened floor of short upper deck, as offered for -200F, also integrated into 747-400F; further developed freight handling system; total cargo volume increased by 41.7 m³ (1,473 cu ft); can carry 110,676 kg (244,000 lb) cargo for 4,300 nm (7,970 km; 4,950 miles);

Boeing 747-400F freighter takes off for first time from Paine Field, Washington, on 7 May 1993

Boeing 747-400F freighter showing short upper deck, nose loading door and optional rear port-side freight door

fuel burn per pound payload 12 per cent lower than 747-200F. Same gross weights as passenger 747-400; max landing weight at optional T-O weight, 302,092 kg (666,000 lb); max zero-fuel weight, 276,691 kg (610,000 lb), can be increased on condition T-O weight is decreased.

ACCOMMODATION: Two-pilot crew, as 747-400. Upward-opening nose cargo door and optional port-side rear cargo door; underfloor cargo doors fore and aft of wing and bulk cargo door aft of rear underfloor door; two crew doors to port. Capacity for 30 pallets on main deck and 32 LD-1 containers plus bulk cargo under floor.

BOEING 747X

New design high-capacity wide-body; existence of market study activity confirmed March 1992. Several versions being considered, including stretched 747-400s (see 747-400 Performance Improvement Package variant). Largest 747X would be a completely new aircraft with wider, circular-section double-deck fuselage seating 612 in three-class configuration (36 first class, 114 business class and 198 economy class on main deck and 264 economy class on upper deck); possible later stretch to increase to 750. Larger wing, carrying four high-bypass turbofans on short pylons; larger tail surfaces; length and wheelbase similar to 747-400 to aid airport compatibility. Initial version max gross weight 545,000 kg (1,201,500 lb); range similar to 747-400.

BOEING 757

TYPE: Medium-range twin-turbofan airliner.

PROGRAMME: New-technology family designated 757/767/777 announced early 1978; 757 has 707/727/737 fuselage cross-section and two large turbofans; Eastern Air Lines and British Airways ordered 21 firm and 24 optioned and 19+18 respectively 13 August 1978; first flight (N757A) 19 February 1982 powered by 166.4 kN (37,400 lb st) Rolls-Royce RB535Cs and designated 757-200; first Boeing airliner launched with foreign engine; FAA certification 21 December 1982; CAA certification 14 January 1983; revenue services began 1 January 1983 (EAL) and 9 February 1983 (BA). First flight of 757 powered by P&W PW2037s 14 March 1984; certificated October 1984 and delivered to Delta; first 757 with RB535E4s delivered to EAL 10 October 1984; first extended range model delivered to Royal Brunei Airlines May 1986; 757 with RB535E4 engines approved FAA EROPS December 1986 (extended to 180 min July 1990); 757 with PW2037/3040 EROPS approved April 1990 (180 min for PW2037 April 1992); Boeing windshear guidance and detection system approved by FAA January 1987. First 757 kept by Boeing for flight test support; used as avionics testbed for Lockheed YF-22. Production rate of 757 reduced from 8.5 per month to seven in June 1993 (instead of September) and then to five in October 1993 (instead of November). Pemco starting conversions of in-service 757s to freighter (15 pallets), combi and a quick-change layout with palletised seating. Discussions under way early 1993 for possible assembly of 757 in China, leading towards full production.

CURRENT VERSIONS: **757-200**: First production passenger airliner; extended range available. *Main description applies to this version, except where indicated.*

757-200PF Package Freighter: Large freight door forward, single crew door and no windows; up to 15 standard 2.24 × 3.18 m (88 × 125 in) cargo pallets on main deck.

757-200 Combi: Mixed cargo/passenger configuration with windows; upward opening cargo door to port (forward) 3.40 × 2.18 m (134 × 86 in); carries up to three 2.24 × 2.74 m (88 × 108 in) cargo containers and 150 passengers; one delivered to Royal Nepal Airlines.

CUSTOMERS: See table on page 441 for orders/deliveries at 31 March 1993.

COSTS: Price of four 757s ordered by Shorouk in November 1992 reported as $240 million.

DESIGN FEATURES: Special Boeing aerofoils; sweepback at quarter-chord 25°; dihedral 5°; incidence 3° 12′.

FLYING CONTROLS: All-speed fully powered outboard ailerons assisted by five flight spoilers on each wing also acting variously as airbrakes and ground spoilers; one additional ground spoiler inboard on each wing; elevators and rudder; double-slotted trailing-edge flaps; full-span leading-edge slats, five sections each wing; variable incidence tailplane. EFIS instruments with engine indication and crew alerting system (EICAS); Collins FCS-700 autopilot flight director system (AFDS), EFIS-700 electronic flight instrument system.

STRUCTURE: Aluminium alloy two-spar fail-safe wing box; centre-section continuous through fuselage; ailerons, flaps and spoilers extensively of honeycomb, graphite composites and laminates; tailplane has full-span light alloy torque boxes; fin has three-spar dual cell light alloy torque box; elevators and rudder have graphite/epoxy honeycomb skins supported by honeycomb and laminated spar and rib assemblies; CFRP wing/fuselage and flap track fairings.

Subcontractors include Hawker de Havilland (wing inspar ribs), Shorts (inboard flaps), CASA (outboard flaps), Boeing Renton (leading-edge slats, main cabin sections), Boeing Helicopters (fixed leading-edges), Boeing Military Airplanes (flight deck), Grumman (overwing spoiler panels), Heath Tecna (wing/fuselage and flap track fairings), Schweizer (wingtips), Vought Aircraft (fin and tailplane, extreme rear fuselage), Rohr Industries (engine support struts), IAI (dorsal fin), Fleet Industries (APU access doors).

LANDING GEAR: Retractable tricycle type, with main and nose units manufactured by Menasco. Each main unit carries a four-wheel bogie, fitted with Dunlop or Goodrich wheels, carbon brakes and tyres. Twin-wheel nose unit, also with Dunlop or Goodrich tyres. All landing gear doors of CFRP/Kevlar. Min ground turning radius 21.64 m (71 ft) at nosewheels, 29.87 m (98 ft) at wingtip.

POWER PLANT: Two 166.4 kN (37,400 lb st) Rolls-Royce 535C, 170 kN (38,200 lb st) Pratt & Whitney PW2037, 178.4 kN (40,100 lb st) Rolls-Royce 535E4, or 185.5 kN (41,700 lb st) Pratt & Whitney PW2040 turbofans, mounted in underwing pods. Fuel capacity 42,597 litres (11,253 US gallons; 9,370 Imp gallons).

ACCOMMODATION: Crew of two on flight deck, with provision for an observer. Five to seven cabin attendants. Nine standard interior arrangements for 178 (16 first class/162 tourist), 186 (16 first class/170 tourist), 202 (12 first class/190 tourist), 208 (12 first class/196 tourist) mixed class passengers, or 214, 220, 223, 224 or 239 all-tourist passengers. First class seats are four-abreast, at 96.5 cm (38 in) pitch; tourist seat pitch is 81 or 86 cm (32 or 34 in), mainly six-abreast, in mixed class arrangements. Large overhead bins of Kevlar provide approximately 0.054 m³ (1.9 cu ft) of stowage per passenger. Choice of two cabin door configurations, with either three passenger doors and two overwing emergency exits on each side (used with 186, 208, 220 and 224 seat interiors), or four doors on each side (used with 178, 202, 214, 223 and 239 seat interiors). All versions have a galley at front on starboard side and another at rear (two on 178 and 186 passenger versions and three on 239 version plus one amidships); toilet at front on port side and three more at rear (186, 202, 208, 220, 224 passengers) or two at rear (239) or amidships (178, 214, 223 passengers). Coat closet at front of first class cabins and 214/220 passenger interiors. Baggage/cargo hold doors on starboard side.

SYSTEMS: AiResearch ECS; General Electric engine thrust management system; Honeywell-Vickers engine driven hydraulic pumps; four Abex electric hydraulic pumps. Hydraulic system maximum flow rate 140 litres (37 US gallons; 30.8 Imp gallons)/min at T-O power on engine driven pumps; 25.4-34.8 litres (6.7-9.2 US gallons; 5.6-7.7 Imp gallons)/min on electric motor pumps; 42.8 litres (11.3 US gallons; 9.4 Imp gallons)/min on ram air turbine. Independent reservoirs, pressurised by air from pneumatic system, maximum pressure 207 bars (3,000 lb/sq in) on primary pumps. Sundstrand electrical power generating system and ram air turbine; and AlliedSignal Garrett GTCP331-200 APU. Wing thermally anti-iced.

AVIONICS: Honeywell inertial reference system (IRS) (first commercial application of laser gyros); IRS provides position, velocity and attitude information to flight deck displays, and the flight management computer system (FMCS) and digital air data computer (DADC) supplied by Honeywell; FMCS provides automatic en route and terminal navigation capability, and also computes and commands both lateral and vertical flight profiles for optimum fuel efficiency, maximised by electronic linkage of the FMCS with automatic flight control and thrust management systems; Boeing windshear detection and guidance system is optional; aircraft for British Airways and Monarch Airlines have Bendix/King ARINC 700 series avionics, including colour weather radar and seven digital comm nav and identification systems; RMI-743 radio distance magnetic indicator (RDMI) and optional radio magnetic indicator (RMI).

DIMENSIONS, EXTERNAL:

Wing span	38.05 m (124 ft 10 in)
Wing chord: at root	8.20 m (26 ft 11 in)
at tip	1.73 m (5 ft 8 in)
Wing aspect ratio	7.8
Length: overall	47.32 m (155 ft 3 in)
fuselage	46.96 m (154 ft 10 in)
Height overall	13.56 m (44 ft 6 in)
Tailplane span	15.21 m (49 ft 11 in)
Wheel track	7.32 m (24 ft 0 in)
Wheelbase	18.29 m (60 ft 0 in)
Passenger doors (two, fwd, port):	
Height	1.83 m (6 ft 0 in)
Width	0.84 m (2 ft 9 in)
Passenger door (rear, port): Height	1.83 m (6 ft 0 in)
Width	0.76 m (2 ft 6 in)
Service door (fwd, stbd): Height	1.65 m (5 ft 5 in)
Width	0.76 m (2 ft 6 in)
Service door (stbd, opposite second passenger door):	
Height	1.83 m (6 ft 0 in)
Width	0.84 m (2 ft 9 in)
Service door (rear, stbd): Height	1.83 m (6 ft 0 in)
Width	0.76 m (2 ft 6 in)
Emergency exits (four, overwing):	
Height	0.97 m (3 ft 2 in)
Width	0.51 m (1 ft 8 in)
Emergency exits, optional (two, aft of wings):	
Height	1.32 m (4 ft 4 in)
Width	0.61 m (2 ft 0 in)

DIMENSIONS, INTERNAL:

Cabin (aft of flight deck to rear pressure bulkhead):	
Length	36.09 m (118 ft 5 in)
Max width	3.53 m (11 ft 7 in)
Max height	2.13 m (7 ft 0 in)
Floor area	116.04 m² (1,249 sq ft)
Passenger section volume	230.50 m³ (8,140 cu ft)
Underfloor cargo volume (bulk loading):	
fwd	19.82 m³ (700 cu ft)
rear	30.87 m³ (1,090 cu ft)

AREAS:

Wings, gross	185.25 m² (1,994.0 sq ft)
Ailerons (total)	4.46 m² (48.0 sq ft)
Trailing-edge flaps (total)	30.38 m² (327.0 sq ft)
Leading-edge slats (total)	18.39 m² (198.0 sq ft)
Flight spoilers (total)	10.96 m² (118.0 sq ft)
Ground spoilers (total)	12.82 m² (138.0 sq ft)
Fin	34.37 m² (370.0 sq ft)
Rudder	11.61 m² (125.0 sq ft)
Tailplane	50.35 m² (542.0 sq ft)
Elevators (total)	12.54 m² (135.0 sq ft)

WEIGHTS AND LOADINGS (with 186 passengers. A: 535E4 engines, B: PW2037s, C: PW2040s):

Operating weight empty: A	57,180 kg (126,060 lb)
B, C	57,039 kg (125,750 lb)
Max basic T-O weight: A, B, C	99,790 kg (220,000 lb)
Max T-O weight (medium-range):	
A, B, C	104,325 kg (230,000 lb)
Max T-O weight (long-range):	
A, B, C	113,395 kg (250,000 lb)
Max landing weight: A, B, C	89,810 kg (198,000 lb)
757-200PF	95,255 kg (210,000 lb)
Max zero-fuel weight: A, B, C	83,460 kg (184,000 lb)
757-200PF	90,720 kg (200,000 lb)
Max wing loading: A, B, C at max basic T-O weight	538.5 kg/m² (110.3 lb/sq ft)
A, B, C at long-range max T-O weight	587.8 kg/m² (120.4 lb/sq ft)
Max power loading:	
at max basic T-O weight:	
A	279.68 kg/kN (2.74 lb/lb st)
B	293.5 kg/kN (2.88 lb/lb st)
C	268.97 kg/kN (2.64 lb/lb st)
at long-range max T-O weight:	
A	317.81 kg/kN (3.12 lb/lb st)
B	333.51 kg/kN (3.27 lb/lb st)
C	305.1 kg/kN (3.00 lb/lb st)

PERFORMANCE: (with 186 passengers; at max basic T-O weight except where indicated):

Max operating speed: A, B, C	Mach 0.86

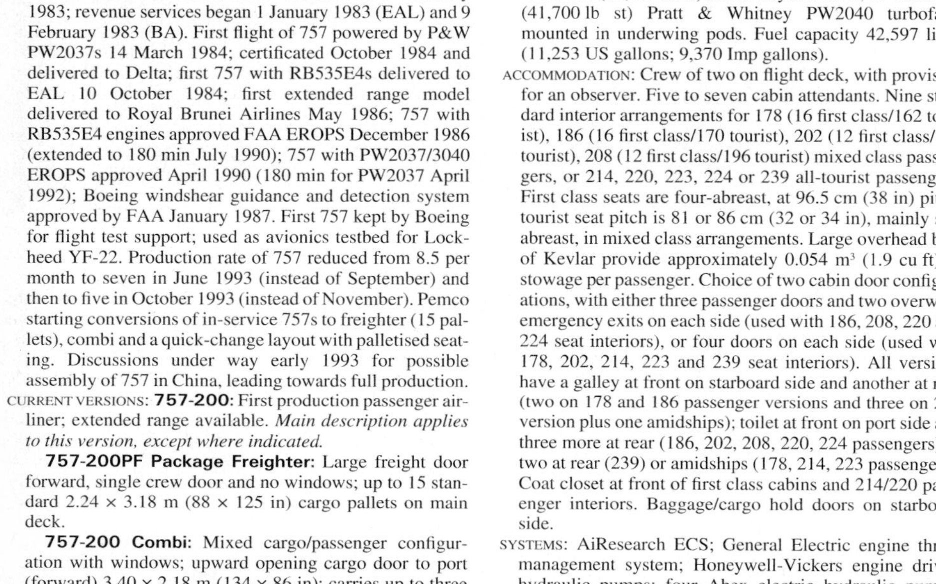

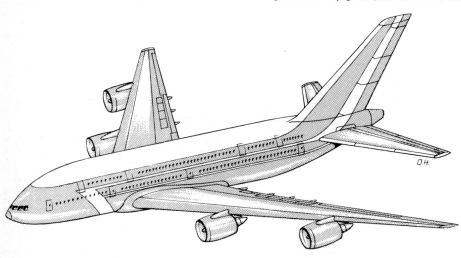

Artist's impression of Boeing 747X ***(Flight International)***

BOEING—AIRCRAFT: USA

Boeing 757-200 for Ladeco takes off from the Boeing airfield at Renton, Washington

Cruising speed: A, B, C　　　　　　　　Mach 0.80
Approach speed at S/L, flaps down, max landing weight:
　A, B, C　　　132 knots (245 km/h; 152 mph) EAS
Initial cruising height: A　　　11,880 m (38,970 ft)
　B, C　　　　　　　　　　　　11,675 m (38,300 ft)
Runway LCN at ramp weight of 100,244 kg (221,000 lb), optimum tyre pressure and subgrade C flexible pavement: H40 × 14.5-19.0 tyres　　　　　　　　　36
T-O field length (S/L, 29°C):
　at max basic T-O weight: A　　1,646 m (5,400 ft)
　　B　　　　　　　　　　　　　1,791 m (5,875 ft)
　　C　　　　　　　　　　　　　1,637 m (5,370 ft)
　at long-range max T-O weight: A　2,134 m (7,000 ft)
　　B　　　　　　　　　　　　　2,792 m (9,160 ft)
　　C　　　　　　　　　　　　　2,118 m (6,950 ft)
Landing field length at max landing weight:
　A　　　　　　　　　　　　　　1,411 m (4,630 ft)
　B, C　　　　　　　　　　　　　1,460 m (4,790 ft)
Range with 186 passengers:
　at max basic T-O weight:
　　A　　　　　2,820 nm (5,226 km; 3,247 miles)
　　B, C　　　　2,980 nm (5,522 km; 3,431 miles)
　at long-range max T-O weight:
　　A　　　　　3,820 nm (7,079 km; 4,399 miles)
　　B, C　　　　4,000 nm (7,408 km; 4,603 miles)
757-200PF, max long-range T-O weight, 22,680 kg (50,000 lb) payload:
　A　　　　　　3,700 nm (6,857 km; 4,261 miles)
　B, C　　　　　3,885 nm (7,200 km; 4,474 miles)
OPERATIONAL NOISE LEVELS (FAR Pt 36 Stage 3):
　T-O, at max basic T-O weight, cutback power:
　　A　　　　　　　　　　　　　　82.2 EPNdB
　　B　　　　　　　　　　　　　　86.2 EPNdB
　　C (estimated)　　　　　　　　84.7 EPNdB
　Approach at max landing weight, 30° flap:
　　A　　　　　　　　　　　　　　95.0 EPNdB
　　B, C　　　　　　　　　　　　　97.7 EPNdB
　Sideline: A　　　　　　　　　　93.3 EPNdB
　　B　　　　　　　　　　　　　　94.0 EPNdB
　　C (estimated)　　　　　　　　94.6 EPNdB

BOEING 767

TYPE: Medium/long-range twin-turbofan airliner.
PROGRAMME: Launched on receipt of United Air Lines order for 30 on 14 July 1978; construction of basic 220-passenger 767-200 began 6 July 1979; first flight (N767BA) 26 September 1981 with P&W JT9D turbofans; first flight fifth aircraft with GE CF6-80A 19 February 1982; 767 with JT9D-7R4D certificated 30 July 1982; with CF6-80A 30 September 1982.
First delivery with JT9D (United Air Lines) 19 August 1982; first delivery with CF6 (Delta) 25 October 1982. 767-200 with JT9D-7R4 or CF6-80A or -80A2 approved for EROPS January 1987; EROPS approval for 767-200 and -300 with PW4000 obtained April 1990. Boeing windshear detection and guidance system FAA approved for 767-200 and -300 February 1987; production to be reduced from five a month to three in October 1993 (instead of four in November).
CURRENT VERSIONS: **767-200**: Basic model. Medium-range variant has reduced fuel; higher gross weight variant certificated June 1983. *Description applies to basic 767-200, except where indicated.*
767-200ER: Extended range version; first flight 6 March 1984; basic -200ER with centre-section tankage and gross weight increased to 156,490 kg (345,000 lb) first delivered to Ethiopian Airlines 23 May 1984; optional higher gross weights are 159,211 kg (351,000 lb), 172,365 kg (380,000 lb) and 175,540 kg (387,000 lb).
767-300: 269-passenger stretched version, with 3.07 m (10 ft 1 in) plug forward of wing and 3.35 m (11 ft) plug aft, and same gross weight as 767-200; strengthened landing gear and thicker metal in parts of fuselage and underwing skin; same flight deck and systems as other 767s; same engine options as 767-200ER; first ordered 29 September 1983. First flight with JT9D-7R4D engines 30 January 1986; certificated with JT9D-7R4D and CF6-80A2 22 September 1986; British Airways ordered 11 in August 1987, later increased to total 25, with Rolls-Royce RB211-524H, for delivery from November 1989.
767-300ER: Extended range, higher gross weight version; development began January 1985; optional gross weights 172,365 kg (380,000 lb), 175,540 kg (387,000 lb) and 181,437 kg (400,000 lb); further increased centre-section tankage. Engine choice CF6-80C2, PW4000, RB211-524H; structural reinforcement; certificated late 1987. Launch customer American Airlines (15), delivered from February 1988.
767ERX/ERY: Boeing's initial 767ERX study of 1992 extended tailplane tankage to give additional 550 nm (1,019 km; 633 miles) range; 767ERY would have wing area and tankage volume extended by increased wing chord, modified front spar, extended span and winglets; fuselage section 48, pressure bulkhead and landing gear would be reinforced; GE, P&W and RR studying 278.0 kN (62,500 lb st) engines; 767ERY could have payload improved by 5,987 kg (13,200 lb), range extended to more than 7,000 nm (12,972 km; 8,060 miles), hot and high take-off performance improved and cruising speed pushed towards Mach 0.84. Boeing says it could produce 767-300ERX in two years and 767-300ERY in about three years from order.
767-300ER Freighter: See separate entry.
767 AWACS: See page 454.
CUSTOMERS: See table on earlier page for orders/deliveries at 31 March 1993; overall total orders 665 at 10 June 1993. Original prototype became 767 Airborne Surveillance Testbed (formerly AOA) for US Army (see 1991-92 *Jane's*). One reconfigured by E-Systems as medevac aircraft for Civil Reserve Air Fleet.
DESIGN FEATURES: Special Boeing aerofoils; quarter-chord sweepback 31° 30′; thickness/chord ratio 15.1 per cent at root, 10.3 per cent at tip; dihedral 6°; incidence 4° 15′.
FLYING CONTROLS: Inboard all-speed and outboard low-speed ailerons supplemented by flight spoilers also acting as airbrakes and lift dumpers; single-slotted, linkage-supported outboard trailing-edge flaps, double-slotted inboard; track-mounted leading-edge slats; variable incidence tailplane; no trim tabs; all control surfaces hydraulically powered; roll and yaw trim through spring feel system; triple digital flight control computers and EFIS; Boeing windshear detection and guidance system optional.
STRUCTURE: Fail-safe structure; CFRP wing spoilers; tailplane and fin contain aluminium honeycomb. Subcontractors include Boeing Helicopters (wing fixed leading-edges); Grumman Aerospace (wing centre-section and adjacent lower fuselage section; fuselage bulkheads); Vought Aircraft (horizontal tail); Canadair (rear fuselage); Alenia (wing control surfaces, flaps and leading-edge slats, wingtips, elevators, fin and rudder, nose radome); Fuji (wing fairings and main landing gear doors); Kawasaki (centre fuselage body panels; exit hatches; wing in-spar ribs); Mitsubishi (rear fuselage body panels; stringers; passenger and cargo doors; dorsal fin).
LANDING GEAR: Hydraulically retractable tricycle type; Menasco twin-wheel nose unit retracts forward; Cleveland

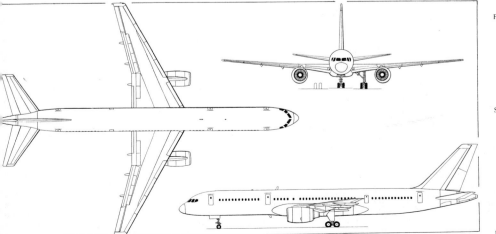

Boeing 757-200 twin-turbofan short/medium-range transport (*Dennis Punnett*)

USA: AIRCRAFT—BOEING

Artist's impression of Boeing 767-300F freighter, the first of which were ordered by United Parcel Service

Pneumatic main gear, comprising two four-wheel bogies, which retract inward; oleo-pneumatic shock absorbers; Bendix wheels and brakes; mainwheel tyres size 45 × 17-20, pressure 12.6 bars (183 lb/sq in); nosewheel tyres size 37 × 14-15, pressure 10.0 bars (145 lb/sq in); steel disc brakes on all mainwheels; electronically controlled anti-skid units.

POWER PLANT: Two high bypass turbofans in pods, pylon-mounted on the wing leading-edges. Alternative engines available for all models are General Electric CF6-80A and Pratt & Whitney JT9D-7R4D, both rated at 213.5 kN (48,000 lb st), and CF6-80A2, JT9D-7R4E and JT9D-7R4E4, rated at 222.4 kN (50,000 lb st). Additionally, 767-200, 767-200ER and 767-300 are available with Pratt & Whitney PW4050 rated at 222.4 kN (50,000 lb st), PW4052 rated at 231.3 kN (52,000 lb st) and General Electric CF6-80C2B2 rated at 233.5 kN (52,500 lb st). General Electric CF6-80C2B4, rated at 257.5 kN (57,900 lb st), available on 767-200ER, 767-300 and 767-300ER. Pratt & Whitney PW4056, rated at 252.4 kN (56,750 lb st), and PW4060 and General Electric CF6-80C2B6 rated at 266.9 kN (60,000 lb st), available only on extended range versions. Rolls-Royce RB211-524G, rated at 269.6 kN (60,600 lb st), available on 767s entering service from early 1990. Fuel in one integral tank in each wing, and in centre tank, with total capacity of 63,216 litres (16,700 US gallons; 13,905 Imp gallons) in 200/300; 767-200ER has additional 14,195 litres (3,750 US gallons; 3,122 Imp gallons) in second centre-section tank, raising total capacity to 77,412 litres (20,450 US gallons; 17,028 Imp gallons). 767-300ER has further expanded wing centre-section tank (optional on -200ER), bringing total capacity to 91,039 litres (24,050 US gallons; 20,026 Imp gallons). Refuelling point in port outer wing.

ACCOMMODATION: Normal operating crew of two on flight deck, with third position optional. Basic accommodation in -200 models for 216 passengers, made up of 18 first class passengers forward in six-abreast seating at 96.5 cm (38 in) pitch, and 198 tourist class in mainly seven-abreast seating at 87 cm (34 in) pitch. Type A inward opening plug doors provided at both front and rear of cabin on each side of fuselage, with Type III emergency exit over wing on each side. Total of five toilets installed, two centrally in main cabin, two aft in main cabin, and one forward in first class section. Galleys situated at forward and aft ends of cabin. Alternative single class layouts provide for 230 tourist passengers, seven-abreast at 86 cm (34 in) pitch; 242 passengers seven-abreast at 81 cm (32 in) pitch; 255 passengers mainly seven-abreast (two-three-two) at 76 cm (30 in) pitch, or eight-abreast (two-four-two) at 81 cm (32 in) pitch. Max seating capacity in -200 models (requiring additional overwing emergency exit) 290 passengers, mainly eight-abreast, at 76 cm (30 in) pitch; capacity in -300 is 290 passengers seven-abreast. Underfloor cargo holds of -200 versions can accommodate, typically, up to 22 LD2 or 11 LD1 containers. 767-300 underfloor cargo holds can accommodate 30 LD2 or 15 LD1 containers. Starboard side forward and rear cargo doors of equal size standard on 767-200 and 767-300, but larger port side forward cargo door standard on 767-200ER and 767-300ER and optional on 767-200 and 767-300, to permit loading of Type 2 pallets, three such pallets being accommodated in -200/200ER and four in -300/300ER. Bulk cargo door at rear on port side. Overhead stowage for carry-on baggage. Cabin air-conditioned, cargo holds heated.

SYSTEMS: AiResearch dual air cycle air-conditioning system. Pressure differential 0.59 bar (8.6 lb/sq in). Electrical supply from two engine driven 90kVA three-phase 400Hz constant frequency AC generators, 115/200V output. 90 kVA generator mounted on APU for ground operation or for emergency use in flight. Three hydraulic systems at 207 bars (3,000 lb/sq in), for flight control and utility functions, supplied from engine driven pumps and a Garrett bleed air powered hydraulic pump or from APU. Maximum generating capacity of port and starboard systems is 163 litres (4 US gallons; 35.8 Imp gallons)/min; centre system 185. litres (49 US gallons; 40.8 Imp gallons)/min, at 196.5 bar (2,850 lb/sq in). Reservoirs pressurised by engine bleed a via pressure regulation module. Reservoir relief valv pressure nominally 4.48 bars (65 lb/sq in). Additiona hydraulic motor driven generator, to provide essenti: functions for extended range operations, standard on 767 200ER and 767-300ER and optional on 767-200 and 767 300. Nitrogen chlorate oxygen generators in passenge cabin, plus gaseous oxygen for flight crew. APU in tailcon to provide ground and in-flight electrical power and press urisation. Anti-icing for outboard wing leading-edge (none on tail surfaces), engine air inlets, air data senso and windscreen.

AVIONICS: Standard avionics include ARINC 700 serie equipment (Bendix/King VOR/marker beacon receiver ILS receiver, radio altimeter, transponder, DME, ADF an RDR-4A colour weather radar in aircraft for All Nippor Britannia and Transbrasil). Collins caution annunciato dual digital flight management systems, and triple digita flight control computers, including FCS-700 flight contro system, EFIS-700 electronic flight instrument system an RMI-743 radio distance magnetic indicator. Honeywel IRS, FMCS and DADC, as described in Boeing 757 entr Options include Boeing's windshear detection and guid ance system.

DIMENSIONS, EXTERNAL:
Wing span	47.57 m (156 ft 1 in
Wing chord: at root	8.57 m (28 ft 1¼ in
at tip	2.29 m (7 ft 6 in
Wing aspect ratio	7.9
Length: overall: 200/200ER	48.51 m (159 ft 2 in
300/300ER	54.94 m (180 ft 3 in
fuselage: 200/200ER	47.24 m (155 ft 0 in
300/300ER	53.67 m (176 ft 1 in
Fuselage: Max width	5.03 m (16 ft 6 in
Height overall	15.85 m (52 ft 0 in
Tailplane span	18.62 m (61 ft 1 in
Wheel track	9.30 m (30 ft 6 in
Wheelbase: 200/200ER	19.69 m (64 ft 7 in
300/300ER	22.76 m (74 ft 8 in
Passenger doors (two, fwd and rear, port):	
Height	1.88 m (6 ft 2 in
Width	1.07 m (3 ft 6 in
Galley service door (two, fwd and rear, stbd):	
Height	1.83 m (6 ft 0 in
Width	1.07 m (3 ft 6 in
Emergency exits (two, each): Height	0.97 m (3 ft 2 in
Width	0.51 m (1 ft 8 in
Cargo doors (two, fwd and rear, stbd):	
Height	1.75 m (5 ft 9 in
Width	1.78 m (5 ft 10 in
* Larger cargo door (fwd, port):	
Height	1.75 m (5 ft 9 in
Width	3.40 m (11 ft 2 in

* Standard on ER models, optional for -200/300

DIMENSIONS, INTERNAL:
Cabin, excl flight deck:	
Length: 200/200ER	33.93 m (111 ft 4 in
300/300ER	40.36 m (132 ft 5 in
Max width	4.72 m (15 ft 6 in
Max height	2.87 m (9 ft 5 in
Floor area: 200/200ER	154.9 m² (1,667 sq ft
300/300ER	184.0 m² (1,981 sq ft

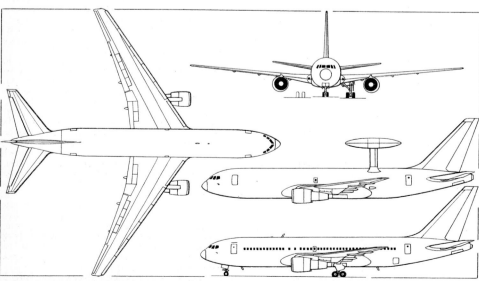

Boeing 767-200 wide-bodied airliner with extra side view of proposed AWACS version (Dennis Punnett)

Boeing 767 of Hungary's Malev airline

Volume: 200/200ER	428.2 m³ (15,121 cu ft)
300/300ER	483.9 m³ (17,088 cu ft)
Volume, flight deck	13.5 m³ (478 cu ft)
Baggage holds (containerised), volume:	
200/200ER	74.8 m³ (2,640 cu ft)
300/300ER	101.9 m³ (3,600 cu ft)
Bulk cargo hold volume:	
all models	12.2 m³ (430 cu ft)
Combined baggage hold/bulk cargo hold volume:	
200/200ER	87.0 m³ (3,070 cu ft)
300/300ER	114.1 m³ (4,030 cu ft)
Total cargo hold volume:	
200/200ER	111.3 m³ (3,930 cu ft)
300/300ER	147.0 m³ (5,190 cu ft)
AREAS:	
Wings, gross	283.3 m² (3,050.0 sq ft)
Ailerons (total)	11.58 m² (124.6 sq ft)
Trailing-edge flaps (total)	36.88 m² (397.0 sq ft)
Leading-edge slats (total)	28.30 m² (304.6 sq ft)
Spoilers (total)	15.83 m² (170.4 sq ft)
Fin	30.19 m² (325.0 sq ft)
Rudder	15.95 m² (171.7 sq ft)
Tailplane	59.88 m² (644.5 sq ft)
Elevators (total)	17.81 m² (191.7 sq ft)

WEIGHTS AND LOADINGS (A: 767-200 basic/JT9D-7R4D engines, B: 767-200 basic/CF6-80A, C: medium-range version/JT9D-7R4D, D: medium-range version/CF6-80A, E: 767-200ER/PW4050, F: 767-200ER/CF6-80C2B2, G: 767-200ER/PW4056, H: 767-200ER/CF6-80C2B4, J: 767-300/PW4050, K: 767-300/CF6-80C2B2, L: 767-300 higher gross weight version/PW4050, M: 767-300 higher gross weight version/CF6-80C2B2, N: 767-300ER/PW4056, P: 767-300ER/CF6-80C2B4, Q: 767-300ER/PW4060):

Manufacturer's weight empty:	
A, C	74,752 kg (164,800 lb)
B, D	74,344 kg (163,900 lb)
E	76,339 kg (168,300 lb)
F	76,249 kg (168,100 lb)
G	76,566 kg (168,800 lb)
H	76,476 kg (168,600 lb)
J, L	79,560 kg (175,400 lb)
K, M	79,379 kg (175,000 lb)
N	80,785 kg (178,100 lb)
P	80,603 kg (177,700 lb)
Q	81,374 kg (179,400 lb)
Operating weight empty: A, C	80,921 kg (178,400 lb)
B, D	80,512 kg (177,500 lb)
E	83,552 kg (184,200 lb)
F	83,461 kg (184,000 lb)
G	83,778 kg (184,700 lb)
H	83,688 kg (184,500 lb)
J, L	87,135 kg (192,100 lb)
K, M	86,953 kg (191,700 lb)
N	89,312 kg (196,900 lb)
P	89,131 kg (196,500 lb)
Q	89,902 kg (198,200 lb)

Max payload (767-200, 216 passengers; 767-200ER, 174 passengers; 767-300, 261 passengers; 767-300ER, 210 passengers): A, B, C, D 19,595 kg (43,200 lb)

E, F, G, H	16,574 kg (36,540 lb)
J, K, L, M	23,677 kg (52,200 lb)
N, P, Q	20,003 kg (44,100 lb)
Max fuel weight:	
A, B, C, D, J, K, L, M	51,131 kg (112,725 lb)
E, F	62,613 kg (138,038 lb)
G, H, N, P, Q	73,635 kg (162,338 lb)
Max T-O weight: A, B	136,078 kg (300,000 lb)
C, D	142,881 kg (315,000 lb)
E, F, J, K	156,489 kg (345,000 lb)
G, H, N, P	175,540 kg (387,000 lb)
L, M	159,211 kg (351,000 lb)
Q	181,437 kg (400,000 lb)
Max ramp weight: A, B	136,985 kg (302,000 lb)
C, D	143,789 kg (317,000 lb)
E, F, J, K	157,396 kg (347,000 lb)
G, H, N, P	175,994 kg (388,000 lb)
L, M	159,664 kg (352,000 lb)
Q	181,890 kg (401,000 lb)
Max zero-fuel weight: A, B	112,491 kg (248,000 lb)
C, D	113,398 kg (250,000 lb)
E, F	114,757 kg (253,000 lb)
G, H	117,934 kg (260,000 lb)
J, K, L, M, N, P	126,098 kg (278,000 lb)
Q	130,634 kg (288,000 lb)
Max landing weight: A, B	122,470 kg (270,000 lb)
C, D	123,377 kg (272,000 lb)
E, F	126,098 kg (278,000 lb)
G, H	129,273 kg (285,000 lb)
J, K, L, M, N, P	136,078 kg (300,000 lb)
Q	145,149 kg (320,000 lb)
Max wing loading: A, B	480.24 kg/m² (98.36 lb/sq ft)
C, D	504.26 kg/m² (103.28 lb/sq ft)
E, F, J, K	552.25 kg/m² (113.11 lb/sq ft)
G, H, N, P	619.53 kg/m² (126.89 lb/sq ft)
L, M	561.87 kg/m² (115.08 lb/sq ft)
Q	640.33 kg/m² (131.15 lb/sq ft)

PERFORMANCE (at max T-O weight except where indicated):
Normal cruising speed, all versions   Mach 0.80
Approach speed at max landing weight:

A, B, C, D	136 knots (252 km/h; 157 mph)
E	138 knots (256 km/h; 159 mph)
F, G, H	140 knots (259 km/h; 161 mph)
J, K, L, M, N, P	141 knots (261 km/h; 162 mph)
Q	145 knots (269 km/h; 167 mph)
Initial cruise altitude: A	11,950 m (39,200 ft)
B	12,100 m (39,700 ft)
C	11,650 m (38,200 ft)
D	11,800 m (38,700 ft)
E	11,215 m (36,800 ft)
F	11,460 m (37,600 ft)
G	10,925 m (35,850 ft)
H	10,850 m (35,600 ft)
J, M	11,250 m (36,900 ft)
K	11,340 m (37,200 ft)
L	11,125 m (36,500 ft)
N, P	10,600 m (34,800 ft)
Q	10,400 m (34,100 ft)
Service ceiling, OEI: A, C	6,525 m (21,400 ft)
B, D	6,430 m (21,100 ft)
E	6,850 m (22,500 ft)
F	7,200 m (23,600 ft)
G	7,250 m (23,800 ft)
H	7,375 m (24,200 ft)
J, L	6,035 m (19,800 ft)
K, M	6,150 m (20,200 ft)
N, P	6,615 m (21,700 ft)
Q	6,550 m (21,500 ft)
T-O field length: A, B	1,798 m (5,900 ft)
C	1,951 m (6,400 ft)
D	1,981 m (6,500 ft)
E	2,347 m (7,700 ft)
F	2,316 m (7,600 ft)
G, H	2,774 m (9,100 ft)
J	2,560 m (8,400 ft)
K	2,469 m (8,100 ft)
L, M	2,652 m (8,700 ft)
N	2,926 m (9,600 ft)
P	2,956 m (9,700 ft)
Q	2,774 m (9,100 ft)
Design range: A	3,160 nm (5,856 km; 3,639 miles)
B	3,220 nm (5,967 km; 3,708 miles)
C	3,795 nm (7,033 km; 4,370 miles)
D	3,850 nm (7,135 km; 4,433 miles)
E	5,365 nm (9,942 km; 6,178 miles)
F	5,410 nm (10,026 km; 6,230 miles)
G	6,770 nm (12,546 km; 7,796 miles)
H	6,805 nm (12,611 km; 7,836 miles)
J	4,000 nm (7,413 km; 4,606 miles)
K	4,020 nm (7,450 km; 4,629 miles)
L	4,230 nm (7,839 km; 4,871 miles)
M	4,260 nm (7,895 km; 4,905 miles)
N	5,740 nm (10,637 km; 6,610 miles)
P	5,760 nm (10,674 km; 6,633 miles)
Q	6,060 nm (11,230 km; 6,978 miles)

OPERATIONAL NOISE LEVELS (FAR Pt 36, Stage 3):

T-O at max basic T-O weight: B	87.1 EPNdB
H	90.4 EPNdB
Approach at max landing weight: B	101.6 EPNdB
H	101.7 EPNdB
Sideline: B	95.4 EPNdB
H	96.6 EPNdB

BOEING 767-300 FREIGHTER

TYPE: Freighter version of 767-300ER.
PROGRAMME: Launched January 1993 by United Parcel Service order; design to be completed second quarter of 1994; first flight second quarter 1995; first UPS delivery October 1995.
CUSTOMERS: UPS ordered 30 (plus 30 options) January 1993.
DESIGN FEATURES: Modifications include reinforced landing gear and internal wing structure; main deck floor strengthened to take 24 containers; no passenger windows; 2.67 × 3.4 m (8 ft 9 in × 11 ft 1¾ in) freight door forward to port; type rating and extensive component commonality with 757 Freighter.
POWER PLANT: In competition early 1993 between Rolls-Royce RB211-524G and variety of P&W JT9D and PW4000 and GE CF6-80A and C versions; thrusts between 213.5 kN (48,000 lb st) and 269.6 kN (60,600 lb st).
WEIGHTS AND LOADINGS:
  Max payload   50,800 kg (112,000 lb)
  Max T-O weight   185,065 kg (408,000 lb)
  Planned alternative T-O weight   192,775 kg (425,000 lb)
PERFORMANCE:
  Range: with 40,823 kg (90,000 lb) payload
    4,000 nm (7,400 km; 4,600 miles)
  with 50,802 kg (112,000 lb) payload
    3,000 nm (5,550 km; 3,450 miles)

BOEING 777-200

TYPE: Long-range, high-capacity twin-turbofan airliner.
PROGRAMME: Formerly known as 767-X, now 777-200; brief details announced 8 December 1989; launch order by United Airlines 15 October 1990 (see Customers); Boeing formal programme launch 29 October 1990; planned rollout Spring 1994, first flight 1 June 1994; basic 777 A Market to be certificated May 1995; 777 B Market to be certificated December 1996; nine development 777s planned, three of which (one with each engine type) will fly 1,000 cycles before certification; flight and maintenance manuals and production simulators to be available well before first deliveries.
CURRENT VERSIONS: **777-200 A Market:** Basic aircraft with max T-O weight of 229,520 kg (506,000 lb) and alternative max T-O weights 233,600 kg (515,000 lb) and 242,670 kg (535,000 lb); max payload 54,930 kg (121,100 lb); twin-aisle cabin layout; 375 to 400 two-class passengers or 305 to 328 three-class passengers or 418 to 440 all-economy depending on choice of nine- or 10-abreast in tourist class and seven- or eight-abreast in business class; range with full passengers 4,050 nm (7,505 km; 4,660 miles).
777-200 B Market: Max T-O weight 263,085 kg (580,000 lb) or 267,620 kg (590,000 lb); max payload 54,660 kg (120,500 lb); same passenger capacity as basic aircraft; range with full passengers 6,350 nm (11,770 km; 7,310 miles).
777 Stretch: Designed to increase passenger capacity and extend range to 7,000 nm (12,972 km; 8,060 miles); fuselage to be stretched by 19 frames adding 10.13 m (33 ft 3 in) in two plugs fore and aft of the wing; seating would increase from 305 to 368 in three-class, 375 to 451 in two-class and 440 to 550 in all-tourist; escape provisions would be increased with a fifth door on each side; T-O gross weight would exceed 272,155 kg (600,000 lb); more powerful engines would be needed; main routes would be Asia-USA, Middle East-Europe and Europe-USA.
CUSTOMERS: First firm order 34 plus options for 34 from United Airlines 15 October 1990 (since modified), to be powered by 325 kN (73,000 lb) Pratt & Whitney PW4073; United version has max T-O weight 234,000 kg (515,880 lb) carrying 363 passengers in two classes for up to 4,200 nm (7,785 km; 4,836 miles); in April 1993, four A Market deliveries for 1996 converted to two B Market in 1998 and two in 1999; 11 A Market to be delivered during 1995; other launch customer is All Nippon Airways with 15 firm plus options for 10 placed 19 December 1990 (PW4073A). Other customers Euralair (two), Thai International (eight, with Rolls-Royce Trent, plus options for six), British Airways (15 firm, with GE90, plus options for 15), Japan Air Lines (10 firm, plus options for 10), Cathay Pacific (11 firm, plus options for 11), Emirates (seven plus seven), Lauda Air (four), CAAC China Southern (six), International Lease Finance (six firm plus options for two) and Continental Airlines (five plus five) by May 1993. Total orders 130 at 29 June 1993, plus options for 94.
COST: Estimated development cost $4 billion (1990); aircraft cost $106 to 129 million (1991).
DESIGN FEATURES: New wing of 31.6° sweepback at quarter-chord incorporates new technology to allow Mach 0.83

450 USA: AIRCRAFT—BOEING

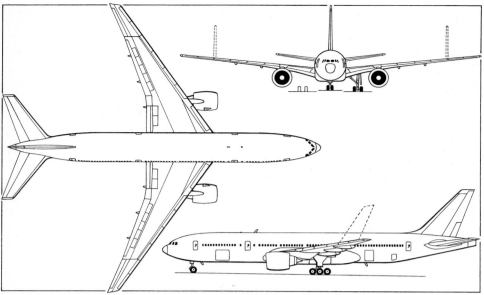

Boeing 777-200 twin-turbofan high-capacity airliner (*Jane's/Mike Keep*)

Boeing 777 A Market/B Market airframes are externally the same

Twin-aisle first class cabin in Boeing 777

cruise in combination with high thickness for economic structure and large internal volume, long span for improved take-off and payload/range and large area for high cruise altitude and low approach speed; no winglet. Boeing wants to achieve 180 min EROPS clearance initial entry into service. Cylindrical fuselage wider than 767 to allow twin aisle seating for from six to 10 abreast toilets and overhead baggage bins designed to allow rapid change of layout. Outer 6.48 m (21 ft 3 in) of each wing can be folded to vertical to reduce gate width requirement airports.

FLYING CONTROLS: Hydraulic fully powered controls (with Teijin Seiki actuators), three GEC-Marconi primary flight control computers (PFCC) and four actuator control electronics (ACE) boxes by Lear Astronics and Teijin Seiki included in GEC-Marconi Avionics full fly-by-wire system; outboard low-speed aileron and inboard all-speed aileron acting as flaperon; five outboard and two inboard spoiler panels on each wing; seven leading-edge slats on each wing; double-slotted inboard flaps and single-slotted outboard flaps; variable-incidence tailplane; large rudder tab. Teijin Seiki, Parker-Bertea and Moog elevator and spoiler actuators; tailplane trim module and hydraulic brake by E-Systems.

Linked control columns and aileron wheels retained for crew familiarity; both back-driven by Rockwell Collins triplex autopilot to let both pilots observe control demands GEC-Marconi fly-by-wire system controls primary, secondary control and high-lift surfaces; full three-axis digital system with direct analog backup; conventional control characteristics with selected enhancement functions; EFI based on liquid crystal flat panel displays. Flight controls integrated with aircraft information management system (AIMS) and triple ARINC 629 databus (see Avionics). Head-up display and enhanced vision system optional.

STRUCTURE: Composites used for moving trailing-edge surfaces and spoiler panels, tail surfaces except leading edges, wing fixed leading-edge, cabin floor beams, engine nacelles, wingroot fairings and main landing gear doors. Toughened materials for high damage resistance and to allow simple low temperature bolted repairs. Metal structure includes thick skins without need for tear straps; no bonding; single-piece fuselage frames; wing top skin and stringers made in new 7055 aluminium alloy; some tail skins and main floor beams in composites of carbon and toughened resins; 10 per cent of structure weight in composites.

Fully digital product definition with all parts created by Dassault/IBM CATIA CAD/CAM and communicated to manufacturing and publications; structure and system integration, tube and cable run design completed before design release. More than 200 design/build teams have ensured that design, fabrication and test have proceeded concurrently for structure and systems.

Centre and rear fuselage barrel sections, tailcone, doors wingroot fairing and landing gear doors made in Japan Wing and tail leading-edges and moving wing parts, landing gear, floor beams, nose landing gear doors, wingtips dorsal fin and nose radome made by Rockwell, Grumman Alenia (Italy), Embraer (Brazil), Short Brothers (UK), Singapore Aerospace Manufacturing, HDH and ASTA (Australia), Korean Air and other subcontractors. Boeing manufactures flight deck and forward cabin, basic wing and tail structures and engine nacelles, and assembles and tests.

LANDING GEAR: Retractable tricycle type (Menasco/Messier Bugatti joint design for main gear); two main legs carrying six-wheel bogies with steering rear axles automatically engaged by nose gear steering angle; twin-wheel steerable nose gear; Goodyear mainwheel tyres H49 × 19-22 with 32 bias ply rating; Michelin radial nosewheel tyres 44 × 18-18 (24 ply); Bendix Carbenix 4000 mainwheel brake arranged so that initial toe-pedal pressure used during taxying applies brakes to alternate sets of three wheels to save brake wear; full toe-pedal pressure applies all six brakes together.

POWER PLANT: Boeing offers choice of Pratt & Whitney Rolls-Royce and General Electric turbofans as follows: for 777 A Market, P&W 327 kN (73,500 lb st) PW4073A or PW4073; R-R 317 kN (71,200 lb st) Trent 870 or 871; GE 331 kN (74,500 lb st) GE90-B3 or -B2. For 777 B Market and 777 Stretch, P&W 366 kN (82,200 lb st) PW4082 or 376 kN (84,600 lb st) PW4084; R-R 366 kN (82,200 lb st) Trent 882 or 375 kN (84,400 lb st) Trent 884; GE 374 kN (84,100 lb st) GE90-B1 or 377 kN (84,700 lb st) GE90-B4. All fuel contained in integral tanks in wing torsion box and wing centre-section, with reserve tank, surge tank and fuel vent and jettison pipes all inboard of wing fold; combined capacity of main, centre and reserve tanks is 117,300 litres (31,000 US gallons; 25,833 Imp gallons) in 777-200 and 169,200 litres (44,700 US gallons; 37,250 Imp gallons) in 777-200 Stretch.

ACCOMMODATION: Two-pilot crew. See under Current Versions for passenger capacities. Internal fuselage width (5.87 m; 19 ft 3 in) designed to give wide choice of twin-aisle seat layouts from six to 10 abreast and easy adaptability of cabin for different combinations of classes and service locations; new pivoting and translating overhead baggage bins allow 0.08 m³ (3.0 cu ft) volume per passenger.

Underfloor compartments will accommodate containers from LD1 to LD6 and LD10 and LD11, as well as 96 in and

Three-class cabin layout and main structural features of Boeing 777 A Market/B Market

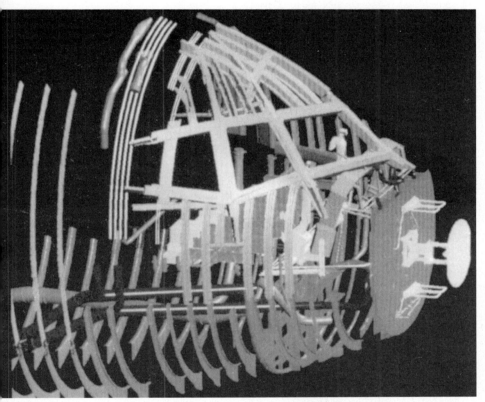

The Boeing 777 is the first aircraft to have been designed completely with computers. Dassault/IBM CATIA 3D picture of structure with ducts, fittings and antennae in place. CATIA can show accurate shapes and dimensions and calculate structural loads

88 in wide pallets; capacity allows 18 LD3s in forward compartment and 14 in rear, plus 16.99 m³ (600 cu ft) bulk cargo, or six 96 in wide pallets in front and 14 LD3s in rear, or six pallets in front and four in rear (latter requires optional 2.69 m; 8 ft 10 in wide rear freight door); mechanised cargo handling system; bulk cargo compartment at rear with separate door holds 16.98 m³ (600 cu ft). Optional underfloor crew rest module with four bunks, two business class seats and stowage space, occupying same floor area as a 96 in wide pallet, requires only an electrical connection and access hatch in cabin floor.

SYSTEMS: Garrett air drive unit using bleed air from engines, APU or ground supply drives central hydraulic system; cabin air supply and pressure control by AiResearch; Sundstrand variable-speed, constant-frequency AC electrical power generating system, with two 120kVA integrated drive generators, one APU-driven generator and ram air turbine system. AlliedSignal Garrett GTCP331-500 APU; Hamilton Standard air-conditioning; Smiths Industries ultrasonic fuel quantity gauging system and electrical load management system; optional wingtip folding by Montek.

AVIONICS: Honeywell airplane information management system (AIMS) integrates flight management, flight and navigation display generators, onboard maintenance, aircraft condition monitoring, communications management, engine data interface and data conversion gateway in two similar cabinets; AIMS results in 20 per cent weight reduction, 30 per cent power reduction and 100 per cent improvement in mean time between unscheduled removals (MTBUR) of avionic system line-replaceable units; ARINC 629 databus links all units to a single twisted wire pair, reducing wire bundles from 600 in Model 767 to 400, connectors from 4,860 to 1,580, wire length from 114 km (71 miles) to 48 km (30 miles) and wire weight from 1,179 kg (2,600 lb) to 658 kg (1,450 lb); GPS and satellite communications are standard equipment. Honeywell air data/inertial reference system (ADIRS), with ring laser gyros. Rockwell Collins maintenance access terminal with liquid crystal display and disk loaders for programming 50 onboard systems; computer controlled cabin management system controls lighting and atmosphere and gives high-quality digital sound at each seat.

DIMENSIONS, EXTERNAL:
Wing span	60.93 m (199 ft 11 in)
Wing span with tips folded	47.32 m (155 ft 3 in)
Wing aspect ratio	8.68
Length overall	63.73 m (209 ft 1 in)
Fuselage: Length	62.78 m (205 ft 11½ in)
Max diameter	6.20 m (20 ft 4 in)
Max height: overall	18.44 m (60 ft 6 in)
to tip of folded wing	14.22 m (46 ft 8 in)
Tailplane span	21.35 m (70 ft 0½ in)
Wheelbase	25.88 m (84 ft 11 in)
Wheel track	10.97 m (36 ft 0 in)
Passenger doors (four port, four stbd):	
Height	1.88 m (6 ft 2 in)
Width	1.07 m (3 ft 6 in)
Max height to sill	5.51 m (18 ft 1 in)
Forward cargo door, stbd: Height	1.70 m (5 ft 7 in)
Width	2.72 m (8 ft 11 in)
Max height to sill	3.05 m (10 ft 0 in)
Rear cargo door, stbd, standard:	
Height	1.88 m (6 ft 2 in)
Width	1.78 m (5 ft 10 in)
Max height to sill	3.40 m (11 ft 2 in)
Rear cargo door, stbd, optional width	2.69 m (8 ft 10 in)
Bulk cargo door, stbd: Height	0.91 m (3 ft 0 in)
Width	1.14 m (3 ft 9 in)
Max height to sill	3.48 m (11 ft 5 in)

DIMENSIONS, INTERNAL:
Cabin: Length	48.97 m (160 ft 8 in)
Max width	5.87 m (19 ft 3 in)
Floor area	279.1 m² (3,004 sq ft)
Max underfloor cargo hold volume	160.16 m³ (5,656 cu ft)

AREAS:
Wings, projected	427.8 m² (4,605.0 sq ft)
Horizontal tail, projected	101.26 m² (1,090 sq ft)
Vertical tail, projected	53.23 m² (573 sq ft)

WEIGHTS AND LOADINGS (A1, A2, A3: 777-200 A Market, at different max T-O weights. B1, B2: 777-200 B Market, at different max T-O weights):
Operating weight empty: A1, A2	135,580 kg (298,900 lb)
A3	135,875 kg (299,550 lb)
B1, B2	138,120 kg (304,500 lb)
Max payload: A1, A2	54,930 kg (121,100 lb)
A3	54,635 kg (120,450 lb)
B1, B2	54,660 kg (120,500 lb)
Max T-O weight: A1	229,520 kg (506,000 lb)
A2	233,600 kg (515,000 lb)
A3	242,670 kg (535,000 lb)
B1	263,085 kg (580,000 lb)
B2	267,620 kg (590,000 lb)
Max ramp weight allowance: A, B	907 kg (2,000 lb)
Max zero-fuel weight:	
A1, A2, A3	190,510 kg (420,000 lb)
B1, B2	192,775 kg (425,000 lb)
Max landing weight: A1	200,035 kg (441,000 lb)
A2, A3	201,850 kg (445,000 lb)
B1	204,115 kg (450,000 lb)
B2	206,385 kg (455,000 lb)

452 USA: AIRCRAFT—BOEING

Model of Boeing 777 showing six-wheel main bogies, which have steerable rear wheels

Max fuel weight: A1, A2, A3 94,210 kg (207,700 lb)
   B1, B2 135,845 kg (299,490 lb)
PERFORMANCE (estimated):
Cruising Mach number 0.83
Runway ACN (Flex, med subgrade, code B) 51.1
Range: A1, 375 two-class passengers, with allowances
   4,050 nm (7,505 km; 4,660 miles)
B2 at optional weight, 305 three-class passengers, with allowances 6,600 nm (12,231 km; 7,590 miles)
A1, max payload, with allowances
   2,300 nm (4,260 km; 2,645 miles)
B2, max payload, with allowances
   4,500 nm (8,340 km; 5,180 miles)

Boeing 777 two-man flight deck has five flat panel displays in horizontal row on main panel and two FMS control and display panels and a multi-function display on the centre console

---

BOEING DEFENSE & SPACE GROUP
PO Box 3999, Seattle, Washington 98124-2499
Telephone: 1 (206) 773 0530
Fax: 1 (206) 773 4261
PRESIDENT: B. Dan Pinick
EXECUTIVE VICE-PRESIDENT: C. G. King
DIRECTOR, PUBLIC RELATIONS AND ADVERTISING:
Peter B. Dakan
MANAGER, PUBLIC RELATIONS: Susan A. L. Bradley

In addition to Military Airplanes and Helicopters Divisions, Defense & Space Group controls Electronic Systems Division, Missiles & Space Division and Product Support Division. Major programme activities are airborne warning and control; partner with Grumman (prime) and Lockheed in A-X programme (see entry for US Navy); partner in F-22 team (see entry for Lockheed F-22A); Inertial Upper Stage rocket motor for Space Shuttle; Avenger air defence system; RAH-66 Comanche combat helicopter (see Boeing-Sikorsky entry, page 459); B-2 bomber (see Northrop entry); and V-22 Osprey tilt-rotor aircraft (see Bell-Boeing entry).

ELECTRONIC SYSTEMS DIVISION (Boeing Defense & Space Group)
PO Box 3999, Seattle, Washington 98124
Telephone: 1 (206) 773 0667
Fax: 1 (206) 773 3900
VICE-PRESIDENT AND GENERAL MANAGER: J. Dempster

BOEING E-3 SENTRY
French designation: **E-3F Système de Détection Aéroporté**
UK designation: **E-3D Sentry AEW. Mk 1**
US designation: **E-3B/C AWACS**
TYPE: Mobile, flexible, survivable, jamming resistant, high capacity radar station and command, control and communications centre; airborne warning and control system (AWACS).
PROGRAMME: Two prototype EC-137Ds used to test competing radars; Westinghouse selected; full-scale development completed 1976. USAF received 34 E-3s by June 1984, including two prototypes.
First production E-3A delivered to 552nd Airborne Warning and Control Wing, Tactical Air Command, at Tinker AFB, Oklahoma, on 24 March 1977; initial operational capability (IOC) April 1978. At various times, E-3s deployed to Iceland, Germany, Saudi Arabia, Sudan, the Mediterranean area, South West Asia and the Pacific and in support of drug enforcement programme. E-3As began to work with NORAD continental air defence on 1 January 1979.
The 552nd Wing has three AWACS squadrons and supporting units. Overseas units include the 960th, 961st and 962nd AWAC squadrons based respectively at NAS Keflavik, Iceland, Kadena AB, Okinawa, Japan, and Elmendorf, Alaska, providing command and control capability to CINCLANT (through Commander, Iceland Defence Force) and CINCPAC.
Production completed 1991; 1,010th and final 707 airframe, an E-3, retained by Boeing for tests and trials; upgrades continue; Boeing to transfer avionics to 767 airframe as successor AWACS (which see).
CURRENT VERSIONS: **Core E-3A:** Initial standard of 23 production USAF Sentries. Equipment detailed in 1987-88 *Jane's*.
E-3B: Block 20 modification updated two EC-137Ds and 22 USAF Core E-3As to E-3B standard by adding ECM-resistant voice communications, a third HF and five more UHF radios (making 12), faster IBM CC-2 computer with larger memory, five more SDCs (making 14), Westinghouse austere maritime surveillance capability to main radar, provision for Have Quick anti-jamming in UHF radios, self-defence, and radio teletype. First E-3B redelivered to USAF 18 July 1984; remaining 23 modified with Boeing kits at Tinker AFB. Under Project Snappy, unspecified sensor installed in seven USAF E-3B/Cs by January 1991 for participation in Gulf War; eight further aircraft modified. *Detailed description applies to USAF E-3B/C, except where indicated.*
US/NATO Standard E-3A: Original standard for USAF aircraft Nos. 26 to 34 of which deliveries began December 1981, and of updated aircraft No. 3. Additions include full maritime surveillance capability, CC-2 computer, additional HF radios, ECM-resistant voice communications, radio teletype, provision for self-defence, and ECM. Eighteen NATO E-3As are to this standard.
E-3C: USAF Block 25 modification of 10 USAF E-3As began 1984. Adds five more SDCs, five more UHF radios and (first use 1 February 1991) Have Quick A-Nets secure communications system.
JE-3C: In May 1987, Boeing received $241.5 million USAF contract for full-scale development and integration into USAF and NATO E-3s of AN/AYR-1 ESM system to detect signals from hostile and friendly targets. First installation in JE-3C temporary test aircraft 73-1674; tested over 1,000 hours in 155 sorties, beginning September 1990; second in NATO E-3A LX-N-90442, re-delivered October 1991; flew 15 trial sorties, early 1992.
KE-3A: Boeing designation for Saudi Arabian tanker

based on E-3 airframe; CFM56 engines; no rotodome or other surveillance equipment; eight built (not included in E-3 production totals).

E-3D Sentry AEW. Mk 1: British government announced order for six, 18 December 1986 and exercised option for seventh in October 1987. CFM56-2A-3 power plants. First flight E-3D (ZH101) 11 September 1989; first flight fully equipped 5 January 1990; flown to UK on 3 November 1990 after combined British-French airworthiness certification and in-flight refuelling trials. Second aircraft (ZH102) flown to Waddington 4 July 1990 for fitting out by a consortium led by British Aerospace (which see) and handed over to RAF as first receipt, 24 March 1991. Final UK E-3D (ZH107), the 1,009th airframe of the Boeing 707 family, first flown 14 June 1991 and delivered to RAF 12 May 1992. Sentry Training Squadron formed at Waddington 1 June 1990. No. 8 Squadron, RAF, formed with E-3Ds at Waddington on 1 July 1991; declared to NATO on 1 July 1992; RAF aircraft are E-3D component, NAEWF. Boeing gave UK 130 per cent industrial offset.

E-3F SDA: France ordered three with CFM56-2A-3 power plants, February 1987; fourth aircraft ordered 1987, but options for two more dropped August 1988. First (No. 201) flown 27 June 1990; delivered UTA Industries at Le Bourget for fitting-out 10 October 1990; to Avord for integration trials and handover 19 December 1990; official delivery to CEAM for military acceptance, 22 May 1991. Delivery of remaining E-3Fs followed 27 July and 11 September 1991 and 15 February 1992. 36e Escadre de Détection Aéroportée formed at Avord 1 March 1990 as operating wing of two squadrons (Escadrons 1/36 'Berry' and 2/36 'Nivernais'). Boeing gave France 130 per cent industrial offset.

Other upgrades: USAF Electronic Systems Division proposed $425 million MSIP over five years for E-3s, adding improved radar detection, passive sensors and other features.

All USAF, NATO and RAF E-3s will in future have Joint Tactical Information Distribution System (JTIDS).

Block 30/35 programme additionally provides USAF E-3s with Boeing/UTL AN/AYR-1 ESM system, Radar System Improvement Programme (RSIP), upgrading of JTIDS to Tactical Digital Information Link-J (TADIL-J), CC-2 computer memory upgrade with VLSI circuitry and bubble memory and ability to use Global Positioning System (GPS). AN/AYR-1 installation, weighing 862 kg (1,900 lb), includes two 'canoe' radomes on sides of forward fuselage; each 3.96 m (13 ft 0 in) long, 0.84 m (2 ft 9 in) wide and protruding 0.46 m (1 ft 6 in). Planned initial operational capability for Block 30/35 in 1993.

All aircraft from No. 25 have a hardpoint for stores under each inner wing.

In 1989, Westinghouse awarded development contract for upgrading AN/APY-1/2 radars of USAF E-3s with new processors, displays and pulse compression to double performance against small targets like cruise missiles. If successful, this will be applied to all 34 E-3B/Cs from 1994 onwards.

Contract for NATO E-3 upgrade placed January 1993; designated Mod Block 1; includes new colour displays, Have Quick secure radios and Link 16 (JTIDS) data link; initial two aircraft under 1993 contract; follow-on order for remaining 16 modifications placed mid-1993; all retrofit work by DASA in Germany.

CUSTOMERS: Total of 68 ordered (USAF 34, NATO 18, Saudi Arabia five, UK seven, France four); last delivered May 1992. First NATO production E-3A flew at Renton 18 December 1980 and delivered to system integrator Dornier at Oberpfaffenhofen 19 March 1981; all 18 delivered to NATO between 22 January 1982 and 25 April 1985. NATO E-3As assigned to Nos. 1, 2 and 3 Squadrons and Training Centre of E-3A component, NATO Airborne Early Warning Force, based at Geilenkirchen, Germany.

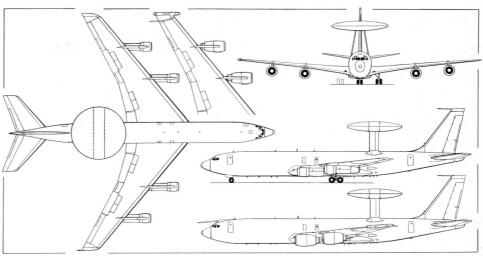

Boeing E-3A Sentry, with lower side view and wing scrap view of Royal Air Force E-3D Sentry AEW. Mk 1 *(Dennis Punnett)*

Boeing E-3A of the Royal Saudi Air Force *(Paul Jackson)*

Up to six NATO E-3As detached to forward operating locations (FOL) at Konya (Turkey), Previza (Greece), Trapani/Birgi (Italy) and Ørland (Norway). NAEWF has NATO Command status and is staffed by personnel from Belgium, Canada, Denmark, Greece, Italy, the Netherlands, Norway, Portugal, Turkey and USA, plus non-aircrew personnel from Luxembourg. ESM systems upgrade under way for 1995 completion. Seven-year E-3A modernisation plan agreed June 1990, subject to funding.

Sale of five E-3s to Royal Saudi Air Force approved October 1981 under Peace Sentinel programme: also included eight **KE-3A** tanker/transports. Selected CFM56-2A-2 with Hispano Suiza thrust reversers for E-3s in 1984. First E-3 handed over 30 June 1986; all 13 E-3s and KE-3s delivered by 24 September 1987.

COSTS: $129 million, flyaway, NATO 1982; $222 million programme unit cost, UK 1987; France $254 million, 1987. E-3D programme cost $1.3 billion (1987); E-3F programme cost $550 million (1987); $294.6 million NATO upgrade (Mod Block 1) initial contract, 1992.

DESIGN FEATURES: Construction details of the E-3 appeared in 1987-88 and earlier *Jane's*.

FLYING CONTROLS: Essentially as for Boeing 707 family.
STRUCTURE: RAF aircraft have additional wing stringers outboard of outer engines because of wingtip pods and trailing-edge HF antennae.
POWER PLANT: Four Pratt & Whitney TF33-PW-100/100A turbofans, each rated at 93.4 kN (21,000 lb st), mounted in pods beneath the wings. Fuel contained in integral wing tanks. Usable fuel 90,528 litres (23,915 US gallons; 19,913 Imp gallons). Provision for in-flight refuelling, with receptacle for boom over flight deck. Four CFM56-2A-2/3 turbofans on French (2A-3), Saudi (2A-2) and UK (2A-3) aircraft; rated at 106.8 kN (24,000 lb st) for take-off; 104.1 kN (23,405 lb st) max continuous. SOGERMA in-flight refuelling probe (in addition to receptacle) on E-3D and E-3F.
ACCOMMODATION: Basic E-3A operational crew of 17 includes a flight crew complement of four plus 13 AWACS specialists, though this latter number can vary for tactical and defence missions. Full crew complement is two pilots, navigator, flight engineer, tactical director, fighter allocator, two weapons controllers, surveillance controller, link manager, three surveillance operators, communications

Boeing E-3D Sentry AEW. Mk 1 of the Royal Air Force escorted by a Tornado F. Mk 3

operator, radar technician, communications technician and computer display technician. E-3B/C have extra four AWACS specialist positions increasing crew to 17, plus four on flight deck. Aft of flight deck, from front to rear of fuselage, are communications, data processing and other equipment bays; multi-purpose consoles; communications, navigation and identification equipment; and crew rest area, galley and parachute storage rack.

SYSTEMS: A liquid cooling system provides protection for the radar transmitter. An air cycle pack system, a draw-through system, and two closed loop ram-cooled environmental control systems ensure a suitable environment for crew and avionics equipment. Electrical power generation has a 600kVA capability. Distribution centre for mission equipment power and remote avionics in lower forward cargo compartment. Rear cargo compartment houses radar transmitter and an APU. External sockets allow intake of power when aircraft is on ground. Two separate and independent hydraulic systems power essential flight and mission equipment; either system can satisfy requirements of both equipment groups in an emergency.

AVIONICS: Elliptical cross-section rotodome of 9.14 m (30 ft) diameter and 1.83 m (6 ft) max depth, mounted 3.35 m (11 ft) above fuselage, comprises four essential elements: a turntable, strut mounted above rear fuselage, supporting rotary joint assembly to which are attached sliprings for electrical and waveguide continuity between rotodome and fuselage; structural centre-section of aluminium skin and stiffener supporting the Westinghouse AN/APY-1 surveillance radar (AN/APY-2 from No. 25 onwards, and in all export E-3s) and IFF/TADIL-C antennae, radomes, auxiliary equipment for radar operation and environmental control of the rotodome interior; liquid cooling of the radar antennae; and two radomes of multi-layer glassfibre sandwich material, one for surveillance radar and one for IFF/TADIL-C array. For surveillance operations rotodome is hydraulically driven at 6 rpm, but during non-operational flights it is rotated at only ¼ rpm, to keep bearings lubricated. Radar operates in E/F-band and can function as both a pulse and/or a pulse Doppler radar for detection of aircraft targets. A similar pulse radar mode with additional pulse compression and sea clutter adaptive processing is used to detect maritime/ship traffic. Radar is operable in six modes: PDNES (pulse Doppler non-elevation scan), when range is paramount to elevation data; PDES (pulse Doppler elevation scan), providing elevation data with some loss of range; BTH (beyond the horizon), giving long-range detection with no elevation data; Maritime, for detection of surface vessels in various sea states; Interleaved, combining available modes for all-altitude longer-range aircraft detection, or for both aircraft and ship detection; and Passive, which tracks enemy ECM sources without transmission-induced vulnerability. Radar antennae, spanning about 7.32 m (24 ft), and 1.52 m (5 ft) deep, scan mechanically in azimuth, and electronically from ground level up into the stratosphere. Heart of the data processing capability of the first 24 aircraft in their original core E-3A form is an IBM 4 Pi CC-1 high-speed computer (see 1987-88 *Jane's* for details). From 25th aircraft, the new and improved IBM CC-2 computer was installed from the start, with a main storage capacity of 665,360 words. Data display and control are provided by Hazeltine high resolution colour situation display consoles (SDC) and auxiliary display units (ADU). The E-3B carries 14 SDCs and two ADUs. Navigation/guidance relies upon two Delco AN/ASN-119 Carousel IV inertial navigation platforms, a Northrop AN/ARN-120 Omega set which continuously updates the inertial platforms, and a Teledyne Ryan AN/APN-213 Doppler velocity sensor to provide airspeed and drift information. Communications equipment provides HF, VHF and UHF channels through which information can be transmitted or received in clear or secure mode, in voice or digital form. Bendix/King weather radar in nose. Identification is based on an Eaton (AIL) AN/APX-103 interrogator set which is the first airborne IFF interrogator to offer complete AIMS Mk X SIF air traffic control and Mk XII military identification friend or foe (IFF) in a single integrated system. Simultaneous Mk X and Mk XII multi-target and multi-mode operations allow the operator to obtain instantaneously the range, azimuth and elevation, code identification, and IFF status, of all targets within radar range. NATO E-3As carry, and USAF aircraft have provisions for, a radio teletype. All aircraft from No. 25 have an inboard underwing hardpoint on each side. There is no current requirement for either USAF or NATO AWACS to carry weapons, but on NATO E-3As these hardpoints can be used to mount additional podded items of ECM equipment. E-3Ds carry Loral 1017 'Yellow Gate' (ARI 18240) ESM pods at the wingtips; AN/AYR-1 ESM retrofitted in USAF and NATO E-3s from 1991 (canoe-shaped pod scabbed to each side of forward fuselage and sensor in tailcone).

DIMENSIONS, EXTERNAL:

Wing span: normal	44.42 m (145 ft 9 in)
E-3D	44.98 m (147 ft 7 in)
Length overall	46.61 m (152 ft 11 in)
Height overall	12.73 m (41 ft 9 in)

WEIGHTS AND LOADINGS (E-3D):

Fuel weight (JP4)	70,510 kg (155,448 lb)
Normal T-O weight	147,417 kg (325,000 lb)
Max T-O weight	150,820 kg (332,500 lb)
Max ramp weight	151,953 kg (335,000 lb)

Artist's impression of AWACS conversion of Boeing 767 airliner

PERFORMANCE:

Max level speed	460 knots (853 km/h; 530 mph)
Service ceiling: TF33	over 8,850 m (29,000 ft)
CFM56	over 10,670 m (35,000 ft)
Max unrefuelled range	more than 5,000 nm (9,266 km; 5,758 miles)
Endurance on station, 870 nm (1,610 km; 1,000 miles) from base	6 h
Max unrefuelled endurance	more than 11 h

BOEING 767 AWACS

TYPE: Airborne warning and control system successor to E-3.

PROGRAMME: Boeing announced, December 1991, definition studies for modified 767-200ER airliner (which see) with Westinghouse AN/APY-2 radar; project sustained by Japanese interest; engine selection October 1992; J-STARS (radar reconnaissance), military transport and tanker versions also possible.

CUSTOMERS: Japanese government purchase decision, December 1992; parliamentary approval given early 1993; total requirement for four; two to be ordered 1993; modification by Boeing Product Support Division at Wichita from January 1995; avionics installation at Seattle from October 1995; seven-month flight-test programme at Seattle from April 1996; mission system equipment installation from October 1996 (No. 2 aircraft first); delivery of first two aircraft in January 1998.

DESIGN FEATURES: See Boeing 767 airliner entry. Additionally, substantial structural modifications (eight frames and nine floor beams) to accommodate rotodome; ventral fins may be added on rear fuselage.

POWER PLANT: Two 273.6 kN (61,500 lb st) General Electric CF6-80CZ turbofans modified for additional electrical generation. In-flight refuelling receptacle.

ACCOMMODATION: Two-man flight crew; up to 18 mission crew.

SYSTEMS: Two 150kVA generators on each engine replace single 90kVA units; total 600kVA.

AVIONICS: Generally as E-3C; capable of accepting latest AWACS upgrades such as cruise missile and UAV detection.

DIMENSIONS, EXTERNAL:

Wing span	47.57 m (156 ft 1 in)
Length overall	48.51 m (159 ft 2 in)
Height overall	15.85 m (52 ft 0 in)
Radome: Diameter	9.14 m (30 ft 0 in)
Thickness	1.83 m (6 ft 0 in)

WEIGHTS AND LOADINGS:

Max T-O weight	171,004 kg (377,000 lb)

PERFORMANCE:

Service ceiling	10,360-13,100 m (34,000-43,000 ft)
Range, unrefuelled	4,500-5,000 nm (8,340-9,266 km; 5,182-5,758 miles)
Endurance:	
at 1,000 nm (1,853 km; 1,150 miles) radius	7 h
at 300 nm (556 km; 345 miles) radius	10 h
with in-flight refuelling	22 h

BOEING E-6A MERCURY (TACAMO II)

TYPE: Long endurance communications relay aircraft carrying US Navy airborne very low frequency (AVLF) system.

PROGRAMME: US Navy contract placed with Boeing Aerospace 29 April 1983 to replace EC-130Q Hercules TACAMO (take charge and move out); first full test flight of prototype (162782) 1 June 1987: two production E-6As ordered FY 1986; first two delivered to VQ-3 squadron, Barber's Point, Hawaii, 2 August 1989.

CUSTOMERS: US Navy, two, three, three and seven production E-6As ordered in FYs 1986-89, making 16; final delivery (actually aircraft No. 10) 28 May 1992; VQ-3 withdrew last EC-130Q in August 1990, having eight E-6As for Pacific area; VQ-4 assigned seven at Patuxent River, Maryland, for Atlantic area; first of these delivered 25 January 1991; VQ-4 operational June 1991. By early 199 continuous TACAMO airborne alert terminated; Strategi Communications Wing One formed Tinker AFB, Oklahoma, 1 May 1992; VQ-3 to Tinker, September 1992 VQ-4 to Tinker January 1993.

DESIGN FEATURES: Substantially as for USAF E-3, but powered by four CFM International turbofans.

FLYING CONTROLS: Substantially as for E-3. Boeing contrac 1991 to Bendix/King to develop digital AFCS.

STRUCTURE: 75 per cent common with E-3, including EM nuclear hardening; radome support structure deleted Additions include wingtip ESM/Satcom pods and H antenna fairings, increased corrosion protection, forward freight door of 707-320C. Local strengthening to over come stresses caused by banked orbit with antenn deployed: flap guide vanes modified; three minor change to be made to cure tail fin flutter.

POWER PLANT: Four 106.8 kN (24,000 lb st) CFM International F108-CF-100 (CFM56-2A-2) turbofans in individual underwing pods, as on some export E-3s. Fue contained in integral tanks in wings, with single-poin refuelling. In-flight refuelling via boom receptacle above flight deck.

ACCOMMODATION: Basic militarised interior sidewalls, ceilings and lighting are similar to those of the E-3A. Interio divided into three main functional areas: forward of wing (flight deck and crew rest area), overwing (five-man mis sion crew), and aft of wings (equipment). Forward crew area, 50 per cent common with that of E-3A, accommo dates a four-man crew on flight deck. Compartmen immediately aft contains dining, toilet, and eight-bunk res area for spare crew carried on extended or remote deploy ment missions. Crew enter by ladder and hatch in floor c this compartment. Overwing compartment with communi cations and other consoles, their operators, and an airborn communications officer (ACO). To the rear is compart ment containing R/T racks, transmitters, trailing wir antennae and their reels. Bale-out door at rear on starboar side.

SYSTEMS: Some 75 per cent of the E-6A's systems are the same as the E-3. Among those retained are the liquid cool ing system for the transmitters, 'draw-through' coolin system for other avionics, the 600kVA electrical powe generation system, APU, liquid oxygen system, and MIL specification hydraulic oil.

AVIONICS: Three Collins AN/ARC-182 VHF/UHF com trans ceivers, all with secure voice capability; five Collin AN/ARC-190 HF com (one transceiver, one receive only) and Hughes Aircraft AN/AIC-32 crew intercom wit secure voice capability. External aerials for Satcom UHF reception in each wingtip pod; fairings beneath each poc house antenna for standard HF reception. Navigation by triplex Litton LTN-90 ring laser gyro-based inertial reference system integrated with a Litton LTN-211 VLF Omega system and duplex Smiths Industries SFM 102 digital/analog flight management computer system (FMCS). Bendix/King AN/APS-133 colour weather radar in nosecone, with capability for short-range terrain mapping, tanker beacon homing, and waypoint display. Honeywell AN/APN-222 high/low-range (0-15,240 m 0-50,000 ft) radio altimeter, and Collins low-range (0-762 m; 0-2,500 ft) radio altimeter, with ILS and GPWS. General Instrument AN/ALR-66(V)4 electronic support measures (ESM), in each wingtip pod, provide information on threat detection, identification, bearing and approximate range. In overwing compartment, overseen by ACO, is a new communications central console, which incorporates ERCS (emergency rocket communications system) receivers, cryptographic equipment, new teletypes, tape recorders, and other communications equipment, all hardened against electromagnetic interference. In each operational area the E-6A links upward with airborne command posts and the Presidential E-4, to satellites, and to the ERCS; and downward to VLF ground stations and the SSBN fleet. Mean time between failures of complete mission avionics is less than 20 h, but the E-6 is able to carry

BOEING—AIRCRAFT: USA   455

Boeing E-6A Mercury prototype pictured during flight testing over Puget Sound, in Washington State

spares, and a spare crew, to permit extended missions of up to 72 h with in-flight refuelling, and/or deployment to remote bases, where it is capable of autonomous operation.

EQUIPMENT: Main VLF antenna is a 7,925 m (26,000 ft) long trailing wire aerial (LTWA), with a 41 kg (90 lb) drogue at the end, which is reeled out from the middle part of the rear cabin compartment through an opening in the cabin floor. The LTWA, with its drogue, weighs about 495 kg (1,090 lb) and creates some 907 kg (2,000 lb) of drag when fully deployed. Acting as a dipole is a much shorter (1,220 m; 4,000 ft) trailing wire aerial (STWA), winched out from beneath the tailcone. At patrol altitude, with the LTWA deployed, the aircraft enters a tight orbit and the wire stalls, causing it to be almost vertical (70 per cent verticality is required for effective sub-sea communications). Signals transmitted through the trailing wire antennae use 200kW of power, and can be received by submerged SSBNs via a towed buoyant wire antenna.

ARMAMENT: None.

DIMENSIONS, EXTERNAL:
Wing span	45.16 m (148 ft 2 in)
Wing aspect ratio	7.20
Length overall	46.61 m (152 ft 11 in)
Height overall	12.93 m (42 ft 5 in)
Wheel track	6.73 m (22 ft 1 in)
Wheelbase	17.98 m (59 ft 0 in)
Forward cargo door: Height	2.34 m (7 ft 8 in)
Width	3.40 m (11 ft 2 in)
Height to sill	3.20 m (10 ft 6 in)

AREAS:
Wings, gross	283.4 m² (3,050.0 sq ft)

WEIGHTS AND LOADINGS:
Operating weight empty	78,378 kg (172,795 lb)
Max fuel	70,305 kg (155,000 lb)
Max T-O weight	155,128 kg (342,000 lb)
Max wing loading	547.4 kg/m² (112.13 lb/sq ft)
Max power loading	363.26 kg/kN (3.56 lb/lb st)

PERFORMANCE (S/L, ISA, estimated):
Dash speed	530 knots (981 km/h; 610 mph)

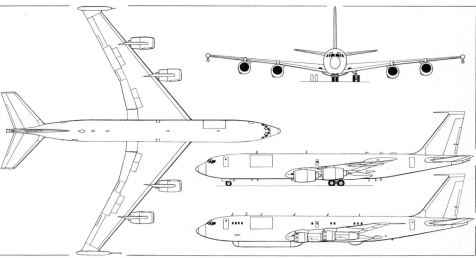
Boeing E-6A Mercury, with additional side view (lower) of E-8C J-STARS *(Dennis Punnett)*

Cruising speed at 12,200 m (40,000 ft)	455 knots (842 km/h; 523 mph)
Patrol altitude	7,620-9,150 m (25,000-30,000 ft)
Service ceiling	12,800 m (42,000 ft)
Critical field length	2,042 m (6,700 ft)
Max effort T-O run	1,646 m (5,400 ft)
Max effort T-O run with fuel for 2,500 nm (4,630 km; 2,875 miles)	732 m (2,400 ft)
Landing run at max landing weight	793 m (2,600 ft)
Mission range, unrefuelled	6,350 nm (11,760 km; 7,307 miles)
Endurance: unrefuelled	15 h 24 min
on-station, 1,000 nm (1,850 km; 1,150 miles) from T-O	10 h 30 min
with one refuelling	28 h 54 min
with multiple refuelling	72 h

BOEING/GRUMMAN E-8C (J-STARS)

See Grumman E-8C (J-STARS) and details in *Jane's Civil and Military Aircraft Upgrades* and *Jane's Battlefield Surveillance Systems*.

BOEING HELICOPTERS (Boeing Defense & Space Group, Helicopters Division)

PO Box 16858, Philadelphia, Pennsylvania 19142
Telephone: 1 (215) 591 3575
Fax: 1 (215) 591 2701
PRESIDENT AND GENERAL MANAGER: Edward J. Renouard
VICE-PRESIDENT, RESEARCH AND ENGINEERING:
  John Diamond
PUBLIC RELATIONS MANAGER: Robert Torgerson

Vertol Aircraft Corporation (formerly Piasecki Helicopter Corporation) purchased in 1960, becoming Vertol Division of Boeing; now Helicopters Division of Boeing Defense & Space Group. Produced more than 2,500 tandem-rotor helicopters for US military services and export. Main production programme, modernisation of early CH-47s to CH-47Ds for US Army. Production of H-46 Sea Knight SR&M improvement kits for US Navy and Marine Corps completed January 1989.

Boeing Helicopters teamed with Bell Helicopter Textron in V-22 Osprey programme (see Bell/Boeing entry). Joined with Sikorsky Aircraft 3 June 1985 to compete for US Army LH (RAH-66 Comanche) light helicopter, with Boeing integrating avionics; declared competition winner in April 1991 (see under separate Boeing/Sikorsky entry for details). All-composite rotor blade for Bell UH-1H developed for US Army; first blade delivered June 1989 to Army Engineering Flight Agency for interchangeability check with Bell-produced blade.

Boeing Helicopters produces fixed leading-edge components for Boeing 737/747/757/767 and metal leading-edge slats for 757.

Workforce 7,000 in 1992. Manufacturing plant at Ridley Township, Pennsylvania, has 325,150 m² (3,500,000 sq ft) of covered floor space; flight test centre at Greater Wilmington, Delaware, has 8,565 m² (92,200 sq ft). An 11,055 m² (119,000 sq ft) development facility and 17,466 m² (188,000 sq ft) office/laboratory/computer centre added at Ridley Township early 1987.

BOEING 107 (H-46)

USN and USMC planning dynamic system upgrade of all H-46s in five-year programme beginning 1994. Refer to *Jane's Civil and Military Aircraft Upgrades*.

BOEING 114 and 414

US Army designations: CH-47 and MH-47 Chinook
Canadian Forces designation: CH-147
Royal Air Force designations: Chinook HC. Mk 1/2
Spanish Army designation: HT.17

TYPE: Tandem-rotor, twin-turbine transport helicopter.

PROGRAMME: Design of all-weather medium transport helicopter for US Army began 1956; first flight of first of five YCH-47As 21 September 1961; for details of CH-47A (354 built for US Army) and CH-47B (108 built for US Army) see 1974-75 *Jane's*. Performance increased in CH-47C by uprated transmissions and 2,796 kW (3,750 shp) T55-L-11A; integral fuel capacity increased to 3,944 litres (1,042 US gallons; 867.6 Imp gallons); first flight 14 October 1967; 270 delivered to US Army from Spring 1968; 182 US Army CH-47Cs retrofitted with composite rotor blades; integral spar inspection system (ISIS) introduced 1973 together with crashworthy fuel system retrofit kit; full details in 1980-81 *Jane's*. Transmissions of some As and Bs upgraded to CH-47C standard.

CURRENT VERSIONS: **CH-47D:** US Army contract to modify one each of CH-47A, B and C to prototype Ds placed 1976; first flight 11 May 1979; first production contract October 1980; first flight 26 February 1982; first delivery 31 March 1982; initial operational capability (IOC) achieved 28 February 1984 with 101st Airborne Div; second multi-year production contract for 144 CH-47Ds awarded 13 January 1989, bringing total CH-47D (and MH-47E) ordered to 472; further two Gulf War attrition replacements authorised; deliveries reached 406 by June 1992; production rate four per month; programme completion due October 1993; US regular Army deliveries mostly completed October 1990 with 17th operating unit (C Company, 228th Aviation Regiment, Fort Wainright, Alaska); remainder of CH-47Ds for Army Reserve and National Guard by late 1993; deliveries to National Guard (Texas) began 1988.

CH-47D update includes strip down to bare airframe, repair and refurbish, fit Textron Lycoming T55-L-712 turboshafts, uprated transmissions with integral lubrication and cooling, composite rotor blades, new flight deck compatible with night vision goggles (NVG), new redundant electrical system, modular hydraulic system, advanced automatic flight control system, improved avionics and survivability equipment, Solar T62-T-2B APU operating hydraulic and electrical systems through accessory gear drive, single-point pressure refuelling, and triple external cargo hooks. Composites account for 10 to 15 per cent of structure. About 300 suppliers involved.

At max gross weight of 22,680 kg (50,000 lb), CH-47D has more than double useful load of CH-47A. Sample loads include M198 towed 155 mm howitzer, 32 rounds of ammunition and 11-man crew, making internal/external load of 9,980 kg (22,000 lb); D5 caterpillar bulldozer weighing 11,225 kg (24,750 lb) on centre cargo hook; US Army Milvan supply containers carried at up to 130 knots (256 km/h; 159 mph); up to seven 1,893 litre (500 US gallon, 416 Imp gallon), 1,587 kg (3,500 lb) rubber fuel blivets carried on three hooks. Four CH-47Ds converted for in-flight refuelling from C-130 at up to 120 knots (222 km/h; 138 mph) by day and night and in moderate turbulence, approved July 1988; graphite fuel boom at lower starboard side contains telescoping aluminium tube and can accept flow rate of 568 litres (150 US gallons; 125 Imp gallons)/min to refuel completely in six minutes; first delivery to US Army July 1988.

CH-47D Special Operations Aircraft: Two battalions of 160th Special Operations Aviation Regiment (at Fort Campbell, Kentucky, and Hunter AAF, Georgia) equipped pending availability of MH-47E with 32 CH-47D SOA fitted with refuelling probes (first refuelling July 1988), thermal imagers, Bendix/King RDR-1300 weather radar, improved communications and navigation equipment, and two pintle-mounted 7.62 mm machine guns. Navigator/commander's station fitted in some SOAs.

MH-47E: Special Forces variant; planned procurement 51, deducted from total 472 CH-47D conversions; prototype development contract 2 December 1987; long-lead items for next 11 helicopters authorised 14 July 1989; firm order for 11, plus option on next 14, awarded 30 June 1991; Lot II (14 helicopters) confirmed 23 June 1992; total 26 firm orders, plus further 25 required. Prototype (88-0267) flew 1 June 1990; delivered 10 May 1991; initial production aircraft (90-0414) flown 1992; first 11 (of 24 intended) originally due to be delivered from November 1992 to 2 Battalion of 160th Special Operations Aviation Regt at Fort Campbell, Kentucky. Later helicopters earmarked for 3 Battalion/160 SOAR at Hunter AAF, Georgia, and 1/245th Aviation Battalion (SOA), Oklahoma National Guard, Lexington (eight and 16 MH-47Es respectively). Following mission software problems, deliveries re-scheduled for November 1993; operational training to begin in May 1994.

Mission profile 5½ hour covert deep penetration over 300 nm (560 km; 345 mile) radius in adverse weather, day or night, all terrain with 90 per cent success probability. Requirements include self deployment to Europe in stages of up to 1,200 nm (1,677 km; 1,042 miles), 44-troop capacity, powerful defensive weapons and ECM. Equipment includes IBM-AlliedSignal Bendix integrated avionics with four-screen NVG compatible EFIS; dual MIL-STD-1553 digital databuses; AN/ASN-145 AHRS; jamming-resistant radios; Rockwell Collins CP1516-ASQ automatic target handoff system; inertial AN/ASN-137 Doppler, Rockwell Collins AN/ASN-149(V)2 GPS/Navstar receiver and terrain referenced positioning navigation systems; Rockwell Collins ADF-149; laser (Perkin-Elmer AN/AVR-2), radar (E-systems AN/APR-39A) and missile (Honeywell AN/AAR-47) warning systems; ITT AN/ALQ-136(V) pulse jammer and Northrop AN/ALQ-162 CW jammer; Tracor M-130 chaff/flare dispensers; Texas Instruments AN/APQ-174A radar with modes for terrain following down to 30 m (100 ft), terrain avoidance, air-to-ground ranging and ground mapping; Hughes AN/AAQ-16 FLIR in chin turret; digital moving map display; uprated T55-L-714 turboshafts with FADEC; increased fuel capacity; additional troop seating (44 max); OBOGS; rotor brake; 272 kg (600 lb) rescue hoist with 61 m (200 ft) usable cable; two M-2 0.50 in window-mounted machine guns (port forward: starboard aft); provisions for Stinger AAMs using FLIR for sighting. This system largely common with equivalent Sikorsky MH-60K (which see).

MH-47E has nose of Commercial Chinook to allow for weather radar, if needed; forward landing gear moved 1.02 m (3 ft 4 in) forward to allow for all-composite external fuel pods (also from Commercial Chinook) that double fuel capacity; Brooks & Perkins internal cargo handling system.

Chinook HC. Mk 1/2: RAF versions (35 of 41 remaining); designation CH47-352; all HC. Mk 1s upgraded to HC. Mk 1B (see 1989-90 and earlier *Jane's*); UK MoD authorised Boeing to update 33 Mk 1Bs to Mk 2, equivalent to CH-47D, October 1989; changes include new automatic flight control system, updated modular hydraulics, stronger transmission, improved T62-T-2B APU, airframe reinforcements, low IR paint scheme, long-range fuel system and standardisation of defensive aids package (IR jammers, chaff/flare dispensers, missile approach warning and machine-gun mountings). Conversion continues from 1991 to July 1995. Chinook HC. Mk 1B ZA718 began flight testing Chandler Evans/Hawker Siddeley dual-channel, FADEC system for Mk 2 in October 1989. Same helicopter to Boeing, March 1991; rolled out as first Mk 2 19 January 1993; to UK for A&AEE trials, March 1993.

HT.17 Chinook: Spanish Army version.

Boeing 234: Commercial version, now out of production; described in 1991-92 *Jane's*.

Boeing 414: Export military version, described in 1985-86 *Jane's*. Now superseded by CH-47D International Chinook (see below).

CH-47D International Chinook: Boeing 414-100 first sold to Japan; Japan Defence Agency ordered two for JGSDF and one for JASDF Spring 1984; first flight (N7425H) January 1986 and, with second machine, delivered to Kawasaki Heavy Industries April 1986 for fitting out; co-production arrangement (see under Japan: Kawasaki **CH-47J**); by Spring 1993, 16 firm orders for JASDF and 36 for JGSDF; final total of at least 54 expected. International Chinook available in four versions with combinations of standard or long range (MH-47E type) fuel tanks and T55-L-712 SSB or T55-L-714.

Advanced Chinook: Chinook upgrade study with 3,729 kW (5,000 shp) class engines, redesigned rotor blades and hubs, and additional (MH-47E) fuel.

CUSTOMERS: Total 732 CH-47A/B/C built and 472 CH-47/MH-47E conversions authorised for US Army by Janua 1989.

Exports (see also rapid reference table) include fi CH-47Cs to Argentina (three air force; two army); Au tralia 12 CH-47Cs with crashworthy fuel system; Cana nine CH-47Cs designated **CH-147**, delivered from Se tember 1974 (details in 1985-86 *Jane's*); Japan (tw CH-47D International Chinooks, plus six kits to initia licenced production); Spanish Army 19, designated HT.1 of which 10 CH-47Cs (one lost) and nine CH-47D Inte national Chinooks delivered to 5th Helicopter Transpo Battalion (Bheltra-V) at Colmenar Viejo, Madrid (last s have Bendix/King RDR-1400 weather radar), original ni being upgraded by Boeing in USA to CH-47D betwe August 1990 and mid-1993, first redelivery to Spain on 2 September 1991; South Korea, total 24 International Chin ooks, comprising 18 for army (six ordered 1987 plus 12 1988; deliveries from 1988) and six for air force (ordere early 1990; delivered by February 1992); Taiwan (thre notionally civil Boeing 234MLR); Thailand, 12 for arm (three ordered August 1988, three early 1990; deliveri 1990 to February 1992); UK RAF 41 (41, including 41, see entry for Chinook HC. Mk 1/2); 12 civilian, which few left operating; two for trials; 45 in kits, compri ing 40 for Italy and five for Japan.

Agusta sold licence-built CH-47Cs to Egypt (15 Greece (10, of which nine to be converted to CH-47D Boeing between March 1992 and 1995; first redelivery October 1993), Iran (68), Italy (41, including 10 CH-47 Plus, to which standard 23 earlier helicopters converte refer Italy), Libya (20), Morocco (nine) and Pennsylvan Army National Guard (11). Kawasaki (which see) has fir orders for 52 (air force 16; army 36) including FY5 (199 funding.

Total Chinook orders, including civil, 1,089. S CH-47Ds, ordered by China January 1989, embargoed t US government; Australia requires four; Netherlan requires six from Agusta (plus seven from Canadia surplus).

CHINOOK PRODUCTION

Customer	Boeing	Boeing kits	Agusta	Kawasaki
Civilian	12			
Argentina	5			
Australia	12			
Canada	9			
Egypt			15	
Greece			10	
Iran		38	30	
Italy		2	39	
Japan	2	6		44[1]
Libya			20	
Morocco			9	
South Korea	24			
Spain	19			
Taiwan	3[2]			
Thailand	6			
UK (RAF)	41			
US Army	732		11	
(Total 1,089)	865	46	134	44
Rebuilds				
Greece	9			
Italy			23	
Spain	9			
UK (RAF)	33			
US Army	472			
(Total 546)	523		23	

[1]Further two expected
[2]Civilian standard

COSTS: $773 million (1989) for conversion of 144 airframes t CH-47D; $81.8 million (placed 1987) for development o MH-47E and conversion of one prototype; $422 millio (1989-91) for 11 MH-47Es and option on further 14. £14 million upgrade for RAF aircraft (33 to Mk 2 standard 1993-95).

DESIGN FEATURES: Two three-blade intermeshing contra rotating tandem rotors; front rotor turns clockwise, viewe from above; rotor transmissions driven by connectin shafts from combiner gearbox, which is driven by rear mounted engines. Classic rotor heads with flapping an drag hinges; manually foldable blades, using Boeing Heli copters VR7 and VR8 aerofoils with cambered leading edges; blades can survive hits from 23 mm HEI and AP rounds; rotor brake optional. Constant cross-section cabi with side doors at front and rear loading ramp that can b opened in flight; underfloor section sealed to give flotatio after water landing; access to flight deck from cabin; mai cargo hook mounting covered by removable floor panel s that load can be observed in flight.

FLYING CONTROLS: Differential fore and aft cyclic for pitc attitude control; differential lateral cyclic pitch (from rud der pedals) for directional control; automatic control t keep fuselage aligned with line of flight. Dual hydrauli rotor pitch-change actuators: secondary hydraulic actu ators in control linkage behind flight deck for autopilo

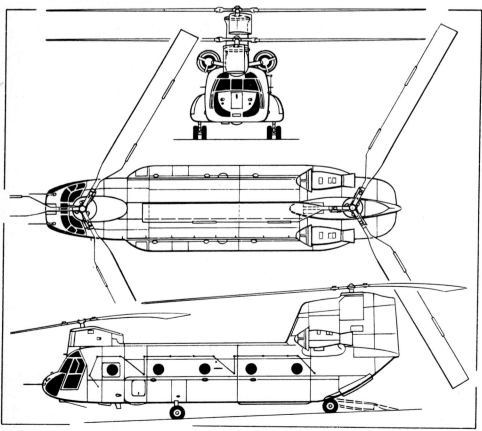

Boeing CH-47D military helicopter. Broken lines show rear loading ramp lowered (*Dennis Punnett*)

autostabiliser input; autopilot provides stabilisation, attitude hold and outer-loop holds.

STRUCTURE: Blades based on D-shaped glassfibre spar, fairing assembly of Nomex honeycomb core and crossply glassfibre skin.

LANDING GEAR: Non-retractable quadricycle type, with twin wheels on each front unit and single wheels on each rear unit. Oleo-pneumatic shock absorbers in all units. Rear units fully castoring; power steering on starboard rear unit. All wheels are size 24 × 7.7-VII, with tyres size 8.50-10-III, pressure 4.62 bars (67 lb/sq in). Two single-disc hydraulic brakes. Provision for fitting detachable wheel-skis.

POWER PLANT: Two Textron Lycoming T55-L-712 turboshafts, pod-mounted on sides of the rear pylon, and each with a standard power rating of 2,237 kW (3,000 shp) and maximum rating of 2,796 kW (3,750 shp). Textron Lycoming T55-L-712 SSB engine has standard power rating of 2,339 kW (3,137 shp) and maximum rating of 3,217 kW (4,314 shp); transmission capacity (CH-47D and MH-47E) 5,593 kW (7,500 shp) on two engines and 3,430 kW (4,600 shp) one engine inoperative; rotor rpm 225. Self-sealing pressure refuelled crashworthy fuel tanks in external fairings on sides of fuselage. Total fuel capacity 3,899 litres (1,030 US gallons; 858 Imp gallons). Oil capacity 14 litres (3.7 US gallons; 3.1 Imp gallons). From January 1991, 100 CH-47Ds fitted with engine air particle separator. Standard in MH-47E and optional in International Chinook are two Textron Lycoming T55-L-714 turboshafts, each with a standard power rating of 3,108 kW (4,168 shp) continuous and emergency rating of 3,629 kW (4,867 shp). MH-47E has 7,828 litres (2,068 US gallons; 1,722 Imp gallons) of fuel in panniers, plus 663 litres (175 US gallons; 146 Imp gallons) in floor tanks; total fuel 8,491 litres (2,243 US gallons; 1,868 Imp gallons). CH-47D SOA and MH-47E have 8.53 m (24 ft 0 in) refuelling probe on starboard side of forward fuselage.

ACCOMMODATION: Two pilots on flight deck, with dual controls. Jump seat for crew chief or combat commander. Jettisonable door on each side of flight deck. Depending on seating arrangement, 33 to 55 troops can be accommodated in main cabin, or 24 litters plus two attendants, or (see under Current Versions) vehicles and freight. Rear loading ramp can be left completely or partially open, or can be removed to permit transport of extra-long cargo and in-flight parachute or free-drop delivery of cargo and equipment. Main cabin door, at front on starboard side, comprises upper hinged section which can be opened in flight, and lower section with integral steps. Lower section is jettisonable. Triple external cargo hook system, as on Commercial Chinook, with centre hook rated to carry max load of 11,793 kg (26,000 lb) and the forward and rear hooks 7,711 kg (17,000 lb) each, or 10,433 kg (23,000 lb) in unison. Provisions are installed for a power-down ramp and water dam to permit ramp operation on water, for forward and rear cargo hooks, internal ferry fuel tanks, external rescue hoist, and windscreen washers.

SYSTEMS: Cabin heated by 200,000 BTU heater/blower. Hydraulic system provides pressure of 207 bars (3,000 lb/sq in) for flying controls. Max flow rate 53.0 litres (14 US gallons; 11.65 Imp gallons)/min. Spherical hydraulic reservoir, volume 5,326 cm³ (325 cu in), pressurised to 1.72 bars (25 lb/sq in). Utility hydraulic system, pressure 231 bars (3,350 lb/sq in), max flow rate 51.5 litres (13.6 US gallons; 11.3 Imp gallons)/min. Piston type reservoir, volume 7,014 cm³ (428 cu in), of which 5,326 cm³ (325 cu in) are usable, pressurised to 3.86 bars (56 lb/sq in). Electrical system includes two 40kVA air-cooled alternators driven by transmission drive system. Solar T62-T-2B APU runs accessory gear drive, thereby operating all hydraulic and electrical systems.

AVIONICS (International CH-47D: US Army CH-47D assumed to be generally similar. Avionics for RAF HC. Mk 1 listed in 1985-86 and earlier editions): Standard avionics include ARC-102 HF com radio, Collins ARC-186 UHF/AM-FM, Magnavox ARC-164 UHF/AM com; C-6533 intercom; Bendix/King AN/APX-100 IFF; APN-209 radar altimeter; AN/ARN-89B ADF; AN/ARN-118 Tacan; AN/ARN-123 VOR/glideslope/marker beacon receiver; and AN/ASN-43 gyromagnetic compass. Flight instruments are standard for IFR, and include an AN/AQU-6A horizontal situation indicator. AFCS maintains helicopter stability, eliminating the need for constant small correction inputs by the pilot to maintain desired attitude. The AFCS is a redundant system using two identical control units and two sets of stabilisation actuators. RAF Chinooks have GEC-Marconi ARI 18228 RWR and (from 1990) Loral AN/ALQ-157 IR jammers, Honeywell AN/AAR-47 missile approach warning equipment and Racal RNS252 Super TANS INS including GPS. Chelton 19-400 satellite communications antenna on some RAF helicopters. Tracor M-130 chaff/flare dispensers in RAF Chinooks.

EQUIPMENT: Hydraulically powered winch for rescue and cargo handling, rearview mirror, plus integral work stands and step for maintenance.

DIMENSIONS, EXTERNAL:
Rotor diameter (each)	18.29 m (60 ft 0 in)
Rotor blade chord (each)	0.81 m (2 ft 8 in)
Distance between rotor centres	11.94 m (39 ft 2 in)
Length: overall, rotors turning	30.14 m (98 ft 10¾ in)

MH-47E prototype showing extended nose and refuelling probe

RAF Chinook HC. Mk 1B, as used by Special Forces in 1991 Gulf War, retrofitted with fore and aft radar warning receiver antennae, chaff dispenser boxes each side of rear fuselage, flare dispensers each side of rear rotor pylon, satellite communications antenna on spine, missile approach warning system on both rotor pylons and attachment points for IR jammer below engine exhausts (*Paul Jackson*)

fuselage: Army CH-47D	15.54 m (51 ft 0 in)
International CH-47D and MH-47E	15.87 m (52 ft 1 in)
Width, rotors folded: CH-47D	3.78 m (12 ft 5 in)
MH-47E	4.78 m (15 ft 8 in)
Height to top of rear rotor head:	
CH-47D	5.77 m (18 ft 11 in)
MH-47E	5.59 m (18 ft 4 in)
Ground clearance, rotors turning:	
front approach	3.33 m (10 ft 11 in)
rear approach: CH-47D	5.78 m (18 ft 11½ in)
MH-47E	5.59 m (18 ft 4 in)
Ground clearance, static, rear approach	4.90 m (16 ft 0¾ in)
Wheel track (c/l of shock absorbers):	
CH-47D	3.20 m (10 ft 6 in)
MH-47E	3.63 m (11 ft 11 in)
Wheelbase: CH-47D	6.86 m (22 ft 6 in)
MH-47E	7.87 m (25 ft 10 in)
Passenger door (fwd, stbd): Height	1.68 m (5 ft 6 in)
Width	0.91 m (3 ft 0 in)
Height to sill	1.09 m (3 ft 7 in)
Rear loading ramp entrance: Height	1.98 m (6 ft 6 in)
Width	2.31 m (7 ft 7 in)
Height to sill	0.79 m (2 ft 7 in)

DIMENSIONS, INTERNAL:
Cabin, excl flight deck: Length	9.30 m (30 ft 6 in)
Width: mean	2.29 m (7 ft 6 in)
at floor	2.51 m (8 ft 3 in)
Height	1.98 m (6 ft 6 in)
Floor area	21.0 m² (226.0 sq ft)
Usable volume	41.7 m³ (1,474 cu ft)

AREAS:
Rotor blades (each)	7.43 m² (80.0 sq ft)
Rotor discs (total)	525.3 m² (5,655 sq ft)

WEIGHTS AND LOADINGS (CH-47D and International Chinook: see tables, MH-47E: as below):
Weight empty	12,210 kg (26,918 lb)
Useful load	12,284 kg (27,082 lb)
Max fuel weight	6,815 kg (15,025 lb)
Max T-O weight	24,494 kg (54,000 lb)

PERFORMANCE (CH-47D and International Chinook: see tables, MH-47E at 22,680 kg; 50,000 lb: as below):
Max level speed	154 knots (285 km/h; 177 mph)
Max cruising speed at S/L	140 knots (259 km/h; 161 mph)
Max rate of climb	561 m (1,840 ft)/min
Service ceiling	3,095 m (10,150 ft)
Hovering ceiling: IGE	2,990 m (9,800 ft)
IGE, ISA + 20°C (68°F)	2,410 m (7,900 ft)
OGE	1,675 m (5,500 ft)
OGE, ISA + 20°C (68°F)	1,005 m (3,300 ft)
Range at S/L	613 nm (1,136 km; 706 miles)

US ARMY CH-47D CHINOOK WEIGHTS AND PERFORMANCE

	Condition 1	Condition 2	Condition 3	Condition 4
Take-off condition				
Altitude	1,220 m (4,000 ft)	Sea level	1,220 m (4,000 ft)	Sea level
Temperature	35°C (95°F)	15°C (59°F)	35°C (95°F)	15°C (59°F)
Empty weight	10,615 kg (23,402 lb)	10,615 kg (23,402 lb)	10,538 kg (23,232 lb)	10,151 kg (22,379 lb)
T-O weight	19,178 kg (42,280 lb)	22,679 kg (50,000 lb)	19,657 kg (43,336 lb)	22,679 kg (50,000 lb)
Payload: external	6,968 kg (15,362 lb)	10,341 kg (22,798 lb)	—	—
internal	—	—	6,308 kg (13,907 lb)	—
Max level speed, S/L, ISA,				
max continuous power, no external load	161 knots (298 km/h; 185 mph)	161 knots (298 km/h; 185 mph)	154 knots (285 km/h; 177 mph)	—
Average cruising speed	120 knots (222 km/h; 138 mph)	132 knots (245 km/h; 152 mph)	134 knots (248 km/h; 154 mph)	138 knots (256 km/h; 159 mph)
Max rate of climb, S/L,				
ISA, intermediate rated power	669 m (2,195 ft)/min	464 m (1,522 ft)/min	640 m (2,100 ft)/min	464 m (1,522 ft)/min
Hovering ceiling OGE, ISA, max power	3,215 m (10,550 ft)	1,524 m (5,000 ft)	2,972 m (9,750 ft)	1,524 m (5,000 ft)
Mission radius	30 nm (55.5 km; 34.5 miles)	30 nm (55.5 km; 34.5 miles)	100 nm (185 km; 115 miles)	—
Ferry range	—	—	—	1,093 nm (2,026 km; 1,259 miles)

Condition 1
T-O weight is gross weight for 61 m (200 ft)/min vertical rate of climb to hover OGE at 1,220 m/35°C (4,000 ft/95°F). External payload is carried outbound only. Fuel reserve is 30 min cruise fuel. Max speed shown is at T-O weight less external payload.

Condition 2
T-O gross weight is max structural T-O weight for which vertical climb capability at S/L, ISA is 271 m (890 ft)/min. Otherwise same as Condition 1.

Condition 3
T-O weight is gross weight for hover OGE at 1,220 m/35°C (4,000 ft/95°F). Radius is with inbound payload 50 per cent of outbound internal payload. Fuel reserve is 30 min cruise fuel. Max speed shown is at T-O weight.

Condition 4
T-O weight is max structural T-O weight. Max ferry range (internal and external auxiliary fuel). Optimum cruise climb to 2,440 m (8,000 ft) and complete cruise at 2,440 m (8,000 ft). Fuel reserve is 10 per cent of initial fuel.

INTERNATIONAL CHINOOK WEIGHTS AND PERFORMANCE

Configuration designation	International CH-47D/-712	International CH-47D/-714	International CH-47D(LR)/-712	International CH-47D(LR)/-714
Power plant	T55-L-712 SSB	T55-L-714	T55-L-712 SSB	T55-L-714
Fuel capacity	3,902 litres (1,030 US gallons; 858 Imp gallons)	3,902 litres (1,030 US gallons; 858 Imp gallons)	7,834 litres (2,068 US gallons; 1,723 Imp gallons)	7,834 litres (2,068 US gallons; 1,723 Imp gallons)
Take-off condition				
Altitude	1,220 m (4,000 ft)	1,220 m (4,000 ft)	1,220 m (4,000 ft)	1,220 m (4,000 ft)
Temperature	35°C (95°F)	35°C (95°F)	35°C (95°F)	35°C (95°F)
Mission T-O weight (hover OGE)	20,094 kg (44,300 lb)	22,426 kg (49,440 lb)	19,695 kg (43,420 lb)	21,918 kg (48,320 lb)
Empty weight	10,670 kg (23,523 lb)	10,693 kg (23,574 lb)	11,016 kg (24,286 lb)	11,039 kg (24,337 lb)
Payload[1]: internal (inbound payload 50% outbound)	6,739 kg (14,857 lb)	8,895 kg (19,610 lb)	6,018 kg (13,268 lb)	8,081 kg (17,816 lb)
Mission radius[1]	100 nm (185 km; 115 miles)	100 nm (185 km; 115 miles)	100 nm (185 km; 115 miles)	100 nm (185 km; 115 miles)
Average cruise speed[1]	132 knots (245 km/h; 152 mph)	133 knots (246 km/h; 153 mph)	135 knots (250 km/h; 155 mph)	137 knots (254 km/h; 157 mph)
Max radius: full fuel (inbound payload 50% outbound)	140 nm (259 km; 161 miles)	130 nm (241 km; 150 miles)	335 nm (621 km; 386 miles)	316 nm (585 km; 363 miles)
Max level speed, S/L, ISA[1], max continuous power, at T-O weight	156 knots (289 km/h; 180 mph)	161 knots (298 km/h; 185 mph)	153 knots (283 km/h; 176 mph)	157 knots (291 km/h; 181 mph)
Hover ceiling, OGE, max power:				
ISA, T-O weight[1]	3,080 m (10,100 ft)	2,415 m (7,930 ft)	3,050 m (10,000 ft)	2,395 m (7,850 ft)
35°C (95°F), 22,680 kg (50,000 lb) gross weight	244 m (800 ft)	1,128 m (3,700 ft)	91 m (300 ft)	844 m (2,900 ft)
Max T-O weight	24,494 kg (54,000 lb)	24,494 kg (54,000 lb)	24,494 kg (54,000 lb)	24,494 kg (54,000 lb)

[1]T-O weight is gross weight for hover OGE at 1,220 m/35°C (4,000 ft/95°F). Radius is with inbound payload 50 per cent of outbound payload. Fuel reserve is 30 min cruise fuel. Max speed shown is at T-O weight.

BOEING 360
Details of this experimental advanced technology demonstrator can be found in the 1992-93 and previous editions of *Jane's*, together with an illustration. Advanced technologies developed as part of Boeing 360 programme helped to support those for CH-47D/MH-47E Chinooks, V-22 Osprey and RAH-66 Comanche.

BELL/BOEING V-22 OSPREY
Boeing is teamed with Bell Helicopter Textron in development of V-22 Osprey tilt-rotor transport. Full description in Bell/Boeing entry in this section.

BOEING/SIKORSKY RAH-66 COMANCHE
Boeing Helicopters formed First Team with Sikorsky in June 1985 to compete for US Army Light Helicopter (LH) programme. Full description in Boeing/Sikorsky entry after Boeing entries.

MILITARY AIRPLANES DIVISION (Boeing Defense & Space Group)
PO Box 3999, Seattle, Washington 98124
Telephone: 1 (206) 655 1198
Fax: 1 (206) 655 7012
VICE-PRESIDENT AND GENERAL MANAGER: Richard Hardy
PUBLIC RELATIONS MANAGER: Peri Widener

LOCKHEED/BOEING F-22A
F-22A design chosen 23 April 1991 to go forward into engineering and manufacturing development (EMD) to produce another nine YF-22 development aircraft and up to 648 production aircraft. Boeing share includes development and construction of wings and aft fuselage; installation of engines, nozzles and APU; operation of Integrated Technology Development Laboratory and Boeing 757 Avionics Flying Laboratory; and training systems development and integration.
For details of F-22A, see entry under Lockheed.

BOEING EX
TYPE: Carrier-compatible maritime surveillance and early warning aircraft.
PROGRAMME: Design study began 1988; present configuration announced late 1992; intended to compete for US Navy

Artist's impression of the Boeing EX joined-wing maritime surveillance and early warning aircraft

E-X requirement for E-2C Hawkeye replacement early next century; wind tunnel testing continuing in 1993-94.
DESIGN FEATURES: Joined-wing diamond planform (high-dihedral front wing, anhedral rear wing) chosen primarily to obtain best results from conformal, active phased-array antennae; radar already tested in one-fifth scale model with satisfactory results; wings rooted at mid-point of front fuselage and single fin, and joined near tips; engines pylon-mounted amidships, with third engine internally for electrical power generation; configuration provides radar with 360° scan and enhanced detection range. Wing sweep angle 40° backward on front leading-edge and 40° forward on rear wing trailing-edge, dictated by radar scan angle. Wings foldable for carrier stowage; two underwing stations for ESM pods, drop tanks or other stores.

POWER PLANT: Two 41.3 kN (9,275 lb st) General Electric TF34-GE-400 turbofans, or more powerful derivative, pod-mounted on sides of centre-fuselage.
ACCOMMODATION: Crew of four (pilot and three sensor operators), all on ejection seats.
SYSTEMS: 1,260 kW (1,690 shp) General Electric T700-GE-401 turboshaft, mounted in fuselage, serves as APU to generate electrical power for radar. Conformal antennae along virtually full span of all four wing surfaces. Wide variety of other sensors could be carried for several alternative missions.
DIMENSIONS, EXTERNAL:
Wing span 19.30 m (63 ft 4 in)
Width, wings folded 8.86 m (29 ft 1 in)
Length overall 15.59 m (51 ft 2 in)

Height overall 5.64 m (18 ft 6 in)
AREAS:
Wings, gross 78.50 m² (845 sq ft)
WEIGHTS AND LOADINGS:
Operating weight empty 16,048 kg (35,380 lb)
Max T-O weight 25,537 kg (56,300 lb)
PERFORMANCE (estimated):
Max level speed Mach 0.76-0.80
On-station patrol speed Mach 0.38
Range: internal fuel only
2,590 nm (4,800 km; 2,982 miles)
with two 1,514 litre (400 US gallon; 333 Imp gallon) external tanks 2,970 nm (5,504 km; 3,420 miles)

PRODUCT SUPPORT DIVISION (Boeing Defense & Space Group)

PO Box 7730, Wichita, Kansas 67277
Telephone: 1 (316) 526 3902
Fax: 1 (316) 523 5369
VICE-PRESIDENT AND GENERAL MANAGER: Lee Gantt
PUBLIC RELATIONS MANAGER: Carolyn Russell

Product Support Division formed 1991 for modification and post-production support of Boeing military products. Programmes include support of B-52 Stratofortress, KC-135 tanker/transport, VC-137, TC-43, C-18, C-22 and E-4B; offensive avionics system of Rockwell B-1B; mission data preparation systems; and all-composites replacement wing for US Navy A-6 Intruder. Contracts for 178 new wing sets for A-6E placed by March 1988; final set delivered May 1992. Additional wings required following cancellation of A-12 programme; Boeing to build further 120 wings for final assembly at Seattle; first delivery January 1995.

BOEING KC-135 STRATOTANKER

Modification continues of the KC-135 with CFM56 turbofans (KC-135R); French C-135FRs to receive underwing refuelling pods. Details in *Jane's Civil and Military Aircraft Upgrades*.

BOEING 707 TANKER/TRANSPORT

Several contractors besides Boeing offer tanker conversions of Boeing 707 airliners in various configurations. Details in *Jane's Civil and Military Aircraft Upgrades*.

BOEING/SIKORSKY

BOEING HELICOPTERS and SIKORSKY AIRCRAFT

Boeing/Sikorsky LH Program Office, Scott Plaza II, Suite 635, Philadelphia, Pennsylvania 19142
Telephone: 1 (215) 591 8820
Fax: 1 (215) 591 8819
RAH-66 PROGRAMME DIRECTOR: Merrick W. Hellyar
VICE-PRESIDENT, RAH-66 PROGRAMME: William W. Walls Jr
PUBLIC AFFAIRS OFFICER: Jack Satterfield

Boeing and Sikorsky began LHX collaboration June 1985; development centre initially Wichita, but transferred to Philadelphia in early 1990.

BOEING/SIKORSKY RAH-66 COMANCHE

TYPE: Two-seat reconnaissance/attack and air combat helicopter.
PROGRAMME: Light Helicopter Experimental (LHX) design concepts requested by US Army 1982; original plan for 5,000 to replace UH-1, AH-1, OH-58 and OH-6; reduced in 1987 to 2,096 scout/attack only, replacing 3,000 existing helicopters; to 1,292 in 1990 (with further 389 possible). LHX request for proposals issued 21 June 1988; 23-month demonstration/validation contracts issued to Boeing/Sikorsky First Team and Bell/McDonnell Douglas Super Team. Boeing/Sikorsky selected 5 April 1991; to build three (reduced from four by 1992 decision) demonstration/validation prototypes in 78-month programme (first flights August 1995, June 1996 and June 1997), plus static test airframe and propulsion system testbed. LHTEC T800 engine specified October 1988. LHX designation changed to LH early 1990, then US Army designation RAH-66 Comanche in April 1991. Original timetable was: 39-month FSD phase with two additional prototypes, starting August 1995; first low-rate production contract due October 1996; initial 72 helicopters to be built by 2002; first full-rate production contract due November 1998; balance of 1,220 built by 2010. IOC in December 1998. Longbow radar installed from production Lot 4 onwards. FY 1993 budget delayed development and production phases indefinitely, anticipating Longbow integration from first aircraft; development and production phases reinstated January 1993; further three prototypes to be built for engineering and manufacturing development phase, in FY 1998-2003; first production procurement in FY 1999; low-rate production of first 24 in FY 2001, followed by 48 in FY 2002, 96 in FY 2003, then 120 per year; IOC in January 2003.
COSTS: $34,000 million programme, including $1,960 million dem/val and $900 million FSD but reduced to $2,240 million dem/val/FSD between 1993-97 by cancellation of three of six planned prototypes; $8.9 million flyaway unit cost (1988 values), increased to $13 million by early 1993. By 1993 (in 1994 dollars, estimated), procurement unit cost $21 million, programme unit cost $27 million.
DESIGN FEATURES: RAH-66 will be lighter, but only slightly smaller, than AH-64 Apache; specified empty weight of 3,402 kg (7,500 lb) increased 22.5 per cent by early 1992, as result of Army add-ons, including allowance for Longbow radar; mission equipment package has maximum commonality with F-22A ATF technology. Design has eight-blade Sikorsky fan-in-fin shrouded tail rotor and five-blade version of DASA/MBB all-composites bearingless main rotor system, and internal weapon stowage. Other members of Boeing/Sikorsky First Team include Boeing Defense and Space Group (flight control computer), General Electric Armament Systems Department, with Giat of France (turreted gun and ammunition feed), Hamilton Standard (flight control computer, wide field-of-view helmet-mounted display system, air data system, environmental control and collective protection system, and air vehicle interface computer), Harris Corporation (3D digital map display, super high speed databus, sensor data distribution network, multi-function controls and

Recent video impression of the Boeing/Sikorsky RAH-66 Comanche showing nose-mounted sensors and gun, midships weapon doors open and tilted fan-in-fin tail

460 USA: AIRCRAFT—**BOEING/SIKORSKY**

Internal arrangement of the Boeing/Sikorsky RAH-66 Comanche

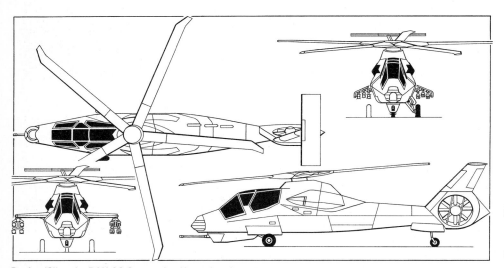

Boeing/Sikorsky RAH-66 Comanche. Upper head-on view shows internal weapons bay doors open; lower head-on view shows eight Stinger (port) and four Hellfire missiles (starboard) on detachable stub-wings
(Jane's/Mike Keep)

Working mockup of RAH-66 front cockpit shows two central multi-function visual displays and central input keyboard flanked by system and weapon information displays. Sidestick cyclic control is on the starboard console

displays), Kaiser Electronics (helmet-mounted display system), Link Flight Simulation (operator training systems), Martin Marietta (electro-optical night navigation and targeting systems), TRW Military Electronics Division with Westinghouse Defense and Electronics (signal and data processors and aircraft survivability equipment). Split torque transmission, obviating need for planetary gearing. T tail unit (upper part folds down for air transportation). Detachable stub-wings for additional weapon carriage and/or auxiliary fuel tanks (EFAMS: external fuel and armament management system). Nose hinges to starboard for access to sensors and ammunition bay. Extremely low radar and infra-red signatures. Eight deployable inside Lockheed C-5 Galaxy with only removal of main rotor; ready for flight 20 minutes after transport lands. Combat turnround time 13 minutes. Fantail disc density, duct depth and taper, space between rotor and vertical strut and rotor tilt largely dictated by aural and other signature factors.

FLYING CONTROLS: Dual triplex fly-by-wire, with sidestick cyclic-pitch controllers and normal collective levers. Main rotor blades removable without disconnecting control system.

STRUCTURE: Largely composite airframe and rotor system. Fuselage built around composite internal box-beam; non-load-bearing skin panels, more than half of which can be hinged or removed for access to interior (eg, weapons bay doors can double as maintenance work platforms). Eight-blade Fantail rear rotor operable with 12.7 mm calibre bullet hits; or for 30 minutes with one blade missing. Main rotor blades and tail section by Boeing, forward fuselage and final assembly by Sikorsky.

LANDING GEAR: Retractable tailwheel type, with single wheel on each unit. Main units can 'kneel' for air transportability.

POWER PLANT: Two LHTEC T800-LHT-800 turboshafts, each rated at 1,002 kW (1,344 shp). Transmission rating 1,532 kW (2,054 shp). Internal fuel capacity 984 litres (260 US gallons; 217 Imp gallons). External tanks totalling 3,483 litres (920 US gallons; 766 Imp gallons) for self-deployment; total fuel capacity 4,467 litres (1,180 US gallons; 982.5 Imp gallons). Main rotor tip speed 196 m (644 ft)/s; 314 rpm.

ACCOMMODATION: Pilot (in front) and WSO in identical stepped cockpits, pressurised for chemical/biological warfare protection. Crew seats resist 11.6 m (38 ft)/s vertical crash landing.

AVIONICS: Martin Marietta night vision pilotage system and Kaiser/Hamilton Standard helmet-mounted display; integrated cockpit, second-generation FLIR targeting, and digital map display. Dual anti-jam VHF-FM and UHF-AM Have Quick tactical communications, VHF-AM, anti-jam HF-SSB, airborne target handover system, Navstar GPS, IFF and radar altimeter. Laser warning and radar warning receivers, RF and IR jammers. Miniaturised version of Longbow radar in one-third of fleet although all to have carriage provision. Maximum avionics commonality required with USAF F-22A ATF programme. Two 15.2 × 20.3 cm (6 × 8 in) flat-screen LCDs in each cockpit (one monochrome for FLIR/TV, one colour for moving map, tactical situation and night operations), plus a 10.2 × 10.2 cm (4 × 4 in) monochrome LCD for fuel and armament information. Three redundant databuses: one low-speed (MIL-STD-1553B), one high-speed and one very-high-speed (fibre-optic based) for signal data distribution.

ARMAMENT: General Electric/Giat three-barrel 20 mm cannon in undernose turret, with up to 500 rounds (320 rds normal for primary mission). Cost-saving option of fixed forward cannon. Side-opening weapons bay door in each side of fuselage, on each of which can be mounted up to three Hellfire or six Stinger missiles or other weapons. Four more Hellfires or eight Stingers can be deployed from multiple carriers under tip of each optional stub-wing, or a 1,741.5 litre (460 US gallon; 383 Imp gallon) auxiliary fuel tank for self-deployment. All weapons can be fired, and targets designated, from push-buttons on collective and sidestick controllers.

DIMENSIONS, EXTERNAL:	
Main rotor diameter	11.90 m (39 ft 0½ in)
Fantail diameter	1.37 m (4 ft 6 in)
Fantail blade chord	0.17 m (6¾ in)
Length: overall, rotor turning	14.28 m (46 ft 10¼ in)
fuselage (excl gun barrel)	13.22 m (43 ft 4½ in)
Height over tailplane	3.39 m (11 ft 1½ in)
Fuselage: Max width	2.29 m (7 ft 6 in)
Width over mainwheels	2.31 m (7 ft 7 in)
Tailplane span	2.82 m (9 ft 3 in)
AREAS:	
Main rotor disc	111.21 m² (1,197.04 sq ft)
Fantail disc	1.47 m² (15.90 sq ft)
WEIGHTS AND LOADINGS (estimated):	
Weight: empty	3,515 kg (7,749 lb)
empty, combat equipped (incl Longbow)	
	4,167 kg (9,187 lb)
Max useful load	1,185 kg (2,612 lb)
T-O weight: primary mission*	4,587 kg (10,112 lb)
max (self-deployability)	7,790 kg (17,174 lb)

* with two crew, full internal fuel, 320 rds gun ammunition, four Hellfires and two Stingers

PERFORMANCE (at 1,220 m; 4,000 ft and 34°C; 95°F, estimated):
Max level (dash) speed	177 knots (328 km/h; 204 mph)
Vertical rate of climb	360 m (1,182 ft)/min
Ferry range with external tanks	
	1,260 nm (2,335 km; 1,451 miles)
Endurance (standard fuel)	2 h 30 min
g limits	+3.5/−1
Masking	1.6 s
180° hover turn to target	4.7 s
Snap turn to target at 80 knots (148 km/h; 92 mph)	3.0 s

BOWERS

PETER M. BOWERS
458 16th Avenue South, Seattle, Washington 98168
Telephone: 1 (206) 242 2582

BOWERS FLY BABY 1-A

TYPE: Single-seat homebuilt.
PROGRAMME: Produced to compete in Experimental Aircraft Association design contest, organised to encourage development of simple, low-cost, easy-to-fly aeroplane that could be built by inexperienced amateurs for recreational flying; first flown 27 July 1960.
CUSTOMERS: Plans available and about 4,300 sets sold by early 1992. Construction of well over 700 Fly Babys known; about 450 flown.
The following description applies to the original single-seat Fly Baby 1-A:
COSTS: Kits: series of three kits available from Classic Flying Machines (15903 Lakewood Blvd, 103 Bellflower, California 90706): first kit $1,750 is for fuselage and is crated in a workbench/jig; second kit $1,500 is for wings, tail and landing gear; third kit $1,500 contains finishing materials except power plant, finish colour and instruments. Plans: $65.
DESIGN FEATURES: Wing section NACA 4412. Non-retractable tailwheel landing gear. Cockpit canopy optional. Several individual amateur builders amended plans during 1973 to allow for construction of two-seat versions. Although Mr Bowers does not recommend conversion, he has made some additions to plans to cater for it.
STRUCTURE: All-wood construction, except for Dacron covering on wings and tail unit.
LANDING GEAR: Non-retractable taildragger; examples flown with floats and skis.
POWER PLANT: One 63.5 kW (85 hp) Teledyne Continental C85. Suited to modified Volkswagen engines of over 1,800 cc. Fuel capacity 60.5 litres (16 US gallons; 13.3 Imp gallons).

DIMENSIONS, EXTERNAL:	
Wing span	8.53 m (28 ft 0 in)
Length overall	5.64 m (18 ft 6 in)
Height, wings folded	1.98 m (6 ft 6 in)
WEIGHTS AND LOADINGS:	
Weight empty	274 kg (605 lb)
Max T-O weight	419 kg (924 lb)
Max power loading	6.60 kg/kW (10.87 lb/hp)

Bowers Fly Baby 1-A built by Mr Don Bell of Seattle, Washington *(Peter M. Bowers)*

PERFORMANCE (at max T-O weight):
Max level speed at S/L	over 104 knots (193 km/h; 120 mph)
Cruising speed	91-96 knots (169-177 km/h; 105-110 mph)
Landing speed	39 knots (73 km/h; 45 mph)
Max rate of climb at S/L	335 m (1,100 ft)/min
T-O and landing run	76 m (250 ft)
Range with max fuel	277 nm (515 km; 320 miles)

BOWERS FLY BABY 1-B

TYPE: Biplane version of Fly Baby 1-A.
PROGRAMME: During 1968 Mr Bowers designed and built set of interchangeable biplane wings for original prototype Fly Baby; first flown 27 March 1969; about 22 more Fly Baby 1-Bs completed and flown.
COSTS: Plans: $65 plus $15 supplement for biplane wings.
DESIGN FEATURES: Biplane wings have same aerofoil and incidence as monoplane, but incorporate aluminium alloy wingtip bows; ailerons fitted to lower wings only; changeover from monoplane to biplane configuration can be accomplished by two people in approximately one hour.
POWER PLANT: Intended to use same engines as monoplane, up to Teledyne Continental O-200; since some biplane builders desired to use 93 kW (125 hp) Textron Lycoming O-290, modification authorised for biplane, whereby wing sweep is decreased by 5° for CG reasons.

DIMENSIONS, EXTERNAL:	
Wing span	6.71 m (22 ft 0 in)
Height overall	2.08 m (6 ft 10 in)
AREAS:	
Wings, gross	13.94 m² (150.0 sq ft)
WEIGHTS AND LOADINGS:	
Weight empty	295 kg (651 lb)
Max T-O weight	440 kg (972 lb)
Max wing loading	31.56 kg/m² (6.48 lb/sq ft)

PERFORMANCE (63.4 kW; 85 hp, at max T-O weight):
Cruising speed	75 knots (140 km/h; 87 mph)
Max rate of climb at S/L	267 m (875 ft)/min

BRANSON

BRANSON AIRCRAFT CORPORATION
4890 Wheeling Street, Denver, Colorado 80239
Telephone: 1 (303) 371 9112
Fax: 1 (303) 371 1813
Telex: 45-4577 BRANSON DVR

PRESIDENT: Carl F. Branson
EXECUTIVE VICE-PRESIDENT: Roger P. Kirwan

Founded in 1966 for special design and custom manufacturing of auxiliary fuel tanks, special interiors and equipment for civil aircraft; offers extended range, weight increase, extended width cargo door and air ambulance modifications for Cessna Citation I, II and S/II; long-range tanks for Learjet 55; large cargo door for Fairchild FH-227 and Fokker F27.

Details in 1991-92 *Jane's;* now transferred to *Jane's Civil and Military Aircraft Upgrades.*

BRANTLY

BRANTLY HELICOPTER INDUSTRIES USA CO LTD
Wilbarger County Airport, 12399 Airport Drive, Vernon, Texas 76384
Telephone: 1 (817) 552 5451
Fax: 1 (817) 552 2703
PRESIDENT: James T. Kimura

Japanese-American businessman James T. Kimura acquired type certificates and manufacturing and marketing rights to Brantly B-2B and 305 from Hynes Aviation Industries March 1989; Brantly Helicopter Industries formed 8 May 1989; 2,787 m² (30,000 sq ft) facility established at Wilbarger County Airport. Production methods being continuously improved. Turbine powered 305 planned.

BRANTLY B-2B

TYPE: Two-seat light helicopter.
PROGRAMME: Maximum production capacity 300 a year.
CUSTOMERS: First B-2B from current production line handed over to Japanese customer 25 August 1990.
COSTS: $120,000 to $135,000 (1991).
DESIGN FEATURES: Three-blade main rotor with flapping hinges close to hub and pitch-change/flap/lag hinges at 40 per cent blade length; symmetrical inboard blade section with 29 per cent thickness/chord ratio, outboard section NACA 0012; outer blade sections quickly removable for compact storage; rotor brake standard; two-blade antitorque rotor. Transmission through automatic centrifugal clutch. Tail rotor drive through flexible couplings and intermediate gearbox. New lightweight starter and alternator replaces generator. Cabin will be streamlined.
FLYING CONTROLS: Conventional direct mechanical control; small fixed tailplane.
STRUCTURE: Inboard blade section has stainless steel leading-edge spar; outboard portion has extruded aluminium spar, polyurethane core with bonded aluminium envelope riveted to spar. All-metal rotor. Fuselage steel tube centre-section with all-metal stressed skin tailcone.
LANDING GEAR: Alternative skid or float gear. Skid type has small retractable wheels for ground handling, fixed tailskid and four shock absorbers with rubber in compression. Inflatable pontoons, which attach to standard skids, are available to permit operation from water.
POWER PLANT: One 134 kW (180 hp) Textron Lycoming IVO-360-A1A flat-four engine, mounted vertically, with dual fan cooling system. Rubber bag type fuel tank under engine, capacity 117 litres (31 US gallons; 25.8 Imp

462 USA: AIRCRAFT—BRANTLY/CALIFORNIA MICROWAVE

Brantly Helicopter Industries B-2B two-seat light helicopter

gallons). Refuelling point on port side of fuselage. Oil capacity 5.7 litres (1.5 US gallons; 1.25 Imp gallons).
ACCOMMODATION: Totally enclosed circular section cabin for two persons seated side by side. Forward hinged door on each side. Dual controls, cabin heater and demisting fan standard. Compartment for 22.7 kg (50 lb) baggage in forward end of tail section.
AVIONICS: Provision for all standard nav/com radios.
EQUIPMENT: Blind-flying instrumentation available as an option, but not certificated for instrument flight. Twin landing lights in nose.

DIMENSIONS, EXTERNAL:
Main rotor diameter	7.24 m (23 ft 9 in)
Main rotor blade chord: inboard	0.225 m (8.85 in)
outboard	0.203 m (8.0 in)
Tail rotor diameter	1.30 m (4 ft 3 in)
Length: overall, rotors turning	8.53 m (28 ft 0 in)
fuselage	6.62 m (21 ft 9 in)
Height overall	2.06 m (6 ft 9 in)
Skid track	1.73 m (5 ft 8½ in)
Passenger doors (each): Height	0.79 m (2 ft 7 in)
Width	0.86 m (2 ft 9¾ in)
Baggage compartment door:	
Mean height	0.25 m (9¾ in)
Length	0.55 m (1 ft 9¾ in)

DIMENSIONS, INTERNAL:
Cabin: Length	1.83 m (6 ft 0 in)
Max width	1.19 m (3 ft 11 in)
Max height	0.99 m (3 ft 3 in)
Floor area	2.60 m² (28.0 sq ft)
Volume	2.78 m³ (98.0 cu ft)
Baggage compartment	0.17 m³ (6.0 cu ft)

AREAS:
Main rotor blades (each)	0.69 m² (7.42 sq ft)
Main rotor disc	41.16 m² (443.0 sq ft)
Tail rotor disc	1.32 m² (14.19 sq ft)

WEIGHTS AND LOADINGS:
Weight empty: with skids	463 kg (1,020 lb)
with floats	481 kg (1,060 lb)
Max T-O weight	757 kg (1,670 lb)
Max disc loading	18.40 kg/m² (3.77 lb/sq ft)

PERFORMANCE (at max T-O weight):
Max level speed at S/L	87 knots (161 km/h; 100 mph)
Max cruising speed (75% power)	78 knots (145 km/h; 90 mph)
Max rate of climb at S/L	580 m (1,900 ft)/min
Service ceiling	3,290 m (10,800 ft)
Hovering ceiling IGE	2,040 m (6,700 ft)
Range with max fuel, with reserves	217 nm (400 km; 250 miles)

BRANTLY 305

TYPE: Five-seat light helicopter.
PROGRAMME: Original first flight January 1964; FAA type approval 29 July 1965; about 44 built during mid-1960s. Improved prototype with redesigned rotor head and new blade aerofoil completed about 30 hours flying by mid-January 1990. New main rotor bearing fitted. Cabin to be streamlined. New production may begin during 1993.
CURRENT VERSIONS: Turbine engines planned for a medevac version.
DESIGN FEATURES: Rotor system and airframe as for B-2B, but enlarged.
FLYING CONTROLS: Conventional direct mechanical.
STRUCTURE: As for B-2B.
LANDING GEAR: Choice of skid, wheel or float gear. Skid type has four oleo struts, two on each side, and small retractable ground handling wheels. Wheel gear has single mainwheels and twin nosewheels, all with oleo-pneumatic shock absorbers. Goodyear mainwheels and tyres size 6.00-6, pressure 2.07 bars (30 lb/sq in); Goodyear nose wheels and tyres size 5.00-5, pressure 1.93 bars (28 lb/sq in). Goodyear single-disc hydraulic brakes on mainwheels.
POWER PLANT: One 227.4 kW (305 hp) Textron Lycoming IVO-540-B1A flat-six engine, mounted vertically, with dual cooling fans. One rubber fuel cell under engine, capacity 163 litres (43 US gallons; 35.8 Imp gallons). Refuelling point in port side of fuselage. Oil capacity 9 litres (2.5 US gallons; 2.1 Imp gallons).
ACCOMMODATION: Two individual seats side by side, with dual controls. Rear bench seat for three persons. Door on each side. Rear compartment for 113 kg (250 lb) of baggage, with downward hinged door on starboard side.
AVIONICS: Bendix/King or Narco radio, to customer specification.
EQUIPMENT: Blind-flying instrumentation available, but helicopter not certificated for instrument flight.

DIMENSIONS, EXTERNAL:
Main rotor diameter	8.74 m (28 ft 8 in)
Main rotor blade chord, constant	0.254 m (10 in)
Tail rotor diameter	1.30 m (4 ft 3 in)
Length: overall, rotors turning	10.03 m (32 ft 11 in)
fuselage	7.44 m (24 ft 5 in)
Height overall	2.44 m (8 ft 1 in)
Wheel track	2.10 m (6 ft 10¼ in)
Wheelbase	2.15 m (7 ft 0½ in)
Passenger doors (each): Height	0.82 m (2 ft 8½ in)
Width	1.02 m (3 ft 3⅞ in)
Baggage compartment door:	
Mean height	0.30 m (1 ft 0¼ in)
Width	0.69 m (2 ft 3 in)

DIMENSIONS, INTERNAL:
Cabin: Length	2.30 m (7 ft 6½ in)
Max width	1.39 m (4 ft 6¾ in)
Max height	1.22 m (4 ft 0½ in)
Baggage compartment	0.47 m³ (16.7 cu ft)

AREAS:
Main rotor blades (each)	0.09 m² (11.79 sq ft)
Tail rotor blades (each)	0.05 m² (0.50 sq ft)
Main rotor disc	59.96 m² (645.4 sq ft)
Tail rotor disc	1.32 m² (14.19 sq ft)

WEIGHTS AND LOADINGS:
Weight empty	816 kg (1,800 lb)
Max T-O and landing weight	1,315 kg (2,900 lb)
Max disc loading	21.92 kg/m² (4.49 lb/sq ft)

PERFORMANCE (at max T-O weight):
Max level speed at S/L	104 knots (193 km/h; 120 mph)
Max cruising speed at S/L	96 knots (177 km/h; 110 mph)
Max rate of climb at S/L	297 m (975 ft)/min
Service ceiling	3,660 m (12,000 ft)
Hovering ceiling IGE	1,245 m (4,080 ft)
Range with max fuel and max payload, 15 min reserves	191 nm (354 km; 220 miles)

Brantly Helicopter Industries 305 five-seat light helicopter

BUSH
BUSH CONVERSIONS INC
Box 431, Udall, Kansas 67146
Telephone: 1 (316) 782 3851
VICE-PRESIDENT: Barbara Williams

Company offers 'taildragger' conversions for Cessna 150/152, 172/Skyhawk, R172K and 175, designed by former Ralph Bolen Inc.
Details in 1991-92 *Jane's*; now appear in *Jane's Civil and Military Aircraft Upgrades*.

CALIFORNIA HELICOPTER
CALIFORNIA HELICOPTER INTERNATIONAL
2935 Golf Course Drive, Ventura, California 93003-7604
Telephone: 1 (805) 644 5800
Fax: 1 (805) 644 5132
Telex: 6831165 CHI UW
PRESIDENT: Gary A. Podolny
VICE-PRESIDENT AND GENERAL MANAGER: W. E. 'Jake' Dangle

Rights to manufacture turbine conversion kits for Sikorsky S-58 and spare parts, plus support of worldwide S-58/S-58T fleet, bought from Sikorsky in 1981 after latter had converted or produced conversion kits for about 146 S-58s. California Helicopter offers dynamic component exchange service for S-58/S-58T.
Details in 1991-92 *Jane's*; now transferred to *Jane's Civil and Military Aircraft Upgrades*.

CALIFORNIA MICROWAVE
CALIFORNIA MICROWAVE INC
(Government Electronics Division)
6022 Variel Avenue, PO Box 2800, Woodland Hills, California 91367
Telephone: 1 (818) 992 8000
Fax: 1 (818) 992 5079
Telex: 910 494 2794
GENERAL MANAGER: R. Medlin

California Microwave's recent products include surveillance conversions of a CASA C-212 and de Havilland DHC-7 to meet US Army's Airborne Reconnaissance-Low requirement (see 1991-92 *Jane's* and *Jane's Civil and Military Aircraft Upgrades*); and a multi-role surveillance information relay adaptation of the Rutan Long-EZ designated CM-30 Optionally Piloted Vehicle (see *Jane's Battlefield Surveillance Systems*).

CAT

COMMUTER AIR TECHNOLOGY

4700 North Airport Drive, Suite 206, Scottsdale, Arizona 85260
Telephone: 1 (602) 951 6288
Fax: 1 (602) 998 1239

PRESIDENT: Keith Nickels
VICE-PRESIDENT, AIRCRAFT MARKETING: Frank Rast

Company converts and markets commuter versions of Beechcraft Super King Air 200, incorporating most Raisbeck King Air modifications. Transregional 250 CATPASS has max zero-fuel weight increase to 4,990 kg (11,000 lb); ST 17 to have 1.22 m (4 ft 0 in) fuselage stretch to accommodate 17 passengers.

Details in *Jane's Civil and Military Aircraft Upgrades.*

CAVENAUGH

CAVENAUGH AVIATION INC

5600 John F. Kennedy Boulevard, Suite 730, Houston, Texas 77302
Telephone: 1 (713) 442 3500
Fax: 1 (713) 442 3559

PRESIDENT: Dudley N. Cavenaugh
DIRECTOR OF MARKETING: Norris D. Allee

Company offers cargo conversions of Mitsubishi MU-2 series (see 1991-92 *Jane's*). Details now in *Jane's Civil and Military Aircraft Upgrades.*

CESSNA

CESSNA AIRCRAFT COMPANY (Subsidiary of Textron Inc)

PO Box 7706, Wichita, Kansas 67277-7706
Telephone: 1 (316) 941 6000
Fax: 1 (316) 941 7812
Telex: 417 400

CHAIRMAN AND CEO: Russell W. Meyer Jr
SENIOR VICE-PRESIDENT, AIRCRAFT MARKETING: Gary W. Hay
DIRECTOR OF PUBLIC RELATIONS: H. Dean Humphrey

Founded by late Clyde V. Cessna 1911; incorporated 7 September 1927; former Pawnee and Wallace aircraft divisions in Wichita consolidated in Aircraft Division mid-1984; acquired by General Dynamics as wholly owned subsidiary 1985; acquisition by Textron announced February 1992.

Owned subsidiaries include McCauley Accessory Division, Dayton, Ohio; Cessna Finance Corporation in Wichita. Sold 49 per cent interest in Reims Aviation of France to Compagnie Française Chaufour Investissement (CFCI) February 1989; Reims continues manufacturing Cessna F406 Caravan II and holds option to build Cessna single-engined aircraft when Cessna restarts production.

Total 177,841 aircraft produced by December 1992; 140 aircraft delivered in 1992, including 62 Caravan Is, 22 Citation IIs, one Citation III, 51 Citation Vs, 10 Citation VIs and 15 Citation VIIs. Total employees 5,477 on 1 January 1993; 2,000th Citation delivered 1993.

CESSNA SUSPENDED PRODUCTION

Since acquisition by Textron, Cessna has considered starting manufacture of single-engined light aircraft. Aircraft types described in mid-1980s *Jane's*.

REIMS-CESSNA F406/CARAVAN II

Turboprop version of Cessna 400 series developed in France by Reims Aviation, which see.

CESSNA 208 CARAVAN I/U-27A

TYPE: Single-turboprop civil and military multi-mission aircraft.
PROGRAMME: First flight of engineering prototype (N208LP) 9 December 1982; first production Caravan I rolled out August 1984; FAA certification October 1984; full production started 1985; amphibian float version certificated March 1986.
CURRENT VERSIONS: **208A**: Basic utility model for passengers or cargo. Commissioned by Federal Express Corporation as **Cargomaster** freighter with special features including T-O weight 3,629 kg (8,000 lb), Bendix/King avionics, no cabin windows or starboard rear door, more cargo tiedowns, additional cargo net, underfuselage cargo pannier of composite materials, 15.2 cm (6 in) vertical extension of fin/rudder, jetpipe deflected to carry exhaust clear of pannier.

208B: Stretched version, developed at request of Federal Express. Commissioned by Federal Express as **Super Cargomaster**; first flight 3 March 1986; certificated October 1986; first delivery to Federal Express 31 October 1986; deliveries completed early 1992. Features include fuselage stretched by 1.22 m (4 ft), payload of 1,587 kg (3,500 lb) and 12.7 m³ (450 cu ft) of cargo volume. Aircraft produced in 1991 given 503 kW (675 shp) P&WC PT6A-114A.

Grand Caravan: Announced at NBAA 1990; stretched to accommodate up to 14 passengers in quick-change interior; powered by 503 kW (675 shp) P&WC PT6A-114A.

U-27A: Military utility/special mission derivative of 208A and 208B versions of Caravan I; announced Spring 1985; roles include cargo, logistic support, paratroop or supply dropping, medevac, electronic surveillance, forward air control, passenger/troop transport, C³I, maritime patrol, SAR, psychological warfare, radio relay/RPV control, military base support, range safety patrol, reconnaissance and fire patrol. Fittings can include six underwing and one centreline hardpoints, observation windows and bubble windows for downward view, centreline reconnaissance pod, 2.8 m³ (84 cu ft) cargo pannier from 208A and two-part electrically actuated upward and downward rolling shutter door with slipstream deflector.
CUSTOMERS: Federal Express Corporation took delivery of 40 208As and 210 208Bs before contract fulfilled in February 1992. Other customers include Royal Canadian Mounted Police (first amphibian float version); Brazilian Air Force (three plus four on order), Liberian Army (one), Royal Thai Army (10).

US State Department using two U-27As with VLF/Omega long-range navigation systems in anti-narcotics campaign. More than 10 per cent of Caravan Is are in military or paramilitary service, some with CIA.

Total of 541 Caravan Is delivered by December 1992.
COSTS: $923,500 (Caravan I); $966,500 (Cargomaster); $1,012,000 (Grand Caravan); $1,055,200 (Super Cargomaster).
DESIGN FEATURES: Claimed as first all-new single-engined turboprop general aviation aircraft; intended to replace de Havilland Canada Beavers and Otters, Cessna 180s, 185s and 206s in worldwide utility role.

Main qualities are high speed with heavy load, compatibility with unprepared strips, economy and reliability with minimum maintenance; can also carry weather radar, air-conditioning and oxygen systems; optional packs for fire-fighting, photography, spraying, ambulance/hearse, border patrol, parachuting and supply dropping, surveillance and government utility missions; optional wheel or float landing gear.

First single-engined aircraft to achieve FAA certification for ILS in Category II conditions (Federal Express aircraft equipped); approval for IFR cargo operations 1989 made France and Ireland first European countries to allow single-engined public transport day/night IFR operation; since approved also in Canada, Denmark and Sweden.

Wing aerofoil NACA 23017.424 at root, 23012 at tip; dihedral 3° from root; incidence 2° 37′ at root, −0° 36′ at tip.
FLYING CONTROLS: Plain mechanical controls; lateral control by small ailerons and slot-lip spoilers ahead of outer section of flaps; aileron trim standard; all tail control surfaces

Cessna Grand Caravan quick-change transport

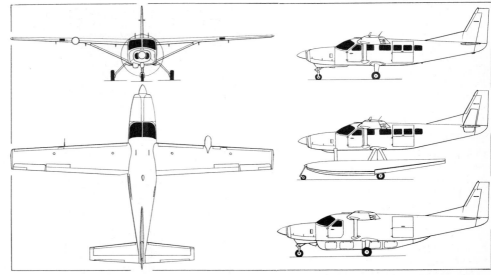

Cessna Caravan I with side views, top to bottom, of 208A Caravan I, 208A amphibian and 208B Super Cargomaster for Federal Express *(Dennis Punnett)*

horn balanced; fixed tailplane with vortex generators above, ahead of elevator; elevator trim tabs; electrically actuated single-slotted flaps occupy more than 70 per cent of trailing-edge and deflect to maximum 30°.

STRUCTURE: Fail-safe two-spar wing; conventional fuselage.

LANDING GEAR: Non-retractable tricycle type, with single wheel on each unit. Tubular spring cantilever main units; oil-damped spring nosewheel unit. Mainwheel tyres size 6.50-10; nosewheel 6.50-8. Oversize tyres, mainwheels 8.50-10, nosewheel 22 × 8.00-8, and extended nosewheel fork, optional. Hydraulically actuated single-disc brake on each mainwheel. Certificated in floatplane and amphibian versions, with floats by Wipline.

POWER PLANT: One Pratt & Whitney Canada PT6A-114 turboprop, flat rated at 447 kW (600 shp) to 3,800 m (12,500 ft), and driving a Hartzell HC-B3MN3/M10083 three-blade constant-speed reversible-pitch and feathering composites propeller. Integral fuel tanks in wings, total capacity 1,268 litres (335 US gallons; 279 Imp gallons), of which 1,257 litres (332 US gallons; 276.5 Imp gallons) are usable.

ACCOMMODATION: Pilot and up to nine passengers or 1,360 kg (3,000 lb) of cargo. Maximum seating capacity with FAR Pt 23 waiver is 14. Cabin has a flat floor with Brownline cargo track attachments for a combination of two- and three-abreast seating, with an aisle between the seats. Forward hinged door for pilot, with direct vision window, on each side of forward fuselage. Airstair door for passengers at rear of cabin on starboard side. Cabin is heated and ventilated. Optional air-conditioning. Two-section horizontally split cargo door at rear of cabin on port side, flush with floor at bottom and with square corners. Upper portion hinges upward, lower portion forward 180°. Electrically operated, flight openable tambour roll-up door with airflow deflecting spoiler optional. In a cargo role cabin will accommodate typically two D-size cargo containers or up to ten 208 litre (55 US gallon; 45.8 Imp gallon) drums.

SYSTEMS: Electrical system is powered by 28V 200A starter/generator and 24V 45Ah lead-acid battery (24V 40Ah nickel-cadmium battery optional). Standby electrical system, with 95A alternator, optional. Hydraulic system for brakes only. Oxygen system, capacity 3.31 m³ (116.95 cu ft), optional. Vacuum system standard. Cabin air-conditioning system optional on c/n 208-00030 onwards. De-icing system, comprising electric propeller de-icing boots, pneumatic wing, wing strut and tail surface boots, electrically heated windscreen panel, heated pitot/static probe, ice detector light and standby electrical system, all optional.

AVIONICS: Standard avionics include Bendix/King Silver Crown package comprising a single nav/com, ADF, transponder and audio console. Optional avionics include Bendix/King RDS-82 colour weather radar in pod on starboard wing leading-edge.

EQUIPMENT: Standard equipment includes sensitive altimeter, electric clock, magnetic compass, attitude and directional gyros, true airspeed indicator, turn and bank indicator, vertical speed indicator, ammeter/voltmeter, fuel flow indicator, ITT indicator, oil pressure and temperature indicator, windscreen defrost, ground service plug receptacle, variable intensity instrument post lighting, map light, overhead courtesy lights (three) and overhead floodlights (pilot and co-pilot), approach plate holder, cargo tiedowns, internal corrosion proofing, vinyl floor covering, emergency locator beacon, partial plumbing for oxygen system, pilot's adjustable fore/aft/vertical/reclining seat with five-point restraint harness, tinted windows, control surface bonding straps, heated pitot and stall warning systems, retractable crew steps (port side), tiedowns and towbar. Optional equipment includes co-pilot's and passenger seats, stowable, folding utility seats, digital clock, fuel totaliser, turn co-ordinator, flight hour recorder, fire extinguisher, dual controls, co-pilot flight instruments, floatplane kit (on c/n 208-00030 onwards), hoisting rings (for floatplane), inboard fuel filling provisions (included in floatplane kit), ice detection light, courtesy lights on wing leading-edges, passenger reading lights, omniflash beacon, rudder gust lock, retractable crew step for starboard side, oversized tyres, electric trim system, oil quick drain valve and fan driven ventilation system.

DIMENSIONS, EXTERNAL (208A):
Wing span 15.88 m (52 ft 1 in)
Wing chord: at root 1.98 m (6 ft 6 in)
  at tip 1.22 m (4 ft 0 in)
Wing aspect ratio 9.6
Length overall: landplane 11.46 m (37 ft 7 in)
Height overall: landplane 4.32 m (14 ft 2 in)
  amphibian (on land) 5.33 m (17 ft 6 in)
Tailplane span 6.25 m (20 ft 6 in)
Wheel track: landplane 3.56 m (11 ft 8 in)
  amphibian 3.25 m (10 ft 8 in)
Wheelbase: landplane 3.54 m (11 ft 7½ in)
  amphibian 4.44 m (14 ft 7 in)
Propeller diameter 2.54 m (8 ft 4 in)
Airstair door: Height 1.27 m (4 ft 2 in)
  Width 0.61 m (2 ft 0 in)
Cargo door: Height 1.27 m (4 ft 2 in)
  Width 1.24 m (4 ft 1 in)

DIMENSIONS, INTERNAL (208A):
Cabin: Length, excl baggage area 4.57 m (15 ft 0 in)
  Max width 1.57 m (5 ft 2 in)
  Max height 1.30 m (4 ft 3 in)
  Volume 9.67 m³ (341.4 cu ft)

AREAS:
Wings, gross 25.96 m² (279.4 sq ft)
Vertical tail surfaces (total, incl dorsal fin) 3.57 m² (38.41 sq ft)
Horizontal tail surfaces (total) 6.51 m² (70.04 sq ft)

WEIGHTS AND LOADINGS (civil 208A. L: landplane, F: floatplane, A: amphibian):
Weight empty: L 1,724 kg (3,800 lb)
  F 2,020 kg (4,454 lb)
  A 2,177 kg (4,799 lb)
Max baggage (all) 147 kg (325 lb)
Max fuel (all) 1,009 kg (2,224 lb)
Max ramp weight: L 3,327 kg (7,335 lb)
  F, A 3,463 kg (7,635 lb)
Max T-O and landing weight, and max zero-fuel weight:
  L 3,311 kg (7,300 lb)
  F, A 3,447 kg (7,600 lb)
Max wing loading: L 127.4 kg/m² (26.1 lb/sq ft)
  F, A 132.8 kg/m² (27.2 lb/sq ft)
Max power loading: L 7.41 kg/kW (12.17 lb/shp)
  F, A 7.71 kg/kW (12.7 lb/shp)

WEIGHTS AND LOADINGS (U-27A. L and A as above):
Weight empty, standard: L 1,752 kg (3,862 lb)
  A 2,233 kg (4,922 lb)
Max ramp weight: L 3,645 kg (8,035 lb)
  A 3,463 kg (7,635 lb)
Max T-O weight: L 3,629 kg (8,000 lb)
  A 3,447 kg (7,600 lb)
Max landing weight: L 3,538 kg (7,800 lb)
  A 3,311 kg (7,300 lb)
Max wing loading: L 139.8 kg/m² (28.6 lb/sq ft)
  A 132.8 kg/m² (27.2 lb/sq ft)
Max power loading: L 8.11 kg/kW (13.33 lb/shp)
  A 7.71 kg/kW (12.67 lb/shp)

PERFORMANCE (civil 208A. L: landplane, F: floatplane, A: amphibian):
Max operating speed (all) 175 knots (325 km/h; 202 mph) IAS
Max cruising speed at 3,050 m (10,000 ft):
  L 184 knots (341 km/h; 212 mph)
  F 159 knots (295 km/h; 183 mph)
  A 153 knots (283 km/h; 176 mph)
Stalling speed, power off:
  L: flaps up 73 knots (135 km/h; 84 mph) CAS
    flaps down 60 knots (111 km/h; 69 mph) CAS
  F, A, landing configuration 58 knots (107 km/h; 67 mph) CAS
Max rate of climb at S/L: L 370 m (1,215 ft)/min
  F 306 m (1,005 ft)/min
  A 290 m (952 ft)/min
Service ceiling: L 8,410 m (27,600 ft)
  F 7,285 m (23,900 ft)
  A 7,010 m (23,000 ft)
Max operating altitude (all) 9,145 m (30,000 ft)
T-O run: L 296 m (970 ft)
T-O run, water: F 468 m (1,535 ft)
  A 469 m (1,540 ft)
T-O to 15 m (50 ft): L 507 m (1,665 ft)
  F, from water 843 m (2,765 ft)
  A, from water 859 m (2,820 ft)
Landing from 15 m (50 ft): L 472 m (1,550 ft)
Landing run: L 197 m (645 ft)
Range with max fuel, at max cruise power, allowances for start, taxi and reserves stated:
  L at 3,050 m (10,000 ft), 45 min 970 nm (1,797 km; 1,117 miles)
  L at 6,100 m (20,000 ft), 45 min 1,275 nm (2,362 km; 1,468 miles)
  F at 3,050 m (10,000 ft), 30 min 898 nm (1,664 km; 1,034 mile
  A at 3,050 m (10,000 ft), 30 min 868 nm (1,608 km; 999 mile
Range with max fuel at max range power, allowances above:
  L at 3,050 m (10,000 ft) 1,115 nm (2,066 km; 1,284 mile
  L at 6,100 m (20,000 ft) 1,370 nm (2,539 km; 1,578 mile
g limits +3.8/−1.

PERFORMANCE (U-27A. L and A as above):
Max cruising speed at 3,050 m (10,000 ft):
  L 184 knots (341 km/h; 212 mph)
  A 163 knots (302 km/h; 188 mph)
Stalling speed in landing configuration:
  L 61 knots (113 km/h; 71 mp
  A 58 knots (107 km/h; 67 mp
Max rate of climb at S/L: L 320 m (1,050 ft)/m
  A 274 m (900 ft)/m
Service ceiling: L 7,770 m (25,500
  A 6,100 m (20,000
T-O run at S/L: L 368 m (1,205
  A, from water 500 m (1,640
T-O to 15 m (50 ft) at S/L: L 674 m (2,210
  A, from water 872 m (2,860
Landing from 15 m (50 ft) at S/L, without propeller reve
  sal: L 505 m (1,655
  A 560 m (1,835
Landing run at S/L, without propeller reversal:
  L 227 m (745
  A 297 m (975
Range at 3,050 m (10,000 ft) at max cruise power, allo
  ances for T-O, climb, cruise, descent, and 45 m
  reserves: L 1,085 nm (2,011 km; 1,249 mile
  A 955 nm (1,770 km; 1,100 mile

CESSNA 525 CITATIONJET

TYPE: Six/seven-seat business jet.

PROGRAMME: Announced at NBAA convention 1989; replace Citation 500 and I (production of which stoppe 1985); first flight of FJ44 turbofans in Citation 500 Ap 1990; first flight of CitationJet (N525CJ) 29 April 199 first flight of second (pre-production) prototype 2 November 1991; FAA certification for single-pilot ope ation received 16 October 1992; first customer delivery 3 March 1993. Anticipated market for 1,000 in 10 years.

CUSTOMERS: Orders for 50 placed at NBAA convention 198 production sold for 18 months after first delivery. Fir orders approx 100 by March 1993; Cessna planned deliver 47 by end of 1993.

COSTS: $2,620,000.

DESIGN FEATURES: Compared with Citation I, CitationJet h fuselage shortened by 0.27 m (10¾ in) and wing spa reduced by 0.57 m (1 ft 10½ in); cabin height increased b 13 cm (5 in) by lowering centre aisle. New supercritic laminar-flow wing aerofoil; high T-tail; two FJ44 turbe fans; trailing link main landing gear.

POWER PLANT: Two 8.45 kN (1,900 lb st) Williams Inte national/Rolls-Royce FJ44 turbofans. Fuel load: see und Weights and Loadings.

AVIONICS: Honeywell SPZ-5000 flight director/autopilot wi two 5 in displays; stick shaker. Bendix/King KLN 8 Loran C nav and RDS 81 colour weather radar standar KLN 90 GPS long-range nav optional.

DIMENSIONS, EXTERNAL:
Wing span 14.26 m (46 ft 9½ i
Wing aspect ratio 9.

Cessna CitationJet six/seven-seat light business jet

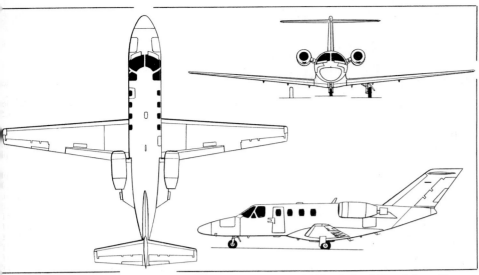

Cessna CitationJet (two Williams International/Rolls-Royce FJ44 turbofans) *(Dennis Punnett)*

Length overall	12.98 m (42 ft 7¼ in)
Height overall	4.18 m (13 ft 8½ in)
Wheel track	3.84 m (12 ft 7¼ in)
Wheelbase	4.54 m (14 ft 10¾ in)
Tailplane span	4.99 m (16 ft 4¾ in)
Crew/passenger door: Height	1.29 m (4 ft 2¾ in)
Width	0.60 m (1 ft 11½ in)

DIMENSIONS, INTERNAL:
Cabin: Length between pressure bulkheads
 4.85 m (15 ft 10¾ in)
 Max width 1.49 m (4 ft 10¾ in)
 Height 1.46 m (4 ft 9½ in)
AREAS:
 Wings, gross 22.30 m² (240.0 sq ft)
 Horizontal tail surfaces (total, incl tab)
  5.03 m² (54.1 sq ft)
 Vertical tail surfaces (total, incl tab) 4.74 m² (51.0 sq ft)
WEIGHTS AND LOADINGS:
 Weight empty 2,794 kg (6,159 lb)
 Max usable fuel weight 1,460 kg (3,220 lb)
 Payload with max fuel 327 kg (721 lb)
 Max T-O weight 4,717 kg (10,400 lb)
 Max ramp weight 4,763 kg (10,500 lb)
 Max zero-fuel weight 3,583 kg (7,900 lb)
 Max landing weight 4,400 kg (9,700 lb)
 Max wing loading 211.6 kg/m² (43.33 lb/sq ft)
 Max power loading 279.3 kg/kN (2.74 lb/lb st)
PERFORMANCE (at max T-O weight except where indicated):
 Max operating speed:
  S/L to 9,300 m (30,500 ft)
   260 knots (482 km/h; 299 mph) CAS
  above 9,300 m (30,500 ft) Mach 0.70
 Max cruising speed at 3,855 kg (8,500 lb) AUW
  380 knots (704 km/h; 438 mph)
 Stalling speed, landing configuration
  81 knots (150 km/h; 94 mph)
 Max rate of climb at S/L 1,050 m (3,450 ft)/min
 Rate of climb at S/L, OEI 326 m (1,070 ft)/min
 Max certificated ceiling 12,500 m (41,000 ft)
 Service ceiling, OEI 7,985 m (26,200 ft)
 T-O balanced field length (FAR Pt 25) 939 m (3,080 ft)
 FAR Pt 25 landing field length at max landing weight
  838 m (2,750 ft)
 Range with one crew, four passengers, 45 min reserves
  1,500 nm (2,780 km; 1,727 miles)

CESSNA 550 CITATION II

TYPE: Six/10-passenger twin-turbofan business jet.
PROGRAMME: Announced 14 September 1976; first flight (N550CC) 31 January 1977; FAR Pt 25 Transport Category certification for two-pilot crew March 1978; phased out in favour of Citation S/II 1984, after 503 Citation IIs delivered. Resumed production announced NBAA convention September 1985.
CURRENT VERSIONS: **550 Citation II**: First version for two-pilot operation. *Data refer to current production 550 Citation II (c/n 0550 and later), unless otherwise indicated.*
 551 Citation II/SP: For single-pilot operation to FAR Part 23 with up to 10 passengers at max T-O weight 5,670 kg (12,500 lb).
CUSTOMERS: Total 660 Citation IIs delivered by 31 December 1992, including 22 in 1992. Production output into 1993 sold.
COSTS: $3,600,375.
DESIGN FEATURES: Citation II 1.14 m (3 ft 9 in) longer than Citation I, greater wing span, increased fuel and baggage capacities. Wing aerofoil NACA 23014 (modified) at centreline, NACA 23012 at wing station 247.95; dihedral 4°; tailplane dihedral 9°.
FLYING CONTROLS: Mechanically actuated ailerons; manual trim tab on port aileron; manual rudder trim; electric elevator trim tab with manual standby; electrically actuated single-slotted flaps; hydraulically actuated airbrake.
STRUCTURE: Two primary, one auxiliary metal wing spars; three fuselage attachment points; conventional ribs and stringers. All-metal pressurised fuselage with fail-safe design providing multiple load paths.
LANDING GEAR: Hydraulically retractable tricycle type with single wheel on each unit. Main units retract inward into the wing, nose gear forward. Free-fall and pneumatic emergency extension systems. Goodyear mainwheels with tyres size 22.0 × 8-10, 10-ply rating, pressure 6.90 bars (100 lb/sq in). Steerable nosewheel (±20°) with Goodyear wheel and tyre size 18.0 × 4.4, 10-ply rating, pressure 8.27 bars (120 lb/sq in). Goodyear hydraulic brakes. Parking brake and pneumatic emergency brake system. Anti-skid system optional. Min ground turning radius about nosewheel 8.38 m (27 ft 6 in).
POWER PLANT: Two Pratt & Whitney Canada JT15D-4B turbofans, each rated at 11.12 kN (2,500 lb st) for take-off, pod-mounted on sides of rear fuselage. Integral fuel tanks in wings, with usable capacity of 2,808 litres (742 US gallons; 618 Imp gallons).
ACCOMMODATION: Crew of two on separate flight deck, on fully adjustable seats, with seat belts and inertia reel shoulder harness. Sun visors standard. Fully carpeted main cabin equipped with seats for six to ten passengers, with toilet in six/eight-seat versions. Main baggage area at rear of cabin. Second baggage area in nose. Total baggage capacity 522 kg (1,150 lb). Cabin is pressurised, heated and air-conditioned. Individual reading lights and air inlets for each passenger. Dropout constant-flow oxygen system for emergency use. Plug type door with integral airstair at front on port side and one emergency exit on starboard side. Doors on each side of nose baggage compartment. Tinted windows, each with curtains. Pilot's storm window, birdproof windscreen with de-fog system, anti-icing, standby alcohol anti-icing and bleed air rain removal system.
SYSTEMS: Pressurisation system supplied with engine bleed air, max pressure differential 0.61 bar (8.8 lb/sq in), maintaining a sea level cabin altitude to 6,720 m (22,040 ft), or a 2,440 m (8,000 ft) cabin altitude to 12,495 m (41,000 ft). Hydraulic system, pressure 103.5 bars (1,500 lb/sq in), with two pumps to operate landing gear and speed brakes. Separate hydraulic system for wheel brakes. Electrical system supplied by two 28V 400A DC starter/generators, with two 350VA inverters and 24V 40Ah nickel-cadmium battery. Oxygen system of 0.62 m³ (22 cu ft) capacity includes two crew demand masks and five dropout constant flow masks for passengers. High capacity oxygen system optional. Engine fire detection and extinguishing systems. Wing leading-edges electrically de-iced ahead of engines; pneumatic de-icing boots on outer wings.

DIMENSIONS, EXTERNAL:
 Wing span 15.76 m (51 ft 8½ in)
 Wing aspect ratio 8.3
 Length overall 14.39 m (47 ft 2½ in)
 Height overall 4.57 m (15 ft 0 in)
 Wheel track 5.36 m (17 ft 7 in)
 Wheelbase 5.55 m (18 ft 2½ in)
DIMENSIONS, INTERNAL:
 Cabin:
  Length, front to rear bulkhead 6.37 m (20 ft 10¾ in)
  Max height 1.46 m (4 ft 9½ in)
 Baggage volume 1.84 m³ (65.0 cu ft)
AREAS:
 Wings, gross 30.00 m² (322.9 sq ft)
 Horizontal tail surfaces (total, incl tab)
  6.56 m² (70.6 sq ft)
 Vertical tail surfaces (total) 4.73 m² (50.9 sq ft)
WEIGHTS AND LOADINGS:
 Weight empty, equipped 3,351 kg (7,388 lb)
 Max fuel weight 2,272 kg (5,009 lb)
 Max T-O weight 6,033 kg (13,300 lb)
 Max ramp weight 6,123 kg (13,500 lb)
 Max zero-fuel weight: standard 4,309 kg (9,500 lb)
  optional 4,990 kg (11,000 lb)
 Max landing weight 5,760 kg (12,700 lb)
 Max wing loading 201.1 kg/m² (41.19 lb/sq ft)
 Max power loading 275.3 kg/kN (2.66 lb/lb st)
PERFORMANCE (at max T-O weight, ISA, except where indicated):
 Max operating speed:
  S/L to 4,265 m (14,000 ft)
   262 knots (486 km/h; 302 mph) IAS
  4,265 m (14,000 ft) to 8,530 m (28,000 ft)
   277 knots (513 km/h; 319 mph) IAS
  8,530 m (28,000 ft) and above Mach 0.705
 Cruising speed at average cruise weight of 4,990 kg (11,000 lb) at 7,620 m (25,000 ft)
  385 knots (713 km/h; 443 mph)
 Stalling speed, clean, at max T-O weight
  94 knots (174 km/h; 108 mph) CAS
 Stalling speed at max landing weight
  82 knots (152 km/h; 95 mph) CAS
 Max rate of climb at S/L 1,027 m (3,370 ft)/min
 Rate of climb at S/L, OEI 322 m (1,055 ft)/min
 Max certificated altitude 13,105 m (43,000 ft)
 Service ceiling, OEI 7,680 m (25,200 ft)
 T-O to 15 m (50 ft) 727 m (2,385 ft)
 T-O balanced field length (FAR Pt 25) 912 m (2,990 ft)
 FAR Pt 25 landing field length at max landing weight
  692 m (2,270 ft)
 Range with max fuel, crew of two and six passengers, allowances for T-O, climb, cruise at 13,105 m (43,000 ft), descent, and 45 min reserves
  1,662 nm (3,080 km; 1,914 miles)
OPERATIONAL NOISE LEVELS (FAR Pt 36):
 T-O 80.1 EPNdB
 Approach 90.5 EPNdB
 Sideline 86.7 EPNdB

CESSNA S550 CITATION S/II
US Navy designation: **T-47A**

TYPE: Six/eight-passenger improved Citation II.
PROGRAMME: Announced 4 October 1983; first flight 14 February 1984; certificated with exemption for single-pilot

Cessna Citation II business jet

466   USA: AIRCRAFT—CESSNA

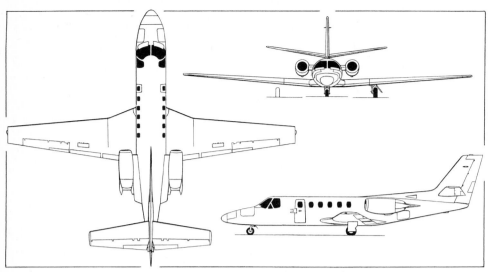

Cessna Citation S/II (Pratt & Whitney Canada JT15D-4B turbofans) *(Dennis Punnett)*

operation July 1984; first delivery late Summer 1984; first delivery Citation S/II ambulance late 1985 to Province of Manitoba, Canada. High capacity brakes introduced as standard and offered as retrofit April 1987, reducing landing distance at max landing weight by 13 per cent.

CURRENT VERSIONS: **S550**: Basic executive transport. *Detailed description applies to this version (c/n 0115 onwards), except where indicated.*

Ambulance: Carries single or double stretchers, up to four medical attendants and large quantities of medical oxygen.

T-47A (Cessna 552): Fifteen Citation S/IIs acquired by US Navy to replace T-39Ds in radar training role; part of five-year programme with three-year option, including provision of aircraft, simulators, maintenance and pilot services for training operators of air-to-air, intercept, air-to-ground and other radars. Differences from S/II include 12.89 kN (2,900 lb st) JT15D-5 turbofans, shorter wing to allow faster climb and Mach 0.733 at 12,200 m (40,000 ft), and Emerson nose-mounted AN/APQ-159 radar. Normal crew includes civilian pilot, Navy instructor, three students. First flight 15 February 1984; FAA certification 21 November 1984. Programme ended September 1991; aircraft returned to manufacturer.

CUSTOMERS: US Navy (see above). Five specially equipped Citation S/IIs delivered to Flight Test Research Institute in Xian for Airborne Remote Sensing Centre of Chinese Academy of Sciences.

DESIGN FEATURES: Improvements, introduced on production line from c/n 506, include new wing aerofoil using Citation III supercritical technology, modified wing/fuselage fairing, extended inboard leading-edge, low drag engine pylon contours, sealed aileron/airbrake gaps, faired flap coves, hydraulically actuated flaps in two sections each side extending further inboard, P&WC JT15D-4B turbofans. New wing reduces cruise drag without sacrificing low speed and short-field capability.

Straight wing; incidence 2° 30′ at centreline, –0° 30′ at station 247.95; dihedral 4°; tailplane dihedral 9°.

Internal refinements include tailcone baggage volume increased to 0.79 m³ (28 cu ft), 12.7 cm (5 in) extra headroom in totally private toilet, soft-touch sound deadening headliners, Citation III-style seats with shoulder harness, lateral seat tracking for better head and elbow room, built-in life jacket stowage, and redesigned sidewall air ducts giving greater insulation and heating and allowing more than 10 per cent extra aisle width.

Options include vanity unit for toilet, refreshment centres in composites, wide door for ambulance/cargo operations, Honeywell EFIS.

FLYING CONTROLS: Mechanically actuated ailerons assisted by geared trim tabs; elevator trim electrically actuated with mechanical standby; mechanical rudder trim; hydraulically actuated Fowler flaps; hydraulically operated airbrakes.

STRUCTURE: As described for Citation II, except ailerons and flaps have graphite composite structure.

LANDING GEAR: As for Citation II except mainwheel tyres have 12-ply rating, pressure 8.27 bars (120 lb/sq in). High capacity brakes manufactured by Aircraft Wheel and Brake Division of Parker Hannifin Corporation.

POWER PLANT: As Citation II, but usable fuel capacity increased to 3,263 litres (862 US gallons; 718 Imp gallons).

ACCOMMODATION: Crew details as for Citation II. Seating for six to eight passengers in main cabin. Standard interior configuration provides for six passenger seats, two forward and four aft facing, each with headrest, seat belt and diagonal inertia reel harness; flushing toilet aft; tracked refreshment centre; forward cabin divider with privacy curtain, aft cabin divider with sliding doors. Passenger service units containing an oxygen mask, air vent and reading light for each passenger. Three separate baggage areas, one in nose section that is externally accessible, one in aft cabin area, and one in tailcone area, with a combined capacity of up to 658 kg (1,450 lb).

SYSTEMS: Pressurisation system as for Citation II but maintains sea level cabin altitude to 6,962 m (22,842 ft), or a 2,440 m (8,000 ft) cabin altitude to 13,105 m (43,000 ft). Hydraulic system as for Citation II. Electrical system as for Citation II, except starter/generators are 300A. TKS Glycol anti-icing system on wing, tail surface de-icing eliminated.

AVIONICS: Standard avionics package comprises Honeywell SPZ-500 integrated flight director/autopilot system, with single-cue command bars, Honeywell C-14D compass system, Honeywell RD-450 (starboard) HSI, dual Collins VHF-22A com transceivers, dual Collins VIR-32 nav receivers with VOR/LOC, glideslope and marker beacon receivers, dual Collins RMI-30, Collins DME-42 with 339F-12 indicator, TDR-90 transponder, Collins ADF-60, and Honeywell Primus 300SL colour weather radar. Optional advanced avionics and instrumentation are available according to customer choice, and include Bendix/King Series III integrated EFIS, nav/com and radar systems.

DIMENSIONS, EXTERNAL:
Wing span over lights	15.90 m (52 ft 2½ in)
Wing chord (mean)	2.06 m (6 ft 9 in)
Wing aspect ratio	7.8
Length overall	14.39 m (47 ft 2½ in)
Height overall	4.57 m (15 ft 0 in)
Wheel track	5.36 m (17 ft 7 in)
Wheelbase	5.55 m (18 ft 2½ in)
Tailplane span	5.79 m (19 ft 0 in)
Cabin door (optional): Height	1.14 m (3 ft 9 in)
Width	0.89 m (2 ft 11 in)

DIMENSIONS, INTERNAL:
Cabin:
Length, front to rear bulkhead	6.37 m (20 ft 10¾ in)
Max height	1.45 m (4 ft 9½ in)
Max width	1.49 m (4 ft 10¾ in)
Baggage capacity (total)	2.27 m³ (80.0 cu ft)

AREAS:
Wings, gross	31.83 m² (342.6 sq ft)
Horizontal tail surfaces (total)	6.48 m² (69.8 sq ft)
Vertical tail surfaces (total)	4.73 m² (50.9 sq ft)

WEIGHTS AND LOADINGS:
Weight empty, equipped	3,655 kg (8,059 lb)
Max baggage weight: cabin/tailcone	272 kg (600 lb)
nose	385 kg (850 lb)
Max ramp weight	6,940 kg (15,300 lb)
Max fuel weight	2,640 kg (5,820 lb)
Max T-O weight	6,849 kg (15,100 lb)
Max landing weight	6,350 kg (14,400 lb)
Max zero-fuel weight	4,990 kg (11,200 lb)
Max wing loading	215.17 kg/m² (44.07 lb/sq ft)
Max power loading	1.42 kg/kN (3.02 lb/lb st)

PERFORMANCE (at max T-O weight, except where indicated):
Max operating speed:
S/L to 2,440 m (8,000 ft)	261 knots (483 km/h; 300 mph) IAS
2,440 m (8,000 ft) to 8,935 m (29,315 ft)	276 knots (511 km/h; 318 mph)
above 8,935 m (29,315 ft)	Mach 0.72

Cruising speed at mid-cruise weight of 5,443 kg (12,000 lb) at 10,670 m (35,000 ft)
403 knots (746 km/h; 463 mph)
Stalling speed: clean, at max T-O weight
94 knots (174 km/h; 108 mph) CAS
at max landing weight
82 knots (152 km/h; 94 mph) CAS
Max rate of climb at S/L	926 m (3,040 ft)/min
Rate of climb at S/L, OEI	262 m (860 ft)/min
Max operating altitude	13,105 m (43,000 ft)
Service ceiling, OEI	7,315 m (24,000 ft)

T-O balanced field length (FAR Pt 25)
987 m (3,240 ft)
FAR 25 landing field length at max landing weight (high capacity brakes)   805 m (2,640 ft)
Range with four passengers, two crew and baggage, zero wind, IFR reserves   1,739 nm (3,223 km; 2,002 miles)
Range with max fuel   1,998 nm (3,701 km; 2,300 miles)

OPERATIONAL NOISE LEVELS (FAR Pt 36):
T-O	78.0 EPNdB
Approach	91.0 EPNdB
Sideline	90.4 EPNdB

CESSNA 560 CITATION V

TYPE: Stretched, eight-passenger version of Citation S/II.

PROGRAMME: First flight of engineering prototype (N560CC) August 1987; announced at NBAA convention 1987; first flight of pre-production prototype early 1988; FAA certification 9 December 1988; first delivery April 1989; 100th aircraft delivered to Cartier, France, April 1991.

CUSTOMERS: Order backlog at end 1992 covered production into 1993; total deliveries 202, including 51 in 1992.

COSTS: $4,950,000.

DESIGN FEATURES: Stretched for full eight-seat cabin and full enclosed toilet/vanity area; seventh cabin window each side; two baggage compartments outside main cabin with total volume 1.16 m³ (41 cu ft) and capacity for 385 kg (850 lb) baggage. Other features include Global GNS-X navigation management system, advanced weather radar, dual transponders, encoding and radio altimeters, thrust reversers, engine synchronisers, 1.8 m³ (64 cu ft) oxygen system, recognition lights, and in-flight telephone.

Cessna 560 Citation V eight-seat business jet

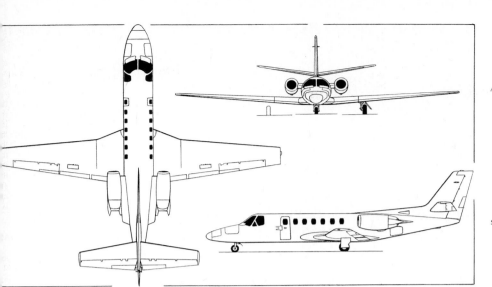

Cessna 560 Citation V (two Pratt & Whitney Canada JT15D-5A turbofans) *(Dennis Punnett)*

FLYING CONTROLS: Category II avionics standard, with integrated Honeywell autopilot and flight guidance system, and EFIS in captain's instrument panel.

POWER PLANT: Two Pratt & Whitney Canada JT15D-5A turbofans, each rated at 12.89 kN (2,900 lb st).

SYSTEMS: De-icing by engine bleed air for inboard wing leading-edges and 'silver' low profile pneumatic boots on outer wings and tailplane.

DIMENSIONS, EXTERNAL: As for Citation S/II except:
Length overall	14.90 m (48 ft 10¾ in)
Tailplane span	6.55 m (21 ft 6 in)
Wheelbase	6.06 m (19 ft 10¾ in)

DIMENSIONS, INTERNAL: As for Citation S/II except:
Cabin: Length, front to rear bulkhead
6.89 m (22 ft 7¼ in)

AREAS: As for Citation S/II except:
Horizontal tail surfaces (total)   7.88 m² (84.8 sq ft)

WEIGHTS AND LOADINGS:
Weight empty, equipped	4,004 kg (8,828 lb)
Max fuel weight	2,640 kg (5,820 lb)
Max T-O weight	7,212 kg (15,900 lb)
Max ramp weight	7,303 kg (16,100 lb)
Max landing weight	6,895 kg (15,200 lb)
Max zero-fuel weight	5,080 kg (11,200 lb)
Max wing loading	226.6 kg/m² (46.41 lb/sq ft)
Max power loading	279.75 kg/kN (2.74 lb/lb st)

PERFORMANCE:
Max operating speed:
2,440 m (8,000 ft) to 8,810 m (28,900 ft)
292 knots (541 km/h; 336 mph) IAS
above 8,810 m (28,900 ft) Mach 0.755
Cruising speed at 10,060 m (33,000 ft)
427 knots (791 km/h; 492 mph)
Stalling speed, clean, at max T-O weight
87 knots (161 km/h; 100 mph)
Max rate of climb at S/L	1,112 m (3,650 ft)/min
Rate of climb at S/L, OEI	360 m (1,180 ft)/min
Max operating altitude	13,700 m (45,000 ft)
Service ceiling, OEI	9,480 m (31,100 ft)
T-O balanced field length (FAR 25)	963 m (3,160 ft)

FAR 25 landing field length, S/L, ISA, at max landing weight   890 m (2,920 ft)
Range with six passengers, two crew, zero wind, at high-speed cruise speed with allowances for T-O, climb, cruise, descent and VFR reserves
1,920 nm (3,558 km; 2,211 miles)

CESSNA 650 CITATION III

Production ended 1992; total deliveries 202, all by late 1991 except one in 1992; superseded by Citation VI and VII (which see). Full details of Citation III in 1991-92 *Jane's*.

CESSNA 650 CITATION VI

TYPE: Six/nine-passenger business jet; simplified, lower-cost version of Citation III.

PROGRAMME: Announced 1990; first aircraft rolled out 2 January 1991; first delivery 1992.

CUSTOMERS: Ten (total deliveries 14) delivered in 1992; production sold into 1993. Two to be delivered to Civil Aviation Administration of China in third quarter of 1993, for navaid flight inspection.

COSTS: $7,850,000.

DESIGN FEATURES: Citation III airframe; retains Garrett TFE731-3B-100 turbofans; empty weight (preliminary) 5,795 kg (12,775 lb). Avionics changes include replacement of Honeywell SPZ-8000 dual digital autopilot/FDS by dual SPZ-650; AHZ-600 AHRS replaced by VG14 and C14D gyros; single AZ-810 digital air data computer for pilot's single altitude encoding servoed altimeter in place of dual ADZ-810; Global/Wulfsberg GNS-X FMS with Loran C, full alphanumeric monochrome CDU. Standard interiors only, choice of three layouts.

FLYING CONTROLS: Pitch axis, variable incidence tailplane and elevator; rudder boosted to counteract asymmetric thrust; hydraulically powered ailerons with manual reversion assisted after 3° movement by outboard spoiler panel; four hydraulically powered spoiler panels on each wing, of which outboard assists aileron, two centre panels act as airbrakes and all four panels used for emergency descent and lift dumping after touchdown. Electrically actuated trailing-edge flaps in three sections each side. Stall strips on inner and outer wings and small fence and turbulators ahead of outer flaps.

STRUCTURE: Conventional light alloy pressurised fuselage of circular section; fail-safe in pressurised area; light alloy tail surfaces; two-spar, fail-safe light alloy wing of bonded and riveted construction; wing built in three sections; flaps of Kevlar and graphite composites.

LANDING GEAR: Hydraulically retractable tricycle type. Main units retract inward into the undersurface of the wing centre-section, nosewheel forward and upward into the nose. Main units of trailing link type, each with twin wheels; steerable nose unit has a single wheel, max steering angle ±70-80°. Oleo-pneumatic shock absorber in each unit. Hydraulically powered nosewheel steering, with an accumulator to provide steering after a loss of normal hydraulic power. Emergency landing gear extension by manual release and free-fall to locked position; pneumatic blowdown system for backup. Mainwheel tyres size 22.0 × 5.75, 10-ply rating, pressure 10.20 bars (148 lb/sq in). Nosewheel tyre size 18.0 × 4.4, 10-ply rating, pressure 8.62 bars (125 lb/sq in). Fully modulated hydraulically powered anti-skid brake system. In the event of hydraulic system failure, an electrically driven standby pump provides pressure for the brakes. Emergency pneumatic brake system. Parking brake. Turning circle based on nosewheel 6.63 m (21 ft 9 in).

POWER PLANT: Two Garrett TFE731-3B-100S turbofans, each rated at 16.24 kN (3,650 lb st) for take-off, pod-mounted on sides of rear fuselage. Hydraulically operated Rohr target type thrust reversers standard. Two independent fuel systems, with integral tanks in each wing; usable capacity 4,183 litres (1,105 US gallons; 920 Imp gallons).

Additional fuel cell behind rear fuselage bulkhead. Single-point pressure refuelling on starboard side of fuselage, to rear of wing trailing-edge. Gravity refuelling point on upper surface of each wing. A boost pump in the port wing fills the fuselage tank when pressure refuelling is not available.

ACCOMMODATION: Crew of two on separate flight deck, and up to nine passengers. Standard interior has six individual seats, with toilet at rear of cabin. The fuselage nose incorporates a radome, high resolution radar, avionics bay and a storage compartment for crew baggage. Electrically heated baggage compartment in rear fuselage with external door on port side. Airstair door forward of wing on port side. Overwing emergency escape hatch on starboard side. Cabin is pressurised, heated and air-conditioned. Windscreen anti-icing by engine bleed air, with alcohol spray backup for port side of the windscreen. Windscreen defogging by warm air, and rain removal by engine bleed air and a mechanically actuated airflow deflector.

SYSTEMS: Environmental control system, with separate control of flight deck and cabin conditions. Direct engine bleed pressurisation system, with nominal pressure differential of 0.67 bar (9.7 lb/sq in), provides 2,440 m (8,000 ft) cabin environment to max certificated altitude and can maintain a sea level cabin environment to approx 7,620 m (25,000 ft). Electrical system includes two 28V 400A DC starter/generators, two 200/115V 5kW three-phase engine driven alternators, two 115V 400Hz solid state static inverters, two 24V 22Ah nickel-cadmium batteries and an external power socket in the tailcone. Hydraulic system of 207 bars (3,000 lb/sq in) powered by two engine driven pressure compensated pumps for operation of spoilers, brakes, landing gear, nosewheel steering and thrust reversers. Hydraulic reservoir with integral reserve and an electrically driven hydraulic pump to provide emergency power. Oxygen system of 1.39 m³ (49 cu ft) capacity with automatic dropout constant-flow oxygen mask for each passenger and a quick-donning pressure demand mask for each crew member. Engine fire detection and extinguishing system. Wing leading-edges de-iced by engine bleed air; tailplane electrically de-iced; fin unprotected. Engine intake anti-icing system.

AVIONICS: Standard avionics include a Honeywell SPZ-650 integrated flight director/autopilot system with AD650A ADI, RD650A HSI and C-14D compass system; Honeywell GH-14 ADI and RD450 HSI with C-14D compass system for co-pilot; AA-300 radio altimeter; dual Collins VHF-22A 720-channel com transceivers, dual VIR-32 nav receivers which include VOR, localiser, glideslope and marker beacon receivers, dual RMI-30, DME-42 DME, and TDR-90 transponder; Honeywell Primus 300SL colour weather radar; Collins ADF-60 ADF; J. E. T. standby attitude gyro; Teledyne angle of attack system; air data computer; dual Avtech audio amplifiers; and Telex microphones, headsets and speakers. A wide range of optional avionics is available including Bendix/King Series III integrated EFIS, nav/com and radar system.

EQUIPMENT: Standard equipment includes dual altimeters, Mach/airspeed indicators, angle of attack indicator, digital clock, instantaneous rate of climb indicators, outside air temperature gauge, crew seats with vertical, fore, aft and recline adjustments, seat belts, shoulder harnesses and inertia reels, six individual passenger seats, three forward and three aft facing with vertical, fore and aft adjustment, lateral tracking and recline adjustments, seat belts and shoulder harnesses, sun visors, flight deck divider with curtain, map case, openable storm windows, electroluminescent and edge-lit instrument panels, stall warning system, cockpit and cabin fire extinguishers, indirect cabin lighting, cabin aisle lights, door courtesy lights, 'Fasten seat belt—No smoking' signs, refreshment centre, cup holders,

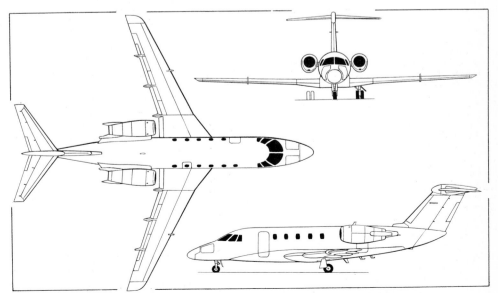

Cessna Citation VI (two Garrett TFE731-3B-100S turbofans) *(Dennis Punnett)*

ashtrays, executive table, aft cabin divider with curtain, emergency exit signs, internal corrosion proofing, emergency battery pack, emergency portable cabin oxygen, navigation and recognition lights, dual landing and taxi lights, dual anti-collision strobe lights, red flashing beacon, dual wing ice lights, lightning protection, static discharge wicks and tiedown provisions.

DIMENSIONS, EXTERNAL:
Wing span	16.31 m (53 ft 6 in)
Wing mean aerodynamic chord	2.08 m (6 ft 9¾ in)
Wing aspect ratio	8.9
Length overall	16.90 m (55 ft 5½ in)
Height overall	5.12 m (16 ft 9½ in)
Tailplane span	5.60 m (18 ft 4½ in)
Wheel track	2.84 m (9 ft 4 in)
Wheelbase	6.50 m (21 ft 4 in)
Cabin door: Width	0.61 m (2 ft 0 in)
Height	1.37 m (4 ft 6 in)

DIMENSIONS, INTERNAL:
Cabin:
Length, front to rear bulkhead	7.01 m (23 ft 0 in)
Length, aft of cockpit divider	5.66 m (18 ft 7 in)
Max width	1.73 m (5 ft 8 in)
Max height	1.78 m (5 ft 10 in)
Baggage capacity (aft)	1.88 m³ (66.4 cu ft)
Crew baggage compartment (nose)	0.17 m³ (6.0 cu ft)

AREAS:
Wings, gross	28.99 m² (312.0 sq ft)
Horizontal tail surfaces (total)	6.26 m² (67.4 sq ft)
Vertical tail surfaces (total)	6.04 m² (65.0 sq ft)

WEIGHTS AND LOADINGS:
Weight empty, standard	5,357 kg (11,811 lb)
Max fuel weight	3,349 kg (7,384 lb)
Max payload	1,583 kg (3,489 lb)
Max T-O weight	9,979 kg (22,000 lb)
Max ramp weight	10,070 kg (22,200 lb)
Max landing weight	9,072 kg (20,000 lb)
Max zero-fuel weight	6,940 kg (15,300 lb)
Max wing loading	344.2 kg/m² (70.51 lb/sq ft)
Max power loading	307.24 kg/kN (3.01 lb/lb st)

PERFORMANCE (at max T-O weight, ISA, except where indicated):
Max operating speed:
S/L to 2,440 m (8,000 ft)
305 knots (565 km/h; 351 mph) IAS
at 11,130 m (36,525 ft)
278 knots (515 km/h; 320 mph) IAS
above 11,130 m (36,525 ft) Mach 0.851
Max cruising speed at 10,670 m (35,000 ft) and 7,257 kg (16,000 lb) cruise weight
472 knots (874 km/h; 543 mph)
Stalling speed: clean, at max T-O weight
125 knots (232 km/h; 144 mph) CAS
flaps and wheels down, at max landing weight
97 knots (515 km/h; 112 mph) CAS
Max rate of climb at S/L	1,127 m (3,700 ft)/min
Rate of climb at S/L, OEI	245 m (805 ft)/min
Time to 13,100 m (43,000 ft)	33 min
Certificated ceiling	15,545 m (51,000 ft)
Ceiling, OEI	7,165 m (23,500 ft)
FAR 25 T-O field length at S/L	1,579 m (5,180 ft)

FAR 25 landing field length at max landing weight
884 m (2,900 ft)
Range, zero wind, with allowances for T-O, climb, descent and 45 min reserves (2 crew, 4 passengers)
2,346 nm (4,348 km; 2,701 miles)
g limits $+3.2/-1$

OPERATIONAL NOISE LEVELS (FAR Pt 36):
T-O	74.0 EPNdB
Approach	85.0 EPNdB
Sideline	81.0 EPNdB

CESSNA 650 CITATION VII

TYPE: More powerful version of Citation III.
PROGRAMME: Announced 1990; engineering prototype first flew February 1991; FAR 25 certification January 1992.
CUSTOMERS: First production aircraft as demonstrator; first customer delivery to golfer Arnold Palmer April 1992; production sold into 1993. Total deliveries 15 (all in 1992).
COSTS: $9,061,625.
DESIGN FEATURES: Citation III airframe with 17.79 kN (4,000 lb st) Garrett TFE731-4R-2S turbofans for improved hot-and-high performance; MMo 0.85.
Specification otherwise as Citation VI except:

WEIGHTS AND LOADINGS:
Weight empty, standard	5,301 kg (11,686 lb)
Max fuel weight	3,350 kg (7,385 lb)
Max T-O weight	10,183 kg (22,450 lb)
Max ramp weight	10,274 kg (22,650 lb)
Max landing weight	9,072 kg (20,000 lb)
Max zero-fuel weight	7,484 kg (16,500 lb)
Max wing loading	351.3 kg/m² (71.96 lb/sq ft)
Max power loading	286.20 kg/kN (2.81 lb/lb st)

PERFORMANCE (at max T-O weight, ISA, except where indicated):
Max operating speed:
S/L to 2,440 m (8,000 ft)
275 knots (509 km/h; 317 mph) IAS
at 11,130 m (36,525 ft)
278 knots (515 km/h; 320 mph) IAS

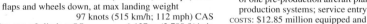

Max cruising speed at 10,670 m (35,000 ft) and 8,165 kg (18,000 lb) cruise weight
478 knots (885 km/h; 550 mph)
Stalling speed, flaps and wheels down, at max landing weight
97 knots (180 km/h; 112 mph)
Rate of climb at S/L, OEI	253 m (830 ft)/min
FAR 25 T-O field length at S/L	1,509 m (4,950 ft)

FAR 25 landing field length at max landing weight
914 m (3,000 ft)
Range, zero wind, with allowances for T-O, climb, cruise, descent and 45 min reserves
2,200 nm (4,077 km; 2,533 miles)
g limits $+3.0/-1$

OPERATIONAL NOISE LEVELS (FAR Pt 36):
T-O (7° flap)	69.3 EPNdB
Approach (full flap)	84.8 EPNdB

CESSNA 750 CITATION X

TYPE: High-speed long-range business jet.
PROGRAMME: Announced October 1990; engine flew on Citation VII testbed (N650) 21 August 1992; first flight delayed three months to September 1993 and two instead of one pre-production aircraft planned to aid integration of production systems; service entry mid-1995.
COSTS: $12.85 million equipped and furnished.
DESIGN FEATURES: High max operating Mach number; US transcontinental and transatlantic range. Forward fuselage/cockpit section derived from Citation VI. Wing sweepback 35-37°.
FLYING CONTROLS: Dual redundant hydraulically powered non-reversible controls; one-piece all-moving horizontal tailplane; speed brakes and spoiler design optimised for drag control.
STRUCTURE: Thick wing skins, milled from solid.
The following data are provisional:
LANDING GEAR: Trailing link main units, each with twin wheels; powered anti-skid brakes; steerable nose unit with twin wheels.
POWER PLANT: Two Allison GMA 3007C turbofans, each rated at 26.69 kN (6,000 lb st) for take-off, pod-mounted on sides of rear fuselage; full authority digital engine controls. Hydraulically operated target type thrust reverser standard. Single point refuelling.
ACCOMMODATION: Crew of two on separate flight deck, and up to 12 passengers; interior custom designed; cabin is pressurised, heated and air-conditioned; 1.79 m (5 ft 10½ in) stand-up headroom; heated and pressurised baggage compartment in rear fuselage with external door. Windscreen electrically heated and demisted.
SYSTEMS: AlliedSignal Garrett GTCP36-150 combined APU/air turbine starter.

DIMENSIONS, EXTERNAL:
Wing span	19.51 m (64 ft 0 in)
Length overall	19.66 m (64 ft 6 in)
Height overall	5.09 m (16 ft 8½ in)

DIMENSIONS, INTERNAL:
Cabin: Length, front to rear pressure bulkhead	8.38 m (27 ft 6 in)
Max width	1.74 m (5 ft 8½ in)
Max height	1.76 m (5 ft 9½ in)
Baggage compartment volume (aft)	2.04 m³ (72.0 cu ft)

WEIGHTS AND LOADINGS:
Weight empty, standard	8,437 kg (18,600 lb)
Max fuel weight	4,990 kg (11,000 lb)
Max T-O weight	14,061 kg (31,000 lb)
Max ramp weight	14,152 kg (31,200 lb)
Max landing weight	13,154 kg (29,000 lb)
Max power loading	263.41 kg/kN (2.58 lb/lb st)

PERFORMANCE (at max T-O weight, ISA, except where indicated):
Max operating Mach No. (MMO) 0.9
Max operating speed (VMO)
350 knots (648 km/h; 403 mph)
Max cruising speed, mid-cruise weight at 11,275 m (37,000 ft) Mach 0.8
Max rate of climb at S/L	1,340 m (4,400 ft)/min
Max operating altitude	15,545 m (51,000 ft)
T-O balanced field length (FAR Pt 25)	1,555 m (5,100 ft)
FAR Pt 25 landing field length	884 m (2,900 ft)

Range, with allowances for T-O, climb, cruise, descent and 45 min reserves 3,300 nm (6,115 km; 3,800 miles)

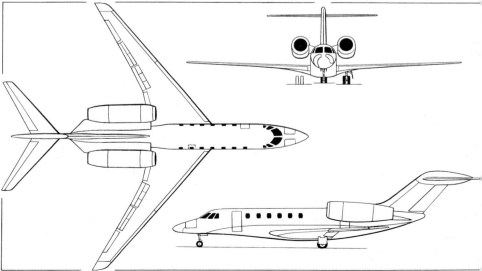

Cessna 750 Citation X before final wing modification introduced in 1992 *(Dennis Punnett)*

Artist's impression of Cessna 750 Citation X business jet

CESSNA 526 JPATS

TYPE: Tandem two-seat jet trainer; contender in USAF/USN joint primary aircraft training system programme.
PROGRAMME: Announced intention of entering JPATS competition with 'all American' aircraft, 1992; selected Williams International as engine partner and FlightSafety International to provide ground based training systems. First flight of engineering prototype expected September/October 1993, second aircraft two months later; certification late March 1994.
DESIGN FEATURES: Based on CitationJet, using similar but shorter laminar flow wings with modified ailerons, redesigned tail, Williams/R-R engines in fuselage, military style canopy; since moved to separate acquisition programme. Wings strengthened for 7g. Fuel capacity 907 kg (2,000 lb) in wings and fuselage. Similar flight control systems; UPC ejection seats; cabin pressure differential 0.34 bar (5.0 lb/sq in).
COSTS: Estimated at under $3 million per aircraft.
POWER PLANT: Two Williams/Rolls-Royce F129 turbofans, derated to 6.67 kN (1,500 lb st) each.
WEIGHTS AND LOADINGS:
Weight empty 2,232 kg (4,920 lb)
Max T-O weight 3,356 kg (7,400 lb)
PERFORMANCE (estimated):
Max level speed 270 knots (500 km/h; 311 mph)
Service ceiling 10,670 m (35,000 ft)
Time to 5,485 m (18,000 ft) 5 min 12 s
Range, with reserves 1,050 nm (1,945 km; 1,209 miles)

CHRYSLER

CHRYSLER TECHNOLOGIES AIRBORNE SYSTEMS INC (CTAS)
PO Box 154580, Waco, Texas 76715-4580
Telephone: 1 (817) 867 4202
Fax: 1 (817) 867 4230
Telex: 163346
PRESIDENT: R. H. Pacey
MARKETING MANAGER, C-27: Wallace Herzog

CTAS services include military and commercial aircraft engineering and modification, including complete flight sciences and structural engineering; major airframe modification; EW and C^3 conversions; Boeing 747 passenger-to-freighter conversions; autopilot/EFIS/EICAS cockpit conversions; prototype engineering; SATCOM design; head-of-state conversions; prime contractor/C-27A intratheatre airlifter. CTAS has unlimited FAA repair station.

CTAS (ALENIA) C-27A SPARTAN

TYPE: Mid-range pressurised STOL transport.
PROGRAMME: US Air Force ordered five Alenia G222-710As (which see in Italian section) from Chrysler on 20 August 1990; further five ordered 4 February 1991; aircraft procured from Alenia by Chrysler and modified for USAF with enhancements including autopilot, INS, new HF/VHF communications and specific mission equipment; first aircraft (I-RAIS) accepted 15 April 1991 and delivered USA 17 April 1991; first flown after C-27A modifications, reserialled 90-0170, 30 July 1991; to USAF 16 August 1991 for 45 days of operational testing at Pope AFB, North Carolina. CTAS provides aircrew academic and flight training, maintenance training and complete logistical support.
CUSTOMERS: US Air Force, 10 plus option on further eight (five in FY 1993 and three in FY 1994); required for short-field operations by USSOUTHCOM in Panama; 90-0171 delivered 310th Airlift Squadron/61st Airlift Group, Howard AFB, Canal Zone, 26 September 1991. Tenth delivered 25 January 1993.
COSTS: First five bought for about $80 million; second five for $72.6 million.
Following description applies to the C-27A, except where indicated. Description of Alenia G222 basic airframe can be found under Alenia, Italy.
POWER PLANT: Two General Electric T64-GE-P4D turboprops, each flat rated at 2,535 kW (3,400 shp) at ISA + 25°C. New turbine inlet temperature indicators with analog pointer and digital readout added to C-27A engine instrument panel.
ACCOMMODATION: Normal crew of three (two pilots and loadmaster); loadmaster's seat by crew entrance door, facing aft in cargo compartment. Standard troop configuration has 34 foldaway sidewall seats with life vest stowage compartment under each seat, and two 20-man liferafts stowed in wing/fuselage fairing. Paratroop version can carry up to 24 fully equipped paratroops and includes door jump platforms and static lines. Cargo version can accept three standard 463L 2.74 m × 2.24 m (108 in × 88 in) pallets, and carry up to 9,000 kg (19,840 lb) of freight. Provision for 135 cargo tiedown points, on 51 cm (20 in) square NATO standard grid. In combat operations, cargo compartment can accommodate one M1038 HMMWV, one M1009 CUCV, one 105 mm howitzer, or six A-22 CDS containers; in aeromedical role, 24 litters and four attendants. Crew and passenger door forward of cabin on left side. Paratroop doors at rear of cargo compartment on left and right sides; emergency exit in latrine, forward cargo compartment on right side. Three emergency exit hatches: one overhead in flight compartment, two in cargo compartment roof, forward and aft of wing. Hydraulically operated rear loading ramp and upward opening door in underside of upswept rear fuselage can be opened in flight for airdrop operations. Six CDS containers (up to 1,000 kg; 2,205 lb each), or a single pallet of up to 5,000 kg (11,023 lb), can be airdropped from rear opening; paratroops jump from either cargo ramp and door opening or from rear side paratroop doors.
SYSTEMS: See G222 in 1987-88 and earlier *Jane's*. Windscreens and quarter-light panels de-iced and demisted electrically; hydraulic wipers and liquid rain repellent system provided for both windscreens. Entire aircraft pressurised and air-conditioned. Normal anti-skid braking provided by No. 2 hydraulic system with maximum braking pressure of 117 bars (1,700 lb/sq in). Emergency braking, without anti-skid, by No. 1 hydraulic system or hydraulic hand pump. Multiple disc brakes by B. F. Goodrich. Redesigned and reinforced flap tracks allow flaps to be lowered at higher airspeeds. Separate oxygen system for aeromedical missions; 12 internal power receptacles in cargo compartment for auxiliary and medical equipment, including eight power receptacles rated at 200V AC, 400Hz and two double fuse receptacles: one 115V AC 60Hz and one 220V AC 60Hz.
AVIONICS: Standard communications equipment includes 7,000-channel UHF radio with SATCOM and Have Quick I/II secure speech capability; two 4,080-channel VHF AM/FM (AN/ARC-186) radios with secure voice capability; VHF FM (AN/ARC-201A SINCGARS) radio; 28,000-channel (AN/ARC-190) HF radio with secure voice, selcal, antijam, and automatic link establishment capability; crew intercom and PA system. Navigation equipment includes LTN-92 INS, with TAS computer, autopilot, flight director, two compasses, two vertical gyros, dual Tacan-VOR/ILS-DME, marker beacon receiver, ADF, two radio magnetic indicators, magnetic compass, radar altimeter, and horizontal situation indicator. Other avionics include IFF/SIF, Bendix/King 1400C weather/search radar, ground proximity warning system, cockpit voice recorder, emergency locator transmitter, and flight data recorder. Davtron 880 digital readout clock on pilot's and co-pilot's yoke in the cockpit.
EQUIPMENT: Loadmaster provided with crew oxygen regulator, quick don mask/goggles, portable oxygen bottle, emergency escape breathing device and life vest. Additional mission handling equipment includes loading winch, pry bar, anti-tilt jack, pallet extenders, and auxiliary ground loading ramps.
WEIGHTS AND LOADINGS:
Weight empty 16,103 kg (35,500 lb)
Max payload 6,740 kg (14,859 lb)
Max useful load 8,472 kg (18,678 lb)
Max fuel weight: JP4 9,348 kg (20,608 lb)
JP5 9,803 kg (21,612 lb)
Max T-O weight 25,800 kg (56,878 lb)
Max landing weight 24,200 kg (53,352 lb)
PERFORMANCE:
Max cruising speed 250 knots (463 km/h; 288 mph)
Airdrop speed 110-140 knots (204-259 km/h; 127-161 mph)
Service ceiling 6,706 m (22,000 ft)
T-O run (1,814 kg; 4,000 lb cargo, ISA + 15°C; 59°F) 457 m (1,500 ft)
Operating range 300 nm (555 km; 345 miles)
Ferry range 1,500 nm (2,779 km; 1,727 miles)

Second US Air Force C-27A Spartan transport version of the Alenia G222, fitted out by CTAS

CIRRUS
CIRRUS DESIGN CORPORATION
Baraboo-Dells Airport, S3440A Highway 12, Baraboo, Wisconsin 53913
Telephone: 1 (608) 356 2266
Fax: 1 (608) 356 2269
PRESIDENT: Alan Klapmeier

Company headquarters being moved from Wisconsin to Duluth International Airport, Minnesota, in mid-1993; plans to concentrate in future on fully certificated factory-built aircraft instead of homebuilt kits, starting with lightplane family based on VK30. First such product is ST-50.

CIRRUS DESIGN CIRRUS VK30
TYPE: Four/five-seat high-performance homebuilt.
PROGRAMME: First flown 11 February 1988; available in prefabricated kit form; 37 kits delivered by May 1993, of which six completed and flown, including Allison 250-B17C turboprop powered version.
COSTS: Airframe kit $56,500; engine kit $36,000 and turbocharged $56,000.
DESIGN FEATURES: Include natural laminar flow wing sections, large Fowler flaps, and a tail mounted pusher propeller. Current model concentrates on the IO-550/TSIO-550 engine. Wing sections based on NASA NLF(1)-0414F.
STRUCTURE: Wings constructed of glassfibre and polyurethane foam, with graphite spar caps. Fuselage has polyurethane foam core, graphite longerons and glassfibre skins. Ailerons and cruise flaps are of graphite/polyurethane/vinylester, and rudder/elevators/tailplane of Kevlar/polyurethane/vinylester.
LANDING GEAR: Retractable tricycle type.
POWER PLANT: One 224 kW (300 hp) Teledyne Continental IO-550-G or 261 kW (350 hp) TSIO-550-B flat-six engine, driving an MT MTV-9 three-blade pusher propeller; or 314 kW (420 shp) Allison 250-B17C turboprop. Fuel capacity 397 litres (105 US gallons; 87.4 Imp gallons).
DIMENSIONS, EXTERNAL:
Wing span 12.04 m (39 ft 6 in)
Wing aspect ratio 12.38
Length overall 7.92 m (26 ft 0 in)

Cirrus Design Corporation Cirrus VK30 four/five-seat homebuilt aircraft

Height overall 3.25 m (10 ft 8 in)
Propeller diameter 1.88 m (6 ft 2 in)
AREAS:
Wings, gross 11.71 m² (126.0 sq ft)
WEIGHTS AND LOADINGS:
Weight empty 1,089 kg (2,400 lb)
Max T-O weight 1,610 kg (3,550 lb)
Max wing loading 137.56 kg/m² (28.17 lb/sq ft)
PERFORMANCE (with TSIO-550-A engine):
Cruising speed, 75% power
more than 260 knots (483 km/h; 300 mph)
Stalling speed, flaps down, engine idling
56 knots (104 km/h; 65 mph)
Max rate of climb at S/L 503 m (1,650 ft)/min
T-O run 412 m (1,350 ft)
Landing run 328 m (1,075 ft)

Range with max fuel, 75% power, with reserves
1,346 nm (2,494 km; 1,550 miles)
g limits +5/−3

CIRRUS DESIGN ST-50
TYPE: Five-passenger business aircraft.
PROGRAMME: Announced 1993; part-funded by Duluth and Minnesota administrations; factory to be built at Duluth Airport and open by end of year; prototype to fly mid-1995; first deliveries 1996.
DESIGN FEATURES: All-composites, single-turboprop pusher design, slightly larger than VK30; pressurised cabin; cruising speed approx 270 knots (500 km/h; 311 mph). No other details known at time of going to press.

CLASSIC
CLASSIC AIRCRAFT CORPORATION
Capital City Airport, Kansing, Michigan 48906
Telephone: 1 (517) 321 7500
Fax: 1 (517) 321 5845
PRESIDENT: Richard S. Kettles
ENGINEERING MANAGER: Robert Edelstein
SALES MANAGER: Donald C. Kettles

CLASSIC WACO CLASSIC YMF SUPER
TYPE: Three-seat sport biplane.
PROGRAMME: Construction began March 1984 under type certificate of original Waco YMF-5; first flight of prototype (N1935B) 20 November 1985; FAA certification 11 March 1986.
CURRENT VERSIONS: **YMF Super:** Larger front cockpit than previously standard F-5 version (see 1992-93 *Jane's*) for commercial joyriding; enlarged forward door; front and rear cockpits 10 cm (4 in) longer; front cockpit 6.3 cm (2½ in) wider.
CUSTOMERS: Total 48 (including F-5s) delivered by 1 January 1993.
COSTS: Standard equipment aircraft $196,000.
DESIGN FEATURES: Wing section Clark Y; dihedral 2°, incidence 0° on upper and lower wings.
FLYING CONTROLS: Ailerons on upper and lower wings; no tabs. Tailplane incidence adjustable by screwjack actuator; ground adjustable trim tab on rudder; elevators.
STRUCTURE: Modern construction techniques, tolerances and materials applied to original design. N type interplane struts; streamlined stainless steel flying and landing wires; all-wood wing with Dacron covering; aluminium ailerons with external chordwise stiffening. Fuselage of 4130 welded steel tubes with internal oiling for corrosion protection; wooden bulkheads; Dacron covering. Braced welded steel tube tail surfaces with Dacron covering.
LANDING GEAR: Non-retractable tailwheel type. Shock absorption by oil and spring shock struts. Steerable tailwheel. Cleveland 30-67F hydraulic brakes on mainwheels only. Cleveland 40-101A mainwheels, tyre size 7.50-10; Cleveland 40-199A tailwheel, tyre size 3.50-4. Mainwheel fairings standard, tailwheel fairing optional. Float and amphibious landing gear optional.
POWER PLANT: One 205 kW (275 hp) Jacobs R-755-B2 air-cooled radial engine (remanufactured), driving a two-blade fixed-pitch wooden propeller. Constant-speed propeller with spinner optional. Engine enclosed with streamline aluminium 'bump' (helmeted) cowling. Fuel contained in two aluminium tanks in upper wing centre-section, total capacity 182 litres (48 US gallons; 40 Imp gallons). Refuelling point for each tank in upper wing surface. Auxiliary tanks, capacity 45 litres (12 US gallons; 10 Imp gallons) each, optional in either or both inboard upper wing panels. Standard oil capacity 15 litres (4 US gallons; 3.33 Imp gallons); with auxiliary fuel tanks 19 litres (5 US gallons; 4.2 Imp gallons).
ACCOMMODATION: Three seats in tandem open cockpits, two side by side in front position, single seat at rear. Dual controls, seat belts with shoulder harness, and pilot's adjustable seat, standard. Front baggage compartment, capacity 11.3 kg (25 lb); rear baggage compartment, volume 0.2 m³ (7.5 cu ft), capacity 34 kg (75 lb).
SYSTEMS: 24V electrical system with battery, alternator and starter for electrical supply to navigation, strobe and rear cockpit lights. Hydraulic system for brakes only.
AVIONICS: Emergency locator transmitter standard. VFR or IFR avionics packages to customer's choice including Bendix/King and Narco nav/com, ADF, DME, transponder and encoding altimeter; Apollo, Arnav, Foster, II Morrow and Northstar Loran C systems; 3M Stormscope WX-8 and 1000 series; Bendix/King KCS 55A slaved compass system with KN 72 VOR/LOC converter; NAT voice-activated intercom; Bendix/King KX 99 and TR 720 hand-held transceivers, and Astrotech LC-2 digital clock.
EQUIPMENT: Toe brakes standard in rear cockpit. Compass, airspeed indicator, turn and bank indicator, rate of climb indicator, sensitive altimeter, recording tachometer, cylinder head temperature gauge and oil pressure and oil temperature gauges standard in rear cockpit. Front cockpit instruments optional. Front and rear windscreens (front removable), front and rear cockpit covers, instrument post lighting, heated pitot, tiedown rings and three-colour paint scheme with choice of two designs, also standard. Optional equipment includes exhaust gas temperature gauge, carburettor temperature gauge, g meter, vacuum or electrically

Classic Waco Classic YMF Super re-creation of the Waco YMF-5 three-seat biplane (*Geoffrey P. Jones*)

driven gyro system, Hobbs meter, outside air temperature gauge, manifold gauge, oil cooler for wooden propeller, ground service plug, landing and taxi lights, front and rear cockpit heaters, flight-approved metal front cockpit cover, map case, glider tow hook, deluxe interior with carpet, leather sidewalls and interior trim, and special exterior paint designs.

DIMENSIONS, EXTERNAL:
Wing span:	
upper: F-5, YMF Super	9.14 m (30 ft 0 in)
lower: F-5, YMF Super	8.18 m (26 ft 10 in)
Length overall: F-5	7.10 m (23 ft 3⅜ in)
YMF Super	7.26 m (23 ft 10 in)
Height overall: F-5	2.57 m (8 ft 5⅜ in)
YMF Super	2.59 m (8 ft 6 in)
Wheelbase: F-5, YMF Super	1.95 m (6 ft 5 in)

Propeller diameter:
F-5, YMF Super   2.44 m (8 ft 0 in)
AREAS:
Wings, gross   21.69 m² (233.5 sq ft)
WEIGHTS AND LOADINGS:
Basic weight empty: F-5   880 kg (1,940 lb)
YMF Super   900 kg (1,985 lb)
Max T-O and landing weight: F-5   1,256 kg (2,770 lb)
YMF Super   1,338 kg (2,950 lb)
Max wing loading: F-5   52.26 kg/m² (10.71 lb/sq ft)
YMF Super   61.67 kg/m² (12.63 lb/sq ft)
Max power loading: F-5   6.21 kg/kW (10.20 lb/hp)
YMF Super   6.53 kg/kW (10.73 lb/hp)
PERFORMANCE (at max T-O weight):
Never-exceed speed (V_{NE}):
F-5, YMF Super   186 knots (344 km/h; 214 mph)

Max level speed at S/L:
F-5, YMF Super   117 knots (217 km/h; 135 mph)
Max cruising speed at S/L:
F-5, YMF Super   104 knots (193 km/h; 120 mph)
Econ cruising speed at 2,440 m (8,000 ft):
F-5, YMF Super   95 knots (177 km/h; 110 mph)
Stalling speed, power off:
F-5   51 knots (94 km/h; 58 mph)
YMF Super   53 knots (97 km/h; 60 mph)
Max rate of climb at S/L:
F-5, YMF Super   235 m (770 ft)/min
T-O run: F-5, YMF Super   152 m (500 ft)
Range, standard fuel, 30 min reserves:
F-5, YMF Super   286 nm (531 km; 330 miles)

COLEMILL
COLEMILL ENTERPRISES INC
PO Box 60627, Cornelia Fort Air Park, Nashville, Tennessee 37206
Telephone: 1 (615) 226 4256
Fax: 1 (615) 226 4702

PRESIDENT: Ernest Colbert

Colemill specialises in performance improvements for single and twin-engined aircraft. Currently offered are Panther Navajo (Piper PA31 and C/R Navajo), Panther II (Piper Chieftain), Panther III (Piper Chieftains), Bearcat (Cessna 310R), Executive 600 (Cessna 310 Models F-Q), President 600 and President II (Beechcraft A55 and B55 Baron), Foxstar Baron (Beechcraft C, D, E and 58 Baron) and Starfire (Beechcraft E33A, F33A, A36, S35 and V35 Bonanza). Details in 1991-92 *Jane's*; now transferred to *Jane's Civil and Military Aircraft Upgrades*.

COMMANDER
COMMANDER AIRCRAFT COMPANY
Wiley Post Airport, 7200 North-West 63rd Street, Bethany, Oklahoma 73008
Telephone: 1 (405) 495 8080
Fax: 1 (405) 495 8383
PRESIDENT: William Boettger
EXECUTIVE VICE-PRESIDENT: Gene Criss
VICE-PRESIDENTS:
Dick Smiley (Operations)
Steve Stoltenborg (Quality Assurance)
Mathew J. Goodman (Sales and Marketing)
Steve Buren (Finance)

Company acquired manufacturing, marketing and support rights for Rockwell Commander 112 and 114 from Gulfstream Aerospace Corporation Summer 1988; spares and support services for existing aircraft and manufacturing based in Oklahoma. A major share offering has been reported to finance expansion, including development of Commander 114B derivatives.

COMMANDER 114B
TYPE: New production version of four-seat, single-engined Commander 114A.
PROGRAMME: Commander 114B certificated 5 May 1992; 25 built in 1992.
CURRENT VERSIONS: **Commander 114B**: Basic version, *as detailed*.
Commander 114TC: Proposed version, to be powered by Textron Lycoming turbocharged engine, driving three-blade McCauley propeller; engineering began 1993.
CUSTOMERS: Factory direct sales for USA and Canada; international dealer network for foreign sales and support.
COSTS: Standard IFR-equipped aircraft $225,000.
DESIGN FEATURES: Improvements in 114B include new cowling design, improved cooling and induction, specially developed McCauley three-blade metal propeller, NACA scoop in dorsal fin, new interior trim and seats, better soundproofing, 28V electrical system, and custom metal instrument panel.
FLYING CONTROLS: Conventional; electric flap control; elevator and rudder trim; dual controls.
LANDING GEAR: Retractable tricycle type; nosewheel (steerable) tyre size 5.00 × 5 (six-ply); mainwheel tyres size 6.00 × 6 (six-ply); dual brakes.
POWER PLANT: One 194 kW (260 hp) Textron Lycoming IO-540-T4B5 flat-six engine, driving a McCauley B3D 32C 419/82NHA-5 three-blade constant-speed metal propeller. Fuel in two integral wing tanks, max capacity 265 litres (70 US gallons; 58.3 Imp gallons), of which 257 litres (68 US gallons; 56.6 Imp gallons) are usable. Max oil capacity 7.6 litres (2 US gallons; 1.67 Imp gallons).
SYSTEMS: 28V DC electrical; hydraulic; cabin heating and ventilation.
AVIONICS: Standard includes Bendix/King KMA 24-03 audio panel, KX 155 digital nav/com, KT 76A transponder, and Terra AT-3000 altitude encoder. Wide range of optional avionics and equipment.

DIMENSIONS, EXTERNAL:
Wing span	9.98 m (32 ft 9 in)
Wing aspect ratio	7.06
Length overall	7.59 m (24 ft 11 in)
Height overall	2.57 m (8 ft 5 in)
Tailplane span	4.10 m (13 ft 5½ in)
Wheel track	3.34 m (10 ft 11½ in)
Wheelbase	2.11 m (6 ft 11 in)
Propeller diameter	1.96 m (6 ft 5 in)
Propeller ground clearance	0.19 m (7½ in)

Commander Aircraft Commander 114B

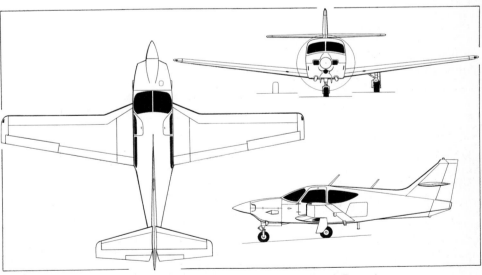
Commander 114B four-seat touring aircraft *(Dennis Punnett)*

DIMENSIONS, INTERNAL:
Cabin: Length	1.91 m (6 ft 3 in)
Max width	1.19 m (3 ft 11 in)
Max height	1.24 m (4 ft 1 in)
Volume	2.83 m³ (100.0 cu ft)
Baggage compartment volume	0.62 m³ (22.0 cu ft)

AREAS:
Wings, gross   14.12 m² (152.0 sq ft)
WEIGHTS AND LOADINGS:
Weight empty   927 kg (2,044 lb)
Baggage capacity   91 kg (200 lb)
Max T-O weight   1,474 kg (3,250 lb)
Max ramp weight   1,479 kg (3,260 lb)
Max wing loading   104.5 kg/m² (21.4 lb/sq ft)
Max power loading   7.62 kg/kW (12.54 lb/hp)
PERFORMANCE (at max T-O weight, ISA, except where indicated):
Max level speed   164 knots (304 km/h; 189 mph)
Cruising speed, 75% power
                                     160 knots (296 km/h; 184 mph)
Stalling speed:
flaps and wheels up   61 knots (113 km/h; 71 mph)
flaps and wheels down   56 knots (104 km/h; 65 mph)
Max rate of climb at S/L   326 m (1,070 ft)/min
Service ceiling   5,120 m (16,800 ft)
T-O run   317 m (1,040 ft)
T-O to 15 m (50 ft)   610 m (2,000 ft)
Landing from 15 m (50 ft)   366 m (1,200 ft)
Landing run   220 m (720 ft)
Range: at 75% power   630 nm (1,167 km; 725 miles)
at 65% power   725 nm (1,343 km; 835 miles)

COMTRAN

COMTRAN INTERNATIONAL INC
1770 Sky Place Boulevard, San Antonio, Texas 78216
Telephone: 1 (512) 821 6301
Fax: 1 (512) 822 7766
Telex: 767438 COMTRAN UD
PRESIDENT: Douglas Jaffe

Company changed name to above; subsidiary of the Jaffe Group, which also includes Jaffe Aircraft and Heli-Air (which see). Comtran Super Q conversion and hush-kitting of Boeing 707 (see 1991-92 *Jane's*) available, but not active; Comtran concentrating on head-of-state conversions of business transports.

DAYTONA

DAYTONA AIRCRAFT CONSTRUCTION INC
1400 Flightline Boulevard, DeLand, Florida 32724
Telephone: 1 (904) 738 7222
Fax: 1 (904) 738 9472

Company established 1990 and acquired design rights to Jamieson D-Series FAA-certificated light aircraft. Standard airframe sold with choice of eight Textron Lycoming engines, various propellers, choice of equipment and fixed or retractable landing gear.

DAYTONA D-SERIES

TYPE: Two/four-seat monoplane with choice of 89.5 to 223.9 kW (120 to 300 hp) engines.
PROGRAMME: Final development underway during 1991, but no subsequent news; full description in 1992-93 *Jane's*.
CURRENT VERSIONS: Engine and equipment choices as listed below, with type designations **D-120** to **D-300** denoting horsepower.
DESIGN FEATURES: Conventional airframe with structure stressed for aerobatics at gross weight of 1,633 kg (3,600 lb).
LANDING GEAR: Option of retractable tricycle gear on all but D-120; air-oleo shock absorbers; nosewheel retracts rearwards and mainwheels inwards with full cover doors; gear operation electric with mechanical standby; Cleveland mainwheels and brakes; all wheels have 6.00-6 tyres; dual pedal-operated brakes.
POWER PLANT: Choice of Textron Lycoming engines of 89.5 kW (120 hp), 119 kW (160 hp), 134 kW (180 hp), 149 kW (200 hp), 186 kW (250 hp), 194 kW (260 hp), 201 kW (270 hp) or 224 kW (300 hp) driving Hartzell or McCauley propellers. Two 117 litre (31 US gallon; 26 Imp gallon) fuel tanks in inner wing; two optional 57 litre (15 US gallon; 12 Imp gallon) tanks in outer wings in D-160 and above; overwing gravity refuelling.
ACCOMMODATION: Dual controls standard; all but D-120 have rear two-seat bench; cabin access via steps aft of wing and two gull-wing doors; air-conditioning optional with D-180 and standard with larger engines.

DIMENSIONS, EXTERNAL:
Wing span 10.97 m (36 ft 0 in)
Wing aspect ratio 7.96
Length overall 7.90 m (25 ft 11 in)
Height overall 2.33 m (7 ft 8 in)
Wheel track 3.36 m (11 ft 0¼ in)
Wheelbase 2.16 m (7 ft 1 in)
AREAS:
Wings, gross 15.13 m² (162.8 sq ft)

Daytona Aircraft Construction D-200 trainer/touring aircraft *(Geoffrey P. Jones)*

WEIGHTS AND LOADINGS:
Weight empty: D-200 567 kg (1,250 lb)
D-300 773 kg (1,704 lb)
Max T-O weight: D-200 1,134 kg (2,500 lb)
D-300 1,636 kg (3,607 lb)
Max wing loading: D-200 74.98 kg/m² (15.36 lb/sq ft)
D-300 108.18 kg/m² (22.16 lb/sq ft)
Max power loading: D-200 7.6 kg/kW (12.5 lb/hp)
D-300 7.3 kg/kW (12.0 lb/hp)
PERFORMANCE:
Max cruising speed:
D-200 175 knots (324 km/h; 201 mph)
D-300 180 knots (333 km/h; 207 mph)
Cruising speed, 75% power:
D-200 165 knots (305 km/h; 190 mph)
D-300 170 knots (315 km/h; 195 mph)
Stalling speed, flaps and gear down:
D-200 51 knots (95 km/h; 59 mph)
D-300 58 knots (107 km/h; 67 mph)
Max rate of climb: D-200 305 m (1,000 ft)/min
D-300 457 m (1,500 ft)/min
Service ceiling: D-200, D-300 5,486 m (18,000 ft)
T-O to 15 m (50 ft): D-200 402 m (1,319 ft)
D-300 366 m (1,200 ft)
Landing from 15 m (50 ft): D-200 372 m (1,220 ft)
D-300 335 m (1,100 ft)
Range, standard fuel, no reserves:
D-200 1,275 nm (2,360 km; 1,466 miles)
D-300 702 nm (1,300 km; 807 miles)
Endurance with optional fuel: D-200 7 h 44 min
D-300 6 h 7 min
g limits: D-200 and D-300 +6/−3

DEE HOWARD

THE DEE HOWARD COMPANY
9610 John Saunders Road, International Airport, PO Box 17300, San Antonio, Texas 78217
Telephone: 1 (512) 828 1341
Telex: 767380

CHAIRMAN: Dee Howard
PRESIDENT AND CEO: Philip Greco
VICE-PRESIDENT, ENGINEERING: David White
VICE-PRESIDENT, MARKETING AND SALES: Roger Munt
DIRECTOR OF MARKETING COMMUNICATIONS: Brian Loflin

Alenia (see Italian section) took initial 60 per cent holding in Dee Howard in Spring 1989.

Company has developed performance improvement package for Learjet 24 and 25; updating United Parcel Service Douglas DC-8s and Boeing 727-100s with Collins EFIS as part of cockpit modernisation programme; 727s also being retrofitted with Rolls-Royce Tay 650 turbofans.

Details given in 1991-92 *Jane's*; now transferred to *Jane's Civil and Military Aircraft Upgrades*.

ENSTROM

THE ENSTROM HELICOPTER CORPORATION
PO Box 490, 2209 North 22nd Street, Twin County Airport, Menominee, Michigan 49858-0490
Telephone: 1 (906) 863 9971
Fax: 1 (906) 863 6821
PRESIDENT AND CEO: Robert M. Tuttle
VICE-PRESIDENT, MANUFACTURING: John E. Hansen
VICE-PRESIDENT, ENGINEERING: Robert L. Jenny
DIRECTOR OF MARKETING: Bill May

Company history since foundation in 1959 detailed in earlier *Jane's*. Acquired by Bravo Investments BVC, Netherlands, January 1980; acquired by group of American investors headed by Mr Dean Kamen and Mr Robert Tuttle September 1984; in early 1990 acquired by investors based in Los Angeles. Total of more than 900 helicopters produced by 1 January 1993.

ENSTROM F28 and 280

TYPE: Three-seat light helicopter.
PROGRAMME: Basic F-28A and 280 described in 1978-79 *Jane's*; replaced by turbocharged F28C and 280C, certificated by FAA 8 December 1975 and last described in 1984-85 *Jane's*; production of these models ceased November 1981; succeeded by F28F and 280F Shark, described in 1985-86 *Jane's*, and 280FX; current models detailed on next page.

Enstrom F28F three-seat light helicopter of the Peruvian Army

CURRENT VERSIONS: **F28F Falcon:** Basic model certificated by FAA January 1981. Recent developments include redesigned main gearbox with heavy wall main rotor shaft (standard on all new F28s and retrofittable to all existing F models); optional lightweight exhaust silencer, reducing noise in hover by 40 per cent and by 30 per cent when flying at 152 m (500 ft) (can be retrofitted to existing F28F, 280F and 280FX); lightweight high efficiency starter motor recently introduced for all models.

Enstrom wet and dry agricultural kit comprises two side-mounted hoppers with quick-fill openings, total capacity 303 litres (80 US gallons; 67 Imp gallons); spraybar 9.04 m (29 ft 8 in) wide, extendable to 11.07 m (36 ft 4 in); manually operated clutch gives positive control of centrifugal pump with capacity for 227 litres (60 US gallons; 50 Imp gallons)/min; dry discharge rate variable from 0 to 272 kg (600 lb)/min; weight of entire quickly removable dispersal system 48 kg (105 lb).

F28F-P Sentinel: Dedicated police patrol version; first delivery October 1986; can be fitted with Locator B, Spectrolabe SX-5 or Carter searchlight and specialised police radio; same specifications and performance as F28F; can also be fitted with FLIR system.

280FX Shark: Certificated to FAA CAR Pt 6 on 14 January 1985. Features include new seats with lumbar support and energy-absorbing foam, new tailplane with endplate fins, tail rotor guard, covered tail rotor shaft, redesigned air inlet system, and completely faired landing gear; optional pneumatic door opener; optional internal tank extends range to 339 nm (627 km; 390 miles).

CUSTOMERS: Chilean Army operates 15 280FXs for primary and instrument training; Peruvian Army has 10 F28Fs for training duties; more than 170 F28F and 280FX helicopters in service worldwide.

DESIGN FEATURES: Three-blade fully articulated head with blades attached by retention pin and drag link; control rods pass inside tubular rotor shaft to swashplate inside fuselage; blade section NACA 0013.5; two-blade teetering tail rotor; blades do not fold. Poly V-belt drive system from horizontally mounted engine to transmission also acts as clutch.

STRUCTURE: Bonded light alloy blades. Fuselage has glassfibre and light alloy cabin section, steel tube centre section frame, and stressed skin aluminium tailboom.

LANDING GEAR: Skids carried on Enstrom oleo-pneumatic shock absorbers. Air Cruiser inflatable floats available optionally.

POWER PLANT: One 168 kW (225 hp) Textron Lycoming HIO-360-F1AD flat-four engine with Rotomaster 3BT5EE10J2 turbocharger. Two fuel tanks, each of 79.5 litres (21 US gallons; 17.5 Imp gallons). Total standard fuel capacity 159 litres (42 US gallons; 35 Imp gallons), of which 151 litres (40 US gallons; 33.3 Imp gallons) are usable. Auxiliary tank, capacity 49 litres (13 US gallons; 10.8 Imp gallons), can be installed in the baggage compartment. Oil capacity 9.5 litres (2.5 US gallons; 2.1 Imp gallons).

ACCOMMODATION: Pilot and two passengers, side by side on bench seat; centre place removable. Fully transparent removable door on each side of cabin. Baggage space aft of engine compartment, capacity 49 kg (108 lb), with external door. Cabin heated and ventilated.

SYSTEMS: Electrical power provided by 12V 70A engine driven alternator; 24V 70A system optional on F28F, standard on 280FX.

AVIONICS: Variety of fits from AR Nav, Bendix/King II Morrow and Northstar.

EQUIPMENT: Standard equipment includes airspeed indicator, sensitive altimeter, compass, outside air temperature gauge, turn and bank indicator, rotor/engine tachometer, manifold pressure/fuel flow gauge, EGT gauge, oil pressure gauge, gearbox and oil temperature gauge, ammeter, cylinder head temperature gauge, and fuel quantity gauge. Eight-light annunciator panel consisting of low rotor rpm, chip detectors (main and tail rotor transmissions), overboost, clutch not fully engaged, low fuel pressure, starter, and low voltage warning lights. Also standard are ground handling wheels with handle, kick-in service steps, floor carpeting, lap belts for all seats, shoulder harnesses for two seats, instrument lighting with dimmer control, position light on each horizontal stabiliser tip, anti-collision strobe light, landing lights, adjustable nose light, soundproofing, main and tail rotor covers and blade tiedowns. Additionally, 280FX has as standard a graphic engine monitor and custom seating.

Optional equipment includes dual controls, cabin heater, fire extinguisher, first aid kit, custom interior and custom paint scheme, floor switch, external power receptacle, third shoulder harness, auxiliary fuel tank, sliding vent windows, eight-day clock, baggage compartment, floats, cargo hook, hardpoints for agricultural equipment or night sign, and wet or wet/dry agricultural dispersal systems. Optional instrumentation includes R. C. Allen attitude gyro and turn co-ordinator, directional gyro, and instantaneous vertical speed indicator. Bose noise-attenuating headsets are optional on all models.

DIMENSIONS, EXTERNAL:
Main rotor diameter	9.75 m (32 ft 0 in)
Tail rotor diameter	1.42 m (4 ft 8 in)
Distance between rotor centres	5.56 m (18 ft 3 in)
Main rotor blade chord	0.24 m (9½ in)

Enstrom 280FX Shark three-seat light helicopter

Enstrom 480/TH-28 four-seat turbine-powered light helicopter

Tail rotor blade chord	0.11 m (4½ in)
Length overall, rotors stationary	8.92 m (29 ft 3 in)
Height to top of rotor head	2.79 m (9 ft 2 in)
Skid track	2.21 m (7 ft 3 in)
Cabin doors (each): Height	1.04 m (3 ft 5 in)
Width	0.84 m (2 ft 9 in)
Height to sill	0.64 m (2 ft 1 in)
Baggage door: Height	0.55 m (1 ft 9½ in)
Width	0.39 m (1 ft 3½ in)
Height to sill	0.86 m (2 ft 10 in)

DIMENSIONS, INTERNAL:
Cabin: Max width: F28F	1.55 m (5 ft 1 in)
280FX	1.50 m (4 ft 11 in)
Baggage compartment volume	0.18 m³ (6.3 cu ft)

AREAS:
Main rotor disc	74.69 m² (804.0 sq ft)
Tail rotor disc	1.66 m² (17.88 sq ft)

WEIGHTS AND LOADINGS (F28F Normal category):
Weight empty, equipped: F28F	712 kg (1,570 lb)
280FX	719 kg (1,585 lb)
Max T-O weight: F28F, 280FX	1,179 kg (2,600 lb)
Max disc loading: F28F, 280FX	15.77 kg/m² (3.23 lb/sq ft)

PERFORMANCE (both versions at AUW of 1,066 kg; 2,350 lb, except where indicated):
Never-exceed speed (V_NE):	
F28F	97 knots (180 km/h; 112 mph)
280FX	102 knots (189 km/h; 117 mph)
Max level speed, S/L to 915 m (3,000 ft):	
F28F	97 knots (180 km/h; 112 mph) IAS
280FX	102 knots (189 km/h; 117 mph) IAS
Max cruising speed:	
F28F	97 knots (180 km/h; 112 mph)
280FX	102 knots (189 km/h; 117 mph)
Econ cruising speed:	
F28F	89 knots (165 km/h; 102 mph)
280FX	93 knots (172 km/h; 107 mph)
Max rate of climb at S/L	442 m (1,450 ft)/min
Certificated operating ceiling	3,660 m (12,000 ft)
Hovering ceiling:	
IGE at AUW of 1,179 kg (2,600 lb)	2,345 m (7,700 ft)
OGE at AUW of 1,066 kg (2,350 lb)	2,650 m (8,700 ft)
Max range, standard fuel, no reserves:	
F28F	228 nm (423 km; 263 miles)
280FX	260 nm (483 km; 300 miles)
Max endurance	3 h 30 min

ENSTROM 480

Military designation: TH-28

TYPE: Four-seat turbine-powered helicopter.

PROGRAMME: Proof-of-concept 280FX, powered by Allison 250 turboshaft, flown December 1988; first flight of definitive wide-cabin 480/TH-28 four-seat prototype (N8631E) in October 1989; second 480/TH-28 (production prototype) flew December 1991; TH-28 model FAA certificated September 1992; four helicopters accumulated over 1,500 flight hours in test programme; FAA certification of 480 anticipated for second half 1993.

CURRENT VERSIONS: **480:** Basic civil version, with four staggered seats or convertible to three-seat training/executive layout; smaller instrument console than TH-28. *Description applies to this version.*

TH-28: Military training/light patrol version; equipped with three crashworthy seats and crashworthy fuel system; capable of training two students simultaneously. Two alternative avionics configurations, one for VFR and one for IFR training. Was competitor in US Army NTH pilot training helicopter programme, but not selected.

CUSTOMERS: First batch of 40 Enstrom 480s sold; order book open for second batch.

COSTS: Production aircraft estimated at less than $725,000 (1993).

DESIGN FEATURES: Three-blade main rotor and dynamic system as for 280FX; 313 kW (420 shp) Allison 250-C20W turboshaft, derated to 212.5 kW (285 shp) for take-off and 191 kW (256 shp) max continuous. Extensive crashworthiness features. Standard engine air inlet particle separator. New widened cabin able to seat three abreast behind pilot in single front seat. Civil cabin layout can be quickly rearranged (see Current Versions).

FLYING CONTROLS: As for 280FX.

DIMENSIONS, EXTERNAL: As for 280FX except:
Distance between rotor centres	5.64 m (18 ft 6 in)

474 USA: AIRCRAFT—ENSTROM/EXPRESS DESIGN

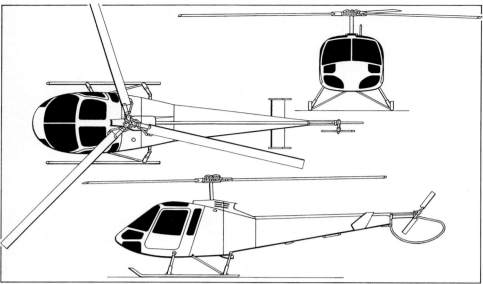

Enstrom 480 light helicopter (Allison 250-C20W turboshaft) *(Jane's/Mike Keep)*

Height to top of rotor head	2.92 m (9 ft 7 in)
Skid track	2.44 m (8 ft 0 in)
DIMENSIONS, INTERNAL:	
Cabin: Max width	1.78 m (5 ft 10 in)
WEIGHTS AND LOADINGS:	
Weight empty	780 kg (1,720 lb)
Max T-O weight	1,292 kg (2,850 lb)
PERFORMANCE (at max T-O weight, ISA, except where indicated):	
Never-exceed speed (V_{NE})	121 knots (225 km/h; 140 mph)
Cruising speed: at AUW of 1,293 kg (2,850 lb)	96 knots (177 km/h; 110 mph)
at AUW of 1,134 kg (2,500 lb)	105 knots (195 km/h; 121 mph)
Max rate of climb at S/L:	
at AUW of 1,293 kg (2,850 lb)	442 m (1,450 ft)/min
at AUW of 1,134 kg (2,500 lb)	482 m (1,583 ft)/min
Service ceiling	3,965 m (13,000 ft)
Hovering ceiling:	
IGE:	
at AUW of 1,293 kg (2,850 lb)	3,050 m (10,000 ft)
at AUW of 1,134 kg (2,500 lb)	4,265 m (14,000 ft)
OGE:	
at AUW of 1,293 kg (2,850 lb)	2,105 m (6,900 ft)
at AUW of 1,134 kg (2,500 lb)	3,660 m (12,000 ft)
Max range	334 nm (620 km; 385 miles)
Endurance	4 h 42 min

EXCALIBUR
EXCALIBUR AVIATION COMPANY
8337 Mission Road, San Antonio, Texas 78214
Telephone: 1 (512) 927 6201
Fax: 1 (512) 927 6287
PRESIDENT: Michael M. Davis

Beechcraft Queen Air 65, A65 and 80 airframes are modified by Excalibur as Queenaire 800 and Queen Air A80s; B80 as Queenaire 8800. Conversion offers improved reliability, speed, range and reduced operating costs. Details in 1991-92 *Jane's*; now transferred to *Jane's Civil and Military Aircraft Upgrades*.

EXPERIMENTAL
EXPERIMENTAL AVIATION
3021 Airport Avenue, Suite 109, Santa Monica, California 90405
Telephone: 1 (310) 391 1943
Fax: 1 (818) 713 8650
CHIEF DESIGNER: Dave Ronneberg

EXPERIMENTAL AVIATION BERKUT
TYPE: Tandem two-seat monoplane.
PROGRAMME: Named after Asian eagle; first flown 11 July 1991.
COSTS: Kit: available, incorporating 27 moulded components and materials.
DESIGN FEATURES: Rear mounted sweptback wings, with winglets. Straight foreplane. Resembles Rutan Long-EZ, of which Dave Ronneberg built seven examples, but features larger cabin and wing strakes with upper camber lift. Intended first for sport and recreational flying; seen to have business applications as well as military. Construction takes about 1,500-2,000 working hours.
FLYING CONTROLS: Ailerons, elevators and rudders; airbrake.
STRUCTURE: Hot wire cut extruded polystyrene foam core, GFRP and CFRP. Rudder cut from GFRP winglets.
LANDING GEAR: Electrohydraulically retractable tricycle landing gear.
POWER PLANT: One 153 kW (205 hp) Textron Lycoming IO-360-B1A flat-four engine, driving a Light Speed Engineering Black Bart pusher propeller. Usable fuel capacity 178 litres (47 US gallons; 39 Imp gallons).
DIMENSIONS, EXTERNAL:
Wing span	8.13 m (26 ft 8 in)
Length overall	5.64 m (18 ft 6 in)
Height overall	2.29 m (7 ft 6 in)
Propeller diameter	1.70 m (5 ft 7 in)

Experimental Aviation Berkut two-seat monoplane *(Craig Schmitman)*

WEIGHTS AND LOADINGS:
Weight empty	469 kg (1,035 lb)
Max T-O weight	907 kg (2,000 lb)
Max power loading	5.93 kg/kW (9.76 lb/hp)

PERFORMANCE:
Max level speed at 1,830 m (6,000 ft)	215 knots (399 km/h; 248 mph)
Cruising speed at 2,440 m (8,000 ft)	208 knots (385 km/h; 239 mph)
Econ cruising speed at 2,440 m (8,000 ft)	187 knots (346 km/h; 215 mph)
Min control speed	58 knots (108 km/h; 67 mph)
Max rate of climb at S/L	610 m (2,000 ft)/min
Service ceiling	9,750 m (32,000 ft)
Range: cruising speed	955 nm (1,770 km; 1,100 miles)
econ cruising speed	1,141 nm (2,114 km; 1,314 miles)

EXPRESS DESIGN
EXPRESS DESIGN INC
Oregon
PRESIDENT: David Ullrich

EDI has acquired assets of former Wheeler Technology Inc (1992-93 *Jane's*) which became bankrupt in late 1990; has resumed production of, and customer support for, former Wheeler Express homebuilt.

EDI (WHEELER) EXPRESS
TYPE: Four-seat, dual-control cross-country homebuilt; conforms to FAR Pt 23.
PROGRAMME: CAD designed by Wheeler Technology Inc as high speed cross-country kitplane, with unusual seating arrangement of one forward and one aft facing seat in rear, behind two front seats with dual controls. Three prototypes built from kits of pre-moulded parts; first flight 28 July 1987. First, larger, production line Express demonstrator

First production-line demonstrator EDI (Wheeler) Express four-seat homebuilt *(Geoffrey P. Jones)*

first flew May 1990, powered by Teledyne Continental engine. Delivered kits incomplete at time of Wheeler bankruptcy; deficiencies made good by EDI, and at least six aircraft now completed. Complete kits now available from Express Design incorporate some modifications, which can also be retrofitted to existing aircraft.

Following details apply to unmodified original Wheeler version:
DESIGN FEATURES: Wing section NASA NFL-1 0215-F (laminar flow).
STRUCTURE: Constructed of composites sandwich material, comprising polyurethane foam core, glassfibre, unidirectional glassfibre tape and vinylester resin.
LANDING GEAR: Non-retractable tricycle type standard. Retractable gear optional.
POWER PLANT: Demonstrator fitted with 156.6 kW (210 hp) Teledyne Continental IO-360-ES1 flat-four engine; 119 or 134 kW (160 or 180 hp) Textron Lycoming engine optional. Fuel capacity 204 litres (54 US gallons; 45 Imp gallons); optional 348 litres (92 US gallons; 76.6 Imp gallons).

DIMENSIONS, EXTERNAL:
Wing span 9.45 m (31 ft 0 in)
Wing aspect ratio 7.63
Length overall 7.62 m (25 ft 0 in)
Height overall 2.13 m (7 ft 0 in)
AREAS:
Wings, gross 11.71 m² (126.0 sq ft)
WEIGHTS AND LOADINGS:
Weight empty 703 kg (1,550 lb)
Baggage capacity, with four 77 kg (170 lb) persons 29 kg (64 lb)
Max T-O weight 1,313 kg (2,895 lb)
Max wing loading 112.18 kg/m² (22.98 lb/sq ft)

PERFORMANCE (149 kW; 200 hp engine, retractable landing gear):
Max level speed at S/L 204 knots (378 km/h; 235 mph)
Max cruising speed at S/L, 75% power and with 45 min reserves 183 knots (338 km/h; 210 mph)
Stalling speed at max T-O weight:
 flaps up 55 knots (101 km/h; 63 mph)
 flaps down 50 knots (92 km/h; 57 mph)
Max rate of climb at S/L 427 m (1,400 ft)/min
Service ceiling 6,100 m (20,000 ft)
Landing run 244 m (800 ft)
Range, 55% power, no reserves
 1,042 nm (1,931 km; 1,200 miles)
g limits +4.4/−2.2
 +8.8/−4.4 ultimate

FAIRCHILD
FAIRCHILD AIRCRAFT INCORPORATED
PO Box 790490, San Antonio, Texas 78279-0490
Telephone: 1 (512) 824 9421
Fax: 1 (512) 820 8656
CHAIRMAN, CEO AND PRESIDENT: Carl A. Albert
EXECUTIVE VICE-PRESIDENT AND CHIEF MARKETING OFFICER:
 D. A. 'Chip' Cipolla
PRESIDENT, GOVERNMENT PROGRAMMES: Hector Cuellar
PRESIDENT, FAIRCHILD AIRCRAFT SERVICES: Ron Stotz
SENIOR VICE-PRESIDENT, ENGINEERING AND PRODUCT DEVELOPMENT: Ronald D. Neal
SENIOR VICE-PRESIDENT, COMMERCIAL MARKETING:
 Chester J. Schickling
SENIOR VICE-PRESIDENTS, SALES AND MARKETING:
 Randolph R. Becker
 Calvin Humphrey
VICE-PRESIDENT, AIRLINE SALES: David E. Norgart
VICE-PRESIDENT, AIRLINE SALES, LATIN AMERICA:
 Juan Gonzalez
VICE-PRESIDENT, CORPORATE COMMUNICATIONS:
 Mark A. Morro

Fairchild Aircraft produces Metroliner series of 19-passenger regional airliners operated by commuter/regional airlines worldwide. Also produces C-26B special mission aircraft; Expediter 23 for cargo/express package operations; and Merlin 23 business aircraft. Workforce 1,250 in December 1992.

FAIRCHILD SA227-CC and -DC METRO 23, MERLIN 23 and EXPEDITER I and 23
Swedish Air Force designation: Tp 88
US military designations: C-26A/B
TYPE: Twin-turboprop 19/20-passenger commuter airliner.
PROGRAMME: Metro III originally SFAR 41 approved; FAR 23 commuter type approval June 1990 for SA227-CC with TPE331-11U-612G engines and SA227-DC with TPE331-12UA-UAR-701Gs. British CAA certification of SA227-AC August 1988; required modifications included dual-redundant stall warning system, dual continuous water/alcohol injection system, modified aileron aerofoil section, externally operable escape hatches. FAR 23 commuter category requirements include all significant CAA certification modifications as required on SA227-AC aircraft.
CURRENT VERSIONS: **SA227-CC and -DC Metro 23:** Current high gross weight aircraft. *Detailed description applies to these versions, except where indicated.*
 Tp 88: Swedish Air Force VIP transport; one delivered.
 C-26A and C-26B: Six C-26As ordered March 1988 as US Air National Guard Operational Support Transport Aircraft (ANGOSTA), later increased to 13, delivered between March 1989 and August 1990; these have quick-change passenger, medevac or cargo interiors. Contract

Fairchild Metro 23 of Air New Zealand link

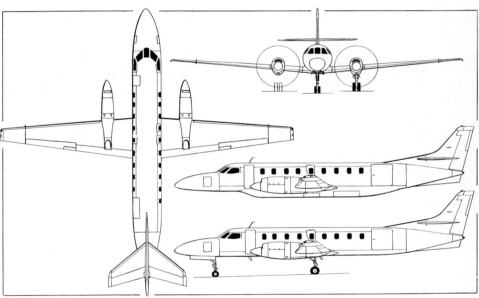
Fairchild Metro 23 commuter airliner with added side view (centre) of Metro 23 EF *(Dennis Punnett)*

Fairchild C-26 (two Garrett TPE331-12UA-UAR-701G turboprops)

476   USA: AIRCRAFT—FAIRCHILD

Fairchild Expediter I with blanked cabin windows, operated on behalf of United Parcel Service

Fairchild Metro 23 cockpit

awarded January 1991 by USAF Aeronautical Systems Division for delivery/logistics support of up to 53 C-26Bs (30 firm, options for 23) over five-year period from January 1992. First 23 C-26Bs delivered to Air and Army National Guard during 1992; first handed over on 2 January and delivered to 128th TFW at Truax Field, Wisconsin, 12 January; first military aircraft qualified for TCAS II and GPS.

Expediter I: All-cargo version; air-conditioning ducts moved to increase cargo volume, reinforced cabin floor, cargo nets and guards; reduced empty weight allowing max payload of more than 2,268 kg (5,000 lb); first 10 to operator SAT-AIR (now Merlin Express) on behalf of United Parcel Service; first of 10 Expediter Is delivered to DHL Worldwide Courier Express April 1985 with structurally reinforced landing gear and wing main spar for max T-O weight 7,257 kg (16,000 lb).

Expediter 23: Increased max T-O weight 7,484 kg (16,500 lb); max payload capability about 2,495 kg (5,500 lb).

Merlin 23: Business aircraft version of Metro 23; 12 and 14 passenger interiors; certificated in FAR 23 commuter category.

CUSTOMERS: More than 950 Metro/Merlin/Expediter/Special Mission aircraft operating in 40 countries; among recent orders were two Metro 23s for Australian regional airline Kendell Airlines. See Current Versions for military operators.

DESIGN FEATURES: Certification to FAR 23 commuter category (Amendment 34) covers ICAO Annex 8; Metro 23 has max T-O weight of 7,484 kg (16,500 lb); changes from Metro III include increased max T-O weight for increased payload, numerous minor changes arising from change from SFAR-41 to FAR 23 commuter category certification basis, continuous alcohol water injection system components relocated from nose bay to wing centre-section and stall avoidance system dual redundant. Certificated for flight into known icing; lightning strike protection claimed to be equal to that of latest commercial jet transports.

Wing section NACA 65_A215 at root, 64_A415 at tip; dihedral 5°; incidence 1° at root, −2° 36′ at tip.

FLYING CONTROLS: Mechanically operated, aerodynamically and mass balanced; manual trim tab in rudder and each aileron; electrically operated variable incidence tailplane; small ventral fin. Hydraulically operated double-slotted trailing-edge flaps.

STRUCTURE: Two-spar fail-safe wing made in one piece; main spar beams have laminated caps (titanium laminations in centre-section); pressurised cylindrical fuselage of 2024 aluminium alloy, flush riveted; glassfibre honeycomb nose cap can contain 0.46 m (18 in) diameter weather radar antenna.

LANDING GEAR: Retractable tricycle type with twin wheels on each unit. Hydraulic retraction, with dual actuators on each unit. All wheels retract forward, main gear into engine nacelles and nosewheels into fuselage. Oleo-pneumatic shock absorber struts. Nosewheel steerable variable authority (±63° max) through rudder pedals or tiller. Free-fall emergency extension, with backup of hand operated hydraulic pump. 18 × 5.5 type VII mainwheels with tubeless tyres, size 19.5 × 6.75-10 ply; nosewheels and low-pressure tubeless tyres, size 18 × 4.40-10 ply, type VII. Tyre pressures at standard T-O weight: nosewheel 4.83 bars (70 lb/sq in), mainwheels 7.31 bars (106 lb/sq in). Aircraft Braking Systems (ABS) self-adjusting hydraulically operated dual rotor disc brakes.

POWER PLANT: Metro 23 (SA227-CC) has two 746 kW (1,000 shp) Garrett TPE331-11U-612G turboprops; Metro 23 (SA227-DC) and C-26 have TPE331-12UA-UAR-701Gs, giving 820 kW (1,100 shp) dry and 820 kW (1,100 shp) with continuous alcohol/water injection system and 746 kW (1,000 shp) max continuous power. McCauley four-blade constant-speed fully feathering reversible-pitch metal propellers; automatic propeller synchronising standard; in-flight windmill start capability. Integral fuel tank in each wing, each with a usable capacity of 1,226 litres (324 US gallons; 270 Imp gallons). Total usable fuel capacity 2,452 litres (648 US gallons; 540 Imp gallons). Refuelling point in each outer wing panel. Automatic fuel heating. Oil capacity 15.1 litres (4 US gallons; 3.3 Imp gallons). Alcohol/water tank in wing centre-section, capacity 53 litres (14 US gallons; 11.7 Imp gallons), with two pumps to pump fluid to engines. Flush mounted fuel vents. Single point rapid defuelling provisions. Negative torque sensing, single red line/autostart, automatic engine temperature limiting and engine fire extinguishing systems.

ACCOMMODATION: Crew of two on flight deck (Metro 23 single pilot approved). Dual controls standard. Bulkhead between cabin and flight deck optional. Standard accommodation for 19-20 passengers seated two-abreast, on each side of centre aisle. High-back, tracking, quickly removable passenger seats standard. Interior convertible to all-cargo or mixed passenger/cargo configuration with movable bulkhead between passenger and cargo sections. Snap-in carpeting. Tiedown fittings for cargo at 0.76 m (30 in) spacing. Integral-step passenger door on port side of fuselage, immediately aft of flight deck. Large cargo loading door on port side of fuselage at rear of cabin, hinged at top. Three window emergency exits, one on the port, two on the starboard side. Forward baggage/avionics compartment in nose, capacity 363 kg (800 lb). Pressurised rear cargo compartment, capacity 385 kg (850 lb). Cabin air-conditioned and pressurised. Electric windscreen de-icing. Two-speed windscreen wipers.

SYSTEMS: AiResearch automatic cabin pressure control system: max differential 0.48 bar (7.0 lb/sq in), providing a sea level cabin altitude to 5,120 m (16,800 ft). Engine bleed air heating, dual air cycle cooling system, with automatic temperature control. Air blower system for on-ground ventilation. Vapour cycle (CFC free) cooling system may be fitted for ground and in-flight operation. Independent hydraulic system for brakes. Dual engine driven hydraulic pumps, using fire resistant MIL-H-83282 hydraulic fluid, provide 138 bars (2,000 lb/sq in) to operate flaps, landing gear actuators and nosewheel steering. Hydraulic system flow rates 30.3 litres (8 US gallons; 6.7 Imp gallons)/min at idle power, both engines; 46.7 litres (12.34 US gallons; 10.27 Imp gallons)/min at T-O and climb power. Air/oil reservoir, pressure 2.27 bars (33 lb/sq in). Electrical system supplied by two 300A 28V DC starter/generators. Fail-safe system with overload and overvoltage protection. Redundant circuits for essential systems. Two 350VA solid state inverters supply 115V and 26V AC. Two 24V 23Ah nickel-cadmium batteries for main services. Engine fire detection system and fire extinguishing system standard. Wing overheat detection system. Oxygen system of 1.39 m^3 (49 cu ft) capacity with flush outlets at each seat; system with capacity of 5.04 m^3 (178 cu ft) optional. Redundant stall avoidance system comprising angle indicator, visual and aural warning. Goodrich automatic, bleed air operated pneumatic de-icing boots on wing and fin; tailplane equipped with abrasion boot. Engine inlet de-icing by bleed air. Electric oil cooler inlet anti-icing. Electric propeller de-icing.

AVIONICS: All commercial avionics systems (Collins, Bendix/King etc) fitted as desired; special/military equipment will be installed on request. Autopilots, global navigation and global positioning, weather/mapping/tracking radars available. Provision for installation of remotely mounted or panel mounted avionics.

EQUIPMENT: Standard equipment includes flight data recorder and cockpit voice recorder; pilot and co-pilot foot warmers; edge lit consoles, pedestal and switch panels; integrally lit instruments; annunciator panel with 48 indicators; internally operated control locks, individual reading lights and air vents for each passenger; heated pitot; heated static sources; baggage compartment, cargo compartment, entrance, map and instrument panel, ice inspection, retractable landing, navigation, rotating beacon and taxi lights; automatic engine start cycle; external power socket; and static wicks.

DIMENSIONS, EXTERNAL:

Wing span	17.37 m (57 ft 0 in)
Wing mean aerodynamic chord	1.84 m (6 ft 0⅓ in)
Wing aspect ratio	10.5
Length overall	18.09 m (59 ft 4¼ in)
Height overall	5.08 m (16 ft 8 in)
Tailplane span	4.86 m (15 ft 11½ in)
Wheel track	4.57 m (15 ft 0 in)
Wheelbase	5.83 m (19 ft 1½ in)
Propeller diameter	2.69 m (8 ft 10 in)
Passenger door (fwd): Height	1.35 m (4 ft 5 in)
Width	0.64 m (2 ft 1 in)
Cargo door (rear): Height	1.30 m (4 ft 3¼ in)
Width	1.35 m (4 ft 5 in)
Height to sill	1.30 m (4 ft 3¼ in)
Forward baggage doors (two, each):	
Height	0.64 m (2 ft 1 in)
Width	0.46 m (1 ft 6 in)
Emergency exits (three, each):	
Height	0.71 m (2 ft 4 in)
Width	0.51 m (1 ft 8 in)

DIMENSIONS, INTERNAL:
Cabin, excl flight deck and rear cargo compartment:
  Length   7.75 m (25 ft 5 in)
  Max width   1.57 m (5 ft 2 in)
  Max height (aisle)   1.45 m (4 ft 9 in)
  Floor area   13.01 m² (140.0 sq ft)
  Volume   13.88 m³ (490.0 cu ft)
Rear cargo compartment (pressurised):
  Length   2.34 m (7 ft 8 in)
  Max width   1.57 m (5 ft 2 in)
  Max height   1.45 m (4 ft 9 in)
  Volume   4.06 m³ (143.5 cu ft)
Nose cargo compartment (unpressurised):
  Length   0.91 m (3 ft 0 in)
  Volume   0.85 m³ (30.0 cu ft)
AREAS:
  Wings, gross   28.71 m² (309.0 sq ft)
  Ailerons (total)   1.31 m² (14.12 sq ft)
  Trailing-edge flaps (total) 3.78 m² (40.66 sq ft)
  Fin, incl dorsal fin   3.40 m² (36.62 sq ft)
  Rudder, incl tab   1.80 m² (19.38 sq ft)
  Tailplane   5.08 m² (54.7 sq ft)
  Elevators   1.98 m² (21.27 sq ft)
WEIGHTS AND LOADINGS (Metro 23):
  Operating weight empty   4,309 kg (9,500 lb)
  Max fuel weight   1,969 kg (4,342 lb)
  Max T-O weight   7,484 kg (16,500 lb)
  Max ramp weight   7,530 kg (16,600 lb)
  Max zero-fuel weight   6,577 kg (14,500 lb)
  Max landing weight   7,110 kg (15,675 lb)
  Max wing loading   261 kg/m² (53.3 lb/sq ft)
  Max power loading   4.56 kg/kW (7.5 lb/shp)
PERFORMANCE (Metro 23 SA227-DC at max T-O weight of 7,484 kg; 16,500 lb, ISA, except where indicated):
  Design diving speed (V_D)
                            311 knots (576 km/h; 358 mph) CAS
  Max operating speed (V_{MO})
                            246 knots (455 km/h; 283 mph) CAS
  Max operating Mach No. (M_{MO})   0.52
  Max cruising speed at 97% rpm, bleed low, at 3,350 m
    (11,000 ft)   293 knots (542 km/h; 337 mph) TAS
  Stalling speed:
    flaps and wheels up
                            103 knots (191 km/h; 118 mph) IAS
    flaps and wheels down
                            89 knots (164 km/h; 102 mph) IAS
  Max rate of climb at S/L, bleed open
                            701 m (2,300 ft)/min
  Rate of climb at S/L, OEI, bleed closed
                            176 m (580 ft)/min
  Service ceiling   7,620 m (25,000 ft)
  Service ceiling, OEI   3,535 m (11,600 ft)
  T-O to 15 m (50 ft), wet power 1,414 m (4,640 ft)
  Landing run   848 m (2,780 ft)
Range, Metro 23 (SA227-DC):

Fairchild MMSA with centreline surveillance pod

  with 19 passengers and baggage, FAA IFR reserves
                            over 1,114 nm (2,065 km; 1,283 miles)
  with 2,268 kg (5,000 lb), FAA IFR reserves
                            over 533 nm (988 km; 614 miles)
  C-26 with 1,315 kg (2,900 lb) payload, 1,969 kg
    (4,342 lb) fuel, VFR reserves
                            1,613 nm (2,990 km; 1,858 miles)

FAIRCHILD SPECIAL MISSION AIRCRAFT

TYPE: Special mission versions of Metro 23; replaced Metro III.
PROGRAMME: Development and production on demand.
CURRENT VERSIONS: **Multi-Mission Surveillance Aircraft (MMSA)**: MMSA provides aircraft not permanently configured for single mission; capable of performing multiple missions while preserving ability to return quickly to passenger, VIP, cargo or airborne ambulance/evacuation configuration. Fairchild and Lockheed Fort Worth Company joined in developing and marketing MMSA as low cost, very capable surveillance system consisting of Metro 23, centreline-mounted surveillance pod, mission dedicated radar system, cabin mounted command control communications and intelligence (C³I) and sensor control console and pilot displays. Electronic reconnaissance systems also available. Baseline surveillance/reconnaissance system currently (1993) comprises electro-optical camera, Loral FLIR and infra-red line scanner, Mitsubishi pilot's FLIR, systems operator console, long-range optical system and air-to-air radar or maritime radar; other options available. GEC-Marconi Seaspray 2000 radar to be considered in Summer 1993.
Airborne early warning: Swedish Defence Materiel Administration (FMV) ordered Fairchild study of airborne early warning (AEW) version of Metro III in 1982, carrying dorsal active array radar antenna; initial wind tunnel testing by LTV in Dallas during 1983; FMV ordered Metro III to test Ericsson PS-890 (now FSR-890) Erieye early 1986; FSR-890 is fixed antenna, electronically scanned E/F band radar scanning to one side at a time over 120° arc; first flight with mockup antenna October 1986; delivered to Sweden after 116 hours aerodynamic and handling tests by Fairchild October 1987; exhibited at Farnborough Air Show 1988; first flight with operating radar January 1991.
Antenna in composites housing approx 8 m (26 ft 3 in) long, mounted on pylons above fuselage with ram air cooling; antenna mounted at incidence −2° to allow for angle of attack at patrol speeds; auxiliary fins on tailplane; enlarged ventral fin; Turbomach T-62T APU, producing 60kVA electrical supply for radar, mounted in E-Systems pod on centreline pylon under wing; periscope sight in flight deck roof; larger emergency exit on port side of cabin. Data link to connect with Swedish STRIL60/90 not selected. FMV has now selected Saab 340B for AEW role.
Loiter speeds include 135-146 knots (250-270 km/h; 155-168 mph) with flaps at 50 per cent, 164-175 knots (305-325 km/h; 189-202 mph) with flaps at 25 per cent; endurance 4-6 h in patrol area 100 nm (185 km; 115 miles) away from base.
Multi-mission C³I and surveillance: Multi-role, multi-mission capability with interchangeable mission configuration units (MCUs), APG-66 radar, and surveillance pod for FLIR/cameras (air-to-air and air-to-ground); capabilities and configurations, and C³I systems, can be tailored to mission requirements; special equipment modification and support are offered.
Other variants include **flight inspection**, **photo reconnaissance**, **electronic intelligence** and **airborne critical care**.
CUSTOMERS: Total 44 delivered in various configurations (Metro III, 23 and C-26). MMSA being evaluated by Poland and Hungary mid-1993.

FREEDOM MASTER
FREEDOM MASTER CORPORATION
150 Hamlin Avenue, Satellite Beach, Florida 32939
Telephone: 1 (407) 459 3200
PRESIDENT: Ronald A. Lueck

FREEDOM MASTER FM-2 AIR SHARK I
TYPE: Four-seat homebuilt sporting amphibian.
PROGRAMME: Prototype first flown 5 April 1985; kits of premoulded parts available.
COSTS: Kit: $39,900.
DESIGN FEATURES: Wing section NASA NLF-0215-F. Optional wing extensions.
STRUCTURE: Wing spar caps of S-2 glassfibre or Kevlar rovings. All aerofoils are vacuum moulded. Wing upper shell and upper fuselage of glassfibre/epoxy and Clark foam plastics sandwich construction, with wing lower surfaces of Kevlar/epoxy. Hull and control surfaces of Kevlar/epoxy with Nomex honeycomb cores. Winglets of glassfibre/epoxy. Tail unit construction similar to wings.
LANDING GEAR: Retractable tricycle type.
POWER PLANT: One 149 kW (200 hp) Textron Lycoming IO-360-C1C flat-four engine; alternative engines up to 194 kW (260 hp). Fuel capacity 416 litres (110 US gallons; 91.6 Imp gallons).
ARMAMENT: Hardpoints for weapons and other stores optional.
DIMENSIONS, EXTERNAL:
  Wing span, standard   11.58 m (38 ft 0 in)
  Wing aspect ratio   10.13
  Length overall   7.54 m (24 ft 9 in)
  Height overall   2.49 m (8 ft 2 in)
  Propeller diameter   1.83 m (6 ft 0 in)
AREAS:
  Wings, gross   13.24 m² (142.5 sq ft)
WEIGHTS AND LOADINGS:
  Weight empty   680 kg (1,500 lb)
  Baggage capacity   54 kg (120 lb)

Freedom Master FM-2 Air Shark I four-seat homebuilt amphibian *(Joel Rieman)*

  Max T-O weight   1,270 kg (2,800 lb)
  Max wing loading   95.94 kg/m² (19.65 lb/sq ft)
  Power loading (200 hp engine) 8.52 kg/kW (14.00 lb/hp)
PERFORMANCE (at max T-O weight):
  Max level speed   191 knots (354 km/h; 220 mph)
  Econ cruising speed at 2,285 m (7,500 ft)
                            148 knots (274 km/h; 170 mph)
  Stalling speed   51 knots (94 km/h; 58 mph)
  Max rate of climb at S/L 564 m (1,850 ft)/min
  Service ceiling   5,640 m (18,500 ft)
  T-O run: land   366 m (1,200 ft)
          water   610 m (2,000 ft)
  Landing run   458 m (1,500 ft)
  Range   1,476 nm (2,735 km; 1,700 miles)
  Endurance   11 h 48 min
  g limits   +10/−4

FREEWING

FREEWING AIRCRAFT CORPORATION
Building 340-1300, University of Maryland, College Park, Maryland 20742
Telephone: 1 (301) 314 7794
Fax: 1 (301) 314 9592/9590
PRESIDENT: Hugh Schmittle
EXECUTIVE VICE-PRESIDENT: Odile Legeay

Company formed as result of sponsored and partially state-funded research and flight testing of Freewing principle at University of Maryland. Four manned prototypes tested in flight or wind tunnel, leading to Mk 5 with commercial prospects as a light aircraft; further Tilt-Body principle developed for use in US Department of Defense UAV Joint Project Office close range and vertical launch and recovery UAV programmes (see *Jane's Battlefield Surveillance Systems* for details).

FREEWING FREEBIRD Mk 5

TYPE: Two-seat light aircraft embodying Freewing principle.
PROGRAMME: Four previous versions tested; first flight Mk 5 with 48.5 kW (65 hp) Rotax 582 two-stroke, 22 March 1993; since re-engined with 74.6 kW (100 hp) Mid-West Aero (Londavia) AE100R twin-chamber rotary engine; company plans production civil Freebird light aircraft development, probably during 1994.
DESIGN FEATURES: Freewing principle is that wing is freely pivoted on its centre of gravity and can behave as a flying wing; at any given airspeed, aerodynamic centre of wing is aft of pivot point and its angle of attack is determined by setting of inboard elevator surfaces; lateral control is by normal outboard ailerons; problem of narrow CG range of flying wing is avoided by fact that fuselage weight is applied only through CG of wing, whatever its angle of attack; fuselage attitude is controlled separately by using what appears to be elevator simply as a trimming surface; because wing is free to maintain constant angle of attack, it damps turbulence, giving smooth ride and limiting structural stress in turbulence; CG limits of fuselage are controlled essentially by available tail trimming effort.

Prototype Mk 5 normally flies in this mode but, in order to allow flap to be used to steepen approach and increase lift at low airspeed (which cannot be done with flying wing), wing of Mk 5 can be locked at incidence of 2° in flight; elevators are then controlled as flaps and trimming surfaces at fuselage tail become normal elevators; mechanical switching ensures that control column always produces normal aircraft reactions.

Wing of Mk 5 has Roncz aerofoil designed for maximum C_L at low airspeed and is reflexed at trailing-edge to counter rotational moment at higher speeds; wing is free to tilt through arc of 35°; chord constant; sweep angle 10°; dihedral 0°; wing of production aircraft will be adapted for greater speed range.
FLYING CONTROLS: Mechanical, with aileron and elevator linkages passing through pivot centre; normal dual control columns operate ailerons and wing trailing-edge elevators in Freewing flight, ailerons and fuselage elevator in fixed-wing flight; normal dual rudder pedals and throttle; locking lever to lock/unlock wing and transfer control functions.
STRUCTURE: Fabric-covered steel tube fuselage and tail; steel tube wing centre section, metal covered; remainder of wing is fabric-covered wood; GFRP engine cowling.
LANDING GEAR: Prototype has non-retractable tailwheel type; production aircraft would have tricycle gear.
POWER PLANT: One 74.6 kW (100 hp) Mid-West Aero AE100R water-cooled twin-chamber rotary engine.

DIMENSIONS, EXTERNAL:
Wing span   10.16 m (33 ft 4 in)
AREAS:
Wings, gross   14.19 m² (152.78 sq ft)
Ailerons (total)   1.40 m² (15.2 sq ft)
Flaps (total)   1.12 m² (12.0 sq ft)
Tailplane   1.07 m² (11.8 sq ft)
Elevators (total)   0.84 m² (9.0 sq ft)
WEIGHTS AND LOADINGS:
Basic weight empty   304 kg (670 lb)
Max T-O and landing weight   476 kg (1,050 lb)
Max wing loading   3.12 kg/m² (6.87 lb/sq ft)
Max power loading   3.95 kg/kW (11.67 lb/hp)
PERFORMANCE:
Never-exceed speed (V_{NE})   78 knots (145 km/h; 90 mph)
Max level speed   70 knots (129 km/h; 80 mph)
Cruising speed, 5,500 rpm   61 knots (113 km/h; 70 mph)
Stalling speed, flaps up   26 knots (48 km/h; 30 mph)
Max rate of climb, pilot only   305 m (1,000 ft)/min
Normal T-O run, flaps up   91 m (300 ft)
Min T-O run, flaps up   61 m (200 ft)
Landing run, flaps up   122 m (400 ft)
Range, no reserves   113 nm (209 km; 130 miles)
Endurance   3 h
g limits   +4/-

Freewing Freebird Mk 5 prototype light aircraft with lockable wing

---

GALAXY

GALAXY GROUP
Van Nuys, California
EXECUTIVES: Gordon Cooper and Pendleton Parrish

Markets turboprop re-engining kits (Allison 250) for piston light twins (Beech Baron and Duke, Cessna 421, Piper Chieftain) and planning similar conversion of Lockheed JetStar I/II with twin Allison GMA 3007s as **Galaxy 700 JetStar**; also developing own-design **Galaxy Star** as twin-Allison turbofan STOL transport of about 10,000 kg (22,043 lb) MTOGW, to operate from 457 m (1,500 ft) unprepared airstrips. Plans to conduct JetStar and Galaxy Star programmes at new facility in Shawnee, Oklahoma.

GARRETT — see AlliedSignal

GENERAL DYNAMICS

3190 Fairview Park Drive, Falls Church, Virginia 22042
Telephone: 1 (703) 876 3000

Military aircraft activities of General Dynamics sold to Lockheed in December 1992, becoming Lockheed Fort Worth on 1 March 1993. Details of F-16 Fighting Falcon appear on Lockheed pages of this edition; other GD aerospace activities outlined in 1992-93 and earlier *Jane's*.

GEVERS

GEVERS AIRCRAFT INC
PO Box 430, Brownsburg, Indiana 46112
Telephone: 1 (317) 852 2735
PRESIDENT, CHIEF ENGINEER: David E. Gevers
VICE-PRESIDENT: Lawrence B. Schmidt
BUSINESS MANAGER: Theresa A. Gevers
MARKETING: Robert L. Glasa

Gevers Aircraft formed 1988 to investigate feasibility of several design innovations for new general aviation aircraft. Although work has begun on a six-seater, four-seat and larger variants are under investigation.

GEVERS GENESIS

TYPE: Six-seat amphibious monoplane.
PROGRAMME: Design began in 1981; radio controlled model flown in 1990; wind tunnel testing and extensive computer aided stability analysis nearing completion in early 1993. Patents being applied for.
DESIGN FEATURES: Goals include greatly improved performance over conventional aircraft, modular and interchangeable subassemblies, twin-engined reliability with single-engined simplicity, multi-purpose landing gear, and manufacturing and construction simplicity. Can be used as training aircraft capable of simulating variety of configurations, and as testbed for aerofoil trials.

Wing has centre-section attached to top of fuselage and two telescopic sections that extend wing span; movable wing sections completely enclosed by centre-section when retracted inwards. Amphibious hull, and T tail above long sweptback fin; several styles of cabin can be accommodated (interchangeable) as upper fuselage area, including those for luxury, economy, military, medical and cargo roles. Wing centre-section aerofoil NACA 66_3-018; dihedral 3°; sweep 3° 17′ at quarter-chord; 0° sweep for wing extensions.
FLYING CONTROLS: Ailerons (inboard droop with extension of flaps), one-piece elevator and rudder; servo tab in rudder and anti-servo in elevator; fixed leading-edge slats on wing extensions and movable slats on centre-section; flight adjustable tailplane incidence.
STRUCTURE: Riveted aluminium alloy wings, tail and lower fuselage; cabin compartment of glassfibre.
LANDING GEAR: Converts in flight for water, land, snow or intermittent snow/hard surface operation; retractable.
POWER PLANT: Two 242.4 kW (325 hp) Textron Lycoming TIO-540 flat-six engines, each driving a Hartzell counter-rotating, two-blade constant-speed feathering and reversible-pitch metal pusher propeller. Total usable fuel capacity 757 litres (200 US gallons; 166 Imp gallons) in two wing mounted bladder type tanks; fuel gravity fed, with electric pumps for backup and aerobatic operation.
ACCOMMODATION: Clamshell type doors on both sides of cabin allow each passenger direct exit. Optional cabin configurations include medical and cargo layouts. During water operations, only upper half of door operational; during non-water operations lower half of door opens to floor level for ease of loading.
SYSTEMS: Pressurised and unpressurised models to be offered.
AVIONICS: Dual VFR and IFR instrument panels, with day and night capability.

DIMENSIONS, EXTERNAL:
Wing span, extended   15.24 m (50 ft 0 in)
Wing chord: at fixed root   3.25 m (10 ft 8 in)
   at fixed tip   2.03 m (6 ft 8 in)
   at extension root and tip   1.12 m (3 ft 8 in)
Wing aspect ratio (extended)   8.3
Length overall   11.58 m (38 ft 0 in)
Height overall   3.76 m (12 ft 4 in)
Tailplane span   4.45 m (14 ft 7 in)
Wheel track   3.20 m (10 ft 6 in)
Wheelbase   6.35 m (20 ft 10 in)
Propeller diameter   1.98 m (6 ft 6 in)
Passenger door: Height   0.965 m (3 ft 2 in)
   Width   2.24 m (7 ft 4 in)
DIMENSIONS, INTERNAL:
Cabin: Length   3.05 m (10 ft 0 in)
   Max width   1.21 m (3 ft 11½ in)

Max height	1.07 m (3 ft 6 in)
Volume	3.74 m³ (132.0 cu ft)
Nose baggage compartment	0.28 m³ (10.0 cu ft)
Rear baggage compartment	0.42 m³ (15.0 cu ft)

AREAS:
Wing extended, gross	28.61 m² (308 sq ft)
Ailerons retracted (total)	1.16 m² (12.50 sq ft)
Ailerons extended (total)	2.79 m² (30.0 sq ft)
Trailing-edge flaps (total)	0.71 m² (7.67 sq ft)
Fin	4.16 m² (44.8 sq ft)
Rudder, incl tab	1.39 m² (15.0 sq ft)
Tailplane	3.81 m² (41.0 sq ft)
Elevator, incl tab	1.90 m² (20.5 sq ft)

WEIGHTS AND LOADINGS (estimated):
Weight empty	1,542 kg (3,400 lb)
Max T-O weight	2,722 kg (6,000 lb)
Max wing loading (extended)	95.1 kg/m² (19.48 lb/sq ft)
Max power loading	5.61 kg/kW (9.23 lb/hp)

PERFORMANCE (estimated):
Max level speed at max T-O weight	267 knots (496 km/h; 308 mph)
Cruising speed at max T-O weight, at S/L:	
75% power	243 knots (450 km/h; 280 mph)
65% power	234 knots (435 km/h; 270 mph)
OEI	208 knots (386 km/h; 240 mph)
Stalling speed: flaps, landing gear and wings retracted	95 knots (175 km/h; 109 mph)
flaps, landing gear and wings extended	55 knots (102 km/h; 63 mph)
Rotation speed	66 knots (123 km/h; 76 mph)
Approach speed	72 knots (132 km/h; 82 mph)
T-O run	183 m (600 ft)
T-O to 15 m (50 ft)	427 m (1,400 ft)
Landing from 15 m (50 ft)	607 m (1,990 ft)
Landing run	71 m (230 ft)

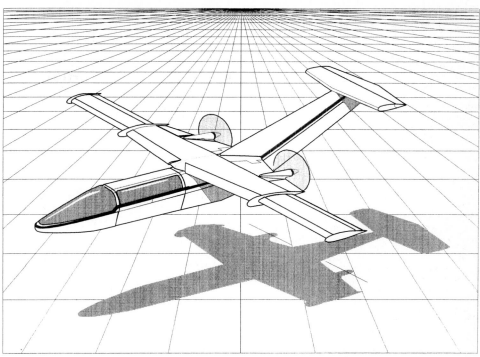

Artist's impression of Gevers Aircraft Genesis with wings extended; shadow represents wings in retracted position

GLOBAL

GLOBAL HELICOPTER TECHNOLOGY INC

1 South Norwood Drive, Hurst, Texas 76053
Telephone: 1 (817) 282 8500
Fax: 1 (817) 282 0322
PRESIDENT: Daniel W. Pettus
VICE-PRESIDENT: Dennis Halwes

Company specialises in modifications to helicopters. Variety of military programmes include Bell UH-1HP High Performance conversion of Bell 205 (with France Aviation) by fitting Textron Lycoming T53-L-703 and Bell 212 drive train.

GLOBAL HELICOPTER HUEY 800

Conversion of Bell UH-1H with 986 kW (1,322 shp) LHTEC T800-LHT-800 engine; first flown 15 June 1992; FAA and military certification in progress. Cost of conversion $800,000. Redeliveries could begin second quarter of 1994. Frahm vibration absorber system optional; integral particle separator and FADEC control; MIL-STD-1553B databus; Hispano-Suiza reduction gearbox.

Claimed to have significantly lower fuel consumption than UH-1 with existing T53-L-13B or other engines currently proposed for refit. Prototype set new point-to-point distance record in Spring 1993 (Los Angeles to Atlanta), approx 1,713 nm (3,175 km; 1,973 miles), exceeding by more than 540 nm (1,000 km; 621 miles) the previous record for helicopters in 3,000-4,500 kg (6,614-9,921 lb) class. Average fuel consumption was 140 kg (308.5 lb)/h, compared with more than 225 kg (496 lb)/h for standard UH-1; approx 110 kg (242 lb) of fuel remained at end of flight.

GREAT PLAINS

GREAT PLAINS AIRCRAFT SUPPLY CO INC

PO Box 304, St Charles, Illinois 60174
Telephone: 1 (708) 464 4178
Fax: 1 (708) 464 4182

Details of current Sonerai I single-seat Formula V racing homebuilt can be found in 1992-93 *Jane's*.

GREAT PLAINS SONERAI II, IIL, II-LT and II-LTS

TYPE: Two-seat, dual-control sporting homebuilt.
PROGRAMME: Prototype of tandem two-seat Sonerai II first flown July 1973; at least 500 since built and flown. Many components, complete kits for fuselage, tail and wings, and materials, available to amateur constructors; estimated building time 850 working hours.

One variant, first flown June 1980, is **Sonerai IIL**, with low-wing instead of mid-wing configuration and 3° dihedral. Another variant, first flown January 1983, is **Sonerai II-LT**, which is similar to Sonerai IIL but has 2,200 cc VW engine standard, tricycle landing gear and larger front cockpit for taller pilot; retrofit kit was produced to allow existing Sonerais to be fitted with tricycle gear. **Sonerai II-LTS** is basically stretched version of Sonerai II-LT, first flown in June 1984.
COSTS: Kit: $5,200. Plans: $99.95.
DESIGN FEATURES: Wing section NACA 64212.
STRUCTURE: Wings of aluminium alloy construction; full span aluminium alloy ailerons. Welded steel tube fuselage and tail unit, fabric covered except for glassfibre engine cowlings.
LANDING GEAR: Non-retractable tailwheel type.
POWER PLANT: One 44.7, 56 or 61 kW (60, 75 or 82 hp) HAPI converted Volkswagen 1,600, 1,834 or 2,200 cc motorcar engine. The 2,200 cc version is standard for the Sonerai II-LT and II-LTS. Fuel capacity 38 litres (10 US gallons; 8.3 Imp gallons) for all versions except Sonerai II-LTS, which has standard capacity of 68 litres (18 US gallons; 15 Imp gallons).

DIMENSIONS, EXTERNAL:
Wing span	5.69 m (18 ft 8 in)
Wing aspect ratio	4.15

Great Plains Sonerai II two-seat homebuilt *(Peter M. Bowers)*

Length overall:	
all, except Sonerai II-LTS	5.74 m (18 ft 10 in)
Sonerai II-LTS	6.20 m (20 ft 4 in)
Propeller diameter	1.32 to 1.37 m (4 ft 4 in to 4 ft 6 in)

AREAS:
Wings, gross	7.80 m² (84.0 sq ft)

WEIGHTS AND LOADINGS (Sonerai II):
Weight empty	227 kg (500 lb)
Utility max T-O weight	431 kg (950 lb)
Max T-O weight	521 kg (1,150 lb)
Max wing loading	66.84 kg/m² (13.69 lb/sq ft)

PERFORMANCE (Sonerai II/IIL with 1,700 cc engine):
Max level speed at S/L	139 knots (257 km/h; 160 mph)
Econ cruising speed at S/L	113 knots (209 km/h; 130 mph)
Stalling speed	38 knots (71 km/h; 44 mph)
Max rate of climb at S/L	152 m (500 ft)/min
T-O run	274 m (900 ft)
Landing run	152 m (500 ft)
Range, with reserves	304 nm (563 km; 350 miles)
g limits: pilot only, aerobatic	±6
max T-O weight, utility	±4.4

GRUMMAN

GRUMMAN CORPORATION
1111 Stewart Avenue, Bethpage, New York 11714
Telephone: 1 (516) 575 0574
Fax: 1 (516) 575 2164
CHAIRMAN OF THE BOARD AND CEO: Renso L. Caporali
PRESIDENT AND COO: Robert J. Myers
SENIOR VICE-PRESIDENT, BUSINESS DEVELOPMENT:
    Thomas J. Kane Jr.
VICE-PRESIDENT, PUBLIC AFFAIRS: Robert P. Harwood

Grumman Aircraft Engineering Corporation incorporated 6 December 1929; Grumman Corporation formed as small holding company 1969 for Grumman Aerospace Corporation, Grumman Allied Industries Inc and Grumman Data Systems Corporation. Ten operating divisions created February 1985 (see 1990-91 and earlier editions), followed by further reorganisations in 1987-88. Corporate structure further consolidated in early 1991 into four operating groups, of which Aircraft Group and Systems Group merged on 1 January 1993:

Aerospace & Electronics Group, including former Aircraft Systems, Aerostructures, St Augustine, Melbourne Systems, Space Electronics, and Space Station Program Support divisions. See following entry.

Data Systems and Services Group, including former Technical Services, Data Systems, and Systems Support divisions (PRESIDENT: Gerald H. Sandler).

Allied Group, including former Allied (Vehicles and Marine) division (PRESIDENT: Leonard Rothenberg).

Aerospace & Electronics Group is prime contractor and installs systems in E-8C J-STARS aircraft, which see. A corporate services division covers legal, purchasing, contract and other services common to all divisions.

Grumman Corporation employees total 21,400; Bethpage plant closed for aircraft assembly, November 1992, but remains for component manufacture; all final assembly and flight test at government-owned plant, Calverton, New York State; aircraft upgrades at Melbourne, Florida.

GRUMMAN AEROSPACE & ELECTRONICS GROUP
PRESIDENT: Al Verderosa

Current products include E-2C Hawkeye and rebuilt versions of A-6 Intruder and F-14 Tomcat for US Navy. Contracted in February 1984 to produce 270 shipsets of Tay turbofan nacelles and thrust reversers for Gulfstream IV and Fokker 100; design and manufacture of complete tail section of V-22 Osprey (see under Bell/Boeing) ordered August 1984; builds composites ailerons, elevators and rudders for McDonnell Douglas C-17A; responsible for engineering and logistic support for Fairchild A-10 Thunderbolt II since October 1987. $20 million study award for A-X naval strike aircraft given to Grumman-led consortium (with Boeing and Lockheed) December 1991 (see US Navy entry).

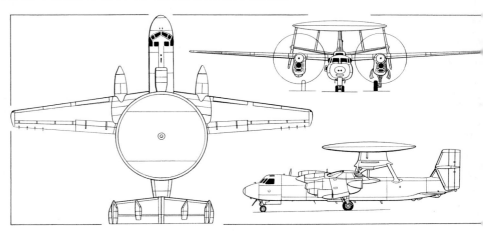

Grumman E-2C Hawkeye twin-turboprop airborne early warning and control aircraft *(Dennis Punnett)*

GRUMMAN HAWKEYE
US Navy designations: E-2C and TE-2C
TYPE: Shipborne and land-based airborne early warning and control aircraft.
PROGRAMME: First flight of first of three prototypes 21 October 1960; total 59 production E-2As, of which 51 updated to E-2B by end 1971 apart from two TE-2A trainers and two converted to E-2C prototypes (see earlier *Jane's*); first flight of E-2C prototype 20 January 1971; production started mid-1971; first flight production aircraft 23 September 1972; total 174 ordered, of which 154 delivered at start of 1993; planned production six per year until 1995. AN/APS-145 radar replaces AN/APS-125, AN/APS-138 and AN/APS-139 in new-built E-2Cs.

Evaluation of Grumman/General Electric AN/APS-145 began 1986; radar tracks more than 2,000 targets and operates at longer ranges, has improved jamming resistance and sharper fully automated/optimised overland detection. AN/APS-145 radar and new main operator displays, IFF, mission computer processor, JTIDS tactical software and upgraded engines form core of current Update Development Program (Groups I and II), now fully developed and being delivered in new production aircraft from October 1991. Major retrofit programme for USN and FMS aircraft is planned, including upgrade of all 18 Group I and minimum of 36 older ('Group 0') aircraft to Group II standard from FY 1995. Second production Group I aircraft (163535 – 123rd E-2C; 102nd USN E-2C) set 20 (broke 14 and established six new) records for time-to-height, altitude and 100 km closed-circuit, 17-19 December 1991.

USN previously considered Advanced Tactical Support (ATS) aircraft programme, including an AEW configuration, to replace E-2C; ATS plans cancelled 1991 and USN considering launch of E-X programme for follow-on AEW, although no funding apparently available in defence budgets prior to 2000; USN accordingly studying potential for major operational upgrades of E-2C beyond provisions of current Update Development Program.
CURRENT VERSIONS: **E-2B**: Withdrawn from USN service.
E-2C: Current service and production version (*as detailed*). USN variants are:

	Quantity	First aircraft
E-2C 'Group 0'	100	158638
E-2C Group I	18	163029
E-2C Group II	21	164108
Total	139	

TE-2C: Training model, based on E-2C.
E-2T: Conversion of E-2B for Taiwan with AN/APS-138 radar and electronic warfare upgrades.
CUSTOMERS: US Navy orders for E-2C totalled 139 by FY 1992, of which 130 delivered by January 1993. Target of 175 cut to 145 in early 1991 and to 139 in January 1992 (excluding two prototypes).

E-2C entered service with VAW-123 at NAS Norfolk, Virginia, November 1973 and went to sea on board USS *Saratoga* late 1974; E-2C issued to 18 other squadrons, including two of Naval Reserve; two TE-2C in service, including 158639 assigned to JTIDS development. Current squadrons are VAW-110 and 112 to 117 at Miramar, California, with VAW-88 of Reserves; VAW-120 to 126 with VAW-78 of Reserves at Norfolk, Virginia. VAW-110 and VAW-120 are training units. Miramar is base of Group II squadron, VAW-113, June 1992. VAW-113 first operational evaluation cruise on USS *Carl Vinson*, 1993. US Navy transferred two to US Coast Guard (operations with CG Air Wing 1 at St Augustine, Florida, began January 1987) and two to US Customs Service for anti-narcotics operations; USCS aircraft to USCG in August 1989 and all returned to USN on 23 October 1991 (due to rotations, nine different aircraft used).

Exports to Israel (four delivered 1978), Japan (received four in 1982 and four in 1984, and ordered three in FY 1989 for 1992-93 delivery and two in FY 1990 for delivery in 1993; all with 601 Squadron, HQ, Misawa), Egypt (accepted five, beginning October 1987; ordered sixth in 1989 for 1993 delivery), Singapore (received four with AN/APS-138 in 1987 for 111 Squadron at Tengah) and France (ordered four in 1993). Taiwan: four E-2B to be converted to E-2T by Grumman in early 1993. Total export orders 35; 21 delivered by January 1993. Singapore requires two more. Japan upgraded radars to AN/APS-145, 1991.
COSTS: $59 million, flyaway, USN FY 1992.
DESIGN FEATURES: Details here apply to E-2C. Hawkeye can cover naval task force in all weathers flying at 9,150 m (30,000 ft) and can detect and assess approaching aircraft in excess of 300 nm (556 km; 345 miles); AN/APS-145 has total radiation aperture control antenna (TRAC-A) to reduce sidelobes to offset jamming; radar sweeps six million cubic mile envelope and simultaneously monitors surface ships; long range, automatic target track initiation and high-speed processing enable each E-2C to track more than 2,000 targets simultaneously and automatically, and control more than 40 intercepts; Randtron Systems AN/APA-171 antenna housed in 7.32 m (24 ft) diameter radome, rotating at 5-6 rpm above rear fuselage; antenna arrays in rotodome provide radar sum and difference signals and IFF.

AN/APS-139 and Allison T56-427 engines form **Group I** update; first operational aircraft (163538) delivered to VAW-112 on 8 August 1989; 18 built; AN/APS-139 can detect cruise missiles at ranges exceeding 100 nm (185 km; 115 miles); also monitors maritime traffic; radar coverage extended by AN/ALR-73 passive detection system (PDS), detecting electronic emitters at twice radar detection range; more detailed description in 1979-80 *Jane's*; AN/APS-145 in **Group II** aircraft from December 1991; other enhancements give Group II 96 per cent expansion in radar volume, 400 per cent extra target tracking capability, 40 per cent more radar and identification range and 960 per cent increase in numbers of targets displayed; Group II adds JTIDS in 1993-94; has GPS provisions.

Conventional airframe with nose-tow catapult attachment, arrester hook and tail bumper; parts of tail made of composites to reduce radar reflection; wings fold hydraulically on skewed hinges to lie parallel to fuselage; wing incidence 4° at root, 1° at tip.
FLYING CONTROLS: Fully powered with artificial feel; tailplane has 11° dihedral; four fins and three double-hinged rudders; long-span ailerons droop automatically when hydraulically operated Fowler flaps are extended; autopilot provides autostabilisation or full flight control.
STRUCTURE: Wing centre-section has three beams, ribs and machined skins; hinged leading-edge provides access to

Grumman E-2C Hawkeye early warning and control aircraft of the US Navy

flying and engine controls. Fuselage conventional light metal. Composites used in parts of tail.

LANDING GEAR: Hydraulically retractable tricycle type. Pneumatic emergency extension. Steerable nosewheel unit retracts rearward. Mainwheels retract forward and rotate to lie flat in bottom of nacelles. Twin wheels on nose unit only. Oleo-pneumatic shock absorbers. Mainwheel tyres size 36 × 11 Type VII 24-ply, pressure 17.9 bars (260 lb/sq in) on ship, 14.5 bars (210 lb/sq in) ashore. Hydraulic brakes. Hydraulically operated retractable tailskid. A-frame arrester hook under tail.

POWER PLANT: Two 3,803 kW (5,100 ehp) Allison T56-A-427 turboprops, driving Hamilton Standard type 54460-1 four-blade fully feathering reversible-pitch constant-speed propellers. These have foam filled blades which have a steel spar and glassfibre shell. T56-A-427 engines provide 15 per cent improvement in efficiency, compared with -425 installed prior to 1989.

ACCOMMODATION: Normal crew of five, consisting of pilot and co-pilot on flight deck, plus ATDS Combat Information Center (CIC) staff of combat information centre officer, air control officer and radar operator. Downward hinged door, with built-in steps, on port side of centre-fuselage and three overhead escape hatches.

SYSTEMS: Pneumatic boot de-icing on wings, tailplane and fins. Spinners and blades incorporate electric anti-icing.

AVIONICS: Randtron AN/APA-171 rotodome (radar and IFF antennae), General Electric AN/APS-145 advanced radar processing system (ARPS) with fully automatic overland/ overwater detection capability (in current production aircraft and scheduled for eventual retrofit in all E-2Cs), improved Hazeltine IFF detector processor, Litton AN/ALR-73 passive detection system, Hazeltine AN/APA-172 control indicator group with Loral enhanced (colour) main display units, Litton OL-77/ASQ computer programmer (L-304) with Loral enhanced high-speed processor, AN/ARC-158 UHF data link, AN/ARQ-34 HF data link, JTIDS Class 2 HP terminal, ASM-440 in-flight performance monitor, AN/AIC-14A intercom, Litton AN/ASN-92 CAINS carrier aircraft inertial navigation system, GPS, GEC-Marconi standard central air data computer, AN/ASN-50 heading and attitude reference system, Collins AN/ARA-50 UHF ADF, AN/ASW-25B ACLS and Honeywell AN/APN-171 (V) radar altimeter.

DIMENSIONS, EXTERNAL:
Wing span	24.56 m (80 ft 7 in)
Wing chord: at root	3.96 m (13 ft 0 in)
at tip	1.32 m (4 ft 4 in)
Wing aspect ratio	9.3
Width, wings folded	8.94 m (29 ft 4 in)
Length overall	17.54 m (57 ft 6¼ in)
Height overall	5.58 m (18 ft 3¾ in)
Diameter of rotodome	7.32 m (24 ft 0 in)
Tailplane span	7.99 m (26 ft 2½ in)
Wheel track	5.93 m (19 ft 5¾ in)
Wheelbase	7.06 m (23 ft 2 in)
Propeller diameter	4.11 m (13 ft 6 in)

AREAS:
Wings, gross	65.03 m² (700.0 sq ft)
Ailerons (total)	5.76 m² (62.0 sq ft)
Trailing-edge flaps (total)	11.03 m² (118.75 sq ft)
Fins, incl rudders and tabs:	
outboard (total)	10.25 m² (110.36 sq ft)
inboard (total)	4.76 m² (51.26 sq ft)
Tailplane	11.62 m² (125.07 sq ft)
Elevators (total)	3.72 m² (40.06 sq ft)

WEIGHTS AND LOADINGS:
Weight empty	17,859 kg (39,373 lb)
Max fuel (internal, usable)	5,624 kg (12,400 lb)
Max T-O weight	24,161 kg (53,267 lb)
Max wing loading	371.5 kg/m² (76.10 lb/sq ft)
Max power loading	3.18 kg/kW (5.22 lb/shp)

PERFORMANCE (at max T-O weight):
Max level speed	338 knots (626 km/h; 389 mph)
Max cruising speed	325 knots (602 km/h; 374 mph)
Cruising speed (ferry)	259 knots (480 km/h; 298 mph)
Approach speed	103 knots (191 km/h; 119 mph)
Stalling speed (landing configuration)	75 knots (138 km/h; 86 mph)
Service ceiling	11,275 m (37,000 ft)
Min T-O run	564 m (1,850 ft)
T-O to 15 m (50 ft)	793 m (2,600 ft)
Min landing run	439 m (1,440 ft)
Ferry range	1,541 nm (2,856 km; 1,775 miles)
Time on station, 175 nm (320 km; 200 miles) from base	4 h 24 min
Endurance with max fuel	6 h 15 min

GRUMMAN INTRUDER
US Navy designations: A-6E and KA-6D

TYPE: Carrier-based all-weather attack aircraft, with tanker variant.

PROGRAMME: 708 new A-6 and EA-6As built; about 375 still in service with US Navy and Marine squadrons and readiness training squadrons. First A-6A entered service in February 1963 with training squadron VA-42; first operational deployment with VA-75, May 1965. Final aircraft (205th new production A-6E, 164385) delivered 3 February 1992.

CURRENT VERSIONS: **A-6A/B/C:** Only those converted to other variants still in service; described in 1978-79 *Jane's*.

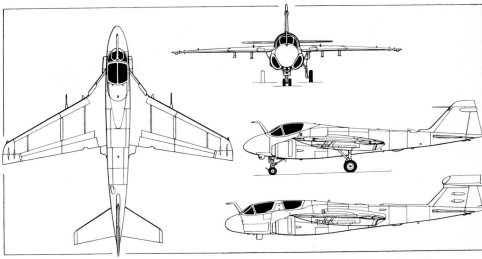
Grumman A-6E/TRAM, with additional side view of EA-6B (bottom) (*Dennis Punnett*)

Grumman A-6E/TRAM Intruders of VA-75 'Sunday Punchers' from USS *Kennedy*

EA-6B Prowler: Advanced electronics development of EA-6A; described separately.

KA-6D: Dedicated tanker; 78 converted from A-6A and 12 from A-6E (see 1980-81 *Jane's*); latest configuration has no weapon capability and can carry five 1,514 litre (400 US gallon; 333 Imp gallon) drop tanks; fitted as probe and drogue receiver as well as tanker; drogue retracts into tunnel under rear fuselage.

A-6E: Began as advanced conversion of A-6A (240 converted) with multi-mode radar and IBM computer first fitted to EA-6B; first flight 10 November 1970; first deployed September 1972; approved for service November 1972. Total 205 new-build A-6Es funded by end FY 1988, of which final 21 have composites wings (see below).

A-6E/TRAM: Current operational version equipped with Target Recognition Attack Multi-Sensor; first flight October 1974; delivery of fully equipped aircraft began 14 December 1978 with 160995; first carrier deployment completed May 1980; all older A-6Es converted to TRAM by 1988; eight Grumman TC-4C (modified Gulfstream I) used for training.

US Navy A-6E squadrons are VA-52, 95, 115, 128 (training), 145, 165, 196 at Whidbey Island, Washington, with two rotated to Atsugi, Japan; VA-34, 35, 36, 42 (training), 55, 65, 75 and 85, at Oceana, Virginia; Reserves VA-205 at Atlanta, Georgia. Marine squadrons are VMA (AW)-224 and 332 at Cherry Point, North Carolina, including one rotated to Atsugi, Japan.

A-6E Harpoon: 50 A-6Es each fitted to carry four McDonnell Douglas Harpoon anti-ship missiles; deployment began 1981; all subsequent new-build and converted aircraft also equipped.

Re-winged A-6E: Last metal-winged A-6E delivered September 1988; final 21 production aircraft fitted with composites wings. Boeing Military Airplanes won competition to produce graphite/epoxy wing for A-6F (new build and retrofit) to overcome fatigue problems in service and give 8,800 hour service life; first flight 3 April 1989. Naval trials completed June 1990. Initial contract for 179 wing sets plus options for 327; wings installed on A-6E after A-6F cancelled; initial order completed May 1992; further 120 to be delivered from Boeing at Seattle from 1993. First re-winged A-6E redelivered to VA-176 on 4 October 1990; rebuilds at Boeing, Wichita (11 only), Grumman (St Augustine), Grumman (Calverton), and USN Depots at Norfolk and Alameda. USN plans 332 re-winged A-6Es of SWIP Block 1 and SWIP Block 1A variants, and ordered 305 retrofit kits in November 1992 for upgrading of aircraft's detection and ranging system, but force reduction plans under consideration in 1993 call for withdrawal of entire A-6 fleet by 2000.

A-6E SWIP: Systems Weapon Integration Program partially applied to last 33 production aircraft. Baseline SWIP (metal wings) and Block 1 SWIP (composites wings) have identical avionics, including provision for AGM-88 HARM, AGM-84 Harpoon and AGM-65 Maverick, and various airframe improvements. SWIP Block 1A covers GPS, AN/ASN-139 INS, GEC-Marconi standard central air data computer on MIL-STD-1553B digital databus, upgrade of Tacan from AN/ARN-84 to AN/ARN-118, HUD, multi-function display, wing fillet modifications to reduce approach speed (AoA increased from 15° to 19°) and thus increase max arresting weight and IDAP (Integrated Defense Avionics Program) combining towed decoy, chaff dispensers and radar warning receiver. Fillet trials A-6 flew August 1991; first of two avionics trials aircraft (152936) on 30 September 1992; second in May 1993; IOC due 1995; total 236 conversions planned to Block 1A.

A-6F and A-6G: Abandoned Intruder developments (see 1990-91 *Jane's*). Five pre-series A-6Fs employed for trials.

CUSTOMERS: US Navy and Marine Corps; see Current Versions.

COSTS: $35.5 million (1988) system unit cost.

DESIGN FEATURES: Wing sweepback 25° at quarter-chord; outer wing panels fold upward more than 90° for stowage; TRAM ball under nose, aft of radome. Survivability improvements incorporated; fire-extinguishing system (halon) inerting fuel tanks; fire suppression in areas around fuselage fuel tanks using dry foam blocks; and self-sealing fuel lines in engine cavities.

FLYING CONTROLS: Hydraulic fully powered controls; slab tailplane; lateral control spoilers ahead of near-full-span Fowler flaperons; split surface airbrakes in trailing-edge outboard of flaperons; sharp-edged leading-edge stall strip next to fuselage with fixed droop section outboard, then long-span leading-edge slats. Fences near wingroot and near tip.

STRUCTURE: Conventional all-metal; graphite/epoxy wing

being retrofitted; aluminium alloy control surfaces and titanium high strength fittings, such as wing-fold.

LANDING GEAR: Hydraulically retractable tricycle type. Twin-wheel nose unit retracts rearward. Single-wheel main units retract forward and inward into air intake fairings. A-frame arrester hook under rear fuselage; nose-tow catapult fitting.

POWER PLANT: Two 41.4 kN (9,300 lb st) Pratt & Whitney J52-P-408 turbojets. Replacement by two 53.4 kN (12,000 lb st) J52-P-409s under study. Max internal fuel capacity 8,873 litres (2,344 US gallons; 1,952 Imp gallons). Provision for up to five external fuel tanks under wing and centreline stations, each of 1,135 litres (300 US gallons; 250 Imp gallons) or 1,514 litres (400 US gallons; 333 Imp gallons) capacity. Removable flight refuelling probe projects upward immediately forward of windscreen.

ACCOMMODATION: Crew of two on Martin-Baker GRU7 ejection seats, which can be reclined to reduce fatigue during low-level operations. Bombardier/navigator slightly behind and below pilot to starboard. Hydraulically operated rearward sliding canopy.

SYSTEMS: AiResearch ECS for cockpit and avionics bay. Dual hydraulic systems for operation of flight controls, leading-edge and trailing-edge flaps, wingtip speed-brakes, landing gear brakes and cockpit canopy, each rated at 119.4 litres (31.5 US gallons; 26.2 Imp gallons) per min, with air/oil separated reservoir, pressurised at 2.76 bars (40 lb/sq in). One electrically driven hydraulic pump provides restricted flight capability by supplying the tailplane and rudder actuators only, each rated at 11.4 litres (3 US gallons; 2.5 Imp gallons) per min, with internally pressurised reservoir, pressurised at 1.08 bars (15.7 lb/sq in). Electrical system powered by two Garrett constant speed drive units that combine engine starting and electric power generation, each delivering 30kVA; retrofit of Sundstrand generators (as in EA-6B) under way. A Garrett ram air turbine, mounted so that it can be projected into the airstream above the port wingroot, provides in-flight emergency electric power for essential equipment.

AVIONICS: Single Norden AN/APQ-148 or AN/APQ-156 simultaneous multi-mode nav/attack radar. IBM and Fairchild nav/attack computer system and interfacing data converter. Conrac Corporation armament control unit. RCA video tape recorder for post-strike assessment of attacks; replacement unit by Precision Echo to be retrofitted. Litton AN/ALR-67 RWR. Radar provides simultaneous ground mapping; tracking and rangefinding of fixed or moving targets; and terrain clearance or terrain following manoeuvres. During 1981-83, it was updated by an improved AMTI (airborne moving target indication) radar to enhance its ability to detect moving targets. IBM AN/ASQ-133 or AN/ASQ-155 solid state digital computer is coupled to A-6E's radar, inertial and Doppler navigational equipment, communications and AFCS. Fairchild signal data converter accepts analog input data from up to 60 sensors, converting data to a digital output that is fed into nav/attack system computer. Conrac armament control unit (ACU) provides all inputs and outputs necessary to select and release weapons. Kaiser AN/AVA-1 multimode display serves as a primary flight aid for navigation, approach, landing and weapons delivery. TRAM package includes undernose precision-stabilised turret, with a sensor package containing both infra-red and laser equipment; INS updated with Litton AN/ASN-92 CAINS; new communications-navigation-identification (CNI) equipment including AN/APX-72 IFF transponder; and automatic carrier landing capability. Sensor package is integrated with multi-mode radar, providing capability to detect, identify and attack a wide range of targets (as well as view the terrain) under adverse weather conditions, and with improved accuracy, using either conventional or laser guided weapons. Bombardier/navigator operates TRAM system by first acquiring target on his radar screen. He then switches to FLIR (forward looking infra-red) system, using an optical zoom to enlarge target's image. After identifying and selecting his targets, bombardier uses a laser designator to mark target with a laser spot, on which his own laser guided weapons, or those from another aircraft, will home. Using TRAM's laser spot detector, A-6E can also acquire a target being illuminated from another aircraft, or designated by a forward air controller on the ground. Some A-6s modified for compatibility with crew night vision goggles.

ARMAMENT: Five weapon attachment points, each with a 1,633 kg (3,600 lb) capacity (max external stores load 8,165 kg; 18,000 lb). Typical weapon loads are twenty-eight 500 lb bombs in clusters of six, or three 2,000 lb general purpose bombs plus two 1,135 litre (300 US gallon; 250 Imp gallon) drop tanks. AIM-9 Sidewinder missiles can be carried for air-to-air use. Harpoon missile capability added to weapons complement of A-6E/TRAM. The HARM missile has been test flown on the A-6E. Up to 20 Brunswick Defense ADM-141 TALD (Tactical Air-Launched Decoy) gliders, or two in addition to normal bomb load. Flight and firing tests have been carried out with the AGM-123A Skipper II, also on an A-6E.

DIMENSIONS, EXTERNAL:

Wing span	16.15 m (53 ft 0 in)
Wing mean aerodynamic chord	3.32 m (10 ft 10¾ in)
Wing aspect ratio	5.3
Width, wings folded	7.72 m (25 ft 4 in)
Length overall	16.69 m (54 ft 9 in)
Height overall	4.93 m (16 ft 2 in)
Tailplane span	6.21 m (20 ft 4½ in)
Wheel track	3.32 m (10 ft 10½ in)
Wheelbase	5.24 m (17 ft 2¼ in)

AREAS:

Wings, gross	49.14 m^2 (528.9 sq ft)
Flaperons (total)	3.81 m^2 (41.0 sq ft)
Trailing-edge flaps (total)	9.66 m^2 (104.0 sq ft)
Leading-edge slats (total)	4.63 m^2 (49.8 sq ft)
Fin	5.85 m^2 (62.93 sq ft)
Rudder	1.52 m^2 (16.32 sq ft)
Tailplane	10.87 m^2 (117.0 sq ft)

WEIGHTS AND LOADINGS:

Weight empty	12,525 kg (27,613 lb)
Fuel load: internal	7,230 kg (15,939 lb)
external (five tanks)	4,558 kg (10,050 lb)
Max external load	8,165 kg (18,000 lb)
Max T-O weight: catapult	26,580 kg (58,600 lb)
field	27,397 kg (60,400 lb)
Max zero-stores weight	20,521 kg (45,242 lb)
Max landing weight: carrier	16,329 kg (36,000 lb)
field	20,411 kg (45,000 lb)
Max wing loading	557. kg/m^2 (114.2 lb/sq ft)
Max power loading	330.88 kg/kN (3.25 lb/lb st)

PERFORMANCE (clean, except where stated):

Never-exceed speed (V_{NE})	700 knots (1,297 km/h; 806 mph)
Max level speed at S/L	560 knots (1,037 km/h; 644 mph)
Cruising speed at optimum altitude	412 knots (763 km/h; 474 mph)
Approach speed	110 knots (204 km/h; 127 mph)
Stalling speed at max landing weight: flaps up	136 knots (253 km/h; 157 mph)
flaps down	98 knots (182 km/h; 113 mph)
Max rate of climb at S/L	2,323 m (7,620 ft)/min
Rate of climb at S/L, OEI	646 m (2,120 ft)/min
Service ceiling	12,925 m (42,400 ft)
Service ceiling, OEI	6,400 m (21,000 ft)
Min T-O run	1,185 m (3,890 ft)
T-O to 15 m (50 ft)	1,390 m (4,560 ft)
Landing from 15 m (50 ft)	774 m (2,540 ft)
Min landing run	521 m (1,710 ft)
Range with max military load	878 nm (1,627 km; 1,011 miles)
Ferry range with max external fuel: tanks retained	2,380 nm (4,410 km; 2,740 miles)
tanks jettisoned when empty	2,818 nm (5,222 km; 3,245 miles)

GRUMMAN PROWLER

US Navy designation: EA-6B

TYPE: Four-seat carrier-borne ECM aircraft.

PROGRAMME: Development contract issued Autumn 1966; externally similar to basic A-6 except longer nose enclosing four-seat cockpit and large pod on fin; first flight 25 May 1968; delivery of first 12 production aircraft started January 1971; last of 170 (164403) delivered 29 July 1991.

CURRENT VERSIONS: **Prototypes:** Five aircraft.

Standard: Following 23 aircraft.

EXCAP: Expanded capability; 25 aircraft.

ICAP-1: Improved capability variant; increased jamming capacity; all aircraft updated to ICAP-2; 45 new-build; 21 production EA-6B modified to ICAP-1 and all production aircraft to 1983 built to this standard; modifications include seven-band onboard tactical jamming system, reduced response time and multi-format display, automatic all-weather carrier landing system (ACLS), new defensive ECM, new communications, navigation and identification (CNI) system.

ICAP-2: Introduced further improved jamming capa ity; now standard; prototype first flight 24 June 1980; fir of 72 production aircraft (161776) delivered 3 Janua 1984; earlier aircraft modified to same standard; exciter each of five external jamming pods can generate signals one of seven different frequency bands instead of one, ar each pod can jam in two frequency bands simultaneousl

Subvariants of ICAP-2 are Block 82 and Block 8 (funded FY 1982 and FY 1986). Following 12 baseline ai craft, 162223 delivered 21 January 1986 as first of 2 Block 82s with HARM missile capability; Block 86 bega 29 July 1988 at 163049, covering final 37 with expande communications system and enhanced signal processing

ADVCAP (advanced capability): Initiated 1983; Litt Industries Amecom Division with Texas Instruments ar ITT contracted to produce new receiver/processor grou for tactical jamming system; ADVCAP EA-6B can car HARM (flight tested during 1989) or tanks on four inboa underwing pylons; additional outboard station under eac wing for ECM pods (total six underwing stations); ce treline station for tank or ECM pod; first flight of prototy (156482) rebuilt to ADVCAP 29 October 1989; retrofit ADVCAP to 102 ICAP-2 Prowlers began 1991 with co tract for one; funding for three planned in FY 1993. Prin feature of ADVCAP is Lockheed-Sanders AN/ALQ-14 communications jammer with underfuselage anten group; last of seven AN/ALQ-149 development mode delivered for flight trials early 1989; AN/ALQ-149 co tains eight assemblies in aircraft equipment bay includir all antennae, receivers, signal recognisers, computers ar controls needed to detect, identify, evaluate and jam hosti communications signals and long-range early warnir radar systems; Sanders has options to supply 95 productic AN/ALQ-149s. Other ADVCAP changes include MFI HUD display system, updated AN/ARN-118 Taca increased emergency RAT capacity, third AN/ARC-18 radio, Band 2/3 transmitter, improved stall/manoeuvrir capability, two additional (making four) AN/ALE-39 cha dispensers, Navstar GPS, disc-based recorder/onboa programme loader, fitting of 53.38 kN (12,000 lb st) J5 P-409 turbojets to boost max landing weight to 21,546 k (47,500 lb), fin height increased by 0.5 m (1 ft 7¾ in recontoured slats and flaps, drooped wing leading-edge new forward fuselage strakes and wingtip speed-bra modification to operate as ailerons in conjunction wi digital flight control system. Aerodynamic changes deve oped as Vehicle Enhancement Program (VEP, former VIP); first flight of VEP June 1992. Full ADVCAP cha ges constitute Block 91; 'export standard' Block 90 iden ical apart from lower (Block 86) level of jamming suite.

Block 2000: Proposed enhancement of ADVCAP wi JTIDS, satellite communications and survivabili measures.

CUSTOMERS: US Navy and Marine Corps, total 170; 12 US squadrons (VAQ-129 for training, 130, 131, 132, 134, 13 136, 137, 138, 139, 140 and 141, all at Whidbey Islan Washington, including one rotated to Atsugi, Japa equipped with Prowler by 1989. First detachment of U Marine Corps Prowler squadron VMAQ-2 began trainin on EA-6B at Cherry Point, North Carolina, in Septemb 1977 and deployed in late 1978. Two additional detac ments became full squadrons on 1 July 1992 — VMAQ- at Iwakuni, Japan and VMAQ-3 at Cherry Point — whi on 1 October 1992, VMAQ-4 elevated from Reserve full-time squadron at Whidbey Island, Washington; at lea one is deployed at all times. Deployment with USN reser units began in June 1989 with conversion of VAQ-309 Whidbey, from EA-6As. VAQ-209 followed at Andrew AFB, 1991. VAQ-35 formed at Whidbey, 1991, in 'ele tronic aggressor' role, complementing EA-6A operat

EA-6B PRODUCTION

Block	Quantity	First aircraft	Remarks
5	1	156478	Prototype
10	1	156479	Prototype
15	1	156480	Prototype
20	1	156481	Prototype
25	1	156482	Prototype
30	12	158029	
37	8	158540	J52-P-408 from 158544
40	3	158649	
45	19	158799	First EXCAP
50	3	159582	
55	3	159585	
60	11	159907	First ICAP-1
70	8	160437	
75	6	160786	
80	6	161115	
85	6	161242	
90	6	161347	
95	6	161774	ICAP-2 from 161776
100	6	161880	
105	8	162223	Block 82 from 162225
110	6	162934	
115	6	163030	
115	6	163044	Block 86 from 163049
120	12	163395	
125	12	163520	
130	9	163884	
135	3	164401	

VAQ-33 at Key West, Florida, although both units due to de-commission in late 1993 and pass aircraft to USN Reserve.
COSTS: $55.7 system unit cost, 1987.
DESIGN FEATURES: Generally as for A-6E; four crew and fin-tip antenna.
FLYING CONTROLS: As A-6E, but modified wingtip speed brakes in ADVCAP (see above).
STRUCTURE: Wings as for A-6E, but reinforced to allow for greater gross weight, fatigue life and 5.5g load factor. Fuselage as for A-6E, but lengthened by 1.37 m (4 ft 6 in).
LANDING GEAR: As for A-6E, except for reinforcement of attachments, A-frame arrester hook, and upgrading of structure to cater for increased gross weight.
POWER PLANT: Two Pratt & Whitney J52-P-408 turbojets, each rated at 49.8 kN (11,200 lb st). ADVCAP retrofit includes 53.38 kN (12,000 lb st) J52-P-409s.
ACCOMMODATION: Crew of four under two separate upward opening canopies. Martin-Baker GRUEA 7 ejection seats for crew. The two additional crewmen are ECM Officers to operate the ALQ-99 equipment from the rear cockpit. Either ECMO can independently detect, assign, adjust and monitor the jammers. The ECMO in the starboard front seat is responsible for communications, navigation, defensive ECM and chaff dispensing.
SYSTEMS: Generally as for A-6E.
AVIONICS: AN/ALQ-99F tactical jamming system, in five integrally powered pods, with a total of 10 jamming transmitters. Each pod covers one of seven frequency bands. Sensitive surveillance receivers in the fin-tip pod for long-range detection of radars; emitter information is fed to a central digital computer (AN/AYK-14 in ICAP-2 aircraft) that processes the signals for display and recording. Detection, identification, direction-finding and jammer-set-on sequence can be performed automatically or with manual assistance from crew. PRB Associates AN/TSQ-142 tactical mission support system. Teledyne Systems AN/ASN-123 navigation system with digital display group.
ARMAMENT: Originally unarmed, but currently capable of carrying Texas Instruments AGM-88A HARM anti-radar missiles underwing. Four underwing hardpoints on ICAP-2 aircraft, six on ADVCAP EA-6B.
DIMENSIONS, EXTERNAL: As for A-6E, except:
Width, wings folded 7.87 m (25 ft 10 in)
Length overall 18.24 m (59 ft 10 in)
Height overall 4.95 m (16 ft 3 in)
Wheelbase 5.23 m (17 ft 2 in)
WEIGHTS AND LOADINGS:
Weight empty 14,321 kg (31,572 lb)
Internal fuel load 6,995 kg (15,422 lb)
Max external fuel load 4,547 kg (10,025 lb)
T-O weight: from carrier in standoff jamming configuration (5 ECM pods) 24,668 kg (54,383 lb)
from field in ferry range configuration (max internal and external fuel) 27,226 kg (60,045 lb)
Max T-O weight, catapult or field 29,483 kg (65,000 lb)
Max zero-fuel weight 17,672 kg (38,961 lb)
Max landing weight, carrier or field 20,638 kg (45,500 lb)
Max wing loading 600.5 kg/m² (123 lb/sq ft)
Max power loading 296.0 kg/kN (2.90 lb/lb st)
PERFORMANCE (A: clean, B: 5 ECM pods):
Never-exceed speed (V_{NE}) 710 knots (1,315 km/h; 817 mph)
Max level speed at S/L:
A 566 knots (1,048 km/h; 651 mph)
B 530 knots (982 km/h; 610 mph)
Cruising speed at optimum altitude:
A, B 418 knots (774 km/h; 481 mph)
Stalling speed:
flaps up, max power:
A 124 knots (230 km/h; 143 mph)
flaps down, max power:
A 84 knots (156 km/h; 97 mph)
Max rate of climb at S/L: A 3,932 m (12,900 ft)/min
B 3,057 m (10,030 ft)/min
Rate of climb at S/L, OEI: A 1,189 m (3,900 ft)/min
Service ceiling: A 12,550 m (41,200 ft)
B 11,580 m (38,000 ft)
Service ceiling, OEI: A 8,930 m (29,300 ft)
T-O run: B 814 m (2,670 ft)
T-O to 15 m (50 ft): A 823 m (2,700 ft)
Landing run: A 579 m (1,900 ft)
B 655 m (2,150 ft)
Range with max external load, 5% reserves plus 20 min at S/L: B 955 nm (1,769 km; 1,099 miles)
Ferry range with max external fuel:
tanks retained 1,756 nm (3,254 km; 2,022 miles)
tanks jettisoned when empty 2,085 nm (3,861 km; 2,399 miles)

GRUMMAN TOMCAT

US Navy designation: F-14
TYPE: Two-seat carrier-based long-range interceptor with attack capability.
PROGRAMME: Won US Navy VFX fighter competition 15 January 1969; first flight of first of 12 development aircraft 21 December 1970; original programme was for 497 Tomcats including 12 development aircraft; programme since extended into 1990s; 28 Tomcat squadrons scheduled by

ADVCAP system test aircraft for Grumman EA-6B Prowler

end 1987, including four reserve and two training squadrons, operating from 12 carriers and Naval Air Stations at Miramar (California), Oceana (Virginia), and Dallas (Texas).

Initial F-14A deployed with USN squadrons VF-1 and VF-2 October 1972; total 557, including 12 development aircraft, delivered to US Navy by April 1987, when production ended; final 102 aircraft (beginning 161597) delivered from FY 1983 powered by improved TF30-P-414A turbofans, having same rating as original 93 kN (20,900 lb st) TF30-P-412A.
CURRENT VERSIONS: **F-14A**: Initial and main version, supplied to US Navy and Iran. Total of 557 built (last aircraft 162711), ending 13 March 1987; 32 reworked to F-14B and 18 to F-14D(R); 385 remained in 1993. Retrofit with Tape 115B (later Tape 116) started May 1991 (see under Avionics) permitting conventional bombing. First two F-14A **'Bomcat'** squadrons, VF-14 and VF-32, began cruise on USS *Kennedy*, 7 October 1992.

F-14MMCAP: (Formerly F-14A++.) Multi-Mission Capability Avionics Program; US Navy proposal, 1992, to upgrade 250 F-14As with improved avionics.

F-14B: Known as F-14A(Plus) until 1 May 1991; second use of F-14B designation; interim improved version re-engined with GE F110-GE-400 turbofans pending introduction of F-14D (see below). Development began July 1984; Grumman prime contractor with General Electric as subcontractor. F110 has 82 per cent parts commonality with F110-GE-100 in USAF F-15s and F-16s; 1.27 m (4 ft 2 in) plug inserted in afterburner section to match engine to F-14A inlet position and airframe contours; only secondary structure requires modification; new engine allows unrestricted throttle handling throughout flight envelope and fewer compressor stalls. NASA scoops for Vulcan cannon; glove vanes eliminated; cockpit modified. Full-scale development used two aircraft, including (first use) F-14B prototype (157986), which made first flight with definitive F110-GE-400 engines 29 September 1986; one prototype was to be upgraded to full F-14D.

F-14B followed F-14A into production; first flight production F-14B (162910) 14 November 1987; 18, 15 and five funded in FY 1986-88; first delivery, to VF-101, 11 April 1988; IOC with two US Navy squadrons achieved early 1989; F-14B production deliveries (last aircraft 163411) completed May 1990. Additionally, 32 F-14As upgraded to F-14B. Further five-seven conversions funded ($143 million) in FY 1992; Grumman to build conversion kits, then compete with Norfolk naval depot for installation contract; FY 1993 funds of $175 million for approximately 12 more conversions. B variant issued to VF-24, 74, 101, 103, 142, 143 and 211.

F-14A/B Upgrade: Total 196 aircraft (all extant F-14Bs plus balance in late production F-14As) to be retrofitted from FY 1994 (30 aircraft; then 25 per year) with some F-14D features, including AN/ALR-67 RWR (to A models), BOL chaff dispenser, AN/AYK-14 mission computer (plus interface to analog 5400 computer), new tactical information display to permit addition of future weapon systems without re-wiring aircraft, MIL-STD-1553B digital databus, and airframe time compliance requirements (TCRs) for 7,500-hour life.

F-14D Super Tomcat: Improved version with AN/APG-71 radar, NACES seats in NVG-compatible cockpits, twin IRST/TV pod (ordered January 1993 for installation before August 1996), JTIDS link (from 1993), AN/ALR-67 RWR (from 1993), plus enhanced missile capability (AIM-120 AMRAAM, AIM-54C Phoenix). Total 37 out of 127 planned new F-14Ds funded (seven, 12 and 18 in FYs 1988-90) before programme cancelled as economy measure in 1989; plans continued for 400 F-14A and F-14B to be remanufactured to F-14D; six **F-14D(R)** conversions funded FY 1990; Grumman working on four conversions, USN at Norfolk NADEP on two; FY 1991-95 funding plans for 12, 18, 20, 24 and 24 all cancelled February 1991, but FY 1991 dozen restored April 1991 (Grumman eight, NADEP four); F-14D programme thus 37 new, 18 rebuilt. Last Grumman redelivery April 1993; last from Norfolk in November 1993. First flight of first of three development aircraft 24 November 1987; 36 months flight testing included a TA-3B Skywarrior. Prototype, 161867, delivered trials squadron VX-4 at Point Mugu, California, May 1990; first production F-14D (163412) rolled out 23 March 1990; last (164604) delivered 20 July 1992. First F-14D(R) (161659) delivered September 1991; two production aircraft to **NF-14D** for permanent test use by VX-4 at Point Mugu, California. Training squadron, VF-124 (first user) October 1990; official acceptance 16 November 1990; first embarked squadrons, VF-11 and VF-31; VF-1 and VF-2 to convert from F-14A in 1993; all F-14D based at Miramar, California. Last F-14D(R) due for redelivery in Spring 1993.

Future upgrades: F-14Ds and F-14A/B Upgrades provide USN with core force of 251 modernised Tomcats for which future planning includes digital flight control system from FY 1996; GPS from FY 1995 (to F-14Ds first); AN/ARC-210 radios from FY 1998; and improved mission recorders from FY 1999. Navy also considering further bombing upgrades (laser designator, FLIR and AN/ALE-50 towed decoy) to offset possible withdrawal of A-6 Intruder by 2000.

Other upgrades recently proposed are:

Quickstrike: Suggested first-stage development of F-14D for extended ground attack potential, providing F-15E Eagle-like capability to the fleet, including standoff weapons; potentially available (132-aircraft proposal) from 1994.

Super Tomcat-21: Company-funded study for US Navy multi-role fighter for next century as economic alternative to Navy version of USAF Advanced Tactical Fighter (ATF). Quickstrike improvements plus new slotted flaps, extended-chord leading-edge slats, enlarged wing glove fairings containing extra fuel, new frameless windscreen. Grumman claims Tomcat-21 can have 90 per cent of ATF capability for 60 per cent of cost. Conversion possible from F-14A.

Attack Super Tomcat-21 (AST-21): Suggested interim replacement for A-12 Avenger II with low-level penetration capability and nuclear as well as conventional armament. Terrain avoidance radar; two extra weapon stations.

ASF-14: Grumman proposal for alternative to Navalised ATF as F-14 evolutionary development including avionics, and possibly power plants, developed for ATF. Could be available by year 2000.
CUSTOMERS: US Navy (see Current Versions). Present F-14 squadrons are (variants in parentheses) are VF-1 (A to D 1993), VF-2 (A to D 1993), VF-11 (D), VF-21 (A), VF-24 (B to A 1992), VF-31 (D), VF-51 (A), VF-111 (A), VF-124 (D, training), VF-154 (A), VF-211 (B to A 1992), VF-213 (A), 301 (Reserve) and 302 (Reserve) at Miramar, California; VF-14 (A), VF-32 (A), VF-33 (A), VF-41 (A), VF-74 (B), VF-84 (A), VF-101 (A/B training), VF-102 (A), VF-103 (B), VF-142 (B) and VF-143 (B) at Oceana, Virginia; VF-201 (A) and VF-202 (A) of the Reserve at Dallas, Texas. In August 1991, 'Top Gun' Fighter Weapons School at Miramar received 'aggressor' F-14As in Sukhoi Su-27 markings.

Iran also acquired 79 F-14As in 1976-78; retained Phoenix missile system, but had slightly different ECM equipment.
COSTS: $984 million fixed-price contract in July 1984 for development of F-14B and F-14D Super Tomcats.
DESIGN FEATURES: Wing sweepback variable from 20° leading-edge to 68°; oversweep of 75° used for carrier stowage without wing fold; wing pivot point 2.72 m (8 ft 11 in) from aircraft centreline; fixed glove has dihedral to

484 USA: AIRCRAFT—GRUMMAN

Grumman F-14B Tomcat of US Navy Squadron VF-74 from USS *Saratoga*

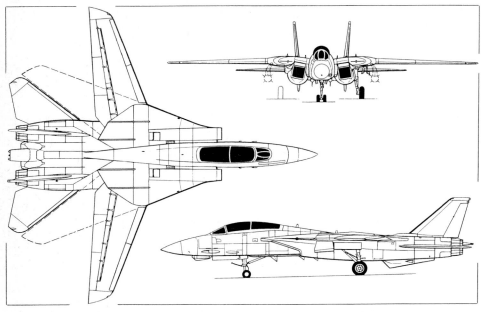

Grumman F-14B Tomcat carrier-based multi-role fighter (*Dennis Punnett*)

Second production F-14D (163413), used by trials squadron VX-4

minimise cross-sectional area and reduce wave drag; small canards on F-14A known as glove vanes extend forward progressively to 15° from inboard leading-edge to balance supersonic trim change and unload tail surfaces.

FLYING CONTROLS: Lateral control by long-span spoilers, ahead of flaps, and tailerons; automatic leading-edge slats assist manoeuvring; strakes emerge from wing glove leading-edge at high airspeeds; automatic wing sweep has manual override; automatic scheduling of control with airspeed; autostabilisation and angle of attack protection; autopilot and automatic carrier landing system (ALCS). Airbrake panel above and below tail, between fins. Twin fins and rudders.

STRUCTURE: Wing carry-through is one-piece electron beam-welded structure of Ti-6A1-4V titanium alloy with 6.71 m (22 ft) span. Fuselage has machined frames, titanium main longerons and light alloy stressed skin; centre fuselage is fuel-carrying box; radome hinges upwards for access to radar; fuel dump pipe at extreme tail; fins and rudders of light alloy honeycomb sandwich; tailplanes have multiple spars, honeycomb trailing-edges and boron/epoxy composites skins.

LANDING GEAR: Retractable tricycle type. Twin-wheel nose unit and single-wheel main units retract forward, main units inward into bottom of engine air intake trunks. Original beryllium brakes were replaced with Goodyear lightweight carbon brakes from Spring 1981. Arrester hook under rear fuselage, housed in small ventral fairing. Nose-tow catapult attachment on nose unit.

POWER PLANT (F-14B/D): Two General Electric F110-GE-400 turbofans rated at 71.56 kN (16,088 lb st) dry and 120.1 kN (27,000 lb st) with afterburning. Garrett ATS200-50 air turbine starter. F110 engine has 43 per cent more reheated thrust and 37 per cent more military thrust (without afterburning) than TF30-P-414A in F-14A; results in 20 per cent more specific excess energy, 30 per cent lower specific fuel consumption in afterburner, 62 per cent greater deck launch intercept radius and 34 per cent more combat air patrol time; can be launched without afterburner; time to 10,670 m (35,000 ft) reduced by 61 per cent and acceleration time by 43 per cent. Integral fuel tanks in outer wings, each with capacity of 1,117 litres (295 US gallons; 246 Imp gallons); between engines in rear fuselage, with capacity of 2,453 litres (648 US gallons; 539 Imp gallons); and forward of wing carry-through structure, capacity 2,616 litres (691 US gallons; 575 Imp gallons); plus two feeder tanks with combined capacity of 1,726 litres (456 US gallons; 380 Imp gallons). Total internal fuel capacity 9,029 litres (2,385 US gallons; 1,986 Imp gallons). An external auxiliary fuel tank can be carried beneath each intake trunk, each containing 1,011 litres (267 US gallons; 222 Imp gallons). Retractable flight refuelling probe on starboard side of fuselage near front cockpit.

ACCOMMODATION: Pilot and naval flight officer seated in tandem on Martin-Baker NACES (or GRU7A in F-14A/B) rocket assisted zero/zero ejection seats, under a one-piece bubble canopy, hinged at the rear and offering all-round view.

AVIONICS: In F-14A, Hughes AN/AWG-9 weapons control system, with ability to detect airborne targets at ranges of more than 65-170 nm (120-315 km; 75-195 miles) according to their size, and ability to track 24 enemy targets and attack six of them simultaneously at varied altitudes and distances. Fairchild AN/AWG-15F fire control set; CP-1066/A central air data computer; CP-1050/A computer signal data converter; AN/ASW-27B digital data link; AN/APX-76(V) IFF interrogator; AN/APX-72 IFF transponder; AN/ASA-79 multiple display indicator group; Kaiser Aerospace AN/AVG-12 vertical and head-up display system. AN/ARC-51 and AN/ARC-15 UHF com; AN/ARR-69 UHF auxiliary receiver; KY-2 cryptographic system; LS-460/B intercom; AN/ASN-92(V) INS; A/A24G39 AHRS; AN/APN-15 beacon augmentor; AN/APN-194(V) radar altimeter; AN/ARA-63A receiver-decoder; AN/ARN-84 micro Tacan; AN/ARA-50 UHF ADF; AN/APR-27/50 radar receiver; AN/APR-25/45 radar warning set. TV optical unit in undernose pod. Northrop Corporation television camera set (TCS) mounted beneath nose is closed-circuit TV system, offering both wide-angle (acquisition) and telescopic (identification) fields of view. TCS automatically searches for, acquires and locks on to distant targets, displaying them on monitors for the pilot and flight officer. Small undernose pod for Sanders AN/ALQ-100/12 deception jamming system, relocated under camera package of aircraft with Northrop TCS. During 1980-81, 49 F-14As were allocated to carry TARPS (tactical air reconnaissance pod system), containing a KS-87B frame camera, KA-99 low altitude panoramic camera, and AN AAD-5 infra-red reconnaissance equipment, on underbelly attachment; by 1993 TARPS capability in 23 F-14As, 1 F-14Bs and 55 F-14Ds.

In F-14D, some 60 per cent of analog avionics made digital, giving new weapons management, navigation, displays and control functions. MIL-STD-1553B digital bus interconnects Litton AN/ALR-67 threat warning and recognition system, joint tactical information distribution system (JTIDS), GE Aerospace Electronic Systems infra-red search and track sensor (IRST) and television camera set (TCS); emphasis on commonality with F/A-18 and later A-6. (Westinghouse/ITT AN/ALQ-165 Airborne Self Protection Jammer cancelled 1992.) New Hughes AN/APG-71 replaces AN/AWG-9 radar with improved ECCM, monopulse angle tracking, digital scan control, target identification and raid assessment. AN/APG-71 features non-co-operative target identification, and ECCM using low-sidelobe antenna and sidelobe blanking guard channel, frequency agility, new high-speed digital signal processor based on AN/APG-70 radar in US Air Force multi-staged improvement programme for F-15. Litton AN/ASN-139 INS; Smiths AN/AYQ-15 stores management system. ECM equipment includes Goodyear AN/ALE-29 and AN/ALE-39 chaff and flare dispensers with integral jammers.

All in-service F-14A Tomcats given Tape 116 computer software addition in Summer 1992 to allow full ground attack with conventional bombs. F-14Ds of VF-11 and VF-31 have Tape G-6 software changes.

ARMAMENT: One General Electric M61A-1 Vulcan 20 mm gun mounted in the port side of forward fuselage, with 675 rounds of ammunition. Four AIM-7 Sparrow air-to-air missiles mounted partially submerged in the underfuselage, or four AIM-54 Phoenix missiles carried on special pallets which attach to the bottom of the fuselage. Two wing pylons, one under each fixed wing section, can carry four AIM-9 Sidewinder missiles or two additional Sparrow or Phoenix missiles with two Sidewinders. F-14D has bombing capability; Rockeye and CBU-59 cluster bombs validated for F-14 December 1992; GBU-16 LGB and Gator mine to follow; AGM-88 HARM ARM and SLAM ASMs planned, but not yet funded.

DIMENSIONS, EXTERNAL:
Wing span: unswept	19.54 m (64 ft 1½ in)
swept	11.65 m (38 ft 2½ in)
overswept	10.15 m (33 ft 3½ in)
Wing aspect ratio	7.28
Length overall	19.10 m (62 ft 8 in)
Height overall	4.88 m (16 ft 0 in)
Tailplane span	9.97 m (32 ft 8½ in)
Distance between fin tips	3.25 m (10 ft 8 in)
Wheel track	5.00 m (16 ft 5 in)
Wheelbase	7.02 m (23 ft 0½ in)

AREAS:
Wings, gross	52.49 m² (565.0 sq ft)
Leading-edge slats (total)	4.29 m² (46.2 sq ft)
Trailing-edge flaps (total)	9.87 m² (106.3 sq ft)
Spoilers (total)	1.97 m² (21.2 sq ft)
Fins (total)	7.90 m² (85.0 sq ft)
Rudders (total)	3.06 m² (33.0 sq ft)

Horizontal tail surfaces (total) 13.01 m² (140.0 sq ft)
'EIGHTS AND LOADINGS (F-14D with F110-GE-400):
Weight empty 18,951 kg (41,780 lb)
Fuel (usable): internal 7,348 kg (16,200 lb)
external 1,724 kg (3,800 lb)
Max external weapon load 6,577 kg (14,500 lb)
T-O weight: fighter/escort mission 29,072 kg (64,093 lb)
fleet air defence mission 33,157 kg (73,098 lb)
max 33,724 kg (74,349 lb)
Max wing loading 642.5 kg/m² (131.59 lb/sq ft)
Max power loading 140.4 kg/kN (1.38 lb/lb st)
:RFORMANCE (F110 engines):
Max level speed Mach 1.88
Max cruising speed Mach 0.72
Carrier approach speed 125 knots (232 km/h; 144 mph)
Service ceiling above 16,150 m (53,000 ft)
Field T-O distance 762 m (2,500 ft)
Field landing distance 732 m (2,400 ft)
Field range with external fuel approx 1,600 nm (2,965 km; 1,842 miles)

Grumman E-8A J-STARS prototype, displaying underfuselage radome

GRUMMAN E-8C (J-STARS)

Grumman Aerospace & Electronics Group is prime con-actor for US Air Force/Army Joint Surveillance and Target ttack Radar Systems; two development E-8As (formerly -18C) served in Gulf War January 1992; one E-8B airframe cquired, but resold before conversion; production aircraft ill be re-engined ex-airline Boeing 707-320s.

Grumman involved with J-STARS since 1985; $523 mil-on (later increased to $643 million – not to exceed) contract warded November 1990 including enhancements to system nd replacement third prototype. Low-rate initial production dvanced procurement contract signed by USAF 24 April 1992; full funding for first lot of two E-8Cs at eventual estimated cost of $450 million. Third prototype at Grumman, Melbourne, in December 1992 after modification at St Augustine; deliveries from November 1995 to March 2001 after conversion at Grumman, Lake Charles, where first two airframes delivered May - June 1992; IOC 1997; total 19 aircraft, including refurbished E-8As.

$1.9 million USAF contract to Grumman-led consortium in 1992 to study multi-mission airborne surveillance technology (MAST) leading to aircraft augmenting existing E-3 AWACS and E-8 J-STARS. NATO considering international J-STARS force, late 1992, not necessarily using Boeing 707 airframe.

Further details in *Jane's Civil and Military Aircraft Upgrades* and *Jane's Battlefield Surveillance Systems.*

;ULFSTREAM AEROSPACE

;ULFSTREAM AEROSPACE CORPORATION

'O Box 2206, Savannah International Airport, Savannah, Georgia 31402-2206
elephone: 1 (912) 964 3000
'ax: 1 (912) 964 3775 or 1 (912) 966 4171
elex: 546470 GULF AERO
HAIRMAN AND CEO: William C. Lowe
RESIDENT AND COO: Fred A. Breidenbach
ENIOR VICE-PRESIDENT, ENGINEERING: Charles N. Coppi
ENIOR VICE-PRESIDENT, MARKETING: Robert H. Cooper
ORPORATE COMMUNICATIONS MANAGER: Henry Ogrodzinsky

Original facilities purchased by Allen E. Paulson from irumman Corporation in 1978; facilities expanded and two iulfstream models introduced when company purchased om Mr Paulson by Chrysler Corporation in 1985; Mr Paul-on and Forstmann Little and Company completed repur-hase from Chrysler Corporation 19 March 1990.

AlliedSignal AiResearch facility at Long Beach Inter-ational Airport, California, bought by Gulfstream in 1986 nd expanded to accommodate outfitting and completion of 7 Gulfstream IVs at a time; Gulfstream also operates sub-ontracting facility in Oklahoma City. Oklahoma Corpor-tion's three facilities cover more than 185,806 m² (2.0 mil-on sq ft) of manufacturing and servicing space; being xpanded by additional 50,168 m² (540,000 sq ft), mainly for iulfstream V production, beginning Autumn 1993 with sec-nd phase starting 1995.

Flight testing to begin late June 1993 of 'Quiet Spey' pro-uction hushkit to enable earlier Gulfstreams II, IIB and III over 440 of which in worldwide service) to comply with 'AR Stage 3 noise regulations.

;ULFSTREAM AEROSPACE GULFSTREAM IV and IV-SP

YPE: Twin-turbofan long-range business transport.

ROGRAMME: Design started March 1983; manufacture of four production prototypes (one for static testing) began 1985; first aircraft (N404GA) rolled out 11 September 1985; first flight 19 September 1985; first flight of second prototype 11 June 1986 and third prototype August 1986; FAA certification 22 April 1987 after 1,412 hours flight testing. Westbound round-the-world flight from Le Bourget, Paris, on 12 June 1987, covering 19,887.9 nm (36,832.44 km; 22,886.6 miles), took 45 h 25 min at average 437.86 knots (811.44 km/h; 504.2 mph) and set 22 world records; eastbound round-the-world flight in N400GA from Houston, Texas, on 26/27 February 1988 covered 20,028.68 nm (37,093.1 km; 23,048.6 miles) in 36 h 8 min 34 s at average 554.15 knots (1,026.29 km/h; 637.71 mph), setting 11 records and bettering United Airlines 747SP circuit flown 30 days before; aircraft carried 3,629 kg (8,000 lb) optional internal long-range tank. In March 1993, Gulfstream IV-SP N485GA set new world speed and distance records in class, at 503.57 knots (933.21 km/h; 579.87 mph) and 5,139 nm (9,524 km; 5,918 miles) respectively, on routine business flight from Tokyo, Japan, to Albuquerque, USA.

:URRENT VERSIONS: **Gulfstream IV:** Current version until 1992; *detailed description applies to this model except where indicated.*

Gulfstream IV-SP: Improved, higher weight version announced at NBAA Convention, Houston, in October 1991; designation applies to all new IVs sold after 6 September 1992; max payload increased by 1,134 kg (2,500 lb) and max landing weight increased by 3,402 kg (7,500 lb), with no increase in guaranteed manufacturer's bare weight empty. Payload/range envelope extended, expanded capability Honeywell SPZ-8000 flight guidance and control system; plus enhancements including GPS, TCAS, T-O performance monitors and turbulence prediction systems.

SRA-4, C-20, Tp 102: Special missions aircraft, described separately.

CUSTOMERS: Total 207 Gulfstream IVs and IV-SPs delivered by end 1992, including 25 in 1992; annual production rate 30 aircraft.

DESIGN FEATURES: Differences from Gulfstream III (1987-88 and earlier *Jane's*) include aerodynamically redesigned wing contributing to lower cruise drag; wing also structurally redesigned with 30 per cent fewer parts, 395 kg (870 lb) lighter and carrying 453 kg (1,000 lb) more fuel; increased tailplane span; fuselage 1.37 m (4 ft 6 in) longer, with sixth window each side; Rolls-Royce Tay turbofans; flight deck with electronic displays; digital avionics; and fully integrated flight management and autothrottle systems.

Advanced sonic rooftop aerofoil; sweepback at quarter-chord 27° 40′; thickness/chord ratio 10 per cent at wing station 50, 8.6 per cent at station 414; dihedral 3°; incidence 3° 30′ at root, −2° at tip; NASA (Whitcomb) winglets.

FLYING CONTROLS: Hydraulically powered flying controls with manual reversion; trim tab in port aileron and both elevators; two spoilers on each wing act differentially to assist aileron and together with third spoiler each side act collectively as airbrakes and lift dumpers; single-slotted Fowler flaps; four vortillons and a single 'tripper' strip under leading-edge of each wing ensure good stalling behaviour.

STRUCTURE: Wing manufactured by Textron Aerostructures. Light alloy airframe except for carbon composites ailerons, spoilers, rudder and elevators, some tailplane parts, some cabin floor structure, and parts of flight deck; winglets of aluminium honeycomb.

LANDING GEAR: Retractable tricycle type with twin wheels on each unit. Main units retract inward, steerable nose unit forward. Mainwheel tyres size 34 × 9.25-16 for IV and H34 × 9.25-18 for IV-SP; pressure 12.07 bars (175 lb/sq in). Nosewheel tyres size 21 × 7.25-10, pressure 7.9 bars (115 lb/sq in). IV has Aircraft Braking Systems air-cooled carbon brakes and IV-SP has Dunlop air-cooled carbon brakes; both use Aircraft Braking Systems anti-skid units and digital electronic brake-by-wire system. Dowty electronic steer-by-wire system.

POWER PLANT: Two Rolls-Royce Tay Mk 611-8 turbofans, each flat rated at 61.6 kN (13,850 lb st) to ISA +15°C. Target type thrust reversers. Fuel in two integral wing tanks, with total capacity of 16,542 litres (4,370 US gallons; 3,639 Imp gallons). Single pressure fuelling point in leading-edge of starboard wing.

ACCOMMODATION: Crew of two plus cabin attendant. Standard seating for 14 to 19 passengers in pressurised and air-conditioned cabin. Galley, toilet and large baggage compartment, capacity 907 kg (2,000 lb), at rear of cabin. Integral airstair door at front of cabin on port side. Baggage compartment door on port side. Electrically heated wraparound windscreen. Six cabin windows, including two overwing emergency exits, on each side.

SYSTEMS: Cabin pressurisation system max differential 0.65 bar (9.45 lb/sq in); dual air-conditioning systems. Two independent hydraulic systems, each 207 bars (3,000 lb/sq in). Maximum flow rate 83.3 litres (22 US gallons; 18.3 Imp gallons)/min. Two bootstrap type hydraulic reservoirs, pressurised to 4.14 bars (60 lb/sq in). Garrett GTCP36-100 APU in tail compartment, flight rated to 12,500 m (41,000 ft) since s/n 1156. Electrical system includes two 36kVA alternators with two solid state 30kVA converters to provide 23kVA 115/200V 400Hz AC power and 250A of regulated 28V DC power; two 24V 40Ah nickel-cadmium storage batteries and external power socket. Wing leading-edges and engine inlets anti-iced.

AVIONICS: Standard items include a Honeywell SPZ-8000 digital AFCS, Honeywell FMZ-800 Phase II flight management system and six 20.3 cm × 20.3 cm (8 in × 8 in) colour CRT displays, two each for primary flight instruments, navigation and engine instrument and crew alerting systems (EICAS); dual fail-operational flight guidance systems including autothrottles; dual air data systems and digital colour radar. System integration is accomplished through a Honeywell avionics standard communications bus (ASCB). Other factory-installed avionics include dual VHF/HF com; dual VOR/LOC/GS and markers; dual DME; dual ADF; dual radio altimeters; dual transponders; dual cockpit audio; dual flight guidance and performance computers; dual laser INS; AHRS; and cockpit voice recorder. The system is designed to provide growth potential for interface with MLS, GPS and VLF Omega in future developments. Flight data recorder. Optional avionics include Racal Satfone satellite communications equipment.

DIMENSIONS, EXTERNAL:
Wing span over winglets 23.72 m (77 ft 10 in)
Wing chord:
at root (fuselage centreline) 5.94 m (19 ft 5⅞ in)
at tip 1.85 m (6 ft 0¾ in)
Wing aspect ratio 5.92
Length overall 26.92 m (88 ft 4 in)
Fuselage: Length 24.03 m (78 ft 10 in)
Max diameter 2.39 m (7 ft 10 in)
Height overall 7.45 m (24 ft 5⅛ in)
Tailplane span 9.75 m (32 ft 0 in)
Wheel track 4.17 m (13 ft 8 in)
Wheelbase 11.61 m (38 ft 1¼ in)
Passenger door (fwd, port): Height 1.57 m (5 ft 2 in)
Width 0.91 m (3 ft 0 in)
Baggage door (rear): Height 0.91 m (2 ft 11¾ in)
Width 0.72 m (2 ft 4½ in)

DIMENSIONS, INTERNAL:
Cabin:
Length, incl galley, toilet and baggage compartment 13.74 m (45 ft 1 in)
Max width 2.24 m (7 ft 4 in)
Max height 1.85 m (6 ft 1 in)
Floor area 22.9 m² (247 sq ft)
Volume 47.77 m³ (1,687 cu ft)
Flight deck volume 3.51 m³ (124.0 cu ft)
Rear baggage compartment volume 4.78 m³ (169.0 cu ft)

AREAS:
Wings, gross 88.29 m² (950.39 sq ft)

486 USA: AIRCRAFT—GULFSTREAM AEROSPACE

Gulfstream Aerospace Gulfstream IV of United Arab Emirates Royal Flight *(Peter J. Cooper)*

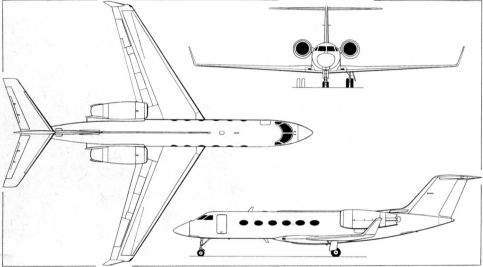

Gulfstream Aerospace Gulfstream IV twin-turbofan business transport *(Dennis Punnett)*

Development prototype of the Gulfstream SRA-4 electronic warfare support aircraft

Ailerons (total, incl tab)	2.68 m² (28.86 sq ft)
Trailing-edge flaps (total)	11.97 m² (128.84 sq ft)
Spoilers (total)	7.46 m² (80.27 sq ft)
Winglets (total)	2.38 m² (25.60 sq ft)
Fin	10.92 m² (117.53 sq ft)
Rudder, incl tab	4.16 m² (44.75 sq ft)
Horizontal tail surfaces (total)	18.83 m² (202.67 sq ft)
Elevators (total, incl tabs)	5.22 m² (56.22 sq ft)

WEIGHTS AND LOADINGS:
Manufacturer's weight empty:
  IV, IV-SP   16,102 kg (35,500 lb)
Typical operating weight empty: IV 19,278 kg (42,500 lb)
  IV-SP   19,277 kg (42,500 lb)
Max payload: IV   1,814 kg (4,000 lb)
  IV-SP   2,948 kg (6,500 lb)
Max usable fuel: IV, IV-SP   13,381 kg (29,500 lb)
Max T-O weight: IV   33,203 kg (73,200 lb)
  IV-SP   33,838 kg (74,600 lb)
Max ramp weight: IV   33,384 kg (73,600 lb)
  IV-SP   34,019 kg (75,000 lb)
Max zero-fuel weight: IV   21,092 kg (46,500 lb)
  IV-SP   22,226 kg (49,000 lb)
Max landing weight: IV   26,535 kg (58,500 lb)
  IV-SP   29,937 kg (66,000 lb)
Max wing loading: IV   375.9 kg/m² (77.02 lb/sq ft)
  IV-SP   383.2 kg/m² (78.49 lb/sq ft)
Max power loading: IV   269.6 kg/kN (2.64 lb/lb st)
  IV-SP   274.7 kg/kN (2.69 lb/lb st)

PERFORMANCE (at max T-O weight, ISA, except where indicated):
Max operating speed
  340 knots (629 km/h; 391 mph) CAS *or* Mach 0.88
Max cruising speed at 9,450 m (31,000 ft)
  505 knots (936 km/h; 582 mph)
Normal cruising speed at 13,715 m (45,000 ft)
  Mach 0.80 (459 knots; 850 km/h; 528 mph)
Stalling speed at max landing weight:
  IV, wheels and flaps up
    122 knots (227 km/h; 141 mph)
  IV-SP, wheels and flaps up
    130 knots (241 km/h; 150 mph)
  IV, wheels and flaps down
    108 knots (200 km/h; 124 mph)
  IV-SP, wheels and flaps down
    115 knots (213 km/h; 133 mph)
Approach speed at max landing weight:
  IV   140 knots (259 km/h; 161 mph)
  IV-SP   149 knots (276 km/h; 172 mph)
Max rate of climb at S/L: IV   1,220 m (4,000 ft)/min
  IV-SP   1,210 m (3,970 ft)/min
Rate of climb at S/L, OEI: IV   337 m (1,105 ft)/min
  IV-SP   314 m (1,030 ft)/min
Max operating altitude   13,715 m (45,000 ft)
Runway PCN   2
FAA balanced T-O field length at S/L:
  IV   1,609 m (5,280 ft)
  IV-SP   1,662 m (5,450 ft)
Landing from 15 m (50 ft): IV   1,032 m (3,386 ft)
  IV-SP   973 m (3,190 ft)
Range:
  with max payload, normal cruising speed and NBAA IFR reserves: IV   3,633 nm (6,732 km; 4,183 miles)
    IV-SP   3,338 nm (6,186 km; 3,843 miles)
  with max fuel, eight passengers, at Mach 0.80 and with NBAA IFR reserves
    4,220 nm (7,820 km; 4,859 miles)

OPERATIONAL NOISE LEVELS (FAR Pt 36):
  T-O: IV   76.8 EPNdB
    IV-SP   77.5 EPNdB
  Approach: IV   91.0 EPNdB
    IV-SP   92.0 EPNdB
  Sideline: IV   87.3 EPNdB
    IV-SP   86.6 EPNdB

GULFSTREAM AEROSPACE SRA-4
Swedish Air Force designation: Tp 102
US military designations: C-20F/G/H

TYPE: Special requirements aircraft version of Gulfstream IV.
PROGRAMME: Development aircraft (N413GA) for electronic warfare support version, integrated by Electrospace Systems Inc (now CTAS: see under Chrysler heading) exhibited at Farnborough Air Show 1988.
CURRENT VERSIONS: **Electronic warfare support:** Development aircraft had forward underfuselage pod for jamming antennae; cabin contains operators' consoles and microwave generator and amplifier rack, modulation generator rack, radio racks, and chaff supply and cutters to simulate EW from adversary aircraft and missiles; aircraft could be used to test and evaluate weapon systems and to develop electronic warfare tactics.
  Electronic surveillance/reconnaissance: Possible sensors include side-looking synthetic aperture radar in belly-mounted pod under forward fuselage, long-range oblique photographic camera (LOROP), ESM, VHF/UHF/HF communications for C³, chaff dispensers, infrared countermeasures in tailcone, SAR equipment and accommodation for operators for each system. Typical mission profile with 1,950 kg (4,300 lb) payload allows 10.5 hours on station at loiter altitudes between 10,670 m and 15,550 m (35,000 ft and 51,000 ft).
  Maritime patrol: Equipment includes high definition surface search radar, forward looking infra-red detection system (IRDS), electronic support measures (ESM), flare marker launch tubes, nav/com and ESM consoles, positions for up to eight observers/console operators, stowage and deployment for survival equipment, and crew rest area. SRA-4 with 1,950-4,173 kg (4,300-9,200 lb) mission payload including 272 kg (600 lb) expendable stores can operate at 600 nm (1,112 km; 690 mile) radius for four to six hours; outbound flight at 12,500 m (41,000 ft) and 454 knots (841 km/h; 523 mph); search at 3,050 m (10,000 ft) but spend one-third of time at 61 m (200 ft); return flight at 13,715 m (45,000 ft).
  ASW: Equipment includes nose radar able to detect periscope and snorkels, FLIR, sonobuoy launchers, acoustic processor, magnetic anomaly detector (MAD) in tail, ESM, torpedo stowage in weapon bay under forward fuselage, and anti-shipping missile carried on each underwing hardpoint. Mission profile with six crew and 2,503 kg (5,518 lb) payload can stay 4.3 hours in hi-lo loitering and manoeuvring at 1,000 nm (1,853 km; 1,151 miles) radius. Mission profile for **anti-shipping** with two missiles allows 1,350 nm (2,502 km; 1,554 miles) outbound flight at high altitude; descent to 61 m (200 ft) for 100 nm (185 km; 115 mile) attack run at 350 knots (649 km/h; 403 mph); launch missile 50 nm (93 km; 57 miles) from target; return to base at 13,715 m (45,000 ft).
  Medical evacuation: Accommodation for 15 stretchers and attendants.
  Priority cargo transport: Cargo door (see below) plus floor-mounted cargo roller system.
  C-20F: US Army administrative transport (one); delivered to Davison Air Command, Andrews AFB, Maryland, 1991.

C-20G: Operational Support Aircraft (OSA) for US Navy and USMC. Convertible interior for passengers and cargo (26 passengers; large cargo door, extra emergency exits). First two for VR-48 at Andrews AFB, Maryland, from March 1994; three more under contract, two for USN Reserve, one for USMC.

C-20H: USAF aircraft, to be based at Andrews AFB, Maryland (one on order).

CUSTOMERS: US services, as detailed above. Swedish Defence Material Administration (FMV) concluded contract 29 June 1992 for three Gulfstream IVs; one as **Tp 102** transport (delivered 24 October 1992 to F16 Wing at Uppsala); remaining two to be modified for electronic intelligence gathering; to be operated by Swedish Air Force; delivery in 1995 and enter operational service 1997, replacing Caravelles. Japanese Civil Aviation Bureau ordered one SRA for flight inspection. Other sales to undisclosed customers.
COSTS: Over $42 million for C-20Gs (1992).
DESIGN FEATURES: Missions include surveillance/reconnaissance, electronic warfare support, maritime patrol, anti-submarine warfare, medical evacuation, priority cargo and administrative transport. Cabin arranged for rapid role changes; upward-opening cargo door 1.60 m (5 ft 3 in) high by 2.11 m (6 ft 11 in) wide can be fitted to starboard ahead of wing to allow for bulky cargo, mission equipment or stretchers.

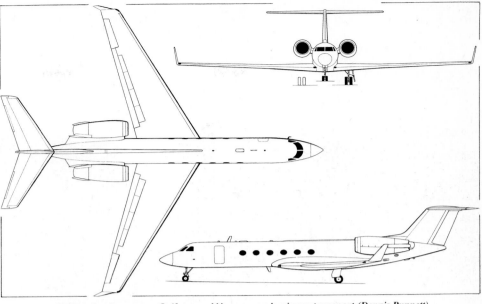

Gulfstream Aerospace Gulfstream V long-range business transport *(Dennis Punnett)*

GULFSTREAM AEROSPACE GULFSTREAM V

TYPE: Twin-turbofan long-range business transport.
PROGRAMME: Study announced at NBAA Convention, Houston, in October 1991; preliminary design and detailed market studies, engine analysis with Rolls-Royce in 1991 followed by exhibition of fuselage mockup at NBAA Convention in September 1992. First flight 1995; initial deliveries in 1996.
CUSTOMERS: At least 10 ordered, with some 12 more deposits held.
COSTS: $29.5 million (1993) for first 24 aircraft.
DESIGN FEATURES: 5,000 nm (9,266 km; 5,757 mile) range with eight passengers at Mach 0.80 cruising speed; T-O and landing distances better than Gulfstream IV.
POWER PLANT: Two 65.30 kN (14,680 lb st) BMW Rolls-Royce BR 710 turbofans.

DIMENSIONS, EXTERNAL:
Wing span over winglets	27.69 m (90 ft 10 in)
Length overall	29.41 m (96 ft 6 in)
Height overall	7.39 m (24 ft 3 in)
Tailplane span	10.72 m (35 ft 2 in)

DIMENSIONS, INTERNAL:
Cabin: Length, incl galley, toilet and baggage compartment 15.57 m (51 ft 1 in)
Max width	2.24 m (7 ft 4 in)
Max height	1.88 m (6 ft 2 in)
Volume	53.94 m³ (1,905 cu ft)
Flight deck: Length	2.06 m (6 ft 9 in)
Volume	4.36 m³ (154.0 cu ft)

WEIGHTS AND LOADINGS (approx):
Operating weight empty	20,638 kg (45,500 lb)
Max payload	2,948 kg (6,500 lb)
Max usable fuel	17,418 kg (38,400 lb)
Max T-O weight	38,600 kg (85,100 lb)
Max landing weight	31,751 kg (70,000 lb)

PERFORMANCE (estimated, at max T-O weight, ISA, except where indicated):
Normal cruising speed at 12,500 m (41,000 ft)
Mach 0.80 (459 knots; 850 km/h; 528 mph)

Mockup of the long-range twin-turbofan Gulfstream V

FAA balanced T-O field length at S/L:
max T-O weight 1,677 m (5,500 ft)
with fuel for 4,000 nm (7,412 km; 4,606 miles)
1,204 m (3,950 ft)
with fuel for 3,000 nm (5,559 km; 3,454 miles)
1,156 m (3,790 ft)
Range with 726 kg (1,600 lb) payload, at normal cruising speed and with NBAA IFR reserves
6,300 nm (11,675 km; 7,254 miles)

GULFSTREAM AEROSPACE GULFSTREAM XT

Extended technology aircraft, currently subject of major market research programme and engineering study; for introduction second half of 1990s. Approximately similar size to Gulfstream II; Mach 0.85 normal cruising speed; max ramp weight 22,679 kg (50,000 lb); payload 907 kg (2,000 lb); range 3,000 nm (5,559 km; 3,454 miles).

GULFSTREAM AEROSPACE TECHNOLOGIES
(Oklahoma Operations)
Wiley Post Airport, Box 22500, Oklahoma City, Oklahoma 73123
Telephone: 1 (405) 789 5000
Telex: 747193
PRESIDENT: John Podger

VICE-PRESIDENT, OPERATIONS: Jerry Cooper
Former Gulfstream Commander Jetprop factory 55,740 m² (600,000 sq ft) now used for subcontracted manufacture and engineering. Single-engined Commander 112 and 114 line sold to Commander Aircraft Company (which see) June 1988. Manufacturing and marketing rights for twin-engined Commanders sold December 1989 to Precision Airmotive of Everett, Washington (see Twin Commander Corporation in this section). Gulfstream Aerospace authorised 2 May 1988 to produce spares for out-of-production Douglas airliners from DC-3 to DC-10. Contract announced 8 September 1989 to design, tool and manufacture 200 wing sets for British Aerospace Jetstream 41 for delivery between 1990 and 2000; first set delivered November 1990.

HAWK
HAWK AIRCRAFT DEVELOPMENT CORPORATION
Clermont County Airport, Batavia, Ohio 45103
PRESIDENT: Hal Shevers

HAWK 72
TYPE: Four-seat all-metal light aircraft.
PROGRAMME: Company announced in January 1992 intention to certificate and manufacture clone of Cessna 172P; discussions have taken place with Ayres for use of production facilities at latter's Albany, Georgia, factory; deliveries to start early 1994.

COSTS: Target $100,000 with basic Bendix/King or Terra avionics.
DESIGN FEATURES: As Cessna 172P final production version, last described in 1985-86 *Jane's*; to retain Textron Lycoming O-320-D2J engine; electrically driven standby attitude indicator as standard; electronic engine monitoring system.

HELI-AIR
HELI-AIR (JAFFE HELICOPTER INCORPORATED)
119 Ida Road, Broussard, Louisiana 70518
Telephone: 1 (318) 837 2376
Fax: 1 (318) 837 2113
Telex: 5101010395
PRESIDENT: Gary J. Villiard
DIRECTOR OF OPERATIONS: David A. Brown
Member of Jaffe Group. Heli-Air Bell 222SP conversion continues.

HELI-AIR (BELL) 222SP
TYPE: Re-engined Bell 222.
PROGRAMME: First flight of Allison-engined Bell 222 test conversion (N5008Q) 10 November 1988; first flight of production conversion 12 January 1990; certificated 21 August 1991.
COSTS: $1.45 million (1992).
DESIGN FEATURES: Heli-Air-developed modification to replace standard 510 kW (684 shp) LTS 101-750C-1 turboshafts in Bell 222A, 222B and 222UT with two 485 kW (650 shp) Allison 250-C30Gs. Performance unchanged except for improved engine-out performance and 36 kg (80 lb) greater useful load.
Details of 222SP conversion in *Jane's Civil and Military Aircraft Upgrades*.

USA: AIRCRAFT—HONDA/KAMAN

HONDA
HONDA MOTOR COMPANY

HONDA UA-5
TYPE: Four/six-seat materials technology research aircraft.
PROGRAMME: Joint project with Mississippi State University, to adapt or develop advanced aerospace composites materials for use in automobile design. Prototype (N3079N) completed 1992; began taxi tests early 1993. Operates in FAA Experimental category; not intended for sale.
DESIGN FEATURES: High-wing monoplane, with forward-swept wing; tall, sweptback fin/rudder and high-mounted horizontal tail surfaces; fuselage reminiscent of Mitsubishi MU-2; two small turbojet or turbofan engines pylon-mounted high on rear fuselage, canted outward at about 30°; retractable tricycle landing gear.

KAMAN
KAMAN AEROSPACE CORPORATION
(Subsidiary of Kaman Corporation)
Old Windsor Road, PO Box No. 2, Bloomfield, Connecticut 06002
Telephone: 1 (203) 242 4461
Fax: 1 (203) 243 6125
Telex: 710 425 3411
PRESIDENT AND CEO: Walter R. Kozlow
VICE-PRESIDENT, ENGINEERING: David J. White
DIRECTOR, PUBLIC RELATIONS: J. Kenneth Nasshan

Founded 1945 by Charles H. Kaman, now Chairman of Board of Kaman Corporation. Developed servo-flap control of helicopter main rotor, initially in counter-rotating two-blade main rotors, and still used in H-2 Seasprite four-blade main rotor. R&D programmes sponsored by US Army, Air Force, Navy and NASA include advanced design of helicopter rotor systems, blades and rotor control concepts, component fatigue life determination and structural dynamic analysis and testing. Kaman has undertaken helicopter drone programmes since 1953; is continuing advanced research in rotary wing unmanned aerial vehicles (UAVs).

Kaman is major subcontractor in many aircraft and space programmes, including design, tooling and fabrication of components in metal, metal honeycomb, bonded and composites construction, using techniques such as filament winding and braiding. Participates in programmes including Grumman A-6 and F-14, Bell/Boeing V-22, Boeing 737, 747, 757 and 767, Sikorsky UH-60 and SH-60, and NASA Space Shuttle Orbiter. Kaman also supplies acoustic engine ducts for P&W JT8D and thrust reversers for GE CF6-80C/E engines.

Kaman designed and, since 1977, has been producing all-composite rotor blades for Bell AH-1 Cobras for US and foreign armies; participated in AH-1 air-to-air Stinger (ATAS) programme; supplied composite-blade kits for Boeing CH-46 Sea Knights.

Prototype ML-30 Magic Lantern mine detection equipment in pods aft of main landing gear of SH-2F

KAMAN SEASPRITE and SUPER SEASPRITE
US Navy designation: SH-2
TYPE: Shipboard ASW helicopter with SAR, observation and utility capability.
PROGRAMME: First flight 2 July 1959; successive versions for US Navy described in earlier *Jane's*; SH-2F put back into production in 1981; from 1967 all single-engined SH-2A/B Seasprites progressively converted to twin-engined UH-2Cs with General Electric T58-GE-8Bs; later modified to Mk I Light Airborne Multi-Purpose System (LAMPS) standard to give small ships ASW, anti-ship surveillance and targeting (ASST), SAR and utility capability; all SH-2s subsequently upgraded to SH-2Fs, each with stronger landing gear, T58-GE-8F engines and improved rotor.

Operational deployment of LAMPS Mk I to HSL squadrons began 7 December 1971; by December 1991, more than 1,000,000 hours flown from ship classes FFG-7, DD-963, DDG-993, CG-47, FFG-1, FF-1052, FF-1040, CG-26, CGN-35, CGN-38 and BB-61; US Coast Guard WMEC and WHEC cutters being equipped to operate SH-2.
CURRENT VERSIONS (in service 1993): **SH-2F**: LAMPS Mk I; 88 delivered between May 1973 and 1982, of which 75 operational February 1989; 16 SH-2Ds converted to SH-2F; first operational unit, HSL-33, deployed to Pacific 11 September 1973; 54 new SH-2Fs ordered FY 1982-86 (18, 18, six, six and six), delivered by December 1989; another six ordered in FY 1987, being completed as SH-2G (see below). SH-2F described in 1987-88 *Jane's*; SH-2F fleet to be upgraded to SH-2G.

For operation in Arabian Gulf since 1987, 16 SH-2Fs augmented standard AN/ALR-66A(V)1 RWR and AN/ALE-39 chaff/flare dispensers by adding Sanders AN/ALQ-144 IR jammers on starboard side of tail rotor driveshaft, Loral AN/AAR-47 missile warning equipment, Collins AN/ARC-182 secure VHF/UHF radios, AN/DLQ-3 missile warning and jamming equipment, and Hughes AN/AAQ-16 FLIR under nose.

SH-2G Super Seasprite: SH-2F upgrade initiated FY 1987; airframe changes included replacing T58 with T700-GE-401 engines and adopting all-composite main rotor blades with 10,000 hour life; fuel consumption improved by over 20 per cent. Avionics improvements include MIL-STD-1553B digital databus, onboard acoustic processor, multi-function raster display, AN/ASN-150 tactical navigation display, and 99-channel sonobuoys. SH-2G qualified for dipping sonar, air-to-surface missiles, FLIR sensors and various guns, rockets and countermeasures.

First flight of SH-2F 161653 as YSH-2G T700 engine testbed, April 1985; first flight with full avionics 28 December 1989; this helicopter delivered 1991, followed by six new-build SH-2Gs ordered FY 1987; no further new production envisaged. June 1987 contract launched conversion programme from SH-2F to SH-2G performed by Kaman; 18 conversion kits on order by January 1993, for completion by June 1994; rebuilds refurbished for further 10,000 flying hours. First production SH-2G (163541) flown March 1990; first delivery, to HSL-84, in December 1992; remaining five to follow in 1993; rebuilds thereafter.
Detailed description applies to SH-2G.

SH-2G can be configured for Magic Lantern podded laser equipment for subsurface mine detection; ML-30 prototype equipment operationally tested in single SH-2F during Gulf War, 1991; two ML-90 prototypes ordered late 1991; delivered to USN, October 1992. Further unknown sensor reportedly under development.
CUSTOMERS: US Navy (see Current Versions). Eight (now four) active Navy LAMPS Mk I squadrons supplemented in 1984 by three Naval Air Reserve squadrons; 24 SH-2Fs transferred to HSL-74, 84 and 94 Reserve squadrons at South Weymouth (Massachusetts), North Island (California), and Willow Grove (Pennsylvania) as new aircraft delivered to HSLs; HSL-84 and -94 receiving eight SH-3Gs each; remaining eight 'G' versions for storage/maintenance. Front-line squadrons are HSL-30, 32 and 34 at Norfolk, Virginia, for Atlantic-based ships; and HSL-33 at North Island, California, for the Pacific; active and Reserve LAMPS Mk I SH-2s will remain in service until 2010. SH-2G offered for-export with various equipment from above; Egypt and Thailand interested. Taiwan offered 12 surplus SH-2Fs by Pentagon, September 1992, in $161 million programme.
COSTS: $12 million per conversion from SH-2F to SH-2G (1993 prices).
DESIGN FEATURES: Main rotor rpm 298; main and tail rotor blades folded manually; nose opens and folds back for shipboard stowage; lateral pylons for torpedoes or tanks; MAD bird in holder extending from starboard side.
FLYING CONTROLS: Main rotor blades fixed on hub; pitch changed by trailing-edge servo flaps.
STRUCTURE: All-metal airframe with flotation hull; titanium main rotor hub.
LANDING GEAR: Tailwheel type, with forward retracting twin mainwheels and non-retractable tailwheel. Liquid spring shock absorbers in main gear legs; oleo-pneumatic shock absorber in tailwheel unit, which is fully castoring for taxying but locked fore and aft for T-O and landing. Main wheels have 8-ply tubeless tyres size 17.5 × 6.25-11, pressure 17.25 bars (250 lb/sq in); tailwheel 10-ply tube-type tyre size 5.00-5, pressure 11.04 bars (160 lb/sq in).
POWER PLANT: Two 1,285 kW (1,723 shp) General Electric T700-GE-401 turboshafts, one on each side of rotor pylon structure. Basic fuel capacity of 1,802 litres (476 US gallons; 396 Imp gallons), including up to two external auxiliary tanks with a combined capacity of 757 litres (200 US gallons; 166.5 Imp gallons). Ship-to-air helicopter in-flight refuelling (HIFR).
ACCOMMODATION: Crew of three, consisting of pilot, co-pilot/tactical co-ordinator, and sensor operator. One passenger with LAMPS equipment installed; four passengers or two litters with sonobuoy launcher removed. Provision for transportation of internal or external cargo. Space for additional troop seats.
SYSTEMS: Include dual 30kVA electrical system and Turbomach T-62 gas turbine APU.
AVIONICS: LAMPS Mk I mission equipment includes Canadian Marconi LN-66HP surveillance radar; General Instruments AN/ALR-66A(V)1 radar warning/ESM; Teledyne Systems AN/ASN-150 tactical management system; dual Collins AN/ARC-159(V)1 UHF radios; Texas Instruments AN/ASQ-81(V)2 magnetic anomaly detector; Computing Devices AN/UYS-503 acoustic processor; Flightline Electronics AN/ARR-84 sonobuoy receiver and AN/ARN-146 on-top position indicator; Tele-Dynamics AN/AKT-22(V)6 sonobuoy data link; 15 DIFAR and DICASS sonobuoys; AN/ALE-39 chaff/flare dispensers; AN/ASQ-188 torpedo presetter. The US Navy plans to retrofit additional self-defence equipment in fleet SH-2Gs, consisting of Hughes AN/AAQ-16 FLIR, Sanders AN/ALQ-144 IR jammers, Loral AN/AAR-47 missile

warning and Collins AN/ARC-182 VHF/UHF secure radio.
EQUIPMENT: Cargo hook for external loads, capacity 1,814 kg (4,000 lb); and folding rescue hoist, capacity 272 kg (600 lb).
ARMAMENT: One or two Mk 46 or Mk 50 torpedoes; eight Mk 25 marine smoke markers. Provision for pintle-mounted 7.62 mm machine-gun in each cabin doorway.
DIMENSIONS, EXTERNAL:

Main rotor diameter	13.41 m (44 ft 0 in)
Main rotor blade chord	0.56 m (1 ft 10 in)
Tail rotor diameter	2.46 m (8 ft 1 in)
Tail rotor blade chord	0.236 m (9.3 in)
Length: fuselage, excl tail rotor	12.34 m (40 ft 6 in)
overall (rotors turning)	16.00 m (52 ft 6 in)
nose and blades folded	11.68 m (38 ft 4 in)
Height: overall (rotors turning)	4.62 m (15 ft 2 in)
blades folded	4.14 m (13 ft 7 in)
Width overall, incl MAD	3.74 m (12 ft 3 in)
Stabiliser span	2.97 m (9 ft 9 in)
Wheel track (outer wheels)	3.30 m (10 ft 10 in)
Wheelbase	5.13 m (16 ft 10 in)
Tail rotor ground clearance	2.12 m (6 ft 11½ in)

AREAS:

Main rotor blades (each)	3.75 m² (40.33 sq ft)
Tail rotor blades (each)	0.295 m² (3.175 sq ft)
Main rotor disc	141.31 m² (1,521.11 sq ft)
Tail rotor disc	4.77 m² (51.32 sq ft)

WEIGHTS AND LOADINGS:

Weight empty	4,173 kg (9,200 lb)
Max T-O weight	6,124 kg (13,500 lb)
Max disc loading	43.31 kg/m² (8.87 lb/sq ft)

PERFORMANCE (ISA):

Max level speed at S/L	138 knots (256 km/h; 159 mph)
Normal cruising speed	120 knots (222 km/h; 138 mph)
Max rate of climb at S/L	762 m (2,500 ft)/min
Rate of climb at S/L, OEI	530 m (1,740 ft)/min
Service ceiling	7,285 m (23,900 ft)
Service ceiling, OEI	4,816 m (15,800 ft)
Hovering ceiling: IGE	6,340 m (20,800 ft)
OGE	5,486 m (18,000 ft)
Max range, 2 external tanks	478 nm (885 km; 500 miles)
Max endurance, 2 external tanks	5 h

Production Kaman SH-2G Super Seasprite in new tactical light grey paint scheme

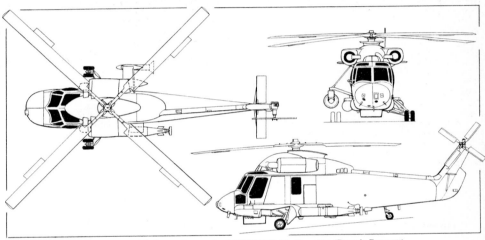

Kaman SH-2G Super Seasprite LAMPS Mk I helicopter (*Dennis Punnett*)

KAMAN K-MAX

TYPE: Civil external lift intermeshing rotor aircraft (K-MAX, K-1200) and military multi-mission intermeshing rotor aircraft (MMIRA).

PROGRAMME: First flight (N3182T) 23 December 1991; 75 hours flown by first public showing 22 March 1992; second and (static) third prototypes under construction; certification expected early 1994; deliveries 1994; initial production batch of six helicopters earmarked for logging operators.

CUSTOMERS: None announced by mid-1993.

COSTS: Initially offered only for lease at $1,000 (1992) per flying hour for 1,000 hours per year.

DESIGN FEATURES: Optimised for forestry support, lifting 2,270 kg (5,000 lb) payload at tree-line altitude of 2,440 m (8,000 ft), OGE. Kaman intermeshing two-blade contra-rotating rotors with separate inclined shafts emerging from common transmission; lifting power increased, because induced drag and downwash of intermeshing rotor system is reduced and power drain of tail rotor system eliminated; blade centreline offset from hub; blade length increased by 15 cm (6 in) after early trials; single drag bearing with drag damper; small trailing-edge flaps set blade pitch; light control loads and low feedback eliminate need for powerful pitch-change rods and levers and hydraulic powered controls; all bending and twisting caused by pitch change accommodated by blade flexing; engine mounted horizontally behind transmission. Minimum overhaul life of all components except engine, 2,500 hours.

FLYING CONTROLS: Pitch and lateral attitude controlled by synchronised cyclic pitch control of both rotors; yaw at low speed controlled by differential collective pitch from foot pedals; differential fore and aft cyclic also applied for yaw control in forward flight. Blade pitch change tabs attached at 75 per cent span of each rotor blade; no hydraulic boost; each rotor has its own swashplate and light mechanical linkage inside driveshaft and through each blade; control mixer box is modular and easily accessible. Aerodynamic rudder for directional trimming augmented by fins at extremities of variable incidence elevators after early flight testing.

STRUCTURE: Light alloy airframe; GFRP and CFRP rotor blades and tabs.

LANDING GEAR: Fixed nosewheel gear designed to lie out of pilot's field of view; impact sustaining suspension with transverse mounting tube for mainwheels; rubber-in-compression suspension for mainwheels; oleo for nosewheel; bear paw plate round each wheel for operation from soft

Original configuration of Kaman K-Max single-seat working helicopter with intermeshing rotors

USA: AIRCRAFT—KAMAN/KING'S

Kaman K-Max prototype in approximate production configuration with elevators deflected downwards

ground and snow; nosewheel swivels and locks; main wheels have individual foot-powered brakes and parking brake.

POWER PLANT: One 1,343 kW (1,800 shp) Textron Lycoming T53-17A-1 turboshaft (civil equivalent of military T53-L-703), flat rated at 1,119 kW (1,500 shp) for take-off up to about 2,315 m (7,600 ft) and 1,007 kW (1,340 shp) continuous up to about 1,980 m (6,500 ft); transmission rating 1,119 kW (1,500 shp). Fuel capacity 871 litres (230 US gallons; 191 Imp gallons) located at aircraft CG; receptacle for hot refuelling with rotors running; dual electric fuel pumps.

ACCOMMODATION: Pilot only, in Simula crash impact absorbing seat with five-point harness; seat and rudder pedals adjustable; heater and windscreen demister; window removable for operation in hot weather; curved windscreen in production version. Tool/cargo compartment 0.74 m (26 cu ft) fitted with 2,268 kg (5,000 lb) stress tiedown rings. Provision for unmanned operation.

SYSTEMS: DC electrical system with starter/generator; no hydraulics.

EQUIPMENT: Pilot-controlled swivelling landing light; standard configuration is for lifting slung loads, but electrical fittings provided for Bambi firefighting bucket, Loadcell and long-line hook gear; kits planned for patrol, agricultural applications and similar missions.

DIMENSIONS, EXTERNAL:
Rotor diameter, each	14.68 m (48 ft 2 in)
Length overall, rotors turning	15.85 m (52 ft 0 in)
Wheel track	3.56 m (11 ft 8 in)

WEIGHTS AND LOADINGS:
Operating weight empty	2,041 kg (4,500 lb)
Max hook capacity	2,721 kg (6,000 lb)
Max fuel weight	699 kg (1,541 lb)
Max T-O weight:	
without jettisonable load	2,721 kg (6,000 lb)
with jettisonable load	over 4,762 kg (10,500 lb)
Max transmission power loading (T-O power)	4.25 kg/kW (7.0 lb/shp)

PERFORMANCE:
Target hovering performance: OGE with 2,268 kg (5,000 lb) slung load and 1.5 h fuel
2,440 m (8,000 ft)

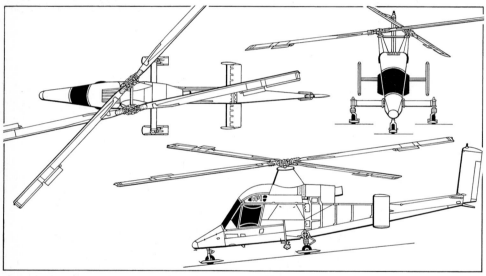

Production version of Kaman K-Max intermeshing rotor helicopter (*Jane's/Mike Keep*)

Detail of Kaman K-Max rotors

KING'S — *see TKEF*

LAKE
LAKE AIRCRAFT INC
606 North Dyer Boulevard, Kissimmee Airport, Kissimmee, Florida 34741
Telephone: 1 (407) 847 9000
Fax: 1 (407) 847 4516
PRESIDENT: Armand E. Rivard
EXECUTIVE VICE-PRESIDENT: Gordon Collins
VICE-PRESIDENT, INTERNATIONAL: Haig Hagopian
VICE-PRESIDENT, MARKETING AND DOMESTIC SALES:
  Kevin T. Tracey

LAKE LA-250 RENEGADE and TURBO 270 RENEGADE
TYPE: Lengthened five-passenger version of LA4-200.
PROGRAMME: LA-250 FAA certificated August 1983.
CURRENT VERSIONS: **LA-250 Renegade:** Standard version. *Described below.*
  Turbo 270 Renegade: Turbocharged version with 186 kW (250 hp) Textron Lycoming TIO-540-AA1AD. Set world altitude record of 7,465 m (24,500 ft) for small amphibians August 1983.
  Special Edition Seafury: Renegade with salt water operating package including lifting rings, stainless steel brake discs, custom interior and survival package. Available since Summer 1990.
DESIGN FEATURES: Single-step all-metal double-sealed boat hull, 1.05 m (3 ft 5 in) longer than LA4-200 with deeper V bottom and additional strakes; retractable water rudder in base of aerodynamic rudder. Tapered wings attached directly to hull sides; wing section NACA 4415 at root, 4409 at tip; dihedral 5° 30'; incidence 3° 15'.
FLYING CONTROLS: Manually operated ailerons, elevators and rudder; ground adjustable aileron trim tabs; outer portion of port elevator separate from inboard and operated hydraulically as trimmer; hydraulically operated slotted flaps.
STRUCTURE: Wing has duralumin leading/trailing-edge torsion boxes separated by single main spar; light alloy monocoque wing floats; hull alodined and zinc chromated inside and out; polyurethane external paint; metal ailerons and tail surfaces.
LANDING GEAR: Hydraulically retractable tricycle type. Consolidated oleo-pneumatic shock absorbers in main gear, which retracts inward into wings. Nosewheel retracts forward. Oleo extension increased compared with LA4-200 for greater ground clearance; wheelbase increased by 0.43 m (1 ft 5 in). Gerdes mainwheels with Goodyear tyres, size 6.00-6, pressure 2.41 bars (35 lb/sq in). Gerdes nosewheel with Goodyear tyre size 5.00-5, pressure 1.38 bars (20 lb/sq in). Gerdes disc brakes; parking brake. Nosewheel free to swivel 30° left/right.
POWER PLANT: One 186 kW (250 hp) Textron Lycoming IO-540-C4B5 flat-six engine in Renegade and Seafury, driving a Hartzell three-blade constant-speed Q-tip metal pusher propeller. Turbocharged TIO-540-AA1AD engine in Turbo Renegade. Standard usable fuel capacity 204 litres (54 US gallons; 45 Imp gallons); optional usable capacity of 340 litres (90 US gallons; 75 Imp gallons).
ACCOMMODATION: Enclosed cabin seating pilot and five passengers. Front and rear seats removable. Front seats have inertia reel harnesses as standard. Dual controls standard; dual brakes for co-pilot optional. Entry via two front-hinged windscreen sections; upward hinged gull-wing cargo door standard. Baggage compartment (larger than in LA4-200) aft of cabin. Windscreen defrosting.
SYSTEMS: Vacuum system for flight instruments. Hydraulic system, pressure 86.2 bars (1,250 lb/sq in), for flaps, horizontal trim and landing gear actuation; handpump provided for emergency operation. Engine driven 12V 60A alternator and 12V 30Ah battery. Janitrol 30,000 BTU heater optional.
AVIONICS: Basic avionics installation includes com and nav antennae, cabin speaker, microphone and circuit breakers. An extensive range of avionics by Bendix/King, Collins and Narco, and autopilots by Brittain and Edo-Aire Mitchell, are available to customers' requirements.
EQUIPMENT: Standard equipment includes full blind-flying instrumentation, electric clock, manifold pressure gauge, outside air temperature gauge, recording tachometer, fuel pressure and quantity indicators, oil pressure and temperature indicators, cylinder head temperature gauge, ammeter, stall warning device, control locks, carpeted floor, four fresh air vents, tinted glass for all windows, dual windscreen defrosters, inertia reel shoulder harness on front seats, shoulder restraining on rear seats, map pocket on front seats, baggage tiedown straps, landing and taxi lights, navigation lights, strobe light, heated pitot, fuselage nose bumper, paddle, cleat, line, full flow oil filter, quick fuel drains, and inboard and outboard tiedown rings. Optional equipment includes hour meter, true airspeed indicator, shoulder harness for rear seats, alternate static source, manual/automatic bilge pump, cabin fire extinguisher, and external metallic paint finish.

DIMENSIONS, EXTERNAL:
  Wing span   11.68 m (38 ft 4 in)
  Wing chord, mean   1.35 m (4 ft 5 in)

Lake Turbo 270 Renegade taxying up a slipway

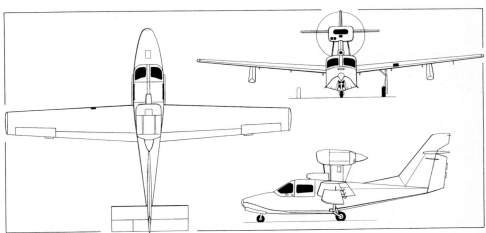
Lake LA-250 Renegade six-seat amphibian *(Dennis Punnett)*

  Wing aspect ratio   8.7
  Length overall   8.64 m (28 ft 4 in)
  Height overall   3.05 m (10 ft 0 in)
  Tailplane span   3.05 m (10 ft 0 in)
  Wheel track   3.40 m (11 ft 2 in)
  Wheelbase   3.13 m (10 ft 3 in)
  Propeller diameter   1.93 m (6 ft 4 in)
DIMENSIONS, INTERNAL:
  Cabin: Length   2.03 m (6 ft 8 in)
    Max width   1.05 m (3 ft 5½ in)
    Max height   1.32 m (3 ft 11½ in)
AREAS:
  Wings, gross   15.79 m² (170.0 sq ft)
  Ailerons (total)   1.16 m² (12.5 sq ft)
  Trailing-edge flaps (total) 2.28 m² (24.5 sq ft)
  Fin   1.25 m² (13.5 sq ft)
  Rudder   0.79 m² (8.5 sq ft)
  Tailplane   1.45 m² (15.6 sq ft)
  Elevators (total)   0.78 m² (8.4 sq ft)
WEIGHTS AND LOADINGS:
  Weight empty, equipped: Renegade 839 kg (1,850 lb)
    Turbo Renegade   875 kg (1,930 lb)
  Max usable fuel   240 kg (528 lb)
  Max ramp, T-O and landing weight 1,383 kg (3,050 lb)
  Max wing loading   87.6 kg/m² (17.94 lb/sq ft)
  Max power loading   7.42 kg/kW (12.2 lb/hp)
PERFORMANCE (at max T-O weight, S/L, ISA):
  Never-exceed speed (VNE):
    Renegade   148 knots (274 km/h; 170 mph)
  Max level speed at 1,980 m (6,500 ft):
    Renegade   139 knots (258 km/h; 160 mph)
  Max cruising speed:
    Renegade, 75% power at 1,980 m (6,500 ft)
                          132 knots (245 km/h; 152 mph)
    Turbo Renegade, 78% power at 6,100 m (20,000 ft)
                          148 knots (274 km/h; 170 mph)
  Stalling speed, power off:
    landing gear and flaps up
                          54 knots (100 km/h; 62 mph) IAS
    landing gear and flaps down
                          48 knots (89 km/h; 56 mph) IAS
  Max rate of climb at S/L   274 m (900 ft)/min
  Service ceiling: Renegade   4,480 m (14,700 ft)
    Turbo Renegade   6,100 m (20,000 ft)
  T-O run: from land   268 m (880 ft)
    from water   381 m (1,250 ft)
  Range with max fuel, 30 min reserves:
    Renegade   900 nm (1,668 km; 1,036 miles)
  Endurance at 78% power: Turbo Renegade   5 h 18 min

LAKE SEAWOLF
TYPE: Single-engined military/maritime surveillance amphibian.
PROGRAMME: Introduced early 1985; prototype Seawolf (N1401G) first shown publicly at Paris Air Show 1985. No subsequent news received, and no known customers. Description in 1992-93 and earlier *Jane's*; illustration in 1991-92 edition.

LANCAIR

LANCAIR INTERNATIONAL INC
2244 Airport Way, Redmond, Oregon 97756
North American dealer: Neico Aviation Inc, at same address.
PRESIDENT, NEICO AVIATION: Lance Neibauer

More than 850 Lancairs now sold to customers in 29 countries.

LANCAIR INTERNATIONAL LANCAIR 235
TYPE: Side by side two-seat sporting and cross-country homebuilt.
PROGRAMME: Prototype Lancair 200 first flew June 1984; **Lancair 235** became model offered in kit form with 88 kW (118 hp) O-235 engine and 0.40 m³ (14.0 cu ft) baggage area; glassfibre was replaced by Kevlar in subsequent form.
COSTS: Believed only now available as prefabricated parts; no kits or plans.
DESIGN FEATURES: Wing section NLF 0215F.
STRUCTURE: Composite airframe of glassfibre, Nomex honeycomb, polyimide Rohacell foam and epoxy resin.
LANDING GEAR: Retractable tricycle type.
POWER PLANT: One 88 kW (118 hp) Textron Lycoming O-235 flat-four engine. Fuel capacity 125-136 litres (33-36 US gallons; 27.5-30 Imp gallons).

DIMENSIONS, EXTERNAL:
Wing span	7.16 m (23 ft 6 in)
Wing aspect ratio	7.27
Length overall	5.99 m (19 ft 8 in)
Width, wings folded	2.54 m (8 ft 4 in)
Height overall	1.85 m (6 ft 1 in)
Propeller diameter	1.47-1.52 m (4 ft 10 in-5 ft 0 in)

AREAS:
Wings, gross	7.06 m² (76.0 sq ft)

WEIGHTS AND LOADINGS:
Weight empty	386 kg (850 lb)
Max T-O weight	635 kg (1,400 lb)
Max wing loading	89.94 kg/m² (18.42 lb/sq ft)
Max power loading	7.22 kg/kW (11.86 lb/hp)

PERFORMANCE:
Max level speed at S/L	195 knots (362 km/h; 225 mph)
Cruising speed	182 knots (338 km/h; 210 mph)
Stalling speed, engine idling	48 knots (89 km/h; 55 mph)
Max rate of climb at S/L	472 m (1,550 ft)/min
Service ceiling	6,100 m (20,000 ft)
T-O run	176 m (575 ft)
Landing run	211 m (690 ft)
Range with max fuel	868 nm (1,609 km; 1,000 miles)

LANCAIR INTERNATIONAL LANCAIR 320
TYPE: As for Lancair 235.
PROGRAMME: Followed Lancair 235, introducing larger fuselage, increased flap effectiveness, and longer landing gear with oleo-pneumatic nose strut. On 8 January 1990 received FAA approval under 51 per cent rule for **Fast-Build** kit, which reduces construction time by more than 700 working hours. Pacific Aeronautical Inc of the Philippines (which see) assembling kits.
COSTS: Kits: $20,450; Fast-Build $26,500. Plans $325.
DESIGN FEATURES: See Programme.
POWER PLANT: One 119 kW (160 hp) Textron Lycoming O-320 flat-four engine. Fuel capacity 163-200 litres (43-53 US gallons; 36-44 Imp gallons). Two optional 38 litre (10 US gallon; 8.3 Imp gallon) auxiliary fuel tanks.

DIMENSIONS, EXTERNAL:
Wing span	7.16 m (23 ft 6 in)
Length overall	6.40 m (21 ft 0 in)
Height overall	2.13 m (7 ft 0 in)
Propeller diameter	1.78 m (5 ft 10 in)

AREAS: As for Lancair 235

WEIGHTS AND LOADINGS:
Weight empty	472 kg (1,040 lb)

Lancair International Lancair 360 *(Austin J. Brown)*

Max T-O weight	764 kg (1,685 lb)
Max wing loading	108.25 kg/m² (22.17 lb/sq ft)
Max power loading	6.42 kg/kW (10.53 lb/hp)

PERFORMANCE:
Max level speed at S/L	226 knots (418 km/h; 260 mph)
Cruising speed at 2,285 m (7,500 ft)	208 knots (386 km/h; 240 mph)
Stalling speed	55 knots (102 km/h; 63 mph)
Max rate of climb at S/L	503 m (1,650 ft)/min
Service ceiling	5,485 m (18,000 ft)
T-O run	214 m (700 ft)
Range, no reserves	1,259 nm (2,333 km; 1,450 miles)
g limits	+9/-4.5

LANCAIR INTERNATIONAL LANCAIR 360
Similar to Lancair 320 but with 134 kW (180 hp) Textron Lycoming O-360 engine. Empty weight 442-483 kg (975-1,065 lb); max level speed 230 knots (426 km/h; 265 mph); max rate of climb at S/L 594 m (1,950 ft)/min.

LANCAIR INTERNATIONAL LANCAIR IV
TYPE: Four-seat homebuilt, with dual controls.
PROGRAMME: Kit deliveries began 1990. On 20 February 1991, prototype set NAA world speed record between San Francisco and Denver of 314.7 knots (583.2 km/h; 362.4 mph).
COSTS: Kit: $45,900.
DESIGN FEATURES: Conventional seating for four persons in airframe of new design; high cruising speeds.
FLYING CONTROLS: Conventional control plus Fowler flaps.
STRUCTURE: Carbonfibre/Kevlar/epoxy airframe, with Nomex honeycomb cores.
LANDING GEAR: As Lancair 320.
POWER PLANT: One 261 kW (350 hp) Teledyne Continental TSIO-550-B1B twin-turbocharged flat-six engine. Fuel capacity 303 litres (80 US gallons; 66.6 Imp gallons).

DIMENSIONS, EXTERNAL:
Wing span	9.19 m (30 ft 2 in)
Wing aspect ratio	9.29
Length overall	7.62 m (25 ft 0 in)
Height overall	2.44 m (8 ft 0 in)
Propeller diameter	1.93 m (6 ft 4 in)

AREAS:
Wings, gross	9.10 m² (98.0 sq ft)

WEIGHTS AND LOADINGS:
Weight empty	794 kg (1,750 lb)
Max T-O weight	1,315 kg (2,900 lb)
Max wing loading	144.5 kg/m² (29.59 lb/sq ft)
Max power loading	5.04 kg/kW (8.29 lb/hp)

PERFORMANCE:
Cruising speed, 75% power	287 knots (531 km/h; 330 mph)
Stalling speed	63 knots (116 km/h; 72 mph)
Max range, 75% power at 7,620 m (25,000 ft), standard fuel, no reserves	1,259 nm (2,333 km; 1,450 miles)
g limits	+9/-4.5 ultimate

LANCAIR INTERNATIONAL LANCAIR ES
TYPE: Four-seat cabin monoplane.
PROGRAMME: Prototype unveiled at Oshkosh Autumn 1992; flight testing continuing early 1993. Certification awaiting new simplified version of FAR Pt 23 governing light aircraft.
DESIGN FEATURES: Low-wing monoplane; hardtop cabin with paired side-by-side seats.
STRUCTURE: All-composites.
LANDING GEAR: Non-retractable type.
POWER PLANT: One 149 kW (200 hp) Teledyne Continental IO-360 flat-six engine; two-blade constant-speed propeller.

WEIGHTS AND LOADINGS:
Max T-O weight	1,315 kg (2,900 lb)

PERFORMANCE:
Max cruising speed	more than 165 knots (306 km/h; 190 mph)
Max range	more than 1,130 nm (2,094 km; 1,301 miles)

LEARJET

LEARJET INC
(Subsidiary of Bombardier Inc)
One Learjet Way, PO Box 7707, Wichita, Kansas 67277
Telephone: 1 (316) 946 2000
Fax: 1 (316) 946 2220
Telex: 417441
PRESIDENT AND CEO: Brian Barents
VICE-PRESIDENT, MARKETING AND SALES: Ted Farid
VICE-PRESIDENT, OPERATIONS: Richard E. Hamlin
VICE-PRESIDENT, PRODUCT SUPPORT: Donald E. Grommesh
VICE-PRESIDENT, CORPORATE AFFAIRS: William G. Robinson
VICE-PRESIDENT, ENGINEERING: William W. Greer

Acquisition of Learjet by Canada's Bombardier announced April 1990 and concluded 22 June 1990 for $75 million; name changed to Learjet Inc; Bombardier assumed responsibility for Learjet's line of credit.

Company originally founded 1960 by Bill Lear Snr as Swiss American Aviation Corporation (SAAC); transferred to Kansas 1962 and renamed Lear Jet Corporation; Gates Rubber Company bought about 60 per cent of company 1967; company renamed Gates Learjet Corporation; 64.8 per cent of company acquired by Integrated Acquisition Inc September 1987 and renamed Learjet Corporation; all manufacturing moved from Tucson, Arizona, to Wichita during 1988, leaving only customer service and modification centre in Tucson; became part of Bombardier Inc June 1990; now part of Bombardier Aerospace Group, North America (which see in Canadian section). Wichita workforce 2,715 in 1992; total 3,315 at all locations.

Learjet is major subcontractor for Martin Marietta Manned Space Systems, Boeing and US Air Force. Has subsidiary to maintain 83 Learjet 35As operating with US Air Force and Air National Guard (as C-21As). More than 1,690 Learjets produced to date.

Learjet bought manufacturing and marketing rights and tooling of Aeronca thrust reversers, for application to Learjet and other aircraft, March 1989.

Learjet projected 27 deliveries in 1993, compared to 25 in 1992.

LEARJET 35A and 36A
US Air Force designation: **C-21A**
TYPE: Light twin-turbofan business jet.
PROGRAMME: First flight of first turbofan Learjet (known as Model 26 and using Garrett TFE731-2s), 4 January 1973; production Models 35 and 36, differing in fuel capacity and accommodation, announced May 1973; FAA certification July 1974; French and UK certification 1979.
CURRENT VERSIONS: **Learjet 35A** and **36A**: Current production models of 35 and 36, with higher standard max T-O weight. *Description applies to these models.*

C-21A: USAF received 80 Learjet 35As on lease in 1984-85 and purchased them for $180 million in September 1986; used as Operational Support Aircraft for priority cargo, medevac and personnel transport, replacing T-39 Sabreliners; four more C-21As bought 1987 by Air National Guard and assigned to Andrews AFB, Maryland.

Special missions versions: Described separately.
CUSTOMERS: Seven 35As delivered 1991, bringing total (both models) by that date to 727 (no more recent total received).

COSTS: Standard 35A $4.94 million (1993); 36A $5.14 million (1993).
DESIGN FEATURES: Softflite package includes full-chord shallow fences bracketing ailerons, with arrowhead energisers on leading-edges and two rows of boundary layer energiser strips between fences. Wing section NACA 64A109 with modified leading-edge; dihedral 2° 30′; incidence 1°; sweepback at quarter-chord 13°.

Century III improvements, Softflite low-speed handling package and engine synchronisers now standard for both models; higher max T-O weight of 8,300 kg (18,300 lb), originally optional, now standard; improvements introduced in 1983 include T/R-4000 thrust reversers with standby hydraulic power reservoir, single-engined reverse capability, quick removal hot section, prevention of reversal at high thrust, reverse available within two seconds of touchdown, throttle retard system and reverse thrust at reduced gas generator rpm. Special interior introduced 1985 offers better leg and headroom, Erda 10-way adjustable seats, stereo and in-flight telephone, lavatory enclosed by doors and electronically controlled washbasin cabinet.
FLYING CONTROLS: Manually actuated flying controls; balance tabs in both ailerons and electrically operated trim tab in port aileron; electrically actuated variable incidence tailplane; electric trim tab in rudder; small ventral fin; hydraulically actuated single-slotted flaps; hydraulically actuated spoilers ahead of flaps.
STRUCTURE: All-metal; eight-spar wing with milled skins; fail-safe fuselage.
LANDING GEAR: Retractable tricycle type, with twin wheels on each main unit and single steerable nosewheel, maximum steering angle 45° either side of centreline. Hydraulic actuation, with backup pneumatic extension. Oleo-pneumatic shock absorbers. Goodyear multiple-disc hydraulic brakes. Pneumatic emergency braking system. Parking brakes. Fully modulated anti-skid system. Min ground turning radius about nosewheel 6.43 m (21 ft 1 in).
POWER PLANT: Two Garrett TFE731-2-2B turbofans, each rated at 15.6 kN (3,500 lb st), pod-mounted on sides of rear fuselage. Fuel in integral wing and wingtip tanks and a fuselage tank, with a combined usable capacity (Learjet 35A) of 3,500 litres (925 US gallons; 770 Imp gallons). Learjet 36A has a larger fuselage tank, giving a combined usable total of 4,179 litres (1,104 US gallons; 919 Imp gallons). Refuelling point on upper surface of each wingtip tank. Fuel jettison system.
ACCOMMODATION: Crew of two on flight deck, with dual controls. Up to eight passengers in Learjet 35A; one on inward facing seat with toilet on starboard side at front, then two pairs of swivel seats which face fore and aft for take-off and landing, with centre aisle, and three on forward-facing couch at rear of cabin. Two forward storage cabinets, one on each side; two folding tables standard. Alternative arrangement, available optionally, places a refreshment area in the middle of the cabin, accessible from fore and aft club seating areas, each for four passengers. Learjet 36A can accommodate up to six passengers, one pair of swivel seats being removed. Toilet and stowage space under front inward facing seat which can be screened from remainder of cabin. Baggage compartment with capacity of 226 kg (500 lb) aft of cabin. Two-piece clamshell door at forward end of cabin on port side, with integral steps built into lower half. Emergency exit on starboard side of cabin. Birdproof windscreens.
SYSTEMS: Environmental control system comprises cabin pressurisation, ventilation, heating and cooling. Heating and pressurisation by engine bleed air, with a max pressure differential of 0.65 bar (9.4 lb/sq in), maintaining a cabin altitude of 1,980 m (6,500 ft) to an actual altitude of 13,715 m (45,000 ft). Freon R12 vapour cycle cooling system supplemented by a ram-air heat exchanger. Flight control system includes dual yaw dampers, dual stick pushers, dual stick shakers and Mach trim. Anti-icing by engine bleed air for wing, engine nacelle leading-edges and windscreen; tailplane anti-iced electrically; electrical heating of pitot heads, stall warning vanes and static ports; and alcohol spray on windscreen and nose radome. Hydraulic system supplied by two engine driven pumps, either pump capable of maintaining the full system pressure of 103.5 bars (1,500 lb/sq in), for operation of landing gear, brakes, flaps and spoilers. Hydraulic system maximum flow rate 15 litres (4 US gallons; 3.33 Imp gallons) per min. Cylindrical reservoir pressurised to 1.38 bars (20 lb/sq in). Electrically driven hydraulic pump for emergency operation of all hydraulic services. Pneumatic system of 124 to 207 bars (1,800 to 3,000 lb/sq in) pressure for emergency extension of landing gear and operation of brakes. Electrical system powered by two 30V 400A brushless generators, two 1kVA solid state inverters to provide AC power, and two 24V 37Ah lead-acid batteries. Oxygen system for emergency use, with crew demand masks and dropout masks for each passenger.
AVIONICS: Standard Collins avionics include dual FIS-84/EHSI-74 flight directors, integrated with J. E. T. FC-530 FCS and dual yaw dampers; dual VHF-22A com transceivers with CTL-22 controls; dual VIR-32 nav receivers with CTL-32 controls; ADF-60 with CTL-62 control; dual DME-42 with IND-42C indicators; dual TDR-90 transponders with CTL-92 controls; ALT-55B radio altimeter with DRI-55 indicator; dual Allen 3137 RMIs; UNS-1B long-range nav system; Honeywell Primus 450 colour weather radar; dual J. E. T. VG-206D vertical gyros; dual J. E. T. DN-104B directional gyros; pilot's IDC electric encoding altimeter with altitude preselect and IDC air data unit; co-pilot's IDC barometric altimeter; dual Teledyne SL2-9157-3 IVSIs; dual marker beacon lamps; dual D. B. audio systems; J. E. T. PS-835D emergency battery and AI-804 attitude gyro; dual Davtron 877 clocks; annunciator package; N_1 reminder; avionics master switch; chip detector; flap preselect; Wulfsberg Flitefone VI; Bendix/King KHF 950 HF; Frederickson Jetcal 5 Selcal; Rosemount air data system and SAS/TAT/TAS indicator.
EQUIPMENT: Standard equipment includes thrust reversers, dual angle of attack indicators, engine synchronisation meter, cabin differential pressure gauge, cabin rate of climb indicator, interstage and turbine temperature gauges, turbine and fan speed gauges, wing temperature indicator, alternate static source, depressurisation warning, engine fire warning lights, Mach warning system, dual stall warning system, fire axe, cabin fire extinguisher, cabin stereo cassette player, EEGO audio distribution system; baggage compartment, courtesy, instrument panel, flood, map, cockpit dome, and reading lights; dual anti-collision, landing, navigation, recognition, strobe, taxi and maintenance lights, wing ice detection light; dual engine fire extinguishing systems with 'systems armed' and fire warning lights, maintenance interphone jack plugs, engine synchronisation system, control lock, external power socket, and lightning protection system.

One of 80 Learjet C-21As sold to US Air Force for operational support missions *(Peter J. Cooper)*

DIMENSIONS, EXTERNAL:
Wing span over tip tanks	12.04 m (39 ft 6 in)
Wing chord: at root	2.74 m (9 ft 0 in)
at tip	1.55 m (5 ft 1 in)
Wing aspect ratio	5.7
Length overall	14.83 m (48 ft 8 in)
Height overall	3.73 m (12 ft 3 in)
Tailplane span	4.47 m (14 ft 8 in)
Wheel track	2.51 m (8 ft 3 in)
Wheelbase	6.15 m (20 ft 2 in)
Passenger door:	
Standard: Height	1.57 m (5 ft 2 in)
Width	0.61 m (2 ft 0 in)
Optional: Height	1.57 m (5 ft 2 in)
Width	0.91 m (3 ft 0 in)
Emergency exit: Height	0.71 m (2 ft 4 in)
Width	0.48 m (1 ft 7 in)

DIMENSIONS, INTERNAL:
Cabin: Length: incl flight deck: 35A	6.63 m (21 ft 9 in)
36A	5.77 m (18 ft 11 in)
excl flight deck: 35A	5.21 m (17 ft 1 in)
36A	4.06 m (13 ft 4 in)
Max width	1.50 m (4 ft 11 in)
Max height	1.32 m (4 ft 4 in)
Volume, incl flight deck: 35A	9.12 m³ (322.0 cu ft)
36A	7.25 m³ (256.0 cu ft)
Baggage compartment: 35A	1.13 m³ (40.0 cu ft)
36A	0.76 m³ (27.0 cu ft)

AREAS:
Wings, gross	23.53 m² (253.3 sq ft)

WEIGHTS AND LOADINGS (35A and 36A):
Weight empty, equipped	4,590 kg (10,119 lb)
Max payload (incl full fuselage tank)	1,534 kg (3,381 lb)
Max T-O weight	8,300 kg (18,300 lb)
Max ramp weight	8,391 kg (18,500 lb)
Max landing weight	6,940 kg (15,300 lb)
Max wing loading	352.5 kg/m² (72.2 lb/sq ft)
Max power loading	266.03 kg/kN (2.61 lb/lb st)

PERFORMANCE (35A and 36A, at max T-O weight except where indicated):
Max operating speed	Mach 0.81
Max level speed at 7,620 m (25,000 ft)	471 knots (872 km/h; 542 mph)
Max cruising speed, mid-cruise weight, at 12,500 m (41,000 ft)	460 knots (852 km/h; 529 mph)
Econ cruising speed, mid-cruise weight, at 13,700 m (45,000 ft)	418 knots (774 km/h; 481 mph)
Stalling speed, wheels and flaps down, engines idling	96 knots (178 km/h; 111 mph) IAS
Max rate of climb at S/L	1,323 m (4,340 ft)/min
Rate of climb at S/L, OEI	390 m (1,280 ft)/min
Service ceiling	12,500 m (41,000 ft)
Service ceiling, OEI	7,620 m (25,000 ft)
T-O balanced field length, FAR Pt 25:	
at 7,711 kg (17,000 lb)	1,323 m (4,340 ft)
at 8,300 kg (18,300 lb)	1,515 m (4,972 ft)
Landing distance, FAR Pt 25, at max landing weight	937 m (3,075 ft)
Range with 4 passengers, max fuel and 45 min reserves:	
35A	2,196 nm (4,069 km; 2,528 miles)
36A	2,522 nm (4,673 km; 2,904 miles)

OPERATIONAL NOISE LEVELS (FAR Pt 36):
T-O	83.9 EPNdB
Approach	91.4 EPNdB
Sideline	86.7 EPNdB

LEARJET 35A/36A SPECIAL MISSIONS VERSIONS

TYPE: Special mission adaptations for Learjet 35A/36A.
CURRENT VERSIONS: **EC-35A**: Used for EW training simulation or as stand-off ECM/ESM platform.
PC-35A: Maritime patrol; equipment includes 360° sea surveillance digital radar, high resolution television, forward-looking infra-red (FLIR), infra-red linescanner

Special missions Learjet U-36A of the Japan Maritime Self-Defence Force

494 USA: AIRCRAFT—LEARJET

(IRLS), electronic support measures (ESM), magnetic anomaly detector, integrated tactical displays, and VLF Omega or other long-range navaids; hardpoint under each wing with Alkan 165B ejector for external stores up to 454 kg (1,000 lb); drop hatch for rescue gear, multi-track digital recorders, homing systems, ASW sonobuoy systems and data annotated hand-held cameras.

RC-35A and RC-36A: Reconnaissance versions; standard installations include long-range oblique photographic cameras (LOROP), side-looking synthetic aperture radars, and surveillance camera system in external pods. Geological versions delivered to China (see below).

UC-35A: Utility versions for transport, navaids calibration, medevac and target towing. Certificated tow systems include Hayes Universal Tow Target System (HUTTS) with or without MTR-101, Flight Refuelling LLHK, MRTT and EMT TGL targets, and Marquardt aerial target launch and recovery tow reel.

U-36A: Extensively modified for Japan Maritime Self-Defence Force (JMSDF). Four delivered for target towing, anti-ship missile simulation and ECM; tip pods extended, in association with Shin Meiwa, to house HWQ-1T missile seeker simulator, AN/ALQ-6 jammer and cameras. Additional equipment includes long-range ocean surveillance radar in underbelly fairing, AN/ALE-43 chaff dispenser, ARS-1-L high-speed tow sleeve with scoring, two-piece windscreen with electrical demisting for low-level missions, expanded underwing stores capability, greater max T-O and landing weights. Further deliveries to Japan expected during 1990s.

CUSTOMERS: 23 customer countries include Argentina, Australia, Bolivia, Brazil, Chile, People's Republic of China (two 36As delivered 1984, plus three 35As with geological equipment and Goodyear SLAR delivered 1985), Finland (three 35A target tugs also equipped for mapping, medevac, pollution control, oblique photography and SAR), Germany (four 35A/36A target tugs), Japan, Mexico, Peru, Saudi Arabia, Sweden, Switzerland, Thailand, UK, USA and Yugoslavia. Nearly 200 Learjet 30 Series aircraft now flying with or for military services, including 83 C-21As (which see).

DESIGN FEATURES: Civilian and paramilitary missions include aerial survey, aeronautical research, airways calibration, ASW, atmospheric research, border patrol, ESM/ECM, geophysical survey, maritime patrol, pilot training, radar surveillance, reconnaissance, search and rescue, and weather modification.

PERFORMANCE (PC-35A):
Operating speed:
 at 11,275-12,500 m (37,000-41,000 ft)
  415 knots (769 km/h; 478 mph)
 at 4,575-7,620 m (15,000-25,000 ft)
  319 knots (590 km/h; 367 mph)
 S/L to 610 m (2,000 ft)
  250 knots (463 km/h; 288 mph)
Max rate of climb at S/L 1,380 m (4,525 ft)/min
Range:
 at high altitude 2,249 nm (4,168 km; 2,590 miles)
 at medium altitude 1,617 nm (2,996 km; 1,862 miles)
 at low altitude 1,060 nm (1,964 km; 1,220 miles)

LEARJET 31A

TYPE: Successor to Learjet 31.
PROGRAMME: Learjet 31 introduced September 1987; first flight of aerodynamic prototype 11 May 1987; first production aircraft (N311DF) used as systems testbed; FAA certification 12 August 1988. Learjet 31A and 31A/ER announced October 1990 to replace Learjet 31.
CURRENT VERSIONS: **Learjet 31:** See 1990-91 *Jane's*.
 Learjet 31A: Current version; *description applies to this model, except where indicated.*
 Learjet 31A/ER: Optional extended range version with 2,627 litres (694 US gallons; 578 Imp gallons) fuel and higher max T-O weight.
CUSTOMERS: 19 Learjet 31s and four 31As delivered in 1992. Total approx 70 Learjet 31s and 31As delivered. Recent new orders include two 31s for Singapore Airlines for cadet pilot and first officer training. Government of Baluchistan (Pakistani state) took delivery of a 31A in early 1993 as VIP transport.
COSTS: Standard Learjet 31A $4.71 million (1993).
DESIGN FEATURES: Learjet 31 (1990-91 *Jane's*) combined fuselage/cabin and power plant of Learjet 35A/36A with wing of Learjet 55; Delta Fins added to eliminate Dutch roll, stabilise aircraft at high airspeeds, induce docile stall and reduce approach speeds and field lengths; stick pusher/puller and dual yaw dampers no longer required for departure; stick shaker and single yaw damper retained for comfort.

Additional features of Learjet 31A include cruise Mach number up to 12,500 m (41,000 ft) increased four per cent to 0.81 and VMO increased from 300 knots (556 km/h; 345 mph) to 325 knots (602 km/h; 374 mph) IAS. Increases mainly benefit descent from high altitudes. Learjet 31A also features integrated digital avionics package.
FLYING CONTROLS: All control surfaces mechanically actuated; ailerons have brush seals and geared tabs; electrically actuated trim tab on port aileron. Electrically actuated tailplane incidence control has separate motors for pilot and co-pilot and single-fault survival protection; aircraft can be manually controlled following tailplane runaway and landed with reduced flap. Rudder has electric trim tab; automatic electric rudder assist servo operates automatically if rudder pedal loads exceed 22.6 kg (50 lb). Full-chord fences bracket the ailerons; airflow between fences corrected by arrowhead energisers on leading-edge, row of round-head screws aft of leading-edge and two rows of energiser strips near ailerons. Single spoiler panel in each wing used as airbrake and lift dumper. Hydraulically actuated flaps extend to 40°. Optional drag parachute mounted on inside of baggage hatch under tail.
STRUCTURE: Multi-spar wing with machined skins.
LANDING GEAR: Retractable tricycle gear; main legs retract inward, nose leg forward; twin mainwheels with anti-skid disc brakes; nosewheel has full-time digital steer-by-wire replacing speed-limited steering of Learjet 31. Max airspeed with gear extended 260 knots (481 km/h; 299 mph); tyre limiting speed 183 knots (339 km/h; 210 mph). Ground turning radius about nosewheel 11.91 m (39 ft 1 in).
POWER PLANT: Two 15.56 kN (3,500 lb st) Garrett TFE731-2 turbofans with digital electronic engine controller giving automatic retention of power settings above 4,575 m (15,000 ft) and special idling control for descent from 15,545 m (51,000 ft). Engine synchroniser fitted. Optional Dee Howard 4000 thrust reverser system weighs 109 kg (240 lb). One integral fuel tank in each wing holds 641 kg (1,413 lb); standard fuselage tank 608 kg (1,340 lb); ER fuselage tank 804 kg (1,773 lb); fuselage fuel transferred by pilots by gravity or pump; single-point pressure refuelling optional.
ACCOMMODATION: Cabin furnishings include a three-seat divan, four Erda 10-way adjustable individual seats, side facing seat with toilet, two folding tables, baggage compartment, forward privacy curtain, overhead panels with reading lights, indirect lighting, air vents and oxygen masks.
SYSTEMS: Hydraulic system operates flaps, landing gear, airbrake, wheelbrakes and thrust reversers; system pressure 69-120.6 bars (1,000-1,750 lb/sq in); pneumatic standby for gear extension and wheelbrakes. Normal cabin pressure differential 0.64 bar (9.4 lb/sq in) with automatic flood engine bleed if cabin altitude exceeds 2,820 m (9,250 ft); pop-out emergency oxygen for passengers and masks for crew. Electrical system based on two starter/generators, two nickel-cadmium batteries and two inverters; both buses can run from one engine; electrics operate tailplane incidence, rudder assister and nosewheel steering. De-icing by bleed air for wing, engine intakes and windscreen; tailplane electrically heated; fin not protected. Alcohol spray for radome to stop shed ice entering engines; controls prevent internal ice and condensation during long descents. Single engine at idle, burning 15.9 kg (35 lb) fuel every 10 minutes, acts as APU.
AVIONICS: Bendix/King integrated digital avionics package.
EQUIPMENT: Throttle-mounted landing gear warning mute and go-around switches, nacelle heat annunciator, engine synchroniser and synchroscope, recognition light, wing ice light, emergency press override switches, transponder ident switch in pilot's control wheel, engine synchroniser and synchroscope, flap preselect, crew lifejackets, cockpit dome lights, cockpit speakers, crew oxygen masks and fire extinguisher are standard.
DIMENSIONS, EXTERNAL:
Wing span 13.33 m (43 ft 8 in)
Length overall 14.83 m (48 ft 8 in)
Height overall 3.73 m (12 ft 3 in)

Learjet 31A twin-turbofan business aircraft

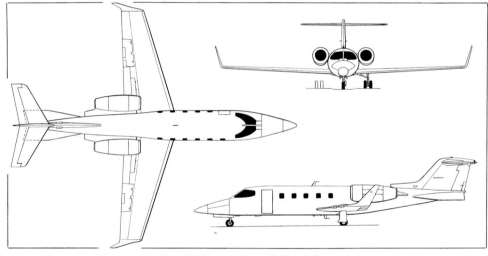

Learjet 31A business aircraft *(Dennis Punnett)*

DIMENSIONS, INTERNAL:
　Cabin: Length:
　　incl flight deck: 31A　　6.63 m (21 ft 9 in)
　　　31A/ER　　6.27 m (20 ft 7 in)
　　excl flight deck: 31A　　5.21 m (17 ft 1 in)
　　　31A/ER　　4.85 m (15 ft 11 in)
　　Max width: 31A, 31A/ER　　1.50 m (4 ft 11 in)
　　Max height: 31A, 31A/ER　　1.32 m (4 ft 4 in)
　　Volume, incl flight deck: 31A　　9.12 m³ (322.0 cu ft)
　　　31A/ER　　8.83 m³ (312.0 cu ft)
　　Baggage compartment: 31A　　1.13 m³ (40.0 cu ft)
　　　31A/ER　　0.85 m³ (30.0 cu ft)
AREAS:
　Wings, gross　　24.57 m² (264.5 sq ft)
WEIGHTS AND LOADINGS:
　Weight empty: 31A　　4,700 kg (10,361 lb)
　　31A/ER　　4,717 kg (10,400 lb)
　Basic operating weight empty: 31A　　4,881 kg (10,761 lb)
　　31A/ER　　4,899 kg (10,800 lb)
　Max payload: 31A　　1,016 kg (2,239 lb)
　　31A/ER　　998 kg (2,200 lb)
　Payload with max fuel: 31A　　846 kg (1,865 lb)*
　　31A/ER　　588 kg (1,297 lb)
　Max fuel weight: 31A　　1,890 kg (4,166 lb)
　　31A/ER　　2,086 kg (4,599 lb)
　Max T-O weight: 31A (standard)　　7,031 kg (15,500 lb)
　　31A (optional), 31A/ER　　7,484 kg (16,500 lb)
　Max ramp weight: 31A (standard)　　7,144 kg (15,750 lb)
　　31A (optional), 31A/ER　　7,597 kg (16,750 lb)
　Max zero-fuel weight　　5,896 kg (13,000 lb)
　Max landing weight　　6,940 kg (15,300 lb)
　Max wing loading: 31A　　286.1 kg/m² (58.60 lb/sq ft)
　　31A/ER　　304.6 kg/m² (62.38 lb/sq ft)
　Max power loading: 31A　　225.32 kg/kN (2.21 lb/lb st)
　　31A/ER　　239.87 kg/kN (2.36 lb/lb st)
* 31A at optional T-O/ramp weights

PERFORMANCE (at max standard T-O weight, S/L, ISA, except where indicated):
　Never-exceed speed (V_{NE})
　　　325 knots (602 km/h; 374 mph) IAS
　Max operating Mach number, up to 12,500 m (41,000 ft)
　　　Mach 0.81
　Cruising speed:
　　at 13,715 m (45,000 ft) 449 knots (832 km/h; 516 mph)
　　at 10,975-12,500 m (36,000-41,000 ft)
　　　463 knots (855 km/h; 532 mph)
　Stalling speed at typical landing weight
　　　93 knots (173 km/h; 107 mph)
　Max rate of climb at S/L: 31A　　1,670 m (5,480 ft)/min
　　31A/ER　　1,555 m (5,100 ft)/min
　Rate of climb at S/L, OEI: 31A　　576 m (1,890 ft)/min
　　31A/ER　　466 m (1,530 ft)/min
　Max certificated ceiling　　15,545 m (51,000 ft)
　Service ceiling, OEI: 31A　　9,510 m (31,200 ft)
　　31A/ER　　8,840 m (29,000 ft)
　T-O balanced field length, FAR Pt 25:
　　31A　　893 m (2,930 ft)
　　31A/ER　　1,000 m (3,280 ft)
　FAR Pt 91 landing distance:
　　31A and 31A/ER　　844 m (2,767 ft)
　Range at econ cruising speed with four passengers, 45 min reserves: 31A　　1,561 nm (2,892 km; 1,797 miles)
　　31A/ER　　1,806 nm (3,346 km; 2,079 miles)
OPERATIONAL NOISE LEVELS (FAR Pt 36):
　T-O: 31A　　79.5 EPNdB
　　31A/ER　　81.0 EPNdB
　Approach: 31A and 31A/ER　　92.6 EPNdB
　Sideline: 31A　　87.2 EPNdB
　　31A/ER　　87.0 EPNdB

LEARJET 45

TYPE: 10/12-seat business jet.
PROGRAMME: Design started September 1992; unveiled at NBAA Convention 20 September 1992; certification planned for 1996 to latest FAR and JAR standards.
COSTS: $5.9 million (June 1992).
DESIGN FEATURES: Designed to combine docile handling characteristics of 31/31A and 60 with exceptional fuel efficiency and good overall performance, and offer increased maintainability and reliability; new larger fuselage, wing and tail unit; exceptional head and shoulder room; wing carry-through spar recessed beneath floor; latest technology systems. Wing designed with NASA.
STRUCTURE: Computervision CADDS 5 digital design system adopted by Learjet for tail design and Shorts for fuselage; concurrent engineering technique used.
LANDING GEAR: Retractable tricycle trailing-link type, for softer and smoother taxying.
POWER PLANT: Two Garrett TFE731-20 turbofans, each rated at 15.57 kN (3,500 lb st); target type thrust reversers; FADEC engine control. Total fuel capacity 3,392 litres (896 US gallons; 746 Imp gallons).
ACCOMMODATION: Two pilots plus eight to 10 passengers. Galley and coat closet.
AVIONICS: Honeywell Primus 1000 with EICAS, dual PFDs and MFDs; heart of system is IC-600 integrated avionics computer, which combines EFIS and EICAS processor, flight director and digital autopilot in single LRU. Primus II nav/com/ident radios in dual configuration; Primus 650 weather radar. Options include Honeywell traffic alert and

Artist's impression of Learjet 45 business jet

Learjet 45 cockpit with Honeywell Primus 1000 integrated avionics system

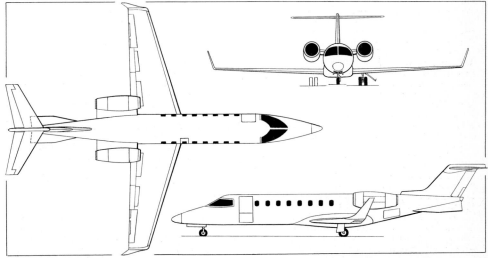

Learjet 45 eight/10-passenger business jet (two Garrett TFE731-20 turbofans) (*Dennis Punnett*)

496 USA: AIRCRAFT—LEARJET/LOCKHEED

collision avoidance system (TCAS II); Primus 870 advanced weather radar with turbulence detection; microwave landing system; dual solid-state fibre-optic gyro-based AHRS.

DIMENSIONS, EXTERNAL:
Wing span	13.35 m (43 ft 9½ in)
Length overall	17.89 m (58 ft 8½ in)
Height overall	4.48 m (14 ft 8½ in)

DIMENSIONS, INTERNAL:
Cabin:	
Length: incl flight deck	7.54 m (24 ft 9 in)
excl flight deck	6.02 m (19 ft 9 in)
Max width	1.55 m (5 ft 1 in)
Max height	1.50 m (4 ft 11 in)
Volume, excl flight deck	14.02 m³ (495 cu ft)
Baggage compartment volume	1.42 m³ (50.0 cu ft)

WEIGHTS AND LOADINGS:
Weight empty	5,307 kg (11,700 lb)
Max payload	1,497 kg (3,300 lb)
Max fuel weight	2,721 kg (6,000 lb)
Max T-O weight	8,845 lb (19,500 lb)
Max ramp weight	8,958 kg (19,750 lb)
Max landing weight	8,709 kg (19,200 lb)
Max zero-fuel weight	6,804 kg (15,000 lb)
Max power loading	284.0 kg/kN (2.79 lb/lb st)

PERFORMANCE (estimated):
Max cruising speed at 11,275 m (37,000 ft)	464 knots (859 km/h; 534 mph)
Econ cruising speed at 13,715 m (45,000 ft)	406 knots (753 km/h; 468 mph)
Max rate of climb at S/L	975 m (3,200 ft)/min
Service ceiling	13,715 m (45,000 ft)
T-O to 11 m (35 ft)	1,280 m (4,200 ft)
Landing from 15 m (50 ft)	911 m (2,990 ft)
Range with four passengers, zero wind, ISA:	
VFR	2,199 nm (4,074 km; 2,532 miles)
NBAA IFR	1,849 nm (3,426 km; 2,129 miles)

LEARJET 55B and 55C

Replaced by Learjet 60; last Learjet 55C (c/n 147) delivered December 1990 to UK. Full details in 1991-92 *Jane's*.

LEARJET 60

TYPE: Medium-range business jet.
PROGRAMME: Announced 3 October 1990 as Learjet 55C successor; first flight of proof-of-concept aircraft with one PW305 turbofan 18 October 1990; flight testing resumed 13 June 1991 with two PW305s and stretched fuselage (more than 300 hours flown by May 1992); first production aircraft first flight (N601LJ) 15 June 1992; certification awarded 15 January 1993; deliveries started immediately.
CUSTOMERS: E-Systems Greenville, Texas, Division ordered two in 1993 for flight inspection missions for FAA; five 60s delivered to customers in January 1993; first entered service with Herman Miller Inc.
DESIGN FEATURES: Largest Learjet; P&WC PW305 engines; T tail; winglets; delta fins.
FLYING CONTROLS: Spoilers can be partially extended to adjust descent rates.
POWER PLANT: Two P&WC PW305 turbofans, each flat rated at 20.46 kN (4,600 lb st) at up to 27°C (80°F). Total fuel capacity 3,560 kg (7,850 lb).
ACCOMMODATION: Two crew and six to nine passengers; gross pressure cabin volume 15.57 m³ (550 cu ft); compared with 55C, main cabin is 0.71 m (2 ft 4 in) longer and rear baggage hold section 0.38 m (1 ft 3 in) longer; full-across aft toilet has flat floor, large mirror, coat closet and external servicing; total 1.67 m³ (59 cu ft) baggage capacity divided between an externally accessible hold (larger than that of Learjet 55C) and internal pressurised, heated compartment that is accessible in flight; galley cabinet has storage for dinnerware, warming oven, cold liquid dispensers and ice storage; entertainment centre; 10-way adjusting seating is standard.
SYSTEMS: Windscreen demisted by electrically heated gold film which also diminishes sun heating in flight and on ground and produces warmth during prolonged flight at high altitude. Full-time digital steer-by-wire nosewheel control operates throughout taxying, take-off and landing.
AVIONICS: Standard fully integrated all-digital Collins Pro Line 4 avionics include EFIS, dual digital air data computers, dual navigation and communications radios, dual automatic heading and attitude reference systems (AHRS), advanced Collins autopilot and long-range navaid as standard; circuit breaker and controls panels redistributed, as in Learjet 31A.

Learjet 60 prototype during first flight with two P&WC PW305 turbofans

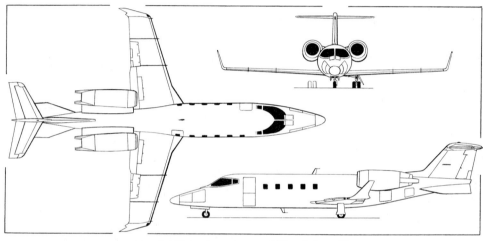
Learjet 60 business transport (*Dennis Punnett*)

DIMENSIONS, EXTERNAL:
Wing span	13.34 m (43 ft 9 in)
Wing chord: at root	2.74 m (9 ft 0 in)
at tip	1.12 m (3 ft 8 in)
Length: overall	17.88 m (58 ft 8 in)
fuselage	17.02 m (55 ft 10 in)
Height overall	4.47 m (14 ft 8 in)
Tailplane span	4.47 m (14 ft 8 in)
Cabin door: Width	0.64 m (2 ft 1 in)
Height to sill	0.69 m (2 ft 3 in)

DIMENSIONS, INTERNAL:
Cabin: Length, incl flight deck	7.04 m (23 ft 1 in)
Max width	1.80 m (5 ft 11 in)
Max height	1.71 m (5 ft 7½ in)

WEIGHTS AND LOADINGS (A: standard, B: optional)
Weight empty: A, B	6,315 kg (13,922 lb)
Max fuel (usable): A, B	3,560 kg (7,850 lb)
Max T-O weight: A	10,319 kg (22,750 lb)
B	10,478 kg (23,100 lb)
Max ramp weight: A	10,432 kg (23,000 lb)
B	10,591 kg (23,350 lb)
Max landing weight: A, B	8,845 kg (19,500 lb)

PERFORMANCE (estimated):
Max cruising speed	463 knots (858 km/h; 533 mph)
Max operating altitude	15,545 m (51,000 ft)
Balanced field length: A	1,582 m (5,190 ft)
B	1,634 m (5,360 ft)
Range with VFR reserves	2,740 nm (5,078 km; 3,155 miles)
Range with four passengers from runway shorter than 1,494 m (4,900 ft)	2,497 nm (4,441 km; 2,760 miles)

Range examples:
can fly New York-Los Angeles against 85 per cent Boeing winds with NBAA reserves
can reach either US coast after taking off from Aspen, Colorado, in ISA +20° with four passengers

LOCK HAVEN
LOCK HAVEN AIRPLANE COMPANY
William T. Piper Memorial Airport, 345 Proctor Street, Lock Haven, Pennsylvania 17745
Telephone: 1 (717) 748 0810

Company announced **Premier Aztec SE** refurbished Piper Aztec E in August 1991; features factory overhauled engines, new Bendix/King avionics, specially designed custom interior and exterior; new one-piece windscreen and cabin windows; cost, IFR-equipped, $225,000.

LOCKHEED
LOCKHEED CORPORATION
4500 Park Granada Boulevard, Calabasas, California 91399-0610
Telephone: 1 (818) 876 2620
Fax: 1 (818) 876 2329
CHAIRMAN, PRESIDENT AND CEO: Daniel M. Tellep
VICE-CHAIRMAN: Vincent N. Marafino
EXECUTIVE VICE-PRESIDENT: Dr Vance D. Coffman

Former Lockheed Aircraft Corporation renamed Lockheed Corporation in September 1977. Activities include design and production of aircraft, electronics, satellites, space systems, missiles, ocean systems, information systems, and systems for strategic defence and for command, control, communications and intelligence. Tactical Military Aircraft Division of General Dynamics (producer of F-16) acquired December 1992 and renamed Lockheed Fort Worth Company.
Lockheed Corporation facilities cover more than 2,991,475 m² (32,200,000 sq ft); total employees 89,000 in 25 US states and worldwide.
Activities are administered by four subsidiary groups: Aeronautical Systems Group (see next entry); Technology Services Group; Electronic Systems Group (comprising Lockheed Sanders Inc, CalComp, Lockheed Commercial Electronics Company, and Lockheed Canada Inc); and Missiles and Space Systems Group (comprising Lockheed Missiles & Space Company Inc, Lockheed Integrated Solutions Company and Lockheed Technical Operations Company).

LOCKHEED AERONAUTICAL SYSTEMS GROUP (LASG)

86 South Cobb Drive, Marietta, Georgia 30063
Telephone: 1 (404) 494 4411
Telex: 542642 LOCKHEED MARA
PRESIDENT: Kenneth W. Cannestra
VICE-PRESIDENT, GROUP BUSINESS OPERATIONS:
    A. G. van Schaick
PUBLIC INFORMATION: J. Brian Johnstone

LASG includes four operating companies:
Lockheed Aeronautical Systems Company
    Next entry
Lockheed Advanced Development Company
    Follows Aeronautical Systems Company
Lockheed Fort Worth Company
    Follows Advanced Development Company
Lockheed Aircraft Service Company
    Follows Lockheed Fort Worth Company

LASG and Lockheed Aeronautical Systems Company have moved aircraft manufacturing from Burbank to Marietta. Lockheed Advanced Development Company autonomous as second component of LASG at Palmdale. Former Burbank Division produced P-3 Orion and made parts of TR-1 and Lockheed Georgia's C-5B. LASG workforce totals approximately 40,000.

LOCKHEED AERONAUTICAL SYSTEMS COMPANY (LASC) (Division of Lockheed Aeronautical Systems Group)

86 South Cobb Drive, Marietta, Georgia 30063
Telephone: 1 (404) 494 4411
Telex: 542642 LOCKHEED MARA
PRESIDENT: Kenneth W. Cannestra
EXECUTIVE VICE-PRESIDENT, AERONAUTICAL SYSTEMS:
    H. Bard Allison

In April 1991, Lockheed won competition to produce F-22 with General Dynamics and Boeing Military Airplanes. Lockheed and Vought are studying advanced tactical surveillance aircraft (ATS), based on combining S-3A airframe with electronically scanned array radar in triangular dorsal radome, to replace E-2C; advanced tactical transport (ATT) for 21st century tactical airlift; and advanced technology tactical transport (ATTT) small, low-cost STOL transport. Other experimental programmes include C-141 with electromechanical instead of hydraulic flying control actuators; and four-year, four-phase investigation of integrated vehicle/propulsion concepts for supersonic cruise.

Other long-term activities at Marietta include production of C-130 Hercules. Re-winging of C-5A completed 1987 and new C-5Bs completed March 1989; new work will consist of P-3C line, transferred from Burbank, and F-22. In December 1991, Marietta was awarded $20 million contract to study potential A-X attack aircraft for US Navy (which see) in consortium also including Boeing and General Dynamics.

LOCKHEED F-22

TYPE: US Air Force advanced tactical fighter (ATF).
PROGRAMME: US Air Force ATF requirement for McDonnell Douglas F-15 Eagle replacement incorporating low observables technology and supercruise (supersonic cruise without afterburning); parallel assessment of two new power plants; request for information issued 1981; concept definition studies awarded September 1983 to Boeing, General Dynamics, Grumman, McDonnell Douglas, Northrop and Rockwell; requests for proposals issued September 1985; submissions received by 28 July 1986; USAF selection announced 31 October 1986 of demonstration/validation phase contractors: Lockheed YF-22 and Northrop YF-23 (see 1991-92 *Jane's*); each produced two prototypes and ground-based avionics testbed; first flights of all four prototypes 1990. Competing engine demonstration/validation programmes launched September 1983; ground testing began 1986-87; flight-capable Pratt & Whitney YF119s and General Electric YF120s ordered early 1988; all four aircraft/engine combinations flown.

Decision of 11 October 1989 extended evaluation phase by six months; draft request for engineering and manufacturing development (EMD) proposals issued April 1990; first artists' impressions released May 1990; final engineering and manufacturing development (EMD) requests issued for both weapon system and engine 1 November 1990; proposals submitted 2 January 1991; F-22 and F119 power plant announced by USAF as winning combination, 23 April 1991; EMD contract given 2 August 1991 for 11 flying prototypes (including two tandem-seat F-22Bs), plus one static and one fatigue test airframes; first flight mid-1996; preliminary design review covering all aspects of the design completed 30 April 1993; final assembly will remain at Marietta rather than moving to Lockheed Fort Worth; critical design review November 1994; pre-production batch of four aircraft to be awarded January 1996; service entry 2002. Minor design changes for production F-22 announced July 1991; see Design Features, Dimensions and Areas. Suggested name of SuperStar rejected in 1991 and F-22 remains unnamed. Ground attack role added May 1993 (see under Armament).

Lockheed teamed with General Dynamics (Fort Worth) and Boeing Military Airplanes to produce two YF-22 prototypes, civil registrations N22YF (with GE YF120) and N22YX (P&W YF119); USAF serial numbers 87-0700 and 87-0701 assigned, but not applied. N22YF rolled out at Palmdale 29 August 1990; first flight/ferry to Edwards AFB 29 September 1990; first air refuelling (11th sortie) 26 October 1990; thrust vectoring in flight 15 November 1990; anti-spin parachute for high angle of attack tests on 34th-43rd sorties; flight testing temporarily suspended 28 December 1990; 43 sorties/52.8 hours. N22YX first flight Palmdale-Edwards 30 October 1990; AIM-9M Sidewinder (28 November 1990) and AIM-120 AMRAAM (20 December 1990) launch demonstrations; achieved Mach 1.8 on 26 December 1990; temporarily grounded after 31 sorties/38.8 hours, 28 December 1990.

Lockheed YF-22 prototype during early flight testing

Model of F-22A in the markings of USAF's 1st Fighter Wing (current F-15 Eagle operator)

Flight test demonstrations included 100°/s roll rate at 120 knots (222 km/h; 138 mph) and supercruise flight in excess of Mach 1.58 without afterburner.

Second (F119-powered) YF-22 taken by road to Palmdale mid-1991; fitted with strain gauges; began further 100 hour test programme 30 October 1991; due for completion April 1992; gathered data on aerodynamic loads, flight control aerodynamic effects, vibration/acoustic fatigue and maximum coefficient of lift; flown by 6511st Test Squadron (F-22 Combined Test Force) of 6510th Test Wing at Edwards AFB; non-fatal crash landing at Edwards 25 April 1992, following pilot-induced oscillations; total 100.4 hours in 70 flights since October 1990; non-flyable, but repaired for use as antenna testbed at Rome Air Development Center, Griffiss AFB, New York.

Lockheed responsible for project control and systems integration; workload shared equally between three partners during dem/val phase; original EMD and production plan called for Lockheed to construct forward fuselage and components, including cockpit, with avionics architecture and functional design, displays, controls, air data system, apertures, edges, tail assembly, landing gear, environmental control system and final assembly. Boeing responsible for wings, wings aft sections, power plant installation, auxiliary power generation system, radar, infra-red search and track system (if fitted in production aircraft) and avionics ground prototype. Avionics flight tested in a modified Boeing 757 (N757A) (first flight 17 July 1989). Boeing providing supporting simulators and training systems. General Dynamics was responsible for mid-fuselage and key systems including electrical, hydraulic, fuel, flight controls and armament; also integrated electronic warfare system (INEWS), integrated communications/navigation/identification avionics (CNI) and INS subsystems. With acquisition of former General Dynamics, Fort Worth, Lockheed now controls 67.5 per cent of programme; programme involves 650 suppliers in 32 US states.

Prototype first flight planned June 1996; low rate production decision in January 1997; first production delivery January 2000; high rate production decision due July 2001.

CURRENT VERSIONS: **F-22A/B**: Single/two-seat production versions for USAF.
    NATF: Projected US Navy variant to replace Grumman F-14 Tomcat; original requirement for 546 had already been cut to 384; concurrent development abandoned in 1991.
CUSTOMERS: US Air Force: two YF-22 demonstrators; planned 11 EMD (test) aircraft reduced to nine (including two tandem-seat) in January 1993; 648 production aircraft; latter to be funded from 1997, beginning with annual batches of four, 12, 12, 24, 36 and 48 in 1997-2002. Original requirement was 750. Peak production, 48 per year.
COSTS: $818 million contracts to both ATF teams, October 1986, for 54-month studies; each airframe team investing own funds (Lockheed/Boeing/GD team investment totalled $675 million); each engine contractor, about $50

498 USA: AIRCRAFT—LOCKHEED

Lockheed YF-22 prototype launching an AIM-9 Sidewinder

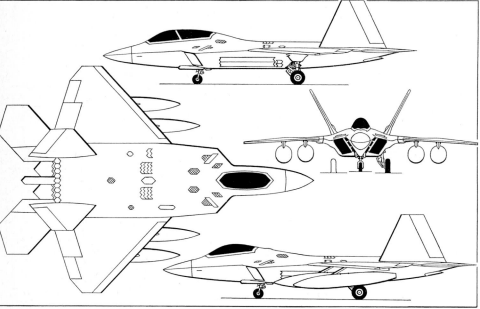

Production configuration of Lockheed F-22A with additional side view (top) of two-seat F-22B
(Jane's/Mike Keep)

Model of F-22A showing upper surfaces and vectoring exhaust nozzles

million; total $3,800 million spent by USAF on both ATFs up to April 1991; programme cost (1992) for 648 aircraft is $13,000 million for development and $47,000 million for production; programme unit cost $147 million (1992); fly-away cost $59.2 million. EMD contract, August 1991, comprised $9,550 million for 11 (subsequently nine) airframes, plus $1,375 million to P&W for 33 (later amended to 27) engines.

DESIGN FEATURES: (F-22) Low observables configuration and construction; stealth/agility trade-off decided by design team; target thrust/weight ratio 1.4 (achieved ratio possibly 1.1 at T-O weight); greatly improved reliability and maintainability for high sortie-generation rates, including 15 minute combat turn-round time; enhanced survivability through 'first-look first-kill' capability; short T-O and landing distances; supersonic cruise and manoeuvring (supercruise) in region of Mach 1.5 without afterburning; internal weapons storage and generous internal fuel; conformal sensors. Wing and horizontal tail leading-edge sweep 42° (both 48° on YF-22); trailing-edge 17° forward, increased to 42° outboard of ailerons (straight trailing edge on YF-22); all-moving five-edged horizontal tail (four-edged elements on YF-22). Vertical tail surfaces (18 per cent larger on YF-22) canted outwards at 28°; leading- and trailing-edge sweep 22.9°; biconvex aerofoil. F-22's wing and stabilator areas same as YF-22, despite re-profiling. F-22 wing taper ratio 0.169; leading-edge anhedral 3.25°; root twist 0.5°; tip twist −3.1°; thickness/chord ratio 5.92 at root, 4.29 at tip; custom-designed aerofoil. Horizontal tails have no dihedral or twist. Sidewinder AAMs stored internally in sides of intake ducts, with AMRAAMs in ventral weapons bay. Diamond-shaped cheek air intakes with highly contoured air ducts; intakes approximately 0.46 m (1 ft 6 in) farther forward on YF-22; single-axis thrust vectoring included on PW119, but specified performance achievable without. Additional production F-22 changes from YF-22 include decreased wingroot thickness, modified camber and twist (increasing anhedral); all 48° plan angles changed to 42°; blunter nose; wheelbase reduced by approx 0.46 m (1 ft 6 in); wheel track reduced by same; revised undercarriage legs and doors; constant chord ailerons; re-profiled cockpit canopy; dorsal airbrake deleted.

FLYING CONTROLS: Triplex, digital, fly-by-wire system using line-replaceable electronic modules to enhance maintainability; thrust vectoring utilised to augment aerodynamic pitch control power and provide firm control even at low speeds and high angles of attack. Technology and control concepts demonstrated throughout the flight envelope during Prototype Air Vehicle test programme, including flight at AoA greater than 60°; wind tunnel testing with models of production aircraft successfully attained AoAs greater than 85°. Ailerons and flaperons occupy almost entire wing trailing-edge; full-span leading-edge flaps; conventional rudders in vertical tail surfaces; slab taileron surfaces; airbrake not included (differential rudder for speed control); sidestick controller. Control surface authorities: leading-edge flaperons 3° up/35° down (5°/37° overtravel); trailing-edge flaperons 20° up/40° down; aileron ±20°; horizontal tail 30° up/25° down; rudder ±30°; speedbrake (rudder) 30° out.

STRUCTURE: Largely metal (aluminium/titanium/steel 33/24/5 per cent) in prototypes. Extensive use of thermoset and thermoplastic composite structures; combined total of 27 to 32 per cent in production aircraft.

LANDING GEAR: Menasco retractable tricycle undercarriage, stressed for no-flare landings of up to 3.05 m (10 ft)/s. Nosewheel tyre 23.5 × 7.5-10; mainwheel tyres 37 × 11.5-18.

POWER PLANT: Two 155 kN (35,000 lb st) class Pratt & Whitney F119-PW-100 advanced technology reheated engines reportedly developed from F100 turbofan. Two-dimensional convergent/divergent exhaust nozzles with thrust vectoring for enhanced performance and manoeuvrability.

ACCOMMODATION: Pilot only on zero/zero ejection seat and wearing tactical life support system with improved g-suits and pressure breathing. Pilot's view over nose −15°.

SYSTEMS: Include Normalair-Garrett OBOGS and Allied-Signal APU.

AVIONICS: F-22 has first fully integrated avionics suite. Hughes common integrated processor (CIP) is core of entire system, supporting Westinghouse radar (air-to-air and navigation), Sanders/General Electric electronic warfare (RF warning and countermeasures) subsystem, and TRW communications-navigation-identification subsystem. Fused situational awareness information is displayed to pilot via four Sanders/Kaiser colour liquid crystal multi-function displays (MFD); MFD bezel buttons provide pilot format control. CIP also contains mission software that uses tailorable mission planning data for sensor emitter management and multi-sensor fusion; mission-specific information delivered to system through Fairchild data transfer equipment that also contains mass storage for default data and air vehicle operational flight programme; stores management system and Litton inertial reference system. General purpose processing capacity of CIP is rated at more than 700 million instructions per second (MIPS) with growth to 2,000 MIPS; signal processing capacity greater than 20 billion operations per second (BOPS) with expansion capability to 50 BOPS; CIP contains more than 300 Mbytes of memory with growth potential to 650 Mbytes. Final integration, as well as integration of the entire suite with non-avionics systems, undertaken at F-22 Avionics Integration Laboratory, Seattle, Washington; airborne integration supported by Boeing 757 flying testbed.

ARMAMENT: Internal long-barrel M61A2 20 mm gun (production F-22). Three internal bays (see Design Features) for AIM-9 Sidewinder (side bays only) and/or AIM-120 AMRAAM AAMs on 'revolutionary weapon racks'. Four underwing stores stations at 317 mm (125 in) and 442 mm (174 in) from centreline of fuselage. Under $6.5 million contract addition on 27 May 1993, weapon bay and avionics to be adapted for delivery of AIM-X missile and 453 kg (1,000 lb) Joint Direct Attack Munition (JDAM); two JDAMs will replace two AIM-120A AMRAAMs in main weapon bay; provision also to be made for two AGM-137A Tri-Service Standoff Attack Missiles (TSSAM) on underwing pylons.

DIMENSIONS, EXTERNAL:

Wing span: YF-22	13.11 m (43 ft 0 in)
F-22	13.56 m (44 ft 6 in)
Wing chord:	
at root (theoretical)	9.85 m (32 ft 3½ in)
at tip (reference)	1.66 m (5 ft 5½ in)
at tip (actual)	1.19 m (3 ft 11 in)
Wing aspect ratio: YF-22	2.20
F-22	2.36
Length overall: YF-22	19.56 m (64 ft 2 in)
F-22	18.92 m (62 ft 1 in)
Height overall: YF-22	5.36 m (17 ft 7 in)
F-22	5.00 m (16 ft 5 in)
Tail span: horizontal surfaces	8.84 m (29 ft 0 in)
vertical surfaces	5.97 m (19 ft 7 in)
Wheelbase	6.04 m (19 ft 9⅝ in)
Weapon bay ground clearance	0.93 m (3 ft 0¾ in)

AREAS:

Wings, gross: YF-22 and F-22	78.0 m² (840.0 sq ft)
Leading-edge flaps (total)	4.76 m² (51.2 sq ft)
Flaperons (total)	5.10 m² (55.0 sq ft)
Ailerons (total)	1.98 m² (21.4 sq ft)
Stabilators (total):	
YF-22 and F-22	12.63 m² (136.0 sq ft)
Vertical tails (total): YF-22	20.25 m² (218.0 sq ft)
F-22	16.54 m² (178.0 sq ft)
Rudders/speedbrakes (total): F-22	5.09 m² (54.8 sq ft)

WEIGHTS AND LOADINGS (estimated):

Weight empty: YF-22	over 13,608 kg (30,000 lb)
Max T-O weight: F-22	27,216 kg (60,000 lb)

PERFORMANCE (YF-22, demonstrated):

Max level speed: supercruise	Mach 1.58
with afterburning	Mach 1.7 at 9,150 m (30,000 ft)
Ceiling	15,240 m (50,000 ft)
g limit	+7.9

PERFORMANCE (F-22A, design target, estimated):

Max level speed at S/L	800 knots (1,482 km/h; 921 mph)
g limit	+9

LOCKHEED/AERMACCHI/ROLLS-ROYCE T-BIRD II

TYPE: Jet primary trainer (JPATS candidate); modified Aermacchi MB-339A.

PROGRAMME: Lockheed received MB-339A as demonstrator May 1992; larger wingtip fuel tanks and small wing fences added; engine hushkit being developed by Aermacchi and Rolls-Royce; choice of Martin-Baker Mk 10 or Mk 16L zero/zero seats to be decided; manufacturing partner expected to be selected mid-1993. Further details and photograph in Aermacchi entry in Italian section.

LOCKHEED 185/285/685/785 ORION

US Navy designation: P-3

CF designations: CP-140 Aurora/CP-140A Arcturus

TYPE: Land-based maritime patrol and ASW aircraft.

PROGRAMME: Lockheed won competition for off-the-shelf ASW aircraft 1958; first flight aerodynamic prototype 19 August 1958; first flight fully equipped YP-3A (YP3V-1) 25 November 1959; details of initial production P-3A and WP-3A in 1978-79 *Jane's*; details of P-3B and EP-3B in 1983-84 *Jane's*; P-3C produced from 1969. P-3A/B are Model 185; P-3C is Model 285. Total 642 P-3s built in California (Burbank, then Palmdale); last delivered (to Canada) May 1991; production line reopened at Marietta, Georgia, August 1991; first P-3C from Marietta (for Korea) due to roll out 19 May 1994. Following cancellation of P-7, USN, Germany and others considering proposed Orion II.

CURRENT VERSIONS: **P-3C**: First flight 18 September 1968; in service 1969; introduced A-NEW system based on Univac computer integrating all ASW information for retrieval, display and transmission of tactical data without routine log-keeping; USN received 118 of P-3C Baseline variant.

P-3C Update I: (Model 285A) Baseline P-3Cs followed from January 1975 by 31 P-3C Update I; new avionics and software included magnetic drum to increase computer memory sevenfold, new versatile computer language, Omega navigation, improved directional acoustic frequency analysis and recording (DIFAR) processing sensitivity, AN/ASA-66 tactical displays for two sensor stations, and improved magnetic tape transport.

Update II: (Model 285A) Applied to 44 aircraft built from August 1977; added infra-red detection system (IRDS) and sonobuoy reference system (SRS); Harpoon missile system incorporated from August 1977. 24 more USN P-3Cs received interim **Update II.5** of 1981 including more reliable navigation and communication systems; IACS submarine communications link; MAD compensation group adaptor; standardised wing pylons; and improved fuel tank vents.

Update III: (Model 285G for USN) Deliveries started May 1984; applied to last 50 USN P-3Cs; includes new IBM Proteus acoustic processor (doubling sonobuoy handling capacity), new sonobuoy receiver replacing DIFAR, improved APU, and higher capacity environmental control system. Baseline P-3C to III retrofit kit first installed in P-3C of VP-31 in 1987 (new designation **IIIR**); fitting of 18 more kits started June 1987; USN planned force of 138 Update III/IIIR aircraft; following cancellation of Update IV, further 109 may be added, giving standard fleet of 247 P-3Cs by 2006. Pakistan Orions, basically Update IIIs, have certain systems replaced by export-standard equipment and thus known as **II.75** versions. *Detailed description applies to P-3C/Update III except where indicated.*

Update IV: Programme cancelled by US Navy on 14 October 1992, but was in full-scale development by Boeing Defense and Space Group for installation from early 1990s; prototype conversion (160292) flew December 1991; equipment stripped out and prototype returned to normal use; originally intended for P-7A LRAACA; all P-3C Update I, II and II.5 were to be retrofitted, equipping VP-8, 10, 11, 23 and 26 at Brunswick, Maine, beginning FY 1994; one P-3C used for aerodynamic and functional testing of Eaton AIL Division AN/ALR-77 ESM system with 36 antennae mounted in four groups at wingtips; installation abandoned in favour of Litton AN/ALR-66(V)5 in same position; other features included Texas Instruments AN/APS-137(V) inverse synthetic aperture radar, Texas Instruments AN/AAS-36 IR detection system, AT&T AN/UYS-2 signal processor and Texas Instruments AN/ASQ-81 MAD. Planned for 109 retrofits.

Outlaw Hunter/OASIS: Single P-3C of VP-19 modified for 1991 Gulf War by USN and Tiburon Systems under Outlaw Hunter programme as over-the-horizon (OTH) targeting platform with ability to maintain battle area overall plot for battle group commanders. Two further P-3Cs, designated OASIS I and II (OTH Airborne Sensor Information System) modified for joint operation.

UP-3D: Kawasaki ECM trainer for JMSDF; planned purchase of two during FYs 1994-95.

UP-3C: Kawasaki-built service trials aircraft; one funded FY 1991; one more required; previously designated **NP-3C**.

EP-3: Elint version of P-3C developed by Kawasaki (which see) for JMSDF (first flight October 1990); two aircraft ordered FY 1987 and 1988; delivered March and November 1991; two more ordered in FYs 1992 and 1993; fifth planned in FY 1994 or 1995; no relation to USN RP-3D.

EP-3E Aries II: Ten P-3As and two EP-3Bs converted to EP-3E Aries I or similar 'Deepwell' (seven) and 'Batrack' (EP-3Bs) standards; radars in large canoe-shaped fairings above and below fuselage and ventral radome forward of wing; avionics believed to include GTE-Sylvania AN/ALR-60 communications intercept and analysis system, Raytheon AN/ALQ-76 noise jamming pod, Loral AN/ALQ-78 automatic ESM system, Magnavox AN/ALQ-108 IFF jammer, Sanders AN/ALR-132 infrared jammer, ARGO Systems AN/ALR-52 instantaneous frequency measuring equipment, Texas Instruments AN/APS-115 frequency agile search radar, Hughes AN/AAR-37 infra-red detector, Loral AN/ASA-66 tactical display, Cardion AN/ASA-69 scan converter and Honeywell AN/ASQ-114 computer. Twelve low-houred P-3Cs selected to replace EP-3E Aries I with USN special reconnaissance squadrons VQ-1 at Agana NAS, Guam, and VQ-2 at Rota, Spain; equipment transferred from original EP-3E but moderate upgrade includes faster processing and standardisation of previous two configurations within Aries I and addition to wingtips of IBM AN/ALR-76 ESM/RWR in pods; first Aries II conversion (156507) completed November 1988, to Patuxent River test centre 21 July 1990; first delivery (157320) to VQ-2 29 June 1991; initial five conversions by Lockheed Aircraft Service Company's Aeromod Center at Greenville, South Carolina, but contract then reassigned to Naval Air Depot, Alameda, 31 July 1991. VQ-2's last aircraft due June 1995.

CP-140A Arcturus: Three P-3s for Canadian Forces;

Two 'Slicks' and a 'Dome'— P-3A(CS) and P-3 AEW&C, respectively — of the US Customs service

USA: AIRCRAFT—LOCKHEED

no ASW equipment; for environmental and fishery patrol replacing CP-121 Trackers; equipment includes Texas Instruments AN/APS-134 Plus radar, Honeywell AN/APN-194 RAWS, Bendix AN/ASW-502 AFCS, Canadian Marconi AN/APN-510 Doppler radar, Litton LN-33 INS and a Leigh AN/ASH-502 flight recorder. Delivered to IMP Group at Halifax, Nova Scotia, for completion; final departure from Palmdale, 30 May 1991; delivery to CF December 1992 to February 1993; based at Greenwood.

P-3 AEW&C: Airborne early warning and control; first flight of prototype (N91LC) converted from Australian P-3B and fitted with empty Randtron AN/APA-171 7.32 m (24 ft) diameter rotodome 14 June 1984; testing of installed General Electric AN/APS-125 radar began 1988; military version would have AN/APS-145 radar as in Grumman E-2C Hawkeye. Other systems would include C^3 system to receive, process and transmit tactical information on HF, UHF, VHF and Satcom channels; AN/ARC-187 satellite communication system; and Collins five-tube colour EFIS-86B flight instruments. General Electric AN/APS-145 radar available from late 1989.

First order from US Customs May 1987 for one plus option for three; first flight US Customs aircraft with AN/APS-125 radar 8 April 1988; aircraft (N145CS ex-N91LC) called *Blue Eagle* delivered to NAS Corpus Christi, Texas, 17 June 1988 and used for anti-narcotics patrol over Caribbean and Gulf of Mexico; retrofitted with AN/APS-138; second ex-RAAF P-3 AEW&C called *Blue Eagle II*, fitted with improved AN/APS-138 radar, delivered to US Customs June 1989; third and fourth (both ex-USN P-3Bs) delivered 26 June 1992 and late 1993. Additional systems include CDC AN/AYK-14 computer with Honeywell 1601M array processor, dual Sanders Miligraphics touch-sensitive colour display screens for digital target data, Hazeltine AN/TPX-54 IFF, dual AN/ARC-182 VHF/UHF com radios, dual AN/ARC-207 HF and dual Wulfsberg VHF/UHF-FM radios. Type nickname 'P-3 Dome'.

P-3H Orion II: Not yet funded by US Navy. Proposed P-3C upgrade to replace cancelled P-7A. Update IV avionics plus new wings and engines (Allison T406 or GE T407) and, possibly, HUDs and flat-panel cockpit displays having high commonality with Hercules II. *See also Addenda.*

EP-3J Orion: Electronic warfare trainer based on conversion of P-3B with AN/USQ-113 communications intrusion and deception system, plus AN/ALQ-167, AN/ALQ-170 and AN/AST-4/6 pods. Modified by Chrysler Technologies. Service with VAQ-33 squadron at Key West; first delivery 17 March 1992; two aircraft (152719 and 152745); to transfer to VP-94 in 1993. Planned Phase 2 modifications to include AN/ALT-40(V) radar jammer.

P-3T/UP-3T: Conversions of USN P-3A for Thailand; five airframes purchased 1992; two for modification to P-3T; one to UP-3T; and two for spares break-down.

P-3W: Australian P-3Cs; have AQS-901 processing system and Barra sonobuoys in place of Proteus and AN/AQA-7 equipment of USN P-3C. Upgrade of 10 with Elta ESM equipment (replacing AN/ALQ-78) begun by AWADI in Australia, 1991; other upgrades planned. P-3W is local designation of P-3C-II.5; earlier P-3C-IIs retain original designation.

CUSTOMERS: US Navy, 552, comprising one YP-3; 157 P-3As, of which 38 modified to UP-3A, seven to EP-3A, six to RP-3A, five to VP-3A (three via WP-3A), 12 to TP-3A, two to EP-3B and 10 to EP-3E; 124 P-3Bs, one converted to NP-3B and one under conversion in 1991 to RP-3B; 267 P-3Cs, 12 intended for EP-3E-II conversion; one RP-3D; two National Oceanographic and Atmospheric Administration WP-3Ds. Final USN P-3C (163295) delivered 17 April 1990. Status of USN Orion force after 1992-93 disbandments shown in table; new total of 18 regular ASW squadrons. Last Reserve P-3A ASW mission 22 March 1990 (152158 of VP-64); last regular USN P-3B ASW mission 11 September 1990 (VP-22).

Total 90 P-3s from Californian production for export to Australia (10 Model 185B P-3Bs, one transferred from US Navy; 10 P-3C-IIs and 10 P-3C-II.5s, both Model 285D); six P-3Bs transferred to Portugal as P-3P, one to New Zealand, two to US Customs (after conversion to AEW&C), Canada (18 CP-140 Auroras, three CP-140A Arcturus), Iran (six Model 685A P-3Fs), Japan (three Model 785A P-3C-IIs from Lockheed; Kawasaki produced 66 P-3C-IIs and has 34 P-3C-IIIs, four EP-3s and one UP-3C on order; P-3C-II to be updated), Netherlands (13 Model 285E P-3C-II.5s), New Zealand (five P-3Bs updated to P-3Ks; plus one ex-Australian P-3K), Norway (five Model 185C P-3Bs, two transferred from US Navy, and four P-3C-IIIs; five P-3Bs transferred to Spain and two converted to P-3N; last conversion re-delivered 21 June 1992) and Pakistan (three P-3C Update II.75s; crew training completed December 1991; aircraft embargoed by US government and stored). Marietta production initially for South Korea (eight P-3C Update IIIs with AN/APS-134 radar; first rollout May 1994; last delivery September 1995). Four USN P-3As transferred to US Customs as UP-3As (alternative designation P-3A(CS) 'Slick') fitted with IR detection system and AN/APG-63 lookdown radar. Chile obtained two P-3As and six UP-3As from USN early in 1993. Others civilianised for various operators, including N406TP temporarily with Allison GMA 2100 turboprop and Dowty Aerospace R373 composite propeller in port outer nacelle, 1990; and five P-3A 'Aerostar' firefighters of Aero Union Corporation. Greece plans to buy five surplus P-3As and one UP-3A for 1994-95 delivery.

COSTS: $600 million for eight P-3Cs for South Korea 1990; $840 million (1990) including engines, training and spares. C$159 million for three semi-complete CP-140A airframes, plus C$59 million for radar, spares, logistics and project management.

US NAVY ORION SQUADRONS

At late 1993

Squadron	Variant	Base
VP-1	P-3C-IIIR	Barber's Point
VP-4	P-3C-IIIR	Barber's Point
VP-5	P-3C-IIIR	Jacksonville
VP-8	P-3C-II.5	Brunswick
VP-9	P-3C-IIIR	Barber's Point
VP-10	P-3C-II.5	Brunswick
VP-11	P-3C-II.5	Brunswick
VP-16	P-3C-IIIR	Jacksonville
VP-17	P-3C-III	Barber's Point
VP-22	P-3C-I/II.5/III/IIIR	Barber's Point
VP-23	P-3C-II	Brunswick
VP-24	P-3C-IIIR	Jacksonville
VP-26	P-3C-II	Brunswick
VP-30	P-3C-II.5/IIIR, TP-3A, VP-3A	Jacksonville
VP-40	P-3C-II.5	Brunswick
VP-45	P-3C-IIIR	Jacksonville
VP-46	P-3C-IIIR	Jacksonville
VP-47	P-3C-III	Barber's Point
VP-49	P-3C-IIIR	Jacksonville
VPU-1	P-3B	Brunswick
VPU-2	P-3B, UP-3A	Moffett Field
VQ-1	EP-3E, UP-3B	Cubi Point
VQ-2	EP-3E-II, P-3B	Rota, Spain
VXN-8	P-3B, RP-3D	Patuxent River
VX-1	P-3C/II/III/IIIR	Patuxent River
VAQ-33	EP-3J, P-3B	Key West

Reserves

VP-60	P-3B	Glenview
VP-62	P-3C-III	Willow Grove
VP-64*	P-3B	Willow Grove
VP-65	P-3C	Point Mugu
VP-66	P-3B	Willow Grove
VP-67*	P-3B	Memphis
VP-68	P-3C-I	Patuxent River
VP-69	P-3C-I	Whidbey Island
VP-90*	P-3B	Glenview
VP-91	P-3C-IIIR	Moffett Field
VP-92	P-3C-II	South Weymouth
VP-93*	P-3B	Selfridge ANGB
VP-94	P-3B	New Orleans

* Planned 1993 disbandment vetoed by Congress

ORION FUNDING

Fiscal Year	P-3A	P-3B USN	P-3B Export	P-3C USN	P-3C Export	P-3D	Remarks
1959	7						
1961	12						
1962	42						
1963	48						
1964	48						
1965		48	5*				* New Zealand
1966		44					
1967		32	15*				* New Zealand 5, Australia 10
1968				25			
1969				23			
1970				23		1*	* RP-3D
1971				12			
1972				24			
1973				12	6*		* Iran (P-3F)
1974				12		2*	* WP-3D
1975				12			
1976				11			
1977				15	10*		* Australia
1978				14	4*		* Canada CP-140
1979				12	14*		* Canada CP-140
1980				12	5*		* Japan 3, Netherlands 2
1981				12	4*		* Netherlands
1982				12	4*		* Netherlands
1983				6	13*		* Netherlands 3, Australia 10
1984				5			
1985				9			
1986				9			
1987				7	4*		* Norway
1989					3*		* Pakistan
1990					3*		* Canada CP-140A
1991					8*		* Korea
	157	124	20	267	78	3	
			144		345		
TOTAL			649				

DESIGN FEATURES: Pressurised cabin. Wing section NACA 0014 (modified) at root, NACA 0012 (modified) at tip; dihedral 6°; incidence 3° at root, 0° 30′ at tip.
FLYING CONTROLS: Hydraulically boosted ailerons, elevators and rudder; fixed tailplane; Lockheed-Fowler trailing-edge flaps.
STRUCTURE: Conventional aluminium alloy with fail-safe box beam wing.
LANDING GEAR: Hydraulically retractable tricycle type, with twin wheels on each unit. All units retract forward, mainwheels into inner engine nacelles. Oleo-pneumatic shock absorbers. Mainwheels have size 40-14 type VII 26-ply tubeless tyres, pressure 7.58-12.41 bars (110-180 lb/sq in) at 36,287 kg (80,000 lb) T-O weight, 12.41 bars (180 lb/sq in) at 57,606 kg (127,000 lb) T-O weight, 13.10 bars (190 lb/sq in) at 61,235 kg (135,000 lb) max normal T-O weight. Nosewheels have size 28-7.7 type VII tubeless tyres, pressure 10.34 bars (150 lb/sq in). Hydraulic brakes. No anti-skid units.
POWER PLANT: Four 3,661 kW (4,910 ehp) Allison T56-A-14 turboprops, each driving a Hamilton Standard 54H60-77 four-blade constant-speed propeller. Fuel in one tank in fuselage and four wing integral tanks, with total usable capacity of 34,826 litres (9,200 US gallons; 7,660 Imp gallons). Four overwing gravity fuelling points and central pressure refuelling point. Oil capacity (min usable) 111 litres (29.4 US gallons; 24.5 Imp gallons) in four tanks
ACCOMMODATION: Normal 10-man crew: pilot, co-pilot, flight engineer and nav/com operator on flight deck; tactical co-ordinator, two acoustic sensor operators, MAD operator, ordnance man and flight technician; up to 13 additional relief crew or passengers. Flight deck has wide-vision windows, and circular windows for observers are provided fore and aft in the main cabin, each bulged to give 180° view. Main cabin is fitted out as a five-man tactical compartment (containing advanced electronic, magnetic and sonic detection equipment), an all-electric galley and large crew rest area.
SYSTEMS: Air-conditioning and pressurisation system supplied by two engine driven compressors. Pressure differential 0.37 bar (5.4 lb/sq in). Hydraulic system, pressure 207 bars (3,000 lb/sq in), for flaps, control surface boosters, landing gear actuation, brakes and bomb bay doors. Three hydraulic pumps, each rated at 30.3 litres (8.0 US gallons; 6.7 Imp gallons)/min at 0-152 bars (0-2,200 lb/sq in), 22.7 litres (6.0 US gallons; 5.0 Imp gallons)/min at 205 bars (2,975 lb/sq in). Class one non-separated air/oil reservoir, Type B pressurised. Electrical system utilises three 60kVA generators for 120/208V 400Hz AC supply. 24V DC supply. Integral APU with 60kVA generator for ground air-conditioning, electrical supply and engine starting. Anti-icing by bleed air on wing and electrical heating on tailplane and fin. Electrically de-iced propeller spinners.
AVIONICS: The AN/ASQ-114 general purpose digital computer is the heart of the P-3C system. Together with the AN/AYA-8 data processing equipment and computer controlled display systems, it permits rapid analysis and utilisation of electronic, magnetic and sonic data. Nav/com system comprises two LTN-72 inertial navigation systems; AN/APN-227 Doppler; AN/ARN-81 Loran A and C; AN/ARN-118 Tacan; two VIR-31A VOR/LOC/GS/MB receivers; AN/ARN-83 LF-ADF; AN/ARA-50 UHF direction finder; AN/AJN-15 flight director indicator for tactical directions; HSI for long-range flight directions; glideslope indicator; on-top position indicator; two AN/ARC-161 HF transceivers; two AN/ARC-143 UHF transceivers; AN/ARC-101 VHF receiver/transmitter; AN/AGC-6 teletype and high-speed printer; HF and UHF secure communication units; AN/ACQ-5 data link communication set and AN/AIC-22 interphone set; AN/APX-72 IFF transponder and AN/APX-76 SIF interrogator. Electronic computer controlled display equipment includes AN/ASA-70 tactical display, AN/ASA-66 pilot's display, AN/ASA-70 radar display and two auxiliary readout (computer stored data) displays.

ASW equipment includes two AN/ARR-72 sonar receivers, replaced in Update III by AN/ARR-78; two AN/AQA-7(V)8 DIFAR (directional acoustic frequency analysis and recording) sonobuoy indicator sets, replaced in Update III by AN/UYS-1 Proteus; hyperbolic fix unit; acoustic source signal generator; time code generator and AN/AQH-4(V) sonar tape recorder; AN/ASQ-81 magnetic anomaly detector; AN/ASA-64 submarine anomaly detector; AN/ASA-65 magnetic compensator; AN/ALQ-78 electronic countermeasures set; AN/APS-115 radar set (360° coverage); AN/ASA-69 radar scan converter; undernose AN/AAS-36 IRDS; KA-74 forward computer assisted camera; KB-18A automatic strike assessment camera with horizon-to-horizon coverage; RO-308 bathythermograph recorder.

Additional items include AN/APN-194 radar altimeter; two AN/APQ-107 radar altimeter warning systems; A/A24G-9 true airspeed computer; AN/ASW-31 automatic flight control system. P-3Cs delivered from 1975 have the avionics/electronics package updated by addition of an extra 393K memory drum and fourth logic unit, Omega navigation, new magnetic tape transport, and an AN/ASA-66 tactical display for the sonar operators. To accommodate the new systems a new operational software computer programme was written in CMS-2 language. GEC-Marconi AQS-901 acoustic signal processing and

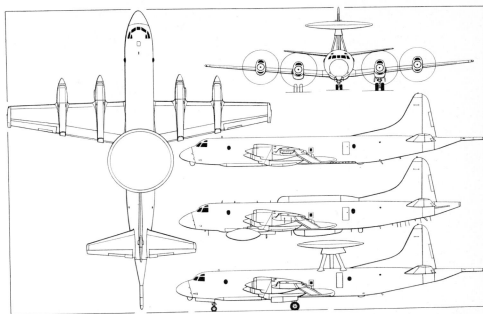
Lockheed P-3 AEW&C, with added side views of EP-3E Aries II (centre) and P-3C Orion (top) *(Dennis Punnett)*

One of three Lockheed P-3C-II Orions ordered by the Pakistan Navy

Lockheed EP-3E Aries II communications interception aircraft of USN squadron VQ-2 *(Paul Jackson)*

display system in RAAF P-3Ws. AN/ALR-66(V)5 passive radar detection system (ESM), to be housed in wingtip pods, is under development for Update IV P-3C by General Instrument, and will also provide targeting data for the aircraft's Harpoon missiles. AN/ALR-66(V)3 installed in Japanese and Norwegian P-3Cs and as retrofit in P-3P and CP-140. Wing span increased by some 0.81 m (2 ft 8 in) to accommodate ESM antennae and receivers. Similar Israeli Elta equipment for Australian retrofit. Loral AN/ALQ-157 IR jammers retrofitted each side of rear fuselage on USN P-3Cs. AN/ALR-66(V)5 replaces Loral AN/ALQ-78A pod on inboard wing pylon.

EQUIPMENT: Searchlight replaces one wing pylon, starboard. Search stores, such as sonobuoys and sound signals, are launched from inside cabin area in the P-3A/B. In the P-3C sonobuoys are loaded and launched externally and internally. Sonobuoys are ejected from P-3C aircraft with explosive cartridge actuating devices (CAD), eliminating the need for a pneumatic system. Australian P-3Ws use SSQ-801 Barra sonobuoys.

ARMAMENT: Bomb bay, 2.03 m wide, 0.88 m deep and 3.91 m long (80 × 34.5 × 154 in), forward of wing, and 10 underwing pylons. Stores can include (weapons bay/underwing, maximum) Mk 46 torpedo 8/0; Mk 50 torpedo 6/0; Mk 54 depth bomb 8/10; B57 nuclear depth charge 3/0; Mk 82 560 lb bomb 8/10; Mk 83 980 lb bomb 8/8; Mk 36 destructor 8/10; Mk 40 destructor 3/8; LAU-68A pod (seven 2.75 in rockets), or LAU-69A (nineteen 2.75 in rockets), or LAU-10A/C (four 5 in rockets), or SUU-44A (eight flares) 0/4; Mk 52 mine 3/8; Mk 55 or Mk 56 mine 1/6; Mk 60 torpedo 0/6; AGM-86 Harpoon anti-ship missile 0/8. Two AIM-9L Sidewinder AAMs underwing for self-defence. Max total weapon load includes six 2,000 lb mines under wings and a 3,290 kg (7,252 lb) internal load made up of two Mk 101 depth bombs, four Mk 44 torpedoes, pyrotechnic pistol and 12 signals, 87 sonobuoys, 100 Mk 50 underwater sound signals (P-3A/B), 18 Mk 3A marine markers (P-3A/B), 42 Mk 7 marine markers, two B. T. buoys, and two Mk 5 parachute flares. Harpoon missiles are standard fit on a proportion of US Navy P-3Cs.

DIMENSIONS, EXTERNAL:
Wing span	30.37 m (99 ft 8 in)
Wing chord: at root	5.77 m (18 ft 11 in)
at tip	2.31 m (7 ft 7 in)
Wing aspect ratio	7.5
Length overall	35.61 m (116 ft 10 in)
Height overall	10.27 m (33 ft 8½ in)
Fuselage diameter	3.45 m (11 ft 4 in)
Tailplane span	13.06 m (42 ft 10 in)
Wheel track (c/l shock absorbers)	9.50 m (31 ft 2 in)
Wheelbase	9.07 m (29 ft 9 in)
Propeller diameter	4.11 m (13 ft 6 in)
Cabin door: Height	1.83 m (6 ft 0 in)
Width	0.69 m (2 ft 3 in)

DIMENSIONS, INTERNAL:
Cabin, excl flight deck and electrical load centre:
Length	21.06 m (69 ft 1 in)
Max width	3.30 m (10 ft 10 in)
Max height	2.29 m (7 ft 6 in)
Floor area	61.13 m² (658.0 sq ft)
Volume	120.6 m³ (4,260 cu ft)

AREAS:
Wings, gross	120.77 m² (1,300.0 sq ft)
Ailerons (total)	8.36 m² (90.0 sq ft)
Trailing-edge flaps (total)	19.32 m² (208.0 sq ft)
Fin, incl dorsal fin	10.78 m² (116.0 sq ft)
Rudder, incl tab	5.57 m² (60.0 sq ft)
Tailplane	22.39 m² (241.0 sq ft)
Elevators, incl tabs	7.53 m² (81.0 sq ft)

WEIGHTS AND LOADINGS (P-3B/C):
Weight empty	27,890 kg (61,491 lb)
Max fuel weight	28,350 kg (62,500 lb)
Max expendable load	9,071 kg (20,000 lb)
Max normal T-O weight	61,235 kg (135,000 lb)
Max permissible weight	64,410 kg (142,000 lb)
Design zero-fuel weight	35,017 kg (77,200 lb)
Max landing weight	47,119 kg (103,880 lb)
Max wing loading	507.0 kg/m² (103.8 lb/sq ft)
Max power loading	4.18 kg/kW (6.87 lb/ehp)

PERFORMANCE (P-3B/C, at max T-O weight, except where indicated otherwise):
Max level speed at 4,575 m (15,000 ft) at AUW of 47,625 kg (105,000 lb)	411 knots (761 km/h; 473 mph)
Econ cruising speed at 7,620 m (25,000 ft) at AUW of 49,895 kg (110,000 lb)	328 knots (608 km/h; 378 mph)
Patrol speed at 457 m (1,500 ft) at AUW of 49,895 kg (110,000 lb)	206 knots (381 km/h; 237 mph)
Stalling speed: flaps up	133 knots (248 km/h; 154 mph)
flaps down	112 knots (208 km/h; 129 mph)
Rate of climb at 457 m (1,500 ft)	594 m (1,950 ft)/min
Time to 7,620 m (25,000 ft)	30 min
Service ceiling	8,625 m (28,300 ft)
Service ceiling, OEI	5,790 m (19,000 ft)
T-O run	1,290 m (4,240 ft)
T-O to 15 m (50 ft)	1,673 m (5,490 ft)
Landing from 15 m (50 ft) at design landing weight	845 m (2,770 ft)
Mission radius (3 h on station at 457 m; 1,500 ft)	1,346 nm (2,494 km; 1,550 miles)
Max mission radius (no time on station) at 61,235 kg (135,000 lb)	2,070 nm (3,835 km; 2,383 miles)
Ferry range	4,830 nm (8,950 km; 5,562 miles)

Max endurance at 4,575 m (15,000 ft):
two engines	17 h 12 min
four engines	12 h 20 min

LOCKHEED VIKING
US Navy designation: S-3

TYPE: Carrier-borne ASW aircraft.
PROGRAMME: Production of 187 S-3As for US Navy ended mid-1978; data in 1991-92 and previous editions of *Jane's*. Additional details of S-3B and ES-3A in *Jane's Civil and Military Aircraft Upgrades*.
CURRENT VERSIONS: **S-3A**: Initial production ASW version (see 1978-79 *Jane's*).
S-3B: Lockheed defined Weapon System Improvement Program, full-scale development ordered 18 August 1981; first flight of first of two development S-3Bs 13 September 1984; installation at Cecil Field NAS, Florida; all 83 planned S-3Bs completed for Atlantic Fleet by December 1991. Further 38 kits being embodied in Pacific Fleet S-3s at NAS North Island, California, from March 1992.
ES-3A: Contract for development of electronic reconnaissance E-3A awarded March 1988. Prototype delivered to USN 9 August 1991; first of 15 'production' kit conversions by Navy at Cecil Field flew 21 January 1992.

LOCKHEED 382 HERCULES
US Air Force designations: C-130, AC-130, DC-130, EC-130, HC-130, JC-130, LC-130, MC-130, NC-130, RC-130 and WC-130
US Navy designations: C-130, DC-130, EC-130, LC-130 and TC-130
US Marine Corps designation: KC-130
US Coast Guard designations: C-130, EC-130 and HC-130
Canadian Forces designations: CC-130 and CC-130T
RAF designations: Hercules C. Mk 1K, C. Mk 1P, W. Mk 2 and C. Mk 3P
Spanish designations: T.10, TK.10 and TL.10
Swedish designation: Tp 84
Export designations: C-130H, C-130H-30, KC-130H, C-130H-MP and VC-130H

TYPE: Tactical transport and multi-mission aircraft.
PROGRAMME: US Air Force specification issued 1951; first production contract for C-130A to Lockheed September 1952; two prototypes, 231 C-130As, 230 C-130Bs and 49 C-130Es manufactured (details in earlier *Jane's*). For late military versions, see below; more than 70 models/variants delivered. Lockheed delivered 2,000th Hercules (C-130H 91-1231 to 165th ALS, Kentucky ANG) 14 May 1992.
CURRENT VERSIONS: **C-130H**: Deliveries started March 196 to Royal New Zealand Air Force; in service with 64 countries. Features include updated avionics, improved wing, new corrosion protection, and Allison T56-A-15 engine flat rated at 3,362 kW (4,508 shp). Can deliver up to 19,052 kg (42,000 lb) by low-velocity air drop, or by low altitude parachute extraction system (LAPES). Total 21 regular transport C-130Hs funded for USAF/ANG/AFRe in FYs 1973-91; further 29 in FY 1993; from 1994, new aircraft have AlliedSignal TCAS II collision avoidance system; Night Vision Instrument System introduced to USAF aircraft from May 1993; others for special roles, a detailed below. In January 1993, Chrysler Technologie Airborne Systems and AlliedSignal Aerospace won cockpit upgrade competition for USAF Hercules (and 14 C-141s) involving four 150 × 200 mm (6 × 8 in) liqui crystal displays; AlliedSignal digital autopilot, Sundstran ground collision avoidance and other avionics improve ments. Total 672 Hercules to be modified, including C-130Es and Hs. *Detailed description applies mainly to C-130H, except where indicated.*

C-130H-MP: Maritime patrol version; one delivered to Indonesian Air Force, three to Royal Malaysian Air Force Max T-O weight 70,310 kg (155,000 lb), max payload 18,630 kg (41,074 lb), and T56-A-15 engines; search time 2 h 30 min at 1,525 m (5,000 ft) at 1,800 nm (3,333 km 2,070 miles) radius or 16 h 50 min at 200 m (370 km; 23 miles) radius. Optional and standard search feature include sea search radar, observer seats and windows, INS Omega navigation, crew rest and galley slide-in module flare launcher, loudspeaker, rescue kit airdrop platform side-looking radar, passive microwave imager, low light TV, infra-red scanner, camera with data annotation an ramp equipment pallet with observer station.

C-130H-30: Stretched version of current production C-130; fuselage lengthened by 4.57 m (15 ft 0 in); troop and cargo capability increased by 40 per cent.

AC-130H Spectre: Gunship version with sideways firing 105 mm recoilless howitzer, 40 mm cannon and tw 20 mm Vulcan guns; infra-red and low light TV sensors and side-looking head-up display for aiming at night while circling target; in-flight refuelling. Conversion by Lockheed Aircraft Service Co. New fire control computers and

USAF HERCULES FUNDING

Fiscal Year	C-130A	C-130B	C-130D	C-130E	HC-130	C-130H	MC-130H	AC-130U	Remarks
1953	7								1 to AC-
1954	20								6 to AC-
1955	48								7 to AC-
1956	84								3 to AC-
1957	60	5	12						
1958		47							
1959		15							
1960		18							
1961		41		16					3 to WC-
1962		10		83					1 to MC-;
									8 to EC-
1963				44					8 to EC-;
									1 to MC-
1964				80*	15†				* 14 to MC-;
									* 3 to WC-;
									† 1 to EC-;
									† 1 to WC-
1965						33			8 to C-;
									6 to WC-
1966						15			
1967						3			
1968				18					
1969				18*	15				* 10 to AC-130H
1970				18					
1972				12					
1973						15			12 to EC-
1974						53			1 to YMC-
1978						8			
1979						8			
1980						8			
1981						6			
1982						8			
1983						8*	1		* 4 to LC-
1984						10	2		
1985						16	2		
1986						18	1		
1987						8	1	1	
1988			2			16	11		
1989			1			20	2	6	
1990						12		5	
1991						16			
1992						14	1		
1993						29			
Total	219	136	12	289	84	273	24	13	**1,050**

Note: Not all conversions are given

navigation and sensors under Special Operations/Forces Improvements (SOFI) being installed. Flight testing began September 1989; first upgrade completed mid-1990; last of current nine due 1993. In service with 16th Special Operations Squadron at Hurlburt Field, Florida.

EC-130H Compass Call: Works with ground-based C³CM to jam enemy command, control and communications. Operated by 41st Electronic Combat Squadron at Davis Monthan AFB, Arizona. (Eight earlier EC-130Es of 7th ACCS being updated by Unisys to ABCCC III standard in $34 million programme, 1990.)

HC-130H: Extended range USAF aircraft for aerial recovery of personnel or equipment and other duties; 43 delivered from October 1964; update announced Spring 1987 includes self-contained navigation, night vision goggles cockpit and new communications equipment; applied to 31 aircraft; 21 of these modified for in-flight refuelling and now designated HC-130P; US Coast Guard ordered 35, final 10 as HC-130H-7 with T56A-7B power plants. Further three HC-130H(N)s funded FYs 1988-90 for 210th ARS, ANG, in Alaska, delivered from 28 November 1990; has helicopter refuelling capability from outset.

JC-130H: Four US Air Force HC-130Hs equipped to recover re-entering space capsules.

DC-130H: Two US Air Force HC-130Hs modified for drone control.

KC-130H: Probe-drogue tanker similar to KC-130R; exported to Argentina (two), Brazil (two), Israel (two), Morocco (two), Saudi Arabia (eight), Spain (five) and Singapore (one).

LC-130H: Similar to LC-130R (which see); four acquired by 139th AS, ANG, at Schenectady.

MC-130H Combat Talon II: Conversion of new-build C-130H for day/night infiltration and exfiltration, resupply of Special Operations Forces, psychological warfare and aerial reconnaissance; terrain-following radar; six-man crew; 25 (including YMC-130H prototype 74-1686, now retired) funded in FYs 1983-90; first flight by E-Systems (see below) at Greenville Spring 1988; flight testing began at Edwards AFB September 1988; MC-130Hs augment MC-130Es at Hurlburt Field; first four delivered 17 October 1991 and temporarily with 8th SOS prior to formation of 15th SOS on 1 October 1992; others to 1st SOS at Kadena, Japan; 7th SOS at Alconbury, UK (beginning with 86-1699 on 10 September 1992); and 1550th ATS (training) at Kirtland AFB, New Mexico (three aircraft).

Equipment includes Emerson Electric AN/APQ-170 precision ground mapping/weather/terrain following and avoidance radar in enlarged radome, inertial navigation, automatic computed air release point, high-speed low level release system, ground acquisition receiver/interrogator, Texas Instruments AN/AAQ-15 infra-red detection system, eight multi-function displays, secure voice UHF/VHF-FM radios, retractable FLIR pod, angle of attack probe, AN/ALQ-8 ECM pod under each wing, and in-flight refuelling. Defensive equipment includes Litton AN/ALR-69 RWR, ITT AN/ALQ-172 detector/jammer, Watkins Johnson WJ-1840 signal detector, Cincinnati Electronics AN/AAR-44 launch warning receiver, Northrop QRC-8402 IR jammer and chaff/flare dispensers. IBM Federal Systems Division is prime contractor for systems integration, with E-Systems as subcontractor for avionics installation and modification.

RC-130H: Unofficial designation for two Moroccan aircraft fitted with SLAR by Flight Systems Inc.

VC-130H: VIP transport.

C-130J: Military version of Hercules II; development of advanced Hercules continuing for foreign sales.

C-130K: RAF version of C-130H; much of avionics and instrumentation made in UK; 66 delivered as **Hercules C. Mk 1** beginning September 1966; one modified by Marshall of Cambridge for RAF Meteorological Research Flight as **Hercules W. Mk 2**. Thirty lengthened by 4.57 m (15 ft), equivalent to commercial L-100-30, and redesignated **Hercules C. Mk 3**; first aircraft modified at Marietta, remaining 29 by Marshall of Cambridge. Six modified to **C. Mk 1K** tankers, of which four received Racal 'Orange Blossom' ESM, which also fitted to one C. Mk 1; 20 C. Mk 3s received AN/APN-169B station-keeping equipment; some have provision for AN/ALQ-157 IR jammers; AN/ALR-66 RWR ordered; all fitted with refuelling probes 1982-1989.

HC-130N: US Air Force search and rescue version of C-130H for recovery of aircrew and space capsules; 15 delivered with helicopter refuelling capability; advanced direction finding equipment.

HC-130P: C-130H modified for refuelling helicopters in flight and recovering parachute-borne payloads; 20 built for USAF; details in 1979-80 *Jane's*. HC-130N/Ps to be upgraded by Rockwell with SOF Improvement package including self-protection aids for refuelling operations in hostile airspace.

EC-130Q: Similar to earlier EC-130G, but with improved equipment and crew accommodation for TACAMO command communication with submarines; 18 built; HF and VLF SIMOP (simultaneous operation). Last aircraft (161531) retired from VQ-4 on 26 May 1992; replaced by Boeing E-6A; two EC-130Qs converted to TC-130Q; three to NASA as transports; remainder to storage.

TC-130Q: Two surplus EC-130Qs with trailing wire aerial removed to permit normal cargo loading via rear doors; wingtip pods retained; first (159348) noted 1990. Two TC-130Gs are similar.

KC-130R: Probe-drogue tanker version of C-130H; 14 delivered to US Marine Corps VMGR-252 and 352; changes from KC-130F (1975-76 *Jane's*) include 3,362 kW (4,508 shp) engines, higher T-O and landing weights, external fuel tanks for additional 10,296 litres (2,720 US gallons; 2,265 Imp gallons) fuel, and removable 13,627 litre (3,600 US gallon; 2,997 Imp gallon) fuel tank in cargo hold (all fuel can be used to increase tanker's range); single-point refuelling of normal and additional tanks from existing filler; operating weight empty 36,279 kg (79,981 lb); max T-O weight 79,378 kg (175,000 lb); can off-load up to 20,865 kg (46,000 lb) of fuel, equivalent to 26,790 litres (7,077 US gallons; 5,893 Imp gallons), at radius of 1,000 nm (1,850 km; 1,150 miles); maximum off-load capability 31,750 kg (70,000 lb), equivalent to 40,765 litres (10,769 US gallons; 8,967 Imp gallons).

LC-130R: C-130H with wheel-ski landing gear for US Navy Squadron VXE-6 in Antarctic; details in 1979-80 *Jane's*.

CC-130T: Canadian Forces' designation for five CC-130s converted by Northwest Industries, 1992-93, as tankers with two FRL Mk 32B hose-drum units beneath wings and 13,627 litre (3,600 US gallon; 2,998 Imp gallon) tank in cargo hold.

KC-130T: Tanker for US Marine Corps (Reserve), able to refuel helicopters and fighters; eight delivered to Marine Aerial Refueller Transport Squadron 234 (VMGR-234), starting November 1983; others to VMGR-452; deliveries two per year since 1990; total 28 required by 1995 (14 per squadron). Similar to KC-130R, but with updated avionics including INS, Omega and Tacan, new autopilot and flight director and solid-state search radar; KC-130Ts delivered in 1984 had Bendix/King AN/APS-133 colour radar, flush antennae and orthopaedically designed crew seats.

KC-130T-30: Stretched KC-130T (similar to C-130H-30) for US Navy; first (164597) delivered to VMGR-452, 29 October 1991; second in November 1991 to VMGR-234.

C-130T: Transport, with secondary refuelling capability, for US Navy; first of six (164762) to VR-54 at New Orleans 20 August 1991; first of six to VR-48 at NARF Andrews, Maryland, October 1992. Total 22 required. VR-53 next to form, at Martinsburg, West Virginia.

AC-130U Spectre: New gunship version of C-130H; details under Rockwell International entry.

EC-130V: Surveillance conversion of USCG HC-130H (1721) with Grumman E-2C Hawkeye's AN/APS-125 radar. Conversion by General Dynamics, Fort Worth; first flight 31 July 1991; to CGAS Clearwater, Florida.

Hercules II: Privately funded development, launched 1991, with Allison GMA 2100 turboprops, six-blade propellers (development contract to Dowty, December 1992) and two-crew cockpit (plus navigator as customer's option). Available from 1996, given launch customer(s); military version designated C-130J; commercial version designated Advanced L-100. Prototype (2,131st production Hercules; c/n 5400) for completion July 1995.

CUSTOMERS: Military and government Hercules operators for C-130H export versions only are given in the table.

COSTS: C-130H $31.5 million (1988) flyaway; $48 million (1987) programme unit cost for AC-130H.

DESIGN FEATURES: Can deliver loads and parachutists over lowered rear ramp and parachutists through side doors; removable external fuel tanks outboard of engines are standard fittings; cargo hold pressurised. Wing section NACA 64A318 at root and NACA 64A412 at tip; dihedral 2° 30′; incidence 3° at root, 0° at tip.

FLYING CONTROLS: All control surfaces boosted by dual hydraulic units; trim tabs on ailerons, both elevators and

MILITARY/GOVERNMENT C-130H/L-100-20/30 DELIVERIES
To end 1992; excluding USA

	C-130H	C-130H-30	KC-130H	Other
Abu Dhabi	6			
Algeria	10	7		
Argentina	5		2	
Australia	12			
Belgium	12			
Bolivia	2			
Brazil	6		2	
Cameroon	2	1		
Canada	14			
Chad	1	1		
Chile	2			
Colombia	2			
Denmark	3			
Dubai		1		
Ecuador	3			
Egypt	22	2		1 (VC-)
France	3	9		
Gabon	1			
Greece	12			
Indonesia	3	7		1 (-MP)
Iran	32			
Israel	10		2	
Italy	14			
Japan	15			
Jordan	4			
South Korea	8	4		
Libya	16			
Malaysia	6	1		3 (-MP)
Morocco	15		2	2 (RC-)
New Zealand	5			
Niger	2			
Nigeria	6	3		
Norway	6			
Oman	3			
Philippines	3			
Portugal	5	1		
Saudi Arabia	35	2	8	1 (VC-)
	3 HS	1 HS		
Singapore	5		1	
Spain	7	1		5
Sudan	6			
Sweden	6			
Taiwan	13			
Thailand	6	6		
Tunisia	2			
UK				66 (130K)
Venezuela	8			
Yemen	2			
Zaïre	7			
Total (514)	371	47	22	74

Key: MP - C-130H-MP maritime patrol
RC - RC-130H electronic reconnaissance
VC - VC-130H VIP transport
HS - C-130/L-100 airborne hospital
K - C-130K United Kingdom variant

Note: Customers and variants as delivered; transfers and conversions not listed.

504 USA: AIRCRAFT—LOCKHEED

Lockheed C-130H of 164th Airlift Squadron of the US Air National Guard over its home town of Mansfield, Ohio

Lockheed C-130H of the Royal Danish Air Force equipped with AN/ALR-56 radar warning receiver system in wingtip pods *(Paul Jackson)*

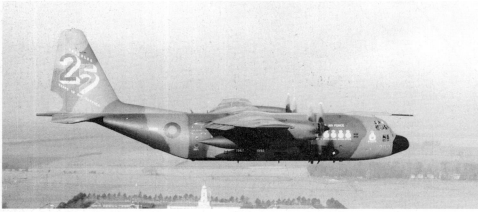

Hercules C. Mk 1 over Royal Air Force College, Cranwell, marking 25 years of service, December 1991
(Paul Jackson)

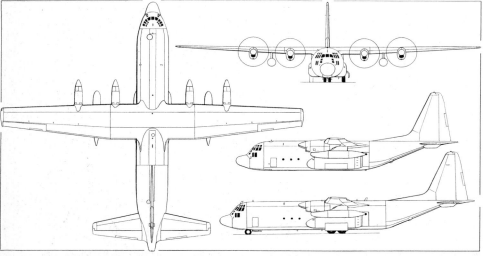

Lockheed C-130H-30 stretched Hercules (RAF C. Mk 3), with upper side view of standard C-130H
(Dennis Punnett)

rudder; elevator tabs have AC main supply and D standby; Lockheed-Fowler trailing-edge flaps; provisi for two removable afterbody ventral strakes.
STRUCTURE: All-metal two-spar wing with integrally stiffene taper-machined skin panels up to 14.63 m (48 ft 0 in) lon
LANDING GEAR: Hydraulically retractable tricycle type. Ea main unit has two wheels in tandem, retracting into fairin built on to the sides of the fuselage. Nose unit has tw wheels and is steerable ±60°. Oleo shock absorbers. Mai wheel tyres size 56 × 20-20, pressure 6.62 bars (96 lb/s in). Nosewheel tyres size 39 × 13-16, pressure 4.14 ba (60 lb/sq in). Goodyear air-cooled multiple disc hydraul brakes with anti-skid units. Retractable combinati wheel-skis available. Min ground turning radius: C-130H 11.28 m (37 ft) about nosewheel and 25.91 m (85 ft) abo wingtip; C-130H-30, 14.33 m (47 ft) about nosewheel and 27.43 m (90 ft) about wingtip.
POWER PLANT: Four 3,362 kW (4,508 shp) Allison T56-A-1 turboprops, each driving a Hamilton Standard type 54H6 four-blade constant-speed fully feathering reversible-pit propeller. Fuel in six internal (four integral and two bla der-type) tanks in wings, with total capacity of 25,81 litres (6,820 US gallons; 5,679 Imp gallons) and tw optional underwing pylon tanks, each with capacity 5,300 litres (1,400 US gallons; 1,166 Imp gallons). Tot fuel capacity 36,416 litres (9,620 US gallons; 8,010 In gallons). Single pressure refuelling point in starboa wheel well. Overwing gravity fuelling. Oil capacity 18 litres (48 US gallons; 40 Imp gallons).
ACCOMMODATION: Crew of four on flight deck, comprisin pilot, co-pilot, navigator and systems manager (fully pe formance qualified flight engineer on USAF aircraft). Pr vision for fifth man to supervise loading. Sleeping bun for relief crew, and galley. Flight deck and main cab pressurised and air-conditioned. Standard complemen for C-130H are as follows: 92 troops, 64 paratroopers, litter patients plus two attendants. Corresponding data f C-130H-30 are 128 troops, 92 paratroopers, and 97 litt patients plus four attendants. Air transport and airdro loads such as Sheridan light armoured vehicle, 19,051 (42,000 lb) when rigged for airdrop, are common to bo C-130H and the C-130H-30; light and medium towed art lery weapons, or variety of wheeled and tracked vehicl and multiple 463L supply pallets (five in C-130H a seven in C-130H-30, plus one on rear ramp for each mode are transportable; C-130H-30 is only airlifter which ca airdrop entire field artillery section (ammo platfor weapon, prime mover, and eight crew jumping over ram in one pass. Hydraulically operated main loading door a ramp at rear of cabin. Paratroop door on each side aft landing gear fairing. Two emergency exit doors standar two additional doors optional on C-130H-30.
SYSTEMS: Air-conditioning and pressurisation system ma pressure differential 0.52 bar (7.5 lb/sq in). Three indepe dent hydraulic systems, utility and booster systems ope ating at a pressure of 207 bars (3,000 lb/sq in), rated at 65 litres (17.2 US gallons; 14.3 Imp gallons)/min for utili and booster systems, 30.3 litres (8.0 US gallons; 6.7 In gallons)/min for auxiliary system. Reservoirs are unpres urised. Auxiliary system has handpump for emergencie Electrical system supplied by four 40kVA AC alternator plus one 40kVA auxiliary alternator driven by APU in po main landing gear fairing. Four transformer-rectifiers f DC power. Current production aircraft incorporate syste and component design changes for increased reliabilit There are differences between the installed components f US government and export versions. Babcock Power L High Volume Mine Layer (HVML) system available as a option, using modular roll-on pallets. Leading-edges wing, tailplane and fin anti-iced by engine bleed air.
AVIONICS: Standard fit specified by US government comprise dual AN/ARC-190 HF com, dual AN/ARC-186 VH com, dual AN/ARC-164 UHF com, AN/AIC-13 PA sy tem, AN/AIC-18 intercom, AN/APX-100 IFF/AIMS AT transponder, dual AN/ARN-118 UHF nav, du AN/ARN-147 VHF nav, self-contained navigation syste (SCNS), dual DF-206A ADF, DF-301E UHF directi finder, emergency locator transmitter (ELT AN/APN-218 Doppler nav, AN/APN-232 combined alt tude radar alt, dual C-12 compass system, dual FD-10 flight director system, either capable of coupling wit FD-109 autopilot, Sundstrand ground proximity warnin system, Kollsman altitude alerter, Westinghouse lo power colour radar (LPCR 130-1) replacing Sperry rad from March 1993, AN/APN-169C station keeping equi ment, A-100A cockpit voice recorder, flight data recorde AN/AAR-47 missile warning system, provisions f AN/ALE-47 flare and chaff dispensing system, provisio for AN/ALQ-157 infra-red countermeasures system, pr visions for KY-58 secure voice, provisions for microwav landing system (Canadian Marconi CMSLA syste ordered for C-130 fleet retrofit, 1991), provisions for UST Satcom system.
ARMAMENT: Fitment of Rockwell Hellfire ASM AC-130H/U studied 1991-92.
DIMENSIONS, EXTERNAL:
Wing span 40.41 m (132 ft 7 i
Wing chord: at root 4.88 m (16 ft 0 i
mean 4.16 m (13 ft 8½ i
Wing aspect ratio 10
Length overall:
all except HC-130H and C-130H-30

RL Mk 32B hose-drum unit fitted to Canadian CC-130T *(Paul Jackson)*

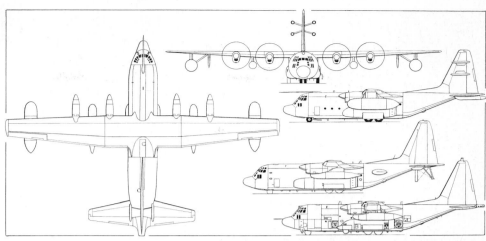

Lockheed EC-130E Rivet Rider/Volant Solo upgrade, with additional side view of (bottom) AC-130U Spectre gunship and (centre) EC-130H Compass Call *(Dennis Punnett)*

	29.79 m (97 ft 9 in)
C-130H-30	34.37 m (112 ft 9 in)
Height overall	11.66 m (38 ft 3 in)
Tailplane span	16.05 m (52 ft 8 in)
Wheel track	4.35 m (14 ft 3 in)
Wheelbase	9.77 m (32 ft 0¾ in)
Propeller diameter	4.11 m (13 ft 6 in)
Main cargo door (rear of cabin):	
Height	2.77 m (9 ft 1 in)
Width	3.12 m (10 ft 3 in)
Height to sill	1.03 m (3 ft 5 in)
Paratroop doors (each): Height	1.83 m (6 ft 0 in)
Width	0.91 m (3 ft 0 in)
Height to sill	1.03 m (3 ft 5 in)
Emergency exits (each): Height	1.22 m (4 ft 0 in)
Width	0.71 m (2 ft 4 in)

DIMENSIONS, INTERNAL:
Cabin, excl flight deck:

Length excl ramp: C-130H	12.50 m (41 ft 0 in)
C-130H-30	17.07 m (56 ft 0 in)
Length incl ramp: C-130H	15.73 m (51 ft 8½ in)
C-130H-30	20.33 m (66 ft 8½ in)
Max width	3.12 m (10 ft 3 in)
Max height	2.81 m (9 ft 2¾ in)
Floor area, excl ramp: C-130H	39.5 m² (425.0 sq ft)
Volume, incl ramp: C-130H	123.2 m³ (4,351.0 cu ft)
C-130H-30	161.3 m³ (5,696.0 cu ft)

AREAS:

Wings, gross	162.12 m² (1,745.0 sq ft)
Ailerons (total)	10.22 m² (110.0 sq ft)
Trailing-edge flaps (total)	31.77 m² (342.0 sq ft)
Fin	20.90 m² (225.0 sq ft)
Rudder, incl tab	6.97 m² (75.0 sq ft)
Tailplane	35.40 m² (381.0 sq ft)
Elevators, incl tabs	14.40 m² (155.0 sq ft)

WEIGHTS AND LOADINGS:
Operating weight empty:

C-130H	34,686 kg (76,469 lb)
C-130H-30	36,397 kg (80,242 lb)
Max fuel weight: internal	20,108 kg (44,330 lb)
external	8,255 kg (18,200 lb)
Max payload: C-130H	19,356 kg (42,673 lb)
C-130H-30	17,645 kg (38,900 lb)
Max normal T-O weight	70,310 kg (155,000 lb)
Max overload T-O weight	79,380 kg (175,000 lb)
Max normal landing weight	70,310 kg (155,000 lb)
Max overload landing weight	79,380 kg (175,000 lb)
Max zero-fuel weight, 2.5g	54,040 kg (119,142 lb)
Wing loading at max normal T-O weight	434.5 kg/m² (89 lb/sq ft)
Power loading at max normal T-O weight	5.23 kg/kW (8.6 lb/shp)

PERFORMANCE (C-130H at max normal T-O weight, unless indicated otherwise):

Max cruising speed	325 knots (602 km/h; 374 mph)
Econ cruising speed	300 knots (556 km/h; 345 mph)
Stalling speed	100 knots (185 km/h; 115 mph)
Max rate of climb at S/L	579 m (1,900 ft)/min
Service ceiling at 58,970 kg (130,000 lb) AUW	10,060 m (33,000 ft)
Service ceiling, OEI, at 58,970 kg (130,000 lb) AUW	8,075 m (26,500 ft)
Runway LCN: asphalt	37
concrete	42
T-O run	1,091 m (3,580 ft)
T-O to 15 m (50 ft)	1,573 m (5,160 ft)
Landing from 15 m (50 ft):	
at 45,360 kg (100,000 lb) AUW	731 m (2,400 ft)
at 58,967 kg (130,000 lb) AUW	838 m (2,750 ft)
Landing run at 58,970 kg (130,000 lb) AUW	518 m (1,700 ft)

Range with max payload, with 5% reserves and allowance for 30 min at S/L
  2,046 nm (3,791 km; 2,356 miles)
Range with max fuel, incl external tanks, 7,081 kg (15,611 lb) payload, reserves of 5% initial fuel plus 30 min at S/L 4,250 nm (7,876 km; 4,894 miles)

New production C-130H in 'proud gray' colour scheme serving Kentucky ANG *(Paul Jackson)*

LOCKHEED L-100 SERIES COMMERCIAL HERCULES

TYPE: Civilian version of C-130.
PROGRAMME: Initial variants described in earlier *Jane's*; current models below.
CURRENT VERSIONS: **L-100-20 (Model 382E):** Fuselage

Proposed RAMTIP (Reliability And Maintainability Technology Insertion Program) two-pilot cockpit for the C-130 includes five Litton Canada colour liquid crystal displays for flight, engine and systems information and four LCD tabular displays

stretched by 2.54 m (8 ft 4 in); certificated 4 October 1968; Allison 501-D22A engines; one L-100-20 was Lockheed HTTB testbed (see 1991-92 *Jane's*); military/government operators listed under C-130 entry.

L-100-30 (Model 382G): Fuselage stretched 2.03 m (6 ft 8 in); military operators listed under C-130 entry; first operator Saturn Airways in December 1970.

L-100-30HS: Hospital version fitted by Lockheed Aircraft Service (which see) with operating theatre, intensive care, advanced anaesthesia and X-ray facilities; five delivered to Saudi Arabia; electrical generators and air-conditioners in underwing pods allow hospital to operate independently for 72 hours.

Advanced L-100 Freighter: Civil equivalent of military Hercules II, with GMA 2100 turboprops; Pemco cargo loading system; able to carry standard M-1/M-2 containers, loaded via standard rear ramp/door or optional 2.74 × 3.51 m (9 ft 0 in × 11 ft 6 in) side cargo door; max cargo payload 20,553 kg (45,312 lb) palletised, 21,625 kg (47,675 lb) bulk; max zero-fuel weight 56,245 kg (124,000 lb); range at MZFW 1,997 nm (3,701 km; 2,300 miles).

CUSTOMERS: See table. Total 112 L-100s delivered by January 1993: 21 standard length, 24 L-100-20s and 67 L-100-30s. Deliveries in 1992 to Frame Air and Ethiopian Airlines (one each).

DESIGN FEATURES: Details of C-130H apply to L-100, except as detailed below. All C-130s and L-100s delivered since April 1984 have two 0.61 × 1.22 m (24 × 48 in) emergency exits which, together with rear personnel doors, allow carriage of 79 passengers; optional additional exit each side allows for 100 passengers; supplemental oxygen provided for passengers; various galley and toilet layouts available.

FLYING CONTROLS: As C-130H.
STRUCTURE: As C-130H.
LANDING GEAR: As for C-130H, except mainwheel tyre pressure 3.24-7.38 bars (47-107 lb/sq in) and nosewheel tyre pressure 4.14 bars (60 lb/sq in). Min ground turning radius: L-100-20, 26.8 m (88 ft); L-100-30, 27.5 m (90 ft).
POWER PLANT: Four 3,362 kW (4,508 shp) Allison 501-D22A turboprops.
DIMENSIONS, EXTERNAL: As for C-130H except:

Length overall: L-100-20	32.33 m (106 ft 1 in)
L-100-30	34.37 m (112 ft 9 in)
Wheelbase: L-100-20	11.30 m (37 ft 1 in)
L-100-30	12.32 m (40 ft 5 in)
Crew door (integral steps): Height	1.14 m (3 ft 9 in)
Width	0.76 m (2 ft 6 in)
Height to sill	1.04 m (3 ft 5 in)

DIMENSIONS, INTERNAL:

Cabin, excl flight deck:	
Length: L-100-20	15.04 m (49 ft 4 in)
L-100-30, excl ramp	17.07 m (56 ft 0 in)
L-100-30, incl ramp	19.93 m (65 ft 4¾ in)
Max height	2.74 m (9 ft 0 in)
Floor area, excl ramp: L-100-20	46.36 m² (499.0 sq ft)
L-100-30	52.30 m² (563.0 sq ft)
Floor area, ramp	9.57 m² (103.0 sq ft)
Volume, incl ramp: L-100-20	150.28 m³ (5,307 cu ft)
L-100-30	171.1 m³ (6,042 cu ft)

WEIGHTS AND LOADINGS:

Operating weight empty:	
L-100-20	34,781 kg (76,680 lb)
L-100-30	35,260 kg (77,736 lb)
Max payload: L-100-20	23,637 kg (52,110 lb)
L-100-30	23,158 kg (51,054 lb)
Max ramp weight	70,670 kg (155,800 lb)
Max T-O weight	70,308 kg (155,000 lb)
Max landing weight	61,235 kg (135,000 lb)
Max zero-fuel weight	58,420 kg (128,790 lb)
Max fuel weight	29,380 kg (64,772 lb)
Max wing loading	433.5 kg/m² (88.8 lb/sq ft)
Max power loading	5.23 kg/kW (8.6 lb/shp)

PERFORMANCE (at max T-O weight, except where indicated):

Max cruising speed at 6,100 m (20,000 ft) at 54,430 kg (120,000 lb) AUW	308 knots (571 km/h; 355 mph)
Landing speed	124 knots (230 km/h; 143 mph)
Max rate of climb at S/L	518 m (1,700 ft)/min
Runway LCN: asphalt	37
concrete	42
FAR T-O field length	1,905 m (6,250 ft)
FAR landing field length at max landing weight	1,478 m (4,850 ft)
Range:	
with max payload, 45 min reserves	1,334 nm (2,472 km; 1,536 miles)
with zero payload	4,830 nm (8,951 km; 5,562 miles)

LOCKHEED L-100 OPERATORS

Operator (in January 1993)	L-100	L-100-20	L-100-30
Advanced Leasing Corp (USA)			1
Air Algérie			3
Angola Air Charter			2
Argentine Air Force			1
China General Aviation Corporation			2
Dubai (UAE) Air Force			1
Ecuador Air Force			1
Endiama (Angola)			1
Ethiopian Airlines			2
Frame Air (Indonesia)			2
Gabon Air Force		1	1
Indonesian Air Force			1
Jamahiriya Air Transport (Libya)			2
Jet Fret (France)			1
Kuwait Air Force			4
Merpati (Indonesia)			2
North West Territories (Canada)			1
Pakistan Air Force	1		
Pelita Air Service (Indonesia)			4
Peru Air Force		5	
Petroléos Mexicanos			1
Philippine Air Force		2	
Safair Freighters			4
Saudi Government			6
Schreiner Airways (Netherlands)			1
Southern Air Transport	4	5	9
Tepper Aviation			2
Transafrik (Sao Tomé)		1	3
Uganda Air Cargo			1
United Arab Airlines (Libya)		2	1
Worldwide Trading (USA)			1
Written off	16	8	7
	21	24	67

Lockheed L-100-30 of Southern Air Transport flying relief missions for the World Food Programme in Somalia *(Paul Jackson)*

OPERATIONAL NOISE LEVELS (FAR Pt 36, Stage 2):

T-O sideline	96.7 EPNdB
T-O flyover: at T-O power	97.8 EPNdB
at cutback power	94.8 EPNdB
Approach flyover	98.1 EPNdB

LOCKHEED HTTB

High technology testbed (HTTB) conversion of L-100-20; data in 1991-92 and previous editions of *Jane's*. Lost in test accident, 3 February 1993; replacement under consideration.

LOCKHEED ADVANCED DEVELOPMENT COMPANY (LADC) (Division of Lockheed Aeronautical Systems Group)

1011 Lockheed Way, Palmdale, California 93599-3740
Telephone: 1 (805) 572 2974
Fax: 1 (805) 572 2209
PRESIDENT: Sherman Mullin
EXECUTIVE VICE-PRESIDENT: Jack S. Gordon
DIRECTOR OF COMMUNICATIONS: James W. Ragsdale

LADC is component of Lockheed Aeronautical Systems Group, functioning autonomously at Palmdale; has specialised in 'black' or covert development programmes, including U-2/TR-1, SR-71 and F-117A; also developing supersonic STOVL fighter. Unconfirmed reports of involvement in Aurora strategic reconnaissance programme (see USAF entry); company semi-officially known as Skunk Works.

ADVANCED STOVL STRIKE FIGHTER (ASTOVL)

PROGRAMME: Competing Lockheed and McDonnell Douglas/BAe (see International section) teams chosen by ARPA and US Navy in March 1993 to receive 30-month concept validation contracts for potential ASTOVL fighter for the US Marine Corps. Lockheed proposal has canard foreplanes and trapezoidal wing with twin vertical fins. Powered by Pratt & Whitney F119, which drives vertically mounted fan for lift. Propulsion and lift system to be tested during 1990s. A prototype might fly in about 2000. See also USAF section and Addenda.

LOCKHEED F-117A

TYPE: Precision attack aircraft with stealth elements, optimised for radar energy dispersion and low IR emission.
PROGRAMME: DARPA requested stealthy fighter proposals from US aerospace industry, 1974; unsolicited Lockheed bid submitted 1975 and selected against Northrop competition, 1976; Northrop design similarly faceted, but with delta wing and single air intake above cockpit; development began with USAF Flight Dynamics Laboratory contract to Lockheed Advanced Development Projects, funded by DARPA under Have Blue programme; two XST (Experimental Stealth Technology) prototypes produced (build Nos. 1001 and 1002), each powered by two 11.12-12.46 kN (2,500-2,800 lb st) GE CJ610s; first flight at Groom Lake, Nevada, December 1977 by William C Park; first prototype crashed 4 May 1978; second XST crashed at Tonopah Test Range 1980; similar to F-117 apart from inward-canted ruddervators. Span 6.86 m (22 ft 6 in); length, excluding nose probe, 11.58 m (38 ft 0 in); height 2.29 m (7 ft 6 in); max T-O weight 5,443 kg (12,000 lb); leading-edge sweep 72° 30′; four-transparency canopy; max speed Mach 0.8; endurance 1 h.

Development and manufacture of operational F-117 under Senior Trend programme began with contract award on 16 November 1978; fabrication started July 1979 of five pre-series aircraft (79-10780 to 784); first delivered to Tonopah 16 January 1981; initial flight 18 June 1981; 10782 crashed 20 April 1982. Planned production of 100 reduced to 59 (Article numbers 785-843) of which 785 (80-0785) crashed on first flight 21 June 1982 and three more lost by end 1992; first aerial refuelling (10780) 18 November 1981; first night flight (10782) 22 March 1982; first weapon drop of BDU-33 (10781) 7 July 1982;

first handover to USAF (00787) 23 August 1982; IOC achieved by 4450th Test Squadron 28 October 1983; 10,000 flying hours achieved September 1986; first picture and designation released 10 November 1988; first operational deployment in Operation Just Cause over Panama, 21 December 1989, when two F-117As each dropped a 2,000 lb laser guided bomb on barracks area at Rio Hato; over 40 F-117As participated in 1991 Gulf War against Iraq, flying some 1,270 missions. Weapon system improvement began with first flight of aircraft modified under Offensive Capability Improvement Program on 1 December 1988; first 'production' re-delivery (20805) 27 November 1990; programme to continue to 2005; improvements installed by 1991 include 'four-dimensional' flight management system (time on waypoint ±1 second) and new cockpit instrumentation with full colour MFDs and digital moving map; Lockheed redelivered 73 F-117 upgrades, including 15 second phase, from Palmdale by early 1992; work continues at one aircraft per month. First phase, beginning 1984, involved replacement of Delco M362F computers by IBM AP-102. Second phase known as Offensive Combat Improvement Program; includes entirely revised cockpit with Honeywell MFDs, three-dimensional flight management system, large moving map or digital situation display, autothrottle and pilot-activated recovery system. Third phase, flight tested 1992, involves new, turret-mounted infra-red acquisition and designation sensor (IRADS) by TI; Honeywell ring laser gyro INS, Collins GPS. Achieved 50,000 flying hours in February 1991.

In 1993, Lockheed made unsolicited offer to US Navy of folding-wing F-117 for carrier-borne operations. Leading-edge sweep would be about 48°.

CURRENT VERSIONS: **F-117A:** *As described.*

CUSTOMERS: USAF; five pre-series plus 59 production; deliveries seven, eight, eight, eight, eight, seven, five, four and three in calendar years 1982-90 (final delivery, of 88-0843, on 12 July 1990).

First F-117A unit, 4450th Tactical Group at Tonopah (122 nm; 225 km; 140 miles north-west of Las Vegas), Nevada, formed 15 October 1979 and equipped with 18 A-7D Corsair IIs until first F-117A arrived; P-Unit (later 4451st Test Squadron) formed June 1981; Q-Unit (later 4452nd TS) began operations 15 October 1982, flying first 4450th TG training operation (00786); Group transferred from direct control of Tactical Air Command to Tactical Fighter Weapons Center at Nellis AFB 1985; Z-Unit (later 4453rd Test and Evaluation Squadron) formed 1 October 1985; first of several public air display appearances made April 1990; unit renamed 37th Tactical Fighter Wing of 12th Air Force 5 October 1989; strength 40 F-117As divided between 415th and 416th TFS; A-7Ds replaced in training and chase duty by eight T/AT-38 Talons of 417th TFTS, also with balance of 16 remaining F-117As; first assignment away from Tonopah was covert deployment to England AFB on 18 June 1990; wing and squadrons deleted 'Tactical' prefix 1 November 1991; transfer to Holloman AFB, New Mexico, began with delivery of 00791 for maintenance familiarisation, 7 January 1992; movement of remaining aircraft began 9 May 1992, when unit became 49th Fighter Wing (7th, 8th and 9th Fighter Squadrons).

COSTS: $6,560 million programme (1990), including $2,000 million R&D, $4,270 million for procurement and $295.4 million for infrastructure. Average unit cost $42.6 million (then-year dollars). *See also Addenda.*

DESIGN FEATURES: Multi-faceted airframe designed to reflect radar energy away from originating transmitter, particularly downward-looking AEW aircraft; vortexes from many sharp edges, including leading-edge of wing, designed to form co-ordinated lifting airflow pattern; wings have 67° 30′ sweepback, much greater than needed for subsonic performance, with aerofoil formed by two flat planes underneath and three on upper surface; forward underwing surface blends with forward fuselage; all doors and access panels have serrated edges to suppress radar reflection; internal weapons bay 4.7 m (15 ft 5 in) long and 1.75 m (5 ft 9 in) wide divided longitudinally by two lengthwise doors hinged on centreline; boom refuelling receptacle on port side of top plate, aft of cockpit. Frontal radar cross-section estimated as 0.01 m² (0.1 sq ft).

FLYING CONTROLS: Four omnidirectional air probes at nose indicate GEC Astronics quadruplex fly-by-wire control system, similar to that of F-16, using two-section elevons and all-moving ruddervators together for control and stability; ruddervators swept about 65° and set at 85° to each other.

STRUCTURE: Material principally aluminium; two-spar wings; fuselage has flat facets mounted on skeletal subframe, jointed without contour blending; surfaces coated with various radar-absorbent materials. Weapons bay doors and landing gear leg doors of composites; nickel alloy honeycomb jetpipes. Ruddervators replaced by new units of thermoplastic graphite composites construction, removing previous speed restriction due to flutter; new ruddervator first flown on 10784, 18 July 1989; last ruddervator fitted 1992. For cost reduction, many components adapted from other aircraft, including Lockheed SR-71, P-3, C-130, L-1011 and F-104.

LANDING GEAR: Tricycle type by Menasco, with single wheels all retracting forward. Loral brakes (steel originally, being

The Lockheed Have Blue stealth prototype, which first flew in December 1977, proved the feasibility of the faceted design

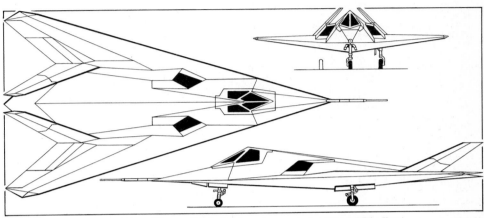

Lockheed Have Blue stealth technology demonstrator *(Jane's/Mike Keep)*

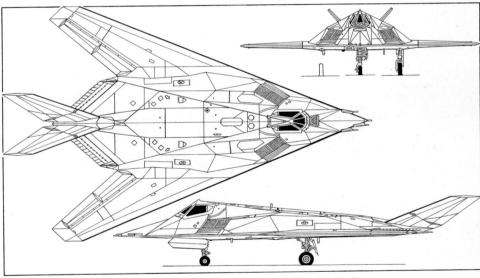

Lockheed F-117A stealth attack aircraft *(Jane's/Mike Keep)*

F-117A engine exhaust, showing vertical guide vanes *(Paul Jackson)*

508 USA: AIRCRAFT—**LOCKHEED**

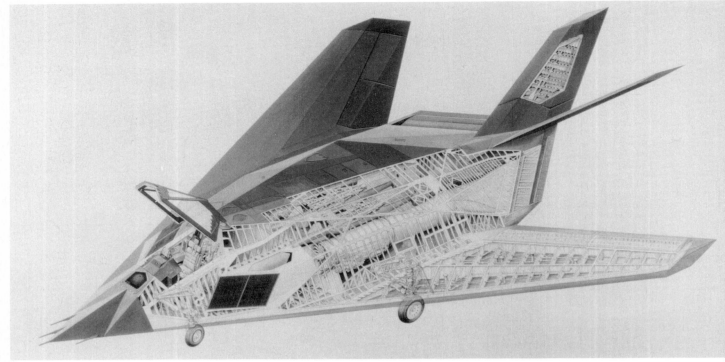

Cutaway drawing of the Lockheed F-117A

replaced by carbon/carbon), wheels (F-15E size) and antiskid system. Goodyear tyres. All doors have serrated edges to suppress radar reflections. Emergency arrester hook with explosively jettisoned cover; Pioneer Aerospace braking parachute (black).

POWER PLANT: Two 48.0 kN (10,800 lb st) class General Electric F404-GE-F1D2 non-augmented turbofans (replacement by F412 under examination, but unlikely). Rectangular overwing air intakes with 2.5 × 1.5 cm (1 × ⅝ in) heated grid for anti-icing and low observability. Auxiliary air intake doors in horizontal surface immediately to the rear. Part of cold air ingested bypasses engine and is mixed with exhaust gases for cooling. Narrow-slot 'platypus' exhausts, designed by Astech/MCI, in rear fuselage, 1.65 m (5 ft 5 in) long and 0.10 m (4 in) high, wi extended lower lip, surrounded by heat tiles of type used Space Shuttle and with 11 vertical, internal guide vane Sundstrand air turbine starter. In-flight refuelling rece tacle in decking aft of cockpit, illuminated for nig refuelling by lamp at apex of cockpit. Optional drop ta on internal weapon pylon.

ACCOMMODATION: Pilot only; McDonnell Douglas ACES zero/zero ejection seat. Five Sierracin/Sylmar Corporati individually framed flat-plate windows, including singl piece windscreen. Transparencies gold-coated for rad dissipation. Canopy hinged to open upward and backwar

SYSTEMS: AlliedSignal environmental control, auxilia power and emergency power systems.

AVIONICS: Forward-looking infra-red (FLIR) sensor, wi dual fields of view, in recessed emplacement, covered fine mesh screen, below windscreen. Retractable dow ward-looking DLIR and laser designator beneath forwa fuselage to starboard of nosewheel bay; FLIR and DLIR Texas Instruments (to be replaced by improved equipme during third-phase retrofit in 1994). HUD based on Kais AN/AVQ-28; large head-down display for FLIR image flanked by two multi-function CRTs. Retractable rad antennae beneath fuselage, ahead of port main landi gear, and on spine. Honeywell radar altimeter, Honeywe SPN-GEANS INS (replaced by Honeywell H-423/E ri laser gyro from August 1991; Rockwell Collins GPS to added); IBM AP-102 mission computer (replacing origin three Delco M362F computers); GEC-Marconi flight co trol computer/navigation interface and autopilot comput (NIAC) system; SLI Avionic Systems Corporati expanded data transfer system and AHRS. Harris Corpo ation digital moving map added as retrofit with full-colo MFDS. Proposed adaptation to carry TARPS reconna sance pod.

ARMAMENT: "Full range of USAF tactical fighter ordnance principally two 2,000 lb bombs: BLU-109B low-lev laser guided or GBU-10/GBU-27 laser guided glide wea ons; alternatively, AGM-65 Maverick or AGM-88 HAR ASMs. Provision for AIM-9 Sidewinder (against AWAC aircraft). Internal carriage on two extensible beams weapon bay. (Only missiles with seeker heads extende below aircraft prior to launch; bombs released from with weapons bay.)

Top view of F-117A showing facets which ensure stealth and vortex lift *(Eric Schulzinger and Denny Lombard)*

Lockheed F-117A of the 49th Fighter Wing *(Eric Schulzinger and Denny Lombard)*

DIMENSIONS, EXTERNAL:
Wing span	13.20 m (43 ft 4 i
Length overall	20.08 m (65 ft 11 i
Height overall	3.78 m (12 ft 5 i

AREAS (estimated):
Wings, gross	105.9 m² (1,140 sq

WEIGHTS AND LOADINGS:
Weight empty (estimated)	13,608 kg (30,000 l
Internal weapons load	2,268 kg (5,000 l
Max T-O weight	23,814 kg (52,500 l

PERFORMANCE (* not confirmed by USAF):
Max level speed	561 knots (1,040 km/h; 646 mp
Normal max operating speed	Mach 0
* T-O speed at normal combat weight	165 knots (306 km/h; 190 mp
* Landing speed	150 knots (227 km/h; 172 mp
Mission radius, unrefuelled, 2,268 kg (5,000 lb) weape load	600 nm (1,112 km; 691 mile
g limit	

OCKHEED FORT WORTH COMPANY (Division of Lockheed Aeronautical Systems Group)

O Box 748, Fort Worth, Texas 76101
elephone: 1 (817) 777 2000
ax: 1 (817) 763 4797
RESIDENT: Gordon R. England
ICE-PRESIDENT, F-16 PROGRAMS: Dain M. Hancock
ICE-PRESIDENT, PROGRAM DEVELOPMENT: A. Dwain Mayfield
IRECTOR OF COMMUNICATIONS: Mike Hatfield
ANAGER, PUBLIC AFFAIRS: Joe W. Stout

General Dynamics' Tactical Military Aircraft Division at ort Worth sold to Lockheed in December 1992; became ockheed Fort Worth Company on 1 March 1993.

Fort Worth activities include production of F-16 Fighting alcon, previously shared development of F-22A ATF (with ockheed and Boeing) and provision of spares, support and odification/update for F-111. Possible re-allocation of F-22 ork-shares between Fort Worth and original partner at ockheed Marietta under consideration in 1993. Fort Worth orkforce being reduced to 5,800 by 1994.

Fokker-built F-16B of the Royal Netherlands Air Force *(Paul Jackson)*

OCKHEED (GENERAL DYNAMICS) F-16 FIGHTING FALCON

YPE: Single-seat and two-seat multi-role fighter.

ROGRAMME: Emerged from YF-16 of US Air Force Lightweight Fighter prototype programme 1972 (details under General Dynamics in 1977-78 and 1978-79 *Jane's*); first flight of prototype YF-16 (72-01567) 2 February 1974; first flight of second prototype (72-01568) 9 May 1974; selected for full-scale development 13 January 1975; day fighter requirement extended to add air-to-ground capability with radar and all-weather navigation; production of six single-seat F-16As and two two-seat F-16Bs began July 1975; first flight of full-scale development aircraft 8 December 1976; first flight of F-16B 8 August 1977.

URRENT VERSIONS: **F/A-16**: Proposed modification of 300 Block 30/32 aircraft for close air support (CAS)/battlefield air interdiction (BAI) in late 1990s; head-steered FLIR, Pave Penny laser ranger and 30 mm cannon pod. 200 F-16Cs to have received CAS/BAI modifications from 1995, including DTS, Navstar GPS and improved data modem. Block 30/32 upgrade abandoned January 1992 in favour of CAS/BAI assignment of Block 40/42 aircraft, having LANTIRN capability; these require more simple modification with ground data link, laser spot-tracker, anti-jam radio, missile approach warner, provision for pilot's night vision goggles and upgrades to LANTIRN pods. Deployment plan envisages 4½ wings of Block 40/42s for night CAS and two wings of Block 30/32s for day CAS. Meanwhile, first dedicated CAS/BAI aircraft are F-16As of ANG's 138th FS at Syracuse, New York; operational 1989, equipment including fixed GPU-5/A 30 mm centreline cannon; first improved data modem installed November 1990; squadron received F-16Cs in 1991.

F-16A: First production version for air-to-air and air-to-ground missions; production for USAF completed March 1985, but still available for other customers; international sales continue; powered since late 1988 (Block 15OCU) by P&W F100-PW-220 turbofan; Westinghouse AN/APG-66 range and angle track radar; first flight of first aircraft (78-0001) 7 August 1978; entered service with 388th TFW at Hill AFB, Utah, 6 January 1979; combat ready October 1980, when named Fighting Falcon; most now serving ANG and AFRES; power plants being upgraded to F100-PW-220E, 1991-1996. Also produced in Europe. Built in Blocks 01, 05, 10 and 15, of which Blocks 01 and 05 retrofitted to Block 10 standard 1982-84; Block 15 retrofitted to OCU standard from late 1987.

Operational Capabilities Upgrade (OCU): USAF/NATO co-operative programme to equip F-16A/B for next-generation BVR air-to-air and air-to-surface weapons; radar and software updated, fire control and stores management computers improved, data transfer unit fitted, combined radar-barometric altimeter fitted, and provision for AN/ALQ-131 jamming pods. Ring laser INS and upgrade from P&W F100-PW-200 to F100-PW-220E planned for 1990s. FMS exports since 1988 to Block 15OCU standard with F-16C features including ring laser INS, AN/ALR-69 RWR, F100-PW-220 power plant and AIM-9P-4 Sidewinder AAM capability.

Mid-Life Update (MLU): Development authorised 3 May 1991 (signature of final partner); US government contract to GD 15 June 1991; originally planned to be applied to 533 aircraft of USAF (130), Belgium (110), Denmark (63), Netherlands (172) and Norway (58) from 1996 in co-development/co-production programme. USAF withdrew 1992, but ordered 223 modular computer retrofit kits from MLU to equip Block 50/52 aircraft. European share re-negotiated on 28 January 1993 to new totals of Belgium 48 plus 24 options; Denmark 61; Netherlands 136; and Norway 56; Letters of Offer and Acceptance finalised mid-1993. Cockpit similar to F-16C/D Block 50 with wide-angle HUD, night vision goggle compatibility, modular mission computer replacing existing three, digital terrain system, AN/APG-66(V2A) fire control radar, Navstar GPS, improved data modem and provision for microwave landing system (MLS). Inlet hardpoints and wiring for FLIR pods will be added to Block 10 aircraft. Options include helmet-mounted display and Hazeltine AN/APX-111 IFF interrogator/transponder; both taken up by Netherlands and Norway. In addition, 150 F-16A/Bs offered to Taiwan in 1992 to be built to Block 15MLU standard. Aircraft for prototype conversion delivered to GD in September 1992, including Danish F-16B ET-204, Netherlands F-16B J-650, Norwegian F-16A 299 and one USAF.

F-16(ADF): Modification of 279 (actually 272 because of pre-conversion attrition) Block 15 F-16A/Bs as USAF air defence fighters to replace F-4s and F-106s with 11 Air National Guard squadrons; ordered October 1986. Modifications include upgrade of AN/APG-66 radar to improve small target detection and provide continuous-wave illumination, provision of AMRAAM data link, improved ECCM, Bendix/King AN/ARC-200 HF/SSB radio (F-16A only), Teledyne/E-Systems Mk XII advanced IFF, provision for Navstar GPS Group A, low altitude warning, voice message unit, night identification light (port forward fuselage of F-16A only), and ability to carry and guide two AIM-7 Sparrow missiles. First successful guided launch of AIM-7 over Point Mugu range, California, February 1989; F-16(ADF) can carry up to six AIM-120 AMRAAM or AIM-9 Sidewinder or combinations of all three missiles; retains internal M61 20 mm gun. GD converted one prototype, then produced modification kits for installation by USAF Ogden Air Logistics Center, Utah, in conjunction with upgrade to OCU avionics standard; first Ogden aircraft, F-16B 81-0817, completed October 1988. Development completed at Edwards AFB during 1990; operational test and evaluation with 57th Fighter Weapons Wing at Nellis AFB, Nevada; first F-16(ADF), 81-0801, delivered to 114th Fighter Training Squadron at Kingsley Field, Oregon, 1 March 1989; 194th Fighter Interceptor Wing, California ANG, Fresno, achieved IOC in 1989, following receipt of first aircraft (F-16B 82-1048) on 13 April 1989; first AIM-7 launch by ANG (159th FS) June 1991. Programme completed early 1992; includes approximately 30 F-16Bs.

F-16B: Standard tandem two-seat version of F-16A; fully operational both cockpits; fuselage length unaltered; reduced fuel.

F-16C/D: Single-seat and two-seat USAF Multinational Staged Improvement Program (**MSIP**) aircraft respectively, implemented February 1980. MSIP expands growth capability to allow for ground attack and beyond-visual-range missiles, and all-weather, night and day missions; **Stage I** applied to Block 15 F-16A/Bs delivered from November 1981 included wiring and structural changes to accommodate new systems; **Stage II** applied to Block 25 F-16C/Ds from July 1984 includes core avionics, cockpit and airframe changes. **Stage III** includes installation of systems as they become available, beginning 1987 and extending up to Block 50/52, including selected retrofits back to Block 25. Changes include Westinghouse AN/APG-68 multi-mode radar with better range, resolution, more operating modes and better ECCM than AN/APG-66; advanced cockpit with better interfaces and up-front controls, GEC-Marconi wide-angle HUD, two multi-function displays, Fairchild mission data transfer equipment and radar altimeter; expanded base of fin giving space for proposed later fitment of AN/ALQ-165 Airborne Self Protection Jamming system (since cancelled); increased electrical power and cooling capacity; structural provision for increased take-off weight and manoeuvring limits; and MIL-STD-1760 weapons interface for use of smart weapons such as AIM-120A AMRAAM and AGM-65D IR Maverick. First AIM-120 operational launch (by any aircraft), 27 December 1992; F-16D (90-0778) of 33rd FS/363rd FW destroyed Iraqi MiG-25.

Common engine bay introduced at **Block 30/32** (deliveries from July 1986) to allow fitting of either P&W F100-PW-220 (Block 32) or GE F110-GE-100 (Block 30) Alternate Fighter Engine. Other changes include computer memory expansion and seal-bonded fuselage fuel tanks. First USAF wing to use F-16C/Ds with F110 engines was 86th TFW at Ramstein AB, Germany, from October 1986. Additions in 1987 included full Level IV multi-target compatibility with AMRAAM (as Block 30B), voice message unit, Shrike anti-radiation missiles (from August), crash survivable flight data recorder and modular common inlet duct allowing full thrust from F110 at low airspeeds.

Software upgraded for full Level IV multi-target compatibility with AMRAAM early 1988. Industry-sponsored development of radar missile capability for several European air forces resulted in firing of AIM-7F and AIM-7M missiles from F-16C in May 1988; capability introduced mid-1991; missiles guided using pulse Doppler illumination while tracking targets in a high PRF mode of the AN/APG-68 radar.

US Air Force F-16C/Ds of 52nd FW at Spangdahlem AB, Germany, have HARM/Shrike capability and operate alongside A-10A Thunderbolt IIs in Wild Weasel defence suppression role.

Block 40/42 Night Falcon (deliveries from December 1988) upgrades include AN/APG-68(V) radar allowing 100 h operation before maintenance, full compatibility with Martin Marietta low altitude navigation and targeting infra-red for night (LANTIRN) pods, four-channel digital flight control system, expanded capacity core computers, diffractive optics HUD, enhanced envelope gunsight, Navstar GPS, improved leading-edge flap drive system, improved cockpit ergonomics, high gross weight landing gear, structural strengthening, and provision for improved EW equipment. LANTIRN gives day/night standoff target identification, automatic target handoff for multiple launch of Mavericks, autonomous laser guided bomb delivery and precision air-to-ground laser ranging. Combat Edge pressure breathing system installed 1990 for higher pilot *g* tolerance.

First Block 40 F-16C/Ds issued in late 1990 to 363rd FW (Shaw AFB, South Carolina); first LANTIRN pods to 36th FS/51st FW at Osan, South Korea, in 1992.

Block 50/52 (deliveries began with F-16C 90-0801 in October 1991 for operational testing) upgrades include F110-GE-129 and F100-PW-229 increased performance engines (IPE), AN/APG-68(V5) radar with advanced programmable signal processor employing VHSIC technology, Have Quick IIA UHF radio, Have Sync VHF anti-jam radio and AN/ALR-56M advanced RWR. Changes planned for 1993 include full integration of HARM/Shrike anti-radiation missiles via Texas Instruments interface, upgraded programmable display generator with digital terrain system (DTS) provisions and scope for digital map capability, ring laser INS (Honeywell H-423 selected 1990) and AN/ALE-47 advanced chaff/flare dispenser. Upgrades being considered for 1994-95 include a digital terrain system with colour map, colour multi-function displays, head-steered FLIR with helmet-mounted display, self-protection jammer and onboard oxygen generating system (OBOGS).

First operational unit is 4th FS of 388th FW at Hill AFB, Utah, from October 1992; others to 480th FS of 52nd FW at Spangdahlem, Germany, replacing Block 30 aircraft from (first delivery) 20 February 1993.

Block 60/62: Projected development to meet possible USAF multi-role fighter requirement, employing technology from F-22A programme.

NF-16D: Variable stability in-flight simulator test aircraft (VISTA) modified from Block 30 F-16D (86-0048) ordered December 1988 to replace NT-33A testbed. Features include vertical surface direct force generators above and below wings, Calspan variable stability flight control system, fully programmable cockpit controls and displays, additional computer suite, permanent flight test data recording system, variable feel centrestick and computer, and safety pilot in rear cockpit. Internal gun, RWR and chaff/flare equipment removed, providing space for Phase II and III growth including additional computer, reprogrammable display generator and customer hardware allowance. 'Israeli-type' bulged spine. First flight 9 April 1992; delivery due July 1992 but aircraft stored after five flights because of funding shortage; proposed to fit axisymmetric thrust vectoring engine nozzle (AVEN) for 60-flight trials programme beginning May 1993.

510 USA: AIRCRAFT—**LOCKHEED**

F-16C of 52nd FW, USAF Europe, armed with AGM-88 HARM anti-radiation missiles *(Paul Jackson)*

F-16N: US Navy supersonic adversary aircraft (SAA) modified from F-16C/D Block 30; selected January 1985; deliveries of 26 aircraft started 1987 and completed 1988; features include AN/APG-66 instead of AN/APG-68 radar, F110-GE-100 engine, deletion of M61 gun, AN/ALR-69 RWR, titanium in lower wing fittings instead of aluminium and cold working of lower wing skin holes to resist greater frequency of high *g*; wingtips fitted only for AIM-9 practice missiles and ACMI AIS pods, but normal tanks and stores on other stations. Four of 26 are two-seat **TF-16N**. F/TF-16Ns serve with 'Top Gun' Fighter Weapons School (eight) and with VF-126 (six) at NAS Miramar, California, VF-45 (six) at NAS Key West, Florida, and VF-43 (six detached from VF-45) at NAS Oceana, Virginia.

FS-X and TFS-X: F-16 derivatives selected by Japan Defence Agency for its FS-X requirement 19 October 1987; details under Mitsubishi in Japan section.

F-16 Recce: Four existing European recce pods, including that for Tornado, demonstrated in flight on F-16 fighters with minimum changes; RNethAF reconnaissance **F-16A(R)** operational since 1983 with Orpheus pods. Specially designed General Dynamics ATARS pod extensively tested with near-real-time reconnaissance capability; ATARS semi-conformal centreline pod on an F-16B (75-0752) housed advanced electro-optical and IR sensors for day/night, all-speed, all-altitude operation and standoff capability; three multi-position sensors carried out work of seven fixed cameras; sensor positioning and real-time viewing in cockpit; in-flight imagery review/manipulation/frame selection followed by digital data linking of selected frames to ground stations. Operators on ground able to analyse, annotate and disseminate imagery and their reports electronically only minutes after images taken. US Air Force wants to replace RF-4C with 108 ATARS F-16s (36 USAF; 72 ANG); conversion by Ogden ALC, 1996-97; designation would be **RF-16**. Full-scale development planned for FY 1992, followed by retrofit of 50-60 Block 30 aircraft in late 1990s.

AFTI/F-16: Modified pre-series F-16A (75-0750) used for US Air Force Systems Command Advanced Fighter Technology Integration (AFTI); first flight 10 July 1982; previous achievements detailed in 1986-87, 1987-88 and 1991-92 *Jane's*; currently with Block 15 horizontal tail surfaces, Block 25 wing and Block 40 avionics. Trials programmes include automatic target designation and attack (1988), night navigation and map displays (1988-89), digital data link and two-aircraft operations (1989), autonomous attack (1989-91) and night attack (1989-92); tested automatic ground collision avoidance system and pilot-activated, low-level pilot disorientation recovery system 1991; LANTIRN pod and Falcon Night FLIR trials, 1992. Total 500 sorties by November 1991.

F-16XL: Two F-16XL prototypes, in flyable storage since 1985, leased from General Dynamics by NASA; first flight of single-seat No. 1, 9 March 1989; NASA modified this aircraft at Dryden with wing glove having laser-perforated skin to smooth airflow over cranked arrow wing in supersonic flight, reducing drag and turbulence and saving fuel. Two-seat No. 2, with GE F110-GE-129 engine, similarly converted early 1992. F-16XL described in 1985-86 *Jane's*.

F-16B-2: Second prototype F-16B (75-0752) converted to private-venture testbed of close air support and night navigation and attack systems; equipment includes F-16C/D HUD, helmet sight or GEC-Marconi Cat's Eyes NVGs, Falcon Eye head-steered FLIR or LANTIRN nav/attack pods, digital terrain system (Terprom), and automatic target handoff system. Alternative nav/attack FLIR pods comprise GEC-Marconi Atlantic and Martin Marietta Pathfinder (LANTIRN derivative). NVG compatible cockpit lighting. Equipment testing continues on AFTI testbed (which see).

Falcon Century: Programme to evaluate and monitor developments and maintain a master plan for F-16s into the next century.

Falcon 21: Company studies for advanced F-16 derivative in 21st century.

CUSTOMERS: See table. Total 3,835 production aircraft ordered by Spring 1993, including planned USAF procurement of 2,203; however, Congress expected to perpetuate procurement at 12 per year for indefinite period, beginning FY 1994.

By July 1992 total of 2,466 delivered from Fort Worth and another 615 from assembly lines in Belgium, Netherlands and Turkey. Combined total 3,117 by December 1992.

COSTS: $18 million, USAF flyaway, FY 1992 prices. $31.5 million for modification of one F-16D Block 30 as NF-16D, which see, placed December 1988; MLU worth $2,000 million, 1991, including $300 million development phase. F-16 ADF programme cost $1 million per aircraft.

DESIGN FEATURES (refers mainly to Block 40 F-16C/D): Cropped delta wings blended with fuselage, with highly swept vortex control strakes along fuselage forebody and joining wings to increase lift and improve directional stability at high angles of attack; wing section NACA 64A-204; leading-edge sweepback 40°; relaxed stability (rearward CG) to increase manoeuvrability; deep wing-roots increase rigidity, save 113 kg (250 lb) structure weight and increase fuel volume; fixed-geometry engine intake; pilot's ejection seat inclined 30° rearwards; single-piece birdproof forward canopy section; two ventral fins below wing trailing-edge. Baseline F-16 airframe life planned as 8,000 hours with average usage of 55.5 per cent in air combat training, 20 per cent ground attack and 24.5 per cent general flying; structured strengthening programme for pre-Block 50 aircraft required during 1990s.

FLYING CONTROLS: Four-channel digital fly-by-wire (analog in earlier variants); pitch/lateral control by pivoting monobloc tailerons and wing-mounted flaperons; maximum rate of flaperon movement 52°/s; automatic wing leading-edge manoeuvring flaps programmed for Mach number and angle of attack; flaperons and tailerons interchangeable left and right; sidestick control column with force feel replacing almost all stick movement.

STRUCTURE: Wing, mainly of light alloy, has 11 spars, five ribs and single-piece upper and lower skins; attached to fuselage by machined aluminium fittings; leading-edge flaps are one-piece bonded aluminium honeycomb and driven by rotary actuators; fin is multi-spar, multi-rib with graphite epoxy skins; brake parachute or ECM housed in fairing aft of fin root; tailerons have graphite epoxy laminate skins, attached to corrugated aluminium pivot shaft and removable full-depth aluminium honeycomb leading-edge; ventral fins have aluminium honeycomb and skins; split speed-brakes in fuselage extensions inboard of tailerons open to 60°. Nose radome by Brunswick Corporation.

LANDING GEAR: Menasco hydraulically retractable type, nose unit retracting rearward and main units forward into fuselage. Nosewheel is located aft of intake to reduce the risk of foreign objects being thrown into the engine during ground operation, and rotates 90° during retraction to lie horizontally under engine air intake duct. Oleo-pneumatic struts in all units. Aircraft Braking Systems mainwheels and brakes; Goodyear or Goodrich mainwheel tyres, size 27.75 × 8.75-14.5, pressure 14.48-15.17 bars (210-220 lb/sq i at T-O weights less than 13,608 kg (30,000 lb). Steerab nosewheel with Goodyear, Goodrich or Dunlop tyre, si 18 × 5.7-8, pressure 20.68-21.37 bars (300-310 lb/sq in) T-O weights less than 13,608 kg (30,000 lb). All but tw main unit components interchangeable. Brake-by-wi system on main gear, with Aircraft Braking Systems ant skid units. Runway arresting hook under rear fuselag Irvin 7.01 m (23 ft 0 in) diameter braking parachute fitte in Greek, Indonesian, Netherlands (retrofit complete December 1992), Norwegian, Turkish and Venezuela F-16s. Landing/taxi lights on nose landing gear door.

POWER PLANT: One 131.6 kN (29,588 lb st) General Electr F110-GE-129, or one 129.4 kN (29,100 lb st) Pratt Whitney F100-PW-229 afterburning turbofan as alternative standard. These Increased Performance Engines (IPI installed from late 1991 in Block 50 and Block 52 aircra Immediately previous standard was 128.9 kN (28,984 st) F110-GE-100 or 105.7 kN (23,770 lb st) F100-PW-22 in Blocks 40/42. (Of 990 F-16Cs and 160 F-16Ds delivere to USAF by December 1991, 489 with F100 and 661 wi F110. IPE variants have half share each in FY 1992 pr curement of 48 F-16s for USAF, following eight reliabili trial installations including six Block 30 aircraft whic flew 2,400 h between December 1990 and Septemb 1992.) F100s of ANG and AFRes F-16A/Bs upgraded -220E standard from late 1991. Fixed geometry intak with boundary layer splitter plate, beneath fuselage. Apa from first few, F110-powered aircraft have intake widene by 30 cm (1 ft 0 in) from 368th F-16C (86-0262); Israe second-batch F-16D-30s have power plants locally mod fied by Bet-Shemesh Engines to F110-GE-110A with pr vision for up to 50 per cent emergency thrust at low leve USAF AVEN (axisymmetric vectoring engine nozzl trials, 1992, involve F100-IPE-94 and F110; possib application to F-16. Standard fuel contained in wing an five seal-bonded fuselage cells which function as tw tanks; see under Weights and Loadings for quantitie Halon inerting system. In-flight refuelling receptacle in to of centre-fuselage, aft of cockpit. Auxiliary fuel can be ca ried in drop tanks: one 1,136 litres (300 US gallons; 25 Imp gallons) under fuselage; 1,402 litres (370 US gallon 308 Imp gallons) under each wing. Optional 2,271 lit (600 US gallon; 500 Imp gallon) underwing tanks.

ACCOMMODATION: Pilot only in F-16C, in pressurised and ai conditioned cockpit. McDonnell Douglas ACES II zer zero ejection seat. Bubble canopy made of polycarbona advanced plastics material. Inside of USAF F-16C/ canopy (and most Belgian, Danish, Netherlands and No wegian F-16A/Bs) coated with gold film to dissipate rad energy. In conjunction with radar-absorbing materials i air intake, this reduces frontal radar signature by 40 pe cent. Windscreen and forward canopy are an integral un without a forward bow frame, and are separated from th aft canopy by a simple support structure which serves als as the breakpoint where the forward section pivots upwar and aft to give access to the cockpit. A redundant safe lock feature prevents canopy loss. Windscreen/canop design provides 360° all-round view, 195° fore and aft, 40 down over the side, and 15° down over the nose. To enab the pilot to sustain high *g* forces, and for pilot comfort, th seat is inclined 30° aft and the heel line is raised. In norm operation the canopy is pivoted upward and aft by electr cal power; the pilot is also able to unlatch the canop manually and open it with a backup handcrank. Emergenc jettison is provided by explosive unlatching devices an two rockets. A limited displacement, force sensing contr stick is provided on the right hand console, with a suitabl armrest, to provide precise control inputs during comba manoeuvres. The F-16D has two cockpits in tanden equipped with all controls, displays, instruments, avionic and life support systems required to perform both trainin and combat missions. The layout of the F-16D second sta tion is similar to the F-16C, and is fully systems operational. A single-enclosure polycarbonate trans parency, made in two pieces and spliced aft of the forwar seat with a metal bow frame and lateral support membe provides outstanding view from both cockpits.

SYSTEMS: Regenerative 12kW environmental control system with digital electronic control, uses engine bleed air fo pressurisation and cooling of crew station and avionic compartments. Two separate and independent hydrauli systems supply power for operation of the primary fligh control surfaces and the utility functions. System pressur (each) 207 bars (3,000 lb/sq in), rated at 161 litres (42.5 U gallons; 35.4 Imp gallons)/min. Bootstrap type reservoirs rated at 5.79 bars (84 lb/sq in). Electrical system powere by engine driven Westinghouse 60kVA main generato and 10kVA standby generator (including ground annun ciator panel for total electrical system fault reporting), wit Sundstrand constant speed drive and powered by a Sund strand accessory drive gearbox. 17Ah battery. Four dedi cated, sealed cell batteries provide transient electrica power protection for the fly-by-wire flight control system An onboard Sundstrand/Solar jet fuel starter is provide for engine self-start capability. Simmonds fuel measurin system. Garrett emergency power unit automatically drives a 5kVA emergency generator and emergency pump to provide uninterrupted electrical and hydraulic power fo control in the event of the engine or primary power system becoming inoperative.

F-16 CUSTOMERS

Operator	Total	Single-seat	Quantity	Two-seat	Quantity	Power plant	First delivery	Squadrons (or base)
Bahrain	12	F-16C-40	8	F-16D-40	4	F110-GE-100	March 1990	(Sheikh Isa)
Belgium	160[1]	F-16A-10	55	F-16B-10	12	F100-PW-200	January 1979	349, 350
		F-16A-15	41	F-16B-15	8	F100-PW-200	October 1982	23, 31
		F-16A-15OCU	40	F-16B-15OCU	4	F100-PW-220	January 1988	1, 2
Denmark	70	F-16A-10	30[1]	F-16B-10	8[1]	F100-PW-200	January 1980	723, 727, 730
		F-16A-15	16[1]	F-16B-15	4[1]	F100-PW-200	May 1982	723, 727, 730
		F-16A-15OCU	8[3]	F-16B-15OCU	4[3]	F100-PW-220	December 1987	726
Egypt	174	F-16A-15	34	F-16B-15	7[2]	F100-PW-200	March 1982	72, 74
		F-16C-32	36	F-16D-32	4	F100-PW-200	August 1986	(Beni Sueif)
		F-16C-40	40	F-16D-40	7	F110-GE-100	October 1991	(Abu Sueir)
		F-16C-50		F-16D-50		F110-GE-129	1994	Total 46[8] (Saqqara)
Greece	80	F-16CG-40	34	F-16DG-40	6	F110-GE-100	November 1988	330, 346
		F-16CG		F-16DG		F110-GE-129	1996	Total 40
Indonesia	12	F-16A-15OCU	8	F-16B-15OCU	4	F100-PW-220	December 1989	3
Israel	210	F-16A-10/15	67	F-16B-10/15	8	F100-PW-200	January 1980	101, plus two
		F-16C-30	51	F-16D-30	24	F110-GE-100A	December 1986	(Hatzor, Ramat David)
		F-16C-40	30	F-16D-40	30	F110-GE-200	July 1991	
Korea, South	160	F-16C-32	30	F-16D-32	10	F100-PW-220	March 1986	161, 162
		F-16C-52		F-16D-52		F100-PW-229		Total 120[9]
Netherlands	213[3]	F-16A-10	47	F-16B-10	13	F100-PW-200	June 1979	322, 323
		F-16A-15	84	F-16B-15	18	F100-PW-200	May 1982	306, 311, 312, 314, 316
		F-16A-15OCU	46	F-16B-15OCU	5	F100-PW-220	February 1988	313, 315
Norway	74	F-16A-10	29[3]	F-16B-10	7[3]	F100-PW-200	January 1980	332, 338
		F-16-15	31[3]	F-16B-15	5[3]	F100-PW-200	June 1982	331, 334
				F-16B-15OCU	2	F100-PW-220	July 1989	331
Pakistan	111	F-16A-15	28	F-16B-15	12	F100-PW-200	January 1983	9, 11, 14
		F-16A-15OCU	60	F-16B-15OCU	11	F100-PW-220	—	Embargoed[10]
Portugal	20	F-16A-15OCU	17	F-16B-15OCU	3	F100-PW-220E	June 1994	201
Singapore	10[7]	F-16A-15OCU	4	F-16B-15OCU	4	F100-PW-220	February 1988	140
		F-16A	2				1993	140
Thailand	36	F-16A-15OCU	12	F-16B-15OCU	6	F100-PW-220	June 1988	103
		F-16A-15OCU	12	F-16B-15OCU	6	F100	1995	
Turkey	240[4]	F-16C-30	35	F-16D-30	9	F110-GE-100	May 1987	141, 142
		F-16C-40	101	F-16D-40	15	F110-GE-100	July 1990	161, 162 plus four
		F-16C-50		F-16D-50		F110-GE-129		Total 80
USAF	2,203	F-16A-10	255	F-16B-10	74	F100-PW-200	August 1978	See note A
		F-16A-15	409[5]	F-16B-15	47[6]	F100-PW-200	September 1981	See note A
		F-16C-25	177	F-16D-25	49	F100-PW-200	July 1984	See note B
		F-16C-30/32	448	F-16D-30/32	53	both (100/220)	July 1986	See note B
		F-16C-40/42	384	F-16D-40/42	78	both (100/220)	December 1988	See note B
		F-16C-50/52		F-16D-50/52		both (129/229)	October 1991	Total 229
US Navy	26	F-16N-30	22	TF-16N-30	4	F110-GE-100	June 1987	43, 45, 126, FWS
Venezuela	24	F-16A-15	18	F-16B-15	6	F100-PW-200	September 1983	161, 162
TOTAL	3,835							

NOTES:

[1] Built by Sabca (Belgium); 221st and last Sabca F-16 (BAF FA-136) delivered 22 October 1991
[2] One built by Fokker (Netherlands)
[3] Built by Fokker; 300th and last Fokker F-16 (RNethAF J-021) delivered 27 February 1992
[4] Two F-16Cs and six F-16Ds built by GD; remainder by TAI
[5] Two built by Fokker
[6] Three built by Sabca
[7] Plus nine leased from USAF
[8] TAI production
[9] 12 built by GD, 36 CKD kits and 72 produced locally by Samsung Aerospace
[10] Six F-16As and three F-16Bs built and stored in USA

NOTE A: Currently operated by: 89, 93, 465 and 466 squadrons of AFRES; 111, 114, 119, 134, 136, 159, 169, 171, 178, 179, 186 and 194 interceptor squadrons, ANG (all F-16ADF); 107, 121, 148, 152, 160, 170, 182, 184, 195 and 198 fighter squadrons, ANG.

NOTE B: Currently operated by: 8th FW, Kunsan, South Korea (35, 80 Sqdns); 31st FW, Homestead AFB, Florida (307, 308, 309 Sqdns); 51st FW, Osan, South Korea (36 Sqdn); 52nd FW, Spangdahlem, Germany (23, 81, 480 Sqdns); 56th FW, McDill AFB, Florida (61, 62, 63 Sqdns); 57th FW, Nellis AFB, Nevada (414 Sqdn, F-16FWS); 58th FW, Luke AFB, Arizona (310, 311, 314, 425 Sqdns); 86th FW, Ramstein, Germany (512, 526 Sqdns); 343rd W, Eielson, Alaska (18 Sqdn); 347th FW, Moody AFB, Georgia (68, 69, 70 Sqdns); 366th W, Mountain Home AFB, Idaho (389 Sqdn); 388th FW, Hill AFB, Utah (4, 34, 421 Sqdns); 432nd FW, Misawa, Japan (13, 14 Sqdns); and also 302, 457, 704 and 706 squadrons, AFRES; 112, 113, 118, 120 (124 in 1994), 125, 127, 131, 138, 149, 157, 161, 163, 174, 175 and 176 squadrons, ANG; Thunderbirds air demonstration team.

REMARKS: Blocks 1 and 5 retrofitted to Block 10 standard 1982-84; Block 15 retrofitted to Block 15OCU (avionics standard only) from 1987. New build F-16As are Block 15OCU from November 1987.

Export programme code-names are: Bahrain - Peace Crown; Egypt - Peace Vector I-IV; Greece - Peace Xenia I-II; Indonesia - Peace Bima-Sena; Israel - Peace Marble I-III; Pakistan - Peace Gate I-IV; Singapore - Peace Carvin I-II; South Korea - Peace Bridge I-II; Thailand - Peace Naresuan I-II; Turkey - Peace Onyx I-II (and CAE-Link simulator, Peace Onyx III); and Venezuela - Peace Delta.

Omitted are two YF-16s and eight pre-production aircraft (six F-16A including one converted to two-seat F-16XL; two F-16B).

RECENT ORDERS: Morocco (second-hand aircraft) notification for 20, July 1991, later postponed; Singapore, two attrition replacements 1992, ex-USAF F-16As; Greece ordered 40 F-16C/Ds, 1992; Taiwan offered 150 F-16A/B-15MLUs in 1992.

INTENTIONS: Singapore to lease nine USAF F-16As (ex-Thunderbirds team) for three years, then buy nine F-16A/Bs; USAF expects Congressional mandate to buy at least 12 in FY 1994 and more thereafter.

AVIONICS: Westinghouse AN/APG-68(V) pulse Doppler range and angle track radar, with planar array in nose. Provides air-to-air modes for range-while-search, uplook search, velocity search, air combat, track-while-scan (ten targets), raid cluster resolution, single target track and (later) high PRF track to provide target illumination for AIM-7 missiles; and air-to-surface modes for ground mapping, Doppler beam sharpening, ground moving target, sea target, fixed target track, target freeze after pop-up, beacon, and air-to-ground ranging. Forward avionics bay, immediately forward of cockpit, contains radar, air data equipment, inertial navigation system, flight control computer, and combined altitude radar altimeter (CARA). Rear avionics bay contains ILS, Tacan and IFF, with space for future equipment. A Dalmo Victor AN/ALR-69 radar warning system is replaced in USAF Block 50/52 by Loral AN/ALR-56M advanced RWR; ALR-56M ordered for USAF Block 40/42 retrofit and (first export) Korean Block 52s. Tracor AN/ALE-40(V)-4 chaff/flare dispensers (AN/ALE-47 in Block 50/52); provision for Westinghouse AN/ALQ-131 jamming pods and planned AN/ALQ-184. Communications equipment includes Magnavox AN/ARC-164 UHF Have Quick transceiver (AN/URC-126 Have Quick IIA in Block 50/52); provisions for a Magnavox KY-58 secure voice system; Collins AN/ARC-186 VHF AM/FM transceiver (AN/ARC-205 Have Sync Group A in Block 50/52); government furnished AN/AIC-18/25 intercom; and SCI advanced interference blanker. Honeywell central air data computer. Litton LN-39 standard inertial navigation system (ring laser Litton LN-93 or Honeywell H-523 in Block 50/52 and current FMS F-16A/B; LN-93 for Egypt, Israel, Indonesia, Korea, Pakistan and Portugal, plus Netherlands retrofit); Gould AN/APN-232 radar altimeter; Collins AN/ARN-108 ILS; Collins AN/ARN-118 Tacan; Teledyne Electronics AN/APX-101 IFF transponder with a government furnished IFF control; government furnished National Security Agency KIT-1A/TSEC cryptographic equipment; Lear Astronics stick force sensors; GEC-Marconi wide-angle holographic electronic head-up display with raster video capability (for LANTIRN) and integrated keyboard; Rockwell GPS/Navstar; General Dynamics enhanced stores management computer; Teledyne Systems general avionics computer; Honeywell multi-function displays; data entry/cockpit interface and dedicated fault display by Litton Canada and General Dynamics/Lockheed, Fort Worth; Fairchild data transfer set; and Astronautics cockpit/TV set. Cockpit and core avionics integrated on two MIL-STD-1553B multiplex buses. Optional equipment includes Collins VIR-130 VOR/ILS and ARC-190 HF radio. Essential structure and wiring provisions are built into the airframe to allow for easy incorporation of future avionics systems under development for the F-16 by the US Air Force. Israeli Air Force F-16s have been extensively modified with Israeli designed and manufactured equipment, as well as optional US equipment, to tailor them to the IAF defence role. This includes Elisra SPS 3000 self-protection jamming equipment in enlarged spines of F-16D-30s and Elta EL/L-8240 ECM in third batch F-16C/Ds, replacing Rapport. Fin-root fairing houses Loral Rapport ECM in Israeli F-16As; Belgian F-16s have Dassault Electronique Carapace ECM from 1992; Dutch aircraft have Orpheus reconnaissance pods; South Korea requires ITT/Westinghouse AN/ALQ-165 ASPJ in 120 F-16C/Ds. Pakistan F-16s carry Thomson-CSF Atlis laser designator pods. Turkish aircraft

512 USA: AIRCRAFT—LOCKHEED

(158 to be modified by 1996) to share 60 LANTIRN pods; LANTIRN also purchased by South Korea and required for second Thailand batch. Enhanced capability LANTIRN (second-generation FLIR) tested by F-16 at Eglin AFB, early 1993. Historical details in 1986-87 and earlier *Jane's*.

ARMAMENT: General Electric M61A1 20 mm multi-barrel cannon in the port side wing/body fairing, equipped with a General Electric ammunition handling system and an enhanced envelope gunsight (part of the head-up display system) and 511 rounds of ammunition. There is a mounting for an air-to-air missile at each wingtip, one underfuselage centreline hardpoint, and six underwing hardpoints for additional stores. For manoeuvring flight at $5.5g$ the underfuselage station is stressed for a load of up to 1,000 kg (2,200 lb), the two inboard underwing stations for 2,041 kg (4,500 lb) each, the two centre underwing stations for 1,587 kg (3,500 lb) each, the two outboard underwing stations for 318 kg (700 lb) each, and the two wingtip stations for 193 kg (425 lb) each. For manoeuvring flight at $9g$ the underfuselage station is stressed for a load of up to 544 kg (1,200 lb), the two inboard underwing stations for 1,134 kg (2,500 lb) each, the two centre underwing stations for 907 kg (2,000 lb) each, the two outboard underwing stations for 204 kg (450 lb) each, and the two wingtip stations for 193 kg (425 lb) each. There are mounting provisions on each side of the inlet shoulder for the specific carriage of sensor pods (electro-optical, FLIR, etc); each of these stations is stressed for 408 kg (900 lb) at $5.5g$, and 250 kg (550 lb) at $9g$. Typical stores loads can include two wingtip mounted AIM-9L/M/P Sidewinders, with up to four more on the outer underwing stations; Rafael Python 3 on Israeli F-16s from early 1991; centreline GPU-5/A 30 mm cannon; drop tanks on the inboard underwing and underfuselage stations; a Martin Marietta Pave Penny laser spot tracker pod along the starboard side of the nacelle; and bombs, air-to-surface missiles or flare pods on the four inner underwing stations. Stores can be launched from Aircraft Hydro-Forming MAU-12C/A bomb ejector racks, Hughes LAU-88 launchers, or Orgen triple or multiple ejector racks. Non-jettisonable centreline GPU-5/A 30 mm gun pods on dedicated USAF ground-attack F-16As. Weapons launched successfully from F-16s, in addition to Sidewinders and AMRAAM, include radar guided Sparrow and Sky Flash air-to-air missiles, French Magic 2 infra-red homing air-to-air missiles, AGM-65A/B/D/G Maverick air-to-surface missiles, HARM and Shrike anti-radiation missiles, Harpoon anti-ship missiles, and, in Royal Norwegian Air Force service, the Penguin Mk 3 anti-ship missile.

F-16C Fighting Falcon of the Bahrain Amiri Air Force *(G. Rolle/J. L. Gaynecoetche)*

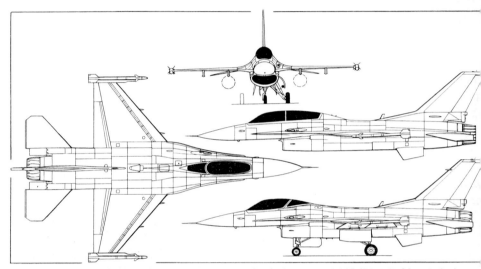

F-16C (GE F110 turbofan) with extra side view (top) of two-seat F-16D (P&W F100 turbofan) *(Dennis Punnett)*

DIMENSIONS, EXTERNAL (F-16C, D):
Wing span: over missile launchers	9.45 m (31 ft 0 in)
over missiles	10.00 m (32 ft 9¾ in)
Wing aspect ratio	3.0
Length overall	15.03 m (49 ft 4 in)
Height overall	5.09 m (16 ft 8½ in)
Tailplane span	5.58 m (18 ft 3¾ in)
Wheel track	2.36 m (7 ft 9 in)
Wheelbase	4.00 m (13 ft 1½ in)

AREAS (F-16C, D):
Wings, gross	27.87 m² (300.0 sq ft)
Flaperons (total)	2.91 m² (31.32 sq ft)
Leading-edge flaps (total)	3.41 m² (36.72 sq ft)
Vertical tail surfaces (total)	5.09 m² (54.75 sq ft)
Rudder	1.08 m² (11.65 sq ft)
Horizontal tail surfaces (total)	5.92 m² (63.70 sq ft)

WEIGHTS AND LOADINGS:
Weight empty:	
F-16C: F100-PW-220	8,273 kg (18,238 lb)
F110-GE-100	8,627 kg (19,020 lb)
F-16D: F100-PW-220	8,494 kg (18,726 lb)
F110-GE-100	8,853 kg (19,517 lb)
Max internal fuel: F-16C	3,104 kg (6,846 lb)
F-16D	2,567 kg (5,659 lb)
Max external fuel (both)	3,066 kg (6,760 lb)
Max external load (both)	5,443 kg (12,000 lb)
Typical combat weight:	
F-16C (F110)	10,780 kg (23,765 lb)
Max T-O weight:	
air-to-air, no external tanks:	
F-16C (F110)	12,331 kg (27,185 lb)
with external load:	
F-16C Block 30/32	17,010 kg (37,500 lb)
F-16C Block 40/42	19,187 kg (42,300 lb)
Wing loading:	
at 12,927 kg (28,500 lb) AUW	464 kg/m² (95.0 lb/sq ft)
at 19,187 kg (42,300 lb) AUW	688 kg/m² (141.0 lb/sq ft)
Thrust/weight ratio (clean)	1.1 to 1

PERFORMANCE:
Max level speed at 12,200 m (40,000 ft) above Mach 2.0
Service ceiling more than 15,240 m (50,000 ft)
Radius of action:
F-16C Block 40, two 907 kg (2,000 lb) bombs, tw Sidewinders, 3,940 litres (1,040 US gallons; 867 Im gallons) external fuel, tanks dropped when empty, h lo-lo-hi 740 nm (1,371 km; 852 mile
F-16C Block 40, four 907 kg (2,000 lb) bombs, tw Sidewinders, 1,136 litres (300 US gallons; 250 Im gallons) external fuel, tanks retained, hi-lo-lo-hi 340 nm (630 km; 392 mile
F-16C Block 40, two Sparrows and two Sidewinder 3,940 litres (1,040 US gallons; 867 Imp gallon external fuel, 2 h 10 min CAP 200 nm (371 km; 230 mile
F-16C Block 40, as immediately above, point interce 710 nm (1,315 km; 818 mile
Ferry range, with drop tanks more than 2,100 nm (3,890 km; 2,415 mile
Max symmetrical design g limit with full internal fuel

LOCKHEED AIRCRAFT SERVICE COMPANY (LAS) (Division of Lockheed Aeronautical Systems Group)

1800 East Airport Drive, Ontario, California 91761-0033
Telephone: 1 (909) 395 2411
PRESIDENT: Harold T. Bowling
EXECUTIVE VICE-PRESIDENT: John S. McLellan
VICE-PRESIDENT, PROGRAMMES: Ned Duffield
VICE-PRESIDENT, OPERATIONS: Robert Chambers
DIRECTOR, PUBLIC RELATIONS: John R. Dailey

LAS is oldest aircraft modification company in the world; has designed, fabricated and installed major aircraft structural modifications and integrated complete avionic systems for US military, foreign governments and commercial customers; has completed conversions on such aircraft as Boeing KC-135 and 707, McDonnell Douglas DC-8 and Lockheed C-130, C-141, L-188 Electra, L-1011 and P-3. Manufactures solid state flight data recorders and tactical reconnaissance pods; operates OMNILOG (worldwide logistics support service) which tracks more than 20 million aircraft components. Lockheed Aeromod Center Inc (LACI) of Greenville, South Carolina, maintains and modifies large military and commercial aircraft; in December 1990 opened 21,739 m² (234,000 sq ft) western area facility for commercial aircraft maintenance and modification at Tucson, Arizona.

In September 1991, Lockheed Commercial Aircraft Center Inc (LCAC) opened at Norton AFB, California, for maintenance and modification of Boeing 747s in partnership with Japan Airlines. Lockheed Support Systems Inc, Arlington, Texas, provides worldwide, low/medium technology operation, maintenance and support services on broad array of fixed and rotary wing aircraft for US government and private industry.

Airod SDN BHD (which see), joint venture between LAS International and Aerospace Industries Malaysia, maintains, modifies and overhauls aircraft at Subang International Airport, Kuala Lumpur. Guangzhou Aircraft Maintenance Engineering Company Limited (GAMECO) is joint venture company formed by Civil Aviation Administration of China-Guangzhou Regional Administration (CAAC-GRA), Lockheed Aircraft Service International (LASI) and Hutchison China Trade Holdings Limited (HCT); new maintenance hangar is the largest in China and able to accommodate or B-747 and two B-737s or four B-737s simultaneously; han; ar and adjoining shop complex occupy 14,000 m² (150,70 sq ft).

LAS specialises in design and application of special mi sion packages for converted C-130s, for electronic warfar command control communication and countermeasure sy tems, and signal intelligence; is currently upgradin AC-130H Spectre gunships into Special Operations Force Improvement (SOFI) configuration; major enhancemen include new computers, integrated weapon sensors, mult mode displays, software and gun mountings. First MOD-9 MC-130E Combat Talon I aircraft completed in Decembe 1991, incorporating enhanced electronic warfare, navigati and cockpit display capabilities. Ten C-130s converted in fully self-contained Rapid Response Mobile Hospital sys tems, including five L-100-30HS; hospital fleet includes ai craft carrying four-wheel drive ambulance and evacuatic aircraft for 52 patients.

LAS P-3 conversions include one US Customs ant narcotics P-3A modified in 1984, including fitting of Hughe

N/APG-63 radar and infra-red detection; Lockheed Aeromod modified three more P-3As using LAS kits; all have inertial navigation and multi-standard com radios; LAS fitted rotodome radar antenna to P-3 AEW&C; two P-3s converted to US Navy VP special mission; conversion of EP-3E Aries II electronic surveillance Orion.

Lockheed L-1011 TriStar conversions have included range, MTOGW and fuel capacity increases; VIP custom interiors; and L-1011 Freighter 2000 programme.

Further data in 1991-92 *Jane's* and in *Jane's Civil and Military Aircraft Upgrades*.

Lockheed AC-130H Spectre gunship redelivered to 16th Special Operations Squadron after SOFI upgrade by LAS

LOPRESTI

LOPRESTI FLIGHT CONCEPTS
50 Seagull Avenue, Vero Beach, Florida 32960
PRESIDENT: LeRoy P. LoPresti
Telephone: 1 (407) 778 5133

Original LoPresti Piper Aircraft Engineering Co (see 1990-91 *Jane's*) formed as Piper subsidiary, but operations suspended December 1990 when Piper abandoned programme; design team retained in LoPresti Speed Merchants (performance improvement kits for Piper small singles and twins); LoPresti Flight Concepts formed October 1991 to acquire rights to SwiftFury and carry through to certification and production; sale of rights foundered late 1992. LoPresti Flight Concepts considering designing new two-seat aircraft from scratch, to sell for $130,000-140,000 plus avionics; this could be available by 1996. Full details of LP-1 SwiftFury can be found in 1992-93 *Jane's*.

LTV

LTV AEROSPACE AND DEFENSE COMPANY
Re-named Vought Aircraft Company (which see) in 1992.

MACHEN

MACHEN INC
South 3608 Davison Boulevard, Spokane, Washington 99204
Telephone: 1 (509) 838 5326
Fax: 1 (509) 838 0831
EXECUTIVE VICE-PRESIDENT: James S. Christy

Output has been conversion of Piper (Ted Smith) Aerostars to Superstar 650, 680 and 700.

Product support for earlier Aerostars now handled by sister company AAC (Aerostar Aircraft Corporation), which has also proposed new twin-turbofan derivative (see Aerostar Aircraft Corporation in this section). Details of all conversions can be found in 1991-92 *Jane's*, and currently in *Jane's Civil and Military Aircraft Upgrades*.

Machen Superstar 650 conversion of the Piper Aerostar 601P

MARSH

MARSH AVIATION COMPANY
060 East Falcon Drive, Mesa, Arizona 85205-2590
Telephone: 1 (602) 832 3770
Fax: 1 (602) 985 2840
Telex: 165 028
PRESIDENT: Floyd D. Stilwell
EXECUTIVE VICE-PRESIDENT: Bill Walker Jr

Rockwell Thrush Commander and Schweizer Super Ag-Cat conversions abandoned because of US product liability laws, but would be supplied to foreign governments if ordered. Fitting of Garrett TPE331-14GR to Grumman HU-16 Albatross begun July 1993.

MARSH TS-2F TURBO TRACKER
TYPE: Grumman S-2 re-engined with Garrett TPE331 turboprops.
PROGRAMME: First flight of Marsh S-2 firefighting conversion (N426DF) 24 November 1986; certificated 19 February 1990; first flight TS-2F3 26 July 1991.
CURRENT VERSIONS: **TS-2F1 Turbo Tracker:** Formerly S-2AT; powered by 932 kW (1,250 shp) TPE331-14 engines; Marsh operating original prototype as firefighter for California Department of Forestry; firefighting retardant tank holds 3,028 litres (800 US gallons; 666 Imp gallons).
TS-2F3 Turbo Tracker: Formerly S-2ET; ASW and maritime patrol version; first flight 26 July 1991; carrier FCLP qualification achieved 14 January 1992; intended 24 per year by 1994. *Details below apply to this version.*
CUSTOMERS: One for California Forestry Department; seven military ASW TS-2F3s on order and 17 to follow in 1993; a few may have Motorola SLAR mounted along top of fuselage or AN/APS-143 in rotating radome.

Prototype Marsh TS-2F1 Turbo Tracker prototype conversion of Grumman S-2 Tracker

COSTS: Conversion with 40kVA AC power, no flight deck avionics or mission equipment, $3.7 million (1992).
DESIGN FEATURES: Equipment for military maritime patrol and ASW includes search radar, FLIR, MAD, sonobuoy launchers, pictorial navigation avionics with automatic search and attack modes, six underwing stores points and torpedo bay. Modern instrumentation, EFIS cockpit and AFCS; modifications improve stall behaviour; empty weight reduced by 907 kg (2,000 lb); airframe life 30 years or 30,000 h; nacelles and fuselage streamlined to achieve 60 knot (111 km/h; 69 mph) increase in cruising speed with 50 per cent lower fuel consumption; take-off and landing runs decreased by 25 per cent and single-engined rate of climb at gross weight increased by 230 m (750 ft)/min.
POWER PLANT: TS-2F3 powered by two 1,227 kW (1,645 shp) Garrett TPE331-15s driving Hartzell five-blade, 3.35 m (11 ft 0 in) diameter reversing propellers; take-off rating 1,313 kW (1,760 shp); negative torque sensing to indicate engine failure; electronic integrated engine controllers with automatic starting; composites cowlings. Max usable fuel 2,850 litres (753 US gallons; 431 Imp gallons).

514   USA: AIRCRAFT—MARSH/McDONNELL DOUGLAS

SYSTEMS: 800A, 28/48V dual DC electrical system and dual 40kVA, 110V AC system with inverter backup; two batteries; engine-bleed ECS; fire detection and halon extinguisher.
WEIGHTS AND LOADINGS:
Weight empty   6,278 kg (13,840 lb)
Max fuel weight   2,288 kg (5,045 lb)
Max T-O weight   13,154 kg (29,000 lb)
Max landing weight   11,113 kg (24,500 lb)
Take-off power loading (TPE331-15s)
                                  5.36 kg/kW (7.59 lb/shp)
PERFORMANCE:
Max level speed at S/L   260 knots (482 km/h; 299 mph)
Cruising speed at 7,010 m (23,000 ft)
                           252 knots (467 km/h; 290 mph)
Min approach speed   76 knots (140 km/h; 87 mph)
Service ceiling   7,315 m (24,000 ft)
T-O run at 7,711 kg (17,000 lb)   142 m (465 ft)
Landing run   243 m (800 ft)
Max range, no reserves   647 nm (1,200 km; 746 miles)

MAULE
MAULE AIR INC
Lake Maule, Route 5, Box 319, Moultrie, Georgia 31768
Telephone: 1 (912) 985 2045
Fax: 1 (912) 890 2402
Telex: 804613 MAULE MOUL
PRESIDENT: Belford D. Maule
SALES MANAGER: Don Merrill

Original Maule Aircraft Corporation formed to manufacture M-4, a four-seat extrapolation of Piper Cub; transferred to Moultrie, Georgia, 1968; production ceased 1975; Maule Air Inc formed 1984 to produce uprated M-5 Lunar Rocket and M-7 Super Rocket; Lunar Rocket discontinued, but variants listed below currently available.

MAULE STAR ROCKET and SUPER ROCKET
TYPE: Four/five-seat light aircraft.
CURRENT VERSIONS: **M-7-235 Super Rocket:** Long wing span five-seater, with choice of 175 kW (235 hp) Textron Lycoming O-540-J1A5D or IO-540-W1A5D engine.
MX-7-160 Star Rocket: As for MX-7-180A (see below) but with 119 kW (160 hp) Textron Lycoming O-320-B2D engine. Developed in 1992. Tricycle landing gear version is **MXT-7-160**, offering slightly lower cruising speed; higher weight empty.
MX-7-180A Star Rocket: Fuselage of M-6 with shorter span wing of M-5 and glassfibre wingtips, greater fuel capacity, ailerons and seven-position flaps of M-7; certificated 9 November 1984; powered by 134 kW (180 hp) Textron Lycoming O-360-C1F/C4F; choice of tricycle or tailwheel landing gear, with latter becoming **MXT-7-180A**, and having height of 2.54 m (8 ft 4 in), weight empty of 644 kg (1,420 lb), and cruising speed (75% power) of 117 knots (217 km/h; 135 mph).
MX-7-235 Super Rocket: As MX-7-180 but with 175 kW (235 hp) Textron Lycoming O-540-J1A5D or IO-540-W1A5D engine.
Turboprop: Maule certificating a version powered by 313 kW (420 shp) Allison 250-C20 flat rated at 268 kW (360 shp); tricycle landing gear.
CUSTOMERS: Total of more than 1,600 produced; recent sales include 16 to Mexican Air Force and Army (10 MXT-7-180 with tricycle landing gears and six M-7-235s).
COSTS: MX-7-160 (1993) $80,000; M-7 and MX-7-235 $110,000 (1993) with tricycle gear and $100,000 (1993) with tailwheel gear; turboprop version $350,000 (1993).
DESIGN FEATURES: USA 35B (modified) wing section; dihedral 1°; incidence 0° 30′; cambered wingtips standard.
FLYING CONTROLS: Ailerons linked to rudder servo tab to allow normal flying with aileron wheel only; trim tab in port elevator; rudder trim by spring to starboard rudder pedal; flap deflection 40° down for slow flight, 24°, 0° and 7° up for improved cruise performance; underfin on floatplane and amphibious versions.
STRUCTURE: All-metal two-spar wing with dual struts and glassfibre tips; fuselage frame of welded 4130 steel tube with Ceconite covering aft of cabin and metal doors and skin round cabin; glassfibre engine cowling.
LANDING GEAR: Non-retractable tailwheel type. Maule oleo-pneumatic shock absorbers in main units. Maule steerable tailwheel. Cleveland mainwheels with Goodyear or McCreary tyres size 17 × 6.00-6, pressure 1.79 bars (26 lb/sq in). Tailwheel tyre size 8 × 3.50-4, pressure 1.03-1.38 bars (15-20 lb/sq in). Cleveland hydraulic disc brakes. Parking brake. Oversize tyres, size 20 × 8.50-6 (pressure 1.24 bars; 18 lb/sq in), and fairings aft of mainwheels optional. Tricycle gear optional on MX-7-180A, becoming **MXT-7-180**. Provisions for fitting optional Edo Model 248B2440 floats or Edo Model 797-2500 amphibious floats, Aqua Model 2400 floats, or Federal Model C2200H or C3000 or FluiDyne 3000 Mk IIIA skis.
POWER PLANT: One flat-four or flat-six engine, driving a Sensenich two-blade constant-speed propeller, as detailed under Current Versions. Three-blade McCauley propeller optional. Two fuel tanks in wings with total usable capacity of 151 litres (40 US gallons; 33 Imp gallons). Auxiliary fuel tanks in outer wings, to provide total capacity of 265 litres (70 US gallons; 58 Imp gallons). Refuelling points on wing upper surface.

Maule M-7-235 Super Rocket five-seat STOL aircraft

DIMENSIONS, EXTERNAL:
Wing span: M-7   10.02 m (32 ft 10¾ in)
         MX-7   9.40 m (30 ft 10 in)
Wing chord, constant: all   1.60 m (5 ft 3 in)
Wing aspect ratio: M-7   6.4
         MX-7   6.0
Length overall:
    MX-7-180, MX-7-235   7.16 m (23 ft 6 in)
Height overall:
    all tailwheel types (landplane) 1.93 m (6 ft 4 in)
    all tricycle types   2.54 m (8 ft 4 in)
    all amphibians   3.20 m (10 ft 6 in)
Propeller diameter: MX-7-180A   1.93 m (6 ft 4 in)
         MX-7-160   1.88 m (6 ft 2 in)
DIMENSIONS, INTERNAL:
Cabin: Width, all   1.07 m (3 ft 6 in)
AREAS:
Wings, gross: M-7   16.44 m² (177.0 sq ft)
         MX-7   15.38 m² (165.6 sq ft)
WEIGHTS AND LOADINGS (L: landplane, F: floatplane, A: amphibian):
Weight empty: MX-7-160 (L)   603 kg (1,330 lb)
         MX-7-180A (L)   613 kg (1,350 lb)
         MX-7-235 (L)   669 kg (1,475 lb)
Max T-O and landing weight:
         MX-7-160 (L)   998 kg (2,200 lb)
         MX-7-180A (L)   1,089 kg (2,400 lb)
         all models (F, A)   1,247 kg (2,750 lb)
Max wing loading:
         M-7 (L)   68.91 kg/m² (14.12 lb/sq ft)
         M-7 (F, A)   75.83 kg/m² (15.54 lb/sq ft)
         MX-7 (L)   70.75 kg/m² (14.49 lb/sq ft)
         MX-7 (F, A)   81.10 kg/m² (16.61 lb/sq ft)
Max power loading:
         MX-7-180A (L)   8.45 kg/kW (13.89 lb/hp)
PERFORMANCE (at max T-O weight, ISA: L, F and A as for Weights and Loadings):
Max level speed at S/L:
    MX-7-235/IO-540 (L)
                             147 knots (273 km/h; 170 mph)
    M-7 (F, A), MX-7 (F, A)
                             130 knots (241 km/h; 150 mph)
    MX-7-180A (L)   134 knots (248 km/h; 154 mph)
Max cruising speed (75% power) at optimum altitude:
    M-7 (L), MX-7 (L)
                             139 knots (257 km/h; 160 mph)
    M-7 (F, A), MX-7 (F, A)
                             125 knots (232 km/h; 144 mph)
    MX-7-160   117 knots (217 km/h; 135 mph)
    MX-7-180A (L)   122 knots (225 km/h; 140 mph)
Stalling speed, flaps down, power off:
    M-7 (L)   31 knots (57 km/h; 35 mph)
    M-7 (F, A), MX-7-235 (L)
                              47 knots (87 km/h; 54 mph)
    MX-7 (L)   35 knots (65 km/h; 40 mph)
Max rate of climb at S/L:
    MX-7-235, M-7 (L)   610 m (2,000 ft)/min
    MX-7-235 (F, A)   381 m (1,250 ft)/min
    M-7-235 (F, A)   411 m (1,350 ft)/min
    MX-7-160   251 m (825 ft)/min
    MX-7-180A (L)   280 m (920 ft)/min
Service ceiling:
    M-7-235 (L), MX-7-235 (L)   6,100 m (20,000 ft)
    MX-7-160   3,960 m (13,000 ft)
    MX-7-180A (L)   4,575 m (15,000 ft)
T-O run: on land, one person/50% fuel:
    MX-7-160   183 m (600 ft)
    MX-7-180A (L)   168 m (550 ft)
    on water: M-7 (F, A)   305 m (1,000 ft)
    MX-7-235 (F, A)   335 m (1,100 ft)
T-O to 15 m (50 ft): on land:
    MX-7-160   360 m (1,180 ft)
    MX-7-180A   351 m (1,150 ft)
    on water: M-7-235 (F, A)   381 m (1,250 ft)
    MX-7-235 (F, A)   412 m (1,350 ft)
Landing from 15 m (50 ft): on land:
    M-7-235 (L), MX-7 (L)   152 m (500 ft)
    M-7-235 (A), MX-7-235 (A)   305 m (1,000 ft)
    on water:
    M-7-235 (F, A), MX-7-235 (F, A)   305 m (1,000 ft)
Landing run on land and water:
    M-7-235 (A), MX-7-235 (A)   244 m (800 ft)
Range with standard fuel:
    M-7-235, MX-7-235/O-540 (L)
                             425 nm (788 km; 490 miles)
    M-7-235 (L), MX-7-235 (L)
                             460 nm (853 km; 530 miles)
    MX-7-160   469 nm (869 km; 540 miles)
    MX-7-180A (L), 30 min reserves
                             434 nm (804 km; 500 miles)
    M-7-235/O-540 (F, A)   286 nm (531 km; 330 miles)
    M-7-235/IO-540 (F, A)   321 nm (595 km; 370 miles)
Range with auxiliary fuel:
    M-7-235/O-540 (L), MX-7-235/O-540 (L)
                             746 nm (1,384 km; 860 miles)
    M-7-235/IO-540 (L), MX-7-235 (L)
                             807 nm (1,496 km; 930 miles)
    MX-7-180A (L)   977 nm (1,810 km; 1,125 miles)
    M-7-235/O-540 (F, A)   608 nm (1,126 km; 700 miles)
    M-7-235/IO-540 (F, A)   668 nm (1,239 km; 770 miles)

MCDONNELL DOUGLAS
MCDONNELL DOUGLAS CORPORATION
Box 516, St Louis, Missouri 63166
Telephone: 1 (314) 232 0232
Fax: 1 (314) 234 3826
Corporate Office
CHAIRMAN AND CEO: John F. McDonnell
PRESIDENT: Gerald A. Johnston
DIRECTOR, INTERNATIONAL PUBLIC RELATIONS:
    Andrew Wilson

Formed 28 April 1967 by merger of Douglas Aircraft Co Inc and the McDonnell Company; encompasses their subsidiaries plus former Hughes Helicopter Company acquired in 1984 and renamed McDonnell Douglas Helicopter Company. Employees totalled about 87,400 worldwide in 1992; total office, engineering, laboratory and manufacturing floor area was then 3.8 million m² (41 million sq ft).

Major operating units of McDonnell Douglas Corporation (MDC) Aerospace Group were reorganised in August 1992 and now are:
Douglas Aircraft Company (DAC)
Follows this entry

McDonnell Douglas Aerospace (MDA)
Follows Douglas entry

MDA administratively divided into MDA East and MDA West and comprise the government aerospace business of McDonnell Douglas Corporation including the C-17 Globemaster III, formerly a Douglas Aircraft programme. MDA-W address is 5301 Bolsa Avenue, Huntington Beach, California 92647. *Telephone:* 1 (714) 896 1300. *Fax:* 1 (714) 896 1308. DAC concentrates on commercial airliners.

First McDonnell Douglas MD-87 for Swiss carrier CTA *(Anton Wettstein)*

DOUGLAS AIRCRAFT COMPANY (Division of McDonnell Douglas Corporation)

HEADQUARTERS: 3855 Lakewood Boulevard, Long Beach, California 90846
Telephone: 1 (310) 593 8611
Fax: 1 (310) 496 8720
Telex: 674357
PRESIDENT: Robert H. Hood Jr
EXECUTIVE VICE-PRESIDENTS:
John D. Wolf
David R. Hinson (Marketing and Business Development)
VICE-PRESIDENTS:
John N. Feren (General Manager, Marketing)
Allen C. Haggerty (General Manager, Design and Technology)
Walter J. Orlowski (General Manager, Development Programs)
PUBLIC RELATIONS MANAGER:
Don Hanson (Commercial Transports)

Douglas Aircraft Company operates plants at Long Beach and Torrance, California (latter to close 1993); Macon, Georgia; Columbus, Ohio; and Salt Lake City, Utah. DC designation superseded by MD in 1983, starting with MD-80 (DC-9 Super 80); DC-8, DC-9 and DC-10 remain unchanged. Douglas delivered 976 DC-9s up to 1982 and its 2,000th twin-jet airliner, an MD-80 to American Airlines, on 11 June 1992. Table shows backlog, orders and deliveries of current production types at 31 March 1993. Plan to sell up to 49 per cent of Douglas to Taiwan Aerospace during 1992 and jointly develop MD-12 did not materialise; in early 1993, Douglas still working to establish an airliner manufacturing alliance, but not based on a specific programme.

Douglas reduced workforce from between 40,000 and 50,000 in 1990, before military C-17 was hived off, to around 30,000 by end of 1992 and expects to stabilise at around 28,000; productivity increased so that man-hours to produce an MD-11 reduced from 300,000 in 1991 to 150,000 by end of 1993 and for a twin-jet from 100,000 to 50,000; MD-90 manufacturing system applied also to MD-80 during 1993; company can therefore remain viable while producing fewer than 30 twin-jets a year, compared with 136 needed in 1989, and around 30 tri-jets a year. Net orders for 1992 shown in table suggest extent of order deferments and cancellations; Douglas says current orders for twin-jets will generate revenue until 1996 and for tri-jets until 1995; company can fund own development of forthcoming derivatives MD-90-50 and MD-11D.

DOUGLAS AIRLINER ORDERS AND DELIVERIES
(backlog up to 31 March 1993)

Order Backlog	Firm	Other*	Total
MD-11	83	116	199
MD-80 Series	89[1]	108	197
MD-90	77[2]	102	179

1992 Orders		Net after changes
MD-11	8	1
MD-80 Series	31	19
MD-90	26	16

1992 Deliveries		Cumulative Total
MD-11	42	85
MD-80 Series	84	1,058
MD-90	0	0

Notes: [1] Up to 60 could be converted to MD-90
[2] Including Trunkliner
*'Other' includes options and reserved positions

MCDONNELL DOUGLAS MD-80 SERIES
TYPE: Twin-turbofan short/medium-range airliner.
PROGRAMME: Began as Super 80 higher capacity variant of DC-9; first flight 18 October 1979; first flight of second and third prototypes (N1002G and N1002W) 6 December 1979 and 29 February 1980 respectively; FAA certification 26 August 1980; first delivery, to Swissair, 12 September 1980. MD-80 Series continues in production with backlog of 234 at end 1992, since reduced; MD-80ADV (see below) being offered early 1993.

CURRENT VERSIONS: **MD-81:** Basic version with maximum seating for 172 passengers; P&W JT8D-209 engines with automatic power reserve; two-man crew; maximum five-abreast passenger seating.

MD-82: Announced 16 April 1979; powered by P&W JT8D-217s for hot and high performance and increased payload/range; same size cabin as MD-81 and -83; first flight 8 January 1981; certificated 31 July 1981 at max T-O weight 66,680 kg (147,000 lb); in service August 1981; same fuel capacity and landing weight as MD-81. Second version powered by JT8D-217As and with higher max T-O weight certificated mid-1982.

Chinese MD-82: Agreement signed 12 April 1985 for assembly by Shanghai Aviation Industrial Corporation (see SAMF in Chinese section) of 25 out of 26 MD-82s ordered by China; another five MD-82s and five MD-83s approved April 1990; US-built first aircraft delivered 30 September 1985; first flight of SAMF assembled MD-82 2 July 1987; in service 4 August 1987; second aircraft delivered 18 December 1987; FAA certificate extended to Chinese-built aircraft 9 November 1987; 27 MD-82s delivered by SAMF by end 1992 (plus five MD-83, see Trunkliner below). Douglas interests vested in McDonnell Douglas Pacific & Asia Ltd; SAMF assembles aircraft and makes tailplane and landing gear doors; Chengdu Aircraft Industrial Corporation is second source for nose sections for China and USA.

MD-83: Extended range version powered by JT8D-219s, announced 31 January 1983; two per cent lower fuel consumption than -217As; two extra fuel tanks in cargo compartment. Passenger capacity same as MD-81 and -82. First flight 17 December 1984; FAA certification 1985; in service Alaska Airlines and Finnair early 1986; on 14 November 1985, Finnair MD-83 made longest MD-80 flight covering 3,404 nm (6,308 km; 3,920 miles) from Montreal to Helsinki in 7 h 26 min; first revenue transatlantic service flown by Transwede between Stockholm and Fort Lauderdale, Florida, with stops at Oslo and Gander. Five MD-83s built in China by SAMF (see MD-82 entry above) completed first quarter 1993; three more to come; all to be sold outside China.

MD-87: Short fuselage version for maximum 139 single-class passengers; fin height increased; powered by JT8D-217Cs with two per cent lower fuel consumption than 217As; other -200 series engines available; first flight 4 December 1986; certificated 21 October 1987; first deliveries to Finnair and Austrian Airlines; optional front and rear cargo compartment auxiliary fuel tanks each hold 2,139 litres (565 US gallons; 470.5 Imp gallons). MD-87 has MD-80 cruise performance improvement package including fillet fairing between engine pylons and fuselage, fairing on APU, improved sealing on horizontal tail, low drag flap hinge fairings, and extended low-drag tailcone; MD-87 also first of series with EFIS, AHRS and HUD as standard.

MD-88: Combines JT8D-219 power plant with EFIS cockpit displays, flight management system, onboard windshear detection system and increased use of composites in structure. Redesigned cabin interior for 142 passengers (14 first/128 coach class) five-abreast; wider aisle; redesigned overhead bins. First flight 15 August 1987; FAA certification 9 December 1987; entered service 5 January 1988 with principal customer Delta Air Lines (125 ordered).

MD-80ADV: Advanced variant of basic MD-80 family being offered to customers early 1993; payload/range capability and P&W JT8D-219 turbofans of MD-83 with improvements offered in later MD-80 variants, including glass cockpit; additional luggage capacity.

MD-80/90T Trunkliner: Agreement worth £1 billion signed 25 June 1992 to produce three MD-82, 17 MD-82T and 20 MD-90-30T Trunkliners in China; assembly by Shanghai Aviation Industrial Corporation (SAIC); Chinese content to be raised progressively to 93 per cent by weight during Trunkliner production; Chengdu Aircraft Industrial Corporation making nose sections for Trunkliners and US-built MD-80/90s; Shenyang will take over assembly of tail surfaces, incorporating SAIC tailplanes and elevators, and Xian will make fuselages and wings; offset work to be offered on IAE V2500 engines. Further discussions planned for 1995 could lead to 130 additional MD-90Ts. MD-82/90Ts will have dual tandem landing gear with four-wheel bogies that will allow aircraft to use all airfields on internal Chinese routes with full passenger and baggage payloads; gear designed jointly by Douglas and Shanghai Aircraft Research Institute of CATIC under separate contract signed in 1990; prototype dual tandem gear to fly in last SAIC MD-82, September 1994; three more MD-82s to be completed before Trunkliner programme starts; first MD-82T to be delivered in 1996 and first MD-90T in 1997.

MD-80 Executive Jets: Corporate and executive versions of MD-83 and MD-87 offered; typically seating 20 passengers; MD-83 maximum range 4,100 nm (7,598 km; 4,721 miles); MD-87 maximum range 4,500 nm (8,339 km; 5,182 miles).

CUSTOMERS: For order backlogs and deliveries, see table; 1,000th delivered 23 March 1992. Order backlog dropped

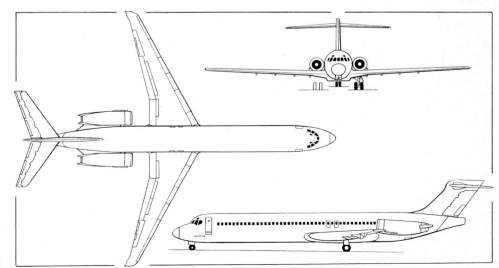
McDonnell Douglas MD-87, a short-fuselage variant of the MD-80 series airliners *(Dennis Punnett)*

USA: AIRCRAFT—McDONNELL DOUGLAS

Recent production MD-80; MD-81, -82, -83 and -88 are externally similar

by 20 firm and 17 'other' between end 1992 and 31 March 1993.

DESIGN FEATURES: MD-80 has DC-9 wing span increased by centre-section plugs and 0.61 m (2 ft 0 in) wingtip extensions; fuselage extended by plugs fore and aft of wing; larger wing holds more fuel; systems improvements include digital integrated flight guidance and control system, 'dial a flap' control for more accurate flap settings, flow-through cooling of avionics compartment, larger capacity APU, recirculation of ventilating air, and advanced digital fuel gauging system. Performance management system similar to that of DC-10 standard from April 1983; optional flight management system giving horizontal and vertical guidance. Other features include increased use of composites, such as Kevlar wing/fuselage fillets introduced 1983. Flight deck changes include advanced attitude and heading reference system, optional Honeywell EFIS, flat LED displays, alternative flight management systems, and Honeywell windshear guidance system (certificated June 1989; now standard on all new MD-80s and retrofittable).

Wing sweepback at quarter-chord 24° 30′; mean thickness/chord ratio 11.0 per cent; dihedral 3°; incidence 1° 15′.

FLYING CONTROLS: Manual ailerons; elevators with assister tabs; electrically actuated variable incidence tailplane; hydraulically actuated rudder with manual standby; automatic landing available; full-span, three-position leading-edge slats; three spoilers per wing, of which outboard two for flight and ground braking and one for lift dumping; hydraulically actuated double-slotted flaps cover 67 per cent of span; one underwing vortillon fence on each wing.

STRUCTURE: All-metal two-spar wing with riveted spanwise stringers; glassfibre trailing-edges on wings, ailerons, flaps, elevators and rudder; detachable wingtips; most of cabin floor made of balsa or Nomex core sandwich; engine pylons by Calcor and fuselage panels by Alenia.

LANDING GEAR: Retractable tricycle type of Cleveland Pneumatic manufacture, with steerable nosewheels (±27° on MD-81/82/87/88; ±25° on MD-83). Hydraulic retraction, nose unit forward, main units inward. Twin Goodyear wheels and tyres on each unit. Mainwheel tyres size 44.5 × 16.5-20, pressure 11.38 bars (165 lb/sq in). Nosewheel tyres size 26 × 6.6-14, pressure 10.34 bars (150 lb/sq in). Goodyear disc brakes. Hydro-Aire Mk IIIA anti-skid units. Douglas ram air brake cooling. Min ground turning radius: MD-81, 82, 83 and 88 about nosewheel 22.43 m (73 ft 7¼ in); MD-87 about nosewheel 19.54 m (64 ft 1¼ in); MD-81, 82, 83 and 88 about wingtip 20.04 m (65 ft 9 in); MD-87 about wingtip 19.63 m (64 ft 5 in).

POWER PLANT: Two Pratt & Whitney JT8D-209 turbofans in MD-81, pod mounted one each side of rear fuselage, and each rated at 82.3 kN (18,500 lb st), with emergency thrust reserve of 3.34 kN (750 lb). MD-82 has JT8D-217s, each rated at 89 kN (20,000 lb st), with emergency thrust reserve of 3.78 kN (850 lb), or -217As of similar rating. MD-83 has JT8D-219 engines of 93.4 kN (21,000 lb st) with thrust reserve of 3.78 kN (850 lb). MD-87 has JT8D-217C engines of 89 kN (20,000 lb st), with an emergency thrust reserve of 3.78 kN (850 lb). MD-88 has 93.4 kN (21,000 lb st) JT8D-219 turbofans with thrust reserve of 3.1 kN (700 lb). Target type thrust reversers. Standard fuel capacity in MD-81/82/87/88 is 21,876 litres (5,779 US gallons; 4,812 Imp gallons); increased in MD-83 (and, optionally, MD-87) to 26,260 litres (6,939 US gallons; 5,778 Imp gallons) by two 2,195 litre (580 US gallon; 483 Imp gallon) auxiliary tanks in cargo compartment. Pressure refuelling point in starboard wing leading-edge. Overwing gravity refuelling points.

ACCOMMODATION: Crew of two and observer on flight deck, plus cabin attendants. Seating arrangements are optional to meet specific airline requirements. Maximum optional seating capacity is for 172 passengers (130 in MD-87). Fully pressurised and air-conditioned. One toilet forward on port side, two at rear of cabin. Provisions for galley at each end of cabin. Passenger door at front of cabin on port side, with built-in electrically operated airstairs, and rear hydraulically operated ventral stairway, are standard. Servicing and emergency exit doors at starboard forward end and port rear end of cabin. Three cargo doors for underfloor holds on starboard side. Overwing emergency exits, two each side.

SYSTEMS: AiResearch dual air cycle air-conditioning and pressurisation system utilising engine bleed air, max differential 0.54 bar (7.77 lb/sq in). Two separate 207 bar (3,000 lb/sq in) hydraulic systems for operation of spoilers, flaps, slats, rudder, landing gear, nosewheel steering, brakes, thrust reversers and ventral stairway. Max flow rate 30.3 litres (8 US gallons; 6.7 Imp gallons)/min. Airless bootstrap type reservoirs, output pressure 2.07 bars (30 lb/sq in). Pneumatic system, for air-conditioning/pressurisation, engine starting and ice protection, utilises 8th or 13th stage engine bleed air and/or APU. Electrical system includes three 40kVA 120/208V three-phase 400Hz alternators, two engine driven, one driven by APU. Oxygen system of diluter demand type for crew on flight deck; continuous flow chemical canister type with automatic mask presentation for passengers. Anti-icing of wing, engine inlets and tailplane by engine bleed air. Electric windscreen de-icing. Thermal anti-icing of leading-edges. APU provides pneumatic and electrical power on ground, and electrical power in flight.

AVIONICS: All-digital avionics, including dual Honeywell integrated flight systems; Honeywell Cat. IIIA autoland; autopilot and stability augmentation; performance management system; speed command with digital full-time autothrottles; thrust rating indicator; dual Honeywell air data systems; colour weather radar. Sundstrand HUD optional.

DIMENSIONS, EXTERNAL (all versions, except as indicated):
Wing span	32.87 m (107 ft 10¼ in)
Wing chord: at root	7.05 m (23 ft 1½ in)
at tip	1.10 m (3 ft 7½ in)
Wing aspect ratio	9.62
Length overall: except MD-87	45.06 m (147 ft 10 in)
MD-87	39.75 m (130 ft 5 in)
Length of fuselage:	
except MD-87	41.30 m (135 ft 6 in)
MD-87	36.30 m (119 ft 1 in)
Fuselage outside diameter	3.61 m (11 ft 10 in)
Height overall: except MD-87	9.19 m (30 ft 2 in)
MD-87	9.50 m (31 ft 2 in)
Tailplane span	12.24 m (40 ft 2 in)
Wheel track	5.08 m (16 ft 8 in)
Wheelbase: except MD-87	22.07 m (72 ft 5 in)
MD-87	19.18 m (62 ft 11 in)
Passenger door (port, fwd):	
Height	1.83 m (6 ft 0 in)
Width	0.86 m (2 ft 10 in)
Height to sill	2.36 m (7 ft 9 in)
Service door (stbd, fwd): Height	1.22 m (4 ft 0 in)
Width	0.69 m (2 ft 3 in)
Height to sill	2.36 m (7 ft 9 in)
Service door (port, rear): Height	1.52 m (5 ft 0 in)
Width	0.69 m (2 ft 3 in)
Height to sill	2.82 m (9 ft 3 in)
Freight and baggage hold doors:	
Height	1.27 m (4 ft 2 in)
Width	1.35 m (4 ft 5 in)
Height to sill: fwd	1.47 m (4 ft 10 in)
centre	1.42 m (4 ft 8 in)
rear	1.65 m (5 ft 5 in)
Rear cargo door (MD-87): Height	1.27 m (4 ft 2 in)
Width	0.91 m (3 ft 0 in)
Heights to sills	as MD-81/82/83/8
Emergency exits (overwing, port and stbd):	
Height	0.91 m (3 ft 0 in)
Width	0.51 m (1 ft 8 in)

DIMENSIONS, INTERNAL:
Cabin, excl flight deck, incl toilets:	
Length	30.78 m (101 ft 0 in)
Max width	3.14 m (10 ft 3¾ in)
Max height	2.06 m (6 ft 9 in)
Floor area	89.65 m² (965.0 sq ft)
Volume	191.9 m³ (6,778 cu ft)
Freight holds (underfloor, MD-81/82):	
fwd	12.29 m³ (434.0 cu ft)
centre	10.65 m³ (376.0 cu ft)
rear	12.54 m³ (443.0 cu ft)
Freight holds (underfloor, MD-83 with extra fuel tanks):	
total	28.7 m³ (1,013 cu ft)
Freight holds (underfloor MD-87):	
total	26.6 m³ (938 cu ft)
with extra fuel tanks	19.7 m³ (695 cu ft)

AREAS:
Wings, gross	115.1 m² (1,239 sq ft)
Ailerons (total)	3.53 m² (38.0 sq ft)
Fin, excl dorsal fin	9.51 m² (102.4 sq ft)
Rudder	6.07 m² (65.3 sq ft)
Tailplane	29.17 m² (314.0 sq ft)

WEIGHTS AND LOADINGS
Operating weight empty: -81	35,329 kg (77,888 lb)
-82, -88	35,369 kg (77,976 lb)
-83 optional fuel	36,145 kg (79,686 lb)
-87 standard fuel	33,237 kg (73,274 lb)
-87 optional fuel	33,965 kg (74,880 lb)
Fuel load:	
-81, -82, -87 standard	17,763 kg (39,162 lb)
-83, -87 optional	21,216 kg (46,773 lb)
Max structural payload:	
-81	18,194 kg (40,112 lb)
-82, -88	19,969 kg (44,024 lb)
-83 optional fuel	19,193 kg (42,314 lb)
-87 standard	17,566 kg (38,726 lb)
-87 optional fuel	16,837 kg (37,120 lb)
Max T-O weight: -81 (-217 engines), -87 standard	63,503 kg (140,000 lb)
-81 (-217A engines), -82, -87 optional, -88 standard	67,812 kg (149,500 lb)
-83 optional	72,575 kg (160,000 lb)
Max zero-fuel weight: -81	53,524 kg (118,000 lb)
-82, -83	55,338 kg (122,000 lb)
-87 standard and optional	50,802 kg (112,000 lb)
Max landing weight:	
-81, -87 standard	58,060 kg (128,000 lb)
-82, -87 optional, -88	58,967 kg (130,000 lb)
-83 optional	63,276 kg (139,500 lb)
Max wing loading:	
-81, -87 standard	534.6 kg/m² (109.5 lb/sq ft)
-82, -87 optional, -88 standard	574.7 kg/m² (117.7 lb/sq ft)
-83, -88 optional	615.0 kg/m² (126.0 lb/sq ft)
Max power loading: -81	385.8 kg/kN (3.78 lb/lb st)
-82, -87 optional, -88	381.0 kg/kN (3.74 lb/lb st)
-83 optional	388.5 kg/kN (3.81 lb/lb st)
-87 standard	356.8 kg/kN (3.50 lb/lb st)

PERFORMANCE (at max T-O weight except where indicated):
Max level speed: all	500 knots (925 km/h; 575 mph)
Max cruising speed: all	Mach 0.8
Normal cruising speed: all	Mach 0.7
FAA T-O field length: -81	1,954 m (6,410 ft)
-82	2,315 m (7,595 ft)
-83	2,462 m (8,075 ft)
-87	1,913 m (6,275 ft)
FAA landing field length, at max landing weight:	
-81	1,451 m (4,760 ft)
-82	1,463 m (4,800 ft)
-83	1,540 m (5,050 ft)
-87	1,451 m (4,760 ft)
Range with max fuel:	
-87 standard	2,980 nm (5,522 km; 3,431 miles)
-87 optional	3,650 nm (6,764 km; 4,203 miles)
Range (-81, -82, -83 with 155 passengers, domestic reserves; -87 with 130 passengers, domestic reserves):	
-81	1,630 nm (3,020 km; 1,877 miles)
-82	2,176 nm (4,032 km; 2,505 miles)
-83	2,618 nm (4,851 km; 3,014 miles)
-87 standard	2,405 nm (4,457 km; 2,769 miles)
-87 optional	2,874 nm (5,326 km; 3,309 miles)

OPERATIONAL NOISE LEVELS (FAR Pt 36):
T-O: -81, -82, -83	90.4 EPNdB
-87 estimated	88.7 EPNdB
Sideline: -81, -82, -83	94.6 EPNdB
-87 estimated	92.8 EPNdB
Approach: -81, -82, -83	93.3 EPNdB
-87 estimated	93.3 EPNdB

Prototype of MD-90-30, eventually to succeed the MD-80

MCDONNELL DOUGLAS MD-90

TYPE: Stretched MD-80 follow-on, powered by IAE V2500 turbofans.
PROGRAMME: Launched 14 November 1989; first flight (N901DC) 22 February 1993; flutter envelope cleared to 395 knots (732 km/h; 454 mph) and Mach 0.895; certification and first deliveries September and October 1994 respectively. See Chinese production under MD-80/90T Trunkliner.
CURRENT VERSIONS: **MD-90-30**: Has MD-80 fuselage lengthened by 1.45 m (4 ft 9 in) ahead of wing; same enlarged tail surfaces as MD-87; powered elevators; 153 two-class passengers, five-abreast; max 172 passengers limited by exit doors and hatches; two IAE V2525-D5 turbofans.
MD-90-30T Trunkliner: 20 ordered for assembly and delivery by Shanghai Aviation Industrial Corporation (SAIC) 25 June 1992; will have dual tandem landing gear; first delivery 1997 (see MD-80/90T Trunkliner entry).
MD-90-50: Extended range version of MD-90-30 for 153 passengers; two IAE V2528-D5 turbofans.
MD-90-55: Similar to MD-90-50 but with extra pair of doors in forward fuselage section to allow max 187 charter class passengers.
CUSTOMERS: Launch customer Delta Air Lines (50 ordered plus 115 on option, later reported as 26 firm and 74 options); other customers reported to include Alaska Airlines (reduced from 20 to 10), GATX, Great China, SAS (six plus six converted from MD-80s) and Japan Air System. For totals, see table on page 515.
DESIGN FEATURES: To be built on MD-80 production line; powered by IAE V2500 engines rated by engine control system for power required by MD-90-30 and MD-90-50; 10 more two-class passengers than MD-80 accommodated by forward fuselage stretch of 1.4 m (4 ft 6 in) to compensate for higher engine weight; better power:weight ratio than MD-80; noise level expected to be 20dB below Stage 3 and with very low emissions; dual tandem main landing gear of Trunkliner variant available for all customers; improved cabin includes larger baggage bins, better lighting and hand rail at bin level.
FLYING CONTROLS: Powered elevators with dual actuators and manual reversion, with servo tabs to cope with increased pitch-axis inertia caused by heavier engines and longer forward fuselage; spoilers for airbrake and lift dumping; flight deck similar to MD-88, but Douglas planning new six-screen layout similar to that of MD-11. Three-position leading-edge slats.
STRUCTURE: Structure broadly as late MD-80, but new manufacturing system (applied also to MD-80) allows both types to be built on same production line and in about half the man-hours of earlier MD-80s; subassemblies contributed by Alenia, Aerospace Technologies of Australia, Aerospatiale, CASA, Chengdu Aircraft Industrial Corporation, Shanghai Aviation Industrial Corporation (SAIC), Shanghai Aircraft Manufacturing Factory (SAMF).
LANDING GEAR: Retractable tricycle type with twin wheels on all units, as MD-80; MD-90-30T Trunkliner will have dual tandem gear with four wheels on each main unit to allow aircraft to use all Chinese runways; dual tandem gear also to be offered as option for other MD-90s.
POWER PLANT: Two 111.21 kN (25,000 lb st) IAE V2525-D5 in MD-90-30; two 124.55 kN (28,000 lb st) V2528-D5 in MD-90-50 and -55; thrust maintained at S/L up to 86°F ambient temperature; power output determined by electronic engine control; cascade thrust reversers for use on ground only; MD-90-30 fuel tankage 22,107 litres (5,840 US gallons; 4,863 Imp gallons); MD-90-50 tankage 28,845 litres (7,620 US gallons; 6,345 Imp gallons) including 6,738 litres (1,780 US gallons; 1,482 Imp gallons) in extra tanks in baggage compartment.

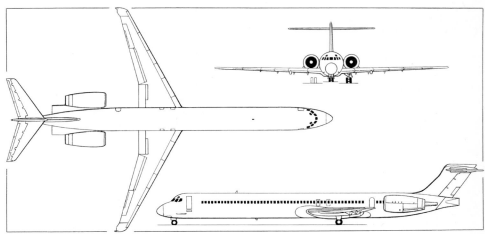
McDonnell Douglas MD-90-30 airliner (two IAE V2525-D5 turbofans) *(Dennis Punnett)*

SYSTEMS: New system elements, compared with MD-80, include variable-speed, constant frequency electrical generation by AlliedSignal, new Garrett 421 kW (565 shp) GTCP131-9D APU to provide greater engine-starting power and 8,000 hour life, carbon wheel brakes saving 181 kg (400 lb) weight, centre wing de-icing system using warmed fuel from engine oil cooler circulated through inboard fuel tanks, new vacuum toilets and new environmental control system providing higher flow rates.
AVIONICS: Honeywell supplies electronic flight instruments, flight management system upgrade for MD-90 and V2500, developed digital flight guidance system (DFGS) with auxiliary control system (ACS), new air data computer, and advanced inertial reference system with ring laser gyros.

DIMENSIONS, EXTERNAL (all versions):
Wing span	32.87 m (107 ft 10 in)
Length overall	46.51 m (152 ft 7 in)
Height overall	9.33 m (30 ft 7¼ in)
Wheel track	5.09 m (16 ft 8½ in)
Wheelbase	23.52 m (77 ft 2 in)

DIMENSIONS, INTERNAL:
Baggage volume (total): -30	36.8 m³ (1,300 cu ft)
-50, -55 with optional fuel	23.3 m³ (822 cu ft)

AREAS:
Wings, gross: all	112.3 m² (1,209 sq ft)

WEIGHTS AND LOADINGS:
Operating weight empty: -30	40,007 kg (88,200 lb)
-50	41,640 kg (91,800 lb)
-55	42,844 kg (94,455 lb)
Space limited payload: -30	17,350 kg (38,250 lb)
-50, -55 with extra tanks	15,195 kg (33,500 lb)
Max T-O weight: -30	70,760 kg (156,000 lb)
-50, -55	78,244 kg (172,500 lb)
Max ramp weight: -30	71,215 kg (157,000 lb)
-50, -55	78,698 kg (173,500 lb)
Max zero-fuel weight: -30	58,965 kg (130,000 lb)
-50, -55	61,235 kg (135,000 lb)
Max landing weight: -30	64,410 kg (142,000 lb)
-50, -55	71,210 kg (150,000 lb)
Max wing loading: -30	614.75 kg/m² (125.91 lb/sq ft)
-50, -55	679.78 kg/m² (139.23 lb/sq ft)
Max power loading: -30	318.14 kg/kN (3.12 lb/lb st)
-50, -55	315.93 kg/kN (3.08 lb/lb st)

PERFORMANCE (estimated, at max T-O weight, ISA, except where indicated):
Cruising speed at 10,670 m (35,000 ft):		
all	437 knots (809 km/h; 503 mph)	Mach 0.76
FAA T-O field length: -30		2,135 m (7,000 ft)
-50, -55		2,347 m (7,700 ft)
FAA landing field length, at max landing weight:		
-30		1,564 m (5,130 ft)
-50, -55		1,628 m (5,340 ft)
Range, with international reserves: (-30 and -50 with 153 passengers, -55 with 187 passengers):		
-30		2,266 nm (4,200 km; 2,610 miles)
-50		3,022 nm (5,600 km; 3,480 miles)
-55		2,700 nm (5,003 km; 3,109 miles)

MCDONNELL DOUGLAS MD-95

TYPE: 95/124-passenger airliner based on shortened MD-87 fuselage, MD-80 wing and MD-88/MD-90 avionics.
PROGRAMME: Announced at Paris Air Show June 1991; initially planned for assembly in China after placing of Trunkliner order; in early 1993, programme responsibility with Douglas and full development not launched; first flight tentatively planned for second half 1995 and deliveries in late 1996; MD-95 could be manufactured outside China.
CURRENT VERSIONS: Choice of two gross weights, as shown below; fuselage length constant, but P&W-powered aircraft has one fewer fuselage frame aft of wing and R-R-powered aircraft has one fewer frame ahead of wing to compensate for different engine weights.
CUSTOMERS: Northwest Airlines participated in early development, but did not become lead customer.
DESIGN FEATURES: About 1 m (3 ft 3 in) longer than DC-9-30; DC-9/MD-80 fuselage cross-section; MD-80 wing planform; systems and avionics are blend of low cost and advanced technology.
FLYING CONTROLS: As MD-90 with vertical and lateral navigation not dependent on radio.
STRUCTURE: Generally as MD-80/MD-90. Subassemblies would be supplied by CASA, Aerospatiale, Alenia, AeroSpace Technologies of Australia, Shanghai Aircraft Manu-

518 USA: AIRCRAFT—McDONNELL DOUGLAS

Artist's impression of the McDonnell Douglas MD-95 95/124-passenger airliner

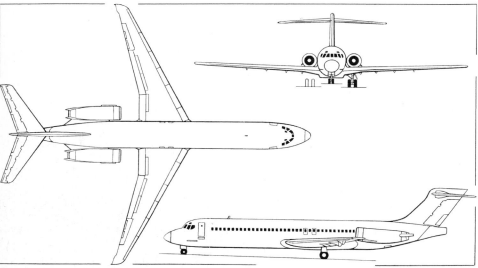

McDonnell Douglas MD-95 airliner (two Pratt & Whitney JT8D-200 series or Rolls-Royce Tay turbofans)
(Dennis Punnett)

Artist's impression of the McDonnell Douglas MD-XX short/medium-range airliner

facturing Factory, Shanghai Aviation Industrial Corporation, and Chengdu Aircraft Industrial Corporation.
POWER PLANT: Choice of two P&W JT8D-216 or Rolls-Royce Tay 671, both flat rated at 73.4 kN (16,500 lb st) up 29°C. Also offered: optional P&W JT8D-218 or Rolls-Royce Tay 670 rated at 80.07 kN (18,000 lb st). Cascade reversers. BMW, Volvo and Textron Lycoming are partners in Tay 670 development and engine might be manufactured in China.
ACCOMMODATION: Alternative cabin layouts for 124 all-tourist at 89, 76 and 74 cm (35, 30 and 29 in) pitch, 12 first class and 93 tourist, or 80 business class and 15 tourist.
DIMENSIONS, EXTERNAL:
Wing span	32.86 m (107 ft 10 in)
Length overall	37.33 m (122 ft 6 in)
Height overall	9.30 m (30 ft 8 in)
Tailplane span	12.25 m (40 ft 2 in)

DIMENSIONS, INTERNAL:
Underfloor cargo volume   23.6 m³ (833 cu ft)

WEIGHTS AND LOADINGS (A: JT8D-216, B: Tay 671, C: JT8D-218, D: Tay 670):
Operating weight empty: A, C	32,477 kg (71,600 lb)
B, D	31,796 kg (70,100 lb)
Max T-O weight: A, B	54,204 kg (119,500 lb)
C, D	58,059 kg (128,000 lb)
Max zero-fuel weight: A, B, C, D	46,266 kg (102,000 lb)
Max landing weight: A, B, C, D	49,895 kg (110,000 lb)

PERFORMANCE (estimated):
Cruising speed: A, B, C, D	Mach 0.7
Initial cruise altitude: A, B	11,275 m (37,000 ft)
FAA T-O field length: A	1,920 m (6,300 ft)
B, C	2,012 m (6,600 ft)
D	2,073 m (6,800 ft)
FAA landing field length: A, B, C, D	1,305 m (4,280 ft)

Range with 105 passengers and baggage:
A	1,600 nm (2,965 km; 1,842 miles)
B	1,785 nm (3,304 km; 2,053 miles)
C	2,210 nm (4,095 km; 2,544 miles)
D	2,371 nm (4,387 km; 2,727 miles)

MCDONNELL DOUGLAS MD-XX

TYPE: Advanced technology medium-range airliner.
PROGRAMME: Preliminary studies begun May 1991. Possible launch 1996 or later; first deliveries before year 2000.
DESIGN FEATURES: Trans-continental, 200/220-passenger transport to improve passenger environment; stretch up to 250 passengers foreseen; Mach 0.8 cruise; range from 3,000 to 4,500 nm (5,560 to 8,339 km; 3,455 to 5,180 miles); provision of office facilities and modern communications techniques. Fuselage cross-section twin-aisle over 5.18 m (17 ft 0 in) wide, 5.40 m (17 ft 8½ in) high; two LD-2 containers side by side under floor; power plant probably two advanced ducted propellers on wing; five year NASA-Douglas study on advanced composite transport will produce a composite wing box of MD-XX size about 1995; main objective is cost reduction. Possible technical features include composite primary and secondary structures in airframe; fly-by-light signalling for control, sensor and data distribution; power-by-wire controls.
DIMENSIONS, EXTERNAL (provisional):
Wing span	39.3 m (128 ft 11¼ in)
Length of fuselage	46.0 m (150 ft 11 in)

MCDONNELL DOUGLAS MD-11

TYPE: Medium/long-range passenger and freight follow-on to DC-10.
PROGRAMME: Revealed at Paris Air Show 1985; British Caledonian ordered nine 3 December 1986; official programme launch 30 December 1986; five-aircraft flight test programme (four with GE engines, one with P&W); first flight 10 January 1990 powered by CF6s; first flight of third prototype powered by P&W PW4460s, 26 April 1990; certificated 8 November 1990; first delivery to Finnair 29 November 1990, entering service 20 December. Planned production 36 in 1991, 40 in 1992, little more than half that number in 1993; 100th MD-11 delivered 30 June 1993. Certification with R-R Trent 650 discontinued.
CURRENT VERSIONS: **MD-11**: Standard passenger version for 323 passengers in two-class layout; max range 7,008 nm (12,987 km; 8,070 miles). Planned production of all variants about two per month in 1993, increasing again from 1995. About 1,814 kg (4,000 lb) excess empty weight and higher than planned engine fuel burn led to range shortfall at introduction into service; MD-11A1 incorporated Phase 1 weight saving of 1,197 kg (2,640 lb) and Phase 2 drag reduction introduced progressively; increased gross weight of 283,721 kg (625,500 lb), greater than original maximum optional gross weight, certificated 4 November 1992; additional fuel offered in one or two 7,472 litre (1,974 US gallons, 1,644 Imp gallons) tanks in underfloor hold have restored most of originally specified range, but not fuel burn per passenger; MD-11A2 version under study end 1992 could have further weight and drag reduction plus a gross weight increase to provide another 100 to 200 nm (185 to 370 km; 115 to 230 miles) range. *Detailed description applies to first three variants listed.*
MD-11 Combi: Mixed cargo/passenger version for four to 10 cargo pallets and 168 to 240 passengers; ranges from 5,210 nm (9,656 km; 6,000 miles) to 6,947 nm

(12,875 km; 8,000 miles). Main deck cargo door at rear on port side. Certificated April 1992 to latest FAA Class C smoke and fire containment requirements.

MD-11CF: Convertible freighter; launched August 1991 with order from Martinair-Holland (four firm orders for delivery 1995, one on option). Main deck cargo door at front on port side. Certification due 1994.

MD-11F: All-freight version.

MD-11B/B1/B2: Initial stretches to MD-11 being discussed with several airlines late 1992; B would have same size fuselage, but gross weight increased to 299,371 kg (660,000 lb) and range increased to 7,300 nm (13,528 km; 8,406 miles); B1 would be slightly stretched, carry 20 more passengers, but sacrifice some range; B2 would have shortened fuselage with 250-passenger capacity, but range increased to nearly 8,000 nm (14,826 km; 9,212 miles).

MD-11C: Tentatively planned increased capacity version with 3.18 m (10 ft 5 in) root plugs in wings to increase span to 58.22 m (191 ft 0 in) and area to 440.45 m² (4,741 sq ft); other changes include four-wheel centre main landing gear, enlarged tail surfaces and varying fuselage lengths. MD-11C would carry 298 passengers for 8,000 nm (14,825 km; 9,200 miles); MD-11C1 would be slightly stretched to carry 329 passengers for 7,300 nm (13,528 km; 8,405 miles); MD-11C2 would be further stretched to carry 353 passengers for 6,900 nm (12,787 km; 7,945 miles). Requires risk-sharing partners for development.

MD-11D: Tentatively planned increased capacity version with underfloor panorama deck in forward freight space seating 43 business class passengers; MD-11D1 would have reduced gross weight 266,258 kg (587,000 lb) and 5,200 nm (9,636 km; 5,988 miles) range; MD-11D2 would have 283,722 kg (625,500 lb) gross weight and 6,100 nm (11,304 km; 7,024 miles) range; all versions would carry 362 passengers; panorama deck requires impact absorbing belly structure, strengthened cabin floor beams, relocated nose landing gear, and two Type B doors replacing forward freight door; windows would be extra large. Increased passenger baggage would occupy eight of 14 LD-3 containers in aft hold. Douglas could develop with own resources.

CUSTOMERS: For orders and deliveries, see table on page 515.
DESIGN FEATURES: Compared with DC-10, MD-11 has winglets above and below each wingtip; tailplane has advanced cambered aerofoil, reduced sweepback and 7,571 litre (2,000 US gallon; 1,665 Imp gallon) fuel trim tank; extended tailcone of chisel profile; two-crew all-digital flight deck; restyled interior; choice of GE CF6-80C2D1F and P&W PW4460 engines. Wing has Douglas aerofoil section; sweepback at quarter-chord 35°; dihedral 6°; incidence at root 5° 51'.

FLYING CONTROLS: Ailerons powered by National Water Lift actuators; electrohydraulically actuated variable incidence tailplane with slotted elevators in two sections each side powered by Parker Hannifin and Teijin Seiki actuators; inboard all-speed ailerons and outboard low-speed ailerons; dual section rudder split into vertical segments; near-full-span leading-edge slats; double-slotted trailing-edge flaps with offset external hinges; five spoilers in groups of four and one on each wing; Category IIIb automatic landing (certificated April 1991) standard.

STRUCTURE: Composites used in virtually all control surfaces, engine inlets and cowlings, and wing/fuselage fillets; wing has two-spar structural box with chordwise ribs and skins with spanwise stiffeners; upper winglet of ribs, spars and stiffened aluminium alloy skin with carbonfibre trailing-edge; lower winglet carbonfibre; inboard ailerons have metal structure with composites skin; outboard ailerons all-composites; inboard flaps composites-skinned metal; outboard flaps all-composites; spoilers aluminium honeycomb and composites skin; tailplane has CFRP trailing-edge; elevators CFRP.

Rear engine inlet duct and fan cowl doors, and nose cowl outer barrels on wing-mounted engines, are of composites construction. Inner surfaces of engine nacelles are acoustically treated.

Suppliers include Alenia (fin, rudder, fuselage panels, winglets), AP Precision Hydraulics (centreline and nose landing gear), AlliedSignal Bendix Landing Systems (mainwheels and carbon brakes), CASA (horizontal tail surfaces), General Dynamics Convair Division (fuselage sections), Embraer (outboard flap sections), Fischer GmbH (composite flap hinge fairings), Pneumo Abex Corporation (main landing gear), Rohr Industries (engine pylons), Honeywell (advanced flight deck and avionics), and Westland Aerospace (flap vane and inlet duct extension rings).

LANDING GEAR: Hydraulically retractable tricycle type, with additional twin-wheel main unit mounted on the fuselage centreline; heaviest proposed variants might have four-wheel centreline bogie; nosewheel and centreline units retract forward, main units inward into fuselage. Twin-wheel steerable nose unit (±70°). Main gear comprises two four-wheel bogies. Oleo-pneumatic shock absorbers in all units. Loral nosewheels and Goodyear tyres size 40 × 15.5-16, pressure 13.44 bars (195 lb/sq in). Main and centreline units have Bendix wheels and Goodyear tyres size 54 × 21-24, pressure 13.79 bars (200 lb/sq in). Bendix carbon brakes with air convection cooling; Loral anti-skid system. Min ground turning radius about nosewheel 26.67 m (87 ft 6 in); about wingtip 35.90 m (117 ft 9½ in).

McDonnell Douglas MD-11 of LTU powered by P&W PW4460 turbofans

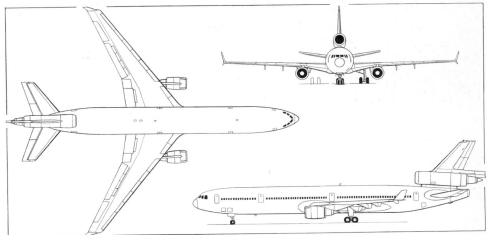

McDonnell Douglas MD-11 medium/long-range transport *(Dennis Punnett)*

MD-11 two-crew flight deck with six large displays and two flight management system panels on console. The third FMS panel at rear is for groundcrew use when testing aircraft avionics

POWER PLANT: Three Pratt & Whitney PW4460 turbofans, each originally rated at 266.9 kN (60,000 lb st), or three General Electric CF6-80C2D1F turbofans, each rated at 273.57 kN (61,500 lb st); P&W thrust later increased to overcome speed and field performance shortcomings; two engines mounted on underwing pylons, the third above the rear fuselage at the base of the fin. Refuelling point in leading-edge of each wing. Standard fuel capacity: MD-11 152,108 litres (40,183 US gallons; 33,459 Imp gallons); MD-11F and Combi 146,305 litres (38,650 US gallons; 32,182 Imp gallons); one or two 7,472 litre (1,974 US gallon; 1,643 Imp gallon) tanks can be added in cargo hold to restore specified range.

ACCOMMODATION: Crew of two. Standard mixed class seating for 323 (MD-11) or 214 (Combi); max passenger capacity 410. Crew door and three passenger doors each side, all eight of which open sliding inward and upward. Two freight holds in lower deck, forward and aft of wing, and one bulk cargo compartment in rear fuselage. Forward freight hold is heated and ventilated; rear freight hold

520 USA: AIRCRAFT—McDONNELL DOUGLAS

heated only. MD-11 Combi has a lower deck cargo door in centre compartment on port side of fuselage for loading of pallets, an upward opening main deck cargo door on port side at rear of cabin, and two additional emergency exit doors, one on each side of passenger cabin immediately forward of the main deck cargo door. MD-11F/CF have port side forward main deck cargo door.

SYSTEMS: Air-conditioning system includes three AiResearch air bearing air cycle units with two automatic digital pressure controllers and electro-manual backup. Cabin max pressure differential 0.59 bar (8.6 lb/sq in). Three independent hydraulic systems for operation of flight controls and braking, with motor/pump interconnects to allow one system to power another. Electrical system comprises three 400Hz, 100/120kVA integrated drive generators, one per engine; one 90kVA generator in APU; 50Ah battery; four transformer-rectifiers to convert AC power to DC; and 25kVA drop-out air driven emergency generator. Pneumatic system, max controlled pressure 0.27-0.41 bar (4-6 lb/sq in) at 230°C, supplies air-conditioning, engine bleed air anti-icing for wing (outer slats) and tailplane leading-edges, galley vent jet pump, and cargo compartment floor heating. Plumbed gaseous oxygen system for crew, using EROS equipment; chemical oxygen generators with automatically deploying masks for passengers. Portable oxygen cylinders for attendants and first aid. De-icing for windscreens, angle of attack sensors, TAT probe and static port plate. AlliedSignal Garrett TSCP700-4E APU.

AVIONICS: Avionics integrator, Honeywell, responsible for flight guidance/flight deck system consisting of 44 line-replaceable units. These include aircraft system controllers (ASC) that performs flight engineering control and monitoring functions, providing automated hydraulic, electrical, environmental and fuel systems; a central fault display system (CFDS); an electronic instrument system (EIS) using six 20 × 20 cm (8 × 8 in) colour CRTs; a flight management system (FMS); an automatic flight system (AFS) featuring Cat. IIIb autoland, windshear detection and guidance computer with escape capability and full-time longitudinal stability augmentation; a laser inertial reference system (IRS); and a digital air data computer (DADC).

DIMENSIONS, EXTERNAL:
Wing span	51.77 m (169 ft 10 in)
Wing chord: at root	10.71 m (35 ft 1¾ in)
at tip	2.73 m (8 ft 11½ in)
Wing aspect ratio	7.5
Length overall: with PW4460	61.24 m (200 ft 11 in)
with CF6-80	61.37 m (201 ft 4 in)
Fuselage: Length	58.65 m (192 ft 5 in)
Max diameter	6.02 m (19 ft 9 in)
Height overall	17.60 m (57 ft 9 in)
Tailplane span	18.03 m (59 ft 2 in)
Wheel track	10.67 m (35 ft 0 in)
Wheelbase	24.61 m (80 ft 9 in)
Crew doors (two, each): Height	1.93 m (6 ft 4 in)
Width	0.81 m (2 ft 8 in)
Passenger doors:	
Height: front pair	1.93 m (6 ft 4 in)
rear six doors	1.93 m (6 ft 4 in)
Width: front pair	0.8 m (2 ft 8 in)
rear six doors	1.07 m (3 ft 6 in)
* Lower deck forward freight door:	
Height	1.68 m (5 ft 6 in)
Width	2.64 m (8 ft 8 in)
Lower deck centre freight door (standard):	
Height	1.68 m (5 ft 6 in)
Width	1.78 m (5 ft 10 in)
Lower deck bulk cargo door: Height	0.91 m (3 ft 0 in)
Width	0.76 m (2 ft 6 in)
Combi main deck cargo door (port, rear):	
Height	2.59 m (8 ft 6 in)
Width	4.06 m (13 ft 4 in)
CF main deck cargo door (port, forward):	
Height	2.59 m (8 ft 6 in)
Width	3.56 m (11 ft 8 in)

* *Centre freight door of Combi also this size*

DIMENSIONS, INTERNAL:
Cabin:
Length, flight deck door to rear bulkhead	46.51 m (152 ft 7¼ in)
Max width	5.71 m (18 ft 9 in)
Max height	2.41 m (7 ft 11 in)
Floor area, incl galleys and toilets	244.7 m² (2,634.0 sq ft)
Volume, incl galleys and toilets	599.3 m³ (21,165 cu ft)
Lower deck freight holds, volume	194 m³ (6,850 cu ft)

AREAS:
Wings, gross	338.9 m² (3,648 sq ft)
Winglets (total)	7.42 m² (80.0 sq ft)
Vertical tail surfaces (total)	56.2 m² (605.0 sq ft)
Horizontal tail surfaces (total)	85.5 m² (920.0 sq ft)

WEIGHTS AND LOADINGS:*
Operating weight empty: -11	131,035 kg (288,880 lb)
-11F	113,630 kg (250,510 lb)
-11 Combi	129,591 kg (285,700 lb)
Weight limited payload: -11	51,058 kg (112,564 lb)
-11F	91,078 kg (200,790 lb)
-11 Combi	65,454 kg (144,300 lb)
Max T-O weight: all	283,725 kg (625,500 lb)
Max zero-fuel weight: -11	181,435 kg (400,000 lb)
-11F	204,700 kg (451,300 lb)
-11 Combi	195,040 kg (430,000 lb)
Max landing weight: -11	195,040 kg (430,000 lb)
-11F	213,870 kg (471,500 lb)
-11 Combi	207,745 kg (458,000 lb)

* *Empty weights with P&W engines about 317 kg (700 lb) lower than with GE engines*

PERFORMANCE:
Max operating Mach number (M_{MO}): all	0.945
Max level speed at 9,449 m (31,000 ft):	
all	Mach 0.87 (511 knots; 945 km/h; 588 mph)
FAA T-O field length, MTOW, S/L, ISA +15°C:	
-11	3,200 m (10,500 ft)
-11F and Combi	3,127 m (10,260 ft)
FAA landing field length, MLW, S/L:	
-11	1,966 m (6,450 ft)
-11F	2,131 m (6,990 ft)
-11 Combi	2,027 m (6,650 ft)
Design range, FAR international reserves:	
-11, 323 passengers (2-class)	6,791 nm (12,566 km; 7,810 miles)
-11F	3,626 nm (6,710 km; 4,170 miles)
-11 Combi, 214 passengers, 6 pallets	6,273 nm (11,609 km; 7,215 miles)
Range, MD-11A1, reported:	
P&W engines	6,803 nm (12,607 km; 7,823 miles)
GE engines	6,869 nm (12,730 km; 7,900 miles)

MCDONNELL DOUGLAS MD-12

TYPE: High capacity long-range airliner.

PROGRAMME: Definitive four-engined layout announced April 1992; development delayed following failure to conclude arrangement with Taiwan Aerospace; new partners or partner group being sought.

CURRENT VERSIONS: Initially offered in Long Range, High Capacity, Freight and Combi versions; later to be offered in Stretch Medium Range, Stretch Long Range and Twin Engined versions.

CUSTOMERS: No firm customers announced by April 1993.

COSTS: Development spending decreased from $36 million in second quarter 1992 to $11 million in last quarter.

DESIGN FEATURES: Full-length two-deck fuselage seating up to 511 passengers in three classes; range would allow New York to Taipei or Singapore to Zurich non-stop.

FLYING CONTROLS: Fly-by-wire control system with all-glass cockpit; normal ailerons assisted by outboard spoiler panels; four elevator sections; rudders divided into fore and aft and upper and lower panels; total seven spoilers per wing for lateral control, airbrakes and lift dumping; six leading edge slat/flap sections per wing; wing sweepback 35°.

LANDING GEAR: Four main legs each to carry one four-wheel bogie; inner legs retract into fuselage, outers into wing and fuselage; twin-wheel nose leg.

POWER PLANT: Four engines; choice of GE CF6-80C2D1 (273.6 kN; 61,500 lb st each) or Pratt & Whitney PW4460 (266.9 kN; 60,000 lb st each) or Rolls-Royce Trent 760 (284.7 kN; 64,000 lb st each).

ACCOMMODATION: Two-pilot flight deck on main deck level; possible cross-qualifying with MD-11; crew rest compartment at front end of upper deck; cabin crew rest area at aft end; main deck has three aisles; Long Range interior seats 430 passengers in three-class layout, including 78 business class in twin-aisle layout on shortened upper deck plus 25 first class and 327 economy class on main deck; High Capacity version seats 95 business class and 65 economy class on full-length twin-aisle upper deck plus 27 first class and 324 economy class on main deck; both decks can accommodate freight in Freight and Combi versions; underfloor freight compartments can hold two LD3 containers abreast; eight passenger doors on main deck and four escape doors on upper deck; one stairway in Long Range version and two in High Capacity.

DIMENSIONS, EXTERNAL:
Wing span	64.92 m (213 ft 0 in)
Length: fuselage	58.82 m (193 ft 0 in)
overall	63.40 m (208 ft 0 in)
Height overall	22.55 m (74 ft 0 in)
Tailplane span	22.55 m (74 ft 0 in)
Wheel track	11.58 m (38 ft 0 in)
Wheelbase	26.82 m (88 ft 0 in)

DIMENSIONS, INTERNAL:
Cabin: Main deck width at floor	7.39 m (24 ft 3 in)
Upper deck width at shoulder	5.20 m (17 ft 1 in)
Height (all decks)	8.50 m (27 ft 11 in)
Underfloor freight volume	126.4 m³ (4,468 cu ft)

Artist's impression of the McDonnell Douglas MD-12

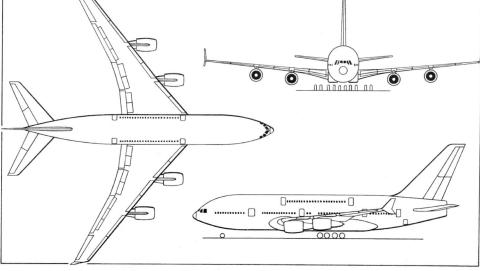

Four-engined two-deck McDonnell Douglas MD-12, announced in April 1992 and temporarily shelved in June 1992 (*Jane's/Mike Keep*)

Main deck freight volume	521.5 m³ (18,420 cu ft)	Max landing weight: LR, HC	291,209 kg (642,000 lb)	
Upper deck freight volume	198.1 m³ (7,000 cu ft)	F	333,393 kg (735,000 lb)	

Main deck freight volume 521.5 m³ (18,420 cu ft)
Upper deck freight volume 198.1 m³ (7,000 cu ft)
AREAS:
Wings, gross 543.0 m² (5,846 sq ft)
Vertical tail surfaces (total) 96.1 m² (1,035 sq ft)
Horizontal tail surfaces (total) 113.8 m² (1,225 sq ft)
WEIGHTS AND LOADINGS (LR: Long Range, HC: High Capacity, F: Freighter):
Operating weight empty: LR 187,633 kg (413,700 lb)
HC 194,683 kg (429,200 lb)
F 173,727 kg (383,000 lb)
Weight limited payload: F only 141,522 kg (312,000 lb)
Max T-O weight: LR, HC, F 430,463 kg (949,000 lb)
Max zero-fuel weight: LR, HC 273,065 kg (602,000 lb)
F 315,249 kg (695,000 lb)
Max landing weight: LR, HC 291,209 kg (642,000 lb)
F 333,393 kg (735,000 lb)
Max wing loading:
LR, HC, F 792.7 kg/m² (162.3 lb/sq ft)
PERFORMANCE (estimated):
Cruising Mach number 0.85
FAA T-O field length (ISA + 30°C):
LR, HC, F 3,078 m (10,100 ft)
FAA landing field length (S/L):
LR, HC 2,575 m (8,450 ft)
F 2,834 m (9,300 ft)
Design range (FAR international reserves):
LR 8,000 nm (14,825 km; 9,200 miles)
HC 7,200 nm (13,343 km; 8,280 miles)
F 4,400 nm (8,154 km; 5,060 miles)

HIGH SPEED COMMERCIAL TRANSPORT

Douglas study suggests as much as 20 per cent of traffic demand in 2000 could be met by High Speed Commercial Transports. In-house and NASA contract research under way for last five years; $8 million five-year NASA Langley study contract awarded 1991; certification not before 2005; in service 2006; 300 passengers; range 5,000 nm (9,266 km; 5,757 miles); potential sales, 500 to 1,500; main operating areas Pacific Rim and North Atlantic.

Douglas also member of international SST study group (see SCTICSG in International section) formed March 1990; current research focusing on 300-passenger Mach 1.6-2.4 aircraft with range of 5,825 nm (10,780 km; 6,700 miles); target dates are first flight 2003, certification no earlier than 2005 and in service 2006.

MCDONNELL DOUGLAS AEROSPACE (Division of McDonnell Douglas Corporation)
Box 516, St Louis, Missouri 63166
Telephone: 1 (314) 232 0232
Fax: 1 (314) 234 3826
EXECUTIVE VICE-PRESIDENT: John P. Capellupo
VICE-PRESIDENT/GENERAL MANAGER F-15:
Peter J. von Minden

VICE-PRESIDENT/GENERAL MANAGER F/A-18: Larry A. Lemke
DIRECTOR OF COMMUNICATIONS: James R. Reed
MDA comprises:
McDonnell Aircraft Company
Follows this entry
McDonnell Douglas Helicopter Company
Follows Aircraft

McDonnell Douglas Aerospace Transport Aircraft
Follows Helicopters
McDonnell Douglas Missile Systems Company
McDonnell Douglas Space Systems Company
McDonnell Douglas Electronic Systems Company
See previous editions of *Jane's*

MCDONNELL AIRCRAFT COMPANY (Division of McDonnell Douglas Aerospace)
Box 516, St Louis, Missouri 63166
Telephone: 1 (314) 232 0232
Fax: 1 (314) 234 3826

Development and production continues to concentrate on F/A-18 Hornet, T-45 Goshawk, F-15 Eagle and AV-8B Harrier II. McDonnell teamed with Vought in study of potential replacement for A-12A Advanced Tactical Aircraft, designated A-X (see US Navy entry). Joined with British Aerospace, February 1992, to study joint advanced STOVL fighter (see International section).

An F/A-18 Hornet (164712) delivered to VFA-137 squadron, USN, on 10 February 1993 was the 10,001st tactical military jet built at St Louis (excluding F-4 and F-15 assembly in Japan), comprising 62 Phantom Is, 895 Banshees, 521 Demons, two XF-85s, two XF-88s, 807 Voodoos, 5,057 Phantom IIs, 1,225 Eagles, 1,152 Hornets, 268 Harrier IIs and 10 production Goshawks.

MCDONNELL DOUGLAS/BAe AV-8B HARRIER II
Details of Harrier II and current improvement programmes are given in International section.

MCDONNELL DOUGLAS/BAe T45TS
US Navy designation: T-45A Goshawk
Details under McDonnell Douglas/BAe entry in International section.

MCDONNELL DOUGLAS F-15C/D EAGLE
Israeli Defence Force name: Baz (Falcon)
TYPE: Twin-turbofan air superiority fighter with secondary attack role.
PROGRAMME: First flight of YF-15 27 July 1972; first F-15C (78-468) 26 February 1979; first F-15D 19 June 1979; P&W F100-PW-220 standard since 1985; last of 894 USAF F-15A/B/C/Ds delivered 3 November 1989; production restarted during 1991 to produce five for Israel and 12 for Saudi Arabia; now concentrated on F-15E (which see).
CURRENT VERSIONS: **F-15A:** Initial single-seat version (see 1981-82 *Jane's*).
F-15B: Initial two-seat operational training version; first flight 7 July 1973.
F-15C: Became standard single-seat production version from June 1979. *Detailed description applies to this version.*
F-15D: Standard two-seat production version from June 1979.
F-15E: See separate entry.
F-15J and DJ: Single- and two-seat versions for JASDF; built by Mitsubishi in Japan (which see) for JASDF.
MSIP: Multi-staged improvement programme, first funded 1983; testing began December 1984; first production MSIP F-15C unveiled 20 June 1985; all other F-15Cs to be retrofitted. MSIP included central computer capacity multiplied by four and processing speed three times as fast; original armament control system panel replaced by single Honeywell multi-purpose colour display which, linked to the Dynamic Controls Corporation AN/AWG-27 programmable armament control system computer, allows for advanced versions of AIM-7, AIM-9, and AIM-120 AMRAAM. Other improvements include Northrop enhanced AN/ALQ-135 internal countermeasures set, Loral AN/ALR-56C RWR, Tracor AN/ALE-45 chaff/flare dispenser, and Magnavox EW warning system. Provision for JTIDS Class 2 terminals; final 43 production F-15C/D. MSIP included AN/APG-70 radar with memory increased to 1,000K and processing speed trebled.
CUSTOMERS: Including prototypes, 1,017 F-15A/Bs and C/Ds produced up to November 1989, when production temporarily halted; 366 F-15As, 58 F-15Bs, 409 F-15Cs and 61 F-15Ds for US Air Force; 19 F-15As (plus four ex-USAF), two F-15Bs, 24 F-15Cs and two F-15Ds for Israel Defence Force; 46 F-15Cs and 16 F-15Ds for Royal Saudi Air Force; and two F-15Js and 12 F-15DJs for Japan ASDF. Further 17 built in 1991-92, of which nine F-15Cs and three F-15Ds for Saudi Arabia between August 1991 and July 1992 under Peace Sun IV programme (in addition, USAF transferred 20 F-15Cs and four F-15Ds as emergency aid September-October 1990); Israel received five F-15Ds under Peace Fox IV programme between 1 May and 26 August 1992, when final two first-generation F-15s (90-0278 and 90-0279) delivered; additional 20 F-15As and five F-15Bs from USAF stocks delivered to Israel between October 1991 and 29 July 1992. Also in 1992, first early production Eagles redesignated **GF-15A/B** and relegated to static instruction at 3700th TTW, Sheppard Technical Training Center. Total US production 1,034, plus F-15E and F-15S; including FY 1993 contracts, 188 JASDF F-15s ordered from Mitsubishi; total commitments 1,222. Air combat claims by F-15s: Israel 56.5; USAF 36; RSAF two.
COSTS: $55.2 million, flyaway, Mitsubishi production in 1990.
DESIGN FEATURES: NACA 64A aerofoil section with conical camber on leading-edge; sweepback 38° 42' at quarter-chord; thickness/chord ratio 6.6 per cent at root, 3 per cent at tip; anhedral 1°; incidence 0°. Twin fins positioned to receive vortex flow off wing and maintain directional stability at high angles of attack. Straight two-dimensional external compression engine air inlet each side of fuselage. Air inlet controllers by Hamilton Standard. Air inlet actuators by National Water Lift.
All F-15Cs fitted to carry two 2,771 litre (732 US gallon; 609 Imp gallon) conformal fuel tanks (CFT) attached to fuselage sides aft of engine air intakes; same load factors as main airframe and removable in 15 minutes; CFT include stub pylons carrying four AIM-7F Sparrows or AIM-120 AMRAAMs.
FLYING CONTROLS: Plain ailerons and all-moving tailplane with dog-tooth extensions, both powered by National Water Lift hydraulic actuators; rudders have Ronson Hydraulic Units actuators; no spoilers or trim tabs; Moog boost and pitch compensator for control column; plain flaps; upward-opening airbrake panel mounted on fuselage between fins and cockpit.
STRUCTURE: Wing based on torque box with integrally machined skins and ribs of light alloy and titanium; aluminium honeycomb wingtips, flaps and ailerons; airbrake panel of titanium, aluminium honeycomb and graphite/epoxy composites skin.
LANDING GEAR: Hydraulically retractable tricycle type, with single wheel on each unit. All units retract forward. Cleveland nose and main units, each incorporating an oleo-pneumatic shock absorber. Nosewheel and tyre by Goodyear, size 22 × 6.6-10, pressure 17.93 bars (260 lb/sq in). Mainwheels by Bendix, with Goodyear tyres size 34.5 × 9.75-18, pressure 23.44 bars (340 lb/sq in). Bendix carbon heat-sink brakes. Hydro-Aire wheel braking skid control system.
POWER PLANT: Two Pratt & Whitney F100-PW-220 turbofans, each rated at 104.3 kN (23,450 lb st), installed, with afterburning for take-off. Internal fuel in structural wing tanks and six Goodyear fuselage tanks, total capacity 7,836 litres (2,070 US gallons; 1,724 Imp gallons). Simmonds fuel gauge system. Optional conformal fuel tanks attached to side of engine air intakes, beneath wing, each containing 2,771 litres (732 US gallons; 609 Imp gallons). Provision for up to three additional 2,309 litre (610 US gallon; 508 Imp gallon) external fuel tanks. Max total internal and external fuel capacity 20,441 litres (5,400 US gallons; 4,496 Imp gallons).
ACCOMMODATION: (F-15A/C) Pilot only, on McDonnell Douglas ACES II zero/zero ejection seat. Stretched acrylic canopy and windscreen.
SYSTEMS: AiResearch air-conditioning system. Three independent hydraulic systems (each 207 bars; 3,000 lb/sq in) powered by Abex engine driven pumps; modular hydraulic packages by Hydraulic Research and Manufacturing Company. Lucas Aerospace 40/50/60kVA generating system for electrical power, with Sundstrand constant speed drive units and ECDEC Corp transformer-rectifiers. The oxygen system includes a Simmonds liquid oxygen indicator. Garrett APU for engine starting, and for the provision of limited electrical or hydraulic power on the ground independently of the main engines.
AVIONICS: General Electric automatic analog flight control system standard. Hughes Aircraft AN/APG-63 X-band pulse Doppler radar (upgraded to AN/APG-70 in final 43 F-15C/Ds of MSIP), equipped since 1980 with a Hughes Aircraft programmable signal processor, provides long-range detection and tracking of small high-speed targets operating at all altitudes down to treetop level, and feeds accurate tracking information to the IBM CP-1075 128K (32K on early F-15C/Ds) central computer to ensure effective launch of the aircraft's missiles or the firing of its internal gun. For close-in dogfights, the radar acquires the target automatically and the steering/weapon system information

New blue and grey camouflage on F-15D Eagle of USAF's 60th FS/33rd FW from Eglin AFB *(Paul Jackson)*

522 USA: AIRCRAFT—McDONNELL DOUGLAS

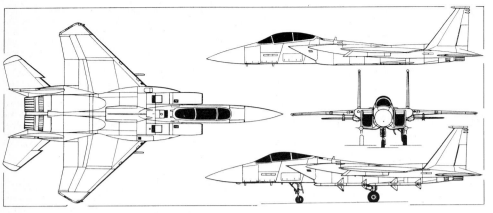

McDonnell Douglas F-15C Eagle single-seat air superiority fighter, with additional side view (top) of two-seat F-15D *(Dennis Punnett)*

Late production F-15D Eagle two-seat trainer for the Israel Defence Force

is displayed on a McDonnell Douglas Electronics AN/AVQ-20 HUD. A Teledyne Electronics AN/APX-101 IFF transponder informs ground stations and other suitably equipped aircraft that the F-15 is friendly. It also supplies data on the F-15's range, azimuth, altitude and identification to air traffic controllers. A Hazeltine AN/APX-76 IFF interrogator informs the pilot if an aircraft seen visually or on radar is friendly. A Litton reply evaluator for the IFF system operates with the AN/APX-76. A Honeywell vertical situation display set, using a CRT to present radar, electro-optical identification and attitude director indicator formats to the pilot, permits inputs received from the aircraft's sensors and the central computer to be visible to the pilot under all light conditions. Honeywell AN/ASK-6 air data computer and AN/ASN-108 AHRS, latter also serving as a backup to the Honeywell CN-1655A/ASN or Litton CN-1655/ASN ring laser gyro INS which provides the basic navigation data and is the aircraft's primary attitude reference. In addition to giving the aircraft's position at all times, the INS provides pitch, roll, heading, acceleration and speed information. Other specialised equipment for flight control, navigation and communications includes a Collins AN/ARN-118 Tacan; Collins HSI to present aircraft navigation information on a symbolic pictorial display; Collins ADF and AN/ARN-112 ILS receivers; Magnavox AN/ARC-164 UHF transceiver and UHF auxiliary transceiver. The communications sets have cryptographic capability. Dorne and Margolin glideslope localiser antenna, and Teledyne Avionics angle of attack sensors. Northrop (Defense Systems Division) Enhanced AN/ALQ-135(V) internal countermeasures set provides automatic jamming of enemy radar signals; Loral AN/ALR-56C RWR; Magnavox AN/ALQ-128 EW warning set; and Tracor AN/ALE-45 chaff dispenser. Bendix tachometer, fuel and oil indicators; Plessey feel trim actuators. IBM CO-1075C very high speed integrated circuit (VHSIC) central computer to be installed in 1994.

ARMAMENT: Provision for carriage and launch of a variety of air-to-air weapons over short and medium ranges, including four AIM-9L/M Sidewinders, four AIM-7F/M Sparrows or eight AIM-120 AMRAAM, and a 20 mm M61A1 six-barrel gun with 940 rounds of ammunition. General Electric lead-computing gyro. A Dynamic Controls Corporation AN/AWG-27 armament control system keeps the pilot informed of weapons status and provides for their management. Three air-to-surface weapon stations allow for the carriage of up to 10,705 kg (23,600 lb) of bombs, rockets or additional ECM equipment.

DIMENSIONS, EXTERNAL:
Wing span	13.05 m (42 ft 9¾ in)
Wing aspect ratio	3.01
Length overall	19.43 m (63 ft 9 in)
Height overall	5.63 m (18 ft 5½ in)
Tailplane span	8.61 m (28 ft 3 in)
Wheel track	2.75 m (9 ft 0¼ in)
Wheelbase	5.42 m (17 ft 9½ in)

AREAS:
Wings, gross	56.5 m² (608.0 sq ft)
Ailerons (total)	2.46 m² (26.48 sq ft)
Flaps (total)	3.33 m² (35.84 sq ft)
Fins (total)	9.78 m² (105.28 sq ft)
Rudders (total)	1.85 m² (19.94 sq ft)
Tailplanes (total)	10.34 m² (111.36 sq ft)

WEIGHTS AND LOADINGS:
Operating weight empty (no fuel, ammunition, pylons or external stores)	12,973 kg (28,600 lb)
Max fuel load: internal (JP4)	6,103 kg (13,455 lb)
CFTs (2, total)	4,315 kg (9,514 lb)
external tanks (3, total)	5,395.5 kg (11,895 lb)
max internal and external	15,814 kg (34,864 lb)
T-O weight (interceptor, full internal fuel, 4 Sparrows and 940 cannon rounds)	20,244 kg (44,630 lb)
T-O weight (incl three 2,309 litre; 610 US gallon; 508 Imp gallon drop tanks)	26,521 kg (58,470 lb)
Max T-O weight with CFTs	30,845 kg (68,000 lb)
Max wing loading	546.1 kg/m² (111.8 lb/sq ft)
Max power loading	147.87 kg/kN (1.45 lb/lb st)

PERFORMANCE:
Max level speed	more than Mach 2.5 (800 knots; 1,482 km/h; 921 mph CAS)
Approach speed	125 knots (232 km/h; 144 mph) CAS
Service ceiling	18,300 m (60,000 ft)
T-O run (interceptor)	274 m (900 ft)
Landing run (interceptor), without braking parachute	1,067 m (3,500 ft)
Ferry range: with external tanks, without CFTs	more than 2,200 nm (4,076 km; 2,533 miles)
with CFTs	2,600 nm (4,818 km; 2,994 miles)
Max endurance: with in-flight refuelling	15 h
unrefuelled, with CFTs	5 h 15 min
Design g limits	+9/−3

MCDONNELL DOUGLAS F-15E/S EAGLE

TYPE: Two-seat dual role attack/air superiority fighter.
PROGRAMME: Demonstration of industry-funded *Strike Eagle* prototype (71-0291) modified from F-15B, including accurate blind weapons delivery, completed at Edwards AFB and Eglin AFB during 1982; product improvements tested in *Strike Eagle*, an F-15C and an F-15D between November 1982 and April 1983, including first take-off at 34,019 kg (75,000 lb), 3,175 kg (7,000 lb) more than F-15C with conformal tanks; new weight included conformal tanks, three other external tanks and eight 500 lb Mk 82 bombs; 16 different stores configurations tested, including 2,000 lb Mk 84 bombs, and BDU-38 and CBU-58 weapons delivered visually and by radar; full programme go-ahead announced 24 February 1984; first flight of first production F-15E (86-0183) 11 December 1986; first delivery to Luke AFB, Arizona, 12 April 1988; first delivery 29 December 1988 to 4th Wing at Seymour Johnson AFB, North Carolina. Small number of F-15Es used for trials with 3246th Test Wing at Eglin AFB, Florida, and 6510th TW (412th TW from October 1992) at Edwards AFB, California; trials include 87-0180 with GE F110-GE-129 engines in place of F100s; P&W F100-PW-229 first flown in F-15E of 6510th TW 2 May 1990. To be modified for full 'Wild Weasel' anti-radar capability with AGM-88 HARM missiles by 1996.

CURRENT VERSIONS: **F-15E**: Basic version, *as detailed*.
F-15F: Proposed single-seat version, optimised for air combat; not built.
F-15H: Proposed export version, lacking specialised air-to-ground capability; supplanted by F-15S.
F-15S: Export version of F-15E, lacking some air-to-air and air-to-ground capabilities; Martin Marietta Sharpshooter replaces LANTIRN targeting pod; Saudi Arabian request for 72 aircraft approved by US government in December 1992; initially designated **F-15XP**; first funds assigned by US government on 23 December 1992; planned deliveries in 1995-98.
Saudi versions comprise 24 optimised for air-to-air missions and 48 optimised for air-to-ground; largely outfitted with F-15C/D systems; AN/APG-70 radar 'de-tuned' to match AN/APG-63 performance, lacking computerised mapping; some ECM deleted; no conformal tanks (or tangential stores carriage). Armament includes AGM-65D/G Maverick, AIM-9M and AIM-9S Sidewinder missiles, and CBU-87 and GBU-10/12 bombs. Saudi programme includes about 154 Pratt & Whitney F110-PW-229 engines.

CUSTOMERS: US funding for originally planned 392 reduced to 200; however, nine funded in FY 1992, comprising three Desert Storm loss replacements and six with proceeds of sale of 24 surplus F-15C/Ds to Saudi Arabia; deliveries to be completed 1994.

F-15E FUNDING

Batch	FY	Quantity	First aircraft
Lot I	1986	8	86-0183
Lot II	1987	42	87-0169
Lot III	1988	42	88-1667
Lot IV	1989	36	89-0471
Lot V	1990	36	90-0227
Lot VI	1991	36	91-0300
Lot VII	1992	9	
Total		209	

Initial USAF unit, 4th Wing, declared operational October 1989, currently with 334, 335 and 336 Squadrons; 461st and 555th Fighter Squadrons of 58th FW at Luke AFB, Arizona, equipped by end 1989 and others to part squadron within 57th Fighter Wing of USAF Weapons School at Nellis AFB, Nevada; 90th FS of 3rd Wing at Elmendorf, Alaska, received first F-15E on 29 May 1991; 391st FS, sole F-15E squadron in multi-type 366th Wing at Mountain Home AFB, Idaho, received first aircraft (re-allocated from early production) 6 November 1991; 492nd and 494th FS of 48th FW at Lakenheath, UK, received first aircraft on 21 February 1992 (29 received by 31 December 1992). $122 million contract placed with McDonnell Douglas by US Government on 18 December 1992 for 72 Saudi aircraft.

COSTS: $35 million, flyaway.

DESIGN FEATURES: Mission includes approach and attack at night and in all weather; main systems include new high resolution, synthetic aperture Hughes AN/APG-70 radar, wide field-of-view FLIR, Martin Marietta LANTIRN navigation (AN/AAQ-13) and targeting (AN/AAQ-14) pods beneath starboard and port air intakes respectively; air-to-air capacity with AIM-7 Sparrow, AIM-9 Sidewinder and AIM-120 AMRAAM retained; rear cockpit has four multi-purpose CRT displays for radar, weapon selection, and monitoring enemy tracking systems; front cockpit modifications include redesigned up-front controls, wide field-of-view HUD, colour CRT multi-function displays for navi-

McDONNELL DOUGLAS—AIRCRAFT: USA

McDonnell Douglas F-15E Eagle (F100-PW-229 engines) of 492nd FS/48th FW, Lakenheath, UK *(Paul Jackson)*

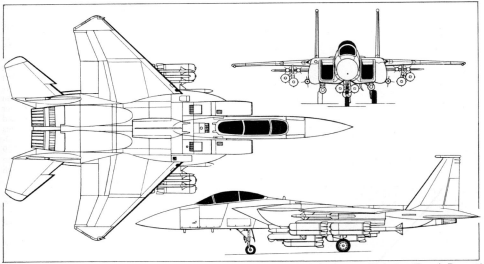

McDonnell Douglas F-15E Eagle equipped for high ordnance payload air-to-ground mission *(Dennis Punnett)*

gation, weapon delivery, moving map, precision radar mapping and terrain following. Engines have digital electronic control, engine trimming and monitoring; fuel tanks are foam-filled; more powerful generators; better environmental control.

Internal fuel capacity slightly reduced to 7,643 litres (2,019 US gallons; 1,681 Imp gallons) to accommodate avionics; conformal tanks have provision for tangential carriage of both air-to-ground and air-to-air weapons; CFT fuel is 2,737 litres (723 US gallons; 602 Imp gallons) per side; in addition up to three external drop tanks.

FLYING CONTROLS: Digital triple-redundant Lear Astronics flight control system capable of automatic coupled terrain following.

STRUCTURE: 60 per cent of normal F-15 structure redesigned to allow 9g and 16,000 hours fatigue life; superplastic forming/diffusion bonding used for upper rear fuselage, rear fuselage keel, main landing gear doors, and some fuselage fairings, plus engine bay structure.

LANDING GEAR: Bendix wheels and Michelin AIR X radial tyres on all units. Nosewheel tyre size 22 × 7.75-9, mainwheel tyres size 36 × 11-18; tyre pressure 21.03 bars (305 lb/sq in) on all units. Bendix five-rotor carbon disc brakes.

POWER PLANT: Originally as F-15C/D, but option of Pratt & Whitney F100-PW-229s or General Electric F110-GE-129s. USAF aircraft 135 onwards (90-0233), built from August 1991, have 129.4 kN (29,100 lb st) Pratt & Whitney F100-PW-229s, which also ordered for 72 Saudi F-15S. CFT and drop tank provision as for F-15C.

ACCOMMODATION: Two crew, pilot and weapon systems officer, in tandem on McDonnell Douglas ACES II zero/zero ejection seats. Single-piece, upward-hinged canopy with increased bird resistance.

SYSTEMS: As F-15C/D, but Lucas Aerospace generating system for electrical power, with Sundstrand 60/75/90kVA constant speed drive units. Litton molecular sieve oxygen generating system (MSOGS) introduced in 1991 to replace the liquid oxygen system. Fuel tanks are foam-filled. Improved environmental control.

AVIONICS: As F-15C/D, except triple redundant Lear Astronics digital flight control system with automatic terrain following standard. Hughes Aircraft AN/APG-70 I-band pulse Doppler radar provides full F-15C air-to-air capability plus high resolution synthetic aperture radar for air-to-ground; terrain following capability provided by Martin Marietta AN/AAQ-13 LANTIRN navigation pod, and FLIR imagery displayed on Kaiser ID-2394/A wide field-of-view HUD; moving map display by Bendix/King RP-341/A remote map reader; Honeywell CN-1655A/ASN ring laser gyro INS for navigation and attitude information; IBM CP-1075C very high speed integrated circuit (VHSIC) central computer introduced in 1992. Rockwell Collins miniature airborne GPS receiver is programmed for installation in 1995; Honeywell digital map system programmed to replace remote map reader in 1996.

ARMAMENT: 20 mm M61A1 six-barrel gun in starboard wing-root, with 512 rds. General Electric lead computing gyro. Provision on underwing (one per wing) and centreline pylons for air-to-air and air-to-ground weapons and external fuel tanks. Wing pylons use standard rail and launchers for AIM-9 Sidewinder and AIM-120 AMRAAM air-to-air missiles; AIM-7 Sparrow and AIM-120 AMRAAM can be carried on ejection launchers on the fuselage or on tangential stores carriers on CFTs. Maximum aircraft load (with or without CFTs) is four each AIM-7 or AIM-9, or up to eight AIM-120. Single or triple rail launchers for AGM-65 Maverick air-to-ground missiles can be fitted to wing stations only. Tangential carriage on CFTs provides for up to six bomb racks on each tank, with provision for multiple ejector racks on wing and centreline stations. Edo BRU-46/A and BRU-47/A adaptors throughout, plus two LAU-106A/As each side of lower fuselage. F-15E can carry a wide variety and quantity of guided and unguided air-to-ground weapons, including Mk 20 Rockeye (26), Mk 82 (26), Mk 84 (seven), BSU-49 (26), BSU-50 (seven), GBU-10 (seven), GBU-12 (15), GBU-15 (two), GBU-24 (five), CBU-52 (25), CBU-58 (25), CBU-71 (25), CBU-87 (25) or CBU-89 (25) bombs; SUU-20 training weapons (three); A/A-37 U-33 tow target (one); and B57 and B61 series nuclear weapons (five). An AN/AXQ-14 data link pod is used in conjunction with the GBU-15; LANTIRN pod illumination is used to designate targets for laser-guided bombs; AGM-130 powered stand-off bomb integrated in 1993; AGM-88 HARM capability in 1996. AN/AWG-27 armament control system.

DIMENSIONS, EXTERNAL: As for F-15C
AREAS: As for F-15C
WEIGHTS AND LOADINGS (F100-PW-220 engines):
Operating weight empty (no fuel, ammunition, pylons or
  external stores) 14,515 kg (32,000 lb)
Max weapon load 11,113 kg (24,500 lb)
Max fuel weight: internal (JP4) 5,952 kg (13,123 lb)
  CFTs (2, total) 4,265 kg (9,402 lb)
  external tanks (3, total) 5,395 kg (11,895 lb)
  max internal and external 15,613 kg (34,420 lb)
Max T-O weight 36,741 kg (81,000 lb)
Max zero-fuel weight 28,440 kg (62,700 lb)
Max landing weight:
  unrestricted 20,094 kg (44,300 lb)
  at reduced sink rates 36,741 kg (81,000 lb)
Max wing loading 650.5 kg/m² (133.2 lb/sq ft)
Max power loading 176.13 kg/kN (1.73 lb/lb st)
PERFORMANCE:
Max level speed at height Mach 2.5
Max combat radius 685 nm (1,270 km; 790 miles)
Max range 2,400 nm (4,445 km; 2,765 miles)

MCDONNELL DOUGLAS F-15S/MTD

TYPE: One-off F-15 STOL/manoeuvring technology demonstrator for USAF.

PROGRAMME: Ordered October 1984; first prototype F-15B (71-0290) modified; first flight 7 September 1988; first flight with two-dimensional vectoring nozzles 10 May 1989; reverse thrust applied in flight 22 March 1990; demonstrated deceleration from Mach 1.6 to Mach 0.7. From May 1991 participated in USAF/MDC integrated controls and avionics for air superiority (ICAAS) programme. Completed assigned programme 12 August 1991 (138th sortie) by demonstrating autonomous night landing capability. Aircraft to NASA/Dryden early 1993 for baseline performance assessment with conventional nozzles; to be fitted with pitch and yaw vectoring nozzles designed by Pratt & Whitney for ACTIVE (Advanced Control Technology for Integrated VEhicles) flight test programme, beginning September 1994. Additional data in 1991-92 *Jane's*.

MCDONNELL DOUGLAS F/A-18 HORNET

US Navy/Marine Corps designations: F/A-18A, B, C, D
Royal Australian Air Force designations: AF-18A and ATF-18A
Canadian Forces designations: CF-188A/B
Spanish Air Force designations: C.15 and CE.15
TYPE: Carrier-borne and land-based attack/fighter.
PROGRAMME: US Navy study of VFAX low cost, lightweight multi-mission fighter accepted Spring 1974; VFAX study terminated August 1974 and replaced by derivative of either General Dynamics YF-16 or Northrop YF-17 lightweight fighter prototypes; McDonnell Douglas proposed F-17 derivative with Northrop as associate; resultant Navy Air Combat Fighter called Hornet accepted in two versions, F-18 fighter and A-18 attack aircraft; single F/A-18 selected to fill both roles; McDonnell Douglas prime contractor and Northrop principal subcontractor for all versions agreed 1985; first Hornet flight (160775) 18 November 1978; 11 development aircraft flying by March 1980; delivery of F/A-18A/B (TF-18A designation dropped) to US Navy and Marines began May 1980 and completed 1987; millionth flying hour achieved 10 April 1990.

Enhancements to Hughes AN/APG-65 radar funded ($65.7 million) May 1990; co-operative venture with USN, Canada and one other Hornet operator; new signal and data processors, upgraded receiver/exciter. Resultant AN/APG-73 received $257 million funding in June 1991 for initial production of 12 in FY 1992; first flight in trials Hornet at St Louis on 15 April 1992; to be installed in production F/A-18s from June 1994.

CURRENT VERSIONS: **F/A-18A:** Single-seater. 371 F/A-18As and 39 two-seat F/A-18Bs (plus 11 prototypes, including two tandem-seat trainers) for USN and USMC as escort fighters to replace F-4s and as attack aircraft replacing A-7s under FYs 1979-85 contracts; first development squadron (VFA-125) formed at NAS Lemoore, California, November 1980; in service 7 January 1983 with Marine Fighter/Attack Squadron 314 at MCAS El Toro, California; first Navy squadron, VFA-113 of Pacific Fleet, October 1983; first Atlantic Fleet squadrons formed NAS Cecil Field, Florida, 1 February 1985; same month VFA-113 and VFA-25 embarked in USS *Constellation*.

US Navy and Marines squadrons with F/A-18A/Bs are tabulated. Electronic warfare training squadron VAQ-34 partly equipped with F/A-18A/Bs from 1 January 1992 at Point Mugu; use AN/ALQ-167 jamming, AN/ALE-41 chaff and AN/AST-4/6 threat simulator pods. 'Aggressor' training aircraft of VFA-127 wear desert camouflage and Iraqi/Libyan insignia. First combat experience by VFA-131, 132, 314 and 323 from USS *Coral Sea* attacking Libyan targets 1986.

F/A-18B: Combat capable two-seater; originally designated TF-18A; internal fuel capacity reduced by 6 per cent; production figures and US operating squadrons, see F/A-18A.

F/A-18C and F/A-18D: Single- and two-seat versions respectively. Purchased from FY 1986 onwards; 138 F/A-18Cs and 30 F/A-18Ds bought under 1986-87 procurements are baseline non-night attack models; overall total of 534 F/A-18C/Ds (including Night Attack – see below) funded between FY 1986 and FY 1993 (batches of 84, 84, 84, 84, 66, 48, 48 and 36); first flight of production F/A-18C (163427) 3 September 1987. *Main description applies to F/A-18C except where indicated*.

Modifications include provision for up to six AMRAAM missiles (two on fuselage and two on each outboard pylon); up to four imaging IR Maverick missiles (one on each wing pylon); provision for AN/ALQ-165 airborne self-protection jammer (ASPJ) interchangeable with AN/ALQ-126B; provision for reconnaissance equipment; upgraded stores management set with 128K memory, Intel 8086 processor, MIL-STD-1553B armament multiplex bus with MIL-STD-1760 weapons interface capability; flight incident recorder and monitoring set (FIRAMS) with integrated fuel/engine indicator, data storage set for recording maintenance and flight incidents data, signal data processor interfacing with fuel system to provide overall system control, enhanced built-in test capability and automatic CG adjustment as fuel is consumed; maintenance status panel isolating faults to card level; and new faster XN-6 mission computer with twice memory of previous XN-5; upgraded to XN-8 from FY 1991; Hughes AN/APG-73 radar (first F/A-18 flight test 15 April 1992) to become standard from June 1994. Small rectangular fence retrofitted to US Navy aircraft above LEX strake just ahead

USA: AIRCRAFT—McDONNELL DOUGLAS

McDonnell Douglas F/A-18Cs of USN squadron VFA-15 'Valions'

of wing leading-edge broadens LEX vortexes, reduces fatigue and improves directional control at angles of attack higher than 45°.

F/A-18C/D Night Attack: First flight of prototype 6 May 1988; one Night Attack F/A-18C (163985) and one D (163986) delivered to Naval Air Test Center, Patuxent River, on 1 and 14 November 1989 respectively; all F/A-18Cs and Ds delivered subsequently (FY 1988 procurement) have all-weather night attack avionics. US Navy squadrons equipped with F/A-18C Night Attack; US Marine Corps 'VMFA-' squadrons also F/A-18C Night Attack; Marines 'VMFA(AW)-' squadrons with F/A-18D Night Attack; deliveries of night-capable C versions to VMFA-312 began 8 August 1991; night-capable Ds to VMFA(AW)-121 from 11 May 1990; third Night D squadron, VMFA(AW)-225, equipped from 1 July 1991, but later to adopt reconnaissance role. Marine Corps needs six squadrons of F/A-18Ds to replace Grumman A-6Es, McDonnell Douglas OA-4s and RF-4Bs in attack, reconnaissance and forward air controller roles; this requires 96 aircraft, of which first 48 authorised in 1990; remainder in F/A-18D(RC) configuration (see below). Navy squadrons unchanged, with two-seaters used as trainers only. 1,000th Hornet was F/A-18D 164237, delivered to VMF(AW)-242 on 22 April 1991. Total 61 Night D versions delivered by 31 December 1992. USN/USMC operating squadron status in table.

Night Attack system includes pilot's night vision goggles, Hughes AN/AAR-50 thermal imaging navigation set (TINS) presenting forward view in Kaiser AN/AVQ-28 raster HUD, colour multi-function displays and Honeywell colour digital moving map; external sensor pods comprise Loral AN/AAS-38 targeting FLIR and TINS; USMC version of F/A-18D has mission-capable rear cockpit with no control column, but two sidestick weapons controllers and two 12.7 cm (5 in) MFDs in addition to colour map display; may be converted to dual control, with stick and throttles, for pilot training.

F/A-18D(RC): Simple reconnaissance version, launched 1982 and first flown 1984, included a twin-sensor package replacing guns in nose; was to be fitted with Martin Marietta Advanced Tactical Airborne Reconnaissance System (ATARS) centreline pod containing Loral AN/UPD-8 high resolution synthetic aperture SLAR supplementing nose-mounted optical and IR sensors; images transmitted in real time by data link and also viewed in rear cockpit; pod has capacity for electro-optical camera; pod, data link and processing equipment flight tested in USMC RF-4B; F/A-18D(RC) can be reconverted to fighter/attack overnight; first aircraft delivered to El Toro, February 1992; ATARS development cancelled mid-1993.

F/A-18E/F: Described separately.

AF-18A and ATF-18A: Royal Australian Air Force versions; decision to purchase 75 announced 20 October 1981; deliveries started 17 May 1985; first flight of ATF-18A assembled by AeroSpace Technologies of Australia (ASTA, which see), 26 February 1985; first flight of Australian manufactured aircraft (ATF-18A, A21-104) 3 June 1985; last of 57 single-seat and 18 two-seat Hornets

HORNET PRODUCTION

FY	Block	US Navy/Marines F/A-18A/C	US Navy/Marines F/A-18B/D	Canada CF-18A	Canada CF-18B	Australia AF-18A	Australia AF-18B	Spain EF-18A	Spain EF-18B	Kuwait KAF-18C	Kuwait KAF-18D	Running Totals
1976	(1)	3										3
	(2)	3	1									7
	(3)	3	1									11
1979	(4)	7	2									20
1980	(5)	3	4									27
	(6)	8	1									36
	(7)	9										45
1981	(8)	11	4		4							64
	(9)	17	4	1	5							91
	(10)	22	3	5	3							124
1982	(11)	17	3	7								151
	(12)	23	1	8	2							185
	(13)	19		6	1							211
1983	(14)	18	3	7	1	3	7					250
	(15)	29	2	6	2	4						293
	(16)	32		7	1	4						337
1984	(17)	24	3	7	2	7			2			382
	(18)	25	4	7	1		5	1	2			427
	(19)	27	1	7	1	3	2		4			472
1985	(20)	24	3	6	2	6		3	1			517
	(21)	26	1	8		5		5	3			565
	(22)	30		8		4	2	9				618
1986	(23)	24[1]	7[2]	8		4	2	9				672
	(24)	21	8		9	4		6				720
	(25)	17	7		6	5		2				757
1987	(26)	25	3			4		5				794
	(27)	26	2			3		5				830
	(28)	25	3			1		7				866
1988	(29)	20[3]	10[4]						2			898
	(30)	16	10						4			928
	(31)	18	10						2			958
1989	(32)	17	7									982
	(33)	23	7									1,012
	(34)	24	6									1,042
1990	(35)	22								5	3	1,072
	(36)	11	10							3	3	1,099
	(37)	13	10							2	2	1,126
1991	(38)	12	4							8		1,150
	(39)	12	4							8		1,174
	(40)	12	4							6		1,196
Subtotal		718[5]	153[6]	98	40	57	18	60	12	32	8	
1992			36									
1993			36									
Subtotal		943		138		75		72		40		1,268
Finland												64[7]
Switzerland												34[8]
TOTAL		943		138		75		72		40		1,366

Notes: [1]Begins F/A-18C production
[2]Begins F/A-18D production
[3]Begins F/A-18C Night Attack production
[4]Begins F/A-18D Night Attack production
[5]Nine prototypes, 371 As, 138 Cs and 200 C Nights
[6]Two prototypes, 39 Bs, 30 Ds and 82 D Nights
[7]Four only funded by 1993
[8]None funded by early 1993

HORNET OPERATING SQUADRONS

Unit	Base	Version	Remarks
Regular Navy			
VFA-15	Cecil Field	C Night	
VFA-22	Lemoore	C Night	
VFA-25	Lemoore	C Night	
VFA-27	Lemoore	A	
VFA-37	Cecil Field	C Night	
VFA-81	Cecil Field	C	
VFA-82	Cecil Field	C	
VFA-83	Cecil Field	C	
VFA-86	Cecil Field	C	
VFA-87	Cecil Field	C Night	
VFA-94	Lemoore	C Night	
VFA-97	Lemoore	A	
VFA-105	Cecil Field	A, B, C, D Night	
VFA-106	Cecil Field	A, C, D Night	Training
VFA-113	Lemoore	C, Night	
VFA-125	Lemoore	A, C, D Night	Training
VFA-127	Fallon	A	Aggressor
VFA-131	Cecil Field	C Night	
VFA-132	Cecil Field	A	
VFA-136	Cecil Field	C Night	
VFA-137	Lemoore	C Night	
VFA-146	Lemoore	C Night	
VFA-147	Lemoore	C Night	
VFA-151	Lemoore	C	
VFA-192	Atsugi, Japan	C	
VFA-195	Atsugi, Japan	C	
VX-4	Point Mugu	A, C, D	
VX-5	China Lake	A, C, D	
Blue Angels	Pensacola	A, B	
NATC	Patuxent River	A, C, D	
NSWC	Fallon	A, B	
NWC	China Lake	A, C, D	
NWEF	Albuquerque	A	
PMTC	Point Mugu	A, B, C	
TPS	Patuxent River	A, B	
Marine Corps			
VMFAT-101	El Toro	A, C, D Night	Training
VMFA-115	Beaufort	A	
VMFA(AW)-121	El Toro	D Night	
VMFA-122	Beaufort	A	
VMFA-212	Kanehoe Bay	C	
VMFA(AW)-224	Beaufort	D Night	1993
VMF(AW)-225	El Toro	D Night	
VMFA-232	Kanehoe Bay	C	
VMFA-235	Kanehoe Bay	C	
VMFA(AW)-242	El Toro	D Night	
VMFA-251	Beaufort	A	
VMFA-312	Beaufort	C Night	
VMFA-314	El Toro	A	
VMFA-323	El Toro	A	
VMFA(AW)-332	Beaufort	D Night	1993
VMFA-451	Beaufort	A	
VMFA(AW)-533	Beaufort	D Night	
Naval Reserve			
VFA-203	Cecil Field	A	
VFA-204	New Orleans	A	
VFA-303	Lemoore	A	
VFA-305	Point Mugu	A	
Marine Corps Reserve			
VMFA-112	Dallas	A	
VMFA-134	El Toro	A	
VMFA-142	Cecil Field	A	
VMFA-321	Washington	A	

Disbanded units are VFA-132 on 1 June 1992; VMFA-333 and VMFA-531 both 31 March 1992; VAQ-34 on 1 October 1993. All F/A-18A operators.

delivered 16 May 1990; Hornet replaced Dassault Mirage IIIO; units are No. 2 OCU, Williamtown, No. 3 Squadron (formed August 1986) and No. 77 Squadron at same base, and No. 75 Squadron, Tindal. Weapons include AIM-9L, AGM-88 HARM, AGM-84 Harpoon and 2,000 lb LGBs; from 1990, remaining 74 aircraft being fitted with F-18C/D type avionics and provision for Loral AN/AAS-38 IR tracking and laser designating pod.

CF-18A and B: Canada's purchase of 138 Hornets (finalised as 98 CF-18As and 40 two-seat CF-18Bs, known respectively as **CF-188A** and **CF-188B**) announced 10 April 1980; first flight of CF-18 29 July 1982; deliveries between 25 October 1982 and September 1988; CAF units were No. 410 OCU and Nos. 416 and 441 Squadrons at CFB Cold Lake, Alberta, 425 and 433 at Bagotville, Quebec, and 439 and 421 Squadrons of No. 4 Fighter Wing/No. 1 Air Division at Baden Sollingen, Germany; last aircraft left Europe 26 January 1993. Unit strengths are six As and 20 Bs at the OCU, 10 As and three Bs in home-based squadrons and 43 As and six Bs previously shared by the German-based squadrons. Differences from US Navy F/A-18 include ILS, in-flight identification spotlight in port side of fuselage, and provision for LAU-5003 19-tube pods for CRV-7 70 mm (2.75 in) high velocity submunition rockets; other weapons are AIM-7M and AIM-9L air-to-air missiles, 500 lb Mk 82 bombs and Hunting BL755 CBUs. Pilot has comprehensive cold weather land survival kit. Upgrade planned for late 1990s.

EF-18A and B: Spanish versions; purchase of 60 single-seat Hornets and 12 two-seaters, known respectively as **C.15** and **CE.15**, under Futuro Avion de Combate y Ataque programme announced 30 May 1983; financial restrictions reduced number from 84 and deliveries then stretched from 36, 24 and 12 from 1986 to 1988 to 11, 26, 15, 12 and eight from 1986 to 1990; maintenance performed in Spain by CASA, which also works on Canadian Hornets in Europe and USN Hornets with 6th Fleet in Mediterranean; first flight Spanish Hornet 4 December 1985; deliveries began 10 July 1986; all 12 trainers delivered by early 1987; armament includes AIM-7F and AIM-9L air-to-air missiles, AGM-84 Harpoon, AGM-88 HARM and free-fall bombs; AIM-120 AMRAAM ordered 1990. First 36 aircraft have Sanders AN/ALQ-126B deception jammers ordered in 1987; final 36 received Northrop AN/ALQ-162(V) systems. Units equipped are Ala de Caza 15 (15th Fighter Wing) formed at Zaragoza December 1985 and operational December 1987, with 30 A and six B shared between Escuadrones 151 and 152; Ala de Caza 12 at Torrejon (Escuadrones 121 and 122) completed re-equipping in July 1990. One attrition replacement required. Agreement in 1992 to upgrade Spanish aircraft to F-18A+/B+ standard, close to F/A-18C/D; of 71 available, 46 to be converted by McDonnell Douglas between September 1992 and March 1994; final 25 by CASA by 1995. Engineering Change Proposal 287 includes later mission and armament computers, databuses, data-storage set, new wiring, pylon modifications and software.

CUSTOMERS: See table of production. Total 1,149 Hornets delivered by January 1993, of which 839 to US forces; USN proposed procurement of 36 in FY 1994 towards total of 1,156 production Hornets required. 1.86 million flying hours to end 1992 (all operators).

In addition to US Navy/Marine Corps, Australia, Canada and Spain (see Current Versions), Switzerland selected 26 F/A-18Cs and eight F/A-18Ds powered by GE F404-GE-402 engines in October 1988, as its Neue Jagdflugzeug to replace F-5Es; purchase ratified by parliament 12 June 1992; confirmed by referendum 6 June 1993; deliveries planned from early 1995. Kuwaiti contract signed September 1988 for 32 F/A-18Cs and eight F/A-18Ds together with AGM-65G Maverick, AGM-84 Harpoon, AIM-7F Sparrow and AIM-9L Sidewinder; first flight 19 September 1991; first three delivered to No. 25 Squadron 25 January 1992; 20 more before year end; final delivery to No 9 Sqdn September 1993; F404-GE-402 power plants. Finland selected seven F/A-18Ds (built by McAir) and 57 F/A-18Cs (for assembly from kits by Valmet); announcement 6 May 1992; letter of offer signed 5 June 1992; -402 engines; initial procurement of four aircraft in 1993. Malaysia to order eight F/A-18Ds.

COSTS: $25 million, flyaway unit cost, 1991. $55,632 million (1991) US programme, 1,167 aircraft.

DESIGN FEATURES: Sharp-edged, cambered leading-edge extensions (LEX), slots at fuselage junction and outward-canted twin fins are designed to produce high agility and docile performance at angles of attack over 50°; wings have 20° sweepback at quarter-chord; wings fold up 90° at inboard end of ailerons, even on land-based F/A-18s; landing gear designed for unflared landings on runways as well as on carriers.

FLYING CONTROLS: Full digital fly-by-wire controls using ailerons and tailerons for lateral control, plus flaps in flaperon form at low airspeeds; leading- and trailing-edge flaps scheduled automatically for high manoeuvrability, fast cruise and slow approach speed; both rudders turned in at take-off and landing to provide extra nose-up trim effort; fly-by-wire returns towards 1g flight if pilot releases controls; lateral and then directional control progressively washed out as angle of attack reaches extreme values;

First KAF-18D Hornet for the air force of Kuwait

526 USA: AIRCRAFT—McDONNELL DOUGLAS

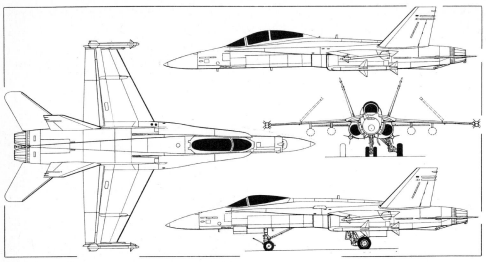

McDonnell Douglas F/A-18C Hornet with additional side view (top) of F/A-18D *(Dennis Punnett)*

F/A-18D Night Attack Hornet of Marine Corps squadron VMFA(AW)-242 'Batmen' firing rockets

Tiger-striped CF-18A Hornet 188764, among last Canadian aircraft to leave Europe, January 1993 *(Paul Jackson)*

height, heading and airspeed holds provided in fly-by-wire system; US Navy aircraft can land automatically using carrier-based guidance system; airbrake panel located on top of fuselage, between fins. Bertea hydraulic actuators for trailing-edge flaps; Hydraulic Research actuators for ailerons; National Water Lift actuators for tailerons.
STRUCTURE: Multi-spar wing mainly of light alloy, with graphite/epoxy inter-spar skin panels and trailing-edge flaps; tail surfaces mainly graphite/epoxy skins over aluminium honeycomb core; graphite/epoxy fuselage panels and doors; titanium engine firewall. Northrop produces rear and centre fuselages; assembly and test at McDonnell Douglas St Louis factory; CASA produces horizontal tail surfaces, flaps, leading-edge extensions, speedbrakes, rudders and rear side panels for all F/A-18s.
LANDING GEAR: Dowty retractable tricycle type, with twin-wheel nose and single-wheel main units. Nose unit retracts forward, mainwheels rearward, turning 90° to stow horizontally inside the lower surface of the engine air ducts. Bendix wheels and brakes. Nosewheel tyres size 22 × 6.6-10, 20 ply, pressure 24.13 bars (350 lb/sq in) for carrier operations, 10.34 bars (150 lb/sq in) for land operations. Mainwheel tyres size 30 × 11.5-14.5, 24 ply, pressure 24.13 bars (350 lb/sq in) for carrier operations, 13.79 bars (200 lb/sq in) for land operations. Ozone nosewheel steering unit. Nose unit towbar for catapult launch. Arrester hook, for carrier landings, under rear fuselage.
POWER PLANT: Two General Electric F404-GE-400 low bypass turbofans initially, each producing approx 71.2 kN (16,000 lb st) with afterburning. F404-GE-402 EPE (Enhanced Performance Engine) standard from early 1992; rated at approx 78.3 kN (17,600 lb st). Self-sealing fuel tanks and fuel lines; foam in wing tanks and fuselage voids. Internal fuel capacity (JP5) approx 6,061 litres (1,600 US gallons; 1,333 Imp gallons). Provision for up to three 1,250 litre (330 US gallon; 275 Imp gallon) external tanks. Canadian Hornets carry three 1,818 litre (480 US gallon; 400 Imp gallon) tanks. Flight refuelling probe retracts into upper starboard side of nose. Simmonds fuel gauging system. Fixed ramp air intakes.
ACCOMMODATION: Pilot only, on Martin-Baker SJU-5/6 ejection seat, in pressurised, heated and air-conditioned cockpit. Upward opening canopy, with separate windscreen, on all versions. Two pilots in F/A-18B and USN F/A-18D; pilot and Naval Flight Officer in USMC F/A-18D.
SYSTEMS: Two completely separate hydraulic systems, each at 207 bars (3,000 lb/sq in). Max flow rate 212 litres (56 US gallons; 46.6 Imp gallons)/min. Bootstrap type reservoir, pressure 5.86 bars (85 lb/sq in). AiResearch air-conditioning system. General Electric electrical power system. Garrett GTC36-200 APU for engine starting and ground pneumatic, electric and hydraulic power. Oxygen system. Fire detection and extinguishing systems.
AVIONICS: Include an automatic carrier landing system (ACLS) for all-weather carrier operations; a Hughes Aircraft AN/APG-65 multi-mode digital air-to-air and air-to-ground tracking radar, with air-to-air modes which include velocity search (VS), range while search (RWS), track while scan (TWS), which can track ten targets and display eight to the pilot, and raid assessment mode (RAM). Honeywell digital moving map display; Smiths Industries multi-purpose colour cockpit display; Collins AN/ARN-118 Tacan, AN/ARC-182 UHF/VHF com and DF-301E UHF/DF; Magnavox AN/ALR-50 and Littor AN/ALR-67 RWRs; GEC-Marconi Type 117 laser desig nator; Goodyear AN/ALE-39 chaff dispenser (AN/ALE-47 from FY 1993); Sanders AN/ALQ-126F ECM; Harris AN/ASW-25 radio data link; Eator AN/ARA-63 receiver/decoder; GEC-Marconi FID 2035 horizontal situation display; Bendix/King HSI; J. E. T ID-1791/A flight director indicator (ITT/Westinghouse AN/ALQ-165 airborne self-protection jammer [ASPJ cancelled in 1992; replacement sought); General Electric quadruple-redundant fly-by-wire flight control system with direct electrical backup to all surfaces and direc mechanical backup to tailerons; two Control Data AN/AYK-14 digital computers; Litton AN/ASN-130A inertial navigation system (plus GPS from FY 1993); two Kaiser multi-function CRTs, central GEC-Marconi-Ben dix/King CRT and Kaiser AN/AVQ-28 HUD; Conrac communications system control; Normalair-Garrett digita data recorder for Bendix/King maintenance recording sys tem; flight incident recording and monitoring system (FIRAMS); Smiths standby altimeter; and Kearflex standby airspeed indicator, standby vertical speed indi cator, and cockpit pressure altimeter. Night Attack F/A-18 has Hughes AN/AAR-50 thermal imaging navigation se and Loral AN/AAS-38 NITE Hawk targeting FLIR, Hon eywell colour digital moving map display (replacing film strip map) and provision for GEC-Marconi Cat's Eyes NVGs.
ARMAMENT: Nine external weapon stations, comprising two wingtip stations for AIM-9 Sidewinder air-to-air missiles two outboard wing stations for an assortment of air-to-air or air-to-ground weapons, including AIM-7 Sparrows AIM-9 Sidewinders, AIM-120 AMRAAMs (launch trials by VX-4 in 1992), AGM-84 Harpoons and AGM-65F Maverick missiles; two inboard wing stations for externa fuel tanks, air-to-ground weapons or Brunswick ADM-141 TALD tactical air-launched decoys; two nacelle fuselage stations for Sparrows or Martin Marietta AN/ASQ-173 laser spot tracker/strike camera (LST/SCAM) o AN/AAS-38 and AN/AAR-50 sensor pods (see Avionics) and a centreline fuselage station for external fuel or weap ons. Air-to-ground weapons include GBU-10 and -12 laser guided bombs, Mk 82 and Mk 84 general purpose bombs and CBU-59 cluster bombs. An M61A1 20 mm six-barre gun, with 570 rounds, is mounted in the nose and has a McDonnell Douglas director gunsight, with a conventional sight as backup.

DIMENSIONS, EXTERNAL:
Wing span	11.43 m (37 ft 6 in)
Wing span over missiles	12.31 m (40 ft 4¾ in)
Wing chord: at root	4.04 m (13 ft 3 in)
at tip	1.68 m (5 ft 6 in)
Wing aspect ratio	3.5
Width, wings folded	8.38 m (27 ft 6 in)
Length overall	17.07 m (56 ft 0 in)
Height overall	4.66 m (15 ft 3½ in)
Tailplane span	6.58 m (21 ft 7¼ in)
Distance between fin tips	3.60 m (11 ft 9½ in)
Wheel track	3.11 m (10 ft 2½ in)
Wheelbase	5.42 m (17 ft 9½ in)

AREAS:
Wings, gross	37.16 m² (400.0 sq ft)
Ailerons (total)	2.27 m² (24.4 sq ft)
Leading-edge flaps (total)	4.50 m² (48.4 sq ft)
Trailing-edge flaps (total)	5.75 m² (61.9 sq ft)
Fins (total)	9.68 m² (104.2 sq ft)
Rudders (total)	1.45 m² (15.6 sq ft)
Tailerons (total)	8.18 m² (88.1 sq ft)

WEIGHTS AND LOADINGS:
Weight empty	10,455 kg (23,050 lb)
Max fuel weight: internal (JP5)	4,926 kg (10,860 lb)
external: F/A-18 (JP5)	3,053 kg (6,732 lb)
CF-18 (JP4)	4,246 kg (9,360 lb)
Max external stores load	7,031 kg (15,500 lb)
T-O weight: fighter mission	16,651 kg (36,710 lb)
attack mission	23,541 kg (51,900 lb)
max	approx 25,401 kg (56,000 lb)
Max wing loading (attack mission)	600.83 kg/m² (123.06 lb/sq ft)
Max power loading (attack mission)	156.80 kg/kN (1.54 lb/lb st)

PERFORMANCE:
Max level speed	more than Mach 1.8
Max speed, intermediate power	more than Mach 1.0
Approach speed	134 knots (248 km/h; 154 mph)
Acceleration from 460 knots (850 km/h; 530 mph) to 920 knots (1,705 km/h; 1,060 mph) at 10,670 m (35,000 ft)	under 2 min
Combat ceiling	approx 15,240 m (50,000 ft)
T-O run	less than 427 m (1,400 ft)
Min wind-over-deck:	
launching	35 knots (65 km/h; 40 mph)
recovery	19 knots (35 km/h; 22 mph)
Combat radius, interdiction, hi-lo-hi	290 nm (537 km; 340 miles)
Combat endurance, CAP 150 nm (278 km; 173 miles) from aircraft carrier	1 h 45 min
Ferry range, unrefuelled	more than 1,800 nm (3,336 km; 2,073 miles)

MCDONNELL DOUGLAS F/A-18E/F HORNET

TYPE: Single/two-seat carrier-based strike/attack and maritime air supremacy aircraft.

PROGRAMME: Proposed 1991 as replacement for cancelled GD/MDC A-12 and follow-on to early F/A-18As and other USN tactical aircraft as they retire; development funding approved by Congress for FY 1992; $3,715 million development contract signed by USN on 7 December 1992; covers seven test aircraft (five Es; two Fs) and three static airframes, plus 7½ years of engineering and support activities. $754 million award to GE for F414 engine development in 1992. First flight due late 1995.

CURRENT VERSIONS: **F/A-18E**: Single-seat.
F/A-18F: Two-seat combat-capable trainer.

CUSTOMERS: US Navy. Seven prototypes; approval (subject to review) for 12, 12 and 18 production aircraft in FYs 1997-99.

COSTS: Development estimated $4.8 billion (1992); $1,089 million in FY 1992 budget and $943 million in FY 1993. Navy plans 1,000 aircraft in $49,000 million programme.

DESIGN FEATURES: Stretched versions of F/A-18C/D; gross landing weight increased by 4,536 kg (10,000 lb); 0.86 m (2 ft 10 in) fuselage plug; wings photometrically increased in size to provide 9.29 m² (100.0 sq ft) extra area and 1.31 m (4 ft 3½ in) span increase; wings 2.5 cm (1 in) deeper at root; larger horizontal tail surfaces; LERX size substantially increased in early 1993 (from total 5.8 m²; 62.4 sq ft to 7.0 m²; 75.3 sq ft - compared with 5.2 m²; 56.0 sq ft on current F/A-18C/Ds), restoring F/A-18C manoeuvre capability at 30-35° AoA; additional 1,637 kg (3,600 lb) of internal fuel and 1,406 kg (3,100 lb) of external fuel; 40 per cent extra range; further two (making 11) weapon hardpoints (Stations 2 and 10, inboard of wingtips, for AAMs and ASMs of up to 520 kg; 1,146 lb); additional survivability measures; air intakes redesigned and slewed to increase mass flow to more powerful F414-400 engines.

FLYING CONTROLS: As F/A-18C/D.
STRUCTURE: As F/A-18C/D.
LANDING GEAR: As F/A-18C/D.
POWER PLANT: Two General Electric F414-GE-400 turbofans, each rated at approx 97.86 kN (22,000 lb st). Internal fuel capacity (JP5 fuel) 8,062 litres (2,130 US gallons; 1,774 Imp gallons). Provision for three 1,818 litre (480 US gallon; 400 Imp gallon) external tanks.
ACCOMMODATION: As F/A-18C/D.
SYSTEMS: High commonality with F/A-18C/D. Leland Electrosystems power generating system; provides 60 per cent more electrical power than F/A-18C. Hamilton Standard air-conditioning; Vickers hydraulic pumps.
AVIONICS: Hughes AN/APG-73 radar; 152 mm (6 in) square tactical situation active matrix display; over 90 per cent avionics commonality with F/A-18C.
ARMAMENT: Full range of USN offensive and defensive ordnance. 'Bring-back' load increased to 4,082 kg (9,000 lb).

DIMENSIONS, EXTERNAL (approx):
Wing span over missiles	13.62 m (44 ft 8½ in)
Width, wings folded	9.32 m (30 ft 7¼ in)
Length overall	18.31 m (60 ft 1¼ in)
Height overall	4.88 m (16 ft 0 in)

AREAS:
Wings, gross	46.45 m² (500.0 sq ft)

WEIGHTS AND LOADINGS:
Weight, empty: design target	13,387 kg (29,574 lb)
specification limit	13,864 kg (30,564 lb)
Max fuel weight: internal (JP5)	6,531 kg (14,400 lb)
external (JP5)	4,436 kg (9,780 lb)
Max external stores load	8,051 kg (17,750 lb)
T-O weight, attack mission	29,937 kg (66,000 lb)
Max wing loading	620.0 kg/m² (127.0 lb/sq ft)
Max power loading	147.1 kg/kN (1.44 lb/lb st)

Artist's impression of the McDonnell Douglas F/A-18E with enlarged LERX

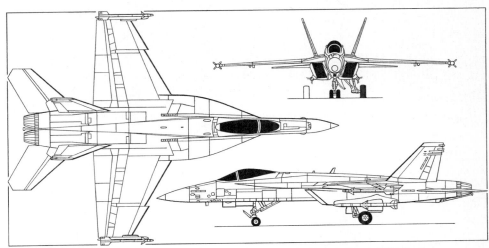

McDonnell Douglas F/A-18E advanced version of the Hornet (*Jane's/Mike Keep*)

PERFORMANCE (estimated):
Max level speed at altitude	more than Mach 1.8
Combat ceiling	15,240 m (50,000 ft)
Min wind-over-deck:	
launching	30 knots (56 km/h; 34.5 mph)
recovery	15 knots (28 km/h; 17.5 mph)

Combat radius specification: interdiction with four 1,000 lb bombs, two Sidewinders and two 1,818 litre (480 US gallon; 400 Imp gallon) external tanks, navigation FLIR and targeting FLIR:
hi-lo-lo-hi	390 nm (722 km; 449 miles)
fighter escort with two Sidewinders and two AMRAAM	410 nm (760 km; 472 miles)

Combat endurance: maritime air superiority, six AAMs, three 1,818 litre (480 US gallon; 400 Imp gallon) external tanks, 150 nm (278 km; 173 miles) from aircraft carrier 2 h 15 min

---

MCDONNELL DOUGLAS HELICOPTER COMPANY (Division of McDonnell Douglas Aerospace)

5000 East McDowell Road, Mesa, Arizona 85205-9797
Telephone: 1 (602) 891 3000
Fax: 1 (602) 891 5599
Telex: 3719337 MD HC C MESA
OTHER WORKS: Culver City, California 90230
Telephone: 1 (213) 305 5000
Telex: 182436 HU HELI C CULV
PRESIDENT: Dean C. Borgman
SENIOR VICE-PRESIDENT, OPERATIONS: Patrick R. McGinnis
SENIOR VICE-PRESIDENT, PROGRAMMES: Chuck Suyo
VICE-PRESIDENTS:
Ervin J. Hunter (Military Programmes)
Andrew H. Logan (Commercial Programmes)
Al Winn (Engineering Division)
MANAGER, COMMUNICATIONS: Ken Jensen

Hughes Helicopters Inc became subsidiary of McDonnell Douglas Corporation 6 January 1984; name changed to McDonnell Douglas Helicopter Company 27 August 1985; teamed with Bell Helicopter Textron for US Army Light Helicopter (LH) competition (unsuccessful); Chain Gun systems manufactured at Culver City.

Main company base at Mesa, Arizona, with 52,955 m² (570,000 sq ft) AH-64 Apache assembly and testing factory and another 123,980 m² (1,334,500 sq ft) completed in 1986; MD 500/530 production line transferred to Mesa 1986-87; workforce 3,800 in mid-1993. Single-engined helicopter deliveries totalled 114 in 1991-92 (57 in each year); 1992 total included 38 MD 520Ns.

Model 300 helicopter design rights sold to Schweizer Aircraft Corporation (which see) at Elmira, New York, 1986, following licence production by Schweizer since 1983; McDonnell Douglas helicopters produced under licence by RACA, Argentina (civil 500D and 500E), Kawasaki, Japan (civil and military 500D), Korean Air (civil and military 500D and 500E, excluding TOW variants, and fuselages for all MD 500s sold worldwide), Agusta, Italy (500D, 500E and 530F civil variants); Agusta licence extended mid-1993 to include MD 520N (first Italian deliveries mid-1994).

MCDONNELL DOUGLAS MD 500/530

TYPE: Single-engined utility helicopter.
PROGRAMME: First flight MD 500E (N5294A) 28 January 1982; first flight MD 530F 22 October 1982.
CURRENT VERSIONS: **MD 500E**: Replaced MD 500D in production 1982; deliveries started December 1982; Allison 250-C20R became optional replacement for standard 250-C20B in late 1988; window area of forward canopy increased in 1991 model. MD 500E introduced many cabin improvements including more space for front and rear seat occupants, lower bulkhead between front and rear seats, T tail and optional four-blade Quiet Knight tail rotor.

MD 530F Lifter: Powered by Allison 250-C30; transmission rating increased from 280 kW (375 shp) to 317 kW (425 shp) from 11 July 1985; diameter of main rotor increased by 0.3 m (1 ft 0 in) and of tail rotor by 5 cm (2 in); cargo hook kit for 907 kg (2,000 lb) external load available; certificated 29 July 1983; first delivery 20 January 1984. *Data below refer to MD 500E and 530F, except where indicated.*

MD 500/530 Defender: Described separately.

CUSTOMERS: More than 4,550 MD 500/530 series produced by early 1993; 57 MD 500/520/530 series helicopters delivered in 1992, including 38 MD 520Ns (which see); 500th MD 500E delivered 1 April 1992. Recent orders include 10 MD 530Fs for Mexican Air Force.

COSTS: Typical MD 500E $700,000; MD 530F $925,000.

DESIGN FEATURES: Fully articulated five-blade main rotor with blades retained by stack of laminated steel straps; blades can be folded after removing retention pins; two-blade tail rotor with optional X-pattern four-blade Quiet Knight tail rotor to reduce external noise; optional high skid landing gear to protect tail rotor in rough country; protective skid on base of lower fin; narrow-chord fin with high-set tailplane and endplate fins introduced with MD 500D. Main rotor rpm (500E/530F) 492/477 normal; main rotor tip speed 207-208 m (680-684 ft)/s; tail rotor rpm, 2,933/2,848.

528 USA: AIRCRAFT—McDONNELL DOUGLAS

McDonnell Douglas MD 500E five/seven-seat utility and executive helicopter

FLYING CONTROLS: Plain mechanical without hydraulic boost. Pilot sits on left instead of normal right-hand seating.

STRUCTURE: Airframe based on two A frames from rotor head to landing gear legs, enclosing rear-seat occupants; front-seat occupants protected within straight line joining rotor hub and forward tips of landing skids; engine mounted inclined in rear of fuselage pod, with access through clam-shell doors; main rotor blades have extruded aluminium spar hot-bonded to wraparound aluminium skin; tail rotor blades have swaged tubular spar and metal skin.

LANDING GEAR: Tubular skids carried on McDonnell Douglas oleo-pneumatic shock absorbers. Utility floats, snow skis and emergency inflatable floats optional.

POWER PLANT: MD 500E powered by 313 kW (420 shp) Allison 250-C20B or 335.6 kW (450 shp) 250-C20R turboshaft, derated in both cases to 280 kW (375 shp) for T-O; max continuous rating 261 kW (350 shp). MD 530F has 485 kW (650 shp) Allison 250-C30 turboshaft, derated to 317 kW (425 shp) for take-off and 261 kW (350 shp) max continuous. MCP transmission rating 261 kW (350 shp). Two interconnected bladder fuel tanks with combined usable capacity of 232 litres (61.3 US gallons; 51 Imp gallons). Self-sealing fuel tank optional. Refuelling point on starboard side of fuselage. Auxiliary fuel system, with 79.5 litre (21 US gallon; 17.5 Imp gallon) internal tank, available optionally. Oil capacity 5.7 litres (1.5 US gallons; 1.2 Imp gallons).

ACCOMMODATION: Forward bench seat for pilot and two passengers, with two or four passengers, or two litter patients and one medical attendant, in rear portion of cabin. Low-back front seats and individual rear seats, with fabric or leather upholstery, optional. Baggage space, capacity 0.31 m³ (11 cu ft), under and behind rear seat in five-seat form. Clear space for 1.19 m³ (42 cu ft) of cargo or baggage with only three front seats in place. Two doors on each side. Interior soundproofing optional.

SYSTEMS: Aero Engineering Corporation air-conditioning system or Fargo pod-mounted air-conditioner optional.

AVIONICS (MD 500E): Optional avionics include dual Bendix/King KY 195 com, KX 175 nav/com, KR 85 ADF, and KT 76 transponder; dual Collins VHF-251 com, VHF-251/351 nav/com, IND-350 nav indicator, ADF-650 ADF, and TDR-950 transponder; intercom system, headsets, microphones; and public address system.

EQUIPMENT: Standard equipment includes basic VFR instruments and night flying lighting. Optional equipment includes shatterproof glass, heating/demisting system, radios and intercom, attitude and directional gyros, rate of climb indicator, nylon mesh seats, dual controls, cargo hook, cargo racks, underfuselage cargo pod, heated pitot tube, extended landing gear, blade storage rack, litter kit, emergency inflatable floats, inflated utility floats and 30 million candlepower Spectrolab SX-16 Nightsun searchlight.

DIMENSIONS, EXTERNAL:
Main rotor diameter: 500E	8.05 m (26 ft 5 in)
530F	8.33 m (27 ft 4 in)
Main rotor blade chord	0.171 m (6¾ in)
Tail rotor diameter: 500E	1.37 m (4 ft 6 in)
530F	1.42 m (4 ft 8 in)
Distance between rotor centres:	
500E	4.67 m (15 ft 4 in)
530F	4.88 m (16 ft 0 in)
Length overall, rotors turning:	
500E	8.61 m (28 ft 3 in)
530F	8.97 m (29 ft 5 in)
Length of fuselage	7.49 m (24 ft 7 in)
Height to top of rotor head (standard skids: for extended skids add 0.27 m; 11¾ in)	2.67 m (8 ft 9 in)
Tailplane span	1.65 m (5 ft 5 in)
Skid track (standard)	1.91 m (6 ft 3 in)
Cabin doors (each): Height	1.13 m (3 ft 8½ in)
Max width	0.76 m (2 ft 6 in)
Height to sill: 500E	0.79 m (2 ft 7 in)
530F	0.76 m (2 ft 6 in)
Cargo compartment doors (each):	
Height	1.12 m (3 ft 8¼ in)
Width	0.88 m (2 ft 10½ in)
Height to sill: 500E	0.71 m (2 ft 4 in)
530F	0.66 m (2 ft 2 in)

DIMENSIONS, INTERNAL:
Cabin: Length	2.44 m (8 ft 0 in)
Max width	1.31 m (4 ft 3½ in)
Max height	1.52 m (5 ft 0 in)

AREAS:
Main rotor blades (each): 500E	0.62 m² (6.67 sq ft)
530F	0.65 m² (6.96 sq ft)
Tail rotor blades (each): 500E	0.063 m² (0.675 sq ft)
530F	0.066 m² (0.711 sq ft)
Main rotor disc: 500E	50.89 m² (547.81 sq ft)
530F	54.58 m² (587.50 sq ft)
Tail rotor disc: 500E	1.53 m² (16.47 sq ft)
530F	1.65 m² (17.72 sq ft)
Fin	0.56 m² (6.05 sq ft)
Tailplane	0.76 m² (8.18 sq ft)

WEIGHTS AND LOADINGS:
Weight empty: 500E	655 kg (1,445 lb)
530F	717 kg (1,580 lb)
Max normal T-O weight: 500E	1,361 kg (3,000 lb)
530F	1,406 kg (3,100 lb)
Max overload T-O weight:	
500E, 530F	1,610 kg (3,550 lb)
Max gross weight, external load:	
530F	1,701 kg (3,750 lb)
Max normal T-O disc loading:	
500E	26.76 kg/m² (5.48 lb/sq ft)
530F	25.78 kg/m² (5.28 lb/sq ft)
Max normal T-O power loading:	
500E	4.87 kg/kW (8.00 lb/shp)
530F	4.44 kg/kW (7.29 lb/shp)

PERFORMANCE (A: 500E, B: 530F, at max normal T-O weight):
Never-exceed speed (V_NE) at S/L:	
A, B	152 knots (282 km/h; 175 mph)
Max cruising speed at S/L:	
A	134 knots (248 km/h; 154 mph)
B	133 knots (246 km/h; 153 mph)
Max cruising speed at 1,525 m (5,000 ft):	
A	132 knots (245 km/h; 152 mph)
B	134 knots (248 km/h; 154 mph)
Econ cruising speed at S/L:	
A	129 knots (239 km/h; 149 mph)
B	131 knots (243 km/h; 151 mph)
Econ cruising speed at 1,525 m (5,000 ft):	
A, B	123 knots (228 km/h; 142 mph)
Max rate of climb at S/L: A	536 m (1,760 ft)/min
B	631 m (2,070 ft)/min
Vertical rate of climb at S/L: A	248 m (813 ft)/min
B	446 m (1,462 ft)/min
Service ceiling: A	4,575 m (15,000 ft)
B	4,875 m (16,000 ft)
Hovering ceiling IGE: ISA: A	2,590 m (8,500 ft)
B	4,360 m (14,300 ft)
ISA+20°C: A	1,830 m (6,000 ft)
B	3,660 m (12,000 ft)
Hovering ceiling OGE: ISA: A	1,830 m (6,000 ft)
B	3,660 m (12,000 ft)
ISA+20°C: A	975 m (3,200 ft)
B	2,970 m (9,750 ft)
Range, 2 min warm-up, standard fuel, no reserves:	
A at S/L	233 nm (431 km; 268 miles)
B at S/L	202 nm (374 km; 232 miles)
A at 1,525 m (5,000 ft)	258 nm (478 km; 297 miles)
B at 1,525 m (5,000 ft)	228 nm (422 km; 262 miles)

MCDONNELL DOUGLAS 500/530 DEFENDER

US Army designations: AH-6, EH-6, MH-6

TYPE: Military derivatives of MD 500/530.

PROGRAMME: Earlier 500MD Scout Defender, TOW Defender, 500MD/ASW Defender and 500MD Defender II described in 1987-88 and earlier Jane's. Except for TOW Defender, military Defenders have same airframe as current civil 500/530; versions available detailed below.

CURRENT VERSIONS: **500MG Defender:** As 530MG, but with 313 kW (420 shp) Allison 250-C20B and MD 500E rotor system.

TOW Defender: Retains round nose of original MD 500; M65 TOW sight in nose; carries four TOW missiles available with Allison 250-C20B, 250-C20R or 250-C30.

Paramilitary MG Defender: Introduced July 1985 as low-cost helicopter suitable for police, border patrol, rescue, narcotics control and internal security use; available in either 500E or 530F configurations.

530MG Defender: Based on MD 530F Lifter; first flight of prototype/demonstrator (N530MG) 4 May 1984; designed mainly for point attack and anti-armour, but also suitable for scout, day and night surveillance, utility, cargo lift and light attack. Integrated crew station with multi-function display allows hands on lever and stick (HOLAS) control of weapon delivery, communications and flight control; HOLAS based on Racal RAMS 3000, designed for all-weather and NOE flight and connected to MIL-STD-1553B digital databus linking the processor interface unit (PIU), control and display unit (CDU) and data transfer unit (DTU); CDU used for flight planning, navigation, frequency selection and subsystem management and has its own monochrome display and keyboard; multi-function display is high definition monochrome tube with symbolic and alphanumeric capability; data input to DTU using ground loader unit inserted in cockpit receptacle.

Other equipment includes Astronics Corporation autopilot; Decca Doppler navigator with Racal Doppler velocity sensor; GEC-Marconi FIN 1110 AHRS; twin Collins VHF/UHF AM/FM radios; Bendix/King HF radio, ADF, VOR, radar altimeter and transponder; Telephonics intercom; SFENA attitude indicator. Options include mast-mounted Hughes TOW sight, FLIR, RWR, IFF, GPWS and laser ranger.

Weapons qualified or tested include TOW 2, FN Herstal pods containing 7.62 mm or 0.50 in machine-gun, and 2.75 in rockets in 7-tube or 12-tube launchers; stores attached by standard 14 in NATO racks. Future armament will include General Dynamics air-to-air Stinger and 7.62 mm McDonnell Douglas Chain Gun. Chaff and flare dispensers with automatic chaff discharge available. Both cyclic sticks have triggers for gun or rocket firing; co-pilot/gunner's visual image display has two handgrips for TOW/FLIR operation.

Nightfox: Introduced 1986 for low-cost night surveillance and military operations; equipment includes FLIR Systems Series 2000 thermal imager and night vision goggles, with same weapons as 530MG; available in both 500MG and 530MG forms.

Following versions of H-6 have been used by the US Army's 160th Special Operations Aviation Regiment at Fort Campbell, Kentucky:

EH-6B: Four conversions of OH-6A; two to AH-6C, two to MH-6B.

MH-6B: Twenty-four conversions of OH-6A, plus two from EH-6B; four to AH-6C; 10 survivors sold in 1991-92.

AH-6C: Eleven conversions of OH-6A, plus six from EH/MH-6. Six lost; others remain.

MH-6C: Three conversions of OH-6A; all remain.

EH-6E: Three new-build; all to MH-6H.

MH-6E: Fifteen new-build; 10 to MH-6H; two to MH-6J; one lost.

AH-6F: Eight new-build; six to AH-6G; two lost.

AH-6G: Four new-build, plus six from AH-6F.

MH-6H: Two new-build (FY 1988), plus 13 conversions from EH/MH-6E; two to MH-6J; two lost.

AH-6J: Twenty or more new-build from FYs 1988-91. Procurement in service by mid-1992; four or more converted from MH-6E/H.

MH-6J: Small number funded in FYs 1990-91.

AH-6F and MH-6E based on MD 500MG; AH-6G and MH-6F/H based on MD 530MG; all can carry 7.62 mm Minigun, machine-gun and rocket pods, with provision for air-to-air Stinger; MH-6E, MH-6F and MH-6H have multi-function displays and FLIR used in association with night vision goggles.

Re-equipment of 160th SOAR with NOTAR variant officially began with four conversions and two new AH-6/MH-6s, of which first pair formally delivered on 1 July 1992. Programme of 39 conversions cancelled soon afterwards. Meanwhile, up to 30 new-build NOTAR H-6s delivered in 1991-92 to unannounced orders. AH-6J/MH-6J have common avionics and multi-role capability with AH-6G armed version and MH-6H personnel transport. Folding tailboom for compact air transport. M134 7.62 mm Minigun; BEI Hydra 70 mm rocket pods; Aerocrafter 0.5 in machine-gun pods. Optional Litton AIM-1 laser marker, Hughes AN/AAQ-16 FLIR. Optional cabin fuel tanks, capacity 110 litres (29 US gallons; 24 Imp gallons) or 236 litres (62.5 US gallons; 52 Imp gallons). Bendix/King KNS600 FMS, Omega/VLF, TACAN, VOR, GPS; Honeywell AN/APN-209 radar altimeter.

CUSTOMERS: Operated in various configurations (incl MD 520N) by US Army and military forces of Colombia, Kenya, Israel, South Korea, Japan and Philippines; 1991 deliveries included 12 military MD 500 series; eight more 500MG Defenders (making total 30) delivered 1992 to Philippine Air Force.

AVIONICS: See Current Versions.
ARMAMENT: See Current Versions.
DIMENSIONS, EXTERNAL:
As for 500E/530F except:
Length of fuselage: 500MD/TOW   7.62 m (25 ft 0 in)
530MG   7.29 m (23 ft 11 in)
Height to top of rotor head:
500MD/TOW   2.64 m (8 ft 8 in)
530MG   2.62 m (8 ft 7 in)
530MG with MMS   3.41 m (11 ft 2½ in)
Height over tail (endplate fins):
500MD/TOW   2.71 m (8 ft 10½ in)
530MG   2.59 m (8 ft 6 in)
Width over skids: 500MD/TOW   1.93 m (6 ft 4 in)
530MG   1.96 m (6 ft 5 in)
Width over TOW pods: 500MD/TOW, 530MG
   3.23 m (10 ft 7¼ in)
Tailskid ground clearance:
500MD/TOW   0.64 m (2 ft 1¼ in)
530MG   0.61 m (2 ft 0 in)
WEIGHTS AND LOADINGS:
Weight empty, equipped:
500MD/TOW   849 kg (1,871 lb)
530MG   898 kg (1,979 lb)
Max T-O weight:
500MD/TOW, normal   1,361 kg (3,000 lb)
500MD/TOW, max overload   1,610 kg (3,550 lb)
530MG, normal   1,406 kg (3,100 lb)
530MG, max overload   1,701 kg (3,750 lb)
Max disc loading:
500MD/TOW   31.64 kg/m² (6.48 lb/sq ft)
530MG   31.16 kg/m² (6.38 lb/sq ft)
PERFORMANCE (at max normal T-O weight, except where indicated):
Never-exceed speed (V$_{NE}$) at S/L:
500MD/TOW, 530MG
   130 knots (241 km/h; 150 mph)
Max cruising speed at S/L:
500MD/TOW, 530MG
   121 knots (224 km/h; 139 mph)
Max cruising speed at 1,525 m (5,000 ft):
500MD/TOW   120 knots (222 km/h; 138 mph)
530MG   123 knots (228 km/h; 142 mph)
Max rate of climb at S/L, ISA:
500MD/TOW   520 m (1,705 ft)/min
530MG   626 m (2,055 ft)/min
Vertical rate of climb at S/L:
500MD/TOW   248 m (813 ft)/min
530MG   445 m (1,660 ft)/min
Service ceiling: 500MD/TOW   4,635 m (15,210 ft)
530MG   over 4,875 m (16,000 ft)
Hovering ceiling IGE:
ISA: 500MD/TOW   2,590 m (8,500 ft)
530MG   4,360 m (14,300 ft)
ISA+20°C: 500MD/TOW   1,830 m (6,000 ft)
530MG   3,660 m (12,000 ft)
500MD/TOW, 35°C   1,340 m (4,400 ft)
530MG, 35°C   2,680 m (8,800 ft)
Hovering ceiling OGE:
ISA: 500MD/TOW   1,830 m (6,000 ft)
530MG   3,660 m (12,000 ft)
ISA+20°C: 500MD/TOW   975 m (3,200 ft)
530MG   2,970 m (9,750 ft)
500MD/TOW, 35°C   732 m (2,400 ft)
530MG, 35°C   2,120 m (6,950 ft)
Range, 2 min warm-up, standard fuel, no reserves:
500MD/TOW at S/L   203 nm (376 km; 233 miles)
530MG at S/L   176 nm (326 km; 202 miles)
500MD/TOW at 1,525 m (5,000 ft)
   227 nm (420 km; 261 miles)
530MG at 1,525 m (5,000 ft)
   200 nm (370 km; 230 miles)
Endurance with standard fuel, 2 min warm-up, no reserves:
500MD/TOW at S/L   2 h 23 min
530MG at S/L   1 h 56 min
500MD/TOW at 1,525 m (5,000 ft)   2 h 35 min
530MG at 1,525 m (5,000 ft)   2 h 7 min

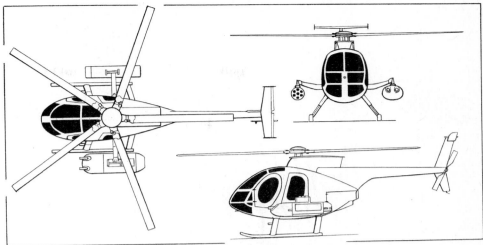

McDonnell Douglas 530MG Defender, with seven-tube rocket launcher and FN Herstal gun-pod *(Dennis Punnett)*

McDonnell Douglas AH-6F of the US Army's 160th Special Operations Aviation Regiment, fitted with Stinger missiles *(Press Office Sturzenegger)*

McDonnell Douglas 530MG Defender of the Colombian Air Force *(Press Office Sturzenegger)*

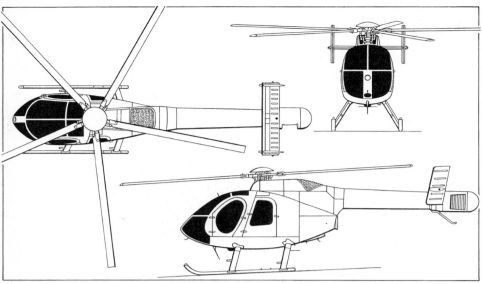

McDonnell Douglas MD 520N five-seat NOTAR helicopter *(Jane's/Mike Keep)*

McDonnell Douglas MD 520N of Fuchs Helikopter of Switzerland

MD 520N flying with twin pop-out floats inflated

MCDONNELL DOUGLAS MD 520N

TYPE: Five-seat light utility helicopter with no tail rotor (NOTAR).

PROGRAMME: First flight OH-6A NOTAR testbed 17 December 1981; programme details in 1982-83 *Jane's*; extensive modifications during 1985 with second blowing slot, new fan, 250-C20B engine and MD 500E nose; flight testing resumed 12 March 1986 and completed in June; retired to US Army Aviation Museum, Fort Rucker, Alabama, October 1990.

Commercial MD 520N and MD 530N NOTAR helicopters announced February 1988 and officially launched January 1989, to be powered by 335.6 kW (450 shp) 250-C20R-2 and 485 kW (650 shp) 250-C30 respectively; first flight of first MD 530N (N530NT) 29 December 1989; first flight 520N (N520NT) 1 May 1990 and first production 520N 28 June 1991; 520N certificated 13 September 1991; first production 520N (N521FB) delivered to Phoenix Police Department 31 October 1991. MD 520N set new Paris-London speed record in September 1992, at 1 h 22 min 29 s.

CURRENT VERSIONS: **MD 520N**: NOTAR version of MD 500, offering more power, higher operating altitude and greater max T-O weight than MD 500E. Transmission rating 280 kW (375 shp) for T-O; max continuous 261 kW (350 shp). Fuel capacity 235 litres (62 US gallons; 51.6 Imp gallons). *Description applies to this version.*

MD 530N: Company "has no plans" to certificate MD 530N; MD 520N fills most of its roles.

Military variants: Also being developed, including retrofit of some US Army AH/MH-6s as MH-6J (which see).

CUSTOMERS: Orders and options for more than 180 MD 520Ns received by March 1992; 48 delivered between 31 October 1991 and May 1993.

COSTS: Original $2.2 million 24-month contract from US Army Applied Technology Laboratory and Defense Advanced Research Projects Agency (DARPA) for modification of OH-6A (65-12917). Typical cost of MD 520N $825,000.

DESIGN FEATURES: NOTAR (no tail rotor) system provides anti-torque and steering control without an external tail rotor, thus eliminating the danger of tail strikes; air emerging through Coanda slots and steering louvres is cool and at low velocity. Significantly quieter than conventional tail rotor helicopters. Main rotor rpm 477; main rotor tip speed 208 m (684 ft)/s; NOTAR system fan rpm 5,388. Emergency floats among options.

FLYING CONTROLS: Unboosted mechanical, as in earlier models. Rotor downwash over tailboom deflected to port by two Coanda-type slots fed with low-pressure air from engine-driven variable-pitch fan in root of tailboom; this counters normal rotor torque; some of fan air is also vented at tail through variable-aperture louvres controlled by pilot's foot pedals, giving steering control in hover and forward flight. Port moving fin on tailplane connected to foot pedals, primarily to increase directional control during autorotation and allow touchdown at under 20 knots (37 km/h; 23 mph); starboard fin independently operated by yaw damper.

STRUCTURE: Same as for MD 500E/530F, except graphite composites tailboom; metal tailplane and fins; new high-efficiency fan with composites blades fitted in production aircraft. NOTAR system components now have twice the lifespan of conventional tail rotor system assemblies.

POWER PLANT: Allison 250-C20R, derated to 317 kW (425 shp) for T-O and 280 kW (375 shp) max continuous.

DIMENSIONS, EXTERNAL:
Rotor diameter	8.33 m (27 ft 4 in)
Length: overall, rotor turning	8.69 m (28 ft 6 in)
fuselage	7.62 m (25 ft 0 in)
Height to top of rotor head:	
standard skids	2.74 m (9 ft 0 in)
extended skids	3.01 m (9 ft 10¾ in)
Height to top of fins	2.22 m (7 ft 4 in)
Tailplane span	2.07 m (6 ft 9½ in)
Skid track	1.98 m (6 ft 6 in)

AREAS:
Rotor disc	54.5 m² (586.8 sq ft)

WEIGHTS AND LOADINGS:
Weight empty: standard	742 kg (1,636 lb)
Max fuel weight	183 kg (404 lb)
Max hook capacity	1,004 kg (2,214 lb)
Max T-O weight: normal	1,519 kg (3,350 lb)
with external load	1,746 kg (3,850 lb)
Max normal disc loading	27.87 kg/m² (5.71 lb/sq ft)
Max normal power loading	5.44 kg/kW (8.93 lb/shp)

PERFORMANCE (at normal max T-O weight, ISA, except where indicated):
Never-exceed speed (V_{NE})	152 knots (281 km/h; 175 mph)
Max cruising speed at S/L	135 knots (249 km/h; 155 mph)
Max rate of climb at S/L: ISA	564 m (1,850 ft)/min
ISA + 20°C	480 m (1,575 ft)/min
Service ceiling	4,320 m (14,175 ft)
Hovering ceiling IGE	2,753 m (9,034 ft)
Hovering ceiling OGE: ISA	1,537 m (5,043 ft)
ISA +20°C	1,292 m (4,241 ft)
Range at S/L	217 nm (402 km; 250 miles)
Endurance at S/L	2 h 24 min

MCDONNELL DOUGLAS MD EXPLORER

TYPE: Eight-seat twin-engined light helicopter.
PROGRAMME: Formerly MDX and then given engineering designations MD 900/901; now known simply as MD Explorer. Announced February 1988; launched January 1989; Hawker de Havilland Limited of Australia designed and manufactures airframe; Canadian Marconi tested initial version of integrated instrumentation display system (IIDS) early 1992; Kawasaki Heavy Industries completed 50 h test of transmission early 1992. Other partners include Aim Aviation (interior), IAI (cowling and seats) and Lucas Aerospace (actuators). First flight 18 December 1992 (N900MD) at Mesa, Arizona; 19 flights (23 h) made by 19 May 1993; aircraft 2 to 7 to be used for certification and demonstration flights; VFR certification and deliveries expected from October 1994.
CURRENT VERSIONS: *Details apply to version with P&WC engines, except where indicated.*
CUSTOMERS: Certificates of interest for about 257 by early 1993 (from over 100 operators), representing planned production until 1997; by then 107 had converted to firm orders; market estimated at 800 to 1,000 in first decade; planned production rate 108 a year by 1996.
COSTS: Target direct operating cost $320 (1991) per hour.
DESIGN FEATURES: NOTAR anti-torque system; all-composite five-blade rotor with bearingless flexbeam retention and pitch case; tuned fixed rotor mast and mounting truss for vibration reduction; replaceable rotor tips; modified A frame construction from rotor mounting to landing skids protects passenger cabin; energy-absorbing seats absorb $30g$ vertically and $18.4g$ horizontally; onboard health monitoring, exceedance recording and blade track/balance.
FLYING CONTROLS: NOTAR tailboom (see details under MD 520N); mechanical engine control from collective-pitch lever is backup for electronic full-authority digital engine control. Automatic stabilisation and autopilot offered for IFR operation.
STRUCTURE: Cockpit, cabin and tail largely carbonfibre; top fairings Kevlar composites; no magnesium. Transmission overhaul life 5,000 hours; glassfibre blades have titanium leading-edge abrasion strip and are attached to bearingless hub by carbonfibre encased glassfibre flexbeams; rotor blades and hub on condition.
LANDING GEAR: Fixed skids; wheels optional.
POWER PLANT: First 128 MD Explorers to be powered by two Pratt & Whitney Canada PW206Bs, each rated at 469 kW (629 shp) for 5 min for T-O; thereafter, will be available powered by two Turbomeca TM319-2 Arrius 2Cs, each rated at 478 kW (641 shp) for 5 min for T-O. Transmission rating 746 kW (1,000 shp) for 5 min for T-O, 671 kW (900 shp) max continuous, and 429 kW (575 shp) for 2½ min OEI. Fuel contained in single tank under passenger cabin, useful capacity 602 litres (159 US gallons; 132.4 Imp gallons); optional 666 litres (176 US gallons; 146 Imp gallons).
ACCOMMODATION: Two pilots or pilot/passenger in front; six passengers in club-type seating in main cabin; rear baggage compartment accessible through rear door. Impact absorbing seats. Cabin can accept long loads reaching from flight deck to rear door.
AVIONICS: Single- or two-pilot IFR avionics with appropriate instrument panels incorporating Canadian Marconi integrated instrumentation display system; electronic display of engine and system data; panel space for weather display in IFR version or FLIR display in law enforcement version.
EQUIPMENT: External cargo hook with 1,361 kg (3,000 lb) capacity.

DIMENSIONS, EXTERNAL:
Rotor diameter	10.31 m (33 ft 10 in)
Length: overall, rotor turning	11.99 m (39 ft 4 in)
fuselage	9.70 m (31 ft 10 in)
Fuselage width	1.63 m (5 ft 4 in)
Height to top of rotor head	3.66 m (12 ft 0 in)
Min fuselage ground clearance	0.38 m (1 ft 3 in)
Tailplane span	2.79 m (9 ft 2 in)
Skid track	2.24 m (7 ft 4 in)
Cabin door width	1.27 m (4 ft 2 in)

DIMENSIONS, INTERNAL:
Volume: Cabin	4.88 m³ (172.5 cu ft)
Baggage (if closed off)	1.46 m³ (51.4 cu ft)

AREAS:
Rotor disc	83.52 m² (899.04 sq ft)

WEIGHTS AND LOADINGS:
Weight empty: standard configuration	1,458 kg (3,215 lb)
Usable fuel load	438-472 kg (965-1,041 lb)
Max internal payload	1,163 kg (2,565 lb)
Max slung load	1,361 kg (3,000 lb)
Max T-O weight: inertial load	2,699 kg (5,950 lb)
slung load	3,035 kg (6,690 lb)
Max disc loading:	
internal load	32.32 kg/m² (6.62 lb/sq ft)
slung load	36.33 kg/m² (7.44 lb/sq ft)
Max power loading:	
internal load	3.62 kg/kW (5.95 lb/shp)
slung load	4.07 kg/kW (6.69 lb/shp)

PERFORMANCE (estimated):
Never-exceed speed (VNE) at S/L, ISA	173 knots (321 km/h; 200 mph)
Max cruising speed at S/L, 38°C (100°F)	148 knots (274 km/h; 170 mph)
Max rate of climb	853 m (2,800 ft)/min
Vertical rate of climb	411 m (1,350 ft)/min
Rate of climb, OEI	305 m (1,000 ft)/min
Service ceiling: both engines	5,485 m (18,000 ft)
OEI	3,200 m (10,500 ft)
Hovering ceiling:	
IGE: ISA	3,960 m (13,000 ft)
ISA + 20°C	2,985 m (9,800 ft)
OGE: ISA	3,385 m (11,100 ft)
ISA + 20°C	2,255 m (7,400 ft)
Hovering ceiling, OEI:	
IGE, ISA at 87% T-O weight	1,220 m (4,000 ft)
with 1,361 kg (3,000 lb) slung load	2,375 m (7,800 ft)
Max range at 1,525 m (5,000 ft), ISA	299-323 nm (555-599 km; 345-372 miles)
Max endurance	3 h 24 min-3 h 36 min

Prototype McDonnell Douglas MD Explorer during its first flight

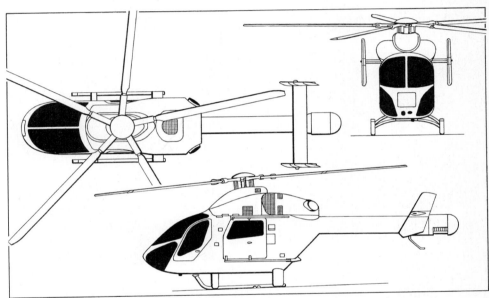
McDonnell Douglas MD Explorer eight-seat commercial helicopter (Jane's/Mike Keep)

MCDONNELL DOUGLAS APACHE

US Army designations: AH-64A, C and D
Israel Defence Force name: Petan (Cobra)

TYPE: Day/night twin-engined attack helicopter.
PROGRAMME: Original Hughes Model 77 entered for US Army advanced attack helicopter (AAH) competition; first flights of two development prototype YAH-64s 30 September and 22 November 1975; details of programme in 1984-85 and earlier *Jane's*; selected by US Army December 1976; named Apache late 1981.
Deliveries started 26 January 1984; 700th delivered December 1991; 720 by April 1992 (including 18 to Israel); 811th and final (plus five flying prototype and pre-production) due for delivery December 1994. Self deployment capability shown by 14th Apache with four 871 litre (230 US gallon; 191 Imp gallon) external tanks, which flew 1,020 nm (1,891 km; 1,175 miles) from Mesa to Santa Barbara with 45 minutes fuel remaining on 4 April 1985; initial operating capability achieved by 3rd Squadron, 6th Cavalry Regiment, July 1986; 29 of 40 planned AH-64A battalions, including six National Guard, combat-ready by December 1992; first combat use (11 AH-64As) in operation Just Cause, Panama, December 1989; used extensively (288) during January/February 1991 Gulf War against Iraq, including first air strike of conflict. First AH-64As issued to Army National Guard in 1987; fourth ArNG unit (1/211 AvRgt in Utah) established 1990; first overseas regiment 2/6 Cavalry Regiment, Illesheim, Germany, September 1987; eighth in Europe (3/4 AvRgt at Finthen) equipped 1990; battalion consists of 18 AH-64As and 13 Bell OH-58 Kiowas; more than 160 AH-64As based in Germany.
Eleven-month programme to integrate air-to-air Stinger began October 1987; four missiles mounted in pairs on wingtips; five firings early 1989; air-to-air development programme included firing two AIM-9 Sidewinders in hover and at 80 knots (148 km/h; 92 mph) at White Sands, New Mexico, November 1987; laser ranging and tracking tests on Bell UH-1 and LTV A-7 flown in 1989; M-230 Chain Gun being improved for air-to-air use; Matra Mistral captive carry tests completed. New missile control system by Base 10 Defense of Trenton, New Jersey, used for two more Stinger firings during 1990; Sidearm anti-radiation missile from AH-64A hit RF emitter on armoured vehicle at US Naval Weapons Test Center 25 April 1988.
CURRENT VERSIONS: **AH-64A:** Production for US Army and export. *Detailed description applies to AH-64A except where indicated.*
AH-64B: Cancelled in 1992. Was planned near-term upgrade of 254 AH-64As with improvements derived from operating experience in 1991 Gulf War, including GPS,

532 USA: AIRCRAFT—McDONNELL DOUGLAS

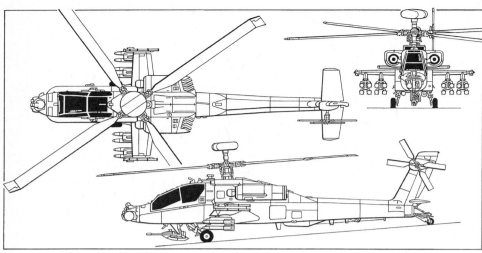

McDonnell Douglas AH-64D Longbow Apache tandem two-seat advanced attack helicopter
(Jane's/Mike Keep)

APACHE PROCUREMENT

US Army

Year	Quantity	First aircraft
FY 1973	3[1]	73-22247
FY 1979	3[2]	79-23257
FY 1982	11	82-23355
FY 1983	48	83-23787
FY 1984	112	84-24200
FY 1985	138	85-25351
FY 1986	116	86-8940
FY 1987	101	87-0407
FY 1988	67	88-0197
	+10	88-0275
FY 1989	72	89-0192
FY 1990	66	90-0280
	+66	90-0415
Subtotal	**813**	

Exports

Israel	18	801 (1990)
Saudi Arabia	12	90-0291 (from April 1993)
Egypt	24	(from 1994)
Greece	12	(from 1995)
UAE	20	(from 1993)
Undisclosed (FMS)	8	
TOTAL	**907**[3]	

Notes: [1]Prototypes; 73-22247 was static test airframe
[2]Pre-production
[3]Plus South Korean and Kuwaiti interest

McDonnell Douglas AH-64A Apache armed with Hellfire ATMs and rocket pods

COSTS: $10 million, 11 month programme to integrate air-to-air Stinger started October 1987. Longbow radar costs $2 million (1990) without supporting modifications compared with $9.96 million flyaway cost of AH-64A. Longbow R&D contract (for four prototypes) $194.6 million. Programme cost (807 aircraft at 1991 values) $1,169 million. Longbow Apache programme unit cost $15.2 million; procurement unit cost $11.3 million (1993 estimates). AH-65C $7.9 million programme and $6.8 million procurement unit costs (1993 estimates).

DESIGN FEATURES: AH-64A is required to continue flying for 30 min after being hit by 12.7 mm bullets coming from anywhere in the lower hemisphere plus 20°; also survives 23 mm hits in many parts; target acquisition and designation system (TADS) and pilot night vision system (PNVS) sensors mounted in nose; low airspeed sensor above main rotor hub; avionics in lateral containers; chin-mounted Chain Gun fed from ammunition bay in centre fuselage; four weapon pylons on stub wings (six when air-to-air capability is installed); engines widely separated, with integral particle separators and built-in exhaust cooling fittings; four-blade main rotor with lifting aerofoil blade section and swept tips; blades can be folded or easily removed; tail rotor consists of two teetering two-blade units crossed at 55° to reduce noise; airframe meets full crash-survival specifications. Two AH-64As will fit in C-141, six in C-5 and three in C-17A.

Main transmission, by Litton Precision Gear Division, can operate for one hour without oil; tail rotor drive, by Aircraft Gear Corporation, has grease lubricated gearboxes with Bendix driveshafts and couplings; gearboxes and shafts can operate for one hour after ballistic damage; main rotor shaft runs within airframe-mounted sleeve, relieving transmission of flight loads and allowing removal of transmission without disturbing rotor; AH-64A has flown aerobatic manoeuvres and is capable of flying at 0.5*g*.

FLYING CONTROLS: Fully powered controls with stabilisation and automatic flight control system; automatic hover hold; tailplane incidence automatically adjusted by Hamilton Standard control to streamline with downwash during hover and to hold best fuselage attitude during climb, cruise, descent and transition.

STRUCTURE: Main rotor blades (by Tool Research and Engineering Corporation, Composite Structures Divisions) tolerant to 23 mm cannon shells, have five U-sections forming spars and skins bonded with structural glassfibre tubes, laminated stainless steel skin and composites rear section; blades attached to hub by stack of laminated steel straps with elastomeric bearings. Teledyne Ryan produces all fuselages, wings, tail, engine cowlings, canopies and avionics containers.

LANDING GEAR: Menasco trailing arm type, with single mainwheels and fully castoring, self-centring and lockable tailwheel. Mainwheel tyres size 8.50-10, tailwheel tyre size 5.00-4. Hydraulic brakes on main units. Main gear is non-retractable, but legs fold rearward to reduce overall height for storage and transportation. Energy absorbing main and tail gears are designed for normal descent rates of up to 3.05 m (10 ft)/s and heavy landings at up to 12.8 m (42 ft)/s. Take-offs and landings can be made at structural design gross weight on terrain slopes of up to 12° (head-on) and 10° (side-on).

POWER PLANT: Two 1,265 kW (1,696 shp) General Electric T700-GE-701 turboshafts, derated for normal operations to provide reserve power for combat emergencies, and with automatic OEI rating of 1,285 kW (1,723 shp); -701C engines from 604th (89-0192) AH-64A onward (1990) giving 1,342 kW (1,800 shp) max continuous and 1,447 kW (1,940 shp) OEI. Engines mounted one on each side of

SINCGARS radios, target handover capability, better navigation, and improved reliability including new rotor blades.

AH-64C: Upgrade of AH-64As to near AH-64D standard, apart from omission of Longbow radar and retention of -701 engines; provisions for optional fitment of both; deliveries from mid-1995. Initial funding for two prototype conversions awarded in September 1992; first flight due January 1994. With exception of AH-64Ds and re-sales, all remaining US Army AH-64As (approximately 540) to be modified.

AH-64D Longbow Apache: Current improvement programme based on Westinghouse mast-mounted Longbow millimetre-wave radar and Martin Marietta Hellfire with RF seeker; includes more powerful GE T700-GE-701C engines, larger generators for 70kVA peak loads, Plessey AN/ASN-157 Doppler navigation, MIL-STD-1553B databus allied to dual 1750A processors, and a vapour cycle cooling system for avionics; early user tests completed April 1990.

Full-scale development programme, lasting 51 months, authorised by Defense Acquisition Board December 1990, but airframe work extended to 70 months to coincide with missile development; supporting modifications being incorporated progressively; first flight of AH-64A (82-23356) with dummy Longbow radome 11 March 1991; first (89-0192) of four AH-64D prototypes flown 15 April 1992; second aircraft flew 13 November 1992; fitted with radar in mid-1993 and due to fly August 1993; Nos. 3 and 4 delivered with radar in Autumn 1993; four AH-64Ds and two AH-64Cs to fly 3,300-hour test programme; production deliveries to start mid-1996. Only 227 conversions from AH-64A. Longbow can track flying targets and see through rain, fog and smoke that defeat FLIR and TV; RF Hellfire can operate at shorter ranges; it can lock-on before launch or launch on co-ordinates and lock-on in flight; Longbow scans through 360° for aerial targets or scans over 270° in 90° sectors for ground targets; mast-mounted rotating antenna weighs 113 kg (250 lb).

Further possible modifications include 'man-print' cockpit with large displays, air-to-air missiles, digital autostabiliser, integrated GPS/Doppler/INS/air data/laser/radar altimeter navigation system, digital communications, faster target handoff system, and enhanced fault detection with data transfer and recording.

CUSTOMERS: US Army 807, of which more than 770 delivered by March 1993; see Programme and Current Versions for details. Confirmed exports totalled 94 by May 1993. Israel ordered 18 in March 1990; first two delivered 12 September 1990 to 'Wasp' Squadron, powered by T700-GE-701s. Deliveries expected to begin in January 1993 to Saudi Arabia (12) and in 1994 to Egypt (24). Further orders from Greece (12) and United Arab Emirates (20), both in December 1991; six deliveries planned to UAE in 1993; plus 14 in 1994. Greek deliveries from 1995. In October 1992, US Army agreed to transfer 24 Apaches to Israel.

Close-up view of Longbow mast-mounted radome

McDonnell Douglas AH-64D Longbow Apache prototype (89-0192) with radome for Longbow mast-mounted radar

fuselage, above wings, with key components armour-protected. Upper cowlings let down to serve as maintenance platforms. Two crash resistant fuel cells in fuselage, combined capacity 1,422 litres (376 US gallons; 313 Imp gallons). 'Black Hole' IR suppression system protects aircraft from heat-seeking missiles: this eliminates an engine bay cooling fan, by operating from engine exhaust gas through ejector nozzles to lower the gas plume and metal temperatures.

ACCOMMODATION: Crew of two in tandem: co-pilot/gunner in front, pilot behind on 48 cm (19 in) elevated seat. Crew seats, by Simula Inc, are of lightweight Kevlar. Teledyne Ryan canopy, with PPG transparencies and transparent acrylic blast barrier between cockpits, is designed to provide optimum field of view. Crew stations are protected by Ceradyne Inc lightweight boron armour shields in cockpit floor and sides, and between cockpits, offering protection against 12.7 mm armour piercing rounds. Sierracin electric heating of windscreen. Seats and structure designed to give crew a 95 per cent chance of surviving ground impacts of up to 12.8 m (42 ft)/s.

SYSTEMS: AiResearch totally integrated pneumatic system includes a shaft driven compressor, air turbine starters, pneumatic valves, temperature control unit and environmental control unit. Parker Bertea dual hydraulic systems, operating at 207 bars (3,000 lb/sq in), with actuators ballistically tolerant to 12.7 mm direct hits. Redundant flight control system for both rotors. In the event of a flying control system failure, the system activates Honeywell secondary fly-by-wire control. Bendix electrical power system, with two 35kVA fully redundant engine driven AC generators, two 300A transformer-rectifiers, and URDC standby DC battery. Garrett GTP 36-55(H) 93 kW (125 shp) APU for engine starting and maintenance checking. DASA (TST) electric blade de-icing.

AVIONICS: Main avionics bays are adjacent to co-pilot/gunner's position, in large fairings on sides of fuselage. Communications equipment includes AN/ARC-164 UHF, AN/ARC-186 UHF/VHF, Tempest C-10414 intercom, KY-28/58/TSEC crypto secure voice, and C-8157 secure voice control. Plessey Electronic Systems AN/ASN-137 lightweight Doppler navigation system, with Litton LR-80 (AN/ASN-143) strapdown AHRS. Doppler system, with AHRS, permits nap-of-the-earth navigation and provides for storing target locations. Avionics fit includes an AN/ARN-89B ADF and an AN/APX-100 IFF transponder with KIT-1A secure encoding. Honeywell digital automatic stabilisation equipment (DASE). Aircraft survivability equipment (ASE) consists of an Aerospace Avionics AN/APR-39 passive RWR, an AN/AVR-2 laser warning receiver, a Sanders AN/ALQ-144 IR jammer, chaff dispensers, and an AN/ALQ-136 radar jammer. Other avionics include Astronautics Corpn HSI, an AN/APU-209 radar altimeter video display unit, remote magnetic indicator, and Pacer Systems omnidirectional, low-airspeed air data system. A Honeywell all-raster symbology generator processes TV data from IR and other sensors, superimposes symbology, and distributes the combination to CRT and helmet-mounted displays in the aircraft. BITE fault detection/location system. Martin Marietta Orlando Aerospace target acquisition and designation sight and AN/AAQ-11 pilot's night vision sensor (TADS/PNVS) comprises two independently functioning, fully integrated systems mounted on the nose. The TADS consists of a rotating turret (±120° in azimuth, +30/−60° in elevation) that houses the sensor subsystems, an optical relay tube in the CPG's cockpit, three electronic units in the avionics bay, and cockpit-mounted controls and displays. It is used principally for target search, detection and laser designation, with the CPG as primary operator (though it can also provide backup night vision to the pilot in the event of a PNVS failure). Once acquired by the TADS, targets can be tracked manually or automatically for autonomous attack with gun, rockets or Hellfire missiles. The TADS daylight sensor consists of a TV camera with narrow (0.9°) and wide angle (4.0°) fields of view; direct view optics (4° narrow and 18° wide angle); a laser spot tracker; and an International Laser Systems laser rangefinder/designator. The night sensor, in the starboard half of the turret, incorporates a FLIR sight with narrow, medium and wide angle (3.1, 10.1 and 50.0°) fields of view. The PNVS consists of a FLIR sensor (30° × 40° field of view) in a rotating turret (±90° in azimuth, +20/−45° in elevation) mounted above the TADS; an electronics unit in the avionics bay; and the pilot's display and controls. It provides the pilot with thermal imaging that permits nap-of-the-earth flight to, from and within the battle area at night or in adverse daytime weather, at altitudes low enough to avoid detection by the enemy. PNVS imagery is displayed on a single monocle in front of one of the pilot's eyes; flight information such as airspeed, altitude and heading is superimposed on this imagery to simplify the piloting task. The monocle is a part of the Honeywell integrated helmet and display sighting system (IHADSS) worn by both crew members.

ARMAMENT: McDonnell Douglas M230 Chain Gun 30 mm automatic cannon, located between the mainwheel legs in an underfuselage mounting with Smiths Industries electronic controls. Normal rate of fire is 625 rds/min of HE or HEDP (high explosive dual purpose) ammunition, which is interoperable with NATO Aden/DEFA 30 mm ammunition. Max ammunition load is 1,200 rds. Gun mounting is designed to collapse into fuselage between pilots in the event of a crash landing. Four underwing hardpoints, with Aircraft Hydro-Forming pylons and ejector units, on which can be carried up to sixteen Hellfire anti-tank missiles or up to seventy-six 2.75 in FFAR (folding fin aerial rockets) in their launchers or a combination of Hellfires and FFAR. Planned modification adds two extra hardpoints for four Stinger, four Mistral or two Sidewinder (including Sidearm anti-radiation variant) missiles. Hellfire remote electronics by Rockwell; Bendix aerial rocket control system; multiplex (MUX) system units by Honeywell. Co-pilot/gunner (CPG) has primary responsibility for firing gun and missiles, but pilot can override his controls to fire gun or launch missiles.

DIMENSIONS, EXTERNAL:
Main rotor diameter	14.63 m (48 ft 0 in)
Main rotor blade chord	0.53 m (1 ft 9 in)
Tail rotor diameter	2.79 m (9 ft 2 in)
Length overall: tail rotor turning	15.54 m (51 ft 0 in)
both rotors turning	17.76 m (58 ft 3⅛ in)
Wing span: clean	5.23 m (17 ft 2 in)
over empty weapon racks	5.82 m (19 ft 1 in)
Height: over tail fin	3.52 m (11 ft 6½ in)
over tail rotor	4.30 m (14 ft 1¼ in)
to top of rotor head	3.84 m (12 ft 7 in)
overall (top of air data sensor)	4.66 m (15 ft 3½ in)
overall, AH-64D	4.90 m (16 ft 1 in)
Distance between c/l of pylons:	
inboard pair	3.20 m (10 ft 6 in)
outboard pair	4.72 m (15 ft 6 in)
Tailplane span	3.40 m (11 ft 2 in)
Wheel track	2.03 m (6 ft 8 in)
Wheelbase	10.59 m (34 ft 9 in)

AREAS:
Main rotor disc	168.11 m² (1,809.5 sq ft)
Tail rotor disc	6.13 m² (66.0 sq ft)

WEIGHTS AND LOADINGS:
Weight empty	5,092 kg (11,225 lb)
Max internal fuel weight	1,157 kg (2,550 lb)
Max external stores weight	771 kg (1,700 lb)
Structural design gross weight	6,650 kg (14,660 lb)
Primary mission gross weight	6,552 kg (14,445 lb)
Design mission gross weight	8,006 kg (17,650 lb)
Max T-O weight	9,525 kg (21,000 lb)
Max disc loading	56.69 kg/m² (11.61 lb/sq ft)

PERFORMANCE (A: at 6,552 kg; 14,445 lb AUW with -701 engines, B: Longbow Apache with -701C engines; ISA except where indicated):
Never-exceed speed (V_{NE})	197 knots (365 km/h; 227 mph)
Max level and max cruising speed	158 knots (293 km/h; 182 mph)
Max vertical rate of climb at S/L:	
A	762 m (2,500 ft)/min
B	771 m (2,530 ft)/min
Service ceiling	6,400 m (21,000 ft)
Service ceiling, OEI: A	3,290 m (10,800 ft)
B	3,804 m (12,480 ft)
Hovering ceiling:	
IGE: A	4,570 m (15,000 ft)
B	5,246 m (17,210 ft)
OGE: A	3,505 m (11,500 ft)
B	4,124 m (13,530 ft)
Max range, internal fuel (30 min reserves)	260 nm (482 km; 300 miles)
Ferry range, max internal and external fuel, still air, 45 min reserves	1,024 nm (1,899 km; 1,180 miles)
Endurance at 1,220 m (4,000 ft) at 35°C	1 h 50 min
Max endurance, internal fuel	3 h 9 min
g limits at low altitude and airspeeds up to 164 knots (304 km/h; 189 mph)	+3.5/−0.5

WEIGHTS FOR TYPICAL MISSION PERFORMANCE (A: anti-armour at 1,220 m/4,000 ft and 35°C, 4 Hellfire and 320 rds of 30 mm ammunition; B: as A, but with 1,200 rds; C: as A, but with 6 Hellfire and 540 rds; D: anti-armour at 610 m/2,000 ft and 21°C, 16 Hellfire and 1,200 rds; E: air cover at 1,220 m/4,000 ft and 35°C, 4 Hellfire and 1,200 rds; F: as E but at 610 m/2,000 ft and 21°C, 4 Hellfire, 19 rockets, 1,200 rds; G: escort at 1,220 m/4,000 ft and 35°C, 19 rockets and 1,200 rds; H: escort at 610 m/2,000 ft and 21°C, 38 rockets and 1,200 rds):

Mission fuel: A	727 kg (1,602 lb)
G	741 kg (1,633 lb)
E	745 kg (1,643 lb)
C	902 kg (1,989 lb)
B	1,029 kg (2,269 lb)
D	1,063 kg (2,344 lb)
H	1,077 kg (2,374 lb)
F	1,086 kg (2,394 lb)
Mission gross weight: A	6,552 kg (14,445 lb)
E	6,874 kg (15,154 lb)
G	6,932 kg (15,282 lb)
B, C	7,158 kg (15,780 lb)
D	7,728 kg (17,038 lb)
F	7,813 kg (17,225 lb)
H	7,867 kg (17,343 lb)

TYPICAL MISSION PERFORMANCE (A-H as above):
Cruising speed at intermediate rated power:	
C	147 knots (272 km/h; 169 mph)
D	148 knots (274 km/h; 170 mph)
F	150 knots (278 km/h; 173 mph)
B	151 knots (280 km/h; 174 mph)
E, H	153 knots (283 km/h; 176 mph)
A	154 knots (285 km/h; 177 mph)
G	155 knots (287 km/h; 178 mph)
Max vertical rate of climb at intermediate rated power:	
B, C	137 m (450 ft)/min
H	238 m (780 ft)/min
F, G	262 m (860 ft)/min
E	293 m (960 ft)/min
D	301 m (990 ft)/min
A	448 m (1,470 ft)/min
Mission endurance (no reserves): A, E, G	1 h 50 min
C	1 h 47 min
D, F, H	2 h 30 min
B	2 h 40 min

MCDONNELL DOUGLAS AEROSPACE TRANSPORT AIRCRAFT (Division of McDonnell Douglas Aerospace)

C-17 Globemaster III programme transferred from Douglas Aircraft Company to McDonnell Douglas Aerospace in 1992.

MCDONNELL DOUGLAS C-17A GLOBEMASTER III

TYPE: Long-range and intra-theatre heavy cargo transport.
PROGRAMME: US Air Force selected McDonnell Douglas to develop C-X cargo aircraft 28 August 1981; full-scale development called off January 1982 and replaced on 26 July 1982 by slow-paced preliminary development order; development and three prototypes (one flying) ordered 31 December 1985; fabrication of first C-17A (T-1/87-0025) began 2 November 1987; first production C-17A contract 20 January 1988; assembly started at Long Beach 24 August 1988; assembly of prototype completed 21 December 1990; first flight 15 September 1991 – also delivery to 6517th Test Squadron/6510th TW, Edwards AFB (unit renamed 417th TS/412th TW on 2 October 1992); 100th hour/35th sortie 17 January 1992; first aerial refuelling 11 April 1992 during 51st sortie (155.8 hours); first stall testing 7 May 1992; 200th hour/65 sorties 8 May 1992; achieved V_{NE} (Mach 0.875/510 knots) May 1992; first opening of cargo door in flight 17 June 1992; 300th hour/73 sorties 13 August 1992; 100th sortie/326 hours 29 August 1992; new software for expansion of flight envelope flown 5 January 1993. Two static test airframes (T2 and T3) commissioned 30 November 1991 and Spring 1992; development to continue until 1993; named Globemaster III on 5 February 1993; peak production target 18 per year; assembly in new 102,200 m² (1.1 million sq ft) facility at Long Beach, California, dedicated 13 August 1987. Feasibility study for hose-drogue tanker/transport combi under way Spring 1991. Total 405 sorties/1,458 hours to 31 May 1993.
CUSTOMERS: US Air Force; original requirement 210; cut to 120 by 1991.

C-17 FUNDING

Batch	FY	Quantity	First aircraft
Proto	—	1	87-0025
Lot I	1988	2	88-0265
Lot II	1989	4	89-1189
Lot III	1990	4	90-0532
—	1991	—	—
Lot IV	1992	4	
Lot V	1993	6	
Total		20 + 1	

Eight requested for FY 1994 (long lead-time funding awarded February 1993).
Initial assignments of production aircraft as below:
P1/88-0265: First flight/delivery 18 May 1992, 417th TS; structural, systems, acoustic, mission systems, unpaved field and air refuelling test aircraft; air load testing began October 1992 flying M60 tank (41,277 kg; 91,000 lb); first C-17 airdrop, 3 May 1993.
P2/88-0266: First flight/delivery 21 June 1992, 417th TS; range/payload trials and tests of additional avionics, including global positioning system, communications/navigation equipment.
P3/89-1189: First flight/delivery 7 September 1992; landed on desert strip at Edwards AFB, 417th TS; electromechanical systems and all-weather trials; Eglin AFB, 2 November for six months of climatic trials; overseas deployments for 12 months, from September/October 1993.
P4/89-1190: First flight 9 December 1992; delivery 20 January 1993 to 417th TS; first C-17 without test instrumentation; initial evaluation and airlift training with USAF Operational Test and Evaluation Center, Kirtland AFB, New Mexico.
P5/89-1191: First flight 31 January 1993; production configuration; delivered for lightning and electromagnetic resistance evaluation at Patuxent River NAS 12 March 1993; 17th ALS/437th AW, Charleston AFB, late 1992.
P6/89-1192: First flight 8 May 1993; first direct delivery to Charleston AFB 14 June; maintenance familiarisation, then 17th ALS/437th AW; first of 12 for squadron; IOC 1994.
COSTS: Initial award $3,387 million for development, placed 31 December 1985. First production contract $603.6 million (1988) for two aircraft; second production contract $757 million for four aircraft (1989); originally expected unit cost $125 million. Programme unit cost now $294 million (1991), or $35,274 million for 120 aircraft. December 1992 estimate of costs for development, plus first six production aircraft, was $7,700 million ($2,000 million above target; $1,130 million over maximum price). Lot III ceiling price $1,200 million; estimated (1993) cost $1,000 million.
DESIGN FEATURES: Externally blown flap system based on McDonnell Douglas YC-15 medium STOL transport prototypes (see 1979-80 *Jane's*), with extended flaps in exhaust flow from engines during take-off and landing; combines load-carrying capacity of C-5 with STOL performance of C-130; required to operate from 915 m (3,000 ft) long and 18.3 m (60 ft) wide runways, complete 180° three-point turn in 25 m (82 ft) and reverse up 1 in 50 gradient when fully loaded using thrust reversers. Structure designed to survive battle damage and protect crew; essential line-replaceable units (LRU) to be replaceable in flight; rear loading ramp. Supercritical wing with 25° sweepback; 2.90 m (9.5 ft) high NASA winglets.
FLYING CONTROLS: First military transport with all-digital FBW control system and two-crew cockpit with sidestick controllers; two full-time, all-function HUDs and four multi-function electronic displays; outboard ailerons and four spoilers per wing; four elevator sections; two-surface rudder split into upper and lower segments; full-span leading-edge slats; two-slot, fixed vane, simple hinged flaps over about two-thirds of trailing-edge; small strakes under tail. Quadruple-redundant General Electric digital fly-by-wire flight control system, with mechanical backup.
STRUCTURE: Major subassemblies produced in new factory at Macon, Georgia; some 50 subcontractors, of which 21 for airframe; subcontractors include Beechcraft (composites winglets), Grumman Aerostructures (composites ailerons, rudder and elevators), Lockheed (wing components up to sixth aircraft only), Vought (vertical and horizontal stabilisers, engine nacelles and thrust reversers), Reynolds Metals Company (wing skins), CC Industries (wing spars and stringers), Kaman Aerospace (wing ribs and bulkheads), Martin Marietta (tailcone), Heath Tecna (wing-to-fuselage fillet), Aerostructures Hamble (composite flap hinge fairing and trailing-edge panels) and Northwest Composites (main landing gear pod panels). C-17 structure is 69.3 per cent aluminium; 12.3 per cent steel alloys; 10.3 per cent titanium and 8.1 per cent composites.
LANDING GEAR: Hydraulically retractable tricycle type, with free-fall emergency extension; designed for sink rate of 4.57 m (15 ft)/s and suitable for operation from paved runways or unpaved strips. Mainwheel units, each consisting of two legs in tandem with three wheels on each leg, rotate 90° to retract into fairings on lower fuselage sides; tyre size 50 × 21-20; pressure 9.52 bars (138 lb/sq in); maximum sink rate 4.6 m (15 ft)/s. Menasco twin-wheel nose leg retracts forwards; tyre size 40 × 16-14; pressure 10.69 bars (155 lb/sq in). AlliedSignal (Bendix) wheels and carbon brakes. Min ground turning radius at outside mainwheels 17.37 m (57 ft 0 in); min taxiway width for three-point turn 27.43 m (90 ft 0 in); wingtip/tailplane clearance 74.24 m (237 ft 0 in).
POWER PLANT: Early aircraft have four Pratt & Whitney F117-PW-100 (PW2040) turbofans, with maximum flat rating of 181.0 kN (40,700 lb st), pylon-mounted in individual underwing pods and each fitted with a directed-flow thrust reverser deployable both in flight and on the ground. Power plant contract open to competition for later C-17As. Provision for in-flight refuelling. Two outboard wing fuel tanks of 21,210 litres (5,603 US gallons; 4,666 Imp gallons) each; two inboard wing fuel tanks of 30,056 litres (7,940 US gallons; 6,612 Imp gallons) each; total capacity 102,610 litres (27,086 US gallons; 22,572 Imp gallons) Plessey fuel pumps.
ACCOMMODATION: Normal flight crew of pilot and co-pilot side by side on flight deck, two observer positions and loadmaster station at forward end of main floor; access to flight deck via downward opening airstair door on port side of lower forward fuselage. Bunks for crew immediately aft of flight deck area; crew comfort station at forward end of cargo hold. Main cargo hold able to accommodate Army wheeled vehicles, including five-ton expandable vans in two rows, or three Jeeps side by side, or up to three AH-64A Apache attack helicopters, with straight-in loading via hydraulically actuated rear loading ramp which forms underside of rear fuselage when retracted. Aircraft fitted with 27 stowable tip-up seats along each side wall and another 48 seats which can be erected along the centreline; optionally up to 48 litters for medical evacuation mission or up to 100 passengers on 10-passenger pallets in addition to 54 sidewall seats. Air delivery capability for nine 463L pallets plus two on ramp; or handling system for 18 463L pallets. Airdrop capability includes single platforms of up to 27,215 kg (60,000 lb), multiple platforms of up to 49,895 kg (110,000 lb), or up to 102 paratroops. Equipped for low altitude parachute extraction system (LAPES) drops. The C-17A will be the only aircraft to airdrop outsize firepower such as the US Army's new infantry fighting vehicle; it will also be able to carry the M1 main battle tank in combination with other vehicles. The cargo handling system includes rails for airdrops and rails/rollers for normal cargo handling. Each row of rails/rollers can be converted quickly by a single loadmaster from one configuration to the other. Cargo tiedown rings, each stressed for 11,340 kg (25,000 lb), all over cargo floor forming grid averaging 74 cm (29 in) square. Three quick-erecting litter stanchions, each supporting four litters, permanently carried. Main access to cargo hold is via rear loading ramp, which is itself stressed for 18,145 kg (40,000 lb) of cargo in flight. Underfuselage door aft of ramp moves upward inside fuselage to facilitate loading and unloading. Paratroop door at rear on each side; two ditching exits overhead, aft of the paratroop doors, and two overhead forward of the wing box.
SYSTEMS: Include AlliedSignal (AiResearch) computer controlled integrated environmental control system and cabin pressure control system; 2,440 m (8,000 ft) equivalent cabin pressure up to 11,280 m (37,000 ft); quad-redundant flight control and four independent 276 bar (4,000 lb/sq in) hydraulic systems; independent fuel feed systems; electrical system; AlliedSignal (Garrett) GTCP331 APU (at front

First prototype McDonnell Douglas C-17A with flaps and landing gear extended

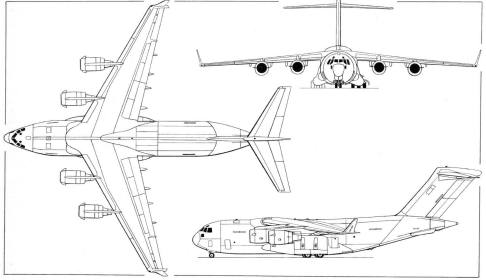

McDonnell Douglas C-17A Globemaster III long-range heavy cargo transport *(Dennis Punnett)*

McDONNELL DOUGLAS—AIRCRAFT: USA

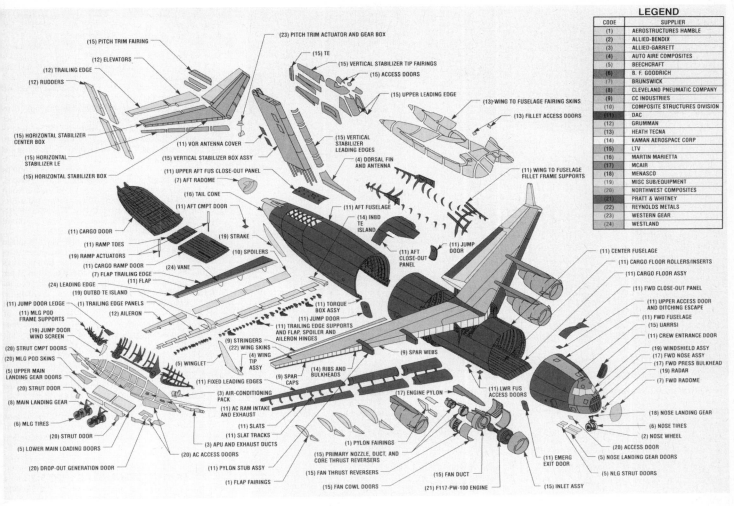

C-17A airframe suppliers

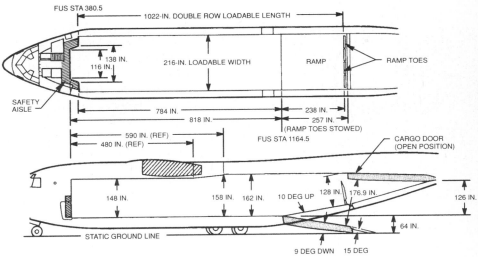

C-17A cargo area dimensions

of starboard landing gear pod), provides auxiliary power for environmental control system, engine starting, and on-ground electronics requirements; onboard inert gas generating system (OBIGGS) for the explosion protection system, pressurised by engine bleed air at 4.1 bars (60 lb/sq in) to produce NEA (nitrogen-enriched air) 4 gas and governed by a Gull Inc system controller; fire suppression system; Pyrotector (Gravnier Inc) smoke detection systems. All phases of cargo operation and configuration change capable of being handled by one loadmaster. Electrical system includes single 90kVA generator per engine and an APU, providing 115/200V, three-phase, 400Hz power; four 200A transformer rectifiers providing 28V DC; single-phase, 1,000VA inverter for ground refuelling and emergency AC; and two 40Ah Ni-Cad batteries for APU starting and emergency DC. Aeromedical equipment provided with 60Hz power.

AVIONICS: Honeywell dual air data computers, with advanced digital avionics and four full-colour multi-function displays (MFDs), two GEC-Marconi full flight regime foldable head-up displays, plus integrated mission and communications keyboards (MCKs) and displays (MCDs). Primary flight data presented on HUD and a selectable mode for the MFD. Horizontal navigation situation, computer-generated flight plan and weather radar overlay selectable on MFD. Station keeping (SKE), engine and flight control configuration data available on MFDs. All frequency tuning for nav/com accomplished from glareshield control panel. MCDs have frequency and channel pre-storage facility and provide for flight plan entry manually or by pre-programmed cassette, permitting insertion of in-flight planning changes without disturbing ongoing navigation. All MCD information for flight and navigation monitoring is presented on the HUD and MFDs. Teledyne Controls warning and caution system. Master warning caution annunciator provides automatic monitoring of all main systems and provides visual alerts on glareshields, aural and voice alerts on intercom. Other equipment includes AlliedSignal (Bendix/King) AN/APS-133(V) weather/mapping radar, Delco Electronics mission computer with MDC software; and electronic control system, Hamilton Standard aircraft and propulsion data management computer, General Dynamics automatic test equipment, and support equipment data acquisition and control system, Sierra Research station keeping equipment, and Telephonic Corporation radio management system. Communications include UHF and Satcom, VHF-AM/FM, HF, secure voice and jam-resistant UHF/VHF/FM, intercom, IFF/SIF and en route army/marine UHF LOS and Satcom hookup; navigation includes four inertial reference units, VOR/DME, TACAN, ILS/marker beacon and UHF-DF; other avionics include ground proximity warning, radar altimeter, cockpit voice recorder, crash position indicator, flight data recorder, diagnostic and integrated test system and airborne integrated data system recorder. Development of defensive electronic systems was authorised in 1988.

DIMENSIONS, EXTERNAL:
Wing span: wings only	50.29 m (165 ft 0 in)
between winglet tips	51.76 m (169 ft 10 in)
Wing aspect ratio	7.165
Length overall	53.04 m (174 ft 0 in)
Height overall	16.79 m (55 ft 1 in)
Fuselage diameter	6.85 m (22 ft 6 in)
Tailplane span	19.81 m (65 ft 0 in)
Wheel track	10.27 m (33 ft 8½ in)
Wheelbase	20.05 m (65 ft 9½ in)

DIMENSIONS, INTERNAL:
Cargo compartment:
Length, incl 6.05 m (19 ft 10 in) rear loading ramp
26.82 m (88 ft 0 in)
Loadable width 5.49 m (18 ft 0 in)
Height under wing 3.76 m (12 ft 4 in)
Max height 4.11 m (13 ft 6 in)
Volume 592 m³ (20,900 cu ft)

AREAS:
Wings, gross	353 m² (3,800 sq ft)
Ailerons (total)	11.83 m² (127.34 sq ft)
Tailplane	79.2 m² (845 sq ft)

WEIGHTS AND LOADINGS (target):
Operating weight empty 122,016 kg (269,000 lb)
Typical payload:
 inter-theatre logistics mission (2.5g load factor)
  54,421 kg (120,000 lb)
 heavy logistics mission (2.25g load factor)
  68,039 kg (150,000 lb)
Max payload 78,108 kg (172,200 lb)
Ramp capacity 18,143 kg (40,000 lb)
Max T-O weight 263,083 kg (580,000 lb)
Max wing loading 745.21 kg/m² (152.63 lb/sq ft)
Max power loading 363.37 kg/kN (3.56 lb/lb st)

First production Globemaster III (88-0265) photographed from a KC-135 during in-flight refuelling

PERFORMANCE (estimated, after USAF requirements reduced in early 1991):
Normal cruising speed at 8,535 m (28,000 ft)
Mach 0.77
Max cruising speed at low altitude
350 knots (648 km/h; 403 mph) CAS
Airdrop speed: at S/L
115-250 knots (213-463 km/h; 132-288 mph) CAS
at 7,620 m (25,000 ft)
130-250 knots (241-463 km/h; 150-288 mph) CAS
Approach speed with max payload
115 knots (213 km/h; 132 mph) CAS
Runway LCN (paved surface) better than 49
T-O field length with 72,575 kg (160,000 lb) payload and fuel for 2,400 nm (4,447 km; 2,763 miles)
2,286 m (7,500 ft)
Landing field length with 75,750 kg (167,000 lb) payload using thrust reversal 915 m (3,000 ft)
Radius, T-O with 36,786 kg (81,100 lb) payload in 975 m (3,200 ft), land in 823 m (2,700 ft), T-O with similar payload in 853 m (2,800 ft) and land in 792 m (2,600 ft) all at load factor of 3g, no in-flight refuelling
500 nm (925 km; 575 miles)
Radius, T-O with 56,245 kg (124,000 lb) payload in 2,012 m (6,600 ft) at load factor of 2.25g, land in 915 m (3,000 ft), T-O with zero payload (load factor of 3g) in 671 m (2,200 ft) and land in 701 m (2,300 ft), no in-flight refuelling 1,900 nm (3,520 km; 2,190 miles)
Range with payloads indicated, with no in-flight refuelling:
72,575 kg (160,000 lb), T-O in 2,286 m (7,500 ft), land in 915 m (3,000 ft), load factor of 2.25g
2,400 nm (4,445 km; 2,765 miles)
68,039 kg (150,000 lb), T-O in 2,320 m (7,600 ft), land in 885 m (2,900 ft), load factor of 2.25g
2,700 nm (5,000 km; 3,110 miles)
54,421 kg (120,000 lb), T-O in 1,830 m (6,000 ft), land in 853 m (2,800 ft), load factor of 2.5g
2,800 nm (5,190 km; 3,225 miles)
self-ferry (zero payload), T-O in 1,128 m (3,700 ft), land in 701 m (2,300 ft), load factor of 2.5g
4,700 nm (8,710 km; 5,412 miles)

MCDONNELL DOUGLAS KDC-10

McDonnell Douglas Aerospace received $32 million US Airforce Foreign Military Sale contract February 1993 to launch development of military boom tanker transport version of DC-10-30CF for Royal Netherlands Air Force; two aircraft acquired from Martinair; total contract value about $100 million. Further conversions expected.

Maiden flight of P3/89-1189 from Long Beach on 7 September 1992

MELEX

MELEX USA INC
1221 Front Street, Raleigh, North Carolina 27609
Telephone: 1 (919) 828 7645
Fax: 1 (919) 834 7290
Telex: 825868 MELEX UF
VICE-PRESIDENT: George Lundy

Subsidiary of Pezetel (PZL) of Poland (which see), responsible for sale and support of PZL Warszawa PZL-104 Wilga and PZL Mielec M-18/M-18A Dromader agricultural aircraft in western hemisphere.

MELEX TURBINE DROMADER

TYPE: Turboprop version of Dromader.
PROGRAMME: First flight of prototype (N2856G) 17 August 1985; STC issued 25 April 1986.
CURRENT VERSIONS: **Turbine Dromader:** Agricultural aircraft.
Water bomber: Under development; could replenish by scooping from water troughs while taxying on land; hydroskis, for replenishing on water without floats, under consideration.
DESIGN FEATURES: Converted by Turbines Inc of Terre Haute, Indiana, in co-operation with Melex USA Inc; 895 kW (1,200 shp) P&WC PT6A-45AG derated to 735 kW (986 shp), driving five-blade propeller; Dromader empty weight reduced by 363 kg (800 lb); normal operating speed increased by 20 knots (37 km/h; 23 mph) and T-O distance reduced by 10 per cent; standard fuel capacity increased to 719 litres (190 US gallons; 185 Imp gallons); hydraulic and electrical systems improved; optional passenger in rearward-facing seat; two-seat pilot trainer available.
FLYING CONTROLS and STRUCTURE: As M-18A Dromader (see Poland).

MELEX WILGA 80-550

TYPE: Teledyne Continental powered version of Polish Wilga 80.
PROGRAMME: Prototype (N7131G) first flown 1992; by March 1993 had completed factory flight tests and was then to apply for STC April/May. Melex to supply kits to factory

Prototype Turbine Dromader conversion with Pratt & Whitney Canada PT6A-45AG engine

MELEX/MONTANA—AIRCRAFT: USA

Polish Wilga 80, re-engined by Melex with a Teledyne Continental IO-550 flat-six engine

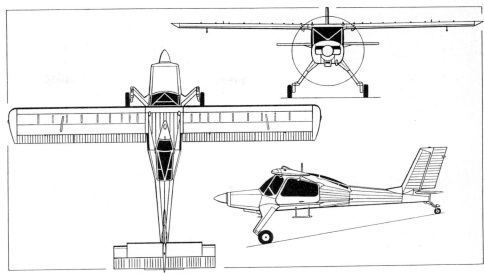
Wilga 80-550 utility aircraft (one Teledyne Continental IO-550) *(Dennis Punnett)*

which will produce Wilga 80s and 80-550s; Melex also continues selling 80s and converts new 80s to 80-550s.
CURRENT VERSIONS: **Wilga 80-550**: General purpose light aircraft.
Blue Shield: Electronic surveillance version, for use by law enforcement agencies and military.
DESIGN FEATURES: Developed from Wilga 80; Teledyne Continental engine; better fuel economy, and intention to add extra fuel capacity; 45.4 kg (100 lb) lighter, and intention to add 100 kg (220 lb) to gross weight; speed increased by 13 knots (24 km/h; 15 mph); shorter T-O run; Cleveland wheel brakes and optional 66 cm (26 in) tundra tyres.
POWER PLANT: One 224 kW (300 hp) Teledyne Continental IO-550 flat-six engine.

MID-CONTINENT
MID-CONTINENT AIRCRAFT CORPORATION
Drawer L, Highway 84 East, Hayti, Missouri 63851
Telephone: 1 (314) 359 0500
Fax: 1 (314) 359 0538

Operator and distributor of Schweizer Ag-Cats and Ayres Thrushes. Details of Ag-Cat re-engined with 895 kW (1,200 hp) Wright R-1820-202A in 1991-92 *Jane's* and now in *Jane's Civil and Military Aircraft Upgrades*.

MOLLER
MOLLER INTERNATIONAL
1222 Research Park Drive, Davis, California 95616
Telephone: 1 (916) 756 5086
Fax: 1 (916) 756 5179
PRESIDENT: Dr Paul S. Moller
DIRECTOR OF MARKETING: Jack G. Allison

Company formed in 1982 to develop Volantor VTOL aircraft, including high power/weight ratio engines and three-axis stabilisation; prototypes called **Discojet**, **XM-4** and **200X** were saucer-shaped with multiple lift fans and centrally mounted cockpit; 200X made 150 flights from 1987 onwards; powered by eight 37.3 kW (50 hp) Wankel engines.

Current model, intended for production, is four-seat **M400 Skycar** (as illustrated). Still under development and yet to make first flight, it will be powered by eight 112 kW (150 hp) Moller rotary engines mounted in pairs in four ducts and driving independent contra-rotating fans. Two sets of cascading vanes at rear of each duct vector thrust vertically for take off, hover and landing; in horizontal flight, vanes move concurrently to act as elevators and differentially as ailerons; four servo motors power each set of vanes. Full time fly-by-wire provides stability augmentation, controlled by sensors and input from pilot's dual joysticks. EFIS flight instrumentation.

Development has cost $28 million, and by early 1993 Moller reportedly had 88 Skycars on order; cost $995,000 (1993). Expected gross weight and max level speed are 1,134 kg (2,500 lb) and 334 knots (620 km/h; 385 mph) respectively.

Moller M400 Skycar in unfinished form

MONTANA
MONTANA COYOTE INC
3302 Airport Road, PO Box 9272, Helena, Montana 59601
Telephone: 1 (406) 449 3556
Fax: 1 (406) 449 3570
EXECUTIVE VICE-PRESIDENT AND OPERATIONS MANAGER: Kenneth D. Probst

MONTANA COYOTE
TYPE: Two-seat STOL homebuilt.
PROGRAMME: First flown in Spring 1991; kits available since Summer 1991. Building time 600-800 hours.
COSTS: Kit: $16,500, including basic VFR and engine instruments, prewelded components, fuel and brake systems, hardware and covering; does not include engine, propeller and paint.
STRUCTURE: Wooden wings, and 4130 steel tube fuselage, rudder and elevators, Dacron covered. Flaps and ailerons have aluminium alloy skins bonded to foam core.

Montana Coyote two-seat STOL homebuilt

Wingtips, cowlings, seats, instrument panel, spinner and fuel tank of GFRP.
LANDING GEAR: Non-retractable tailwheel type, with brakes. Optional floats, amphibious floats and wheel/skis.
POWER PLANT: One engine of 74.5-149 kW (100-200 hp), weighing 159 kg (350 lb) or under. Fuel capacity 151 litres (40 US gallons; 33.3 Imp gallons).
DIMENSIONS, EXTERNAL:
Wing span 11.58 m (38 ft 0 in)
Wing aspect ratio 8.71
Length overall 7.62 m (25 ft 0 in)
Height overall 1.98 m (6 ft 6 in)
Propeller diameter 1.88 m (6 ft 2 in)
AREAS:
Wings, gross 15.40 m² (165.75 sq ft)
WEIGHTS AND LOADINGS:
Weight empty 408-476 kg (900-1,050 lb)
Baggage capacity 72.5 kg (160 lb)
Max T-O weight 839 kg (1,850 lb)
Max wing loading 54.49 kg/m² (11.16 lb/sq ft)
PERFORMANCE:
Max level speed 104 knots (193 km/h; 120 mph)
Max cruising speed at 1,525 m (5,000 ft)
87 knots (161 km/h; 100 mph)
Econ cruising speed at 1,525 m (5,000 ft)
82 knots (153 km/h; 95 mph)
Stalling speed, power off 31 knots (57 km/h; 35 mph)
Max rate of climb at S/L 381 m (1,250 ft)/min
Service ceiling 4,575 m (15,000 ft)
T-O and landing run 107 m (350 ft)
Range 434 nm (804 km; 500 miles)
Endurance 6 h
g limits +5/−2

MOONEY
MOONEY AIRCRAFT CORPORATION
PO Box 72, Louis Schreiner Field, Kerrville, Texas 78028
Telephone: 1 (210) 896 6000
Fax: 1 (210) 896 6818
CHAIRMAN AND PRESIDENT: Alexandre Couvelaire
CEO: Jacques Esculier
VICE-PRESIDENT, SALES: Jeffrey T. Dunbar
VICE-PRESIDENT, MANUFACTURING: C. Keith Russey

Original Mooney company formed in Wichita, Kansas, 1948; produced single-seat M-18 Mite until 1952; later history recorded in 1987-88 *Jane's*. Alexandre Couvelaire, President of Euralair/Avialair Paris, France, and Michel Seydoux, President of MSC, jointly acquired Mooney in 1985; Mooney and Aerospatiale (Socata) announced joint development of TBM 700 June 1987 (see under Socata in French section), but Mooney withdrew in Spring 1991. Three types listed remain in production; other types, including proposed Enhanced Flight Screener, discontinued in 1991.

MOONEY MSE
TYPE: Four-seat touring aircraft, formerly M20J.
PROGRAMME: First flight of original Mooney 201 June 1976; certificated September 1976.
CURRENT VERSIONS: **MSE**: Standard version, *to which detailed description applies.*
201ATS: Advanced trainer version of 201SE (Special Edition) produced 1989; main differences from standard aircraft include 14V DC electrics, 70A 12V alternator, dual brakes, hard-wearing interior, larger oil access door, instructor and pupil approach chart holders, white instrument panel, standby vacuum system, three strobe lights, high visibility paint scheme, Bendix/King IFR radio package, and David Clark four-position intercom system.
CUSTOMERS: Total of more than 1,800 201s delivered by early 1992; 35 MSE and five ATS produced in 1992; 41 MSE and six ATS planned for 1993.
COSTS: MSE $191,260; ATS $163,975.
DESIGN FEATURES: High-efficiency touring aircraft, originally designed by Mooney brothers; wing section NACA 63₂-215 at root, 64₁-412 at tip; dihedral 5° 30′; incidence 2° 30′ at root, 1° at tip; wing swept forward 2° 29′.
FLYING CONTROLS: Manually operated; sealed gap, differentially operated ailerons; fin and tailplane integral so that both tilt, varying tailplane incidence for trimming; no trim tabs; electrically actuated single-slotted flaps.
STRUCTURE: Single-spar wing with auxiliary spar out to mid-position of flaps; wing and tail surfaces covered with stretch formed wraparound skins. Steel tube cabin section covered with light alloy skin; semi-monocoque rear fuselage with extruded stringers and sheet metal frames.
LANDING GEAR: Electrically retractable levered suspension tricycle type with airspeed safety switch bypass. Nosewheel retracts rearward, main units inward into wings. Rubber disc shock absorbers in main units. Cleveland mainwheels, size 6.00-6, and steerable nosewheel, size 5.00-5. Tyre pressure, mainwheels 2.07 bars (30 lb/sq in), nosewheel 3.38 bars (49 lb/sq in). Cleveland hydraulic single-disc brakes on mainwheels. Parking brake.
POWER PLANT: One 149 kW (200 hp) Textron Lycoming IO-360-A3B6D flat-four engine, driving a McCauley B2D34C214/90DHB-16 two-blade constant-speed metal propeller. Two integral fuel tanks in wings, with combined usable capacity of 242 litres (64 US gallons; 53.3 Imp gallons). Refuelling points in wing upper surface. Oil capacity 7.5 litres (2 US gallons; 1.7 Imp gallons).
ACCOMMODATION: Cabin accommodates four persons in pairs on individual vertically adjusting seats with reclining back, armrests and lumbar support. Dual controls standard. Overhead ventilation system. Cabin heating and cooling system, with adjustable outlets and illuminated control. One-piece wraparound windscreen. Tinted Plexiglas windows. Rear seats removable for freight stowage. Rear seats fold forward for carrying cargo. Single door on starboard side. Compartment for 54 kg (120 lb) baggage behind cabin, with access from cabin or through door on starboard side.
SYSTEMS: Hydraulic system for brakes only. Electrical system includes 70A alternator, 28V 70Ah battery, voltage regulator and warning lights, together with protective circuit breakers. Windscreen defrosting system standard.

Mooney MSE four-seat light aircraft (Textron Lycoming IO-360-A3B6D engine)

AVIONICS: A complete range of digital IFR avionics, including autopilot and weather avoidance systems, is available as an option.
EQUIPMENT: Standard equipment includes 'Greystone' instrument panel with blind-flying instruments, engine indicators, FT101 digital fuel totaliser, fuel sight gauges in wings, two electric fuel quantity gauges, electric OAT gauge, CHT and EGT gauges, alternate static source, and panel/cabin lighting, navigation lights, landing/taxi light, three high intensity strobe lights, grey tinted windscreen and cabin windows, seat belts and shoulder harnesses for all seats, assist straps and baggage straps, hatrack, multiple cabin fresh air vents, cargo tiedowns, wing jackpoints and external tiedowns, towbar, fuel tank quick drains and fuel sampler cup, auxiliary power plug, heated pitot tube, epoxy polyimide anti-corrosion treatment, and overall external polyurethane paint finish. Optional equipment includes export altimeter with millibar subscale, co-pilot's toe brakes, and deluxe control wheels.
DIMENSIONS, EXTERNAL:
Wing span 11.00 m (36 ft 1 in)
Wing chord, mean 1.50 m (4 ft 11¼ in)
Wing aspect ratio 7.45
Length overall 7.52 m (24 ft 8 in)
Height overall 2.54 m (8 ft 4 in)
Tailplane span 3.58 m (11 ft 9 in)
Wheel track 2.79 m (9 ft 2 in)
Wheelbase 1.82 m (5 ft 11½ in)
Propeller diameter 1.88 m (6 ft 2 in)
Propeller ground clearance 0.24 m (9½ in)
DIMENSIONS, INTERNAL:
Cabin: Length 2.90 m (9 ft 6 in)
  Max width 1.10 m (3 ft 7½ in)
  Max height 1.13 m (3 ft 8½ in)
Baggage door: Width 0.53 m (1 ft 9 in)
  Height 0.43 m (1 ft 5 in)
Baggage compartment volume 0.38 m³ (13.5 cu ft)
AREAS:
Wings, gross 16.24 m² (174.8 sq ft)
Ailerons (total) 1.06 m² (11.4 sq ft)
Trailing-edge flaps (total) 1.66 m² (17.9 sq ft)
Fin 0.73 m² (7.92 sq ft)
Rudder 0.58 m² (6.23 sq ft)
Tailplane 1.99 m² (21.45 sq ft)
Elevators (total) 1.11 m² (12.05 sq ft)
WEIGHTS AND LOADINGS:
Weight empty 809 kg (1,784 lb)
Max T-O and landing weight 1,243 kg (2,740 lb)
Max wing loading 76.5 kg/m² (15.67 lb/sq ft)
Max power loading 8.34 kg/kW (13.7 lb/hp)
PERFORMANCE (at max T-O weight):
Never-exceed speed (V$_{NE}$) 196 knots (364 km/h; 226 mph)
Max level speed at S/L 175 knots (325 km/h; 202 mph)
Max cruising speed, 75% power at 2,470 m (8,100 ft)
168 knots (311 km/h; 193 mph)

Mooney MSE four-seat touring aircraft (Dennis Punnett)

Econ cruising speed, 55% power at 2,470 m (8,100 ft)
  152 knots (282 km/h; 175 mph)
Stalling speed:
  flaps up 63 knots (117 km/h; 73 mph) IAS
  wheels and flaps down
    53 knots (98 km/h; 61 mph) CAS
Max rate of climb at S/L 314 m (1,030 ft)/min
Service ceiling 5,670 m (18,600 ft)
T-O to 15 m (50 ft) 463 m (1,517 ft)
Landing from 15 m (50 ft) 491 m (1,610 ft)
Landing run 235 m (770 ft)
Range: 55% power, no reserves
  1,059 nm (1,962 km; 1,219 miles)
  75% power, no reserves
  951 nm (1,762 km; 1,095 miles)

MOONEY TLS

TYPE: Four-seat turbocharged light aircraft, formerly M20M.
PROGRAMME: Announced 2 February 1989 as TLS (Turbocharged Lycoming Sabre); certificated 1989.
CUSTOMERS: 21 produced in 1992; 38 planned for 1993.
COSTS: Price $287,315 (1993).
DESIGN FEATURES: Stretched fuselage; turbocharged and intercooled engine.
POWER PLANT: One 201.3 kW (270 hp) turbocharged and intercooled Textron Lycoming TIO-540-AF1A flat-six engine, driving a McCauley three-blade metal propeller. Two integral fuel tanks in inboard wing leading-edges, with a combined capacity of 363 litres (96 US gallons; 80 Imp gallons), of which 341 litres (90 US gallons; 75 Imp gallons) are usable. Two-piece nose cowling is of composite glassfibre/graphite construction.
ACCOMMODATION: All seats have centre and side armrests (removable in rear seats) and European-style headrests. Rear seats moved aft by 10 cm (4 in) to increase legroom. Pilot and co-pilot seats have inertia reel shoulder harnesses.
SYSTEMS: Oxygen system, capacity 3.26 m³ (115 cu ft), with masks and overhead outlets, standard.
AVIONICS: Full range of Bendix/King avionics packages available as options, including EFIS, EHI-40 coupled to KLN 88 Loran C with moving map display, flight director, altitude and vertical speed preselect, R/Nav, and weather avoidance systems.
EQUIPMENT: Standard equipment includes attitude indicator, IFR directional gyro, fuel flow indicator, annunciator panel with press-to-test, electric/manual elevator trim and electric rudder trim with console- or panel-mounted LED indicators, avionics master switch, forward centre console, console-mounted chart holder, pilot's and co-pilot's map lights, high speed electric starter, dual batteries, console-mounted weight-and-balance computer, chrome-plated collapsible towbar, cigarette lighter and ashtrays, cabin, baggage door and ignition locks, and speedbrakes.
DIMENSIONS, EXTERNAL:
Wing span 11.00 m (36 ft 1 in)
Length overall 8.15 m (26 ft 9 in)
Height overall 2.51 m (8 ft 3 in)
Propeller diameter 1.91 m (6 ft 3 in)
DIMENSIONS, INTERNAL:
Cabin: Length 3.20 m (10 ft 6 in)

Mooney TLS four-seat turbocharged light aircraft

Floor area 3.53 m² (38.0 sq ft)
Volume 3.88 m³ (137 cu ft)
Baggage compartment volume 0.64 m³ (22.6 cu ft)
WEIGHTS AND LOADINGS:
Weight empty 913 kg (2,012 lb)
Max T-O weight 1,451 kg (3,200 lb)
Max wing loading 89.35 kg/m² (18.3 lb/sq ft)
Max power loading 7.21 kg/kW (11.85 lb/hp)
PERFORMANCE (at max T-O weight, ISA, except where indicated):
Max cruising speed:
  at 3,960 m (13,000 ft) 200 knots (371 km/h; 230 mph)
  at 7,620 m (25,000 ft) 223 knots (413 km/h; 257 mph)
Stalling speed:
  flaps and wheels up 65 knots (121 km/h; 75 mph)
  flaps and wheels down 60 knots (111 km/h; 69 mph)
Max rate of climb at S/L 375 m (1,230 ft)/min
Certificated ceiling 7,620 m (25,000 ft)
Range with max fuel 1,070 nm (1,983 km; 1,232 miles)

MUSTANG

MUSTANG AERONAUTICS INC

PO Box 1685, Troy, Missouri 48099
Telephone: 1 (313) 362 4295
Fax: 1 (313) 588 6788
PRESIDENT: Chris Tieman

Mustang Aeronautics Inc is new owner of M-II Mustang II, formerly marketed by Bushby Aircraft Inc. Details of current Bushby Midget Mustang in 1992-93 *Jane's*.

MUSTANG AERONAUTICS M-II MUSTANG II

TYPE: Two-seat, dual-control sporting homebuilt.
PROGRAMME: Side by side two-seat derivative of Midget Mustang; first flight 9 July 1966.
CUSTOMERS: About 1,200 being built by amateurs in early 1993, at which time about 300 completed.
DESIGN FEATURES: Description applies to de luxe model, stressed for +6g; empty weight quoted includes IFR instrumentation and nav/com equipment; can also be operated as aerobatic aircraft in 'Sport' configuration, identical to de luxe model except that electrical system, radio, additional IFR instrumentation, soundproofing and upholstery, wheel fairings, and some baggage capacity deleted; 'Sport' model has empty weight of 431 kg (950 lb), T-O weight of 567 kg (1,250 lb) and stressed for +9g. Wing section NACA 64A212 at root, NACA 64A210 at tip.
STRUCTURE: All-aluminium alloy flush riveted.
LANDING GEAR: Standard version has non-retractable tailwheel gear. Alternative non-retractable tricycle gear.
POWER PLANT: Normally one 134 kW (180 hp) Textron Lycoming IO-360; also can be fitted with 119 kW (160 hp) O-320. Provision for other engines including a 93 kW (125 hp) Textron Lycoming O-290. Fuel capacity 94.6 litres (25 US gallons; 21 Imp gallons). Optional integral wing fuel tanks, each with a capacity of 68 litres (18 US gallons; 15 Imp gallons). Provision for wingtip tanks.
DIMENSIONS, EXTERNAL:
Wing span 7.37 m (24 ft 2 in)
Wing aspect ratio 6.01
Length overall 5.94 m (19 ft 6 in)
Height overall 1.60 m (5 ft 3 in)
Propeller diameter 1.83 m (6 ft 0 in)
AREAS:
Wings, gross 9.02 m² (97.12 sq ft)
WEIGHTS AND LOADINGS:
Weight empty, equipped, 134 kW (180 hp) engine and tailwheel landing gear 454 kg (1,000 lb)
Baggage capacity 34 kg (75 lb)
* Max T-O weight 726 kg (1,600 lb)
* *Except for countries that restrict max wing loading to 73.2 kg/m² (15 lb/sq ft), where T-O weight of 658 kg (1,450 lb) applies*
PERFORMANCE (134 kW; 180 hp engine, tailwheel landing gear, at max T-O weight):
Max level speed at S/L 195 knots (362 km/h; 225 mph)
Max cruising speed at 2,285 m (7,500 ft)
  187 knots (346 km/h; 215 mph)

Mustang Aeronautics M-II Mustang II

Stalling speed: flaps up 53 knots (96 km/h; 60 mph)
  flaps down 51 knots (94 km/h; 58 mph)
Max rate of climb at S/L 670 m (2,200 ft)/min
Service ceiling 6,400 m (21,000 ft)
T-O run 137 m (450 ft)
Landing run 168 m (550 ft)
Range: with standard fuel, 75% power
  460 nm (853 km; 530 miles)
  with optional wing tanks
  1,120 nm (2,076 km; 1,290 miles)

NASA
NATIONAL AERONAUTICS AND SPACE ADMINISTRATION (Office of Aeronautics)

600 Independence Avenue SW, Washington, DC 20546
Telephone: 1 (202) 453 2693
Fax: 1 (202) 426 4256
Telex: 89530 NASA WSH
ASSOCIATE ADMINISTRATOR: Dr Wesley Harris
DIRECTOR FOR AERONAUTICS: Cecil C. Rosen
DIRECTOR FOR NASP OFFICE: Vincent Rausch

NASA operates fleet of 112 aircraft, comprising 34 dedicated to research and 70 to programme support; flew 29,000 hours in 1991, including 24,600 for R&D and programme support; employs 137 pilots, including 45 astronauts. Main operating bases are:
Ames Research Center, Moffett Field, California
Dryden Flight Research Facility, Edwards AFB, California
Langley Research Center, Hampton, Virginia
Lewis Research Center, Cleveland, Ohio
Other significant aircraft operations bases are:
Johnson Space Center, Houston, Texas
Wallops Flight Facility, Wallops Island, Virginia

NATIONAL AERO-SPACE PLANE (X-30)

TYPE: Reusable spacecraft; development of manned X-30 abandoned 1993 (see Addenda).
PROGRAMME: Announced 4 February 1986 by President Reagan as aerospace plane that "could shrink travel times between Washington, DC, and Tokyo . . . to less than two hours"; NASA and US Department of Defense then initiated joint National Aero-Space Plane (NASP) research programme. Objective is to develop technology for single-stage-to-orbit flight with air-breathing primary propulsion and horizontal take-off and landing. Enabling technologies will also allow high-speed cruise within atmosphere. Since ground-based test facilities cannot adequately simulate all the effects of flight environment above approximately Mach 8, NASP programme is focused on X-30 flight research vehicle to fly to orbital speed.

NASA and DoD jointly manage team of five major contractors: Pratt & Whitney and Rocketdyne work on propulsion; General Dynamics, McDonnell Douglas and Rockwell International conduct airframe and vehicle integration tasks; contractor team, formed 1990, has focused on new, single concept for the X-30. Technology development tasks being accomplished by contractors and by NASA and DoD laboratories; US Air Force continues as the lead agency. Large scale tests have proven fuselage tank and hydrogen-cooled engine structures at crucial (Mach 16) conditions; ground-based scramjet tests simulated flight conditions up to Mach 4 for large scale engine models and up to Mach 8 for ⅓-scale models; key scramjet components were tested up to Mach 18.

Technology readiness reviewed in 1993; decision then made to abandon third phase, to build and flight test two X-30 vehicles; further details in Addenda. Programme options include rocket-boosted flight test of vehicle components to precede X-30, using SR-71 as launch platform for unmanned craft.

NASP-derived vehicles (NDVs) have civil and military applications, including flexible, efficient access to Earth orbit for denser and higher priority payloads carried by conventional rocket-powered boosters; NDV attributes include reusability, widened launch and recovery windows for orbital flight, plus recovery-to-airport abort modes.
COSTS: $2,000 million up to 1999, including $700 million from industry.
DESIGN FEATURES: Twin-fin lifting body with small, compound-delta wings; propulsion by three to five scramjets in underfuselage pod, plus single booster rocket of 222-333 kN (50,000-75,000 lb) thrust for entry into orbit and re-entry; two-person crew compartment. Basic vehicle work-split gives McDonnell Douglas the centre-fuselage, overall aerodynamics, stability and control, and vehicle thermal control systems; GD responsible for rear fuselage and tail, crew compartment, airframe and airframe/engine integration; Rockwell (North American Aircraft) for forward fuselage, wings, vehicle management system and subsystems; Rocketdyne leads on engine system integration; Pratt & Whitney for power plant controls. X-30 expected to be some 45-60 m (148-197 ft) long, with T-O gross weight in region of 110,000-140,000 kg (242,500-308,650 lb).
STRUCTURE: Aerodynamic heating produces temperatures and heat loads on ascent that exceed those of US Space Shuttle on descent; high-temperature, lightweight metal-matrix composites (fibre-reinforced titanium) used for airframe primary structure; carbon-carbon panels and active cooling with internal fluid flows protect airframe's hottest spots; graphite-epoxy tanks hold slush-hydrogen fuel. Lower surface of lifting-body fuselage contoured to condition inflows for scramjets used at speeds above Mach 6.

NASA TRIALS PROGRAMMES

Recent developments summarised hereunder:
Boeing 747SP: Engineering work under way to build Stratospheric Observatory for Infra-Red Astronomy (SOFIA) with 2.50 m (98 in) telescope; to replace Lockheed C-141A (NASA714) based at Ames; partner is German Ministry of Science (BMFT); cruise altitude 12,500-13,700 m (41,000-45,000 ft); to fly 120 sorties (8 hours each) per year for 20 years.
General Dynamics F-16XL: Acquired two aircraft, 1991-92, for advanced research into laminar flow with applications in High-Speed Civil Transport programme. (See Lockheed entry, this edition.) First (single-seat) aircraft (NASA849/75-0749) flown at Dryden from 3 May 1990 before transfer to Langley; second (two-seat) (NASA846/75-0747) for trials 1992-96 at Dryden; first flight 29 September 1992 with passive (no suction) glove on starboard wing; due to begin trials with active glove on port wing in June 1995.
Grumman X-29: Both stood-down from use at Dryden by September 1991 (see 1990-91 and earlier *Jane's*) after 374 flights. New series of high AoA trials undertaken in mid-1992 by second aircraft (NASA049/82-0049); both currently in flyable storage at Dryden; total over 450 sorties since 1984; last sortie (by NASA049) 18 October 1992.
Lockheed EC-130Q: Two ex-US Navy delivered to Wallops in December 1991 for modification to remote sensing platforms, replacing NP-3A Orion (NASA428).
Lockheed P-3B: Entered service at Wallops, 1992.
Lockheed SR-71: Two SR-71As (NASA832/64-17971 and NASA844/64-17980) and one SR-71B (NASA831/64-17956) acquired in 1991 in support of high speed research and support to X-30 NASP. Based at Dryden. Flights include crew proficiency in NASA831; began series of high altitude ultraviolet spectronomy missions March 1993.
McDonnell Douglas F/A-18: Includes F/A-18A High Angle of Attack Research Vehicle (HARV) (NASA840/161251) with thrust vectoring system; based at Dryden; first flight 16 January 1991 at Dryden; first flight with vectoring 15 July 1991; reached 79° AoA in sustained flight, 1992; trials continue to 1994.
Rockwell OV-10A: Entered service at Langley late 1992 for wake vortex studies.
Rockwell/MBB X-31: Described in International section; NASA acquired both aircraft, April 1992; based at Dryden in international test programme; achieved 70° AoA in sustained flight, 1992.
Sikorsky UH-60A: Acquired for Ames in 1991 as Rotorcraft Aircrew Systems Concept Airborne Laboratory (RASCAL; NASA750); investigating increased agility for helicopters and improving automatic terrain avoidance; heavily instrumented rotor blades.
McDonnell Douglas F-15A: (NASA835/71-0287) Used as testbed for integrated digital electronic engine and flight control systems; demonstrated first self-repairing flight control systems. Currently flown in programme to improve engine performance, fuel efficiency and lifetime (performance seeking controls), and for studies of engine-only control research. Goals include landing on engine controls only (aerodynamic surface locked); made first touch-and-go landing with aerodynamic controls locked and using thrust vectoring for pitch attitude and lateral control in April 1993.
Lockheed C-130B Hercules: Earth Resources Aircraft Program at Ames operates C-130B (N707NA/58-0712) to acquire data for earth research; carries variety of sensors for Federal, state, university and industrial investigations.

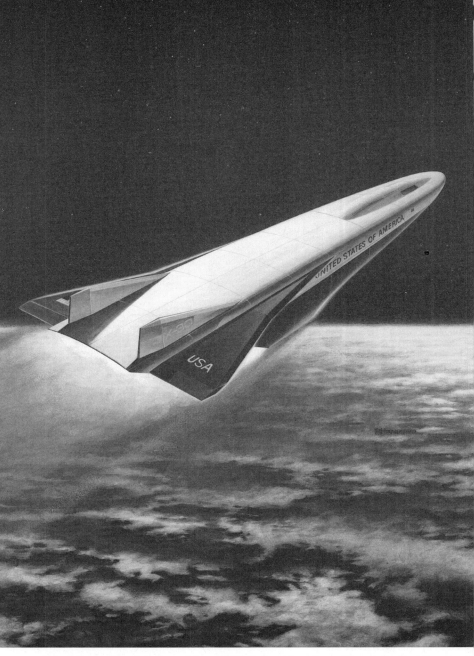

Artist's impression of the (now defunct) X-30 National Aero-Space Plane in Earth orbit

Calibration of RASCAL Sikorsky Black Hawk after modification by NASA

NASA Dryden Research Center F/A-18 with thrust deflector paddles as part of high angle of attack research

NASA has kept this SR-71B trainer in flying condition to maintain crew familiarity

One of the two F-16XLs has been used to investigate laminar flow at supersonic speed by means of a glove on the port wing

Lockheed C-130B Hercules modified by NASA for Earth resources exploration

542   USA: AIRCRAFT—NASA/NORTHROP

Artist's impression of SOFIA Boeing 747, modified by NASA to carry infra-red telescope

NEWCAL
NEWCAL AVIATION INC
14 Riser Road, Little Ferry, New Jersey 07643
Telephone: 1 (201) 440 1990
Fax: 1 (201) 440 8961
CONTACT: Perry Nifirous

NEWCAL (DE HAVILLAND CANADA) DHC-4T CARIBOU
TYPE: Turboprop conversion of DHC-4 STOL utility transport.
PROGRAMME: Prototype conversion made first flight 16 November 1991; crashed 27 August 1992 on take-off from NewCal's Gimli, Manitoba, Canada base; programme continuing; certification expected mid-1993.
CUSTOMERS: 20 Caribous available for conversion early 1992, incl NewCal's own fleet of seven; six options then held for DHC-4Ts (IAT, Mexair and another operator).
COSTS: Approx $4.5 million conversion cost (excluding cos of airframe) in 1992.
DESIGN FEATURES: Replaces 1,081 kW (1,450 hp) P&W R-2000 piston radials with two flat rated 1,062 kW (1,42 shp) P&WC PT6A-67R turboprops and new four-blad propellers; MTOGW unchanged at 12,927 kg (28,500 lb) but basic empty weight reduced by 794 kg (1,750 lb) t 7,484 kg (16,500 lb), resulting in 1,474 kg (3,250 lb increase in max payload to 5,443 kg (12,000 lb); servic ceiling raised to more than 8,840 m (29,000 ft), and speed improved by 10 per cent.

NORDAM
NORDAM AIRCRAFT MODIFICATION DIVISION
624 East 4th Street, Tulsa, Oklahoma 74120
Telephone: 1 (918) 560 5560
Fax: 1 (918) 560 5528
Telex: 158105

MARKETING DIRECTOR: Jack Arehart
Received FAA STC 26 June 1992 to modify Boeing 737-200s to meet FAA Stage 3/ICAO Chapter 3 noise levels; reduces 75dBA footprint by 69 per cent at 52,390 kg (115,500 lb) and allows most 737-200s to use 40° flaps during approach. Aircraft began commercial service with Lufthansa and Air New Zealand July 1992; over 100 firm orders by January 1992; cost $3 million per aircraft.

NORTHROP
NORTHROP CORPORATION
1840 Century Park East, Los Angeles, California 90067-2199
Telephone: 1 (310) 553 6262
Fax: 1 (310) 552 3109 or 552 4104
CHAIRMAN, PRESIDENT AND CEO: Kent Kresa
CORPORATE VICE-PRESIDENT AND CHIEF PUBLIC AFFAIRS OFFICER: Les Daly

Company formed 1939 by John K. Northrop to produce military aircraft; activities extended to missiles, target drones, electronics, space technology, communications, support services and commercial products; name changed from Northrop Aircraft Inc to Northrop Corporation in 1959; company now organised in Aircraft, B-2 (formerly Advanced Systems), and Electronics Systems Divisions, plus Northrop Advanced Technology and Design Center (established 1991) and subsidiary Northrop Worldwide Aircraft Services Inc. Worldwide workforce nearly 34,000 in early 1993; reduced by 2,400 during year.

Aircraft Division (nearly 13,000 personnel) designs and manufactures fighters, produces fuselage sections for McDonnell Douglas F/A-18 (1,000th set delivered October 1990) and major subassemblies for commercial aircraft including Boeing 747 fuselages, of which 1,000th set delivered in 1993. Northrop is prime contractor for AGM/MGM-137 Tri-Service Stand-Off Attack Missile and builds UAVs including BQM-74. B-2 Division manages a numbe of programmes and is prime contractor for B-2 advance technology stealth bomber. Northrop Worldwide Aircraf Services Inc, Lawton, Oklahoma, provides technical an support services to US armed forces and customs service Electronics Systems Division at Rolling Meadows, Illinois Norwood, Massachusetts, and Hawthorne, California, prod uces strategic and tactical navigation and guidanc equipment, passive sensor and tracking systems, electroni countermeasures (including AN/ALQ-135 internal counter measures for USAF F-15 and AN/ALQ-162 for three arme services), rate and rate-integrating gyros and strapdown guid ance systems.

AIRCRAFT DIVISION
One Northrop Avenue, Hawthorne, California 90250
Telephone: 1 (310) 332 1000
Fax: 1 (310) 332 3396
CORPORATE VICE-PRESIDENT AND AIRCRAFT DIVISION GENERAL MANAGER: Wallace C. Solberg
MANAGER, AIRCRAFT BUSINESS DEVELOPMENT: Jim Murphy

Principal subcontractor for F/A-18 (see McDonnell Douglas) and produces Boeing 747 main fuselage, upper deck, cargo door and passenger doors. Boeing 747 and other civilian work assigned to separate operating division from 2 January 1993. Research projects involve advanced simulators and composite materials. Agreement with Embraer, 1992, to promote EMB-312 Super Tucano for USAF's JPATS competition (which see). Continues support of Northrop F-5/T-38 family of fighters/trainers.

NORTHROP XST
Unbuilt design for stealth fighter offered to meet DARPA Have Blue requirement, 1974-75, but revealed only in 1992. Had many points of similarity to winning Lockheed XST (which compare), apart from dorsal air intake.

Artist's impression of the now-declassified Northrop XST, unsuccessful contending design for Have Blue
(Jane's/Mike Keep)

B-2 DIVISION

8900 East Washington Boulevard, Pico Rivera, California 90660-3737
Telephone: 1 (310) 942 3000
PRESIDENT AND GENERAL MANAGER: Oliver C. Boileau
VICE-PRESIDENT, BUSINESS AND ADVANCED SYSTEMS DEVELOPMENT: Ralph D. Crosby Jr

NORTHROP B-2

TYPE: Low-observable strategic penetration bomber.
PROGRAMME: Development of high level bomber started 1978; contract placed by USAF Aeronautical Systems Division October 1981; design modified for low altitude operation 1983; KC-135 testbed for B-2 avionics flying at Edwards AFB since January 1987; six B-2s assigned to trials; all but second will be refurbished for operational service; two static airframes also funded, of which structural test airframe exceeded ultimate (150 per cent) load test before fracture at 160 per cent, December 1992. 3,600 hour test programme planned, of which 26 per cent concerned with low observables (LO). First prototype B-2 (82-1066) rolled out USAF Plant 42 at Palmdale 22 November 1988; first flight (and delivery to Edwards AFB) 2 h 20 min 17 July 1989; first flight refuelling (from KC-10A) 8 November 1989. Block 1 of test programme, up to 13 June 1990, comprised 67 hours in 16 flights, covering aerodynamic performance and airworthiness. Block 2 LO testing began on 17th flight, 23 October 1990, after lay-up for preparation; further LO modifications, late 1992; trials completed and placed in storage by March 1993. Second B-2 (82-1067) first flew Palmdale to Edwards, 19 October 1990; loads test and envelope expansion; non-standard LO configuration; plans for use as permanent testbed reversed in April 1993, and aircraft to be refurbished for USAF (replacing No. 1, which to remain in trials configuration). Third B-2 (82-1068) flew 18 June 1991; first with full mission avionics, including AN/APQ-181 radar; fourth (82-1069) also for avionics, LO and weapons testing, flew 17 April 1992 and made first B-2 bomb drop on 4 September 1992 (inert 2,000 lb Mk 84 from 6,100 m; 20,000 ft). Fifth (82-1070) for weapons, LO and climatic testing, flown 5 October 1992; received LO improvements in early 1993. Sixth (82-1071), flown 2 February 1993, for operational testing (this first occasion on which new B-2 not delivered to Edwards on first flight). Total 215 sorties/1,000 hours up to February 1993 by B-2 fleet, including first night flight on 4 June 1992 and first night refuelling on 2 July 1992. All aircraft operated by 6520th Test Squadron of 6510th TW, Edwards AFB (re-designated 420th TS/412th TW on 2 October 1992).

CURRENT VERSIONS: **B-2A Block 30**: Envisaged service standard; five of six trials aircraft eventually to be upgraded from Block 10, although early production aircraft to be delivered in Block 20 configuration.

CUSTOMERS: USAF originally wanted 133 B-2s including prototype, but budget cuts reduced this to 76 by 1991. Six prototypes (five eventually for USAF) funded 1982; followed by further three whilst programme still 'black' (possibly FY 1988), three in FY 1989, two in FY 1990 and two in FY 1991, for total of 16 (15 USAF); programme frozen by Congress October 1991, at 16 B-2s for delivery before end 1995; USAF seeking minimum of 20 operational bombers, including one with conditional FY 1992 funds and four requested for FY 1993; last mentioned five conditionally approved in FY 1993 budget; funds to be released when B-2 achieves 'stealthiness' target. B-2 operating units to be 393rd and 750th Bomb Squadrons of 509th Bomb Wing (formed 1 April 1993) at Whiteman AFB, Missouri, where extensive, purpose designed facilities being built; Oklahoma Air Logistics Center is primary B-2 depot facility.

COSTS: $2 billion contract to start B-2 production 19 November 1987; programme cost $64,700 million (then-year prices) for 75 aircraft; currently estimated $44,400 million for 20 B-2s or $41,800 million for 15 (1992); unit programme cost (20 aircraft) $2,220 million each; or flyaway, $1,020 million per aircraft; absolute ceiling of $44,500 million imposed on programme.

DESIGN FEATURES: Blended flying wing, with straight leading-edges, swept at 33°; centre and tip sections have sharp, strongly under-cambered fixed leading-edges; two dielectric panels underwing outboard of flight deck cover dual radar antennae; 'double-W' trailing-edge incorporating elevons and drag rudders outboard of engines; two side by side weapons bays in lower centrebody each have small, drop-down spoiler panels ahead of doors, generating vortexes to ensure clean weapon release; engines fed by S-shaped air ducts; irregular-shaped air intakes feed engines, with three-pointed splitter plates ahead of inlets which remove boundary layer and provide secondary airflow for cooling and IR emissions control; upper lip of intake has single point; two auxiliary air inlet doors mounted on top of intake trunks remain open on ground and in slow-speed flight; two V-shaped overwing exhausts set well forward of trailing-edge; titanium on wing surface behind engine outlet; provision for mixing chloro-fluorosulfonic acid with exhaust gases to eliminate visible contrails; wingtips and leading-edges have dielectric covering of aerofoil section to mask radar-dissipating sawtooth construction.

Northrop assisted by Boeing Defense and Space Group, General Electric and Vought; planned 80,000 hours of testing aircraft's components include 24,000 hours wind tunnel tests, 44,000 hours avionics testing and 6,000 hours full-scale 'plastic bird' control system tests; flight testing to total 3,600 hours with six aircraft; all locations in airframe stored in CAD/CAM three-dimensional database used for machine tool, robot and tooling reference; prototype built on production tooling to accuracy of ±6.3 mm (¼ in) from tip to tip; nearly 900 new materials and processes developed; 131 Northrop subcontractors, plus 69 for Boeing, 52 for General Electric and 32 for Vought, announced by late 1989.

FLYING CONTROLS: Four control surfaces on each outer body/wing section, totalling approx 15 per cent of wing area combining aileron, elevator, rudder and flap functions; two inner pairs act in unison as elevons for slow-speed flight control; outboard elevon pair employed in normal flight; outermost surfaces, split horizontally, function as fast-moving drag rudders and speed-brakes, remaining partly deployed at most times (latter set at ±45° on landing approach); for normal flight, lower halves of drag rudders deploy to 90°, followed by upper halves to extent required to execute manoeuvre; no high-lift devices; beaver tail behind centre fuselage acts as pitch-axis trimming surface and, with elevons, as gust alleviation system; elevons deflect at up to 100°/s; four groups of four pressure sensors around nose section indicate quadruplex full fly-by-wire flying control system.

STRUCTURE: Mostly of composites (extensively graphite/epoxy), radar absorbent honeycomb structure and skin; outer skin of materials and coatings designed to reduce radar reflection and heat radiation. Northrop builds forward centre section including cockpit, integrates and assembles aircraft; Boeing Military Airplanes produces aft centre section and outboard sections; Vought produces intermediate fuselage sections and aluminium and titanium structural components and composites parts; combined workforce is 19,000, including 4,300 at Boeing and 2,000 at Vought.

LANDING GEAR: Tricycle type, adapted from Boeing 757/767. Inward retracting four-wheel main bogies have large trapezoidal door of thick cross-section. Rearward-retracting two-wheel nose unit has small door with sawtooth edges and large rear door, also used for crew access. Two landing lights on nosewheel leg. Landing gear limiting speed 224 knots (415 km/h; 258 mph).

POWER PLANT: Four 84.5 kN (19,000 lb st) General Electric F118-GE-110 non-afterburning turbofans mounted in pairs within wing structure, each side of weapons bay. In-flight refuelling receptacle in centrebody spine.

ACCOMMODATION: Two crew, with upward firing ejection seats: pilot to port, mission commander/instructor pilot to starboard. Provision for third member. Both forward positions have conventional control columns. Flight, engine, sensor and systems information presented on nine-tube EFIS display. Either crew member capable of flying complete mission, although data entry panels biased towards weapon systems officer on starboard seat. Four flight deck windows.

Northrop B-2 prototype (four General Electric F118 turbofans)

544   USA: AIRCRAFT—NORTHROP/ORLANDO

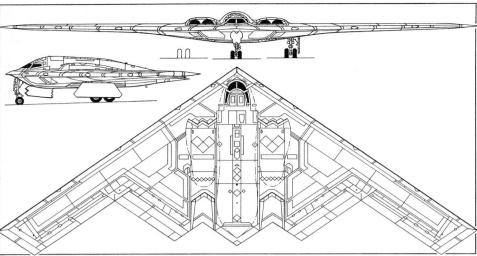

Northrop B-2 strategic penetration bomber (*Jane's/Mike Keep*)

SYSTEMS: Hydraulic system for flying controls operates at 27 bars (4,000 lb/sq in). Garrett APU outboard of port engine bay, covered by triangular door flush with wing surface.
AVIONICS: Hughes AN/APQ-181 low-probability-of-intercept (LPI) J-band covert strike radar, having 21 modes including terrain following and terrain avoidance. IBM Federal Systems AN/APR-50 (ZSR-63) RWR; ZSR-62 defensive aids equipment (role unspecified). Rockwell VLF/L receiver. Rockwell Collins TCN-250 Tacan, VIR-130 ILS and ICS-150X intercom. To ease crew workload, three-position selector switch in cockpit activates/deactivates appropriate equipment for 'take-off' (transfer mission data tape, checklist and appropriate flight control mode), 'go to war' (flight controls in 'stealthy' mode, weapons ready and radio emitters switched off) and 'land' (reactivate systems and perform checklist).
ARMAMENT: Boeing advanced applications rotary launcher each of two side by side weapons bays in lower central body. Total capacity of 16 AGM-69 SRAM II, AGM-129 ACMs or 16 AGM-137 TSSAMs (Tri-Service Standoff Attack Missiles). Alternative weapons include B61 tactical/strategic or 16 B83 strategic free-fall nuclear bombs; 80 Mk 82 1,000 lb bombs; 16 Mk 84 2,000 bombs; 36 M117 750 lb fire bombs; 36 CBU-87/89/97/ cluster bombs; and 80 Mk 36 1,000 lb or Mk 62 sea mines.

DIMENSIONS, EXTERNAL:
  Wing span   52.43 m (172 ft 0 in)
  Length overall   21.03 m (69 ft 0 in)
  Height overall   5.18 m (17 ft 0 in)
  Wheel track   12.20 m (40 ft 0 in)
AREAS (estimated):
  Lower surface   over 464.5 m² (5,000 sq
WEIGHTS AND LOADINGS:
  Weight empty   45,360-49,900 kg (100,000-110,000 lb)
  Max weapon load   22,680 kg (50,000 lb)
  Max internal fuel capacity
                       81,650-90,720 kg (180,000-200,000 lb)
  Normal T-O weight   168,433 kg (371,330 lb)
  Max T-O weight   181,437 kg (400,000 lb)
  Max wing loading   390.6 kg/m² (80.0 lb/sq ft)
  Max power loading   536.80 kg/kN (5.26 lb/lb)
PERFORMANCE:
  Approach speed   140 knots (259 km/h; 161 mph)
  Service ceiling   15,240 m (50,000 ft)
  Range with eight SRAMs and eight B83 bombs, totalling 16,919 kg (37,300 lb), at max T-O weight:
    hi-hi-hi   6,300 nm (11,675 km; 7,255 miles)
    hi-lo-hi (1,000 nm; 1,853 km; 1,152 miles at low level)
                           4,400 nm (8,154 km; 5,067 miles)
  Range with eight SRAMs and eight B61 bombs, totalling 10,886 kg (24,000 lb), at 162,386 kg (358,000 lb) T-O weight:
    hi-hi-hi   6,600 nm (12,231 km; 7,600 miles)
    hi-lo-hi (1,000 nm; 1,853 km; 1,152 miles at low level)
                           4,500 nm (8,339 km; 5,182 miles)
  Range:
    with one aerial refuelling
                           over 10,000 nm (18,532 km; 11,515 miles)

Northrop B-2 prototype in steeply banked flight

ORLANDO
ORLANDO HELICOPTER AIRWAYS INC
2774 Carrier Avenue, Building 141, Sanford, Florida 32773-9378
Telephone: 1 (407) 841 3480 and 323 1756
Fax: 1 (407) 330 2647
Telex: 52 9450
PRESIDENT: Fred P. Clark

Founded at Sanford, Florida, 1964 to remanufacture, sell and operate Sikorsky helicopters; holds large parts stocks for Sikorsky S-52, S-55/H-19 and S-58/H-34, and holds FAA type certificates for all H-19 and H-34 models; affiliate company Rotorparts Inc holds world's largest stocks of S-55 parts and supplies components for S-55, S-58, S-61 and S-62; Orlando has remanufactured more than 100 Sikorsky helicopters; plans to convert military surplus S-61, S-62 and S-64, re-engine helicopters with turbines, and experiment with alternative gasoline/alcohol/propane fuels.

Programme manufacture of Orlando OHA-S-55 Bearcats in China by Guangzhou Orlando Helicopters inactive since mid-1989 as a result of US trade restrictions; details in 1991-92 and earlier editions.

Orlando Helicopter Airways Aggressor conversion of the Sikorsky H-19/S-55

ORLANDO/SIKORSKY S-55/H-19

Further details of all models given in 1991-92 and earlier Jane's; can now be found in *Jane's Civil and Military Aircraft Upgrades*.
TYPE: Remanufactured Sikorsky S-55/H-19.
CURRENT VERSIONS: **Vistaplane:** Passenger/air ambulance seating eight passengers or 11 in high-density version; six stretchers and two attendants; optional cabin floor viewing window.
OHA-S-55 Heli-Camper: VIP model sleeping four in fully fitted camper interior.
OHA-S-55 Nite-Writer: Aerial advertising model with 12.2 × 2.4 m (40 × 8 ft) Sky Sign Inc computerised electronic billboard.
OHA-S-55 Bearcat: Carries Orlando-developed dry and spray dispensing systems, with quick conversion to fertilisers and seeds.
OHA-S-55 Heavy Lift: External load model for logging, firefighting and construction; useful load 1,361 kg (3,000 lb).
OHA-AS-55/QS-55 Aggressor: Orlando modified 15 S-55s to look like Mil Mi-24 'Hind-Es' for US Army Missile Command; used for training and as targets; dummy pilots in Mil cockpits when droned, or with pilot (when not droned) sitting high up and looking out through windows above Mi-24 type oil cooler fairings.
OHA-AT-55 Defender: Low cost multi-role military helicopter, for use in troop transport, armoured assault or medevac duties; powered by 626 kW (840 shp) Garrett TPE331-3 turboshaft or Wright R-1300-3 radial.
COSTS: Standard Bearcat $330,000; Aggressor under $1 million; Defender under $1 million (1992).

ORLANDO/SIKORSKY S-58/H-34

Further details of all models can be found in the 1991-92 Jane's or, currently, in *Jane's Civil and Military Aircraft Upgrades*.

Orlando Helicopter Airways OHA-AT-55 Defender

TYPE: Remanufactured Sikorsky S-58/H-34.
CURRENT VERSIONS: **Agricultural and Heavy Lift:** Powered by 1,137 kW (1,525 hp) Wright Cyclone R-1820-84; equipped as equivalent OHA-S-55 versions.
Heli-Camper: Equipped as OHA-S-55 Heli-Camper with additional features.
Orlando Airliner: High-density 18-passenger version of Sikorsky twin-turbine S-58T (see California Helicopters entry in this section).
COSTS: Standard Orlando Airliner $1.3 million (1992).

PEMCO

PEMCO AEROPLEX INC (Subsidiary of Precision Standard Inc)
PO Box 2287, Birmingham, Alabama 35201-2287
Telephone: 1 (205) 591 7870
Fax: 1 (205) 592 6306
Telex: (810) 733 3687

Former Hayes International Corporation acquired and renamed by Precision Standard Inc of Birmingham, Alabama, September 1988; aircraft maintenance, modification, repair and cargo conversions; besides those below, company has converted Convair 240, 340, 580, 640, Douglas DC-6, DC-8, Gulfstream I, Boeing 727-100/200. See *Jane's Civil and Military Aircraft Upgrades*.

PEMCO AEROPLEX BAe 146-200 CARGO CONVERSION

Appointed by BAe to make first conversion of 146-200 (see UK section) to all-cargo configuration; installed port side upward opening 3.30 m × 1.98 m (10 ft 10 in × 6 ft 6 in) cargo door aft of wing; applied strengthening and fitted out cabin. Converted aircraft can operate as all-cargo or passenger carrier. Pemco Aeroplex obtained FAA supplemental type certificate and makes all conversions, but BAe markets; company has installed Mode-S, TCAS and windshear customer-ordered modifications to BAe 146; 146-200 and 146-300 also converted to QC quick change.

PEMCO AEROPLEX BOEING 737-300 FREIGHTER/QC

Only post-delivery quick change and freighter conversions of Boeing 737-300 offered worldwide. Programme certificated in France, Germany, Singapore, Sweden and USA. Stronger floor-beam anchorages; additional fuselage doublers to accept new 3.55 m × 2.28 m (11 ft 8 in × 7 ft 6 in) cargo door. Since launch of programme in 1990, contracted to convert 39 737s, of which 16 completed and redelivered by early 1993.

PEMCO AEROPLEX BOEING 747-100 CARGO CONVERSION

Ex-TWA Boeing 747-100 converted to all-cargo configuration; 3.12 m × 3.48 m (10 ft 3 in × 11 ft 5 in) cargo door port side aft of wing; 9g bulkhead aft of flight deck; cargo compartment floor beams replaced; upper deck modified; windows blanked off; new interior cabin liner; provision for electric cargo handling system. Main deck can carry 29 cargo containers, with another 13 on lower deck; total payload 91,625 kg (202,000 lb). FAA supplemental type certificate gained April 1988; two modified in 1989; three more modified in 1990; sixth completed in 1991.

PEMCO AEROPLEX BOEING 757QC

Pemco Aeroplex presently developing Boeing 757 passenger to freighter/QC conversion programme through Data Licensing Agreement with Boeing Company. Conversion will be available from Spring 1993, and Pemco Aeroplex expected to initially certificate modification in Europe and USA.

PEMCO AEROPLEX LOCKHEED L-1011F TRISTAR

Ex-Eastern Air Lines L-1011-1 converted to all-cargo configuration for Aircraft Sales Company; 2.84 m × 4.32 m (9 ft 4 in × 14 ft 2 in) cargo door fitted port side forward of wing; 9g bulkhead aft of flight deck; windows blanked off; tracks and rollers on upper and lower decks; max payload 54,430 kg (120,000 lb); first flight 7 May 1987; FAA type approved; further conversions to order.

PEMCO AEROPLEX MCDONNELL DOUGLAS DC-9 CARGO CONVERSION

All-cargo conversion of DC-9; 2.06 m × 3.45 m (6 ft 9 in × 11 ft 4 in) cargo door port side aft of wing; 9g net barrier; cockpit emergency exit; smoke detection system; overhead storage rack/bin modification; altered interior panels and modifications to air-conditioning and oxygen systems; 15,876 kg (35,000 lb) cargo capacity in main cabin without operating restrictions.

PIASECKI

PIASECKI AIRCRAFT CORPORATION
Second Street West, Essington, Pennsylvania 19029-0360
Telephone: 1 (215) 521 5700
Fax: 1 (215) 521 5935
PRESIDENT: Frank N. Piasecki
VICE-PRESIDENTS:
 Donald N. Meyers (Engineering)
 Frederick W. Piasecki (Technology)
 John W. Piasecki (Contracts)
DIRECTOR, MILITARY REQUIREMENTS: Joseph P. Cosgrove

Formed in 1955 by Frank Piasecki, former Chairman of the Board and President of Piasecki Helicopter Corporation, now Helicopters Division, Boeing Defense and Space Group.

Current projects include modified AH-64A attack helicopter with vectored thrust ducted propeller and wing designed by Piasecki; US Marine Corps medium lift replacement (MLR) concept exploration program; US certification of Swidnik W-3A Sokół (see Polish section) and modification for US market.

Piasecki AH-64A Apache vectored thrust combat agility demonstrator (VTCAD): Aviation Applied Technology Directorate, US Army Aviation and Troop Command, Fort Eustis, Virginia, awarded $4.34 million contract to Piasecki in Spring 1991 for R&D on compound helicopter with Piasecki vectored thrust ducted propeller (VTDP) and wing, in air-to-air combat role. Work includes wind tunnel testing of VTDP, preliminary design of modified Apache and real-time piloted simulation. Design objectives to increase max

Model of AH-64A Apache VTCAD with PiAC vectored thrust ducted propeller and wing

PIPER

PIPER AIRCRAFT CORPORATION
2926 Piper Drive, PO Box 1328, Vero Beach, Florida 32960
Telephone: 1 (407) 567 4361
Fax: 1 (407) 778 2144
PRESIDENT AND COO: Chuck Suma
VICE-PRESIDENT, MARKETING AND SALES:
  Warner Hartlieb
DIRECTOR, ENGINEERING: Elliott Nichols
MARKETING CO-ORDINATOR: Renee While

Piper purchased by Lear Siegler Inc effective 1 March 1984; Lock Haven and Piper facilities in Pennsylvania closed second half 1984; Lakeland, Florida, plant phased out October 1985. Activities concentrated at Vero Beach and new 12,077 m² (130,000 sq ft) plant completed October 1986 for Cheyenne IIIA and 400 production (transferred from Lakeland).

Company acquired by Mr M. Stuart Millar 12 May 1987 and became subsidiary of Romeo Charlie Inc; LoPresti Piper Advanced Engineering Group set up as subsidiary December 1987; Piper North Corporation subsidiary formed November 1989 at Lock Haven to re-establish aircraft manufacturing at former headquarters under loan guarantee agreement with Commonwealth of Pennsylvania. Increasing cash flow problems in second half of 1990 resulted in closure of both subsidiaries and reduced production. Sale of type certificates and manufacturing rights for number of out-of-production models were negotiated August 1990 and March 1991. Aerospatiale (Socata) negotiations to buy Piper finally broken off 22 March 1991, reportedly because of product liability uncertainties. Voluntarily filed for protection under Chapter 11 of US Bankruptcy Code July 1991; negotiations announced October 1991 with The Cyrus Eaton Group International Ltd, originally to have been concluded January 1992 but US Bankruptcy Court extended reorganisation plan filing deadline to 31 July 1992 following rival bid by Duncan Investments of Denmark. Negotiations were not concluded and take-over did not happen. In late 1991, Piper International reportedly formed to hold all eight Piper aircraft type certificates and have aircraft manufactured in East Europe, Russia or even China. Principal creditor Congress Financial Corporation made further loan early 1992 to complete aircraft from outstanding order book; this was repaid that September.

During 1992, Piper delivered 85 aircraft, a substantial increase over previous year; estimates for 1993 were for 120 aircraft, of which 24 delivered by early April. Workforce increased to about 330 by early 1993. On 5 February 1993, company given further 90 days to plan its recovery from Chapter 11, and Congress Financial Corporation offered further line of credit. Letter of Intent (LoI) signed by Piper 8 April 1993 to sell its assets to Pilatus of Switzerland for $36 million, linked with reorganisation plan that would permit resumption of production of Super Cub, Warrior II, Archer II, Arrow, Dakota, Saratoga II HP and Seneca III; Pilatus would gain marketing base and potential production capability in USA for PC-12. However, Piper creditors rejected original plan on 7 May; revised plan to be submitted, but outcome undeclared at time of going to press; in any event, Pilatus would have to contest any alternative bids for Piper in order to accomplish takeover.

Piper designs are manufactured in Argentina by Chincul SA (which see) and in Brazil by Embraer subsidiary Neiva (which see); PZL Mielec in Poland (which see) produces Seneca II as M-20 Mewa.

PIPER SUSPENDED PRODUCTION
Piper's small singles and piston twins, last fully described in 1990-91 *Jane's*, included the following:

Piper PA-18-150 Super Cub: Tandem two-seat light aircraft; original FAA Type Approval received 18 November 1949; rights acquired by WTA Inc 1982 but Piper resumed production 1988; 55 were delivered in 1989 and 1990, and 27 were on order on 1 January 1991, but only one delivered in 1992. Further 19 could be delivered in 1993 if production resumed.

Piper PA-28-161 Warrior II and Cadet: Development of four-seat PA-28 Cherokee Cruiser; longer span wings with tapered outer panels introduced, with 112 kW (150 hp) engine, on 1972 Warrior I; 119 kW (160 hp) engine introduced on 1976 Warrior II. Cadet two/four-seat version optimised for training, introduced 1988. Total 2,985 PA-28-161s sold by 1 January 1991; further 28 built in 1992.

Piper PA-28-181 Archer II: Introduced October 1972 as Cherokee Challenger, successor to PA-28 Cherokee 180; became Cherokee Archer 1974; redesignated Cherokee Archer II in 1976; tapered wings of Warrior II introduced 1978. Total of 9,894 Cherokee 180s/Archer IIs delivered by 1 January 1991; 20 built in 1992.

Piper PA-28R-201 Arrow and PA-28R-201T Turbo Arrow: Derived from Cherokee Arrow II, as retractable landing gear version of Archer II with more powerful engine; tapered wings of Archer II introduced 1977; turbo version flew December 1976; T tailed version introduced 1979 as Arrow IV but production of low-tail version later resumed. Total 33 Arrows and 10 Turbo Arrows delivered in 1989; further five Arrows delivered in 1992.

Piper PA-28-236 Dakota: Introduced 1978 as more powerful version of Warrior/Archer/Arrow range, with 175 kW (235 hp) Textron Lycoming O-540-J3A5D engine and increased fuel capacity. Total 742 sold by 1 January 1991, and two more in 1992.

Piper PA-32-301 Saratoga and PA-32R-301 Saratoga SP/II HP: Announced December 1979 as replacements for six-seat Cherokee Six 300 and T tail Lance; Saratoga SP has retractable landing gear. Total of 819 Saratogas delivered by 1 January 1993, including 55 in 1990 and four in 1992. Deliveries began in May 1993 of **II HP**, a significantly upgraded SP with cabin class interior, new instrument panel layout to enhance pilot proficiency, aerodynamic modifications to improve performance, and new exterior styling.

Piper PA-34-220T Seneca III: PA-34 Seneca six-seat twin-engined light aircraft announced September 1971; redesignated Seneca II 1975 and PA-34-220T Seneca III, with more powerful engines, in 1981. Total of 4,477 Senecas delivered by 1 January 1993, including 50 in 1990 and 13 in 1992. Variants of earlier Seneca II produced in Poland as **M-20 Mewa** (see entry for WSK-PZL Mielec).

Piper PA-44-180 Seminole: Four-seat twin-engine light aircraft first flown May 1976; production of basic Seminole and Turbo Seminole began 1978; production terminated 1982 after 361 Seminoles and 87 Turbo Seminoles built; basic Seminole restored to production 1988 and a further 2? delivered by 1 January 1991; one more built in 1992.

PIPER PA-42-720 CHEYENNE IIIA
TYPE: Seven/10-passenger corporate and commuter airline transport and advanced trainer.
PROGRAMME: Announced as Cheyenne III; production prototype flew 18 May 1979; certification received early 1980 89 built before replaced by Cheyenne IIIA; certificated March 1983.
CURRENT VERSIONS: **Cheyenne IIIA:** Standard version, to which detailed description applies.
  Customs High Endurance Tracker (CHET): Aircraft fitted with special sensors delivered to US Drug Enforcement Administration 15 February 1984; used for variety of day and night surveillance and identification missions; further eight ordered March 1985, delivered monthly interval from September 1985. AN/APG-66 radar and Texas Instruments ventral FLIR turret.
CUSTOMERS: Advanced trainers delivered to Lufthansa (seven with Collins five-tube EFIS on flight decks configured to resemble airline's Airbus A310s); other sales include to Alitalia (three), Austrian Airlines, Turkish Air League (two), All Nippon Airways (four) and CAAC of China (four); CAAC received only Cheyenne IIIA built in 199? (one); available at one year's notice.
DESIGN FEATURES: Cheyenne III differences from Cheyenne I and II include increased wing span, lengthened fuselage, T tail and more powerful PT6A-41 engines; Cheyenne IIIA introduced PT6A-61 engines, increased max cruising speed, higher certificated ceiling, and improvements to interior layout, air-conditioning and electrical system. Wing section NACA 63_2A415 (modified) at root, NACA 63_1A212 at tip; dihedral 5°; incidence 1° 30′; no sweepback.
FLYING CONTROLS: Manually actuated controls; servo tab in rudder; anti-servo tab on elevator.
LANDING GEAR: Hydraulically retractable tricycle type with single wheel on each unit. Main units retract inward, nosewheel aft. Pneumatic blow-down system for emergency extension, with manually operated hydraulic system as backup. Piper oleo-pneumatic shock absorbers. Cleveland mainwheels with tyres size 6.50-10, 12-ply Type III, pressure 6.90 bars (100 lb/sq in). Cleveland steerable nosewheel with tyre size 17.5 × 6.25, 10-ply rating Type III, pressure 4.83 bars (70 lb/sq in). Goodrich hydraulically operated disc brakes. Parking brake.
POWER PLANT: Two Pratt & Whitney Canada PT6A-61 turboprops, each flat rated at 537 kW (720 shp) and driving a Hartzell three-blade constant-speed feathering and reversible-pitch metal propeller with Q-tips. Automatic propeller feathering system and synchrophaser optional. Each wing has four interconnected fuel cells and a tip tank, with a combined total capacity of 2,127 litres (562 US gallons 468 Imp gallons), of which 2,120 litres (560 US gallons 466 Imp gallons) are usable.
ACCOMMODATION: Pilot and co-pilot on four-way adjustable seats with armrests, headrests, shoulder safety belts with inertia reels, and stowage for oxygen mask beneath seats Certificated for single pilot operation. Dual controls standard. Cabin seats six to nine passengers.
SYSTEMS: AiResearch pressurisation system with max differential of 0.43 bar (6.3 lb/sq in), maintaining a cabin altitude of 3,050 m (10,000 ft) to a height of 10,060 m (33,000 ft) Environmental control system, combining the functions of heater, air-conditioner and dehumidifier. Hydraulic system supplied by dual engine driven pumps. Pneumatic system and vacuum system supplied by engine bleed air. Electrical system includes two 28V 300A engine driven generators and 24V 40Ah storage battery. Oxygen system of 0.62 m³ (22 cu ft) capacity with ten outlets. Pneumatic wing and tailplane de-icing boots, electric anti-icing of engine air intakes, heated pitots, electric propeller de-icing, and windscreen heating.
AVIONICS: Standard Bendix/King avionics package includes 4 in EHSI, KFC-325 autopilot with flight director and altitude preselect, RDS-81 radar, dual slaved compass systems and audio panels. Alternative Collins package with same capabilities.
DIMENSIONS, EXTERNAL:
  Wing span over tip tanks 14.53 m (47 ft 8 in)
  Wing chord: at root 3.12 m (10 ft 3 in)
    at tip 0.97 m (3 ft 2 in)

Piper Cheyenne IIIA used for advanced civil pilot training by China's CAAC

Wing aspect ratio	7.8
Length overall	13.23 m (43 ft 4¾ in)
Height overall	4.50 m (14 ft 9 in)
Tailplane span	6.65 m (21 ft 10 in)
Wheel track	5.72 m (18 ft 9 in)
Wheelbase	3.23 m (10 ft 7¼ in)
Propeller diameter	2.41 m (7 ft 11 in)
Distance between propeller centres	5.38 m (17 ft 8 in)
Passenger door: Height	1.12 m (3 ft 8 in)
Width	0.73 m (2 ft 5 in)
Baggage doors:	
Nose: Height	0.51 m (1 ft 8 in)
Width	0.66 m (2 ft 2 in)
Utility door (aft): Height	0.76 m (2 ft 6 in)
Width	0.43 m (1 ft 5 in)
Nacelle locker doors: Height	0.86 m (2 ft 10 in)
Width	0.61 m (2 ft 0 in)

DIMENSIONS, INTERNAL:

Cabin (incl flight deck and rear baggage area):	
Length	6.99 m (22 ft 11 in)
Max width	1.30 m (4 ft 3 in)
Max height	1.32 m (4 ft 4 in)
Volume	approx 9.91 m³ (350.0 cu ft)
Baggage compartment volume:	
nose	0.46 m³ (16.25 cu ft)
rear	0.88 m³ (31.0 cu ft)
nacelle locker (two, each)	0.16 m³ (5.6 cu ft)

AREAS:

Wings, gross	27.22 m² (293.0 sq ft)
Ailerons (total)	1.25 m² (13.5 sq ft)
Trailing-edge flaps (total)	3.17 m² (34.1 sq ft)
Fin	4.26 m² (45.81 sq ft)
Rudder, incl tab	1.60 m² (17.25 sq ft)
Tailplane	4.39 m² (47.3 sq ft)
Elevators, incl tab	2.26 m² (24.3 sq ft)

WEIGHTS AND LOADINGS:

Basic weight empty	3,101 kg (6,837 lb)
Max T-O weight	5,080 kg (11,200 lb)
Max ramp weight	5,119 kg (11,285 lb)
Max zero-fuel weight	4,241 kg (9,350 lb)
Max landing weight	4,685 kg (10,330 lb)
Max wing loading	186.6 kg/m² (38.22 lb/sq ft)
Max power loading	4.73 kg/kW (7.78 lb/shp)

PERFORMANCE (at max T-O weight, except where indicated):

Max level speed at average cruise weight of 4,127 kg (9,100 lb) 305 knots (565 km/h; 351 mph)
Cruising speed at max cruise power, at average cruise weight of 4,127 kg (9,100 lb):
  at 6,700 m (22,000 ft) 305 knots (565 km/h; 351 mph)
  at 7,620 m (25,000 ft) 302 knots (560 km/h; 348 mph)
  at 9,460 m (31,000 ft) 293 knots (543 km/h; 337 mph)
  at 10,670 m (35,000 ft) 282 knots (523 km/h; 325 mph)
Minimum single-engine control speed (V$_{MC}$)
  91 knots (169 km/h; 105 mph)
Stalling speed, engines idling, at 5,080 kg (11,200 lb):
  flaps and gear up 100 knots (186 km/h; 115 mph) IAS
  flaps and gear down
    87 knots (162 km/h; 100 mph) IAS
Rotation speed 93 knots (172 km/h; 107 mph) IAS
Approach speed 109 knots (202 km/h; 126 mph) IAS
Max rate of climb at S/L 725 m (2,380 ft)/min
Rate of climb at S/L, OEI 191 m (625 ft)/min
Service ceiling 10,925 m (35,840 ft)
Service ceiling, OEI 6,460 m (21,200 ft)
T-O run 465 m (1,525 ft)
T-O to 15 m (50 ft) 726 m (2,380 ft)
Landing from 15 m (50 ft) 928 m (3,043 ft)
Landing from 15 m (50 ft) with propeller reversal
  788 m (2,586 ft)
Landing run 583 m (1,914 ft)
Landing run with propeller reversal 444 m (1,457 ft)
Accelerate/stop distance 1,025 m (3,363 ft)
Range with max fuel, allowances for taxi, T-O, climb, descent, 45 min reserves at max range power, ISA:
  max cruising power at:
    6,700 m (22,000 ft)
      1,372 nm (2,542 km; 1,580 miles)
    7,620 m (25,000 ft)
      1,510 nm (2,798 km; 1,739 miles)
    9,460 m (31,000 ft)
      1,850 nm (3,428 km; 2,130 miles)
    10,670 m (35,000 ft)
      2,060 nm (3,817 km; 2,372 miles)
  max range power at:
    6,700 m (22,000 ft)
      1,803 nm (3,341 km; 2,076 miles)
    7,620 m (25,000 ft)
      1,945 nm (3,604 km; 2,240 miles)
    9,460 m (31,000 ft)
      2,200 nm (4,077 km; 2,533 miles)
    10,670 m (35,000 ft)
      2,270 nm (4,207 km; 2,614 miles)

PIPER PA-42-1000 CHEYENNE 400

TYPE: Seven/10-passenger light business transport and advanced trainer.
PROGRAMME: Announced September 1982 as Cheyenne IV and later known as Cheyenne 400LS; first flight of Cheyenne 400 prototype (N400PT) 23 February 1983; first flight of second prototype 23 June 1983; FAA certification

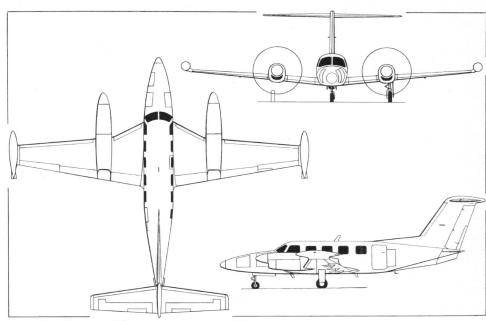

Piper Cheyenne 400 twin-turboprop seven/10-passenger business transport *(Dennis Punnett)*

Piper Cheyenne 400 (Garrett TPE331-14A/14B turboprops)

received 13 July 1984; deliveries commenced July 1984.
CUSTOMERS: First delivery to Garrett Turbine Engine Company 26 July 1984; 1,000th aircraft of Cheyenne series was Cheyenne 400 delivered to customer in Connecticut.
DESIGN FEATURES: Similar to Cheyenne III but flush riveted throughout; wings adapted for TPE331 engines and strengthened inboard of wheel wells to accept new main landing gear; outer wing panels modified to incorporate integral fuel tanks and carry wingtip tanks; fuselage strengthened for increased pressurisation; new multi-ply stretched acrylic windows; minor modifications to tail unit to cater for higher speeds.
LANDING GEAR: Hydraulically retractable tricycle type, with single wheel on each unit. Main units retract inward into wings, nosewheel rearward. Redesigned by comparison with Cheyenne IIIA, for improved ground attitude and increased landing weight. Mainwheels and tyres size 6.50-10, 12-ply rating, pressure 7.58 bars (110 lb/sq in). Steerable nosewheel, size 6.00-6, with 17.5 × 6.25-6 tyre, 10-ply rating, pressure 5.52 bars (80 lb/sq in). Hydraulically actuated dual disc brakes, with multiple brake pads, on each mainwheel.
POWER PLANT: Two 1,226.5 kW (1,645 shp) Garrett TPE331-14A/14B 'handed' (left/right rotating) turboprops, each flat rated at 746 kW (1,000 shp) and driving a Hartzell four-blade constant-speed reversible-pitch metal propeller; propeller synchrophaser standard. Fuel system as for Cheyenne IIIA, except usable fuel capacity is 2,158 litres (570 US gallons; 474.5 Imp gallons).
ACCOMMODATION: Although certificated for single pilot operation, provision for crew of two side by side on separate flight deck. Standard cabin has six seats, toilet with solid divider and door, and walk-in baggage area (capacity 136 kg; 300 lb); nose baggage compartment, capacity 136 kg (300 lb).
SYSTEMS: Bleed air from both engines for heating, cooling and pressurisation. Max pressure differential 0.51 bar (7.5 lb/sq in), maintaining a cabin altitude of 3,040 m (9,980 ft) at a height of 12,500 m (41,000 ft). Independent emergency bleed air pressurisation system. Automatic dropout oxygen masks.
AVIONICS: As for Cheyenne IIIA except standard autopilot is Bendix/King KFC 400 with air data system.
DIMENSIONS, EXTERNAL: As for Cheyenne IIIA, except:

Height overall	5.18 m (17 ft 0 in)
Propeller diameter	2.69 m (8 ft 10 in)
Nose baggage doors:	
Fwd: Height	0.30 m (1 ft 0 in)
Width	0.60 m (2 ft 0 in)
Rear: Height	0.66 m (2 ft 2 in)
Width	0.51 m (1 ft 8 in)

DIMENSIONS, INTERNAL: As for Cheyenne IIIA, except:

Cabin: Max height	1.42 m (4 ft 8 in)
Baggage compartment volume:	
nose	0.48 m³ (17.0 cu ft)
rear	0.88 m³ (31.0 cu ft)

WEIGHTS AND LOADINGS:

Weight empty, standard	3,412 kg (7,522 lb)
Max usable fuel	1,732 kg (3,819 lb)
Max T-O weight	5,466 kg (12,050 lb)
Max ramp weight	5,504 kg (12,135 lb)
Max landing weight	5,035 kg (11,100 lb)
Max zero-fuel weight	4,536 kg (10,000 lb)
Max wing loading	200.8 kg/m² (41.13 lb/sq ft)
Max power loading	2.23 kg/kW (3.66 lb/shp)

PERFORMANCE (at max T-O weight except where indicated):
Max operating speed
  Mach 0.62 (246 knots; 455 km/h; 283 mph EAS)
Cruising speed at max cruise power at AUW of 4,536 kg (10,000 lb):
  at 7,620 m (25,000 ft) 358 knots (663 km/h; 412 mph)
  at 12,500 m (41,000 ft) 335 knots (621 km/h; 386 mph)

548　USA: AIRCRAFT—PIPER/QUESTAIR

Minimum single-engine control speed (V_{MC})
　　99 knots (183 km/h; 114 mph) IAS
Stalling speed, engines idling:
　flaps and landing gear up
　　93 knots (172 km/h; 107 mph) IAS
　flaps and landing gear down
　　84 knots (156 km/h; 97 mph) IAS
Max rate of climb at S/L　　1,071 m (3,515 ft)/min
Rate of climb at S/L, OEI　　314 m (1,030 ft)/min
Service ceiling　　12,130 m (39,800 ft)
Service ceiling, OEI　　7,650 m (25,095 ft)
T-O run　　606 m (1,987 ft)
T-O to 15 m (50 ft)　　861 m (2,825 ft)
Landing from 15 m (50 ft) at max landing weight
　　860 m (2,820 ft)
Landing run　　427 m (1,400 ft)
Accelerate/stop distance　　1,113 m (3,650 ft)
Range at max cruise power at 10,670 m (35,000 ft), with allowances for start, taxi, T-O, climb, descent and 45 min reserves at max range power:
　with 8 passengers 1,167 nm (2,162 km; 1,343 miles)
　with 3 passengers 1,718 nm (3,183 km; 1,978 miles)
Max range, at max range power at 12,500 m (41,000 ft), allowances as above:
　with 8 passengers 1,547 nm (2,867 km; 1,781 miles)
　with 3 passengers 2,283 nm (4,230 km; 2,629 miles)

Piper PA-46-350P Malibu Mirage cabin class, pressurised, single-piston-engined aircraft

PIPER PA-46-350P MALIBU MIRAGE
TYPE: Six-seat light business transport.
PROGRAMME: Announced November 1982; FAA certification of original PA-46-310P Malibu received September 1988; production deliveries began November 1983; 402 built before replaced by PA-46-350P Malibu Mirage October 1988; Piper delivered 24 Malibu Mirages in 1990; total orders had reached 200 by May 1990 and combined deliveries 518 by Spring 1991; FAA temporary ban on IMC flights and Special Certification Review begun early 1991; returned to full operation February 1992.
CUSTOMERS: 10 delivered in 1992.
DESIGN FEATURES: Textron Lycoming TIO-540-AE2A introduced on Malibu Mirage; other changes include dual 70A 28V alternators, split bus electrical system; redesigned flight deck; additional options include computerised fuel management system and pilot's electrically heated windscreen. Wing dihedral 4.5°.
FLYING CONTROLS: Conventional ailerons; horn balanced elevators and rudder; trim tab in elevator; electrically operated trailing-edge flaps.
STRUCTURE: Cantilever high aspect ratio all-metal wings; light alloy fuselage, fail-safe construction in pressurised area; light alloy tail surfaces.
LANDING GEAR: Hydraulically retractable tricycle type with single wheel on each unit; main units retract inward into wingroots, nosewheel rearward, rotating 90° to lie flat under baggage compartment.
POWER PLANT: One 261 kW (350 hp) Textron Lycoming TIO-540-AE2A turbocharged and intercooled flat-six engine, driving a Hartzell two-blade constant-speed propeller. Fuel system capacity 462 litres (122 US gallons; 101.6 Imp gallons), of which 454 litres (120 US gallons; 100 Imp gallons) are usable. Oil capacity 11.5 litres (3 US gallons; 2.5 Imp gallons).
ACCOMMODATION: Pilot and five passengers in pressurised, heated and ventilated cabin; unpressurised baggage compartment in nose, and pressurised space at rear of cabin, each with capacity of 45 kg (100 lb).
SYSTEMS: Pressurisation max differential 0.38 bar (5.5 lb/sq in), to provide a cabin altitude of 2,400 m (7,900 ft) to a height of 7,620 m (25,000 ft). Hydraulic system pressure 107 bars (1,550 lb/sq in). Dual engine driven vacuum pumps standard. Standard electrical system has two 70A/28V alternators; 24V 10Ah battery; full icing protection optional.
AVIONICS: Standard Bendix/King IFR avionics package.
DIMENSIONS, EXTERNAL:
Wing span　　13.11 m (43 ft 0 in)
Wing aspect ratio　　10.57

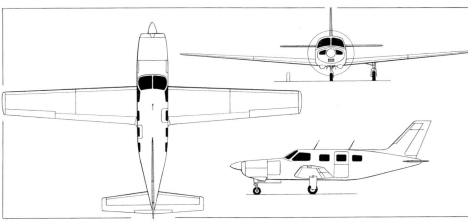

Piper PA-46-350P Malibu Mirage (Textron Lycoming TIO-540-AE2A) *(Dennis Punnett)*

Length overall　　8.72 m (28 ft 7¼ in)
Height overall　　3.51 m (11 ft 6 in)
Tailplane span　　4.42 m (14 ft 6 in)
Wheel track　　3.75 m (12 ft 3½ in)
Wheelbase　　2.44 m (8 ft 0 in)
Propeller diameter　　2.03 m (6 ft 8 in)
Passenger door (port, rear):
　Height　　1.17 m (3 ft 10 in)
　Width　　0.61 m (2 ft 0 in)
Baggage door (port, nose):
　Height　　0.58 m (1 ft 11 in)
　Width　　0.48 m (1 ft 7 in)
DIMENSIONS, INTERNAL:
Cabin: Length, instrument panel to rear pressure bulkhead　　3.76 m (12 ft 4 in)
　Max width　　1.26 m (4 ft 1½ in)
　Max height　　1.19 m (3 ft 11 in)
Baggage compartment volume:
　nose　　0.37 m³ (13.0 cu ft)
　rear cabin　　0.57 m³ (20.0 cu ft)
AREAS:
Wings, gross　　16.26 m² (175.0 sq ft)
WEIGHTS AND LOADINGS:
Weight empty, standard (estimated)　　1,265 kg (2,790 lb)
Max T-O weight　　1,950 kg (4,300 lb)
Max ramp weight　　1,958 kg (4,318 lb)
Max zero-fuel weight　　1,860 kg (4,100 lb)
Max landing weight　　1,860 kg (4,100 lb)
Max wing loading　　119.9 kg/m² (24.6 lb/sq ft)
Max power loading　　7.47 kg/kW (12.3 lb/hp)

PERFORMANCE:
Max level speed at mid-cruise weight
　　232 knots (430 km/h; 267 mph)
Cruising speed at optimum altitude, mid-cruise weight:
　high speed cruise power
　　225 knots (417 km/h; 259 mph)
　normal cruise power　215 knots (398 km/h; 247 mph)
　econ cruise power　199 knots (369 km/h; 229 mph)
　long range cruise power
　　168 knots (311 km/h; 193 mph)
Stalling speed:
　flaps and wheels up　71 knots (132 km/h; 82 mph)
　flaps and wheels down　60 knots (111 km/h; 69 mph)
Max rate of climb at S/L　　371 m (1,218 ft)/min
Max certificated ceiling　　7,620 m (25,000 ft)
T-O run　　467 m (1,530 ft)
T-O to 15 m (50 ft)　　724 m (2,375 ft)
Landing from 15 m (50 ft)　　599 m (1,964 ft)
Landing run　　311 m (1,018 ft)
Range with max fuel, allowances for start, T-O, climb, descent, plus 45 min reserves, at optimum altitude:
　high speed cruise power
　　990 nm (1,834 km; 1,140 miles)
　normal cruise power
　　1,056 nm (1,957 km; 1,216 miles)
　econ cruise power　1,184 nm (2,194 km; 1,363 miles)
　long range cruise power
　　1,450 nm (2,687 km; 1,669 miles)

QUESTAIR

QUESTAIR INC
7700 Airline Road, PO Box 18946, Greensboro, North Carolina 27419
Telephone: 1 (919) 668 7890 or (800) 852 3889
Fax: 1 (919) 668 7960
CEO: Mark Chambers
DIRECTOR, SALES AND MARKETING: Charles S. Haupt
　Questair was purchased by Chicago businessman Bob McLallen. Details of M20 Venture in 1992-93 *Jane's*.

QUESTAIR SPIRIT
TYPE: Side by side 2+1 seat sporting and cross-country homebuilt.
PROGRAMME: Construction of prototype began in 1990; first flight February 1991. Kits are available, with engine included at customer's request.
COSTS: Kit: $24,900 (1992).

Questair Spirit two/three-seat sporting and cross-country monoplane

DESIGN FEATURES: Non-retractable landing gear derivative of Venture, powered by Teledyne Continental IO-360 engine but with Textron Lycoming O-360 as optional power plant.
LANDING GEAR: Non-retractable tricycle type.
POWER PLANT: One 156.6 kW (210 hp) Teledyne Continental IO-360-ES flat-six engine. Alternatively 134 kW (180 hp) Textron Lycoming O-360. Fuel capacity 212 litres (56 US gallons; 46.6 Imp gallons).
DIMENSIONS, EXTERNAL:
Wing span 8.38 m (27 ft 6 in)
Wing aspect ratio 10.43
Length overall 4.95 m (16 ft 3 in)
Height overall 2.34 m (7 ft 8 in)
Propeller diameter 1.73 m (5 ft 8 in)
AREAS:
Wings, gross 6.74 m² (72.50 sq ft)
WEIGHTS AND LOADINGS:
Weight empty 465 kg (1,025 lb)
Max T-O weight 771 kg (1,700 lb)
Max wing loading 114.5 kg/m² (23.45 lb/sq ft)
Max power loading 4.92 kg/kW (8.10 lb/hp)
PERFORMANCE:
Max level speed at 2,135 m (7,000 ft)
 195 knots (361 km/h; 225 mph)
Max cruising speed at 2,135 m (7,000 ft)
 185-195 knots (343-361 km/h; 213-225 mph)
Stalling speed 56 knots (104 km/h; 65 mph)
Max rate of climb at S/L
 457-610 m (1,500-2,000 ft)/min
Range, with reserves 1,030 nm (1,908 km; 1,186 miles)
Endurance 6 h 10 min
g limits +6/−3

RAISBECK

RAISBECK ENGINEERING INC
7675 Perimeter Road South, Boeing Field International, Seattle, Washington 98108
Telephone: 1 (206) 763 2000
Fax: 1 (206) 767 7726
CEO: James D. Raisbeck
VICE-PRESIDENT, SERVICE AND SALES: Robert P. Steinbach
DIRECTOR, TECHNICAL SERVICES: Michael P. Tougher

Develops, certificates and sells general aviation advanced technology modifications to improve performance, safety and productivity of business aircraft.

Company holds 25 STCs from FAA for upgrade systems for entire Beech King Air/Super King Air range: these are grouped into **TFPS** (Hartzell/Raisbeck Quiet Turbofan Propeller Systems) with four-blade Hartzell propellers to reduce external noise, and cabin and flight deck noise levels reduced by 7-10 dBA; **RARS** (Ram Air Recovery System), developed in association with P&WC for Super King Air 200 to have more complete sealing of the engine nacelle inlet section and new fixed turning vane; **SFES** (Short Field Enhancement System; Enhanced Performance Leading-Edges) for Super King Air 200/B200, featuring new composites constructed wing leading-edges, intercooler ducting, wing-to-fuselage fairings and flush-mounted Goodrich de-icing boots, combined with Hartzell Quiet Turbofan Propeller conversion to reduce T-O speed and FAA T-O run substantially; **DABS** (Dual Aft Body Strakes) that substitute dual ventral strakes for original single strakes on all King Air models to reduce rear cabin area vibration; **HFGD** (High-Flotation Gear Doors) for King Air F90/100s and Super King Air 200s; **NWLS** (Nacelle Wing Locker Systems) certificated to FAA FAR Pt 135 for King Air 90/100 and Super King Air 200/300, to carry 272 kg (600 lb) and increase baggage volume by 0.45 m³ (16 cu ft); and **SESF/CESF** (Stainless Exhaust Stack Fairing/Composite Exhaust Stack Fairings) for all models.

In late 1992 it was estimated that some 130 King Airs/Super King Airs were adding Raisbeck systems each year. Further details in 1992-93 *Jane's* and current *Jane's Civil and Military Aircraft Upgrades*. A further STC covers Hartzell/Raisbeck Quiet Turbofan Propellers with optional 71 per cent cruise rpm for de Havilland Canada DHC-6-300 Twin Otter.

RAM

RAM AIRCRAFT CORPORATION
7505 Airport Drive, Waco Regional Airport, PO Box 5219, Waco, Texas 76708
Telephone: 1 (817) 752 8381
Fax: 1 (817) 752 3307
Telex: 1 800 445 9713 (USA)
PRESIDENT: Jack M. Riley Jr (Engineering)
INTERNATIONAL SALES:
 Doug Mackay (South America)
 Bob Neal (Europe and Pacific)
SALES MANAGER: David Seesing
PRESS RELATIONS: Chuck Morrow

RAM specialises in modification of selected single- and twin-engined general aviation aircraft for improved performance and efficiency. All modifications FAA approved by award of STC. Export STC modification kits available with new engines or RAM-remanufactured 100 per cent balanced versions of Teledyne Continental TSIO-520 engine (further details in 1990-91 *Jane's*). RAM Cessna 310, 320, 340 and 414A conversions designated Series I (310 hp) and Series III (325 hp) with Teledyne Continental camshafts, and Series II and Series IV respectively with RAM economy camshafts developed by Crane Cams Inc of Daytona Beach, Florida; these give easier starting, increased manifold pressure and smoother idling, together with claimed 3-5 per cent reduction in fuel consumption at cruise power. Refurbishing, repairing and maintenance also offered.

RAM/Cessna 172 and T210 STC to install 119 kW (160 hp) Textron Lycoming O-320-D2G or uprated O-320-E2D. Similar upgrade available for **Piper PA-28-140** and **PA-28-151** with O-320-D3G. Replacement of TSIO-520-C/H engine by 231 kW (310 hp) RAM remanufactured TSIO-520-M/R in **RAM/Cessna T206/210**.

New or RAM remanufactured and uprated TSIO-520 engines of various models installed in **Cessna 310, Cessna 340/340A, Cessna 401/402A/402B/402C, Cessna 414, Cessna 421C, Beech Baron 58P/58TC**, in most cases to allow operation at increased max T-O weight and to offer increased useful load; some models with Hartzell Q-tip propellers; **Cessna 414AW** and **Cessna 421CW** have winglets. **Cessna 414AW Series V** has 261 kW (350 hp) liquid-cooled TSIO-550A Voyager engines driving new McCauley three-blade propellers. **Series VI** announced 1991; 250 kW (335 hp) TSIO-520 engines installed in Cessna 340/340A/414; vortex generators, intercooler scoops and landing gear STC; max T-O weight and useful load increased.

Further details of RAM conversions can be found in 1991-92 and earlier *Jane's*, and currently in *Jane's Civil and Military Aircraft Upgrades*.

RANS

RANS INC
4600 Highway 183 Alternate, Hays, Kansas 67601
Telephone: 1 (913) 625 6346
Fax: 1 (913) 625 2795
PRESIDENT: Randy Schlitter

Details of S-4 Coyote, S-6 Coyote II, S-9 Chaos, S-10 Sakota, S-11 Pursuit, S-12 Airaile and S-14 Airaile can be found in 1992-93 *Jane's*.

RANS S-7 COURIER
TYPE: Two-seat STOL homebuilt.
PROGRAMME: Offering standard kits (300-500 working hours to assemble) or a quick-build kit (150 working hours, approximately). Prototype first flew October 1985. Estimated building time from kit is 500-700 working hours.
COSTS: Kit: $12,600.
DESIGN FEATURES: Follows general configuration of Coyote (see 1992-93 *Jane's*); fitted with dual controls and suited to training, recreational flying, agricultural spraying etc.
FLYING CONTROLS: Conventional control, with flaps.
STRUCTURE: Fabric covered, with aluminium alloy wings, and welded 4130 steel fuselage and tail unit. Glassfibre engine cowling.
LANDING GEAR: Non-retractable tailwheel type. Optional floats and skis.
POWER PLANT: One 48.5 kW (65 hp) Rotax 582, with 2.58:1 reduction gear. Optional 67 kW (90 hp) AMW 636. Fuel capacity 49 litres (13 US gallons; 10.8 Imp gallons).
DIMENSIONS, EXTERNAL:
Wing span 8.92 m (29 ft 3 in)
Wing aspect ratio 5.67
Length overall 6.40 m (21 ft 0 in)
Height overall 1.91 m (6 ft 3 in)
Propeller diameter 1.73 m (5 ft 8 in)

RANS S-7 Courier homebuilt aircraft

AREAS:
Wings, gross 14.03 m² (151.0 sq ft)
WEIGHTS AND LOADINGS:
Weight empty, equipped 204 kg (450 lb)
Max T-O weight 465 kg (1,025 lb)
Max wing loading 33.14 kg/m² (6.79 lb/sq ft)
PERFORMANCE (two persons):
Cruising speed 69 knots (129 km/h; 80 mph)
Stalling speed, power off:
 flaps up 31 knots (57 km/h; 35 mph)
 flaps down 26 knots (49 km/h; 30 mph)
Max rate of climb at S/L 229 m (750 ft)/min
Service ceiling 4,115 m (13,500 ft)
T-O run 54 m (175 ft)
Landing run to safe turn speed 93 m (303 ft)
Range 208 nm (386 km; 240 miles)
Endurance 3 h
g limits +6/−3

RILEY

RILEY INTERNATIONAL CORPORATION
2206 Palomar Airport Road, Suite B-2, Carlsbad, California 92008
Telephone: 1 (619) 438 9089
Fax: 1 (619) 438 0578
PRESIDENT: Jack M. Riley

Company formed to continue marketing Riley conversions of production aircraft. Riley holds total of 120 STCs for conversions developed over 39 years. Current models: **Riley Rocket P-210** conversion of Cessna P-210 Pressurised Centurion offered with remanufactured 231 kW (310 hp) turbocharged Teledyne Continental TSIO-520-AF and numerous avionics, equipment and performance enhancement changes; **Riley Sky Rocket Cessna P-337** offered with similar basic conversion and optional features.

Further details in 1991-92 and earlier *Jane's*; can also now be found in *Jane's Civil and Military Aircraft Upgrades*.

ROBINSON
ROBINSON HELICOPTER COMPANY
24747 Crenshaw Boulevard, Torrance, California 90505
Telephone: 1 (310) 539 0508
Fax: 1 (310) 539 5198
Telex: 18-2554
PRESIDENT: Franklin D. Robinson
MARKETING DIRECTOR: Tim Goetz

In late 1992 had 420 employees; new 24,619 m² (265,000 sq ft) plant for R44 production scheduled for completion late 1993; plans to add at least 150 employees 1993 and work up to 800 when R44 in full production.

ROBINSON R22 BETA
TYPE: Light two-seat helicopter.
PROGRAMME: Design began 1973; first flight 28 August 1975; first flight of second R22 early 1977; FAA certification 16 March 1979; UK certification June 1981; deliveries began October 1979; R22 Alpha certificated October 1983; R22 Beta announced 5 August 1985; 2,000th R22 delivered 22 November 1991; 222 delivered in 1992, 67 per cent for export.
CURRENT VERSIONS: **R22 Beta:** Standard version from c/n 501 onwards; increased horsepower. *Main description applies to this version except where indicated.*
R22 Mariner: Has floats and ground wheels; first delivered for operation from tuna fishing boats off Mexico and Venezuela; 21 delivered in 1992.
R22 Police: Special communications and other equipment including removable port-side controls, 70A alternator, searchlight, loudspeaker, siren and transponder.
R22 IFR: Equipped with flight instruments and radio to allow training for helicopter IFR flying (see under Avionics).
External load R22: Hook kit certificated for 181 kg (400 lb) produced by Classic Helicopter Corporation, Boeing Field, Seattle, Washington; weighs 2.3 kg (5 lb); used for slung load training; never-exceed speed with load in place limited to 75 knots (139 km/h; 86 mph).
R22 Agricultural: Equipped with Apollo Helicopter Services DTM-3 spray system; FAA approved December 1991. Low drag belly tank contains 151 litres (40 US gallons; 33.3 Imp gallons); boom length 7.31 m (24 ft 0 in); tank frame attached to landing gear mounting points with four bolts and wing nuts; installation of entire system requires no tools and can be completed by one person in approximately five minutes. System cost $6,000 (1992) including installation kit.

CUSTOMERS: Production rate eight per week in early 1993; total production 2,260 by 31 November 1992, when outstanding orders totalled approx 25-30 aircraft. Ten delivered to Turkish Army for training, early 1992. Forty ordered for Buenos Aires Police late 1992, for delivery at four per month from January 1993; comprises 27 Betas, 10 Mariners and three IFRs.
COSTS: $120,000 typically.
DESIGN FEATURES: Horizontally mounted piston engine drives transmission through multiple V belts and sprag-type overrunning clutch; main and tail gearboxes use spiral bevel gears; maintenance-free flexible couplings of proprietary manufacture used in both main and tail rotor drives. Two-blade semi-articulated main rotor, with tri-hinged under-slung rotor head to reduce blade flexing, rotor vibration and control force feedback, and an elastic teeter hinge stop to prevent blade-boom contact when starting or stopping rotor in high winds; blade section NACA 63-015 (modified); two-blade tail rotor on port side; rotor brake standard.
FLYING CONTROLS: All mechanical; cyclic stick mounted between pilots with hand grips on swing arm for comfortable access from either seat.
STRUCTURE: All-metal bonded blades with stainless steel spar and leading-edge, light alloy skin and light alloy honeycomb filling; cabin section of steel tube with metal and plastics skinning; full monocoque tailboom.
LANDING GEAR: Welded steel tube and light alloy skid landing gear, with energy absorbing crosstubes. Twin float/skid gear on Mariner with additional tailplane surface on lower tip of fin.
POWER PLANT: One 119 kW (160 hp) Textron Lycoming O-320-B2C flat-four engine (derated to 97.5 kW; 131 hp for T-O), mounted in the lower rear section of the main fuselage, with cooling fan. Light alloy main fuel tank in upper rear section of the fuselage on port side, usable capacity 72.5 litres (19.2 US gallons; 16 Imp gallons). Optional auxiliary fuel tank, capacity 39.75 litres (10.5 US gallons; 8.7 Imp gallons). Oil capacity 5.7 litres (1.5 US gallons; 1.25 Imp gallons).
ACCOMMODATION: Two seats side by side in enclosed cabin, with inertia reel shoulder harness. Curved two-panel windscreen. Removable door, with window, on each side. Police version has observation doors with bubble windows, which are also available as options on other models. Baggage space beneath each seat. Cabin heated and ventilated.
SYSTEMS: Electrical system, powered by 12V DC alternator, includes navigation, panel and map lights, dual landing lights, anti-collision light and battery. Second battery optional.
AVIONICS: A Bendix/King KY 197 VHF radio is standard optional avionics include a KN 53 nav receiver, Apollo II Bendix/King or Northstar Loran C, KT 76A transponder and KR 87 ADF. IFR trainer standard equipment is same as for basic Beta. Avionics include AIM 305-1AL DVF artificial horizon, Bendix/King KEA 129 encoding altimeter Astronautics DC turn co-ordinator, Bendix/King KCS 55A HSI, KR 87 ADF, KX 165 nav/com digital display radio KT 76A transponder and KR 22 marker beacon receiver and Astro Tech LC-2 digital clock. IFR trainer optional avionics include Bendix/King KN 63 DME.
EQUIPMENT: Standard equipment includes rate of climb indicator, sensitive altimeter, quartz clock, hour meter, low rotor rpm warning horn, temperature and chip warning lights for main gearbox and chip warning light for tail gearbox.

DIMENSIONS, EXTERNAL:
Main rotor diameter	7.67 m (25 ft 2 in)
Tail rotor diameter	1.07 m (3 ft 6 in)
Main rotor blade chord	0.18 m (7.2 in)
Distance between rotor centres	4.39 m (14 ft 5 in)
Length overall (rotors turning)	8.76 m (28 ft 9 in)
Fuselage: Length	6.30 m (20 ft 8 in)
Max width	1.12 m (3 ft 8 in)
Height overall	2.67 m (8 ft 9 in)
Skid track	1.93 m (6 ft 4 in)

DIMENSIONS, INTERNAL:
Cabin: Max width	1.12 m (3 ft 8 in)

AREAS:
Main rotor blades (each)	0.70 m² (7.55 sq ft)
Tail rotor blades (each)	0.037 m² (0.40 sq ft)
Main rotor disc	46.21 m² (497.4 sq ft)
Tail rotor disc	0.89 m² (9.63 sq ft)
Fin	0.21 m² (2.28 sq ft)
Stabiliser	0.14 m² (1.53 sq ft)

WEIGHTS AND LOADINGS:
Weight empty (without auxiliary fuel tank)	379 kg (835 lb)
Fuel weight: standard	52 kg (115 lb)
auxiliary	28.6 kg (63 lb)
Max T-O and landing weight	621 kg (1,370 lb)
Max zero-fuel weight	569 kg (1,255 lb)
Max disc loading	13.43 kg/m² (2.75 lb/sq ft)

PERFORMANCE (at max T-O weight):
Never-exceed speed (V_{NE}) without sling load	102 knots (190 km/h; 118 mph)
Max level speed	97 knots (180 km/h; 112 mph)
Cruising speed, 75% power at 2,440 m (8,000 ft)	96 knots (177 km/h; 110 mph)
Econ cruising speed	82 knots (153 km/h; 95 mph)
Max rate of climb at S/L	366 m (1,200 ft)/min
Rate of climb at 1,525 m (5,000 ft)	323 m (1,060 ft)/min
Service ceiling	4,265 m (14,000 ft)
Hovering ceiling IGE	2,125 m (6,970 ft)
Range with auxiliary fuel and max payload, no reserves	319 nm (592 km; 368 miles)
Endurance at 65% power, auxiliary fuel, no reserves	3 h 20 min

ROBINSON R44
TYPE: Four-seat light helicopter with dual controls.
PROGRAMME: Development began 1986; first flight 31 March 1990; prototype (N44RH) and second aircraft had flown more than 200 hours by end January 1992; first flight of third R44 March 1992; certification completed 10 December 1992; expected in production 1993 at new factory at present site.
CUSTOMERS: First 25 supplied only to customers in south western USA for easy recall if modification required $15,000 deposits for 36 taken on first day of sales 22 March 1992; order backlog exceeded 115 by 30 November 1992 first production R44 (c/n 0005) to be delivered to customer at Heli-Expo 1993.
COSTS: Standard price $235,000 (1993).
DESIGN FEATURES: New design incorporating some proven concepts of R22; designed to requirements of FAR Pt 27 features for comfort and safety include electronic throttle governor to reduce pilot workload by controlling rotor and engine rpm during normal operations, rotor brake advanced warning devices, automatic clutch engagement to simplify and reduce start-up procedure and reduce chance of overspeed, T-bar cyclic control, and crashworthy features including energy-absorbing landing gear and lap/shoulder strap restraints designed for high forward g loads. High reliability and low maintenance; patented rotor design with tri-hinge (see R22); low noise levels.
FLYING CONTROLS: Conventional, with Robinson central cyclic stick; rpm governor; rotor brake standard; left-hand collective lever and pedals removable.
LANDING GEAR: Fixed skids.
POWER PLANT: One 194 kW (260 hp) Textron Lycoming O-540 flat-six engine, derated to 165 kW (225 hp) at T-O 153 kW (205 hp) continuous.
ACCOMMODATION: Four persons seated 2 + 2. Concealed baggage compartments.
SYSTEMS: Heating and ventilation.

Robinson R22 Agricultural with Apollo DTM-3 spray system

Robinson R22 Beta two-seat helicopters awaiting delivery

AVIONICS: Wide selection of optional avionics and instruments.
EQUIPMENT: Standard equipment includes auxiliary fuel system, heater and Bendix/King KY 197 VHF radio with intercom for all four occupants.

DIMENSIONS, EXTERNAL:
Main rotor diameter	10.06 m (33 ft 0 in)
Tail rotor diameter	1.47 m (4 ft 10 in)
Height overall	3.28 m (10 ft 9 in)
Skid track	2.18 m (7 ft 2 in)

DIMENSIONS, INTERNAL:
Cabin: Max width	1.28 m (4 ft 2½ in)

AREAS:
Main rotor disc	79.46 m² (855.3 sq ft)
Tail rotor disc	1.70 m² (18.35 sq ft)

WEIGHTS AND LOADINGS:
Weight empty, standard	635 kg (1,400 lb)
Fuel weight: standard	86 kg (190 lb)
auxiliary	50 kg (110 lb)
Max T-O and landing weight	1,088 kg (2,400 lb)
Max disc loading	13.70 kg/m² (2.81 lb/sq ft)

PERFORMANCE:
Cruising speed, 75% power	113 knots (209 km/h; 130 mph)
Max rate of climb at S/L	305 m (1,000 ft)/min
Service ceiling	4,270 m (14,000 ft)
Hovering ceiling:	
IGE at 1,088 kg (2,400 lb)	1,860 m (6,100 ft)
OGE	1,370 m (4,500 ft)
Max range, no reserves	approx 347 nm (643 km; 400 miles)

Robinson R44 four-seat light helicopter

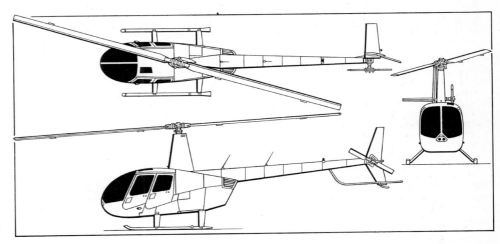

Robinson R44 light helicopter (one Textron Lycoming O-540 flat-six) *(Jane's/Mike Keep)*

ROCKWELL INTERNATIONAL

ROCKWELL INTERNATIONAL CORPORATION
2201 Seal Beach Boulevard, PO Box 4250, Seal Beach, California 90740-8250
Telephone: 1 (310) 797 3311
CHAIRMAN AND CEO: Donald R. Beall
EXECUTIVE VICE-PRESIDENTS AND COOs:
 Kent M. Black
 Sam F. Iacobellis

North American Aviation founded 1928 and manufactured aircraft from 1934; merged with Rockwell-Standard Corporation of Pittsburgh, Pennsylvania (which manufactured Aero Commander aircraft), 22 September 1967, forming North American Rockwell Corporation; present name adopted 1973. Four major businesses comprise aerospace, automotive, electronics and graphics. Aerospace includes production of military aircraft, manned and unmanned space systems, rocket engines, advanced space-based surveillance systems, and high energy laser and other directed energy programmes. Electronics includes industrial automation equipment and systems, avionics products and systems, and related communications technologies primarily used in commercial and military aircraft; commercial telecommunications systems and products; and defence electronics systems and products for precision guidance and control, for tactical weapons and for command, control, communications and intelligence.

NORTH AMERICAN AIRCRAFT
201 North Douglas Street, El Segundo, California 90245
Telephone: 1 (213) 647 1000
PRESIDENT: John J. Pierro
Palmdale Facility:
2825 East Avenue P, Palmdale, California 93550
Telephone: 1 (805) 273 6000
VICE-PRESIDENT AND GENERAL MANAGER: C. W. Bright
Tulsa Facility:
2000 North Memorial Drive, Tulsa, Oklahoma 74158
Telephone: 1 (918) 835 3111

VICE-PRESIDENT AND GENERAL MANAGER:
 William P. Swiech

In addition to programmes detailed, North American Aircraft is working on X-30 National Aero-Space Plane (NASP—see NASA entry); Rockwell teamed with DASA in X-31A programme and to offer Ranger 2000 for JPATS training aircraft requirement (see International section). Rockwell has Lockheed as partner in one of five consortia studying potential A-X naval strike aircraft.

ROCKWELL (LOCKHEED) AC-130U SPECTRE

TYPE: Aerial gunship.
PROGRAMME: Development of new gunship version of Lockheed C-130 Hercules launched 6 July 1987 with $155,233,489 contract to North American Aircraft Operations; prototype (87-0128) funded in FY 1986, six production aircraft in FY 1989, five in FY 1990 and one in FY 1992 — last mentioned contract awarded 31 December 1992, increasing total to 13.

AC-130U FUNDING

Batch	FY	Quantity	First aircraft
—	1987	1	87-0128
Lot I	1989	6	89-0509
Lot II	1990	5	90-0163
Lot III	1992	1	92-

Prototype delivered to Rockwell as standard C-130 transport, 28 July 1988; first post-conversion flight on 20 December 1990 was also ferry to Edwards AFB, California, for trials with 6510th Test Wing; work on remaining AC-130Us began January 1991; by January 1993, six aircraft complete, of which three with 418th Test Squadron/412th TW (ex-6510th TW) at Edwards AFB; operational unit to be 16th Special Operations Squadron at Hurlburt Field, Florida; deliveries from 1994. Armament, firing to port, consists of (front to rear) General Electric GAU-12/U 25 mm six-barrel Gatling gun with 3,000 rounds, Bofors 40 mm gun, and a 105 mm gun based on US Army howitzer; addition of Rockwell Hellfire ASMs under consideration 1992; guns can be slaved to Hughes AN/APQ-180 (modified AN/APG-70) digital fire control radar, Texas Instruments AN/AAQ-117 FLIR or GEC-Marconi all-light-level television (ALLTV), for night and adverse weather attack on ground targets; sideways-facing

Third production Lockheed Hercules conversion by Rockwell to AC-130U Spectre gunship

552 USA: AIRCRAFT—ROCKWELL INTERNATIONAL/ROGERSON

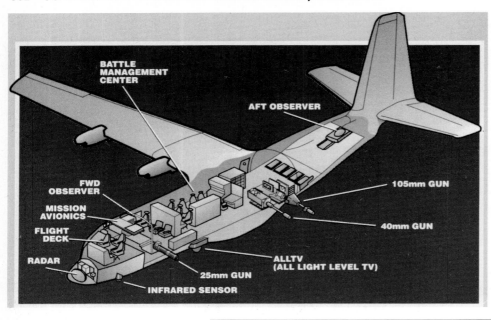

Weapon and sensor locations in the AC-130U gunship

HUD for visual aiming. Attack method is to circle target at altitude firing into apex of turn on ground, but guns can now be trained, relieving pilot of absolute precision flying; flight path is also less predictable; can fire on two targets simultaneously. AC-130U can refuel in flight and fly escort, surveillance, search, rescue and armed reconnaissance/interdiction missions.

Prone observer's position on rear ramp; starboard side observer's window aft of flight deck, and battle management centre in cabin with seven positions at monitoring consoles and four IBM AP-102 computers; crew totals 13 including flight crew and loaders. Defensive aids believed similar to those in MC-130H; modified fuel tank pylons contain IR countermeasures; total of 300 chaff bundles and 90 MJU7 or 180 M206 flares in three launchers under fuselage; ITT Avionics AN/ALQ-172 jammer in base of fin; Loral AN/ALR-56M RWR; other equipment includes combined INS and Navstar GPS, triple MIL-STD-1553B digital databuses and Spectra ceramic armour protection.

Full details of C-130H in Lockheed entry in this section.

AEROSPACE BUSINESSES
2201 Seal Beach Boulevard, PO Box 4250, Seal Beach, California 90740-8250
Telephone: 1 (310) 797 3311
EXECUTIVE VICE-PRESIDENT AND COO: Sam F. Iacobellis
Rocketdyne Division
6633 Canoga Avenue, Canoga Park, California 91303
Telephone: 1 (818) 710 6300
PRESIDENT: Robert D. Paster

Space Systems Division
12214 Lakewood Boulevard, Downey, California 90241
Telephone: 1 (810) 922 2111
PRESIDENT: Robert G. Minor
MANAGER, EXTERNAL COMMUNICATIONS AND MEDIA RELATIONS: Janet L. Dean

Group includes Downey, Seal Beach and Palmdale facilities and contractual support groups in main national space centres; work includes development and fabrication of manned and unmanned space systems; Rockwell group produced five Shuttle Orbiter spacecraft for NASA, last described fully in 1992-93 *Jane's*.

ROCKY MOUNTAIN
ROCKY MOUNTAIN HELICOPTERS INC
PO Box 1337, Provo, Utah 84603
Telephone: 1 (801) 375 1124
Fax: 1 (801) 375 6712
CHAIRMAN AND CEO: James Burr
PRESIDENT AND COO: J. Russell Spray
DIRECTOR OF MARKETING: Orin 'Spike' Kinghorn
DIRECTOR OF MARKETING, MEDICAL: David L. Dolstein
VICE-PRESIDENT, HUMAN RESOURCES: Donald G. Andrews

ROCKY MOUNTAIN HELICOPTERS ALLSTAR AS 350 CONVERSION
Programme to re-engine Eurocopter AS 350 with 485 kW (650 shp) Allison 250-C30M turboshaft in place of standard Arriel or LTS101; supplemental type certificate purchased from Soloy 1990, since when weight has been reduced by 85 kg (187 lb) partly by use of composites in cowling; first conversion completed 1990 and certification then expected April 1991; conversion kit without engine costs $65,000; engine and installation by Rocky Mountain Helicopters $290,000; sold as conversion kit.

ROCKY MOUNTAIN HELICOPTERS LAMA CONVERSION
Design for replacing fixed shaft Turbomeca Artouste IIIB in Eurocopter Lama with 671 kW (900 shp) P&WC PT6B-35H free turbine turboshaft started early 1992; re-engined Lama (N220 RM) was introduced at Heli-Expo 93 in February 1993, after 8 February first flight; engine change expected to reduce maintenance, fuel consumption, weight and noise; conversion, which takes five hours, carried out by AeroProducts International, a subsidiary of RMH.

ROGERSON
ROGERSON HILLER CORPORATION (Subsidiary of Rogerson Aircraft Corporation)
2140 West 18th Street, Port Angeles, Washington 98362
Telephone: 1 (206) 452 6891
Fax: 1 (206) 457 9039
PRESIDENT: Norman B. Hirsh
VICE-PRESIDENT, BUSINESS DEVELOPMENT: James Brown

Formed January 1973 as Hiller Aviation Inc to acquire from Fairchild Industries the design rights, production tooling and spares of Hiller 12E helicopters and support worldwide Hiller fleet; manufactured new 12Es and 12E4s and revived five-seat FH-1100 after 1980; Hiller Aviation became wholly owned subsidiary of Rogerson Aircraft April 1984; renamed Hiller Helicopters and began deliveries of redesignated UH-12E and RH-1100 July 1984; since renamed Rogerson Hiller; moved to new facility at Port Angeles, Washington, November 1985; L. O. M. Corporation, which produces rotor blades, acquired September 1985 and renamed Aerobond; rotor blades manufactured at Port Angeles. Rogerson produces underfloor fuel tanks for Boeing 737-300, -400 and -500, and participates in Lockheed-led development and test of flight refuelling hosereel unit for USAF.

Production of UH-12E relaunched during 1991 and development of RH-1100S continued in 1992 but company was put up for sale in Spring 1993, with RH-1100S development placed in abeyance.

HILLER RH-1100
TYPE: Five-seat utility helicopter.
PROGRAMME: Original FH-1100 was refined development of OH-5A; 246 built by Fairchild before production ended 1974. Development of Rogerson Hiller versions restarted early 1992.
CURRENT VERSIONS: **RH-1100C:** Original civil version, derived from FH-1100; prototype used for development of 1100S, but 1100C still potentially available for production under licence. Max T-O weight 1,292 kg (2,850 lb).
RH-1100S: Proposed uprated civil version with new seven-seat cabin section, more powerful power plant and lighter gross weight; development resumed early 1992, but suspended Spring 1993. Description in 1992-93 *Jane's*.

HILLER UH-12E HAULER
TYPE: Three/four-seat utility helicopter.
PROGRAMME: Resumed deliveries of original Hiller UH-12E production rate one per month early 1992. Proposed 1991 (as **UH-12ET**) for US Army New Training Helicopter (NTH) programme, with Soloy/Allison 250-C20B turboshaft, but unsuccessful.
CUSTOMERS: Deliveries of aircraft equipped for agricultural work to Government of India began May 1986; more recent orders from Hungary (three) and Taiwan (four).
DESIGN FEATURES: Blades interchangeable individually, bolted to forks which are retained at rotor head by tension-torsion bars; no blade folding. Main rotor blade thickness/chord ratio 12 per cent; main rotor/engine rpm ratio 1:8.66; tail rotor/engine rpm ratio 1:1.44. Mechanical drive through two-stage planetary main transmission; bevel gear drive to auxiliaries; tail rotor gearbox.
FLYING CONTROLS: Teetering two-blade main rotor manually controlled through stabiliser bar carrying aerodynamic paddles (servo rotor). Horizontal stabiliser incidence ground adjustable.
STRUCTURE: Airframe is light alloy platform with engine on steel tube mounting and beaded aluminium alloy sheet tailboom with no internal stiffeners; horizontal stabiliser with steel tube spar, aluminium alloy ribs and skin; horizontal stabiliser replaced on four-seat version by inverted-V surfaces of similar construction, mounted forward of tail rotor.
LANDING GEAR: Wide track light alloy tube skids carried on spring steel cross members. Optional extended legs. Ground handling wheels standard. Optional 'zip-on' pontoons can be attached above the skids to permit water or land operations, but require extended landing gear.
POWER PLANT: One 253.5 kW (340 hp) Textron Lycoming VO-540-C2A flat-six engine, installed vertically and derated to 227.5 kW (305 hp). Engine muffler optional. Single bladder fuel cell, capacity 174 litres (46 US gallons;

Hiller RH-1100C prototype, UH-12E Hauler and mockup of RH-1100S

38 Imp gallons), mounted in lower portion of rear fuselage, beneath engine. Two optional auxiliary fuel tanks, mounted in fuselage on each side of engine; capacity 76 litres (20 US gallons; 16.6 Imp gallons) each. Oil capacity 12.5 litres (3.3 US gallons; 2.7 Imp gallons).
ACCOMMODATION: Three persons side by side on bench seat. Seat belts with provision for shoulder harness. Dual controls optional. Forward-hinged door on each side, with sliding window. Baggage compartment immediately aft of engine. Heater/defroster optional.
SYSTEMS: Electrical system includes a 72A alternator, nickel-cadmium battery, and battery temperature monitor.
AVIONICS: A range of optional avionics is available.
EQUIPMENT: External cargo hook and various spraygear systems optional.

DIMENSIONS, EXTERNAL:
Main rotor diameter	10.80 m (35 ft 5 in)
Main rotor blade chord (constant)	0.33 m (1 ft 1 in)
Servo rotor diameter	3.05 m (10 ft 0 in)
Tail rotor diameter	1.68 m (5 ft 6 in)
Distance between rotor centres	6.17 m (20 ft 3 in)
Length: overall, rotors turning	12.41 m (40 ft 8½ in)
fuselage	8.69 m (28 ft 6 in)
Height to top of rotor head	3.08 m (10 ft 1¼ in)
Skid track	2.29 m (7 ft 6 in)
Cabin doors (standard, each):	
Height	1.13 m (3 ft 8½ in)
Max width	0.81 m (2 ft 8 in)
Height to sill	0.58 m (1 ft 11 in)

DIMENSIONS, INTERNAL:
Cabin: Length	1.52 m (5 ft 0 in)
Max width	1.50 m (4 ft 11 in)
Max height	1.35 m (4 ft 5 in)
Floor area	1.16 m² (12.5 sq ft)

AREAS:
Tail rotor blades (each)	0.094 m² (1.01 sq ft)
Main rotor disc	91.97 m² (990.0 sq ft)
Tail rotor disc	2.57 m² (27.7 sq ft)

WEIGHTS AND LOADINGS:
Weight empty	798 kg (1,759 lb)
Max T-O weight	1,406 kg (3,100 lb)
Max disc loading	15.28 kg/m² (3.13 lb/sq ft)
Max power loading	6.18 kg/kW (10.16 lb/hp)

PERFORMANCE (at max T-O weight, except where indicated):
Never-exceed (V_{NE}) and max level speed	83 knots (154 km/h; 96 mph)
Cruising speed	78 knots (145 km/h; 90 mph)
Max rate of climb at S/L	393 m (1,290 ft)/min
Vertical rate of climb at S/L	225 m (740 ft)/min
Service ceiling:	
at AUW of 1,270 kg (2,800 lb)	4,575 m (15,000 ft)
at max T-O weight	2,255 m (7,400 ft)
Hovering ceiling IGE:	
at AUW of 1,270 kg (2,800 lb)	3,170 m (10,400 ft)
at max T-O weight	2,315 m (7,600 ft)
Hovering ceiling OGE:	
at AUW of 1,270 kg (2,800 lb)	2,070 m (6,800 ft)
at max T-O weight	1,155 m (3,800 ft)
Range, 30 min reserves:	
standard fuel	150 nm (278 km; 173 miles)
auxiliary fuel	316 nm (585 km; 364 miles)
Endurance with auxiliary fuel, 30 min reserves	4 h 1 min

ROTORWAY
ROTORWAY INTERNATIONAL
300 South 25th Avenue, Phoenix, Arizona 85009
Telephone: 1 (602) 278 8899
Fax: 1 (602) 278 7657
PRESIDENT: John Netherwood

All assets of former RotorWay Aircraft Inc were purchased by Englishman John Netherwood, and new company established on 1 June 1990, now trading as RotorWay International. Product line centres on Exec 90, the former Exec with some 23 modifications.

ROTORWAY INTERNATIONAL EXEC 90
TYPE: Side by side two-seat homebuilt helicopter.
PROGRAMME: RotorWay International currently manufactures Exec 90 helicopter kits for final assembly by amateur builders, powered by company's own R. I. 2.66 litre (162 cu in) liquid-cooled engine; kit comes complete and only requires avionics and paint; in addition to regular kit, there is Quick Kit with many components prefinished (reducing building time from 500 working hours to 250); some foreign distributors also sell helicopter fully assembled and test flown for delivery. RotorWay offers complete flight orientation and maintenance training, and comprehensive customer service programme. Exec 90 currently marketed in USA and 33 other countries.
COSTS: Kit: $47,000 standard; $53,000 for Quick Kit.
DESIGN FEATURES: Asymmetrical aerofoil section two-blade main rotor. All-metal aluminium alloy blades attached to aluminium alloy teetering rotor hub by retention straps. Teetering tail rotor, with two blades each comprising steel spar and aluminium alloy skin. Elastomeric bearing rotor hub system with dual push/pull cable controlled swashplate for cyclic pitch control.
STRUCTURE: Blades as detailed under Design Features. Basic 4130 steel tube airframe structure, with wrap-around glass-fibre fuselage/cabin enclosure. Aluminium alloy monocoque tailboom.

RotorWay International Exec 90 two-seat homebuilt helicopter

LANDING GEAR: Basic twin skid type.
POWER PLANT: One 113.3 kW (152 hp) RotorWay International R. I. engine with dual electronic ignition. Standard fuel capacity 64.4 litres (17 US gallons; 14.2 Imp gallons).

DIMENSIONS, EXTERNAL:
Main rotor diameter	7.62 m (25 ft 0 in)
Length of fuselage	6.71 m (22 ft 0 in)
Height to top of main rotor	2.44 m (8 ft 0 in)

WEIGHTS AND LOADINGS:
Weight empty	420 kg (925 lb)
Crew weight	181 kg (400 lb)
Max T-O weight	646 kg (1,425 lb)

PERFORMANCE (at max T-O weight):
Never-exceed (V_{NE}) and max level speed	100 knots (185 km/h; 115 mph)
Normal cruising speed	82 knots (153 km/h; 95 mph)
Max rate of climb at S/L	305 m (1,000 ft)/min
Service ceiling	3,050 m (10,000 ft)
Hovering ceiling, with two persons:	
IGE	2,135 m (7,000 ft)
OGE	1,525 m (5,000 ft)
Range with max fuel at optimum cruising power	156 nm (290 km; 180 miles)
Endurance with max fuel at optimum cruising power	2 h

SABRELINER
SABRELINER CORPORATION
18118 Chesterfield Airport Road, Chesterfield, Missouri 63005-1121
Telephone: 1 (314) 537 3660
Fax: 1 (314) 537 9053/9587
Telex: 44-7227
OTHER WORKS: Lambert-St. Louis International Airport; Spirit of St. Louis Airport, Missouri; Perryville Municipal Airport, Missouri; Neosho, Missouri; Independence, Kansas
CHAIRMAN AND CEO: F. Holmes Lamoreux
VICE-PRESIDENT, GOVERNMENT OPERATIONS: Reuben D. Best
VICE-PRESIDENT, COMMERCIAL MARKETING: Karl R. Childs
VICE-PRESIDENT, ENGINEERING: Bob D. Hanks

Sabreliner Corporation formed July 1983 after Wolsey & Co bought Sabreliner Division of Rockwell International; supports about 600 Sabreliners still in service; refurbishes and upgrades Lockheed T-33 and AT-33 for sale in Central and South America; two Sabreliner 40R and 24 AT-33s delivered to Ecuadorian Air Force; Sabreliner 40 and 60 installation of overhauled P&W JT12s and fitted with Collins EFIS and new interior and exterior; contract completed for logistics support of US Navy Grumman TC-4Cs based at Whidbey Island (Washington), Oceana (Virginia) and Cherry Point (North Carolina) after awarded early 1985; five-year contract for worldwide support of US Navy and Marine Corps CT-39 Sabreliner awarded April 1989; installed service life extension to three T-39As and five T-39Bs, first of which was redelivered Spring 1988. Details now in *Jane's Civil and Military Aircraft Upgrades*.

Excalibur programme, for Sabreliner 40 and 60 with 10,000 flying hours, introduced early 1989; refurbished aircraft became Sabreliner 40EX and 60EX and have 5,000 more airframe hours; alternative modifications give 30,000 hour/15,000 mission and 30,000 hour/30,000 mission life extension. Service life extension contract for 644 Cessna T-37Bs of USAF Air Training Command awarded August 1989; modification includes replacement of wing carry-through structure, lower front wing spars, tail mounting structure and tailplane; after three Sabreliner-prepared prototypes, SLEP kits will be installed by US Air Force.

Sabreliner rebuilds Textron Lycoming T53 turboshafts for US Army Bell UH-1s and supports airborne and ground-based training systems for US Navy's Undergraduate Naval Flight Officer (UNFO) programme at NAS Pensacola, Florida. Preliminary designs of Model 85 derivative completed Autumn 1990; investment partner being sought.

Sabreliner contracted by US Navy to supply 17 T-39N radar trainers for use in UNFO programme; based on Sabreliner 40, STC approved June 1991. First aircraft flew 16 May 1991 and was delivered 28 June; all 17 now in service and fly approx 1,000 hours each month.

SADLER
SADLER AIRCRAFT CORPORATION
8225 East Montebello, Scottsdale, Arizona 85250
Telephone: 1 (602) 994 4631
Fax: 1 (602) 481 0574
PRESIDENT: William G. Sadler
MARKETING DIRECTOR: F. Brent Stewart

Formerly American Microflight Inc; developing series of low cost close support and patrol/reconnaissance derivatives.

SADLER A-22 LASA
TYPE: Light armed surveillance aircraft.
PROGRAMME: Design started June 1986; construction began January 1987; first flight of prototype (N22AB) 8 October 1989.
CURRENT VERSIONS: **A-22 LASA:** Standard version, with retractable landing gear.
T-22: Pod with two side by side seats and dual controls, fitted to same basic airframe.
CUSTOMERS: Procurement of 20 reported.

COSTS: Programme cost $2.5 million (1989); average equipped cost of production aircraft $150,000.
DESIGN FEATURES: Wings unswept; constant chord with Hoerner droop tips; outer panels fold upward in two hinged sections for transportation or storage. Wings use NACA 63₃-A218 laminar flow aerofoil section; 18 per cent thickness/chord ratio; 3° dihedral; 4° incidence.
FLYING CONTROLS: Plain ailerons; twin rudders operating in unison; elevator with electrically operated trim tab; four-position electrically operated trailing-edge flaps.

USA: AIRCRAFT—SADLER/SCALED

Prototype Sadler A-22 LASA single-seat light armed surveillance aircraft

STRUCTURE: Aluminium alloy wing with shear web main spar, tubular capstrips and spreaders, and aluminium alloy skin; Kevlar composite fuselage pod; twin aluminium alloy tailbooms; tail surfaces of aluminium alloy construction and skin.

LANDING GEAR: Retractable tricycle landing gear. Fully swivelling nosewheel with steering through differential braking of mainwheels.

POWER PLANT: One 224 kW (300 hp) water-cooled Chevrolet V-6 all-aluminium fuel injected engine, modified for aircraft use, with custom forged crankshaft, stainless steel forged connecting rods, forged aluminium pistons, stainless steel inlet and exhaust valves, and a heavy duty reduction drive to a two- or four-blade composites pusher propeller. Standard fuel capacity 75.7 litres (20 US gallons; 16.6 Imp gallons). Turbocharged version available for high altitude operations.

ACCOMMODATION (A-22): Pilot only, in enclosed cockpit, with bullet resistant Lexan canopy.

EQUIPMENT: Demonstrated with stabilised video camera, low light level TV/infra-red sensor; video display in cockpit; real time video downlink. In light ground attack role, BEI Defense Systems M146 armament management system control panel for BEI Hydra-70 rocket systems with seven-tube launchers.

ARMAMENT: Four underwing hardpoints for up to 454 kg (1,000 lb) of stores; provision for 7.62 mm machine gun in each inboard wing section.

DIMENSIONS, EXTERNAL:
Wing span	6.70 m (22 ft 0 in)
Wing chord, constant	1.27 m (4 ft 2 in)
Wing aspect ratio	5.28
Length overall	4.78 m (15 ft 8 in)
Height overall	1.14 m (3 ft 9 in)
Height, wings folded	2.13 m (7 ft 0 in)
Width, wings folded	2.24 m (7 ft 4 in)
Propeller diameter	1.42 m (4 ft 8 in)

AREAS:
Wings, gross	8.51 m² (91.6 sq ft)

WEIGHTS AND LOADINGS:
Weight empty	386 kg (850 lb)
Max payload	454 kg (1,000 lb)
Max T-O weight	975 kg (2,150 lb)
Max wing loading	114.6 kg/m² (23.47 lb/sq ft)
Max power loading	4.35 kg/kW (7.17 lb/hp)

PERFORMANCE (estimated at max T-O weight at S/L):
Max level speed at S/L	165 knots (306 km/h; 190 mph)
Stalling speed, flaps down	62 knots (115 km/h; 72 mph)
Max rate of climb at S/L	610 m (2,000 ft)/min
T-O run	153 m (500 ft)
Landing run	183 m (600 ft)
Range with standard fuel	260 nm (483 km; 300 miles)

SADLER FANJET

TYPE: Fanjet version of A-22.

DESIGN FEATURES: In joint co-operation programme with engine manufacturer; same basic configuration and size but with redesigned wings and fuselage pod.

SADLER OPV SERIES

TYPE: Optionally piloted patrol and observation aircraft.
PROGRAMME: Design of OPV series started January 1982; first model flew 27 August 1987. Since 1987 manned and unmanned tests conducted with autopilot, command uplink and autonomous guidance via GPS navigation.
CURRENT VERSIONS: **OPV18-50 and OPV18-100**: Wing span 5.49 m (18 ft 0 in).
OPV22-50 and OPV22-100: Wing span 6.71 m (22 ft 0 in); -50 models available with a 38 kW (51 hp) Rotax 532.2V two-stroke; -100 models have 71 kW (95 hp) Teledyne Continental GR-36 rotary engine.
COSTS: Programme cost $1.2 million (1988); manned version costs from $33,000 for basic OPV18-50 to $100,000 for fully equipped OPV22-100; high volume cost for UAVs from $10,000.
DESIGN FEATURES: Similar to A-22 except outer wings fold upward in single hinged sections; fixed landing gear when piloted, UAV is rail or zero-length launched and parachute recovered.

A-22 LASA description applies, except as follows:

DIMENSIONS, EXTERNAL:
Wing span: OPV18	5.49 m (18 ft 0 in)
OPV22	6.71 m (22 ft 0 in)
Wing aspect ratio: OPV18	4.32

AREAS:
Wings, gross: OPV18	6.97 m² (75.0 sq ft)

WEIGHTS AND LOADINGS (A: OPV18-50, B: OPV22-50, C: OPV18-100, D: OPV22-100):
* Max payload: A	45 kg (100 lb)
B	91 kg (200 lb)
C	136 kg (300 lb)
D	181 kg (400 lb)
Max T-O weight: A	329 kg (725 lb)
B	397 kg (875 lb)
C	454 kg (1,000 lb)
D	521 kg (1,150 lb)
Max wing loading: A	47.21 kg/m² (9.67 lb/sq ft)
C	65.08 kg/m² (13.33 lb/sq ft)
Max power loading: A	8.66 kg/kW (14.22 lb/hp)
B	10.45 kg/kW (17.16 lb/hp)
C	6.39 kg/kW (10.53 lb/hp)
D	7.34 kg/kW (12.11 lb/hp)

* UAV versions can carry additional 45-68 kg (100-150 lb) payload

PERFORMANCE (at max T-O weight):
Max level speed at S/L:	
A	100 knots (185 km/h; 115 mph)
B	102 knots (189 km/h; 117 mph)
C	130 knots (241 km/h; 150 mph)
D	128 knots (237 km/h; 147 mph)
Stalling speed, flaps down:	
A	47 knots (87 km/h; 54 mph)
B	40 knots (74 km/h; 46 mph)
C	56 knots (104 km/h; 65 mph)
D	46 knots (86 km/h; 53 mph)
Max rate of climb at S/L: A	183 m (600 ft)/min
B	213 m (700 ft)/min
C	229 m (750 ft)/min
D	244 m (800 ft)/min
T-O run: A	244 m (800 ft)
B	213 m (700 ft)
C	176 m (575 ft)
D	153 m (500 ft)
Landing run: A	213 m (700 ft)
B	176 m (575 ft)
C	244 m (800 ft)
D	122 m (400 ft)
Range with standard fuel:	
A	282 nm (523 km; 325 miles)
B	325 nm (603 km; 375 miles)
C	260 nm (482 km; 300 miles)
D	304 nm (563 km; 350 miles)

SCALED

SCALED COMPOSITES INC

Hangar 78, Mojave Airport, Mojave, California 93501
Telephone: 1 (805) 824 4541
Fax: 1 (805) 824 4174
PRESIDENT: Elbert L. (Burt) Rutan
VICE-PRESIDENT AND GENERAL MANAGER:
  Michael W. Melvill

Scaled Composites Inc bought by Beech Aircraft Corporation June 1985; sold back to Burt Rutan November 1988 and integrated in joint venture with Wyman-Gordon Company of Worcester, Massachusetts; Scaled will produce composite aerospace structures for Wyman-Gordon and continues to provide R&D facilities to individuals and companies; several projects developed for Beechcraft retained by Scaled. Scaled associated with Wyman-Gordon Composites (formerly KDI of Buena Park, California) to offer series production capacity.

Past projects have included NASA AD-1 oblique wing research aircraft (see 1981-82 *Jane's*); Fairchild/Ames 62 per cent scale New Generation Trainer (see 1982-83 *Jane's*); M115-6.85 SCAT 1 85 per cent scale demonstrator for the Beech 2000 Starship (see Beech Aircraft in this section); prototype Aviation Composites Mercury canard microlight (see Sport Aircraft section 1987-88 *Jane's*); Model 120-9E proof-of-concept Predator 480 agricultural aircraft and ATAC entry in 1985-86 *Jane's*); Rutan 81 Catbird (see Scaled entry in 1989-90 *Jane's*); full-scale demonstrator of California Microwave CM-44 manned surrogate UAV (see *Jane's Battlefield Surveillance Systems*); production run of Teledyne Ryan Aeronautical Scarab Mach 0.8 ground-launched reconnaissance UAV; composite wings and tail surfaces of Orbital Sciences/Hercules Pegasus air-launched space booster; sail for Americas Cup catamaran; Model 133-4.62 proof-of-concept AT³ (see 1992-93 *Jane's*); Model 143 Triumph (see 1991-92 *Jane's*); Pond Racer PR-01 (see 1992-93 *Jane's*); and subscale radar cross-section models of aircraft such as Sukhoi Su-25. Current programmes include Freewing Scorpion and RAPTOR UAVs.

Scaled 2,787 m² (30,000 sq ft) factory, next door to Rutan Aircraft Factory, has space and facilities for short-run production and flight testing.

RUTAN 151 ARES

TYPE: Agile Response Effective Support (ARES) low cost military and paramilitary aircraft.
PROGRAMME: Rutan made design study of US Army Low Cost Battlefield Attack Aircraft (LCBAA) 1981; ARES design began 1985; first flight (N151SC) 19 February 1990; 178 hours flown by December 1991; speeds demonstrated to 305 knots IAS (656 km/h; 351 mph) in level flight, 405 knots TAS (750 km/h; 466 mph) at 7,620 m (25,000 ft); more than 1,000 nm (1,850 km; 1,150 miles) flown on internal fuel at 7,620 m (25,000 ft); live firing tests with GAU-12/U cannon November 1991, funded by US Air Force. ARES is still flown regularly, and Scaled Composites is working towards selling complete programme with ultimate goal of production.
CURRENT VERSIONS: ARES design can be produced in different sizes, with different armaments and in two-seat form.
CUSTOMERS: Potentially Pacific Rim nations; US Navy and Air Force interest.
COSTS: Initial development as private venture; no outside funding. Production aircraft would cost $1 million. Development prototypes could be supplied for $5 million each.
DESIGN FEATURES: Offset fuselage layout, with engine intake to port and gun to starboard, to prevent gun gases interfering with engine operation; General Electric five-barrel GAU-12/U 25 mm gun installed in 'focused depression' in starboard side of fuselage so that gun blast impinges on forward fuselage and counteracts recoil and cockpit canopy is shielded from blast; engine installed off-axis in rear fuselage, with curved duct from circular intake on port side; jet-pipe curved to align efflux with fuselage axis; these curves hide compressor and turbine from radar; fuselage offset 76 mm (3.0 in) to port of wing centreline; fins and rudders mounted on short tailbooms, which shield jet efflux. Inner wing swept at 50°, outer wing at about 15°; foreplanes swept about 10° forward. Natural angle of attack limiting has been demonstrated; airflow separates on canard and thereby limits angle of attack to something less than stall angle of main wing.

Rutan offering to produce prototypes for trials by potential customers in two years; potential roles include anti-helicopter, close air support, forward air control, operational trainer, armed border patrol and anti-narcotics.
FLYING CONTROLS: Elevators on foreplanes; ailerons on wings; airbrake/flaps on inner wings. Built-in aerodynamic angle of attack protection.
STRUCTURE: Extensive use of carbonfibre/epoxy over foam/PVC cores.
LANDING GEAR: Hydraulically retractable tricycle type, with single wheel on each unit. Main units retract rearward into tailbooms, nosewheel forward into fuselage.

POWER PLANT: One 13.12 kN (2,950 lb st) Pratt & Whitney JT15D-5 turbofan with electronic fuel control unit, offset 8° to port of centreline, with thrust line corrected by curved jetpipe. Curved air intake on port side of centre fuselage only. Maximum internal fuel capacity 771 kg (1,700 lb). Provision for external fuel tank, capacity 227 kg (500 lb).
ACCOMMODATION: Pilot only, on Universal Propulsion Company SIIS-3ER ejection seat. Canopy hinged at rear. Two-seat version planned for training missions.
SYSTEMS: Oxygen system standard.
ARMAMENT: One internal 25 mm General Electric GAU-12/U five-barrel rapid-firing cannon on starboard side of fuselage, with 220 rounds. Provision for carriage of two AIM-9L Sidewinder or four AIM-92 Stinger air-to-air missiles on external stores hardpoints.

DIMENSIONS, EXTERNAL:
Wing span	10.67 m (35 ft 0 in)
Wing aspect ratio	6.5
Foreplane span	5.84 m (19 ft 2 in)
Foreplane aspect ratio	10.7
Length overall	8.97 m (25 ft 5¼ in)
Height overall	3.00 m (9 ft 10 in)

AREAS:
Wings, gross	17.49 m² (188.3 sq ft)
Foreplanes (total)	3.19 m² (34.3 sq ft)

WEIGHTS AND LOADINGS:
Weight empty, unarmed	1,308 kg (2,884 lb)
Max internal fuel weight	771 kg (1,700 lb)
Max T-O weight: unarmed	2,179 kg (4,804 lb)
with GAU-12/U and 220 rds	2,767 kg (6,100 lb)
Max wing loading	158.17 kg/m² (32.40 lb/sq ft)
Max power loading	210.90 kg/kN (2.07 lb/lb st)

PERFORMANCE (estimated):
Max limiting Mach number	0.65
Max level speed at 10,670 m (35,000 ft)	375 knots (695 km/h; 432 mph)

TOYOTA USA/RUTAN LIGHT AIRCRAFT

TYPE: Six-seat private aircraft.
PROGRAMME: Reportedly being flight tested; said to be inspired by Toyota effort to enter light aircraft manufacturing using motorcar engines. Toyota USA's private aircraft maintenance fixed base at Long Beach, California, acquired in 1989, called Toyota Aviation USA, represents Toyota exploration of light aircraft business. Prototype first flew late 1991 (N191SC); all-composites structure and powered by Toyota Lexus engine.

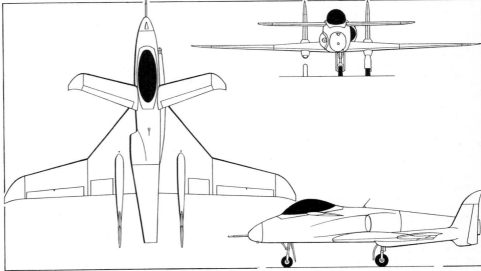

Photograph and three-view drawing (*Jane's/Mike Keep*) of the Rutan 151 ARES turbofan powered agile combat aircraft

SCHAFER
SCHAFER AIRCRAFT MODIFICATIONS INC
Route 10, Box 301, Madison Cooper Airport, Waco, Texas 76708
Telephone: 1 (817) 753 1551
Fax: 1 (817) 753 8416
Telex: 795902 SCHAFER CFTO
PRESIDENT: Earl Schafer
EXECUTIVE VICE-PRESIDENT AND DIRECTOR OF MARKETING:
R. B. Stevens

Company formed 1977; originally manufactured auxiliary fuel tanks and modifications for Cessna 300 and 400 series aircraft; since 1979, has developed own modification programmes.

Comanchero: Conversion of Piper Pressurised Navajo; re-engined with 559 kW (750 shp) P&WC PT6A-135 turboprops; useful load increased by 345 kg (760 lb).

Comanchero 500A: Similar conversion of Piper Chieftain, with 410 kW (550 shp) P&WC PT6A-20s.

Comanchero 500B: Powered by 533 kW (715 shp) P&WC PT6A-27s.

Neiva Comanchero/NE-821 Carajá: As Comanchero 500B but conversions by Neiva (see Brazilian section).

Comanchero 750: Piper Cheyenne II re-engined with 462 kW (620 shp) P&WC PT6A-28s.

DC-3-65TP Cargomaster: Conversion of Douglas DC-3 with 1,062 kW (1,424 shp) P&WC PT6A-65AR turboprops; fuselage stretched 1.02 m (3 ft 4 in) forward of wing to retain CG range.

Details, given in 1991-92 and earlier *Jane's*, can now be found in *Jane's Civil and Military Aircraft Upgrades*.

Schafer Comanchero 500 conversion of Piper Chieftain

SCHWEIZER
SCHWEIZER AIRCRAFT CORPORATION
PO Box 147, Elmira, New York 14902
Telephone: 1 (607) 739 3821
Fax: 1 (607) 796 2488
Telex: 932459 SCHWEIZER BIGF
PRESIDENT: Paul H. Schweizer
EXECUTIVE VICE-PRESIDENTS:
Leslie E. Schweizer
W. Stuart Schweizer
VICE-PRESIDENT: Michael D. Oakley
DIRECTOR, MARKETING: Cole Hedden
MANAGER, MARKETING AND COMMUNICATIONS:
Barbara J. Tweedt

Established 1939 to produce sailplanes; present models described in appropriate section of earlier editions; from mid-1957 to 1979 Schweizer also made Grumman (later Gulfstream American) Ag-Cat under subcontract; all rights to Ag-Cat purchased January 1981; delivery of Ag-Cat Super-B, since supplemented by turboprop version, started October 1981; Ag-Cat marketing and support based at Elmira factory.

Schweizer acquired rights for sole US manufacture of Hughes 300 light helicopter 13 July 1983; Schweizer supporting earlier Hughes 300s and US Army TH-55 trainers; first Elmira-built 300C completed June 1984; Schweizer purchased US rights for whole 300C programme from McDonnell Douglas Helicopter Company (formerly Hughes Helicopters) 21 November 1986.

Schweizer subcontracts include work for Bell Helicopter, Boeing, Sikorsky and others; company is involved in design and prototyping, and in projects to develop heavy lift vehicles, aerial applicators for pheromones, centrifuges and spatial disorientation trainers.

SCHWEIZER SA 2-37A
US military designation: RG-8A
TYPE: Two-seat quiet special missions aircraft.
PROGRAMME: First flight of SA 2-37A (N9237A) in 1986; three RG-8As (85-0047, 85-0048 and 86-0404) bought by US Army; one (85-0048) lost in accident; remaining two transferred to US Coast Guard for anti-narcotics operations; prototype fitted with Hughes AN/AAQ-16 and

Schweizer RG-8A two-seat drug interdiction aircraft of the US Coast Guard *(Press Office Sturzenegger)*

other manufacturers' thermal imaging systems. Full description in 1991-92 and earlier *Jane's*.

CUSTOMERS: US Army and US Coast Guard.
DESIGN FEATURES: Modification of Schweizer SGM 2-37 motor glider, but with slightly greater wing span, drooped leading-edge and leading-edge fences on outer wing panels to improve stall, much more powerful engine with large exhaust silencers on fuselage sides, three-blade quiet propeller, fuselage modified to accept bulged canopy and larger engine, streamlined fairings and hydraulic parking brake on mainwheels; more than treble standard fuel capacity, but optional extra tank also available; removable underfuselage skin and hatches give access to 1.84 m³ (65 cu ft) payload bay behind cockpit, in which pallets holding LLLTV, FLIR or camera payloads can be quickly removed and installed; other engines and larger payloads available for surveillance, basic and advanced training, operator training, glider and banner towing, and priority cargo delivery. Certificated to FAR Pt 23 for day and night IFR; inaudible when overflying at 'quiet mode' speed at about 610 m (2,000 ft) using 38.8 kW (52 hp) from its 175 kW (235 hp) Textron Lycoming IO-540 engine.

Wing section Wortmann FX-61-163 at root and FX-60-126 (modified) at tip; outer wing panels and horizontal tail can be removed for transport.

POWER PLANT: One 175 kW (235 hp) Textron Lycoming IO-540-W3A5D flat-six engine, driving a McCauley three-blade constant-speed propeller. Standard fuel capacity of 196.8 litres (52 US gallons; 43.3 Imp gallons), increasable optionally to 253.6 litres (67 US gallons; 55.8 Imp gallons). Shadin fuel flow system.

ACCOMMODATION: Seats for two persons side by side under two-piece upward opening canopy, hinged on centreline. Dual controls, seat belts and inertia reel harnesses standard. Compartment aft of seats enlarged to accommodate pallet containing up to 340 kg (750 lb) of sensors or other equipment.

DIMENSIONS, EXTERNAL:
Wing span	18.745 m (61 ft 6 in)
Wing aspect ratio	18.97
Fuselage length	8.46 m (27 ft 9 in)
Height overall (tail down)	2.36 m (7 ft 9 in)
Wheel track	2.79 m (9 ft 2 in)
Wheelbase	5.99 m (19 ft 8 in)
Propeller diameter	2.18 m (7 ft 2 in)

AREAS:
Wings, gross	18.52 m² (199.4 sq ft)

WEIGHTS AND LOADINGS:
Weight empty	918 kg (2,025 lb)
Max mission payload	340 kg (750 lb)
Max T-O weight	1,587 kg (3,500 lb)
Max wing loading	85.65 kg/m² (17.55 lb/sq ft)
Max power loading	9.06 kg/kW (14.89 lb/hp)

PERFORMANCE (at max T-O weight):
Max permissible diving speed (V$_D$)	176 knots (326 km/h; 202 mph) CAS
Cruising speed at 1,525 m (5,000 ft):	
75% power	138 knots (256 km/h; 159 mph)
65% power	129 knots (239 km/h; 148 mph)
Approach speed	117 knots (217 km/h; 135 mph) CAS
Quiet mode speed	70-80 knots (130-148 km/h; 80-92 mph)
Optimum climbing speed	77 knots (142 km/h; 88 mph) CAS
Stalling speed:	
airbrakes open	71 knots (132 km/h; 82 mph)
airbrakes closed	67 knots (124 km/h; 77 mph)
Max rate of climb at S/L	292 m (960 ft)/min
Service ceiling	5,490 m (18,000 ft)
T-O run (S/L, ISA): hard surface	387 m (1,270 ft)
grass	533 m (1,750 ft)
T-O to 15 m (50 ft) (S/L, ISA):	
hard surface	612 m (2,010 ft)
grass	759 m (2,490 ft)
Landing from 15 m (50 ft) (S/L, ISA):	
hard surface	680 m (2,230 ft)
grass	732 m (2,400 ft)
Best glide ratio	20
g limits	+6.6/−3.3

SCHWEIZER AG-CAT SERIES

TYPE: Agricultural biplane.
PROGRAMME: First flight of original Grumman Ag-Cat 27 May 1957; first deliveries 1959; production resumed in October 1981 with improved G-164B designated Ag-Cat Super-B, later joined by Ag-Cat Super-B Turbine.
CURRENT VERSIONS: **Ag-Cat Super-B/600 (G-164B):** Basic model powered by 447.5 kW (600 hp) P&W R-1340 radial engine, driving Pacific Propeller Type 12D40/AG100 two-blade constant-speed metal propeller; improvements in Super-B include 40 per cent more hopper capacity and 0.97 m (3 ft 2 in) wider stainless steel gatebox and bottom loader as standard equipment; upper wing raised 20 cm (8 in) to improve pilot's view and increase load carrying capability, operating speed and climb rate. *Detailed description applies to this version, except where indicated.*
Ag-Cat 450B (G-164B): Generally similar alternative version, available to special order; powered by 335.5 kW (450 hp) P&W R-985 radial engine; usable fuel 242 litres (64 US gallons; 53.3 Imp gallons); hopper volume 1.23 m³ (43.5 cu ft) and capacity 1,230 litres (325 US gallons; 271 Imp gallons).
Ag-Cat Super-B Turbine (G-164B): Similar to basic Super-B, but powered by choice of 373 kW (500 shp) P&WC PT6A-11AG, 507 kW (680 shp) PT6A-15AG or 559 kW (750 shp) PT6A-34AG turboprop engines. Airframe now being offered to customers to install own engine; PT6A-15, PT6A-20, PT6A-27, PT6A-28 or PT6A-34 acceptable; new air induction filter; improved oil cooling system; fuel capacity increased to 454 litres (120 US gallons; 100 Imp gallons).
CUSTOMERS: Total 2,628 Ag-Cats built by Schweizer under contract, including 1,730 G-164As, 832 G-164Bs, 44 G-164Cs and 22 G-164Ds. Ag-Cat Super-B Turbine also manufactured under licence by Ethiopian Airlines Technical Services Division in Addis Ababa (which see) since December 1986.
DESIGN FEATURES: Wing section NACA 4412; dihedral 3°; incidence 6°; uncowled piston engine.
FLYING CONTROLS: Mechanical, with trim tab on port elevator; ground adjustable tabs on one aileron, rudder and starboard elevator; no flaps.
STRUCTURE: Each wing has two aluminium spars and sheet skin wrapped over whole upper surface and back to front spar on under surface; remainder of underside fabric covered; leading-edges in five panels to simplify repair; glass-fibre wingtips; light alloy ailerons; wire-braced fabric covered tail surfaces on steel tube frame; welded steel tube fuselage with removable side panels.
LANDING GEAR: Non-retractable tailwheel type. Cantilever spring steel legs. Cleveland mainwheels with tyres size 8.50-10, 8-ply, pressure 2.42 bars (35 lb/sq in). Steerable tailwheel with tyre size 12.4-4.5, pressure 3.45 bars (50 lb/sq in). Cleveland heavy duty air-cooled hydraulic disc brakes. Parking brake.
POWER PLANT: One Pratt & Whitney nine-cylinder air-cooled radial engine with Pacific Propeller constant-speed propeller, or Pratt & Whitney Canada PT6A turboprop, as detailed under Current Versions. Fuel tanks in upper wing, with combined usable capacity of 302 litres (80 US gallons; 66.6 Imp gallons). Single-point refuelling on upper surface of upper wing centre-section. Oil capacity 32.2 litres (8.5 US gallons; 7.1 Imp gallons).
ACCOMMODATION: Single seat in enclosed cockpit. Reinforced fairing aft of canopy for turnover protection. Canopy side panels open outward and down, canopy top upward and to starboard, to provide access. Baggage compartment. Cockpit pressurised against dust ingress and ventilated by ram air. Safety padded instrument panel. Air-conditioning by J. B. Systems optional.
SYSTEMS: Hydraulic system for brakes only. Optional electrical system with 24V alternator, navigation and/or strobe lights, external power socket, and electric engine starter.
EQUIPMENT: Radio optional. Standard equipment includes control column lock, instrument glareshield, seat belt and shoulder harness, tinted windscreen, stall warning light, refuelling steps and assist handles, tiedown rings, and urethane paint external yellow finish.

Schweizer Ag-Cat Super-B Turbine agricultural aircraft with wide inter-plane gap

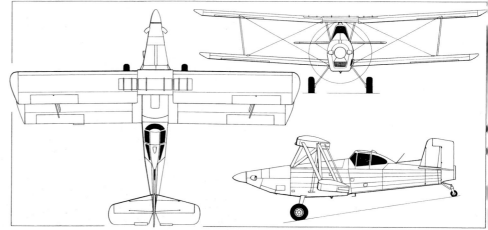

Schweizer Ag-Cat Super-B Turbine agricultural aircraft *(Dennis Punnett)*

Forward of cockpit, over CG, is a glassfibre hopper, capacity 1,514 litres (400 US gallons; 333 Imp gallons), for agricultural chemicals (dry or liquid); distributor beneath fuselage. Low-volume, ULV or high-volume spray system, with leading- or trailing-edge booms. Emergency dump system for hopper load; can be used also for water bomber operations.

DIMENSIONS, EXTERNAL (A: Super-B/600, B: Super-B Turbine, C: 450B):

Wing span: upper	12.92 m (42 ft 4½ in)
lower	12.36 m (40 ft 6¾ in)
Wing chord, constant	1.47 m (4 ft 10 in)
Wing aspect ratio: upper wing	8.7
biplane, effective mean	5.5
Length overall: A	7.44 m (24 ft 5 in)
B	8.41 m (27 ft 7¼ in)
C	7.54 m (24 ft 9 in)
Height overall: A, C	3.51 m (11 ft 6 in)
B	3.68 m (12 ft 1 in)
Tailplane span	3.96 m (13 ft 0 in)
Wheel track	2.44 m (8 ft 0 in)
Wheelbase	5.59 m (18 ft 4 in)
Propeller diameter (max)	2.74 m (9 ft 0 in)
Propeller ground clearance: A	0.27 m (10.8 in)
B	0.37 m (1 ft 2¾ in)
Cockpit door: Height	0.53 m (1 ft 9 in)
Width	0.64 m (2 ft 1 in)
Height to sill	0.71 m (2 ft 4 in)

DIMENSIONS, INTERNAL:

Cockpit: Length	1.27 m (4 ft 2 in)
Max width	0.76 m (2 ft 6 in)
Max height	1.14 m (3 ft 9 in)
Hopper volume	1.51 m³ (53.5 cu ft)

AREAS:

Wings, gross	36.48 m² (392.7 sq ft)
Ailerons (total)	2.92 m² (31.4 sq ft)
Fin	1.67 m² (17.97 sq ft)
Rudder	1.12 m² (12.0 sq ft)
Tailplane	2.12 m² (22.8 sq ft)
Elevators	2.06 m² (22.2 sq ft)

WEIGHTS AND LOADINGS:

Weight empty equipped, spray and duster versions:	
A	1,656 kg (3,650 lb)
B	1,429 kg (3,150 lb)
C	1,508 kg (3,325 lb)
Certificated gross weight	2,358 kg (5,200 lb)
Max T-O weight (CAM.8)	3,184 kg (7,020 lb)
Max wing loading	87.42 kg/m² (17.91 lb/sq ft)
Max power loading: A	7.12 kg/kW (11.71 lb/hp)
B	5.70-8.54 kg/kW (9.36-14.04 lb/shp)
C	9.49 kg/kW (15.6 lb/hp)

PERFORMANCE (A: Super-B/600, B: Super-B Turbine with PT6A-15AG engine, C: 450B):

Never-exceed speed (VNE)	136 knots (252 km/h; 157 mph)
Working speed: A, C	100 knots (185 km/h; 115 mph)
B	113 knots (209 km/h; 130 mph)
Stalling speed, power off	56 knots (103 km/h; 64 mph)
T-O run	120 m (394 ft)
T-O to 15 m (50 ft) at 2,358 kg (5,200 lb) certificated gross weight: A	320 m (1,050 ft)
B	274 m (900 ft)
C	396 m (1,300 ft)
Landing from 15 m (50 ft)	407 m (1,333 ft)
Landing run	157 m (513 ft)
Range with max fuel	172 nm (318 km; 198 miles)
Design *g* limits, all versions	+4.2/–1

SCHWEIZER 300C

TYPE: Three-seat light utility helicopter.

PROGRAMME: First flight August 1969; first flight Hughes production model December 1969; FAA certification May 1970 (basic Hughes 300 described in 1976-77 *Jane's*). Production of 300C transferred from Hughes Helicopters to Schweizer July 1983; first flight Schweizer 300C June 1984; Schweizer bought entire programme November 1986.

CURRENT VERSIONS: **300C**: Standard civil version, *to which main description applies*.

300C Sky Knight: Special police version; options include safety mesh seats with inertia reel shoulder harnesses, public address/siren system, searchlight, integrated communications, heavy duty 28V 100A electrical system, cabin heater, night lights with strobe beacons, cabin utility light, fire extinguisher, first aid kit and map case.

TH-300C: Military training version.

CUSTOMERS: Hughes produced 2,800 of all versions, including TH-55A Osage for US Army, before transfer to Schweizer. $4.9 million US Army order for 30 300Cs and spares placed in late 1985; two batches of 24 TH-300C training helicopters delivered to Royal Thai Army between March 1986 and mid-1989; 250th Schweizer 300C delivered late 1989. Recent Sky Knight customers include police departments of Baltimore, Maryland; Columbus, Ohio; Hillsborough County Sheriff's Office, Florida; Warren, Michigan; Lake County, Indiana; and Kansas City, Missouri.

DESIGN FEATURES: Fully articulated three-blade main rotor; fully interchangeable blades; blade section NACA 0015; elastomeric dampers; two-blade teetering tail rotor; limited blade folding; no rotor brake; multiple V-belt and pulley reduction gear/drive system between horizontally mounted engine and transmission, with electrically controlled belt-tensioning system instead of clutch; braced tubular tailboom.

Schweizer 300C Sky Knight light utility helicopter of the Minnesota State Patrol

FLYING CONTROLS: Mechanical, with first pilot on left; electric cyclic trim; separate dihedral tailplane and fin.

STRUCTURE: Main rotor blades bonded with constant-section extruded aluminium spar, wraparound skin and trailing-edge; tail rotor blades have steel tube spar and glassfibre skin; steel tube cabin section with light alloy, stainless steel and Plexiglas skin.

LANDING GEAR: Skids carried on oleo-pneumatic shock absorbers. Replaceable skid shoes. Two ground handling wheels with 0.25 m (10 in) balloon tyres, pressure 4.14-5.17 bars (60-75 lb/sq in). Available optionally on floats made of polyurethane coated nylon fabric, 4.70 m (15 ft 5 in) long and with a total installed weight of 27.2 kg (60 lb).

POWER PLANT: One 168 kW (225 hp) Textron Lycoming HIO-360-D1A flat-four engine, derated to 142 kW (190 hp), mounted horizontally aft of seats. Two aluminium fuel tanks, total capacity 185.5 litres (49 US gallons; 40.8 Imp gallons), mounted externally aft of cockpit. Crash resistant fuel tank optional. Oil capacity 9.5 litres (2.5 US gallons; 2.1 Imp gallons).

ACCOMMODATION: Three persons side by side on sculptured and cushioned bench seat, with shoulder harness, in Plexiglas enclosed cabin. Carpet and tinted canopy standard. Forward hinged, removable door on each side. Dual controls optional. Baggage capacity 45 kg (100 lb). Exhaust muff heating and ventilation kits available.

SYSTEMS: Standard electrical system includes 24V 70A alternator, 24V battery, starter and external power socket.

AVIONICS: Optional avionics include Bendix/King KY 196A com transceiver and headsets, KR 86 ADF and KT 76A transponders.

EQUIPMENT: Standard equipment includes map case, first aid kit, fire extinguisher, engine hour meter, and main rotor blade tiedown kit. Optional equipment includes amphibious floats, litter kits, cargo racks with combined capacity of 91 kg (200 lb), external load sling of 408 kg (900 lb) capacity, Simplex Model 5200 agricultural spray or dry powder dispersal kits, Sky Knight law enforcement package, instrument training package, throttle governor, start-up overspeed control unit, night flying kit, dual controls, all-weather cover, heavy duty skid plates, single or dual exhaust mufflers, door lock and dual oil coolers.

DIMENSIONS, EXTERNAL:

Main rotor diameter	8.18 m (26 ft 10 in)
Main rotor blade chord	0.171 m (6¾ in)
Tail rotor diameter	1.30 m (4 ft 3 in)
Distance between rotor centres	4.66 m (15 ft 3½ in)
Length overall, rotors turning	9.40 m (30 ft 10 in)
Height: to top of rotor head	2.66 m (8 ft 8⅝ in)
to top of cabin	2.19 m (7 ft 2 in)
Width: rotor partially folded	2.44 m (8 ft 0 in)
cabin	1.30 m (4 ft 3 in)
Skid track	1.99 m (6 ft 6½ in)
Length of skids	2.51 m (8 ft 3 in)
Passenger doors (each): Height	1.09 m (3 ft 7 in)
Width	0.97 m (3 ft 2 in)
Height to sill	0.91 m (3 ft 0 in)

AREAS:

Main rotor blades (each)	0.70 m² (7.55 sq ft)
Tail rotor blades (each)	0.08 m² (0.86 sq ft)
Main rotor disc	52.5 m² (565.5 sq ft)
Tail rotor disc	1.32 m² (14.2 sq ft)
Fin	0.23 m² (2.5 sq ft)
Horizontal stabiliser	0.246 m² (2.65 sq ft)

WEIGHTS AND LOADINGS:

Weight empty	474 kg (1,046 lb)
Max T-O weight: Normal category	930 kg (2,050 lb)
external load	975 kg (2,150 lb)
Max disc loading: Normal category	17.67 kg/m² (3.62 lb/sq ft)
Max power loading: Normal category	6.55 kg/kW (10.8 lb/hp)

PERFORMANCE (at max Normal T-O weight, ISA):

Never-exceed speed (VNE) at S/L	91 knots (169 km/h; 105 mph)
Max cruising speed	82 knots (153 km/h; 95 mph)
Speed for max range, at 1,220 m (4,000 ft)	67 knots (124 km/h; 77 mph)
Max rate of climb at S/L	229 m (750 ft)/min
Service ceiling	3,110 m (10,200 ft)
Hovering ceiling: IGE	1,800 m (5,900 ft)
OGE	840 m (2,750 ft)
Range at 1,220 m (4,000 ft), 2 min warm-up, max fuel, no reserves	194 nm (360 km; 224 miles)
Max endurance at S/L	3 h 24 min

SCHWEIZER 330

TYPE: Turbine powered three/four-seat light helicopter.

PROGRAMME: Announced 1987; first flight in public (N330TT) 14 June 1988; entered in US Army New Training Helicopter (NTH) competition, but excluded from contest; now intended for commercial sale; FAA certification received September 1992.

CUSTOMERS: More than 20 ordered by early 1992; 1993 production run of 18 sold.

COSTS: Base $433,775 for civil aircraft (1993).

DESIGN FEATURES: Civil roles include law enforcement, scout/observation, aerial photography, light utility, agricultural spraying and personal transport; uses turbine fuel rather

Prototype Schweizer 330 (Allison 250-C20 turboshaft)

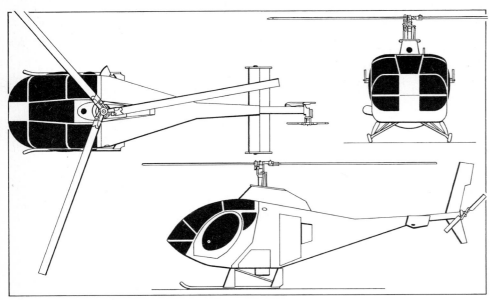

Schweizer 330 three/four-seat light helicopter (Jane's/Mike Keep)

than scarcer Avgas; extremely low flat rating of engine promises outstanding hot and high performance; third occupant seat allows second student to observe instruction; three sets of flying controls can be fitted; streamlined external envelope and large tailplane with endplate fins added during development; rotor rpm 471.
FLYING CONTROLS: Similar to 300C; see Design Features.
STRUCTURE: Generally similar to 300C.
POWER PLANT: One Allison 250-C20 turboshaft capable of 313.2 kW (420 shp). Max fuel capacity 227 litres (60 US gallons; 50 Imp gallons).

DIMENSIONS, EXTERNAL:
Main and tail rotor diameters: As for Model 300C
Length overall, rotors turning 9.40 m (30 ft 10 in)
Height overall 2.89 m (9 ft 5¾ in)
Skid track 2.08 m (6 ft 10 in)

DIMENSIONS, INTERNAL:
Cabin: Width at seat 1.72 m (5 ft 7¾ in)
Height at seat 1.35 m (4 ft 5¼ in)

WEIGHTS AND LOADINGS:
Weight empty 508 kg (1,120 lb)
Normal T-O weight 998 kg (2,200 lb)

PERFORMANCE (at Normal T-O weight):
Max cruising speed 108 knots (200 km/h; 124 mph)
Normal cruising speed 100 knots (185 km/h; 115 mph)
Hovering ceiling: IGE 5,485 m (18,000 ft)
OGE 4,265 m (14,000 ft)
Max range at 1,220 m (4,000 ft), no reserves
269 nm (498 km; 309 miles)
Max endurance, no reserves 4 h 12 min

SEQUOIA

SEQUOIA AIRCRAFT CORPORATION
2000 Tomlynn Street, PO Box 6861, Richmond, Virginia 23230
Telephone: 1 (804) 353 1713
Fax: 1 (804) 359 2618
PRESIDENT: Alfred P. Scott

SEQUOIA 300 SEQUOIA

TYPE: Side by side two-seat utility and aerobatic homebuilt.
PROGRAMME: First flown 26 April 1992; no plans or kits available; entire programme for sale 1993.
DESIGN FEATURES: Wing section NACA 64_2A215 at root, NACA 64A210 at tip.
STRUCTURE: Wings, ailerons, slotted flaps and tail unit of flush riveted aluminium alloy construction. Wings have glassfibre tips. Welded steel tube fuselage, covered entirely with lightweight shell of glassfibre/PVC foam/glassfibre sandwich, attached to tubing with glassfibre and epoxy resin.
LANDING GEAR: Retractable tricycle type.
POWER PLANT: One 224 kW (300 hp) Textron Lycoming TIO-540-S1AD turbocharged engine in prototype. Other Textron Lycoming engines of 175-224 kW (235-300 hp) may be used. Fuel capacity 291.5 litres (77 US gallons; 64 Imp gallons). Provision for tip tanks of approximately 75 litres (20 US gallons; 16.7 Imp gallons) capacity each; and two underwing attachment points for additional fuel tanks, radar or stores of up to 272 kg (600 lb) combined weight.

DIMENSIONS, EXTERNAL:
Wing span 9.14 m (30 ft 0 in)
Wing aspect ratio 6.92
Length overall 7.62 m (25 ft 0 in)
Height overall 2.90 m (9 ft 6 in)
Propeller diameter 2.03 m (6 ft 8 in)

AREAS:
Wings, gross 12.08 m² (130.0 sq ft)

WEIGHTS AND LOADINGS:
Weight empty 816 kg (1,800 lb)
Baggage capacity 45 kg (100 lb)
Max T-O weight: Utility 1,270 kg (2,800 lb)
Aerobatic 1,088 kg (2,400 lb)
Max wing loading: Utility 105.16 kg/m² (21.54 lb/sq ft)
Max power loading (prototype): Utility
5.68 kg/kW (9.33 lb/hp)

PERFORMANCE (estimated):
Max level speed at S/L 195 knots (362 km/h; 225 mph)
Max cruising speed at 2,440 m (8,000 ft)
185 knots (343 km/h; 213 mph)
Stalling speed: clean 75 knots (139 km/h; 86 mph)
flaps and wheels down 60 knots (111 km/h; 69 mph)
Max rate of climb at S/L 664 m (2,180 ft)/min
Service ceiling 7,620 m (25,000 ft)
T-O run 457 m (1,500 ft)
Landing run 548 m (1,800 ft)
Range at max cruising speed, with 45 min reserves
868 nm (1,609 km; 1,000 miles)

SEQUOIA FALCO F.8L

TYPE: Side by side two-seat, dual control homebuilt; provision for child's seat in baggage space.
PROGRAMME: Sequoia Aircraft markets plans and kits to build improved version of Falco F.8L high-performance monoplane, designed in Italy by Ing Stelio Frati of General Avia (which see) and first flown on 15 June 1955.

Sequoia 300 Sequoia

Sequoia Falco F.8L homebuilt aircraft

COSTS: Kit: $60,000. Plans: $400.
DESIGN FEATURES: NACA $64_2212.5$ wing section at root, NACA 64_2210 at tip.
STRUCTURE: Entire airframe has plywood covered wood structure, with overall fabric covering and glassfibre wing fillets. Optional metal control surfaces.
LANDING GEAR: Retractable tricycle type.
POWER PLANT: One 119 kW (160 hp) Textron Lycoming IO-320-B1A is standard for kit-built aircraft. Optional engines for which Sequoia offers installation kits are the 112 kW (150 hp) IO-320-A1A and 134 kW (180 hp) IO-360-B1E. Fuel capacity 151 litres (40 US gallons; 33.3 Imp gallons). Optional 7.6 litre (2 US gallon; 1.7 Imp gallon) header tank to permit inverted flight.

DIMENSIONS, EXTERNAL:
Wing span 8.00 m (26 ft 3 in)
Wing aspect ratio 6.40
Length overall 6.50 m (21 ft 4 in)
Height overall 2.29 m (7 ft 6 in)

AREAS:
Wings, gross 10.00 m² (107.6 sq ft)

WEIGHTS AND LOADINGS (119 kW; 160 hp engine):
Weight empty 550 kg (1,212 lb)
Payload with max fuel 194 kg (428 lb)

Baggage capacity	40 kg (88 lb)	PERFORMANCE (119 kW; 160 hp engine):		Max rate of climb at S/L	347 m (1,140 ft)/min
Max aerobatic weight	748 kg (1,650 lb)	Max level speed at S/L	184 knots (341 km/h; 212 mph)	Service ceiling	5,790 m (19,000 ft)
Max T-O weight	853 kg (1,880 lb)	Cruising speed		Range at econ cruising speed	
Max wing loading	85.31 kg/m² (17.47 lb/sq ft)		more than 174 knots (322 km/h; 200 mph)		868 nm (1,609 km; 1,000 miles)
Max power loading	7.17 kg/kW (11.75 lb/hp)	Stalling speed, flaps and wheels down			
			54 knots (100 km/h; 62 mph)		

SIERRA
SIERRA INDUSTRIES
Garner Municipal Airport, PO Box 5184, Uvalde, Texas 78802-5184
Telephone: 1 (210) 278 4381
Fax: 1 (210) 278 7649
PRESIDENT: Mark Huffstutler

Sierra Industries acquired assets of R/STOL Systems Inc 24 September 1986; assets included about 100 supplemental type certificates for STOL and performance modifications to 40 types of Beech, Cessna and Piper light and business aircraft (see tables in 1990-91 and earlier *Jane's*).

In-Wing Weather Radar: RCA Weather Scout radar installed in reinforced wing leading-edge bay in Cessna 182, various Cessna 206 models and Cessna 201.

Cessna Auxiliary Fuel Systems: 204 litres (54 US gallons; 45 Imp gallons) additional fuel tankage in Cessna 206 and 207.

Cessna 210 Landing Gear Door Modification: Mainwheel doors removed and replaced by metal fairings to minimise door-fouling problems; 8.6 kg (19 lb) increase in useful load.

Safety and Performance Conversions: Available for Beech A36 Bonanza and single- and smaller twin-engined Cessna and Piper aircraft; Hi-Lift full span wing leading- and trailing-edge modifications improve landing and take-off distances, give maximum resistance to stall and high manoeuvrability at low speeds.

Sierra/Cessna Eagle Series
Offered as **Eagle:** Cessna Citation 500 and Citation I with inner wing thickness increased, tips extended; fuel capacity increased by 392 kg (865 lb).
Eagle SP: Similar conversion of Cessna Citation 501.
Longwing: Features wingtip extension of Eagle without wingroot thickening; fuel capacity increased by 54 kg (120 lb).
Sierra/Citation Eagle 400: Basic Eagle modifications; 11.12 kN (2,500 lb st) JT15D-4s replace JT15D-1s or 1As.
Details can now be found in *Jane's Civil and Military Aircraft Upgrades*.

SIKORSKY
SIKORSKY AIRCRAFT, DIVISION OF UNITED TECHNOLOGIES CORPORATION
6900 Main Street, Stratford, Connecticut 06601-1381
Telephone: 1 (203) 386 4000
Fax: 1 (203) 386 7300
Telex: 96 4372
OTHER WORKS: Troy, Alabama; South Avenue, Bridgeport, Connecticut; Shelton, Connecticut; West Haven, Connecticut; Development Flight Test Center, West Palm Beach, Florida
PRESIDENT: Eugene Buckley
EXECUTIVE VICE-PRESIDENT, OPERATIONS AND ENGINEERING:
 Raymond P. Kurlak
SENIOR VICE-PRESIDENTS:
 G. C. Kay (Administration)
 Robert R. Moore (Production Operations)
VICE-PRESIDENTS, INTERNATIONAL PROGRAMMES:
 Robert M. Baxter (International and S-76 Commercial Business)
 David L. Powell (Japan)
 Edward M. Francis (Korea)
 Clement C. Peterson (Middle East)
 James J. Satterwhite (Turkey)
VICE-PRESIDENT, RESEARCH AND ENGINEERING:
 Dr Kenneth M. Rosen
VICE-PRESIDENT, GOVERNMENT BUSINESS DEVELOPMENT:
 Gary F. Rast
VICE-PRESIDENT, COMMUNICATIONS: James H. Lynch
MANAGER OF PUBLIC RELATIONS: William S. Tuttle

Founded as Sikorsky Aero Engineering Corporation by late Igor I. Sikorsky; has been division of United Technologies Corporation since 1929; began helicopter production in 1940s; has produced more than 7,800 rotating wing aircraft, further 1,800 Sikorsky helicopters built by foreign licensees. Delivered 126 military helicopters in 1992. Workforce 11,500 in 1993.

Headquarters and main plant at Stratford, Connecticut; main current programmes include UH-60 Black Hawk and derivatives, CH-53E Super Stallion and S-76 series. Sikorsky and Boeing Helicopters won US Army RAH-66 Comanche light helicopter demonstration/validation order on 5 April 1991 (see under Boeing/Sikorsky in this section).

Sikorsky licensees include Westland of the UK, Agusta of Italy, Eurocopter of France and Germany, Mitsubishi of Japan and Pratt & Whitney Canada Ltd. Sikorsky and Embraer of Brazil signed agreement in Summer 1983 to transfer technology covering design and manufacture of composites components. Sikorsky and CASA of Spain signed MoU in June 1984 covering long-term helicopter industrial co-operation programme; CASA builds tail rotor pylon, tailcone and stabiliser components for H-60 and S-70, with first CASA S-70 components delivered to Sikorsky January 1986.

Westland plc (see UK section) shareholders approved joint Fiat/Sikorsky plan involving financial and technical support and minor equity participation in Westland on 12 February 1986; Westland then licensed to manufacture S-70 series.

Plan to establish an S-76 production line in South Korea announced February 1988; Daewoo-Sikorsky Aerospace Ltd (DSA) formed with facility at Changwon to produce civil and military S-76/H-76. Licence to manufacture S-70 in Japan granted to Mitsubishi Heavy Industries in 1988.

BOEING/SIKORSKY RAH-66 COMANCHE
Boeing/Sikorsky submission for US Army LH, now RAH-66 Comanche, selected to proceed into demonstration/validation phase on 4 April 1991. Details under Boeing/Sikorsky in this section.

SIKORSKY S-80/H-53E
US Marine Corps and Navy designations: CH-53E Super Stallion and MH-53E Sea Dragon
TYPE: Three-engined heavy-lift helicopter.
PROGRAMME: Phase I development funding for CH-53E Super Stallion allocated 1973; first flight of first of two prototypes 1 March 1974; first flight of first production prototype 8 December 1975; first flight of second production prototype March 1976; deliveries to US Marine Corps started 16 June 1981; 142 CH-53Es and 46 MH-53Es funded up to and including FY 1992; further 16 CH and four MH funded in FY 1993; combined totals of 16 and 18 planned for FYs 1994 and 1995; total requirement is for 177 CH-53Es.
CURRENT VERSIONS: **CH-53E Super Stallion:** Used by US Marine Corps for amphibious assault, carrying heavy equipment and armament, and recovering disabled aircraft; also used by US Navy for vertical onboard delivery and recovery of damaged aircraft from aircraft carriers; went into operation in Mediterranean area Summer 1983 with HC-4 at Sigonella, Italy; also serves as part of HC-2 at Norfolk, Virginia, and with Marines squadrons HMH-361, HMH-465 and HMH-466 at Tustin, California, and HMH-461 and HMH-464 at New River, North Carolina, plus HMT-302 for training at Tustin. *Detailed description applies to CH-53E, but applicable also to MH-53E, except where indicated.*

Planned improvements for CH-53E include composites tail rotor blades, uprated GE T64-416 engines, Omega navigation system, ground proximity warning system, flight crew night vision system, improved internal cargo handling system, missile alerting system, chaff/flare dispensers, nitrogen fuel inerting system, and facility for refilling hydraulic system inside cargo compartment. CH-53E will possibly be equipped with self-defence airto-air missiles; initial trials of AIM-9 Sidewinder conducted at Naval Air Test Center, Patuxent River, Maryland. Total 131 delivered by January 1993.

Full-scale development of helicopter night vision system (HNVS) for CH-53E began June 1986, in co-operation with Northrop Electro-Mechanical Division; HNVS includes Martin Marietta pilot night vision system (PNVS) and Honeywell integrated helmet and display sighting system (IHADSS) from Bell AH-1S surrogate trainer; HNVS will allow low level operations in night and adverse weather; HNVS ground testing began 1988; operational evaluation began August 1989. Smaller-scale capability authorised 1993, with contract to EER Systems for installation of Hughes AN/AAQ-16B FLIR, Teledyne Ryan AN/APN-217 Doppler and Rockwell Collins GPS 3A; total 24 upgrades to follow.

MH-53E Sea Dragon: Airborne mine countermeasures helicopter able to tow through water hydrofoil sledge carrying mechanical, acoustic and magnetic sensors; early history in 1982-83 *Jane's*; nearly 3,785 litres (1,000 US gallons; 833 Imp gallons) extra fuel carried in enlarged sponsons made of composites; improved hydraulic and electrical systems; minefield, navigation and automatic flight control system with automatic towing and approach and departure from hover modes. First flight of first preproduction MH-53E 1 September 1983; first delivery to US Navy 26 June 1986; in operational service with HM-14 at Norfolk, Virginia, 1 April 1987; another MH-53E delivered to HM-12 (now AMCM training unit) in Spring 1987 was 100th H-53E; 32 delivered to USN by December 1991, including some with HM-15 at Alameda, California, for Pacific Fleet; none delivered in 1992; first carrier deployment by HM-15 on board USS *Tripoli*, 9 December 1989. Also partly equips Marines squadron HMH-461 at New River, North Carolina. Total requirement is 56. Delivery due April 1994 of MH-53E retrofitted with upgraded avionics package by EER Systems, comprising two 15.5 cm (6 in) horizontal situation display colour screens, Fairchild mission data loader and Rockwell Collins GPS 3A; upgrade of entire MH-53E fleet planned.
CUSTOMERS: US Navy and Marine Corps (see Programme and Current Versions). Japan has acquired helicopters similar to MH-53E (see S-80M entry). Over 690 of Sikorsky S-65/S-80/H-53 series built by 1993.

SEA DRAGON and SUPER STALLION DELIVERIES

	USN/USMC	JAPAN
1980	2	
1981	14	
1982	24	
1983	20	
1984		
1985	10	
1986	9	
1987	12	
1988	13	1
1989	12	3
1990	14	2
1991	6	
1992	8	4
Totals	**162**	**10**

COSTS: $24.36 million (1992) projected average unit cost.
DESIGN FEATURES: Fully articulated seven-blade main rotor; blade twist 14°; hydraulic powered blade folding for main rotor; tail pylon folds hydraulically to starboard; fourblade tail rotor on pylon canted 20° to port to derive some lift from tail rotor and extend CG range; cranked, strutbraced tailplane; rotor brake standard; fuselage stressed for 20g vertical and 10g lateral crash loads.

Sikorsky CH-53E Super Stallion heavy lift helicopter of the US Marine Corps, in Saudi Arabia for Operation Desert Storm, 1991 (*Paul Jackson*)

560　USA: AIRCRAFT—SIKORSKY

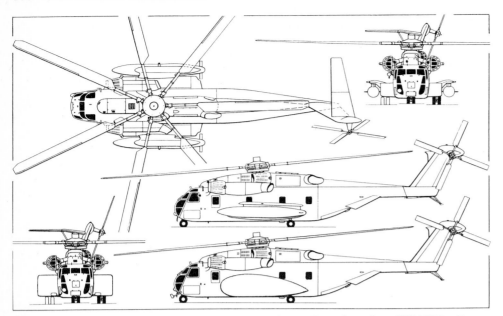

Sikorsky CH-53E Super Stallion, with additional lower side view and lower front view of MH-53E Sea Dragon
(Dennis Punnett)

FLYING CONTROLS: Fully powered, with autostabilisation and autopilot. See also Current Versions and Avionics.
STRUCTURE: Fuselage has watertight primary structure of light alloy, steel and titanium; glassfibre/epoxy cockpit section; extensive use of Kevlar in transmission fairing and engine cowlings; main rotor blades have titanium spar, Nomex honeycomb core and glassfibre/epoxy composites skin; titanium and steel rotor head; Sikorsky blade inspection method (BIM) sensors detect blade spar cracks occurring in service; tail rotor of aluminium; pylon and tailplane of Kevlar composites.
LANDING GEAR: Retractable tricycle type, with twin wheels on each unit. Main units retract into rear of sponsons on each side of fuselage. Fully castoring nosewheels.
POWER PLANT: Three General Electric T64-GE-416 turboshafts, each with a max rating of 3,266 kW (4,380 shp) for 10 min, intermediate rating of 3,091 kW (4,145 shp) for 30 min and max continuous power rating of 2,756 kW (3,696 shp). Retrofit planned with T64-GE-419s. Transmission rated at 10,067 kW (13,500 shp) for take-off. Self-sealing bladder fuel cell in forward part of each sponson, each with capacity of 1,192 litres (315 US gallons; 262 Imp gallons). Additional two-cell unit, with capacity of 1,465 litres (387 US gallons; 322 Imp gallons), brings total standard internal capacity to 3,849 litres (1,017 US gallons; 847 Imp gallons). (Total internal capacity of MH-53E is 12,113 litres; 3,200 US gallons; 2,664 Imp gallons.) Optional drop tank outboard of each sponson of CH-53E, total capacity 4,921 litres (1,300 US gallons; 1,082 Imp gallons). (MH-53E can carry seven internal range extension tanks, total capacity 7,949 litres; 2,100 US gallons; 1,748 Imp gallons.) Forward extendable probe for in-flight refuelling. Alternatively, aircraft can refuel by hoisting hose from surface vessel while hovering.
ACCOMMODATION: Crew of three. Main cabin of CH-53E will accommodate up to 55 troops on folding canvas seats along walls and in centre of cabin. Door on forward starboard side of main cabin. Hydraulically operated rear loading ramp. Typical freight loads include seven standard 1.02 × 1.22 m (3 ft 4 in × 4 ft) pallets. Single-point central hook for slung cargo, capacity 16,330 kg (36,000 lb).
SYSTEMS: Hydraulic system, with four pumps, for collective, cyclic pitch/roll, yaw and feel augmentation flight control servo mechanisms; engine starters; landing gear actuation; cargo winches; loading ramp; and blade and tail pylon folding. System pressure 207 bars (3,000 lb/sq in), except for engine starter system which is rated at 276 bars (4,000 lb/sq in). (Separate hydraulic system in MH-53E to power AMCM equipment.) Electrical system includes three 115V 400Hz 40-60kVA AC alternators, and two 28V 200A transformer-rectifiers for DC power. Solar APU.
AVIONICS: Hamilton Standard automatic flight control system, using two digital onboard computers and a four-axis autopilot. Retrofit test flown late 1993, comprising four Canadian Marconi CM A-2082 15 cm (6 in) square colour displays, tied with GPS, Doppler and AHARS; installation by Teledyne Ryan.
EQUIPMENT: MH-53E equipment includes Westinghouse AN/AQS-14 towed sonar, AN/AQS-17 mine neutralisation device, AN/ALQ-141 electronic sweep, and Edo AN/ALQ-166 towed hydrofoil sled for detonating magnetic mines.
DIMENSIONS, EXTERNAL (CH-53E and MH-53E):
Main rotor diameter	24.08 m (79 ft 0 in)
Main rotor blade chord	0.76 m (2 ft 6 in)
Tail rotor diameter	6.10 m (20 ft 0 in)
Length overall: rotors turning	30.19 m (99 ft 0½ in)
rotor and tail pylon folded	18.44 m (60 ft 6 in)
Length of fuselage	22.35 m (73 ft 4 in)
Width of fuselage	2.69 m (8 ft 10 in)
Width overall, rotor and tail pylon folded:	
CH-53E	8.66 m (28 ft 5 in)
MH-53E	8.41 m (27 ft 7 in)
Height: to top of main rotor head	5.32 m (17 ft 5½ in)
tail rotor turning	8.97 m (29 ft 5 in)
rotor and tail pylon folded	5.66 m (18 ft 7 in)
Wheel track (c/l of shock struts)	3.96 m (13 ft 0 in)
Wheelbase	8.31 m (27 ft 3 in)

DIMENSIONS, INTERNAL (CH-53E and MH-53E):
Cabin:
Length (rear ramp/door hinge to fwd bulkhead)	9.14 m (30 ft 0 in)
Max width	2.29 m (7 ft 6 in)
Max height	1.98 m (6 ft 6 in)

AREAS (CH-53E and MH-53E):
Main rotor disc	455.38 m² (4,901.7 sq ft)
Tail rotor disc	29.19 m² (314.2 sq ft)

WEIGHTS AND LOADINGS:
Weight empty: CH-53E	15,072 kg (33,228 lb)
MH-53E	16,482 kg (36,336 lb)
Internal payload (100 nm; 185 km; 115 miles radius):	
CH-53E	13,607 kg (30,000 lb)
External payload (50 nm; 92.5 km; 57.5 miles radius):	
CH-53E	14,515 kg (32,000 lb)
Max external payload: CH-53E	16,330 kg (36,000 lb)
Useful load, influence sweep mission:	
MH-53E	11,793 kg (26,000 lb)
Max T-O weight (CH-53E and MH-53E):	
internal payload	31,640 kg (69,750 lb)
external payload	33,340 kg (73,500 lb)
Max disc loading:	
internal payload	69.47 kg/m² (14.23 lb/sq ft)
external payload	73.21 kg/m² (14.99 lb/sq ft)
Max power loading:	
internal payload	3.14 kg/kW (5.17 lb/shp)
external payload	3.31 kg/kW (5.44 lb/shp)

PERFORMANCE (CH-53E and MH-53E, ISA, at T-O weight of 25,400 kg; 56,000 lb):
Max level speed at S/L	170 knots (315 km/h; 196 mph)
Cruising speed at S/L	150 knots (278 km/h; 173 mph)
Max rate of climb at S/L, 11,340 kg (25,000 lb) payload	762 m (2,500 ft)/min
Service ceiling at max continuous power	5,640 m (18,500 ft)
Hovering ceiling at max power:	
IGE	3,520 m (11,550 ft)
OGE	2,895 m (9,500 ft)
Self-ferry range, unrefuelled, at optimum cruise condition for best range:	
CH-53E	1,120 nm (2,075 km; 1,290 miles)

SIKORSKY S-80E and S-80M

Export versions of CH/MH-53E, available as **S-80E** basic heavy transport helicopter and **S-80M** mine countermeasures helicopter; S-80E has single-point cargo hook and VFR equipment; options include troop seats, cabin soundproofing, internal cargo winch, 272 kg (600 lb) external hoist, two-point external cargo attachment, automatic blade and tail folding, two 2,460.5 litre (650 US gallon; 541 Imp gallon) external tanks, ground-to-air or in-flight refuelling equipment, engine air particle separators, and wide selection of com/nav equipment.

Japan allocated funding for 11 S-80Ms (four in FY 1986, two in FY 1987, four in FY 1989 and one in FY 1991; further one required); first S-80M-1 (8623) completed 13 January 1989 and handed over to Japan Defence Agency 30 November 1989; initially to 51 Squadron at Atsugi for trials; 10 delivered by January 1993.

SIKORSKY S-70A

US Army designations: UH-60A, UH-60L and UH-60Q Black Hawk, EH-60C and MH-60K
US Air Force designations: UH-60A, HH-60G, MH-60G Pave Hawk
US Marine Corps designation: VH-60N

TYPE: Infantry squad transport helicopter; also adapted for other roles.
PROGRAMME: UH-60A declared winner of US Army Utility Tactical Transport Aircraft System (UTTAS) competition against Boeing Vertol YUH-61A 23 December 1976; first flight of first of three YUH-60A competitive prototypes 17 October 1974; early development history recorded in 1982-83 and earlier *Jane's*. For retrofitted improvements, see UH-60L below.
CURRENT VERSIONS: **UH-60A Black Hawk**: Initial production version, designed to carry crew of three and 11 troops; also can be used without modification for medevac, reconnaissance, command and control, and troop supply; cargo hook capacity 3,630 kg (8,000 lb); one UH-60A can be carried in C-130, two in C-141 and six in C-5.

Medevac kits delivered from 1981; missile qualification completed June 1987, with day and night firing of Hellfire in various flight conditions; airborne target handover system (ATHS) qualified; cockpit lighting suitable for night vision goggles fitted to production UH-60s since November 1985 and retrofitted to those built earlier; US Army began testing Honeywell Volcano mine dispensing system July 1987; Volcano container is disposable and dispenses 960 Gator anti-tank and anti-personnel mines; usage monitor to measure certain rotor loads installed in 30 UH-60As; wire strike protection added to UH-60s and EH-60s during 1987; accident data recorders also fitted. Total 985 built before production change to UH-60L in 1989. *Detailed description applies to UH-60A/L except where indicated*.

Enhanced Black Hawk: Incorporates active and passive self-defence systems, retrofitted by Corpus Christi Army Depot, Texas, to new build UH-60A/Ls; first 15 delivered to US Army in South Korea November 1989. Equipment includes Tracor AN/ARN-148 Omega navigation receiver, Motorola AN/LST-5B satellite UHF communications transceiver, Bendix/King AN/ARC-199 HF-SSB, and AEL AN/APR-44(V)3 specific threat RWR complementing existing AN/APR-39 general threat RWR;

Sikorsky S-80M (export version of MH-53E) in Japanese service

BLACK HAWK DELIVERIES

	US Army	USAF	Philippines	SOA	China	Taiwan	Jordan	Brunei	Westland	Australia	Saudi Arabia	Colombia	Turkey	Korea Co-prod	Korea FMS	Korea RoKAF	Japan	Egypt	Bahrain	Mexico	DEA	Morocco	Hong Kong
1978	2																						
1979	36																						
1980	67																						
1981	119																						
1982	126																						
1983	126																						
1984	123		2		9																		
1985	120				15	11																	
1986	102					3	1	2	1												1		
1987	99	11					2			1+(13)											4		
1988	82	16								7+(11)	2	5	6										
1989	72	18								13+(14)	10	5											
1990	72	26								17	1		6	3+(13)			1+(2)	2	1				
1991	49			1						1	4			4	3	3				2			
1992	40	7									4			(24)								2	2
TOTALS	1,235	78	2	1	24	14	3	2	1	39	21	10	12	7	3	3	1	2	1	2	5	2	2

Notes: Kits in parentheses; NOT to be included in totals (appear also under year of completion)
DEA - Drug Enforcement Agency (USA)
SOA - Special Operations Aircraft

M134 Minigun can be fitted on each of two pintle mounts, replacing M60 machine gun.

JUH-60A: At least seven used temporarily for trials.

GUH-60A: At least 20 grounded airframes for technical training.

HH-60D Night Hawk: One prototype completed for abandoned USAF combat rescue variant.

EH-60A: (Designation **EH-60C** reserved, but not adopted by US Army.) Prototype YEH-60A ordered in October 1980 to carry 816 kg (800 lb) Quick Fix IIB battlefield ECM detection and jamming system. TRW Electronic Systems Laboratories was prime contractor for AN/ALQ-151(V)2 ECM kit, with installation by Tracor Aerospace; four dipole antennae on fuselage and deployable whip antenna; hover infra-red suppressor subsystem (HIRSS) standard. YEH-60A first flight 24 September 1981; order for Tracor Aerospace to modify 40 UH-60As to EH-60A standard under $51 million contract placed October 1984; first delivery July 1987 as part of US Army Special Electronics Mission Aircraft (SEMA) programme; 66 funded by FY 1987; programme completed 1989. Intercepts/locates AM, FM, CW and SSB radio emissions from upper HF to mid-VHF ranges over bandwidths of 8, 30 or 50 kHz; jams VHF communications with Fairchild AN/TLQ-17A; jams radar with ITT AN/ALQ-136(V)2 pulsed and Northrop AN/ALQ-162(V)2 CW transmitters. Protective systems of UH-60A/L (M-130 chaff/flare and AN/ALQ-144 IR jammer) augmented by Lockheed Sanders AN/ALQ-156(V)2 missile plume detector and Litton AN/APR-39(V)2 RWR. Used by regular Army and Guard (D Coy, 135th Aviation of Kansas ArNG). Contract for conversion of 32 EH-60s to Advanced Quick Fix due 1995 for 1997 deployment; improvements to include AN/APR-39A(V)2 RWR, new PRC-118 wideband datalink, AEL/Sanders TACJAM-A ECM subsystem, IBM Communications High Accuracy Locating System - Exploitable (CHALS-X) and UH-60L engines and gearbox for increase in max weight from 7,845 kg (17,295 lb) to 10,206 kg (22,500 lb).

MH-60A: About 30 modified for Army 160th Special Operations Aviation Regt; believed to be fitted with FLIR, Omega/VLF navigation, multi-function displays, auxiliary fuel tanks and door-mounted Minigun; interim equipment, pending MH-60K, but replaced by MH-60L in late 1990 and passed to 1/245 Avn, Oklahoma ArNG.

VH-60N: Nine for US Marine Corps Executive Flight Detachment of squadron HMX-1 at Quantico, Virginia, to replace UH-1Ns; deliveries started November 1988; known as VH-60A until redesignated 3 November 1989; two standard UH-60As also supplied. Additional equipment includes more durable gearbox, weather radar, SH-60B-type flight control system and ASI, cabin soundproofing, VIP interior, cabin radio operator station, EMP hardening and extensive avionics upgrading.

HH/MH-60G Pave Hawk: Replaced US Air Force HH-60D Night Hawk rescue helicopters, which were not funded (see 1987-88 *Jane's*); converted from UH-60A/L, including 10 originally delivered to 55th Aerospace Rescue and Recovery Squadron (now Special Operations Squadron) at Eglin AFB, Florida, in 1982-83, initially remaining as UH-60As; all progressively fitted by Sikorsky Support Services at Troy, Alabama, with aerial refuelling probe, 443 litre (117 US gallon; 97.5 Imp gallon) internal auxiliary fuel tank and fuel management panel; then to Pensacola NAD for mission avionics and modified instrument panel. Total 98 funded up to FY 1991, plus five in FY 1992 budget; all designated MH-60G until 1 January 1992, when 82 in combat rescue role redesignated HH-60G, balance of 16 remaining as MH-60G for special operations units. All have Doppler/INS, electronic map display, Tacan, RDR-14 lightweight weather/ground-mapping radar, secure HF, and Satcom; MH-60G additionally has ESSS (see Armament paragraph) for weapons and additional fuel carrying capability, plus door-mounted 0.50 in machine-guns and Pave Low III FLIR as fitted in MH-53H/J; individual aircraft configurations vary, as only 10 per cent of fleet is to full 'G' standard. Issued to 55th SOS (first operational MH-60G unit, March 1992); 1551st FTS, Kirtland AFB, New Mexico; 38th RqS (Rescue Squadron), Osan, Korea; 66th RqS, Nellis AFB, Nevada (formed 1 March 1991); 56th RqS, Keflavik, Iceland (from December 1991); plus 304th RqS, Portland, Oregon (AFRes); 102nd RqS, Suffolk County, New York ANG; 129th RqS, Moffett Field NAS, California ANG; 210th RqS, Kulis, Alaska ANG; 39th RqS at Misawa, Japan (1992); 71st SOS, Davis Monthan AFB, Arizona (1992); 48th RqS, Holloman AFB, New Mexico 1993; 71st SOS, Davis-Monthan AFB, Arizona (1992), AFRes; 301st RqS, Homestead AFB (temporarily Patrick AFB), Florida (1992), AFRes; and 41st RqS, Patrick AFB, Florida (October 1992). Re-equipment planned of 33rd RqS, Kadena, Okinawa; plus 203 RqS at Hickam, Hawaii.

MH-60K: US Army special operations aircraft (SOA); prototype ordered in January 1988; first flight 10 August 1990. US Army funded two batches of 11 with options for

Sikorsky HH-60G Pave Hawk air rescue helicopter

Sikorsky MH-60K special operations helicopter

another 38; first production aircraft complete, February 1992; trials at Patuxent River and Edwards AFB before intended first deliveries in June 1992 to 160th Special Operations Aviation Group (part of 160 SOA Regiment). Planned assignments to 'C' and 'D' Companies of 1 Battalion/160 SOA Group at Fort Campbell, Kentucky; 3/160 Avn at Hunter AAF, Georgia; and 1/245 Avn, Oklahoma ArNG. Deliveries delayed by software problems with special operations equipment; first 10 accepted in 1992 in non-operational state; remaining 12 in storage; start of training by 160 SOA Group planned for February 1994. Features include provision for additional 3,141 litres (829 US gallons; 691 Imp gallons) of internal and external fuel (see Power Plant), plus flight refuelling capability, integrated avionics system with electronic displays, Hughes AN/AAQ-16 FLIR, Texas Instruments AN/APQ-174A terrain following/terrain avoidance radar, T700-GE-701C engines and uprated transmission, external hoist, wire-strike protection, rotor brake, tiedown points, folding tailplane, AFCS similar to that of SH-60B, strengthened pintle mounts for 0.50 in machine guns, provision for Stinger missiles, missile warning receiver, pulse radio frequency jammer, CW radio jammer, laser detector, chaff/flare dispensers, and IR jammer.

SH-60B Seahawk: US Navy ASW/ASST helicopter, described separately.

SH-60F Seahawk: US Navy carrier-borne inner-zone ASW helicopter to replace SH-3D Sea King. See Seahawk entry.

HH-60H and HH-60J Jayhawk: Search and rescue/special warfare helicopters; see Seahawk entry.

UH-60J: Designation of Japanese-built S-70A-12.

UH-60L: Replaced UH-60A in production for US Army from October 1989 (aircraft 89-26180 onwards); prototype first flight 22 March 1988; first delivery 7 November 1989 to Texas ArNG. Powered by T700-GE-701C engines with uprated 2,535 kW (3,400 shp) transmission. Current production aircraft fitted with hover infra-red suppression system (HIRSS) to cool exhaust in hover as well as forward flight; older UH-60s retrofitted.

MH-60L: Similar to MH-60A; for 160th SOAR, US Army; converted late 1990; to be replaced by MH-60K.

UH-60M: Proposed enhanced version for US Army. Cancelled early 1989 in favour of UH-60L.

UH-60P: South Korean Army version of UH-60L (S-70A-18) with minor avionics modifications to meet local requirements; first of three UH-60Ls delivered by Sikorsky 10 December 1990; balance of 80 UH-60Ps to be assembled locally by Korean Air with increasing indigenous content, in $500 million, five-year programme.

UH-60Q: 'Dustoff' medical evacuation version for US Army; includes OBOGS, patient monitoring equipment, rescue hoist, dual mode IR/white searchlight, navigation (GPS, TACAN, Doppler and INS), communications and survivability upgrades and improved litter arrangement. Prototype flown 31 January 1993 (conversion of UH-60A by Serv-Air Inc at Richmond, Kentucky). Delivered to Tennessee ArNG at Lovell Field, Chattanooga, on 12 March 1993. Evaluation for 12 months from September 1993; production decision 1994; requirement for 120.

Exports comprise: S-70A-1: FMS deal for Royal Saudi Land Forces Army Aviation Command; 12 delivered January to April 1990 to squadron based at King Khaled Military City; modified to **Desert Hawk** and one (delivered December 1990) fitted with VIP interior; Desert Hawk has 15 troop seats, blade erosion protection using polyurethane tape and spray-on coating, Racal Jaguar 5 frequency-hopping radio, provision for searchlights, internal auxiliary fuel tanks, and external hoist.

S-70A-1L: Medical evacuation version for Saudi Arabia; infra-red filtered searchlight, rescue hoist, improved AN/ARC-217 HF com, AN/ARN-147 VOR/ILS, AN/ARN-149 ADF, air-conditioning and provision for six stretchers; eight delivered from December 1991; further eight required.

S-70A-5: Two for Philippine Air Force.

S-70A-9: Royal Australian Air Force; 39 S-70A-9s to replace Bell UH-1s; deliveries ran from October 1987 to 1 February 1991; first S-70A-9 completed by Sikorsky, remainder assembled by Hawker de Havilland in Australia; aircraft transferred to Australian Army in February 1989, but RAAF continues to maintain them.

S-70A-11: Three to Jordan in 1987.

S-70A-12: Japanese UH-60J built by Mitsubishi; 13 for air force and eight for navy on firm order, including FY 1992 funding.

S-70A-14: Two VIP for Brunei, delivered 1986.

S-70A-16: Reserved for Westland Helicopters (see UK section).

S-70A-17: Turkish Jandarma ordered six in September 1988; deliveries completed December 1988; further six (including two VIP) delivered from late 1990 to Turkish National Police. See S-70A-28.

S-70A-18: Korea (see UH-60P).

S-70A-19: Reserved for Westland of UK.

S-70A-21: Two VIP versions to Egypt, 1990.

S-70A-22: Korean VIP version.

S-70A-24: Two UH-60Ls for Mexico. Delivered 1991.

S-70A-25/26: Moroccan Gendarmerie ordered two

Sikorsky UH-60L making first trial lift of vehicle-mounted Avenger air defence system, 18 August 1992

US ARMY BLACK HAWK FUNDING

Fiscal Year	Quantity	First aircraft*	Remarks
1972	(3)	72-21650	
1977	15	77-22714	
1978	56	78-22961	
1979	92	79-23263	
1980	94	80-23416	
1981	85	81-23547	
1982	102	82-23660	⎫
1983	96	83-23837	⎬ MYP1, 18 December 1981
1984	96	84-23933	⎭
1985	96	85-24387	⎫
1986	96	86-24483	⎬ MYP2, 31 October 1984
1987	104[1]	87-24579	⎭
1988	72	88-26000	⎫
1989	72	89-26133	⎬ MYP3, 11 January 1988
1990	72	90-26218	⎭
1991	48[2]	91-26319	
1992	60	92-26402	⎫
1993	60		⎪
1994	60		⎬ MYP4, 28 April 1992
1995	60		⎪
1996	60		⎭
	1,496(+3)		
Supplement	13		Gulf War attrition batch
	1,509(+3)		

Notes: [1]Planned 96; additional eight result from multi-year procurement (MYP) savings
[2]Planned 40; plus eight from MYP savings
*Serial numbers not always consecutive; include transfers to USAF and exports

Black Hawks with different seating arrangements in 1991; delivered October 1992; began operations 11 November 1992; fitted with colour weather radar.

S-70A-27: Hong Kong. Two delivered 16 December 1992 to Royal Hong Kong Auxiliary Air Force; unit became Government Flying Service 1 April 1993. Fitted with FLIR and searchlight. Requirement for further four in 1995.

S-70A-28: Turkish follow-on batch; 95 ordered 8 December 1992, of which first five to Jandarma on 4 January 1993, followed by 40 to armed forces by end of 1993; balance of 50 co-produced in Turkey by TUSAS (which see).

Direct transfers include two UH-60Ls to Bahrain, early 1991; five UH-60Ls delivered to Colombian Air Force in July 1988 for anti-narcotics operations; five more sold February 1989; 10 for Israel in 1993.

Reserved designations as yet not ordered comprise S-70A-2 for Germany; -3 Spain; -4 Switzerland; -6 Thailand (transport); -7 Peru; -8 Brazil; -10 Israel; -13 not used; -15 Sweden; -20 Thailand (VIP); -23 Algeria.

Morocco ordered two VIP Black Hawks 1991.

S-70C: Commercial version, described separately and under SH-60B Seahawk.

CUSTOMERS: Total 1,496 funded by US Army to end of FY 1996, including EH-60As and diversions to USAF, Bahrain, Colombia, Saudi Arabia and Philippines; see table; 1,000th accepted 17 October 1988; US Army Black Hawks in service in Hawaii, Korea, Panama, Germany and with Army National Guard and Army Reserve.

US Army loaned 12 UH-60As to the US Drug Enforcement Agency, augmenting five bought direct from Sikorsky.

See Current Versions for export models and details.
COSTS: $5.87 million (1992) US Army unit cost.
DESIGN FEATURES: Represented new generation in technology for performance, survivability and ease of operation when introduced to replace UH-1 as US Army's main squad-

carrying helicopter; adapted to wide variety of other roles, including several maritime applications. Four-blade main rotor; one-piece forged titanium rotor head with elastomeric blade retention bearings providing all movement and requiring no lubrication; hydraulic drag dampers; bifilar self-tuning vibration absorber above head; blades have 18° twist, and tips swept at 20°; thickness and camber vary over the length of blades, based on Sikorsky SC-1095 aerofoil; blades tolerant up to 23 mm hits and spar tubes pressurised with gauges to indicate loss of pressure following structural degradation. Transmission rating 2,109 kW (2,828 shp) in UH-60A, uprated to 2,535 kW (3,400 shp) in models with T700-GE-701C engines.

Two pairs of tail rotor blades fastened in cross-beam arrangement, mounted to starboard; tail rotor pylon tilted to port to produce lift as well as anti-torque thrust and to extend permissible CG range; fixed fin large enough to allow controlled run-on landing following loss of tail rotor.

FLYING CONTROLS: Rotor pitch control powered by two independent hydraulic systems; Hamilton Standard AFCS with digital three-axis autopilot provides speed and height control and coupled modes. Full-time autostabilisation includes feet-off heading hold cancelling torque-induced yaw at all airspeeds and during hover; positive fuselage attitude control provided by electrically driven variable incidence tailplane moving from +34° in hover to −6° during autorotation; angle is controlled by combined sensing of airspeed, collective-lever position, pitch attitude rate and lateral acceleration.

STRUCTURE: Main blade spar is formed and welded into oval titanium tube, with Nomex core, graphite trailing-edge, and covered by glassfibre/epoxy skin; titanium leading-edge abrasion strip and Kevlar tip. Cross-beam composites tail rotor, eliminating all rotor head bearings. Light alloy airframe designed to retain 85 per cent of its flight deck and passenger space intact after vertical impact at 11.5 m (38 ft)/s, lateral impact at 9.1 m (30 ft)/s, and longitudinal impact at 12.2 m (40 ft)/s; also withstands simultaneous 20g forward and 10g downward impact; glassfibre and Kevlar used for cockpit doors, canopy, fairings and engine cowlings; glassfibre/Nomex floors; tailboom folds to starboard and main rotor mast can be lowered for transport/storage.

LANDING GEAR: Non-retractable tailwheel type with single wheel on each unit. Energy absorbing main gear with a tailwheel which gives protection for the tail rotor in taxying over rough terrain or during a high-flare landing. Axle assembly and main gear oleo shock absorbers by General Mechatronics. Mainwheel tyres size 26 × 10.00-11, pressure 8.96-9.65 bars (130-140 lb/sq in); tailwheel tyre size 15 × 6.00-6, pressure 6.21-6.55 bars (90-95 lb/sq in).

POWER PLANT: Two 1,210 kW (1,622 shp) intermediate rating General Electric T700-GE-700 turboshafts initially. From late 1989 (UH-60L), two T700-GE-701C engines, each developing 1,342 kW (1,800 shp). (T700-GE-701A engines with max T-O rating of 1,285 kW; 1,723 shp optional in export models.) Transmission rating 2,535 kW (3,400 shp). Two crashworthy, bulletproof fuel cells, with combined usable capacity of 1,361 litres (360 US gallons; 299.5 Imp gallons), aft of cabin. Single-point pressure refuelling, or gravity refuelling via point on each tank. Auxiliary fuel can be carried internally in one of several optional arrangements, or externally by the ESSS system. Two external tanks each of 870.5 litres (230 US gallons; 191.5 Imp gallons); up to two internal tanks, each of 700 litres (185 US gallons; 154 Imp gallons).

ACCOMMODATION: Two-man flight deck, with pilot and co-pilot on armour protected seats. A third crew member is stationed in the cabin at the gunner's position adjacent to the forward cabin windows. Forward hinged jettisonable door on each side for access to flight deck area. Main cabin open to cockpit to provide good communication with flight crew and forward view for squad commander. Accommodation for 11 fully equipped troops, or 14 in high density configuration. Eight troop seats can be removed and replaced by four litters for medevac missions, or to make room for internal cargo. An optional layout is available to accommodate a maximum of six litter patients. Cabin heated and ventilated. External cargo hook, having a 3,630 kg (8,000 lb) lift capability, enables UH-60A to transport a 105 mm howitzer, its crew of five and 50 rounds of ammunition. Rescue hoist of 272 kg (600 lb) capacity optional. Large rearward sliding door on each side of fuselage for rapid entry and exit. (Executive interiors for 7-12 passengers available for the S-70A.)

SYSTEMS: Solar 67 kW (90 hp) T-62T-40-1 APU; Garrett or Sundstrand APU. An optional winterisation kit provides a second hydraulic accumulator installed in parallel with the APU hydraulic start accumulator, maintaining engine start capability at low ambient temperatures; Bendix 30/40kVA and 20/30kVA electrical power generators; 17Ah nickel-cadmium battery. Engine fire extinguishing system. Rotor blade de-icing system standard on US Army aircraft, optional for export. Electric windscreen de-icing.

AVIONICS: Com equipment comprises E-Systems AN/ARC-186 VHF-FM, GTE Sylvania AN/ARC-115 VHF-AM, Magnavox AN/ARC-164 UHF-AM, Collins AN/ARC-186(V) VHF-AM/FM, Bendix/King AN/APX-100 IFF transponder, Magnavox TSEC/KT-28 voice security set, and intercom. Nav equipment comprises Emerson AN/ARN-89 ADF, Bendix/King

Sikorsky S-70A-27 Black Hawk of Hong Kong Government Flying Services

Sikorsky UH-60L in South Korean markings represents the UH-60P to be co-produced under licence by Korean Air

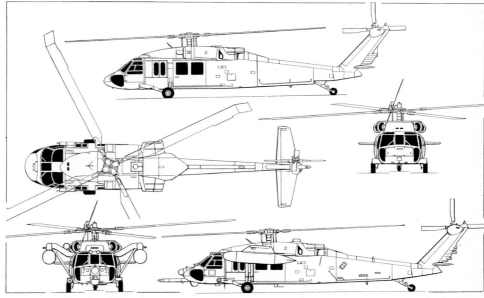

Sikorsky UH-60L Black Hawk combat assault helicopter with additional lower side view and lower front view of MH-60K special operations variant (*Dennis Punnett*)

564 USA: AIRCRAFT—SIKORSKY

Sikorsky S-70A-28 Black Hawk before delivery to Turkey in January 1993

Moroccan Gendarmerie S-70A Black Hawk with nose-mounted colour weather radar

Sikorsky S-70A-1L medical evacuation helicopter operated by the Royal Saudi Land Forces

AN/ARN-123(V)1 VOR/marker beacon/glideslope receiver, Honeywell AN/ASN-43 gyro compass, Plessey Electronic Systems AN/ASN-128 Doppler, and Honeywell AN/APN-209(V)2 radar altimeter. E-Systems Melpar/Memcor AN/APR-39(V)1 RWR and Sanders AN/ALQ-144 IR countermeasures set. Tracor M-130 chaff/flare dispenser. Hamilton Standard AFCS with digital three-axis autopilot. Additional avionics and self-protection equipment installed in Enhanced Black Hawk, as described under Current Versions.

MH-60K avionics include Texas Instruments AN/APQ-174A terrain following/terrain avoidance radar, Honeywell AN/AAR-47 missile warning system, Northrop AN/ALQ-136 pulse radio frequency jammer, Northrop AN/ALQ-162 CW radio jammer, E-Systems AN/APR-39A and -44 pulse/CW warning receivers, Perkin-Elmer AN/AVR-2 laser detector, M-130 chaff/flare dispensers, and Lockheed Sanders AN/ALQ-144 IR jammer.

HH/MH-60G avionics include Bendix/King RDR-1400C (AN/APN-239) radar, Rockwell Collins AN/ASN-149 GPS, Litton or Honeywell SNU-84-1 ringlaser gyro INS, GEC-Marconi AN/ASN-137 Doppler, Hughes AN/AAQ-16 FLIR, Sanders AN/ALQ-144 IR jammer, chaff/flare dispensers (M-130 being replaced by AN/ALE-40, but AN/ALE-47 to follow) and AN/URC-108 Satcom.

ARMAMENT: New production UH-60As and Ls from c/n 431 onward incorporate hardpoints for an external stores support system (ESSS). This consists of a combination of fixed provisions built into the airframe and four removable external pylons from which fuel tanks and a variety of weapons can be suspended. Able to carry more than 2,268 kg (5,000 lb) on each side of the helicopter, the ESSS can accommodate two 870 litre (230 US gallon; 191.5 Imp gallon) fuel tanks outboard, and two 1,703 litre (450 US gallon; 375 Imp gallon) tanks inboard. This allows the UH-60A to self deploy 1,200 nm (2,220 km; 1,380 miles) without refuelling. The ESSS also enables the Black Hawk to carry Hellfire laser-guided anti-armour missiles, gun or M56 mine dispensing pods, ECM packs, rockets and motorcycles. Up to 16 Hellfires can be carried externally on the ESSS, with another 16 in the cabin to provide the capability to land and reload. (Laser designation provided by Bell OH-58 helicopter or ground troops.) Two pintle mounts in cabin, adjacent to forward cabin windows on each side, can each accommodate a 0.50 in calibre General Electric GECAL 50 or 7.62 mm six-barrel Minigun.

DIMENSIONS, EXTERNAL:
Main rotor diameter	16.36 m (53 ft 8 in)
Main rotor blade chord	0.53 m (1 ft 8¾ in)
Tail rotor diameter	3.35 m (11 ft 0 in)
Length overall: rotors turning	19.76 m (64 ft 10 in)
rotors and tail pylon folded	12.60 m (41 ft 4 in)
Length of fuselage:	
UH-60A/MH-60G, excl flight refuelling probe	15.26 m (50 ft 0¾ in)
MH-60G, incl retracted refuelling probe	17.38 m (57 ft 0¼ in)
Width: UH-60A fuselage	2.36 m (7 ft 9 in)
MH-60G with auxiliary tanks	5.46 m (17 ft 11 in)
Max depth of fuselage	1.75 m (5 ft 9 in)
Height: overall, tail rotor turning	5.13 m (16 ft 10 in)
to top of rotor head	3.76 m (12 ft 4 in)
in air transportable configuration	2.67 m (8 ft 9 in)
Tailplane span	4.38 m (14 ft 4½ in)
Wheel track	2.705 m (8 ft 10½ in)
Wheelbase	8.83 m (28 ft 11¾ in)
Tail rotor ground clearance	1.98 m (6 ft 6 in)
Cabin doors (each): Height	1.37 m (4 ft 6 in)
Width	1.75 m (5 ft 9 in)

DIMENSIONS, INTERNAL:
Cabin: Volume	11.61 m³ (410.0 cu ft)

AREAS:
Main rotor blades (each)	4.34 m² (46.70 sq ft)
Tail rotor blades (each)	0.41 m² (4.45 sq ft)
Main rotor disc	210.15 m² (2,262 sq ft)
Tail rotor disc	8.83 m² (95.03 sq ft)
Tailplane	4.18 m² (45.0 sq ft)
Vertical stabiliser	3.00 m² (32.3 sq ft)

WEIGHTS AND LOADINGS:
Weight empty: UH-60A	5,118 kg (11,284 lb)
UH-60L	5,216 kg (11,500 lb)
Payload: internal, UH-60A/L	1,197 kg (2,640 lb)
underslung, UH-60A/L	3,629 kg (8,000 lb)
Mission T-O weight: UH-60A	7,708 kg (16,994 lb)
Max alternative T-O weight:	
UH-60A	9,185 kg (20,250 lb)
UH-60L	9,979 kg (22,000 lb)
Max disc loading:	
UH-60A at mission T-O weight	36.68 kg/m² (7.51 lb/sq ft)
UH-60A/S-70A at max alternative T-O weight	47.49 kg/m² (9.73 lb/sq ft)
Max power loading:	
UH-60A at mission T-O weight	3.66 kg/kW (6.01 lb/shp)
UH-60A/S-70A at max alternative T-O weight	4.73 kg/kW (7.78 lb/shp)

PERFORMANCE (UH-60A at mission T-O weight, except where

indicated):
Max level speed at S/L   160 knots (296 km/h; 184 mph)
Max level speed at max T-O weight
                         158 knots (293 km/h; 182 mph)
Max cruising speed:
  UH-60A   139 knots (257 km/h; 160 mph)
  UH-60L   150 knots (278 km/h; 173 mph)
Single-engine cruising speed at 1,220 m (4,000 ft), and
  35°C (95°F)   105 knots (195 km/h; 121 mph)
Vertical rate of climb at 1,220 m (4,000 ft) and 35°C
  (95°F): UH-60A   125 m (411 ft)/min
  UH-60L   239 m (785 ft)/min
Service ceiling   5,790 m (19,000 ft)
Hovering ceiling: IGE at 35°C 2,895 m (9,500 ft)
  OGE, ISA   3,170 m (10,400 ft)
  OGE at 35°C   1,705 m (5,600 ft)
Range with max internal fuel at max T-O weight, 30 min
  reserves: UH-60A   319 nm (592 km; 368 miles)
  UH-60L   315 nm (584 km; 363 miles)
Range with external fuel tanks on ESSS pylons:
  with two 870 litre (230 US gallon; 191.5 Imp gallon)
    tanks   880 nm (1,630 km; 1,012 miles)
  with two 870 litre (230 US gallon; 191.5 Imp gallon) and
    two 1,703 litre (450 US gallon; 375 Imp gallon)
    tanks   1,200 nm (2,220 km; 1,380 miles)
Endurance: UH-60A   2 h 18 min
  UH-60L   2 h 6 min
  MH-60G with max fuel   4 h 51 min

SIKORSKY S-70B
US Navy designations: SH-60B Seahawk, SH-60F and HH-60H
US Coast Guard designation: HH-60J Jayhawk
Spanish Navy designation: HS.23

TYPE: ASW, and anti-ship surveillance and targeting helicopter.

PROGRAMME: Naval development of Sikorsky UTTAS (UH-60A Black Hawk) utility helicopter; won US Navy LAMPS Mk III competition for shipboard helicopter in 1977; first flight of first of five prototypes (161169) 12 December 1979; development details in 1982-83 *Jane's*; first 18 SH-60Bs authorised FY 1982.

CURRENT VERSIONS: **SH-60B**: Initial production version for ASW/ASST, *as detailed under main description*.

XSH-60J: Japan Maritime Self-Defence Force (JMSDF) placed $27 million order for two **S-70B-3**s for installation of Japanese avionics and mission equipment; first flights 31 August and early October 1987; 1,007 hour test programme by Japan Defence Agency Technical Research and Development Institute between 1 June 1989 and 7 April 1991 to evaluate largely Japanese avionics for SH-60J, but AN/APS-124 radar.

SH-60J: Mitsubishi (which see) is manufacturing SH-60J Seahawk for JMSDF to replace Sikorsky SH-3 Sea Kings by mid-1990s; total 80 required; 47 funded including FY 1992.

UH-60J: Japan Self-Defence Forces acquiring UH-60J version of Mitsubishi SH-60J for search and rescue. Sikorsky-built prototype (N7267D), plus two CKD kits, delivered late 1990. JASDF requires 43, 13 funded up to FY 1992; eight funded for JMSDF to FY 1992.

S-70B: Royal Australian Navy selected Seahawk, designated **S-70B-2**, for role adaptable weapon system (RAWS) full-spectrum ASW helicopter with autonomous operating capability; order for eight confirmed 9 October 1984; eight more ordered May 1986; operates from RAN 'Adelaide' class (FFG-7) guided missile frigates; first flight (N7265H, now N24-001) at West Palm Beach 4 December 1987; 14 originally planned to be assembled in Australia, but announced late 1988 Sikorsky would assemble first eight; ASTA in Australia (which see) designated to assemble remainder in early 1989; first S-70B-2 handed over in USA 12 September 1989; formal acceptance at Nowra, NSW, 4 October 1989; Seahawk Introduction and Transition Unit became HS-816 at Nowra, 1 June 1991; formed first ship's flight, September 1991; squadron formally commissioned 23 July 1992. S-70B-2 has substantially different avionics to USN version: THORN EMI Super Searcher radar (capable of tracking 32 surface targets) and Rockwell Collins advanced integrated avionics including cockpit controls and displays, navigation receivers, communications radios, airborne target hand-off data link and tactical data system (TDS); final delivery 11 September 1991.

Spanish Navy received six **S-70B-1**s from December 1988 (designated **HS.23**) for operation from four FFG-7 frigates by Escuadrilla 010 at Rota; similar to USN SH-60B, but with Bendix/King AN/AQS-13F dipping sonar; further six ordered 1991.

SH-60F: CV Inner Zone ASW helicopter, known as CV-Helo, for close-in ASW protection of aircraft carrier groups; $50.9 million initial US Navy contract for full-scale development and production options placed 6 March 1985; replacing SH-3H Sea King; Seahawk prototype modified as SH-60F test aircraft; first flight 19 March 1987; initial fleet deployment with USS *Nimitz* in 1991.

SH-60F has all LAMPS Mk III avionics, fairings and equipment removed, including cargo hook and RAST system main and tail probes, but installation provisions

Sikorsky SH-60B Seahawk LAMPS Mk III helicopter with parent ship USS *Crommelin* (FFG-37)

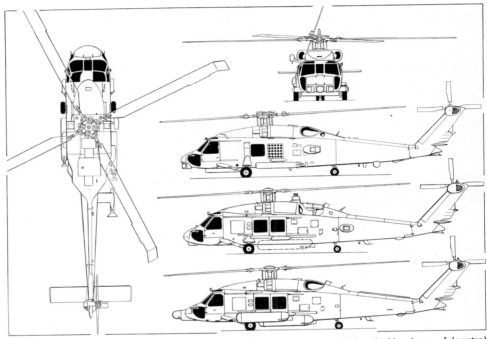
Sikorsky SH-60B Seahawk twin-turbine ASW/ASST helicopter with additional side views of (centre) HH-60H and (bottom) HH-60J Jayhawk (*Dennis Punnett*)

Sikorsky SH-60F Ocean Hawk of HS-2, assigned to aircraft carrier USS *Nimitz*

566 USA: AIRCRAFT—SIKORSKY

Sikorsky HH-60H combat strike-rescue/special warfare support helicopter

Sikorsky HH-60J Jayhawk rescue and law enforcement helicopter for the US Coast Guard

Sikorsky S-70B Seahawk operated by the Spanish Armada

retained. Replaced by integrated ASW mission avionics including Allied Signal (Bendix Oceanics) AN/AQS-13F dipping sonar, MIL-STD-1553B databus, dual Teledyne Systems AN/ASN-150 tactical navigation computers and AN/ASM-614 avionics support equipment, automatic flight control system with quicker automatic transition and both cable and Doppler autohover, tactical data link with other aircraft, communications control system, multi-function keypads and displays for each of four crew members, internal/external fuel system and extra weapon station to port allowing carriage of three Mk 50 homing torpedoes, provision for surface search radar, FLIR, night vision equipment, passive ECM, MAD, air-to-surface missile capability, sonobuoy data link, chaff/sonobuoy dispenser, attitude and heading reference system (AHRS), Navstar GPS, fatigue monitoring system and increase of max T-O weight to 10,659 kg (23,500 lb); secondary missions include SAR and plane guard.

HH-60H: US Navy ordered five (increased to 25 including seven in FY 1993) for strike-rescue/special warfare support (HCS), designated HH-60H, in September 1986; first flight (163783) 17 August 1988; accepted by USN 30 March 1989; in service with HCS-4 at Norfolk, Virginia, January 1990; final delivery July 1991, completing HCS-5 at Point Mugu, California although long-term requirement is 42; both squadrons are Reserves, and regularly operate alongside SH-60F units; missions are to recover four-man crew at 250 nm (463 km; 288 miles) from launch point or fly 200 nm (371 km; 230 miles) and drop eight SEALs from 915 m (3,000 ft).

Close derivative of SH-60F, with same T700-GE-401C engines and HIRSS as SH-60B/F; equipment includes General Instrument AN/APR-39A(XE)2 RWR, Hughes AN/AVR-2 laser warning receiver, Honeywell AN/AAR-47 missile plume detector, Loral AN/ALE-47 chaff/flare dispenser, AN/ALQ-144 IR jammer, night vision goggle cockpit and two cabin-mounted M60D 7.62 mm machine guns; provision for weapon pylons; required to operate from decks of FFG-7, DD-963, CG-47 and larger vessels, as well as unprepared sites. Cubic AN/ARS-6 personnel locator system to be installed from FY 1991. Some equipped with Seahawk-type Indal RAST (recovery assist, secure and traverse) equipment. Armament development authorised October 1991 for installation of Hellfire ASM, 70 mm (2.75 in) rockets and forward-firing guns.

HH-60J Jayhawk: Ordered in parallel with HH-60H; adapted for US Coast Guard medium range recovery (MRR) role; 40 funded, including six in FY 1991 and one each in FYs 1992-93; seven more required. First flight (USCG 6001) 8 August 1989; first delivery to USCG (6002) at Elizabeth City CGAS) 16 June 1990; subsequently to Mobile, Traverse City, San Fransisco, Cape Cod, Sitka, Kodiak and (December 1992) Clearwater CGAS. When carrying three 455 litre (120 US gallon; 100 Imp gallon) external tanks, HH-60J can fly out 300 nm (556 km; 345 miles) and return with six survivors in addition to four-man crew, or loiter for 1 h 30 min when investigating possible smugglers; other duties include law enforcement, drug interdiction, logistics, aids to navigation, environmental protection and military readiness; compatible with decks of 'Hamilton' and 'Bear' class USCG cutters. Equipment includes Bendix/King RDR-1300C search/weather radar, AN/ARN-147 VOR/ILS, KDF 806 direction finder, GPS, Tacan, VHF/UHF-DF, TacNav, dual U/VHF-FM radios, HF radio, IFF, V/U/HF IFF crypto computers, NVG-compatible cockpit, rescue hoist and external cargo hook.

S-70B-6: Hybrid SH-60B/F for Greece; five (plus three options) selected December 1991 and ordered 17 August 1992 for MEKO 200 frigates; armament includes NFT Penguin ASMs; avionics include AN/AQS-18(V)-3 dipping sonar, AN/APS-143(V3) radar and AN/ALR-66(V)-2 ESM; deliveries from fourth quarter of 1994.

S-70C(M)-1 Thunderhawk: Delivery began July 1990 to Taiwanese Navy of 10 SH-60B Seahawks; given S-70C designations; shipboard deployment from 1993 aboard six FFG-7 frigates. Equipment includes Bendix Oceanics AN/AQS-18(V) dipping sonar, Telephonics AN/APS-128PC radar and Litton AN/ALR-606(V)-2 ESM integrated with radar antenna; no MAD.

CUSTOMERS: Total US Navy requirement 260 SH-60Bs, plus five prototypes; 181 on order, including 12 in FY 1993; first flight production Seahawk 11 February 1983; deliveries at two a month; first squadron was HSL-41 at NAS North Island, San Diego, California; operational deployment began 1984; 11 US Navy squadrons operating by March 1991 (HSLs 41, 43, 45, 47 and 49 at NAS North Island; 40, 42, 44, 46 and 48 at NAS Mayport, Florida; and 51 at Atsugi, Japan, formed 1 October 1991); SH-60Bs deployed in 'Oliver Hazard Perry' (FFG-7) class frigates, 'Spruance' class and Aegis equipped destroyers, 'Ticonderoga' class guided missile cruisers. US Navy requires 150 SH-60Fs; total 82 on order, comprising seven pre-series plus 18 each in FYs 1988, 1989 and 1991, 12 in FY 1992 and nine in FY 1993; two used for operational evaluation; five more delivered by September 1989; in service with HS-2, 4, 6 and 10 squadrons NAS North Island, California, and HS-3 at Jacksonville, Florida; last-mentioned equipped from 27 August 1991 as first East Coast squadron.

Exported to Australia, Greece, Japan and Spain (see Current Versions). S-70B-4 and -5 are derivatives of

SEAHAWK DELIVERIES

	SH-60B series USN	SH-60B series Spain	SH-60B series Japan	SH-60B series Australia	SH-60F series USN	SH-60F series Taiwan	HH-60H	HH-60J
1983	9							
1984	27							
1985	24		2					
1986	24							
1987	22				2			
1988	16	6		(14)				
1989	8			4	9		8	
1990	6			6	19	2	4	6
1991	6			6	18	8	6	10
1992	6				17			12
Totals	**148**	**6**	**2**	**16**	**65**	**10**	**18**	**28**

Notes: Kits in parentheses; NOT to be included in totals (appear also under year of completion)

SH-60F and HH-60H, respectively; not taken up.

COSTS: $20.25 million (1992) USN programme unit cost.

DESIGN FEATURES: SH-60B Seahawk designed to provide all-weather detection, classification, localisation and interdiction of surface ships and submarines, either controlled through data link from mother ship or operating independently; secondary missions include SAR, vertical replenishment, medevac, fleet support and communications relay.

New features include more powerful navalised GE T700-GE-401 engines, additional fuel, sensor operator's station, port-side internal launchers for 25 sonobuoys, pylon on starboard side of tailboom for MAD bird, lateral pylons for two torpedoes or external tanks, chin-mounted ESM pods, sliding cabin door, rescue hoist, electrically actuated blade folding, rotor brake, folding tail, short-wheelbase tailwheel landing gear with twin tailwheels stressed for lower crash impact, DAF Indal RAST recovery assist, secure and traversing for haul-down landings on small decks and moving into hangar, hovering in-flight refuelling system, and emergency flotation system; pilots' seats not armoured. SH-60B gives 57 minutes more listening time on station and 45 minutes more ship surveillance and targeting time than LAMPS Mk I.

For operation in Gulf during Iran-Iraq war, 25 SH-60Bs fitted with upper and lower Lockheed Sanders AN/ALQ-144 IR jammers, Tracor AN/ALE-39 chaff/flare dispensers, Honeywell AN/AAR-47 electro-optical missile warning, and a single 7.62 mm machine gun in door, for a weight penalty of 169 kg (369.5 lb); seven Seahawks fitted with Ford Aerospace AN/AAS-38 FLIR on port weapon pylon with instantaneous relay to parent ship.

First Block I SH-60B update, introduced in production Lot 9, delivered from October 1991, includes provision for NFT AN/AGM-119 Penguin anti-ship missile, Mk 50 advanced lightweight torpedo, Rospatch AN/ARR-84 99-channel sonobuoy receiver (replacing ARR-75), Chelton AN/ARC-182 V/UHF FM radio and Rockwell/Collins Class 3A Navstar GPS; 115 Penguin-capable Seahawks to come from retrofitting back to Lot 5. Block II development, including airborne low-frequency sonar and inverse synthetic aperture radar, began 1991 with preliminary design contract award to IBM; for retrofitting from 1999; new sonar decision announced December 1991 in favour of Hughes/Thomson Sintra FLASH (folding light acoustic sonar for helicopters); 158 to replace AN/AQS-13F in SH-60F; plus 185 for installation in SH-60B; associated AT&T AN/UYS-2 enhanced modular signal processor; max weight increased to 10,659 kg (23,500 lb).

FLYING CONTROLS: As for UH-60.

STRUCTURE: Basically as for UH-60 plus marine corrosion protection; single cabin door, starboard side, narrower than UH-60.

POWER PLANT: Two 1,260 kW (1,690 shp) General Electric T700-GE-401 turboshafts in early aircraft; 1,417 kW (1,900 shp) T700-401C turboshafts introduced in 1988 and on HH-60H/J. Transmission rating 2,535 kW (3,400 shp). Internal fuel capacity 2,233 litres (590 US gallons; 491 Imp gallons). Hovering in-flight refuelling capability. Two 455 litre (120 US gallon; 100 Imp gallon) auxiliary fuel tanks on fuselage pylons optional (three on HH-60J). Hover infra-red suppressor subsystem (HIRSS) exhaust cowling fitted to HH-60H.

ACCOMMODATION: Pilot and airborne tactical officer/backup pilot in cockpit, sensor operator in specially equipped station in cabin. Dual controls standard. Sliding door with jettisonable window on starboard side. Accommodation heated, ventilated, air-conditioned.

SYSTEMS: Generally as for UH-60A.

AVIONICS: Com equipment comprises Collins AN/ARC-159(V)2 UHF and AN/ARC-174(V)2 HF, Hazeltine AN/APX-76A(V) and Bendix/King AN/APX-100(V)1 IFF transponders, TSEC/KY-75 voice security set, TSEC/KG-45(E-1) com security, Telephonics OK-374/ASC com system control group and Sierra Research AN/ARQ-44 data link and telemetry (Rockwell Collins DHS-901 in Australian Seahawks). Nav equipment comprises Collins AN/ARN-118(V) Tacan, Honeywell AN/APN-194(V) radar altimeter, Teledyne Ryan AN/APN-217 Doppler, and Collins AN/ARA-50 UHF DF. Mission equipment includes Sikorsky sonobuoy launcher, Edmac AN/ARR-75 and R-1651/ARA sonobuoy receiving sets (AN/ARR-84 receiver in Australian Seahawks and for USN Block 1 upgrade), Texas Instruments AN/ASQ-81(V)2 MAD (CAE AN/ASQ-504[V] internal MAD in Australian Seahawks), Raymond MU-670/ASQ magnetic tape memory unit, Astronautics IO-2177/ASQ altitude indicator, Fairchild AN/ASQ-164 control indicator set and AN/ASQ-165 armament control indicator set, Texas Instruments AN/APS-124 search radar (under front fuselage) (THORN EMI SuperSearcher in Australia; Telephonics AN/APS-128PC in Taiwan), IBM AN/UYS-1(V)2 Proteus acoustic processor (Computing Devices UYS-503 for Australia) and CV-3252/A converter display, Control Data AN/AYK-14 (XN-1A) digital computer, and Raytheon AN/ALQ-142 ESM (in chin-mounted pods). SH-60F has Bendix Oceanics AN/AQS-13F dipping sonar; AN/AQS-18 in Taiwanese S-70s. During 1991 Gulf War, pod-mounted Hughes AN/AAQ-16 FLIR fitted to five SH-60Bs and Texas Instruments AN/AAQ-17 FLIR deployed on one SH-60B; GEC-Marconi Sea Owl IR turret evaluated later in 1991. Australian Seahawks fitted with AN/ALE-47 chaff/flare dispensers, AN/AAR-47 missile detectors and AN/AAQ-16 FLIR for Gulf War. See also Current Versions for variants other than SH-60B.

EQUIPMENT: External cargo hook (capacity 2,722 kg; 6,000 lb) and rescue hoist (272 kg; 600 lb) standard.

ARMAMENT: Includes two Mk 46 torpedoes and (IOC 1993) NFT AN/AGM-119B Penguin Mk 2 Mod 7 anti-shipping missiles. Block I upgrade integrates Penguin and Honeywell Mk 50 Advanced Lightweight Torpedo from 1993. HH-60H has two pintle-mounted M60D machine-guns but to receive Rockwell AGM-114 Hellfire ASMs, 70 mm (2.75 in) rocket pods and forward-firing guns.

DIMENSIONS, EXTERNAL: As UH-60A except:

Length overall, rotors and tail pylon folded:	
SH-60B	12.47 m (40 ft 11 in)
HH-60H	12.51 m (41 ft 0⅝ in)
HH-60J	13.13 m (43 ft 0⅞ in)
Length of fuselage: HH-60J	15.87 m (52 ft 1 in)
Width, rotors folded	3.26 m (10 ft 8½ in)
Height: to top of rotor head	3.79 m (12 ft 5⅜ in)
overall, tail rotor turning	5.18 m (17 ft 0 in)
overall, pylon folded	4.04 m (13 ft 3¼ in)
Wheelbase	4.83 m (15 ft 10 in)
Tail rotor ground clearance	1.83 m (6 ft 0 in)
Main/tail rotor clearance	6.6 cm (2⅝ in)

AREAS: as UH-60A

WEIGHTS AND LOADINGS:

Weight empty: SH-60B ASW	6,191 kg (13,648 lb)
HH-60H	6,114 kg (13,480 lb)
HH-60J	6,086 kg (13,417 lb)
Useful load: HH-60J	3,551 kg (7,829 lb)
Internal payload: HH-60H	1,860 kg (4,100 lb)
Mission gross weight:	
SH-60B ASW	9,182 kg (20,244 lb)
SH-60B ASST	8,334 kg (18,373 lb)
Max gross weight:	
SH-60B Utility, HH-60H	9,926 kg (21,884 lb)
HH-60J	9,637 kg (21,246 lb)
Max disc loading:	
SH-60B Utility, HH-60H	47.24 kg/m² (9.67 lb/sq ft)
HH-60J	45.86 kg/m² (9.39 lb/sq ft)
Max power loading:	
SH-60B Utility, HH-60H	3.92 kg/kW (6.44 lb/shp)
HH-60J	3.80 kg/kW (6.25 lb/shp)

PERFORMANCE:

Cruising speed at S/L:	
HH-60H	147 knots (272 km/h; 169 mph)
HH-60J	146 knots (271 km/h; 168 mph)
Dash speed at 1,525 m (5,000 ft), tropical day:	
SH-60B	126 knots (234 km/h; 145 mph)
Vertical rate of climb at S/L, 32.2°C (90°F):	
SH-60B	213 m (700 ft)/min
Vertical rate of climb at S/L, 32.2°C (90°F), OEI:	
SH-60B	137 m (450 ft)/min

SIKORSKY S-70C

TYPE: Commercial or military H-60 for non-FMS customers.

PROGRAMME: See Customers below.

CUSTOMERS: Delivery of 24, designated **S-70C-2**, with undernose radar to People's Republic of China completed December 1985 but offered for sale in 1992 due to spares embargo; 14 supplied to Republic of China Air Force, Taiwan; one each for Westland plc (designated Westland **WS-70L**) and Rolls-Royce in UK; R-R aircraft used as testbed for R-R Turbomeca RTM 322 turboshafts; all based on Black Hawk. **S-70C(M)-1** designation assigned to 10 Seahawks for Taiwan (see SH-60B entry).

DESIGN FEATURES: Certificated to FAR Pt 21.25; roles include utility transport, external lift, maritime and environmental survey, forestry and conservation, mineral exploration and heavy construction support; powered by General Electric CT7-2C or -2D engines; options include de-icing kit for main and tail rotors, cargo hook for 3,630 kg (8,000 lb) loads, rescue hoist, aeromedical evacuation kit, and winterisation kit. Combined transmission rating (continuous) 2,334 kW (3,130 shp).

FLYING CONTROLS: As for UH-60.

STRUCTURE: As for UH-60.

POWER PLANT: Two 1,212 kW (1,625 shp) General Electric CT7-2C or 1,285 kW (1,723 shp) CT7-2D turboshafts, or equivalent military T700s. Rolls-Royce Turbomeca RTM 322 in demonstrator G-RRTM. Maximum fuel capacity 1,370 litres (362 US gallons; 301 Imp gallons).

ACCOMMODATION: Flight deck crew of two, with provision for 12 passengers in standard cabin configuration and up to 19 passengers in high density layout. Forward hinged door on each side of flight deck for access to cockpit area. Large rearward sliding door on each side of main cabin.

DIMENSIONS, INTERNAL:

Cabin: Length	3.84 m (12 ft 7 in)
Max width	2.34 m (7 ft 8 in)
Max height	1.37 m (4 ft 6 in)
Floor area	8.18 m² (88.0 sq ft)
Volume	10.96 m³ (387.0 cu ft)
Baggage compartment volume	0.52 m³ (18.5 cu ft)

WEIGHTS AND LOADINGS:

Weight empty	4,607 kg (10,158 lb)
Max external load	3,630 kg (8,000 lb)
Max T-O weight	9,185 kg (20,250 lb)
Max disc loading	43.73 kg/m² (8.96 lb/sq ft)
Max power loading	3.94 kg/kW (6.47 lb/shp)

PERFORMANCE (ISA, at max T-O weight):

Never-exceed speed (V_{NE})	195 knots (361 km/h; 224 mph)

Sikorsky S-70C commercially available derivative of the UH-60 series

568   USA: AIRCRAFT—SIKORSKY

Max level speed at S/L	157 knots (290 km/h; 180 mph)
Cruising speed at S/L	145 knots (268 km/h; 167 mph)
Max rate of climb at S/L	615 m (2,020 ft)/min
Service ceiling	4,360 m (14,300 ft)
Service ceiling, OEI	1,095 m (3,600 ft)
Hovering ceiling OGE	1,204 m (3,950 ft)

Range at 135 knots (250 km/h; 155 mph) at 915 m (3,000 ft) with max standard fuel, 30 min reserves
255 nm (473 km; 294 miles)
Range, max fuel, no reserves
297 nm (550 km; 342 miles)

SIKORSKY S-76
TYPE: Twin-turbine helicopter.
PROGRAMME: Announced 19 January 1975; four prototypes begun May 1976; first flight (N762SA) 13 March 1977; certification to FAR Pt 29 in November 1978; deliveries started early 1979; delivery of Mk II began 1 March 1982; Mk II modification kits available for earlier S-76s.
CURRENT VERSIONS: **S-76**: Original transport version delivered before 1 March 1982 (see earlier editions).
  S-76A+: Retrofit of Turbomeca Arriel 1S to S-76A.
  S-76 Mark II: Standard version from March 1982; no longer current. Details in 1991-92 and earlier *Jane's*.
  S-76 Utility: More basic version of S-76 Mk II with sliding doors (one each side), dual controls and floors stressed for 976 kg/m² (200 lb/sq ft); options include fixed landing gear with low pressure tyres, crash-resistant fuel tanks, baggage compartment auxiliary tank for 416 litres (100 US gallons; 91.5 Imp gallons), armoured crew seats, cargo hook, rescue hoist, engine air particle separators, and provision for stretchers. Philippine Air Force received 17 military S-76 Utility, of which 12 AUH-76s fitted for COIN, troop/logistic support and medevac, two others for SAR, and three as 12 or eight passenger transports.
  S-76B: Powered by P&W PT6B-36 engines; described separately; current production model. One S-76B flown with Fantail anti-torque system as part of RAH-66 Comanche development programme; see H-76 Eagle.
  S-76C: Powered by Turbomeca Arriel 1S1 engines described separately; current production model.
  H-76 and H-76N Eagle: Military and naval developments of S-76B; described separately.
  Shadow: One-off modification (N765SA) of S-76 Sikorsky Helicopter Advanced Demonstrator and Operator Workload fitted with nose-mounted single-seat cockpit for US Army Rotorcraft Technology Integration (ARTI) programme; main cabin used for safety pilots and test operators; first flight 24 June 1985; trials of full cockpit began Spring 1986; now has fly-by-wire side-arm control stick, voice interactive system, remote map reader, FLIR HUD with programmable symbol generator, visually coupled helmet-mounted display and dual CRT displays with touch-sensitive screens. Later modifications include upgraded engines and dynamic components, NVG compatible cockpit and cabin, reconfigurable evaluation cockpit, and high visibility paint for night and NOE flying.
  Partners in programme include Bendix/King, Kaiser, Litton, Northrop, Pacer Systems, Rockwell Collins, Plessey Electronic Systems and Hamilton Standard.
  Used 1990 in Boeing/Sikorsky LH First Team programme to flight test NVPS (night vision pilotage system) developed by Martin Marietta.
CUSTOMERS: Total 389 all-civil versions delivered by 3 December 1992; total includes 283 S-76s (incl Mark II), 17 S-76A+ conversions, 71 S-76Bs and 18 S-76Cs. Japan Maritime Safety Agency selected S-76C in advanced SAR configuration, with first delivery anticipated Winter 1993.

Sikorsky S-76A+ SAR helicopter of former Royal Hong Kong Auxiliary Air Force, with GEC-Marconi MRTS FLIR turret under nose, searchlight under cabin, sliding doors and cargo hook

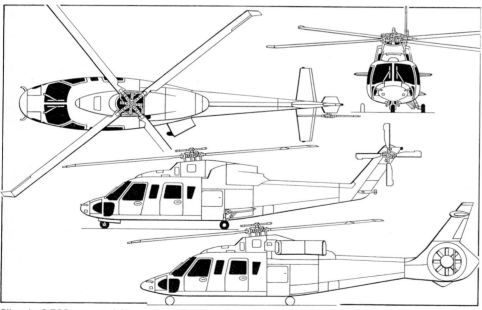
Sikorsky S-76C commercial transport helicopter with additional side view (bottom) of S-76B experimentally fitted with fan-in-fin tail of the RAH-66 Comanche *(Jane's/Mike Keep)*

Sikorsky Helicopter Advanced Demonstrator of Operator Workload (Shadow), based on an S-76B

SIKORSKY S-76B
TYPE: PT6B-36A-engined S-76 Mark II.
PROGRAMME: Initiated October 1983; first flight (N3123U) 22 June 1984; first flight of reworked ground test S-76B February 1985; certificated in FAA Category A early 1987; UK certification testing began July 1987; increase in max T-O weight to 5,307 kg (11,700 lb) approved by FAA in June 1988.
CUSTOMERS: S-76Bs operating in China, Germany, Japan, Netherlands, South Korea, UK and USA; four delivered in 1992; total 71 delivered by 31 December 1992.
DESIGN FEATURES: Meets FAR Pt 29 with Category A IFR intended for offshore support, business transport, medical evacuation and general utility use; technology and aerodynamics based on those of UH-60 Black Hawk.
  Four-blade main rotor with high twist and varying section and camber based on Sikorsky SC-1095; tapered blade tip has 30° leading-edge sweep; fully articulated rotor head with single elastomeric bearings; hydraulic drag dampers; dual bifilar vibration absorber assemblies above rotor head; four-blade cross-beam tail rotor on port side; transmission rating 969 kW (1,300 shp); rotor brake optional.
  Experimental higher harmonic control (HHC) system, to damp vibration through control actuators at rotor head, tested Spring 1985; demonstrated 90 per cent vibration reduction. Ice protection kit weighing 68 kg (150 lb), of which 45 kg (100 lb) could be removed in Summer, also under development. Optional manual blade folding to be introduced in 1993.
  Additional power of P&WC PT6B-36A engines and raised transmission rating increases take-off power in S-76B by 48 per cent; useful load increased 51 per cent in hot and high conditions; max T-O weight increased by 635 kg (1,400 lb); tail rotor pylon area reduced by 15 per cent and engine exhaust reconfigured.
  Emergency medical service installation includes multiple-position pivoting primary patient litter, a second litter, and track-mounted seats for four attendants, forward and rear oxygen systems, and dual access to external power on ground; cabin volume 4,000 litres (141 cu ft).
FLYING CONTROLS: Dual powered hydraulic controls with autostabilisation and autopilot; releasable spring-centring trim system for both cyclic and collective controls; full IFR instrument panel with optional EFIS and weather radar.
STRUCTURE: Main rotor blades have formed and welded titanium oval-section tubular main spar with Nomex honeycomb aerofoil core, glassfibre composite outer skin and titanium/nickel leading-edge abrasion strips. Fuselage contains extensive Kevlar and honeycomb components.
LANDING GEAR: Hydraulically retractable tricycle type, with single wheel on each unit. Nosewheel retracts rearward, main units inward into rear fuselage; all three wheels are enclosed by doors when retracted. Mainwheel tyres size 14.5 × 5.5-6, pressure 11.38 bars (165 lb/sq in); nosewheel tyre size 13 × 5.00-4, pressure 9.31 bars (135 lb/sq in). Hydraulic brakes; hydraulic mainwheel parking brake. Non-retractable tricycle gear, with low pressure tyres, optional.
POWER PLANT: Two Pratt & Whitney Canada PT6B-36A turboshafts; five minute T-O rating 732 kW (981 shp); 30 minute OEI rating 771 kW (1,033 shp); three-minute OEI rating, S/L, 90°F, 686 kW (920 shp); max continuous rating at 1,524 m (5,000 ft) and 30°C (77°F) 504 kW (676 shp). Transmission rating 1,118 kW (1,500 shp). Standard

Sikorsky S-76B (Pratt & Whitney Canada PT6B-36A turboshafts)

fuel capacity 1,064 litres (281 US gallons; 234 Imp gallons); optional auxiliary tank capacity 208 litres (55 US gallons; 46 Imp gallons).

ACCOMMODATION: Two pilots and 12-13 passengers. Three four-abreast rows of seats, floor mounted at a pitch of 79 cm (31 in). A number of executive layouts are available, including a four-passenger 'office in the sky' configuration. Executive versions have luxurious interior trim, full carpeting, special soundproofing, radio telephone, and co-ordinated furniture. Dual controls optional. Two large doors on each side of fuselage, hinged at their forward edges; sliding doors available optionally. Baggage hold aft of cabin, with external door each side of fuselage. Cabin heated and ventilated. Windscreen demisting and dual windscreen wipers. Windscreen heating and external cargo hook optional.

SYSTEMS: Hydraulic system at pressure of 207 bars (3,000 lb/sq in) supplied by two pumps driven from main gearbox. Hydraulic system maximum flow rate 15.9 litres (4.2 US gallons; 3.5 Imp gallons) per minute. Bootstrap reservoir. Pump head pressure 3.45 bars (50 lb/sq in). In VFR configuration, electrical system comprises two 200A DC starter/generators and a 24V 17Ah nickel-cadmium battery. In IFR configuration, system comprises gearbox driven 7.5kVA generator, and a 115V 600VA 400Hz static inverter for AC power. 34Ah battery optional. Engine fire detection and extinguishing system.

AVIONICS: Wide range of optional avionics available, according to configuration, including full IFR system based on Honeywell SPZS-7000 digital AFCS with EFIS, or Hamilton Standard AFCS, VHF nav receivers, transponder, compass system, weather radar, long-range navaid, radar altimeter, ADF, DME, and ELT and sonic transmitters.

EQUIPMENT: Standard equipment includes provisions for dual controls; cabin fire extinguishers; cockpit, cabin, instrument, navigation and anti-collision lights; landing light; external power socket; first aid kit; and utility soundproofing. Optional equipment includes air-conditioning, cargo hook, rescue hoist, emergency flotation gear, engine air particle separators and litter installation.

DIMENSIONS, EXTERNAL:	
Main rotor diameter	13.41 m (44 ft 0 in)
Main rotor blade chord	0.39 m (1 ft 3½ in)
Tail rotor diameter	2.44 m (8 ft 0 in)
Tail rotor blade chord	0.16 m (6½ in)
Length: overall, rotors turning	16.00 m (52 ft 6 in)
fuselage	13.43 m (44 ft 0¾ in)
Height overall, tail rotor turning	4.41 m (14 ft 5¾ in)
Tailplane span	3.05 m (10 ft 0 in)
Width of fuselage	2.13 m (7 ft 0 in)
Wheel track	2.44 m (8 ft 0 in)
Wheelbase	5.00 m (16 ft 5 in)
Tail rotor ground clearance	1.97 m (6 ft 5¾ in)
DIMENSIONS, INTERNAL:	
Passenger cabin: Length	2.46 m (8 ft 1 in)
Max width	1.93 m (6 ft 4 in)
Max height	1.35 m (4 ft 5 in)
Floor area	4.18 m² (45.0 sq ft)
Volume	5.78 m³ (204 cu ft)
Baggage compartment volume	1.08 m³ (38 cu ft)
AREAS:	
Main rotor disc	141.26 m² (1,520.53 sq ft)
Tail rotor disc	4.67 m² (50.27 sq ft)
Tailplane	2.00 m² (21.5 sq ft)
WEIGHTS AND LOADINGS:	
Weight empty, standard equipment	3,012 kg (6,641 lb)
Max T-O weight	5,307 kg (11,700 lb)
Max disc loading	37.6 kg/m² (7.69 lb/sq ft)
Max power loading	4.75 kg/kW (7.80 lb/shp)
PERFORMANCE (at max T-O weight, ISA):	
Max level speed	155 knots (287 km/h; 178 mph)
Cruising speed	145 knots (269 km/h; 166 mph)
Max rate of climb at S/L	502 m (1,650 ft)/min
Max operating altitude	4,575 m (15,000 ft)
Service ceiling, OEI	1,980 m (6,500 ft)
Hovering ceiling OGE	1,143 m (3,750 ft)
Range at 139 knots (257 km/h; 160 mph) at 915 m (3,000 ft) with standard fuel:	
no reserves	357 nm (661 km; 410 miles)
30 min reserves	288 nm (533 km; 331 miles)

SIKORSKY S-76C

TYPE: Arriel 1S1-engined S-76B.

PROGRAMME: Announced June 1989; first flight 18 May 1990; FAA certification and first deliveries April 1991.

CUSTOMERS: Two ordered for Royal Hong Kong Auxiliary Air Force and eight for Spanish Air Force (for IFR training at Ala 78 helicopter school, Granada/Armilla); deliveries to Spain began November 1991; Bond Helicopters, Scotland, ordered seven for delivery from 1993; Malaysian Helicopter Services ordered three for 1992 delivery plus options for three, deliverable in 1993. Eight delivered in 1992, as four to Spain, three to Malaysia and one executive to US customer; 18 total deliveries by 31 December 1992; 1993 deliveries include two to Spain and three to Bond.

DESIGN FEATURES: Same airframe and power train as S-76B, but powered by two Turbomeca Arriel 1S1 turboshafts; initial certification for max T-O weight 5,171 kg (11,400 lb); useful load 299 kg (660 lb) greater than S-76 for same range; in typical UK CAA IFR offshore configuration, S-76C can fly 12 passengers for 220 nm (407 km; 253 miles) or 10 passengers for 320 nm (593 km; 368 miles) with 45 min reserves.

POWER PLANT: Two Turbomeca Arriel 1S1 turboshafts, each rated at 539 kW (723 shp) for take-off and max continuous, 591 kW (792 shp) max contingency OEI for 2½ minutes and 498 kW (668 shp) for normal cruise. Transmission and fuel details as S-76B.

ACCOMMODATION: As S-76B.

DIMENSIONS, EXTERNAL: As S-76B

DIMENSIONS, INTERNAL: As S-76B

AREAS: As S-76B

WEIGHTS AND LOADINGS:	
Weight empty, standard equipment	2,849 kg (6,282 lb)
Max T-O weight	5,171 kg (11,400 lb)
Max disc loading	36.61 kg/m² (7.50 lb/sq ft)
Max power loading	4.63 kg/kW (7.60 lb/shp)
PERFORMANCE (at max T-O weight, ISA, except where indicated):	
Max level speed at S/L	155 knots (287 km/h; 178 mph)
Normal cruising speed at S/L	145 knots (269 km/h; 166 mph)
Max rate of climb at S/L, T-O power	445 m (1,460 ft)/min
Service ceiling: two engines	3,505 m (11,800 ft)
single engine at 30 min power	630 m (2,070 ft)
Fuel consumption at 138 knots (255 km/h; 159 mph) at 915 m (3,000 ft), 3,261 kg (10,700 lb) gross weight	268 kg/h (593 lb/h)
Range at 138 knots (255 km/h; 159 mph) at 915 m (3,000 ft), with standard fuel:	
no reserves	430 nm (798 km; 494 miles)
30 min reserves	366 nm (678 km; 421 miles)

SIKORSKY H-76 EAGLE

TYPE: Military armed utility derivative of S-76B.

PROGRAMME: First flight (N3124G) February 1985; weapon firing from four-station pitch-compensated armament pylon at Mojave, California, early 1987.

CURRENT VERSIONS: **H-76**: Standard military armed version, *as described in detail.*

H-76N: Naval version, announced 1984; no known customers so far. Details in 1991-92 and earlier *Jane's.*

Fantail Demonstrator: Fitted with Boeing/Sikorsky First Team Fantail anti-torque system for RAH-66 Comanche (which see) and flown 6 June 1990; eight-blade tail rotor 1.20 m (3 ft 11¼ in) diameter, generating 737 kg (1,624 lb) max thrust at 610 m (2,000 ft) and 35°C (95°F).

DESIGN FEATURES: Airframe modifications include armoured crew seats, sliding cabin doors, heavy duty floor, optical sight above instrument panel, self-sealing high-strength fuel tanks, and door-mounted weapons; main transmission, intermediate and tail rotor gearboxes uprated; main rotor hub and shaft strengthened; broader tail rotor blade chord; dual spars in tail pylon; fuselage side skin thickened to resist weapon blast; Strobex rotor blade tracker optional.

H-76 can be equipped for troop transport/logistic support, as gunship, or for airborne assault, air observation post, combat SAR, evacuation, ambulance and conventional SAR. Can carry either mast-mounted or roof-mounted sight, plus HUD, laser ranger and integrated armament management system; also provision for self-protection system including radar warning, infra-red jammer, and chaff/flare dispensers; high-clearance landing gear and Honeywell SPZ-7000 dual digital autopilot available.

Pitch-compensated armament pylon has faired leading-edge, giving 3-4 knots speed increase; weapons tested include Giat 20 mm cannon pod, 7.62 mm and 12.7 mm machine-gun pods, VS-MD-H mine dispenser and 70 mm (2.75 in) rockets; H-76 could carry 16 air-to-air missiles or eight AAMs and two cannon pods; integrated armament management system allows weapons to be selected from collective lever as well as from panel; system ready lights and sideslip trim ball on HUD.

External/internal dimensions generally as for S-76B; *details below show differences from civil S-76s:*

POWER PLANT: As for S-76B, except fuel is contained in two high strength, optionally self-sealing, tanks located below the rear cabin, with a total capacity of 993 litres (262.4 US gallons; 218.4 Imp gallons). Gravity refuelling point on each side of fuselage. Engine fire detection and extinguishing systems. Engine air particle separator optional.

ACCOMMODATION: Pilot and co-pilot, plus varying troop/passenger loads according to role. Armoured pilots seats optional. Ten fully armed troops can be transported, or seven troops when configured as an airborne assault vehicle with multi-purpose pylon system (MPPS) and one 7.62 mm door gun installed. For evacuation use the cabin can be equipped with 12 seats or, in emergency, all seats can be removed and 16 persons can be airlifted sitting on the cabin floor. For SAR use the cabin will accommodate three patients on litters, or six persons lying prone on the floor and on the rear cabin raised deck. The standard medevac layout provides for three litters and a bench seat for two medical attendants.

SYSTEMS: Generally as for S-76B, except electrical system has a 17 or optional 34Ah battery. Engine ice protection by bleed air anti-icing system.

AVIONICS: Typical avionics include VHF-20A VHF transceiver, AN/ARC-186 VHF-AM/FM com, 719A UHF com, ADF-60A ADF, DF-301E UHF DF, VIR-30A VOR with ILS, glideslope and marker beacon receivers, DME-40 DME, TDR-90 transponder and dual RMI-36 RMI, all by Collins; course deviation indicators, ELT, Andrea A301-61A intercom, cabin speaker system and loudhailer. Targeting equipment can include FLIR, Saab-Scania reticle sight, TOW roof sight or TOW mast-mounted sight and laser rangefinder.

EQUIPMENT: Typical equipment includes dual controls and instrumentation, stability augmentation system, dual 5 in VGIs, Allen RCA-26 standby self-contained attitude indicator, Collins ALT-50A radio altimeter, soundproofing, 'Fasten seat belt—No smoking' sign, first aid kit, two cabin fire extinguishers, external power socket, provisions for optional emergency flotation system, and provisions for installation of cargo hook with certificated capacity of 1,497 kg (3,300 lb) and rescue hoist of 272 kg (600 lb) capacity. Standard lighting includes cockpit, cabin and

Sikorsky S-76C twin-turboshaft helicopter of Malaysian Helicopter Services

570 USA: AIRCRAFT—SIKORSKY

Sikorsky H-76 Eagle military utility helicopter (Pratt & Whitney Canada PT6B-36 turboshafts)

Height overall	4.62 m (15 ft 2 in)
Fuselage: Max height	2.29 m (7 ft 6 in)
Max width	2.16 m (7 ft 1 in)
Wheel track	3.48 m (11 ft 5 in)
Wheelbase	6.88 m (22 ft 7 in)

DIMENSIONS, INTERNAL:
Cabin:

Max width	1.83 m (6 ft 0 in)
Max height	1.83 m (6 ft 0 in)
Volume	16.88 m³ (596 cu ft)
Baggage volume	4.76 m³ (168 cu ft)

WEIGHTS AND LOADINGS:

Manufacturer's weight empty	6,228 kg (13,730 lb)
Max T-O weight	10,079 kg (22,220 lb)

PERFORMANCE:

Max cruising speed	160 knots (296 km/h; 184 mph)
Best range speed	140 knots (259 km/h; 161 mph)
Service ceiling	4,575 m (15,000 ft)
Service ceiling, OEI	2,135 m (7,000 ft)
Hovering ceiling: IGE	3,290 m (10,800 ft)
OGE	1,980 m (6,500 ft)

Range, at best range speed with 10% reserves
400 nm (741 km; 460 miles)

instrument lights, navigation lights, anti-collision strobe light, and a battery operated cabin emergency light.

ARMAMENT: One 7.62 mm machine gun can be pintle-mounted in each doorway and fired with or without the MPPS system installed. Pintles incorporate field of fire limiters and will accept FN Herstal or Maramount M60D machine guns. The MPPS can be installed on the cabin floor, providing the capability to carry and deploy pods containing single or twin 7.62 mm machine guns, 0.50 in machine guns, 2.75 in and 5 in rocket pods, Mk 66 2.75 in rockets, Oerlikon 68 mm rockets, mines, Hellfire, TOW, Sea Skua and Stinger missiles, and Mk 46 torpedoes (see also Design Features).

WEIGHTS AND LOADINGS:

Basic weight empty	2,545 kg (5,610 lb)
Weight empty, equipped (typical)	3,030 kg (6,680 lb)
Max fuel weight	792 kg (1,745 lb)
Max T-O weight	5,171 kg (11,400 lb)

PERFORMANCE: Similar to S-76B, but range highly variable according to loading and mission

SIKORSKY S-92

TYPE: Development (Growth Hawk) of S-70 Black Hawk/Seahawk.

PROGRAMME: Announced March 1992; market evaluation co-ordinated with Mitsubishi Corporation and Mitsubishi Heavy Industries had expected to lead to production decision by end 1992, first flight 1994 and deliveries 1996; launch postponed until world market improves, but investment, design and market study continuing during 1993.

CURRENT VERSIONS: **S-92C**: Civil version, *to which details below apply.*
S-92M: Military and/or naval version, not necessarily tied to launch of civil version. Folding tailboom and main blades on naval variant.

CUSTOMERS: Offshore oil support, civil passenger transport, cargo transport and SAR seen as main S-92C market; possible USMC interest in S-92M as CH-46 replacement.

COSTS: Launch conditional on achieving acceptable price.

DESIGN FEATURES: Basically Black Hawk rotor head with bifilar vibration damper; inclined crossbeam four-blade tail rotor to starboard of tail pylon; improved main rotor blades with compound swept (30°) and drooped (20°) tips; high-set tailplane to port of pylon; SH-60B Seahawk heavy duty transmission; 44 per cent commonality with Black Hawk/Seahawk in rotors and drive train; to be designed to FAA/CAA/JAR standards for Category A/JAR Group 1 OEI performance.

LANDING GEAR: Retractable tricycle; main units retract rearward into sponsons; nosewheel retracts rearwards under flight deck.

POWER PLANT: Initially two GE CT7-6Xs, but later alternative would be Rolls-Royce Turbomeca RTM 322. Engine ratings: 1,305 kW (1,750 shp) for T-O, 1,081 kW (1,450 shp) continuous, 1,641 kW (2,200 shp) OEI 30 seconds, 1,305 kW (1,750 shp) OEI 30 minutes. Fuel (2,195 litres; 580 US gallons; 483 Imp gallons) in sponsons; crash-safe suction fuel system.

ACCOMMODATION: Two-pilot crew on separate flight deck; 19-22 passengers in three-abreast configuration with windows for each seat row; rear loading ramp optional (capacity for three LD3 containers); passenger door to starboard at front of cabin; escape hatch at rear seat row; internal flotation, emergency lighting and raft.

SYSTEMS: Dual hydraulic and electrical systems.

AVIONICS: Full IFR system with Honeywell six-screen EFIS system, integrated navigation and flight control avionics; health and usage monitoring system.

DIMENSIONS, EXTERNAL:

Main rotor diameter	16.36 m (53 ft 8 in)
Tail rotor diameter	3.35 m (11 ft 0 in)
Length overall, rotors turning	20.19 m (66 ft 3 in)
Fuselage length	16.51 m (54 ft 2 in)
Tailplane span	2.92 m (9 ft 7 in)

Mockup of Sikorsky S-92C proposed 19/22-passenger transport

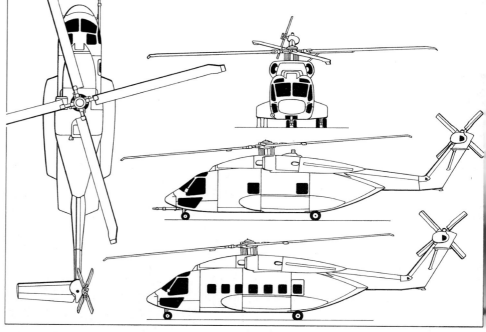

Sikorsky S-92M military helicopter with additional side view (bottom) of civil S-92C (*Jane's/Mike Keep*)

SIKORSKY TILT-ROTOR

One-sixth scale wind tunnel test model of possible 30-passenger tilt-rotor design built 1991. Features include twin three-blade rotors able to retract from full 2.44 m (8 ft) diameter (max lift for T-O and hover) to 1.60 m (5 ft 3 in) diameter for reduced drag in forward flight.

SIKORSKY S-2K

Feasibility studies for very large civil transport (VLCT) helicopter which could carry up to 100 passengers; year 2000 is target date for introduction; Sikorsky believes development work being undertaken for new Army Advanced Cargo Aircraft should produce refinements in helicopter rotor drive and other systems to satisfy civil requirements for reliability, comfort, operating economics and safety in large passenger carrying helicopters.

SKYSTAR

SKYSTAR AIRCRAFT CORPORATION

Nampa Municipal Airport, 100 North Kings Road, Nampa, Idaho 83687
Telephone: 1 (208) 466 1711
Fax: 1 (208) 466 7194
PRESIDENT: Philip L. Reed III
DIRECTOR, SALES AND MARKETING: Larry E. Robb

SkyStar Aircraft Corporation has taken over Kitfox programme from former Denney Aerocraft Company.

SKYSTAR KITFOX IV

TYPE: Side by side two-seat, dual control homebuilt.
PROGRAMME: Prototype Kitfox first flown 7 May 1984. New Vixen derivative is to be certificated under new FAA primary aircraft category.
CURRENT VERSIONS: **Kitfox IV:** Also referred to as Kitfox IV-1200. *Detailed description applies to this version.*
  Speedster: Higher performance version through optional package (see Design Features).
  Vixen: Non-retractable tricycle landing gear version of Kitfox, with steerable nosewheel, and Rotax 912 engine under new smooth engine cowlings. To be FAA certificated under primary aircraft category.
  XL: Fourth variant of basic Kitfox IV; Rotax 503 engine; max T-O weight 431 kg (950 lb). Flight testing to VLA criteria completed early 1993.
CUSTOMERS: Some 2,000 Kitfox aircraft ordered, and 850 built.
COSTS: Kits: start at $14,995 including engine; Kitfox with Speedster package (Rotax 912) is approximately $25,000 including engine.
DESIGN FEATURES: Designed to have exceptional short-field performance. In mid-1991, Kitfox IV introduced, featuring new wing with laminar flow section and all-metal hinge brackets for full-span flaperons; windscreen material thickened and given increased slope and, with other changes, resulted in substantial increase in cruising speed; stalling and landing speeds decreased, former due to clever use of flaperons; differential aileron control provides outstanding stability and roll control; can have 0.14 m³ (5 cu ft) storage space behind seat; wing folding standard. A **Speedster** optional package is available, which increases cruising speed to approximately 109 knots (202 km/h; 125 mph) with Rotax 912 engine; included are cropped wings of about 8.84 m (29 ft) span with new wingtips, revised tail unit, wheel fairings, airscoop and more. Optional agricultural spray and underfuselage cargo pods.
STRUCTURE: Wings of aluminium alloy tubing with aluminium inserts, plywood ribs and drooped glassfibre tips, with Stits Poly-Fiber covering overall. Steel tube fuselage and tail unit, with Poly-Fiber covering.
LANDING GEAR: Non-retractable tailwheel type, with hydraulic disc brakes. Optional composite floats, amphibious floats and skis.
POWER PLANT: One 38.8 kW (52 hp) Rotax 503 or 48.5 kW (65 hp) Rotax 582 LC two-cylinder two-stroke engine, with 3:1 reduction gear. Optional 59.7 kW (80 hp) Rotax 912 flat-four. Fuel capacity 36 litres (9.5 US gallons; 8 Imp gallons) standard. Two 22.7 litre (6 US gallon; 5 Imp gallon) or two 49 litre (13 US gallon; 10.8 Imp gallon) optional fuel tanks in wings.

DIMENSIONS, EXTERNAL:
  Wing span   9.75 m (32 ft 0 in)
  Length overall   5.38 m (17 ft 8 in)
  Height overall   1.73 m (5 ft 8 in)
  Propeller diameter   1.73 m (5 ft 8 in)
AREAS:
  Wings, gross   12.16 m² (130.84 sq ft)
WEIGHTS AND LOADINGS:
  Weight empty   216 kg (475 lb)
  Pilot weight range   68-113 kg (150-250 lb)
  Max recommended pilot and passenger weight
                                181 kg (400 lb)
  Max T-O weight   544 kg (1,200 lb)
  Max wing loading   44.77 kg/m² (9.17 lb/sq ft)
PERFORMANCE (two crew, unless stated otherwise):
  Max level speed   109 knots (201 km/h; 125 mph)
  Cruising speed   91 knots (169 km/h; 105 mph)
  Stalling speed   25 knots (47 km/h; 29 mph)
  Max rate of climb at S/L, at max T-O weight
                                396 m (1,300 ft)/min
  Service ceiling, pilot only   6,100 m (20,000 ft)
  T-O run, pilot only   26 m (85 ft)
  Landing run   46 m (150 ft)
  Range: standard fuel   195 nm (362 km; 225 miles)
    max fuel   825 nm (1,528 km; 950 miles)
  Endurance, standard fuel   3 h
  g limits, pilot only   +6/−3

SkyStar Aircraft Kitfox IV two-seat homebuilt

SNOW

SNOW AVIATION INTERNATIONAL INC

228 South Third Street, Columbus, Ohio 43207
Telephone: 1 (614) 443 2711
Fax: 1 (614) 443 2861
PRESIDENT: Harry T. Snow Jr
EXECUTIVE VICE-PRESIDENT: William G. Ferguson
VICE-PRESIDENT, ENGINEERING: Arthur E. Eckles
VICE-PRESIDENT, SALES: Emil E. Kluever

In addition to continued development of its SA-204C/SA-210TA transport, Snow is leading a group modifying Lockheed C-130 Hercules transports with a modern technology cockpit and upgraded systems (see *Jane's Civil and Military Aircraft Upgrades*).

SNOW AVIATION SA-204C and SA-210TA

TYPE: Transport with STOL and rough-field capability.
CURRENT VERSIONS: **SA-204C:** Civil version; could carry 64 passengers or 11,340 kg (25,000 lb) of cargo (seats on pallets for fast removal).
  SA-210TA: Military version; would accommodate 61 combat-equipped troops, or 38 paratroops, or 40 litters

Artist's impression of Snow Aviation SA-210TA STOL-C/AT Cargo/Assault Transport

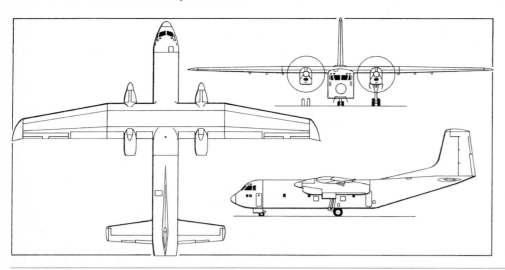

with attendants, or four 463L military pallets in all-cargo configuration, with max payload of 10,886 kg (24,000 lb).
PROGRAMME: Design began 1986, with major redesign 1987 and 1991; construction of prototype to begin late 1993 with first flight of prototype anticipated Spring 1994 and first production aircraft one year later. Certification under FAR Pt 25/JAR 25 expected 1995. No orders by early 1993.
COSTS: Standard aircraft $9.65 million civil and $10.25 million military (1993).
POWER PLANT: To be powered by two Allison GMA 2100/SA turboprop engines, each flat rated at 3,132 kW (4,200 shp) driving Dowty-Rotal/Hamilton Standard six-blade constant-speed propellers; max fuel capacity 9,342 litres (2,468 US gallons; 2,055 Imp gallons).
WEIGHTS AND LOADINGS:
Max T-O weight (estimated)   29,620 kg (65,300 lb)
PERFORMANCE (estimated):
Cruising speed at 3,050 m (10,000 ft)   260 knots (482 km/h; 299 mph)

Snow Aviation SA-210TA *(Jane's/Mike Keep)*

SOLOY

SOLOY CORPORATION
450 Pat Kennedy Way SW, Olympia, Washington 98501-7298
Telephone: 1 (206) 754 7000
Fax: 1 (206) 943 7659
PRESIDENT: Joe E. Soloy
EXECUTIVE VICE-PRESIDENT/GENERAL MANAGER:
   Scott Smith
VICE-PRESIDENTS:
   Nick Parkinson (Engineering)
   Tim Koester (Marketing)

Leader in field of engineering and certificating turbine engine installations for aircraft originally powered by reciprocating engines; develops and manufactures turbine conversion kits for Bell 47 and Hiller UH-12E helicopters and Cessna 206 and 207 light aircraft. Patented Dual Pac combining gearbox technology, now applied in contract development of Tridair Gemini ST twin-engined version of Bell 206L-3 LongRanger III; also used in PT6 Dual Pac twin-engine/single propeller power plant being developed jointly with Pratt & Whitney Canada; ENAER T-35 Turbo Pillán (which see) developed and tested by Soloy under contract to Empresa Nacional de Aeronautica, Chile.

Further details of conversions given in *Jane's* up to and including 1991-92 edition; can now be found in *Jane's Civil and Military Aircraft Upgrades.*

SOLOY TURBINE PAC and DUAL PAC CONVERSIONS

Turbine Pac: Based on 313 kW (420 shp) Allison 250-C20S, with robust 522 kW (700 shp) reduction gearbox, frame and associated systems manufactured by Soloy. Distinctive characteristics are slow turning propeller (1,810 rpm), downward ejecting exhaust, and compressor inlet at rear of engine. Details of various applications, including more powerful versions, in 1991-92 and earlier *Jane's.*

Dual Pac: Initiated as joint development with Pratt & Whitney Canada; first model with two 522 kW (700 shp) PT6B turboshaft engines driving single propeller through combining gearbox rated for output of 1,268 kW (1,700 shp) at 1,700 rpm to accommodate other near term applications requiring higher power. Patented drive train features independent operation of redundant gearing, freewheeling, governing and lubrication systems, either engine to provide power independently to propeller. Designed to qualify single propeller aircraft as multi-engined for FAR Pt 135 IFR operations; provides twin-engined safety, nacelle free wings, centreline thrust and reduced noise signature. Pusher and four-engine (twin Dual Pac) applications planned; version for rotary wing aircraft designed (see Tridair Gemini ST in this section).

SPORT

SPORT AIRCRAFT INC
44211 Yucca Avenue, Unit A, Lancaster, California 93534
Telephone: 1 (805) 949 2312
OWNER: Phil Tucker

SPORT AIRCRAFT (SUNDERLAND) S-18

TYPE: Side by side two-seat, dual control homebuilt; provision for child's jump seat.
PROGRAMME: Originating from Thorp T-18 Tiger; modified S-18 offered in plans form by Sunderland Aircraft, while Sport Aircraft Inc made kits and prefabricated components available. Following closure of Sunderland Aircraft, Sport Aircraft took over sale of plans.
COSTS: Kit: $13,500 without engine, propeller, instruments, avionics and canopy (1992). Plans: $195 (1992).
DESIGN FEATURES: Wing section LDS-4-212.
FLYING CONTROLS: Conventional control, plus flaps. Optional electric aileron and pitch trim.
STRUCTURE: Aluminium alloy construction.
LANDING GEAR: Non-retractable tailwheel type, with brakes.
POWER PLANT: One Textron Lycoming engine in 100-134 kW (135-180 hp) category; typically 112 kW (150 hp) O-320. Optional Ford V6 motorcar engine modification by Blanton. Fuel capacity 110 litres (29 US gallons; 24.1 Imp gallons). Optional two integral tanks in wing leading-edges, each 47.3 litres (12.5 US gallons; 10.4 Imp gallons).
DIMENSIONS, EXTERNAL:
Wing span	6.35 m (20 ft 10 in)
Wing aspect ratio	5.05
Length overall	5.89 m (19 ft 4 in)
Height overall	1.55 m (5 ft 1 in)
Propeller diameter	1.73 m (5 ft 8 in)

AREAS:
Wings, gross	8.0 m² (86.0 sq ft)

WEIGHTS AND LOADINGS (104.4 kW; 140 hp engine):
Weight empty	419 kg (923 lb)
Baggage capacity	36 kg (80 lb)
Max T-O weight	726 kg (1,600 lb)
Max wing loading	90.8 kg/m² (18.6 lb/sq ft)
Max power loading	6.95 kg/kW (11.43 lb/hp)

Sport Aircraft S-18 two-seat homebuilt (112 kW; 150 hp Textron Lycoming O-320-E2A engine)

PERFORMANCE (112 kW; 150 hp Textron Lycoming):
Max level speed at S/L	159 knots (295 km/h; 183 mph)
Max cruising speed (75% power) at 2,590 m (8,500 ft)	156 knots (290 km/h; 180 mph)
Econ cruising speed at 3,050 m (10,000 ft)	139 knots (257 km/h; 160 mph)
Stalling speed, flaps up and engine idling	55 knots (102 km/h; 63 mph)
Max rate of climb at S/L	366 m (1,200 ft)/min
Service ceiling	6,100 m (20,000 ft)
Range: with max fuel	460 nm (853 km; 530 miles)
with max payload	868 nm (1,609 km; 1,000 miles)
Endurance	3 h 12 min
g limits	+5/−2.5

STARPAC

STAR OF PHOENIX AIRCRAFT
Casa Grande, Arizona
COO: Charles Kallmann

STARPAC PHOENIX FLYER

TYPE: Updated model of Thorp T211.
PROGRAMME: StarPAC has acquired rights to T211 from former Thorp Aero Inc, which has ceased trading. Two T211s almost finished by StarPAC by Spring 1993, with further 10 under construction; deliveries began July 1993; production 20 per month by 1994.
CURRENT VERSIONS: **AeroSPORT:** Mid-range model.
AeroSPORT-XL: Uprated engine.
Trainer: Base version, with 74.6 kW (100 hp) Teledyne Continental O-200A; *details apply to this version.*
COSTS: $60,000-70,000.

DESIGN FEATURES: Improved version with aerodynamic glass-fibre cowling; glassfibre wing and tailplane tips; new electrical system; larger wheels with Cleveland brakes.
FLYING CONTROLS: Plain ailerons; all-moving tailplane; wide-span manually operated three-position trailing-edge flaps.
STRUCTURE: Light alloy fuselage; wing with main spar, false spar and ribbed skin; tailplane and rudder have ribbed skins.

LANDING GEAR: Non-retractable tricycle type; oleo-pneumatic shock absorber in each leg; all three wheels size 5.00-5; mainwheels with Cleveland brakes.
POWER PLANT: One 74.6 kW (100 hp) Continental O-200A flat-four engine, driving a Sensenich two-blade fixed-pitch propeller. Fuel cell aft of cabin, capacity 79.5 litres (21 US gallons; 17.5 Imp gallons).
ACCOMMODATION: Two seats side by side beneath rearward sliding transparent canopy. Baggage compartment behind seats; capacity 36 kg (80 lb).
SYSTEMS: Electrical system supplied by 12V 20A generator and 12V 2Ah storage battery.
AVIONICS: Single nav/com; ADF; transponder with Mode C.

DIMENSIONS, EXTERNAL:
Wing span 7.62 m (25 ft 0 in)
Length overall 5.49 m (18 ft 0 in)
Height overall 1.92 m (6 ft 3½ in)
Propeller diameter 1.70 m (5 ft 7 in)
AREAS:
Wings, gross 9.75 m² (105.0 sq ft)
WEIGHTS AND LOADINGS:
Weight empty 340 kg (750 lb)
Max T-O and landing weight 576 kg (1,270 lb)
Max wing loading 59.1 kg/m² (12.1 lb/sq ft)
Max power loading 7.73 kg/kW (12.7 lb/hp)

PERFORMANCE (at max T-O weight):
Max level speed at S/L 137 knots (254 km/h; 158 mph)
Cruising speed at 75% power 105 knots (194 km/h; 121 mph)
Stalling speed, flaps down 39 knots (73 km/h; 45 mph)
Max rate of climb at S/L 228 m (750 ft)/min
Service ceiling 5,790 m (19,000 ft)
T-O run 91 m (300 ft)
Landing run 61 m (200 ft)
Range at 55% power, with max fuel reserves 347 nm (640 km; 400 miles)

STODDARD-HAMILTON

STODDARD-HAMILTON AIRCRAFT INC
8701 58th Avenue North-East, Arlington, Washington 98223
Telephone: 1 (206) 435 8533
Fax: 1 (206) 435 9525
PRESIDENT: Theodore E. Setzer
MARKETING AND SALES MANAGER: Bill Sprague

Glasair I and II have been superseded by II-S and III. Details of Glasair I in 1989-90 *Jane's*. Brief details of GlaStar appeared in early 1992.

STODDARD-HAMILTON GLASAIR II-S

TYPE: Side by side two-seat, dual control homebuilt.
PROGRAMME: Glasair II-S (stretched) available in RG (retractable landing gear), FT (non-retractable tricycle gear) and TD (taildragger) forms to supersede earlier Glasair series.
COSTS: Kits: $24,500 for RG; $17,500 for FT; $16,900 for TD, all without engine, propeller, instruments and avionics.
DESIGN FEATURES: Wing section NASA LS(I)-0413. Upswept Hoerner style trailing-edges.
STRUCTURE: Glassfibre and foam composite construction.
LANDING GEAR: Three types available, as detailed above. Brakes.
POWER PLANT: One 119-134 kW (160-180 hp) Textron Lycoming O-360 series engine. Fuel capacity 182 litres (48 US gallons; 40 Imp gallons). Auxiliary tanks of 42 litres (11 US gallons; 9.2 Imp gallons) capacity in optional wingtip extensions.

DIMENSIONS, EXTERNAL (all versions):
Wing span, standard 7.10 m (23 ft 3½ in)
Wing aspect ratio 6.67
Length overall 6.16 m (20 ft 2½ in)
Height overall 2.07 m (6 ft 9½ in)
AREAS:
Wings, gross 7.55 m² (81.3 sq ft)
WEIGHTS AND LOADINGS:
Weight empty: RG 601 kg (1,325 lb)
FT 567 kg (1,250 lb)
TD 544 kg (1,200 lb)
Baggage capacity: all versions 36 kg (80 lb)
Max T-O weight: RG, FT 953 kg (2,100 lb)
TD 907 kg (2,000 lb)
Max wing loading: RG, FT 126.1 kg/m² (25.83 lb/sq ft)
TD 120.1 kg/m² (24.60 lb/sq ft)
PERFORMANCE (RG and TD with 180 hp engine, FT with 160 hp):
Max level speed at S/L:
RG 217 knots (402 km/h; 250 mph)
FT 207 knots (383 km/h; 238 mph)
TD 208 knots (385 km/h; 239 mph)
Econ cruising speed at 2,440 m (8,000 ft):
RG 189 knots (351 km/h; 218 mph)
FT 178 knots (330 km/h; 205 mph)
TD 179 knots (332 km/h; 206 mph)
Stalling speed, flaps down, power off:
RG 43 knots (80 km/h; 50 mph)
FT, TD 42 knots (78 km/h; 49 mph)

Service ceiling: all versions approx 5,790 m (19,000 ft)
Max rate of climb at S/L:
all versions 732 m (2,400 ft)/min
Range:
RG, standard fuel 1,125 nm (2,085 km; 1,296 miles)
RG, with auxiliary fuel 1,383 nm (2,563 km; 1,593 miles)
FT, TD, standard fuel 1,042 nm (1,931 km; 1,200 miles)
FT, TD, with auxiliary fuel 1,281 nm (2,373 km; 1,475 miles)
g limits at AUW of 708 kg (1,560 lb):
all versions +6/−4 limit
+9/−6 ultimate

STODDARD-HAMILTON GLASAIR III

PROGRAMME: Introduced 1987, retaining a similar configuration to the earlier models but designed to offer exceptional performance, constructional simplicity and economical kit price. Construction takes approximately 1,800 working hours.
COSTS: Kit: $33,500 without engine, propeller, instruments and avionics.
DESIGN FEATURES: Larger and wider fuselage for increased baggage space, payload capacity and comfort, also improving the longitudinal and directional stability and thereby making it a better cross-country and IFR aircraft. Thicker windscreen to improve protection against bird strikes at higher speeds, and additional glassfibre laminates, integral longerons, and lay-up schedule which provides structurally stronger and torsionally stiffer fuselage. Wing section LS(I)-0413; strengthened; carries more fuel than previous models. NACA style air vents provide cabin ventilation. Under development is turbocharging system to be offered as retrofit kit option for IO-540-K engined Glasair IIIs. Projected max cruising speed with system is 284 knots (526 km/h; 327 mph) at 5,485 m (18,000 ft).

Stoddard-Hamilton Glasair III (background) and Glasair II-S RG (foreground)

LANDING GEAR: Retractable.
POWER PLANT: One 224 kW (300 hp) Textron Lycoming IO-540-K1H5. Fuel capacity in wings 201 litres (53 US gallons; 44 Imp gallons). Fuselage header tank, capacity 30 litres (8 US gallons; 6.7 Imp gallons). Optional tanks in wingtip extensions, total capacity 41.6 litres (11 US gallons; 9.2 Imp gallons).

DIMENSIONS, EXTERNAL:
Wing span, standard 7.10 m (23 ft 3½ in)
Length overall 6.52 m (21 ft 4¼ in)
Height overall 2.29 m (7 ft 6 in)
AREAS:
Wings, gross 7.55 m² (81.3 sq ft)
WEIGHTS AND LOADINGS:
Weight empty 703 kg (1,550 lb)
Baggage capacity 45 kg (100 lb)
Max T-O weight, without wingtip extensions 1,089 kg (2,400 lb)
Max wing loading 144.13 kg/m² (29.52 lb/sq ft)
Max power loading 4.86 kg/kW (8.00 lb/hp)
PERFORMANCE (standard wings, except where indicated):
Max level speed at S/L 252 knots (467 km/h; 290 mph)
Cruising speed: 75% power at 2,440 m (8,000 ft) 243 knots (451 km/h; 280 mph)
50% power at 5,335 m (17,500 ft) 223 knots (414 km/h; 257 mph)
Stalling speed: pilot only, flaps and wheels up 65 knots (119 km/h; 74 mph)
flaps down, at max T-O weight 70 knots (129 km/h; 80 mph)
Max rate of climb at S/L 732 m (2,400 ft)/min
Service ceiling approx 7,315 m (24,000 ft)
Range at 55% power:
standard fuel 1,112 nm (2,061 km; 1,281 miles)
with tip tanks 1,313 nm (2,433 km; 1,512 miles)
g limits at AUW of 962 kg (2,120 lb) +6/−4 limit
+9/−6 ultimate

SUPER 580

SUPER 580 AIRCRAFT COMPANY
(Division of Flight Trails Inc)
2192 Palomar Airport Road, Carlsbad, California 92008
Telephone: 1 (619) 438 3600
Fax: 1 (619) 753 1531
Telex: 140414
PRESIDENT: Ted Vallas
SENIOR VICE-PRESIDENT: James Coleman

Licensed by Allison Gas Turbine Division of General Motors for turboprop conversion and remanufacturing of Convair 340/440/580 series transports into Super 580s.

Now included in *Jane's Civil and Military Aircraft Upgrades*.

SUPER 580 AIRCRAFT SUPER 580

TYPE: Re-engined Convair 340/440/580.
PROGRAMME: First Super 580 conversion by Hamilton Aviation, Tucson, Arizona; first flight 21 March 1984; FAA approval to CAR 4b by means of STC; delivered to The Way International June 1984. Super 580 ST stretch for 78 passengers or nine LD3 containers investigated 1984.
DESIGN FEATURES: Replaces piston engines of 340/440 or Allison 501-D13 of 580 with Allison 501-D22G turboprops, each flat rated at 2,983 kW (4,000 shp) and driving Hamilton Standard four-blade, constant-speed, feathering and reversing propeller; compared with 580, Super 580 has 2.5 per cent better sfc, improved engine-out performance, 13 per cent longer range, 40 per cent greater hot day payload, 40 per cent lower operating costs, and cruising speeds up to 325 knots (602 km/h; 374 mph).

SWEARINGEN

SWEARINGEN AIRCRAFT INC
1234 99th Street, San Antonio, Texas 78214
Telephone: 1 (512) 921 0055
Fax: 1 (512) 921 0198
CHAIRMAN AND CEO: E. J. Swearingen
PRESIDENT AND COO: John R. Novak
SENIOR VICE-PRESIDENT: James R. Hyslop
VICE-PRESIDENTS:
Joseph E. Karwowski (Procurement and Manufacturing programmes)
Robert N. Buckley
DIRECTORS:
John C. Maurer (International Marketing)
Thomas F. Connelly (US Sales)
Robert A. Kromer (Technical Marketing Operations)

'Ed' Swearingen is well known for designing Merlin and Metro commuter and business aircraft, and for engineering such aircraft as Piper Twin Comanche and Lockheed 731 JetStar II; also built prototype SA-32T for Jaffe (1991-92 *Jane's*). Now developing Swearingen SJ30 small business jet, which will be built, flight tested and delivered from new facility being established in Martinsburg, West Virginia.

SWEARINGEN SJ30

TYPE: Twin-turbofan pressurised business jet.
PROGRAMME: Announced 30 October 1986 as SA-30 Fanjet; Gulfstream Aerospace, Williams International and Rolls-Royce announced they were joining programme in October 1988 and aircraft renamed Gulfstream SA-30 Gulfjet; Gulfstream withdrew from programme 1 September 1989; place taken by Jaffe Group of San Antonio, Texas, and aircraft renamed Swearingen/Jaffe SJ30; now solely a Swearingen project; first flight of prototype (N30SJ) 13 February 1991; second (pre-production) aircraft expected 1993; certification expected April 1994; over 300 flight test hours flown by December 1992. Planned output six per month by mid-1994, 10-12 a month by 1997.
CUSTOMERS: Three distributors appointed in USA; 10 distributors appointed for international sales.
COSTS: Standard aircraft $2.595 million (1989 dollars).
DESIGN FEATURES: Tapered, 30° sweptback wing of computer-designed section with integral fuel tanks in torsion box.
FLYING CONTROLS: Mechanical, with electrically actuated variable incidence tailplane, rudder and aileron trim tabs; two large outward-canted ventral fins under tail; slotted Fowler trailing-edge flaps actuated electrically; hydraulically actuated full span leading-edge slats; single electro-hydraulically actuated airbrake/lift dumper panel on each wing ahead of flap.
STRUCTURE: All-metal with chemically milled skins on fuselage.
LANDING GEAR: Retractable tricycle type, with twin wheels on each unit. Trailing-link oleo-pneumatic suspension on main units. Hydraulic actuation, main units retracting inward and rearward into fuselage, nose unit forward. Electrically steerable nose unit retracts forward.
POWER PLANT: Two 8.45 kN (1,900 lb st) Williams International FJ44 turbofans, pod-mounted on pylons on sides of rear fuselage. Fuel in three integral tanks, one in each wing and one in rear fuselage; combined capacity 1,893 litres (500 US gallons; 416 Imp gallons). Single-point refuelling.
ACCOMMODATION: Pilot and one passenger (or co-pilot) on flight deck. Main cabin separated by a bulkhead; four chairs in facing pairs, each with adjustable reclining backs and retractable armrests, plus two foldaway tables; optional refreshment centre at front, and toilet, washbasin and storage cabinets at rear. Other layouts for up to seven passengers optional. Lengthened cabin for two extra seats in production version. Airstair passenger door at front on port side. Baggage compartment aft of main cabin, with external access via port-side door aft of wing. Two-piece birdproof electrically heated wraparound windscreen.
SYSTEMS: Cabin pressurised to 0.69 bar (10.0 lb/sq in) on prototype (0.83 bar; 12.0 lb/sq in on second and production aircraft), and heated by engine bleed air; cooled by a freon-cycle system. Hydraulic system (207 bars; 3,000 lb/sq in) for actuation of leading-edge slats, airbrake/lift dumpers, and landing gear extension/retraction. Two 300A engine driven starter/generators and static inverters. Redundant frequency-wild alternators provide power for windscreen heating. Wing and tailplane have TKS liquid de-icing systems; inlets de-iced by engine bleed air.
AVIONICS: Bendix/King Gold Crown III dual IFR avionics standard, with colour weather radar and autopilot/flight director.

DIMENSIONS, EXTERNAL:
Wing span	11.07 m (36 ft 4 in)
Wing aspect ratio	8.00
Length overall	12.98 m (42 ft 7¼ in)
Height overall	4.24 m (13 ft 11 in)

DIMENSIONS, INTERNAL:
Cabin: Length:	
between pressure bulkheads	5.23 m (17 ft 2 in)
passenger section	3.61 m (11 ft 10 in)
Max width	1.43 m (4 ft 8½ in)
Max height	1.31 m (4 ft 3½ in)
Volume	8.95 m³ (316.22 cu ft)

AREAS:
Wings, gross	15.33 m² (165.0 sq ft)

WEIGHTS AND LOADINGS:
Weight empty, equipped	2,817 kg (6,210 lb)
Fuel weight	1,519 kg (3,350 lb)
Max ramp weight	4,762 kg (10,500 lb)
Max zero-fuel weight	3,447 kg (7,600 lb)
Max T-O weight	4,717 kg (10,400 lb)
Max landing weight	4,309 kg (9,500 lb)
Max wing loading	291.5 kg/m² (59.7 lb/sq ft)
Max power loading	278.9 kg/kN (2.74 lb/lb st)

PERFORMANCE (estimated, at max T-O weight except where indicated):
Max operating speed (V_{MO})	Mach 0.82 (470 knots; 871 km/h; 541 mph)
Max cruising speed	445 knots (824 km/h; 512 mph)
Long-range cruising speed	413 knots (765 km/h; 475 mph)
Stalling speed at max landing weight	81 knots (150 km/h; 93 mph)
Max rate of climb at S/L	1,195 m (3,920 ft)/min
Max operating altitude	13,100 m (43,000 ft)
FAA T-O balanced field length	1,015 m (3,330 ft)
FAA landing distance	762 m (2,500 ft)
Range at Mach 0.72 (413 knots; 765 km/h; 475 mph):	
NBAA VFR reserves	2,076 nm (3,845 km; 2,389 miles)
NBAA IFR reserves	1,730 nm (3,204 km; 1,991 miles)

Swearingen SJ30 (two Williams Rolls-Royce FJ44 turbofans)

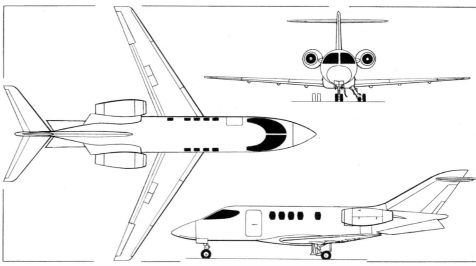

Swearingen SJ30 business jet (*Dennis Punnett*)

TAYLORCRAFT

TAYLORCRAFT AIRCRAFT

PO Box 480, 820 East Bald Eagle Street, Lock Haven, Pennsylvania 17745
Telephone: 1 (717) 748 8262
Fax: 1 (717) 748 5031
PRESIDENT: John P. Polychron
CHIEF ENGINEER: Darrel C. Romick
GENERAL MANAGER: Fred C. Burlingame

Company re-formed 1 April 1968 mainly for product support, but also manufactured several Taylorcraft types (see 1989-90 *Jane's*); company bought by group of former Piper employees 9 July 1985 and moved to Lock Haven, Pennsylvania; Taylorcraft bought manufacturing rights for Edo floats early 1986; production ceased August 1986 and company filed for Chapter 11 protection late 1986.

Taylorcraft acquired by Aircraft Acquisition Corporation with partners in November 1989; six new models, all based on same basic airframe, announced; since April 1992, company independent of Aircraft Acquisition Corporation and renamed Taylorcraft Aircraft; ceased production in October 1992.

TAYLORCRAFT F22 SERIES

TYPE: Two/three-seat trainer and sporting aircraft.
PROGRAMME: Construction began 29 December 1989; rolled out mid-January 1990; for production, see below.
CURRENT VERSIONS: **F22 Classic:** Basic aircraft powered by 88 kW (118 hp) Textron Lycoming O-235-L2C driving Sensenich fixed-pitch propeller; tailwheel landing gear; production starting.
F22A Tractor: F22 with tricycle landing gear; production starting.
F22B Ranger: Powered by 134 kW (180 hp) Textron Lycoming O-360-A4M driving Sensenich fixed-pitch propeller; tailwheel landing gear; certification expected July 1992.
F22C Trooper: F22B with tricycle landing gear; certificated 3 April 1992.
F22S: F22B on Edo 2000 floats; certification expected Autumn 1992.
COSTS: Basic prices (1992): F22, $62,900; F22A, $65,900; F22B, $69,900; F22C, $72,900; F22S, $89,650.
DESIGN FEATURES: Similar to Taylorcraft F-21B (see 1989-90 *Jane's*), but with flaps, larger door, new exhaust, choice of tricycle or tailwheel landing gear. Two-spar wing with dual struts either side and jury struts; NACA 23012 aero-

Taylorcraft F22 two/three-seat light aircraft

foil section; dihedral 1°; gross weight of all versions being raised to 907 kg (2,000 lb).
FLYING CONTROLS: Frise wide-span ailerons; trim tab in port elevator; split flaps; IFR instruments optional.
STRUCTURE: Wings have aluminium I-beam spars with stamped metal ribs and Dacron covering; steel tube fuselage, with aluminium alloy skinned underside and Dacron elsewhere; wire braced steel tube tail surfaces, with Dacron covering.
LANDING GEAR: Choice of non-retractable tailwheel or tricycle. Two side Vs and half-axles. Mainwheels fitted with 6.00-6 4-ply tyres; Scott swivelling tailwheel with 8 in pneumatic tyre. Nosewheel has 5.00-5 4-ply tyre. Dual Cleveland hydraulic toe brakes standard. Wheel fairings optional. Edo or Aqua aluminium floats and Aero M1500 or M2000 skis optional.
POWER PLANT: One 88 kW (118 hp) Textron Lycoming O-235-L2C flat-four engine, driving a Sensenich 72CK-O-50 two-blade fixed-pitch metal propeller, or one 134 kW (180 hp) Textron Lycoming O-360-A4M, driving a Sensenich 76EM8S5. One fuel tank in each wing, with combined usable capacity of 151 litres (40 US gallons; 33.3 Imp gallons). Oil capacity 5.7 litres (1.5 US gallons; 1.2 Imp gallons).
ACCOMMODATION: Two adjustable seats, with shoulder harness, side by side. Folding rear seat optional, to accommodate one passenger. Dual controls and toe brakes standard.

Metal door with hinged window each side. Baggage compartment aft of seats, standard capacity 91 kg (200 lb), and side windows. Accommodation heated and ventilated.
SYSTEMS: Electrical system powered by 12V 60A engine driven alternator, with 12V battery; navigation lights and stall warning. Engine driven vacuum pump when optional blind-flying instrumentation installed.
AVIONICS: Standard avionics include single com transceiver, transponder with Mode C encoder and intercom. Terra, Narco and Bendix/King avionics packages, incl IFR, optional.

DIMENSIONS, EXTERNAL:
Wing span 10.99 m (36 ft 0 in)
Wing chord, constant 1.60 m (5 ft 3 in)
Length overall 6.90 m (22 ft 8½ in)
Height overall: with tailwheel 1.98 m (6 ft 6 in)
 with nosewheel 2.31 m (7 ft 7 in)
Tailplane span 3.05 m (10 ft 0 in)
Wheel track 1.98 m (6 ft 6 in)
Propeller diameter: O-235 1.83 m (6 ft 0 in)
 O-360 1.92 m (6 ft 4 in)
Passenger door: Height 0.95 m (3 ft 1½ in)
 Width 0.76 m (2 ft 5¾ in)
 Height to sill 0.60 m (1 ft 11½ in)
DIMENSIONS, INTERNAL:
Cabin: Length 2.68 m (8 ft 9½ in)
 Max width 1.01 m (3 ft 3¾ in)
 Max height 1.17 m (3 ft 10¼ in)
Floor area 2.14 m² (23.0 sq ft)
Volume 2.47 m³ (87.4 cu ft)
Baggage compartment:
 Volume 1.36 m³ (48.1 cu ft)
AREAS:
Wings, gross 17.07 m² (183.71 sq ft)
Ailerons (total) 1.86 m² (20.0 sq ft)
Trailing-edge flaps (total) 3.72 m² (40.0 sq ft)
Fin 0.49 m² (5.3 sq ft)
Rudder 0.59 m² (6.3 sq ft)
Tailplane 1.21 m² (13.0 sq ft)
Elevators (total, incl tab) 0.99 m² (10.66 sq ft)
WEIGHTS AND LOADINGS:
Minimum weight empty 494 kg (1,090 lb)
Max T-O weight, all versions: Normal* 794 kg (1,750 lb)
 Utility 626 kg (1,380 lb)
Max wing loading 46.48 kg/m² (9.52 lb/sq ft)
Max power loading: F22/22A 9.02 kg/kW (14.83 lb/hp)
 F22B, F22C 5.92 kg/kW (9.7 lb/hp)
* to be increased to 907 kg (2,000 lb)
PERFORMANCE (at max T-O weight, except where indicated):
Never-exceed speed (V_{NE})
 128 knots (238 km/h; 148 mph)
Max level speed 108 knots (201 km/h; 125 mph)
Max cruising speed at 2,745 m (9,000 ft):
 F22/22A, F22B, F22C 104 knots (193 km/h; 120 mph)
Econ cruising speed at 3,810 m (12,500 ft):
 F22/22A 95 knots (175 km/h; 109 mph)
 F22B, F22C 104 knots (193 km/h; 120 mph)
Stalling speed: flaps up 42 knots (78 km/h; 48 mph)
 flaps down 36 knots (66 km/h; 41 mph)
Max rate of climb at S/L: F22/22A 229 m (750 ft)/min
 F22B, F22C 457 m (1,500 ft)/min
Service ceiling 5,485 m (18,000 ft)
T-O run: F22/22A 220 m (720 ft)
 F22B, F22C 122 m (400 ft)
T-O to 15 m (50 ft): F22/22A 348 m (1,140 ft)
 F22B, F22C 198 m (650 ft)
Landing from 15 m (50 ft) 221 m (725 ft)
Landing run 149 m (487 ft)
Range, 75% power with 30 min reserves:
 F22/22A 555 nm (1,028 km; 639 miles)
 F22B, F22C 378 nm (699 km; 435 miles)
g limits:
 Normal category, max T-O weight +3.8/−1.52
 Utility category, at 626 kg (1,380 lb) AUW
 +4.4/−1.76

TCM

TELEDYNE CONTINENTAL MOTORS
PO Box 90, Mobile, Alabama 36601
Telephone: 1 (205) 438 3411
Fax: 1 (205) 438 3411 ext 179
DIRECTOR OF OEM MARKETING: Tim Archer
 Company is developing a Voyager liquid-cooled engine installation for later Beechcraft Bonanzas. An equivalent programme for the Piper Chieftain is in abeyance.
 Bonanza: For late-production A36 and B36TC Bonanzas; replaces standard plain and turbocharged 212.5 kW (285 hp) and 224 kW (300 hp) engines with turbocharged liquid-cooled 224 kW (300 hp) Voyager T-550 engine driving a new three-blade McCauley propeller; new front cowling; radiator and oil cooler inside engine compartment; propeller arc moved forward about 7.6 cm (3 in); first flight of proof-of-concept prototype installation in A36 Bonanza (N9127Q) Summer 1989; if an STC is obtained, TCM would offer retrofits.

THORP — see STARPAC

THURSTON

THURSTON AEROMARINE CORPORATION
4 Ledge Road, Cumberland Foreside, Maine 04110
Telephone: 1 (207) 829 6108
PRESIDENT: David B. Thurston
 Company owns Type Certificate A15EA and related data for original TSC-1A Teal amphibian and, following cessation of operations by International Aeromarine Corporation of Sanford, Florida, during 1991, design data and all rights to TA16 Seafire reverted to Thurston Aeromarine Corporation. Plans to develop eight-seat Seamaster twin-engined amphibian.

THURSTON TSC-1A3 TEAL III
TYPE: Two-seat amphibian.
PROGRAMME: Construction of prototype started February 1991; first flight September 1991; production aircraft were to be available during 1992; plans include licence manufacture by Canadian Amphibians Ltd of Ontario, Canada.
CURRENT VERSIONS: **Teal III:** Basic model, as described below.
COSTS: Standard aircraft $125,000 (1992).
DESIGN FEATURES: Development of TSC-1A Teal with wing span increased by 1.22 m (4 ft 0 in); horizontal tail surfaces increased in area; pylon mount similar to TA16 Seafire. 134 kW (180 hp) engine in place of original 112 kW (150 hp) unit. Hull structure revised to take tricycle landing gear with nosewheel steering; max T-O weight increased to 1,043 kg (2,300 lb). Wing section NACA 4415.
FLYING CONTROLS: Conventional ailerons; elevator and rudder both with trim tabs.
STRUCTURE: All-metal D-spar wing; all-metal semi-monocoque fuselage with GFRP foredeck and cabin top skins; cantilever all-metal T tail.
LANDING GEAR: Retractable tricycle type.
POWER PLANT: One 134 kW (180 hp) Textron Lycoming O-360-A1F6D flat-four engine driving a Hartzell HC-C2YK-1BF/F7666A-2 tractor propeller. Two leading-edge fuel tanks, total capacity 174 litres (46 US gallons; 38.3 Imp gallons).
ACCOMMODATION: Enclosed cabin seating two side by side; baggage space behind seats.
DIMENSIONS, EXTERNAL:
Wing span 10.97 m (36 ft 0 in)
Wing aspect ratio 7.32
Length overall 7.27 m (23 ft 10¼ in)
Height overall 2.79 m (9 ft 2 in)
Propeller diameter 1.88 m (6 ft 2 in)
AREAS:
Wings, gross 16.44 m² (177.0 sq ft)
WEIGHTS AND LOADINGS:
Weight empty 680 kg (1,500 lb)
Max T-O weight 1,043 kg (2,300 lb)
Baggage weight 113 kg (250 lb)
Max wing loading 63.42 kg/m² (12.99 lb/sq ft)
Max power loading 7.78 kg/kW (12.78 lb/hp)
PERFORMANCE:
Max level speed 101 knots (187 km/h; 116 mph) IAS
Max cruising speed at 2,285 m (7,500 ft)
 104 knots (193 km/h; 120 mph)
Cruising speed at 75% power at 2,285 m (7,500 ft)
 97 knots (180 km/h; 112 mph)
Stalling speed, no flaps, power off
 48 knots (89 km/h; 55 mph)
Max rate of climb at S/L 244 m (800 ft)/min
Service ceiling 4,875 m (16,000 ft)
T-O run: land 305 m (1,000 ft)
 water 366 m (1,200 ft)
Landing run: land 213 m (700 ft)
 water 275 m (900 ft)
Range with standard fuel 434 nm (804 km; 500 miles)
Endurance 5 h 30 min
g limit 5.7

THURSTON TA16 SEAFIRE
TYPE: Four-seat amphibian.
PROGRAMME: Construction of prototype (N16SA) started June 1980; first flight 10 December 1981; original development by International Aeromarine continued by Thurston; FAR Pt 23 type certification had been expected September 1992; initial deliveries planned for first quarter 1993.
CURRENT VERSIONS: **Seafire:** Basic version, as described below.
COSTS: Standard aircraft approximately $235,000 (1992).
DESIGN FEATURES: Wing section NACA 64_2A215; dihedral 3°; incidence 4°.
FLYING CONTROLS: Thurston ailerons with ground adjustable trim tab; single-slotted trailing-edge flaps; elevators with bungee trim system; conventional rudder. Dual controls.
STRUCTURE: All-metal constant chord wings; all-metal single-step planing hull with retractable water rudder; cantilever all-metal tail surfaces.
LANDING GEAR: Hydraulically retractable tricycle type. Main units retract inward; steerable nosewheel retracts forward to close opening in hull, which needs no closure doors. Oleo-pneumatic shock absorbers. Parker-Hannifin aluminium alloy wheels, all three size 6.00-6, with tyre size 17.5 × 6.30, 6-ply rating. Tyre pressures: mainwheels 2.76 bars (40 lb/sq in), nosewheel 2.07 bars (30 lb/sq in). Parker-Hannifin dual-pad disc brakes. Toe brakes. Parking brake. Wheel landing gear designed to meet Canadian DoT snow-ski load conditions.
POWER PLANT: One 186 kW (250 hp) Textron Lycoming O-540-A4D5 flat-six engine, pylon mounted and braced from upper surface of hull directly over wing, driving Hartzell two-blade constant-speed metal tractor propeller. Fuel tank in leading-edge of each wing, with combined capacity of 340 litres (90 US gallons; 75 Imp gallons). Refuelling point on upper surface of each wing. Oil capacity 11.5

Thurston TA16 Seafire four-seat amphibian prototype

litres (3 US gallons; 2.5 Imp gallons). Engine air intake incorporates filter and automatic inlet door which opens if main duct becomes blocked by ice or debris.

ACCOMMODATION: Pilot and three passengers in pairs in enclosed cabin, with two-section rearward sliding canopy. Forward section slides aft over rear canopy, and both may then be rotated to either side of hull, or removed, to facilitate loading/unloading of bulky items. All glass tinted. Adjustable seats with belts and shoulder harness. Space for baggage or freight at rear of cabin. Accommodation heated and ventilated.

SYSTEMS: Electrical system powered by 24V 70A engine driven alternator; 24V 37Ah battery. Electrically driven hydraulic pump provides system pressure of 69 bars (1,000 lb/sq in) for actuation of landing gear. Oxygen system optional.

AVIONICS: Standard Narco package; optional nav/com equipment to customer requirements.

DIMENSIONS, EXTERNAL:
Wing span	11.28 m (37 ft 0 in)
Wing chord, constant	1.52 m (5 ft 0 in)
Wing aspect ratio	7.40
Length overall	8.28 m (27 ft 2 in)
Length of hull	7.42 m (24 ft 4 in)
Height overall	3.28 m (10 ft 9 in)
Tailplane span	3.05 m (10 ft 0 in)
Wheel track	4.01 m (13 ft 2 in)
Wheelbase	3.28 m (10 ft 9 in)
Propeller diameter	2.03 m (6 ft 8 in)

DIMENSIONS, INTERNAL:
Cabin: Length	2.13 m (7 ft 0 in)
Max width	1.01 m (3 ft 4 in)
Max height	1.22 m (4 ft 0 in)
Floor area	1.86 m² (20.0 sq ft)
Volume	2.26 m³ (80.0 cu ft)

AREAS:
Wings, gross	17.00 m² (183.0 sq ft)
Ailerons (total)	1.11 m² (12.0 sq ft)
Trailing-edge flaps (total)	2.51 m² (27.0 sq ft)
Fin	2.34 m² (25.2 sq ft)
Dorsal fin	0.12 m² (1.30 sq ft)
Rudder	0.67 m² (7.2 sq ft)
Tailplane	1.95 m² (21.0 sq ft)
Elevators	1.45 m² (15.6 sq ft)

WEIGHTS AND LOADINGS:
Weight empty, equipped	885 kg (1,950 lb)
Max fuel weight	245 kg (540 lb)
Max T-O and landing weight	1,451 kg (3,200 lb)
Max wing loading	85.44 kg/m² (17.5 lb/sq ft)
Max power loading	7.80 kg/kW (12.8 lb/hp)

PERFORMANCE (at max T-O weight):
Never-exceed speed	160 knots (298 km/h; 185 mph) IAS
Max level speed at 2,135 m (7,000 ft)	152 knots (281 km/h; 175 mph) IAS
Max cruising speed, 75% power at 2,135 m (7,000 ft)	139 knots (257 km/h; 160 mph)
Econ cruising speed, 67% power at 2,135 m (7,000 ft)	135 knots (249 km/h; 155 mph)
Stalling speed, engine idling:	
flaps up	63 knots (117 km/h; 73 mph)
flaps down	52 knots (97 km/h; 60 mph)
Max rate of climb at S/L	323 m (1,060 ft)/min
Service ceiling	5,485 m (18,000 ft)
T-O run: land	198 m (650 ft)
water	259 m (850 ft)
T-O to 15 m (50 ft): land	305 m (1,000 ft)
water	366 m (1,200 ft)
Landing from 15 m (50 ft):	
land or water	366 m (1,200 ft)
Range with max fuel	868 nm (1,609 km; 1,000 miles)

THURSTON TA19 SEAMASTER

TYPE: Eight-seat transport amphibian.

PROGRAMME: Design studies completed 1991; construction of prototype expected to begin early 1993; first flight expected 1995, with production aircraft available 1997.

CURRENT VERSIONS: **Seamaster**: Basic version, as described below.

COSTS: Standard aircraft approximately $1.2 million.

DESIGN FEATURES: Being developed as modern turboprop replacement for Grumman Goose and Mallard amphibians. Wing section GA 40U-A215.

FLYING CONTROLS: Droop ailerons, conventional rudder and elevators; double-slotted flaps.

STRUCTURE: Aluminium alloy main structure; GFRP secondary structure (wingtips, forward cabin top, bow deck).

LANDING GEAR: Retractable tricycle type with steerable nosewheel; underwing floats; skis optional.

POWER PLANT: Two 335.5 kW (450 shp) Allison 250-B17 turboprops, each driving a Hartzell three-blade constant speed propeller. Four interconnected fuel tanks in each wing, total capacity 1,363 litres (360 US gallons; 300 Imp gallons).

ACCOMMODATION: Two pilots on flight deck; eight passengers in main cabin.

DIMENSIONS, EXTERNAL:
Wing span	16.15 m (53 ft 0 in)
Wing aspect ratio	8.3
Length overall	11.89 m (39 ft 0 in)
Height overall	4.27 m (14 ft 0 in)
Propeller diameter	2.24 m (7 ft 4 in)

AREAS:
Wings, gross	31.12 m² (335.0 sq ft)

WEIGHTS AND LOADINGS:
Weight empty	2,204 kg (4,860 lb)
Max T-O weight	3,901 kg (8,600 lb)
Fuel weight	1,111 kg (2,450 lb)
Max wing loading	125.34 kg/m² (25.67 lb/sq ft)
Max power loading	5.81 kg/kW (9.56 lb/hp)

PERFORMANCE (at max T-O weight):
Max level speed at S/L	191 knots (354 km/h; 220 mph)
Cruising speed, 75% power at 2,285 m (7,500 ft)	174 knots (322 km/h; 200 mph)
Econ cruising speed, 50% power at 2,285 m (7,500 ft)	152 knots (281 km/h; 175 mph)
Stalling speed	52 knots (97 km/h; 60 mph)
Max rate of climb at S/L	457 m (1,500 ft)/min
Max rate of climb at S/L, OEI	152 m (500 ft)/min
Service ceiling	6,100 m (20,000 ft)
T-O run: land	427 m (1,400 ft)
water	564 m (1,850 ft)
Landing run: land	366 m (1,200 ft)
water	427 m (1,400 ft)
Range with max standard fuel, 50% power, 30 min reserves	1,163 nm (2,156 km; 1,340 miles)
Range with max payload, same power and reserves as above	955 nm (1,770 km; 1,100 miles)
Endurance with max standard fuel	9

TKEF
THE KING'S ENGINEERING FELLOWSHIP

Municipal Airport, Orange City, Iowa 51041
Telephone: 1 (712) 737 4444
Fax: 1 (712) 737 3344
PRESIDENT: Carl A. Mortenson
MANUFACTURING: Angel Aircraft Corporation (same address)
MARKETING: TradeLink Inc, Roanoke, Texas
Telephone: 1 (817) 491 2442
Fax: 1 (817) 430 4807

Angel, developed by The King's Engineering Fellowship (TKEF) through donations and designed by Carl Mortenson, follows earlier Evangel (described in 1974-75 *Jane's*).

KING'S 44 ANGEL

TYPE: Specialised light twin-engined utility aircraft.

PROGRAMME: Design started November 1972; prototype construction began January 1977; prototype (N44KE) built on production tooling; first flight 13 January 1984; structural testing began 1990; certification to FAR Pt 23 (Normal category) for day, night, VFR and IFR conditions gained 20 October 1992. Production began Spring 1993; c/n 002 to be completed late Summer 1993; production goals of four aircraft in 1993, 10 in 1994, 18 in 1995 and 24 in 1996.

COSTS: Standard aircraft, IFR-equipped $585,000 (1993).

DESIGN FEATURES: Designed for missionary aviation but not exclusively; other commercial uses include air taxi, air ambulance, air observation/patrol, fishery/pipeline/border inspection, tracking, mining/oil/rubber/forestry operations, ranching and firefighting leader. Design goals include low manufacturing costs, STOL capability, operation from soft and rough fields, easy repair in field, and easy loading of bulky cargo. Vortex generator on wing outboard panels for enhanced lift. Crashworthiness built and tested in key structures; seats dynamically tested to absorb 20g vertically and 26g horizontally; cabin structure includes areas of double-wall and tested for overturn impact survivability.

Wing section NACA 23018-23010 with modified leading-edge; sweepback 15° 36′ at leading-edge, 11° at quarter-chord; dihedral 5° 24′; incidence 3° at root, –0° 22′ at tip.

FLYING CONTROLS: Manual (cable) actuated. Lateral control by multiple small-plate spoilers immediately forward of flaps; 'trimmerons' outboard of flaps to assist during single-engined flight; elevators and rudder, both with trim tabs; almost full-span hydraulically actuated semi-Fowler flaps deflecting to maximum 37°.

STRUCTURE: Riveted aluminium alloy and welded 4130/4340 steel tube; wing has built-up capstrip spars and 19 die-formed ribs each side; broad-chord fin and rudder with large dorsal fin; CFRP spinners; GFRP/epoxy in some areas.

LANDING GEAR: Retractable tricycle type. Electrohydraulic retraction, mainwheels inward into wingroots, nosewheel rearward. Emergency extension by handpump or gravity. Cleveland main wheel tyres size 8.50-10 (pressure 2.41 bars; 35 lb/sq in), McCreary or Goodyear nosewheel tyre size 8.50-6 (pressure 1.03 bars; 15 lb/sq in). Cleveland disc brakes. Min turning circle 5.11 m (16 ft 9 in).

POWER PLANT: Two 224 kW (300 hp) Textron Lycoming IO-540-M1C5 flat-six engines, mounted on top of the wings inboard and each driving a Hartzell three-blade constant-speed feathering pusher propeller. Two wing fuel tanks, gravity flow cross-fed, with total capacity 844 litres (223 US gallons; 185.7 Imp gallons). Refuelling point in top of wing. Oil capacity 22.7 litres (6 US gallons; 5 Imp gallons).

ACCOMMODATION: Enclosed cabin seating up to eight persons including pilot. Five seats can be removed for carrying cargo, including four 208 litre (55 US gallon; 46 Imp gallon) drums. Rearmost bench seat is fixed. Four large windows and one smaller circular window on each side of cabin. Horizontally divided clamshell door on port side at

front of cabin; emergency exit on starboard side. Heating and window air vents standard. Compartment for 90 kg (200 lb) of baggage at rear of fuselage, with door on port side.
SYSTEMS: Hydraulic system, with electric pump, for landing gear and flap actuation. Electrical system with two alternators includes 12V DC battery in nose.
AVIONICS: IFR. Bendix/King avionics, including twin com/nav transceivers, glideslope, dual ADF, transponder, ELT. Weather radar, GPS, Loran C and HF com optional.

DIMENSIONS, EXTERNAL:
Wing span	12.16 m (39 ft 10¾ in)
Wing aspect ratio	7.06
Wing chord: at root	2.59 m (8 ft 6 in)
at tip	1.00 m (3 ft 3½ in)
Length overall	10.21 m (33 ft 6 in)
Height overall	3.51 m (11 ft 6 in)
Tailplane span	5.30 m (17 ft 4½ in)
Wheel track	3.95 m (12 ft 11¾ in)
Wheelbase	4.67 m (15 ft 4 in)
Propeller diameter	1.93 m (6 ft 4 in)
Propeller ground clearance at T-O	0.23 cm (9.01 in)
Cabin door (port, fwd): Height	0.91 m (3 ft 0 in)
Width	1.04 m (3 ft 5 in)
Height to sill	0.91 m (3 ft 0 in)
Baggage door: Height	0.41 m (1 ft 4 in)
Width	0.61 m (2 ft 0 in)
Height to sill	1.17 m (3 ft 10 in)
Emergency exit door (stbd, fwd):	
Height	0.48 m (1 ft 7 in)
Width	1.04 m (3 ft 5 in)
Max height	1.32 m (4 ft 4 in)

DIMENSIONS, INTERNAL:
Cabin, incl flight deck:	
Length	3.51 m (11 ft 6 in)
Max width	1.07 m (3 ft 6 in)
Max height	1.14 m (3 ft 9 in)
Volume	2.38 m³ (84.0 cu ft)
Baggage compartment volume (aft)	0.28 m³ (10.0 cu ft)

AREAS:
Wings, gross	20.94 m² (225.4 sq ft)
Ailerons	0.39 m² (4.20 sq ft)
Trailing-edge flaps	4.05 m² (43.60 sq ft)
Spoilers	0.37 m² (4.00 sq ft)
Rudder, incl tab	1.41 m² (15.15 sq ft)
Tailplane	6.23 m² (67.02 sq ft)
Elevators, incl tabs	2.76 m² (29.67 sq ft)

WEIGHTS AND LOADINGS:
Weight empty	1,760 kg (3,880 lb)
Max fuel weight	612 kg (1,350 lb)
Max T-O and landing weight	2,631 kg (5,800 lb)
Max wing loading	125.6 kg/m² (25.73 lb/sq ft)
Max power loading	5.87 kg/kW (9.67 lb/hp)

PERFORMANCE:
Never-exceed speed (V_{NE})	209 knots (387 km/h; 240 mph)
Max level speed at S/L	180 knots (333 km/h; 207 mph)
Cruising speed, 65% power at 3,500 m (11,500 ft)	169 knots (313 km/h; 195 mph)
Stalling speed:	
power off, flaps down	57 knots (106 km/h; 66 mph)
power on	51 knots (95 km/h; 59 mph)
Max rate of climb at S/L	434 m (1,423 ft)/min
Rate of climb, OEI	75 m (245 ft)/min
Service ceiling	5,975 m (19,605 ft)
Service ceiling, OEI	1,654 m (5,425 ft)
T-O run	201 m (658 ft)
T-O to 15 m (50 ft)	387 m (1,270 ft)
Landing from 15 m (50 ft)	325 m (1,065 ft)
Landing run	178 m (584 ft)
Range with max fuel, VFR reserves:	
65% power	1,263 nm (2,340 km; 1,454 miles)
45% power	1,605 nm (2,974 km; 1,848 miles)
35% power	1,720 nm (3,187 km; 1,980 miles)
Endurance, VFR reserves: with 8 occupants	3 h
with max fuel and 3 occupants:	
65% power	7 h 54 min
45% power	11 h 18 min
35% power	13 h 6 min

Prototype TKEF Angel light twin for missionary and commercial roles

TRADEWIND
TRADEWIND TURBINES CORPORATION
PO Box 31930, 4105 Tradewind Road, Amarillo, Texas 79120
Telephone: 1 (806) 376 5203
Fax: 1 (806) 376 9725
PRINCIPAL: Joe C. Boyd
CEO: J. A. Whittenburg III

Company acquired STC in October 1989 for Allison Gas Turbine Division/Soloy-developed STC for re-engining of Beech A36 Bonanza; Teledyne Continental 212.5 kW (285 hp) piston engine replaced by 313 kW (420 shp) Allison 250-B17D or 336 kW (450 shp) -B17F2; fuel capacity increased to 424 litres (112 US gallons; 93 Imp gallons) in two tanks in wings and two Osborne wingtip tanks; optional Bendix/King EFIS.

Details, in 1991-92 *Jane's*, now shown in *Jane's Civil and Military Aircraft Upgrades*.

TRIDAIR
TRIDAIR HELICOPTERS INC
3000 Airway Avenue, Costa Mesa, California 92626
Telephone: 1 (714) 540 3000
Fax: 1 (714) 540 1042
PRESIDENT: Douglas Daigle

TRIDAIR (BELL) 206L-3ST and 206L-4ST GEMINI ST
TYPE: Twin-engined conversion of Bell 206L LongRanger.
PROGRAMME: Announced at Helicopter Association International (HAI) show 1989; first flight of prototype N700TH 16 January 1991; second prototype (N100TH) first flew 15 September 1992, featuring redundant electrical systems and modified fuel system; FAA issued Supplemental Type Certificate (provisional) 25 February 1993, allowing installation of two Allison 250-C20R engines in Bell 206L-1/-3/-4; final certification expected June 1993; third prototype based on new Bell 206L-4. Initial customer installations carried out at Soloy's Olympia plant; six Approved Installation Centers organised by March 1993, to begin conversions from early 1994. See also *Jane's Civil and Military Aircraft Upgrades*.
CURRENT VERSIONS: **206L-3ST**: *As described in detail below.*
206L-4ST: Based on 206L-4 airframe; higher internal gross weight of 2,018 kg (4,450 lb).
CUSTOMERS: Tridair ordered 25 new airframes from Bell Helicopter Canada for delivery at rate of two per month from November 1990; Bell Helicopter announced in March 1992 that it will build new **206LT TwinRangers** in its Canadian factory (which see) with first delivery in third quarter 1993; Tridair to specialise in conversions. 15 new-build and eight conversions ordered by June 1992; 50 conversions expected by end of 1993; approved installation centres being established (see Programme).
COSTS: Modification kit $609,500; new converted aircraft $1,479,500.
DESIGN FEATURES: Replaces single 485 kW (650 shp) Allison 250-C30P with two 335 kW (450 shp) 250-C20Rs with max continuous rating of 276 kW (370 shp) each; fuel capacity increased to 416 litres (110 US gallons; 91.6 Imp gallons); Soloy Dual Pac combining gearbox with individual freewheels concentrates engine outputs into original single input drive to transmission; transmission rating 325 kW (435 shp); current STC limitations are single-engine

Tridair Helicopters (Bell) 206L-3ST Gemini ST prototype

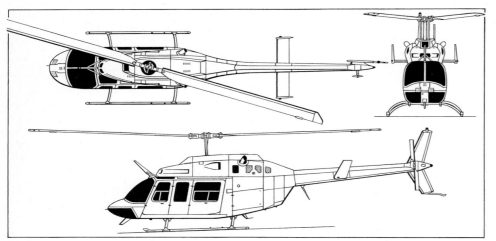

operation restricted to emergency conditions only, max gross weight limited to current internal gross weight of 206L-3 at 1,882 kg (4,150 lb), flight in falling and blowing snow not authorised; flight test reports being generated which should expand STC to include internal gross weight of 1,928 kg (4,250 lb) for 206L-3ST and 2,018 kg (4,450 lb) for 206L-4ST; snow testing was scheduled to be completed by end of March 1993.

WEIGHTS AND LOADINGS:
Weight empty	1,175 kg (2,590 lb)
Max fuel weight	340 kg (750 lb)
Max external load	907 kg (2,000 lb)
Max T-O weight (internal load)	1,928 kg (4,250 lb)

PERFORMANCE (estimated):
Never-exceed speed (V_{NE})	130 knots (241 km/h; 150 mph)
Max cruising speed	117 knots (217 km/h; 135 mph)
Max rate of climb at S/L	472 m (1,550 ft)/min
Service ceiling	6,100 m (20,000 ft)
Hovering ceiling: IGE	4,815 m (15,800 ft)
OGE	1,675 m (5,500 ft)
Range with max payload and max internal fuel	347 nm (643 km; 400 miles)

Tridair Helicopters (Bell) 206L-3ST Gemini ST (two Allison 250-C20Rs) *(Jane's/Mike Keep)*

TURBOTECH
TURBOTECH INC
2215 Parrott Way, Kelso, Washington 98666
Telephone: 1 (206) 423 7699
Fax: 1 (206) 423 7695
PRESIDENT: Nancy J. Soukup

Turbotech installs more powerful and more reliable engines in various aircraft; company holds more than 30 STCs for multiple conversions; also undertakes one-off engine changes and add-on turbo packages with FAA certification; recent examples include Grumman Widgeon fitted with 261 kW (350 hp) Textron Lycoming O-540 engines.

STCs obtained for two Cessna 185 modifications in 1982, to provide full performance to service ceiling; these covered alternative installation of 261 kW (350 hp) Textron Lycoming TIO-540-J2BD or addition of a Garrett turbocharger to 224 kW (300 hp) Continental IO-520-D fitted in standard Cessna 185, which gave full performance to ceilings of 7,620 m (25,000 ft) and 7,315 m (24,000 ft) respectively; Cessna 206 also approved with TIO-540-J2BD turbocharged engine. Gross weight increased by 93 kg (250 lb) with installation of wingtip tanks, approved July 1983.

Other developments include installation in Grumman AA-5B Tiger of a 134 kW (180 hp) Textron Lycoming O-360-A4K; 172 kW (230 hp) Teledyne Continental O-470 installed in Stinson 108. Current feasibility studies for Cessna 180, 182, 207 and 210 engine conversions; turbocharged Textron Lycoming O-540-L3C5D, with intercooler, automatic wastegate and magneto pressurisation, being installed in Cessna 182RG; latter also to be available for TR 182.

Turbotech Cessna 185 conversion

Turbotech Grumman Widgeon with 261 kW (350 hp) Textron Lycoming O-540 engines

TWIN COMMANDER
TWIN COMMANDER AIRCRAFT CORPORATION
19003 59th Drive NE, Arlington, Washington 98223
Telephone: 1 (206) 435 9797
Fax: 1 (206) 435 1112
PRESIDENT: Dan Montgomery

Purchased product support and manufacturing rights to former Commander series of piston and turboprop twins from Gulfstream Aerospace, December 1989 (production had been suspended since 1984); Twin Commander initially manufacturing and distributing spares, but has been negotiating with groups outside USA to manufacture new aircraft.

UNC
UNC HELICOPTER (part of UNC Inc Aviation Services Division)
1101 Valentine Extension, Ozark, Alabama 36360
Telephone: 1 (205) 774 2529
Fax: 1 (205) 774 6756
VICE-PRESIDENT, MARKETING, UNC INC: Robert Marchetti

Provides rotary-wing line and intermediate maintenance for US military helicopters; basic and advanced pilot training for US and NATO helicopter pilots; and refurbishment, re-engineering and certification of military and commercial helicopters. Is currently prime contractor on US Army's UH-1H retirement programme, for which it is overhauling 250 of these helicopters for redelivery as foreign military sales; and has teamed with GE Aircraft Engines to develop and market Ultra Huey for offer to US Army National Guard and potential foreign customers.

UNC (BELL) ULTRA HUEY
TYPE: Upgraded/modified UH-1H, principally with General Electric T700-GE-701C engine.
PROGRAMME: Demonstrator (N700UH) first flown 21 August 1992; completed hot and high trials in Alamosa, Colorado; undertook marketing tour of US National Guard/Army bases; UNC looking to launch customer with some 100 helicopters to modify.
DESIGN FEATURES: Replacement of 1,044 kW (1,400 shp) T53-L-13B with 1,342 kW (1,800 shp) General Electric T700-GE-701C (modified civil CT7-2A); transmission uprated from 865 kW (1,160 shp) to 962 kW (1,290 shp); new Lucas Western speed reduction gearbox; main/tail rotor blades, thicker skinned tailboom/rotor pylon and landing gear from Bell 212; UH-1N anti-torque system; Bendix King EFIS nav/com and Canadian Marconi avionics; wire strike protection; max T-O weight increased from 4,309 kg (9,500 lb) to 4,763 kg (10,500 lb); NV-compatible cockpit lighting. Results in better hot and high performance, increased range/payload, and 20 per cent lower sfc.

USAF
UNITED STATES AIR FORCE MATERIEL COMMAND

Aeronautical Systems Division, Wright-Patterson AFB, Dayton, Ohio 45433-6503
Telephone: 1 (513) 255 3334
PUBLIC AFFAIRS OFFICER: Capt Jamie S. Scearse

Formed 1 July 1992, incorporating former Logistics and Systems Commands; duties include management of procurement programmes.

ADVANCED TACTICAL AIRCRAFT (ATA)
PROGRAMME: USAF plans adoption of Navy's A-X (which see) as replacement for General Dynamics F-111, Lockheed F-117 and McDonnell Douglas F-15E. ATA joint concept definition and demonstration/validation phase to begin in 1994.

JOINT STEALTH STRIKE AIRCRAFT (JSSA) and JOINT ATTACK FIGHTER (JAF)
PROGRAMME: 1993 concepts for multi-service, dual role aircraft, representing high (JSSA) and low-medium (JAF) mix. Fall-back programmes in event of cancellation of F-X, F-22, MRF or F/A-18E.

MULTI-ROLE FIGHTER (MRF)
PROGRAMME: Replacement for F-16 Fighting Falcon, to which A-10A and Navy's F/A-18 may be added; currently on 'indefinite hold' but USAF continuing to fund F-16 production at low rate to hold open option of F-16 based MRF.

JOINT PRIMARY AIRCRAFT TRAINING SYSTEM (JPATS)
TYPE: Primary/basic pilot trainer.
PROGRAMME: USAF and USN issued Joint-Service Operational Requirements Document, November 1991; Department of Defense go-ahead of 19 January 1993 re-timed programme; draft RFP 1 March; final RFP 1 August (aircraft only; associated ground training package now separate); selection October 1993; work start May 1994; USAF requirement 417 aircraft (originally to have been 495), with FY 1998 IOC; USN requires 347, with 2000 IOC; carrier compatibility not needed.
COSTS: $6,000 million programme.
DESIGN FEATURES: Tandem, stepped cockpits specified November 1991, eliminating Promavia Jet Squalus, Teledyne Ryan (with revised Fairchild Republic T-46 design) and Saab 2060. Minimum 250 knots (463 km/h; 288 mph) IAS cruising speed; 270 knots (500 km/h; 311 mph) objective; canopy resistance to 1.8 kg (4 lb) bird at objective speed; pressurised accommodation; provision for blind-flying hood in front cockpit; zero height/60 knots (111 km/h; 69 mph) ejection seats.

Contenders include Rockwell International/MBB Ranger 2000, Beech/Pilatus PC-9 Mk II, Cessna/Flight Safety trainer version of CitationJet, Grumman/Agusta S.211, Vought/FMA IA 63 Pampa 2000, Lockheed/Aermacchi/Rolls-Royce MB-339 'T-Bird II' and Northrop/Embraer EMB-312H Super Tucano.
STRUCTURE: Life of 24 years/19,000 flying hours; +6/−3g.
POWER PLANT: Turboprop or jet power.
AVIONICS: To include GPS microwave landing system, collision warning system and provision for HUD.

ENHANCED FLIGHT SCREENER (EFS)
PROGRAMME: Pre-selection trainer. Chosen Slingsby Firefly designated T-3A; objections of losing contenders overruled 1992; programme proceeding. Refer to Slingsby in UK section.

ADVANCED SURVEILLANCE AND TRACKING TECHNOLOGY/AIRBORNE RADAR DEMONSTRATOR (ASTT/ARD)
PROGRAMME: Part of AFMC's Aeronautical Systems Division programme to develop new AWACS aircraft for use in 21st century; wind tunnel tests conducted late 1990; flight test demonstrator not expected before 1996. Current proposals envisage Boeing 747 platform aircraft, with GE Aerospace (now Martin Marietta) advanced L-band phased-array radar installed in very large E-Systems dorsal radome (40 × 14.5 × 3.8 m; 131.2 × 47.6 × 12.5 ft), called a 'synergistic plandome', supported by multiple-strut structure above fuselage.

STRATEGIC RECONNAISSANCE AIRCRAFT
Some disinformation, and minimal verifiable fact, are available on the 'black' programmes which have followed the Lockheed U-2, A-12/SR-71 and F-117. Data below have been gathered from many uncorroborated sources and do not conform to *Jane's* usual standards of accuracy.

'Lockheed Aurora': Name accidentally revealed, 1983, in connection with USAF manned reconnaissance assets; press reports, 1988, of Mach 5 recce aircraft under development; linked to Lockheed test site at Groom Lake, Nellis range, Nevada; reportedly cancelled by Congress in July 1990. Since June 1991, sonic booms of Mach 3-4 aircraft, about 27 m (90 ft) in length and heading towards southern Nevada, tracked by seismological stations; radio intercepts reveal operating altitude of at least 20,420 m (67,000 ft) for unknown aircraft approaching Edwards AFB, April 1992; ex-SR-71 aircrew believed posted to Groom Lake since 1990. Deep rumbling/booming engine note associated with new aircraft reported night flying at Beale AFB, California (9th Recce Wing; former SR-71 operator) since late 1991. Aircraft apparently has pulse detonation wave engine which, under varying conditions, produces distinctive 'donuts-on-a-rope' or 'string of sausages' condensation trails. Two contenders for Aurora title appeared in 1992:

1. 'X-30 Aurora'. Artist's conception, based on known performance and operational parameters of Mach 8 aircraft produced design similar to X-30 NASP (see three-view drawing and NASA entry); one retrospective claim of 1989 sighting. Length 27 m (90 ft). Alleged in late 1992 that X-30 is merely camouflage for Aurora programme.

2. 'XB-70 Aurora'. Forward fuselage resembling front of

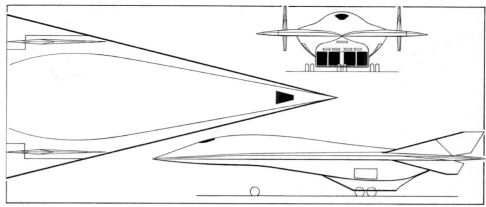

Conjectural Aurora design derived from basic requirements of Mach 8 flight; length 27 m (90 ft) *(Jane's/Mike Keep)*

Alleged Aurora design based on sighting reports during 1992; length 60 m (200 ft) *(Jane's/Mike Keep)*

Alleged 'Northrop TR-3A' reconnaissance aircraft *(Jane's/Mike Keep)*

Lockheed SR-71, but with fatter, more rounded chines, noted being loaded into C-5 transport at Lockheed Burbank in January 1992. Aircraft with this front section, but otherwise resembling North American (Rockwell) XB-70 of 1960s, reported landing at Lockheed's Helendale private airfield July 1992; 60 m (200 ft) long and with raised rear section implying provision for carriage of 'pick-a-back' aircraft (as per SR-71 predecessor A-12(M) with D-21A reconnaissance drone); see artist's impression; airliner crew reported airmiss near Los Angeles, August 1992, with F-16 size aircraft having SR-71 shape forward fuselage and indistinct tail surfaces; possibly after launch from Aurora.

'Northrop TR-3A': Flying wing of approximately 20 m (66 ft) span first reported over California in 1990; originally believed scale demonstrator of GD/McDonnell A-12; late associated with real-time intelligence gathering for Lockheed F-117 attack force. Possibly developed from Northrop THA (Tactical High Altitude Penetrator), prototype of which firs flew mid-1981. No recent reports.

US ARMY
UNITED STATES ARMY AVIATION SYSTEMS COMMAND
400 Goodfellow Boulevard, St Louis, Missouri 63120-1798
Telephone: 1 (314) 263 1599

NEW TRAINING HELICOPTER (NTH)
PROGRAMME: Eleven-week flight evaluation undertaken in Autumn 1992 of Eurocopter AS 350, Bell 206TH, Enstrom TH-28 and Grumman/Schweizer TH-330 as potential replacement for 270 ageing Bell UH-1 Iroquois (Huey) helicopters at Fort Rucker training establishment. Former lease option now eliminated in favour of direct purchase of 157 three-seat (two pupils and instructor) commercial light helicopters and option on further 157. Final offers submitted by 16 February 1993; Bell contender (to be known as TH-67 Creek) announced winner March 1993; initial funding for 102 in FY 1993 (including 30 with FY 1992 funds) and 55 to follow. (See Bell entry in Canadian section).

ADVANCED CARGO AIRCRAFT (ACA)
PROGRAMME: Planned replacement of existing Boeing CH-47D Chinooks; full scale engineering development 1998; IOC 2002; 11,800 kg (26,014 lb) payload with full fuel; 18,000 kg (39,683 lb) 'zero range' lift. Sikorsky to offer new design; Boeing offering interim CH-47F Advanced Chinook with MH-47E fuel tanks, refuelling probe, Boeing type rotors and advanced avionics; esti mated non-recurring cost $200-300 million (1992) and fly away cost $15-20 million.

UH-X
PROGRAMME: Replacement for Bell UH-1H Iroquois utilit helicopter; 1,279 required for procurement from FY 1998 specification currently undefined; contenders includ Sikorsky UH-60X Advanced Black Hawk, Eurocopter LTV AS 565 Panther 800 (see this edition) and improved re-engined UH-1s (see Bell Helicopter, UNC and Globa Helicopters).

USN
UNITED STATES NAVAL AIR SYSTEMS COMMAND
Jefferson Plaza 1, Washington, DC 20361-0001
Telephone: 1 (262) 693 2260

NAVAIR has over 46,000 military and civilian employees and an annual budget of $13,000 million.

A-X
TYPE: Carrier-borne strike aircraft with air-to-air capabilities.
PROGRAMME: Following January 1991 cancellation of General Dynamics/McDonnell Douglas A-12 naval attack aircraft, Tentative Operational Requirement issued for substitute aircraft to replace Grumman A-6 aboard USN carriers; USAF expecting to adopt same aircraft as ATA, replacing F-111, F-117 and F-15E. Five consortia each awarded $20 million contracts 30 December 1991 for 'trade studies and risk-reduction activities':
Grumman, with Boeing and Lockheed; offering new design.
General Dynamics (later absorbed by Lockheed), with McDonnell Douglas and Northrop; A-12 derivative.
McDonnell Douglas with Vought; new design.
Lockheed, with Boeing and General Dynamics (later absorbed by Lockheed); F-22 variant.
Rockwell with Lockheed; new design.
Original plan called for submission of studies by September 1992; solicitation for advanced development proposals in October 1992; contractor selection May 1993; demonstration/validation phase 1994 and full-scale development in 1996. Programme delayed by two years following late 1992 Congressional mandate for competitive prototyping; concurrently, improved air-to-air performance added to requirements, prompting unofficial designation change to A/F-X; in-service date 2007; between 400 and 500 A-Xs required. Performance requirements include 5,443 kg (12,000 lb) ordnance load; carriage of all USAF/USN air-launched weapons except AIM-54 Phoenix; endurance 2 h; and range of 600 nm (1,112 km; 691 miles). By mid-1993, survival of A-X programme was in doubt.

NAVAL ADVANCED TACTICAL FIGHTER (NATF)
Plans for a naval adaptation of the winning contender for the USAF ATF competition were terminated in early 1991 by removal of funding until after 1997. IOC for a low-observables replacement for the Grumman F-14 Tomcat planned in 2005-2008, requiring concept definition for a new NATF to begin around 1999.

E-X
TYPE: Shipboard AEW aircraft.
PROGRAMME: Replacement for Grumman E-2C Hawkeye; consideration of upgraded E-2, adaptation of Lockheed S-3 Viking or all-new Grumman ATS (Advanced Tactical System) and Boeing 'joined wing' aircraft.

STOVL STRIKE FIGHTER (SSF)
TYPE: Carrier-borne fighter.
PROGRAMME: Requirement defined by USN in April 1992 project managed by ARPA; RFP deadline 5 Decembe 1992; proposals received from Lockheed Fort Worth (ex General Dynamics), Lockheed, Grumman, Boeing an McDonnell Douglas/British Aerospace; two to be selecte for 30 month concept validation phase; ARPA then to sel ect most promising design; demonstrator to fly 2000 fighter in service 2010. Potential replacement for F/A-1 Hornet and AV-8B Harrier II.

First decision, March 1993 awarded Phase 2 concep validation contracts to Lockheed (see this section) an McDonnell Douglas (see International section). Forme uses shaft-driven fan for lift during STO/VL phase; latte relies on gas-coupled fan; both fans are mounted forwarc rear element of thrust coming from engine nozzle deflec tion; fan doors on both aircraft close for normal forwar flight. Lockheed team ($32.9 million contact) include Pratt & Whitney (F119 engine), Allison, Rolls-Royce an Hercules; McDonnell Douglas ($27.7 million) teame with BAe and General Electric (F120 engine). One non flying airframe from each team to be pylon-mounted a NASA/Ames for engine trials and later tested in wind tur nel at same facility. One design to be selected for Phase production of two flying prototypes, powered by singl afterburning engine in class of Lockheed F-22's powe plant: 200 kN (45,000 lb st) in conventional flight; 165 kl (37,000 lb st) during lift phase.
DESIGN FEATURES: Rear-mounted Advanced Tactical Fighte engine driving horizontally mounted lift fan in forwar fuselage either by shaft or by gas drive.

VALSAN
VALSAN INC
3010 Westchester Avenue, Purchase, New York 10577
Telephone: 1 (914) 694 3600
Fax: 1 (914) 694 6062
CHAIRMAN: Robert E. Wagenfeld
PRESIDENT, MARKETING: Walter H. Johnson

VALSAN 727RE 'QUIET 727'
TYPE: Re-engined/quietened Boeing 727-100 and -200.
PROGRAMME: Prototype conversion of 727 (Sterling Airways 727-200 OY-SAS), to meet FAR Pt 36 Stage 3 and ICAO Annex 16 Chapter 3 noise limits, completed June 1988; first flight 12 July 1988; STC awarded October 1988; certification of first customer 727-100 conversion July 1990; additional winglet system conversions certificated 1992.
CUSTOMERS: Customers include Federal Express (15 ordered, 46 options), Texas Air Leasing (10 ordered, 30 options), Sterling Airways (two ordered, 13 options), IAL Aviation Services (10 ordered, 20 options), International Aircraft Services (10 ordered, 20 options), Triangle Aircraft Services (one), Electra Aviation (10 ordered, 20 options), North-west Holding (one), Amway (one), Baker (one), Northern Lakes Equity (one), Northern Lakes Financial (one), Sun Country Airlines (two) and Airline of the Americas (two).
Details in 1991-92 *Jane's* and now in *Jane's Civil and Military Aircraft Upgrades*.

Valsan 'Quiet 727' conversion of the Boeing 727-200 for Federal Express

VAN'S
VAN'S AIRCRAFT INC
PO Box 160, North Plains, Oregon 97133
Telephone: 1 (503) 647 5117
PRESIDENT: Richard VanGrunsven

Details of the current single-seat Van's RV-3 sporting homebuilt, available in kit and plan forms, can be found in the 1992-93 *Jane's*.

VAN'S RV-4
TYPE: Tandem two-seat sporting homebuilt.
PROGRAMME: First flight of prototype 21 August 1979. Plans and kits available to homebuilders. By early 1993, over 3,400 sets of plans sold, with about 1,600 aircraft under construction and 460 RV-4s flying.
COSTS: Kit: $8,890. Plans: $165. Information pack: $8.
DESIGN FEATURES: Wing section Van's Aircraft 135.
POWER PLANT: One 112 kW (150 hp) Textron Lycoming O-320-E1F flat-four engine. Fuel capacity 121 litres (32 US gallons; 26.6 Imp gallons).
DIMENSIONS, EXTERNAL:
Wing span 7.01 m (23 ft 0 in
Wing aspect ratio 4.8

Length overall	6.21 m (20 ft 4½ in)
Height overall	1.60 m (5 ft 3 in)
Propeller diameter	1.73 m (5 ft 8 in)

AREAS:
Wings, gross	10.22 m² (110.0 sq ft)

WEIGHTS AND LOADINGS:
Weight empty	404 kg (890 lb)
Baggage capacity	13.6 kg (30 lb)
Max T-O weight	680 kg (1,500 lb)
Max wing loading	66.58 kg/m² (13.64 lb/sq ft)
Max power loading	6.07 kg/kW (10.00 lb/hp)

PERFORMANCE (at max T-O weight, and prior to prototype's aerodynamic clean-up):
Max level speed at S/L	175 knots (323 km/h; 201 mph)
Econ cruising speed, 55% power at 2,440 m (8,000 ft)	142 knots (264 km/h; 164 mph)
Stalling speed	47 knots (87 km/h; 54 mph)
Max rate of climb at S/L	503 m (1,650 ft)/min
Service ceiling	5,945 m (19,500 ft)
T-O run	137 m (450 ft)
Landing run	130 m (425 ft)
Range with max fuel, 55% power	695 nm (1,287 km; 800 miles)

Van's RV-4 built by Raymond DeRoche of Sainte-Ruffine, France

VAN'S RV-6

TYPE: Side by side two-seat sporting homebuilt.
PROGRAMME: Prototype first flew June 1986. Plans and kits available. By early 1993, more than 2,700 sets of plans sold, with about 800 aircraft under construction and 120 flying.
COSTS: Kit: $9,615. Plans: $195. Information pack: $8.
DESIGN FEATURES: Basically side by side two-seat derivative of RV-4.
POWER PLANT: One 112-134 kW (150-180 hp) Textron Lycoming flat-four engine. Fuel capacity 140 litres (37 US gallons; 30.8 Imp gallons).

DIMENSIONS, EXTERNAL: As RV-4 except:
Length overall	6.16 m (20 ft 2½ in)

AREAS: As RV-4
WEIGHTS AND LOADINGS:
Weight empty	431 kg (950 lb)
Baggage capacity	27.2 kg (60 lb)
Max T-O weight	726 kg (1,600 lb)
Max wing loading	71.04 kg/m² (14.55 lb/sq ft)

PERFORMANCE (at max T-O weight):
Max level speed at S/L	175 knots (323 km/h; 201 mph)
Econ cruising speed, 55% power at 2,440 m (8,000 ft)	146 knots (270 km/h; 168 mph)
Stalling speed	47 knots (87 km/h; 54 mph)
Max rate of climb at S/L	503 m (1,650 ft)/min
Service ceiling	5,945 m (19,500 ft)
T-O run	160 m (525 ft)
Landing run	153 m (500 ft)
Range with max fuel (55% power)	803 nm (1,488 km; 925 miles)

VAN'S RV-6A

TYPE: Tricycle landing gear variation of RV-6.
PROGRAMME: Prototype first flew July 1988. Plans and kits available. AIEP of Nigeria (which see) has developed version known as Air Beetle, intended as military screening trainer. By early 1993, 25 RV-6As had flown and about 800 were under construction.
COSTS: Kit: $10,215. Plans: $205. Information pack: $8.
POWER PLANT: Similar to RV-6; fuel capacity 144 litres (38 US gallons; 31.6 Imp gallons).

DIMENSIONS, EXTERNAL: As RV-6 except:
Length overall	6.07 m (19 ft 11 in)
Height overall	2.13 m (7 ft 0 in)

AREAS: As RV-6
WEIGHTS AND LOADINGS: As RV-6 except:
Weight empty	451 kg (995 lb)

PERFORMANCE (two persons):
Max level speed	174 knots (322 km/h; 200 mph)
Econ cruising speed, 55% power at 2,440 m (8,000 ft)	144 knots (267 km/h; 166 mph)
Stalling speed	48 knots (89 km/h; 55 mph)
Max rate of climb at S/L	427 m (1,400 ft)/min
Service ceiling	4,970 m (16,300 ft)
T-O run	92 m (300 ft)
Landing run	153 m (500 ft)
Range, 55% power at 2,440 m (8,000 ft)	760 nm (1,408 km; 875 miles)

VARDAX

VARDAX CORPORATION
3025 Eldridge Avenue, Bellingham, Washington 98225
Telephone: 1 (206) 671 7817
Fax: 1 (206) 671 7820
PRESIDENT: Dara Wilder

VARDAX VAZAR DASH 3

TYPE: Re-engined DHC-3 Otter.
PROGRAMME: Prototype (N9707B) completed 1986; 13 conversions completed by December 1991, when three more in progress. Aircraft operating in North America, Europe and Antarctica. Conversion takes six weeks. Further details can be found in *Jane's Civil and Military Aircraft Upgrades*.
DESIGN FEATURES: Re-engined with 559 kW (750 shp) P&WC PT6A-135 turboprop, giving improved performance and increases useful load by 331 kg (730 lb); additional 159 litre (50.4 US gallon; 42 Imp gallon) fuel tank; thermoformed bubble windows. Can operate on wheels, skis or choice of Edo 7170 or 7490 amphibious floats. Details of standard DHC-3 in 1967-68 *Jane's* and of Vazar Dash 3 changes in 1990-91 *Jane's*.

VAT

VERTICAL AVIATION TECHNOLOGIES INC
PO Box 2527, Sanford, Florida 32773
Telephone: 1 (407) 322 9488
Fax: 1 (407) 330 2647
PRESIDENT: Bradley G. Clark
CONSULTANT ENGINEER: Ralph Alex

Vertical Aviation Technologies Inc is FAA approved repair facility for various Sikorsky helicopters. In 1984 development began to modify four-seat Sikorsky S-52-3 former production helicopter (of which about 50 originally produced) into kit for assembly by individuals, corporations or military, with all tooling and fixtures completed by 1988. In 1990 development began for new version of Sikorsky S-55 named Elite. Company also remanufactures Sikorsky S-58T.

VAT HUMMINGBIRD

TYPE: Four-seat helicopter.
CUSTOMERS: Ten Hummingbird kits sold at $86,000 including engine by May 1993.
DESIGN FEATURES: Based on previously type certificated Sikorsky S-52-3. VAT replaced standard Sikorsky S-52-3 power plant with 224 kW (300 hp) liquid-cooled, aluminium block Chevrolet motorcar engine. Newly manufactured airframes and tailcones identical to S-52-3, except for nose section. Assembly takes about 1,000 working hours and kit includes original Sikorsky components. Currently under development are five-blade rotor system, Allison turbine power plant, supercharged liquid-cooled power plant, and military/law enforcement variant.

DIMENSIONS, EXTERNAL:
Main rotor diameter	10.06 m (33 ft 0 in)
Length overall, rotors turning	12.11 m (39 ft 9 in)
Length of fuselage	9.27 m (30 ft 5 in)
Height to top of rotor head	2.62 m (8 ft 7 in)
Height overall	2.87 m (9 ft 5 in)

AREAS:
Main rotor disc	79.46 m² (855.3 sq ft)

WEIGHTS AND LOADINGS:
Weight empty	771 kg (1,700 lb)
Max payload	453 kg (1,000 lb)
Max T-O weight	1,225 kg (2,700 lb)

PERFORMANCE (three-blade rotor):
Cruising speed	85 knots (157 km/h; 98 mph)
Max rate of climb at S/L	366 m (1,200 ft)/min
Service ceiling	3,353 m (11,000 ft)
Range	approx 304 nm (563 km; 350 miles)

Vertical Aviation Technologies Hummingbird kit-built version of modified Sikorsky S-52-3 helicopter, with components

VAT ELITE

TYPE: Highly modified version of Sikorsky S-55.
PROGRAMME: Certificating five-blade rotor system. Various configurations include law enforcement, EMS, VIP, agricultural spraying and sightseeing.
CUSTOMERS: Second Elite being manufactured in May 1993. Price $850,000 in VIP form.
DESIGN FEATURES: New streamline engine doors; new style cockpit and windscreen; composite transmission cowlings; new tail fairings; installation of Garrett TSE331 turboshaft engine. Can be equipped with up to 10 passenger seats, rescue hoist, cargo sling, air-conditioning, and other equipment. New five-blade rotor system will increase performance at altitude and increase max speed, plus reduce noise levels.
POWER PLANT: One Garrett TSE331, derated from 626 kW (840 shp) to 522 kW (700 shp). Usable fuel capacity 681 litres (180 US gallons; 150 Imp gallons).

WEIGHTS AND LOADINGS:
Weight empty 2,132 kg (4,700 lb)
Max T-O weight: standard 3,266 kg (7,200 lb)
restricted 3,583 kg (7,900 lb)

PERFORMANCE:
Never-exceed speed (V$_{NE}$) 99 knots (183 km/h; 114 mph)
Hovering ceiling: IGE 3,050 m (10,000 ft)
OGE 2,040 m (6,700 ft)
Range, 20 min reserves 282 nm (523 km; 325 miles)

Vertical Aviation Technologies Elite version of Sikorsky S-55

VECTOR

VECTOR AIRCRAFT COMPANY
PO Drawer 3350, 165 Scott Avenue, Suite 102, Morgantown, West Virginia 26505
Telephone: 1 (304) 291 2376
Fax: 1 (304) 292 1902
PRESIDENT: Darus H. Zehrbach
CHIEF ENGINEER: Michael E. Renforth
GENERAL MANAGER: Lee Anne Demus

Aircraft Acquisition Corporation, owners of Helio Aircraft Corporation, Taylorcraft Aircraft Corporation and New Technik Corporation, formed Taylor Kits Corporation in February 1990 to market aircraft, registered under the FAA 51 per cent owner built rule, based on the original C. G. Taylor designs. Company now renamed Vector Aircraft.

All kits are equal to certificated production aircraft. All kits use 'Modular Mode' construction sequence, with four major subassemblies shipped in sequence to save cost and storage space.

First aircraft offered was Taylorcraft F21B, with many of original hard to fabricate parts replaced by glassfibre and advanced composites components. Wooden spar augmented with optional glassfibre/Kevlar unit for better vibration damping, energy absorption characteristics and slightly decreased weight. Original aircraft designed for 48.5 kW (65 hp) Franklin engine, but kit flown with engines from 67 to 93.2 kW (90 to 125 hp); updated kit with new wing and tail surfaces offers 7 to 10 per cent higher speeds.

VECTOR AIRCRAFT T-CRAFT
TYPE: Side by side two-seat cabin homebuilt.
DESIGN FEATURES: Strut braced high-wing monoplane. NACA 23012 wing section.
FLYING CONTROLS: Modified Frise type ailerons. Conventional tail unit; trim tab on port elevator.
STRUCTURE: Fabric covered wings have wood and glassfibre/Kevlar spars, and aluminium ribs. Warren truss fuselage structure of 4130 steel tubing, with aluminium alloy and fabric covering. Glassfibre nose cowling. Wire braced tail unit of steel tube and fabric construction.
LANDING GEAR: Non-retractable tailwheel type, with hydraulic disc brakes and optional mainwheel fairings and floats.
POWER PLANT: One engine of 67-119 kW (90-160 hp), with Rotax and Textron Lycoming engines recommended. Fuel capacity 91-152 litres (24-40 US gallons; 20-33.3 Imp gallons).

DIMENSIONS, EXTERNAL:
Wing span 10.97 m (36 ft 0 in)
Length overall 6.78 m (22 ft 3 in)
Height overall 2.08 m (6 ft 10 in)

WEIGHTS AND LOADINGS (88 kW; 118 hp engine):
Weight empty 476 kg (1,050 lb)
Max payload 312 kg (687.5 lb)
Max T-O weight 794 kg (1,750 lb)

PERFORMANCE (88 kW; 118 hp engine):
Max level speed at 1,145 m (3,750 ft) 108 knots (200 km/h; 124 mph)
Econ cruising speed at 1,525 m (5,000 ft) 89 knots (164 km/h; 102 mph)
Stalling speed, power off 39 knots (71 km/h; 44 mph)
Max rate of climb at S/L 230 m (756 ft)/min
Service ceiling 5,335 m (17,500 ft)
T-O run 116 m (380 ft)
Landing run 122 m (400 ft)
Range, no reserves 943 nm (1,747 km; 1,086 miles)

VECTOR AIRCRAFT TWIN-T
TYPE: Two/three-seat twin-engined homebuilt.
DESIGN FEATURES: Strut braced high-wing monoplane. NACA 23012 wing section.
FLYING CONTROLS: Fowler trailing-edge flaps; trim tabs on port elevator and rudder.
STRUCTURE: Fabric covered wings have glassfibre spars and aluminium ribs. Modified Frise type ailerons and slotted flaps of aluminium/glassfibre. Warren truss 4130 steel tube fuselage structure with non-structural glassfibre skins. 4130 steel tube tail unit with glassfibre skins.
LANDING GEAR: Non-retractable tricycle type, with hydraulic disc brakes and glassfibre wheel fairings.
POWER PLANT: Two engines of 67-119 kW (90-160 hp) each. Four fuel tanks, total capacity 182-303 litres (48-80 US gallons; 40-66.6 Imp gallons).

DIMENSIONS, EXTERNAL:
Wing span 10.97 m (36 ft 0 in)
Wing aspect ratio 7.00
Length overall 7.51 m (24 ft 7½ in)
Height overall 3.07 m (10 ft 1 in)

AREAS:
Wings, gross 17.19 m^2 (185.0 sq ft)

WEIGHTS AND LOADINGS (two 119 kW; 160 hp engines):
Weight empty 639.5 kg (1,410 lb)
Max payload 373 kg (823 lb)
Max T-O weight 1,020 kg (2,250 lb)
Max wing loading 59.38 kg/m^2 (12.16 lb/sq ft)

PERFORMANCE (estimated, with two 88 kW; 118 hp engines):
Max level speed at 1,525 m (5,000 ft) 129 knots (240 km/h; 149 mph)
Econ cruising speed at 1,525 m (5,000 ft) 109 knots (201 km/h; 125 mph)
Stalling speed, flaps down, power off 37 knots (68 km/h; 42 mph)
Max rate of climb at S/L 511 m (1,675 ft)/min
Service ceiling 6,100 m (20,000 ft)
T-O run 70 m (227 ft)
Landing run 60 m (196 ft)
Range with max fuel, with IFR reserves 661 nm (1,224 km; 761 miles)

Vector Aircraft Twin-T homebuilt (*Jane's*/Mike Keep)

Prototype Vector Aircraft T-Craft (88 kW; 118 hp Textron Lycoming engine)

VIKING

VIKING AIRCRAFT LTD
PO Box 20791, Carson City, Nevada 89721
Telephone: 1 (702) 884 4716
Fax: 1 (702) 883 9158
OWNER: Patrick Taylor

VIKING DRAGONFLY

TYPE: Side by side two-seat, dual control sporting homebuilt.
PROGRAMME: Prototype first flew 16 June 1980. Plans are available, together with preformed engine cowling and canopy. Also, kits of prefabricated component parts, requiring no complex jigging or tooling, available. In this form aircraft known as 'Snap' Dragonfly. It is estimated that kits save builder more than 700 working hours.
CURRENT VERSIONS: **Dragonfly Mark I:** Original configuration, with non-retractable mainwheels at tips of foreplane.
  Dragonfly Mark II: In parallel production, for operation from unprepared strips and narrow taxiways. Main landing gear in short non-retractable cantilever units under wings, with individual hydraulic toe brakes, and increased foreplane and elevator areas; wheel track 2.44 m (8 ft).
  Dragonfly Mark III: Flight tested in 1985; non-retractable tricycle landing gear.
CUSTOMERS: By February 1993, more than 100 kits delivered, together with 1,800 sets of plans; some 70 kit built and more than 300 plans built aircraft flying.
COSTS: Kit: $12,000. Plans: $259 in USA; $300 foreign.
DESIGN FEATURES: Wing section Eppler 1213. Foreplane GU25 section.
STRUCTURE: Composites wing, foreplane and tail unit structures of styrene foam, glassfibre, carbonfibre and epoxy. Semi-monocoque fuselage, formed (not carved) from 12.5 mm (½ in) thick urethane foam, with strips of 18 mm (¾ in) foam bonded along edges to allow large-radius external corners. Fuselage covered with glassfibre inside and out.
LANDING GEAR: See Current Versions. Brakes fitted.
POWER PLANT: One 44.5 kW (60 hp) 1,835 cc modified Volkswagen motorcar engine; 1,600 cc engine, rated at 33.5 kW (45 hp), optional. Fuel capacity 56.8 litres (15 US gallons; 12.5 Imp gallons).

Viking Aircraft Dragonfly Mark II

DIMENSIONS, EXTERNAL:
Wing span	6.71 m (22 ft 0 in)
Wing aspect ratio	9.18
Foreplane span	6.40 m (21 ft 0 in)
Length overall	5.79 m (19 ft 0 in)
Height overall: Mk I	1.22 m (4 ft 0 in)
Propeller diameter	1.32 m (4 ft 4 in)

AREAS:
Wings, gross	4.90 m² (52.75 sq ft)

WEIGHTS AND LOADINGS:
Weight empty	283 kg (625 lb)
Baggage capacity	22.6 kg (50 lb)
Max T-O weight	522 kg (1,150 lb)
Max wing loading	106.53 kg/m² (21.80 lb/sq ft)

PERFORMANCE:
Max level speed at 2,285 m (7,500 ft), 75% power	143 knots (266 km/h; 165 mph)
Cruising speed at S/L	104-146 knots (193-270 km/h; 120-168 mph)
Stalling speed	42 knots (78 km/h; 48 mph)
Max rate of climb at S/L	259 m (850 ft)/min
Service ceiling	5,640 m (18,500 ft)
T-O run	213 m (700 ft)
Range	434 nm (804 km; 500 miles)
Endurance	4 h 15 min
g limits	+4.4/−2

VOLMER

VOLMER AIRCRAFT
Box 5222, Glendale, California 91201
Telephone: 1 (818) 247 8718
PRESIDENT: Volmer Jensen

Other current products include foot-launched VJ-23E Swingwing and VJ-24W Sunfun (see Microlights tables in 1992-93 *Jane's*).

VOLMER VJ-22 SPORTSMAN

TYPE: Side by side two-seat, dual-control homebuilt amphibian.
PROGRAMME: First flight of prototype 22 December 1958, plans available.
CUSTOMERS: Total of 889 sets of plans sold by January 1993, and more than 100 Sportsman amphibians flying. Some have tractor propellers, but this modification is not recommended by Mr Jensen.
COSTS: Plans: $250.
DESIGN FEATURES: Wings from standard Aeronca Chief or Champion.
STRUCTURE: Wings have wooden spars, light alloy ribs and fabric covering, and carry stabilising floats under the tips. Plans of specially designed wing, with wooden ribs and spars, available. Flying-boat hull of wooden construction, coated with glassfibre. Steel tube tail unit, fabric covered.
LANDING GEAR: Retractable tailwheel type. Water rudder.
POWER PLANT: One 63.5 kW (85 hp) Teledyne Continental C85, 67 kW (90 hp) or 74.5 kW (100 hp) Teledyne Continental O-200-B. Fuel capacity 76 litres (20 US gallons; 16.7 Imp gallons).

DIMENSIONS, EXTERNAL:
Wing span	11.12 m (36 ft 6 in)
Wing aspect ratio	7.61
Length overall	7.32 m (24 ft 0 in)
Height overall	2.44 m (8 ft 0 in)

AREAS:
Wings, gross	16.3 m² (175.0 sq ft)

Volmer VJ-22 Sportsman two-seat homebuilt amphibian

WEIGHTS AND LOADINGS (63.5 kW; 85 hp):
Weight empty	454 kg (1,000 lb)
Max T-O weight	680 kg (1,500 lb)
Max wing loading	41.85 kg/m² (8.57 lb/sq ft)
Max power loading	10.71 kg/kW (17.65 lb/hp)

PERFORMANCE (63.5 kW; 85 hp, at max T-O weight):
Max level speed at S/L	83 knots (153 km/h; 95 mph)
Max cruising speed	74 knots (137 km/h; 85 mph)
Stalling speed	39 knots (72 km/h; 45 mph)
Max rate of climb at S/L	183 m (600 ft)/min
Service ceiling	3,960 m (13,000 ft)
Range with max fuel, no reserves	260 nm (480 km; 300 miles)

VOLPAR

VOLPAR AIRCRAFT CORPORATION
7701 Woodley Avenue, Van Nuys, California 91406
Telephone: 1 (818) 994 5023
Fax: 1 (818) 988 8324
Telex: 651482 VOLPAR B VAN
PRESIDENT: Andy Savva
VICE-PRESIDENTS:
  Robert C. Dunigan (Maintenance and Engineering)
  Joel Fogelson (Marketing)
  Frank V. Nixon (Development)
Volpar Aircraft Corporation acquired by Gaylord Holdings of Switzerland in early 1990. Following Chapter 11 bankruptcy protection, a $7 million cash injection from a foreign investor has enabled work on Falcon PW300-F20 programme to resume. Further details of programmes can be found in *Jane's Civil and Military Aircraft Upgrades*.

VOLPAR FALCON PW300-F20

TYPE: Re-engined Dassault Falcon 20.
PROGRAMME: Announced at Paris Air Show 1989. Modified Falcons will meet FAR Pt 36 Stage 3 noise requirements and cruise 50 knots (93 km/h; 58 mph) faster at 12,500 m (41,000 ft); estimated max range with NBAA IFR reserves 2,600 nm (4,818 km; 2,994 miles). Baseline testing with CF700-2D2 engines completed at Van Nuys August 1990; prototype then grounded for PW305 installation (first flight 12 February 1991). To be marketed by Advanced Falcon Aircraft Partnership of Greenwich, Connecticut.
COSTS: $3.8 million, including thrust reversers and new paint.
DESIGN FEATURES: Replaces GE CF700 engines on Falcon 20s with 23.24 kN (5,225 lb st) P&WC PW305 turbofans with variable inlet guide vanes for improved high altitude performance; Rohr Industries thrust reversers standard.

VOLPAR T-33V

TYPE: Modernised Lockheed T-33.
PROGRAMME: Upgraded and re-engined Lockheed T-33, being developed during 1990 in collaboration with William F. Chana Associates; Allison J33 would be replaced by P&WC PW300 turbofan, flat rated at 21.13 kN (4,750 lb st); aircraft weight reduced by 499 kg (1,000 lb) and fuel consumption reduced by up to two-thirds.

Volpar estimated market for 250 modified T-33Vs and would also supply modification kits; about 1,000 T-33s thought to be in service worldwide.

VOUGHT

VOUGHT AIRCRAFT COMPANY

9314 West Jefferson Boulevard, PO Box 655907, Dallas, Texas 75265-5907
Telephone: 1 (214) 266 2543
Fax: 1 (214) 266 4982
PRESIDENT AND CEO: Gordon L. Williams
VICE-PRESIDENT, CORPORATE COMMUNICATIONS: Lynn J. Farris

Original Lewis & Vought Corporation founded 1917; became Chance Vought Corporation, 31 December 1960; merged with Ling-Temco Electronics, 31 August 1961 to form Ling-Temco-Vought Inc (later The LTV Corporation); following file for bankruptcy (Chapter 11) in July 1986, LTV aerospace/defence operations reorganised into two groups 29 September 1986, as LTV Aircraft Products Group and LTV Missiles and Electronics Group; latter included Missiles Division at Grand Prairie, Texas, AM General Division at South Bend, Indiana, and Sierra Research Division at Buffalo, New York. AM General Division and Sierra Research Division subsequently sold, and LTV Aircraft Products Group redesignated Aircraft Division.

LTV Corporation announced plans in May 1991 to dispose of Aircraft and Missiles Division belonging to LTV Aerospace and Defense Company to raise cash to help LTV emerge from Chapter 11 bankruptcy. Early in 1992, LTV reached agreement with Martin Marietta Corporation and Lockheed Corporation to sell the two divisions; this followed by series of bankruptcy court hearings and offers from other bidders including Thomson-CSF of France, The Carlyle Group (a Washington investment group) and Loral Corporation. Bankruptcy court approved $214 million sale of Aircraft Division to The Carlyle Group and its 49 per cent minority partner, Northrop Corporation, and sale of the Missiles Division for $261 million to Loral Corporation, 13 August 1992; sale finalised on 31 August, at which time Aircraft Division became Vought Aircraft Company and Missiles Division became Loral Vought Systems Corporation; now no formal relationship between Vought Aircraft Company and Loral Vought Systems Corporation.

Current output includes about one-third of B-2 airframe by weight, aft fuselage and tail surfaces of Boeing 747, tailplane of Boeing 767, tail surfaces of Boeing 757, engine nacelles and tail sections of McDonnell Douglas C-17A and engine nacelles for Canadair Challenger and Regional Jet; teaming agreement with American Eurocopter Corporation and LHTEC for possible production of Panther 800, a T800-powered version of AS 565 Panther for US Army; Vought and FMA of Argentina teamed to promote IA 63 Pampa 2000 for USAF/USN JPATS programme; Vought is partner in McDonnell Douglas consortium offering potential A-X design to US Navy (teaming announced 16 July 1991).

VOUGHT (LTV) VOUGHT A-7 CORSAIR II

Details of the YA-7F/A-7 Plus appeared in the 1991-92 *Jane's*; earlier versions are described in the 1979-80, 1983-84 and 1989-90 editions. Rebuilds and upgrades continue; see *Jane's Civil and Military Aircraft Upgrades*.

VOUGHT (FMA) PAMPA 2000

TYPE: Tandem-seat jet primary trainer ('missionised' version of FMA IA 63, which see, for JPATS competition).
PROGRAMME: Teaming for JPATS began with LTV/FMA agreement of May 1990 to offer Pampa in this USAF/USN programme; September 1991 agreement with AlliedSignal to maximise latter's content (Garrett TFE731-2 engine, Bendix/King avionics and AirResearch environmental control system); UNC Aviation Services joined team April 1992 as contractor logistics support partner for aircraft portion of programme; Loral Corporation became ground-based training partner May 1992. Vought using one prototype and one production IA 63 as JPATS development (see FMA entry in Argentine section); first flight in fully modified JPATS configuration 26 May 1993; second aircraft due to fly June 1993. If successful in competition, Vought will build whole Pampa airframe in its Dallas, Texas factory; US content would be more than 90 per cent.

VOUGHT (EUROCOPTER) PANTHER 800

TYPE: Re-engined Eurocopter AS 565 Panther.
PROGRAMME: Being developed for offer to US Army and ArNG as off-the-shelf replacement for UH-1 Iroquois. Vought teamed with American Eurocopter, LHTEC and IBM Corporation; potential market for 500-1,000; Vought would provided manufacturing and marketing capability in USA. Prototype (N6040T) began tethered hover trials March 1992 and made first flight 12 June 1992 at Eurocopter Grand Prairie, Texas.
POWER PLANT: Two LHTEC T800 turboshafts, with FADEC, each derated 657 kW (881 shp). Integral particle separator on each engine. Transmission ratings 1,030.5 kW (1,382 shp) max, 974.5 kW (1,307 shp) for 30 min, 909 kW (1,219 shp) max continuous for hover, and 854.5 kW (1,146 shp) max continuous for forward flight speeds above 75 knots (139 km/h; 86 mph). Main rotor rpm 365.
WEIGHTS AND LOADINGS:
  Operating weight empty   2,994 kg (6,600 lb)
  Max T-O weight   4,250 kg (9,370 lb)
PERFORMANCE (at max T-O weight):
  Max level speed at S/L, ISA
     143 knots (265 km/h; 165 mph)
  Hovering ceiling OGE, ISA + 10°C   1,980 m (6,500 ft)
  Max range at S/L, ISA, standard fuel
     420 nm (778 km; 483 miles)

WAG-AERO

WAG-AERO INC

PO Box 181, 1216 North Road, Lyons, Wisconsin 53148
Telephone: 1 (414) 763 9586
Fax: 1 (414) 763 7595
PRESIDENT: Richard H. Wagner
MARKETING SUPERVISOR: Mary Pat Henningfield

WAG-AERO SPORT TRAINER

TYPE: Four different modern versions of the Piper J-3.
PROGRAMME: Sport Trainer first flew 12 March 1975.
CURRENT VERSIONS: **Sport Trainer:** Basic two-seat sporting aircraft following original design, with wooden main spar and ribs, light alloy leading-edge and fabric covering. The fuselage and tail unit are of welded steel tube with fabric covering. Can be powered by any flat-four Continental, Franklin or Textron Lycoming engine of between 48.5 and 93 kW (65 and 125 hp). *Description applies to this version.*

Acro Trainer: Differs from standard version by having strengthened fuselage, shortened wings (8.23 m; 27 ft), modified lift struts, improved wing fittings and rib spacing, and new leading-edge.

Observer: Replica of L-4 military liaison aircraft.
Super Sport: Structural modifications to accept engines of up to 112 kW (150 hp), making it suitable for glider towing, bush operations, or operation as floatplane.
COSTS: Plans: reportedly $65; cost of building estimated at about $18,000.
DIMENSIONS, EXTERNAL:
  Wing span   10.73 m (35 ft 2½ in)
  Wing aspect ratio   6.94
  Length overall   6.82 m (22 ft 4½ in)
  Height overall   2.03 m (6 ft 8 in)
AREAS:
  Wings, gross   16.58 m² (178.5 sq ft)
WEIGHTS AND LOADINGS:
  Weight empty   327 kg (720 lb)
  Max T-O weight   635 kg (1,400 lb)
  Max wing loading   38.3 kg/m² (7.84 lb/sq ft)
PERFORMANCE (at max T-O weight):
  Max level speed at S/L   89 knots (164 km/h; 102 mph)
  Cruising speed   82 knots (151 km/h; 94 mph)
  Stalling speed   34 knots (63 km/h; 39 mph)
  Max rate of climb at S/L   149 m (490 ft)/min
  Service ceiling   over 3,660 m (12,000 ft)
  T-O run   114 m (375 ft)
  Range: at cruising speed with standard fuel (45.5 litres; 12 US gallons; 10 Imp gallons)
     191 nm (354 km; 220 miles)
  with auxiliary fuel (98.5 litres; 26 US gallons; 21.6 Imp gallons)
     395 nm (732 km; 455 miles)

WAG-AERO WAG-A-BOND

TYPE: Side by side two-seat homebuilt; replica of Piper PA-15 Vagabond.
PROGRAMME: Prototype Wag-A-Bond completed by Wag-Aero in May 1978.
DESIGN FEATURES: Two versions (**Classic** and **Traveler**). Traveler is modified and updated version of Vagabond with port and starboard doors, overhead skylight window, extended sleeping deck (conversion from aircraft to camper interior taking about two minutes and accommodating two persons), extended baggage area, engine of up to 85.7 kW (115 hp), and provision for full electrical system.
STRUCTURE: All-wood wing and aluminium alloy aileron structures. Welded steel tube and flat plate fuselage structure, and steel tube tail unit. Complete airframe is fabric covered.
LANDING GEAR: Non-retractable tailwheel type. Optional skis.
POWER PLANT: Traveler can be powered by a Textron Lycoming engine of 80.5-85.7 kW (108-115 hp). Classic can be powered by a Teledyne Continental engine of 48.5-74.5 kW (65-100 hp). Fuel capacity: Traveler 98.5 litres (26 US gallons; 21.6 Imp gallons), Classic 45.5 litres (12 US gallons; 10 Imp gallons).
DIMENSIONS, EXTERNAL:
  Wing span   8.32 m (29 ft 3½ in)
  Wing aspect ratio   5.82
  Length overall   5.66 m (18 ft 7 in)
  Height overall   1.83 m (6 ft 0 in)
AREAS:
  Wings, gross   13.70 m² (147.5 sq ft)

UK registered Wag-Aero Acro Trainer *(PFA)*

WEIGHTS AND LOADINGS:
Weight empty: Traveler 329 kg (725 lb)
Classic 290 kg (640 lb)
Baggage capacity:
Traveler 27 kg (60 lb)
Classic 18 kg (40 lb)
Max T-O weight: Traveler 658 kg (1,450 lb)
Classic 567 kg (1,250 lb)
Max wing loading: Traveler 47.99 kg/m² (9.83 lb/sq ft)
Classic 41.35 kg/m² (8.47 lb/sq ft)
PERFORMANCE:
Max level speed: Traveler 118 knots (219km/h; 136 mph)
Classic 91 knots (169 km/h; 105 mph)
Cruising speed: Traveler 108 knots (200 km/h; 124 mph)
Classic 83 knots (153 km/h; 95 mph)
Stalling speed: Traveler, Classic 39 knots (73 km/h; 45 mph)

Max rate of climb at S/L: Traveler 259 m (850 ft)/min
Classic 190 m (625 ft)/min

WAG-AERO 2+2 SPORTSMAN
TYPE: Four-seat homebuilt.
PROGRAMME: Plans and material kits available.
COSTS: Plans: reportedly $89; cost of building estimated at about $22,000.
DESIGN FEATURES: Based on Piper PA-14 Family Cruiser. True four-seater, with option of hinged rear fuselage decking to provide access to baggage and rear seat areas. The rear seat itself can be removed for cargo or litter.
FLYING CONTROLS: Include upper and lower spoilers.
STRUCTURE: Similar construction to Wag-A-Bond, with glass-fibre tips. Alternatively, drawings and materials provided to modify standard PA-12, PA-14 or PA-18 wings. Pre-welded fuselage structure also available.
POWER PLANT: Engine of 93-149 kW (125-200 hp). Usable fuel capacity 148 litres (39 US gallons; 32.5 Imp gallons).

DIMENSIONS, EXTERNAL:
Wing span 10.90 m (35 ft 9 in)
Wing aspect ratio 7.34
Length overall 7.12 m (23 ft 4½ in)
Height overall 2.02 m (6 ft 7½ in)
AREAS:
Wings, gross 16.18 m² (174.12 sq ft)
WEIGHTS AND LOADINGS:
Weight empty 490 kg (1,080 lb)
Max T-O weight 998 kg (2,200 lb)
Max wing loading 61.67 kg/m² (12.63 lb/sq ft)
PERFORMANCE (typical; actual data depend on engine fitted):
Max level speed 112 knots (207 km/h; 129 mph)
Cruising speed 108 knots (200 km/h; 124 mph)
Stalling speed 33 knots (62 km/h; 38 mph)
Max rate of climb at S/L 244 m (800 ft)/min
Service ceiling 4,510 m (14,800 ft)
Range at cruising speed 582 nm (1,078 km; 670 miles)

WHITE LIGHTNING
WHITE LIGHTNING AIRCRAFT CORPORATION
Box 497, Walterboro, South Carolina 29488-0497
Telephone: 1 (803) 549 1800
PRESIDENT: Howell C. Jones Jr

WHITE LIGHTNING AIRCRAFT WHITE LIGHTNING
TYPE: Four-seat homebuilt.
PROGRAMME: Prototype first flew 8 March 1986. In same year it established several world speed records in Classes C1b and C1c. By 15 January 1992, 39 assembled aircraft and 32 kits had been delivered; 13 of latter had by then been flown. Construction from kit takes about 1,000 working hours for the first-time builder.
COSTS: Kit: $33,000. Many optional kits, including electrical for $1,110.
DESIGN FEATURES: Options include electrical, various engine installation, light, strobe light and interior kits, plus other items. Wing section NACA 66_2-215.
STRUCTURE: Wings have graphite tubular 'wet' main spar, and glassfibre/epoxy front and rear spars and ribs, all pre-cast in the lower glassfibre/epoxy skin. Glassfibre/epoxy fuselage moulded in upper and lower halves. Tail unit has spars and ribs of fin and tailplane pre-cast into one skin of each.
LANDING GEAR: Retractable tricycle type, with brakes.
POWER PLANT: Prototype has 156.6 kW (210 hp) Teledyne Continental IO-360-CB flat-six engine. Usable fuel capacity 257 litres (68 US gallons; 56.6 Imp gallons).

DIMENSIONS, EXTERNAL:
Wing span 8.43 m (27 ft 8 in)
Wing aspect ratio 8.60
Length overall 7.11 m (23 ft 4 in)
Height overall 2.18 m (7 ft 2 in)
Propeller diameter 1.85 m (6 ft 1 in)
AREAS:
Wings, gross 8.27 m² (89.0 sq ft)
WEIGHTS AND LOADINGS:
Weight empty 635 kg (1,400 lb)
Max baggage capacity 154 kg (340 lb)
Max T-O weight 1,088 kg (2,400 lb)
Max wing loading 131.7 kg/m² (26.97 lb/sq ft)
Max power loading 6.95 kg/kW (11.43 lb/hp)

White Lightning Aircraft White Lightning four-seat composites homebuilt *(Howard Levy)*

PERFORMANCE (at max T-O weight):
Max level speed at S/L 243 knots (450 km/h; 280 mph)
Max cruising speed at 2,440 m (8,000 ft) 230 knots (426 km/h; 265 mph)
Econ cruising speed at 3,350 m (11,000 ft) 221 knots (410 km/h; 255 mph)
Stalling speed: flaps up 78 knots (145 km/h; 90 mph)
flaps down 58 knots (108 km/h; 67 mph)
Max rate of climb at S/L 579 m (1,900 ft)/min
Service ceiling 7,010 m (23,000 ft)
T-O run, full flap 381 m (1,250 ft)
Landing run 397 m (1,300 ft)
Range with max fuel 1,389 nm (2,575 km; 1,600 miles)
Endurance 6 h 10 min
g limits +4.4/−2 utility

WIPAIRE
WIPAIRE INC
8520 River Road, Inver Grove Heights, Minnesota 55076
Telephone: 1 (612) 451 1205
Fax: 1 (612) 451 1786
CHAIRMAN: Robert D. Wiplinger
PRESIDENT: Robert L. Nelson
SALES AND MARKETING: James J. Jacobsen

Wipaire Inc is manufacturer of Wipline amphibious and seaplane floats. Company also modifies Cessna 185, 206 and 208, and de Havilland Canada DHC-2 Beaver and DHC-6 Twin Otter aircraft. It holds more than 32 STCs for modifications, including structural rebuilds, structural modifications, engine modifications and installations, and float conversions. **4000 Float** is certificated for the Cessna 185 and 206; **6000 Float** for de Havilland Canada DHC-2; **8000 Float** for Cessna Caravan; and **13000 Float** for de Havilland Canada DHC-6 Twin Otter.

Other conversions include fitting 224 kW (300 hp) Teledyne Continental IO-550D engine and McCauley three-blade propeller to Cessna A185E/F; fitting 224 kW (300 hp) IO-550F or TSIO-520-M and McCauley Black Mac three-blade propeller to Cessna 206 as **Boss I 206**, while also offering co-pilot's composite door, 46 cm (18 in) wing extensions (each), Monarch fuel tanks, custom upholstery and avionics, and other work that includes overhaul; modification of Beavers to **Super Beavers**, including 0.71 m (2 ft 4 in) rearward extension of cabin, addition of 0.85 m × 0.27 m (2 ft 9½ in × 10½ in) baggage door, two extra Panaview windows each side, tinted forward skylight windows, articulated seats with inertia-reel belts, forward- or rearward-facing three-person centre bench seat with underneath stowage and cabin soundproofing, plus optional electrically actuated flaps, customised instrument panel, Digiflow fuel-metering system, 3M Stormscope, S-Tech Series 50 autopilot with electric trim, altitude hold and flight director, Hartzell three-blade constant-speed propeller and more, plus refurbishment (most modifications also available to Mk III Turbo-Beaver). Seaplane kits (amphibious float installation optional) for Cessna 208I Caravan and de Havilland Canada DHC-6 Twin Otter include auxiliary finlets on horizontal tailplane outboard panels, and all necessary structural reinforcements.

Wipaire modified Cessna 208 Caravan with 8000 floats

UZBEKISTAN

CHKALOV

TASHKENTSKOYE APOiCh (Tashkent Aircraft Production Corporation named after V. P. Chkalov)
61 Vorovski Street, 700016 Tashkent
Telephone: 7 (3712) 32 11 67 or 33 65 32
Fax: 7 (3712) 68 03 18
Telex: 11475 TAPO SU

Founded in 1932, this large manufacturing centre had built approximately 1,100 aircraft by Autumn 1992, including Li-2s, Il-14s, An-8s, Ka-22s, An-12s, An-22s and Il-76s. It has repaired MiG and Sukhoi fighters and Tu-22M bombers. About 120 Il-76s have been exported. It is preparing to build Il-114 twin-turboprop transports at the rate of five to seven a month.

YUGOSLAVIA (FORMER)

New state of Yugoslavia proclaimed 27 April 1992, following secession of Bosnia-Hercegovina (which see for details of Soko company; but Soko aircraft described under this entry).

UTVA

UTVA—SOUR METALNE INDUSTRIJE, RO FABRIKA AVIONA
Jabučki Put bb, YU-26000 Pančevo
Telephone: 38 (13) 515383, 512584
Fax: 38 (13) 519859
Telex: 13250 UTVA YU
MANAGING DIRECTOR: Jovo Opsenica
PRODUCTION MANAGER: Dragan Nikodinović
MARKETING MANAGER: Radivoj Perišić

Utva (Sheldrake) Aircraft Industry formed at Zemun, 5 June 1937; to Pančevo 1939; post-war production of Trojka, 212, 213, Aero 3, Utva 56, 60, 65P and 66; currently 1,500 employees and 85,500 m² (920,320 sq ft) of covered floor area. Subcontract work included flaps and weapon pylons for Soko/Avioane Orao/IAR-93; centre and rear fuselage sections, tail surfaces and gun pod for Super Galeb; components for FLS Aerospace Optica; Boeing 737 rib assemblies; Boeing 747 assemblies; Boeing 757 wingtips and floor supports; tools for Israel Aircraft Industries, McDonnell Douglas and CIS aircraft industry; external trade prohibited by UN embargo 1992.

Following termination of aircraft production at Soko on 1 May 1992 (see Bosnia-Hercegovina), some machinery and jigs from G-4 Super Galeb and Gazelle production lines removed to 'new' Yugoslavia; near-term prospects for return to production at Utva extremely remote, but these aircraft and their developments are listed under this entry for continuity, pending end to the civil war in former Yugoslavia. Utva primarily involved with aircraft maintenance and munitions production during civil war, development of Lasta being allowed to lapse.

UTVA-75

No Utva-75 light aircraft believed manufactured in recent years; Utva-75A and Utva-75AG11 agricultural sprayer last described in 1991-92 *Jane's*.

UTVA LASTA 1 (SWALLOW)

TYPE: Tandem two-seat, primary/basic trainer and light attack aircraft.
PROGRAMME: Designed by VTI (which see) for basic instrument, navigation and armament training; programme launched April 1983; construction of first prototype began July 1983; first flight (54001) 2 September 1985; second prototype (54002) flew July 1986; construction of pre-series production aircraft began May 1989, but ceased shortly before planned first flight in early 1992.
CURRENT VERSIONS: **Lasta 1:** Pre-series batch for Yugoslav federal air force (RV-PVO).
Lasta 2: Definitive production version; described separately.
CUSTOMERS: RV-PVO only; two prototypes. Six pre-series not built. Details in 1992-93 *Jane's*.

Aircraft of former Yugoslavia in new national markings include this Slovenian Utva-75

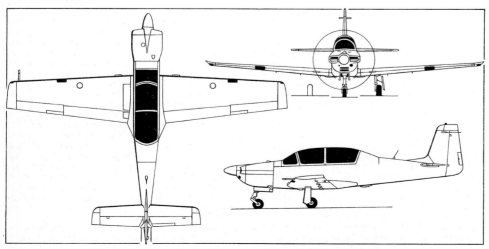

Utva Lasta 2 (intended production configuration) *(Jane's/Mike Keep)*

UTVA LASTA 2 (SWALLOW)

TYPE: Tandem-seat, primary/basic trainer and light attack aircraft.
PROGRAMME: Significantly redesigned version of Lasta 1 (see above); programme launched January 1989; construction of first prototype (54003) begun September 1990; first revealed in model form at 1991 Paris Air Show; first flight planned mid-1992 but not achieved; planned total two prototypes and one static test airframe.
CURRENT VERSIONS: **Lasta 2:** Compared with pre-series Lasta 1, has reduced weight, shorter fuselage, increased wing span with upturned tips, two-blade propeller, no aerodynamic balancing of rear control surfaces, no underfuselage strake, reduced ventral fin, trailing link undercarriage, twin doors for nosewheel bay and single-piece canopy.
CUSTOMERS: Yugoslav federal air force (RV-PVO): two prototypes and 50 production aircraft on order; deliveries suspended.
DESIGN FEATURES: Designed to meet FAR Pt 23; cost-efficient lead-in trainer to G-4 Super Galeb, with similar cockpit design; comparatively high wing loading with clear stall warning; fin and rudder optimised for high rudder efficiency when spinning. Leading-edge sweep 3° 10'; wing section NACA 64_2215 at root, NACA 64_2212 at tip; dihedral 4°; incidence 4°; twist 5°. Upturned wingtips.
FLYING CONTROLS: Mechanical control via pushrods; Frise ailerons with servo tabs; horn balanced elevators and rudder; small wedge-type stall strip on each wing; long dorsal fin and shallow ventral fin; no fences or vortillons; trim and servo tab in starboard aileron; servo tab in port aileron; trim tab in rudder and starboard elevator; all trim tabs electrically actuated; electrically operated single-slotted Fowler flaps with externally hinged axis.
STRUCTURE: Conventional light alloy construction; two-spar wing with machine spars and chemically milled stressed skin and integral fuel tanks; GFRP wingtips; fuselage reinforced by stringers and frames.
LANDING GEAR: Hydraulically retractable type of Prva Petoletka-Trstenik (PPT) design and manufacture; trailing link design; single wheels throughout; nosewheel retracts rearward (doors each side of nosewheel bay), main units inward; oleo-pneumatic shock absorbers; ±20° castoring nosewheel. Dunlop France tyres, size 15 × 6.00-6 (pressure 4.2 bars; 60.9 lb/sq in) on mainwheels and size 5.00-4 (pressure 3.5 bars; 50.75 lb/sq in) on nosewheel; PPT single-disc hydraulic brakes on each mainwheel; parking brake. Min ground turning radius 2.10 m (6 ft 11 in).
POWER PLANT: One 224 kW (300 hp) Textron Lycoming AEIO-540-L1B5D flat-six engine, driving Hartzell HC-C2YR-4CF/FC 8475-6 constant-speed, two-blade propeller. Two integral wing fuel tanks, each 88 litres (23.2 US gallons; 19.4 Imp gallons) usable; 5 litre (1.3 US gallon; 1.1 Imp gallon) collector tank in fuselage; total usable fuel 181 litres (47.8 US gallons; 39.8 Imp gallons); total fuel capacity 187 litres (49.4 US gallons; 41.1 Imp gallons). Gravity refuelling point on top surface of each wing. Oil capacity 15.14 litres (4.0 US gallons; 3.3 Imp gallons).
ACCOMMODATION: Tandem seats, with rear raised 76 mm (3 in); seats adjustable vertically. Single, framed canopy, hinged to starboard. One-piece windscreen hinges to starboard for access to instruments. Cockpits heated and ventilated. Baggage compartment aft of rear seat, volume 0.05 m³ (1.77 cu ft).
SYSTEMS: Dual hydraulic systems: No. 1 driven by 24V electric motor, flow 2.5 litres (0.66 US gallon; 0.55 Imp gallon)/min at 8 bars (116 lb/sq in), for landing gear, with auxiliary hand-pump in front cockpit and unpressurised tank; No. 2 hydraulic system, with separate tank, for brakes. 28V DC electrical system with 2kW alternator and 17Ah nickel-cadmium battery; two static inverters, total 50VA, provide 115V/26V, 400Hz AC power. Windscreen demisting and de-icing by hot air. Optional oxygen system.
AVIONICS: Collins VHF-22 VHF com, Collins VIR-32 VOR/ILS/marker, Collins DME-42, Collins ADF-462, SFIM CG-90 gyromagnetic compass and SFENA H-140 gyro horizon; standard blind flying instrumentation by Teleoptik and SATAIR in both cockpits. PK1 reflector gunsight.
ARMAMENT: Two underwing hardpoints for bombs of up to 120 kg, L-57-07 rocket pods (each seven 57 mm FFARs) or twin 7.62 mm machine-gun pods (500 rds/gun).

DIMENSIONS, EXTERNAL:
Wing span	8.89 m (29 ft 2 in)
Wing chord: at root	1.71 m (5 ft 7¼ in)
at tip	0.94 m (3 ft 1 in)
Wing aspect ratio	6.73
Length: overall	7.61 m (24 ft 11¾ in)
fuselage, excl propeller	6.61 m (21 ft 8¼ in)
Height overall: empty, equipped	3.10 m (10 ft 2 in)
max T-O weight	2.60 m (8 ft 6⅜ in)
Tailplane span	3.46 m (11 ft 4¼ in)
Wheel track	2.24 m (7 ft 4¼ in)
Wheelbase	1.70 m (5 ft 7 in)
Propeller diameter	1.98 m (6 ft 6 in)
Propeller ground clearance	0.27 m (10⅝ in)

AREAS:
Wings, gross	11.4 m² (122.7 sq ft)
Ailerons (total)	1.92 m² (20.67 sq ft)
Trailing-edge flaps (total)	2.10 m² (22.60 sq ft)
Winglets	0.40 m² (4.31 sq ft)
Fin (excl dorsal fin)	0.64 m² (6.89 sq ft)
Rudder (incl tab)	0.61 m² (6.57 sq ft)
Tailplane	1.15 m² (12.38 sq ft)
Elevators (incl tab)	0.82 m² (8.82 sq ft)

WEIGHTS AND LOADINGS:
Weight empty, equipped	881 kg (1,942 lb)
Max payload	252 kg (556 lb)
Max fuel weight, usable	130 kg (287 lb)
T-O weight, training mission	1,191 kg (2,626 lb)
Max T-O and landing weight	1,443 kg (3,181 lb)
Max zero-fuel weight	1,313 kg (2,895 lb)
Max wing loading:	
training mission	104.5 kg/m² (21.40 lb/sq ft)
armed mission	126.6 kg/m² (25.93 lb/sq ft)
Max power loading:	
training mission	5.32 kg/kW (8.75 lb/hp)
armed mission	6.44 kg/kW (10.60 lb/hp)

PERFORMANCE (at 1,191 kg; 2,626 lb):
Never-exceed speed (VNE)	264 knots (490 km/h; 304 mph)
Max level speed at S/L	178 knots (330 km/h; 205 mph)
Max cruising speed, 75% power at 1,830 m; (6,000 ft)	168 knots (312 km/h; 194 mph)
Econ cruising speed, 65% power at 1,830 m; (6,000 ft)	154 knots (285 km/h; 177 mph)
Stalling speed, engine idling:	
flaps up	61 knots (113 km/h; 71 mph)
flaps down	69 knots (128 km/h; 80 mph)
Max rate of climb at S/L	540 m (1,770 ft)/min
Service ceiling	5,400 m (17,725 ft)
T-O run	350 m (1,149 ft)
T-O to 15 m (50 ft)	600 m (1,969 ft)
Landing from 15 m (50 ft)	650 m (2,133 ft)
Landing run	350 m (1,149 ft)
Range with max fuel, 10% reserves	519 nm (962 km; 597 miles)

Model of Lasta 2 *(Paul Jackson)*

First Soko G-4 Super Galeb of the Myanmar Air Force

OPTICA and NAC 6 FIELDMASTER

Assembly lines were established by Utva for Optica and Fieldmaster (see under FLS Aerospace and Croplease in United Kingdom section). Five Fieldmasters said to be under construction in 1990; none completed.

(SOKO) G-4 SUPER GALEB (SEAGULL)

TYPE: Two-seat basic/advanced trainer and ground attack.
PROGRAMME: Replacement for G-2 Galeb and Lockheed T-33; designed by VTI (which see); programme launched October 1973; construction of first prototype begun May 1975; first flight (23004) 17 July 1978; second prototype (23005) 17 December 1979; third airframe for ground testing; first of six pre-production aircraft flew 17 December 1980; first full production, 1983. G-4M to have new avionics; planned first flight July 1992, but abandoned. Mostar factory abandoned 1 May 1992; some Super Galeb airframes left uncompleted; jigs transferred to Utva, but no evidence of re-started production.
CURRENT VERSIONS: **G-4 PPP**: First prototype and six pre-series aircraft (23601-23606); fixed tailplane with inset elevators and no anhedral.
 G-4: Second prototype and all full production aircraft; all-moving tailplane with anhedral. Retrofit planned with provision for AAMs (probably AA-8 and AA-11) and ASMs (probably AS-7 and AS-9), improved electronic equipment and better power plant de-icing. *Main description applies to G-4, except where indicated.*
 G-4M: Revised avionics and nav/attack systems, including gyro platform, IFF, flight data recorder, indigenous electronic sight. HUD, multi-function displays and wingtip missile rails. Payload increased by 405 kg (893 lb). Intended to undertake portion of weapon training syllabus previously performed by front-line aircraft.
CUSTOMERS: Yugoslavia; 150 required; issued to Vojno-Vazduhoplovna Akademija (Air Force Academy) at Zemunik (to Udbina 1991) for first 60 hours of pilot training up to streaming, advanced school at Pula (also moved 1991) for further 120 hours for interceptor student stream, advanced flying school at Titograd providing alternative 120 hour ground attack course, and Letece Zvezde (Flying Stars) aerobatic team of Academy (first public display 20 May 1990). Units of 'new' Yugoslav air force (formed 19 May 1992) include 249 'Cobras' Squadron of 105 Fighter-Bomber Regiment. Six handed over to Myanmar (Burma) from January 1991; training in Yugoslavia early 1991 before delivery; second six to Myanmar early 1992.
DESIGN FEATURES: Optimised for easy transition of pupils to more advanced combat aircraft; reduces training costs; good manoeuvring characteristics at high Mach numbers and high altitude. Sweptback wing, 26° at leading-edge, 45° at root, 22° at quarter-chord; wing section NACA 64A211; dihedral 1°; no incidence or twist. Sweptback vertical fin; all-moving horizontal tail surfaces have 10° anhedral for optimum rudder efficiency when spinning.
FLYING CONTROLS: Mechanical control with non-linear transmission; dual hydraulic fully powered controls for tailplane and ailerons; directional control fully mechanical with direct linkage. Ailerons, without aerodynamic balance (except G-4 PPP); elevator with internal aerodynamic balance on G-4 PPP; rudder has aerodynamic balance and trim tab. Small wedge-type stall strips on wings; fences at two-thirds of wing semi-span; single ventral and dorsal fins; directional trim by electromechanical tab in rudder; tab in elevator on G-4 PPP. Artificial feel provided by loading simulators for longitudinal and lateral planes (built-in spring for additional stick loading in longitudinal plane on G-4 PPP). Door-type airbrake under rear fuselage. Single-slotted trailing-edge flaps (double slotted on G-4 PPP) operated by electrically controlled hydraulic actuators.
STRUCTURE: Single-piece, two-spar, all-metal wing with integrally machined skin panels inboard and chemically milled skin outboard; shallow boundary layer fence above each wing, forward of inboard end of aileron; wings attached to fuselage at six points; entire trailing-edge of conventional all-metal sealed ailerons and flaps; integral fuel tank in wingroot. All-metal, semi-monocoque fuselage; rear detachable for engine access. Optional composites tail unit.
LANDING GEAR: Prva Petoletka-Trstenik (PPT) hydraulically retractable tricycle type, with single wheel on each unit. Nosewheel retracts forward, main units inward into wings and fuselage; max steering angle ±90°. Oleo-pneumatic shock absorber in each leg. Hydraulically steerable nose unit optional (not on current aircraft). Trailing link main units. Mainwheels fitted with Miloje Zakić tyres, size 615 × 225-10PR, pressure 4.5 bars (65 lb/sq in), and PPT hydraulic brakes. Nosewheel has tyre size 6.50-5.5 TC, pressure 4.2 bars (61 lb/sq in). Brake parachute container at base of rudder. Provision for attaching two assisted take-off rockets under centre-fuselage. Operable from grass at normal T-O weight. Min ground turning radius 4.70 m (15 ft 5 in).
POWER PLANT: One licence-built Rolls-Royce Viper Mk 632-46 turbojet, rated at 17.8 kN (4,000 lb st). Fuel in two rubber fuselage tanks, each 506 litres (134 US gallons; 111.3 Imp gallons); collector tank, 57 litres (15.0 US gallons; 12.5 Imp gallons); and two integral wing tanks, each 348 litres (92 US gallons; 76.5 Imp gallons); total internal fuel 1,764 litres (466 US gallons; 388 Imp gallons). Provision for two underwing auxiliary tanks, on inboard pylons, total capacity 737 litres (195 US gallons; 162 Imp gallons); G-4M also has centreline tank of 449 litres (119 US gallons; 99 Imp gallons). Max fuel capacity: G-4, 2,501 litres (660 US gallons; 550 Imp gallons); G-4M, 2,950 litres (779 US gallons; 649 Imp gallons). Oil capacity 7.38 litres (1.95 US gallons; 1.6 Imp gallons). Gravity refuelling system standard; pressure system optional.
ACCOMMODATION: Crew of two in tandem on Martin-Baker zero/zero Mk 10Y ejection seats (zero height/90 knot Mk J8 optional, but not on current aircraft), with ejection through the individual sideways hinged (to starboard) canopy over each seat. Rear seat raised by 25 cm (10 in). Cockpit pressurised and air-conditioned.
SYSTEMS: Engine compressor bleed air used for pressurisation (max differential, 0.21 bar; 3.0 lb/sq in), air-conditioning, anti-g suit, windscreen anti-misting and anti-icing systems, and to pressurise fuel tanks. Engine face and (optional) intake lip de-icing. Dual hydraulic systems, pressure 206 bars (2,988 lb/sq in), for flying control servos, flap and airbrake actuators, landing gear retraction and extension, and wheel brakes. Hydraulic system flow rate 45 litres (12 US gallons; 10 Imp gallons)/min for main system, 16 litres (4.2 US gallons; 3.5 Imp gallons)/min for flight control system. Emergency Abex electric pump, providing 4.6 litres (1.2 US gallons; 1.0 Imp gallons)/min to one aileron. Electrical system supplied by 9kW 28V DC generator, with 36Ah nickel-cadmium battery for ground/emergency power and self-contained engine starting. Two static inverters, total output 600VA, provide 115V/26V 400Hz AC power; G-4M total output 900VA. Gaseous oxygen system, adequate for two crew for 2 h 30 min, comprises two bottles, each 7.3 litres (0.26 cu ft) at nominal 150 bars (2,175 lb/sq in).
AVIONICS: Dual controls and full blind-flying instrumentation in each cockpit. Standard nav/com equipment comprises EAS type ER4.671D or Rudi Čajavec RCE 163 Kondor

588　YUGOSLAVIA: AIRCRAFT—UTVA

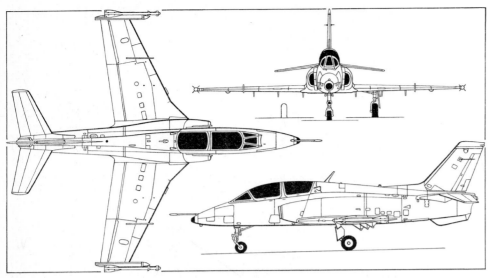

Soko G-4 Super Galeb with wingtip 'Aphid' AAMs (Rolls-Royce Viper Mk 632 turbojet) *(Jane's/Mike Keep)*

Soko G-4 Super Galeb jet training and light attack aircraft in blue/white/red colours of the Letece Zvezde aerobatic team

Model of Soko G-4M Super Galeb with underwing tanks and Maverick ASMs, plus wingtip AA-8 'Aphids' *(Paul Jackson)*

VHF com radio, GEC-Marconi AD 370B or Iskra VARK-01 ADF, Collins VIR-30 VOR/ILS, Iskra 75R4 VOR marker beacon receiver, Collins DME 40, TRT AHV-6 radio altimeter. SFIM CG-90 gyro-compass, SFENA H-140 artificial horizon, GEC-Marconi ISIS type D-282 weapon sight and Iskra SD-1 radar warning receiver. Optional M. Kobac UHF. G-4M has RCE 163 VHF, M. Kobac UHF, VARK-01, VIR-30, 75R4, DME-40 and AHV-6 plus IFF, SFIM SHARP gyro platform, Electronic Industry RPL flight data recorder, Zrak ENP-MG4 HUD, Rudi Čajavec ENS-MG4 sight, SD-1 RWR and optional chaff/flare dispensers. Development of a photo reconnaissance/infra-red linescan pod and night illumination system, and selection of an alternative off-the-shelf reconnaissance pod, deferred pending identification of customer requirement.

ARMAMENT: Removable ventral gun pod containing 23 mm GSh-23L twin-barrel rapid fire cannon with 200 rds; optional replacement on G-4M by wet pylon of 400 kg (882 lb) capacity. Two attachments under each wing: inboard pylons (all wet) stressed to 350 kg (772 lb) on G-4 and 500 kg (1,102 lb) on G-4M; outboard pylons to 250 kg (551 lb) on G-4 and 350 kg (772 lb) on G-4M. G-4M wingtip launch rails for R-60 (AA-8 'Aphid') IR AAMs; provision for Hughes AGM-65B Maverick ASM on outboard pylons. Total weapon load capability, with centreline gun pod, 1,280 kg (2,822 lb), or 1,800 kg (3,968 lb) for G-4M. In addition to standard high explosive bombs and napalm pods, typical Yugoslav stores include S-8-16 cluster bombs, each with eight 16 kg fragmentation munitions; KPT-150 expendable containers, each with up to 40 anti-personnel or 54 anti-tank bomblets; L-57-16MD pods, each with sixteen 57 mm rockets; L-128-04 pods, each with four 128 mm rockets; adaptors for twin 5 in HVAR rockets, single 57 mm VRZ-57 training rockets; SN-3-050 triple carriers for 50 kg bombs; SN-3-100 triple carriers for 100 kg bombs; KM-3 pods each containing a single 12.7 mm (0.50 in) gun; SAM Z-80 towed target system; and auxiliary fuel tanks on the inboard attachments.

DIMENSIONS, EXTERNAL:
Wing span: G-4	9.88 m (32 ft 5 in)
G-4M over launchers	10.05 m (33 ft 0 in)
G-4M over missiles	10.31 m (33 ft 9⅞ in)
Wing chord:	
at root, with apex	3.69 m (12 ft 1¼ in)
at root, without apex, at centreline	2.86 m (9 ft 4⅝ in)
at tip	1.20 m (3 ft 11¼ in)
Wing aspect ratio: G-4	4.73
Length: overall	12.25 m (40 ft 2¼ in)
without pitot	11.35 m (37 ft 2⅞ in)
fuselage	11.02 m (36 ft 1⅞ in)
Height overall	4.30 m (14 ft 1¼ in)
Tailplane span	3.97 m (13 ft 0¼ in)
Wheel track	3.49 m (11 ft 5½ in)
Wheelbase	4.15 m (13 ft 7½ in)

AREAS:
Wings, gross	19.5 m² (209.9 sq ft)
Ailerons (total)	1.358 m² (14.62 sq ft)
Trailing-edge flaps (total)	3.34 m² (35.95 sq ft)
Airbrake	0.438 m² (4.71 sq ft)
Fin	2.39 m² (25.73 sq ft)
Rudder	0.57 m² (6.14 sq ft)
Horizontal tail surfaces (total)	4.669 m² (50.26 sq ft)
Elevators, including tab (G-4 PPP only)	1.00 m² (10.76 sq ft)

WEIGHTS AND LOADINGS (G-4 PPP as for G-4, unless otherwise stated):
Weight empty, equipped: G-4	3,172 kg (6,993 lb)
G-4M clean	3,403 kg (7,502 lb)
G-4M with missile rails	3,435 kg (7,573 lb)
Max payload: G-4	2,053 kg (4,526 lb)
G-4M	2,458 kg (5,419 lb)
Max fuel weight: internal: G-4	1,307 kg (2,881 lb)
G-4M	1,376 kg (3,034 lb)
external: G-4	575 kg (1,268 lb)
G-4M	925 kg (2,039 lb)
T-O weight, training mission:	
G-4 PPP	4,639 kg (10,227 lb)
G-4	4,708 kg (10,379 lb)
G-4M	4,971 kg (10,959 lb)
Max T-O weight: G-4	6,300 kg (13,889 lb)
G-4M	6,400 kg (14,110 lb)
Max landing weight: G-4 PPP	4,639 kg (10,227 lb)
G-4	4,708 kg (10,379 lb)
G-4M	4,971 kg (10,959 lb)
Max wing loading: training mission:	
G-4 PPP	238.0 kg/m² (48.75 lb/sq ft)
G-4	241.0 kg/m² (49.36 lb/sq ft)
G-4M	255.0 kg/m² (52.23 lb/sq ft)
combat mission:	
G-4 PPP/G-4	323.0 kg/m² (66.15 lb/sq ft)
G-4M	328.0 kg/m² (67.18 lb/sq ft)
Max power loading: training mission:	
G-4 PPP	260.6 kg/kN (2.56 lb/lb st)
G-4	264.5 kg/kN (2.59 lb/lb st)
G-4M	279.3 kg/kN (2.74 lb/lb st)
combat mission:	
G-4 PPP/G-4	353.9 kg/kN (3.47 lb/lb st)
G-4M	359.6 kg/kN (3.52 lb/lb st)

PERFORMANCE (G-4 at 4,708 kg; 10,379 lb, G-4M at 4,971 kg; 10,959 lb):
Never-exceed speed (VNE) Mach 0.9
Max level speed:
  at 10,000 m (32,800 ft): G-4 Mach 0.81
  G-4M Mach 0.80
  at 4,000 m (13,120 ft):
    G-4 491 knots (910 km/h; 565 mph)
    G-4M 486 knots (900 km/h; 559 mph)
Max cruising speed at 6,000 m (19,700 ft) and 95% rpm
    456 knots (845 km/h; 525 mph)
Econ cruising speed at 6,000 m (19,700 ft)
    297 knots (550 km/h; 342 mph)
Stalling speed, power on:
  flaps up: G-4 119 knots (220 km/h; 137 mph)
  G-4M 121 knots (225 km/h; 140 mph)
  flaps down: G-4 97 knots (180 km/h; 112 mph)
  G-4M 100 knots (185 km/h; 115 mph)
Max rate of climb at S/L: G-4 1,860 m (6,100 ft)/min
  G-4M 1,800 m (5,900 ft)/min
Service ceiling: G-4 12,850 m (42,160 ft)
  G-4M 12,500 m (41,010 ft)
T-O run: G-4 572 m (1,877 ft)
  G-4M 600 m (1,969 ft)
T-O to 15 m (50 ft): G-4 900 m (2,953 ft)
  G-4M 950 m (3,117 ft)
Landing from 15 m (50 ft):
  without drag-chute: G-4 1,065 m (3,494 ft)
  G-4M 1,130 m (3,707 ft)
  with drag-chute: G-4 690 m (2,264 ft)
  G-4M 750 m (2,461 ft)
Landing run: G-4 815 m (2,674 ft)
  G-4M 860 m (2,822 ft)
Range at 11,000 m (36,090 ft), 10% reserves:
  max internal fuel:
    G-4 1,025 nm (1,900 km; 1,180 miles)
    G-4M 971 nm (1,800 km; 1,118 miles)
  max internal and external fuel:
    G-4 1,349 nm (2,500 km; 1,553 miles)
    G-4M 1,565 nm (2,900 km; 1,802 miles)
Range at 11,000 m (36,090 ft) with cannon and four 277 kg (610 lb) Hunting BL-755 CBUs:
    G-4 701 nm (1,300 km; 807 miles)
    G-4M, plus two AA-8s
    647 nm (1,200 km; 745 miles)

(SOKO) G-5

Development programme (undertaken by VTI which see) for advanced jet trainer, largely based on G-4, but with modern cockpit, similar to F-16; Garrett TFE731-4 turbofan; and designed with assistance of Lockheed Fort Worth (formerly General Dynamics); US participation withdrawn early 1992.

(SOKO) GAZELLE

TYPE: Light military utility helicopter.
PROGRAMME: Licenced manufacture by Soko since 1971 (see Eurocopter in International section); first delivery 1974; licence obtained for SA 342L1. Production terminated by dissolution of former Yugoslavia in May 1992; production assets to Utva, but no evidence of re-started manufacture.
CURRENT VERSIONS: **SA 341H Partizan**: Original version; out of production.
  SA 342L1 GAMA (GAzelle MAljutka): Anti-tank version with 9M14M Maljutka (AT-3 'Sagger') missiles; outrigger weapons pylons, trained in elevation by electric motor, including standard B-500 weapon racks with 14-inch hook spacing; normal armament, four 9M14M anti-tank missiles and two 9M32M Strela (SA-7 'Grail') IR homing missiles for anti-air/self-defence; alternatively, two L-57-07 launchers, each with seven 57 mm FFARs; APX M-334-25 gyro-stabilised weapon sight above cabin; PKV reflector sight for unguided ordnance (latter compatible with NVGs).
  SA 342L1 HERA (HElikopter RAdio): Reconnaissance version; APX M-334-25 sight with laser rangefinder, combined with Racal Doppler 80/RNS 252 navigation system, gyromagnetic compass and radio altimeter; data relayed directly to artillery ground post; photographic recce with Hasselblad Mk 70 system, comprising camera, autowinder and changeable lenses from 100 mm f3.5 to 350 mm f5.6; NBC contamination assessment, including fully automatic RDIV-2 equipment relaying radiation readings to ground station.
CUSTOMERS: 132 SA 341Hs produced, in addition to 21 from French production; total Gazelle production was planned to reach 300.

Soko-manufactured SA 342L GAMA Gazelle, armed with AT-3 'Sagger' and SA-7 'Grail' missiles

VAZDUHOPLOVNO TEHNICKI INSTITUT (VTI)

VTI RATNO VAZDUHOPLOVSTVO I PROTIV-VAZDUSNA OBRANA

Niška bb, Žarkovo, YU-11133 Beograd
Telephone: 38 (11) 510 735
Fax: 38 (11) 505 448
Telex: 11892
DIRECTOR: Maj-Gen Sava Pustinja, MSc

Aeronautical Technical Institute of Air Force & Anti-Aircraft Defence (VTI RV I PVO) founded 1946; responsible for design of G-2 Galeb, Jastreb, Orao (in collaboration with Romania), G-4 Super Galeb and Lasta prior to their manufacture by Soko and Utva. Institute has laboratories for aerodynamic testing, stress testing, avionics and flight controls; specific research includes high/low-speed aerodynamic investigation, static and fatigue testing, aircraft engine development, instruments, avionics and other airborne equipment, weapons, materials, fuels and lubricants. Most recent major military programme was Novi Avion. Efforts reportedly switched to civilian aircraft design in 1992.

NOVI AVION (NEW AIRCRAFT)

TYPE: Single-engined, single-seat multi-role combat aircraft and two-seat operational trainer.
PROGRAMME: Indigenous successor to MiG-21; design definition completed by VTI in early 1990, assisted by several Western aviation companies, notably Dassault of France; discussions with European and US firms on possible collaborative airframe development and possible use of Rolls-Royce, General Electric or Pratt & Whitney power plants; plans for manufacture by Soko overtaken by Bosnian declaration of independence, early 1992; progress halted by civil war and international embargo; long-term prospects uncertain; military requirements for aircraft reportedly cancelled.
COSTS: Estimated development expenditure $150-200 million per year throughout 1990s.
DESIGN FEATURES: Mid-positioned, delta wings with 46° leading-edge sweep and 3° trailing-edge forward sweep; canard foreplanes with 45° leading-edge sweep; single fin with inset rudder; no horizontal tail; 'Rafale-type' twin, kidney-shaped air intakes for high AoA performance; low-set cockpit. Rates of turn: sustained, 20°/s; instantaneous, 30°/s.
FLYING CONTROLS: Digital FBW control system; two-section elevons form entire rear surface of wing; movable canards.
STRUCTURE: Widespread use of carbon composites, Kevlar, aluminium and aluminium-lithium alloys, titanium and other advanced materials.
LANDING GEAR: Tricycle, with single wheels; all retracting forward into fuselage.
POWER PLANT: One engine in 100 kN (22,480 lb st) class. Internal fuel capacity 3,300 litres (871 US gallons; 726 Imp gallons); external tanks for further 5,800 litres (1,531 US gallons; 1,276 Imp gallons).

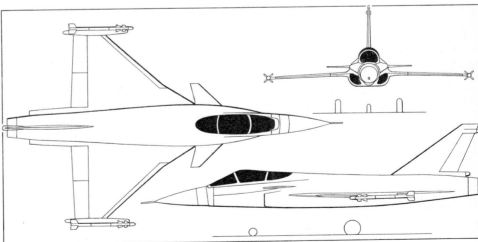

Vazduhoplovno Tehnicki Institut (VTI) design for Novi Avion *(Jane's/Mike Keep)*

Model of the proposed Novi Avion combat aircraft

ACCOMMODATION: Pilot only, under three-piece canopy; 17° downward view over nose.
AVIONICS: Nose-mounted radar; digital flight control system; multi-function nav/attack system; secure communications and advanced cockpit displays.
ARMAMENT: Eleven hardpoints, including two for wingtip infra-red AAMs and one for cannon up to 30 mm; maximum individual bomb size, 1,000 kg; precision guided bombs and missiles.

DIMENSIONS, EXTERNAL:
Wing span (without missiles)	8.00 m (26 ft 3 in)
Length overall	13.75 m (45 ft 1⅜ in)
Height overall	4.87 m (15 ft 11¾ in)
Foreplane span	3.55 m (11 ft 7¾ in)
Wheel track	2.42 m (7 ft 11¼ in)
Wheelbase	4.20 m (13 ft 9⅜ in)

AREAS:
Wings, gross	30.00 m² (322.9 sq ft)

WEIGHTS AND LOADINGS:
Weight empty, equipped	6,247 kg (13,772 lb)
Max fuel weight: internal	2,565 kg (5,655 lb)
external	4,500 kg (9,921 lb)
Max T-O weight	13,400 kg (29,542 lb)
Max wing loading	446.67 kg/m² (91.48 lb/sq ft)
Max power loading	134 kg/kN (1.31 lb/lb st)

PERFORMANCE:
Max level speed	Mach 1.85
Max rate of climb	15,000 m (49,200 ft)/min
T-O and landing run (close air support)	750 m (2,461 ft)
Tactical radius (close air support)	216-539 nm (400-1,000 km; 248.5-621 miles)
Ferry range	2,031 nm (3,765 km; 2,339 miles)

LIGHTER THAN AIR

> The inclusion of balloons produced primarily for sport or recreational flying has been discontinued. Last full coverage of such types appeared in the 1991-92 edition, with a partial update in the 1992-93 *Jane's*.

CANADA

21ST CENTURY
21ST CENTURY AIRSHIPS INC
110 Pony Drive, Newmarket, Ontario L3Y 4W2
Telephone: 1 (416) 898 6274
Fax: 1 (416) 898 7245
PRESIDENT AND CEO: Hokan Colting
   Company President is experienced pilot of hot-air balloons and airships and former head of Colting Balloons Ltd, formed in Ireland in 1975; has lived and worked in Canada since 1980.

21ST CENTURY AIRSHIP PROTOTYPES
TYPE: Non-rigid helium airships.
CURRENT VERSIONS: **No. 1:** Spherical envelope of 10.06 m (33 ft) diameter, powered by two 17.9 kW (24 hp) engines; max speed 16 knots (30 km/h; 18 mph); several hundred flights made over three-month period.
   No. 2: No details known at time of going to press.
   No. 3: Basically cigar-shaped, double-skinned envelope with 2.13 m (7 ft) long nose; outer (load-bearing) skin of lightweight polyester, inner skin of Kevlar reinforced Mylar film. Reached altitude of 2,270 m (7,450 ft) on 8 November 1992 on power of backup engine only, establishing nine world records in FAI Classes BA2-BA10.
   No. 4: Spherical envelope of 15.85 m (52 ft) diameter, designed to carry eight passengers and reach speed of 29 knots (54 km/h; 33 mph) on power of two 59.7 kW (80 hp) engines.
DESIGN FEATURES: Control system eliminates need for conventional fins and rudders, and reduces ground crew requirement to one or two persons; very manoeuvrable at low speeds.
FLYING CONTROLS: New (patent pending) method for steering and altitude control utilising variable and deflected thrust from two of three engines.
POWER PLANT: Three piston engines, third of which operates independently and is intended for backup only.

CHINA, PEOPLE'S REPUBLIC

HADG
HUAHANG AIRSHIP DEVELOPMENT GROUP
PO Box 307, Jingmen, Hubei 434535
Telephone: 86 (7267) 32525
Fax: 86 (7267) 32551
Telex: 4282
PRESIDENT: Li Hong Chou
INFORMATION: Dong Xue Ping
   Nearly a decade ago, China initiated design and construction of its own airships. Leading manufacturer in this field is HADG, which launched its first passenger-carrying non-rigid, the FK-4, on 10 August 1990. (For description and illustration, see 1992-93 and earlier editions.)
   To meet present domestic market needs, and to promote its own products, in 1992 HADG developed new, remotely controlled airship, the FK-6.

HADG FK-6
TYPE: Remotely controlled non-rigid airship.
PROGRAMME: Developed and first flown 1992.
DESIGN FEATURES: Expected to be used mainly for aerial photography and advertising. Non-rigid envelope and ballonets deflate automatically in emergency.
FLYING CONTROLS: Eight-channel remote controller actuates control surfaces, engine intakes and emergency system.
STRUCTURE: Envelope of lightweight, high-strength/low-permeability glue-coated fabric, with most seams heat-sealed for maximum strength. Gondola is single-keel structure with thin walls.

Huahang (HADG) FK-6 remotely controlled non-rigid airship

POWER PLANT: Two YH-40 piston engines (power not stated), with twin thrust-vectoring ducted propulsors.
DIMENSIONS, EXTERNAL:
Envelope: Length overall	14.00 m (45 ft 11¼ in)
Max diameter	4.50 m (14 ft 9¼ in)
Height overall, incl gondola	5.00 m (16 ft 5 in)

DIMENSIONS, INTERNAL:
Envelope volume	88.0 m³ (3,108 cu ft)

WEIGHTS AND LOADINGS:
Max payload	15 kg (33 lb)
Max T-O weight	108 kg (238 lb)

PERFORMANCE:
Max level speed	27 knots (50 km/h; 31 mph)
Max endurance	3 h

FRANCE

AERAZUR
AERAZUR (Member company of the Groupe Zodiac)
Division Equipements Aéronautiques
58 boulevard Galliéni, F-92137 Issy-les-Moulineaux Cedex
Telephone: 33 (1) 45 54 92 80
Fax: 33 (1) 45 57 99 85
Telex: 270 887 F
CEO: Jean Louis Gerondeau
INTERNATIONAL MARKETING MANAGER: Jean-Pierre Fetu
   Aerazur began building lighter than air craft before Second World War, and in 1960s manufactured world's largest non-rigid kite balloons (up to 15,000 m³; 529,720 cu ft), used for tests in the atmosphere of French nuclear weapons. It designs and manufactures envelopes for UK Skyship series, and is associated in designing envelope for Sentinel 5000 airship for US Navy.

GERMANY

GEFA-FLUG
GESELLSCHAFT ZUR ENTWICKLUNG UND FÖRDERUNG AEROSTATISCHER FLUGSYSTEME mbH
Weststrasse 14, D-5100 Aachen
Telephone: 49 (241) 874026/27
Fax: 49 (241) 875206
MANAGING DIRECTOR: Karl Ludwig Busemeyer
   Gefa-Flug established 1975 to operate advertising and passenger-carrying hot-air balloons; has R&D department to develop remotely controlled airships for manned and unmanned civil and environmental research applications such as aerial photogrammetry and pollution monitoring. Photogrammetry for archaeological research has been undertaken in Egypt, France, Germany, Greece, Israel, Oman, Pakistan, Syria, Turkey, the UK, Yugoslavia and elsewhere.
   Company currently operates more than half a dozen hot-air balloons, four hot-air airships and a number of remotely controlled aerostats, and has sold about 100 systems (mostly unmanned and remotely controlled) worldwide. Fleet includes two AS 80 GD hot-air airships (see 1991-92 *Jane's*), a modified version of the Thunder & Colt AS 80 Mk II. Future market seen as survey projects for environmental research; Gefa-Flug also developing manned family of civil hot-air aerostats, and developing scientific measuring equipment in partnership with German Mining Technology, a partly government-funded body.

WDL
WESTDEUTSCHE LUFTWERBUNG THEODOR WÜLLENKEMPER KG
(WDL Flugdienst GmbH and WDL Luftschiff GmbH)

Flughafen Essen-Mülheim, D-4330 Mülheim/Ruhr
Telephone: 49 (208) 378080
Fax: 49 (208) 3780833 (Management)
49 (208) 3780841 (Operations)
Telex: 856810
OWNER: Theodor Wüllenkemper

WDL resumed airship construction in 1987 with first batch of new design known as WDL 1B. In addition to airship activities, WDL operates six Fokker F27 transports on passenger/cargo flights within Europe. It is reportedly planning an 86 m (282 ft) long, 10,000 m³ (353,147 cu ft), 16-passenger airship to follow WDL 1B.

WDL 1B

TYPE: Passenger-carrying non-rigid airship.
PROGRAMME: First flight (D-LDFP *Asahi*) 30 August 1988; first delivery (to Mitsui, Japan) mid-September 1988.
CUSTOMERS: Four completed by early 1991 (latest information received), including two for Mitsui.
DESIGN FEATURES: World's largest fully certificated non-rigid; improved gondola compared with WDL 1; computer-controlled coloured advertising graphics and 9,000-bulb night sign on *Asahi*.
FLYING CONTROLS: Two ballonets (40 per cent air-filled) and 300 litres (79.25 US gallons; 66 Imp gallons) water ballast; ballonet valves actuated by digital computer which indicates gas temperature and pressure.
POWER PLANT: Two 157 kW (210 hp) Teledyne Continental IO-360-CD flat-six engines, each driving a two-blade propeller. Total fuel capacity (two tanks) 400 litres (105.7 US gallons; 88 Imp gallons).

WDL 1B seven-passenger non-rigid (two Teledyne Continental IO-360-CD flat-six engines)

ACCOMMODATION: Pilot and up to seven passengers in new-design gondola.
AVIONICS: Instrumentation improved compared with WDL 1.
DIMENSIONS, EXTERNAL:
Envelope: Length overall 59.90 m (196 ft 6¼ in)
Max diameter 16.40 m (53 ft 9¾ in)
Gondola: Length overall 7.60 m (24 ft 11¼ in)
Width: top 2.10 m (6 ft 10¾ in)
bottom 1.30 m (4 ft 3¼ in)
Height: excl landing gear 2.53 m (8 ft 3½ in)
incl landing gear 3.68 m (12 ft 1 in)
DIMENSIONS, INTERNAL:
Envelope volume 7,200 m³ (254,265 cu ft)
WEIGHTS AND LOADINGS:
Envelope weight 2,000 kg (4,409 lb)
Max payload 1,180 kg (2,601 lb)
Weight empty 5,100 kg (11,243 lb)
PERFORMANCE:
Never-exceed speed (VNE) 48 knots (90 km/h; 60 mph)
Max manoeuvring speed 35 knots (65 km/h; 40 mph)
Max endurance 22 h

ZEPPELIN
LUFTSCHIFFBAU ZEPPELIN GmbH

LZ-Gelande, Postfach 2540, D-7990 Friedrichshafen 1
Telephone: 49 (7541) 202-0
Fax: 49 (7541) 202 250
Telex: 734323 ZMET D
MANAGING DIRECTOR: Dipl Phys Heinrich Kollmann
CHIEF DESIGNER: Hans Hagenlocher

LZ N05

TYPE: Rigid helium airship demonstrator and pilot trainer.
PROGRAMME: Study group for revival of modern rigid airships formed 1989; concluded that combination of vectored thrust and new constructional approach offered best solution, resulting in NT (new technology) programme; 10 m (32 ft 10 in) long, remotely controlled proof-of-concept model tested extensively in 1991; design definition of N05 demonstrator completed late 1992; prototype go-ahead expected 1993, completion 1995. Intended to be precursor of product line of commercial airships for such applications as scientific research, environmental monitoring, TV missions and tourism.
DESIGN FEATURES: Internal 'prism' structure modified and optimised from Stuttgart University proposal of 1975 and British patent of 1924; multiple gas cells increase safety features; placement of vectored thrust units near CG and also at rear enhances manoeuvrability and control at low speeds, further improving safety and reducing ground crew requirements (and thereby operating costs).
STRUCTURE: Internal primary structure of aluminium tubular longerons and carbonfibre tubular frames, braced in Warren girder pattern, providing internal prism-shaped structure accommodating for less than 15 per cent of gross weight; this frame and pressurised hull are both load-carrying structures to enhance safety factor. Airship designed to remain airworthy in event of breakdown of differential pressure between envelope and atmosphere.
POWER PLANT: To be decided (piston, turboprop and diesel options under consideration).
ACCOMMODATION: Crew of two and six passengers.
DIMENSIONS, EXTERNAL:
Hull: Length overall 62.80 m (206 ft 0½ in)
Max diameter 12.90 m (42 ft 4 in)

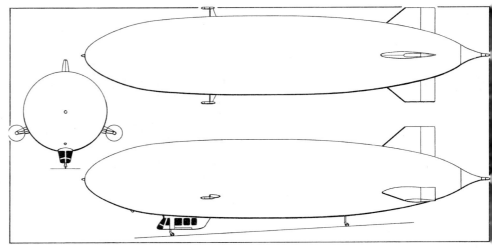

Zeppelin LZ N05 prototype of proposed new family of rigid airships (*Jane's/Mike Keep*)

DIMENSIONS, INTERNAL:
Hull volume 5,400 m³ (190,700 cu ft)
WEIGHTS AND LOADINGS:
Useful load 1,300 kg (2,866 lb)
Max T-O weight 5,380 kg (11,860 lb)
PERFORMANCE (estimated):
Max level speed 76 knots (140 km/h; 87 mph)
Cruising speed 62 knots (115 km/h; 71 mph)
Pressure ceiling 2,500 m (8,200 ft)
Endurance at 38 knots (70 km/h; 43 mph):
with max payload 18 h
with reduced payload 36 h

LZ N30

TYPE: Passenger carrying rigid helium airship.
PROGRAMME: Envisaged as passenger transport and/or surveillance platform.

DIMENSIONS, EXTERNAL:
Hull: Length overall 110.00 m (360 ft 10¾ in)
Max diameter 22.50 m (73 ft 10 in)
DIMENSIONS, INTERNAL:
Hull volume 30,000 m³ (1,059,440 cu ft)
WEIGHTS AND LOADINGS:
Useful load 15,000 kg (33,069 lb)
Max T-O weight 30,000 kg (66,138 lb)
PERFORMANCE (estimated):
Max level speed 76 knots (140 km/h; 87 mph)
Cruising speed 67 knots (125 km/h; 78 mph)
Pressure ceiling 3,000 m (9,850 ft)
Endurance at 38 knots (70 km/h; 43 mph):
with max payload 23 h
with reduced payload 82 h

RUSSIA

DKBA
DESIGN BUREAU AUTOMATICA

1 Letnaya Sreet, 141700 Dolgoprudny, Moscow Region
Telephone: 7 (095) 4088909
Fax: 7 (095) 4087511
CHIEF DESIGNER: Petr Petrovich Dementyev

DKBA DS-3

TYPE: Multi-purpose semi-rigid airship.
PROGRAMME: First revealed in form of model at 1989 Paris Air Show; was originally planned to make first flight in 1992, but has been delayed due to lack of funding; had not flown by mid-1993, but development continuing.
DESIGN FEATURES: Helium-filled semi-rigid, with cruciform tail surfaces.
STRUCTURE: Internally rigged hull of rubber coated Lavsan fabric.
POWER PLANT: Twin engines of unknown type; ducted propellers, one each side of gondola, can be vectored 120° upward and 120° downward.
ACCOMMODATION: Gondola accommodates crew of two and up to eight passengers; up to 3,000 kg (6,614 lb) in cargo

hold or up to 2,000 kg (4,409 lb) bulky loads on external sling.

DIMENSIONS, EXTERNAL:
Hull: Length overall 62.00 m (203 ft 5 in)
Max diameter 15.00 m (49 ft 2½ in)
DIMENSIONS, INTERNAL:
Hull volume 8,040 m³ (283,930 cu ft)
WEIGHTS AND LOADINGS:
Structural weight empty 4,500 kg (9,921 lb)
PERFORMANCE (estimated):
Max level speed 54 knots (100 km/h; 62 mph)
Max cruising speed 43 knots (80 km/h; 50 mph)
Max operating altitude 3,000 m (9,850 ft)
Range at cruising speed of 38 knots (70 km/h; 43 mph)
1,080 nm (2,000 km; 1,243 miles)
Endurance at cruising speed of 38 knots (70 km/h; 43 mph) 29 h

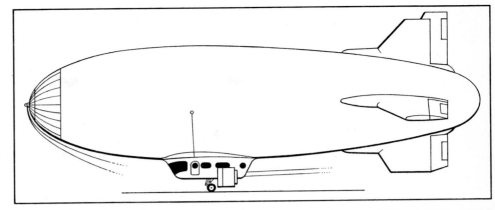

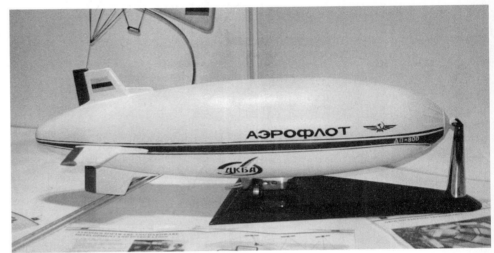

Preliminary drawing (*Jane's/Mike Keep*) and model photo (*Mark Lambert*) of the DKBA DS-3 passenger/cargo airship

THERMOPLANE
THERMOPLANE DESIGN BUREAU, MOSCOW AVIATION INSTITUTE

4 Volokolamskoe Highway, 125871 Moscow
Telephone: 7 (095) 1584127
Fax: 7 (095) 1582977
DIRECTOR OF MAI: Academician Yuri A. Ryzhov
CHIEF DIRECTOR, THERMOPLANE DESIGN BUREAU:
Yuri G. Ishkov

Thermoplane Design Bureau formed at MAI in 1980 to study requirement for large, unballasted airship, ecologically friendly, able to transport heavy or bulky cargoes in Siberian winter conditions, yet needing minimal ground support bases, hangars, mooring masts and personnel.

TDB ALA-40

TYPE: Proof-of-concept demonstrator for unique 'two-gas' airship.
PROGRAMME: Began 1985; originally funded by former Soviet government and first reported 1989, when preliminary design work completed; objective to develop ecologically clean heavy-lift airships to transport bulky cargo (eg for construction, logging or mining industries) in remote areas; three prototypes being built, with funding in 1992 being provided by joint venture complex of Russian timber, oil and other industries; buoyancy, static and dynamic ground testing of first prototype 1992 at Aviastar factory in Ulyanovsk; first flight expected Summer 1993, followed by 02 later same year and 03 in Spring 1994.
DESIGN FEATURES: Disc-shaped envelope with elliptical cross-section and unique two-gas lift system; disc configuration claimed to offer smaller size, lower weight, greater strength, and more even distribution of stress loads, than conventional cigar shape; Mi-2 helicopter fuselage adapted as gondola.
FLYING CONTROLS: Centre portion of envelope is occupied by helium- or hydrogen-filled spheres, remainder being filled with natural gas; hot air bled from main propulsion engines heats natural gas, as required, to augment basic lift provided by spheres; smaller engines fore and aft generate additional lift and control in conjunction with horizontal and vertical control surfaces.
STRUCTURE: Composites structure and skin (thermoplastics skin, based on Terlon, and carbonfibre tapes).
POWER PLANT: Two 112 kW (150 hp) piston engines for primary propulsion; propellers mounted on extension shafts from envelope/gondola junction; additional small engines on centreline (one forward, one aft) to supplement main lift and control systems.
ACCOMMODATION: Modified Mi-2 helicopter fuselage serves as gondola for test crew and equipment.

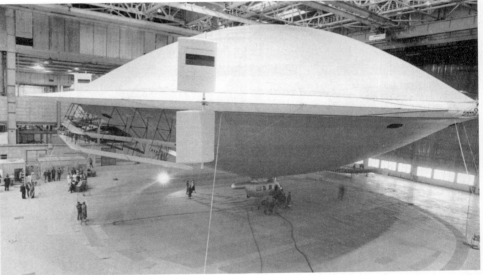

The ALA-40 proof-of-concept Thermoplane under test at Ulyanovsk in early 1992 (*Tass*)

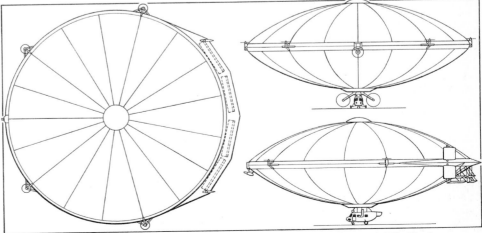

Three-view drawing of the ALA-40 proof-of-concept Thermoplane (*Jane's/Mike Keep*)

DIMENSIONS, EXTERNAL:
Envelope: Diameter 40.00 m (131 ft 3 in)
Max depth 16.00 m (52 ft 6 in)
Height: overall 19.514 m (64 ft 0¼ in)
to envelope c/l 11.514 m (37 ft 9¼ in)
to base of tail unit 7.662 m (25 ft 1¾ in)
to base of envelope 3.00 m (9 ft 10 in)
DIMENSIONS, INTERNAL:
Envelope volume: helium 5,800 m^3 (204,825 cu ft)
hot air 4,860 m^3 (171,629 cu ft)
total 10,660 m^3 (376,455 cu ft)
WEIGHTS AND LOADINGS:
Weight empty 6,150 kg (13,558 lb)
Fuel weight 200 kg (441 lb)
Max payload 2,150 kg (4,740 lb)
Max T-O weight 8,500 kg (18,739 lb)
PERFORMANCE (estimated):
Max level speed 59 knots (110 km/h; 68 mph)
Cruising speed 43 knots (80 km/h; 50 mph)
Pressure ceiling 2,000 m (6,560 ft)

TDB ALA-100 and ALA-300

TYPE: Projected intermediate size Thermoplanes.
ACCOMMODATION: Crew of 12 and 50 passengers in ALA-100; 16 crew and 80 passengers in ALA-300.
DIMENSIONS, EXTERNAL:
Length overall: 100 146.0 m (479 ft 0 in)
300 195.0 m (639 ft 9¼ in)
Envelope diameter: 100 138.0 m (452 ft 9 in)
300 184.0 m (603 ft 8 in)
Envelope max depth: 100 56.0 m (183 ft 8¾ in)
300 70.0 m (229 ft 8 in)
Cargo platform diameter: 100 25.0 m (82 ft 0¼ in)
300 30.0 m (98 ft 5 in)
Cargo platform height: 100 5.0 m (16 ft 4¾ in)
300 6.0 m (19 ft 8¼ in)
WEIGHTS AND LOADINGS:
Max payload: 100 100,000 kg (220,462 lb)
300 300,000 kg (661,387 lb)
PERFORMANCE (estimated):
Cruising speed: 100 32 knots (60 km/h; 37 mph)
300 38 knots (70 km/h; 43 mph)
Range: 100, 300 2,698 nm (5,000 km; 3,106 miles)

TDB ALA-600

TYPE: Projected 600-tonne Thermoplane.
PROGRAMME: First flight planned for late 1995, service entry by 2000.
ACCOMMODATION: Crew of four per shift (16 total), plus 150 passengers or equivalent cargo.
DIMENSIONS, EXTERNAL:
Length overall 210.0 m (688 ft 11¾ in)
Envelope: Diameter 198 m (649 ft 7¼ in)
Max depth 82.0 m (269 ft 2½ in)
Cargo platform: Diameter 35.0 m (114 ft 10 in)
Height 8 m (26 ft 3 in)
WEIGHTS AND LOADINGS:
Fuel weight 190,000 kg (418,878 lb)
Max payload 600,000 kg (1,322,770 lb)
Max T-O weight 1,200,000 kg (2,645,545 lb)
PERFORMANCE (estimated):
Max level speed 97-108 knots (180-200 km/h; 112-124 mph)
Cruising speed 76 knots (140 km/h; 87 mph)
Pressure ceiling 7,000 m (22,965 ft)
Range 2,698 nm (5,000 km; 3,106 miles)

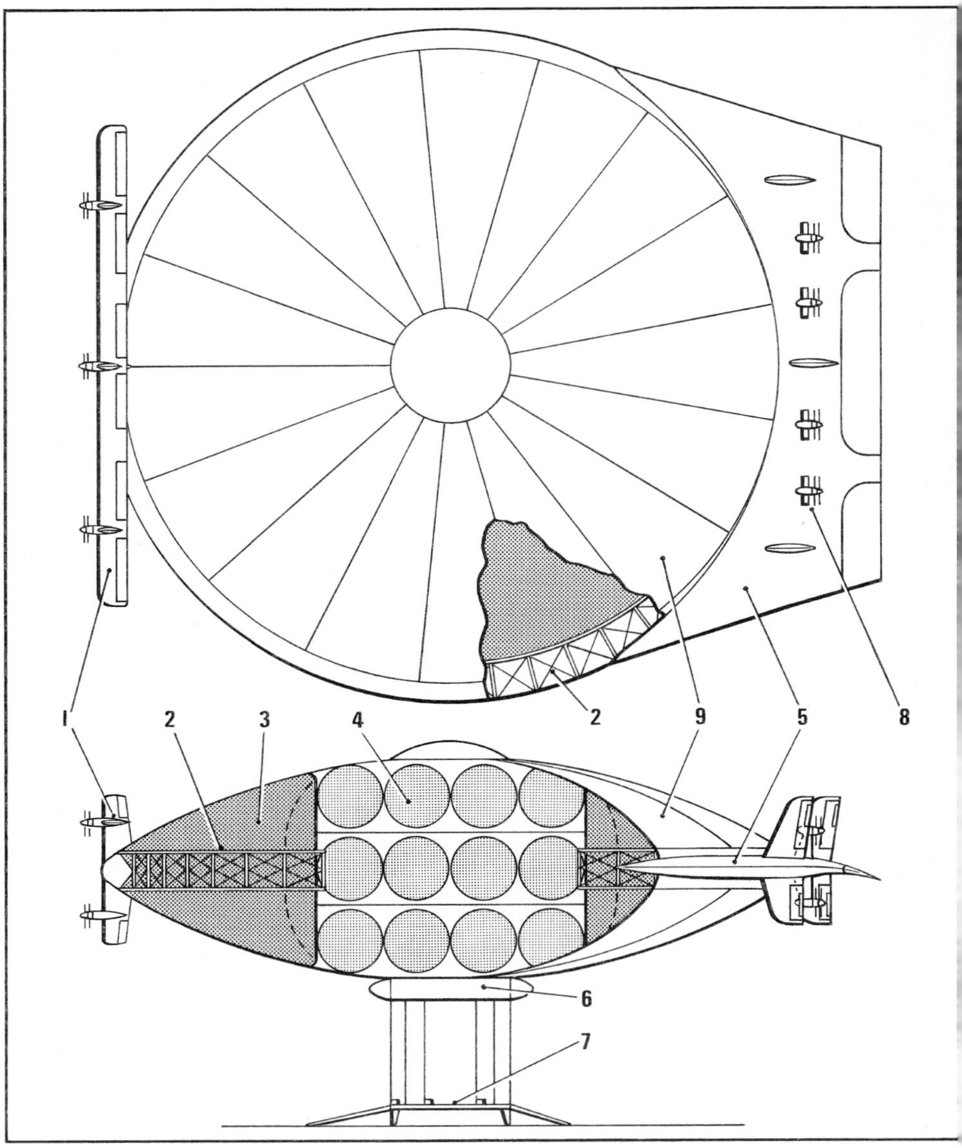

Structural details of the Thermoplane ALA-600:
1. Front vertical and horizontal stabilisers
2. Internal hull construction
3. Hot air/natural gas volume
4. Hydrogen or helium spheres
5. Rear stabiliser
6. Fuselage module
7. Cargo platform
8. Engines
9. Hull skin

UKRAINE

AEROS

NPP AEROSTATIC TECHNICS

31A Zamarstynivska Street, 290000 Lvov
Telephone: 7 (322) 524423
Fax: 7 (322) 340017
CHAIRMAN: Igor Pasternak
PRESIDENTS:
Gennady Verba
Evgeny Zlatkyn
BUSINESS DEVELOPMENT, ADVANCED PROGRAMMES:
Andrew Albeschenko
US OFFICE: Worldwide Aeros Corporation, 885 Second Avenue at One Dag Hammarskjold Plaza, 7th Floor, New York, NY 10017
Telephone: 1 (212) 702 4834
Fax: 1 (212) 207 9888
COMMUNICATIONS: Irina Svirid

Company specialises in developing and manufacturing airships and tethered arostats for such applications as atmospheric research, advertising and surveillance.

AEROS-50

TYPE: Non-rigid helium airship.
PROGRAMME: Prototype nearing completion April 1993; first flight scheduled for September 1993.
DESIGN FEATURES: Designed for pilot training, promotional and pleasure flights, ecological research and cartography. 12-gore envelope with 2.96 fineness ratio; cruciform tailfins, gas-filled from envelope; airship said to be capable of automatic docking, but details not provided.
FLYING CONTROLS: Single ballonet, air-filled via scoop (with diaphragm valve) aft of main engine propeller duct; valving of gas optionally automatic or manual; cable operated tail control surfaces; water ballast tank in gondola.
STRUCTURE: Envelope made of diagonally duplicated, rubberised cotton-based fabric; nose reinforced by 12 battens; tail control surfaces made of composites; gondola has tubular aluminium frame with composites skin, and is suspended from envelope by four catenary curtains; three aluminium skids under gondola.
POWER PLANT: One 33.6 kW (45 hp) CZ-400 piston engine, driving a two-blade fixed-pitch ducted pusher propeller. Fuel capacity 25 litres (6.6 US gallons; 5.5 Imp gallons).
DIMENSIONS, EXTERNAL:
Length overall 20.80 m (68 ft 3 in)
Width overall 8.00 m (26 ft 3 in)
Height overall 9.10 m (29 ft 10¼ in)
Envelope: Length 20.70 m (67 ft 11 in)
Max diameter 7.00 m (23 ft 0 in)
Gondola: Length 2.51 m (8 ft 2¾ in)
Max width 1.20 m (3 ft 11¼ in)
Max height 1.50 m (4 ft 11 in)
Propeller diameter 1.20 m (3 ft 11¼ in)
DIMENSIONS, INTERNAL:
Envelope volume 497.0 m^3 (17,551 cu ft)
Ballonet volume 70.0 m^3 (2,472 cu ft)
AREAS:
Envelope surface, total 380.5 m^2 (4,096 sq ft)
Tail surfaces (four, total) 12.8 m^2 (137.8 sq ft)
WEIGHTS AND LOADINGS: No details received
PERFORMANCE:
Max level speed 38 knots (70 km/h; 43 mph)
Econ cruising speed 27 knots (50 km/h; 31 mph)
Pressure ceiling 620 m (2,035 ft)
Max rate of climb 600 m (1,970 ft)/min
Max rate of descent 300 m (985 ft)/min
Range at econ cruising speed, 20% reserves 216 nm (400 km; 248 miles)
Max endurance 10 h

AEROS-500

TYPE: Projected 20-passenger rigid airship.
PROGRAMME: Design development under way 1993 for completion mid-1994; planned to form production basis for manufacture in California, USA, by end 1994.

DESIGN FEATURES: Designed for all-weather operation as passenger transport and ecological laboratory with high safety factor; lifting gas is helium/hydrogen mix; hull is of elliptical/paraboloid shape with fineness ratio 3.0; skin anti-icing by warm engine bleed air; tail-fins in X configuration; automatic T-O and landing without ground crew claimed, but no details given.

FLYING CONTROLS: Hull incorporates two ballonets (one forward, one aft); rudders located in slipstream of main propeller.

STRUCTURE: Hull has four-longeron/eight-rib metal frame and 0.15 mm duralumin sheet skin; inner gas envelope comprises multiple polythene cells, with provision for longitudinal warm-air ducts between these and metal skin; gondola/passenger cabin is of composites construction.

POWER PLANT: Main propulsion from 895 kW (1,200 hp) piston engine driving rear mounted, reversible-pitch ducted pusher propeller; laterally mounted 373 kW (500 hp) piston engine driving 360° vectorable ducted tractor propeller on each side on equatorial line of 'midships section.

ACCOMMODATION: Flight crew of two, plus up to four cabin attendants; max seating for 20 passengers; provision for in-flight refreshment and film projection.

DIMENSIONS, EXTERNAL:
Hull: Length 54.0 m (177 ft 2 in)
Max diameter 18.0 m (59 ft 0¾ in)
Height overall 23.6 m (77 ft 5 in)
Width over tail-fins 22.0 m (72 ft 2 in)
DIMENSIONS, INTERNAL:
Hull volume 8,500 m³ (300,175 cu ft)
Ballonet volume (two, total) 4,600 m³ (162,447 cu ft)
WEIGHTS AND LOADINGS: No details received
PERFORMANCE (estimated):
Max level speed 108 knots (200 km/h; 124 mph)
Cruising speed 65 knots (120 km/h; 75 mph)
Pressure ceiling 4,500 m (14,760 ft)
Max range 809 nm (1,500 km; 932 miles)
Endurance 18 h

UNITED KINGDOM

AAC
ADVANCED AIRSHIP CORPORATION LTD
Airship Facility, Jurby, Isle of Man

AAC went into receivership on 1 February 1993; offered for sale March 1993.

AAC ANR
ANR prototype was about 90 per cent complete in mid-1991 (envelope, tail-fins and part of gondola) when lack of continued funding caused work to be suspended. Description and illustration can be found in 1992-93 *Jane's*.

AHA
ADVANCED HYBRID AIRCRAFT LTD
Gorse Bank, Brookhill Road, Ramsey, Isle of Man
Telephone: 44 (624) 812023
Fax: 44 (624) 812023
MANAGER: Bruce N. Blake

AHA HORNET LV

TYPE: Recreational hybrid buoyant aircraft (50 per cent buoyant).

PROGRAMME: Design of Hornet LV (leisure variant) initiated October 1991; next stage will be to develop prototype (not for certification) and undertake six-month US demonstration tour, followed by set-up of production lines in USA and UK; initial work will concentrate on kits or fully assembled examples of single-seat sports version. This would then be followed by two-seat LV, and later by certificated two-or-more-seat AW (aerial work) version. Funding being sought in UK and Australia early 1993.

CUSTOMERS: Aimed specifically at US sport flying market; response to survey among EAA members expected to encourage funding for programme, leading to sales of up to 100 LVs in USA during first five years. Market interest also received from Denmark, France, Germany, India, Italy and UK.

DESIGN FEATURES: Single-seat gondola/cabin is essentially microlight 'podule' with fixed tricycle landing gear. Hornet can be taxied like a microlight, mast-moored by one person, and remain on ground when vacated by pilot. Some 67.45 m² (726 sq ft) of advertising area (10.06 × 3.35 m; 33 × 11 ft panel each side) is available on envelope.

POWER PLANT: Two 18 kW (24 hp) König SC 430 three-cylinder radial engines, mounted on tilting stub-wings that can be vectored upward 30° for climb.

DIMENSIONS, EXTERNAL:
Envelope: Length overall 15.24 m (50 ft 0 in)
Max diameter 3.81 m (12 ft 6 in)
Height overall (incl gondola) 4.88 m (16 ft 0 in)
Wing span 6.86 m (22 ft 6 in)
Wheel track 2.13 m (7 ft 0 in)
Wheelbase 2.74 m (9 ft 0 in)
Propeller diameter 1.37 m (4 ft 6 in)
Distance between propeller centres 4.95 m (16 ft 3 in)

DIMENSIONS, INTERNAL:
Envelope volume 121.8 m³ (4,300 cu ft)
Ballonet volume (two, total) 12.18 m³ (430 cu ft)
AREAS:
Envelope surface, total 145.86 m² (1,570 sq ft)
WEIGHTS AND LOADINGS:
Weight empty 152.4 kg (336 lb)
Weight empty at mast (heaviness) 30 kg (66 lb)
Max normal heaviness 122.5 kg (270 lb)
Gross aerostatic lift at 5% ballonet inflation 122.5 kg (270 lb)
Disposable load 92.5 kg (204 lb)
Max T-O weight 250 kg (540 lb)
PERFORMANCE (estimated):
Max level speed 51 knots (95 km/h; 59 mph)
Min flying speed 23 knots (43 km/h; 27 mph)
Max rate of climb at S/L, wings vectored 30° 610 m (2,000 ft)/min

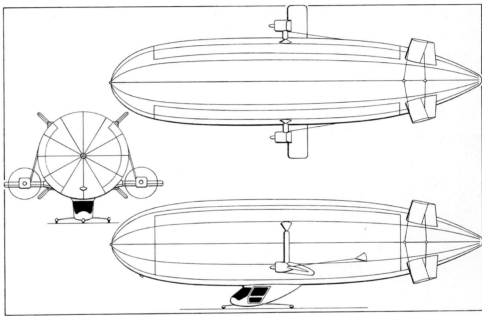

General arrangement of the AHA single-seat Hornet LV (*Jane's*/Mike Keep)

CAMERON
CAMERON BALLOONS LTD
St John's Street, Bedminster, Bristol BS3 4NH
Telephone: 44 (272) 637216
Fax: 44 (272) 661168
Telex: 444825 GASBAG G
MANAGING DIRECTOR:
D. A. Cameron, BSc (Aero Eng), MIE, MRAeS
SALES DIRECTOR: Philip Dunnington, BA

Cameron Balloons, which celebrated its 21st birthday in 1991, holds CAA, FAA, French CNT and German Musterzulassungsschein type certificates for its balloons. It is now world's largest manufacturer of special-shape and conventional hot-air balloons, and by 1 January 1993 had produced nearly 3,000 from main factory in Bristol and smaller unit at Harrogate. A sister company in Dexter, Michigan, USA, had then produced more than 700.

Details of Cameron's conventional-shape balloon range last appeared in the 1992-93 *Jane's*.

Cameron also designs and produces hot-air airships, being first company to develop a craft of this type. Production hot-air airships now include pressurised DP series. Helium filled airships have included DG-19 single-seater and DG-25 two-seater, both now superseded by single-seat DG-14.

CAMERON DP SERIES

TYPE: Pressurised hot-air airships; figures in designations indicate volume in thousands of cubic feet.

PROGRAMME: DP prototype (G-BMEZ) made first flight April 1986; DP series first/second/third in World Hot-Air Airship Championships 1990; DP 70 first in British and European Championships 1991; DP 80 holds current world endurance record.

CURRENT VERSIONS: **DP 60**: Single/two-place.
DP 70: As DP 60, but larger envelope; two-seater.
DP 80: Two-seater, with better performance in hot/high conditions.
DP 90: Larger-envelope version of DP 80.

CUSTOMERS: By 1 January 1993 Cameron had completed, in addition to DP 50 prototype, 22 DP series airships; DP 60/70 customer countries comprise Belgium, Chile, Luxembourg, Switzerland, UK, USA and Venezuela; DP 90s sold to customers in Australia, Brazil, Germany and Switzerland.

DESIGN FEATURES: Differs from earlier hot-air airships in utilising single engine for both propulsion and pressure control; Cameron self-regulating pressure control system; can be flown unpressurised in event of engine failure and landed like hot-air balloon. Twin silencers minimise engine noise level; ergonomically designed cockpit.

STRUCTURE: All-aluminium gondola. Envelope incorporates twin pressure relief valves and Hyperlast fabric top.

POWER PLANT: One 21 kW (28 hp) König SD 570 four-cylinder two-stroke engine standard for DP 60/70, optional for DP 80/90, driving a three-blade König shrouded pusher propeller. Standard engine for DP 80/90 is 47 kW (63 hp) Rotax 582 UL. Fuel capacity 22.7 litres (6 US gallons; 5 Imp gallons) of 40:1 two-stroke mixture for König engine; Rotax runs on unleaded automotive fuel.

ACCOMMODATION: Gondola seats one or two persons with full-harness seat belts. Full height polycarbonate windscreen.

EQUIPMENT: Instrumentation comprises manometer, rev counter, fuel contents gauge, voltmeter, variometer, altimeter and thermistor. 12V 30Ah battery standard. Propane fuel in two Worthington, Cameron 599 or Cameron 426 tanks, according to model, with respective capacities of 77.3, 86.4 or 109 litres (20.4, 22.8 or 28.8 US gallons; 17, 19 or 24 Imp gallons) per tank. Manual piezo-electric or electronic ignition.

UK: LIGHTER THAN AIR—CAMERON

Cameron DP 80 commissioned by a US balloon museum

Gondola of the Cameron DP series

Height overall, incl gondola	8.23 m (27 ft 0 in)
Banner area (each side): Length	9.50 m (31 ft 2 in)
Height	2.59 m (8 ft 6 in)

DIMENSIONS, INTERNAL:
Envelope volume 396.4 m³ (14,000 cu ft)

ROZIERE BALLOONS

Rozières are combination helium/hot-air balloons in which small propane burner maintains helium efficiency at night when solar heating is absent, providing greatly extended flight duration without expending helium or sand ballast. Cameron manufactures five types: the R-15, R-42, R-60, R-77 and R-225 (classes AM-3/5/7/7/11 respectively: see 1992-93 *Jane's*). An R-60 (illustrated in 1991-92 *Jane's*) made first UK/USSR balloon flight in October 1990. Another R-60 completed east-west Atlantic crossing (Canary Islands-Venezuela) in February 1992; five Rozières were produced for 1992 transatlantic race, and several more under construction late 1992 for other long-distance flights, including two attempts to make first round-the-world flight.

DESIGN FEATURES: Helium-filled, with main and tail ballonets, 0.305 m (12 in) manual gas valves, main and overpressure patch-type rip panels and ball type mast connection. Major use expected to be for advertising, but could also find applications for surveillance or as camera platform; can be moored to portable mast, but in adverse weather could be hangared without loss of expensive helium that deflation would involve.
STRUCTURE: Gondola has welded steel frame, aluminium skin and three-wheel landing gear.
POWER PLANT: One 21 kW (28 hp) König SD 570 four-cylinder two-stroke engine, driving a six-blade ducted fan with pitch adjustment, flow straighteners and 40° thrust vectoring. Electric starting. Fuel capacity 22.7 litres (6 US gallons; 5 Imp gallons).
ACCOMMODATION: Pilot only.
DIMENSIONS, EXTERNAL:

Envelope: Length overall	19.00 m (62 ft 4 in)
Max diameter	6.86 m (22 ft 6 in)

Cameron R-77 transatlantic Rozière balloon

DIMENSIONS, EXTERNAL:
Envelope: Length overall: DP 60	30.48 m (100 ft 0 in)
DP 70	32.31 m (106 ft 0 in)
DP 80	33.83 m (111 ft 0 in)
DP 90	35.05 m (115 ft 0 in)
Height: DP 60	13.72 m (45 ft 0 in)
DP 70	14.63 m (48 ft 0 in)
DP 80	15.24 m (50 ft 0 in)
DP 90	15.54 m (51 ft 0 in)
Max width: DP 60	11.28 m (37 ft 0 in)
DP 70	11.89 m (39 ft 0 in)
DP 80	12.19 m (40 ft 0 in)
DP 90	12.80 m (42 ft 0 in)
Max display area per side: DP 60	86.96 m² (936 sq ft)
DP 70	95.97 m² (1,033 sq ft)
DP 80	91.97 m² (990 sq ft)
DP 90	99.03 m² (1,066 sq ft)

DIMENSIONS, INTERNAL:
Envelope volume: DP 60	1,699.0 m³ (60,000 cu ft)
DP 70	1,982.2 m³ (70,000 cu ft)
DP 80	2,265.3 m³ (80,000 cu ft)
DP 90	2,548.5 m³ (90,000 cu ft)

WEIGHTS AND LOADINGS:
Gondola (all models)	195 kg (430 lb)
Envelope: DP 60	146 kg (322 lb)

Useful passenger load (S/L, 15°C):
DP 60	149 kg (328 lb)
DP 70	222 kg (489 lb)
DP 80	285 kg (628 lb)
DP 90	359 kg (791 lb)
Max T-O weight: DP 60: one person	469 kg (1,034 lb)
two persons	546 kg (1,204 lb)
DP 70: one person	485 kg (1,069 lb)
two persons	562 kg (1,239 lb)

PERFORMANCE:
Max level speed: DP 60 15 knots (28 km/h; 17 mph)

CAMERON DG-14

TYPE: Non-rigid helium airship.
PROGRAMME: First flight (G-BRDU) October 1989; trials complete 1990; certificated to BCAR 31, 20 February 1991.

Newest and smallest Cameron helium airship, the single-seat DG-14, about to land

SLINGSBY

SLINGSBY AVIATION LTD

Ings Lane, Kirkbymoorside, North Yorkshire YO6 6EZ
Telephone: 44 (751) 32474
Fax: 44 (751) 31173
Telex: 57597 SLINAV G

MANAGING DIRECTOR: Dr David Holt
MARKETING DIRECTOR, AIRCRAFT: John C. Dignan
AIRSHIP PROJECT MANAGER: John Bewley

Slingsby Aviation was principal subcontractor to Airship Industries (UK) Ltd, supplying gondolas, nosecones, fins and control surfaces made in composites materials for Skyship range; following collapse of Airship Industries in September 1990, Slingsby acquired design and manufacturing rights for Skyship 500/500HL/600 range and is now CAA and FAA type certificate holder for these airships.

By end of 1992, six Skyship 500s and nine 600s had been completed by Airship Industries and Slingsby Aviation; two 500s have been converted to HL standard.

SKYSHIP 500HL

TYPE: Non-rigid helium airship (converted Skyship 500).
PROGRAMME: First flight (G-SKSB) 30 July 1987.
CUSTOMERS: Two conversions completed: in service with Airship International of Orlando, Florida, USA.
DESIGN FEATURES: Combines modified Skyship 500 gondola with larger envelope of Skyship 600, providing ability to operate with greater payload in hotter climates and at higher altitudes than Skyship 500.
POWER PLANT: Two 152 kW (204 hp) Porsche 930/01 non-turbocharged engines; fuel capacity 545 litres (144 US gallons; 120 Imp gallons).
DIMENSIONS, EXTERNAL: As for Skyship 600 except:

Gondola: Length overall	9.24 m (30 ft 3½ in)
Max width	2.41 m (7 ft 10¾ in)

DIMENSIONS, INTERNAL: As for Skyship 600 except:

Gondola: Cabin length	4.20 m (13 ft 9½ in)
Cabin height	1.96 m (6 ft 5 in)

WEIGHTS AND LOADINGS:

Gross disposable load	2,190 kg (4,829 lb)

PERFORMANCE:

Max level speed	50 knots (93 km/h; 57 mph)
Cruising speed	up to 47 knots (87 km/h; 54 mph)
Pressure ceiling	3,050 m (10,000 ft)
Endurance at 35 knots (65 km/h; 40 mph)	17 h

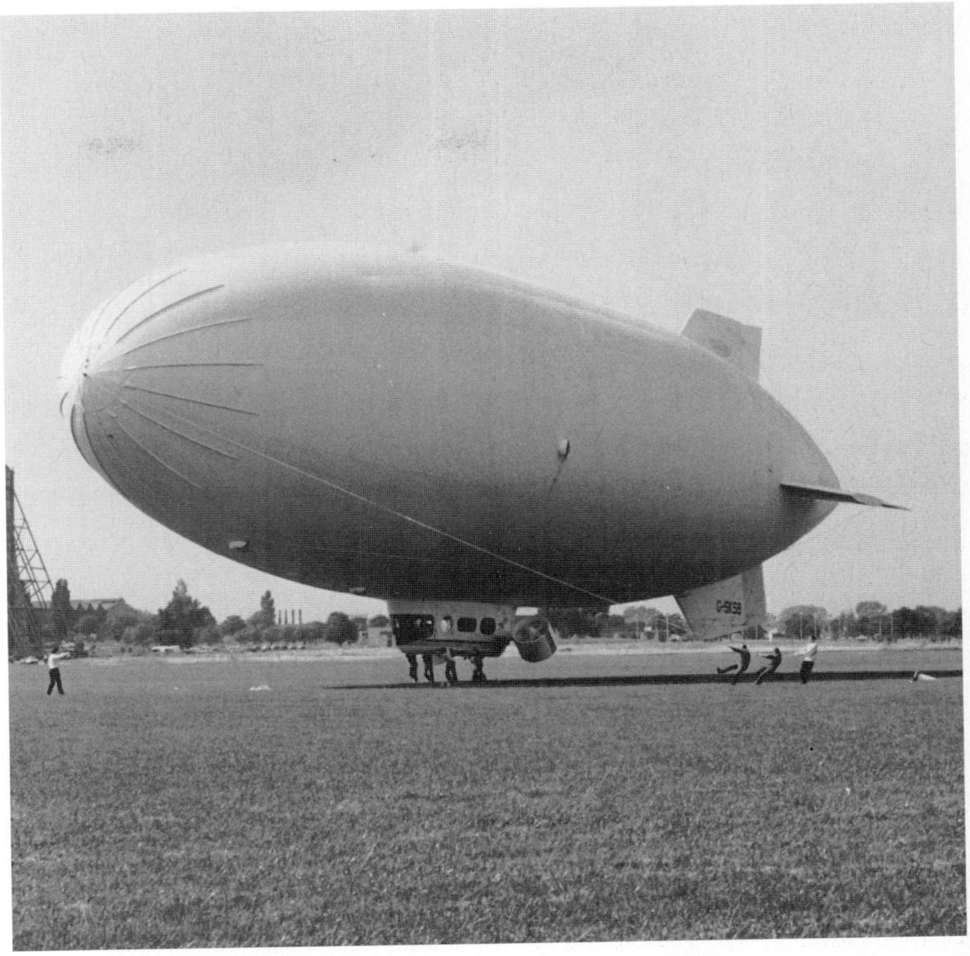

Skyship 500HL (G-SKSB) heavy-lift airship at its Cardington base

SKYSHIP 600

TYPE: Non-rigid helium airship.
PROGRAMME: First flight 6 March 1984; special category C of A awarded by UK CAA 1 September 1984; aerial work certification received Spring 1986; full passenger-carrying C of A 8 January 1987, initiating Skycruise aerial sightseeing service over London, San Francisco, Munich and Sydney in 1987, and over Paris in 1988. First US FAA type certificate awarded to an airship for civil use was issued to Skyship 600 May 1989.
CUSTOMERS: Total of nine built.
DESIGN FEATURES: Non-rigid envelope, with one ballonet forward and one aft (together 26 per cent of total volume); ballonet air intake aft of each propulsor unit. Cruciform tail unit.
FLYING CONTROLS: Vectored thrust propulsion (see Power Plant); differential inflation of ballonets for static fore and aft trim; cable operated rudders and elevators, each with spring tab; ballast in box below crew seats; disposable water ballast in tanks at rear.
STRUCTURE: Envelope manufactured, by Aerazur division of Zodiac group in France, from single-ply polyester fabric, coated with titanium dioxide loaded polyurethane to reduce ultraviolet degradation; polyvinylidene chloride film bonded onto inner coating of polyurethane on inside of envelope minimises loss of helium gas. Four parabolic arch load curtains, carrying multiple Kevlar 29 gondola suspension cables. Nose structure is domed disc, moulded from GFRP and carrying fitting by which airship is moored to its mast. Each tail surface attached to envelope at root and braced by wires on each side; all four surfaces constructed from interlocking ribs and spars of Fibrelam with GFRP skins.

One-piece moulded gondola of Kevlar reinforced plastics, with flooring and bulkheads of Fibrelam panels; those forming engine compartment at rear faced with titanium for fire protection.
LANDING GEAR: Single two-wheel assembly with double tyres, mounted beneath rear of gondola.
POWER PLANT: Two 190 kW (255 hp) Porsche 930/01 six-cylinder air-cooled and turbocharged piston engines mounted in rear of gondola. Each drives a ducted propulsor consisting of a Hoffmann five-blade reversible-pitch propeller within an annular duct of carbonfibre reinforced GFRP. Each propulsor can be rotated about its pylon attachment to gondola through an arc of 210°, 90° upward and 120° downward, vectored thrust thus providing both V/STOL and in-flight hovering ability. Fuel tank, capacity 682 litres (180 US gallons; 150 Imp gallons), at rear of engine compartment. Auxiliary fuel tanks optional. Engine modifications include provision of automatic mixture control, fuel injection and electronic ignition.
ACCOMMODATION: Seats for pilot and co-pilot, with dual controls, although airship is designed to be flown by single pilot. Max capacity 13 passengers in addition to pilot.
SYSTEMS: 28V electrical system, supplied by engine driven alternators.
AVIONICS: Include Bendix/King Silver Crown series dual nav/com, ADF, Omega, VOR/ILS and weather radar.
DIMENSIONS, EXTERNAL:

Envelope: Length overall	59.00 m (193 ft 7 in)
Max diameter	15.20 m (49 ft 10½ in)
Height overall, incl gondola and landing gear	20.30 m (66 ft 7¼ in)
Width over tailfins	19.20 m (63 ft 0 in)
Gondola: Length overall	11.67 m (38 ft 3½ in)
Max width	2.56 m (8 ft 4¾ in)

DIMENSIONS, INTERNAL:

Envelope volume	6,666 m³ (235,400 cu ft)
Ballonet volume (two, total)	1,733 m³ (61,200 cu ft)
Gondola: Cabin length	6.89 m (22 ft 7¼ in)
Cabin height	1.92 m (6 ft 3½ in)
Cabin floor area (usable)	12.00 m² (130.0 sq ft)

WEIGHTS AND LOADINGS (design):

Gross disposable load	2,343 kg (5,165 lb)

PERFORMANCE:

Max level speed	58 knots (107 km/h; 67 mph)
Cruising speed (70% power)	52 knots (96 km/h; 60 mph)
Pressure ceiling	3,050 m (10,000 ft)
Still air range at 40 knots (74 km/h; 46 mph), without auxiliary tanks	550 nm (1,019 km; 633 miles)

THUNDER & COLT

THUNDER & COLT LTD

Maesbury Road, Oswestry, Shropshire SY10 8HA
Telephone: 44 (691) 670644
Fax: 44 (691) 670617
Telex: 35503 COLT G

MANAGING DIRECTOR: Malcolm Gillespie
TECHNICAL DIRECTOR: Mark Broome
PRODUCTION DIRECTOR: Paul Dickinson
MARKETING DIRECTOR: Chris Kirby
PUBLIC RELATIONS MANAGER: Carolle Doyle

Thunder & Colt produces a range of hot-air airships, and the GA 42 helium-filled airship. Company also well known for range of hot-air balloons, from one-person types up to large (8,495 m³; 300,000 cu ft) passenger-carrying models, as detailed in 1991-92 *Jane's*, and produced world's largest hot-air balloon, the 73,624 m³ (2.6 million cu ft) *Virgin Otsuka Pacific Flyer*. Its hot-air balloons currently (early 1993) hold all of world's ultimate distance, duration and altitude records. Company stated in 1992 to be seeking purchaser for its airship division.

COLT AS 80 Mk II and AS 105 Mk II

TYPE: Hot-air airship.
PROGRAMME: AS 80 Mk II developed and certificated 1988; production continuing.
CURRENT VERSIONS: **AS 80 Mk II**: *Description applies to this version except where indicated.*

AS 105 Mk II: As AS 80 except for larger envelope.
CUSTOMERS: Total of 18 AS 80 Mk II and AS 105 Mk II built by 1 January 1993, including two **AS 80 GD** for Gefa-Flug of Germany (which see).
DESIGN FEATURES: Envelope fineness ratio 2.5; cruciform tail surfaces with rudder on each vertical fin; twin catenary load suspension system distributes loads of gondola weight and power plant forces evenly into envelope.
FLYING CONTROLS: Burners located inside envelope and operated electrically, with manual override; pilot lights fitted with electric spark and piezo-electric ignition, specially modified to operate directly underneath inflation fan without blowing out; steering by rudder on each vertical fin, operated by cables connecting them to gondola.
STRUCTURE: Envelope made from HTN90K high-tenacity and high temperature resistant fabric; engine drives generator supplying electric fan used to pressurise envelope; additional air for envelope pressurisation via scoop located in propeller slipstream. Gondola is stainless steel tube spaceframe with aluminium skin panels; windscreen, of polycarbonate sheet, forms partially enclosed cockpit.
LANDING GEAR: Non-retractable tricycle type.
POWER PLANT: One 38.8 kW (52 hp) Rotax 462 two-cylinder water-cooled engine, driving a two-blade pusher propeller.
ACCOMMODATION: Pilot and one passenger in tandem; second seat can be replaced by auxiliary fuel tank.

598 UK/USA: LIGHTER THAN AIR—THUNDER & COLT/ABC

Colt AS 80 Mk II single/two-person hot-air airship

AVIONICS: Standard instrumentation comprises altimeter, variometer, envelope temperature and pressure gauges, propane volume gauge, petrol level gauge, voltmeter, engine water temperature and engine rpm gauge. A 720-channel VHF com transceiver option, with pilot-passenger intercom, is available.

DIMENSIONS, EXTERNAL:
Envelope: Length overall: 80	31.00 m (101 ft 8½ in)
105	34.00 m (111 ft 6½ in)
Max diameter: 80	12.40 m (40 ft 8¼ in)
105	13.87 m (45 ft 6 in)
Gondola: Length overall: 80, 105	3.84 m (12 ft 7¼ in)
Max width: 80, 105	1.75 m (5 ft 9 in)
Height overall: 80, 105	1.80 m (5 ft 11 in)
Propeller diameter: 80, 105	1.57 m (5 ft 2 in)

DIMENSIONS, INTERNAL:
Envelope volume: 80	2,123.8 m³ (75,000 cu ft)
105	2,973.3 m³ (105,000 cu ft)

WEIGHTS AND LOADINGS:
Envelope: 80	190 kg (419 lb)
105	213 kg (469 lb)
Gondola, empty: 80, 105	210 kg (463 lb)
Propane fuel: 80, 105	60-110 kg (132-242 lb)
Gross lift: 80	600 kg (1,323 lb)

PERFORMANCE:
Max level speed: 80, 105	20 knots (37 km/h; 23 mph)
Max endurance: 80, 105	2 h 30 min

COLT AS 56
TYPE: Single-seat hot-air airship.
PROGRAMME: Winner at first World Hot-Air Airship Championships 1988.
DESIGN FEATURES: Generally similar to those of AS 80/105 above.
POWER PLANT: One 21 kW (28 hp) König SC 570 four-cylinder two-stroke driving a two-blade pusher propeller.

DIMENSIONS, EXTERNAL:
Envelope: Length overall	28.00 m (91 ft 10½ in)
Max diameter	11.20 m (36 ft 9 in)
Gondola: Length overall	2.75 m (9 ft 0¼ in)
Max width	1.50 m (4 ft 11 in)
Height overall	1.40 m (4 ft 7 in)
Propeller diameter	1.32 m (4 ft 4 in)

DIMENSIONS, INTERNAL:
Envelope volume	1,585.75 m³ (56,000 cu ft)

WEIGHTS AND LOADINGS:
Envelope	120 kg (265 lb)
Gondola	110 kg (242 lb)
Propane fuel	40 kg (88 lb)
Max T-O weight	365 kg (805 lb)

PERFORMANCE:
Max level speed	15 knots (28 km/h; 17 mph)
Max endurance	1 h 30 min

COLT AS 261
Largest hot-air airship ever flown; first flight 1989; described and illustrated in 1991-92 and earlier *Jane's*. No further examples completed.

COLT GA 42
TYPE: Non-rigid helium airship.
PROGRAMME: Joint project between Thunder & Colt and sister company Airborne Industries (Thunder & Colt undertook design and production of gondola and design of envelope; Airborne Industries responsible for envelope manufacture); prototype (G-MATS) made first flight 2 September 1987; type certificate issued May 1989 after intensive testing and modification; certificated to BCAR CAP 471 Section Q in aerial work category.
CUSTOMERS: Seven built by January 1993 for customers in Finland, France, UK and USA.
DESIGN FEATURES: Envelope shape and stability maintained by two catenary curtains which transfer weight loading to top of envelope; cruciform tail surfaces; gondola supported by seven catenary cables each side (three forward, four aft); quick-release mooring by hard nose with 12 battens.
FLYING CONTROLS: Two air-filled ballonets (one forward, one aft, with max fill ratio of 25 per cent), inflated by engine driven Porsche axial fan and controlled manually or (at pre-set levels) by automatic pressure relief valve; helium pressure relief valve operates automatically, but can be actuated manually in emergency; movable control surfaces on all except upper vertical tail-fin. Fly-by-wire flight control system, using three linear DC actuators; twin analog backup circuits.
STRUCTURE: Envelope made of polyurethane coated polyimide, with ultra-violet blockage, hydrolysis proofed. Tail-fins fabricated from GFRP composites and attached to hull by rigging wires; aluminium nose battens; all-aluminium riveted gondola.
LANDING GEAR: Fully castoring, rubber sprung single wheel.
POWER PLANT: One 74.5 kW (100 hp) Teledyne Continental O-200-B flat-four engine, driving a Hoffmann four-blade fixed-pitch wood/composite pusher propeller. Fuel capacity 130 litres (34.3 US gallons; 28.6 Imp gallons).
ACCOMMODATION: Side by side bucket seats for crew of two, with full harnesses, on fully enclosed and heated flight deck; 180° single-curvature polycarbonate windscreen, plus observation window in roof for viewing airship interior.
SYSTEMS: Electrical systems powered by 25Ah wet cell battery, charged via engine mounted alternators capable of delivering 60A at 14V DC. Ample battery power available to supply control systems for landing in event of alternator failure.
AVIONICS: VHF nav/com, ADF, transponder, VOR, and ADF indicators. Instrument panel roof-mounted in order not to obstruct all-round field of view.

DIMENSIONS, EXTERNAL:
Envelope: Length overall	27.50 m (90 ft 2¾ in)
Max diameter	9.20 m (30 ft 2¼ in)
Height overall between fin-tips	11.20 m (36 ft 9 in)
Banner area (each side):	
Length	14.70 m (48 ft 2¾ in)
Height	7.50 m (24 ft 7¼ in)
Gondola: Length overall	3.44 m (11 ft 3½ in)
Max width	1.54 m (5 ft 0½ in)
Height: excl wheel	1.55 m (5 ft 1 in)
incl wheel	2.14 m (7 ft 0¼ in)

DIMENSIONS, INTERNAL:
Envelope volume	1,189.3 m³ (42,000 cu ft)
Ballonet volume (two, total)	approx 311.5 m³ (11,000 cu ft)

WEIGHTS AND LOADINGS:
Max static heaviness	100 kg (220 lb)
Design gross lift	1,100 kg (2,425 lb)

PERFORMANCE:
Cruising speed: max	40 knots (74 km/h; 46 mph)
normal	32 knots (60 km/h; 37 mph)
econ	27 knots (50 km/h; 31 mph)
Max rate of climb	396 m (1,299 ft)/min
Max rate of descent	456 m (1,496 ft)/min
Pressure ceiling (ISA)	2,900 m (9,515 ft)
Max endurance	7 h 30 min
Pitch angle limits	+30°/−25°

Two examples of the two-place, 42,000 cu ft Colt GA 42 helium airship

UNITED STATES OF AMERICA

ABC
AMERICAN BLIMP CORPORATION
1900 North-East 25th Avenue (Suite 5), Hillsboro, Oregon 97124
Telephone: 1 (503) 693 1611
Fax: 1 (503) 681 0906
PRESIDENT: James Thiele
EXECUTIVE VICE-PRESIDENT: Charles Ehrler
CHIEF ENGINEER: Rudy Bartel

ABC A-60 PLUS LIGHTSHIP
TYPE: Five-seat non-rigid helium airship.
PROGRAMME: A-50 prototype (see 1990-91 *Jane's*) made first flight 9 April 1988; first flight of production A-60 (1991-92 *Jane's*) June 1989; A-60 certificated by FAA 18 May 1990 for day/night and VFR/IFR flight; first delivery (to Virgin Lightships Ltd) November 1990; A-60 Plus certificated Autumn 1991 by FAA; CAA approval, and APU upgrade, expected 1993.
CURRENT VERSIONS: **A-60:** Initial production version, described in 1991-92 *Jane's*. A-60 Plus upgrade retrofit kit available.
A-60 Plus: Current version from 1991; payload increased by 181 kg (400 lb) compared with A-60; Light-sign message board option from 1992. *Description applies to this version.*
CUSTOMERS: Three A-60s delivered to Virgin Lightships 1990-91, and one leased to Japan Airship Services Novem-

ber 1992; two A-60 Plus ordered by Virgin (first delivery 14 April 1993).
DESIGN FEATURES: Conventional-shape envelope; cruciform tail surfaces; gondola suspended by 12 catenary cables each attached to external patches, eliminating need for internal cables, gas-tight fittings and bellow sleeves. Can be inflated without a net by attaching ballasted gondola before adding helium; gondola can be removed from inflated Lightship without a net by ballasting catenary cables.
FLYING CONTROLS: Single internal ballonet; rudder or elevator on each of four tail-fins; primary flight controls for left-hand front seat only.
STRUCTURE: Outer envelope skin of Dacron/Mylar with separate urethane film inner gas-tight bladder and single ballonet. All structural attachments to sewn outer bag, such as nose mooring, fin base and guy wires, car catenary and handling lines, made with webbing reinforcements sewn directly to envelope.
LANDING GEAR: Single twin-wheel unit beneath gondola; small tailwheel at base of lower vertical fin.
POWER PLANT: Twin 60 kW (80 hp) Limbach L 2000 engines, pusher-mounted to enhance propulsive efficiency and reduce noise. Standard capacity of rear-mounted fuel tank is 280 litres (74 US gallons; 61.6 Imp gallons); can be refuelled in the field without mooring.
ACCOMMODATION: Two single seats and three-person rear bench seat in gondola.
SYSTEMS: APU capable of delivering 2.2kW at 110V certificated December 1990; to be upgraded to 2.5kW in 1993. Used mainly to power internal illumination lights, but can be used also for other airborne equipment.
EQUIPMENT: Certificated for gyro-stabilised TV camera mount, complete with microwave downlink for live broadcast; can be operated simultaneously with APU.

DIMENSIONS, EXTERNAL:
Envelope: Length overall 39.01 m (128 ft 0 in)
Max diameter 10.01 m (32 ft 10 in)
Fineness ratio 3.80
Gondola: Length overall 3.96 m (13 ft 0 in)
Width 1.52 m (5 ft 0 in)
Height overall 2.90 m (9 ft 6 in)
Propeller diameter 1.52 m (5 ft 0 in)
DIMENSIONS, INTERNAL:
Envelope volume 1,925.5 m³ (68,000 cu ft)
Gondola: Cabin length 2.74 m (9 ft 0 in)
Cabin height 1.83 m (6 ft 0 in)
AREAS:
Tail-fins (four, total) 42.74 m² (460.0 sq ft)
WEIGHTS AND LOADINGS:
Total weight empty 1,216 kg (2,680 lb)
Max buoyancy 1,814 kg (4,000 lb)
Max dynamic lift 113 kg (250 lb)
Max useful lift, ISA at 656 m (2,000 ft)
 680 kg (1,500 lb)
Max gross weight 1,993 kg (4,394 lb)
PERFORMANCE:
Max level speed 46 knots (85 km/h; 53 mph)
Max rate of ascent 457 m (1,500 ft)/min
Service ceiling 2,225 m (7,300 ft)
Max rate of descent 396 m (1,300 ft)/min
Min T-O distance 112 m (366 ft)
Max range at 35 knots (65 km/h; 40 mph)
 521 nm (965 km; 600 miles)
Max endurance at 35 knots (65 km/h; 40 mph) 15 h

ABC A-120 LIGHTSHIP
Conceptual study under way early 1993 for nine-passenger development with 3,398 m³ (120,000 cu ft) envelope and stretched gondola. Power plant to be decided, but performance expected to be comparable to that of A-60 Plus.

American Blimp Corporation A-60 Plus Lightship on its first flight

Gondola and power plant of the ABC Lightship

AEROSTAR
AEROSTAR INTERNATIONAL INC
(Subsidiary of Raven Industries Inc)
Box 5057, Sioux Falls, South Dakota 57117-5057
Telephone: 1 (605) 331 3500
Fax: 1 (605) 331 3520
VICE-PRESIDENT, SALES: Larry Manderscheid

EARTHWINDS
Larry Newman, who crossed Atlantic in helium balloon *Double Eagle II* in 1978 and Pacific in *Double Eagle V* three years later, hopes to attempt non-stop round-the-world flight in unique double-balloon *Earthwinds*. Complete system, devised by Aerostar, comprises upper helium balloon and lower air-filled balloon, between which is an all-composites, pressurised three-place gondola made by Scaled Composites Inc. Crew members are to be Richard Branson of UK, who part-funded project, and Maj Gen Vladimir Dzhanibekov, a Russian cosmonaut.

Height of helium balloon is approx 51.8 m (170 ft) and that of anchor balloon 30.5 m (100 ft); when fully inflated, complete system will have overall height of approx 106.7 m (350 ft). Air balloon is to act as 'flexible ballast', avoiding need to vent helium when it expands in Sun's heat by lightly pressurising lower sphere; latter would be deflated each night to maintain *Earthwinds* combination at constant altitude of about 10,670 m (35,000 ft). Two 9 kW (12 hp) liquid-cooled engines in gondola provide power to operate onboard electrics and avionics, control balloons' helium and air, and pressurise, heat and ventilate accommodation. Including 2,650 litres (700 US gallons; 583 Imp gallons) of fuel for these engines, total empty equipped weight of complete system is 8,900 kg (19,620 lb). Com/nav avionics, provided by Bendix/King and NASA, include GPS, HF and VHF radio, and satellite navigation and tracking.

Take-off is to be made from Akron, Ohio, and flight is expected to take between 12 and 20 days, travelling mainly between 45th and 50th parallels.

BARNES
BARNES AIRSHIPS

BARNES WHISPERSHIP
First flown in 1991, the Whispership is a single/two-place recreational non-rigid with an X-fin tail unit and pusher propeller. Known details in 1992-93 *Jane's*; all requests for further information have failed to elicit response.

MEMPHIS
MEMPHIS AIRSHIPS INC
Isle-A-Port Airport, 1720 Harbor Avenue, PO Box 13037, Memphis, Tennessee 38113
Telephone: 1 (901) 775 0386
Fax: 1 (901) 775 0917
PRESIDENT: Steve Garner

EXECUTIVE VICE-PRESIDENT: Jim Groce

Memphis Airships formed 1981 as producer of small unmanned advertising airships and fan supported inflatables. Ultrablimp small air recreation vehicle described and illustrated in 1987-88 *Jane's*, Dreamfinder and EXP-II in 1986-87 and 1990-91 *Jane's* respectively. Latest known product was Zephyr 200 (1992-93 *Jane's*), but no news of this since early 1991.

TCOM
TCOM LIMITED PARTNERSHIP
7115 Thomas Edison Drive, Columbia, Maryland 21046
Telephone: 1 (410) 312 2400
Fax: 1 (410) 312 2455
Telex: 198105
VICE-PRESIDENT, MARKETING: Stephen Silvoy

TCOM TETHERED AEROSTATS
TCOM, a Westinghouse subsidiary until becoming independent in late 1989, has produced some 50 radar-carrying tethered aerostats for land- and ship-based surveillance, air defence, drug interdiction and similar tasks since the early 1970s. A powered tether system and low helium leakage enable them to operate for extended periods. Principal types are as follows:

STARS (small tethered aerostat relocatable system): Transportable tactical system, mainly for border and coastal surveillance; can be mounted on flat-bed trailer, deployed from ship or truck, or towed by helicopter. Length 25.60 m (84 ft), volume 736.2 m³ (26,000 cu ft), payload 125 kg (275 lb); sensor line-of-sight range 60 nm (111 km; 69 miles) from altitude of 750 m (2,460 ft). Principal customer: US Coast Guard (five STARS/Superstars).

Superstars: Larger version of STARS, 31 m (101.7 ft) long, with volume of 1,557.4 m³ (55,000 cu ft). US Army has four, known as **SASS** (small aerostat surveillance system) sea-based; two to US Customs Service; eight supplied to South Korea. Normal sensor is modified Westinghouse AN/APG-66 radar.

LASS (low altitude surveillance system): Aerostat is 71.02 m (233 ft) long and has volume of 10,335.6 m³ (365,000 cu ft). One to US Coast Guard; exports include one each to Kuwait (since lost) and Saudi Arabia. Normal sensor is modified Westinghouse AN/TPS-63 radar.

Further details can be found in *Jane's Battlefield Surveillance Systems*.

THOMPSON
JAMES THOMPSON, AIAA
1700 Citizens Plaza, Louisville, Kentucky 40202
Telephone: 1 (502) 566 0501
Fax: 1 (502) 589 0148

THOMPSON AIRSHIP
TYPE: Sport/advertising non-rigid helium airship.
PROGRAMME: Protracted construction on opportunity basis (see earlier editions), but by January 1993 nearly all components completed and airship transferred to Oregon for final assembly, rigging and flight test; hoped to make first flight late April 1993.
DESIGN FEATURES: See accompanying illustration. Inverted Y tail surfaces (all three angles 120°); no catenary curtains (cables attached to five finger patches on each side of envelope suspend gondola slightly forward of centre of buoyency to compensate for pitch-up). Gondola stabilised by three cables attached to propeller shroud.
FLYING CONTROLS: Two 56.6 m³ (2,000 cu ft) air ballonets, fillable unequally to provide pitch trim; elevators on lower pair of tail surfaces operate differentially for roll control; 81.6 kg (180 lb) of lead shot or sand ballast in gondola.
STRUCTURE: Tail unit is fabric covered aluminium tube structure. Gondola has steel tube frame, with glassfibre and urethane foam skin panels.
POWER PLANT: One 1,200 cc Honda liquid-cooled engine, with 2.2:1 toothed-belt reduction drive to a two-blade shrouded wooden pusher propeller.
ACCOMMODATION: Pilot and one passenger in gondola.
DIMENSIONS, EXTERNAL:
Envelope: Length overall	24.91 m (81 ft 9 in)
Max diameter	7.91 m (25 ft 11¼ in)
Fineness ratio	3.15

DIMENSIONS, INTERNAL:
Envelope volume: helium	695.0 m³ (24,544 cu ft)
ballonets (two, total)	97.9 m³ (3,456 cu ft)
design total	792.9 m³ (28,000 cu ft)

WEIGHTS AND LOADINGS:
Envelope	244 kg (538 lb)
Gondola	187 kg (412 lb)
Ballast	81.6 kg (180 lb)
Max T-O weight	696 kg (1,534 lb)

PERFORMANCE (estimated):
Design speed	30 knots (55 km/h; 34 mph)

Model of the Thompson two-person airship

ULITA
ULITA INDUSTRIES INC (Manufacturing Division)
PO Box 412, Sheboygan, Wisconsin 53082-0412
Telephone and Fax: 1 (414) 458 2842
PRESIDENT AND CHAIRMAN: Thomas S. Berger

Ulita has undergone organisational and financial restructuring in recent years, resulting in termination of UM20-48 and UM25-64 programmes (see 1992-93 *Jane's*). Current hardware development programme is UM10-23 Cloud Cruiser; concept development and detail design of UM30-71 are being continued.

ULITA UM10-23 CLOUD CRUISER
TYPE: Non-rigid helium airship.
PROGRAMME: Prototype under construction for completion mid-1993; enlarged slightly from original 637.2 m³ (22,503 cu ft) UM10-22 design described in 1992-93 and earlier editions.
DESIGN FEATURES: Optimised for ARV (air recreational vehicle) market as low-cost light utility craft; to be available initially in kit form only, but certification being considered. Envelope fineness ratio 3.6; twin horizontal and vertical stabilisers, each with movable control surface, attached to ends of internally mounted inverted V structure.
FLYING CONTROLS: Two internal ballonets (one forward, one aft) for pitch trim; twin elevators and rudders for pitch and yaw control.
LANDING GEAR: Landing wheel, mounted on outrigger, on each side of gondola; small wheel at base of each tail-fin.

Model of the Ulita UM10-23 Cloud Cruiser air recreational vehicle airship

POWER PLANT: Two 18 kW (24 hp) König SC 340 three-cylinder two-stroke engines, each driving a three-blade fixed-pitch ducted propeller, mounted on outrigger from gondola and rotatable to provide horizontal or vertical thrust. Main fuel tank in gondola, with provisions for auxiliary tanks on outriggers. Main fuel tank capacity 60.6 litres (16 US gallons; 13.3 Imp gallons); increasable with auxiliary tanks to 98.5 litres (26 US gallons; 21.6 Imp gallons).
ACCOMMODATION: Side by side seating, with dual control wheels, for pilot and one pupil/passenger. Gondola ventilated, but not heated.
SYSTEMS: 12V DC battery and alternators.

DIMENSIONS, EXTERNAL:	
Length overall	25.48 m (83 ft 7 in)
Width overall	8.15 m (26 ft 8¾ in)
Height overall	8.97 m (29 ft 5 in)
Envelope: Length overall	24.99 m (82 ft 0 in)
Max diameter	6.95 m (22 ft 9½ in)
Gondola: Length: at top	3.12 m (10 ft 3 in)
overall	3.44 m (11 ft 3½ in)
Width: at top	1.32 m (4 ft 4 in)
at floor	1.04 m (3 ft 5 in)
overall (incl propulsors)	4.33 m (14 ft 2½ in)
Height: excl landing gear	1.71 m (5 ft 7¼ in)
incl landing gear	2.12 m (6 ft 11½ in)
Propeller diameter (each)	1.07 m (3 ft 6 in)
DIMENSIONS, INTERNAL:	
Envelope volume	658.4 m³ (23,250 cu ft)
Ballonet volume (two, total)	144.8 m³ (5,115 cu ft)
WEIGHTS AND LOADINGS:	
Weight empty	427 kg (941 lb)
Fuel (standard)	32.7 kg (72 lb)
Crew of 2	154 kg (340 lb)
Payload	32 kg (70 lb)
Total gross lift	645 kg (1,423 lb)
PERFORMANCE (estimated):	
Max level speed	39 knots (72 km/h; 45 mph)
Econ cruising speed	26 knots (48 km/h; 30 mph)
Pressure ceiling	2,440 m (8,000 ft)

ULITA UM30-71

TYPE: Non-rigid helium airship.

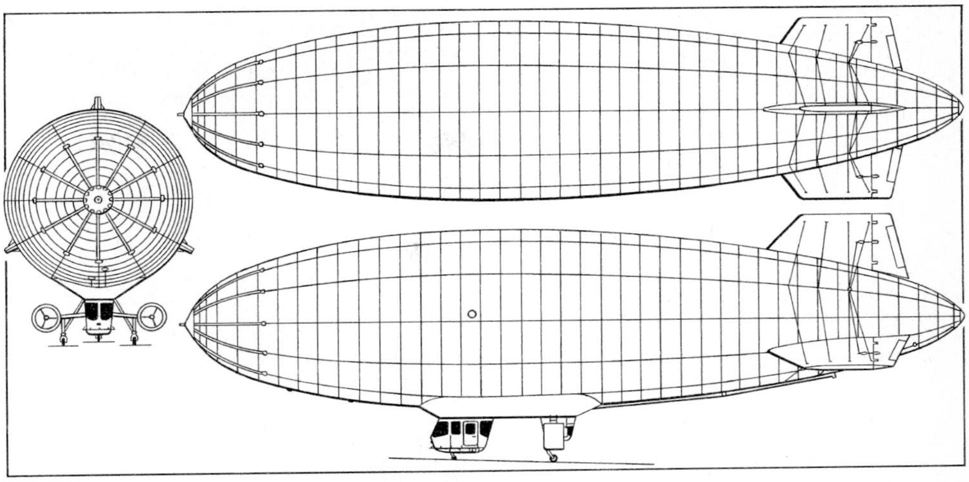

General appearance of the projected Ulita UM30-71 *(Jane's/Mike Keep)*

PROGRAMME: Long-term project; detail design continuing in 1993.
DESIGN FEATURES: See accompanying drawing.
POWER PLANT: One 149 kW (200 hp) Textron Lycoming IO-360 flat-four engine, driving a pair of five-blade, reversible-pitch ducted propellers in vectoring units.
ACCOMMODATION: Pilot and up to three passengers.

DIMENSIONS, EXTERNAL:	
Length overall	41.45 m (136 ft 0 in)
Width overall	9.94 m (32 ft 7¼ in)
Height overall	13.65 m (44 ft 9½ in)
Envelope: Length overall	41.15 m (135 ft 0 in)
Max diameter	9.87 m (32 ft 4½ in)
Gondola: Length: at top	7.16 m (23 ft 6 in)
overall	7.84 m (25 ft 8½ in)
Width: at top	1.71 m (5 ft 7¼ in)
at floor	1.31 m (4 ft 3½ in)
overall (incl propulsors)	7.01 m (23 ft 0 in)
Height: excl landing gear	2.50 m (8 ft 2½ in)
incl landing gear	3.05 m (10 ft 0 in)
Propeller diameter (each)	1.30 m (4 ft 3 in)
DIMENSIONS, INTERNAL:	
Envelope volume	2,022.2 m³ (71,413 cu ft)
Ballonet volume (two, total)	404.4 m³ (14,283 cu ft)
WEIGHTS AND LOADINGS:	
Weight empty	1,386 kg (3,055 lb)
Fuel (standard)	163 kg (360 lb)
Payload	231 kg (510 lb)
Total gross lift	1,857 kg (4,095 lb)

UPSHIP

UPSHIP DIRIGIBLE AIRSHIPS

Route 1, Box 386B, Jackson, Georgia 30233
Telephone: 1 (706) 775 4952
Fax: 1 (706) 775 2919
DIRECTOR: Jesse Blenn

UPSHIP 100-001

TYPE: Semi-rigid helium airship.
PROGRAMME: Design started 1989; original 750-001 (see 1992-93 *Jane's*) now enlarged as 100-001 proof-of-concept vehicle; progress in 1992 included full-scale gondola and tail-fin mockups; patent applications and further detail design work planned for 1993; hopes to begin full time construction of 100-001 in 1995.
CUSTOMERS: Markets foreseen in tourism, advertising, scientific research, cargo, resource management, and minimum-impact logging operations, where UPships can be built and operated at competitive price.
DESIGN FEATURES: Aim is to develop helium airships able to utilise full potential of buoyant flight; semi-rigid configuration chosen as offering better low-speed control and more design flexibility than a non-rigid; simple but effective control system claimed to give design efficient control at all speeds and to eliminate major expense of need for ground crew. Conventional-shape envelope, with inverted V tail surfaces.
FLYING CONTROLS: Pitch and yaw control by ruddervators; ballonet for pitch trim.

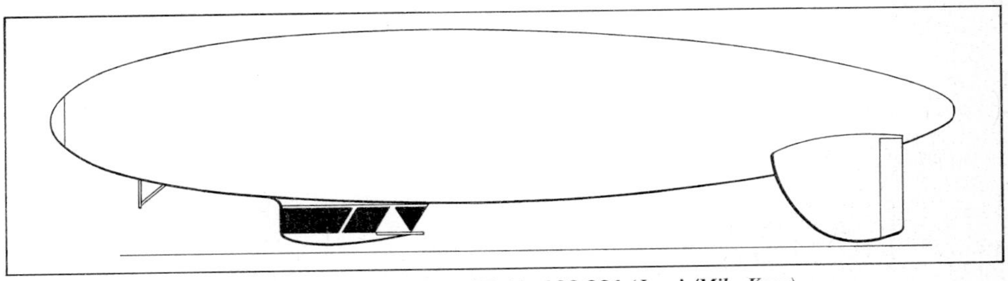

Preliminary drawing of the UPship 100-001 *(Jane's/Mike Keep)*

STRUCTURE: Triangular-section aluminium girder keel; outer envelope of sewn polyester fabric, with three-compartment inner helium bag of Mylar film; tail-fins and ruddervators have balsa and carbonfibre framework with polyester fabric covering.
POWER PLANT: Three 3.7 kW (5 hp) Honda GX 140 piston engines, each driving a mahogany propeller. Fuel capacity 66 litres (17.5 US gallons; 14.6 Imp gallons).
ACCOMMODATION: Pilot and one passenger, seated in tandem.

DIMENSIONS, EXTERNAL:	
Envelope: Length overall	30.48 m (100 ft 0 in)
Max diameter	6.10 m (20 ft 0 in)
Height overall, incl gondola	7.52 m (24 ft 8 in)
Propeller diameter	1.02 m (3 ft 4 in)
DIMENSIONS, INTERNAL:	
Envelope volume	620.8 m³ (21,925 cu ft)
Ballonet volume (20%)	124.17 m³ (4,385 cu ft)
WEIGHTS AND LOADINGS (approx):	
Weight empty	363 kg (800 lb)
Useful lift	181 kg (400 lb)
Total lift	544 kg (1,200 lb)
PERFORMANCE (estimated):	
Max level speed	29 knots (53 km/h; 33 mph)
Cruising speed, 70% power	25 knots (47 km/h; 29 mph)
Ceiling above T-O: normal	610 m (2,000 ft)
max	1,830 m (6,000 ft)
Range at cruising speed, two engines, 20% reserves	510 nm (946 km; 588 miles)
Endurance, conditions as above	20 h
Fuel consumption (cruising speed, two engines)	2.23 litres (0.59 US gallon; 0.49 Imp gallon)/h

WAI

WESTINGHOUSE AIRSHIPS INC

PO Box 17193 (MS 1780), Baltimore, Maryland 21203
Telephone: 1 (410) 993 1734
Fax: 1 (410) 765 2950
ODM PROGRAMME MANAGER: James F. Luck
VICE-PRESIDENT, BUSINESS DEVELOPMENT:
E. Judson Brandreth Jr

Following collapse of Airship Industries (UK) Ltd in 1990, Westinghouse Electric Corporation acquired that company's 50 per cent share in their WAI joint venture company, together with Sentinel 1000/5000 and all military/defence applications of SK 500/500HL/600 Skyships.

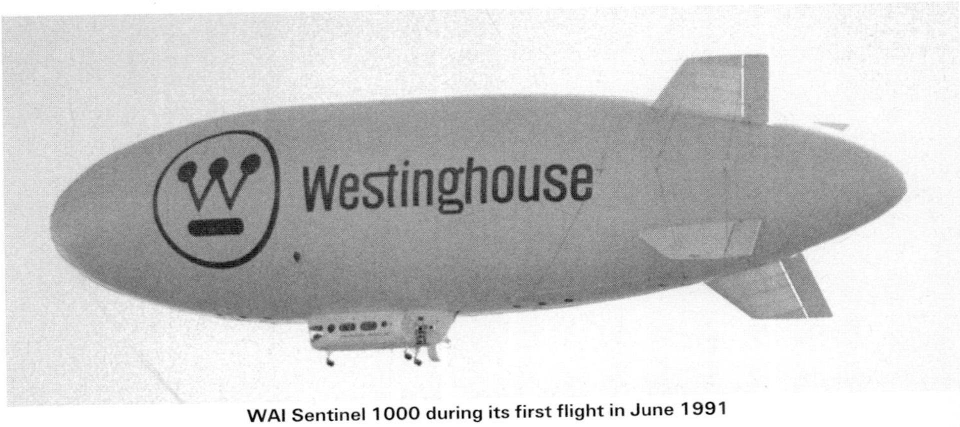

WAI Sentinel 1000 during its first flight in June 1991

WAI SENTINEL 1000

TYPE: Non-rigid helium airship.
PROGRAMME: Initiated by Airship Industries (UK) and taken over after collapse by WAI; first flight 26 June 1991; total

277 hours in 83 flights by 1 January 1993, including first flight with production-standard fly-by-light system 11 April 1992; certification expected Summer 1993.
DESIGN FEATURES: Derivative of Skyship 600 (see under Slingsby in UK part of this section); larger envelope (currently the largest non-rigid airship flying); higher performance engines; upgraded avionics (by Ferranti International); substantial increase in payload; improved flight control through autopilot/autostabilisation elements of YEZ-2A flight control system. Has been designated official trials vehicle for US Navy YEZ-2A programme.
FLYING CONTROLS: GEC-Marconi optical fibre (fly-by-light) flight control system.
STRUCTURE: Generally similar to that of Skyship 600.
POWER PLANT: Two 175 kW (235 hp) Porsche engines.

DIMENSIONS, EXTERNAL:
  Envelope: Length overall   67.50 m (221 ft 5½ in)
  Max diameter   16.70 m (54 ft 9½ in)
  Height overall (incl gondola and landing gear)
     20.20 m (66 ft 3¼ in)
  Gondola   As for Skyship 600
DIMENSIONS, INTERNAL:
  Envelope volume   10,700 m³ (377,868 cu ft)
  Ballonet volume (total)   2,405 m³ (84,932 cu ft)
  Gondola   As for Skyship 600
WEIGHTS AND LOADINGS (design):
  Gross disposable load   4,200 kg (9,260 lb)
PERFORMANCE (estimated):
  Max level speed   58 knots (107 km/h; 67 mph)
  Cruising speed (70% power) 50 knots (92 km/h; 57 mph)
  Still air range at 36 knots (67 km/h; 41 mph) with long-endurance tanks
     2,400 nm (4,448 km; 2,764 miles)

General configuration of the YEZ-2A Sentinel 5000

WAI SENTINEL 5000
US Navy designation: YEZ-2A

TYPE: AEW non-rigid helium airship.
PROGRAMME: Naval Airship Program (NAP) initiated by Naval Air Systems Command 1985 to investigate suitability of airships in AEW role; early history of programme in previous editions of *Jane's*; $168.9 million contract to WAI June 1987, selecting Sentinel 5000 as basis for operational development model (ODM) airship. In October 1988, NAP evolved from an exclusively US Navy programme to one by Defense Advanced Research Projects Agency (DARPA); since FY 1990 has been funded under DoD's Air Defense Initiative. Programme still focuses primarily on demonstration of long-range, long-endurance surveillance system capable of seeing low-observable sea-skimming cruise missiles; other potential applications include OTH (over the horizon) targeting, drug surveillance and interdiction functions.

Original Navy requirement was for independent airborne early warning system capable of operating with surface attack groups anywhere in the world. ODM vehicle will have unrefuelled endurance of 2-3 days; by refuelling and replenishing from surface units within a task force, a mission capability of some 30 days is intended. Designated USN missions are surveillance and targeting, AEW and communications. Current ARPA contract continues until May 1994 and includes critical design review of ODM; recently renewed USN interest expected to produce funding to build ODM for first flight 1996 and production start in 1998.
CUSTOMERS: US Navy; original NASC contract included options for up to five more Sentinels after ODM.
DESIGN FEATURES: Will be largest non-rigid ever built; general configuration similar to Skyship 600 and Sentinel 1000, but much larger, with considerably different gondola; four tail-fins in X configuration.
FLYING CONTROLS: GEC-Marconi computer controlled, optically signalled fly-by-light flight control system, using two MIL-STD-1750A standard computers and a MIL-STD-1553B databus, providing redundant fail-safe actuation of movable tail surfaces (two on each tail-fin) via fibre optic signalling; automatic stability and good handling characteristics throughout flight regime; hands-off

Ground test vehicle for the YEZ-2A Sentinel 5000 gondola and propulsion system, at Weeksville, North Carolina

autopilot control for cruising, hovering and mooring; and automatic speed control.
STRUCTURE: Envelope made of single-ply Dacron weave material, laminated with Tedlar and Mylar and bonded with Hytrel.
POWER PLANT: Unpressurised engine room aft contains CODAG (combined diesel and gas turbine) propulsion system, comprising internally mounted pair of 1,394 kW (1,870 hp) ducted propulsion CRM diesel units, with thrust vectoring, plus a 1,394 kW (1,870 shp) General Electric CT7-9 turboprop, mounted on gondola centreline at rear, to provide additional power when higher dash speed required. Substantial winching bay between engine room aft and pressurised accommodation section forward, from where refuelling, re-storing and personnel transfers will be conducted through large opening in base of gondola.
ACCOMMODATION: Crew of 10-15 in multi-deck pressurised gondola. Upper deck provides living accommodation for crew, including double cabins, showers, separate wardrobe galley and small gymnasium. Mission avionics, control information centre, galley, replenishment stores and refuelling equipment on lower main deck. Below these two is smaller flight deck containing flight controls and instrumentation.

DIMENSIONS, EXTERNAL:
  Envelope: Length overall   129.54 m (425 ft 0 in)
  Max diameter   32.00 m (105 ft 0 in)
  Height overall, incl gondola and landing gear
     46.33 m (152 ft 0 in)
  Gondola: Length overall   25.91 m (85 ft 0 in)
  Max width   5.08 m (16 ft 8 in)
  Max height   7.32 m (24 ft 0 in)
DIMENSIONS, INTERNAL:
  Envelope volume   70,792 m³ (2,500,000 cu ft)
PERFORMANCE (estimated):
  Max level speed (3 engines)
     88 knots (163 km/h; 101 mph)
  Operating height range   S/L to 3,050 m (10,000 ft)
  Pressure ceiling   4,270 m (14,000 ft)
  Max unrefuelled endurance at 40 knots (74 km/h; 46 mph) at 1,525 m (5,000 ft)   more than 60 h
  Mission capability   30 days

AERO ENGINES

This section includes all available details of engines of the aircraft which are featured in this annual. Engines of homebuilts, and propulsion systems used purely for UAVs, targets, missiles and unmanned spaceflight are no longer included. Readers are referred for these subjects to *Jane's Battlefield Surveillance Systems*, *World Unmanned Aircraft* and *Jane's Space Directory*.

AUSTRALIA

HDHV
HAWKER DE HAVILLAND VICTORIA LTD
Box 779H, GPO Melbourne, Victoria 3001
Telephone: 61 (3) 647 6111
Fax: 61 (3) 646 3431
Telex: AA 30721

HDHV makes components for engines and airframes. It makes F404 blades and seals, and assembles and tests engines for the RAAF. It makes CF6-50 and -80 rings for GE, and CFM56 rings, and supports RAAF Viper engines.

AUSTRIA

ROTAX
BOMBARDIER-ROTAX GmbH
Post Box 5, A-4623 Gunskirchen
Telephone: 43 (7246) 271-0
Fax: 43 (7246) 370
Telex: 25 546 BRG K A
PRODUCT MANAGER: Josef Fürlinger

This company is one of the world's largest producers of light piston engines. In the past 10 years it has sold 50,000 engines for aircraft propulsion. Details of the main models were given in the 1992-93 *Jane's*. The types described below are used in a growing range of kit-built and factory-built light aircraft, an increasing proportion of which are being officially certificated under the European JAR Very Light Aircraft and US FAA Primary Aircraft regulations.

ROTAX 912 A
This engine was specially designed for use in aircraft. It was added to the range in 1988. 912 A3 certificated engine with hydraulic governor for constant-speed propeller and 912 UL DCDI A3 for experimental aircraft introduced in mid-1993.
TYPE: Four-cylinder four-stroke piston engine, certificated to FAR 22. Dry sump forced lubrication. Electric starter. Reduction gear 2.273:1.
CYLINDERS: Air-cooled barrels, liquid-cooled heads. Bore 79.5 mm (3.13 in). Stroke 61.0 mm (2.40 in). Capacity 1,211.2 cc (73.912 cu in). Compression ratio 9.0:1.
IGNITION: Breakerless dual-condenser discharger.
OPTIONAL EQUIPMENT: Vacuum pump, engine mount, oil and coolant radiators, air filter and engine instruments.

Rotax 582 twin-cylinder piston engine

FUEL: Leaded or unleaded RON 95 or Avgas 100 LL.
WEIGHT, DRY:
   With carburettors, intake silencer and exhaust system
                               65.2 kg (143.7 lb)
RATING: 59.0 kW (80 hp) at 5,500 rpm

ROTAX 582
Adapted Skidoo engine frequently used in very light aircraft.
TYPE: Two-cylinder two-stroke rotary-valve piston engine.
CYLINDERS: In-line in water-cooled block. Bore 76.0 mm (2.99 in). Stroke 64.00 mm (2.52 in). Capacity 580.7 cc (35.44 cu in).
IGNITION: Dual breakerless.

Rotax 912 A four-cylinder piston engine

WEIGHT, DRY:
   With carburettors and exhaust system   33.8 kg (74.5 lb)
RATING: 47.0 kW (63 hp) at 6,500 rpm

ROTAX 532
Adapted Skidoo engine used in many very light aircraft.
TYPE: Two-cylinder two-stroke rotary-valve piston engine. Oil 1:50 mix. Fuel RON 95.
CYLINDERS: In-line in water-cooled block. Bore 72.00 mm (2.83 in). Stroke 64.0 mm (2.52 in). Capacity 521.1 cc (31.799 cu in).
IGNITION: Dual breakerless.
WEIGHT, DRY:
   With carburettors   33.0 kg (72.8 lb)
RATING: 48.0 kW (64.0 hp) at 6,500 rpm

BELGIUM

TECHSPACE
TECHSPACE AERO SA (Subsidiary of Fabrique Nationale Nouvelle Herstal)
Route de Liers 121, B-4411 Herstal
Telephone: 32 (41) 784671
Fax: 32 (41) 785207 (general)
     32 (41) 786739 (sales & marketing)
Telex: B 41223 FABNA
DIRECTOR AND GENERAL MANAGER: P. Bourgeois
BUSINESS DEVELOPMENT: J. C. Morin

Previously known as FN Moteurs, this company began jet engine production in 1949. Today's activity is equally distributed in three sectors: fighter engine production, medium/large transport and space propulsion production, and depot maintenance for air forces. Major programmes are:

Production and assembly of P&W F100 fan and engine core modules, and for assembly and test of complete engines.

Responsible for four major parts of GE F110: fan discs 1 and 3, HP turbine disc, and fan stator case.

Member of consortium (RR, MTU, SNECMA, TA) producing Tyne 21 and 22 engines for Atlantique and Transall, with a 9.5 per cent share.

After having produced parts for the JT8D and JT9D-7R4, TA signed a partnership agreement with Pratt & Whitney for a 3 per cent share in the PW4000 series. TA is responsible for the HP compressor case and various other components.

Since 1972, in association with SNECMA, TA has developed and produced the lubrication modules of all CFM56 versions. TA has increased its participation up to 10 per cent of the SNECMA share (5 per cent overall) in the -5A and -5C versions, to power respectively the Airbus A320 and A340. TA is responsible for the lubrication and shop modules and other stator or rotor parts. TA is also performing endurance testing of complete engines.

TA has signed a co-operation agreement with Turbomeca, and is developing several critical parts of the TM 333.

BRAZIL

CELMA
CELMA-CIA ELECTROMECANICA
PO Box 90341, 25669-900 Petrópolis, RJ
Telephone: 55 (242) 43 4962
Fax: 55 (242) 42 3684
Telex: (021) 21271
PRESIDENT: Alexandre Gonçalves Silva

TECHNICAL DIRECTOR: Carlos A. R. Pereira
This company of 1,050 people has facilities totalling 45,000 m² (485,000 sq ft) in which it overhauls many kinds of engine (GE, P&W, P&WC, CFM and RR) and accessories. It shares in the production of components for the Rolls-Royce Spey 807 turbofan.

CANADA

P&WC
PRATT & WHITNEY CANADA (Subsidiary of United Technologies Corporation)

1000 Marie Victorin, Longueuil, Quebec J4G 1A1
Telephone: 1 (514) 677 9411
Fax: 1 (514) 647 3620
Telex: 05 267509
PRESIDENT AND CEO: L. D. Caplan
EXECUTIVE VICE-PRESIDENT: G. P. Ouimet
VICE-PRESIDENTS:
  C. Lloyd (Marketing and Customer Support)
  F. Osborne (Communications)
  B. H. Sanders (Engineering)
PUBLIC RELATIONS DIRECTOR: Louise Fleischmann

Pratt & Whitney Canada is owned 97 per cent by United Technologies Corporation, Connecticut, USA, and is the P&W Group member responsible for engines for general aviation and regional transport. P&WC employs some 8,500 persons. By 1 January 1993 it had delivered 37,787 engines. Link with Klimov (Russia) June 1993.

P&WC JT15D

Designed to power business aircraft, small transports and training aircraft, the JT15D first ran on 23 September 1967. Initial application was the twin-engined Cessna Citation. Up to 1976 Cessna used the **JT15D-1**. Late that year it announced the Citation I powered by the **JT15D-1A** and the Citation II powered by the **JT15D-4**. During 1983 the D-1A was replaced by the **D-1B**.

Other twin-engined business jets powered by the JT15D-4 are the Aérospatiale Corvette and Mitsubishi Diamond I. TBO is 3,500 h for the JT15D-1/D-1A, and 3,000 h for the JT15D-4.

JT15D-4B. Altitude optimised variant for the Citation S/II.
JT15D-4C. Oil system for sustained inverted flight, and an electronic fuel control. Powers Agusta S.211.
JT15D-4D. Flat rated for improved hot/high performance for the Diamond IA.
JT15D-5. Growth version. A new fan with higher pressure ratio and flow, plus an improved boost stage and HP compressor, are combined to produce 25 per cent more altitude cruise thrust, with a 3 per cent improvement in sfc. HP turbine blades and electronic fuel control are also improved. Powers Cessna T-47A, Diamond II and Beechjet.
JT15D-5A. Hydromechanical fuel control. Selected for Citation V.
JT15D-5B. Dash-5A modified to suit Beech T-1A Jayhawk.
JT15D-5C. Oil system for sustained inverted flight. Powers Agusta S.211A.

By 1993 total deliveries of all JT15D engines had reached 4,543. Operating time totalled 14,563,546 h.

The following description relates to the JT15D-1B, except where indicated:
TYPE: Two-shaft turbofan.
FAN: Single-stage axial with 28 solid titanium blades with part-span shrouds. Mass flow, 34 kg (75 lb)/s; bypass ratio about 3.3; fan pressure ratio 1.5.
COMPRESSOR: Single stage titanium centrifugal. Overall pressure ratio about 10:1. (D-4 and D-5 have axial boost stage between fan and compressor.)
COMBUSTION CHAMBER: Annular reverse flow type. Spark igniters at 5 and 7 o'clock (viewed from rear).
FUEL SYSTEM: Pump delivering at 44.8 bars (650 lb/sq in). D-1, 4, 4B, 4D, 5A have DP-L2 hydromechanical control; 4C, 5 and 5C have JFC 118 or 119 electronic system.
FUEL GRADES: JP-1, JP-4, JP-5 to CPW 204.
TURBINE: Single-stage HP with 71 solid blades; two-stage LP, first stage cast integrally with 61 blades and second carrying 55 blades in fir tree roots.
LUBRICATION SYSTEM: Integral oil system, with gear type pump delivering at up to 5.52 bars (80 lb/sq in). Capacity, 9.0 litres (2.4 US gallons; 2.0 Imp gallons).
OIL SPECIFICATION: PWA521 Type II, CPW 202.
STARTING: Air turbine starter or electric starter/generator.
DIMENSIONS:
  Diameter: JT15D-1 691 mm (27.2 in)
  JT15D-4 686 mm (27.0 in)
  Length overall: JT15D-1 1,506 mm (59.3 in)
  JT15D-4 1,600 mm (63.0 in)
WEIGHT, DRY:
  JT15D-1, 232.5 kg (514 lb)
  JT15D-1A, -1B 235 kg (519 lb)
  JT15D-4 253 kg (557 lb)
  JT15D-4B 258 kg (568 lb)
  JT15D-4C 261 kg (575 lb)
  JT15D-4D 255 kg (560 lb)
  JT15D-5, -5A 291.5 kg (632 lb)
  JT15D-5B 291.7 kg (643 lb)
  JT15D-5C 302 kg (665 lb)
PERFORMANCE RATINGS:
  T-O: JT15D-1, -1A, -1B 9.79 kN (2,200 lb st)
  JT15D-4, -4B, -4C, -4D 11.12 kN (2,500 lb st)
  JT15D-5, -5A, -5B 12.9 kN (2,900 lb st)
  JT15D-5C 14.19 kN (3,190 lb st)
Max continuous:
  JT15D-1, -1A, -1B 9.3 kN (2,090 lb st)
  JT15D-4, -4B, -4C, -4D 10.56 kN (2,375 lb st)
  JT15D-5, -5A, -5B 12.9 kN (2,900 lb st)
  JT15D-5C 14.19 kN (3,190 lb st)
SPECIFIC FUEL CONSUMPTION (T-O):
  JT15D-1, -1A, -1B 15.30 mg/Ns (0.540 lb/h/lb st)
  JT15D-4, -4B, -4C, -4D 15.92 mg/Ns (0.562 lb/h/lb st)
  JT15D-5, -5A, 5B 15.61 mg/Ns (0.551 lb/h/lb st)
  JT15D-5C 16.23 mg/Ns (0.573 lb/h/lb st)

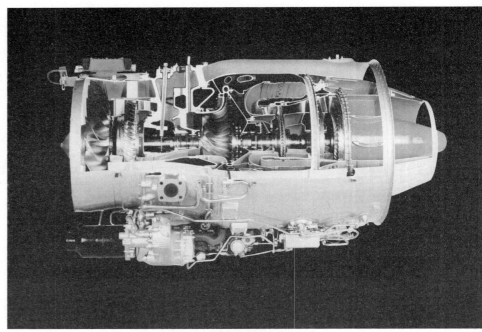

Cutaway P&WC JT15D-5 turbofan

Cutaway P&WC PW305 turbofan

P&WC PW305

This high BPR turbofan has been optimised for advanced intercontinental business aircraft. MTU of Germany is responsible for the LP turbine section, as a partner sharing 25 per cent of the estimated development cost of C$500 million.

Transport Canada certification was achieved in mid-1990. FAA certification and CAA validation followed in mid-1991, and in 1992 127 engines had flown 8,962 h. The PW305 powers the BAe 1000, Learjet 60 and Volpar F20 re-engine.
TYPE: Two-shaft turbofan.
FAN: Single-stage, overhung ahead of front bearing. Pointed rotating spinner. Bypass ratio (S/L, T-O) 4.5.
COMPRESSOR: Four axial stages followed by one centrifugal. Core pressure ratio 15.
COMBUSTION CHAMBER: Annular, fed around periphery by ring of separate curved pipes ducting air from diffuser case of centrifugal compressor.
FUEL GRADES: JP-1, JP-4, JP-5 to CPW 204.
CONTROL SYSTEM: Dowty and Smiths Industries full-authority digital, with dual channels.
HP TURBINE: Two axial stages, the first having air-cooled blades.
LP TURBINE: Three axial stages joined via centre stage disc to fan shaft. Two main shaft bearings.
LUBRICATION SYSTEM: Integral oil system with gear-type pump. Capacity 8 litres (2.11 US gallons; 1.76 Imp gallons).
OIL SPECIFICATION: CPW 202.
STARTING: Electric starter/generator.
DIMENSIONS (estimated):
  Diameter 970.3 mm (38.2 in)
  Length 2,062 mm (81.2 in)
WEIGHT, DRY (installed): 450 kg (993 lb)
PERFORMANCE RATINGS (ISA):
  T-O, S/L 23.24 kN (5,225 lb st) or flat rated at 21.13 kN (4,750 lb st) to 25°C
  Cruise at Mach 0.8 at 12,200 m (40,000 ft)
    5.03 kN (1,132 lb)
SPECIFIC FUEL CONSUMPTION (estimated):
  T-O as above 11.53 mg/Ns (0.407 lb/h/lb st)
  Cruise as above 19.40 mg/Ns (0.685 lb/h/lb)

P&WC PW100

The PW100 is a free turbine turboprop consisting of turbomachine and reduction gearbox modules connected by a rque-measuring driveshaft and integrated structural intake ase.

Flight development of the PW100 began in February 1982. rincipal versions are as follows:

PW118. T-O rated at 1,411 ekW; 1,342 kW (1,892 ehp; ,800 shp) at 1,300 propeller rpm to 33°C. Selected for MB-120 Brasilia. Certificated March 1986.

PW118A. T-O rated at 1,411 ekW; 1,342 kW (1,892 ehp; ,800 shp) at 1,300 propeller rpm to 42°. Selected for MB-120 Brasilia. Certificated June 1987.

PW119B. T-O rated at 1,702 ekW; 1,626 kW (2,282 ehp; ,180 shp) at 1,300 propeller rpm to 23°C. Selected for Dornier 328.

PW120. T-O rated at 1,566 ekW; 1,491 kW (2,100 ehp; ,000 shp) at 1,200 propeller rpm to 27.7°C. Selected for Aérospatiale/Alenia ATR 42 and Snow SA-210TA. Certificated December 1983.

PW120A. T-O rated at 1,566 ekW; 1,491 kW (2,100 ehp; ,000 shp) at 1,200 propeller rpm to 29°C. Selected for Dash 8-100. Certificated September 1984.

PW121. T-O rated at 1,679 ekW; 1,603 kW (2,252 ehp; ,150 shp) at 1,200 propeller rpm to 26°C. Selected for Dash 8-100 and ATR 42. Certificated February 1987.

PW123. T-O rated at 1,866 ekW; 1,775 kW (2,502 ehp; ,380 shp) at 1,200 propeller rpm to 35°C. Selected for Dash 8-300. Certificated June 1987.

PW123AF. T-O rated at 1,866 ekW; 1,775 kW (2,502 ehp; 2,380 shp) at 1,200 propeller rpm to 35°C. Selected for CL-215/415.

PW123B. T-O rated at 1,958 ekW; 1,864 kW (2,626 ehp; 2,500 shp) at 1,200 propeller rpm to 30°C. Selected for Dash 8-300.

PW124. Growth version, with T-O rating of 1,880 ekW; 1,790 kW (2,522 ehp; 2,400 shp) to 34.4°C. **PW124A** is identical but has higher max continuous power rating.

PW124B. PW124 with PW123 turbomachinery to suit four-blade propeller at 1,200 rpm, same rating. Selected for ATR 72. Certificated May 1988.

PW125B. Growth PW124 with T-O rating of 1,958 ekW; 1,864 kW (2,626 ehp; 2,500 shp) at 1,200 propeller rpm to 30°C. Powers Fokker 50. Certificated May 1987.

PW126. Growth engine, max contingency 2,078 ekW; 1,978 kW (2,786 ehp; 2,653 shp) at 1,200 propeller rpm to 32.4°C. Powers BAe ATP. Certificated May 1987.

PW126A. Growth 124A with T-O rating of 2,084 ekW; 1,985 kW (2,795 ehp; 2,662 shp) at 1,200 propeller rpm to 29.2°C. Selected for BAe ATP. Certificated June 1989.

PW127. T-O rated at 2,148 ekW; 2,051 kW (2,880 ehp; 2,750 shp) at 1,200 propeller rpm to 31.6°C. Selected for ATR 72 and ATR 42.500.

PW127B. T-O rated at 2,147 ekW; 2,051 kW (2,880 ehp; 2,750 shp) at 1,200 propeller rpm to 30°C. Selected for hot/high-performance Fokker 50.

PW127C. T-O rated at 2,148 ekW; 2,051 kW (2,880 ehp; 2,750 shp) at 1,200 propeller rpm to 30.2°C. Selected for XAC Y7-200A and, as **PW127D**, for late BAe ATPs and Jetstream 61.

By December 1992 deliveries of PW100 engines had reached 2,794. Operating time reached 14,619,590 h.

The following description applies to all PW100 series engines:

TYPE: Free turbine turboprop.

PROPELLER DRIVE: Twin-layshaft gearbox with propeller shaft offset above turbomachine. Max propeller speed 1,200 rpm.

AIR INTAKE: S-bend duct. A secondary duct forms a flowing bypass to prevent foreign object ingestion.

COMPRESSOR: Two centrifugal impellers in series, each driven by its own turbine. Air guided through ring of curved pipes from LP diffuser to HP entry.

COMBUSTION CHAMBER: Annular reverse flow type, with 14 air blast fuel nozzles around periphery and two spark igniters.

FUEL SYSTEM: Hydromechanical control and electronic power management.

FUEL GRADES: JP-1, JP-4, JP-5 to PWA Spec 522.

TURBINES: Single-stage HP with 47 air-cooled blades. Single-stage LP with 53 solid blades. Two-stage power turbine, first with 68 blades and second with 74, all with shrouded tips.

ACCESSORY DRIVES: Pads driven by HP compressor, for starter/generator, hydromechanical fuel control and hand turning. Pads on reduction gearbox for alternator, hydraulic pump, propeller control module, overspeed governor and electric auxiliary pump. Electric torque signal and auto power augmentation.

LUBRICATION SYSTEM: One pressure pump and two scavenge pumps, all driven off HP rotor. Integral tank, capacity 9.44 litres (2.5 US gallons, 2.08 Imp gallons).

OIL SPECIFICATION: CPW202 or PWA521 Type II.

STARTING: Electric starter/generator.

DIMENSIONS:
Length: PW118, 118A, 119A 2,057 mm (81 in)
others 2,134 mm (84 in)
Width: PW118-121 635 mm (25 in)
others 660 mm (26 in)
Height: PW118-121 787 mm (31 in)
others 838 mm (33 in)

WEIGHT, DRY:
PW118 391 kg (861 lb)
PW118A 394 kg (866 lb)
PW119A 415.5 kg (916 lb)

Cutaway P&WC PW100 turboprop

PW120 417.8 kg (921 lb)
PW120A 423 kg (933 lb)
PW121 425 kg (936 lb)
PW123, 123B, 123AF 450 kg (992 lb)
others 481 kg (1,060 lb)

PERFORMANCE RATINGS (S/L, static):
T-O: See under model listings
Max cruise:
PW118 1,188 ekW; 1,128 kW
(1,593 ehp; 1,512 shp) at 1,300 rpm to 20°C
PW118A 1,188 ekW; 1,127 kW
(1,594 ehp; 1,512 shp) at 1,300 rpm to 29.4°C
PW119B 1,339 ekW; 1,267 kW
(1,820 ehp; 1,734 shp) at 1,300 rpm to 22°C
PW120 1,271 ekW; 1,207 kW
(1,704 ehp; 1,619 shp) at 1,200 rpm to 15°C
PW120A 1,296 ekW; 1,231 kW
(1,738 ehp; 1,651 shp) at 1,200 rpm to 15°C
PW121 1,330 ekW; 1,268 kW
(1,784 ehp; 1,700 shp) at 1,200 rpm to 15°C
PW123, 124 1,593 ekW; 1,514 kW
(2,136 ehp; 2,030 shp) at 1,200 rpm to 22.2°C
PW124A 1,635 ekW; 1,553 kW
(2,192 ehp; 2,083 shp) at 1,200 rpm to 26°C
PW124B 1,593 ekW; 1,514 kW
(2,136 ehp; 2,030 shp) at 1,200 rpm to 22.2°C
PW125B 1,623 ekW; 1,513 kW
(2,203 ehp; 2,030 shp) at 1,200 rpm to 22.2°C
PW126 1,635 ekW; 1,553 kW
(2,192 ehp; 2,083 shp) at 1,200 rpm to 26.3°C
PW126A 1,632 ekW; 1,553 kW
(2,190 ehp; 2,081 shp) at 1,200 rpm to 27.2°C
PW127 1,667 ekW; 1,591 kW
(2,236 ehp; 2,134 shp) at 1,200 rpm to 22.7°C
PW127B 1,424 ekW; 1,351 kW
(1,911 ehp; 1,812 shp) at 1,200 rpm to 34.7°C
PW127C 1,667 ekW; 1,589 kW
(2,237 ehp; 2,132 shp) at 1,200 rpm to 20.5°C

SPECIFIC FUEL CONSUMPTION:
T-O rating:
PW118 84.2 μg/J (0.498 lb/h/ehp)
PW118A 85.2 μg/J (0.504 lb/h/ehp)
PW119B 82.8 μg/J (0.490 lb/h/ehp)
PW120, 120A 82.0 μg/J (0.485 lb/h/ehp)
PW121 80.4 μg/J (0.476 lb/h/ehp)
PW123, 123AF 79.4 μg/J (0.470 lb/h/ehp)
PW124, 124A, 124B 79.1 μg/J (0.468 lb/h/ehp)
PW123B, 125B, 126 78.2 μg/J (0.463 lb/h/ehp)
PW126A 77.9 μg/J (0.461 lb/h/ehp)
PW127, 127B, 127C 77.6 μg/J (0.459 lb/h/ehp)

P&WC PT6A

US military designations: **T74** (separate entry) and **T101** (PT6A-45R)

The PT6A is a free turbine turboprop, built in many versions since November 1959. By late 1992 a total of 30,015 had logged over 208 million hours with 5,237 customers.

Current versions of the PT6A are as follows:

PT6A-11. Flat rated at 394 ekW; 373 kW (528 ehp; 500 shp) at 2,200 propeller rpm to 42°C. Piper Cheyenne I and IA, T-1040, original Claudius Dornier Seastar and Harbin Y-12 I.

PT6A-110. Flat rated at 374 ekW; 354 kW (502 ehp; 475 shp) at 1,900 propeller rpm to 38°C. Dornier 128-6.

PT6A-11AG. Flat rated at 394 ekW; 373 kW (528 ehp; 500 shp) at 2,200 propeller rpm to 42°C. Can use diesel fuel. Ayres Turbo-Thrush, Schweizer Turbo Ag-Cat and Weatherly 620 TP.

PT6A-112. Flat rated at 394 ekW; 373 kW (528 ehp; 500 shp) at 1,900 propeller rpm to 56°C. Cessna Conquest I, Reims-Cessna F 406 Caravan II, Dornier Composite Seastar CD2 and AAC Turbine P-210 conversion.

PT6A-114. Flat rated at 471 ekW; 447 kW (632 ehp; 600 shp) at 1,900 propeller rpm to 54.4°C. Cessna Caravan I, with single exhaust.

PT6A-114A. Flat rated at 529 ekW; 503 kW (709 ehp; 675 shp) at 1,900 propeller rpm to 46°C. Cessna Caravan 208B.

PT6A-15AG. Flat rated at 533 ekW; 507 kW (715 ehp; 680 shp) at 2,200 propeller rpm to 22°C. Can use diesel fuel. Ayres Turbo-Thrush, Frakes Turbo-Cat, Schweizer Turbo Ag-Cat D, Air Tractor AT-400, and prototypes of IAR-827TP and IAR-825TP Triumf.

PT6A-21. Flat rated at 432.5 ekW; 410 kW (580 ehp; 550 shp) at 2,200 propeller rpm to 33°C. Mates A-27 power unit with A-20A gearbox. Beechcraft King Air C90.

PT6A-25. Flat rated at 432.5 ekW; 410 kW (580 ehp; 550 shp) at 2,200 propeller rpm to 33°C. Oil system for sustained inverted flight. Beechcraft T-34C and NAC Firecracker.

PT6A-25A. Some castings of magnesium alloy instead of aluminium alloy. Pilatus PC-7.

PT6A-25C. Flat rated at 584 ekW; 559 kW (783 ehp; 750 shp) at 2,200 propeller rpm to 31°C. A-25 with A-34 hot end and A-27 first stage reduction gearing. Embraer EMB-312.

PT6A-27. Flat rated at 553 ekW; 507 kW (715 ehp; 680 shp) at 2,200 propeller rpm to 22°C, attained by increase in mass flow, at lower turbine temperatures than in PT6A-20. Hamilton Westwind II/III (Beech 18) conversions, Beechcraft 99 and 99A, and U-21A and U-21D, DHC-6 Twin Otter 300, Pilatus/Fairchild Industries PC-6/B2-H2 Turbo-Porter, Frakes Aviation (Grumman) Mallard conversion, Let L-410A Turbolet, Saunders ST-27A conversion, Embraer EMB-110 Bandeirante (early), Harbin Y-12 II, SAC Spectrum-One, and Schafer Comanchero 500B/Neiva Carajá.

PT6A-28. Similar to PT6A-27, this has an additional cruise rating of 486 ekW (652 ehp) available to 21°C and max cruise up to 33°C. Beechcraft King Air E90 and A100, and 99A, Piper Cheyenne II, Avtek 400 and Embraer Xingu I.

PT6A-34. Flat rated at 584 ekW; 559 kW (783 ehp; 750 shp) at 2,200 propeller rpm to 31°C, this version has air-cooled nozzle guide vanes. IAI 102/201 Arava, Saunders ST-28, Frakes Aviation (Grumman) Mallard conversion, Airmaster Avalon 680, Omni Turbo Titan, Spectrum-One, Embraer EMB-110P1/P2 and EMB-111, and Carajá.

PT6A-34B. Aluminium alloy replaces magnesium in major castings. Beechcraft T-44A.

PT6A-34AG. Agricultural, certificated on diesel fuel. Frakes conversion of Ag-Cat and Ayres Turbo-Thrush. PZL-106AT/BT Turbo-Kruk, Schweizer Turbo Ag-Cat and Croplease Fieldmaster.

PT6A-135. Flat rated at 587 ekW; 559 kW (787 ehp; 750 shp) at 1,900 rpm. Changed drive ratio to reduce noise; higher cycle temperatures. JetCrafters Taurus, Embraer 121A1 Xingu II, Piper Cheyenne IIXL, Advanced Aircraft Turbine P-210 and Regent 1500, Airmaster Avalon 680, and Schafer Comanchero/Comanchero 750 conversions.

PT6A-135A. Higher thermodynamic ratings. Avtek 400, Beech F90-1, Dornier Composite Seastar and OMAC Laser 300.

PT6A-36. Flat rated at 586 ekW; 559 kW (786 ehp; 750 shp) at 2,200 rpm to 36°C. Similar to -34 but higher cycle temperatures. IAI 101B/202 Arava and Beechcraft C99.

PT6A-41. Higher mass flow, air-cooled nozzle guide vanes and two-stage free turbine. T-O rating of 673 ekW; 634 kW (903 ehp; 850 shp) at 2,000 propeller rpm to 41°C. Thermodynamic power 812 ekW (1,089 ehp). Beechcraft Super King Air 200 and C-12, and Piper Cheyenne III.

PT6A-41AG. For agricultural aviation. Frakes Turbo-Cat and Schweizer Turbo Ag-Cat.

PT6A-42. A-41 with increase in cruise performance. Beechcraft Super King Air B200.

PT6A-45A. A-41 with gearbox to transmit higher powers at reduced speeds. Rated at 916 ekW; 875 kW (1,229 ehp; 1,173 shp) at 1,700 rpm to 8°C, or to 21°C with water injection. Shorts 330 and Mohawk 298.

PT6A-45B. A-45A with increased water injection.

PT6A-45R. A-45B with reserve power rating and deleted water system.

T101-CP-100. A-45R for Shorts C-23A.

PT6A-50. A-41 with higher ratio reduction gear for quieter operation. T-O 875.5 ekW; 835 kW (1,174 ehp; 1,120 shp) with water at 1,210 propeller rpm up to 34°C. DHC-7.

PT6A-60A. A-45B with jet flap intake and increased mass flow for high altitude cruise. Rated at 830 ekW; 783 kW (1,113 ehp; 1,050 shp) at 1,700 rpm to 25°C. Beech Super King Air 300 and OMAC Laser 300A.

PT6A-61. A-60 gas generator matched with A-41 power section with 2,000 rpm gearbox. T-O rating 673 ekW; 634 kW (903 ehp; 850 shp) to 46°C. Cheyenne IIIA.

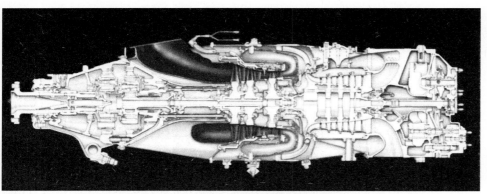

Longitudinal section of P&WC PT6A-67 turboprop

PT6A-62. Flat rated at 708 kW (950 shp). Pilatus PC-9.
PT6A-64. A-67 gas generator with A-61 gearbox. Flat rated at 522 kW (700 shp) at 2,000 rpm to 63.5°C. TBM 700.
PT6A-65B. A-65R without reserve rating. Flat rated at 875.5 ekW; 820 kW (1,174 ehp; 1,100 shp) at 1,700 rpm to 51°C. Beechcraft 1900 and C-12J, and CASA Aviocar C-212P.
PT6A-65R. A-45R with four-stage compressor with jet flap intake, fuel control and fuel dump. Improved hot end and exhaust duct. Reserve power 1,087 ekW; 1,026 kW (1,459 ehp; 1,376 shp) at 1,700 rpm to 28°C. Alternative T-O at 975 ekW; 917 kW (1,308 ehp; 1,230 shp) at 1,700 rpm to 24°C. Shorts 360.
PT6A-65AR. Reserve power 1,125 ekW; 1,062 kW (1,509 ehp; 1,424 shp) at 1,700 rpm to 27.7°C. Shorts 360 and AMI DC-3.
PT6A-66. Flat rated at 674 ekW; 534 kW (905 ehp; 850 shp) at 2,000 rpm to 62.2°C. Piaggio Avanti with opposed rotation gearbox.
PT6A-67. Flat rated at 870 ekW; 820 kW (1,167 ehp; 1,100 shp) at 1,700 rpm to 60°C. Beechcraft RC-12K.
PT6A-67A. Flat rated at 948 ekW; 894 kW (1,272 ehp; 1,200 shp) at 1,700 rpm to 51°C. Beech Starship 2000.
PT6A-67AF. Flat rated at 1,125 ekW; 1,061 kW (1,509 ehp; 1,424 shp) at 1,700 rpm to 37.2°C. Conair Turbo Firecat.
PT6A-67B. A-67 modified for medium altitudes. Flat rated at 895 kW (1,200 shp) at 1,700 rpm to 52°C. Pilatus PC-12.
PT6A-67CF. Growth version of A-67AF. Flat rated at 1,231 kW (1,650 shp) at 1,700 rpm to 37.8°C. Brazilian S-2 Turbine Tracker.
PT6A-67D. A-67B with A-67R gearbox. Flat rated at 909 kW (1,219 shp) at 1,700 rpm to 39°C. Beech 1900D.
PT6A-67R. A-67 with reserve power rating for commuter aircraft. Flat rated at 1,125 ekW; 1,061 kW (1,509 ehp; 1,424 shp) at 1,700 rpm to 37.2°C. Shorts 360-300 and Basler Turbo 67.

The following data apply generally to the PT6A series:
TYPE: Free turbine axial-plus-centrifugal turboprop.
PROPELLER DRIVE (all models up to and including PT6A-41): Two-stage planetary. Ratio 15. Higher ratio gears for A-45R, -50, -60, -65 and -67.
AIR INTAKE: Annular at rear with screen. Aircraft-supplied alcohol or inertial anti-icing.
COMPRESSOR: Three axial stages, plus single centrifugal (-65 series, four axial stages). PT6A-27: pressure ratio 6.7, mass flow 3.1 kg (6.8 lb/s). PT6A-65: pressure ratio 10, mass flow 4.3 kg (9.5 lb/s).
COMBUSTION CHAMBER: Annular reverse flow, with 14 simplex burners.
FUEL SYSTEM: Bendix DP-F2 pneumatic automatic fuel control. A-50 has DP-F3 with starting spill valve and motive flow systems; A-60 series (except -62) have Woodward 83212 hydromechanical system.
FUEL GRADE: JP-1, JP-4, JP-5, MIL-J-5624. Gasolines (MIL-G-5572) grades 80/87, 91/98, 100/130 and 115/145 for up to 150 h during any overhaul period. Agricultural use, diesel.
TURBINES: Up to A-34 two single-stage axial; HP (58 blades) drives compressor, and LP (41 shrouded blades) drives output shaft. A-41 onward have two-stage LP turbine. All blades have fir tree root.
ACCESSORIES: Pads on accessory case (rear of engine) for starter/generator, hydraulic pump, aircraft accessory drive, vacuum pump and tachometer generator. Pad on reduction gear for propeller governor, constant speed unit and tachometer.
LUBRICATION SYSTEM: One pressure and four scavenge elements driven by gas generator. Integral oil tank 8.75 litres (2.3 US gallons).
OIL SPECIFICATION: CPW202, PWA521 Type II (7.5 cs vis) (MIL-L-23699, MIL-L-7808 for military engines).
STARTING: Electric starter/generator on accessory case.
DIMENSIONS:

Max diameter	483 mm (19 in)
Length, excl accessories:	
PT6A-11 to -36	1,575 mm (62 in)
PT6A-41, 42, 61	1,701 mm (67 in)
PT6A-45, 60A	1,829 mm (72 in)
PT6A-50	2,133 mm (84 in)
PT6A-62, -66	1,778 mm (70 in)
PT6A-65B, -67	1,880 mm (74 in)
PT6A-65R, -65AR	1,905 mm (75 in)

WEIGHT, DRY:

PT6A-11	148.8 kg (328 lb)
PT6A-11AG	149.7 kg (330 lb)
PT6A-110, -112	151.5 kg (334 lb)
PT6A-114	158.8 kg (350 lb)
PT6A-15AG, -21, -27, -28	148.8 kg (328 lb)
PT6A-25	160.1 kg (353 lb)
PT6A-25A	155.6 kg (343 lb)
PT6A-25C	151.9 kg (335 lb)
PT6A-34, -36	150.1 kg (331 lb)
PT6A-34B, -135, -135A	156.0 kg (344 lb)
PT6A-34AG	150.1 kg (331 lb)
PT6A-41, -42	182.8 kg (403 lb)
PT6A-41AG	186.9 kg (412 lb)
PT6A-45A	196.8 kg (434 lb)
PT6A-45R	203.2 kg (448 lb)
PT6A-50	275.3 kg (607 lb)
PT6A-60A	215.5 kg (475 lb)
PT6A-61	194.6 kg (429 lb)
PT6A-62	205.9 kg (454 lb)
PT6A-64	207.0 kg (456 lb)
PT6A-65B, -65R	218.2 kg (481 lb)
PT6A-65AR	220.4 kg (486 lb)
PT6A-66	213.2 kg (470 lb)
PT6A-67, -67A	229.5 kg (506 lb)
PT6A-67AF	241.0 kg (532 lb)
PT6A-67CF	253.0 kg (558 lb)
PT6A-67R, -67B, -67D	233.5 kg (515 lb)

PERFORMANCE RATINGS (S/L, static):
T-O: See under model listings
Max continuous:

PT6A-110	374 ekW; 354 kW (502 ehp; 475 shp) at 1,900 rpm (to 38°C)
PT6A-11, 11AG	394 ekW; 373 kW (528 ehp; 500 shp) at 2,200 rpm (to 42°C)
PT6A-112	394 ekW; 373 kW (528 ehp; 500 shp) at 1,900 rpm (to 56°C)
PT6A-114	471 ekW; 447 kW (632 ehp; 600 shp) at 1,900 rpm (to 54.4°C)
PT6A-15AG, -27, -28	533 ekW; 507 kW (715 ehp; 680 shp) at 2,200 rpm (to 22°C)
PT6A-21	432.5 ekW; 410 kW (580 ehp; 550 shp) at 2,200 rpm (to 33°C)
PT6A-25, -25A	432.5 ekW; 410 kW (580 ehp; 550 shp) at 2,200 rpm (to 33°C)
PT6A-25C	584 ekW; 559 kW (783 ehp; 750 shp) at 2,200 rpm (to 31°C)
PT6A-34	584 ekW; 559 kW (783 ehp; 750 shp) at 2,200 rpm (to 30°C)
PT6A-135	587 ekW; 559 kW (787 ehp; 750 shp) at 1,900 rpm (-135 to 29°C, -135A to 34°C)
PT6A-36	586 ekW; 559 kW (786 ehp; 750 shp) at 2,200 rpm (to 36°C)
PT6A-41	673 ekW; 634 kW (903 ehp; 850 shp) at 2,000 rpm (to 41°C)
PT6A-42	674 ekW; 634 kW (904 ehp; 850 shp) at 2,000 rpm (to 41°C)
PT6A-45A, -45B, -45R	798 ekW; 761 kW (1,070 ehp; 1,020 shp) at 1,700 rpm (to: -45A, 26.7°C; -45B, 29°C; -45R, 33°C)
PT6A-50	762 ekW; 725.5 kW (1,022 ehp; 973 shp) at 1,210 rpm (to 32°C)
PT6A-60A	830 ekW; 783 kW (1,113 ehp; 1,050 shp) at 1,700 rpm (to 25°C)
PT6A-61	673 ekW; 634 kW (902 ehp; 850 shp) at 2,000 rpm (to 46°C)
PT6A-62	751 ekW; 708 kW (1,008 ehp; 950 shp) at 2,000 rpm (to 37°C)
PT6A-64	557 ekW (747 ehp) at 2,000 rpm (to 63.5°C)
PT6A-65B	875 ekW; 820 kW (1,174 ehp; 1,100 shp) at 1,700 rpm (to 45°C)
PT6A-65R	931 ekW; 875 kW (1,249 ehp; 1,173 shp) at 1,700 rpm (to 38°C)
PT6A-66	675 ekW; 634 kW (905 ehp; 850 shp) at 2,000 rpm (to 62°C)
PT6A-67	870 ekW; 820 kW (1,167 ehp; 1,100 shp) at 1,700 rpm (to 54.4°C)
PT6A-67B	949 ekW (1,272 ehp) at 1,700 rpm (to 52°C)
PT6A-67CF	1,298 ekW (1,740 ehp) at 1,700 rpm (to 37.8°C)
PT6A-67D	1,009 ekW (1,353 ehp) at 1,700 rpm (to 39°C)
PT6A-67R	965 ekW; 910 kW (1,294 ehp; 1,220 shp) at 1,700 rpm (to 48°C)

Max cruise rating:

PT6A-110	374 ekW; 354 kW (502 ehp; 475 shp) at 1,900 rpm (to 19°C)
PT6A-11	394 ekW; 373 kW (528 ehp; 500 shp) at 2,200 rpm (to 37°C)
PT6A-11AG	394 ekW; 373 kW (528 ehp; 500 shp) at 2,200 rpm (to 36°C)
PT6A-112	394 ekW; 373 kW (528 ehp; 500 shp) at 1,900 rpm (to 48°C)
PT6A-114	471 ekW; 447 kW (632 ehp; 600 shp) at 1,900 rpm (to 31.1°C)
PT6A-15AG	as PT6A-2
PT6A-21, -25, -25A	432.5 ekW; 410 kW (580 ehp; 550 shp) at 2,200 rpm (to 47.2°C; 33°C; 33°C)
PT6A-25C	545 ekW; 522 kW (731 ehp; 700 shp) at 2,200 rpm (to 19°C)
PT6A-27	486 ekW; 462 kW (652 ehp; 620 shp) at 2,200 rpm (to 21°C)
PT6A-28	486 ekW; 462 kW (652 ehp; 620 shp) at 2,200 rpm (to 33°C)
PT6A-34	545 ekW; 522 kW (731 ehp; 700 shp) at 2,200 rpm (to 19°C)
PT6A-135	548 ekW; 522 kW (735 ehp; 700 shp) at 1,900 rpm (-135 to 37°C, -135A to 41°C)
PT6A-36	548 ekW; 522 kW (735 ehp; 700 shp) at 2,200 rpm (to 19°C)
PT6A-41	673 ekW; 634 kW (903 ehp; 850 shp) at 2,000 rpm (to 28°C)
PT6A-42	674 ekW; 634 kW (904 ehp; 850 shp) at 2,000 rpm (to 33°C)
PT6A-45	749 ekW; 713 kW (1,004 ehp; 956 shp) at 1,425 rpm (to 15°C)
PT6A-50	706 ekW; 671 kW (947 ehp; 900 shp) at 1,020-1,160 rpm (to 23°C)
PT6A-60A	791 ekW; 746 kW (1,061 ehp; 1,000 shp) at 1,700 rpm (to 28°C)
PT6A-61	as max continuous but to 43.8°C
PT6A-62	712 ekW; 671 kW (955 ehp; 900 shp) at 2,000 rpm (to 32°C)
PT6A-64	557 ekW (747 ehp) at 2,000 rpm (to 58°C)
PT6A-65B, -65R	762 ekW; 713 kW (1,022 ehp; 956 shp) at 1,425 rpm (to 27°C)
PT6A-65AR	762 ekW; 713 kW (1,022 ehp; 956 shp) at 1,425 rpm (to 29°C)
PT6A-66	675 ekW; 634 kW (905 ehp; 850 shp) at 2,000 rpm (to 56°C)
PT6A-67, -67A	792 ekW; 746 kW (1,062 ehp; 1,000 shp) at 1,700 rpm (to 50°C)
PT6A-67AF	808 ekW; 760 kW (1,083 ehp; 1,020 shp) at 1,700 rpm (to 37°C)
PT6A-67B	792 ekW (1,062 ehp) at 1,700 rpm (to 45°C)
PT6A-67CF	868 ekW (1,163 ehp) at 1,425 rpm (to 36°C)
PT6A-67D	837 ekW (1,122 ehp) at 1,700 rpm (to 34°C)
PT6A-67R	808 ekW; 761 kW (1,083 ehp; 1,020 shp) at 1,425 rpm (to 35°C)

SPECIFIC FUEL CONSUMPTION:
At T-O rating:

PT6A-110	111.0 μg/J (0.657 lb/h/ehp)
PT6A-11, -11AG	109.4 μg/J (0.647 lb/h/ehp)
PT6A-112	107.6 μg/J (0.637 lb/h/ehp)
PT6A-114	108.2 μg/J (0.640 lb/h/ehp)
PT6A-15AG, -27, -28	101.8 μg/J (0.602 lb/h/ehp)
PT6A-21, -25, -25A	106.5 μg/J (0.630 lb/h/ehp)
PT6A-25C, -34, -34B, -34AG	100.6 μg/J (0.595 lb/h/ehp)
PT6A-135, -135A	98.9 μg/J (0.585 lb/h/ehp)
PT6A-36	99.7 μg/J (0.590 lb/h/ehp)
PT6A-41, -61	99.9 μg/J (0.591 lb/h/ehp)
PT6A-42	101.5 μg/J (0.601 lb/h/ehp)
PT6A-45A, -45B	93.5 μg/J (0.554 lb/h/ehp)
PT6A-45R	93.4 μg/J (0.553 lb/h/ehp)
PT6A-50	94.6 μg/J (0.560 lb/h/ehp)
PT6A-60A	92.6 μg/J (0.548 lb/h/ehp)
PT6A-62	95.8 μg/J (0.567 lb/h/ehp)
PT6A-64	118.8 μg/J (0.703 lb/h/ehp)
PT6A-65B	90.6 μg/J (0.536 lb/h/ehp)
PT6A-65R	86.5 μg/J (0.512 lb/h/ehp)
PT6A-65AR	86.0 μg/J (0.509 lb/h/ehp)
PT6A-66	104.6 μg/J (0.620 lb/h/ehp)
PT6A-67	92.4 μg/J (0.547 lb/h/ehp)
PT6A-67A	92.8 μg/J (0.549 lb/h/ehp)
PT6A-67AF, -67R	87.9 μg/J (0.520 lb/h/ehp)
PT6A-67B	93.3 μg/J (0.552 lb/h/ehp)
PT6A-67CF	83.7 μg/J (0.495 lb/h/ehp)
PT6A-67D	89.6 μg/J (0.530 lb/h/ehp)

OIL CONSUMPTION:
Max 0.091 kg (0.20 lb)/h

P&WC T74

T74 is a US designation for military versions of the PT6A turboprop and PT6B turboshaft.

T74-CP-700. US Army PT6A-20. More than 300 delivered to Beech for 129 U-21A. Inertial separator system.

T74-CP-702. Rated at 580 ekW (778 ehp) and retrofitted Beechcraft U-21.

P&WC PT6B/PT6C

The PT6B is the commercial turboshaft version of the PT6A and has a lower ratio reduction gear. Current versions are:

PT6B-36. Reverse drive 6,409 rpm gearbox. T-O rating 732 kW (981 shp) to 15°C, with 2½-min contingency 770 kW (1,033 shp) to 15°C. Sikorsky S-76B.

PT6B-36A. Identical to -36 but with different ratings.

PT6C. Direct drive from the power turbine.

DIMENSIONS:
Max diameter: PT6B-36, -36A	495 mm (19.5 in)
Length, excl accessories:	
PT6B-36, -36A,	1,504 mm (59.2 in)

WEIGHT, DRY:
PT6B-36	169 kg (372 lb)
PT6B-36A	171 kg (378 lb)

PERFORMANCE RATINGS:
T-O: See under model listings
Max cruise, continuous:	
PT6B-36	640 kW (870 shp) to 15°C
PT6B-36A	652 kW (887 shp) to 15°C

SPECIFIC FUEL CONSUMPTION:
At T-O rating:	
PT6B-36	100.5 µg/J (0.594 lb/h/shp)
PT6B-36A	98.2 µg/J (0.581 lb/h/shp)

OIL CONSUMPTION:
Max	0.091 kg (0.20 lb)/h

P&WC PT6T TWIN-PAC

US military designation: T400 (separate entry)

First run in July 1968, the Twin-Pac comprises two PT6 turboshaft engines side by side and driving a combining gearbox.

PT6T-3. T-O rating 1,342 kW (1,800 shp). For Bell and Agusta-Bell 212 and California/Sikorsky S-58T.

In these applications, shaft power is limited by the transmission. In the Model 212 the 1,342 kW (1,800 shp) PT6T-3 is restricted to a T-O rating of 962 kW (1,290 shp) and 843 kW (1,130 shp) for continuous power. In the S-58T the limits are 1,122 kW (1,505 shp) at T-O and 935 kW (1,254 shp) for continuous operation.

PT6T-3B. PT6T-3 with some T-6 hardware and improved single-engine performance. Bell 212 and 412.

PT6T-3BE. PT6T-3B with upgraded combining reduction gearbox and modified torque control unit. For Bell 412HP.

PT6T-3D. Improved engine for 412HP; certification due September 1993.

PT6T-6. Improved compressor-turbine nozzle guide vanes and rotor blades. S-58T and AB 212. By 1991 22,754,776 equivalent PT6 hours had been flown by PT6T engines in 3,204 helicopters in 83 countries.

The following features differ from the PT6:

TYPE: Coupled free turbine turboshaft.

SHAFT DRIVE: Combining gearbox comprises three separate gear trains, two input and one output, each contained within an individual sealed compartment and all interconnected by driveshafts. Overall reduction ratio 5.

AIR INTAKES: Additional inertial particle separator to reduce ingestion. High frequency compressor noise suppressed.

FUEL SYSTEM: As PT6 with manual backup, and dual manifold for cool starts. Automatic power sharing and torque limiting.

Cutaway P&WC PW206A free turbine turboshaft

FUEL GRADES: JP-1, JP-4 and JP-5.

ACCESSORIES: Starter/generator and tachogenerator on accessory case at front of each power section. Other drives on gearbox, including power turbine governors and tachogenerators, and provision for blowers and aircraft accessories.

OIL SPECIFICATION: PWA Spec 521. For military engines, MIL-L-7808 and -23699.

STARTING: Electrical, with cold weather starting down to −54°C.

DIMENSIONS:
Length	1,702 mm (67.0 in)
Width	1,118 mm (44.0 in)
Height	838 mm (33.0 in)

WEIGHT, DRY (standard equipment):
PT6T-3	292 kg (645 lb)
PT6T-3B, -3BE, -6	298 kg (657 lb)

PERFORMANCE RATINGS:
T-O (5 min):
Total output, at 6,600 rpm:	
PT6T-3, -3B, -3BE, -3D	1,342 kW (1,800 shp)
PT6T-6	1,398 kW (1,875 shp) (to 21°C)
Single power section only, at 6,600 rpm:	
PT6T-3, -3B	671 kW (900 shp)
PT6T-6, -3BE (2½ min)	764 kW (1,025 shp)
30 min power (single power section), at 6,600 rpm:	
PT6T-3B, -3BE, -6	723 kW (970 shp)

Cruise A:
Total output, at 6,600 rpm:	
PT6T-3, -3B, -3BE	932 kW (1,250 shp)
PT6T-6	1,014 kW (1,360 shp)
Single power section only, at 6,600 rpm:	
PT6T-3, -3B	466 kW (625 shp)
PT6T-6	500 kW (670 shp)

Cruise B:
Total output, at 6,600 rpm:	
PT6T-3, -3B, -3BE	820 kW (1,100 shp)
PT6T-6	891 kW (1,195 shp)
Single power section only, at 6,600 rpm:	
PT6T-3, -3B	410 kW (550 shp)
PT6T-6	440 kW (590 shp)
Ground idle, at 2,200 rpm	44.7 kW (60 shp) max

SPECIFIC FUEL CONSUMPTION:
At 2½-min rating (single power section):	
PT6T-3B, -3D	100.7 µg/J (0.596 lb/h/shp)
PT6T-6	101.6 µg/J (0.602 lb/h/shp)

OIL CONSUMPTION:
Max (for both gas generators)	0.18 kg (0.4 lb)/h

P&WC T400

Military version of the PT6T Twin-Pac. Deliveries totalled 799 engines, completed 1978. Details in *Jane's* to 1992-93.

P&WC PW200

Development announced 1983. The basic design is flexible and is planned to permit increased power and reduced sfc without dimensional change. The **PW206A** was certificated in December 1991 to power the McDonnell Douglas Explorer. By 1993, 11 had been delivered. The **PW206B**, due for certification in October 1994, will power the Eurocopter EC 135.

TYPE: Free-turbine turboshaft.

AIR INTAKE: Inwards amidships through mesh screen.

COMPRESSOR: Single-stage centrifugal of machined titanium. Pressure ratio 8.

COMBUSTION CHAMBER: Reverse-flow annular with 12 air blast nozzles. Two capacitor discharge igniters.

FUEL SYSTEM: Hydromechanical with digital electronics powered by dedicated alternator.

FUEL GRADE: JP-1, JP-4, JP-5 or a range of gasolines.

TURBINES: Single-stage axial compressor and power turbines with blades held in dovetail slots. Cold junction temperature sensing.

GEARBOX: 206A: front mounted combined reduction and accessory gearbox with 6,000 rpm output; includes phase shift torquemeter and drives for starter/generator, hydraulic pump and alternator. 206B: similar but bevel accessory gearbox with 5,898 rpm output.

DIMENSIONS:
Length: 206A	912 mm (35.9 in)
206B	1,042 mm (41.0 in)
Width	500 mm (19.7 in)
Height: 206A	566 mm (22.3 in)
206B	627 mm (24.7 in)

PERFORMANCE RATINGS:
206A: 2½ min	476 kW (638 shp)
30 min and T-O	463 kW (621 shp)
Max continuous	410 kW (550 shp)
206B:	
2½ min, 30 min and T-O	414 kW (555 shp)
Max continuous	398 kW (534 shp)

SPECIFIC FUEL CONSUMPTION:
206A: 2½ min	93.3 µg/J (0.552 lb/h/shp)
30 min and T-O	93.6 µg/J (0.554 lb/h/shp)
Max continuous	95.3 µg/J (0.564 lb/h/shp)
206B:	
2½ min, 30 min and T-O	95.3 µg/J (0.564 lb/h/shp)
Max continuous	96.1 µg/J (0.569 lb/h/shp)

P&WC PT6B-36 turboshaft

CHINA, PEOPLE'S REPUBLIC

AIC
AVIATION INDUSTRIES OF CHINA
PRESIDENT: Zhu Yuli

Replaced former Ministry of Aero-Space Industry 1993 as industrial ruling body for Chinese aviation activities.

CAREC
CHINA NATIONAL AERO-ENGINE CORPORATION
67 Jiao Nan Dajie, Beijing 100712
Telephone: 86 (1) 4013322 ext 5401

Cables: 9696
INTERNATIONAL MARKETING:
CATIC (China National Aero-Technology Import and Export Corporation)
5 Liangguochang Road, East City District (PO Box 647), Beijing 100010
Telephone: 86 (1) 4017722
Fax: 86 (1) 4015381
Telex: 22318 AEROT CN
PRESIDENT: Liu Guomin
EXECUTIVE VICE-PRESIDENT: Tang Xiaoping

In the past 30 years China has enormously expanded it aero-engine industry. Today there are eight major engin manufacturing centres, five factory-managed design inst tutes and four engine research and design institutes. Ove 48,000 engines of 25 types have been manufactured for th air force and navy, 756 engines of 10 types manufactured fo CAAC, and smaller numbers for export. Nine types of larg solid and liquid rocket engines have also been manufacture Most factories are seeking foreign orders or partners.

The following survey is by alphabetical order in English c the design or manufacturing organisation.

BUAA
BEIJING UNIVERSITY OF AERONAUTICS AND ASTRONAUTICS
37 Xue Yuan Road, Beijing 100083
Telephone: 86 (1) 201 7251

Telex: 222700 BUAA CN

This university's propulsion department is producing the **WP11** turbojet, for various manned and unmanned applications. It is derived from the Turbomeca Marboré II.

CEC
CHENGDU ENGINE COMPANY
PO Box 77, Chengdu, Sichuan 610067
Telephone: 86 (28) 443628
Fax: 86 (28) 442470
Telex: 60142 CET CN
Cables: 4721

GENERAL MANAGER: Duan Changping

This company was formed in October 1958, and is also known as CEF (Chengdu Aero-Engine Factory). Most of the staff and resources came from the Shenyang Overhaul Factory, and the first task was to produce the **RD-500K** (RR Derwent derivative) turbojet for a cruise missile.

Today CEC has a site area of 137 ha (338.5 acres) and a workforce of almost 20,000. It produces the **WP6** turboje (see LM), the **WP13** turbojet (see LMC) and components fo the Pratt & Whitney **JT8D** turbofan, including combustio liners. In October 1988 SNECMA announced that it wa assisting CATIC to develop the improved **WP13G** an **WP14** for later F-7 versions. New annular combustio chambers will be produced.

CLXMW
CHANGZHOU LAN XIANG MACHINERY WORKS
PO Box 37, Changzhou, Jiangsu 213123
Telephone: 86 (596) 602095, 602097
Cables: 5046

DIRECTOR: Tian Taiwu

Also known as JHEF (Jiangxi Helicopter Engine Factory), this works has a payroll of over 5,000. Its main product is the **WZ6** turboshaft and the **WZ6G** industrial derivative. Rated at 1,145 kW (1,536 shp) and with a dry weight of 300 k (661 lb), the WZ6 is derived from the Turbomeca Turm IIIC. Work began in 1975, testing occupied 1980-82 an WZ6 engines first flew in a Z-8 helicopter in 1986.

DEMC
DONGAN ENGINE MANUFACTURING COMPANY
PO Box 51, Harbin, Heilongjiang 150066
Telephone: 86 (451) 802120
Fax: 86 (451) 802266
Telex: 87131 HDEC CN
Cables: 0021
GENERAL MANAGER: Song Jingang

Also known as HEF (Harbin Engine Factory), this establishment was founded in 1948 and employs more than 10,000. Its first product was the 1,268 kW (1,700 hp) **HS7**, a 14-cylinder radial piston engine based on the Soviet Shvetsov ASh-82V. In parallel, in 1957-59, a few ASh-21 engines were made, but only the HS7 went into production, for the Z-5 helicopter. In the late 1950s there was a need for a better engine for the Il-12, Il-14, Tu-2 and Curtiss C-46, with better altitude performance. The result, produced from 1962 until 1980, was the **HS8**, which combined the main body and supercharger of the HS7 with the reduction gear of the ASh-82T. The HS8 is rated at 1,380 kW (1,850 hp).

A larger effort began in 1968 when development of the **WJ5** was transferred from ZEF. A derivative of the Ivchenko (ZMKB Progress) AI-24A rated at 1,901 ekW (2,550 ehp) was developed. Among problems solved were broaching of the turbine disc and welding of the annular combustion chambers. Following 5,678 test hours the WJ5 was certificated in January 1977. From 1969 an even larger project was development of the **WJ5A**, to meet the needs of the SH-5. The turbine section was almost completely redesigned, with air-cooled rotor blades, and many other parts changed in design or material. The first 5A went on test in late 1970, and certification was gained in January 1980. This modernise engine is rated at 2,350 ekW (3,150 ehp) up to 30°C. In 197 work began on modifying the original WJ5 to enable the Y-series of aircraft to have better hot/high performance. Th result, certificated in 1982, is the **WJ5A I**, basically a 5A fla rated at 2,162.5 ekW (2,900 ehp) to 38°C.

In 1988 an agreement was reached with General Electric o the USA for development of the **WJ5E** with reduced sfc. Th prototype passed initial static tests in 1990, with sfc reduce from 98.7 µg/J (0.584 lb/h/shp) to 89.4 µg/J (0.529 lb/h/shp) The WJ5E is the engine of the Y7-200B.

In 1966 Harbin was assigned the task of modifying th WJ5 into the **WZ5** turboshaft engine for the Z-6 helicopter After a prototype had been run this engine was transferred t ZARI (which see).

GADRI
GUIZHOU AERO-ENGINE DESIGN AND RESEARCH INSTITUTE
Supported by many companies, GADRI developed the **WP7B** fighter engine which is included in the entry for LMC.

GEF — *see under LMC*

HEF — *see under DEMC*

JHEF — *see under CLXMW*

LM
LIMING ENGINE MANUFACTURING CORPORATION
PO Box 424 (6 Dongta St, Dadong District), Shenyang, Liaoning 110043
Telephone: 86 (24) 443139
Fax: 86 (24) 732221
Cables: 4104
Telex: 80025 LMMCS CN
GENERAL MANAGER: Yan Guangwei

With site area exceeding 100 ha (247.1 acres), more than 200,000 m² (2,152,850 sq ft) covered area, and a workforce of over 20,000, Liming (Daybreak) is one of the largest and most experienced aero-engine centres in China. Alternatively known as SEF (Shenyang Aero-Engine Factory), or just as The New Factory, it was set up on the basis of the old overhaul factory in 1954-57. The first product was the **WP5** turbojet, a licensed version of the Soviet Klimov VK-1F (Rolls-Royce Nene derivative). Production of the WP5 was achieved in June 1956. By 1957 the WP5A, based on the VK-1A, was in production, but in 1963, following increased demand to power the H-5, production was transferred to XEF (see XAE).

In 1956 SADO (Shenyang Aero-Engine Design Office) was established, later being restyled SARI. This undertook the design of the first Chinese turbojet, the **PF1A**, based on the WP5 but smaller and rated at 15.7 kN (3,527 lb st) to power the JJ-1 trainer. Design was complete in 1957 and the PF1A powered the prototype JJ-1 in July 1958, but later the requirement for the JJ-1 was dropped.

WP6A afterburning turbojet

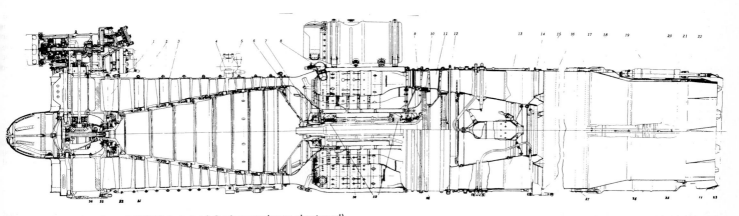

Longitudinal section through WP6A turbojet (afterburner shown shortened)
(1) variable inlet stator, *(2)* stator vanes, *(3)* compressor case, *(4)* air bleed band, *(5)* bleed actuating cylinder, *(6)* rear load relief cavity, *(7)* centre bearing, *(8)* starting igniter, *(9)* stage 1 nozzle, *(10)* turbine rotor, *(11)* stage 2 nozzle, *(12)* quick-release ring, *(13)* diffuser, *(14)* quick-release ring, *(15)* front flange, *(16)* case, *(17)* shroud, *(18)* bracket, *(19)* actuating cylinder, *(20)* adjustable flap flange, *(21)* actuator and rod heat shield, *(22)* nozzle adjusting ring, *(23)* copper plate, *(24)* flap, *(25)* centring pin, *(26)* cylinder cowl, *(27)* clamp strip, *(28)* rear bearing, *(29)* oil jet, *(30)* flame tube, *(31)* compressor, *(32)* front load relief cavity, *(33)* front bearing, *(34)* front case

LM WP6

In early 1958 Soviet documents arrived for licence production of the Tumansky (Soyuz) RD-9BF-811. This supersonic axial turbojet with afterburner, with 2,521 component parts, proved a major challenge, but the Shenyang **WP6** was first tested at the end of 1958. Tests were not successful, but following improvements in quality control trial production restarted in 1961. Subsequently several thousand, with progressive upgrades, have been made for the J-6, JJ-6 and Q-5. Much further work led to the **WP6A** for the Q-5 I attack aircraft, with a variable inlet stator stage and increased turbine temperature. This engine was certificated in August 1973. A further variant was the **WP6B** for the J-12, not made in quantity.

The following is a description of the WP6A:

INLET: Cast assembly with four de-iced radial struts, one housing drive to accessory section above and projecting ahead of inlet. Central fixed bullet and front bearing.

COMPRESSOR: Welded ring construction with one row of variable inlet stators and nine rows of rotor blades. T-O airflow 46.2 kg (101.85 lb)/s. Pressure ratio 7.44.

COMBUSTION CHAMBER: Can-annular type with 12 flame tubes, each terminating in a section of turbine inlet periphery. Spill type burners. Two igniters fed from starting tank.

TURBINE: Two-stage type with blades inserted into large flat discs, driving compressor via tubular shaft.

AFTERBURNER: Constant diameter type with main starting burner in turbine rear cone and single ring of fuel nozzles and gutter flameholders around rear of cone. Ten adjustable nozzle flaps positioned by four rams.

DIMENSIONS:
Length	5,483 mm (215.9 in)
Max height	950 mm (37.4 in)
Diameter	668 mm (26.3 in)

WEIGHT, DRY: 725 kg (1,598 lb)

PERFORMANCE RATINGS (S/L, static):
Afterburner: WP6A	36.78 kN (8,267 lb st)
WP6B	39.72 kN (8,929 lb st)
Max dry: WP6A	29.42 kN (6,614 lb st)
Normal: WP6A	24.03 kN (5,401 lb st)
WP6B	24.51 kN (5,511 lb st)

SPECIFIC FUEL CONSUMPTION (WP6A):
Afterburner	0.163 kg/h/N (45.24 mg/Ns; 1.597 lb/h/lb st)
Max	0.1 kg/h/N (27.76 mg/Ns; 0.980 lb/h/lb st)
Normal	0.099 kg/h/N (27.48 mg/Ns; 0.970 lb/h/lb st)

WP7C afterburning turbojet (afterburner not fitted)

LM WP7

To meet the needs of the J-7 fighter programme the Shenyang factory produced the Tumansky (Soyuz) R-11F-300 afterburning turbojet under licence. Designated **WP7**, this Mach 2 engine was an even greater challenge as many key Soviet documents were not supplied, and 1,097 documents had errors or omissions. Eventually, using indigenous materials, the first WP7 went on test in October 1965. Dry and afterburning ratings were 38.2 kN (8,598 lb st) and 56.4 kN (12,676 lb st) respectively. By 1970 production of this engine was transferred to LMC. Shenyang continued to introduce design improvements, and a stall flutter problem was solved by using 24 larger blades in the first stage of the compressor instead of 31, while other changes were made to the HP turbine disc, bearing lubrication and afterburner nozzle flap design.

In 1964 a decision had to be made on how to power the J-8 fighter. No engine in the 120 kN (12 tonne) class could be produced in time, but Rong Ke, deputy director of Beijing Aeronautical Materials Institute, undertook to produce air-cooled blades within a year to allow two uprated WP7 engines to be used. In May 1965 the resulting engine was authorised as the **WP7A**. After prolonged testing against forged blades with three large cooling holes the decision was taken to use nine-hole cast blades. Dry and afterburning ratings were established at 43.14 kN (9,698 lb st) and 59 kN (13,265 lb st) respectively. These engines powered the J-8 on

WS6 afterburning turbofan

LM WS6

Also in 1964 a meeting was held to select an engine for new indigenous fighter. The choice fell on an augmente turbofan, the **WS6**. In collaboration with SARI, drawing were produced in 1964-66, and by 1969 two prototyp engines had been made. Then the cultural revolution delaye the programme by about 10 years, but after 1978 further pr gress was possible, and eight more engines were built. A tw spool engine with air-cooled HP blades, it overcame mar problems and reached design figures in 1980. Augmente thrust was increased to 122.1 kN (27,450 lb st) in 1982, b by then the associated fighter programme had been cancelle

TYPE: Two-shaft augmented turbofan for superson applications.
FAN: Three stages, First stage transonic. No inlet guide vane Mass flow 155 kg (342 lb)/s. Bypass ratio 1.
COMPRESSOR: 11 stages, with variable inlet vanes and fift stage bleed. Pressure ratio 6.78 at 9,400 rpm. Overa pressure ratio 14.44.
COMBUSTION CHAMBER: Can-annular.
TURBINES: Two-stage HP, inlet temperature 1,177°C. Tw stage LP.
AFTERBURNER: Five circular gutters and six fuel-injectic zones. Max temperature 1,527°C.
NOZZLE: Outer nozzle only, multi-flap type.
DIMENSIONS:
    Length   4,654 mm (183.2 i
    Diameter (nozzle)   1,370 mm (53.94 i
WEIGHT, DRY   2,100 kg (4,630 lb
THRUST RATINGS (S/L):
    Max T-O   122.1 kN (27,445 lb s
    Max dry   71.1 kN (15,991 lb s
SPECIFIC FUEL CONSUMPTION:
    Max T-O   64.01 mg/Ns (2.26 lb/h/lb s
    Max dry   17.56 mg/Ns (0.62 lb/h/lb s

its first flight in July 1969, and were certificated in June 1982. Subsequently LMC developed the **WP7B** (see LMC) and **WP7C**. The latter powers J-7 and J-7 II aircraft. Thrust ratings are: max 60.6 kN (13,623 lb st), max dry 42.65 kN (9,588 lb st) and normal 33.54 kN (7,540 lb st). TBO is 300 h. Latest known variant is the **WP7F**, which powers the J-7E; dry rating is 44.13 kN (9,921 lb st) and with afterburning 63.74 kN (14,330 lb st).

LMC
LIYANG MACHINERY CORPORATION
PO Box 5, Pingba, Guizhou 561102
Telephone: 86 (34) 551779, 523311
Telex: 66044 LYMCG CN
Cables: 4099, 4101 PINGBA
GENERAL MANAGER: Hu Wenqin

With a covered area of 750,000 m² (8,073,200 sq ft), and a workforce exceeding 10,000, LMC is also known as GEF (Guizhou Aero-Engine Factory). The associated GADRI undertook the development of the **WP7B** afterburning turbojet, the programme being transferred to LMC because Shenyang (LM) was overloaded.

LMC WP7B

This engine powers the J-7 fighter and JJ-7 trainer. The main change in the 7B concerned the structure and length of the afterburner. Severe problems were encountered with the cast air-cooled turbine blades, which tended to crack, with the high reject rate of castings, and with burning of the rear fuselage caused by excessive afterburner wall temperatures. The 7B was eventually certificated in 1978 and succeeded the WP7 in production in 1980. Further changes enabled TBO to be increased from 100 h to 200 h. Guizhou then eliminated the separate petrol (gasoline) starter and its tank and supply system, which reduced weight and maintenance and improved air starting. The engine is known as the **WP7B (M batch)** and entered production in 1982. A further modification, in the **WP7B (BM)**, reduces weight by 17 kg (37.5 lb), enabling the F-7M to add drop tanks.

The following refers to the WP7B (BM):
TYPE: Two-shaft turbojet with afterburner.
INLET: No inlet guide vanes, first LP compressor stage overhung ahead of front roller bearing.

LMC WP13A II turbojet (afterburner not fitted)

COMPRESSOR: Three-stage LP compressor with pressure rati of 2.74. Three-stage HP compressor giving overall press ure ratio of 8.1. All blades inserted into discs carried o short tubular shafts.
COMBUSTION CHAMBER: Can-annular, with ten flame tube Nos. 1 and 6 being of a different pattern and incorporatin torch igniters. Air-film liners coated on both sides wit ceramic material.
TURBINES: Single-stage HP with 96 inserted shrouded blade Single-stage LP with shrouded blades. Outlet gas tempera ture 1,083°K (810°C).
AFTERBURNER: Multiple gutters and double-wall liner. Multi flap nozzle driven by four hydraulic rams. Up to 40 operation permitted in each 200 h overhaul period.
PERFORMANCE RATINGS (S/L, static):
    Max afterburner   59.82 kN (13,448 lb st
    Max dry   43.15 kN (9,700 lb s
SPECIFIC FUEL CONSUMPTION (as above):
    Max afterburner   56.37 mg/Ns (1.99 lb/h/lb
    Max dry   28.61 mg/Ns (1.01 lb/h/lb s

LMC WP13

The next major production of LMC was the **WP13** Though this has some features in common with the Gavrilo R-13 it is a Chinese development, based on experience wit the WP7. A two-spool afterburning turbojet, it was develope to power the J-7 III and J-8 II. Compared with the WP7 th

WP7B(BM) afterburning turbojet for the F-7M Airguard fighter

rflow is increased, work per stage improved, pressure ratio ised (the HP spool having five stages) and surge margin oubled. New titanium alloys were used for the compressor scs and blades, and in a major development two more new tanium alloys were used for the cast compressor casings. WP13 development began in 1978, and it was decided to ake the engine a 50/50 joint project with CEC. Both factors tested engines, and certification was gained in 1985. Further development introduced air-cooled HP turbine blades d modifications to the combustion chamber and afterurner, the afterburner of the **WP13A II** being longer.

MENSIONS:
Length overall:
- WP13 — 4,600 mm (181.1 in)
- WP13A II — 5,150 mm (202.75 in)

Diameter: both — 907 mm (35.71 in)
Max height: WP13 — 1,085 mm (42.72 in)

WEIGHT, DRY:
- WP13 — 1,211 kg (2,670 lb)
- WP13A II — 1,201 kg (2,648 lb)

PERFORMANCE RATINGS (S/L, static):
Max afterburner:
- WP13 — 64.73 kN (14,550 lb st) at 11,156 LP rpm
- WP13A II — 65.9 kN (14,815 lb st)

Max dry:
- WP13 — 40.21 kN (9,039 lb st) at 11,156 LP rpm
- WP13A II — 42.7 kN (9,590 lb st)

SPECIFIC FUEL CONSUMPTION (as above):
Max afterburner:
- WP13 — 63.73 mg/Ns (2.25 lb/h/lb st)
- WP13A II — 62.32 mg/Ns (2.20 lb/h/lb st)

Max dry:
- WP13 — 27.19 mg/Ns (0.96 lb/h/lb st)
- WP13A II — 28.04 mg/Ns (0.99 lb/h/lb st)

OVERHAUL LIFE:
- WP13 — 500 h (total service life 1,500 h)
- WP13A II — 300 h (including up to 90 h in afterburner)

AC
HENYANG AERO-ENGINE COMPANY

This was formed in 1979 from the merger of SEF and ARI. Today it is part of LM.

AMP
HANGHAI AERO-ENGINE MANUFACTURING PLANT

O Box 600, 600 Guangzhong Road, Shanghai 200083
elephone: 86 (21) 6650644
ax: 86 (21) 6651482
elex: 33136 SHAIR CN
ables: 5834

DIRECTOR: Shen Huansheng

This factory was built in 1971-74 and was originally the SAF (Shanghai Aero-Engine Factory). It has a covered area of 56,191 m² (604,855 sq ft) and workforce of 2,000. Apart from various non-aero engines and components its main development effort was the **WS8** turbofan, to power the Y-10. Work proceeded quickly and the first engine went on test in June 1975. Eight engines were built, one running a 1,000 h test, one a 150 h certification test and one was flight tested, making eight flights totalling 22 h. A front-fan engine with a short bypass duct, the WS8 was rated at 80 kN (18,000 lb st). About 17 per cent was titanium, and new techniques included Ni-Cd anti-corrosion coating, graphite varnish of titanium parts and aluminised siliconising of turbine blades. The Y-10 was not put into production and the engine had no other application.

ARI
HENYANG AVIATION ENGINE RESEARCH INSTITUTE

O Box 428, Shenyang, Liaoning 110015
elephone: 86 (24) 820057
ax: 86 (24) 820673
elex: 80055 SARI CN
ables: 4391 (national), SARI (international)

DIRECTOR: Hai Yide

Shenyang Aero-engine Research Institute (SARI), founded in 1961, has nearly 3000 employees. It is responsible for research, design and development of large and intermediate size turbojet and turbofan engines and their components and systems. In the 1960s SARI modified the **WP7** engine into the **WP7A**, and transferred the engine to Liming Corporation in Shenyang and Liyang Corporation for serial production for F-7 and J-8 aircraft. From 1965 SARI developed the **WS6**, a high-thrust turbofan with afterburner. Ten WS6 demonstrator engines were built. All performance goals were achieved, but the WS6 was not put into production (see under LM).

SARI is now developing new types of turbojet and turbofan engines.

EF — *Original abbreviation of SAC (see LM)*

MPMC
OUTH MOTIVE POWER AND MACHINERY COMPLEX

O Box 211, Zhuzhou, Hunan 412002
elephone: 86 (733) 21151
ax: 86 (733) 24220
elex: 995002 CHENF CN
ables: 2820

AEC (South Aero-Engine Company)
ddress as above
ENERAL MANAGER: Wu Shenduo

With a covered area of nearly 300,000 m² (3,230,000 sq ft) d workforce of over 10,000, SMPMC is one of the larger AS establishments. Until 1983 its aero-engine division, AEC, was known as ZEF (Zhuzhou Aero-Engine Factory), d in 1951 was set up as the first aero-engine factory in hina. Its first product was the Soviet Shvetsov **M-11FR** raal piston engine rated at 119 kW (160 hp), the first three eing completed in July 1954. Mass production followed. To eet the needs of the Y-5 (licensed An-2) ZEF began in September 1956 to work on the **HS5** (licensed ASh-62IR). Over 600 of these 746 kW (1,000 hp) radial piston engines were produced by 1986, some being installed in CAAC Li-2s. Lacking the chosen Praga Doris B engine to power the CJ-6 trainer, the Soviet Ivchenko (ZMKB Progress) AI-14R radial piston engine was produced as the **HS6**. Rated at 191 kW (260 hp), the HS-6 entered production in June 1962, about 700 being produced. To improve performance, especially at altitude, ZEF increased rpm, compression ratio and supercharger gear ratio. The result, in 1963, was the **HS6A**, with T-O power increased to 212.5 kW (285 hp). About 3,000 were made by 1986. In 1975 the engine was again modified to power the Y-11; rpm were increased and the reduction gear strengthened. The resulting **HS6D**, with power of 224 kW (300 hp), was certificated in August 1980. The **HS6E**, for the NAMC Haiyan, has a further increased compression ratio and modifed exhaust valves and reduction gear, raising output to 261 kW (345 hp). In 1990 the simplified **HS6K** was certificated at 298 kW (400 hp) and is intended as the future engine of the N-5A. Experimental models, in the 1963-70 period, were the turbocharged **HS6B** and the **HS6C** for helicopters, used in the 701 and Yan'an II helicopters.

Work on gas turbines began in January 1965 in support of the development by BIAA (Beijing Institute of Aeronautics and Astronautics) of the **WP11**. This simple turbojet, rated at 8.3 kN (1,874 lb st), powered the WZ-5 unmanned reconnaissance vehicle. The WP11 first ran in June 1971 and was certificated in 1980, manufacture then being transferred to ZEF. In September 1965 ZEF was selected to develop the **WJ5** turboprop, but this was transferred in 1968 to HEF (see DEMC). In 1969 ZEF was ordered to develop the **WJ6** turboprop, based on the Soviet Ivchenko AI-20M, to power the Y-8. Testing started in 1970, but various problems delayed certification until January 1977. Further changes (for example to compressor vane angle, igniter and lubrication clearances) resulted in TBO being raised in stages from 300 h to 3,000 h. T-O power was 3,169 ekW (4,250 ehp) and weight 1,200 kg (2,645 lb). In 1977 work began on the **WJ6A** to power the Y-8C with a pressure cabin and greater payload. By using air-cooled blades and raising the rpm this engine was successfully run in 1983 at 3,393 ekW (4,550 ehp).

In 1980 ZEF began the assembly and test of the **WZ8** (Turbomeca Arriel 1C) for the Z-9 helicopter. ZEF gives the output as 522 kW (700 shp) for a weight of 118 kg (260 lb). An all-Chinese WZ8 ran in 1985, and resulted in major technical upgrades at Zhuzhou (the high-voltage power unit was the only imported part). As part of the offset, 40 accessory gearboxes were supplied to France.

ZEF also produces solid rocket motors for air-to-air missiles.

AE
IAN AERO-ENGINE CORPORATION

O Box 13, Xian, Shaanxi 710021
elephone: 86 (29) 61951, 63366, 722600
ax: 86 (29) 72 2721
elex: 70102 XIARO CN
ables: 5411
ENERAL MANAGER: Wang Xin Yan

Construction of this factory began in August 1958, as XEF Xian Aero-Engine Factory), and today its workforce xceeds 16,000. Its first task, undertaken in collaboration ith HEF and SEF, was to develop and produce the **WP8** rbojet. Based on the Soviet Mikulin RD-3M, this large gine is rated at 93.17 kN (20,944 lb st) and weighs 3,133 kg ,906 lb). It posed the greatest challenge to the embryonic hinese industry, especially as many questions were left answered by the licensor (for example, the composition d work processes for the nozzle guide vanes). In 1963 the hole WP8 programme was transferred to XEF, and the gine was certificated and put into production in January 967.

In 1963 the increased demand for **WP5A** engines for the -5 resulted in series production being transferred to XEF om SEF. In 1965, together with Chengdu Aircraft Corporion, XEF developed the modified **WP5D** for the JJ-5

WP8 single-shaft turbojet

WS9 two-shaft augmented turbofan

trainer. In 1966 it developed the **WP5B** and produced it in quantity for Soviet-made MiG-15bis fighters. In 1976 it developed and series-produced the **WP5C** to power Soviet-made MiG-17 all-weather fighters. XEF also developed and produced WQJ-1 turbo-starters and WDZ-1 APUs.

In 1972 Premier Zhou Enlai called for a major upgrade in Chinese aeronautical quality and technology. As a result negotiations with Rolls-Royce began in May 1972, leading to a licence to produce the Spey 202 augmented turbofan signed on 13 December 1975. The work was assigned to XEF, the engine being designated **WS9**. After a great effort, of major benefit to the industry, four WS9 engines were successfully tested in China and Britain in 1979-80.

Since 1980 XEF has started business relations with Rolls-Royce, Pratt & Whitney (USA and Canada), GE, Textron Lycoming and Allison. Numerous types of engine parts are produced for these companies, as well as numerous non-aero products.

ZAC
ZHUZHOU AERO-ENGINE COMPANY

This organisation, apparently an alternative or earlier title for SAEC (see SMPMC entry), was formed in 1983 by merging ZEF and ZARI. Its main task so far has been to produce the **WJ9** turboprop to power a production version of the Y-12. This engine, rated at 507 kW (680 shp), is based on the WZ8 (Arriel 1C), but with a new reduction gear, accessory drives and fuel and lubrication systems. The first engine ran in 1986 and met the design figures. A production WJ9 would probably also power agricultural and other special-purpose aircraft.

ZARI
ZHUZHOU AERO-ENGINE RESEARCH INSTITUTE

This organisation was established in 1968. From the start it has been tasked with R&D of shaft-drive gas-turbine engines, starting with the **WZ5** (see DEMC). The poor payload and lack of twin-engine safety of the Z-5 and Z-6 led to a new helicopter, the Z-7, in 1970. This was powered by the twin-packaged **WZ5A**, developed at ZARI. Contingency power of each power section was 1,737 kW (2,330 shp). Design was finished in 1971, but the cultural revolution caused a delay of eight years. The WZ5A then reached its design targets, but was cancelled upon the abandonment of the Z-7. ZARI was merged with ZEF in 1983 to form ZAC.

ZEF
ZHUZHOU AERO-ENGINE FACTORY

Established in 1951 as China's first aero-engine factory (see SMPMC entry), ZEF was amalgamated with ZARI in 1983 to form ZAC.

CZECH REPUBLIC

OMNIPOL
OMNIPOL CO LTD
Nekázanka 11, CR-112 21 Prague 9
Telephone: 42 (2) 21401111
Fax: 42 (2) 21402241
Telex: 121297 OMPO

Omnipol is now a joint stock company responsible for exporting some products of the Czech aviation industry and for supplying information on those products which are available for export.

AEROTECHNIK
CR-68604 Uherské Hradiste, Kunovice
Telephone: 42 (632) 5122
Fax: 42 (632) 5128
Telex: 60380
TECHNICAL DEPT: Ing E. Parma

This company is now producing the long-established inverted four-cylinder Walter Mikron piston engine, of 2,440 cc (149 cu in) capacity. The factory at Mor. Trebová calls it the **Mikron IIIS (A)**. It powers the Aerotechnik L-13SW Vivat motor glider. Since 1989 the **Mikron IIISE (AE)** has also been produced, with electric starter and alternator. This increases weight from 65.9 kg (145.3 lb) to 70.0 kg (154 lb). A silencer may be fitted.
PERFORMANCE RATINGS (both versions):
  T-O   48.5 kW (65 hp) at 2,600 rpm
  Cruise   35.5 kW (47.6 hp) at 2,350 rpm

Aerotechnik Mikron IIISE (AE) piston engine

AVIA-HSA
AVIA-HAMILTON STANDARD AVIATION LTD
Jinonicka 329, CR-15000 Prague 5, Letňany
Telephone: 42 (2) 858 5643
Fax: 42 (2) 322 742
Telex: Prague AVIA C 121 475
PRESIDENT: Mark Petru

This company is at present engaged in series production of piston engines, propellers and spare parts. Formed in 1992, it is the successor to the original Avia KP. These engines are now FAA approved in the USA, where they are marketed by LPE (which see).

AVIA M 137 A

Designed to power light aerobatic aircraft, the 134 kW (180 hp) M 137 A piston engine is a modification of the M 337 with fuel and oil systems for aerobatic operation and without a supercharger. It powers the Zlin 42 M and Z 526 F. The **M 137 AZ** is a modified version, with an inverted air intake at the rear so that a dust filter can be incorporated. Details are as M 337, with the following differences:
CRANKSHAFT: No oil holes for propeller control.
FUEL SYSTEM: Type LUN 5150 pump; system designed for sustained aerobatics.
STARTER: LUN 2131 electric.
DIMENSIONS:
  Length   1,344 mm (52.9 in)

AVIA-HSA/MOTORLET—ENGINES: CZECH REPUBLIC

Avia M 137 A six-cylinder air-cooled piston engine

Width	443 mm (17.44 in)
Height	630 mm (24.80 in)
WEIGHT (incl starter):	141.5 kg (312 lb)
PERFORMANCE RATINGS:	
T-O	134 kW (180 hp) at 2,750 rpm
Max continuous	119 kW (160 hp) at 2,680 rpm
Max cruise	104.5 kW (140 hp) at 2,580 rpm
SPECIFIC FUEL CONSUMPTION:	
At T-O rating	91.26 µg/J (0.540 lb/h/hp)
At max cruise rating	81.96 µg/J (0.485 lb/h/hp)

AVIA M 337 A

The M 337 A six-cylinder air-cooled supercharged engine powers several types of light aircraft that were built in Czechoslovakia, including the L-200D Morava, Zlin 43 and Zlin 726K. It can be supplied with hubs for fixed-pitch or controllable-pitch propellers. The fully aerobatic **M 337 AK** is fitted to the Zlin 142.

TYPE: Six-cylinder inverted inline air-cooled, ungeared, supercharged and with direct fuel injection.
CYLINDERS: Bore 105 mm (4.13 in). Stroke 115 mm (4.53 in). Swept volume 5.97 litres (364.31 cu in). Compression ratio 6.3 : 1. Steel cylinders with cooling fins machined from solid. Cylinder bores nitrided. Detachable cylinder heads are aluminium alloy castings.
PISTONS: Aluminium alloy with graphited surfaces.
CONNECTING RODS: Two split big ends bolted together.
CRANKSHAFT: Forged from chrome vanadium steel, machined all over. Nitrided crankpins.
CRANKCASE: Heat treated magnesium alloy (Elektron).
IGNITION: Shielded type. Two plugs per cylinder.
LUBRICATION: Dry sump pressure feed type. The M 337 AK has a system for sustained inverted operation.
SUPERCHARGER: Centrifugal, ratio 7.4 : 1.
FUEL SYSTEM: Low pressure injection system with nozzles in front of inlet valves. The M 337 A has a unified fuel injection pump.
FUEL GRADE: Minimum 72-78 octane.
STARTING: Electric, engaged by an electromagnet.
ACCESSORIES: One 600W 28V dynamo. Electric tachometer. Propeller control unit. Mechanical tachometer on oil pump. Hydraulic pump to special order.
PROPELLER DRIVE: Direct left hand tractor.
DIMENSIONS:

Length, excl propeller boss	1,410 mm (55.51 in)
Width	472 mm (18.58 in)
Height	628 mm (24.72 in)
WEIGHT, DRY:	148 kg (326.3 lb)
PERFORMANCE RATINGS:	
T-O	157 kW (210 hp) at 2,750 rpm
Max cruise at 1,200 m (3,940 ft)	112 kW (150 hp) at 2,400 rpm
SPECIFIC FUEL CONSUMPTION:	
At T-O rating	100.6 µg/J (0.595 lb/h/hp)
At max cruise rating at 1,200 m (3,940 ft)	72.7 µg/J (0.430 lb/h/hp)

MOTORLET

MOTORLET A. S.
CR-158 01 Prague 5, Jinonice
Telephone: 42 (2) 521119
Fax: 42 (2) 526060
GENERAL MANAGER: Ing Josef Krča

Motorlet operates the main aero engine establishment in the Czech Republic, based on the former Walter factory. The Walter name continues as a trademark. The shareholding company listed above is a member of the Aero Group.
In 1991 Motorlet signed an agreement with GE for future joint development and manufacture. GE supplies CT7-9 engines for the L-610 transport.

WALTER M 601

The M 601 was designed to power the L-410 transport. It drives an Avia V 508 propeller.
The first version, rated at 410 ekW (550 ehp), ran in October 1967. Development of the **M 601 B**, of increased diameter, started during 1968. The Let L-410M powered by M 601 B engines was in Aeroflot service by early 1979. The M 601B powers the L-410UVP. In 1982 a further variant, the **M 601 D**, entered production. This gives increased power and can be operated to longer TBO. It powers the PZL-106 BT-601.
By 1985 the **M 601 Z** had entered production to power the Z 137T agricultural aircraft. It drives an auxiliary piston compressor for the spraying/dusting installation. TBO is 1,500 h. The **M 601 E** powers the L-410UVP-E and has been selected to power the Myasishchev M-101. It has a TBO of 2,000 h without hot-section inspection. It drives a VJ 8.508E five-blade propeller, and an alternator for anti-icing windscreens and propeller blades. It is certificated to JAR and FAR Pt 33 in the Czech Republic.
The **M 601 T** is an aerobatic version for the PZL-130 Turbo-Orlik. A new M 601 version to have a TBO of 3,000 to 4,000 h is under development. The **M601 F** is intended for Western markets, initially with FAA and Canadian certification. It powers the L-420.
Two M601 turboprops rated at 662 kW (888 shp) power the Raketa-2 (not to be confused with the 2.2) Ekranoplan.
TYPE: Free turbine turboprop.
PROPELLER DRIVE: Reduction gear at front of engine with drive from free turbine. Reduction ratio 14.9.
AIR INTAKE: Annular, at rear (reverse flow engine).
COMPRESSOR: Two axial stages of stainless steel, plus single centrifugal stage of titanium. Pressure ratio (601 B) 6.4, (601 D) 6.55, (601 E, F, T) 6.65, at 36,660 rpm gas generator speed. Air mass flow (601 B) 3.25 kg (7.17 lb)/s, (601 D) 3.55 kg (7.83 lb)/s, (601 E, F, T) 3.6 kg (7.94 lb)/s.
COMBUSTION CHAMBER: Annular combustor with rotary fuel injection and low-voltage ignition.
COMPRESSOR TURBINE: Single stage with solid blades; inlet temperature 952°C.
POWER TURBINE: Single stage.
FUEL SYSTEM: Low pressure regulator, providing gas generator and power turbine speed controls.
FUEL GRADE: PL6, RT, TS-1 and Jet A-1 kerosene.
LUBRICATION SYSTEM: Pressure gear-pump circulation. Integral oil tank.
OIL SPECIFICATION: B3V synthetic oil or Aeroshell 500, 555, 560.
STARTING: LUN 2132-8 8kW electric starter/generator.
DIMENSIONS:

Length: 601 D	1,658 mm (65.27 in)
601 B, E, Z, F, T	1,675 mm (65.94 in)
Width	590 mm (23.23 in)
Height	650 mm (25.59 in)
WEIGHT, DRY:	
601 B, D	193 kg (425.5 lb)
601 Z	197 kg (434.3 lb)
601 E	200 kg (441 lb)
601 F	203 kg (448 lb)
601 T	202 kg (445 lb)
PERFORMANCE RATINGS (T-O):	
601 B	515 kW (691 shp)
601 D	540 kW (724 shp)
601 E, T	560 kW (751 shp)
601 Z	382 kW (512 shp)
601 F	580 kW (778 shp)
SPECIFIC FUEL CONSUMPTION (T-O):	
601 B	110.8 µg/J (0.656 lb/h/ehp)
601 D	110.5 µg/J (0.654 lb/h/ehp)

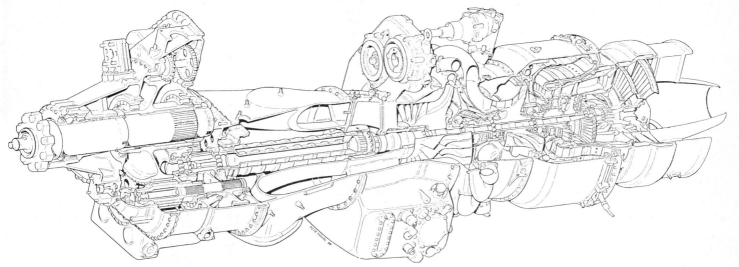

Cutaway drawing of the Walter M 602 turboprop

CZECH REPUBLIC/FRANCE: ENGINES—MOTORLET/JPX

Walter M 601 E turboprop

TYPE: Three-shaft turboprop.
PROPELLER DRIVE: Double spur reduction, ratio 12.58.
AIR INTAKE: At front, S duct from chin inlet passing up behind reduction gear.
COMPRESSORS: LP centrifugal, pressure ratio 4.15 at 25,000 rpm. HP centrifugal, pressure ratio 3.133 at 31,489 rpm. Overall pressure ratio 13. Mass flow 7.33 kg (16.16 lb)/s.
COMBUSTION CHAMBER: Short annular reverse flow with 14 simplex nozzles and low-voltage semiconductor ignition.
COMPRESSOR TURBINES: Single-stage HP, single-stage LP.
POWER TURBINE: Two-stage, 16,600 rpm.
FUEL SYSTEM: LP electrohydraulic regulator and electronic limiter.
FUEL GRADE: T-1, TS-1, RT, Jet A-1.
LUBRICATION SYSTEM: Pressure gear-pump circulation, integral oil tank and cooler.
OIL SPECIFICATION: B3V, AeroShell 500, 550.
STARTING: LUN 5362-8 pneumatic.
DIMENSIONS:
  Length   2,669 mm (105.08 in)
  Width   753 mm (29.65 in)
  Height   872 mm (34.33 in)
WEIGHT, DRY (equipped):   570-580 kg (1,257-1,279 lb)
PERFORMANCE RATINGS (S/L):
  T-O   1,360 kW (1,824 shp)
  Max continuous   1,200 kW (1,608 shp)
  Cruise   700 kW (938 shp)
SPECIFIC FUEL CONSUMPTION:
  T-O   94.5 µg/J (0.559 lb/h/shp)

601 E, T	109.7 µg/J (0.649 lb/h/ehp)
601 F	106.9 µg/J (0.633 lb/h/ehp)
601 Z	135.8 µg/J (0.804 lb/h/ehp)

WALTER M 602

This engine was developed to power the L-610, which first flew on 28 December 1988. By 1993 no production decision had been reached, though 34 engines had run.

PS
POVAŽSKÉ STROJÁRNE
Aero Engine and Tools Division
CR-01734 Považská Bystrica
Telephone: 42 (822) 23037
Fax: 42 (822) 23778
Telex: 075233, 075316
DIVISION EXECUTIVE MANAGER: Jan Pirič
CHIEF DESIGNER: E. Jiři Bednář

PS-AED succeeded ZVL, a major new design bureau which also included the VUM research institute. It partnered the Progress/Lotarev (ZMKB) bureau in the former Soviet Union (which see under Ukraine) in designing the DV-2 described under ZMKB. The DV-2 is produced by both partners, as noted under PS-ZMKB in the International section. PS is building the DV-2 and continues to develop it.

PS-ZMKB DV-2 two-shaft turbofan

EGYPT

AOI
AOI ENGINE FACTORY (EF)
PO Box 12, Helwan
Telephone: 20 (2) 781157, 781404 and 781088
Fax: 20 (2) 781236
Telex: 23138 ENFAC UN

CHAIRMAN: Dr Mohamed El Semary

Among other work, EF assembles and tests the Larzac 04 and PT6A-25E. Since 1980 EF has carried out complete overhaul, repair and test of APUs and the SNECMA Atar 9C and Larzac, GE CT64 and Allison 501. Plans for 1993 include manufacture and overhaul of various gas turbines.

FRANCE

JPX
SARL JPX
Z. I. Nord, BP 13, F-72320 Vibraye
Telephone: 33 43 93 61 74
Fax: 33 43 93 62 71
Telex: 722151 F

Details of smaller JPX engines were last given in the 1992-93 edition of *Jane's*.

JPX 4T60/A

Unlike previous JPX engines, this is a four-stroke with four opposed air-cooled cylinders with bore 93.0 mm (3.66 in) and stroke 75.4 mm (2.97 in). Capacity 2,050 cc (125 cu in). Compression ratio 8.2. Overall length 650 mm (25.6 in). Width 805 mm (31.7 in). Weight, dry (without propeller hub) 73.0 kg (161 lb). Developed from the Volkswagen VW126A of smaller capacity, the 4T60/A has an electric starter and alternator, and is rated at 47.8 kW (65 hp) at 3,200 rpm, and 42.0 kW (57 hp) at 2,500 rpm. Using 4-star motor fuel or 100LL fuel, this engine has flown over 400,000 h in the Robin ATL and Jodel D 18 and D 19, with TBO of 1,000 h. The **4T60/AES** for ultralight aircraft weighs 61 kg (134 lb).

JPX 4TX 75

Designed with the benefit of 4T60 experience, the 4TX 75 has four cylinders of 95 mm (3.74 in bore) and 82 mm (3.23 in) stroke, capacity being 2,325 cc (141.9 cu in). Compression ratio is 8.7, and fuel can be 100LL or 4-star motor fuel. Dry weight is 75 kg (165 lb) and take-off power 56 kW (75 hp) at 2,800 rpm. JAR/E and FAR Pt 33 certification was due in 1993.

JPX 4TX 75

JPX T620

JPX has announced a family of small single-shaft turbojets, of simple design with a centrifugal compressor and inward-radial turbine. The largest, the T620, has a rating of 0.196 kN (44.06 lb st) at 80,000 rpm, and is suitable for light aircraft. Weight without the electronic fuel control is 6.2 kg (13.67 lb).

JPX T620 turbojet

MICROTURBO
MICROTURBO SA

This company no longer produces or supports engines for manned aircraft. The last entry was in the 1992-93 *Jane's*.

SNECMA
SOCIETE NATIONALE D'ETUDE ET DE CONSTRUCTION DE MOTEURS D'AVIATION

2 boulevard du Général Martial Valin, F-75724 Paris Cedex 15
Telephone: 33 (1) 40 60 80 80
Fax: 33 (1) 40 60 81 02
Telex: 600 700 SNECMA
CHAIRMAN AND CEO: Gérard Renon
SENIOR VICE-PRESIDENT:
  Pierre Alesi (Engineering and Production)
VICE-PRESIDENTS:
  Jean Bonnet (Military Programmes and Marketing)
  Dominique Hedon (Commercial Programme Management and Marketing)
COMMUNICATION MANAGER: Jean-Claude Nicolas

More than 5,200 Atar turbojets have been produced for Mirage fighters. SNECMA is now producing the M53 turbojet and developing the M88. It is also participating in international collaborative programmes, as described hereunder.

Today, SNECMA heads about 26,000 persons formed around its major subsidiaries: SEP (rocket motors, remote sensing and composites), Hispano-Suiza (aeronautical equipment, nuclear and robotic equipment), Messier-Bugatti (landing gears) and Sochata (engine repair).

SNECMA Atar 9K50 afterburning turbojet

SNECMA ATAR 9K50

The Atar is a single-shaft military turbojet, first run in 1946 and subsequently developed and cleared for flight at Mach numbers greater than 2. The final version was the Atar 9K50, which powers all production Mirage F1 and Mirage 50 versions.

DIMENSIONS:
  Diameter   1,020 mm (40.2 in)
  Length overall   5,944 mm (234 in)
WEIGHT, DRY:
  Complete with all accessories   1,582 kg (3,487 lb)
PERFORMANCE RATINGS:
  With afterburner   70.6 kN (15,870 lb st) at 8,400 rpm
  Without afterburner   49.2 kN (11,055 lb st)
SPECIFIC FUEL CONSUMPTION:
  With afterburner   55.5 mg/Ns (1.96 lb/h/lb st)
  Without afterburner   27.5 mg/Ns (0.97 lb/h/lb st)
OIL CONSUMPTION: 1.5 litres (3.2 US pints; 2.64 Imp pints)/h

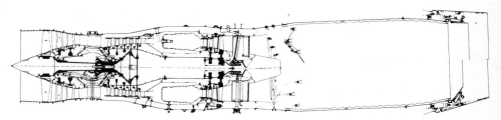

Longitudinal section through the SNECMA M53-P2 augmented bypass turbojet

SNECMA M53

The M53 is a single-shaft bypass turbojet capable of propelling fighter aircraft to Mach 2.5, without any throttle limitations over the flight envelope. Its modular construction allows easier maintenance.

It includes a three-stage fan, five-stage compressor (pressure ratio 9.8 at 10,600 rpm), annular combustion chamber, two-stage turbine and an afterburner equipped with a multi-flap variable nozzle. The control system is monitored by an Elecma electronic computer. The following versions are in service:

M53-5. Produced in 1980-85 as the initial engine of the Mirage 2000.

M53-P2. Designed to power the Mirage 2000 from 1985; current production version.

DIMENSIONS:
  Length, overall   5,070 mm (199.6 in)
  Max diameter   1,055 mm (41.5 in)

SNECMA M53-P2 augmented bypass turbojet

WEIGHT, DRY:
  M53-5   1,470 kg (3,240 lb)
  M53-P2   1,500 kg (3,307 lb)
PERFORMANCE RATINGS:
  Max with afterburner:
    M53-5   88.2 kN (19,830 lb st)
    M53-P2   95.0 kN (21,355 lb st)
  Max without afterburner:
    M53-5   54.4 kN (12,230 lb st)
    M53-P2   64.3 kN (14,455 lb st)
SPECIFIC FUEL CONSUMPTION (without afterburner):
    M53-5   24.64 mg/Ns (0.87 lb/h/lb st)
    M53-P2   25.55 mg/Ns (0.90 lb/h/lb st)

SNECMA M88

The M88 is a family of advanced augmented turbofans, built around the same core, with thrust ranging from 75 kN (16,860 lb) to 105 kN (23,600 lb). In a six-year demonstration programme an uprated core with 1,850°K turbine entry temperature completed its first simulated altitude tests in February 1987.

M88-2. This is the basic engine of the family, under development since 1987 for the Rafale ACT/ACM aircraft for the French Air Force and Navy. The first engine went on test on 27 February 1989, afterburner operation began in April 1989 and one engine flew in the Rafale A in February 1990. Two engines were installed in the Rafale D to continue flight tests from the beginning of 1991.

The following relates to the basic M88-2:
TYPE: Two-shaft augmented turbofan.
COMPRESSOR: Variable inlet guide vanes, three-stage LP (fan), six-stage HP, with three variable stator stages.
COMBUSTOR: Annular, with aerodynamic injection.
TURBINES: Single-stage HP, single-stage LP.
ACCESSORIES: Full-authority digital control system.
DIMENSION:
  Length   3,540 mm (139.0 in)
WEIGHT, DRY: 897 kg (1,970 lb)
PERFORMANCE RATINGS (installed):
  With afterburner   72.9 kN (16,400 lb st)
  Without afterburner   48.7 kN (10,950 lb st)
SPECIFIC FUEL CONSUMPTION:
  With afterburner   49.9 mg/Ns (1.76 lb/h/lb st)
  Without afterburner   24.9 mg/Ns (0.88 lb/h/lb st)

GE/SNECMA GE90

SNECMA participates in the General Electric GE90 with a share of 25 per cent in all phases of the programme.

GE/SNECMA/MTU/VOLVO/FIAT CF6

As a continuation of the CF6-50 and CF6-80A programmes, SNECMA participates in the General Electric CF6-80C2, in the derived LM6000 industrial gas turbine with a share of 10 per cent, and in the CF6-80E1 with a share of 20 per cent.

CFM INTERNATIONAL CFM56

This programme is covered under CFM in the International part of this section.

SNECMA/TURBOMECA LARZAC

This appears under Turbomeca-SNECMA GRTS.

SNECMA M88-2 augmented turbofan

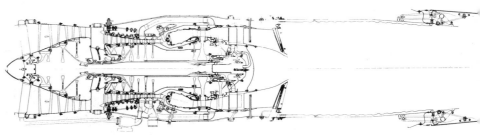

Longitudinal section through SNECMA M88 augmented turbofan

KLIMOV/SNECMA

To power future Russian trainers, an engine derived from the Larzac is being developed jointly with Klimov Corporation.

RR/SNECMA/MTU/TA TYNE

Under Rolls-Royce licence, the Tyne 22 turboprop for the Transall C-160 was put back into production in October 1977. The similar Tyne 21 is in production for the Dassault Atlantique 2. These 4,549 kW (6,100 ehp) engines were last described in the 1975-76 *Jane's*.

TURBOMECA
SOCIETE TURBOMECA
F-64511 Bordes
Telephone: 33 59 32 84 37
Fax: 33 59 53 15 12, 33 59 53 20 93
Telex: 560928
PARIS OFFICE: 1 rue Beaujon, F-75008 Paris
Telephone: 33 (1) 40 75 01 78
Fax: 33 (1) 40 75 01 79
Telex: 650347
CHAIRMAN: Mme Sonia Meton
PRESS RELATIONS MANAGER: Jacques Millepied

By 1 January 1993 more than 27,000 Turbomeca aircraft engines had been delivered to customers in 125 countries. Approximately 14,000 more engines have been built under licence by what are today Rolls-Royce plc in the UK, Teledyne CAE in the USA, ENMASA in Spain, Hindustan Aeronautics in India, Bet-Shemesh in Israel, SMPMC in China and factories in Romania and former Yugoslavia.

A European Small Engines Co-operation Agreement signed in April 1985 joins Turbomeca, MTU of Germany and Rolls-Royce of the UK in promoting three complementary new engines: the Turbomeca TM 333, MTU-Turbomeca-RR MTR 390 and Rolls-Royce Turbomeca RTM 322. Other European small engine makers may join the collaboration, in which each partner may share in engines sold to its own government.

In March 1989 a UK marketing company was formed, Turbomeca Ltd. Address as for Rolls-Royce Turbomeca (see International part of this section); General Manager St John Williamson.

Total covered floor area for Turbomeca's three plants at Bordes, Mézières and Tarnos is 140,487 m² (1,512,188 sq ft). The company employs about 3,962 people.

ROLLS-ROYCE TURBOMECA ADOUR

This turbofan appears in the International part of this section.

TURBOMECA-SNECMA LARZAC

This turbofan appears under Turbomeca-SNECMA GRTS.

MTU-TURBOMECA-RR MTR 390

This appears in the International part of this section.

ROLLS-ROYCE TURBOMECA RTM 322

See International part of this section.

TURBOMECA ARRIUS

Previously known as the TM 319, this turboshaft is compact, with just a single centrifugal compressor, reverse-flow combustor, single-stage HP and single-stage LP turbines, and FADEC control.

The initial rating is 340 kW (456 shp). The first was run on the bench on 21 February 1983. Full authority digital electronic control is supplied by Elecma.

Production deliveries for the AS 355/555 began in 1987 and reached 192 by 1993. The same gas generator is used in the Arrius 1D turboprop.

Arrius 1. Basic version to which data below apply. Twin TM 319s at a T-O rating of 358 kW (480 shp) are an alternative power plant for the Eurocopter BO 108 (EC 155).

Arrius 2. Growth version, max T-O rating 473 kW (634 shp). Optional for McDonnell Douglas Explorer.
DIMENSIONS:
  Length   782 mm (30.78 in)
  Width   360 mm (14.2 in)
  Height   540 mm (21.26 in)
WEIGHT, DRY: 87 kg (192 lb)
PERFORMANCE RATINGS (Arrius 1, ISA, S/L):
  Max contingency   408 kW (547 shp)
  Max T-O   380 kW (509 shp)
  Max continuous   295 kW (395 shp)

Turbomeca Arrius free turbine turboshaft

Turbomeca Arrius 1D free turbine turboprop

Turbomeca Arriel 1S free turbine turboshaft

TURBOMECA ARRIUS 1D
The turboprop version is fully aerobatic. The gas generator and power turbine modules are identical with those of the Arrius 1. The first Arrius 1D ran on 11 September 1985. Flight testing in an Epsilon began in November 1985, followed by a Valmet L-90 TP in December 1987. The Arrius 1D is flying in the Socata Oméga.

DIMENSIONS:
Length 826 mm (32.52 in)
Width 476 mm (18.74 in)
Height 590 mm (23.22 in)
WEIGHT, DRY (bare): 111 kg (245 lb)
PERFORMANCE RATINGS (ISA, S/L):
T-O 313 kW (420 shp)
Cruise (6,100 m; 20,000 ft) 179 kW (240 shp)

TURBOMECA ARRIEL
This turboshaft powers the single-engined AS 350 Ecureuil/AS 550 Fennec, twin-engined AS 365/565 Dauphin/Panther, Agusta A 109K and 109K2, and Sikorsky S-76A+ and S-76C and Eurocopter/Kawasaki BK 117 C1.

The Arriel has modular construction. The first complete engine ran on 7 August 1974.

There are 13 qualified production versions, differing essentially only in power rating:

Arriel 1A, 1B. T-O rating 478 kW (641 shp); power AS 350B/BA and AS 365C.
Arriel 1C. T-O rating 492 kW (660 shp); powers AS 365N.
Arriel 1D. T-O rating 510 kW (684 shp); powers AS 350B and L1.
Arriel 1C1, 1M, 1K, 1S. All have T-O rating of 522 kW (700 shp); power AS 365F and N1, A 109K and S-76A+.
Arriel 1C2. T-O rating 550 kW (738 shp); powers AS 365N2.
Arriel 1D1. T-O rating 558 kW (749 shp); powers AS 350B2. and AS 550U2. Selected for Kamov Ka-128.
Arriel 1M1. T-O rating 558 kW (749 shp); powers AS 565UA and AS 565MA.
Arriel 1S1. T-O rating 539 kW (723 shp); powers S-76C.
Arriel 1E. T-O rating 528 kW (708 shp); powers BK 117 C1.
Arriel 2. Growth version with single-stage gas generator turbine. To provide at least 16 per cent more power than 1S. Candidate engine for EC 120, Dauphin and S-76.

By 31 December 1992 a total of 3,220 Arriels had been delivered, with production continuing at the rate of 180 per year. These engines had then flown 4.8 million hours in 79 countries. These totals do not include engines produced under licence in China.

The following relates to the Arriel 1A, 1B:
TYPE: Single-shaft free turbine turboshaft.
COMPRESSOR: Single-stage axial and supersonic centrifugal. Pressure ratio 9.
COMBUSTION CHAMBER: Annular, with flow radially outwards and then inwards. Centrifugal fuel injection.
GAS GENERATOR TURBINE: Two integral cast axial stages with solid blades.
POWER TURBINE: Single axial stage with inserted blades.
REDUCTION GEAR: Light alloy gearbox, containing two stages of helical gears, giving drive at 6,000 rpm to front and rear. Hydraulic torquemeter.
ACCESSORY DRIVES: Main pad provides for optional 12,000 rpm alternator; other drives for oil pumps, tachometer generator, governor and starter.
LUBRICATION SYSTEM: Independent circuit through gear pump and metallic cartridge filter.
OIL SPECIFICATION: AIR 3512 or 3513A.
STARTING: Electric starter or starter/generator.

DIMENSIONS:
Length, excl accessories 1,090 mm (42.91 in)
Height overall 569 mm (22.40 in)
Width 430 mm (16.93 in)
WEIGHT, DRY:
With all engine accessories 120 kg (265 lb)
PERFORMANCE RATINGS: See variants list
SPECIFIC FUEL CONSUMPTION:
T-O 96.8 µg/J (0.573 lb/h/shp)

TURBOMECA ASTAZOU TURBOSHAFT
This turboshaft series is derived from the Astazou II turboprop. Variants are:

Astazou IIA. Rated at 390 kW (523 shp) for SA 318C.
Astazou IIIA. Derived from IIA but with revised turbine to match power needs of SA 341G. Produced jointly by Turbomeca and Rolls-Royce, with 1,008 delivered.
Astazou XIVB and XIVF. For SA 319B; XIVB is civil and XIVF military. Flat rated to 441 kW (591 shp) (1 h) up to 4,000 m (13,125 ft) or +55°C.
Astazou XIVH. For SA 342J/L, to remove altitude and temperature limitations; 1,146 delivered.
Astazou XVIIIA. Higher gas temperature. Powers AS 360C.
Astazou XX. Fourth axial compressor stage added. Designed for operation in hot and high countries.

By 1993 a total of 2,253 Astazous had been delivered.
The following description relates to the Astazou III, except where indicated:
TYPE: Single-shaft axial-plus-centrifugal turboshaft.
REDUCTION GEAR: Two-stage epicyclic having helical primary and straight secondary gears. Ratio 7.039 : 1 (XIVB/F, 7.345; XVIIIA, 7.375).
COMPRESSOR: Single-stage axial (IIA, III), two-stage axial (XIV, XVIII) or three-stage axial (XX) followed by single-stage centrifugal. Mass flow 2.5 kg (5.5 lb)/s.
COMBUSTION CHAMBER: Reverse flow annular with centrifugal injector using rotary atomiser. Two ventilated torch igniters.
TURBINE: Three-stage axial with blades integral with discs.
ACCESSORIES: Five drive pads on casing forming rear of air intake.
FUEL SYSTEM: Automatic constant speed control.
LUBRICATION SYSTEM: Pressure type with gear type pumps. Oil tank of 8 litre (17 US pints; 14 Imp pint) capacity.
STARTING: Electrical, automatic.

DIMENSIONS:
Length overall: Astazou IIA 1,272 mm (50.0 in)
Astazou III, XIVB/F 1,433 mm (56.3 in)
Astazou XIVH 1,470 mm (57.9 in)
Astazou XVIIIA 1,327 mm (52.2 in)
Astazou XX 1,529 mm (60.22 in)
Height: Astazou IIA 458 mm (18.0 in)
Astazou III, XIVH 460 mm (18.1 in)
Astazou XVIIIA 698 mm (27.48 in)
Astazou XX 721 mm (28.4 in)
Width: Astazou IIA 480 mm (18.8 in)
Astazou III, XIVH 460 mm (18.1 in)
WEIGHTS:
Equipped: Astazou III 147 kg (324 lb)
Astazou III (suffix 2) 150 kg (330 lb)
Astazou XIVB/F 166 kg (366 lb)
Astazou XIVH 160 kg (353 lb)
Astazou XVIIIA 155 kg (341 lb)
Astazou XX 195 kg (430 lb)
PERFORMANCE RATINGS:
Max power: Astazou IIA 390 kW (523 shp)
Astazou III 441 kW (592 shp)
Astazou III (suffix 2) 481 kW (645 shp)
Astazou XX 749 kW (1,005 shp)
One hour: Astazou XIVB/F 441 kW (591 shp)
Astazou XVIIIA 651 kW (873 shp)
 maintained at sea level to 40°C
Max continuous: Astazou IIA 353 kW (473 shp)
Astazou III 390 kW (523 shp)
Astazou III (suffix 2) 441 kW (592 shp)
Astazou XIVB/F 405 kW (543 shp)
Astazou XIVH flat rated in SA 341 at 440.7 kW
 (591 shp) to 55°C or 4,000 m (13,125 ft)
Astazou XVIIIA 600 kW (805 shp)
Astazou XX 675 kW (905 shp)
SPECIFIC FUEL CONSUMPTION:
At max power rating:
Astazou IIA 105.3 µg/J (0.623 lb/h/shp)
Astazou III 108.7 µg/J (0.643 lb/h/shp)
Astazou III (suffix 2) 109.9 µg/J (0.650 lb/h/shp)
Astazou XIVB/F 105.5 µg/J (0.624 lb/h/shp)
Astazou XVIIIA 91.3 µg/J (0.540 lb/h/shp)
Astazou XX 85.9 µg/J (0.508 lb/h/shp)

TURBOMECA ARTOUSTE III
The Artouste IIIB is a single-shaft turboshaft with two-stage axial-centrifugal compressor and three-stage turbine. Pressure ratio 5.2. Mass flow 4.3 kg/s (9.5 lb/s) at 33,300 rpm. Built under licence in India by Hindustan Aeronautics.

The IIIB powers the SA 315B and SA 316B/C. The uprated IIID has a reduction gear giving 5,864 rpm at the driveshaft (instead of 5,773) and in revised equipment. The IIID powers the SA 316C; data are for this version. A total of 2,525 Artouste III engines had been built by 1988.
DIMENSIONS:
Length 1,815 mm (71.46 in)
Height 627 mm (24.68 in)
Width 507 mm (19.96 in)
WEIGHT, DRY: 178 kg (392 lb)
PERFORMANCE RATING (T-O, maintained up to 55°C at S/L or up to 4,000 m; 13,125 ft): 440 kW (590 shp)
SPECIFIC FUEL CONSUMPTION: 126.2 µg/J (0.747 lb/h/shp)

TURBOMECA TM 333
This turboshaft was launched in July 1979 to power the AS 365 and other helicopters in the 4,000 kg (8,800 lb) class, including the Indian ALH. French certification of the 1A version was obtained on 11 July 1986.

TM 333 1A. Basic version, composed of a gas generator module, free turbine module and reduction gear module.
TM 333 1M. For military AS 565.
TM 333 2B. Growth version with single crystal HP turbine, giving T-O rating of 747 kW (1,001 shp). Selected for HAL (India) ALH. First flight 20 August 1992.

The TM 333 is one of three new engines included in the European Small Engines Co-operation Agreement. Data are for the 2B:
TYPE: Free turbine turboshaft.
COMPRESSOR: Variable inlet guide vanes, two-stage axial compressor, single-stage centrifugal.
COMBUSTION CHAMBER: Annular, reverse flow.
GAS GENERATOR TURBINE: Single-stage with uncooled inserted blades.
POWER TURBINE: Single-stage axial with uncooled inserted blades.
GEARBOX: Two stages to give drive at 6,000 rpm to front output shaft.
LUBRICATION: Independent system. Oil passes through gear pump and metallic cartridge filter.
FUEL SYSTEM: Microprocessor numerical control.

TURBOMECA TURMO

The Turmo free turbine engine is in service in both turboshaft and turboprop versions. A total of 2,012 have been delivered.

Current variants are as follows:

Turmo IIIC$_5$, IIIC$_6$, IIIC$_7$. For SA 321F/G/H/Ja. Total production 549.

Turmo IIIE$_6$. Higher turbine temperature.

Turmo IVA. Civil engine derived from IIIC$_4$, with contingency rating of 1,057 kW (1,417 shp). The IVB is a military version.

TYPE: Free turbine turboshaft.
REDUCTION GEAR: IIIC$_3$, C$_5$ and E$_3$ fitted with rear mounted reduction gear; IIIC$_4$ direct drive.
COMPRESSOR: Single-stage axial followed by single-stage centrifugal. Pressure ratio 5.9 on IIIC$_3$. Mass flow 5.9 kg (13 lb)/s.
COMBUSTION CHAMBER: Reverse flow annular with centrifugal fuel injector using rotary atomiser. Two ventilated torch igniters.
GAS GENERATOR TURBINE: Two-stage axial.
POWER TURBINE: Two-stage axial unit in IIIC$_3$, C$_5$ and E$_3$, and single stage in IIIC$_4$.
ACCESSORIES: Pads for oil pump, fuel control, electric starter, tacho-generator and, on IIIC$_4$, oil cooler fan.
FUEL SYSTEM: Fuel control for gas generator on IIIC$_3$, C$_5$ and E$_3$, with speed limiter for power turbine on E$_3$. Constant-speed system on IIIC$_4$ power turbine.
FUEL GRADE: AIR 3405 for IIIC$_4$.
LUBRICATION SYSTEM: Pressure type with oil cooler and 13 litre (27.5 US pint; 23 Imp pint) tank.
OIL SPECIFICATION: AIR 3155A, or AIR 3513, for IIIC$_4$.
STARTING: Automatic system with electric starter motor.
DIMENSIONS:
  Length:
    Turmo IIIC$_3$, C$_5$ and E$_3$ 1,975.7 mm (78.0 in)
    Turmo IIIC$_4$ 2,184 mm (85.5 in)
    Turmo IIID$_3$ 1,868 mm (73.6 in)
  Width:
    Turmo IIIC$_3$, C$_5$ and E$_3$ 693 mm (27.3 in)
    Turmo IIIC$_4$ 637 mm (25.1 in)
    Turmo IIID$_3$ 934 mm (36.8 in)
  Height:
    Turmo IIIC$_3$, C$_5$ and E$_3$ 716.5 mm (28.2 in)
    Turmo IIIC$_4$ 719 mm (28.3 in)
    Turmo IIID$_3$ 926 mm (36.5 in)
WEIGHT, DRY:
  Turmo IIIC$_3$ and E$_3$, fully equipped 297 kg (655 lb)
  Turmo IIIC$_5$, IIIC$_6$ and IIIC$_7$ 325 kg (716 lb)
  Turmo IIIC$_4$, equipped engine   225 kg (496 lb)
  Turmo IIID$_3$, basic engine   365 kg (805 lb)
PERFORMANCE RATINGS:
  T-O: Turmo IIIC$_3$, D$_3$ and E$_3$ 1104 kW (1,480 shp)
  Turmo IIIE$_6$ 1,181 kW (1,584 shp)
  Max contingency:
    Turmo IIIC$_4$ at 33,800 gas generator rpm
                                         1,032 kW (1,384 shp)
  Turmo IIIC$_6$ at 33,550 gas generator rpm
                                         1,156 kW (1,550 shp)
  Turmo IIIC$_7$ at 33,800 gas generator rpm
                                         1,200 kW (1,610 shp)
  Turmo IVA at 33,950 gas generator rpm
                                         1,057 kW (1,417 shp)
  Turmo IVC at 33,800 gas generator rpm
                                         1,163 kW (1,560 shp)
  T-O and intermediate contingency:
    Turmo IIIC$_5$ 1,050 kW (1,408 shp)
SPECIFIC FUEL CONSUMPTION:
  At T-O rating:
    Turmo IIIC$_3$ and E$_3$ 101.9 µg/J (0.603 lb/h/shp)
    Turmo IIID$_3$ 104.1 µg/J (0.616 lb/h/shp)
  At max contingency rating:
    Turmo IIIC$_4$, C$_5$, C$_6$, C$_7$ and IV
                               106.8 µg/J (0.632 lb/h/shp)
    Turmo IVA   106.3 µg/J (0.629 lb/h/shp)

DIMENSIONS:
  Length, including accessories   1,045 mm (41.1 in)
  Height overall   712 mm (28.0 in)
  Width   454 mm (17.9 in)
WEIGHT, DRY: 156 kg (345 lb)
PERFORMANCE RATINGS:
  Max contingency   788 kW (1,057 shp)
  T-O   747 kW (1,001 shp)
  Max continuous   663 kW (889 shp)
SPECIFIC FUEL CONSUMPTION:
  Max contingency   88 µg/J (0.523 lb/h/shp)
  T-O   89.4 µg/J (0.529 lb/h/shp)
  Max continuous   91.7 µg/J (0.543 lb/h/shp)

Turbomeca TM 333 2B free turbine turboshaft

TURBOMECA MAKILA

This turboshaft powers the AS 332. Derived partly from the Turmo, it incorporates rapid-strip modular construction; three axial stages of compression plus one centrifugal; centrifugal atomiser; two-stage gas generator turbine with cooled blades; two-stage free power turbine; and lateral exhaust.

Makila 1A. Certificated 1980. OEI rating 1,310 kW (1,757 shp).

Makila 1A1. Certificated 1984. OEI rating 1,400 kW (1,877 shp).

Makila 1A2. Under development. OEI rating (30 s) 1,569 kW (2,103 shp), (2 min) 1,433 kW (1,921 shp).

By 1 January 1993 deliveries of Makila engines had reached 1,112.

DIMENSIONS:
  Length, intake face to rear face   1,395 mm (54.94 in)
  Max diameter   514 mm (20.25 in)
WEIGHT, DRY: Basic   210 kg (463 lb)
  Equipped   243 kg (535 lb)
PERFORMANCE RATINGS (ISA, S/L):
  Max contingency: see model listings
  Cruise: 1A, 1A1, 1A2   700 kW (939 shp)
SPECIFIC FUEL CONSUMPTION:
  Max contingency: 1A   83.9 µg/J (0.496 lb/h/shp)
    1A1   81.4 µg/J (0.481 lb/h/shp)
  Cruise: 1A   97.7 µg/J (0.578 lb/h/shp)
    1A1   95.0 µg/J (0.562 lb/h/shp)

Turbomeca Makila 1A free turbine turboshaft

TURBOMECA-SNECMA
GROUPEMENT TURBOMECA-SNECMA (GRTS)

2 blvd du Gén Martial Valin, F-75725 Paris Cedex 15
Telephone: 33 (1) 40 60 80 80
Fax: 33 (1) 40 60 81 02

Groupement Turbomeca-SNECMA is a company formed jointly by Société Turbomeca and SNECMA to manage the Larzac turbofan launched in 1968.

TURBOMECA-SNECMA LARZAC

In February 1972 the **Larzac 04** turbofan was selected for a joint Franco-German programme to provide propulsion for the Alpha Jet (see under Dassault/Dornier in 1992-93 International part of Aircraft section). In addition to the two French engine partners, two German companies, MTU and KHD (now BMW RR), shared in production and development. A total of 1,264 engines were assembled in the two countries. The Larzac was adopted early in 1993 as the power plant of the MiG AT twin-engined advanced trainer.

Current versions are as follows:

Larzac 04-C6. Two-stage fan, four-stage HP compressor, annular combustion chamber, single-stage HP turbine with cooled blades and single-stage LP turbine. Maximum airflow

Turbomeca-SNECMA Larzac 04-C20 two-shaft turbofan

28 kg (62 lb)/s, pressure ratio 10.6 and BPR 1.13. First production delivery September 1977.
Larzac 04-C20. Growth version with increased mass flow compressor and higher temperature HP turbine. Pressure ratio 11.3 and BPR 1.04. First run March 1982; first flight December 1982; production deliveries from September 1984.

DIMENSIONS:
Overall length of basic engine	1,179 mm (46.4 in)
Overall diameter	602 mm (23.7 in)

WEIGHT, DRY: 302 kg (666 lb)
T-O THRUST (S/L, static):
Larzac 04-C6	13.19 kN (2,966 lb)
Larzac 04-C20	14.12 kN (3,175 lb)

SPECIFIC FUEL CONSUMPTION:
Larzac 04-C6	20.1 mg/Ns (0.71 lb/h/lb st)
Larzac 04-C20	20.95 mg/Ns (0.74 lb/h/lb st)

GERMANY

BMW RR
BMW ROLLS-ROYCE GmbH
Hohenmarkstrasse 60-70, Post Box 1246, D-6370 Oberursel
Telephone: 49 (61) 71 500 591
Fax: 49 (61) 71 500 633
Telex: 410727
CHAIRMAN: Albert Schneider
PUBLIC AFFAIRS: Holger Lapp

This joint venture company was established in July 1990 by BMW AG of Munich (50.5 per cent) and Rolls-Royce plc of London (49.5 per cent). It is located at the facilities of the former KHD Luftfahrttechnik, with over 1,000 employees and capital of DM30 million. It is regarded as a German company.

It has taken over activities of the former KHD, including the Tornado secondary power system, T117 propulsion for the CL-289 UAV, T118 APU demonstrator and T128 low-cost turbojet. It has a 20 per cent share in the Rolls-Royce Tay programme, 5 per cent in the Rolls-Royce Trent, and 5 per cent in the SNECMA contribution to the CFM56-5.

Both partners are supporting the BR700 series, a new family of turbofans. These will go into production at a new factory being built in Brandenburg, 20 miles south of Berlin (former East Germany). Up to 45 per cent of the cost (estimated at £450 million) could be covered by a 'reunification loan'.

BMW ROLLS-ROYCE BR 700
Basic core, used in other engines of the family. A new design, launched during the course of 1991, was to run in 1992. Ten-stage compressor with inlet guide vanes and next four stators variable. High-intensity smokeless annular combustor. Two-stage turbine. Together with the three currently projected derived engines, to be developed, manufactured, sold and supported by BMW Rolls-Royce.

BR 710. Described separately.
BR 715. Turbofan for regional jet transports. Fan diameter 1,346 mm (53 in), plus two-stage LP booster, driven by three-stage LP turbine. Ratings 62.3-80.1 kN (14,000-18,000 lb st).
BR 720. Turbofan for larger jet aircraft. Fan diameter 1,397 mm (55 in), plus three-stage booster, driven by four-stage turbine. Ratings 80.1-97.9 kN (18,000-22,000 lb st).

BR 710
This is the baseline engine of the family, and the first in timing. Components have been tested and the first complete engine is to run in September 1994. The initial application is the Gulfstream GV, for which an order for 200 engines was placed on 8 September 1992. The GV is to fly in late 1995 and engine certification is due in August 1996. The second application is the Canadair Global Express. Certification is scheduled for February 1997, with the first delivery to Canadair late that year.
TYPE: Two-shaft turbofan.
CONFIGURATION: Single-stage wide-chord fan, 10-stage HP compressor, two-stage HP and LP turbines.
EMISSIONS: Noise significantly below FAR 36 Stage 3, NO$_x$ 30 per cent below ICAO, with advanced combustor designed for further 30 per cent reduction.
FAN DIAMETER: 1,219 mm (48.0 in)
RATING (T-O, ISA + 20°C):
GV	66.28 kN (14,900 lb st)
Global Express	65.30 kN (14,680 lb st)

Cutaway drawing of BR 710 installation in Global Express

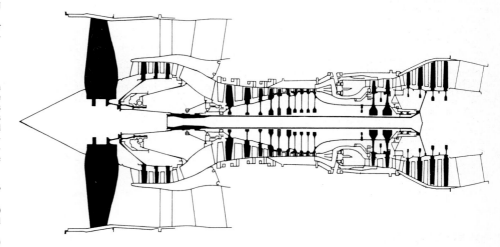
Longitudinal section of the BR 720 two-shaft turbofan

HIRTH
GÖBLER-HIRTHMOTOREN GmbH
Post Box 20, Max Eyth Strasse 10, D-7141 Benningen
Telephone: 49 (71) 44 6074
Fax: 49 (71) 44 5415
Telex: 7 264 530 GHIR D

This company produces small piston engines for microlights and other aircraft in the range 2.61-82 kW (3.5-110 hp). Details of the 16.2 kW (22 hp) F 263 R 53 and 17 kW (23 hp) F 22 were last given in the 1992-93 *Jane's*.

HIRTH F 23A
Cylinders 72 mm (2.835 in) bore and 64 mm (2.52 in) stroke, giving capacity of 521 cc (31.79 cu in). Compression ratio 10.5, using 50:1 fuel mix. Equipped weight 24.0 kg (52 lb), silencer adding 4.5 kg (10 lb). Max power 30 kW (40 hp) at 5,500 rpm.

HIRTH 2701 R 03
Cylinders 70 mm (2.756 in) bore and 64 mm (2.52 in) stroke, giving capacity of 493 cc (30.08 cu in). Compression ratio 11, using 50:1 fuel mix. Weight with fan and recoil starter 32.8 kg (72.5 lb). Max power 32 kW (43 hp) at 6,750 rpm.

HIRTH 2702 R 03
Cylinders as F 23A. Weight with fan and recoil starter 35.0 kg (77 lb). Max power 31 kW (41.6 hp) at 5,500 rpm.

HIRTH 2703
Similar to 2702 but twin carburettors. Max power 44 kW (59 hp) at 6,200 rpm.

HIRTH 2704
Again enlarged, to 625 cc (38.1 cu in). Max power 38 kW (51 hp) at 4,500 rpm.

HIRTH F 30
Four-cylinder engine tested in two sizes, the smaller (1,042 cc) being a twinned 2702 but lighter. Both can have direct or geared drive.

1,042 cc (63.58 cu in) version: Weight dry 36.0 kg (79.4 lb). Max power 48.5 kW (65 hp) at 4,500 rpm, 82 kW (110 hp) at 6,500 rpm.

LIMBACH

LIMBACH FLUGMOTOREN GmbH

Kotthausener Strasse 5, D-5330 Königswinter 21, Sassenberg
Telephone: 49 (2244) 2322 and 3031
Fax: 49 (2244) 6976
Telex: 889574 plm d
GENERAL MANAGER: Friedrich Welbhoff
SALES MANAGER: Pierre Schmitt
In USA: **J. T. Consulting**
3114 N Sunset, Sand Springs, Oklahoma 74063
Telephone: 1 (918) 241 5468
Fax: 1 (918) 245 6910

This company manufactures four-stroke and two-stroke piston engines for light aeroplanes and powered gliders. Full details appeared in the 1992-93 *Jane's*.

LIMBACH SL 1700

Several variants of this engine have been certificated. The following description refers to the SL 1700E:
TYPE: Four-cylinder opposed air-cooled four-stroke piston engine.
CYLINDERS: Bore 88 mm (3.46 in). Stroke 69 mm (2.71 in). Capacity 1,680 cc (102.51 cu in). Compression ratio 8.
INDUCTION: Stromberg-Zenith 150CD carburettor.
FUEL GRADE: 90 octane.
IGNITION: Single Slick 4230 magneto feeding one Bosch WB 240 ERT 1 plug in each cylinder.
STARTING: One Fiat 0.37 kW (0.5 hp) starter (EA, EAI, one Bosch 0.3 kW; 0.4 hp).
ACCESSORIES: Ducellier 250W alternator; APG 17.09.001 fuel pump.
DIMENSIONS:
  Length overall   618 mm (24.3 in)
  Width overall   764 mm (30.1 in)
  Height overall   368 mm (14.5 in)
WEIGHT, DRY: 73 kg (161 lb)
PERFORMANCE RATINGS:
  T-O   51 kW (68 hp) at 3,600 rpm
  Continuous   45.5 kW (61 hp) at 3,200 rpm

LIMBACH L 1800

Basically an L 2000 with reduced stroke.
CYLINDERS: Bore 90 mm (3.54 in). Stroke 69 mm (2.7 in). Capacity 1,756 cc (107 cu in). Compression ratio 7.5.
FUEL GRADE: 100LL or Mogas.
WEIGHT, DRY: about 70 kg (154 lb)
PERFORMANCE RATINGS (S/L):
  T-O   49.2 kW (66 hp) at 3,600 rpm
  Cruise   41 kW (55 hp) at 2,600 rpm

LIMBACH L 2000

This family is based on the SL 1700 with increased bore and stroke:
L 2000EO1. As 1700E1. Installed in Fournier RF-5, RF-9 and RF-10, IAR IS-28M2 and Aerotechnik L-13L Vivat motor gliders.
L 2000EA1. As 1700EA1. Installed in Scheibe SF-25C Falke 2000 and SF-36.
L 2000EB1. As 1700EB1. Installed in Grob G 109, Valentin Taifun and Hoffmann Dimona.
L 2000DA2. Compression ratio reduced to 8.5. JAR 22E certificated with TBO 1,000 h. Installed in Robin ATL II and Stiletto T-9. Certificated to JAR 22E.
Details as for SL 1700, except for following:
CYLINDERS: Bore 90 mm (3.54 in). Stroke 78.4 mm (3.09 in). Capacity 1,994 cc (120.26 cu in). Compression ratio 8.7 (EAI, 8.9).
FUEL GRADE: 100LL or Mogas.
WEIGHT, DRY (with all accessories):
  L 2000EOI   70 kg (154 lb)
  L 2000EAI   69 kg (152 lb)
  L 2000EBI   71.5 kg (157.5 lb)
PERFORMANCE RATINGS:
  T-O: all models   59 kW (80 hp) at 3,400 rpm
  Continuous:
    L 2000EOI, EAI   52 kW (70 hp) at 3,000 rpm
    L 2000EBI   53 kW (72 hp) at 3,000 rpm

Limbach L 2400EO3X four-stroke piston engine

LIMBACH L 2400

Similar to L 2000 but larger. JAR 22 certificated.
L 2400EB1A. Installed in G 109A and G 109B.
L 2400EB1B. Installed in Taifun.
L 2400EB1C. Optional on HK 36 Super Dimona.
L 2400EB1D. Installed in Stemme S 10.
L 2400DAB3X. Version with dual magneto ignition. Weight with all accessories and silencer 88 kg (194 lb). T-O power 68.6 kW (92 hp) at 3,200 rpm.
L 2400EO3X. Single ignition, single carburettor.
CYLINDERS: Bore 97 mm (3.82 in). Stroke 82 mm (3.23 in). Capacity 2,424 cc (147.91 cu in). Compression ratio 8.5.
INDUCTION: Twin Stromberg-Zenith 150 CD-3 carburettors.
FUEL GRADE: Minimum 96 RON.
IGNITION: Slick 4230 magneto feeding single Bosch WB 240 ERT 1 plugs.
ACCESSORIES: 1.4 kW starter, 14-V 33 or 55 A generator, fuel pump and tachometer. Provision for Hoffmann or Mühlbauer variable-pitch propeller.
WEIGHT, DRY: 82 kg (181 lb)
PERFORMANCE RATINGS:
  L 2400EB1A, B and C:
    T-O (5 min)   65 kW (87 hp) at 3,200 rpm
    Continuous   63 kW (84.5 hp) at 3,000 rpm
  L2400EB1D:
    T-O   70 kW (94 hp) at 3,400 rpm

MTU

MOTOREN- UND TURBINEN-UNION MÜNCHEN GmbH

Dachauer Str 665, Munich-Allach (Postal address, Post Box 500640, D-8000 Munich 50)
Telephone: 49 (89) 1489 0
Fax: 49 (89) 1489 4303
Telex: 529 500-15 MT D

MTU München, a subsidiary of Deutsche Aerospace AG, is Germany's largest aero engine company. It produces engines for most classes of aircraft. In March 1990 it signed an agreement which gives Pratt & Whitney a firm foothold in Europe and makes MTU United Technologies' 'preferred partner worldwide'.

PRATT & WHITNEY PW4084

MTU is a partner with 12.5 per cent share. It is reponsible for the LP turbine. The engine will power the Airbus A330 and Boeing 777.

GENERAL ELECTRIC CF6

MTU has approximately a 12 per cent share in the manufacture of the CF6-50, approximately an 8 per cent share of the CF6-80A/A1 for the A310 and 767 and a 9 per cent share of the CF6-80C2 for the A300-600, 747 and 767. MTU makes HP turbine parts.

PRATT & WHITNEY PW2000

MTU is a partner, with FiatAvio, in the PW2037 and 2040. It is responsible for the LP turbine, under a 21.2 per cent share.

IAE V 2500

MTU has a 12.1 per cent share in IAE (see International part of this section).

PRATT & WHITNEY JT8D-200

MTU has a 12.5 per cent share, being largely responsible for the LP turbine.

EUROJET EJ200

MTU has a 33 per cent share in this engine, described in the International part of this section.

TURBOMECA-SNECMA LARZAC

MTU has a 23 per cent share (see under France).

TURBO-UNION RB199

MTU has a 40 per cent share in this engine, described in the International part of this section.

ROLLS-ROYCE TYNE

MTU has a 28 per cent share in about 170 Tyne turboprops for the Transall. MTU supports all Tyne 21 engines (Atlantique) and Tyne 22 (Transall), as well as Tynes used by civil operators.

ALLISON 250-C20B

MTU licence built more than 700 engines, designated 250-MTU-C20B, for the PAH-1 and VBH military variants of the Eurocopter BO 105 helicopter. MTU is converting C20Bs into C20Rs and is supporting engines used by civil operators.

RTM 322

Under an agreement of October 1991, MTU has a 15 per cent share in the production and further development of this engine.

MTU-TURBOMECA-RR MTR 390

Details of this three-nation helicopter engine are given in the International part of this section.

P&WC PW305

MTU has a 25 per cent share in this Canadian turbofan. One of its responsibilities is the LP turbine.

ZOCHE

MICHAEL ZOCHE

Keferstrasse 13, D-8000 Munich 40
Telephone: 49 (89) 34 45 91
Fax: 49 (89) 34 24 51
Telex: 523 402 ZOCHE D

This company's diesel aero engines incorporate tungsten counterweights and full aerobatic pressure lubrication. Both engines have a propeller governor and four accessory drive pads. The weights given include starter, alternator, governor, vacuum pump and turbocharger.

ZOCHE ZO 01A

CYLINDERS: Four, arranged at 90°. Bore 95 mm (3.74 in). Stroke 94 mm (3.70 in). Capacity 2,665 cc (162.6 cu in). Compression ratio 17.
FUEL GRADES: Diesel No. 2, JP-4 or Jet A.

Zoche ZO 02A eight-cylinder diesel engine

DIMENSIONS:
  Length   720 mm (28 in)
  Height and Width   530 mm (20.9 in)
  Diameter   640 mm (25.2 in)
WEIGHT, DRY: 84 kg (185 lb)
PERFORMANCE RATING: 110 kW (150 hp) at 2,500 rpm
SPECIFIC FUEL CONSUMPTION (above rating):
                      65.0 µg/J (0.3855 lb/h/hp)

ZOCHE ZO 02A

This is a double-row eight-cylinder engine using 01A cylinders. It is 935 mm (36.8 in) long, weighs 118 kg (259 lb) and has a T-O rating of 220 kW (300 hp).

INDIA

GTRE
GAS TURBINE RESEARCH ESTABLISHMENT
Suranjan Das Road, Post Bag 9302, C. V. Raman Nagar, Bangalore 560 093
Telephone: 91 (812) 570698
Telex: 0845 2438 GTRE IN
DIRECTOR: Dr R. Krishnan

Established in 1959, the GTRE is one of 45 R&D establishments administered by the DRDO (Defence Research & Development Organisation). By far its biggest challenge is the design and development of a new engine for fighter aircraft.

GTRE GTX
This engine is planned as the power plant of the production LCA (Light Combat Aircraft). Although influenced by existing engines, the GTX is a completely Indian project, and is being developed in the following versions:

GTX37-14U. This afterburning turbojet was the first designed in India. First run in 1977, it has a three-stage LP compressor and seven-stage HP compressor, both driven by single-stage turbines. It is flat rated to ISA +30°C at 44.5 kN (10,000 lb st) dry and 64.3 kN (14,450 lb st) with full reheat. A few engines will continue running to support later variants.

GTX37-14UB. Turbofan version with bypass ratio of 0.215. Max thrust 88.9 kN (19,990 lb st) with a larger frontal area.

GTX-35. Advanced turbojet with five-stage HP compressor, new annular combustor and increased turbine temperature. Offered required thrust for LCA, but higher fuel consumption due to higher thrust levels.

Kaveri. Improved turbofan planned as engine for LCA. Earlier designation of GTX-35VS changed to Kaveri with a redesigned core compressor of six stages, updated full-authority digital engine control of Indian design and advanced exhaust nozzle. Core to run in December 1993 and engine in March 1994.
LP COMPRESSOR: Three stages, with transonic blading. Pressure ratio 3.38.
HP COMPRESSOR: Six stages with some variable stators. Pressure ratio 6.5. Overall pressure ratio 22.
COMBUSTION CHAMBER: Annular with air-blast atomisers.
HP TURBINE: Heavily loaded single-stage with DS cooled blades. Maximum entry gas temperature 1,427°C. Later to have thermal barrier coating.
LP TURBINE: Single-stage, cooled.
CONTROL SYSTEM: FADEC being developed in India and of Indian design.
PERFORMANCE RATINGS (flat rated to ISA + 20°C, S/L):
Max: dry 51.3 kN (11,530 lb st)
with afterburning 80.2 kN (18,030 lb st)

HAL
HINDUSTAN AERONAUTICS LTD
PO Box 5150, 15/1 Cubbon Road, Bangalore 560 001
Telephone: 91 (812) 76901
OFFICERS: see Aircraft section

The Bangalore Engine and Koraput Divisions of HAL constitute the main aero engine manufacturing elements of the Indian aircraft industry.

BANGALORE COMPLEX (Engine Division)
Adour 811 engines are manufactured under Rolls-Royce Turbomeca licence. The Orpheus 701 and Dart 536-2T are manufactured under licence from Rolls-Royce. The Artouste IIIB is made under licence from Turbomeca.

KORAPUT DIVISION
This Division was established to manufacture under Soviet government licence the Tumansky R-11 afterburning turbojet. With help from the Soviet Union, the first engine was run in early 1969. In 1977 production switched to the Tumansky (Soyuz) R-25 for the MiG-21bis, followed in 1984 by the R-29B for the MiG-27M.

INTERNATIONAL PROGRAMMES

BMW ROLLS-ROYCE
See under Germany in this section.

CFM
CFM INTERNATIONAL SA
2 boulevard du Général Martial Valin, F-75015 Paris, France
Telephone: 33 (1) 40 60 81 90
Fax: 33 (1) 40 60 81 47
Sales Engineering: F-75016 Paris
Telephone: 33 (1) 44 14 54 15
Fax: 33 (1) 44 14 55 54
CFM International Inc
Interstate Hwy 75, POB 15514, Cincinnati, Ohio 45215, USA
Telephone: 1 (513) 563 4180
Fax: 1 (584) 612 638
CHAIRMAN AND CEO: F. Avanzi
MARKETING DIRECTOR: Richard B. Shaffer

CFM International, a joint company, was formed by SNECMA (France) and General Electric (USA) in 1974 to provide management for the CFM56 programme and a single customer interface.

GE is responsible for design integration, the core engine and the main engine control. The core engine is derived from that of the F101 turbofan developed for the US military. SNECMA is responsible for the low-pressure system, gearbox, accessory integration and engine installation.

CFM INTERNATIONAL CFM56
US military designation: F108

In the late 1960s SNECMA and General Electric (now GE Aircraft Engines) concluded that a large market existed for a high bypass ratio engine in the ten ton class (97.9-106.8 kN; 22,000-24,000 lb st). The first CFM56 demonstrator ran at GE's Evendale plant on 20 June 1974.

The CFM56 designation covers a family of engines from 82.3 to 151.25 kN (18,500-34,000 lb st). By February 1993 a total of 6,000 engines had been delivered, against orders for more than 8,000, and 2,020 aircraft had logged over 45 million flight hours. The following are current versions:

CFM56-2. Certificated 8 November 1979, under FAR Pt 33 and JAR-E, at 106.8 kN (24,000 lb st); but a 97.9 kN (22,000 lb) T-O rating is used to re-engine the DC-8-60 to Super 70 standard. Scheduled operations began on 24 April 1982 and 110 aircraft in service with 16 operators have logged more than 7 million engine flight hours. Engine-caused shop visit rate is 0.16 and the dispatch reliability is 99.77 per cent. The CMF56-2 exists in several variants (CFM56-2C1 to -2C6), differing by their ratings or configuration details.

CFM56-2A. Certificated 6 June 1985 at 106.8 kN (24,000 lb st), flat rated to 35°C (95°F), the CFM56-2A-2 and 2A-3 powers the US Navy E-6 communications aircraft, the Royal Saudi Air Force E-3 and KE-3, the E-3 for the UK and France, and the Boeing E-8A for the US Air Force and Army. These applications require long duration oiltank, reverser and gearbox for two high capacity integrated drive generators.

CFM56-2B. Certificated 25 June 1982 at 97.9 kN (22,000 lb st), flat rated to 32°C (90°F), the CFM56-2B1 was selected by the US Air Force for its KC-135A tanker re-engining programme on 22 January 1980. First flight of a KC-135R took place on 4 August 1982 and production **F108-CF-100** engines power KC-135R aircraft delivered from late 1983. CFM56-2B1 also powers the C-135CFR tankers of the French Armée de l'Air. Total early 1993 was 335 aircraft.

CFM56-3B1. Derivative of CFM56-2, rated at 89.00 kN (20,000 lb st), flat rated to 30°C (86°F), with smaller fan. Powers Boeing 737-300 and -500. First ran in March 1982. US and French certification granted 12 January 1984. Entered airline service December 1984. Rerated at 82.3 kN (18,500 lb st), powers 737-500 which entered service February 1990.

CFM56-3B2. Certificated at 97.90 kN (22,000 lb st), flat rated to 30°C (86°F), on 20 June 1984. For 737-300 and 737-400 with improved payload/range from short, hot, high airfields. 737-400 entered service September 1988.

CFM56-3C1. Rated at 104.5 kN (23,500 lb st) for 737-400. Certificated December 1986. Offered as common engine for all 737 models at 82.3-104.5 kN (18,500-23,500 lb st). Over 3,000 Dash-3 engines delivered.

By December 1992 more than 169 customers had ordered 5,774 CFM56-3 engines. Over 1,250 Boeing 737s in service had logged more than 24 million engine flight hours. Engine-caused shop visit rate (12 month rolling averages) was 0.08, and dispatch reliability 99.96 per cent.

CFM56-5A1. Launched September 1984, for A320. Has fan diameter of -2, with improved aerodynamics in all LP and HP components, advanced clearance control features and FADEC. Nominal rating of 111.21 kN (25,000 lb st), flat rated to 30°C (86°F). Certificated 27 August 1987 and

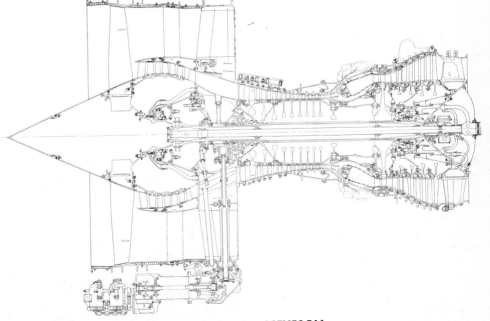

Longitudinal section of CFM56-5A1

622 INTERNATIONAL: ENGINES—CFM/EUROJET

CFM International CFM56-5C two-shaft turbofan

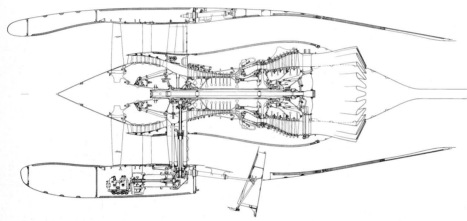

Longitudinal section of CFM56-5C complete nacelle, showing (lower half) reverser in operation

entered service April 1988. Outstanding reliability demonstrated by the first 280 A320s in service with 20 operators resulted in 120-min ETOPS certification.

CFM56-5A3. Rated at 117.9 kN (26,500 lb st). To meet specific airline requirements for A320.

CFM56-5B2. High-performance derivative with core of 5C4. Rated at 137.9 kN (31,000 lb st) for A321. First run 25 October 1991 at Villaroche. Advanced combustor to minimise NO$_x$ emissions. Certification due May 1993, entry to service due early 1994.

CFM56-5B4. Version of -5B derated to 117.9 kN (26,500 lb st); option for A320.

CFM56-5C2. Advanced fan, new four-stage LP compressor, active clearance control HP spool, new turbine section (5-stage LP, new frame, modulated clearance, new aerodynamics throughout), integrated mixer nozzle and FADEC. Rated at 138.8 kN (31,200 lb st). Powers A340, certificated 31 December 1991, service entry February 1993.

CFM56-5C3. Growth version rated at 144.6 kN (32,500 lb st). Certificated 31 December 1991 at 950°C TET. **5C3/F** certificated 1 March 1993 at 965°C TET. To enter service late 1993. **5C3/G** to be certificated 1994 at 975°C.

CFM56-5C4. Growth version to be certificated late 1994 at 151.25 kN (34,000 lb st) at 975°C TET to provide increased range/payload capability for A340.

TYPE: Two-shaft subsonic turbofan.

FAN: Single-stage axial. Forged titanium disc holding (CFM56-2) 44 titanium blades, each with a tip shroud to form a continuous ring; (CFM56-3) 38 titanium blades, each with part-span shroud; (CFM56-5) 36 titanium blades, each with part-span shroud. Mass flow (-2A2) 370 kg (817 lb)/s, (-3B2) 312 kg (688 lb)/s, (-5A1) 386 kg (852 lb)/s, (-5B2) 434 kg (956 lb)/s, (-5C2) 466 kg (1,027 lb)/s. Bypass ratio (-2) 6, (-3) 5, (-5A) 6, (-5B) 5.5, (-5C) 6.6.

LP COMPRESSOR: Three axial stages (4 on -5B and -5C), on titanium drum bolted to fan disc. A ring of bleed doors allows core airflow to escape into fan duct at low power settings.

HP COMPRESSOR: Nine-stage rotor with three stages of titanium blades and remainder of steel. Stator vanes are steel, with first four stators variable. Overall pressure ratio (-2) 2_ class, (-3C) 30.6, (-5A) 31.3, (-5B) 35, (-5C2/-5C3, max climb) 37.4.

COMBUSTION CHAMBER: Machined ring, fully annular, with advanced film cooling.

HP TURBINE: Single-stage with air-cooled stator and rotor air foils, advanced materials on -5. HP system carried in two bearings.

LP TURBINE: Four-stage (5 on -5C) with tip shrouds.

EXHAUST UNIT (FAN): Constant diameter duct of sound absorbent construction. Outer cowl and engine cowl form convergent plug nozzle, with airframe mounted reverser.

EXHAUST UNIT (CORE): Fixed-area with convergent nozzle, mixer on -5C.

ACCESSORY DRIVE: (CFM56-2 and -5) Gearbox in front sump transmits drive from front of HP spool to transfer gearbox on underside of fan case. Air starter at transfer gearbox (-2) or accessory gearbox (-5). (CFM56-3) Side mounted accessory drive gearbox with transfer gearbox; air starter pad on accessory gearbox.

CONTROL SYSTEM: Hydromechanical with electronic trim (-2, -3); FADEC (-5).

LUBRICATION: Dry sump system.

DIMENSIONS:

Length, excl spinner:
CFM56-2	2,430 mm (95.7 in)
CFM56-3	2,360 mm (93.0 in)
CFM56-5A1, -5A3	2,422 mm (95.4 in)
CFM56-5B1, -5B2	2,601 mm (102.4 in)
CFM56-5C	2,616 mm (103 in)

Fan diameter:
CFM56-2, -5A, -5B	1,735 mm (68.3 in)
CFM56-3	1,524 mm (60.0 in)
CFM56-5C	1,836 mm (72.3 in)

WEIGHT, DRY:
CFM56-2A2	2,187 kg (4,820 lb)
CFM56-2B1	2,119 kg (4,671 lb)
CFM56-2C series	2,102 kg (4,635 lb)
CFM56-3B1	1,940 kg (4,276 lb)
CFM56-3B2, -3C	1,951 kg (4,301 lb)
CFM56-5A1	2,257 kg (4,975 lb)
CFM56-5A3	2,266 kg (4,995 lb)
CFM56-5B1, -5B2	2,381 kg (5,250 lb)
CFM56-5C	2,492 kg (5,494 lb)

PERFORMANCE RATINGS:

Max T-O: see under model listings

Cruise, installed, 10,670 m (35,000 ft), Mach 0.8, ISA:
CFM56-2A2	23.08 kN (5,188 lb)
CFM56-2B1	22.10 kN (4,969 lb)
CFM56-2C series	23.82 kN (5,356 lb)
CFM56-3B1	20.67 kN (4,650 lb)
CFM56-3B1 rerate	19.57 kN (4,400 lb)
CFM56-3B2	22.42 kN (5,040 lb)
CFM56-3C	23.88 kN (5,370 lb)
CFM56-5A1, -5A3	22.23 kN (5,000 lb)
CFM56-5B1, -5B2	25.98 kN (5,840 lb)
CFM56-5C (with mixer)	30.74 kN (6,910 lb)

SPECIFIC FUEL CONSUMPTION (cruise, as above):
CFM56-2A2	18.72 mg/Ns (0.661 lb/h/lb)
CFM56-2B1	18.61 mg/Ns (0.657 lb/h/lb)
CFM56-2C series	18.44 mg/Ns (0.651 lb/h/lb)
CFM56-3 (all)	18.55 mg/Ns (0.655 lb/h/lb)
CFM56-5A1, -5A3, -5B1, -5B2	16.87 mg/Ns (0.596 lb/h/lb)
CFM56-5C (with mixer)	16.06 mg/Ns (0.567 lb/h/lb)

CFM INTERNATIONAL CFM88

This proposed turbofan is based on the core of the SNECMA M88-2 fighter engine, with an added seventh HP compressor stage. The LP spool would have a single fan stage. The CFM88 is seen as leading to a family of engines from 53.38 kN (12,000 lb st) to 88.97 kN (20,000 lb st), for 50/120-seat regional transports, business jets and military transports.

EUROJET

EUROJET TURBO GmbH

Arabellastrasse 13, D-8000 Munich 81, Germany
Telephone: 49 (89) 9210050
Fax: 49 (89) 92100539 and 92100562
Telex: 5212124 EJET D
MANAGING DIRECTOR: M. J. Roberts

Formed in August 1986 by a consortium of Fiat Aviazione (now FiatAvio) of Italy, MTU-München GmbH of (then) West Germany, Rolls-Royce of the UK and Sener (now ITP) of Spain, this company became operational on 1 January 1987 as EUROJET Turbo GmbH. It was established to co-ordinate the design, development and manufacture of the EJ200 engine for the European Fighter Aircraft (now Eurofighter 2000).

The workshare was proportional to the expected aircraft requirement of the four nations, and was agreed as follows: Fiat (21 per cent), LP turbine and shaft, interstage support, reheat system, gearbox, oil system, and participation in the intermediate casing; MTU (33 per cent), LP compressor, HP compressor, participation in HP turbine, and FADEC (full authority digital electronic control) design responsibility; Rolls-Royce (33 per cent), combustion system, HP turbine, intermediate casing, and participation in LP and HP compressors, LP turbine, interstage support, reheat system and nozzle; ITP (13 per cent), nozzle, jetpipe, exhaust diffuser and bypass duct.

Engine build and test during development and production is at each partner's facilities. Each partner provides comprehensive support for engines of its own national air force. The original requirement of the four nations was for 760 aircraft, including about 2,000 engines. The contract for full scale development was signed in November 1988. In the same month the first Design Verification Engine began running at MTU. By the end of 1992 the EJ200 had run more than 2,000 h and was due to begin flight testing in 1993.

EUROJET EJ200

This engine is an advanced turbofan designed for Mach numbers of about 2. It is fully modular, and allows for on-condition maintenance with built-in engine health monitoring and test equipment. Low maintenance and life-cycle cost along with high reliability have been prime design criteria. The total number of aerofoils is only approximately 60 per cent of those used in the RB199. The EJ200 programme has achieved all technical milestones on time and cost. This includes 150 h of accelerated mission endurance testing at

EUROJET/IAE—ENGINES: INTERNATIONAL

EJ200 FSD engine for the Eurofighter 2000 fighter aircraft

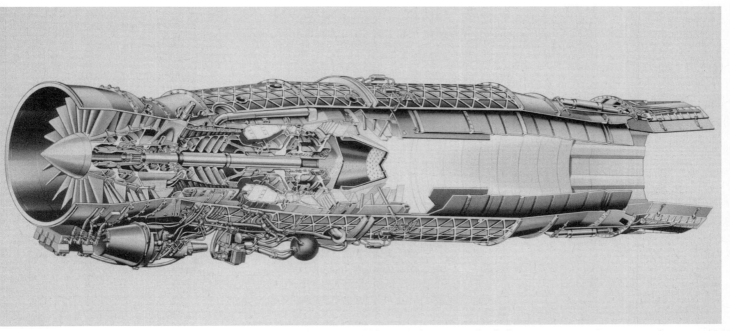

Internal arrangement of EUROJET EJ200 two-shaft augmented turbofan

full temperature rating. Entry into service is expected to be in 2000.
TYPE: Two-shaft augmented turbofan.
LP COMPRESSOR: Three stages, with 3D transonic blades of robust large-chord section. No inlet guide vanes. Third stage blisks. Overhung ahead of high-capacity ball bearing and forward roller bearing. No inlet guide vanes. Bypass ratio about 0.4. Pressure ratio 4-plus.
HP COMPRESSOR: Five stages, with first-stage variable inlet guide vanes and blisk rotor stage. Shaft supported between front ball and rear roller bearings. Overall pressure ratio more than 25.
COMBUSTOR: Fully annular, with vaporising burners.
HP TURBINE: Single-stage, with low density air-cooled single crystal blades.
LP TURBINE: Single-stage, with single crystal blades. Both turbine bearings in single interstage support frame.
EXHAUST SYSTEM: High efficiency augmentor of burn-then-mix type, with fully variable convergent-divergent nozzle.
ACCESSORIES: Central gearbox driven via tower shaft in interstage support. Full authority digital electronic control. Integrated health monitoring system, produced by consortium led by Dornier, with Smiths Industries, Tecnost and Ceselsa. Rotating-tank oil system to give artificial positive gravity at all times.
DIMENSIONS: Smaller throughout than RB199.
WEIGHT: In 1,000 kg (2,200 lb) class.
PERFORMANCE RATING: In 90 kN (20,000 lb st) class.

IAE
INTERNATIONAL AERO ENGINES AG
OFFICES: Corporate Center II, 628 Hebron Avenue, Glastonbury, Connecticut 06033, USA
Telephone: 1 (203) 652 1800
Fax: 1 (203) 659 1410
Telex: 4436031 INTLAERO
PRESIDENT: Roland P. Dilda
EXECUTIVE VICE-PRESIDENT: R. Goatmon
VICE-PRESIDENT, BUSINESS: Phillip Aldred
VICE-PRESIDENT, MARKETING: C. R. Nelson
PUBLIC RELATIONS MANAGER: Robert Nuttall

IAE is a management company set up on 14 December 1983 to direct the entire programme for the V2500 turbofan worldwide. The following shareholders signed a 30-year agreement: Rolls-Royce, UTC (P&W), Japanese Aero Engines (IHI, KHI, MHI), MTU and FiatAvio. Responsibility for each module is allocated according to shareholding. Overall engineering direction is delegated to Pratt & Whitney. Rolls-Royce manages the nacelle programme. Engines are assembled and tested by Pratt & Whitney and Rolls-Royce.

JAEC is responsible for the fan (derived from 535E4) and LP compressor, Rolls-Royce for the HP compressor, Pratt & Whitney for the combustor and HP turbine, MTU for the LP turbine and Fiat for the gearbox. Turbines, gearbox and FADEC use PW2000 technology.

IAE V2500
V2500-A1. In service on the A320. IAE supplies the complete package, including the nacelle (by Rohr/Shorts). Testing of the engine began in December 1985. A flight programme on a Boeing 720B in Canada was completed in 35 h in Spring 1988, and every ingestion and fan-blade-off test was passed first time (believed an industry record). The first pair of propulsion systems was delivered to Airbus Industrie in March 1988, and the V2500 was certificated in June 1988.

The first V2500-powered A320 made its first flight on 28 July 1988. The aircraft entered service with Adria Airways in May 1989. Awarded 120-min EROPS approval January 1992. Shop visit rate (12 months to May 1992) was 0.03 per 1,000 h, an industry record.

624　INTERNATIONAL: ENGINES—IAE/MTR

IAE V2500 two-shaft turbofan

Subsequent development to higher thrust has been achieved by increasing the core airflow and aerodynamic changes. Pressure ratio of the LP compressor is increased by adding a fourth stage. All engines in the A (Airbus) and D (Douglas) series have common fans and cores, and fit into a common nacelle which for the MD-90 is modified for fuselage side mounting. A5/D5 engines all certificated November 1992.

V2522-A5. Rated at 99.79 kN (22,000 lb st). Four-stage LP compressor, as in all A5 and D5 versions. Bypass ratio 4.9. Pressure ratio 25.3. For A319.
V2527-A5. Flat rated at 117.88 kN (26,500 lb st). Bypass ratio 4.75. Pressure ratio 28.6. For A320. Flight testing from 1992; first delivery (to United) late 1993.
V2525-D5. Flat rated at 111.25 kN (25,000 lb st). Bypass ratio 4.8. Pressure ratio 27.7. For MD-90-30. Flight testing from February 1993; first delivery (to Delta) 1994.
V2528-D5. Rated at 124.55 kN (28,000 lb st). Bypass ratio 4.7. Pressure ratio 30.4. For MD-90-50.
V2530-A5. Rated at 133.4 kN (30,000 lb st). Bypass ratio 4.6. Pressure ratio 32.5. For A321. Flight testing from March 1993; first delivery (Lufthansa) 1994.
V2535. Provisional designation for future growth engine to be rated in the 155.7 kN (35,000 lb st) class, with increased diameter fan and revised LP compressor and turbine.

By March 1993 a total of 33 customers had specified the V2500 for 750 aircraft, comprising 511 A320/321s and 239 MD-90s.

The primary features of the V2500-A5/D5 are as follows:
TYPE: Two-spool subsonic turbofan.
FAN: Single-stage with wide-chord shroudless blading. Diameter 1,600 mm (63.0 in). Pressure ratio 1.70. Bypass ratio 4.6. Mass flow 384 kg (848 lb)/s.
LP COMPRESSOR: Four stages, bolted to rear of fan to boost inlet to core. (Three stages in A1 version.)
HP COMPRESSOR: Ten stages of blading supported by a drum rotor. Inlet guide and first three vane stages variable. Overall pressure ratio 29.4.
COMBUSTOR: Annular segmented construction eliminates hoop stresses and provides low emissions and uniform exit temperatures.
HP TURBINE: Two stages of air-cooled single-crystal blading in powder metallurgy discs. Active tip clearance control.
LP TURBINE: Five stages of uncooled blading in welded and bolted rotor. Active clearance control.
GEARBOX: Modular unit, fan-case mounted.
CONTROL SYSTEM: Full authority digital electronic control (FADEC) to provide command outputs for engine fuel flow, stator vane angle, bleed modulation, turbine and exhaust case cooling, oil cooling, ignition and reverser functions. Supplied by Hamilton Standard.
NACELLE: Full length nacelle with reverser. Cowl load sharing to minimise case deflections. Acoustically treated.
DIMENSIONS:
　Length (flange to flange)　　　　　　3,200 mm (126 in)
　Fan diameter　　　　　　　　　　　　1,600 mm (63 in)
WEIGHT, DRY (with original single-stage LP compressor):
　Bare engine　　　　　　　　　　　　　2,370 kg (5,224 lb)
　Complete power plant, incl nacelle　3,311 kg (7,300 lb)
PERFORMANCE RATINGS (installed):
　T-O, S/L, ISA　133.4 kN (30,000 lb st) to ISA + 15°C
　Cruise Mach 0.8, 10,670 m (35,000 ft)
　　　　　　　　　　　　　　　　　　25.6 kN (5,752 lb)
SPECIFIC FUEL CONSUMPTION (Cruise Mach 0.8, 10,670 m;
　35,000 ft, installed)　　　　　16.26 mg/Ns (0.575 lb/h/lb)

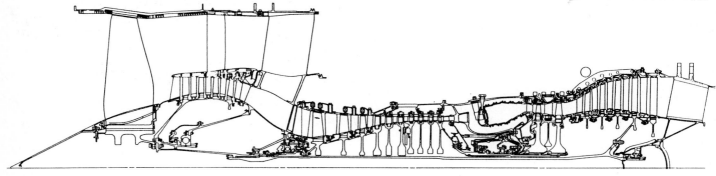

Longitudinal section of IAE V2500-A5/D5 two-shaft turbofan

MTR
MTU TURBOMECA ROLLS-ROYCE GmbH
Arabellastrasse 13, D-8000 Munich 81, Germany
Telephone: 49 (89) 910 2017
Fax: 49 (89) 910 1543
MANAGING DIRECTOR: H. J. Tenter
TECHNICAL DIRECTOR: J. L. Boucon
PROGRAMME DIRECTOR: H. Mohnkopf
COMMERCIAL DIRECTOR: B. Lory
MARKETING DIRECTOR: R. Sanderson

This company is owned equally by the three participants. It was set up in 1989 to produce and subsequently support the MTR 390 engine, and to act as contractor for customers.

MTR 390
This turboshaft engine was selected to power the Franco-German Tiger anti-tank and Gerfaut escort/support helicopters. The engine is suitable for military and civil applications in helicopters and fixed-wing aircraft in the form of single and twin installations.

The main characteristics of the modular engine are: ample emergency power reserve for OEI operation, high alternating output shaft power capability, low fuel consumption under part load, good acceleration, low life-cycle cost, easy handling and simple maintenance. It has a high performance FADEC and an engine monitoring system for flight and maintenance crew support.

Design studies have been extended to derivatives for other applications. A 6,000 rpm drive version (MTR 390L) and a direct-drive version (MTR 390T) have been defined.

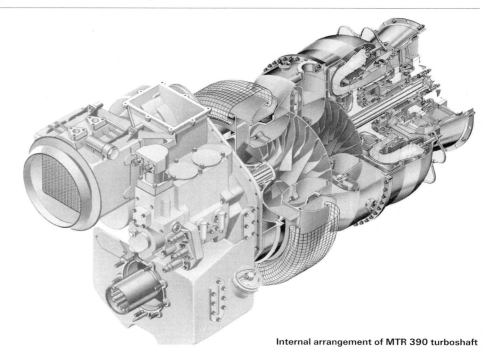

Internal arrangement of MTR 390 turboshaft

First run took place at MTU in December 1989, and the first flight engines for the first Tiger prototypes were delivered in September 1990. First flights in a Panther testbed and the prototype Tiger took place on schedule in 1991. By the end of 1992 more than 3,000 hours had been run, including 500 on flight test.

TYPE: Free turbine turboshaft.
COMPRESSOR: Two centrifugal stages for erosion and FOD resistance. Mass flow 3.2 kg (7.05 lb)/s. Pressure ratio 13.
COMBUSTION CHAMBER: Annular reverse-flow, with airblast fuel injectors for low emissions.
HP TURBINE: Single-stage gas generator turbine with high performance blade cooling, single-crystal blades and powder-metal disc.
LP TURBINE: Two-stage free power turbine with shrouded blades.
GEARBOX: Reduces the speed of the power turbine to the output shaft speed of 8,000 rpm. The accessory gearbox in the upper part provides the support and drive for the front and top mounted engine equipment.
CONTROL SYSTEM: A FADEC, with engine monitoring system.
LUBRICATION SYSTEM: Integral oil system, with engine-mounted tank and oil cooler with fan.
DIMENSIONS:
Length overall 1,078 mm (42.4 in)
Width overall 442 mm (17.4 in)
Height overall 682 mm (26.9 in)
WEIGHT, DRY: 169 kg (372.6 lb)
PERFORMANCE RATINGS (uninstalled, ISA, S/L):
Super contingency (OEI, 20 s) 1,160 kW (1,556 shp)
T-O 958 kW (1,285 shp)
Max continuous 873 kW (1,171 shp)
SPECIFIC FUEL CONSUMPTION:
T-O 76.05 µg/J (0.450 lb/h/shp)
Max continuous 76.90 µg/J (0.455 lb/h/shp)

PS-ZMKB

PS: see under Czech Republic
ZMKB: see under Ukraine
The DV-2 turbofan is a joint programme by these two organisations. It is described under ZMKB Progress in the Ukraine section.
Development began in 1978 in parallel with that of a new trainer to replace the L-39. Under an intergovernment agreement, the former Czechoslovakia was responsible for the aircraft and the former Soviet Union (then at the Lotarev MKB) for the engine. In 1982-87, 16 prototype DV-2 engines were tested. Series production began in 1989 at what are today PS Aero Engine division and ZMKB Progress. Current development includes life extension and a FADEC.

ROLLS-ROYCE/SATURN

Rolls-Royce: see under UK
Lyulka: see under Saturn/Lyulka in Russia
It was announced on 16 May 1990 that a preliminary agreement had been signed between Rolls-Royce, the Lyulka (Saturn) engine design bureau and the Sukhoi aircraft design bureau which is intended to lead to the joint development of an engine for a supersonic business jet. By May 1991 the two engine teams were to have determined the engine configuration and main characteristics. Further progress now depends on a business evaluation, finance, and the approval of the two governments to launch the programme.

ROLLS-ROYCE TURBOMECA

ROLLS-ROYCE TURBOMECA LIMITED
4/5 Grosvenor Place, London SW1X 7HH, UK
Telephone: 44 (71) 235 3641
Fax: 44 (71) 245 6385
Telex: 918944

This joint company was formed in June 1966 to control design, development and production programmes for the Adour two-shaft turbofan.
In 1980 Rolls-Royce Turbomeca launched the RTM 321 turboshaft demonstrator, leading to the RTM 322.

ROLLS-ROYCE TURBOMECA ADOUR
US military designation: F405

The Adour was designed for the SEPECAT Jaguar. The whole engine is simple and robust and of modular design.
Bench testing began at Derby on 9 May 1967. Engines for Jaguars were assembled at Derby (RR) and Tarnos (Turbomeca) from parts made at single sources in Britain and France. Turbomeca makes the compressors, casings and external pipework.
Following selection of the Adour for the Mitsubishi T-2 trainer and F-1 fighter/support aircraft, Ishikawajima-Harima Heavy Industries produced the Adour from 1970 under a licence agreement. In 1972 a non-afterburning Adour was selected to power the British Aerospace Hawk advanced trainer. More than 2,500 engines have been produced, including licence manufacture in Japan, India and Finland. Flight hours exceed 4 million.
Currently produced versions of the engine are as follows:

Mk 151. Non-afterburning version for Hawk. Internal components and certification temperatures identical to Mk 102 and Mk 801A. Qualified in 1975.
Mk 801A. Japanese designation TF40-IHI-801A. For Mitsubishi T-2 and F-1. Qualified in 1972. (See Ishikawajima-Harima in Japanese section.)
Mk 804. Uprated engine for Jaguar International. Rating with full afterburner at Mach 0.9 at S/L, ISA, increased by 27 per cent. Qualified in 1976.
Mk 811. Uprated version for Jaguar International. Revised compressor aerodynamics and hot-end improvements. Assembled by Hindustan Aeronautics, with increasing Indian manufactured content.
Mk 815C. Mk 804 uprated to Mk 811 performance level by conversion at overhaul.
Mk 851. Non-afterburning version of Mk 804 for export Hawk.
Mk 861. Non-afterburning version of Mk 811, first deliveries 1981.
Mk 861-49. US designation **F405-RR-400**. Derated version of Mk 861, for prototype McDD/BAe T-45A Goshawk for US Navy. Certificated 1988.
Mk 871. Uprated version for BAe Hawk Series 100 and 200. Certificated late 1990.
F405-RR-401. US version of Mk 871 with minor changes for production T-45A Goshawk.

The following refers to non-afterburning versions, except where indicated:
TYPE: Two-shaft turbofan for subsonic aircraft.
FAN: Two-stage. Full length bypass duct. Bypass ratio, 0.75-0.80.
COMPRESSOR: Five stages. Overall pressure ratio 11.
COMBUSTION CHAMBER: Annular, with 18 air spray fuel nozzles and two igniter plugs. Lucas engine fuel system.
HP TURBINE: Single-stage, air-cooled.
LP TURBINE: Single-stage. Squeeze-film bearings.
DIMENSIONS:
Length overall:
  Mks 102, 801A, 804, 811 2,970 mm (117 in)
  Mks 151, 851, 861, 861-49, 871 1,956 mm (77 in)
Inlet diameter (all) 559 mm (22 in)
Max width (all) 762 mm (30 in)
Max height (all) 1,041 mm (41 in)
WEIGHT, DRY:
Mk 102, 801A 704 kg (1,552 lb)
Mk 104, 804 713 kg (1,571 lb)
Mk 151 553 kg (1,220 lb)
Mk 851 568 kg (1,252 lb)
Mk 861 577 kg (1,273 lb)
Mk 811 738 kg (1,627 lb)
Mk 871 603 kg (1,330 lb)
PERFORMANCE RATINGS (S/L T-O):
Mk 102, 801A 32.5 kN (7,305 lb st)*
Mk 104 35.1 kN (7,900 lb st)*
Mk 151, 851 23.1 kN (5,200 lb st)
Mk 804 35.8 kN (8,040 lb st)*
Mk 861 25.4 kN (5,700 lb st)
Mk 861-49 24.2 kN (5,450 lb st)
Mk 811 37.4 kN (8,400 lb st)*
Mk 871 26.2 kN (5,900 lb st)
*With afterburner
SPECIFIC FUEL CONSUMPTION (Mk 102):
S/L static, dry 21 mg/Ns (0.74 lb/h/lb st)
Mach 0.8, 11,890 m (39,000 ft) 27 mg/Ns (0.955 lb/h/lb)

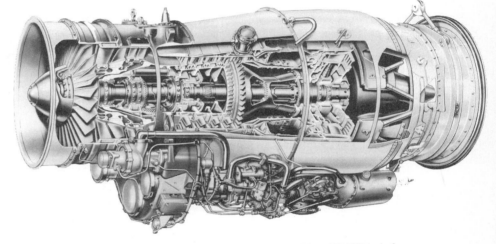

Cutaway drawing of Rolls-Royce Turbomeca Adour Mk 871 turbofan

ROLLS-ROYCE TURBOMECA RTM 322

Rolls-Royce and Turbomeca combined their extensive experience in helicopter gas turbines to produce the RTM 322 family of engines. Since 1986, Piaggio of Italy has been a 10 per cent participant. The launch engine is the RTM 322-01 turboshaft, which is conservatively rated at 1,566 kW (2,100 shp) with easy growth potential to 2,237 kW (3,000 shp).
The family, which will include turboprop and turbofan derivatives, is configured to combine simple design, reliability, low fuel consumption, light weight and low cost of ownership. A turboprop using the RTM 322-01 core would produce 1,193-1,491 kW (1,600-2,000 shp), with potential for growth to 2,088 kW (2,800 shp). It is therefore suitable for aircraft in the 35-70 seat range.
The turboshaft itself has full-authority digital electronic control, availability of different output drive configurations, a choice of three starting systems, and options for an inlet particle separator and infra-red suppressor. Combined with engine mounts configured for compatibility with a number of existing airframes, these features give the unit a wide range of potential civil and military applications in the 7 to 15 tonne class. Examples are: EH 101, Sikorsky S-92, Black Hawk and Seahawk series, European NH 90, Westland WS-70 and AH-64 Apache. The engine has been studied by the US Army as a potential growth power plant for the Black Hawk and Apache, and during 1987 the US Navy carried out an operability study in an SH-60B. In 1985 UTC (Pratt & Whitney) signed a licence agreement for US and Canadian markets.
In 1988 a major competition was held between Rolls-Royce Turbomeca and General Electric for the production engine contract of all UK EH 101 helicopters. In September 1988 the Minister of Defence Procurement announced that Rolls-Royce Turbomeca had won this competition for approximately 500 engines, as it provided 'the best value for

Rolls-Royce Turbomeca RTM 322-01 turboshaft

money'. The first flight on the EH 101 was planned for early 1993. Another application is the twin-engined Kamov Ka-62R.

The first complete RTM 322-01 ran on 4 February 1985. Over 1,790 kW (2,400 shp) has been demonstrated. A total of over 10,000 hours running have been completed, which includes 1,100 hours of flight development in the S-70C (from 14 June 1986) and then in the SH-60B. UK military certification was completed in October 1988, and was followed by civil certification in May 1992.

The following particulars apply to the RTM 322-01 turboshaft:

COMPRESSOR: Three-stage axial and single-stage centrifugal.
COMBUSTOR: Annular reverse flow. Ignition by Lucas Aerospace exciter.
TURBINES: Two-stage gas generator turbine. Cooling is applied to the 1st and 2nd stage stators and 1st stage rotor. The 2nd stage rotor is made of single crystal material and is uncooled. Two-stage power turbine with drive to front or rear.
DIMENSIONS:
Length overall 1,171 mm (46 in)
Diameter 604 mm (23.8 in)
WEIGHT, DRY: 240 kg (538 lb)
PERFORMANCE RATINGS (S/L):
Max contingency 1,724 kW (2,312 shp)
Max T-O 1,566 kW (2,100 shp)
Typical cruise 940 kW (1,260 shp)
SPECIFIC FUEL CONSUMPTION:
Cruise (as above) 81 μg/J (0.48 lb/shp/h)

TURBO-UNION

TURBO-UNION LTD
PO Box 3, Filton, Bristol BS12 7QE, UK
Telephone: 44 (272) 791234
Fax: 44 (272) 797575
Telex: 44185 RR BSLG
MUNICH OFFICE: Arabellastrasse 13, D-8000 Munich 81, Germany
Telephone: 49 (89) 9242 1
Fax: 49 (89) 915870
Telex: 524151 TUD
CHAIRMAN: C. H. Green
MANAGING DIRECTOR: Kurt Münzenmaier

Formed in 1969 as a European engine consortium comprising Rolls-Royce plc (40 per cent) of the United Kingdom, MTU Motoren- und Turbinen-Union München GmbH (40 per cent) of Germany and FiatAvio SpA (20 per cent) of Italy. The consortium was established to design, develop, manufacture and support the RB199 turbofan for the Panavia Tornado aircraft.

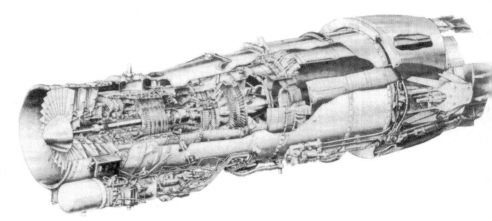

Cutaway drawing of Turbo-Union RB199 Mk 104 three-shaft augmented turbofan

TURBO-UNION RB199

The RB199 is a three-spool turbofan offering low fuel consumption for long-range dry cruise and approximately 100 per cent thrust augmentation with full afterburner for short take-off, combat manoeuvre and supersonic acceleration. An integral thrust reverser system is available. It was the first military engine with FADEC without hydromechanical backup.

In-service experience of over 2 million flying hours, at low level in the most arduous conditions, has proven the resilience of the RB199 to birdstrike and foreign object damage (FOD). This is a direct result of the relatively short, rigid rotating assemblies held between the small bearing spans in a three-spool layout.

Over 2,400 engines have been produced, with the present engine family comprising:

Mk 103. Standard production engine, with integral thrust reverser, for Panavia Tornado IDS variants.

Mk 104. Identical to the Mk 103 other than the jetpipe, which is extended by 360 mm (14 in) to provide up to 10 per cent greater thrust and reduced specific fuel consumption. The Mk 104 is the standard production engine for Tornado ADV variants.

Mk 104D. The power plant for the BAe experimental aircraft programme (EAP) advanced technology demonstrator.

Mk 104E. Selected as the interim engine for the European Fighter Aircraft (EFA) programme, the RB199 will be used to conduct flight test work on early EFA prototypes.

Mk 105. Similar to the Mk 103, the Mk 105 incorporates an increased mass flow LP compressor producing higher pressure ratios, and single-crystal HP turbine blades. In addition to a 10 per cent thrust increase, these improvements also give significant reductions in life-cycle cost. In service as the power plant for the Tornado ECR.

The following description refers to the Mk 105:

TYPE: Three-shaft turbofan with afterburner and reverser.
LP COMPRESSOR: Three-stage axial of titanium alloy. Casing of three bolted sections. Rotor of three discs welded together. Rotor blades secured by dovetail roots, all with snubbers. Mass flow approx 74.6 kg (164 lb)/s. Bypass ratio about 1:1.
IP COMPRESSOR: Three stages of titanium alloy. Rotor has welded discs in which blades are secured by dovetails.
HP COMPRESSOR: Six-stage; material changes from titanium at front to heat resisting alloy at rear, except stator blades are heat resisting steel throughout. Rotor discs secured by ten through-bolts, carrying blades by dovetail roots. Bevel drive to gearbox. Overall pressure ratio greater than 23.
BYPASS DUCT: Fabricated in titanium.
COMBUSTION CHAMBER: Annular flame tube fabricated from nickel alloy, bolted at rear end between outer casing, forged and chemically milled in nickel-iron alloy, and inner casing of nickel alloy. Carries 13 double-headed fuel vaporisers which give combustion without visible smoke. Two igniter plugs. Hot-streak injector for afterburner ignition.
HP TURBINE: Shrouded single stage. Entry temperature over 1,327°C. Rotor blades and stator vanes air-cooled.
IP TURBINE: Shrouded single stage. Air-cooled stator vanes and rotor blades.
LP TURBINE: Two-stage with shrouded hollow uncooled rotor blades.
AFTERBURNER: Front end of titanium fabricated jetpipe carries afterburner in which bypass air and core gas burn concurrently, without a mixing section. For core flow, two gutter flameholders fed by upstream atomisers. For bypass flow, reverse colander with radial extensions, each containing vaporising primary burner, between which multiple jets inject remainder of afterburner fuel. Fully modulated augmentation.
NOZZLE: Variable area, short petal, convergent nozzle operated by shroud actuated by four screwjacks, driven by fourth stage HP air motor via flexible shafting. Each of 14 master and 14 secondary petals is precision cast in cobalt alloy which minimises friction.
REVERSER: External two bucket type driven via flexible shaft by motor using HP air. In stowed position outer skins form aircraft profile. Deployment takes 1 s at any thrust setting from idle to max dry.
ACCESSORY DRIVES: Accessory gearbox on underside of intermediate casing (quick attach/detach coupling) carries hydromechanical portions of main and afterburner fuel systems, oil tank and pump, and output shaft to aircraft gearbox carrying KHD gas turbine starter/APU.
FUEL SYSTEM: Electronic main engine control unit uses signals from pilot's lever and power plant sensors. Afterburner fuel from engine driven vapour core pump.
DIMENSIONS:
Length overall: Mk 103 3,251 mm (128 in)
Mk 104 3,607 mm (142 in)
Mk 105 3,302 mm (130 in)
Intake diameter: Mks 103, 104 719 mm (28.3 in)
Mk 105 752 mm (29.6 in)
WEIGHT, DRY (excl reverser):
Mk 103 965 kg (2,107 lb)
Mk 104 976 kg (2,151 lb)
Mk 105 980 kg (2,160 lb)
PERFORMANCE RATINGS (S/L, ISA):
Max dry: Mks 103, 104 40.48 kN (9,100 lb st)
Mk 105 42.95 kN (9,656 lb st)
Max afterburning: Mk 103 71.17 kN (16,000 lb st)
Mk 104 72.95 kN (16,400 lb st)
Mk 105 74.73 kN (16,800 lb st)

WILLIAMS-ROLLS

See under Williams (USA)

ISRAEL

BSEL
BET-SHEMESH ENGINES LTD
Mobile Post Haela, Bet-Shemesh 99000
Telephone: 972 (2) 911661-6
Fax: 972 (2) 911970 or 915117

Bet-Shemesh Engines is owned by Ormat Turbines and Pratt & Whitney. Its 300 employees produce Turbomeca Marboré VI and Bet-Shemesh Sorek 4 turbojets (not included as it powers missiles and UAVs) and parts for the F100, J52, J79 and JT8. Support is provided for the Allison 250, F100, J79, Marboré and PW1120. There is an important investment casting division. BSEL is one of the few companies with experience of the F100 upgrading programme to the 220-E version.

IAI
ISRAEL AIRCRAFT INDUSTRIES LTD
Ben-Gurion International Airport, 70100
Telephone: 972 (3) 97131111
Fax: 972 (3) 972290
Telex: ISRAV-IL 371133

The Engine Overhaul Plant is part of IAI's Bedek Aviation Division. It produces J79-J1E engines under GE licence and performs extensive overhaul and maintenance of civil and military engines. Details of the J79 can be found in the 1978-79 edition of *Jane's*.

ITALY

ALFA ROMEO AVIO
ALFA ROMEO AVIO SpA
I-80038 Pomigliano D'Arco, Naples
Telephone: 39 (81) 8430111
Telex: 710083 ARAVIO
CHAIRMAN: Gen Fulvio Ristori
MANAGING DIRECTOR: Ing Filippo De Luca

Alfa Romeo Avio was prime contractor for the manufacture, under General Electric licence, of the J85, J79 and T58. It manufactures CF6 combustors and JT9D components, and assembles PT6T engines for the AB 212. Under GE licence it is responsible for the hot section of the T64-P4D, co-produced with FiatAvio, and participates in the RB199. The company is a partner in Italian production of the Rolls-Royce Spey 807. In November 1988 it became a 6.4 per cent partner in the Rolls-Royce Tay programme.

In February 1986 it began deliveries of GE T700-401 engines for EH 101 prototypes. It supplies components for T700 engines fitted to American helicopters and is developing new versions. It is also involved, with FiatAvio, in the development of the GE CT7-6, aimed at the EH 101, NH 90 and a new version of the A 129.

ALFA ROMEO AVIO AR.318
Alfa Romeo Avio developed this simple turboprop to cover powers from 298 to 596 kW (400-800 shp). Prototype, rated at 453 kW (608 shp) for T-O, flown in King Air. Full description in 1991-92 *Jane's*. No production application reported.

ARROW
ARROW ENGINEERING SRL
Via Badiaschi 25, I-29100 Piacenza
Telephone: 39 (523) 41932-42271
Fax: 39 (523) 41340
CHAIRMAN: Guido Polidoro
ENGINEERING MANAGER: Attilio Palladini

This company produces modular air-cooled two-stroke piston engines for microlights, ultralights and homebuilts. Smaller engines were described in the 1992-93 *Jane's*. The following are the two largest models.

ARROW GP 1000
This is the flat-four version, using the same cylinders as in the smaller engines, with bore 74.6 mm (2.94 in) and stroke 57.0 mm (2.24 in). Two 40 mm Bing carburettors are fitted. Capacity 996 cc (60.77 cu in). Helical gear reduction of 0.387 ratio.
DIMENSIONS:
Length	521 mm (20.45 in)
Width	490 mm (19.29 in)
Height	458 mm (18.03 in)
WEIGHT, READY TO RUN:	65.0 kg (119 lb)
PERFORMANCE RATING:	89.5 kW (120 hp) at 6,800 rpm
FUEL CONSUMPTION:	8-14 litres (2.11-3.70 US gallons; 1.76-3.08 Imp gallons)/h

ARROW GP 1500
This is the flat-six version of the GP 1000, using the same cylinders. Three 40 mm Bing carburettors are fitted. Capacity 1,495 cc (91.17 cu in). Helical gear reduction of 0.387 ratio.
DIMENSIONS:
Length	821 mm (32.32 in)
Width	490 mm (19.29 in)
Height	458 mm (18.03 in)
WEIGHT, READY TO RUN:	87.5 kg (192.90 lb)
PERFORMANCE RATING:	134 kW (180 hp) at 6,800 rpm
FUEL CONSUMPTION:	10-14 litres (2.38-3.69 US gallons; 2.20-3.08 Imp gallons)/h

Arrow GP 1500 flat-six engine

CRM
CRM
Via Manzoni 12, I-20121 Milan
Telephone: 39 (02) 708326

This company is famous for high-speed diesel engines for patrol boats and yachts, derived from the Isotta-Fraschini Asso 1000 W-18 aero engine of 1928. One version has been selected by the US Navy for the Sentinel 5000 airship.

CRM 18D/SS
This engine is a four-stroke turbocharged diesel. It has 18 cylinders in W formation, three banks of six, with pre-combustion chambers.
CYLINDERS: Water-cooled. Bore 150 mm (5.91 in). Stroke 180 mm (7.09 in). Capacity 57,260 cc (3,495 cu in). Compression ratio 16.25. Two turbochargers.
DIMENSIONS:
Length	3,370 mm (132.68 in)
Width	1,350 mm (53.15 in)
Height	1,304 mm (51.34 in)
WEIGHT, DRY:	about 1,700 kg (3,745 lb)
PERFORMANCE RATING:	1,380 kW (1,850 hp) at 2,100 rpm

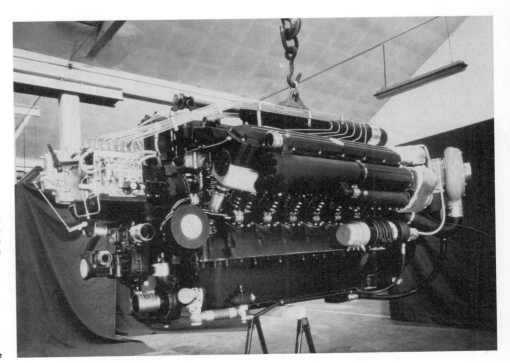

CRM 18D/SS 18-cylinder diesel engine

FIAT

FIATAVIO
Via Nizza 312, I-10127 Turin
Telephone: 39 (11) 69311
Telex: 221320 FIATAV
MANAGING DIRECTOR: P. Torricelli

FiatAvio's main aircraft engine programmes now concern the IAE V2500, Turbo-Union RB199, Rolls-Royce Spey 807 and Viper 600, Pratt & Whitney PW2037/2040 and PW4000, and General Electric GE90, CF6 (including CF6-80C2), CT7 and T64. It is a partner in EUROJET Turbo. FiatAvio makes the FA150-Argo APU for the AMX, transmissions for Eurocopter France helicopters and many other aviation products.

EUROJET TURBO EJ200
Fiat is responsible for 21 per cent of this programme. Its duties include design, development and manufacture of the LP turbine and shaft, rear bearing support, gearbox, oil tank and pump, afterburner and some accessories including the nozzle control unit and part of the DECU (FADEC) software.

TURBO-UNION RB199
Fiat holds 20 per cent of the shares of Turbo-Union Ltd. Its responsibility is the LP turbine and shaft, exhaust diffuser, jetpipe and nozzle.

IAE V2500
The V2500 is produced by the IAE consortium. Fiat is responsible for the accessory gearbox, oil tank and pumps, exhaust case and No. 5 bearing compartment.

ROLLS-ROYCE SPEY 807
This turbofan is produced for the AMX, under a Rolls-Royce licence to the Italian government, by FiatAvio (prime contractor in Italy) and CELMA (prime contractor in Brazil).

PRATT & WHITNEY PW2000 and PW4000
Since 1974 FiatAvio has been responsible for design and production of the accessory drive gearbox for these Pratt & Whitney engines.

GENERAL ELECTRIC GE90
FiatAvio is a risk-sharing partner on this engine.

GENERAL ELECTRIC CF6
FiatAvio produces components for the CF6 for GE and SNECMA. For GE the company supplies accessory gearboxes, inlet gearboxes and shafts. SNECMA is supplied with gearbox components and shafts for CF6-50 engines. FiatAvio is collaborating with GE on CF6-80C/C2 engines.

GENERAL ELECTRIC T64-P4D
This turboprop powers most versions of the Alenia G222. Under a licence agreement between GE and the Italian government, the engine is manufactured in Italy, with FiatAvio as prime contractor.

GENERAL ELECTRIC T700/CT7
Parts of the T700 are made by FiatAvio, while for the EH 101 helicopter the CT7-6 is being developed by GE, FiatAvio and Alfa Romeo Avio.

ROLLS-ROYCE VIPER 600
Development of this turbojet was undertaken in collaboration with Rolls-Royce. For most versions, components rearward of the compressor (except turbine discs and blades) are FiatAvio's responsibility. However, the Mk 632-43 is licensed to Piaggio.

PIAGGIO

INDUSTRIE AERONAUTICHE E MECCANICHE RINALDO PIAGGIO SpA
Via Cibrario 4, I-16154, Genoa
Telephone: 39 (10) 600 41
Fax: 39 (10) 603378
Telex: 270695 AERPIA I
WORKS AND OFFICERS: see Aircraft section

The Aero Engine Division of Piaggio manufactures the following engines under licence agreements: Rolls-Royce Viper 11, 526, 540 and 632-43 turbojets; Textron Lycoming T53-L-13, T55-L-11 and -712 turboshafts; and Rolls-Royce 1004 turboshaft. Piaggio also participates in co-production under licence of the Rolls-Royce Spey 807 turbofan and has joined Rolls-Royce Turbomeca in development and production of the RTM 322-01 turboshaft. The Engine Division also develops and produces IR suppression devices.

VM

VM MOTORI SpA
Via Ferrarese 29, I-44042 Cento (Fe)
Telephone: 39 (51) 908511
Fax: 39 (51) 908517
Telex: 511642

VM Motori specialises in high speed lightweight diesel engines. Following automotive production, the company entered the aeronautical field with a range of horizontally opposed engines. All are air-cooled four-stroke compression ignition engines, cooled by a propylenic glycol mixture and burning Jet A-1, JP-4, JP-5, JP-8 or similar fuel with direct injection by camshaft-driven plunger pumps. Each engine is turbocharged for operation to 8,850 m (29,000 ft). The engines are fully modular and are being offered with an initial TBO of 3,000 h. Cylinder size is 130 mm (5.1 in) by 110 mm (4.33 in) and compression ratio 18. Specific fuel consumption (econ cruise) is 106.4 µg/J (0.63 lb/h/hp).

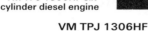

VM TPJ 1306HF six-cylinder diesel engine

VM TPJ 1304HF
Four cylinders, capacity 5.84 litres (356 cu in). Weight, dry with electrical system, 185 kg (408 lb). Maximum power 154 kW (206 hp) at 2,640 rpm.

VM TPJ 1306HF
Six cylinders, capacity 8.76 litres (535 cu in). Weight, dry with electrical system, 243 kg (536 lb). Maximum power 235 kW (315 hp) at 2,640 rpm.

VM TPJ 1308HF
Eight cylinders, capacity 11.68 litres (713 cu in). Weight, dry with electrical system, 298 kg (657 lb). Maximum power 316 kW (424 hp) at 2,640 rpm.

JAPAN

IHI

ISHIKAWAJIMA-HARIMA JUKOGYO KABUSHIKI KAISHA (Ishikawajima-Harima Heavy Industries Co Ltd)
Shin Ohtemachi Bldg 2-1, Ohtemachi 2-chome, Chiyoda-ku, Tokyo 100
Aero-Engine and Space Operations (ASO):
ADDRESS: As above
Telephone: 81 (3) 3224 5333
Fax: 81 (3) 3244 5398
Telex: 22232 IHIHQT J
PRESIDENT, ASO: Shozo Ojimi
GENERAL MANAGER, BUSINESS PLANNING:
  Kuniyasu Yamanaka

IHI's Aero-Engine & Space Operations specialises in the development and manufacture of aero-engines, space-related equipment, and land/marine gas turbines, as well as maintenance and repair. It has three plants and 3,750 employees. The number of jet engines so far produced totals about 3,500.

IHI began production of the J3 turbojet using Japan's own technology in 1959. This was followed by the licensed production of the J79, T64, T58 and TF40 (Adour) engines. In recent years, the F100, T56, F3 and T700 have been added to the product line.

The company has been involved in numerous engine development projects including the national project for the FJR710 and the Japan-Britain joint project for the RJ500. Currently, as the leader of a Japanese consortium, IHI is participating in the IAE V2500.

IHI is actively involved in many aspects of spaceflight.

IHI F3
Development of this turbofan began in 1976, with funding by the JDA's Technical Research & Development Institute. The Phase 1 **XF3-1** form has a single-stage fan with bypass ratio of 1.9, five-stage transonic compressor, 12-burner combustor and single-stage HP and LP turbines. Rating is 11.79 kN (2,650 lb st).

F3-IHI-30 two-shaft turbofan

In 1977 JDA contracted with IHI for the **XF3-20**, with reduced bypass ratio and higher turbine temperature to give a rating of 16.28 kN (3,660 lb st). This was followed by the **XF3-30**, which in 1982 was selected by the JASDF as the engine for the T-4 trainer. XF3-30 qualification was completed in March 1986. The engine is now redesignated **F3-IHI-30**, and the first production engine was delivered to JDA on 17 December 1987.
TYPE: Two-shaft turbofan.
FAN: Two-stage axial. No inlet guide vanes. Mass flow 34 kg (75 lb)/s. Pressure ratio 2.6. Bypass ratio 0.9.
COMPRESSOR: Five stages. First two stators variable. Overall pressure ratio 11.
COMBUSTION CHAMBER: Annular, with 12 duplex fuel nozzles.
HP TURBINE: Single-stage, air-cooled rotor blades.
LP TURBINE: Two-stage, tip shrouded.
FUEL SYSTEM: Hydromechanical, with electronic supervisor.
DIMENSIONS:
  Length   1,340 mm (52.76 in)
  Inlet diameter   560 mm (22.0 in)
WEIGHT, DRY: 340 kg (750 lb)
PERFORMANCE RATING (T-O, S/L):
                              16.37 kN (3,680 lb st) class
SPECIFIC FUEL CONSUMPTION:
                              19.83 mg/Ns (0.7 lb/h/lb st)

KAWASAKI
KAWASAKI JUKOGYO KABUSHIKI KAISHA (Kawasaki Heavy Industries Ltd)
1-18 Nakamachi-dori 2-chome, Chuo-ku, Kobe 650-91
Telephone: 81 (78) 371 9530
Jet Engine Division: World Trade Centre, 4-1 Hamamatsu-cho 2-chome, Minato-ku, Tokyo
Telephone: 81 (3) 3435 2111
Fax: 81 (3) 3578 3519
OFFICERS: see Aircraft section

In 1967 KHI started manufacturing T53 turboshafts. Deliveries of the resulting KT5311A, KT5313B and T53-K-13B engines totalled 324 by 1989. KHI now licence-builds the T53-K-703 for AH-1S HueyCobras (58 by 1989) and T55-K-712 for CH-47J Chinooks (72 by 1989). Kawasaki shared in parts manufacturing for the Adour, F100 and JT8D and IHI-assembled T56. It is a member of the IAE consortium (see under International heading), and is a risk-sharing partner on the PW4000 and RR Trent.

KHI is producing the small KJ12 turbojet of 1.47 kN (331 lb st) and is developing a helicopter turboshaft engine in the 597-746 kW (800-1,000 shp) class.

MITSUBISHI
MITSUBISHI JUKOGYO KABUSHIKI KAISHA (Mitsubishi Heavy Industries Ltd)
HEAD OFFICE: 5-1, Marunouchi 2-chome, Chiyoda-ku, Tokyo 100
Telephone: 81 (3) 3212 3111
NAGOYA AEROSPACE SYSTEMS: 10 Oye-cho, Minato-ku, Nagoya 455
Telephone: 81 (52) 611 2111
Fax: 81 (52) 612 3763
NAGOYA GUIDANCE AND PROPULSION: 1200 O-aza, Higashi Tanaka, Komaki
Telephone: 81 (568) 79 2111
OFFICERS: see Aircraft section

Between January 1973 and June 1981, under licence agreement with Pratt & Whitney, MHI delivered 72 JT8D-M-9 turbofans. MHI entered into a risk- and revenue-sharing agreement on the JT8D-200 in 1984, and on the PW4000 in 1989. In collaboration with IHI and Kawasaki, MHI participates in the V2500 (see IAE in the International part of this section).

NAL
NATIONAL AEROSPACE LABORATORY
7-44-1 Jindaijihigashi-machi, Chofu City, Tokyo 182
Telephone: 81 (422) 47 5911
Fax: 81 (422) 48 5888
DIRECTOR-GENERAL: Kazuaki Takashima
DIRECTOR OF AERO-ENGINE DIVISION: Hiroyuki Nouse

The NAL is a government establishment responsible for research and development. In 1971 the Ministry of International Trade and Industry (MITI) funded a high bypass ratio turbofan development programme. This engine, the FJR710, is still being used in experimental testing. A description appeared in the 1991-92 *Jane's*.

POLAND

PZL
PEZETEL FOREIGN TRADE ENTERPRISE
Al. Stanów Zjednoczonych 61, PL-04-028 Warsaw 50
Telephone: 48 (22) 108001
Fax: 48 (22) 132356 and 132835
Telex: 814651

For details of the organisation and activities of the Polish aircraft and aero engine industry, see the PZL entry in the Aircraft section.

IL
INSTYTUT LOTNICTWA (Aviation Institute)
HEADQUARTERS: Al. Krakowska 110/114, PL-02-256 Warsaw-Okecie
Telephone: 48 (22) 460993 and 460171
Fax: 48 (22) 464 432
Telex: 813 537
MANAGING DIRECTOR: Roman Czerwiński, MScEng
CHIEF CONSULTANT FOR SCIENTIFIC AND TECHNICAL CO-OPERATION: Jerzy Grzegorzewski, MScEng

The Aviation Institute is concerned with aeronautical research and testing. It can construct prototypes to its own design.

IL D-18A
This completely new engine was first run on 16 April 1992.
TYPE: Two-shaft turbofan.
AIR INTAKE: Direct pitot intake without inlet guide vanes.
FAN: Two-stage axial, with steel blades and stators. EB-welded steel rotor carried in ball and roller bearings. Pressure ratio 2.07. Bypass ratio 0.7.
COMPRESSOR: Five-stage compressor on HP shaft. Steel blades and stators. Rotor consists of two parts bolted together, both EB-welded, connected with HP turbine by Hirth coupling. Overall pressure ratio 8.
COMBUSTION CHAMBER: Annular type with 18 integral vaporisers, six starting atomisers and two high-energy igniters.
FUEL SYSTEM: Full-authority digital electronic control.
TURBINE: Single-stage HP and single-stage LP, both carried in roller bearings between rotors. Gas temperature 900°C before HP turbine.
JET PIPE: Plain fixed convergent nozzles for both core gas and bypass flow.
ACCESSORY DRIVES: Accessory gearbox driven by power off-take from front of HP shaft.
LUBRICATION SYSTEM: Integral oil system with vane pumps. Oil/fuel heat exchanger.
OIL SPECIFICATION: Synthetic, type SDF.
MOUNTING: Two main pads on intermediate casing. One rear strut on either side of centreline.
STARTING: 9 kW (12 hp), 27V starter-generator driven by aircraft battery or ground power unit.
DIMENSIONS:
  Length   1,940 mm (76.37 in)
  Width   750 mm (29.52 in)
  Height   900 mm (35.43 in)
WEIGHT, DRY: 380 kg (837.7 lb)
PERFORMANCE RATING:
  T-O   17.65 kN (3,968 lb st)
SPECIFIC FUEL CONSUMPTION:
  At T-O rating   20.96 mg/Ns (0.74 lb/h/lb st)

IL D-18A two-shaft turbofan

POLAND: ENGINES—IL/WSK-PZL RZESZÓW

IL K-15
This turbojet was announced in Summer 1988. It is being produced by WSK-PZL Rzeszów, and is a candidate engine for the I-22 Iryda.
TYPE: Single-shaft turbojet.
COMPRESSOR: Six stages. Rotor blades and shrouded stator blades of stainless steel. Two blow-off valves. Pressure ratio 5.3. Mass flow 23 kg (50.7 lb)/s.
COMBUSTION CHAMBER: Short annular type, with 18 vaporising burners and six starting atomisers. Electric ignition.
FUEL SYSTEM: Hydromechanical, with electronic blow-off valve control and overspeed and overtemperature limiters.
FUEL GRADE: Kerosene PSM-2 or TS-1.
TURBINE: Single-stage. Disc attached by Hirth coupling.
LUBRICATION SYSTEM: Self-contained recirculatory system except total-loss for rear bearing. Fully aerobatic.
OIL SPECIFICATION: Type SDF synthetic.
ACCESSORY DRIVES: Gearbox at bottom of intake casing driven by bevel gear from front of compressor.
STARTING: 27V starter/generator in nose bullet.
DIMENSIONS:
   Length overall   1,560 mm (61.42 in)
   Width   725 mm (28.54 in)
   Height   892 mm (35.12 in)
WEIGHT, DRY: 320 kg (705 lb)
PERFORMANCE RATINGS:
   T-O   14.7 kN (3,305 lb st) at 15,900 rpm
   Max continuous   11.5 kN (2,585 lb st) at 15,025 rpm
SPECIFIC FUEL CONSUMPTION:
   At T-O rating   28.49 mg/Ns (1.006 lb/h/lb st)

IL K-15 single-shaft turbojet

IL SO-3
This single-shaft turbojet was an improved version of the SO-1. Both engines were produced to power the TS-11 trainer, deliveries totalling 30 SO-1 engines and 580 SO-3. A description appeared in the 1991-92 *Jane's*.

WSK-PZL KALISZ
WYTWÓRNIA SPRZETU KOMUNIKACYJ-NEGO-PZL KALISZ
ul. Czestochowska 140, PL-62-800 Kalisz
Telephone: 48 (62) 77351
Fax: 48 (62) 37084
Telex: 046 2231
GENERAL DIRECTOR: Wlodzimierz Jerzyk

In 1952 the Soviet Union transferred responsibility for manufacture and service support of Soviet air-cooled radial piston engines to the WSK (transport equipment manufacturing centre) at Kalisz. Current production is centred on the following, plus the TWD-10B turboprop (described under WSK-PZL Rzeszów), which was supplied to the USSR. Kalisz also overhauls the TWD-10B.

PZL (IVCHENKO) AI-14R
The original 260 hp AI-14R version of this nine-cylinder air-cooled radial engine was produced in very large quantities, in the Soviet Union, China and Poland. Subsequent versions are:

AI-14RA. Rated at 191 kW (256 hp). Piston compressor drive and pneumatic starter. For Yak-12, Yak-18, PZL-101A Gawron and PZL-104 Wilga 35.
AI-14RA-KAF. RA with carburettor further aft, for Wilga 80.
AI-14RD. Rated at 206 kW (276 hp). Electric starter. For PZL-104 Wilga and PZL-130 Orlik.
AI-14RDP. With pneumatic starter. For PZL-104 Wilga and PZL-130 Orlik.
K8-AA. Rated at 246 kW (330 hp). Direct drive and pneumatic starter. Aerobatic. For PZL-130 Orlik until piston engine abandoned in favour of turboprop.

The following description refers to the AI-14RA:
TYPE: Nine-cylinder air-cooled radial.
CYLINDERS: Bore 105 mm (4.125 in). Stroke 130 mm (5.125 in). Capacity 10.16 litres (620 cu in). Compression ratio 5.9.
FUEL GRADE: 91 to 100 octane.
PROPELLER DRIVE: Planetary gears, ratio 0.787.
DIMENSIONS:
   Length   956 mm (37.63 in)
   Diameter   985 mm (38.78 in)
WEIGHT, DRY: 200 kg (441 lb)
PERFORMANCE RATINGS:
   T-O   191 kW (256 hp) at 2,350 rpm

PZL K8-AA (M-14Pm) nine-cylinder piston engine

Rated   162 kW (217 hp) at 2,050 rpm
SPECIFIC FUEL CONSUMPTION:
   T-O   95-104.3 μg/J (0.562-0.617 lb/h/hp)

PZL (SHVETSOV) ASz-62R
Power plant of the An-2 transport biplane, the ASz-62R was developed in the Soviet Union as the ASh-62. Current versions are:

ASz-62IR-16. Centrifugal oil filter. For An-2 and PZL-106 Kruk.
ASz-62IR-M18. Hydraulic airframe pump drive. For PZL M-18 Dromader.
ASz-62IR-M18/DHC-3. As M18 plus vacuum pump. For DHC-3 Otter.
K9-AA. Uprated engine designed at Kalisz, 860 kW (1,170 hp) at 2,300 rpm. Electric starter and hydraulic airframe pump. For PZL M-24 Super Dromader.
K9-BA. First of B-series, with improved cylinders and piston rings. Provision for feathering propeller, for C-47/DC-3 conversion.
K9-BB. Improved engine for M-24 Super Dromader.

PZL ASz-62IR nine-cylinder piston engine

K9-BC. Improved engine for An-2 and PZL-106 Kruk.
TYPE: Nine-cylinder air-cooled radial.
CYLINDERS: Bore 155 mm (6.10 in). Stroke 174 mm (6.85 in). Capacity 29.87 litres (1,823 cu in). Compression ratio 6.4.
FUEL GRADE: 91 to 100 octane.
PROPELLER DRIVE: Planetary gears, ratio 0.687.
The following relates to the IR-16:
DIMENSIONS:
   Length overall   1,130 mm (44.50 in)
   Diameter   1,375 mm (54.13 in)
WEIGHT, DRY:
   Without power take-off   580 kg (1,279 lb)
PERFORMANCE RATINGS:
   T-O   735 kW (985 hp) at 2,200 rpm
   Rated power   603 kW (809 hp) at 2,100 rpm
SPECIFIC FUEL CONSUMPTION:
   T-O   112 μg/J (0.661 lb/h/hp)

WSK-PZL RZESZÓW
WYTWÓRNIA SPRZETU KOMUNIKACYJ-NEGO-PZL RZESZÓW
ul. Hetmańska 120, PL-35-078 Rzeszów, PO Box 340
Telephone: 48 (17) 46100
Fax: 48 (17) 42625
Telex: 0633353
GENERAL MANAGER: Eng Tadeusz Cebulak

Current production at WSK Rzeszów is centred on the TWD-10B turboprop supplied to PZL for the An-28 (also produced by WSK-PZL Kalisz, which see); the GTD-350 turboshaft, together with WR-2 reduction gear for Mi-2 helicopters; the tropicalised SO-3 turbojet (see under IL); and the PZL-3S piston engine for agricultural aircraft.

TWD-10B
The Soviet designed OMKB/Glushenkov TVD-10B turboprop (see Mars under Russia), rated at 716 kW (960 shp), is made under licence in Poland for the An-28 STOL light transport built at WSK-PZL Mielec.
TYPE: Free turbine turboprop.
AIR INTAKE: Three radial struts, inlet guide vanes and starter de-iced by bleed air from combustion chamber.
COMPRESSOR: Six axial stages and one centrifugal. Stage 1 has front bearing journal, bolted to stages 2 to 6 which are pinched by compressor shaft used as tie bolt. Blades in dovetail roots. Pressure ratio 7.4. Mass flow 4.6 kg (10.14 lb)/s at 29,600 rpm.
COMPRESSOR CASING: Forward upper and lower halves in titanium; rear section welded from sheet steel and containing anti-surge bleed valve and radial diffuser.
COMBUSTION CHAMBER: Annular with centrifugal burner, and two starting units each with semiconductor igniter and auxiliary burner.
FUEL SYSTEM: Comprises supply pump, filter, pump governor, acceleration control, signalling block and thermocorrector.
FUEL GRADE: T-1, T-2, TS-1, RT, PSM-2, Jet A-1.
COMPRESSOR TURBINE: Two-stage axial. Blades held by fir tree roots. Inlet guide vanes of hollow sheet with air cooling. Casing has ceramic liner.

POWER TURBINE: Single-stage axial, blades with fir tree roots, held in front roller and rear ball bearing.
REDUCTION GEARS: Single-stage spur high speed gear to accessory box and propeller gear; under the high speed gear is the feathering pump oil tank. Accessory box drives 16kW alternator, propeller tachometer and reduction gear oil pump, with propeller brake. Upper driveshaft to single-stage planetary reduction gear.
ENGINE ACCESSORIES: Box contains oil centrifuge; and drives tachometer, oil pump, fuel pump and pump governor. The starter is on the front with a claw clutch.
OIL SYSTEM: Closed and pressurised. Gas generator and reduction gear pumps. Oil tank capacity 16 litres (4.2 US gallons; 3.5 Imp gallons).
OIL GRADE: Oil mixture: 25 per cent MK-22 or MS-20, 75 per cent MK-8 or MS-8P.
DIMENSIONS:
  Length without airframe jetpipe   2,060 mm (81.1 in)
  Width   555 mm (21.9 in)
  Height   900 mm (35.4 in)
WEIGHT, DRY: Bare   230 kg (507 lb)
  Complete engine   300 kg (661 lb)
PERFORMANCE RATINGS:
  T-O   754 kW (1,011 shp)
  Nominal   613 kW (823 shp)
  Max cruise   543 kW (728 shp)
SPECIFIC FUEL CONSUMPTION (T-O):
                                     96.4 µg/J (0.570 lb/h/shp)

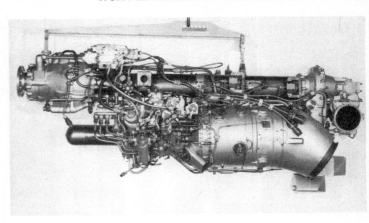

OMKB TWD-10B licensed to PZL Rzeszów

PZL-10W

This is the helicopter version of the TWD-10B, two of which power the PZL Swidnik Sokół. It uses the same gas generator, with the following differences:
TYPE: Free turbine turboshaft engine.
POWER TURBINE: Single-stage axial, with blades held in fir tree roots. Speed maintained at 22,490 rpm.
FUEL SYSTEM: Pump governor provides automatic operation at selected constant helicopter rotor speeds, as well as control of anti-surge bleed valve, maintaining constant fuel flow in starting cycle, limiting shaft speed and gas temperature and automatic switch-off of faulty engine and selection of emergency power on remaining unit.
FUEL GRADE: T1, T2, TS-1, RT, PSM-2a, Jet A-1.
OIL SYSTEM: Closed and pressurised, with one delivery and four-section scavenge pump. Oil tank airframe-mounted, normal capacity 14 litres (3.7 US gallons; 3.08 Imp gallons).
OIL GRADE: B-3W synthetic, Castrol 599 or 5000, Aeroshell or Elf turbine oil, or oil mixture 25 per cent MK-22 or MS-20 and 75 per cent MK-8 or MS-8P.
ACCESSORIES: Integral cast box drives 5kW starter, starter unit, electronic temperature limiter, vibration sensor, tachometer generator, oil pumps and centrifuge, phase torquemeter, de-icing valve, power-turbine speed limiter and operation time counter.
DIMENSIONS:
  Length with jetpipe   1,875 mm (73.8 in)
  Width   740 mm (29.0 in)
WEIGHT, DRY: 141 kg (310 lb)
PERFORMANCE, RATINGS (ISA):
  2.5 min   846 kW (1,134 shp)
  30 min   736 kW (987 shp)
  T-O   662 kW (888 shp)
  Continuous   574 kW (770 shp)
SPECIFIC FUEL CONSUMPTION:
  T-O   101.11 µg/J (0.60 lb/h/shp)

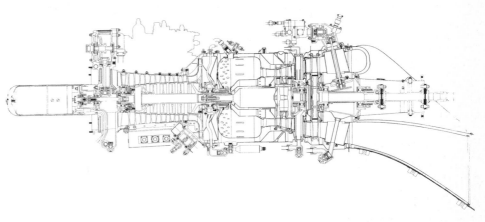

Longitudinal section through PZL-10W turboshaft

PZL GTD-350

The **GTD-350** is a helicopter turboshaft. In the Mi-2, the drive is taken from the rear. Though developed by the Isotov bureau in the former Soviet Union, it is in production only in Poland. PZL Rzeszów has developed a new version rated at 331 kW (444 shp) and designated **GTD-350P**. Technical life of the GTD-350 is 4,000 h.
TYPE: Axial/centrifugal-flow free turbine turboshaft.
AIR INTAKE: Annular. Automatic de-icing of inlet guide vanes and bullet by air bleed.
COMPRESSOR: Seven axial stages and one centrifugal, all of steel. Pressure ratio 6.05. Mass flow 2.19 kg (4.83 lb)/s at 45,000 rpm.
COMBUSTION CHAMBER: Reverse-flow type with air supply through two tubes. Centrifugal duplex single-nozzle burner. Semiconductor igniter plug.
FUEL SYSTEM: NR-40TA pump governor; RO-40TA power turbine governor; DS-40 controlling bleed valves; and electromagnetic starting valve.
FUEL GRADE: TS-1, TS-2 or Jet A-1.
COMPRESSOR TURBINE: Single stage. Shrouded blades with fir tree roots. Temperature before turbine 940°C (GTD-350P, 985°C).
POWER TURBINE: Two-stage constant speed (24,000 rpm). Shrouded blades with fir tree roots. Discs bolted together. Turbine stators integrally cast.
REDUCTION GEARING: Two sets of gears, with ratio of 0.246 : 1, in magnesium alloy casing. Output speed 5,900 rpm.
LUBRICATION SYSTEM: Closed type. Gear type pump with one pressure and four scavenge units. Cooler and tank, capacity 12.5 litres (3.30 US gallons; 2.75 Imp gallons).

PZL GTD-350 free turbine turboshaft

OIL GRADE: B3-W (synthetic), Castrol 98 or 5000, Elf Turbojet II or Shell Turbine Oil-500.
ACCESSORIES: STG3 3kW starter/generator, NR-40TA governor pump, D1 tachometer and oil pumps driven by gas generator. RO-40TA speed governor, D1 tachometer and centrifugal breather driven by power turbine.
STARTING: STG3 starter/generator suitable for operation at up to 4,000 m (13,125 ft) altitude.
DIMENSIONS:
  Length overall   1,385 mm (54.53 in)
  Max width   520 mm (20.47 in)
  Width (with jetpipes)   626 mm (24.65 in)
  Max height   630 mm (24.80 in)
  Height (with jetpipes)   760 mm (29.9 in)
WEIGHT, DRY:
  Less jetpipes and accessories   139.5 kg (307 lb)
PERFORMANCE RATINGS:
  T-O rating (6 min) at 96% max gas generator rpm:
    GTD-350   298 kW (400 shp)
    GTD-350P   331 kW (444 shp)
  Nominal rating (1 h) at 90% gas generator rpm:
    GTD-350   238.5 kW (320 shp)
    GTD-350P   261 kW (350 shp)
  Cruise rating (I)
    212.5 kW (285 shp) at 87.5% gas generator rpm
  Cruise rating (II)
    175 kW (235 shp) at 84.5% gas generator rpm

SPECIFIC FUEL CONSUMPTION:
  T-O   136 µg/J (0.805 lb/h/shp)
  Nominal   146 µg/J (0.861 lb/h/shp)
  Cruise (I)   154 µg/J (0.913 lb/h/shp)
  Cruise (II)   165 µg/J (0.978 lb/h/shp)
OIL CONSUMPTION:
  Max   0.3 litre (0.63 US pint; 0.53 Imp pint)/h

PZL-3S

Derived from Soviet AI-26W via LiT-3. Applications include PZL-106A Kruk, IAR-827A and conversions of Grumman/Schweizer Ag-Cat A, B and C, Thrush Commander, DHC-2 Beaver and DHC-3 Otter.
TYPE: Seven-cylinder air-cooled radial.
CYLINDERS: Bore 155.5 mm (6.12 in). Stroke 155 mm (6.1 in). Capacity 20.6 litres (1,265 cu in). Comp ratio 6.4.
PISTONS: Forged aluminium.
INDUCTION SYSTEM: Float type carburettor. Mechanically driven supercharger.
FUEL GRADE: Aviation gasoline, minimum 91 octane.
LUBRICATION: Gear type oil pump. Oil grade Aero Shell 100 or other to MIL-L-6082.
PROPELLER DRIVE: Direct. Provision for constant-speed US-132000A propeller.
ACCESSORIES: ANG 6423 Prestolite alternator and two output shafts, one 20 kW (27 hp) (26 kW; 35 hp max) for spraying pump and the other 3.7 kW (5 hp).

POLAND/RUSSIA: ENGINES—WSK-PZL RZESZÓW/TURBOMECANICA

Cutaway PZL-3S seven-cylinder piston engine

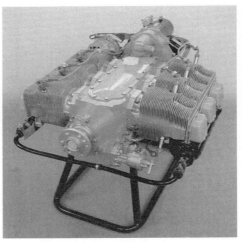

PZL-F 6A-350C1R six-cylinder piston engine

PZL-F 4A-235B31 four-cylinder piston engine

PZL-F 2A-120C1 two-cylinder piston engine

STARTING: Electric.
DIMENSIONS:
Diameter 1,267 mm (49.88 in)
Length 1,177 mm (46.34 in)
WEIGHT, DRY: 411 kg (906 lb)
PERFORMANCE RATINGS:
Max T-O 447 kW (600 hp) at 2,200 rpm
Max continuous 410 kW (550 hp) at 2,050 rpm
Cruise (75 per cent) 310 kW (415 hp) at 2,000 rpm
SPECIFIC FUEL CONSUMPTION:
T-O, max continuous 105 μg/J (0.61 lb/h/hp)
Cruise 86 μg/J (0.51 lb/h/hp)

PZL-3SR

This is the geared version of the PZL-3S. Applications include the PZL-106AR and BR and M-21. The following are the main differences:

PROPELLER DRIVE: Planetary gear of 0.7 ratio. Provision for constant-speed propeller, Type US-133000.
DIMENSION:
Length 1,271 mm (50.06 in)
WEIGHT, DRY: 446 kg (983 lb)

PZL-F ENGINES

In 1975 Pezetel acquired rights to manufacture and market the entire range of air-cooled piston engines formerly produced by the Franklin Engine Company (Air-cooled Motors) of the USA. These engines, known as PZL-F, are being produced by WSK-PZL Rzeszów.

Current applications include the SZD-45-2 Ogar F motor glider (2A-120C1), PZL-126 Mrówka (2A-120C1), PZL-110 Koliber (4A-235B31), and PZL Mielec M-20 Mewa (6A-350C1R). The 2A-120C1, 4A-235B31 and 6A-350C1R each have a Polish CACA certificate.

All models are of the horizontally opposed type, with cylinders of 117.48 mm (4.625 in) bore and 88.9 mm (3.5 in) stroke. All have direct drive and operate on 100/130 grade fuel. Accessories normally include electric starter, alternator and fuel pump. Other details are tabulated.

Engine model	Cylinder arrangement	Capacity cc (cu in)	Compression ratio	Max T-O rating at S/L kW (hp) at rpm	Overall dimensions mm (in) length	width	height	Weight, dry kg (lb)
2A-120C1	2 horiz	1,916 (117)	8.5	45 (60) at 3,200	630 (24.8)	795 (31.3)	515 (20.3)	69.1 (152)
4A-235B31	4 horiz	3,850 (235)	8.5	87 (116.7) at 2,800	750 (29.5)	795 (31.3)	637 (25.1)	103 (227)
6A-350C1R	6 horiz	5,735 (350)	10.5	153 (205) at 2,800	960 (37.8)	795 (31.3)	641 (25.25)	150 (331)

ROMANIA

AEROSTAR
AEROSTAR SA
9 Condorilor St, R-5500 Bacau
Telephone: 40 (931) 41885
Fax: 40 (931) 41885

OFFICERS: See Aircraft section

Among a diverse range of other products (see Aircraft section), this reorganised factory is manufacturing under licence the Russian VMKB (Vedeneyev) M-14P and M-14V26 piston engines, and the Soyuz/Tumansky RU-19A-300 gas turbine.

Aerostar-built M-14P nine-cylinder radial

TURBOMECANICA
INTREPRINDEREA TURBOMECANICA BUCURESTI
244 Bd Pacii, Sector 6, R-77826 Bucharest
Telephone: 40 (1) 760 6150
Fax: 40 (1) 769 4687
Telex: 10151 TURMO R

GENERAL DIRECTOR: Serban Ciorapciu

Founded in 1975, this factory produces under licence the Rolls-Royce Viper 632-41 and 633-47 and Spey 512-14DW, Turbomeca Turmo IVC, and helicopter gearboxes and rotor heads.

RUSSIA

Since 1989 the entire industry of the former Soviet Union has been undergoing great changes. Of course, the most profound changes have reflected the breakup of the Soviet Union into separate republics. As in the Aircraft Section, engine centres are now listed under the name of the republic in which they are situated. Most are in this section, but several will be found under Ukraine. In addition, the entire management of the industry has been revolutionised. The changes are generally intended to increase commercial viability, and they are superimposed on changes of name caused by the replacement of deceased or retired chief constructors by their successors. For nearly 70 years the procedure was for new engines to be designed at a KB (construction bureau) always identified by the name of the bureau head, or chief constructor. Their work had the benefit of support from the laboratories of the Central Institute of Aviation Motors. Once cleared for production, an engine would be assigned to a factory for series production. The factory (or factories) remained anonymous, though its GAZ (state aviation factory) number might become known (eg, see introduction to Klimov). Today the production plants are being named, many even have a chief designer, and they are taking display stands at exhibitions alongside the KBs. Meanwhile, confusion is increased further by the fact that some of the KBs have received names in honour of former chief designers or based on the city where they are located. As in the case of aircraft, each bureau no longer has guaranteed large home orders but has to fend for itself. This is reflected in the pages that follow. One immediate result is the formation of **ASSAD** in March 1992. This Union of Aviation Engine Producers is intended to assist strategic planning throughout the Independent States, and develop foreign economic links. ASSAD has been established by Viktor Chuiko, formerly Deputy Minister of the Soviet Aviation Industry. It is open to membership by major foreign (mainly engine) companies.

AO 'AVIADVIGATEL'
PERM SCIENTIFIC AND PRODUCTION ENTERPRISE 'AIRCRAFT ENGINES'

93 Komsomolsky Prospect (PO Box 624), Perm 614600
Telephone: 7 (83422) 452019
Fax: 7 (83422) 459777
Telex: 134802 LAVA SU
Cables: Perm 10 LAVA
GENERAL DESIGNER AND DIRECTOR: Yuri E. Reshetnikov
VICE-PRESIDENT, CHIEF DESIGNER: Mikhail I. Kuzmenko

This large design bureau, until 1990 named MKB engine design bureau, was founded 53 years ago by Shvetsov (which see), who was followed in 1953-89 by Soloviev. The engines created by Shvetsov and Soloviev powered 26 types of production aeroplanes and helicopters. Total flying time of their gas-turbine engines exceeds 42 million hours. These are fitted to the Tu-124, Tu-134, Tu-154M, Il-76, Il-62M, MiG-31, and Mi-6 and Mi-10 helicopters. These are flown by 22 countries.

The PS-90A, Soloviev's last design, is fitted to the Tu-204 and Il-96-300. The engines designed by 'Aviadvigatel' are manufactured by two large-scale production plants in Perm and Rybinsk. Many of the new designs are based on the PS-90A core: the PS-90A12, PS-90A14, D-100, D-110, and the PS-92 operating on liquefied gas. Attention is also being paid to propfans with contra-rotating fans, with a bypass ratio of about 15. Another area of design is industrial gas-turbine plant for gas pumping and power generation.

D-20P

The D-20P was the first turbofan in the USSR to enter scheduled airline service. A two-shaft engine, it was qualified in 1962 to power the Tu-124 at a rating of 52.96 kN (11,905 lb st). A full description last appeared in *Jane's* in 1987-88.

D-25V

D-25V is the turboshaft which powers the Mi-6 and Mi-10 and -10K helicopters. The helicopter power plant comprises two D-25V engines, identical except for handed jetpipes, and an R-7 gearbox. The latter has four stages of large gearwheels providing an overall ratio of 69.2. The R-7 is 2,795 mm (110.04 in) high, 1,551 mm (61.06 in) wide and 1,852 mm (72.91 in) long. Its dry weight is 3,200 kg (7,054 lb).

The D-25V is flat rated to maintain rated power to 3,000 m (10,000 ft) or to temperatures up to 40°C at S/L.

The **D-25VF** is uprated to 4,847 kW (6,500 shp), and operates at higher turbine gas temperatures. It powered the prototype V-12.

The following details apply to the basic D-25V:
TYPE: Single-shaft turboshaft with free power turbine.
COMPRESSOR: Nine-stage axial, with fixed inlet guide vanes and blow-off valves. Pressure ratio 5.6 at T-O power, 9,950 rpm.
COMBUSTION CHAMBER: Can-annular with 12 flame tubes.
FUEL GRADE: T-1, TS-1 to GOST 10227-62 (DERD.2494, MIL-F-5616).
TURBINE: Single-stage compressor turbine, overhung behind rear roller bearing. Two-stage power turbine, overhung on end of rear output shaft. Normal power turbine rpm, 7,800-8,300; maximum 9,000.
LUBRICATION: Pressure circulation at 3.5-4.5 bars (50-64 lb/sq in). Separate systems for gas generator and for power turbine, transmission and gearbox.
OIL GRADE: Gas generator, MK-8 to GOST 6457-53 or GOST 982-56. Power turbine and gearbox, mixture (75-25 Summer, 50-50 Winter) of MK-22 or MS-20 to GOST 1013-49 and MK-8 or 982-56.
ACCESSORIES: SP3-12TV electric supply and starting system; fuel supply to separate LP and HP systems; airframe accessories driven off upper and lower portions of R-7 gearbox.
STARTING: The SP3-12TV comprises an STG-12TM starter/generator on each engine, igniter unit, two spark plugs with cooling shrouds, two contactors, solenoid air valve, pressure warning, PSG-12V control panel and electrohydraulic cutout switch of the TsP-23A centrifugal governor.
DIMENSIONS:
Length overall, bare	2,737 mm (107.75 in)
Length overall with transmission shaft	5,537 mm (218.0 in)
Width	1,086 mm (42.76 in)
Height	1,158 mm (45.59 in)

WEIGHT, DRY:
With engine mounted accessories	1,325 kg (2,921 lb)

PERFORMANCE RATINGS:
T-O	4,157 kW (5,575 shp)
Rated power	3,552 kW (4,764 shp)
Cruise (1,000 m; 3,280 ft, 135 knots; 250 km/h; 155 mph)	2,983 kW (4,000 shp)

SPECIFIC FUEL CONSUMPTION:
T-O, as above	107.0 µg/J (0.633 lb/h/shp)
Cruise, as above	98.7 µg/J (0.584 lb/h/shp)

D-30

The D-30 has powered the Tu-134 since 1966. Since 1972 the Tu-134A has been powered by the **D-30 II** with reverser. Since 1982 the Tu-134A-3 has been powered by the **D-30 III** with an LP zero stage, providing the existing ratings at reduced gas temperature and maintained to ISA + 25°C. The Myasishchev M-55 was powered by two **D-30V12** engines each rated at only 49 kN (11,015 lb st). The Bartini (Beriev) VVA-14 was powered by four **D-30N** engines, the front pair for starting the air cushion. The gas generator (HP section and combustor) was the basis of the D-30F6 described later.
TYPE: Two-shaft turbofan (bypass turbojet).
FAN: Four-stage axial (LP compressor) (III, five-stage). First stage has shrouded titanium blades held in disc by pinned joints. Pressure ratio (T-O rating, 7,700 rpm, S/L, static), 2.65. Mass flow 126.8 kg (279.5 lb)/s. Bypass ratio 1.
COMPRESSOR: Ten-stage axial (HP compressor). Drum and disc construction, largely of titanium. Pressure ratio (T-O rating, 11,600 rpm, S/L, static), 7.1. Overall pressure ratio, 17.65.
COMBUSTION CHAMBER: Can-annular with 12 flame tubes fitted with duplex burners.
FUEL GRADE: T-1 and TS-1 to GOST 10227-62 (equivalent to DERD.2494 or MIL-F-5616).
TURBINE: Two-stage HP turbine. First stage has cooled blades in both stator and rotor. LP turbine also has two stages. All blades shrouded and bearings shock mounted.
JETPIPE: Main and bypass mixer with curvilinear ducts. D-30-II engine of Tu-134A fitted with twin-clamshell reverser.
LUBRICATION: Open type, with oil returned to tank.
OIL GRADE: Mineral oil MK-8 or MK-8P to GOST 6457-66 (equivalent to DERD.2490 or MIL-O-6081B).
ACCESSORIES: Automatic ice protection system, fire extinguishing for core and bypass flows, vibration detectors on casings, oil chip detectors and automatic limitation of exhaust gas temperature to 620°C at take-off or when starting and to 630°C in flight (5 min limit). Shaft driven accessories driven via radial bevel gear shafts in centre casing, mainly off HP spool. D-30-II carries constant speed drives for alternators.
STARTING: Electric DC starting system incorporating STG-12TVMO starter/generators.
DIMENSIONS:
Length overall	3,983 mm (156.8 in)
Base diameter of inlet casing	1,050 mm (41.3 in)

WEIGHT, DRY: 1,550 kg (3,417 lb)
PERFORMANCE RATINGS:
T-O	66.68 kN (14,990 lb st)
Long-range cruise rating, 11,000 m (36,000 ft) and Mach 0.75	12.75 kN (2,866 lb)

SPECIFIC FUEL CONSUMPTION:
T-O	17.22 mg/Ns (0.608 lb/h/lb st)
Cruise, as above	22.38 mg/Ns (0.79 lb/h/lb)

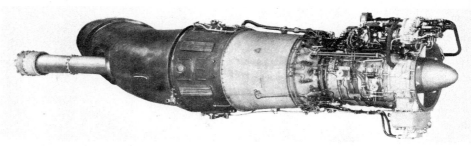
Aviadvigatel (Soloviev) D-25V free turbine turboshaft

Aviadvigatel (Soloviev) D-30 III two-shaft turbofan with reverser

Aviadvigatel (Soloviev) D-30KU-154 two-shaft turbofan with reverser

D-30K

Despite its designation, this turbofan is much larger and more powerful than the D-30. The basic **D-30KU**, to which the specification details apply, replaced the NK-8-4 as power plant of the Il-62M in 1974. The **D-30KU II**, or **KU-154**, is configured to suit the Tu-154M and is derated to 104 kN (23,380 lb st) up to ISA + 15°C. The **D-30KP**, rated at 117.7 kN (26,460 lb st), up to ISA + 15°C, was the original engine of all versions of the Il-76. Clamshell reversers are fitted to all four engines of this aircraft, and to the outer engines of the Il-62M. These reversers are airframe assemblies incorporated in the nacelle. In 1980 the KP was replaced in production by the **KP II** which maintains full power to ISA + 23°C. The **D-30KPV**, rated at 117.7 kN (26,460 lb st) and without reverser, powers the Beriev A-40 Albatross.

The following refers to the original D-30KU:
TYPE: Two-shaft turbofan, with mixer and reverser.
FAN (LP COMPRESSOR): Three stages, mainly of titanium alloy. First-stage rotor blades with part-span snubbers. Mass flow, 269 kg (593 lb)/s at 4,730 rpm (87.9 per cent), with bypass ratio of 2.42.
HP COMPRESSOR: Eleven stages, first two having part-span snubbers. Guide vanes turn 30° over 7,900-9,600 rpm, while air is bled from fifth and sixth stages. Overall pressure ratio (S/L, static) 20 at HP speed of 10,460 rpm (96 per cent).
COMBUSTION CHAMBER: Can-annular type with 12 flame tubes. Each tube comprises hemispherical head and eight short sections welded with gaps for dilution air. Single swirl type main/pilot burner centred in each tube. Igniter plugs in two tubes.
FUEL GRADE: T-1, TS-1, GOST-10227-86, A-1 (D1655/63t), DERD.2494 or 2498, Air 3405/B or 3-GP-23e.
TURBINES: Two-stage HP turbine with cooled blades in both stages. Second-stage rotor blades tip shrouded. Max gas temperature 1,122°C. Four-stage LP turbine with shrouded blades.
LUBRICATION: Closed type. Fuel/oil heat exchanger and centrifugal air separator with particle warning.
OIL GRADE: MK-8 or MK-8P to GOST 6457-66 (mineral) or BNII NP-50-1-4F to GOST 13076-67 (synthetic).
ACCESSORIES: Front and rear drive boxes under engine carry all shaft driven accessories. Differential constant speed drive to alternator and air turbine starter.
STARTING: Pneumatic starter fed by ground supply, APU or cross-bleed.
DIMENSIONS:
Length with reverser	5,700 mm (224 in)
Inlet diameter	1,455 mm (57.28 in)

Maximum diameter of casing	1,560 mm (61.4 in)
WEIGHT, DRY:	
With reverser	2,668 kg (5,882 lb)
Without reverser	2,318 kg (5,110 lb)
PERFORMANCE RATINGS (ISA):	
T-O	107.9 kN (24,250 lb st) to 21°C
Cruise at 11,000 m (36,000 ft) and Mach 0.8	
	27 kN (6,063 lb)
SPECIFIC FUEL CONSUMPTION:	
At T-O rating	14.11 mg/Ns (0.498 lb/h/lb st)
Cruise, as above	20.82 mg/Ns (0.735 lb/h/lb)

PS-90A (D-90A)

This high bypass ratio turbofan is not derived from any existing engine. It is assembled from 11 modules, and is designed for long life, high reliability and low fuel burn.

Bench testing began in 1984. Flight testing was in progress in 1987 with an engine replacing the starboard inner D-30KP in an Il-76. Certification was completed in 1991. The PS-90A powers the Il-96-300, first flown on 28 September 1988 with engines derated to 132.4 kN (29,762 lb st), and the Tu-204, first flown on 2 January 1989 with fully rated PS-90As. This engine is the first to have a designation reflecting the name of the General Designer (Soloviev).

In 1990 Mr Reshetnikov said that the sfc then achieved was 16.85 mg/Ns (0.595 lb/h/lb st), while the specification figure was 0.58. He said modifications intended to achieve the target included modifying blade profiles, and possibly adding a third stage to the LP compressor. For reliability reasons the variable HP stators will be changed from hydraulic to pneumatic operation. Many parts, including the nose spinner and titanium honeycomb nozzle, will be changed to composite material.

Since 1991 development has been in collaboration with the CFM partners GE and SNECMA. The French company would redesign the engine aerodynamically, while GE would undertake an integrity analysis. The resulting **PSKh-90** (PSX-90) would have T-O rating of 17.5 tonnes (171.6 kN; 38,580 lb st) with sfc of 16.03 mg/Ns (0.566 lb/h/lb).

TYPE: Two-shaft turbofan with mixer and fan reverser.
FAN: Single-stage, with 33 titanium blades, with snubbers. Hub/tip ratio 0.34. Bypass ratio (T-O) 4.6, (cruise) 4.8.
LP COMPRESSOR: Two-stage booster bolted to rear of fan.
HP COMPRESSOR: 13-stage spool with variable inlet guide vanes and first two stators. Overall pressure ratio (cruise) 35.5. Speed (max) 11,820 rpm.
COMBUSTION CHAMBER: Can-annular with 12 flame tubes with vaporising burners and two igniters.
HP TURBINE: Two stages, with advanced blades cooled by air passed through cold heat exchanger. Entry gas temperature 1,640°K (1,367°C).
LP TURBINE: Four stages.
JETPIPE: Mixer combines core and bypass flows to single nozzle.
CONTROL SYSTEM: Full-authority digital electronic.
REVERSER: Multiple blocker doors close off fan duct as translating mid-section of cowl moves to rear, to uncover all-round reverser cascades. No core reverser.

DIMENSIONS:	
Fan diameter	1,900 mm (74.8 in)
Length overall	4,964 mm (195.4 in)
WEIGHT, DRY:	2,950 kg (6,503 lb)
PERFORMANCE RATINGS (ISA):	
T-O, S/L	156.9 kN (35,275 lb st) to 30°C
Cruise at 11,000 m (36,000 ft) and Mach 0.8	
	34.32 kN (7,716 lb)
SPECIFIC FUEL CONSUMPTION:	
Cruise, as above	16.85 mg/Ns (0.595 lb/h/lb)

D-30F6

This large supersonic engine was designed from 1972 expressly for the MiG-31. Requirements included Mach 2.83 cruise at 11,000-21,000 m (36,100-68,900 ft) with the lowest possible fuel consumption, and Mach 1.25 at S/L. The engine comprises seven interchangeable modules. By 1992 it had attained an outstanding standard of reliability in several hundred thousand flight hours.

An unaugmented derivative rated at 49 kN (11,023 lb st), powers the subsonic high-altitude Myasishchyev M-55.

TYPE: Two-shaft augmented turbofan (bypass turbojet).
LP COMPRESSOR: Five stages, fixed inlet guide vanes. Mass flow 150 kg (331 lb)/s. Pressure ratio 3. Bypass ratio 0.57.
HP COMPRESSOR: 10 stages, first row variable stators and bypass doors behind stages 4 and 5. Pressure ratio 7.05. Overall pressure ratio 21.15.
COMBUSTION CHAMBER: Can-annular with 12 interlinked flame tubes.
TURBINES: Two-stage HP, with entry temperature 1,367°C. Cooling air bled from HP stages 5, 10, cooled in heat exchanger in bypass duct. Two-stage LP.
AFTERBURNER: High-volume, with four flameholder rings.
NOZZLE: Multi-flap con-di type with large variable area and cooling flows. Flow stabilised by auxiliary valve plates in divergent petals.
CONTROL SYSTEM: FADEC mounted on airframe.
AUXILIARIES: Independent gas-turbine APU under compressor used for starting (one per engine). Independent lubrication system.

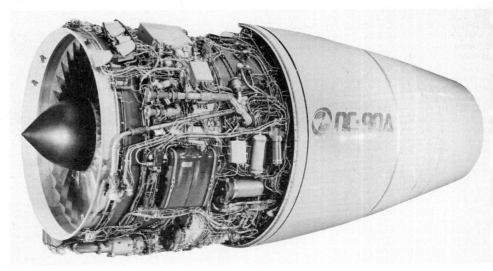

Aviadvigatel (Soloviev) PS-90A two-shaft turbofan

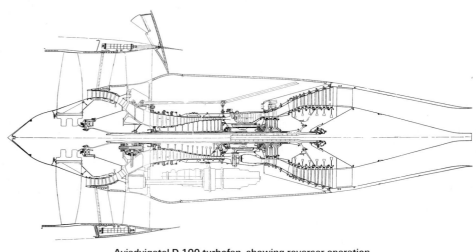

Aviadvigatel D-100 turbofan, showing reverser operation

DIMENSIONS:	
Inlet diameter	1,020 mm (40.2 in)
Length	7,040 mm (277.2 in)
WEIGHT, DRY:	2,416 kg (5,326 lb)
PERFORMANCE RATINGS (S/L static):	
Dry	93.2 kN (20,944 lb st)
Augmented	151.9 kN (34,170 lb st)
Frontal (max)	186.1 kN (41,843 lb st)
Power/frontal area	18,900 kg/m^2 (3,871 lb/sq ft)
SPECIFIC FUEL CONSUMPTION:	
Dry	20.4 mg/Ns (0.72 lb/h/lb st)
Augmented	53.8 mg/Ns (1.9 lb/h/lb st)

D-100/110/112

The D-100 and D-110 are turbofans for the 1990s, based on the PS-90A core. The **D-100** would have a fan with a diameter of 2,350 mm (92.52 in), the bypass ratio being about 8. Overall pressure ratio would be 35, TET 1,310°C, dry weight 3,350-3,550 kg (7,385-7,826 lb), T-O thrust 19-20 tonnes (186.3-196.1 kN; 41,890-44,090 lb st) and sfc (cruise, 11 km; 36,000 ft, M 0.8) of 14.73 mg/Ns (0.52 lb/h/lb).

Perm (Aviadvigatel) D-30F6 augmented two-shaft bypass turbojet

The **D-110** is a more advanced engine with a geared fan of 2,670 mm (105 in) diameter, giving a bypass ratio of 12 to 13. Overall pressure ratio would be 31, TET 1,282°C, dry weight 3,520 kg (7,760 lb), T-O thrust 20 tonnes (196.1 kN; 44,090 lb st) and sfc (cruise, same conditions) 14.73 mg/Ns (0.52 lb/h/lb).

An accompanying illustration shows the two engines together with an alternative gearless birotative propfan, the **D-112**. This has some features resembling the Garrett ATF3, with the same double reverse-flow arrangement, accessories being at the rear and the core jet being turned 180° to escape into the fan duct. This enables existing gas generator modules to be used for engines of super high bypass ratio, typically 15, as the fan shafts are not passed through the core.

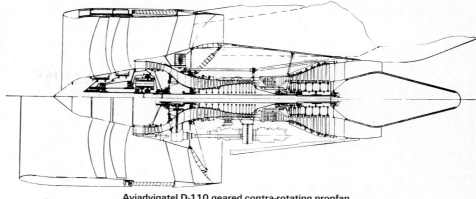

Aviadvigatel D-110 geared contra-rotating propfan

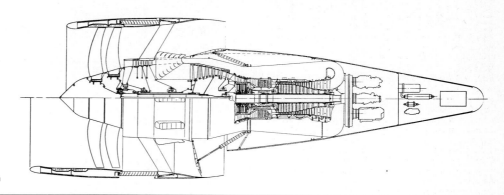

Aviadvigatel D-112 reverse-flow gearless propfan

AVIADVIGATEL — *see AO 'Aviadvigatel'*

BAKANOV

This designer, at Voronezh, is working on four completely new piston engines for general aviation.

BAKANOV M-16

Largest of the new engines, this is intended to replace the M-14P radial. It has eight air-cooled cylinders in double-X form (tandem pairs of cylinders forming a 90° X seen from the front). Maximum rating is 224 kW (300 hp). Applications include the Myasishchev Skif, Yak-56, Yak-57 and Yak-58.

BAKANOV M-17

An X-4 engine produced by taking half an M-16. To be rated in 1993 at 112 kW (150 hp), the M-17 could power the Be-103 and Yak-112. It is also a candidate for the Myasishchev SL.

BAKANOV M-18

This engine has two cylinders, smaller than those of the M-16 and M-17. To be rated at 30 kW (40 hp), it will power microlights.

BAKANOV M-19

This engine has four cylinders of the type used in the M-18. It will be rated at up to 59.7 kW (80 hp) for various light aircraft.

CHERNYSHOV
MOSCOW MACHINE-BUILDING PRODUCTION ASSOCIATION NAMED FOR V. V. CHERNYSHOV
121362 Moscow

Telephone: 7 (095) 491 57 22
Telex: ELIKA

This series production factory makes the RD-33 and TV7-117, both designed by Klimov Corporation.

GALIGUSOV

V. I. Galigusov was a Chief Designer in the Rybinsk KB; see RKBM.

GAVRILOV

V. Gavrilov was Chief Designer of the R-13-300 afterburning turbojet described under Soyuz (Tumansky).

GLUSHENKOV — *see Mars*

ISOTOV — *see Klimov*

IVCHENKO — *see ZMKB Progress under Ukraine*

KHACHATUROV

K. Khachaturov was Chief Designer of the R-27, R-29 and R-35 families of engines, within the Mikulin/Tumansky bureau (see Soyuz).

KKBM
KUIBYSHEV ENGINE DESIGN BUREAU
GSPE Trud, 443026 Samara 36
Telephone: 7 (8846) 250 02 28
Fax: 7 (8846) 250 39 79

This is the bureau formerly named for N. D. Kuznetsov. Engines begun under his direction but constructed later, and all work on LH$_2$ (liquid hydrogen) and LNG (liquid natural gas) are listed under Samara. See also Trud. Details of the P-020 and P-065 small piston engines appeared in the 1992-93 *Jane's*.

NK-6

For historical interest the tandem version of this turboprop, the **2TV-2F**, powered the eight-engined Tu-95-I, first prototype 'Bear'.

NK-12M

Designed at Kuibyshev in 1947-52 as the **TV-12**, under N. D. Kuznetsov and former German engineers, the NK-12M is the most powerful turboprop in the world. The **NK-12M** developed 8,948 ekW (12,000 ehp). The **NK-12MV** is rated at 11,033 ekW (14,795 ehp) and powered the Tu-114, driving four-blade contra-rotating propellers of 5.6 m (18 ft 4 in) diameter. As the **NK-12MA**, rated at 11,185 kW (15,000

RUSSIA: ENGINES—KKBM/KLIMOV CORPORATION

KKBM NK-12MV turboprop *(Piotr Butowski)*

shp), it powers the An-22, with propellers of 6.2 m (20 ft 4 in) diameter. A further application is in the Tupolev Tu-95/-142 bomber and its derivatives, and Tu-126, all usually powered by the NK-12MV. A single **NK-12MK** provides propulsion for the Orlyonok family of Ekranoplans. Tandem engines provide cruise propulsion for the Sukhoi S-90-200.

The NK-12M has a 14-stage axial-flow compressor. Pressure ratio varies from 9 to 13 according to altitude, and variable inlet vanes and blow-off valves are necessary. A can-annular type combustion system is used: each flame tube is mounted centrally on a downstream injector, but all tubes merge at their maximum diameter to form an annular secondary region. The single turbine is a five-stage axial. Mass flow is 65 kg (143 lb)/s.

The casing is made in four portions, from sheet steel, precision welded. An electric control for variation of propeller pitch is incorporated, to maintain constant speed.

DIMENSIONS:
Length 6,000 mm (236.2 in)
Diameter 1,150 mm (45.3 in)
WEIGHT, DRY: 2,350 kg (5,181 lb)
PERFORMANCE RATINGS (NK-12MV):
T-O 11,033 ekW (14,795 ehp)
Nominal power 8,826 ekW (11,836 ehp) at 8,300 rpm
Idling speed 6,600 rpm

NK-16

This powerful turboprop, derived from the NK-12 and rated at 13,795 kW (18,500 shp), powered the Tu-96 of 1956.

NK-8

The NK-8 was developed through a number of variants, the most powerful of which is the NK-144. Basic versions are the 99.1 kN (22,273 lb st) **NK-8-4**, later uprated to 103 kN (23,150 lb st), which originally powered the Il-62, and the 93.2 kN (20,950 lb st) **NK-8-2** which was the original engine of the Tu-154. The **NK-8-2U** powered the Tu-154B-2. The NK-8-4 remains in service with several Il-62 (not Il-62M) operators, including LOT. It led to the NK-86 described later. Twin **NK-8-4K** engines provide starting power for the Orlyonok family of Ekranoplans.

TYPE: Two-shaft turbofan.
FAN: Two-stage axial, with anti-flutter sweptback blades on first rotor stage. Pressure ratio 2.15 at 5,350 rpm. Bypass ratio 1.02 (NK-8-2, 1.00).
COMPRESSOR: Two IP stages on fan shaft. Six-stage HP compressor. Construction almost wholly of titanium. Core pressure ratio, 10.8 at 6,950 HP rpm (NK-8-2, 10 at 6,835 rpm).
COMBUSTION CHAMBER: Annular, with 139 burners.
FUEL GRADE: T-1 and TS-1 to GOST 10227-62 or T-7 to GOST 12308-66 (equivalent to Avtur 50).

TURBINE: Single-stage HP turbine, two-stage LP turbine, all with shrouded rotor blades, air-cooled discs and hollow nozzle blades (stators). Gas temperature, not over 870°C (1,143°K) ahead of turbine, not over 670°C (NK-8-2, 650°C) downstream.
JETPIPE: Mixer leads bypass flow into common jetpipe which may be fitted with blocker/cascade type reverser giving up to 48 per cent (NK-8-2, 45 per cent) reverse thrust, and noise suppressor.
LUBRICATION: Continuous pressure feed and recirculation. Pressure not less than 2.28 bars (33 lb/sq in).
OIL GRADE: Mineral oil MK-8 or MK-8P to GOST 6457-66 (DERD.2490 or MIL-O-6081B).
ACCESSORIES: These include automatic flight deck warning of vibration, ice and fire. All grouped beneath fan duct casing. RTA-26-9-1 turbine temperature controller by Smiths Industries.
STARTING: HP spool driven by constant-speed drive type PPO-62M, or started pneumatically by air from TA-6 APU from ground hose or by air bleed (NK-8-2, pneumatic only).

DIMENSIONS:
NK-8-4: Length, no reverser 5,100 mm (201 in)
NK-8-2: Length, with reverser 5,288 mm (208.19 in)
Length, no reverser 4,762 mm (187.48 in)
Diameter 1,442 mm (56.8 in)
WEIGHT, DRY:
NK-8-4: no reverser 2,100 kg (4,629 lb)
with reverser 2,400 kg (5,291 lb)
NK-8-2: no reverser 2,100 kg (4,629 lb) ma
with reverser 2,350 kg (5,180 lb) ma
PERFORMANCE RATINGS:
NK-8-4: T-O rating 103.0 kN (23,150 lb st)
Cruise rating at 11,000 m (36,000 ft) and 458 knot
(850 km/h; 530 mph) 27.0 kN (6,063 lb)
NK-8-2: T-O rating 93.2 kN (20,950 lb st)
NK-8-2U: T-O rating 103.0 kN (23,150 lb st)
SPECIFIC FUEL CONSUMPTION:
At cruise rating at 11,000 m (36,000 ft) and 458 knot
(850 km/h; 530 mph):
NK-8-4 22.1 mg/Ns (0.78 lb/h/lb)
NK-8-2 21.53 mg/Ns (0.76 lb/h/lb)

NK-20

Large two-shaft augmented turbofan qualified 1966 a 196.1 kN (44,090 lb st) max T-O.

NK-22

Large two-shaft augmented turbofan qualified 1962 a 119.6 kN (26,895 lb st) dry and 156.9 kN (35,275 lb st) max

NK-25

Large two-shaft augmented tubofan qualified about 197 at 186.4 kN (41,900 lb st) dry and 245.2 kN (55,115 lb s max.

NK-44

Projected two-shaft turbofan, 392 kN (88,180 lb st).

NK-86

This turbofan of 127.5 kN (28,660 lb st) is closely relate to the NK-8 series. Four power the Il-86 with combine reversers and noise attenuators.

NK-144

Brief details of this SST engine last appeared in the 1991-92 *Jane's*.

KKBM NK-86 two-shaft turbofan *(Flight International)*

KLIMOV CORPORATION
ST PETERSBURG NPO IM. KLIMOV

13 Kantemirovskaya Street, St Petersburg 184100
Telephone: 7 (245) 00 35
Fax: 7 (245) 33 55
Telex: 121282 JET SU
GENERAL DESIGNER: Alexander Alexandrovich Sarkisov
CHIEF DESIGNER: Piotr D. Gavra
DEPUTY GENERAL MANAGER: Andrei P. Listratov

The great design bureau and factory at what used to be Leningrad, the former Factory No. 117, was a major centre for high-power piston engines developed under Vladimir Yakovlyevich Klimov. In 1946 it was selected to build the Rolls-Royce Nene turbojet, later developed as the Klimov VK-1. Klimov was succeeded by his deputy, Sergei Pietrovich Isotov, who developed gas turbines mainly for helicopters but who in 1968 moved into the field of fighters. Isotov died in 1983 and was succeeded by Vladimir Styepanov (who retired early) and Alexander Sarkisov, but today the bureau has been renamed for its founder, and the '117' is continued in its engine designations.

Engines designed by NPO Klimov are manufactured at Perm (Sverdlov) and Zaporozhye (Motorostroitel), except for the TV7-117 which is in production at the Chernishov factory (Moscow) and Mars (Omsk), and the RD-33, made by Che nishov. June 1993 link with P&W Canada.

TV2-117

The power plant of the Mi-8 comprises two **TV2-117A** engines coupled through a VR-8A gearbox. The complet package incorporates a control system (separate from the control system of each gas generator) which maintains de sired rotor speed, synchronises the power of both engine and increases the power of the remaining engine if the othe should fail.

TV2-117TG. Qualified to operate on all normal gas-turbine fuels, and on gasoline (petrol), benzine, diesel oil, liquefied natural gas, propane or butane gas. Flown on Mi-8TG, ratings unchanged, and selected as interim engine for Mi-38. A foreign production facility is sought.
TYPE: Free turbine helicopter turboshaft.
COMPRESSOR: Ten-stage axial. Inlet guide vanes and stators of stages 1, 2 and 3 are variable. Pressure ratio 6.6 at 21,200 rpm.
COMBUSTION CHAMBER: Annular, with eight burner cones.
FUEL GRADE: T-1 or TS-1 to GOST 10227-62 specification (Western equivalents, DERD.2494, MIL-F-5616).
TURBINE: Two-stage axial compressor turbine with solid blades. Two-stage free power turbine.
OUTPUT SHAFT: Conveys torque from the free turbine to the overrunning clutch of the main gearbox (VR-8A) and also to the speed governor. Max output speed 12,000 rpm; main rotor speed 192 rpm.
ACCESSORIES: Engine control system includes fuel, hydraulic, anti-icing, gas temperature restriction, engine electric supply and starting, and monitoring systems. Up to 1.8 per cent of the mass flow can be used to heat the intake and other parts liable to icing. Fire extinguishant can be released by the pilot.
LUBRICATION: Pressure circulation type. Oil is scavenged from the five main bearings by the lower pump, returned through the air/oil heat exchanger and thence to the tank.
OIL GRADE: Synthetic, Grade B-3V to MRTU 38-1-157-65 (nearest foreign substitute Castrol 98 to DERD.2487).
STARTING: The SP3-15 system comprises DC starter/generator, six storage batteries, control panel, ground supply receptacle, and control switches and relays; airframe mounted except the GS-18TP starter/generator. The ignition unit comprises a control box, two plugs, solenoid valve, and switch. The starting fuel system comprises an automatic unit on the NR-40V pump, constant-pressure valve, and two igniters.
DIMENSIONS:
Length overall	2,835 mm (111.5 in)
Width (without jetpipe)	547 mm (21.5 in)
Height	745 mm (29.25 in)

WEIGHT, DRY:
Engine, without generator etc	338 kg (745 lb)
VR-8A gearbox, less entrapped oil	745 kg (1,642 lb)

PERFORMANCE RATINGS:
Max	1,250 kW (1,677 shp)
T-O (S/L, static)	1,118 kW (1,500 shp)
Max continuous	895 kW (1,200 shp)
Cruise (122 knots; 225 km/h; 140 mph at 500 m; 1,640 ft)	746 kW (1,000 shp)

SPECIFIC FUEL CONSUMPTION:
T-O, as above	102.4 µg/J (0.606 lb/h/shp)
Cruise, as above	115.4 µg/J (0.683 lb/h/shp)

TV3-117

This second-generation turboshaft has been produced in very large numbers. Bench testing began in 1974, the first flight was in 1976 and series production began in 1978.
TV3-117BK. Electronic control. Rated at 1,618 kW (2,170 shp). Powers some Ka-27s and Ka-28.
TV3-117MT. 1,434 kW (1,923 shp). Powers Mi-8T/TB/TBK, -14, -17, -24.
TV3-117V. 1,633 kW (2,190 shp). Powers some Ka-27s, 29 and -32.
TV3-117VK. Electronic control. Rated at 1,618 kW (2,170 shp). Powers Ka-50.
TV3-117VM. Electronic control. Rated at 1,545 kW (2,070 shp) to 3,600 m (11,810 ft). Powers Mi-17-1VA, -25, 28 and -35.
Data below refer mainly to the MT:
TYPE: Free turbine turboshaft.
COMPRESSOR: Ten-stage axial. Inlet guide vanes and first three stators variable. Pressure ratio 7.5.
COMBUSTION CHAMBER: Annular, improved version of TV2-117.
TURBINES: Two-stage gas generator turbine, improved from TV2-117. Two-stage power turbine.
OUTPUT: As TV2-117 but more compact jetpipe.
STARTING: BK, MT, pneumatic air turbine; V, VK, VM, electric.
DIMENSIONS:
Length	2,085 mm (82.1 in)
Width	640 mm (25.2 in)
Height	725 mm (28.5 in)

WEIGHT, DRY: 285 kg (628 lb)
PERFORMANCE RATINGS (S/L, max T-O): See variants
SPECIFIC FUEL CONSUMPTION:
Max T-O, TV3-117V	96.3 µg/J (0.57 lb/h/shp)

TV7-117

Described as a third-generation engine, the TV7-117 has a modular core incorporating advanced features and materials, and envisaged as the basis for various jet and shaft engines. The initial turboprop version was flight tested on two Il-76s and the prototype Il-114. Drives Stupino SV-34 propeller with six composite blades. To have been certificated August 1992, with series production at Chernyshov and Mars (Omsk).

TV7-117C. Turboprop version, selected (as **TV7-117-3**) to power the Il-114, produced in collaboration with Polish industry.
TVD-117E. Turboprop tailored to Ekranoplan propulsion, rated at 1,840 kW (2,467 shp). Three power the Raketa 2.2.
TV7-117V. Turboshaft version. Flat rated at 1,728 kW (2,318 shp) to 2,700 m (8,860 ft); OEI contingency 2,610 kW (3,500 shp) at S/L only. Powers Mi-38. Subject of agreement with Eurocopter (see Mil in aircraft section).
L-3000. Uprated version for Chinese aircraft at present powered by WJ5.
A growth core with two centrifugal stages is under development. Output will be in the 2,985 kW (4,000 shp) class, and will be unaffected by sand or dust.
The following description refers to the basic Il-114 engine:
TYPE: Free turbine turboprop.
COMPRESSOR: Annular ram inlet around reduction gear tapers to entry to five-stage axial compressor, with variable inlet guide vanes and next two stators, followed by centrifugal stage on same shaft. Pressure ratio 16.
COMBUSTION CHAMBER: Annular folded reverse flow. Minimum pollution with wide range of fuels, including LNG and LPG.
TURBINES: Two-stage gas generator turbine with cooled blades. Entry temperature 1,242°C. Two-stage power turbine.
REDUCTION GEAR: Planetary type, with new tooth profiles and anti-vibration mountings.
CONTROL SYSTEM: Full-authority electronic, with separate automatic control for ground and flight operation.
STARTING: Pneumatic air turbine, mass flow 0.2 kg (0.44 lb)/s.
DIMENSIONS:
Length	2,143 mm (84.37 in)
Width	886 mm (34.88 in)
Height	940 mm (37.0 in)

WEIGHT, DRY: 520 kg (1,146 lb)
PERFORMANCE RATINGS:
Max T-O
    1,839 kW (2,466 shp) to 35°C and 250 m (820 ft)
Cruise (6,000 m; 19,685 ft at 270 knots; 500 km/h; 311 mph)   1,342 kW (1,800 shp)
SPECIFIC FUEL CONSUMPTION:
Cruise, as above	67.09 µg/J (0.397 lb/h/shp)

Klimov (Isotov) TV2-117A free turbine turboshaft

Klimov TV3-117 free turbine turboshaft

Klimov TV7-117-3 (117C) turboprop with six-blade propeller for Ilyushin Il-114 *(Flight International)*

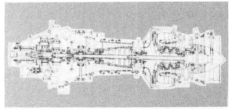

Longitudinal section of Klimov TV7-117V helicopter turboshaft

Klimov RD-33 augmented bypass turbojet

Klimov RD-60A two-shaft turbofan (Piotr Butowski)

RD-33

The contract for this important fighter engine was won competition with two other engine bureaux, and it w designed before Isotov's death. Detail design began in 196 meeting the requirements of the MiG-29, and the first ben run took place in 1972. Deliveries were initiated in 1976, a mass production at Chernyshov and Omsk followed in 198 It powers all known MiG-29 versions and is offered as t engine of an upgraded MiG-21.

RD-33K. Uprated version, with greater airflow a increased turbine entry temperature; 86.0 kN (19,335 lb s Powers MiG-29K and -29M.

RD-33/191. Hot-end improvements to extend TBO fro 800 to 1,200 h. Adopted by Luftwaffe.

The following refers to the original production RD-33:
TYPE: Two-shaft afterburning bypass turbojet (low ra turbofan).
LP COMPRESSOR: Four stages. Front bearing carried in fou strut nose, but no inlet guide vanes. Bypass ratio 0.4.
HP COMPRESSOR: Nine stages. Overall pressure ratio 20.
COMBUSTION CHAMBER: Annular with air-blast fuel nozzl giving generally smokeless combustion of a range of fuel
TURBINES: Single-stage HP turbine with single-crystal cool blades. Entry gas temperature 1,397°C. Single-stage L turbine.
AFTERBURNER: Combustion in both core and bypass flow Nozzle with fully variable area and profile in primary a secondary flows. Outer nozzle has 24 flaps. Vectoring no zles under preliminary development by CIAM and Soyu
ACCESSORIES: Oil tank, hydromechanical fuel control and au iliaries grouped above engine to reduce cross-sectio Closed lubrication system functions under all positive negative g-loads. Multipurpose self-diagnostic system.
DIMENSIONS:
Length 4,127 mm (162.5 i
Max diameter 1,000 mm (39.37 i
WEIGHT, DRY:
Bare engine 1,055 kg (2,326 l
Complete power plant 1,217 kg (2,683 l
PERFORMANCE RATINGS (S/L):
Max augmented 81.4 kN (18,300 lb s
Max dry 49.4 kN (11,110 lb s
SPECIFIC FUEL CONSUMPTION:
Max augmented 59.48 mg/Ns (2.10 lb/h/lb s
Max dry, S/L 21.8 mg/Ns (0.77 lb/h/lb s

KLIMOV RD-60A

In 1990 the RKBM (Novikov) RD-60 was developed in the RD-60A two-shaft turbofan for use in a future advance trainer to replace the Czech L-29 and L-39, for which five ai frame OKBs were competing in 1992. In order to reduce cos and forge international links, this engine has now bee replaced by one differing only in detail from the Frenc Turbomeca-SNECMA Larzac, an engine dating from 1968

KOBCHYENKO — *see Mars and Soyuz*

KOLIESOV — *see RKBM*

KUZNETSOV
Kuibyshev

Kuznetsov was deputy to General V. Ya. Klimov during the Second World War. In the late 1940s his own bureau developed large turboprops and turbofans. The NK-4 was transferred to Ivchenko, and is described under ZMKB Progress in the Ukraine section as the AI-20. Other engines, see KKBM and Samara/Trud.

LOTAREV — *see ZMKB Progress under Ukraine*

LYULKA — *see Saturn*

MARS
OMSK AIRCRAFT ENGINE DESIGN BUREAU 'MARS'
644021 Omsk-21
Telephone: 7 (3812) 33 49 81, 33 00 84
Telegraph: Omsk-21 MARS
Teletype: Omsk-3274 MARS
Telex: 133112 + MARS SU

This bureau was formerly headed by Glushenkov. Development, in partnership with CIAM (Central Institute for Aviation Motors), began in 1957. The first product was the 224 kW (300 hp) GTD-1, followed by the GTD-5 and -5M, which continue in production. The first major engine for aircraft propulsion was the GTD-3, produced from 1964 as a twin package plus common reduction gear for helicopters. In 1970 came the TVD-10 turboprop. Apart from APUs, the latest Mars engines are the TVD-20 and TV-O-100.

The associated Omsk (OMKB) production factory produces several types of engine, including the RD-33 and TV7-117.

TVD-10
Military designation: GTD-3

This free turbine engine was developed to power the Kamov Ka-25 helicopter. The civil turboshaft was licensed to Poland for production, and a description appears under WSK-PZL Rzeszów. Ka-25 engines were made in the Soviet Union, the **GTD-3F** being rated at 671 kW (900 shp) and the **GTD-3BM** at 738 kW (990 shp).

The **TVD-10B** is the turboprop version selected to power the An-28 and Roks-Aero T-101, and is a candidate engine for the T-106 and T-501. This engine is produced in Poland as the **TWD-10B**, as described under WSK-PZL Rzeszów. The Omsk design bureau is developing the **TVD-10BA**, rated a 790 kW (1,060 shp) for the An-28A for Polar operations.

TV-O-100

This engine, designed under Kobchyenko (see Soyuz), wa developed to provide a modern core in the 537 kW (720 shp class. Its initial application is in the single-turbine Ka-12 helicopter. Like the helicopter, the engine was to have bee produced in Romania (by Turbomecanica), but this arrange ment was cancelled in 1992. Future development of th engine to 619 kW (830 shp), with pressure ratio about 10. and turbine entry temperature of 1,077°C, may be carried ou in collaboration with that country. A derated version (53 kW; 710 shp) is a candidate engine for the Ka-118. Th **TVD-100** turboprop version could be a candidate for th twin-engined Sukhoi S-86 and Roks-Aero T-610. It coul

ive tractor or pusher propellers. CIAM, the national aero-
gine research organisation, is testing a heat exchanger with
ich specific fuel consumption may be reduced by 15-20
r cent.
PE: Free-turbine turboshaft. Modular construction.
MPRESSOR: Inlet above engine leads via large dust/sand
extractor to two-stage axial compressor, with inlet and
intermediate stator stages variable, and single centrifugal
stage. Pressure ratio 9.2.
MBUSTION CHAMBER: Annular folded reverse flow.
RBINES: Single-stage gas generator turbine with uncooled
blades. Turbine entry temperature 1,027°C. Single-stage
power turbine.
NTROL SYSTEM: Dual-channel electronic.
TPUT SHAFT: Central quill shaft drives gear train at front,
with triple spur gears to 6,000 rpm output shaft at top.
MENSIONS:
Length   1,275 mm (50.2 in)
Width   780 mm (30.7 in)
Height   735 mm (28.9 in)
IGHT, DRY (equipped):   160 kg (353 lb)
RFORMANCE RATINGS (S/L):
Maximum   537 kW (720 shp)
Cruise   343 kW (460 shp)
ECIFIC FUEL CONSUMPTION:
Cruise   109.2 µg/J (0.646 lb/h/shp)

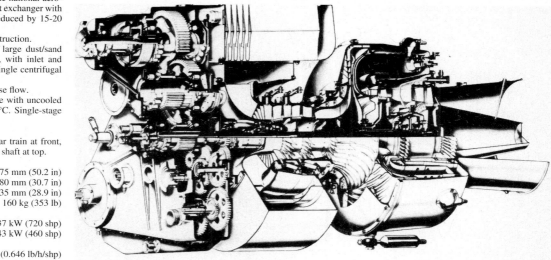

Cutaway drawing of OMKB Mars TV-O-100 turboshaft

TVD-20

This turboprop was to be fitted to the production An-3, and
s been selected to power the Roks-Aero T-101V. It uses a
rivative of the TVD-10B gas generator, with a zero stage
 a total of seven) on the compressor and a second stage on
 power turbine.
MPRESSOR: Seven axial stages plus one centrifugal. Inlet
faces aft.
MBUSTION CHAMBER: As TVD-10B.
RBINES: Two-stage gas-generator turbine. Two-stage free
power turbine.

PROPELLER DRIVE: Quill shaft leads forward from power tur-
bine to drive two-stage gearbox with accessory drives
above. Drive to AV-17 reversible-pitch propeller incorpor-
ates brake for safe loading of chemicals with engine
running.
ACCESSORIES: Two 27 V electric starters, provision for dusting
or spraying pumps.
DIMENSIONS:
Length   1,770 mm (69.7 in)
Width   850 mm (33.5 in)
Height   845 mm (33.3 in)
WEIGHT, DRY: 285 kg (628 lb)
PERFORMANCE RATING (S/L):
T-O   1,029 kW (1,380 shp)
SPECIFIC FUEL CONSUMPTION:
T-O as above   85.4 µg/J (0.506 lb/h/shp)

ETSKHVARICHVILI — *see Soyuz*

MIKULIN

A. A. Mikulin was a leading designer of engines from
16. The bureau is today called Soyuz.

NPK — *see Soyuz*

OTOROSTROITEL — *see ZMKB Progress under Ukraine*

OVIKOV — *see RKBM*

MSK, OMKB — *see Mars*

ERM — *see AO 'Aviadvigatel'*

ROGRESS — *see ZMKB Progress under Ukraine*

KBM
KBM (RYBINSK ENGINE-BUILDING DESIGN OFFICE)

2903 Rybinsk, Jaroslavskaya obl.
lephone: 7 (86370) 4 31 44
lex: Start
ANAGING DIRECTOR AND CHIEF DESIGNER:
Aleksandr S. Novikov
This large design office has produced more than 50 types
 engine under four General Designers (V. A. Dobrynin, P.
 Kolesov, V. I. Galigusov and A. S. Novikov). It has a large
search effort on non-metallic materials, especially carbon
mposites, silicon nitride and silicon carbide.

KOLESOV VD-7

This large single-shaft turbojet was designed for the M-6
d ZMS (NATO 'Bison').
VD-7B. Powered all production versions of M-6, ZM and
MS, rated at 127.5 kN (28,660 lb st). A different version
s used as the outboard engines of the M-50 (NATO 'Boun-
r'). Ten (eight at the front for starting, two on the tail for
uise) powered the giant KM and *Lun* Ekranoplans in
67-72.
VD-7M. Version designed for supersonic flight, with afterrner, for prototype Tu-105 and Tu-22 (NATO 'Blinder')
d inboard positions of M-50; max rating 156.9 kN
5,275 lb).
RD-7M-2. Developed engine for the production Tu-22;
ax rating 161.8 kN (36,376 lb st).

RKBM Kolesov VD-7 turbojet (without afterburner)

KOLESOV VD-19

Development of VD-7 with reduced diameter, to power Tu-28-80 prototype interceptors. Max rating with afterburner 100 kN (22,480 lb st).

KOLESOV RD-36-35

This was the first Soviet lift jet to go into production.
RD-36-35FVR. Lift engine of Yak-36M (Yak-38), in twin installation.
RD-36-35PR. Lift engine of Bartini VVA-14, two batteries of six.
TYPE: Single-shaft turbojet for operation in vertical attitude.
COMPRESSOR: Six stages. Airflow 45.3 kg (100 lb)/s.
COMBUSTION CHAMBER: Annular, with 30 burners.
TURBINE: Single-stage, with air-impingement starting.
NOZZLE: Fixed area and direction (usually 15° backward), no provision for vectoring.
WEIGHT, DRY: 201.5 kg (444 lb)
PERFORMANCE RATING: 29.9 kN (6,725 lb st)

KOLESOV RD-36-51

Despite the similarity of designation, this engine has no connection with the RD-36-35.
RD-36-51A. Designed for Tu-144D, cruising at Mach 2.2. *Following description applies to this version.*
RD-36-51V. Simplified version with fixed nozzle, rated at only 68.6 kN (15,430 lb st). Powers Myasishchev M-17 Stratosfera.
TYPE: Single-shaft turbojet.
COMPRESSOR: 14-stage axial with variable inlet vanes and first five and last five stator stages.
COMBUSTION CHAMBER: Tubo-annular with 16 burners.
TURBINE: Three-stage axial with cooled blades. Entry temperature 1,160°C.
NOZZLE: Laval type with adjustable spike.
ACCESSORIES: Airframe mounted, driven via tower shaft and gearbox.
DIMENSIONS:
Length 5,228 mm (205.8 in)
Diameter 1,415 mm (55.7 in)
WEIGHT, DRY: 4,125 kg (9,094 lb)
PERFORMANCE RATING (T-O): 196.12 kN (44,090 lb st)
SPECIFIC FUEL CONSUMPTION (M 2.2 cruise, hi-alt):
24.93 mg/Ns (0.88 lb/h/lb)

GALIGUSOV RD-38

This is the lift engine of the Yak-38M. Turbine entry gas temperature increased to 1,097°C.
WEIGHT, DRY: 231 kg (509 lb)
PERFORMANCE RATING: 31.87 kN (7,165 lb st)

Cutaway RKBM RD-36-35FVR lift turbojet *(Flight International)*

RKBM RD-36-51A turbojet *(Flight International)*

NOVIKOV RD-41

This lift engine was developed for the Yak-141 supersonic V/STOL aircraft. It is a fresh design, though owing something to the RD-36-35FVR. Design began in 1982, and nearly 30 engines were produced before work was halted in late 1991.
TYPE: Single-shaft lift turbojet.
COMPRESSOR: Seven stages. Made of titanium alloy with some composites. Airflow 55.3 kg (121.9 lb)/s.
COMBUSTION CHAMBER: Minimum length annular, with fuel/air mixing taking place upstream. Material (except flame tube) titanium.
TURBINE: Single stage with high-nickel blades inserted in titanium disc. Entry gas temperature 1,207°C.
NOZZLE: Axi-symmetric, with two hydraulic rams on each side (four per engine), operating on aircraft fuel, to vector nozzle ±12.5° fore/aft.
CONTROL SYSTEM: Three engines of Yak-141 are interlinked by three-channel FADEC. During take-off and landing no pilot inputs to engines are needed. Auxiliary hydromechanical control.
DIMENSIONS:
Length 1,594 mm (62.75 in)
Inlet diameter 635 mm (25.0 in)
WEIGHT, DRY: 290 kg (639 lb)
PERFORMANCE RATING (ISA, S/L): 40.21 kN (9,040 lb st)

NOVIKOV RD-60

This turbojet is a derivative of the RD-36-35FVR modified for operation in the horizontal attitude, with a long-life lubrication system and other changes. It is the take-off booster fitted to the prototype Beriev A-40.
PERFORMANCE RATING: 29.85 kN (6,581 lb st)

RKBM PROJECT

In 1990 RKBM revealed plans for a range of engines based on a core already developed. This single-shaft core has a nine-stage compressor driven by a single-stage turbine. Mass flow is 22.5 kg (49.6 lb)/s, pressure ratio 12.4 and TET 1,274°C. Three derived engines would be: a helicopter turboshaft rated at 4,480-5,970 kW (6,000-8,000 shp), a high bypass ratio turbofan in the 12 tonne (117.5 kN, 26,455 lb st) class, and an advanced propfan, with three core booster stages, giving a thrust of 12-15 tonnes (117.5-147.1 kN, 26,455-33,070 lb st).

RKBM RD-41 lift turbojet *(Piotr Butowski)*

TVD-1500

This new gas turbine has been developed as a core suitable to power turboshaft, turboprop and turbofan engines. Features include minimal number of parts, advanced materials (new titanium alloys, new refractory materials and new composites), full-authority digital control and modular construction. Prototypes are running of the shaft and propeller versions. There is no immediate plan to build a turbofan, but this would be in the 7.85-9.81 kN (1,764-2,205 lb st) class.

The basic engine powers the Antonov An-38. Certification

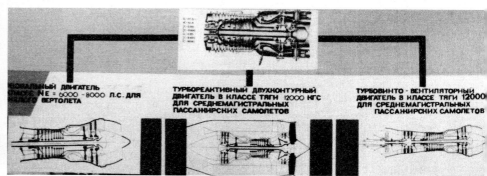

RKBM exhibit showing future engines based on a common core *(Flight International)*

n this application is scheduled for 1995. The **TVD-1500S** turboprop would power the Sukhoi S-80. The **TVD-1500SKh** would power the Antonov SKhS agricultural aircraft. The **TVD-1500I** pusher propfan would power the Ilyushin Il-Kh. Two **RD-600S (TVD-1500A)** turboshaft engines would be installed in the rear fuselage of the Myasishchev Yamal amphibian, driving (via shafts) a pusher propeller behind the rudder. The **TVD-1500V** turboshaft (also described as **RD-600**) will power the Kamov V-62 helicopter, a near-term programme, and the later Ka-52.

TYPE: Free turbine turboshaft or turboprop.
COMPRESSOR: Three axial stages with inlet guide vanes and both stators variable, followed by one centrifugal. Pressure ratio 14.4. Electrochemically machined blades, electron-beam welded rotors and precision-cast casing.
COMBUSTION CHAMBER: Annular reverse-flow. Multi-fuel capability.
TURBINES: Two-stage compressor turbine with monocrystal or directionally solidified blades. Entry temperature 1,267°C. Two-stage power turbine with shaft passing through the engine to front drive.
REDUCTION GEAR: On turboprop, two-stage spur gear followed by single-stage planetary.

DIMENSIONS:
Width	620 mm (24.4 in)
Height	760 mm (29.9 in)
Length: Turboprop	1,965 mm (77.4 in)
Turboshaft	1,250 mm (49.2 in)

WEIGHT, DRY:
Turboprop	240 kg (529 lb)

PERFORMANCE RATINGS (S/L):
T-O	970 kW (1,300 shp)
Contingency (TVD-1500V)	1,156 kW (1,550 shp)
Cruise (7,500 m; 24,600 ft, Mach 0.65)	559 kW (750 shp)

SPECIFIC FUEL CONSUMPTION:
Cruise, as above	63.2 µg/J (0.374 lb/h/hp)

RKBM TVD-1500 turboprop

DN-200

Designated for Diesel Novikov, this unusual piston engine is hoped to be the most economical in the world. It is being designed to replace American engines in light aircraft, though beginning with the Yak-112. Five single-cylinder test engines had been built by 1991 to assist in reaching the target fuel consumption. The first DN-200 ran in 1991, and certification is due in 1994-95.

TYPE: Two-stroke liquid-cooled diesel piston engine.
CYLINDERS: Opposed pistons driving crankshafts along each side. Bore 72 mm (2,835 in). Stroke of each piston 72 mm (2.835 in). Capacity 1,759 cc (107.3 cu in).

WEIGHT, DRY: 165 kg (364 lb)

PERFORMANCE RATING (S/L):
T-O	149 kW (200 hp)
Cruise	119 kW (160 hp)

SPECIFIC FUEL CONSUMPTION:
T-O	45.5 µg/J (0.27 lb/h/hp)

RKBM DN-200 diesel engine

RYBINSK — see RKBM

SALYUT

This name (good health) is that of the MMPO (production factory) making the Saturn/Lyulka AL-31F.

SAMARA
SAMARA STATE SCIENTIFIC & PRODUCTION ENTERPRISE

443026 Samara 36 (Kuibyshev)
CHIEF DESIGNER: Vladimir N. Orlov
DEPUTY CHIEF DESIGNER: Valentin Osipov

This company, also known as SSPE Trud (labour or toil), is the successor to the large KB formerly headed by Kuznetsov (which see), and it has retained his NK initials on engines whose design was begun under his direction. See also KKBM.

SAMARA NK-88

Samara/Trud has been working since about 1968 on the use of LH/LNG (liquid hydrogen and liquefied natural gas) for various gas-turbine engines. A particularly important development programme was the conversion of an NK-8-2 engine (see KKBM) to burn either of the two new liquids, and of the tankage, piping and control system of a Tu-154 to accept such an engine. The result was the Tu-155 aircraft, which on 15 April 1988 became the first aircraft to fly with an engine burning liquid hydrogen. This was fed to the modified engine, designated an **NK-88**, installed in the right-hand (not centre) position. After 12 flights the aircraft was converted to feed LNG to the test engine, thereafter flying to Nice, Hanover and Berlin. Samara would like to test these fuels on an A310 with JT9D engines, followed by PW4000s.

SAMARA NK-92

This is a variant of the NK-93 described below. It could be used on the projected twin-engined 150/220-seat Il-90.

SAMARA NK-93

This is the most powerful propfan known to be under development anywhere in the world. Parametric studies began in 1985, and confirmed a 7 per cent advantage in propulsive efficiency for a two-stage contra-rotating ducted propfan over open-rotor and ducted single-stage propfans. The NK-93 entered preliminary design in 1988. The first complete engine, incorporating both the gas generator and the propfans, went on test in December 1989. By early 1992 five complete engines had run, with about 15 more expected to be required to complete certification. Samara/Trud describe the NK-93 as similar in concept to the MTU/Pratt & Whitney CRISP, but much later in the stage reached; possible co-operation is being discussed.

The immediate applications envisaged for the NK-93 include the Il-96M and Tu-204. Flight development is scheduled to begin in 1994, with a single engine mounted on an Il-76 testbed aircraft. Certificated series produced NK-93 engines are planned for 1997.

TYPE: Three-shaft geared counter-rotating shrouded propfan.
FAN: Two stages, counter-rotating, same directions as Tu-95 ('Bear') propellers. Front fan (40 per cent power) eight blades, rear (60 per cent power) ten blades. Blades swept 30°, pitch range 110°. Prototype blades solid magnesium, production blades (by Stupino propeller factory) solid sparless graphite-epoxy composite retained by short steel root slotted into disc. Blade length 1,050 mm (41.34 in). Mass flow (cruise) 1,000 kg (2,205 lb)/s. Bypass ratio 17.
GEARBOX: Planetary, transmitting 22,370 kW (30,000 shp) through seven satellite pinions. Designed for service life of 20,000 h.
COMPRESSORS: Seven-stage LP, titanium discs and blades. Eight-stage HP, first five titanium, last three steel. Pressure ratio 37.
COMBUSTION CHAMBER: Fully annular with vaporising burners. Being studied for use of LNG (liquefied natural gas) fuel.
TURBINES: Single-stage HP with cooled single-crystal blades drives HP compressor. Single-stage IP drives LP compressor. Three-stage LP drives propfan gearbox.

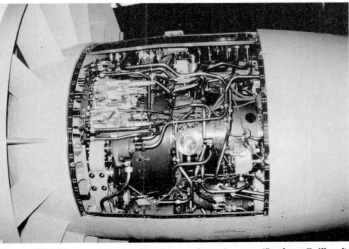
Samara NK-93 in complete nacelle with propfan exit vanes (*Stephane Guilbaud*)

Samara NK-93 propfan assembly (*Stephane Guilbaud*)

Plastic-covered cutaway Samara NK-321 augmented turbofan (*Piotr Butowski*)

DIMENSIONS:
Length	about 5,500 mm (216.5 in)
Inlet diameter	1,455 mm (57.28 in)
Propfan diameter	2,900 mm (114.2 in)
Shroud external diameter	3,150 mm (124.0 in)

WEIGHT, DRY: 3,650 kg (8,047 lb)
PERFORMANCE RATING (ISA, S/L):
Max T-O 176.5 kN (39,683 lb st)
SPECIFIC FUEL CONSUMPTION (cruise 11,000 m; 36,090 ft, M 0.75): 13.89 mg/Ns (0.49 lb/h/lb)

SAMARA NK-321

This carefully planned long-range bomber engine is the power plant of the Tupolev Tu-160 ('Blackjack'). It was designed entirely by the Samara/Trud team, and was derived from the NK-32 gas generator. Design began around 1977, and the first prototype NK-321 went on test in 1980. It has been in production since 1986, and is still being made at a modest rate.

TYPE: Three-shaft bypass augmented turbojet (low-bypass-ratio turbofan).
COMPRESSORS: Three-stage LP (fan); five-stage IP; seven-stage HP. Designed for maximum efficiency and highest overall pressure ratio. First stage also designed for minimal radar reflectivity of any radiation managing to reach it. Materials titanium, steel and (final stages) high-nickel alloy. Mass flow 365 kg (805 lb)/s. Pressure ratio (T-O) 28.4. Bypass ratio 1.4.
COMBUSTION CHAMBER: Annular, vaporising burners, no visible smoke and near-perfect uniformity of temperature at HP turbine face.
TURBINES: Single-stage HP, diameter about 1,000 mm (39.34 in), cooled blades of single-crystal material. Entry gas temperature 1,630°K (1,357°C). Single-stage IP, L blades. Two-stage LP, DS blades.
AFTERBURNER: Designed for peak efficiency and maximu thrust for lowest gas temperature to minimise IR signatu No visible smoke. Downstream of multi-lobe flow mix Fully variable con/di nozzle.
CONTROL SYSTEM: Electrical, with hydromechanical backu Studies in progress for later switch to FADEC.
DIMENSIONS:
Length about 6,000 mm (236
Inlet diameter 1,460 mm (57.5
WEIGHT, DRY: about 3,400 kg (7,496
PERFORMANCE RATINGS (ISA, S/L):
Max T-O 245 kN (55,077 lb
Max dry 137.2 kN (30,843 lb

SATURN
NPO SATURN

13 Kasatkin St, 129301 Moscow
Telephone: 7 (095) 283 09 13
Fax: 7 (095) 286 7 566
PRESIDENT: Dr Viktor M. Chepkin

This bureau was previously named for its famous founder, Arkhip M. Lyulka. In addition to the engines listed below, this bureau is collaborating with Rolls-Royce on the engine to power the Gulfstream/Sukhoi SSBJ (see International section). Also known as Saturn/Lyulka.

AL-7
Service designation: TRD-31

The first AL-7 ran on the bench in late 1952 and the first production version was cleared for use in 1954 at a design rating of 63.74 kN (14,330 lb st). The main innovation in this engine was supersonic flow through the first two stages of the compressor. In the mid-1950s the afterburning **AL-7F** powered the initial production Su-7, and the Tu-98, Il-54 and La-250. Later versions were the **AL-7RV** and **AL-7PB** without afterburner, the latter powering the Be-10. By 1963 production had been transferred to an uprated afterburning version designated **AL-7F-1** to power the Su-7BM and later versions and the Su-9 series. The final production version was the **AL-7F-2**.

TYPE: Single-shaft turbojet, with or without afterburner.
COMPRESSOR: Nine-stage axial (eight stages in original AL-7 design). First two stages widely separated axially, with variable stators ahead of first two stages. Mass flow 114 kg (251 lb)/s. Pressure ratio 9.1.
COMBUSTION CHAMBER: Annular type with perforated inner flame tube. Multiple downstream fuel injectors inserted through cups in forward face of liner.
TURBINE: Two-stage axial-flow type. Both wheels overhung behind rear bearing; front disc bolted to flange on hollow tubular driveshaft. Entry temperature 927°C.
AFTERBURNER (AL-7F): Comprises upstream diffuser and downstream combustion section. Pilot combustor includes single nozzle ring and flame holder; main spray ring and gutter flame holder assembly downstream at greater radius. Refractory liner in combustion section. Variable-area nozzle, with 24 hinged flaps.
ACCESSORIES: Fuel pump and control unit, oil pumps, hydraulic pump, electric generator, tachometer and other items grouped into quickly replaceable packages beneath compressor casing.

DIMENSIONS (AL-7F-1):
Length	6,810 mm (268 in)
Max diameter	1,250 mm (49.2 in)

WEIGHT, DRY:
AL-7F-1	2,010 kg (4,431 lb)
AL-7F-2	2,103 kg (4,636 lb)

PERFORMANCE RATINGS:
AL-7	63.4 kN (14,250 lb st)
AL-7F, with afterburner	88.25 kN (19,840 lb st)
AL-7PB, RV	63.74 kN (14,330 lb s
AL-7-1	68.65 kN (15,432 lb s
AL-7F-1, with afterburner	99.12 kN (22,282 lb s
AL-7F-2, with afterburner	97.08 kN (21,825 lb s

SPECIFIC FUEL CONSUMPTION:
AL-7, max continuous rating
25.49 mg/Ns (0.90 lb/h/lb s

AL-21

This important engine, in various **AL-21F-3** versions, h been produced in thousands for such aircraft as the MiG-2 Su-24 and Su-17/20/22. Developed in the early 1970s an cleared for production in 1970-74, it is noteworthy for i major advances in the design of the compressor, combusti chamber, turbine and afterburner.

TYPE: Single-shaft turbojet, with afterburner.
COMPRESSOR: 14-stage axial. Inlet frame carries front bearin with hot-air de-icing of fixed inlet vanes. Variable stato downstream pivoted to casing and central bullet. Variab stators ahead of first five stages. Parallel HP section wi independently scheduled variable stators ahead of stag 10, 11, 12, 13 and 14. Mass flow 104 kg (229 lb)/s. Pres ure ratio 14.75.
COMBUSTION CHAMBER: Can-annular type, with 12 flame tub each with duplex downstream burner.
TURBINE: Three stages with air-cooled first stage blades. Ent temperature 1,112°C. Rotor assembly supported by re

bearing held in eight-strut rear frame and drives via cone coupled to rear of compressor shaft.
TERBURNER: Rapid combustion downstream of three injector/flameholder rings. Corrugated insulated liner with ceramic coating. Fully variable nozzle with 24 flaps driven by six rams. Full-regime control giving smooth lightup and modulated power over whole range from flight idle to max afterburner.

IGHT, DRY: 1,720 kg (3,792 lb)
RFORMANCE RATINGS (S/L):
Military power 76.5 kN (17,200 lb st) at 8,400 rpm
Max afterburner 110.5 kN (24,800 lb st)
CIFIC FUEL CONSUMPTION (S/L):
Military power 22.6 mg/Ns (0.80 lb/h/lb st)

AL-31

This was A. Lyulka's last and greatest engine. With afterrner, as the **AL-31F**, it powers all the initial production verns of the Su-27, for which it entered series production in 84. Slightly modified engines might power the prototype khoi S-21 SSBJ. Engines designated **R-32** (the Service signation for AL-31 engines) were fitted to the P-42, an -27 development aircraft, which set rate of climb records in 86. A high proportion of the engine is titanium or stainless el. Production by Salyut MMPO.

AL-31FM. Developed version offering higher thrust and ter fuel economy. Powers initial version of Su-35. T-O ing 130.4 kN (29,320 lb st). Data for AL-31F:
PE: Two-shaft augmented turbofan.
N: Inlet has 23 variable guide vanes. Four fan stages slotted into discs. Bypass ratio 0.6.
MPRESSOR: Variable inlet guide vanes followed by ninestage HP spool with first three stators variable. Easy field replacement of damaged blades. Mass flow 110 kg (243 lb)/s. Overall pressure ratio 23.
MBUSTION CHAMBER: Annular, with 28 downstream burners fed from inner manifold. Auto continuous ignition during missile launch.
RBINES: Single-stage HP with cooled blades, using air/air heat exchanger in bypass duct. Entry gas temperature up to 1,427°C. Two-stage LP with cooled blades. Both turbines have active tip clearance control.
PIPE: Short mixer section to combine core and bypass flows upstream of afterburner.
TERBURNER: Two flameholder rings downstream of multiple radial spray bars. Interlinked primary and secondary multi-flap nozzles are angled about 5° downwards.
NTROL SYSTEM: Hydromechanical full-regime control giving smooth power from flight idle to max afterburner in all manoeuvre conditions. Auto elimination of surge 'at Mach numbers 2-2.5 when normal, flat and inverted spins occur'. Linked via software to Su-27 fly-by-wire flight control system. All accessories grouped above engine.
MENSIONS:
Length 4,950 mm (195 in)
Diameter 1,220 mm (48.0 in)
IGHT, DRY: 1,530 kg (3,373 lb)
RFORMANCE RATINGS:
Max augmented 122.6 kN (27,560 lb st)
Max dry 79.43 kN (17,857 lb st)
ECIFIC FUEL CONSUMPTION:
Max dry 18.98 mg/Ns (0.67 lb/h/lb st)

AL-34

This turboprop in the 410 kW (550 shp) class was reported 1990 to have been selected to power the Sukhoi S-86 ecutive aircraft (rated at 522 kW; 700 shp), Myasishchev el (rated at 373 kW; 500 shp) and the Roks-Aero T-601 lity transport. In the S-86, the AL-34 would drive a pusher ntraprop.

AL-35

Two AL-35F engines, each rated at 133 kN (29,900 lb st), wer the Su-35 (see AL-31FM).

Saturn (Lyulka) AL-31F two-shaft augmented bypass turbojet

Saturn (Lyulka) AL-31F sectioned display engine *(Piotr Butowski)*

SAT-41

This completely new engine is being developed to power advanced tactical fighters of the late 1990s. It will provide for long-term sustained supersonic flight, STOL, enhanced combat manoeuvrability and economical operation with low specific weight. Features include a very small parts count (for example by using few stages of compression), an advanced combustion chamber with 'effective mixture formation and cooling', a highly loaded turbine with single-crystal blades incorporating 'a new cooling concept', and a multimode variable nozzle.

SATURN RTWD-14

The RTWD-14 is the primary source of shaft power on the Buran spacecraft. Three are installed, two active and one on standby. These engines run on the liquid oxygen/liquid hydrogen main propellants.

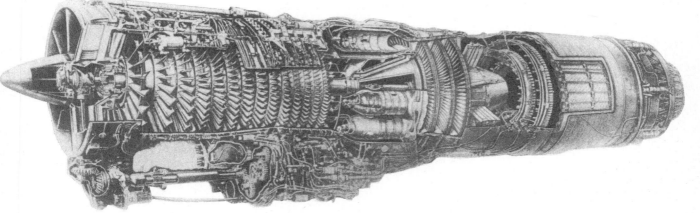

Cutaway drawing of Saturn (Lyulka) AL-21F-3 single-shaft afterburning turbojet

SHVETSOV

FOUNDER OF BUREAU: Arkadiya Dmitrievich Shvetsov

A. D. Shvetsov was responsible for the M-11 five-cylinder air-cooled radial engines made in enormous numbers in 1928-59 at powers from 74.6 kW (100 hp) to 149 kW (200 hp). He later developed larger radials, the most important being the ASh-82, which at ratings up to 1,491 kW (2,000 hp) powered Lavochkin fighters, the Tu-2 and Tu-4 bombers, Il-12 and -14 transports and Mi-4 and Yak-24 helicopters. The ASh-62 was made in large numbers at ratings in the 750 kW (1,000 hp) class for the Li-2 and An-2. The ASh-62M agricultural version, developed by Vedeneyev, and all An-2 engines after 1952, were transferred to Poland (see WSK-PZL Kalisz and WSK-PZL Rzeszów).

Another important engine was the 545 kW (730 hp) ASh-21, fitted to the Yak-11. All these have been described in earlier editions. The ASh-62 and ASh-82 have been produced under licence in China, as described in appropriate Chinese entries in this section. Today the bureau is PNPP Perm, which see for later engines.

SOLOVIEV — see AO 'Aviadvigatel'

SOYUZ

Luzhnetskaya nab. 2/4, Moscow 119048
Telephone: 7 (95) 33 242 2862
Teletype: 207022 Kabina
GENERAL DESIGNER: Vassili K. Kobchyenko
DEPUTY GENERAL DESIGNER: Alexander Girnov

This design bureau is today known as the **MNPK**, Moscow Scientific Production Corporation 'Soyuz' (Alliance). The longest established engine KB in Russia, it was founded in 1933 by A. A. Mikulin, who was General Constructor during the design of the AM-3 (see below). He was succeeded, successively, by Tumansky, Favorski (briefly) and now by Kobchyenko. The latter headed the design of the TV-O-100, which is described under Mars. The names of several Chief Designers appear in the entries listing their engines.

AM-3
Service designation: TRD-3

The AM-3M single-spool turbojet was developed under the leadership of P. F. Zubets from the Mikulin M-209 (civil RD-3 or AM-3) which powered the Tu-88 prototype bombers and was adapted for the USSR's first jet transport, the Tu-104. The usual engine of Tu-16s is the **RD-3M-500**. This engine was produced at the Xian factory in China as the WP8 (see under XAE, China).

It has a simple configuration, with an eight-stage compressor, annular combustion system with 14 flame tubes, and a two-stage turbine. The compressor casing is made in front, centre and rear portions, the front casing housing a row of inlet guide vanes. The inlet bullet houses an S-300M gas turbine starter, developing 75 kW (100 hp) at 31,000-35,000 rpm. Pressure ratio is 6.4; temperature after turbine 720°C.

DIMENSIONS:
Length overall	5,340 mm (210.23 in)
Diameter	1,400 mm (55.12 in)

PERFORMANCE RATINGS (T-O):
RD-3	85.8 kN (19,285 lb st)
RD-3D	85.3 kN (19,180 lb st)
RD-3M-500	93.19 kN (20,950 lb st)

TUMANSKY RD-9

First axial turbojet of wholly Soviet design to be placed in production, this engine was designed under Tumansky in the Mikulin bureau in 1950-51 and received the CIAM designation **AM-5**. The afterburning **AM-5F** was selected for the MiG-19 and the unaugmented engine for the Yak-25. After Mikulin's removal in 1956 the engine was redesignated **RD-9**, the afterburning type being called **RD-9B**. In 1957 the **RD-9BM** entered production, with a hydraulically actuated three-step nozzle. All variants have a nine-stage compressor, can-annular combustor and two-stage turbine; pressure ratio at take-off is 7.14. Production in the Soviet Union tapered off in 1958 in favour of the R-11. In China, production of the RD-9BM began in 1960 and has continued for the J-6 and Q-5. Diameter of all models is 813 mm (32.0 in). Maximum T-O rating of the BM is 32.36 kN (7,275 lb st).

A detailed description and drawing will be found under the designation WP6 in the Chinese (LM) part of this section.

TUMANSKY R-11
Service designation: TRD-37

The first R-11 turbojets (known as the R-37 series by the armed forces) ran in 1953 and entered production in 1956. The R-11 is a two-spool engine, with accessories grouped on the underside of the compressor. It was one of the first engines to have an overhung first rotor stage without inlet guide vanes. Other features include a can-annular combustion chamber with ten flame tubes, which in some versions provides large air bleed couplings for flap blowing, and, in most installations, a large afterburner with multi-flap variable nozzle. All early versions (but not the Chinese derivative) are started on petrol (gasoline) fed from a separate tank, and gaseous oxygen can be fed to the burners to increase relight altitude from 8 to 11.9 km (26,250 to 39,000 ft). Considerably more than 20,000 of all versions have been delivered from Soviet plants, to power many types of aircraft including versions of the MiG-21 and Yak-28. Many more have been produced by Hindustan Aeronautics in India, and by the Chinese as the LEMC WP7. Known Soviet versions are:

R-11. Initial production series, rated at 38.25 kN (8,600 lb st) dry and 50.0 kN (11,240 lb st) with afterburner. Three-stage transonic LP compressor and three-stage HP, each driven by single-stage turbine.

R-11AF2-300. Fitted with enlarged afterburner and new nozzle. Powered speed-record Ye-66 in 1959. Dry thrust unchanged but maximum increased to 58.4 kN (13,120 lb st) Powers most Yak-28 versions.

R-11F. Original engine uprated, with small afterburner; dry thrust unchanged, maximum 56.4 kN (12,676 lb st).

R-11FS, F2S. Suffix S signifies provision for flap blowing. Final version was the **R-11F2S-300**. This powered not only various MiG-21s but also initial versions of the Sukhoi T-58, with dry and maximum ratings of 38.2 kN (8,588 lb st) and 60.8 kN (13,668 lb st). The F2S-300 was until at least 1986 still in production in India and China. Details will be found in the entry on the Chinese WP7.

GAVRILOV R-13

This two-spool turbojet supplanted the R-11 as the MiG-21 engine in the late 1960s, before giving way to the R-25. It also powered the Su-15M and -15TM. It has an advanced afterburner, and a new five-stage HP compressor handling a greater airflow. An unaugmented version powers the Su-25 (see R-195 below). The **PDM** fitted to the record breaking Ye-66B of 1974 may have been an R-13. The standard ratings for the **R-13-300** are 41.55 kN (9,340 lb st) dry and 64.72 kN (14,550 lb st) with afterburner. Fuller details will be found in the entry on the Chinese WP13.

GAVRILOV R-195

This tough and simple turbojet was developed from the R-13 to power the Su-25 and Su-28 family of attack and training aircraft. It has also been selected to power the Sukhoi S-54. Originally designated **R-95**, it was further developed into the R-195 before entering production in about 1979. The entire engine was designed to offer considerable resistance to 23 mm gunfire, and to continue to operate after suffering damage in eight places.

Tumansky R-11F2S-300 two-shaft turbojet, with afterburner removed

Gavrilov R-195 two-shaft turbojet *(Flight International)*

TYPE: Two-shaft turbojet.
LP COMPRESSOR: Three stages. No inlet guide vanes or variable stators.
HP COMPRESSOR: Five stages. No variable stators, but auto bleed valves.
COMBUSTION CHAMBER: Can-annular with multiple duplex burners. Cleared for kerosene, diesel oil and MT petrol.
TURBINES: Single-stage HP, single-stage LP.
JETPIPE: No afterburner, simple fixed-area nozzle.
DIMENSIONS:
 Length 3,300 mm (130 in)
 Diameter 914 mm (36 in)
WEIGHT, DRY: 990 kg (2,183 lb)
PERFORMANCE RATING:
 T-O 44.13 kN (9,921 lb st)

TUMANSKY R-15

This single-shaft afterburning turbojet was developed from 1959 for supersonic applications.

R-15K. Original version developed to power K-20 cruise missile.

R-15B-300. Original production engine for MiG-25P, PU, RB, RBK, RBS, RBT, RBV and RU. Retained in MiG-25BM, RBF and RBSh.

R-15BD-300. Developed engine with greater thrust and TBO of 1,000 h. Fitted to MiG-25PD and PDS.

R-15BF2-300. Uprated engine fitted to first Ye-155M.

TYPE: Single-shaft turbojet with afterburner. Mainly steel construction.
COMPRESSOR: Five stages with fixed inlet vanes. Pressure ratio 7.
COMBUSTION CHAMBER: Annular, 18 vaporising burners for T-6 fuel (freezing point -62.2°C, flash point 54.4°C), plus water/methanol injection.
TURBINE: Single stage.
AFTERBURNER: Three spray rings light in succession.
PERFORMANCE RATING (ISA, S/L):
 Max dry: R-15B-300 73.5 kN (16,525 lb st)
  R-15BD-300 86.24 kN (19,387 lb st)
 Max afterburner: R-15B-300 100.1 kN (22,500 lb st)
  R-15BD-300 110.0 kN (24,700 lb st)
  R-15BF2-300 132.3 kN (29,740 lb st)

TUMANSKY RU-19

Service designation: TRD-29

This single-shaft turbojet has been used to power aircraft and as an APU and emergency booster (in the An-26). When used as an APU its thrust is reduced depending upon the shaft power extracted. The fully rated **RU-19-300** has a thrust of 8.83 kN (1,985 lb st). This is installed in the An-24RV and An-24RT. The **RU-19A-300** is rated at 7.85 kN (1,765 lb st) and is installed in the An-26 and An-30. When providing full electrical power the residual thrust is only 2.16 kN (485 lb st).

Cutaway Tumansky R-15B-300 single-shaft afterburning turbojet *(Piotr Butowski)*

Gavrilov R-25-300 afterburning turbojet *(Flight International)*

An RU-19 was selected as the APU/booster for the MAI-Tashkent Semurg, the rating in this application being 3.1 kN (697 lb st). Another RU-19, equipped as a turbojet and fully rated, powers the La-17 target (see RPVs and Targets in 1977-78 *Jane's*). The RU-19 powered the Yak-30 and Yak-32 jet trainers. The engine is built under licence by Aerostar of Romania.

METSKHVARICHVILI R-21

N. Metskhvarichvili was Chief Designer of the R-21, fitted to the Mikoyan Ye-8. Derived from the R-11F, it had an LP compressor increased in diameter from 772 to 845 mm (30.4 to 33.3 in), and an afterburner nozzle increased in diameter from 902 to 987 mm (35.5 to 38.9 in). Weight rose from 1,165 to 1,250 kg (2,568 to 2,755 lb). Dry and max thrusts were respectively 46.06 and 70.56 kN (10,350 and 15,860 lb st).

GAVRILOV R-25

This two-spool turbojet was a new design, and its compressor of high pressure ratio confers a markedly lower sfc than the R-13. The R-25 has redesigned accessory systems but is installationally interchangeable with the R-13. The first LP compressor stage has 21 titanium blades of large chord. The afterburner has two stages (the R-11 and R-13 having one) and so can be used in combat at high altitudes. The

Aerostar-built RU-19A-300 APU/turbojet

Two views of the Khachaturov R-27V-300 vectored-thrust turbojet (box above left nozzle is added for demonstration purposes) *(Flight International)*

R-25-300 powers the Su-15bis and MiG-21bis, with ratings of 40.26 kN (9,050 lb st) dry and 69.62 kN (15,650 lb st) with afterburner. No longer made, except under licence by HAL of India for MiG-21bis.

TUMANSKY R-266/R-31
These designations were used for the engines of the R-15 family used for FAI-homologated record purposes.

KHACHATUROV R-27
This engine was a natural growth development from the R-11, -13, -25 series. It entered production in 1970 to power the initial series versions of the MiG-23.

R-27F2M-300. Engine of MiG-23MF, MS and UB. Fitted with afterburner and water injection.

R-27V-300. Afterburner replaced by bifurcated nozzles each rotating through about 100° (cycle time 8 s), driven by two hydraulic motors linked by cross-shaft with spring synchronisation. Provides lift/cruise propulsion of Yak-38.

TYPE: Two-shaft turbojet.
LP COMPRESSOR: Five stages. No inlet guide vanes or variable stators. Circulation bleed over first-stage rotor.
HP COMPRESSOR: Six stages. No variable stators.
COMBUSTION CHAMBER: Annular, burners fed from inner manifold. Water injection with max afterburner.
TURBINES: Single-stage HP with air-cooled blades, single-stage LP.
JETPIPE: Afterburner with profiled multi-flap primary and secondary nozzles. R-27V, plenum chamber and two vectoring nozzles.
DIMENSIONS:
  Length: R-27   4,850 mm (191 in)
          R-27V   3,706 mm (145.9 in)
  Diameter   1,012 mm (39.84 in)
WEIGHT, DRY:
  R-27   1,500 kg (3,300 lb)
  R-27V   1,350 kg (2,976 lb)
PERFORMANCE RATINGS:
  R-27F2M-300 max T-O   98 kN (22,045 lb st)
  R-27F2M-300 max dry   68.64 kN (15,430 lb st)
  R-27V-300 max T-O   65.70 kN (14,770 lb st)

KHACHATUROV R-28
Previously designated R-27VM-300, this is the lift/cruise engine of the Yak-38M and -39; the airflow is increased.
PERFORMANCE RATING:
  Max T-O   68.0 kN (15,300 lb st)

KHACHATUROV R-29
This augmented turbojet is simpler than the corresponding American F100, with fewer compressor stages and a lower pressure ratio; but it is more powerful and costs much less. Different subtypes are fitted to all current MiG-23 and MiG-27 versions for Warsaw Pact front-line use, and to the Su-22. In all these aircraft water injection is used on take-off, the MiG-23MF water tank having a capacity of 28 litres (7.4 US gallons; 6.2 Imp gallons).

The following versions have been identified:

R-29-300. Original full-rated production engine for MiG-23MF and related versions.

R-29PN. This replaced the R-29B as the standard engine of non-export MiG-23 aircraft.

Khachaturov R-35 augmented turbojet *(Piotr Butowski)*

R-29B-300. Simplified engine with small afterburner and short two-position nozzle for subsonic low-level operation. Fitted to all MiG-27 versions, with fixed or variable inlet ducts.

R-29BS-300. This is the engine of the Su-22 (export Su-17 versions).
COMPRESSOR: Five-stage LP, six-stage HP. Overall pressure ratio (29B) 12.4, (29-300) 13.1. Mass flow (29B) 105 kg (235 lb)/s, (R-29-300) 110 kg (242.5 lb)/s.
COMBUSTION CHAMBER: Annular, vaporising burners.
TURBINES: HP has single stage with air-cooled blades; max 8,800 rpm. Single-stage LP; max 8,500 rpm.
AFTERBURNER: Fuel rings with separate light-up give modulated fully variable augmentation. Fully variable nozzles differ in different installations (see variants).
DIMENSIONS (R-29-300):
  Length   4,960 mm (195 in)
  Max diameter   912 mm (35.9 in)
WEIGHT, DRY:
  R-29B   1,760 kg (3,880 lb)
  R-29-300   1,880 kg (4,145 lb)
PERFORMANCE RATINGS:
  Max S/L unaugmented   78.45 kN (17,635 lb st)
  Min with afterburner   97.1 kN (21,825 lb st)
  Max T-O, wet (R-29B-300)   112.8 kN (25,350 lb st)
  Max T-O, wet (R-29-300)   122.3 kN (27,500 lb st)

KHACHATUROV R-35
A team led by Khachaturov produced this advanced development of the R-29B. The **R-35-300** powers the MiG-23ML, MLA and MLD. Mass flow and pressure ratio as for R-29-300, but dry weight is 1,765 kg (3,891 lb) and maximum thrust 83.85 kN (18,850 lb st) dry and 127.46 kN (28,660 lb st) with maximum afterburner.

KOBCHYENKO R-79
This engine was originally developed as the engine of the Sukhoi T-4. After 14 Mach 3 qualified engines had been built for this programme a further 12 were built to a new design as the (lift/cruise) power plant of the Yak-141. Though development of this supersonic carrier-based VTOL aircraft was suspended in 1991, development of the R-79 is continuing according to Mr Kobchyenko, "with funds from sources outside the Russian government".

By 1993 the 12 revised engines had run about 3,500 h including over 500 h in flight. Engine fitted to the Yak-141 is designated **R-79V-300**. Work is also proceeding on the upgraded **R-79M**, rated at about 176.2 kN (39,600 lb st), with FADEC control, new combustion chamber, and (for STOL applications) a fixed axi-symmetric nozzle limited to vector angles of 20° up and 20° down. The nozzle has a fixed shroud and is powered by fuel-pressure rams. It could power the Sukhoi S-37.

TYPE: Two-shaft augmented turbofan with vectoring nozzle.
COMPRESSOR: Five-stage LP (fan). Mass flow 120 kg (265 lb)/s. Six-stage HP. Bypass ratio 1.0. HP bleed from two stages to provide aircraft hover control power.
COMBUSTION CHAMBER: Annular, with vaporising burners fed from an inner manifold, giving very low emissions. Non traditional double-zone design.
TURBINES: Single-seat HP with air-cooled single-crystal blades. Max entry gas temperature exceeds 1,600°C. HP/LP turbines rotate in different directions.
AFTERBURNER: Fuel burner rings just behind LP turbine light up in sequence to give fully modulated variable augmentation. Can be used with the nozzle in the 95° position for hovering flight.
NOZZLE: Convergent/divergent type with variable primary and secondary profile and area. Connected to bypass duct periphery by two tapering-wedge pipe sections which

Khachaturov R-29-300 afterburning turbojet *(Piotr Butowski)*

Vectoring tailpipe and nozzle of Kobchyenko R-79 augmented turbofan *(Piotr Butowski)*

rotate in opposite directions to vector nozzle from 0° (forward flight) to 63° (STO) and 95° (VL and hovering).
CONTROL SYSTEM: Three-channel electronic, with duplicated hydromechanical units. Automatically varies engine thrust to trim aircraft in pitch, supplies modulated air to roll and yaw control jets, supplies bleed air to start lift engines, and controls main engine fuel flow and drives to main hydraulic and electric power.
DIMENSIONS:
Inlet (fan) diameter 1,100 mm (43.31 in)
Max diameter (external) 1,716 mm (67.56 in)
Length 5,229 mm (205.87 in)
WEIGHT, DRY: about 2,750 kg (6,063 lb)
PERFORMANCE RATINGS (ISA, S/L):
Max dry 107.63 kN (24,200 lb st)
Max afterburner 152.0 kN (34,170 lb st)
Max (full aircraft control bleed) 137.3 kN (30,864 lb st)
SPECIFIC FUEL CONSUMPTION:
Max dry, as above 18.70 mg/Ns (0.66 lb/h/lb st)

SOYUZ GTE-400

This new turboshaft, rated at 298 kW (400 shp), has been selected for the twin-engined version of the Ka-118.

SOYUZ TVD-450

This turboprop may use the same core as the GTE-400. A twinned version, driving through a coupling gearbox to a single propeller, powers the Sukhoi S-86, as an alternative to the Saturn AL-34. Each power section is rated at 336 kW (450 shp).

SOYUZ R123-300

Projected engine for low cost propulsion of light high-subsonic aircraft.
TYPE: Two-shaft turbofan.
FAN: Single stage. Bypass ratio about 6.
COMPRESSOR: Two axial stages, with variable inlet vanes, and single centrifugal.
COMBUSTION CHAMBER: Annular folded reverse-flow.
TURBINES: Single-stage HP, single-stage LP.
JETPIPE: Mixer leads to combined nozzle.
PERFORMANCE RATING: To be agreed, probably in 13.4-14 kN (3,000-3,150 lb st) class.

Kobchyenko R-79 augmented, vectored turbofan (nozzle horizontal) *(Piotr Butowski)*

SOYUZ TV116-300

Now under development, on basis of a proven experimental gas generator, this engine is intended for light aeroplanes and helicopters.
TYPE: Turboprop or turboshaft with single-shaft gas generator.
COMPRESSOR: Two-stage centrifugal.
COMBUSTION CHAMBER: Annular folded reverse-flow.
TURBINES: Two-stage gas generator, with cooled blades. Two-stage power turbine with drive shaft to front of engine and suitable reduction gearbox for application.
ACCESSORIES: Bevel drive to tower shaft at 12 o'clock position above inlet.
DIMENSIONS:
Length (without gearbox) 800 mm (31.5 in)
Diameter 420 mm (16.54 in)
WEIGHT, DRY (without gearbox): 150 kg (331 lb)
PERFORMANCE RATING (ISA, S/L):
Max T-O 805 kW (1,080 shp)
SPECIFIC FUEL CONSUMPTION:
Max T-O 78.2 µg/J (0.463 lb/h/shp)

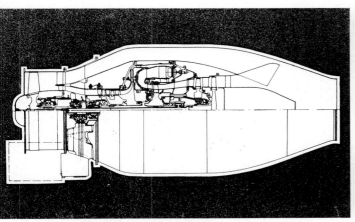

Longitudinal section of Soyuz R123-300 (provisional)

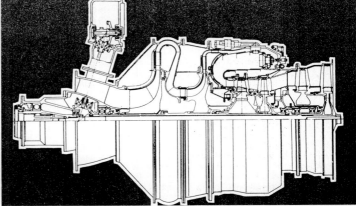

Longitudinal section of Soyuz TV116-300 as bare turbomachine (provisional)

SVERDLOV

This is the name of the production factory at Perm, the principal source of helicopter engines of Klimov design.

TRUD

SSPE TRUD
443026 Samara 36

Meaning labour or toil, this is the name of the production factory on the Samara river at Kuibyshev for Kuznetsov engines. See Samara and KKBM.

TUMANSKY

Academician Sergei Konstantinovich Tumansky, who died in 1973, was a noted designer of piston engines. His RD-9 axial turbojet went into production in 1953. On Mikulin's disgrace in 1956, Tumansky was appointed head of the bureau. Production of his subsequent engines easily exceeds that of any other family of aircraft gas turbines in the post-1955 era. His engines are produced by Soyuz and UMKPA (which see).

UMKPA

CHIEF DESIGNER: Alexei Ryzhov
The Ufa Engine-Manufacturing Production Association is a principal source of engines designed by the Tumansky KB. Their catalogue and exhibit material makes no mention of Tumansky, but lists engines as 'products'. **Product 25-11** (69.2 kN; 15,555 lb st combat rating) is the R-25-300; **Product 55B** (117 kN; 26,300 lb st combat rating) is the R-35F-300; **Product 95-I** (64.5 kN; 14,500 lb st 'first reheat rating') is the R-13-300; and **Product 95-III** (T-O 40.2 kN; 9,037 lb st) is the R-195.

VAZ

The Togliatti (Fiat) car plant has designed two RC (rotary combustion) Wankel type engines for surface applications, and has now produced a fresh design for aircraft.

VAZ-413

This twin-rotor engine is rated at 103 kW (138 hp). Two power the Volga-2 Ekranoplan.

D-200/VAZ-430

This liquid-cooled twin-rotor aircraft engine has been developed from the VAZ-4304 automotive engine designed by the RPD NTS VAZ bureau. Its first application is the Mi-34V helicopter, in which the Vedeneyev M-14V radial is replaced by two rotary engines. Mil general designer Marat Tishchenko said "We could not find a gas turbine small enough". The Mi-34V was flying, complete with new rotor, in 1993 and the D-200 is now also offered for light utility aircraft. It runs on 91 to 92 octane Mogas. TBO is set at 600 h, rising quickly to 1,000 h.
IGNITION: Two spark plugs per rotor, each with its own ignition coil.

WEIGHT, DRY:	150 kg (330 lb)
PERFORMANCE RATINGS (S/L):	
T-O	162 kW (217 hp)
Contingency (Mi-34V)	198.5 kW (266 hp)

VAZ D-200 twin-chamber rotary engine for helicopters and light aircraft *(Mark Lambert)*

VEDENEYEV

FORMER GENERAL DESIGNER: Ivan M. Vedeneyev

This designer was responsible for improvement of the AI-14 piston engine designed by the Ivchenko bureau. He also developed the ASh-62M, produced in Poland as the ASz-62M, from Shvetsov's ASh-62IR. His engines are now produced by the VMKB (see below).

VMKB

Engines of Vedeneyev design are now attributed to the MKB (motor construction bureau) at Voronezh (hence VMKB). They are built under licence by Aerostar of Romania.

VMKB M-14V-26

Derived from the AI-14 engines for fixed-wing aircraft, the M-14V-26 powers the Ka-26 helicopter. In this installation the stub wing carries an engine on each tip. Beneath the rotor an R-26 gearbox combines the power of the engines and distributes it equally between the two coaxial main rotors. The same engine in a revised installation powers the initial single-engined Mi-34 helicopter. Production of the M-14 was transferred to the Romanian aerospace industry.

The engine has forced cooling by an axial fan driven via a friction clutch and extension shaft ahead of the main output bevel box at 1.452 times crankshaft speed. The planetary gearbox has a ratio of 0.309 and incorporates friction and ratchet clutches. The central R-26 gearbox has a ratio of 0.34; it also drives the generator, hydraulic pump, oil pump and tachometer generator.

DIMENSIONS:	
Length	1,145 mm (45.08 in)
Diameter	985 mm (38.78 in)
WEIGHT, DRY:	245 kg (540 lb)
PERFORMANCE RATINGS:	
T-O	239 kW (320 hp) at 2,800 rpm
Max continuous I	205 kW (275 hp) at 2,450 rpm
Max continuous II	142 kW (190 hp) at 2,350 rpm
Cruise I	142 kW (190 hp) at 2,350 rpm
Cruise II	108 kW (145 hp) at 2,350 rpm
SPECIFIC FUEL CONSUMPTION:	
At cruise ratings	77.7 µg/J (0.46 lb/h/hp)

VMKB M-14P

For fixed-wing applications, the M-14P was used with direct drive in the original version of the MAI Kvant, at a T-O rating of 242 kW (325 hp). Currently, the Kvant has an M-14 II of 268 kW (360 hp). The same rating is quoted for the M-14P and M-14PT fitted to the Su-26, -26M, -26MX and -29, and the Yak-18T, -50, -52, -53, -55 and -58, in each case with a controllable-pitch propeller. An M-14P will also power the prototype Myasishchev Gjel, and similar engines have been selected for the Roks-Aero T-401, T-407 and T-433. The M-14PF, rated at 294 kW (395 hp), powers the Su-29T and 29U.

VMKB M-18

This engine has been produced for homebuilts, microlights and powered flexwings.
TYPE: Two-stroke piston engine with two horizontally opposed air-cooled cylinders.

WEIGHT, DRY:	
Geared	28 kg (61.7 lb)
Ungeared	13.7 kg (30.2 lb)
PERFORMANCE RATING (T-O):	
Geared	40 kW (53.6 hp)
Ungeared	30 kW (40.2 hp)

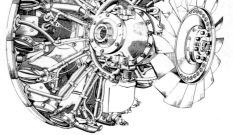

Drawing of VMKB M-14V-26 radial piston engine, with cooling fan and right-angle drive

VOLSZHSKY

This design bureau has produced the following small piston engine.

BAZ-21083

Apart from the T-O rating of 59.7 kW (80 hp), nothing is yet known of this engine, of which two will drive pusher propellers in the Phoenix-Aviatechnica LKhS-4 biplane.

ZAPOROZHYE — *see ZMKB Progress under Ukraine*

SINGAPORE

SA
SINGAPORE AEROSPACE LTD
ADDRESSES: See Aircraft section

SA manufactures engine parts for General Electric and Turbomeca. It has 2 per cent of the PW4000 programme and is sole source of eleven high-precision parts, mainly in the compressor section.

SOUTH AFRICA

ATLAS
ATLAS AVIATION (A member of Simera Division of Denel (Pty) Ltd)
PO Box 11, Atlas Road, Kempton Park 1620, Transvaal
Telephone: 27 (11) 927 9111
Fax: 27 (11) 395 1103
Telex: 724403
CEO: J. J. Eksteen

Atlas is manufacturing the Rolls-Royce Viper 540 turbojet under sublicence from Piaggio of Italy, for use in Atlas Impala attack trainers.

SPAIN

ITP
INDUSTRIA DE TURBO PROPULSORES SA
Edificio 300, Parque Tecnológico, E-48016 Zamudio (Vizcaya)
Telephone: 34 (94) 489 21 00
Fax: 34 (94) 489 21 93
COMMERCIAL OFFICE: Travesfa Costa Brava 6 (Carretara Colmenar Viejo), E-28034 Madrid
Telephone: 34 (91) 384 80 00
Fax: 34 (91) 384 80 01

ITP was formed in 1989 as Spain's aero-engine company to participate in the design, development, manufacture, sale and support of gas turbine engines. Shareholders are CST (4 per cent), Rolls-Royce plc (45 per cent) and Turbo 2000 SA (51 per cent), the latter being in turn owned by CASA and Bazan (Spanish government) and Sener (private sector). The Ajalvir plant near Madrid, acquired from CASA, overhauls turboshaft, turboprop and turbojet engines and accessories (GE, PWC, Garrett, SNECMA, TM, Textron Lycoming and Allison). The newly constructed Zamudio plant near Bilbao manufactures components for civil and military engines. Deliveries began December 1991. ITP is the Spanish participant (13 per cent) in the EJ200 engine programme to power the Eurofighter 2000. Active plans include participation in international civil programmes (Trent), marine propulsion and industrial power projects.

SENER
SENER-INGENIERIA Y SISTEMAS SA
Severo Ochoa (PTM), E-28760 Tres Cantos, Madrid
Telephone: 34 (1) 807 7000
Fax: 34 (1) 807 7201

Sener, engaged in NATO programmes and ESA space activities, was to have been the Spanish partner in the consortium that will build the EJ200 engine of the Eurofighter 2000. This role has now been assumed by ITP, in which Sener is a shareholder, as briefly referred to in previous entry and in the International part of this section.

SWEDEN

FLYGMOTOR
VOLVO FLYGMOTOR AB
S-461 81 Trollhättan
Telephone: 46 (520) 94000
Fax: 46 (520) 34010
Telex: 420 40 VOLFA S

Volvo Flygmotor produces aircraft engines and space propulsion components. Since 1980 it has been a risk and revenue sharing partner with General Electric on the CF6-80A and -80C for wide-body transports. Volvo Flygmotor also participates in development and production of the Garrett TFE731-5 turbofan and TPE331-14/15 turboprop. The company also has agreements with Pratt & Whitney on the JT8D-200, PW2000 and V2500, and with Rolls-Royce on the Tay, making combustors for the Tay 610-650 and compressor intermediate casings for the Mk 670.

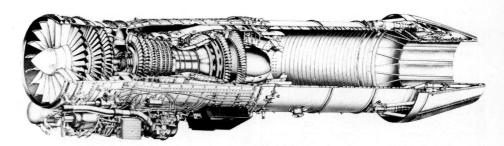

Cutaway drawing of Flygmotor RM12 two-shaft augmented turbofan

FLYGMOTOR RM8
The RM8 is a Swedish military version of the Pratt & Whitney JT8D turbofan which Flygmotor developed to power the Saab 37 Viggen. Following the **RM8A** for the AJ 37, SF 37, SH 37 and SK 37, the **RM8B** was developed for the JA 37. The major change to improve functional stability at high altitude involved replacing the first stage of the LP compressor by a third stage on the fan. Flygmotor delivered 207 RM8A and 173 RM8B engines. Details appeared in the 1992-93 *Jane's*.

FLYGMOTOR RM12
The RM12 is a version of the F404 developed jointly by General Electric and Volvo Flygmotor to power the JAS 39 Gripen. General Electric retains rights to the design, and supplies approximately 60 per cent by value of parts.

Volvo Flygmotor is a partner in all F404 applications. Volvo Flygmotor will supply parts to General Electric similar to those parts that it manufactures for the RM12.

The RM12 thrust improvement has been achieved by increasing the turbine inlet temperature by up to 105°C and by increasing fan airflow. The fan meets more stringent bird strike requirements, and this has required changes to the control system, with built-in redundancy to ensure get-home power. Increased temperature and pressure have required changes to hot section materials. RM12 testing started at GE in June 1984, and the Gripen first flew on 9 December 1988. Three production engines were delivered by 1993, in which year delivery began of 140 Gripen aircraft.

The following are features of the RM12:
FAN: Variable first-stage stator. Airflow 68 kg (150 lb)/s. Bypass ratio 0.28.
WEIGHT, DRY: 1,050 kg (2,315 lb)
PERFORMANCE RATINGS:
Max T-O: dry 54 kN (12,140 lb st)
augmented 80.5 kN (18,100 lb st)

TURKEY

TEI
TUSAŞ ENGINE INDUSTRIES
Muttalip Mevkii Mrk, PK 610 Eskisehir
Telephone: 90 (22) 222 030
Fax: 90 (22) 222 057
Telex: 35356 tmco tr
MANAGING DIRECTOR: F. M. Erureten

TEI was formed in January 1985 as a joint venture between the Turkish government (53.78 per cent) and US General Electric (46.2 per cent). The factory produces the F110 fighter engine, and had delivered 139 engines by the end of 1992. TEI also produces numerous other aircraft engine components, particularly for GE but also for Pratt & Whitney, SNECMA and CFM.

UKRAINE

ZMKB PROGRESS
ZAPOROZHYE MACHINE-BUILDING DESIGN BUREAU PROGRESS
Zaporozhye 330064
Telephone: 7 (612) 650327
Fax: 7 (612) 654697
Telex: 127464 Bolt SU
GENERAL DESIGNER: Fyodor Mikhailovich Muravchenko
CHIEF DESIGNER: Vladimir Ivanovich Kolesnikov
COMMUNICATION ADVISER: Alexander Vladimirovich Gavrishenko
Telephone: 7 (612) 656105

Zaporozhye Machine-Building Design Bureau was founded by Ivchenko in May 1945. From 1968 it was headed by Lotarev, and since 1988 by Muravchenko.

Until 1991 ZMKB Progress was a member of the group of enterprises of the USSR Aviation Industry Ministry. Now it is a state enterprise of Ukraine.

ZMKB Progress carries out the complete cycle of creation of gas-turbine engines, the cycle comprising the design, development, manufacture of the development batch and certification, as well as supervision of series production and operation. Series production of engines designed and developed by ZMKB Progress is carried out by the Motor Sich enterprise (formerly, Zaporozhye Industrial Association Motorostroitel).

Not including Ivchenko's AI-14 and AI-26 piston engines (see PZL Kalisz and Rzeszów in Polish section), 17 types of ZMKB Progress engines are in service on 42 types of aircraft. These engines are in service in 40 countries. The total of more than 30,000 ZMKB Progress aircraft gas-turbine engines have accumulated over 250,000,000 h in service.

AI-20
This turboprop was the first gas-turbine engine designed by Zaporozhye collective, headed by Ivchenko. AI-20 was developed in 1955-57 as an alternative to the NK-4 by Kuznetsov design bureau for the Il-18 aircraft, mainly by Zaporozhye designers but using the experience of Mikulin design bureau in developing the AM-3M engine, and the experience of CIAM specialists. Mass production of AI-20 began in 1958 for Il-18 and An-10A. After that several modifications appeared:

AI-20K. Installed on Il-18V, An-10A and An-12. Rated at 2,983 ekW (4,000 ehp) at S/L, ISA. Produced at SMPMC in China as WJ6.

AI-20M. Known as AI-20 Series 6. Initial T-O rating 3,126 ekW (4,192 ehp), later increased to 3,169 ekW (4,250 ehp). Fitted to An-12 and Il-18/20/22/38. Differs from AI-20 by shrouded turbine blades.

AI-20DK. Known as AI-20D Series 3. Rated at 3,124 ekW (4,190 ehp), navalised engine fitted to An-8 and Beriev Be-12.

AI-20DM. Known as AI-20D Series 4. Rated at 3,760 ekW (5,042 ehp). Fitted to An-8 and Be-12.

AI-20D Series 5. Developed for An-32 with 3,760 ekW (5,042 ehp), to operate from −60°C to +55°C with automatic variation of propeller pitch.

Service life and TBO of different modifications depend on service conditions and are: service life from 6,000 h (AI-20D Series 5) to 20,000 h (AI-20M); TBO 7,000 h (AI-20M).

The following description refers to the AI-20M:
TYPE: Single-shaft turboprop.
COMPRESSOR: Axial, ten stages, with four bypass valves, which are used at starting and transient ratings. Pressure ratio from 7.6 at T-O (ground) to 9.2 (cruise). Mass flow 20.7 kg (45.6 lb)/s. Stator casing of sheet stainless steel.
COMBUSTION CHAMBER: Annular with ten burner cones and two pilot burners and igniter plugs. The casing is one of the load carrying elements of the engine.
TURBINE: Axial, three stages. Rotor blades shrouded at inner and outer ends and installed in pairs in slots of air-cooled discs. First guide vanes and discs are cooled by secondary air from combustion chamber. Maximum entry temperature is 900°C at S/L (937°C for AI-20DM). Rotor speed 12,300 rpm, except 10,400 at ground idle.
REDUCTION GEAR: Planetary type, two-stage, incorporating a six-cylinder torquemeter and negative-thrust transmitter (IKM-type). Reduction ratio 0.08732.
LUBRICATION: Pressure-feed type with full recirculation.
OIL GRADE: 75 per cent MK-8 or MK-8P to GOST-6457-66 (DERD.2490 or MIL-O-6081B) and 25 per cent MS-20 or MK-22 to GOST 21743-76 (DERD.2472 or MIL-O-6082B).
STARTING: Two electric starter/generators. Type STG-12-TMO-1000, supplied from ground or TG-16 or AI-8 APU.
FUEL GRADE: T-1, TS-1, T2, RT, to GOST-10227-86 (DERD.2492, JP-1 to MIL-F-56616).
DIMENSIONS:
Length 3,096 mm (121.89 in)
Width 842 mm (33.15 in)
Height 1,180 mm (46.46 in)
WEIGHT, DRY: 1,040 kg (2,293 lb)
PERFORMANCE RATINGS:
T-O (S/L, static) 3,169 ekW (4,250 ehp)
Max cruise (340 knots; 630 km/h; 391 mph at 8,000 m; 26,250 ft) 1,938 ekW (2,600 ehp)
SPECIFIC FUEL CONSUMPTION:
T-O 89.42 µg/J (0.529 lb/h/ehp)
Cruise, as above 73.3 µg/J (0.434 lb/h/ehp)

AI-24
This turboprop powers the An-24 and its derivatives. There are four main versions:

AI-24 (Series 1). Was in production in 1961-64.

AI-24 (Series 2). Installed on An-24A, An-24B, An-24V, An-24T and An-24PV. The An-24A, -24B, -24V and -24T have provision for water injection. Production began in 1964.

AI-24P. Multi-fuel version for Ekranoplan cruise propulsion. Rated at 1,840 kW (2,467 shp). One powers the SM-6 series and two power Meteor 2.

AI-24T. Powers the An-24A and An-24B. These aircraft of Series 1 are without provision for water injection. Production began in 1967.

AI-24VT. Powers the An-26 and An-30. Production began in 1971.
TYPE: Single-shaft turboprop.
COMPRESSOR: Ten-stage axial. Pressure ratio, (AI-24T) 7.85. Mass flow 14.4 kg (31.7 lb)/s. Speed (Series 2) 15,100 rpm, (T, VT) 15,800 rpm.
COMBUSTION CHAMBER: Annular, with eight simplex burners and two starting units, each comprising a body, pilot burner and igniter plug.
FUEL GRADE: T-1, TS-1 to GOST 10227-86 (DERD.2494 or MIL-F-5616).
TURBINE: Three-stage axial with solid blades. Rotor/stator sealing effected by soft inserts mounted in grooves in nozzle assemblies. Entry temperature 877°C.
REDUCTION GEAR: Planetary type, incorporating hydraulic torquemeter and electromagnetic negative thrust transmitter for propeller autofeathering. Type AV-72 propeller (AI-24T drives AV-72T propeller). Ratio 0.08255.
LUBRICATION: Pressure circulation system.
OIL GRADE: 75 per cent GOST 6457-66 or MK-8 (DERD.2490 or MIL-O-6081B) and 25 per cent MS-20 or MS-22 (DERD.2472 or MIL-O-6082B).
ACCESSORIES: Mounted on front casing are starter/generator, alternator, aerodynamic probe, ice detector, negative-thrust feathering valve, torque transmitter, oil filter, propeller speed governor and centrifugal breather. Below casing are oil unit, air separator, LP and HP fuel pumps and drives to hydraulic pump and tachometer generators.
STARTING: Electric STG-18TMO starter/generator supplied from ground power or from TG-16 APU.
DIMENSIONS:
Length overall 2,346 mm (92.36 in)
Width 677 mm (26.65 in)
Height 1,075 mm (42.32 in)
WEIGHT, DRY: 600 kg (1,323 lb)
PERFORMANCE RATINGS:
T-O: AI-24 (Series 1) 1,901 ekW (2,550 ehp)
AI-24T, VT 2,074 ekW (2,781 ehp)
Cruise rating at 243 knots (450 km/h; 280 mph) at 6,000 m (19,685 ft):
AI-24 (Series 2) 1,208 ekW (1,620 ehp)
AI-24T 1,253 ekW (1,680 ehp)
SPECIFIC FUEL CONSUMPTION:
At cruise rating:
AI-24 (Series 2) 86.0 µg/J (0.509 lb/h/ehp)
AI-24T, VT 87.5 µg/J (0.518 lb/h/ehp)
OIL CONSUMPTION: 0.85 kg (1.87 lb)/h

AI-25
This bypass engine powers the Yak-40 passenger aircraft. Series production began in 1967.
TYPE: Two-shaft bypass engine. Bypass ratio 2.2.
FAN: Three-stage axial subsonic; drum-and-disc with pin-jointed blades. Pressure ratio 1.7 at 10,560 rpm.
COMPRESSOR: Eight-stage, drum-and-disc rotor. Pressure ratio 4.7 at 16,640 rpm. Total pressure ratio at T-O rating (S/L) 8.
COMBUSTION CHAMBER: Annular, with 12 centrifugal fuel nozzles.
HP TURBINE: Single-stage, shrouded, with cooled nozzle guide vanes. Maximum rotor inlet temperature 933°C.
LP TURBINE: Two-stage, shrouded, cooled discs.
CONTROL SYSTEMS: Automatic hydromechanical, main and emergency.
STARTING: SV-25 air starter.
OIL SYSTEMS: Autonomous, recirculation type.
DIMENSIONS:
Length 1,993 mm (78.46 in)
Width 820 mm (32.28 in)
Height 896 mm (35.28 in)
WEIGHT, DRY: 320 kg (705 lb)
PERFORMANCE RATINGS:
T-O (S/L) 14.71 kN (3,307 lb)
Long-range cruise (296 knots; 550 km/h; 342 mph at 6,000 m; 20,000 ft) 4.34 kN (976 lb)
SPECIFIC FUEL CONSUMPTION:
T-O 15.86 mg/Ns (0.56 lb/h/lb st)
Cruise, as above 22.09 mg/Ns (0.78 lb/h/lb)

ZMKB Progress (Ivchenko) AI-20M single-shaft turboprop (*Jacques Marmain*)

ZMKB Progress (Ivchenko) AI-24 single-shaft turboprop (*Jacques Marmain*)

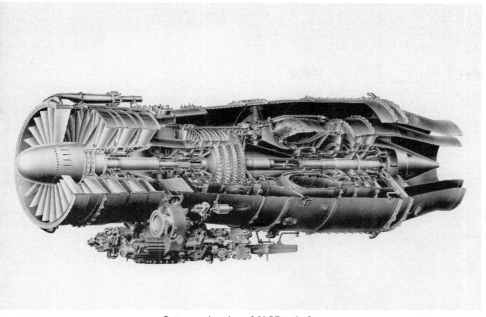

Cutaway drawing of AI-25 turbofan

AI-25TL

This modified engine powers the L-39 trainer. Series production began in 1973. Bypass ratio 2. Differs from the AI-25 by:

COMPRESSOR: Nine-stage. Total pressure ratio at maximum continuous rating (S/L, ISA) 9.5.
HP TURBINE: Single-stage, cooled, maximum rotor inlet temperature 1,037°C.
EXHAUST UNIT: Extension pipe, common nozzle.
CONTROL SYSTEM: Hydromechanical. Duplicated by a standby and an autonomous emergency system.
WEIGHT, DRY: 350 kg (772 lb)
PERFORMANCE RATING:
T-O (S/L, ISA) 16.87 kN (3,792 lb st)
SPECIFIC FUEL CONSUMPTION:
T-O 16.43 mg/Ns (0.58 lb/h/lb st)

DV-2

This small turbofan was designed by Progress and developed jointly with ZVL (now PS, which see under Czech Republic) to replace the AI-25TL as the engine of the Aero L-59 (previously L-39 MS) trainer. The engine was designed to fit the existing engine bay, so that L-39s can be re-engined. The same core is to be used in several engines:

DV-2. Basic turbofan. The description below applies to this engine. Powers L-59 and selected to power Yak-130, Il-108 business jet and China/Pakistan K-8 single-engined trainer/attack aircraft.
DV-2B. Larger fan, with bypass ratio 2. T-O rating 25 kN (5,610 lb st); sfc 15.17 mg/Ns (0.536 lb/h/lb st); dry weight about 500 kg (1,102 lb).
DV-2F. Afterburning engine for supersonic applications. Bypass ratio 1.46; maximum thrust 37.5 kN (8,422 lb st); sfc 61.1 mg/Ns (2.16 lb/h/lb st); dry weight about 600 kg (1,323 lb).
DV-12. Turboshaft for helicopters and tilt-rotor aircraft. Large curved jetpipe to side and rear gearbox and output shaft. Max contingency power 5,595 kW (7,500 shp); sfc 68.1 µg/J (0.403 lb/h/shp); dry weight 675 kg (1,488 lb).
DV-22. Turbofan for transport aircraft. Mass flow 122 kg (269 lb)/s; bypass ratio 5; T-O thrust 32.32 kN (7,265 lb st) to 30°C; sfc 10.5 mg/Ns (0.37 lb/h/lb st); dry weight 700 kg (1,543 lb). Planned for later Il-108.
DV-32. Turboshaft for helicopters and tilt-rotor aircraft. As DV-12 but front drive. Same weight.
DV-X. Projected core for pusher turboprop or propfan. T-O rating 4,425 kW (5,932 shp) plus 2.4 kN (540 lb st) to 30°C; sfc 75.4 µg/J (0.446 lb/h/shp).

TYPE: Two-shaft turbofan.
FAN: Single stage, overhung, supersonic, with 15 titanium blades of large chord, without snubbers, hub/tip ratio 0.37, bypass ratio 1.46. Diameter 645 mm (25.4 in). Mass flow 49.5 kg (109 lb)/s. Pressure ratio 2.3.
LP COMPRESSOR: Two stages, rotating with fan.
HP COMPRESSOR: Seven stages. Two-position IGVs, stators cantilevered stages 3-6. Three bypass valves. Overall pressure ratio 13.5.
COMBUSTION CHAMBER: Annular, 16 fuel nozzles, giving low emissions.
HP TURBINE: Single stage, convective cooling, max gas temperature 1,147°C.
LP TURBINE: Uncooled, two-stage.
STARTING: Air turbine.
CONTROL SYSTEM: Main electronic-hydromechanical, with digital block, plus reserve and emergency hydromechanical systems. System developed jointly with Yugostroy (Czech), which makes production system.
LUBRICATION: Autonomous, recirculating, aerobatic.
DIMENSIONS:
Length 1,721 mm (67.75 in)
Width 994 mm (39.1 in)
Height 1,037 mm (40.8 in)
WEIGHT, DRY: 450 kg (992 lb)
PERFORMANCE RATING (S/L, T-O): 21.58 kN (4,852 lb st)
SPECIFIC FUEL CONSUMPTION: 16.7 mg/Ns (0.59 lb/lb st)

D-36

The D-36 was the first Soviet engine with a high bypass ratio. Bench tests began in 1971, flight tests in 1974, and series production in 1977.

It is the base engine for a whole family: the **D-136** turboshaft; **D-236** propfan demonstrator; **D-336** industrial engine; and **D-436K/T, D-436T1/T2** turbofans. The D-36 was also used as a model for attaining design objectives of the D-18T. The D-36 powers the Yak-42 (three), and An-72 and An-74 and the Kometa 2 and Vykhr 2 Ekranoplans (all two each).

TYPE: Three-shaft turbofan.
CONFIGURATION: Three-shaft, with minimum number of bearings (6) and without intershaft bearings. The bearings are resilient and resilient/damping. Their oil cavities have contact radial face seals. Made of 12 modules. Bypass ratio 6.3.
FAN: Single-stage, transonic; 29 titanium blades with part-span shrouds; 48 outlet guide vanes (the number of blades and vanes chosen for minimum noise). Blade containment by winding Kevlar fibre on the fan casing. Maximum speed 5,400 rpm. Mass flow 255 kg (562 lb)/s.
IP COMPRESSOR: Six stages with inlet guide vanes adjusted on the bench, then fixed in position. Three blow-off valves. Discs and rotor blades of titanium, stator vanes of steel. Maximum speed 10,500 rpm.
HP COMPRESSOR: Seven-stage, with adjustable inlet guide vanes. Three blow-off valves. Rotor blades and discs of two aft stages of steel. Maximum speed 14,170 rpm. Pressure ratio (overall) 18.7.
INTERMEDIATE CASE: Cast magnesium alloy. Forms connecting duct from LPC to HPC and bypass duct, and used for locating HPC front bearing and mounting LPC and HPC casings. Inside is a drive from HP rotor to accessory gearbox on intermediate case under nacelle cowl. The front engine mount is attached to the case.
COMBUSTION CHAMBER: Annular, with 24 burners and 2 igniters. Integral with HPT nozzle guide vanes (they form a single module). Combustion chamber case made by explosion stamping. Flame tube elements rolled and welded into one unit.
HP TURBINE: Single stage. Max inlet 1,137°C. Rotor blades tip-shrouded, convective-film cooling system, attached by fir-tree with two blades in each groove. Nozzle vanes have convective cooling.
IP TURBINE: Single stage. Uncooled, tip-shrouded, rotor blades. Nozzles guide vanes cooled by 3rd stage HPC air.
TURBINE SUPPORT HOUSING: Module located between HPT and IPT rotors and combined with wide IPT nozzle vanes. Outer casing attached to inner casing by spokes passing through wide hollow nozzle vanes. Inner casing carries rear bearings of HP and IP rotors. Oil supplied through hollow nozzle vanes.
FAN (LP) TURBINE: Three-stage with tip-shrouded blades, air-cooled discs.
EXHAUST UNIT: Consists of rear-bearing housing and main duct nozzle. Rear engine mount attached to housing.

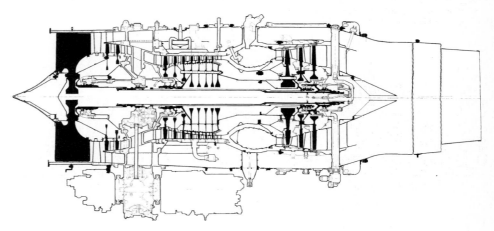

Longitudinal section through DV-2 turbofan

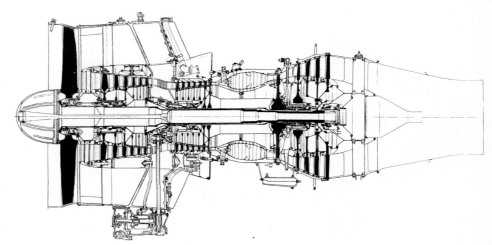

Longitudinal section through D-36 turbofan

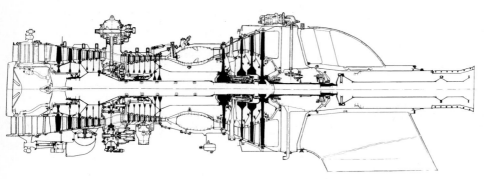

Longitudinal section through D-136 turboshaft

ZMKB Progress D-236 propfan (*Jacques Marmain, Aviation Magazine International, Paris*)

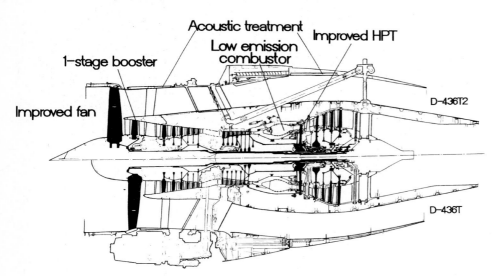

Design differences between ZMKB Progress D-436T and D-436T2

MOUNTING: Universal: under and over the wing, in the fuselage and on both sides without changing engine.
CONTROL SYSTEM: Hydromechanical, with inner redundancy and electronic unit controlling gas temperature and rotor speeds. Compressor blow-off valves controlled by independent pneumonic (bubble memory) system located near the valves. Engine compressor air is the operating medium. The D-36 is provided with sensors sufficient for FADEC control. Testing with DSIC FADEC began August 1992.
STARTING: By air-turbine starter, on accessory gearbox, from ground source, APU or operating engine. Inflight starting can be assisted by starter.
COWLING: On Yak-42, nacelle with short fan duct. On An-72 and An-74 nacelle has common nozzle of fan and core ducts.
DIMENSIONS:
Length 3,470 mm (136.6 in)
Diameter 1,373 mm (54.06 in)
WEIGHT, DRY: 1,100 kg (2,425 lb)
PERFORMANCE RATINGS (ISA):
T-O static 63.74 kN (14,330 lb st)
Max cruise at 8,000 m (26,250 ft) at Mach 0.75
15.7 kN (3,527 lb)
SPECIFIC FUEL CONSUMPTION:
T-O 10.195 mg/Ns (0.360 lb/h/lb st)
Max cruise, as above 18.4 mg/Ns (0.65 lb/h/lb)

D-136

This turboshaft engine was designed for the Mi-26 transport helicopter. Bench testing started in 1977, and series production in 1982. It is composed of ten modules, each of which (except for main module) can be removed and replaced without disturbing neighbouring modules on installed engine. Five gas-generator modules are identical with those of D-36.
TYPE: Two-spool free turbine turboshaft.
GAS GENERATOR: As D-36 but with redesigned intermediate case between LPC and HPC (no fan duct). Accessory gearboxes at top and bottom on intermediate case. Speed (max): LPC 10,950 rpm, HPC 14,170 rpm.
FREE TURBINE: Two-stage with uncooled nozzle guide vanes and tip-shrouded blades, with air-cooled discs. Combined as separate module with support housing, outer and inner casings connected by 11 wide load-bearing struts. Inner casing contains front roller and rear ball bearings of free turbine. Struts carry oil pipes and speed sensor drive.
DRIVE SHAFT: Flexibly mounted shaft at rear, transmits torque from free-turbine rotor to helicopter transmission by splines.
EXHAUST NOZZLE: Curved to side.
STARTING: By air turbine starter.
CONTROL SYSTEM: Hydromechanical speed governor of free turbine with power synchroniser of both Mi-26 engines. Free turbine speed maintained at 8,300 rpm, changed by pilot command in range of ±300 rpm. Electronic control of gas temperature and speed of free turbine and gas-generator rotors. Compressor blow-off valves controlled by self-contained pneumonic system. On Mi-26, engine is equipped with particle separator.
DIMENSIONS:
Length 3,715 mm (146.26 in)
Width 1,403 mm (55.24 in)
Height 1,133 mm (44.61 in)
WEIGHT DRY: 1,050 kg (2,315 lb)
PERFORMANCE RATINGS (ISA, S/L):
Max contingency 8,500 kW (11,400 shp)
T-O 74.57 kW (10,000 shp)
Max climb 6,338 kW (8,500 shp)
SPECIFIC FUEL CONSUMPTION: 77.09 µg/J (0.456 lb/h/shp)

D-236

This propfan demonstrator was designed on the base of the D-36 core, but it incorporates a reduction gearbox to drive the SV-36 contra-rotating propfan. Design began in 1979. The aim is to create a demonstrator for studying the geared propfan. Bench tests began in 1985, and the tests in the Il-76 flying testbed began in 1987.
The Yakovlev design bureau has been working on propfan integration since 1987. The Yak-42E-LL flying testbed with a D-236 replacing one D-36 first flew in March 1991.
CONFIGURATION: Three-shaft, the propfan being driven by a separate turbine.
GAS GENERATOR: Generally as D-136.
REDUCTION GEAR: Planetary, single-stage, ratio 5.64; contra-rotating output shafts, ratio of outer to inner shaft torque is 41 to 59 per cent.
PROPFAN TURBINE: Three-stage, shrouded rotor blades.
CONTROL SYSTEM: D-36 type, hydromechanical with electronic unit limiting gas temperature and rotor speeds. Independent pneumonic (bubble memory) system controlling compressor blow-off valves.
PROPFAN: SV-36, by Stupino Design Bureau of Machine-Building. Blades of composites (fibrecarbonglass) without metal elements. Front propeller 8, rear propeller 6 blades. Rotation speed, T-O 1,100 rpm, cruising ratings 960 rpm, ground idle 500-600 rpm. Front and rear propeller speeds maintained equal at all ratings. Blade angle control, digital electronic with hydraulic backup. At max cruise rating (as below) propeller efficiency 0.87; at take-off rating thrust to power ratio 0.94. Propfan diameter 4.20 m (165.4 in).
PERFORMANCE RATINGS:
T-O (S/L, ISA) 8,090 kW (10,850 shp)
Max cruise (11,000 m; 36,100 ft, Mach 0.7)
4,730 ekW (6,340 ehp)
SPECIFIC FUEL CONSUMPTION:
T-O, as above 78.1 µg/J (0.462 lb/h/shp)
Cruise, as above 60.67 µg/J (0.359 lb/h/shp)

D-436

A growth version of the D-36. The initial version, in accordance with preliminary requirements of the Antonov and Yakovlev design bureaux for further development of the An-72, An-74 and Yak-42, was designated **D-436K**.
This first ran in 1985, when the total service time of the D-36 was 1 million hours. In 1987 the D-436K was certificated in accordance with USSR airworthiness regulations.
Later, the **D-436T** and **D-436M** versions, for the Tu-334-1 and Yak-42M aircraft respectively, were developed in accordance with more precise requirements. They differ from the D-436K in the accessory gearbox design and mix of accessories, the provision of the D-18T type reverser on the D-436T and some changes in the aerodynamics of components. The D-436M has been selected to power the Be-200.
The D-436T has been bench-tested since the beginning of 1990. The following description refers to this version:
TYPE: Three-shaft turbofan.
CONFIGURATION: Six rotor bearings without intershaft bearings. Bearings are resilient and resilient/damping.
FAN: As D-36, with improved performance. Fan speed increased to 5,850 rpm.
COMPRESSORS AND INTERMEDIATE CASING: Modules same as D-36. Speeds and pressure ratio increased.
COMBUSTION CHAMBER: Redesigned for increased gas temperature, reduced emissions and easier high-altitude start. 18 single-orifice nozzles, some with pneumatic atomisers.
HP TURBINE: As D-18T.
IP AND FAN TURBINES: Almost identical to D-36.
REVERSER: As D-18T.
CONTROL SYSTEM: As D-36.
STARTING: Pneumatic, as D-36.
MOUNTING: Universal, as D-36.
Taking into consideration the requirements of Western aircraft manufacturers for regional and medium-range aircraft, the **D-436T1/T2** modification was proposed in 1991. D-436T1 and D-436T2 are practically the same in design, but differ in performance and date of certification. Changes in D-436T1/T2 compared with D-436T are:
FAN: D-18T type but higher performance. Diameter as D-436, but speed increased.
COMPRESSOR: Compressor booster stage driven by fan shaft. Total pressure ratio of core increased, while HP and IP rotor speeds remain unchanged.
COMBUSTION CHAMBER: Changed to ensure moderate gas temperature rise, and reduction of emissions to meet most strict requirements.
HP TURBINE: Aerodynamics changed. Three-dimensional airfoils of nozzle guide vanes. Blade and disc cooling system improved.
LP TURBINE: Some improvement in nozzle guide vane aerodynamics.
FAN TURBINE: Disc strengthened for increased speed.

D-436 VARIANTS

	D-436K	D-436T	D-436T1	D-436T2
Fan diameter mm (in)	1,373 (54.06)	1,373 (54.06)	1,373 (54.06)	1,373 (54.06)
Weight, dry kg (lb)	1,250 (2,756)	1,250 (2,756)	1,450 (3,197)	1,450 (3,197)
Take-off rating				
Conditions	S/L, static, +15°C		S/L, static, +30°C	
Thrust (ideal) kN (lb)	73.53 (16,535)	73.53 (16,535)	75.0 (16,865)	80.39 (18,078)
Max HP rotor inlet temperature °K (°C)	1,470 (1,197)	1,470 (1,197)	1,483 (1,210)	1,520 (1,247)
Max cruise rating				
	8,000 m (26,250 ft) 0.75 M, ISA		11,000 m (36,100 ft) 0.75 M, ISA	
Thrust (ideal) kN (lb)	18.63 (4,189)	12.75 (2,866)	14.71 (3,307)	15.3 (3,439)
SFC mg/Ns (lb/h/lb)	18.4 (0.65)	17.8 (0.63)	17.28 (0.61)	17.28 (0.61)
Bypass ratio	6.2	6.0	4.95	4.9
Pressure ratio	21.0	21.9	25.2	26.2

ACOUSTIC LINING: Area of acoustic panels in bypass duct and air intake is twice that of D-36. Noise level meets latest requirements.
CONTROL SYSTEM: Sensors on engine allow a FADEC system.
COWLING: Nacelle of 3/4 type, as D-18T.

D-18T

The D-18T is a large HBPR turbofan. It powers the An-124 (Ruslan) and An-225 (Mriya). This engine incorporates design and technological improvements achieved during design, development and operation of the D-36 engine.

D-18T development began in 1979 by the starting of a single-spool core engine. A two-spool core engine was developed, and the first run of a full-scale engine was accomplished in 1980. First flight on the Il-76T flying testbed was carried out in 1982. On 24 December 1982 the An-124 made its first flight.

Development of a derivative engine based on the D-18T was initiated in 1991. This engine, the **D-18TM**, with thrust increased to 245-294 kN (55,115-66,138 lb), has been selected to power the An-218. It will probably be equipped with a bigger fan, 3D turbine with single-crystal blades and FADEC control system.

TYPE: Three-shaft turbofan.
CONFIGURATION: Each of the three rotors is carried on two bearings (total 6). The bearings are resilient and resilient/damping. Oil cavities of the fan and IPC bearings are provided with labyrinth seals, others with radial-face contact seals. The engine comprises 17 modules.
FAN: Single-stage, supersonic, with 33 titanium blades with part-span shrouds. Stator has 56 carbon-glass plastic vanes, with epoxide matrix and inner and outer titanium shrouds. Leading-edge protected by stainless-steel strip. Containment of separated blade provided by winding Kevlar-type fibre on the inside surface of the case. Blades attached by fir-tree roots in disc slots. Speed (max) 3,450 rpm. Mass flow 765 kg (1,687 lb)/s. Bypass ratio 5.6.
IP COMPRESSOR: Seven-stage, transonic, with variable inlet guide vanes and 8 blow-off valves on case. Titanium blades and steel vanes. Speed (max) 5,900 rpm.
HP COMPRESSOR: Seven-stage with adjustable inlet guide vanes. Blades of first four stages of titanium, remainder of steel. Speed (max) 9,100 rpm. Overall pressure ratio (T-O) 25, (cruise) 27.5.
INTERMEDIATE CASE: Intended to form transition path from IPC to HPC and fan duct as well as attachment of HPC bearing and IPC and HPC cases. Drive from HP rotor to accessory gearbox in lower part of case. Front engine mounts attached to inner case. Aluminium outer shell riveted. Inner shell and struts of titanium.
COMBUSTION CHAMBER: High-temperature, annular, with 22 main fuel nozzles and 2 igniters. Integrated with HPT NGVs (a separate module). Case consists of outer and inner shells. Cooling air (for IPT) passes between them. Combustion chamber specially modified for low emissions.
HP TURBINE: Single-stage with tip-shrouded blades with convection-film aircooling. Blades mounted by fir-tree roots (a pair in each slot). Turbine support housing with IP and HP rotor bearings. Vanes air-cooled.
IP TURBINE: Single-stage with tip-shrouded blades with convective aircooling.
FAN TURBINE (LPT): Four-stage with uncooled blades with tip shrouds. Rotor of drum-disc type. Outer case cooled by air from fan duct.
EXHAUST UNIT: Comprises rear support case with LPT bearing and rear engine mount, and core nozzle. Struts shaped to untwist turbine outlet flow.
REVERSER: Attached to rear flange of intermediate case, with 12 doors pulled inwards and blocking fan duct by axial movement of translating cowl section, which simultaneously opens peripheral cascade rings. Control and drives are hydromechanical, using engine oil.

ZMKB D-27 propfan

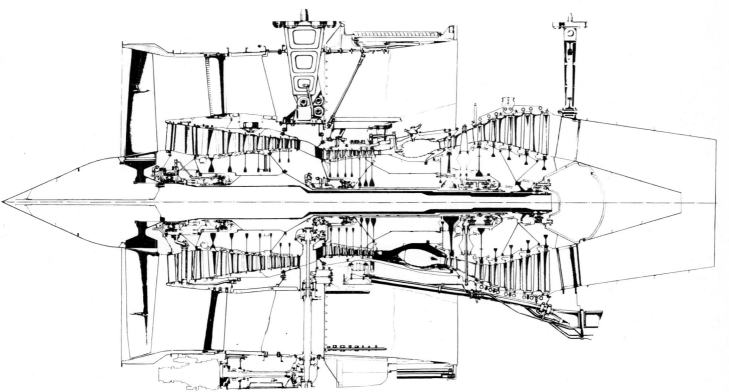

Longitudinal section through D-18T turbofan

UKRAINE/UK: ENGINES—ZMKB PROGRESS/EMDAIR

ZMKB D-27 propfan on outdoor testbed with SV-27 propeller

LUBRICATION SYSTEM: Self-contained continuous circulation under pressure.
CONTROL SYSTEM: Similar to D-36. Self-contained pneumonic system controls blow-off valves and HPC inlet guide vanes. Engine control provides automatic starting and maintaining given rating over complete operating range.
ACCESSORIES: Accessory gearbox drives constant-speed generator and two hydraulic pumps.

DIMENSIONS:
Length 5,400 mm (212.6 in)
Fan diameter 2,330 mm (91.73 in)
WEIGHT, DRY: 4,100 kg (9,039 lb)
PERFORMANCE RATINGS:
T-O (S/L, ISA+13°C) 229.8 kN (51,660 lb st)
Max cruise (11,000 m; 36,100 ft, Mach 0.75, ISA)
  47.67 kN (10,716 lb)

SPECIFIC FUEL CONSUMPTION:
T-O 10.195 mg/Ns (0.360 lb/h/lb st)
Cruise, as above 16.142 mg/Ns (0.570 lb/h/lb)

D-27

Since 1985 work has been under way to create a gas generator (named later 27) with high cycle parameters intended to create a family of medium-size engines (similar to the D-36 family). These will have to meet modern Western requirements.

The design envisages the creation of engines of different configuration with minimum changes to the core.

Having acquired experience with the D-236 demonstrator, ZMKB Progress began to design a propfan. Using the 27 gas generator, the D-27 propfan is being designed with SV-27 contra-rotating tractor propfans developed by Stupino Design Bureau of Machine Building. This engine is intended for cargo and passenger aircraft, with improved take-off and landing performance.

In comparison with the SV-36, the SV-27 has blades of greater sweep, higher efficiency at cruising ratings and improved acoustics. Rotational speed: (T-O) 1,000 rpm, (cruise) 850 rpm.

Take-off rating is 10,350 kW, 10,440 ekW (13,880 shp, 14,000 ehp); the tractor installation is used on the An-70T and is specified for the An-180. The propfan Yak-46 specifies the theoretically more efficient pusher installation, timed more than a year later (1995-96); the Yakovlev Corporation describe this as having a thrust of 109.8 kN (24,690 lb). Sfc at max cruise (11 km, M 0.7, ISA) is 8.1 mg/Ns (0.286 lb/h/ehp).

D-227

This propfan has been named as the probable power plant of the Tu-334 TVDD. They would be installed with pusher propellers giving a thrust of 78.5-88.25 kN (17,650-19,840 lb st).

UNITED KINGDOM

ALVIS
ALVIS UAV ENGINES LTD
Lynn Lane, Shenstone, Lichfield WS14 0DT
Telephone: 44 (543) 481819
Fax: 44 (543) 481393

In 1992 Alvis plc backed a management buy-out of the UAV engine business of Norton Motors. Though unmanned air vehicles are no longer featured in this annual (see *Jane's Battlefield Surveillance Systems*), this announcement is made to assist readers. The manned aircraft Norton engines are the property of MWAE (which see).

EMDAIR
EMDAIR LTD
Harbour Road, Rye, East Sussex TN31 7TH
Telephone: 44 (797) 223460
Fax: 44 (797) 224615
Telex: 957116 FOR EMDAIR

Activities centre on lightweight four-stroke flat-twin engines. These air-cooled engines run on Avgas and have four-valve heads, electronic management systems, central and side spark plugs, direct injection, alternator and electric starter. They are intended to combine high specific output at low rpm with low sfc, low costs and long life. Each engine is available in A (direct drive) and B (geared, torsional damping and extension shaft) versions.

EMDAIR CF 077A
Crankshaft rotation counterclockwise, seen from the front. An exhaust muffler (silencer) is optional. Conforms to JAR 22.
CYLINDERS: Bore 95.0 mm (3.74 in). Stroke 88.9 mm (3.50 in). Capacity 1,261 cc (76.91 cu in). Compression ratio 9.5.
DIMENSIONS:
Length 400 mm (15.75 in)
Width 698 mm (27.5 in)
Height 447 mm (17.75 in)
WEIGHT, DRY (with starter): 47.27 kg (104 lb)
PERFORMANCE RATINGS:
Max 44.8 kW (60 hp) at 3,600 rpm
Max cruise 33.6 kW (45 hp) at 3,090 rpm
FUEL CONSUMPTION:
Max cruise
  13.25 litres (3.5 US gallons; 2.9 Imp gallons)/h

EMDAIR CF 077B
This version powers the Verilite Sunbird, designed to FAR Pt 23. Geared output with torsional vibration damping, reducing propeller speed to 2,600 rpm.

EMDAIR CF 092A
CYLINDERS: Bore 104.0 mm (4.094 in). Stroke 88.9 mm (3.50 in). Capacity 1,511 cc (92.17 cu in). Compression ratio 9.5. The **092B** is a geared version.
DIMENSIONS:
Length 401 mm (15.8 in)
Width 698 mm (27.5 in)
Height 447 mm (17.75 in)
WEIGHT, DRY (with starter): 51.0 kg (112 lb)
PERFORMANCE RATINGS:
Max 52.2 kW (70 hp) at 3,600 rpm
75 per cent cruise 39.1 kW (52.5 hp) at 3,075 rpm
FUEL CONSUMPTION:
Max cruise
  14.4 litres (3.8 US gallons; 3.2 Imp gallons)/h

EMDAIR CF 112
The **112B** has an epicyclic gearbox giving a propeller speed of 2,500 rpm at a crankshaft speed of 3,600 rpm. The following relates to the **112A**:
CYLINDERS: Bore 110 mm (4.35 in). Stroke 96.0 mm (3.78 in). Capacity 1,834 cc (112 cu in). Compression ratio 9.5.
DIMENSIONS:
Length 412.75 mm (16.25 in)
Width 711.2 mm (28.0 in)
Height 425 mm (16.75 in)

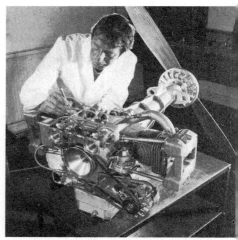

Emdair CF 092B two-cylinder piston engine

WEIGHT, DRY (with starter): 57.28 kg (128 lb)
PERFORMANCE RATINGS:
Max 63.4 kW (85.0 hp) at 3,600 rpm
75 per cent cruise 47.5 kW (63.75 hp) at 3,270 rpm
FUEL CONSUMPTION:
Max cruise
  15.5 litres (4.1 US gallons; 3.4 Imp gallons)/h

MWAE
MID WEST AERO ENGINES
Hangar 38, Staverton Airport, Gloucestershire GL51 6SR
Telephone: 44 (452) 857456
Fax: 44 (452) 856519
TECHNICAL DIRECTOR: Peter Watts

This company has taken over the Hewland piston engine fitted to the Super2, as well as single- and twin-rotor engines for manned aircraft previously developed by Norton. Details of the **MW90** twin-rotor engine appeared in the 1992-93 Jane's.

MWAE75
This engine was designed by the Hewland company. It powers the ARV-1 Super2. The AE45 is a two-cylinder version. Certificated JAR-E and BCAR-C.
TYPE: Three-cylinder inline two-stroke.
CYLINDERS: Liquid-cooled. Capacity 748 cc (45.5 cu in).
WEIGHT, DRY:
    With geared drive, starter and generator   50 kg (110 lb)
RATING: 57.4 kW (77 hp) at 6,750 rpm

MWAE100R
TYPE: Wankel-type twin-rotor.
ROTORS: Air-cooled (housing water/glycol), total capacity 558 cc (35.9 cu in).
WEIGHT, DRY: 52.0 kg (114.6 lb)
RATING: 74.6 kW (100 hp) at 7,000 rpm

NORTON
NORTON MOTORS LTD
See under Alvis and MWAE.

ROLLS-ROYCE
ROLLS-ROYCE plc
65 Buckingham Gate, London SW1E 6AT
Telephone: 44 (71) 222 9020
Fax: 44 (71) 233 1733
Telex: 918091
MAIN LOCATIONS:
PO Box 31, Moor Lane, Derby DE24 8BJ
Telephone: 44 (332) 242424
Fax: 44 (332) 249936/7
PO Box 3, Filton, Bristol BS12 7QE
Telephone: 44 (272) 791234
Fax: 44 (272) 797575
CHAIRMAN: Sir Ralph Robins
CEO: Dr Terry Harrison
DIRECTOR OF PUBLIC AFFAIRS: Michael Farlam

Rolls-Royce civil and military gas turbines are used by more than 310 airlines, 110 armed forces and 700 executive customers.

The main aircraft gas turbine activities are at Derby, Glasgow, Bristol and Coventry. Aircraft gas turbine techniques are applied to industrial and marine uses at Ansty. Employees total 53,000.

In April 1985 Rolls-Royce joined with Turbomeca and MTU in the European Small Engines Co-operation Agreement. This involves the Rolls-Royce Turbomeca RTM 322, MTU-Turbomeca-RR MTR 390, and Turbomeca TM 333. Further details are given under Turbomeca (France) and in the International part of this section. The company is a partner in IAE, EUROJET, Turbo-Union, Rolls-Royce Turbomeca and Williams-Rolls Inc.

In 1990, BMW Rolls-Royce was formed, which see in German part of this section.

ROLLS-ROYCE TURBOMECA ADOUR and RTM 322
See the International part of this section.

TURBO-UNION RB199
See the International part of this section.

EUROJET EJ 200
See the International part of this section.

IAE V2500
See the International part of this section.

ROLLS-ROYCE RB211
The designation RB211 applies to a family of three-shaft turbofans of high bypass ratio and high pressure ratio, ranging in thrust from 166.4 kN (37,400 lb st) to over 445 kN (100,000 lb st). These ratings include the derived 535 and Trent, which are described separately. For all applications Rolls-Royce retains responsibility for the complete propulsion system.

The **RB211-22B**, fitted to the L-1011-1 and -100 TriStar, is flat rated at 186.8 kN (42,000 lb st) to 29°C. Certificated in February 1972 by the CAA and in April 1972 by the FAA. Production ceased in 1982 when over 670 engines had been delivered. By 1993 engine flight hours in service exceeded 23 million.

The **RB211-524** series of engines was developed from the RB211-22B and covers a range of thrusts from 222.4 kN (50,000 lb st) to 269.6 kN (60,600 lb st). The -524 entered airline service in 1977 with the L-1011 and 747. By the beginning of 1992 over 900 engines had been delivered, and service experience of over 19 million hours achieved.

The **RB211-524B**, which powers the L-1011-200, L-1011-500 and 747, is certificated at 222.4 kN (50,000 lb st) to 28.9°C. A 524B4 completed 24,267 h on a Delta L-1011 without removal, an industry record.

The **RB211-524C** offers increased thrust ratings for the 747. It entered service in April 1980 with a rating of 229.1 kN (51,500 lb st).

The **RB211-524D4** is an improved engine rated at 235.75 kN (53,000 lb st) for the 747.

The **RB211-524G** incorporates advanced features proven on the 535E4, such as the wide-chord fan, 3D aerodynamics, directionally solidified HP and IP turbine blades and integrated mixer nozzle. Another new feature is a full-authority digital control system. The engine can operate at G rating (258 kN; 58,000 lb st) or H rating (269.6 kN; 60,600 lb st) and powers the 747-400 and 767-300, with 180 minute EROPS clearance.

The **RB211-524H**, rated at 269.6 kN (60,600 lb st) entered service on the Boeing 747-400 and 767-300 in February 1990, and is mechanically identical to the 524G.

The following description relates to the RB211-524G/H:
TYPE: Three-shaft axial turbofan. Overall pressure ratio 33.
FAN: Single-stage overhung, driven by LP turbine. Composite nosecone, 24 hollow wide-chord blades in titanium alloy, controlled diffusion outlet guide vanes. Aluminium casing, with Armco containment ring. Mass flow 728 kg (1,604 lb)/s. Bypass ratio 4.3.

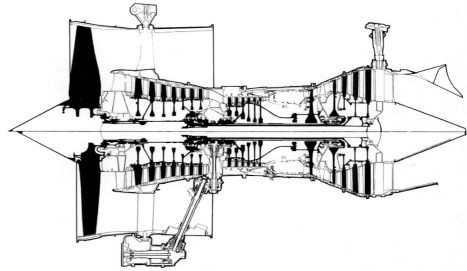

Longitudinal sections of the Rolls-Royce RB211-524G (upper half) and RB211-524D4 Upgrade (lower)

Rolls-Royce RB211-524G three-shaft turbofan

UK: ENGINES—ROLLS-ROYCE

IP COMPRESSOR: Seven-stage, driven by IP turbine. Two drums, one of titanium discs welded together and the other of welded steel discs, bolted to form one rotor, carrying titanium blades. Aluminium and steel casings carry steel stator blades. Single-stage titanium variable inlet guide vanes.

HP COMPRESSOR: Six-stage, driven by HP turbine. Welded titanium discs, single steel disc and welded nickel alloy discs bolted together carrying titanium, steel and nickel alloy blades. Steel casing carries steel and Nimonic stator blades.

COMBUSTION CHAMBER: Fully annular, with steel outer casings and nickel alloy combustor. Downstream fuel injection by 18 airspray burners with annular atomisers. High energy igniter plugs in Nos 8 and 12 burners.

HP TURBINE: Single-stage, with directionally solidified nickel alloy rotor blades, both convection- and film-cooled, mounted in nickel alloy disc by fir tree roots.

IP TURBINE: Single-stage, with directionally solidified nickel alloy rotor blades fir tree mounted in nickel alloy disc.

LP TURBINE: Three-stage, with nickel alloy rotor blades fir tree mounted in steel discs.

EXHAUST NOZZLE: Integrated nozzle with deep-chute forced mixer.

ACCESSORY DRIVES: Radial drive from HP shaft to gearbox on fan casing. Accessories include integrated drive generator and aircraft hydraulic pumps.

LUBRICATION SYSTEM: Continuous circulation dry sump system supplying oil to four bearing chambers with a combination of ball and roller bearings. 27 litre (57.6 US pint; 48 Imp pint) oil tank integral with gearbox.

DIMENSIONS:
Length overall:
RB211-22B, -524C2   3,033 mm (119.4 in)
RB211-524B4, -524D4   3,106 mm (122.3 in)
RB211-524G, -524H   3,175 mm (125.0 in)
Fan diameter:
RB211-22B, -524C2   2,154 mm (84.8 in)
RB211-524B4, -524D4   2,180 mm (85.8 in)
RB211-524G, -524H   2,192 mm (86.3 in)

WEIGHT, DRY:
RB211-22B   4,171 kg (9,195 lb)
RB211-524B4, B4 Improved   4,452 kg (9,814 lb)
RB211-524C2   4,472 kg (9,859 lb)
RB211-524D4, D4 Upgrade, -524G/H   4,479 kg (9,874 lb)

PERFORMANCE RATINGS:
T-O: see model listings
Cruise at 10,670 m (35,000 ft) and Mach 0.85 (uninstalled):
RB211-22B   42.2 kN (9,495 lb)
RB211-524B4, B4 Improved   48.9 kN (11,000 lb)
RB211-524C2   51.1 kN (11,490 lb)
RB211-524D4 (all models)   50.0 kN (11,230 lb)
RB211-524G, -524H   52.5 kN (11,813 lb)

SPECIFIC FUEL CONSUMPTION (cruise):
RB211-22B   17.79 mg/Ns (0.628 lb/h/lb)
RB211-524B4   17.56 mg/Ns (0.620 lb/h/lb)
RB211-524B4 Improved   17.16 mg/Ns (0.606 lb/h/lb)
RB211-524C2   18.21 mg/Ns (0.642 lb/h/lb)
RB211-524D4   17.48 mg/Ns (0.617 lb/h/lb)
RB211-524D4 Upgrade (1987)   17.02 mg/Ns (0.601 lb/h/lb)
RB211-524G, -524H   16.15 mg/Ns (0.570 lb/h/lb)

ROLLS-ROYCE TRENT

This is the most powerful Rolls-Royce aircraft engine. Its detailed engineering design began in 1988 to meet the propulsion requirements of wide-bodied aircraft planned for the 1990s, the Airbus A330 and Boeing 777. To meet the requirements for these aircraft a range of Trent engines is being developed.

Trent Designation	Take-off Thrust
768	300.3 kN (67,500 lb st)
772	316.3 kN (71,100 lb st)
870	333.2 kN (74,900 lb st)
877	356.2 kN (80,070 lb st)
884	384.8 kN (86,500 lb st)

The Trent is a low risk derivative of the RB211-524, employing advances in design techniques and material. It incorporates the first increase in diameter of the fan since the entry into service of the RB211-22B in April 1972, even though power at this size has increased by nearly 45 per cent due to design and material improvements.

Another change in the Trent design is that, whilst the successful modular design of the RB211 has been retained, the order of module removal has been altered for increased ease of maintenance. With the RB211 design the IP compressor and intermediate casing are removed from the front of the fan case assembly, with all other modules being removed rearwards. On the Trent all modules are removed rearwards, facilitating complete gas generator removal from the fan case and further improving maintenance flexibility.

The Trent 700 for the A330 was launched in April 1989. For this application the Trent 700 will be certificated in December 1993 and cleared to operate at both 768 and 772 ratings. The Trent 772 will enable the A330s to fly long distances, for example from the US west coast to Europe. The Trent 800, designed to meet the thrust requirements of the Boeing 777, will be certificated at the thrust rating of the Trent 884 in February 1995. The Trent has growth potential to meet all thrust requirements of derivative and future aircraft.

The following outlines major differences between the Trent and the RB211-524G/H:

FAN: Single stage with 26 wide chord blades of hollow titanium. Blades made by superplastic forming and diffusion bonding to form integral canted spars running from root to tip.

IP COMPRESSOR: Eight stages, giving increased core airflow. Variable inlet guide vanes and first two stator stages.

HP COMPRESSOR: Six-stage design offering greater efficiency with improved tip clearance control based on V2500 technology.

COMBUSTION CHAMBER: Based on latest technology optimised for reduced emissions.

HP TURBINE: Fitted with blading of the latest three-dimensional design, with directionally solidified blades.

IP TURBINE: Increased annulus design with single-crystal uncooled blades.

LP TURBINE: Trent 700 has four stages with three-dimensional blading. Trent 800 five-stage.

GEARBOX: Single-piece casting with gear and bearings individually replaceable without disturbance of adjacent gears. Separate oil tank and filler.

REVERSER: Four-door type on 700; cascade type on 800.

MOUNTING: Trent 700 core mounted. Trent 800 front fan case mount, rear core mount with thrust struts to front of core. Designed for enhanced rigidity, reduced distortion of the fan case and reduced engine weight.

ENGINE CONTROL: Full authority digital engine control (FADEC). Gives improved fuel consumption, better engine control and reduces pilot work load by interacting with aircraft computers.

DIMENSIONS:
Length overall: 700   3,912 mm (154 in)
800   4,369 mm (172 in)
Fan diameter: 700   2,474 mm (97.4 in)
800   2,794 mm (110 in)

WEIGHT, DRY:
700   4,785 kg (10,550 lb)
800   5,957 kg (13,133 lb)

PERFORMANCE RATINGS:
T-O: See above

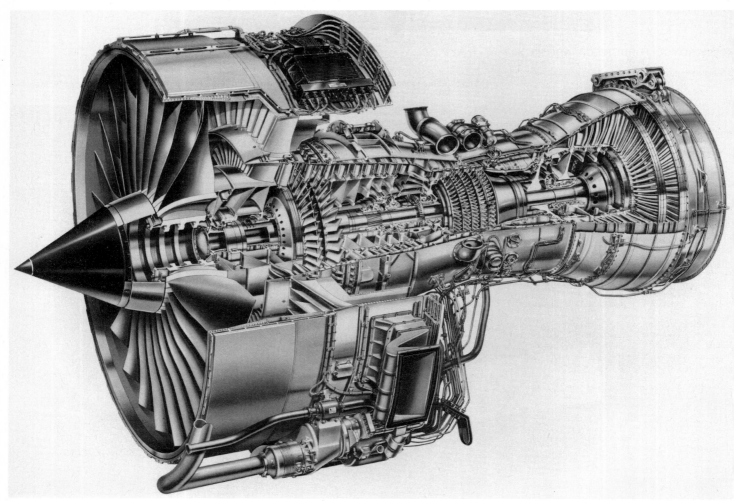

Cutaway drawing of Rolls-Royce Trent

Cruise (10,670 m; 35,000 ft, Mach 0.82 [0.83 for Trent 800]):
768, 772   64.8 kN (14,570 lb)
870, 877, 884   72.0 kN (16,190 lb)
SPECIFIC FUEL CONSUMPTION (cruise):
700   15.66 mg/Ns (0.553 lb/h/lb)
800   15.78 mg/Ns (0.557 lb/h/lb)

ROLLS-ROYCE 535

The **535C** was launch engine for the Boeing 757. It has an HP module based on the RB211-22B, six-stage IP compressor without variable stator vanes, and a scaled down -524 fan. Fan airflow is 18 per cent lower than that of the -22B, and core airflow 12 per cent lower. The 535C entered service on 1 January 1983.

The **535E4** is an advanced version offering increased thrust, together with reduced fuel consumption relative to the 535C. The E4 allowed 757s powered by it to be cleared for 120 minute extended-range operations (EROPS) in December 1986, extended to 180 minutes in 1990.

An uprated 535E4 entered service in August 1989. Its T-O rating of 191.7 kN (43,100 lb st) enables the 757 to carry heavier payloads from the most noise-sensitive airports. By 1993 over 40 operators had exceeded 6,000,000 h on the 535, including 150,000 EROPS.

The following description relates to the 535E4:
TYPE: Three-shaft turbofan.
FAN: Single-stage, with only 22 wide chord blades, without snubbers, each of activated diffusion bonded titanium skins on titanium honeycomb core. Fan case of Rohrbond with Kevlar containment. Mass flow 522 kg (1,150 lb)/s. Bypass ratio 4.3 (535C, 4.4).
IP COMPRESSOR: Six stages of controlled diffusion design. No variable vanes.
HP COMPRESSOR: Six stages of end-bend blading, with stage -4, -5 and -6 discs in titanium super alloy. Low expansion casing for improved tip clearance control. Overall engine pressure ratio 25.8 (535C, 21.1).
COMBUSTION CHAMBER: Annular, 18 airspray nozzles, flexible liner mountings, heatshields and thermal barrier coatings.
HP TURBINE: Single-stage. Rotor blades, directionally solidified, cast with HP leading-edge cooling, HP and LP internal air cooling passages both with triple pass system. Nozzle guide vanes with curved stacking, highly cooled and with thermal barrier coating on platforms.
IP TURBINE: Single stage. Cooled NGVs with multi-lean stacking for improved airflow onto high aspect ratio blades.
LP TURBINE: Three stages. All turbine casings double wall and cooled.
JETPIPE: Core and bypass flows mixed in integrated nozzle.
REVERSER: Fan reverser only. Jacks move translating cowl to rear, blocker deals seal fan duct and uncover cascade vanes. Over expansion reduces core thrust.
GEARBOX: Mounted under fan case, driven from HP spool.
DIMENSIONS:
Length: 535C   3,010 mm (118.5 in)
535E4   2,995 mm (117.9 in)
Inlet diameter: 535C   1,877 mm (73.9 in)
535E4   1,892 mm (74.5 in)
WEIGHT, DRY:
535C   3,309 kg (7,294 lb)
535E4   3,295 kg (7,264 lb)
PERFORMANCE RATINGS (note: flexible T-O ratings involving considerable derating are used in operation):
T-O (S/L, ISA): 535C   166.4 kN (37,400 lb st)
535E4   178.4-192 kN (40,100-43,100 lb st)
Max climb (10,670 m; 35,000 ft, Mach 0.80):
535C   40.1 kN (9,023 lb)
535E4   41.4 kN (9,300 lb)
Max cruise (10,670 m; 35,000 ft, Mach 0.80):
535C   37.6 kN (8,453 lb)
535E4   38.7 kN (8,700 lb)
SPECIFIC FUEL CONSUMPTION (cruise, as above):
535C   18.30 mg/Ns (0.646 lb/h/lb)
535E4   16.94 mg/Ns (0.598 lb/h/lb)

ROLLS-ROYCE RB168 SPEY

The RB168 military versions of the Spey are as follows:
Mk 101. Developed from civil Mk 505. Fitted to BAe Buccaneer.
Mk 202/203. Supersonic fighter engine with augmentation. Fitted to Phantom FG.1/FGR.2 and to Chinese H-7 (B-7) prototypes.
Mk 250/251. Fully marinised. Fitted to BAe Nimrod MR.1, R.1 and MR.2.
Mk 807. Mk 101 rotors within Mk 555 structure. Produced under licence in Italy with Brazilian participation for AMX aircraft.
DIMENSIONS:
Length: Mk 101   2,911 mm (114.6 in)
Mk 202, 203   5,204 mm (204.9 in)
Mk 250, 251   2,972 mm (117.0 in)
Mk 807   2,456 mm (96.7 in)
Diameter   825 mm (32.5 in)
WEIGHT, DRY:
Mk 101   1,121 kg (2,471 lb)
Mk 202, 203   1,857 kg (4,093 lb)
Mk 250, 251   1,243 kg (2,740 lb)
Mk 807   1,096 kg (2,417 lb)

Rolls-Royce 535E4 three-shaft turbofan

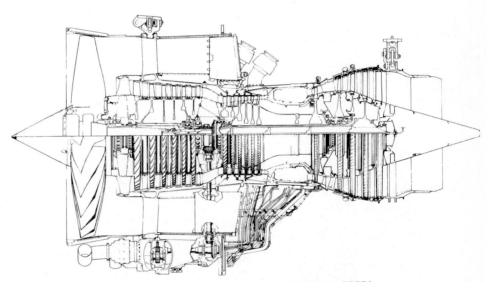

Longitudinal section and cutaway of Rolls-Royce 535E4

Rolls-Royce Spey Mk 807 two-shaft turbofan

658 UK: ENGINES—ROLLS-ROYCE

PERFORMANCE RATINGS (max T-O):
- Mk 101, 807 — 49.1 kN (11,030 lb st)
- Mk 202, 203: dry — 54.5 kN (12,250 lb st)
- augmented — 91.2 kN (20,515 lb st)
- Mk 250, 251 — 53.3 kN (11,995 lb st)

ROLLS-ROYCE TAY

The Tay turbofan is designed around the core and external gearbox of the RB183 Mk 555. The LP system has been tailored to complement this by maintaining core inlet and outlet conditions similar to those of the original engine. The widechord fan and three-stage IP compressor are driven by a three-stage LP turbine which uses the latest proven technology. The cold bypass air and hot exhaust are combined in a forced mixer. The bypass duct is carbonfibre composite.

The initial production versions are the **Tay 611** and **620**. The **Tay 650** gives a 9 per cent increase in maximum take-off thrust and a 15 per cent increase in maximum continuous, climb, and cruise thrusts, achieved by a small increase in fan diameter and an advanced HP turbine.

The initial versions received certification from the CAA in June 1986. Over 650 of these had accumulated 1.5 million hours in service by 1 January 1993. Certification of the Tay 650 was achieved ahead of schedule in June 1988. The Tay easily meets all current emission standards, and enables the Fokker 100, Gulfstream IV and re-engined 727-100 to comply with FAR Pt 36 Stage 3 noise requirements.

Tay 610-8. Selected for Gulfstream IV. Certificated on 26 June 1986. All replaced by **Tay 611** (see data).

Tay 620-15. Selected for Fokker 100. Entered service April 1988. Proposed for Fokker 70.

Tay 650-14. Selected to re-engine BAe One-Eleven.
Tay 650-15. Specified for higher-performance versions of Fokker 100. Entered service October 1989.
Tay 651-54. Selected to re-engine 727-100; in service December 1992.

TYPE: Two-shaft turbofan.
FAN: Single-stage with wide chord blades. Diameter (620), 1,118 mm (44 in), (650) 1,143 mm (45 in).
LP COMPRESSOR: New design with three stages on fan shaft.
HP COMPRESSOR: 12-stage axial (RB183 Mk 555).
COMBUSTION CHAMBER: Tubo-annular with 10 flame tubes, each with one burner.
FUEL SYSTEM: As RB183 Mk 555 but with improved fuel control unit.
HP TURBINE: All Mks except 650 and 670, two stages as RB183 Mk 555. Tay 650 and 670, advanced two-stage design.
LP TURBINE: New design with three stages.
BYPASS DUCT: Carbonfibre composite.
MIXER: Forced deep chute type with 12 lobes.
DIMENSIONS:
- Length: Tay 611, 620, 650, 651 — 2,405 mm (94.7 in)
- Inlet diameter: Tay 611, 620 — 1,118 mm (44.0 in)
- Tay 650 — 1,138 mm (44.8 in)

WEIGHT, DRY:
- Tay 611 — 1,339 kg (2,951 lb)
- Tay 620 — 1,445 kg (3,185 lb)
- Tay 650 — 1,515 kg (3,340 lb)
- Tay 651 — 1,533 kg (3,380 lb)

PERFORMANCE RATING (T-O):
- Tay 611, 620-15 — 61.61 kN (13,850 lb st) to 30°C
- Tay 650, 651 — 67.17 kN (15,100 lb st) to 30°C

SPECIFIC FUEL CONSUMPTION (cruise):
- Tay 611 (13,100 m; 43,000 ft, Mach 0.8) — 20.1 mg/Ns (0.71 lb/h/lb)
- Tay 620 and 650 (9,145 m; 30,000 ft, Mach 0.73) — 19.5 mg/Ns (0.69 lb/h/lb)
- Tay 651 (10,669 m; 35,000 ft, Mach 0.78) — 19.5 mg/Ns (0.69 lb/h/lb)

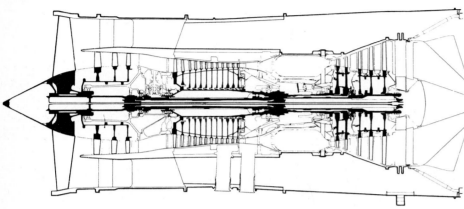

Longitudinal section of Rolls-Royce Tay two-shaft turbofan

Rolls-Royce Tay two-shaft turbofan

ROLLS-ROYCE TYNE

The 4,549 ekW (6,100 ehp) Tyne two-shaft turboprop is being produced by a consortium comprising SNECMA and MTU (prime contractors carrying out assembly and test), Rolls-Royce and TA. The Mk 21 engine powers the Atlantique and the Mk 22 the Transall.

ROLLS-ROYCE VIPER

This turbojet remains in production. More than 5,000 are in civil and military operation.

Current versions are as follows:

Viper 11 (Mk 200 Series). Single-shaft seven-stage axial compressor driven by single-stage turbine. Mass flow 20 kg (44 lb)/s. Type tested at 11.12 kN (2,500 lb st) and powers Jindivik Mk 3 drone, Jet Provost T.4 and 5, SOKO Galeb and HJT-16 Kiran Mk I/IA.

A Viper 11 version, the 22-1, was built under licence in Italy by Piaggio and HDH of Australia for the Aermacchi MB-326.

Viper 500 Series. Development with zero stage on compressor. Major applications include early HS 125 (Mks 521, 522) and PD-808 executive aircraft (Mk 526), and Strikemaster (Mk 535), MB-326 (Mk 540) and Jastreb (Mk 531) training and light combat aircraft. Mk 540 built under licence by Piaggio.

Viper 600 Series. Eight-stage compressor driven by two-stage turbine; annular vaporising combustion chamber. Take-off rating 16.7 kN (3,750 lb st) civil and 17.8 kN (4,000 lb st) military. Agreement signed with Fiat (Italy) in July 1969 for technical collaboration (see FiatAvio).

The civil Viper 601 powers the BAe 125-600; the military 632 is fitted to the G-4 Super Galeb, MB-326K and MB-339. The 632-41 powers the Orao/IAR-93A and IAR-99. The 632 is built under licence in Italy, Romania and former Yugoslavia. The 633 has an afterburner of the two-gutter type, with hot streak ignition; rating 22.3 kN (5,000 lb st).

The **Viper 680** is the latest variant for the MB-339. It produces 10 per cent more thrust than the Viper 632.

The following details apply to the Viper 600 Series:
TYPE: Single-shaft axial turbojet.
COMPRESSOR: Eight-stage. Steel drum type rotor with disc assemblies. Magnesium alloy casing with blow-off valve.
COMBUSTION CHAMBER: Short annular type with 24 vaporising burners and six starting atomisers. Electric ignition.
FUEL SYSTEM: Hydromechanical, with pump, barometric control and air/fuel control.
TURBINE: Two-stage axial. Shrouded blades attached to discs by fir tree roots and locking strips.
ACCESSORY DRIVES: Gearbox driven from front of compressor by bevel gear.
LUBRICATION SYSTEM: Self contained recirculatory system. Military version fully aerobatic.
STARTING: 24V starter/generator.
DIMENSIONS:
- Length (flange to flange):
  - Viper 531, 535, 632, 680 — 1,806 mm (71.1 in)
- Max width — 749.3 mm (29.5 in)
- Height — 901.7 mm (35.5 in)

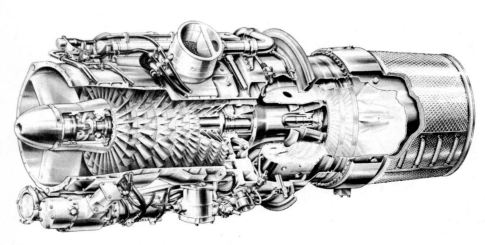

Cutaway drawing of Rolls-Royce Viper 632 single-shaft turbojet

ROLLS-ROYCE—ENGINES: UK   659

Rolls-Royce Pegasus 11-61 (F402-RR-408) vectored-thrust turbofan

WEIGHT, DRY:
Viper 11	281 kg (620 lb)
Viper 531, 535, 540	358 kg (790 lb)
Viper 601, 632	376.5 kg (830 lb)
Viper 680	379 kg (836 lb)

PERFORMANCE RATINGS (T-O):
Viper 11	11.12 kN (2,500 lb st)
Viper 531	13.9 kN (3,120 lb st)
Viper 535, 540	14.9 kN (3,360 lb st)
Viper 601	16.7 kN (3,750 lb st)
Viper 632	17.66 kN (3,970 lb st)
Viper 680	19.39 kN (4,360 lb st)

SPECIFIC FUEL CONSUMPTION (T-O):
Viper 11	30.3 mg/Ns (1.07 lb/h/lb st)
Viper 500 series	28.3 mg/Ns (1.00 lb/h/lb st)
Viper 601	26.9 mg/Ns (0.95 lb/h/lb st)
Viper 632	27.5 mg/Ns (0.97 lb/h/lb st)

ROLLS-ROYCE PEGASUS
USMC designation: F402

The Pegasus is a two-shaft turbofan designed for short take-off/vertical landing (STOVL) applications. It powers all versions of the Harrier attack aircraft. The Pegasus provides both lift and propulsive thrust through four swivelling exhaust nozzles which vector engine thrust from horizontal, for conventional (wing-borne) flight, to vertical, for jet-borne flight. To minimise aircraft control problems in jet-borne flight, thrust is divided between the engine nozzles to ensure the resultant thrust passes through a fixed point irrespective of nozzle angle; LP and HP spools rotate in opposite directions, to minimise gyroscopic effects; and HP bleed air is used for aircraft stabilisation.

The Pegasus entered service in the BAe Harrier in 1969 as the **Pegasus 11 Mk 103** (US designation **F402-RR-402**). A marinised version, designated **Mk 104**, is fitted to the BAe Sea Harrier.

The **Pegasus 11-21**, developed for the McDonnell Douglas/BAe Harrier II, has the US designation **F402-RR-406** and UK designations **Mk 105** (GR. Mk 5 & 7) and **Mk 106** (Sea Harrier). Delivery of 11-21 engines began in December 1984. Since 1986 Pegasus 11-21 engines for the USMC's AV-8B and UK's GR.5 & 7 have been fitted with digital engine control systems (DECS) in place of the earlier hydromechanical systems.

The latest production version of the Pegasus engine is the **11-61**, designated **F402-RR-408** by the US Marine Corps. To enhance the combat effectiveness of the Harrier and reduce the cost of engine ownership, 11-61 has increased lift thrust, up to 15 per cent more at high ambient temperatures, and twice the hot-end life of its predecessor. Maintenance activity is significantly reduced by incorporation of proven digital engine control and engine monitoring systems, improved inspection facilities and modular construction. The Pegasus 11-61 entered service in 1990 with the US Marine Corps, and has been specified by the Italian and Spanish navies for their Harrier II Plus aircraft.

Pegasus engines are currently in service with the Royal Air Force and Royal Navy, the US Marine Corps and the navies of India, Italy and Spain. Total Pegasus experience exceeds 1 million hours, with over 1,000 engines manufactured.

The following details apply to the Pegasus 11-61 (F402-RR-408):
TYPE: Two-shaft vectored-thrust turbofan.
FAN: Three-stage, overhung ahead of front bearing. Titanium alloy blades with advanced circumferential snubbers.
HP COMPRESSOR: Eight-stage with titanium alloy rotor blades and discs.
COMBUSTION SYSTEM: Annular with 'T-shaped' fuel vaporisers.
FUEL SYSTEM: Digital engine control system (DECS) comprising duplicated full-authority digital engine control units (DECU) and a hydromechanical fuel metering unit (FMU). FMU includes a cockpit-selectable manual fuel control system (MFCS) with manual back-up.
TURBINES: Two-stage cooled HP turbine incorporating single-crystal blading. Two-stage uncooled LP turbine with single-crystal material in first stage.
THRUST NOZZLES: Two fabricated steel zero-scarf front (cold) nozzles and two Nimonic rear (hot) nozzles rotated simultaneously by bleed-air motor under pilot command.
LUBRICATION SYSTEM: Self-contained with fuel-cooled oil cooler.
STARTING: Gas turbine starter/APU driving through engine gearbox.
DIMENSIONS:
Width, incl nozzles	2,510 mm (98.8 in)
Length, incl nozzles	3,485 mm (137.2 in)
Diameter, intake	1,222 mm (48.1 in)

WEIGHT, DRY (incl nozzles): 1,896 kg (4,180 lb)
PERFORMANCE RATING (max lift thrust, S/L static):
Pegasus 11-61   105.1 kN (23,620 lb)

ROLLS-ROYCE ADVANCED STOVL
Rolls-Royce is reported to be in partnership with Pratt & Whitney on a derivative of the F119 and with General Electric on a derivative of the F120 (1990-91 *Jane's*), as candidate engines for a future ASTOVL aircraft. Funding is provided by the US Advanced Research Projects Agency (see under US Air Force in main Aircraft section).

MTU-TURBOMECA-RR MTR 390
See the International part of this section.

ROLLS-ROYCE GEM
The Gem was developed for the Westland Lynx helicopter. Subsequent applications are the Westland 30 and advanced Lynx versions, and Agusta A 129. The Gem 41 and 60 have been civil certificated.

Choice of a two-spool gas generator gives fast response to power demand without the need for a complex control system. There are seven major modules, each assembled, tested and released as an interchangeable unit.

Provision is made for in-flight and on-ground condition monitoring systems. Features include access ports for intrascope inspection of each LP compressor stage, HP compressor, combustor, LP turbine and power turbine, and mountings for vibration pickups.

The following versions have been announced:
Gem Mk 1001. Engine change unit for Lynx of British services and French Navy. Rated at 671 kW (900 shp).
Gem 2. Export military, rated at 671 kW (900 shp).
RR 1004. For Agusta A 129. Direct drive in place of reduction gearbox and electronic instead of hydromechanical control. Production under licence by Piaggio.
Gem 41 series. Modified compressor to increase mass flow by about 10 per cent plus small increase in TET.
Gem 42 series. Same ratings as Gem 41, but improved reliability and power retention.
Gem Mk 510. Civil Gem 41 for Westland 30.
Gem 60 series. Further increase in mass flow of approximately 26 per cent over Gem 41, derived from LP compressor incorporating new blades; TET also increased.
Gem Mk 531. Civil Gem 60 for Westland 30.

The following description relates to the Gem Mk 1001:
TYPE: Free turbine turboshaft.
SHAFT DRIVE: Single-stage double-helical reduction gear with rotating planet cage carried by ball bearing at front and roller bearing at rear. Alternative of direct drive output at 27,000 rpm or gearbox giving governed output speed of 6,000 rpm.
LP COMPRESSOR: Four-stage axial.
HP COMPRESSOR: Single-stage centrifugal. Overall pressure ratio 12.0.
COMBUSTION CHAMBER: Annular reverse flow with air atomiser fuel sprays. High energy ignition.
HP TURBINE: Single-stage close-coupled to HP impeller.
LP TURBINE: Single-stage with shrouded blades.
POWER TURBINE: Two-stage axial with shrouded blades.
ACCESSORY DRIVES: Bevel gear on front of HP shaft drives starter/generator, fuel pump, oil cooler fan and other accessories.
FUEL SYSTEM: Plessey fluidics automatic control, and power matching for multi-engine installation. Hamilton Standard electronic control fitted to RR 1004 and Gem 60.

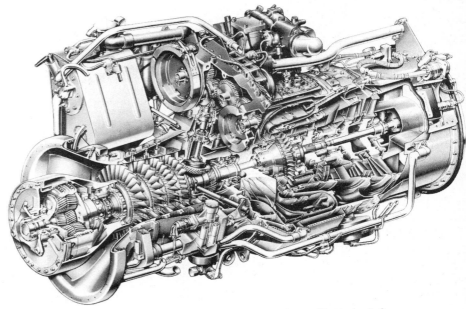

Cutaway drawing of Rolls-Royce Gem 42 free turbine turboshaft

GEM ENGINE RATINGS, kW (shp) ISA S/L STATIC

Designation	Date In Service	Emergency (20 s)	One Engine Inoperative			Normal Twin Operation		
			Max Contingency (2½ min)	Intermediate Contingency (60 min)	Max (T-O) (30 min)	Max (T-O) (5 min)	Max Continuous	
Gem 2, Mk 1001	1976	N/A	671 (900)	619 (830)	N/A	619 (830)	559 (750)	
RR 1004	1990	759 (1,018)	704 (944)	657 (881)	657 (881)	N/A	615 (825)	
Gem 41-1, Mk 1014	1978	N/A	835 (1,120)	790.5 (1,060)	N/A	746 (1,000)	664 (890)	
Gem 42, Mk 1017	1987	N/A	835 (1,120)	790.5 (1,060)	N/A	746 (1,000)	664 (890)	
Gem 60-3/1 Mk 530	1983	N/A	897 (1,203)	844 (1,132)	N/A	844 (1,132)	821.75 (1,102)	

N/A: Not applicable

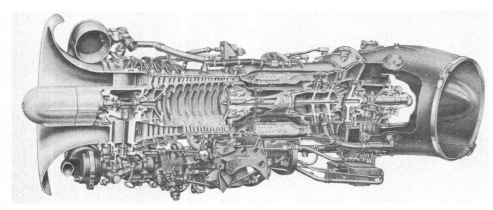

Cutaway drawing of Rolls-Royce Gnome H.1400-1 free turbine turboshaft

LUBRICATION SYSTEM: Engine mounted oil tank and cooler. Magnetic chip detectors. Oil filter in accessory wheelcase.
DIMENSIONS:
  Length overall　　　　　　　　　1,099 mm (43.2 in)
  Width overall　　　　　　　　　　　575 mm (22.6 in)
  Height overall　　　　　　　　　　596 mm (23.5 in)
WEIGHT, DRY:
  Gem 1001, 2　　　　　　　　　　　178 kg (393 lb)
  Gem 41, 42　　　　　　　　　　　 183 kg (404 lb)
  RR 1004　　　　　　　　　　　　　163 kg (360 lb)
  Gem 60　　　　　　　　　　　　　 185 kg (407 lb)
PERFORMANCE RATINGS: see table
SPECIFIC FUEL CONSUMPTION:
  50 per cent max T-O:
    Except Gem 60　　　　　110 µg/J (0.65 lb/h/shp)
    Gem 60　　　　　　　　103 µg/J (0.61 lb/h/shp)

ROLLS-ROYCE GNOME
Gnome is the name given to the General Electric T58 turboshaft which Rolls-Royce manufactures. More than 2,300 have been delivered in the following versions:
H.1000. Initial version, rated at 783 kW (1,050 shp) for Whirlwind and Agusta-Bell 204B.
H.1200. Rated at 932 kW (1,250 shp). Used in Agusta-Bell 204B, Boeing 107 and some Kawasaki KV107-II-5s. Coupled version for Wessex comprises two H.1200s driving through a coupling gearbox.
H.1400. Rated at 1,044 kW (1,400 shp). Based on the H.1200, with modified compressor to increase airflow.
H.1400-1. Rated at 1,145 kW (1,535 shp). Increased gas generator speed and improved gas generator turbine blades. In production for Sea King and Commando.
H.1400-1T. Tropical model; turbine nozzle adjusted for better high ambient performance.

The following description refers to the H.1400-1:
TYPE: Free turbine turboshaft.
COMPRESSOR: Ten-stage axial. Variable inlet guide vanes and first three stator rows. Mass flow 6.26 kg (13.8 lb)/s.
COMBUSTION CHAMBER: Annular with 16 Simplex injectors, eight on each of two sets of manifolds. One Lodge high energy igniter.
FUEL SYSTEM: Lucas hydromechanical controlled by HSDE computer.
FUEL GRADE: DERD.2452, 2453, 2454, 2486, 2494 and 2498 (NATO F44, F34, F40, F35 and F43).
GAS-PRODUCER TURBINE: Two-stage.
POWER TURBINE: Single-stage free turbine.
REDUCTION GEAR: Optional double helical gear providing reduction from nominal 19,500 rpm power turbine speed to 6,600 rpm at left or right output shaft.
ACCESSORY DRIVES: Fuel and lubrication systems mounted beneath compressor casing. Power take-off shaft up to 100 shp.
LUBRICATION: Scavenged gear pumps. Serck oil cooler.
STARTING: Rotax electric starter in nose bullet.
DIMENSIONS (H.1400-1):
  Length　　　　　　　　　　　　　1,392 mm (54.8 in)
  Max height　　　　　　　　　　　　549 mm (21.6 in)
  Max width　　　　　　　　　　　　　577 mm (22.7 in)
WEIGHT, DRY:
  H.1400-1　　　　　　　　　　　　　148 kg (326 lb)
PERFORMANCE RATINGS (at power turbine shaft):
  Max contingency (2½ min; multi-engine aircraft only):
    H.1400-1　　　　　　　　　　1,238 kW (1,660 shp)
    H.1400-1T　　　　　　1,092 kW (1,465 shp) to 45°C
  Max one-hour (single engine):
    H.1400-1　　　　　　　　　　1,145 kW (1,535 shp)
    H.1400-1T　　　　　　1,030 kW (1,380 shp) to 45°C
  Max continuous:
    H.1400-1　　　　　　　　　　　932 kW (1,250 shp)
    H.1400-1T　　　　　　　783 kW (1,050 shp) to 45°C
SPECIFIC FUEL CONSUMPTION:
  At max contingency rating:
    H.1400-1　　　　　　　102.75 µg/J (0.608 lb/h/shp)
    H.1400-1T (30°C)　　　105.8 µg/J (0.626 lb/h/shp)

UNITED STATES OF AMERICA

ALLISON
ALLISON GAS TURBINE DIVISION, GENERAL MOTORS CORPORATION
PO Box 420, Indianapolis, Indiana 46206-0420
Telephone: 1 (317) 230 2000
GENERAL MANAGER: Dr F. Blake Wallace

Allison Gas Turbine Division has 5,700 employees producing gas turbine engines and components for aircraft, vehicular and industrial/marine applications.

A major research effort is in hand to develop a lift fan for the future Lockheed ASTOVL aircraft driven by hot gas from a P&W/RR F119-derived engine. Collaboration on the T800 engine with Garrett (AlliedSignal) is described under LHTEC on a later page.

ALLISON 250
US military designations: T63 and T703

The Allison 250 is a small turboshaft/turboprop. Deliveries exceed 25,000, logging over 82 million hours.

A development contract for the T63 military version was received by Allison in June 1958. Details of early versions last appeared in the 1978-79 *Jane's*; the following are current models:
T63-A-720. Military turboshaft engine with hot-end improvements, increasing T-O rating to 313 kW (420 shp), for the Bell OH-58C.
T703-A-700. Military turboshaft engine corresponding to 250-C30R, for Bell OH-58D.
250-B17. Uprated version of B15 turboprop. The **B17B** operates at 17°C higher turbine gas temperature with hot-end improvements which maintain full power at high ambient temperatures. Produced from 1974 for Turbostar 402 conversions, Turbostar 414, ASTA Nomad N22 and 24, SIAI-Marchetti SM.1019E and various agricultural aircraft. **B17C** introduced improved gearbox allowing use of 313 kW (420 shp) on T-O. Produced for Nomad N22 and 24, SF.260TP and 600TP, Turbostar 402/414, Allison Bonanza, Bobby Aerostar, Par Turbine Aerostar, LoPresti Piper SwiftFury, AASI Jetcruzer prototype, Glasair III, Advanced Airship ANR, BN-2T Turbine Islander, Composite Eagle, AP.68TP Viator and Maule MX-7. **B17D** produced for SF.260TP, Aucán, Redigo, Fuji T-5, HTT-34 and Mentor 420. **B17F**, with new compressor as in -C20R for increased T-O power, introduced in 1985 for BN-2T, SF.600TP, HX-1, Redigo, Turbine P-210 and A36 Bonanza.
250-C20B. Introduced 1974, and rated at 313 kW (420 shp). For Bell and Agusta-Bell 206B JetRanger III and B LongRanger, MD 500D, BO 105CBS and E, Kitty Hawk, FH-1100 and UH-12E, Ka-226, RFB Fantrainer 400 and Bell 47G conversions.
250-C20F. For AS 355 Ecureuil 2/TwinStar.
250-C20J. For Bell and Agusta-Bell 206B JetRanger III.
250-C20R. Derivative of C20B with new axial-centrifugal compressor. **C20R/1** with redundant overspeed system for twin-engine applications certificated September 1986 for A 109A Mk II and Gemini ST. **C20R/2** for single-engine helicopters certificated early 1987 for MD 500ER, JetRanger III, LongRanger and MD 520N.
250-C20S. Turboprop for Soloy Turbine Pac conversions of Cessna 185, 206 and 207. **C20W** powers Schweizer TH-330 and Enstrom 480.
250-C28. Major redesign with single centrifugal compressor only, with increased airflow. Reduced noise and emissions, and minimal infra-red signature. New main gearbox. Certificated December 1977.
250-C28B. With particle separator; 2½-min rating of 410 kW (550 shp). Powers Bell LongRanger I.
250-C28C. Improved model with plain inlet; 2½-min rating of 410 kW (550 shp). Powers BO 105 LS.
250-C30. Advanced single-stage compressor and dual ignition. Initial rating 485 kW (650 shp), with a 2½ min rating of 522 kW (700 shp). Certification completed March 1978. Produced for S-76 and MD 530G.
250-C30G. Produced for Heli-Air Bell 222; C30G2 for Bell 230.
250-C30L. With digital control, produced for Bell 406CS.
250-C30M. Produced for AS 350.
250-C30P. Produced for LongRanger III.
250-C30R. With digital control, produced for AHIP OH-58D.
250-C30S. Produced for S-76.
TYPE: Light turboshaft or turboprop.
COMPRESSOR: C20, C20B and B17B have six axial stages and one centrifugal. B17F and C20R have four axial stages and one centrifugal. C28 and C30 models have single-stage centrifugal compressor only. Pressure ratio: C20B, B17, 7.2; C28B/C, C30, 8.4. Mass flow: C20B, B17C, 1.56 kg (3.45 lb)/s; C28B/C, 2.02 kg (4.45 lb)/s; C30, 2.54 kg (5.6 lb)/s.
COMBUSTION CHAMBER: Single can type at rear. Single duplex fuel nozzle in rear face. One igniter on C20B, B17C, C28B/C and C30L/M/P/R. Dual igniters on C30 and C30S; optional on C28B/C.

Allison 250-B17C free turbine turboprop

Allison 250-C20R free turbine turboshaft

TURBINES: Two-stage gas producer turbine and two-stage free power turbine. Integrally cast rotor blades and wheels.
CONTROL SYSTEM: Pneumatic-mechanical system (B17C, hydromechanical; C20B, C28, C30/30M/P/S, pneumatic-mechanical; C30L, C30R, supervisory electronic).
FUEL: Primary fuels are ASTM-A or A-1 and MIL-T-5624, JP-4, JP-5 and diesel fuel.
LUBRICATION: Dry sump.
OIL SPECIFICATION: MIL-L-7808 and MIL-L-23699.
DIMENSIONS:
Length: B17C	1,143 mm (45.0 in)
C20B, R	985 mm (38.8 in)
C28C	1,201 mm (47.3 in)
C30	1,041 mm (41.0 in)
Width: B17C, C20B	483 mm (19.0 in)
C28, C30	557 mm (21.94 in)
Height: B17C	572 mm (22.5 in)
C20B	589 mm (23.2 in)
C28, C30	638 mm (25.13 in)
WEIGHT, DRY: B17C	88.4 kg (195 lb)
C20B	71.5 kg (158 lb)
C20R	76.0 kg (168 lb)
C28	99.3 kg (219 lb)
C28B, C	104 kg (230 lb)
C30	109.3 kg (240 lb)

PERFORMANCE RATINGS (S/L, ISA):
T-O:
B17C, C20B, C20J, C20S (5 min)	313 kW (420 shp)
B17F, C20R (5 min, to 26.7°C)	335 kW (450 shp)
C20F	313 kW (420 shp)
C28, 28B, 28C (30 min)	373 kW (500 shp)
C30 (5 min)	485 kW (650 shp)
Max continuous: B17C (cruise)	275 kW (369 shp)
C20B	276 kW (370 shp)
C20R	283 kW (380 shp)
C28	368 kW (494 shp)
C30	415 kW (557 shp)
Cruise B (75 per cent): B17C	206 kW (277 shp)
C20B	207 kW (278 shp)
C20R	236 kW (317 shp)
C28	274 kW (367 shp)
C30	312 kW (418 shp)

SPECIFIC FUEL CONSUMPTION:
At T-O rating: B17C	111 µg/J (0.657 lb/h/shp)
C20B	110 µg/J (0.650 lb/h/shp)
C20R	103 µg/J (0.608 lb/h/shp)
C28	102.5 µg/J (0.606 lb/h/shp)
C30	100 µg/J (0.592 lb/h/shp)
At cruise B rating: B17C	120.8 µg/J (0.715 lb/h/shp)
C20B	120 µg/J (0.709 lb/h/shp)
C20R	112.5 µg/J (0.666 lb/h/shp)
C28	112 µg/J (0.664 lb/h/shp)
C30	111 µg/J (0.657 lb/h/shp)

ALLISON T56 and 501
US military designation: T56

Current versions of the T56 are as follows:

T56-A-14. Rated at 3,661 ekW (4,910 ehp). Generally similar to T56-A-15, but seven-point suspension and detail changes. Powers the P-3B and C Orion.

T56-A-15. Rated at 3,661 ekW (4,910 ehp). Introduced air-cooled turbine blades. Powers current C-130.

T56-A-423 and **A-16.** Rated at 3,661 ekW (4,910 ehp). Powers US Navy versions of the C-130.

T56-A-425. Rated at 3,661 ekW (4,910 ehp). Powers C-2A Greyhound and early E-2C Hawkeye.

T56-A-427. Increased power and 13 per cent improvement in specific fuel consumption. Digital electronic supervisory control and modified propeller drive to maintain 1,106 rpm output with 14,239 power section rpm. Powers E-2C delivered since 1987.

501-D22A. Rated at 3,490 ekW (4,680 ehp). Commercial version of T56-A-15. Powers Lockheed L-100; -D22A2 powers CV-580.

501-D22G. Powers Super-580 and Kelowna CV-5800.

Including the Allison 501 commercial engines, production of these engines reached 15,240 by 1 January 1993.

The following details apply to the T56-A-15:
TYPE: Axial-flow turboprop.
PROPELLER DRIVE: Combination spur/planetary gear type. Overall gear ratio 13.54. Power section rpm 13,820. Weight of gearbox approximately 249 kg (550 lb) with pads on rear face for accessory mounting.
COMPRESSOR: Fourteen-stage axial flow. Pressure ratio 9.5. Mass flow 14.70 kg (32.4 lb)/s. Constant speed 13,820 rpm.
COMBUSTION CHAMBER: Six stainless steel can-annular type perforated combustion liners within one-piece stainless steel outer casing. Two ignitors in diametrically opposite combustors.
FUEL SYSTEM: High pressure type. Bendix control system. Water/alcohol augmentation system available.
FUEL GRADE: MIL-J-5624, JP-4 or JP-5.
TURBINE: Four-stage. Rotor assembly consists of four stainless steel discs, with first stage having hollow air-cooled blades, secured by fir tree roots. Gas temperature before turbine 1,076°C.
ACCESSORY DRIVES: Accessory pads on rear face of reduction gear housing at front end of engine.
LUBRICATION SYSTEM: Low pressure. Dry sump. Pesco dual-element oil pump. Normal oil supply pressure 3.8 bars (55 lb/sq in).
OIL SPECIFICATION: MIL-L-7808.
MOUNTING: Three-point suspension.
STARTING: Air turbine, gearbox mounted.
DIMENSIONS:
Length (all current versions)	3,708 mm (146 in)
Width (all current versions)	686 mm (27 in)
Height: A-15, A-16, A-425, D-22A	991 mm (39 in)
A-14	1,118 mm (44 in)

WEIGHT, DRY:
A-14	855 kg (1,885 lb)
A-15	828 kg (1,825 lb)
A-16	835 kg (1,841 lb)
A-425	860 kg (1,895 lb)
D22A	832 kg (1,834 lb)

PERFORMANCE RATINGS (S/L, ISA, static):
T-O: A-14, A-15, A-16, A-423, A-425
3,661 ekW; 3,424 kW (4,910 ehp; 4,591 shp)
A-427   3,915 kW (5,250 shp)
501-D22A   3,490 ekW (4,680 ehp)
Normal: A-14, A-15, A-16, A-425, D22A
3,255 ekW; 3,028 kW (4,365 ehp; 4,061 shp)
SPECIFIC FUEL CONSUMPTION:
At max rating:
A-14, A-15, D22A   84.67 µg/J (0.501 lb/h/ehp)
At normal rating:
A-14, A-15   87.4 µg/J (0.517 lb/h/ehp)
OIL CONSUMPTION:
A-14, A-15
1.3 litres (0.35 US gallon; 0.29 Imp gallon)/h

ALLISON GMA 1107
US military designation: T406-AD-400

The T406 turboshaft, developed for the tilt-rotor V-22 Osprey, is the basis for the GMA 1107 turboshaft, GMA 2100 turboprop and GMA 3007 turbofan. The last two are described separately. All rest on 170 million T56 hours built into the core.

The T406-AD-400 is a free turbine, front drive 4,588 kW (6,150 shp) turboshaft incorporating high efficiency components and reduced maintenance features required for operation in the V-22 Osprey. It features six rows of variable compressor stators, dual full-authority digital electronic fuel controls, self-contained oil system capable of engine operation in the vertical position and modular construction. The T406 has completed its flight rating tests and 21 engines have been delivered for the V-22 flight test programme. Full production was scheduled for 1993.

The T406 is applicable to other helicopter and tilt-rotor aircraft as the **GMA 1107**. This engine is undergoing FAA certification.

TYPE: Axial flow turboshaft.
COMPRESSOR: Fourteen-stage axial flow, with variable inlet guide vanes and first five stator rows. Pressure ratio 16.7. Mass flow 16.1 kg (35.5 lb)/s.
COMBUSTOR: Annular convection film-cooled, with 16 airblast-type fuel nozzles providing smoke-free operation. Dual capacitor discharge ignition.
CONTROL SYSTEM: Full-authority digital.
FUEL GRADE: MIL-T-5624; grades JP-4, JP-5 and MIL-T-83188; JP-8.
TURBINE: Gas generator turbine has two axial stages with air-cooled single-crystal blading; both stages overhung to the rear of the gas generator thrust bearing. Power turbine has two axial stages on a straddle-mounted shaft which runs the entire length of the engine. Film-damped bearings eliminate the need for a centre bearing.
OUTPUT: Power turbine forward shaft drives a torquemeter assembly which is directly coupled to the V-22 rotor gearbox. The torque tube housing serves as the main engine mount.
ACCESSORY DRIVES: An engine accessory gearcase is mounted beneath the air inlet housing. It provides for engine starter, generator, oil pump and fuel pump metering unit drives.
LUBRICATION SYSTEM: Self contained, featuring positive scavenging sumps, 3-micron filtration, quantitative debris monitor and a bottom-mounted, all-attitude oil reservoir with service scuppers on each side of the engine.
OIL GRADE: MIL-L-7808 or MIL-L-23699.
DIMENSIONS:
Length overall, without gearbox	1,958 mm (77.08 in)
Length from inlet flange	1,521 mm (59.88 in)
Width	671 mm (26.40 in)
Height	864 mm (34.00 in)

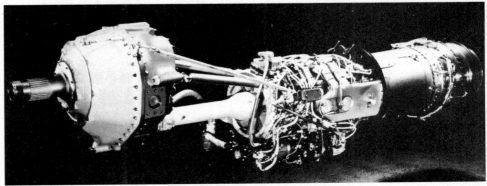

Allison T56-A-427 single-shaft turboprop

Allison T406-AD-400 free turbine turboshaft

WEIGHT, DRY:	440.4 kg (971 lb)
PERFORMANCE RATINGS:	
Max power (S/L)	4,586 kW (6,150 shp) to 43°C
Max cont power (static 1,219 m; 4,000 ft)	
	4,391 kW (5,888 shp) to 25°C
SPECIFIC FUEL CONSUMPTION:	
Max cont power	71.0 µg/J (0.42 lb/h/shp)

ALLISON GMA 2100

The GMA 2100 turboprop is being developed for the new generation of high-speed regional aircraft as well as military transports and maritime patrol aircraft. The engine combines the T406 power section with a new propeller reduction gearbox based on the highly successful T56/501 design. The GMA 2100 was scheduled for certification in early 1993. It has been selected by Saab (derated to 3,096 kW; 4,152 shp) to power their Saab 2000 and by IPTN to power their N-250-100 aircraft.

The first GMA 2100 went on test in June 1988. A prototype GMA 2100 engine successfully completed flight testing on a P-3A aircraft in late 1990. The prototype consisted of a T406 power section, the new reduction gearbox and a flange-mounted Dowty Aerospace six-bladed propeller. Six flight test and six production engines had been delivered by January 1993, when full production began.

TYPE: Free turbine, axial flow turboprop.
COMPRESSOR: Fourteen-stage, axial flow, with variable inlet guide vanes and first five stator rows.
COMBUSTOR: Annular design with 16 airblast fuel nozzles.
CONTROL SYSTEM: Full-authority digital (FADEC), co-ordinating engine and propeller functions.
HP TURBINE: Two-stage axial design with air-cooled single-crystal blading.
POWER TURBINE: Two uncooled stages on straddle-mounted shaft.
PROPELLER DRIVE: Completely new design saving 68 kg (150 lb) weight, with life of 30,000 h. Accessories grouped on rear face.
DIMENSIONS:
Diameter overall	1,151 mm (45.3 in)
Length overall	2,743 mm (108 in)
WEIGHT, DRY:	702.2 kg (1,548 lb)
PERFORMANCE RATINGS:	
S/L, T-O	4,474 kW (6,000 shp)
SPECIFIC FUEL CONSUMPTION:	
S/L, T-O	69.31 µg/J (0.41 lb/h/shp)

Allison GMA 2100 free turbine turboprop

ALLISON GMA 3007

The GMA 3007 turbofan is being developed to power regional airliners and medium/large business jets. The engine utilises the high-pressure spool from the T406/GMA 2100 with a new low-pressure spool. The GMA 3007 was selected in 1990 by Embraer to power the EMB-145, and by Cessna for the Citation X. Growth versions are planned with ratings exceeding 53.4 kN (12,000 lb st).

The first GMA 3007 went on test in July 1991. Additional development during 1992 included full altitude calibration, acoustic tests, initial endurance testing, and initiation of flight testing on a Cessna Citation VII. The five active engines in the programme had accumulated over 1,000 h through 1992. Certification is due in early 1994.

TYPE: Two-shaft subsonic turbofan.
FAN: Single-stage, direct drive featuring wide chord, clapperless blades. Bypass ratio of 5.0. Blades replaceable on aircraft. Diameter 978 mm (38.5 in).
HP COMPRESSOR: Fourteen-stage axial flow, with variable inlet guide vanes and first five stator rows, all steel. Overall pressure ratio 23.
COMBUSTOR: Annular design with 16 airblast fuel nozzles. Dual capacitor-discharge ignition.
CONTROL SYSTEM: Full-authority digital (FADEC).
HP TURBINE: Two-stage axial with air-cooled single-crystal blading.
LP TURBINE: Three-stage axial uncooled design.
BYPASS DUCT: Single-piece full-length composite with provisions for thrust reverser.
DIMENSIONS:
Diameter overall	1,105 mm (43.5 in)
Length	2,705 mm (106.5 in)
WEIGHT, DRY (with bypass duct):	717.1 kg (1,581 lb)
PERFORMANCE RATINGS:	
S/L, T-O	32.03 kN (7,200 lb st) to 30°C
SPECIFIC FUEL CONSUMPTION:	
S/L, T-O	11.05 mg/Ns (0.39 lb/h/lb st)

Allison GMA 3007 turbofan

CFE

CFE COMPANY

111 South 34th Street, PO Box 62332, Phoenix, Arizona 85082-2332
Telephone: 1 (602) 231 3285
Fax: 1 (602) 231 7722

This company was formed jointly by Garrett Engine Division (now AlliedSignal Propulsion) and General Electric Aircraft Engines in June 1987. It is managing all phases of the development, manufacture, marketing and support of the CFE738 turbofan.

CFE738

This turbofan is being produced to power regional airliners and large business jet aircraft. The CFE738 is being designed to the latest airline standard technology, with modular construction for 'on wing' maintenance. Its core is essentially that of the GE27, developed under the US Army's MTDE (Modern Technology Demonstrator Engine) programme and part of the T407/GLC38 programme. Engine cores are shipped from GE to Phoenix complete with the engine control system. Garrett/AlliedSignal is responsible for the fan, LP turbine and accessory gearbox, and for engine assembly and test. Complete engine testing for certification programmes will be shared equally by the two partners. Deliveries of the first version, CFE738-1, were due in 1993 after 4,000 hours of testing.

In April 1990 the CFE738 was selected to power the Dassault Falcon 2000. The thrust was announced as 26.69 kN (6,000 lb).

The following data relate to the CFE738-1. Growth versions are planned with thrust ratings exceeding 31.1 kN (7,000 lb).

CFE/GARRETT—ENGINES: USA

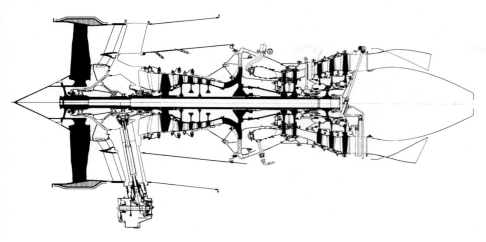

Longitudinal section through CFE738 two-shaft turbofan

TYPE: Two-shaft subsonic turbofan.
FAN: Single stage with 28 inserted titanium blades with part-span dampers and rotating pointed spinner. Front end of LP shaft held in large-capacity ball bearing. Mass flow 109 kg (240 lb)/s. Pressure ratio 1.7. Bypass ratio 5.3.
HP COMPRESSOR: Five axial stages followed by one centrifugal. First three stator stages variable. Overall pressure ratio 35.
COMBUSTION CHAMBER: Centrifugal diffuser leads into annular chamber with 15 fuel injectors.
HP TURBINE: Two stages with cooled blades.
LP TURBINE: Three stages.
JETPIPE: Fixed mixer assembly with 20 chutes for combining the hot and cold flows from core and bypass duct. Provision for reverser.
DIMENSIONS:
Diameter overall 902 mm (35.5 in)
Length 1,735 mm (68.3 in)
WEIGHT, DRY: 596.5 kg (1,315 lb)
PERFORMANCE RATINGS (ideal installation):
S/L, T-O 26.69 kN (6,000 lb st) to 30°C (86°F)
Cruise, 12,200 m (40,000 ft), Mach 0.8
6.07 kN (1,365 lb)
SPECIFIC FUEL CONSUMPTION:
S/L, T-O 10.54 mg/Ns (0.372 lb/h/lb st)
Cruise (as above) 18.27 mg/Ns (0.645 lb/h/lb)

DYNA-CAM

Details of the barrel-type axial-piston Dyna-Cam engine appeared in the 1992-93 *Jane's*.

GARRETT
ALLIEDSIGNAL PROPULSION ENGINE COMPANY
111 South 34th Street, PO Box 52181, Phoenix, Arizona 85072-2181
Telephone: 1 (602) 231 1000
Fax: 1 (602) 231 7722
PRESIDENT: James A. Robinson

AlliedSignal (Garrett) has been called the world's largest producer of small gas turbines. At the end of 1984 it announced its collaboration with Allison on the T800-LHT-800 turboshaft for the LH helicopter programme (now RAH-66 Comanche). This is mentioned under LHTEC.

The CFE738 turbofan, being developed jointly by Allied-Signal and General Electric, is to be found under CFE.

GARRETT ATF3
US military designation: F104-GA-100

The ATF3 was the first engine to combine three-spool design with a reverse-flow combustion system and turbines, and mixed-flow exhaust.

The arrangement of components allows the fan design to be determined largely independently of the gas generator, and permits operation at optimum fan speed. Omission of inlet guide vanes, mixing of the exhaust with the fan airflow, and double reversal of the airflow, enable the ATF3 to offer reductions in noise and IR signature.

In May 1976 it was announced that the **ATF3-6A** had been selected by Dassault to power the Falcon 200 business jet. FAA certification was achieved on 24 December 1981, and Falcon 200 deliveries began in 1983. It also powers the HU-25A Guardian.

TYPE: Three-shaft axial-flow turbofan.
LOW PRESSURE (FAN) SYSTEM: Single-stage titanium fan, driven by three-stage IP turbine. Bypass ratio 2.8 at take-off. Mass flow 73.5 kg (162 lb)/s.
INTERMEDIATE PRESSURE SYSTEM: Five-stage titanium compressor, driven by two-stage LP turbine. Airflow is delivered to rearward facing HP compressor via eight tubes feeding into annular duct. Core airflow 18.15 kg (40 lb)/s.
HIGH PRESSURE SYSTEM: Single titanium centrifugal compressor, driven by single-stage HP turbine. IP airflow enters the impeller from the rear. Overall pressure ratio (T-O) 21, (high altitude cruise) 25.
COMBUSTION SYSTEM: Reverse-flow annular type.
TURBINES: Single-stage HP, three-stage IP and two-stage LP turbines drive, respectively, the HP, fan (LP) and IP compressors. IP and LP turbines have shrouded blades. Air-cooled HP rotor blades. Exhaust gases turned 180° through eight sets of cascades to mix with fan bypass flow.
FUEL SYSTEM: Electromechanical, incorporating solid state computer. Manual emergency backup system.
ACCESSORY DRIVES: Three drive pads on rear-mounted gearbox driven by HP shaft, providing for hydraulic pump, starter/generator and one spare.
EXHAUST SYSTEM: Mixed fan and turbine exhaust discharged via annular nozzle surrounding combustion section.
LUBRICATION SYSTEM: Hot tank integral with gearbox.
STARTING: Electric or pneumatic.
DIMENSIONS:
Length 2,591 mm (102.0 in)
Max diameter 853 mm (33.6 in)
WEIGHT, DRY: 510 kg (1,125 lb)
PERFORMANCE RATINGS (uninstalled):
T-O (ISA, S/L, static) 24.20 kN (5,440 lb st)

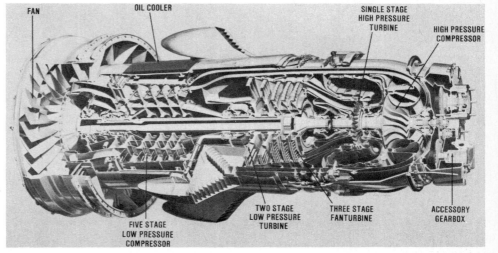

Cutaway drawing of Garrett ATF3-6A three-shaft turbofan

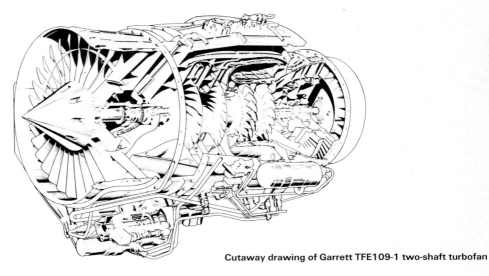

Cutaway drawing of Garrett TFE109-1 two-shaft turbofan

Cruise (12,200 m; 40,000 ft at Mach 0.8)
4.69 kN (1,055 lb)
SPECIFIC FUEL CONSUMPTION:
At T-O rating (S/L, ISA static)
14.33 mg/Ns (0.506 lb/h/lb st)
At cruise (as above) 23.51 mg/Ns (0.83 lb/h/lb)

GARRETT TFE109
US military designation: F109-GA-100

Based on the core of the T76/TPE331 turboprops, with performance improvements, this turbofan was selected in July 1982 as the engine of the US Air Force's Fairchild T-46A. This was subsequently cancelled, but the engine was certificated and production engines delivered.

TFE109-1. Basic engine, as described below. Powers Jet Squalus 1300.
TFE109-2. Rated (continuous) at 6.7 kN (1,500 lb st). Selected for twin-engined Promavia ATTA 3000.
TFE109-3. Rated (continuous) at 7.12 kN (1,600 lb st). Selected for Jet Squalus 1600.

The core of the TFE109 is used in the T800 helicopter engine, described under LHTEC.

TYPE: Two-shaft turbofan.
FAN: Single-stage, with 28 blades with part-span shrouds

(snubbers) inserted in self-de-icing rotating spinner.
COMPRESSOR: Tandem two-stage centrifugal with titanium impellers. LP and HP bleeds.
COMBUSTION CHAMBER: Annular reverse flow type, with piloted air blast nozzles for low emissions.
HP TURBINE: Two-stage axial, with single-pass cooling.
LP TURBINE: Two-stage axial, with tip shrouds.
CONTROL SYSTEM: Full-authority digital with hydromechanical backup.
ACCESSORIES: Mounted on one-piece gearbox, under intermediate case, with drive from HP shaft taken through one of five main aerofoil struts spaced at 72°.
DIMENSIONS:
Length 1,130 mm (44.495 in)
Diameter 597 mm (23.51 in)
WEIGHT, DRY: 199 kg (439 lb)
PERFORMANCE RATINGS (ISA):
T-O (unlimited) 5.92 kN (1,330 lb st)
Max continuous (9,145 m; 30,000 ft, Mach 0.5)
  1.78 kN (400 lb)
SPECIFIC FUEL CONSUMPTION:
Max T-O, as above 11.10 mg/Ns (0.392 lb/h/lb st)

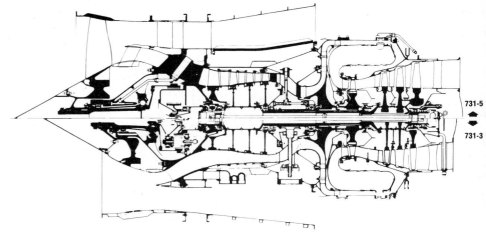

Longitudinal section of TFE731-5 (top half) and TFE731-3 (bottom half) geared turbofans

GARRETT TFE731

Announced in April 1969, the TFE731 is a two-spool geared turbofan designed for business jet aircraft. Use of a geared fan confers flexibility in operation and yields optimum performance at up to 15,545 m (51,000 ft).

TFE731-2. First production model. Deliveries for the Falcon 10 took place in August 1972.

TFE731-3. Increased turbine temperature. First delivered in 1974 for re-engined JetStar II. Also selected for the Learjet 54/56, Cessna Citation III, Dassault Falcon 50, BAe 125-700, IAI Westwind 1 and 2 and Astra, Rockwell International's Sabreliner 65A, CASA C-101 and Argentine IA 63.

TFE731-4. Powers Citation VII. Certificated late 1991.

TFE731-5. Higher bypass ratio fan driven by new LP turbine. Certificated in November 1983 for BAe 125-800 and CASA C-101.

TFE731-5A. Mixer nozzle reducing sfc and raising standard thrust to that of the APR rating. Certificated in December 1984 for Dassault Falcon 900, and offered as retrofit for Falcon 20. Volvo Flygmotor has a 5.6 per cent share of the 731-5 production programme.

TFE731-5B. Uprated version, certificated in 1990 for Falcon 900B and Falcon 20 retrofit.

TFE731-20. First of new series scheduled for certification 1995. Changes include smaller -5 fan, new HP compressor, HP turbine and gearbox; FADEC. Selected for Learjet 45.

TYPE: Turbofan with two shafts and geared front fan.
FAN: Single-stage axial titanium fan, with inserted blades. The fan shaft is connected directly to the planetary gearbox ring gear. Mass flow, sea level static, TFE731-2, 51.25 kg (113 lb)/s; -3, 53.7 kg (118.3 lb)/s; -5, 64.86 kg (143 lb)/s; -20, 55.8 kg (123 lb)/s. Bypass ratio, -2, 2.66; -3, 2.80; -5, 3.48; -20, 3.70.
COMPRESSOR: Four-stage LP, followed by centrifugal HP on separate shaft running at higher speeds. Overall pressure ratio (S/L, static): -2, 14.0; -3, 14.6; -20, 14.3.
COMBUSTION CHAMBER: Annular reverse flow type, with 12 nozzles injecting tangentially. Meets EPA/FAA emission requirements.
FUEL SYSTEM: Hydro-electronic, with single lever control; -20, FADEC, hydromechanical backup.
TURBINES: Single-stage HP and three-stage LP. Average HP inlet gas temperature, S/L, max T-O, -2, 860°C; -3, 907°C; -5, 952°C; -20, 941°C.
ACCESSORY DRIVES: Pads provided for hydraulic pump, starter/generator or starter motor and alternators. Pads on rear side of drive fuel control and oil pump.
DIMENSIONS:
Length overall: -2, -3 1,520 mm (59.83 in)
  -3A, -3B 1,440 mm (59.70 in)
  -4 1,477 mm (58.15 in)
  -5 1,665 mm (65.54 in)
  -5A, -5B 2,314 mm (91.10 in)
  -20 1,515 mm (59.65 in)
Intake diameter 716 mm (28.20 in)
Width: -2, -3, -3A, -3B, -4, -20 869 mm (34.20 in)
  -5, -5A, -5B 858 mm (33.79 in)
Height overall:
  -2, -3, -3A, -3B, -4, -20 1,000 mm (39.36 in)
  -5, -5A, -5B 1,029 mm (40.52 in)
WEIGHT, DRY:
  -2 337 kg (743 lb)
  -3 342 kg (754 lb)
  -3A 352 kg (775 lb)
  -3B 349 kg (769 lb)
  -4 373 kg (822 lb)
  -5 387 kg (852 lb)
  -5A 402 kg (884 lb)
  -5B 408 kg (899 lb)
  -20 379 kg (836 lb)
PERFORMANCE RATINGS:
Max T-O (S/L): -2 15.57 kN (3,500 lb st) to 22.2°C
  -3, -3A 16.46 kN (3,700 lb st) to 24.4°C
  -3B 16.24 kN (3,650 lb st) to 21.1°C
  -4 18.15 kN (4,080 lb st) to 24.4°C
  -5 19.15 kN (4,304 lb st) to 23°C
  -5A 20.02 kN (4,500 lb st) to 23°C
  -5B 21.13 kN (4,750 lb st) to 23°C
  -20 15.57 kN (3,500 lb st) to 33.9°C
Max T-O (APR, auto performance reserve):
  -3, -3A, -3B 17.13 kN (3,850 lb st)
  -5, -5A 20.02 kN (4,500 lb st)
  -4, -5B: as above
  -20 16.24 kN (3,650 lb st)
Cruise (12,200 m; 40,000 ft at Mach 0.8):
  -2 3.36 kN (755 lb)
  -3 3.64 kN (817 lb)
  -3A 3.73 kN (838 lb)
  -3B 3.75 kN (844 lb)
  -4 4.13 kN (929 lb)
  -5 4.25 kN (955 lb)
  -5A 4.39 kN (986 lb)
  -5B 4.68 kN (1,052 lb)
  -20 3.90 kN (876 lb)
SPECIFIC FUEL CONSUMPTION:
Cruise (as above): -2 23.08 mg/Ns (0.815 lb/h/lb)
  -3 23.65 mg/Ns (0.835 lb/h/lb)
  -3A 23.30 mg/Ns (0.823 lb/h/lb)
  -3B 23.11 mg/Ns (0.816 lb/h/lb)
  -4 22.55 mg/Ns (0.796 lb/h/lb)
  -5 22.72 mg/Ns (0.802 lb/h/lb)
  -5A 21.84 mg/Ns (0.771 lb/h/lb)
  -5B 21.41 mg/Ns (0.756 lb/h/lb)
  -20 20.62 mg/Ns (0.728 lb/h/lb)

GARRETT TFE1042-70

US military designations: Dry F124-GA-100, afterburning F125-GA-100

This engine was developed by the ITEC (International Turbine Engine Co) as the power plant of the AIDC Ching-kuo IDF (Indigenous Defensive Fighter). Fully modular, it has matured as a family of engines to be marketed by Allied-Signal. The basic engine, the 1042-70, had by 1993 completed over 13,000 hours of testing. Initial flight release was achieved in February 1989, with full qualification in September 1991; production deliveries began a month later.

The following are existing and planned versions:

TFE1042-70. Basic version, in production by ITEC for Ching-kuo IDF. Description below applies to this version.

F125-GA-100. Similar to TFE1042-70. Being marketed at 28.47 kN (6,400 lb st) intermediate thrust and 41.81 kN (9,400 lb st) maximum.

F125X. Growth version for 1996. Maximum rating 54.49 kN (12,250 lb st).

F125XX. Growth engine for 1999. Maximum thrust 71.17 kN (16,000 lb st).

F124-GA-100. A non-afterburning version of the F125, the F124 is being proposed as an alternative engine for the T-45A Goshawk. Designed as a 'drop in' replacement engine, the F124 has an engine mounted accessory gearbox (EMAG), and a hydromechanical tertiary control unit (TCU) as a backup to the dual FADEC. F124 engine development testing began in 1990. Airflow 42.7 kg (94.1 lb)/s, length 1,925 mm (75.8 in), dry weight 575 kg (1,268 lb), T-O rating 28.02 kN (6,300 lb st).

F124X. F124 growth engine for 1996. Maximum thrust 36.12 kN (8,120 lb st).

F124XX. F124 dry growth engine for 2000. Maximum thrust 51.16 kN (11,500 lb st).

Data for 1042-70:
TYPE: Two shaft augmented turbofan.
FAN: Three stages with rotating spinner. Maximum airflow 43.29 kg (95.4 lb)/s. Bypass ratio 0.3.
COMPRESSOR: Four axial stages followed by one centrifugal.
COMBUSTION CHAMBER: Annular.
TURBINES: Single-stage HP and LP turbine.
AUGMENTOR: Reheat in bypass and core flows. Three mechanical actuators drive 10-flap variable nozzle.
DIMENSIONS:
Inlet diameter 605 mm (23.8 in)
Length 2,880 mm (134.0 in)
Maximum nozzle diameter 782 mm (30.8 in)
WEIGHT, DRY: 603 kg (1,330 lb)
PERFORMANCE RATINGS:
Intermediate 26.80 kN (6,025 lb st)
Maximum augmented 41.15 kN (9,250 lb st)
SPECIFIC FUEL CONSUMPTION:
TFE1042-70 (intermediate) 23.79 mg/Ns (0.84 lb/h/lb st)

Garrett 1042-70 augmented turbofan

GARRETT TPE331

US military designation: T76

Based upon experience with APUs, this was the first Garrett engine for aircraft propulsion. By January 1993 deliveries of all versions had passed 13,000.

The following are major versions:

TPE331 series I, II. FAA certificated in February 1965. Rated at 451 ekW; 429 kW plus 0.33 kN (605 ehp; 575 shp plus 75 lb st). Redesignated **TPE331-25/61** and **-25/71** and produced until 1970. Powers MU-2 (A to E models), Porter, Jet Liner, Super Turbo 18, FU-24, Hawk Commander and 680, and Turbo Beaver.

TPE331-1 series. Certificated December 1967 at 526 ekW; 496 kW plus 0.44 kN (705 ehp; 665 shp plus 100 lb st). Powers MU-2 (F and G), Turbo-Porter and AU-23A Peacemaker, CJ600, Turboliner, Interceptor 400, Turbo Commander and (customer option) Thrush Commander, Merlin IIB and Fletcher 1284, Turbo Thrush and Turbo Ag-Cat.

TPE331-2 series. Certificated in December 1967 at 563 ekW; 533 kW plus 0.45 kN (755 ehp; 715 shp plus 102 lb st). Powers Skyvan, CASA 212 pre-series, Turbo Goose and Turbo Beaver.

TPE331-3 series. Certificated in March 1970 at 674 ekW; 626 kW plus 0.71 kN (904 ehp; 840 shp plus 159 lb st). Uprated gas generator with increased airflow and pressure ratio, but same turbine temperature as in original TPE331. Powers Merlin III, IV and Metro, and Jetstream III.

TPE331-5/6 series. The -5 was certificated in March 1970; this matches the gas generator of the -3 with the 715 shp gearbox, and is flat rated to 2,135 m (7,000 ft). Powers MU-2, King Air B100 (-6), CASA 212 (-5), Dornier 228 (-5), and Commander 840/900. The -5 designation indicates output speed of 1,591 rpm; the -6 has an output speed of 2,000 rpm.

TPE331-8. Matches compressor and gearbox of -3 with new turbine section. Thermodynamic power of 676 ekW; 645 kW (905 ehp; 865 shp) plus 0.47 kN (105 lb st), but flat rated at 533 kW (715 shp) to 36°C. Certification was received in November 1976. Powers Conquest II.

TPE331-9. Thermodynamic rating 645 kW (865 shp).

TPE331-10. Rated at 746 kW (1,000 shp). Certificated January 1978. Powers Marquise and Solitaire, Commander 980/1000, Merlin IIIC, CASA 212-200 and Jetstream 31.

TPE331-11. Certificated 1979. Higher gearbox limit; wet rating 820 kW (1,100 shp). Powers Metro III.

TPE331-12. Same size as -10 but offers 834 kW (1,100 shp). Certificated December 1984. Powers Jetstream Super 31 and Metro 23. The **TPE331-12B** powers the Shorts Tucano. Rolls-Royce makes 30 per cent by value of engines for Shorts, and supports RAF engines in service.

TPE331-14/15. Scaled-up models, with thermodynamic power in the 1,227 kW (1,645 shp) class. The -14 was certificated in April 1984 and is flat rated at 746 kW (1,000 shp) for the Cheyenne 400. TPE331-15AW powers re-engined S-2 Tracker.

TPE331-14GR/HR (clockwise/anti-clockwise) handed engine, flat rated at 1,462 kW (1,960 shp). Powers Jetstream 41 and candidate for An-38.

T76. Military engine, with gas generator similar to TPE331-1 but with front end inverted, to give inlet above spinner. All models power OV-10 Bronco.

Except for the TPE331-14/15, all versions are of similar frame size, and the following data apply generally to all:

TYPE: Single-shaft turboprop.
PROPELLER DRIVE: Two-stage reduction gear, one helical spur and one planetary, with overall ratio of 20.865 or 26.3.
COMPRESSOR: Tandem two-stage centrifugal made from titanium. Mass flow, 2.61 kg (5.78 lb)/s for -25/61, -25/71; 2.81 kg (6.2 lb)/s for -1; 2.80 kg (6.17 lb)/s for -2 and T76; 3.52 kg (7.75 lb)/s for -5; and 3.54 kg (7.8 lb)/s for -3. Pressure ratio 8.0 for -25/61, -25/71; 8.34 for -1; 8.54 for -2 and T76; 10.37 for -5 and -3.
COMBUSTION CHAMBER: Annular, with capacitor discharge igniter plug on turbine plenum.
FUEL SYSTEM: Woodward or Bendix control. Max fuel pressure 41.4 bars (600 lb/sq in).
FUEL GRADE (TPE331): ASTM designation D1655-64T types Jet A, Jet B and Jet A-1; MIL-F-5616-1, Grade JP-1.
TURBINE: Three-stage axial. In early models, blades cast integrally with disc. In -10, -11, -12 first-stage disc with inserted blades. In -14/15 inserted blades in all three stages. Inlet gas temperature, 987°C for -25/61, -25/71; 993°C for T76, 1,005°C for all other models.
ACCESSORIES: AND 20005 Type XV-B tachometer generator, AND 20002 Type XII-D starter/generator, AND 20010 Type XX-A propeller governor and AND 20001 Type XI-B hydraulic pump.
LUBRICATION SYSTEM: Medium pressure dry sump system. Normal oil supply pressure 6.90 bars (100 lb/sq in).
OIL SPECIFICATION: MIL-L-23699-(1) or MIL-L-7808.
STARTING: Pad for 399A starter/generator.
DIMENSIONS (approx):
Length overall:
TPE331   1,092 to 1,333 mm (43-52.5 in)

Cutaway drawing of Garrett TPE331-14 turboprop

T76	1,118 mm (44 in)
Width: TPE331	533 mm (21 in)
T76	483 mm (19 in)
Height: TPE331	660 mm (26 in)
T76	686 mm (27 in)

WEIGHT, DRY:
TPE331-25/61, 71	152 kg (335 lb)
TPE331-1, -2	152.5 kg (336 lb)
T76	155 kg (341 lb)
TPE331-3	161 kg (355 lb)
TPE331-5	163 kg (360 lb)
TPE331-8	168 kg (370 lb)
TPE331-10	172 kg (380 lb)
TPE331-11	182 kg (400 lb)
TPE331-12	176 kg (387 lb)
TPE331-14/-14GR/-15	256 kg (565 lb)

PERFORMANCE RATINGS:
T-O   see under model listings
Military (30 min): T76-G-410/411
         533 kW; 563 ekW (715 shp; 755 ehp)
Normal: T76-G-410/411
         485 kW; 514.5 ekW (650 shp; 690 ehp)
Max cruise (ISA, 3,050 m; 10,000 ft and 250 knots; 463 km/h; 288 mph):
TPE331-25/61, 71	332 kW (445 shp)
TPE331-1	404 kW (542 shp)
TPE331-2, T76	430 kW (577 shp)
TPE331-3, -5	530 kW (710 shp)

SPECIFIC FUEL CONSUMPTION:
At T-O rating:
TPE331-25/61, 71	111.5 µg/J (0.66 lb/h/shp)
TPE331-1	102.2 µg/J (0.605 lb/h/shp)
TPE331-2	99.4 µg/J (0.588 lb/h/shp)
TPE331-3	99.7 µg/J (0.59 lb/h/shp)
TPE331-5	105.8 µg/J (0.626 lb/h/shp)
TPE331-8	96.7 µg/J (0.572 lb/h/shp)
TPE331-10	94.6 µg/J (0.560 lb/h/shp)
TPE331-11	94.3 µg/J (0.558 lb/h/shp)
TPE331-12	92.8 µg/J (0.549 lb/h/shp)
TPE331-14/-15	84.8 µg/J (0.502 lb/h/shp)
T76-G-410/411	101.4 µg/J (0.60 lb/h/shp)

OIL CONSUMPTION:
Max   0.009 kg (0.02 lb)/h

GARRETT TPF351

This is Garrett's first free turbine turboprop. It is being developed initially as the **TPF351-20** for pusher installation in the Embraer/FAMA CBA-123. The engine is modular, and arranged for clockwise (CW) or counterclockwise (CCW) propeller rotation (both forms of gearbox are shown in the drawing).

TYPE: Free turbine pusher turboprop.
AIR INLET: Ram inlet at front of nacelle with duct passing above accessory gearbox.
COMPRESSOR: Tandem two-stage centrifugal. Mass flow 6.35 kg (14.0 lb)/s. Pressure ratio 13.3.
COMBUSTION CHAMBER: Annular reverse flow with dual-channel ignition system.
FUEL SYSTEM: FADEC with mechanical backup fuel control unit and integral electrical shutoff valve.
TURBINES: Two-stage gas generator turbine. Three-stage power turbine; inter-turbine temperature 805°C (1,481°F); EGT 571°C (1,061°F).
JET PIPE: Bifurcated with diagonal outlet on each side (depicted in drawing as above and below).
PROPELLER DRIVE: Epicyclic with drive to auxiliary gearbox.
DIMENSIONS:
Length overall	1,954 mm (76.94 in)
Width	606 mm (23.84 in)
Height	838 mm (33.0 in)

WEIGHT, DRY:
CW	340.2 kg (750.1 lb)
CCW	347.0 kg (765.1 lb)

PERFORMANCE RATING (ISA, S/L, T-O):
Thermodynamic   1,566 kW (2,100 shp), torque-limited
                                           to 1,081 kW (1,450 shp)
SPECIFIC FUEL CONSUMPTION (T-O):
                         83.66 µg/J (0.495 lb/h/shp)

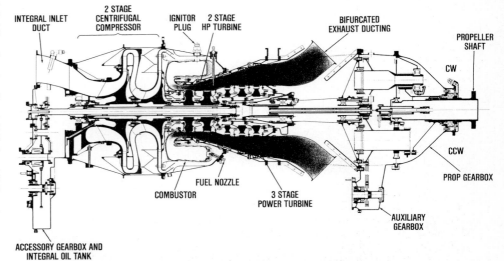

Longitudinal section of Garrett TPF351-20 free turbine pusher turboprop

GENERAL ELECTRIC
GE AIRCRAFT ENGINES
One Neumann Way, Evendale, Ohio 45215-6301
Telephone: 1 (513) 243 6136
Fax: 1 (513) 786 1568
Telex: 212078 GEAEG UR
PRESIDENT AND CEO: B. H. Rowe

Current products of GE Aircraft Engines include the F103, F110, F118, F404, F414, T64, T700 and TF34 for military use, and the GE90, CF6, CF34, CT7 and GLC38 for the commercial and general aviation market. A new turbofan is being developed in partnership with AlliedSignal (Garrett) and appears under CFE. In partnership with SNECMA of France a company was formed to develop and market the CFM56 turbofan, as described in the International part of this section under CFM International.

Details of the CF700, CJ610, F101, J79, J85, T58 and TF39 engines, which are no longer in production, can be found in earlier editions of *Jane's*.

General Electric CF34-3A1 two-shaft turbofan

GENERAL ELECTRIC F404
In May 1975 the US Navy selected the McDonnell Douglas/Northrop team to develop its F/A-18 Hornet, powered by two F404-GE-400 engines.

First F404 engine test took place in December 1976. Preliminary flight rating test took place in May 1978, first F/A-18 flight in November 1978 and MQT (model qualification test) in June 1979. The first production delivery took place in December 1979.

The following are current versions of the F404:

F404-GE-400. Original production engine for all F/A-18 models. Also powers Grumman X-29 and Rockwell/MBB X-31. Max rating 71.2 kN (16,000 lb st).

F404-GE-402. Enhanced Performance Engine (EPE) in the 79 kN (17,700 lb st) class; combines a 2,000-h hot section with up to 25 per cent increase in thrust, achieved through higher temperature, increased fan speed and improved design and materials in the turbines and afterburner. Delivery 1991 (F/A-18 for Kuwait) and 1992 (US Navy).

F404-GE-100D. A 49 kN (11,000 lb st) non-afterburning derivative of F404-GE-400 with modified control system for Singapore Air Force A-4SU Super Skyhawk.

F404-GE-F1D2. A non-afterburning turbojet derivative of the F404-GE-400 powering the US Air Force F-117A.

F404/RM12. An 80 kN (18,100 lb st) F404 derivative. Redesigned fan handles 10 per cent greater airflow and has increased resistance to foreign-object damage. Increased turbine inlet temperature provides operating margin for lower right hand corner of flight envelope. Modified control system for JAS 39 Gripen. Interim engine for Indian Light Combat Aircraft and proposed for re-engining Chinese F-7 and J-8. Candidate for former Yugoslav Novi Avion and other programmes. See Volvo Flygmotor, Sweden.

F404-GE-400D. Unaugmented version of F404-GE-400 rated at 48.0 kN (10,800 lb st). Powered Grumman A-6F prototype.

The following description applies to the F404-GE-402:
TYPE: Two-shaft augmented low bypass ratio turbofan (turbojet with continuous bypass bleed).
FAN: Three-stage. Bypass ratio 0.27. Mass flow 66.2 kg (146 lb)/s.
HP COMPRESSOR: Seven-stage. Overall pressure ratio, 26 : 1 class.
COMBUSTION CHAMBER: Single-piece annular.
HP TURBINE: Single-stage with air-cooled blades.
LP TURBINE: Single-stage.
EXHAUST SYSTEM: Close coupled afterburner. Convergent-divergent exhaust nozzle with hydraulic actuation.
CONTROL SYSTEM: Electrical-hydromechanical.
DIMENSIONS:
Length overall 4,030 mm (158.8 in)
Max diameter 880 mm (34.8 in)

WEIGHT, DRY: 1,016 kg (2,240 lb)
PERFORMANCE RATING:
Max T-O 79 kN (17,700 lb st)

GENERAL ELECTRIC F414
The F414 is the engine for the US Navy's F/A-18E/F Hornet multi-mission aircraft. Derived from the F404, the **F414-GE-400** is rated at 97.86 kN (22,000 lb st). Production qualification is planned for 1996, and shipments are due in 1998.

GENERAL ELECTRIC ASTOVL
Using the F120 variable-geometry fighter engine (see 1990-91 *Jane's*) as a basis, General Electric in partnership with Rolls-Royce is developing a candidate engine for the McDonnell Douglas Advanced STOVL combat aircraft. It will have a shaft drive to a separate lift fan.

GENERAL ELECTRIC TF34
This turbofan won a 1965 US Navy competition. In August 1972 the **TF34-GE-2**, the initial variant for the Lockheed S-3A Viking, completed its model qualification test (MQT) and entered fleet service in February 1974. In January 1975 GE began shipment of the **TF34-GE-400A/B**, which replaced the GE-2 as S-3A engine. In 1970 the TF34 was selected to power the A-10A attack aircraft as the **TF34-GE-100**, with a long fan duct and side mountings. In 1974 a third version was selected to provide auxiliary (thrust) power for the Sikorsky S-72 research aircraft.
TYPE: Two-shaft high bypass ratio turbofan.
FAN: Single-stage fan has blades forged in titanium. Mass flow (TF34-400A/B) 153 kg (338 lb)/s at 7,365 rpm with pressure ratio 1.5. Bypass ratio 6.2.
COMPRESSOR: 14-stage axial on HP shaft. Inlet guide vanes and first five stators variable. First nine rotor stages titanium, remainder high nickel alloy. Performance at max S/L rating, core airflow 21.3 kg (47 lb)/s at 17,900 rpm with pressure ratio 14 : 1, overall pressure ratio 21.
COMBUSTION CHAMBER: Annular Hastelloy liner and front dome, with 18 carburetting burners.
TURBINE: Two-stage HP; four-stage LP with tip-shrouded blades. EGT 1,225°C maximum.
FUEL SYSTEM: Hydromechanical with electronic amplifier. Fuel grade JP-4 or JP-5.
DIMENSIONS:
Max diameter:
TF34-GE-400A/B 1,321 mm (52.0 in)
TF34-GE-100 1,245 mm (49.0 in)
Basic length (both) 2,540 mm (100.0 in)
WEIGHT, DRY:
TF34-GE-400A/B 670 kg (1,478 lb)
TF34-GE-100 653 kg (1,440 lb)
PERFORMANCE RATING (max T-O; S/L static):
TF34-GE-400A/B 41.3 kN (9,275 lb st)
TF34-GE-100 40.3 kN (9,065 lb st)
SPECIFIC FUEL CONSUMPTION (max T-O, S/L static):
TF34-GE-400A/B 10.3 mg/Ns (0.363 lb/h/lb st)
TF34-GE-100 10.5 mg/Ns (0.370 lb/h/lb st)

GENERAL ELECTRIC CF34
The CF34 is a commercial adaptation of the TF34. Total airflow at take-off power with automatic power reserve (APR) is 151 kg (332 lb)/s. Bypass ratio is 6.3.
CF34-1A. Certificated August 1982. Powers Challenger 601-1A. Rated at 38.48 kN (8,650 lb st).
CF34-3A. Powers Challenger 601-3A. Rated at 40.66 kN (9,140 lb st) with APR, or 38.48 kN (8,650 lb st) without.
CF34-3A1. Powers Canadair RJ. Data below for this:
DIMENSIONS:
Length overall 2,616 mm (103.0 in)
Max diameter (at mounts) 1,245 mm (49.0 in)
PERFORMANCE RATINGS (S/L, static):
T-O (APR) 41.0 kN (9,220 lb st)
T-O (Normal) 38.8 kN (8,729 lb st)
SPECIFIC FUEL CONSUMPTION (as above):
T-O (Normal) 10.11 mg/Ns (0.357 lb/h/lb st)

GENERAL ELECTRIC GE90
On 16 January 1990 GE announced that it was developing a high-bypass turbofan with thrust in the range 333-444 kN (75,000-100,000 lb st) capable of powering all new and derivative widebody aircraft that may enter the market in the mid-1990s. The first GE90 to run reached the record thrust of 468 kN (105,400 lb st) on 3 April 1993.

The GE90 is scheduled for certification at 387 kN (87,000 lb st) in May 1994, and entry into service at 350.1 kN (78,700 lb st) on the Boeing 777 in third quarter 1995.

This new design is based on a compressor scaled directly from GE's Energy Efficient Engine programme, a joint GE/NASA-funded project aimed at establishing a technology base for engines of the 1990s. It features the largest fan yet built, with composite wide-chord blades.

The large fan diameter increases the bypass ratio to 9, higher than that of any engine yet developed. The higher bypass ratio will help achieve a 10 per cent improvement in specific fuel consumption compared to today's large turbofan engines, while providing reduced noise levels.

Improvements to the combustor design will result in significantly improved emissions (oxides of nitrogen).

The engine features a fan case mount with an integrally stiffened system that bypasses thrust loads around the short, stiff core. This system, in conjunction with the straddle-mounted core, will deliver improved performance retention and reduced airline operating costs.

The compressor permits a shorter, lighter, stiffer engine with a high overall cycle pressure ratio that also contributes to

General Electric F404-GE-100D unaugmented turbofan

GENERAL ELECTRIC—ENGINES: USA 667

General Electric GE90 engineering mockup

compressor variable bypass doors. CF6-80, electronic trimming.
FUEL GRADES: Fuels conforming to ASTM-1655-65T, Jet A, Jet A1 and Jet B, and MIL-T-5624G2 grades JP-4 or JP-5 are authorised, but Jet A is primary specification.
LUBRICATION SYSTEM: Dry sump centre-vented nominal pressure is 2.07-6.21 bars (30-90 lb/sq in).
STARTING: Air turbine starter mounted on the front of the accessory gearbox at the through shaft.
NOISE SUPPRESSION: Acoustic panels integrated with fan casing, fan front frame and thrust reverser.
DIMENSIONS:
  Max height (over gearbox)   2,675 mm (105.3 in)
  Length overall (cold):
    CF6-50 series   4,394 mm (173.0 in)
    CF6-80A   3,998 mm (157.4 in)
WEIGHT, DRY:
  Basic engine:
    CF6-50C2   3,960 kg (8,731 lb)
    CF6-50E, -E1   3,851 kg (8,490 lb)
    CF6-50E2   3,977 kg (8,768 lb)
    CF6-80A, -80A2   3,854 kg (8,496 lb)
    CF6-80A3   3,819 kg (8,420 lb)
  Reverser:
    CF6-50E   962 kg (2,121 lb)
PERFORMANCE RATINGS:
  Max T-O, uninstalled, ideal nozzle: See under model listings
  Max cruise at 10,670 m (35,000 ft), Mach 0.85, flat rated to ISA + 10°C, uninstalled, real nozzle:
    CF6-50C2, -E2   50.3 kN (11,300 lb)
    CF6-80A, -80A1   45.9 kN (10,320 lb)
SPECIFIC FUEL CONSUMPTION:
  At T-O thrust, as above:
    CF6-50E   10.65 mg/Ns (0.376 lb/h/lb st)
    CF6-50C2, -E2   10.51 mg/Ns (0.371 lb/h/lb st)
    CF6-80A   9.74 mg/Ns (0.344 lb/h/lb st)
OIL CONSUMPTION: 0.9 kg (2.0 lb)/h

the significantly improved specific fuel consumption. The double-dome combustor was designed specifically for low NO_x emissions. Under low thrust, the outer combustor nozzles are used; as thrust is increased, the inner set of main nozzles begins operation. The high-pressure turbine features single-crystal blades, and the turbine discs are boltless, with smooth side plates for low windage losses.

GE's partners include SNECMA, FiatAvio and IHI. Among the GE90's features are:
FAN: Low speed, low pressure ratio with bypass ratio of approximately 9; composite, unshrouded wide-chord rotor blades; structural outlet guide vanes. Blade/vane quantities and spacing optimised for low acoustics.
LP COMPRESSOR: Three stages; low speed; moderate pressure ratio; low noise features.
HP COMPRESSOR: Ten stages; pressure ratio 23; derived from GE/NASA Energy Efficient Engine (E^3) technology demonstrator. Overall pressure ratio over 45.
COMBUSTOR: Double annular design for low NO_x emissions.
HP TURBINE: Two stages with monocrystal blades and powdered metal discs.
LP TURBINE: Six stages with relatively low stage loading for improved efficiency, low noise features.
FUEL SYSTEM: FADEC control of fuel flow, variable geometry and active clearance control systems.
DIMENSIONS:
  Fan diameter   3,124 mm (123 in)
  Length, flange to flange   4,902 mm (193 in)

GENERAL ELECTRIC CF6
US military designation (CF6-50E): F103-GE-100

On 11 September 1967 General Electric announced the commitment of corporate funding for development of the CF6 turbofan for wide-body transports. The CF6-6D for the DC-10-10 was selected by United and American on 25 April 1968. Details of this and other early versions appeared in the 1987-88 Jane's. The following are current versions:

CF6-50C2/E2. Rated 233.5 kN (52,500 lb st) to 30°C. Certification 1978. Military -50C2 powers KC-10; -50E2 powers Boeing E-4.

CF6-80A/A3. Improved sfc and performance retention, with length and weight reduced by elimination of turbine mid frame and reduction in combustor and diffuser length. Engine rated at 213.5 kN (48,000 lb st) as the -80A/A1, and 222.4 kN (50,000 lb st) as the -80A2/A3. Fitted to 767 and A310. Programme launched November 1977, first engine ran October 1979 and certification October 1981. Production split between GE and SNECMA. The -80A and -80C2 on the 767 were first to receive FAA approval (January 1989) for 180 min EROPS operations.

CF6-80C2. Described separately.

The following data relate to the CF6-50C2/E2, with -80 differences noted:
TYPE: Two-shaft high bypass ratio turbofan.
FAN: Single-stage with three-stage LP compressor both driven by LP turbine. The 38 fan rotor blades have anti-vibration shrouds at two-thirds span. Blades, discs, spool of titanium; exit guide vanes of aluminium; fan frame and shaft of steel; spinner and fan case of aluminium alloy. Total airflow 591 kg (1,303 lb)/s, bypass ratio 5.7. CF6-80A/A1 has better efficiency and birdstrike resistance, with Kevlar containment in fan case. Fan diameter 2,195 mm (86.4 in). Mass flow, CF6-80A/A1, 651 kg (1,433 lb)/s; -80A2/A3, 663 kg (1,460 lb)/s. Bypass ratio, -80A/A1, 4.7; -80A2/A3, 4.6.
LP COMPRESSOR: Three core booster stages; 12 bypass doors maintain flow matching between fan/LP and core by opening at low power settings, closed during take-off and cruise.
HP COMPRESSOR: Fourteen-stage with inlet guide vanes and first six stator rows having variable incidence. Core airflow 125 kg (276 lb)/s. CF6-80A series incorporate bore cooling for blade/casing clearance control, and one-piece steel casing with insulated aft stages and short diffuser section. Overall pressure ratio (T-O), 29.13 (-50C), 30.1 (-50E), 28.0 (-80A/A1), 29.0 (-80A2/A3).
COMBUSTOR: Fully annular. CF6-80A has rolled ring combustor, 152 mm (6.0 in) shorter, mounted at aft flange.
HP TURBINE: Two-stage air-cooled, TET 1,330°C. CF6-80A has no turbine mid-frame and eliminates one main bearing, and HP case has active clearance control.
LP TURBINE: Four-stage constant tip diameter with nominal 871°C inlet temperature. Rotor blades tip-shrouded and not air-cooled. CF6-80A, new turbine with active clearance control.
THRUST REVERSER (FAN): Rear portion of fan outer cowl translates aft on rotating ballscrews to uncover cascade vanes. Blocker doors (16) flush-mounted in cowl on link arms hinged in inner cowl, rotate inwards to expose cascade vanes and block fan duct. CF6-80A1/A3 (A310) similar; 767 reverser by Boeing.
ACCESSORY DRIVE: Inlet gearbox in forward sump transfers energy from the core. Transfer gearbox on bottom of fan frame with starter, fuel pump, main engine control, lubrication pump and tachometer. Pads for aircraft hydraulic pumps, constant speed drive and alternator. CF6-80A gearbox in environmental enclosure on core; -80A1 on fan case.
FUEL SYSTEM: Hydromechanical, schedules acceleration and deceleration fuel flow, variable stator vane position and LP

GENERAL ELECTRIC CF6-80C2

This engine is a major redesign for higher thrust and improved sfc, based on the CF6-80A1/A3 but with a 2,362 mm (93 in) diameter fan. It has a four-stage LP compressor and LP turbine redesigned aerodynamically with 5½ stages. The first CF6-80C2 ran in May 1982, and exceeded 276 kN (62,000 lb st) corrected thrust. Flight test on an A300 took place between August and December 1984, leading to certification on 28 June 1985. The engine entered revenue service on 5 October 1985. Programme sharing agreements have been signed with SNECMA of France, MTU of Germany, Volvo Flygmotor of Sweden and FiatAvio of Italy. Applications are shown in table on this page.

The CF6-80C2 differs from earlier CF6 engines in the following features:
FAN: Single-stage, with integrally mounted four-stage booster (LP compressor). Mainly titanium except for steel mid-fan shaft, aluminium spinner and blade-containment shroud of layers of Kevlar around aluminium case. Eighty composite exit guide vanes canted for better aerodynamic efficiency. Mass flow 802 kg (1,769 lb)/s; bypass ratio 5.05.
LP COMPRESSOR: Four stages with blades and vanes mounted orthogonally, with dovetail offset from centre of pressure to reduce bending.
HP COMPRESSOR: 14-stage, with inlet guide vanes and first five stator rows with variable incidence. Blades in stages 1-5 titanium, 6-14 steel; vanes all steel. One-piece steel casing with insulated aft stages. Core airflow 154 kg (340 lb)/s. Overall pressure ratio 30.4.
COMBUSTOR: Annular, rolled ring construction, aft-mounted with film cooling.
HP TURBINE: Two-stage. Stage one blades directionally solidified. Casing with active and passive clearance control. No midframe.
LP TURBINE: Five stages, with cambered struts in rear frame to reduce exit swirl, effectively producing another half-stage. Rear hub heated by exhaust gas to reduce thermal stress.

CF6-80C2 and -80E1 MODELS

Model	Ideal nozzle	Real nozzle	Application
		Thrust	
CF6-80C2A2	238 kN (53,500 lb)	233.5 kN (52,460 lb) to 43.9°C	A310-200, -300
CF6-80C2A3	268 kN (60,200 lb)	262 kN (58,950 lb) to 30°C	A300-600
CF6-80C2A5	272.5 kN (61,300 lb)	267.5 kN (60,100 lb) to 30°C	A300-600R
CF6-80C2A8	262.4 kN (59,000 lb)	257.4 kN (57,860 lb) to 35°C	A310-300
CF6-80C2B1	252 kN (56,700 lb)	247.6 kN (55,670 lb) to 30°C	747-200, -300
CF6-80C2B1F	257.6 kN (57,900 lb)	253.1 kN (56,900 lb) to 32.2°C	747-400
CF6-80C2B2	233.5 kN (52,500 lb)	229.4 kN (51,570 lb) to 32.2°C	767-200ER, -300ER
CF6-80C2B2F	234.5 kN (52,700 lb)	229.8 kN (51,650 lb) to 32.2°C	767-300ER
CF6-80C2B4	257.5 kN (57,900 lb)	252.9 kN (56,850 lb) to 32.2°C	767-200ER, -300ER
CF6-80C2B4F	258 kN (58,100 lb)	253.3 kN (56,950 lb) to 32.2°C	767-300ER
CF6-80C2B6	270 kN (60,800 lb)	265.6 kN (59,700 lb) to 30°C	767-300ER
CF6-80C2B6F, B7F	270 kN (60,800 lb)	265.8 kN (59,750 lb) to 30°C	767-300ER
CF6-80C2D1F	273.5 kN (61,500 lb)	271.2 kN (60,960 lb) to 30°C	MD-11
CF6-80E1A2	300.3 kN (67,500 lb)	286.9 kN (64,500 lb) to 30°C	A330
CF6-80E1A3	320.3 kN (72,000 lb)	303.6 kN (68,250 lb) to 30°C	A330

USA: ENGINES—GENERAL ELECTRIC

General Electric CF6-80C2 two-shaft turbofan

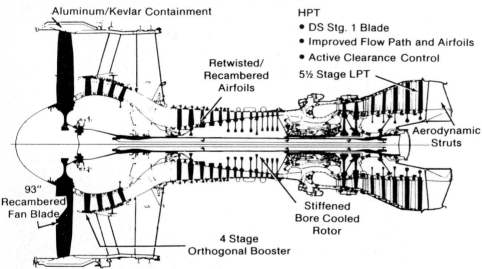
Features of General Electric CF6-80C2 two-shaft turbofan

General Electric F110-GE-129 two-shaft augmented turbofan

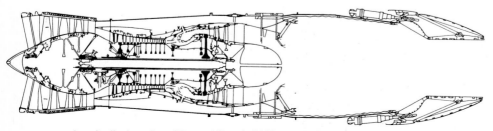
Longitudinal section of General Electric F110 two-shaft augmented turbofan

FUEL SYSTEM: FADEC or hydromechanical fan speed control with electronic supervision; one throttle position corresponds to each engine rating in all flight conditions.
DIMENSION:
Length 4,087 mm (160.9 in)
WEIGHT, DRY: 4,144 kg (9,135 lb)
PERFORMANCE RATINGS (uninstalled, ideal nozzle):
Max T-O See table
Max cruise (10,670 m; 35,000 ft, Mach 0.85)
  50.4 kN (11,330 lb)
SPECIFIC FUEL CONSUMPTION:
T-O, as above 9.32 mg/Ns (0.329 lb/h/lb st)

GENERAL ELECTRIC CF6-80E1

This engine is a major redesign for higher thrust, based on the CF6-80C2 but with a 2,438 mm (96 in) diameter fan. It has a four-stage LP compressor and LP turbine redesigned aerodynamically with 5½ stages. The first CF6-80E1 ran in December 1990 at over 320.3 kN (72,000 lb st) corrected thrust. Flight tested on an A300 early 1992; certification was due March 1993. The engine was to enter revenue service in 1993 on the A330.

The CF6-80E1 development is being shared with SNECMA of France, MTU of Germany, Volvo Flygmotor of Sweden, and FiatAvio of Italy.

The CF6-80E1 differs from the CF6-80C2 in the following features:
FAN: Diameter increased; number of blades reduced from 38 to 34.
LP COMPRESSOR: Flow capacity increased 9 per cent at a 12 per cent increased pressure ratio.
HP COMPRESSOR: High-temperature alloys in last stage. Overall pressure ratio: E1A2, 32.6; E1A3, 34.6.
COMBUSTOR: No change.
HP TURBINE: High-temperature alloys and improved cooling.
LP TURBINE: High-temperature alloys, improved cooling changes and aerodynamic changes.
FUEL SYSTEM: Surface-mounted FADEC; a higher-capacity system with an externally valved staging system (50 per cent of nozzles can be turned off) for low thrust operability.
DIMENSIONS:
Length: Engine 4,405 mm (173.5 in)
  Propulsion system 7,356 mm (289.6 in)
WEIGHT, DRY:
Engine 4,820 kg (10,627 lb)
Propulsion system 6,556 kg (14,453 lb)
PERFORMANCE RATINGS: See table
SPECIFIC FUEL CONSUMPTION (Max T-O):
E1A2 9.26 mg/Ns (0.327 lb/h/lb st)
E1A3 9.63 mg/Ns (0.340 lb/h/lb st)

GENERAL ELECTRIC F110

The **F110** (previously F101 DFE) is a fighter engine derivative of the F101. The first ran in late 1979. In early 1984 the USAF selected the F110 to power future F-16 aircraft.

The following are current versions of the F110:
F110-GE-100. Initial USAF engine, also selected by Israel (100 engines) and Turkey (177 engines). Other orders have been received from Bahrain, Egypt and Greece. Delivery of production F-16C/D aircraft with F110 engines began in mid-1986. The US Navy selected the Dash-100 to power its F-16Ns for the adversary role in its Top Gun programme.

F110-GE-400. Powers F-14A (Plus) and F-14D Tomcat for the US Navy. First production aircraft with this engine delivered to Navy in November 1987. Tomcats powered with this engine show a significant improvement in fuel consumption and the ability to catapult launch without use of an afterburner, resulting in a 61 per cent time-to-climb reduction and a 62 per cent improvement in mission range. Ratings are: maximum 120.2 kN (27,000 lb st), dry 71.2 kN (16,000 lb st).

F110-GE-129. Rated at 129 kN (29,000 lb st), the GE IPE (Improved Performance Engine) is the successor to the F110-GE-100. Through the use of improved design and materials, higher operating temperatures, speeds and pressures, the GE IPE increases thrust levels by as much as 30 per cent in certain areas of the flight envelope, while retaining more than 80 per cent parts commonality. The IPE's digital electronic control has 50 per cent fewer parts than previous controls, and offers substantially improved reliability. The IPE first flew in an F-16C/D in August 1988, and a USAF field service evaluation programme began in the latter half of 1990. Service entry was achieved in 1991. The engine has also been ordered by Turkey and selected to power Japan's FS-X fighter.

The following refers to the F110-GE-129:
FAN: Three stages. Bypass ratio 0.76. Airflow 122 kg (270 lb)/s.
HP COMPRESSOR: Nine stages. Overall pressure ratio, 31 : 1 class.
COMBUSTOR: Annular scroll.
HP TURBINE: Single stage with air-cooled blades.
LP TURBINE: Two stages, drives fan.
AUGMENTOR: Close coupled mixed-flow linear thrust afterburner.
EXHAUST NOZZLE: Convergent/divergent exhaust nozzle with hydraulic actuation.
DIMENSIONS:
Length 4,626 mm (182.3 in)
Diameter 1,180 mm (46.5 in)
PERFORMANCE RATINGS (S/L):
T-O 129 kN (29,000 lb st)
Max dry 75.7 kN (17,000 lb)

GENERAL ELECTRIC F118

This unaugmented turbofan has been developed under USAF contract to meet the demanding propulsion requirements of the Northrop B-2A bomber. An 84.52 kN (19,000 lb) class derivative of the F101 and F110 engines, the non-afterburning **F118-GE-100** employs new long-chord fan technology with the compressor and turbine used on the F110. The F118 has a higher airflow capacity and higher air pressure ratio than the F110, resulting in higher thrust. The engine was fully qualified in 1987, has successfully powered the B-2A in the USAF flight test programme since July 1989, and has accumulated more than 6,000 test hours to date. A feature is the use of common F101/F110 production tooling for low cost.

The accompanying photograph shows the actuation ring for the trailing flaps of the inlet guide vanes, a large inlet bleed-air de-icing pipe, and the ribbed bypass duct. The cooled jets issue across what appear to be carbon areas above the B-2A wing.

In Autumn 1989 a Lockheed U-2 began a flight test programme powered by a derivative of the F118, designated **F118-GE-101**. This engine is lighter and shorter but delivers more dry thrust than the J75-P-13B normally fitted.

General Electric F118-GE-100 two-shaft unaugmented turbofan

GENERAL ELECTRIC GLC38

The GLC38 is a member of the T407/GLC38/CFE738 engine family, which includes turboprop, turboshaft, turbofan and unducted fan (UDF) versions. The **T407-GE-400** turboprop was selected to power the Lockheed P-7A ASW aircraft, a programme terminated in 1990. The **GLC38** is the commercial turboprop engine, which is the subject of a joint development agreement with Textron Lycoming. The following refers to the basic GLC38 power plant (including the propeller gearbox):

TYPE: Two-shaft modular turbine engine.
COMPRESSOR: Five axial stages followed by one centrifugal stage.
COMBUSTION CHAMBER: Short annular type with 15 injectors.
TURBINES: Two-stage air-cooled compressor turbine. Three-stage power turbine.
DIMENSIONS:
Max envelope length 2,972 mm (117 in)
Max envelope diameter 838 mm (33 in)
PERFORMANCE RATING (S/L, T-O):
T407 4,475 kW (6,000 shp) class
SPECIFIC FUEL CONSUMPTION:
"15-30 per cent better than engines in its class"

General Electric T407/GLC38 basic core

GENERAL ELECTRIC T64

The T64 was developed initially for the US Navy. It is available as a turboshaft or turboprop. Current versions are:
T64-GE-100. -7A with improved turbine. Powers CH-53C and MH-53J. Also being upgraded by kit from -7A.
T64-GE-413. Powers CH-53D.
T64-GE-415. Improved combustion liner and turbine cooling. Powers RH-53D.
T64-GE-416. As -415. Powers CH/MH-53E.
T64-GE-416A. As -416, improved turbine.
T64-GE-419. As -416A, with integral fuel/oil heat exchanger, and OEI emergency power on a 32.2°C day. To power CH/MH-53E from 1993.
CT64-820-4. Turboprop, powers DHC-5D.
T64/P4D. Turboprop, powers G222 and C-27A. Production by Fiat, supported by Alfa Romeo Avio from 1975.
TYPE: Free turbine turboshaft/turboprop.
COMPRESSOR: Fourteen-stage axial-flow, single-spool steel rotor for -820/-1/2/3, titanium and steel compressor for -100, -413, -415, -416, -416A, -419, -P4D and CT64-820-4. Inlet guide vanes and first four stages of stator vanes variable, air mass flow per second: -100, -413, -415, -416, -416A, -419, 13.3 kg (29.4 lb); -820-4, 11.9 kg (26.2 lb); P4D, 12.2 kg (27.0 lb). Pressure ratio: -820-4, 12.5; -100, -413, -415/-416/-416A, -419, 14.0; P4D, 13.0.
COMBUSTION CHAMBER: Annular type. Double fuel manifold feeds twelve duplex type fuel nozzles.
GAS GENERATOR TURBINE: Two-stage, coupled directly to compressor rotor by spline connection. Engines rated 3,228 kW (4,330 shp) or over have air-cooled first-stage blades.
POWER TURBINE: Two-stage, independent of gas generator.
REDUCTION GEAR: Remotely mounted for turboprop, offset and accessible for inspection and replacement. Ratio 13.44.
STARTING: Mechanical, airframe supplied.
DIMENSIONS:
Length: T64-GE-100, -413, -415, -416, -416A, -419
  2,006 mm (79 in)
 T64/P4D, CT64-820-4 2,793 mm (110 in)
Width: T64-GE-100, -413, -415, -416, -416A, -419
  660 mm (26.0 in)
 T64/P4D, CT64-820-4 683 mm (26.9 in)
Height: T64-GE-100, -413, -415, -416, -416A, -419
  825 mm (32.5 in)
 T64/P4D, CT64-820-4 1,168 mm (46 in)
WEIGHT, DRY:
 T64-GE-100, -413, -415, -416, -416A 327 kg (720 lb)
 T64-GE-419 343 kg (755 lb)

General Electric GLC38 turboprop mockup

General Electric T64-GE-415/416 turboshaft

670 USA: ENGINES—GENERAL ELECTRIC

CT64-820-4	520 kg (1,145 lb)
T64/P4D	538 kg (1,188 lb)

PERFORMANCE RATINGS (max rating (S/L)):
T64-GE-100	3,229 kW (4,330 shp) to 29.4°C
T64-GE-413	2,927 kW (3,925 shp)
T64-GE-415, -416, -416A	3,266 kW (4,380 shp)
T64-GE-419	3,542 kW (4,750 shp)
CT64-820-4	2,336 kW (3,133 shp)
T64/P4D	2,535 kW (3,400 shp)

SPECIFIC FUEL CONSUMPTION (max rating (S/L)):
T64-GE-100, -413, -415, -416, -416A, -419	79.4 µg/J (0.47 lb/h/shp)
CT64-820-4, T64/P4D	81 µg/J (0.48 lb/h/shp)

General Electric T700-401 free turbine turboshaft

General Electric CT7-9 free turbine turboprop

Full scale model of General Electric CT7-6 turboshaft *(Brian M. Service)*

GENERAL ELECTRIC T700

The T700 was selected in 1971 to power the US Army's utility tactical transport aircraft system (UTTAS). The first T700 went to test in 1973, and in 1976 it was the first turboshaft to pass current US military qualification standards. The engine went into production in 1978. Of the following models, only the Step 2 growth derivatives are in production.

T700-GE-700. First production model delivered from early 1978. Following description refers to this version, except where otherwise noted. Powers UH-60A Black Hawk.

T700-GE-701. First-step growth derivative. Upgrade of 10 per cent. Powers AH-64 Apache.

T700-GE-401. Navalised first-step derivative. Powers SH-60B Seahawk, SH-2G Super Seasprite and AH-1W SuperCobra.

T700-GE-701C. Second-step growth derivative. Provides 20 per cent more power than -700. Went into production in 1989 for Black Hawk and Apache. Offered for future S-70/WS-70 Black Hawk derivatives for international sales.

T700-GE-401C. Navalised second-step growth derivative. Chosen by US Navy for future Seahawks and derivatives. First production engines delivered in 1988 to power HH-60H; Royal Australian Navy also a launch customer.

GE alliances on T700 engines include: agreements with European Gas Turbines of Great Britain that result in a UK manufactured T700; a licence agreement with Alfa Romeo Avio (with additional participation of FiatAvio) for manufacture and assembly of T700/CT7 engines in Italy; a licence agreement with Australia to assemble T700s; another with Japan to build and test T700s for Black Hawks and Seahawks; and a licence agreement with KAL for manufacture and test of T700 engines for Black Hawks in South Korea.

TYPE: Ungeared free turbine turboshaft engine.
INTAKE: Annular, with anti-iced separator designed to remove 95 per cent of sand, dust and foreign object ingestion. Extracted matter discharged by blower driven from accessory gearbox.
COMPRESSOR: Combined axial/centrifugal. Five axial stages and single centrifugal stage mounted on same shaft. Each axial stage is one-piece 'blisk' (blades plus disc) in AM355 steel highly resistant to erosion. Inlet guide vanes and first two stator stages variable. Pressure ratio about 15. Mass flow about 4.5 kg (10 lb)/s at 44,720 rpm.
COMBUSTION CHAMBER: Fully annular. Central fuel injection to maximise acceptance of contaminated fuel and give minimal smoke generation. Ignition power from separate winding on engine mounted alternator serves dual plugs.
TURBINE: Two stage gas generator (HP) turbine. Rated speed (S/L, ISA, max T-O), 44,720 rpm. Two stage free power turbine, with tip shrouded blades and segmented nozzles. Output speed, 21,000 rpm.
CONTROLS: Hydromechanical control can be replaced in less than 12 minutes. Electrical control provides multi-engine speed and torque matching.
ACCESSORIES: Grouped at top of engine, together with engine control system. Integral oil tank, plus emergency mist lubrication. Torque sensor provides signal to electronic control.

DIMENSIONS:
Length overall	1,168 mm (46.0 in)
Width	635 mm (25 in)
Height overall	584 mm (23 in)

WEIGHT, DRY (with particle separator):
T700-700	198 kg (437 lb)
T700-401	197 kg (434 lb)
T700-701C	207 kg (456 lb)
T700-401C	208 kg (458 lb)

PERFORMANCE RATINGS (ISA, S/L, static):
T700-700: intermediate	1,210 kW (1,622 shp)
continuous	987 kW (1,324 shp)
T700-701: contingency	1,285 kW (1,723 shp)
intermediate	1,262 kW (1,692 shp)
continuous	1,171 kW (1,570 shp)
T700-401: contingency	1,285 kW (1,723 shp)
intermediate	1,260 kW (1,690 shp)
continuous	1,072 kW (1,437 shp)
T700-701C: maximum	1,409 kW (1,890 shp)
intermediate	1,342 kW (1,800 shp)
continuous	1,239 kW (1,662 shp)
T700-401C: contingency	1,447 kW (1,940 shp)
intermediate	1,342 kW (1,800 shp)
continuous	1,239 kW (1,662 shp)

SPECIFIC FUEL CONSUMPTION (ISA, S/L, static):
T700-700: continuous	79.41 µg/J (0.470 lb/h/shp)
T700-701: continuous	78.73 µg/J (0.466 lb/h/shp)
T700-401: continuous	79.60 µg/J (0.471 lb/h/shp)
T700-701C: continuous	77.56 µg/J (0.459 lb/h/shp)
T700-401C: continuous	77.56 µg/J (0.459 lb/h/shp)

GENERAL ELECTRIC CT7

Commercial engine based on the T700. Certification in April 1977. The **CT7-2A** turboshaft powers the Bell 214ST. The **CT7-2D** and **-2D1** power S-70 Black Hawks. The CT7-2D is the commercial equivalent of the -701/-401. The CT7-2D1 is the commercial equivalent of the -701C/-401C and received FAA certification in June 1989.

The **CT7-6** is a Step 3 growth engine, co-developed by Alfa Romeo Avio, FiatAvio and GE for the EH 101 and other helicopter applications, such as the NH 90. First delivery for the EH 101 took place in April 1988; first flight of a CT7-6 powered EH 101 took place in September 1988. The marinised **CT7-6A** powers the Italian Navy's EH 101 prototype. The CT7-6/-6A has received FAA certification, as well as RAI and BCAA validation.

The same core is used in the CT7 turboprop, which has a remote propeller gearbox. This engine received FAA certification in August 1983 and is in production as the **CT7-5A** for the Saab 340 and as the **CT7-7A** for the Airtech CN-235.

CT7 turboprop growth engines include the **CT7-9B** and **-9C** flat rated at 1,395 kW (1,870 shp). Increased power is obtained by improvements in aerodynamics, materials and turbine cooling. These engines power later versions of the

N-235 and Saab 340. The -9D is the power plant for the Let -610G.

Data below are for the current turboshaft versions; see adjacent table for turboprop models.

DIMENSIONS:
Length 1,168 mm (46.0 in)
Diameter (max envelope) 635 mm (25.0 in)

WEIGHT, DRY: CT7-2A 202 kg (444 lb)
CT7-2D/2B 201 kg (442 lb)
CT7-6 209 kg (460 lb)

PERFORMANCE RATINGS (S/L, static, 15°C):
Contingency (2½ min OEI):
CT7-2A, 2B, -2D 1,287 kW (1,725 shp)
CT7-6 1,491 kW (2,000 shp)
T-O (5 min) and en route contingency (30 min):
CT7-2A, 2B, -2D 1,212 kW (1,625 shp)
CT7-6 1,491 kW (2,000 shp)

CT7 turboprops

	-5A	-7A	-9B	-9C	-9D
Length mm (in)	2,438 (96)	2,438 (96)	2,438 (96)	2,438 (96)	2,438 (96)
Max diameter mm (in)	737 (29)	737 (29)	737 (29)	737 (29)	737 (29)
Weight, dry kg (lb)	349.3 (770)	351.5 (775)	360.6 (795)	360.6 (795)	360.6 (795)
PERFORMANCE RATINGS:					
S/L T-O kW (shp)	1,294 (1,735)	1,268 (1,700)	1,305 (1,750)	1,305 (1,750)	1,305 (1,750)
APR auto power reserve kW (shp)	—	—	1,394 (1,870)	1,394 (1,870)	1,447 (1,940)
Max cruise at 4,575 m (15,000 ft) ekW (ehp)	978 (1,312)	978 (1,312)	1,052 (1,411)	1,118 (1,499)	1,052 (1,411)

SPECIFIC FUEL CONSUMPTION:
Max continuous:
CT7-2A, -2B 79.9 μg/J (0.473 lb/h/shp)
CT7-6 (15°C) 79.4 μg/J (0.470 lb/h/shp)
CT7-6 (35°C) 85.9 μg/J (0.508 lb/h/shp)

LHTEC

LIGHT HELICOPTER TURBINE ENGINE COMPANY

The Paragon Building, Suite 400, 12400 Olive Boulevard, St Louis, Missouri 63141

MEMBER COMPANIES: **Allison Gas Turbine Division**
PO Box 420, Indianapolis, Indiana 46206-0420
AlliedSignal Propulsion Engines
PO Box 52181, Phoenix, Arizona 85072-2181

These two companies have combined their resources to develop the T800-LHT-800 975 kW (1,300 shp) class turboshaft engine for the multi-service RAH-66 Comanche programme. The CTS800 is a version for commercial applications.

LHTEC T800-LHT-800

This fully metric, high power density helicopter engine stems from programmes that the two partners started prior to 1983. AlliedSignal's core engine based on the F109 had extensive running time. Allison's engine was developed under the Army's Advanced Technology Demonstrator Engine (ATDE) programme. The blending of these two engines produced the T800 technology prototypes, which exceeded 940 kW (1,260 shp) in December 1984, with lower than specified fuel consumption. After fierce competition, LHTEC was selected by the US Army in October 1988 to continue the engine development and subsequent production. By February 1993 civil **CTS800** and military T800 engines had run more than 116,000 hours of testing.

The CTS800/T800 is designed for unprecedented reliability, maintainability and supportability, tilt-rotor compatibility and low specific fuel consumption. Its basic design philosophy includes module/LRU removal and replacement times of less than 15 minutes using six basic hand tools, and on-condition maintenance.

Twin CTS800s have been flying in an Agusta A 129 since 1988. In 1992 CTS800s flew in a UH-1, Lynx, Coast Guard HH-65 and Panther 800. The CTS800/T800 received FAA certification and military qualification in 1993. Production deliveries for the RAH-66 began in 1993.

In December 1992 the US Army awarded a growth contract to increase power to over 1,044 kW (1,400 shp).

Cutaway drawing of LHTEC T800 free turbine turboshaft

Turboprop, ground power and armoured-vehicle derivatives are expected.
INLET: Annular, with solids extracted by particle separator of over 97 per cent efficiency.
COMPRESSOR: Two centrifugal stages in tandem. One-piece titanium impellers.
COMBUSTION CHAMBER: Annular reverse flow.
COMPRESSOR TURBINE: Two axial stages with single-pass cooling.
POWER TURBINE: Two axial stages with tip shrouds.
LUBRICATION SYSTEM: Self-contained, with tank of 4.1 litres (1.08 US gallons, 0.90 Imp gallon), air/oil heat exchanger.
OIL GRADES: MIL-L-7808, MIL-STD-23699.
FUEL GRADES: MIL-T-5624, MIL-T-83133, JP-4, JP-5, JP-8 (emergency DF-A, DF-1, DF-2).
CONTROL SYSTEM: FADEC.
ACCESSORIES: Mounted above engine with drive from core. Self-contained electrical system.
REDUCTION GEAR: Drive at 23,000 rpm to front or rear. Reduction gearbox gives output at 6,000-6,600 rpm.
DIMENSIONS:
Length 800 mm (31.5 in)
Width 549 mm (21.6 in)
Height 658 mm (25.9 in)
WEIGHT, DRY: T800 143 kg (315 lb)
CTS800 with gearbox 163 kg (360 lb)
PERFORMANCE RATINGS:
Contingency (2 min) 1,007 kW (1,350 shp)
T-O (5 min) 985.8 kW (1,322 shp)
T-O (30 min) 895 kW (1,200 shp)
Continuous 764 kW (1,025 shp)
FUEL FLOW:
Contingency 285 kg (630 lb)/h
5 min 274 kg (603 lb)/h
30 min 256 kg (565 lb)/h
Continuous 219 kg (483 lb)/h

LPE

LIGHT POWER ENGINE CORPORATION

265 Scott Ave, Suite 102, PO Box 3350, Morgantown, West Virginia 26505
Telephone: 1 (304) 291 2376
Fax: 1 (304) 292 1902

This company is marketing a wide range of ZM water-cooled 90° V-8 piston engines. Entirely of original design, these have capacities of 400, 500 or 600 cu in (6.56, 8.2 or 9.84 litres), and are available upright or inverted, with direct fuel injection, with or without reduction gear or turbocharger. Direct-drive engines weigh 168-202 kg (370-446 lb) and are rated at 171.5-283 kW (230-380 hp) at 2,700 rpm. Geared engines weigh from 195-225 kg (430-496 lb) and are rated at 313-447 kW (420-600 hp) at 4,300 or 4,500 rpm.

LPE is the US distributor for Avia engines (which see under Czech Republic).

NELSON

NELSON AIRCRAFT CORPORATION

PO Box 454, 8075 Pennsylvania Ave, Irwin, Pennsylvania 15642
Telephone: 1 (412) 863 5900
PRESIDENT: Charles R. Rhoades
DEVELOPMENT DEPARTMENT: 420 Harbor Drive, Naples, Florida 33940
Telephone: 1 (813) 263 1670
Fax: 1 (813) 434 5993

Nelson Aircraft Corporation, among its many industrial activities, produces to order the Nelson H-63 four-cylinder two-cycle air-cooled engine, which is certificated by the FAA as a power unit for single-seat helicopters, and is also available as a power plant for propeller driven aircraft.

NELSON H-63

US military designation: YO-65

Developed originally as a power unit for single-seat helicopters, the H-63 is available in two versions, as follows:
H-63C. Basic helicopter power unit for vertical installation. Battery/electronic ignition and direct drive.
H-63CP. Basically as H-63C, but without clutch, fan and shroud. Intended primarily for installation in horizontal position, with direct drive to propeller.

Nelson H-63C four-cylinder two-stroke engine

Full details of these engines appeared in the 1992-93 *Jane's*.

POWER RATINGS:
T-O: H-63C 32 kW (43 hp) at 4,000 rpm
H-63CP 35.8 kW (48 hp) at 4,400 rpm
Max continuous: H-63C 32 kW (43 hp) at 4,000 rpm
H-63CP 33.6 kW (45 hp) at 4,000 rpm

Nelson H-63CP four-cylinder engine for fixed-wing aircraft

672 USA: ENGINES—PRATT & WHITNEY

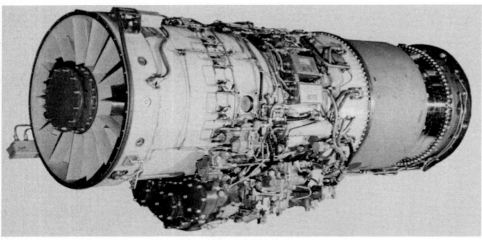

Pratt & Whitney J52-P-408 two-shaft turbojet

PRATT & WHITNEY
UNITED TECHNOLOGIES PRATT & WHITNEY

HEADQUARTERS: 400 Main Street, East Hartford, Connecticut 06108
Telephone: 1 (203) 565 4321
Fax: (PR) 1 (203) 565 8896
Government Engines and Space Propulsion
PO Box 109600, West Palm Beach, Florida 33410-9600
Telephone: (PR) 1 (407) 796 7885
Fax: (PR) 1 (407) 796 7258
PRESIDENT: Karl K. Krapek

Pratt & Whitney Aircraft was formed in 1925. Today Pratt & Whitney is the world's largest producer of gas turbine engines. Excluding P & W Canada (which see) it had by 1993 delivered nearly 68,000 aircraft gas turbines, including more than 27,000 airline jet engines.

PRATT & WHITNEY JT8
US military designation: J52

The J52 is a two-spool turbojet, with 12 compressor stages, a can-annular combustion system fed by 36 dual orifice injectors and single-stage HP and LP turbines. The P-408 has two-position inlet guide vanes and air-cooled first-stage turbine vanes and blades. New production P-409 offers higher thrust and increased life.

J52-P-6A, 6B, 8A, 8B. Rated at 37.8 kN (8,500 lb st) (6A, 6B) or 41.4 kN (9,300 lb st) (8A, 8B). Powers A-4 and A-6.

J52-P-408/P-408A. Rated at 49.8 kN (11,200 lb st). Powers A-4F, A-4M, some export A-4 versions, EA-6B.

J52-P-409. Rated at 53.4 kN (12,000 lb st). Powers EA-6B ADVCAP, available for upgraded A-6E.

Data below are for the P-408:
DIMENSIONS:
Diameter   814.3 mm (32.06 in)
Length   3,020 mm (118.9 in)
WEIGHT, DRY: 1,052 kg (2,318 lb)

PRATT & WHITNEY JT8D

This turbofan was developed to power the Boeing 727. Military versions have been developed in Sweden by Volvo Flygmotor (see RM8 in that company's entry).

The JT8D entered commercial service on 1 February 1964. It has since become the most widely used commercial jet engine, almost 12,000 having logged more than 407 million flight hours by January 1993 (over 434 million including -200 series).

The following are current versions:
JT8D-9, -9A. Develops 64.5 kN (14,500 lb st) to 28.9°C at S/L. Specified for 727-100, -100C and -200, 737-200, -200C and T-43A, DC-9-20, -30, -40, C-9A, C-9B and VC-9C, Caravelle 12 and Kawasaki C-1. Produced under licence in Japan (see Mitsubishi).
JT8D-11. Develops 66.7 kN (15,000 lb st) to 28.9°C at S/L. Specified for DC-9-20, -30 and -40.
JT8D-15. Develops 69 kN (15,500 lb st) to 28.9°C. Powers Mercure, Advanced 727 and 737, and DC-9.
JT8D-15A. In 1982 new components in the Dash-15 resulted in a 5.5 per cent reduction in cruise fuel consumption. The same parts in the Dash-17 produce the **JT8D-17A**, and when fitted to the Dash-17R the **JT8D-17AR**.
JT8D-17. Develops 71.2 kN (16,000 lb st) to 28.9°C. Powers Advanced Boeing 727 and 737, and DC-9.
JT8D-17R. Normal T-O rating 72.95 kN (16,400 lb st) but has capability of providing 4.448 kN (1,000 lb) additional thrust in the event of significant thrust loss on any other engine.
JT8D-200 Series. Described separately.
TYPE: Two-spool turbofan.
FAN: Two-stage front fan. Mass flow: -9, -9A, 145 kg (319 lb)/s; -11, -15, 146 kg (322 lb)/s; -17, 147 kg (324 lb)/s; -17R, 148 kg (326 lb)/s. Bypass ratio: -9, -9A, 1.04; -11, 1.05; -15, 1.03; -17, 1.02; -17R, 1.00.
LP COMPRESSOR: Six-stage, integral with fan.
HP COMPRESSOR: Seven-stage. Overall pressure ratio: -9, -9A, 15.9; -11, 16.2; -15, 16.5; -17, 16.9; -17R, 17.3.
COMBUSTION CHAMBER: Can-annular type with nine cylindrical flame tubes, each with a single Duplex burner.
HP TURBINE: Single-stage. Solid blades in -9, air-cooled in -1 and later.
LP TURBINE: Three-stage.
DIMENSIONS:
Diameter   1,080 mm (42.5 i
Length   3,137 mm (123.5 i
WEIGHT, DRY:
JT8D-9, -9A   1,532 kg (3,377 l
JT8D-11   1,537 kg (3,389 l
JT8D-15   1,549 kg (3,414 l
-15A   1,576 kg (3,474 l
JT8D-17   1,556 kg (3,430 l
-17A   1,577 kg (3,475 l
JT8D-17R   1,585 kg (3,495 l
-17AR   1,588 kg (3,500 l
PERFORMANCE RATINGS:
T-O thrust (S/L, static)   see model descriptio
Max cruise thrust (10,670 m; 35,000 ft at Mach 0.8):
JT8D-9, -9A   18.2 kN (4,100 l
JT8D-11   17.6 kN (3,950 l
JT8D-15, -15A   18.2 kN (4,100 l
JT8D-17, -17R, -17A, -17AR   18.9 kN (4,240 l
SPECIFIC FUEL CONSUMPTION:
T-O rating:
JT8D-9, 9A   16.85 mg/Ns (0.595 lb/h/lb s
JT8D-11   17.56 mg/Ns (0.620 lb/h/lb s
JT8D-15   17.84 mg/Ns (0.630 lb/h/lb s
-15A   16.63 mg/Ns (0.587 lb/h/lb s
JT8D-17   18.27 mg/Ns (0.645 lb/h/lb s
-17A   17.05 mg/Ns (0.602 lb/h/lb s
JT8D-17R   18.55 mg/Ns (0.655 lb/h/lb s
-17AR   17.31 mg/Ns (0.611 lb/h/lb s
Max cruise rating, as above:
JT8D-9, -9A   22.86 mg/Ns (0.807 lb/h/l
JT8D-11   23.14 mg/Ns (0.817 lb/h/l
JT8D-15   22.97 mg/Ns (0.811 lb/h/l
JT8D-17, -17R   23.37 mg/Ns (0.825 lb/h/l

PRATT & WHITNEY JT8D-200 SERIES

This reduced noise derivative of the JT8D combines the HP compressor, HP turbine spool and combustion section of the JT8D-9 with advanced LP technology. It offers increase thrust with reduced noise and specific fuel consumption. The fan has increased diameter. The new six-stage LP co pressor, integral with the fan, offers increased pressure rati The LP turbine has 20 per cent greater annular area an achieves a higher efficiency. Surrounding the engine is a ne bypass duct. The exhaust system includes a 12 lobe mixe FAA certification of the JT8D-209 was awarded in Jur 1979. Over 2,500 engines have been delivered of the follow ing series:

JT8D-209. Rated at 82.2 kN (18,500 lb st) to 25°C, an 85.6 kN (19,250 lb st) following loss of thrust on any othe engine. Entered service in October 1980, powering th MD-81.
JT8D-217. Rated at 88.96 kN (20,000 lb st), and 92.7 kN (20,850 lb st) following loss of thrust on any other engin Powers MD-82.
JT8D-217A. T-O thrust available to 28.9°C or up to 1,52 m (5,000 ft). Powers MD-82.
JT8D-217C. Incorporates JT8D-219 performanc improvements to reduce sfc. Powers MD-82 and -87.
JT8D-219. Rated at 93.4 kN (21,000 lb st), with a reserv thrust of 96.5 kN (21,700 lb st). Powers MD-83 and othe MD-80 aircraft.
TYPE: Two-spool turbofan.
FAN: Single-stage front fan has 34 titanium blades, with par span shrouds. Mass flow: -209, 213 kg (469 lb)/s; -21 (all), 219 kg (483 lb)/s; -219, 221 kg (488 lb)/s. Bypas ratio: -209, 1.78; -217 (all), 1.73; -219, 1.77.
LP COMPRESSOR: Six-stage axial, integral with fan.
HP COMPRESSOR: Seven-stage axial. Overall pressure rati -209, 17.1; -217 (all), 18.6; -219, 19.2.
COMBUSTION CHAMBER: Nine can-annular low-emissions bu ners with aerating fuel nozzles.
HP TURBINE: Single-stage. Air-cooled blades in -217, -217C and -219.
LP TURBINE: Three-stage.
DIMENSIONS:
Diameter   1,250 mm (49.2 in
Length   3,911 mm (154 in
WEIGHT, DRY:
JT8D-209   2,056 kg (4,533 lb
JT8D-217, -217A   2,072 kg (4,524 lb
JT8D-217C, -219   2,092 kg (4,612 lb
PERFORMANCE RATINGS:
T-O (S/L static):   see model description
Max cruise thrust (10,670 m; 35,000 ft at Mach 0.8):
JT8D-209   22.0 kN (4,945 lb
JT8D-217, -217A, -217C   23.31 kN (5,240 lb
JT8D-219   23.35 kN (5,250 lb
SPECIFIC FUEL CONSUMPTION:
Max cruise rating, as above:
JT8D-209   20.50 mg/Ns (0.724 lb/h/lb
JT8D-217, -217A   21.32 mg/Ns (0.753 lb/h/lb
JT8D-217C   20.84 mg/Ns (0.736 lb/h/lb
JT8D-219   20.87 mg/Ns (0.737 lb/h/lb

Pratt & Whitney JT8D-219 two-shaft turbofan

PRATT & WHITNEY PW4000

The PW4000 is a third generation turbofan for wide-body transports. Ratings range from 222.4 kN (50,000 lb st) to 373.7 kN (84,000 lb st). The first engine achieved 275 kN (61,800 lb st) during initial testing in April 1984. First flight test on an A300B took place on 31 July 1985. Certificated July 1986 at 249 kN (56,000 lb st) and in 1988 at 266.9 kN (60,000 lb st). Entered service on Pan Am A310 on 20 June 1987. The last two numbers denote thrust (thus, the initially certificated 56,000 lb st engine is the PW4056). Programme sharing agreements have been signed with Techspace Aero (Belgium), FiatAvio (Italy), Norsk Jet Motors (Norway), Kawasaki (Japan), Samsung (South Korea), Eldim (Netherlands), Mitsubishi (Japan) and Singapore Aircraft Industries.

Fuel consumption was initially reduced seven per cent compared with the JT9D-7R4. There are about half as many parts, promising reductions in maintenance cost exceeding 25 per cent. HP compressor pressure ratio is increased by 10 per cent and the HP rotor operates at 27 per cent higher speed.

The PW4000 incorporates single-crystal turbine blades, aerodynamically enhanced aerofoils, and a full-authority digital electronic engine control (FADEC). The PW4000 was the first FADEC engine approved for EROPS operations beyond 180 min from alternates.

For the A330 the **PW4168** has a fan diameter increased to 2,535 mm (99.8 in) and an extra stage on the LP compressor and LP turbine. Initial A330 deliveries will be at 302.5 kN (68,000 lb st) in 1994. For the Boeing 777 the **PW4084** will have a fan of 2,845 mm (112 in) diameter (using new hollow titanium blades without part-span clappers), six-stage LP compressor and seven-stage LP turbine. The first PW4084 achieved 400.3 kN (90,000 lb st) in August 1992. Initial 777 deliveries in 1995 will be at 343 kN (77,000 lb st).

Pratt & Whitney is evaluating ADP (Advanced Ducted Propeller) derivatives. These could provide over 444.8 kN (100,000 lb st) for new large aircraft by 2000.

The following data apply to the basic PW4000, except where indicated:

TYPE: Two-shaft turbofan.
FAN: Single-stage. Titanium alloy hub retains 38 titanium alloy blades with aft part-span shrouds. Diameter 2,377 mm (93.6 in). Data for 249 kN (56,000 lb st) rating: mass flow 773 kg (1,705 lb)/s. Fan pressure ratio 1.7. Bypass ratio 4.85.
LP COMPRESSOR: Four stages with controlled diffusion aerofoils.
HP COMPRESSOR: Eleven stages with first four vane rows variable. Clearance control accomplished via rotor/case thermal matching. Overall pressure ratio at 251.5 kN rating, 30.0.
COMBUSTOR: Annular, forged nickel alloy roll-ring with double-pass cooling. 24 air-blast anti-coking injectors. Segmented Floatwall burner liner to be introduced in 1993.
HP TURBINE: Two stages with air-cooled blades cast as single crystal (PWA 1480) in first row and directional crystal (PWA 1422) in second row, retained in double-hub nickel alloy rotor with active clearance control. Vane aerofoils thermal barrier coated.
LP TURBINE: Four stages with active clearance control.
CONTROL SYSTEM: Full authority digital electronic with dual channel computer.
DIMENSIONS:
Length: PW4000   3,371 mm (132.7 in)
  PW4168   4,143 mm (163.1 in)
  PW4084   4,868 mm (191.7 in)
Fan case diameter: PW4000   2,463 mm (96.98 in)
  PW4168   2,718 mm (107.0 in)
  PW4084   3,048 mm (120.0 in)
WEIGHT, DRY:
Basic PW4000   4,264 kg (9,400 lb)
PW4168 (complete system)   6,509 kg (14,350 lb)
Basic PW4084   6,335 kg (13,965 lb)
RATINGS: See table
SPECIFIC FUEL CONSUMPTION (ISA, ideal nozzle, Mach 0.8, 10,670 m; 35,000 ft):
PW4056   15.21 mg/Ns (0.537 lb/h/lb)

Pratt & Whitney PW4084 turbofan

PW4000 MODELS

Model	Thrust (ideal nozzle)	Application
PW4152	231.3 kN (52,000 lb) to 42.2°C	A310-300
PW4156	249.1 kN (56,000 lb) to 30°C	A300-600/A310-300
PW4158	258.0 kN (58,000 lb) to 30°C	A300-600R
PW4168	302.5 kN (68,000 lb) to 30°C	A330
PW4052	232.2 kN (52,200 lb) to 33.3°C	767-200/-200ER
PW4056	252.4 kN (56,750 lb) to 33.3°C	767-300/-300ER/747-400
PW4060	266.9 kN (60,000 lb) to 33.3°C	767-300ER/747-400
PW4062	275.8 kN (62,000 lb) to 30°C	767-300ER/747-400
PW4074	329.2 kN (74,000 lb) to 30°C	777
PW4077	342.5 kN (77,000 lb) to 33.3°C	777
PW4084	373.7 kN (84,000 lb) to 30°C	777
PW4460	266.9 kN (60,000 lb) to 30°C	MD-11
PW4462	275.8 kN (62,000 lb) to 30°C	MD-11

Ratings are S/L static, no bleed or power extraction

PRATT & WHITNEY JT9D

This was the first of the new era of large, high bypass ratio turbofans on which the design of wide-body commercial transports rests. First run was in December 1966. The first flight of the Boeing 747 was on 9 February 1969.

Current versions include:

JT9D-3A. Water injection rating of 200.8 kN (45,150 lb) to 26.7°C. Powers 747-100 and -200B.

JT9D-7. Higher thrust version with air-cooled DS HP turbine blades; powers 747-200B, C, F and SR.

JT9D-7A. Aerodynamic improvements; powers 747-200 and 747SP.

JT9D-7F, -7J. Improved DS blades, 7J with shower-head cooled first stage, giving -7F T-O rating without water injection.

JT9D-7Q, -7R. Described later.

JT9D-20, -20J. D-7A and D-7J with accessory gearbox under fan case. Powers DC-10-40.

JT9D-59A, -70A. Fan diameter approximately 25.4 mm (one inch) larger, with re-profiled blades; LP compressor has zero (fourth) stage and is redesigned; burners recontoured, No. 3 bearing carbon seal added, HP blades directionally solidified, and HP annulus larger. Both configured for common nacelle for 747 (-70A) or DC-10 and A300B (-59A).

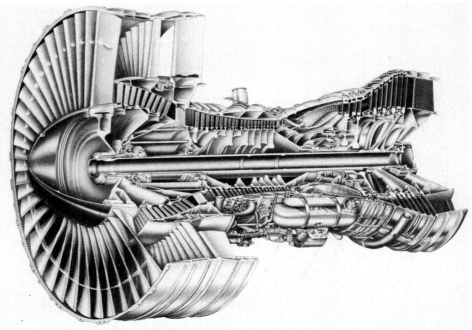

Cutaway drawing of Pratt & Whitney JT9D-7R4 two-shaft turbofan

USA: ENGINES—PRATT & WHITNEY

Pratt & Whitney PW2000 series two-shaft turbofan

JT9D-7Q, -7Q3. As -59A and -70A but configured like -7 for 747-200 nacelle. Thrust 236-249 kN (53,000-56,000 lb).

JT9D-7R4 Series. Seven models (7R4D to 7R4H1), with larger fan with wide chord blades, zero stage on LP compressor, improved combustor, H1 has single crystal HP turbine blades, increased diameter LP turbine, supervisory electronic fuel control and many smaller changes. 7R4D, 7R4E and 7R4E4 for Boeing 767; 7R4D1 and 7R4E1 for A310; -7R4G2 for 747-300; -7R4H1 for the A300B-600.

Deliveries exceed 3,265 and flight time by early 1993 was in excess of 118 million h. Full details appeared in the 1992-93 *Jane's*.

DIMENSIONS:
JT9D-3A, -7, -7A, -7F, -7J, -20:
  Diameter   2,427 mm (95.56 in)
  Length (flange to flange)   3,255 mm (128.15 in)
JT9D-59A, -70A, -7Q, -7Q3:
  Diameter   2,464 mm (97.0 in)
  Length   3,358 mm (132.2 in)
JT9D-7R4D to H: Diameter   2,463 mm (96.98 in)
  Length   3,371 mm (132.7 in)

WEIGHT, DRY:
Guaranteed, including standard equipment:
  JT9D-3A   3,905 kg (8,608 lb)
  JT9D-7, -7A, -7F, -7J   4,014 kg (8,850 lb)
  JT9D-20   3,833 kg (8,450 lb)
  JTRD-20J   3,883 kg (8,560 lb)
  JT9D-59A   4,146 kg (9,140 lb)
  JT9D-70A   4,153 kg (9,155 lb)
  JT9D-7Q   4,216 kg (9,295 lb)
  JT9D-7R4D, E, E4   4,039 kg (8,905 lb)
  JT9D-7R4D1, E1   4,029 kg (8,885 lb)
  JT9D-7R4G2   4,143 kg (9,135 lb)
  JT9D-7R4H1   4,029 kg (8,885 lb)

PERFORMANCE RATINGS (ideal nozzle):
T-O, dry: JT9D-3A   193.9 kN (43,600 lb st) to 26.7°C
  JT9D-7   202.8 kN (45,600 lb st) to 26.7°C
  JT9D-7A   205.7 kN (46,250 lb st) to 26.7°C
  JT9D-7F   213.5 kN (48,000 lb st) to 26.7°C
  JT9D-7J, -20J   222.4 kN (50,000 lb st) to 30°C
  JT9D-20   206.0 kN (46,300 lb st) to 28.9°C
  JT9D-59A, -70A, -7Q   236.0 kN (53,000 lb st) to 30°C
  JT9D-7R4D, D1   213.5 kN (48,000 lb st) to 33°C
  JT9D-7R4E, E1   222.4 kN (50,000 lb st) to 33°C
  JT9D-7R4E4   222.4 kN (50,000 lb st) to 45.6°C
  JT9D-7R4G2   243.4 kN (54,750 lb st) to 30°C
  JT9D-7R4H1   249.0 kN (56,000 lb st) to 30°C
T-O, wet: JT9D-3A   200.8 kN (45,150 lb st) to 26.7°C
  JT9D-7   210.0 kN (47,200 lb st) to 30°C
  JT9D-7A   212.4 kN (47,750 lb st) to 30°C
  JT9D-7F   222.4 kN (50,000 lb st) to 30°C
  JT9D-20   220.0 kN (49,400 lb st) to 30°C
Max cruise, 10,670 m (35,000 ft) at Mach 0.85:
  JT9D-3A, -7   45.4 kN (10,200 lb)
  JT9D-7A   48.2 kN (10,830 lb)
  JT9D-7F, -7J   49.2 kN (11,050 lb)
  JT9D-20, -20J   47.5 kN (10,680 lb)
  JT9D-59A, -70A, -7Q   53.2 kN (11,950 lb)
  JT9D-7R4D, D1   50.0 kN (11,250 lb)
  JT9D-7R4E, E1   52.0 kN (11,700 lb)
  JT9D-7R4G2, H1   54.5 kN (12,250 lb)

SPECIFIC FUEL CONSUMPTION (ideal nozzle):
Max cruise, ISA + 10°C, Mach 0.85 at 10,670 m (35,000 ft):
  JT9D-3A   17.67 mg/Ns (0.624 lb/h/lb)
  JT9D-7   17.55 mg/Ns (0.620 lb/h/lb)
  JT9D-7A   17.69 mg/Ns (0.625 lb/h/lb)
  JT9D-7F, -7Q, -59A, -70A   17.87 mg/Ns (0.631 lb/h/lb)
  JT9D-20, -20J   17.67 mg/Ns (0.624 lb/h/lb)
  JT9D-7R4D, D1   17.42 mg/Ns (0.615 lb/h/lb)
  JT9D-7R4E, E1   17.55 mg/Ns (0.620 lb/h/lb)
  JT9D-7R4G2   18.10 mg/Ns (0.639 lb/h/lb)
  JT9D-7R4H1   17.79 mg/Ns (0.628 lb/h/lb)

PRATT & WHITNEY PW2000
US military designation: F117

The PW2000 is a third generation turbofan upon which major changes have been made since 1972. In mid-1980 was scaled up to be compatible with the Boeing 757-200, 170.1 kN (38,250 lb st). The first model was given the designation **PW2037**, the last two digits denoting thrust in thousands of pounds. The first engine test run took place December 1981. FAA certification was achieved in December 1983, and the first flight was made on the prototype 757 on 14 March 1984. FAA 120 min EROPS received December 1989; 180 min received April 1992.

Companies participating are MTU of Germany and FiatAvio of Italy. Pratt & Whitney bears 70.8 per cent of the programme, MTU 21.2 per cent and FiatAvio 4 per cent. A collaboration agreement between these companies was signed July 1977. In 1987 Volvo of Sweden joined the programme as a 4 per cent manufacturing partner.

Current applications are in versions of the Boeing 757 and McDonnell Douglas C-17A. The PW2037-powered 757 was certificated on 25 October 1984 and entered revenue service on 1 December. An uprated engine, the **PW2040**, was certificated in January 1987 and entered service in September 1987. The C-17A is powered by four **F117-PW-100** engines similar to the PW2040. The F117 was certificated in December 1988 and was to enter USAF service in mid-1993.

Four **PW2337** engines rated at 170.1 kN (38,250 lb st) began flight testing in the Ilyushin Il-96M in April 1993, with certification scheduled for 1995.

More than 570 PW2000 engines have been delivered accumulating 5.2 million flight hours.

TYPE: Two-shaft turbofan of high bypass ratio.
FAN: Single-stage. Titanium forged hub, with 36 inserted titanium alloy blades with part-span shrouds. Tip diameter 1,994 mm (78.5 in). Mass flow 608 kg (1,340 lb)/s. Pressure ratio 1.7. Bypass ratio 6.0.
LP COMPRESSOR: Four stages, with controlled diffusion aerofoils with thick leading- and trailing-edges.

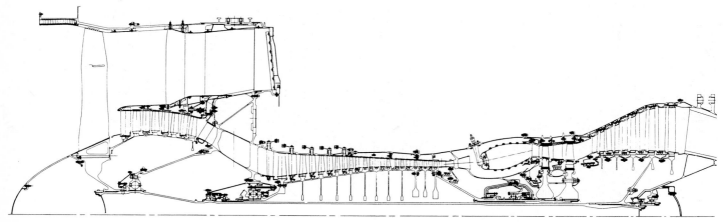

Longitudinal section of Pratt & Whitney PW2000 two-shaft turbofan

HP COMPRESSOR: Twelve stages, with controlled diffusion aerofoils. Variable vanes on first five stages and active clearance control on last eight stages. Overall cruise pressure ratio 31.8.
COMBUSTION CHAMBER: Annular, with flame tube fabricated in nickel alloy. Single-pipe fuel nozzles.
HP TURBINE: Two stages with air-cooled blades cast as single crystals in PW 1480 and PW 1484 alloy. Rotors with active clearance control. Both discs of PW1100 nickel based powder.
LP TURBINE: Five stages, with active clearance control.
CONTROL SYSTEM: Full authority digital electronic with two redundant computers.
DIMENSIONS:
Length 3,729 mm (146.8 in)
Fan case diameter 2,154 mm (84.8 in)
WEIGHT, DRY:
PW2000 3,311 kg (7,300 lb)
F117 3,220 kg (7,100 lb)
PERFORMANCE RATING (T-O, S/L):
PW2037 170.1 kN (38,250 lb st)
PW2040 185.5 kN (41,700 lb st)
SPECIFIC FUEL CONSUMPTION (ideal nozzle, cruise at Mach 0.8 at 10,670 m; 35,000 ft): 15.95 mg/Ns (0.563 lb/h/lb)

PRATT & WHITNEY F100

Family of afterburning turbofan fighter engines. Versions listed below for the F-15 and F-16 are now leading to a 160.1 kN (36,000 lb st) version for possible use in the MRF (Multi-Role Fighter) and Navy AX (Advanced Attack Aircraft). Nearly 6,000 have flown nearly 10 million hours since 1974.

F100-PW-100. Powered all early versions of F-15.
F100-PW-200. Modified for early versions of F-16.
F100-PW-220. Redesigned to incorporate new technology including digital control and later material and heat-transfer concepts for extended life and improved safety and maintainability. Entered Alternate Fighter Engine competition for combined future F-15 and F-16 production.
F100-PW-220E. Conversion of earlier engines by kit providing all advanced features of PW-220. USAF plans to upgrade all PW-100/-200 engines during 1990s.
F100-PW-229. Most advanced production version, incorporating further new technology.
F100-PW-220P. Being developed to incorporate Dash-229 technology into earlier engines, notably advanced fan, augmentor fuel management and digital control.

The following refers to the PW-229:
TYPE: Two-shaft axial turbofan with mixed-flow afterburner (augmentor).
FAN: Damage-tolerant three-stage with mid-span shrouds on Stages 1 and 2. Inlet diameter 886.5 mm (34.9 in). Bypass ratio 0.36.
COMPRESSOR: Damage-tolerant 10-stage, with variable stators on first three stages. Overall pressure ratio 32.
COMBUSTOR: Smokeless annular with Floatwall film cooling, 24 air blast nozzles and continuous capacitor discharge ignition.
HP TURBINE: Two stages with single-crystal air-cooled blades. Inspection interval 4,300 cycles (7-9 years).
LP TURBINE: Two stages with uncooled directionally solidified blades. Inspection interval 4,300 cycles.
AFTERBURNER: 11-segment mixed flow with advanced fuel management and high-energy ignition.
NOZZLE: Variable-area convergent/divergent with balanced-beam multi-flap construction.
CONTROL SYSTEM: FADEC with integrated diagnostics and aircraft control system interactive capability.
DIMENSIONS:
Length 4,855 mm (191.15 in)
Diameter overall 1,181 mm (46.5 in)
WEIGHT, DRY:
PW-220 1,451 kg (3,200 lb)
PW-220E 1,429 kg (3,151 lb)

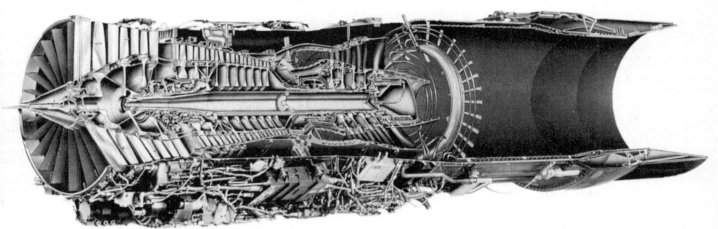

Cutaway drawing of Pratt & Whitney F100-PW-229 two-shaft augmented turbofan

Cutaway drawing of Pratt & Whitney F100-PW-220 two-shaft augmented turbofan

Pratt & Whitney F119, with 2D (two-dimensional) nozzle and fairing as fitted in YF-22

676 USA: ENGINES—PRATT & WHITNEY/SOLOY

PW-229	1,681 kg (3,705 lb)
PW-220P	1,526 kg (3,365 lb)

PERFORMANCE (S/L static):
Military (dry):

PW-220, 220E	65.26 kN (14,670 lb st)
PW-229	79.18 kN (17,800 lb st)
PW-220P	74.29 kN (16,700 lb st)

Max (afterburner):

PW-220, 220E	106.0 kN (23,830 lb st)
PW-229	129.45 kN (29,100 lb st)
PW-220P	120.1 kN (27,000 lb st)

PRATT & WHITNEY F119

This advanced augmented turbofan was selected by the US Air Force over the rival General Electric F120 to power the Lockheed F-22A Advanced Tactical Fighter in April 1991. Basic requirements were the simplest and most robust design for maximum reliability and maintainability, supercruise (supersonic persistence) capability without afterburner and a 2D (two-dimensional) nozzle incorporating limited thrust vectoring. The rival companies began formal development of their competing designs in 1983. From the outset Pratt & Whitney chose to offer a mature low-risk engine, even if in some respects it might appear to be less advanced than the rival F120. This philosophy was followed when, as the result of weight growth in the competing ATF aircraft, the thrust requirements were upgraded in early 1988. By this time the design had long been fixed and the first YF119 demonstrator was about to run. The decision was taken not to change the engine, but to meet the extra thrust requirement by a subsequent slight increase in fan diameter, leaving the rest of the engine unaltered.

Bench testing began in December 1988. In the Summer of 1990, F119 engines made 65 flights totalling 153 h, with no stalls or shutdowns. By April 1991, engines on ground test had run over 3,000 h, including 1,500 with the 2D nozzle. Production engines are due for delivery from 1997, the full programme total exceeding 1,500 engines priced at over $12 billion.

In partnership with Rolls-Royce, a derivative of the F119 is being developed as a candidate engine for the Lockheed ASTOVL combat aircraft, with gas drive to an Allison lift fan.

TYPE: Two-shaft low-bypass ratio augmented turbofan.
FAN: Three stages of snubberless wide-chord blades.
COMPRESSOR: Multi-stage spool with integrally bladed rotor turning in opposition to the fan.
COMBUSTION CHAMBER: High-intensity short annular with smokeless burning.
TURBINES: Single-stage contra-rotating HP and LP turbines.
CASING: Integrally stiffened bypass duct and inner compressor casing both split for easy maintenance access.
AUGMENTOR: Single spray ring for combustion downstream of both bypass duct and core.
NOZZLE: Two-dimensional (rectangular) with external flaps to meet aircraft aerodynamic and LO (low observables) requirements and internal flaps to match engine operating parameters. Hydraulic control to ±20°. No reversing capability.
ACCESSORIES: Grouped for immediate access from below. All LRUs mounted one-deep. FADEC control.
DIMENSIONS: Generally similar to PW1129
PERFORMANCE RATING (S/L): 155.6 kN (35,000 lb st) class

ROCKETDYNE
ROCKETDYNE DIVISION OF ROCKWELL INTERNATIONAL

6633 Canoga Avenue, Canoga Park, California 91303
Telephone: 1 (818) 710 6300
Fax: 1 (818) 710 2866
Telex: 698478
PRESIDENT: R. D. Paster

Rocketdyne is devoted primarily to rocket engines for the US Air Force and the National Aeronautics and Space Administration. It was established as a separate division in November 1955.

ROCKETDYNE SSME

On 13 July 1971 Rocketdyne was selected by NASA to develop the main engine for the Shuttle Orbiter. Three of these engines provide a total of 6,523 kN (1,466,400 lb) vacuum thrust, and power the vehicle to near orbit.

Flight certification was achieved in December 1980. Another cycle in the FPL (full power level) certification programme was completed in April 1983. Certification required four test series of 5,000 s each, involving engine power levels normally at 104 per cent. The Shuttle programme began its operational phase with the STS-5 launch of the *Columbia* on 11 November 1982. A total of 24 launches, all successful, had been accomplished prior to the loss of *Challenger* on 28 January 1986.

Testing is being continued to extend the life of the HP turbomachinery, increase performance margin by reducing flow losses, and investigate other modifications that could allow uprating or life extension.

SSME undergoing FPL test

COMBUSTION CHAMBER: Channel wall construction with regenerative cooling by the hydrogen fuel. Concentric element injector.
TURBOPUMPS: Two low pressure pumps boost the inlet pressures for two high pressure pumps. Dual pre-burners provide turbine drive gases to power the high-pressure pumps. Hydrogen pump discharge pressure is 485.4 bar (6,445 lb/sq in) at 35,080 rpm; it develops 51,528 kW (69,100 hp).
CONTROLLER: Honeywell digital computer provides closed loop control, in addition to data processing and signal conditioning for control, checkout, monitoring engine status and maintenance data acquisition.
CONTROLS: Hydraulic actuation. Dual redundant self monitoring servo actuators respond to signals from the controller to position the ball valves. A pneumatic system provides backup for engine cut-off.
MAINTENANCE: Airline type maintenance procedure for on-the-vehicle servicing. Planned life between overhauls 55 flights.
DIMENSIONS:

Length	4,242 mm (167 in)
Diameter at nozzle exit	2,388 mm (94 in)

PERFORMANCE:

S/L thrust (one engine, 104%)	1,751.7 kN (393,800 lb)
Vacuum thrust	2,174.3 kN (488,800 lb)
Specific impulse	453
Chamber pressure	215.5 bars (3,126 lb/sq in)
Throttling ratio	1.6
Expansion ratio	77.5

ROTORWAY
ROTORWAY INTERNATIONAL

300 South 25th Avenue, Phoenix, Arizona 85009
Telephone: 1 (602) 278 8899
Fax: 1 (602) 278 7657
Telex: 683 5059

RotorWay International produces its own liquid-cooled aircraft power plant.

ROTORWAY RI 162

TYPE: Horizontally opposed, water-cooled, four-cylinder, four-stroke piston engine. Crankshaft vertical.
CYLINDERS: Offset left and right for plain connecting rods side by side. Capacity 2.66 litres (162 cu in). Compression ratio 9.6.

INDUCTION: Single two-barrel downdraught carburettor with integral manifold heating.
IGNITION: Dual direct-fire electronic, with redundant coils, sensors, timing processors and plugs.
FUEL GRADE: Autogas 92 octane or Avgas 100LL.
WEIGHT, DRY (with starter): 77.1 kg (170 lb)
PERFORMANCE RATING: 113 kW (152 hp) at 4,400 rpm

RotorWay RI 162 four-cylinder helicopter engine

SOLOY
SOLOY CORPORATION

450 Pat Kennedy Way SW, Olympia, Washington 98502
Telephone: 1 (206) 754 7000
Fax: 1 (206) 943 7659
PRESIDENT: Joe I. Soloy

Further details can be found in the US Aircraft section.

SOLOY TURBINE PAC

The Soloy Turbine Pac is an FAA Supplemental Type Certificate approved turboprop engine assembly, rated at 312 kW (418 shp) with a propeller rpm range of 1,450 to 1,810. Its Allison 250-C20S turboshaft engine is combined with Soloy's propeller gearbox and other components to produce a turboprop configured for single-engined aircraft. Its high thrust line and rear inlet suit it particularly to bush aircraft. The engine assembly includes propeller governing and overspeed systems, and a self-contained lubrication system. Customised models are available in pusher configuration and can utilise the 485 kW (650 shp) Allison 250-C30 engine for tractor or pusher configurations.

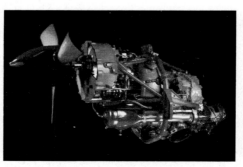

Soloy 206 Turbine Pac free turbine turboprop

SOLOY DUAL PAC INC

Address as Soloy Corporation

SOLOY DUAL PAC

The first production model of the Soloy Dual Pac multi-engine system utilises two Allison 250-C30S turboshaft engines, each rated at 522 kW (700 shp). The Soloy combining gearbox is rated at 1,119 kW (1,500 shp) to accommodate possible future power increases for the Allison engine. In 1992 a Dual Pac Cessna Caravan conversion was offered using a pair of the original PT6A-114A engines. The Dual Pac's redundancy, separation, and isolation of engine and drive train systems allows it to satisfy FAA requirements for designation as a twin engine, and Dual Pac powered aircraft to be defined as multi-engined. Patented free-wheeling units at the final stage drive train provide automatic disengagement in the event of an engine shutdown, with no adverse effect on aircraft drag or thrust symmetry.

The Dual Pac is designed for tractor or pusher configuration, and is planned for use in both single-propeller and multi-propeller aircraft.

TCM
TELEDYNE CONTINENTAL MOTORS
(Aircraft Products)

PO Box 90, Mobile, Alabama 36601
Telephone: 1 (205) 438 3411
Fax: 1 (205) 432 2922
Telex: 505519
PRESIDENT: Bryan L. Lewis
DIRECTOR, PUBLIC RELATIONS: Susan Brane

TCM produces piston, rotary combustion and turboprop engines for general aviation.

TELEDYNE CONTINENTAL O-200 SERIES

The O-200-A is a four-cylinder horizontally opposed air-cooled engine. It is fitted with a single updraught carburettor, dual magnetos and starter and generator. The O-200-B is designed for pusher installation.

For other details, see table.

Teledyne Continental IO-240 four-cylinder air-cooled engine

REPRESENTATIVE TCM HORIZONTALLY OPPOSED ENGINES

Engine Model	No. of Cylinders	Bore and Stroke mm (in)	Capacity litres (cu in)	Power Ratings kW (hp) at rpm Take-off	Power Ratings kW (hp) at rpm M.E.T.O.	Comp. Ratio	Dry Weight* kg (lb)	Length mm (in)	Width mm (in)	Height mm (in)	Octane Rating
O-200-A	4	103.2×98.4 (4¹/₁₆×3⅞)	3.28 (201)	74.5 (100) at 2,750	74.5 (100) at 2,750	7.0	99.8 (220)	725 (28.53)	802 (31.56)	589 (23.18)	80/87
IO-240	4	112.7×98.4 (4⁷/₁₆×3⅞)	3.94 (240)	93.2 (125) at 2,800	93.2 (125) at 2,800	8.5	113 (250)	749 (29.5)	839 (33.03)	592 (23.31)	100/100LL
IO-360-ES	6	112.7×98.4 (4⁷/₁₆×3⅞)	5.9 (360)	157 (210) at 2,800	157 (210) at 2,800	8.5	158.8 (350)	923 (36.32)	839 (33.05)	598 (23.52)	100/100LL
IO-360-KB	6	112.7×98.4 (4⁷/₁₆×3⅞)	5.9 (360)	145.5 (195) at 2,600	145.5 (195) at 2,600	8.5	148.3 (327)	864 (34.03)	841 (33.11)	781 (30.74)	100/130
TSIO-360-GB-C, D	6	112.7×98.4 (4⁷/₁₆×3⅞)	5.9 (360)	168 (225) at 2,800	168 (225) at 2,800	7.5	136 (300)	910† (35.84)	838 (33.03)	603 (23.75)	100/130
TSIO-360-EB LIO-360-FB	6	112.7×98.4 (4⁷/₁₆×3⅞)	5.9 (360)	149 (200) at 2,575	149 (200) at 2,575	7.5	175 (385)	1,437 (56.58)	795 (31.30)	671 (26.44)	100/130
TSIO-360-GB, LB	6	112.7×98.4 (4⁷/₁₆×3⅞)	5.9 (360)	156.5 (210) at 2,700	156.5 (210) at 2,700	7.5	175 (386)	902 (35.52)	795 (31.30)	699 (27.53)	100/130
TSIO-360-KB	6	112.7×98.4 (4⁷/₁₆×3⅞)	5.9 (360)	164 (220) at 2,800	164 (220) at 2,800	7.5	178 (392)	1,437 (56.58)	795 (31.30)	672 (26.44)	100/130
TSIO-360-MB	6	112.7×98.4 (4⁷/₁₆×3⅞)	5.9 (360)	156.5 (210) at 2,700	156.5 (210) at 2,700	7.5	186.9 (412)	1,087 (42.78)	873 (34.37)	834 (32.82)	100LL
O-520-D	6	133×101.6 (5¼×4)	8.5 (520)	224 (300) at 2,850	212.5 (285) at 2,700	8.5	208.2 (459)	949 (37.36)	901 (35.46)	604 (23.79)	100/130
O-520-L	6	133×101.6 (5¼×4)	8.5 (520)	224 (300) at 2,850	212.5 (285) at 2,700	8.5	211.7 (466.7)	1,039 (40.91)	852 (33.56)	591 (23.25)	100/130
O-520-M, -MB	6	133×101.6 (5¼×4)	8.5 (520)	212.5 (285) at 2,700	212.5 (285) at 2,700	8.5	188 (415)	1,189 (46.80)	852 (33.56)	518 (20.41)	100/130
TSIO-520-AF	6	133×101.6 (5¼×4)	8.5 (520)	231 (210) at 2,700	212.5 (285) at 2,600	7.5	197.8 (436.15)	1,039 (40.91)	852 (33.56)	598 (23.54)	100/130
TSIO-520-B, -BB	6	133×101.6 (5¼×4)	8.5 (520)	213 (285) at 2,700	213 (285) at 2,700	7.5	219 (483)	1,490 (58.67)	852 (33.56)	516 (20.32)	100/130
TSIO-520-BE	6	133×101.6 (5¼×4)	8.5 (520)	231 (310) at 2,600	231 (310) at 2,600	7.5	?	1,083 (42.64)	1,079 (42.5)	851 (33.5)	100LL
TSIO-520-CE	6	133×101.6 (5¼×4)	8.5 (520)	242.5 (325) at 2,700	242.5 (325) at 2,700	7.5	237 (527)	1,039 (40.91)	852 (33.56)	597 (23.54)	100LL
TSIO-520-C	6	133×101.6 (5¼×4)	8.5 (520)	212.5 (285) at 2,700	212.5 (285) at 2,700	7.5	208 (458)	1,040† (40.91)	852 (33.56)	509 (20.04)	100/130
TSIO-520-E, -EB	6	133×101.6 (5¼×4)	8.5 (520)	224 (300) at 2,700	224 (300) at 2,700	7.5	219 (483)	1,010† (39.75)	852 (33.56)	527 (20.74)	100/130
TSIO-520-J, N, -JB, -NB	6	133×101.6 (5¼×4)	8.5 (520)	231 (310) at 2,700	231 (310) at 2,700	7.5	221.3 (487.8)	997 (39.25)	852 (33.56)	516 (20.32)	100/130
TSIO-520-L, -LB	6	133×101.6 (5¼×4)	8.5 (520)	231 (310) at 2,700	231 (310) at 2,700	7.5	244.5 (539)	1,286 (50.62)	852 (33.56)	508 (20.02)	100/130
TSIO-520-M, P, R	6	133×101.6 (5¼×4)	8.5 (520)	231 (310) at 2,700	212.5 (285) at 2,600	7.5	198 (436)	1,040† (40.91)	852 (33.56)	598 (23.54)	100/130

*With accessories; †Not including turbocharger;
‡ weight 220 kg (486 lb)

REPRESENTATIVE TCM HORIZONTALLY OPPOSED ENGINES Continued

Engine Model	No. of Cylinders	Bore and Stroke mm (in)	Capacity litres (cu in)	Power Ratings kW (hp) at rpm Take-off	Power Ratings kW (hp) at rpm M. E. T. O.	Comp. Ratio	Dry Weight* kg (lb)	Dimensions Length mm (in)	Dimensions Width mm (in)	Dimensions Height mm (in)	Octane Rating
TSIO-520-T	6	133×101.6 (5¼×4)	8.5 (520)	231 (310) at 2,700	231 (310) at 2,700	7.5	193.4 (426.3)	970 (38.2)	852 (33.56)	819 (32.26)	100/130
TSIO-520-UB	6	133×101.6 (5¼×4)	8.5 (520)	224 (300) at 2,700	224 (300) at 2,700	7.5	191.6 (422.5)	1,136 (44.5)	852 (33.56)	733 (28.86)	100/130
TSIO-520-VB	6	133×101.6 (5¼×4)	8.5 (520)	242.5 (325) at 2,700	242.5 (325) at 2,700	7.5	207.2 (456.7)	997 (39.25)	852 (33.56)	518 (20.41)	100/130
TSIO-520-WB	6	133×101.6 (5¼×4)	8.5 (520)	242.5 (325) at 2,700	242.5 (325) at 2,700	7.5	188.75 (416.1)	1,286 (50.62)	852 (33.56)	509 (20.02)	100/130
GTSIO-520-D, H	6	133×101.6 (5¼×4)	8.5 (520)	280 (375) at 3,400	280 (375) at 3,400	7.5	250 (550.4)	1,081 (42.56)	880 (34.04)	680 (26.78)	100/130
GTSIO-520-F, K	6	133×101.6 (5¼×4)	8.5 (520)	324 (435) at 3,400	324 (435) at 3,400	7.5	272.0 (600)	1,426 (56.12)	880 (34.04)	664 (26.15)	100/130
GTSIO-520-L, M, N	6	133×101.6 (5¼×4)	8.5 (520)	280 (375) at 3,350	280 (375) at 3,350	7.5	228 (502)‡	1,114 (43.87)	880 (34.04)	671 (26.41)	100/130
LTSIO-520-AE	6	133×101.6 (5¼×4)	8.5 (520)	186.5 (250) at 2,400	186.5 (250) at 2,400	8.5	172.2 (379.6)	967 (38.07)	846 (33.29)	543 (21.38)	100/130
Voyager 550	6	133×108 (5¼×4¼)	9.0 (550)	261 (350) at 2,700	261 (350) at 2,700	7.5	228.6 (504)				100/100LL
Voyager T-550	6	133×108 (5¼×4¼)	9.0 (550)	224 (300) at 2,500	224 (300) at 2,500	7.5	204 (450)				100LL
Voyager GT-550	6	133×108 (5¼×4¼)	9.0 (550)	298 (400) at 2,700	298 (400) at 2,700	7.5	249.5 (550)				100LL
IO-550-B	6	133×108 (5¼×4¼)	9.0 (550)	224 (300) at 2,700	224 (300) at 2,700	8.5	207.9 (462)	964 (37.97)	852 (33.56)	694 (27.32)	100LL
IO-550-C, E	6	133×108 (5¼×4¼)	9.0 (550)	224 (300) at 2,700	224 (300) at 2,700	8.5	196.4 (433)	1,100.1 (43.31)	852 (33.56)	502 (19.78)	100LL
IO-550-F, L	6	133×108 (5¼×4¼)	9.0 (550)	224 (300) at 2,700	224 (300) at 2,700	7.5	191.9 (423)	1,039 (40.91)	852 (33.56)	516 (20.32)	100LL
IO-550-G	6	133×108 (5¼×4¼)	9.0 (550)	209 (280) at 2,500	209 (280) at 2,500	7.5	210.9 (465)	1,189 (46.80)	852 (33.56)	518 (20.41)	100LL
TSIO-550B	6	133×108 5¼×4¼	9.0 (550)	261 (350) at 2,700	261 (350) at 2,700	7.5	257 (566)	1,086 (42.75)	1,072 (42.20)	853 (33.60)	100/100LL

*With accessories; †Not including turbocharger;
‡N weight 220 kg (486 lb)

TELEDYNE CONTINENTAL IO-240

This four-cylinder engine stems from a design by former licensee Rolls-Royce. For details, see table.

TELEDYNE CONTINENTAL IO-520 SERIES

These engines are similar to the earlier IO-360, but with cylinders of larger bore. They are fitted with an alternator driven either by a belt or by a face gear on the crankshaft. The TSIO-520 series are turbocharged. In 1981 TCM announced a lightweight series of engines with magnesium replacing aluminium in some areas, modified camshaft and cylinder heads (with parallel valves or inclined valves of larger diameter), and a range of turbocharging options. The GTSIO-520 series is geared.

Teledyne Continental IO-550 six-cylinder air-cooled engine

TELEDYNE CONTINENTAL IO-550

In 1984 this series of fuel-injected engines was introduced, similar to the IO-520 but with greater stroke. Initial applications were the Beechcraft Baron and Bonanza.

TELEDYNE CONTINENTAL VOYAGER 550

Largest of the new range of liquid-cooled engines, the Voyager 550 offers the same advantages as its smaller predecessors. The T-550 was designed for Bonanza conversions and the geared GT-550 for the Chieftain.

Teledyne Continental TSIO-550B six-cylinder liquid-cooled engine

Teledyne Continental Voyager T-550 six-cylinder liquid-cooled engine

TELEDYNE CAE

TELEDYNE CAE DIVISION OF TELEDYNE INC
1330 Laskey Road, Toledo, Ohio 43612-0971
Telephone: 1 (419) 470 3000
Fax: 1 (419) 470 3386
Telex: EASYLINK 6 288 4828
PRESIDENT: Robert S. Van Huysen

Teledyne CAE produces small gas turbine engines for training aircraft, missiles and UAVs. See 1987-88 *Jane's* for further details.

TELEDYNE CAE 352
US military designation: J69

One version of this simple turbojet is used in manned aircraft:

J69-T-25A (Teledyne CAE Model 352-5A). Long life version; powers Cessna T-37B.

DIMENSIONS (nominal):
Length overall 899 mm (35.39 in)
Width 566 mm (22.30 in)
WEIGHT, DRY: 165 kg (364 lb)
PERFORMANCE RATINGS:
Max T-O thrust 4.56 kN (1,025 lb) at 21,730 rpm
SPECIFIC FUEL CONSUMPTION:
Max T-O 32.30 mg/Ns (1.14 lb/h/lb st)

TEXTRON LYCOMING

TEXTRON LYCOMING (Turbine Engine Division)
550 Main Street, Stratford, Connecticut 06497
Telephone: 1 (203) 385 2000
Fax: 1 (203) 385 3255
PRESIDENT: David G. Assard

Textron Lycoming produces several families of engines, including the T53, T55, LF 500 and LT 101, with turboshaft, turboprop, turbofan, vehicular and marine variants. In June 1987 Textron Lycoming and General Electric agreed joint development of gas turbines for aero and ground applications. The aero engine is the T407/GLC38, described under General Electric.

TEXTRON LYCOMING ALF 502

The ALF 502 was launched in 1969, primarily for commercial and executive aircraft. The core is the T55, and construction is modular.

Current versions are as follows:

ALF 502L. First commercial version, FAA certificated in February 1980. Powers Canadair Challenger 600 in ALF 502L-2 form. L-2A, L-2C and L-3 certificated 1982-3.

ALF 502R. Reduced rating, FAA certificated January 1981 as R-3 to power BAe 146. Improved R-3A, R-4 and R-5 certificated 1982-3. R-6 certificated 1984.

TEXTRON LYCOMING ALF 502, LF 500, LTC1/T53 and LTC4/T55 ENGINES

Manufacturer's and civil designation	Military designation	Type*	T-O Rating kN (lb st) or max kW (hp)	SFC μg/J; ‡mg/Ns (lb/h/hp; ‡lb/h/lb st)	Weight, Dry less tailpipe kg (lb)	Max diameter mm(in)	Length overall mm (in)	Remarks
T5311A	—	ACFS	820 kW (1,100 shp)	115 (0.68)	225 (496)	584 (23)	1,209 (47.6)	Bell 204B
T5313B	—	ACFS	1,044 kW (1,400 shp)	98 (0.58)	245 (540)	584 (23)	1,209 (47.6)	Bell 205A
T5317A	—	ACFS	1,119 kW (1,500 shp)	99.7 (0.59)	256 (564)	584 (23)	1,209 (47.6)	Bell 205A-1
—	T53-L-13B	ACFS	1,044 kW (1,400 shp)	98 (0.58)	245 (540)	584 (23)	1,209 (47.6)	Advanced UH-1H, AH-1G
—	T53-L-703	ACFS	1,106 kW (1,485 shp)	101.4 (0.60)	247 (545)	584 (23)	1,209 (47.6)	Bell AH-1S TOW/Cobra
LTC1K-4K	—	ACFS	1,156 kW (1,550 shp)	98.7 (0.584)	234 (515)	584 (23)	1,209 (47.6)	Bell XV-15
—	T53-L-701	ACFP	1,082 ekW (1,451 ehp)	101.4 (0.60)	312 (688)	584 (23)	1,483 (58.4)	Grumman OV-1D, AIDC (Taiwan) T-CH-1
—	YT55-L-9	ACFP	1,887 ekW (2,529 ehp)	102.7 (0.608)	363 (799)	615 (24.2)	1,580 (62.2)	Piper Enforcer
—	T55-L-7C	ACFS	2,125 kW (2,850 shp)	101.4 (0.60)	267 (590)	615 (24.2)	1,118 (44)	
T5508D (LTC4B-8D)	—	ACFS	2,186 kW (2,930 shp) flat rated at 1,678 kW (2,250 shp)	100.1 (0.592) 106.0 (0.628)	274 (605)	610 (24)	1,118 (44)	Bell 214A, 214B
—	T55-L-712†	ACFS	2,796 kW (3,750 shp)	89.6 (0.53)	340 (750)	615 (24.2)	1,181 (46.5)	Boeing CH-47D
AL5512	—	ACFS	3,039 kW (4,075 shp)	89.6 (0.53)	355 (780)	615 (24.2)	1,118 (44)	Boeing 234
—	T55-L-714	ACFS	3,629 kW (4,867 shp)	85.0 (0.503)	363 (800)	615 (24.2)	1,181 (46.5)	Boeing MH-47E
ALF 502R-3	—	ACFF	29.8 kN (6,700 lb)	‡11.64 (‡0.411)	576 (1,270)	1,059 (41.7)	1,443 (56.8)	BAe 146
ALF 502R-3A	—	ACFF	31.0 kN (6,968 lb)	‡11.55 (‡0.408)	576 (1,270)	1,059 (41.7)	1,443 (56.8)	BAe 146
ALF 502R-5	—	ACFF	31.0 kN (6,968 lb)	‡11.55 (‡0.408)	583 (1,283)	1,059 (41.7)	1,443 (56.8)	BAe 146
ALF 502R-6	—	ACFF	33.36 kN (7,500 lb)	‡11.73 (‡0.415)	589 (1,298)	1,059 (41.7)	1,487 (58.56)	BAe 146
ALF 502L/L-2	—	ACFF	33.36 kN (7,500 lb)	‡12.1 (‡0.428)	589 (1,298)	1,059 (41.7)	1,487 (58.56)	Canadair Challenger 600
ALF 502L-2A	—	ACFF	33.36 kN (7,500 lb)	‡11.70 (‡0.414)	589 (1,298)	1,059 (41.7)	1,487 (58.56)	
ALF 502L-3	—	ACFF	34.74 kN (7,800 lb)	‡11.73 (‡0.415)	589 (1,298)	1,059 (41.7)	1,487 (58.56)	
LF 507-1F	—	ACFF	31.14 kN (7,000 lb)	‡11.50 (‡0.406)	628 (1,385)	1,059 (41.7)	1,487 (58.56)	Avro Int. RJ
LF 507-1H	—	ACFF	31.14 kN (7,000 lb)	‡11.50 (‡0.406)	628 (1,385)	1,059 (41.7)	1,487 (58.56)	Avro Int. RJ

*ACFS axial plus centrifugal, free turbine shaft; ACFP axial plus centrifugal, free turbine propeller; ACFF axial plus centrifugal, free turbine fan
†Applies to T55-L-11A, C**, D, E** and 712**, those designated ** having 2½ min contingency rating of 3,357 kW (4,500 shp)

680 USA: ENGINES—TEXTRON LYCOMING

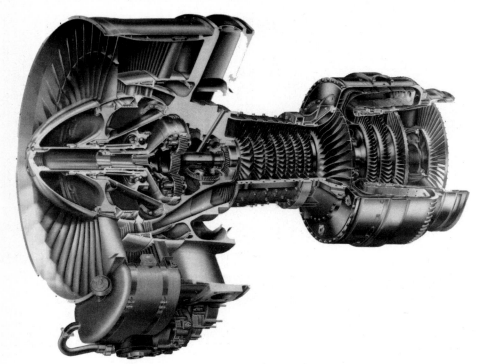

Cutaway drawing of Textron Lycoming LF 507 geared turbofan

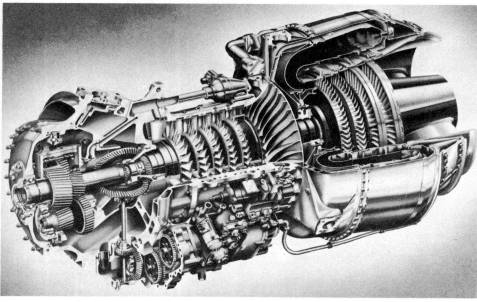

Cutaway drawing of Textron Lycoming T53-L-13B free turbine turboshaft

Cutaway drawing of Textron Lycoming T55-L-714 (military) or AL5512 (commercial) free turbine turboshaft

By 1992 a total of 184 ALF 502L engines had flown 605,000 h in the Challenger, while 1,050 ALF 502R engines had flown 6,600,000 h in the BAe 146. The ALF 502 and LF 507 operate on-condition.

TYPE: High bypass ratio, two-shaft geared turbofan.
FAN MODULE: Cast frame includes four engine mounts 90° apart, and may carry reverser. Fan rotor blades are base and part-span shrouded. Mounted directly behind rotor (6,700 lb st engines) is a single or (7,500 lb st engines) two stages of compression. Anti-icing of LP compressor inlet by bleed air. Accessory gearbox on fan frame takes HP shaft power. Reduction gear couples LP turbine to fan. Bypass ratio: 502R-3, 5.71; 502R-5, 5.6; 502L, 5.0.
COMPRESSOR: HP compressor has seven stages and single centrifugal. Acceleration bleed between stages 6 and 7 operated by main fuel control. Overall pressure ratio: R-3, 11.6; R-5, 12.0; L-2, 13.6.
TURBINE: HP has two air-cooled stages. LP has two stages. All rotor blades base shrouded; LP tip shrouded.
COMBUSTION CHAMBER: One-piece annular combustor wraps around turbine. Atomising nozzles inserted through outer chamber at rear. Disconnecting permits removal of combustor/turbine module, providing access to HP turbine.
ACCESSORY DRIVES: Accessory gearbox carries main fuel control, oil pump and filter, tachometer (if required) and provisions for customer accessories.
DATA: See table on page 679.

TEXTRON LYCOMING LF 500

In September 1988 Textron Lycoming announced plans for a family of commercial turbofans based on the ALF 502R, using the same core. The engines will range from 31.14 kN (7,000 lb st) to 62.28 kN (14,000 lb st), the most powerful versions requiring an increase in fan diameter. The **LF 507-1H**, the first engine in the family, was certificated in October 1991. The **LF 507-1F** was certificated in March 1992 and entered airline service with the Avro RJ85 in April 1993.

Features include an additional supercharger stage, machined combustor liner, cast fourth nozzle, reduced turbine inlet temperature, steel compressor casing, improved reduction gear lubrication and FADEC. Data, see table on page 679.

TEXTRON LYCOMING LTC1
US military designation: T53

The T53 was developed under a joint US Air Force/Army contract. More than 19,000 have logged over 46 million hours since 1956.

Licences for manufacture of the T53 are held by BMW Rolls-Royce in Germany, Piaggio in Italy, Kawasaki in Japan, and in Taiwan.

Current versions are as follows:
T53-L-13. Uprated L-11. Redesigned 'hot end' and initial stages of compressor. Four turbine stages, compared with two in earlier models, and variable inlet guide vanes combined with redesigned first two compressor stages. Atomising combustor to facilitate operation on a wider range of fuels. Powers Bell UH-1M and UH-1H and AH-1F HueyCobra. The **T5313A** commercial version has been superseded by the **T5313B**.
T53-L-701. Turboprop incorporating 'split power' reduction gear.
T53-L-703. Improved durability L-13. Thermodynamic rating 1,343 kW (1,800 shp).
LTC1K-4K. Direct drive L-13 suitable for operation from 105° nose up to 90° nose down.
T5317A. Improvements over L-13 include improved cooling of first gas producer turbine nozzle plus air-cooled blades in first turbine rotor.

The following details apply to the T53-L-13 and L-701:
TYPE: Free turbine turboshaft.
COMPRESSOR: Five axial stages followed by single centrifugal stage. Variable inlet guide vanes. Pressure ratio 7.2. Mass flow 5.53 kg (12.2 lb)/s at 25,150 gas producer rpm.
COMBUSTION CHAMBER: Annular reverse flow, with 22 atomising injectors.
FUEL CONTROL SYSTEM: Chandler Evans TA-2S and TA-7 system with one dual fuel pump, 41.4 bars (600 lb/sq in). Interstage air bleed control.
FUEL GRADE: ASTM A-1, MIL-J-5624, MIL-F-26005A, JP-1, JP-4, JP-5, CITE.
TURBINE: First two stages, driving compressor, use hollow air-cooled stator vanes and cored-out cast steel rotor blades. Second two stages, driving reduction gearing, have solid steel blades.
ACCESSORIES: Electric starter or starter/generator. Bendix-Scintilla TGLN high energy ignition unit.
LUBRICATION: Recirculating system, with gear pump, 4.83 bars (70 lb/sq in).
OIL GRADE: MIL-L-7808, MIL-L-23699.
DATA: See table on page 679.

TEXTRON LYCOMING LTC4
US military designation: T55

This engine is based on the T53 with higher mass flow. Total operating time by 1992 on more than 4,000 engines was over 6 million hours.

Current versions are as follows:
LTC4B-8D. Modified T55-L-7C.

T5508D. Commercial version of LTC4B-8D.

T55-L-11 (LTC4B-11B) series. Uprated L-7, with variable inlet guide vanes and two-stage compressor turbine.

T55-L-712. Improved L-11D. Wide chord compressor blades without inlet guide vanes, and one-piece rotor.

AL5512. Commercial L-712, with engine out contingency rating of 3,250 kW (4,355 shp).

T55-L-714. Growth version with cooled gas-generator turbine blades, FADEC and improved torquemeter. OEI rating (contingency) 3,780 kW (5,069 shp); max continuous 3,108 kW (4,168 shp).

The following description applies to the T55-L-714:

TYPE: Free turbine turboshaft.
COMPRESSOR: Seven axial stages followed by single centrifugal stage. Pressure ratio 9.3. Mass flow 13.19 kg (29.08 lb)/s.
COMBUSTION CHAMBER: Annular reverse flow. Twenty-eight atomizing nozzles.
FUEL SYSTEM: Chandler Evans FADEC type EMC-32T-2, consisting of hydromechanical unit with gear type pump and DECU.
FUEL GRADE: MIL-J-5624L grade JP-4, JP-5, MIL-T-83133 grade JP-8 or CITE.
TURBINE: Gas generator turbine has two stages with cooled blades and cooled shrouds to give tip clearance control. Two-stage power turbine.
ACCESSORIES: Electric starter or starter/generator, or air or hydraulic starter. Bendix-Scintilla TGLN high energy ignition unit. Four igniter plugs.
LUBRICATION: Recirculating. Integral tank and cooler.
OIL GRADE: MIL-L-7808, MIL-L-23699.
DATA: See table on page 679.

TEXTRON LYCOMING LT 101
US military designation: YT702-LD-700

The LT 101 is designed for low life cycle costs and fuel efficiency. Each engine comprises an accessory reduction gearbox, gas generator and combustor/power turbine module. The engine has a single axial compressor stage followed by a single centrifugal stage, a reverse flow annular combustor, a single-stage gas generator turbine, and a single-stage power turbine. Front gearboxes provide output speeds of 1,925, 6,000 or 9,545 rpm. The 6,000 rpm gearbox has both forward and aft drives. The engine has either a scroll or radial inlet. Mass flow is 2.03 kg (4.8 lb)/s, and pressure ratio 8.5.

Current production versions include turboshaft (LTS) and turboprop (LTP) models, with max power in the 459 to 548 kW (615 to 735 shp) range. All are certificated under FAR Pt 33 for 2,400 h TBO or on-condition maintenance. LT 101 operating experience now totals more than 3 million hours, mainly in helicopters.

TEXTRON LYCOMING LTS 101 ENGINES

Engine Model	Performance Rating (T-O, S/L) kW (shp)	SFC µg/J (lb/h/shp)	Weight, Dry kg (lb)	Length mm (in)	Diameter mm (in)
LTS 101-600A-2	459 (615)	96.5 (0.571)	115 (253)	785 (30.9)	599 (23.6)
LTS 101-600A-3	459 (615)	98.4 (0.582)	120 (265)	785 (30.9)	599 (23.6)
LTS 101-650B-1	410 (550)	97.5 (0.577)	124 (273)	790 (31.1)	645 (25.4)
LTS 101-650C-2/C-3/C-3A	447 (600)	96.7 (0.572)	109.5 (241)	787 (31.0)	574 (22.6)
LTS 101-750B-1	410 (550)	97.5 (0.577)	124 (273)	822 (32.36)	627 (24.7)
LTS 101-750B-2	510 (684)	96.3 (0.570)	123 (271)	822 (32.36)	627 (24.7)
LTS 101-750C-1	510 (684)	97.5 (0.577)	110.5 (244)	790 (31.1)	574 (22.6)

TEXTRON LYCOMING LTS 101

The **LTS 101-600A-2** is a 6,000 rpm power plant for the Eurocopter AS 350D AStar. The **650C-2/C-3** is a 9,545 rpm power plant for the Bell 222. The **650B-1**, a 6,000 rpm engine with a radial inlet for the Eurocopter/Kawasaki BK117A. The **750B-1** powers the BK117B. The **600A-3** powers the AS 350D Mk 3. A growth version, the **750C-1**, powers the Bell 222B and 222UT. The **750B-2**, with radial inlet, powers the AS 366/HH-65A Dolphin.
DATA: See table above.

TEXTRON LYCOMING LTP 101

The LTP 101 turboprop incorporates a free power turbine, provisions for tractor or pusher installation, hydraulic propeller governor, radial screened inlet and anti-icing protection. Output speed is 1,700–1,950 rpm. The **LTP 101-600A-1A** and **-700A-1A** power the P.166-DL3, Air Tractor and Cresco agricultural aircraft, Riley Cessna 421 and Page Turbo Thrush and Ag-Cat. It has flown in Piper Brave, Turbine Islander and Dornier 128-6 prototypes.

DIMENSIONS:
Length   914 mm (36.0 in)
Diameter   533 mm (21.0 in)
WEIGHT, DRY: 152 kg (335 lb)
PERFORMANCE RATINGS (T-O, S/L):
LTP 101-600   462 ekW (620 ehp)
LTP 101-700   522 ekW (700 ehp)
SPECIFIC FUEL CONSUMPTION (T-O, S/L):
LTP 101-600, -700   93 µg/J (0.55 lb/h/ehp)

Left to right: Textron Lycoming LTS 101-750B-1 turboshaft, LTP 101-700A-1A turboprop and LTS 101-750C-1 turboshaft

---

RECIPROCATING ENGINE DIVISION
652 Oliver St, Williamsport, Pennsylvania 17701
Telephone: 1 (717) 327 7041
Fax: 1 (717) 327 7022

This division is the world's largest producer of piston engines for general aviation.

TEXTRON LYCOMING O-235 SERIES

Four cylinders of 111 mm (4⅜ in) bore and 98.4 mm (3⅞ in) stroke. The high compression O-235-N is the most recent production version of the O-235, used in several trainers. It requires 100 octane fuel.

TEXTRON LYCOMING O-320 and IO-320 SERIES

Cylinder bore increased to 130 mm (5⅛ in). The O-320 is an engine in the 112-119 kW (150-160 hp) class. Both carburetted and fuel injected versions are produced in low and high compression models for use with 80/87 or 100 octane minimum grade fuels, respectively. Fully aerobatic models are available.

Textron Lycoming IO-360 four-cylinder piston engine

USA: ENGINES—TEXTRON LYCOMING

Textron Lycoming O-540 six-cylinder piston engine

Textron Lycoming IO-720 eight-cylinder piston engine

Textron Lycoming TIO-540-AE2A turbocharged piston engine

TEXTRON LYCOMING O-540 and IO-540 SERIES

The O-540 is a direct drive, six-cylinder version of the four-cylinder O-360. It is available in low and high compression versions, and the VO-540 is a helicopter power plant with crankshaft vertical. Fuel injected IO-540 models are manufactured with ratings of 186-224 kW (250-300 hp). An aerobatic version is available.

TEXTRON LYCOMING TIO-540 SERIES

This is a turbocharged version of the fuel injected IO-540, some of which have tuned induction. It is manufactured for unpressurised and pressurised aircraft, and several models incorporate an intercooler.

TEXTRON LYCOMING TIO-541 SERIES

Although the displacement of this turbocharged, six-cylinder series is the same as that of the TIO-540, the TIO-541 and geared TIGO-541 are totally redesigned. The TIO-541-E is rated at 283 kW (380 hp) and the geared TIGO-541-E at 317 kW (425 hp). A double scroll blower is available to provide cabin pressurisation.

TEXTRON LYCOMING IO-720 SERIES

This eight-cylinder version of the IO-540 is used at ratings from 280 to 298 kW (375 to 400 hp).

TEXTRON LYCOMING O-360 and IO-360 SERIES

The O-360 series is basically the same as the O-320 except for an increase in stroke to 111 mm (4⅜ in). Like the O-320, this engine is manufactured with low or high compression, with carburettor or fuel injection. The various models include aerobatic capability, a specific design for helicopters and turbocharged versions. The IO-360-A has fuel injection, tuned induction and high output cylinders, while the IO-360-B has continuous flow port injection and standard cylinders. In the TIO-360-C the turbocharger is pilot controlled.

TEXTRON LYCOMING HORIZONTALLY OPPOSED PISTON ENGINES

Engine Model*	No. of Cylinders	Rated output at Sea Level kW (hp) at rpm	Capacity litres (cu in)	Compression Ratio	Fuel grade Minimum	Weight, Dry kg (lb)	Length Overall mm (in)	Width Overall mm (in)	Height Overall mm (in)	Gear Ratio†
O-235-C	4	86 (115) at 2,800	3.85 (233)	6.75	80/87	97.5 (215)	751 (29.56)	812 (32.00)	569 (22.40)	D
O-235-L, M	4	88 (118) at 2,800 78 (105) at 2,400	3.85 (233)	8.5	100	98 (218)	738 (29.05)	812 (32.00)	569 (22.40)	D
O-235-N, P	4	87 (116) at 2,800	3.85 (233)	8.1	100	98 (218)	738 (29.05)	812 (32.00)	569 (22.40)	D
O-320-A, E	4	112 (150) at 2,700	5.2 (319.8)	7.0	80/87	110 (243)	751 (29.56)	819 (32.24)	584 (22.99)	D
(H)O-320B2C	4	119 (160) at 2,700	5.2 (319.8)	8.5	91/96	115 (255)	751 (29.56)	819 (32.24)	584 (22.99)	D
O-320-D	4	119 (160) at 2,700	5.2 (319.8)	8.5	91/96	114 (253)	808 (31.82)	819 (32.24)	488 (19.22)	D
AEIO-320-D	4	119 (160) at 2,700	5.2 (319.8)	8.5	100	123 (271)	780 (30.70)	819 (32.24)	589 (23.18)	D
AEIO-320-E	4	112 (150) at 2,700	5.2 (319.8)	7.0	80/87	117 (258)	738 (29.05)	819 (32.24)	584 (22.99)	D
(L)IO-320B, C	4	119 (160) at 2,700	5.2 (319.8)	9.0	100	117.5 (259)	853 (33.59)	830 (32.68)	488 (19.22)	D
(L)O-360-A	4	134 (180) at 2,700	5.92 (361)	8.5	91/96	120 (265)	808 (31.82)	848 (33.37)	488 (19.22)	D
O-360-F	4	134 (180) at 2,700	5.92 (361)	8.5	100	122 (269)	808 (31.81)	859 (33.38)	507 (19.96)	D
TO-360-C, F	4	157 (210) at 2,575 to 3,050 m (10,000 ft)	5.92 (361)	7.3	100	154 (343)	876 (34.50)	921 (36.25)	534 (21.02)	D
IO-360-A	4	149 (200) at 2,700	5.92 (361)	8.7	100	133 (293)	757 (29.81)	870 (34.25)	491 (19.35)	D
IO-360-B	4	134 (180) at 2,700	5.92 (361)	8.5	100	122 (268)	757 (29.81)	848 (33.37)	631 (24.84)	D
LIO-360-C	4	149 (200) at 2,700	5.92 (361)	8.7	100	134 (298)	855 (33.65)	870 (34.25)	495 (19.48)	D
TIO-360-C	4	210 (282) at 2,575 to 3,050 m (10,000 ft)	5.92 (361)	7.3	100	158 (348)	910 (35.82)	921 (36.25)	550 (21.65)	D

*Model designation code: AE, Aerobatic engine; G, Geared; H, Helicopter; I, Fuel injected; L, Left-hand rotation crankshaft; O, Opposed cylinders; S, Supercharged; T, Turbocharged. †D, Direct drive.

TEXTRON LYCOMING HORIZONTALLY OPPOSED PISTON ENGINES Continued

HIO-360-D1A	4	142 (190) at 3,200 to 1,280 m (4,200 ft)	5.92 (361)	10.0	100	146 (321)	894 (35.23)	904 (35.62)	495 (19.48)	D
(L)HIO-360-F1AD	4	142 (190) at 3,050	5.92 (361)	8.0	100	133 (293)	797 (31.36)	870 (34.25)	507 (19.97)	D
AEIO-360-A	4	149 (200) at 2,700	5.92 (361)	8.7	100	139 (307)	780 (30.70)	870 (34.25)	492 (19.35)	D
AEIO-360-B	4	134 (180) at 2,700	5.92 (361)	8.5	91/96	125 (277)	738 (29.05)	848 (33.37)	631 (24.84)	D
O-540-B	6	175 (235) at 2,575	8.86 (541.5)	7.2	80/87	166 (366)	945 (37.22)	848 (33.37)	624 (24.56)	D
O-540-E	6	194 (260) at 2,700	8.86 (541.5)	8.5	91/96	167 (368)	976 (38.42)	848 (33.37)	624 (24.56)	D
O-540-A	6	186 (250) at 2,700	8.86 (541.5)	8.5	91/96	161 (356)	976 (38.42)	848 (33.37)	624 (24.56)	D
O-540-J	6	175 (235) at 2,400	8.86 (541.5)	8.5	100	162 (357)	989 (38.93)	848 (33.37)	519 (20.43)	D
IO-540AA1A5	6	201 (270) at 2,700	8.86 (541.5)	7.3	80/87	217 (479)	999 (39.34)	880 (34.70)	498 (19.60)	D
IO-540-S	6	224 (300) at 2,700	8.86 (541.5)	8.7	100	200 (441)	997 (39.24)	880 (34.70)	498 (19.60)	D
IO-540-C	6	186 (250) at 2,575	8.86 (541.5)	8.5	91/96	170 (375)	976 (38.42)	848 (33.37)	622 (24.46)	D
IO-540-K	6	224 (300) at 2,700	8.86 (541.5)	8.7	100	201 (443)	999 (39.34)	870 (34.25)	498 (19.60)	D
IO-540-T4A5D	6	194 (260) at 2,700	8.86 (541.5)	8.5	100	187 (412)	989 (38.93)	848 (33.37)	546 (21.50)	D
IO-540-W1A5	6	175 (235) at 2,400	8.86 (541.5)	8.5	100	166 (367)	989 (38.93)	848 (33.37)	492 (19.35)	D
AEIO-540-D	6	194 (260) at 2,700	8.86 (541.5)	8.5	91/96	174 (386)	999 (39.34)	848 (33.37)	621 (24.46)	D
AEIO-540-L	6	224 (300) at 2,700	8.86 (541.5)	8.0	100	202 (445)	989 (38.93)	870 (34.25)	622 (24.46)	D
TIO-540-AF1A	6	201 (270) at 2,575 to 4,575 m (15,000 ft)	8.86 (541.5)	7.3	100	223 (491)	1,022 (40.24)	848 (33.37)	727 (28.62)	D
TIO-540-AE2A	6	261 (350) at 2,600	8.86 (541.5)	7.3	100	249 (549)	1,067 (42.02)	1,182 (46.52)	705 (27.75)	D
TIO-540-C	6	186 (250) at 2,575 to 4,575 m (15,000 ft)	8.86 (541.5)	7.2	100	205 (456)	1,026 (40.38)	848 (33.37)	770 (30.33)	D
TIO/LTIO-540-F	6	242 (325) at 2,575 to 4,575 m (15,000 ft)	8.86 (541.5)	7.3	100	233 (514)	1,304 (51.34)	870 (34.25)	570 (22.42)	D
TIO/LTIO-540-J	6	261 (350) at 2,575 to 4,575 m (15,000 ft)	8.86 (541.5)	7.3	100	235 (518)	1,308 (51.50)	870 (34.25)	573 (22.56)	D
TIO-540-S	6	224 (300) at 2,700 to 3,660 m (12,000 ft)	8.86 (541.5)	7.3	100	228 (502)	1,004 (39.56)	915 (36.02)	667 (26.28)	D
TIO/LTIO-540-U	6	261 (350) at 2,500 to 4,575 m (15,000 ft)	8.86 (541.5)	7.3	100	248 (547)	1,204 (47.40)	870 (34.25)	574 (22.59)	D
TIO/LTIO-540-V	6	269 (360) at 2,600 to 5,485 m (18,000 ft)	8.86 (541.5)	7.3	100	248 (547)	1,352 (53.21)	886 (34.88)	621 (24.44)	D
TIO-541-E	6	283 (380) at 2,900 to 4,575 m (15,000 ft)	8.86 (541.5)	7.3	100	270 (596)	1,282 (50.70)	905 (35.66)	640 (25.17)	D
TIGO-541-E	6	317 (425) at 3,200 to 4,575 m (15,000 ft)	8.86 (541.5)	7.3	100	319 (704)	1,462 (57.57)	885 (34.86)	575 (22.65)	0.667
IO-720-A, B, D	8	298 (400) at 2,650	11.84 (722)	8.7	100	258 (568)	1,179 (46.41)	870 (34.25)	573 (22.53)	D

*Model designation code: AE, Aerobatic engine; G, Geared; H, Helicopter; I, Fuel injected; L, Left-hand rotation crankshaft; O, Opposed cylinders; S, Supercharged; T, Turbocharged. †D, Direct drive.

THIOKOL

THIOKOL CORPORATION
2475 Washington Blvd, Ogden, Utah 84401-2398
Telephone: 1 (801) 629 2270
Fax: 1 (801) 629 2420
SOLID PROPELLANT ROCKET MOTOR PLANTS: Elkton, Maryland; Huntsville, Alabama; Promontory and Brigham City, Utah
CHAIRMAN AND CEO: U. E. Garrison
PRESIDENT: John R. Myers
DIRECTOR, COMMUNICATION AND MARKETING:
Gilbert R. Barley

Established in 1929, Thiokol produced the first American synthetic rubber. In 1943, the discovery by Thiokol of liquid polymer, a new type of synthetic rubber, paved the way for the 'case-bonded' rocket. The company's polysulphide liquid polymer proved to be the catalyst for production of efficient solid motors, as well as the development of the world's largest.

THIOKOL SHUTTLE SSRM

The NASA Shuttle solid rocket motors (SSRM) are manufactured and cast in four segments at the main plant in Utah. The SSRMs are then transported to the Vehicle Assembly Building at Kennedy Space Center (KSC), Florida, where they are stacked vertically on either side of the external tank on top of the mobile platform launcher. All three are then joined to the orbiter (Space Shuttle).

During the first two minutes of flight the two Thiokol boosters provide the majority of thrust. A flexible bearing nozzle driven by two actuators provides thrust-vector control for the boosters. The boosters separate after burnout and fall into the Atlantic Ocean about 130 miles from KSC. Two retrieval ships recover the fallen boosters and transport them to port where they are cleaned, disassembled, and shipped back to Utah for refurbishment and reloading. The loaded boosters are then shipped back to KSC.

Thiokol booster motors have flown on every manned Space Shuttle flight. With the signing of the 1991 Buy 3 contract, the booster motor will continue to boost Space Shuttle through the 1990s.

DIMENSIONS:
Length   38.47 m (126.2 ft)
Diameter   3,708 mm (146 in)
WEIGHTS:
Propellant   503,478 kg (1,109,980 lb)
Loaded motor   569,883 kg (1,256,378 lb)
THRUST:
Average vacuum   11,521 kN (2,590,000 lb)
BURN TIME: 123 s

WILLIAMS

WILLIAMS INTERNATIONAL
2280 West Maple Road, PO Box 200, Walled Lake, Michigan 48390
Telephone: 1 (313) 624 5200
Fax: 1 (313) 669 1577
VICE-PRESIDENT, BUSINESS DEVELOPMENT:
Raymond C. Preston
PUBLIC RELATIONS: Christina J. Pearce
Telephone: 1 (313) 960 2409
Fax: 1 (313) 669 3790

Details of the engines manufactured by Williams for unmanned applications can be found in the 1987-88 *Jane's*.

WILLIAMS-ROLLS FJ44
US military designation: F129

Rolls-Royce joined Williams as a partner on the FJ44 in 1989. Williams is the design authority, with Rolls-Royce providing its expertise for certain components as well as assisting in product support. In production, the majority of manufacturing and final assembly is being accomplished at the Williams facility in Ogden, Utah, while Rolls-Royce manufactures the LP turbine shaft and the three turbine rotors. The engine is marketed and supported worldwide by Williams-Rolls Inc, a joint business arrangement between Williams International and Rolls-Royce.

The FJ44, with its low acquisition and maintenance cost, low specific fuel consumption, and high reliability, made feasible a new category of light business jets and trainers. It completed FAA certification on schedule in March 1992, with the first production engines being delivered in June 1992. The Cessna CitationJet is in production, and other air-

Williams-Rolls FJ44 two-shaft turbofan

craft either in development or certification include the Swearingen SJ30, the Promavia twin-engine ATTA trainer, the Promavia single-engine Squalus trainer, and the AASI Stratocruzer business jet. All FJ44 engines are assembled at and shipped from Williams' Ogden, Utah facility.

Intended applications include the H&M 303 19-passenger trijet, a twin-FJ44 version of the AASI Jetcruzer and a twin-FJ44 version of the Machen Aerostar.

Williams International was to announce during 1993 details of its military version of the FJ44, the **F129**. This engine has been selected by Cessna for its JPATS twin-engine trainer that was to fly in 1993.

FAN: Single-stage wide-chord axial. Bypass ratio 3.28.
COMPRESSOR: Single-stage axial on LP shaft followed by a single-stage centrifugal on HP shaft.
COMBUSTION CHAMBER: Annular radial outflow.
TURBINES: Single-stage axial HP turbine. Two-stage axial LP turbine. All three rotors with inserted blades.
CONTROL SYSTEM: Hydromechanical.
NOZZLE: Full-length fan duct. Common fan and core exhaust.
DIMENSIONS:
  Length   1,024 mm (40.3 in)
  Max diameter   602 mm (23.7 in)
WEIGHT, DRY: 202 kg (445 lb)
PERFORMANCE RATINGS:
  T-O (S/L) to 22°C   8.45 kN (1,900 lb st)
  Max continuous (11,000 m; 36,090 ft; Mach 0.7)
                                                            2.25 kN (506 lb)
SPECIFIC FUEL CONSUMPTION:
  T-O (S/L)   13.45 mg/Ns (0.475 lb/h/lb lb)
  Max continuous (as above)   21.47 mg/Ns (0.758 lb/h/lb)

YUGOSLAVIA (FORMER)

Aero-engine manufacture in Yugoslavia had a long tradition. Over 1,200 engines of indigenous or licensed design were built in the country between 1923 and 1941.

The group of aero-engine makers was integrated within the Business Association of the Yugoslav Aerospace Industry. They manufactured under collaborative or counter-purchase arrangements. The group also performed modifications, upgrades, repair and overhaul of engines and their components. There was also a Business Association of Air Depots. The fate of the engine factories in the war was not known as this edition went to press. The *Jane's* enquiries for the 1993-94 update were all returned undeliverable because of international sanctions or military blockade.

DMB
DVADESETPRVI MAJ BEOGRAD
Ul. Oslobodjenja 1, YU-11090 Belgrade
Telephone: 38 (11) 593685 and 593982

Fax: 38 (11) 593982 and 594985
Telex: 71084 DMB YU
 Manufactured Turbomeca turboshaft and turboprop and parts of RR Viper turbojet.

MOMA STANOJLOVIĆ
MOMA STANOJLOVIĆ AIR DEPOT
YU-11273 Batajnica
Telephone: 38 (11) 618987
Fax: 38 (11) 619483

Telex: 11675 VZ MOMA YU
 The Moma Stanojlović Air Depot used to co-operate with DMB in the manufacture of the Turbomeca Astazou III and Astazou XIV turboshaft engines and overhaul military engines.

ORAO
ORAO AIR DEPOT
YU-71163 Rajlovac
Telephone: 38 (71) 455444

Fax: 38 (71) 455383
Telex: 41230
 Depot overhauled Viper turbojets and made parts for GE, RR and SNECMA engines.

ADDENDA

AIRCRAFT

ARGENTINA

AERO BOERO (page 1)
AERO BOERO 180
TYPE: Two-seat pre-selection aircraft.
CURRENT VERSIONS: **AB 180 PSA**: Modified AB 180; *details as for AB 180 RVR except as follows:*
POWER PLANT: Fuel capacity 176 litres (46.5 US gallons; 38.7 Imp gallons).
ACCOMMODATION: Tandem seats for instructor and pupil. Baggage compartment aft of seats, with external access.
SYSTEMS: 12V 35Ah battery.
AVIONICS: Include Bendix/King KX 155 VHF radio, VOR and ATC transponder.

DIMENSIONS, EXTERNAL:
Wing span   10.90 m (35 ft 9¼ in)
DIMENSIONS, INTERNAL:
Baggage compartment volume   0.21 m³ (7.42 cu ft)
AREAS:
Wings, gross   17.28 m² (186.0 sq ft)
WEIGHTS AND LOADINGS:
Weight empty   640 kg (1,411 lb)
Max T-O weight   890 kg (1,962 lb)
Max wing loading   51.50 kg/m² (10.55 lb/sq ft)
Max power loading   6.64 kg/kW (10.90 lb/hp)
PERFORMANCE:
Never-exceed speed (V_{NE})   132 knots (245 km/h; 152 mph)
Cruising speed   107 knots (198 km/h; 123 mph)
Landing speed   46 knots (84 km/h; 53 mph)
Stalling speed: flaps up   45 knots (82 km/h; 51 mph)
    flaps down   43 knots (79 km/h; 49 mph)
Max rate of climb at S/L   312 m (1,023 ft)/min
T-O run   85 m (279 ft)
Landing run   88 m (289 ft)

Aero Boero 180 PSA tandem-seat pre-selection aircraft

BRAZIL

EMBRAER (page 13)

EMBRAER EMB-120 BRASILIA
CURRENT VERSIONS: **EMB-120**: Initial production version, with 1,118.5 kW (1,500 shp) PW115 engines and Hamilton Standard 14RF four-blade propellers (replaced at early stage by higher output PW115s of 1,193 kW; 1,600 shp).
EMB-120RT (Reduced Take-off): Current standard production version; 1,342 kW (1,800 shp) PW118 engines for better field performance; most early aircraft now retrofitted to RT standard. Also, from late 1986, in hot/high version with PW118As which maintain max output up to ISA + 15°C at S/L.
Improved Brasilia: As described on page 14; provisional designation EMB-120X.
EMB-120 Cargo: All-cargo version with 4,000 kg (8,818 lb) capacity; floor and sidewall protection, fire detection system, smoke curtain between flight deck and cargo cabin, and cargo restraint net.
EMB-120 Combi: Mixed-configuration version with quick-release seats, 9g movable rear bulkhead and cargo restraint net; retains toilet and galley; typical capacity 30 passengers and 700 kg (1,543 lb) of cargo or 19 passengers and 1,100 kg (2,425 lb) cargo.
EMB-120ER (Extended Range): As described on page 14; introduced early 1992; available as new-build or retrofit.
EMB-120QC (Quick Change): Convertible in 40 minutes from 30-passenger layout to 3,500 kg (7,716 lb) all-cargo configuration with floor and sidewall protection, fire detection system, smoke curtain aft of flight deck, 9g removable rear bulkhead and cargo restraint net; conversion from cargo to passenger interior takes 50 minutes. First customer (14 May 1993 delivery) Total Linhas Aéreas of Brazil.

EMBRAER EMB-145
PROGRAMME: Additional partners announced by 10 June 1993 were Sonaca of Belgium (rear fuselage section 1, including engine pylons and cargo door, plus centre-fuselage section 1); ENAER of Chile (fin, tailplane and elevators); C & D Interiors (passenger cabin and cargo compartment interiors); Honeywell Business and Commuter Aviation Systems Division, USA (complete digital avionics suite, based on its Primus 1000 system); Auxiliary Power International Corporation, USA (18.6 kW; 25 shp APS-500 APU); Hamilton Standard, USA (air-conditioning and bleed air systems); B. F. Goodrich, USA (wheels and brakes); Crane Hydro-Aire Division, USA (brake-by-wire control system); and ABG-SEMCA of France (pressurisation system).

Single prototype (for first flight early 1995) will be joined by three pre-series aircraft for development flight testing and certification; CTA and FAA type approval expected first quarter 1996, with simultaneous European JAA certification; pre-series aircraft will later be refurbished for customer delivery.
CURRENT VERSIONS: **EMB-145**: As described on pages 16-17.
EMB-145ER (Extended Range): Under simultaneous development for full-passenger payload range of 1,180 nm (2,187 km; 1,359 miles); max T-O weight increased to 20,600 kg (45,415 lb), but payload and other max operating and landing weights unchanged; max cruising speed at 95 per cent of MTOGW reduced to 416 knots (771 km/h; 479 mph); FAR T-O field length at MTOGW 1,760 m (5,775 ft), FAR landing field length at typical landing weight 1,500 m (4,922 ft).
CUSTOMERS: Letters of intent from 10 airlines by 10 June 1993, covering more than 120 EMB-145s.

EMBRAER EMB-312 TUCANO
PROGRAMME: First flight of second pre-series EMB-312F 8 April 1993; both due for delivery to French CEV at Mont-de-Marsan July 1993 for one-year evaluation; delivery schedule for first 20 production Fs July 1994-July 1995; second batch (28 aircraft) to follow July 1995-July 1996; final 30 in two blocks of 15 until May 1998.
CUSTOMERS: Total Tucano deliveries more than 570 by June 1993.

CANADA

CANADAIR (page 26)
CANADAIR CHALLENGER
CURRENT VERSIONS: **601-3R**: Introduced May 1993; first delivery (to J. C. Penney Co of Dallas, Texas) July 1993; GE CF34-3A1 engines (same ratings as -3As in 601-3A); tailcone fuel tank of 601-3A/ER standard, conferring same intercontinental range; improved environmental control and pitot static systems. Data generally as for 601-3A/ER except balanced T-O field length (S/L, ISA) 1,844 m (6,050 ft).

604: Proposed next-generation version; enlarged cabin, uprated CF34-3B engines, new avionics suite, 4,000 nm (7,413 km; 4,606 mile) range, improved reliability, lower life-cycle costs; first flight/certification/delivery start planned for 1995.

CZECH REPUBLIC

LET (page 66)
Joint venture agreement concluded, subject to Czech government approval, between Fairchild Aircraft of USA and Czech holding company Aero SA; Fairchild to invest more than $100 million in modernising Let factory and acquiring majority shareholding; Fairchild to assist in marketing **L-420** (improved L-410 with M 601, P&WC or Garrett turboprops) and L-610G; Let to undertake subcontract manufacture of Metro 23. FAA certification for L-420 expected before end of 1993.

FRANCE

DASSAULT AVIATION (page 75)
MIRAGE 2000

French requirements have again been modified within the overall total of 318 and now stand at 126 Mirage 2000Cs, 27 Mirage 2000Bs, 75 Mirage 2000Ns and 90 Mirage 2000Ds.

First six 2000Ds, assigned to EC 1/3 at Nancy/Ochey, achieved IOC (temporarily at Mont-de-Marsan) on 1 August 1993, with equipment including Thomson PDLCT laser-designation pods.

FOURNIER
AVIONS FOURNIER

c/o Tours Aviation, Tours St Symphorien Airport, F-37100 Tours, France
Telephone: 33 47 51 25 64
Fax: 33 47 54 29 49

New René Fournier aircraft now produced by Tours Aviation; Fournier RF-6 adopted and converted as T67 by Slingsby in UK; RF-9 now being produced by ABS in Germany.

FOURNIER RF-47

TYPE: Two-seat light aircraft.
PROGRAMME: First flight 9 April 1993; certification expected late 1993; to be produced also in kits.
DESIGN FEATURES: Designed as very light club trainer and pleasure aircraft with clearance for spinning, but not aerobatics.
FLYING CONTROLS: Controls specially tailored in Fournier tradition for precise response, light forces and excellent handling characteristics for high-grade training and personal pleasure; fixed tailplane with trim tab on elevator; slotted flaps.
STRUCTURE: First aircraft has wooden fuselage and wood and fabric wings; second aircraft, intended for certification, will have wing spar and other components in new wood/CFRP/wood sandwich, which halves the weight of the main spar, but increases strength factor by half.
LANDING GEAR: Prototype has tricycle landing gear with steering by brakes, but tailwheel landing gear will be offered as an option.
POWER PLANT: Prototype has 67 kW (90 hp) Sauer modification of Volkswagen engine running at maximum 2,700 rpm, and silencer, but other engines will be optional.
ACCOMMODATION: Two seats side by side under single-piece canopy swinging up and rearwards; fresh-air vents.

DIMENSIONS, EXTERNAL:
Wing span 10.00 m (32 ft 9¾ in)
Wing aspect ratio 9.43
Length overall 6.25 m (20 ft 6 in)
Height overall 2.10 m (6 ft 10¾ in)

Fournier RF-47 very light trainer and club aircraft (*Mark Lambert*)

AREAS:
Wings, gross 10.6 m² (114.1 sq ft)
WEIGHTS AND LOADINGS:
Manufacturer's weight empty 380 kg (838 lb)
Max T-O weight 600 kg (1,323 lb)
Max wing loading 56.60 kg/m² (11.59 lb/sq ft)
Max power loading 8.94 kg/kW (14.70 lb/hp)
PERFORMANCE:
Never-exceed speed (V_NE)
 145 knots (270 km/h; 167 mph)
Max level speed 119 knots (220 km/h; 137 mph)
Max cruising speed 103 knots (190 km/h; 118 mph)
Normal cruising speed 97 knots (180 km/h; 112 mph)
Stalling speed, flaps 40° 39 knots (72 km/h; 45 mph)
Max rate of climb at S/L 240 m (787 ft)/min
Service ceiling 4,000 m (13,125 ft)
T-O run 260 m (853 ft)
Endurance, according to power setting 4 to 6 h
Max cross-wind for T-O/landing
 22 knots (41 km/h; 25 mph)
g limits, Utility category +4.4/−2.2

GERMANY

GROB (page 103)
GROB STRATO 2C

Reported in April 1993 that engine installation is being reversed to pusher configuration; this improves efficiency, smooths airflow over the wing and leaves the forward fuselage clear for air sampling.

Artist's impression showing new pusher layout of power plant

RFB (page 104)
FANTRAINER 800

TYPE: Two-seat military trainer.
PROGRAMME: Announced at Paris Air Show 1993 as more powerful version of Fantrainer 600.
CUSTOMERS: Unnamed customer expected to place order end 1993.
DESIGN FEATURES: Same airframe and turboshaft/shrouded fan propulsion system as Fantrainer 600; same airframe and adaptable cockpit section, but added rocket escape system.
FLYING CONTROLS: As earlier Fantrainers.
STRUCTURE: Forward fuselage based on GFRP keel structure carrying all main loads with side shells forming cockpit contour and canopy layout; all-composites wings autoclaved for long life (Royal Thai Air Force airframes have suffered no deterioration from damp or heat); rear fuselage and tail surfaces metal stressed skin.
LANDING GEAR: Retractable tricycle type, with main gear built in single removable assembly; nosewheel retracts upwards and forwards into nose fairing.
POWER PLANT: One Allison 250-C30 turboshaft, with output

uprated to as much as 597 kW (800 shp) by end 1993 with no increase in size or weight; engine drives five-blade wooden fan turning in close-fitting ring duct; fan designed to absorb up to 746 kW (1,000 hp); engine exhaust protects fan from icing. Fuel: see under Weights and Loadings.

ACCOMMODATION: Two pilots in tandem on fixed seats; UPC rocket extraction system can save pilots in zero/zero conditions; firing of extraction rockets separated by 0.5 s; environmental control system optional. Cockpit and canopy contours can be adapted to simulate operational aircraft.

SYSTEMS: As Fantrainer 600.
AVIONICS: Customer choice.
DIMENSIONS, EXTERNAL:
Wing span 9.74 m (31 ft 11½ in)
Length overall 9.48 m (31 ft 1¼ in)
Height overall 3.16 m (10 ft 4½ in)
Wheel track 1.94 m (6 ft 4½ in)
Wheelbase 3.89 m (12 ft 9¼ in)
WEIGHTS AND LOADINGS (A: Aerobatic, U: Utility):
Weight empty: A, U 1,180 kg (2,601 lb)
Weight of pilots with parachutes 200 kg (441 lb)
Other payload: A 64 kg (141 lb)
U 156 kg (344 lb)
Max fuel weight, internal: A 176 kg (388 lb)
U 384 kg (846 lb)
Max fuel weight, external: U 400 kg (882 lb)
Max T-O weight: A 1,600 kg (3,527 lb)
U 2,310 kg (5,093 lb)
Max landing weight: A 2,000 kg (4,41 lb)
Max power loading: A 2.68 kg/kW (4.41 lb/hp)
U 3.87 kg/kW (6.37 lb/hp)
PERFORMANCE (at max Aerobatic T-O weight):
Max level speed at 4,575 m (15,000 ft)
259 knots (480 km/h; 298 mph)
Cruising speed at 4,575 m (15,000 ft)
243 knots (450 km/h; 280 mph)
Stalling speed 61 knots (113 km/h; 71 mph)
Max rate of climb at S/L 900 m (2,952 ft)/min
Rate of climb at 4,575 m (15,000 ft)
780 m (2,559 ft)/min
T-O and landing run 250 m (820 ft)
Range at 3,050 m (10,000 ft), max internal fuel, 45 min reserves
561 nm (1,040 km; 646 miles)
Endurance at 3,050 m (10,000 ft), max internal fuel, 45 min reserves
4.1 h
g limits: A +6/−3
U +4.4/−1.76

RFB Fantrainer 800 prototype (one uprated Allison 250-C30 turboshaft)

RFB MFI-10C VIPAN/PHÖNIX

TYPE: Military/civil utility aircraft.
PROGRAMME: Revival of Swedish Malmö Flygindustri MFI-10 Vipan, last described in 1971-72 *Jane's*. RFB plans initially to sell the military aircraft with tailwheel landing gear.
CURRENT VERSIONS: **MFI-10C Vipan:** Original military utility aircraft built and flown by Malmö Flygindustri in June 1962; powered by Textron Lycoming IO-360-A.
Phönix: Civil four-seater development with tricycle landing gear proposed by RFB.
DESIGN FEATURES: Originally designed for Swedish Defence Board and Royal Swedish Aero Club as observation and casevac aircraft and tourer; precision airframe gives high efficiency.
FLYING CONTROLS: Frise ailerons, variable incidence tailplane; trim tab in rudder; split flaps.
STRUCTURE: Central cabin has steel tube structure and doors covering full length of cabin without pillar; nearly all of remainder in aluminium honeycomb sandwich giving high strength with very low weight; single-spar wing with single bracing struts and prominent jury struts.

MFI-10 Vipan utility aircraft design, being revived by Rhein Flugzeugbau in Germany *(Kenneth Munson)*

LANDING GEAR: Fixed; steerable tailwheel; mainwheels on single laminated GFRP bow.
POWER PLANT: One 149 kW (200 hp) Textron Lycoming IO-360-A driving three-blade constant-speed propeller; two fuel tanks in wingroots hold total of 200 litres (52.8 US gallons; 44 Imp gallons).
DIMENSIONS, EXTERNAL:
Wing span 10.61 m (34 ft 9¾ in)
Length overall 7.92 m (25 ft 11¾ in)
Height overall 2.13 m (6 ft 11¾ in)
AREAS:
Wings, gross 15.7 m² (169.0 sq ft)
WEIGHTS AND LOADINGS:
Weight empty 650 kg (1,433 lb)
Max T-O weight 1,173 kg (2,586 lb)
Max wing loading 74.71 kg/m² (15.30 lb/sq ft)
Max power loading 7.87 kg/kW (12.93 lb/hp)

PERFORMANCE (N: Normal, U: Utility):
Never-exceed speed (V_{NE}):
N, U 163 knots (302 km/h; 187 mph)
Max level speed at S/L:
N, U 132 knots (245 km/h; 152 mph)
Cruising speed, 75% power at 2,300 m (7,545 ft):
N, U 124 knots (230 km/h; 143 mph)
Touchdown speed: N 46 knots (85 km/h; 53 mph)
U 41 knots (76 km/h; 47 mph)
Max rate of climb at S/L: N 228 m (748 ft)/min
U 318 m (1,043 ft)/min
Service ceiling: N 4,800 m (15,750 ft)
U 5,750 m (18,865 ft)
T-O run: N 155 m (509 ft)
U 115 m (378 ft)
Range, 70% power at S/L:
N, U 593 nm (1,100 km; 684 miles)

INDONESIA

IPTN (page 113)
IPTN N-250-100

PROGRAMME: IPTN has decided to launch with 64/68-seat sized model instead of 50/54-seat (though first prototype will remain configured as 50-seater); plugs of 0.50 m (1 ft 7¾ in) forward and 1.00 m (3 ft 3¼ in) aft of wing; accordingly has redesignated initial version N-250-100. IPTN now says first flight early 1995, then three N-250-100 development aircraft (plus two static test); first production aircraft first flight mid-1996, service entry (Indonesian regional airlines) mid-1997. Passenger, cargo and combi versions to be offered. *Following details apply to N-250-100.*
ACCOMMODATION: 64 passengers at 81 cm (32 in) pitch or 68 at 76 cm (30 in).
DIMENSIONS, EXTERNAL:
Wing span 28.00 m (91 ft 10¼ in)
Length overall 28.153 m (92 ft 4½ in)
Fuselage: Length 26.774 m (87 ft 10 in)
Max diameter 2.90 m (9 ft 6¼ in)
Height overall 8.785 m (28 ft 9¾ in)
Tailplane span 9.04 m (29 ft 8 in)
Wheel track (c/l of shock struts) 4.10 m (13 ft 5½ in)
Wheelbase 10.253 m (33 ft 7¾ in)
WEIGHTS AND LOADINGS:
Max fuel weight 4,200 kg (9,259 lb)
Max T-O weight 24,800 kg (54,675 lb)
Max landing weight 24,600 kg (54,234 lb)

INTERNATIONAL

AIRBUS INDUSTRIE (page 114)
AIRBUS A319

The programme was firmly launched during the Paris Air Show 1993; leasing company ILFC had already ordered six plus two options in December 1992; first flight expected Autumn 1995; flight testing at Toulouse; final assembly and furnishing by Deutsche Aerospace Airbus at Hamburg; entry into service expected in Spring 1996; development cost $275 million; sales expected to reach 450 by 2011.

A319 is 3.73 m (12 ft 3 in) shorter than A320 and seats 130, but has the longest range of the A320 family and of the category at 2,700 nm (5,000 km; 3,105 miles); crew cross-qualification between A319, A320, A321, A340 and A330 will radically simplify training and currency of experience for long-range and short-range crews.

ATR (page 131)

ATR 42-500

TYPE: Uprated ATR 42 with 'new look' interior.
PROGRAMME: Announced at Paris Air Show 1993; will become baseline version of ATR 42; certification expected in March 1995.
CUSTOMERS: Launch orders expected to come from South America during 1993.
DESIGN FEATURES: More powerful engines (see below); reinforced wings to allow greatly increased cruising speed of 304 knots (563 km/h; 350 mph) and higher weights; could depart at max gross weight from Denver, Colorado, in 30°C (86°F) ambient; all systems improvements of ATR 72, including flight management computers, to be incorporated.
POWER PLANT: To be powered by two P&WC PW127 turboprops (as ATR 72), derated from 2,051 kW (2,750 shp) to the 1,790 kW (2,400 shp) of the PW124, giving high power reserve; ATR 72-210 nacelles; propellers have four unswept blades; fuel capacity unchanged at 4,500 kg (9,921 lb).
ACCOMMODATION: Seating for 46 to 50 passengers at 76 cm (30 in) pitch; completely new interior with new ceiling and sidewalls, indirect lighting, more sound damping and active sound control; call buttons and reading lights relocated; overhead bins lengthened to 2 m (6 ft 6¼ in) to accommodate skis, golf clubs and fishing equipment carried as hand baggage.
SYSTEMS: New systems transferred from ATR 72.
WEIGHTS AND LOADINGS:
Operating weight empty 10,980 kg (24,207 lb)

Model of ATR 42-500 (two P&WC PW127s) *(Mark Lambert)*

Max payload 5,620 kg (12,390 lb)
Max T-O weight 18,500 kg (40,785 lb)
Max power loading (with power reserve) 4.51 kg/kW (7.42 lb/hp)

PERFORMANCE:
Cruising speed 304 knots (563 km/h; 350 mph)
T-O field length, S/L 1,140 m (3,740 ft)
Max range 1,000 nm (1,853 km; 1,151 miles)

PANAVIA (page 171)

PANAVIA TORNADO

German Air Force IDS: Mid-life update planned in two parts: KWA (Kampfwertanpassungsprogramm: Combat Efficiency Enhancement Programme) is first stage, involving addition of FLIR and GPS, plus associated software (change to Ada language), avionics and display changes; KWE (Kampfwerterhaltungsprogramm: Combat Efficiency Upgrade Programme) is later element, concentrating on electronic warfare capability and involving new defensive aids computer, plus missile approach warning and enhanced radar warning equipment.

Royal Air Force GR. Mk 4: Prototype Mk 4 ZD708 first flew 29 May 1993 (XZ631, planned as initial trials aircraft, is only partial conversion). RAF expects having to forego installation of GEC-Marconi terrain-referenced navigation and FLIR for financial reasons, but these fitted in prototypes.

Tornado ECR: Italian prototype MM7079 flew on 20 July 1992.

Maiden flight of the Tornado GR. Mk 4 prototype, Warton, 29 May 1993

SCTICSG (page 178)

TYPE: Second-generation supersonic airliner programme (Supersonic Commercial Transport International Co-operative Study Group).
PROGRAMME: Work on second-generation SST begun by Aerospatiale and French national institute ONERA in June 1989; SNECMA and Rolls-Royce began joint engine studies also in 1989; MTU and FiatAvio joined them later; Aerospatiale and BAe started joint study in April 1990; SCT (supersonic commercial transport) Group started with Boeing, McDonnell Douglas and Deutsche Aerospace in May 1990 with nickname Group of Five; objective to study market and environmental problems; Alenia, Tupolev and Japan Aircraft Development Corporation (Mitsubishi, Kawasaki and Fuji) joined during 1991; P&W and GE working together on engine problems; Phase 1 market, certification, noise and environmental study completed by mid-1993 and Phase 2 launched to study technical and economic feasibility; new SST in service 2005.

CURRENT VERSIONS: **Alliance:** Existence vigorously denied to this book in early 1993, but confirmed at Paris Air Show 1993 as defining French Aerospatiale/ONERA/SNECMA study for Concorde-style 250- to 300-passenger SST with range of 5,935 nm (11,000 km; 6,835 miles); double delta wing with high aspect ratio outer panels.

Advanced Supersonic Transport: British study for Concorde-style SST combining low aspect ratio ogival wing and canard control surfaces.

Tu-244: Tupolev designation for its own SST study as successor to original Tu-144; model shown at Paris Air Show 1993 (see photo page 694).

ASCT: Advanced Supersonic Commercial Transport; generic name for various US and Japanese studies.

DESIGN FEATURES: Cruising speed Mach 2.05 chosen to allow cruising at lower altitudes, where emissions are not so critical; fare surcharge not more than 10 to 20 per cent for three-class layout. Typical dimensions: fuselage length 88 m (288 ft 8 in), wing span 45 m (147 ft 7 in).
POWER PLANT: Four dual cycle engines meeting normal FAR 36 Stage 3 noise levels; nitrous oxide emission reduced by 80 per cent; dual cycle concept under study by SNECMA designated Mid Tandem Fan promises 15 to 20 dB noise reduction at take-off; Rolls-Royce and P&W participating with SNECMA; fuel consumption 0.043 kg/seat/km (0.175 lb/seat/nm).

ISRAEL

IAI (page 181)

IAI (MIKOYAN) MiG-21-2000

IAI announced teaming agreement with Aerostar of Romania in June 1993 to offer MiG-21 upgrade and associated training and ground support facilities on international market. Romanian Air Force MiG-21bis displayed at Paris Air Show had Elta EL/M-2032 lookdown/shootdown radar with Doppler beam sharpening (replacing former Soviet 'Jaybird'); El-Op HUD, monochrome HDD and colour tactical display; new one-piece windscreen; Martin-Baker ejection seat; and single mission computer compatible with a MIL-STD-1553 databus. Possible weapons could include Rafael Python 3 infra-red AAMs and Mk 82 laser-guided bombs. Other possible options in upgrade package could include Elbit DASH (display and sight helmet), increased internal/external fuel capacity and airframe life extension.

Elbit has $300 million 1993 contract to upgrade Romanian Air Force MiG-21s, though precise avionics fit and number of aircraft involved (reportedly 100) not revealed.

IAI MiG-21-2000 upgrade of Romanian Air Force MiG-21bis *(Mark Lambert)*

IAI (MARSH/GRUMMAN) S-2UP TRACKER

Customer revealed as Argentine Navy ($30 million contract to upgrade one S-2E and provide kits for customer to upgrade five more). Bedek Aviation Division's Shaham Plant teamed with Marsh Aviation Co (see entry in main US section); upgrade includes refit with Garrett TPE331-15 turboprops and five-blade propellers; improved air-conditioning and oxygen systems; wing-mounted high-candlepower searchlight; new autopilot; modernisation of other flight systems; payload and take-off weights, and performance, increased without loss of carrier launch and retrieval capability.

IAI has also developed a higher level S-2E upgrade package incorporating specified radar plus ASW, EW and/or a day/night reconnaissance system, all controlled through one multi-function display. Argentine Navy upgrade apparently did not include these features.

Prototype S-2UP (modified S-2E) upgraded by IAI and Marsh Aviation for the Argentine Navy *(Mark Lambert)*

ITALY

GENERAL AVIA (page 196)
GENERAL AVIA F.22/C PINGUINO

TYPE: Two-seat utility aircraft.
PROGRAMME: Extrapolation of F.22 Pinguino family (which see) of light aircraft; prototype flying early 1993; certification expected by end 1993; batch production started.
DESIGN FEATURES: Same basic airframe as rest of family, but with more powerful Textron Lycoming IO-360 engine; retractable landing gear optional.
POWER PLANT: One 134 kW (180 hp) Textron Lycoming IO-360-A1A driving Hartzell HC-C2YK-1BF constant-speed two-blade metal propeller. Fuel capacity 160 litres (42.3 US gallons; 35.2 Imp gallons).
DIMENSIONS, EXTERNAL: As F.22/A and /B
WEIGHTS AND LOADINGS:
Basic weight empty 585 kg (1,290 lb)
Max T-O weight: Aerobatic 850 kg (1,874 lb)
Utility 900 kg (1,984 lb)
Max wing loading: Aerobatic 78.56 kg/m² (16.09 lb/sq ft)
Utility 83.18 kg/m² (17.04 lb/sq ft)
Max power loading: Aerobatic 6.34 kg/kW (10.41 lb/hp)
Utility 6.71 kg/kW (11.02 lb/hp)
PERFORMANCE (at Utility category max T-O weight):
Max level speed 178 knots (330 km/h; 205 mph)
Stalling speed, flaps down 54 knots (100 km/h; 63 mph)
Max rate of climb at S/L 540 m (1,772 ft)/min
Service ceiling 6,100 m (20,000 ft)
T-O run 200 m (656 ft)
Landing run 240 m (788 ft)
Max range, standard fuel 755 nm (1,400 km; 870 miles)

PARTENAVIA (page 197)
PARTENAVIA PD.93 IDEA

TYPE: Four-seat trainer and utility aircraft.
PROGRAMME: Announced at Paris Air Show 1993; first flight expected beginning of 1994; company claims that commonality with P.68 would allow certification in mid-1994; to be manufactured in Italy.
DESIGN FEATURES: Pascale design begun before AerCosmos took over Partenavia; incorporates cantilever wing, rear fuselage and tail surfaces of P.68 light twin; wing span 1 m (3 ft 3¼ in) less than P.68; designed to latest version of FAR Pt 23 Utility category; roles include training, civilian parachuting, aerial photography, ambulance, fire patrol and aerial work.
FLYING CONTROLS: Slab tailplane with anti-balance/trim tab; wide-span slotted flaps.
STRUCTURE: Same as P.68 with new materials to improve maintainability and reliability.
LANDING GEAR: Initially fixed tricycle with spring leaf main legs and oleo-pneumatic nosewheel strut; retractable landing gear to be offered later.
POWER PLANT: One 149 kW (200 hp) Textron Lycoming IO-360-A1B6 driving a three-blade constant-speed GFRP-bladed propeller. Fuel tanks between spars at root of each wing contain total 250 litres (66 US gallons; 55 Imp gallons).
ACCOMMODATION: Four-seat cabin with rear baggage space; door next to pilot's seat at front; rear cabin door to starboard; floor hatch behind pilots.
AVIONICS: Provision for IFR avionics.
DIMENSIONS, EXTERNAL:
Wing span 11.00 m (36 ft 1 in)
Wing chord, constant 1.55 m (5 ft 1 in)
Length overall 8.00 m (26 ft 3 in)
Height overall 3.20 m (10 ft 6 in)
AREAS:
Wing area 17.05 m² (183.5 sq ft)
WEIGHTS AND LOADINGS:
Weight empty, equipped 770 kg (1,698 lb)
Max T-O and landing weight 1,250 kg (2,756 lb)
Max wing loading 73.31 kg/m² (15.02 lb/sq ft)
Max power loading 8.39 kg/kW (13.78 lb/hp)
PERFORMANCE:
Never-exceed speed (V_{NE})
 200 knots (370 km/h; 230 mph)
Max cruising speed, 75% power at 2,286 m (7,500 ft)
 137 knots (254 km/h; 158 mph)
Econ cruising speed, 55% power at 3,658 m (12,000 ft)
 122 knots (226 km/h; 140 mph)
Stalling speed: flaps up 57 knots (104 km/h; 65 mph)
 flaps down 47 knots (87 km/h; 54 mph)
Max rate of climb at S/L 289 m (950 ft)/min
Service ceiling 5,485 m (18,000 ft)
Range, 45 min reserves:
 at 75% power, 2,286 m (7,500 ft)
 755 nm (1,400 km; 870 miles)
 at 55% power, 3,660 m (12,000 ft)
 944 nm (1,750 km; 1,087 miles)
g limits +4.4/−1.76

Artist's impression of Partenavia PD.93 Idea utility aircraft based on P.68 components

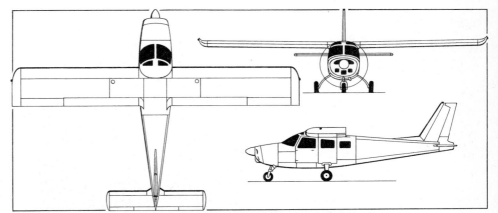

Partenavia PD.93 Idea (Textron Lycoming IO-360-A1B6) *(Jane's/Mike Keep)*

PARTENAVIA PD 90 TAPETE AIR TRUCK

TYPE: Twin-turboprop light utility transport.
PROGRAMME: Announced at Paris Air Show 1993; programme not firmly launched at that time.
DESIGN FEATURES: Twin-boom layout allows integrated loading ramp/airstair; intended missions include 18-passenger commuter freighter with 1,800 kg (3,968 lb) payload, ambulance, parachuting, tactical transport, maritime surveillance and firefighting.
FLYING CONTROLS: Fixed T-tail with elevator and trim tab inset; flaps inboard and outboard of tail booms.
STRUCTURE: Conventional light metal, stressed skin.
LANDING GEAR: Retractable tricycle type with twin wheels on each main unit.
POWER PLANT: Two 559 kW (750 shp) P&WC PT6 turboprops, driving three-blade feathering, reversible-pitch metal propellers. Fuel capacity 1,810 litres (478 US gallons; 398 Imp gallons).
ACCOMMODATION: Rectangular cabin; five windows each side; crew access on port side; side door at starboard rear; integral airstair/ramp openable in flight.

DIMENSIONS, EXTERNAL:
Wing span	18.0 m (59 ft 0¾ in)
Length overall	13.4 m (43 ft 11½ in)
Height overall	4.7 m (15 ft 5 in)

DIMENSIONS, INTERNAL:
Cabin: Length	5.00 m (16 ft 4¾ in)
Max width	2.00 m (6 ft 6¾ in)
Max height	1.70 m (5 ft 7 in)

AREAS:
Wings, gross	37.71 m² (405.9 sq ft)

WEIGHTS AND LOADINGS:
Weight empty, standard	3,600 kg (7,937 lb)
Max payload	1,790 kg (3,946 lb)
Max T-O and landing weight	6,300 kg (13,889 lb)
Max ramp weight	6,340 kg (13,977 lb)
Max zero-fuel weight	5,550 kg (12,236 lb)
Max wing loading	167.1 kg/m² (34.22 lb/sq ft)
Max power loading	5.64 kg/kW (9.26 lb/shp)

PERFORMANCE:
Max level speed at 4,270 m (14,000 ft)	240 knots (445 km/h; 276 mph)
Max cruising speed at 3,660 m (12,000 ft)	232 knots (430 km/h; 267 mph)
Stalling speed: flaps up	83 knots (154 km/h; 96 mph)
flaps down	71 knots (130 km/h; 81 mph)
Max rate of climb at S/L: both engines	579 m (1,900 ft)/min
OEI	122 m (400 ft)/min
Service ceiling: both engines	9,145 m (30,000 ft)
OEI	4,270 m (14,000 ft)
T-O to 15 m (50 ft)	450 m (1,477 ft)
Landing from 15 m (50 ft)	600 m (1,969 ft)
Range, at 3,050 m (10,000 ft), 45 min reserves: long-range cruise with max fuel	1,057 nm (1,960 km; 1,217 miles)
max cruise with max payload	418 nm (775 km; 481 miles)

Partenavia PD 90 Tapete Air Truck utility transport project (Mark Lambert)

JAPAN

ISHIDA (page 205)
ISHIDA TW-68

Reported June 1993 that Ishida Foundation support for programme has been withdrawn and existing funding due to expire at end of month. Separate TW-68 support company, for which buyer being sought, formed by Ishida Aerospace Research.

MITSUBISHI (page 209)
MITSUBISHI FS-X

The following details have appeared in the Japanese aviation press:
DESIGN FEATURES: Wing leading-edge sweepback 33° 12'; incidence 2° 30'.
POWER PLANT: Max internal fuel capacity 4,675 litres (1,235 US gallons; 1,028 Imp gallons) (reduced to 3,978 litres; 1,051 US gallons; 875 Imp gallons in TFS-X). Max external fuel capacity (both) 5,678 litres (1,500 US gallons; 1,249 Imp gallons).

DIMENSIONS, EXTERNAL:
Wing span: over missile rails	11.13 m (36 ft 6¼ in)
excl missile rails	10.80 m (35 ft 5¼ in)
Length overall	15.27 m (50 ft 1¼ in)
Height overall	4.97 m (16 ft 3¾ in)
Tailplane span	6.05 m (19 ft 10¼ in)
Wheelbase	4.05 m (13 ft 3½ in)

AREAS:
Wings, gross	33.17 m² (357.0 sq ft)
Leading-edge flaps (total)	4.70 m² (50.59 sq ft)
Flaperons (total)	3.96 m² (42.63 sq ft)
Horizontal tail surfaces (total)	7.05 m² (75.89 sq ft)

WEIGHTS AND LOADINGS:
Weight empty	9,525 kg (21,000 lb)
Max fuel load: internal*	3,642 kg (8,029 lb)
external	4,422 kg (9,750 lb)
Max T-O weight with external stores	22,100 kg (48,722 lb)

* 3,099 kg (6,832 lb) in TFS-X

NETHERLANDS

FOKKER (page 215)
FOKKER 50

CURRENT VERSIONS: **Fokker 60 Utility:** New designation for stretched utility version of Fokker 50. In production for Singapore; RNethAF expected to order four during 1993.

FOKKER 70
CUSTOMERS: Launch customers Sempati Air (10, plus five options) and Pelita Air (five).

POLAND

PZL WARSZAWA-OKECIE (page 240)
PZL KOLIBER II

Export version of Koliber 150 (which see); powered by Textron Lycoming 112 kW (150 hp) O-320-E2A; offered in Normal and Utility category of FAR 23 with optional full IFR equipment for instrument training; weights and performance differ slightly from those given in main Polish section. Cadmus Corporation in Northfield, Illinois, is sole distributor for USA.

AIRCRAFT: ADDENDA

ROMANIA

IAR (page 252)

Prototype IAR-46 two-seat light aircraft developed from the IS-28M2 motor glider *(Kenneth Munson)*

RUSSIA

AERORIC
AERORIC RESEARCH AND PRODUCTION ENTERPRISE
2 Sibirskaya Street, 86 Nizhny Novgorod, 603600 Russia
Telephone/Fax: 7 (8312) 441 965
CHIEF DESIGNER: Dr V. P. Morozov

This enterprise designs light aircraft for production at the Sokol State Aircraft Building Plant, Nizhny Novgorod.

AERORIC DINGO
TYPE: Light multi-purpose amphibious aircraft.
PROGRAMME: First seen in mockup form at MosAeroshow '92; prototype under construction 1993; first flight scheduled early 1994.
CUSTOMERS: Orders for 300 claimed mid-1993.
DESIGN FEATURES: Basically conventional low-wing twin-boom pusher engined configuration; unique air cushion landing system, permitting operation from hard ground, water, snow, ice, mud and marshy surfaces; able to overcome ground obstructions up to 30 cm (1 ft) high and cross ditches 90 cm (3 ft) wide.
LANDING GEAR: Air cushion formed under wing, contained at sides by air bladders, at front and rear by flexible flat flaps attached to wing by hinges and retracted during flight.
POWER PLANT: One Pratt & Whitney Canada PT6 turboprop; three-blade pusher propeller. Air cushion generated by TVA-200 turboshaft.
ACCOMMODATION: Main cabin, aft of flight deck, accommodates seven persons on bench seats; alternative configurations for air ambulance, photography, forest firefighting, mail and freight carrying.

Mockup of AeroRIC Dingo with air cushion landing gear *(Mark Lambert)*

DIMENSIONS, EXTERNAL:
Wing span 14.33 m (47 ft 0 in)
Wing aspect ratio 8.43
Length overall 12.80 m (42 ft 0 in)
Height overall 3.95 m (13 ft 0 in)
AREAS:
Wings, gross 24.33 m² (261.9 sq ft)
WEIGHTS AND LOADINGS:
Weight empty 2,160 kg (4,760 lb)
Max payload 845 kg (1,860 lb)
Max fuel 600 kg (1,320 lb)
Max T-O weight 3,600 kg (7,940 lb)
Max wing loading 148.0 kg/m² (30.3 lb/sq ft)
PERFORMANCE (estimated):
Max level speed 167 knots (310 km/h; 192 mph)
Max cruising speed at 1,980 m (6,500 ft) 148 knots (275 km/h; 170 mph)
T-O speed 70 knots (130 km/h; 81 mph)
Landing speed 62 knots (115 km/h; 72 mph)
Service ceiling 3,350 m (11,000 ft)
Design range 460 nm (850 km; 530 miles)
Ferry range 810 nm (1,500 km; 930 miles)

AVIATIKA
AVIATIKA JOINT STOCK COMPANY
33A Leningradsky Prospekt, 125284 Moscow
Telephone: 7 (095) 945 56 54
Fax: 7 (095) 945 29 00
PRESIDENT: Igor B. Pjankov

Aviatika aircraft are manufactured by Moscow Aircraft Production Group (MAPO), responsible for MiG-29 production; both MAI-890 and MAI-890U exhibited 1993 Paris Air Show. Glider and autogyro derivatives under development.

AVIATIKA MAI-890
TYPE: Single- and two-seat light multi-purpose biplane.
CURRENT VERSIONS: **MAI-890** single-seater; **MAI-890U** two-seater.
DESIGN FEATURES: Strut- and wire-braced biplane; conventional tail surfaces carried on tubular boom; engine mounted under trailing-edge of upper wing; slight sweep-back on all wings; dihedral on lower wings only. Semi-aerobatic and capable of flying in rough air conditions.
FLYING CONTROLS: Full-span ailerons on lower wings; large-area rudder and elevators, each with ground adjustable tab.
STRUCTURE: Aircraft grade aluminium and titanium alloys, alloy steels; fabric covering unaffected by sun's radiation or atmospheric precipitation. Designed in accordance with FAR 23.
LANDING GEAR: Non-retractable tricycle type; single wheel on each unit; cantilever spring main legs. Optional skis or floats.
POWER PLANT: One 48 kW (64 hp) Rotax 582 piston engine; two-blade pusher propeller. Standard fuel 50 litres (13.2 US gallons; 11 Imp gallons); provision for 55 litre (14.5 US gallon; 12 Imp gallon) auxiliary tank on MAI-890.

Aviatika MAI-890U two-seat light biplane *(Brian M. Service)*

ACCOMMODATION: Open cockpit standard; doors optional, one jettisonable; seats side by side in MAI-890U; provision for backpack parachute, as alternative to BRS ballistically deployed parachute system for crew and aircraft.
EQUIPMENT: Provision for up to 120 kg (265 lb) payload on four attachments, under engine mountings (60 kg; 132 lb), or underbelly (100 kg; 220 lb), or at lower wingtips (each 45 kg; 99 lb), including agricultural dusting and spraygear.
DIMENSIONS, EXTERNAL (A: MAI-890, B: MAI-890U):
Wing span, upper: A, B 8.11 m (26 ft 7½ in)
Length overall, incl pitot: A 5.32 m (17 ft 5½ in)
B 5.50 m (18 ft 0½ in)
Height overall: A, B 2.25 m (7 ft 4¾ in)
AREAS:
Wings, gross, total: A, B 14.29 m² (153.8 sq ft)
WEIGHTS AND LOADINGS:
Weight empty: A 205 kg (452 lb)
B 235 kg (518 lb)
Max T-O weight: A 315 kg (695 lb)
B 450 kg (992 lb)
PERFORMANCE:
Max level speed: A 75 knots (140 km/h; 87 mph)
B 67 knots (125 km/h; 77 mph)
Cruising speed: A 48-65 knots (90-120 km/h; 56-74 mph)
B 48-59 knots (90-110 km/h; 56-68 mph)
T-O speed: A 34 knots (63 km/h; 39 mph)
B 39 knots (72 km/h; 45 mph)
Landing speed: A 33 knots (60 km/h; 38 mph)
B 37 knots (68 km/h; 43 mph)
Max rate of climb at S/L: A 336 m (1,102 ft)/min
B 180 m (590 ft)/min
Service ceiling: A 5,500 m (18,045 ft)
B 4,000 m (13,125 ft)
T-O run: A 50 m (165 ft)
B 80 m (265 ft)
Landing run: A 85 m (280 ft)
B 110 m (360 ft)
Max range:
standard fuel: A, B 145 nm (270 km; 167 miles)
with auxiliary tank: A 295 nm (550 km; 340 miles)

BERIEV (page 255)

BERIEV Be-32
NATO reporting name: Cuff
TYPE: Twin-turboprop multi-purpose light transport.
PROGRAMME: Development of Be-30, first flown in prototype form on 3 March 1967; small number of Be-30s built, but programme terminated when Aeroflot ordered Let L-410As from Czechoslovakia. Hard currency shortage revived programme 1993, with modestly upgraded version known as Be-32; one of original Be-30s exhibited at 1993 Paris Air Show, as Be-32 demonstrator; production to be centred at Irkutsk.
DESIGN FEATURES: Conventional cantilever high-wing monoplane; three-section wings, with anhedral on outer panels; semi-monocoque fuselage of rectangular section; swept-back vertical tail surfaces; engines at tip of centre-section each side.
FLYING CONTROLS: Conventional three-axis; double-slotted flaps.
STRUCTURE: All-metal; spars and skin panels of wing torsion box are mechanically and chemically milled profile pressings; detachable bonded leading-edge; half of wings and most of tail unit covered with thin honeycomb panels stiffened with stringers; fuselage covered mainly with bonded and spot welded chemically milled panels.
LANDING GEAR: Tricycle type; single wheel on each unit; nosewheel retracts forward, mainwheels rearward into engine nacelles; low-pressure tyres. Optional floats and skis.
POWER PLANT: Two 1,011 kW (1,356 shp) Mars (Omsk) TVD-20 turboprops; three-blade propellers; four integral wing fuel tanks. Garrett engines optional.
ACCOMMODATION: Basic seating for two crew and 14-17 passengers in pairs, with centre aisle; other versions include a seven-passenger business transport, cargo version with 3.64 m² (39.2 sq ft) cargo floor for 2,000 kg (4,410 lb) freight, transport with sidewall seats for 12 paratroops or 17 troops, and ambulance for nine stretcher patients, six seated casualties and attendant. Carry-on baggage compartment on starboard side, aft of cabin seating, opposite forward-hinged door and airstairs; toilet to rear.
SYSTEMS: Three-phase AC electrical system, with two 16kW 200V generators. Wing and tail unit hot air de-icing, using engine bleed air. Cabin heated and ventilated.
AVIONICS: Com/nav radio, autopilot, and system for automatic approach to altitude of 50 m (165 ft) standard; nav system with roller-map display optional. New Bendix/King avionics optional.

Refurbished Beriev Be-30, displayed as Be-32 demonstrator (*Brian M. Service*)

DIMENSIONS, EXTERNAL:
Wing span 17.00 m (55 ft 9¼ in)
Length overall 15.70 m (51 ft 6 in)
Height overall 5.52 m (18 ft 1½ in)
Tailplane span 6.36 m (20 ft 10½ in)
Wheel track 5.20 m (17 ft 0¾ in)
Wheelbase 4.75 m (15 ft 7 in)
DIMENSIONS, INTERNAL:
Passenger cabin: Length 5.66 m (18 ft 7 in)
Width 1.50 m (4 ft 11 in)
Height 1.82 m (5 ft 11 in)
AREAS:
Wings, gross 32.0 m² (345 sq ft)
WEIGHTS AND LOADINGS:
Max T-O weight 7,000 kg (15,430 lb)
PERFORMANCE (estimated):
Max cruising speed 259 knots (480 km/h; 298 mph)
Range: with 17 passengers 350 nm (650 km; 404 miles)
with 7 passengers 863 nm (1,600 km; 994 miles)

MiG (page 279)

MIKOYAN MiG-AT
TYPE: Two-seat advanced jet trainer and light attack aircraft.
PROGRAMME: Basic details on page 287. Since written, MiG-AT selected as one of two finalists in Russian competition to replace Aero L-29 and L-39 Albatros; design refined, as shown in accompanying photograph of model, following October 1992 agreement under which two prototypes will be powered by Larzac engines supplied by SNECMA of France; first flight scheduled for first quarter 1995, service entry 1996; Russian requirement for 700 trainers in this category; whether or not MiG-AT wins competition, will be offered for export.
DESIGN FEATURES: Basically as three-view drawing on page 288; wingroot leading-edges now sweptforward, with engine air intakes overwing; wingtips more square; deeper ventral fin; tailcone now comprises two front-hinged door type airbrakes. Onboard simulation of manoeuvring target, meteorological conditions and system failures via HUD.
LANDING GEAR: Retractable tricycle type; single wheel on each unit; wide-track main units retract inward, nosewheel rearward; high-efficiency brakes; operation practicable from unpaved surfaces of bearing ratio 6 kg/cm² (85.3 lb/sq in).
POWER PLANT: Two Turbomeca-SNECMA Larzac 04-R20 turbofans; each 14.12 kN (3,175 lb st).
ACCOMMODATION: Two crew in tandem, on zero/zero ejection seats; rear seat raised to improve occupant's forward view; birdproof canopy.
AVIONICS: Multi-functional central computer, with all data integrated through MIL-STD-1553B databus; two multi-functional CRT displays with buttons; HUD with input from colour video and TV camera; laser rangefinder;

Model of MiG-AT with Larzac 04-R20 turbofans (*Mark Lambert*)

AIRCRAFT: ADDENDA 693

HSI/ADI; automatic control system; air data system; INS; Tacan; ILS; radar warning receivers; IFF.
ARMAMENT (optional): Four underwing hardpoints for up to 2,000 kg (4,410 lb) of guided and unguided missiles, guns and bombs.
DIMENSIONS, EXTERNAL:
Wing span 10.60 m (34 ft 9½ in)
Wing aspect ratio 5.35
Length overall 11.145 m (36 ft 6¾ in)
Height overall 4.27 m (14 ft 0 in)
Wheel track 3.80 m (12 ft 5¾ in)
AREAS:
Wings, gross 21.00 m² (226.0 sq ft)
WEIGHTS AND LOADINGS:
Fuel weight 1,000 kg (2,205 lb)
Normal T-O weight 4,620 kg (10,185 lb)
Max T-O weight 6,800 kg (14,990 lb)
PERFORMANCE (estimated):
Max level speed 460 knots (850 km/h; 528 mph)
T-O speed 97 knots (180 km/h; 112 mph)
Landing speed 92 knots (170 km/h; 106 mph)
Service ceiling 15,000 m (49,200 ft)
Normal range 647 nm (1,200 km; 745 miles)
g limits +8.0/−3.0

MIL (page 289)

MIL Mi-26

TYPE: Twin-turbine multi-purpose heavy-lift helicopter.
PROGRAMME: Details of wide range of available versions given at 1993 Paris Air Show.
CURRENT VERSIONS: **Mi-26**: Basic military transport (page 295).
Mi-26A: With PNK-90 integrated flight/nav systems for automatic approach and descent to critical decision point, and other tasks.
Mi-26T: Basic civil freight transport, generally as military Mi-26. Variants include **Firefighting Mi-26**, able to dispense 7,500 litres (1,980 US gallons; 1,650 Imp gallons) fire retardant from one or two vents; **Geological Survey Mi-26** towing seismic gear, with tractive force of 10,000 kg (22,045 lb) or more, at 97-108 knots (180-200 km/h; 112-124 mph) at 55-100 m (180-330 ft) for up to three hours.
Mi-26MS: Medical evacuation version, with life support section for four casualties and two medics, surgical section for one casualty and three medics, pre-operating section for two casualties and two medics, ambulance section for five stretcher patients, three seated casualties and two attendants; laboratory; and amenities section with toilet, washing facilities, food storage and recreation unit.
Mi-26P: Transport for 63 passengers, basically four-abreast in airline type seating, with centre aisle; toilet, galley and cloakroom aft of flight deck.
Mi-26TM: Flying crane, with gondola for pilot/sling supervisor under fuselage aft of nosewheels or under rear loading ramp; rear gondola can accommodate pilot and trainee.
Mi-26TZ: Tanker with 14,040 litres (3,710 US gallons; 3,088 Imp gallons) additional fuel and 1,040 litres (275 US gallons; 228 Imp gallons) lubricants, dispensed through four hoses for aircraft, ten hoses for ground vehicles.
Mi-26M: Upgrade under development in 1993; all-GFRP main rotor blades of new aerodynamic configuration, new ZMKB Progress D-127 turboshafts, and modified integrated flight/nav system. Transmission rating unchanged, but full payload capability maintained under 'hot and high' conditions, OEI safety improved, hovering and service ceilings increased, and greater max payload for crane operations.

MIL Mi-52

TYPE: Three-seat light helicopter.
PROGRAMME: Announced, and model displayed, at 1993 Paris Air Show; first flight scheduled 1994; production at small plant in Moscow, adjoining OKB.
DESIGN FEATURES: Smallest helicopter yet developed by Mil; configuration shown in accompanying photograph of model. Four-blade main rotor, two-blade tail rotor.
LANDING GEAR: Non-retractable tricycle type; single wheel on each unit; cantilever spring legs; fairing over each wheel.
POWER PLANT: One rotary (Wankel type) engine, probably related to VAZ-430 in Mi-34 VAZ.
ACCOMMODATION: Enclosed cabin for pilot, in front, and two passengers on rear bench seat; two doors on each side.
WEIGHTS AND LOADINGS:
Max T-O weight 1,150 kg (2,535 lb)

Two-man gondola under rear ramp of Mi-26TM crane helicopter (*Mark Lambert*)

Model of Mil Mi-52 three-seat light helicopter (*Brian M. Service*)

SUKHOI (page 308)

SUKHOI Su-30

TYPE: Two-seat twin-engined multi-role combat aircraft.
PROGRAMME: Two prototypes built at Irkutsk, as Su-27PU (page 313); first flight 30 December 1989; production as long-range fighter for PVO; first two production Su-30s, without military equipment, built for Zhukovsky aerobatic team; military IOC imminent mid-1993.
CURRENT VERSIONS: **Su-30**: Basic long-range interceptor, designed for missions of 10 hours or more, including group missions as described for Su-27PU.
Su-30M: As basic Su-30 but equipped for multi-role duties. Export version is **Su-30MK**.
DESIGN FEATURES: Two-seat development of Su-27/27UB, with new avionics; without advanced radar, foreplanes, flight control system for static instability and new power plant of Su-35.
POWER PLANT: As Su-27, but flight refuelling probe standard.
AVIONICS: Include new navigation system based on Loran, Omega and Mars; two identical cockpits; capable of Su-27UB training roles; fire control system able to engage two air-to-air targets simultaneously. Provision for fitting foreign-made airborne and weapon systems at customer's request.
ARMAMENT: Twelve hardpoints for more than 8,000 kg (17,635 lb) of stores, including anti-radiation missiles, TV and laser homing missiles and bombs. Conventional bombs, rockets, other munitions and 30 mm gun as Su-27.
DIMENSIONS, EXTERNAL: As Su-27 except:
Height overall 6.357 m (20 ft 10¼ in)
WEIGHTS AND LOADINGS:
Normal T-O weight 24,000 kg (52,910 lb)
Max T-O weight 33,000 kg (72,750 lb)
PERFORMANCE:
Max level speed at height Mach 2.0
T-O run 550 m (1,805 ft)
Landing run 670 m (2,200 ft)
Combat range:
 internal fuel 1,620 nm (3,000 km; 1,865 miles)
 with one in-flight refuelling
  2,805 nm (5,200 km; 3,230 miles)

Sukhoi Su-27UB demonstrator representing Su-30MK multi-role combat aircraft (*Mark Lambert*)

First seen at the 1993 Paris Air Show, the Kh-59M (NATO AS-18) cruise missile is one of the weapons carried by the Su-30M. Powered by a ventral turbofan, and guided by a Granit 7TM1 TV camera in the nose, the 920 kg (2,028 lb) Kh-59M has a range of more than 65 nm (120 km; 75 miles) (*Brian M. Service*)

SUKHOI Su-32

TYPE: Tandem two-seat civil and military primary trainer.
PROGRAMME: Announced mid-1993; first flight planned 1994.
DESIGN FEATURES: Conventional low-wing monoplane; see accompanying three-view drawing.
LANDING GEAR: Retractable tricycle type; single wheel on each unit; main units retract inward, nosewheel rearward.
POWER PLANT: Choice of one 265 kW (355 hp) Vedeneyev M-14P radial or 324 kW (435 hp) Teledyne Continental GTSIO-520 piston engine; three-blade propeller.
ACCOMMODATION: Two seats in tandem; rear seat raised; unique Zvezda ejection system, based on K-36 but ejecting crew through canopy on diverging trajectories, without seats.

DIMENSIONS, EXTERNAL:
Wing span   8.50 m (27 ft 10¾ in)
Length overall   7.85 m (25 ft 9 in)

WEIGHTS AND LOADINGS:
Weight empty   850 kg (1,874 lb)
Max T-O weight   1,300 kg (2,865 lb)

PERFORMANCE (estimated. A: with M-14P, B: with GTSIO-520):
Max level speed: A   200 knots (370 km/h; 230 mph)
              B   275 knots (510 km/h; 317 mph)
Landing speed: A, B   60 knots (110 km/h; 69 mph)
Rate of climb at S/L: A   810 m (2,650 ft)/min
                    B   990 m (3,250 ft)/min
Max range: A, B   810 nm (1,500 km; 932 miles)

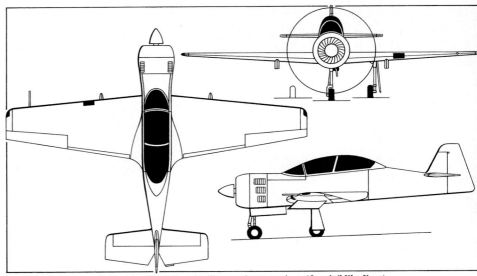

Sukhoi Su-32 civil and military primary trainer (*Jane's/Mike Keep*)

TUPOLEV (page 319)

TUPOLEV Tu-330

TYPE: Twin-turbofan freight transport.
PROGRAMME: Announced Spring 1993 as replacement for Antonov An-12; first flight scheduled 1995.
DESIGN FEATURES: See accompanying three-view drawing. Wing design basically similar to Tu-334; engine pylons as Tu-204; rear-loading ramp.
LANDING GEAR: Retractable tricycle type; twin-wheel nose unit; each main unit three pairs of wheels in tandem.
POWER PLANT: Two Aviadvigatel PS-90 turbofans. Rolls-Royce, General Electric and Pratt & Whitney turbofans under consideration for export sales.

DIMENSIONS, EXTERNAL:
Wing span   47.47 m (155 ft 9 in)
Length overall   39.60 m (130 ft 0 in)
Height overall   14.60 m (47 ft 10 in)
Tailplane span   18.33 m (60 ft 0 in)
Wheel track   5.00 m (16 ft 5 in)

WEIGHTS AND LOADINGS:
Design payload   25,000-30,000 kg (55,100-66,135 lb)

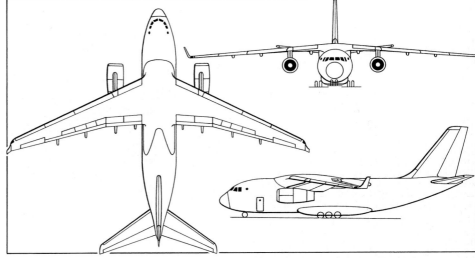

Tupolev Tu-330 twin-turbofan freight transport (*Jane's/Mike Keep*)

TUPOLEV TRANSPORT CONCEPTS

 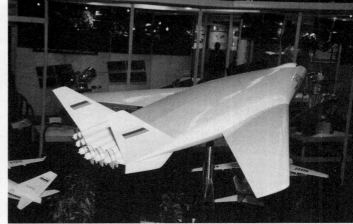

Among Tupolev concepts displayed in model form at 1993 Paris Air Show were Tu-244 (left), a supersonic airliner successor to Concorde, and Tu-404, an 800-passenger super-widebody powered by six pusher propfans (*Brian M. Service*)

YAKOVLEV (page 331)

YAKOVLEV Yak-3

TYPE: Reproduction single-seat piston-engined fighter.
PROGRAMME: Prototype of original wooden-skinned Yak-3 flew 1943; deliveries to Soviet air forces began July 1944, totalled 4,848 (of 36,737 Yakovlev single-engined Second World War fighters); described often as most agile monoplanes of 1939-45 period. Metal-skinned reproductions now available, with Western power plant and uprated instrumentation; new-build 'prototype' displayed 1993 Paris Air Show; 20 being built, to meet orders.
DESIGN FEATURES: Precise reproduction in metal of Second World War airframe, except for repositioned carburettor air intake above engine cowling to suit changed engine. Conventional cantilever low-wing configuration, with tapered, round-tipped wings.
POWER PLANT: Reconditioned 925 kW (1,240 hp) Allison V-1710 twelve-cylinder V liquid-cooled piston engine replaces original 1,240/1,300 hp VK-105PF-2; three-blade propeller.

DIMENSIONS, EXTERNAL:
Wing span   9.20 m (30 ft 2¼ in)
Length overall   8.49 m (27 ft 10¼ in)

AREAS:
Wings, gross   14.85 m² (159.8 sq

WEIGHTS AND LOADINGS:
Max T-O weight   2,697 kg (5,945 lb)
Max wing loading   181.6 kg/m² (37.15 lb/sq
Max power loading   2.91 kg/kW (4.80 lb/

PERFORMANCE:
Max level speed:
  at height   350 knots (648 km/h; 402 mph)
  at S/L   307 knots (570 km/h; 354 mph)
Time to turn 360° at 1,000 m (3,280 ft)   19

AIRCRAFT: ADDENDA 695

Reproduction Yakovlev Yak-3 fighter with Allison engine *(Brian M. Service)*

Yakovlev Yak-18T four-seat light aircraft *(Paul Duffy)*

Prototype Yak-54 displayed at 1993 Paris Air Show (see page 334) *(Mark Lambert)*

YAKOVLEV Yak-18T

TYPE: Four-seat multi-purpose light aircraft.
PROGRAMME: Prototype first flew Summer 1967 as extensively redesigned cabin version of veteran Yak-18 trainer; produced subsequently in large series at Smolensk, including trainer, ambulance, communications and light freight versions; last described in 1989-90 *Jane's*; production now resumed to meet new contracts, one from Philippine Air Force.
DESIGN FEATURES: Conventional cantilever low-wing monoplane; three-section two-spar wings; semi-monocoque fuselage of basically square section; strut- and wire-braced tail surfaces; wing section Clark YH, thickness/chord ratio 14.5 per cent at root, 9.3 per cent at tip; dihedral on outer panels only.
STRUCTURE: All-metal construction, except for fabric covering on parts of outer wing panels, ailerons and tail surfaces.
LANDING GEAR: Tricycle type; single wheel on each unit; pneumatic retraction, mainwheels inward into centre-section, nosewheel rearward; oleo-nitrogen shock absorbers; castoring nosewheel is non-steerable; tyres size 500 × 150 mm on mainwheels, 400 × 150 mm on nosewheel; pneumatic brakes on mainwheels.
POWER PLANT: One 265 kW (355 hp) Vedeneyev M-14P radial piston engine; two-blade variable-pitch metal propeller; wingroot fuel tanks, capacity 208 litres (55 US gallons; 45.75 Imp gallons).
ACCOMMODATION: Cabin seats four persons in pairs, with forward-hinged jettisonable door each side; rear bench seat removable for freight carrying; ambulance configuration accommodates pilot, stretcher patient and medical attendant; baggage compartment aft of rear seat.
SYSTEMS: Pneumatic system for landing gear and flaps; electrical system includes red panel lighting, nav and landing lights, and fintip anti-collision beacon.
AVIONICS: UHF radio, intercom, radio compass, radio altimeter and flight recorder standard.

DIMENSIONS, EXTERNAL:
Wing span 11.16 m (36 ft 7¼ in)
Length overall 8.35 m (27 ft 4¾ in)
AREAS:
Wings, gross 18.75 m² (201.8 sq ft)
WEIGHTS AND LOADINGS (A: instructor and one pupil, B: four persons):
Max payload: A 306 kg (675 lb)
B 436 kg (960 lb)
Max T-O weight: A 1,500 kg (3,307 lb)
B 1,650 kg (3,637 lb)
Max wing loading: A 80 kg/m² (16.4 lb/sq ft)
B 88 kg/m² (18.0 lb/sq ft)
Max power loading: A 5.66 kg/kW (9.32 lb/hp)
B 6.22 kg/kW (10.25 lb/hp)
PERFORMANCE (at max T-O weight):
Max level speed: A, B 159 knots (295 km/h; 183 mph)
Max cruising speed: B 135 knots (250 km/h; 155 mph)
Max rate of climb at S/L: B 300 m (985 ft)/min
Service ceiling: A, B 5,500 m (18,000 ft)
T-O run: A 330 m (1,085 ft)
B 400 m (1,315 ft)
Landing run: A 400 m (1,315 ft)
B 500 m (1,640 ft)
Range with max fuel, with reserves:
A 350 nm (650 km; 403 miles)
B 485 nm (900 km; 560 miles)

YAKOVLEV Yak-42 and Yak-142

Additional current versions:
Yak-42A: Experimental version displayed at 1993 Paris Air Show; similar to Yak-42D but with Western avionics. Production version, with Bendix/King avionics, to be designated **Yak-142**.
Yak-42B: As Yak-42A but Bendix/King nav system and flight deck CRTs.
Yak-42T: Freighter, with 2.50 m × 2.00 m (8 ft 2½ in × 6 ft 6¾ in) cargo door in upper deck; max freight load 12,000 kg (26,455 lb), including standard eight containers in underfloor holds.

YAKOVLEV Yak-77

TYPE: Twin-turbofan executive/regional transport.
PROGRAMME: Announced Spring 1993, as replacement for Yak-48 programme; first flight scheduled 1996.
DESIGN FEATURES: General configuration and dimensions similar to those of now-abandoned Yak-48 (see page 334). TsAGI high-efficiency wing section, with compound sweepback; rear-mounted turbofans; design range more than double that of Yak-48.
POWER PLANT: Two Allison GMA 3000 series turbofans.
ACCOMMODATION: Two crew side by side on flight deck; main cabin able to accommodate eight business passengers or 30 in regional transport configuration, with changes only to cabin arrangements.

DIMENSIONS, EXTERNAL:
Wing span 21.50 m (70 ft 6 in)
WEIGHTS AND LOADINGS:
Payload, executive version 900 kg (1,985 lb)
Max T-O weight 25,000 kg (55,115 lb)
PERFORMANCE (estimated):
Econ cruising speed at 12,200 m (40,000 ft) Mach 0.75

YAKOVLEV Yak-130

TYPE: Two-seat advanced jet trainer.
PROGRAMME: Yak-UTS (page 337) now designated Yak-130; one of two designs (other MiG-AT) selected as competitors for Russian contract to replace L-29 and L-39; now being developed in partnership with Aermacchi of Italy; first flight scheduled before mid-1994; tender now includes Penza (Russia)/CAE Electronics (Canada) simulators and Yak-54 for screening and primary training.
DESIGN FEATURES: Modified version of quadruplex digital fly-by-wire control system of Yak-141; prototype to be stable; 5 per cent longitudinal instability planned for production to reproduce handling characteristics of MiG-29/Su-27.
POWER PLANT: Two Klimov-modified ZMKB Progress DV-2S turbofans; each 21.58 kN (4,852 lb st). Full-authority digital fuel control system.

YAKOVLEV Yak-242

TYPE: Twin-turbofan medium-range airliner.
PROGRAMME: Replaces three-engined Yak-42M (page 333); first of two prototypes to fly 1994; certification scheduled mid-1996.
CUSTOMERS: Aeroflot requirement for 200 aircraft in this category, stated some years ago, still firm.
DESIGN FEATURES: Configuration generally as Yak-42M, except for deletion of third engine from base of fin.
POWER PLANT: Two Aviadvigatel PS-90A12 turbofans; each 118 kN (26,525 lb st).
ACCOMMODATION: Crew as Yak-42M; 130-180 passengers; typically 162 passengers in three cabins, 12 four-abreast in first class cabin at front, remainder six-abreast, all with centre aisle. Doors at front and rear of cabin, with service doors opposite; two emergency exits each side, over wing.
AVIONICS: Bendix/King flat-panel flight deck displays.
WEIGHTS AND LOADINGS:
Max T-O weight 63,000 kg (138,890 lb)
PERFORMANCE (estimated):
Typical range, with 162 passengers
2,320 nm (4,300 km; 2,670 miles)

Model of Yakovlev Yak-242 twin-turbofan airliner *(Brian M. Service)*

SOUTH AFRICA

AEROTEK (page 338)

Prototype of the Aerotek Hummingbird two-seat observation aircraft *(Kenneth Munson)*

SWEDEN

IG JAS (page 348)

First two production JAS 39A Gripens. On 8 June 1993 aircraft 02, in markings of 7 Wing, was first to be handed over to the Swedish Air Force

SWITZERLAND

PILATUS (page 358)

PILATUS PC-7

Reported during Paris Air Show in June 1993 that South Africa was to buy 60 PC-7s for SAAF Harvard replacement programme. Designated **PC-7 Mk 2**, they will differ from standard version in having Martin-Baker CH-11A ejection seats and two (instead of six) underwing stations, plumbed only for auxiliary fuel tanks.

PILATUS PC-12

PROGRAMME: Second prototype (HB-FOB) made first flight 28 May 1993; 01 and 02 had logged some 350 and 10 hours' flying respectively by start of Paris Air Show on 10 June.

UKRAINE

ANTONOV (page 366)

Antonov An-32P Firekiller at 1993 Paris Air Show
(*Brian M. Service*)

UNITED KINGDOM

JETSTREAM HOLDINGS LTD (page 379)

BAe JETSTREAM 51

TYPE: Up to 52-seat turboprop transport, developed from Jetstream 61.
PROGRAMME: Market definition study began June 1993, following nine months of customer research.
POWER PLANT: Two 2,983 kW (4,000 shp) turboprops.
ACCOMMODATION: Up to 52 passengers, four-abreast, at 81 cm (32 in) pitch.
PERFORMANCE (estimated):
  Max level speed   360 knots (667 km/h; 414 mph)
  Cruising altitude   7,620-9,450 m (25,000-31,000 ft)

BAe JETSTREAM 71

TYPE: Up to 78-seat turboprop transport, developed from Jetstream 61.
PROGRAMME: As for Jetstream 51.
DESIGN FEATURES: Could operate 350 nm (648 km; 403 mile) sector with max passenger load after T-O from 1,200 m (3,940 ft) airfield at S/L, ISA.
POWER PLANT: As for Jetstream 51.
ACCOMMODATION: Up to 78 passengers; pitch as for Jetstream 51.
PERFORMANCE (estimated):
  Max level speed   350 knots (648 km/h; 403 mph)
  Cruising altitude   as for Jetstream 51

UNITED STATES OF AMERICA

BOEING (page 440)

BOEING 737-X

Boeing announced 29 June 1993 it has authorised the offer for sale of the 737-X family (see under Boeing in USA section); full launch to follow if sufficient orders received; first deliveries expected 1997; aircraft to be offered as a family of three with the smallest and largest having two-class seating respectively for 108 and 157 passengers. Capacity of the middle-sized 737-X would be about the same as that of the existing 737-300.

Main innovations are to be a modified wing with greater chord and span and greater fuel capacity, giving US transcontinental range; power plant would be two CFMI CFM56-3XS turbofans in new nacelles; cruising speed would be increased to Mach 0.8. Otherwise, 737-X structure and systems would be kept very close to those of existing 737s.

LOCKHEED AERONAUTICAL SYSTEMS COMPANY (page 497)

P-3H ORION II

Marketing of this proposed next-generation Orion was abandoned in mid-1993; P-3C Update III remains available for export; Lockheed offering new-built 'green' P-3Cs to US Navy as alternative to proposed refurbishment of corrosion-affected aircraft (from which avionics would be transferred).

LOCKHEED ADVANCED DEVELOPMENT COMPANY (page 506)

F-117A FUNDING

Year	Quantity	Serial Numbers
FY 1979	5	79-10780 to 79-10784
FY 1980	7	80-0785 to 80-0791
FY 1981	7	81-10792 to 81-10798
FY 1982	8	82-0799 to 82-0806
FY 1983	7	83-0807 to 83-0808 and 83-0815 to 83-0819
FY 1984	9	84-0809 to 84-0812 and 84-0824 to 84-0828
FY 1985	11	85-0813 to 85-0814, 85-820 and 85-0829 to 85-0836
FY 1986	7	86-0821 to 86-0823 and 86-0837 to 86-0840
FY 1988	3	88-0841 to 88-0843
	64	

Artist's impression of a possible configuration of an advanced STOVL fighter (ASTOVL) by Lockheed Advanced Development Company. Lockheed is working on preparatory development for ARPA and the US Navy, using a developed P&W F119 engine, Rolls-Royce powered lift technology and an Allison shaft-driven lift fan

NASA (page 540)

X-30 NASP

Design of the manned X-30 single-stage-to-orbit aerospacecraft was abandoned in May 1993 in response to budget cuts and uncertainties regarding the maturity of technology required to construct a demonstrator. The NASP effort has now switched to trials in 1996-97 of hypersonic test vehicles mounted on Minuteman boosters.

RAYTHEON COMPANY (see also BEECH, page 423)
141 Spring Street, Lexington, Massachusetts 02173
Telephone: 1 (617) 862 6600
PRESIDENT: Max E. Bleck

RAYTHEON (BAe) 125 CORPORATE 800 and 1000

PROGRAMME: BAe sold Corporate Jets to US Raytheon company for £250 million in 1993; included are Corporate 800, Corporate 1000, BAe's Hawarden executive jet facility near Chester, Hatfield engineering and support centre, and Arkansas Aerospace Inc of Little Rock, USA. Final assembly facility at Chester to receive $1 million upgrade. Raytheon contracted BAe Airbus division at Chester to continue supplying airframe kits for at least three years.

AERO ENGINES

CANADA

P&WC (page 604)

P&WC PW300

Study of this turboprop began in 1992. It would be based on an uprated version of the PW305 turbofan core, a reduction gear using PW100 geometry and a FADEC. Starting by battery would enable an APU to be eliminated. Thermodynamic power would be in the 3,729 kW (5,000 hp) class, but the PW300 will be flat rated at around 2,983 kW (4,000 shp) to suit the needs of the market for fast aircraft in the 70/80-seat size. Proof-of-concept testing was to begin in 1993, with the first demonstrator scheduled to run in early 1994.

INDEXES

(Items in italics refer to the ten previous editions)

AIRCRAFT

On this page are aircraft generally identified only by numbers. Most were previously listed as 'Model 707' and so on under M

T-1A-2 Sport Trainer (Great Lakes) ...... *(1987-88)* 424	285 Orion (Lockheed) 499	737-400 (Boeing) 442
GCBC Scout (Champion) ...... *(1985-86)* 392	300 Sequoia (Sequoia) 538	*737-500 (Boeing)* 443
H-3 (Piasecki) ...... *(1989-90)* 476	*300C (Hughes)* ...... *(1984-85)* 419	*737X (Boeing)* 443, 697
H-4 (Piasecki) ...... *(1991-92)* 475	300C (Schweizer) 557	*747 (Boeing)* ...... *(1990-91)* 375
A (Air & Space) 412	*301 (Bell)* ...... *(1990-91)* 366	*747 CRAF (Boeing)* ...... *(1990-91)* 390
Tech 4 (New Technik) ...... *(1991-92)* 456	*301 (Mooney)* ...... *(1985-86)* 469	*747, Discontinued models (Boeing)* 443
A (Helio) ...... *(1987-88)* 435	*303 (Cessna)* ...... *(1985-86)* 382	*747 Freighter conversions (Boeing)* ...... *(1991-92)* 372
(Bellanca) ...... *(1985-86)* 348	305 (Brantly) 462	*747SP (Boeing)* ...... *(1985-86)* 378
(RAF) ...... *(1985-86)* 612	330 (Schweizer) 557	747-100 cargo conversion (Pemco) 545
(Learjet) ...... *(1990-91)* 430	*340A/340A II/340A III (Cessna)* ...... *(1984-85)* 370	*747-100 conversion (IAI)* 184
(RAF) ...... *(1985-86)* 612	*340A/340A (RAM/Cessna)* ...... *(1991-92)* 466	*747-200F Freighter (Boeing)* ...... *(1990-91)* 377
A (Learjet) 494	340B (Saab) 352	747-200 Freighter conversion (IAI) 184
(RAF) ...... *(1985-86)* 612	350 Super King Air (Beech) 429	*747-300 (Boeing)* ...... *(1990-91)* 377
(RAF) ...... *(1985-86)* 612	358, EL-1 (Gavilan) 62	747-400 (Boeing) 443
A/36A (Learjet) 492	*360 (Boeing)* ...... *(1992-93)* 353	747-400F (Boeing) 445
A/36A special missions versions (Learjet) 493	*360 (Hillman)* ...... *(1983-84)* 404	747X (Boeing) 446
P, Lightning (Beech) ...... *(1984-85)* 309	*360 (Shorts)* ...... *(1986-87)* 311	750 Citation X (Cessna) 468
Defiant (Rutan) ...... *(1985-86)* 613	*360 Advanced (Shorts)* ...... *(1987-88)* 324	757 (Boeing) 446
, Angel (TKEF) 576	*369D (Kawasaki/Hughes)* ...... *(1984-85)* 160	757-200 (Boeing) 446
(Learjet) 495	382 Hercules (Lockheed) 502, 505	*757 LRAACA (Boeing)* ...... *(1988-89)* 371
(Gates Learjet) ...... *(1986-87)* 409	*385 TriStar conversions (Lockheed)* ...... *(1991-92)* 430	757QC (Pemco) 545
B (Learjet) ...... *(1991-92)* 419	*400 (Avtek)* ...... *(1986-87)* 335	767 (Boeing) 447
C (Learjet) ...... *(1991-92)* 419	*400 Beechjet (Beech)* ...... *(1988-89)* 338	767-200ER (Boeing) 447
Baron (Beech) 425	*400 Combat Twin (Bell)* ...... *(1987-88)* 22	767-300 (Boeing) 447
Baron (Ram/Beech) ...... *(1991-92)* 466	*400 TwinRanger (Bell)* ...... *(1985-86)* 340	767-300ER (Boeing) 447
P Baron (Beech) ...... *(1987-88)* 354	400A (Avtek) 419	767-300 Freighter (Boeing) 449
TC (Beech) ...... *(1984-85)* 342	*400A (Bell)* ...... *(1987-88)* 22	*767 Airborne Surveillance Testbed (Boeing)* ...... *(1991-92)* 366
Predator (Rutan) ...... *(1983-84)* 481	400A Beechjet (Beech) 432	*767 AOA (Boeing)* ...... *(1990-91)* 371
(Learjet) 496	*400A Interceptor (Prop-Jets)* ...... *(1985-86)* 490	767 AWACS (Boeing) 454
(RAF) ...... *(1985-86)* 613	*400C Aurora (MTC)* ...... *(1988-89)* 445	*767-X (Boeing)* ...... *(1990-91)* 384
Grizzly (Rutan) ...... *(1985-86)* 503	*400M Aurora (MTC)* ...... *(1987-88)* 477	777-200 (Boeing) 449
(Rutan) ...... *(1983-84)* 480	*401/402A/402B/402C (RAM/Cessna)* ...... *(1991-92)* 466	785 Orion (Lockheed) 499
Apache (McDonnell Douglas) ...... *(1984-85)* 439	*402C (Cessna)* ...... *(1991-92)* 385	*7J7 (Boeing)* ...... *(1988-89)* 364
Skipper (Beech) ...... *(1984-85)* 306	406-5 (Reims-Cessna) ...... *(1984-85)* 75, 372	*7J7-YXX (International)* ...... *(1992-93)* 89, 354
Catbird (Scaled Composites) ...... *(1989-90)* 498	406 (AHIP) (Bell) 436	*800J (Aerofan)* ...... *(1986-87)* 329
(Sabreliner) ...... *(1987-88)* 501	*406 CS (Bell)* ...... *(1992-93)* 345	*1300 Commuter (Beech)* ...... *(1990-91)* 355
-B55 (Beech) ...... *(1983-84)* 303	*406LM Long Ranger (Bell)* ...... *(1985-86)* 340	*1400 Commuterliner (Skytrader)* ...... *(1988-89)* 488
(De Vore) ...... *(1986-87)* 399	*410 (Teledyne Ryan)* ...... *(1990-91)* 522	*1900 Exec-Liner (Beech)* ...... *(1991-92)* 353
King Air (Beech) ...... *(1983-84)* 308	*412 (Agusta-Bell)* ...... *(1986-87)* 151	*1900C Airliner (Beech)* ...... *(1991-92)* 353
Sunbird (Verilite) ...... *(1990-91)* 525	*412 (Bell)* ...... *(1986-87)* 370	1900D (Beech) 430
(Boeing) ...... *(1991-92)* 366	*412 (Bell)* ...... *(1989-90)* 20	*2000 (Morrisey)* ...... *(1989-90)* 470
(Spratt) ...... *(1983-84)* 583	*412 (Nurtanio/Bell)* ...... *(1985-86)* 99	2000 (Saab) 354
4 (Boeing) 455	412 HP (Bell) 23, 191	2000 Starship 1 (Beech) 431
4B Commander (Commander) 471	*412 SP (Agusta-Bell)* ...... *(1991-92)* 157	2000A Starship (Beech) 431
3-3.62 (Scaled Composites) ...... *(1989-90)* 498	412 SP (Bell) 23	*2100 (Lear Fan)* ...... *(1984-85)* 429
3-4.62 (Scaled Composites) ...... *(1992-93)* 498	414 (Boeing) 455	*2150A Kachina (Varga)* ...... *(1983-84)* 497
2 (ZLIN) 69	*414 (RAM/Cessna)* ...... *(1991-92)* 466	*2150A-TG, Kachina (Hibbard/Varga)* ...... *(1983-84)* 402
3 Triumph (Scaled Composites) ...... *(1991-92)* 473	*414A Chancellor (Cessna)* ...... *(1985-86)* 376	*2180 Kachina (Varga)* ...... *(1983-84)* 498
5 (Myers) ...... *(1991-92)* 455	*414AW (RAM/Cessna)* ...... *(1991-92)* 466	*5800 (KFC)* ...... *(1992-93)* 29
1 ARES (Scaled Composites/Rutan) 554	*414AW Series V (Ram/Cessna)* ...... *(1991-92)* 466	
2 (Cessna) ...... *(1991-92)* 385	*421 Golden Eagle (Cessna)* ...... *(1991-92)* 385	**A**
2 Aerobat (Cessna) ...... *(1984-85)* 359	*421C and 421CW (RAM/Cessna)* ...... *(1991-92)* 466	
2 (RAM/Cessna) ...... *(1991-92)* 465	*425 Conquest I (Cessna)* ...... *(1991-92)* 385	A-1 (AMX) 128
5 Orion (Lockheed) 499	430 (Bell) 25	*A1 Flamingo-Trainer (FUS)* ...... *(1983-84)* 86
5 Skywagon (Cessna) ...... *(1991-92)* 385	*440 (Bell)* ...... *(1987-88)* 22	A-1 Husky (Aviat) 417
0 Super King Air (Beech) 428	*440E Commuter (Moller)* ...... *(1986-87)* 467	*A-3 (Cranfield)* ...... *(1987-88)* 314
0T (Beech) ...... *(1989-90)* 353	*441 Conquest II (Cessna)* ...... *(1991-92)* 385	*A-4 conversions (SAI)* ...... *(1987-88)* 215
0X (Moller) 537	442 (Bell) 25	*A-4 Skyhawk upgrade (RNZAF)* ...... *(1991-92)* 184
1 (Mooney) ...... *(1985-86)* 467	*480 (Advanced Technology)* ...... *(1985-86)* 321	*A-4PTM Skyhawk (Grumman/McDonnell Douglas)* ...... *(1986-87)* 424
1 (Mooney) ...... *(1988-89)* 443	480 (Enstrom) 473	A-4S-1 Super Skyhawk (SA) 338
1AT Advanced Trainer (Mooney) ...... *(1991-92)* 453	*480 Predator (ATAC)* ...... *(1986-87)* 335	A-5 (NAMC) 51
1SE (Mooney) ...... *(1991-92)* 453	497 (Romaero) 253	A-5, NAMC, modernisation (Alenia) 195
4B (Fuji-Bell) ...... *(1984-85)* 157	*500 (Cessna)* ...... *(1984-85)* 376	*A-5K Kong Yun (NAMC)* ...... *(1989-90)* 40
4-2 (Fuji-Bell) ...... *(1989-90)* 178	500 Citation, Upgrade Programme (AlliedSignal) 415	A-5M (NAMC) 53
5 (Agusta-Bell) ...... *(1988-89)* 151	500/520/530 (KA/McDonnell Douglas) 214	*A-6 Intruder (Grumman)* ...... *(1987-88)* 427
5 (Bell) ...... *(1991-92)* 358	500/530 Defender (McDonnell Douglas) 528	A-6E Intruder (Grumman) 481
5 (Mooney) ...... *(1988-89)* 444	500D (Kawasaki/McDonnell Douglas) 208	A-6E/TRAM (Grumman) 481
5A-1, Advanced (Fuji-Bell) ...... *(1992-93)* 157	*500MD/ASW Defender (McDonnell Douglas)* ...... *(1985-86)* 461	*A-6F Intruder 11 (Grumman)* ...... *(1990-91)* 418
5B, Advanced (Fuji-Bell) 204	*500ME (Hughes)* ...... *(1983-84)* 406	*A-6G (Grumman)* ...... *(1990-91)* 418
5 UH-1HP Huey II (Bell) 434	*500MF (Hughes)* ...... *(1983-84)* 406	*A-7 Corsair II (LTV)* ...... *(1989-90)* 445
6, Cessna, Turbine Pac (Soloy) ...... *(1991-92)* 490	*501 Citation I/SP (Cessna)* ...... *(1984-85)* 377	*A-7 Corsair II update programmes (LTV)* ...... *(1991-92)* 430
6B Jet Ranger III (Agusta-Bell) 190	503A (Air Tractor) ...... *(1990-91)* 341	*A-9B-M Quail (AAMSA)* ...... *(1983-84)* 161
6B Jet Ranger III (Bell) ...... *(1992-93)* 17	525 CitationJet (Cessna) 464	*A-10A (Fairchild Republic)* ...... *(1987-88)* 414
6 B-3 Jet Ranger III (Bell) 20	550 Citation II (Cessna) 465	*A10B (HDH)* ...... *(1986-87)* 7
6L Texas Ranger (Bell) ...... *(1985-86)* 341	560 Citation V (Cessna) 466	*A-12 (Lockheed)* ...... *(1992-93)* 402
6L-1 Long Ranger II (Bell) ...... *(1983-84)* 317	*600A (Piper)* ...... *(1984-85)* 478	*A-12A (General Dynamics/McDonnell Douglas)* ...... *(1992-93)* 382
6L-3 Long Ranger III (Bell) ...... *(1992-93)* 18	*602P (Piper)* ...... *(1985-86)* 488	*A20 (AAC)* ...... *(1985-86)* 6
6L-3ST Gemini ST (Tridair) 577	*650 Citation III (Cessna)* ...... *(1991-92)* 390	A-22 LASA (Sadler) 553
6L-L4 LongRanger IV (Bell) 21, 365	650 Citation VI (Cessna) 467	A-36 Bonanza (Beech) 424
6L-4ST Gemini ST (Tridair) 577	650 Citation VII (Cessna) 468	A-36 Halcón (CASA) 347
6LT TwinRanger (Bell) 22	*660 Citation VII (Cessna)* ...... *(1991-92)* 370	A-36 Halcón (ENAER) 43
7, Cessna, Turbine Pac (Soloy) ...... *(1991-92)* 490	*670 Citation IV (Cessna)* ...... *(1990-91)* 400	A-40 Albatross (Beriev) 255
8 Caravan I/U-27A (Cessna) 463	*680 Rotor (Aerofan)* ...... *(1990-91)* 366	A-50 Mainstay (Beriev) 259
8/208B Caravan Fire Fighter (Aero Union) ...... *(1991-92)* 338	685 Orion (Lockheed) 499	*A 109A Mk II (Civil versions) (Agusta)* ...... *(1991-92)* 154
9 HueyCobra (Modernised versions) (Bell) 434	*700P (Piper)* ...... *(1985-86)* 488	*A 109A Mk II (Military, Naval & Law Enforcement versions) (Agusta)* ...... *(1991-92)* 154
9 Improved SeaCobra (Bell) 435	*700R (Aerofan)* ...... *(1986-87)* 329	A 109C/CM/EOA (Agusta) 187
9 SeaCobra (Bell) ...... *(1990-91)* 361	707 (Boeing) 441	*A 109 EOA (Agusta)* ...... *(1991-92)* 154
9 SuperCobra (Bell) 435	*707 Tanker/Transport (Boeing)* 372	A 109K/KM/KN/K2 (Agusta) 188
2 (Agusta-Bell) ...... *(1990-91)* 161	707/720 conversions (IAI) 184	*A 109KN (Agusta)* 155
2 Twin Two-Twelve (Bell) 22	707TT (Alenia) 194	*A 109 MAX (Agusta)* ...... *(1990-91)* 158
2ASW (Agusta-Bell) 190	*712 FSW (Grumman)* ...... *(1990-91)* 420	*A 129 developed versions (Agusta)* ...... *(1991-92)* 161
4ST SuperTransport (Bell) ...... *(1990-91)* 364	*720 AAT (Boeing)* ...... *(1991-92)* 372	*A 129, LAH Tonal (JEH)* ...... *(1990-91)* 132
2 (Bell) ...... *(1990-91)* 365	*727 (Boeing)* ...... *(1983-84)* 327	A 129 Mangusta (Agusta) 189
2 conversion (Heli-Air) ...... *(1991-92)* 412	*727, Boeing, conversion (Lockheed)* ...... *(1987-88)* 451	*A-132 Tangará (Aerotec)* ...... *(1983-84)* 10
2 SP (Heli-Air) 487	*727 RE Quiet 727 (Valsan)* ...... *(1991-92)* 495	*A-135 Tangará II (Aerotec)* ...... *(1988-89)* 9
0 (Bell) 24	*727 update (Dee Howard)* ...... *(1991-92)* 396	A 139 Utility (Agusta) 190
1 (Mooney) ...... *(1985-86)* 468	731 Falcon 20B Retrofit Programme (AlliedSignal) 415	*A300 (Airbus)* ...... *(1984-85)* 99
4 (Boeing) ...... *(1991-92)* 369	*737 (Boeing)* ...... *(1983-84)* 328	*A300B2-600 (Airbus)* ...... *(1983-84)* 101
9 HueyCobra (Bell) ...... *(1983-84)* 320	737-100 (Boeing) 441	A300-600 (Airbus) 114
52TSE (Mooney) ...... *(1991-92)* 454	737-200 (Boeing) 441	
57 TLS (Mooney) ...... *(1991-92)* 454	737-300 (Boeing) 441	
80 (Enstrom) 472	737-300QC conversions (Pemco) 545	

INDEXES: AIRCRAFT—A

A300-600 Convertible (Airbus)	116
A300-600 Freighter (Airbus)	117
A300-600 ST Super Transporter (SATIC)	178
A300C (Airbus) .... (1986-87)	105
A300C4/F4 (Airbus) .... (1983-84)	102
A300F (Airbus) .... (1986-87)	105
A310 (Airbus)	117
A319 (Airbus)	121, 687
A320 (Airbus)	118
A321 (Airbus)	121
A330 (Airbus)	125
A340 (Airbus)	123
A340-300 Combi (Airbus) .... (1992-93)	97
A350 (Airbus) .... (1991-92)	114
A-750 (Airmaster) .... (1986-87)	331
A-1200 (Airmaster) .... (1988-89)	322
AA-23 STOL 180 (Taylorcraft) .... (1991-92)	492
AA-23A1 Ranger (Taylorcraft) .... (1991-92)	492
AA-23B Aerobat (Taylorcraft) .... (1991-92)	492
AA-23B1 RS (Taylorcraft) .... (1991-92)	492
AA200 Orion (CRSS)	112
AA300 Rigel (CRSS)	112
AA330 Theta (CRSS)	112
AAA (AAA)	114
AAA (see Advanced Amphibious Aircraft)	114
AAC (see Aerostar Aircraft Corporation)	410
AAC (see Aircraft Acquisition Corporation) .... (1992-93)	323
AAC (see Hawker de Havilland Ltd) .... (1985-86)	7
AAI (see American Aviation Industries) .... (1990-91)	336
AAMSA (see Aeronautica Agricola Mexicana SA) ... (1984-85)	164
AASI (see Advanced Aerodynamics and Structures Inc)	410
AB 115 Trainer (Aero Boero)	1
AB 115/150 (Aero Boero) .... (1983-84)	1
AB 115 BS (Aero Boero) .... (1983-84)	1
AB 150 Ag (Aero Boero) .... (1983-84)	1
AB 150 RV (Aero Boero) .... (1983-84)	1
AB 180 Ag (Aero Boero) .... (1988-89)	1
AB 180 PSA (Aero Boero)	685
AB 180 RV (Aero Boero) .... (1988-89)	1
AB 180 RVR (Aero Boero) .... (1988-89)	1
AB 180 SP (Aero Boero) .... (1988-89)	1
AB 260 Ag (Aero Boero) .... (1988-89)	2
AC-4 Andorinha (Super Rotor) .... (1992-93)	16
AC-05 Pijao (Aviones de Colombia)	61
AC-130 (Lockheed)	502
AC-130U Spectre (Rockwell/Lockheed)	551
ACA (BAe) .... (1990-91)	309
ACAC (see American Champion Aircraft Corporation)	415
ACAP, Model D292 (Bell) .... (1988-89)	347
ACAP programme (Bell) .... (1988-89)	347
ACAP programme (Sikorsky/US Army) .... (1986-87)	512
A. C. E. (see A. C. E. International) .... (1990-91)	57
ACS (see Aero-Club Der Schweiz)	205
ACT (Dassault-Breguet) .... (1984-85)	65
ACT Jaguar (BAe) .... (1985-86)	126
ACT (see Aviation Composite Technology) .... (1992-93)	174
ACX (Dassault-Breguet) .... (1984-85)	65
AD-1 Oblique-wing aircraft (NASA) .... (1983-84)	445
ADA (see Aeronautical Development Agency)	106
ADAL (see Aeronautical Development Associates Ltd) .... (1983-84)	248
ADDAX/LCRV (Williams) .... (1985-86)	315
ADOCS programme (Boeing Vertol) .... (1986-87)	386
AE 206 Mistral (Aviasud)	74
AE 207 Mistral Twin (Aviasud)	75
AE 209 Albatros (Aviasud)	75
AE 210 Alizé	74
Ae 270 (AERO)	65
AERO (see Aero Vodochody Akciova Spolecnost)	63
AEROTEK (see Aeronautical Systems Technology)	338
AF-18A (McDonnell Douglas)	523
AFTI/F-16 (General Dynamics) .... (1986-87)	413
AG-5B Tiger (American General)	416
AG-6 (Aerostar)	249
AGV (Aerospatiale) .... (1990-91)	58
AH-1 (4B)W Viper (Bell) .... (1992-93)	343
AH-1E HueyCobra (Bell)	434
AH-1F HueyCobra (Bell)	434
AH-1G HueyCobra (Bell) .... (1990-91)	361
AH-1J SeaCobra (Bell)	435
AH-1P HueyCobra (Bell)	434
AH-1Q HueyCobra (Bell) .... (1990-91)	361
AH-1S (Fuji-Bell)	205
AH-1S HueyCobra (Bell)	434
AH-1T Improved SeaCobra (Bell)	435
AH-1W Cobra Venom (Bell)	436
AH-1W Super Cobra (Bell)	435
AH-6 (McDonnell Douglas)	528
AH-58D Warrior (Bell) .... (1989-90)	362
AH-64 Apache (McDonnell Douglas)	531
AH-64A Apache Vectored Thrust Combat Agility Demonstrator (Piasecki)	545
AHIP Army Helicopter Improvement Program (Bell)	436
AIA (see Aviation Industries of Australia)	4
AIC (see Ames Industrial Corporation) .... (1984-85)	300
AIC (see Aviation Industries of China)	44
AICSA (see Aero Industrial Colombiana SA)	60
AIDC (see Aero Industry Development Center)	363
AIEP (see Aeronautical Industrial Engineering and Project Management Co Ltd)	224
AIM (see Aerospace Industries of Malaysia) ..... (1988-89)	174
AIPI (see Aerotech Industries Phillipines Inc)	226
AIRTECH (see Aircraft Technology Industries)	126
AISA (see Aeronautica Industrial SA)	343
AJ 37 Viggen (Saab) .... (1984-85)	196
AJS 37 Viggen (Saab)	351
ALR (see Arbeitsgruppe für Luft- und Raumfahrt) ... (1987-88)	227
AMI (see Aero Modifications International) ...... (1992-93)	327
AMIN (see Aero Maroc Industries) .... (1984-85)	165
AMIT Fouga (AIA) .... (1991-92)	149
AMX (AMX International)	128
An-2 Colt (Antonov)	366
An-2 Antek (PZL Mielec)	236
An-3 (Antonov) .... (1992-93)	280
An-3M (WSK-PZL Warszawa-Okecie/Antonov) ..... (1986-87)	193
An-12 Cub (Antonov)	366
An-22 Antheus (Antonov) .... (1991-92)	231
An-24 Coke (Antonov) .... (1991-92)	232
An-26 Curl (Antonov)	280
An-28 Cash (PZL Mielec)	231
An-30 Clank (Antonov) .... (1991-92)	233
An-32 Cline (Antonov)	367
An-38 (Antonov)	369
An-70 (Antonov) .... (1991-92)	235
An-70T (Antonov)	369
An-72 Coaler (Antonov)	370
An-72P Coaler (Antonov)	371
An-74 Coaler (Antonov)	370
An-74 Variant Madcap (Antonov) .... (1991-92)	236
An-77 (Antonov)	369
An-124 Condor (Antonov)	372
An-180 (Antonov)	374
An-218 (Antonov)	374
An-225 Cossack (Antonov)	374
An-400 (Antonov) .... (1984-85)	211
AOI (see Arab Organisation for Industrialisation)	72
AP 68TP-300 (Partenavia) .... (1988-89)	162
AP 68TP-600 Viator (Partenavia)	198
ARA 3600 (Promavia)	12
ARIA, EC-18B (Boeing) .... (1988-89)	370
ARTB (Lockheed) .... (1991-92)	429
ARTI (Bell) .... (1988-89)	348
ARTI (Boeing) .... (1986-87)	386
ARV (see Island Aircraft Ltd) .... (1988-89)	279
ARV-1 Super2 (Aviation Scotland)	376
ARV-1 Super2 (Island Aircraft) .... (1990-91)	315
AS-32T Turbo Trainer (FFA) .... (1983-84)	192
AS-61 (Agusta-Sikorsky)	191
AS-61N1 (Agusta-Sikorsky) .... (1990-91)	163
AS-61R (Agusta-Sikorsky)	191
AS100 (Aerospatiale) .... (1991-92)	58
AS 202 Bravo (FFA) .... (1986-87)	214
AS 202/18A Bravo (FFA)	357
AS 330 Puma (Aerospatiale) .... (1991-92)	58
AS 330 Puma conversions (Atlas)	342
AS 332 Super Puma (Eurocopter)	146
AS 332 Super Puma Mk II (Aerospatiale) .... (1991-92)	60
AS 350 AllStar conversion (Rocky Mountain)	552
AS 350 Ecureuil/AStar (Eurocopter)	150
AS 351 (Aerospatiale) .... (1984-85)	56
AS 355 Ecureuil 2/TwinStar (Eurocopter)	151
AS 365F Dauphin 2 (Aerospatiale) .... (1988-89)	64
AS 365N2 Dauphin 2 (Eurocopter)	152
AS 365X DGV Dauphin (Eurocopter)	154
AS 365X FBW Dauphin (Eurocopter)	154
AS 366 Dauphin 2 (Aerospatiale) .... (1991-92)	66
AS 532 Cougar (Eurocopter)	146
AS 532 Cougar Mk II (Eurocopter)	148
AS 550 Fennec (Eurocopter)	150
AS 555 Ecureuil 2/TwinStar (Aerospatiale) ...... (1990-91)	63
AS 555 Fennec (Eurocopter)	151
AS 565 Dauphin 2 (Aerospatiale) .... (1990-91)	64
AS 565 Panther (Helibras)	20
AS 565 Panther, army/air force versions (Eurocopter)	153
AS 565 Panther, naval versions (Eurocopter)	154
ASF, air-superiority fighter (USSR) .... (1990-91)	293
ASH-3H (Agusta-Sikorsky)	191
ASI (see Aerodynamics & Structures Inc) .... (1990-91)	343
ASP-XJ (Baldwin) .... (1989-90)	348
AST (see Aviaspetrans Consortium)	255
ASTA (see AeroSpace Technologies of Australia Pty Ltd)	5
ASTOR (Airborne Stand-off Radar) Programme (RAE) .... (1989-90)	325
ASTOR Defender (Pilatus Britten-Norman) .... (1992-93)	310
ASTOVL programme (NASA) .... (1988-89)	447
ASTRA Hawk (Cranfield) .... (1988-89)	296
AT² (Scaled Composites) .... (1992-93)	438
AT-3 Tzu-Chung (AIDC)	364
AT-9 (Avocet) .... (1988-89)	326
AT-46A (Fairchild Republic) .... (1986-87)	403
AT-301/301B Air Tractor (Air Tractor) .... (1988-89)	322
AT-302 Air Tractor (Air Tractor) .... (1985-86)	320
AT-400 Turbo Air Tractor (Air Tractor)	413
AT-400A (Air Tractor) .... (1988-89)	323
AT-401 Air Tractor (Air Tractor)	412
AT-401A Air Tractor (Air Tractor) .... (1992-93)	324
AT-402 Turbo Air Tractor (Air Tractor)	413
AT-500 (Air Tractor) .... (1985-86)	320
AT-501 (Air Tractor) .... (1992-93)	325
AT-502 (Air Tractor)	413
AT-503/503A (Air Tractor)	414
AT-802 (Air Tractor)	414
AT-2000 (Aermacchi/Deutsche Aerospace)	114
ATA (General Dynamics/McDonnell Douglas) ... (1992-93)	382
ATA (USAF)	579
ATAC (see Advanced Technology Aircraft Co Inc) ... (1986-87)	335
ATB (Northrop) .... (1987-88)	479
ATF (Lockheed)	497
ATF (Northrop/McDonnell Douglas) .... (1991-92)	457
ATF (USAF) .... (1991-92)	495
ATF-18A (McDonnell Douglas)	523
ATF 580S Turbo Flagship (Allison) .... (1987-88)	346
ATI-2 Fantrainer (RFB) .... (1984-85)	90
ATL (FMA) .... (1985-86)	5
ATL (Robin) .... (1991-92)	83
ATL, AIRCRAFT TECHNOLOGY LTD (China) ... (1987-88)	38
ATL2, Atlantique2 (Dassault Aviation)	81
ATM 42L (ATR) .... (1989-90)	121
ATOPS (NASA) .... (1990-91)	470
ATP (BAe) .... (1992-93)	295
ATP, Jetstream (BAe)	382
ATR (see Avions de Transport Régional)	131
ATR 42 (ATR)	131
ATR 42-500 (ATR)	688
ATR 52C (ATR)	133
ATR 72 (ATR)	133
ATR 82 (ATR)	134
ATTA 3000 (Promavia/MiG)	175
AT-TC-3 (AIDC) .... (1984-85)	203
ATSF (Aerospatiale)	74
ATTAS, VFW 614 (MBB) .... (1989-90)	10
AU-24A (Helio) .... (1991-92)	4
AUH-76 (Sikorsky) .... (1984-85)	50
AV-8A (BAe) .... (1989-90)	30
AV-8B (McDonnell Douglas/BAe)	16
AV-8S Harrier (BAe) .... (1991-92)	31
AW1-2, Fantrainer (RFB) .... (1984-85)	9
AX (CASA)	34
A-X (USN)	58
ACA INDUSTRIES INC (USA) .... (1992-93)	32
ACE AIRCRAFT MANUFACTURING CO (USA) .... (1984-85)	55
A. C. E. INTERNATIONAL (France) .... (1990-91)	5
Acey Deucy Mk II Biplane (Powell) .... (1984-85)	59
Acro-Husky (Aviat)	4
Acro-Sport I (EAA) .... (1983-84)	55
Acro-Sport II (EAA) .... (1983-84)	55
ADAMS INDUSTRIES INC (USA) .... (1983-84)	29
Adnan 1 (IAF)	18
ADVANCED AERODYNAMICS AND STRUCTURES INC (USA)	4
Advanced Airborne Command Post (Boeing) ... (1985-86)	36
ADVANCED AIRCRAFT CORPORATION (USA) .. (1992-93)	32
ADVANCED AMPHIBIOUS AIRCRAFT (International)	11
Advanced Cargo Aircraft (US Army)	58
Advanced Electronic Countermeasures Programme (Boeing) .... (1989-90)	39
Advanced Light Helicopter (HAL/Eurocopter)	10
Advanced STOVL Strike Fighter (Lockheed)	50
Advanced Surveillance and Tracking Technology/ Airborne Radar Demonstrator (USAF)	57
Advanced Tactical Aircraft (General Dynamics/ McDonnell Douglas) .... (1991-92)	405, 49
Advanced Tactical Fighter (McDonnell Douglas/ Northrop) .... (1991-92)	45
Advanced Tactical Fighter ATF (USAF)	57
ADVANCED TECHNOLOGY AIRCRAFT CO INC (USA) .... (1986-87)	33
Advanced Technology Tactical Transport (Beech) .. (1988-89)	34
Advanced Transport Systems Research Vehicle (NASA) .... (1986-87)	46
AERITALIA/AERMACCHI/EMBRAER (International) .... (1986-87)	10
AERITALIA—SOCIETA AEROSPAZIALE ITALIANA SpA (Italy) .... (1990-91)	15
AERMACCHI/DEUTSCHE AEROSPACE (International)	11
AERMACCHI SpA (Italy)	18
Aerobat (Reims/Cessna) .... (1986-87)	7
Aerobat Model 152 (Cessna) .... (1984-85)	35
Aerobat Model AA-23B (Taylorcraft) .... (1991-92)	49
AERO BOERO SA (Argentina)	4
Aerobot (Moller) .... (1989-90)	46
AERO-CLUB DER SCHWEIZ (Switzerland) .... (1985-86)	20
AERO CLUBE DE RIO CLARO (Brazil) .... (1986-87)	1
AERO CZECH AND SLOVAK AEROSPACE INDUSTRY LTD (Czech Republic)	63
AERO COMPOSITE TECHNOLOGY INC (USA)	41
AERODIS AMERICA INC (USA) .... (1992-93)	32
AERODYNAMICS & STRUCTURES INC (USA) ..... (1990-91)	34
AERODYNE SYSTEMS ENGINEERING LTD (USA) .... (1987-88)	34
AEROFAN AIRCRAFT MANUFACTURING CO (USA) .... (1986-87)	33
AERO GARE (USA) .... (1984-85)	55
AERO INDUSTRIAL COLOMBIANA SA (Colombia)	6
AERO INDUSTRY DEVELOPMENT CENTER (Taiwan)	36
Aero Knight Twister (Payne) .... (1985-86)	60
AERO MAROC INDUSTRIE (Morocco) .... (1984-85)	16
AERO MERCANTIL SA (Colombia) .... (1991-92)	4
AERO MOD GENERAL (USA) .... (1987-88)	34
AERO MODIFICATIONS INTERNATIONAL INC (USA) .... (1992-93)	32
AERONAUTICA AGRICOLA MEXICANA SA (Mexico) .... (1984-85)	16
AERONAUTICA INDUSTRIAL SA (Spain)	34
AERONAUTICAL DEVELOPMENT AGENCY (India)	10
AERONAUTICAL DEVELOPMENT ASSOCIATES LTD (UK) .... (1983-84)	24
AERONAUTICAL INDUSTRIAL ENGINEERING AND PROJECT MANAGEMENT COMPANY LTD (Nigeria)	22
AERONAUTICAL SYSTEMS TECHNOLOGY (South Africa)	33
AERONAUTICA MACCHI SpA (Italy)	18
AERORIC RESEARCH AND PRODUCTION ENTERPRISE (Russia)	69
AEROSPACE CONSORTIUM INC, THE (Canada)	2
AEROSPACE INDUSTRIES OF MALAYSIA (Malaysia) .... (1988-89)	17
Aero-Space Plane (NASA)	54
AEROSPACE TECHNOLOGIES OF AUSTRALIA PTY LTD (Australia)	5
AEROSPATIALE SNI (France)	7
Aero Sportster (Moser) .... (1985-86)	60
AEROSTAR AIRCRAFT CORPORATION (USA)	41
AEROSTAR SA (Romania)	24
Aerostar 600A (Piper) .... (1984-85)	47
Aerostar 602P (Piper) .... (1985-86)	48
Aerostar 700P (Piper) .... (1985-86)	48
Aerostar 3000 (AAC)	41
Aerostar P-3A (Aero Union) .... (1992-93)	33
Aero Subaru (Fuji) .... (1986-87)	16
AERO-TECH (USA) .... (1985-86)	57
AEROTECH INDUSTRIES PHILIPPINES INC (Philippines)	22
AEROTECHNIK (Czechoslovakia) .... (1984-85)	4
AEROTEC S/A INDUSTRIA AERONAUTICA (Brazil) .... (1988-89)	9
AERO TRADER LTD (New Zealand) .... (1988-89)	17
AERO UNION CORPORATION (USA)	41
AERO VODOCHODY AKCIOVA SPOLECNOST (Czech Republic)	63
AERO VODOCHODY NARODNI PODNIK (Czechoslovakia) .... (1990-91)	49

A - B—AIRCRAFT: INDEXES 701

ft-body strake system (Raisbeck) ... (1991-92) 465
g-Cat B-Plus (Schweizer) ... (1983-84) 482
g-Cat Series (Schweizer) ... 556
g-Cat Super B Turbine (EAL) ... (1989-90) 56
g-Chuspi IAP-002 (Indaer Peru) ... (1992-93) 173
gcraft (Taylorcraft) ... (1991-92) 492
g Husky (Cessna) ... (1991-92) 385
gile Falcon (General Dynamics) ... (1991-92) 403
GRICOPTEROS LTDA (Colombia) ... (1984-85) 42
gricultural Remote Control Helicopter (DHI) ... 213
GRO-COPTEROS LTDA (Colombia) ... 60
gro-Turbo, Zlin 37T (Moravan) ... (1983-84) 47
gro Turbo, Z 137T (Zlin) ... 71
gTrainer, (Cessna/Aviones de Colombia) ... 61
g Truck (Cessna) ... (1991-92) 385
GUSTA SpA ... 187
ggressor (Orlando) ... 545
iglon, R 1180 (Robin) ... (1984-85) 77
ileron-rudder interconnect programme (NASA) ... (1986-87) 469
IR AMERICA INC (USA) ... (1986-87) 330
IR & SPACE AMERICA INC (USA) ... 412
ir Bearing Fan (Sadlier) ... (1989-90) 6
ir Beetle (AIEP) ... 224
irbuggy (WHE) ... (1984-85) 299
AIRBUS INDUSTRIE (International) ... 114
AIRCORP PTY LTD (Australia) ... 4
Aircraft 101 (Tupolev) ... (1985-86) 262
AIRCRAFT ACQUISITION CORPORATION (USA) ... (1992-93) 323
AIRCRAFT DESIGNS (BEMBRIDGE) LTD (UK) .. (1984-85) 256
AIRCRAFT TECHNOLOGY INDUSTRIES (International) ... 126
AIRESEARCH AVIATION COMPANY (USA) ... (1986-87) 330
Air Force One VC-25A (Boeing) ... (1991-92) 372
Airfox, IAR-317 (IAR) ... (1991-92) 205
Airguard, F-7M (CAC) ... 44
AIRMASTER INC (USA) ... (1989-90) 341
AIROD SENDIRIAN BERHAD (Malaysia) ... 214
Airone F.220 (General Avia) ... 197
Air Sentry (Fairchild) ... (1985-86) 401
Air Shark 1 (Freedom Master) ... 477
AIRTECH CANADA (Canada) ... 20
AIR TRACTOR INC (USA) ... 412
Airtrainer CT4 (PAC) ... 222
Airtruk (PL-12 (Transavia) ... (1991-92) 6
Ajeet (HAL) ... (1984-85) 93
Ajeet Trainer (HAL) ... (1988-89) 98
AKASAMITRA HOMEBUILT AIRCRAFT ASSOCIATION (Indonesia) ... (1986-87) 100
Albatros AE 209 (Aviasud) ... 75
Albatros L-39 (AERO) ... 63
Albatross A-40 (Beriev) ... 255
Albatross, G-111 (Grumman/Resorts International) ... (1986-87) 416
Albatross Tanker conversion (Aero Union) ... (1991-92) 338
ALENIA (Italy) ... 194
Alizé (Dassault Aviation) ... (1991-92) 74
Alizé AE 210 (Aviasud) ... 74
Alliance (SCTICSG) ... 688
ALLIEDSIGNAL AVIATION SERVICES (USA) ... 415
ALLISON GAS TURBINE DIVISION, GMC (USA) . (1990-91) 341
Alouette III (Aerospatiale) ... (1990-91) 59
Alouette III, IAR-316B (ICA) ... (1989-90) 213
Alouette III SA 316B (HAL) ... 109
Alouette III, SA 316B (HAL/Aerospatiale) ... (1987-88) 106
Alpha (Robinson) ... (1985-86) 493
Alpha Jet (Dassault/Dornier) ... 137
Alpha series (Robin) ... (1984-85) 76
Alpha Sport (Robin) ... (1983-84) 76
Alpha XH-1 (Atlas) ... (1987-88) 216
AMECO-HAWK INTERNATIONAL (USA) ... (1992-93) 327
AMERICAN AVIATION INDUSTRIES (USA) ... (1990-91) 336
AMERICAN CHAMPION AIRCRAFT CORPORATION (USA) ... 415
AMERICAN EUROCOPTER CORPORATION (USA) ... 416
AMERICAN GENERAL AIRCRAFT CORPORATION (USA) ... 416
AMES INDUSTRIAL CORPORATION (USA) ... (1984-85) 300
Amphibian Air Car Jr (Spencer) ... (1984-85) 600
AMS/OIL (USA) ... (1984-85) 559
AMX INTERNATIONAL LTD (International) ... 128
ANDERSON AIRCRAFT CORPORATION (USA) ... (1984-85) 560
Andorinha AC-4 (Super Rotor) ... (1992-93) 16
Angel (TKEF) ... 576
Antek, An-2 (PZL-Mielec) ... 236
Antheus, An-22 (Antonov) ... (1991-92) 231
ANTONOV DESIGN BUREAU (Ukraine) ... 366
Apache (McDonnell Douglas) ... 531
Apache, sea-going (McDonnell Douglas) ... (1989-90) 467
Apache 1 (ACT) ... (1991-92) 186
APOLLO HELICOPTER SERVICES (USA) ... (1987-88) 347
ARAB ORGANISATION FOR INDUSTRIALISATION ... 72
Arava (IAI) ... (1989-90) 149
ARBEITSGEMEINSCHAFT TRANSALL (International) ... (1989-90) 146
ARBEITSGRUPPE FÜR LUFT- UND RAUMFAHRT (Switzerland) ... (1987-88) 227
Archer II (Piper) ... (1990-91) 477
ARCTIC AIRCRAFT COMPANY (USA) ... 416
Arctic Tern (Arctic) ... 416
Arcturus CP-140A (IMP) ... 38
Arcturus CP-140A (Lockheed) ... 499
ARES (Scaled Composites/Rutan) ... 554
AROCET INC (USA) ... (1988-89) 326
Arrow (Piper) ... (1990-91) 478
ARV AVIATION LTD (UK) ... (1987-88) 294
ARVIN/CALSPAN ADVANCED TECHNOLOGY CENTER (USA) ... (1983-84) 349
AStar AS 350 (Eurocopter) ... 150
Astra, 1125 (IAI) ... 182
Asuka (Kawasaki) ... (1988-89) 167
Asuka (NAL) ... (1991-92) 175
Atlant VM-T (Myasishchev) ... (1991-92) 273
Atlantic 1 Modernisation (Alenia) ... (1991-92) 163
Atlantic 1 Modernisation Programme (Dornier) ... (1989-90) 94
Atlantique 2 (Dassault Aviation) ... 81
ATLAS AIRCRAFT CORPORATION OF SOUTH AFRICA (PTY) LIMITED (South Africa) ... (1991-92) 210, 753
ATLAS AVIATION (PTY) LIMITED (South Africa) ... 339
Atlas Military Trainer (Atlas) ... (1991-92) 210
Attack Helicopter (Cardoen) ... (1987-88) 35
Aucán (ENAER) ... (1988-89) 33
Aurora (MTC) ... (1988-89) 445
Aurora CP-140 (Lockheed) ... 499
AUSTRALIAN AIRCRAFT CONSORTIUM PTY LTD (Australia) ... (1985-86) 6
AUSTRALIAN AUTOGYRO CO, THE (Australia) ... 5
Australian basic trainer (AAC) ... (1983-84) 6
Australian F/A-18A (McDonnell Douglas) ... (1986-87) 452
Autogyro (Everett) ... (1989-90) 316
Avalon 680 (Airmaster) ... (1986-87) 331
AVALON AVIATION (Canada) ... (1989-90) 18
Avalon Twin Star 800 (Airmaster) ... (1985-86) 319
Avalon Twin Star 1000 (Airmaster) ... (1988-89) 322
Avanti P.180 (Piaggio) ... 199
Avenger II A-12A (General Dynamics/McDonnell Douglas) ... (1992-93) 382
AVIASPETRANS CONSORTIUM (Russia) ... 255
AVIASUD ENGINEERING SA (France) ... 74
AVIAT INC (USA) ... 417
AVIATIKA JOINT STOCK COMPANY (Russia) ... 691
AVIATION COMPOSITE TECHNOLOGY (Philippines) ... (1992-93) 174
AVIATION INDUSTRIES OF AUSTRALIA PTY LTD (Australia) ... 4
AVIATION INDUSTRIES OF CHINA (China) ... 44
AVIATION SCOTLAND LTD (UK) ... 376
AVIATSIONNYI KOMPLEKS IMIENI S. V. ILYUSHINA (Russia) ... 259
AVIATSIONNYI NAUCHNO-PROIZVODST-VENNIY KOMPLEKS-ANPK MiG IMIENI A. I. MIKOYANA (Russia) ... 279
AVIATSIONNYI NAUCHNO-TEKHNICHESKIY KOMPLEKS IMIENI A. N. TUPOLEVA (Russia) ... 319
AVID AIRCRAFT INC (USA) ... 418
Avid Flyer Mark IV (Avid) ... 418
AVIOANE SA (Romania) ... 250
Aviocar, C-212 Series 200 (CASA) ... (1987-88) 217
Aviocar, C-212 Series 300 (CASA) ... 345
Aviocar, C-212P (CASA) ... (1991-92) 214
Aviocar, NC-212-200 (IPTN) ... 113
Aviocar Special Mission Versions (CASA) ... 346
Aviojet, C-101 (CASA) ... 347
AVIOLIGHT (Italy) ... (1992-93) 153
AVIONES DE COLOMBIA SA (Colombia) ... 60
Avion Liviano (ENAER) ... (1989-90) 33
AVIONS DE TRANSPORT REGIONAL (International) ... 131
AVIONS MUDRY et CIE (France) ... 86
AVIONS PIERRE ROBIN (France) ... 89
AVIONS PIERRE ROBIN INC (Canada) ... (1985-86) 29
AVIOTECHNICA LTD (International) ... 134
AVRO INTERNATIONAL AEROSPACE (International) ... 135
AVSTAR INC (USA) ... (1991-92) 343
AVTEK CORPORATION (USA) ... 419
AYMET AEROSPACE INDUSTRIES (Turkey) ... 365
AYRES CORPORATION (USA) ... 420
Ayres Thrush modification (Serv-Aero) ... (1984-85) 497
Aztec, amphibious (Wipaire) ... (1991-92) 496

B

B Scamp (Agricopteros/Aerosport) ... (1984-85) 42
B-1B Lancer (Rockwell) ... (1990-91) 491
B-2 (Northrop) ... 543
B2-16 Bushmaster (Aircorp) ... 4
B-2B (Brantly) ... 461
B-2B (Brantly-Hynes) ... (1983-84) 347
B-2B Brantly (Naras) ... (1989-90) 107
B2-N Bushmaster (Aircorp) ... (1991-92) 3
B-6 (XAC) ... (1992-93) 44
B-7 (XAC) ... 58
B-8MJ (Bensen) ... (1983-84) 543
B-18 Hover-Gyro (Bensen) ... (1984-85) 561
B36TC, Turbo Bonanza (Beech) ... 424
B-52 Stratofortress (Boeing) ... (1991-92) 371
B60 Duke (Beech) ... (1983-84) 305
B100 King Air (Beech) ... (1983-84) 308
B200 Super King Air (Beech) ... 427
B200T Maritime Patrol (Beech) ... 428
BA-14 Starling (FFV Aerotech) ... (1990-91) 218
BA-14B (MFI) ... 350
BAC (see Buchanan Aircraft Corporation) ... 6
BAC 1-11 (Dee Howard) ... (1991-92) 396
BAC 111-400 (Dee Howard) ... (1986-87) 398
BAC-204 (BAC) ... 6
BAC/BAe One-Eleven (Romaero) ... 253
BAe (see British Aerospace plc) ... 377
BAe Corporate 800 (BAe) ... 377
BAe Corporate 1000 (BAe) ... 378
BAe 125 Series 700 (BAe) ... (1983-84) 251
BAe 125 Series 700-11 (BAe) ... (1991-92) 301
BAe 125-800 (BAe) ... 204
BAe 125 Series 900 (BAe) ... (1989-90) 300
BAe 132 (BAe) ... (1990-91) 301
BAe 146 (BAe) ... 383
BAe 146 CC Mk 2 (BAe) ... 383
BAe 146 NRA (BAe) ... (1992-93) 296
BAe 146 Military Variants (BAe) ... (1992-93) 298
BAe 146 Series 100 (BAe) ... 383
BAe 146 Series 200 (BAe) ... 383
BAe 146 Series 300 (BAe) ... 383
BAe 146 Series 350 (BAe) ... (1988-89) 285
BAe 146-200 cargo conversion (Pemco) ... 545
BAe 146MSL (BAe) ... (1988-89) 286
BAe 146MT (BAe) ... (1988-89) 286
BAe 146-QC Convertible (BAe) ... 383
BAe 146-QT Quiet Trader (BAe) ... 383
BAe 748 (AEW) (HAL) ... (1987-88) 107
BAe 748 Coastguarder (BAe) ... (1984-85) 261
BAe 748 Series 2B (BAe) ... (1983-84) 253
BAe 748 Series 2B Super (BAe) ... (1983-84) 254
BAe 748 Water Bomber Conversion (Cranfield) ... (1987-88) 314
BAe 748(M) (HAL) ... (1984-85) 94
BAe RJ70/RJ80 (BAe) ... (1992-93) 296
BAF (see British Air Ferries) ... 391
BD-4 (Bede) ... (1992-93) 504
BD-10 (Bede) ... 422
Be-12 Tchaika (Beriev) ... (1991-92) 239
Be-32 Cuff (Beriev) ... 692
Be-40 Albatross (Beriev) ... 255
Be-42 Mermaid (Beriev) ... 257
Be-103 (Beriev) ... 258
Be-200 (Beriev) ... 257
BGP1 Biplane (Plumb) ... (1983-84) 536
BHEL (see Bharat Heavy Electricals Ltd) ... 106
BHK (see Bell Helicopter Korea) ... 212
BK 117 (Eurocopter/Kawasaki) ... 159
BK 117 (Nurtanio/MBB) ... (1983-84) 97
BK 117 conversion (Heli-Air) ... (1991-92) 412
BK 117 Improvement Programme (MBB/Kawasaki) ... (1984-85) 113
BK 117A (MBB/Kawasaki) ... (1984-85) 115
BK 117A-3M (MBB/Kawasaki) ... (1987-88) 123
BK 117M (MBB) ... (1990-91) 97
BN-2A Mk III Tri-Commutair (IAC/Pilatus Britten-Norman) ... (1984-85) 424
BN-2B Islander (Pilatus Britten-Norman) ... 396
BN-2T Special Role Turbine Defenders (Pilatus Britten-Norman) ... (1992-93) 310
BN-2T Special Role Turbine Islander (Pilatus Britten-Norman) ... (1992-93) 310
BN-2T Turbine Islander (Pilatus Britten-Norman) ... 397
BN 109 (NTT) ... (1988-89) 128
BO 105 (Eurocopter) ... 154
BO 105 LS (Eurocopter Canada) ... 38
BO 105/PAH-1/VBH/BSH-1 (Eurocopter) ... 155
BO 108 (Eurocopter) ... (1992-93) 117
BO 108 (MBB) ... (1984-85) 89
BOR-4 (USSR) ... (1991-92) 298
BOR-5 (USSR) ... (1991-92) 298
BR2000 (Bromon) ... (1990-91) 392
BX-50A (Berger-Helicopter) ... (1984-85) 549
BX-110 (Berger-Helicopter) ... (1983-84) 531
BX-110A (Berger) ... (1984-85) 549

Baaz (MiG) ... 283
Backfire, Tu-22M (Tupolev) ... 324
Badger, Tu-16 (Tupolev) ... 319
Baghdad 1 (IAF) ... 181
Bahadur (HAL) ... 111, 280
BALDWIN AIRCRAFT INTERNATIONAL (USA) ... (1990-91) 347
B & B AVIATION (USA) ... (1983-84) 296
Bandeirante (EMBRAER) ... 14
BARNEY OLDFIELD AIRCRAFT CO (USA) ... (1985-86) 578
Baron 58 (Beech) ... 425
Baron 58 (RAM/Beech) ... (1991-92) 466
Baron 58P (Beech) ... (1987-88) 354
Baron 58TC (Beech) ... (1985-86) 328
Baron Model 95-B55 (Beech) ... (1983-84) 303
Baron Model E55 (Beech) ... (1984-85) 311
BASLER TURBO CONVERSIONS INC (USA) ... 422
Battlefield Lynx (Westland) ... 409
Bayern II (Maiss) ... (1985-86) 562
Baz F-15 (McDonnell Douglas) ... 521
Bazant (WSK-PZL Swidnik/Mil) ... (1985-86) 183
BEACHNER, CHRIS (USA) ... (1985-86) 576
Beagle H-5 (Harbin) ... (1986-87) 34
Bear, Tu-95 (Tupolev) ... 320
Beaver conversions (Wipaire) ... (1991-92) 496
Beaver, DHC-2 (Modern Wing) ... (1988-89) 31
Beaver, DHC-2/PZL-3S (Airtech Canada) ... (1991-92) 17
BEECH AIRCRAFT CORPORATION (USA) ... 423
Beechcraft 99 conversions (ENAER Chile) ... (1987-88) 37
Beechjet 400A (Beech) ... 432
Beech Research & Development Projects (Beech) .. (1988-89) 340
BEETS, GLENN (USA) ... (1983-84) 560
Belfast conversion (Marshall) ... (1985-86) 292
BELL/BOEING (USA) ... 438
BELLANCA AIRCRAFT ENGINEERING INC (USA) ... (1985-86) 348
BELLANCA INC (USA) ... (1989-90) 371
BELL HELICOPTER CANADA (Canada) ... (1987-88) 21
BELL HELICOPTER KOREA INC (Korea, South) ... 212
BELL HELICOPTER TEXTRON (Canada) ... 20
BELL HELICOPTER TEXTRON INC (USA) ... 434
BEN-AIR LTD (UK) ... (1983-84) 267
Bengali, HN 500 (Nicollier) ... (1983-84) 521
BENGIS AIRCRAFT COMPANY (PTY) LTD (South Africa) ... (1992-93) 264
Berkut (Experimental Aviation) ... 474
BERIEV (Russia) ... 255
BERIEV DESIGN BUREAU (USSR) ... (1991-92) 239, 754
Beriev new flying-boat (Beriev) ... (1989-90) 244
Beta Bird (Hovey) ... (1985-86) 595
Beta, R22 (Robinson) ... 550
Beta, XTP-1 (Atlas) ... (1987-88) 204
BEZZOLA, GION (Switzerland) ... (1983-84) 531
BHARAT HEAVY ELECTRICALS LTD (India) ... 106
Bionic Bat (MaCready) ... (1985-86) 444
Bird (Taylor) ... (1985-86) 626
Bison M-3 (Myasishchev) ... (1991-92) 273
Black Hawk (Sikorsky) ... 365, 560
'Blackbirds' (Lockheed) ... (1983-84) 417
BLACKHOLE INVESTMENTS LTD (Canada) ... (1988-89) 21
Blackjack, Tu-160 (Tupolev) ... 326
Blinder, Tu-22 (Tupolev) ... 323
BOEING CANADA LTD (Canada) ... (1991-92) 20
BOEING COMMERCIAL AIRPLANE GROUP (USA) ... 440
BOEING COMPANY, THE (USA) ... 440
BOEING DEFENSE & SPACE GROUP (USA) ... 452
BOEING FABRICATION AND SERVICES COMPANY OF CANADA (Canada) ... (1986-87) 20
BOEING HELICOPTERS (USA) ... 455
BOEING MILITARY AIRPLANES (USA) ... (1992-93) 348
Boeing Orders and Deliveries (Boeing) ... 441

702 INDEXES: AIRCRAFT—**B** – **C**

BOEING/SIKORSKY (USA) 459
BOEING VERTOL COMPANY (USA) (1986-87) 381
BOMBARDIER AEROSPACE GROUP (Canada) 25
Bonanza 400 (Smith) (1988-89) 488
Bonanza A36 (Beech) 424
Bonanza F33A/C (Beech) 423
Bonanza V35B (Beech) (1986-87) 338
BOND, JOHN (USA) (1984-85) 562
BOUNSALL, EDDIE (USA) (1984-85) 562
BOWERS, PETER M (USA) 461
Boxer, LDA-500 (Lockspeiser) (1986-87) 298
Boxer, LDA-1000 (Lockspeiser) (1986-87) 299
BOYETTE, ERNIE (USA) (1983-84) 544
BRANSON AIRCRAFT CORPORATION (USA) 461
BRANTLY HELICOPTER INDUSTRIES USA CO LTD (USA) 461
BRANTLY-HYNES HELICOPTER INC (USA) (1983-84) 347
Brasilia, EMB-120 (EMBRAER) 14
Bravo, AS 202 (FFA) (1986-87) 214
Bravo, AS202/18A (FFA) 357
BREDANARDI COSTRUZIONI AERONAUTICHE SpA (Italy) (1989-90) 171
Breezy Model RLU-1 (RLU) (1983-84) 578
Brewer (Yakovlev) (1990-91) 290
BRISTOL AEROSPACE LTD (Canada) 25
BRITISH AEROSPACE plc (UK) 377
BRITISH AEROSPACE AIRBUS LTD (UK) 377
BRITISH AEROSPACE AIRCRAFT GROUP (UK) (1984-85) 256
BRITISH AEROSPACE CORPORATE JETS LTD (UK) 377
BRITISH AEROSPACE DEFENCE LTD (UK) 385
BRITISH AEROSPACE DYNAMICS GROUP (UK) (1984-85) 257
BRITISH AEROSPACE REGIONAL AIRCRAFT LTD (UK) 383
BRITISH AIR FERRIES (UK) 391
BROMON AIRCRAFT CORPORATION (USA) (1990-91) 392
Bronco OV-10 (Rockwell) (1990-91) 490
Bronco, OV-10D Plus (Rockwell) (1991-92) 468
BROOKLANDS AIRCRAFT LTD (UK) (1990-91) 312
Buccaneer update (BAe) (1990-91) 310
BUCHANAN AIRCRAFT CORPORATION LTD (Australia) 5
Buffalo (Boeing Canada) (1988-89) 21
Bulldog, SA-3-120 Series 120 (BAe) (1983-84) 258
Bull Thrush S2R-R1820/510 (Ayres) (1990-91) 345
BURKHART GROB LUFT- UND RAUMFAHRT (Germany) 100
BUSH CONVERSIONS INC (USA) 462
Bushmaster B2-16 (Aircorp) 4
Bushmaster B2N (Aircorp) (1991-92) 3
Businessliner (Cessna) (1985-86) 383

C

C-1 modifications (Kawasaki) (1988-89) 167
C-2A Greyhound (Grumman) (1989-90) 417
C4M Kudu (Atlas) (1983-84) 184
C-5 Galaxy (Lockheed) (1989-90) 443
C-8A Buffalo (DH Canada) (1983-84) 26
C-9A Nightingale (Douglas) (1985-86) 454
C-9B Skytrain II (Douglas) (1985-86) 454
C-12 Super King Air (Beech) 428
C-12J (Beech) (1991-92) 353
C.14 (Dassault Aviation) (1991-92) 68
C.15 (McDonnell Douglas) 523
C-17 (McDonnell Douglas) (1985-86) 459
C-17A Globemaster III (McDonnell Douglas) 534
C-18A (Boeing) (1986-87) 380
C-19A (Boeing) (1990-91) 390
C-20 Gulfstream III (Gulfstream Aerospace) (1987-88) 432
C-20F/G/H, SRA-4 (Gulfstream Aerospace) 486
C-21A (Learjet) 492
C22J (Agusta-Caproni Vizzola) (1990-91) 168
C22R (Caproni Vizzola) (1985-86) 154
C-23A Sherpa (Shorts) (1992-93) 312
C-23B Sherpa (Shorts) (1992-93) 312
C-25A (Boeing) (1986-87) 380
C-26A/B (Fairchild) 475
C-27 (USAF) (1991-92) 495
C-27A Spartan (Alenia) 195
C-27A Spartan (Chrysler) 469
C-29A (BAe) 377
C-47TP Super Dakota (SAAF) (1992-93) 265
C90A King Air (Beech) (1991-92) 349
C90B King Air (Beech) 426
C-95 (EMBRAER) (1991-92) 10
C-97 (EMBRAER) (1989-90) 12
C99 Airliner (Beech) (1987-88) 360
C-101 Aviojet (CASA) 347
C-101BB Aviojet (Indaer Chile/CASA) (1983-84) 33
C-122S (Chadwick) (1988-89) 379
C-123T (Mancro/Fairchild) (1983-84) 428
C-130 (Lockheed) 502
C-130 Aerial spray system (Aero Union) (1991-92) 338
C-130 AEW (Lockheed) (1986-87) 442
C-130 Air Tanker conversion (Aero Union) (1990-91) 339
C-130 Auxiliary Fuel System (Aero Union) (1991-92) 338
C-130 Conversions (IAI) 184
C-130 Conversions (Lockheed) (1991-92) 430
C-130 Fireliner (Aero Union) (1991-92) 338
C-141 StarLifter (Lockheed) (1983-84) 423
C-160 (Second series) (Transall) (1989-90) 146
C-160S (Transall) (1984-85) 126
C-160SE (Transall) (1984-85) 126
C180 (Dynac) (1990-91) 405
C 206L-III (Cardoen) (1990-91) 29
C-212 Series 200 Aviocar (CASA) (1987-88) 217
C-212 Series 300 Aviocar (CASA) 345
C-212 Series 300 Aviocar - Special Mission Versions (CASA) 346
C-212P Aviocar (CASA) (1991-92) 214
C235XP (Dynac) (1990-91) 405
CAC (see Chengdu Aircraft Industrial Corporation) 44
CAC (see Commercial Airplane Company) 203
CAC (see Commuter Aircraft Corporation) (1983-84) 348
CAF (see Changhe Aircraft Factory) 47
CAF, counter-air fighter (USSR) (1990-91) 293
CAI (see Ciskei Aircraft Industries) (1991-92) 48
CAI (see Composites Aircraft Industries) (1987-88) 217
CAMC (see Changhe Aircraft Manufacturing Corporation) (1989-90) 35
CA-05 Christavia Mk 1 (Elmwood) 37
CAP (see Club Aviazione Popolare) (1983-84) 528
CAP 10 B (Mudry) 86
CAP 21 (Mudry) 87
CAP '92' (Mudry) (1991-92) 80
CAP 230 (Mudry) (1991-92) 79
CAP 231 (Mudry) 87
CAP 231 EX (Mudry) 88
CAP X (Mudry) (1987-88) 84
CAP X4 (Mudry) (1991-92) 80
CASA (see Construcciones Aeronauticas SA) 343
CASA 3000 (CASA) 344
CASA AX (CASA) 344
CASTOR Islander (Pilatus Britten-Norman) (1986-87) 307
CAT (see Commuter Air Technology) 463
CATIC, CHINA NATIONAL AERO-TECHNOLOGY IMPORT AND EXPORT CORPORATION (China) 44
CB 206L-III (Metalnor) (1992-93) 32
CBA-123 Vector (EMBRAER/FMA) 143
CC-130 Hercules (Lockheed) 502
CC-132 (DH Canada) (1986-87) 25
CC-138 (Boeing Canada) (1988-89) 21
CC-142 (de Havilland) 36
CC-144 (Canadair) 26
CC-150 Polaris (Airbus) 117
CCE (see Colani/Composite Engineering) (1989-90) 89
CCE-208 (Colani/Composite Engineering) (1989-90) 89
CD2 Seastar (Dornier/Composite) (1991-92) 94
CE.15 (McDonnell Douglas) 523
CE-144A (Canadair) 26
CF-5A Upgrade programme (Bristol Aerospace) (1991-92) 23
CFM (see Cook Flying Machines) 391
CF-18 Hornet (McDonnell Douglas) (1983-84) 433
CF-116 (Bristol Aerospace) 25
CF-188A/B (McDonnell Douglas) 523
CH-3 (Sikorsky) (1983-84) 485
CH-8 Christavia Mk 4 (Elmwood) 37
CH-34 Super Puma (Aerospatiale) (1991-92) 58
CH-46 Sea Knight (Boeing) (1991-92) 366
CH-47 Chinook (Boeing) 455
CH-47C Chinook (Agusta/Boeing) 194
CH-47D Weights and Performance (Boeing) 458
CH-47J Chinook (Kawasaki/Boeing) 208
CH-50 (Eurocopter) 150
CH 50 Mini Z (Zenair) (1984-85) 526
CH-53D Yasur 2000 (IAI) 184
CH-53E Super Stallion (Sikorsky) 559
CH-55 (Eurocopter) 151
CH-60 (Zenair) (1985-86) 543
CH-100, Mono Z (Zenair) (1983-84) 510
CH-113 Labrador (Boeing of Canada) (1985-86) 19
CH-113A Voyageur (Boeing of Canada) (1985-86) 19
CH-118 Iroquois (Bell) (1990-91) 360
CH-124 (Sikorsky) (1983-84) 485
CH-135 (Bell) 20
CH-136 (Bell) (1987-88) 365
CH-139 (Bell) 20
CH-146 Griffon (Bell) 23
CH-147 (Boeing) 455
CH-148 Petrel (EHI) 140
CH-149 Chimo (EHI) 140
CH 200 (Zenair) 40
CH 250 (Zenair) 40
CH 300 Tri-Z (Zenair) 40
CH 400 (Zenair) (1985-86) 30
CH 601 (Zenair) 40
CH 601HD (Zenair) 40
CH 601HDS (Zenair) 40
CH 701 STOL (Zenair) 41
CH 701-AG STOL (Zenair) 41
CH 2000 Trainer (Zenair) 41
CHK-91 Chang-Gong 91 (KA) 213
CHKALOV (see Tahkentskoye Apoichi) 586
CJ-6A (NAMC) 54
CL-1 Zipper (Cleary) (1985-86) 394
CL-215 (Canadair) (1991-92) 26
CL-215T (Canadair) 30
CL-415 (Canadair) 31
CL-600 Challenger (Canadair) (1983-84) 21
CL-601 Challenger (Canadair) (1983-84) 21
CL-601 RJ (Canadair) (1989-90) 25
CMC (see Chichester-Miles Consultants Ltd) 392
CN-235 (Airtech) 126
CNA (see Centrul National Aeronautic) (1989-90) 209
CNIAR (see Centrul National al Industriei Aeronautice Române) (1991-92) 203
CP-140 Aurora (Lockheed) 499
CP-140A Arcturus (IMP) 38
CP-140A Arcturus (Lockheed) 499
CRSS (see P. T. Cipta Restu Sarana Svaha) 112
C-SAM Gulfstream III (Gulfstream Aerospace) (1984-85) 412
CSH-2 Rooivalk (Atlas) 340
CT4 Airtrainer (PAC) 222
CT4D Airtrainer (PAC) (1989-90) 190
CT-134A Musketeer II (Beech) (1984-85) 306
CT-142 (de Havilland) 36
CTA (see Centro Técnico Aeroespacial) (1990-91) 10
CV 5800 (KFC) 39
Cabri G2 (Guimbal) 85
Cadet (Piper) (1990-91) 477
Cadet (Robin) (1992-93) 69
CAGNY, RAYMOND DE (France) (1990-91) 67
CALIFORNIA HELICOPTER INTERNATIONAL (USA) 462
CALIFORNIA MICROWAVE INC (USA) 462
CALSPAN CORPORATION (see Arvin Calspan Advanced Technology Center) (1983-84) 349
CALYPSO AIRWAYS (USA) (1985-86) 3
Camber Il-86 (Ilyushin) 2
CAM Special (Carothers) (1985-86) 5
CAMMACORP (USA) (1986-87) 3
CANADAIR GROUP, BOMBARDIER INC (Canada) (1992-93)
Canary Hawk (Karp) (1983-84) 5
Candid, Il-76 (Ilyushin) 2
Canguro, SF.600 TP (SAAC) 1
Canguro, SF.600 TP (SIAI-Marchetti) (1990-91) 1
Canso/Turbo-Canso water bomber (Avalon) (1983-84)
CAPRONI VIZZOLA COSTRUZIONI AERONAUTICHE SpA (Italy) (1989-90) 1
Captain (Russell) (1985-86) 61
Carajá (EMBRAER/Piper) (1989-90) 1
Carajá NE-821 (Neiva) (1992-93) 1
Caravan I/U-27A (Cessna) 4
Caravan II, F406 (Reims/Cessna)
Caravan, Fire Fighter (Aero Union) (1991-92) 3
CARDOEN, INDUSTRIAS, LTDA (Chile) (1991-92)
Careless, Tu-154 (Tupolev) 3
Cargo aircraft ground mobility system (Boeing) (1991-92) 3
CargoLifter (Bensen) (1985-86) 3
Cargoliner (Cavenaugh) (1991-92) 3
Caribou DHC-4T (Newcal) 5
Carioca (EMBRAER/Piper) (1983-84) 1
CAROTHERS, CHUCK (USA) (1985-86) 5
Cash, An-28 (PZL Mielec) 2
CASTOR Islander (Pilatus Britten-Norman) (1986-87) 3
Catbird (Scaled Composites) (1989-90) 4
CATH (Eurocopter) (1987-88) 1
Cava (Atlas) (1991-92) 21
CAVENAUGH AVIATION INC (USA) 4
CELAIR (PTY) LTD (South Africa) (1992-93) 2
CENTRAIR SA (France) (1988-89)
CENTRO TECNICO AEROESPACIAL (Brazil) (1990-91) 1
CENTRUL NATIONAL AERONAUTIC (Romania) (1989-90) 2
CENTRUL NATIONAL AL INDUSTRIEI AERONAUTICE ROMANE (Romania) (1991-92) 2
Centurion (Cessna) (1991-92) 3
Centurion conversion (Air America) (1984-85) 3
Century V Eagle (Sierra) (1989-90) 5
Century 600 (Colemill) (1985-86) 3
CEPPATELLI, GIANFRANCO (Italy) (1984-85) 5
Cessna 210 landing gear door modification (Sierra) (1991-92) 4
Cessna 340/340A (Air America) (1986-87) 3
CESSNA AIRCRAFT COMPANY (USA) 4
Cessna auxiliary fuel systems (Sierra) (1991-92) 4
Cessna suspended production (Cessna) 4
CHADWICK HELICOPTERS INC (USA) (1989-90) 4
Challenger 600 (Canadair) 26, 6
Challenger 601 (Canadair)
Challenger 604 (Canadair) 6
CHAMPION AIRCRAFT COMPANY INC (USA) (1986-87) 3
Chancellor (Cessna) (1985-86) 3
Chandra, RTAF-4 (RTAF) (1983-84) 1
Charger (Tupolev) (1983-84) 24
Chang-Gong 91 CHK-91 (KA) 21
CHANGHE AIRCRAFT FACTORY (China) 4
CHANGHE AIRCRAFT MANUFACTURING CORPORATION (China) (1989-90) 3
Charlie, P.66C-160 (Partenavia) (1984-85) 15
Charlie Trainer P.66T (Partenavia) (1984-85) 15
Cheetah (Atlas) 34
Cheetah (HAL/Aérospatiale) (1987-88) 10
Cheetah, SA315B (HAL) 10
Chegoong-Ho (KA) (1988-89) 17
CHENGDU AIRCRAFT INDUSTRIAL CORPORATION (China) 4
Cherokee Arrow Trainer (Chincul/Piper) (1983-84)
Cherokee Chief TG180, Piper (Rays) (1985-86) 61
Cherokee PA-28 conversion (Isham/Piper) (1984-85) 42
Chetak (HAL) 10
Cheyenne I (Piper) (1983-84) 45
Cheyenne IA (Piper) (1986-87) 48
Cheyenne II (Piper) (1983-84) 45
Cheyenne IIXL, (Piper) (1986-87) 48
Cheyenne IIIA (Piper) 54
Cheyenne IV (Piper) (1983-84) 46
Cheyenne 400 (Piper) 54
CHICHESTER-MILES CONSULTANTS LTD (UK) 39
Chieftain (Piper) (1986-87) 48
Chimo Ch-140 (EHI) 14
CHINCUL S. A. C. A. I. F. I. (Argentina)
Ching-Kuo (AIDC) 36
Chinook 114 (Boeing) 45
Chinook 414 (Boeing) 45
Chinook CH-47 (Kawasaki/Boeing) 208, 45
Chinook CH-47C (Agusta/Boeing) 194, 45
Chinook Commercial (Boeing) (1991-92) 36
Chinook HC Mk1B (Boeing) (1989-90) 38
Chinook HC Mk 1/2 (Boeing) 45
Chinook production (Boeing) 45
Chinook, weights and performance (Boeing) 45
Chiricahua UV-20A (Pilatus) 35
Christavia Mk 1 (Elmwood) 3
Christavia Mk 4 (Elmwood) 3
CHRISTEN INDUSTRIES INC (USA) (1990-91) 40
Chrysalis, DHR G-1 (Wood) (1983-84) 50
CHRYSLER TECHNOLOGIES AIRBORNE SYSTEMS INC (USA) 46
Chujiao-6 (NAMC) (1989-90) 4
Chujiao-6A (NAMC) 5
Chuji Jiaolianji-6 (NAMC) (1989-90) 4
Chuji Jiaolianji-6A (NAMC) 5
Chung Cheng (AIDC) (1987-88) 23
Chuspi, IAP-001 (Indaer Peru) (1992-93) 17
CIPTA RESTU SARANA SYAHA (Indonesia) 11
CIRRUS DESIGN CORPORATION (USA) 47
Cirrus VK 30 (Cirrus) 47
CISKEI AIRCRAFT INDUSTRIES (PTY) LTD (Ciskei) (1991-92) 4
Citabria (Champion) (1985-86) 39
Citation, Air ambulance equipment (Branson) (1991-92) 38

C – E—AIRCRAFT: INDEXES 703

itation, Extended range fuel system (Branson) ...... (1991-92) 382
itation, Extended width cargo door (Branson) ...... (1991-92) 382
itation I (Cessna) ...... (1984-85) 376
itation I/SP Model 501 (Cessna) ...... (1984-85) 377
itation II 550 (Cessna) 465
itation II Weight increase (Branson) ...... (1991-92) 382
itation III (Cessna) ...... (1991-92) 390
itation IV (Cessna) ...... (1990-91) 400
itation V 560 (Cessna) 466
itation VI 650 (Cessna) 467
itation VII 650 (Cessna) 468
itation VII Model 660 (Cessna) ...... (1991-92) 391
itationJet 525 (Cessna) 464
itation S/II (Cessna) 465
itation X 750 (Cessna) 468
itation 500 Upgrade Programme (AlliedSignal) 415
IVIL AVIATION DEPARTMENT (India) ...... (1991-92) 100
lank, An-30 (Antonov) ...... (1991-92) 233
LASSIC AIRCRAFT CORPORATION (USA) 470
lassic, Il-62 (Ilyushin) ...... (1989-90) 246
lassic, Il-62M/MK (Ilyushin) ...... (1989-90) 247
LAUDIUS DORNIER SEASTAR GmbH & Co KG (West Germany) ...... (1990-91) 90
LEARY AIRCRAFT CORPORATION (USA) ...... (1985-86) 394
line, An-32 (Antonov) 367
lobber, Yak-42 (Yakovlev) 332
LUB AVIAZIONE POPOLARE (Italy) ...... (1983-84) 528
lub Sprint (FLS) 395
LUTTON, ERIC (UK) ...... (1983-84) 532
oaler, An-72, An-74 (Antonov) 370
oaler, An-72P (Antonov) 371
Coastguarder (BAe) ...... (1984-85) 260
Cobra 2000 (Bell) ...... (1983-84) 321
Cobra Venom AH-1W (Bell) 436
Cock, An-22 (Antonov) ...... (1991-92) 231
Codling (Yakovlev) ...... (1983-84) 245
Coke (XAC) ...... (1991-92) 46
Coke, An-24 (Antonov) ...... (1991-92) 232
COLANI/COMPOSITE ENGINEERING (West Germany) ...... (1989-90) 89
COLEMILL ENTERPRISES INC (USA) 471
Collegiate (Rankin) ...... (1987-88) 493
Colt (SAP) ...... (1991-92) 44
Colt, An-2 (Antonov) 366
Colt, An-2 (PZL Mielec) 236
Comanche (LoPresti Piper) ...... (1989-90) 487
Comanche RAH-66 (Boeing/Sikorsky) 459
Comanchero (Schafer) ...... (1991-92) 475
Comanchero 500 (Schafer) ...... (1991-92) 475
Comanchero 750 (Schafer) ...... (1991-92) 475
Combat Scout 406 CS (Bell) ...... (1992-93) 345
Commander 112 (Commander) ...... (1991-92) 394
Commander 114B (Commander) 471
COMMANDER AIRCRAFT COMPANY (USA) 471
Commander Jetprop 840 (Gulfstream Aerospace) .. (1984-85) 413
Commander Jetprop 900 (Gulfstream Aerospace) .. (1984-85) 414
Commander Jetprop 980 (Gulfstream Aerospace) .. (1984-85) 414
Commander Jetprop 1000 (Gulfstream Aerospace) ...... (1984-85) 414
Commander Jetprop 1200 (Gulfstream Aerospace) ...... (1984-85) 415
Commander Jetprop Special Mission Aircraft (Gulfstream Aerospace) ...... (1984-85) 415
Commando (Westland) 405
COMMERCIAL AIRPLANE COMPANY 203
Commercial Chinook (Boeing) ...... (1991-92) 369
Commercial Hercules (Lockheed) 505
COMMONWEALTH AIRCRAFT CORPORATION LIMITED (Australia) ...... (1985-86) 6
Commuter IIA/IIB (International Helicopters) ...... (1983-84) 562
COMMUTER AIRCRAFT CORPORATION (USA) . (1983-84) 348
COMMUTER AIR TECHNOLOGY 463
COMPOSITE AIRCRAFT CORPORATION (USA) . (1987-88) 409
COMPOSITE AIRCRAFT INDUSTRIES (South Africa) ...... (1987-88) 217
COMPOSITE INDUSTRIES LTD (Australia) ...... (1987-88) 7
COMTRAN LTD (USA) 472
CONAIR AVIATION LTD (Canada) 33
CONDOR AERO INC (USA) ...... (1983-84) 550
Condor, An-124 (Antonov) 372
Condor F.20 TP (General Avia) ...... (1990-91) 170
CONDOR SA (Romania) ...... (1991-92) 203
Condor, Type 526 AFM (Aerotechnik/Zlin) ...... (1984-85) 44
Conestoga (Skytrader) ...... (1988-89) 487
Conquest I (Cessna) ...... (1991-92) 385
Conquest II (Cessna) ...... (1991-92) 385
CONSTRUCCIONES AERONAUTICAS SA (Spain) 343
Convair 5800 (Kelowna) 39
COOK FLYING MACHINES (UK) 391
Coot-A, Il-20 (Ilyushin) 259
Coot-B, Il-22 (Ilyushin) 260
Corisco (EMBRAER/Piper) ...... (1984-85) 17
Corisco Turbo (EMBRAER/Piper) ...... (1988-89) 15
Cormoran (CCE) ...... (1989-90) 89
Cormorano, S.700 (SIAI-Marchetti) ...... (1984-85) 151
Corporate 77-32 (Boeing) ...... (1987-88) 376
Corporate 77-33 (Boeing) ...... (1987-88) 378
Corporate 77-43 (Boeing) ...... (1986-87) 369
Corporate 77-52 (Boeing) ...... (1987-88) 384
Corporate 77-62 (Boeing) ...... (1987-88) 385
Corporate 800 (BAe) 377, 698
Corporate 1000 (BAe) 378
Corsair II (LTV) ...... (1989-90) 445
Corsair II Update Programmes (LTV) ...... (1991-92) 430
Corsair, modified (Riley) ...... (1983-84) 472
Cossack An-225 (Antonov) 374
COSTRUZIONI AERONAUTICHE GIOVANNI AGUSTA SpA (Italy) ...... (1989-90) 158
Cougar AS 532 (Eurocopter) 146
Cougar Mk II AS 532 (Eurocopter) 148
Courier 600, 700, 800, 900 (Helio) ...... (1991-92) 412
Courier 700 and 800 (Helio) ...... (1987-88) 435
Courier S-7 (Rans) 549
COX AIRCRAFT CORPORATION (USA) ...... (1983-84) 372
Coyote (Montana) 537

CRANFIELD AERONAUTICAL SERVICES LIMITED (UK) ...... (1991-92) 315
CRANFIELD INSTITUTE OF TECHNOLOGY, COLLEGE OF AERONAUTICS (UK) ...... (1991-92) 315
Credible Hawk (Sikorsky) ...... (1987-88) 513
Creek TH-67 (Bell) 20
Cresco 08-600 (PAC) 222
CROPLEASE PLC (UK) 393
Crusader (Cessna) ...... (1985-86) 393
Crusader (Dassault Aviation) ...... (1991-92) 74
Crusty (Tupolev) ...... (1984-85) 248
Cryoplane (DASA) 96, 138
Cub (SAC) ...... (1991-92) 41
Cub, An-12 (Antonov) 366
Cuesta (EMBRAER) 19
Cuff, Be-32 (Beriev) 692
Curl (XAC) ...... (1991-92) 47
Curl, An-26 (Antonov) 366
Curucaca (IPE) ...... (1992-93) 16
Cutlass (Cessna) ...... (1984-85) 360
Cutlass RG (Cessna) ...... (1991-92) 385
Cygnet, SF-2A (Sisler) ...... (1983-84) 580

D

D2 Special, 100 (Denight) ...... (1984-85) 571
D.92 Bébé (Jodel) ...... (1985-86) 551
D-260/D295 Senior Aerosport (D'Apuzzo) ...... (1983-84) 552
D292 (ACAP) (Bell) ...... (1988-89) 347
D-500 (Egrett) ...... (1991-92) 104
D-3300 (McDonnell Douglas) ...... (1983-84) 442
DA-2A (Davis) ...... (1985-86) 588
DA-5A (Davis) ...... (1983-84) 552
DA-6 (Davis) ...... (1983-84) 553
DASA (see Deutsche Aerospace AG) 95
DASA 80/130 passenger airliner (Deutsche Airbus) ...... (1991-92) 88
DC-3 AMI Jet Prop (Professional Aviation) 343
DC-3 conversion, Turbo-67 (Basler) ...... (1991-92) 346
DC-3 Spraying System (Aero Union) ...... (1984-85) 300
DC-3 Turbo Conversions (Basler) 422
DC-3 Turbo Express (USAC) ...... (1987-88) 528
DC-3-65TP Cargomaster (Schafer/AMI) ...... (1991-92) 475
DC-3 65TPs (AMI) ...... (1992-93) 327
DC-3/2000 (Airtech Canada) ...... (1990-91) 18
DC-8 freighter conversion (Alenia) 195
DC-8 Series 71/72/73 (McDonnell Douglas) ...... (1991-92) 435
DC-8 Super 71/72/73 (Cammacorp) ...... (1986-87) 387
DC-8 update (Dee Howard) ...... (1991-92) 396
DC-9 (McDonnell Douglas) ...... (1985-86) 454
DC-9 cargo conversion (Pemco) 545
DC-10 (McDonnell Douglas) ...... (1991-92) 442
DC-10 Series 30CF (Douglas) ...... (1990-91) 455
DC-10 Series 30F (McDonnell Douglas) ...... (1990-91) 455
DC-130 Hercules (Lockheed) 502
DHC-2 Beaver (Modern Wing) ...... (1988-89) 31
DHC-2 Mk III Turbo-Beaver (DH Canada) ...... (1984-85) 24
DHC-2/PZL-3S Beaver (Airtech Canada) ...... (1991-92) 17
DHC-3 Otter (DH Canada) ...... (1983-84) 24
DHC-3/1000 Otter Airtech Canada ...... (1991-92) 17
DHC-3/PZL-3S Otter (Airtech) ...... (1983-84) 20
DHC-3-T (Cox) ...... (1983-84) 372
DHC-4T Caribou (NewCal) 542
DHC-5 Buffalo (DH Canada) ...... (1983-84) 24
DHC-5/C-8A augmentor wing jet/STOL research aircraft (DH Canada) ...... (1983-84) 26
DHC-5D Buffalo (Boeing Canada) ...... (1988-89) 21
DHC-5E Transporter (DH Canada) ...... (1984-85) 24
DHC-6 Twin Otter Series 300 (Boeing Canada) ...... (1989-90) 21
DHC-6 Twin Otter Series 300M (DH Canada) ...... (1985-86) 25
DH-6-300 MR Military Twin Otter (DH Canada) ... (1985-86) 25
DHC-7 ARL (California Microwave) ...... (1991-92) 384
DHC-7 Dash 7 (Boeing Canada) ...... (1989-90) 21
DHC-7 Dash 7R (DH Canada) ...... (1984-85) 29
DHC-7 Dash 7IR (Boeing Canada) ...... (1988-89) 21
DHC-8 Dash 8 Series 100 (de Havilland) 34
DHC-8 Dash 8 Series 200 (de Havilland) 34
DHC-8 Dash 8 Series 300 (de Havilland) 36
DHC-8 Dash 8 Series 400 (de Havilland) 37
DHC-8 Dash 8M (de Havilland) 36
DHI (see Daewoo Heavy Industries) 213
DHR G-1 Chrysalis (Wood) ...... (1983-84) 509
DMAV (see Dual Mode Air Vehicle) ...... (1992-93) 373
Do 24TT (Dornier) ...... (1985-86) 87
Do 28 D-2 Skyservant (Dornier) ...... (1987-88) 93
DR 400 Dauphin (Robin) 89
DR 400 Remo (Robin) 91
DR 400/100 Cadet (Robin) ...... (1992-93) 69
DR 400/160 Major (Robin) 90
DR 400/180 Régent (Robin) 90
DR 400/180R Remorqueur (Robin) ...... (1988-89) 81
DR 400/200R (Robin) 91
D-Series light aircraft (Daytona) 472
DV 20 Katano (HOAC) 11
DYNAC (see Dynac International Corporation) ...... (1990-91) 405

DAEDALUS RESEARCH INC (USA) ...... (1988-89) 383
DAEWOO HEAVY INDUSTRIES CO LTD (South Korea) .. 213
Dakota (Piper) ...... (1990-91) 479
DASA/AEROSPATIALE/ALENIA (International) ...... (1991-92) 120, 749
DASA/TUPOLEV (International) ...... (1991-92) 750
Dash 7, DHC-7 (Boeing Canada) ...... (1989-90) 21
Dash 7IR, DHC-7 (Boeing Canada) ...... (1986-87) 27
Dash 7R, DHC-7 (DH Canada) ...... (1984-85) 29
Dash 8 Series 100 DHC-8 (de Havilland) 34
Dash 8 Series 200 (de Havilland) 35
Dash 8 Series 300, DHC-8 (de Havilland) 36
Dash 8 Series 400, DHC-8 (de Havilland) 37
Dash 8M, DHC-8 (de Havilland) 36
DASSAULT AVIATION (France) 75
DASSAULT/DORNIER (International) 137
DATWYLER, MDC MAX, AG (Switzerland) ...... (1991-92) 220
Dauphin 2 AS 365F (Aerospatiale) ...... (1990-91) 64

Dauphin 2 AS 365N2 (Eurocopter) 152
Dauphin 2 AS 366 (Aerospatiale) ...... (1991-92) 66
Dauphin 2 AS 565 (Aerospatiale) ...... (1990-91) 64
Dauphin 2, SA 365M (Aerospatiale) ...... (1985-86) 57
Dauphin 2, SA 365N (Aerospatiale) ...... (1988-89) 63
Dauphin 2, SA 365N1 (Aerospatiale) ...... (1989-90) 65
Dauphin 2, SA 365S (Aerospatiale) ...... (1989-90) 66
Dauphin 2, SA 366 (Aerospatiale) ...... (1989-90) 66
Dauphin AS 365X DGV (Eurocopter) 154
Dauphin AS 365X FBW (Eurocopter) 154
Dauphin, DR 400 (Robin) 89
Dauphin, SA 360C (Aerospatiale) ...... (1984-85) 57
Dauphin X 380 (Eurocopter) 154
DAYTONA AIRCRAFT CONSTRUCTION INC (USA) 472
DE CHEVIGNY/WILSON (International) 137
Decathlon (Champion) ...... (1985-86) 391
DEE HOWARD COMPANY, THE (USA) 472
Defender (McDonnell Douglas) 528
Defender (Pilatus Britten-Norman) 397
Defender 4000 (Pilatus Britten-Norman) 398
Defender, Battlefield Surveillance (Pilatus Britten-Norman) .. 397
Defender, Border Patrol (Pilatus Britten-Norman) 398
Defender, ELINT (Pilatus Britten-Norman) 398
Defender, Maritime (Pilatus Britten-Norman) 398
Defiant (Rutan) ...... (1985-86) 613
Defiant 300 (UDRD) ...... (1990-91) 191
Defiant 500 (UDRD) ...... (1992-93) 175
Defiant 1000 (UDRD) 227
DE HAVILLAND AIRCRAFT OF CANADA LTD, THE (Canada) ...... (1987-88) 28
DE HAVILLAND INC (Canada) 34
DELAERO BUSINESS AND COMMERCIAL AVIATION LTD (Russia) ...... (1992-93) 198
Delphin (Myasishchev) 302
Delta, P.66D (Aviolight) ...... (1991-92) 163
DENIGHT, RONALD A (USA) ...... (1984-85) 571
DEUTSCHE AEROSPACE AG (Germany) 95
DEUTSCHE AEROSPACE AIRBUS GmbH (Germany) 96
DEUTSCHE AEROSPACE AIRBUS/TUPOLEV (International) 138
DE VORE AVIATION CORPORATION (USA) ...... (1986-87) 399
Diamond I (MAI) ...... (1983-84) 427
Diamond IA (MAI) ...... (1985-86) 446
Diamond IA Long-range tank (Branson) ...... (1991-92) 382
Diamond II (MAI) ...... (1985-86) 447
Dingo (Aeroric) 691
Dino (Ganzavia) ...... (1991-92) 100
Discojet (Moller) 537
Dolphin, HH-65A (Aerospatiale) ...... (1991-92) 66
Doppelraab, RE 2 (Ehmann) ...... (1984-85) 544
DORNA, H. F. (Iran) ...... (1991-92) 144
DORNIER COMPOSITE AIRCRAFT GmbH (Germany) ...... (1992-93) 78
DORNIER LUFTFAHRT GmbH (Germany) 96
Dornier 128-6 (Dornier) ...... (1985-86) 85
Dornier 228 (Dornier) ...... (1991-92) 88
Dornier 228 (HAL) 110
Dornier 228 Maritime Patrol (Dornier) 97
Dornier 228 Maritime Pollution Surveillance (Dornier) 97
Dornier 228 Photogrammetry/Geo-Survey (Dornier) 98
Dornier 228 Reconnaissance (Dornier) ...... (1990-91) 94
Dornier 228 Sigint (Dornier) ...... (1990-91) 94
Dornier 228-212 (Dornier) 96
Dornier 328 (Dornier) 98
Dornier 328S, Stretched (Dornier) 99
DOUGLAS AIRCRAFT COMPANY (USA) 515
DOUGLAS/CATIC (International) 138
Dragon (Phalanx) ...... (1987-88) 482
Dragonfly (Viking) 583
Draken J35J (Saab-Scania) ...... (1991-92) 220
Dreadnought (Sanders) ...... (1984-85) 493
Dromader, M-18 (PZL Mielec) 233
Dromader Mikro (WSK-PZL Mielec) ...... (1983-84) 176
Dromader Mini, M-21 (PZL Mielec) ...... (1991-92) 193
Dromader Super, M-24 (PZL Mielec) ...... (1991-92) 180
Dromander Water Bomber (Melex) ...... (1991-92) 465
Dual aft-body stroke system (Raisbeck) ...... (1991-92) 465
DUAL MODE AIR VEHICLE INC (USA) ...... (1992-93) 373
Dual Pac conversions (Soloy) 572
Duchess 76 (Beech) ...... (1984-85) 310
Duke (Beech) ...... (1983-84) 305
Dyna-cam engine project (Piper) ...... (1988-89) 459
DYNAC INTERNATIONAL CORPORATION (USA) ...... (1990-91) 405

E

E-2B (Grumman) ...... (1991-92) 406
E-2C (Grumman) 480
E-3B/C AWACS (Boeing) 452
E-3 Sentry (Boeing) 452
E-3D Sentry AEW Mk 1 (BAe/Boeing) 391, 452
E-3F (Boeing) 452
E-4 (Boeing) ...... (1985-86) 363
E-6A Mercury/Tacamo II (Boeing) 454
E-8 (Boeing) 441
E-8 J-STARS (Boeing/Grumman) ...... (1991-92) 366
E-9A (de Havilland) 36
E.25 Mirlo (CASA) 347
E.26 Tamiz (CASA) ...... (1989-90) 219
E.26 Tamiz (ENAER) 41
E47 Elf (Ward) ...... (1985-86) 571
E55 Baron (Beech) ...... (1984-85) 311
E-86C J-STARS (Grumman) 485
E-266 (MiG) ...... (1987-88) 259
E-266M (Mikoyan) ...... (1984-85) 227
E400 (Gyroflug) ...... (1988-89) 91
E401 (Gyroflug) ...... (1987-88) 98
EA-6 Intruder (Grumman) ...... (1987-88) 427
EA-6A Intruder (Grumman) ...... (1990-91) 416
EA-6B Prowler (Grumman) 482
EA7 Optica (Edgley) ...... (1985-86) 289
EA230 Ultimate (ACS) ...... (1985-86) 205

704 INDEXES: AIRCRAFT—**E – F**

EAA (see Eagle Aircraft Australia) (1991-92) 4
EAA (see Experimental Aircraft Association, Inc) (1984-85) 572
EAA Biplane (EAA) (1983-84) 555
EAI (see Eagle Aircraft International) 6
EAL (see Ethiopian Airlines S. C.) 72
EAP (BAe) (1990-91) 309
EAV-8B (McDonnell Douglas/BAe) (1986-87) 120
EBI Rotor Buggy (Boyette) (1983-84) 544
EC-1 (Kawasaki) (1988-89) 167
E. C.2 Easy Too (Clutton-Tabenor) (1983-84) 550
EC-18B (Boeing) (1988-89) 370
EC-18B ARIA (USAF/Boeing) (1989-90) 525
EC-18C (Boeing) (1988-89) 392
EC-18D (CTAS) (1991-92) 392
EC-20F (Grumman Aerospace) (1988-89) 400
EC-24A (Electrospace Systems) (1989-90) 406
EC-95 (EMBRAER) (1991-92) 10
EC 120 (Eurocopter/CATIC/SA) 158
EC-130 Hercules (Lockheed) 502
EC 135 (Eurocopter) 156
EC-137D (Boeing) (1987-88) 388
EC Super Five (Eurocopter) 155
EF-111A (Grumman/General Dynamics) (1989-90) 421
EF-111A update (Grumman) (1991-92) 410
EFA (see European Fighter Aircraft) (1992-93) 122
EFS (Mooney) (1991-92) 454
EFS (USAF) (1992-93) 456
EGHS (see Eau Gallie High School) (1985-86) 590
EH-1H Iroquois (Bell) (1990-91) 360
EH-6 (McDonnell Douglas) 528
EH-60C (Sikorsky) 560
EH 101 (EHI) 140
EHI (see EH Industries Limited) 140
EL-1 (Gavilan) 62
EM (see Elicotteri Meridionali SpA) (1988-89) 155
EMB-110 Bandeirante (EMBRAER) 14
EMB-110P1 (EMBRAER) (1984-85) 10
EMB-110P1/41 (EMBRAER) (1984-85) 11
EMB-110P1SAR (EMBRAER) (1983-84) 11
EMB-110P2 (EMBRAER) (1984-85) 11
EMB-110P2/41 (EMBRAER) (1984-85) 11
EMB-110P3 (EMBRAER) (1983-84) 11
EMB-111 (EMBRAER) 14
EMB-120 Brasilia (EMBRAER) 14, 685
EMB-121A Xingu I (EMBRAER) (1983-84) 15
EMB-121A1 Xingu II (EMBRAER) (1987-88) 16
EMB-121B Xingu III (EMBRAER) (1987-88) 17
EMB-123 (EMBRAER) (1987-88) 17
EMB-145 Jetliner (EMBRAER) 16, 685
EMB-201A Ipanema (EMBRAER) (1992-93) 15
EMB 202 Ipanema (EMBRAER) 19
EMB-312 Tucano (EMBRAER) 17, 685
EMB-312H Tucano (EMBRAER) 18
EMB-326GB Xavante (EMBRAER/Aermacchi) (1984-85) 15
EMB-710 Carioca (EMBRAER/Piper) (1983-84) 18
EMB-711 ST Corisco Turbo (EMBRAER/Piper) (1988-89) 15
EMB-711T Corisco (EMBRAER/Piper) (1984-85) 17
EMB-712 Tupi (EMBRAER/Piper) (1988-89) 15
EMB-720D Minuano (EMBRAER/Piper) 19
EMB-721 Sertanejo (EMBRAER/Piper) (1985-86) 16
EMB-810D Cuesta (EMBRAER/Piper) 19
EMB-820 Navajo (EMBRAER/Piper) (1989-90) 15
EMB N-821 (EMBRAER/Piper) (1985-86) 16
EMBRAER (see Empresa Brasileira de Aeronáutica SA) 13
ENAER (see Empresa Nacional de Aeronáutica de Chile) 41
ENET (see Escuela Nacional de Educacion Tecnica No 1) (1990-91) 2
EOS/SFA (Frederick-Ames) (1984-85) 575
ER-2 (Lockheed) (1989-90) 436
E-Tan (IAC) (1991-92) 228
EUROFAR (see European Future Advanced Rotorcraft 160
EX (Boeing) 458
E-X (USN) 580
EX101 Firefly (Nutting) (1985-86) 605
Exec 80 (RotorWay) 533

Eagle (Composite) (1985-86) 395
Eagle (Eagle Aircraft) (1983-84) 372
Eagle I (Christen) (1984-85) 568
Eagle I/II/III (Composite/Windecker) (1985-86) 395
Eagle 220 (Eagle) (1983-84) 372
Eagle 300 (Celair) (1992-93) 264
Eagle 300 (Eagle) (1983-84) 372
Eagle series, Cessna, R/STOL mods (Sierra) (1991-92) 479
Eagle, F-15C/D (McDonnell Douglas) 521
Eagle, F-15DJ (Mitsubishi/McDonnell Douglas) (1991-92) 174
Eagle, F-15E/S (McDonnell Douglas) 522
Eagle, F-15J (Mitsubishi/McDonnell Douglas) 209
Eagle H-76 (Sikorsky) 569
Eagle, Windecker (Composite) (1983-84) 371
Eagle X (Composite Industries) (1987-88) 7
Eagle XTS (EAI) 6
EAGLE AIRCRAFT AUSTRALIA (Australia) (1991-92) 4
EAGLE AIRCRAFT COMPANY (USA) (1984-85) 386
EAGLE AIRCRAFT INTERNATIONAL (Australia) 6
Easy Too, E. C.2 (Clutton-Tabenor) (1983-84) 550
EAU GALLIE HIGH SCHOOL (USA) (1985-86) 590
EBERSHOFF AIRPLANE COMPANY (USA) (1984-85) 573
Echo P92 (Tecnam) 202
ECTOR AIRCRAFT COMPANY INC (USA) (1989-90) 406
Ecureuil/AStar AS 350 (Eurocopter) 150
Ecureuil 2/TwinStar AS 355 (Eurocopter) 151
EDGLEY AIRCRAFT LTD (UK) (1985-86) 289
EDWARDS, WILLIAM (USA) (1983-84) 583
EGRETT (International) 138
Egrett-1 (Egrett) (1991-92) 121
Egrett II (Egrett) 138
EH INDUSTRIES LIMITED (International) 140
EHMANN, ROLF (Germany) (1984-85) 544
EIDGENOSSISCHES FLUGZEUGWERK (Switzerland) 363
EKIN (ENGINEERING) CO LTD, W. H. (UK) (1984-85) 299
ELECTROSPACE SYSTEMS INC (USA) (1990-91) 406
Elf, E47 (Ward) (1985-86) 571
ELICOTTERI MERIDIONALI SpA (Italy) 194
Elite (VAT) 582
ELMWOOD AVIATION (Canada) 37
EMIND (Yugoslavia) (1991-92) 497
EMBRAER/FMA (International) 143
EMPRESA BRASILEIRA DE AERONAUTICA SA (Brazil) 13
EMPRESA NACIONAL DE AERONAUTICA DE CHILE (Chile) 41
Enforcer (Piper) (1985-86) 488
Enhanced Eagle, F-15E (McDonnell Douglas) (1983-84) 432
Enhanced Flight Screener (USAF) (1992-93) 456
ENSTROM HELICOPTER CORPORATION, THE (USA) 472
Epervia (Epervia Aviation) 11
EPERVIA AVIATION (Belgium) 11
Epsilon (Aerospatiale) (1988-89) 56
Epsilon, TB 30 (SOCATA) (1990-91) 87
Equator (Equator) (1985-86) 87
Equator, P-300/-350/-400/-420/-450/550 (Equator) (1984-85) 85
EQUATOR AIRCRAFT GESELLSCHAFT FUR FLUGZEUGBAU mbH ULM (West Germany) (1987-88) 96
ESCOLA ENGENHARIA DE SAO CARLOS (Brazil) (1986-87) 17
ESCUELA NACIONAL DE EDUCACION TECNICA No 1 (Argentina) (1990-91) 2
Eshet (EAL) 72
Esqualo-180 (Moura) (1983-84) 505
Esquilo, HA-1 (Eurocopter) 150
Esquilo, HB 350 (HELIBRAS) 19
Esquilo, HB 355F2 (HELIBRAS) 19
Esquilo, TH-50 (Eurocopter) 150
Esquilo, UH-12 (Eurocopter) 150
Esquilo, UH-12B (Eurocopter) 151
Esquilo, VH-55 (Eurocopter) 151
ETHIOPIAN AIRLINES CORPORATION (Ethiopia) (1988-89) 53
ETHIOPIAN AIRLINES S. C. (Ethiopia) 72
EUROCOPTER CANADA LTD (Canada) 37
EUROCOPTER SA (International) 145
EUROCOPTER/CATIC/SA 158
EUROCOPTER/KAWASAKI (International) 159
Eurofar (Eurofar) 160
Eurofighter 2000 (Eurofighter) 161
EUROFIGHTER JAGDFLUGZEUG GmbH (International) 161
EUROFLAG srl (International) 162
EURO-HERMESPACE SA (International) 163
EUROPATROL (International) 163
European Fighter Aircraft (Eurofighter) (1992-93) 122
EUROPEAN FUTURE ADVANCED ROTORCRAFT (International) 160
EUROPEAN FUTURE LARGE AIRCRAFT GROUP (International) (1990-91) 131
Eurotrainer 2000 (FFT) (1991-92) 97
Eurotrainer 2000A (FFT) (1992-93) 80
Evader (Skytrader) (1988-89) 487
EVER (Egrett) 140
EVERETT R. J. ENGINEERING LTD (UK) (1989-90) 316
EVERGREEN AIR CENTER INC (USA) (1987-88) 413
EXCALIBUR AVIATION COMPANY (USA) 474
Executive 600 (Colemill) (1991-92) 393
Executive commuter, Mark III (JetCraft) (1989-90) 427
Executive Mark I (JetCraft) (1990-91) 424
Expediter 1 (Fairchild) 475
Expediter 23 (Fairchild) 475
Experimental Aircraft 105 (USSR) (1991-92) 298
EXPERIMENTAL AIRCRAFT ASSOCIATION, INC. (USA) (1984-85) 572
EXPERIMENTAL AVIATION (USA) 474
Explorer (ACAC) 415
Explorer (De Chevigny/Wilson) 137
Explorer MD (McDonnell Douglas) 531
EXPRESS DESIGN INC (USA) 474
Express, Wheeler (Express Design) 474
Extender (McDonnell Douglas) (1991-92) 442
Extra 230 (Extra) (1990-91) 99
Extra 260 (Extra) (1991-92) 95
Extra 300 (Extra) 100
Extra 400 (Extra) 100
EXTRA-FLUGZEUGBAU (Germany) 100

F

F-1 (Mitsubishi) (1987-88) 182
F1 Mirage (Dassault Aviation) (1992-93) 58
F-4 Phantom, modernised (Boeing/McDonnell Douglas) (1986-87) 380
F-4 Phantom II (McDonnell Douglas) (1991-92) 435
F-4 Phantom NWDS (Boeing) (1990-91) 391
F4B-2/P-12C Fighter replica (Aero Tech) (1985-86) 575
F-4EJKai (Mitsubishi) 209
F-4F ICE Programme (DASA) 99
F-5 Plus upgrade (IAI) 184
F-5 Upgrade Programmes (Bristol Aerospace) 25
F-5 Waco Classic (Classic) (1992-93) 371
F-5E Tiger II (KA) (1988-89) 174
F-5E Tiger II (Northrop) (1986-87) 470
F-5E/RF-5E conversion (SA) 388
F-5F Tiger II (KA) (1988-89) 174
F-5F Tiger II (Northrop) (1986-87) 470
F-5G Tigershark (Northrop) (1986-87) 472
F-5T (China) (1983-84) 33
F-6 (Shenyang) (1987-88) 46
F-7 (CAC) 44
F-8 (SAC) 56
F-8 (Shenyang) (1987-88) 46
F-8 oblique-wing demonstrator (NASA) (1988-89) 447
F-8 II (SAC) (1989-90) 43
F.8L Falco (Sequoia) 558
F-14 airflow experiments (NASA) (1986-87) 469
F-14 Tomcat (Grumman) 483
F-15 Advanced Technology Eagle (McDonnell Douglas) (1984-85) 449
F-15, HIDEC (NASA) (1990-91) 469
F-15C/D Eagle (McDonnell Douglas) 521
F-15DJ (Mitsubishi/McDonnell Douglas) (1991-92)
F-15E Eagle (McDonnell Douglas)
F-15F Eagle (McDonnell Douglas)
F15F (Procaer/General Avia) (1985-86)
F-15H Eagle (McDonnell Douglas) (1992-93)
F-15J (Mitsubishi/McDonnell Douglas)
F-15S Eagle (McDonnell Douglas)
F-15S/MTD (McDonnell Douglas) (1992-93)
F-15 upgrade (IAI)
F-16 (SSA)
F-16 Customers, table of (Lockheed)
F-16 Fighting Falcon (Lockheed)
F-16/79 (General Dynamics) (1986-87)
F-16/101 (General Dynamics) (1983-84)
F-16C/D Fighting Falcon (TAI)
F-16E (General Dynamics) (1983-84)
F-16XL (General Dynamics) (1986-87)
F-19 (Lockheed) (1986-87)
F-20 Tigershark (Northrop) (1986-87)
F.20 TP Condor (General Avia) (1990-91)
F-21 (Taylorcraft) (1987-88)
F-21A (Taylorcraft) (1984-85)
F-21A Kfir (IAI) (1989-90)
F-21B (Taylorcraft) (1989-90)
F-22 (Lockheed)
F-22 (Taylorcraft) (1989-90)
F.22 Series (Taylorcraft)
F22A Classic 118 (Taylorcraft) (1991-92)
F22A Pinguino (General Avia
F.22/B Pinguino (General Avia)
F.22/C Pinguino (General Avia)
F.22/R Pinguino-Sprint (General Avia)
F27 Firefighter (Conair)
F27 Firefighting, Fokker (Conair) (1986-87)
F27 Friendship (Fokker) (1987-88)
F27 Friendship ARAT (SECA) (1991-92)
F-27 Large cargo door (Branson) (1991-92)
F27 Maritime (Fokker) (1985-86)
F-28 (Enstrom)
F28 Fellowship (Fokker) (1987-88)
F33A/C Bonanza (Beech)
F90 King Air (Beech) (1983-84)
F90-1, King Air (Beech) (1987-88)
F-104CCV (MBB) (1985-86)
F-104S (Aeritalia/Lockheed) (1984-85)
F-104S ASA (Alenia/Lockheed) (1991-92)
F-106B storm hazards tests (NASA) (1987-88)
F-111 (BAe)
F-111 upgrade (Rockwell International) (1991-92)
F-111G (USAF/General Dynamics) (1991-92)
F-117A (Lockheed)
F 152 Aerobat (Reims) (1986-87)
F 172 Skyhawk/100 (Reims) (1986-87)
F.220 Airone (General Avia)
F406 Caravan II (Reims/Cessna)
F1300 NGT Jet Squalus (General Avia) (1986-87)
F1300 NGT Jet Squalus (Promavia) (1992-93)
F.1300 Squalo (General Avia) (1985-86)
F.3500 Sparviero (General Avia) (1990-91)
F/A-18 high-alpha research vehicle (NASA) (1991-92)
F/A-18 Hornet (McDonnell Douglas)
F/A-18A, B, C, D Hornet (McDonnell Douglas)
F/A-18E/F Hornet (McDonnell Douglas)
F/A-18L (McDonnell Douglas) (1985-86)
FA 152 Aerobat (Reims) (1986-87)
FA-200 Aero Subaru (Fuji) (1986-87)
FAC (see Fuerza Aérea Chilena) (1983-84)
FAMA (see Fabrica Argentina de Materiales Aeroespaciales) (1989-90)
FFA-2000 (FFA) (1989-90)
FH-227 Large cargo door (Branson) (1991-92)
FH-1100A/B (Hiller) (1983-84)
FH-1100C (Hiller) (1984-85)
FIMA (Future International Military/Civil Airlifter) (1989-90)
FJX (Shorts) (1989-90)
FM-2 Air Shark 1 (Freedom Master)
FMA (see Fabrica Militar de Aviones)
FRC 225 (Fairchild) (1983-84)
FRWE (Australia) (1991-92)
FSW demonstrator (Grumman) (1989-90)
FS-X (Mitsubishi) 209, 6
FT-5 (CAC)
FT-6 (State Aircraft Factories) (1984-85)
FT-7 (GAIC)
FU24-954, Fletcher (PAC)
FUS (see Flugzeug-Union-Sud GmbH) (1984-85)
F + W (see Swiss Federal Aircraft Factory)

FABRICA ARGENTINA DE MATERIALES AEROESPACIALES (Argentina) (1989-90)
FABRICA MILITAR DE AVIONES SA (Argentina)
FAIRCHILD AIRCRAFT INCORPORATED (USA)
FAIRCHILD INDUSTRIES INC (USA) (1987-88)
FAIRCHILD REPUBLIC COMPANY (USA) (1987-88)
Fajr (IRGC) (1991-92)
Falco F.8L (Sequoia)
Falcon 20, 731 Engine retrofit programme (Garrett) (1991-92)
Falcon 20B, 731 Engine Retrofit Programme (AlliedSignal)
Falcon 50 (Dassault Aviation)
Falcon 900 (Dassault Aviation) (1992-93)
Falcon 900B (Dassault Aviation)
Falcon 2000 (Dassault Aviation)
Falcon A (Munninghoff) (1985-86)
Falcon, F-28F (Enstrom)
Falcon PW300-F20 (Volpar)
Fan Commander (American Aviation) (1985-86)
Fanjet (Sadler)
Fan Ranger (Rockwell International/DASA) (1992-93)
FanStar (AAI) (1989-90)
Fanstar 200T (Fanstar) (1990-91)
FANSTAR PARTNERS (USA) (1990-91)
Fantan (NAMC)
Fantrainer 400/600 (RFB) (1991-92)

F -H —AIRCRAFT: INDEXES 705

antrainer 400/600 (RTAF-RFB) (1991-92) 228
antrainer 800 (RFB) 686
antrainer 1000 (RFB) (1990-91) 103
armer-C (Shenyang) (1987-88) 46
armer-D (Shenyang) (1987-88) 46
ARRINGTON AIRCRAFT CORPORATION (USA) (1991-92) 401
ellowship, F28 (Fokker) (1987-88) 187
encer, Su-24 (Sukhoi) 309
ennec AS 550 (Eurocopter) 150
ennec AS 555 (Eurocopter) 151
FA FLUGZEUGWERKE ALTENRHEIN AG (Switzerland) 357
FT GESELLSCHAFT FUR FLUGZEUG- UND FASERVERBUND-TECHNOLOGIE GmbH (Germany) (1992-93) 79
FV AEROTECH (Sweden) (1991-92) 215
ibra 8 (Fibra) (1986-87) 207
IBRA AB (Sweden) (1987-88) 221
iddler (Tupolev) (1989-90) 287
ieldmaster NAC 6 (Croplease) 393
ighter (AIDC) (1988-89) 218
ighting Falcon F-16 (Lockheed) 509
ighting Falcon F-16C/D (TAI) 365
inback (SAC) 56
inesse, S-1275 (Schapel) (1983-84) 482
irebar (Yakovlev) (1990-91) 290
irecat (Conair) 33
irecracker (NAC) (1988-89) 298
irecracker, NDN 1 (NDN) (1983-84) 272
irefighter (Conair) 33
irefly 160, T67M (Slingsby) (1989-90) 330
irefly, EX 101 (Nutting) (1985-86) 605
irefly T67 (Slingsby) 401
irefly T67M200 (Slingsby) 402
irefly T67M260 (Slingsby) 402
irefly T67M Mk II (Slingsby) 401
iremaster (Croplease) 393
irestar tanker conversion (Aero Union) (1991-92) 338
ishbed (HAL) (1983-84) 96
ishbed, MiG-21 (CAC) (1991-92) 33
ishbed, MiG-21 (MiG) (1992-93) 214
ishpot (Sukhoi) (1983-84) 230
itter-A (Sukhoi) (1988-89) 257
itter-C, D, E, F, G, H, J and K (Sukhoi) 308
LAGLOR, K (USA) (1985-86) 592
laglor Sky Scooter (Headberg) (1985-86) 594
lagon Su-15 (Sukhoi) (1992-93) 236
laming, PZL-105 (PZL Warszawa-Okecie) 241
lamingo III TM-20B (Bengis) (1992-93) 264
lamingo, T-433 (ROKS-Aero) 306
lamingo-Trainer (FUS) (1984-85) 86
lanker, Su-27 (Sukhoi) 313
LIGHT REFUELLING LTD (UK) (1985-86) 290
loatmaster (GAF) (1985-86) 7
logger-A, B, C, E, F, G, H and K (MiG) 279
logger-D and J (MiG) 280
logger-J (HAL) 111
LS AEROSPACE LTD (UK) 394
LUG- und FAHRZEUGWERKE AG (Switzerland) (1986-87) 214
LUGZEUG-UNION-SUD GmbH (West Germany) (1984-85) 86
ly Baby 1-A (Bowers) 461
ly Baby 1-B (Bowers) 461
MV RADAR (Sweden) (1989-90) 220
okker 50 (Fokker) 215
okker 50 Hot and High (Fokker) (1991-92) 181
okker 50 Maritime and Surveillance versions (Fokker) (1991-92) 181
okker 50 Special Mission Versions (Fokker) 217
okker 50-200 (Fokker) (1990-91) 185
okker 60 Utility (Fokker) 690
okker 70 (Fokker) 219, 690
okker 100 (Fokker) 216
okker 130 (Fokker) 220
OKKER, NV KONINKLIJKE NEDERLANDSE VLIEGTUIGFABRIEK (Netherlands) 215
oo Fighter (Stewart) (1984-85) 602
ORBES, DAVID (USA) (1983-84) 602
orger (Yakovlev) 331
orward Swept Wing Demonstrator (Grumman) (1990-91) 420
oxbat, MiG-25 (MiG) 281
oxhound, MiG-31 (MiG) 286
oxstar Baron (Colemill) (1991-92) 394
RAKES AVIATION INC (USA) (1989-90) 411
REDERICK-AMES RESEARCH CORPORATION (USA) (1984-85) 575
reebird Mk 5 (Freewing) 478
reedom Fighter (Northrop) (1992-93) 426
REEDOM MASTER CORPORATION (USA) 477
reelance (NAC) (1989-90) 318
reestyle, Yak-141 (Yakovlev) 336
REEWING AIRCRAFT CORPORATION (USA) 478
R GROUP PLC (UK) (1986-87) 298
riendship, F27 (Fokker) (1987-88) 185
riendship ARAT, F27 (SECA) (1991-92) 84
rogfoot Su-25 (Sukhoi) 310
rogfoot Su-28 (Sukhoi) 310
UERZA AEREA CHILENA (Chile) (1983-84) 32
UJI HEAVY INDUSTRIES LTD (Japan) 203
ulcrum, MiG-29 (MiG) 283
ully enclosed landing gear doors (Raisbeck) (1991-92) 465
US (see Flugzeug-Union-Sud GmbH) (1984-85) 86
uture Large Aircraft (Euroflag) 162
uture SST (SCTICSG) (1991-92) 141

G

2 Cabri (Guimbal) 85
-4 Super Galeb (UTVA) 587
-5 (UTVA) 589
110 (Grob) (1984-85) 86
G 111 (Grob) (1983-84) 87
G-111 Albatross (Grumman/Resorts International) (1986-87) 416
G 112 (Grob) (1985-86) 88
G 115 (Grob) 101
G 115T (Grob) 101
G 116 (Grob) (1990-91) 100
G159C Gulfstream I-C (Gulfstream Aerospace) (1983-84) 395
G-164 C-T Turbo Cat (Marsh) (1991-92) 432
G222 (Alenia) 195
G 520 (Egrett) 138
G.802 Orion (Grinvalds) (1983-84) 516
G802/1 Spur (Operation Ability) (1989-90) 319
G1159C (Gulfstream Aerospace) (1988-89) 400
GA (see Gippsland Aeronautics) 7
GA-55 LightWing (Hughes) 9
GA-200 (GA) 7
GAC COMMANDER (see Gulfstream Aerospace Corporation, Commander Division) (1983-84) 399
GAF (see Government Aircraft Factories) (1986-87) 6
GAIC (see Guizhou Aviation Industry Corporation) 47
GAIGC (see Guizhou Aviation Industry Group Company) (1989-90) 36
GA-K-22 Dino (Ganzavia) (1991-92) 100
GB-2 Retro (Bezzola) (1983-84) 531
GF 200 (Grob) 102
GN, Autogyro (AISA) (1983-84) 185
GOHL (see Guangzhou Orlando Helicopters Ltd) (1992-93) 35
GP-180 (Gates/Piaggio) (1985-86) 114
GR-582 LightWing (Hughes) 8
GR-912 LightWing (Hughes) 9
GROB (see Burkhart Grob Luft- und Raumfahrt) 100
GTP-350 Gemini (Daedalus) (1987-88) 410

Gabier (SOCATA) (1985-86) 81
GafHawk 125 (Hawk) (1989-90) 425
GafHawk 125-200 (Ameco-Hawk) (1991-92) 341
Gajaraj, Il-76 (Ilyushin) 261
Galaxy 700 JetStar (Galaxy) 478
Galaxy, C-5 (Lockheed) (1989-90) 443
GALAXY GROUP (USA) 478
Galaxy Star (Galaxy) 478
Galérien (SOCATA) (1983-84) 78
Galopin (SOCATA) (1983-84) 78
GANZAVIA GT (Hungary) (1991-92) 100
Gardian (Dassault-Breguet) (1986-87) 71
Gardian 2 (Dassault Aviation) (1991-92) 75
Gardian 50 (Dassault Aviation) (1991-92) 76
GARRETT GENERAL AVIATION SERVICES DIVISION (USA) (1992-93) 378
Gasior PZL-140 (PZL Warszawa-Okecie) (1990-91) 204
GATES LEARJET CORPORATION (USA) (1987-88) 418
GATES/PIAGGIO (International) (1985-86) 114
Gavião HB 315B (Helibras) 19
Gavilan (Aero Mercantil) (1991-92) 48
GAVILAN SA, EL (Colombia) 62
Gazelle (UTVA) 589
Gazelle SA 342 (Eurocopter) 149
G/B AIRCRAFT (USA) (1983-84) 560
Gee Bee Y Replica (Flaglor) (1985-86) 592
Gem-260 (Miller) (1985-86) 601
Gemini GTP-350 (Daedalus) (1987-88) 410
Gemini ST (Global) (1991-92) 406
Gemini ST (Tridair) 577
Gemsbok (Atlas) (1991-92) 211
GENERAL AVIA COSTRUZIONI AERONAUTICHE SRL (Italy) 196
General Aviation stall/spin flight research (NASA) (1987-88) 478
GENERAL DYNAMICS CORPORATION (USA) 478
GENERAL DYNAMICS/MCDONNELL DOUGLAS (USA) (1992-93) 382
Genesis (Gevers) 478
Geofizika M.55 (Myasishchev) 301
Gepal Mk IV 550 (AMIN) (1983-84) 161
GEPARD SENSOR TECHNOLOGIES SYSTEMS AG (Switzerland) (1991-92) 222
Gerfaut (Eurocopter) 156
Geronimo (Seguin) (1988-89) 473
GEVERS AIRCRAFT INC (USA) 478
G. F. GESTIONI INDUSTRIALI (Italy) 187
GIPPSLAND AERONAUTICS PTY LTD (Australia) 7
Glasair II-S (Stoddard-Hamilton) 573
Glasair III (Stoddard-Hamilton) 573
GLASFASER ITALIANA SRL (Italy) (1990-91) 171
Global Express (Canadair) 30
GLOBAL HELICOPTER TECHNOLOGY INC (USA) 479
Globemater III C-17A (McDonnell Douglas) 534
Golden Eagle (Cessna) (1991-92) 385
GOODE, RICHARD (UK) (1983-84) 534
Goshawk T-45A (McDonnell Douglas/BAe) 168, 385
GOVERNMENT AIRCRAFT FACTORIES (Australia) (1986-87) 6
Gratch T-101 (ROKS-Aero) 303
GREAT LAKES AIRCRAFT COMPANY INC (USA) (1988-89) 393
GREAT PLAINS AIRCRAFT SUPPLY CO INC (USA) 479
Greyhound, C-2A (Grumman) (1989-90) 417
Griffon (Agusta-Bell) 191
Griffon CH-146 (Bell) 23
Grillon 120 (Sellet-Pelletier) (1991-92) 85
GRINVALDS, JEAN (France) (1983-84) 516
Gripen, JAS 39 (IG JAS) 348
Grizzly, Model 72 (Rutan) (1985-86) 503
GRUMMAN AEROSPACE CORPORATION (USA) (1985-86) 413
GRUMMAN CORPORATION (USA) 480
GUANGZHOU ORLANDO HELICOPTERS LTD (China) (1992-93) 35
Guardian A-750 (Airmaster) (1986-87) 331
Guardian A-1200 (Airmaster) (1988-89) 322
Guardian HU-25 (Dassault Aviation) (1991-92) 74
Guardian HU-25A (Dassault-Breguet) (1985-86) 69
Guerrier (SOCATA) (1987-88) 84
GUIMBAL, BRUNO (France) 85
GUIZHOU AVIATION INDUSTRY CORPORATION (China) 47
GUIZHOU AVIATION INDUSTRY GROUP COMPANY (China) (1989-90) 36
Gulfjet (Gulfstream Aerospace) (1989-90) 424
GULFSTREAM AEROSPACE CORPORATION (USA) 485
GULFSTREAM AEROSPACE CORPORATION, COMMANDER DIVISION (USA) (1983-84) 399
GULFSTREAM AEROSPACE TECHNOLOGIES (USA) 487
Gulfstream I-C, G159C (Gulfstream Aerospace) (1983-84) 395
Gulfstream II-B (Gulfstream Aerospace) (1986-87) 424
Gulfstream III (Gulfstream Aerospace) (1987-88) 432
Gulfstream III C-SAM (Gulfstream Aerospace) (1984-85) 412
Gulfstream III Maritime (Gulfstream Aerospace) (1984-85) 412
Gulfstream III Special Mission Versions (Gulfstream Aerospace) (1987-88) 432
Gulfstream IV (Gulfstream Aerospace) 485
Gulfstream IV-B (Gulfstream Aerospace) (1986-87) 426
Gulfstream IV-SP (Gulfstream Aerospace) 485
Gulfstream V (Gulfstream Aerospace) 487
Gulfstream VI-SU (Sukhoi/Gulfstream) (1990-91) 144
Gulfstream SRA-1 (Gulfstream Aerospace) (1984-85) 412
Gulfstream XT (Gulfstream Aerospace) 487
Gull (EGHS) (1985-86) 590
Guri (ENET) (1989-90) 753
GYROFLUG-FFT (West Germany) (1989-90) 96
Gzel M-101 (Myasishchev) 302

H

H-1 (Hamilton) (1989-90) 425
H-2 (Hynes) (1987-88) 436
H-5 (HAMC) (1988-89) 39
H-5 (Harbin) (1986-87) 45
H-5 (Hynes) (1987-88) 436
H-6 (XAC) (1992-93) 44
H-7 (XAC) (1991-92) 45
H-40 (RFB) 104
H-53E (Sikorsky) 559
H-76 Eagle (Sikorsky) 569
H-76N (Sikorsky) (1991-92) 489
H-550A Stallion (Helio) (1991-92) 413
HA-1 Esquilo (Eurocopter) 150
HA-2M Sportster (Hollmann) (1985-86) 595
HAFFS (See Heliborne Aerial FireFighting System) (1991-92) 338
HAI (see Hellenic Aerospace Industry) 105
HAL (see Hindustan Aeronautics Limited) 107
HAMC (see Harbin Aircraft Manufacturing Corporation) 48
HAP/PAH-2/HAC-3G (Eurocopter) (1986-87) 116
HB-23 Hobbyliner (HB Aircraft) (1991-92) 7
HB-23 Scanliner (HB-Aircraft) (1991-92) 7
HB-202 (HB-Aircraft) (1991-92) 8
HB 315B Gaviao (Helibras) 19
HB 350 Esquilo (Helibras) 19
HB 355F2 Esquilo (Helibras) 19
HC-130 Hercules (Lockheed) 502
HD.21 (Aerospatiale) (1991-92) 58
HDH (see Hawker de Havilland Ltd) 8
HE.10B (Bell) (1990-91) 360
HEFFS (see Helicopter Emergency Fire Fighting System) (1987-88) 342
HELIBRAS (see Helicipteros do Brasil S/A) 19
HH-1H Iroquois (Bell) (1990-91) 360
HH-2 (Kaman) (1990-91) 425
HH-3 (Sikorsky) (1983-84) 485
HH-3A (Sikorsky) (1983-84) 485
HH-3F Pelican (Agusta/Sikorsky) (1991-92) 157
HH-60 Night Hawk (Sikorsky) (1987-88) 513
HH-60D Night Hawk (Sikorsky) (1983-84) 488
HH-60G Pave Hawk (Sikorsky) 560
HH-60H (Sikorsky) 565
HH-60J (Mitsubishi) (1988-89) 171
HH-60J Jayhawk (Sikorsky) 565
HH-65A Dolphin (Aerospatiale) (1991-92) 66
HHC helicopter (McDonnell Douglas) (1985-86) 463
HIDEC F-15 (NASA) (1990-91) 469
HJ-5 (Harbin) (1986-87) 34
HJT-16 Kiran Mk I (HAL) (1983-84) 93
HJT-16 Kiran Mk II (HAL) (1990-91) 105
HKP-4C (Kawasaki-Boeing) (1984-85) 159
Hkp 10 (Aerospatiale) (1991-92) 58
HM-1 (Helibras/Eurocopter) 20, 153
HMP (see Huabei Machinery Plant) (1988-89) 39
HN 500 Bengali (Nicollier) (1983-84) 521
HOTOL (BAe) (1992-93) 303
HPT-32 (HAL) 110
HS 9 Biplane (Schoenenberg) (1983-84) 526
HS.23 (Sikorsky) 565
HS 125 Series 700 (BAe) (1983-84) 251
HS 748 (BAe 748) Coastguarder (BAe) (1984-85) 261
HS 748 (BAe 748) Super (BAe) (1986-87) 282
HS 748 (HAL BAe) (1984-85) 94
HS 748 ASWAC (HAL) 110
HS 748 Military Transport (BAe) (1983-84) 254
HS 748 Series 2B (BAe) (1983-84) 253
HT.17 (Boeing) 455
HT.21 (Aerospatiale) (1991-92) 58
HTH (see Helitrans Hybrid-Flugzeugbau GmbH) (1985-86) 89
HTT-34 (HAL) (1991-92) 104
HTTB (Lockheed) (1991-92) 429
HU-1H (Fuji-Bell) (1992-93) 157
HU.10B (Bell) (1990-91) 360
HU-16B Albatross Tanker conversion (Aero Union) (1991-92) 338
HU-25 Guardian (Dassault Aviation) (1991-92) 74
HU-25A Guardian (Dassault-Breguet) (1985-86) 69
HVS (HVS) (1984-85) 544
HVS (see Hutter/Villinger/Schule) (1984-85) 544
HX-1 (Hamilton) (1991-92) 412
HXT-2 (Hamilton) (1991-92) 412
HZ-5 (Harbin) (1986-87) 34

706 INDEXES: AIRCRAFT—**H - J**

Haitun Z-9 (HAMC) 51
Haiyan (NAMC) .... (1991-92) 41
Halcón, A-36 (ENAER) 43
Halcón, T-36 (CASA) 347
Halcón, T-36 (ENAER) 43
Halo, Mi-26 (MIL) 295
HAMILTON AEROSPACE (USA) .... (1991-92) 412
HARBIN AIRCRAFT MANUFACTURING CORPORATION (China) 48
Harbin/Nanchang (Mil) Mi-8 (State Aircraft Factories) .... (1984-85) 38
Harke Mi-10 (MIL) .... (1991-92) 266
Harrier (BAe) 389
Harrier II, (McDonnell Douglas/BAe) 164
Harrier II Plus (McDonnell Douglas/BAe) 167
Harrier III (McDonnell Douglas/BAe) 167
Harrier GR. Mk 3 (BAe) .... (1991-92) 311
Harrier GR. Mk 5, 5A and 7, and T. Mk 10 (McDonnell Douglas/BAe) 164
Harrier T. Mk 4/4A (BAe) .... (1991-92) 311
Harrier T. Mk 4N/8N (BAe) 389
Harrier T. Mk 60 (BAe) 389
Hauler UH-12E (Rogerson Hiller) 552
Have Blue, XST (Lockheed) 502
Havoc, Mi-28 (MIL) 296
Hawk 72 (Hawk) 487
Hawk 200 Series (single seat) (BAe) 388
HAWK AIRCRAFT DEVELOPMENT CORPORATION (USA) 487
Hawk customers (BAe) 388
HAWK INTERNATIONAL (USA) .... (1989-90) 425
Hawk T. Mk 1 and 1A (BAe) 385
Hawk, two-seat versions (BAe) 385
Hawk, US Navy (McDonnell Douglas/BAe) .... (1984-85) 116
Hawk VTXTS (McDonnell Douglas/BAe) .... (1984-85) 450
Hawker 731 (AiResearch) .... (1986-87) 330
HAWKER DE HAVILLAND LTD (Australia) 8
HAWKER DE HAVILLAND VICTORIA LTD (Australia) .... (1989-90) 5
Hawkeye (Grumman) 480
HAYES INTERNATIONAL CORPORATION (USA) .... (1988-89) 402
Haze, Mi-14 (MIL) 291
HB-AIRCRAFT INDUSTRIES LUFTFAHRZEUG AG (Austria) .... (1992-93) 9
Headwind (Stewart) .... (1984-85) 602
Heintz Zenith-CH 400 (Zenair) .... (1985-86) 30
HELI-AIR (USA) 487
Heliborne Aerial Fire Fighting System (Aero Union) .... (1991-92) 338
Heli-Camper (Orlando) 545
Helicop-Jet (Helicop-Jet Canada) .... (1988-89) 30
HELICOP-JET PROJECT MANAGEMENT (Canada) .... (1989-90) 29
HELICOP-JET PROJECT MANAGEMENT (France) .... (1986-87) 75
Helicopter Emergency Fire Fighting System (Aero Union) .... (1987-88) 342
HELICOPTEROS DO BRASIL S/A (Brazil) 19
HELIO AIRCRAFT CORPORATION (USA) 389
HELIO AIRCRAFT LTD (USA) .... (1987-88) 435
Heli-Stat (Piasecki) see Lighter Than Air section
Helitankers (Conair) 33
HELITRANS HYBRID-FLUGZEUGBAU GmbH (West Germany) .... (1985-86) 89
Helitruck (HTH) .... (1985-86) 89
Helix-A and D (Kamov) 272
Helix-B (Kamov) 273
Helix-C (Kamov) 274
HELLENIC AEROSPACE INDUSTRY (Greece) 105
HELWAN (Egypt) .... (1983-84) 48
HEMET VALLEY FLYING SERVICE (USA) .... (1989-90) 426
Hercules (Lockheed) 502
Hercules C. Mk 1, Mk 2 (Lockheed) .... (1989-90) 439
Hercules C. Mk 1K, C. Mk 1P, W. Mk 2 and C. Mk 3P (Lockheed) 502
Hercules, commercial (Lockheed) 505
Hercules conversions (IAI) 184
Hercules conversion (IPTN) 113
Hercules conversions (Marshall) .... (1987-88) 315
Hercules, Receiver (Marshall) .... (1990-91) 316
Hercules T.10, TK.10, TL.10 (Lockheed) 502
Hercules, Tanker (Marshall) .... (1990-91) 316
Hermès (Euro-Hermespace) .... (1992-93) 124
Hermit (Mil) 297
HIBBARD AVIATION (USA) .... (1984-85) 417
High-alpha research vehicle F/A-18 (NASA) .... (1991-92) 456
High Speed Commercial Transport (McDonnell Douglas) 521
HILLER AVIATION INC (USA) .... (1983-84) 402
HILLER HELICOPTERS (USA) .... (1985-86) 425
HILLMAN HELICOPTER ASSOCIATES (USA) .... (1984-85) 419
Hind, Mi-24 (Mil) 294
Hind-D, Mi-25 (Mil) 295
Hind look-alike (Orlando) .... (1990-91) 474
HINDUSTAN AERONAUTICS LIMITED (India) 107
Hip (State Aircraft Factories) .... (1983-84) 38
Hip, Mi-8 (Mil) 289
Hip-G, Mi-9 (Mil) 291
Hip-H, Mi-17 (Mil) 292
Hip-H, Mi-171 (Mil) 292
Hip-K, Mi-17P (Mil) 292
HOAC AUSTRIA FLUGZEUGWERK WIENER NEUSTADT GmbH (Austria) 11
Hobbyliner (HB-Aircraft) .... (1991-92) 7
Hobbyliner, HB-Aircraft (CAI) .... (1991-92) 48
HOFFMANN, WOLF, FLUGZEUGBAU AG (Germany) .... (1992-93) 82
Hokum, Ka-50 (Kamov) 275
Hokum, Ka-136 (Kamov) .... (1991-92) 254
Holiday Knight Twister SKT-1250 (Payne) .... (1985-86) 608
HOLLMANN TECHNOLOGIES INC (USA) .... (1985-86) 595
HONDA MOTOR COMPANY (USA) 488
Hong-5 (Harbin) .... (1986-87) 34
Hong-6 (XAC) .... (1992-93) 44
Hong-7 (XAC) .... (1991-92) 45
Hongzhaji-5 (Harbin) .... (1986-87) 34
Hongzhaji-6 (XAC) .... (1992-93) 44
Hongzhaji-7 (XAC) .... (1991-92) 45
Hongzhaji Jiaolianji-5 (Harbin) .... (1986-87) 34
Hongzhaji Zhenchaji-5 (Harbin) .... (1986-87) 34
Hongzhen-5 (Harbin) .... (1986-87) 34
Hoodlum-A, Ka-26 (Kamov) 271
Hoodlum-B, Ka-126 (Kamov) 277
Hook, Mi-6 (Mil) 289
Hook, Mi-22 (Mil) 289
Hoplite, Mi-2 (PZL Swidnik) 237
Hormone, Ka-25 (Kamov) 271
Hornet 2000 (McDonnell Douglas) .... (1990-91) 451
Hornet, F/A-18 (McDonnell Douglas) 523
Hornet F/A-18E/F (McDonnell Douglas) 527
Hornet, M79 (Aerodyne) .... (1987-88) 341
Hornet production, table (McDonnell Douglas) 524
Horus (MBB) .... (1991-92) 94
'Hotter' (Vardax/Serv-Aero) .... (1989-90) 526
Hound (State Aircraft Factories, China) .... (1985-86) 38
Hover-Gyro, B-18 (Bensen) .... (1984-85) 561
HOWARD HUGHES ENGINEERING PTY LTD (Australia) 8
HUABEI MACHINERY PLANT (China) .... (1988-89) 39
Huey II, UH-1HP (Bell) 434
Huey 800 (Global) 479
HueyCobra (Bell) .... (1990-91) 361
HueyCobra, Modernised Versions (Bell) 434
Humlan 2 (Elkstrom) .... (1983-84) 530
Hummingbird (Aerotek) 338
Hummingbird (VAT) 581
HUNTING FIRECRACKER AIRCRAFT LTD (UK) .... (1985-86) 290
Husky A-1 (Aviat) 417
HUTTER/VILLINGER/SCHULE (West Germany) .... (1984-85) 544
Hybrid Aircraft (Vivian Associates) .... (1992-93) 29
HYNES (USA) .... (1989-90) 426
Hytex (MBB) .... (1991-92) 94

I

I-22 Iryda (WSK-PZL Mielec) 228
IA 58A Pucará (FMA) 3
IA 58C Pucará (FAMA) .... (1989-90) 3
IA 63 Pampa (FMA) 3
IA 66 (FMA) .... (1988-89) 2
IAC (see International Aeromarine Corporation) .... (1990-91) 424
IAC (see International Aircraft Company) .... (1991-92) 228
IAC (see International Aircraft Corporation) .... (1987-88) 437
IAF (see Iraqi Air Force) 181
IAI (see Israel Aircraft Industries) 181
IAI-101 Arava (IAI) .... (1984-85) 130
IAI-101B (IAI) .... (1986-87) 135
IAI-102 Arava (IAI) .... (1986-87) 135
IAI-201 Arava (IAI) .... (1986-87) 135
IAI-202 Arava (IAI) .... (1986-87) 135
IAI 1124 Westwind (IAI) .... (1986-87) 136
IAI 1124 Westwind I (IAI) .... (1986-87) 136
IAI 1124A Westwind 2 (IAI) .... (1988-89) 140
IAI 1124N Sea Scan (IAI) .... (1986-87) 136
IAI 1125 Astra (IAI) 182
Iak-52 (Aerostar) 248
IAM (see Industrie Aeronautiche Meridionali) .... (1989-90) 166
IAP-001 Chuspi (Indaer Peru) .... (1992-93) 173
IAP-002 Ag-Chuspi (Indaer Peru) .... (1992-93) 173
IAP-003 Urpi (Indaer Peru) .... (1992-93) 174
IAR-28MA (ICA) .... (1989-90) 212
IAR-46 (IAR) 252
IAR-93 (SOKO/Avioane) 179
IAR-99 Şoim (Avioane) 250
IAR-109 Swift (Avioane) 251
IAR-316B Alouette III (ICA) .... (1989-90) 213
IAR-317 Airfox (IAR) .... (1991-92) 205
IAR-330 Puma 2000 (IAR) 253
IAR-330L Puma (IAR) 252
IAR-503A (INAv) 248
IAR-705 (Avioane) 251
IAR-823 (ICA) .... (1989-90) 212
IAR-825TP Triumf (ICA) .... (1990-91) 210
IAR-827A (ICA) .... (1989-90) 212
IAR-827TP (ICA) .... (1983-84) 183
IAR-828 (IAR) 252
IAR-828TP (ICA) .... (1988-89) 200
IAR-831 Pelican (ICA) .... (1990-91) 210
IAR SA (Romania) 252
IAv BACAU (see Intreprinderea de Avioane Bacau) .... (1990-91) 207
IAv BUCURESTI (see Intreprinderea de Avioane Bucuresti) .... (1990-91) 208
IAv CRAIOVA (see Intreprinderea de Avioane Craiova) .... (1991-92) 207
ICA (see Intreprinderea de Constructii Aeronautice) .... (1990-91) 210
ICE Programme (MBB) 99
IG JAS (see Industrigruppen JAS AB) 348
IJ-22 Orao (Soko/Avioane) 179
IL (see Instytut Lotnictwa) 228
Il-18 (Ilyushin) .... (1983-84) 205
IL-18 DORR (Ilyushin) .... (1989-90) 244
Il-20 (Ilyushin) 259
Il-22 (Ilyushin) 260
Il-38 May (Ilyushin) 260
Il-62 (Ilyushin) .... (1989-90) 246
Il-62M (Ilyushin) .... (1989-90) 247
Il-62MK (Ilyushin) .... (1989-90) 247
Il-76 (Ilyushin) 261
Il-76 Command Post (Ilyushin) 263
Il-76 Refuelling tanker (Ilyushin) .... (1988-89) 234
Il-78M (Ilyushin) 264
Il-86 (Ilyushin) 265
Il-86 Command Post (Ilyushin) 266
Il-90 (Ilyushin) .... (1991-92) 246
Il-96 Twin-engined version (Ilyushin) .... (1990-91) 248
Il-96-300 (Ilyushin) 26
Il-96-350 (Ilyushin) .... (1990-91) 24
Il-96M (Ilyushin) 26
Il-96T (Ilyushin) 26
Il-102 (Ilyushin) 26
Il-103 (Ilyushin) 26
Il-108 (Ilyushin) 26
Il-114 (Ilyushin) 27
ILYUSHIN (see Aviatsionnyi Kompleks Imieni S. V. Ilyushina) 25
Il-X (Ilyushin) .... (1991-92) 24
INAv (see Institutul de Aviatie) 24
INDAER PERU (see Industria Aeronautica Del Peru SA) 22
IPAI-26 Tuca (São Carlos) .... (1984-85)
IPAI-27 Jipe Voador (São Carlos) .... (1983-84) 1
IPAI-28 Super Surubim (Rio Claro) .... (1984-85)
IPAI-29 Tira Prosa (São Carlos) .... (1983-84)
IPAI-30 (São Carlos) .... (1984-85)
IPE (see Industria Paranaense Estruturas) .... (1992-93)
IPE 04 (IPE) .... (1989-90)
IPE 06 Curucaca (IPE) .... (1992-93)
IPTN (see Industri Pesawat Terbang Nusantara) 11
IRGC (see Islamic Revolutionary Guards Corps) .... (1991-92) 14

IAR SA (Romania) 25
IAROM SA (Romania) 24
Idea PD.93 (Partenavia) 68
ILYUSHIN (Russia) 25
ILYUSHIN DESIGN BUREAU (USSR) .... (1991-92) 241, 75
IML GROUP (New Zealand) .... (1984-85) 17
IMP AEROSPACE LTD (Canada) ....
Impala Mk 2 (Atlas) .... (1986-87) 20
Improved Fouga (IAI) .... (1985-86) 13
Improved SeaCobra (Bell) 43
INDAER CHILE (see Industria Aeronautica de Chile) .... (1983-84)
INDUSTRIA AERONAUTICA DE CHILE (Chile) .... (1983-84)
INDUSTRIA AERONAUTICA DEL PERU SA (Peru) 22
INDUSTRIA AERONAUTICA NEIVA S/A (Brazil) .... (1992-93)
INDUSTRIA METALURGICA DEL NORTE LTDA (Chile) .... (1992-93)
INDUSTRIA PARANAENSE DE ESTRUTURAS (Brazil) .... (1992-93)
INDUSTRIAS CARDOEN LTDA (Chile) .... (1991-92)
INDUSTRIE AERONAUTICHE E MECCANICHE RINALDO PIAGGIO SpA (Italy) 1
INDUSTRIE AERONAUTICHE MERIDIONALI (Italy) .... (1989-90) 1
INDUSTRIGRUPPEN JAS AB (Sweden) 3
INDUSTRI PESAWAT TERBANG NUSANTARA (Indonesia) 1
INGENIEURBURO PROF. DIP LING C DORNIER JR (West Germany) .... (1983-84)
INSTITUTUL DE AVIATIE SA (Romania) 2
INSTYTUT LOTNICTWA (Poland) 2
Interceptor 400A (Prop-Jets) .... (1985-86) 4
Intercity 748 (BAe) .... (1983-84) 2
INTERNATIONAL AEROMARINE CORPORATION (USA) .... (1990-91) 4
INTERNATIONAL AIRCRAFT COMPANY (Thailand) .... (1991-92) 2
INTERNATIONAL AIRCRAFT CORPORATION (USA) .... (1987-88) 4
INTERNATIONAL HELICOPTERS INC (USA) .... (1983-84) 5
Interstate Privateer (Arctic) 4
Interstate S1B2 Arctic Tern (Arctic) 4
INTRAN USA INC (USA) .... (1984-85) 4
INTREPRINDEREA DE AVIOANE BACAU (Romania) .... (1990-91) 2
INTREPRINDEREA DE AVIOANE BUCURESTI (Romania) .... (1990-91) 2
INTREPRINDEREA DE AVIOANE CRAIOVA (Romania) .... (1991-92) 2
INTREPRINDEREA DE CONSTRUCTII AERONAUTICE (Romania) .... (1990-91) 2
Intruder II (Grumman) .... (1990-91) 4
Intruder A-6E (Grumman) 48
In-wing weather radar (Sierra) .... (1991-92) 4
Ipanema EMB-201A (EMBRAER) .... (1992-93)
Ipanema EMB-202 (EMBRAER) ....
IRAQI AIR FORCE (Iraq) 1
Iroquois (Bell) .... (1991-92) 3
Iryda I-22 (WSK-PZL Mielec) 2
Iryda M-95 (PZL Mielec) 2
Iryda M-97 (PZL Mielec) 2
ISHAM AIRCRAFT (USA) .... (1984-85) 4
ISHIDA AEROSPACE RESEARCH INC (Japan) 2
Iskierka, M-26 (PZL Mielec) 2
Iskra TS-11R (PZL Mielec) 2
Iskra-bis, TS-11 (WSK-PZL Mielec) .... (1988-89) 1
ISLAMIC REVOLUTIONARY GUARDS CORPS (Iran) .... (1991-92) 1
ISLAND AIRCRAFT LTD (UK) .... (1990-91) 3
Islander (PADC/Pilatus Britten-Norman) 2
Islander (Romaero) 2
Islander AL. Mk 1 (Pilatus Britten-Norman) 3
Islander BN-2B (Pilatus Britten-Norman) 3
Islander CC. Mks 2/2A (Pilatus Britten-Norman) 3
ISRAEL AIRCRAFT INDUSTRIES LTD (Israel) 1

J

J-5 Janowski (Marko-Elektronik) .... (1984-85) 5
J-6 (Shenyang) .... (1987-88)
J-7 (CAC) ....
J-7 III (SAC) .... (1989-90)
J-8 (SAC) ....
J-8 (Shenyang) .... (1987-88)
J-8 II (SAC) .... (1989-90)
J-9 (CAC) ....
J-9 (CATIC) .... (1988-89)
J-10 (Jackson) .... (1984-85) 5
J-12 (NAMC) .... (1989-90)

J – M—AIRCRAFT: INDEXES 707

-22 Orao (SOKO/Avioane) 179
32E Lansen (Saab-Scania) (1987-88) 225
35J Draken (Saab-Scania) (1991-92) 220
A 37 Viggen (Saab-Scania) (1991-92) 219
ADC (see Japan Aircraft Development Corporation) 206
AS 39 Gripen (IG JAS) 348
C-2D Tri-Coupé (Coupe-Aviation) (1983-84) 514
C-130 Hercules (Lockheed) 502
D1HW 1.7 Headwind (Stewart) (1984-85) 602
D2FF Foo Fighter (Stewart) (1984-85) 602
EH (see Joint European Helicopter) (1991-92) 129
F (Eurofighter) (1989-90) 129
H-7 (XAC) 58
J-5 (CAC) 44
J-6 (State Aircraft Factories) (1984-85) 35
-7 (GAIC) 47
Magnum (O'Neill) (1986-87) 475
PATS (Cessna) 469
S-1 (Anger/Selcher) (1985-86) 616
-STARS (Boeing) (1989-90) 392
VX (Bell/Boeing Vertol) (1987-88) 398
W-1 (ACA) (1991-92) 335
W-2 (ACA) (1991-92) 335
W-3 (ACA) (1991-92) 335
Z-6 (Shenyang) (1986-87) 39

ABIRU AIRCRAFT PTY LTD (Australia) 9
ACKSON, L. A. (USA) (1984-85) 579
ACOBS JETS INC (USA) (1983-84) 410
AFFE AIRCRAFT INC (USA) (1992-93) 389
aguar International (HAL/SEPECAT) 108
aguar International (SEPECAT) 179
anowski J-5 (Marko-Electronik) (1984-85) 547
APAN AIRCRAFT DEVELOPMENT CORPORATION (Japan) 206
APAN AMATEUR BUILT AIRCRAFT LEAGUE (Japan) (1984-85) 547
AVELIN AIRCRAFT COMPANY INC (USA) (1987-88) 437
avelin, Cessna P210 (Air America) (1986-87) 330
ayhawk HH-60J (Sikorsky) 565
ayhawk T-1A (Beech) 432
ETCRAFT USA INC (USA) (1990-91) 424
et Cruiser, Mark II (JetCraft) (1990-91) 424
etCruzer 450 (AASI) 410
etCruzer 500P (AASI) 410
etCruzer 650 (AASI) 411
etCruzer ML-2 (AASI) 410
et Cruzer 620 (AASI) (1991-92) 335
etliner, EMB-145 (EMBRAER) 16
et Prop DC-3 AMI (Professional Aviation) 343
etRanger III Model 206B (Bell) (1992-93) 17
etRanger III Model 206B-3 (Bell) 20
et Squalus (General Avia) (1986-87) 158
et Squalus, F1300 NGT (Promavia) 12
etStar laminar flow control (NASA) (1986-87) 469
etstream 31 (BAe) (1989-90) 297
etstream 41 (BAe) 380
etstream 51 (BAe) 697
etstream 61 (BAe) 383
etstream 71 (BAe) 697
etstream ATP (BAe) 382
ETSTREAM HOLDINGS (UK) 379
etstream Super 31 (BAe) 379
etstream T. Mk 3 (BAe) (1988-89) 280
et Wasp II (Texas Helicopter) (1983-84) 496
ian-6 (Shenyang) (1987-88) 46
ian-7 (CAC) 44
ian-8 (Shenyang) (1987-88) 46
ian-8 (SAC) 56
ian-8 II (SAC) (1989-90) 43
ian Hong-7 (XAC) 58
ianjiao-5 (CAC) 44
ianjiao-6 (China) (1984-85) 35
ianjiao-7 (GAIC) 47
ianjiji-6 (Shenyang) (1987-88) 46
ianjiji-7 (CAC) 47
ianjiji-8 (SAC) 56
ianjiji-8 (Shenyang) (1987-88) 46
ianjiji-8 II (SAC) (1989-90) 43
ianjiji Hongzhaji-7 (XAC) 58
ianjiji Jiaolianji-5 (CAC) 44
ianjiji Jiaolianji-6 (China) (1984-85) 35
ianjiji Jiaolianji-7 (GAIC) 47
ipe Voador, IPAI-27 (São Carlos) (1983-84) 19
oint Attack Fighter (USAF) 579
OINT EUROPEAN HELICOPTER (International) (1991-92) 129
oint Primary Aircraft Training System (USAF) 579
oint Stealth Strike Aircraft (USAF) 579
olly Green Giant (Sikorsky) (1983-84) 485
ONES, STANLEY (USA) (1983-84) 563
ORDAN AEROSPACE (Jordan) (1991-92) 177
OSES, CHARLES RUPERT (Brazil) (1983-84) 505
URCA, MARCEL (France) 86

K

-1 Flamingo-Trainer (FUS) (1983-84) 86
-8 Karakorum 8 (NAMC/PAC) 169
A (see Korean Air) 213
A-6 (Grumman) (1987-88) 427
A-6D (Grumman) 481
a-25 (Kamov) 271
a-26 (Kamov) 271
a-26 Jet control testbed (Kamov) (1991-92) 250
a-27 (Kamov) 272
a-28 (Kamov) 272
a-29 (Kamov) 273
a-32 (Kamov) 274
a-34? (Kamov) (1991-92) 254
a-50 Werewolf (Kamov) 275
a-62 (Kamov) 276
a-118 (Kamov) (1992-93) 213
Ka-126 Hoodlum-B (Kamov) 277
Ka-128 (Kamov) 278
Ka-136 Hokum (Kamov) (1991-92) 254
Ka-226 (Kamov) 278
KBHC (see Korea Bell Helicopter Company) (1992-93) 165
KC-10A (McDonnell Douglas) (1991-92) 442
KC-130 (Lockheed) 502
KC-135A Stratotanker (Boeing) (1991-92) 371
KDC-10 (McDonnell Douglas) 536
KE-3A (Boeing) (1991-92) 372
KF02 (WSK-PZL Krosno) (1988-89) 182
KFC (see Kelowna Flightcraft Air Charter) 39
KF-X (SSA/General Dynamics) (1991-92) 179
KM-2 Kai (Fuji) 203
KM-2D (Fuji) (1988-89) 166
KM-180 (SACAB) (1987-88) 226
KM C-180 (SACAB) (1987-88) 226
K-Max (Kaman) 489
KS-3A Viking (Lockheed) (1986-87) 439
KTX-1 Korint (DHI) 213
KTX-2 (SSA) 214
KV-107II (Kawasaki/Boeing) (1984-85) 159
KV-107IIA (Kawasaki) (1991-92) 173

Kachina, Model 2150A-TG (Hibbard/Varga) (1983-84) 402
Kachina, Model 2150A (Varga) (1983-84) 497
Kachina, Model 2180 (Varga) (1983-84) 498
Kaczor, PZL-107 (PZL Warszawa-Okecie) 244
KAMAN AEROSPACE CORPORATION (USA) 488
KAMOV (Russia) 271
KAMOV DESIGN BUREAU (USSR) (1991-92) 249
Kania (PZL Swidnik) 238
Karakorum 8 (NAMC/PAC) 169
KARP, LARRY (USA) (1983-84) 564
Katana (HOAC) 11
Katana T30 (Terzi) 203
KATRAN (International) 163
Katran (Katran) 163
KAWASAKI JUKOGYO KABUSHIKI KAISHA (Japan) 206
Kawka, PZL-107 (WSK-PZL Warszawa-Okecie) (1983-84) 171
KELEHER, JAMES J. (USA) (1984-85) 580
KELOWNA FLIGHTCRAFT GROUP LTD (Canada) 39
Kfir (IAI) 182
Kiddy Hawk (Calypso) (1985-86) 373
King Air B100 (Beech) (1983-84) 308
King Air C90/C90A update (Raisbeck) (1991-92) 456
King Air C90A (Beech) (1991-92) 349
King Air C90B (Beech) 426
King Air conversion (Frakes) (1989-90) 411
King Air Exec-Liner (Beech) (1987-88) 360
King Air F90 (Beech) (1983-84) 307
King Air F90-1 (Beech) (1987-88) 356
Kingbird, F27 (Fokker) (1985-86) 171
King Cat (Mid-Continent) (1991-92) 452
KING'S ENGINEERING FELLOWSHIP, THE (USA) (1992-93) 391
Kiowa (Bell) 436
Kiowa Warrior (Bell) 436
Kiran Mk I/IA (HAL) (1983-84) 93
Kiran Mk II (HAL) (1990-91) 105
Kitfox (Skystar) 571
KOC HOLDING (Turkey) (1991-92) 229
Koliber II (PZL Warszawa-Okecie) 690
Koliber 150, PZL-110 (PZL Warszawa-Okecie) 244
Koliber 235, PZL-111 (PZL Warszawa-Okecie) 245
Kong Yun (NAMC) (1991-92) 40
KOREA BELL HELICOPTER COMPANY (Korea) (1992-93) 165
KOREAN AIR (Korea) 213
Korean Indigenous Trainer (DHI) (1992-93) 164
Korint KTX-1 (DHI) 213
Kruk, PZL-106B (PZL-Warszawa-Okecie) 242

L

L-2M Tech Two (New Technik) (1992-93) 456
L-8 (NAMC/PAC) (1988-89) 136
L-19 (Ector) (1983-84) 373
L-39 Albatros (Aero) 63
L-39 MS (Aero) (1991-92) 51
L-59 (Aero) 64
L-70 Militrainer (Valmet) (1989-90) 57
L-80 TP (Valmet) (1985-86) 50
L-90 TP Redigo (Valmet) 73
L-100 conversions (IAI) 184
L-100 series Commercial Hercules (Lockheed) 505
L-270 (Aero) (1991-92) 51
L-410 UVP-E (Let) 66
L-420 (Let) 685
L-602 (Let) (1983-84) 46
L-610 (Let) 67
L-1011 (Lockheed) (1989-90) 438
L-1011 conversions (Lockheed) (1989-90) 445
L-1011F (Hayes-Lockheed) (1988-89) 402
L-1011F (Pemco) 545
L-1011 TriStar conversions (Lockheed) (1991-92) 430
LA-4-200 Amphibian (Lake) (1991-92) 415
LA-250 Renegade (Lake) 491
LAS (see Lockheed Aircraft Service Company) (1986-87) 434
LC-130 Hercules (Lockheed) 502
LCPT (Rushjet/General Technologies/Aerosud) (1992-93) 136
LCRA (NAL) (1988-89) 100
LDA-500 Boxer (Lockspeiser) (1986-87) 298
LDA-1000 Boxer (Lockspeiser) (1986-87) 299
LET (see Let Akciova Spolecnost) 66
LF 2 (HOAC) (1992-93) 9
LH (Bell) (1990-91) 366
LHX. (Boeing/Sikorsky) (1991-92) 381
LKhS-4 (Phöenix-Aviatechnica) 303
LP-1 SwiftFury (LoPresti) (1992-93) 404
LRAACA (Lockheed) (1991-92) 426
LSA (Jabiru) 9
LT-1 Swati (BHEL) 106
Labrador, CH-113 (Boeing Canada) (1985-86) 19
LAH Tonal (JEH) (1990-91) 132
LAKE AIRCRAFT INC (USA) 491
LAKE AMPHIBIAN INC (USA) (1985-86) 431
Lama SA 315B (HAL) 109
Lama SA 315B (HAL/Aerospatiale) (1987-88) 106
Lama conversion (Rocky Mountain) 552
Lancair IV (Lancair) 492
Lancair 235 (Lancair) 492
Lancair 320 (Lancair) 492
Lancair 360 (Lancair) 492
Lancair ES (Lancair) 492
Lancer B-1B (Rockwell) (1990-91) 491
Lansen J 32E (Saab-Scania) (1987-88) 225
Lapwing (Snipe) (1985-86) 305
Lark 1B (Keleher) (1984-85) 580
LASA A-22 (Sadler) 553
Laser 300 (Omac) (1990-91) 473
Laser Jet (Machen) (1985-86) 445
LASHER, C. W. (USA) (1983-84) 366
Lasta 1 (UTVA) 586
Lasta 2 (UTVA) 586
Lavi (IAI) (1988-89) 137
Lavi Technology Demonstrator (IAI) (1991-92) 146
Lear Fan 2100 (Lear Fan) (1984-85) 276, 429
LEAR FAN LIMITED (USA) (1985-86) 433
LEAR FAN LTD (UK) (1984-85) 276
LEARJET INC (USA) 492
Learjet 25D (Gates Learjet) (1986-87) 406
Learjet 25G (Dee Howard/Gates) (1986-87) 398
Learjet 25G (Gates Learjet) (1986-87) 407
Learjet 31 (Learjet) (1990-91) 430
Learjet 31A (Learjet) 494
Learjet 35A/36A (Learjet) 492
Learjet 35A/36A Special missions versions (Learjet) 493
Learjet 45 (Learjet) 495
Learjet 55 (Gates Learjet) (1986-87) 409
Learjet 55 Long-range tanks (Branson) (1991-92) 382
Learjet 55B (Learjet) (1991-92) 419
Learjet 55C (Learjet) (1991-92) 419
Learjet 60 (Learjet) 496
LEG-AIR CORPORATION (USA) (1985-86) 599
LEICHTFLUGZEUGE-ENTWICKLUNGEN DIPL ING HERMANN MYLIUS (West Germany) (1985-86) 92
Leonardo, M. B.3 (Militi) (1983-84) 528
Leopard (CMC) 392
Leshii SL-90 (Aviotechnica) 134
LET AKCIOVA SPOLECNOST (Czech Republic) 66
LET KONCERNOVY PODNIK (Czechoslovakia) (1991-92) 51, 747
Lift Activator Disc (VTOL Aircraft) (1987-88) 10
Light Aircraft (Toyota/Rutan) 555
Light Attack Helicopter (HAL) 107
Light Combat Aircraft (ADA) 106
Light Combat Aircraft (HAL) (1990-91) 105
Light Helicopter (Bell) 25
Lightning Model 38P (Beech) (1984-85) 309
Light Sport Aircraft (Jabiru) 9
Light Transport Aircraft (HAL) 111
LightWing GA-55 (Hughes) 9
LightWing GR-582 (Hughes) 8
LightWing GR-912 (Hughes) 9
LOCK HAVEN AIRPLANE COMPANY (USA) 496
LOCK HAVEN RE-MAN CENTER (USA) (1987-88) 441
Lockheed F-104/Shuttle Insulation Tests (NASA) (1983-84) 445
LOCKHEED ADVANCED DEVELOPMENT COMPANY (USA) 506
LOCKHEED AERONAUTICAL SYSTEMS COMPANY (USA) 497
LOCKHEED AERONAUTICAL SYSTEMS GROUP (USA) 497
LOCKHEED AIRCRAFT SERVICE COMPANY (USA) 512
Lockheed Aurora (USAF) 579
LOCKHEED-CALIFORNIA COMPANY (USA) (1986-87) 435
LOCKHEED CORPORATION (USA) 496
LOCKHEED FORT WORTH COMPANY (USA) 508
LOCKHEED-GEORGIA COMPANY (USA) (1986-87) 439
LOCKSPEISER AIRCRAFT LTD (UK) (1987-88) 314
Long-EZ, RAF Model 61 (Rutan) (1985-86) 613
LongRanger II (Bell) (1983-84) 317
LongRanger III (Bell) (1992-93) 18
LongRanger IV (Bell) 21
LongRanger IV (Aymet) 365
LOPRESTI FLIGHT CONCEPTS (USA) 513
LOPRESTI PIPER AIRCRAFT ENGINEERING COMPANY (USA) (1991-92) 465
LOVAUX LIMITED (UK) (1991-92) 316
LTV AEROSPACE AND DEFENSE COMPANY (USA) (1992-93) 404
Lunar Rocket (Maule) (1985-86) 448
LUNDS TEKNISKE (Norway) 224
LUSCOMBE AIRCRAFT LTD (UK) (1987-88) 315
Lynx (Westland) 405
Lynx-3 (Westland) (1987-88) 337

M

M-1 (Super Rotor) (1992-93) 17
M-4 (Myasishchev) (1991-92) 273
M-5 Lunar Rocket (Maule) (1985-86) 448
M-6 Super Rocket (Maule) (1983-84) 429
M-7 Starcraft (Maule) (1991-92) 434
M-11 Mustang (Mustang) 539
M-12 Tchaika (Beriev) (1991-92) 239
M-17 Mystic (Myasishchev) 301
M-18 Dromader (PZL Mielec) 233
M-20 Mewa (PZL Mielec) 234
M20J (Mooney) (1985-86) 467
M20K (Mooney) (1985-86) 468
M20L PFM Pegasus (Mooney) (1990-91) 445
M-21 Dromader Mini (PZL Mielec) (1991-92) 193
M-24 Dromader Super (PZL Mielec) (1991-92) 180

M-26 Iskierka (PZL Mielec) 235
M-55 (Myasishchev) 301
M79 Hornet (Aerodyne) (1987-88) 341
M-95 Iryda (PZL Mielec) 231
M-97 Iryda (PZL Mielec) 231
M-101 Gzhel (Myasishchev) 302
M-103 Skif (Myasishchev) 302
M400 Skycar (Moller) 537
MA-2 Mamba (AIA) 4
MAC (see Melbourne Aircraft Corporation) (1991-92) 5
MAFFS (see Modular Aerial Fire Fighting System (Aero Union) (1991-92) 338
MAI (see Mitsubishi Aircraft International) (1986-87) 446
MAI (see Moscow Aviation Institute) (1985-86) 233
MAI-890 (Aviatika) 691
MAS (see Ministry of Aero-Space Industry) (1992-93) 32
M. B.3 Leonardo (Militi) (1983-84) 528
M. B.4 (Militi) (1983-84) 528
M. B.339 Uprated version (Aermacchi) (1984-85) 139
MB-339A (Aermacchi) 185
MB-339B (Aermacchi) (1989-90) 156
MB-339C (Aermacchi) 186
MB-339D (Aermacchi) (1988-89) 146
MB-339K (Aermacchi) (1989-90) 157
MB-339RM (Aermacchi) (1990-91) 154
MBB (see Messerschmitt-Bölkow-Blohm GmbH) (1992-93) 77
MC-130 Hercules (Lockheed) 502
MD3 Swiss Trainer (MDB) 357
MD-9XX (McDonnell Douglas) (1989-90) 456
MD-11 (McDonnell Douglas) 518
MD-11 freight door (Marshall) (1991-92) 317
MD-11X (McDonnell Douglas) (1985-86) 459
MD-12 (McDonnell Douglas) 520
MD-12X (McDonnell Douglas) (1991-92) 444
MD-80 Series (McDonnell Douglas) 515
MD-81 (McDonnell Douglas) 515
MD-82 (McDonnell Douglas) 515
MD-82 (SAMF/McDonnell Douglas) 57
MD-83 (McDonnell Douglas) 515
MD-83 (SAMF/McDonnell Douglas) 57
MD-87 (McDonnell Douglas) 515
MD-88 (McDonnell Douglas) 515
MD-89 (McDonnell Douglas) (1985-86) 455
MD-90 (McDonnell Douglas) 517
MD-90 UHB programme (McDonnell Douglas) (1989-90) 456
MD-90-30T (McDonnell Douglas) 515
MD-90-30T (SAMF/McDonnell Douglas) 57
MD-91 (McDonnell Douglas) (1989-90) 456
MD-91V (McDonnell Douglas) (1989-90) 456
MD-92 (McDonnell Douglas) (1989-90) 456
MD-92V (McDonnell Douglas) (1989-90) 456
MD-94X (McDonnell Douglas) (1987-88) 463
MD-95 (Douglas/CATIC) 138
MD-95 (McDonnell Douglas) 517
MD-100 (McDonnell Douglas) (1984-85) 455
MD 500D (Kawasaki/McDonnell Douglas) 208
MD 500/530 (McDonnell Douglas) 527
MD-500/530 civil versions (McDonnell Douglas) (1990-91) 446
MD 500/530 Defender (McDonnell Douglas) 528
MD 520N NOTAR (McDonnell Douglas) 530
MD 530N (McDonnell Douglas) 530
MD 900 MDX (McDonnell Douglas) (1990-91) 462
MD Explorer (McDonnell Douglas) 531
MDX (McDonnell Douglas) (1991-92) 448
MD-XX (McDonnell Douglas) 518
METALNOR (see Industria Metalurgica del Norte) (1992-93) 32
MF-85 Series (Ruschmeyer) 350
MFI (see Malmo Forsknings & Innovations AB) 350
MFI-10C Vipan (RFB) 687
MFI-11 (MFI) 350
MFI-18 (MFI) (1985-86) 202
MFT/LF (Embraer) (1991-92) 10
MH-6 (McDonnell Douglas) 528
MH-47 Chinook (Boeing Helicopters) 455
MH-53E Sea Dragon (Sikorsky) 559
MH-53J Pave Low III enhanced (Sikorsky) (1991-92) 480
MH-60G Pave Hawk (Sikorsky) 560
MH-60K (Sikorsky) 560
Mi-2 (PZL Swidnik) 237
Mi-2B (PZL Swidnik) (1991-92) 195
Mi-2R (WSK-PZL Swidnik/Mil) (1984-85) 179
Mi-6 (Mil) 289
Mi-8 (Harbin/Nanchang) (1984-85) 38
Mi-8 (Mil) 289
Mi-9 (Mil) 291
Mi-10 (Mil) (1991-92) 266
Mi-10K (Mil) (1991-92) 266
Mi-14 (Mil) 291
Mi-17 (Mil) 292
Mi-18 (Mil) (1983-84) 225
Mi-22 (Mil) 289
Mi-24 (Mil) 294
Mi-25 (Mil) 295
Mi-26 (Mil) 295, 693
Mi-28 (Mil) 296
Mi-30 (Mil) (1989-90) 272
Mi-32 (Mil) (1989-90) 272
Mi-34 (Mil) 297
Mi-34 VAZ (Mil) 298
Mi-35 (Mil) 298
Mi-38 (Mil) 298
Mi-40 (Mil) 299
Mi-46K (Mil) 300
Mi-46T (Mil) 299
Mi-52 (Mil) 693
Mi-54 (Mil) 300
Mi-171 (Mil) 292
MiG-18-50 (MiG) 288
MiG-21 (MiG) (1992-93) 214
MiG-21bis (HAL) (1987-88) 108
MiG-21 upgrade (IAI) 184
MiG-21 upgrade (HAL) 111
MiG-21, upgraded (MiG) (1990-91) 256
MiG-21-2000 (IAI) 688
MiG-23 (MiG) 279
MiG-23 upgrade (AIA) 184
MiG-24 (MiG) (1991-92) 256
MiG-25 (MiG) 281
MiG-27 (MiG) 280
MiG-27M (HAL) 111
MiG-29 (MiG) 283
MiG-31 (MiG) 286
MiG-33 (MiG) 287
MiG 105, Experimental Aircraft (MiG) (1991-92) 298
MiG fighter, new (MiG) (1991-92) 263
MiG-AT (MiG) 287, 692
MiG SVB (MiG) 288
MIT (see Massachusetts Institute of Technology) (1985-86) 466
M. J.5 Sirocco (Jurca) 86
M. J.20 Tempete (Jurca) (1983-84) 518
M. J.52 Zephyr (Jurca) 86
ML74 Wasp (Aerodyne) (1987-88) 341
ML79 Hornet (Aerodyne) (1987-88) 341
MP-18 Dragon (Phalanx) (1987-88) 482
MP.205 Busard (Lefebvre) (1985-86) 555
MPC75 (MPC Aircraft) (1991-92) 134
MSE (Mooney) 538
MSSA (Pilatus Britten-Norman) 398
MT20 Military Trainer (Mooney) (1985-86) 468
MTC (see MOSHIER TECHNOLOGIES CORPORATION (USA) (1988-89) 445
MU-2 (MAI) (1985-86) 445
MU-2 (Mitsubishi) (1986-87) 170
MU-2 Express conversions (MU-2) (1991-92) 455
MU-300 (Mitsubishi) (1987-88) 181
MX-7 Starcraft (Maule) (1991-92) 434
MY 102 Tornado (Mylius) (1985-86) 92

MACAVIA INTERNATIONAL (USA) (1991-92) 431
MacCREADY, DR PAUL B. (USA) (1985-86) 444
MACERA, SILVANO (Italy) (1984-85) 546
MACFAM WORLD TRADERS (Canada) (1984-85) 524
MACHEN INC (USA) 513
Madcap, An-74 (Antonov) (1991-92) 236
Maestro (Yakovlev) (1990-91) 290
Magnum (Avid) 419
Magnum (O'Neill) (1986-87) 475
MAGNUM AIRCRAFT (USA) (1991-92) 432
Maiden (Sukhoi) (1983-84) 230
Mail, Be-12 (Beriev) (1991-92) 239
Mainstay, A-50 (Beriev) 259
MAISS, ULRICH (Germany) (1985-86) 562
Major, DR 400/160 (Robin) 90
Malibu (Piper) (1988-89) 458
Malibu Mirage (Piper) 548
MALMO FORSKNINGS & INNOVATIONS AB (Sweden) 350
Mamba MA-2 (AIA) 4
MANCRO AIRCRAFT COMPANY (USA) (1983-84) 428
Mangusta, A 129 (Agusta) 189
Maritime, F27 (Fokker) (1985-86) 171
Maritime Defender (Pilatus Britten-Norman) 398
Maritime Enforcer, F27 (Fokker) (1985-86) 171
Maritime Patrol B200T (Beech) 428
Mark I (Spitfire) (1984-85) 508
Mark II (Spitfire) (1984-85) 508
Mark II Jet Cruiser (JetCraft) (1990-91) 424
Mark III Executive Commuter (JetCraft) (1989-90) 427
Mark IV (Spitfire) (1983-84) 494
Mark IV System (Raisbeck/Western) (1983-84) 469
Mark VI System (Raisbeck) (1988-89) 459
Mark VII System (Raisbeck) (1988-89) 460
MARKO-ELEKTRONIK COMPANY (Poland) (1984-85) 547
MARMANDE AERONAUTIQUE (France) (1985-86) 74
Marquise (MAI) (1985-86) 445
MARSH AVIATION COMPANY (USA) 513
MARSHALL OF CAMBRIDGE (ENGINEERING) LTD (UK) (1992-93) 307
Mascot, HJ-5 (Harbin) (1986-87) 34
MASSACHUSETTS INSTITUTE OF TECHNOLOGY (USA) (1985-86) 466
Matador VA.1 (BAe) (1991-92) 311
Matador II, VA.2 (McDonnell Douglas/BAe) 164
MAULE AIR INC (USA) 514
Mavi Isik 78-XA (THvK/KIBM) (1983-84) 197
Mavi Isik-B (THvK/KIBM) (1983-84) 197
Mavi Isik-G (THvK/KIBM) (1983-84) 197
Maxi Cat (Aero Mod) (1984-85) 300
May, Il-38 (Ilyushin) 260
MBB/AEROSPATIALE (International) (1983-84) 110
MBB/CATIC (International) (1987-88) 128
MBB HELICOPTER CANADA LTD (Canada) (1991-92) 29
MBB HELICOPTER CORPORATION (USA) (1991-92) 434
MBB/KAWASAKI (International) (1991-92) 129
MCDONNELL AIRCRAFT COMPANY (USA) (1992-93) 406
MCDONNELL DOUGLAS AEROSPACE (USA) 521
MCDONNELL DOUGLAS AEROSPACE TRANSPORT AIRCRAFT (USA) 534
MCDONNELL DOUGLAS/BAe (International) 164
MCDONNELL DOUGLAS CORPORATION (USA) 514
MCDONNELL DOUGLAS HELICOPTER COMPANY (USA) 527
MDB FLUGTECHNIK AG (Switzerland) 357
MELBOURNE AIRCRAFT CORPORATION PTY LTD (AUSTRALIA) (1991-92) 5
MELEX USA (USA) 536
Mercury E-6A (Boeing) 454
MERIDIONALI SpA ELICOTTERI (Italy) (1985-86) 150
Merlin, EH101 (EHI) 140
Merlin IIIC, SA227-TT (Fairchild) (1983-84) 379
Merlin IVC (SA227-AT) (Fairchild) (1991-92) 400
Merlin V (Fairchild) (1989-90) 410
Merlin 23 (Fairchild) 475
Merlin 200 Aerobot (Moller) (1987-88) 474
Merlin 300 (Fairchild Aircraft) (1988-89) 388
Merlin 300 Aerobot (Moller) (1988-89) 443
Merlin 400 Aerobot (Moller) (1989-90) 468
Merlin 400SP (Fairchild) (1987-88) 416
MERLYN PRODUCTS INC (USA) (1989-90) 467
Mermaid, A-40 (Beriev) 255
Mermaid, B-42 (Beriev) 25
MESSERSCHMITT-BÖLKOW-BLOHM GmbH (Germany) (1992-93) 7
Meteorite (Wallis) (1986-87) 31
Meteorite 2 (Wallis) (1987-88) 33
Metro III (Fairchild Aircraft) (1991-92) 39
Metro IIIA (Fairchild) (1986-87) 40
Metro V (Fairchild) (1990-91) 41
Metro VI (Fairchild) (1987-88) 41
Metro 23 (Fairchild) 47
Metro 23 Special Mission Aircraft (Fairchild) 47
Metro 25 (SA227-DC) (Fairchild Aircraft) (1990-91) 41
Mewa, M-20 (PZL Mielec) 234
MEYER, CLAIR O (USA) (1983-84) 56
Microjet 200B (Microjet) (1990-91) 80
MICROJET SA (France) (1990-91) 80
Midas, Il-78 (Ilyushin) 264
MID-CONTINENT AIRCRAFT CORPORATION (USA) 53
MiG (Russia) 27
MIKOYAN DESIGN BUREAU (USSR) (1991-92) 25
MIL DESIGN BUREAU (Russia) (1992-93) 224
MILITI, BRUNO (Italy) (1983-84) 52
MILLER, J. W. AVIATION INC (USA) (1985-86) 601
MILLER, WILLIAM Y (USA) (1985-86) 602
Miltrainer, L-70 (Valmet) (1989-90) 57
Mini Z, CH 50 (Zenair) (1984-85) 52
Mini-Esqualo (Brazil) (1985-86) 53
MINISTRY OF AERO-SPACE INDUSTRY (China) (1992-93) 32
Mini-Swat, S-350 (Schapel) (1983-84) 481
MINTY, E. R. (Australia) (1983-84) 9
Minuano (EMBRAER/Piper) 1
Mirage 3 NG (Dassault-Breguet) (1989-90) 68
Mirage III (Dassault-Breguet) (1988-89) 65
Mirage III series (Dassault) (1992-93) 58
Mirage III/5 upgrade (IAI) 183, 184
Mirage IV-P (Dassault-Breguet) (1988-89) 66
Mirage 5 (Dassault-Breguet) (1988-89) 65
Mirage 50 (Dassault-Breguet) (1988-89) 65
Mirage 50 Upgrade (ENAER) (1988-89) 33
Mirage 50M (Dassault-Breguet) (1989-90) 68
Mirage 2000 (Dassault Aviation) 75, 686
Mirage 4000 (Dassault-Breguet) (1988-89) 70
Mirage F1 (Dassault Aviation) (1992-93) 58
Mirage advanced technology update programmes (Dassault Aviation) (1991-92) 67
Mirage Improvement Programme (F + W) 363
Mirage modifications (IAI) (1987-88) 144
Mirage safety improvement programme (SABCA) 13
Mirlo, E.25 (CASA) 347
Mission Adaptive Wing (NASA/USAF/Boeing) (1988-89) 371
Missionmaster (GAF) (1985-86) 7
Mistral AE-206 (Aviasud) 74
Mistral Twin (Aviasud) 75
MITSUBISHI AIRCRAFT INTERNATIONAL INC (USA) (1986-87) 446
MITSUBISHI JUKOGYO KABUSHIKI KAISHA (Japan) 209
Mod AH-1S (Bell) (1984-85) 328

Note: Aircraft previously identified as 'Model xxx' are now listed as numbers only at the beginning of the index.

MODERN WING AIRCRAFT LTD (Canada) (1988-89) 31
MOD SQUAD, THE, INC (USA) (1988-89) 442
Modular Aerial Fire Fighting System (Aero Union) (1991-92) 338
Mohawk (Grumman) (1983-84) 391
Mohawk, Grumman, modified (PIA) (1988-89) 450
Mohawk OV-1 updates (Grumman) (1990-91) 420
Mojave (Piper) (1986-87) 482
Moller 400 Volantor (Moller) (1991-92) 423
MOLLER INTERNATIONAL (USA) 537
Molniya-1 (Molniya) 300
MOLNIYA SCIENTIFIC AND INDUSTRIAL ENTERPRISE (Russia) 300
MONARCH AVIATION INC (USA) (1987-88) 474
Monarch B (MIT) (1985-86) 466
Monex (Monnett) (1983-84) 568
MONG/MORTENSEN (USA) (1985-86) 602
Mong/Mortensen Biplane (1985-86) 602
Mongol MiG-21 (MiG) (1992-93) 214
MONNETT EXPERIMENTAL AIRCRAFT INC (USA) (1985-86) 602
Montalva (Super Rotor) (1992-93) 17
MONTANA COYOTE INC (USA) 537
MOONEY AIRCRAFT CORPORATION (USA) 538
MORAVAN AKCIOVÁ SPOLECNOST (Czech Republic) 69
MORAVAN NARODNI PODNIK (Czechoslovakia) (1991-92) 54
MORRISEY AIRCRAFT COMPANY (USA) (1989-90) 470
MORRISEY COMPANY, THE (USA) (1984-85) 585
MORTENSEN, DAN (see Mong/Mortensen) (1985-86) 602
MOSCOW AVIATION INSTITUTE (USSR) (1985-86) 233
MOSER, E. H., AIRCRAFT COMPANY (USA) (1985-86) 604
MOSHIER TECHNOLOGIES CORPORATION (USA) (1988-89) 445
MOSKOVSKII MASHINOSTROITELNYY ZAVOD IMIENI A. I. MIKOYANA (Russia) (1992-93) 214
MOSKOVSKII MASHINOSTROITELNYY ZAVOD 'SKOROST' IMIENI A. S. YAKOVLEVA (Russia) 331
MOSKOVSKII VERTOLETNAY ZAVOD IMIENI M. L. MILYA (Russia) 289
Mosquito P.86 (Aviolight) (1990-91) 170
Moss Tu-126 (Tupolev) (1990-91) 284
Moujik (Sukhoi) (1988-89) 257
Mountaineer (Ector) (1988-89) 384
MPC AIRCRAFT GmbH (International) (1992-93) 128
Mriya, An-225 (Antonov) 374
Mrówka, PZL-126 (PZL Warszawa-Okecie) 245
MU-2 MODIFICATIONS INC (USA) (1992-93) 424
MUDRY et CIE, AVIONS (France) 86
Multi-Role Fighter (USAF) 579
MUNNINGHOFF, IVAN (USA) (1985-86) 604
Mushshak (PAC) 225
Musketeer II, CT-134A (Beech) (1984-85) 306
MUSTANG AERONAUTICS INC (USA) 539

M – P —AIRCRAFT: INDEXES 709

ustang (Venture Aviation) (1984-85) 611
ustang II M-II (Mustang) 539
YASISHCHEV DESIGN BUREAU (Russia) 301
YERS AVIATION (USA) (1991-92) 455
YLIUS (see Leichtflugzeuge-Entwicklungen Dip Ing Hermann Mylius) (1985-86) 92
ystére-Falcon 10MER (Dassault) (1984-85) 70
ystére-Falcon 20 (Dassault Aviation) (1991-92) 74
ystére-Falcon 20-5 (Dassault-Breguet) (1989-90) 76
ystére-Falcon 20 G (Dassault-Breguet) (1985-86) 69
ystére-Falcon 100 (Dassault Aviation) (1991-92) 75
ystére-Falcon 200 (Dassault Aviation) (1991-92) 74
ystére-Falcon 900 (Dassault Aviation) (1992-93) 64
ystére Falcon 2000 (Dassault-Breguet) (1989-90) 79
ystic, M-17 (Myasishchev) 301

N

-5A (NAMC) 53
22C Nomad (GAF) (1986-87) 6
24 Nomad (GAF) (1985-86) 7
-228 (IPTN) (1989-90) 108
-250 (IPTN) 113
-250-100 (IPTN) 687
-260 (IPTN) (1989-90) 108
-442 (NTT) (1989-90) 138
-821 Corajá (Embraer/Piper) (1985-86) 16
-821 Corajá (NEIVA) (1985-86) 17
-2130 (IPTN) 113
AC (see Norman Aeroplane Company Ltd, The) (1990-91) 317
AC 1 (NAC) (1989-90) 318
AC 6 Fieldmaster (Croplease) 393
AL (see National Aeronautical Laboratory) 111
AL (see National Aerospace Laboratory) 210
AMC (see Nanchang Aircraft Manufacturing Company) 51
AMC/PAC (International) 169
AS-332 Super Puma (IPTN) 113
ASA (see National Aeronautics and Space Administration) 540
ASA Trials Programmes (NASA) 540
ATF, Naval Advanced Tactical Fighter (USN) 580
BELL-412 (IPTN) 113
BK-117 (IPTN) (1991-92) 105
BO 105 (IPTN) 113
C-130 Hercules (Lockheed) 502
C-212-200 Aviocar (IPTN) 113
DN (see NDN Aircraft Ltd) (1984-85) 279
DN 1 Firecracker (NDN) (1983-84) 272
DN 1T Turbo Firecracker (NDN) (1983-84) 272
DN 6 Fieldmaster (NDN) (1984-85) 279
E-821 Carajá (EMBRAER) (1989-90) 15
E-821 Carajá (NEIVA) (1992-93) 16
EIVA (see Industria Aeronáutica Neiva S/A) (1992-93) 16
GT (Next Generation Trainer) (Fairchild) (1987-88) 414
H 90 (NH Industries) 170
IPPI (see Nihon Hikoki Kabushiki Kaisha) 211
SA-300 Puma (Nurtanio) (1984-85) 97
TT (see New Transport Technologies Ltd) (1990-91) 138
V/STOL (Skytrader) (1989-90) 517
WI (see Northwest Industries Limited) 39

amcu (ENAER) 43
ammer (IAI) 184
ANCHANG AIRCRAFT MANUFACTURING COMPANY (China) 51
ARAS AVIATION PVT LTD (India) (1989-90) 107
ASH AIRCRAFT LTD (UK) 396
ATIONAL AERONAUTICAL LABORATORY (India) 111
ATIONAL AERONAUTICS AND SPACE ADMINISTRATION (USA) 540
ATIONAL AEROSPACE LABORATORY (Japan) 210
ational Aero-Space Plane (NASA) 540
ATIONAL AIRCRAFT FACTORIES (China) (1989-90) 34
avajo (EMBRAER/Piper) (1989-90) 15
avajo (Piper) (1983-84) 455
avajo C/R (Piper) (1984-85) 471
aval Advanced Tactical Fighter, NATF (USN) 580
DN AIRCRAFT LTD (UK) (1984-85) 279
ew Brave, PA-36 (WTA/Piper) (1989-90) 528
EWCAL AVIATION INC (USA) 542
EW TECHNIK INC (USA) (1992-93) 426
ew Training Helicopter (US Army) 580
EW TRANSPORT TECHNOLOGIES LTD (International) (1990-91) 138
ew Tupolev Bomber (see Tu-(160?)) (1989-90) 286
ext Generation Trainer (NGT) (Fairchild) (1987-88) 414
GT Turboprop Trainer (Atlas) 339
H INDUSTRIES SARL (International) 170
ICOLLIER, HENRI (France) (1984-85) 539
ight Falcon (General Dynamics) 509
ightfox (McDonnell Douglas) 528
ight Hawk (Sikorsky) (1987-88) 513
ightRanger (Global) (1991-92) 405
ight Stalker (Grumman/Dassault-Breguet) (1989-90) 421
IHON HIKOKI KABUSHIKI KAISHA (Japan) 211
imrod (BAe) (1988-89) 293
imrod AEW Mk 3 (BAe) (1987-88) 310
ite-Writer (Orlando) 545
omad (ASTA) (1988-89) 4
omad N22/24 (GAF) (1985-86) 6
omad N22C (GAF) (1986-87) 6
ong 5 (NAMC) 53
ongye Feiji 5 (NAMC) 53
ORDAM (USA) 542
ORMAN AEROPLANE COMPANY LTD, THE (UK) (1990-91) 317
ORTH AMERICAN AIRCRAFT (USA) 551
ORTH AMERICAN SPACE OPERATIONS (USA) (1989-90) 494
ORTHROP CORPORATION (USA) 542
orthrop TR-3A (USAF) 580
ORTHWEST INDUSTRIES LIMITED (Canada) 39
ovi Avion (VTI) 589
uri upgrade, S-61A (Airod) 214

NUTTING, D. (USA) (1985-86) 605
NYGE-AERO AB (Sweden) (1987-88) 221

O

O-2ST (Robertson/Brico) (1984-85) 485
O2-337 Sentry (Summit) (1989-90) 521
OA7 (Optica) (1986-87) 304
OA7-300 Optica (FLS) 394
OA-46A (Fairchild Republic) (1986-87) 403
OGMA (see Oficinas Gerais de Material Aeronáutico) 247
OH-6D (Kawasaki/McDonnell Douglas) 208
OH-58 (Bell) (1987-88) 365
OH-58D Kiowa (Bell) 436
OH-X (Kawasaki) 208
OPV Series (Sadler) 554
OV-1 (Grumman) (1983-84) 391
OV-1 updates (Grumman) (1990-91) 420
OV-1A modified (PIA) (1988-89) 450
OV-10 (Rockwell) (1990-91) 490
OV-10D Plus Bronco (Rockwell) (1991-92) 468

Oblique-wing demonstrator F-8 (NASA) (1988-89) 447
Observer P.68 (Partenavia) (1991-92) 165
Ocean Hawk SH-60F (Sikorsky) (1992-93) 447
OFICINAS GERAIS DE MATERIAL AERONAUTICO (Portugal) 247
OMAC 1 (Omac) (1983-84) 450
OMAC INC (USA) (1990-91) 473
Omega TB30 (SOCATA) (1989-90) 87
Omega TB31 (SOCATA) 92
OMNIPOL COMPANY LIMITED (Czech Republic) 63
OMNI TITAN CORPORATION (USA) (1987-88) 486
OMNI WELD (USA) (1985-86) 605
One-Eleven (BAe) (1984-85) 264
One-Eleven (Romaero) 253
One-Eleven, Auxiliary Fuel System (Cranfield) (1988-89) 296
One-Eleven, Tay (BAe) (1988-89) 287
One-Eleven 2400 (Dee Howard) (1991-92) 396
One-Eleven 2500 (Dee Howard) (1991-92) 396
O'NEILL AIRPLANE COMPANY (USA) (1987-88) 480
OPERATION ABILITY LTD (UK) (1991-92) 318
Optica (Optica) (1986-87) 304
OPTICA AVIATION (UK) (1986-87) 304
Optica EA7 (Edgley) (1985-86) 289
Optica OA7-300 (FLS) 394
Optica Scout (Brooklands) (1987-88) 311
Optica Scout Mk II (Lovaux) (1991-92) 316, 501
Orao J22 (SOKO/Avioane) 179
ORCA AIRCRAFT LTD (UK) (1990-91) 317
Orel T-602 (ROKS-Aero) 307
Orion (Lockheed) 499
Orion, AA200 (CRSS) 112
Orion air tanker conversion (Aero Union) (1990-91) 339
Orion II P-3H (Lockheed) 697
Orion, Lockheed P-3B, upgrade (Boeing) (1985-86) 364
Orion, operation ability modified (Operation Ability) (1989-90) 316
Orion P-3C (Kawasaki/Lockheed) 207
ORLANDO HELICOPTER AIRWAYS INC (USA) 544
Orlik, PZL-130 (PZL Warszawa-Okecie) (1991-92) 201
Oryx, AS 330 (Atlas) 342
Osprey V-22 (Bell/Boeing) 391, 438
Otter, DHC-3/1000 (Airtech Canada) (1991-92) 17

P

P1 Bandeirante (EMBRAER) (1984-85) 11
P2 Bandeirante (EMBRAER) (1984-85) 12
P2V-5 Tanker conversion (Evergreen) (1987-88) 413
P-3 conversions (Lockheed) (1991-92) 430
P-3 Orion (Lockheed) 499
P3 Rattler (Luscombe) (1985-86) 291
P-3A Aerostar (Aero Union) (1991-92) 338
P-3A Orion airtanker conversion (Aero Union) (1990-91) 339
P3B Viper (Luscombe) (1985-86) 291
P-3B Orion upgrade (Boeing) (1985-86) 364
P-3C Orion (Kawasaki/Lockheed) 207
P-3C update IV (Boeing) (1991-92) 366
P-3H Orion II (Lockheed) 697
P-7A (Lockheed) (1991-92) 426
P-9D (McDonnell Douglas) (1988-89) 431
P-51B Mustang scale replica (Meyer) (1983-84) 567
P.66C-160 Charlie (Partenavia) (1984-85) 153
P.66D Delta (Aviolight) (1991-92) 163
P.66T Charlie Trainer (Partenavia) (1984-85) 153
P.68 Observer (Partenavia) (1991-92) 165
P.68 Observer 2 (Partenavia) 197
P.68C (Partenavia) 197
P.68TC (Partenavia) 197
P.70 Acey Deucy (Powell) (1985-86) 610
P.86 Mosquito (Aviolight) (1990-91) 170
P92 Echo (Tecnam) 202
P.95 (EMBRAER) 14
P120 (CATIC) (1987-88) 38
P120L (Eurocopter/CATIC/SA) (1992-93) 119
P132 (BAe) (1992-93) 296
P.134 (BAe) 378
P.135 (BAe) 378
P.166-DL3-MAR (Piaggio) (1983-84) 152
P.166-DL3SEM (Piaggio) 201
P.180 Avanti (Piaggio) 199
P.210 Centurion Conversion (Air America) (1986-87) 330
P.210 Rocket (Riley) (1991-92) 467
P.210 Turbine (Advanced Aircraft) (1991-92) 336
P-300 Equator (Equator) (1984-85) 85
P-337 Sky Rocket (Cessna/Riley) (1991-92) 467
P-350 Equator (Equator) (1984-85) 85
P-400 Equator (Equator) (1984-85) 85
P-420 Turbo Equator (Equator) (1984-85) 85
P.450 Equator (Equator) (1984-85) 85

P-550 Turbo Equator (Equator) (1984-85) 85
PA-18-150 Super Cub (Piper) (1992-93) 429
PA-18-150 Super Cub (WTA/Piper) (1987-88) 532
PA-24 Comanche (LoPresti Piper) (1989-90) 487
PA-28 Cherokee conversion (Isham/Piper) (1984-85) 426
PA-28-140/-151 (RAM/Piper) (1991-92) 465
PA-28-161 Cadet (Piper) (1990-91) 477
PA-28-161 Warrior II (Piper) (1990-91) 476
PA-28-181 Archer II (Piper) (1990-91) 477
PA-28-236 Dakota (Piper) (1990-91) 479
PA-28R-201 Arrow (Piper) (1990-91) 478
PA-28R-300 Pillan (Indaer Chile/Piper) (1983-84) 32
PA-28R-300XBT (Piper) (1985-86) 480
PA-28RT-201T Turbo Arrow (Piper) (1990-91) 478
PA-28RT-201T Turbo Arrow IV (Piper) (1988-89) 452
PA-31 Navajo (Piper) (1983-84) 455
PA-31-325 Navajo C/R (Piper) (1984-85) 471
PA-31-350 Chieftain (Piper) (1986-87) 481
PA-31-350 T-1020 (Piper) (1985-86) 488
PA-31P-350 Mojave (Piper) (1986-87) 482
PA-31T (Piper-Chincul) (1989-90) 2
PA-31T Cheyenne II (Piper) (1984-85) 474
PA-31T-1 Cheyenne IA (Piper) (1986-87) 483
PA-31T-2 Cheyenne IIXL (Piper) (1986-87) 484
PA-31T-3 T-1040 (Piper) (1985-86) 489
PA-32-301 Saratoga (Piper) (1992-93) 429
PA-32-301T Turbo Saratoga (Piper) (1984-85) 469
PA-32R-301 Saratoga SP (Piper) 546
PA-32R-301T Turbo Saratoga SP (Piper) (1988-89) 454
PA-34-220T Seneca III (Piper) (1990-91) 480
PA-36 New Brave (WTA/Piper) (1989-90) 528
PA-38-112 Tomahawk II (Piper) (1984-85) 477
PA-42-720 Cheyenne IIIA (Piper) 546
PA-42-1000 Cheyenne 400 (Piper) 547
PA-44-180 Seminole (Piper) 546
PA-46-310P Malibu (Piper) (1988-89) 458
PA-46-350P Malibu Mirage (Piper) 548
PA-48 Enforcer (Piper) (1985-86) 488
PA-60 Aerostar 600A (Piper) (1984-85) 478
PA-60-700P Aerostar 700P (Piper) (1985-86) 488
PAC (see Pacific Aerospace Corporation Limited) 221
PAC (see Pakistan Aeronautical Complex) 225
PACE (see Goode, Richard) (1983-84) 534
PACI (see Philippine Aircraft Company Inc) 226
PADC (see Philippine Aerospace Development Corporation) 227
PAH-2 Tiger (Eurocopter) (1991-92) 118
PAH-2/HAC (MBB/Aérospatiale) (1983-84) 110
PAI (see Pacific Aeronautical Inc) 227
PC-6 Turbo-Porter (Pilatus) 358
PC-6 Turbo-Porter Agricultural (Pilatus) (1986-87) 216
PC-7 Turbo-Trainer (Pilatus) 359, 697
PC-7/CH Turbo-Trainer (Pilatus) 359
PC-9 (Pilatus) 360
PC-9 Mk II (Pilatus) 361
PC-9/A Pilatus (HDH) (1992-93) 7
PC-12 (Pilatus) 362, 697
PD 90 Tapete Air Truck (Partenavia) 690
PD. 93 Idea (Partenavia) 689
PDQ (see PDQ Aircraft Products) (1983-84) 574
PFM (Mooney) (1990-91) 467
PIA (see Pittsburgh Institute of Aeronautics) (1988-89) 450
PIC (see Promavia International Corporation) 39
PIK-23 Towmaster (Valmet) (1985-86) 50
PL-3 Guri (ENET) (1989-90) 753
PL-12 550T (Transavia) (1986-87) 8
PL-12 M300 (Transavia) (1985-86) 8
PL-12-U (Transavia) (1991-92) 6
PPA-1, Agricultural aircraft (Emind) (1991-92) 497
PPA-2, Agricultural aircraft (Emind) (1991-92) 497
PR-01 Pond Racer (Scaled Composites) (1992-93) 439
PRC-105 (Lockheed) (1987-88) 445
PROCAER (see Progetti Costruzioni Aeronautiche) (1983-84) 153
PS-5 (HAMC) 48
PT-6 (NAMC) (1989-90) 41
PT-6A (NAMC) 54
PT NURTANIO (see PT Industri Pesawat Terbang Nurtanio) (1985-86) 98
PTS-2000 (Aermacchi/Deutsche Aerospace) 114
PZL (see Zrzeszinie Wytworcow Sprzetu Lotniczego Il Silnikowego PZL) (1991-92) 188
PZL-104 Wilga 35 and 80 (PZL Warszawa-Okecie) 241
PZL-105 Flaming (PZL Warszawa-Okecie) 241
PZL-106AT Turbo-Kruk (WSK PZL-Okecie) (1985-86) 187
PZL-106B Kruk (PZL Warszawa-Okecie) 242
PZL-106BT Turbo-Kruk (PZL Warszawa-Okecie) 243
PZL-107 Kaczor (PZL Warszawa-Okecie) 244
PZL-107 Kawaka (WSK-PZL Warszawa-Okecie) (1983-84) 171
PZL-110 Koliber 150 (PZL Warszawa-Okecie) 244
PZL-111 Koliber 235 (PZL Warszawa-Okecie) 245
PZL-118 Mewa (WSK PZL-Mielec) (1986-87) 184
PZL-126 Mrówka (PZL Warszawa-Okecie) 245
PZL-130 Turbo Orlik (PZL Warszawa-Okecie) (1991-92) 201
PZL-130T Turbo-Orlik (PZL Warszawa-Okecie) 246
PZL-140 Gasior (PZL Warszawa-Okecie) (1990-91) 204
PZL-230F Skorpion (PZL Warszawa-Okecie) 247
PZL SWIDNIK SA (see Zygmunta Pulawskiego-PZL Swidnik) 237
PZL WARSZAWA-OKECIE (see Panstwowe Zaklady Lotnicze Warszawa-Okecie) 240

PACIFIC AERONAUTICAL INC (Philippines) 227
PACIFIC AEROSPACE CORPORATION LIMITED (New Zealand) 221
PAKISTAN AERONAUTICAL COMPLEX (Pakistan) 225
Palomino (Palomino) (1984-85) 587
PALOMINO AIRCRAFT ASSOCIATES (USA) (1985-86) 606
Pamela (Tucker) (1983-84) 590
Pampa, IA 63 (FMA) 3
Pampa 2000 (Vought) 584
PANAVIA AIRCRAFT GmbH (International) 171
Panda (GOHL) (1992-93) 35
PANSTWOWE ZAKLADY LOTNICZE WARSZAWA-OKECIE (Poland) 240
Pantera 50C (ENAER) 43

710 INDEXES: AIRCRAFT—**P - S**

Panther II (Colemill) ... *(1991-92)* 393
Panther 800 (Vought) ... 584
Panther AS565 Helibras (Eurocopter) ... 20
Panther AS565, army/air force versions (Eurocopter) ... 153
Panther AS565, naval versions (Eurocopter) ... 154
Panther Navajo (Colemill) ... *(1991-92)* 393
Panther Navajo/winglets (Colemill) ... *(1985-86)* 394
Panther SA 365K (Aerospatiale) ... *(1989-90)* 66
Panther SA 365M (Aerospatiale) ... *(1988-89)* 64
Paris Jet IV (Jacobs Jets) ... *(1983-84)* 410
Parkot Biplane (Parkot) ... *(1985-86)* 564
PARKOT, ROMAN (Poland) ... *(1985-86)* 564
PARTENAVIA COSTRUZIONI AERONAUTICHE SpA (Italy) ... 197
Pathfinder (Piasecki) ... *(1989-90)* 476
Pathfinder III (Piasecki) ... *(1991-92)* 461
Patroller (Maule) ... *(1985-86)* 448
Paulistinha 56 (CTA) ... *(1989-90)* 10
Pave Hawk (Sikorsky) ... 560
Pawnee conversion (Chincul/Piper) ... *(1983-84)* 3
PDQ AIRCRAFT PRODUCTS (USA) ... *(1983-84)* 574
PEACOCK AIRCRAFT LEASE & OVERHAUL LTD (Canada) ... *(1983-84)* 31
Pegasus, M20L (Mooney) ... *(1987-88)* 476
Pelican, AS-61R (Agusta-Bell) ... 191
Pelican, IAR-831 (ICA) ... *(1990-91)* 210
PEMCO AEROPLEX (USA) ... 545
Peregrine (Gulfstream Aerospace) ... *(1984-85)* 415
Performance 2000 (Cagny) ... *(1989-90)* 67
Petan (McDonnell Douglas) ... 531
Petrel 42/72 (ATR) ... *(1990-91)* 122
Petrel (Nash) ... 396
Petrel CH-148 (EHI) ... 140
PEZETEL FOREIGN TRADE ENTERPRISE LTD (Poland) 228
Pezetel Thrush S2R-R3S (Ayres) ... *(1984-85)* 304
PHALANX ORGANISATION INC (USA) ... *(1988-89)* 450
Phalcon AEW (IAI) ... 184
Phantom, new wing (BAe) ... *(1990-91)* 311
Phantom 2000 (IAI) ... 184
Phantom F-4, modernised (Boeing/McDonnell Douglas) ... *(1986-87)* 380
PHILIPPINE AEROSPACE DEVELOPMENT CORPORATION (Philippines) ... 227
PHILIPPINE AIRCRAFT COMPANY INC (Philippines) ... 226
Phillicopter Mk 1 (Vectol Helicopters) ... 11
Phoenix F-22 (Taylorcraft) ... *(1989-90)* 524
PHÖENIX-AVIATECHNICA (Russia) ... 303
PIAGGIO SpA, INDUSTRIE AERONAUTICHE E MECCANICHE, RINALDO (Italy) ... 199
PIASECKI AIRCRAFT CORPORATION (USA) ... 545
Pijao AC-05 (Aviones de Colombia) ... 61
PILATUS BRITTEN-NORMAN LTD (UK) ... 396
PILATUS FLUGZEUGWERKE AG (Switzerland) ... 358
Pillán (Piper) ... *(1985-86)* 480
Pillán, PA-28R-300 (Indaer. Chile) ... *(1983-84)* 32
Pillán T-35 (ENAER) ... 41
Pillán, T35C (CASA/Enaer) ... *(1985-86)* 202
Pinguino F.22 (General Avia) ... 196, 689
Pinguino-Sprint, F.22/R (General Avia) ... 196
PIPER AIRCRAFT CORPORATION (USA) ... 546
PIPER NORTH CORPORATION (USA) ... *(1991-92)* 464
Piper Suspended Production (Piper) ... 546
Piranha 6 (ALR) ... *(1987-88)* 227
Pirania (IL) ... 228
PITTS AEROBATICS (USA) ... *(1983-84)* 467, 574
PITTSBURGH INSTITUTE OF AERONAUTICS (USA) ... *(1988-89)* 450
PLUMB, BARRY G (UK) ... *(1983-84)* 536
Pober P-9 Pixie (EAA) ... *(1983-84)* 556
Polaris CC-150 (Airbus) ... 117
Pole Star (HAMC) ... 51
Pond Racer (Scaled Composites) ... *(1992-93)* 439
PORSCHE AVIATION PRODUCTS INC (USA) ... *(1988-89)* 459
POWELL, JOHN C. (USA) ... *(1985-86)* 610
Predator, Model 59 (Rutan) ... *(1983-84)* 481
Predator, Model 480 (ATAC) ... *(1986-87)* 335
Premier Aztec SE (Lock Haven) ... 496
President 600 (Colemill) ... *(1991-92)* 394
Pressurised Centurion (Cessna) ... *(1991-92)* 385
Privateer S-4 (Arctic) ... 417
PROCAER (see Progetti Costruzioni Aeronautiche Srl) ... *(1985-86)* 161
PROFESSIONAL AVIATION SERVICES (South Africa) ... 343
PROGETTI COSTRUZIONI AERONAUTICHE Srl (Italy) ... *(1985-86)* 161
'Project Stealth' (Lockheed) ... *(1983-84)* 415
PROMAVIA SA (Belgium) ... 12
PROMAVIA INTERNATIONAL CORPORATION (Canada) ... 39
PROMAVIA/MiG (International) ... 175
Propfan test assessment programme (NASA/Lockheed) ... *(1989-90)* 444
Prop-Jet, XP-99 (Smith) ... *(1985-86)* 518
Prop-jet Bonanza (Allison) ... *(1990-91)* 341
Prop-Jet Bonanza (Tradewind) ... *(1991-92)* 493
PROP-JETS INC (USA) ... *(1986-87)* 488
Prowler EA-6B (Grumman) ... 482
P. T. CIPTA RESTU SARANA SVAHA (Indonesia) ... 112
PT INDUSTRI PESAWAT TERBANG NURTANIO (Indonesia) ... *(1985-86)* 98
Pucará IA58A (FMA) ... 3
Pucará IA58C (FAMA) ... *(1989-90)* 3
Pucará IA66 (FAMA) ... *(1988-89)* 2
Pulsar (Partenavia) ... *(1984-85)* 156
Puma (Nurtanio/Agrospatiale) ... *(1984-85)* 97
Puma AS330 conversions (Atlas) ... 342
Puma, IAR-330L (IAR) ... 252
Puma, Persian Gulf War (Westland) ... *(1991-92)* 334

Q

Q-5 (NAMC) ... 51
Q-5K (NAMC) ... *(1991-92)* 40
Q-5M (NAMC) ... *(1989-90)* 40
Qiang-5 (NAMC) ... 51
Qiangjiji-5 (NAMC) ... 51
Quail (Aerosport) ... *(1983-84)* 539
Quail A9B-M (AAMSA) ... *(1983-84)* 161
QUASAR, P. JEAN and P. MONTEL (France) ... *(1985-86)* 561
Quaser 200 (Quaser) ... *(1985-86)* 561
Queenaire 800/8800 (Excalibur) ... *(1991-92)* 398
QUESTAIR INC (USA) ... 548
Questor (Omni) ... *(1985-86)* 605
Quiet 727 (Valsan) ... *(1991-92)* 495
Quiet Trader, BAe 146 (Avro) ... 136
Quiet Turbofan propeller systems (Raisbeck) ... *(1991-92)* 465

R

R-1 (Rivers) ... *(1985-86)* 615
R-11 (Joses) ... *(1983-84)* 505
R22 Alpha (Robinson) ... *(1985-86)* 493
R22 Beta (Robinson) ... 550
R22HP (Robinson) ... *(1983-84)* 477
R44 (Robinson) ... 550
R90-230RG (Ruschmeyer) ... 104
R92 (Regioliner) ... *(1992-93)* 134
R-95 (EMBRAER) ... *(1991-92)* 10
R122 (Regioliner) ... *(1992-93)* 134
R172K Skyhawk conversions (Isham/Cessna) ... *(1984-85)* 426
R 235 Gabier (SOCATA) ... *(1985-86)* 81
R 235 Guerrier (SOCATA) ... *(1988-89)* 84
R 1180 Aiglon (Robin) ... *(1984-85)* 77
R 2000 Alpha series (Robin) ... *(1984-85)* 77
R 2112 Alpha (Robin) ... *(1983-84)* 76
R 2160 Alpha Sport (Robin) ... *(1983-84)* 76
R 3000 series (Robin) ... 91
RACA (see Representaciones Aero Commerciales Argentinas) SA ... *(1984-85)* 5
RAE (see Royal Aerospace Establishment) ... *(1989-90)* 325
RAF (see Rutan Aircraft Factory) ... *(1985-86)* 612
RAH-66 Comanche (Boeing/Sikorsky) ... 459
RC-130 Hercules (Lockheed) ... 502
RE 2 Doppelraab (Ehmann) ... *(1984-85)* 544
RF-4EJ (Mitsubishi) ... 209
RF-5E Tigereye (Northrop) ... *(1986-87)* 472
RF-19 (Lockheed) ... *(1988-89)* 411
RF-47 (Fournier) ... 686
RFB (see Rhein-Flugzeugbau GmbH) ... 104
RG-8A (Schweizer) ... 555
RH-1100 (Hiller) ... *(1985-86)* 426
RH-1100 (Rogerson Hiller) ... 552
RJ 70 (Avro) ... 136
RJ 85 (Avro) ... 136
RJ 100 (Avro) ... 136
RJ 115 (Avro) ... 136
RJ 120 (Avro) ... *(1992-93)* 710
RJX (Avro) ... 136
RLU (see Roloff, Charles, Robert Liposky & Carl Unger) ... *(1983-84)* 578
RLU-1 Breezy (RLU) ... *(1983-84)* 578
RNZAF (see Royal New Zealand Air Force) ... *(1991-92)* 184
RTAF (SWDC) (see Royal Thai Air Force) ... *(1992-93)* 279
RTAF-4 Chandra (RTAF) ... *(1983-84)* 196
RTAF-5 (RTAF) ... *(1987-88)* 233
RU-21J (Beech) ... *(1985-86)* 330
RV-1 Mohawk (Grumman) ... *(1990-91)* 420
RV-1D (Grumman) ... *(1983-84)* 391
RV-4 (Van's) ... 580
RV-6 (Van's) ... 581
RV-6A (Van's) ... 581
Racer (Statler/Beck) ... *(1983-84)* 584
Rafale (Dassault Aviation) ... 79
RAISBECK ENGINEERING INC (USA) ... 549
Rallye series (Socata) ... *(1985-86)* 81
RAM AIRCRAFT CORPORATION (USA) ... 549
RAM AIRCRAFT MODIFICATIONS INC (USA) ... *(1983-84)* 469
Ram air recovery system (Raisbeck) ... *(1991-92)* 465
Ram-M (USSR) ... *(1989-90)* 295
Ranger (Grob) ... *(1985-86)* 88
Ranger (Luscombe) ... *(1986-87)* 299
Ranger (UAT) ... *(1988-89)* 311
Ranger 2000 (Rockwell International/DASA) ... 177
Ranger Model AA-23A1 (Taylorcraft) ... *(1991-92)* 482
RANKIN AIRCRAFT (USA) ... *(1987-88)* 493
RANS INC (USA) ... 549
Rat'ler (Helio) ... *(1987-88)* 435
Rattler, P3 (Luscombe) ... *(1985-86)* 291
Raven (Grumman/General Dynamics) ... *(1989-90)* 421
Raven, EF-111A, update (Grumman) ... *(1991-92)* 410
RAY'S AVIATION (USA) ... *(1985-86)* 615
RAYTHEON COMPANY (USA) ... 698
Redigo (Valmet) ... 73
Regent 1500 (Advanced Aircraft) ... *(1991-92)* 336
Régent, DR 400/180 (Robin) ... 90
Regianne 2000, scale (Tesori) ... *(1983-84)* 508
REGIOLINER GmbH (International) ... *(1992-93)* 134
Regional Jet (Canadair) ... 28
Regional Jetliner (Avro) ... 135
REIMS AVIATION SA (France) ... 88
Remo 180, DR 400 (Robin) ... 91
Remorqueur, DR 400/180R (Robin) ... *(1988-89)* 81
Renegade 1 (Lasher) ... *(1983-84)* 566
Renegade LA-250 (Lake) ... 491
REPRESENTACIONES AEROCOMMERCIALES ARGENTINAS SA (Argentina) ... *(1984-85)* 5
Retro, GB-2 (Bezzola) ... *(1983-84)* 531
Reusable Space Plane (USSR) ... *(1989-90)* 295
RHEIN-FLUGZEUGBAU GmbH (Germany) ... 104
Rigel AA300 (CRSS) ... 112
RILEY AIRCRAFT MANUFACTURING (USA) ... *(1983-84)* 470
RILEY INTERNATIONAL CORPORATION (USA) ... 549
RIO CLARO, AERO CLUBE DE (Brazil) ... *(1986-87)* 17
RIVERS (see STONE, C. RIVERS) ... *(1985-86)* 615
ROBERTSON AIRCRAFT CORPORATION (USA) . *(1984-85)* 484
Robin 200 (Robin) ... 89
ROBIN, AVIONS PIERRE (France) ... 89
ROBINSON HELICOPTER COMPANY INC (USA) ... 550
Rocket 340 (Riley) ... *(1983-84)* 471
Rocket P-210 (Riley) ... *(1991-92)* 467
Rocket Power 414 (Riley) ... *(1983-84)* 471
ROCKWELL INTERNATIONAL CORPORATION (USA) . 551
ROCKWELL INTERNATIONAL/ DEUTSCHE AEROSPACE (International) ... 176
ROCKY MOUNTAIN HELICOPTERS INC (USA) ... 552
ROGERSON HILLER CORPORATION (USA) ... 552
ROKS-AERO CORPORATION (Russia) ... 303
ROLOFF, CHARLES, ROBERT LIPOSKY & CARL UNGER (USA) ... *(1983-84)* 578
ROMAERO SA (Romania) ... 253
Rombac 1-11 (Romaero) ... 253
Rooivalk, CSH-2 (Atlas) ... 340
Rooivalk XH-2 (Atlas) ... *(1990-91)* 213
Rotor Buggy, EBI (Boyette) ... *(1983-84)* 544
ROTORWAY INTERNATIONAL (USA) ... 553
ROYAL AEROSPACE ESTABLISHMENT (UK) ... *(1989-90)* 325
ROYAL NEW ZEALAND AIR FORCE (New Zealand) ... *(1991-92)* 184
ROYAL THAI AIR FORCE (Thailand) ... *(1992-93)* 279
R/STOL SYSTEMS (USA) ... *(1986-87)* 499
RUSCHMEYER LUFTFAHRTTECHNIK (Germany) ... 104
RUSJET/GENERAL TECHNOLOGIES/AEROSUD (International) ... *(1992-93)* 136
RUSSELL, Michael (USA) ... *(1985-86)* 616
Rutan 68 (Ams/Oil) ... *(1984-85)* 559
RUTAN AIRCRAFT FACTORY INC (USA) ... *(1985-86)* 503, 612

S

S-1 Pitts Special (Christen) ... *(1987-88)* 407
S1 Solar Wind (Venture) ... *(1985-86)* 542
S1B2 Arctic Tern (Arctic Aircraft) ... 416
S-1D Pitts Special (Christen) ... *(1987-88)* 407
S-1E Special (Pitts) ... *(1983-84)* 574
S-1T Pitts Special (Aviat) ... 418
S-2A Pitts Special (Christen) ... *(1987-88)* 407
S-2B Pitts (Aviat) ... 418
S-2E (Pitts) ... *(1983-84)* 574
S-2E Tracker upgrade (IMP) ... 38
S-2K (Sikorsky) ... 571
S2R Turbo-Thrush (Ayres) ... 421
S2R-R1340, Thrush (Ayres) ... 420
S2R-R1820/510 Thrush (Ayres) ... 421
S2R-T Turbo Thrush (Marsh) ... *(1991-92)* 432
S-2S Pitts (Aviat) ... 418
S-2T Turbo Tracker (Grumman) ... *(1991-92)* 410
S-2T Turbo Tracker (Marsh) ... *(1991-92)* 432
S-2UP Tracker (IAI) ... 689
S-3 Viking (Lockheed) ... 502
S-4 Privateer (Arctic) ... 417
S-7 Courier (Rans) ... 549
S10gsm (Gepard) ... *(1991-92)* 222
S-18 (Sport Aircraft) ... 572
S-21 (Sukhoi) ... 316
S-21-G supersonic business jet (Sukhoi/Gulfstream) ... *(1992-93)* 139
S-51 (Sukhoi) ... 317
S-52 civil conversion (Orlando) ... *(1988-89)* 449
S-52-3 (VAT) ... *(1990-91)* 525
S-54 (Sukhoi) ... 317
S-55/H-19 (Orlando) ... *(1991-92)* 459
S-58/H-34 (Orlando) ... *(1991-92)* 460
S-58T (California Helicopter) ... *(1991-92)* 384
S-61 (Mitsubishi/Sikorsky) ... *(1991-92)* 174
S-61A Nuri Upgrade (Airod) ... 214
S-61A/B (Sikorsky) ... *(1983-84)* 485
S-61R (Agusta-Sikorsky) ... *(1986-87)* 152
S-61R (Sikorsky) ... *(1983-84)* 485
S-61/SH-3 (Sikorsky) ... *(1991-92)* 480
S-65/MH-53J Pave Low III Enhanced (Sikorsky) ... *(1991-92)* 480
S-69 (Sikorsky) ... *(1983-84)* 486
S-70A (Sikorsky) ... 560
S-70B (Sikorsky) ... 565
S-70C (Sikorsky) ... 567
S-72XI X-Wing project (Sikorsky) ... *(1990-91)* 516
S-75 ACAP (Sikorsky/US Army) ... *(1986-87)* 512
S-76 (Sikorsky) ... 568
S-76 Shadow (Sikorsky) ... *(1991-92)* 489
S-76A+ (Sikorsky) ... *(1991-92)* 487
S-76B (Sikorsky) ... 568
S-76C (Sikorsky) ... 569
S-80 (Sukhoi) ... 317
S-80E (Sikorsky) ... 560
S-80/H-53E (Sikorsky) ... 559
S-80M (Sikorsky) ... 560
S-84 (Sukhoi) ... 318
S-86 (Sukhoi) ... 318
S-92 (Sikorsky) ... 570
S-99 (Sukhoi) ... *(1992-93)* 246
S-185 (Schapel) ... *(1986-87)* 505
S.211 (Agusta) ... 193
S.226 (SIAI-Marchetti) ... *(1983-84)* 168
S312 Tucano (Shorts) ... 399
S-350 Mini-Swat (Schapel) ... *(1983-84)* 481
S-525 Super Swat (Schapel) ... *(1983-84)* 481
S550 Citation S/11 (Cessna) ... 465
S.700 Cormorano (SIAI/Marchetti) ... *(1984-85)* 151
S-1080 Thunderbolt (Schapel) ... *(1983-84)* 481
S-1275 Finesse (Schapel) ... *(1983-84)* 482
SA (see Singapore Aerospace Ltd) ... 338
S. A. III Swalesong (Coates) ... *(1985-86)* 568
SA 2-37A (Schweizer) ... 555
SA-3-120 Bulldog Series 120 (BAe) ... *(1984-85)* 264
SA30 Fanjet (Swearingen) ... *(1988-89)* 491
SA-30 Gulfjet (Grumman Aerospace) ... *(1989-90)* 424
SA-32T (Jaffe) ... *(1991-92)* 413
SA-204C (Snow Aviation) ... 571

S—AIRCRAFT: INDEXES 711

SA-210AT (Snow Aviation) 571
SA227 Metro 23 (Fairchild) 475
SA227-AC Metro III (Fairchild Aircraft) (1991-92) 398
SA227-AT Merlin IVC (Fairchild Aircraft) (1991-92) 400
SA227-DC Metro 23 (Fairchild Aircraft) 475
SA227-TT Merlin IIIC (Fairchild) (1983-84) 379
SA227 TT/41 (Fairchild Aircraft) (1988-89) 388
SA-228-AE (Fairchild Aircraft) (1990-91) 410
SA 315B Lama (HAL) 109
SA 315B Lama (HAL/Aerospatiale) (1987-88) 106
SA 316B Alouette III (Aerospatiale) (1990-91) 59
SA 316B Alouette III (HAL/Aerospatiale) 109
SA 319B Alouette III (Aerospatiale) (1990-91) 59
SA 321 Super Frelon (Aerospatiale) (1983-84) 52
SA 330 Puma (Nurtanio/Aerospatiale) (1983-84) 97
SA 341 Gazelle (Aerospatiale/Westland) (1984-85) 54
SA 342 Gazelle (Eurocopter) 149
SA 360 Dauphin (Aerospatiale) (1984-85) 57
SA 360 C Dauphin (Aerospatiale) (1984-85) 57
SA 365F Dauphin 2 (Aerospatiale) (1989-90) 64
SA 365K Panther (Aerospatiale) (1989-90) 66
SA 365M Dauphin 2 (Aerospatiale) (1985-86) 57
SA 365M Panther (Aerospatiale) (1988-89) 64
SA 365N Dauphin 2 (Aerospatiale) (1988-89) 63
SA 365N1 Dauphin 2 (Aerospatiale) (1989-90) 65
SA 365S Dauphin 2 (Aerospatiale) (1989-90) 66
SA 366 Dauphin 2 (Aerospatiale) (1989-90) 66
SA-550 Spectrum-One (Spectrum) (1984-85) 493
SA-882 Flying Wing (Schapel) (1986-87) 505
SA-981 Swat (Schapel) (1983-84) 481
SAAC (see Sammi Agusta Aerospace Corporation) (1992-93) 165
SAAF (see South African Air Force) (1992-93) 265
SABCA (see Société Anonyme Belge de Constructions Aéronautiques) 13
SAC (see Shaanxi Aircraft Corporation) 54
SAC (see Shenyang Aircraft Company) 56
SAC (see Spectrum Aircraft Corporation) (1984-85) 493
SACAB (see Scandinavian Aircraft Construction AB (Sweden) (1988-89) 212
SADLIER (see VTOL Industries Australia Ltd) (1989-90) 6
SAH-1 (Orca) (1990-91) 317
SAI (see Singapore Aircraft Industries) (1988-89) 203
SAIC (see Shanghai Aviation Industrial Corporation) (1988-89) 43
SAMF (see Shanghai Aircraft Manufacturing Factory) 57
SAP (see Shijiazhuang Aircraft Plant) 57
SATIC (see Special Aircraft Transport International Company) 178
SB-4 (Seabird) (1990-91) 6
SB5 Sentinel (Seabird) (1991-92) 6
SB7L Seeker (Seabird) 10
SC.7 Skyvan Series 3 and 3M (Shorts) (1986-87) 312
SC-95 (EMBRAER) (1991-92) 10
SC 01B Speed Canard (FFT) (1992-93) 79
SCTICSG (see Supersonic Commercial Transport International Co-operation Study Group) 178, 688
SD27 (Sivel) 201
SE-86 (CAI) (1987-88) 217
SECA (see Société d'Exploitation et de Constructions Aeronautiques) (1991-92) 84
SEPECAT (see Société Europeenne de Production de l'Avion E. C. A. T.) 179
SF-2A Cygnet (Sisler) (1983-84) 580
SF 37 Viggen (Saab) (1984-85) 196
SF.260 (Agusta) 191
SF.260TP (Agusta) 192
SF 340A (Saab-Scania) (1986-87) 208
SF.600TP Canguro (PADC) 194
SF.600TP Canguro (SIAI-Marchetti) (1990-91) 167
SGAC (see Shenzhen General Aviation Company) 57
SH-2 Seasprite (Kaman) 488
SH-3 Sea King (Sikorsky) (1983-84) 485
SH-3D (Agusta-Sikorsky) (1984-85) 145
SH-3D/TS (Agusta-Sikorsky) 191
SH-3H (Agusta-Sikorsky) 191
SH-5 (HAMC) 48
SH 37 Viggen (Saab) (1984-85) 196
SH-60B Sea Hawk (Sikorsky) 563
SH-60F Ocean Hawk (Sikorsky) 565
SH-60J (Mitsubishi/Sikorsky) 210
SJ30 (Swearingen) 574
SK 37 Viggen (Saab) (1984-85) 196
SKT-1250, Holiday Knight Twister (Payne) (1985-86) 608
SL (Myasishchev) (1992-93) 234
SL 90 Leshii (Aviotechnica) 134
SOCATA (see Société de Construction d'Avions de Tourisme et d'Affaires) 92
SOGEPA (see Societé de Gestion de Participations Aeronautiques) 95
SOKO (see Vazduhoplovna Industrija Soko) 13
SONACA (see Société Nationale de Construction Aérospatiale SA) 13
SP-2H Firestar tanker conversion (Aero Union) (1991-92) 338
SR-71 (Lockheed) (1992-93) 402
SRA-4 (Gulfstream Aerospace) 486
SS-2A (Shin Meiwa) 212
SSA (see Samsung Aerospace) 214
SSBJ (Sukhoi/Gulfstream) (1992-93) 139
SSF (McDonnell Douglas/BAe) 167
SST (Grumman Aerospace) (1989-90) 425
SST-X001 (JADC) 206
ST-17 (CAT) (1991-92) 384
ST-50 (Cirrus) 470
ST-220 (Akasamitra) (1985-86) 98
ST1 700 Conestoga (Skytrader) (1990-91) 516
ST1 700 MD Evader (Skytrader) (1990-91) 516
STAF (see Shaanxi Transport Aircraft Factory) (1988-89) 43
STARPAC (see Star of Phoenix Aircraft) 572
STOL CH 701 (Zenair) 41
STOL CH 701-AG Zenair (Zenair) 41
STOVL Prototype (Sadleir) 9
STOVL Strike Fighter (McDonnell Douglas) 169
STOVL technology development programmes (NASA) (1990-91) 470
Su-7B (Sukhoi) (1988-89) 257
Su-9 (Sukhoi) (1983-84) 230
Su-11 (Sukhoi) (1983-84) 230
Su-15 (Sukhoi) (1992-93) 236
Su-17 (Sukhoi) 308
Su-20 (Sukhoi) 308
Su-21 (Sukhoi) (1987-88) 274
Su-22 (Sukhoi) 308
Su-24 (Sukhoi) 309
Su-25 (Sukhoi) 310
Su-25T (Su-34) (Sukhoi) 311
Su-26 (Sukhoi) (1985-86) 253
Su-26M (Sukhoi) 312
Su-27 (Sukhoi) 313
Su-28 (Sukhoi) 310
Su-29 (Sukhoi) 315
Su-30 (Sukhoi) 693
Su-31 (Sukhoi) (1992-93) 243
Su-31T/U (Sukhoi) 315
Su-32 (Sukhoi) 694
Su-33 (Sukhoi) 316
Su-34 (Sukhoi) 311
Su-35 (Sukhoi) 316
Su-37 (Sukhoi) (1992-93) 244
SW-4 (PZL Swidnik) 240
SW 70 (Sikorsky-Westland) (1987-88) 338
SAAB/FAIRCHILD AIRCRAFT (International) (1985-86) 124
SAAB-SCANIA AB 351
Saab 35 Draken (Saab-Scania) (1991-92) 220
Saab 114 (Saab) (1983-84) 190
Saab 340A (Saab) (1989-90) 220
Saab 340B (Saab) 352
Saab 2000 (Saab) 354
Saab 2110 Gripen (Saab) (1983-84) 189
Sabre 2 (CAC) (1988-89) 36
Sabre Super Searcher (Sabreliner) (1987-88) 501
SABRELINER CORPORATION (USA) 553
SABRELINER DIVISION, ROCKWELL INTERNATIONAL (USA) (1983-84) 480
SADLEIR (VTOL Industries) (Australia) 9
SADLER AIRCRAFT CORPORATION (USA) 553
Safari An-28 (PZL Mielec) (1992-93) 178
Safety and performance conversions (Sierra) (1991-92) 478
Salon, Mi-8 (Mil) (1983-84) 223
SAMMI AGUSTA AEROSPACE CORPORATION LTD (South Korea) (1992-93) 165
SAMSUNG AEROSPACE (South Korea) 214
Samurai 350 (Merlyn) (1988-89) 442
SANDBERG, JOHN (USA) (1984-85) 596
SANDERS AIRCRAFT INC (USA) (1984-85) 493
Sanger (DASA) 99
Saratoga (Piper) 546
Saratoga SP (Piper) 546
SAO CARLOS (see Escola De Engenharia de Sao Carlos) (1986-87) 17
SARCUP programme (Boeing Canada/CAF) (1985-86) 19
Scale Regianne 2000 (Tesori) (1983-84) 508
SCALED COMPOSITES INC (USA) 554
Scamp Model B (Agricopteros/Aerosport) (1984-85) 42
SCANDINAVIAN AIRCRAFT CONSTRUCTION AB (Sweden) (1988-89) 212
Scanliner (HB-Aircraft) (1991-92) 7
Scanliner, HB-Aircraft (CAI) (1991-92) 48
SCENIC AIRLINES INC (USA) (1987-88) 502
SCHAFER AIRCRAFT MODIFICATIONS INC (USA) 555
SCHAPEL AIRCRAFT COMPANY (USA) (1986-87) 505
SCHOENENBERG, HEINRICH (West Germany) (1984-85) 544
SCHWEIZER AIRCRAFT CORPORATION (USA) 555
Scorpion 133 (RotorWay) (1984-85) 596
Scout (ACAC) 416
Scout, 8GCBC (Champion) (1985-86) 392
Scout A (Skytrader) (1988-89) 488
Scout Defender (Hughes) (1985-86) 461
Scoutmaster (Lovaux) (1991-92) 316
Scout-STOL (Skytrader) (1988-89) 487
Sea Bird (SEFA Asia) 227
SEABIRD AVIATION AUSTRALIA PTY LTD (Australia) 10
SeaCobra (Bell) (1990-91) 361
Sea Dragon MH-53E (Sikorsky) 559
Seafire, TA16 (Thurston) 575
Sea Harrier (BAe) 389
Sea Harrier FRS. Mks 1 and 2 (BAe) 389
Sea Harrier FRS. Mk 51 (BAe) 389
Sea Hawk (Aero Gare) (1984-85) 558
Seahawk SH-60B (Sikorsky) 565
Sea Hawker (Leg Air) (1985-86) 599
SEAIR PACIFIC PTY LTD (Australia) (1991-92) 6
Sea King (Sikorsky) (1991-92) 480
Sea King (Westland) 403
Sea King conversion (1) (IMP) (1991-92) 28
Sea King conversion (2) 38
Sea Knight (Boeing) (1991-92) 366
Seamaster TA-19 (Thurston) 576
SEAPLANES INC (USA) (1985-86) 507
SeaRanger, TH-57 (Bell) (1987-88) 365
Searchmaster (GAF) (1985-86) 7
Sea Scan (IAI) (1986-87) 136
Seasprite (Flight Dynamics) (1984-85) 574
Seasprite (Kaman) 488
Seastar (Ingenieurburo Dornier) (1983-84) 88
Seastar CD2 (Dornier Composite) (1991-92) 94
Sea Thrush (Speciality Aircraft Sales) (1985-86) 30
Sea Warrior (Agusta) 192
Seawolf (Lake) 491
Seeker SB7L (Seabird) 10
SEFA ASIA INC (Philippines) 227
SEGUIN AVIATION INC (USA) (1988-89) 473
SELCHER, JOHN L. (USA) (1985-86) 616
SELLET/PELLETIER HELICOPTERE (France) (1991-92) 85
Seminole (Piper) 546
Semurg (MAI) (1984-85) 222
Seneca III (EMBRAER/Piper) (1992-93) 16
Seneca III, PA-34-220T (Piper) (1990-91) 480
Senior Aero Sport (D'Apuzzo) (1983-84) 552
Sentinel, F-27 (Fokker) (1985-86) 171
Sentinel, F28F-P (Enstrom) 472
Sentinel, SB5 (Seabird) (1991-92) 6
Sentry 02-337 (Summit) (1989-90) 521
Sentry, E-3 (Boeing) 452
Sentry E-3D, AEW Mk I (BAe) (1992-93) 303
Sentry E-3D, AEW Mk I (Boeing) 452
SEQUOIA AIRCRAFT CORPORATION (USA) 558
Sequoia 300 (Sequoia) 558
Sertanejo (EMBRAER/Piper) (1985-86) 16
SERV-AERO ENGINEERING INC (USA) (1984-85) 497
SEYEDO SHOHADA PROJECT, DEFENCE INDUSTRIES (Iran) 181
SHAANXI AIRCRAFT COMPANY (China) 54
SHAANXI TRANSPORT FACTORY (China) (1988-89) 43
Shadow (Sikorsky) (1991-92) 489
Shadow II (CFM) 392
Shadow Series C (CFM) 391
Shadow Series C-D (CFM) 391
Shahal (IAI) 182
Shahbaaz (PAC) 225
Shama (Wells) (1985-86) 634
Shamsher (HAL) 108
SHANGHAI AIRCRAFT MANUFACTURING FACTORY (China) 57
SHANGHAI AVIATION INDUSTRIAL CORPORATION (China) (1988-89) 43
Shark (Enstrom) (1985-86) 396
Shennong-1 (CATIC) (1988-89) 34
SHENYANG AIRCRAFT CORPORATION (China) 56
Shenyang JJ-5 (China) (1983-84) 33
Shenyang/Tianjin JJ-6 (State Aircraft Factories) (1984-85) 35
SHENZHEN GENERAL AVIATION COMPANY LTD (China) 57
Sherpa (Shorts) 399
Sheriff (Aircraft Designs/Britten) (1983-84) 248
SHESTAVIN, STANISLAV (USSR) (1984-85) 550
SHIJIAZHUANG AIRCRAFT PLANT (China) 57
SHINMAYWA INDUSTRIES LTD (Japan) 211
SHIN MEIWA INDUSTRY CO LTD (Japan) (1992-93) 163
Shoestring (Condor) (1983-84) 550
SHORT BROTHERS PLC (UK) 399
Shorts 330 (Shorts) (1992-93) 311
Shorts 330-UTT (Shorts) (1992-93) 312
Shorts 360-300 (Shorts) (1991-92) 323
Shorts 360-300F (Shorts) (1991-92) 323
Short-field enhancement system (Raisbeck) (1991-92) 465
SHOWA HIKOKI KOGYO KABUSHIKI KAISHA (Japan) (1983-84) 160
Shui Hong (China) (1983-84) 38
Shuihong-5 (HAMC) 48
Shuishang Hongzhaji 5 (HAMC) 48
Shuttle Spacecraft (Rockwell International) 434
SIAI-MARCHETTI SpA (Italy) (1989-90) 166
Sierra 200 (Beech) (1984-85) 306
SIERRA INDUSTRIES (USA) 559
Sierra safety and performance conversions for Beech, Cessna and Piper aircraft. Tables (1990-91) 502
SIKORSKY AIRCRAFT, DIVISION OF UNITED TECHNOLOGIES CORPORATION (USA) 559
Silhouette (Lunds Tekniske) 224
Silver (Agusta-Sikorsky) (1990-91) 163
SINGAPORE AEROSPACE LTD (Singapore) 338
SINGAPORE AIRCRAFT INDUSTRIES PTE LTD (Singapore) (1988-89) 203
SingleTwin (Aerofan) (1986-87) 329
Sirocco (Jurca) 86
SISLER AIRCRAFT COMPANY (USA) (1983-84) 580
SIVEL srl (Italy) 201
Skif (Myasishchev) 302
Skipper, Model 77 (Beech) (1984-85) 306
Skorpion PZL-230 (PZL Warszawa-Okecie) 247
Skybolt F-7P (CAC) (1990-91) 33
Sky Dancer (Bond) (1984-85) 562
Skyfarmer (Transavia) (1992-93) 8
Skyfox (Boeing) (1987-88) 393
Skyfox (PAC) 226
Skyfox (Skyfox Aviation) 10
SKYFOX AVIATION (Australia) 10
SKYFOX CORPORATION (USA) (1985-86) 517
Skyhawk (Cessna) (1991-92) 385
Skyhawk/100, F172 (Reims/Cessna) (1986-87) 78
Skyhawk A-4PTM (Grumman/McDonnell Douglas) (1986-87) 424
Skyhawk R172K conversions (Isham/Cessna) (1984-85) 426
Skyhawk upgrade (IAI) 184
Skyhawk upgrade (RNZAF) (1991-92) 184
Skyhook (The Australian Autogyro Co) 5
Skyhook Mini Chopper (Minty) (1983-84) 9
Sky Knight (Schweizer) 557
Skylane (Cessna) (1991-92) 385
Skylane RG (Cessna) (1991-92) 385
Skyrocket (Bellanca) (1985-86) 348
Sky Rocket P-337 (Riley) (1991-92) 467
Skyservant, Do 28 D2 (Dornier) (1987-88) 93
SKYSTAR AIRCRAFT CORPORATION (USA) 571
SKYTRADER CORPORATION (USA) (1990-91) 516
Skyvan SC.7 Series 3 (1987-88) 325
Skyvan SC.7 Series 3M (Shorts) (1989-90) 307
Skywagon (Cessna) (1991-92) 385
SLINGSBY AVIATION LIMITED (UK) 401
SMITH AERO INC, MIKE (USA) (1988-89) 488
SNIPE AIRCRAFT DEVELOPMENTS LTD (UK) (1985-86) 305
SNOW AVIATION INTERNATIONAL INC (USA) 571
SOCATA/MOONEY (International) (1987-88) 140
SOCIÉTÉ ANONYME BELGE DE CONSTRUCTIONS AÉRONAUTIQUES (Belgium) 13
SOCIÉTÉ DE CONSTRUCTION D'AVIONS DE TOURISME ET D'AFFAIRES (France) 92
SOCIÉTÉ D'EXPLOITATION ET DE CONSTRUCTIONS AÉRONAUTIQUES (France) (1991-92) 84
SOCIETE DE GESTION DE PARTICIPATIONS AERONAUTIQUES (France) 95
SOCIÉTÉ EUROPÉENNE DE PRODUCTION DE L'AVION E. C. A. T. (International) 179

712 INDEXES: AIRCRAFT—**S – T**

SOCIÉTÉ NATIONALE DE CONSTRUCTION AEROSPATIALE SA (Belgium) 13
Şoim, IAR-99 (Avioane) 250
SOKO/AVIOANE (International) 179
SOKO/CNIAR (International) (1990-91) 143
SOKO/IAv CRAIOVA (International) (1991-92) 141
Sokól T-401 (ROKS-Aero) 305
Sokól W-3 (PZL Swidnik) 238
Solar Wind (Venture) (1985-86) 542
Solitaire (MAI) (1985-86) 445
SOLOY CORPORATION (USA) 572
Sonerai II (Great Plains) 479
Sonerai IIL (Great Plains) 479
Sonerai II-LT (Great Plains) 479
Sonerai II-LTS (Great Plains) 479
Sootless exhaust stack fairings (Raisbeck) (1991-92) 465
SOUR VAZDUHOPLOVNA INDUSTRIJA SOKO (Yugoslavia) (1988-89) 498
SOUTH AFRICAN AIR FORCE (South Africa) (1992-93) 265
SOUTHERN CLASSIC AEROPLANES (USA) (1985-86) 620
Southern Classic Biplane (Southern Classic) (1985-86) 620
SOVIET AEROSPACE INDUSTRIES ASSOCIATION (Russia) (1992-93) 235
Spaceplane studies (NAL) 210
Space Shuttle (USSR) (1991-92) 297
Space transportation system (Rockwell) (1992-93) 434
Sparrow (Nyge-Aero) (1986-87) 208
Sparrow Hawk (Aero Dynamics) (1985-86) 574
Spartacus, AP 68TP-300 (Partenavia) (1988-89) 162
Spartacus, Maritime Patrol Version (Partenavia) (1984-85) 156
Spartacus Series 300 (Partenavia) (1984-85) 155
Spartan, C-27A (Alenia) 195, 469
Sparviero (Macera) (1984-85) 546
Special (Denight) (1984-85) 571
SPECIAL AIRCRAFT TRANSPORT INTERNATIONAL COMPANY (International) 178
Special Mission Aircraft (Fairchild) 477
Special S-1 (Pitts) (1983-84) 467
Special S-1, Pitts (Christen) (1987-88) 407
Special, S-1D/-1E (Pitts) (1983-84) 574
Special S-1T Pitts (Aviat) 418
Special S-2A, Pitts (Christen) (1987-88) 407
Special, S-2E/-2S (Pitts) (1983-84) 574
SPECIALTY AIRCRAFT SALES (Canada) (1986-87) 30
Spectre, AC-130U (Rockwell/Lockheed) 551
SPECTRUM AIRCRAFT CORPORATION (USA) (1984-85) 493
Spectrum-One (Spectrum) (1984-85) 493
Speed Canard SC OIB (FFT) (1992-93) 79
SPEZIO (USA) (1983-84) 583
Spiral project, 50-50 (USSR) (1991-92) 298
Spirit (Pace) (1983-84) 534
Spirit (Questair) 548
Spirit 750 (Advanced Aircraft) (1985-86) 316
SPITFIRE HELICOPTER COMPANY LTD (USA) (1986-87) 519
SPITFIRE HELICOPTERS INTERNATIONAL (Spain) (1985-86) 202
Spitfire Mark I (Spitfire) (1984-85) 508
Spitfire Mark II (Spitfire) (1984-85) 508
Spitfire Mark IV (Spitfire) (1983-84) 494
SPORT AIRCRAFT INC (USA) 572
Sportsman VJ-22 (Volmer) 583
Sportster, HA-2M (Hollmann) (1985-86) 595
Sport Trainer (Great Lakes) (1987-88) 424
Sport Trainer (Wag-Aero) 584
SPRATT & COMPANY INC (USA) (1983-84) 583
Sprint (FLS) 395
Spur (Operation Ability) (1989-90) 319
Squalo, F.1300 (General Avia) (1985-86) 156
Stainless exhaust stack fairings (Raisbeck) (1990-91) 486
Stalker (Stoddard-Hamilton) (1990-91) 518
Stallion (Helio) (1991-92) 413
Starcraft (Maule) (1991-92) 434
Starfighter (Aeritalia/Lockheed) (1984-85) 134
Starfire Bonanza (Colemill) (1991-92) 394
StarLifter, C-141 (Lockheed) (1983-84) 423
Starling, BA-14 (FFV Aerotech) (1990-91) 218
STAR OF PHOENIX AIRCRAFT (USA) 572
Star Rocket (Maule) 514
Starship 1 (Beech) 431
Starship 2000 (Beech) 431
Starship 2000A (Beech) 431
STATE AIRCRAFT FACTORIES (China) (1986-87) 33
Statesman (BAe) (1992-93) 296
Stationair 6 (Cessna) (1990-91) 394
Stationair 8 (Cessna) (1984-85) 365
Stealth Fighter (Lockheed) (1990-91) 432
Stellux 3esse3 Trenzo (Stellux) (1988-89) 165
STELLUX AIRCRAFT CORPORATION (Italy) (1989-90) 177
STEWART AIRCRAFT CORPORATION (USA) (1985-86) 623
Stiletto (MFI) 351
Stiletto T-9 (Terzi) 202
Stiletto, T-9R (Terzi Aerodin) (1992-93) 156
Stinson 108-3 (Univair) (1987-88) 528
STODDARD-HAMILTON AIRCRAFT INC (USA) 573
STONE, C. RIVERS (USA) (1985-86) 615
STOVL Strike Fighter (USN) 580
STRATEGIC BUSINESS UNIT HELICOPTERS (Germany) (1991-92) 91
STRATEGIC BUSINESS UNIT MILITARY AIRCRAFT (Germany) (1991-92) 93
STRATEGIC BUSINESS UNIT REGIONAL AIRCRAFT (Germany) (1991-92) 88
STRATEGIC BUSINESS UNIT SPACE TRANSPORTATION AND PROPULSION SYSTEMS (Germany) (1991-92) 94
Strategic Reconnaissance Aircraft (USAF) 579
Strato I (Egrett) 138
Strato II (Egrett) (1991-92) 121
Strato 2C (Grob) 103, 686
Stratocruzer 1250-ER (AASI) 411
Stratofortress, B-52 (Boeing) (1991-92) 371
Stratosfera M-17 (Myasishchev) 301
Stratotanker, KC-135A (Boeing) (1991-92) 371
Streak Shadow (CFM) 392
Stretch 580 (KFC) (1991-92) 28
Stretched Gazelle (Aérospatiale) (1984-85) 54
Strikemaster, total sales (BAe) (1989-90) 312
SUKHOI DESIGN BUREAU AVIATION SCIENTIFIC-INDUSTRIAL COMPLEX (Russia) 308
SUKHOI DESIGN BUREAU (USSR) (1991-92) 275
SUKHOI/GULFSTREAM (International) (1992-93) 139
SUMMIT AVIATION INC (USA) (1989-90) 521
Sunbird (De Vore) (1986-87) 399
Sunbird (Verilite) (1990-91) 525
Sundowner 180 (Beech) (1984-85) 306
Sup' Air (Centrair) (1987-88) 70
Super2 (Island Aircraft) (1990-91) 315
Super2 ARV-1 (Aviation Scotland) 376
Super-7 (CAC) 46
Super 71, 72, 73 (McDonnell Douglas) (1986-87) 453
Super 310 (Riley) (1990-91) 488
Super 340 (Riley) (1990-91) 488
Super 580 (Super 580 Aircraft) 573
SUPER 580 AIRCRAFT COMPANY (USA) 573
Super 580 Falcon (Super 580 Aircraft) (1989-90) 522
Super 580 ST (Super 580 Aircraft) (1990-91) 519
Super 748 (BAe) (1989-90) 301
Super Ag Max (Aero Mod) (1987-88) 342
Super Beaver (Wipaire) (1991-92) 496
Super Bug (Bensen) (1983-84) 542
SuperCobra (Bell) 436
Super Courier (Helio) (1991-92) 413
Super Cub PA-18-150 (Piper) 546
Super Cub, PA-18-150 (WTA/Piper) (1987-88) 532
Super Dakota C-47TP (SAAF) (1992-93) 265
Super Decathlon (ACAC) 415
Super Decathlon (Champion) (1985-86) 391
Super Etendard (Dassault Aviation) (1991-92) 74
Super Frelon (Aérospatiale) (1983-84) 52
Super Galeb (UTVA) 587
Super Gyro-Copter (Bensen) (1985-86) 578
Super Hornet (McDonnell Douglas) (1987-88) 460
Super Huey (Bell) (1991-92) 358
Super JetRanger (Global) (1989-90) 415
Super King Air B200 (Beech) 427
Super King Air 200/B200 US Military versions (Beech) 428
Super King Air 300 (Beech) (1991-92) 353
Super King Air 300 LW (Beech) 429
Super King Air 350 (Beech) 429
Super Lynx (Westland) 409
Super Mirage 4000 (Dassault-Breguet) (1986-87) 68
Super Mountaineer (Ector) (1985-86) 396
Super Mustang T-100 (Avstar) (1990-91) 344
Super Phantom (IAI) 184
Super Puma AS 332 (IPTN) (1989-90) 109
Super Puma AS 332 (Eurocopter) 146
Super Puma Mk II AS 332 L2 (Eurocopter) 148
Super Puma, NAS-332 (IPTN) 113
Super Q (Comtran) (1991-92) 395
Super Ranger (Luscombe) (1986-87) 300
Super Ranger (UAT) (1988-89) 312
Super Rocket (Maule) 514
SUPER ROTOR INDUSTRIA AERONAUTICA LTDA, M. M. (Brazil) (1992-93) 16
Super Seasprite (Kaman) 488
Super Skyhawk A-4S-1 (SA) 338
Supersonic business jet (Sukhoi/Gulfstream) (1992-93) 139
SUPERSONIC COMMERCIAL TRANSPORT INTERNATIONAL CO-OPERATION STUDY GROUP (International) 178, 688
Supersonic STOVL fighter (USN) (1992-93) 457
Super Stallion CH-53E (Sikorsky) 559
Superstar I (Machen) (1985-86) 444
Superstar II (Machen) (1985-86) 444
Superstar III (Machen) (1985-86) 445
Superstar 650 (Machen) (1991-92) 431
Superstar 680 (Machen) (1991-92) 431
Superstar 700 (Machen) (1991-92) 431
Superstar F-22 (Lockheed) (1992-93) 396
Super Surubim, IPAI-28 (Rio Claro) (1984-85) 18
Super Swat S-525 (Schapel) (1983-84) 481
SuperTransport, Model 214ST (Bell) (1990-91) 364
Super Transporter (Airbus) 178
Super Tucano (Embraer) 18
Surveiller (Boeing) (1987-88) 376
Sutlej, An-32 (Antonov) 367
Swalesong (Coates) (1985-86) 568
Swat, SA-981 (Schapel) (1983-84) 481
Swati, LT-1 (BHEL) 106
SWEARINGEN AIRCRAFT INC (USA) 573
Swift, IAR-109 (Avioane) 251
SwiftFire (LoPresti Piper) (1990-91) 486
Swiftfury (LoPresti) (1992-93) 404
SwiftFury (LoPresti Piper) (1990-91) 485
SWISS FEDERAL AIRCRAFT FACTORY (F + W) (Switzerland) 363
Swiss Trainer MD3 (MDB) 357

T

T1 Flamingo-Trainer (FUS) (1983-84) 86
T-1A Jayhawk (Beech) 432
T-2 (Mitsubishi) (1989-90) 183
T-2CCV (Mitsubishi) (1986-87) 171
T-4 (Kawasaki) 207
T-5 (Fuji) 203
T-9 (Stoddard-Hamilton) (1990-91) 518
T-9 Stiletto (Terzi Aerodin) 202
T-9R Stiletto (Terzi Aerodin) (1992-93) 156
T.16 (Dassault Aviation) 82
T.17 (Boeing) (1991-92) 372
T.18 (Dassault Aviation) 83
T-18 Tiger (Thorp) (1985-86) 627
T-27 (EMBRAER) 17
T-30 Katana (Glasfaser) (1990-91) 171
T30 Katana (Terzi) 203
T-33V (Volpar) 584
T-34 Turbo Mentor (Marsh) (1983-84) 429
T-34C (Beech) (1989-90) 349
T-34C, Laminar Flow Jetstar (NASA) (1986-87) 469
T-35 Pillán (ENAER) 41
T-35C (CASA Enaer) (1985-86) 200
T-35DT Turbo Pillán (ENAER) 42
T-35T (ENAER) (1988-89) 33
T-35TX (ENAER Chile) (1987-88) 36
T-36 Halcón (CASA) 347
T-36 Halcón (ENAER) 43
T-43A (Boeing) (1990-91) 372
T-45 (McDonnell Douglas/BAe) (1984-85) 116, 450
T45 Turbine Dromader (Melex) (1991-92) 451
T45 Turbine Dromader (WSK-PZL Mielec) 233
T-45A Goshawk (McDonnall Douglas/BAe) 168, 385
T-45TS (McDonnell Douglas/BAe) (1987-88) 132
T-46A (Fairchild) (1987-88) 414
T-47A Citation S/II (Cessna) 465
T67 Firefly (Slingsby) 401
T67M Firefly 160 (Slingsby) (1989-90) 330
T67M Mk II Firefly (Slingsby) 401
T67M200 Firefly (Slingsby) 402
T67M260 Firefly (Slingsby) 402
T68 (Slingsby) (1983-84) 282
T69 (Slingsby) (1983-84) 282
T-100 Super Mustang (Avstar) (1990-91) 344
T-101 Gratch (ROKS-Aero) 303
T-106 (ROKS-Aero) 304
T206/210 (RAM/Cessna) (1991-92) 465
T207 (RAM/Cessna) (1985-86) 491
T211 (Adams/Thorp) (1983-84) 292
T-211 (Thorp) (1992-93) 454
T-300A Skyfarmer (Transavia) (1992-93) 8
T303 Crusader (Cessna) (1985-86) 382
T310 (RAM/Cessna) (1991-92) 465
T-400 Beechjet 400A (Beech) 432
T-400 Skyfarmer (Transavia) (1992-93) 8
T-401 Sokol (ROKS-Aero) 305
T-407 (ROKS-Aero) 305
T-433 Flamingo (ROKS-Aero) 306
T-501 (ROKS-Aero) 306
T-602 Orel (ROKS-Aero) 307
T-610 Voyage (ROKS-Aero) 307
T-1020 (Piper) (1985-86) 488
T-1040 (Piper) (1985-86) 489
TA-7 Corsair II, Vought (LTV) (1989-90) 445
TA16 Seafire (Thurston) 575
TA19 Seamaster (Thurston) 576
TAAL (see Taneja Aerospace and Aviation) 111
TAC (see Taiwan Aerospace Corporation) 365
TACAMO II, E-6 (Boeing) 454
TAH-1F HueyCobra (Bell) 434
TAI (see Tusas Aerospace Industries Inc) 365
TAI (see Tusas Havacilik ve Uzay Sanayii As) (1987-88) 234
TAP (see Thompson Aero Products) (1989-90) 523
TAV-8A (BAe) (1989-90) 308
TAV-8B (McDonnell Douglas/BAe) 164
TAV-8S Harrier (BAe) (1991-92) 311
TB 9 Tampico (SOCATA) 93
TB 10 Tobago (SOCATA) 93
TB 11 Tobago (SOCATA) (1983-84) 78
TB 16 (Socata) (1985-86) 83
TB 20/21 Trinidad (SOCATA) 93
TB 30 Epsilon (SOCATA) (1990-91) 87
TB 30 Omega (SOCATA) (1989-90) 87
TB 31 Omega (SOCATA) 92
TB 200 Tobago XL (SOCATA) 93
TBM 700 (SOCATA) 94
TC-130 Hercules (Lockheed) 502
TCA, NATO (Dornier) (1989-90) 94
T-CH-1 (AIDC) (1988-89) 217
TCM (see Teledyne Continental Motors) 575
TD Technology Demonstrator (IAI) 182
TE-2C Hawkeye (Grumman) 480
TF-25 (Robertson) (1984-85) 486
TG-01 (Tsilefski) (1983-84) 526
TG-10 (Venga) (1989-90) 30
TG 180 Piper Cherokee Chief (Ray's) (1985-86) 615
TH-1S HueyCobra (Bell) 434
TH-28 (Enstrom) 473
TH-50 Esquilo (Eurocopter) 150
TH-57 SeaRanger (Bell) (1987-88) 365
TH-67 Creek (Bell) 21
THvK (1984-85) 205
TKEF (see The King's Engineering Fellowship) 576
TLS (Mooney) 539
TM (see TM Aircraft) (1985-86) 628
TM-5 (TM) (1985-86) 628
TM 20B Flamingo III (Bengis) (1992-93) 264
TNT (New Technology Wing) testbed aircraft (Dornier) (1983-84) 82
Tp 84 Hercules (Lockheed) 502
Tp 88 (Fairchild) 475
Tp 89 (CASA) (1991-92) 213
Tp 100 (Saab) 352
Tp 101 Super King Air B200 (Beech) 427
Tp 102 SRA-4 (Gulfstream Aerospace) 486
TP-400 (Piper) (1989-90) 486
TP-600 (Piper) (1989-90) 486
TR-1 (Lockheed) (1989-90) 436
TR-3A Northrop (USAF) 580
TRANSAVIA (see Transfield Construction Pty Ltd) 231
TS-2F Turbo Tracker (Marsh) 513
TS-11 Iskra-bis (WSK-PZL Mielec) (1988-89) 184
TS-11R Iskra (PZL Mielec) 231
TSC-1A3 Teal III (Thurston) 575
Tsu-Chiang, AT-3 (AIDC) (1991-92) 227
TSz-1 (Shestavin) (1984-85) 550
TT30 (Westland) (1987-88) 337
TT300 (Westland) (1987-88) 337
TTH 90 (NH 90) 170
TTTS (Tanker/Transport Training System (USAF) (1990-91) 524
Tu-16 (Tupolev) 319
Tu-22 (Tupolev) 323
Tu-22M (Tupolev) 324
Tu-28P (Tupolev) (1989-90) 287
Tu-95 (Tupolev) 320
Tu-126 (Tupolev) (1990-91) 284

T – V—AIRCRAFT: INDEXES 713

Tu-128 (Tupolev) (1989-90) 287
Tu-134 (Tupolev) (1984-85) 248
Tu-134BSh (Tupolev) 327
Tu-134UBL (Tupolev) 327
Tu-142 (Tupolev) 320
Tu-144 (Tupolev) (1983-84) 241
Tu-144D (Tupolev) (1983-84) 142
Tu-154 (Tupolev) (1988-89) 270
Tu-154A/B (Tupolev) (1985-86) 263
Tu-154C (Tupolev) (1990-91) 288
Tu-154M (Tupolev) 327
Tu-154S (Tupolev) 327
Tu-155 (Tupolev) (1990-91) 288
Tu-160 (Tupolev) 326
Tu-(160?) (Tupolev) (1989-90) 286
Tu-164 (Tupolev) (1984-85) 250
Tu-204 (Tupolev) 328
Tu-330 (Tupolev) 694
Tu-334 (Tupolev) 330
TUPOLEV (see Aviatsionnyi Nauchno-Tekhnicheskiy Kompleks Imieni A. N. Tupoleva) 319
TUSAS (see Turk Ucak Sanayii Anonim Ortakligi) .. (1985-86) 213
TW-68 (Ishida) 205, 690
TX-1 (Mooney) (1985-86) 468

Tacamo II, E-6A (Boeing) 454
Taifun 11S (Valentin) (1988-89) 96
Taifun 12E (Valentin) (1987-88) 103
TAGANROG AVIATSIONNYI NAUCHNO-TEKHNICHESKIY KOMPLEKS IMIENI G. M. BERIEVA (Russia) 255
'Taildragger' conversions, Bolen (Bush) (1991-92) 384
TAIWAN AEROSPACE CORPORATION (Taiwan) 365
Tamiz, E.26 (CASA) (1989-90) 219
Tamiz, E.26 (ENAER) 41
Tampico Club TB 9 (SOCATA) 93
TANEJA AEROSPACE AND AVIATION LTD (India) 111
Tangará, A-132 (Aerotec) (1983-84) 10
Tangará II (Aerotec) (1988-89) 9
Tanker/Transport Training System (USAF) (1990-91) 524
TASHKENT AIRCRAFT PRODUCTION CORPORATION (CHKALOV) (Russia) 586
Taurus (Spitfire/WSK-PZL Swidnik) (1984-85) 508
Taurus (WSK-PZL Swidnik) (1985-86) 184
Taurus modification (Swearingen) (1988-89) 490
TAYLOR, MRS. JOHN F. (UK) (1985-86) 570
TAYLORCRAFT AIRCRAFT (USA) 574
T-Bird II (Lockheed/Aermacchi/Rolls-Royce) 186, 499
TBM SA (International) (1991-92) 143
Tchaika (Beriev) (1991-92) 239
T-Craft (Vector) 582
Teal III, TSC-1A3 (Thurston) 575
Tech 2 (New Technik) (1991-92) 456
Tech 4 (New Technik) (1991-92) 456
TECNAM srl (Italy) 202
TEHNOIMPORTEXPORT FOREIGN TRADE COMPANY (Romania) (1990-91) 207
TELEDYNE CONTINENTAL MOTORS (USA) 575
TELEDYNE RYAN AERONAUTICAL (USA) (1991-92) 493
TERZI AERODIN (Itay) 202
TESORI, ROBERT (Canada) (1983-84) 508
Texas Gem (Miller) (1985-86) 601
TEXAS HELICOPTER CORPORATION (USA) (1984-85) 510
Texas Ranger, Model 206L (Bell) (1984-85) 326
THE KING'S ENGINEERING FELLOWSHIP (USA) 576
Theta AA330 (CRSS) 112
THOMPSON AERO PRODUCTS (USA) (1989-90) 523
THORP T211 AIRCRAFT INC (USA) (1988-89) 493
THORP AERO INC (USA) (1992-93) 454
THORP ENGINEERING COMPANY (USA) (1985-86) 627
Thorpe T211 (Intran) (1984-85) 425
Thrush, Ayres conversion (Aero Mod) (1987-88) 342
Thrush Commander/R-1820 (Serv-Aero) (1984-85) 497
Thrush modification (Serv-Aero/Ayres) (1984-85) 497
Thrush S2R-600 (Ayres) (1990-91) 345
Thrush S2R-R1340 (Ayres) 420
Thrush S2R-R1820/510 (Ayres) 421
Thunderbird 211 (Mod Squad) (1988-89) 442
Thunderbolt II (Fairchild Republic) (1987-88) 414
Thunderbolt, S-1080 (Schapel) (1983-84) 481
THURSTON AEROMARINE CORPORATION (USA) 575
Tiger (Eurocopter) 156
Tiger (Thorp) (1985-86) 627
Tiger II (KA) (1988-89) 174
Tiger II (Northrop) (1992-93) 426
Tiger AG-5B (American General) 416
Tiger, AS332 (Aerospatiale) (1991-92) 59
Tigereye, RF-5E (Northrop) (1986-87) 472
Tigershark, F-20 (Northrop) (1986-87) 472
Tigre (Eurocopter) 156
Tilt-rotor (Agusta) (1987-88) 161
Tilt-rotor (Sikorsky) (1991-92) 452
Tilt-rotor aircraft (USSR) (1988-89) 277
Tira Prosa, IPAI-29 (Sao Carlos) (1983-84) 19
TM AIRCRAFT (USA) (1985-86) 628
Tobago, TB 10 (SOCATA) 93
Tobago, TB 11 (SOCATA) (1983-84) 78
Tobago XL, TB 200 (SOCATA) 93
Tomahawk II, PA-38-112 (Piper) (1984-85) 477
Tomahawk SP (Aero Boero/Piper) (1985-86) 1
Tomcat F-14 (Grumman) 483
Tonopah Low (Forbes) (1983-84) 558
Tornado 2000 (Panavia) (1992-93) 133
Tornado ADV (Panavia) 174, 688
Tornado ECR (Panavia) 173, 688
Tornado F. Mk 2, 2A and 3 (Panavia) 174
Tornado GR. Mk 1, 1A and 4 (Panavia) 171, 688
Tornado IDS (Panavia) 171
Tornado IDS Mid-Life improvement programme (Panavia) (1990-91) 140
Tornado, MY 102 (Mylius) (1985-86) 92
Towmaster PIK-23 (Valmet) (1985-86) 50
Tracker S-2E upgrade (IMP) 38
Tracker upgrading (IAI) 689
TRADEWIND TURBINES CORPORATION (USA) 577
TRAGO MILLS LTD (UK) (1988-89) 311
Trainer (Aero Boero) 1
Trainer (Skyfox) (1985-86) 517
Trainer Cargo Aircraft, NATO (Dornier) (1989-90) 94
Transall C-160 (International) (1991-92) 144
TRANSALL, ARBEITSGEMEINSCHAFT (International) (1989-90) 146
TRANSFIELD CONSTRUCTION PTY LTD, TRANSAVIA DIVISION (Australia) 11
Transporter, DHC-5E (Canada) (1984-85) 24
Transregional 250 (CAT) (1991-92) 384
Trenzo (Stellux) (1989-90) 177
Tri-Commutair, BN-2A Mk III (IAC/Pilatus Britten-Norman) (1984-85) 424
TriCoupé, JC-2D (Coupé Aviation) (1983-84) 514
TRIDAIR HELICOPTERS INC (USA) 577
Trinidad TB20/21 (SOCATA) 93
TriStar cargo conversions (Pemco) 545
TriStar conversions (Lockheed) (1991-92) 430
TriStar modification, civil (Marshall) (1991-92) 317
TriStar tanker conversion (Marshall) (1991-92) 317
Tri-Tail Bonanza (Smith) (1987-88) 522
Triumf, IAR-825TP (ICA) (1990-91) 210
Triumph (Scaled Composites/Rutan) (1991-92) 473
Tri-Z CH-300 (Zenair) 40
TSILEFSKI, G (France) (1983-84) 526
Tsunami (Sandberg) (1984-85) 596
Tuca, IPAI-26 (Sao Carlos) (1984-85) 19
Tucano EMB-312 (EMBRAER) 17
Tucano EMB-312H (EMBRAER) 18
Tucano T. Mk I (Shorts) 399
TUCKER, L. G. (USA) (1983-84) 590
Tupi, EMB-712 (EMBRAER/Piper) (1988-89) 15
TUPOLEV (Russia) 319
TUPOLEV DESIGN BUREAU (USSR) (1991-92) 284, 756
Turbine 207 (MacAvia) (1991-92) 431
Turbine 404 (Riley) (1983-84) 471
Turbine Bonanza (Allison) (1988-89) 324
Turbine Chancellor (Riley) (1984-85) 484
Turbine Defender (Pilatus Britten-Norman) (1992-93) 309
Turbine Defender, Special Role (Pilatus Britten-Norman) (1992-93) 310
Turbine Dromader (Melex) (1991-92) 451
Turbine Dromader (Melex) 536
Turbine Dromader T45 (WSK-PZL Mielec) 233
Turbine Eagle 421 (Riley) (1990-91) 489
Turbine Islander BN-2T (Pilatus Britten-Norman) 397
Turbine Islander, Special Role (Pilatus Britten-Norman) (1992-93) 310
Turbine Malibu (Piper) (1989-90) 486
Turbine Malibu (Riley) (1984-85) 484
Turbine Mentor (Allison) (1990-91) 341
Turbine Mentor 34C (Beech) (1989-90) 349
Turbine P-210 (Advanced Aircraft) (1991-92) 336
Turbine P-210 (Riley) (1983-84) 470
Turbine Pac conversions (Soloy) 572
Turbine Rocket 421 (Riley) (1988-89) 462
Turbo-67 DC-3 conversion (Basler) (1991-92) 346
Turbo 270 Renegade (Lake) 491
Turbo 350 Bonanza (Machen) (1985-86) 445
Turbo Air Tractor (Air Tractor) 413
Turbo-Albatros (Frakes) (1987-88) 417
Turbo Arrow (Piper) (1990-91) 478
Turbo Arrow IV (Piper) (1988-89) 452
Turbo-Beaver, DHC-2 Mk III (de Havilland Canada) (1984-85) 24
Turbo Bonanza, B36 TC (Beech) 424
Turbo-Canso (Avalon) (1989-90) 18
Turbo Cat, G-162 C-T (Marsh) (1991-92) 432
Turbo Centurion (Cessna) (1991-92) 385
Turbo Equator (Equator) (1985-86) 87
Turbo Firecat (Conair) 33
Turbo Flagship ATF 580S (Allison) (1987-88) 346
Turbo-Kruk PZL-106BT (PZL Warszawa-Okecie) 243
Turbo Mentor, T-34 (Marsh) (1983-84) 429
Turbo Mooney 231 (Mooney) (1984-85) 457
Turbo Orlik (Airtech) (1988-89) 18
Turbo Orlik (PZL Warszawa-Okecie) 246
Turbo Otter (Cox) (1983-84) 372
Turbo-Panda, Y-12 (China) (1984-85) 41
Turbo Pillán (Enaer Chile) (1985-86) 32
Turbo Pillán, T-35FDT (ENAER) 42
Turbo-Porter PC-6 (Pilatus) 358
Turbo-Porter Agricultural (Pilatus) (1986-87) 216
Turbo S-2 conversion (Marsh/Grumman) (1990-91) 444
Turbo Saratoga (Piper) (1984-85) 469
Turbo Saratoga SP (Piper) (1988-89) 454
Turbo Sea Thrush, Terr-Mar (Ayres) (1991-92) 345
Turbo Skylane (Cessna) (1984-85) 361
Turbo Skylane RG (Cessna) (1991-92) 385
Turbo-Stationair 6 (Cessna) (1990-91) 394
Turbo-Stationair 8 (Cessna) (1984-85) 365
TURBOTECH INC (USA) 578
Turbo-Thrush S-2R (Ayres) 421
Turbo Thrush, S2R-T (Marsh) (1991-92) 432
Turbo Titan (OMNI) (1987-88) 480
Turbo Tracker S-2T(Grumman) (1991-92) 410
Turbo Tracker S-2T (Marsh) (1991-92) 432
Turbo Tracker TS-2F (Marsh) 513
Turbo Trainer, AS 32T (FFA) (1983-84) 192
Turbo-Trainer, PC-7 (Pilatus) 359
Turbo Trainer, SA-32T (Jaffe) (1991-92) 413
Turbolet (LET) (1989-90) 50
TURK HAVA KUVVETLERI, KAYSERI HAVA IKMAL BAKIM MERKEZI KOMUTANLIGI (Turkey) (1984-85) 205
TURK UCcAK SANAYII ANONIM ORTAKLIGI (Turkey) (1985-86) 213
TURKISH AIR FORCE (Turkey) (1984-85) 205
TUSAŞ AEROSPACE INDUSTRIES INC (Turkey) 365
TUSAS HAVACILIK VE UZAY SANAYII AS (Turkey) (1987-88) 234
TWIN COMMANDER AIRCRAFT CORPORATION (USA) 578
Twin Equator (Equator) (1985-86) 87
Twin Otter Series 300 (Boeing Canada) (1989-90) 21
TwinRanger (Bell) (1985-86) 340
TwinRanger (Bell Helicopter Textron) (1988-89) 20
TwinRanger (Luscombe) (1986-87) 300
Twin Ranger (UAT) (1988-89) 312
TwinRanger, 206LT (Bell) 22
Twin Ranger, new (Bell) (1992-93) 344
TwinStar (Eurocopter) 151
Twin Star 1000 (Airmaster) (1988-89) 322
Twin-T (Vector) 582
Twin Tech (New Technik) (1991-92) 456
Twin Two-Twelve (Bell) (1987-88) 370
Twin Two-Twelve (Bell) 22
Type 10A (Jones) (1983-84) 563
Type 30 (Westland) (1987-88) 337
Type 75 (UTVA) (1987-88) 534
Type 78 (UTVA) (1985-86) 529
Type 101 (Tupolev) (1985-86) 262
Type 128-2 (Dornier) (1983-84) 81
Type 128-6 (Dornier) (1987-88) 93
Type 231 (Mooney) (1985-86) 468
Type 300 (Fairchild) (1985-86) 402
Type 301 (Mooney) (1985-86) 469
Type 328 (Dornier) (1987-88) 95
Type 340 (Saab-Fairchild) (1985-86) 124
Type 360 Advanced (Shorts) (1987-88) 324
Type 400 (Fairchild) (1984-85) 394
Type 440 (Saab-Fairchild) (1985-86) 125
Type 460P (Wren) (1988-89) 496
Type 526 AFM Condor (Aerotechnik/Zlin) (1984-85) 44
Type 731 Hawker (AiResearch) (1986-87) 330
Type 731 Hawker (Garrett) (1987-88) 417
Type 748(M) (HAL/BAe) (1984-85) 94
Tzu-Chung AT-3 (AIDC) 364
Tzukit (IAI) (1991-92) 149

U

U-2 (Lockheed) (1983-84) 415
U-2 (NASA/Lockheed) (1983-84) 445
U-2R (Lockheed) (1989-90) 436
U-26A (Cessna) (1990-91) 394
U-27A Caravan (Cessna) 463
U-36A Learjet (Gates Learjet) (1986-87) 407
U-125 (BAe) 204, 377
UA-5 (Honda) 488
UAT (see United Aerospace Technologies Ltd) (1988-89) 311
UDRD (see Universal Dynamics Research and Development) 227
UH-1 (Bell) (1991-92) 358
UH-1D/H/V (Bell) (1990-91) 360
UH-1H (Fuji-Bell) 204
UH-1HP (Global) (1992-93) 382
UH-1HP Huey II (Bell) 434
UH-1N (Bell) 22
UH-2 Seasprite (Kaman) (1984-85) 427
UH-12 (Hiller) (1985-86) 426
UH-12 Esquilo (Eurocopter) 150
UH-12B Esquilo (Eurocopter) 151
UH-12E Hauler (Rogerson Hiller) 552
UH-46 Sea Knight (Boeing) (1991-92) 366
UH-60A Black Hawk (Sikorsky) 560
UH-60A Pave Hawk (Sikorsky) 560
UH-60J (Mitsubishi/Sikorsky) 210
UH-60L Black Hawk (Sikorsky) 365, 560
UH-60Q Black Hawk (Sikorsky) 560
UHB MD-90 Programme (McDonnell Douglas) (1989-90) 456
UHCA (Airbus) (1992-93) 97
UH-X (US Army) 580
US-1A (Shin Meiwa) 211
USAC (see United States Aircraft Corporation) (1987-88) 528
USAF (see United States Air Force Materiel Command) 579
US ARMY (see United States Army Aviation Systems Command) 580
USN (see United States Naval Air Systems Command) 580
UV-18A (Boeing Canada) (1988-89) 21
UV-18B (Boeing Canada) (1988-89) 21
UV-20A Chiricahua (Pilatus) 358
UV-23 (Skytrader) (1989-90) 517

Ultimate EA 230 (ACS) (1985-86) 205
Ultra High Capacity Aircraft (Airbus) 126
Ultra Huey (UNC) 578
Ultra Imp (Aerocar) (1983-84) 539
Ultralite Test Vehicle (Flight Dynamics) (1985-86) 403
UNC HELICOPTER (USA) 578
UNITED AEROSPACE TECHNOLOGIES LTD (UK) (1988-89) 311
UNITED STATES AIRCRAFT CORPORATION (USA) (1987-88) 528
UNITED STATES AIR FORCE MATERIEL COMMAND (USA) 579
UNITED STATES ARMY AVIATION SYSTEMS COMMAND (USA) 580
UNITED STATES NAVAL AIR SYSTEMS COMMAND (USA) 580
UNIVAIR AIRCRAFT CORPORATION (USA) (1988-89) 494
UNIVERSAL DYNAMICS RESEARCH AND DEVELOPMENT (Philippines) 227
Up-gun AH-1S (Bell) (1987-88) 369
Urpi, IAP-003 (Indaer Peru) (1992-93) 174
UTVA-SOUR METALNE INDUSTRIJE, RO FABRIKA AVIONA (Yugoslavia) 586
UTVA-75A (Utva) (1991-92) 500
UTVA-75AG11 (Utva) (1991-92) 501
UTVA-75A21 (Utva) (1990-91) 529
UTVA-78 (Utva) (1985-86) 529

V

V-1-A Vigilante (Ayres) (1991-92) 345
V-2 (Mil) 289
V-8 (Mil) 289
V-8 Special (Beachner) (1985-86) 576

714 INDEXES: AIRCRAFT—V - Z

V-14 (Mil) 291
V-22 Osprey (BAe) 391
V-22 Osprey (Bell/Boeing) 438
V35B Bonanza (Beech) (1986-87) 338
V-62 (Kamov) (1990-91) 254
V2000 Starfire (Vulcan) (1990-91) 526
VA.1 Matador (BAe) (1991-92) 311
VA.2 Matador II (McDonnell Douglas/BAe) 164
VAAC Harrier (Cranfield) (1987-88) 314
VAT (see Vertical Aviation Technologies) 581
VC-9C (McDonnell Douglas) (1985-86) 454
VC10 Tankers (BAe) (1992-93) 303
VC-25A (Boeing) (1991-92) 372
VC-97 (EMBRAER) 14
VC-130H Hercules (Lockheed) 502
VC-137 (Boeing) 441
VFW 614 ATTAS (MBB) (1989-90) 101
VH-3 (Sikorsky) (1983-84) 485
VH-55 Esquilo (Eurocopter) 151
VH-60N (Sikorsky) 560
VI-SU Gulfstream (Sukhoi/Gulfstream) (1990-91) 145
VJ-22 Sportsman (Volmer) 583
V-Jet (Williams) (1986-87) 525
VK30 Cirrus (Cirrus) 470
VLA-1 (Nyge-Aero) (1986-87) 208
VM-T Atlant (Myasishchev) (1991-92) 273
VTI (Vazduhoploxno Tehnicki Institut) 589
VU-9 (EMBRAER) (1984-85) 15
VZLU (see Vyzkumny a Zkusebni Letecky Ustav) (1983-84) 46

Vajra (Dassault Aviation) 75
VALENTIN FLUGZEUGBAU GmbH (West Germany) (1988-89) 96
Valiant (Luscombe) (1985-86) 290
VALMET AVIATION INDUSTRIES (Finland) 73
VALSAN INC (USA) 580
VAN'S AIRCRAFT INC (USA) 580
VARDEX CORPORATION (USA) 581
VARGA AIRCRAFT CORPORATION (USA) (1984-85) 511
VariEze, RAF Model 31/33 (Rutan) (1985-86) 612
VariViggen, RAF Model 27/32 (Rutan) (1985-86) 612
Vazar Dash 3 (Vardex) (1991-92) 496
VAZDUHOPLOVNA INDUSTRIJA SOKO DD (Yugoslavia) 13
VAZDUHOPLOVNO TECHNICKI INSTITUT (Yugoslavia) 589
Vector, CBA-123 (EMBRAER/FMA) 143
VECTOR AIRCRAFT COMPANY (USA) 582
VEETOL HELICOPTERS PTY LTD (Australia) 11
Veltro 2 (Aermacchi) (1987-88) 156
VENGA AIRCRAFT INC (Canada) (1990-91) 29
Venom Mk 2 (Wallis) (1987-88) 331
Ventura, C22J (Augusta Caproni Vizzola) (1990-91) 168
VENTURE AVIATION INC (USA) (1984-85) 611
VERILITE AIRCRAFT CO (USA) (1990-91) 525
VERTICAL AVIATION TECHNOLOGIES (USA) 581
VERTOLETYI NAUCHNO-TEKHNICHESKIY KOMPLEKS IMIENI N. I. KAMOVA (Russia) 271
Very Large Commercial Transport (Airbus) 126
Viator AP.68TP-600 (Partenavia) 198
Victor (Partenavia) (1984-85) 153
Viggen, JA37 (Saab-Scania) (1991-92) 219
Viggen AJS 37 (Saab) 351
Vigilante V-1-A (Ayres) (1991-92) 345
Viking (Bellanca) (1986-87) 363
Viking (Lockheed) 502
VIKING AIRCRAFT LTD (USA) 583
Vindicator (Wallis) (1987-88) 331
Vinka (Valmet) (1989-90) 57
VINTEN LTD, W (UK) (1986-87) 316
VINTEN/WALLIS (see W. Vinten Ltd) (1983-84) 282
Vipan MFI-10C (RFB) 687
Viper AH-1(4B)W (Bell) (1992-93) 343
Viper, P3B (Luscombe) (1985-86) 291
Viper I (Thunder Wings) (1984-85) 608
Viscount-life extension (BAF) 391
Vistaliner (Scenic) (1987-88) 502
Vistaplane (Orlando) 545
VIVIAN ASSOCIATES LIMITED, LR (Canada) (1992-93) 29
Volantor (Moller) (1991-92) 452
VOLMER AIRCRAFT (USA) 583
VOLPAR AIRCRAFT CORPORATION (USA) 583
VOUGHT AIRCRAFT COMPANY (USA) 584
VOUGHT CORPORATION (USA) (1983-84) 498
Voyage T-610 (ROKS-Aero) 307
Voyager (Voyager) (1987-88) 529
VOYAGER AIRCRAFT INC (USA) (1987-88) 529
Voyageur, CH-113A (Boeing Canada) (1985-86) 19
VTOL AIRCRAFT PTY LTD (Australia) (1991-92) 7
VTOL airliner studies (NAL) 211
VTOL INDUSTRIES AUSTRALIA LTD (Australia) 9
VULCAN AIRCRAFT CORPORATION (USA) (1990-91) 526
VYZKUMNY A ZKUSEBNI LETECKY USTAV (Czechoslovakia) (1983-84) 46

W

W-3 Sokól (PZL Swidnik) 238
WA-116 (Wallis) (1992-93) 316
WA-116 Venom (Wallis) (1987-88) 331
WA-116-T (Wallis) (1992-93) 316
WA-116/W (Wallis) (1983-84) 283
WA-116X (Wallis) (1992-93) 316
WA-117/R-R (Wallis) (1986-87) 317
WA-117/R-R MoD operations (Wallis) (1991-92) 327
WA-118/M Meteorite (Wallis) (1986-87) 317
WA-120/R-R (Wallis) (1987-88) 331
WA-121 (Wallis) (1992-93) 316
WA-121/M Meteorite 2 (Wallis) (1987-88) 331
WA-122/R-R (Wallis) (1987-88) 332
WA-122/R-R (Wallis) (1992-93) 316
WA-201 (Wallis) (1992-93) 316
WC-130 Hercules (Lockheed) 502
WHE (see Ekin, W. H. (Engineering) Co Ltd) (1984-85) 299
WM-2 (Miller) (1985-86) 602
WS 70 (Westland) (1991-92) 333
WS-70 (Westland/Sikorsky) (1989-90) 335
WS 70L (Westland) 410
WSK-PZL KROSNO (Poland) (1988-89) 182
WSK-PZL MIELEC (see Wytwornia Sprzetu Komunikacyj-nego-PZL Mielec) 228
WSK-PZL SWIDNIK (see Wytwornia Sprzetu Komunikacyj-nego Im. Zygmunta Pulawskiego-PZL Swidnik) (1992-93) 181
WSK PZL-WARSZAWA-OKECIE (see Wytwornia Sprzetu Komunikacyj-nego PZL-Warszawa-Okecie) (1989-90) 203

Waco Classic F5 (Classic) (1992-93) 371
Waco Classic YMF Super (Classic) 470
Wag-a-Bond (Wag Aero) 584
WAG-AERO INC (USA) 584
WALLIS AUTOGYROS LTD (UK) 402
WARD, M. (UK) (1985-86) 571
Warrior (Agusta) 192
Warrior II (Piper) (1990-91) 476
Warrior AH-58D (Bell) (1989-90) 362
Wasp (Texas Helicopter) (1983-84) 496
Wasp II (Texas Helicopter) (1983-84) 496
Wasp II (Williams) (1985-86) 524
Wasp ML74 (Aerodyne) (1987-88) 341
WEATHERLEY AVIATION COMPANY INC (USA) (1983-84) 500
WELLS, EUGENE W. (USA) (1985-86) 634
Werewolf, Ka-50 (Kamov) 275
Westair 204 (Western) (1985-86) 634
WESTERN AIRCRAFT CORPORATION (USA) (1985-86) 634
WESTERN WINGS (USA) (1987-88) 530
Westland 30 (Westland) (1987-88) 317
WESTLAND GROUP PLC (UK) 402
WESTLAND HELICOPTERS LIMITED (UK) 402
Westwind 1 (IAI) (1986-87) 136
Westwind 2, 1124A (IAI) (1988-89) 140
Whisper Jet, Mark III (JetCraft) (1987-88) 438
White Lightning (White Lightning) 585
WHITE LIGHTNING AIRCRAFT CORPORATION (USA) 585
Wildfire (Statler/Beck) (1984-85) 602
Wilga 35, PZL-104 (PZL Warszawa-Okecie) 241
Wilga 80, PZL-104 (PZL Warszawa-Okecie) 241
Wilga 80-550 (Melex) 536
WILLIAMS, DAVID (UK) (1985-86) 315
WILLIAMS INTERNATIONAL (USA) (1987-88) 530
WING AIRCRAFT COMPANY (USA) (1983-84) 500
Wing lockers (Raisbeck) (1991-92) 465
Wingtip vortex turbine research (NASA) (1988-89) 447
WIPAIRE (USA) 585
WIPLINE INC (USA) (1985-86) 524
WOOD, THOMAS M (Canada) (1983-84) 509
Wren, AS 202/18A (FFA) 357
WREN AIRCRAFT INC (USA) (1988-89) 496
WTA INC (USA) (1991-92) 497
WYTWORNIA SPRZETU KOMUNIKACYJ NEGO-PZL KROSNO (Poland) (1988-89) 182
WYTWORNIA SPRZETU KOMUNIKACYJ NEGO-PZL MIELEC (Poland) 228
WYTWORNIA SPRZETU KOMUNIKACYJ NEGO-PZL WARSZAWA-OKECIE (Poland) (1989-90) 203
WYTWORNIA SPRZETU KOMUNIKACYJ NEGO Im. ZYGMUNTA PULAWSKIEGO-PZL SWIDNIK (Poland) (1992-93) 181

X

X.001 (Xausa) (1984-85) 546
X4 (Robin) 92
X-29A (Grumman) (1990-91) 420
X-30 (NASA) 540, 698
X-31A EFM (Rockwell International/Deutsche Aerospace) 176
X 380 Dauphin (Eurocopter) 154
XAC (see Xian Aircraft Manufacturing Company) 57
XCH-62 (Boeing) (1986-87) 381
XH-1 (Atlas) (1987-88) 216
XH-2, Rooivalk (Atlas) (1990-91) 213
XH-59 (Sikorsky) (1983-84) 486
X-Jet (Williams) (1986-87) 525
XM-4 (Moller) 537
XP-99 Prop-Jet (Smith) (1985-86) 518
XR Learjet (Dee Howard) (1991-92) 396
XSH-60J (Mitsubishi-Sikorsky) (1987-88) 182
XST (Northrop) 542
XST Have Blue (Lockheed) 506
XT-4 (Kawasaki) (1985-86) 162
XTP-1 (Atlas) (1989-90) 215
XV-15 (Bell) (1989-90) 368
XAUSA, RENATO (Italy) (1984-85) 546
Xavante (EMBRAER/Aermacchi) (1984-85) 15
XIAN AIRCRAFT MANUFACTURING COMPANY (China) 57
Xingu II (EMBRAER) (1987-88) 16
Xingu III (EMBRAER) (1987-88) 17
Xuanfeng (China) (1985-86) 38
X-Wing project (Sikorsky) (1990-91) 516

Y

Y-5B (SAP) 57
Y7 (XAC) 58
Y-7 Derivative (XAC) 60
Y-7H (XAC) 60
Y-8 Yunshuji-8 (SAC) 54
Y-9 (State Aircraft Factories) (1983-84) 39
Y-10 (Shanghai) (1987-88) 45
Y-11 (Harbin) (1986-87) 34
Y-11B (I) (HAMC) 49
Y-12 (HAMC) (1989-90) 37
Y-12 (II) (HAMC) 50
Y-14-100 (XAC) (1989-90) 47
YAH-64 (McDonnell Douglas) (1985-86) 463
Yak-3 (Yakovlev) 694
Yak-18T (Yakovlev) 695
Yak-28 (Yakovlev) (1990-91) 290
Yak-36M (Yakovlev) 331
Yak-38 (Yakovlev) 331
Yak-40 (Yakovlev) (1983-84) 245
Yak-40TL (Skorost/Yakovlev) 331
Yak-41 (Yakovlev) (1992-93) 257
Yak-42 (Yakovlev) 332, 695
Yak-42M (Yakovlev) 333
Yak-44 (Yakovlev) (1992-93) 259
Yak-46 propfan (Yakovlev) 333
Yak-46 turbofan (Yakovlev) 334
Yak-48 (Yakovlev) 334
Yak-50 (Yakovlev) (1989-90) 294
Yak-52 (Condor) (1991-92) 203
Yak-52 (Yakovlev) 334
Yak-53 (Yakovlev) (1989-90) 294
Yak-54 (Yakovlev) 334
Yak-55 (Yakovlev) (1989-90) 295
Yak-55M (Yakovlev) 335
Yak-56 (Yakovlev) (1992-93) 260
Yak-58 (Yakovlev) 335
Yak-77 (Yakovlev) 695
Yak-112 (Yakovlev) 335
Yak-130 (Yakovlev) 695
Yak-141 Freestyle (Yakovlev) 336
Yak-142 (Yakovlev) 695
Yak-144 (Yakovlev) 337
Yak-242 (Yakovlev) 699
Yak-UTS (Yak 130) (Yakovlev) 337
YAKOVLEV (see Moskovskii Mashinostroitelnyy Zavod 'Skorost' Imieni A. S. Yakovleva (Russia) 331
Ye-133 (Mikoyan) (1987-88) 259
YF-22A (USAF) (1990-91) 433, 523
YF-23A (Northrop) (1991-92) 457
YF-23A (USAF) (1990-91) 523
YMF Super (Classic) 470
YS-11E (Nippi) 211
YSX (JADC) 206
YSX-75B (JADC) (1992-93) 158
YT-17 (Aerotec) (1983-84) 10

YAKOVLEV (Russia) 331
YAKOVLEV DESIGN BUREAU (USSR) (1991-92) 293, 756
Yamal (Aviaspetstrans) 255
Yasur 2000 (IAI) 184
Yun-5 (SAP) 57
Yun-7 (XAC) 58
Yun-8 (SAC) 54
Yun-10 (Shanghai) (1987-88) 45
Yun-11 (Harbin) (1986-87) 34
Yun-11B (HAMC) 49
Yun-12 (II) (HAMC) 50
Yun-14 (XAC) (1989-90) 47
Yunshuji-5 (SAP) 57
Yunshuji-7 (XAC) 58
Yunshuji-8 (SAC) 54
Yunshuji-10 (Shanghai) (1987-88) 45
Yunshuji-11 (Harbin) (1986-87) 34
Yunshuji-11B (I) (HAMC) 49
Yunshuji-12 (HAMC) 50
Yunshuji-14 (XAC) (1989-90) 47

Z

Z-5 (State Aircraft Factories) (1985-86) 38
Z-6 (HAMC) (1988-89) 39
Z-8 (CAF) 47
Z-9 Haitun (HAMC/Eurocopter) 51
Z 37T Agro Turbo (Zlin) 71
Z 50 (Zlin) 70
Z 50 LS (Zlin) (1986-87) 49
Z 61 L (Zlin) (1990-91) 55
Z 90 (Zlin) 71
Z 137T Agro Turbo (Zlin) 71
Z 142 (Zlin) 69
Z 143 (Zlin) 70
Z 242L (Zlin) 69
ZLAC (see Zhongmeng Light Aircraft Company) 60
ZLIN (see Moravan Akciová Spolećnost) 69

Zafar 300 (Seyedo Shohada) (1992-93) 139
ZENAIR LTD (Canada) 39
Zenith CH 200 (Zenair) 40
Zenith CH 250 (Zenair) 40
Zenith CH 2000 (Zenair) 41
Zephyr (Jurca) 86
Zhi-5 (State Aircraft Factories) (1985-86) 38
Zhi-6 (Harbin) (1986-87) 34
Zhi-8 (CAF) 47
Zhi-9 (HAMC) 51
Zhishengji-5 (State Aircraft Factories) (1985-86) 38
Zhishengji-6 (Harbin) (1986-87) 35
Zhishengji-8 (CAF) 47
Zhishengji-9 (HAMC) 51
ZHONGMENG LIGHT AIRCRAFT CORPORATION (China) 60
Zipper (Cleary) (1985-86) 394
ZRZESZINIE WYTWORCOW SPRZETU LOTNICZEGO I SILNIKOWEGO PZL (Poland) (1991-92) 188
ZYGMUNTA PULAWSKIEGO-PZL SWIDNIK (Poland) 237

PRIVATE AIRCRAFT

(Incorporating Sport, Homebuilt and Microlight Aircraft)

Private aircraft retained in the book are now listed in the main Aircraft index. This index mainly covers entries in the preceding ten editions, but aircraft and manufacturers now covered in the main Aircraft index are indicated thus *

A

A-1 (CCMI) (1992-93) 554
A-2-S Barracuda (Hobbystica) (1992-93) 559
A-2-2 Barracuda (Hobbystica) (1992-93) 559
A-5 (Aeroprakt) (1987-88) 611
A-6 (Byelyi) (1987-88) 612
A6M5 Zero (WAR) (1988-89) 612
A-10 Silver Eagle (Mitchell Wing) (1991-92) 599
A-11M (Aeroprakt) (1986-87) 610
AA-200 (Aerodis) (1992-93) 498
ABC (Dimanchev/Valkanov) (1992-93) 464
ACE (see Aircraft Composite Engineering) (1985-86) 639
ACT (see Aircraft Composite Technology) (1983-84) 652
AD-100, Traveller (NAI) (1987-88) 563
AFG (see Allgäuer UL-Flugzeugbau GmbH) (1987-88) 592
AG 02 (Gatard) (1987-88) 574
AG 02 Sp (Gatard) (1988-89) 522
AG-7 (IAC) (1986-87) 674
AG-38A Terrier (Mitchell Wing) (1991-92) 560
AG-38-A War Eagle (Mitchell Wing) (1991-92) 560
AG Bearcat (Litecraft) (1986-87) 682
AG Zipper (Zenair) (1987-88) 561
AILE (see Air Industrie Loisir Engineering) (1986-87) 563
A. I. R.'s SA (see Air International Services) (1992-93) 555
AJS 2000 (Sky-Craft) (1986-87) 609
AM-DSI (Leader's International) (1986-87) 679
AM-DSII (Leader's International) (1987-88) 676
AMF-S14 Miranda (Falconar) (1992-93) 467
AMF-S14H Miranda (Falconar) (1992-93) 552
AN-20 (Neukom) (1983-84) 646
AN-20B (Neukom) (1984-85) 674
AN-20B (Richter) (1983-84) 643
AN-20K, Piccolo (Borowski) (1987-88) 594
AN-21R (Neukom) (1984-85) 674
AN-22 (Albatros) (1987-88) 593
AN-22 (Neukom) (1984-85) 674
AN-22 (Point) (1985-86) 660
ARV-IK Golden Hawk (Falconar) (1992-93) 552
ASA 200 (Feugray) (1986-87) 574
ASL (see Club d'Avions Super Légers du Cher) (1992-93) 555
ASL 18 La Guehe (ASL) (1992-93) 555
AT-9 (Stoddard Hamilton) (1988-89) 602
ATI (see Advanced Technologies) (1991-92) 540
ATOI (Terrade) (1987-88) 590
ATOL (Martenko) (1991-92) 511
ATTL-1 (Euronef) (1992-93) 464
AUI (see Aviazione Ultraleggera Italiana) (1992-93) 558

ABE, KEIICHI (Japan) (1984-85) 673
ABERNATHY, BOB (USA) (1986-87) 622
Acapella (Option Air Reno) (1986-87) 692
Accipiter II (Pagen) (1984-85) 728
ACE AIRCRAFT COMPANY (USA) (1992-93) 494
ACES HIGH LIGHT AIRCRAFT (Canada) (1992-93) 551
Acey Deucy, Powell P.70 (Acey Deucy) (1986-87) 622
ACEY DEUCY SPORT AIRCRAFT INC (USA) (1986-87) 622
Acroduster 1 SA-700 (Aerovant) (1992-93) 499
Acroduster Too SA-750 (Stolp) (1992-93) 539
Acro Sport I (Acro Sport) (1992-93) 494
Acro Sport II (Acro Sport) (1992-93) 494
ACRO SPORT INC (USA) (1992-93) 494
Acro X (Catto) (1986-87) 648
Acro-Zenith CH 150 (Zenair) (1992-93) 472
ADVANCED AVIATION INC (USA) (1992-93) 495, 562
ADVANCED COMPOSITE TECHNOLOGY (USA) (1987-88) 628
ADVANCED TECHNOLOGIES (USA) (1991-92) 540
Adventure (Mead Engineering) (1986-87) 685
ADVENTURE AIR (USA) (1992-93) 496
Adventurer (Adventure) (1992-93) 496
Adventurer 2 + 2 (Adventure) (1992-93) 496
Adventurer 4 Place (Adventure) (1992-93) 496
Aeolus (Aeolus) (1985-86) 665
Aeolus (Poisestar) (1983-84) 650
AEOLUS AVIATION (UK) (1985-86) 665
AERMAS (see Automobiles Martini) (1992-93) 554
Aermas 386 (Aermas) (1992-93) 554
Aero (Bowen) (1984-85) 718
Aerobatic-Speedwings (Avid Aircraft) (1992-93) 504
AEROCAR INC (USA) (1992-93) 496
*AERO COMPOSITE TECHNOLOGY INC (USA)
AÉRO DELTA SERVICE (France) (1986-87) 562
AERO DESIGNS INC (USA) (1992-93) 498
AERODIS AMERICA INC (USA) (1992-93) 498
AERODIS, SARL (France) (1991-92) 512
AERO DYNAMICS LTD (USA) (1990-91) 560
AERODYNE SYSTEMS INC (USA) (1985-86) 674
Aero-Fox (Lite-Flite) (1987-88) 679
AERO FLIGHT (USA) (1990-91) 560
AERO INNOVATIONS INC (USA) (1987-88) 631
Aerokart 4315 (Aerokart) (1986-87) 562
Aerokart 4320 (Aerokart) (1986-87) 563
AÉROKART, SOCIÉTÉ (France) (1986-87) 562
AEROLIGHT FLIGHT DEVELOPMENT INC (USA) (1984-85) 686
AEROLITES INC (USA) (1992-93) 498
AERO MIRAGE (USA) (1991-92) 535
AERONAUTIC 2000 (France) (1991-92) 587
Aeroplane XP (UFM of Kentucky) (1984-85) 718
AÉROPLUM (France) (1986-87) 563
Aeropower 28 (Hughes) (1989-90) 534
AEROPRAKT (USSR) (1987-88) 611
AEROSERVICE ITALIANA SRL (Italy) (1992-93) 558
AERO-SOL (USA) (1992-93) 562
AEROSPORT LTD (USA) (1992-93) 499
Aerostar (Goldwing) (1986-87) 667
Aerostat 340 (Aerolight) (1983-84) 652
AÉROSTRUCTURE (France) (1984-85) 663
AERO TECH AVIATION LTD (Canada) (1983-84) 633
AEROTECH DYNAMICS CORPORATION (USA) (1986-87) 631
AERO-TECH INDUSTRIES INC (USA) (1992-93) 499
AEROTECH INTERNATIONAL LTD (UK) (1991-92) 593
AEROTIQUE AVIATION INC (USA) (1986-87) 631
AEROVANT AIRCRAFT CORPORATION (USA) (1992-93) 499
AERO VENTURES (USA) (1985-86) 674
AERO VISIONS INTERNATIONAL (USA) (1991-92) 536
Ag Bearcat (Aerolites) (1992-93) 498
Ag-Hawk (CGS Aviation) (1992-93) 563
AG Master (Hutchinson) (1989-90) 577
Agricol (ULAC) (1985-86) 664
Agri-Delta (Agri-Delta) (1983-84) 638
AGRI-PLANE (see La Culture de l'An 2000 SA) (1986-87) 563
AgStar (Pampa's Bull) (1992-93) 550
AILES DE K, L DE KALBERMATTEN (Switzerland) (1985-86) 663
AILE VOLANTE B. DANIS (France) (1987-88) 572
Airaile S-12 (Rans) (1992-93) 532
Airaile S-14 (Rans) (1992-93) 532
AIR AND SPACE (USA) (1992-93) 500
Air Arrow (Mountain Valley) (1991-92) 506
AIRBORNE WING DESIGN (USA) (1984-85) 686
Aircamper B4 (Pietenpol) (1988-89) 590
Aircamper GN-1 (Grega) (1992-93) 518
AIR-CEFELEC AÉRONAUTIQUE (France) (1983-84) 638
Airco DH2 (Redfern) (1992-93) 532
AIR COMMAND INTERNATIONAL (USA) (1992-93) 500, 562
AIR COMPOSITE (France) (1992-93) 474
AIRCRAFT COMPOSITE ENGINEERING (Australia) (1985-86) 639
AIRCRAFT COMPOSITE TECHNOLOGY (USA) (1983-84) 652
AIRCRAFT DESIGNS INC (USA) (1992-93) 500
AIRCRAFT DEVELOPMENT INC (USA) (1992-93) 562
AIRCRAFT DYNAMICS INC (USA) (1992-93) 501
AIR CREATION (France) (1992-93) 554
Aircross (Distri Snap) (1988-89) 522
Air Cub (Chuck's) (1987-88) 650
Air-Dinghy (Polaris Motor) (1988-89) 538
Air-Dinghy 1+1 (Polaris Motor) (1988-89) 538
AIR INDUSTRIE LOSIER ENGINEERING SA (France) (1986-87) 563
AIR INTERNATIONAL SERVICE SA (France) (1992-93) 555
AIRMASS—CIRRUS INDUSTRIES (USA) (1983-84) 653
Air Master (Aerolights) (1992-93) 498
AIR NOSTALGIA (USA) (1992-93) 502
AIRPLANE FACTORY INC, THE (USA) (1989-90) 616
Airplume (Croses, Emilien) (1984-85) 666
Airplume (Croses, Yves) (1992-93) 555
AIR SERVICE (Hungary) (1992-93) 557
*Air Shark I, FM-2 (Freedom Master)
AIRTECH CANADA (Canada) (1991-92) 585
AKAFLIEG MÜNCHEN (see Flugtechnische Forschungsgruppe an der Technischen Universität München) (1991-92) 522
Akro Model Z (Laser) (1988-89) 504
ALBARDE (France) (1986-87) 564
Albatros (Vulcan) (1992-93) 557
Albatros DVa (Ryder) (1992-93) 534
ALBATROS ULTRA FLIGHT GmbH (West Germany) (1987-88) 592
Albertan (Ultimate Aircraft) (1987-88) 554
ALDERFER GYROCHOPPER AIRCRAFT CORPORATION (USA) (1988-89) 556
Alex-I (Parmesani) (1987-88) 604
ALIFERRARI (Italy) (1992-93) 558
Allegro (A. I. R's SA) (1992-93) 555
ALLGÄUER UL-FLUGZEUGBAU GmbH (West Germany) (1987-88) 592
ALPAIR ULM (Italy) (1987-88) 602
ALPHA (Poland) (1992-93) 488
ALPHA AVIATION SUPPLY CO (USA) (1992-93) 502
Alto 833 (ULM Aérotechnic) (1983-84) 641
Alto-Stratus (Stratus) (1983-84) 690
ALTURAIR (USA) (1992-93) 502
American (American Air Jet) (1991-92) 538
AMERICAN AEROLIGHTS INC (USA) (1983-84) 653
AMERICAN AIRCRAFT INC (USA) (1992-93) 563
AMERICAN AIR JET INC (USA) (1991-92) 538
AMERICAN AIR TECHNOLOGY (USA) (1987-88) 638
AMERICAN MICROFLIGHT INC (USA) (1986-87) 634
AMF MICROFLIGHT LTD (UK) (1992-93) 490
ANDERSON EARL (USA) (1987-88) 639
ANDREASSON, BJÖRN (Sweden) (1992-93) 489
ANGLIN ENGINEERING (USA) (1992-93) 502
ANGLIN SPECIAL AERO PLANES INC (USA) (1984-85) 688
Antares SB1 (Soyer/Barritault) (1992-93) 484
ANTONOV (Ukraine) (1992-93) 560
APCO AVIATION (West Germany) (1987-88) 593
Apollo (Product MZ) (1987-88) 600
Aquitain (Lanot) (1986-87) 580
Aquitain 2 + 2 (Lanot) (1983-84) 640
Ara, PM-2 (MDK Krakow) (1983-84) 644
Arco (Reuter) (1986-87) 596
*ARCTIC AIRCRAFT COMPANY (USA)
*Arctic Tern, Interstate S1B2 (Arctic)
Ariete (Tesori) (1987-88) 553
Arrow (Arrow Aircraft) (1992-93) 551
Arrow, P-3 (UP) (1987-88) 726
ARROW AIRCRAFT CO (Canada) (1992-93) 465, 551
Arthur (Dohet) (1987-88) 573
Ascender (Freedom Fliers) (1987-88) 665
Ascender (Pterodactyl) (1983-84) 670
Ascender II+ (Coldfire) (1991-92) 596
Ascent I (Brown) (1989-90) 566
Asterix (Michel) (1987-88) 582
Asterix 001 (Nickel & Foucard) (1992-93) 480
Astra (Astra International) (1983-84) 654
ASTRA INTERNATIONAL (USA) (1983-84) 654
ATLANTIS AVIATION INC (USA) (1984-85) 688
Atlas (Birdman) (1983-84) 634
Ausra VK-8 (Kensgaila) (1992-93) 487
AUSTIN AEROSPACE CORPORATION (USA) (1987-88) 640
AUSTRALIAN AEROLITE (Australia) (1985-86) 639
AUSTRALITE INC (USA) (1992-93) 503
Austrostrike (Steinbach-Delta) (1987-88) 543
Autan M. J.53 (Jurca) (1992-93) 478
Autogyro 18A (Air and space) (1992-93) 500
AUTOMATED AIRCRAFT TOOLING (USA) (1992-93) 503
AUTOMOBILES MARTINI SA (France) (1992-93) 554
Avenger (Airborne Wing) (1983-84) 653
AviaStar (Pampa's Bull) (1992-93) 550
*AVIASUD ENGINEERING SA (France)
AVIATECHNICA, CO-OPERATIVE DESIGN OFFICE (Russia) (1992-93) 488
AVIATIKA LTD (Russia) (1992-93) 489
AVIAT INC (USA) (1992-93) 503
AVIATION COMPOSITES (UK) (1988-89) 544
AVIAZIONE ULTRALEGGERA ITALIANA (Italy) (1992-93) 558
Avid Aerobat (Avid Aircraft) (1992-93) 504
*AVID AIRCRAFT INC (USA)
Avid Amphibian (Avid Aircraft) (1992-93) 504
Avid Commuter (Avid Aircraft) (1992-93) 504
Avid Flyer (Avid Aircraft) (1991-92) 540
*Avid Flyer Mark IV (Avid Aircraft)
Avid Landplane (Avid Aircraft) (1992-93) 504
Avion (Avion) (1983-84) 655
Avion (Ken Brock) (1991-92) 596
AVIONS A. GATARD (France) (1989-90) 544
AVIONS JODEL SA (France) (1992-93) 476
AVIONS MUDRY ET CIE (France) (1983-84) 641
AVIONS YVES GARDAN (France) (1986-87) 575

B

B1-RD (Powers Bashforth) (1992-93) 568
B1-RD (Robertson) (1984-85) 712
B1-RD Instructor (Robertson) (1984-85) 712
B1-RD Two-seater (Powers Bashforth) (1992-93) 528
B2-RD (Robertson) (1983-84) 672
B4 Aircamper (Pietenpol) (1988-89) 590
B-8 Gyro Glider (Bensen) (1986-87) 638
B-8M Gyro Copter (Bensen) (1986-87) 638
B-8MW Hydro-Copter (Bensen) (1986-87) 638
B-8V Gyro Copter (Bensen) (1986-87) 638
B-8W Hydro Glider (Bensen) (1986-87) 638
B-10 (Mitchell Wing) (1991-92) 600
B-80 Gyro Glider (Bensen) (1986-87) 639
B-80D Gyro Glider (Bensen) (1986-87) 639
B-85 Rotax (Bensen) (1986-87) 640
B-532 King Cobra B (Advanced Aviation) (1992-93) 495
BA-4B (Andreasson) (1992-93) 489
BA-12 Slandan (Andreasson, MFI) (1992-93) 560
BA-14 (MFI) (1990-91) 612
BA 83 (Ikarusflug) (1990-91) 609
BAC (see Brutsche Aircraft Corporation) (1992-93) 563
BD-4 (Bede) (1992-93) 504
BD-4 Modification (Ward) (1987-88) 732
BD-5 (Alturair) (1992-93) 502
*BD-10 (Bede)
B-EC-9 Paras-Cargo (Croses) (1986-87) 570
BF-3 (CUP) (1992-93) 558
BF-10 (Bebe) (1992-93) 505
BG-1 (Bredelet and Gros) (1987-88) 568
BKFI (Nord) (1989-90) 550
BLI-KEA (Hinz) (1991-92) 522
BOAC (see Barney Oldfield Aircraft Company) (USA) (1992-93) 505
BW-20 Giro Kopter (BW Rotor) (1992-93) 563
BX-2 Cherry (Brändli) (1992-93) 490
BX-111 Giro-Hili (Berger) (1987-88) 610
BX-200 (Bristol) (1991-92) 543

Baby Ace Model D (Ace Aircraft) (1992-93) 494
Baby Bird (Stitts) (1987-88) 712
'Baby' Lakes (BOAC) (1992-93) 505
Balade ST 80 (Stern) (1992-93) 484
Baladin (Paumier) (1992-93) 481
Baladin NP2 (Parent) (1992-93) 481
Balans (Kolodziej) (1984-85) 725, 730
Balerit (Mignet) (1992-93) 556
Banchee (International Ultralite) (1985-86) 692
B & B AIRCRAFT COMPANY (USA) (1984-85) 689
B & F TECHNIK VERTRIEBS GmbH (Germany) (1992-93) 556
B & G AIRCRAFT COMPANY INC (USA) (1985-86) 678
Bandit (Solar Wings) (1992-93) 561
Banty (Butterfly Aero) (1992-93) 563
BARBERO (France) (1986-87) 565
BARDOU, ROBERT (France) (1986-87) 565
BARNETT ROTORCRAFT (USA) (1992-93) 504

716 INDEXES: PRIVATE AIRCRAFT—**B** - **C**

BARNEY OLDFIELD AIRCRAFT COMPANY
(USA) (1992-93) 505
Barnstormer (Fisher) (1983-84) 660
Baroudeur (Aéronautic 2000) (1991-92) 587
Barracuda (Barracuda) (1992-93) 504
BARRACUDA (see Bueth Enterprises Inc) (1992-93) 504
Barracuda A-2-2 (Hobbystica) (1992-93) 559
Barracuda A-2-S (Hobbystica) (1992-93) 559
Bathtub Mk I (Kimbrel/Dormoy) (1988-89) 578
Bausalz (LO-Fluggeratebau) (1992-93) 557
BEACHY, ERNEST (USA) (1986-87) 637
Bearcat (Aerolites) (1992-93) 498
Bearcat (Litecraft) (1986-87) 682
Bearcat (Ultra Classics) (1984-85) 719
BEARD, ROBERT (USA) (1987-88) 641
BEAUJON AIRCRAFT CO (USA) (1986-87) 637
BEAUJON ULTRALIGHTS (USA) (1983-84) 656
Beaver RX-28 (Spectrum) (1988-89) 510
Beaver RX-35 (Spectrum) (1988-89) 510
Beaver RX-35 Floater (Spectrum) (1988-89) 510
Beaver RX-550 Series (Spectrum) (1988-89) 511
BEAVER RX ENTERPRISES LTD (Canada) (1992-93) 552
Bébé Series (Jodel) (1992-93) 476
*BEDE JET CORPORATION (USA
BEDE-MICRO AVIATION INC (USA) (1992-93) 505
BEDSON (see GORDON BEDSON AND
ASSOCIATES) (1986-87) 537
BEIJING UNIVERSITY OF AERONAUTICS AND
ASTRONAUTICS (China) (1992-93) 553
BELIOVKIN (Latvia) (1992-93) 559
BELL, F. M. ULTRALITE AIRCRAFT INC (USA) (1986-87) 637
BENSEN AIRCRAFT CORPORATION (USA) (1986-87) 638
BERGER-HELICOPTER (Switzerland) (1987-88) 610
*Berkut (Experimental Aviation)
BERNARD BROC ENTERPRISE (France) (1992-93) 555
Beryl (Piel) (1992-93) 482
Beta Bird (Hovey) (1991-92) 598
BEZZOLA-VILS, GION (Switzerland) (1987-88) 610
Bidulm 44 (Cosmos) (1987-88) 570
Bifly (Lascaud) (1992-93) 556
BIGUÁ (Argentina) (1991-92) 502
Biguá amphibian (Biguá) (1991-92) 502
BINDER AVIATIK GmbH (West Germany) (1989-90) 609
Biplane (Ostrowski) (1985-86) 662
Biplum (Guerpont) (1988-89) 522
Bird TA-2/3 (Taylor) (1992-93) 541
BIRDMAN AIRCRAFT INTERNATIONAL (USA) (1985-86) 679
BIRDMAN ENTERPRISES 1986 LTD (Canada) (1990-91) 605
BIRDMAN SPORTS LTD (UK) (1983-84) 688
Blackbird (Black) (1987-88) 642
BLACK, FRANK E. (USA) (1987-88) 642
BLACK, JAMES G. (USA) (1983-84) 656
BLALOCK, KEN (USA) (1987-88) 642
Blue Max (Blue Max) (1992-93) 550
BLUE MAX ULTRALIGHT (Australia) (1992-93) 550
Blue Sky-3 (Korean Air) (1992-93) 559
Bobcat (First Strike) (1992-93) 514
Boomerang (Fisher) (1983-84) 660
Boondocker (For Two Enterprises) (1992-93) 565
Boondocker 42 (For Two Enterprises) (1992-93) 565
Boredom Fighter (Wolf) (1988-89) 614
BOROWSKI, LTB (West Germany) (1987-88) 594
BOUDEAU, MICHEL (France) (1986-87) 566
BOUNSALL AIRCRAFT (USA) (1987-88) 642
Bouvreuil (Pottier) (1988-89) 530
Bowen Aero (Bowen Aero) (1984-85) 726
BOWEN AERO (USA) (1984-85) 726
*BOWERS, PETER M. (USA)
Boxmoth (Thomson) (1983-84) 678
Boxmoth B (Thomson) (1986-87) 724
BRÄNDLI, MAX (Switzerland) (1992-93) 490
Bravo OM-1 (Morrisey) (1987-88) 686
BREDELET AND GROS (France) (1987-88) 568
BREEN AVIATION (UK) (1983-84) 646
Breezer (Beachy) (1986-87) 637
BRIFFAUD, GEORGES (France) (1986-87) 566
BRISTOL, URIEL (Bristol) (1991-92) 543
BROC ENTERPRISE, BERNARD (France) (1992-93) 555
Brok-1M Garnys (LAK) (1992-93) 559
BROKAW AVIATION INC (USA) (1992-93) 507
Bronco (Thalhofer) (1987-88) 598
BROWN, MICHAEL (USA) (1990-91) 567
BRÜGGER MAX (Switzerland) (1992-93) 490
BRUTSCHE AIRCRAFT CORPORATION
(USA) (1992-93) 563
BRYAN, COURTNEY (USA) (1986-87) 644
Buccaneer SX (Advanced Aviation) (1992-93) 496
Buccaneer II, XA/650 (Advanced Aviation) (1992-93) 496
Buckeye (Coldfire) (1991-92) 597
Buckeye (Freedom Fliers) (1987-88) 666
Buckeye B (Coldfire) (1992-93) 564
Buckeye 503B (Coldfire) (1992-93) 564
BUCK SPORT AIRCRAFT INC (USA) (1987-88) 646
'Buddy Baby' Lakes (BOAC) (1992-93) 506
BUETH ENTERPRISES, WB, INC (USA) (1992-93) 504
Bull (Windryder) (1987-88) 735
Bullet (Brokaw) (1992-93) 507
Bullet 2100 (Aerocar) (1989-90) 560
Bumble Bee (Aircraft Designs) (1992-93) 500
Bumble Camel (Smith, Ronald) (1992-93) 553
BURKE AIRCRAFT (USA) (1987-88) 646
Busard (Lefebvre) (1985-86) 555
Bushbird (Fletcher) (1987-88) 548
BUSHBY AIRCRAFT INC (USA) (1992-93) 507
Bushmaster (Sylvaire) (1987-88) 553
Bushmaster I Mk2 (Snowbird) (1992-93) 553
Bushmaster II (Morgan) (1986-87) 690
Bushranger (Seabird) (1986-87) 540
Butterfly (Butterfly) (1984-85) 659
Butterfly (Butterfly Aero) (1992-93) 563
BUTTERFLY AERO (USA) (1992-93) 563
BUTTERFLY COMPANY (Belgium) (1985-86) 642
BUTTERWORTH, G. N. (USA) (1988-89) 561
BW ROTOR COMPANY INC (USA) (1992-93) 563
BYELYI (USSR) (1987-88) 612

C

C22 (Ikarus) (1992-93) 556
C22 HP (Ikarus) (1992-93) 556
C32 (Ikarus) (1989-90) 610
*CA-05 Christavia Mk 1 (Elmwood)
CA-61/-61R Mini Ace (Cvjetkovic) (1992-93) 510
CA-65 (Cvjetkovic) (1992-93) 510
CA-65A (Cvjetkovic) (1992-93) 510
CB-1 Biplane (Hatz) (Kelly) (1992-93) 520
CC-01 (Chudzik) (1987-88) 568
C. C. C. (Mathews) (1986-87) 684
CCMI (see Central China Mechanical Institute) (1992-93) 554
*CFM (see Cook Flying Machines)
CGS (see CGS Aviation Inc) (1987-88) 650
CGS Hawk (Micronautics) (1989-90) 607
*CH-8 Christavia Mk 4 (Elmwood)
CH 150 (Zenair) (1992-93) 472
CH 180 (Zenair) (1992-93) 472
*CH 250 (Zenair)
*CH 300 Tri-Z (Zenair)
CH 600 (Zenair) (1991-92) 510
*CH 601 (Zenair)
*CH 601HD (Zenair)
*CH 601HDS (Zenair)
*CH 701 STOL (Zenair)
*CH 701-AG STOL (Zenair)
CH 801 Super STOL (Zenair) (1992-93) 474
CH 2000 (Zenair) (1992-93) 474
CJ-1 (Corby) (1988-89) 503
CJ-3D Cracker Jack (Wood Wing) (1992-93) 548
CJC.01 (Chaboud) (1987-88) 568
CL-1 Ersatz Junkers (Haigh) (1986-87) 669
CMV (see Construction Machines Volantes) (1987-88) 569
C. P. 70/750/751 Beryl (Piel) (1992-93) 482
C. P. 80 (Piel) (1988-89) 529
C. P. 90 Pinocchio (Piel) (1992-93) 482
C. P. 150 Onyx (Piel) (1992-93) 482
C. P. 300 series, Emeraude (Piel) (1992-93) 481
C. P. 600 series, Super Diamant (Piel) (1992-93) 482
C. P. 1320 (Piel) (1992-93) 482
CSRI-01 Tse Tse (Souquet) (1987-88) 589
CUP (see Centro Ultraleggeri Partenopeo) (1992-93) 558
CX-901 (Command-Aire) (1984-85) 692

Cabin Starduster (Stolp) (1992-93) 540
CABRINHA ENGINEERING INC (USA) (1992-93) 508
Cadi Z001 (Composite Aircraft) (1992-93) 510
CALAIR CORPORATION (Australia) (1991-92) 502
CALDER (USA) (1984-85) 690
CALVEL PLASTIQUE AERONAUTIQUE, JACQUES
(France) (1984-85) 664
CAMPBELL-JONES, MICHAEL (UK) (1983-84) 646
CANA AIRCRAFT COMPANY (Singapore) (1987-88) 607
Canadian Skyrider (Aero Tech) (1983-84) 633
CANADIAN ULTRALIGHT MANUFACTURING LTD
(Canada) (1992-93) 552
CANAERO DYNAMICS AIRCRAFT INC (Canada) (1988-89) 508
Canelas Ranger (Martin Uhia) (1992-93) 560
Capella (Flightworks) (1992-93) 516
Capella XS (Flightworks) (1992-93) 516
Capena (Pena) (1987-88) 584
CARLSON AIRCRAFT INC (USA) (1992-93) 508, 563
CAROTHERS, CHUCK AIR SHOW (USA) (1986-87) 646
Carrera (Advanced Aviation) (1992-93) 496
Carrera 180 (Advanced Aviation) (1990-91) 558
Carrera 182 (Advanced Aviation) (1992-93) 496
CASCADE ULTRALIGHTS INC (USA) (1988-89) 562
CASSAGNERES E (USA) (1992-93) 508
CASSUTT (USA) (1988-89) 562
Cassutt IIIM (National) (1992-93) 527
Catfish (Eaves) (1987-88) 660
CATTO AIRCRAFT (USA) (1986-87) 648
Cavalier SA 102.5 (MacFam) (1992-93) 524
CAVU Flyer (Peters) (1985-86) 702
C—CON GmbH (Germany) (1992-93) 556
Cekady (Cowan) (1991-92) 504
Celebrity (Fisher Aero) (1992-93) 516
Celerity (Mirage Aircraft) (1992-93) 526
CENTRAIR SN (France) (1992-93) 555
CENTRAL CHINA MECHANICAL INSTITUTE
(China) (1992-93) 554
CENTRO ULTRALEGGERI PARTENOPEO
(Italy) (1992-93) 558
Centurion (Sport Aviation) (1983-84) 690
CESLOVAS (Lithuania) (1992-93) 559
CGS AERONAUTICS (USA) (1990-91) 616
CGS AVIATION INC (USA) (1992-93) 563
CHABOUD, CLAUDE (France) (1987-88) 568
Challenger (Quad City) (1992-93) 568
Challenger (Ulm Pyrénées) (1986-87) 591
Challenger II (Quad City) (1992-93) 568
Challenger II Special (Quad City) (1992-93) 530
Challenger Special (Quad City) (1992-93) 530
Channel Wing (Smith) (1986-87) 710
Chaos S-9 (Rans) (1992-93) 532
Charger MA-5 (Marquart) (1992-93) 526
CHARGUS GLIDING CO (UK) (1984-85) 676
Chaser S (Mc²) (1992-93) 567
CHASLE, YVES (France) (1992-93) 555
Cherry BX-2 (Brändli) (1992-93) 490
Chevvron 2-32C (AMF) (1992-93) 490
Chicken Hawk (Atlantis) (1984-85) 688
Chickinox (Dynali SA) (1992-93) 551
Chinook 1-S (Canadian Ultralight) (1992-93) 552
Chinook 2-S plus 2 (Canadian Ultralight) (1992-93) 552
Chinook, WT-11 (Birdman) (1986-87) 544
Choucas (Bardou) (1986-87) 565
*Christavia Mk 1 (Elmwood)
*Christavia Mk 4 (Elmwood)
CHRISTEN INDUSTRIES INC (USA) (1990-91) 568
CHUCK'S AIRCRAFT COMPANY (USA) (1992-93) 564
CHUDZIK (France) (1987-88) 568
CICARÉ, AUGUSTO (Argentina) (1991-92) 502
CIRCA REPRODUCTIONS (Canada) (1992-93) 466
*CIRRUS DESIGN CORPORATION (USA)
*Cirrus VK 30 (Cirrus)
Citabriette N3-2C (Mosler) (1991-92) 561
Classic (Cosy) (1992-93) 485
Classic (Fisher) (1986-87) 664
Classic (Fisher) (1992-93) 515
CLASSIC AIRCRAFT REPLICAS INC (USA) (1992-93) 509
Classic FP-404 (Fisher) (1989-90) 572
Clipper (Bell, F. M.) (1986-87) 638
Clipper (Worldwide Ultralite) (1984-85) 722
Clipper II (Sunrise) (1992-93) 540
Clipper Super Sport (Sunrise) (1992-93) 571
Clipper Ultralight (Sunrise) (1992-93) 571
Cloudbuster (Cloudbuster) (1985-86) 681
CLOUDBUSTER ULTRALIGHTS INC (USA) (1985-86) 681
Cloud Dancer (Rodger) (1984-85) 712
Cloud Dancer (US Aviation) (1992-93) 572
CLOUD DANCER AEROPLANE WORKS INC
(USA) (1987-88) 651
Club (Jodel) (1992-93) 476
CLUB D'AVIONS SUPERS LEGERS DU CHER
(France) (1992-93) 555
CLUTTON-TABENOR, Eric Clutton (USA) (1987-88) 651
CMI (Iraq) (1991-92) 522
Coach (Ehroflug) (1991-92) 589
Coach IIS (Ehroflug) (1992-93) 560
COATES, J. R. (UK) (1986-87) 612
Cobra (Advanced Aviation) (1992-93) 495
Cobra (Cobra Helicopters) (1992-93) 509
Cobra B (Advanced Aviation) (1992-93) 562
Cobra Carrera (Advanced Aviation) (1990-91) 558
Cobra HP (Advanced Aviation) (1983-84) 652
COBRA HELICOPTERS (USA) (1992-93) 509
Cobra Mk 1 (Romain & Sons) (1992-93) 561
COLANI, LUIGI (Switzerland) (1986-87) 608
COLDFIRE SYSTEMS INC (USA) (1992-93) 564
Colibri 2 MB-2 (Brügger) (1992-93) 490
COLLINS AERO (USA) (1992-93) 509
COLOMBAN, MICHEL (France) (1992-93) 475
COLUMBIA PLASTICS INC (USA) (1983-84) 658
Comet (UP) (1984-85) 728
Comet 135 (UP) (1983-84) 690, 694
Comet 165 (UP) (1984-85) 730
Comet 185 (UP) (1984-85) 730
COMMAND-AIRE AIRCRAFT (USA) (1985-86) 682
Commander 447 (Air Command) (1992-93) 562
Commander 500 (Air Command) (1992-93) 500
Commander 503 (Air Command) (1992-93) 500
Commander Elite 532 (Air Command) (1992-93) 500
Commander Elite 532 Two-seater (Air Command) (1992-93) 500
Commander Elite Sport 532 (Air Command) (1991-92) 536
Commander Elite Sport 582 (Air Command) (1992-93) 500
Commander Sport 447 (Air Command) (1992-93) 562
Commander Sport 503 (Air Command) (1992-93) 500
Commander Tandem 1000 (Air Command) (1992-93) 500
COMMISSION INTERNATIONALE DE VOL LIBRE
(France) (1983-84) 623
Commuter II-A (Cobra) (1992-93) 509
Commuter IIA (Tamarind) (1986-87) 721
Commuter IIB (Tamarind) (1986-87) 721
COMPETITION AIRCRAFT (USA) (1983-84) 658
COMPOSITE AIRCRAFT DESIGN INC (USA) (1992-93) 510
COMPOSITE CONSTRUCTIONS GmbH
(Germany) (1991-92) 589
COMPOSITE INDUSTRIES LTD (Australia) (1987-88) 537
CONDOR (UFM) (1992-93) 488
Condor II (Condor) (1985-86) 682
Condor III (Condor) (1985-86) 682
Condor III + 2 (Condor) (1985-86) 682
Condor 2090 (Agri-Plane) (1986-87) 563
CONDOR AIRCRAFT (USA) (1985-86) 682
CONTROT (France) (1986-87) 568
Convertible (Firebolt) (1992-93) 514
*COOK FLYING MACHINES (UK)
CO-OPERATIVE DESIGN OFFICE AVIATECHNICA
(Russia) (1992-93) 488
CONSTRUCTION MACHINES VOLANTES
(France) (1987-88) 569
Corben Jr Ace (Acro Sport) (1988-89) 550
CORBY, JOHN C. (Australia) (1988-89) 503
Cormorano (Aliferrari) (1992-93) 558
Corsair F4U (Air Nostalgia) (1992-93) 502
Corsair F4U (WAR) (1988-89) 612
Cosmagri (Cosmos) (1987-88) 571
Cosmos (Skylines) (1984-85) 714
Cosmos Dragster (Skylines) (1986-87) 709
COSMOS SARL (France) (1992-93) 555
COSY EUROPE (Germany) (1992-93) 485
Cougar Model 1 (Acro Sport) (1992-93) 494
COUNTRY AIR INC (USA) (1990-91) 570
Country Eagle XL (Aeroservice Italiana) (1988-89) 536
COUPE, AVIONS, JACQUES (France) (1992-93) 475
*Courier S-7 (Rans)
COWAN, PETER (Canada) (1991-92) 504
Coyote (Advanced Aviation) (1983-84) 652
*Coyote (Montana)
Coyote (Rans) (1988-89) 592
Coyote II S-6 (Rans) (1991-92) 601
Coyote II S-6ES (Rans) (1992-93) 569
Coyote S-4 (Rans) (1992-93) 569
Co Z DEVELOPMENT CORPORATION (USA) (1992-93) 510
Cozy (Co Z Europe) (1989-90) 550
Cozy Mk IV (Co Z) (1992-93) 510
COZZA/SCHOLL AIRCRAFT INC (USA) (1984-85) 692
Cracker Jack CJ-3D (Wood Wing) (1992-93) 548
Craft 200 (Craft Aerotech) (1992-93) 510
CRAFT AEROTECH (USA) (1992-93) 510
Cricri MC 15 (Colomban) (1992-93) 475
Criquet EC-6 (Croses) (1992-93) 476
Cropstar (Mainair Sport) (1992-93) 561
Crossbow (Cunion) (1984-85) 726
CROSES, EMILIEN (France) (1984-85) 532, 666
CROSES, EMILIEN (France) (1992-93) 476
CROSES, YVES (France) (1992-93) 555

C – F—PRIVATE AIRCRAFT: INDEXES 717

Cross-Country (Polaris) (1992-93) 559
Cruiser (Falconar) (1992-93) 468
Crystall (Kuibyshev) (1992-93) 560
CS Flash (Mainair Sports) (1987-88) 616
CSN (USA) (1992-93) 510
Cubmajor (Falconar) (1992-93) 467
Cuby I (Aces High) (1992-93) 551
Cuby I (Magal) (1991-92) 586
Cuby II (Aces High) (1992-93) 551
Cuby II (Magal) (1991-92) 586
CUELLAR, JESSE (USA) (1986-87) 651
Cuellar Monoplane (Cuellar) (1986-87) 651
Culex (Fisher) (1986-87) 664
Culex (Fisher Aero) (1992-93) 515
Culite (Aero Visions) (1991-92) 536
Currie Wot (PFA) (1988-89) 546
Curtiss P-40C (Thunder Wings) (1988-89) 606
CUSTOM AIRCRAFT CONVERSIONS (USA) (1984-85) 692
CUSTOM FLIGHT COMPONENTS (Canada) (1992-93) 466
Cutlass (Skyhook) (1984-85) 688, 692
CUVEILIER, ROLAND (France) (1986-87) 571
CVJETKOVIC, ANTON (USA) (1992-93) 510
Cyclone V300 (Windryder) (1992-93) 548
Cygnet SF-2A (Hapi) (1991-92) 553
Cygnus 21P (Gyro 2000) (1992-93) 518
Cygnus 21T (Gyro 2000) (1990-91) 576
Cygnus 21TX (Gyro 2000) (1992-93) 518

D

D-01, Ibis (Delemontez) (1986-87) 572
D.2 (Jones, Mason) (1986-87) 546
D 2 INC (USA) (1992-93) 510
D.9 Bébé Series (Jodel) (1992-93) 476
D.11 (Jodel) (1992-93) 476
D.18 (Jodel) (1992-93) 476
D.19 (Jodel) (1992-93) 477
D31 (PFA) (1988-89) 547
D.112 Club (Jodel) (1992-93) 476
D.113 (Jodel) (1992-93) 476
D.119 (Jodel) (1992-93) 476
D.150 Mascaret (Jodel) (1992-93) 477
D-201 (D'Apuzzo) (1992-93) 511
DA-2A (Davis) (D2) (1992-93) 510
DA-2B (DZ) (1992-93) 510
DA-5A (Davis) (D2) (1992-93) 511
DA-7 (Davis) (1992-93) 512
DD-100 (Delmotte) (1987-88) 573
DD-200 (Delmotte) (1987-88) 573
DK-1 Der Kricket (FLSZ) (1992-93) 516
DK-3 (Janowski) (1988-89) 540
DL 260 (Lesage) (1986-87) 582
DP01 (Prost) (1984-85) 671
DX-1 (Davis Wing) (1986-87) 654

Daki 530Z/Ikenga (Gyro 2000) (1992-93) 518
DALLACH FLUGZEUGE GmbH (Germany) (1992-93) 556
DANIS (see Aile Volante B. Danis) (1987-88) 572
D'APUZZO, NICHOLAS E (USA) (1992-93) 511
Dart M. J.3 (Jurca) (1992-93) 478
Das Ultralighterfighter (Airplane Factory) (1989-90) 616
DAVENPORT, BRAD (USA) (1986-87) 653
DAVIS, LEEON D (USA) (1992-93) 512
DAVIS WING LTD (USA) (1992-93) 512
DEDALUS Srl (Italy) (1992-93) 486
DEGRAW, RICHARD R (USA) (1986-87) 654
DELEMONTEZ-DESJARDINS (France) (1986-87) 572
DELMOTTE (France) (1987-88) 573
Delta (Nordpunkt) (1992-93) 490
Delta Agro (Beliovkin) (1992-93) 559
Delta Bird (Hovey) (1991-92) 598
Delta Dart (Ritec) (1990-91) 609
Delta Dart II (C-Con) (1992-93) 556
Delta Hawk (Hovey) (1991-92) 598
Delta JD-2 (Dyke Aircraft) (1992-93) 512
DELTA TECHNOLOGY (USA) (1987-88) 656
Demoiselle (Ultra Efficient Products) (1992-93) 572
DENNEY AEROCRAFT COMPANY (USA) (1992-93) 512
Der Jäger D. IX (White) (1988-89) 613
Der Kricket, DK-1 (FLSZ) (1992-93) 516
DESIGNABILITY (1984-85) 678
DESIGNABILITY LTD (UK) (1983-84) 647
Diamant (Franz) (1987-88) 594
Diana (Hoffmann) (1986-87) 595
DIEHL AERO-NAUTICAL (USA) (1991-92) 597
DIMANCHEV, GEORGI/VALKANOV, VESELIN (Bulgaria) (1992-93) 464
DiMASCIO (USA) (1986-87) 656
Dipper W-7 (Collins Aero) (1992-93) 509
Discovery (Progress) (1992-93) 529
DISTRI SNAP SARL (France) (1988-89) 522
Djins 300 (Ulm Technic Acromarine) (1984-85) 668
DJURIC, VLADIMIR (West Germany) (1986-87) 594
DMITRIYEV, V (1987-88) 612
DOBROCINSKI, TADEUSZ (Poland) (1984-85) 673
DOHET, RAMYOND (France) (1987-88) 573
Don Quixote J-1B (Alpha) (1992-93) 502
DORNA, H. F. CO (Iran) (1992-93) 486
DOUBLE EAGLE AIRCRAFT (USA) (1986-87) 656
Double Eagle TU-10 (Mitchell Aerospace) (1991-92) 600
Doublestar (Starflight) (1983-84) 676
Dragon (Dragon) (1984-85) 678
*Dragonfly (Viking)
DRAGON LIGHT AIRCRAFT CO LTD (UK) (1985-86) 666
Dragoon (Pegasus) (1987-88) 620
Dragster 25 (Cosmos) (1987-88) 570
Dragster 25 (Skylines/Cosmos) (1986-87) 709
Dragster 34 (Cosmos) (1987-88) 570
Dragster 34 (Skylines/Cosmos) (1986-87) 709
Dream (NAC) (1986-87) 690
Drifter/ARV Series (Maxair) (1992-93) 526
DRIGGERS, MICHAEL (USA) (1987-88) 658
Druine D31 Turbulent (PFA) (1988-89) 547

Dual (American Air) (1987-88) 638
Dual Star (Pioneer) (1985-86) 702
Dual Striker Trike (Flexi-Form) (1986-87) 612
Dual Trainer Raven (Hornet) (1986-87) 614
Duet (Designability) (1983-84) 647
Duet (Jordan Aviation) (1984-85) 680
DURAND ASSOCIATES INC (USA) (1992-93) 512
Durand Mk V (Durand) (1992-93) 512
DURUBLE, ROLAND (France) (1986-87) 572
DYKE AIRCRAFT (USA) (1992-93) 512
DYNALI SA (Belgium) (1992-93) 551
Dynamic Trick (Polaris) (1988-89) 538

E

EA-1 Kingfisher (Anderson) (1986-87) 636
EA-1 Kingfisher (Anderson-Warner) (1989-90) 600
EAC-3 Pouplume (Croses) (1991-92) 513
EC-6 Criquet (Croses) (1992-93) 476
EC-8 Tourisme (Croses) (1991-92) 514
E-Racer (Sierra Delta) (1992-93) 535
EZ-1 (Aircraft Development) (1992-93) 562
E. Z Bird (Gyro 2000) (1992-93) 566
EZE-Max (T. E. A. M.) (1989-90) 621

Eagle (American Aerolights) (1983-84) 653
Eagle (Ultrasport) (1984-85) 672
Eagle 2-PLC (Aeroservice Italiana) (1992-93) 558
Eagle 2-PLC (Ultrasport) (1986-87) 603
Eagle II (Aviat) (1992-93) 503
Eagle II (Christen) (1990-91) 568
EAGLE PERFORMANCE (USA) (1992-93) 564
EAGLE WINGS AIRCRAFT (USA) (1988-89) 567
Eagle X (Composite Industries) (1987-88) 537
Eagle XL (Aeroservice Italiana) (1992-93) 558
Eagle XL (American Aerolights) (1983-84) 653
Eagle XL (Ultrasport) (1986-87) 602
Eagle XL 2-place (American Aerolights) (1983-84) 653
Eagle XL Country (Aeroservice Italiano) (1992-93) 558
EARLY BIRD AIRCRAFT COMPANY (USA) (1992-93) 564
EARTHSTAR AIRCRAFT INC (USA) (1992-93) 513, 564
EASTERN ULTRALIGHTS INC (USA) (1984-85) 694
EASTWOOD AIRCRAFT PTY LTD (Australia) (1992-93) 550
EAVES, LEONARD (USA) (1987-88) 660
EBERSCHOFF AND COMPANY (USA) (1988-89) 568
Eclair VI (CMV) (1985-86) 652
Eclipse (Phase III) (1984-85) 710
Ecnelight (Columbia Plastics) (1983-84) 658
Edelweiss (Duruble) (1986-87) 572
Edelweiss 150 (Duruble) (1986-87) 572
EGRETTE, Association (France) (1986-87) 573
EICH, JAMES P (USA) (1991-92) 549
EIPPER INDUSTRIES (USA) (1988-89) 568
EKSTRÖM, STAFFAN W (Sweden) (1988-89) 541
Electric Powered Glider (Gigax) (1984-85) 674
Elete (Rotorway) (1990-91) 590
El Moscardon (Farge) (1987-88) 537
*ELMWOOD AVIATION (Canada)
Emeraude (Piel) (1992-93) 481
EPA Costruzioni Aeronautiche (Italy) (1992-93) 558
Epervier (Epervier) (1992-93) 551
*Epervier ARV (Epervier)
*EPERVIER AVIATION SA (Belgium)
Epsilon (Distri Snap) (1988-89) 522
ERPALS INDUSTRIE (see Etudes et réalisations de prototypes pour l'aviation légère et sportive) (1986-87) 574
Ersatz Junkers CL-1 (Haigh) (1986-87) 669
E Sporty (Falconar) (1992-93) 467
Esqualo II (MIM) (1987-88) 544
ETS NION AÉRONAUTIQUES (France) (1992-93) 556
ETUDES ET RÉALISATIONS DE PROTOTYPES POUR L'AVIATION LÉGÈRE ET SPORTIVE (France) (1986-87) 574
EURONEF SA (Belgium) (1992-93) 464
Europlane (Stern) (1992-93) 484
EURO WING (UK) (1983-84) 648
EVANS AIRCRAFT (USA) (1992-93) 513
EVERGREEN ULTRALITE INC (USA) (1986-87) 661
Excel (Alpair) (1987-88) 602
Excelsior (St Croix) (1992-93) 570
Exec (RotorWay) (1990-91) 590
*Exec 90 (RotorWay)
Executive (Hornet) (1987-88) 614
*EXPERIMENTAL AVIATION (USA)
Explorer (Advanced Aviation) (1992-93) 496
*Express (Wheeler)
Exp-Spirito (UFM) (1983-84) 680
E. Z. Bird (Gyro 2000) (1992-93) 518

F

F-1B (Polaris) (1992-93) 559
F2-Foxcat (Aeroservice Italiana) (1992-93) 558
F2 Foxcat (Ultrasport) (1986-87) 603
F2-Foxcat 2 PLC (Aeroservice Italiana) (1992-93) 558
F2 Twin Foxcat (Ultrasport) (1986-87) 603
F3 (CUP) (1992-93) 558
F.3 (Féré) (1986-87) 574
F4U Corsair (Air Nostalgia) (1992-93) 502
F4U Corsair (WAR) (1988-89) 612
*F.8L Falco (Sequoia)
F9 (Falconar) (1992-93) 467
F10 (Falconar) (1992-93) 467
F11A (Falconar) (1992-93) 467
F12 (Falconar) (1992-93) 468
FA-1 (Fulton) (1991-92) 528
FB-1 (Tirith) (1986-87) 621
FB-2 (Tirith) (1986-87) 621
FK6 (MBB) (1987-88) 596
FK9 (Funk) (1991-92) 589
FK9-503 (B & F Technik) (1992-93) 556
FLSZ (see Flight Level Six-Zero) (1992-93) 516

*FM-2 Air Shark (Freedom Master)
FP-101 (Fisher) (1992-93) 565
FP-202 Koala (Fisher) (1992-93) 565
FP-303 (Fisher) (1992-93) 565
FP-404 (Fisher) (1992-93) 515
FP-404 Classic (Fisher) (1992-93) 565
FP-404 Classic II (Fisher) (1992-93) 565
FP-505 Skeeter (Fisher) (1992-93) 565
FP-606 Sky Baby (Fisher) (1992-93) 565
FSRW-1 (Smith) (1987-88) 541
Fw 190A Focke-Wulf (Thunder Wings) (1988-89) 606

Fair-Fax Trikes (Riemann) (1987-88) 596
*Falco F.8L (Sequoia)
Falcon (American Aerolights) (1983-84) 654
Falcon (American Aircraft) (1992-93) 563
Falcon (Blalock) (1987-88) 642
Falcon, AG-38 (Mitchell Wing) (1985-86) 698
Falcon B.1 (SCWAL) (1992-93) 551
Falcon B.2 (SCWAL) (1992-93) 551
Falcon XP (American Aircraft) (1987-88) 639
FALCONAR AVIATION (Canada) (1992-93) 467, 552
FARGGE, JUAN DE LA (Argentina) (1987-88) 537
Farmate (SV Aircraft) (1992-93) 550
Faucheux (Danis) (1987-88) 572
FÉRÉ, RENÉ (France) (1986-87) 574
FEUGRAY, G (France) (1986-87) 574
Fiesta (Aero Innovations) (1987-88) 631
FIGHTER ESCORT WINGS (USA) (1992-93) 514
Firebird FB-1 (Tirith) (1986-87) 621
Firebird FB-2 (Tirith) (1986-87) 621
Firebird Flyer (Competition Aircraft) (1983-84) 658
FIREBIRD LEICHTFLUGZEUGBAU GmbH (West Germany) (1984-85) 669
FIREBIRD SCHWIEGER & JEHLE (West Germany) (1986-87) 594
Firebolt (Starfire) (1987-88) 710
FIRE BOLT (USA) (1992-93) 514
Firebolt Convertible (Starfire) (1991-92) 572
Firefly (Payne) (1987-88) 690
Firefly (Statler) (1986-87) 715
Firestar KX (Kolb) (1992-93) 566
FIRST STRIKE AVIATION INC (USA) (1992-93) 514
FISHER AERO CORPORATION (USA) (1992-93) 515
FISHERCRAFT INC (USA) (1989-90) 574
FISHER FLYING PRODUCTS INC (USA) (1992-93) 514, 565
Flamingon (Kolodziej) (1983-84) 692
Flash (Pegasus) (1987-88) 618
Flash-3 (Wolff) (1991-92) 524
FLETCHER, DANIEL & RICHARD (Canada) (1987-88) 548
FLEXI-FORM SKYSAILS (UK) (1986-87) 612
FLICK AVIATION (USA) (1987-88) 664
Flick Monoplane (Flick Aviation) (1987-88) 664
FLIGHT 95 (Australia) (1987-88) 538
FLIGHT DYNAMICS INC (USA) (1986-87) 664
FLIGHT LEVEL SIX-ZERO INC (USA) (1992-93) 516
FLIGHT RESEARCH (UK) (1986-87) 612
FlightStar (Pampa's Bull) (1991-92) 583
FlightStar (Pioneer) (1985-86) 703
FLIGHTWORKS CORPORATION (USA) (1992-93) 516
Flitzer Z-1 (Sky Craft) (1992-93) 492
Floater (Beaver) (1992-93) 552
FLOW STATE DESIGN (USA) (1992-93) 565
FLUGTECHNISCHE FORSCHUNGSGRUPPE an der TECHNISCHEN UNIVERSITÄT MÜNCHEN (Germany) (1991-92) 522
*Fly Baby 1-A (Bowers)
*Fly Baby 1-B (Bowers)
Flycycle (Kennedy Aircraft) (1991-92) 555
Flyer (Kolb) (1983-84) 665
Flying-Boat (Polaris Motor) (1987-88) 605
Flying Flea (Falconar Aviation) (1992-93) 468, 552
FLYLITE (UK) (1986-87) 613
Focke-Wulf 190 (WAR) (1988-89) 612
Focke-Wulf Fw 190 (Thunder Wings) (1988-89) 606
Fokker D. VII (Squadron Aviation) (1988-89) 600
FOKKER DR 1 (Redfern) (1992-93) 532
Fokker Dr 1 Replica Triplane (Sands) (1988-89) 596
Fokker D. VII Replica (Iseman) (1986-87) 675
Fokker-Lite (Tact-Avia) (1986-87) 720
Forger (Chargus) (1984-85) 677
Forger (Pegasus) (1987-88) 618
FORSGREN, LYLE (USA) (1987-88) 665
FOR TWO ENTERPRISES (USA) (1992-93) 565
FOTON (USSR) (1991-92) 527
Fourtouna M. J.14 (Jurca) (1992-93) 478
Fox (Gigax) (1984-85) 674
Fox-C22 (Ikarus Deutschland) (1988-89) 533
Fox-D (Ikarus Deutschland) (1989-90) 610
Foxbat (Manta) (1987-88) 680
Foxcat (Ultrasport) (1986-87) 603
FRANCE AÉROLIGHTS SA (France) (1987-88) 574
FRANZ (see Ul-Bau-Franz) (1987-88) 594
Fred Series 3 (Clutton-Tabenor) (1987-88) 651
Freedom 28 (BAC) (1992-93) 563
Freedom 48 (BAC) (1988-89) 558
FREEDOM FLIERS INC (USA) (1987-88) 572
Freedom Machine KB-2 (Brock) (1992-93) 506
*FREEDOM MASTER CORPORATION (USA)
FREE FLIGHT AVIATION PTY LTD (Australia) (1986-87) 538
FREE SPIRIT AIRCRAFT COMPANY LTD (USA) (1988-89) 573
Free Spirit Mark II (Cabrinha) (1992-93) 508
FRIEBE LUFTFAHRT-BEDARF (West Germany) (1983-84) 642
FRIEDMAN, JOSEPH (USA) (1990-91) 575
Frolov Monoplane (1987-88) 612
FROLOV, V (USSR) (1987-88) 612
FULMAR (France) (1983-84) 640
FULMAR ULTRALIGHT DIFFUSION (Belgium) (1985-86) 642
FULTON AIRCRAFT LTD (UK) (1991-92) 528
FUM (Australia) (1987-88) 538
Fun Fly (Swiss Aerolight) (1992-93) 560
Fun GT 447 (Air Creation) (1992-93) 554
Fun Javelin (1986-87) 538

718 INDEXES: PRIVATE AIRCRAFT—**F - K**

FUNK, OTTO and PETER (Germany) .... (1991-92) 589
Funplane 1 (UPM) .... (1990-91) 610
Fun Racer 447 (Air Creation) .... (1992-93) 555
Fury (Two Wings) .... (1992-93) 571
Fury II (Isaacs) .... (1988-89) 544

G

G11 (Moto-Delta) .... (1983-84) 640
G-802 Orion (Aerodis) .... (1991-92) 512
*GA55 LightWing (Hughes)
GA-55 Aeropower LightWing (Hughes) .... (1990-91) 532
GB-3 Vetro (Bezzola) .... (1987-88) 610
G. B.10 Pou-Push (Briffaud) .... (1986-87) 566
GL06 (Landray) .... (1987-88) 579
GMD 330 (Goodwin) .... (1985-86) 667
GN-1 Aircamper (Grega) .... (1992-93) 518
GP3 Osprey II (Osprey Aircraft) .... (1992-93) 527
GP4 (Osprey Aircraft) .... (1992-93) 528
GQ-01 Monogast (Quaissard) .... (1986-87) 590
GR-2 Whisper (Grove) .... (1987-88) 668
*GR-582 LightWing (Hughes)
*GR-912 LightWing (Hughes)
GT1000 GT (Rotary Air Force) .... (1992-93) 470
GT, Quicksilver (Eipper) .... (1988-89) 569
GW-1 Windwagon (Watson) .... (1992-93) 546
GY-120 (Gardan) .... (1986-87) 575

Gambit (SciCraft) .... (1992-93) 486
GANZER, DAVID, W (USA) .... (1987-88) 666
GARDAN (see Avions Yves Gardan) .... (1986-87) 575
GARDNER, DEGA APPLIED TECHNOLOGY (UK) .... (1987-88) 614
Garnis (LAK) .... (1988-89) 543
GATARD (see Avions A. Gatard) .... (1989-90) 544
GEISER, JEAN MARC (France) .... (1984-85) 668
Gemini (Ganzer) .... (1987-88) 666
Gemini (Thruster Aircraft) .... (1987-88) 542
Gemini 2 (Mainair Sports) .... (1987-88) 616
Gemini Flash 2 (Mainair Sports) .... (1987-88) 616
Gemini Flash 2 Alpha (Mainair Sports) .... (1992-93) 561
GEMINI INTERNATIONAL INC (USA) .... (1991-92) 597
GENERAL GLIDERS (Italy) .... (1992-93) 558
Gerfan (Air Composite) .... (1992-93) 474
Gidra (Aeroprakt) .... (1986-87) 610
GIGAX, HANS (Switzerland) .... (1985-86) 663
Gil (Kolodziej) .... (1984-85) 692
Giro-Heli (Berger) .... (1987-88) 610
Giro Kopter (BW Rotar) .... (1992-93) 563
Glasair II (Stoddard-Hamilton) .... (1989-90) 594
*Glasair II-S (Stoddard-Hamilton)
*Glasair III (Stoddard-Hamilton)
Glasair FT (Stoddard-Hamilton) .... (1989-90) 594
Glasair RG (Stoddard-Hamilton) .... (1989-90) 594
Glasair TD (Stoddard-Hamilton) .... (1989-90) 594
Glastar (Stoddard-Hamilton) .... (1992-93) 539
Gnatsum M. J.7, M. J.77 (Jurca) .... (1992-93) 478
GOLDEN AGE AIRCRAFT COMPANY (USA) .... (1985-86) 688
GOLDEN CIRCLE AIR (USA) .... (1992-93) 565
Golden Hawk (Falconar) .... (1992-93) 467, 552
Golden Interceptor (Eagle Performance) .... (1992-93) 564
Goldwing (Goldwing) .... (1986-87) 667
GOLDWING LTD (USA) .... (1986-87) 667
GOODWIN PROPELLERS (UK) .... (1985-86) 667
GORDON BEDSON AND ASSOCIATES (Australia) .... (1986-87) 537
Gotz 50 (Heinrich Zettle) .... (1992-93) 557
Grach (Leningrad Central) .... (1982-83) 634
Graflite (Lundy) .... (1991-92) 558
*GREAT PLAINS AIRCRAFT SUPPLY CO INC (USA)
GREENWOOD AIRCRAFT INC (USA) .... (1986-87) 668
GREGA, JOHN W (USA) .... (1992-93) 518
GROEN BROTHERS AVIATION INC (USA) .... (1992-93) 518
Ground Troop Army Trainers (Rotec) .... (1987-88) 703
GROVE AIRCRAFT COMPANY (USA) .... (1987-88) 668
GROVER AIRCRAFT CORPORATION (USA) .... (1986-87) 669
Gryf ULMI (Tepelne) .... (1992-93) 554
GUERPONT (France) .... (1988-89) 522
Guppy SNS-2 (Sorrell) .... (1992-93) 536
Guri (Pazmany) .... (1988-89) 588
Gypsy Trainer (R. D. Aircraft) .... (1992-93) 569
GYRO 2000 INC (USA) .... (1992-93) 518, 566
Gyrochopper II (Alderfer) .... (1988-89) 556
Gyrochopper III (Alderfer) .... (1988-89) 556
Gyro Copter (Bensen) .... (1986-87) 638
Gyrocopter (Wombat) .... (1992-93) 493
Gyro Glider (Bensen) .... (1986-87) 638
Gyroplane, JE-2 (Eich) .... (1991-92) 549

H

H-3 Pegasus (Howland) .... (1992-93) 566
H-39 Diana (Hoffmann) .... (1986-87) 595
HA-2M Sportster (Aircraft Designs) .... (1992-93) 500
HFL-UL (Höhenflug) .... (1984-85) 670
H. M.81 (Midwest) .... (1985-86) 697
H. M.81 Tomcat (Waspair) .... (1983-84) 683
HM-290E Flying Flea (Falconar) .... (1992-93) 552
HM-293 (Falconar) .... (1992-93) 552
HM-360 (Falconar) .... (1992-93) 552
HM-380 (Falconar) .... (1992-93) 552
HM-1000 Balerit (Mignet) .... (1992-93) 556
HN 433 Menestral (Nicollier) .... (1992-93) 480
HN 434 Super Menestral (Nicollier) .... (1992-93) 480
HN 600 Week-End (Nicollier) .... (1992-93) 481
HN 700 Menestrel II (Nicollier) .... (1992-93) 480
HX-321 (Hamilton) .... (1988-89) 574

HAIGH, JOSEPH (USA) .... (1986-87) 669
Half Pint (Medway Microlights) .... (1988-89) 546
HALLMARK AIRCRAFT CORPORATION (USA) .... (1984-85) 700
HAMILTON AEROSPACE (USA) .... (1988-89) 574
Hamlet (Aeroprakt) .... (1986-87) 610
HAPI ENGINES INC (USA) .... (1991-92) 553
HARMON (USA) .... (1987-88) 668
Harrier (Wills Wing) .... (1982-83) 657
HATZ AIRPLANE SHOP (USA) .... (1989-90) 576
Hawk (Austin) .... (1986-87) 636
Hawk (CGS Aeronautics) .... (1990-91) 616
Hawk (Kestrel) .... (1992-93) 468
Hawk Arrow (CGS Aviation) .... (1992-93) 564
Hawk II Arrow (CGS Aviation) .... (1992-93) 564
Hawk Classic (CGS Aviation) .... (1992-93) 564
Hawk II Classic (CGS Aviation) .... (1992-93) 564
Hawk Series (Groen) .... (1992-93) 518
Hawk Tandem (CGS Aeronautics) .... (1990-91) 616
HEADBERG AVIATION INC (USA) .... (1992-93) 519
Helicopter (BW Rotor) .... (1992-93) 563
HELICRAFT INC (USA) .... (1992-93) 519, 566
Helios (Aéro Delta) .... (1986-87) 562
Heron (Fulmar) .... (1984-85) 659
HEWA-TECHNICS (West Germany) .... (1990-91) 550
HFL-FLUGZEUGBAU GmbH (Germany) .... (1992-93) 556
Hidra (Aeroprakt) .... (1987-88) 611
HIGHCRAFT AERO-MARINE (USA) .... (1986-87) 670
High Gross STOL (Avid Aircraft) .... (1992-93) 504
Hi-Max (T. E. A. M.) .... (1992-93) 571
Hi-Nuski Huski (Advanced Aviation) .... (1983-84) 652
HINZ, LUCIA AND BERNHARD (Germany) .... (1991-92) 522
Hiperbipe SNS-7 (Sorrell) .... (1992-93) 536
Hiperlight EXB (Sorrell) .... (1992-93) 570
Hiperlight SNS-8 (Sorrell) .... (1992-93) 570
HIPP'S SUPERBIRDS INC (USA) .... (1992-93) 520, 566
Hirondelle (Chasle) .... (1992-93) 555
Hirondelle, PGK-1 (Western Aircraft) .... (1992-93) 472
HISTORICAL AIRCRAFT CORPORATION (USA) .... (1992-93) 520
Hitchiker (Labahan) .... (1991-92) 583
HIWAY HANG GLIDERS LTD (UK) .... (1984-85) 726
HLAMOT DESIGN (Hungary) .... (1992-93) 558
Hlamot-M (Hlamot Design) .... (1992-93) 558
HOBBYSTICA AVIO (Italy) .... (1992-93) 559
HOFFMANN FLUGZEUGBAU KG, WOLF (West Germany) .... (1986-87) 595
HOLCOMB, JERRY (USA) .... (1991-92) 554
Hollman Bumble Bee (Aircraft Designs) .... (1992-93) 500
Hollman Sportster (Aircraft Designs) .... (1992-93) 500
Honcho II (Magnum) .... (1992-93) 567
HONGXING MACHINERY FACTORY (China) .... (1985-86) 650
Horizon (Aero Visions) .... (1991-92) 536
Horizon 1 (Fisher Aero) .... (1992-93) 516
Horizon 2 (Fisher Aero) .... (1992-93) 516
Hornet (Free Flight) .... (1986-87) 538
Hornet (SR-1) .... (1985-86) 709
Hornet 130S (Free Flight) .... (1983-84) 633
HORNET MICROLIGHTS LTD (UK) .... (1991-92) 594
HORNET MICROLIGHTS (TEMPLEWARD LTD) (UK) .... (1987-88) 614
Hornet R (Hornet) .... (1991-92) 594
Hornet RS (Hornet) .... (1991-92) 594
Hornet RSE (Hornet) .... (1991-92) 594
HOSKINS, CRAIG (USA) .... (1986-87) 671
HOVEY, ROBERT W. (USA) .... (1991-92) 598
HOWARD HUGHES ENGINEERING PTY LTD (Australia) .... (1992-93) 463
HOWATHERM GmbH & Co KG (West Germany) .... (1986-87) 596
HOWLAND AERO DESIGNS (USA) .... (1992-93) 566
HUABEI MACHINERY FACTORY (China) .... (1987-88) 562
Huber 101-1 Aero (Huber) .... (1985-86) 690
HUBER, JAMES M (USA) .... (1985-86) 690
*HUGHES, HOWARD, ENGINEERING PTY LTD (Australia)
HUMBERT (France) .... (1987-88) 574
Humlan (Ekström) .... (1988-89) 541
HUMMEL AIRCRAFT (USA) .... (1988-89) 576
Hummel Bird (Hummel) .... (1988-89) 576
Hummel Bird (Morry Hummel) .... (1992-93) 566
Hummer (Maxair) .... (1985-86) 696
Hummingbird (Degraw) .... (1986-87) 654
Hummingbird (Gemini) .... (1984-85) 698
*Hummingbird (VAT)
Hummingbird 103 (Gemini) .... (1991-92) 597
Hummingbird Longwing (Gemini) .... (1991-92) 597
HUNTAIR LTD (UK) .... (1985-86) 667
Hurricane (Windryder) .... (1992-93) 548
Hurricane 100 (Windryder) .... (1992-93) 548
HUSKY MANUFACTURING LTD (Canada) .... (1991-92) 505
HUTCHINSON AIRCRAFT COMPANY (USA) .... (1989-90) 577
Hybred R (Medway Microlights) .... (1992-93) 561
Hydro-Copter (Bensen) .... (1986-87) 638
Hydro-Glider (Bensen) .... (1986-87) 638
Hydrolight (Diehl) .... (1988-89) 566
Hydroplum (Tisserand/Hydroplum) .... (1992-93) 476
Hydroplum II/Petrel (Tisserand/Hydroplum) .... (1989-90) 544
HYDROPLUM SARL (France) .... (1992-93) 476

I

I-66L San Francesco (Iannotta) .... (1992-93) 486
IAC (see International Aeromarine Corporation) .... (1986-87) 674
IAC (see International Aircraft Corporation) .... (1986-87) 674
IPE-06 (Industria Paranaense) .... (1991-92) 585

Iannotta 940 Zefiro (Iannotta) .... (1988-89) 538
IANNOTTA DOTTING ORLANDO (Italy) .... (1992-93) 486
Ibis (Delemontez) .... (1986-87) 572
Ibis RJ-03 (Junqua) .... (1992-93) 477
IKARUS-COMCO (West Germany) .... (1984-85) 670
IKARUS DEUTSCHLAND COMCO GmbH (Germany) .... (1992-93) 556
IKARUSFLUG (Germany) .... (1992-93) 556
Ikenga/Cygnus 21P (Gyro 2000) .... (1992-93) 518
Ikenga/Cygnus 21T (Gyro 2000) .... (1990-91) 576
Ikenga/Cygnus 21TX (Gyro 2000) .... (1992-93) 518
Imp (Aerocar/Brown) .... (1986-87) 643
INAV Ltd (USA) .... (1986-87) 674
INDUSTRIA PARANAENSE DE ESTRUTURAS LTDA (Brazil) .... (1991-92) 585
Instructor (Robertson) .... (1984-85) 712
INSTYTUT LOTNICTWA (Poland) .... (1992-93) 559
Intermezzo (Aéroplum) .... (1986-87) 563
INTERNATIONAL AEROMARINE CORPORATION (USA) .... (1986-87) 674
INTERNATIONAL AIRCRAFT CORPORATION (USA) .... (1986-87) 674
INTERNATIONAL ULTRALIGHTS INC (USA) .... (1985-86) 692
INTERNATIONAL ULTRALITE AVIATION INC (USA) .... (1985-86) 692
*Interstate Privateer (Arctic)
*Interstate S1B2 Arctic Tern (Arctic)
Intruder SNS-10 (Sorrell) .... (1992-93) 570
Invader (Hornet) .... (1986-87) 614
Invader Mk III-B (Ultra Efficient Products) .... (1992-93) 572
Invader Updated (Hornet) .... (1986-87) 614
ISAACS, JOHN O (UK) .... (1988-89) 544
ISEMAN BOB (USA) .... (1986-87) 675
Italzair (Italzair) .... (1986-87) 602
Italzair (Nike Aeronautica) .... (1988-89) 538
ITALZAIR (see Nike Aeronautica) .... (1986-87) 602

J

J-1B Don Quixote (Alpha) .... (1992-93) 502
J-2 Polonez (Janowski) .... (1992-93) 559
J2G2 (Black) .... (1983-84) 656
J-3 Kitten (Anglin) .... (1984-85) 688
J-3 Kitten (Grover) .... (1985-86) 689
J-3 Kitten (Hipp's Superbirds) .... (1992-93) 566
J-4 Sportster (Grover) .... (1985-86) 689
J-4 Sportster (Hipp's Superbirds) .... (1992-93) 566
J 4B (Barnett Rotorcraft) .... (1992-93) 504
J 4B-2 (Barnett Rotorcraft) .... (1992-93) 504
J-5 Janowski (Marko-Elektronik) .... (1986-87) 605
J-5 Marco (Alpha) .... (1992-93) 488
J-5 Marco (Hewa-Technics) .... (1990-91) 550
J-5 Marco (Marko-Elektronik) .... (1986-87) 605
J-6 Karatoo (Skyway Aircraft) .... (1991-92) 570
JB-1000 (Leader's International) .... (1992-93) 566
JC-01 (Coupé-Aviation) .... (1992-93) 475
JC-3 (Coupé-Aviation) .... (1992-93) 476
JC-24C Weedhopper C (Weedhopper) .... (1983-84) 683
JC-200 (Coupé-Aviation) .... (1992-93) 476
JD-2 Delta (Dyke Aircraft) .... (1992-93) 512
JE-2 Gyroplane (Eich) .... (1991-92) 549
J. H.03 Le Courlis (Erpals Industrie) .... (1986-87) 574
JN-1 (Jim Peris) .... (1992-93) 567
J. T.1 Monoplane (Taylor) .... (1992-93) 492
J. T.2 Titch (Taylor) .... (1992-93) 493
JT-5 (Tervamaki) .... (1988-89) 516
Ju 87B2 replica (Langhurst) .... (1987-88) 676

*Jabiru (Jabiru Aircraft)
*JABIRU AIRCRAFT (AUSTRALIA)
Jackeroo (Winton) .... (1986-87) 543
Jaeger (Eagle Wings) .... (1988-89) 567
JANOWSKI, JAROSLOW (Poland) .... (1992-93) 559
Javelin (Flight 95) .... (1987-88) 538
JAVELIN AIRCRAFT COMPANY INC (USA) .... (1992-93) 520
Jeep (Fum) .... (1987-88) 538
Jenny (Cloud Dancer) .... (1987-88) 65
Jenny (Early Bird) .... (1992-93) 564
Jet Hawk II (Saunders) .... (1987-88) 704
Jet Wing ATV (Flight Designs) .... (1984-85) 696
JINZHOU MICROLIGHT HELICOPTER COMPANY (China) .... (1986-87) 560
JODEL, AVIONS SA (France) .... (1992-93) 476
JONES, MASON (Canada) .... (1986-87) 546
JORDAN AVIATION LTD (UK) .... (1985-86) 668
Jr Ace, Pober (Acro Sport) .... (1992-93) 495
JULIAN, CHRIS (UK) .... (1991-92) 529
Junior Ace Model E (Ace Aircraft) .... (1992-93) 494
Juno (Thor) .... (1988-89) 512
JUNQUA/ANDREAZZA, ROGER and JEAN-YVES (France) .... (1986-87) 576
JUNQUA, ROGER (France) .... (1992-93) 477
Junster I (MacFam) .... (1992-93) 524
Junster II (MacFam) .... (1992-93) 524
*JURCA, MARCEL (France)

K

K-236 (Klampfl) .... (1992-93) 464
KB-2 Freedom Machine (Brock) .... (1992-93) 506
KB-3 (Ken Brock) .... (1992-93) 563
KDA (see Kany, Dreyer and Aupy) .... (1983-84) 640
Kh-14A (Dmitriev) .... (1987-88) 612
KR-1 (Rand Robinson) .... (1992-93) 531
KR-2 (Rand Robinson) .... (1992-93) 531
KR-100 (Rand Robinson) .... (1992-93) 531
KSML-IM Nyirseg-2 (Air Service) .... (1992-93) 558
KSML-P1 Nemere (Air Service) .... (1992-93) 557
KSML-P2 Nyirseg (Air Service) .... (1992-93) 557
KX 400 (Kolb) .... (1985-86) 692

KALER CORPORATION, THE (USA) .... (1985-86) 692
Kanion (Sznapki) .... (1983-84) 687, 692
KANY, DREYER and AUPY (France) .... (1983-84) 640
Karatoo J-6 (Skyway Aircraft) .... (1991-92) 570
Kasia DK-3 (Marganski) .... (1988-89) 540
Kasperwing 1-80 (Cascade) .... (1988-89) 562
*Katana T30 (Terzi)
Katta (Yakovlev) .... (1991-92) 528
Kelly-D (Kelly) .... (1992-93) 520
Kelly-D11 (Kelly) .... (1989-90) 578
KELLY, DUDLEY R (USA) .... (1992-93) 520

K – M—PRIVATE AIRCRAFT: INDEXES 719

KEN BROCK MANUFACTURING INC (USA) ... *(1992-93)* 506, 563
KENNEDY AIRCRAFT COMPANY INC (USA) ... *(1991-92)* 555
KENSGAILA'S AIRCRAFT ENTERPRISE, V (Lithuania) ... *(1992-93)* 487
Kestrel (Helicraft) ... *(1992-93)* 566
Kestrel Hawk (Sun Fun) ... *(1987-88)* 552
KESTREL SPORT AVIATION (Canada) ... *(1992-93)* 468
KIMBERLEY, GARETH J. (Australia) ... *(1992-93)* 550
KIMBREL, MICHAEL G. (USA) ... *(1991-92)* 598
King Cobra (Advanced Aviation) ... *(1985-86)* 673
King Cobra B (Advanced Aviation) ... *(1992-93)* 495
King Fisher (Anderson) ... *(1986-87)* 636
Kingfisher (Anderson-Warner) ... *(1989-90)* 600
Kis (Tri-R) ... *(1992-93)* 542
Kitfox (Denney Aerocraft) ... *(1990-91)* 572
Kitfox III (Denney Aerocraft) ... *(1991-92)* 548
Kitfox IV (Denny Aerocraft) ... *(1992-93)* 512
Kit Hawk (Kestrel) ... *(1992-93)* 468
Kitten (Anglin) ... *(1984-85)* 688
Kitten (Grover) ... *(1985-86)* 689
KLAMPFL (Austria) ... *(1992-93)* 464
Knight Twister (Payne) ... *(1985-86)* 607
Koala (Fisher) ... *(1986-87)* 662
Koala FP-202 (Fisher) ... *(1992-93)* 565
Kodiak (Sequoia) ... *(1991-92)* 569
KOLB AIRCRAFT INC (USA) ... *(1992-93)* 521, 566
Kolibri (AFG) ... *(1987-88)* 592
KOLODZIEJ, ZDZISLAW (Poland) ... *(1984-85)* 725, 730
KONSUPROD GmbH & Co KG (Germany) ... *(1992-93)* 557
Kopykat (Kopykat) ... *(1992-93)* 566
KOPYKAT (USA) ... *(1992-93)* 566
KOREAN AIR (South Korea) ... *(1992-93)* 559
KOROLEV, Y (USSR) ... *(1987-88)* 612
KUIBYSHEV, PETER ALMURZIN (Russia) ... *(1992-93)* 560

L

L5 (Lucas) ... *(1992-93)* 479
L6 (Lucas) ... *(1992-93)* 480
L7 (Lucas) ... *(1992-93)* 480
L10 (Lucas) ... *(1992-93)* 480
LA-4 Vortex (Langebro) ... *(1992-93)* 490
L. A4a Luton Minor (PFA) ... *(1988-89)* 547
LACO-125 (Laven) ... *(1991-92)* 556
LACO-145 (Laven) ... *(1991-92)* 556
LACO-145A Special (Laven) ... *(1992-93)* 522
LAK (see Litovskaya Aviatsionnaya Konstruktsiya) ... *(1991-92)* 593
LF-1 (Forsgren) ... *(1987-88)* 665
LG 150 (Leglaive-Gautier) ... *(1987-88)* 580
LK-2 Sluka (Letov) ... *(1992-93)* 554
LK-3 Nova (Letov) ... *(1992-93)* 554
LM-1 (Light Miniature) ... *(1992-93)* 522
LM-1U (Light Miniature) ... *(1992-93)* 522
LM-1X (Light Miniature) ... *(1992-93)* 522
LM-2P (Light Miniature) ... *(1991-92)* 556
LM-2U (Light Miniature) ... *(1992-93)* 523
LM-2X (Light Miniature) ... *(1992-93)* 523
LM-3U (Light Miniature) ... *(1992-93)* 523
LM-3X (Light Miniature) ... *(1992-93)* 523
LM-5X (Classic) ... *(1992-93)* 509
LNB 11 (Cuvelier) ... *(1986-87)* 571
LO-120 (LO-Fluggeratebau) ... *(1992-93)* 557
LO-120 (Vogt) ... *(1991-92)* 590
LO-125 (LO-Fluggeratebau) ... *(1992-93)* 557
LO-150 (Vogt) ... *(1991-92)* 590
LP-01 Sibylle (Lendepergt) ... *(1992-93)* 479

LABAHAN, ROBERT (Australia) ... *(1991-92)* 583
LACO (USA) ... *(1988-89)* 580
Lacroix 2L12 (Cuvelier) ... *(1986-87)* 571
Lacroix LNB-11 (Cuvelier) ... *(1986-87)* 571
Ladybird (Campbell-Jones) ... *(1983-84)* 646
Lady Bug (McAsco) ... *(1987-88)* 549
La Guêpe (ASL) ... *(1992-93)* 555
Lala 1 (Djuric) ... *(1986-87)* 594
*Lancair IV (Lancair International)
Lancair 200 (Neico Aviation) ... *(1991-92)* 562
*Lancair 235 (Lancair International)
*Lancair 320 (Lancair International)
*Lancair 360 (Lancair International)
*LANCAIR INTERNATIONAL INC (USA)
Lancer MA-4 (Marquart) ... *(1992-93)* 525
Lancer Mk IV (Pacific Kites) ... *(1984-85)* 686
LANDRAY, GILBERT (France) ... *(1987-88)* 579
LANGEBRO AVIATION (Sweden) ... *(1992-93)* 490
LANGHURST, LOUIS F. (USA) ... *(1987-88)* 676
LANOT-AVIATION (France) ... *(1986-87)* 580
LASCAUD, ETS D (France) ... *(1992-93)* 556
Laser (Kolb) ... *(1992-93)* 522
LASER AEROBATICS (Australia) ... *(1988-89)* 504
LATÉCOÈRE (see Société Industrielle Latécoère SA) ... *(1986-87)* 580
Latécoère 225 (Latécoère) ... *(1986-87)* 580
Laughing Gull (Earthstar) ... *(1990-91)* 572
Laughing Gull II (Earthstar) ... *(1990-91)* 572
LAVEN, JOE (USA) ... *(1992-93)* 522
LAVORINI, NEDO (Italy) ... *(1987-88)* 604
Lazair (Ultraflight) ... *(1989-90)* 607
Lazair II (Ultraflight) ... *(1989-90)* 607
LEADER'S INTERNATIONAL INC (USA) ... *(1992-93)* 566
LEADING EDGE AIR FOILS INC (USA) ... *(1991-92)* 567
Le Courlis (Erpals Industrie) ... *(1986-87)* 574
Lederlin 380-L (Lederlin) ... *(1992-93)* 478
LEDERLIN, FRANÇOIS (France) ... *(1992-93)* 478
LEE, G, CIRCA REPRODUCTIONS (Canada) ... *(1985-86)* 644
LEFEBVRE, ROBERT (France) ... *(1986-87)* 581
Legato (Automated Aircraft) ... *(1992-93)* 503
LEGLAIVE-GAUTIER (France) ... *(1987-88)* 580
LEICHTFLUGZEUG GmbH & Co KG (Germany) ... *(1992-93)* 557
LENDEPERGT, PATRICK (France) ... *(1992-93)* 479
Leone 7T7, T8 (AUI) ... *(1988-89)* 536
Leone T7 (AUI) ... *(1992-93)* 558
Le Pelican (Ultravia) ... *(1988-89)* 513
LESAGE, CHRISTIAN ... *(1986-87)* 582
Le Solitaire (Cuvelier) ... *(1986-87)* 571
Lespace (Proust/Sigur) ... *(1986-87)* 589
LETOV LTD (Czechoslovakia) ... *(1992-93)* 554
LETOV, RUDY (Czechoslovakia) ... *(1990-91)* 607
LIGETI AERONAUTICAL PTY LTD (Australia) ... *(1991-92)* 583
LIGHT AERO INC (USA) ... *(1989-90)* 579
Light Flyer (Freedom Fliers) ... *(1984-85)* 698
Light Flyer (Pterodactyl) ... *(1983-84)* 672
LIGHT MINIATURE AIRCRAFT INC (USA) ... *(1992-93)* 522
Lightning Bug (Lightning Bug Aircraft) ... *(1992-93)* 523
LIGHTNING BUG AIRCRAFT CORPORATION (USA) ... *(1992-93)* 523
Lightning DS (Ultra Sports) ... *(1984-85)* 684
Lightning P-38 (Mitchell Wing) ... *(1991-92)* 560
Lightning Sport Copter (Vancraft) ... *(1992-93)* 572
LightWing Aeropower 28 (Hughes) ... *(1989-90)* 534
LightWing Aeropower GA-55 (Hughes) ... *(1990-91)* 532
*LightWing GA-55 (Hughes)
*LightWing GR-582 (Hughes)
*LightWing GR-912 (Hughes)
LightWing R 55 (Hughes) ... *(1989-90)* 534
Lion Cub (MFS) ... *(1987-88)* 608
LITECRAFT INC (USA) ... *(1986-87)* 682
LITE-FLITE INC (USA) ... *(1987-88)* 679
LITOVSKAYA AVIATSIONNAYA KONSTRUKTSIYA (USSR) ... *(1991-92)* 593
LITTLE AIRPLANE COMPANY, THE (USA) ... *(1985-86)* 694
Little Bi (Simpson Midwest) ... *(1986-87)* 709
Little Dipper replica (R. D. Aircraft) ... *(1991-92)* 568
LOEHLE ENTERPRISES (USA) ... *(1992-93)* 524, 567
LO-FLUGGERATEBAU GmbH (Germany) ... *(1992-93)* 557
Lomac Trick (Polaris) ... *(1991-92)* 592
Lone Ranger (Striplin) ... *(1983-84)* 677
Lone Ranger (Striplin) ... *(1992-93)* 540
Lone Ranger Ultralight (Striplin) ... *(1992-93)* 571
LoneStar (Star Aviation) ... *(1991-92)* 572
LONG, JIM (USA) ... *(1983-84)* 666
Longnose Pelican (Ultravia) ... *(1988-89)* 513
Long Wing Super Pup (Preceptor) ... *(1992-93)* 529
LOPRESTI BROTHERS AIRCRAFT INC (USA) ... *(1987-88)* 680
LOTUS CARS LTD (Lotus) ... *(1985-86)* 668
Lotus Microlight (Lotus) ... *(1985-86)* 668
LUCANT, MOTO (France) ... *(1987-88)* 580
LUCAS, EMILE (France) ... *(1992-93)* 479
Lucky (T. E. D. A.) ... *(1991-92)* 592
*LUNDS TEKNISKE (Norway)
LUNDY, BRIAN (USA) ... *(1991-92)* 558
Luton L. A.4a Minor (PFA) ... *(1988-89)* 547

M

M1 (Firebird) ... *(1986-87)* 594
M-1 Midget Mustang (Bushby) ... *(1992-93)* 507
M-1 Slavutitch (Antonov) ... *(1992-93)* 560
M-1A Midget Mustang (Bushby) ... *(1992-93)* 507
*M-11 Mustang II (Bushby)
M20 Venture (Questair) ... *(1992-93)* 530
M-80 (Morin) ... *(1992-93)* 480
MA-4 Lancer (Marquart) ... *(1992-93)* 525
MA-5 Charger (Marquart) ... *(1992-93)* 526
MAC Mattison (see Hallmark) ... *(1984-85)* 703
MAI-89 (Foton) ... *(1991-92)* 527
MAI-890 (Aviatika) ... *(1992-93)* 489
MB-2 Colibri 2 (Brügger) ... *(1992-93)* 490
MB.10 (Boudeau) ... *(1986-87)* 566
MBA (see Micro Biplane Aviation) ... *(1985-86)* 669
MBB (see Messerschmitt-Bölkow-Blohm GmbH) ... *(1987-88)* 596
MC 15 Cricri (Colomban) ... *(1992-93)* 475
MC 100 (Colomban) ... *(1990-91)* 542
MEG-2XH (Helicraft) ... *(1992-93)* 566
MFI (see Malmo Forsknings & Innovations AB) ... *(1992-93)* 560
MFI-9 HB (Andreasson) ... *(1992-93)* 490
MFS (see Microlight Flight Systems Pty Ltd) ... *(1989-90)* 613
MIM (see Mauricio Impelizieri P. Moura) ... *(1987-88)* 544
MIT (see Massachusetts Institute of Technology) ... *(1987-88)* 684
M. J.2 Tempête (Jurca) ... *(1992-93)* 477
M. J.3 Dart (Jurca) ... *(1992-93)* 478
M. J.3H Dart (Jurca) ... *(1987-88)* 577
M. J.4 Shadow (Jurca) ... *(1992-93)* 478
*M. J.5 Sirocco (Jurca)
M. J.5 Sirocco Sport Wing (Jurca) ... *(1992-93)* 478
M. J.7 Gnatsum (Jurca) ... *(1992-93)* 478
M. J.7S Solo (Jurca) ... *(1987-88)* 578
M. J.8 (Jurca) ... *(1992-93)* 478
M. J.9 One-Oh-Nine (Jurca) ... *(1992-93)* 478
M. J.10 Spit (Jurca) ... *(1988-89)* 526
M. J.12 Pee-40 (Jurca) ... *(1988-89)* 526
M. J.14 Fourtouna (Jurca) ... *(1992-93)* 478
M. J.22 Tempête (Jurca) ... *(1992-93)* 477
M. J.51 Sperocco (Jurca) ... *(1992-93)* 478
*M. J.52 Zephyr (Jurca)
M. J.53 Autan (Jurca) ... *(1992-93)* 478
M. J.70 (Jurca) ... *(1992-93)* 478
M. J.77 Gnatsum (Jurca) ... *(1992-93)* 478
M. J.80 One-Nine-Oh (Jurca) ... *(1992-93)* 478
M. J.90 (Jurca) ... *(1992-93)* 478
M. J.100 (Jurca) ... *(1988-89)* 526
MM-1 (Bushby) ... *(1990-91)* 567
MP-16 (Fulmar) ... *(1984-85)* 659
M. P. Unik (Pomergau) ... *(1987-88)* 550
MT5 (VPM SnC) ... *(1986-87)* 604
MT7 (VPM SnC) ... *(1986-87)* 604
Mü 30 (Akaflieg München) ... *(1988-89)* 531
MW4 (Whittaker) ... *(1983-84)* 651
MW5 Sorcerer (Whittaker) ... *(1992-93)* 562
MW5, Whittaker (Aerotech) ... *(1986-87)* 612
MW5K Sorcerer (Aerotech) ... *(1991-92)* 593
MW6 Merlin (Whittaker) ... *(1992-93)* 562
MW7 (Whittaker) ... *(1992-93)* 562
MX Sport (Quicksilver) ... *(1992-93)* 569
MX Sprint (Quicksilver) ... *(1992-93)* 569
MZF1 (Nord) ... *(1989-90)* 550

MACAIR INDUSTRIES INC (Canada) ... *(1992-93)* 468
MACFAM (USA) ... *(1992-93)* 524
Mach 01 (France Aérolights) ... *(1986-87)* 574
Mach .07 (Beaujon) ... *(1986-87)* 637
Mach 20 (France Aérolights) ... *(1986-87)* 574
Maestro (A. I. R's SA) ... *(1992-93)* 555
MAGAL HOLDINGS LTD (Canada) ... *(1991-92)* 586
Magnum V8 (O'Niél) ... *(1992-93)* 527
MAGNUM INDUSTRIES (USA) ... *(1992-93)* 525, 567
MAINAIR SPORTS LTD (UK) ... *(1992-93)* 561
Majorette (Falconar) ... *(1992-93)* 467
MALCOLM AIRPLANE COMPANY (USA) ... *(1992-93)* 567
MALMO FORSKNINGS & INNOVATIONS AB (Sweden) ... *(1992-93)* 560
Mangoos (Flow State Design) ... *(1992-93)* 565
MANTA PRODUCTS INC (USA) ... *(1987-88)* 680
Manx (Moshier Technologies) ... *(1991-92)* 561
Maranda (Falconar) ... *(1992-93)* 467, 552
Marco J-5 (Alpha) ... *(1992-93)* 488
Marco J-5 (Hewa-Technics) ... *(1990-91)* 550
MARGANSKI, EDWARD (Poland) ... *(1988-89)* 540
MARKGRAFLICH BADISCHE VERWALTUNG (Germany) ... *(1992-93)* 557
MARKO-ELEKTRONIK COMPANY (Poland) ... *(1986-87)* 605
Mariah, T-100A (Turner) ... *(1983-84)* 680
Mariah, T-100B (Turner) ... *(1985-86)* 712
Mariah, T-100D (Turner) ... *(1992-93)* 571
Mariner (Two Wings) ... *(1992-93)* 572
MARQUART, ED (USA) ... *(1992-93)* 525
MARQUIOND R (France) ... *(1987-88)* 582
MARTENKO FINLAND OY (Finland) ... *(1992-93)* 474
MARTIN UHIA (Spain) ... *(1992-93)* 560
Mascaret D.150 (Jodel) ... *(1992-93)* 477
MASSACHUSETTS INSTITUTE OF TECHNOLOGY (USA) ... *(1987-88)* 684
MATTHEWS, LYLE (USA) ... *(1986-87)* 684
MATTISON AIRCRAFT COMPANY (USA) ... *(1983-84)* 666
Maverick (Skytek) ... *(1990-91)* 532
MAXAIR AIRCRAFT CORPORATION (USA) ... *(1992-93)* 526
Maya (Percy) ... *(1992-93)* 550
Mc² (Microlights) (USA) ... *(1992-93)* 567
McASCO AIRCRAFT DIVISION (Canada) ... *(1987-88)* 548
MDK KRAKOW-NOWA HUTA (Poland) ... *(1983-84)* 644
MEAD ENGINEERING COMPANY (USA) ... *(1986-87)* 685
Meadowlark (Meadowlark) ... *(1986-87)* 686
MEADOWLARK ULTRALIGHT CORPORATION (USA) ... *(1986-87)* 686
Medvegalis (Yuodinas) ... *(1992-93)* 559
MEDWAY MICROLIGHTS (UK) ... *(1992-93)* 561
MEM REPULOGEPES SZOLGALAT (Hungary) ... *(1989-90)* 611
Menestrel II HN 700 (Nicollier) ... *(1992-93)* 480
Menestrel HN 433 (Nicollier) ... *(1992-93)* 480
Mercure (Danis) ... *(1987-88)* 572
Mercury (Aviation Composites) ... *(1988-89)* 544
Mercury (Inav/Aviation Composites) ... *(1986-87)* 675
Merganser (Merganser) ... *(1987-88)* 682
MERGANSER AIRCRAFT CORPORATION (USA) ... *(1988-89)* 585
Merlin (Macair) ... *(1992-93)* 468
Merlin (Mainair) ... *(1985-86)* 668
Merlin GT (Malcolm Airplane) ... *(1992-93)* 567
MESSERSCHMITT-BÖLKOW-BLOHM GmbH (West Germany) ... *(1987-88)* 596
MICHEL (France) ... *(1987-88)* 582
Michelob Light Eagle (MIT) ... *(1987-88)* 684
MICRO BIPLANE AVIATION (UK) ... *(1985-86)* 669
MICROFLIGHT AIRCRAFT LTD (UK) ... *(1991-92)* 594
Micro-Imp (Aerocar) ... *(1992-93)* 497
Microlight (Lotus) ... *(1985-86)* 668
MICROLIGHT FLIGHT SYSTEMS PTY LTD (South Africa) ... *(1989-90)* 613
MICRONAUTICS INC (Canada) ... *(1989-90)* 607
Micro-Star (Friebe) ... *(1983-84)* 642
Micro Star (Gygax) ... *(1983-84)* 646
Microtug Tow Trike (Mainair) ... *(1985-86)* 669
Midget V (Abe) ... *(1983-84)* 644
Midget 6 (Abe) ... *(1984-85)* 673
Midget Mustang (Bushby) ... *(1992-93)* 507
MIDLAND ULTRALIGHTS LTD (UK) ... *(1987-88)* 617
MIDWEST MICROLIGHTS (USA) ... *(1985-86)* 696
Mifeng-II (Beijing University) ... *(1992-93)* 554
Mifeng-2 (Beijing University) ... *(1992-93)* 553
Mifeng-3 (Beijing University) ... *(1992-93)* 553
Mifeng-3C (Beijing University) ... *(1992-93)* 553
Mifeng-4 and 4A (Beijing University) ... *(1992-93)* 553
Mifeng-5 (Beijing University) ... *(1992-93)* 553
Mignet (Lavorini) ... *(1987-88)* 604
MIGNET (see Société d'exploitation des Aéronefs Henri Mignet) ... *(1988-89)* 527
Mikro-2M (Shcheglov) ... *(1987-88)* 612
Milan (Howatherm) ... *(1986-87)* 596
MILLER AIR SPORTS INC (USA) ... *(1988-89)* 585
MINER, EARL L (USA) ... *(1987-88)* 682
Mini-500 (Revolution Helicopter) ... *(1992-93)* 533
Mini Ace (Cvjetkovic) ... *(1992-93)* 510
Mini Coupe (Buck Sport) ... *(1986-87)* 644
Minihawk (Falconar) ... *(1992-93)* 467
Mini-Imp (Aerocar) ... *(1992-93)* 496
Mini-Jumbo (Say, Maxwell) ... *(1986-87)* 549
Mini Master (Powers Bashforth) ... *(1992-93)* 528
MiniMAX (T. E. A. M.) ... *(1992-93)* 571
MINIMAX T. E. A. M. (USA) ... *(1985-86)* 697
Minimum (NST) ... *(1987-88)* 596
Minimum-Trike (NST) ... *(1987-88)* 596
Mini-Spit (Ailes de K) ... *(1985-86)* 663
Ministar (NST) ... *(1987-88)* 596
Minor, L. A. 4a (PFA) ... *(1988-89)* 547
Mirac (Mainair) ... *(1985-86)* 669
Mirage (Ultralight Flight) ... *(1983-84)* 681
Mirage II (Mirage) ... *(1987-88)* 684
MIRAGE AIRCRAFT INC (USA) ... *(1992-93)* 526
MIRA SLOVAK AVIATION (USA) ... *(1987-88)* 684

720 INDEXES: PRIVATE AIRCRAFT—**M - P**

Missile (Moyes) ... *(1984-85)* 724, 730
Mister America (Harmon) ... *(1987-88)* 668
Mistral II, BA83 (Ikarusflug) ... *(1990-91)* 609
Mistral 53 (Ikarusflug) ... *(1992-93)* 556
Mistral 462 (Aviasud) ... *(1992-93)* 474
Mistral 532 (Aviasud) ... *(1992-93)* 474
*Mistral, Twin-engined (Aviasud)
MITCHELL AEROSPACE COMPANY (USA) ... *(1992-93)* 526
MITCHELL AIRCRAFT CORPORATION (USA) ... *(1983-84)* 667
MITCHELL WING (see Mitchell Aerospace Company) ... *(1992-93)* 526
Mobycoptère (Pauchard) ... *(1987-88)* 584
Model 75 Gemini (Ganzer) ... *(1987-88)* 666
*Model 300 Sequoia (Sequoia)
Model 302 Kodiak (Sequoia) ... *(1991-92)* 569
Model 380 ATTL-1 (Euronef) ... *(1992-93)* 464
Model A (Partnerships) ... *(1984-85)* 708
Model A, Standard Ritz (Loehle) ... *(1988-89)* 581
Model C (Cosmos Sarl) ... *(1992-93)* 555
Model C (Meadowlark) ... *(1986-87)* 686
Model D, Baby Ace (Ace Aircraft) ... *(1992-93)* 494
Model E, Junior Ace (Ace Aircraft) ... *(1992-93)* 494
Model RF (Double Eagle) ... *(1986-87)* 656
Model S-12-E (Spencer) ... *(1992-93)* 536
Model 'S' Sidewinder (Smyth) ... *(1992-93)* 535
Model T (Worldwide Ultralite) ... *(1984-85)* 722
Model W-8 Tailwind (Wittman) ... *(1992-93)* 548
Model W-10 Tailwind (Wittman) ... *(1992-93)* 548
Model Z (Laser Aerobatics) ... *(1988-89)* 504
Mohawk (Warpath) ... *(1983-84)* 682
Monogast (Quaissard) ... *(1986-87)* 590
Monoplane (Frolov) ... *(1987-88)* 612
Monoplane (Lucant) ... *(1987-88)* 580
Monoplane, J. T.1 (Taylor) ... *(1992-93)* 492
Mono Z-CH 100 (Zenair) ... *(1992-93)* 472
*MONTANA COYOTE INC (USA)
MONTGOMERIE GYROCOPTERS, JIM (UK) ... *(1992-93)* 492
Mooney Mite (Mooney Mite) ... *(1986-87)* 689
MOONEY MITE AIRCRAFT CORPORATION (USA) ... *(1986-87)* 689
MORGAN AIRCRAFT (USA) ... *(1986-87)* 690
MORIN, A (France) ... *(1992-93)* 480
MORREY HUMMEL (USA) ... *(1992-93)* 566
Morrisey 2000 (Morrisey) ... *(1992-93)* 527
MORRISEY AIRCRAFT CORPORATION (USA) ... *(1987-88)* 686
MORRISEY BILL (USA) ... *(1992-93)* 527
MOSHIER TECHNOLOGIES (USA) ... *(1991-92)* 561
MOSLER AIRFRAMES & POWER PLANTS (USA) ... *(1991-92)* 561, 600
MOTO-DELTA (see Geiser, Jean Marc) ... *(1984-85)* 668
Moto du Cicl (Humbert) ... *(1987-88)* 574
MOTOPLANE (France) ... *(1983-84)* 641
MOUNTAIN VALLEY AIR (Canada) ... *(1991-92)* 506
MOURA, MAURICIO IMPELIZIERI P (Brazil) ... *(1987-88)* 544
Mouse (Nostalgair) ... *(1987-88)* 687
Mr East (Mathews) ... *(1986-87)* 684
MUDRY ET CIE, AVIONS (France) ... *(1983-84)* 641
MUP (Russia) ... *(1992-93)* 560
Mup-02M (Mup) ... *(1992-93)* 560
MURPHY, A (Ireland) ... *(1986-87)* 599
MURPHY AVIATION LTD (Canada) ... *(1992-93)* 468
Mustang (Cobra Helicopters) ... *(1992-93)* 509
Mustang (Ulm Progress) ... *(1987-88)* 590
*Mustang 11 (Bushby)
Mustang 5151 and 5151RG (Loehle) ... *(1992-93)* 524
Mustang P-51D (Falconar) ... *(1992-93)* 468
Mustang P-51D/TF (Fighter Escort) ... *(1992-93)* 514
MUSTANG HELICOPTER COMPANY (USA) ... *(1992-93)* 527

N

N2 Mouse (Nostalgair) ... *(1987-88)* 687
N3 Pup (Nostalgair) ... *(1987-88)* 688
N3 Pup (Preceptor) ... *(1992-93)* 529, 568
N3-2 Pup (Preceptor) ... *(1992-93)* 529
N3-2C Citabriette ... *(1992-93)* 529
N3-C (Mosler) ... *(1991-92)* 561
N3-C (Preceptor) ... *(1992-93)* 529
NAC (see National Aircraft Corporation) ... *(1986-87)* 690
NAI (see Nanjing Aeronautical Institute) ... *(1988-89)* 516
NF.001 (Michel) ... *(1987-88)* 582
NP2 Baladin (Parent) ... *(1992-93)* 481
NST (see Norbert Schwarze Maschinenbau) ... *(1987-88)* 596

NANJING AERONAUTICAL INSTITUTE (China) ... *(1988-89)* 516
NASH, PAUL L (USA) ... *(1984-85)* 707
NATIONAL AERONAUTICS & MANUFACTURING COMPANY (USA) ... *(1992-93)* 527
NATIONAL AIRCRAFT CORPORATION (USA) ... *(1986-87)* 690
NEICO AVIATION INC (USA) ... *(1991-92)* 562
NELSON, ROBERT (USA) ... *(1984-85)* 707
Nemere (Air Sevice) ... *(1992-93)* 557
NESSA AIRCRAFT COMPANY (Canada) ... *(1992-93)* 470
Nessall (Nessa Aircraft) ... *(1992-93)* 470
NEUKOM SEGELFLUGZEUGBAU, ALBERT (Switzerland) ... *(1985-86)* 663
Nexus (Goldwing) ... *(1984-85)* 688
NICKEL & FOUCARD, RUDY, JOSEPH (France) *(1992-93)* 480
NICOLLIER, AVIONS H (France) ... *(1992-93)* 480
Nieuport 11 (Circa Reproductions) ... *(1992-93)* 466
Nieuport 11 (Leading Edge Air Foils) ... *(1992-93)* 567
Nieuport 12 (Circa Reproductions) ... *(1992-93)* 466
Nieuport 12 (Leading Edge Air Foils) ... *(1992-93)* 567
Nieuport 17/24 (Redfern) ... *(1992-93)* 532
Nifty (Morrisey) ... *(1992-93)* 527
NIKE AERONAUTICA SRL (Italy) ... *(1992-93)* 486
Nimbus Minimot (APCO Aviation) ... *(1987-88)* 593
Ninja series (CMV) ... *(1987-88)* 569
NION (see Ets Nion Aéronautiques) ... *(1992-93)* 556
NIPPER KITS AND COMPONENTS LTD (UK) ... *(1992-93)* 492
Nipper Mk IIIb (Nipper) ... *(1992-93)* 492
NOBLE HARDMAN AVIATION LTD (UK) ... *(1989-90)* 556, 614
Nomad (Flight Research) ... *(1985-86)* 666
Nomad 425 (Flight Research) ... *(1986-87)* 612
Nomad II (Magnum) ... *(1992-93)* 567
Nomad Tribe (Flight Research) ... *(1986-87)* 612
NORBERT SCHWARZE MASCHINENBAU (West Germany) ... *(1987-88)* 596
NORD, FLUGZEUGBAU NORD (West Germany) ... *(1989-90)* 550
NORDPUNKT AG (Switzerland) ... *(1992-93)* 490
Norseman (Husky) ... *(1991-92)* 505
NORTHSTAR LTD (USA) ... *(1986-87)* 690
NOSTALGAIR INC (USA) ... *(1987-88)* 687
Nova (Aircraft Designs) ... *(1991-92)* 538
Nova (Letov) ... *(1992-93)* 554
Nova 2000 (Nova Air) ... *(1985-86)* 701
NOVA AIR INC (USA) ... *(1985-86)* 701
NOVADYNE AIRCRAFT INC (Canada) ... *(1990-91)* 536
Nugget (Davenport) ... *(1986-87)* 653
Nuwaco T-10 (Aircraft Dynamics) ... *(1992-93)* 501
NUWACO AIRCRAFT COMPANY (USA) ... *(1989-90)* 584
Nyirseg (Air Service) ... *(1992-93)* 557

O

O&O Special (Wittman) ... *(1987-88)* 735
OM-1 (Morrisey) ... *(1987-88)* 686

Observer (Swiss Aerolight) ... *(1992-93)* 560
Omega (UPM) ... *(1990-91)* 610
Omega II (UPM) ... *(1990-91)* 610
OMNI-WELD INC (USA) ... *(1984-85)* 708
One-Nine-Oh (Jurca) ... *(1992-93)* 478
One-Oh-Nine M. J.9 (Jurca) ... *(1992-93)* 478
O'NIEL AIRPLANE COMPANY (USA) ... *(1992-93)* 527
Onyx C. P. 150 (Piel) ... *(1992-93)* 482
Opteryx (South and Rogers) ... *(1983-84)* 675
Opteryx III (Opteryx) ... *(1986-87)* 692
OPTERYX ULTRALIGHT AIRCRAFT (USA) ... *(1986-87)* 692
Optimum 88 (Optimum) ... *(1992-93)* 551
OPTIMUM AIRCRAFT GROUP (Bulgaria) ... *(1992-93)* 551
OPTION AIR RENO (USA) ... *(1986-87)* 692
Orion (Skyhook Sailwings) ... *(1984-85)* 682
Orion G-802 (Aerodis) ... *(1991-92)* 512
Orlik RO-7 (Orlinski) ... *(1992-93)* 488
ORLINSKI, ROMAN (Poland) ... *(1992-93)* 488
Osprey II GP3 (Osprey Aircraft) ... *(1992-93)* 527
OSPREY AIRCRAFT (USA) ... *(1992-93)* 527
Ostrowski Biplane (Ostrowski) ... *(1985-86)* 662
OSTROWSKI, JERZY (Poland) ... *(1985-86)* 662

P

P-3 (Up) ... *(1987-88)* 726
P-4 (Up) ... *(1987-88)* 726
P-9 Pixie (Acro Sport) ... *(1992-93)* 494
P-38 Lightning (Mitchell Wing) ... *(1991-92)* 560
P-40 (DiMasco) ... *(1986-87)* 656
P-40 (Loehle) ... *(1992-93)* 524
P-40C Curtiss (Thunder Wing) ... *(1988-89)* 606
P-47 Thunderbolt (WAR) ... *(1988-89)* 612
P.50 Bouvreuil (Pottier) ... *(1988-89)* 530
P-51D Mustang (Falconar) ... *(1992-93)* 468
P.60 Minacro (Pottier) ... *(1992-93)* 482
P.70 Powell Acey Deucy (Acey Deucy) ... *(1986-87)* 622
P.70S (Pottier) ... *(1992-93)* 482
P.80S (Pottier) ... *(1987-88)* 588
P.100TS (Pottier) ... *(1987-88)* 588
P.105TS (Pottier) ... *(1987-88)* 588
P.110TS (Pottier) ... *(1987-88)* 588
P.170S (Pottier) ... *(1992-93)* 483
P.180S (Pottier) ... *(1992-93)* 483
P.200 Series (Pottier) ... *(1992-93)* 483
P. A.8 (Parmesani) ... *(1987-88)* 604
PFA (see Popular Flying Association) ... *(1992-93)* 492
PGK-1 Hirondelle (Western Aircraft) ... *(1992-93)* 472
PL.2 (Pazmany) ... *(1992-93)* 528
PL.3 (Pazmany) ... *(1988-89)* 588
PL.4A (Pazmany) ... *(1992-93)* 528
PL-9 Stork (Pazmany) ... *(1992-93)* 528
PM-2 (Paraplane) ... *(1991-92)* 567
PM-2 Ara (MDK Krakow) ... *(1983-84)* 644
PT-2 Sassy (ProTech Aircraft) ... *(1991-92)* 564
PT-2B Prostar (ProTech Aircraft) ... *(1992-93)* 530
PUL 9 (Nike) ... *(1992-93)* 486
P. U. P. (Matthews) ... *(1986-87)* 684
PX-2000 (Aile) ... *(1986-87)* 564
PZL P.11c (Historical Aircraft) ... *(1992-93)* 520

PACIFIC KITES (New Zealand) ... *(1984-85)* 686
Padre (Morrisey) ... *(1987-88)* 686
PAGEN, DENNIS (USA) ... *(1984-85)* 728
Pago Jet (Pago Jet) ... *(1991-92)* 600
PAGO JET (USA) ... *(1991-92)* 600
PAMPA'S BULL SA (Argentina) ... *(1992-93)* 463, 550
Panther (Rotec) ... *(1992-93)* 569
Panther (Ultra Sports) ... *(1984-85)* 684
Panther Plus (Rotec) ... *(1992-93)* 534
Panther 2 Plus (Rotec) ... *(1992-93)* 534
Panther XL (Pegasus) ... *(1987-88)* 619
Panther XL (Ultra Sports) ... *(1984-85)* 684
Papillon (Broc) ... *(1992-93)* 555
Papillon (Brugger) ... *(1984-85)* 674
Parafan (Centrair) ... *(1991-92)* 555
Parafan II (Centrair) ... *(1990-91)* 541
Parakeet (Rose) ... *(1992-93)* 534
Paraplane (Markgraflich) ... *(1992-93)* 557
PARAPLANE CORPORATION (USA) ... *(1991-92)* 567
Paraplane PM-2 (Paraplane) ... *(1991-92)* 562
Paras-Cargo, B-EC-9 (Croses) ... *(1986-87)* 570
Parascender I (Powercraft) ... *(1992-93)* 568
Parascender II (Powercraft) ... *(1992-93)* 568
Parasol (Aerotique) ... *(1986-87)* 631
PARENT, NORBERT (France) ... *(1992-93)* 481
PARKER, CALVIN Y (USA) ... *(1988-89)* 588
PARMESANI SANDRO (Italy) ... *(1987-88)* 604
PARSONS, BILL (USA) ... *(1991-92)* 562
PARTNERSHIPS LIMITED INC (USA) ... *(1985-86)* 702
Pathfinder II (Huntair) ... *(1984-85)* 679
Patrilor 3 (Aile) ... *(1986-87)* 563
PAUCHARD, GIRONEF (France) ... *(1987-88)* 584
PAUMIER, NORBET (France) ... *(1992-93)* 481
PAUP AIRCRAFT (USA) ... *(1984-85)* 708
PAYNE, VERNON W (USA) ... *(1987-88)* 690
PAZMANY AIRCRAFT CORPORATION (USA) ... *(1992-93)* 528
P-Craft (Paup) ... *(1984-85)* 708
Pee-40 (Jurca) ... *(1988-89)* 526
Pegasus (Those Flying Machines) ... *(1983-84)* 679
Pegasus II (TFM) ... *(1986-87)* 723
Pegasus Q (Pegasus Transport) ... *(1991-92)* 594
Pegasus Q 462 (Solar Wings) ... *(1992-93)* 561
Pegasus Quastar (Solar Wings) ... *(1992-93)* 561
Pegasus Supra (TFM) ... *(1986-87)* 723
Pegasus Titan (TFM) ... *(1986-87)* 724
PEGASUS TRANSPORT SYSTEMS LTD (UK) ... *(1991-92)* 594
Pegasus XL (Solar Wings) ... *(1992-93)* 561
Pegasus XL-R (Pegasus Transport) ... *(1991-92)* 594
Pelican Club (Ultravia) ... *(1989-90)* 539
Pelican Club GS, PL and S (Ultravia) ... *(1992-93)* 471
Pelican Club UL (Ultravia) ... *(1992-93)* 553
PENA, LOUIS (France) ... *(1987-88)* 584
Penetrater (Ultra Efficient Products) ... *(1992-93)* 572
PERCY, GRAHAM J (Australia) ... *(1992-93)* 550
Perigee (Holcomb) ... *(1991-92)* 554
PERIS, JIM (USA) ... *(1992-93)* 567
PERSONAL PLANES INC (USA) ... *(1983-84)* 670
PETERS, WILLIAM R (USA) ... *(1985-86)* 702
Petit Breezy (Matthews) ... *(1983-84)* 666
Petrel (S. M. A. N.) ... *(1992-93)* 484
Petrel (Tisserand/Hydroplum) ... *(1989-90)* 544
Phantom (Motoplane) ... *(1983-84)* 641
Phantom (Phantom Sport) ... *(1992-93)* 567
Phantom (Ultralight Flight) ... *(1987-88)* 725
Phantom F11 (Phantom Sport) ... *(1992-93)* 568
PHANTOM SPORT PLANES INC (USA) ... *(1992-93)* 567
PHASE III INC (USA) ... *(1984-85)* 710
Phoenix (Nova Air) ... *(1985-86)* 701
Phoenix (Phoenix Aviation) ... *(1989-90)* 585
Phoenix (Phoenix Development) ... *(1986-87)* 695
Phoenix (Soleair) ... *(1983-84)* 651
Phoenix (Tube Works) ... *(1987-88)* 722
PHOENIX AVIATION (USA) ... *(1989-90)* 585
PHOENIX DEVELOPMENT & SPECIALITIES INC (USA) ... *(1986-87)* 695
Phoenix SG-1 (Westwind) ... *(1987-88)* 623
Phoenix SL (Bryan) ... *(1986-87)* 644
PHÖNIX-AVIATECHNICA (Russia) ... *(1992-93)* 489
Photon (Pegasus) ... *(1987-88)* 620
Piccolo (Technoflug) ... *(1989-90)* 610
Piccolo, AN-20K (Borowski) ... *(1987-88)* 594
PIEL, AVIONS CLAUDE (France) ... *(1992-93)* 481
PIETENPOL, BUCKEYE ASSOCIATION (USA) ... *(1992-93)* 528
PIETENPOL, DONALD (USA) ... *(1988-89)* 590
PINAIRE ENGINEERING INC (USA) ... *(1992-93)* 568
Pinocchio C. P. 90 (Piel) ... *(1992-93)* 482
Pintail (Little Airplane) ... *(1985-86)* 694
PIONEER INTERNATIONAL AIRCRAFT INC (USA) ... *(1985-86)* 703
Pipistrelle 2B (Southdown Aerostructure) ... *(1986-87)* 620
Pipistrelle 2C (Southdown Aerostructure) ... *(1987-88)* 622
PITTS (see Christen Industries Inc) ... *(1987-88)* 650
Pixie (Skyhook Sailwings) ... *(1984-85)* 682
Pixie, P-9 Pober (Acro Sport) ... *(1992-93)* 494
POINT AVIATION FLUGZEUGBAU GmbH (West Germany) ... *(1985-86)* 660
POISESTAR LTD (UK) ... *(1983-84)* 650
POLARIS SRL (Italy) ... *(1992-93)* 559
Polliwagen (Polliwagen) ... *(1989-90)* 586
POLLIWAGEN INC (USA) ... *(1989-90)* 586
Polonez J-2 (Janowski) ... *(1992-93)* 559
POLTERGEIST AIRCRAFT DIVISION (USA) ... *(1986-87)* 697
POMERGAU, MICHAEL (Canada) ... *(1987-88)* 550
Pontresina (Colani) ... *(1986-87)* 608
Pony (Tesori Aircraft) ... *(1986-87)* 552
Poppy (Dedalus) ... *(1992-93)* 486
POPULAR FLYING ASSOCIATION, THE (UK) ... *(1992-93)* 492
POTTIER, JEAN (France) ... *(1992-93)* 482
Pou (Control) ... *(1986-87)* 568
Pouplume EAC-3 (Croses) ... *(1991-92)* 513
Pou-Push (Briffaud) ... *(1986-87)* 566
POUR LE MERITE ULTRALIGHTS (Australia) ... *(1987-88)* 540
Poussin AG 02 (Gatard) ... *(1987-88)* 574
Poussin AG 02 Sp (Gatard) ... *(1988-89)* 522
POWERCRAFT CORPORATION (USA) ... *(1992-93)* 568
Powergyro (Bensen) ... *(1986-87)* 640
POWERS BASHFORTH AIRCRAFT CORPORATION (USA) ... *(1992-93)* 528, 568
PRECEPTOR AIRCRAFT (USA) ... *(1992-93)* 529, 568
Predator (Mustang Helicopter) ... *(1992-93)* 527
PRESCOTT AERONAUTICAL CORPORATION (USA) ... *(1989-90)* 586
PREVOT (France) ... *(1986-87)* 589
*Privateer S-4 (Arctic)
PRODUCT MZ (Hungary) ... *(1987-88)* 600
PROGRESS AERO INC ... *(1992-93)* 529
Project 102 (Windsoar) (Birdman) ... *(1983-84)* 634
Prop-Copter (Driggers) ... *(1985-86)* 683
Prop Kopter (BW Rotor) ... *(1992-93)* 563
Prospector (Bounsall) ... *(1987-88)* 642
PROST, D (West Germany) ... *(1984-85)* 670
Prostar PT-2B (ProTech Aircraft) ... *(1992-93)* 530
PROTECH AIRCRAFT INC (USA) ... *(1992-93)* 530
PROUST, CHRISTIAN/SIGUR (France) ... *(1986-87)* 589
Prowler (Prowler Aviation) ... *(1992-93)* 530
PROWLER AVIATION LTD (USA) ... *(1992-93)* 530
PTERODACTYL (USA) ... *(1984-85)* 710
PTERODACTYL LTD (USA) ... *(1983-84)* 670
Ptiger (Freedom Fliers) ... *(1984-85)* 697
Ptraveler (Freedom Fliers) ... *(1984-85)* 696

P – S—PRIVATE AIRCRAFT: INDEXES 721

Pulsar (Aero Designs) .... (1992-93) 498
Puma (Ultrasports) .... (1983-84) 651
Pup (Nostalgair) .... (1987-88) 688
Pup Series (Preceptor) .... (1992-93) 529
Pursuit S-11 (Rans) .... (1992-93) 532
Pusher (Prescott) .... (1989-90) 586

Q

Q (Pegasus) .... (1991-92) 594
Q2 Quickie (Quickie Aircraft) .... (1986-87) 700
Q-200 Quickie (Quickie Aircraft) .... (1986-87) 700
QM-1, Quicksilver (Eipper) .... (1984-85) 694
QM-2, Quicksilver (Eipper) .... (1984-85) 694

Qingting 5 (Shijiazhuang) .... (1989-90) 608
Qingting W-5 series (SAP) .... (1992-93) 554
Qingting W6 (SAP) .... (1991-92) 554
QUAD CITY ULTRALIGHTS (USA) .... (1983-84) 672
QUAD CITY ULTRALIGHT AIRCRAFT CORPORATION (USA) .... (1992-93) 530, 568
Quail (Aerosport) (D 2) .... (1992-93) 511
QUAISSARD, G (France) .... (1986-87) 590
Quartz GT 462 (Air Creation) .... (1991-92) 588
Quartz GT 503 (Air Creation) .... (1992-93) 554
*QUESTAIR INC (USA)
Questor (Nostalgair) .... (1987-88) 688
Questor (Omni-Weld) .... (1984-85) 708
Quickie (Quickie Aircraft) .... (1986-87) 699
QUICKIE AIRCRAFT CORPORATION (USA) .... (1986-87) 699
Quickie Q2 (Quickie Aircraft) .... (1986-87) 700
Quickie Q2 Modification (Tri Q) .... (1988-89) 606
Quicksilver (Eipper) .... (1988-89) 568
QUICKSILVER ENTERPRISES INC (USA) .... (1992-93) 568
Quicksilver GT (Quicksilver) .... (1992-93) 568
Quicksilver GT 400 (Quicksilver) .... (1991-92) 601
Quicksilver GT 500 (Quicksilver) .... (1992-93) 569
Quicksilver MXL (Quicksilver) .... (1990-91) 619
Quicksilver MXL II (Quicksilver) .... (1990-91) 619
Quicksilver MXL II Sport (Quicksilver) .... (1992-93) 569
Quicksilver MXL II Sprint (Quicksilver) .... (1992-93) 569
Quicksilver QM-1 (Eipper) .... (1984-85) 694
Quicksilver QM-2 (Eipper) .... (1984-85) 694

R

R55 LightWing (Hughes) .... (1989-90) 534
RB-1 (Birdman) .... (1984-85) 690
RB-1 Lightning (Ultra Sports) .... (1984-85) 684
RB-2 (Roberts Ultralight) .... (1991-92) 602
RB 60 (Barbero) .... (1986-87) 565
RC 412-11 (Cabrinha) .... (1990-91) 568
RD-02A Edelweiss (Duruble) .... (1986-87) 572
RD-03 Edelweiss 150 (Duruble) .... (1986-87) 572
RD-30V (Vidor) .... (1987-88) 606
RD-95 (Vidor) .... (1987-88) 606
Re 2002 scale (Tesori) .... (1987-88) 553
RJ-02 (Junqua) .... (1987-88) 576
RJ-03 Ibis (Junqua) .... (1992-93) 477
RM-02 (Marquoind) .... (1987-88) 582
RO-7 Orlik (Orlinski) .... (1992-93) 488
RV-3 (Van's Aircraft) .... (1992-93) 543
*RV-4 (Van's Aircraft)
*RV-6 (Van's Aircraft)
*RV-6A (Van's Aircraft)
RX-28 (Beaver) .... (1989-90) 606
RX-35 (Beaver) .... (1992-93) 552
RX-550 (Beaver) .... (1992-93) 552
RX-650 (Beaver) .... (1992-93) 552

Racer (Air Creation) .... (1988-89) 518
Rally 2B (Rotec) .... (1992-93) 570
Rally 3 (Rotec) .... (1992-93) 570
Rally Champ (Rotec) .... (1992-93) 570
Rally Sport (Rotec) .... (1992-93) 570
RAM AIRCRAFT COMPANY (Costa Rica) .... (1987-88) 564
RAND ROBINSON ENGINEERING INC (USA) .... (1992-93) 531
Ranger (T. E. D. A.) .... (1991-92) 592
RANGER AVIATION (USA) .... (1983-84) 672
Range Rider (Anglin) .... (1992-93) 502
*RANS INC (USA)
Rapier (Mainair) .... (1985-86) 668
Raven (Hornet) .... (1987-88) 614
Raven II (Custom Aircraft Conversions) .... (1984-85) 692
Razor (Mainair Sports) .... (1989-90) 614
R. D. AIRCRAFT (USA) .... (1992-93) 532, 569
Rebel (Murphy) .... (1992-93) 470
REDFERN WD (USA) .... (1992-93) 532
REIMANN (see Ul-Bau Reimann) .... (1987-88) 596
Reliant (Hipp's Superbirds) .... (1992-93) 566
Reliant II (Hipp's Superbirds) .... (1992-93) 520
Reliant II microlight (Hipp's Superbirds) .... (1991-92) 598
Reliant SX (Hipp's Superbirds) .... (1992-93) 520
Renegade II (Murphy) .... (1992-93) 468
Renegade Spirit (Murphy) .... (1992-93) 469
Replica Junkers Ju 87B-2 'Stuka' (Langhurst) .... (1987-88) 676
REPLICA PLANS (Canada) .... (1992-93) 470
Resurgam (Bedson) .... (1986-87) 537
REUTER, DIPL-ING ADOLF K (West Germany) .... (1986-87) 596
REVOLUTION HELICOPTER CORPORATION INC (USA) .... (1992-93) 533
RICHARD WARNER AVIATION INC (USA) .... (1992-93) 546
RICHTER INGENIEURBÜRO, KLAUS J (West Germany) .... (1983-84) 642
RITEC GmbH (West Germany) .... (1990-91) 609
RITZ AIRCRAFT CO (USA) .... (1984-85) 712
Ritz Standard A1 (Falconar) .... (1992-93) 553
Ritz Standard Model A (Loehle) .... (1989-90) 618
ROBERTSON AIRCRAFT CORPORATION (USA) .... (1985-86) 705
ROBERTS SPORTS AIRCRAFT (USA) .... (1989-90) 594
ROBERTS ULTRALIGHT AIRCRAFT (USA) .... (1991-92) 602
ROCHELT, GÜNTER (West Germany) .... (1983-84) 644
ROCOURT, MICHAEL (Belgium) .... (1987-88) 544
RODGER, ERWIN (USA) .... (1984-85) 712

ROMAIN, J AND SONS (UK) .... (1992-93) 561
Romibutter (Rocourt) .... (1987-88) 544
ROSE (USA) .... (1992-93) 534
ROTARY AIR FORCE INC (Canada) .... (1992-93) 470
Rotax Gyro-Copter (Bensen) .... (1986-87) 640
ROTEC ENGINEERING INC (USA) .... (1992-93) 534, 570
Rotor Lightning Sport Copter (Vancraft) .... (1988-89) 609
*ROTORWAY INTERNATIONAL (USA)
Rouseabout (Seabird) .... (1985-86) 640
Rouseabout Mk 5 (Seabird) .... (1986-87) 540
Royal (Ulm Pyrénées) .... (1986-87) 591
RUTAN AIRCRAFT FACTORY INC (USA) .... (1985-86) 503
Ruty (Albarde) .... (1986-87) 564
RYDER INTERNATIONAL CORPORATION (USA) .... (1992-93) 534

S

S-1 series (Pitts) .... (1986-87) 696
S-2 series (Pitts) .... (1986-87) 696
S-2B Upside-Down Pitts (Hoskins) .... (1986-87) 671
S-4 Coyote (Rans) .... (1992-93) 569
S-4 Saunders .... (1985-86) 616
S-4A Jet Hawk II (Saunders) .... (1987-88) 704
S-6 Coyote II (Rans) .... (1991-92) 601
S-6 Supermarine Replica (Flight Dynamics) .... (1986-87) 664
S-6ES Coyote II (Rans) .... (1992-93) 569
*S-7 Courier (Rans)
S-9 Chaos (Rans) .... (1992-93) 532
S 10 (Free Flight) .... (1986-87) 538
S-10 Sakota (Rans) .... (1992-93) 532
S-11 Pursuit (Rans) .... (1992-93) 532
S-12 Airaile (Rans) .... (1992-93) 532
S-14 Airaile (Rans) .... (1992-93) 532
*S-18 (Sunderland/Sport Aircraft)
S-18 T (Sunderland) .... (1987-88) 718
S-18 T-18C-W (Thorp) .... (1986-87) 724
S-51D (Stewart) .... (1988-89) 601
SA-60 (Silhouette) .... (1989-90) 591
SA 101 Super Starduster (Stolp) .... (1992-93) 540
SA 102.5 Cavalier (MacFam) .... (1992-93) 524
SA 103 (McAsco) .... (1986-87) 547
SA 104 (McAsco) .... (1986-87) 547
SA 105 (MacFam) .... (1988-89) 582
SA 106 (McAsco) .... (1987-88) 549
SA-300 Starduster Too (Stolp) .... (1992-93) 539
SA-500 Starlet (Stolp) .... (1992-93) 539
SA-700 Acroduster 1 (Aerovant) .... (1992-93) 499
SA-750 Acroduster Too (Stolp) .... (1992-93) 539
SA-900 V-Star (Stolp) .... (1992-93) 540
SAP (see Shijiazhuang Aircraft Plant) .... (1991-92) 554
SB1 Antares (Soyer/Barritault) .... (1992-93) 484
SB-1 (Seabird) .... (1986-87) 540
SB-2 (Seabird) .... (1986-87) 540
SB-3 (Seabird) .... (1986-87) 540
S. E.5A (Replica Plans) .... (1992-93) 470
SF-2A Cygnet (Hapi) .... (1991-92) 553
SG-1 Phoenix (Westwind) .... (1987-88) 623
SL-1 (Nelson) .... (1984-85) 707
SL-1 Star-Lite (Star-Lite) .... (1989-90) 593
S. M. A. N. (see Société Morbihannaise d'Aero Navigation) .... (1992-93) 484
SNS-2 Guppy (Sorrell) .... (1992-93) 536
SNS-7 Hiperbipe (Sorrell) .... (1992-93) 536
SNS-8 Hiperlight (Sorrell) .... (1992-93) 570
SNS-8 Hiperlight EXP (Sorrell) .... (1992-93) 570
SNS-9 EXP II (Sorrell) .... (1992-93) 536
SNS-10 Intruder (Sorrell) .... (1992-93) 570
SP (see Sinaya Ptitsa Klub Deltaplaneristov Rostov-Don) .... (1985-86) 665
SP-1 Spunt (Czechoslovakia) .... (1990-91) 607
SP-8UT (SP) .... (1984-85) 676
SPM (SP) .... (1983-84) 646
SR-1 (SR-1) .... (1985-86) 709
SR-1 ULTRA SPORT AVIATION (USA) .... (1985-86) 709
SR-10 (Mitchell Wing) .... (1987-88) 684
ST 80 Balade (Stern) .... (1992-93) 484
ST 87 Europlane (Stern) .... (1992-93) 484
*STOL CH 701 (Zenair)
*STOL CH 701-AG (Zenair)
SV-1 (Skywise) .... (1989-90) 605
SV-2 (Skywise) .... (1991-92) 583
SV-8 (Veenstra) .... (1984-85) 658
SV9 (Veenstra) .... (1985-86) 642
SV 11B (SV Aircraft) .... (1992-93) 550
SV 14 (SV Aircraft) .... (1987-88) 542
SVCIAF (American Microflight) .... (1986-87) 635
SX II Racer 503 (Air Creation) .... (1992-93) 555
SX-300 (Swearingen) .... (1987-88) 718
SX GT 462 (Air Creation) .... (1991-92) 588
SX GT 503 (Air Creation) .... (1991-92) 588
SX GT 503S (Air Creation) .... (1992-93) 554
SX GT 582ES (Air Creation) .... (1992-93) 555
SX Racer (Air Creation) .... (1991-92) 588

Sabre (Danis) .... (1983-84) 640
Sabre (Skyhook) .... (1982-83) 637
Sabre 1K (Pour le Merite) .... (1987-88) 540
Sabre 1R (Pour le Merite) .... (1987-88) 540
Sabre A (Skyhook) .... (1983-84) 688, 692
Sabre B (Skyhook) .... (1983-84) 688, 692
Sabre C (Skyhook) .... (1983-84) 688, 692
Sadler Vampire (American Microflight) .... (1986-87) 634
Sadler Vampire (Skywise) .... (1991-92) 583
Sadler Vampire Trainer (American Microflight) .... (1986-87) 635
Safari G-TB1 (Air Creation) .... (1988-89) 518
Saja (CMI) .... (1991-92) 522
Sakota S-10 (Rans) .... (1992-93) 532
Samolet (Aviatechnica) .... (1992-93) 488
Samourai (Aéronautic 2000) .... (1983-84) 637
Samson Replica (Pitts) (Wolf) .... (1986-87) 737
SANDBERG, JOHN R (USA) .... (1987-88) 704
SANDER VEENSTRA (Australia) .... (1985-86) 642
Sand Piper (B & B) .... (1984-85) 689
SANDS, RON INC (USA) .... (1988-89) 596

San Francesco I-66L (Iannotta) .... (1992-93) 486
SAPHIR AMERICA (USA) .... (1991-92) 602
Sapphire (Winten) .... (1992-93) 551
SARL LA MOUETTE (France) .... (1983-84) 686
Sassy, PT-2 (ProTech Aircraft) .... (1991-92) 564
SAUNDERS, DON (USA) .... (1987-88) 704
SAURIER FLUG SERVICE (Germany) .... (1992-93) 557
Sauterelle (EMC2) .... (1982-83) 624
SAY, MAXWELL A (Canada) .... (1986-87) 549
Scamp A (Aerosport) .... (1992-93) 499
Sceptre I (Roberts Sport) .... (1989-90) 589
Sceptre II (Roberts Sport) .... (1989-90) 589
SCHEIBE FLUGZEUGBAU GmbH (West Germany) .... (1984-85) 671
Schlacro Mü 30 (Akaflieg München) .... (1988-89) 531
SCHLITTER, RANDY (USA) .... (1984-85) 711
SCICRAFT LTD (Israel) .... (1992-93) 486
Scorcher (Mainair Sports) .... (1992-93) 561
Scorpion (Cozza/Scholl) .... (1984-85) 692
Scout (Gardner) .... (1987-88) 614
Scout (Topa) .... (1986-87) 725
Scout Mk III (Wheeler) .... (1992-93) 550
Scwal 100 (Scwal) .... (1992-93) 464
SCWAL SA (Belgium) .... (1992-93) 464, 551
SEABIRD AVIATION PTY LTD (Australia) .... (1987-88) 540
SEABIRD ULTRALIGHT AIRCRAFT (Australia) .... (1985-86) 640
Sea Fury, Hawker (WAR) .... (1988-89) 612
SeaHawk (CGS Aviation) .... (1992-93) 564
*Sea Hawker (Aero Composite)
Sea Panther (Rotec) .... (1992-93) 534
Seawind (SNA) .... (1992-93) 536
Seawind 2000 (Seawind) .... (1989-90) 538
Seawind 2500 (Seawind) .... (1990-91) 536
SEAWIND INTERNATIONAL INC (Canada) .... (1990-91) 536
*Sequoia Model 300 (Sequoia)
*SEQUOIA AIRCRAFT CORPORATION (USA)
Series 2000 autogyro (Rotary Air Force) .... (1992-93) 470
SEVILLE AIRCRAFT INC (Canada) .... (1988-89) 510
Seville Two Place (Seville) .... (1988-89) 510
Shadow (CFM) .... (1989-90) 544
Shadow (Helicraft) .... (1992-93) 519
Shadow (MFS) .... (1989-90) 613
Shadow I 377 (Evergreen) .... (1986-87) 661
*Shadow II (CFM)
Shadow II 503 (Evergreen) .... (1986-87) 661
Shadow M. J.4. (Jurca) .... (1992-93) 478
Shadow Series B and B-D (CFM) .... (1990-91) 554
*Shadow Series C and C-D (CFM)
Sharkfire (Lopresti) .... (1987-88) 680
SHAW, IVAN (UK) .... (1987-88) 622
SHCHEGLOV, V (USSR) .... (1987-88) 612
Sherpa I (Ikarus Deutschland) .... (1989-90) 610
Sherpa II (Ikarus Deutschland) .... (1989-90) 610
SHIJIAZHUANG AIRCRAFT PLANT (China) .... (1991-92) 554
Shrike (Kaler) .... (1985-86) 692
Sibylle LP-01 (Lendepergt) .... (1992-93) 479
Sidewinder (International Ultralights) .... (1985-86) 692
Sidewinder, Model S (Smyth) .... (1992-93) 535
SIERRA DELTA SYSTEMS (USA) .... (1992-93) 535
Sierra Whitehawk 185 (Beck Productions) .... (1984-85) 726, 730
*Silhouette (Lunds Tekniske)
Silhouette (Skyhook Sailwings) .... (1984-85) 682
Silhouette I (Silhouette) .... (1989-90) 591
SILHOUETTE AIRCRAFT INC (USA) .... (1989-90) 591
Silver Eagle II (Eagle Performance) .... (1992-93) 564
Silver Eagle A-10 (Mitchell Wing) .... (1991-92) 599
Silver Cloud (Ranger) .... (1983-84) 672
Silverwing Custom (Preceptor Aircraft) .... (1992-93) 568
SIMPSON MIDWEST ULTRALIGHTS (USA) .... (1986-87) 709
SINAYA PTITSA KLUB DELTAPLANERISTOVE ROSTOV-DON (USSR) .... (1985-86) 665
Sirio 2 (Sirio) .... (1986-87) 602
SIRIO SRL (Italy) .... (1986-87) 602
Sirius (Nion Aeronautics) .... (1991-92) 589
Sirius C (Nion Aeronautiques) .... (1992-93) 556
Sirocco (Aviasud) .... (1992-93) 474
Sirocco 377GB (Midland Ultralights) .... (1987-88) 617
*Sirocco M. J. 5 (Jurca)
Sirocco Sport Wing M. J.5 (Jurca) .... (1992-93) 478
SIX-CHUTER INC (USA) .... (1992-93) 570
Skeeter (Aero Ventures) .... (1985-86) 674
Skeeter (Fisher) .... (1986-87) 664
Skeeter FP505 (Fisher) .... (1988-89) 571
Skybaby (Skyhigh) .... (1986-87) 709
Skybaby (Stits) .... (1987-88) 712
Skybaby FP606 (Fisher) .... (1989-90) 573
Sky Bird (BW Rotor) .... (1992-93) 563
Skybolt (Ebershoff) .... (1988-89) 568
Skybolt (Steen Aero Lab) .... (1992-93) 538
Skybolt (Steen-Stolp) .... (1990-91) 594
Skybolt Mk II (Stolp) .... (1987-88) 716
SKY CRAFT LTD (UK) .... (1992-93) 492
SKY-CRAFT SA (Switzerland) .... (1986-87) 609
Sky Cycle (BW Rotor) .... (1992-93) 563
Skycycle (Kennedy Aircraft) .... (1991-92) 555
Skycycle (R. D. Aircraft) .... (1992-93) 569
Skycycle II (R. D. Aircraft) .... (1992-93) 569
Skycycle '87 (Carlson) .... (1992-93) 508
Skycycle Gipsy series (R. D. Aircraft) .... (1992-93) 569
Skye-Ryder (Six-Chuter) .... (1992-93) 570
SKYE TRECK (Canada) .... (1983-84) 634
Skyfly CA-65 (Cvjetkovic) .... (1992-93) 510
Skyfox (Calair) .... (1991-92) 502
*Skyfox (Skyfox Aviation)
*SKYFOX AVIATION (Australia)
SKYHIGH ULTRALIGHTS INC (USA) .... (1986-87) 709
SKYHOOK SAILWINGS LTD (UK) .... (1984-85) 682, 726
SKY KING INTERNATIONAL (Canada) .... (1985-86) 644
Sky Kopter (BW Rotor) .... (1992-93) 563
Skylark (Airtech) .... (1991-92) 585
Skylark 1 (Helicraft) .... (1991-92) 566
SKYLINES ENTERPRISES (USA) .... (1986-87) 709
Skyote (Skyote) .... (1986-87) 709
SKYOTE AEROMARINE LTD (USA) .... (1986-87) 709
Sky Pup (Sport Flight) .... (1992-93) 570
Skyraider (Worldwide Ultralite) .... (1984-85) 722

722 INDEXES: PRIVATE AIRCRAFT—**S – T**

Sky Ranger (EPA) (1992-93) 558
Sky Ranger (Striplin) (1983-84) 677
Sky Ranger series (Striplin) (1992-93) 540
Sky-Rider (Kimberley) (1992-93) 550
Sky Scooter (Flaglor/Headberg) (1992-93) 519
Skyseeker I/III (Skye Treck) (1983-84) 634
SKYSEEKER AIRCRAFT CORPORATION (Canada) (1985-86) 645
Skyseeker Mk III (Skyseeker) (1985-86) 645
Sky Sport (BW Rotor) (1992-93) 563
SKYTEK AUSTRALIA PTY LTD (Australia) (1991-92) 502
Skywalker (Leichtflugzeug) (1987-88) 596
Sky Walker (Sterner) (1984-85) 716
Skywalker II (Leichtflugzeug) (1992-93) 557
SKYWAY AIRCRAFT (USA) (1991-92) 570
SKYWISE ULTRAFLIGHT PTY LTD (Australia) (1991-92) 503, 583
Sky-Wolff (Wolff) (1991-92) 524
Slandan BA-12 (Andreasson/MFI) (1992-93) 560
Slavutitch M-1 (Antonov) (1992-93) 560
SMITH, ALLAN C, DEVELOPMENTS (Australia) (1987-88) 506
SMITH, RONALD M. B. (Canada) (1992-93) 553
SMITH, S & H, AIRCRAFT (USA) (1991-92) 570
SMITH, WILLIAM M (USA) (1986-87) 710
SMYTH, JERRY (USA) (1992-93) 535
SNA INC (USA) (1991-92) 536
Sno-Bird (Ultralight Aircraft) (1991-92) 604
Sno Bird 503S (SnoBird) (1992-93) 570
Sno Bird 582S (SnoBird) (1992-93) 570
Sno Bird 636D (SnoBird) (1992-93) 570
Sno Bird Aircraft Inc (USA) (1992-93) 570
Snoop (Eastern) (1984-85) 694
SNOWBIRD AEROPLANE CO. LTD, THE (UK) (1992-93) 561
SNOWBIRD AVIATION (Canada) (1992-93) 553
Snowbird Mk III (Noble Hardman) (1987-88) 618
Snowbird Mk IV (Noble Hardman) (1989-90) 556
Snowbird Mk IV (Snowbird Aeroplane) (1992-93) 561
SOCIÉTÉ AÉROKART (France) (1986-87) 562
SOCIÉTÉ D'EXPLOITATION DES AÉRONEFS HENRI MIGNET (France) (1992-93) 556
SOCIÉTÉ INDUSTRIELLE LATÉCOÈRE SA (France) (1986-87) 580
SOCIÉTÉ MAVIC (France) (1986-87) 582
SOCIÉTÉ MORBIHANNAISE D'AÉRO NAVIGATION (France) (1992-93) 484
Solair I (Rochelt) (1983-84) 644
SOLAR WINGS AVIATION LTD (UK) (1992-93) 561
SOLEAIR AVIATION (UK) (1983-84) 651
Solo (American Air) (1987-88) 638
Solo M. J. 7S (Jurca) (1987-88) 578
SOLO WINGS (South Africa) (1992-93) 560
Somethin' Special (Ultralite Soaring) (1984-85) 721
Sonerai I (Great Plains) (1992-93) 517
Sonerai I (Inav) (1986-87) 674
*Sonerai II, IIL, II-LT, II-LTS (Great Plains)
Sonerai II, III, II-LT and II-LTS (Inav) (1986-87) 674
Sonic (Spectra) (1983-84) 690
Sonic Spitfire (Sunrise) (1991-92) 603
Sooper-Coot Model A (Aerocar) (1992-93) 496
Sophie (Egrette) (1986-87) 573
Sopwith Triplane (Circa Reproductions) (1992-93) 466
Sorcerer Aerotech (Whittaker) (1986-87) 615
SORRELL AIRCRAFT COMPANY LTD (USA) (1992-93) 536
SOUQUET (France) (1987-88) 589
SOUTH COAST AIR CENTRE PTY LTD (Australia) (1985-86) 640
SOUTH, DON AND ROGERS, TOM (USA) (1983-84) 675
SOUTHDOWN AEROSTRUCTURE (UK) (1988-89) 548
SOUTHERN AERO SPORTS (UK) (1983-84) 651
SOUTH/RODGERS (USA) (1984-85) 715
SOYER/BARRITAULT, CLAUDE/JEAN (France) (1992-93) 484
Space Walker (Country Air) (1990-91) 570
Spacewalker I (Anglin) (1992-93) 502
Spacewalker II (Anglin) (1992-93) 502
Sparrow (B & G) (1985-86) 678
Sparrow (Carlson) (1992-93) 563
Sparrow II (Carlson) (1992-93) 508
Sparrow-ette (Carlson) (1992-93) 508
Sparrow Hawk (Aero Dynamics) (1985-86) 574
Sparrow Hawk Mk II (Aero Dynamics) (1990-91) 560
Sparrow Sport Special (Carlson) (1992-93) 508
Special (Laven) (1991-92) 556
Special (Pitts) (1986-87) 696
Special (Pitts-Ultimate Aircraft) (1991-92) 508
Special (Whatley) (1988-89) 612
Special I (Cassutt) (1988-89) 562
Special II (Cassutt) (1988-89) 562
Special O & O (Wittman) (1987-88) 735
SPECTRA AIRCRAFT CORPORATION (USA) (1983-84) 690
Spectrum (Microflight Aircraft) (1991-92) 594
Spectrum (Spectrum Aircraft) (1992-93) 562
SPECTRUM AIRCRAFT (UK) (1992-93) 562
SPECTRUM AIRCRAFT INC (Canada) (1988-89) 510
SPENCER AMPHIBIAN AIRCRAFT INC (USA) (1992-93) 536
Sperocco M. J.51 (Jurca) (1992-93) 478
*Spirit (Questair)
Spirit (Ultralite Soaring) (1984-85) 720
Spirit 1 (Spirit One) (1986-87) 712
SPIRIT ONE CORPORATION (USA) (1986-87) 712
Spit (Jurca) (1988-89) 526
Spitfire (Isaacs) (1988-89) 544
Spitfire (Worldwide Ultralite) (1984-85) 722
Spitfire I (Bell, F. M.) (1986-87) 637
Spitfire II (Bell, F. M.) (1986-87) 638
Spitfire II (Sunrise) (1989-90) 596
Spitfire II (Worldwide Ultralite) (1984-85) 722
Spitfire II Elite (Sunrise) (1992-93) 540, 571
Spitfire Mk IX (Thunder Wings) (1988-89) 606
Spitfire series (Sunrise) (1992-93) 571
Split Tail (Nash) (1984-85) 707
Sport (Partnerships) (1984-85) 708
Sport (Ulm Pyrénées) (1986-87) 591
*SPORT AIRCRAFT INC (USA)
SPORT AVIATION (USA) (1983-84) 690
SPORT FLIGHT ENGINEERING INC (USA) (1992-93) 570
Sport Helicopter (Star Aviation) (1992-93) 538
Sport Parasol (Loehle Aviation) (1992-93) 567
Sport Racer (Sport Racer) (1992-93) 537
SPORT RACER INC (USA) (1992-93) 537
Sportsman (Winton) (1986-87) 543
Sportsman 2+2 (Wag-Aero) (1992-93) 546
*Sportsman VJ-22 (Volmer)
Sportster (Grover) (1985-86) 689
Sportster (Poltergeist) (1986-87) 697
Sportster HA-2M (Aircraft Designs) (1992-93) 500
*Sport Trainer (Wag-Aero)
Sportwing D-201 (D'Apuzzo) (1992-93) 511
Sprite (Murphy) (1986-87) 599
SQUADRON AVIATION INC (USA) (1988-89) 600
Staaker Z-1 Flitzer (Sky Craft) (1992-93) 492
Staggerbipe Mk I (WAACO) (1987-88) 542
Stalker T-9 (Stoddard-Hamilton) (1991-92) 573
Stallion (Aircraft Designs) (1992-93) 501
Stallion, CX-901 (Command-Aire) (1984-85) 692
Standard (Hornet) (1987-88) 614
Standard Model A Ritz (Loehle) (1988-89) 581
STAR AVIATION INC (USA) (1992-93) 538
Starcruiser Gemini (Davis Wing) (1992-93) 512
Starduster Too SA-300 (Stolp) (1992-93) 539
STARFIRE AVIATION INC (USA) (1991-92) 572
STAR FLIGHT AIRCRAFT (USA) (1992-93) 570
STARFLIGHT MANUFACTURING (USA) (1984-85) 676
Star Lance I (Nelson) (1984-85) 707
Starlet (Corby) (1988-89) 503
Starlet (CSN) (1992-93) 510
Starlet SA-500 (Stolp) (1992-93) 539
STAR-LITE AIRCRAFT INC (USA) (1989-90) 593
Star-Lite SL-1 (Star-Lite) (1989-90) 593
Starship (Ultimate Hi) (1981-82) 628, 632
Starship Gemini (Davis Wing) (1988-89) 566
S. T. A. R.-Trike (Wisch & Nist) (1987-88) 599
STATLER, WILLIAM H (USA) (1986-87) 714
Statoplan Poussin (Gatard) (1988-89) 522
ST CROIX AIRCRAFT (USA) (1992-93) 537, 570
ST CROIX PROPELLERS (USA) (1984-85) 714
STEEN AERO LAB INC (USA) (1992-93) 538
STEINBACH DELTA (Austria) (1987-88) 543
STERN, RÉNE (France) (1992-93) 484
STERNER AIRCRAFT (USA) (1984-85) 716
STEWART 51 INC (USA) (1992-93) 538
STEWART AIRCRAFT CORPORATION (USA) (1988-89) 601
Stimul-10 (Korolyev) (1987-88) 612
Stinger (SR-1) (1983-84) 675
Stingray (Ace) (1985-86) 639
Sting Ray (Eaves) (1986-87) 659
STITS, DON (USA) (1987-88) 712
*STODDARD-HAMILTON AIRCRAFT INC (USA)
STOLP STARDUSTER CORPORATION (USA) (1992-93) 539
Stork PL-9 (Pazmany) (1992-93) 528
Stratos (HFL) (1986-87) 594
Stratos (Ligeti Aeronautical) (1991-92) 583
Stratos (Stratos) (1984-85) 716
Stratos 300 (HFL-Flugzeugbau) (1992-93) 556
STRATOS AVIATION (USA) (1984-85) 716
STRATUS UNLIMITED INC (USA) (1983-84) 690
Streaker (Abernathy) (1986-87) 622
*Streak Shadow (CFM)
Striker (Flexi-Form) (1985-86) 666
Striker Tri-Flyer (Mainair) (1985-86) 668
STRIPLIN AIRCRAFT CORPORATION (USA) (1992-93) 540
SUMMIT AIRCRAFT CORPORATION (USA) (1987-88) 716
SUN AEROSPACE CORPORATION (USA) (1983-84) 677
SUN AEROSPACE GROUP INC (USA) (1990-91) 594
Sunburst (Airmass) (1983-84) 653
SUNDERLAND AIRCRAFT (USA) (1987-88) 718
Sundowner Convertible (Ultra-Fab) (1985-86) 713
SUN FUN ULTRALIGHT AVIATION (Canada) (1987-88) 552
Sunfun VJ-24W (Volmer) (1992-93) 572
Sunny (Tandem Aircraft) (1991-92) 590
Sun Ray (Sun Aerospace) (1983-84) 677
Sun Ray 100 (Sun Aerospace) (1990-91) 594
Sun Ray 200 (Sun Aerospace) (1990-91) 594
Sunrise (Personal Planes) (1983-84) 670
Sunrise II (Dallach Flugzeuge) (1992-93) 556
SUNRISE ULTRALIGHT MANUFACTURING COMPANY (USA) (1992-93) 540, 571
Super Ace, Pober (Acro Sport) (1992-93) 495
Super Acro Sport (Acro Sport) (1992-93) 494
Super Acro-Zénith CH 180 (Zenair) (1992-93) 472
Super Adventurer (Adventure) (1992-93) 496
Super Adventurer 2 + 2 (Adventure) (1992-93) 496
'Super Baby' Lakes (BOAC) (1992-93) 506
Super Cat (First Strike) (1992-93) 514
Super Cavalier (Macfam) (1988-89) 582
Super Diamant (Piel) (1992-93) 482
Super Emeraude (Piel) (1992-93) 481
Super Hawk (Hovey) (1991-92) 598
Super Honcho (Magnum) (1992-93) 525
Super Kingfisher W-1 (Richard Warner) (1992-93) 546
Super Kitten (Grover) (1985-86) 689
Super Kitten (Hipp's Superbirds) (1992-93) 520
Super Koala (Fisher) (1992-93) 514
Supermarine S.6 Replica (Flight Dynamics) (1986-87) 664
Supermarine Spitfire Mk IX (Thunder Wings) (1988-89) 606
Super-Menestral HN 434 (Nicollier) (1992-93) 480
Super Mono-Fly (Teman) (1984-85) 717
Super Pelican (Ultravia) (1988-89) 513
Super Prospector (Bounsall) (1987-88) 643
Super Pup (Preceptor) (1992-93) 529
Super Scout (Flylite) (1986-87) 613
Super Skybolt (Stolp) (1987-88) 715
Super-Sport (ULM Pyrénées) (1986-87) 591
Super Sportster (Hipp's Superbirds) (1992-93) 520
Super Star (Starflight) (1983-84) 676
Super Starduster (Stolp) (1992-93) 540
Super STOL CH 801 (Zenair) (1992-93) 474
Superwing (Mitchell Wing) (1988-89) 586
Super Zodiac (Zenair) (1986-87) 558
Super Zodiac CH 601HDS (Zenair) (1992-93) 473
Supreme (Hornet) (1984-85) 678
SV AIRCRAFT PTY LTD (Australia) (1992-93) 550
Swallow A (Swallow) (1985-86) 711
SWALLOW AEROPLANE COMPANY, THE (USA) (1985-86) 710
Swallow B (Swallow) (1985-86) 710
SWEARINGEN AIRCRAFT CORPORATION (USA) (1987-88) 718
SWICK AIRCRAFT (USA) (1992-93) 540
Swingwing VJ-23E (Volmer) (1992-93) 572
SWISS AEROLIGHT (Switzerland) (1992-93) 560
SYLVAIRE MANUFACTURING LTD (Canada) (1987-88) 553
SZNAPKI, ZDZISLAW (Poland) (1983-84) 687

T

T-1 (Thor) (1988-89) 512
T-2 (Antonov) (1992-93) 560
T-2 (Thor) (1988-89) 512
T-4 (Antonov) (1992-93) 560
T-IV (Canaero Dynamics) (1986-87) 544
T7 Leone (AUI) (1992-93) 558
T8 Leone (AUI) (1988-89) 536
T-9 Stalker (Stoddard-Hamilton) (1991-92) 573
T-10 Nuwacot (Aircraft Dynamics) (1992-93) 501
*T30 Katana (Terzi)
T-77 (Turner) (1987-88) 723
T-88 (Thruster) (1989-90) 534
T-100A Mariah (Turner) (1983-84) 680
T-100B Mariah (Turner) (1985-86) 712
T-100D Mariah (Turner) (1992-93) 571
T-110 (Turner) (1991-92) 576
T.250, Vortex (Chargus) (1984-85) 677
T300 (Thruster Aircraft) (1992-93) 550
T.440, Forger (Chargus) (1984-85) 677
T.440, Titan (Chargus) (1984-85) 676
T500 (Thruster Aircraft) (1992-93) 550
T.503 (Pegasus) (1987-88) 618
TA (Teratorn) (1984-85) 718
TA-2/3 Bird (Taylor) (1992-93) 541
TA-3 (T. E. A. M.) (1991-92) 603
TA16 (IAC) (1986-87) 674
TA16 Trojan (Thurston) (1991-92) 576
TC-2 (Aero Mirage) (1991-92) 535
TD-13 Wazka (Dobrocinski) (1984-85) 673
TE-F3 (Temple-Wing) (1984-85) 672
TE-F3A (Temple Wing) (1987-88) 598
T-M Scout (Gardner) (1987-88) 614
TM-5 (TM Aircraft) (1985-86) 628
TR-7 (IAC) (1986-87) 674
TR-200 (Tech-Aero) (1992-93) 542
TR 260 (Feugray) (1986-87) 574
TST (Thruster) (1988-89) 506
TU-10 Double Eagle (Mitchell Wing) (1991-92) 600
TX1000 (Starflight) (1984-85) 716

T-Bird I (Golden Circle Air) (1992-93) 565
T-Bird II (Golden Circle Air) (1992-93) 565
T-Bird III (Golden Circle Air) (1992-93) 565
T-Bird Z2000 (Golden Circle Air) (1992-93) 566
T-Craft (Taylor Kits) (1992-93) 542
T. E. A. M. (see Tennessee Engineering and Manufacturing Inc) (1992-93) 571
TEDA (see Tecnologia Europea Divisione Aeronautica) (1991-92) 592
TFM (see Those Flying Machines) (1986-87) 723
TACT-AVIA CORPORATION (USA) (1986-87) 720
Tailwind W-8 (Wittman) (1992-93) 548
Tailwind W-10 (Wittman) (1992-93) 548
TAMARIND INTERNATIONAL LTD (USA) (1986-87) 721
TAMIS MOTOR INDUSTRIES (Turkey) (1992-93) 560
TANDEM AIRCRAFT KG (Germany) (1991-92) 590
Tandem Trainer (Parsons) (1991-92) 562
Tandem Two-seater (Air Command) (1991-92) 537
TAYLOR AERO INC (USA) (1992-93) 541
Taylor Bullet 2100 (Aerocar) (1989-90) 560
TAYLOR KITS CORPORATION (USA) (1992-93) 541
TAYLOR, T (UK) (1992-93) 492
T Clip-Wing Taylorcraft (Swick) (1992-93) 540
TECH-AERO (USA) (1992-93) 542
TECHNOFLUG LEICHTFLUGZEUBAU GmbH (West Germany) (1989-90) 610
TECNOLOGIA EUROPEA DIVISIONE AERONAUTICA Srl (Italy) (1991-92) 592
TEENIE COMPANY (USA) (1992-93) 542
Teenie Two (Parker) (1988-89) 588
TEMAN AIRCRAFT INC (USA) (1984-85) 717
Tempête (Jurca) (1992-93) 477
TEMPLE, PROF DIPL-ING BERNHARD E. R. J. DE (West Germany) (1987-88) 598
TENNESSEE ENGINEERING AND MANUFACTURING INC (USA) (1992-93) 571
Ten Series (Ultimate Aircraft) (1986-87) 552
TEPELNE ISOLACE BUDOV (Czechoslovakia) (1992-93) 554
TERATORN AIRCRAFT INC (USA) (1990-91) 596
Termite (Smith) (1991-92) 570
TERRADE, A (France) (1987-88) 300
Terrier AG-38A (Mitchell Wing) (1991-92) 560
TERVAMÄKI, JUKKA (Finland) (1988-89) 516
*TERZI AERODINE (Italy)
TESORI AIRCRAFT LTD (Canada) (1986-87) 552
TESORI, ROBERT (Canada) (1987-88) 553
Texan Chuckbird (Chuck's Aircraft) (1992-93) 564
Texan Gem 260 (Miller) (1988-89) 585
TFM INC (see Those Flying Machines) (1983-84) 679
THALHOFER LUDWIG (West Germany) (1987-88) 598
The Fun Machine (Loehle) (1989-90) 618
The Minimum (Saphir) (1991-92) 602
THOMPSON AIRCRAFT (USA) (1987-88) 721
THOR AIR (Canada) (1988-89) 512
THORP (USA) (1986-87) 724
Thort T-1 (Thor Air) (1988-89) 512
THOSE FLYING MACHINES, TFM, INC (USA) (1986-87) 723
Thruster (Thruster Aircraft) (1986-87) 542
Thruster (Ultralight) (1984-85) 658
Thruster 447 (Thruster) (1985-86) 641
THRUSTER AIRCRAFT (AUSTRALIA) PTY LTD (Australia) (1992-93) 550

T – Z —PRIVATE AIRCRAFT: INDEXES 723

Thruster TST (Thruster) (1985-86) 641
Thruster Utility (Thruster) (1985-86) 641
Thunderbolt (WAR) (1988-89) 612
Thunder Gull (Earthstar) (1990-91) 573
Thunder Gull J (Earthstar) (1992-93) 564
Thunder Gull JX (Earthstar) (1992-93) 513
THUNDER WINGS (USA) (1988-89) 606
THURSTON AEROMARINE CORPORATION (USA) (1992-93) 542
Tierra (Teratorn) (1983-84) 678
Tierra I (Teratorn) (1990-91) 596
Tierra II (Teratorn) (1990-91) 596
Tierra UL (Teratorn) (1986-87) 722
Tiger Cub (MBA) (1984-85) 682
TI-Ram (Ram) (1987-88) 564
TIRITH MICROPLANE LTD (UK) (1987-88) 622
TISSERAND, CLAUDE (France) (1987-88) 590
Titan (Chargus) (1984-85) 676
Titan T.440 (Pegasus) (1985-86) 670
Titch J. T.2 (Taylor) (1992-93) 493
TM AIRCRAFT (USA) (1985-86) 628
Tomcat (Midwest) (1985-86) 697
Tomcat (Waspair) (1983-84) 682
TOPA AIRCRAFT COMPANY (USA) (1986-87) 725
Toucan (Earthstar) (1985-86) 684
Toucan (Long) (1983-84) 666
Toucan Series II (Canaero Dynamics) (1988-89) 508
Toucan Series II (Novadyne) (1990-91) 536
Toucan T-IV (Canaero Dynamics) (1986-87) 544
Tourisme EC-8 (Croses) (1991-92) 514
Tragonfly (Miner) (1987-88) 682
Traveller (NAI) (1987-88) 563
Traveller WH-I (Wendt) (1986-87) 734
TriCX (Firebird) (1986-87) 594
Trident T-3 (Summit) (1987-88) 716
Tri-Flyer 330 (Mainair) (1985-86) 668
Tri-Flyer Sprint (Mainair) (1985-86) 668
Trike (Delta Wing) (1984-85) 726, 730
Trike (LEAF) (1984-85) 702
Trike (Skyhook Sailwings) (1983-84) 688
Tri-Pacer (Ultrasports) (1983-84) 651
TRI Q DEVELOPMENT COMPANY (USA) (1988-89) 606
TRI-R TECHNOLOGIES (USA) (1992-93) 542
TriStar (Starflight) (1983-84) 676
*Tri-Z CH 300 (Zenair)
Trojan (IAC) (1986-87) 674
Trojan TA16 (Thurston) (1991-92) 576
Trolslandon BA-12, Andreasson (MFI) (1985-86) 662
Tse Tse (Souquet) (1987-88) 589
Tsunami (Sandberg) (1987-88) 704
TUBE WORKS (USA) (1987-88) 722
Tug (Pegasus) (1987-88) 619
Tukan (Aero-Sol) (1992-93) 562
Turbulent D31 (PFA) (1988-89) 547
TURNER AIRCRAFT INC (USA) (1992-93) 542, 571
Twin Foxcat (Ultrasport) (1986-87) 603
TwinStar (Kolb) (1987-88) 578
TwinStar Mark II (Kolb) (1990-91) 578
TwinStar Mark III (Kolb) (1992-93) 521
Twin-T (Taylor Kits) (1992-93) 542
Twin-Varieze (Shaw) (1987-88) 622
Two Easy (Beard) (1987-88) 641
Two Place (Seville) (1988-89) 510
TWO WINGS AVIATION (USA) (1992-93) 542, 571
Type 200/300/400 (MAC) (1983-84) 666
Type 4315 (Aérokart) (1986-87) 562
Type 4320 (Aérokart) (1986-87) 563
Type 4320-A83 (Aérokart) (1983-84) 636
Tyro (South Coast) (1984-85) 657
Tyro Mk 2 (Eastwood Aircraft) (1992-93) 550

U

U-2 Superwing (Mitchell Aerospace) (1991-92) 600
U76 (Brugger) (1984-85) 674
UFM (see Ultra Ligeros Franco Mexicanos) (1992-93) 488
UFM (see Ultralight Flying Machines) (1983-84) 680
UFM of Ky Aeroplane XP (Loehle) (1989-90) 618
ULM 1 Gryf (Tepelne) (1992-93) 554
ULM-1B Moskito (Konsuprod) (1992-93) 557
ULM 811 (Air Cefelec) (1983-84) 638
ULM 812 (Air Cefelec) (1983-84) 638
UP (see UP Sports Ultralite Products) (1986-87) 728
UR-1 (Mira Slovak) (1987-88) 684
UW-7 (Weller) (1986-87) 598

UFM OF KENTUCKY (USA) (1984-85) 718
ULAC (see Ultra Light Aircraft Corporation) (1985-86) 664
UL-BAU-FRANZ (Germany) (1987-88) 594
UL-BAU-REIMANN (Germany) (1987-88) 596
Uli (Scheibe) (1984-85) 671
Ulm (Prevot) (1986-87) 589
ULM AÉROTECHNIC (France) (1983-84) 641
ULM LEDUC (France) (1985-86) 656
Ulm Leduc 104 (Ulm Leduc) (1985-86) 656
ULM PROGRESS (France) (1987-88) 590
ULM PYRÉNÉES (France) (1986-87) 591
ULM TECHNIC AEROMARINE SARL (France) (1984-85) 668
Ultavia (Ultavia) (1984-85) 719
ULTAVIA AIRCRAFT LTD (USA) (1984-85) 719
ULTIMATE AIRCRAFT (Canada) (1992-93) 470
Ultimate Aircraft 10 Dash 100/300 (Ultimate Aircraft) (1992-93) 470
Ultimate Aircraft 20 Dash 300 (Ultimate Aircraft) (1992-93) 471
ULTIMATE HI (USA) (1984-85) 719
Ultra (Scheibe) (1984-85) 671
Ultra-Aire I (Pinaire) (1992-93) 568
Ultra-Aire II (Pinaire) (1987-88) 692
Ultrabat (Australite) (1992-93) 503
ULTRA CLASSICS (USA) (1984-85) 719
ULTRA EFFICIENT PRODUCTS INC (USA) (1992-93) 572
ULTRA-FAB (USA) (1985-86) 713
ULTRAFLIGHT SALES LTD (Canada) (1989-90) 607
Ultra Gull (Earthstar) (1990-91) 573
ULTRALEICHT-FLUGGERÄTEBAU GmbH (Germany) (1992-93) 557
ULTRA LIGEROS FRANCO MEXICANOS (Mexico) (1992-93) 488
ULTRA LIGHT AIRCRAFT CORPORATION (Switzerland) (1985-86) 664
ULTRALIGHT AIRCRAFT INC (USA) (1991-92) 604
ULTRALIGHT AIRCRAFT INDUSTRIES (AUST) PTY LTD (Australia) (1985-86) 641
ULTRALIGHT AVIATION (Australia) (1984-85) 658
ULTRALIGHT AVIATION CENTRE LTD (UK) (1982-83) 615
ULTRALIGHT FLIGHT INC (USA) (1987-88) 725
ULTRALITE SOARING INC (USA) (1985-86) 714
Ultramental (Burke) (1986-87) 644
Ultra Pup (Preceptor) (1992-93) 529, 568
ULTRA SAIL INC (USA) (1990-91) 622
ULTRASPORT SRL (Italy) (1986-87) 602
ULTRA SPORTS LTD (UK) (1984-85) 683
Ultrastar (Kolb) (1988-89) 578
ULTRAVIA AERO INC (Canada) (1992-93) 471, 553
Unik (Pomergau) (1987-88) 550
UP INC (USA) (1987-88) 726
UPM GmbH (West Germany) (1990-91) 610
Upside-Down Pitts S-2B (Hoskins) (1986-87) 671
UP SPORTS (ULTRALITE PRODUCTS) (USA) (1986-87) 728
US AVIATION (USA) (1992-93) 572
US MOYES INC (USA) (1984-85) 728

V

V-8 Special (V-8 Special) (1992-93) 542
V-8 Special SXS (V-8 Special) (1992-93) 542
*VAT (see Vertical Aviation Technologies Inc)
*VJ-22 Sportsman (Volmer)
VJ-23E Swingwing (Volmer) (1992-93) 572
VJ-24W Sunfun (Volmer) (1992-93) 572
VK-8 Ausra (Kengsaila) (1992-93) 487
*VK30 Cirrus (Cirrus)
VP-1 (Evans) (1992-93) 513
VP-2 (Evans) (1988-89) 570
V-Star SA-900 (Stolp) (1992-93) 540
VTOL (BIAA) (1987-88) 562

V-8 SPECIAL (USA) (1992-93) 542
Vagabond (Saurier Flug Service) (1990-91) 609
Vagabond II (Saurier Flug Service) (1992-93) 557
Vagrant II (FLSZ) (1990-91) 574
Vampire Sadler (Skywise) (1988-89) 505
Vancraft (Vancraft Copters) (1992-93) 543
VANCRAFT COPTERS (USA) (1992-93) 543, 572
Vancroft (Vanek) (1986-87) 729
VAN LITH, JEAN (France) (1987-88) 591
Van Lith VIII (Van Lith) (1987-88) 591
*VAN'S AIRCRAFT INC (USA)
VANEK, CHUCK (USA) (1986-87) 729
VECTOR (1984-85) 721
Vector 610 (Aerodyne Systems) (1984-85) 685
Vector 610 (Sky King) (1985-86) 644
Vector 627 (Aerodyne Systems) (1984-85) 685
Vector 627 (Sky King) (1985-86) 644
VEENSTRA (see Sander Veenstra) (1985-86) 642
Velocity (Velocity Aircraft) (1992-93) 544
VELOCITY AIRCRAFT (USA) (1992-93) 544
VENTURE FLIGHT DESIGN INC (Canada) (1986-87) 554
Venture M.20 (Questair) (1992-93) 530
VERNER, VLADA (Czechoslovakia) (1987-88) 564
VERTICAL AVIATION TECHNOLOGIES INC (USA)
Vert-X (Driggers) (1987-88) 658
Vetro (Bezzola) (1987-88) 610
VIDOR, GUISEPPE (Italy) (1987-88) 606
Viking (Northstar) (1986-87) 690
*VIKING AIRCRAFT LTD (USA)
VOGT, HELMUT (Germany) (1991-92) 590
Volksplane VP-1 (Evans) (1992-93) 513
*VOLMER AIRCRAFT (USA)
Volucelle (Junqua) (1987-88) 576
Vortex (Chargus) (1984-85) 677
Vortex (Pegasus) (1992-93) 490
Vortex LA-4 (Langebro) (1985-86) 670
Vortex T.250 (Pegasus) (1985-86) 670
VPM Sn C (Italy) (1986-87) 604
VULCAN UL-AVIATION (Germany) (1992-93) 557

W

W-1 Super Kingfisher (Richard Warner) (1992-93) 546
W-02 (Verner) (1987-88) 564
W-5 and W-5A (SAP) (1992-93) 554
W-5B (SAP) (1992-93) 554
W-6 Qingting (SAP) (1991-92) 554
W-7 Dipper (Collins Aero) (1992-93) 509
W8 Tailwind (Wittman) (1992-93) 548
W10 Tailwind (Wittman) (1992-93) 548
W-11 Boredom Fighter (Wolf) (1988-89) 614
WAACO (see West Australian Aircraft Company) (1988-89) 506
WAC (see Wolfe Aviation Company) (1984-85) 721
WAR (see War Aircraft Replicas) (1988-89) 612
WAT (WAC) (1984-85) 721
WH-1 Traveller (Wendt) (1986-87) 734
WSK-PZL WARSZAWA-OKECIE (see Wytwornia Sprzetu Komunikacyjnego-PZL Warszawa-Okecie) (1983-84) 687
WT-11 Chinook (Birdman) (1986-87) 544
WT-11 Chinook 2S (Birdman) (1985-86) 644
WW-1 Der Jager D IX (White) (1988-89) 613

Waco II (Golden Age) (1985-86) 688
*Wag-a-Bond (Wag-Aero)
*WAG-AERO INC (USA)
WAR AIRCRAFT REPLICAS (USA) (1988-89) 612
WARD, RAY (USA) (1987-88) 732
War Eagle AG-38-A (Mitchell Wing) (1991-92) 560
WARNER AVIATION INC (USA) (1989-90) 600
WARPATH AVIATION (USA) (1983-84) 682
WASPAIR (see Midwest) (1984-85) 722
WASPAIR CORPORATION (USA) (1983-84) 682
WATSON WINDWAGON COMPANY (USA) (1992-93) 546
Wazka TD-13 (Dobrocinski) (1984-85) 673
WEEDHOPPER (see Nova Air) (1984-85) 722
Weedhopper (Weedhopper) (1992-93) 572
Weedhopper C, JC-24C (Weedhopper) (1983-84) 683
WEEDHOPPER INC (USA) (1992-93) 572
WEEDHOPPER OF UTAH INC (USA) (1983-84) 683
Weedhopper Two-seater (Weedhopper) (1992-93) 572
Week-end HN 600 (Nicollier) (1992-93) 481
Wel-1 (Wiweko) (1986-87) 598
WELLER, ROMAN (West Germany) (1986-87) 598
WENDT AIRCRAFT ENGINEERING (USA) (1986-87) 734
Westair 204 (Western) (1985-86) 634
WEST AUSTRALIAN AIRCRAFT COMPANY (Australia) (1988-89) 506
WESTERN AIRCRAFT CORPORATION (USA) (1985-86) 634
WESTERN AIRCRAFT SUPPLIES (Canada) (1992-93) 472
Westland Whirlwind Mark II (Butterworth) (1987-88) 648
WESTWIND CORPORATION LTD (UK) (1988-89) 548
WEWYNE (USSR) (1991-92) 528
WHATLEY, VASCOE Jr (USA) (1988-89) 612
WHEELER AIRCRAFT (SALES) PTY LTD, RON (Australia) (1992-93) 550
*WHEELER TECHNOLOGY INC (USA)
Whing Ding II (WD-11) (Hovey) (1991-92) 598
Whisper (Grove) (1987-88) 668
Whitehawk 185 (Beck) (1983-84) 688
*White Lightning (White Lightning)
*WHITE LIGHTNING AIRCRAFT CORPORATION (USA)
WHITE, MARSHALL E (USA) (1988-89) 613
WHITTAKER, MICHAEL W. J. (UK) (1992-93) 562
Wichawk (Javelin) (1992-93) 520
WICKS AIRCRAFT SUPPLY (USA) (1992-93) 547
WIERZBOWSKI, PAWEL (Poland) (1983-84) 687
Wildente (Ultraleicht) (1992-93) 557
Wildente 2 (Ultraleicht) (1992-93) 557
WIND DANCER (USA) (1987-88) 734
Wind Dancer (Wind Dancer) (1987-88) 734
Windlass Trike (Solo Wings) (1992-93) 560
Wind Rider (Aerotech) (1986-87) 631
WindRyder (WindRyder) (1989-90) 601
WINDRYDER ENGINEERING INC (USA) (1992-93) 548
Windwagon (Watson) (1992-93) 546
WINTON AIRCRAFT (Australia) (1986-87) 543
WINTON, SCOTT (Australia) (1992-93) 551
WISCH & NIST GmbH (West Germany) (1987-88) 599
Witch (Greenwood) (1986-87) 668
WITTMAN, S. J. (USA) (1992-93) 548
WIWEKO, SOEPONO (Indonesia) (1986-87) 598
Wizard (Ultralite Soaring) (1984-85) 720
WOLF, DONALD S (USA) (1988-89) 614
WOLF, STEVE (USA) (1986-87) 737
WOLFE AVIATION COMPANY (USA) (1984-85) 721
WOLFF, ATELIERS PAUL (Luxembourg) (1991-92) 524
Wombat Gyrocopter (Julian) (1991-92) 529
WOMBAT GYROCOPTERS (UK) (1992-93) 493
WONDER MUDRY AVIATION (France) (1984-85) 548
WOOD WING SPECIALITY (USA) (1992-93) 582
WORLD WIDE RACING SERVICES (UK) (1986-87) 622
WORLDWIDE ULTRALITE INDUSTRIES INC (USA) (1984-85) 722
Wren (Advanced Composite) (1987-88) 628
Wren (Calder) (1983-84) 656
Wren (Wren) (1985-86) 715
WREN AVIATION INC (USA) (1985-86) 715

X

X99 (Ulac) (1985-86) 664
XA/650 Buccaneer II (Advanced Aviation) (1992-93) 496

XC 280 Stiletto (Star Flight) (1992-93) 570
XC Series (Star Flight) (1992-93) 570
XL (Pegasus) (1988-89) 546
XL Panther (Pegasus) (1987-88) 619
XL Panther (Ultra Sports) (1984-85) 684
XP Aeroplane, UFM of Ky (Loehle) (1989-90) 618
XP Falcon (American Aircraft) (1987-88) 639
XP-Talon (Aero-Tech) (1992-93) 499
XTC Hydrolight (Diehl) (1991-92) 597

Y

YC-100 Series (Chasle) (1992-93) 555
YC-210 (Chasle) (1987-88) 568

YAKOVLEV (USSR) (1991-92) 528
Yarrow Arrow (Arrow Aircraft) (1992-93) 465
Yuca (Tamis Motor) (1992-93) 560
Yunhe No 1 (Jinzhou) (1986-87) 560
YUODINAS, KINTAUTAS (Lithuania) (1992-93) 559

Z

Z-1 Flitzer (Sky Craft) (1992-93) 492
Z-CH 100 Mono (Zenair) (1992-93) 472
Z-MAX (T. E. A. M.) (1992-93) 571

Zarus (Kolodziej) (1983-84) 692
Zeffir (Calvel) (1984-85) 664
Zefiro 940 (General Gliders) (1992-93) 558
Zem-80 (Instytut Lotnictwa) (1992-93) 559
Zem-90 (Instytut Lotnictwa) (1992-93) 559
*ZENAIR LTD (Canada)
ZENITH-AVIATION (France) (1987-88) 592
*Zenith-CH 200 (Zenair)
*Zenith-CH 250 (Zenair)
*Zenith CH 2000 (Zenair)
Zephir C. P.80 (Piel) (1988-89) 529
Zephyr II (Euro Wing) (1983-84) 648
*Zéphyr M. J. 52 (Jurca)
Zero, Mitsubishi (WAR) (1988-89) 612
ZETTLE, HEINRICH (Germany) (1992-93) 557
Zipper (Mainair Sports) (1987-88) 616
Zipper (Zenair) (1987-88) 560
Zipper II (Zenair) (1987-88) 561
Zipper RX (Zenair) (1987-88) 560
Zippy Sport (Fishercraft) (1989-90) 574
Zodiac CH 600 (Zenair) (1991-92) 510
Zodiac CH 601 (Zenair) (1992-93) 472
Zodiac CH 601HD (Zenair) (1992-93) 473

SAILPLANES

These are no longer included in the book. The index below covers the preceding ten editions

A

A-21S (Caproni Vizzola) (1984-85) 639
A-21SJ (Caproni Vizzola) (1984-85) 639
AFH 22 (Akaflieg Hannover) (1984-85) 628
AFH 24 (Akaflieg Hannover) (1992-93) 582
AFH 26 (Akaflieg Hannover) (1992-93) 582
AK-2 (Akaflieg Karlsruhe) (1985-86) 735
AK-5 (Akaflieg Karlsruhe) (1992-93) 582
AK-8 (Akaflieg Karlsruhe) (1992-93) 582
AMT-100 Ximango (Aeromot) (1992-93) 576
ANB (KAPU) (1986-87) 765
ARPLAM (see All Reinforced Plastic Mouldings) (1990-91) 623
ASH 25 (Schleicher) (1992-93) 590
ASH 25 E (Schleicher) (1992-93) 590
ASH 25 MB (Schleicher) (1990-91) 642
ASH 26 E (Schleicher) (1992-93) 590
ASK 13 (Jubi) (1992-93) 584
ASK 13 (Schleicher) (1988-89) 632
ASK 21 (Schleicher) (1992-93) 588
ASK 23 (Schleicher) (1992-93) 588
ASW 19 (Schleicher) (1986-87) 756
ASW 19 Club (Schleicher) (1985-86) 742
ASW 19 B (Schleicher) (1985-86) 742
ASW 20 (Schleicher) (1986-87) 756
ASW 20 B (Schleicher) (1992-93) 588
ASW 20 BL (Schleicher) (1992-93) 588
ASW 20 C (Schleicher) (1992-93) 588
ASW 20 CL (Schleicher) (1992-93) 588
ASW 20 L (Schleicher) (1986-87) 756
ASW 22 (Schleicher) (1986-87) 756
ASW 22-2 (Schleicher) (1985-86) 744
ASW 22 B (Schleicher) (1992-93) 588
ASW 22 BE (Schleicher) (1992-93) 588
ASW 24 (Schleicher) (1992-93) 590
ASW 24 E (Schleicher) (1992-93) 590
ASW 24 TOP (Schleicher) (1992-93) 590
ASW 27 (Schleicher) (1992-93) 590
ATS-1 Ardhra (Civil Aviation Department) (1992-93) 594

AB-RADAB (Sweden) (1992-93) 600
Acro IAR-35 (IAR) (1992-93) 596
ADVANCED AVIATION INC (USA) (1992-93) 600
AERO CLUBE DE RIO CLARO (Brazil) (1986-87) 740
AEROMOT-AERONAVES e MOTORES SA (Brazil) (1992-93) 576
AEROMOT-INDUSTRIA MEGANICO- METALURGICA (Brazil) (1990-91) 623
AERONAUTIQUE SERVICE (France) (1990-91) 626
AERONAVES E MOTORES SA (Brazil) (1989-90) 623
AEROSTRUCTURE SARL (France) (1987-88) 744
AEROTECHNIK (Czechoslovakia) (1992-93) 578
Aiglon J. P. 15-36 A/AR (CARMAM) (1983-84) 602
AKADEMISCHE FLIEGERGRUPPE BERLIN eV (Germany) (1992-93) 578
AKADEMISCHE FLIEGERGRUPPE BRAUNSCHWEIG eV (Germany) (1991-92) 614
AKADEMISCHE FLIEGERGRUPPE DARMSTADT eV (Germany) (1991-92) 614
AKADEMISCHE FLIEGERGRUPPE ESSLINGEN eV (West Germany) (1985-85) 734
AKADEMISCHE FLIEGERGRUPPE HANNOVER eV (Germany) (1992-93) 582
AKADEMISCHE FLIEGERGRUPPE KARLSRUHE eV (Germany) (1992-93) 582
AKADEMISCHE FLIEGERGRUPPE MÜNCHEN eV (West Germany) (1983-84) 606
AKADEMISCHE FLIEGERGRUPPE STUTTGART eV (Germany) (1992-93) 582
AKAFLIEG BERLIN (see Akademische Fliegergruppe Berlin eV) (1992-93) 578
AKAFLIEG BRAUNSCHWEIG (see Akademische Fliegergruppe Braunschweig eV) (1991-92) 614
AKAFLIEG DARMSTADT (see Akademische Fliegergruppe Darmstadt eV) (1991-92) 614
AKAFLIEG ESSLINGEN (see Akademische Fliegergruppe Esslingen eV) (1985-86) 734
AKAFLIEG HANNOVER (see Akademische Fliegergruppe Hannover eV) (1992-93) 582
AKAFLIEG KARLSRUHE (see Akademische Fliegergruppe Karlsruhe eV) (1991-92) 582
AKAFLIEG MÜNCHEN (see Flugtechnische Forschungsgruppe eV) (1986-87) 748
AKAFLIEG STUTTGART (see Akademische Fliegergruppe Stuttgart eV) (1992-93) 582
ALANNE, PENTTI (Finland) (1985-86) 730
ALEXANDER SCHLEICHER GmbH & CO (Germany) (1992-93) 588
ALL REINFORCED PLASTICS MOULDINGS (Belgium) (1990-91) 623
ANTONOV, OLEG K, DESIGN BUREAU (UKRAINE) (1992-93) 600
APPLEBAY INC (USA) (1985-86) 753
Aqua Glider (Explorer) (1983-84) 624
Araponga (Rio Claro) (1986-87) 740
Ardhra, ATS-1 (Civil Aviation Department) (1992-93) 594
Arrow (UP Sports) (1984-85) 652
Astir (Grob) (1983-84) 608
Astir II (Grob) (1983-84) 608
Atlas 21m (La Mouette) (1992-93) 578
AUTO-AERO KÖZLEKEDÉSTECHNIKAI VALLALAT (Hungary) (1992-93) 594
AVIATECHNIKA (Hungary) (1986-87) 759
AVIONS RENÉ FOURNIER (France) (1992-93) 578
AVIONS POTTIER (France) (1987-88) 746

B

B4-PC11AF (Nippi/Pilatus) (1984-85) 640
B 12T (Akaflieg Berlin) (1989-90) 630
B-13 (Akaflieg Berlin) (1992-93) 578
BIAA (see Beijing Institute of Aeronautics & Astronautics) (1984-85) 621
BK-7 Lietuva (LAK) (1983-84) 623
BL-1 Kea (Hinz) (1990-91) 634
BMW-1 (Wilkes) (1984-85) 652
Bro-IIM Zile (LAK/Oshkinis) (1985-86) 752
Bro-16 (LAK/Oshkinis) (1984-85) 644
Bro-17U Utochka (LAK/Oshkinis) (1984-85) 644
BRO-20 Pukyalis (LAK) (1984-85) 644
BRO-21 Vituris (LAK) (1984-85) 644
BRO-23KR Garnys (Oshkinis) (1992-93) 600

Bakcyl PW-3 (PW) (1992-93) 594
BEIJING INSTITUTE OF AERONAUTICS & ASTRONAUTICS (China) (1984-85) 621
Blanik (LET) (1989-90) 626
Blanik, L-13A (Let) (1984-85) 622
Blue Wren, T-5 (Todhunter) (1987-88) 739
BOULAY-FERRIER (see Boulay, Jacques and Ferrier, Hubert) (1984-85) 624
BOULAY, JACQUES AND FERRIER, HUBERT (France) (1984-85) 624
Brawo, SZD-48-3M (SZD) (1986-87) 761
BRDITSCHKA HB, GmbH & CO KG (Austria) (1987-88) 739
BRYAN AIRCRAFT INC (USA) (1985-86) 753
BURKHART GROB LUFT- UND RAUMFAHRT GmbH (Germany) (1992-93) 582

C

C.38 (CARMAM) (1983-84) 602
CB-7 (CETEC) (1989-90) 624
CETEC (see Central Technical Foundation of Minas Gerais) (1989-90) 624
CT6 (Lightwing) (1989-90) 650
CTA (see Centro Tecnico Aeroespacial) (1984-85) 619

Calif Series (Caproni Vizzola) (1984-85) 639
CALVEL, JACQUES (France) (1984-85) 624
CANARD AVIATION AG (Switzerland) (1989-90) 648
Canard SC (Canard) (1989-90) 648
CAPRONI VIZZOLA COSTRUZIONI AERONAUTICHE SpA (Italy) (1984-85) 639
Carbon Dragon (Maupin) (1992-93) 600
CARMAM (France) (1984-85) 624
CELAIR MANUFACTURING AND EXPORT (PTY) LTD (South Africa) (1992-93) 600
Celstar GA-1 (Celair) (1992-93) 600
CENTRAIR, SA (France) (1992-93) 578
CENTRAL TECHNICAL FOUNDATION OF MINAS GERAIS (Brazil) (1989-90) 624
CENTRO TECNICO AEROESPACIAL (Brazil) (1984-85) 619
CHENGDU AIRCRAFT CORPORATION (China) (1991-92) 608
Chibis PPO-1 (Mikoyan) (1992-93) 600
Chronos (Sarl La Mouette) (1992-93) 578
Cirrus G/81 (Jastreb/Schempp-Hirth) (1984-85) 654
CISKEI AIRCRAFT INDUSTRIES (PTY) LTD (Ciskei) (1988-89) 619
CIVIL AVIATION DEPARTMENT (India) (1992-93) 594
Cloudster, ST-100 (Ryson) (1984-85) 650
Club III, G102 (Grob) (1984-85) 630
Club IIIb, G102 (Grob) (1987-88) 750
Club 101 (Centrair) (1992-93) 578
Companion (Lightwing) (1989-90) 650
Condor (Boulay-Ferrier) (1984-85) 624
Condor (Hollmann) (1984-85) 647

D

D-4 Straton (Olsansky) (1989-90) 626
D-39b (Akaflieg Darmstadt) (1985-86) 734
D-40 (Akaflieg Darmstadt) (1987-88) 747
D 41 (Akaflieg Darmstadt) (1991-92) 614
D 77 Iris (Issoire) (1984-85) 626
DG-101 (Glaser-Dirks) (1985-86) 736
DG-101 Elan (Elan/Glaser-Dirks) (1989-90) 654
DG-200 (Glaser-Dirks) (1985-86) 736
DG-202 (Glaser-Dirks) (1985-86) 736
DG-300 Elan (Elan/Tovarna) (1992-93) 602
DG-300 Elan (Glaser-Dirks) (1985-86) 736
DG-400 (Glaser-Dirks) (1992-93) 582
DG-500 (Glaser-Dirks) (1984-85) 630
DG-500 Elan (Elan/Glaser-Dirks) (1987-88) 772
DG-500 Elan (Glaser-Dirks) (1992-93) 582
DG-600 (Glaser-Dirks) (1992-93) 582
DG-800 (Glaser-Dirks) (1992-93) 582
DWLKK (see Doswiadczalne) Warsztaty Lotni ze Konstrukcji Kompozytowych) (1992-93) 594

Delphin, SF-34 (Scheibe) (1986-87) 753
Dimona H-36 (Hoffmann) (1990-91) 634
Discus (Schempp-Hirth) (1992-93) 584
DOKTOR FIBERGLAS (URSULA HÄNLE) (Germany) (1992-93) 582
DOSWIADCZALNE WARSZTATY LOTNICZE KONSTRUKCJI KOMPOZYTOWYCH (Poland) (1992-93) 594
DSK AIRCRAFT CORPORATION (USA) (1983-84) 624

E

E-14 (Akaflieg Esslingen) (1985-86) 734
E-14 (FAG Esslingen) (1987-88) 748
E 78 Silène (Issoire) (1985-86) 732
EEL (see Entwicklung und Erprobung von Leichflugzeugen) (1987-88) 748
ENSMA (see École Nationale Supérieure de Mécanique et d'Aérotechnique) (1983-84) 603
ES 65 (Schneider) (1988-89) 617
ESAG (see Eksprimentine Sportines Aviacijos Gamykla) (1991-92) 632

ÉCOLE NATIONALE SUPÉRIEURE DE MÉCANIQUE ET D'AÉROTECHNIQUE (France) (1983-84) 603
EKAPRIMENTINE SPORTINES AVIACIJOS GAMYKLA (USSR) (1991-92) 632
ELAN TOVARNA ŠPORTNEGAORODJA N. SOL. O (Yugoslavia) (1992-93) 602
Elan, DG-101 (Elan/Glaser-Dirks) (1989-90) 654
Elan, DG-300 (Elan Tovarna) (1992-93) 602
Elan, DG-500 (Elan/Glaser-Dirks) (1992-93) 582
Elfe M 17 (Parodi) (1985-86) 739
ENTWICKLUNG UND ERPROBUNG VON LEICHFLUGZEUGEN (West Germany) (1987-88) 748
EXPLORER AIRCRAFT COMPANY INC (USA) (1983-84) 624

F

FAG ESSLINGEN (see Flugtechnische Arbeitsgemeinschaft) (1987-88) 748
FS-31 (Akaflieg Stuttgart) (1984-85) 628
Fs-32 (Akaflieg Stuttgart) (1992-93) 582
FVA (see Flugwissenschaftliche Vereinigung Aachen (1920) EV) (1983-84) 608
FVA-20 F. B. Schmetz (FVA) (1983-84) 608

FACTORY SPORTINE (Lithuania) (1992-93) 594
FALCONAR AVIATION LTD (Canada) (1992-93) 576
Falke (Scheibe) (1985-86) 740
Falke 82 (Scheibe) (1984-85) 634
Falke 85 SF-25C (Scheibe) (1985-86) 740
Falke 86 SF-25C (Scheibe) (1986-87) 752
Falke 87 SF-25C (Scheibe) (1987-88) 753
Falke 88 SF-25C (Scheibe) (1989-90) 635
Falke 90, SF-25C (Scheibe) (1991-92) 620
Falke 92 SF-25C (Scheibe) (1992-93) 584
Fauvel AV36 (Falconar) (1992-93) 576
Fauvel AV361 (Falconar) (1992-93) 576
FLIGHT DYNAMICS INC (USA) (1984-85) 647
FLUGTECHNISCHE ARBEITSGEMEINSCHAFT (West Germany) (1987-88) 748
FLUGTECHNISCHE FORSCHUNGSGRUPPE (West Germany) (1986-87) 748
FLUGWISSENSCHAFTLICHE VEREINIGUNG AACHEN (1920) eV (West Germany) (1983-84) 608
FOURNIER AVIATION (France) (1983-84) 603

G

G 102 Series III (Grob) (1987-88) 750
G 103 Twin II (Grob) (1987-88) 750
G 103C Twin III (Grob) (1992-93) 582
G 109 (Grob) (1983-84) 609
G 109B (Grob) (1992-93) 582
GA-1 Celstar (Celair) (1992-93) 600

Gapa, PW-2 (PDWLKK) (1992-93) 594
Gapa D PW-2D (DWLKK) (1992-93) 594
Garnys, BRO-23KR, (Oshkinis) (1992-93) 600
GENERAL GLIDERS (Italy) (1992-93) 594
GLASER-DIRKS FLUGZEUGBAU GmbH (Germany) (1992-93) 582
GLASFASER ITALIANA SrL (Italy) (1983-84) 618
Gobé, R-26S (Auto-Aero) (1989-90) 642
Gobé R-26SU (Auto-Aero) (1992-93) 594
Gradient, VSO 10 (VSO) (1991-92) 610
GROB (see Burkhart Grob Luft- Und Raumfahrt) (1992-93) 582

H

H-36 Dimona (Hoffmann) (1990-91) 634
H-36D (Hoffmann) (1988-89) 628
H-38 Observer (Hoffmann) (1987-88) 752
H 101 Salto 'Hänle' (Doktor Fiberglas) (1992-93) 582
HB (see HB Aircraft Industries Luftfahrzeug AG) (1988-89) 617
HB-21 Hobbyliner (Brditschka) (1985-86) 727
HB-21/2400 Hobbylifter (Brditschka) (1985-86) 727
HB-22 Hobbyliner (Brditschka) (1983-84) 597
HB-22/2400 Hobbylifter (Brditschka) (1983-84) 597
HB-23/2000 Hobbyliner (Brditschka) (1985-86) 727
HK-36 Super Dimona (HOAC) (1992-93) 576
HK-36R Super Dimona (HOAC) (1992-93) 576
HP-18 (Bryan/Schreder) (1985-86) 754
HP-19 (Bryan/Schreder) (1983-84) 624
HP-20 (Bryan/Schreder) (1983-84) 624
HP-21 (Bryan/Schreder) (1985-86) 754
HU-1, Seagull (Shenyang) (1992-93) 576
HU-2 Petrel 650B (Shenyang) (1992-93) 576

H – S—SAILPLANES: INDEXES 725

HB-AIRCRAFT INDUSTRIES LUFTFAHRZEUG AG (Austria) (1988-89) 617
HINZ, LUCIA AND BERNARD (West Germany) (1990-91) 634
HOAC FLUGZEUGWERK WIENER NEUSTADT (Austria) (1992-93) 576
HOFFMAN FLUGZEUGBAU KG, WOLF (West Germany) .... (1 990-91) 634
HOLLMANN (USA) (1984-85) 647
HAVRDA, HORST (Germany) (1991-92) 616

I

IAR-35 Acro (ICA) (1992-93) 596
IPE (see Industria Paranaense de Estruturas) (1990-91) 623
IPE 02 Nhapecam (IPE) (1983-84) 598
IPE 02b Nhapecam II (IPE) (1992-93) 576
IPE-03 (IPE) (1985-86) 728
IPE-04 (IPE) (1991-92) 608
IPE-05 Quero-Quero II (IPE) (1990-91) 624
IPE-05 Quero-Quero III (IPE) (1992-93) 576
IPE-08 (IPE) (1992-93) 576
IS-28B2 (IAR) (1992-93) 596
IS-28M2 (ICA) (1984-85) 642
IS-28M2A (IAR) (1992-93) 596
IS-28MA (ICA) (1983-84) 620
IS-29D2 (IAR) (1992-93) 596
IS-30 (IAR) (1991-92) 632
IS-30A (ICA) (1984-85) 643
IS-32 (IAR) (1991-92) 632

IAR SA (Romania) (1992-93) 596
INAV LTD (USA) (1988-89) 643
INDUSTRIA PARANAENSE DE ESTRUTURAS (Brazil) (1990-91) 623
IPE-INDUSTRIA PROJETOS E ESTRUTURAS AERONAUTICAS LTDA (Brazil) (1992-93) 576
Iris, D 77 (Issoire) (1984-85) 626
ISSOIRE-AVIATION SA (France) (1991-92) 610

J

J5 Janowski (Marco-Elektronik Company) (1985-86) 746
J. P.15-36 A Aiglon (CARMAM) (1983-84) 602
J. P.15-36 A R Aiglon (CARMAM) (1983-84) 602

Jantar 2B (SZD) (1990-91) 645
Jantar 15, SZD-52 (SZD) (1986-87) 762
Jantar Standard 2, SZD-48 (SZD) (1983-84) 619
Jantar Standard 3 (SZD) (1992-93) 596
Janus (Schempp-Hirth) (1992-93) 588
JASTREB FABRICA AVIONA I JEDRILICA (Yugoslavia) (1992-93) 602
Jian Fan, X-7 (Chengdu) (1991-92) 608
Jian Fan, X-9 (Shenyang) (1992-93) 576
JUBI GmbH SPORTFLUGZEUGBAU (Germany) .. (1992-93) 584
Junior (SZD) (1992-93) 596
Junior 1 (Antonov) (1992-93) 600

K

KAI (see Kazan Aviation Institute) (1984-85) 644
KAI-50 (KAI) (1984-85) 644
KAPU (see Kuibyshevski Aviation Institute and Production Unit) (1987-88) 766
KN-1 (Knechtel) (1985-86) 738
KR-1B (Rand Robinson) (1984-85) 650
KR-03A Puchatek (WSK) (1992-93) 596
KW 1 b 2 Quero-Quero II (IPE) (1990-91) 623
KW 1 GB (IPE) (1990-91) 623

KAZAN AVIATION INSTITUTE (USSR) (1984-85) 644
Kit-Club 14-34 (Pottier) (1984-85) 626
Kit-Club 15-34 (Pottier) (1986-87) 745
Kiwi (TWI) (1992-93) 594
KLOTZ (Germany) (1991-92) 616
KNECHTEL, ING WILHELM (West Germany) (1985-86) 738
Kosava-2-S (Jastreb) (1985-86) 760
Krokus, SZD-52 (SZD) (1986-87) 762
KUFFNER (see Leichtflugzeugbau Werner Kuffner) (1987-88) 752
KUIBYSHEVSKI AVIATION INSTITUTE AND PRODUCTION UNIT (1987-88) 766
Kuznechik PPO-2 (Mikoyan) (1992-93) 600

L

L-1 (Arplam) (1989-90) 623
L6-7 (Lucas) (1988-89) 623
L6FS Mouse (Lightwing) (1987-88) 767
L-13A Blanik (Let) (1984-85) 622
L-13SE Vivat (Aerotechnik) (1992-93) 578
L-13SL Vivat (Aerotechnik) (1992-93) 578
L-13SW Vivat (Aerotechnik) (1989-90) 626
L-23 Blanik (LET) (1989-90) 626
L-23 Super Blanik (LET) (1992-93) 578
LAK (see Litovskaya Aviatsionnaya Konstruktsiya) (1987-88) 766
LAK-2 (LAK) (1985-86) 752
LAK-5 Nyamunas (Esag and Ssaktb) (1987-88) 766
LAK-8 (Esag and Ssaktb) (1990-91) 650
LAK-9 Lietuva (LAK) (1983-84) 623
LAK-10 Lietuva (Esag and Ssaktb) (1983-84) 623
LAK-11 Nida (Esag and Ssaktb) (1989-90) 648
LAK-12 Lietuva (Factory Sportine) (1992-93) 594
LAK-12 Lietuva 2R (Factory Sportine) (1992-93) 594
LAK-14 Strazdas (LAK) (1987-88) 767
LAK-15 (Esag and Ssaktb) (1990-91) 650
LAK-16 (Esag) (1991-92) 632
LAK-16M (Factory Sportine) (1992-93) 594
LAK-17 (Factory Sportine) (1992-93) 594
LS4 (Rolladen-Schneider) (1992-93) 584
LS6 (Rolladen-Schneider) (1992-93) 584
LS7 (Rolladen-Schneider) (1992-93) 584

LEICHTFLUGZEUGBAU WERNER KUFFNER (West Germany) (1987-88) 752
LET KONCERNOVY PODNIK (Czechoslovakia) ... (1992-93) 578
Leuvense (Arplam) (1989-90) 623
Libelle (Lutz) (1989-90) 634
Lietuva, LAK-9 (LAK) (1983-84) 623
Lietuva, LAK-10 (LAK) (1983-84) 623
Lietuva LAK-12 (Factory Sportine) (1992-93) 594
Lietuva 2R LAK-12 (Factory Sportine) (1992-93) 594
LIGHTWING RESEACH (UK) (1989-90) 650
Lite-N-Nite (Michna) (1991-92) 634
LITOVSKAYA AVIATSIONNAYA KONSTRUKTSIYA (USSR) (1987-88) 766
LUCAS, EMILE (France) (1988-89) 623
LUNDS TEKNISKE (Norway) (1992-93) 594
Lutin 80 (Aerostructure) (1987-88) 744
LUTZ, JÜRGEN (West Germany) (1989-90) 634
LYNCH, JOHN F. (Australia) (1988-89) 617

M

M2, Milomei (Meier) (1983-84) 610
MG-1 (Civil Aviation Department) (1986-87) 760
Mü 28 (Akaflieg München) (1985-86) 736

MARCO-ELEKTRONIK COMPANY (Poland) (1985-86) 760
Marianne Type 201 (Centrair) (1984-85) 625
Marianne (Centrair) (1992-93) 578
MARSDEN, DR DAVID J (Canada) (1984-85) 620
MAUPIN, JIM (USA) (1992-93) 600
MEIER, MICHEL-LORENZ (West Germany) (1983-84) 610
MICHNA (USA) (1991-92) 634
MIKOYAN (Russia) (1992-93) 600
Milomei M2 (Meier) (1983-84) 610
Mini-Moni (Monnett) (1985-86) 756
Minimus, ST-11 (Stralpes Aero) (1983-84) 605
Mistral-C (Valentin) (1988-89) 635
Model 10 (VSO) (1985-86) 730
Model 77 Solitaire (Rutan) (1983-84) 628
Model 77-6 Solitaire (Rutan) (1985-86) 756
Moka 1 (Klotz) (1991-92) 616
Monarch (Marske) (1992-93) 600
Monerai (INAV) (1987-88) 768
Monerai P (Monnett) (1985-86) 756
Monerai S (Monnett) (1985-86) 755
Monerai-Max (Monnett) (1985-86) 756
Moni (INAV) (1987-88) 768
Moni (Monnett) (1985-86) 756
MONNETT EXPERIMENTAL AIRCRAFT INC (USA) (1985-86) 755
Motorblanik (Alanne) (1983-84) 601
Motorlerche (Alanne) (1984-85) 623
Mouse, L6FS (Lightwing) (1987-88) 767

N

NIPPI (see Nihon Hikoki Kabushiki Kaisha) (1984-85) 640

NATIONAL AIRCRAFT FACTORIES (China) (1989-90) 624
Nhapecam, IPE 02 (IPE) (1983-84) 598
Nhapecan II, IPE-02b (IPE) (1992-93) 676
Nida (Esag and Ssaktb) (1989-90) 648
NIHON HIKOKI KABUSHIKI KAISHA (Japan) (1984-85) 640
Nimbus 2C (Schempp-Hirth) (1983-84) 612
Nimbus 3 (Schempp-Hirth) (1992-93) 584
Nimbus 3D (Schempp-Hirth) (1992-93) 584
Nimbus 4 (Schempp-Hirth) (1992-93) 588
Nyamunas, (Esag and Ssaktb) 1987-88 766

O

O2b (IPE) (1985-86) 728
O3 (IPE) (1986-87) 740

Observer (Hoffmann) (1987-88) 752
OLŠANSKY, OLDŘICH (Czechoslovakia) (1989-90) 626
OMNIPOL FOREIGN TRADE CORPORATION (Czechoslovakia) (1985-86) 730
OŠKINIS (Russia) (1992-93) 600

P

PE-80367 Urubu (CTA) (1983-84) 598
PG-1 Aqua Glider (Explorer) (1983-84) 624
PIK-20E2F (Issoire) (1991-92) 610
PIK-20E2F (17-metre version) (Issoire) (1984-85) 626
PIK-30 (Issoire) (1991-92) 610
PPO-1 Chibis (Mikoyan) (1992-93) 600
PPO-2 Kuznechik (Mikoyan) (1992-93) 600
PW (see Politechnika Warszawaska) (1992-93) 594
PW-2 Gapa (PW) (1991-92) 628
PW-2D Gapa (DWLKK) (1992-93) 594
PW-3 Bakcyl (PW) (1992-93) 594
PW-4 (PW) (1992-93) 596
PW-5 (PW) (1992-93) 596

PARODI MOTORSEGLERTECHNIK (West Germany) (1985-86) 739
Pathfinder (Electra) (1985-86) 739
Pégase (Centrair) (1988-89) 621
Pégase A and B (Centrair) (1992-93) 578
Pégase BC (Centrair) (1986-87) 743
Pégase C (Centrair) (1986-87) 743
Pégase CC (Centrair) (1986-87) 743
Pégase P (Centrair) (1984-85) 625
Pégase Club (Centrair) (1985-86) 732
Petrel 550 (Shenyang) (1989-90) 625
Petrel 650 (Shenyang) (1989-90) 625
Petrel 650B (Shenyang) (1992-93) 576
Piccolo (Technoflug) (1991-92) 626
Piccolo B (Technoflug) (1992-93) 590
Pilatus B4-PCIIAF (Nippi) (1984-85) 640
Pioneer II (Marske) (1992-93) 600
Platypus (Schneider) (1988-89) 617
POLITECHNIKA WARSZAWSKA (Poland) (1992-93) 594
POTTIER, AVIONS (France) (1987-88) 746
PRZEDSIEBIORSTWO DOSWIADCZALNO-PRODUKCYJNE SZYBOWNICTWA (Poland) (1992-93) 596
Puchacz (SZD) (1992-93) 596
Puchatek KR-03A (WSK) (1992-93) 596
Pukyalis, BRO-20 (LAK) (1984-85) 644

Q

Qian Jin, X-10 (Shenyang) (1992-93) 576
Quero-Quero II, KW 1 b 2 (IPE) (1990-91) 623
Quero-Quero III, IPE-05 (IPE) (1992-93) 576

R

R-26SU Gobé (Auto-Aero) (1992-93) 594
RF 5 (Fournier) (1992-93) 578
RF-10 (Aerostructure) (1986-87) 742
RS-15 (Bryan) (1985-86) 753

RAND ROBINSON ENGINEERING INC (USA) (1984-85) 650
RILEY & CO (SPORTAVIA) (Australia) (1983-84) 597
RIO CLARO (see Aéro Clube de Rio Claro) (1986-87) 740
ROLLADEN-SCHNEIDER FLUGZEUGBAU GmbH (Germany) (1992-93) 584
Rooster (Lightwing) (1985-86) 753
RUTAN AIRCRAFT FACTORY INC (USA) (1986-87) 769
RYSON AVIATION CORPORATION (USA) (1985-86) 756

S

S-1 Swift (Swift) (1992-93) 596
S-2A (Strojnik) (1992-93) 602
S 10 (Stemme) (1992-93) 590
S10VC (Stemme) (1992-93) 590
SB-13 (Akaflieg Braunschweig) (1991-92) 614
SC Canard (1985-86) 750
SCM (Canard) (1985-86) 750
SF-25C 2000 (Scheibe) (1985-86) 740
SF-25C Falke 82 (Scheibe) (1984-85) 634
SF-25C Falke 86 (Scheibe) (1985-86) 740
SF-25C Falke 86 (Scheibe) (1986-87) 752
SF-25C Falke 87 (Scheibe) (1987-88) 753
SF-25C Falke 88 (Scheibe) (1989-90) 635
SF-25C Falke 90 (Scheibe) (1991-92) 620
SF-25C Falke 92 (Sheibe) (1992-93) 584
SF-25E Super-Falke (Scheibe) (1992-93) 584
SF-28A Tandem-Falke (Scheibe) (1990-91) 636
SF-34 (Scheibe) (1989-90) 636
SF-34B (Scheibe) (1992-93) 584
SF-36 (Scheibe) (1987-88) 754
SGM 2-37 (Schweizer) (1992-93) 602
SGS 1-36 Sprite (Schweizer) (1992-93) 602
SGS 2-33A (Schweizer) (1992-93) 602
SH-2H (Havrda) (1991-92) 616
SL-2P (Esag and Ssaktb) (1988-89) 642
SSAKTB (see Specialus Sportines Avicijos Konstravimo - Technologinis Biuras) (1991-92) 632
ST-11 (Stralpes Aero) (1992-93) 578
ST-12 (Stralpes Aero) (1986-87) 746
ST-14 (Stralpes Aero) (1986-87) 746
ST-15 (Stralpes Aero) (1992-93) 578
ST-100 Cloudster (Ryson) (1984-85) 650
SY-5S (BIAA) (1984-85) 621
SZD (see Przedsiebiorstow Doswiadczalno- rodukcyjne Szybownictwa) (1992-93) 596
SZD-42-2 Jantar 2B (SZD) (1990-91) 645
SZD-48-1 Jantar Standard 2 (SZD) (1983-84) 619
SZD-48-2 Jantar Standard 2 (SZD) (1983-84) 619
SZD-48-3 Jantar Standard 3 (SZD) (1992-93) 596
SZD-48-3M Brawo (SZD) (1986-87) 761
SZD-50-3 Pushacz (SZD) (1992-93) 596
SZD-51-1 Junior (SZD) (1992-93) 596
SZD-51-3 Junior (SZD) (1985-86) 748
SZD-52 Jantar 15 (SZD) (1986-87) 762
SZD-52 Krokus (SZD) (1986-87) 762
SZD-55 (SZD) (1992-93) 596
SZD-56 (SZD) (1992-93) 596
SZD-59 (SZD) (1992-93) 596

Salto H 101 'Hänle' (Doktor Fiberglas) (1992-93) 582
SARL LA MOUETTE (France) (1992-93) 578
SCHEIBE FLUGZEUGBAU GmbH (Germany) (1992-93) 584
SCHEMPP-HIRTH FLUGZEUGBAU GmbH & Co KG (Germany) (1992-93) 584
SCHLEICHER GmbH & Co, ALEXANDER (Germany) (1992-93) 588
SCHNEIDER PTY LTD, EDMUND (Australia) (1988-89) 617
SCHWEIZER AIRCRAFT CORPORATION (USA) (1992-93) 602
Seagull, HU-1 (Shenyang) (1992-93) 576
SHENYANG SAILPLANE FACTORY (China) (1992-93) 576
Sierra (Advanced Aviation) (1992-93) 600

INDEXES: SAILPLANES—S - Z/HANG GLIDERS—A - M

Entry	Edition	Page
Silène, E 78 (Issoire)	(1985-86)	732
Silhouette (Lunds Tekniske)	(1992-93)	594
Silhouette (Silhouette)	(1988-89)	646
SILHOUETTE AIRCRAFT INC (USA)	(1988-89)	646
SIREN, SA (France)	(1987-88)	746
SOKO VAZDUHOPLOVNA INDUSTRIJA RADNA ORGANIZACIJA VAZDUHOPLOVSTVO (Yugoslavia)	(1985-86)	760
SOLE-77 (Jastreb)	(1987-88)	774
Solitaire, Model 77 (Rutan)	(1985-86)	756
SPECIALUS SPORTINES AVICIJOS KONSTRAVIMO - TECHNOLOGINIS BIURAS (USSR)	(1991-92)	632
Sprite, SGS 1-36 (Schweizer)	(1992-93)	602
Standard III, G102 (Grob)	(1985-86)	737
Standard Astir II (Grob)	(1983-84)	608
Standard Cirrus 75-VTC (Jastreb)	(1992-93)	602
Standard Cirrus G (Jastreb)	(1985-86)	760
Standard Cirrus G/81 (Jastreb)	(1992-93)	602
STEMME GmbH & Co KG (Germany)	(1992-93)	590
STRALPES AÉRO SARL (France)	(1992-93)	578
Straton (Olsansky)	(1989-90)	626
Stratus 500 (Stratus)	(1992-93)	590
STRATUS UNTERNEHMENSBERATUNG GmbH (Germany)	(1992-93)	590
Strazdas, LAK-14 (LAK)	(1987-88)	767
STROJNIK, PROF. ALEX (USA)	(1992-93)	602
SUNDERLAND, GARY (Australia)	(1983-84)	597
Super Blanik L-23 (LET)	(1992-93)	578
Super Dimona, HK-36 (HOAC)	(1992-93)	608
Super Dimona HK-36R (HOAC)	(1992-93)	576
Super-Falke SF-25E (Scheibe)	(1992-93)	584
SWIFT LTD (Poland)	(1992-93)	596
Swift S-1 (Swift)	(1992-93)	596

T

Entry	Edition	Page
T-5 Blue Wren (Todhunter)	(1987-88)	739
TG-7A (Schweizer)	(1990-91)	654
TZ-14 (IPE)	(1990-91)	624
Taifun (Valentin)	(1983-84)	616
Taifun 17E II (TWI)	(1992-93)	590
Taifun 17E-90 (Valentin)	(1986-87)	758
Tandem-Falke SF-28A (Scheibe)	(1990-91)	636
TECHNICAL CENTRE, CIVIL AVIATION DEPARTMENT (India)	(1992-93)	594
TECHNOFLUG LEICHTFLUGZEUBAU GmbH (Germany)	(1992-93)	590
TODHUNTER, R. W. (Australia)	(1987-88)	739
TWI FLUGZEUGGESELLSCHAFT mbH (Germany)	(1992-93)	590
Twin II, G 103 (Grob)	(1986-87)	750
Twin III G 103C (Grob)	(1987-88)	750
Type 101 (Centrair)	(1988-89)	621
Type 101 Club (Centrair)	(1988-89)	621
Type 201, Marianne (Centrair)	(1984-85)	625
Type 1100 Windex (Radab)	(1985-86)	750
Type 2001 Marianne (Centrair)	(1988-89)	622
Type 2001M Marianne M (Centrair)	(1988-89)	622

U

Entry	Edition	Page
ULF-1 (EEL)	(1987-88)	748
ULS (PW)	(1985-86)	746
UP SPORTS (see Ultralite Products)	(1984-85)	652
Uavg-15 (Marsden)	(1984-85)	620
ULTRALITE PRODUCTS (USA)	(1984-85)	652
Urubu, PE-80367 (CTA)	(1983-84)	598
Utochka, Bro-17U (LAK/Oshkinis)	(1984-85)	644

V

Entry	Edition	Page
VSO (see Vyvojová Skupina Orlican)	(1991-92)	610
VSO 10 Gradient (VSO)	(1991-92)	610
VUK-T (Jastreb)	(1992-93)	602
VALENTIN GmbH and Co FLUGZEUGBAU (West Germany)	(1990-91)	642
VALENTIN GmbH GERÄTE UND MASCHINENBAU (West Germany)	(1985-86)	744
Valiant (Schleicher)	(1985-86)	742
Ventus (Schempp-Hirth)	(1992-93)	588
Vesper (CETEC)	(1989-90)	624
Viking TX Mk 1 (Grob)	(1986-87)	750
Vituris, BRO-21 (LAK)	(1984-85)	644
Vivat L-13E (Aerotechnik)	(1992-93)	578
Vivat L-135L (Aerotechnik)	(1992-93)	578
Vivat L-13SW (Aerotechnik)	(1989-90)	626
VYVOJOVÁ SKUPINA ORLICAN (Czechoslovakia)	(1991-92)	610

W

Entry	Edition	Page
WK-1 (Kuffner)	(1984-85)	652
WK-1b (Kuffner)	(1987-88)	752
WSK (see Wytwornia Sprzetu Komunikacyjnego)	(1992-93)	596
WILKES, RAYMOND S. (USA)	(1984-85)	652
Windex 1100 (Radab)	(1985-86)	750
Windex 1200 (Radab)	(1990-91)	649
Windex 1200C (Radab)	(1992-93)	600
Windrose (Maupin)	(1992-93)	600
Woodstock One (Maupin)	(1992-93)	600
WYTWORNIA SPRZETU KOMUNIKACYJNEGO (Poland)	(1992-93)	596

X

Entry	Edition	Page
X-5A (State Aircraft Factories)	(1985-86)	728
X-7 Jian Fan (Chengdu)	(1991-92)	608
X-9 Jian Fan (Shenyang)	(1992-93)	576
X-10 Qian Jin (Shenyang)	(1992-93)	576
X-11 (State Aircraft Factories)	(1984-85)	622
XL-113 (Aerotechnik)	(1992-93)	610
Ximango (Aeromot)	(1992-93)	576

Z

Entry	Edition	Page
Z-16 (IPE)	(1990-91)	623
Zefiro 940A (General Gliders)	(1992-93)	594
Zia (Applebay)	(1985-86)	753
Zile, Bro-11M (LAK/Oshkinis)	(1985-86)	752

HANG GLIDERS

Hang gliders are no longer listed in this book. Those listed below appeared in the 1992-93 and preceding seven editions

A

Entry	Edition	Page
AOK (see Akademicki Osrodek Konstrukcyjny)	(1992-93)	604
Ace (Pegasus)	(1987-88)	778, 782
Ace/Ace RX (small) (Pegasus/Solar Wings)	(1992-93)	604
Ace/Ace RX (medium) (Pegasus/Solar Wings)	(1992-93)	604
Ace/Ace RX (large) (Pegasus/Solar Wings)	(1992-93)	604
Ace Fun (Pegasus/Solar Wings)	(1992-93)	604
Ace Sport (Pegasus/Solar Wings)	(1992-93)	604
Ace Supersport (Pegasus/Solar Wings)	(1992-93)	604
AEROTEC SARL (France)	(1992-93)	604
Airfex (Finsterwalder)	(1992-93)	604
AIRWAVE GLIDERS LTD (UK)	(1992-93)	604
AKADEMICKI OSRODEK KONSTRUKCYJNY (Poland)	(1992-93)	604
Alfa (Vega)	(1992-93)	604
ANTONOV (see O. K. Antonov Design Bureau)	(1992-93)	604
Atlas (La Mouette)	(1987-88)	776
Atlas 14 (La Mouette)	(1992-93)	604
Atlas 16 (La Mouette)	(1992-93)	604
Atlas 18 (La Mouette)	(1992-93)	604
Attack Duck (Wills Wing)	(1986-87)	779
Ayres (Polaris)	(1992-93)	604
Azur (La Mouette)	(1985-86)	716

B

Entry	Edition	Page
BECK PRODUCTIONS, CRAIG (USA)	(1985-86)	718
Bergfex (Finsterwalder)	(1992-93)	604
Black Magic 22 (Airwave)	(1992-93)	604
Black Magic 24 (Airwave)	(1992-93)	604
Black Magic 27 (Airwave)	(1992-93)	604
Blade (Pegasus)	(1987-88)	778
Breez (Progressive)	(1985-86)	720

C

Entry	Edition	Page
Calypso (Airwave)	(1992-93)	604
Chiron 22 (Steinbach-Delta)	(1992-93)	604
Chiron 25 (Steinbach-Delta)	(1992-93)	604
Chiron 28 (Steinbach-Delta)	(1992-93)	604
Classic (Firebird)	(1992-93)	604
Cobra (La Mouette)	(1992-93)	604
Comet 2B (UP)	(1987-88)	780
Comet 2B 165 (UP)	(1992-93)	604
Comet 2B 185 (UP)	(1992-93)	604
Compact (La Mouette)	(1992-93)	604
Condor 5+ (Steinbach-Delta)	(1992-93)	604
CUNION, EVERARD (UK)	(1986-87)	776

D

Entry	Edition	Page
Delta (Polaris)	(1986-87)	774
Delta 13 (Polaris)	(1986-87)	778
Delta 16 (Polaris)	(1986-87)	778
Delta 19 (Polaris)	(1986-87)	778
Delta 21 (Polaris)	(1986-87)	778
DELTA WING AVIATION (USA)	(1992-93)	604
Demon (Flight Designs)	(1985-86)	718
Divine 100 (Steinbach-Delta)	(1992-93)	604
Divine 200 (Steinbach-Delta)	(1992-93)	604
DRACHEN STUDIO KECUR GmbH (Germany)	(1992-93)	604
Dream (Delta Wing)	(1987-88)	779
Dream 165 (Delta Wing)	(1992-93)	604
Dream 175 (Delta Wing)	(1992-93)	604
Dream 185 (Delta Wing)	(1992-93)	604
Duck (Wills Wing)	(1986-87)	779
Duck 160 (Wills Wing)	(1986-87)	779
Duck 180 (Wills Wing)	(1986-87)	780

E

Entry	Edition	Page
E Stratus Series (AOK)	(1992-93)	604
E-4 (AOK)	(1987-88)	782
Ephyr (Optimum)	(1992-93)	604
Euro III (Steinbach-Delta)	(1992-93)	604

F

Entry	Edition	Page
F-14 (AOK)	(1992-93)	604
FR 16 (Polaris)	(1992-93)	604
Favorit (Optimum)	(1991-92)	604
FINSTERWALDER GmbH (Germany)	(1992-93)	604
FIREBIRD SCHWEIGER KG (Germany)	(1992-93)	604
Flair (Rochelt)	(1992-93)	604
Fledge III (Manta)	(1986-87)	777
FLIGHT DESIGNS INC (USA)	(1985-86)	718
FREE FLIGHT MICROLIGHTS AND HANG GLIDERS NZ (New Zealand)	(1986-87)	775
Fun Air (Steinbach-Delta)	(1992-93)	604
Funfex (Finsterwalder)	(1992-93)	604

G

Entry	Edition	Page
GT (Thalhofer)	(1992-93)	604
GTR (Moyes)	(1987-88)	776
GTR 148 (Moyes)	(1991-92)	640
GTR 162 (Moyes)	(1991-92)	640
GTR 175 (Moyes)	(1991-92)	640
GTR 210 (Moyes)	(1991-92)	640
Gamma 177 (Polaris)	(1986-87)	774, 778
Gemini (UP)	(1987-88)	780
Gemini M 134 (UP)	(1992-93)	604
Gemini M 164 (UP)	(1992-93)	604
Gemini M 184 (UP)	(1992-93)	604
Genesis (Pacific Airwave)	(1992-93)	604
Glidezilla 155 (UP)	(1992-93)	604
Gryps (Polaris)	(1992-93)	604
GUNTER ROCHELT (Germany)	(1992-93)	604

H

Entry	Edition	Page
HP 11 (Wills Wing)	(1989-90)	660
HP 170 (Wills Wing)	(1986-87)	779, 780
HP AT 145 (Wills Wing)	(1992-93)	604
HP AT 158 (Wills Wing)	(1992-93)	604
Hermes (La Mouette)	(1987-88)	776
Hermes 13 (La Mouette)	(1988-89)	650
Hermes 14 (La Mouette)	(1988-89)	650
Hermes 15 (La Mouette)	(1988-89)	650
Hermes 16 (La Mouette)	(1988-89)	650

I

Entry	Edition	Page
IKARUSFLUG BODENSEE (West Germany)	(1985-86)	716
INSTYTUT LOTNICTWA (Poland)	(1992-93)	604

J

Entry	Edition	Page
Joker (Thalhofer)	(1992-93)	604

K

Entry	Edition	Page
K2 (Airwave)	(1992-93)	604
K3 (Airwave)	(1992-93)	604
KOLECKI NEW AVIATION ENGINEERING (Sweden)	(1987-88)	778
Komposit UT (Onpo Technologia)	(1992-93)	604

L

Entry	Edition	Page
LA MOUETTE, SARL (France)	(1992-93)	604
Lancer Mk IV (Free Flight)	(1985-86)	717
Lightflex (Finsterwalder)	(1992-93)	604

M

Entry	Edition	Page
MX (Vega)	(1990-91)	658
MX II (Vega)	(1992-93)	604
Magic IV (Airwave)	(1987-88)	778
Magic IV Series (Airwave)	(1990-91)	660

M – Z—HANG GLIDERS/**A**—LIGHTER-THAN-AIR: INDEXES 727

Magic IV FR (Airwave) .... *(1987-88)* 778
Magic 6 (Airwave) .... *(1992-93)* 604
Magic Kiss (Airwave) .... *(1990-91)* 660
Magic Kiss 154 (Pacific Airwave) .... *(1992-93)* 604
MANTA PRODUCTS INC (USA) .... *(1986-87)* 777
Mars (Moyes) .... *(1987-88)* 776
Mars 150 (Moyes) .... *(1992-93)* 604
Mars 170 (Moyes) .... *(1991-92)* 604
Mars 190 (Moyes) .... *(1991-92)* 604
Minifex II (Finsterwalder) .... *(1988-89)* 650
Minifex M2 (Finsterwalder) .... *(1992-93)* 604
Mission 170 (Moyes) .... *(1992-93)* 604
Motolotnia 80 White Eagle (Kolecki) .... *(1987-88)* 778
MOYES DELTA GLIDERS PTY LTD (Australia) .... *(1992-93)* 604
Mystic (Delta Wing) .... *(1987-88)* 779
Mystic 155 (Delta Wing) .... *(1992-93)* 604
Mystic 166 (Delta Wing) .... *(1992-93)* 604
Mystic 177 (Delta Wing) .... *(1992-93)* 604

N

New Wave 15 (Firebird) .... *(1992-93)* 604
New Wave 16 (Firebird) .... *(1992-93)* 604
Nimbus (Swiss Aerolight) .... *(1992-93)* 604

O

O. K. ANTONOV DESIGN BUREAU (Ukraine) .... *(1992-93)* 604
ONPO TECHNOLOGIA (Russia) .... *(1992-93)* 604
Opit (Optimum) .... *(1992-93)* 604
OPTIMUM AIRCRAFT GROUP (Bulgaria) .... *(1992-93)* 604

P

PACIFIC AIRWAVE INC (USA) .... *(1992-93)* 604
PACIFIC WINDCRAFT LTD (USA) .... *(1988-89)* 658
Parafun (Steinbach) .... *(1992-93)* 604
Parafun 1 (Steinbach) .... *(1992-93)* 604
Para Safe (Steinbach) .... *(1992-93)* 604
PEGASUS TRANSPORT SYSTEMS LTD/ SOLAR WINGS LTD (UK) .... *(1992-93)* 604
Perfex (Finsterwalder) .... *(1992-93)* 604
Phoenix Dream (Delta Wing) .... *(1985-86)* 718
Phoenix Streak (Delta Wing) .... *(1985-86)* 718
POLARIS SRL (Italy) .... *(1992-93)* 604
Proair (Progressive) .... *(1985-86)* 720
Profil (La Mouette) .... *(1987-88)* 776
Profil 15 (La Mouette) .... *(1988-89)* 650
Profil 17 (La Mouette) .... *(1988-89)* 650
PROGRESSIVE AIRCRAFT COMPANY (USA) .... *(1986-87)* 777
Prostar II (Progressive) .... *(1985-86)* 720

Q

Quattro (Firebird) .... *(1988-89)* 650
Quattro-S (Firebird) .... *(1992-93)* 604
Quattro-S Piccolo (Firebird) .... *(1992-93)* 604

R

R-10 Stratus (Warsaw Technical University) .... *(1986-87)* 775
R-11 Stratus (Warsaw Technical University) .... *(1986-87)* 775
R-12 Stratus (Warsaw Technical University) .... *(1986-87)* 775
R-15 Stratus (Warsaw Technical University) .... *(1986-87)* 775
Racer GP (Pegasus) .... *(1987-88)* 778
Rapace 15 (Aerotec) .... *(1992-93)* 604
Rapace 16 (Aerotec) .... *(1992-93)* 604
ROCHELT, GÜNTER (Germany) .... *(1992-93)* 604
Rumour (Pegasus/Solar Wings) .... *(1992-93)* 604

S

S4-Plus (Pegasus) .... *(1987-88)* 778
S4-Racer (Pegasus) .... *(1987-88)* 778
SB-1 (AOK) .... *(1987-88)* 778, 782
SB-1 WARS (AOK) .... *(1992-93)* 604
SP (Steinbach-Delta) .... *(1992-93)* 604
SP Vario (Steinbach-Delta) .... *(1992-93)* 604
Schneid Air (Rochelt) .... *(1992-93)* 604
SEEDWINGS (USA) .... *(1992-93)* 604
Sensor 510-B (Seedwings) .... *(1987-88)* 780
Sensor 510-160 VG (Seedwings) .... *(1992-93)* 604
Sensor 510-165 (Seedwings) .... *(1988-89)* 652
Shadow (Flight Designs) .... *(1985-86)* 718
Sierra (Firebird) .... *(1986-87)* 774
Sierra 155 (Firebird) .... *(1986-87)* 778
Sierra 175 (Firebird) .... *(1986-87)* 778
SKRAIDYKLES (Bulgaria) .... *(1991-92)* 639
Skyhawk (Wills Wing) .... *(1987-88)* 780
Skyhawk 168 (Wills Wing) .... *(1991-92)* 642
Skyhawk 188 (Wills Wing) .... *(1991-92)* 642
Slavutitch Sport (Antonov) .... *(1992-93)* 604
Slavutitch-UT (Antonov) .... *(1992-93)* 604
Solar Ace (Pegasus) .... *(1987-88)* 778
Solar Storm (Pegasus) .... *(1987-88)* 778
Spectrum 144 (Wills Wing) .... *(1992-93)* 604
Spectrum 165 (Wills Wing) .... *(1992-93)* 604
Spirit (Firebird) .... *(1988-89)* 650
Spirit New Wave (Firebird) .... *(1987-88)* 777
Spit (Polaris) .... *(1992-93)* 604
Sport 150 (Wills Wing) .... *(1990-91)* 660
Sport AT 150 (Wills Wing) .... *(1991-92)* 642
Sport 167 (Wills Wing) .... *(1990-91)* 660
Sport AT 167 (Wills Wing) .... *(1991-92)* 642
Sport 180 (Wills Wing) .... *(1990-91)* 660
Sport AT 180 (Wills Wing) .... *(1991-92)* 642
Sport American 167 (Wills Wing) .... *(1991-92)* 642
STEINBACH-DELTA FUN-AIR (Austria) .... *(1992-93)* 604
Stratus (Warsaw Technical University) .... *(1986-87)* 775
Stratus E Series (AOK) .... *(1992-93)* 604
Streak (Delta Wing) .... *(1987-88)* 779
Streak 130 (Delta Wing) .... *(1992-93)* 604
Streak 160 (Delta Wing) .... *(1992-93)* 604
Streak 180 (Delta Wing) .... *(1992-93)* 604
Sunfun, VJ-24 (Volmer) .... *(1992-93)* 604
Super 13 (Polaris) .... *(1987-88)* 782
Super 16 (Polaris) .... *(1987-88)* 782
Superfex (Finsterwalder) .... *(1992-93)* 604
Super Sport 143 (Wills Wing) .... *(1992-93)* 604
Super Sport 153 (Wills Wing) .... *(1992-93)* 604
Super Sport 163 (Wills Wing) .... *(1992-93)* 604
Swing (Thalhofer) .... *(1992-93)* 604
Swingwing, VJ-23 (Volmer) .... *(1992-93)* 604
SWISS AEROLIGHT (Switzerland) .... *(1992-93)* 604

T

Tempest (Pegasus) .... *(1987-88)* 778
THALHOFER TEAM (Germany) .... *(1992-93)* 604
Topfex (Finsterwalder) .... *(1992-93)* 604
Touring 13 (Polaris) .... *(1992-93)* 604
Touring 15 (Polaris) .... *(1992-93)* 604
Tropi (Drachen) .... *(1987-88)* 776
Tropi 16 (Drachen) .... *(1992-93)* 604
Tropi 17 (Drachen) .... *(1992-93)* 604
Typhoon (Pegasus) .... *(1987-88)* 778

U

Uno (Firebird) .... *(1992-93)* 604
Uno Jumbo (Firebird) .... *(1992-93)* 604
Uno Piccolo (Firebird) .... *(1992-93)* 604
UP INC (ULTRALITE PRODUCTS) (USA) .... *(1992-93)* 604
US MOYES INC (USA) .... *(1985-86)* 720

V

VJ-23 Swingwing (Volmer) .... *(1992-93)* 604
VJ-24 Sunfun (Volmer) .... *(1992-93)* 604
Vampyre Mk II (Free Flight) .... *(1985-86)* 717
Vega 16/PR (Vega) .... *(1992-93)* 604
VEGA-DRACHENBAU PAWEL WIERZBOWSKI (Austria) .... *(1992-93)* 604
Vision (Pacific Windcraft) .... *(1987-88)* 779
Vision 148 (Pacific Windcraft) .... *(1988-89)* 652
Vision 175 (Pacific Windcraft) .... *(1988-89)* 652
Vision 194 (Pacific Windcraft) .... *(1988-89)* 652
Vision Eclipse (Pacific Windcraft) .... *(1987-88)* 782
Vision Eclipse 17 (Pacific Windcraft) .... *(1988-89)* 652
Vision Mk IV 17 (Pacific Airwave) .... *(1992-93)* 604
Vision Mk IV 19 (Pacific Airwave) .... *(1992-93)* 604
Vision Eclipse 19 (Pacific Windcraft) .... *(1988-89)* 652
VOLMER AIRCRAFT (USA) .... *(1992-93)* 604

W

Wars (AOK) .... *(1987-88)* 778
WARSAW TECHNICAL UNIVERSITY (Poland) .... *(1986-87)* 775
WILLS WING INC (USA) .... *(1992-93)* 604
WILLS WINGS DEUTSCHLAND (Germany) .... *(1992-93)* 604
Windspiel 2 (Ikarusflug) .... *(1985-86)* 716
World Cup 90-5 School Glider (Firebird) .... *(1986-87)* 774, 778
WSK-PZL WARSZAWA-OKECIE (Poland) .... *(1985-86)* 717

X

XC-155 (Moyes) .... *(1992-93)* 604
XS-142 (Moyes) .... *(1992-93)* 604
XS-155 (Moyes) .... *(1992-93)* 604
XS-169 (Moyes) .... *(1992-93)* 604
XT-145 (Moyes) .... *(1992-93)* 604
XT-165 (Moyes) .... *(1992-93)* 604

Z

Z-87 (IL) .... *(1992-93)* 604
Z-90 (IL) .... *(1992-93)* 604
Zeta-80B (IL) .... *(1987-88)* 778
Zeta-84 (IL) .... *(1987-88)* 778, 782
Zeta 87 (IL) .... *(1992-93)* 642

LIGHTER-THAN-AIR

All airships and those balloons that are used commercially or have special technical interest are listed here, as well as balloons and airships which appeared during the past decade

A

A-1 (Aviatik Brno) .... *(1987-88)* 786
A-1 Albatross (Boland) .... *(1985-86)* 770
A-2 Rover (Boland) .... *(1990-91)* 672
A-3 (Piccard) .... *(1983-84)* 705
A-4 (Piccard) .... *(1983-84)* 705
A-5 (Piccard) .... *(1983-84)* 705
A50 (Adams) .... *(1991-92)* 658
A-50, Lightship (ABC) .... *(1990-91)* 671
A50S (Adams) .... *(1991-92)* 659
A55 (Adams) .... *(1991-92)* 659
A55S (Adams) .... *(1991-92)* 652
A-60 Lightship (ABC) .... *(1991-92)* 659
A60 (Adams) .... *(1991-92)* 659
A-60 Plus Lightship (ABC) 598
A60S (Adams) .... *(1991-92)* 659
A-75, Lightship (ABC) .... *(1989-90)* 670
A-105 (Cameron) .... *(1991-92)* 659
A-120 (Cameron) .... *(1991-92)* 659
A-120 Lightship (ABC) 599
A-140 (Cameron) .... *(1991-92)* 659
A-160 (Cameron) .... *(1992-93)* 611
A-180 (Cameron) .... *(1991-92)* 660
A-210 (Cameron) .... *(1991-92)* 660
A-250 (Cameron) .... *(1991-92)* 660
A-300 (Cameron) .... *(1991-92)* 660
A-375 (Cameron) .... *(1992-93)* 611
A-530 (Cameron) .... *(1991-92)* 660
AA-3 (Ballonfabrik) .... *(1992-93)* 606
AA-4 (Ballonfabrik) .... *(1992-93)* 606
AA-5 (Ballonfabrik) .... *(1992-93)* 606
AA-6 (Ballonfabrik) .... *(1992-93)* 606
AAC (see Advanced Airship Corporation) 595
AB (Adams) .... *(1991-92)* 659
AB-1 (Aviatik Brno) .... *(1987-88)* 800
AB-2 (Aerotechnik) .... *(1991-92)* 658
AB-3 (Aerotechnik) .... *(1991-92)* 658
AB-8 (Aerotechnik) .... *(1991-92)* 659
AB-10 (Aerotechnik) .... *(1991-92)* 660
ABC (see American Balloon Company) .... *(1986-87)* 801
ABC (see American Blimp Corporation (USA)) 598
ADA (see Airship Developments Australia) .... *(1992-93)* 605
ADA-1200 (ADA) .... *(1991-92)* 645
AEROS (see NPP Aerostatic Technics) 594
AG-60 (ILC) .... *(1987-88)* 807
AHA (See Advanced Hybrid Aircraft) 595
ALA-40 (Thermoplane) 593
ALA-500 (MAI) .... *(1992-93)* 608
ANR (Advanced Airship) .... *(1992-93)* 608
ANR (Wren) .... *(1987-88)* 793
AS-42 Colt (Thunder and Colt) .... *(1985-86)* 768
AS 42 Mk II, Colt (Thunder & Colt) .... *(1987-88)* 792
AS 56 (Thunder & Colt) 598
AS 76 (Thunder & Colt) .... *(1988-89)* 669
AS 80 Mk II (Thunder & Colt) 597
AS 80 GD (Gefa-Flug) .... *(1991-92)* 646
AS-90 Colt (Thunder and Colt) .... *(1985-86)* 767
AS 90 Mk II (Thunder & Colt) .... *(1989-90)* 669
AS-105 Colt (Thunder & Colt) .... *(1987-88)* 792
AS 105 Mk II (Thunder & Colt) 597
AS 261 (Thunder & Colt) .... *(1991-92)* 651
AV-1 (Aviatik Brno) .... *(1987-88)* 786
AX-4 (Air Service) .... *(1992-93)* 607
AX-5 (Boland) .... *(1987-88)* 806
AX5-40M (Flamboyant) .... *(1991-92)* 658
AX-6 (Fantasy) .... *(1992-93)* 605
AX-6 (Kavanagh) .... *(1992-93)* 605
AX6-56 (Flamboyant) .... *(1992-93)* 658
AX6-56M (Flamboyant) .... *(1991-92)* 658
AX-7 (Boland) .... *(1987-88)* 806
AX-7 (Fantasy) .... *(1992-93)* 605
AX-7 (Kavanagh) .... *(1992-93)* 605
AX7-65 (Flamboyant) .... *(1991-92)* 658
AX7-77M (Flamboyant) .... *(1991-92)* 658
AX-8 (Boland) .... *(1987-88)* 806
AX-8 (Fantasy) .... *(1992-93)* 605
AX-8 (Kavanagh) .... *(1992-93)* 605
AX8-85 (Flamboyant) .... *(1991-92)* 659
AX8-105 (Flamboyant) .... *(1991-92)* 659
AX-9 (Boland) .... *(1987-88)* 806
AX9 (Galaxy) .... *(1992-93)* 614
AX-9 (Kavanagh) .... *(1992-93)* 605
AX-10 (Boland) .... *(1987-88)* 806
AX-10 (Kavanagh) .... *(1992-93)* 605
AX10-150 (Flamboyant) .... *(1991-92)* 660
AX-11 (Boland) .... *(1987-88)* 806
AX-11 (Kavanagh) .... *(1992-93)* 605
AX 11-240 (Flamboyant) .... *(1991-92)* 660
ADAMS' BALLOON LOFT INC (USA) .... *(1992-93)* 613
ADVANCED AIRSHIP CORPORATION LTD (UK) 595
ADVANCED HYBRID AIRCRAFT (UK) 595
AERAZUR (France) 591
AEROLIFT INC (USA) .... *(1991-92)* 652

728 INDEXES: LIGHTER-THAN-AIR—**A – M**

Aeros-50 (Aeros) 594
Aeros-500 (Aeros) 594
AEROSTAR INTERNATIONAL INC (USA) 599
Aerostat, 56,000 cu ft, sea-based (RCA) (1987-88) 808
Aerostat, 250,000 cu ft (ILC) (1987-88) 807
AEROTECHNIK (Czechoslovakia) (1992-93) 606
AEROTEK CORPORATION (USA) (1988-89) 664
Air Chair (Thunder & Colt) (1985-86) 781
AIRBORNE INDUSTRIES LIMITED (UK) (1987-88) 802
AIR SERVICE REPULOGEPES SZOLGALAT (Hungary) (1992-93) 607
Airship (Thompson) (1992-93) 615
AIRSHIP DEVELOPMENTS AUSTRALIA PTY LTD (Australia) (1992-93) 605
AIRSHIP INDUSTRIES (UK) LTD (1991-92) 648
Albatross (ADA) (1991-92) 645
AIRSHIP USA INC (USA) (1986-87) 655
Albatross, A-1 (Boland) (1985-86) 770
America GZ-20A (Goodyear) (1987-88) 796
America N3A (Goodyear) (1984-85) 738
Americana (ABC) (1986-87) 801, 807
AMERICAN BALLOON COMPANY (USA) (1986-87) 801
AMERICAN BLIMP CORPORATION (USA) 598
AVIAN BALLOON COMPANY (USA) (1992-93) 614
AVIATIK CLUB BRNO (Czechoslovakia) (1987-88) 786, 800

B

B 800/2-Ri (Ballonfabrik) (1991-92) 658
BA-3 (Skyrider) (1990-91) 673
BIAA (see Beijing Institute of Aeronautics and Astronautics) (1988-89) 656
BUAA (see Beijing University of Aeronautics and Astronautics) (1991-92) 645
Baby Snake (Boland) (1990-91) 678
BALLONFABRIK SEE- UND LUFTAUSRÜSTUNG GmbH und Co KG (Germany) (1992-93) 606
BALLOONS SERVICE DI BONANNO LETTERIO (Italy) (1991-92) 647
BALLOON WORKS, THE (USA) (1992-93) 614
BALLONS CHAIZE (France) (1992-93) 606
BARNES AIRSHIPS (USA) 599
BEIJING INSTITUTE OF AERONAUTICS AND ASTRONAUTICS (China) (1988-89) 656
BEIJING UNIVERSITY OF AERONAUTICS AND ASTRONAUTICS (China) (1991-92) 645
BOLAND BALLOON (USA) (1990-91) 672, 677
Bumerang (USSR) (1985-86) 765

C

C-56 (Kavanagh) (1991-92) 658
C-65 (Kavanagh) (1991-92) 658
C-77 (Kavanagh) (1991-92) 658
CAC (see Commercial Airship Company) (1987-88) 788
CS.1600 (Chaize) (1991-92) 658
CS.1800 (Chaize) (1991-92) 658
CS.2000 (Chaize) (1991-92) 658
CS.2200 (Chaize) (1991-92) 658
CS.3000 (Chaize) (1991-92) 659
CS.4000 (Chaize) (1991-92) 659
Calibre, 32 (Avian) (1986-87) 802
CALIFORNIA AIRSHIPS (USA) (1987-88) 796
CAMERON BALLOONS LTD (UK) 595
CENTURY AIRSHIPS, 21st (Canada) 591
Century Airships prototypes (Century Airships) 591
CHAIZE (see Ballons Chaize) (1992-93) 606
Circus Maximus (Boland) (1987-88) 805
Cloud Cruiser UM10-22 (Ulita) (1992-93) 615
Cloud Cruiser UM10-23 (Ulita) 600
Cloudhopper Junior (Thunder & Colt) (1988-89) 670
Cloudhopper Midi (Thunder & Colt) (1991-92) 658
Cloudhopper Super (Thunder & Colt) (1991-92) 658
COLT BALLOONS LTD (UK) (1984-85) 736
Concept AX-7 (Cameron) (1992-93) 611
Columbia, GZ 20A (Goodyear) (1987-88) 796
COMMERCIAL AIRSHIP COMPANY (New Zealand) (1987-88) 788
CycloCrane (AeroLift) (1990-91) 672

D

D-38 (Cameron) (1987-88) 791
D-50 (Cameron) (1988-89) 662
D-77 (Kavanagh) (1991-92) 658
D-84 (Kavanagh) (1991-92) 659
D-90 (Kavanagh) (1991-92) 659
D-96 (Cameron) (1989-90) 668
D-105 (Kavanagh) (1991-92) 659
DG-14 (Cameron) 596
DG-19 (Cameron) (1986-87) 788
DG-25 (Cameron) (1987-88) 792
DKBA (See Design Bureau Automatica) 592
DP series (Cameron) 595
DS-3 (DKBA) 592
DEMENTYEV (Russia) 607
Desert Fox (Boland) (1987-88) 805
DESIGN BUREAU AUTOMATICA (Russia) 592
Dimples (Boland) (1987-88) 805
Dino 3 (Aérazur) (1986-87) 784
DragonFly 56 (Balloon Works) (1992-93) 614
DragonFly 65 (Balloon Works) (1992-93) 614
DragonFly 77 (Balloon Works) (1992-93) 614
DragonFly 90 (Balloon Works) (1992-93) 614
DragonFly 90DB (Balloon Works) (1992-93) 614
DragonFly 105 (Balloon Works) (1992-93) 614
DragonFly 560 (Balloon Works) (1991-92) 658
DragonFly 650 (Balloon Works) (1991-92) 659
DragonFly 770 (Balloon Works) (1991-92) 659
DragonFly 900 (Balloon Works) (1991-92) 659
DragonFly 1050 (Balloon Works) (1991-92) 659
Dreamfinder (Memphis) (1986-87) 794
Drifter (Hoverair) (1986-87) 803, 806
Dino 3 (Aérazur) (1987-88) 786
Dinosaure (Aérazur) (1988-89) 657

E

E-120 (Kavanagh) (1991-92) 659
E-140 (Kavanagh) (1991-92) 659
E-160 (Kavanagh) (1991-92) 660
E-200 (Kavanagh) (1991-92) 660
E-210 (Kavanagh) (1991-92) 660
E-240 (Kavanagh) (1991-92) 660
E-260 (Kavanagh) (1991-92) 660
EXP-11 (Memphis) (1990-91) 673
Earthwinds (Aerostar) 599
ENDEAVOUR (UK) (1987-88) 802

F

FK-4 (HADG) (1992-93) 605
FK-6 (HADG) 591
Falcon II (Avian) 614
Fantasy 6 (Fantasy) (1991-92) 658
Fantasy 7 (Fantasy) (1991-92) 658
Fantasy 8-90 (Fantasy) (1991-92) 659
Fantasy 8-105 (Fantasy) (1991-92) 659
FANTASY SKY PROMOTIONS INC (Canada) (1992-93) 605
FireFly 5 (Balloon Works) (1983-84) 705
FireFly 6 (Balloon Works) (1983-84) 706
FireFly 6B (Balloon Works) (1983-84) 706
FireFly 7 (Balloon Works) (1983-84) 706
FireFly 8 (Balloon Works) (1991-92) 659
FireFly 42 (Balloon Works) (1992-93) 614
FireFly 56 (Balloon Works) (1983-84) 706
FireFly 65 (Balloon Works) (1992-93) 614
FireFly 77 (Balloon Works) (1992-93) 614
FireFly 90 (Balloon Works) (1992-93) 614
FireFly 90DF (Balloon Works) (1992-93) 614
FireFly 105 (Balloon Works) (1992-93) 614
FireFly 140 (Balloon Works) (1991-92) 659
FireFly 650 (Balloon Works) (1991-92) 659
FireFly 770 (Balloon Works) (1991-92) 659
FireFly 900 (Balloon Works) (1991-92) 659
FireFly 1050 (Balloon Works) (1991-92) 659
Fitzgerald (Boland) (1987-88) 805
FLAMBOYANT BALLOONS (PTY) LTD (South Africa) (1992-93) 608
Football (Boland) (1990-91) 678
Fred II (Boland) (1983-84) 705

G

GA 42 (Thunder & Colt) 598
GAC-20 (Grace) (1985-86) 772
GB-1 (Thunder & Colt) (1991-92) 658
GB55 (Yost) (1985-86) 780
GEFA FLUG (See Gesellschaft zur Entwicklung und Förderung Aerostatischer Flugsysteme) 591
GZ-20A America (Goodyear) (1986-87) 792
GZ-20A Columbia (Goodyear) (1987-88) 796
GZ-22 (Goodyear Aerospace) (1990-91) 672
GadFly 56 (Balloon Works) (1992-93) 614
Gadfly 560 (Balloon Works) (1991-92) 658
Galaxy 7 (Galaxy) (1992-93) 614
Galaxy 8 (Galaxy) (1992-93) 614
GALAXY BALLOONS INC (USA) (1992-93) 614
GESELLSCHAFT ZUR ENTWICKLUNG UND FÖRDERUNG AEROSTATISCHER FLUGSYSTEME mbh (Germany) 591
GOODYEAR AEROSPACE CORPORATION (USA) (1991-92) 652
GOODYEAR TIRE & RUBBER COMPANY, THE (USA) (1988-89) 665
GRACE (see US Airship) (1986-87) 796
GRACE AIRCRAFT CORPORATION (USA) (1985-86) 772

H

H-34 Skyhopper (Cameron) (1991-92) 658
HACDC (see Hangzhou Airship Comprehensive Development Company) (1988-89) 657
HADG (Huahang Airship Development Group) 591
HAPP (Magnus) (1988-89) 656
HI-SPOT (LMSC) (1983-84) 701
HANGZHOU AIRSHIP COMPREHENSIVE DEVELOPMENT COMPANY (China) (1988-89) 657
Heli-Stat (Piasecki) (1989-90) 673
Helium filled airships (Cameron) (1989-90) 669
High Altitude Platforms (ILC) (1986-87) 793
High Altitude System (RCA) (1987-88) 808
High Fred (Boland) (1985-86) 781
Hornet LV (AHA) 595
Hot-Air Airship (Gefa-Flug) (1988-89) 658
Hot-Air Airship (MEM RSZ) (1987-88) 787
HOVERAIR LTD (USA) (1986-87) 803
Hybrid heavy-lift vehicles (ILC) (1986-87) 793
HYSTAR (Canada) (1988-89) 655
Hystar 103 (Hystar) (1988-89) 655
HUAHANG AIRSHIP DEVELOPMENT GROUP (China) 591

I

ILC DOVER (USA) (1987-88) 797, 807
I-PAAM (Bonanno) (1990-91) 679

K

K 630/1-Ri (Ballonfabrik) (1991-92) 658
K 780/2-Ri (Ballonfabrik) (1991-92) 658
K 945/2-Ri (Ballonfabrik) (1991-92) 658
K 1050/3-Ri (Ballonfabrik) (1991-92) 658
K 1260/3-Ri (Ballonfabrik) (1991-92) 658
K 1360/4-Ri (Ballonfabrik) (1991-92) 658
K 1680/4-Ri (Ballonfabrik) (1991-92) 658
KAVANAGH BALLOONS PTY LTD (Australia) (1992-93) 605
Kenwood (Gefa-Flug) (1988-89) 657

L

LD (Adams) (1991-92) 658
LD-S (Adams) (1991-92) 658
LMSC (see Lockheed Missiles & Space Company) (1983-84) 701
LTA (Magnus) (1986-87) 783
LTA 20-01 (Van Dusen) (1983-84) 696
LTA 20-1 (Magnus) (1988-89) 656
LTA 20-1 (Van Dusen) (1983-84) 696
LTA 20-16 (Magnus) (1986-87) 783
LTA 20-16 (Van Dusen) (1984-85) 732
LTA 20-60 (Van Dusen) (1984-85) 732
LUA (ADA) (1986-87) 782
LUA-1 (Ulita) (1987-88) 798
LUA-2 (Ulita) (1987-88) 799
LUA-6 (ADA) (1985-86) 762
LZ NO5 (Zeppelin) 592
Large-volume aerostat (ILC Dover) (1983-84) 701
LETTERIO BONANNO BALLOONS SERVICE DI (Italy) (1988-89) 658
Levity (Boland) (1990-91) 679
Lightship A-50 (ABC) (1990-91) 671
Lightship A-60 (ABC) (1991-92) 652
Lightship A-60 Plus (ABC) 598
Lightship A-120 (ABC) 599
LIN HA SCIENCE INSTITUTE (China) (1985-86) 762
LOCKHEED MISSILES & SPACE COMPANY INC (a subsidiary of Lockheed Corporation) (USA) (1983-84) 701
LTA SYSTEMS LTD (Canada) (1983-84) 697

M

MAI (see Moscow Aviation Institute) 608
Mark VII-S (TCOM) (1987-88) 809
MEM (see Mém Repülógépes Szologát (1989-90) 677
MG-300 (Aerostar) (1991-92) 658
MG-300 (Raven) (1989-90) 678
MG-1000 (Aerostar) (1991-92) 658
MG-1000 (Raven) (1989-90) 678
MLA-24-A (Spacial) (1985-86) 764
MLA-32-A (Spacial) (1991-92) 647
Magnum IX (Avian) (1992-93) 614
Magnum Super Sport (Avian) (1992-93) 614
MAGNUS AEROSPACE CORPORATION (Canada) (1989-90) 655
Mai (USSR) (1988-89) 659
Manned Demonstrator Vehicle (Magnus) (1989-90) 663
MÉM REPÜLÓGÉPES SZOLOGAT (Hungary) (1989-90) 677
MEMPHIS AIRSHIPS INC (USA) 600
MIA POW (Boland) (1990-91) 679
Mifeng-6 (BUAA) (1991-92) 645
MIKE ADAMS BALLOON LOFT INC (USA) (1985-86) 777
Mirage 56 (Fantasy) (1992-93) 605
Mirage 65 (Fantasy) (1992-93) 605
Model 25M (TCOM Corporation) (1987-88) 808
Model 31Z (Thunder & Colt) (1991-92) 658
Model 32 Calibre (Avian) (1986-87) 806
Model 42 (ThunderColt) (1985-86) 781
Model 42 Series 1 (Thunder & Colt) (1991-92) 658
Model 42A (Thunder & Colt) (1991-92) 658
Model 56 (ThunderColt) (1985-86) 781
Model 56 Series 1 (Thunder & Colt) (1991-92) 658
Model 56A (Thunder & Colt) (1991-92) 658
Model 65 (ThunderColt) (1985-86) 782
Model 65 Series 1 (Thunder & Colt) (1991-92) 658
Model 69A (Thunder & Colt) (1991-92) 659
Model 77 (ThunderColt) (1985-86) 782
Model 77 Series 1 (Thunder & Colt) (1991-92) 659
Model 77A (Thunder & Colt) (1991-92) 659
Model 84 (ThunderColt) (1985-86) 782
Model 84 Series 1 (Thunder & Colt) (1991-92) 659
Model 90 (ThunderColt) (1985-86) 782
Model 90 Series 1 (Thunder & Colt) (1991-92) 659
Model 90 Series 2 (Thunder & Colt) (1991-92) 659
Model 90A (Thunder & Colt) (1991-92) 659
Model 97-34J (Piasecki) (1986-87) 794
Model 105 (ThunderColt) (1985-86) 782
Model 105 Series 1 (Thunder & Colt) (1991-92) 659
Model 105 Series 2 (Thunder & Colt) (1991-92) 659
Model 105A (Thunder & Colt) (1991-92) 659
Model 120A (Thunder & Colt) (1991-92) 659
Model 140 (Thunder & Colt) (1988-89) 671

M – Z—LIGHTER-THAN-AIR: INDEXES 729

Model 140 Series 2 (Thunder & Colt) ... (1991-92) 659
Model 160 Series 1 (Thunder & Colt) ... (1991-92) 660
Model 160A (Thunder & Colt) ... (1991-92) 660
Model 180 Series 1 (Thunder & Colt) ... (1991-92) 660
Model 180A (Thunder & Colt) ... (1991-92) 660
Model 240A (Thunder & Colt) ... (1991-92) 660
Model 280A (Thunder & Colt) ... (1987-88) 811
Model 300A (Thunder & Colt) ... (1991-92) 660
Model 365B/H (TCOM Corporation) ... (1987-88) 809
Model 400A (Thunder & Colt) ... (1991-92) 660
MOSCOW AVIATION INSTITUTE (Russia) ... (1992-93) 608
Mr. Gunsnook (Boland) ... (1990-91) 680
Mr Snake (Boland) ... (1986-87) 802
Mrs Snake (Boland) ... (1990-91) 679

N

N-31 (Cameron) ... (1991-92) 658
N-42 (Cameron) ... 611
N-56 (Cameron) ... (1991-92) 658
N-65 (Cameron) ... (1991-92) 658
N-77 (Cameron) ... (1991-92) 658
N-90 (Cameron) ... (1991-92) 659
N-105 (Cameron) ... (1991-92) 659
N-120 (Cameron) ... (1992-93) 611
N-133 (Cameron) ... (1991-92) 659
N-145 (Cameron) ... (1991-92) 660
N-160 (Cameron) ... (1992-93) 611
N-180 (Cameron) ... (1991-92) 660
N-850 (Cameron) ... (1991-92) 660
NCS ULD 1 (Endeavour) ... (1987-88) 802
NZ-1 (NZA) ... (1987-88) 788
NZA (see New Zealand Airships Ltd) ... (1987-88) 788

NAVAL AIR SYSTEMS COMMAND (USA) ... (1986-87) 796
Naval Airship Program (US Navy) ... (1986-87) 796
NEW ZEALAND AIRSHIPS LTD ... (1987-88) 788
NPP AEROSTATIC TECHNICS (UKRAINE) ... 594

O

O-31-O-77 (Cameron) ... (1991-92) 658
O-84 (Cameron) ... (1991-92) 659
O-90 (Cameron) ... (1992-93) 611
O-105 (Cameron) ... (1991-92) 659
O-120 (Cameron) ... (1991-92) 659
O-140 (Cameron) ... (1991-92) 659
O-160 (Cameron) ... (1991-92) 660

P

PTB 50 (Airborne) ... (1986-87) 799

Pacific Flyer (Thunder & Colt) ... (1992-93) 611
Peaches (Boland) ... (1987-88) 805
PIASECKI AIRCRAFT CORPORATION (USA) ... (1989-90) 673
Piccard (Piccard) ... (1983-84) 705
PICCARD BALLOONS LTD, DON (USA) ... (1983-84) 704
Piccolo (Boland) ... (1983-84) 705
Puppy Chow (Boland) ... (1990-91) 678

R

R-225 (Cameron) ... (1987-88) 802
RBX-7 (Bonanno) ... (1991-92) 647
RS-1 (Wren) ... (1987-88) 793
RS$_3$-03 (MÉM RS$_3$) ... (1983-84) 702, 706
RS$_3$-04 (MÉM RS$_3$) ... (1983-84) 702
RSZ (see Repulogepes Szologat) ... (1991-92) 657
RSZ-03 (MÉM RSZ) ... (1987-88) 811
RSZ-03/1 (RSZ) ... (1991-92) 659
RSZ-04 (MÉM RSZ) ... (1987-88) 811
RSZ-04/1 (RSZ) ... (1991-92) 659
RSZ-05 (RSZ) ... (1991-92) 658
RSZ-06 (MÉM RSZ) ... (1989-90) 678
RSZ-06/1 (Air Service) ... (1992-93) 607
RX-6 (Aerostar) ... (1991-92) 658
RX-7 (Aerostar) ... (1991-92) 659
RX-8 (Aerostar) ... (1991-92) 659
RX-100 (Cameron) ... (1992-93) 611

Rainbow-Stars (Boland) ... (1990-91) 678
RAVEN INDUSTRIES INC (USA) ... (1989-90) 677
RCA SERVICE COMPANY (USA) ... (1987-88) 808
Red Baron (Boland) ... (1990-91) 680
REPULOGEPES SZOLOGAT (Hungary) ... (1991-92) 657
Riff-Raff (Boland) ... (1987-88) 805
Rover, A-2 (Boland) ... (1990-91) 672
Roziere balloons (Cameron) ... 596
Roziere R-15 (Cameron) ... (1992-93) 610
Roziere R-42 (Cameron) ... (1992-93) 610
Roziere R-60 (Cameron) ... (1992-93) 610
Roziere R-77 (Cameron) ... (1992-93) 610
Roziere R-225 (Cameron) ... (1992-93) 610

S

S-40A to S-50A (Aerostar) ... (1991-92) 658
S-52A to S-66A (Aerostar) ... (1991-92) 659
S-71A (Aerostar) ... (1991-92) 660
S-77A (Aerostar) ... (1991-92) 660
SPACIAL (see Servicios Publicitarios Aereos Construccion y Engenieria de Aeronaves Ligeras SA de CV) (Mexico) ... (1992-93) 607
SS-103 (Solo) ... (1989-90) 678
SSP-6000 (ILC) ... (1987-88) 807

Sentinel 1000 (WAI) ... 601
Sentinel 5000 (WAI) ... 602
SERVICIOS PUBLICITARIOS AEREOS CONSTRUCCION Y ENGENIERIA DE AERONAVES LIGERAS SA de CV (Mexico) ... (1992-93) 607
Sky Chariot (Maxi) (ThunderColt) ... (1985-86) 781
Sky Chariot (Midi) (ThunderColt) ... (1985-86) 781
Sky Chariot (Mini) (ThunderColt) ... (1985-86) 780
Skyhawk (Avian) ... (1992-93) 614
SKYRIDER AIRSHIPS INC (USA) ... (1991-92) 653
Skyship 500 (Slingsby) ... (1991-92) 649
Skyship 500HL (Slingsby) ... 597
Skyship 600 (Slingsby) ... 597
Skyship 7000 (Airship Industries) ... (1985-86) 766
Skystar (Cameron) ... (1989-90) 668
SLINGSBY AVIATION LIMITED (UK) ... 597
Sparrow (Avian) ... (1992-93) 614
Spirit of Lake Garda (Boland) ... (1987-88) 805
SOLO SYSTEM INC (USA) ... (1989-90) 677
Stars (TCOM Corporation) ... (1987-88) 808

T

TCOM 25M (TCOM) ... (1987-88) 808
TCOM 365B/H (TCOM) ... (1987-88) 809
TCOM CORPORTION (USA) ... (1987-88) 808
TZ-1 (BIAA) ... (1988-89) 657

TCOM LIMITED PARTNERSHIP (USA) ... 600
Tethered Aerostats (TCOM) ... 600
Thermoplane (MAI) ... (1992-93) 608
THERMOPLANE DESIGN BUREAU (Russia) ... 593
Thompson Airship (Thompson) ... 600
THOMPSON, JAMES, AIAA (USA) ... 600
THUNDER & COLT LTD (UK) ... 597
THUNDER BALLOONS (UK) ... (1984-85) 740
THUNDERCOLT BALLOONS LTD (UK) ... (1983-84) 662, 667
Tianzhou 1 (BIAA/Hongwei) ... (1988-89) 657
Toluca (Spacial) ... (1991-92) 647
Tour d'Argent (Boland) ... (1990-91) 679
Turbo 8 (Avian) ... (1992-93) 614

U

ULD-1 (Cameron) ... (1987-88) 802
ULD-1 (Endeavour) ... (1987-88) 802
UM10-22 Cloud Cruiser (Ulita) ... (1992-93) 615
UM10-23 Cloud Cruiser (Ulita) ... 600
UM20-48 (Ulita) ... (1992-93) 616
UM25-64 (Ulita) ... (1992-93) 616
UM30-71 (Ulita) ... 601
USA 100 (Aerotek) ... (1988-89) 664
US-LTA (see US Lighter Than Air Corporation) ... 616
US-LTA 138-S (US-LTA) ... 600

ULITA INDUSTRIES INC (USA) ... 600
ULITA MANUFACTURING INC (USA) ... (1985-86) 773
Ultrablimp (Memphis) ... (1987-88) 797
Ultralight (Boland) ... (1990-91) 678
Upship 100-001 (Upship) ... 601
Upship 750-001 (Upship) ... (1992-93) 616
UPSHIP PROJECT DIRIGIBLE AIRSHIPS ... 601
Ural-3 (USSR) ... (1983-84) 697
US AIRSHIP CORPORATION (USA) ... (1986-87) 796
US LIGHTER THAN AIR CORPORATION (USA) ... 615

V

VAN DUSEN COMMERCIAL DEVELOPMENT (CANADA) LTD (Canada) ... (1984-85) 732
Virgin Atlantic Flyer (Thunder & Colt) ... (1987-88) 803
Viva-20 (Cameron) ... (1992-93) 610
Viva-31 (Cameron) ... (1992-93) 610
Viva-42 (Cameron) ... (1992-93) 610
Viva-56 (Cameron) ... (1992-93) 610
Viva-65 (Cameron) ... (1992-93) 610
Viva-77 (Cameron) ... (1992-93) 610
Viva-90 (Cameron) ... (1992-93) 610

W

WAI (see Westinghouse-Airship Industries) ... 601
WDL (see Westdeutsche Luftwerbung Theodor Wullenkemper) ... 592
WDL 1B (WDL) ... 592

WATSON MODEL WORKS (USA) ... (1987-88) 796
WESTDEUTSCHE LUFTWERBUNG THEODOR WULLENKEMPER (Germany) ... 592
WESTINGHOUSE-AIRSHIP INC (USA) ... 601
Whispership (Barnes) ... (1992-93) 614
White Dwarf (California Airships) ... (1986-87) 792

X

X-8 (Bonanno) ... (1987-88) 801
Xihu (Lin Hai) ... (1985-86) 762
Xihu-1 (HACDC) ... (1987-88) 786
XXUS (Boland) ... (1985-86) 780

Y

YEZ-2A (WAI) ... 602

Yankee Doodle (ABC) ... (1986-87) 801, 806
YOST, ED (USA) ... (1985-86) 780

Z

Zephyr 200 (Memphis) ... 614
ZEPPELIN, LUFTSCHIFFBAU, GmbH (Germany) ... 592

AERO ENGINES

A

A650 (Alturdyne) (1992-93) 681
AI-14R (WSK-PZL Kalisz) 630
AI-20 (ZMKB Progress) 650
AI-24 (ZMKB Progress) 650
AI-25 (ZMKB Progress) 650
AI-25TL (ZMKB Progress) 651
AIC (see Ames Industrial Corporation) (1985-86) 888
AL-7 (Saturn) 642
AL-21 (Saturn) 642
AL-31 (Saturn) 643
AL-34 (Saturn) 643
AL-35 (Saturn) 643
ALF 502 (Textron Lycoming) 679
AM-3 (Soyuz) 644
AM-5 (Soyuz) (1991-92) 705
AMI (see AeroMotion Inc) (1992-93) 681
APW34 (AVCO/P & W) (1985-86) 895
AR.318 (Alfa Romeo Avio) (1992-93) 644
ARC (see Atlantic Research Corporation) (1984-85) 859
ARTJ 140 (Alfa Romeo) (1985-86) 862
ASSAD (Russia) 632
ASTOVL (General Electric) 666
ASW 24 TOP (F&W) (1990-91) 696
ASz-62R (WSK-PZL Kalisz) 630
ASz-621 R (PZL) (1983-84) 787
ATF-3 (Garrett) 663
AVCO/P & W (USA) (1985-86) 895
AVIA-HSA (see Avia-Hamilton Standard Aviation Ltd) 612

Adour (Rolls-Royce Turbomeca) 625
Advanced Apogee Motor (CSD/SEP) (1984-85) 865
AERO BONNER LTD (UK) (1985-86) 878
AEROJET-GENERAL (USA) (1992-93) 678
AEROJET LIQUID ROCKET COMPANY (USA) (1983-84) 808
AEROJET STRATEGIC PROPULSION COMPANY (USA) (1985-86) 887
AEROJET TACTICAL SYSTEMS COMPANY (USA) (1985-86) 887, 888
AEROJET TECHSYSTEMS CO (USA) 678
AEROMOTION INC (USA) 681
AeroMotion Twin (AMI) (1992-93) 681
AEROSTAR SA (Romania) 632
AEROTECHNIK (Czech Republic) 612
Aerotor 90 (Norton) (1988-89) 710
Air breathing propulsion (CSD) (1992-93) 681
ALFA ROMEO AVIO SpA (Italy) 627
Algol III (CSD) (1984-85) 864
ALLISON GAS TURBINE DIVISION, GENERAL MOTORS CORPORATION 660
ALLISON/MTU (International) (1989-90) 699
Altair III (Morton Thiokol) (1984-85) 879
ALTURDYNE (USA) (1992-93) 681
ALVIS UAV ENGINES LTD (UK) 654
AMES INDUSTRIAL CORPORATION (USA) (1985-86) 888
Antares III (Morton Thiokol) (1984-85) 879
AOI ENGINE FACTORY (135) (Egypt) 614
AO AVIADVIGATEL (Russia) 633
Arbizon (Turbomeca) (1984-85) 815
Arriel (Turbomeca) 617
Arrius (Turbomeca) 616
Arrius 1D (Turbomeca) 617
ARROW (Italy) ENGINEERING SRL 627
Artouste III (Turbomeca) 617
Astazou Turboprop (Turbomeca) (1991-92) 676
Astazou Turboshaft (Turbomeca) 617
Astrobee F (Aerojet) (1983-84) 809
Atar (SNECMA) (1986-87) 886
Atar 9K50 (SNECMA) 615
ATELIERS JPX (France) (1991-92) 673
ATLANTIC RESEARCH CORPORATION (USA) (1984-85) 859
ATLAS AVIATION (South Africa) 649
AVCO LYCOMING/PRATT & WHITNEY (USA) (1987-88) 945
AVCO LYCOMING TEXTRON STRATFORD DIVISION (USA) (1987-88) 941
AVCO LYCOMING WILLIAMSPORT DIVISION (USA) (1987-88) 943
AVIA-HAMILTON STANDARD AVIATION LTD (Czech Republic) 612
AVIA KUNCERNOVY PODNIK (Czech Republic) 628
AVIA NARODNI PODNIK (Czechoslovakia) (1986-87) 883

B

BAZ-21083 (Volszhsky) 648
BR 700 (BMW-RR) 619
BR 710 (BMW-RR) 619
BSEL (see Bet-Shemesh Engines) 627
BUAA (see Beijing University of Aeronautics and Astronautics) 608

Baby Mamba (Dreher) (1983-84) 818
BAJ VICKERS LTD (UK) (1984-85) 846
BAKANOV (Russia) 635
BEIJING UNIVERSITY OF AERONAUTICS AND ASTRONAUTICS (China) 608
BELL AEROSPACE TEXTRON (USA) (1984-85) 863
BET-SHEMESH ENGINES LTD (Israel) 627
BMW ROLLS-ROYCE GmbH 619
BOMBARDIER-ROTAX GmbH (Austria) 603
BONNER (see Aero Bonner Ltd) (1985-86) 878
Booster separation motor (CSD) (1987-88) 946
BORZECKI, JOZEF (Poland) (1986-87) 903
BPD SETTORE DIFESA E SPAZIO (Italy) (1984-85) 828
Butalane (SNPE) (1985-86) 830
Butalite (SNPE) (1984-85) 815

C

CAC (see Commonwealth Aircraft Corporation Ltd) (1986-87) 876
CAM (see Canadian Airmotive Inc) (1992-93) 619
CAM 100 (CAM) 710
CAREC (see China National Aero-Engine Corporation) 608
CATIC (see China National Aero-Technology Import and Export Corporation) (1992-93) 623
CEC (see Chengdu Engine Company) 608
CF 077A (Emdair) 654
CF 077B (Emdair) 654
CF 092A (Emdair) 654
CF6 (General Electric) 667
CF6-80C2 (General Electric) 667
CF6-80E1 (General Electric) 668
CF34 (General Electric) 666
CF 100A (Emdair) (1988-89) 710
CF 112 (Emdair) 654
CF 150A (Emdair) (1986-87) 916
CF700 (General Electric) (1986-87) 940
CFE738 (CFE) 662
CFM56 (CFM International) 621
CFM88 (CFM International) 622
CJ610 (General Electric) (1986-87) 940
CLXMW (see Changzhou Lan Xiang Machinery Works) 608
CRM 18D/SS (Italy) 627
CSD (see Chemical Systems Division) (1992-93) 681
CSTM (see Centro Studi Trasporti Missilistici) (1984-85) 828
CT7 (General Electric) 670
CT58 (General Electric) (1988-89) 725

CAMERA REYNAUD & CIE, SCS (France) (1992-93) 630
CANADIAN AIRMOTIVE INC (Canada) (1992-93) 619
Castor II (Morton Thiokol) (1984-85) 878
Castor IV (Morton Thiokol) (1984-85) 878
CELMA-CIA ELECTROMECANICA (Brazil) 603
CENTRO STUDI TRASPORTI MISSILISTICI (Italy) (1984-85) 828
CFE COMPANY (USA) 662
CFM INTERNATIONAL SA (International) 621
CHANGZHOU LAN XIANG MACHINERY WORKS (China) 608
CHEMICAL SYSTEMS DIVISION (USA) (1992-93) 681
CHENGDU ENGINE COMPANY (China) 608
CHERNYSHOV (Russia) 635
CHINA NATIONAL AERO-ENGINE CORPORATION (China) 608
CHINA NATIONAL AERO-TECHNOLOGY IMPORT AND EXPORT CORPORATION (China) (1992-93) 623
CHOTIA (see Weedhopper of Utah Inc) (1984-85) 863
Civil Spey RB163 (Rolls-Royce) 674
COMMONWEALTH AIRCRAFT CORPORATION LIMITED (Australia) (1986-87) 876
CONTINENTAL (see Teledyne Continental Motors) (1985-86) 918
CRM (Italy) 627
CURTISS-WRIGHT CORPORATION WOODIDGE FACILITY (USA) (1985-86) 896
CUYUNA ENGINE CO (USA) (1987-88) 947

D

D-15 (MKB) (1991-92) 697
D-18A (II) 629
D-18T (ZMKB Progress) 653
D-20P (Soloviev) (1987-88) 923
D-25V (AO Aviadvigatel) 633
D-27 (ZMKB Progress) 654
D-30 (AO Aviadvigatel) 633
D-30F6 (AO Aviadvigatel) 634
D-30K (AO Aviadvigatel) 633
D-36 (ZMKB Progress) 651
D-90A (Soloviev) (1991-92) 698
D-100/110/112 (AO Aviadvigatel) 634
D-136 (ZMKB Progress) 652
D-227 (ZMKB Progress) 654
D-236 (ZMKB Progress) 652
D-236T (Lotarev) (1986-87) 911
D-336 (Progress) (1991-92) 702
D-436 (ZMKB Progress) 652
DEMC (see Dongan Engine Manufacturing Company) 608
DN-200 (RKBM) 641
DV-2 (ZMKB Progress) 651

Dart (Rolls-Royce) (1987-88) 933
Delta (TRW) (1984-85) 892
DETROIT DIESEL ALLISON DIVISION, GENERAL MOTORS CORPORATION (USA) (1983-84) 810
DONGAN ENGINE MANUFACTURING COMPANY (China) 608
DRAGON ENGINE CORPORATION (USA) (1987-88) 974
DREHER ENGINEERING COMPANY (USA) (1984-85) 866
Dual Pac (Soloy) 676
DVADESETPRIVI MAJ BEOGRAD (Yugoslavia) 684
DYNA-CAM INDUSTRIES (USA) (1992-93) 681

E

EJ-200 (Eurojet) 622

ECOLE NATIONAL DES INGENIEURS DE SAINT-ETIENNE (France) (1985-86) 847
EMDAIR LTD (UK) 654
EMG ENGINEERING COMPANY (USA) (1992-93) 682
ENISE (see Ecole National des Ingénieurs de Saint-Etienne) (1985-86) 847
Epictete (SNPE) (1985-86) 850
ETJ1081 (Garrett) (1987-88) 948
EUROJET TURBO GmbH (International) 622
EX-104 (Morton Thiokol) (1984-85) 878
EHMANN, ROLF (West Germany) (1984-85) 818

F

F3 (IHI) 628
F22 (Hirth) (1992-93) 636
F23A (Hirth) (1992-93) 636
F30 (Hirth) (1992-93) 636
F100 (Pratt & Whitney) 675
F100 EMD (Pratt & Whitney) (1984-85) 885
F100-PW-229 (Pratt & Whitney) (1990-91) 749
F100-PW-229 (Pratt & Whitney) 675
F101 (General Electric) (1987-88) 953
F103-GE-100 (General Electric) 667
F104-GA-100 (Garrett) 663
F107 (Williams) (1987-88) 974
F108 (CFM International) 621
F109-GA-100 (Garrett) 663
F110 (General Electric) 668
F112 (Williams) (1987-88) 974
F113 (Rolls-Royce) 674
F117 (Pratt & Whitney) 674
F118 (General Electric) 669
F119 (Pratt & Whitney) 676
F120 (General Electric) (1991-92) 726
F124-GA-100 (Garrett) 664
F125-GA-100 (Garrett) 664
F129 (Williams) 683
F263 R 53 (Hirth) (1992-93) 636
F402, Pegasus (Rolls-Royce) 659
F404 (General Electric) 666
F404, Pegasus (Rolls-Royce) (1989-90) 725
F404J (Flygmotor/General Electric) (1983-84) 790
F405 (Rolls-Royce Turbomeca) 625
F412 (General Electric) (1991-92) 724
F414 (General Electric) 685
FA150 (Fiat) (1985-86) 863
FAM (see France Aéro Moteurs) 630
F + E (see Fischer & Entwicklungen) (1992-93) 635
FE.260AG (Fieldhouse) (1987-88) 927
FE.525A(F) (Fieldhouse) (1987-88) 927
FE.525AG (Fieldhouse) (1987-88) 927
FJ44 (Williams-Rolls) 683
FJR710 (NAL) (1991-92) 689
FNM 1 (see FN Moteurs SA) (1992-93) 619
FW-5 (CSD) (1983-84) 817

FABRIQUE NATIONALE HERSTAL SA (Belgium) (1988-89) 676
FEDERAL DIRECTORATE OF SUPPLY AND PROCUREMENT (Yugoslavia) 705
FIATAVIO (Italy) 628
FIELDHOUSE ENGINES LTD (UK) (1987-88) 927
FISCHER + ENTWICKLUNGEN (Germany) (1992-93) 635
FISHER RESEARCH CORPORATION (USA) (1992-93) 682
FLYGMOTOR (see Volvo Flygmotor AB) 667
FN MOTEURS SA (Belgium) (1992-93) 619
FRANCE AERO MOTEURS SARL (France) 630

G

G2P (see Groupement pour les Gros Propulseurs à Poudre) (1984-85) 810
G8-2 (Gluhareff-EMG) (1992-93) 682
G25B (Zenoah) (1986-87) 902
GADRI (see Guizhou Aero-Engine Design and Research Institute) 608
GDL (USSR) (1984-85) 836
GE27 (General Electric) (1989-90) 736
GE36, UDF (General Electric) (1989-90) 693, 735
GE37 (General Electric) (1990-91) 740
GE38 (General Electric) (1989-90) 735
GE90 (General Electric) 666
GK (SECA) (1987-88) 885
GLC38 (General Electric) 669
GMA 140 TK (Scoma) (1992-93) 63
GMA 500 (Allison) (1984-85) 858
GMA 501 (Allison) 661
GMA 2100 (Allison) 662
GMA 3000 (Allison) (1990-91) 734
GMA 3007 (Allison) 662
GP 1000 (Arrow) 627
GP 1500 (Arrow) 627
GR-18 (TCM/Continental) (1990-91) 752
GR-36 (TCM/Continental) (1990-91) 752
GT250 (Arrow) (1992-93) 644
GT500 (Arrow) (1992-93) 644
GT654 (Arrow) (1992-93) 644
GT1000 (Arrow) (1992-93) 644
GTD-3 (Mars) 638
GTD-350 (WSK-PZL-Rzeszów) 631
GTE-400 (Soyuz) 647
GTRE (see Gas Turbine Research Establishment) 621
GTX (GTRE) 621

GARRETT ENGINE DIVISION (USA) 682
GAS TURBINE RESEARCH ESTABLISHMENT (India) 621
GAVRILOV (Russia) 651
GE AIRCRAFT ENGINES (USA) 685
Gem (Rolls-Royce) 659
GENERAL ELECTRIC (USA) 666
GLUSHENKOV (USSR) (1990-91) 712

G – N—ENGINES: INDEXES 731

Gnome (Rolls-Royce) 660
GRETH, GERRY (USA) .... *(1992-93)* 690
GROUPEMENT POUR LES GROS PROPULSEURS A POUDRE (France) .... *(1984-85)* 810
GROUPEMENT TURBOMECA -SNECMA (GRTS) (France) 635
GUIZHOU AERO-ENGINE DESIGN AND RESEARCH INSTITUTE (China) 608

H

H-63 (Nelson) 671
HAL (see Hindustan Aeronautics Ltd.) 621
HARM TU-780 (Morton Thiokol) .... *(1987-88)* 960
HDHV (see Hawker de Havilland Victoria Ltd) 603
HM7 (SEP) .... *(1984-85)* 813
HP 60Z (Parodi) .... *(1984-85)* 821
HS-5 (CATIC) .... *(1989-90)* 686
HS-6 (CATIC) .... *(1989-90)* 686
HS7 (DEMC) 608
HS8 (DEMC) 608
HS-26 (CATIC) .... *(1989-90)* 686
HS.260A (Fieldhouse) .... *(1987-88)* 927
HS.525A (Fieldhouse) .... *(1987-88)* 927

Harpoon (Morton Thiokol) .... *(1984-85)* 878
Harpoon booster motor (Aerojet) .... *(1984-85)* 856
HAWKER DE HAVILLAND VICTORIA LTD (Australia) .. 603
HAWKER SIDDELEY CANADA INC (Canada) .... *(1991-92)* 662
Hawk motor (Aerojet) .... *(1984-85)* 856
Hellfire, TX 657 (Morton Thiokol) .... *(1987-88)* 960
Hellfire, TX 773 (Morton Thiokol) .... *(1987-88)* 960
HELWAN (Egypt) .... *(1985-86)* 847
HINDUSTAN AERONAUTICS LTD (India) 621
HIRTHMOTOREN, GÖBLER, GmbH (Germany) ... *(1992-93)* 636
Huosai-16 (CATIC) .... *(1987-88)* 890
Hybrid propulsion (CSD) .... *(1987-88)* 946
Hydrazine Thruster (SEP) .... *(1984-85)* 813

I

IAE (see International Aero Engines AG) 623
IAI (see Israel Aircraft Industries Ltd) 627
IAME (see Ital-American Motor Engineering) .... *(1992-93)* 645
IHI (see Ishikawajima-Harima Jukogyo Kabushiki Kaisha) 628
IL (see Instytut Lotnictwa) 629
IL-144 engine (USSR) .... *(1988-89)* 709
IMAER (see Indústria Mecânica E Aeronautica Ltda) .... *(1992-93)* 619
Imaer 1000 (Imaer) .... *(1992-93)* 619
Imaer 2000 (Imaer) .... *(1992-93)* 619
IO-240 (TCM) 678
IO-320 series (Textron Lycoming) 681
IO-360 series (Textron Lycoming) 682
IO-520 series (TCM) 678
IO-540 Series (Textron Lycoming) 682
IO-550 (TCM) 679
IO-720 Series (Textron Lycoming) 681
IOL-200 (TCM) .... *(1986-87)* 959, 960
IOL-300 (TCM) .... *(1986-87)* 959, 960
IPSM-II (Morton Thiokol) .... *(1984-85)* 878
ISM (CSD) .... *(1984-85)* 865
ITP (see Industria de Turbopropulsion) 649
ITS-90 (IHI) .... *(1990-91)* 707
IUS (CSD) .... *(1988-89)* 719

INDUSTRIA DE TURBO PROPULSORES (Spain) 649
INDÚSTRIA MECÂNICA E AERONAUTICA LTDA (Brazil) .... *(1992-93)* 619
INSTYTUT LOTNICTWA (Poland) 629
IN-TECH INTERNATIONAL, INC (USA) .... *(1992-93)* 690
INTERNATIONAL AERO ENGINES AG (International) 623
INTREPRINDEREA TURBOMECANICA BUCURESTI (Romania) 632
ISAYEV (USSR) .... *(1984-85)* 837
ISHIKAWAJIMA-HARIMA JUKOGYO KABUSHIKI KAISHA (Japan) 628
Isolane (SNPE) .... *(1985-86)* 850
ISOTOV (USSR) .... *(1990-91)* 712
ISRAEL AIRCRAFT INDUSTRIES LTD (Israel) 627
ITAL-AMERICAN MOTOR ENGINEERING (Italy) *(1992-93)* 645
IVCHENKO (USSR) .... *(1990-91)* 712

J

J52 (Pratt & Whitney) 672
J69 (Teledyne CAE) 679
J75 (Pratt & Whitney) .... *(1987-88)* 961
J79 (General Electric) .... *(1986-87)* 939
J79-IAI-J1E (IAI) .... *(1992-93)* 644
J85 (General Electric) .... *(1986-87)* 939
J400 (Williams) .... *(1987-88)* 974
J402-CA-400 (Teledyne CAE) .... *(1987-88)* 970
J402-CA-401 (Teledyne CAE) .... *(1987-88)* 970
J402-CA-700 (Teledyne CAE) .... *(1987-88)* 970
J402-CA-702 (Teledyne CAE) .... *(1987-88)* 970
J403-MT-400 (Microturbo) .... *(1987-88)* 895
JB 2 × 250 (Borzecki) .... *(1986-87)* 903
JPX 4T60/A (JPX) 614
JPX 4TX 75 (JPX) 614
JT3D (Pratt & Whitney) .... *(1987-88)* 961
JT4 (Pratt & Whitney) .... *(1987-88)* 961
JT8 (Pratt & Whitney) 672
JT8D (Pratt & Whitney) 672
JT8D-200 Series (Pratt & Whitney) 672
JT9D (Pratt & Whitney) 673
JT15D (P&WC) 604
JTF10A (Pratt & Whitney) .... *(1991-92)* 734
JTF22 (Pratt & Whitney) .... *(1990-91)* 749
JVX (General Electric) .... *(1983-84)* 827

JANOWSKI, JAROSLAW (Poland) .... *(1987-88)* 914
JAVELIN AIRCRAFT COMPANY INC (USA) .... *(1992-93)* 690
JPX, SARL (France) 630

K

K-15 (IL) 630
K-100A (SPP) .... *(1990-91)* 751
KFM 104 (IAME) .... *(1986-87)* 900
KFM 105 (IAME) .... *(1986-87)* 900
KFM 107 (IAME) .... *(1984-85)* 829
KFM 107 Maxi (IAME) .... *(1992-93)* 645
KFM 112 (IAME) .... *(1992-93)* 645
KJ12 (Kawasaki) .... *(1992-93)* 647
KKBM (see Kuibyshev Engine Design Bureau) 635

KAWASAKI JUKOGYO KABUSHIKI KAISHA (Japan) 629
KHACHATUROV (USSR) .... *(1990-91)* 713
KHD LUFTFAHRTTECHNIK GmbH (West Germany) .... *(1990-91)* 697
KLIMOV CORPORATION (Russia) 636
KOLIESOV (USSR) .... *(1990-91)* 715
KOMATSU ZENOAH COMPANY (Japan) .... *(1986-87)* 902
KONIG MOTORENBAU (Germany) .... *(1992-93)* 636
KOPTCHYENKO (USSR) .... *(1990-91)* 715
KOSBERG (USSR) .... *(1984-85)* 840
KUIBYSHEV ENGINE DESIGN BUREAU (USSR) 635
KUZNETSOV (Russia) 638

L

L 90E (Limbach) .... *(1992-93)* 636
L 275E (Limbach) .... *(1992-93)* 636
L 550E (Limbach) .... *(1992-93)* 636
L 1800 (Limbach) 620
L 2000 (Limbach) 620
L 2400 (Limbach) 620
LE-5 (Mitsubishi) .... *(1985-86)* 830
LEM 2fm 17 (Offmar) .... *(1992-93)* 645
LF500 (Textron Lycoming) 680
LHTEC (Allison Gas Turbine Division) .... *(1985-86)* 908
LHTEC (see Light Helicopter Turbine Engine Company) 671
LJ95 (Teledyne CAE) .... *(1985-86)* 918
LM (see Liming Engine Manufacturing Corporation) 608
LMC (see Liyang Machinery Corporation) 610
LO-360 (Avco Lycoming) .... *(1987-88)* 944
LPE (see Light Power Engine Corporation 671
LR89-NA-5 (Rocketdyne) .... *(1984-85)* 887
LR101-NA-7 (Rocketdyne) .... *(1984-85)* 887
LR105-NA-5 (Rocketdyne) .... *(1984-85)* 887
LT101 (Textron Lycoming) 681
LTC1 (Textron Lycoming) 680
LTC4 (Textron Lycoming) 680
LTP 101 (Textron Lycoming) 681
LTS 101 (Textron Lycoming) 681
LTSIO-360 (TCM) .... *(1987-88)* 747

Larzac (Turbomeca/SNECMA) 618
Liftjet (Koliesov) .... *(1989-90)* 713
Liftjet (USSR) .... *(1988-89)* 709
LIGHT HELICOPTER TURBINE ENGINE COMPANY (USA) 671
LIGHT POWER ENGINE CORPORATION (USA) 671
LIMBACH FLUGMOTOREN GmbH (Germany) 620
LIMBACH MOTORENBAU (West Germany) .... *(1985-86)* 855
LIMING ENGINE MANUFACTURING CORPORATION (China) 608
LIYANG MACHINERY CORPORATION (China) 610
LOTAREV ZMKB (USSR) .... *(1990-91)* 716
LOTAREV/ZVL (International) 642
LOTUS CARS LTD (UK) .... *(1987-88)* 927
LTV AEROSPACE AND DEFENCE (USA) .... *(1984-85)* 876
LYCOMING (see Avco Lycoming) .... *(1985-86)* 893
LYULKA (USSR) .... *(1991-92)* 697

M

M-14P (VMKB) 648
M-14V-26 (VMKB) 648
M-16 (Bakanov) 635
M-17 (Bakanov) 635
M-18 (Bakanov) 635
M-18 (VMKB) 648
M-19 (Bakanov) 635
M49 Larzac (SNECMA/ Turbomeca) .... *(1985-86)* 772, 775, 777
M53 (SNECMA) 615
M88 (SNECMA) 616
M 137 (Avia) .... *(1985-86)* 846
M 137A (Avia-HSA) 612
M 337 (Avia) .... *(1985-86)* 846
M 337A (Avia-HSA) 613
M 601 (Motorlet) 613
M 602 (Motorlet) 614
MA-3 (Rocketdyne) .... *(1984-85)* 887
MA-5 (Rocketdyne) .... *(1984-85)* 887
MA-196 (Marquardt) .... *(1988-89)* 728
MA-225XAA (Marquardt) .... *(1985-86)* 908
MAS (see Ministry of Aero-Space Industry) 608
MB-4-80 (Mudry) .... *(1992-93)* 631
MBB (see MBB/ERNO Space Division) .... *(1984-85)* 820
Mk 36 Mod 9 (Morton Thiokol) .... *(1985-86)* 909
MKB (see Motorostroitelnoye Konstruktorskoye Buro) .... *(1991-92)* 697
MNPK Design Bureau (see Soyuz) 662
MS-1500 (Pieper) .... *(1992-93)* 637
MTM 385-R (MTU/Turbomeca) .... *(1987-88)* 907
MTR (see MTU TURBOMECA ROLLS-ROYCE GmbH) 624
MTR 390 (MTR) 624
MTU (see Motoren- und Turbinen-Union München GmbH) .. 620
MUDRY (see Moteurs Mudry-Buchoux (Avions Mudry et Cie)) 631
MWAE (see Mid West Aero Engines) 655
MWAE 75 (MWAE) 655
MWAE 90 (MWAE) .... *(1992-93)* 672
MWAE100R (MWAE) 655

MACHEN INC (USA) .... *(1985-86)* 908
Mage (SEP) .... *(1984-85)* 613
Makila (Turbomeca) 618
Malemute (Morton Thiokol) .... *(1984-85)* 878
MARQUARDT COMPANY, THE (USA) .... *(1992-93)* 691
MARS, OMSK AIRCRAFT ENGINE DESIGN BUREAU (Russia) 638
Maverick Motor (Aerojet) .... *(1987-88)* 937
Maverick, TX-481 (Morton Thiokol) .... *(1987-88)* 960
Maverick, TX-633 (Morton Thiokol) .... *(1987-88)* 960
MBB (West Germany) .... *(1984-85)* 820
MBB/ERNO SPACE DIVISION (West Germany) .... *(1984-85)* 820
Merlyn (In-Tech) .... *(1992-93)* 690
Merlyn (Machen) .... *(1985-86)* 908
MICROTURBO INC (USA) .... *(1992-93)* 692
MICROTURBO SA (France) .... *(1992-93)* 630
MID WEST AERO ENGINES (UK) 655
Mikron IIIS (A) (Aerotechnik) 628
MIKULIN/SOYUZ (Russia) 655
Military Spey (Rolls-Royce) 675
MINISTRY OF AERO-SPACE INDUSTRY (China) 608
Minuteman motor (Aerojet) .... *(1984-85)* 856
MITSUBISHI JUKOGYO KABUSHIKI KAISHA (Japan) ... 629
Model 225 (Allison) .... *(1988-89)* 717
Model 250 (Allison) 660
Model 280 (Allison) .... *(1986-87)* 928
Model 352 (Teledyne CAE) 679
Model 356 (Teledyne CAE) 679
Model 365 (Teledyne CAE) .... *(1985-86)* 918
Model 370 (Teledyne CAE) .... *(1987-88)* 970
Model 370-1 (Teledyne CAE) .... *(1987-88)* 970
Model 372-2 (Teledyne CAE) .... *(1987-88)* 970
Model 372-11A (Teledyne CAE) .... *(1987-88)* 960
Model 373 (Teledyne CAE) .... *(1987-88)* 970
Model 440/555 (Teledyne CAE) .... *(1983-84)* 839
Model 455 (Teledyne CAE) .... *(1987-88)* 970
Model 501 (Allison) 661
Model 501-M62 (Allison) .... *(1988-89)* 718
Model 535 (Rolls-Royce) 657
Model 578DX (PW-Allison) .... *(1991-92)* 736
Model 912-B46 and B51 (Rolls-Royce/Allison) .... *(1984-85)* 824
Model 912-B52 (Allison) .... *(1987-88)* 941
Model 1000 (IMAER) 619
Model 2000 (IMAER) 619
Model 4318F (Northrop) .... *(1987-88)* 960
Model 8096 Agena engine (Bell) .... *(1984-85)* 863
MOLLER INTERNATIONAL (USA) .... *(1992-93)* 691
MOMA STANOJLOVIC AIR DEPOT (Yugoslavia) 684
MORTON THIOKOL INC (USA) .... *(1989-90)* 739
MOTEURS MUDRY-BUCHOUX (AVIONS MUDRY et CIE) (France) .... *(1992-93)* 631
MOTOREN- UND TURBINEN-UNION MÜNCHEN GmbH (Germany) 620
MOTORLET A. S. LTD (Czech Republic) 613
MOTORLET NC, ZAVOD JANA SVERMY (Czechoslovakia) .... *(1988-89)* 683
MOTOROSTROITELNOYE KONSTRUKTORSKOYE BURO (USSR) .... *(1991-92)* 697
MTU/MOTOREN- UND TURBINEN-UNION MÜNCHEN GmbH (West Germany) .... *(1985-86)* 892
MTU TURBOMECA ROLLS-ROYCE GmbH (International) 624
MTU/TURBOMECA SARL (International) .... *(1985-86)* 858

N

NAL (see National Aerospace Laboratory) 629
NGL (see Normalair-Garrett Ltd) .... *(1987-88)* 928
NK-6 (KKBM) 635
NK-8 (KKBM) 636
NK-12M (KKBM) 635
NK-16 (KKBM) 636
NK-20 (KKBM) 636
NK-22 (KKBM) 636
NK-25 (KKBM) 636
NK-86 (KKBM) 636
NK-144 (KKBM) 636
NK-86 (KKBM) 652
K-88 (SAMARA) 641
NK-92 (SAMARA) 641
NK-93 (SAMARA) 641
NK-144 (KKBM) .... *(1991-92)* 695
NK-321 (SAMARA) 642
NPT (see Noel Penny Turbines Ltd) .... *(1991-92)* 710
NPT 301 (Noel Penny) .... *(1991-92)* 710
NPT 401B (Noel Penny) .... *(1989-90)* 721
NPT 754 (Noel Penny) .... *(1991-92)* 710
NR 602 (Norton) .... *(1990-91)* 726
NR 612 (Norton) .... *(1991-92)* 710
NR 622 (Norton) .... *(1991-92)* 709
NR 642 (Norton) .... *(1991-92)* 709
NR 731 (Norton) .... *(1991-92)* 709
NR 801 (Norton) .... *(1991-92)* 709

NATIONAL AEROSPACE LABORATORY (Japan) 629
NATIONAL AIRCRAFT ENGINE FACTORIES (China) .... *(1989-90)* 686
NELSON AIRCRAFT CORPORATION (USA) 671
Nitramite (SNPE) .... *(1985-86)* 850
NOEL PENNY TURBINES LTD (UK) .... *(1991-92)* 710
NORMALAIR-GARRETT LIMITED (UK) .... *(1987-88)* 928
NORTHROP CORPORATION, VENTURA DIVISION (USA) .... *(1987-88)* 960
NORTON MOTORS LTD (UK) .... *(1991-92)* 709
NOVIKOV (USSR) .... *(1990-91)* 720

732 INDEXES: ENGINES—**O - T**

O

O-100-3 (Northrop) (1987-88) 960
O-160 (Avco Lycoming) (1986-87) 933
O-200 Series (TCM) 677
O-235 Series (Textron Lycoming) 681
O-320 Series (Textron Lycoming) 681
O-360 Series (TCM) 697
O-360 Series (Textron Lycoming) 682
O-470 Series (TCM/Continental) (1990-91) 752
O-520 Series (TCM) 697
O-540 Series (Textron Lycoming) 682
O-550 (TCM) 698
OMKB (see Omsk Aircraft Engine Design Bureau) (1991-92) 699
OMS engine (Aerojet) (1992-93) 678

Odin (Rolls-Royce) (1984-85) 853
OFFMAR AVIO srl (Italy) (1992-93) 645
OMNIPOL FOREIGN TRADE CORPORATION (Czech Republic) 612
OMSK AIRCRAFT ENGINE DESIGN BUREAU MARS (Russia) 654
ORAO AIR DEPOT (Yugoslavia) 684
ORAO AIR FORCE DEPOT (Yugoslavia) (1991-92) 744
ORENDA (see Hawker Siddeley Canada Inc) (1991-92) 662

P

P-020 (KKBM) 652
P-065 (KKBM) 652
P41 (Norton) (1987-88) 928
P60 (Norton) (1987-88) 928
P64 (Norton) (1987-88) 928
P.80 (Piper) (1987-88) 929
P.80 (Piper TTL) (1984-85) 854
P.80/2 (Piper) (1987-88) 929
P.80/2 (Piper TTL) (1984-85) 855
P.100 (Piper) (1987-88) 930
P.100 (Piper TTL) (1984-85) 855
P.200 (Piper) (1987-88) 930
P.200 (Piper TTL) (1984-85) 855
P&WC (see Pratt & Whitney Canada) 619
PAL 640 (JPX) (1987-88) 894
PAL 1300 (JPX) (1987-88) 894
PDM (Tumansky) (1985-86) 877
PF1A (LM) 624, 626
PFM 3200 (Porsche) (1990-91) 698
PLT34 (Avco Lycoming) (1987-88) 943
PRV V6 (FAM) (1987-88) 894
PS-90A (AO Aviadvigatel) 634
PT6A (P&WC) 605
PT6B/PT6C (P&WC) 607
PT6T Twin-Pac (P&WC) 607
PUL 212 (JPX) (1992-93) 630
PUL 425 (JPX) (1992-93) 630
PW11XX (Pratt & Whitney) (1985-86) 916
PW100 (P&WC) 604
PW200 (P&WC) 607
PW205B (P&WC) (1987-88) 890
PW 206A/B (P&WC) 623
PW209T (P&WC) (1987-88) 890
PW300 (P&WC) (1989-90) 683, 697
PW305 (P&WC) 604
PW400 (P&WC) (1987-88) 887
PW1120 (Pratt & Whitney) (1991-92) 735
PW1128 (Pratt & Whitney) (1984-85) 885
PW1129 (Pratt & Whitney) (1990-91) 749
PW2000 (Pratt & Whitney) 674
PW3000 (Pratt & Whitney) (1988-89) 734
PW4000 (Pratt & Whitney) 673
PW5000 (Pratt & Whitney) (1991-92) 736
PS (see Povazske Strojarne) 614
PZL (see Pezetel Foreign Trade Enterprise) 629
PZL (see Polskie Zaklady Lotnicze) (1991-92) 690
PZL-3R (WSK-PZL-Rzeszów) (1985-86) 868
PZL-3S (WSK-PZL-Rzeszów) 631
PZL-3SR (WSK-PZL-Rzeszów) 632
PZL-10 (WSK-PZL-Rzeszów) (1983-84) 788
PZL-10S (WSK-PZL-Rzeszów) (1984-85) 833
PZL-10W (WSK-PZL-Rzeszów) 631
PZL-F engines (WSK-PZL Rzeszów) 632
PZL-TVD-10S (WSK-PZL Rzeszów) (1985-86) 867

PARODI, MOTORSEGLERTECHNIK (Germany) (1992-93) 637
Patriot (Morton Thiokol) (1984-85) 878
Pave Tiger (Cuyuna) (1987-88) 947
Peacekeeper motors (Aerojet) (1984-85) 856
Peacekeeper Stage-I (Morton Thiokol) (1984-85) 878
Pedro (Morton Thiokol) (1984-85) 879
Pegasus (Rolls-Royce) 659
PEZETEL FOREIGN TRADE ENTERPRISE (Poland) 629
PIAGGIO, INDUSTRIE AERONAUTICHE E MECCANICHE RINALDO SpA (Italy) 646
PIEPER MOTORENBAU GmbH (Germany) (1992-93) 637
PIPER FM LTD (UK) (1987-88) 929
PIPER LTD (UK) (1985-86) 880
PIPER-TARGET TECHNOLOGY LTD (UK) (1984-85) 854
PNPP AVIADVIGATEL (Russia) 655
Polaris A3R motor (Aerojet) (1984-85) 856
POLSKIE ZAKLADY LOTNICZE (Poland) (1991-92) 690
PORSCHE, DR ING h c F, PORSCHE AG (Germany) (1991-92) 680
POVAZSKE STROJARNE (Czech Republic) 614
PPO AVIADVIGATEL (USSR) (1991-92) 699
PRATT & WHITNEY (see United Technologies Pratt & Whitney) 672
PRATT & WHITNEY CANADA (Canada) (1992-93) 619
Product 25-11 (UMKPA) 647
Product 55B (UMKPA) 647
Product 95-1 (UMKPA) 647
Product 95-111 (UMKPA) 647
PROGRESS (Ukraine) 667
PROGRESS/LOTAREV/ZVL (International) (1992-93) 642
PS-ZMKB (International) 625
PULCH, OTTO (West Germany) (1984-85) 822
PW-ALLISON (USA) (1992-93) 696

R

R-1E (Marquardt) (1992-93) 691
R-4D (Marquardt) (1984-85) 877
R-6C (Marquardt) (1984-85) 877
R-11 (Soyuz) 644
R-13 (Soyuz) 644
R-15 (Soyuz) 645
R-18 (TCM) (1986-87) 961
R-21 (Soyuz) 645
R-25 (Soyuz) 645
R-27 (Soyuz) 646
R-28 (Soyuz) 646
R-29 (Soyuz) 646
R-30 (Marquardt) (1984-85) 877
R-32 (Saturn) 662
R-32 (Tumansky) (1988-89) 709
R-33 (Tumansky) (1988-89) 709
R-35 (Soyuz) 646
R-36 (Norton) (1988-89) 710
R-40A (Marquardt) (1992-93) 691
R-79 (Soyuz) 646
R-123-300 (Soyuz) 647
R-195 (Soyuz) 644
R-266 (Tumansky) (1984-85) 845
R-266/R-31 (Soyuz) 664
RB.163 Spey (Rolls-Royce) 657
RB.168-62 and -66 Spey (Rolls-Royce/Allison) (1984-85) 824
RB.183 (Rolls-Royce) (1992-93) 675
RB.199 (Turbo-Union) 626
RB.211 (Rolls-Royce) 655
RB.401 (Rolls-Royce) (1984-85) 852
RB545 (Rolls-Royce) (1990-91) 730
RB550 (Rolls-Royce) (1990-91) 730
RB580 (Rolls-Royce) (1989-90) 724
RC engines (Avco Lycoming) (1986-87) 932
RD-3M-500 (Mikulin) (1986-87) 912
RD-9 (Soyuz) 644
RD-33 (Klimov) 638
RD-36-35 FVR (RKBM) 640
RD-36-51 (RKBM) 640
RD-38 (RKBM) 640
RD-41 (RKBM) 640
RD-60 (RKBM) 640
RD-60A (Klimov) 638
RD-100/101/103/107/108/111 (GDL) (1984-85) 836
RD-119/214/216/219/253 (GDL) (1984-85) 837
RD-500K (CEC) 624
RFB (see Rhein-Flugzeugbau GmbH) (1984-85) 822
R1 162 (Rotorway) 676
RJ500 (Rolls-Royce) (1983-84) 801
RKBM (see Rybinsk Engine-Building Design Office) 639
RKBM Project (RKBM) 640
RL10 (Pratt & Whitney) (1984-85) 886
RM8 (Flygmotor) 649
RM12 (Flygmotor) 649
RM-1000-A (Retimotor) (1987-88) 885
RM-2000 (Retimotor) (1987-88) 885
RRX1 (Flygmotor) (1984-85) 836
RRX5 (Flygmotor) (1984-85) 836
RS-27 (Rocketdyne) (1984-85) 887
RS-34 (Rocketdyne) (1984-85) 887
RSRM (Thiokol) 704
RTM 321 (Rolls-Royce/Turbomeca) (1983-84) 774, 781, 806
RTM 322 (Rolls-Royce Turbomeca) 625
RTWD-14 (Saturn) 643
RU-19 (Soyuz) 645
RW-100 (Rotor Way) (1984-85) 888
RW-133 (Rotor Way) (1983-84) 838

Ramjets (Marquardt) (1992-93) 691
Ram rocket (MBB) (1984-85) 820
RECTIMO AVIATION SA (France) (1992-93) 631
REFRIGERATION EQUIPMENT WORKS (Poland) (1992-93) 648
RETIMOTOR ENGENHARIA LTDA (Brazil) (1987-88) 885
RHEIN-FLUGZEUGBAU GmbH (West Germany) (1984-85) 822
Rita II (SEP) (1984-85) 813
ROCKETDYNE DIVISION OF ROCKWELL INTERNATIONAL (USA) 676
Rockets (Marquardt) (1988-89) 728
ROLLASON AIRCRAFT & ENGINES LTD (UK) (1983-84) 801
ROLLS-ROYCE plc (UK) 655
ROLLS-ROYCE/ALLISON (International) (1984-85) 824
ROLLS-ROYCE LYULKA (International) (1991-92) 684
ROLLS-ROYCE SATURN (International) 625
ROLLS-ROYCE TURBOMECA LIMITED (International) 625
ROTAX (see Bombardier-Rotax) 603
Rotax 532 (Rotax) 603
Rotax 582 (Rotax) 603
Rotax 912A (Rotax) 603
ROTORWAY INTERNATIONAL (USA) 676
RYBINSK ENGINE-BUILDING DESIGN OFFICE (Russia) 639

S

SA (see Singapore Aerospace Ltd) 649
SAC (see Shenyang Aero-Engine Company) 611
SACMA (France) (1985-86) 849
SAI (see Singapore Aircraft Industries) (1990-91) 710
SAMP (see Shanghai Aero-Engine Manufacturing Plant) 611
SARI (see Shenyang Aviation Engine Research Institute) 611
SARV retro (Morton Thiokol) (1985-86) 879
SAT-41 (Saturn) 643
SC 430 (König) (1992-93) 636
SD 570 (König) (1992-93) 636
SD (SNPE) (1985-86) 850
SE 1800 EIS (Sauer) (1992-93) 638
SECA (see Société d'Entreprises Commerciales et Aéronautiques SA) (1987-88) 885
SEP (see Société Européenne de Propulsion) (1984-85) 812
SF 930 (König) (1991-92) 678
SG 95 (RFB) (1984-85) 822
SL 1700 (Limbach) 620
SMPMC (see South Motive Power and Machinery Complex) 611
SNECMA (see Société Nationale d'Étude et de Construction de Moteurs d'Aviation) 615
SNIA BPD SpA (Italy) (1985-86) 863
SNPE (see Société Nationale des Poudres et Explosifs) (1987-88) 897
SO-1 (IL) (1991-92) 690
SO-3 (IL) 630
SPP (see Sport Plane Power Inc) (1990-91) 751
SR-114-TC-1 (Morton Thiokol) (1987-88) 960
SRM (Morton Thiokol) (1989-90) 739
SS 2100 H1S (Sauer) (1992-93) 638
SSME (Rocketdyne) 696
SSRM (Thiokol) 683
SS-RPV (Microturbo/Turbomeca) (1984-85) 811
SSUS-A motor (Morton Thiokol) (1987-88) 959
ST 2500 H1S (Sauer) (1992-93) 638
STAR-6B (Morton Thiokol) (1984-85) 879
STAR-13 (Morton Thiokol) (1984-85) 879
STAR-17 (Morton Thiokol) (1984-85) 879
STAR-17A (Morton Thiokol) (1984-85) 879
STAR-20 (Morton Thiokol) (1984-85) 879
STAR-24 (Morton Thiokol) (1984-85) 879
STAR-27 (Morton Thiokol) (1984-85) 879
STAR-30 (Morton Thiokol) (1984-85) 879
STAR-31 (Morton Thiokol) (1984-85) 879
STAR-37 (Morton Thiokol) (1984-85) 879
STAR-37FM (Morton Thiokol) (1984-85) 879
STAR-37XF (Morton Thiokol) (1984-85) 879
STAR-48 (Morton Thiokol) (1984-85) 879
STM (Vought) (1983-84) 844
Su-15 (Lyulka) (1987-88) 922

SALYUT (Russia) 641
SAMARA STATE SCIENTIFIC & PRODUCTION ENTERPRISE (Russia) 641
Satellite engines (Aerojet) (1984-85) 856
Saturn 500 (Janowski) (1987-88) 914
SATURN NPO (Russia) 642
SAUER MOTORENBAU GmbH (Germany) (1992-93) 638
SCOMA-ÉNERGIE (France) (1992-93) 631
Segmented motor (CSD) (1984-85) 863
SENER-INGENIERIA Y SISTEMAS SA (Spain) 649
SHANGHAI AERO-ENGINE MANUFACTURING PLANT (China) 611
SHENYANG AERO-ENGINE COMPANY (China) 611
SHENYANG AVIATION ENGINE RESEARCH INSTITUTE (China) 611
SHVETSOV (Russia) 644
Sidewinder motor (Aerojet) (1987-88) 937
SINGAPORE AEROSPACE LTD (Singapore) 649
Slat (LTV) (1984-85) 876
SOCIETA PER AZIONI ALFA-ROMEO (Italy) (1982-83) 742
SOCIETE D'ENTREPRISES COMMERCIALES ET AERONAUTIQUES SA (Belgium) (1987-88) 885
SOCIETE EUROPEENNE DE PROPULSION (France) (1984-85) 812
SOCIETE NATIONALE DES POUDRES ET EXPLOSIFS (France) (1987-88) 897
SOCIETE NATIONALE D'ETUDE ET DE CONSTRUCTION DE MOTEURS D'AVIATION (France) 615
SOCIETE TURBOMECA (France) 616
SOLOY CONVERSIONS LTD (USA) (1990-91) 751
SOLOY CORPORATION (USA) 676
SOLOY DUAL PAC INC (USA) 676
SOLOVIEV MKB (USSR) (1990-91) 721
Sorek 4 (Bet-Shemesh) (1988-89) 696
SOUTH MOTIVE POWER AND MACHINERY COMPLEX (China) 611
SOYUZ (Russia) 644
Space Shuttle OMS engine (Aerojet) (1992-93) 678
Space Shuttle SSRM (Thiokol) 683
Space Shuttle SSME (Rocketdyne) 696
Sparrow/Shrike Skipper motors (Aerojet) (1987-88) 937
Spey, civil (Rolls-Royce) 674
Spey, military (Rolls-Royce) 675
Spey 807 (Fiat) 687
SPORT PLANE POWER INC (USA) (1990-91) 751
Stamo 1000 (Pieper) (1985-86) 856
Stamo MS 1500 (Pieper) 637
Standard missile motor (Aerojet) (1984-85) 856
STATE AIRCRAFT ENGINE FACTORIES (China) (1987-88) 890
Statolite (SNPE) (1985-86) 850
Styx (SEP) (1984-85) 813
Super Sapphire (Bonner) (1985-86) 878
SVERDLOV factory (Russia) 647

T

T53 (Textron Lycoming) 680
T55 (Textron Lycoming) 680
T56 (Allison) 661
T58 (General Electric) (1988-89) 724
T63 (Allison) 660
T64 (General Electric) 669
T70 (TTL) (1987-88) 937
T74 (P&WC) 607
T76 (Garrett) 665
T101 (P&WC) 605
T117 (KHD) (1987-88) 901
T317 (KHD) (1984-85) 819
T400 (P&WC) (1992-93) 623
T406 (Allison) (1988-89) 717

T - Z —ENGINES: INDEXES

Entry	Year	Page
406-AD-400 (Allison)		661
407 (General Electric)	(1990-91)	741
407-GE-400 (General Electric)	(1989-90)	735
620 (JPX)		615
700 (General Electric)		670
701-AD-700 (Allison)	(1988-89)	718
703 (Allison)		660
800-APW-800 (Textron Lycoming/Pratt & Whitney)	(1988-89)	742
800-LHT-800 (LHTEC)		671
CM (see Teledyne Continental Motors)		677
E495-TC700 (Thunder)	(1986-87)	962
E-M-29-8 (Morton Thiokol)	(1984-85)	879
E-M-236 (Morton Thiokol)	(1984-85)	879
E-M-364 (Morton Thiokol)	(1984-85)	879
E-M-416 Tomahawk (Morton Thiokol)	(1984-85)	878
E-M-458 (Morton Thiokol)	(1984-85)	879
E-M-479 (Morton Thiokol)	(1984-85)	879
E-M-521 (Morton Thiokol)	(1984-85)	879
E-M-604 (Morton Thiokol)	(1984-85)	879
E-M-616 (Morton Thiokol)	(1984-85)	879
E-M-640 Altair III (Morton Thiokol)	(1984-85)	879
E-M-696 (Morton Thiokol)	(1984-85)	879
E-M-697 (Morton Thiokol)	(1984-85)	879
E-M-700 (Morton Thiokol)	(1984-85)	879
E-M-707 (Morton Thiokol)	(1984-85)	878
E-M-711 (Morton Thiokol)	(1984-85)	879
E-M-714 (Morton Thiokol)	(1984-85)	879
E-M-762 Antares III (Morton Thiokol)	(1984-85)	879
E-M-764 (Morton Thiokol)	(1984-85)	879
E-M-783 (Morton Thiokol)	(1984-85)	879
E-M-790 (Morton Thiokol)	(1984-85)	879
TEI (see Tusas Engine Industries)		649
TF30 (Pratt & Whitney)	(1991-92)	734
TF33 (Pratt & Whitney)	(1987-88)	961
TF34 (General Electric)		666
TF39 (General Electric)	(1988-89)	722
TF41 Spey (Rolls-Royce/Allison)	(1984-85)	824
TFA (Microturbo)		631
TFE76 (Garrett)	(1988-89)	720
TFE109 (Garrett)		663
TFE731 (Garrett)		664
TFE1042 (Garrett/Volvo Flygmotor)	(1985-86)	898
TFE1042 (Garrett)		664
TIO-540 Series (Textron Lycoming)		682
TIO-541 Series (Textron Lycoming)		682
TJD-76C Baby Mamba (Dreher)	(1983-84)	818
TJD-76D and E (Dreher)	(1983-84)	818
TM 333 (Turboméca)		617
TO-360 (Avco Lycoming)	(1987-88)	944
TOP (F + E)	(1992-93)	635
TP 319 Arrius (Turboméca)	(1991-92)	675
TP-500 (TCM)		698
TPE331 (Garrett)		665
TPF351 (Garrett)		665
TPJ 1304HF (VM Motori)		628
TPJ 1306HF (VM Motori)		628
TPJ 1308HF (VM Motori)		628
TR-201 (TRW)	(1984-85)	892
TRB (Microturbo)	(1992-93)	630
TRD-3 (Soyuz)		644
TRD-29 (Soyuz)		645
TRD-31 (Saturn)		661
TRD-37 (Soyuz)		644
TRI 60 (Microturbo)	(1987-88)	895
TRI 80 (Microturbo)	(1987-88)	895
TRS 18 (Microturbo)		631
TRS 18-046 (Microturbo)	(1986-87)	885
TRS 18-056 (Microturbo)	(1984-85)	811
TRS 18-075 (Microturbo)	(1984-85)	811
TRS 18-076 (Microturbo)	(1984-85)	811
TRS 20/22 (Microturbo)	(1984-85)	811
TRW (see TRW Defense and Space Systems Group)	(1984-85)	892
TTL (see Piper Target Technology)	(1984-85)	854
TTL (see Target Technology Ltd)	(1987-88)	937
TU-758 (Morton Thiokol)	(1984-85)	842
TU-780 (Morton Thiokol)	(1987-88)	960
TU-876 (Morton Thiokol)	(1984-85)	878
TV116-300 (Soyuz)		647
TV2-117 (Klimov)		638
TV3-117 (Klimov)		637
TV7-117 (Klimov)		637
TVD-10 (Mars)		638
TVD-10B (WSK-PZL Rzeszow)	(1990-91)	709
TVD-20 (Mars)		639
TVD-450 (Soyuz)		647
TVD-850 (Isotov)	(1987-88)	918
TVD-1500 (Glushenkov)	(1990-91)	712
TVD-1500 (RKBM)		640
TV-0-100 (Mars)		638
TWD-10B (WSK-PZL-Rzeszow)	(1982-83)	630
TX-175 (Thiokol)	(1984-85)	800
TX-354 Castor II (Morton Thiokol)	(1984-85)	878
TX-481 (Morton Thiokol)	(1987-88)	960
TX-486 (Morton Thiokol)	(1984-85)	878
TX-526 Castor IV (Morton Thiokol)	(1984-85)	878
TX-632 (Morton Thiokol)	(1987-88)	960
TX-633 (Morton Thiokol)	(1987-88)	960
TX-657 (Morton Thiokol)	(1987-88)	960
TX-683 (Morton Thiokol)	(1987-88)	960
TX 773 (Morton Thiokol)	(1987-88)	960
TACTICAL SYSTEMS CO (ATSC)	(1987-88)	937
TARGET TECHNOLOGY LTD (UK)	(1987-88)	937
Tay (Rolls-Royce)		658
TECHSPACE AERO SA (Belgium)		603
TELEDYNE CAE DIVISION OF TELEDYNE INC (USA)		679
TELEDYNE CONTINENTAL MOTORS (USA)		677
TEXTRON LYCOMING (USA)		679
TEXTRON LYCOMING/PRATT & WHITNEY (USA)	(1988-89)	742
THERMO-JET STANDARD INC (USA)	(1992-93)	703
THIOKOL CORPORATION (USA)		683
Three cylinder (Pulch)	(1984-85)	822
THUNDER ENGINES INC	(1986-87)	962
Titan III engines (Aerojet)	(1984-85)	865
Tomahawk (CSD)	(1984-85)	878
Tomahawk (Morton Thiokol)	(1984-85)	878
Tornado (Fisher)	(1991-92)	721
Trent (Rolls-Royce)		656
Trident motors (Morton Thiokol)	(1984-85)	878
TRUD factory (Russia)		647
TRW DEFENSE AND SPACE SYSTEMS GROUP (USA)	(1984-85)	892
Tu-144D engine (Koliesov)	(1989-90)	713
Tu-160 engine (ZMKB)	(1990-91)	725
TUMANSKY (Russia)		647
TUMSAS (Turkey)	(1987-88)	917
Turbine Pac (Soloy)		676
Turbo 90 (CAM)		619
TURBOMECANICA (see Intreprinderea Turbomecanica Bucuresti)		632
TURBOMECA-SNECMA GROUPEMENT GRTS (France)		618
TURBOMECA, SOCIETE (France)		616
TURBO-UNION LTD (International)		626
Turmo (Turbomeca)		618
TUSAS ENGINE INDUSTRIES (Turkey)		649
Twin-Pac PT6T (P&WC)	(1987-88)	889
Tyne (Rolls-Royce)		658
Type 003 (Pulch)	(1984-85)	822
Type 1.7-50 (Dragon)	(1987-88)	947
Type 4T60/A (JPX)		614
Type 18D/55 (CRM)	(1986-87)	645
Type 57 (USSR)	(1986-87)	915
Type 151 (NPT)	(1987-88)	929
Type 171 (NPT)	(1987-88)	929
Type 215 (Cuyuna)	(1987-88)	947
Type 225 (Lotus)	(1987-88)	927
Type 301 (NPT)	(1991-92)	710
Type 331 (NPT)	(1987-88)	929
Type 352 (Teledyne CAE)		700
Type 356 (Teledyne CAE)		700
Type 365 (Teledyne CAE)	(1985-86)	918
Type 370 (Teledyne CAE)	(1987-88)	970
Type 370-1 (Teledyne CAE)	(1987-88)	970
Type 372-2 (Teledyne CAE)	(1987-88)	970
Type 372-11A (Teledyne CAE)	(1987-88)	970
Type 373 (Teledyne CAE)	(1987-88)	970
Type 401 (NPT)	(1984-85)	813
Type 401B (NPT)	(1989-90)	721
Type 402 (SEP)	(1984-85)	813
Type 403 (SEP)	(1984-85)	813
Type 430 (Cuyana)	(1987-88)	947
Type 440/555 (Teledyne CAE)	(1983-84)	839
Type 450 (Lotus)	(1987-88)	927
Type 455 (Teledyne CAE)	(1984-85)	863
Type 460 (Chotia)	(1983-84)	839
Type 490 (Teledyne CAE)	(1988-89)	718
Type 501-M62 (Allison)		673
Type 535 (Rolls-Royce)		673
Type 600N WAEL (NGL)	(1987-88)	928
Type 754 (NPT)	(1991-92)	710
Type 902 (SEP)	(1984-85)	889
Type 904 (SEP)	(1984-85)	889
Type 2701 R 03 (Hirth)	(1992-93)	636
Type 2702 R 03 (Hirth)	(1992-93)	636
Type 2703 (Hirth)	(1992-93)	636
Type 2704 (Hirth)	(1991-92)	698
Type R (MKB)		631
UDF (General Electric)	(1991-92)	726
UL11-02 (Cuyuna)	(1987-88)	947
UMKPA (Russia)		647
UNITED TECHNOLOGIES PRATT & WHITNEY (USA)		672
UNITED TURBINES (UK) LTD (UK)	(1992-93)	678
UP ARROW AIRCRAFT (USA)	(1992-93)	704
V2500 (IAE)		623
VAZ-413 (VAZ)		648
VAZ-430 (VAZ)		648
VD-7 (RKBM)		639
VD-19 (RKBM)		640
VR-35 (Flygmotor)	(1984-85)	836
VAZ car plant (Russia)		648
VEDENEYEV (Russia)		648
Viking (SEP)	(1984-85)	812
Viper (Rolls-Royce)		658
Viper 600 (Fiat/Rolls-Royce)		687
VM MOTORI SpA (Italy)		628
VMKB bureau (Russia)		648
VOLSZHSKY (Russia)		648
VOLVO FLYGMOTOR AB (Sweden)		667
VOUGHT CORPORATION (USA)	(1983-84)	844
Voyager 200 (TCM)	(1990-91)	752
Voyager 300 (TCM)	(1990-91)	752
Voyager 370 (TCM)	(1990-91)	752
Voyager 550 (TCM)		679

W

Entry	Year	Page
W 5/33 (Westmayer)	(1992-93)	618
W40/50; 42/55; 65/80; 70/85 (Weslake)	(1984-85)	855
WAEL 342 (NGL)	(1987-88)	928
WAEL 600N (NGL)	(1987-88)	928
WAM 274 (NGL)	(1985-86)	879
WAM 342 (NGL)	(1985-86)	880
WAM 684 (NGL)	(1983-84)	800
WJ5 (DEMC)		608
WJ6 (SMPMC)		627
WJ9 (ZAC)	(1989-90)	628
WP-2 (CATIC)	(1989-90)	686
WP5 (LM)	(1992-93)	624
WP6 (LM)		609
WP7 (LM)		609
WP-7A (Chengdu)	(1989-90)	686, 687
WP7B (LMC)		610
WP8 (XAE)		623
WP-11 (BUAA)		610
WP13 (LMC)		624
WP13G (CEC)		624
WP14 (CEC)		624
WR2 (Williams)	(1987-88)	974
WR19-A7 (Williams)	(1987-88)	974
WR24 (Williams)	(1987-88)	974
WR34 (Williams)	(1985-86)	922
WR44 (Williams)	(1984-85)	893
W55 (SARI)		626
WS6 (LM)		610
WS8 (SAMP)		626
WS-9 (XAE)		627
WSK-PZL-KALISZ (see Wytwórnia Sprzetu Komunikacyjnego-PZL Kalisz)		630
WSK-PZL-RZESZOW (see Wytwórnia Sprzetu Komunikacyjnego-PZL Rzeszow)		630
WTS34 (Williams)	(1986-87)	963
WTS34-16 (Williams)	(1987-88)	974
WZ5 (ZARI)		628
WZ-6 (CLXMW)		624
WZ8 (SMPMC)		627
Walter M 601 (Motorlet)	(1988-89)	683
WEEDHOPPER OF UTAH INC (USA)	(1984-85)	863
WESLAKE & COMPANY (UK)	(1985-86)	887
WESTERMAYER, OSKAR (Austria)	(1992-93)	618
WILLIAMS INTERNATIONAL (USA)		683
WILLIAMSPORT (USA)		702
WYTWÓRNIA SPRZETU KOMUNIKACYJNEGO-PZL KALISZ (Poland)		630
WYTWÓRNIA SPREZETU KOMUNIKACYJNEGO-PZL RZESZÓW (Poland)		630

X

Entry	Year	Page
XAE (see Xian Aero-Engine Corporation)		611
XF3 (IHI)	(1986-87)	901
XG-40 (Rolls-Royce)	(1992-93)	676
XG50C (Zenoah)	(1986-87)	902
XIAN AERO-ENGINE CORPORATION (China)		611

Y

Entry	Year	Page
YH40 (YMF)	(1992-93)	627
YH280 (YMF)	(1992-93)	627
YJ69-Y-406 (Teledyne CAE)	(1988-89)	739
YLR89-NA-7 (Rocketdyne)	(1984-85)	887
YLR101-NA-15 (Rocketdyne)	(1984-85)	887
YLR105-NA-7 (Rocketdyne)	(1992-93)	887
YMF (see Yuhe Machine Factory)		671
YO-65 (Nelson)		681
YT702-LD-700 (Textron Lycoming)		
YUHE MACHINE FACTORY (China)	(1992-93)	627
YMF (see Yuhe Machine Factory)		627
YO-65 (Nelson)		692
YT702-LD-700 (Textron Lycoming)		701
YUHE MACHINE FACTORY (China)		627

Z

Entry	Year	Page
ZAC (see Zhuzhou Aero-Engine Company)		612
ZARI (see Zhuzhou Aero-Engine Research Institute)		612
ZEF (see Zhuzhou Aero-Engine Factory)		612
ZMKB PROGRESS (Ukraine)		650
ZO 01A (Zoche)		620
ZO 02A (Zoche)		620
ZVL (see Zavody Na Vyrobu Lozisk)	(1992-93)	630
ZAPOROZHYE MACHINE-BUILDING DESIGN BUREAU PROGRESS (Ukraine)		650
ZAVODY NA VYROBU LOZISK (Czechoslovakia)		630
ZENOAH (see Komatsu Zenoah Company)	(1986-87)	902
ZHUZHOU AERO-ENGINE COMPANY (China)		612
ZHUZHOU AERO-ENGINE FACTORY (China)		612
ZHUZHOU AERO-ENGINE RESEARCH INSTITUTE (China)		612
ZOCHE, MICHAEL (Germany)		620